The European Garden Flora is the definitive manual for the accurate identification of cultivated ornamental plants. Designed to meet the highest scientific standards, the vocabulary has nevertheless been kept as uncomplicated as possible so that the work is fully accessible to the informed gardener and landscape architect as well as to the professional botanist. Although based on Europe, the work is also intended to serve as a reference on plants cultivated throughout the world.

Comprehensive keys are provided at the level of family, genus and species, and line diagrams are included to illustrate important diagnostic features of critical taxa. Reference is made to useful illustrations and taxonomic accounts, and a small amount of guidance on cultivation is provided for each genus.

This volume is the fourth in a series of six. It contains accounts of 32 families of dicotyledons, including large, horticulturally important genera such as *Rosa*, *Saxifraga* and *Sedum*. The first three volumes cover the ferns and their allies, the conifers, the monocotyledons and the first 49 families of dicotyledons

THE EUROPEAN GARDEN FLORA

Associated publications

Some of the groups of plants covered in the European Garden Flora lend themselves to special treatment in separate volumes. The first of these 'spin-off' publications is *The Orchid Book* edited by James Cullen (1992). This is a fully updated and revised version of the Orchidaceae which was published in volume II of the Flora in 1984. A number of photographs and line drawings complement the text. It is intended that books on certain other groups will be published over the next few years.

THE EUROPEAN GARDEN FLORA

*A manual for the identification of plants cultivated
in Europe, both out-of-doors and under glass*

VOLUME IV

Dicotyledons (Part II)

edited by

J. Cullen, J.C.M. Alexander, A. Brady, C.D. Brickell,
P.S. Green, V.H. Heywood, P.-M. Jörgensen, S.L. Jury,
S.G. Knees, A.C. Leslie, V.A. Matthews, N.K.B. Robson,
S.M. Walters, D.O. Wijnands and P.F. Yeo

sponsored by

The Royal Botanic Garden, Edinburgh
The Royal Horticultural Society, London
The Stanley Smith Horticultural Trust, Cambridge

CAMBRIDGE
UNIVERSITY PRESS

Published by the Press Syndicate of the University of Cambridge
The Pitt Building. Trumpington Street, Cambridge CB2 1RP
40 West 20th Street, New York, NY 10011–4211, USA
10 Stamford Road, Oakleigh, Melbourne 3166, Australia

First published 1995

Printed in Great Britain
by The Bath Press, Avon

A catalogue record for this book is available from the British Library

Library of Congress cataloguing in publication data

The European garden flora

"Sponsored by the Royal Horticultural Society" – Cover v. 3
Includes bibliographical references.
Includes indexes.
Contents: v. 1. Pteridophyta, Gymnospermae,
Angiospermae – Monocotyledons (part I)—2. Monocotyledons
(part II)—v. 3–4. Dicotyledons (part I–II).
1. Plants, Ornamental—Europe—Identification.
2. Fruit—Europe—Identification. 3. Nuts—Europe—
Identification. I. Walters, S. M. (Stuart Max) II. Royal
Horticultural Society (Great Britain)
SB406.93.E85E97 1984   635.9′094   83–7655
ISBN 0 521 24859 0 (v. 1)
ISBN 0 521 25864 2 (v. 2)

ISBN 0 521 42095 4

CONTENTS

MAPS AND FIGURES

Map 1. Mean minimum January isotherms for Europe (hardiness codes)   xvii

FIGURES

ORGANISATION AND ADVISERS

Editorial Committee

J.C.M. Alexander (Secretary), Royal Botanic Garden, Edinburgh

A. Brady, National Botanic Gardens, Glasnevin, Dublin

C.D. Brickell, Royal Horticultural Society, London

J. Cullen (Chairman), Cambridge

P.S. Green, Royal Botanic Gardens, Kew

V.H. Heywood, Department of Plant Sciences, University of Reading

P.-M. Jørgensen, University of Bergen, Norway

S.L. Jury, Department of Plant Sciences, University of Reading

S.G. Knees (Research Associate), Edinburgh

A.C. Leslie, Royal Horticultural Society, Wisley

V.A. Matthews, Royal Horticultural Society, London

N.K.B. Robson, Natural History Museum, London

S.M. Walters, Cambridge

D.O. Wijnands, Wageningen, The Netherlands

P.F. Yeo, University Botanic Garden, Cambridge

Advisers

Mrs B.M. Allen, Cádiz, Spain

Professor C.D.K. Cook, Zürich, Switzerland

Professor H. Ern, Berlin, Germany

Dr H. Heine, Mannheim, Germany

The European Garden Flora project suffered a double blow shortly before this volume went to press with the deaths of Aidan Brady and Onno Wijnands. The other members of the editorial committee would like to record their gratitude for their considerable contribution to both the style and the content of the European Garden Flora since its inception in 1979.

EDITORS AND CONTRIBUTORS TO VOLUME IV

The various sections of this volume were edited at the following institutions:

Royal Botanic Garden, Edinburgh: *Eucryphiaceae, Dipterocarpaceae, Caryocaraceae, Nepenthaceae, Droseraceae, Papveraceae, Capparidaceae, Resedaceae, Moringaceae, Hamamelidaceae, Crassulaceae, Saxifragaceae* in part, *Cunoniaceae, Bruniaceae, Rosaceae, Chrysobalanaceae, Leguminosae, Krameriaceae.*

National Botanic Gardens, Glasnevin, Dublin: *Actinidiaceae, Sarraceniaceae, Cephalotaceae, Pittosporaceae, Byblidaceae, Roridulaceae.*

Natural History Museum, London: *Dilleniaceae, Paeoniaceae, Ochnaceae, Theaceae, Marcgraviaceae, Guttiferae.*

Royal Horticultural Society's Garden, Wisley: *Platanaceae, Saxifragaceae* in part.

University of Reading: *Cruciferae.*

Contributors

J. Akeroyd (Hindolveston, Norfolk)
J.C.M. Alexander (Royal Botanic Garden, Edinburgh)
P.G. Barnes (Royal Horticultural Society, Wisley)
S.T. Buczacki (Stratford-upon-Avon, Warwickshire)
M. Cheek (Royal Botanic Gardens, Kew)
J.Y. Clark (Bracknell, Berkshire)
A. Coombes (Hillier Arboretum, Hampshire)
J. Cullen (Cambridge)
U. Eggli (Zürich, Switzerland)
H. Ern (Berlin, Germany)
A.R. Ferguson (National Botanic Gardens, Glasnevin, Dublin)
D. Fräenz (Palmengarten, Frankfurt, Germany)
C. Fraile (Royal Botanic Garden, Edinburgh)
J. Fryer (Froxfield, Hampshire)
M.F. Gardner (Royal Botanic Garden, Edinburgh)
R. Gornall (University of Leicester)
Z. Gowler (University of Edinburgh)
P.S. Green (Royal Botanic Gardens, Kew)
N. Groendijk-Wilders (Wageningen, The Netherlands)
E.H. Hamlet (Royal Botanic Garden, Edinburgh)
F.K. Hibberd (Peasmarsh, East Sussex)
R.D. Hyam (Royal Botanic Garden, Edinburgh)
B. Hylmö (Bjuv, Sweden)
M. Jepp (Madang, Papua New Guinea)
S.L. Jury (University of Reading)
C.J. King (University Botanic Garden, Cambridge)
S.G. Knees (Royal Botanic Garden, Edinburgh)
J.M. Lees (formerly at Royal Botanic Garden, Edinburgh)
A.C. Leslie (Royal Horticultural Society, Wisley)

H. McAllister (University of Liverpool Botanic Garden, Ness)
D.R. McKean (Royal Botanic Garden, Edinburgh)
J.B. Martinez-Laborde (Madrid, Spain)
V.A. Matthews (Royal Horticultural Society, London)
H.S. Maxwell (Royal Botanic Garden, Edinburgh)
R.D. Meikle (Minehead, Somerset)
D.M. Miller (Royal Horticultural Society, Wisley)
C.M. Mitchem (Edinburgh)
E.C. Nelson (National Botanic Gardens, Glasnevin, Dublin)
U. Oster (Frankfurt, Germany)
D.A.H. Rae (Royal Botanic Garden, Edinburgh)
J.A. Ratter (Royal Botanic Garden, Edinburgh)
J.E. Richardson (Royal Botanic Garden, Edinburgh)
M. de Ridder (Wageningen, The Netherlands)
N.K.B. Robson (Natural History Museum, London)
G.D. Rowley ('Cactusville', Reading)
S.A. Spongberg (Arnold Arboretum, Harvard University, USA)
L. Springate (University of Reading)
W.T. Stearn (Natural History Museum, London)
N.P. Taylor (Royal Botanic Gardens, Kew)
M.C. Tebbitt (Royal Botanic Garden, Edinburgh)
T. Upson (University of Reading)
F.T. de Vries (Wageningen, The Netherlands)
S.M. Walters (Cambridge)
M.C. Warwick (Royal Botanic Garden, Edinburgh)
C. Whitefoord (Natural History Museum, London)
A.C. Whiteley (Royal Horticultural Society, Wisley)
D.O. Wijnands (Wageningen, The Netherlands)
P.F. Yeo (University Botanic Garden, Cambridge)

ACKNOWLEDGEMENTS

During the writing of this volume, The European Garden Flora project has received substantial support from the following:

(a) The Council of the Royal Horticultural Society: financial support.

(b) The Stanley Smith Horticultural Trust: financial support.

(c) The Royal Botanic Garden Edinburgh: financial support, staff time, services and general support for the secretariat.

(d) The other institutions to which members of the Editorial Committee belong: staff time, support and services.

(g) The Humphrey Whitbread First Charitable Trust: financial support.

The Editorial Committee gratefully acknowledge all this generous help.

Particular thanks are also due to Hazel Hamlet (Edinburgh), Douglas McKean (Edinburgh) and Suzanne Maxwell (Edinburgh) for editorial support. Mark Tebbitt, James Richardson, Cymon Cox and Chris Whitehouse provided intensive short-term assistance with editing and writing accounts, and Valerie Muirhead gave technical assistance with the final manuscript. We are grateful to Peter Barnes, Urs Eggli, Frances Hibberd, Christopher Hogg, Debbie Maizells, Suzanne Maxwell, Susan Oldfield, Sally Rae, Margaret Tebbs, Femke de Vries and Maureen Warwick for the preparation of illustrations.

INTRODUCTION

Amenity horticulture (gardening, landscaping, etc.) touches human life at many points. It is a major leisure activity for very large numbers of people, and is a very important means of improving the environment. The industry that has grown up to support this activity (the nursery trade, landscape architecture and management, public parks, etc.) is a large one, employing a considerable number of people. It is clearly important that the basic material of all this activity, i.e. plants, should be readily identifiable, so that both suppliers and users can have confidence that the material they buy and sell is what it purports to be.

The problems of identifying plants in cultivation are many and various, and derive from several sources which may be summarised as follows:

(a) Plants in cultivation have originated in all parts of the world, many of them from areas whose wild flora is not well known or documented. Many have been introduced, lost and re-introduced under different names.

(b) Plants in gardens are growing under conditions to which they are not necessarily well adapted, and may therefore show morphological and physiological differences from the original wild stocks.

(c) All plants that become established in cultivation have gone through a process of selection, some of it conscious (selection of the 'best' variants, etc.), some of it unconscious (by methods of cultivation and particularly propagation), so that, again, the populations of a species in cultivation may differ significantly from the wild populations.

(d) Many garden plants have been 'improved' by hybridisation (deliberate or accidental), and so, again, differ from the original stocks.

(e) Finally, and perhaps most importantly, the scientific study of plant classification (taxonomy) has concentrated mainly on wild plants, largely ignoring material in gardens.

Nevertheless, the classification of garden plants has a long and distinguished history. Many of the Herbals of pre-Linnaean times (i.e. before 1753) consist partly or largely of descriptions of plants in gardens, and this tradition continued, and perhaps reached its peak in the late eighteenth and early nineteenth centuries—the period following the publication of Linnaeus's major works, when exploration of the world was at its height. This is the period that saw the founding of *Curtis's Botanical Magazine* (1787) and the publication of J.C. Loudon's *Encyclopaedia of plants* (1829 and many subsequent editions).

The further development of plant taxonomy, from about the middle of the nineteenth century to the present, has seen an increasing divergence between garden and scientific taxonomy, leading on the one hand to such works as the *The Royal Horticultural Society's dictionary of gardening* (1951 and reprints), based on G. Nicholson's *Illustrated dictionary of gardening* (1884–8), *The new Royal Horticultural Society's dictionary of gardening* (1992), and the very numerous popular, usually illustrated works on garden plants available today, and, on the other hand, to the Floras, Revisions and Monographs of scientific taxonomy.

Despite this divergence, a number of plant taxonomists realised the importance of the classification and identification of cultivated plants, and produced works of considerable scientific value. Foremost among these stands L.H. Bailey, editor of *The standard cyclopedia of horticulture* (1900, with several subsequent reprints and editions), author of *Manual of cultivated plants* (1924, edn 2, 1949) and founder of the journals *Gentes Herbarum* and *Baileya*. Other important workers in this field are T. Rumpler (*Vilmorin's Blumengärtnerei*, 1879). L. Dippel (*Handbuch der Laubholzkunde*, 1889–93), A. Voss and A. Siebert (*Vilmorin's Blumengärtnerei*, edn 3, 1894–6), C.K. Schneider (*Illustriertes Handbuch der Laubholzkunde*, 1904–12), A. Rehder (*Manual of cultivated trees and shrubs*, 1927, edn 2, 1947), J.W.C. Kirk (*A British garden flora*, 1927), F. Enke (*Parey's Blumengärtnerei*, edn 2, 1958), B.K. Boom (*Flora Cultuurgewassen*, 1959 and proceeding), V.A. Avrorin and M.V. Baranova (*Decorativn'ie Travyanist'ie Rasteniya Dlya Otkritogo Grunta SSSR*, 1977) and R. Mansfeld (*Verzeichnis landwirtschaftlicher und gärtnerischer Kulturpflanzen*, edn 2, 1986).

The present Flora, which, of necessity, is based on original taxonomic studies by many workers, attempts to provide a scientifically accurate and up-to-date means for the identification of plants cultivated for amenity in Europe (i.e. it does not include crops, whether horticultural or agricultural, or garden weeds), and to provide what are currently thought to be their correct names, together with sufficient synonymy to make sense of catalogues and other horticultural works. The needs of the informed amateur gardener have been borne in mind at all stages of the work, and it is hoped that the Flora will meet his needs just as much as it meets the needs of the professional plant taxonomist. The details of the format and use of the Flora are explained in section 2 below (p. xiv).

In writing the work, the Editorial Committee has been fully aware of the difficulties involved. Some of these have been outlined above; others derive from the fact that herbarium material of cultivated plants is scanty and usually poorly annotated, so that material of many species is not available for checking the use of names, or for comparative purposes. Because of these facts, attention has been drawn to numerous problems which cannot be solved but can only be adverted to. The solution of such problems requires much more taxonomic work.

The form in which contributions appear is the responsibility of the Editorial Committee. The vocabulary and the technicalities of plant description are therefore not necessarily those used by the contributors.

1. SELECTION OF SPECIES

The problem of determining which species are in cultivation is complex and difficult, and has no complete and final answer. Many species, for instance, are grown in botanic gardens but not elsewhere; others, particularly orchids, succulents and some alpines, are to be found in the collections of specialists but are not available generally. Yet others have been in cultivation in the past but are now lost, or perhaps linger in a few collections, unrecorded and unpropagated. Further problems arise from the fact that the identification of plants in collections is not always as good as it might be, and some less well-known species probably appear in published lists under the names of other, well-known species (and vice versa).

The Flora attempts to cover all those species that are likely to be found in general collections (i.e. excluding botanic gardens and specialist collections) in Europe, whether they are grown out-of-doors or under glass. In order to produce a working list of such species, a compilation of all European nursery catalogues available to us was made in 1978 by Margaret McDonald, a vacation student working at the Royal Botanic Garden, Edinburgh. Since then, numerous additions have been made. This list (known as the 'Commercial List'), which includes well over 12 000 specific names, forms the basis of the species included here. Since 1987 the annual production by the Hardy Plant Society of the *Plant Finder* has made the process of ascertaining which plants are on sale very much simpler. Similar publications on the continent of Europe – Erhardt, A. & W., *Pflanzen Einkaufsführer* (1990); Pereire, A. & Bonduel, P., *Où trouver vos plantes* (1992) and Van der Laar, H.J., *Naamlijst van Houtige Gewassen* (1985–1989) – have made the process of scanning catalogues no longer necessary. Another work of great benefit to cultivated plant taxonomy is *Index Hortensis* by Piers Trehane; volume I covering herbaceous perennials was published in 1989; other groups will be covered in subsequent volumes. This work gives the nomenclaturally correct names and synonyms for plants in cultivation and also has extensive lists of cultivars. In addition to the 'Commercial List', *The Plant Finder* and *Index Hortensis*, several works on the flora of gardens have been consulted, and the species covered by them have been carefully considered for inclusion. These works are: *The Royal Horticultural Society's dictionary of gardening*, edn 2 (1956, supplement revised 1969); *The new Royal Horticultural Society's dictionary of gardening* (1992); Encke, F. (editor), *Parey's Blumengärtnerei* (1956); Boom, B.K., *Flora der Cultuurgewassen van Nederland* (1959 and proceeding); Bean, W.J., *Trees & shrubs hardy in the British Isles* (edn 8, 1970–81); Krüssmann, G., *Handbuch der Laubgehölze* (edn 2, 1976–8); *Manual of cultivated broad-leaved trees & shrubs* (English edition, translated by M.E. Epp, G.S. Daniels, editor, 1986); Encke, F., Buchheim, G. and Seybold, S. (editor) *Zander's Handwörterbuch der Pflanzennamen* (edn 13, 1984), and, since 1986, Mansfeld, R., *Verzeichnis landwirtschaftlicher und gärtnerischer Kulturpflanzen* (edn 2, 1986) and Jelitto, L. and Schacht, W., *Die Freiland Schmuckstauden* (1963; edn 3, edited by Jelitto, L., Schacht, W., and Fessler, A. 1985), *Hardy herbaceous perennials* revised english edition (1989). Most of the names included in these works are covered by the present Flora, though some have been rejected as referring to plants no longer in general cultivation.

As well as the works cited above, several relating to plants in cultivation in North America have also been consulted: Rehder, A., *Manual of cultivated trees and shrubs* (edn 2, 1947) and *Bibliography of cultivated trees and shrubs* (1949); Bailey, L.H., *Manual of cultivated plants* (edn 2, 1949); *Hortus Third* (edited by the staff of the L.H. Bailey Hortorium, Cornell University, 1976).

The contributors have also drawn on their own experience, as well as that of the family editors, European advisers and other experts, in deciding which species should be included.

Most species have a full entry, being keyed, numbered and described as set out under section 2c below (p. xv). A few, less commonly cultivated species are not keyed or numbered individually, but are described briefly under the species to which they are most likely to key out in the formal key. The system of asterisks used in volumes I and II has been misinterpreted and has now been abandoned.

2. USE OF THE FLORA

a. *The taxonomic system followed in the Flora.* Plants are described in this work in a taxonomic order, so that similar genera and species occur close to each other, rendering comparison of descriptions more easy than in a work where the entries are alphabetical. The families (and higher groups) follow the Engler and Prantl system as set out in H. Melchior's edition (edn 12, 1964) of *Syllabus der Pflanzenfamilien*. The assignment of genera to families also usually follows the *Syllabus*.

The order of the species within each genus has been a matter for the individual author's discretion. In general, however, some established revision of the genus has been followed, or, if no such revision exists, the author's own views on similarity and relationships have governed the order used.

b. *Nomenclature*. The arguments for using Latin names for plants in popular as well as scientific works are often stated and widely accepted, particularly for Floras such as this, which cover an area in which several languages are spoken. Latin names have therefore been used at every taxonomic level. A concise outline of the taxonomic hierarchy and how it is used can be found in C. Jeffrey's *An introduction to plant taxonomy* (1968, edn 2, 1982). Because of the difficulties of providing vernacular names in all the necessary languages (not to say dialects), they have not been included. S. Priszter's *Trees and shrubs of Europe, a dictionary in eight languages* (1983) is a useful source for the vernacular names of woody plants.

Many horticultural reference works omit the authority which should follow every Latin plant name. Knowledge of this authority prevents confusion between specific names that may have been used more than once within the same genus, and makes it possible to find the original description of the species (using *Index Kewensis*, which lists the original references for all Latin names for higher plants published since 1753). In this Flora authorities are therefore given for all names at or below genus level. These are unabbreviated to avoid the obscure contractions which mystify the lay reader, and on occasions the professional botanist. In most cases we have not thought it necessary to include the initials or qualifying words and letters which often accompany author names, e.g. A. Richard, Reichenbach filius, fil. or f. (the exceptions involve a few, very common surnames).

In scientific taxonomic literature, the authority for a plant name sometimes consists of two names joined together by *ex* or *in*. Such formulae have not been used here: the authority has been shortened in accordance with *The international code of botanical nomenclature*, e.g. *Capparis lasiantha* R. Brown ex de Candolle becomes *Capparis lasiantha* de Candolle; *Viburnum ternatum* Rehder in Sargent becomes *Viburnum ternatum* Rehder. The abbreviations *hort.* and *auct.*, which sometimes stand in place of the authority after Latin names, have not been used in this work as they are often obscure or misleading. The situations described by them can be clearly and unambiguously covered by the terms *invalid*, *misapplied* or *Anon*. *Invalid* implies that the name in question has not been *validly* published in accordance with the Code of Nomenclature, and therefore cannot be accepted. *Misapplied* refers to names which have been applied to the wrong species in gardens or in literature. *Anon*. is used with validly published names for which there is no apparent author.

Gardeners and horticulturists complain bitterly when long-used and well-loved names are replaced by unfamiliar ones. These changes are unavoidable if *The international code of botanical nomenclature* is adhered to. Taxonomic research will doubtless continue to unearth earlier names and will also continue to realign or split up existing groups, as relationships are further investigated. However, the previously accepted names are not lost; in this work they appear as synonyms, given in brackets after the currently accepted name; they are also included in the index. Dates of publication are not given either for accepted names or for synonyms.

c. *Descriptions and terminology*. Families, genera and species included in the Flora are mostly provided with full-length descriptions. Shorter, diagnostic descriptions are, however, used for genera or species which differ in only a few characters from others already fully described, e.g.:

3. P. vulgaris Linnaeus. Like *P. officinalis* but leaves lanceolate and corolla red . . .

This implies that the description of *P. vulgaris* is generally similar to that of *P. officinalis* except in the characters mentioned; it should not be assumed that plants of the two species will necessarily look very like each other. Additional species (see p. xiv), subspecies, varieties, formae and cultivars (see p. xvi) are described very briefly and diagnostically.

Unqualified measurements always refer to length (though 'long' is sometimes added in cases where confusion might arise); similarly, two measurements separated by a multiplication sign indicate length and breadth respectively.

The terminology has been simplied as far as is consistent with accuracy. The technical terms which, inevitably, have had to be used are explained in the glossary (p. 551). Technical terms restricted to particular families or genera are defined in the observations following the family or genus description, and are also referred to in the glossary.

d. *Informal keys*. For most genera containing 5 to 20 species (and for most families containing 5 to 20 genera) an informal key is given; this will not necessarily enable the user to identify precisely every species included, but will provide a guide to the occurrence of the more easily recognised characters. A selection of these characters is given, each of which is followed by the entry numbers of those species which show that character. In some cases, where only a few species of a genus show a particular character, the alternative states are not specified, e.g.:

Leaves. Leathery: **18,19**. This means that only species **18** and **19** in the particular genus have leathery leaves; the other species may have leaves of various textures, but they are not leathery.

e. *Formal keys*. For every family containing more than one genus, and for every genus containing more than one full-entry species, a dichotomous key is provided. This form of key, in which a series of decisions must be made between pairs of contrasting character-states, should lead the user step by step to an entry number followed by the name of a species. The reader should then check this identification with the description of that species: in some cases, other less commonly cultivated species may be mentioned under the description of the full-entry species: the brief descriptions of these should also be scanned, so that a final identification can be made. A key to all the families of the Dicotyledons to be included in the Flora is provided (p. 3); this has been modified from the version in volume III and improved versions will be printed in volumes V and VI.

f. *Horticultural information*. Notes on the cultural requirements and methods of propagation are generally included in the observation to each genus; more rarely, such information is given in the observations under the family description. These notes are generally brief and very generalised, and merely provide guidance.

Reference to general works on gardening is necessary for more detailed information.

g. *Citation of literature.* References to taxonomic books, articles and registration lists are cited for each family and genus, as appropriate. No abbreviations are used in these citations (though very long titles have been shortened). The citation of a particular book or article does not necessarily imply that it has been used in the preparation of the account of the particular genus or family in this work. A list giving somewhat fuller bibliographical details of all books cited in volumes I–III is to be found on pp. 429–439 of volume III. A similar list for volumes IV–VI will be published in volume VI.

h. *Citation of illustrations.* References to good illustrations are given for each species (or subspecies or variety); the names under which they were originally published (which may be different from those used here) are not normally given. The illustrations may be coloured or black and white, and may be drawings, paintings or photographs. Up to four illustrations per species have been given, and an attempt has been made to choose from widely available, modern works. Where no illustrations are cited, they either do not exist, as far as we know, or those that do are considered to be of doubtful accuracy or of very restricted availability.

In searching for illustrations, use has been made of *Index Londinensis* (1929–31, supplement 1941), R.T. Isaacson's *Flowering plant index of illustration and information* (1979) and an extensive index compiled over the last 15 years at the Royal Botanic Garden, Edinburgh.

Several pages of figures of diagnostic plant parts are included with various groups in the Flora, and should be particularly helpful when plants are being identified by means of the keys. Some of these are original, others have either been redrawn from various sources or are photocopies of leaves.

i. *Geographical distribution.* The wild distribution, as far as it can be ascertained, is given in italics at the end of the description of each species (or subspecies or variety). The choice and spelling of place names in general follows *The Times Atlas*, Comprehensive edition (1983 reprint), except:

(1) Well-established English forms of names have been used in preference to unfamiliar vernacular names, e.g. Crete instead of Kriti, Naples instead of Napoli;

(2) New names or spellings will be adopted as soon as they appear in readily available works of reference.

j. *Hardiness* (see map, p. xvii). For every species a hardiness code is given. This gives a tentative indication of the lowest temperatures that the species can withstand:

G2 – needs a heated glasshouse even in south Europe.
G1 – needs a cool glasshouse even in south Europe.

H5 – hardy in favourable areas; withstands 0 to –5 °C minimum.
H4 – hardy in mild areas; withstands –5 to –10 °C minimum.
H3 – hardy in cool areas; withstands –10 to –15 °C minimum.
H2 – hardy almost everywhere withstands –15 to –20 °C minimum.
H1 – hardy everywhere; withstands –20 °C and below.

The map of mean January minima (p. xvii) shows the isotherms corresponding to these codes.

k. *Flowering time.* The terms spring, summer, autumn and winter have been used as a guide to flowering times in cultivation in Europe. It is not possible to be more specific when dealing with an area extending from northern Scandinavia to the Mediterranean. In cases where plants do not flower in cultivation, or flower rarely, or whose time of flowering is not recorded, no flowering time is given.

l. *Subspecies, varieties and cultivars.* Subspecies and varieties are included, where appropriate. This is done in various ways, depending on the number of such groups; all are self-explanatory.

No attempt has been made to describe the range of cultivars of any species, either partially or comprehensively. The former is scarcely worth doing, the latter virtually impossible. Reference to individual, commonly grown cultivars is, however, made in various ways:

(1) If a registration list of cultivars exists, it is cited in the 'Literature' paragraph (see section 2g) following the description of the genus.

(2) If a particular cultivar is very widely grown, it may be referred to, either in the description of the species to which it belongs (or most resembles), or in the observations to that species.

(3) If, in a particular species, cultivars are numerous and fall into reasonably distinct groups based on variation in some obvious character, then these groups may be referred to, together with an example of each, in the observations to the species.

m. *Hybrids.* Many hybrids between species (interspecific hybrids) and some between genera (intergeneric hybrids) are in cultivation, and some of them are widely grown. Commonly cultivated interspecific hybrids are, where possible, included as though they were species. Their names, however, include a multiplication sign indicating their hybrid origin; the names of the parents (when known or presumed) are also given. Other hybrids which are less frequently grown are mentioned in the observations to the parent species they most resemble. In some genera where the number of hybrids is very large, only a small selection of those most commonly grown is mentioned.

Map 1. Mean minimum January isotherms for Europe
(hardiness codes). (After Krüssmann, *Handbuch der Laubgehölze,*
1960 and *Mitteilungen der Deutsche Dendrokigische Gesellschaft*
75:, 1983.) Corrected from European Garden Flora volume
II (1984).

ANGIOSPERMAE

Plants herbaceous or woody. Seedling leaves (cotyledons) 1 or 2. Stamens and ovary borne in unisexual or bisexual flowers which generally have protective and/or pollinator-attracting envelopes (perianth, often composed of differentiated sepals and petals). Ovules borne inside closed ovaries; seeds enclosed, until ripe, inside a fruit.

The flowering plants, as generally understood, to which most garden plants belong; there are about 300 000 species arranged in about 11 000 genera in about 400 families; some 15 000–20 000 species are in general cultivation. The group is divided into 2 large classes, the Monocotyledons (53 families in all), covered by volumes I (1986) and II (1984) and the Dicotyledons (about 350 families in all) covered in volume III (1989), this volume and subsequent volumes V and VI.

KEY TO CLASSES

1a. Cotyledon 1, terminal; leaves usually with parallel veins, sometimes these connected by cross-veinlets; leaves without stipules, opposite only in some aquatic plants; flowers usually with parts in 3s; mature root system wholly adventitious   **Monocotyledons** (volumes I & II)

b. Cotyledons usually 2, lateral; leaves usually net-veined, with or without stipules, alternate, opposite or whorled; flowers usually with parts in 2s, 4s or 5s or parts numerous; primary root (tap-root) usually persistent, branched   **Dicotyledons**

1

DICOTYLEDONS
(Part II)

Plants herbaceous or woody. Primary root (tap-root) often persisting, enlarging and branched. Seedling leaves (cotyledons) usually 2, lateral. Leaves alternate, opposite or whorled, rarely absent, veins usually forming a branched network. Parts of the flower usually in 2s, 4s or 5s, or numerous.

KEY TO FAMILIES

Families included in this volume are numbered and provided with page numbers; those already published in volume III are also indicated; the remaining families will be covered in volumes V and VI.

KEY TO GROUPS

1a. Petals present, free from each other at their bases (rarely united above the base), usually falling individually, or petals absent   2

b. Petals present, all united into a tube at their bases, sometimes shortly so, falling as a complete corolla   10

2a. Flowers unisexual and without petals, at least the males borne in catkins which are usually deciduous; plants always woody   **Group A**

b. Flowers with or without petals, unisexual or bisexual, never in catkins; plants woody or not   3

3a. Ovary consisting of 2 or more carpels which are completely free from each other   **Group B** (p. 4)

b. Ovary consisting of a single carpel or of 2 or more carpels which are united to each other wholly or in part (rarely the bodies of the carpels more or less free but the styles united)   4

4a. Perianth of 2 or more whorls, more or less clearly differentiated into calyx and corolla (calyx rarely very small and obscure; excluding aquatic plants with minute, quickly deciduous petals and branch-parasites with opposite, leathery leaves)   5

b. Perianth of a single whorl (which may be petal-like) or perianth completely absent, more rarely the perianth of 2 or more whorls but the segments not differing from whorl to whorl   8

5a. Stamens more than twice as many as the petals   **Group C** (p. 4)

b. Stamens twice as many as the petals or fewer   6

6a. Ovary partly or fully inferior   **Group D** (p. 6)

b. Ovary completely superior   7

7a. Placentation axile, apical, basal or free-central   **Group E** (p. 6)

b. Placentation parietal or marginal   **Group F** (p. 8)

8a. Stamens borne on the perianth or ovary inferior (perianth of female flowers sometimes very small)   **Group G** (p. 9)

b. Stamens free from the perianth; ovary superior or naked (i.e. not surrounded by a perianth)   9

9a. Flowers unisexual   **Group H** (p. 10)

b. Flowers bisexual   **Group I** (p. 10)

10a. Ovary partly or fully inferior   **Group J** (p. 11)

b. Ovary completely superior   11

11a. Corolla radially symmetric   **Group K** (p. 11)

b. Corolla bilaterally symmetric   **Group L** (p. 13)

Group A

1a. Stems jointed; leaves reduced to whorls of scales   **XXXIV. Casuarinaceae** III:14

b. Stems not jointed; leaves not as above   2

2a. Leaves pinnate   3

b. Leaves simple and entire, toothed or lobed (sometimes deeply so)   4

3a. Leaves without stipules; fruit a nut   **XXXVI. Juglandaceae** III:17

b. Leaves with stipules; fruit a legume   **CXIII. Leguminosae** (p. 463)

4a. Leaves opposite, evergreen, entire; fruit berry-like   **Garryaceae**

b. Leaves alternate, deciduous or evergreen; fruit not berry-like   5

5a. Ovules many, parietal; seeds many, cottony-hairy; male catkins erect with the stamens projecting beyond the bracts or hanging and with fringed bracts   **XXXVII. Salicaceae** III:20

b. Ovules solitary or few, not parietal; seeds few, not cottony-hairy; male catkins not as above   6

6a. Leaves dotted with aromatic glands   **Myricaceae**

b. Leaves not gland-dotted   7

7a. Styles 3, each often branched; fruit splitting into 3
    mericarps; seeds with appendages   **Euphorbiaceae**

b. Styles 1–3, not branched; fruit and seeds not as above   8

8a. Plant with milky sap   **XLII. Moraceae** III:86

b. Plants with clear sap   9

9a. Male catkins compound, i.e. each bract with 2 or 3 flowers
    attached to it; styles 2   **XXXVII. Betulaceae** III:45

b. Male catkins simple, i.e. each bract subtending a single
    flower; styles 1 or 3–6   **XXXIX. Fagaceae** III:59

Group B

1a. Trees with bark peeling off in plates; palmately lobed
    leaves and unisexual flowers in hanging, spherical heads
    CI. Platanaceae (p. 165)

b. Combination of characters not as above   2

2a. Perianth-segments and stamens borne independently below
    the ovary, or perianth absent   3

b. Perianth-segments and stamens borne on a rim or cup
    which is borne below the ovary   23

3a. Aquatic plants with peltate leaves and 3 sepals
    LXXXVII. Nymphaeaceae III:400

b. Terrestrial plants, or, if aquatic then without peltate leaves
    and with more than 3 sepals   4

4a. Herbs, succulent shrubs or shrubs with yellow wood or
    climbers with bisexual flowers and opposite leaves   5

b. Trees or shrubs which are neither succulent nor with
    yellow wood, if climbers then with unisexual flowers and
    alternate leaves   10

5a. Perianth absent   **LXXIX. Saururaceae** III:406

b. Perianth present   6

6a. Leaves succulent; stamens in 1 or 2 whorls
    CIII. Crassulaceae (p. 170)

b. Leaves not succulent; stamens spirally arranged, not
    obviously in whorls   7

7a. Petals fringed; fruits formed from each carpel borne on a
    common gynophore   **XCIX. Resedaceae** (p. 164)

b. Petals (when present) not fringed, but sometimes modified
    for nectar-secretion; fruits formed from each carpel not
    borne on a common gynophore   8

8a. Leaves opposite or whorled; flowers small, stalkless, in
    axillary clusters; ovule 1, placentation basal
    XLIX. Phytolaccaceae III:130

b. Combination of characters not as above   9

9a. Sepals differing among themselves, green; stamens ripening
    from the inside of the flower outwards, borne on a nectar-
    secreting disc   **LXXXIV. Paeoniaceae** (p. 17)

b. Sepals all similar, green or petal-like; stamens ripening
    from the outside of the flower inwards; nectar-secreting
    disc absent   **LXXIII. Ranunculaceae** III:325

10a. Leaves simple   11

b. Leaves compound   21

11a. Sepals and petals 5   12

b. Sepals and petals not 5   14

12a. Leaves opposite; stamens 5–10   **Coriariaceae**

b. Leaves alternate; stamens more numerous   13

13a. Anthers opening by pores

    LXXXVII. Ochnaceae (p. 28)

b. Anthers opening by slits   **LXXXIII. Dilleniaceae** (p. 15)

14a. Unisexual climbers   15

b. Erect trees or shrubs, flowers usually bisexual   16

15a. Carpels many; seeds not U-shaped
    LXIV. Schisandraceae III:317

b. Carpels 3 or 6; seeds usually U-shaped
    LXXVI. Menispermaceae III:398

16a. Stamens with a truncate connective which overtops the
    anthers; fruit usually a fleshy aggregate fruit; endosperm
    convoluted   **LXI. Annonaceae** III:315

b. Connective of stamens not as above; fruit not as above;
    endosperm not convoluted   17

17a. Carpels spirally arranged on a long receptacle; stipules
    large, united, early deciduous, leaving a ring-like scar
    LXI. Magnoliaceae III:302

b. Carpels in 1 whorl; stipules absent, minute or united to the
    leaf-stalk, not leaving a ring-like scar   18

18a. Petals present   19

b. Petals absent   20

19a. Sepals free, overlapping, more than 6; ovules solitary in
    each carpel   **LXV. Illiciaceae** III:318

b. Sepals 2–6, united, or if free, then edge-to-edge in bud;
    ovules more than 1 in each carpel
    LX. Winteraceae III:314

20a. Leaves in whorls; flowers bisexual; sepals minute or absent
    LXXI. Eupteleaceae III:324

b. Leaves opposite or alternate; flowers unisexual; sepals 4
    LXXII. Cercidiphyllaceae III:324

21a. Unisexual climbers, or erect shrubs with blue fruits;
    perianth-parts in 3s   **LXXV. Lardizabalaceae** III:396

b. Erect shrubs, fruits not blue; perianth-parts not in 3s   22

22a. Flowers showy, bisexual; leaves not aromatic
    LXXXIV. Paeoniaceae (p. 17)

b. Flowers inconspicuous, unisexual; leaves aromatic
    Rutaceae

23a. Leaves modified into insectivorous pitchers
    CIV. Cephalotaceae (p. 244)

b. Leaves not modified into insectivorous pitchers   24

24a. Flowers unisexual; leaves evergreen
    LXVI. Monimiaceae III:319

b. Flowers bisexual; leaves usually deciduous   25

25a. Stamens all fertile; perianth-whorls with 4–9 segments;
    leaves usually alternate   **CXI. Rosaceae** (p. 323)

b. Inner stamens sterile; perianth-whorls with more than 9
    segments; leaves opposite
    LXVII. Calycanthaceae III:320

Group C

1a. Perianth and stamens borne independently below the
    superior ovary   2

b. Perianth and stamens either borne on the edge of a rim
    or cup which is itself borne below the superior ovary, or
    borne on the top or the sides of the (partly or fully) inferior
    ovary   30

2a. Placentation axile or free-central   3
 b. Placentation marginal or parietal   20
3a. Placentation free-central; sepals usually 2
                              LII. Portulacaceae III:171
 b. Placentation axile; sepals more than 2   4
4a. Leaves all basal, tubular, forming insectivorous pitchers;
    style peltately dilated   **XCIII. Sarraceniaceae** (p. 78)
 b. Leaves not as above; style not dilated   5
5a. Leaves alternate   6
 b. Leaves opposite   17
6a. Anthers opening by terminal pores   7
 b. Anthers opening by longitudinal slits   9
7a. Shrubs with simple leaves without stipules, often covered
    with stellate hairs; stamens inflexed in bud; fruit a berry
                              LXXXVI. Actinidiaceae (p. 26)
 b. Combination of characters not as above   8
8a. Ovary deeply lobed, borne on an enlarged receptacle or
    gynophore; petals not fringed
                              LXXXVII. Ochnaceae (p. 28)
 b. Ovary not lobed, not borne as above; petals often fringed
                              Elaeocarpaceae
9a. Inner whorl of perianth-segments tubular or bifid, nectar-
    secreting; fruit a group of partly to fully united follicles
                              LXXIII. Ranunculaceae III:325
 b. Combination of characters not as above   10
10a. Leaves with translucent, aromatic glands   **Rutaceae**
 b. Leaves without translucent, aromatic glands   11
11a. Large tropical trees; sepals 5, all, or 2 or 3 of them
    enlarged and wing-like in fruit; carpels 3
                              LXXXVIII. Dipterocarpaceae (p. 28)
 b. Combination of characters not as above   12
12a. Stipules absent; leaves evergreen
                              LXXXIX. Theaceae (p. 29)
 b. Stipules present; leaves usually deciduous   13
13a. Filaments free; anthers 2-celled   14
 b. Filaments united into a tube, at least around the ovary,
    often also around the styles; anthers often 1-celled   15
14a. Disc absent; stamens more than 15; leaves simple
                              Tiliaceae
 b. Disc present, conspicuous; stamens 15; leaves dissected
                              Zygophyllaceae
15a. Styles divided, several; stipules often persistent; carpels 5 or
    more   **Malvaceae**
 b. Style 1, capitate or lobed, stigmas 1–several; stipules
    usually deciduous; carpels 2–5   16
16a. Stamens in 2 whorls, those of the outer whorl usually
    sterile   **Sterculiaceae**
 b. Stamens in several whorls, all fertile   **Bombacaceae**
17a. Sepals united, falling as a unit; fruit separating into boat-
    shaped units   **LXXXV. Eucryphiaceae** (p. 23)
 b. Sepals and fruit not as above   18
18a. Small trees; stamens with brightly coloured filaments
    which are at least twice as long as the petals, the anthers
    forming a ring   **XC. Caryocaraceae** (p. 43)
 b. Combination of characters not as above   19

19a. Leaves simple, without stipules, often with translucent
    glands; stamens often united in bundles
                              XCII. Guttiferae (p. 44)
 b. Leaves pinnate, without translucent glands; stamens not
    united in bundles   **Zygophyllaceae**
20a. Aquatic plants with cordate leaves; style and stigmas
    forming a disc on top of the ovary
                              LXXVII. Nymphaeaceae III:400
 b. Combination of characters not as above   21
21a. Carpel 1 with marginal placentation   22
 b. Carpels 2 or more, placentation parietal   23
22a. Leaves bipinnate or modified into phyllodes, with stipules
                              CXIII. Leguminosae (p. 463)
 b. Leaves various but not as above, without stipules
                              LXXIII. Ranunculaceae III:325
23a. Leaves opposite   24
 b. Leaves alternate   26
24a. Styles numerous; floral parts in 3s
                              XCVI. Papaveraceae (p. 103)
 b. Styles 1–5; floral parts in 4s or 5s   25
25a. Style 1; stamens not united in bundles; leaves without
    translucent glands   **Cistaceae**
 b. Styles 3–5, free or variously united; stamens united in
    bundles (sometimes apparently free); leaves with
    translucent glands   **XCII. Guttiferae** (p. 44)
26a. Small trees with aromatic bark; filaments united
                              LXIII. Canellaceae III:317
 b. Herbs, shrubs or trees, bark not aromatic; filaments free
                                                                 27
27a. Trees; leaves with stipules; anthers opening by pore-like
    slits   **Bixaceae**
 b. Herbs or shrubs; leaves usually without stipules; anthers
    opening by longitudinal slits   28
28a. Sepals 2 or rarely 3, quickly deciduous
                              XCVI. Papaveraceae (p. 103)
 b. Sepals 4–8, persistent in flower   29
29a. Ovary closed at its apex, borne on a gynophore; none of
    the petals fringed   **XCVII. Capparidaceae** (p. 128)
 b. Ovary open at its apex, not borne on a gynophore; at least
    some of the petals fringed **XCIX. Resedaceae** (p. 164)
30a. Flowers unisexual; leaf-base oblique   **Begoniaceae**
 b. Flowers bisexual; leaf-base not oblique   31
31a. Placentation free-central; ovary partly inferior
                              LII. Portulacaceae III:171
 b. Placentation not free-central; ovary either completely
    superior or completely inferior   32
32a. Aquatic plants with cordate leaves
                              LXXVII. Nymphaeaceae III:400
 b. Terrestrial plants; leaves various   33
33a. Carpels 1 or 3, excentrically placed at the top of, the
    bottom of, or within, the tubular perigynous zone
                              CXII. Chrysobalanaceae (p. 462)
 b. Carpels and perigynous zone not as above   34
34a. Stamens united in bundles on the same radii as the petals;
    staminodes often present; plants usually rough with
    stinging hairs   **Loasaceae**

b. Combination of characters not as above   35
35a. Sepals 2, united, falling as a unit; plant herbaceous
                              XCVI. Papaveraceae (p. 103)
b. Sepals 4 or 5, usually free, not falling as a unit; trees or
    shrubs   36
36a. Stamens united in several rings or sheets   **Lecythidaceae**
b. Stamens not as above   37
37a. Carpels 8–12, one above the other   **Punicaceae**
b. Carpels fewer, side by side   38
38a. Leaves with stipules   39
b. Leaves without stipules   40
39a. Leaves opposite or in whorls; plant woody
                              CVI. Cunoniaceae (p. 318)
b. Leaves alternate; plants woody or herbaceous
                              CXI. Rosaceae (p. 323)
40a. Leaves with translucent aromatic glands; style 1
                                                  Myrtaceae
b. Leaves without translucent aromatic glands; styles usually
    more than 1   **CV. Saxifragaceae** (p. 244)

Group D

1a. Petals and stamens numerous; plant succulent   2
b. Petals and stamens each fewer than 10; plants usually not
    succulent   3
2a. Stems succulent, often very spiny, leaves absent, very
    reduced or falling early   **LVII. Cactaceae** III:202
b. Stems and leaves succulent, spines usually absent
                              LI. Aizoaceae III:133
3a. Anthers opening by terminal pores   4
b. Anthers opening by longitudinal slits   5
4a. Filaments with a knee-like joint below the anthers; leaves
    with 3 conspicuous main veins from the base
                                          Melastomataceae
b. Filaments straight; leaves with a single main vein
                                          Rhizophoraceae
5a. Placentation parietal, placentas sometimes intrusive   6
b. Placentation axile, basal, apical or free-central   7
6a. Climbing herbs with tendrils; flowers unisexual
                                          Cucurbitaceae
b. Erect herbs or shrubs, if climbing then without tendrils;
    flowers usually bisexual   **CV. Saxifragaceae** (p. 244)
7a. Stamens on the same radii as the petals; trees or shrubs
    with simple leaves   **Rhamnaceae**
b. Stamens not on the same radii as the petals; plants
    herbaceous or woody, leaves simple to compound   8
8a. Flowers borne in umbels, sometimes condensed into heads
    or layers of whorls; leaves usually compound   9
b. Flowers not in umbels; leaves usually simple   10
9a. Fruit splitting into 2 mericarps; flowers usually bisexual;
    petals overlapping in bud; usually herbs without stellate
    hairs   **Umbelliferae**
b. Fruit a berry; flowers often unisexual; petals edge-to-edge
    in bud; plants mostly woody; often with stellate hairs
                                          Araliaceae
10a. Style 1   11
b. Styles more than 1, often 2, divergent   19

11a. Floating aquatic herb; leaf-stalks inflated   **Trapaceae**
b. Terrestrial herbs, trees or shrubs; leaf-stalks not inflated
                                                    12
12a. Small, low shrubs with scale-like, overlapping leaves;
    flowers in heads   **CX. Bruniaceae** (p. 323)
b. Trees, shrubs or herbs with expanded leaves; flowers not
    usually in heads   13
13a. Ovary 1-celled with 2–5 ovules; fruit leathery or drupe-
    like, 1-seeded   **Combretaceae**
b. Ovary usually with 2–5 cells, ovules various; fruit not as
    above   14
14a. Ovule solitary in each cell of the ovary   15
b. Ovules 2–numerous in each cell of the ovary   18
15a. Petals edge-to-edge in bud; flower usually bisexual   16
b. Petals overlapping in bud; flowers often unisexual   17
16a. Stamens with swollen, hairy filaments; petals recurved
                                          Alangiaceae
b. Stamens without swollen, hairy filaments; petals not
    recurved   **Cornaceae**
17a. Flowers in heads subtended by 2 conspicuous, white
    bracts; ovary 6–10-celled   **Davidiaceae**
b. Flowers various, but not as above; ovary 1-celled
                                          Nyssaceae
18a. Sap milky; petals 5; ovary 3-celled   **Campanulaceae**
b. Sap watery; petals 2 or 4; ovary usually 4-celled
                                          Onagraceae
19a. Trees or shrubs; hairs often stellate; fruit a few-seeded
    woody capsule   **CII. Hamamelidaceae** (p. 166)
b. Herbs or shrubs; hairs simple or absent; fruit various, not a
    woody capsule   **CV. Saxifragaceae** (p. 244)

Group E

1a. Perianth bilaterally symmetric   2
b. Perianth radially symmetric (the stamens sometimes not
    radially symmetric due to deflexion)   14
2a. Anthers cohering, covering the ovary like a cap
                                          Balsaminaceae
b. Anthers free, not as above, filaments sometimes united
                                                    3
3a. Anthers opening by terminal pores   4
b. Anthers opening by longitudinal slits   5
4a. Stamens 8, filaments united for at least half their length;
    fruit without barbed bristles   **Polygalaceae**
b. Stamens 3 or 4, filaments free; fruit covered in barbed
    bristles   **CXIV. Krameriaceae** (p. 550)
5a. Plants herbaceous   6
b. Plants woody (trees, shrubs or climbers)   10
6a. Leaves with stipules   7
b. Leaves without stipules   8
7a. Carpel 1; fruit a legume, sometimes 1-seeded
                              CXIII. Leguminosae (p. 463)
b. Carpels 5; fruit a capsule or berry, or splitting into
    mericarps   **Geraniaceae**
8a. Sepals, petals and stamens borne on a rim, cup or tube
    which is borne below the ovary; leaves not peltate   9
b. Sepals, petals and stamens borne independently below the

ovary (rarely the petals and stamens somewhat united at the base); leaves usually peltate **Tropaeolaceae**

9a. Leaves opposite **Lythraceae**

b. Leaves alternate or all basal **CV. Saxifragaceae** (p. 244)

10a. Stamens as many as or fewer than petals, borne on the same radii as the petals **Sabiaceae**

b. Stamens more than the petals, if as many or fewer then not on the same radii as the petals 11

11a. Carpel 1 with its style arising from near its base **CXII. Chrysobalanaceae** (p. 462)

b. Carpels 2 or more, styles not as above 12

12a. Leaves opposite, palmate; sepals united at the base **Hippocastanaceae**

b. Leaves alternate, usually pinnate; sepals free 13

13a. Stipules large, borne within the bases of the leaf-stalks **Melianthaceae**

b. Stipules absent, or, if present, minute, not borne as above **Sapindaceae**

14a. Anthers opening by terminal pores 15

b. Anthers opening by longitudinal slits, or by flaps 21

15a. Leaves with 3 more or less parallel veins from the base; each filament with a knee-like bend below the anther **Melastomataceae**

b. Leaves with 1 main vein; filaments straight or arched 16

16a. Low shrubs; leaves and stems covered in conspicuous, stalked glandular hairs on which insects are often caught 17

b. Shrubs or rarely low shrubs, not glandular-hairy as above 18

17a. Carpels 2 **CVIII. Byblidaceae** (p. 322)

b. Carpels 3 **CIX. Roridulaceae** (p. 323)

18a. Low shrub with unisexual flowers; stamens 4; petals 4, each usually 2 or 3-lobed, rarely a few unlobed **Elaeocarpaceae**

b. Combination of characters not as above 19

19a. Ovary lobed, consisting of several rounded humps, the style arising from the depression between them **LXXXVII. Ochnaceae** (p. 28)

b. Ovary not lobed as above, style terminal 20

20a. Carpels 3; style divided above into 3 branches; nectar-secreting disc absent **Clethraceae**

b. Carpels 4 or 5 (rarely 3); style not divided; nectar-secreting disc usually present **Ericaceae**

21a. Placentation free-central (ovary sometimes with septa below), or basal 22

b. Placentation axile or apical 26

22a. Stamens as many as petals and on the same radii as them 23

b. Stamens more or fewer than the petals, if as many then not on the same radii as them 25

23a. Anthers opening by flaps; stigma 1 **LXXIV. Berberidaceae** III:370

b. Anthers opening by longitudinal slits; stigmas more than 1 24

24a. Sepals 5; ovule 1, basal on a long, curved stalk; stipules absent **Plumbaginaceae**

b. Sepals 2 or rarely 3; ovules usually numerous, rarely 1, then not on a long, curved stalk; stipules usually present **LII. Portulacaceae** III:171

25a. Ovary lobed, consisting of several humps, the style arising from the depression between them; leaves pinnatisect **Limnanthaceae**

b. Ovary not lobed, style terminal; leaves simple, entire **LIV. Caryophyllaceae** III:177

26a. Petals and stamens both numerous; plants with succulent leaves and stems **LI. Aizoaceae** III:138

b. Combination of characters not as above 27

27a. Small, hairless annual herbs growing in water or on wet mud; seeds pitted **Elatinaceae**

b. Combination of characters not as above 28

28a. Sepals, petals and stamens borne on a rim, cup or tube which is inserted below the ovary 29

b. Sepals, petals and stamens inserted individually below the ovary 34

29a. Stamens as many as the petals and borne on the same radii as them **Rhamnaceae**

b. Stamens more or fewer than the petals or if as many then not borne on the same radii as them 30

30a. Style 1 31

b. Styles more than 1, often 2, divergent 32

31a. Calyx-tube not prominently ribbed; seeds with arils; mostly trees, shrubs or climbers **Celastraceae**

b. Calyx-tube prominently ribbed; seeds without arils; mostly herbs **Lythraceae**

32a. Fruit an inflated, membranous capsule; leaves mostly opposite, compound **Staphyleaceae**

b. Combination of characters not as above 33

33a. Trees or shrubs; hairs often stellate; anthers often opening by flaps; fruit a few-seeded, woody capsule **CII. Hamamelidaceae** (p. 166)

b. Herbs or shrubs; hairs simple or absent; anthers opening by longitudinal slits; fruit a non-woody capsule **CV. Saxifragaceae** (p. 244)

34a. Leaves with translucent, aromatic glands **Rutaceae**

b. Leaves without translucent, aromatic glands 35

35a. Flowers with a well-developed disc (usually nectar-secreting) below and/or around the ovary 36

b. Flowers without a disc, nectar secreted in other ways 46

36a. Stamens as many as and on the same radii as the petals 37

b. Stamens more or fewer than the petals, if as many then not on the same radii as them 38

37a. Climbers with tendrils; stamens free **Vitaceae**

b. Erect shrubs without tendrils; stamens with the filaments united at least at the base **Leeaceae**

38a. Resinous trees or shrubs 39

b. Herbs, shrubs or trees, not resinous (sometimes aromatic) 40

39a. Ovules 2 per cell; fruit a drupe or capsule **Burseraceae**

b. Ovules 1 per cell; fruit a drupe **Anacardiaceae**

40a. Plant herbaceous 41

 b. Plant woody (tree, shrub or climber)   42
41a. Petals long-clawed, united above the base; leaves not fleshy
                                                      Stackhousiaceae
 b. Petals entirely free, not long-clawed; leaves fleshy
                                                      Zygophyllaceae
42a. Flowers, or at least some of them, functionally unisexual
     (i.e. apparent anthers not producing pollen or ovary
     containing no ovules)   43
 b. Flowers bisexual   44
43a. Leaves alternate; ovary with 2–5 carpels, not flattened
                                                      Simaroubaceae
 b. Leaves opposite; ovary with 2 (rarely 3) carpels, flattened
     in the plane of the septum   **Aceraceae**
44a. Leaves entire or toothed; stamens 4 or 5, emerging from
     the disc; seeds with arils   **Celastraceae**
 b. Combination of characters not as above   45
45a. Leaves without stipules, not fleshy; filaments of the
     stamens united into a tube   **Meliaceae**
 b. Leaves with stipules, fleshy; filaments of the stamens free
                                                      Zygophyllaceae
46a. Plant herbaceous   47
 b. Plant woody (tree, shrub or climber)   49
47a. Leaves always simple; ovary 6–10-celled by the
     development of 3–5 secondary septa between the original
     3–5 septa during development of the flower   **Linaceae**
 b. Leaves lobed or compound; secondary septa absent 48
48a. Leaves with stipules   **Geraniaceae**
 b. Leaves without stipules   **Oxalidaceae**
49a. Filaments of the stamens united below   50
 b. Filaments of the stamens free   54
50a. Plant succulent, spiny; stamens 8 with woolly filaments;
     plants unisexual   **LVIII. Didiereaceae** III:301
 b. Combination of characters not as above   51
51a. Stamens 2   **Oleaceae**
 b. Stamens 3 or more   52
52a. Leaves without stipules   **XLV. Olacaceae** III:118
 b. Leaves with stipules though these are sometimes quickly
     deciduous   53
53a. Stipules persistent, borne between the bases of the leaf-
     stalks; petals with appendages   **Erythroxylaceae**
 b. Stipules quickly deciduous, not borne between the bases of
     the leaf-stalks; petals without appendages   **Sterculiaceae**
54a. Stamens 8–10   55
 b. Stamens 3–6   57
55a. Petals long-clawed, often fringed or toothed; stamens 10;
     some or all of the sepals with nectar-secreting glands
     outside   **Malpighiaceae**
 b. Petals neither clawed nor fringed nor toothed; stamens
     8–10; sepals without nectaries outside   56
56a. Ovules 1 per cell; sepals united at the base
                                          XLV. Olacaceae III:118
 b. Ovules many per cell; sepals free   **Stachyuraceae**
57a. Staminodes present in flowers which also contain fertile
     stamens   58
 b. Staminodes absent from flowers with fertile stamens,
     present only in functionally female flowers   59

58a. Carpels 2, ovary containing a single apical ovule; stipules
     present, borne within the bases of the leaf-stalks
                                                      Corynocarpaceae
 b. Carpels 2–4, each cell of the ovary containing 1 or 2
     ovules; stipules absent   **Cyrillaceae**
59a. Trees with opposite, pinnate leaves; twigs tipped with large
     dark buds; fruit a samara   **Oleaceae**
 b. Combination of characters not as above   60
60a. Sepals united at the base   61
 b. Sepals free from each other at the base   62
61a. Carpels 2 or rarely 3, 1 or 2 of them sterile, the ovary
     containing 2 apical ovules   **Icacinaceae**
 b. Carpels 3–many, all fertile, the ovary containing 1 or 2
     ovules per cell   **Aquifoliaceae**
62a. Ovules 1 per cell; petals 3 or 4   **Cneoraceae**
 b. Ovules many per cell; petals 5
                                 CVII. Pittosporaceae (p. 320)

Group F
1a. Sepals, petals and stamens borne on a rim or cup which is
     inserted below the ovary   2
 b. Sepals, petals and stamens inserted individually below the
     ovary   4
2a. Trees; leaves bi- or tripinnate; flowers bilaterally symmetric;
     stamens 5, of different lengths   **C. Moringaceae** (p. 165)
 b. Combination of characters not as above   3
3a. Flower-stalks slightly united to the leaf-stalks so that the
     flowers appear to be borne on the latter; petals twisted
     (each overlapped by, and overlapping, 1 other) in bud;
     carpels 3   **Turneraceae**
 b. Flower-stalks not united to leaf-stalks; petals not as above
     in bud; carpels 2 or 4   **CV. Saxifragaceae** (p. 244)
4a. Perianth bilaterally symmetric   5
 b. Perianth radially symmetric   9
5a. Ovary of 1 carpel with marginal placentation
                                 CXIII. Leguminosae (p. 463)
 b. Ovary of 2 or more carpels with parietal placentation 6
6a. Ovary open at the apex; some or all of the petals fringed
                                 XCIX. Resedaceae (p. 164)
 b. Ovary closed at the apex; no petals fringed   7
7a. Petals and stamens 5; carpels 3   **Violaceae**
 b. Petals and stamens 4 or 6; carpels 2   8
8a. Ovary borne on a stalk (gynophore); stamens projecting
     beyond the petals   **XCVII. Capparidaceae** (p. 128)
 b. Ovary not borne on a stalk; stamens not projecting beyond
     the petals   **XCVI. Papaveraceae** (p. 103)
9a. Petals and stamens numerous   **LI. Aizoaceae** III:133
 b. Petals and stamens fewer than 7   10
10a. Stamens alternating with multifid staminodes
                                 CV. Saxifragaceae (p. 244)
 b. Stamens not alternating with multifid staminodes 11
11a. Leaves insect-catching by means of stalked, glandular hairs
                                 XCV. Droseraceae (p. 87)
 b. Leaves not insect-catching   12

12a. Climbers with tendrils; ovary and stamens borne on a common stalk (androgynophore); corona present
Passifloraceae

b. Combination of characters not as above   13

13a. Petals 4, the inner pair 3-fid; sepals 2
XCVI. Papaveraceae (p. 103)

b. Petals not as above; sepals 4 or 5   14

14a. Stamens 6, 4 longer and 2 shorter; carpels 2; fruit with a secondary septum   **XCVIII. Cruciferae** (p. 129)

b. Stamens 4–10, all more or less equal; carpels 2–5; fruit without a secondary septum   15

15a. Petals each with a scale-like appendage at the base of the blade; leaves opposite   **Frankeniaceae**

b. Petals without appendages; leaves alternate or all basal
16

16a. Stipules present   17

b. Stipules absent   18

17a. Stamens 10; flowers in dense, cylindric panicles
CVI. Cunoniaceae (p. 318)

b. Stamens 5; flowers not as above   **Violaceae**

18a. Leaves scale-like, alternate   **Tamaricaceae**

b. Leaves normally developed, usually all basal   **Pyrolaceae**

Group G

1a. Aquatics or rhubarb (Rheum)-like marsh plants with cordate leaves   2

b. Terrestrial plants, not as above   4

2a. Stamens 8, 4 or 2; leaves deeply divided or cordate
Haloragaceae

b. Stamen 1; leaves undivided, not cordate   3

3a. Leaves whorled; fruit small, indehiscent, dry, 1-seeded, not lobed   **Hippuridaceae**

b. Leaves opposite; fruit 4-lobed, up to 4-seeded
Callitrichaceae

4a. Trees or shrubs   5

b. Herbs, climbers or parasites   17

5a. Plant covered with scales; fruit enclosed in the berry-like, persistent, fleshy perianth   **Elaeagnaceae**

b. Plant not covered with scales; fruit not as above   6

6a. Stamen 1, or 1 whole stamen flanked by 2 half-stamens; leaves opposite   **LXXXI. Chloranthaceae** III:415

b. Stamens not as above; leaves usually alternate   7

7a. Stamens on radii alternating with the sepals
Rhamnaceae

b. Stamens on the same radii as the sepals   8

8a. Ovary 2-celled, partly inferior; stellate hairs often present; fruit a woody capsule   **CII. Hamamelidaceae** (p. 166)

b. Combination of characters not as above   9

9a. Stamens 4, situated at the top of the spoon-shaped, petal-like perianth-segments which split apart as the flower opens   **XLIV. Proteaceae** III:105

b. Combination of characters not as above   10

10a. Ovary inferior   11

b. Ovary superior   14

11a. Placentation parietal   **CV. Saxifragaceae** (p. 244)

b. Placentation axile or basal   12

12a. Styles 3–6; fruit a nut surrounded by a scaly cupule
XXXIX. Fagaceae III:59

b. Style 1; fruit not as above   13

13a. Stamens 4 or 5; placentation basal
XLVI. Santalaceae III:119

b. Stamens 5–10; placentation axile   **Cornaceae**

14a. Leaves aromatic, dotted with translucent glands; anthers opening by flaps   **LXVI. Lauraceae** III:321

b. Leaves neither aromatic nor gland-dotted; anthers not opening by flaps   15

15a. Stamens 2, or 8–10 borne at different levels in the perianth-tube; leaves simple, entire   **Thymelaeaceae**

b. Stamens not as above; leaves lobed or compound   16

16a. Inflorescence borne on the shoots of the current year; fruit a schizocarp of 2 (rarely 3) samaras   **Aceraceae**

b. Inflorescence borne on old wood; fruit a legume
CXIII. Leguminosae (p. 463)

17a. Branch parasites with green, forked branches or with small, scale-like leaves joined in pairs
XLVII. Loranthaceae III:120

b. Plants not parasitic, as above   18

18a. Perianth absent; flowers in spikes
LXXIX. Saururaceae III:406

b. Perianth present; flowers usually not in spikes   19

19a. Leaf-base oblique; ovary inferior, 3-celled   **Begoniaceae**

b. Leaf-base not oblique; ovary not as above   20

20a. Ovary superior   21

b. Ovary inferior   26

21a. Carpel 1, ovule 1, apical; perianth tubular
Thymelaeaceae

b. Combination of characters not as above   22

22a. Carpels 3 (rarely 2), ovule 1, basal; perianth persistent in fruit; leaves usually alternate, entire   23

b. Combination of characters not as above   24

23a. Leaves without stipules; stamens 5
LIII. Basellaceae III:176

b. Leaves with stipules united into a sheath (ochrea); stamens usually 6–9   **XLVIII. Polygonaceae** III:121

24a. Leaves alternate, usually lobed or compound
CXI. Rosaceae (p. 323)

b. Leaves opposite, usually entire   25

25a. Ovule 1, fruit a nut; stipules scarious or rarely absent
LIV. Caryophyllaceae III:117

b. Ovules numerous; fruit a capsule; stipules absent
Lythraceae

26a. Leaves pinnate; ovary open at the apex   **Datiscaceae**

b. Leaves not pinnate; ovary closed at apex   27

27a. Ovary 6-celled; perianth 3-lobed or tubular and bilaterally symmetric   **LXXXII. Aristolochiaceae** III:416

b. Combination of characters not as above   28

28a. Ovules 1–5; seed 1   29

b. Ovules and seeds numerous   30

29a. Perianth-segments thickening in fruit; leaves alternate
LV. Chenopodiaceae III:195

b. Perianth-segments not as above; leaves opposite or alternate   **XLVI. Santalaceae** III:119

30a. Styles 2; placentation parietal
    CV. Saxifragaceae (p. 244)
  b. Style 1; placentation axile   **Onagraceae**

Group H
1a. Aquatic herb; leaves divided into thread-like segments
    LXVI. Ceratophyllaceae III:466
  b. Terrestrial plants; leaves not as above   2
2a. Trailing, heather-like shrublet; fruit a berry
    Empetraceae
  b. Combination of characters not as above   3
3a. Flowers in racemes or spikes; fruit a berry or drupe-like; leaves entire, alternate, without stipules; carpels more than 5   **XLIX. Phytolaccaceae** III:129
  b. Combination of characters not as above   4
4a. Ovary 3-celled; styles 3   5
  b. Ovary 1-, 2- or 4-celled; styles 1 or 2   7
5a. Leaves with sheathing, membranous stipules; perianth-segments 6; fruit a nut   **XLVIII. Polygonaceae** III:121
  b. Combination of characters not as above   6
6a. Fruit schizocarpic; sap often milky; deciduous or evergreen herbs, trees or shrubs or stem-succulents; styles usually divided; seeds usually with appendages   **Euphorbiaceae**
  b. Fruit a capsule splitting through the cells; sap not milky; evergreen shrubs; styles undivided; seeds black and shiny, without appendages   **Buxaceae**
7a. Resinous trees or shrubs; leaves simple or pinnate; flowers with a nectar-secreting disc; stamens 3–10; fruit 1-seeded, drupe-like   **Anacardiaceae**
  b. Combination of characters not as above   8
8a. Stamens 2   **Oleaceae**
  b. Stamens more than 2   9
9a. Leaves forming insectivorous pitchers
    XCIV. Nepenthaceae (p. 80)
  b. Leaves not forming insectivorous pitchers   10
10a. Plants aromatic, dioecious; stamens 3–18, filaments united; ovary of 1 carpel containing a single, basal ovule
    XII. Myristicaceae III:317
  b. Combination of characters not as above   11
11a. Placentation parietal; stamens numerous; fruit a berry or capsule   **Flacourtiaceae**
  b. Combination of characters not as above   12
12a. Trees, shrubs or climbers, if herbaceous then flowers sunk in a fleshy receptacle; ovules apical   13
  b. Combination of characters not as above   16
13a. Ovules 4, of which only 1 develops; flowers in axillary racemes   **Daphniphyllaceae**
  b. Ovule 1; flowers not in axillary racemes   14
14a. Sap watery; fruit a drupe   **XL. Ulmaceae** III:79
  b. Sap milky; fruit a syncarp or group of samaras   15
15a. Perianth present; fruit frequently a syncarp of drupes or achenes united with the flat to flask-shaped receptacle
    XLII. Moraceae III:86
  b. Perianth absent; fruit a samara
    XLI. Eucommiaceae III:85
16a. Stinging hairs present or plant rough to the touch;

stamens touch-sensitive, inflexed in bud; leaves often with cystoliths; seed with a straight embryo
    XLIII. Urticaceae III:102
  b. Stinging hairs absent; stamens neither touch-sensitive nor inflexed in bud; cystoliths absent; seed often with a curved embryo   17
17a. Perianth scarious; stamens usually with the filaments united below   **LVI. Amaranthaceae** III:199
  b. Perianth greenish or absent; stamens with free filaments
    18
18a. Leaves all opposite; fruit splitting into 2 mericarps
    Euphorbiaceae
  b. Leaves alternate, at least above; fruit not as above   19
19a. Ovary with cross-walls, containing 4 ovules; leaves leathery   **Buxaceae**
  b. Ovary 1-celled, without cross-walls, containing 1 ovule; leaves not leathery   20
20a. Leaves with stipules; ovule apical   **XLII. Moraceae** III:86
  b. Leaves without stipules (sometimes stem succulent and continuous with the leaves); ovule basal
    LV. Chenopodiaceae III:195

Group I
1a. Flowers in racemes or spikes; fruit a berry or drupe-like; leaves entire, alternate, without stipules
    XLIX. Phytolaccaceae III:124
  b. Combination of characters not as above   2
2a. Trees or trailing, heather-like shrublets, rarely aromatic shrubs   3
  b. Herbs, climbers or non-aromatic shrubs   9
3a. Trailing, heather-like shrublet; fruit a drupe
    Empetraceae
  b. Trees or aromatic shrubs; fruit a drupe, samara, nut or capsule   4
4a. Stamens numerous; ovary with 5 or more cells   5
  b. Stamens 12 or fewer; ovary with up to 4 cells   7
5a. Leaves in whorls   **LXX. Trochodendraceae** III:324
  b. Leaves not in whorls   6
6a. Sepals 4; flowers in hanging spikes
    LXIX. Tetracentraceae III:324
  b. Sepals not 4; flowers in cymes   **Tiliaceae**
7a. Leaves evergreen with translucent, aromatic glands; anthers opening by flaps   **LXVIII. Lauraceae** III:321
  b. Leaves usually deciduous, without translucent, aromatic glands; anthers opening by slits   8
8a. Stamens 2; leaf-base not oblique   **Oleaceae**
  b. Stamens 4–8; leaf-base oblique   **XL. Ulmaceae** III:72
9a. Perianth absent; flowers borne (and often sunk) in a fleshy spike; leaves well developed, often fleshy
    LXXX. Piperaceae III:407
  b. Combination of characters not as above   10
10a. Leaves with stipules which are usually united into a sheath; fruit usually a 3-sided nut
    XLVIII. Polygonaceae III:121
  b. Leaves without stipules; fruit not a 3-sided nut   11

11a. Sepals falling as the flower opens; herbs with palmately-
    lobed leaves and orange sap
                    XCVI. Papaveraceae (p. 103)
  b. Combination of characters not as above   12
12a. Ovary of 1 carpel; fruit 1-seeded; perianth usually petal-
    like, bracts sometimes calyx-like
                    L. Nyctaginaceae III:131
  b. Ovary of 2 or more carpels; fruit 1-many-seeded; perianth
    not petal-like   13
13a. Ovary open at the apex; placentation parietal
                    XCIX. Resedaceae (p. 164)
  b. Ovary closed at the apex; placentation basal, free-central
    or axile   14
14a. Ovule solitary, basal   15
  b. Ovules numerous, axile or free-central   16
15a. Perianth green, membranous or absent; stamens free
                    LV. Chenopodiaceae III:195
  b. Perianth scarious; stamens often united below
                    LVI. Amaranthaceae III:199
16a. Placentation axile; leaves alternate
                    CV. Saxifragaceae (p. 244)
  b. Placentation basal or free-central; leaves usually opposite
                    17
17a. Sepals free; stamens on the same radii as or more
    numerous than the sepals   **LIV. Caryophyllaceae** III:177
  b. Sepals united; stamens as many as and alternating with
    the sepals   **Primulaceae**

Group J

1a. Leaves needle-like or scale-like; plants small, heather-like
    shrublets   **CX. Bruniaceae** (p. 323)
  b. Combination of characters not as above   2
2a. Ovary divided into cells but placentation parietal; leaves
    very succulent   **LI. Aizoaceae** III:133
  b. Combination of characters not as above   3
3a. Inflorescence a head subtended by an involucre of bracts;
    ovule always solitary   4
  b. Inflorescence and ovule not as above   5
4a. Each flower surrounded by a cup-like involucel; stamens 4,
    free; ovule apical   **Dipsacaceae**
  b. Involucels absent; stamens 5, their anthers united into a
    tube; ovule basal   **Compositae**
5a. Stamens 2; stamens and style united into a touch-sensitive
    column; leaves linear   **Stylidiaceae**
  b. Combination of characters not as above   6
6a. Leaves alternate or all basal   7
  b. Leaves opposite or appearing whorled   15
7a. Anthers opening by pores; fruit a berry or drupe
                    Ericaceae
  b. Anthers opening by slits; fruit various   8
8a. Climbers with tendrils and unisexual flowers; stamens 1–5;
    placentation parietal; fruit berry-like   **Cucurbitaceae**
  b. Combination of characters not as above   9
9a. Stamens 10-many; plants woody   10
  b. Stamens 4 or 5; plants mainly herbaceous   12

10a. Leaves gland-dotted, smelling of eucalyptus; corolla
    completely united, unlobed, falling as a whole
                    Myrtaceae
  b. Combination of characters not as above   11
11a. Hairs stellate or scale-like; stamens in 1 series; anthers
    linear   **Styracaceae**
  b. Hairs absent or not as above; stamens in several series;
    anthers broad   **Symplocaceae**
12a. Stigma surrounded by a sheath   **Goodeniaceae**
  b. Stigma not surrounded by a sheath   13
13a. Stamens as many as, and on the same radii as the petals
                    Primulaceae
  b. Stamens not as above   14
14a. Stamens 2 or 4, borne on the corolla; sap not milky
                    Gesneriaceae
  b. Stamens 5 or more, free from the corolla; sap usually
    milky   **Campanulaceae**
15a. Placentation parietal; stamens 2 or 4 and paired
                    Gesneriaceae
  b. Placentation axile or apical; stamens 1 or more, if 4 then
    not paired   16
16a. Stamens 1–3; ovary with 1 ovule   **Valerianaceae**
  b. Stamens 4 or more; ovary usually with 2 or more ovules
                    17
17a. Leaves divided into 3 leaflets; flowers in a few-flowered
    head   **Adoxaceae**
  b. Leaves simple or rarely pinnate; inflorescence various,
    usually not as above   18
18a. Stipules usually borne between the bases of the leaf-stalks
    (sometimes looking like the leaves); ovary usually 2-celled;
    flowers usually radially symmetric; fruit capsular, fleshy or
    schizocarpic   **Rubiaceae**
  b. Stipules usually absent, when present not as above; ovary
    usually with 3 cells (occasionally with 2–5 cells),
    sometimes only 1 cell fertile; flowers often bilaterally
    symmetric; fruit a berry or drupe   **Caprifoliaceae**

Group K

1a. Stamens 2   **Oleaceae**
  b. Stamens more than 2   2
2a. Carpels several, free; plants succulent
                  CIII. Crassulaceae (p. 170)
  b. Carpels united, or, if the bodies of the carpels are more
    or less free, the styles united, rarely the ovary of 1 carpel;
    plants usually not succulent   3
3a. Corolla scarious, 4-lobed; stamens 4, projecting from the
    corolla; leaves often all basal and with parallel veins
                    Plantaginaceae
  b. Combination of characters not as above   4
4a. Central flowers of the inflorescence abortive, their bracts
    forming nectar-secreting pitchers; petals completely united,
    falling as the flower opens   **XCI. Marcgraviaceae** (p. 43)
  b. Combination of characters not as above   5
5a. Stamens more than twice as many as the petals   6
  b. Stamens up to twice as many as the petals   13

6a. Leaves evergreen, divided into 3 leaflets; filaments brightly coloured, at least twice as long as the petals
XC. Caryocaraceae (p. 43)

  b. Leaves simple, entire or lobed, deciduous or evergreen; filaments not as above   7

7a. Leaves with stipules; filaments of stamens united into a tube around the ovary and style   **Malvaceae**

  b. Leaves without stipules; filaments free   8

8a. Anthers opening by pores
LXXXVI. Actinidiaceae (p. 26)

  b. Anthers opening by longitudinal slits   9

9a. Leaves with translucent, aromatic glands; calyx cup-like, unlobed   **Rutaceae**

  b. Leaves without translucent, aromatic glands; calyx not as above   10

10a. Placentation parietal; leaves fleshy   **Fouquieriaceae**

  b. Placentation axile; leaves not fleshy   11

11a. Sap milky; ovules 1 per cell   **Sapotaceae**

  b. Sap not milky; ovules 2 or more per cell   12

12a. Ovules 2 per cell; flowers usually unisexual   **Ebenaceae**

  b. Ovules many per cell; flowers bisexual
LXXXIX. Theaceae (p. 29)

13a. Stamens as many as the petals and on the same radii as them   14

  b. Stamens more or fewer than the petals, if of the same number then not on the same radii as them   20

14a. Tropical trees with milky sap and evergreen leaves
Sapotaceae

  b. Tropical or temperate trees, shrubs, herbs or climbers with watery sap and deciduous leaves   15

15a. Placentation axile   16

  b. Placentation basal or free-central   17

16a. Climbers with tendrils; stamens free   **Vitaceae**

  b. Erect shrubs without tendrils; stamens with their filaments united below   **Leeaceae**

17a. Trees or shrubs; fruit a berry or drupe   18

  b. Herbs (occasionally woody at the extreme base); fruit a capsule   19

18a. Leaves with translucent glands; anthers opening towards the centre of the flower; staminodes absent   **Myrsinaceae**

  b. Leaves without translucent glands; anthers opening towards the outside of the flower; staminodes 5
Theophrastaceae

19a. Sepals 2, free   **LII. Portulacaceae** III:171

  b. Sepals more than 4, united   **Primulaceae**

20a. Flower compressed, with 2 planes of symmetry; stamens in 2 bundles   **XCVI. Papaveraceae** (p. 103)

  b. Combination of characters not as above   21

21a. Leaves bipinnate or replaced by phyllodes; carpel 1, fruit a legume   **CXIII. Leguminosae** (p. 463)

  b. Combination of characters not as above   22

22a. Anthers opening by pores   23

  b. Anthers opening by longitudinal slits, or pollen in masses (pollinia)   24

23a. Stamens free from the corolla-tube, often twice as many as petals   **Ericaceae**

  b. Stamens attached to the corolla-tube, as many as petals
Solanaceae

24a. Leaves alternate or all basal; carpels never 2 and almost free with a single, terminal style   25

  b. Leaves opposite or whorled, alternate only when carpels 2 and almost free with a single, terminal style   41

25a. Plant woody, leaves usually evergreen; stigma not stalked, borne directly on top of the ovary   **Aquifoliaceae**

  b. Combination of characters not as above   26

26a. Prostrate herbs with milky sap and stamens free from the corolla-tube   **Campanulaceae**

  b. Combination of characters not as above   27

27a. Ovary 5-celled   28

  b. Ovary 2-, 3- or 4-celled   30

28a. Placentation parietal; soft-wooded tree   **Caricaceae**

  b. Placentation axile; herbs   29

29a. Leaves fleshy; anthers 2-celled; fruit often deeply lobed, schizocarpic   **Nolanaceae**

  b. Leaves leathery; anthers 1-celled; fruit a capsule or berry
Epacridaceae

30a. Ovary 3-celled   31

  b. Ovary 2- or rarely 4-celled   32

31a. Dwarf, evergreen shrublets; 5 staminodes usually present; some petals overlapping on both sides in bud
Diapensiaceae

  b. Herbs or climbers with tendrils; staminodes absent; petals each overlapping 1 other and overlapped by 1 other in bud   **Polemoniaceae**

32a. Stamens with the filaments united into a tube; flowers in heads; stigma surrounded by a sheath   **Brunoniaceae**

  b. Combination of characters not as above   33

33a. Flowers in spirally coiled cymes or the calyx with appendages between the lobes; style terminal or arising from between the lobes of the ovary   34

  b. Flowers not in spirally coiled cymes, calyx without appendages; style terminal   35

34a. Style terminal; fruit a capsule, usually many-seeded
Hydrophyllaceae

  b. Style arising from the depression between the 4 lobes of the ovary; fruit of up to 4 nutlets, more rarely fruit a 1–4-seeded drupe   **Boraginaceae**

35a. Placentation parietal   36

  b. Placentation axile   37

36a. Corolla-lobes edge-to-edge in bud; leaves either of 3 leaflets or simple, cordate or peltate, hairless; aquatic or marsh plants   **Menyanthaceae**

  b. Corolla-lobes overlapping in bud; leaves never as above; plants not aquatic, not occurring in marshes
Gesneriaceae

37a. Ovules 1 or 2 in each cell of the ovary   38

  b. Ovules 3-many in each cell of the ovary   40

38a. Arching shrub with small purple flowers in clusters on last year's wood   **Buddleiaceae**

  b. Combination of characters not as above   39

39a. Sepals free; corolla-lobes each overlapping 1 other and overlapped by 1 other, and infolded in bud; twiners, herbs or dwarf shrubs **Convolvulaceae**

b. Sepals united; corolla-lobes not as above in bud; trees or shrubs **Boraginaceae**

40a. Corolla-lobes folded, edge-to-edge or overlapping 1 other and overlapped by 1 other in bud; septum of ovary oblique **Solanaceae**

b. Corolla-lobes variously overlapping, but not as above in bud; septum of ovary horizontal **Scrophulariaceae**

41a. Trailing, heather-like shrublet **Ericaceae**

b. Plant not as above 42

42a. Milky sap usually present; fruit usually of 2 almost free 'follicles' and seeds with silky appendages 43

b. Milky sap absent; fruit a capsule or fleshy; seeds without silky appendages 44

43a. Pollen granular; corona absent; corolla-lobes each overlapping 1 other and overlapped by 1 other in bud **Apocynaceae**

b. Pollen usually in pollinia; corona usually present; corolla-lobes as above or edge-to-edge in bud **Asclepiadaceae**

44a. Herbs; flowers in spirally coiled cymes **Hydrophyllaceae**

b. Herbs or shrubs; flowers not in spirally coiled cymes 45

45a. Placentation parietal; carpels 2 46

b. Placentation axile; carpels 2, 3 or 5 47

46a. Leaves compound; epicalyx present **Hydrophyllaceae**

b. Leaves simple; epicalyx absent **Gentianaceae**

47a. Stamens fewer than corolla-lobes **Verbenaceae**

b. Stamens as many as corolla-lobes 48

48a. Carpels 5; shrubs with leaves with spiny margins **Desfontainiaceae**

b. Carpels 2 or 3; herbs or shrubs, leaves without spiny margins 49

49a. Leaves without stipules; carpels 3; corolla-lobes each overlapping 1 other and overlapped by 1 other in bud; plants herbaceous **Polemoniaceae**

b. Leaves with stipules (often reduced to a ridge between the leaf-bases); carpels usually 2; corolla-lobes variously overlapping or edge-to-edge in bud; plants usually woody 50

50a. Corolla usually 5-lobed; stellate and/or glandular hairs absent **Loganiaceae**

b. Corolla 4-lobed; stellate and/or glandular hairs present **Buddleiaceae**

Group L

1a. Stamens more numerous than the corolla-lobes, or anthers opening by pores 2

b. Stamens as many as corolla-lobes or fewer, anthers not opening by pores 6

2a. Anthers opening by pores; leaves undivided; ovary of 2 or more united carpels 3

b. Anthers opening by slits; leaves dissected or compound; ovary of 1 carpel 5

3a. The 2 lateral sepals petal-like; filaments united **Polygalaceae**

b. No sepals petal-like; filaments free 4

4a. Shrubs with alternate or apparently whorled leaves; stamens 5–25 **Ericaceae**

b. Herbs with opposite leaves; stamens 5 **Gentianaceae**

5a. Leaves pinnate or of 3 leaflets; perianth not spurred **CXIII. Leguminosae** (p. 463)

b. Leaves laciniate; perianth spurred **LXXIII. Ranunculaceae** III:325

6a. Stamens as many as corolla-lobes; bilateral symmetry weak 7

b. Stamens fewer than corolla-lobes; bilateral symmetry pronounced 12

7a. Stamens on the same radii as the petals; placentation free-central **Primulaceae**

b. Stamens on different radii from the petals; placentation axile 8

8a. Leaves of 3 leaflets, with translucent, aromatic glands; stamens 5, the 2 upper fertile, the 3 lower sterile **Rutaceae**

b. Combination of characters not as above 9

9a. Ovary of 3 carpels; ovules many **Polemoniaceae**

b. Ovary of 2 carpels; ovules 4 or many 10

10a. Flowers in spirally coiled cymes; fruit of up to 4 one-seeded nutlets **Boraginaceae**

b. Flowers not in spirally coiled cymes; fruit a many-seeded capsule 11

11a. Corolla-lobes each overlapping 1 other and overlapped by 1 other in bud; stamens 5, equal; leaves opposite; climber **Loganiaceae**

b. Corolla-lobes overlapping in bud, but not as above; stamens 4, or 5 and unequal; leaves usually alternate **Scrophulariaceae**

12a. Placentation axile; ovules 4 or many 13

b. Placentation parietal, free-central, basal or apical; ovules many or 1 or 2 20

13a. Ovules numerous but not in vertical rows in each cell 14

b. Ovules 4 or more numerous but then in vertical rows in each cell 16

14a. Seeds winged; mainly trees, shrubs or climbers with opposite, pinnate, palmate or rarely simple leaves **Bignoniaceae**

b. Seeds usually wingless; mainly herbs or shrubs with simple leaves 15

15a. Corolla-lobes variously overlapping in bud; septum of ovary horizontal; leaves opposite or alternate **Scrophulariaceae**

b. Corolla-lobes usually folded, edge-to-edge or overlapping 1 other and overlapped by 1 other in bud; septum oblique; leaves alternate **Solanaceae**

16a. Leaves all alternate, usually with blackish, resinous glands; plants woody **Myoporaceae**

b. At least the lower leaves opposite or whorled, not glandular; plant herbaceous or woody 17

17a. Fruit a capsule; ovules 4–many, usually in vertical rows in each cell 18

b. Fruit not a capsule; ovules 4, side by side 19

18a. Leaves all opposite, often prominently marked with
cystoliths; flower-stalks without swollen glands at the base;
capsule opening elastically, seeds usually on hooked stalks
Acanthaceae

b. Upper leaves alternate, cystoliths absent; flower-stalks with
swollen glands at the base; capsule not elastic, seeds not
on hooked stalks   **Pedaliaceae**

19a. Style arising from the depression between the 4 lobes of
the ovary, or, if terminal then corolla with a reduced upper
lip; fruit usually of 4 one-seeded nutlets; calyx and corolla
often 2-lipped   **Labiatae**

b. Style terminal; corolla with well-developed upper lip; fruit
usually a berry or drupe; calyx often more or less radially
symmetric   **Verbenaceae**

20a. Ovules 4–many; fruit a capsule, rarely a berry or drupe
21

b. Ovules 1 or 2; fruit indehiscent, often dispersed in the
persistent calyx   27

21a. Ovules 4, side by side   **Verbenaceae**

b. Ovules many   22

22a. Placentation free-central; corolla spurred
Lentibulariaceae

b. Placentation parietal or apical; corolla not spurred, rarely
swollen at the base   23

23a. Leaves scale-like, never green; root parasites   24

b. Leaves green, expanded; free-living plants   25

24a. Placentas 4; calyx laterally 2-lipped   **Orobanchaceae**

b. Placentas 2; calyx 4-lobed   **Scrophulariaceae**

25a. Seeds winged; mainly climbers with opposite, pinnately
divided leaves   **Bignoniaceae**

b. Combination of characters not as above   26

26a. Capsule with a long beak separating into 2 curved horns;
plant sticky-velvety   **Martyniaceae**

b. Capsule without beak or horns; plants velvety or variously
hairy or hairless   **Gesneriaceae**

27a. Flowers in heads surrounded by an involucre of bracts;
ovule 1   **Globulariaceae**

b. Flowers not in heads as above; ovary 2-celled, ovules 1 in
each cell, often only 1 maturing   **Scrophulariaceae**

LXXXIII. DILLENIACEAE

Trees, shrubs, woody climbers or perennial herbs, rarely woody at base. Leaves spirally arranged (rarely opposite), simple (rarely pinnatifid or 3-lobed), entire or toothed; stipules absent or united to leaf-stalk. Flowers solitary or in variously cyme- or raceme-like inflorescences, usually bisexual. Sepals 3–20, usually 5, free, spiral, overlapping, persistent. Petals 2–5, free, overlapping, often crumpled in bud, white or yellow to orange, occasionally pink, falling, or sometimes absent. Stamens 3 to numerous, free or in 3–15 bundles with filaments free or partly united, all fertile or some staminodial; anthers opening by slits or rarely pores. Ovary superior, of 1–20 carpels in 1–2 rows, free or sometimes partly united with each other or with receptacular apex, with 1 to numerous ovules in each carpel. Fruit dry and splitting or sometimes not splitting and enclosed in a more or less fleshy calyx; seeds with or without a more or less fringed or lobed aril.

A family of 10–12 genera and about 350 species occurring throughout the tropics and in temperate Australia, but with only one genus in continental Africa.

Literature: Gilg, E. & Werdermann, E., Dilleniaceae in Engler, A. & Prantl, K., *Die natürlichen Pflanzenfamilien*, edn 2, **21**: 7–36 (1925).

1a. Trees or shrubs; leaves closely feather-veined; carpels 4–20, partly united; flowers (cultivated species) 10–20 cm across   **1. Dillenia**
  b. Shrubs or woody-based herbs, sometimes climbing; leaves with lateral veins few or absent; carpels 1–10, free; flowers 6 cm or less across   **2. Hibbertia**

1. DILLENIA Linnaeus
N.K.B. Robson

Trees or sometimes shrubs, usually evergreen, with leaf-scars half to completely encircling stem. Leaves simple, feather-veined with 4–90 pairs of lateral veins, margin entire to wavy or toothed; leaf-stalk with or without narrow wings; leaves of saplings and young trees larger than and sometimes different from those of older branches. Flowers solitary and terminal or axillary, or in axillary raceme-like cymes, racemes or clusters, bisexual. Sepals 4–18, enlarging in fruit. Petals 5 or absent, white or yellow, or occasionally pink. Stamens numerous, free, sometimes outer sterile or with group of free or more or less coherent staminodes; anthers opening by a long terminal pore. Ovary of 4–20 carpels, partly coherent and adherent to conical apex of receptacle; ovules 4–80 per carpel. Young fruit enclosed by enlarged fleshy sepals (pseudocarp); mature carpels spreading, star-like and splitting or remaining indehiscent within sepals. Seeds with or without arils.

A genus of 60 species in Madagascar and from India to Australia and Fiji. Almost all are lowland tropical and therefore greenhouse subjects in Europe.

Literature: Hoogland, R.D., A revision of the genus Dillenia. *Blumea* **7**: 1–145 (1952).

1a. Leaf-stalk unwinged; leaves with 20–70 pairs of lateral veins; flowers solitary, 15–20 cm across
    1. indica
  b. Leaf-stalk winged; leaves with 8–18 pairs of lateral veins; flowers 1–3, 10–12.5 cm across
    2. philippinensis

1. D. indica Linnaeus (*D. speciosa* Thunberg). Illustration: Botanical Magazine, 5016 (1857); Beddome, The Flora Sylvatica for Southern India **1**: pl. 103 (1869).

Tree to 30 m, bark peeling in small hard scales; branches with adpressed silky-hairs when young, leaf-scars forming a half ring. Leaf-stalks 2.5–7.5 cm, not winged; blade mostly 15–30 × 6–12 cm, oblong to elliptic-oblong, apex slightly acuminate to obtuse, base wedge-shaped to rounded, margin slightly to distinctly toothed, lateral veins 20–70 on each side. Flower solitary, terminal, 15–20 cm across. Sepals 5. Petals 5, white with green veins, 7–9 × 5–6.5 cm. Stamens in 2 groups, outer *c.* 550, inner *c.* 25, all white. Carpels 14–20. Pseudocarp indehiscent, yellowish-green, spherical, 8–10 cm across. Seeds without arils. *India, Burma & S China (Yunnan) to Sumatra, Java & Borneo.* G2.

Although frequently planted in warmer regions as an ornamental tree, *D. indica* is confined in Europe to greenhouses, where it apparently rarely flowers.

2. D. philippinensis Rolfe. Illustration: Brown, Products of Philippine forests **2**: f. 62, 63 (1921).

Tree to 17 m, bark peeling in thin plates; branches sometimes spreading-hairy when young, leaf-scars forming complete ring. Leaf-stalks 3.5–5 cm, winged; blade mostly 10–16 × 7–12 cm, ovate or elliptic to oblong or lanceolate, apex slightly acuminate to rounded, base wedge-shaped to rounded, margin slightly toothed or wavy, lateral veins 8–18 on each side. Flowers 1–3, terminal, 10–12.5 cm across. Sepals 5. Petals 5, white, *c.* 6 × 4 cm. Stamens in 2 groups, outer *c.* 230, yellowish white, inner *c.* 40, purplish. Carpels 10–12. Pseudocarp indehiscent, 4–5 × 5–6 cm. Seeds with arils. *Philippines.* G2.

2. HIBBERTIA Andrews
N.K.B. Robson

Trees or shrubs, often straggling or twining, sometimes heather-like (ericoid), evergreen, with leaf-scars encircling less than half of stem. Leaves simple, obscurely feather-veined with up to 20 pairs of lateral veins or 1-veined, margin entire or wavy to toothed; leaf-stalk unwinged or absent, very rarely leaf-like (phylloclade). Flowers solitary or in one-sided, raceme-like cymes, terminal and often apparently also axillary, bisexual. Sepals 5, not enlarging in fruit. Petals 5, sometimes 3 or 4, yellow or rarely white or pink. Stamens 3 to numerous, free or in 2–5 bundles,

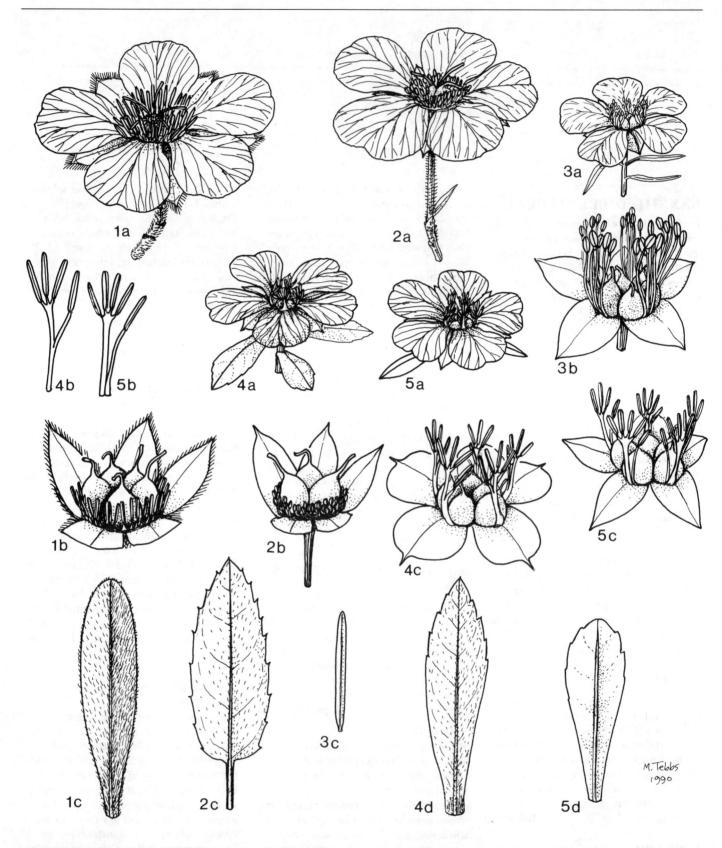

Figure 1. Diagnostic details of *Hibbertia* species. 1a, 1b, Flower and floral parts of *H. scandens*. 1c, Leaf of *H. scandens*. 2a, 2b, Flower and floral parts of *H. dentata*. 2c, Leaf of *H. dentata*. 3a, 3b, Flower and floral parts of *H. procumbens*. 3c, Leaf of *H. procumbens*. 4a, 4b, 4c, Flower and floral parts of *H. tetrandra*. 4d, Leaf of *H. tetrandra*. 5a, 5b, 5c, Flower and floral parts of *H. cuneiformis*. 5d, Leaf of *H. cuneiformis*.

each with 2–6 stamens, sometimes with staminodes or all confined to one side of ovary; anthers opening by a slit or pore. Ovary of 1–5 (rarely to 10 or more) carpels, free or slightly coherent basally; ovules 1–15 per carpel. Fruit of splitting follicles. Seeds with arils, 1–4 per follicle.

A genus of 122 species (when treated, as here, to include *Candollea* Labillardière) in Madagascar, New Guinea, Fiji, New Caledonia and Australia. In northern Europe the Australian species are usually plants for the cool greenhouse.

1a. Stamens and staminodes all free except sometimes at the very base; carpels each with 4–8 ovules; stems trailing to climbing or prostrate   2
  b. Stamens (all fertile) in 5 bundles with filaments of all or most in each bundle completely united; carpels with 2 or 3 ovules; stems more or less erect   4
2a. Leaves 1.5 mm wide, linear; stems procumbent to prostrate, much branched; flowers to 2.5 cm across   **3. procumbens**
  b. Leaves 1.5–4 cm wide, elliptic or oblong to obovate; stems trailing or twining; flowers 3–5 cm across   3
3a. Carpels 5–8; sepals green, 1.5–2.5 cm; leaves entire or almost so   **1. scandens**
  b. Carpels 3; sepals magenta, 0.8–1.2 cm; leaves with toothed or sinuous margins   **2. dentata**
4a. Petals 1.8–2.2 cm, 1.2–1.4 times length of sepals; leaves oblanceolate to spathulate, margin coarsely to obscurely toothed   **4. tetrandra**
  b. Petals 1–1.5 cm, shorter than or up to 1.2 times length of sepals; leaves obovate or obtriangular to oblong, margin entire or with few teeth near apex   **5. cuneiformis**

1. H. scandens (Willdenow) Dryander (*H. volubilis* (Ventenat) Andrews; *Dillenia scandens* Willdenow; *D. volubilis* Ventenat; *D. speciosa* misapplied). Figure 1(1), p. 16. Illustration: Botanical Magazine, 449 (1799); Nees von Esenbeck & Sinning, Sammlung schnöblühender Gewächse, pl. 89 (1831).

Shrub, stems woody, trailing or twining and climbing to *c.* 3 m; young branches silky-hairy. Leaf-stalks *c.* 1.5 cm; blade flat or almost so, 4–9 × 1.5–3.5 cm, elliptic or oblong-elliptic to obovate, almost acuminate to rounded, base narrowing, margin entire or very shallowly toothed towards apex. Flowers solitary, terminal,

almost stalkless; upper leaves bract-like. Sepals green, 1.5–2.5 cm, ovate, acuminate, silky-hairy outside. Petals 5, yellow, 1.5–2.5 cm, obovate or obovate-oblong, entire. Stamens free, very numerous, all fertile. Carpels 5–8, hairless, with 6–8 ovules. *Australia (Queensland, New South Wales)*. H5. Summer.

2. H. dentata R. Brown. Figure 1(2), p. 16. Illustration: Edwards's Botanical Register **4**: pl. 282 (1818); Botanical Magazine, 2338 (1822); The Garden **34**: pl. 659 (1888).

Shrub, stems woody at the base, more slender than those of *H. scandens*, trailing or twining; young branches downy or hairless. Leaves with stalks 0.5–1 cm; blade flat, 5–9 × 2–4 cm, elliptic-oblong to oblong, acute or obtuse, base rounded to almost cordate, margin toothed or sinuous, sometimes downy when young. Flowers solitary, terminal but often appearing lateral, on stalks 0.5–2 cm; upper leaves reduced. Sepals magenta, 0.8–1.2 cm, ovate, acuminate to obtuse, downy or silky-hairy beneath. Petals 5, yellow to orange, 1.5–2.5 cm, obovate to oblong, mucronate to shallowly notched. Stamens free, very numerous, the outer staminodal. Carpels 3, hairless, with 4–8 ovules. *Australia (New South Wales, Victoria)*. H5. Summer.

3. H. procumbens (Labillardière) de Candolle (*H. angustifolia* Salisbury; *Dillenia procumbens* Labillardière). Figure 1(3), p. 16. Illustration: Labillardière, Plantae Novae Hollandiae, **2**: pl. 156 (1806).

Shrub, stems woody, prostrate, spreading and much branched, 10–30 cm; young branches downy. Leaves almost stalkless; blade flat or slightly grooved above, 5–15 × 0.5–1.5 mm, linear, almost acute to rounded, base narrowly wedge-shaped or parallel-sided, margin entire, hairless or with scattered soft hairs. Flowers solitary, terminal and terminating short lateral branches, stalkless; upper leaves bract-like. Sepals *c.* 1 cm, ovate to oblong, mucronate, hairless. Petals 5, yellow, 0.8–1.5 cm, obovate, very shallowly notched. Stamens in 4 or 5 groups each of 5 stamens, all fertile, with filaments free. Carpels 4 or 5, hairless, with 4–6 ovules. *Australia (Victoria, Tasmania)*. H5. Summer.

4. H. tetrandra (Lindley) Gilg (*Candollea tetrandra* Lindley; *C. latifolia* Steudel). Figure 1(4), p. 16. Illustration: Edwards's Botanical Register **29**: pl. 50 (1843).

Shrub, stems woody, erect, to 2.1 m, angular, shortly downy. Leaves stalkless, flat, 1.8–6.5 × 0.6–0.8 cm, oblanceolate to almost spathulate, obtuse to acute, shortly acuminate, base narrowed and encircling stem, margin coarsely to obscurely toothed, hairless. Flowers solitary, terminal, stalkless; upper leaves transitional to sepals. Sepals 1–1.5 cm, ovate-oblong, mucronate to shortly acuminate, hairless. Petals 5, yellow, 1.8–2.2 cm, triangular-obovate, more or less deeply notched. Stamens in 5 bundles, each of 3–5 stamens, all fertile, one with filament free or partly united, the others with filaments completely united. Carpels 5, hairless, with 2 or 3 ovules. *Australia (Western Australia)*. H5. Late spring.

This species may not be distinct from *H. cuneiformis*.

5. H. cuneiformis (Labillardière) Smith (*H. obcuneata* Salisbury; *Candollea cuneiformis* Labillardière). Figure 1(5), p. 16. Illustration: Botanical Magazine, 2711 (1827); Fairall, West Australian plants in cultivation, 163 (1970).

Shrub, stems woody, erect but with tendency to trail or twine, to 2 m, shortly downy. Leaves stalkless, flat, 2–3 × 0.5–1.1 cm, obovate or obtriangular to oblong, obtuse to acute, base narrowed and encircling stem, margin entire or with few teeth near apex, hairless. Flowers solitary, terminal, stalkless; upper leaves transitional to sepals. Sepals 0.8–1.2 cm, ovate to oblong, acuminate, hairless. Petals 5, yellow, 1–1.5 cm, triangular-obovate, notched. Stamens and carpels as in *H. tetrandra*. *Australia (Western Australia)*. H5. Late spring.

LXXIV. PAEONIACEAE

Perennial herbs with fleshy roots, or soft-wooded shrubs. Leaves alternate, compound, at least the lower ones twice divided into 3 or bipinnately divided into pinnately veined leaflets, segments and lobes. Flowers bisexual, radially symmetric, large, solitary and terminal or few from leaf axils. Floral bracts 1–12, leafy, immediately below calyx; sepals 5, free, persistent, overlapping, unequal. Petals 5–9, or more (in garden variants). Stamens numerous; anthers opening by longitudinal slits. Carpels 1–8, free, on a fleshy disc which sometimes extends upwards into fleshy lobes or a sheath; ovules numerous, lateral, in 2 rows. Fruit of 1–5 diverging follicles, each with several seeds, opening by a slit on the upper side. Seeds large, the fertile ones black, infertile red.

A family of one genus or, if *Glaucidium* (III: 326) is included, as by Melville (1983), two genera. This family has often been included in Ranunculaceae. The roots of herbaceous paeonies may be tapering and carrot-like as in *P. mascula*, or from a slender string-like root swollen into a tuber, as in *P. officinalis* and *P. peregrina*.

1. PAEONIA Linnaeus
W.T. Stearn
Description as for the family.

A genus of about 30 species in north-west Africa, southern Europe, northern Asia (from Caucasus to China and Japan) and western North America. Although some species stand out from others by distinctive characters, e.g., *P. tenuifolia* by its intricately divided leaves with linear segments, *P. lutea* by its tall shrubby habit and yellow petals, *P. emodi* by its single carpel, nevertheless in the groups typified by *P. mascula* and *P. officinalis* the taxa intergrade or are distinguishable on leaf characters which, though evident in the living state to the eye when growing, are almost impossible to describe diagnostically and intelligibly; hence reference must often be made to illustrations. The basic leaf pattern of the lower leaves is with 3 sets of 3 broad undivided leaflets, 9 in all, which become cleft into narrower divisions and lobes in a threefold or pinnate manner (Figure 2) reaching final intricacy in the finely divided leaf with very narrow ultimate segments of *P. tenuifolia* (Figure 3).

Literature: Stern, F.C., *A study of the Genus Paeonia* (1946); Kemularia-Natadze, L.M., Kavkazskie prestabiteli roda Paeonia, *Trudy Tbilisskogo Botanicheskogo Instituta* **21**: 1–51 (1961); Melville, R., The affinity of Paeonia, *Kew Bulletin* **38**: 87–105 (1983); Stearn, W.T. & Davis, P.H., *Peonies of Greece* (1984); Smith, G. *World of flowers* part 2 (1984). Haw, S.G. & Lauener, L.A., A review of the infraspecific taxa of Paeonia suffruticosa Andrews, *Edinburgh Journal of Botany* **47**(3): 273–81 (1990).

1a. Plants with persistent, erect, or woody, ascending stems 30–200 cm   2
  b. Plants herbaceous, without persistent woody stems, dying back in autumn to ground buds   5
2a. Lateral leaflets with rounded or wedge-shaped base; carpels at first wholly enclosed in a sheath; petals often 5–9 cm wide
                          1. suffruticosa
  b. Lateral leaflets with narrowly wedge-shaped base; carpels with fleshy non-enclosing lobes at base; petals 2–3 cm wide   3
3a. Bracts and sepals together 13–17; petals dark red   **2. delavayi**
  b. Bracts and sepals together 5–8; petals yellow, white or red   4
4a. Plant to 2.5 m, branched from a single rootstock; ultimate lobes of leaf divisions mostly 1.5–3 cm wide; petals yellow   **3. lutea**
  b. Plant to 50 cm, usually branched, spreading underground into wide masses; ultimate lobes of leaf divisions mostly 5–15 mm wide; petals red, yellow or white
                          4. potaninii
5a. Leaves with very numerous (to 60) linear segments 1–2 mm wide, the uppermost leaf almost immediately below the flower   **17. tenuifolia**
  b. Leaves with 40 or fewer segments, 5 mm or more wide   6
6a. Carpel solitary; petals white
                          13. emodi
  b. Carpels 2–8 (rarely 1); petals various   7
7a. Lower leaves divided to axis into 9–23 distinct leaflets   8
  b. Lower leaves divided into numerous divisions (25 or more), usually themselves lobed   20
8a. Flower-stem with 2 or more flowers
                          12. lactiflora
  b. Flower-stem always with one flower
                                             9
9a. Petals yellow or yellowish   10
  b. Petals red, purplish, pink or white
                                            11
10a. Leaflets covered beneath with inconspicuous very short curved or erect hairs; filaments yellowish
                    5. mlokosewitschii
  b. Leaflets covered beneath with numerous longer hairs; filaments red   **6. wittmanniana**
11a. Carpels hairless   12
  b. Carpels hairy   15
12a. Terminal leaflet obovate to broadly obovate   **9. obovata**
  b. Terminal leaflet narrowly or broadly elliptic to ovate or broadly ovate   13
13a. Carpels 5–8, purple; leaflets purple or purplish beneath
                       7. cambessedesii
  b. Carpels 2–4, green; leaflets green beneath but sometimes with purple or red veins   14
14a. Carpels 2, tapered gradually above into the stigma   **11. coriacea**
  b. Carpels 3–5, with stalkless stigma
                    8. mascula subsp. **russii**
15a. Lower leaves with 17–30 narrowly elliptic, hairless leaflets
                          10. broteroi
  b. Lower leaves with 9–16 leaflets   16
16a. Petals white
              8. mascula subsp. **hellenica**
  b. Petals purple or rose   17
17a. Leaflets broadly ovate or almost spherical, the apex rounded, the margin strongly wavy
              8. mascula subsp. **triternata**
  b. Leaflets ovate to broadly ovate or narrowly elliptic to broadly elliptic, the apex acute, margins flat   18
18a. Stem hairy; leaflets 12–16 on lower leaves, hairy beneath
              8. mascula subsp. **arietina**
  b. Stem hairless; leaflets 9–12 on lower leaves, hairless or hairy beneath   19
19a. Leaflets green beneath when young, hairless or with few hairs; plant 35–60 cm
              8. mascula subsp. **mascula**
  b. Leaflets purplish when young and with long hairs beneath; plant 20–40 cm
              8. mascula subsp. **russii**
20a. Segments of lower leaves 1.5–5 cm wide, without minute bristles on veins above, not toothed at apex
                          15. officinalis
  b. Segments of lower leaves narrower, usually less than 1.5 cm wide (species **14**, **19**, **21**) or without minute bristles along veins above (species **16**)   21
21a. Leaves densely hairy beneath with profuse, very short hairs
                          16. humilis
  b. Leaves hairless or sparsely hairy beneath   22
22a. Leaves without minute bristles along veins above; segments narrowly lanceolate, usually less than 1 cm wide; carpels less than 1.5 cm, almost or quite hairless, purple   **18. × smouthii**
  b. Leaves with minute bristles along veins above; other characters not as above   23
23a. Leaf-segments coarsely toothed at apex; flowers bowl-shaped with ascending deep red, very concave

petals; innermost sepal without apical appendage　　**14. peregrina**

b. Leaf-segments deeply lobed; flowers saucer-shaped with spreading, flatter petals; innermost sepal with very short, conical projection　　24

24a. Stem 1–3 flowered; petals purple, pink or white; carpels greenish　　**21. veitchii**

b. Stem always single-flowered; petals deep red; carpels purplish　　25

25a. Sepals mucronate, pollen fertile　　**19. anomala**

b. Sepals rounded, pollen defective　　**20. × hybrida**

1. P. suffruticosa Haworth (*P. papaveracea* Andrews; *P. moutan* Sims; *P. arborea* Donn). Figure 3(1), p. 22. Illustration: Botanist's Repository **6**: 373 (1804); Botanical Magazine, 1154 (1808), 2175 (1820); Haworth-Booth, The Moutan or Tree Peony, tt. 1–11 (1963); The Garden **112**: 27 (1987).

Deciduous, soft-wooded, well-branched shrub to 2 m, with ascending hairless branches. Leaves doubly pinnate, with lower leaflets elliptic or ovate, some entire, mostly cut from ¼ to ½ into 3 lobes, sparsely hairy beneath. Flowers 12–30 cm across, saucer-shaped; petals white, pink, red, purple or lilac with maroon blotches at the base. Filaments purplish at base, white above. Carpels 5, hairy, enclosed at first in a sheathing outgrowth from the disc. *China*. H3. Spring–summer.

Subsp. **suffruticosa** has 9 (rarely more) ovate to broadly ovate leaflets, more or less hairless leaf-stalks, single or double, rarely blotched flowers, in the full range of colour described above, and a purple carpel sheath.

Subsp. **spontanea** Rehder differs from subsp. *suffruticosa* in its broader, sometimes almost circular leaflets, shortly hairy leaf-stalks, and smaller, unblotched, single, pink to purple (rarely white) flowers.

Subsp. **rockii** Haw & Lauener, from Gansu in China has about 19–31 leaflets, white petals with a basal blotch, and a whitish carpel sheath; also known as 'Rock's Variety' or 'Joseph Rock'.

Haw & Lauener (1990) wrongly thought the author *P. suffruticosa* to be Andrews.

Hybrids between *P. lutea* var. *lutea* and *P. suffruticosa* raised by V. Lemoine et Fils at Nancy and resembling *P. suffruticosa* but with yellow, red-tinged petals come under the collective name **P. × lemoinei** Rehder.

2. P. delavayi Franchet. Illustration: Gardeners' Chronicle, **53**: 405 (1913); Stern, Study of Paeonia, 44 (1946).

Deciduous, soft-wooded shrub to 2 m, with ascending hairless branches. Leaves doubly pinnate with lower leaflets entire, lobed or toothed, the terminal leaflet deeply 3-lobed, all acute or acuminate and hairless. Bracts and sepals together 13–17, with long leafy bracts immediately below the calyx. Flowers 5–9 cm across, cup-shaped; petals incurved, dark red. Filaments dark red. Carpels often 5, hairless. *W China*. H3. Summer.

3. P. lutea Franchet (*P. delavayi* var. *lutea* (Franchet) Finet & Gagnepain). Illustration: Botanical Magazine, 7788 (1901), n.s., 209 (1953); Flora and Sylva **1**: 231 (1903); Journal of the Royal Horticultural Society **72**: 394 (1947); Morley & Everard, Wild flowers of the world, t. 90c (1970).

Deciduous soft-wooded shrub to 2.5 m, with ascending, hairless branches, leafless below current years growth. Leaves twice divided into 3, with leaflets hairless, deeply cut into 5–12 acute lobes. Bracts and sepals together 5–10. Flowers 5–12 cm across, petals spreading, sulphur yellow. Filaments yellow to reddish brown. Carpels 2–4, hairless. *W China (Xizang)*. H3. Spring.

Var. **lutea**. Figure 3(6) p. 22. Flowers *c*. 5 cm across, cup-shaped, hidden amongst the leaves and 3–4 carpels. *W China*.

Var. **ludlowii** F.C. Stern & G. Taylor, differs in its saucer-shaped flowers to 9 cm across, held on long erect stalks above the leaves and 1–3 carpels. *SE Xizang*.

P. delavayi and *P. lutea* hybridise in gardens and are said to intergrade in western China.

4. P. potaninii Komarov (*P. delavayi* var. *angustifolia* Rehder & Wilson). Illustration: Gardeners' Chronicle, **53**: 403 (1913); New Flora & Silva **3**: 279 (1931); National Horticultural Magazine **13**: 220 (1934); Stern, Study of Paeonia, 48, t. facing 50 (1946).

Deciduous shrub 30–80 cm, with erect unbranched or sparingly branched shoots, spreading underground into wide low masses. Leaves doubly pinnate with leaflets hairless and deeply divided into 3–5 narrow lobes themselves often 3-toothed and acuminate. Bracts and sepals together 5–7. Flowers saucer-shaped, 5–6 cm across. Filaments red or greenish. Carpels

2 or 3, hairless. *W China*. H3. Early summer.

Var. **potaninii**. Figure 3(5), p. 22. Petals dark maroon red.

Forma **alba** (Bean) F.C. Stern. Petals white.

Var. **trollioides** (Stapf) F.C. Stern (*P. trollioides* Stapf). Petals yellow.

5. P. mlokosewitschii Lomakin. Figure 2(1), p. 20. Illustration: Botanical Magazine, 8173 (1908); Stern, Study of Paeonia, t. facing 54 (1946).

Stem hairless, 30–60 cm. Leaves mostly 3; lower leaves biternate with 9 broadly elliptic, ovate or obovate acute or obtuse leaflets sparsely hairy below, the hairs very short and curved. Flowers 8–12 cm across, bowl-shaped, petals concave, pale yellow. Filaments yellow. Carpels 2–4, usually 2, with densely felted hairs. *C Caucasus*. H3. Spring.

Hybridised with *P. mascula* subsp. *triternata*, this has produced the hybrid named **P. × chamaeleon** Troitzky.

6. P. wittmanniana Lindley. Illustration: Edwards's Botanical Register **32**: t. 9 (1846); Annales de la Société Royale d'Agriculture et de botanique de Gand, **2**: t. 64 (1846); Botanical Magazine, 6645 (1882), 9249 (1931); Stern, Study of Paeonia, 58, t. facing 59 (1946).

Stem hairless, to 1 m. Leaves mostly 3. Lower leaves twice divided into 3, with 9 broadly elliptic or broadly ovate acute leaflets hairy beneath, the hairs long, to 1.5 mm. Flowers 9–12 cm, bowl-shaped, petals concave, pale citron-yellow. Filaments red. Carpels 2–4, hairless or with densely felted hairs. *Caucasus*. H3. Spring.

Var. **wittmanniana** (*P. wittmanniana* var. *nudicarpa* Schipczinsky). Carpel hairless.

Var. **tomentosa** Lomakin (*P. tomentosa* (Lomakin) N. Busch; *P. wittmanniana* F.C. Stern). Carpels with densely felted hairs.

Hybrids between *P. wittmanniana* and cultivars of *P. lactiflora* raised by G. Arends, Ronsdorf, and V. Lemoine, Nancy, have received the collective name **P. × arendsii** Bergmans.

7. P. cambessedesii (Willkomm) Willkomm (*P. corallina* var. *cambessedesii* Willkomm). Illustration: Botanical Magazine, 8161 (1907); Stern, Study of Paeonia, t. facing 62 (1946).

Stem hairless, 30–45 cm. Leaves 3–6.

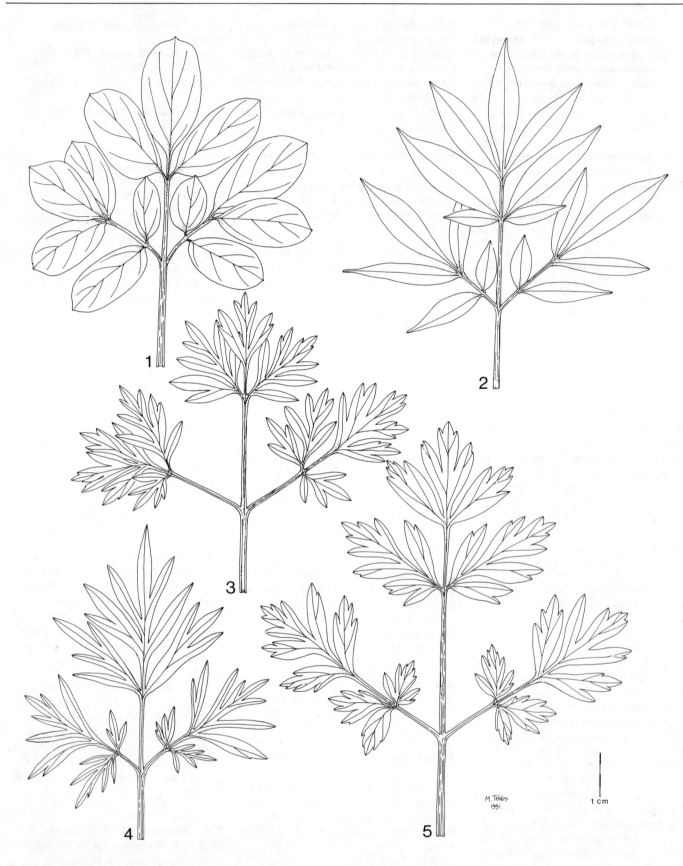

Figure 2. Leaves of *Paeonia* species. 1, *P. mlokosewitschii*.
2, *P. broteroi*. 3, *P. humilis* var. *humilis*. 4, *P. anamola* var.
anomola. 5, *P. peregrina*.

Lower leaves twice divided into 3, with 9 lanceolate, elliptic or ovate hairless acute leaflets, dark green above, purple or purplish beneath. Flowers 6–10 cm across, petals outspread slightly wavy, deep rose. Filaments red. Carpels 5–8, hairless, purple. *Balearic Isles (Majorca)*. H3. Late spring.

8. P. mascula (Linnaeus) Miller (*P. officinalis* var. *mascula* Linnaeus).

Stems hairless or sparsely hairy, 20–75 cm. Leaves 3–7. Lower leaves twice divided into 3, with ovate, elliptic or broadly obovate leaflets entire or sometimes with a few leaflets 2 or 3 lobed, hairless or hairy beneath. Flowers saucer-shaped, 7–13 cm across, petals outspread red, purplish, pink or white. Filaments red, purplish, pink or white. Carpels 2–5, with densely felted hairs. *Mediterranean Europe to SE Asia*. H3. Late Spring.

This collective species as currently accepted exhibits much diversity, being an assemblage of subspecies often and justifiably treated as species.

Subsp. **mascula** (*P. corallina* Retzius). Illustration: Sowerby & Smith, English Botany **22**: t. 1513 (1806); Stern, Study of Paeonia, 68, 69 (1946); Stearn & Davis, Peonies of Greece, t. facing 80 (1984). Stem 35–60 cm. Leaflets green and hairless or sparsely hairy beneath, acute. Petals purplish red. *S Europe, W Asia*.

Subsp. **arietina** (G. Anderson) Cullen & Heywood (*P. arietina* G. Anderson). Illustration: Botanical Register **10**: t. 819 (1824); National Horticultural Magazine **13**: 217 (1934); Stern, Study of Paeonia, 82 (1946). Stem 45–75 cm, sparsely hairy. Leaves green and hairy beneath, many leaflets partly split into 2 narrow acute lobes, making up to 15 lobes. Petals purplish red. *N Italy, N Balkan Peninsula, W Asia*.

Subsp. **hellenica** Tzanoudakis (*P. flavescens* C. Presl). Illustration: Stearn & Davis, Peonies of Greece 98, 99, 100, t. facing 100 (1984). Stem 30–60 cm, hairless. Leaflets green and hairless or sparsely hairy beneath, entire, acute or acuminate. Petals white. *Greece, Sicily*.

Subsp. **triternata** (Boissier) Stearn & Davis (*P. daurica* Andrews). Illustration: Andrews, Botanist's Repository **7**: t. 486 (1807); Botanical Magazine, 1441 (1812); National Horticultural Magazine **13**: 231 (1934); Stern, Study of Paeonia, t. facing 70 (1946); Stearn & Davis, Peonies of Greece, 109, 110 (1984). Stem hairless, 35–60 cm. Leaflets green and sparsely

hairy beneath, entire, with wavy margins and rounded tip. Petals purplish rose. *N Balkan Peninsula, Crimea, Turkey*.

Subsp. **russi** (Bivona) Cullen & Heywood (*P. russi* Bivona; *P. corallina* var. *pubescens* Moris; *P. mascula* var. *russi* (Bivona) Gürke; *P. revelieri* Jordan). Illustration: Jordan & Fourreau, Icones ad Floram Europae **2**: t. 322 (1903); Stern, Study of Paeonia, 64 (1946); Stearn & Davis, Paeonies of Greece, 37, 46, 59, 88, 90, t. facing 90 (1984). Stem hairless, 25–45 cm. Leaflets purplish (when young) and sparsely or densely hairy beneath, entire, acute. Petals pinkish mauve. *Mediterranean islands (Corsica, Sardinia, Sicily, Ionian Islands & adjacent Greek mainland)*.

P. bakeri Lynch. Illustration: Stern, Study of Paeonia, t. facing 85 (1946). Known only in gardens, resembling *P. mascula* subsp. *arietina* in appearance and hairiness but with leaflets broadest below the middle, i.e. ovate or broadly ovate, with rounded, almost truncate or broadly wedge-shaped base; *P. arietina* has leaflets broadest at the middle, i.e. narrowly to broadly elliptic, with wedge-shaped base.

9. P. obovata Maximowicz. Illustration: Gardeners' Chronicle, **57**: 290, f. 94 (1915); Botanical Magazine, 8867 (1916); National Horticultural Magazine **13**: 227 (1934).

Stem hairless, 40–60 cm. Leaves usually 3. Lower leaves twice divided into 3, with broadly elliptic lateral leaflets and a terminal usually obovate one, all acute or shortly acuminate, hairless above, sparsely or densely hairy beneath with long hairs. Flowers 7–10 cm across, petals white or rose-purple. Filaments greenish white or purple. Carpels 2–5, hairless. *E Asia*. H2. Early summer.

Var. **obovata**. Leaves sparsely hairy beneath. Flowers to 7 cm across; petals red. Filaments greenish white. *SE Siberia, Manchuria, E China, Sakhalin, Japan*.

Var. **japonica** Makino (*P. japonica* (Makino) Miyabe & Takeda). Leaves sparsely hairy beneath. Flowers *c.* 7 cm across; petals white. Filaments dark purple. *Japan*.

Var. **willmottiae** (Stapf) F.C. Stern (*P. willmottiae* Stapf). Leaves densely hairy beneath. Flowers to 10 cm across; petals white. Filaments purple. *C China*.

10. P. broteroi Boissier & Reuter (*P. lusitanica* Miller ?). Figure 2(2), p. 20. Illustration: Stern, Study of Paeonia, t. facing 86 (1946).

Stem hairless, 30–40 cm. Leaves 3–5.

Lower leaves twice divided into 3 with basically 9 leaflets usually cut deeply into 2 or 3 segments thus making up to 23 elliptic or broadly elliptic hairless acute segments or leaflets. Flowers 8–10 cm across, bowl-shaped, petals somewhat concave, rose. Filaments yellow. Carpels 2–4, with densely felted hairs. *Portugal, W & S Spain*. H3. Late spring–summer.

11. P. coriacea Boissier. Illustration: Stern, Study of Paeonia, 89 (1946).

Stem hairless, to 40 cm. Leaves 5–6. Lower leaves twice divided into 3 with 9 broadly elliptic, lanceolate or ovate, hairless acute leaflets, some almost divided in two. Flowers 7–15 cm across, petals outspread, rose. Filaments red. Carpels 2–4, hairless. *Spain, Morocco*. H2. Late spring.

12. P. lactiflora Pallas (*P. albiflora* Pallas; *P. cultorum* Bergmans). Illustration: Botanical Magazine, 1756, 1768 (1815), 2888 (1829); Edwards's Botanical Register **1**: t. 42 (1815), **6**: t. 485 (1820), **8**: t. 630 (1822), **17**: t. 1436 (1831).

Stem hairless, 50–90 cm, usually with 2 or more flowers. Leaves 5–7. Lower leaves twice divided into 3 with 9 entire or 2-lobed narrowly elliptic or lanceolate leaflets with sparse minute hairs along veins above, hairless or sparsely hairy beneath, margins rough with minute short stiff hairs. Flower 7–10 cm across, petals white or pink. Filaments yellow. Carpels 3–5, hairless or hairy. *E Siberia, inner Mongolia, Manchuria, China (Shanxi)*. H2. Summer.

This species has been cultivated in China for some 2500 years and since its introduction into European gardens late in the 18th century has produced yet more cultivars.

13. P. emodi Royle. Illustration: Botanical Magazine, 5719 (1868); Coventry, Wild flowers of Kashmir **1**: t. 10 (1923); Stern, Study of Paeonia, t. facing 94 (1946).

Stem hairless, 30–75 cm, with 2–4 flowers. Leaves 5–7. Lower leaves twice divided into 3, with *c.* 9 narrowly elliptic, acuminate segments, wedge-shaped and united by threes at bases, hairless except for minute hairs along main veins above. Flowers 8–12 cm across, petals outspread, white petals. Filaments yellow. Carpel 1 (very rarely 2), hairy (var. **emodi**) or hairless (var. **glabrata** J.D. Hooker & Thomson). *W Himalaya*. H3. Spring.

14. P. peregrina Miller (*P. decora* G. Anderson; *P. romanica* Brandza). Figure

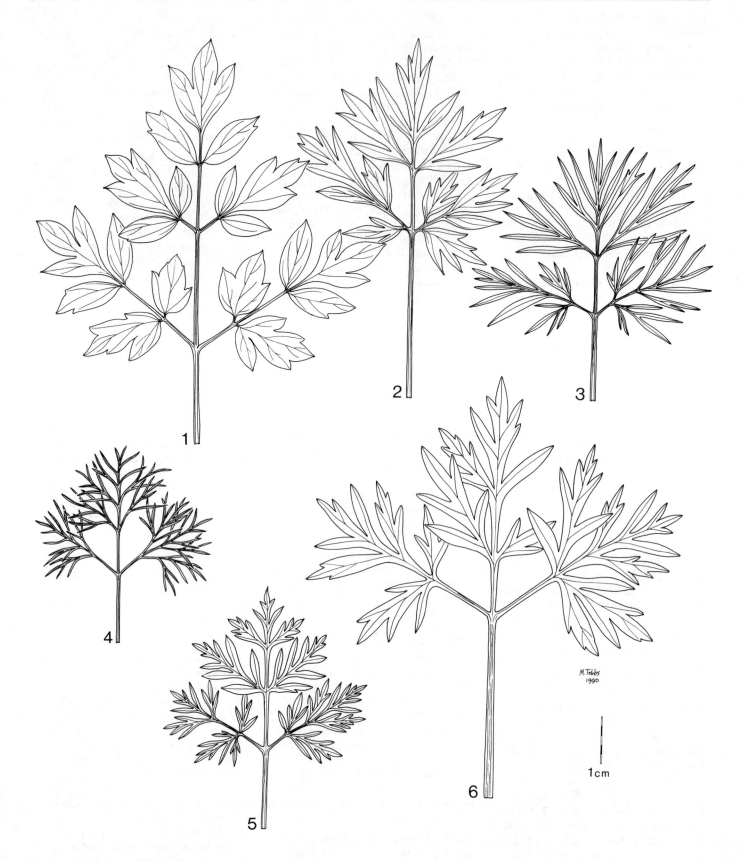

Figure 3. Leaves of *Paeonia* species. 1, *P. suffruticosa.*
2, *P. veitchii* var. *veitchii.* 3, *P.* × *smouthii.* 4, *P. tenuifolia.*
5, *P. potaninii* var. *potaninii.* 6, *P. lutea* var. *lutea.*

2(5), p. 20. Illustration: Sweet, British Flower Garden **1**: t. 70 (1824); Botanical Magazine, 8742 (1918); National Horticultural Magazine **13**: 218 (1934); Stearn & Davis, Peonies of Greece, figs 12, 13, 21, t. 3 (1984).

Stem hairless, 30–80 cm. Leaves 6–12, mostly 8. Lower leaves twice divided into 3 with 15–17 principal divisions, some cut further into 2–3 segments lobed or coarsely toothed, above usually with very minute bristles along the main veins, otherwise hairless, beneath hairless or with sparse hairs. Flowers 10–12 cm across, bowl-shaped, with deep red concave petals. Filaments red. Carpels 1–4, usually 2, with densely felted hairs. *Italy, Balkan Peninsula (Romania to Greece), N Turkey.* H3. Late spring–early summer.

15. P. officinalis Linnaeus (*P. feminea* (Linnaeus) Miller; *P. banatica* Rochel; *P. monticola* Jordan; *P. villarsii* Jordan). Illustration: Botanical Magazine, 3175 (1832); Reichenbach, Icones Florae Germanicae **4**: t. 126 (1840); Stearn & Davis, Peonies of Greece, 24, 25, 27, 30 (1984); Catasagna & Polani, Flora proteta dell'Italia settentrionale, t. facing 180 (1985).

Stem sparsely hairy or hairless, 35–60 cm. Leaves *c.* 5. Lower leaves twice divided into 3, with lateral divisions consisting of 3 leaflets, two narrowly elliptic, the terminal one cut into 3 segments, all together about 20 acute lobes, sparsely hairy or hairless beneath. Flowers 9–13 cm across, petals outspread, red or rose. Filaments red. Carpels 2–3, with densely felted hairs. *France, Switzerland, N Italy, former Yugoslavia, Hungary.* H3. Summer.

Cultivated in Europe during and since the Middle Ages, producing several garden forms, of which the most commonly grown is the deep red fully double var. **rubra** Sabine (*P. festiva* Tausch). The above description refers to subsp. **officinalis**, with all leaf divisions deeply cut; subsp. **banatica** (Rochel) Sóo has less-divided leaves.

16. P. humilis Retzius (*P. officinalis* subsp. *humilis* (Retzius) Cullen & Heywood). Illustration: Botanical Magazine, 1050 (1807), 3431 (1835); Jordan & Fourreau, Icones ad Floram Europae **2**: t. 320, 321 (1903); National Horticultural Magazine **13**: 225 (1934); Stern, Study of Paeonia, 105, 107 (1946).

Stem hairless or hairy, 25–50 cm. Leaves 5–6. Lower leaves twice divided into 3 or bipinnatisect with the segments deeply cut into up to 8 lobes, with in all *c.* 55 lobes, densely hairy beneath. Flowers 9–13 cm across, petals purple. Filaments red. Carpels 2–5, usually 3, hairless or with densely felted hairs. *S Europe.* H3. Summer.

Var. **humilis** (*P. modesta* Jordan). Figure 2(3) p. 20. Carpels hairless or with a few hairs. *Spain, S France.*

Var. **villosa** (Huth) F.C. Stern (*P. paradoxa* G. Anderson; *P. officinalis* subsp. *villosa* (Huth) Cullen & Heywood). Carpels with densely felted hairs. *S France, Italy.*

17. P. tenuifolia Linnaeus. Figure 3(4), p. 22. Illustration: Botanical Magazine, 926 (1806); Sweet, British Flower Garden **7**: t. 345 (1836); National Horticultural Magazine **13**: 229, 230 (1934); Stern, Study of Paeonia, t. facing 110 (1946).

Stem hairless, 25–60 cm. Leaves 9–11. Lower leaves intricately divided into very numerous linear segments 3 mm or less wide, hairless. Flowers 6–8 cm across, broadly bowl-shaped, petals, red concave. Filaments yellow. Carpels 2–4, usually 3, with densely felted hairs. *SE Europe (former Yugoslavia to S Russia).* H3. Spring.

There is a double-flowered red variant 'Plena' and a single pink variant 'Rosea'.

18. P. × smouthii Van Houtte (*P. lactiflora × P. tenuifolia*). Figure 3(3), p. 22. Illustration: Gardeners' Chronicle, **87**: 425, f. 181 (1930).

Stem hairless, 50–55 cm. Leaves 5–7. Lower leaves twice divided into 3, the divisions deeply cut into 15–18 very narrowly elliptic acute segments, 3–10 mm wide, hairless except for minute bristles along veins above. Flowers 1 or 2, bowl-shaped, 6–8 cm across, petals red, *c.* 5 cm. Filaments yellow. Carpels 2–4, mostly 3, dark purple, hairless or slightly hairy. *Garden origin.* H3. Late spring.

19. P. anomala Linnaeus. Illustration: Andrews, Botanist's Repository **8**: t. 514 (1807); Botanical Magazine, 1745 (1815); Stern, Study of Paeonia, 114, t. facing 113 (1946).

Stem hairless, 60–90 cm. Lower leaves twice divided into 3 but the 9 major divisions themselves deeply cut into 3–5 narrow acute segments with very minute bristles along the veins above, hairless beneath. Flowers 7–9 cm across, petals outspread, red. Filaments yellow. Carpels 3–5, mostly 3, hairless or with densely felted hairs. *E Europe to W Asia.* H2. Early summer.

Var. **anomala** (*P. laciniata* Pallas; *P.*

anomala var. *nudicarpa* Huth). Figure 2(4), p. 20. Carpels hairless. *E European Russia to W Siberia.*

Var. **intermedia** (C.A. Meyer) O. & B. Fedtschenko (*P. intermedia* C.A. Meyer). Carpels with densely felted hairs. *NW European Russia (Kola Peninsula) to Turkestan.*

20. P. × hybrida Pallas (*P. anomala* var. *anomala × P. tenuifolia*; *P. bergiana* invalid).

Stem to 60 cm. Flowers 9–10 cm across; petals deep red. Filaments reddish. Carpels 2–4, densely felted hairy. H3. Late spring to early summer.

This hybrid first arose before 1788 in the St Petersburg Botanic Garden and before 1935 in the Hortus Bergianus, Stockholm, in both gardens when *P. tenuifolia* and *P. anomala* var. *anomala* were being grown together, and it has also been made by deliberate crossing in USA. It is almost indistinguishable from wild *P. anomala* var. *intermedia* except for defective pollen and the innermost sepal rounded not mucronate.

21. P. veitchii Lynch (*P. beresowskii* Komarov). Illustration: Gardeners' Chronicle, **46**: 2 (1909); National Horticultural Magazine **13**: 214, 233 (1934); Stern, Study of Paeonia, 115, t. facing 115 (1946).

Stem hairless, 20–50 cm, usually with 2 or 3 flowers. Leaves 4 or 5. Lower leaves twice divided into 3, with the 3 main divisions each with 3 leaflets deeply cut into 3–5 acuminate lobes having minute bristles along the veins above, hairless or with bristly hairs beneath. Flowers 5–9 cm across, petals outspread, purple, rose or white. Filaments yellow. Carpels 2–5, with dense felted hairs. *W China (Gansu, Sichuan, Shaanxi).* H2–3.

Var. **veitchii**. Figure 3(2), p. 22. Leaflets hairless beneath. Petals light purple or white (forma **alba**).

Var. **woodwardii** Bergmans (*P. woodwardii* Cox). Leaflets with bristly hairs beneath. Petals rose.

LXXXV. EUCRYPHIACEAE

Evergreen trees and shrubs (*E. glutinosa* usually deciduous in cultivation). Leaves opposite, simple or pinnate, entire or toothed, leaf-stalk stipules falling early. Flowers bisexual, fragrant, solitary in the leaf axils. Flower-stalks with bracts, usually at the base; inflorescence-stalk usually not evident. Sepals 4, forming a cap over the petals which falls as the flower opens. Petals usually 4, white, rarely pink. Stamens numerous with pink anthers. Ovary superior. Fruit a woody capsule with numerous winged seeds.

A family of only one genus from Australia and South America. All are best given a moist but well-drained, peaty soil in a sunny position but need to be cool and shaded at the base. *E. cordifolia* and *E. × nymansensis* are lime-tolerant but others require an acid soil. In cool areas protection from cold winds is necessary and they grow well sheltered amongst other trees and shrubs but are wind-resistant in mild areas. *E. glutinosa* is the most tolerant of cold, dry conditions. Propagation is by seed or cuttings.

Literature: Bausch, J., A revision of the Eucryphiaceae, *Kew Bulletin* **8**: 317–349 (1938); Dress, W.J., A review of the genus *Eucryphia*, *Baileya* **4**: 116–127 (1956); Wright, D., *Eucryphia, Hoheria* and *Plagianthus, The Plantsman* **5**: 167–185 (1983).

1. EUCRYPHIA Cavanilles
A.J. Coombes
Description as for family.

A genus of 5 species from SE Australia and Chile.

1a. All leaves on mature plants simple (seedlings and vigorous shoots sometimes have leaves with 3 leaflets)   2
  b. Leaves on adult plants compound or with compound and simple leaves mixed   6
2a. Leaves scalloped or with forward-pointing teeth; ovary hairless   **1. cordifolia**
  b. Leaves entire or nearly so; ovary downy, sometimes sparsely so   3
3a. Leaves mostly more than 5 × 2 cm, entire or with few teeth; ovary sparsely downy   **2. 'Penwith'**
  b. Leaves mostly less than 5 × 2 cm, entire; ovary downy   4
4a. Leaves 3–5 cm, more than 3 times as long as wide; flowers 3–5 cm across, cup-shaped at first, opening flat; bracts 2   **3. lucida**
  b. Leaves 2.5 cm or less, not more than 2.5 times as long as wide; flowers 1.8–2.5 cm across, cup-shaped, not opening flat; bracts 2–4   5
5a. Leaves 1–1.5 cm; flowers usually 2 cm across or less; bracts 4   **4. milliganii**

  b. Leaves 1.5–2.5 cm; flowers 2–2.5 cm across; bracts 2–4   **5. × hybrida**
6a. Leaves glaucous beneath, leaflets entire; inflorescence-stalk evident, bracts inserted in middle of inflorescence-stalk   7
  b. Leaves green or grey-green beneath, at least the larger leaves and leaflets toothed; inflorescence-stalk not evident, bracts inserted at base of flower-stalk   8
7a. Leaflets 11–15, lateral leaflets at least 5 times as long as wide, all distinctly mucronate   **7. moorei**
  b. Leaflets usually 3–5, sometimes up to 9, lateral leaflets up to 3 times as long as wide, rounded or shortly mucronate, leaves sometimes simple   **6. × hillieri**
8a. Leaves all pinnate   **8. glutinosa**
  b. At least some leaves simple   9
9a. Leaves and leaflets sharply toothed   **9. × nymansensis**
  b. Leaves and leaflets entire or few toothed at apex   **10. × intermedia**

1. E. cordifolia Cavanilles. Figure 4(1), p. 25. Illustration: Botanical Magazine, 8209 (1908); The Plantsman **5**: 170 (1983); Phillips & Rix, Shrubs, 224 (1989).

Columnar tree to 20 m; young shoots with densely felted hairs. Leaves leathery, oblong, to 8 × 4 cm, toothed at the margin, conspicuously so on young plants, grey and downy beneath. Bracts 6; flowers *c.* 5 cm across; petals occasionally 6. *Chile.* H5. Late summer–autumn.

2. E. 'Penwith' (*E. cordifolia* × *E. lucida*). Figure 4(2), p. 25. Illustration: Thomas, Gardens of the National Trust, t. 175 (1979); Phillips & Rix, Shrubs, 225 (1989).

Tree to 10 m or more. Young shoots with dense spreading hairs. Leaves leathery, oblong, wavy-edged, to 6.5 × 3 cm, entire or with a few teeth towards the apex, glaucous beneath, some with 3 leaflets on vigorously growing young plants. Bracts 4; flowers *c.* 5 cm across. H4. Summer.

Often wrongly attributed to *E. × hillieri*.

3. E. lucida (Labillardière) Baillon (*E. billardieri* Spach). Figure 4 (3), p. 25. Illustration: Botanical Magazine, 7200 (1891); RHS Dictionary, 788 (1956); Curtis, The endemic Flora of Tasmania, **1**: t. (1967); Galbraith, Wild flowers of South-East Australia, pl. 65 (1977).

Columnar tree or shrub to 10 m. Young shoots adpressed downy. Leaves simple, oblong to oblong-lanceolate, entire,

glaucous beneath, to 5 × 1.5 cm; seedlings and sometimes leaves on vigorous shoots with 3 leaflets. Bracts 2; flowers 3–5 cm across. *Australia (Tasmania).* H4. Summer.

'Pink Cloud' is a variant with pink flowers.

4. E. milliganii J.D. Hooker (*E. lucida* var. *milliganii* (J.D. Hooker) Summerhayes). Figure 4(9), p. 25. Illustration: Curtis, The endemic Flora of Tasmania, **1**: t. 34 (1967); Brickell, The RHS gardeners' encyclopedia of plants & flowers, 105 (1989); Phillips & Rix, Shrubs, 225 (1989).

Columnar shrub to 6 m similar to *E. lucida*. Leaves oblong, usually no more than 10–15 × 5–7 mm, glaucous beneath; seedlings and sometimes leaves on vigorous shoots with 3 leaflets. Bracts 4; flowers cup-shaped, *c.* 2 cm across. *Australia (Tasmania).* H4. Summer.

5. E. × hybrida Bausch (*E. lucida* × *E. milliganii*). Figure 4(10), p. 25. Shrub, intermediate between the parents in leaf size, flower size and number of bracts. Generally most resembling *E. milliganii*. Leaves 1.5–2 × 0.8–1.2 cm. Flowers 2–2.5 cm across. *Australia (Tasmania).* H4. Summer.

Occasionally grown as *E. milliganii*.

6. E. × hillieri Ivens (*E. lucida* × *E. moorei*). Figure 4(7), p. 25. Illustration: The Journal of the Royal Horticultural Society **78**: 343 (1953).

Tree to 10 m. Leaves usually pinnate, sometimes simple on mature plants, with 3–5 or to 9 leaflets, glaucous beneath, rounded or shortly mucronate at the apex, mucro 2 mm or less; terminal leaflet largest, narrowly oblong-elliptic, to 6 × 1.5 cm, lateral leaflets oblong, to 3 × 1 cm. Inflorscence-stalk evident; bracts 4, inserted in the middle of the flower-stalk; flowers to 3.5 cm across. H5. Summer.

'Winton' is the most commonly grown cultivar.

7. E. moorei F. Mueller. Figure 4(6), p. 25. Illustration: The Journal of the Royal Horticultural Society **78**: 343 (1953); Phillips & Rix, Shrubs, 225 (1989).

Conical tree to 15 m. Leaves pinnate, usually glaucous beneath; leaflets 11–13 or sometimes more, decreasing in size towards leaf-base, entire, mucronate, mucro to 4 mm. Terminal leaflet to 7 × 1.5 cm. Inflorescence-stalk evident; bracts 4, inserted in the middle of the flower-stalk; flowers *c.* 2.5 cm across. *Australia (New South Wales).* H5. Summer.

The least hardy species.

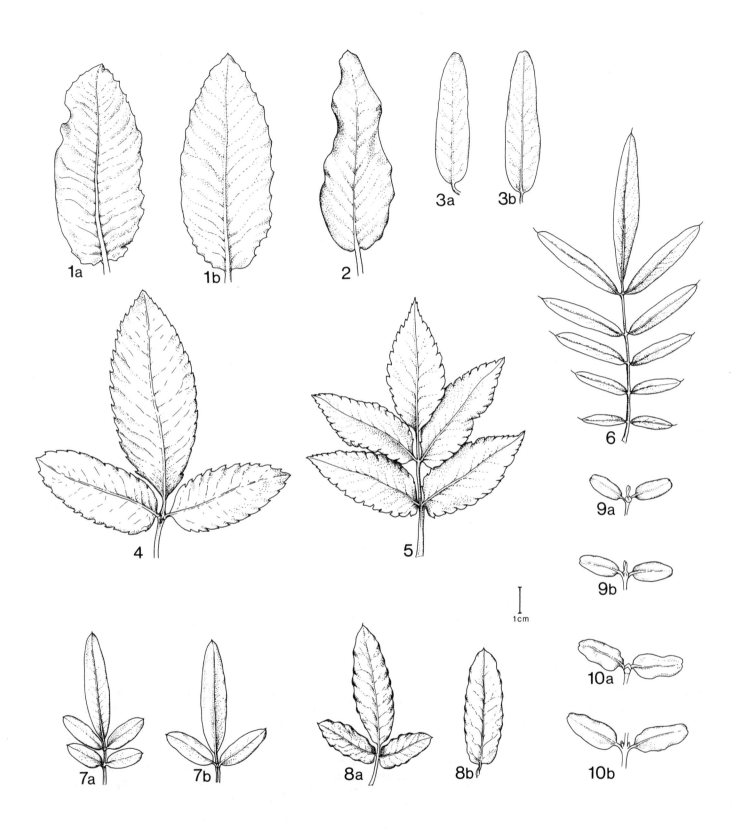

Figure 4. Leaves of *Eucryphia* species. 1a, 1b, *E. cordifolia.*
2, *E. 'Penwith'.* 3a, 3b, *E. lucida.* 4, *E. × nymansensis.*
5, *E. glutinosa.* 6, *E. moorei.* 7a, 7b, *E. × hillieri.*
8a, 8b, *E. × intermedia.* 9a, 9b, *E. milliganii.* 10a, 10b, *E. hybrida.*

8. E. glutinosa (Poeppig & Endlicher) Baillon (*E. pinnatifolia* Gay). Figure 4(5), p. 25. Illustration: Botanical Magazine, 7067 (1889); The Plantsman **5**: 172 (1983); Brickell, The RHS gardeners' encyclopedia of plants & flowers, 63 (1989); Phillips & Rix, Shrubs, 224 (1989).

Deciduous or semi-evergreen tree to 10 m. Leaves pinnate; leaflets 3–5, margin with forward-pointing teeth, to 5 × 3 cm, green and downy on both sides; terminal leaflet stalked, lateral leaflets stalkless or nearly so. Bracts 4; flowers *c.* 6 cm across. *Chile.* H4. Late summer.

The hardiest species; leaves turn orange-red in autumn. Some plants have double or semi-double flowers (Plena group).

9. E. × nymansensis Bausch (*E. cordifolia* × *E. glutinosa*). Figure 4(4), p. 25. Illustration: Thomas, Gardens of the National Trust, pl. 15 (1979); Nelson & Walsh, An Irish florilegium, pl. 22 (1983); The Plantsman **5**: 176 (1983); Phillips & Rix, Shrubs, 224–225 (1989).

Columnar tree to 15 m. Leaves green and thinly hairy on both sides, simple or compound, all sharply toothed. Simple leaves elliptic-oblong to lanceolate, to 6 cm; compound leaves with usually 3 leaflets, terminal one largest. Bracts 4 or 5; flowers to 7.5 cm across. H4. Late summer–autumn.

Several cultivars are grown of which 'Nymansay' is the best known; 'Mount Usher' is also grown.

10. E. × intermedia Bausch (*E. glutinosa* × *E. lucida*). Figure 4(8), p. 25. Illustration: Botanical Magazine n.s., 534 (1969); Phillips & Rix, Shrubs, 224 (1989).

Tree to 10 m, shoots adpressed downy. Leaves green or slightly grey-green but not glaucous beneath, nearly hairless, simple or compound, leaves and leaflets entire or with a few teeth towards the apex. Simple leaves oblong, to 6 × 2.5 cm, compound leaves with usually 3 leaflets, terminal leaflet 1.5–2 times longer than lateral ones. Bracts 4; flowers *c.* 4 cm across. H4. Late summer–autumn.

'Rostrevor' is the best known cultivar.

LXXXVI. ACTINIDIACEAE

Trees, shrubs or rampant vines, usually dioecious. Leaves alternate, simple, evergreen or deciduous. Inflorescences axillary. Flowers usually with 4 or 5 sepals and petals, numerous stamens and a superior ovary with many ovules. Fruit a berry or capsule.

A family of three genera and about 300 species, mainly from tropical and subtropical Asia, with a few members in Australia and America. *Saurauia* is included, although sometimes placed in the separate family Saurauiaceae, or in Dilleniaceae.

1a. Vines or climbing shrubs; inflorescence a cyme with a few flowers; flowers unisexual 2
  b. Trees or shrubs; not vines or climbers; inflorescence a panicle, with 1–500 flowers; flowers bisexual but functionally unisexual **3. Saurauia**
2a. Stamens 10; ovary with 5 cells; style solitary **2. Clematoclethra**
  b. Stamens numerous; ovary with numerous cells; styles numerous **1. Actinidia**

1. ACTINIDIA Lindley
A.R. Ferguson & E.C. Nelson

Deciduous, usually dioecious vines, climbing and clinging by spirally twisting stems. Leaves usually toothed. Flowers unisexual or very infrequently bisexual, solitary or in cymes. Sepals 5. Petals 5. Stamens numerous. Ovary with numerous cells; styles numerous. Fruit a juicy berry.

A genus of about 50 species in eastern Asia, ranging from Korea southwards into the Indonesian archipelago; the centre of diversity is south-western China. While about 10 species are listed in botanical collections, only about 5 are widely cultivated in European gardens, and the identification of others may be very doubtful. *A. chinensis* was probably not cultivated in Europe until the late 1980s; all plants previously grown under this name belong to *A. deliciosa*.

Literature: Ferguson, A.R., Kiwifruit: a botanical review; *Horticultural Reviews* **6**: 1–64 (1984); McMillan-Browse, P., Some notes on actinidias and their propagation, *The Plantsman* **6**: 167–180 (1984).

1a. At least some leaves marked with white patches towards apex 2
  b. Leaves entirely green without white markings 3
2a. Stem pith in plate-like layers; leaf-base cordate; variegation white or pink **5. kolomikta**
  b. Stem pith solid, white; leaf-base rounded; variegation white or yellow **8. polygama**
3a. Leaves with stellate hairs 4
  b. Leaves hairless, or if hairy, stellate hairs absent 5
4a. Fruit spherical, softly hairy, becoming hairless **4. chinensis**
  b. Fruit ellipsoid, with persistent stiff brown hairs **3. deliciosa**
5a. Undersides of leaves glaucous, without bristles **6. melanandra**
  b. Undersides of leaves green, or hairy or with bristles 6
6a. Bristles on leaf-stalks and main veins below **1. arguta**
  b. Bristles absent 7
7a. Fruit purple, not spotted **2. purpurea**
  b. Fruit green, spotted **7. callosa**

1. A. arguta (Siebold & Zuccarini) Miquel (*A. giraldii* Diels; *A. polygama* misapplied; *A. volubilis* misapplied). Illustration: Botanical Magazine, 7497 (1896); Hayashi, Woody plants of Japan, 480 (1985).

Vigorous vine; shoots hairless, lenticels present but indistinct except on sunward side, green and on sunward side red; pith in plate-like layers. Leaves green on both surfaces, obovate to 15 × 10 cm, cordate to rounded at base, tapering to apex, margins with fine bristle-teeth, veins beneath with bristles and tufts of hairs in axils; leaf-stalks red, to 5 cm, sharply recurved at base, with bristles. Flowers fragrant, cup-shaped, in clusters of 2 or 3, *c.* 2 cm across. Petals white, tinged green, concave. Anthers purple. Fruit green-yellow. *E Asia.* H5–G1. Spring–summer.

Variants of *A. arguta* are numerous, including var. **cordifolia** (Miquel) Dunn, leaves with heart-shaped base; leaf-stalks red, and var. **rufa** (Siebold & Zuccarini) Maximowicz (*Trochostigma rufa* Siebold & Zuccarini), leaf-stalks hairless, inflorescence hairy. Plants wrongly labelled *A. giraldii* Diels may also belong here.

2. A. purpurea Rehder (*A. arguta* var. *purpurea* (Rehder) Liang).

Similar to *A. arguta* but leaves bright green; leaf-stalks hairless, bristles absent from stalks and main veins. Fruit purple. *E Asia.* H5–G1. Spring–summer.

Some plants (raised from Wilson 1314) have a remarkable abruptly pointed leaf apex.

3. A. deliciosa (Chevalier) Liang & Ferguson (*A. latifolia* Merrill var. *deliciosa* Chevalier; *A. chinensis* Planchon var. *deliciosa* (Chevalier) Chevalier; *A. chinensis* var. *hispida* Liang). Illustration: Botanical Magazine, 8538 (1914); Hayashi, Woody plants of Japan, 481 (1985); Taylor, Climbing plants, 35 (1987); Phillips & Rix, Shrubs, 86 (1989).

Rampant vine; young stems often translucent red, cloaked with red hairs, pith in plate-like layers. Leaves heart-shaped to ovate, to 15 × 15 cm, rough with star-shaped white and simple red hairs beneath especially on veins, becoming more or less hairless above; leaf-stalk hairy, length variable. Flowers in cymes. Sepals ovate, to 1.5 cm, with brown hairs. Petals to 3 cm, cream-white, larger in female flowers, ovate to elliptic. Fruit ellipsoid, covered with stiff brown hairs, pulp green, black seeds numerous. *China*. H5–G1. Spring.

This is the commercial kiwi fruit, formerly called Chinese gooseberry in Britain. It was originally considered to be *A. chinensis* but is now placed in a different, distinct species. Only var. **deliciosa** is represented in cultivation, and good fruiting clones of it are now marketed ('Hayward' and 'Bruno' are recommended); the best male (pollinator) clones are 'Matua' and 'Tomuri'. Many old plants (of unrecorded provenance) fruit well in western Europe as far north as Northern Ireland; some of these are bisexual. The young shoots will be damaged by late spring frosts.

4. A. chinensis Planchon. Illustration: Hooker, Icones Plantarum **16**: 1593 (1887); Revista di Fruticoltura **52**(10):9 (1990).

Similar to *A. deliciosa*, but distinguished by its spherical fruit initially covered with soft hairs, becoming almost hairless at maturity. *China*. H5. Summer.

Not in cultivation in western Europe except perhaps in recently established scientific and experimental collections.

5. A. kolomikta (Maximowicz & Ruprecht) Maximowicz. Illustration: Beckett, Climbing plants, 66 (1983); Hayashi, Woody plants of Japan, 480 (1985); Brickell, RHS gardeners' encyclopaedia, 167 (1989); Phillips & Rix, Shrubs, 86 (1989).

To 7 m; stems dark, hairless (occasionally with coarse hairs when young), pith brown, in plate-like layers. Leaves to 15 × 12 cm, ovate, base cordate, apex acuminate, variegated with white and pink, hairless except on veins beneath; leaf-stalk 2–3 cm. Flowers in clusters of 1–3; flower-stalks 0.6–1 cm. Petals *c.* 0.5 × 1 cm. Anthers yellow. Ovary hairless. Fruit spherical, *c.* 2 cm across. *E Asia*

(Russia, Siberia, Korea, Japan, W China). H4. Summer.

Some authors suggest the variegation is especially prominent in male plants.

Var. **kolomikta** has leaves to 10 × 8 cm, flower- and leaf-stalks hairless.

Var. **gagnepainii** (Nakai) Li has leaves *c.* 15 × 12.5 cm, flower- and leaf-stalks with densely felted rusty hairs.

6. A. melanandra Franchet.

Almost hairless vine; stems with prominent, almost linear lenticels, glaucous, grey-red; pith in plate-like layers. Leaves *c.* 15 × 6 cm, obovate to oblanceolate, base tapering, heart-shaped or rounded, tapering towards apex, blade margin with fine, cream, cartilaginous teeth, dark green with paler impressed veins above, glaucous beneath and very sparsely hairy with tufts of brown hairs in vein-axils; leaf-stalk sharply curved upwards at base, translucent rose-pink, to 6 cm. Flowers white, 3–5 in cymes, or solitary, *c.* 2.5 cm across. Ovary hairless. Fruit ovoid, *c.* 2.5 cm, red-brown, hairless, glaucous. *China (Sichuan, Hubei, Yunnan)*. H4. Summer.

7. A. callosa Lindley.

Like *A. melanandra*, but leaves and shoots with prominent lenticels. Leaves not glaucous beneath. Flowers in 1–5 flowered cymes, fragrant. Petals white. Anthers yellow. Ovary hairless. Fruit green, spotted. *N India, Bhutan, Nepal, China*. H5. Summer.

8. A. polygama (Siebold & Zuccarini) Maximowicz (*A. volubilis* (Siebold & Zuccarini) Miquel). Illustration: Hayashi, Woody plants of Japan, 480 (1985); Revista di Fruticoltura **52**(10): 12 (1990).

Like *A. kolomikta*, but pith white and solid, and leaves variegated with white or yellow (variegation is not more prevalent in male plants). Flowers occasionally solitary. Fruit to 2.5 cm across, spherical yellow, hairless without lenticels. *Japan, China, Korea*. H5. Summer.

Var. **polygama** has leaves with sparse coarse hairs on veins and yellow anthers.

Var. **lecomtei** (Nakai) Li has hairless leaves and brown anthers.

2. CLEMATOCLETHRA (Franchet) Maximowicz
E.C. Nelson

Deciduous shrubs, with twining stems resembling *Actinidia*. Flowers in cymes. Sepals 5, not deciduous. Petals 5. Stamens 10. Ovary with 5 cells surmounted by a single style; stigma not lobed. Fruit a small berry.

This genus contains perhaps 10 species confined to China; few are in cultivation in Europe and their classification and nomenclature must be considered highly speculative. Cultivation as for *Actinidia*; propagation by seed and by cuttings.

1a. Shoots with brown bristles
3. scandens
 b. Shoots hairless, or perhaps downy when young, without bristles 2
2a. Leaves *c.* 10 cm, not glaucous beneath; lenticels prominent
2. lasioclada
 b. Leaves less than 8 cm, glaucous beneath; lenticels minute
1. actinioides

1. C. actinioides Maximowicz (*C. integrifolia* Maximowicz). Illustration: Botanical Magazine, 9439 (1936).

To 20 m, hairless; shoots grey-brown, lenticels minute. Leaves to 7 × 4 cm, ovate to heart-shaped, apex acuminate, base cordate, green above, glaucous beneath and occasionally with hairs in vein-axils; leaf-stalk to 6 cm. Inflorescences with 1–3 flowers, axillary. Sepals pink, *c.* 0.4 cm across, margins hairy, persisting in fruit and then brown. Petals *c.* 0.8 × 0.5 cm, white. Style to 0.7 cm, persistent in fruit. Berry black or red-black, to 0.7 cm across, seeds 5. *China (Gansu, Sichuan)*. H4. Summer.

2. C. lasioclada Maximowicz.

To 8 m; young shoots downy but rapidly becoming hairless, glaucous grey-brown to red brown; lenticels very prominent. Leaves ovate to heart-shaped, blade to 10 cm, margins with bristle-tipped teeth, and downy hairs in vein-axils beneath, base tapering to cordate; leaf-stalk to 6 cm, red. Inflorescences with 2–7 flowers. Petals white. Fruit black, *c.* 1 cm across. *China*. H5. Summer.

Var. **grandis** (Hemsley) Rehder, with large leaves, is reported in cultivation (Wilson 886); it is doubtful if it is worth maintaining as a variety.

This species clasps supports by the leaf-stalks which when attached have a very tight grip; later the growing stems may also twine around a support.

3. C. scandens (Franchet) Maximowicz (*Clethra scandens* Franchet).

To 10 m; shoots with brown bristles. Leaves ovate to elliptic, apex pointed, base rounded or tapering, 5–12 × 2–7 cm, margins with bristle-tipped teeth, glaucous and downy beneath with bristles on veins on both sides; leaf-stalk to 3 cm, covered with bristles. Inflorescence of 3–6 flowers; flowers less than 1 cm across; petals white. Fruit red, *c.* 1 cm across. *China.* H4. Early summer.

3. SAURAUIA Willdenow
E.C. Nelson

Trees and shrubs. Leaves spirally arranged. Flowers bisexual. Sepals 3–8, green, fused at base. Petals 3–9, white or rarely pink, fused at base. Stamens numerous; anthers yellow. Ovary superior, usually with 5 cells; ovules numerous; styles 3–5. Fruit a juicy berry, seeds small, numerous.

A genus of 300 species distributed throughout tropical Asia and America. The fruits are sometimes eaten. The flowers of **S. purgans** B.L. Burt make a crying noise when opening. The correct spelling of the generic name is *Saurauia* as it is conserved under the *International Code of Botanical Nomenclature*; '*Saurauja*' should not be used.

Literature: Soejarto, D.D., Revision of South American *Saurauia* (Actinidiaceae). *Fieldiana Botany* **2** (n.s.) (1980).

1a. Flowers creamy-white
         3. tomentosa
  b. Flowers pink   2
2a. Leaves with brown felt and scales below   **1. nepalensis**
  b. Leaves with bristles **2. subspinosa**

1. S. nepalensis de Candolle. Illustration: Wallich, Plantae Asiaticae Rariores **2**: t. 178 (1831); Brandis, Indian trees, 62 (1921).

Tree or shrub; wood soft; shoots with felt of hairs and deciduous brown scales. Leaves to 40 cm, with 25–30 pairs of secondary veins, margins toothed, underside covered with brown felt and scales, bright green above, crowded at tips of branches. Flowers pink, in panicles. Styles 4 or 5, free. Fruit sticky, edible. *Himalaya.* G1. Summer.

2. S. subspinosa Anthony. Illustration: Botanical Magazine, 9472 (1937).

Shrub to 3 m, shoots with pale brown bristles. Leaves to 32 × 10 cm, oblanceolate, with 8–12 pairs of secondary veins, margins toothed, hairless, underside with bristles on midrib and secondary veins and sparsely scattered between; leaf-stalk with bristles. Flowers pale pink with deep red throat, in erect pyramidal panicles to 10 cm. Ovary hairless; styles 5 (rarely 3 or 4), fused below middle. Fruit *c.*

1 cm across, red, hairless. *Burma.* H5–G1. Summer.

A most attractive plant, but sensitive to frost. Plants in cultivation in Britain today result from Forrest 25084.

3. S. tomentosa (Kunth) Sprengel. Illustration: Rhodora **65**: 14 (1963).

Tree to 30 m. Leaves *c.* 35 × 12 cm, obovate to elliptic, acuminate, tapered or rounded at base, coarsely hairy, margins finely toothed, dark green or brown above, grey or brown beneath. Inflorescence to 25 cm with 25–75 flowers. Flowers creamy-white, *c.* 2 cm across. Sepals densely hairy. Petals often deeply cut, to 1.3 × 1 cm. Ovary hairless; styles 5. Fruit green with red tint. *Venezuela, Colombia, Ecuador.* G2. Summer.

LXXXVII. OCHNACEAE

Evergreen and deciduous trees, shrubs or more rarely herbs. Leaves alternate, simple, with closely set secondary parallel veins, margins often toothed. Stipules entire, fringed or pectinate. Flowers bisexual, radially symmetric in racemes, panicles or bundles. Sepals 4 or 5, often leathery, persistent, usually enlarging and brightly coloured in fruit. Petals 4 or 5, pale yellow, white or pink, soon falling. Stamens 10-many, filaments persistent, anthers linear, basifixed, opening by longitudinal slits or apical pores. Ovary either unlobed and 1-celled or deeply lobed with style attached to gynoecium by a stalk. Fruit a capsule, or a few black drupes on an enlarged receptacle.

A family of 460 species in 37 genera, only a few of which are regularly cultivated. **Lophira lanceolata** Keay (*L. alata* misapplied, not Gaertner) was formerly cultivated as a hothouse plant but is probably no longer in cultivation in Europe. The seeds are rich in oil and the timber is durable and used in wet constructions. It may have been introduced as a botanical curiosity.

Literature: Engler, Die Pflanzenwelt Afrikas, **3**(2): 478–494 (1921).

OCHNA Schreber
S.G. Knees

Evergreen trees or shrubs. Flowers yellow, solitary or in racemes. Sepals 5, overlapping, persistent; petals 5–10, stamens numerous, ovaries with 3–10 lobes, each developing into a single-seeded drupe, these surrounding a central disc; seeds black.

A genus of 86 species in tropical and

warm temperate Africa and South-East Asia. Plants require a minimum winter temperature of 13 °C and a compost of sandy peat and fibrous loam. Propagation is usually from cuttings of semi-ripened wood taken in late summer and rooted in a frame containing moist sand with bottom heat.

1a. Leaf-margin with fine forward-pointing teeth; petals less than 1 cm
         1. serrulata
  b. Leaf-margin entire or slightly wavy; petals more than 2 cm **2. kirkii**

1. O. serrulata (Hochstetter) Walpers (*O. japonica* misapplied; *O. multiflora* misapplied; *O. serratifolia* misapplied; *O. atropurpurea* invalid, not de Candolle). Illustration: Botanical Magazine, 9402 (1935); RHS dictionary, 1391 (1951); Marshall Cavendish encyclopaedia of gardening, 1352 (1967); Hortus III, 775 (1976); Trees of Southern Africa, 609 (1981).

Shrub 1–2.5 m, branches covered with rough lenticels. Leaves narrowly elliptic, 5–6 × 0.7–1 cm, glossy above, margins with fine, forward-pointing teeth. Flowers 1-several, on short branches. Sepals broadly elliptic, yellowish green at first, becoming bright red at maturity to 1.8 cm; petals obovate, *c.* 8 mm, soon falling. Fruits 5 or 6, ellipsoid, *c.* 1 cm, black, on a hard dome-shaped, red receptacle to 2.5 cm across. *Southern Africa.* G1. Spring–summer.

Perhaps not distinct from **O. atropurpurea** de Candolle, although most of the plants grown in cultivation under this name are probably *O. serrulata*.

2. O. kirkii D. Oliver. Illustration: Graf, Exotica series 4, **2**: 1628 (1985).

Shrub or small tree to 5 m. Leaves oblong–elliptic to narrowly obovate, 8–10 cm, rounded at tip, wedge-shaped to cordate at base, margins entire or slightly wavy, usually with long fine hairs. Flowers in short panicles on short lateral branches. Sepals red, flat and spreading in fruit. Petals *c.* 2.5 cm. Carpels 12. Fruits cylindric, slightly flattened. *E Africa (Tanzania, Mozambique, Kenya).* G2. Spring–summer.

LXXXVIII. DIPTEROCARPACEAE

Trees, usually resinous. Leaves spirally arranged, simple, leathery; stipules present.

Flowers usually bisexual, regular, nodding, scented, usually in axillary panicles, racemes or rarely cymes; sepals 5, all or 2 or 3 enlarged and wing-like in fruit, sometimes edge-to-edge in fruit, but usually overlapping, often with basal tube, persistent; petals 5, twisted; stamens 5–110 in 1–3 whorls or more or less irregular, filaments distinct or more or less basally united, anthers dehiscing longitudinally; ovary inferior, of 2–5 fused carpels, placentation axile. Fruit dry indehiscent, 1-seeded.

A family of 16 genera and 530 species from SE Asia. Although many are valuable timber-producing species it is unlikely that any are cultivated in Europe except in specialist collections of economic or curiosity plants.

LXXXIX. THEACEAE

Trees and shrubs, occasionally lianas. Leaves evergreen or occasionally deciduous, stalked, simple, margins entire or toothed; spirally arranged, usually 2-ranked, without stipules. Flowers usually large, bisexual and regular, axillary and solitary, but sometimes small, unisexual, terminal, in racemes or panicles; occasionally fragrant. Bracteoles 2–12, sometimes intergrading with the sepals, deciduous or persistent. Sepals 4 or 5, basally united, overlapping. Petals 4–many, sometimes basally united, overlapping. Stamens usually numerous, free in 1–5 series, in 5 bundles on a fleshy disc, irregularly or basally united, attached to the base of the corolla; anthers versatile or basifixed. Ovary usually superior, rarely partly inferior, carpels 3–5 rarely 2–10, basally united. Styles free or partially united. Fruit a capsule, sometimes woody, or a berry. Seeds 1-many, sometimes winged.

A family of about 28 genera and 520 species; mainly tropical, a few warm temperate.

1a. Anthers versatile; fruit a capsule, sometimes woody   2
 b. Anthers basifixed; fruit a berry, or rarely a fleshy capsule   6
2a. Deciduous plants; capsule dehiscing apex and base   **1. Franklinia**
 b. Evergreen or rarely deciduous plants; capsule usually dehiscing at apex only; if plant deciduous and capsule splitting down ribs, then

winter buds silvery hairy and seeds 2–4 per cell   3
3a. Deciduous or rarely evergreen plants; capsule without persistent central axis   **2. Stewartia**
 b. Plants evergreen; capsule with persistent central axis   4
4a. Seeds unwinged   **3. Camellia**
 b. Seeds winged   5
5a. Stamens (outer whorl at least) 5, grouped on fleshy disk; capsule cylindric or narrowly ovoid
                                                 4. Gordonia
 b. Stamens free, in 3–5 series; capsule almost spherical   **5. Schima**
6a. Receptacle and sepals enlarged
                                                  6. Visnea
 b. Receptacle and sepals not enlarged
                                                                    7
7a. Fruit a fleshy capsule, yellow, turning red; leaves clustered at ends of branches   **7. Ternstroemia**
 b. Fruit a berry, red, turning purple-black; leaves 2-ranked   8
8a. Flowers 1–1.5 cm, bisexual; bracteoles deciduous   **8. Cleyera**
 b. Flowers 2–7 mm, unisexual; bracteoles persistent   **9. Eurya**

1. FRANKLINIA Marshall
C. Whitefoord
Deciduous shrubs or small trees to 5–7 (rarely 10) m; erect with short trunk; branches stout; bark thin, smooth, grey to reddish brown. Young growth densely silky-hairy. Winter buds naked, reddish brown-hairy. Leaves shortly stalked, blade 6–15 × 3.8–5 cm, obovate-oblong, apex blunt or acute, membranous, bright glossy green above, veins impressed, downy beneath; margin finely saw-toothed; coloured red in autumn. Flowers bisexual, axillary, solitary, almost stalkless, cupped, 7–9 cm, fragrant. Bracteoles 2, beneath calyx, minute, deciduous. Sepals 5, unequal, almost circular, white silky-hairy, persistent. Petals snow-white, 5, shortly united, obovate, margin wavy, membranous. Stamens yellow, numerous, free, attached to base of petals; anthers versatile. Ovary almost spherical, ridged, hairy, 5-celled. Styles 5, united; stigma disc-shaped. Fruit a woody capsule, almost spherical, with persistent central axis, splitting from above and below, the valves remaining attached. Seeds 6–8 per cell, flat, angular, unwinged.

A genus of one species, from N America, which is probably extinct in the wild. *Franklinia* requires well-drained acid soil with constant moisture. It needs long,

hot summers to flower well. In cooler areas it is chiefly valued for the brilliant autumnal colouring of its leaves. It may be propagated by cuttings or by seed. *Franklinia alatamaha* has been crossed with *Gordonia lasiantha*, but the progeny are susceptible to disease.

1. F. alatamaha Marshall (*Gordonia alatamaha* (Marshall) Sargent; *G. pubescens* L'Héritier). Figures 5(1), p. 30 and 7(1), p. 34. Illustration: Sargent, The silva of North America **1**: t. 22 (1891); Krüssmann, Manual of cultivated broad-leaved trees & shrubs **2**: 84 (1986).
 N America (Georgia). H5. Late summer.

2. STEWARTIA Linnaeus
S.A. Spongberg
Trees or shrubs, deciduous, rarely evergreen. Bark smooth, often mottled, peeling and flaking or closely fissured and non-flaking. Leaves stalked, the stalks winged, sometimes enclosing the winter leaf-buds. Leaf-blades papery to leathery; more or less hairless above, hairy beneath at least along the veins; margin toothed. Winter leaf-buds laterally compressed with 1–several overlapping scales, rarely naked. Flowers bisexual, axillary, rarely terminal, solitary, rarely 2 or 3; cup-shaped or flattish, occasionally fragrant, stalked. Bracteoles 2 (rarely 1) beneath the calyx, usually leafy, almost opposite and persistent. Sepals 5, rarely 6, basally joined, usually persistent in fruit. Petals 5, rarely 6–8, snow or cream-white, occasionally stained red outside, basally united in a short tube, margin wavy, outer surface usually finely silky-hairy. Stamens numerous, filaments basally joined, attached to corolla base; anthers versatile. Ovary superior, conical or almost spherical, hairless or hairy, 5- rarely 4- or 6-celled; styles 5 (occasionally 4 or 6), united for most of their length, rarely free, usually persistent in fruit. Fruit a woody capsule, almost spherical to ovoid, strongly angled and ribbed, often beaked, lacking a central axis, splitting between the walls, usually from the apex, or rarely beneath, leaving the apex intact. Seeds 2–4 per cell, usually flattened-obovoid, or disc-shaped, dull-surfaced, narrowly winged, or rarely ovoid, smooth, shiny and unwinged.

A genus of 15 to 20 species, occurring in Korea, Japan, Thailand, Laos, Cambodia, Vietnam and SW China, also in SE USA; about 7 species are grown in Europe. Hybrids sometimes occur, one of which, *S. × henryae*, is included here. The evergreen

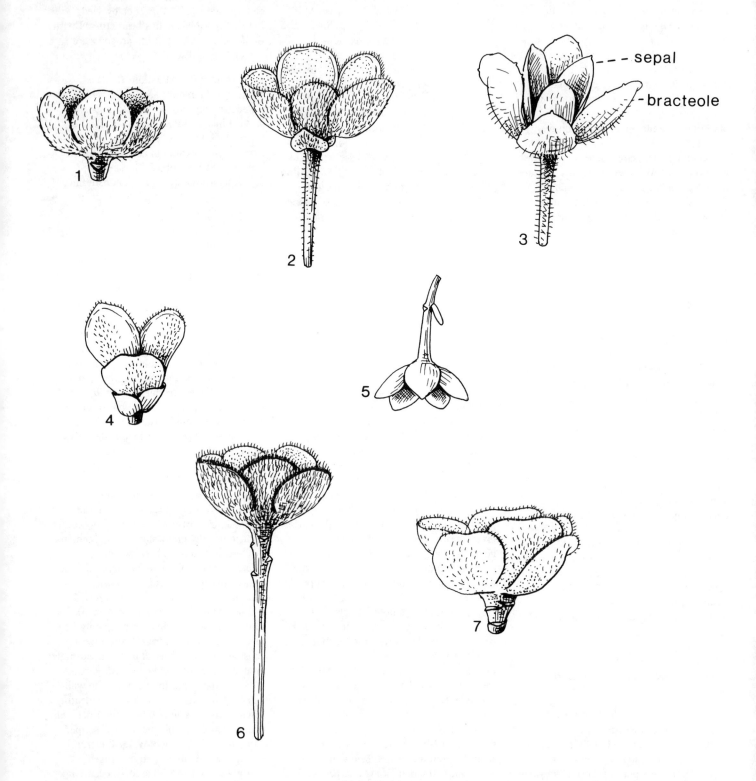

Figure 5. Calyces and bracteoles of Theaceae (× 2). 1, *Franklinia alatamaha*. 2, *Stewartia pseudocamellia*. 3, *S. sinensis*. 4, *Camellia saluensis*. 5, *C. sinensis*. 6, *Gordonia lasianthus*. 7, *G. axillaris*.

species are sometimes placed in *Hartia* Dunn; *S. pteropetiolata* is the only one in European cultivation. They are extremely ornamental, valuable for their bark in winter and the bright autumn leaf colour, as well as for their white flowers. In Europe the flowering season is late summer. Although most species are fairly hardy, they prefer a sheltered position, and deep, well-drained, lime-free soil. They resent disturbance. Stewartias are best propagated from fresh seed, germinating in 2 years, but stratification will shorten the period to about 7 months. Seed is placed in a polythene bag containing equal parts of sand and moss peat, kept for 4 months in warm conditions followed by 3 months cold, and then sown in spring. Cuttings taken in late summer under glass may succeed. The genus *Stewartia* is often misspelt *Stuartia*.

Literature: Spongberg, S.A., unpublished manuscript on *Stewartia* in cultivation in Europe; Spongberg, S.A. A review of the deciduous-leaved species of *Stewartia* (Theaceae), *Journal of the Arnold Arboretum* **55**: 182–214 (1974); Spongberg, S.A. & Fordham, A.J., Stewartias—small trees & shrubs for all seasons, *Arnoldia* **35**: 165–180 (1975); Ye, Chuangxing, A taxonomy of the genera *Stewartia* & *Hartia*, *Acta Scientiarum Naturalium Universitatis Sunyatseni* **4**: 108–116 (1982).

1a. Leaves evergreen; buds naked, enclosed by winged stalks; bracteoles deciduous   **1. pteropetiolata**
 b. Leaves deciduous; buds with 1–several scales, sometimes enclosed by winged stalks; bracteoles persistent   2
2a. Styles free; stalks winged, the wings enclosing lateral and terminal buds; bracteoles 1   **2. ovata**
 b. Styles united for most of their length; stalks narrowly winged, not enclosing buds; bracteoles 2   3
3a. Filaments purplish, anthers bluish; capsule splitting below intact apex, margins of split folding outwards   **3. malacodendron**
 b. Filaments whitish, anthers yellow, orange or violet; capsule splitting from apex   4
4a. Bracteoles equal to, or longer than calyx; bark smooth or fissured; young branches usually terete and straight   5
 b. Bracteoles conspicuously shorter than calyx; bark smooth, mottled;

young branches usually compressed and zigzag   **9. pseudocamellia**
5a. Ovaries and capsules almost spherical, hairless, or downy only at very base   6
 b. Ovaries and capsules conical, ovoid or cylindric, downy over entire surface   7
6a. Ovaries and capsules hairless; seeds 2 per cell; bark on older branches smooth and mottled   **4. serrata**
 b. Ovaries and capsules downy only at very base; seeds 4 per cell; bark on older branches finely fissured   **5. rostrata**
7a. Sepal-apex acute, entire   8
 b. Sepal-apex rounded, with marginal hairs   **8. × henryae**
8a. Bracteoles ovate, very nearly as long as sepals; styles 6–8 mm; seeds 7–9 mm   **6. sinensis**
 b. Bracteoles oblong, much longer than sepals; styles 3–4 mm; seeds 5–6 mm   **7. monadelpha**

1. S. pteropetiolata W.C. Cheng (*Hartia sinensis* Dunn). Illustration: Botanical Magazine n.s., 510 (1967).

Evergreen shrub or small tree to *c.* 14 m, erect; bark silvery-grey, finely fissured; young growth white-silky-hairy at first. Winter buds naked, concealed within winged stalks. Leaves 5.6–12.8 × 2.2–4.9 cm, elliptic to oblong or oblanceolate, margin widely and shallowly toothed, thinly leathery, dark, glossy, with impressed venation above, paler and conspicuously dark-veined beneath. Flowers solitary or sometimes paired, 3–4 cm; stalk 1–2 cm; bracteoles 2, narrow, deciduous; stalks and bracteoles densely long-hairy. Sepals ovate to broadly lanceolate, margin toothed, silky-hairy at base; persistent in fruit. Petals white, obovate to oblong, densely silky-hairy on back. Filaments whitish, united for half their length; anthers yellow. Ovary almost spherical, hairless. Styles 5, united. Capsule ovoid to conical, splitting at apex. Seeds discoid, narrowly winged. *SW China (Guangxi, Yunnan)*. H5. Summer.

The leaves are used locally for making a tea.

2. S. ovata (Cavanilles) Weatherby (*Malacodendron ovatum* Cavanilles; *S. pentagyna* L'Héritier). Illustration: Botanical Magazine, 3918 (1842); Journal of the Arnold Arboretum **55**: 190 (1974); Krüssmann, Manual of cultivated broad-leaved trees & shrubs **3**: 365 (1986).

Deciduous shrub or small tree to 6 m,

erect and bushy; bark greyish-brown, non-flaking, finely fissured; young growth often red-tinged. Winter buds minute, silvery-hairy, with 1 scale, concealed within winged leaf-stalks. Leaves mostly 7–15 × 4.5–7 cm; ovate to ovate-lanceolate, margin widely and minutely saw-toothed, hairless above, often densely hairy along veins beneath. Flowers solitary, 6–8 cm, cup-shaped; stalk stout, to 4 mm; bracteole 1, leafy, oblong, silky-hairy, persistent. Sepals oblong-lanceolate, margin long-hairy, persistent, spreading to erect in fruit. Petals 5 or 6, cream-white, occasionally stained bright red outside, obovate to almost circular. Filaments white, yellowish or rose, shortly united; anthers yellow or orange. Ovary conical, densely silky-hairy; styles 5, free. Capsule ovoid, splitting from apex. Seeds 2 per cell, flat-obovoid, narrowly winged. *SE USA*. H3. Summer.

Forma **grandiflora** (Bean) Kobuski differs from the species in its flowers with 5–8 petals and purple filaments.

3. S. malacodendron Linnaeus (*S. virginica* Cavanilles; *Malachodendron monogynum* Dumont de Courset). Figure 7(2), p. 34. Illustration: Botanical Magazine, 8145 (1907); Krüssmann, Manual of broad-leaved trees & shrubs **3**: 365 (1986).

Deciduous shrub or small tree to 7 m; bark silver-grey, non-flaking; young growth softly hairy. Winter buds minute, densely silvery-hairy, scales 2. Leaves mostly 5.5–11 × 2.5–5 cm, ovate to elliptic, margin minutely saw-toothed, hairless above, thinly hairy beneath, densely hairy along the midrib. Flowers solitary, 6.5–9 cm; stalks to 5 mm; bracteoles 2, leafy, ovate to almost circular, persistent. Sepals ovate to almost circular, margin hairy, persistent and reflexed in fruit. Petals snow-white, obovate to almost circular, outer surface silky-hairy towards the base. Filaments purple, shortly united; anthers bluish. Ovary almost spherical, densely silky-hairy; styles 4 or 5, united. Capsule almost spherical, often broader than long, splitting down ribs, apex remaining intact. Seeds 2–4 per cell, ovoid, angular, smooth, shiny, unwinged. *SE USA*. H3. Late spring–early summer.

4. S. serrata Maximowicz. Illustration: Botanical Magazine, 8771 (1918).

Small deciduous tree to *c.* 10 m; bark smooth, reddish-brown, and flaking; young growth often reddish and hairy. Winter buds with 2 reddish-brown scales. Leaves

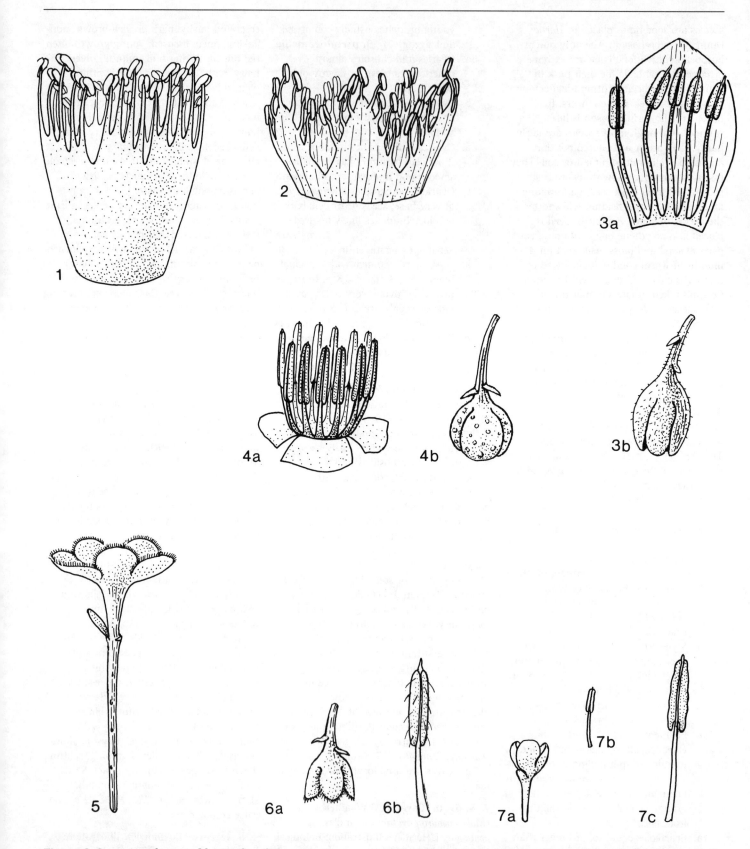

Figure 6. Stamens, calyces and bracteoles of Theaceae.
1, *Camellia saluensis* (stamens × 3). 2, *Gordonia lasianthus* (stamens × 3). 3a, *Visnea mocanera* (stamens × 6). 3b, *V. mocnera* (calyx and bracteoles × 2). 4a, *Ternstroemia gymnanthera* (stamens × 6). 4b, *T. gymnanthera* (calyx and bracteoles × 2). 5, *Schima wallichii* (calyx and bracteoles × 2). 6a, *Clevera japonica* (calyx and bracteoles × 2). 6b, *C. japonica* (stamens × 6). 7a, *Eurya japonica* (stamens × 18).

2.5–7.5 × 1.4–2.6 cm, ovate to lanceolate or oblanceolate, marginal teeth fine and incurved, often long-hairy, hairless above, hairless or hairy along the midrib beneath, sometimes with tufts of hair in vein-axils. Flowers solitary, 5–6.5 cm, cup-shaped; stalk to 5 mm; bracteoles 2, leafy, lanceolate, margin wavy and finely toothed, hairless, eventually deciduous. Sepals lanceolate, like bracteoles often reddish, hairless except for margins, spreading or reflexed in fruit, becoming woody at base. Petals cream-white, sometimes stained red outside, obovate to almost circular, finely silky-hairy on back. Filaments whitish, united for 1/3 of their length; anthers yellow. Ovary amost spherical, hairless. Styles 5, united for most of their length, ending in short stigmatic arms. Capsule almost spherical, shortly beaked, splitting at apex. Seeds 2 per cell, flat-obovoid, narrowly winged. *Japan (Honshu, Shikoku, Kyushu Islands)*. H3. Summer.

5. S. rostrata Spongberg (*S. sinensis* misapplied). Illustration: Journal of the Arnold Arboretum **55**: 199 (1974); Arnoldia **35**: 170, 173 (1975); Dansk Dendrologisk Årsskrift **5**(4): 29 (1981).

Deciduous shrub or small tree to 10 m, branching from the base; bark finely fissured, slate-grey. Winter buds with 2 or 3 finely downy scales. Leaves mostly 6–10.5 × 2.5–4.5 cm, ovate to elliptic, margin bearing fine-pointed teeth, hairless above, sparsely long-hairy beneath especially along the raised midrib, and tufts of hair in vein-axils. Flowers solitary, 5–6 cm; stalks 5–7 mm; bracteoles 2, leafy, ovate, margin wavy, persistent. Sepals ovate to obovate, margin wavy, often strongly reflexed in fruit, becoming woody at base. Petals white, obovate to almost circular, outer surface finely silky-hairy. Filaments whitish, united for 1/3 of their length; anthers yellow. Ovary almost spherical, densely silky-hairy at base, hairless above. Styles 5 or 6, united for most of their length, ending in short stigmatic arms. Capsule spherical or almost so, strongly ribbed, splitting from beaked apex. Seeds 4 per cell, flat-obovoid, notched at base, narrowly winged. *SE China (Zheijang, Jiangxi, Hunan)*. H2. Summer.

S. rostrata was formerly confused with *S. sinensis*, and some plants may be in cultivation under that name. It has been recognised and described only recently.

6. S. sinensis Rehder & Wilson (*S.*

gemmata Chien & Chang). Figures 5(3), p. 30 and 7(3), p. 34. Illustration: Botanical Magazine, 8778 (1918); Krüssmann, Manual of cultivated broad-leaved trees & shrubs **3**: pl. 130 (1986).

Deciduous shrub or small tree to 20 m; bark smooth, reddish-brown, peeling and flaking in large patches; young growth finely downy. Winter buds with 4 or 5 silvery overlapping scales. Leaves 3–10.5 × 1.4–4.5 cm, ovate, elliptic or oblanceolate, apex acuminate, margin finely saw-toothed, both surfaces sparsely and finely hairy, becoming hairless. Flowers solitary, 4–5 cm, fragrant; stalk usually 1–1.6 cm; bracteoles 2, leafy, ovate, margin saw-toothed, persistent. Outer 2 sepals similar to bracteoles, inner 3 much narrower; all spreading or erect in fruit, becoming woody at the base. Petals white, obovate to almost circular, outer surface silky-hairy. Filaments whitish, united for a 1/3 of their length; anthers yellow. Ovary cylindric, densely long-hairy; styles 5 or 6, united for most of their length. Capsule ovoid, splitting at apex. Seeds 2 per cell, flat-obovoid, very narrowly winged. *C China (Zhejiang, Anhui, Hubei)*. H2. Summer.

This species is grown more for its remarkably beautiful bark than for its flowers.

7. S. monadelpha Siebold & Zuccarini (*S. sericea* Nakai). Illustration: Morris Arboretum Bulletin **15**: 46 (1964); Kurata, Illustrated important forest trees of Japan **1**: 180, t. 90 (1971); Krüssmann, Manual of cultivated broad-leaved trees & shrubs **3**: 365 (1986).

Deciduous tree to 25 m, more usually a shrub in cultivation; bark smooth, mottled reddish-brown, thin and peeling. Young growth downy at first. Winter buds with 3–7 finely hairy scales. Leaves 4.2–10 × 1.6–3.4 cm, ovate, elliptic or lanceolate, margin finely saw-toothed, teeth rounded, incurved, both surfaces downy at first, often with tufts of hair in vein-axils beneath. Flowers solitary, rarely 2 together, 2.5–4 cm, spreading; stalk usually 7–10 mm; bracteoles 2, leafy, oblong, margin finely toothed, persistent. Sepals ovate to almost circular, erect or slightly reflexed in fruit. Petals white, obovate to almost circular, silky-hairy on the back. Filaments whitish, shortly united; anthers violet. Ovary conical, 5-angled, densely long-hairy; styles 5, united for most of their length, with short stigmatic arms. Capsule ovoid, splitting from apex. Seeds 2 per cell, irregularly obovoid, base

often notched, very narrowly winged. *Japan (Honshu, Kyushu & Shikoku)*. H2. Summer.

8. S. × henryae Li. Illustration: Morris Arboretum Bulletin **15**: 15 (1964).

Deciduous small tree to *c.* 10 m; bark smooth, peeling. Branchlets mostly zigzag, slightly compressed. Winter buds with 3 or 4 silky-hairy scales. Leaves mostly 5.8–8 × 3–3.5 cm, ovate to elliptic, margin finely saw-toothed, teeth rounded and long-hairy. Flowers solitary, *c.* 5 cm; stalk 1.3–2 cm; bracteoles 2, oblong to oblong-ovate, persistent. Sepals almost circular, margin finely saw-toothed, outer surface silky-hairy, erect or slightly reflexed in fruit, eventually deciduous. Petals white, obovate to almost circular, outer surface silky-hairy. Filaments whitish, united for a 1/3 of their length; anthers yellow. Ovary conical, long-hairy; styles 5, united for most of their length, ending in stigmatic arms. Capsule ovoid, splitting from the apex. Seeds 2 per cell; irregularly ovoid, narrowly winged, base and apex notched. *Garden origin*. H2. Summer.

A putative hybrid between *S. pseudocamellia × S. monadelpha*. It more nearly resembles *S. pseudocamellia*.

9. S. pseudocamellia Maximowicz (*S. grandiflora* Carrière; *S. koreana* Rehder; *S. pseudocamellia* var. *koreana* (Rehder) Sealy). Figure 5(2), p. 30. Illustration: Botanical Magazine, 7045 (1889); Morris Arboretum Bulletin **15**: 47 (1964); Kayata, Illustrated important forest trees of Japan **2**: 128, t. 64 (1974); Krüssmann, Manual of cultivated broad-leaved trees & shrubs **3**: 365 (1986).

Deciduous small tree to 20 m; bark smooth, mottled, peeling and flaking in thin strips. Winter buds with 2–4 finely downy scales. Leaves mostly 3.5–11.5 × 2.5–8 cm, broadly ovate to elliptic, margin shallowly saw-toothed. Flowers solitary, 5–6 cm, cup-shaped; stalk 1–4 cm; bracteoles 2, broadly kidney-shaped, much smaller than sepals, persistent. Sepals almost circular, densely silky-hairy outside, enclosing the maturing fruit, later erect, reflexed or partially deciduous. Petals white, almost circular or obovate, margins wavy, densely silky-hairy outside. Filaments whitish, united for 1/4 of their length; anthers bright yellow. Ovary conical, 5 or 6 angled, densely long-hairy; styles 5 or 6, united for most of their length, ending in short stigmatic arms. Capsule ovoid, strongly ribbed, splitting from the beaked apex. Seeds 4 per cell, ovoid, obovoid or oblong, essentially

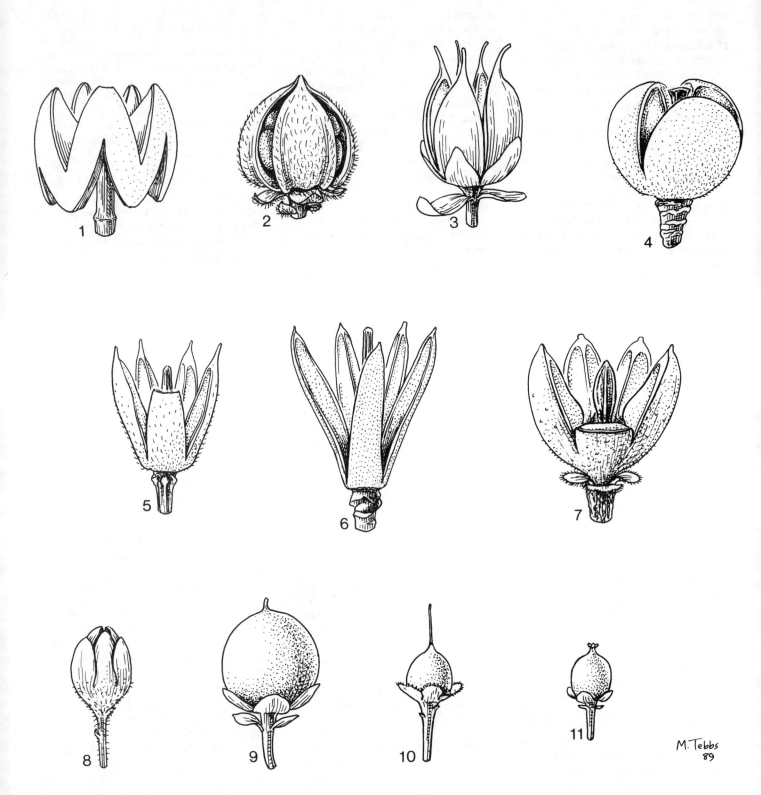

Figure 7. Fruits of Theaceae (× 2). 1, *Franklinia alatamaha*.
2, *Stewartia malacodendron*. 3, *S. sinensis*. 4, *Camellia saluensis*.
5, *Gordonia lasianthus*. 6, *G. axillaris*. 7, *Schima wallichii*.
8, *Visnea mocanera*. 9, *Ternstroemia gymnanthera*.
10, *Cleyera japonica*. 11, *Eurya japonica*.

wingless. *South Korea & Japan (Shikoku, Kyushu, Honshu)*. H2. Summer.

Plants sometimes recognised as var. *koreana* (Rehder) Sealy differ in having very slightly larger, flatter flowers and larger leaves. *S. pseudocamellia* is probably the best species for European gardens, although not the most spectacular, it is better suited to the climate than the more striking American species.

3. CAMELLIA Linnaeus
C. Whitefoord

Evergreen trees and shrubs, erect, occasionally weeping. Leaves usually leathery, occasionally papery, usually toothed, glossy deep green above, paler beneath; sometimes bearing small black or brown dots (cork-warts). Flowers almost terminal at the ends of new shoots, arising between the scales of axillary vegetative buds, solitary or clustered, stalked or almost stalkless, bisexual; occasionally fragrant. Bracteoles 2–8; sepals 5 or 6, sometimes intergrading with bracteoles and then termed perules. Petals 5–14, or in horticultural varieties sometimes more, usually basally united, white, pink or red, rarely yellow. Stamens usually numerous, of unequal length, rarely hairy, the outer filaments partially, often irregularly united, often forming a cup, basally attached to the corolla; anthers versatile. Sepals and stamens sometimes petal-like. Ovary superior, often hairy. Fruit a 1–5-celled capsule, leathery or woody, splitting from the apex, with a persistent axis. Seeds irregularly shaped, generally rounded, unwinged.

A genus formerly considered to have about 80 species, but recently estimated to contain nearer 200, with several thousand cultivars. It occurs in Tibet, India, Assam, Nepal, Sikkim, Bhutan, Burma, Thailand, Laos, Cambodia, Vietnam, China, Korea, Japan, Indonesia and the Philippines. *C. sinensis* is cultivated for tea, *C. oleifera* and *C. sasanqua* provide tea-oil, used for culinary and cosmetic purposes. The more usual species cultivated in Europe are included in the key, but most Camellias planted for ornament are hybrids or cultivars of *C. japonica*, *C. sasanqua* and *C. reticulata*. There is great interest in recently discovered species, particularly those with yellow flowers. *Camellias* prefer a peaty soil (pH 4.5–6.5) but will tolerate neutral well-drained loam with additional peat and leafmould at planting time. They appreciate an annual organic mulch. Camellias thrive in light woodland, but often flower more freely as wall plants. Although many species are hardy, they need shelter because their flowers are susceptible to damage from frost, wind, rain and early morning sun. Their shiny leaves are resistant to pollution; they are very suitable for urban gardens. Having a compact fibrous root-system, they do well in pots and tubs. A few species are grown under glass. Camellias are propagated by seed, cuttings or by grafting. They bloom during the cooler months from October to May.

Literature: Hume, H., *Camellias: kinds & culture* (1951); Hume, H., *Camellias in America* (1946), revised edition (1955); Sealy, J.R., unpublished manuscript on Camellia in cultivation in Europe, *A Revision of the Genus Camellia* (1958); Urquhart, B.L., *The Camellia* **1**: (1956), **2**: (1960); Macoboy, S., *The colour dictionary of Camellias* (1981); Chang Hung Ta, *Camellias* (1984).

1a. Flowers almost stalkless, perulate; bracteoles and sepals intergrading 2
  b. Flowers distinctly stalked; bracteoles clearly distinct from the sepals 11
2a. Perules deciduous, falling as the flower expands 3
  b. Perules persistent, at least until the flower has expanded fully 5
3a. Petals falling as the flower expands; flowers to 3 cm wide **1. kissi**
  b. Petals remaining after the flower has expanded fully; flowers 2.5–15 cm wide 4
4a. Leaves thinly leathery, blunt; stamens loosely spreading **2. sasanqua**
  b. Leaves stiffly leathery, acute; stamens compact and erect **3. oleifera**
5a. Filaments orange, spreading; petals white 6
  b. Filaments white, cream or red, more or less united in a cup or tube; petals white, pink or red 7
6a. Flowers 7–10 cm wide; styles 3, free to the base **4. crapnelliana**
  b. Flowers 12–15 cm wide; styles 5, united for most of their length **5. granthamiana**
7a. Filaments red; petals red; flowers 6 cm wide **6. hongkongensis**
  b. Filaments white or cream; petals pink or white or, if red, then flowers 3–4 cm wide 8

8a. Petals red; flowers 3–4 cm wide; leaves with scattered cork-warts beneath **7. japonica**
  b. Petals pink or white; flowers 2.5–5 cm wide; leaves without cork-warts beneath 9
9a. Leaves narrowly oblong to oblong-elliptic; flowers 2.5–5 cm wide **8. saluenensis**
  b. Leaves elliptic to broadly elliptic; flowers 5.5–15 cm wide 10
10a. Flowers 5.5–7 cm wide; leaf apex usually acute **9. pitardii** var. **yunnanica**
  b. Flowers 7–15 cm wide; leaf apex often blunt **10. reticulata**
11a. Flower-stalks 0.6–1.7 cm, nodding, bearing scars of fallen bracteoles 12
  b. Flower-stalks 3–6.5 mm, straight, bearing persistent bracteoles 13
12a. Flower-stalks 6–10 mm; petals 7 or 8; styles 3 **11. sinensis**
  b. Flower-stalks 1.2–1.7 cm; petals 11; styles 5 (rarely 4) **12. taliensis**
13a. Free part of stamens silky-hairy 14
  b. Free part of stamens hairless 15
14a. Bracteoles and sepals usually narrowly lanceolate; petals wavy-edged **13. salicifolia**
  b. Bracteoles and sepals usually widely ovate; petals flat **14. caudata**
15a. Flowers conical (funnel-shaped); petals pink **15. rosiflora**
  b. Flowers cup-shaped or spreading; petals white or sometimes tinged with pink or lilac outside 16
16a. Stamens shortly united basally, loosely spreading **16. cuspidata**
  b. Stamens united in a tube, erect 17
17a. Flowers cup-shaped, 2 cm wide; petals wholly white **17. tsaii**
  b. Flowers spreading, 3–4 cm wide; petals white, sometimes lilac-tinged outside **18. fraterna**

1. C. kissi Wallich (*C. drupifera* misapplied; *C. caduca* Brandis; *C. iniquicarpa* (Kochs) Cohen Stuart; *Thea bachmaensis* Gagnepain; *T. iniquicarpa* Kochs; *T. brachystemon* Gagnepain; *T. sasanqua* var. *kissi* (Wallich) Pierre). Illustration: Sealy, A Revision of the Genus Camellia, 197 (1958); Chang, Camellias, 52, figs. 3 & 4 (1984).

Shrub or tree to 13 m; branchlets hairy or hairless; bark deciduous. Leaves extremely variable, usually 5.5–11.5 × 1.7–3.5 cm, sometimes 4–5.5 × 1–2 cm, narrowly elliptic, oblong, oblanceolate to broadly ovate, apex blunt or acute; margin finely toothed towards the apex; thinly or thickly leathery, hairless, sometimes hairy at first. Flowers solitary or paired, 2.5–3 cm, spreading, fragrant. Perules rounded, strongly concave, sometimes velvety, falling before the flower expands. Petals white, 7 or 8, free, ovate, slightly notched, inner petals narrower, outer petals intergrading with perules, soon falling. Stamens spreading; filaments hairless, pale orange where shortly united, anthers yellow. Ovary densely silky-hairy. Styles 3, united basally or for most of their length, hairless or hairy. *Nepal, India, Sikkim, Vietnam, Laos, Cambodia, China (Yunnan, Guangxi, Guangdong, Hainan).* G1.

Although available, this species is seldom cultivated. The petals fall so easily that it is hardly ornamental.

2. C. sasanqua Thunberg (*Thea sasanqua* (Thunberg) Cels; *T. sasanqua* var. *thunbergii* Pierre; *Sasanqua vulgaris* Nees; *S. odorata* Rafinesque). Illustration: Botanical Magazine, 9591 (1940); Sealy, A Revision of the Genus Camellia, 203 (1958).

Shrub or tree to 5 m or more; branchlets very slender, sometimes hairy, densely leafy. Leaves 3–6 × 1.5–3 cm, larger in cultivation, elliptic to broad elliptic, apex blunt, marginal teeth rounded, thinly leathery, midrib and veins raised on both surfaces. Flowers solitary, c. 4.5 cm, spreading, slightly fragrant. Perules mostly hairy or velvety, falling as petals unfold. Petals white, 6–8, oblanceolate, rounded, deeply notched, margins often recurved. Stamens loosely spreading, yellow; filaments shortly united, hairless. Ovary densely silky-hairy. Styles 3, free or united for c. ½ their length. *S Japan.* H4. Late autumn–early winter.

Although hardy, this species is not as reliable as *C. japonica*, the flowers being more susceptible to damage. It flowers profusely; an ideal plant for the cold greenhouse. There are many cultivated variants and hybrids with white or pink, single or double flowers to 7.5 cm, with larger, thicker leaves. *C. sasanqua* × *C. japonica* produced **C. × vernalis** (Makino) Makino, the Spring Camellia (which now includes *C. × hiemalis* Nakai, the Cold Camellia, which is later-flowering). *C. × maliflora* is a sterile hybrid of garden origin, cultivated in China since the early 19th century. It was first thought to be a variety of *C. sasanqua* 'Palmer's Double' and later a double variant of *C. rosiflora*, but its ancestry remains obscure.

3. C. oleifera Abel (*C. drupifera* Loureiro; *C. sasanqua* misapplied; *C. sasanqua* var.

oleosa misapplied; *C. oleosa* misapplied; *Sasanqua oleifera* (Abel) Rafinesque; *Thea sasanqua* misapplied; *T. podogyna* Léveillé). Illustration: Edwards's Botanical Register, 942 (1825); Botanical Magazine, n.s., 221 (1954); Sealy, A Revision of the Genus Camellia, 206 (1958).

Shrub or tree to 7 m, stiffly erect; branchlets mostly hairy. Leaves 3.5–8 × 2–4 cm, oblong-elliptic or obovate, apex short, often blunt, margin finely toothed; stiff, leathery, midrib raised on both sides. Flowers solitary or paired, 4–6 cm, spreading, fragrant. Perules concave, margin broad, membranous, densely silky-hairy, falling as petals unfold. Petals white, 5–7, oblanceolate, deeply notched, twisted. Stamens erect, compact, yellow; filaments shortly united. Ovary densely white-hairy. Styles 3, united for most of their length. *N India, Burma, S China, SE Asian islands.* H5. Winter–early spring.

This species is widely cultivated throughout southern China for tea-oil, extracted from the seeds. It has been confused with *C. sasanqua*. The first introduction was a double form known as *C. sasanqua*, 'Lady Banks' Camellia. Fortune's Yellow Camellia ('Jaune') has anemone-form flowers, white with yellow petal-like stamens.

4. C. crapnelliana Tutcher (*Thea crapnelliana* (Tutcher) Rehder). Illustration: Sealy, A Revision of the Genus Camellia, 153 (1958); Chang, Camellias, 46 (1984).

Tree, 5–7 m, with smooth brick-red bark. Young branches stout, hairless. Leaves 8–16.5 × 4–5 cm, elliptic to obovate-elliptic, stiffly leathery, hairless, abruptly pointed, apex often blunt; marginal teeth minute, blunt and widely spaced; prominent venation and scattered cork-warts beneath. Flowers terminal, solitary, 7–10 cm. Perules 7 or 8, hard, cup-shaped, tawny-hairy, persistent. Petals white, 6–8, obovate, often notched, crumpled and wavy-edged. Stamens numerous; filaments united for c. ⅓ their length; anthers bright orange. Ovary densely hairy; styles 3, free to the base. *China (Hong Kong).* G1. Spring.

This striking species is known only from a single wild tree. It is remarkable for its red bark and white flowers with their conspicuous stamens. It is seldom cultivated in Europe or America, although it thrives in some Australian gardens.

5. C. granthamiana Sealy. Illustration: Sealy, A Revision of the Genus Camellia, 166 (1958); Botanical Magazine, n.s., 597

(1971); Chang, Camellias, 30 (1984).

Shrub or tree to 3 m; branchlets hairy at first. Leaves 7–10 × 2–4 cm, oblong-elliptic, apex short, bluntish; margin shallowly toothed, leathery, veins deeply impressed above, raised with scattered cork-warts beneath. Flowers solitary, spreading, 12–15 cm. Perules 12 or more, concave, leathery, hairy, persistent in fruit. Petals white, 8, obovate, deeply notched, crumpled. Stamens densely clustered, bright orange, drooping; filaments shortly united. Ovary densely hairy. Styles 5, united for over ½ their length, hairy, with recurved hairless stigmatic arms. *China (Guangdong & Hong Kong).* G1. Autumn–early winter.

This is the species with the largest, most spectacular flowers. It is commonly grafted on to *C. japonica*. Like *C. crapnelliana*, this species is known from only one wild tree.

6. C. hongkongensis Seemann (*C. japonica* misapplied; *Thea hongkongensis* (Seemann) Pierre). Illustration: Sealy, A Revision of the Genus Camellia, 170 (1958); Chang, Camellias, 30 (1984).

Shrub or slender tree to 10 m; branchlets thick, hairless. Leaves 7–12.5 × 2–4 cm, lanceolate to oblong-elliptic, apex short, often blunt, margin widely and minutely toothed, leathery, metallic bluish-brown when unfolding. Flowers solitary, c. 6 cm wide, almost stalkless, conical. Perules concave, stiffly leathery, velvety, persistent in fruit. Petals light crimson 6 or 7, broadly obovate, shallowly notched. Stamens erect, filaments red, united to ⅔ their length in a wide fleshy tube, basally strongly attached to corolla; anthers yellow. Ovary densely hairy. Styles 3, free. *China (Guangdong & Hong Kong).* G1. Winter.

7. C. japonica Linnaeus (*C. mutabilis* Paxton; *Thea japonica* (Linnaeus) Baillon; *T. camellia* Hoffmannsegg; *T. hozanensis* Hayata). Illustration: Botanical Magazine, 42 (1788); Sealy, A Revision of the Genus Camellia, 177 (1958); Chang, Camellias, 96, figs. 4,5 (1984).

Shrub or tree 7–10 m, much branched; branchlets hairless. Leaves 5–9 × 3–5 cm, broad-elliptic to elliptic-oblong, apex short, usually blunt; margin finely and bluntly toothed; stiffly leathery, hairless, scattered cork-warts beneath. Flowers solitary or paired, 3–4 cm wide, conical. Perules cup-shaped, velvety, persistent until the flower opens fully. Petals red, 5 or 6, obovate, more or less deeply notched, fleshy, outer petals silky-velvety outside.

Stamens numerous; filaments cream-coloured, united for ½–⅔ their length in a wide fleshy tube, basally attached to the corolla for 0.8–1.7 cm; anthers yellow. Ovary hairless. Styles 3, united for *c.* ¾ their length. *Japan (Ryukyu Islands), China (Sichuan, Shandong, Jiangxi).* H3. Spring.

This was the first *Camellia* species to reach Western gardens, in the early 18th century. Flowers vary in size, shape, form and colour in the thousands of cultivars derived from this extremely variable species. Although hardy, they may not always receive enough warmth in the growing season to set buds. *C. japonica* crossed with *C. saluenensis* gave rise to the hardy and free-flowering *C. × williamsii* hybrids; and crossed with *C. sasanqua* produced *C. × vernalis*.

Subsp. **rusticana** (Honda) Kitamura differs mainly in its low spreading, shrubby habit. *NW Japan (Honshu).* Often used in hybridising, it has been crossed with varieties of *C. japonica* to produce the Higo Camellias, a style of camellia traditionally cultivated in Japan. These bear single flowers with a conspicuous boss of flaring stamens and are included here with *C. japonica* hybrids.

8. C. saluenensis Bean (*C. pitardii* var. *lucidissima* (Léveillé) Rehder; *C. pitardii*, misapplied; *Thea camellia* var. *lucidissima* Léveillé; *T. pitardii* (Cohen Stuart) Rehder in part; *T. pitardii* var. *lucidissima* (Léveillé) Rehder). Figures 5(4), p. 30, 6(1), p. 32, and 7(4), p. 34. Illustration: Sealy, A Revision of the Genus Camellia, frontispiece, 184 (1958).

Shrub, 2–3 m, compact, densely leaved; branchlets hairy at first. Leaves 3–7 × 1.5–2.5 cm, narrowly oblong to oblong-elliptic, apex usually blunt; margin finely and regularly toothed, teeth black-tipped, hard, midrib raised, venation impressed above, midrib and venation raised beneath. Flowers solitary or paired, 2.5–5 cm wide, somewhat funnel-shaped at first. Outer perules minute, innermost *c.* 2 × 1.5 cm, usually persistent until petals unfold. Petals white to deep rose, 6 or 7, broadly obovate, notched. Stamens erect; filaments united for most of their length, forming a fleshy cream-coloured cup, attached to the corolla for *c.* 1 cm. Ovary densely white-hairy. Styles 3, united for most of their length. *China (Yunnan).* H5. Autumn–spring.

Although generally hardy, young plants need protection at first. This species was crossed with *C. japonica*, resulting in the

hardy *C. × williamsii* hybrids, perhaps the most suitable camellias for European gardens.

9. C. pitardii var. **yunnanica** Sealy (*Thea speciosa* Handel-Mazzetti, in part; *T. pitardii* (Cohen Stuart) Rehder, in part). Illustration: Botanical Magazine n.s., 633 (1972).

Tree to 7 m, open-branched; branchlets and leaves often hairy. Leaves 6–11 × 2–3.5 cm, elliptic to oblanceolate-elliptic, apex usually acute; margin regularly toothed, thinly leathery; venation visible on both surfaces. Flowers solitary or paired, 5.5–7 cm wide, conical. Perules 9 or 10, velvety, persistent until the petals unfold. Petals pink or white, 5 or 6, obovate, notched, the outer petals shorter, harder and velvety near the apex. Stamens erect; filaments united for ½–⅔ their length, basally strongly attached to the petals. Ovary densely hairy. Styles 3, united except apically. *W China (Yunnan, Sichuan, Guizhou).* H5. Spring.

This variety crossed with *C. japonica* 'Tiffany' produced the cultivar 'El Dorado'.

10. C. reticulata Lindley (*C. spectabilis* Bentham; *Desmitus reticulata* (Lindley) Rafinesque; *Thea reticulata* (Lindley) Kochs). Illustration: Botanical Magazine, 2784 (1827), 9397 (1935); Sealy, A Revision of the Genus Camellia, 182 (1958); Chang, Camellias, 93 (1984).

Shrub or tree to 15 m, loosely branched; branchlets stout, rigid, hairless. Leaves 8–11 × 3–5.5 cm, broad-elliptic, slenderly, sometimes abruptly, pointed, apex often blunt; margin finely and regularly toothed; midrib raised beneath; venation conspicuous on both sides. Flowers solitary, 7–15 cm. Perules persistent until petals unfold, densely silky-velvety. Petals rose-pink, 5–7, obovate to broadly spatula-shaped, notched. Stamens numerous; filaments yellowish, united for ½–⅔ their length, attached to the corolla for *c.* 1 cm. Ovary densely white-hairy. Styles 3, united for most of their length. *China (Yunnan).* H3. Late winter–early spring.

The variants first introduced, such as 'Captain Rawes', had broad-elliptic dull dark green leaves, conspicuously net-veined and semi-double flowers *c.* 15 cm; these were obtained from Chinese gardens, where they had been cultivated for centuries. Variants later discovered in the wild were single-flowered, *c.* 9 cm; some with dull dark green leaves, others with narrower shiny leaves. *C. reticulata* Wild Type, is a large compact shrub, hardier and of a better constitution than

the cultivars. About 105 cultivars of *C. reticulata* are grown in China; of these 44 have reached the West. Although they are successfully grown in America, Australia and New Zealand, they are not so suitable for European gardens. *C. reticulata* has been crossed with several other species, giving rise to numerous hybrids with spectacular flowers, often large, richly coloured and iridescent. **C. × heterophylla** Hu (probably *C. reticulata × japonica*), with large crimson single flowers, is well-known in cultivation.

Var. **rosea** (Makino) Makino (*C. uraku* Kitamura; *C. wabiske* forma *uraku* (Kitamura) Kitamura) and varieties **albo-rosea** Kitamura, **campanulata** Kitamura and **wabiske** (Makino) Kitamura (*C. wabiske* vars.), form the Wabiske Camellias, a group of cultivated camellias of obscure hybrid origin with small single thimble-like flowers, seldom cultivated in the West.

11. C. sinensis (Linnaeus) O. Kuntze (*C. thea* Link; *Thea sinensis* Linnaeus; *T. bohea* Linnaeus; *T. viridis* Linnaeus; *T. cantonensis* Loureiro; *T. cochinchinensis* Loureiro; *T. oleosa* Loureiro). Figure 5(5), p. 30. Illustration: Botanical Magazine, 998 (1807), 3148 (1832); Urquhart, Booth & Jones, The Camellia, **1**: t. 1 (1956); Sealy, A Revision of the Genus Camellia, 114 (1958).

Shrub or small tree, 1–6 m, densely bushy; branchlets usually hairless. Leaves 5–14 × 2–5 cm, lanceolate-elliptic, apex blunt; marginal teeth blunt, black-tipped and incurved; leathery. Flowers solitary or clustered, nodding, shallowly cup-shaped, 2–3.5 cm. Flower-stalk 0.6–1 cm. Bracteoles 2 or 3, soon falling. Sepals 5 or 6, unequal, sometimes velvety, persistent in fruit. Petals white, rarely pink, 7 or 8, broadly ovate. Stamens numerous, spreading; filaments shortly united, white; anthers yellow. Ovary densely white-hairy. Styles 3, united for *c.* ⅔ of their length. *China, Taiwan, Japan. Widely cultivated in Nepal, India, Thailand, Cambodia, Vietnam, and Malaysia.* H5. Spring.

This is the tea plant, a geographically variable species.

Var. **sinensis** is a smaller, hardier, more shrubby plant with smaller leaves and many more flowers. It is more interesting than decorative. *China (W Yunnan, Xizang).*

Var. **assamica** (Masters) Kitamura is the tropical form, a tree with larger, thinner leaves. *S China, Vietnam.*

12. C. taliensis (W.W. Smith) Melchior (*Thea taliensis* W.W. Smith). Illustration:

Botanical Magazine, 9684 (1948); Sealy, A Revision of the Genus Camellia, 128 (1958).

Shrub or small tree 2–7 m, loosely branched; branchlets hairless. Leaves 9–15.3 × 3.5–6 cm, elliptic to broad-elliptic, apex short, blunt; margin widely and bluntly toothed; venation visible both sides. Flowers single or clustered, nodding, 5–6.5 cm, spreading. Flower-stalk 1.2–1.7 cm. Bracteoles 2 or 3, soon falling. Sepals 5, unequal, persistent, sometimes hairy. Petals cream-white, c. 11, broadly ovate to obovate, rounded. Stamens numerous, spreading; filaments united for less than 1/2 their length, yellow; anthers darker. Ovary densely white-hairy. Styles 4 or 5, united for most of their length. *China (Yunnan)*. H5. Winter.

This species is closely related to *C. sinensis*, but differs mainly in its larger flowers with longer stamens and 4 or 5 styles.

13. C. salicifolia Bentham (*Camellia salicifolia* var. *longisepala* Keng; *Thea salicifolia* (Champion) Seemann; *T. salicifolia* var. *warburgii* Kochs; *Camelliastrum salicifolium* (Champion) Nakai). Illustration: Sealy, A Revision of the Genus Camellia, 99 (1958).

Shrub or small tree, branches slender, weeping; branchlets densely tawny-hairy, later hairless. Leaves 5–10 × 1–2.5 cm, narrowly oblong to elliptic-oblong, tapered to an acute apex; margin widely and shallowly toothed, the teeth incurved, black-tipped; firmly papery, smooth above, midrib hairy, with raised midrib and venation beneath. Flowers solitary, 1–2 cm, spreading, slightly scented. Flower-stalk c. 5 mm, densely hairy. Bracteoles 4 or 5, sepals 5; all usually narrowly lanceolate, densely hairy, persistent. Petals 5 or 6, ovate or obovate, margin wavy, backs velvety, white. Stamens erect; filaments united for 2/3 their length in a narrow tube, the free part silky-hairy. Ovary minute, densely hairy. Styles 3, united nearly to the apex. *S China (Guangxi, Fujian, Guangdong, Hong Kong), Taiwan*. G1. Winter.

This species is distinct in its weeping habit and long narrow, willow-like leaves, copper-coloured when young. It flowers profusely, but the small flowers are largely hidden by the leaves.

14. C. caudata Wallich (*C. axillaris* Griffith; *C. gracilis* Hemsley; *C. buisanensis* Sasaki; *Camellia caudata* var. *gracilis* (Hemsley) Yamamoto; *Thea caudata* (Wallich) Seemann; *T. gracilis* (Hemsley) Hayata; *Camelliastrum caudatum* (Wallich) Nakai; *C. gracile* (Hemsley) Nakai; *C. buisanense* (Sasaki) Nakai). Illustration: Sealy, A Revision of the Genus Camellia, 104 (1958).

Shrub or small tree to 7 m; branchlets very slender, downy at first, becoming hairless. Leaves 5–11 × 1–2.5 cm, elliptic to oblong-elliptic, apex long-acuminate, usually acute; margin finely, and sometimes bluntly, toothed; papery, sparsely hairy becoming hairless, the midrib remaining slightly downy. Flowers solitary or paired, c. 2 cm, spreading. Flower-stalk stout, 3–4 mm. Bracteoles 3 or 4, narrowly to widely ovate, hairy, persistent. Sepals 5, widely ovate, with a narrow membranous margin, densely silky-hairy, persistent. Petals white, 5, broadly obovate to almost circular, slightly hairy on the back. Stamens erect, filaments united for over 1/2 their length, white, the free part densely long-hairy; anthers yellow. Ovary densely hairy; styles 3, united except at the apex, hairy. *India (Assam) Bhutan, Burma, Vietnam, China (Zhejiang, Hainan) & Taiwan*. G1. Winter–spring.

15. C. rosiflora Hooker (*Thea rosaeflora* (Hooker) O. Kuntze). Illustration: Botanical Magazine, 5044 (1858); Sealy, A Revision of the Genus Camellia, 71 (1958).

Shrub or small tree with weeping branches; branchlets downy. Leaves 3.5 × 2–2.5 cm, elliptic to broadly elliptic, apex narrow, blunt-tipped; margin widely toothed; thinly leathery, sometimes variegated. Flowers solitary or clustered, c. 3.5 cm, conical. Flower-stalk stout, 4–9 mm, covered by bracteoles. Bracteoles 6–8, silky-hairy, persistent. Sepals 5 or 6, concave, silky-hairy. Petals rose-pink, 6–9, obovate, rounded to obtuse. Stamens erect; filaments united for 1/2 their length, hairless, white; anthers yellow. Ovary hairless. Styles 3, united for most of their length. *E China (Hubei, Sichuan, Jiangsu, Zhejiang)*. G1. Spring.

This species resembles *C. × maliflora* Lindley (*Thea maliflora* (Lindley) Seeman), but the leaves are larger, thicker and more widely toothed, and the flowers single.

16. C. cuspidata (Kochs) Wright (*Thea cuspidata* Kochs; *T. rosaeflora* var. *glabra* Kochs). Illustration: Botanical Magazine, 9277 (1932); Sealy, A Revision of the Genus Camellia, 56 (1958).

Shrub to 4 m, erect; branchlets slender, hairless. Leaves 5–9 × 1.5–3 cm, narrow or broad-elliptic to ovate, apex long, slender and blunt; margin finely toothed, the teeth gland-tipped, glossy, sometimes tinged brownish-purple, finely dotted beneath, copper-tinted when young. Flowers solitary (clustered in the cultivated forms), c. 3 cm, spreading. Flower-stalk c. 3 mm. Bracteoles 4, small, persistent. Sepals 5, cup-shaped. Petals white, 6 or 7, rounded, sometimes notched. Stamens numerous; filaments hairless, white, basally united, spreading; anthers golden. Ovary hairless. Styles 3, united for most of their length. *China (widespread, Shaanxi southwards)* H3. Spring.

17. C. tsaii Hu (*Thea tsaii* (Hu) Gagnepain; *T. fusiger* Gagnepain). Illustration: Hooker's Icones Plantarum, t. 3430 (1943); Sealy, A Revision of the Genus Camellia, 76 (1958); Chang, Camellias, 176 (1984).

Shrub or tree, 5–10 m; branchlets pendent, more or less densely hairy. Leaves 6–9.5 × 2–3 cm, oblong to oblong-elliptic, apex long, slender; margin minutely toothed, wavy-edged, softly papery, midrib impressed above, sparsely hairy with venation and midrib raised beneath; copper-coloured when young. Flowers solitary, borne in profusion, c. 2 cm wide, cup-shaped; fragrant. Flower-stalk 4–5 mm, hairless. Bracteoles 4 or 5, persistent. Sepals 5, small, rounded, silky-hairy. Petals white, flushed pink outside, 5, obovate, rounded, inner petals sometimes notched. Stamens strongly attached to the corolla; filaments united for 1/3–1/2 their length in a fleshy tube, hairless, white; anthers brown. Ovary hairless, pale green. Styles 3, shortly parted at the apex. *Burma, N Vietnam, China (Yunnan)*. G1. Spring.

This species resembles *C. cuspidata* but is easily distinguished by its hairy branchlets, softer leaves and longer flower-stalks. **C. lutchuensis** Ito is a hardier Japanese species closely related to *C. tsaii*, with more strongly scented flowers. Crossing *C. lutchuensis* with *C. japonica* ssp. *rusticana* resulted in the hybrid 'Fragrant Pink', which is hardy in a sunny position. Its scent is only perceptible indoors.

18. C. fraterna Hance (*Thea fraterna* (Hance) O. Kuntze; *T. rosaeflora* var. *pilosa* Kochs; *Theopsis fraterna* (Hance) Nakai). Illustration: Sealy, A Revision of the Genus Camellia, 67 (1958).

Shrub 1–5 m; branchlets densely and coarsely hairy, becoming hairless. Leaves 4–8 × 1.5–3.5 cm, elliptic or oblong-elliptic, apex slender, blunt; margin widely and bluntly toothed, the teeth incurved and black-tipped; leathery, midrib hairy, coarsely hairy beneath at first. Flowers solitary, 3–4 cm, spreading, fragrant. Flower-stalk *c.* 3–4 mm, stout, coarsely hairy. Bracteoles 5, persistent; sepals 5, cup-shaped, both densely and coarsely hairy. Petals white, sometimes flushed lilac outside, 5 or 6, obovate, rounded or notched. Stamens few, erect; filaments hairless, white, united for ½–⅔ their length; anthers orange. Ovary hairless; styles 3, united almost to the apex. *E China.* G1. Late spring.

Crossing with a *C. japonica* cultivar, 'Akebono', produced the hybrid 'Tiny Princess', a small shrub with small pink semi-double flowers.

Hybrids and Cultivars.
The hybrids and cultivars listed are some of those which are frequently mentioned in *Camellia* literature or have received horticultural awards. For recent introductions, specialist catalogues and the publications of *Camellia* societies should be consulted.

Flower sizes.
Very large, over 12.5 cm wide.
Large, 10–12.5 cm wide.
Medium, 7.5–10 cm wide.
Small, 5–7.5 cm wide.

Flower shapes and forms.
Single: one row of not more than 8 regular, irregular or loose petals, and conspicuous stamens.
Semi-double: one or more rows of regular, irregular or loose petals and conspicuous stamens.
Anemone-form: one or more rows of large outer petals, flat or wavy, the centre a mass of intermingled stamens and petal-like stamens.
Paeony-form: a deep rounded flower of mixed petals, petal-like stamens and sometimes stamens.
Double: overlapping petals, cupped, showing stamens in the centre.
Formal-double: many rows of petals, fully overlapping, with no stamens.
Informal-double: many rows of petals, not fully overlapping, with no stamens.

C. japonica cultivars.
These camellias are fairly hardy (H3), but those with larger or paler flowers will need protection from wind and frost. Most *C. japonica* cultivars flower from spring to early summer.

'Adolphe Audusson': strong growth, a dense round bush; flowers large, red,
semi-double.
'Alba Plena' ('Alba Grandiflora'): slow growing, erect, bushy; flowers large, white, formal-double. Probably the best double-white for general planting.
'Alba Simplex': erect, with leaves *c.* 13 cm; flowers large, white, single, opening flat, with conspicuous stamens.
'Althaeiflora': flowers large, dark red, paeony-form.
'Apollo': vigorous and open; flowers medium-sized, rose-red, sometimes white-blotched, semi-double, with a conspicuous boss of stamens. Leaf tips twisted. One of the most satisfactory camellias for open ground, of better habit than 'Adolphe Audusson'.
'Arejishi' ('Are-jishi', 'Arajishi'): growth vigorous and open; leaves large, thick, coarsely toothed. Flowers medium-sized, rich red, dense, paeony-form, with wavy petals.
'Blood of China': vigorous and compact; flowers large, deep salmon-red, semi-double to loose paeony-form. Fragrant.
'Chandleri Elegans' = 'Elegans'
'Contessa Lavinia Maggi': vigorous growth; flowers large, white or pink broadly striped cerise, formal-double. Considered the best variegated camellia for the garden.
'Coquetii': growth slow, erect, bushy; flowers medium-sized, soft red, double, but usually showing some stamens.
'Devonia' ('Devoniensis'): growth vigorous, erect; flowers medium-sized, white, single, rather cup-shaped.
'Donckelaeri': growth slow and bushy, rather erect; flowers large, soft red, marbled pink and white, semi-double. Free flowering and formerly excellent, but plants now weakened by virus infection.
'Elegans' ('Chandleri Elegans'): growth vigorous and spreading; flowers large, rosy pink, anemone-form. A reliable cultivar for general planting.
'Fimbriata Superba' = 'Fred Sander'.
'Fred Sander': growth strong, dense, erect; flowers large, crimson, semi-double, petals wavy and fringed, with many yellow stamens. A sport of 'Tricolor'.
'Gauntlettii' ('Grandiflora Alba', 'Lotus', 'Sode Kakushi', 'Sode-gashuki'): growth vigorous with spreading branches, but with a weak constitution; flowers very large, white, semi-double.
'Gloire de Nantes': growth medium, compact and erect; flowers large, rose pink, semi-double. A very reliable
cultivar with a long flowering season.
'Grandiflora Alba' = 'Gauntlettii'
'Grandiflora Rosea' = 'Lady Clare'
'Jupiter': growth vigorous; narrowly erect; flowers medium-sized, scarlet sometimes blotched with white, single or semi-double with conspicuous stamens. It is free-flowering in the open and sometimes sets seed.
'Konron-Koku' ('Kouron-jura', 'Kuro-Tsubaki', 'Black Camellia'): growth medium, semi-erect; flowers medium, very dark blackish red, semi-double or formal-double. A reliable shrub for the open ground.
'Lady Clare' ('Akashigata', 'Grandiflora Rosea'): vigorous; low and spreading, loose; leaves large. Flowers large, deep pink, semi-double. An old cultivar, but one of the best. It needs some protection.
'Lady de Saumarez': vigorous, compact; flowers medium, bright red, spotted white, semi-double. A sport of 'Tricolor'. A reliable cultivar.
'Lady Vansittart': growth slow; compact, upright habit; leaves wavy. Flowers medium-sized, white, striped and blotched rose-pink, semi-double with wavy edged petals. A good plant for small gardens.
'Lotus' = 'Gauntletii'
'Magnoliiflora' ('Hagaromo'): growth medium, compact; leaves light green. Flowers medium-sized, blush-pink, semi-double.
'Mathotiana' ('Mathotiana Rubra'): growth compact and upright; flowers large, crimson, double to formal-double.
'Mathotiana Alba': vigorous, open habit, erect; flowers large, white, formal-double, occasionally spotted pink; easily damaged, needs protection.
'Mathotiana Rosea': growth vigorous, compact, erect; flowers large, clear pink, formal-double. Needs protection.
'Mathotiana Rubra' = 'Mathotiana'.
'Rubescens Major': growth bushy and compact; flowers large, rose-red or crimson with darker veins, informal double. Very reliable in open ground.
'Sieboldii' = 'Tricolor'.
'Tricolor' ('Sieboldii'): compact, spreading habit; flowers medium-sized, white with pink stripes, slightly cupped, semi-double. Very reliable in open ground.

C. reticulata hybrids and cultivars. Although very spectacular, many of the hybrids developed in America, New Zealand and Australia are not reliably hardy in European gardens. Most of the *C. reticulata* hybrids raised in Europe have *C. saluenensis* in their parentage, often through crossing with *C.* × *williamsii*; these are hardier than the hybrids derived from *C. reticulata* × *C. japonica*. These hybrids need more sunshine than most Camellias, and prefer a lighter, better drained soil. H5–G1.

'Barbara Hillier' (probably *C. reticulata* × *C. japonica*, this clone is often placed under *C.* × *heterophylla* Hu). Flowers large, satin-pink, single.

'Buddha' (*C. reticulata* × *C. pitardii* var. *yunnanica*). Vigorous, erect; flowers very large, rose-pink, semi-double.

'Captain Rawes': flowers very large, carmine rose-pink, semi-double.

'Crimson Robe' ('Dataohong'): flowers very large, carmine, semi-double with wavy, crinkled petals; leaves often edged with cream and yellow.

'Flore Pleno' = 'Robert Fortune'

'Innovation' (*C. reticulata* 'Crimson Robe' × *C. williamsii* 'William's Lavender): flowers large, light crimson, paeony-form; weatherproof.

'Leonard Messel' (*C. reticulata* × *C. williamsii* 'Mary Christian'): a large shrub; flowers large, clear pink, semi-double.

'Pagoda' = 'Robert Fortune'

'Purple Gown' ('Tzepao', 'Zipao', a Chinese cultivar): flowers large to very large, dark purple-red, flecked with white or wine-red, double to paeony-form.

'Robert Fortune' ('Pagoda', 'Flore Pleno', 'Songzilin' or 'Pine-Cone-Scale'): growth compact; flowers large, deep crimson, double.

'Semi-Plena' = 'Captain Rawes'.

'Tali Queen' ('Dali Cha'): the correct name for the Chinese cultivar distributed in the West as 'Noble Pearl'). Growth medium, erect; flowers very large, turkey-red to deep pink, semi-double with wavy petals.

'Tzepao' = 'Purple Gown'

C. cuspidata hybrids and cultivars.
These are small to medium-sized shrubs with slender branches and lanceolate leaves, generally resembling *C. cuspidata*. H3.

'Cornish Snow' (*C. cuspidata* × *C. saluenensis*): medium-sized spreading shrub, with copper-tinted young leaves; flowers small, white flushed pink outside, single, borne abundantly along the slender branches.

'Cornish Spring' (*C. cuspidata* × *C. japonica*): flowers medium-sized, light pink, single.

C. saluenensis hybrids.
C. saluenensis is fairly tolerant of low temperatures, and its hybrid progeny are even hardier.

'Francie L.' (*C. saluenensis* × *C. reticulata* 'Buddha', or possibly *C.* × *williamsii* 'Apple Blossom'): a loose shrub, best trained on a wall; flowers large, rosy crimson, semi-double.

'Grand Jury' (*C. saluenensis* × 'Salutation'): spreading habit; flowers large, apricot-pink, paeony-form; fairly weatherproof. Spring–early summer.

'Inspiration' (*C. saluenensis* × *C. reticulata*, or possibly a variant of *C.* × *williamsii*): flowers large, deep mauve-pink, semi-double.

C. × williamsii hybrids (*C. japonica* × *C. saluenensis* cultivars). These hybrids are generally the best all-round garden camellias for cultivation in Europe; they are hardier than *C. saluenensis* and flower regularly. Although they generally resemble *C. japonica* in habit, the flowers are more like those of *C. saluenensis*. They bear many flowers, and unlike those of *C. japonica*, the flowers fall as they wither.

'Bow Bells': flowers bright rose, semi-double; long flowering season.

'C.F. Coates' (the Fishtail Camellia): leaves 3-lobed at the apex; flowers medium-sized, deep rose, single.

'Debbie' (*C. saluenensis* × *C. japonica* 'Debutante'): flowers medium to large, rosy pink, semi-double to paeony-form.

'Donation' (*C. saluenensis* × *C. japonica* 'Donckelaeri'): vigorous, erect habit; flowers large, soft rosy pink with darker veins, semi-double.

'Elizabeth Rothschild' (*C. saluenensis* × *C. japonica* 'Adolphe Audusson'): flowers medium, soft pink, semi-double.

'Elsie Jury' (*C. saluenensis* × *C. japonica* 'Pukura'): narrow habit; flowers large, deep pink, anemone to paeony-form.

'Francis Hanger' (*C. saluenensis* × *C. japonica* 'Alba Simplex'): growth erect, leaves strongly wavy; flowers medium-sized, white, single.

'J.C. Williams': growth loose; flowers medium, phlox-pink, single.

'Lady Gowrie': growth vigorous, compact; flowers large, pink, semi-double.

'Mary Christian': growth vigorous, erect; flowers small, light pink, single, cup-shaped.

'St. Ewe': tall, bushy habit; flowers medium, phlox-pink, single, cupped.

C. sasanqua hybrids.
These camellias are quite hardy, but their flowers are liable to weather damage. They are best grown against a wall, and need a sunny position to flower well. Leaves small, elliptic; flowers single, semi-double to double, with conspicuous flaring stamens, the petals often ruffled, with shaded margins. H4. Autumn–early spring.

'Narumi-gata' (*C. sasanqua* var. *oleifera* misapplied). Flowers large, creamy white, shaded pink margins, single; slightly fragrant.

'Fukuzu-tsumi' is similar, but flowers later than the previous cultivar.

4. GORDONIA Ellis
C. Whitefoord

Evergreen trees and shrubs. Leaves usually shortly stalked, blade leathery, hairless, margin usually toothed. Winter buds naked, silky-hairy. Flowers usually at branch tips, usually solitary, rarely 2 or 3, sometimes shortly racemose with reduced leaves, bisexual, stalked, sometimes fragrant. Bracteoles 2–12, beneath calyx, sometimes intergrading with sepals, deciduous. Sepals 4 or 5, unequal, persistent. Petals 5–7, shortly united, outer petals somewhat transitional to sepals. Stamens numerous, the outer whorl at least on cup-shaped 5-lobed disc at base of corolla; filaments sometimes irregularly united; anthers versatile. Ovary almost spherical; styles 5, united; stigma lobed. Fruit a woody capsule, ovoid or cylindric, 3–6-celled, splitting from apex to base, central axis persistent. Seeds flat, apically winged.

A genus of 70 species from South-East Asia and the southeastern USA; few are suited to European gardens. They are mostly planted for ornament, but the wood of *G. lasianthus* was used formerly for cabinet-making, and its bark for tanning. Gordonias need well-drained lime-free sandy or peaty soil; they are propagated by seed, layering or from cuttings placed under glass.

1a. Flowers very shortly stalked; petals notched   2
  b. Flowers long-stalked; petals entire   3
2a. Bracteoles 5–12; sepals hard; young growth grey; leaves leathery
           1. axillaris
  b. Bracteoles 4–6; sepals herbaceous; young growth carmine; leaves thickly leathery   **2. chrysandra**

3a. Bracteoles 4–7; sepals densely
   white-velvety   **3. lasianthus**
 b. Bracteoles 2; sepals hairless outside,
   fringed   **4. sinensis**

1. G. axillaris (Ker-Gawler) D. Dietrich
(*Camellia axillaris* Ker-Gawler; *G. anomala*
Sprengel; *Polyspora axillaris* (Ker-Gawler)
Sweet). Figures 5(7), p. 30 and 7(6),
p. 34. Illustration: Botanical Magazine,
2047 (1819), 4019 (1843); Bean, Trees &
shrubs hardy in the British Isles, edn 8, **2**:
pl. 44 (1973).

Large shrub or small tree, 7–10 m;
young growth grey, finely silky-hairy.
Leaves 7–16 × 2–6 cm, elliptic-oblong
to oblanceolate, apex usually blunt, dark
shining green, margin entire or shallowly
toothed towards apex, hairless above,
sparsely silky-hairy beneath at first.
Flowers very shortly stalked, 7.5–13 cm,
fragrant. Bracteoles 5–12, covering stalk,
intergrading with sepals. Sepals almost
circular, hard, becoming woody, finely
silky-hairy. Petals cream-white, widely
obovate, deeply notched, silky towards
base. Filaments yellow; anthers orange-
yellow. Ovary silky-hairy. Capsule
cylindric, 5-furrowed. *Vietnam, China &
Taiwan.* H5. Winter–spring.

2. G. chrysandra Cowan. Illustration:
RHS dictionary of gardening **2**: 912
(1951); Botanical Magazine n.s., 285
(1956).

Large shrub, 3–10 m. Young growth
carmine, finely hairy at first. Leaves
6.5–12.5 × 2.5–3.5 cm, obovate to
lanceolate, apex acute, blunt or rounded,
thickly leathery, glossy, margin slightly
toothed towards apex, hairless above,
sparsely silky-hairy beneath at first.
Flowers very shortly stalked, 7.5–10 cm,
fragrant. Bracteoles 4–6, covering stalk,
intergrading with sepals. Sepals almost
circular, herbaceous, minutely white-
hairy at first. Petals cream-white, widely
obovate, deeply notched, silky-hairy
towards base. Stamens deep yellow. Ovary
densely yellow-velvety. Capsule narrowly
ovoid, 5-furrowed apically. *China (Yunnan).*
G1. Winter–early spring.

G. chrysandra differs from *G. axillaris*
mainly in its smaller flowers and smaller,
thicker leaves.

3. G. lasianthus (Linnaeus) Ellis
(*Hypericum lasianthus* Linnaeus). Figures
5(6), p. 30, 6(2), p. 32 and 7(5) p. 34.
Illustration: Botanical Magazine, 668
(1803).

Shrub or tree to 25 m, usually much

smaller, erect. Leaves 5–15 × 2.5–4.5 cm,
narrowly elliptic to oblanceolate, apex
blunt or acute, thickly leathery, glossy,
hairless, margin shallowly toothed towards
the apex. Autumn colouring scarlet.
Flowers 5–8 cm, fragrant; stalks 3–7.5
cm, usually slender, widening towards the
apex. Bracteoles 4–7, small, well-spaced.
Sepals almost circular, fringed, densely
white-velvety outside. Petals white,
obovate, concave, margins wavy, often
fringed, silky on the back. Stamens yellow,
much shorter than the petals. Ovary white-
silky-hairy. Capsule ovoid, sharply pointed,
silky-hairy. *SE USA.* H5–G1. Late summer.

This species needs long hot summers to
flower well.

4. G. sinensis Hemsley & Wilson.

Tree to 15 m, usually a shrub in
cultivation; young branches grey-brown.
Leaves 10–13.5 × 4–6.5 cm, elliptic or
oblong to oblanceolate, apex acute,
leathery, glossy, margin widely toothed.
Flowers 5–6.5 cm, on stalks *c.* 4 cm, stout,
widening towards the apex. Bracteoles 2.
Sepals almost circular, hairless outside,
fringed with white hairs. Petals white,
obovate, silky-hairy at the base. Stamens
yellow, much shorter than the petals.
Ovary silky-hairy. *China (Sichuan).* G1 or
H5. Winter–early spring.

5. SCHIMA Blume
C. Whitefoord
Evergreen trees or shrubs to 47 m; bark
whitish. Young growth silky-hairy at first.
Leaves shortly stalked, blades mostly 5–17
× 3–8 cm, oblong, lanceolate, ovate or
obovate, usually acuminate; apex acute,
rarely blunt; margin entire to saw-toothed;
papery, thinly or occasionally thickly
leathery, glossy, red-veined, often reddish
beneath and reddish when young. Flowers
bisexual, axillary, usually solitary,
sometimes crowded, racemose, 1.5–7 cm,
fragrant; stalks 1–6 cm, slender or stout,
usually thickened at the apex, bearing
2 deciduous bracteoles. Sepals 5 or 6,
almost circular, silky-hairy, fringed, much
smaller than the petals, persistent. Petals
scarlet in bud, later white, 5 or 6, obovate,
rounded, basally united, hairy outside.
Stamens yellow, numerous, in 3–5 series,
free, attached to corolla base; anthers
versatile. Ovary almost spherical, densely
silky-hairy, 5 (occasionally 6- or 7)-celled.
Styles 5 or 6, united; stigma lobed. Fruit a
woody capsule, red turning black, almost
spherical, splitting apically to about half-
way, central axis persistent. Seeds 3 per

cell, kidney-shaped, winged nearly all
round.

A genus formerly thought to contain
about 15 species, but now reduced to one
extremely variable species with several
geographically distinct subspecies and
varieties. *S. wallichii* occurs in subtropical
or warm temperate regions from the
eastern Himalayas to southwest China,
Japan, Indonesia and Malaysia. *Schima*
requires sheltered woodland conditions
in lime-free soil. It is hardy only in the
mildest and dampest of European gardens,
but succeeds in a conservatory. It may be
propagated by seed or cuttings.

Literature: Bloembergen, S., A critical
study in the complex-polymorphous genus
Schima (Theaceae), *Reinwardtia* **2**:
133–183 (1952).

1. S. wallichii (de Candolle) Korthals (*S.
argentea* E. Pritzel; *S. khasiana* Dyer; *S.
noronhae* Blume; *S. crenata* Korthals; *S.
superba* Gardner & Champion; *S. forrestii*
Airy-Shaw; *S. sericea* Airy-Shaw, *Cleyera
mertensiana* Siebold & Zuccarini; *Gordonia
wallichii* de Candolle). Figures 6(5), p. 32
and 7(7), p. 34. Illustration: Botanical
Magazine, 4539 (1850), 9558 (1939);
Bean, Trees & shrubs hardy in the British
Isles, edn 8, **4**: pl. 38 (1980); Krüssmann,
Manual of cultivated broad-leaved trees &
shrubs, **3**: 310 (1986).

*India & Nepal, China, Taiwan, Indo-China,
Japan (Ryukyu Islands, Ogaswara-gunto),
Malaysia & Indonesia.* H5–GI. Late summer.

Subsp. **noronhae** var. **superba**
(Gardner & Champion) Bloembergen (*S.
argentea* E. Pritzel). This is the hardiest
variant in cultivation. H5.

Subsp. **wallichii** var. **khasiana** (Dyer)
Bloembergen, (*S. khasiana* Dyer).
Illustration: Botanical Magazine, n.s., 143
(1951). This is a variant with large,
toothed, thinly leathery leaves. G1.

6. VISNEA Linnaeus filius
C. Whitefoord
Evergreen trees, 6–8 m, trunk slender,
branching shortly, bark grey. Branchlets
reddish, angular, warty and sparsely hairy.
Leaves shortly stalked, usually 4–6 ×
1.5–2.5 cm, elliptic, lanceolate or
oblanceolate, apex blunt or notched,
margin entire to widely and shallowly
toothed, leathery, glossy, venation obscure,
midrib reddish beneath, sparsely long-
hairy beneath at first. Flowers solitary
or clustered, nodding, 6–10 mm; Flower-
stalks slender, 0.4–1.5 cm; bracteoles

2, minute, well-spaced. Sepals 5, ovate, fleshy, sparsely hairy outside. Petals 5, cream-white, broadly obovate, nearly free. Stamens *c.* 20; attached to the base of the corolla; anthers basifixed. Ovary ovoid, densely hairy, usually 3-celled; styles usually 3, free. Fruit red, turning purple-black, receptacle and erect sepals becoming greatly enlarged, almost spherical, obscurely angular. Seeds 2 per cell, ovoid, unwinged.

A genus of one species from Madeira and the Canary Islands.

1. V. mocanera Linnaeus filius. Figures 6(3), p. 32 and 7(8), p. 34. Illustration: Hooker, W.J., Icones Plantarum, **3**: t. 253 (1840); Revue Horticole: 212 (1882); Kunkel, Flora de Gran Canaria, **1**: 34 (1971); Kunkel, Arboles y arbustos de las islas Canarias, **1**: 105 (1981).

Madeira, Canary Islands. G1–H5. Summer.

An attractive tree often cultivated in the gardens of its native islands. Syrup extracted from the fruit is used locally in food and medicine. Except in the Mediterranean area it is probably better suited to greenhouse conditions. It is easily propagated by cuttings.

7. TERNSTROEMIA Linnaeus filius
C. Whitefoord
Evergreen trees and shrubs. Leaves stalked, usually leathery, margins usually entire, clustered at the ends of branches. Flowers axillary, solitary, stalked, bisexual, or unisexual. Bracteoles 2, almost opposite, minute, persistent. Sepals 5. Petals 5, rarely 6, free or basally united. Stamens 15–300, attached to the base of the corolla, in 2 or more series, filaments united basally, anthers hairless, basifixed. Ovary 2- or 3-celled; styles 2 or 3, united for most of their length, the base at least persistent. Fruit a fleshy capsule. Seeds few, large, unwinged.

A genus of about 85 species, widely distributed in the tropics. One species only is grown in Europe. It needs similar conditions to *Camellia* species: well-drained fertile, acid soil and a sheltered site, or under glass in less favourable areas. It is propagated by seed and cuttings.

1. T. gymnanthera (Wight & Arnott) Sprague (*T. japonica* misapplied; *Cleyera gymnanthera* Wight & Arnott). Figures 6(4), p. 32 and 7(9), p. 34. Illustration: Wight, Icones plantarum indiae orientalis, **1**: t. 47 (1838).

Shrub or small tree to 3 m. Young growth reddish-brown, warty, hairless.

Leaf-stalks short, red. Leaves *c.* 9 × 4.5 cm, elliptic to oblanceolate, apex abruptly narrowed, blunt or rounded, margin entire, thickly leathery, dark green and glossy. Flowers solitary, nodding, often in axils of fallen scales on lower naked part of shoot, 0.8–1.5 cm, fragrant. Flower-stalks 1–2 cm, thickened apically, bearing 2 triangular persistent bracteoles. Sepals almost circular, fleshy, margin sometimes minutely glandular, persistent in fruit. Petals greenish yellow to white, oblong to obovate, cupped, fleshy, united at base, margin thin, sometimes glandular, sometimes notched or tattered. Ovary ovate, hairless. Styles 2 or 3, united for most of their length, base stout, persistent, with large stigmatic crests. Capsule almost spherical or broadly ovoid, yellow, turning red and later splitting irregularly. *Nepal, India, China, Taiwan, S Japan, South Korea.* G1–H5. Late summer.

In cultivation this is a slow-growing shrub with a bushy habit, mostly grown for its foliage, although both flowers and fruit are attractive.

'Variegata' has leaves marbled with grey and margins cream-white, turning pink.

8. CLEYERA Thunberg
C. Whitefoord
Trees or shrubs, evergreen or deciduous. Leaves 2-ranked, stalked, leathery, rarely papery, margins entire or toothed. Flowers solitary or clustered, sometimes on short axillary shoots; bisexual, stalked; stalks thickened at the apex, bearing 2 or 3 minute well-spaced deciduous bracteoles. Sepals 5, unequal. Petals 5, shortly united at base. Stamens 25–30, attached to the base of the corolla; anthers basifixed, bristly-hairy. Ovary 2- or 3-celled. Styles 3 or 4, united for most of their length, usually persistent in fruit. Fruit a spherical or ovoid berry, many seeded. Seeds curved, unwinged, with scanty endosperm.

A genus of 17 species, 1 from Asia and 16 from Mexico to Panama and the West Indies. They are large slow-growing shrubs, grown for their neat habit and foliage, which is reddish when young. They need ordinary, moist, well-drained soil and a sheltered site, but are probably better suited to greenhouse culture; they are propagated by seed or cuttings.

Literature: Kobuski, C.E., Studies in the Theaceae 2, *Cleyera, Journal of the Arnold Arboretum* **18**: 118–129 (1937), Studies in the Theaceae 7, The American species of the genus *Cleyera, Journal of the Arnold Arboretum* **22**: 395–416 (1941).

1a. Leaves firmly papery; margins
　　toothed　　　　　　**1. theoides**
　b. Leaves leathery; margins entire　**2.**
　　　　　　　　　　　　japonica

1. C. theoides (Swartz) Choisy (*C. panamensis* (Standley) Kobuski; *Eroteum theaeoides* Swartz; *Freziera theoides* (Swartz) Swartz; *Eurya theoides* (Swartz) Blume; *E. panamensis* Standley). Illustration: Botanical Magazine, 4546 (1850).

Evergreen tree to 18 m, but usually a shrub to 2 m in cultivation. Young growth minutely hairy. Leaves stalked, blades 3–10 × 2–4.5 cm, obovate to elliptic, sometimes acuminate, apex acute; margin bluntly toothed; firmly papery, very dark green above. Flowers solitary, rarely 2 or 3, nodding, *c.* 1 cm; stalks 1–2 cm. Sepals almost circular, hairless, minutely fringed. Petals cream-white, obovate, rounded, reflexed at the apex. Stamens 25–30; anthers reddish. Ovary ovoid, hairless. Berry purple-black, spherical, 3-celled; seeds 6–8. *Mexico to Panama, Jamaica.* G1–H5. Late summer.

2. C. japonica Thunberg (*C. japonica* var. *kaempferiana* (de Candolle) Sealy; *C. ochnacea* de Candolle; *C. ochnacea* var. *kaempferiana* de Candolle; *Eurya ochnacea* (de Candolle) Szyszylowicz; *Ternstroemia japonica* Thunberg; *Tristylium ochnaceum* (de Candolle) Merrill). Figures 6(6), p. 32 and 7(10), p. 34.

Evergreen tree to 12 m, or in cultivation shrub 1–3 m; bark smooth, grey-brown. Young growth hairless. Leaf-blades 5–7 × 2–3 cm, narrowly or broadly oblong or obovate, often acuminate, apex usually acute, margin usually entire, leathery, glossy above, hairless. Flowers numerous, solitary or in clusters of 2 or 3, nodding, mostly *c.* 1.5 cm, fragrant; stalks 0.5–1 cm. Sepals ovate to almost circular, fringed, much smaller than the petals. Petals waxy, yellowish outside, white inside, narrowly oblong, nearly free, later reflexing. Stamens with white anthers. Ovary almost spherical, hairless, with 2 or 3 cells. Styles 2 or 3, united for most of their length, persistent. Berry spherical to ovoid, red ripening to black, many-seeded. *Nepal, NE India (Assam), Burma, China, Taiwan, Korea, Japan.* G1–H5. Summer.

Var. **wallichiana** (de Candolle) Sealy (*C. ochnacea* var. *wallichiana* de Candolle). Illustration: Botanical Magazine, 9606 (1940). A shrub to 6 m, with most parts larger, leaves 6–9 × 3–4 cm; flowers solitary or clustered 2–6; stalks to 2.5 cm; stamens numerous, filaments white, anthers orange with white bristles. *Nepal, China (Yunnan, Sichuan)*. G1–H5. Late summer.

'Tricolor' (*C. japonica tricolor* Nicholson; *C. japonica foliis variegatis* Verschaffelt; *C. fortunei* J.D. Hooker; *C. japonica* var. *tricolor* W. Miller; *C. japonica* forma *tricolor* (Nicholson) Kobuski; *Eurya latifolia variegata* Verschaffelt). Illustration: Botanical Magazine, 7434 (1895). A variant with dark shining leaves strikingly marbled with grey; margin cream, sometimes flushed pink. It is often sterile. *Introduced from Japan, origin unknown.* G1–H5. Late summer.

9. EURYA Thunberg
C. Whitefoord

Evergreen shrubs and small trees. Leaves very shortly stalked, papery or leathery, margin toothed. Flowers small, unisexual, solitary or clustered, sometimes on short axillary shoots, shortly stalked, fleshy. Bracteoles 2, minute, persistent. Sepals 5. Petals 5, united *c.* ⅓. Stamens 5–25, anthers basifixed, hairless. Ovary 2–5-celled; styles 2 or 3, united for most of their length or nearly free, persistent. Fruit a many-seeded berry; seeds unwinged.

A genus of about 70 species in tropical and warm Asia and the western Pacific. Some are locally used as timbers. *Eurya* differs from *Cleyera* in having the sexes on separate plants, smaller flowers with the petals more united, and hairless anthers. They are slow-growing shrubs appreciated for their neat habit and glossy foliage. They need shelter and well-drained ordinary soil and are propagated by seed or cuttings.

1a. Leaves usually 3.5–8 cm; plant hairless   **1. japonica**
 b. Leaves usually 2.5–3.5 cm; young growth downy   **2**
2a. Leaves narrowly obovate; apex rounded, usually notched; margin usually curved downwards and inwards, bluntly toothed, thickly leathery   **2. emarginata**
 b. Leaves elliptic to obovate, narrowed to a short blunt or acute apex; margin usually flat, finely toothed, papery   **3. chinensis**

1. E. japonica Thunberg (*E. japonica* var. *pusilla* Blume; *E. pusilla* Blume; *E. latifolia* C. Koch). Figures 6(7), p. 32 and 7(11), p. 34. Illustration: Botanical Magazine, n.s., 588 (1970); Krüssmann, Manual of cultivated broad-leaved trees & shrubs **2**: 66 (1986).

Shrub or small tree to 10 m or more, but usually under 3 m in cultivated plants; branches stiffly spreading, hairless. Leaves usually 3.5–8 × 1–3 cm, narrowly elliptic to oblanceolate, usually acuminate, apex blunt or notched, margin bluntly toothed, leathery, dark green above, paler below. Flowers 1–5, on very short stalks, 5–7 mm, female slightly larger than the male, with an objectionable scent. Sepals almost circular, fleshy. Petals pale greenish-yellow, fleshy, broadly ovate, notched. Stamens 10–12, lacking in female flowers, or represented by 1 or 2 staminodes. Ovary (abortive in male flowers) ovoid, 3-celled; styles 3, basally united. Berry spherical, purplish-black, many-seeded. *Japan (Ryukyu Islands), Korea, Taiwan.* G1–H5. Winter–spring.

2. E. emarginata (Thunberg) Makino (*E. chinensis* misapplied; *Ilex emarginata* Thunberg). Illustration: Krüssmann, Manual of cultivated broad-leaved trees & shrubs, **2**: pl. 27 (1986).

Shrub or small tree. Young growth densely brown-hairy. Leaves usually 2.5–3.5 × 1–1.5 cm, narrowly obovate, apex rounded, usually notched, margin curved downwards and inwards, bluntly toothed towards apex, thickly leathery, glossy. Flowers 1–4, clustered, 2–3 mm, very shortly stalked. Sepals fleshy, almost circular, notched. Petals pale yellowish green, oblong. Ovary ovoid; styles united. Berry spherical, black. *Japan (Ryukyu Islands), Korea, China (Zhejiang, Fujian).* G1–H5. Winter–spring.

Var. **microphylla** Makino ('*Microphylla*'). A small spreading, densely branched shrub. Leaves to 9 × 2.5–70 mm, obovate-circular, notched. *Horticultural origin, Japan.* G1–H5. Winter–spring.

3. E. chinensis R. Brown (*E. parvifolia* Gardner; *E. japonica* var. *parvifolia* Thwaites).

Small tree or shrub. Young growth hairy. Leaves usually 2.5–3.5 × 1–1.5 cm, elliptic to obovate, narrowed to a short blunt or acute apex, margin evenly and finely saw-toothed, flat, papery. Flowers solitary or 2–5 clustered, shortly stalked. Male flowers *c.* 5 mm, female *c.* 2 mm. Sepals circular. Male petals broadly obovate, female petals oblong. Styles 3, shortly united. Berry spherical, purplish black. *Sri Lanka, China (Guangdong, Fujian, Guangxi, Yunnan), Taiwan.* G1–H5. Winter–spring.

XC. CARYOCARACEAE

Trees, rarely shrubs. Leaves opposite or spirally arranged; leaflets 3–5, stipules absent. Flowers in terminal racemes, bisexual, radially symmetric; sepals 5 or 6, united at base or reduced and lobed; petals 5 or 6, sometimes united at base, united above; stamens numerous, united at base in a short ring or in 5 bundles opposite the petals, inner stamens often without anthers, filaments brightly coloured, at least twice as long as petals; ovary of 4–20 united carpels, each containing one ovule, placentation axile. Fruit a drupe; seeds kidney-shaped.

A family of 2 genera and 24 species from tropical America especially Amazonia. It is unlikely that species of this family are cultivated outside specialist collections.

XCI. MARCGRAVIACEAE

Climbing and mostly epiphytic shrubs, sometimes with clinging aerial roots, rarely becoming tree-like, hairless. Leaves spirally arranged or in 2 ranks, simple, entire, sometimes of two different shapes, without stipules. Flowers in terminal racemes or raceme-like umbels, bisexual or sometimes sterile, radially symmetric; bracts more or less united to flower-stalk, sometimes modified to form a glandular sac, spur or pitcher. Sepals 4–7, free or more or less united. Petals 4 or 5, free or partly to completely united (then forming a cap or calyptra), pink or white to cream, falling early. Stamens 3–40, free or slightly united, sometimes united with base of petal; anthers opening by a longitudinal slit. Ovary superior, eventually with 2–12 cells; axile placentas with numerous ovules; stigmas 2–12, mostly stalkless, radiating. Fruit a leathery or fleshy berry (sometimes splitting at base); seeds numerous, small.

A family of 5 genera and about 108–113 species in tropical America. In Europe they require hothouse conditions.

1. MARCGRAVIA Choisy
N.K.B. Robson

Climbing epiphytic shrubs with two forms of shoot. Climbing shoots sterile, with clinging aerial roots and 2-ranked, stalkless leaves. Mature shoots pendent, fertile, with larger, spirally-ranked, often stalked leaves, able to revert to climbing form. Leaves evergreen, leathery. Inflorescence stalked, pendent, umbel-like; bracts of central, sterile flowers modified as nectar-containing sacs or pitchers. Sepals 4–6, in alternate pairs, free. Petals 4 or 5, forming a cap and falling early. Stamens 10–50, with filaments united at base. Ovary with 4–12 cells; stigmas stalkless, radiating. Berry spherical.

A genus of about 60 species in tropical America. Only one or two species are commonly cultivated. **M. trinitatis** Presl (*M. rectiflora* Triana & Planchon) and, possibly, the Jamaican **M. brownei** (Triana & Planchon) Krug & Urban are also sometimes grown. Both differ from the following species in having flowers inserted in line with the flower-stalk, and in *M. trinitatis* the inflorescence is erect as well as the flowers.

1a. Leaves very narrowly acuminate; stamens 20–30   **1. umbellata**
  b. Leaves acute to shortly acuminate; stamens 9–16   **2. brachysepala**

1. M. umbellata Linnaeus. Illustration: Martius, Flora Brasiliensís **12**(1): pl. 42 (1878); Fournet, Flore illustrée des phanérogames de Guadeloupe et de Martinique, 541 (1978); Howard, Flora of the Lesser Antilles **5**: 304 (1989).

Climbing shrub to 12 m. Climbing stems with adherent roots; leaves stalkless, *c.* 3 × 2.5 cm, broadly ovate to rounded. Fertile pendent stems with leaves stalked, blade 5–13 × 2–7.5 cm, elliptic-oblong to oblong, very narrowly acuminate, base wedge-shaped to rounded. Inflorescence with flower-stalks radiating at right-angles to inflorescence-stalk; flowers facing downwards. Pitcher-like bracts 3–4.5 cm, on stalks *c.* 1 cm. Fertile flowers 20–30, at right angles to stalks, 2–3.5 cm. Sepals 4, 3–3.5 mm, crescent-shaped. Corolla 'cap' greenish-cream, 8–10 mm. Stamens 20–30. Ovary with 8 cells. Berry red. *West Indies (Lesser Antilles), Colombia to Brazil.* G2. Winter or spring.

2. M. brachysepala Urban. Illustration: Hooker, Exotic Flora, t. 160 (1825), showing the inflorescence upside-down.
Like *M. umbellata*, but leaves on fertile stems 5–11 × 2–4 cm, apex acute to shortly acuminate. Flowers with stamens, 9–16. Ovary with 5–11 cells. *Jamaica.* G2.

Cultivated material could be of either species, which are very similar and often confused.

XCII. GUTTIFERAE

Trees, shrubs or herbs, rarely climbers, mostly containing a yellow to red or clear resinous latex, sometimes with simple or stellate hairs. Leaves usually opposite, rarely whorled or alternate, simple, entire or rarely with margin gland-fringed, nearly always without stipules, usually with glands and/or resin canals that are often translucent. Flowers bisexual or unisexual, radially symmetric, terminal or axillary, solitary or in cymes or false racemes. Sepals 2–6 or more, free or more or less united, overlapping, rarely not distinct from petals. Petals 3–6 or more, free, overlapping, usually in the same direction. Stamens basically in two whorls of 3–5 bundles each of 3 to numerous stamens, the outer whorl usually sterile or absent, the inner (opposite petals) free or partly or wholly united, sometimes the ring of the bundles wholly merged, the stamens then appearing free; ovary superior, 1–12-celled, with 1–many ovules on axile, parietal, basal or rarely apical placentas. Styles 2–12, free or more or less united, or apparently, single. Fruit a capsule opening between cells (rarely also within cells), berry or drupe. Seeds without endosperm, sometimes winged or with arils or with a fleshy outer coat, the embryo sometimes with reduced or without cotyledons.

A family of about 48 genera and more than 1000 species, almost confined to tropical regions apart from the large genus *Hypericum* (sometimes included in the separate family *Hypericaceae*), which is almost cosmopolitan, being absent only from extremely dry or extremely cold regions and most of the lowland tropics. Many *Hypericum* species are used as decoratives, fruits of some species of *Garcinia* and *Mammea* are eaten, and species in many genera are used for pharmaceutical purposes or for their wood.

Literature: Engler, A., Guttiferae. *Die natürlichen Pflanzenfamilien* edn 2, **21**: 154–237 (1925).

1a. Style(s) long, slender; flowers usually bisexual   2
  b. Style(s) absent or very short and thick; flowers usually unisexual   3
2a. Fruit a dry or rarely fleshy capsule; seeds numerous; styles free or partly united; petals yellow or rarely red; mostly shrubs or herbs   **1. Hypericum**

  b. Fruit a drupe; seed solitary; styles completely united, apparently single; inner perianth-segments white; trees   **2. Calophyllum**
3a. Styles free; fruit dehiscent, a leathery capsule; seeds numerous, white with a scarlet aril   **4. Clusia**
  b. Styles united or absent; fruit indehiscent, a fleshy, leathery or woody berry or drupe; seeds 1–13, without arils but sometimes with a fleshy seedcoat   4
4a. Sepals completely united in bud, splitting into 2 or 3 sepals; stamens free or basally united; sterile stamen bundles absent   **3. Mammea**
  b. Sepals free or basally united, 2–5; stamens free or variously united; sterile stamen bundles often present   **5. Garcinia**

1. HYPERICUM Linnaeus
N.K.B. Robson
Small trees, shrubs or herbs, evergreen or deciduous, with pale (translucent or amber) and often dark (black or occasionally red) glands (and/or canals), hairless or sometimes with simple hairs on stem, leaves and/or sepals. Stems usually 2–6-lined when young, eventually terete. Leaves opposite or sometimes in whorls of 3–5, stalkless or shortly stalked, entire or rarely gland-fringed, with pale and/or dark glandular dots, streaks or lines. Flowers bisexual, solitary and terminal or in terminal and sometimes axillary cymes or sometimes with flowers replaced by flowering branches. Sepals 4 or 5, free or partly united, unequal or equal, entire, glandular hairy or toothed, with pale and/or dark glands as in leaves, persistent or rarely falling. Petals 4 or 5, yellow, the parts visible in bud often tinged red, very rarely wholly carmine red or white, usually asymmetric and with a lateral projection, persistent or falling. Stamens usually numerous, in 4 or 5 equal bundles opposite petals or 2 bundles, rarely 1 bundle, united and larger ('stamen bundles 3, rarely 4'), or all bundles merged (stamens 'free' or 'irregular'); filaments in each bundle rarely united more than half way and then bundles alternating with small bilobed scales; anthers small, oblong to elliptic, with an amber or dark gland on connective. Ovary 3–5-celled, or 1-celled with 2–5 placentae bearing 2–many ovules; styles 2–5, long, free or sometimes partly to completely united; stigmas small. Fruit a 2–5-celled capsule splitting between the cells, rarely fleshy and tardily dehiscent

or indehiscent. Seeds small, usually curved, cylindric, sometimes keeled or laterally winged.

A genus of over 400 species, almost cosmopolitan except for lowland tropical, arctic, high altitudinal and desert regions, but relatively rare in cool temperate parts of the Southern Hemisphere. The Eurasian shrubs thrive in any good garden soil, as do some of the North American ones (e.g. *H. frondosum* & *H. prolificum*). Other North American shrubs, however, probably prefer damp conditions (e.g. *H. galioides* & *H. fasciculatum*). Some alpine, mediterranean and SW Asian shrublets and herbs grow naturally in chalk or limestone and thus benefit from (though apparently do not all require) soil with a high pH (e.g. *H. coris* & *H. ericoides*). Like most alpines, these require open sites; most other species (shrubs and herbs) can tolerate a certain amount of shade, but they all flower better in full sun. Other herbs (e.g. *H. elodes* & *H. anagalloides*) are marsh or aquatic plants.

The most important characters to observe are the form and distribution of black glands (when present), the number and disposition of floral parts, and the behaviour of the outer floral whorls after flowering. Variation in the glands of the capsule is also important in some groups; and leaf shape and/or venation can be important at species level. The word 'intramarginal' is used to described veins or glands borne just within the leaf-margins.

Literature: Robson, N.K.B., Studies in the genus *Hypericum* L. (Guttiferae) Parts 1–3, 7–8. *Bulletin of the British Museum (Natural History), Botany* **5**: 291–355 (1977); **8**: 55–226 (1981); **12**: 163–325 (1985); **16**: 1–106 (1987); **20**: 1–151 (1990). Part 3 (1985) contains the accounts of most of the Old-World shrubs in cultivation; Keller, R., *Hypericum* in *Die natürlichen Pflanzenfamilien*, edn 2: **21** 175–183 (1925); Stefanoff, B., *Sistematicheski i geografiski prouchvaniya verchu mediterransko-orientaliskite vidovye na roda Hypericum* L., *Godislinsa na Agronomo-lesovidniya fakultet pri Universitet Sofia* **10**: 19–57 (1932), **11**: 139–186 (1933), **12**: 69–100 (1934); Adams, P., Studies in the Guttiferae I, A synopsis of *Hypericum* section *Myriandra*. *Contributions from the Gray Herbarium of Harvard University*, No. 189 (1962).

1a.	Petals and sometimes stamens falling after flowering, normally withering (except in Section *Myriandra*) 2
b.	Petals and stamens persistent after flowering, withering 56
2a.	Black or red glands absent; leaves in opposite pairs 3
b.	Black or red glands present on sepal and sometimes petal margins; leaves in whorls of 3 (Section *Coridium* in part) **85. empetrifolium**
3a.	Stamens in 3–5 bundles; styles free or partly united, usually spreading; sepals and petals normally 5 4
b.	Stamens in a continuous ring; styles wholly or partly adpressed, apparently united; sepals and petals 4 or 5 (Section *Myriandra*) 44
4a.	Styles 4 or 5; if stem lines present, then those from between leaves weaker or absent 5
b.	Styles 3; stem lines from below leaf midrib weaker or absent (Section *Androsaemum*) 40
5a.	Bracteoles adpressed to sepals; styles 4 or 5; stem and leaves glandular-warty (Section *Psorophytum*) **2. balearicum**
b.	Bracteoles not adpressed to sepals; styles 5; stem and leaves smooth (Section *Ascyreia*) 6
6a.	Leaves (at least upper) stalkless, without visible dense net-veining beneath; styles free 7
b.	Leaves almost stalkless or shortly stalked or, if stalkless, then visibly densely net-veined beneath and/or with styles more or less united 8
7a.	Styles 1.8–2.5 times length of ovary; petal projection acute; leaves elliptic-oblong to oblong, base shallowly cordate to auriculate or rounded **3. cordifolium**
b.	Styles 1–1.5 times length of ovary; petal projection obtuse to rounded; leaves oblong-lanceolate to broadly ovate, base almost cordate to rounded **4. augustinii**
8a.	Stems prostrate to ascending or pendent, forming low mats or clumps; styles almost equalling ovary **5. reptans**
b.	Stems erect to straggling or, if creeping and rooting, then styles 1.5–3 times length of ovary 9
9a.	Styles straight or gradually outcurving; leaves not densely net-veined 10
b.	Styles abruptly curved or flexed below apex or, if straight, then leaves densely net-veined 11
10a.	Sepals erect in bud and fruit, rounded or rarely obscurely mucronate, 4–5 mm; stamens 0.4–0.6 times length of petals; leaves lanceolate-elliptic **6. tenuicaule**
b.	Sepals spreading in bud and fruit, mucronate or rarely rounded, 7–11 mm; stamens 0.25–0.4 times length of petals; leaves ovate-lanceolate to triangular-ovate **7. lobbii**
11a.	Leaves densely net-veined beneath or styles more or less united; styles at least 1.5 times length of ovary 12
b.	Leaves loosely or not visibly net-veined beneath; styles free, less than 1.5 times length of ovary 14
12a.	Styles free 13
b.	Styles united almost to apex **10. monogynum**
13a.	Leaves all stalkless; anthers yellow; sepals elliptic-oblong to narrowly ovate; stems more or less spreading but not creeping or rooting **8. oblongifolium**
b.	Leaves (at least lower) shortly stalked; anthers reddish; sepals broadly elliptic to obovate or almost circular; stems creeping and rooting **9. calycinum**
14a.	Leaves with distinct but loose net-veining beneath; styles 1–1.5 times length of ovary; sepals broad; anthers usually orange or reddish (hybrids of *H. calycinum*) 15
b.	Leaves without visible net-veining or, if visibly net-veined, then styles shorter than ovary or sepals narrowly oblanceolate to linear; anthers pale to deep yellow 17
15a.	Anthers brick-red; stamens 0.5–0.75 times length of petals; green leaves usually mucronate; stems low-arching **25. × moserianum**
b.	Anthers orange; stamens 0.35–0.5 times length of petals; green leaves not or apiculate; stems high-arching or spreading to ascending 16
16a.	Leaves triangular-lanceolate, acute to obtuse; petals golden yellow, not red-tinged; branches arching to spreading to 1.75 m **21. × 'Hidcote'**
b.	Leaves oblong-ovate, nearly mucronate to rounded; petals deep golden yellow, sometimes tinged red outside; branches spreading-ascending to 70 cm **37. × dummeri**

17a. Leaves in one plane or if in four rows, then broadest at middle; stems usually arching, sometimes frond-like   18

b. Leaves in four rows, broadest below middle; stems not or slightly arching, never frond-like   30

18a. Sepals spreading to reflexed in bud and fruit; styles shorter than (or rarely equalling) ovary   19

b. Sepals outcurved to erect in bud and fruit or, if spreading then styles 1.5–2 times length of ovary   21

19a. Leaves elliptic to oblanceolate, almost stalkless; stems persistently 4-lined; young fruits deep wine-red
11. subsessile

b. Leaves narrowly lanceolate to broadly ovate, shortly stalked; stems soon terete; fruits brown, not colouring red   20

20a. Sepals acute or almost so; ovary and capsule narrowly ellipsoid to narrowly ovoid-conic; leaves triangular-lanceolate to broadly ovate   **12. leschenaultii**

b. Sepals mucronate to rounded; ovary and capsule ovoid-ellipsoid to ovoid; leaves oblong-lanceolate to ovate   **13. × 'Rowallane'**

21a. Leaves with distinct, usually continuous, intramarginal veins, all or mostly narrowly elliptic to narrowly oblong; sepals acute to acuminate; stamens 0.75–0.9 times length of petals
14. acmosepalum

b. Leaves without continuous intramarginal veins, relatively broader or broader below middle; sepal apex and relative stamen length various   22

22a. Sepals without or with very narrow translucent margin; stems arching or spreading to pendent but not frond-like, not or scarcely 2-edged when young   23

b. Sepals with marked translucent margin; stem erect to arching or spreading and then often frond-like (i.e. with branches in one plane), markedly 2-edged at least when young (*H. patulum* group)   28

23a. Styles 0.35–1 times ovary; sepals usually ovate-lanceolate to obovate or spatula-shaped, acute to rounded   24

b. Styles 1.5–2 times length of ovary; sepals lanceolate to linear, acute or shortly awned   27

24a. Leaves narrowly elliptic to rather broadly oblong-elliptic; flower buds acute to obtuse; stamens 0.6–0.7 times length of petals; stems rather slender, arching or spreading
15. lagarocladum

b. Leaves broader below middle; flower-buds obtuse to rounded; stamens 0.25–0.4 times length of petals; stems slender to stout but not spindly, arching or spreading   25

25a. Sepals acute to obtuse or rarely rounded-mucronate, not becoming ribbed; stamens 40–50 per bundle   26

b. Sepals rounded or very rarely rounded-mucronate, often becoming ribbed; stamens 60–80 per bundle   **19. hookerianum**

26a. Leaves mucronate-obtuse to rounded, elliptic-oblong to ovate-lanceolate; sepals ovate, oblong or spatula-shaped; styles 0.7–1 times length of ovary   **18. addingtonii**

b. Leaves acute to mucronate-obtuse, lanceolate; sepals ovate-lanceolate or narrowly elliptic-oblong to spatula-shaped; styles *c.* 0.6 times length of ovary
20. × cyathiflorum

27a. Sepals erect in bud and fruit, ovate-oblong to lanceolate or elliptic; petals obovate   **16. wilsonii**

b. Sepals spreading in bud and fruit, narrowly oblong-lanceolate to linear; petals oblanceolate
17. dyeri

28a. Stems erect to arching, eventually 2-lined or terete or persistently 4-lined; sepals acute or mucronate to rounded, usually entire   29

b. Stems spreading, soon 2-lined; sepals obtuse to rounded or slightly and usually mucronate, margin finely toothed or with marginal hairs   **24. patulum**

29a. Sepals entire, rounded; stem branching, frond-like; branchlets markedly compressed, 2-sided
23. uralum

b. Sepals not both entire and rounded; stem branching not or weakly frond-like; branchlets compressed but 4-sided   **22. henryi**

30a. Leaves with main lateral veins (usually conspicuous) joining to form loops and/or with venation rather densely netted; sepals broadest at middle, entire; immature capsules green   31

b. Leaves with main lateral veins (usually obscure) not joining, venation not netted; sepals usually broadest below middle, often toothed; immature capsules turning red   33

31a. Sepals erect in bud and fruit, usually acute to rounded; leaves lanceolate-oblong to broadly elliptic or almost circular, obtuse to notched; flowers deeply cup-shaped, buds usually obtuse to rounded
28. bellum

b. Sepals outcurved to recurved in bud and fruit, usually acute; leaves lanceolate to triangular-ovate, usually acute to acuminate; flowers stellate to shallowly cupped, buds acute or apiculate   32

32a. Petals obovate-oblanceolate; flowers stellate; stamens *c.* 0.6 times length of petals; leaves narrowly lanceolate, glaucous beneath, stalk 0.5–2 mm   **26. maclarenii**

b. Petals broadly obovate to almost circular, flowers cup-shaped; stamens 0.35–0.4 times length of petals; leaves ovate to triangular-lanceolate, paler but not glaucous beneath, stalk 2–4 mm
27. choisianum

33a. Sepals lanceolate to narrowly elliptic or oblanceolate, acute or acuminate; petals with projection acute to obtuse or absent   34

b. Sepals ovate to circular or obovate, acute to rounded; petals with or without obtuse to rounded projection   38

34a. Stamens 0.65–0.8 times length of petals; petals spreading or reflexed
29. kouytchense

b. Stamens 0.3–0.6 times length of petals; petals more or less incurved   35

35a. Styles equalling or longer than ovary; stamens *c.* 0.6 times length of petals; inflorescence solitary or with 3–14 flowers   36

b. Styles shorter than ovary; stamens 0.3–0.4 times length of petals; inflorescence solitary or rarely with 2–4 flowers   37

36a. Ovary 4–5 mm; styles *c.* 1.2 times as long as ovary, straight or flexuous; inflorescence branches relatively slender; leaves with dense gland dots on lower surface
31. stellatum

b. Ovary 5–6.5 mm; styles equalling or slightly longer than ovary, outcurved; inflorescence branches relatively stout; leaves usually without or with few gland dots on lower surface   **32. lancasteri**

37a. Petals 3–3.5 cm; flowers shallowly cup-shaped; leaves elliptic-oblong to lanceolate, obtuse to rounded; styles ¾ length of ovary or more
   30. 'Eastleigh Gold'

   b. Petals 1.5–2 cm; flowers deeply cup-shaped; leaves triangular-ovate to lanceolate; styles ½ length of ovary   **33. curvisepalum**

38a. Sepals nearly apiculate to rounded; stems soon terete   **36. forrestii**

   b. Sepals acute to obtuse; stems more or less persistently 4-lined   39

39a. Styles equalling or shorter than ovary; stamen bundles 0.5–0.7 times length of petals; leaf apex usually obtuse or apiculate
   34. beanii

   b. Styles exceeding ovary; stamen bundles 0.75–0.85 times length of petals; leaf apex usually rounded
   35. pseudohenryi

40a. Sepals shrivelling and deciduous before fruit ripens; styles 3–5 times as long as ovary; leaves (when crushed) usually smelling of goats
   42. hircinum

   b. Sepals persistent at least until fruit ripens; styles 0.5–3 times as long as ovary; leaves (when crushed) very rarely smelling of goats   41

41a. Styles 1–3 times as long as ovary; petals 1.5–4 times as long as sepals; ripe fruit dry, rarely shining, dehiscent   42

   b. Styles half as long as ovary; petals equalling or shorter than sepals; ripe fruit fleshy, dark brown to black, shining, indehiscent
   40. androsaemum

42a. Petals 1.5–2 times as long as sepals; fruit coloured bright red during maturation, dehiscing tardily and incompletely; outer sepals broadly ovate
   1. × inodorum

   b. Petals 3–4 times as long as sepals; fruit not coloured red during maturation, dehiscing completely; outer sepals narrowly oblong to elliptic   43

43a. Inflorescence branches widely spreading; capsule leathery, ovoid to ellipsoid; leaves broadly triangular-ovate to oblong; petals 1.6–3 cm   **38. grandifolium**

   b. Inflorescence branches narrowly ascending; capsule papery, broadly cylindric-ellipsoid; leaves narrowly ovate to triangular-lanceolate; petals 1–1.8 cm   **39. foliosum**

44a. Sepals 5, unequal or equal; petals 5 or rarely 4; styles 3–5, wholly adpressed in flower   45

   b. Sepals 4, markedly unequal, or 2; petals 4; styles 2–4, apically separate in flower   55

45a. Sepals falling in fruit; stamens deciduous with or soon after petals; leaves and sepals usually with basal transverse groove   46

   b. Sepals persistent in fruit; stamens persistent long after petals; leaves and sepals without basal groove   53

46a. Leaves and sepals with basal groove; inflorescence leafy (bracts leaf-like)   47

   b. Leaves and sepals without basal groove; inflorescence naked (bracts much reduced)   **96. nudiflorum**

47a. Mature leaf with expanded blade, usually over 2 mm wide, lower surface visible; sepals usually broadened, not linear   48

   b. Mature leaves and sepals needle-like, only midrib visible beneath
   94. fasciculatum

48a. Capsule 7–15 mm; flowers 1.5–4.5 cm, terminal inflorescence with 1–9 flowers   49

   b. Capsule 4.5–7 mm; flowers 0.9–1.7 cm, terminal inflorescence with 3–25 flowers   51

49a. Flowers solitary or in terminal clusters only, or rarely 1–3 in axils of leaf pairs immediately below, inflorescence corymb-like   50

   b. Flowers in terminal 1–7-flowered axillary clusters from 2 leaf pairs below, inflorescence cylindric
   89. prolificum

50a. Leaves 0.8–2.2 cm wide; styles 3; capsule 1.2–1.5 cm; shrub to 1.35 m   **88. frondosum**

   b. Leaves 3–10 mm wide; styles mostly 5; capsule 7-11 mm; shrub 0.6–1 m   **90. kalmianum**

51a. Inflorescence broadly pyramidal to spherical-cylindric or corymb-like, from 1–3 nodes; sepals spreading to reflexed after flowering   52

   b. Inflorescence narrowly cylindric, from 4–5 nodes; sepals erect after flowering   **93. galioides**

52a. Styles mostly 4 or 5; capsule markedly lobed   **91. lobocarpum**

   b. Styles mostly 3; capsule not or scarcely lobed   **92. densiflorum**

53a. Flowers solitary or rarely 3–5, terminal; stems ascending, mat-forming   **95. buckleii**

   b. Flowers in 3 to numerous-flowered terminal dichasia; stems erect   54

54a. Leaves linear-oblong to lanceolate; capsule triangular-ovoid; shrub
   97. cistifolium

   b. Leaves ovate to elliptic or oblanceolate; capsule ellipsoid to spherical; rhizomatous herb
   98. ellipticum

55a. Styles 3 or 4; leaves elliptic-oblong or ovate, usually more than 1 cm wide; stem not or little branched, erect   **99. crux-andreae**

   b. Styles 2; leaves narrowly oblong to oblanceolate, usually less than 6 mm wide; stem sparsely to much-branched, erect or decumbent and mat-forming   **100. hypericoides**

56a. Stamens in a continuous or irregular narrow ring; black or red glands absent from all parts of plant; stem lines or wings 4; herbs (Section *Brathys*)   57

   b. Stamens in bundles (5 equal or 3 or 4 unequal); black or red glands often present; stems lines or wings absent or 1–6; trees, shrubs or herbs   63

57a. Leaves subulate or scale-like, adpressed; capsule 2–3 times as long as sepals; much branched erect annual   **117. gentianoides**

   b. Leaves flat or recurved, spreading; capsule shorter than or up to 1–3 times as long as sepals; erect to prostrate perennial or annual, branching various   58

58a. Stem and leaves hairy
   118. setosum

   b. Stem and leaves hairless   59

59a. Leaf-bases decurrent, forming a V, margin becoming revolute or leaf linear; stems usually erect and unbranched below inflorescence
   119. gramineum

   b. Leaf-bases cordate to wedge-shaped, not decurrent or forming a V, margin plane or slightly recurved, leaf-blade relatively broader; stems erect to prostrate, branching various   60

60a. Stigmas narrowly to scarcely head-like; stems more or less diffuse and rooting; capsule equalling or shorter than sepals
   123. anagalloides

- b. Stigmas broadly head-like; stems erect to diffuse and rooting; capsule longer to shorter than sepals   **61**
- 61a. Leaves lanceolate to narrowly oblong-elliptic; capsule conic-ellipsoid   **120. majus**
- b. Leaves ovate to broadly elliptic, or obovate; capsule cylindric to spherical   **62**
- 62a. Capsule equalling or exceeding sepals; inflorescence-branching dichasial; uppermost stem-internode shorter than others or absent   **121. mutilum**
- b. Capsule shorter than sepals or, if longer, then inflorescence-branching sympodial; uppermost stem internode not shorter than others   **122. japonicum**
- 63a. Styles adpressed, at least at base, or more or less united   **64**
- b. Styles free, spreading or ascending, not adpressed   **66**
- 64a. Styles and stamen bundles 4 or 5; leaves free   **65**
- b. Styles and stamen bundles 3; leaves perfoliate (Section *Bupleuroides*)   **45. bupleuroides**
- 65a. Tree or shrub; black or reddish glands present at least on leaves and anthers (Section *Campyloporus*)   **1. revolutum**
- b. Erect perennial herb; black or reddish glands absent (Section *Roscyna*)   **44. ascyron**
- 66a. Stamen bundles 5, equal; styles 3 or 4; shrub to 1.5 m, without black glands, but dark red marginal petal glands and sometimes yellowish marginal sepal glands present (Section *Inodora*)   **43. xylosteifolium**
- b. Stamen bundles 3 or 4, unequal, sometimes indistinct; styles 3; trees, shrubs or perennial to annual herbs with or without black glands   **67**
- 67a. Seeds more or less flattened at one end or with a caruncle; trees, shrubs or shrublets without black glands or at least without a regularly spaced intramarginal row of black glands in leaves   **68**
- b. Seeds tapered or rounded at both ends; shrubs to herbs almost always with black glands, usually forming a regularly spaced intramarginal row in leaves   **72**
- 68a. Leaves with densely netted venation; style bases not touching; bark whitish   **69**

- b. Leaves without visible net venation; style bases touching; bark brownish   **71**
- 69a. Shrub or small tree, 1–4 m; leaves more or less narrowly lanceolate to elliptic, acute; older stem-nodes not swollen (Section *Webbia*)   **101. canariense**
- b. Shrublet to 50 cm; leaves broadly ovate or elliptic to nearly circular, obtuse to slightly notched; older stem-nodes more or less swollen (Section *Arthrophyllum*)   **70**
- 70a. Sepals acute to shortly acuminate, lanceolate; leaf-venation not or scarcely prominent   **102. nanum**
- b. Sepals acute to rounded, oblong to ovate; leaf-venation more or less prominent on both sides   **103. vacciniifolium**
- 71a. Branches straggling; black glands on leaves, flowers and sometimes stems; young stems red; petals red-tipped in bud; flowers spreading, styles of one length   **104. pallens**
- b. Branches more or less erect; black glands absent; young stems green; petals not red-tipped in bud; flowers apparently tubular, styles of 2 lengths on different plants   **105. aegypticum**
- 72a. Sepal marginal glands flat-topped or lengthening along margin; stems, when herbaceous, terete; plant usually more or less hairy   **73**
- b. Sepal marginal glands absent or round-topped or spherical; stems, when herbaceous, lined; plant hairless   **83**
- 73a. Sepal marginal glands red; leaves and stems with dense whitish hairs; leaves without intramarginal black glands; flowers apparently tubular (Section *Elodes*)   **116. elodes**
- b. Sepal marginal glands black; leaves and stems hairless or with more or less dense hairs; leaves with intramarginal black glands; flowers stellate (Section *Adenosepalum*)   **74**
- 74a. Shrub; leaves with stalkless marginal glands   **106. glandulosum**
- b. Shrub or herb; leaves entire or with basal glandular fringes   **75**
- 75a. Shrub; stem curled-hairy; leaves hairless; sepals entire or with stalkless marginal glands   **107. reflexum**
- b. Herb; stem curled-hairy or hairless; sepals glandular-toothed or with stalked marginal glands   **76**
- 76a. Plant hairless; sepals, bracts and

base of upper leaves with long glandular hairs on margin   **112. elodeoides**
- b. Plant usually partly hairy; leaves entire; sepals and bracts with glandular marginal hairs or with stalkless glands   **77**
- 77a. Bracts with densely glandular auricles; sepals with long teeth or short marginal hairs   **78**
- b. Bracts not auriculate but often with longer-stalked glands towards base; sepals almost entire or with short teeth or marginal hairs   **81**
- 78a. Leaves with rough very short stiff hairs beneath or rarely wholly hairless; stems hairless   **111. montanum**
- b. Leaves hairy on both sides; stems hairy   **79**
- 79a. Stem and leaves shortly densely hairy to velvety; stems erect, not rooting at nodes   **108. annulatum**
- b. Stem and leaves sparsely hairy; stems decumbent or procumbent, rooting at nodes   **80**
- 80a. Leaves stalkless, 1.2–4.5 cm   **114. delphicum**
- b. Leaves shortly stalked, 0.8–1.5 cm   **115. athoum**
- 81a. Inflorescence hairless; sepals more or less acute to rounded   **82**
- b. Inflorescence hairy; sepals acute or shortly awned   **113. tomentosum**
- 82a. Sepals and sometimes petals with superficial black glands   **109. atomarium**
- b. Sepals and petals without superficial black glands   **110. lanuginosum**
- 83a. Leaves (but not always bracts) without intramarginal black glands (apical black glands or intramarginal translucent glands sometimes present)   **84**
- b. Leaves with intramarginal black glands and/or stalked marginal glands or glandular teeth   **100**
- 84a. Low shrubs or perennial herbs; leaves all in whorls of 3 or 4; (Section *Coridium* in part)   **85**
- b. Habit various; leaves all or nearly all in opposite pairs   **86**
- 85a. Leaves smooth, 0.4–1.8 cm; sepals with stalkless marginal glands   **86. coris**
- b. Leaves papillose, 1.5–3.5 mm; sepal margin without glands or with glandular-hairs   **87. ericoides**
- 86a. Sepals entire, without marginal or intramarginal glands   **87**

b. Sepals with marginal hairs, fringed, finely toothed or entire with intramarginal glands   90

87a. Sepals broadly overlapping; inflorescence with 1–5 flowers (Section *Drosocarpium*, in part)   88

b. Sepals not or only slightly overlapping; inflorescence usually with numerous flowers   89

88a. Dwarf hairless shrub; leaves glaucous   **59. olympicum**

b. Hairy perennial with woody base; leaves green   **60. cerastoides**

89a. Petals clawed, not red-tinged; stems not rooting (Section *Hirtella* in part)   **74. elongatum**

b. Petals not clawed, red-tinged; stems rooting and branching at base (Section *Taeniocarpium* in part)   **77. linarioides**

90a. Petals clawed, not red-tinged or veined (but rarely wholly red), with pellucid dots or short streaks only; stems not creeping or rooting, herbaceous (Section *Hirtella* in part)   91

b. Petals not clawed, usually red-tinged or veined, with pellucid lines and sometimes streaks and dots; stems creeping, often rooting and/or woody (Section *Taeniocarpium* in part)   93

91a. Inflorescence broad, almost corymb-like; petals orange to red   **76. capitatum**

b. Inflorescence long, narrow; petals yellow   92

92a. Sepal margin without glands; capsule 0.8–1.3 cm; leaves usually apiculate   **74. elongatum**

b. Sepal margin continuously glandular; capsule 5–8.5 mm; leaves obtuse   **75. hyssopifolium**

93a. Leaves and usually stems sparsely hairy, curled-hairy or covered in a bloom   94

b. Leaves and stems hairless   96

94a. Leaves ovate, broadly elliptic-oblong or lanceolate, 2–6 cm, sparsely hairy; stems 35–100 cm   **78. hirsutum**

b. Leaves linear or narrowly elliptic to oblong, 0.5–1.5 cm, sparsely hairy or covered in a bloom; stems 10–35 cm   95

95a. Stems sparsely hairy; leaves stiffly hairy; sepals 3–4 mm, margin glandular-toothed to shortly glandular-hairy   **79. kotschyanum**

b. Stems hairless or very shortly curled-hairy; leaves covered in bloom or shortly hairy; sepals 4–7 mm, margin glandular-hairy to glandular-fringed   **80. confertum**

96a. Inflorescence cylindric to narrowly pyramidal; leaf-stalks absent or not jointed; stems (except base) erect or ascending   97

b. Inflorescence corymb-like to broadly pyramidal; leaf-stalks jointed at base; stems diffuse   98

97a. Anthers pink or orange; leaves ovate-cordate to oblong   **81. pulchrum**

b. Anthers yellow; leaves usually linear to linear-lanceolate   **77. linarioides**

98a. Leaves elliptic to oblanceolate or nearly spherical, margin wavy   **82. crenulatum**

b. Leaves ovate to circular, margin plane   99

99a. Leaves 5–20 mm, glaucous beneath only; petals 8–16 mm   **83. nummularium**

b. Leaves 2–7 mm, glaucous on both sides; petals 6–9 mm   **84. fragile**

100a. Leaves all with marginal glands and auricles (Section *Drosocarpium* in part)   101

b. Leaves entire (except sometimes the uppermost), without auricles   102

101a. Leaf-margin glandular toothed with yellowish glands; black glands absent from whole plant   **73. orientale**

b. Leaf-margin and sometimes upper surface with black or orange-brown glandular hairs; black glands also present elsewhere on plant   **72. adenotrichum**

102a. Petals with at least some black glandular hairs (Section *Drosocarpium* in part)   103

b. Petals entire or indented, often with stalkless or nearly marginal black glands (Section *Hypericum*)   115

103a. Sepals entire or rarely eroded, without marginal or intramarginal black glands (sometimes with superficial ones)   104

b. Sepals gland-fringed or hairy, if entire, then with intramarginal black glands   105

104a. Hairless dwarf shrub; capsule erect   **59. olympicum**

b. Hairy perennial herb; capsule pendent   **60. cerastoides**

105a. Capsule valves with longitudinal vittae, usually continuous; plant neither glaucous nor hairy   106

b. Capsule valves with vesicles and/or interrupted vittae or almost smooth; plant usually glaucous or hairy   107

106a. Sepals more or less equal, with glandular-hairy margin and with numerous superficial black glands; leaves on flowering stems linear to lanceolate; petals 8–12 mm   **70. linariifolium**

b. Sepals unequal, often entire, with few or no superficial black glands; leaves on flowering stems ovate to lanceolate or oblong; petals 4–8 mm   **71. humifusum**

107a. Leaves hairy or surface more or less markedly papillose; stems 2-lined   108

b. Leaves hairless; stems terete   109

108a. Stems and leaves hairy; stems almost erect to ascending; petals and sepals 5   **61. origanifolium**

b. Stems hairless; leaves more or less markedly papillose; stems prostrate; petals and sepals usually 4   **62. kelleri**

109a. Leaves densely and conspicuously net-veined; capsule with black glands   **68. richeri**

b. Leaves loosely or indistinctly net-veined; capsule with yellowish glands or almost without glands   110

110a. Sepal margin fringed, without glands   **65. barbatum**

b. Sepal margin glandular-hairy, gland-toothed or entire   111

111a. Petals with apical and marginal black glands only or glands absent   112

b. Petals with black glands over whole surface   114

112a. Stem erect or basally decumbent, stout; leaves 1.2–6 cm; capsule with prominent glands   113

b. Stems ascending or prostrate, slender; leaves 0.5–1.4 cm; capsule almost smooth or with rather faint glands   **69. trichocaulon**

113a. Capsule ovoid with linear and long glands; sepals obtuse to almost acute, erect in fruit   **66. perfoliatum**

b. Capsule narrowly pyramidal with rounded glands; sepals acute or acuminate, spreading or deflexed in fruit   **67. montbretii**

114a. Capsule narrowly pyramidal with faint glands or almost smooth;

leaves ovate-lanceolate to linear; sepals narrowly ovate to ovate-elliptic   **63. rumeliacum**
  b. Capsule ovoid with very prominent orange glands; leaves triangular-lanceolate to narrowly oblong; sepals narrowly lanceolate to oblong   **64. spruneri**
115a. Stems with 2–4 definite lines or wings   116
  b. Stems terete or slightly 2-lined   121
116a. Anther gland amber; sepals ovate, acute or acuminate; flowers 2–3.5 cm   **46. concinnum**
  b. Anther gland black; sepals either ovate and obtuse to rounded or narrower; flowers 5–35 mm   117
117a. Stems 2-lined   118
  b. Stems usually 4-lined or 4-winged   119
118a. Sepal margin entire; capsule with dorsal linear and lateral long or vesicular glands   **48. perforatum**
  b. Sepal margin with fine black glandular teeth or with stalkless black glands; capsule with only linear glands   **51. elegans**
119a. Sepals obtuse, apex entire or with eroded teeth; petals black-gland-dotted or streaked; leaves often without pellucid gland-dots   **47. maculatum**
  b. Sepals acute or acuminate, apex entire; petals without or with few black gland-dots; leaves densely pellucid gland-dotted   120
120a. Petals at least 7.5 mm, usually red-tipped outside; inflorescence loose; leaves narrowly ovate to elliptic or oblong, margin wavy   **49. undulatum**
  b. Petals not more than 8 mm, not red-tipped outside; inflorescence more or less dense; leaves broadly oblong or elliptic to ovate or circular, margin plane   **50. tetrapterum**
121a. Leaf pairs perfoliate; capsule glands vesicular   **52. sampsonii**
  b. Leaf pairs free; capsule glands linear   122
122a. Leaves with gland-dots all or almost all black, usually dense; inflorescence dense   123
  b. Leaves with gland-dots (except round margin) all or mostly translucent, dense to sparse or sometimes almost absent; inflorescence loose   125
123a. Sepals and/or petals with black glandular streaks or lines as well

as dots; leaves mostly ovate to lanceolate or elliptic-oblong   124
  b. Sepals and petals with black glandular dots only; leaves mostly oblong to elliptic   **56. punctatum**
124a. Leaves lanceolate to elliptic-oblong; sepals usually entire   **53. erectum**
  b. Leaves ovate or rarely ovate-oblong; sepals with some glandular marginal hairs   **57. formosum**
125a. Stems erect from creeping and rooting base; leaves 1–7 cm; flowers 1–3 cm   126
  b. Stems prostrate, radiating, rooting at lower nodes; leaves 0.3–1.4 cm; flowers 6–9 mm   **54. yakusimense**
126a. Ovary and capsule almost spherical; leaves densely dotted with translucent glands; styles 6–12 mm   **55. graveolens**
  b. Ovary and capsule narrowly ovoid; leaves scarcely dotted with translucent glands; styles 2–8 mm   **58. scouleri**

Section **Campylosporus** (Spach) Keller. Trees or shrubs, evergreen, with or without black glands. Floral whorls all in 5s. Petals and stamens persistent after flowering. Stamen bundles free. Styles more or less coherent.

1. H. revolutum Vahl.
  Shrub or small tree, *c.* 3 m in cultivation. Leaves 15–45 × 2–12 mm, narrowly elliptic to narrowly oblong or oblanceolate, with parallel veins and cross-veins and black intramarginal gland dots. Flowers solitary, 3.5–8 cm, stellate. Sepals 5–18 mm, narrowly ovate to circular, acute to rounded, margin usually irregularly glandular-hairy. Petals orange-yellow to golden yellow, tinged red-orange, 1.5–4.3 cm, obovate-oblanceolate. Stamens *c.* 0.5 times length of petals. Styles 4–12 mm, 0.8–1.6 times length of ovary, coherent for *c.* four-fifths of their length. *Saudi Arabia to South Africa, Cameroon, Fernando Po.* H5–G1. Summer.
  Subsp. **keniense** (Schweinfurth) N. Robson (*H. keniense* Schweinfurth). Illustration: Bamps, Flore du Congo, Guttiferae, 8 (1970). Leaves with 1 or 2 pairs of cross-veins not interrupting parallel veins; flower-stalk 5–8 mm, petals tinged red outside, stamens *c.* 40 per bundle. *Zaire, Uganda, Kenya, Tanzania.*
  Subsp. **revolutum** (*H. lanceolatum* misapplied; *H. leucoptychodes* A. Richard). Illustration: Flowering Plants of South Africa **20**: t. 787 (1940); Bamps, Flore du Congo, Guttiferae, 8 (1970); Dale &

Greenway, Kenya trees & shrubs, 236 (1961). Leaves with 3–8 pairs of cross-veins interrupting parallel veins, flower-stalk 1–5 mm, petals tinged orange outside, stamens 30–35 per bundle. *Distribution as for the species.*
  H. revolutum is usually unwilling to flower outside in Britain but can do so in a cool greenhouse. It is doubtful if the true subsp. *keniense* is in cultivation in Europe; the plants so labelled are from a somewhat intermediate population of subsp. *revolutum* on Mt Kilimanjaro.

Section **Psorophytum** (Spach) Nyman. Shrubs or small trees, evergreen, without black glands. Floral whorls all in 5s, or styles and capsule valves sometimes 4. Bracteoles adpressed to sepals. Petals and stamens deciduous after flowering. Stamen bundles free. Styles free.

2. H. balearicum Linnaeus. Illustration: Botanical Magazine, 137 (1791); Barceló, Flora de Mallorca **3**: f. 987 (1979); Bulletin of the British Museum (Natural History), Botany **12**: 204 (1985).
  Shrub or tree to *c.* 1 m, forming rounded or spreading bush in cultivation. Stem and leaves glandular-warty. Leaves 6–15 × 3–7 mm, ovate to oblong, rounded, margin more or less wavy. Flowers solitary, 1.5–4 cm, stellate. Bracteoles and sepals 5–7 mm, broadly ovate-elliptic, rounded. Sepals reflexed in fruit, entire or apically toothed. Petals golden yellow, slightly red-tinged outside, 1–2 cm, narrowly obovate-oblanceolate. Stamens *c.* 0.8 times length of petals. Styles 6–13 mm, 2–2.5 times length of ovary, almost erect. *Balearic Islands.* H5. Summer.
  The glandular-warty stems and leaves of *H. balearicum* are unique in the genus and make it immediately recognisable.

Section **Ascyreia** Choisy. Shrubs (rarely trees) or shrublets, evergreen or deciduous, without black glands. Floral whorls all in 5s. Bracteoles not adpressed to sepals. Petals and stamens deciduous after flowering. Stamen bundles free. Styles free to almost completely united.

3. H. cordifolium Choisy (*H. bracteatum* Wallich). Illustration: Wallich, Plantae Asiaticae Rariores, **3**: t. 220 (1831).
  Shrub 1–1.3 m, branches erect, arching or pendent. Leaves 18–62 × 5–23 mm, elliptic-oblong to oblong, mucronate or shortly acuminate, base rounded to shallowly cordate-auriculate, more or less glaucous, thinly leathery, not net-

veined. Inflorescence rounded-pyramidal to cylindric. Flowers 3–5 cm, stellate. Sepals 8–13 mm, usually lanceolate, acute to apiculate, margin entire, erect to ascending in bud and fruit. Petals bright yellow, sometimes tinged red, 1.5–2.2 cm, narrowly obovate to oblanceolate. Stamens 0.65–0.75 times length of petals. Styles 7.5–12 mm, 1.8–2.5 times length of ovary, free, erect. *Nepal*. G1. Summer.

4. H. augustinii N. Robson (*H. leschenaultii* misapplied). Figure 8(1), p. 52. Illustration: Bulletin of the British Museum (Natural History), Botany **12**: 220 (1985).

Shrub 70–130 cm, bushy, branches erect or arching. Leaves stalkless or (upper) stem-clasping, 3–7.5 × 1–4.4 cm, broadly ovate to oblong-lanceolate, acute to rounded and apiculate, base rounded to nearly cordate, glaucous (more so beneath), leathery, not net-veined. Inflorescence almost corymb-like. Flowers 1–13, stellate to shallowly cupped, 4–6.6 cm. Sepals 7–15 mm, ovate to broadly oblong or elliptic, almost apiculate to rounded, entire or apically eroded, erect in bud, spreading in fruit. Petals pale to bright golden yellow, not tinged red, 2–3.6 cm, obovate. Stamens *c.* 0.5 times length of petals. Styles 6–8 mm, *c.* 1.2 times length of ovary, free, erect to gradually divergent. *China (S Yunnan)*. H5. Late summer–autumn.

This striking species with pale-green leaves is marginally hardy outside in southern Britain and Ireland, but it usually succumbs to late spring frosts.

5. H. reptans Dyer. Illustration: Quarterly Bulletin of the Alpine Garden Society **9**: 178 (1937).

Shrublet, prostrate or ascending to 30 cm, forming clumps or mats to 1 m across, sometimes pendent, with branches rooting. Leaves shortly stalked, 2.2 cm × 2–9 mm, larger and broader up stem, elliptic to oblanceolate or obovate, obtuse to rounded, base wedge-shaped, paler beneath, leathery, obscurely net-veined. Flowers solitary, 2–3 cm, deeply cupped. Sepals 6 mm–1.4 cm, oblong to obovate or oblanceolate, obtuse to rounded, entire, reflexed in bud, spreading in fruit. Petals deep golden yellow, sometimes tinged red, 1.1–1.8 cm, broadly obovate. Stamens 0.25–0.35 times length of petals. Styles 2.5–4.5 mm, about equalling ovary, free, sharply outcurved. Capsule fleshy, brick-coloured. *E Himalaya (Nepal to Yunnan)*. H4. Middle–late summer.

6. H. tenuicaule Dyer.

Shrub *c.* 1.5 m, with branches arching to ascending. Leaves stalked, 15–58 × 4–19 mm, lanceolate to elliptic, acute to rounded, base wedge-shaped to narrowed, glaucous beneath, leathery, not net-veined. Flowers 1–7, 1.5–3 cm, deeply cupped. Sepals 4–5 mm, oblong to spathulate, rounded, entire or eroded with minute teeth, erect in bud and fruit. Petals bright yellow, sometimes tinged red, 1–1.3 cm, obovate to almost circular. Stamens *c.* 0.5 times length of petals. Styles 3.5–6 mm, 0.9–1.2 times length of ovary, free, wholly erect or apically outcurved. *E Nepal, Sikkim, W Bhutan*. H4. Summer.

This species has been confused with *H. uralum* (p. 54) but can be recognised by the divergent, not frond-like, branching, the terete stem internodes and the usually erect styles almost equalling or exceeding the ovary.

7. H. lobbii N. Robson (*H. oblongifolium* misapplied; *H. hookerianum* in part, misapplied). Illustration: Botanical Magazine, 4949 (1856).

Shrub 1.2–2 m, branches erect. Leaves stalked, 20–45 × 9–25 mm, ovate-lanceolate to triangular-ovate, obtuse to rounded, apiculate, base broadly wedge-shaped to shortly narrowed, glaucous beneath, thinly leathery, not net-veined. Flowers 1–24, 3–5 cm, shallowly cupped. Sepals 7–11 mm, ovate-oblong to broadly elliptic, usually apiculate, usually finely eroded, more or less spreading in bud and fruit. Petals golden yellow, not tinged red, 2–2.6 cm, broadly obovate. Stamens 0.35–0.4 times length of petals. Styles 4.5–5 mm, *c.* 0.9 times length of ovary, free, almost erect, outcurved towards apex. *NE India (Khasi Hills)*. H4. Late summer.

8. H. oblongifolium Choisy (*H. cernuum* D. Don). Illustration: Choisy, Prodromus d'une monographie de la famille des Hypéricinées, t. 4 (1821).

Shrub 45–240 cm, branches spreading or drooping. Leaves stalkless, 3–9.3 × 1–4.2 cm, elliptic to oblong or ovate-oblong, obtuse or apiculate to rounded, base wedge-shaped to rounded, paler beneath, sometimes glaucous, thickly papery, conspicuously net-veined. Flowers 1–8, 3.5–7.5 cm, stellate. Sepals 5–8 mm, narrowly ovate to elliptic-oblong, acute to rounded, entire or with minute teeth at apex, erect in bud, ascending in fruit. Petals bright yellow to yellow-orange, not red-tinged, 2–3 cm, obovate to oblanceolate. Stamens *c.* 0.7 times length

of petals. Styles 9–14 mm, *c.* 2–3 times length of ovary, free, more or less erect. *W Himalaya*. G1. Summer.

This shrub with spreading branches can resemble a tall, several-flowered *H. calycinum*. It is, however, much less hardy and, at least when young, seems unable to survive outside in Western Europe without winter protection. It may no longer be in cultivation.

9. H. calycinum Linnaeus. Figure 8(8), p. 52. Illustration: Botanical Magazine, 146 (1796); Smith, English Botany **29**: t. 2017 (1809).

Shrub 20–60 cm, evergreen, with creeping, branching stolons and erect, usually unbranched stems. Leaves stalkless or shortly stalked, 4.5–10.4 × 1.5–4.5 cm, oblong to elliptic or narrowly ovate, obtuse or apiculate, base wedge-shaped to rounded, paler but not glaucous and densely net-veined beneath, leathery. Flowers 1–3, stellate. Sepals 1–2 cm, unequal, broadly elliptic to almost circular or obovate, rounded, entire or minutely toothed, erect in bud, more or less ascending in fruit. Petals bright yellow, not red-tinged, 2.5–4 cm, obovate to oblanceolate. Stamens *c.* 0.75 times length of petals. Styles 1.2–2 cm, 1.5–3 times length of ovary, free, erect or spreading. *SE Bulgaria, NW & NE Turkey*. H4. Summer.

This species is well known in cultivation for its large flowers and its ability to tolerate dry shade, but it flowers better in the open. Its propensity to creep is useful (e.g. for covering slopes or margins), but it often creeps further than is desirable. It has recently become susceptible to rust (*Melampsora hypericorum*).

10. H. monogynum Linnaeus (*H. chinense* Linnaeus; *H. salicifolium* Sieber & Zuccarini; *Norysca chinensis* (Linnaeus) Spach; *H. chinense* var. *salicifolium* (Sieber & Zuccarini) Choisy; *H. monogynum* var. *salicifolium* (Sieber & Zuccarini) André; *Norysca chinensis* var. *salicifolia* (Sieber & Zuccarini) Kimura). Figure 8(2), p. 52. Illustration: Botanical Magazine, 334 (1796); Nakai & Honda, Nova Flora Japonica, **10**: 105 (1951); Bean, Trees and shrubs hardy in the British Isles, edn 8, **2**: 419 (1973); Lancaster, Travels in China, 384 (1989).

Shrub 50–130 cm, semi-evergreen, bushy or more usually with branches loose and spreading. Leaves stalkless, 2–11 × 1–4 cm, oblanceolate or elliptic to oblong, acute to rounded, base wedge-shaped to rounded, paler beneath, not glaucous,

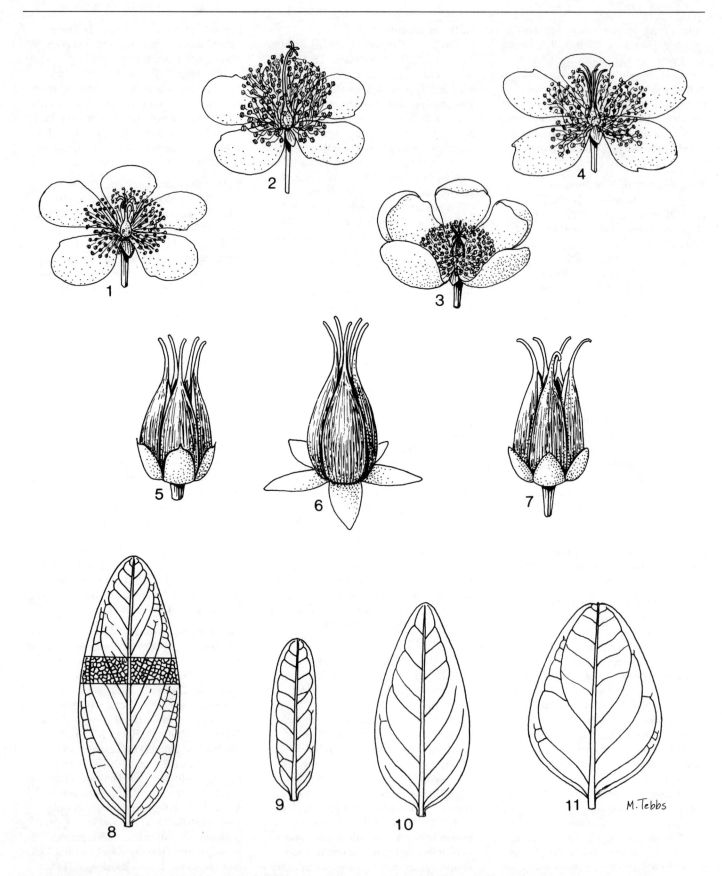

Figure 8. Diagnostic details of *Hypericum* species. Flowers (× 1). 1, *H. augustinii*. 2, *H. monogynum*. 3, *H. hookerianum*. 4, *H. kouytchense*. Fruits (× 2). 5, *H. patulum*. 6, *H. lancasteri*. 7, *H. beanii*. Leaves. 8, *H. calycinum*. 9, *H. acmosepalum*. 10, *H. addingtonii*. 11, *H. bellum*.

thinly leathery, densely net-veined. Inflorescence loose, corymb-like, with 1–30 flowers. Flowers 3–6.5 cm, stellate. Sepals 4–13 m, lanceolate to narrowly oblong or elliptic to oblanceolate, acute to rounded, entire, more or less spreading in bud and fruit. Petals golden to lemon yellow, not red-tinged, 2–3.4 cm, triangular-obovate. Stamens almost equalling petals. Styles 1.2-2 2 cm, *c.* 3.5–5 times length of ovary, united almost to the apex. *SE China, Taiwan.* H5. Summer.

The variant cultivated in Europe recently has been the hardier '*H. salicifolium*', with narrowly elliptic to oblanceolate, acute leaves and a loose inflorescence.

11. H. subsessile N. Robson. Illustration: Lancaster, Travels in China, 247 (1989).

Shrub 1–1.5 m or more, evergreen, branches strong, erect, arching. Leaves almost stalkless, 35–65 × 7–20 mm, narrowly elliptic to oblanceolate, acute to rounded-mucronate, base wedge-shaped, paler to glaucous beneath, thinly leathery, loosely net-veined. Flowers 1–8, terminal, 3.5–4.5 cm, stellate to shallowly cupped. Sepals 1–1.8 cm, unequal, leaf-like, ovate to elliptic, acute or acuminate, entire, outcurved in bud, reflexed in fruit. Petals bright yellow, sometimes red-tinged outside, 1.7–2 cm, oblanceolate-obovate. Stamens *c.* 0.7 times length of petals. Styles 5–6 mm, 0.8–0.9 times length of ovary, free, outcurved near apex. *China (Yunnan, Sichuan).* H4. Summer.

The large purplish-red immature capsules are striking.

12. H. leschenaultii Choisy (*H. triflorum* Blume). Illustration: Botanical Magazine, 9160 (1925); Graf, Tropica, edn 2. **1**: 524 (1981).

Shrub 50–250 cm, evergreen, usually with branches more or less spreading. Leaves shortly stalked, 2.5–8 × 1–3.7 cm, triangular-lanceolate to ovate, acute to rounded-apiculate, base broadly wedge-shaped to rounded, paler to glaucous beneath, thinly leathery, loosely net-veined. Inflorescence almost corymb-like. Flowers 3.5–7 cm, shallowly cupped, in groups of 1–10. Sepals 7–22 mm, unequal, narrowly oblong, elliptic to oblanceolate, margin entire, acute, spreading to recurved in bud and fruit. Petals deep golden yellow, not red-tinged, 2–4.5 cm. Stamens *c.* 0.35 times length of petals. Styles 3.5–7 mm, 0.5–0.9 times length of ovary, free, sharply outcurved. *Indonesia.* H5. Summer.

A much misunderstood species that is rare in cultivation.

13. H. × 'Rowallane' Armytage Moore (probably *H. leschenaultii × H. hookerianum* 'Charles Rogers'; *H. hookerianum* 'Rowallane' F. Schneider). Illustration: Gardeners' Chronicle, **147**: 255 (1960); Bean, Trees and shrubs hardy in the British Isles, edn 8, **2**: t.51 (1973).

Shrub to 3 m, branches erect, gradually outcurving. Leaves shortly stalked, 2.7–6.7 × 1–3.3 cm, ovate to oblong-ovate or oblong-lanceolate, apiculate-obtuse, base wedge-shaped to cordate, paler or glaucous beneath, thinly leathery to thickly papery, not net-veined. Flowers 5–7.5 cm, shallowly cupped, in groups of 1–3. Sepals 1–1.4 cm, oblanceolate to ovate or spathulate, mucronate or rounded, entire or minutely toothed, spreading in bud and fruit. Petals deep golden yellow, not red-tinged, 3–4 cm, obovate to circular. Stamens 0.25–0.35 times length of petals. Styles 6–7.5 mm, *c.* 0.7 times length of ovary, free, diverging. *Garden origin.* H5. Late summer–autumn.

The suggested parentage has not yet been confirmed but seems very likely.

14. H. acmosepalum N. Robson (*H. oblongifolium* misapplied; *H. patulum* var. *oblongifolium* misapplied; *H. kouytchense* misapplied; *H. patulum* var. *henryi* misapplied; *H. henryi* misapplied). Figure 8(9), p. 52. Illustration: Journal of the Royal Horticultural Society **95**: f. 238 (1970); Lancaster, Travels in China, 147 (1989).

Shrub 60–200 cm, bushy, branches strong, erect, gradually outcurving. Leaves shortly and broadly stalked, 1.8–6 cm × 6–20 mm, mostly oblong or elliptic-oblong, obtuse to rounded, markedly paler to glaucous beneath, thinly leathery to thickly papery, not net-veined but with conspicuous intramarginal veins. Flowers 1–6, terminal, 3–5 cm, stellate. Sepals 5–11 mm, ovate to narrowly lanceolate, acute, entire or minutely toothed, outcurved in bud and fruit. Petals deep yellow, sometimes red-tinged, 1.6–2.5 cm, obovate. Stamens 0.75–0.95 times length of sepals. Styles 3–8 mm, equalling or slightly longer than ovary, free, almost erect. *S China (Yunnan, Sichuan, Guizhou, Guangxi).* H4. Summer.

15. H. lagarocladum N. Robson.
Shrub 50–150 cm, diffuse, branches arching or spreading, slender. Leaves narrowly and shortly stalked, 1.8–4.5 cm, narrowly elliptic to broadly oblong-elliptic, acute to rounded, base wedge-shaped, paler beneath, not glaucous, thickly papery, not net-veined. Flowers 1–3 or more, 3–4.5 cm, stellate to shallowly cupped. Sepals 6–10 mm, ovate or oblong-ovate to lanceolate, acute to obtuse, entire or minutely toothed, erect or almost so in bud and fruit. Petals golden yellow, not red-tinged, 1.8–2.3 cm, obovate. Stamens 0.6–0.7 times length of petals. Styles 4–7 mm, 0.5–0.85 times length of ovary, free, almost erect to gradually outcurved. *S China (Yunnan, Sichuan, Guizhou, Hunan).* H5. Summer.

16. H. wilsonii N. Robson (*H. kouytchense* misapplied). Illustration: Botanical Magazine, 9345 (1934).

Shrub 50–100 cm, spreading, branches pendent to somewhat prostrate. Leaves shortly stalked, 2.3–6 cm, elliptic to lanceolate or ovate-lanceolate, acute to rounded, base broadly wedge-shaped to rounded, paler or glaucous beneath, thickly papery, not or obscurely net-veined. Flowers 4–6 cm, stellate, in groups of 1–22. Sepals 7–10 mm, lanceolate or narrowly elliptic, acutely acuminate to shortly awned, entire, erect in bud and fruit. Petals golden yellow, 2–2.5 cm, obovate. Stamens 0.35–0.5 times length of petals. Styles 7–9 mm, 1.5–1.8 times length of ovary, free, erect with outcurved apex. *Central China.* H4. Summer.

17. H. dyeri Rehder (*H. lysimachioides* Dyer). Illustration: Flora of West Pakistan **32**: f. 1A–D (1973).

Shrub 60–120 cm, branches widely spreading. Leaves shortly stalked, 1–6 cm, oblong-lanceolate or lanceolate to ovate, acute to rounded, base wedge-shaped to rounded, markedly glaucous beneath, papery, very loosely net-veined. Inflorescence corymb-like or broadly pyramidal. Flowers 1.5–3.5 cm, stellate. Sepals 4–12 mm, narrowly oblong-lanceolate to linear, acute, spreading in bud and fruit. Petals bright yellow, not tinged red, 1–1.8 cm, oblanceolate. Stamens 0.7–0.85 times length of petals. Styles 4–7 mm, 1.5–2 times length of ovary, free, erect or gradually divergent. *Nepal to Pakistan.* H5–G1. Late summer.

Has been confused with *H. stellatum* and may no longer be cultivated in Europe.

18. H. addingtonii N. Robson (*H. leschenaultii* misapplied). Figure 8(10), p. 52. Illustration: Bulletin of the British Museum (Natural History), Botany **12**: 252 (1985).

Shrub 1.5–2 m, spreading to 2.5 m wide, branches arching or spreading.

Leaves shortly stalked, 2–8.5 cm, elliptic-oblong to ovate-lanceolate, apiculate-obtuse to rounded, base wedge-shaped, paler beneath, not glaucous, thickly papery, not net-veined. Flowers 3–6.5 cm, shallowly cupped, in groups of 1–5. Sepals 7–10 mm, ovate to oblong-ovate or oblong to spathulate, acute to rounded-apiculate, entire or minutely toothed, erect in bud and fruit. Petals golden yellow, not red-tinged, 2–3.2 cm, broadly obovate to circular. Stamens *c.* 0.4 times length of petals. Styles 4.5–7 mm, 0.7–1 times length of petals, free, almost erect with outcurving apex. *SW China (NW Yunnan).* H4. Summer.

All plants grown outside as *H. leschenaultii* (except in Ireland and SW England) have proved to belong to this species.

19. H. hookerianum Wight & Arnott (*H. oblongifolium* misapplied; *H. patulum* var. *oblongifolium* misapplied). Figure 8(3), p. 52. Illustration: Wallich, Plantae asiaticae rariores **3:** t. 244 (1832); Wight, Icones plantae ex India orientali **3:** t. 959 (1845); Fyson, Flora of Nilgiri and Pulney hill-tops **2:** 29 (1915).

Shrub 30–210 cm, bushy, round-topped, branches erect to spreading. Leaves with stalk to 4 mm, blade 1.7–7.8 cm, narrowly lanceolate to oblong-lanceolate or rarely to broadly ovate, acute to rounded, base narrowly wedge-shaped to cordate, paler or glaucous beneath, thickly papery, not net-veined. Flowers 3–6 cm, deeply cupped, in groups of 1–5. Sepals 5–10 mm, elliptic to obovate or circular, rounded, usually entire, erect in bud and fruit. Petals deep golden to pale yellow, 1.5–3 cm, broadly obovate to circular. Stamens 0.25–0.35 times length of petals. Styles 2–4 mm, 0.35–0.7 times length of ovary, free, gradually outcurved. *S India, Himalaya, NW Thailand to Bangladesh, China (NW Yunnan).* H4–5. Summer.

Shows continuous variation over its wide range. 'Charles Rogers' ('Rogersii'; *rogersii* invalid) has persistently 4-angled stems, oblong-ovate to broadly ovate rounded leaves, flowers 4–6 cm on stout stalks, with circular to obovate petals; sepals markedly ribbed.

20. H. × cyathiflorum N. Robson (probably *H. addingtonii* × *H. hookerianum*; *H.* 'Lawrence Johnston'; *H. patulum* 'Gold Cup'; *H. hookerianum* 'Gold Cup'; *H. beanii* 'Gold Cup'). Illustration: Journal of the Royal Horticultural Society **95:** f. 236 (1970).

Shrub to *c.* 1.5 m, rather bushy, branches arching or widely spreading. Leaves shortly and broadly stalked, 3–7.5 cm, lanceolate, acute to apiculate, base wedge-shaped, paler beneath, not glaucous, thickly papery, not net-veined. Inflorescence corymb-like. Flowers 4–5 cm, cupped. Sepals 6–9 mm, ovate-lanceolate to oblong or spathulate, acute to apiculate-obtuse, entire or minutely toothed, erect in bud and fruit. Petals golden yellow, 2–3 cm, broadly oblong-obovate. Stamens 0.3–0.4 times length of petals. Styles 4.5–5 mm, *c.* 0.6 times length of ovary, free, erect with outcurving apex. *Known only in cultivation.* H4. Summer.

This hybrid was apparently introduced to cultivation by Jackman's about 1952 as *H. patulum* 'Gold Cup', but its origins are obscure.

'Gold Cup' is the only known variant.

21. H. × 'Hidcote' (probably *H. × cyathiflorum* 'Gold Cup' × *H. calycinum*; *H. patulum* 'Hidcote Variety'; *H. patulum* 'Hidcote'; *H. hookerianum* 'Hidcote'; *H.* 'Hidcote Gold'). Illustration: Gardeners' Chronicle **147:** 227, 254 (1960); Journal of the Royal Horticultural Society **95:** 244 (1970).

Shrub to 1.75 m, bushy, branches arching to spreading. Leaves shortly stalked, 3–6 cm, triangular-lanceolate, acute to obtuse or slightly mucronate, base broadly wedge-shaped, paler beneath, not glaucous, thickly papery, conspicuously loosely net-veined. Inflorescence corymb-like. Flowers 3.5–6.5 cm, cupped. Sepals 8–11 mm, ovate-oblong to broadly elliptic, rounded, margin finely hairy or finely toothed towards the apex, erect in bud and fruit. Petals golden yellow, 1.5–3.5 cm, obovate. Stamens 0.3–0.5 times length of petals, anthers orange. Styles 8–10 mm, 1–1.5 times length of ovary free, erect or erect with outcurved apex. *Known only in cultivation.* H3. Summer.

Probably the most popular shrubby *Hypericum*. Completely sterile.

22. H. henryi Léveillé & Vaniot (*H. patulum* misapplied).

Shrub 50–300 cm, bushy, branches erect to arching, not or weakly frond-like. Leaves stalkless or shortly stalked, 1–4 cm, narrowly elliptic or lanceolate to ovate, acute to obtuse or rounded, base narrowed or wedge-shaped to rounded, densely glaucous beneath, thickly papery, not net-veined. Inflorescence corymb-like. Flowers 1.5–5.2 cm, very shallowly to deeply cupped, in groups of 1–7. Sepals 4–9 mm, oblong or elliptic to circular or obovate, acute to rounded, entire or finely toothed and translucent, usually erect in bud and fruit. Petals golden to pale yellow, sometimes red-tinged, 8–25 mm, narrowly to broadly obovate. Stamens *c.* 0.5 times length of petals. Styles 2.5–6 mm, 0.7–1.2 times as long as ovary, free, erect with outcurved apex. *E Asia from China to N Sumatra.* H5–G1. Late summer–early autumn.

Subsp. **hancockii** N. Robson. Stems erect, not frond-like, internodes sometimes becoming 2-lined. Leaves narrowly elliptic or lanceolate to ovate-oblong. Sepals (at least outer) broadly elliptic or broadly oblong to circular, obtuse to rounded, entire. *China (S Yunnan) to N Sumatra.* H5–G1.

Subsp. **henryi** (*H. patulum* misapplied). Stems slender, erect to arching or spreading, not or weakly frond-like, internodes persistently 4-lined. Leaves usually ovate-lanceolate to broadly ovate. Sepals broadly elliptic or oblong to ovate or circular, apiculate-obtuse to rounded, entire to finely toothed. *China (C & E Yunnan, C Guizhou).* H5.

Subsp. **uraloides** (Rehder) N. Robson (*H. uraloides* Rehder; *H. eudistichum* invalid). Stems erect to arching, sometimes frond-like towards apex, internodes persistently 4-lined. Leaves narrowly elliptic or lanceolate to ovate-lanceolate. Sepals elliptic or narrowly oblong to oblanceolate, acute or rarely obtuse, entire. *China (N & W Yunnan, W Guizhou, S Sichuan), E Burma.* H5.

True subsp. *hancockii* appears to be rare in cultivation in Europe, but a variant intermediate between it and subsp. *henryi* is sometimes grown.

23. H. uralum D. Don (*H. patulum* misapplied; *H. patulum* var. *uralum* (D. Don) Koehne; *H. hookerianum* Wight & Arnott 'Buttercup'). Illustration: Botanical Magazine, 2375 (1823); Journal of the Royal Horticultural Society **95:** f. 243 (1970).

Shrub 30–200 cm, bushy, branches erect to arching, often frond-like. Leaves 1–4 cm with short flat stalk, lanceolate or older ones ovate, acute to rounded-apiculate, base wedge-shaped, densely glaucous beneath, rather thickly papery, not net-veined. Inflorescence corymb-like, often with lateral flowering branches. Flowers 1.5–3 cm, deeply cupped. Sepals 3.5–9 mm, oblong to elliptic or oblong-spathulate, rounded, margin entire and translucent, erect in bud and fruit. Petals golden to deep yellow, not red-tinged, 9–18 mm, broadly obovate to circular. Stamens 0.25–0.5 times length of petals. Styles 2.5–4.5 mm, 0.6–1 times length of ovary, free, erect, divergent towards apex or wholly outcurved. *China (Xizang), NW Burma, India to Pakistan.* H4. Summer.

24. H. patulum Thunberg (*Norysca patula* (Thunberg) Voigt; *Eremanthe patula* (Thunberg) K. Koch). Figure 8(5), p. 52. Illustration: Botanical Magazine, 5693 (1868); Nakai & Honda, Nova Flora Japonica, **10**: 100 (1951); Plants of the World (Weekly Asahi Encyclopedia), 1511 (1977).

Shrub 30–150 cm in cultivation, bushy, branches arching to spreading, sometimes weakly frond-like. Leaves shortly stalked, 1.5–6 cm, lanceolate or oblong-lanceolate to oblong-ovate, obtuse to rounded, apiculate, base wedge-shaped to abruptly narrowed, rather glaucous beneath, thickly papery, not net-veined. Inflorescence corymb-like, sometimes with lateral flowering branches. Flowers 2.5–4 cm, cupped. Sepals 5–10 mm, broadly ovate to elliptic, circular or obovate to spathulate, apex obtuse to rounded or slightly notched, usually apiculate, margin finely toothed to hairy and translucent, often reddish, erect in bud and fruit. Petals golden yellow, not red-tinged, 1.2–2.8 cm, oblong-obovate to broadly obovate. Stamens 0.4–0.5 times length of petals. Styles 4–5.5 mm, 0.75–0.95 times length of ovary, free, erect with outcurved apex. *China (N Guizhou, Sichuan); introduced further east and into Taiwan and Japan).* H5. Summer–early autumn.

25. H. × moserianum André (*H. patulum × H. calycinum*). Illustration: Gardeners' Chronicle, **147**: 227 (1960).

Shrub 30–70 cm, semi-evergreen, branches spreading or arching, not frond-like. Leaves shortly stalked, 2.2–6 cm × 7–36 mm, oblong-lanceolate to oblong-ovate, acute to rounded-apiculate, base wedge-shaped to rounded, paler beneath, not glaucous, thinly leathery, loosely net-veined. Inflorescence corymb-like without lateral flowering branches. Flowers 4.5–6 cm, stellate or slightly cupped. Sepals 7–10 mm, broadly oblong-elliptic to rounded, margin hairy, translucent, often with reddish zone, erect in bud, spreading in fruit. Petals bright yellow, 2–3 cm, obovate. Stamens 0.5–0.75 times length of petals, anthers reddish. Styles 8–11 mm, 1–1.5 times length of ovary, free, erect with outcurved apex. *Garden origin.* H3.

A variegated variant, 'Tricolor' (*H. × moserianum* var. *tricolor* Maumené), is popular as a foliage plant, the flowers being much smaller than usual.

26. H. maclarenii N. Robson. Illustration: The Garden **112**: 561 (1987); Lancaster, Travels in China, 307 (1989).

Shrub 75–100 cm, branches erect to spreading. Leaves shortly stalked, 2.5–4 cm, narrowly lanceolate, acute, base wedge-shaped, densely glaucous beneath, thickly papery, not net-veined, with wavy intramarginal vein. Flowers 4–5 cm, stellate, in groups of 1–6. Sepals 7–11 mm, narrowly elliptic, acute to acuminate, entire, outcurved to spreading in bud and fruit. Petals golden yellow, sometimes red-tinged, 2–2.5 cm, obovate-oblanceolate. Stamens *c.* 0.6 times length of petals. Styles 6–8 mm, 0.85–1 times length of ovary, free, outcurved towards apex. *China (Sichuan).* H4. Summer.

27. H. choisianum N. Robson (*H. hookerianum* var. *leschenaultii* misapplied). Illustration: Flora of West Pakistan, **31**: f. 1E–H (1973).

Shrub 1–2 m, bushy, branches erect to arching. Leaves with stalk 2–4 mm, blade 2.5–8.8 cm, triangular-lanceolate to ovate, acute or acuminate to obtuse or rounded, base broadly wedge-shaped to cordate, paler beneath, not glaucous, thickly papery, usually markedly net-veined. Flowers 4–7 cm, shallowly to deeply cupped, in groups of 1–7. Sepals 7–18 mm, narrowly to very broadly elliptic, or larger and leaf-like, acute or apiculate to obtuse, entire, spreading to recurved in bud and fruit. Petals deep golden yellow, sometimes red-tinged, 1.6–3 cm, broadly obovate to circular. Stamens 0.35–0.4 times length of petals. Styles 3–5 mm, 0.35–0.7 times length of ovary, free, outcurved towards apex. *China (Yunnan), N Burma, Bhutan to Pakistan.* H4. Summer.

A variable species with at least two distinct variants in cultivation.

28. H. bellum H.L. Li. Figure 8(11), p. 52.

Shrub 30–150 cm, bushy, branches erect to arching. Leaves with stalk 0.5–3 mm, blade 1.5–7.8 cm, oblong-lanceolate to broadly ovate or circular, obtuse to rounded or indented, often apiculate, base broadly wedge-shaped to cordate, paler or glaucous beneath, thickly papery, not net-veined. Flowers 1–7, terminal, 2.5–6 cm, cupped. Sepals 3–12 mm, broadly elliptic to narrowly oblong or obovate, acute to

rounded, margin entire or finely toothed at apex, often translucent, erect in bud and fruit. Petals golden to butter-yellow or pale yellow, not red-tinged, 1.5–3 cm. Stamens 0.35–0.6 times length of petals. Styles 3–8 mm, free, erect to divergent, outcurved towards apex. *China (Yunnan, Sichuan, Xizang), N Burma, India.* H4. Summer.

Subsp. **bellum**. Illustration: Journal of the Royal Horticultural Society **95**: f. 239 (1970). Stems *c.* 1 m. Leaves oblong-ovate to circular, margin often wavy. Flowers 2.5–3.5 cm. Sepals 3–9 × 2.5–6 mm, narrowly elliptic to obovate, rounded or rarely almost apiculate, sometimes finely toothed. Petals golden or pale yellow. *China (Yunnan, Sichuan, Xizang), India.*

Subsp. **latisepalum** N. Robson. Illustration: Lancaster, Travels in China, 226 (1981). Stems *c.* 1.5 m. Leaves oblong-lanceolate to broadly ovate, margin plane. Flowers mostly 4–6 cm. Sepals 8–13 × 5–8 mm, broadly elliptic, acute to obtuse, often apiculate, entire. Petals always golden yellow. *China (Yunnan), N Burma.*

29. H. kouytchense H. Léveillé (*H. penduliflorum* invalid; *H. patulum* var. *grandiflorum* invalid; *H. patulum* 'Sungold'; *H. patulum* 'Laplace'; *H. patulum* 'Summergold'). Figure 8(4), p. 52. Illustration: Gardeners' Chronicle **147**: 255 (1960); Journal of the Royal Horticultural Society **95**: ff. 240–241 (1970).

Shrub 1–1.8 m, bushy, branches arching or pendent. Leaves shortly stalked, 2–5.8 cm, elliptic to ovate or lanceolate, acute to obtuse or rarely rounded-apiculate, base narrowed or wedge-shaped to rounded, paler but not or scarcely glaucous beneath, thickly papery, not net-veined. Flowers 4–6.5 cm, stellate, in groups of 1–11. Sepals 7–15 mm, narrowly ovate to lanceolate, acute to acuminate, entire, erect in bud and fruit. Petals bright golden yellow, not red-tinged, 2.4–4 cm, obovate-oblong to obovate. Stamens 0.65–0.8 times length of petals. Styles 8–10 mm, 1.2–1.35 times length of ovary, free, erect slightly outcurved at apex. Young fruits red. *China (Guizhou).* H4. Summer.

30. H. × 'Eastleigh Gold' (*H. beanii* 'Eastleigh Gold'].

Shrub to *c.* 1 m, loose, branches spreading or drooping. Leaves shortly stalked, 2.5–5.1 cm, elliptic-oblong to lanceolate, obtuse to rounded, base broadly wedge-shaped to rounded, paler or somewhat glaucous beneath, thickly papery, not net-veined. Flowers 5–6.5 cm, very shallowly cupped, in groups of 1–4. Sepals 1–1.2 cm, narrowly oblong or lanceolate, acute to acuminate, margin entire, outcurved in bud and fruit. Petals golden yellow, not red-tinged, 3–3.5 cm, oblong-obovate. Stamens 0.3–0.4 times length of petals. Styles 5.5–6.5 mm, 0.75–0.9 times length of ovary, free, erect, outcurved towards apex. *Garden origin*. H4. Summer.

31. H. stellatum N. Robson (*H. lysimachioides* misapplied; *H. dyeri* misapplied). Illustration: Journal of the Royal Horticultural Society **95**: f. 237 (1970); Bulletin of the British Museum (Natural History), Botany **12**: 280 (1985).

Shrub 1–2.5 m, branching spreading to pendent. Leaves shortly stalked, 2–5.5 cm, oblong-lanceolate to narrowly ovate, acute to rounded-apiculate, base wedge-shaped to rounded, paler or densely glaucous beneath, thickly papery, not net-veined. Flowers 2.5–4 cm, stellate, in groups of 1–14. Sepals 8–13 mm, narrowly lanceolate, acute, entire, widely spreading to recurved, reddish in bud and fruit. Petals golden yellow, sometimes red-tinged, 1.2–2 cm, obovate. Stamens *c.* 0.6 times length of petals. Styles 6–9.5 mm, 1.2–1.5 times length of ovary, free, usually flexuous and twisted. *China (NE Sichuan)*. H4. Summer.

32. H. lancasteri N. Robson. Figure 8(6), p. 52. Illustration: Bulletin of the British Museum (Natural History), Botany **12**: 280 (1985); Lancaster, Travels in China, 147 (1989).

Shrub 30–100 cm, rounded, branches erect to spreading. Leaves shortly stalked, 3–6 cm, oblong to triangular-lanceolate, acute to rounded, base wedge-shaped to rounded, paler or densely glaucous beneath, thickly papery, not net-veined. Flowers forming a loose inflorescence, 3–5.5 cm, stellate to slightly cupped. Sepals 8–11 mm, lanceolate to ovate or oblong, acute to acuminate, entire, purplish-tipped in bud, widely spreading or recurved in bud and fruit. Petals golden yellow, not red-tinged, 1.7–2.8 cm, oblong-ovate. Stamens *c.* 0.6 times length of petals. Styles 5–7 mm, 1–1.2 times length of ovary, free, outcurved but not twisted. *China (N Yunnan, S Sichuan)*. H5. Summer.

33. H. curvisepalum N. Robson. Illustration: Bulletin of the British Museum (Natural History), Botany **12**: 280 (1985); Lancaster, Travels in China, 147 (1989).

Shrub 30–120 cm, branches spreading to pendent, young growth purplish-red. Leaves shortly stalked, 2–4 cm, triangular-lanceolate to triangular-ovate, acute to rounded, base rounded to shallowly cordate, glaucous beneath, thickly papery. Flowers 2–4 cm, deeply cupped, in groups of 1–3. Sepals 8–14 mm, ovate to lanceolate or narrowly elliptic, acute to acuminate or apiculate-obtuse, entire, spreading to recurved, purplish-red in bud and fruit. Petals deep yellow, 1.2–2.2 cm, broadly ovate to circular. Stamens 0.35–0.7 times length of petals. Styles 3–4 mm, *c.* 0.5 times length of ovary, free, spreading. *China (Yunnan, Sichuan, Guizhou)*. H5. Summer.

34. H. beanii N. Robson (*H. patulum* var. *henryi* Bean; *H. pseudohenryi* misapplied). Figure 8(7), p. 52. Illustration: Journal of the Royal Horticultural Society **95**: f. (1970); Yearbook of the International Dendrological Society 1980, 124 (1981); Lancaster, Travels in China, 147 (1989).

Shrub 60–200 cm, bushy, branches erect or arching. Leaves shortly stalked, 2.5–6.5 cm, narrowly elliptic or oblong-lanceolate to ovate, acute to obtuse or sometimes rounded, base wedge-shaped to rounded, paler or glaucous beneath, thickly papery to thinly leathery, not net-veined. Flowers 3–4.5 cm, stellate to deeply cupped, in groups of 1–14. Sepals 6–14 mm, oblong-ovate or broadly elliptic, acute to obtuse, entire or minutely toothed and translucent, erect to spreading in bud and fruit. Petals golden yellow, 1.5–3.3 cm, oblong-obovate to circular. Stamens 0.5–0.7 times length of petals. Styles 4–9 mm, 0.65–1.1 times length of ovary, free, erect, outcurved near apex. *China (E Yunnan, Guizhou)*. H4. Summer.

35. H. pseudohenryi N. Robson (*H. patulum* var. *henryi* misapplied; *H. henryi* misapplied). Illustration: Bulletin of the British Museum (Natural History), Botany **12**: 285 (1985); Lancaster, Travels in China, 310 (1989).

Shrub 70–170 cm, bushy, branches erect to arching. Leaves shortly stalked, 2–8 cm, ovate-oblong to lanceolate-oblong, apiculate-obtuse or usually rounded, base narrowly to broadly wedge-shaped, pale or somewhat glaucous beneath, thickly papery, not net-veined, with wavy marginal vein. Flowers 3–5.5 cm, stellate or slightly cupped, in groups of 1–25. Sepals 6–13 mm, ovate-oblong, acute to obtuse, entire or minutely toothed and translucent, erect to outcurved in bud and

fruit. Petals golden yellow, not red-tinged, 1.6–3 cm, obovate. Stamens 0.75–0.85 times length of petals. Styles 5.5–11 mm, *c.* 1.1 times length of ovary, free, erect or divergent. Young fruits red. *China (Yunnan, Sichuan)*. H4. Summer.

36. H. forrestii (Chittenden) N. Robson (*H. patulum* var. *forrestii* Chittenden; *H. patulum* forma *forrestii* (Chittenden) Rehder; *H. patulum* 'Rothschild's Form'). Illustration: Gardeners' Chronicle **72**: 235 (1922); **147**: 226 (1960); Journal of the Royal Horticultural Society **95**: f. 235 – as H. beanii, f. 242 (1970); Lancaster, Travels in China, 293 (1989).

Shrub 30–150 cm, bushy, branches erect. Leaves 2–6 cm, lanceolate or triangular-ovate to broadly ovate, obtuse to rounded, base broadly wedge-shaped to rounded, paler beneath, thickly papery, not net-veined. Flowers 2.5–6 cm, deeply cupped, in groups of 1–20. Sepals 6–9 mm, ovate or broadly elliptic to circular, rounded or rarely apiculate, entire or finely toothed and often translucent, erect in bud and fruit. Petals golden yellow, not red-tinged 1.8–3 cm, broadly obovate. Stamens 0.4–0.6 times length of petals. Styles 4–7 mm, 0.7–1 times length of ovary, free, erect, outcurved near apex. Young fruits green. *China (Yunnan, Sichuan), NE Burma*. H4. Summer.

37. H. × dummeri N. Robson (*H. forrestii* × *H. calycinum*).

Shrub to 70 cm, branches erect from ascending or creeping but not rooting base. Leaves shortly stalked, 3.5–4.8 cm, oblong-ovate, almost apiculate to rounded, base wedge-shaped, paler or somewhat glaucous beneath, thinly leathery, loosely net-veined. Flowers *c.* 5.5 cm, shallowly cupped, in groups of 1–4. Sepals 6–9 mm, broadly oblong to obovate or almost circular, rounded, margins finely toothed, translucent, erect in bud and fruit. Petals deep golden yellow, sometimes red-tinged, *c.* 3 cm, narrowly obovate. Stamens *c.* 0.4 times length of petals, anthers deep orange. Styles *c.* 8 mm, *c.* 1.2 times length of ovary, free, erect, gradually outcurved. *Garden origin*. H4. Summer.

This hybrid has a lower, more spreading habit than *H. forrestii* but is less invasive than *H. calycinum*. It sometimes produces variegated foliage, especially early in the season. Plants with red-tinged petals have been named 'Peter Dummer'.

Section **Androsaemum** (Duhamel) Godron. Shrubs, deciduous, without black glands. Floral whorls 5 (outer) or 3 (styles). Bracteoles not adpressed to sepals. Petals and stamens deciduous after flowering. Stamen bundles free. Styles free.

38. H. grandifolium Choisy (*Androsaemum webbianum* Spach; *H. elatum* misapplied). Illustration: Webb & Berthelot, Phytographia Canariensis **1**: t. 4E (1836); Schaeffer, Plants of the Canary Islands, 144 (1963); Kunkel, Flora de Gran Canaria **4**: t. 164 (1979).

Shrub 50–200 cm, bushy, branches erect or ascending. Leaves stem-clasping, 3–9 cm, broadly triangular-ovate to oblong-ovate, obtuse to rounded, base cordate to rounded, papery, dense net-veining prominent above. Inflorescence branches widely spreading, flowers 2.5–4.5 cm, stellate, in groups of 1–13. Sepals 6–8 mm, lanceolate to narrowly oblong, acute, entire, spreading to reflexed in flower, erect in fruit. Petals golden yellow, slightly red-tinged, 1.6–3 cm, narrowly oblong-lanceolate. Stamens *c.* 0.9 times length of petals. Styles 7.5–17 mm, 2–3 times length of ovary, free, erect, divergent towards apex. *Canary Islands, Madeira.* H5. Summer–autumn.

A good cool greenhouse plant, but can be grown outside in sheltered locations if protected from frost.

39. H. foliosum Aiton. Illustration: Annual Report of the Missouri Botanical Garden **8**: t. 19 (1897).

Shrub 50–100 cm, bushy, branches erect or spreading. Leaves stalkless, sometimes stem-clasping, rather dense, 3.5–6 cm, narrowly ovate to triangular-lanceolate, obtuse to rounded, base rounded to cordate, slightly paler beneath, not glaucous, papery, net-veining prominent on both sides. Flowers 1–9, 2–3.5 cm, stellate, inflorescence branches narrowly ascending. Sepals 3–6 mm, triangular-lanceolate to oblong-elliptic, acute to obtuse, entire, spreading in flower, deflexed in fruit. Petals golden yellow, 1–1.8 cm, oblanceolate. Stamens equalling or slightly exceeding petals. Styles 5–10 mm, 1.5–2.5 times length of ovary, free, erect or narrowly divergent near apex. *Azores.* H5. Summer.

40. H. androsaemum Linnaeus (*H. bacciferum* Lamarck; *Androsaemum officinale* Allioni; *A. vulgare* Gaertner). Illustration: Reichenbach, Icones Florae Germanicae et Helveticae **6**: t. 352 (1844); Syme, English botany, edn 3, t. 264 (1864); Ross-Craig, Drawings of British plants **6**: t. 6 (1952).

Shrub 30–70 cm, bushy, branches erect from branching base. Leaves stalkless, sometimes stem-clasping, 4–15 cm, oblong to broadly ovate, rounded, base truncate to cordate, paler beneath, not glaucous, papery, net-veining prominent on both sides. Inflorescence branches narrowly ascending. Flowers 1.5–2.5 cm, stellate or cupped. Sepals 6–15 mm, oblong to broadly ovate, rounded, entire, spreading in flower, deflexed in fruit. Petals golden yellow, 6–12 mm, obovate. Stamens *c.* 0.9–1.1 times length of petals. Styles 2–2.5 mm, *c.* 0.5 times length of ovary, free, erect, with upper half sharply outcurved. *W Europe, Mediterranean area to N Iran.* H3. Summer.

This well known and sometimes weedy species is easily distinguished by its short petals and styles, persistent sepals and persistently fleshy fruits. These usually become black, but sometimes remain reddish-brown, especially in variants where the flowers are relatively larger and the leaves red-tinged. Variants with yellowish leaves ('Aureum') or leaves variegated pink and white (forma **variegatum** McClintock & Nelson or 'Mrs Gladys Brabazon') are in cultivation.

41. H. × inodorum Miller (*H. androsaemum × H. hircinum*; *H. elatum* Aiton; *H. anglicum* Bertoloni; *H. multiflorum* invalid; *H. × persistens* 'Elatum' Schneider; *Androsaemum pyramidale* invalid; *A. parviflorum* Spach; *H. × urberuagae* P. & S. Dupont). Illustration: Watson, Dendrologia Britannica **2**: t. 85 (1825); Syme, English botany, edn 3, t. 265 (1865).

Shrub 60–200 cm, bushy, branches erect from branching base. Leaves stalkless or almost so, sometimes stem-clasping, 3.5–11 cm, oblong-lanceolate to broadly ovate, acute to rounded, base rounded to cordate, somewhat paler beneath, not glaucous, papery, net-veining prominent on both sides. Inflorescence branches narrowly ascending. Flowers 1.5–3 cm, stellate or cupped. Sepals 5–9.5 mm, oblong or oblong-lanceolate to broadly ovate, entire, deflexed in flower and fruit. Petals golden yellow, 8–15 mm, oblanceolate to narrowly obovate. Stamens 1.2–1.3 times length of petals. Styles 6–16 mm, 1–2.5 times length of ovary, free, erect, narrowly divergent above. *Spain to Italy, Corsica, but usually originating in cultivation and widely naturalised.* H3. Summer.

This hybrid varies considerably in height, leaf- and flower-size, depending largely on the identity of the *H. hircinum* parent. With *H. hircinum* subsp. *majus*, the plant can reach 2 m and the flowers are relatively large; with subsp. *hircinum* or subsp. *cambessedesii*, the plant is rarely over 1.5 m and the flowers are relatively small ('*H. multiflorum*'). It can be distinguished from its parents by the long styles and persistent sepals. Cultivars with yellow leaves showing different forms of variegation include 'Summergold', 'Ysella', 'Goudelsje', 'Hysan' and 'Beattie's Variety'. 'Elstead' ('Elstead Variety'; *H. × persistens* 'Elstead' Schneider) is a selected variant (or back-cross to *H. androsaemum*) with large fruits that become pinkish-red, not the usual cerise, during maturation, see Botanical Magazine, n.s. 376 (1962). *H. × inodorum*, like the parent *H. androsaemum*, seems to be especially susceptible to *Hypericum* rust (*Melampsora hypericorum*).

42. H. hircinum Linnaeus (*Androsaemum hircinum* (Linnaeus) Spach).

Shrub 50–150 cm, bushy, branches erect to spreading or pendent. Leaves stalkless or almost so, 2–7.5 cm, broadly ovate to triangular-lanceolate, acute to rounded, base shortly narrowed or wedge-shaped to cordate, somewhat paler beneath, rarely glaucous, papery, net-veining prominent on both sides. Flowers 1–20, 2–4 cm, stellate. Sepals 2–8 mm, lanceolate to narrowly ovate, acute or shortly acuminate, entire, curving in flower, deciduous in fruit. Petals bright to golden yellow, 1.1–2.1 cm, oblanceolate to narrowly obovate. Stamens 1–1.2 times length of petals. Styles 1–2.4 cm, 3–5 times length of ovary, free, erect, narrowly divergent above. *France & Spain to the E Mediterranean, Morocco & Saudi Arabia (Asir).* H4. Summer.

A variable species that comprises five subspecies, four of which are in cultivation.

Subsp. **majus** (Aiton) N. Robson (*H. hircinum* var. *majus* Aiton; *H. hircinum* subsp. *hircinum* misapplied). Illustration: Zohary, Flora Palaestina **1**: t. 324 (1966); Bulletin of the British Museum (Natural History), Botany, **12**: 308 (1985). Plant 50–150 cm. Leaves 3–7.5 cm, narrowly ovate to triangular-lanceolate, usually acute to obtuse, margin plane, goat-scented. Petals 1.3–2 cm. Styles 1.3–2.4 cm. Capsule 8–14 mm. *W Europe, (excluding Balearic Islands. Corsica, Sardinia), Sicily, Crete, Rhodes, E Mediterranean area, SW Arabia.*

Subsp. **cambessedesii** (Barceló)

Sauvage (*H.hircinum* var. *minus* Aiton in part; *Androsaemum cambessedesii* Nyman invalid; *H. cambessedesii* Barceló; *H. hircinum* var. *cambessedesii* (Barceló) Ramos; *H. minus* invalid). Illustration: Marès & Vigineix, Catalogue raisonné des plantes de Baleares, t. 3 (1880); Bulletin of the British Museum (Natural History), Botany **12**: 308 (1985). Plant 20–100 cm. Leaves 2.2–4.8 cm, lanceolate to triangular-lanceolate, acute to obtuse, margin plane, goat-scented. Petals 1–1.5 cm. Capsule 8–14 mm. *Balearic Islands.*

Subsp. **hircinum** (*H. hircinum* var. *minus* Aiton in part; *H. hircinum* var. *obtusifolium* Choisy; *H. hircinum* var. *pumilum* Watson; *H. hircinum* subsp. *obtusifolium* (Choisy) Sauvage). Illustration: Watson, Dendrologica Britannica **2**: t. 87 (1825); Bulletin of the British Museum (Natural History), Botany **12**: 308 (1985). Plant to 1 m. Leaves 2.5–4.5 cm, broadly ovate, obtuse to rounded, margin plane, rarely goat-scented. Petals 1.5–1.8 cm. Capsule 6–9 mm. *Corsica & Sardinia.*

Subsp. **albimontanum** (Greuter) N. Robson (*H. hircinum* var. *albimontanum* Greuter). Illustration: Sibthorp & Smith, Flora Graeca **8**: t. 773 (1833); Bulletin of the British Museum (Natural History), Botany **12**: 308 (1985). Plant 50–100 cm. Leaves 3–4.5 cm, broadly ovate to lanceolate, obtuse to rounded, margin wavy, goat-scented. Petals 1.8–2 cm, deeper yellow than those of other subspecies. Capsule 5–8 mm. *S Greece, Crete, Andros, Cyprus.*

Section **Inodora** Stefanoff. Shrub, deciduous, without black glands. Floral whorls in 5s (outer) or 3 or 4 (styles). Bracteoles not adpressed to sepals. Petals and stamens persistent. Stamen bundles free. Styles free.

43. H. xylosteifolium (Spach) N. Robson (*H. inodorum* Willdenow not Miller; *Androsaemum xylosteifolium* Spach; *H. ramosissimum* Ledebour). Illustration: Jaubert & Spach, Illustrationes plantarum orientalium **1**: t. 38 (1842); Bulletin of the British Museum (Natural History), Botany **12**: 316 (1985).

Shrub to 1.5 m, bushy, branches erect to spreading. Leaves with stalk 4–10 mm, blade 1.5–2.7 cm × 8–26 mm, oblong or elliptic to lanceolate or ovate, obtuse to rounded, base broadly wedge-shaped to rounded, paler beneath, not glaucous, thickly papery, conspicuously net-veined. Inflorescence broadly pyramidal to corymb-like. Flowers 1.5–3 cm, stellate.

Sepals 4–12 mm, very narrowly oblong, elliptic to oblanceolate or spathulate, acute, entire or with stalked yellowish glands, erect in bud, spreading in fruit. Petals golden yellow, 8–15 mm, narrowly oblanceolate to narrowly obovate. Stamens about equalling petals. Styles 7–9 mm, 2.5–3 times length of ovary, free, erect, narrowly divergent above. *NE Turkey, Georgia.* H4. Summer.

This species tends to spread and form thickets; it flowers better if pruned. The flowers are individually rather small, but when flowering well they can be effective *en masse.*

Section **Roscyna** (Spach) R. Keller. Perennial herbs, without black glands. Floral whorls all in 5s. Bracteoles not adpressed to sepals. Petals and stamens persistent. Stamen bundles free. Styles free or united for up to ⅘ of their length.

44. H. ascyron Linnaeus (*H. pyramidatum* Aiton; *H. gebleri* Ledebour; *Roscyna americana* Spach; *H. ocymoides* Loddiges invalid; *H. ascyron* var. *longistylum* Maximowicz; *H. ascyron* var. *brevistylum* Maximowicz; *H. ascyron* var. *americanum* (Spach) Y. Kimura; *H. ascyron* var. *vilmorinii* Rehder; *H. ascyron* forma *vilmorinii* (Rehder) Rehder). Figure 9(1,5,9), p. 59. Illustration: Gartenflora, t. 1381 (1892); Botanical Magazine, 8557 (1914); House, Wild flowers of New York, **1**: t. 130a (1918); Plants of the World (Weekly Asahi Encyclopedia), 1509 (1977).

Perennial herb, 50–150 cm, stems usually erect, 4-lined. Leaves stalkless, 4–10 cm × 7–35 mm, narrowly oblong to lanceolate or ovate, acute to obtuse, base wedge-shaped to cordate and stem-clasping, rather paler beneath, thickly papery, net-veining loose. Inflorescence corymb-like to narrowly pyramidal. Flowers 3–8 cm, stellate. Sepals 5–15 mm, ovate-lanceolate or obovate, obtuse or apiculate to acute or rounded, entire, becoming deflexed. Petals golden or sometimes pale yellow, 1.5–4 cm. Stamens 0.4–0.7 times length of petals. Styles 4–15 mm, 0.5–2 times length of ovary, free or united to ⅘ of their length, free part outcurving. *China, Taiwan, Japan, Korea, former USSR (Altai Mts to S Kamchatka), USA & adjacent Canada.* H4. Summer.

A variable species; 'forma *vilmorinii*' has large flowers and broad leaves; '*H. gebleri*' has small flowers and narrow leaves.

Section **Bupleuroides** Stefanoff. Perennial herb, without or with minute black or reddish glands at sepal and petal margins only. Floral whorls 5s (outer) or 4 or 3 (styles). Petals and stamens persistent. Stamen bundles partly united. Styles free, adpressed below.

45. H. bupleuroides Grisebach.
Perennial herb, 45–75 cm, erect, with stems usually terete. Leaves in perfoliate pairs, 7–12 × 4–8 cm, ovate or elliptic-ovate, rounded-apiculate, paler beneath, thinly papery, net-veining dense. Inflorescence broadly pyramidal. Flowers 2.7–4 cm, stellate. Sepals 3–4.5 mm, free, ovate to elliptic-oblong, rounded, entire or with minute black or reddish stalkless marginal glands. Petals golden yellow, 1.5–2 cm, narrowly oblanceolate. Stamens 0.75–0.9 times length of petals. Styles 1.2–1.4 cm, c. 2 times length of ovary, 0.35–0.4 adpressed. *NE Turkey, Georgia.* H4. Summer.

The perfoliate leaves make this species unmistakable.

Section **Hypericum**. Perennial or rarely annual herbs, with black glands on leaves, petals and anthers (except *H. concinnum*), and sometimes sepals and stems. Leaves entire, not auricled. Floral whorls in 5s (outer) or 4 or 3 (styles). Petals and stamens persistent. Stamen bundles partly united. Styles free, spreading. Capsule with linear and/or vesicular glands.

46. H. concinnum Bentham (*H. bracteatum* Kellogg not D. Don; *H. seleri* R. Keller). Illustration: Parsons, The wild flowers of California, 167 (1909); Rickett, Wild flowers of the United States **5**: 161 (1971).

Perennial herb, 15–33 cm; stems wiry, erect or ascending, rarely rooting, 2–4-lined with branches stiffly erect, forming small bushes. Leaves stalkless to shortly stalked, 1.3–3.2 cm × 1–8 mm, narrowly elliptic or narrowly oblong to linear, acute, base wedge-shaped, greyish-green, thinly leathery, obscurely veined. Inflorescence crowded, corymb-like to cylindric. Flowers 2–3.5 cm or more, stellate. Sepals 6–9 mm, broadly to narrowly ovate, acute to acuminate, entire or finely irregularly toothed, without or with few black glands. Petals bright yellow, 1–1.5 cm, oblong to obovate, black glands marginal, stalkless or impressed. Stamens c. 0.8 times length of petals. Styles 6–9 mm, 2–2.7 times length of ovary. Capsule glands linear. *USA (N California).* H4. Summer.

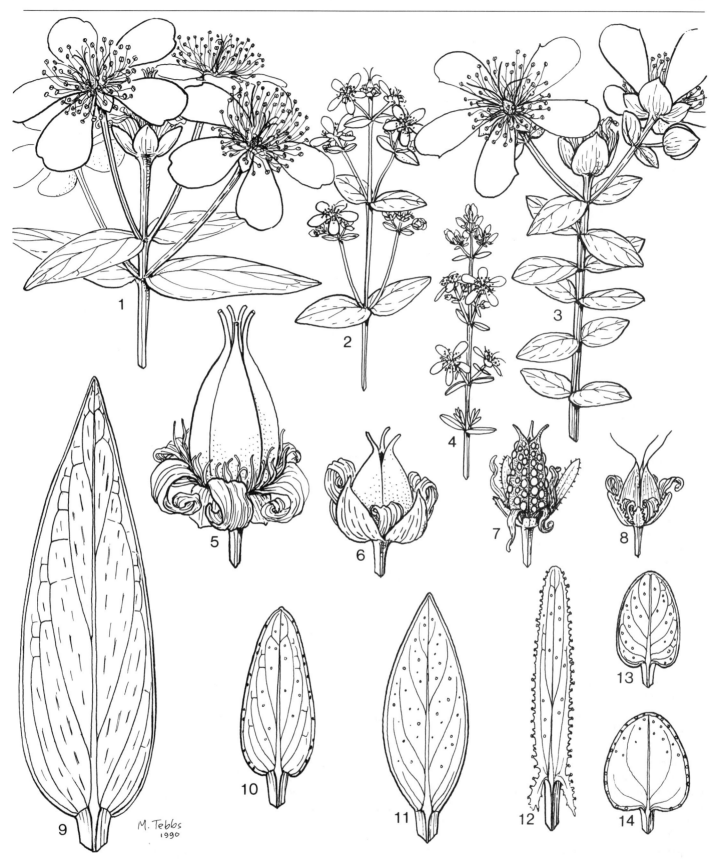

Figure 9. Diagnostic details of *Hypericum* species. Flowers (× 1). 1, *H. ascyron*. 2, *H. scouleri*. 3, *H. olympicum*. 4, *H. linarioides*. Fruits (× 2). 5, *H. ascyron*. 6, *H. olympicum*. 7, *H. montbretti*. 8, *H. coris*. Leaves. 9, *H. ascyron*. 10, *H. scouleri*. 11, *H. olympicum*. 12, *H. orientale*. 13, *H. pulchrum*. 14, *H. nummularium*.

This species superficially resembles a long- and narrow-leaved *H. olympicum*.

47. H. maculatum Crantz (*H. quadrangulum* misapplied).

Perennial herb, 15–100 cm; stems erect from rooting base, 2–4-lined, with branches ascending to spreading. Leaves stalkless, 1–5 cm × 9–20 mm, elliptic or oblong to ovate, rounded, base wedge-shaped, paler beneath, thickly papery, densely net-veined, translucent glands few or none. Flowers numerous, forming corymb-like to narrowly pyramidal inflorescence, 2–3 cm, stellate. Sepals 4–6 × 1–2 mm, free, broadly ovate to oblong, rounded, entire or finely and regularly toothed, without glands. Petals bright yellow, 9–12 mm, oblong-oblanceolate, without marginal glands, with or rarely without numerous superficial black glandular streaks and/or dots. Stamens 0.9–1 times length of petals. Styles 4–6 mm, 1.3–1.6 times length of ovary. Capsule glands linear. *Europe to Siberia*. H3. Summer.

Subsp. **maculatum** (*H. fallax* Grimm; *H. obtusum* Moench, invalid). Illustration: Reichenbach, Icones Florae Germanicae et Helveticae **6**: t. 343, 5178 (1841). Inflorescence-branching narrow (30°). Leaves densely net-veined, usually without translucent glands. Sepals broadly ovate, entire. Petals covered with black dots or short streaks or rarely (var. **immaculatum** (Murbeck) Jordanov & Kozhuharov) without black glands. *Scotland, Ardennes, Massif Central, Pyrenees and from Scandinavia, Germany and the Alps eastward.*

Subsp. **obtusiusculum** (Tourlet) Hayek (*H. dubium* Leers; *H. quadrangulum* subsp. *obtusiusculum* Tourlet. Illustration: Ross-Craig, Drawings of British plants **6**: t. 8 (1952). Inflorescence-branching wide (50°). Leaves usually relatively loosely net-veined, often with some translucent glands. Sepals narrowly ovate to oblong, apex finely and irregularly toothed. Petals with black streaks and lines. *NW Europe, Alps.*

48. H. perforatum Linnaeus (*H. officinarum* Crantz; *H. vulgare* Lamarck). Illustration: Reichenbach, Icones Florae Germanicae et Helveticae **6**: t. 343, f. 5177 (1844); Ross-Craig, Drawings of British plants **6**: t. 7 (1952).

Perennial herb, 10–110 cm; stems erect or decumbent from rooting base, 2-lined, with branches ascending. Leaves stalkless, 8–30 × 1–13 mm, narrowly ovate or lanceolate or elliptic-oblong to linear, obtuse to rounded, base wedge-shaped, paler beneath, thickly papery, not or very loosely net-veined, translucent glands numerous, relatively large. Inflorescence corymb-like to cylindric. Flowers 1.5–3.5 cm, stellate. Sepals 3–7 × 0.5–2 mm, oblong-elliptic to lanceolate or linear, acute or acuminate to shortly awned, almost entire, without glands. Petals bright yellow, 8–18 mm, oblanceolate, with few marginal black glands and sometimes with superficial black streaks. Stamens 0.65–0.85 times length of petals. Styles 5–6 mm, 2.5–3 times length of ovary. Capsule glands linear and laterally vesicular. *Europe to C China, N Africa, W Himalaya; introduced into N America & S temperate regions.* H3. Summer.

Extreme forms have been named var. **latifolium** Koch (large flowers, broad leaves), var. **angustifolium** de Candolle (smallish flowers, narrow leaves) and var. **microphyllum** de Candolle (small flowers, small leaves), but these all seem to merge. Plants growing in very dry habitats (walls, chalk, etc.) can be confused with var. *angustifolium*, but their sepals are usually longer (4–5 mm rather than 3–4 mm). Hybrids between *H. perforatum* and *H. maculatum* (**H. × desetangsii** Lamotte) are common and variable, showing all intermediate forms between those of the parents.

49. H. undulatum Willdenow (*H. decipiens* H. Watson). Illustration: Syme, English botany, edn 3, t. 270 (1864); Butcher & Strudwick, Further illustrations of British plants, f. 108 (1930); Ross-Craig, Drawings of British plants **6**: t. 10 (1952).

Perennial herb, 15–100 cm; stems erect from decumbent rooting base, narrowly 4-winged, with branches spreading to ascending. Leaves stalkless, 7–40 × 8–17 mm, narrowly ovate to elliptic or oblong, rounded, base cordate to rounded and stem-clasping, papery, margin wavy, paler beneath, not glaucous, translucent glands numerous, relatively medium-sized to small. Inflorescence corymb-like to cylindric. Flowers 1.2–1.7 cm, stellate. Sepals 3.5–5.5 mm, lanceolate, acute or acuminate, entire, without glands, with 3–14 superficial black dots. Petals bright yellow, red-tinged, 7.5–10 mm, elliptic, without or with a few marginal or superficial black dots. Stamens *c.* 0.75 times length of petals. Styles 3–4 mm, 1–1.3 times length of ovary. Capsule glands linear. *Wales & SW England to Morocco, Madeira, Azores.* H3. Summer.

50. H. tetrapterum Fries (*H. quadrangulum* Linnaeus, invalid; *H. acutum* Moench, invalid). Illustration: Reichenbach, Icones Florae Germanicae et Helveticae **6**: t. 344, f. 5179 (1844); Ross-Craig, Drawings of British plants **6**: t. 9 (1952).

Perennial herb, 10–100 cm; stems erect from decumbent rooting base or rarely wholly decumbent, rather narrowly 4-winged, branches ascending. Leaves stalkless, 1–4 × 0.7–2.4 cm, ovate or elliptic-oblong to circular, rounded, base rounded to cordate and stem-clasping, margin plane, thinly papery, somewhat paler but not glaucous beneath, with translucent glands numerous, relatively small. Inflorescence corymb-like to cylindric. Flowers 1–1.5 cm, stellate. Sepals 3.5–5 mm, lanceolate to narrowly oblong, acute to acuminate, entire, without or with 1 or 2 superficial black dots and rarely an apical dot. Petals rather pale yellow, 5–8 mm, oblanceolate without or with 1–4 black dots just within the margin. Stamens *c.* 0.9 times length of petals. Styles 2–3.5 mm, 0.8–1.4 times length of ovary. Capsule glands linear. *Europe, N Africa. W Asia, Madeira.* H3. Summer.

Like *H. undulatum, H. tetrapterum* grows best in damp or wet soil.

51. H. elegans Willdenow (*H. kohlianum* Sprengel). Illustration: Reichenbach, Icones Florae Germanicae & Helveticae **6**: t. 350, f. 5190 (1844).

Perennial herb, 15–55 cm; stems erect or decumbent from rooting base, 2-lined, branches ascending. Leaves stalkless, 1–3 cm × 3–6 mm, lanceolate or oblong to linear-oblong, acute to acuminate, base cordate and stem-clasping, margin plane, paler beneath, not glaucous, with numerous translucent glands. Inflorescence broadly pyramidal to cylindric. Flowers 1.5–2 cm, stellate. Sepals 4–5 mm, lanceolate to narrowly oblong, acute to acuminate, margin with stalkless or shortly stalked black glands, occasionally with superficial black dots. Petals bright yellow, 1–1.3 cm, oblanceolate, with only marginal black dots. Stamens 0.9 times length of petals. Styles 4–6 mm, 2 times length of ovary. Capsule glands linear. *C Europe & Balkan Peninsula to Siberia.* H3. Summer.

52. H. sampsonii Hance (*H. electrocarpum* Maximowicz; *H. assamicum* S.N. Biswas). Illustration: Nakai & Honda, Nova Flora Japonica **10**: 128 (1951); Li et al., Flora of Taiwan **2**: 644 (1976).

Perennial herb 20–80 cm; stems solitary or few, erect, terete, with branches erect to ascending. Leaves in perfoliate pairs, blade 3–8 × 1–3.5 cm, rounded, pale or glaucous beneath, thickly papery, net-veining loose. Inflorescence loose, corymb-like to cylindric. Flowers 6–15 mm, stellate with cupped base. Sepals 3–10 mm, oblong or elliptic-oblong to spathulate, rounded, entire. Petals bright yellow, 4–13 mm, elliptic-oblong, with stalkless marginal black glands and rarely superficial black dots and streaks. Stamens *c.* 0.8 times length of petals. Styles 1.5–3 times length of ovary. Capsule glands vesicular, ovoid or elongate, *S Japan to C Burma, NE India.* H5. Summer.

53. H. erectum Murray. Illustration: Nakai & Honda, Nova flora Japonica, No. **10**: 141 (1951); Plants of the World (Weekly Asahi Encyclopedia), 1506 (1977).

Perennial herb, 15–70 cm; stems solitary or few, erect to ascending from rooting base, terete, usually unbranched. Leaves stalkless, 2–5.5 × 0.5–1.7 cm, triangular-lanceolate to elliptic-oblong, obtuse to rounded, based cordate and stem-clasping to rounded, paler beneath, thickly papery, net-veining dense, with only black gland dots. Inflorescence corymb-like to pyramidal. Flowers *c.* 1–1.5 cm, stellate. Sepals 2.5–6 mm, narrowly lanceolate to ovate-lanceolate, acute to obtuse, entire. Petals bright yellow, not red-tinged, 7–9 mm, obovate to oblong-obovate, 1.5–3 times length of sepals, with stalkless or immersed marginal black glands and superficial black dots and streaks or lines. Stamens almost as long as petals. Styles 2.5–3 mm, 1–1.2 times length of ovary. Capsule glands linear. *E Asia from Sakhalin to Taiwan.* H4. Summer.

Plants cultivated under the name *H. erectum* 'Gemo' belong to *H. prolificum*.

54. H. yakusimense Koidzumi (*H. pseudopetiolatum* var. *yakusimense* (Koidzumi) Y. Kimura; *H. kiusianum* var. *yakusimense* (Koidzumi) Kato; *H. yakusimanum* invalid). Illustration: Memoirs of the Faculty of Science and Agriculture, Taihoku Imperial University **2**, Botany No. 4: t. 7, f. 2 (1934).

Perennial or annual herb, 3–8 cm, densely tufted with numerous branches, prostrate or ascending and rooting. Leaves usually shortly stalked, 3–8 × 1–3 mm, narrowly oblong to oblanceolate or obovate, rounded, base wedge-shaped, paler beneath, thickly membranous, with translucent or black superficial gland dots.

Flowers 1–3, terminal, 6–7 mm, stellate. Sepals 2.5–3 × 0.6–1.3 mm, narrowly oblong, obtuse, margin entire, without superficial black glands. Petals bright yellow, red-tinged, 6–7 mm, narrowly oblong, with few stalkless marginal and sometimes superficial black glands. Stamens *c.* 0.7 times length of petals. Styles *c.* 1.5 mm, about equalling ovary. Capsule glands linear. *Japan (Ryukyu Islands: Yakushima).* H4. Summer.

Having no superficial black glands on the leaves and petals, the variant in cultivation in Europe is forma **lucidum** Y. Kimura.

55. H. graveolens Buckley. Illustration: Torreya **27**: 85 (1927); Rickett, Wild flowers of the United States **2**(1): 209 (1967).

Perennial herb, 30–65 cm; stems erect, slightly 2-lined, with branches ascending. Leaves stalkless, 3–6.5 × 1.5–2.5 cm, acute to rounded, base broadly wedge-shaped to cordate and stem-clasping, paler beneath, thickly papery, densely net-veined, with translucent and sometimes black superficial gland dots. Inflorescence corymb-like. Flowers 2–3 cm, stellate. Sepals 5–11 mm, lanceolate, acuminate, margin entire, black glands absent. Petals bright yellow, 1–1.8 cm, obovate to elliptic, with few or no (marginal, stalkless) black glands. Stamens about equalling petals. Styles 6–10 mm, 2–2.5 times length of ovary. Capsule glands linear. *SE USA (North Carolina, Tennessee).* H4. Summer.

H. graveolens and *H. punctatum* appear to hybridise in nature (**H. mitchellianum** Rydberg is intermediate) and may be expected to do so in cultivation.

56. H. punctatum Lamarck (*H. maculatum* Walter not Crantz; *H. corymbosum* Mühlenberg). Illustration: Reichenbach, Iconographia botanica exotica, t. 88 (1827); Rickett, Wild flowers of the United States **1**: 153 (1966); National Museum of Natural Sciences, Ottawa, Publications in Botany, **11**: 31 (1981).

Perennial herb 14–105 cm; stems erect, slightly 2-lined or terete, black-dotted, branches ascending. Leaves usually stalkless, 2–6 × 1–2.6 cm, triangular-lanceolate to oblong-elliptic or oblanceolate, apiculate-obtuse to rounded, base broadly wedge-shaped to cordate and stem-clasping, paler beneath, thinly papery, densely net-veined, with translucent and sometimes black superficial gland dots. Inflorescence corymb-like to

pyramidal, with many flowers. Flowers 8–14 mm, stellate. Sepals 1.5–4 mm, lanceolate to ovate-elliptic, acute to rounded, entire, black glands absent or numerous, superficial. Petals deep to golden yellow, 3–7 mm, elliptic, with (rarely without) dense black glands, marginal and superficial (dots and streaks). Stamens about equalling petals. Styles 1–4 mm, about equalling ovary. Capsule glands linear. *E USA & SE Canada.* H4. Summer.

57. H. formosum Kunth. Illustration: Humboldt, Bonpland & Kunth, Nova Genera et Species plantarum **5**: t. 460 (1822).

Perennial herb, to 95 cm; stems erect, terete, with branches erect. Leaves stalkless, 2.5–5 × 1.7–2.5 cm, broadly ovate to oblong, rounded, base rounded to cordate and stem-clasping, paler beneath, thinly papery, rather densely net-veined, with dense black superficial gland dots. Inflorescence corymb-like to pyramidal. Flowers 1.5–2 cm, stellate. Sepals 5–6 mm, narrowly lanceolate, acute, margin with small stalkless glands, superficial black glands dense, dots and streaks. Petals golden yellow, red-tinged, 1–1.2 cm, oblanceolate with stalkless marginal and superficial black gland dots. Stamens about equalling petals. Styles *c.* 4.5 mm, 1.5 times length of ovary. Capsule glands linear. *WC Mexico.* H5. Summer.

Most, if not all records of *H. formosum* in cultivation in Europe refer to *H. scouleri*; the true *H. formosum* is quite distinct (see key). It may still be grown in the more clement regions, e.g. Spain.

58. H. scouleri J.D. Hooker (*H. formosum* var. *scouleri* (Hooker) Coulter; *H. formosum* subsp. *scouleri* (Hooker) Hitchcock). Figure 9(2,10), p. 59. Illustration: National Museum of Natural Sciences, Ottawa, Publications in Botany, **11**: 29 (1981); Rickett, Wild flowers of the United States **6**(1): 199 (1976).

Perennial herb, 5–60 cm; stems erect or decumbent and rooting, terete, with branches erect. Leaves stalkless, 1–2.8 cm × 6–15 mm, elliptic-oblong to ovate or elliptic, rounded, base rounded or cordate and clasping, paler beneath, thickly papery, not net-veined, superficial black glands absent. Inflorescence broadly to narrowly pyramidal. Flowers 1–2 cm, stellate. Sepals 3–6 mm, elliptic to oblong, rounded, entire, black glands absent or few, marginal and/or superficial. Petals golden yellow, often red-tinged, 7–12 mm, obovate to oblong-oblanceolate, with stalkless marginal black glands only. Stamens 0.7–0.9 times length of petals. Styles 2–6 mm, 1.2–1.5 times length of ovary, erect. Capsule glands linear. *SW Canada, W USA, C Mexico*. H4. Summer.

Subsp. **scouleri**, with stems taller and branched, elliptic-oblong to narrowly ovate leaves, usually axillary leaf clusters, and numerous flowers, occurs at lower altitudes and more southern localities and is a herbaceous border plant.

Subsp. **nortoniae** (Jones) Gillett, with stems shorter, decumbent and rooting, broadly ovate to elliptic leaves, no axillary leaf clusters, and solitary flowers or simple cymes, occurs at higher altitudes and is more suited for the rock garden. In the wild, intermediate forms link it with subsp. *scouleri*.

Section **Drosocarpium** (broad sense). Dwarf shrubs or perennial or annual herbs, with black glands on leaves, sepals (usually), petals and anthers, and sometimes on stems. Leaves rarely gland-fringed and auricled. Floral whorls in 5s (rarely 4s) or 3 or 2 (styles). Petals and stamens persistent. Petal marginal glands stalkless or immersed. Stamen bundles partly united. Styles free, spreading. Capsule with linear to vesicular glands.

59. H. olympicum Linnaeus. Figure 9(3,6,11), p. 59.

Dwarf shrub, 10–55 cm; stems erect to decumbent or rarely prostrate, not rooting, branches ascending. Leaves stalkless, 5–38 × 1–12 mm, oblong to elliptic or lanceolate to linear, acute to obtuse or rarely rounded, base wedge-shaped to rounded, not paler beneath, glaucous, thinly leathery, not net-veined, marginal black glands sometimes absent, superficial black glands absent. Flowers 2–6 cm, stellate. Sepals 9–16 mm, broadly ovate or broadly elliptic-lanceolate, entire, without or with 1 or 2 apical black glands. Petals golden or lemon-yellow, sometimes red-tinged, 1.4–3 cm, oblong to oblanceolate, without or with few (usually marginal) black glands. Stamens *c*. 0.8 times length of petals. Styles 1.4–2 cm, 4–5 times length of ovary. Capsule smooth. *Greece, former Yugoslavia, Bulgaria, NW & S Turkey*. H4. Summer.

Forma **olympicum** (*H. olympicum* var. *latifolum* Sims not Stefanoff; *H. olympicum* forma *majus* Haussknecht). Illustration: Botanical Magazine, 1867 (1817); Smith, Exotic botany **2**: t. 96 (1805–7); Goulandris, Wild flowers of Greece, 33 (1968); Strid, Wild flowers of Mt Olympus,

t. 79, f. 5 (1980). Stems usually erect in cultivation. Leaves 1–3.8 cm × 2–12 mm, elliptic-oblong to lanceolate. Petals always golden yellow, 1.7–3 cm. *N Greece, NW Turkey*.

Forma **uniflorum** Jordanov & Kozhuharov (*H. olympicum* var. *grandiflorum* invalid; *H. olympicum* 'Sunburst'; *H. olympicum* forma *prostratum* Jordanov & Kozhuharov; *H. polyphyllum* misapplied). Illustration: The Plantsman **1**(4): frontispiece (1980); Upward, An illustrated guide to alpines, 74 (1983) – as 'Citrinum'. Stems erect to decumbent. Leaves 8–23 × 5–13 mm, broadly elliptic to obovate. Petals golden or lemon yellow, 2–2.5 cm. *Greece, Bulgaria*.

'Sunburst' is a particularly robust, large-flowered clone. Forma *uniflorum* 'Citrinum' (*H. olympicum citrinum*) is the correct name for the clone with lemon-coloured flowers.

Forma **minus** Haussknecht (*H. polyphyllum* misapplied; *H. repens* misapplied; *H. fragile* misapplied). Illustration: Synge, Some good garden plants: 107, f. 121 (1962); Hay & Synge, The dictionary of garden plants in colour, 12 (1969). Stems decumbent to straggling. Leaves 6–15 × 1–2 mm, narrowly elliptic-oblong. Petals golden or lemon yellow, 1.4–1.8 cm. *S Greece*.

Most plants grown as *H. polyphyllum* belong to forma *uniflorum*. Plants with lemon-coloured flowers with narrow leaves (*H. polyphyllum citrinum* invalid, *H. repens citrinum* invalid) are correctly named forma *minus* 'Sulphureum'. Straggling to prostrate clones are usually grown as *H. fragile* or *H. repens*. True **H. polyphyllum** Boissier & Balansa, from southern Turkey, does not appear to be in cultivation yet.

60. H. cerastoides (Spach) N. Robson (*Campylopus cerastoides* Spach; *H. rhodoppeum* Frivaldsky). Illustration: Jaubert & Spach. Illustrationes plantarum orientalium **1**: t. 15 (1842); Gardeners' Chronicle, **92**: 224 (1932); Hay & Synge, The dictionary of garden plants in colour, 12 (1969).

Perennial herb, 7–27 cm; stems decumbent or ascending, sometimes rooting. Leaves stalkless, 8–30 × 2–13 mm, oblong to elliptic or ovate, rounded or rarely obtuse to acute, base rounded, slightly paler beneath, thickly papery, not net-veined, without or with few black glands, downy. Flowers 1–5, 2–4.5 cm, stellate. Sepals 5–12 mm, unequal, broadly ovate, elliptic to oblong or lanceolate, acute to rounded, entire, without black

glands, softly downy. Petals bright yellow, 9–21 mm, obovate to oblong, with marginal black glands. Stamens 0.5–0.8 times length of petals. Styles 3–4 mm, 1–1.6 times length of ovary. Capsule nodding, with glands linear. *S Bulgaria, NE Greece, NW Turkey*. H3. Summer.

The above description applies to subsp. **ceratoides** from NW Turkey. The erect-stemmed subsp. **meuselianum** Hagemann, from Greece and Bulgaria is apparently not in cultivation.

61. H. origanifolium Willdenow. Illustration: Jaubert & Spach, Illustrationes plantarum orientalium **1**: 6.16 (1842).

Perennial herb, 5–37 cm; stems erect to ascending, not usually rooting, shortly whitish-downy. Leaves stalkless, 5–30 × 4–9 mm, elliptic-oblong to ovate or obovate, acute to obtuse or rarely rounded, base wedge-shaped to rounded, not paler beneath, downy, thickly papery, not net-veined, black glands sometimes superficial. Inflorescence corymb-like to broadly pyramidal. Flowers 1.5–2.5 cm, stellate. Sepals 3–5 mm, narrowly oblong to spathulate, usually obtuse to acute, margin glandular hairy, usually with superficial black glands, finely downy or hairless. Petals golden yellow, red-veined, 9–15 mm, obovate to oblanceolate, sometimes with marginal black glands, with superficial black or amber glands. Stamens *c*. 0.6 times length of petals. Styles 3.5–6 mm, 1.3–2 times length of ovary. Capsule glands linear and vesicular. *W Turkey*. H4. Summer.

62. H. kelleri Baldacci. Illustration: Pacific Horticulture **38**(1): 19 (1977).

Perennial herb, 1–10 cm; stems prostrate, forming mats, creeping and rooting, glaucous. Leaves shortly stalked, 2–6 × 0.8–2 mm, oblong or elliptic, obtuse to acute, base wedge-shaped, not paler beneath, glaucous, papery, net-veined, with few, irregularly spaced marginal black glands, surface wavy, papillose or almost smooth. Flowers single, stellate. Sepals usually 4, 3–4 mm, oblong-elliptic to lanceolate, acute, margin glandular hairy to almost entire. Petals usually 4, golden yellow, red-tinged, 6–7 mm, elliptic, with few scattered black dots. Stamens *c*. 0.6 times length of petals. Styles *c*. 3 mm, about equalling ovary. Capsule glands linear and vesicular. *W Crete*. H5. Summer.

63. H. rumeliacum Boissier.
Perennial herb, 5–40 cm; stems erect to procumbent, rarely rooting, glaucous.

Leaves stalkless, 6–35 × 1–11 mm, linear-oblong or lanceolate to broadly ovate, acute to rounded, base wedge-shaped to rounded, uppermost sometimes black-glandular hairy, thickly papery, scarcely paler beneath, glaucous, obscurely net-veined, without superficial black glands. Flowers 2–3.5 cm, stellate, in groups of 1–15. Sepals 4.5–8.8 mm, narrowly ovate to lanceolate or elliptic, acute to obtuse, margin glandular hairy to fringed, with superficial black dots and streaks. Petals deep yellow, sometimes red-tinged, 1.2–1.8 cm, oblanceolate to elliptic, with black dots scattered over whole surface. Stamens *c.* 0.6 times length of petals. Styles 5– 7 mm, 1.7–2.3 times length of ovary. Capsule with faint, long or rounded vesicles or almost smooth. *S Balkan Peninsula, N Greece.* H4. Summer.

Subsp. **rumeliacum**. Illustration: Journal of the Faculty of Pharmacy, University of Istanbul, **17**: 136 (1981). Stems few, erect or ascending. Leaves linear-oblong to broadly oblong-lanceolate. Flowers 5–15, 2–3 cm. Sepals narrowly ovate to lanceolate, glandular-fringed with teeth spreading. Petals not red-tinged. *S Balkan Peninsula.*

Subsp. **apollinis** (Boissier & Heldreich) Robson & Strid (*H. apollinis* Boissier & Heldreich). Illustration: Stems numerous, ascending or procumbent. Leaves ovate-elliptic. Flowers 1–5, 2.5–3.5 cm. Sepals ovate-elliptic, glandular-hairy to fringed with teeth forward-pointing. Petals usually red-tinged. *N Greece, S Albania.*

64. H. spruneri Boissier.

Perennial herb, 30–60 cm, stems erect or decumbent sometimes rooting, glaucous. Leaves stalkless, 2–6 cm × 5–17 mm, triangular-lanceolate to narrowly elliptic or oblong, obtuse or rounded, base cordate and stem-clasping to rounded, uppermost sometimes with glandular auricles or margin glandular-hairy or toothed, paler beneath, glaucous, thickly papery, not net-veined. Inflorescence corymb-like to broadly pyramidal. Flowers 1.5–2.5 cm, stellate. Sepals 5–6 mm, narrowly lanceolate to oblong, acute, margin glandular-hairy or toothed, with numerous superficial black gland dots. Petals golden yellow, sometimes red-tinged, 1.1–1.4 cm, oblanceolate with black dots scattered over whole surface. Stamens *c.* 0.7 times length of petals. Styles *c.* 7 mm, *c.* 2.3 times length of ovary. Capsule glands round, vesicular. *SE Italy, Greece, Albania & former Yugoslavia.* H5. Summer.

65. H. barbatum Jacquin. Illustration: Sowerby & Smith, English Botany **28**: t. 1986 (1809); Reichenbach, Icones Florae Germanicae et Helveticae **6**: t. 349 (1844).

Perennial herb, 10–45 cm; stems erect or decumbent, not rooting, glaucous. Leaves stalkless or shortly stalked, 6–40 × 2–10 mm, lanceolate to linear-oblong or elliptic-oblong, acute, margin plane or revolute, base wedge-shaped to rounded, glaucous, thickly papery, not net-veined. Flowers 5–20, 1.5– 2.5 cm, stellate. Sepals 3.5–6 mm, lanceolate, acute, fringed without glands, with numerous superficial black dots or streaks. Petals golden yellow, sometimes red-veined, 1–1.5 cm, broadly oblanceolate, with black dots over whole surface or rarely near apex only. Stamens *c.* 0.7 times length of petals. Styles 2.5–4 mm, 1.6–2.5 times length of ovary. Capsule glands vescicular or almost absent. *Austria to Greece, S Italy.* H5. Summer.

66. H. perfoliatum Linnaeus (*H. ciliatum* Lamarck). Illustration: Desfontaines, Choix de Plantes de Corollaire des Instituts de Tournefort, t. 53 (1808); Coste, Flore de France **1**: t. 676 (1901); Bonnier, Flore complète **2**: t. 105 (1913).

Perennial herb, 15–75 cm; stems erect, sometimes from decumbent base, not rooting, glaucous. Leaves stalkless, 1.3–6 cm × 5–28 mm, broadly ovate to triangular or linear-lanceolate, usually rounded, base cordate and stem-clasping, uppermost sometimes black glandular-hairy, somewhat paler beneath, sometimes glaucous, papery, obscurely net-veined, without superficial black glands. Inflorescence corymb-like to broadly pyramidal. Flowers 1.2–2.5 cm, stellate. Sepals 3.5–5 mm, oblong, acute to rounded, densely and irregularly glandular-toothed or hairy, with superficial black glands mostly in 2 vertical rows. Petals golden yellow, not red-tinged, 9–14 mm, oblanceolate, sometimes with superficial black dots or streaks towards apex. Stamens *c.* 0.65 times length of petals. Styles 6–8 mm, 1.7–2 times length of ovary. Capsule glands linear and vesicular. *Mediterranean area (except SE), Canary Islands, Madeira.* H5. Summer.

67. H. montbretii Spach. Figure 9(7), p. 59. Illustration: Jaubert & Spach, Illustrationes plantarum orientalium **1**: t. 32 (1842).

Perennial herb, 15–60 cm; stems erect or decumbent, not rooting, sometimes glaucous. Leaves stalkless, 1.5–5.5 cm

× 7– 8 mm, broadly ovate to oblong or triangular-lanceolate, obtuse to rounded, margin of upper sometimes sparsely glandular-hairy, base cordate, stem-clasping, paler beneath, glaucous, papery, sometimes net-veined, rarely with a few superficial black glands. Inflorescence corymb-like. Flowers 1.5–2.5 cm, stellate. Sepals *c.* 3.8 mm, lanceolate to narrowly oblong, acute or acuminate, margin regularly glandular-hairy, usually with a few superficial black glands, reflexed in fruit. Petals golden yellow, 8–14 mm, oblanceolate, with or without black dots towards apex. Stamens *c.* 0.65 times length of petals. Styles 4.5–6 mm, 1.5–2.2 times length of ovary. Capsule glands prominently vesicular. *SE Balkan Peninsula, W Turkey.* H5. Summer.

68. H. richeri Villars.

Perennial herb, 10–50 cm; stems erect from creeping and rooting base, glaucous. Leaves stalkless, 1–5.5 cm × 5–25 mm, ovate to triangular or elliptic, acute to obtuse, base wedge-shaped to cordate and stem-clasping, all entire, paler beneath, glaucous, thinly papery, densely and conspicuously net-veined, without superficial black or translucent glands. Inflorescence dense, corymb-like. Flowers 2–4.5 cm, stellate. Sepals 3.5–10 mm, ovate to lanceolate or elliptic, acute to acuminate, entire or margin variously glandular-hairy to fringed, with numerous superficial black streaks and dots. Petals golden yellow, tinged red, 1–2.5 cm, clawed, oblanceolate, with black dots scattered over whole surface. Stamens 0.5–0.65 times length of petals. Styles 5–7 mm, 1–2 times length of ovary. Capsule glands long or round, vesicular, black and sometimes orange. *Mts of Europe (Cantabrians to Carpathians).* H4. Summer.

Subsp. **burseri** (de Candolle) Nyman (*H. fimbriatum* var. *burseri* de Candolle; *H. burseri* (de Candolle) Spach). Illustration: Bonnier, Flore complète **2**: t. 106 (1913). Leaves usually obtuse, base stem-clasping. Sepals acute, glandular-toothed to glandular-hairy. Petals 1–2.5 cm. *Pyrenees, Cantabrian Mts.*

Subsp. **richeri** (*H. fimbriatum* Lamarck). Illustration: Bonnier, Flore complète **2**: t. 106 (1913). Leaves usually acute, base usually cordate and stem-clasping, rounded. Sepals acuminate, glandular-fringed. Petals 1–1.7 cm. *W Alps, Jura, Apennines.*

Subsp. **grisebachii** (Boissier) Nyman (*H. alpinum* Waldstein & Kitaibel; *H. alpigenum* Kitaibel; *H. grisebachii* Boissier; *H. richeri*

subsp. *alpigenum* (Kitaibel) E. Schmid).
Illustration: Waldstein & Kitaibel,
Descriptiones et icones plantarum rariorum
Hungariae **3**: t. 265 (1812); Javorka &
Csapody, Iconographia Florae Hungaricae,
337 (1932). Leaves usually obtuse, base
rounded to wedge-shaped. Sepals
acuminate, glandular-hairy to entire.
Petals 1–1.8 cm. *SE Alps, Balkan Peninsula,
Carpathians.*

69. H. trichocaulon Boissier & Heldreich
(*H. repens* misapplied). Illustration:
Sibthorp & Smith, Flora Graeca **8**: t. 775
(1833); Hooker's Icones Plantarum t. 3499
(1951).

Perennial herb, 5–45 cm; stems
procumbent, ascending or pendent,
sometimes rooting, glaucous. Leaves
stalkless or shortly stalked, 5–14 × 4–6
mm, ovate-oblong to elliptic or linear,
obtuse, base rounded, paler beneath,
glaucous, not obviously net-veined,
sometimes with a few superficial black
glands. Flowers solitary or sometimes
2–3, 2–2.5 cm, stellate. Sepals *c.* 6 mm,
oblong, acute, margin glandular-toothed
to entire, with scattered superficial black
dots and streaks. Petals golden yellow,
red-tinged, 1–1.2 cm, obovate, with black
dots scattered towards apex. Stamens
0.65–0.8 times length of petals. Styles *c.*
6 mm, *c.* 2 times length of ovary. Capsule
glands faint, linear and long-vesicular or
almost absent. *Crete.* H5. Summer.

70. H. linariifolium Vahl. Illustration:
Reichenbach, Icones Florae Germanicae et
Helvetica **6**: t. 350, f. 5190b (1844); Syme,
English Botany, edn 3, **2**: t. 272 (1864);
Bonnier, Flore complète **2**: t. 106 (1913);
Ross-Craig, Drawings of British plants **6**: t.
12 (1952).

Perennial herb, 20–50 cm; stems erect
or decumbent and rooting, terete. Leaves
stalkless, 5–35 × 0.5–4 mm, narrowly
oblong to lanceolate or linear, obtuse to
rounded, margin recurved, base rounded
and stem-clasping to wedge-shaped, paler
beneath, papery, not glaucous or net-
veined, without superficial black or
translucent glands. Inflorescence corymb-
like to broadly pyramidal. Flowers *c.* 1.5
cm, stellate. Sepals 2–6 mm, lanceolate to
ovate, acute to obtuse, margin glandular-
hairy, with numerous superficial black dots
and usually streaks. Petals golden yellow,
red-tinged, 8–12 mm, oblanceolate, rarely
with superficial black streaks. Stamens *c.*
0.7 times length of petals. Styles 5–6 mm,
c. 3 times length of ovary. Capsule glands
linear. *W Europe, Madeira.* H4. Summer.

71. H. humifusum Linnaeus. Illustration:
Reichenbach, Icones Florae Germanicae et
Helveticae **6**: t. 44, f. 5176 (1844); Syme,
English botany, edn 3, **2**: t. 271 (1864),
Ross-Craig, Drawings of British plants **6**: t.
11 (1952).

Short-lived perennial or annual herb,
3–30 cm; stems decumbent or prostrate,
rooting, 2-lined. Leaves shortly stalked,
blade 3–20 × 2–6 mm, upper oblong to
lanceolate, lower obovate to lanceolate,
rounded, base wedge-shaped, paler
beneath, not glaucous, papery, not
obviously net-veined, without superficially
black glands, usually with translucent
glands. Flowers 8–12 mm, forming a loose
inflorescence, of 1–20, stellate. Sepals
(sometimes 4) unequal, 3–5 mm, oblong
to lanceolate or ovate, obtuse to acute,
gland-toothed or entire, usually with a few
superficial black dots. Petals (sometimes
4) golden or bright yellow, usually red-
tinged, 4–6 mm, elliptic, very rarely with
superficial black dots. Stamens *c.* 0.8 times
length of petals. Styles 1–2 mm, *c.* 1.5
times length of ovary. Capsule glands
linear. *Europe, Madeira, Azores.* H3.
Summer.

72. H. adenotrichum Spach. Illustration:
Jaubert & Spach, Illustrationes plantarum
orientalium **1**: t. 20 (1842).

Perennial herb, 7–32 cm; stems erect
or decumbent or prostrate, sometimes
rooting. Leaves stalkless with gland-fringed
auricles, 7–26 × 2–5 mm, oblong or
oblanceolate to linear, rounded, margin
black glandular-fringed, base wedge-
shaped, paler beneath, not glaucous,
thickly papery, not net-veined.
Inflorescence corymb-like to shortly
cylindric. Flowers 1.5–2.5 cm, stellate.
Sepals 6–8 mm, narrowly oblong, acute to
rounded, margin black glandular-fringed,
without superficial black glands. Petals
golden yellow, red-tinged, 9–15 mm,
oblong-oblanceolate, with few superficial
black dots near apex. Stamens *c.* 0.7 times
length of petals. Styles *c.* 8 mm, *c.* 2 times
length of ovary. Capsule glands linear.
Turkey. H4. Summer.

73. H. orientale Linnaeus (*H.
ptarmicifolium* Spach; *H. tournefortii* Spach;
H. jaubertii Spach; *H. orientale* var.
ptarmicifolium (Spach) Boissier; *H. orientale*
var. *tournefortii* (Spach) Boissier; *H.
orientale* var. *jaubertii* (Spach) Boissier; *H.
decussatum* Kunze). Figure 9(12), p. 59.
Illustration: Jaubert & Spach, Illustrationes
plantarum orientalium **1**: t. 17–19 (1842).

Perennial herb, 7–45 cm; stems erect or

decumbent and sometimes rooting. Leaves
stalkless with gland-fringed auricles, 10–40
× 2–10 mm, narrowly oblong, elliptic to
oblanceolate or linear, rounded to acute,
margin with amber glandular saw-like
teeth or hairs, base wedge-shaped, slightly
paler beneath, not glaucous, thickly
papery, not net-veined. Inflorescence
corymb-like to shortly cylindric. Flowers
1.5–3 cm, stellate. Sepals 6–8 mm, ovate
or narrowly oblong to obovate, obtuse to
rounded, margin with amber glandular
teeth, black glands absent. Petals golden
yellow, not red-tinged, 1–1.8 cm, oblong-
oblanceolate, without superficial black
glands. Stamens *c.* 0.6 times length of
petals. Styles 8–10 mm, 2–3.3 times length
of ovary. Capsule glands linear. *Turkey,
former USSR (Georgia, Azerbaijan).* H3.
Summer.

H. orientale is very variable; but the
variation (even in cultivation) from erect
broad-leaved and large-flowered (*H.
jaubertii*) to decumbent, narrow-leaved
and small-flowered (*H. ptarmicifolium*) is
continuous.

Section **Hirtella** Stefanoff. Perennial herbs,
usually with basal sterile shoots, usually
with amber to red or black glands on
stems (sometimes on emergences), and
black glands on sepals and petals
(marginal and sometimes superficial) and
occasionally leaves. Anther glands amber.
Floral whorls in 5s or 3s. Petals and
stamens persistent. Petals usually clawed,
with marginal glands on hairs or stalkless,
translucent glands short. Stamen bundles
partly united. Styles free, spreading.
Capsule glands linear.

74. H. elongatum Ledebour (*H.
hyssopifolium* var. *elongatum* (Ledebour)
Ledebour; *H. hyssopifolium* subsp. *elongatum*
(Ledebour) Woronow). Illustration:
Ledebour, Icones plantae rossicae **5**: t. 486
(1834).

Perennial herb, 10–70 cm; stems erect,
sometimes with faint amber glands. Leaves
(main stem) stalkless, 8–32 × 1–3 mm,
narrowly oblong or elliptic to linear-
lanceolate or linear, acute to mucronate
or rounded, margin usually recurved,
base wedge-shaped to parallel-sided, paler
beneath, sometimes glaucous, papery, not
net-veined, without black glands.
Inflorescence narrowly cylindric to
pyramidal. Flowers *c.* 20 cm, 1.5–2.5 cm,
stellate. Sepals unequal or equal, 2.5–4
mm, broadly ovate or oblong to lanceolate
or narrowly elliptic, all entire or some
or all (in one flower) with irregular or

regular stalkless marginal glands. Petals golden yellow, sometimes red-tinged or veined, 7–18 mm, oblanceolate, clawed, without or with few superficial black dots. Stamens *c*. 6 times length of petals. Styles 4–5 mm, 1.3–1.5 times length of ovary. *Mediterranean area to Siberia (Altai Mts)*. H3. Summer.

Only subsp. **elongatum**, with at least some entire sepals, is in cultivation.

75. H. hyssopifolium Villars. Illustration: Reichenbach, Icones Florae Germanicae et Helveticae **6**: t. 351, f. 5190c (1844); Bonnier, Flore complète **2**: t. 105 (1913).

Perennial herb, 20–60 cm; stems erect, without glands. Leaves (main stem) stalkless, 15–27 × 1–6 mm, narrowly elliptic-oblong to linear, rounded, margin often recurved, base narrowly wedge-shaped, papery, paler beneath, not glaucous or net-veined, without black glands. Inflorescence narrowly cylindric 5–19 cm. Flowers numerous, 1–1.5 cm, stellate. Sepals almost equal, 2.5–4.5 mm, oblong to elliptic, rounded or rarely obtuse, margin with stalkless glands to shortly glandular-hairy. Petals golden yellow, red-tinged, 7–10 mm, obovate, scarcely clawed, without superficial black glands. Stamens 0.8–1 times length of petals. Styles 3–3.5 mm, *c*. 1.5 times length of ovary. *European mts*. H4. Summer.

H. hyssopifolium occurs more often in horticultural catalogues than in gardens; but the true plant is cultivated in NW Europe and presumably also near its native areas in the south. The plate in Botanical Magazine, 3277 (1833) under this name represents *H. lydium* Boissier, which appears now to be cultivated in Europe.

76. H. capitatum Choisy (*H. rubrum* Hochstetter; *H. laeve* Boissier & Haussknecht var. *rubrum* (Hochstetter) Boissier). Illustration: Botanical Magazine, 8773 (1918).

Perennial herb, 15–50 cm; stems erect or decumbent at the base, sometimes with numerous small amber glands. Leaves (main stem) almost stalkless, 8–28 × 1.5–3.5 mm narrowly oblong to linear, apiculate or rounded, margin sometimes recurved, base rounded, paler beneath, glaucous, not net-veined, without black glands. Inflorescence dense, very broadly pyramidal to corymb-like, 3–10 cm. Flowers 8–12 mm, stellate. Sepals equal, 1.5–2.5 mm, narrowly oblong to ovate-lanceolate, acute to obtuse or rarely rounded, margin with stalkless glands or gland-toothed. Petals wholly orange or

blood-red to crimson, 5–7 mm, obovate-oblanceolate, clawed, without superficial black glands. Stamens about equalling petals. Styles 3–4 mm, *c*. 2 times length of ovary. *S Turkey*. H5. Summer.

The above description and synonymy refer to the orange- to red-flowered var. **capitatum**. The yellow-flowered var. **luteum** N. Robson (*H. laeve* Boissier & Haussknecht) is not in cultivation. Var. *capitatum* has been introduced to Britain twice (before 1914 and in 1967) but is thought to have been lost to cultivation. It may still be grown in New Zealand. Such a striking species merits a place in gardens or greenhouses.

Section **Taeniocarpium** Jaubert & Spach. Perennial herbs, sometimes woody at the base, without basal sterile shoots, rarely with amber glands on stem, with black glands on sepals, petals and bracteoles (margins only) and sometimes at leaf apex. Anther gland amber. Floral characters as in section *Hirtella* except petals not clawed, usually red-tinged or red- or orange-veined, and with elongate translucent glands.

77. H. linarioides Bosse (*H. repens* invalid; *H. alpestre* Steven; *H. polygonifolium* Ruprecht). Figure 9(4), p. 59. Illustration: Jaubert & Spach, Illustrationes plantarum orientalium **1**: t. 26 (1842).

Perennial herb, 5–33 cm; stems erect or ascending from rooting and branching base. Leaves (main stem) stalkless, 5–30 × 1.5–8 mm, narrowly oblong or elliptic to linear, obtuse, margin often recurved, base wedge-shaped, usually paler beneath, glaucous, thickly papery, not net-veined, without apical black gland. Inflorescence narrowly cylindric to spike-like, to l4 cm. Flowers 1.2–2.2 cm, stellate. Sepals 2–2.5 mm, oblong, acute to rounded, entire or with short-stalked glands towards apex. Petals golden yellow, often red-tinged or veined, 5–12 mm, broadly elliptic. Stamens 0.65–0.7 times length of petals. Styles *c*. 4 mm, *c*. 2 times length of ovary. *S Balkan Peninsula to Iran*. H4. Summer.

78. H. hirsutum Linnaeus. Illustration: Reichenbach, Icones Florae Germanicae et Helveticae **6**: t. 349, f. 5189 (1844); Syme, English Botany, edn 3, **2**: t. 274 (1864); Ross-Craig, Drawings of British plants **6**: t. 14 (1952).

Perennial herb, 35–100 cm; stems erect from decumbent rooting base, downy. Leaves shortly stalked, blade 2–6 cm × 8–20 mm, oblong to elliptic or lanceolate, rounded, margin usually plane, base

wedge-shaped, paler beneath, not glaucous, thinly papery, rather densely net-veined, stiffly hairy. Inflorescence loose, cylindric to narrowly pyramidal, to *c*. 3.5 cm. Flowers *c*. 1.5 mm, stellate. Sepals 3–5 mm, narrowly oblong, acute, margin gland-toothed or glandular-hairy. Petals rather pale yellow, sometimes red-veined, 8–10 mm. Stamens 0.6–0.7 times length of petals. Styles 4–6 mm, *c*. 2 times length of ovary. *W Europe to W China*. H2. Summer.

79. H. kotschyanum Boissier. Illustration: Bouloumoy, Flore du Liban et de la Syrie **2**: t. 72 (1930).

Perennial herb with woody base, 10–30 cm; stems erect or ascending, shortly grey-hairy. Leaves (main stem) stalkless, 5–15 × 1–2.5 mm, narrowly oblong to linear-lanceolate, rounded, margin recurved, base wedge-shaped, paler beneath, not glaucous, papery, not net-veined, stiffly and shortly greyish-hairy. Inflorescence narrowly cylindric to pyramidal. Flowers 1.4–2.2 cm, stellate. Sepals 3–4 mm, broadly oblong to ovate, rounded, 3–5-ribbed, margin gland-toothed or glandular-hairy. Petals golden yellow, often red-veined, 7–11 mm, oblanceolate. Stamens *c*. 0.9 times length of petals. Styles 4–6 mm, 1.5–2 times length of ovary. *Mts of S Turkey*. H5. Summer.

80. H. confertum Choisy.
Perennial herb, 10–35 cm; stems erect or ascending from rooting and branching base, hairless or hairy. Leaves stalkless, 7–20 × 1–4 mm, lanceolate to oblong-linear, acute to apiculate-obtuse, margin recurved, base wedge-shaped to rounded, paler beneath, bloomed to very shortly hairy, not glaucous or net-veined. Inflorescence narrowly pyramidal to cylindric or spike-like. Flowers 1.7–2.5 mm, stellate. Sepals 4–7 mm, lanceolate or oblong to ovate, acute to rounded, 3–5-ribbed, margin glandular-hairy to gland-fringed. Petals golden yellow, red-tinged or veined, 7–16 mm, elliptic-obovate. Stamens 0.5–0.9 times length of petals. Styles *c*. 6 mm, 2–3 times length of ovary. *Turkey, Syria, Lebanon, Cyprus*. H5. Summer.

Subsp. **confertum** (*H. saturejifolium* Jaubert & Spach). Illustration: Jaubert & Spach, Illustrationes plantarum orientalium **1**: t. 28 (1842). Stems downy to bloomed. Sepals lanceolate to narrowly oblong, acute, margin with long glandular hairs or gland-fringed. *W Turkey*.

Subsp. **stenobotrys** (Boissier) Holmboe (*H. stenobotrys* Boissier). Illustration:

Holmboe, Studies on the vegetation of Cyprus, 127 (1914). Stems hairless. Sepals oblong to ovate, usually obtuse, margin glandular-hairy. *S Turkey, Syria, Lebanon, Cyprus.*

81. H. pulchrum Linnaeus. Figure 9(13), p. 59. Illustration: Reichenbach, Icones Flora Germanicae et Helveticae **6**: t. 347, f. 5185 (1844); Syme, English Botany, edn 3, **2**: t. 273 (1864); Ross-Craig, Drawings of British plants **6**: t. 13 (1952).

Perennial herb, 3–90 cm; stems erect or ascending, sometimes budding from roots but stems not rooting. Leaves (main stem) stalkless, 6–20 × 3–12 mm, broadly ovate to oblong, obtuse, margin slightly recurved, base cordate, stem-clasping, paler beneath, not glaucous, thinly leathery, not net-veined, without apical black gland. Inflorescence loose, cylindric to narrowly pyramidal, *c.* 17 cm. Flowers *c.* 1.5 cm, stellate. Sepals 2.5–4 mm, ovate to broadly elliptic or oblong, obtuse to rounded, margin with regular, stalkless or shortly stalked black glands. Petals golden or rarely pale yellow, red-tinged, 7–9 mm, elliptic. Stamens *c.* 0.9 times length of petals; anthers orange to pinkish red. Styles 3.5–4 mm, 2 times length of ovary. *NW Europe.* H2. Summer.

82. H. crenulatum Boissier.
Perennial herb with woody base, 5–30 cm; stems ascending or procumbent, often straggling, not rooting. Leaves shortly stalked, blade 3–12 × 2–7 mm, elliptic to oblanceolate or circular, rounded, margin markedly wavy, base broadly wedge-shaped to cordate, slightly paler beneath, glaucous, thickly papery, not net-veined. Inflorescence corymb-like. Flowers *c.* 1.5 cm, stellate. Sepals 2.5–3 mm, elliptic to oblong, obtuse to rounded, margin with stalkless black glands or with black glandular teeth. Petals golden yellow, red-tinged, 5–9 mm, oblong-elliptic. Stamens *c.* 0.8 times length of petals. Styles *c.* 3 mm, *c.* 1.2 times length of ovary. *Turkey.* H4. Summer.

83. H. nummularium Linnaeus. Figure 9(14), p. 59. Illustration: Reichenbach, Icones Florae Germanicae et Helveticae **6**: t. 346, f. 5184 (1844); Bonnier, Flore complète **2**: t. 105 (1913).

Perennial herb, 8–30 cm; stems erect to decumbent and diffuse, creeping and rooting. Leaves shortly stalked, 5–20 × 3–16 mm, broadly ovate to circular, obtuse, margin plane, base rounded to shallowly cordate, glaucous beneath, thinly

leathery, not net-veined, with 2 apical black glands. Inflorescence corymb-like. Flowers 1.5–3 cm, stellate. Sepals 4–6 mm, oblong or elliptic to obovate, rounded, margin with fine black glandular teeth. Petals bright yellow, sometimes red-veined, 8–16 mm, obovate to elliptic. Stamens *c.* 0.8 times length of petals. Styles 4.5–6 mm, 1.8–2.5 times length of ovary. *Pyrenees & N Spain, SW Alps.* H4. Summer.

84. H. fragile Boissier.
Perennial herb, 4–16 cm; stems erect or straggling, jointed at nodes, not rooting, glaucous. Leaves shortly stalked, blade 2–7 × 2–3 mm, ovate to oblong or circular, rounded, margin slightly thickened, base broadly wedge-shaped to rounded, glaucous on both sides, leathery, not net-veined, usually without apical black gland. Inflorescence corymb-like. Flowers 1–1.5 cm, stellate. Sepals 3–4 mm, oblong to elliptic, acute, margin black glandular hairy. Petals bright yellow, red-tinged, oblanceolate, 6–9 mm, Stamens *c.* 0.65 times length of petals. Styles *c.* 2 mm, equalling ovary. *Greece.* H5. Summer.

This name appears in horticultural contexts more frequently than the plant itself, which is largely confined to alpine collections. Almost all references to *H. fragile* on garden labels, lists and catalogues refer to *H. olympicum*.

Section **Coridium** Spach. Low shrubs, shrublets or perennial herbs with woody base and leaves 3- or 4-whorled, without basal sterile shoots or stem glands, with black glands confined to sepal and sometimes petal margin or rarely absent. Anther gland amber. Floral characters as in section *Taeniocarpium* except that petals and stamens are sometimes deciduous and capsule glands are often vesicular.

85. H. empetrifolium Willdenow.
Low shrub to 60 cm; stems erect, tufted with stiffly erect branching, or decumbent and cushion-like with multi-layered branching, or prostrate and rooting. Leaves stalkless, in whorls of 3, 2–12 × 0.7–2 mm, linear or rarely narrowly elliptic, rounded, margin recurved, base wedge-shaped, paler beneath, not glaucous, leathery, not net-veined. Inflorescence cylindric to narrowly pyramidal. Flowers 1–2 cm, stellate. Sepals 1.5–2.5 mm, oblong to elliptic, rounded, margin with stalkless black glands. Petals golden yellow, oblong-elliptic, 5–10 mm, deciduous. Stamens 0.75–0.8 times length of petals, deciduous. Styles 3–4 mm, 2

times length of ovary. Capsule glands oblique, vesicular. *E Mediterranean area.* H4–5. Summer.

Subsp. **empetrifolium**. Illustration: Botanical Magazine, 178 (1796) – as H. coris; 6764 (1884). Low shrub, erect. Leaves 7–11 mm. Flowers 8–40. Petals 5. H5–G1.

Subsp. **oliganthum** (Rechinger filius) Hagemann (*H. empetrifolium* var. *oliganthum* Rechinger; *H. empetrifolium* var. *prostratum* invalid). Illustration: Sibthorp & Smith, Flora Graeca **8**: t. 774 (1833) – as H. coris. Cushion-like dwarf shrub, decumbent or ascending, branching multi-layered. Leaves 4.5–6 mm. Flowers 4–7. Petals 4–6. *Crete.* H4. Summer.

This is the subspecies commonly cultivated in Western Europe in alpine collections. It survives outside in favourable localities.

Subsp. **tortuosum** (Rechinger) Hagemann (*H. empetrifolium* var. *tortusoum* Rechinger). A dwarf shrub, prostrate, with branching single-layered, leaves 2–3.2 mm, flowers 1–4 and petals 5, from higher altitudes in Crete than subsp. *oliganthum*, does not appear to be in general cultivation.

86. H. coris Linnaeus. Figure 9(8), p. 59. Illustration: Reichenbach, Icones Florae Germanica et Helvetica **6**: 351, f. 5191 (1844); Botanical Magazine, 6563 (1881); Bonnier, Flore complète **2**: t. 105 (1913).

Dwarf shrub or perennial herb with woody base, 10–45 cm; stems erect and tufted or ascending. Leaves stalkless, in whorls of 4 (rarely 3), 4–18 × 0.5–1 mm, linear, shortly apiculate or rounded, margin recurved, base wedge-shaped, glaucous beneath and sometimes above, leathery, not net-veined. Inflorescence broadly pyramidal to shortly cylindric. Flowers 1.5–2 cm, stellate. Sepals 3.5–4 mm, narrowly oblong, rounded, margin with fine black gland-tipped teeth or with stalkless black glands. Petals golden yellow, sometimes red-veined, 9–12 mm, oblanceolate, persistent. Stamens *c.* 0.8 times length of petals, persistent. Styles 6–8 mm, *c.* 2 times length of ovary. Capsule glands linear and oblique-vesicular. *SE France, Switzerland, Italy.* H4. Summer.

Requires open situations and basic soils in the garden or alpine house. It seldom exceeds 15 cm in cultivation.

87. H. ericoides Linnaeus (*H. robertii* Battandier; *H. ericoides* subsp. *robertii*

(Battandier) Maire & Wilczek; *H. ericoides* subsp. *maroccanum* Maire & Wilczek). Illustration: Cavanilles, Icones et descriptiones plantarum **2**: t. 122 (1793); Botanical Magazine, n. s., 36 (1948).

Dwarf shrub, 2–25 cm; stems erect to decumbent with very short internodes, branching strict. Leaves stalkless, in whorls of 4 or 5, 15–35 × 5–6 mm, linear-lanceolate, obtuse to acute and shortly apiculate, margin recurved, base rounded, glaucous, densely papillose, leathery, not net-veined. Inflorescence broadly pyramidal or corymb-like. Flowers *c.* 0.8 cm, stellate. Sepals 2.5–3 mm, elliptic to oblong-oblanceolate, acuminate to acute, margin with short black glandular hairs or entire. Petals golden yellow, 5–6 mm, oblong-elliptic, persistent. Stamens *c.* 0.4 times length of petals, persistent. Styles 3–5 mm, *c.* 2.8 times length of ovary. Capsule glands linear. *Spain, Tunisia, Morocco.* H5. Summer.

The above description refers to the Spanish plant (subsp. **ericoides**), apparently the only subspecies in cultivation.

Section **Myriandra** (Spach) R. Keller. Shrubs or perennial herbs, without black glands. Floral whorls rarely all in 5s, or styles and capsule lobes 2–4; sepals, petals and stamen 'bundles' sometimes 4. Petals and sometimes sepals and stamens deciduous after flowering. Stamen bundles completely united (stamens 'free'). Styles usually adpressed in flower and often in fruit, thus appearing single. Capsule glands not prominent.

88. H. frondosum Michaux (*H. aureum* Bartram; *H. amoenum* Pursh). Figure 10(1,9), p. 68. Illustration: Botanical Magazine, 8498 (1913).

Shrub 60–133 cm; stems erect, forming a rounded bush. Leaves stalkless, 25–65 × 8–22 mm, oblong or sometimes elliptic to oblong-lanceolate or oblanceolate, apiculate-obtuse to rounded, margin plane or recurved, base broadly to narrowly wedge-shaped, bluish or yellowish green, paler or somewhat glaucous beneath, thickly papery, densely net-veined. Inflorescence almost corymb-like. Flowers 1–7, 2.5–4.5 cm. Sepals 4–5, very unequal, leaf-like, 6–20 mm, persistent and enlarging in fruit, ovate or oblong to elliptic to spathulate, rounded or apiculate-obtuse, margin plane or recurved. Petals 4 or 5, golden yellow, 1.2–2.5 cm, obovate to oblanceolate, becoming somewhat incurved-deflexed. Stamens 0.5–0.75 times

length of petals, forming a powder-puff-like cushion. Styles 3, 4–6 mm, *c.* 0.7 times length of ovary. *SE USA (Georgia, Alabama, Tennessee & Kentucky)*. H4. Mid–late summer.

This, the largest-flowered species in section *Myriandra*, makes a fine show with its 'powder-puff' stamens, is hardy in W Europe, and deserves to be more widely grown. Size of flower and leaf, as well as inflorescence form, differentiate it from *H. prolificum*, when both are well grown; but hybrids between the species occur quite frequently in cultivation. A large-flowered clone of Canadian origin is now being grown as 'Sunburst'.

89. H. prolificum Linnaeus (*Myriandra prolifica* (Linnaeus) Spach; *M. spathulata* Spach; *H. spathulatum* (Spach) Steudel). Figure 10(2,4,10), p. 68. Illustration: Watson, Dendrologia Britannica **2**: t. 88 (1825); Garden & Forest **3**: 526 (1890); National Museum of Natural Sciences, Ottawa, Publications in Botany, No. **11**: 10 (1981).

Shrub 20–200 cm, with stems erect to ascending, forming a rounded or irregular bush. Leaves stalkless or shortly stalked, blade 3–7 cm × 6–15 mm, narrowly oblong to elliptic or oblanceolate, rounded-apiculate to acute, margin plane or recurved, base wedge-shaped to attenuate, paler or somewhat glaucous beneath, papery, densely net-veined. Inflorescence cylindric, flowers 1.5–3 cm, in groups of 7–21. Sepals 5, unequal, 4–8 mm, enlarging in fruit, elliptic to obovate or oblanceolate to spathulate, obtuse to acute, margin plane or recurved. Petals 5, golden yellow, 7–15 mm, obovate to oblanceolate or spathulate, becoming incurved-deflexed. Stamens 0.7–0.85 times length of petals. Styles usually 3, 4–6 mm, 1–1.3 times length of ovary. *C & E USA, Canada (S Ontario)*. H4. Summer.

Plants intermediate in form between *H. prolificum* and *H. frondosum* have been named **H. × vanfleetii** Rehder.

90. H. kalmianum Linnaeus. Illustration: Botanical Magazine, 8491 (1913); Garden & Forest **3**: 113 (1890); National Museum of Natural Sciences, Ottawa, Publications in Botany, No. **11**: 8 (1981).

Shrub 14–100 cm, with stems erect, forming a slender, rounded or flat-topped bush. Leaves stalkless, 1.5–4.5 cm × 3–10 mm, narrowly oblong to oblanceolate or linear, rounded to apiculate or obtuse, margin recurved, base narrowly wedge-shaped or narrowed, paler or glaucous

beneath, papery, densely but obscurely net-veined beneath. Inflorescence rounded-corymb-like. Flowers 2–3.5 cm, in groups of 1–7. Sepals 5, almost equal (or rarely 4 unequal), 4–9 mm, enlarging in fruit, elliptic or oblong to obovate, obtuse to acute, margin recurved. Petals 4 or 5, golden yellow, 8–15 mm, obovate to oblong, becoming incurved-deflexed. Stamen 0.65–0.75 times length of petals. Styles usually 5, 3–4 mm, 0.65–0.75 times length of ovary. *USA and Canada (Great Lakes)*. H4. Mid–late summer.

91. H. lobocarpum Gattinger (*H. densiflorum* var. *lobocarpum* (Gattinger) Svenson). Illustration: Garden & Forest **10**: 453 (1897).

Shrub 90–200 cm, with stems erect, forming large clumps. Leaves stalkless or shortly stalked, blade 3.5–5 cm × 3–14 mm, narrowly oblong to lanceolate or linear, rounded-apiculate to obtuse, margin recurved, base narrowly wedge-shaped to attenuate, dull but not bluish or yellowish green, paler or glaucous beneath, papery, venation obscure. Inflorescence spherical-cylindric to broadly pyramidal. Flowers 1–1.5 cm. Sepals 5, equal, 3.5–4.5 mm, not enlarging, narrowly elliptic to narrowly oblong or spathulate, apiculate or acute, margin recurved. Petals 5, golden yellow, 6–8 mm, obovate-oblanceolate, becoming incurved-deflexed. Stamens *c.* 0.85 times length of petals. Styles usually 4 or 5, 2–3 mm, 0.8–0.85 times length of ovary; capsule lobed. *SE USA (S Carolina & Alabama)*. H5. Late summer.

Hybrids between *H. lobocarpum* and *H. prolificum* (**H. × dawsonianum** Rehder) are intermediate in form between the parents.

92. H. densiflorum Pursh (*H. prolificum* var. *densiflorum* (Pursh) A. Gray; *H. glomeratum* Small; *H. nothum* Rehder). Figure 10(3,5,11), p. 68. Illustration: Garden & Forest **3**: 527 (1890); Rickett, Wild flowers of the United States **1**(1): 153 (1966).

Shrub 60–300 cm; stems erect, forming dense bushes. Leaves stalkless, 20–45 × 2–7 mm, very narrowly elliptic-oblong or oblanceolate to linear, rounded-apiculate to acute, margin recurved, base narrowly wedge-shaped to tapered, paler and often glaucous beneath, papery, venation dense but sometimes obscure. Inflorescence broadly pyramidal to cylindric. Flowers 1–1.7 cm. Sepals 5, equal, 4–6 mm, not enlarging, narrowly oblong-lanceolate or oblanceolate to spathulate, mucronate to

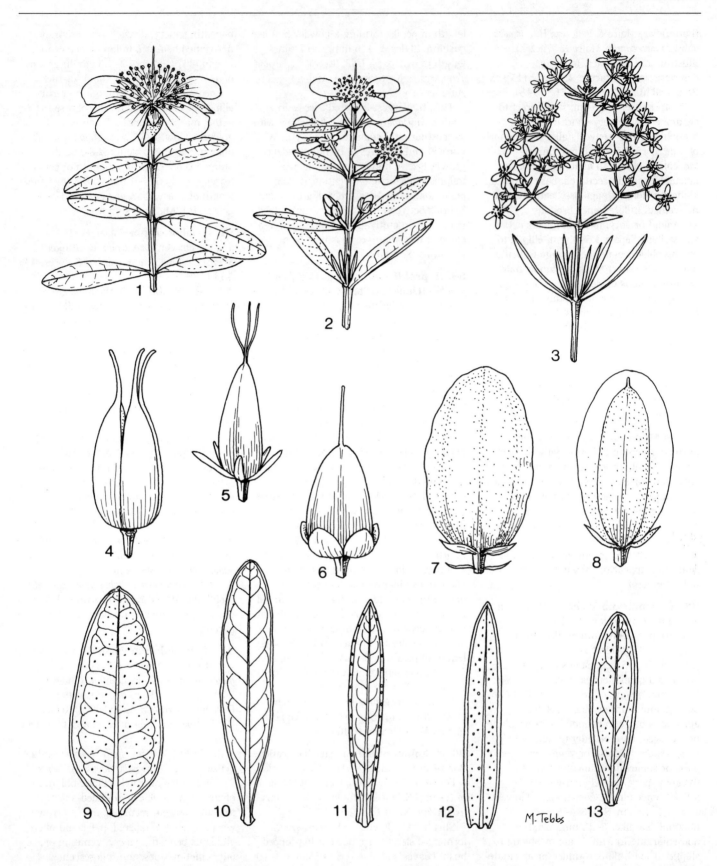

Figure 10. Diagnostic details of *Hypericum* species.
Inflorescences. 1, *H. frondosum*. 2, *H. prolificum*. 3, *H. densiflorum*.
Fruits (× 6). 4, *H. prolificum*. 5, *H. densiflorum*. 6, *H. cistiofolium*.
7, 8, *H. hypericoides*. Leaves (× 2). 9, *H. frondosum*.
10, *H. prolificum*. 11, *H. densiflorum*. 12, *H. cistiflorum*.
13, *H. hypericoides*.

acute, margin recurved. Petals 5, golden yellow, 6–9 mm, obovate-oblanceolate, becoming somewhat deflexed. Stamens *c.* 0.8 times length of petals. Styles usually 3 or 4, 2–3 mm, 0.7–0.85 times length of ovary; capsule not lobed. *E USA (Georgia, Alabama to New Jersey).* H4. Mid–late summer.

A low (to 60 cm) densely branched shrub with small leaves and small flowers that has been grown in Europe as **H. × arnoldianum** Rehder is probably *H. densiflorum × H. lobocarpum.*

93. H. galioides Lamarck (*H. axillare* Lamarck; *H. ambiguum* Elliott; *H. galioides* var. *ambiguum* (Elliott) Chapman). Illustration: Garden & Forest **10**: 433 (1897); Gardeners' Chronicle, **24**: 301 (1898); Rickett, Wild flowers of the United States **2**(1): 213 (1967).

Shrub 50–150 cm, stems erect, forming rounded clumps. Leaves stalkless (sometimes apparently stalked), 5–37 × 1–7 mm, very narrowly oblong-elliptic or oblanceolate to linear, rounded to acute, margin recurved, base tapered, paler but not glaucous beneath, papery, venation sometimes obscure. Inflorescence narrowly cylindric. Flowers 9–14 mm. Sepals 5, equal, 3.5–6.5 mm, not enlarging, oblanceolate to spathulate or linear, apiculate-obtuse to acute, margin recurved. Petals 5, golden yellow, 5–9 mm, obovate-oblanceolate, becoming somewhat deflexed. Stamens *c.* 0.85 times length of petals. Styles 3, 2.5–4 mm, 1.2–1.35 times length of ovary. *SE USA (N Carolina to Texas).* H5. Mid–late summer.

94. H. fasciculatum Lamarck (*H. aspalathoides* Willdenow; *H. galioides* var. *aspalathoides* (Willdenow) Torrey & Gray; *H. galioides* var. *fasciculatum* (Lamarck) Svenson). Illustration: Rhodora **42**: 13, t. 587, f. 1 (1940); Rickett, Wild flowers of the United States **2**(1): 213 (1967).

Shrub 1–1.5 m, with stems erect, forming dense bushes; bark spongy. Leaves stalkless, 1–18 × 0.4–0.5 mm, those in axillary clusters almost as long, linear-subulate, apiculate, margins completely recurved obscuring most of lower surface, base parallel, glaucous beneath, leathery, 1-veined. Inflorescence corymb-like. Flowers 1.5–1.8 cm. Sepals 5, equal, 6–8 mm, not enlarging, linear-subulate, acute, margin completely recurved. Petals 5, golden yellow, 7–9 mm, obovate-oblanceolate, becoming somewhat deflexed. Stamens *c.* 0.75 times length of petals. Styles 3, *c.* 3 mm, about equalling ovary.

SE USA (N Carolina to Texas). H5–G1. Summer.

95. H. buckleiyi M.A. Curtis (*H. buckleiyi* invalid). Illustration: Garden & Forest **4**: 581 (1891); Rickett, Wild flowers of the United States **2**(1): 213 (1967).

Dwarf shrub, 5–45 cm, with stems erect or ascending from decumbent, rooting base, forming low compact mats. Leaves stalkless or almost so, 4–5 × 2–12 mm, oblong or elliptic to oblanceolate or obovate, rounded, margin plane, base wedge-shaped, paler but not glaucous beneath, papery, venation obscure. Flowers 2–2.5 cm, terminal, in groups of 1–5. Sepals 5, equal, 4–5 mm, enlarging somewhat in fruit, broadly elliptic to spathulate, obtuse, plane. Petals 5, golden yellow, 6–11 mm, oblanceolate, becoming deflexed. Stamens *c.* 0.8 times length of petals. Styles 3, 2.5–4 mm, 0.85–1 times length of ovary. *E USA (S Appalachian Mts).* H4. Summer.

96. H. nudiflorum Michaux. Illustration: Rickett, Wild flowers of the United States **2**(1): 211 (1967).

Shrub 1–2 m, with stems erect, woody and much branched below, rather brittle, forming a narrow bush. Leaves stalkless or almost so, 2.2–7 cm × 6–24 mm, lanceolate or elliptic to oblong, obtuse to rounded, margin plane, base wedge-shaped, paler but not glaucous beneath, papery, not net-veined. Inflorescence flattish, corymb-like. Flowers 1.2–1.5 cm. Sepals 5, equal, 1.5–2.5 mm, not enlarging, becoming deflexed, oblong-lanceolate, acute, margin plane. Petals 5, pale yellow, 6.5–8 mm, becoming deflexed. Stamens 0.6–0.75 times length of petals. Styles 3, 2.5–3 mm, 0.8–1 times length of ovary. *SE USA (Virginia to Alabama).* H5. Summer.

97. H. cistifolium Lamarck (*H. opacum* Torrey & Gray). Figure 10(6, 12), p. 68. Illustration: Garden & Forest **5**: 305 (1892); Rickett Wild flowers of the United States **3**(1): 213 (1967); Bell & Taylor, Florida wild flowers and roadside plants, t. 146 (1982).

Shrub 30–100 cm, with stems usually single, erect, sometimes branching strictly above. Leaves stalkless, 15–40 × 4–8 mm, narrowly lanceolate to linear-elliptic or narrowly oblong to oblanceolate, obtuse to rounded, margin recurved, base broadly wedge-shaped to rounded, paler and sometimes glaucous beneath, thinly leathery, venation obscure. Inflorescence a corymb-like cyme. Flowers 9–12 mm.

Sepals 5, unequal, 2.5–4 mm, not enlarging, obovate to oblong, obtuse to rounded, margin plane. Petals 5, golden yellow, 5–6.5 mm, obovate-oblanceolate, becoming deflexed. Stamens *c.* 0.5 times length of petals. Styles 3, 1.8–2 mm, 0.8–1 times length of ovary. *SE USA.* H5. Summer.

A plant for marshy places in favoured localities.

98. H. ellipticum J.D. Hooker. Illustration: House, Wild flowers of New York **1**: t. 130B (1918); Rickett, Wild flowers of the United States **2**(1): 211 (1967).

Perennial herb, 15–40 cm, with stems erect from creeping slender rhizomes, often unbranched. Leaves stalkless, 12–30 × 5–14 mm, oblong to elliptic or oblanceolate, rounded, margin plane, base wedge-shaped to rounded, paler but not glaucous beneath, papery, netted venation visible beneath. Inflorescence a loose, flattish, corymb-like cyme. Flowers 1–1.5 cm. Sepals 4 or 5, unequal, 4.5–6 mm, obovate to elliptic or narrowly oblong, rounded to acute, plane. Petals 4 or 5, golden yellow, 6–9 mm, obovate to oblong-oblanceolate, not becoming deflexed. Stamens *c.* 0.6 times length of petals. Styles 3, 2–3 mm, 0.8–1 times length of ovary. *SE Canada, E USA (Newfoundland to Tennessee).* H3. Summer

Like *H. cistifolium, H. ellipticum* is a plant for the marshy garden, but it is less easily controllable.

99. H. crux-andreae (Linnaeus) Crantz (*Ascyrum crux-andreae* Linnaeus; *Ascyrum stans* Michaux; *Hypericum stans* (Michaux) Adams & Robson). Illustration: Ventenat, Jardin de Malmaison **2**: t. 90 (1805); Rickett, Wild flowers of the United States **2**(1): 209 (1967).

Shrub 20–100 cm, with stems erect, simple or sparsely branched. Leaves stalkless, 12–36 × 6–16 mm, elliptic-oblong to obovate, obtuse to acute, margin slightly recurved, somewhat glaucous beneath, leathery, venation obscure. Inflorescence narrowly cylindric, flowers 2–3 cm, in groups of 1–9. Sepals 4, very unequal, plane; outer 9–20 mm, broadly ovate to circular, acute, inner 7–14 mm, lanceolate, acute. Petals 4, golden to apricot yellow, 1.1–1.8 cm, obovate, not becoming deflexed. Stamens 0.45–0.5 times length of petals. Styles 3 (rarely 4), 1.5–2 mm, 0.5 times length of ovary, distinct, spreading above. *E USA (New York to Texas).* H5. Summer.

The most showy of the 4-petalled Hypericums, it deserves to be more widely grown.

100. H. hypericoides (Linnaeus) Crantz (*Ascyrum hypericoides* Linnaeus). Figure 10(7,8,13), p. 68. Illustration: Deam, Shrubs of Indiana, 239 (1932); Rickett, Wild flowers of the United States **2**(1): 209 (1967).

Shrub 3–15 cm, stems erect to diffuse, simple or sparsely branched or with numerous basal branches. Leaves stalkless, 5–34 × 1–8 mm, linear to oblanceolate, obtuse to rounded, margin slightly recurved, paler but not glaucous beneath, thickly papery, venation obscure. Inflorescence narrowly pyramidal to narrowly cylindric. Flowers 1.5–2.5 cm. Sepals 4, very unequal, plane; outer 5–12.5 mm, broadly ovate to elliptic, obtuse to acute, inner minute or scarcely developed. Petals 4, golden yellow, 7–12 mm, narrowly oblong-elliptic, not reflexed. Stamens *c.* 0.4 times length of petals. Styles 2, *c.* l mm, *c.* 0.4 times length of ovary, distinct, spreading above. *E USA, Mexico to Honduras, Greater Antilles, Bahamas, Bermuda.* Summer. H4–Gl.

Subsp. **hypericoides** is erect and lowland; subsp. **multicaule** (Willdenow) N. Robson (*Ascyrum multicaule* Willdenow; *Hypericum hypericoides* var. *multicaule* (Willdenow) Adams & Robson; *H. stragulum* Adams & Robson) is decumbent to prostrate and upland. The latter is likely to be hardier in NW Europe, but only subsp. *hypericoides* is known to have been cultivated.

Section **Webbia** (Spach) R. Keller. Trees or shrubs without black glands. Floral whorls in 5s or 3 (styles). Petals and stamens persistent. Sepals with minute amber marginal glands. Stamen bundles partly united. Styles free, spreading, separating basally in fruit. Capsule glands linear, obscure.

101. H. canariense Linnaeus (*H. floribundum* Aiton; *H. canariense* var. *floribundum* (Linnaeus) Bornmüller; *Webbia canariensis* (Linnaeus) Webb & Berthelot; *W. floribunda* (Aiton) Spach; *W. platysepala* Spach. Figure 11(1,6,11), p. 71. Illustration: Botanical Cabinet **10**: t. 953 (1842); Reichenbach, Iconographia botanica exotica **1**: t. 95 (1827); Webb & Berthelot, Phytographia Canariensis: t. 4B-D (1836); Schaeffer, Pflanzen der Kanarischen Inseln, edn 2, 147 (1967).

Shrub or small tree, 1–4 m, erect to

somewhat spreading, with branches ascending. Leaves stalkless, 2–6.5 cm × 4–l5 mm, narrowly elliptic to oblong-elliptic, acute, margin plane, base narrowly wedge-shaped, paler but not glaucous beneath, papery, densely net-veined towards margin beneath. Inflorescence broadly to narrowly pyramidal. Flowers 2–3 cm, stellate. Sepals 3–4.5 mm, variable in shape from broadly obovate-oblong and obtuse or rounded to lanceolate and acute, minutely amber-gland-toothed. Petals bright yellow, 1.2–1.7 cm, narrowly oblanceolate. Stamens *c.* 0.7 times length of petals. Styles 8–14 mm, 2.7–4.7 times length of ovary. *Canary Islands, Madeira.* Summer. H5.

Section **Arthrophyllum** Jaubert & Spach. Low to dwarf shrubs with older stem nodes swollen and black glands sometimes on bract and sepal margins. Otherwise as in section *Webbia*.

102. H. nanum Poiret. Figure 11(3,8,13), p. 71. Illustration: Jaubert & Spach, Illustrationes plantarum orientalium **1**: t. 23 (1842).

Dwarf shrub, 5–50 cm, erect, rounded, much branched, with branches erect. Leaves stalkless, 8–20 × 4–17 mm, ovate to broadly elliptic, obtuse to rounded, base wedge-shaped to narrowing, somewhat paler beneath, glaucous, thickly papery, densely net-veined. Flowers 1–9, terminal, 1–2 cm, stellate. Sepals 2–3 mm, oblong-lanceolate or elliptic to ovate, acute, entire. Petals yellow, 7–11 mm, oblanceolate. Stamens *c.* 0.5–0.6 times length of petals. Styles 8–10 mm, *c.* 4 times length of ovary. *Syria, Lebanon, N Israel.* Summer. H5.

H. nanum is a popular alpine-house plant in northern Europe, but could probably be grown outside in the south.

103. H. vacciniifolium Hayek & Siehe.

Dwarf shrub, 8–20 cm, erect or decumbent, rounded, much branched, branches erect. Leaves stalkless or almost so, 9–18 × 4–10 mm, broadly elliptic-oblong to oblanceolate, obtuse to rounded, margin plane, base rounded or wedge-shaped to narrowing, slightly paler or glaucous beneath, thickly papery, densely and prominently net-veined. Flowers 1–3, terminal and on axillary shoots, *c.* 2 cm, stellate. Sepals 2–4 mm, ovate to oblong, entire. Petals yellow, 1–1.2 cm, oblong-oblanceolate. Stamens *c.* 0.8 times length

of petals. Styles 8–10 mm, *c.* 3–4 times length of ovary. *S Turkey.* Summer. H5

Section **Triadenioides** Jaubert & Spach. Low, dense or straggling shrubs or shrublets without swollen stem nodes, with small leaves. Black glands absent or, when present, not confined to margin. Otherwise as in section *Webbia*.

104. H. pallens Banks & Solander (*H. cuneatum* Poiret). Illustration: Jaubert & Spach, Illustrationes plantarum orientalium **1**: t. 25 (1842).

Dwarf shrublet, with stems 5–25 cm, straggling, bright red, sometimes with a few black glands. Leaves stalkless, 3–25 × 2–9 mm, oblong or elliptic to oblanceolate or obovate, obtuse or rounded, margin plane, base wedge-shaped, paler beneath, glaucous, thinly papery, obscurely veined, with few marginal and/or superficial black glands. Flowers 1–3, terminal, 1.4–2.5 cm, stellate. Sepals 2–6 mm, obovate to linear, rounded to acute, entire or with marginal and usually a few superficial black glands. Petals pale yellow, red-tinged, 6–12 mm, oblong, usually with a few black or reddish glands. Stamens *c.* 0.75 times length of petals; anther gland black. Styles 4–5 mm, *c.* 15 times length of ovary. *S Turkey, W Syria.* Summer. H5.

An attractive alpine-house plant with flower bud tips sealing-wax-red.

Section **Adenotrias** (Jaubert & Spach) R. Keller. Usually low shrubs with small leaves, as in section *Triadenioides*, without black glands but with flowers apparently tubular and styles of 2 different lengths, with petal ligules, 'lodicules' alternating with stamen bundles, stamen filaments united for *c.* ⅔ of their length, and petals sometimes deciduous.

105. H. aegypticum Linnaeus (*H. webbii* Steudel; *Triadenia aegyptica* (Linnaeus) Boissier; *Hypericum maritimum* Sieber; *Triadenia maritimum* (Sieber) Boissier). Figure 11(2,7,12), p. 71. Illustration: Edwards's Botanical Register **3**: t. 196 (1817); Botanical Magazine, 6481 (1880).

Shrub to *c.* 50 cm in cultivation, erect, decumbent, spreading or rarely prostrate, densely branched. Leaves stalkless, 3–16 × 1.5–4 mm, narrowly oblong to elliptic, acute, margin plane, base wedge-shaped, uniformly coloured, glaucous, leathery, veins not visible. Flowers 7–9 mm, apparently tubular with spreading margin, solitary, terminal and often on crowded short branches, forming a cylindric false raceme. Sepals 4–6 mm, oblong, obtuse

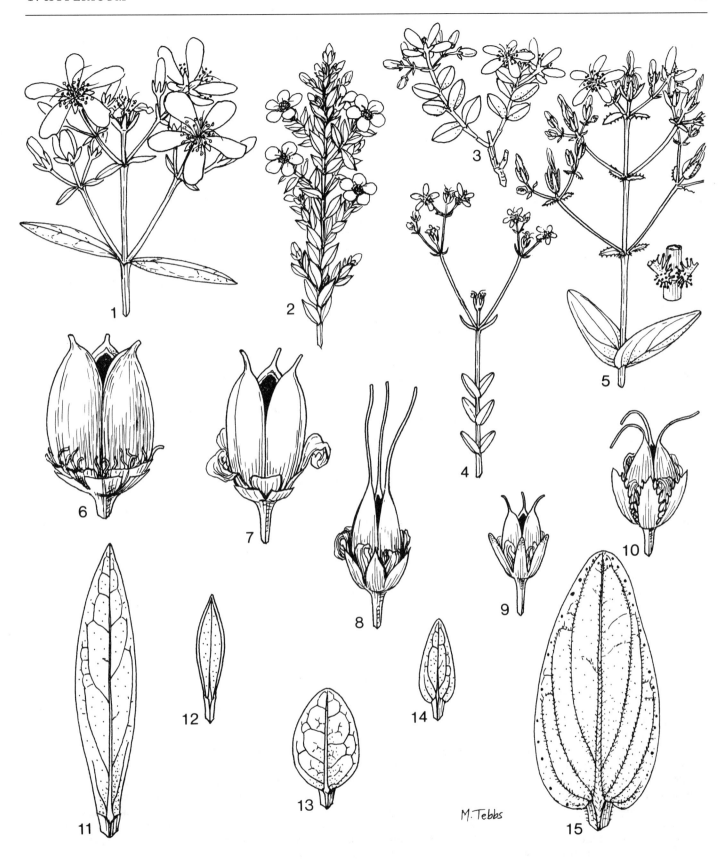

Figure 11. Diagnostic details of *Hypericum* species.
Inflorescences (× 1). 1, *H. canariense*. 2, *H. aegypticum*.
3, *H. nanum*. 4, *H. japonicum*. 5, *H. annulatum*. Fruits (× 5).
6, *H.canariense*. 7, *H. aegypticum*. 8, *H. nanum*. 9, *H. japonicum*.

10, *H. annulatum*. Leaves (× 2). 11, *H. canariense*.
12, *H. aegypticum*. 13, *H. nanum*. 14, *H. japonicum*.
15, *H. annulatum*.

or rarely acute, entire. Petals rather pale yellow, 8–14 mm, oblanceolate. Stamens *c.* 0.9 (short-styled) or 0.6 (long-styled) times length of petals. Styles 1–3 mm, 0.5–0.7 times length of ovary (short-styled) or 4–6 mm, 2–3 times length of ovary (long-styled). *S Morocco, Malta, Sardinia, Libya (Cyrenaica), Greece (Ionian Islands, Peloponnesus), Crete.* Spring–summer or until frost. H5.

Section **Adenosepalum** Spach. Shrubs or perennial herbs, often hairy, with black or amber to reddish glands on leaves (marginal or inframarginal and sometimes elsewhere), sepals (marginal), anthers and sometimes petals (usually marginal only). Floral whorls in 5s (outer) or 3 (styles). Petals and stamens persistent. Petal marginal glands present or absent. Stamen bundles partly united. Styles free, spreading. Capsule glands linear.

106. H. glandulosum Aiton. Illustration: Webb & Berthelot, Phytographia Canariensis **1**: t. 3 (1835).

Shrub, 25–300 cm, spreading, branches ascending, usually hairless. Leaves stalkless, 2.3–5.5 cm × 6–17 mm, elliptic to oblanceolate, margin plane with stalkless black glands, base wedge-shaped to narrowed, paler but not glaucous beneath, papery, scarcely net-veined. Inflorescence broad, rounded, corymb-like. Flowers 1.5–2 cm, stellate. Sepals 5–8 mm, lanceolate, acute, with stalkless marginal glands. Petals pale yellow, red-tinged, 9–11 mm, narrowly oblanceolate. Stamens *c.* 0.7 times length of petals. Styles 5–6 mm, 2–3 times length of ovary. *Canary Islands, Madeira.* Summer. G1.

107. H. reflexum Linnaeus filius. Illustration: Reichenbach, Iconographia Botanica Exotica, t. 86 (1827); Kunkel, Flora de Gran Canaria **4**: t. 165 (1979).

Shrub to *c.* 50 cm, spreading, branches ascending, usually densely white-felted; plant otherwise hairless. Leaves crowded, 4-ranked, stalkless, 9–25 × 4–9 mm, narrowly oblong to triangular-lanceolate or elliptic, acute, margin plane, entire, base cordate to rounded, stem-clasping, scarcely paler beneath, not glaucous, papery, densely net-veined. Inflorescence corymb-like to hemispherical. Flowers 1.5–2 cm, stellate. Sepals 5–7 mm, narrowly lanceolate, acute or acuminate, margin shortly irregularly glandular-hairy. Petals yellow, sometimes red-tinged, 8–10 mm, narrowly oblanceolate. Stamens 0.7–0.9 times length of petals. Styles 7–8 mm,

2.3–3.2 times length of ovary. *Canary Islands.* Summer. G1.

108. H. annulatum Moris (*H. degenii* Bornmüller; *H. atomarium* subsp. *degenii* (Bornmüller) Hayek; *H. atomarium* misapplied). Figure 11(5,10,15), p. 71. Illustration: Moris, Flora Sardoa, t. 22 (1837); Webbia **22**: 275 (1967).

Perennial herb, 20–65 cm in cultivation; stems wholly erect or base decumbent but not rooting, shortly whitish-hairy except inflorescence axis. Leaves stalkless, 1.5–5.5 cm × 8–25 mm, ovate to elliptic, rounded, margin plane, base cordate and stem-clasping, somewhat paler beneath, not glaucous, papery, venation obscure, shortly whitish hairy. Bracts with densely glandular auricles. Inflorescence broadly pyramidal to corymb-like. Flowers 1.5–2 cm, stellate. Sepals 4–6 mm, lanceolate to narrowly oblong, acute to obtuse, margin with long glandular hairs. Petals bright yellow, not red-tinged, 9–12 mm, oblanceolate. Stamens *c.* 0.8 times length of petals. Styles 5–6 mm, 2 times length of ovary. *Sardinia, Balkan Peninsula, Saudi Arabia, NE & E Africa.* H4. Summer.

H. annulatum is much more variable in habit and hairiness in Africa than in Europe. The above description applies to the Balkan variant (*H. degenii*), which is the only one known to be cultivated.

109. H. atomarium Boissier.

Similar to *H. annulatum* but with inflorescence more nearly cylindric (not corymb-like), bracts without auricles; sepals black gland-dotted or -streaked; petals sometimes with a few superficial black glands. *S Greece, W Turkey.* H5. Summer.

Most plants cultivated as *H. atomarium* belong to *H. annulatum*.

110. H. lanuginosum Lamarck (*H. gracile* Boissier; *H. lanuginosum* var. *gracile* (Boissier) Boissier). Illustration: Holmboe, Studies on the Vegetation of Cyprus, 128 (1914); Boulloumoy, Flore du Liban et de la Syrie, t. 71 (1930).

Similar to *H. annulatum*, but inflorescence as in *H. atomarium* and bracts without auricles; sepals 2–5 mm, broadly ovate to oblong or rarely lanceolate, obtuse to rounded; petals rarely red-veined. *S Turkey to N Israel, Cyprus.* H5. Summer.

Most records of *H. lanuginosum* in cultivation turn out to refer to *H. annulatum*, but the true plant has been grown.

111. H. montanum Linnaeus. Illustration: Reichenbach, Icones Florae Germanicae et Helveticae **6**: t. 347 (1844). Syme, English botany, edn 3, **2**: t. 275 (1864); Ross-Craig, Drawings of British plants **6**: t. 15 (1952).

Perennial herb, 20–80 cm; stems erect, hairless. Leaves stalkless, 2–7 × 1–3 cm, ovate to lanceolate or oblong-elliptic, apex rounded, margin plane, base rounded to cordate and stem-clasping, somewhat paler beneath, not glaucous, papery, densely net-veined, usually rough beneath. Inflorescence congested, shortly cylindric to corymb-like. Flowers 15–25 mm, stellate. Bracts with glandular-hairy auricles. Sepals 4–6 mm, narrowly oblong, acute, margin glandular-hairy. Petals pale yellow, 8–12 mm, elliptic. Stamens *c.* 0.65 times length of petals. Styles 3–5 mm, 1.2–1.35 times length of ovary. *W & C Europe, C Russia, Transcaucasia, Morocco.* H4. Summer.

112. H. elodeoides Choisy (*H. napaulense* Choisy).

Perennial herb, 15–73 cm; stems wholly erect or base shortly creeping and rooting, hairless. Leaves stalkless, 10–36 × 2–17 mm, ovate-lanceolate or oblong-linear, usually acute, margin plane, base rounded to cordate and stem-clasping, paler beneath, not glaucous, papery, densely net-veined, uppermost auricled with base and auricles glandular-hairy. Inflorescence cylindric to corymb-like. Flowers 1–1.7 cm, stellate. Sepals 5–9 mm, narrowly oblong-lanceolate to linear, acute, margin glandular-hairy. Petals pale yellow, 7–12 mm, broadly oblanceolate. Stamens *c.* 0.8 times length of petals. Styles 5–7 mm, 1.5–2.5 times length of ovary. *Himalaya to W China.* H5. Summer.

113. H. tomentosum Linnaeus. Illustration: Reichenbach, Icones Florae Germanicae et Helveticae **6**: t. 346, f. 5183 (1844); Bonnier, Flore complète, **2**: t. 104 (1912).

Perennial herb, 10–90 cm; stems decumbent, rooting, densely whitish hairy. Leaves stalkless, 5–22 × 2–11 mm, ovate to oblong, obtuse to rounded, margin plane, base rounded to cordate and stem-clasping, slightly paler beneath, not glaucous, papery, not net-veined, densely whitish hairy. Bracts not auricled. Inflorescence corymb-like or sometimes with branches all down stem and then cylindric. Flowers 8–16 mm, stellate. Sepals 3–6 mm, ovate or elliptic to broadly lanceolate, acute or shortly awned, margin and usually apex glandular-hairy, hairy outside. Petals bright yellow, 6–11 mm, oblanceolate. Stamens *c.* 0.8 times length of petals. Styles 5–7 mm, *c.* 2.4 times length of ovary. *Morocco, W Mediterranean area.* H5. Summer.

114. H. delphicum Boissier & Heldreich.

Perennial herb, 11–45 cm; stems ascending, branching and rooting, bristly-hairy below inflorescence. Leaves stalkless, 1.2–3.5 × 1–3 cm, ovate, rounded, margin plane, base rounded to cordate, stem-clasping, slightly paler beneath, not glaucous, papery, densely net-veined, bristly-hairy. Inflorescence dense, shortly cylindric to corymb-like inflorescence. Flowers 1.2–2 cm, stellate. Sepals 3–5 mm, narrowly oblong, acute, margin glandular-hairy. Petals bright yellow, 8–10 mm, oblanceolate. Stamens *c.* 0.9 times length of petals. Styles *c.* 6 mm, 3 times length of ovary. *Greece (W Aegean Islands).* H5. Summer.

115. H. athoum Boissier & Orphanides.

Perennial herb like *H. delphicum*, but smaller in all parts and with hairs softer and less dense. Stems 10–20 cm, numerous, diffuse, slender, hairy below inflorescence. Leaves shortly stalked, 8–17 × 5–11 mm, broadly elliptic to ovate, rather sparsely hairy. Flowers 1–7, 7, 8–10 mm, from terminal node. Sepals 3–5 mm, narrowly oblong to lanceolate, glandular-fringed. Petals pale yellow veined red, 6–7 mm. Styles *c.* 5 mm. *Greece, N Aegean Islands.* H5. Summer.

A very suitable species for the alpine house.

Section **Elodes** (Adanson) W. Koch. Marsh or aquatic herb, hairy, without black glands but with red glands on sepals and bracts (marginal). Floral whorls in 5s (outer) or 3 (styles). Petals and stamens persistent. Petal marginal glands absent. Petals with ligule. Stamen bundles united, with filaments united above middle, alternating with 'lodicules'. Styles free, erect. Capsule glands linear.

116. H. elodes Linnaeus (*Elodes palustris* Spach). Illustration: Reichenbach, Icones Florae Germanicae et Helveticae, **6**: t. 342, f. 5182 (1844); Syme, English botany, edn 3, **3**: t. 276 (1864); Ross-Craig, Drawings of British plants, **6**: t. 16 (1952).

Perennial herb, 10–30 cm; stems erect from creeping rooting stolon-like base, densely whitish hairy (except in deep water). Leaves stalkless, 5–30 × 5–20 mm, broadly ovate or elliptic to circular,

rounded, margin plane, base rounded to cordate and stem-clasping, slightly paler beneath, not glaucous, papery, obscurely net-veined, whitish hairy. Sepals 2–4 mm, hairless, elliptic, acute or obtuse, margin glandular hairy. Petals bright yellow, 7–11 mm, oblanceolate. Stamens *c.* 0.6 times length of petals. Styles *c.* 4 mm, *c.* 3 times length of ovary. *W Europe, Azores.* H3. Summer.

H. elodes is useful for a bog garden or shallow pond, but requires the soil or water to be nutrient-rich and not extremely acid.

Section **Brathys** (Linnaeus filius) Choisy. annual herbs, deciduous, without black glands. Floral whorls all in 5s or styles 4–2. Petals and stamens persistent. Stamen bundles free or usually in a continuous or interrupted ring. Styles free, spreading above touching bases.

117. H. gentianoides (Linnaeus) Britton, Sterns & Poggenburg (*Sarothra gentianoides* Linnaeus; *H. nudicaule* Walter; *H. sarothra* Michaux). Illustration: Rickett, Wild flowers of the United States **2**(1): 213 (1967); National Museum of Natural Sciences, Ottawa, Publications in Botany, **11**: 12 (1981).

Annual herb, 7–60 mm; stems erect, branches numerous, strict, hairless. Leaves stalkless, 1–4 × 0.4–0.6 mm, narrowly triangular-subulate to linear, acute, margin incurved, base parallel-sided or almost so, thickly papery. Inflorescence pyramidal, with 1–24-flowered branches from nearly all nodes. Flowers 3–5 mm, stellate. Sepals 1.5–2.5 mm, lanceolate to narrowly oblong, acute. Petals orange yellow to bright yellow, tinged red, 2–4 mm, oblong. Stamens 0.5–0.65 times length of petals. Styles 3, 0.8–1.2 mm, *c.* 0.8 times length of ovary. Capsule narrowly cylindric-conic. *Canada (S Ontario), E USA.* H4. Summer.

118. H. setosum Linnaeus (*Ascyrum villosum* Linnaeus; *H. villosum* (Linnaeus) Crantz; *H. pilosum* Walter). Illustration: Bulletin of the British Museum (Natural History), Botany **20**: 69 (1990).

Perennial or annual herb, 20–80 cm; stems erect, usually unbranched below inflorescence, roughly or rather softly hairy. Leaves stalkless, 4–15 × 2–7 mm, narrowly ovate or lanceolate to narrowly oblong-elliptic, apex acute to obtuse, margin recurved, base parallel-sided to rounded and clasping, not paler beneath or glaucous or net-veined, thinly leathery, roughly or rather softly hairy.

Inflorescence cylindric to corymb-like. Flowers 5–11 mm, stellate. Sepals 2.5–5 mm, ovate-lanceolate to obovate, acute, margin bristly hairy, elsewhere sparsely hairy or hairless. Petals deep yellow, not red-tinged, 4–7 mm, obovate. Stamens 0.65–0.75 times length of petals. Styles 3 or 4, 1.5–2 mm, 1–1.3 times length of ovary. *SE USA.* H5. Summer.

119. H. gramineum G. Forster (*Ascyrum involutum* Labillardière; *H. involutum* (Labillardière) Choisy). Illustration: Burbridge & Gray, Flora of the Australian Capital Territory, 259 (1970).

Perennial or annual herb, 2.5–72 cm; stems erect or decumbent but not rooting, hairless. Leaves stalkless, 4–25 × 1.2–8 mm, lanceolate or ovate-lanceolate to oblong or linear, obtuse to rounded, margin plane or recurved, base usually rounded or cordate and stem-clasping, paler and glaucous beneath, papery, not net-veined. Inflorescence obconic to narrowly ellipsoid. Flowers 5–15 mm, stellate. Sepals 2.8–9 mm, lanceolate to narrowly elliptic or oblong, acute, entire. Petals pale yellow to orange, 5–10 mm, obovate to oblanceolate. Stamens 0.4–0.6 times length of petals. Styles 3, 0.7–1.8 mm, 0.5–0.9 times length of ovary. *Bhutan to Taiwan & S to Australia & New Zealand.* H5–G1. Summer.

120. H. majus (A. Gray) Britton (*H. canadense* Linnaeus var. *majus* A. Gray).

Perennial herb, 5–70 cm; stems erect, not or sparsely branched, hairless. Leaves stalkless, 10–45 × 2–12 mm, lanceolate to narrowly oblong-elliptic, acute to rounded, margin plane, base narrowly wedge-shaped to cordate and stem-clasping, not paler or glaucous beneath, papery, densely net-veined. Inflorescence compact, cylindric to corymb-like. Flowers 6–7 mm, stellate. Sepals 3.5–6.5 mm, lanceolate to narrowly elliptic, acute, margin entire. Petals golden yellow, sometimes red-veined, 3.5–6 mm, oblanceolate. Stamens *c.* 0.7 times length of petals. Styles 3, 0.6–1 mm, 0.3–0.5 times length of ovary. *S Canada, N USA.* H3. Summer.

121. H. mutilum Linnaeus (*H. quinquenervium* Walter: *H. parviflorum* Willdenow). Illustration: Correll & Correll, Aquatic and wetland plants of SW United States **2**: 1132 (1975); National Museum of Natural Sciences, Ottawa, Publications in Botany, **11**: 18 (1981).

Perennial or annual herb, 15–60 cm; stems erect or decumbent but not rooting, hairless. Leaves stalkless, 5–27 × 1–15 mm, ovate or ovate-triangular to elliptic-oblong, obtuse to rounded, margin plane, base usually rounded, paler beneath, not glaucous, membranous, loosely and obscurely net-veined. Inflorescence cylindric. Flowers 3–5 mm, stellate. Sepals 2–4 mm, lanceolate or linear-lanceolate to narrowly oblong, usually acute to obtuse, margin entire. Petals golden yellow, 1.7–3.5 mm, oblong. Stamens c. 0.8 times length of petals. Styles 3, c. 0.5 mm, c. 0.5 times length of ovary. *E Canada, E USA*. H4. Summer.

The plant in cultivation in Europe belongs to subsp. **mutilum**, with terminal stem-node very short or absent, leaves usually ovate to elliptic-oblong, paler beneath and bracts subulate.

122. H. japonicum Murray (*H. pusillum* Choisy; *Brathys laxa* Blume; *H. laxum* (Blume) Koidzumi; *Sarothra japonica* (Murray) Y. Kimura; *S. laxa* (Blume) Y. Kimura. Figure 11(4,9,14), p. 71. Illustration: Labillardière, Novae Hollandiae plantarum species **2**: t. 175 (1806); Nakai & Honda, Nova flora Japonica, no. **10**: 242 (1951); Li et al., Flora of Taiwan **2**: 634 (1976).

Annual herb, 2–50 cm; stems erect to decumbent or prostrate and rooting, hairless. Leaves stalkless, 2–18 × 1–10 mm, broadly ovate to circular to oblong or elliptic to oblanceolate, obtuse to rounded, margin plane, base cordate and stem-clasping to wedge-shaped, paler or glaucous beneath, membranous, not net-veined. Inflorescence diffuse. Flowers 4–8 mm, stellate. Sepals 2.5–5 mm, narrowly oblong to elliptic or obovate, acute to rounded, entire. Petals pale yellow to orange, 1.7–5 mm, obovate to oblong or elliptic. Stamens 0.4–0.8 times length of petals. Styles 2–3, 0.4–1 mm, 0.4–0.6 times length of ovary. *E Asia south to SE Australia & New Zealand*. H4–5. Summer.

123. H. anagalloides Chamisso & Schlechtendal (*H. bryophytum* Elmer; *H. tapetoides* A. Nelson). Illustration: Mason, A Flora of the marshes of California, 574 (1957); Rickett, Wild flowers of the United States **4**(1): 195 (1971); National Museum of Natural Sciences, Ottawa, Publications in Botany, **11**: 14 (1981).

Perennial or annual herb, 3–23 cm; stems decumbent or ascending, diffuse, rooting. Leaves stalkless, 3–13 × 1.5–8.5 mm, ovate or circular to elliptic, oblong or oblanceolate, rounded, margin plane, base narrowly wedge-shaped to cordate, not

paler beneath or glaucous, membranous, net-veining sometimes visible. Flowers 3–8 mm, stellate. Sepals (sometimes 4) 2–4 mm, narrowly elliptic-oblong to oblanceolate or spathulate, acute to rounded, entire. Petals (sometimes 4) golden yellow to salmon-orange, 1.7–5 mm, oblanceolate. Stamens c. 0.7 times length of petals. Styles, 3, 0.5–1.5 mm, 0.5–1.5 times length of ovary. *W North America*. H4. Summer.

2. CALOPHYLLUM Linnaeus
N.K.B. Robson

Trees or shrubs, evergreen with milky-yellow or clear latex, and simple hairs at least on buds. Leaves paired, usually stalked, entire, leathery, with closely parallel lateral veins alternating with latex canals. Flowers bisexual or rarely unisexual, in terminal and/or axillary cymes or false racemes or rarely axillary and solitary. Perianth-segments 4–16 in alternating pairs, outer 1–3 pairs rarely distinct from inner, white, symmetric, margin entire, fringed with hairs, deciduous. Stamens in 4 obscure bundles or usually apparently free, numerous, deciduous; filaments slender, anthers ovate to linear-oblong, without connective gland. Ovary l-celled, with basal placenta bearing 1 ovule; style elongate; stigma expanded, often peltate. Fruit a drupe with fleshy to fibrous mesocarp. Seed solitary, large.

A genus of about 187 species, 179 in the Old World (mainly Indo-Malaysia, including *C. inophyllum*) and about 8 in the New World. All species require hot-house treatment and are usually propagated by cuttings of half-ripe shoots with bottom heat.

Literature: Stevens, P.F., A revision of the Old World species of *Calophyllum* (Guttiferae). *Journal of the Arnold Arboretum* **61**: 117–699 (1980).

1a. Perianth-segments 8–13, 'petals' larger than 'sepals'; flowers bisexual, c. 2.5 cm; fruit 2.5–5 cm
1. inophyllum
  b. Perianth-segments 4–9, 'petals' absent or smaller than 'sepals'; flowers apparently bisexual and male, 0.8–1.2 cm, fruit c.1.5 cm
2. brasiliense

1. C. inophyllum Linnaeus. Illustration: Bamps, Robson & Verdcourt, Flora of Tropical East Africa, Guttiferae, 4 (1978).

Tree 7–35 m, broader than high, with spreading crown; latex cream-coloured; twigs hairless. Leaf-blades 8–23 × 4.5–12

cm. Flowers 5–15, c. 2.5 cm, bisexual, in axillary false racemes 7–15 cm. Perianth-segments 8–13, outer 2 ('sepals') 5.5–10 mm, ovate to almost circular, next 2 ('sepals') 9–15 mm, elliptic, inner 4–9 4–16 mm, obovate to elliptic or oblong, all reflexing. Stamens yellow to orange. Ovary pink, 1.5–3.5 mm, spherical; style 2.5–9 mm, flexuous. Drupe 1.7–2.2 cm, spherical. *Coasts from E Africa & Madagascar to Polynesia*. G2.

2. C. brasiliense Cambessèdes (*C. antillanum* Britton; *C. calaba* invalid). Illustration: Martius, Flora Brasiliensis **12**(1): t. 80 (1888).

Tree 12–30 m, with spreading crown; latex whitish; twigs minutely hairy when young. Leaves 5–12 cm, obtuse to notched, base wedge-shaped. Flowers 3–15, 9–12 mm, unisexual (females apparently bisexual), in axillary false racemes 2–5 cm. Perianth-segments 4–9, outer 2 ('sepals') 3–4 mm, elliptic, next 2 ('sepals') and sometimes up to 5 more ('petals') 4–5 mm, obovate-circular, all reflexing. Male flowers with c. l6 orange stamens and sometimes an ovary rudiment. Female ('bisexual') flowers with 8–12 sterile stamens; ovary c. 1.5 mm, spherical to ellipsoid, style 1.5 mm, bent. Drupe 1.5–2 cm, spherical. *C & N South America, West Indies*. G2.

3. MAMMEA Linnaeus
N.K.B. Robson

Trees, evergreen, with yellow latex, hairless. Leaves paired, stalked, entire, leathery, densely net-veined between parallel lateral veins, with numerous translucent glandular dots and streaks in the vein interstices (areoles). Flowers bisexual (or female) and male, the bisexual and female solitary, axillary, the male in axillary clusters usually in axils of fallen leaves. Calyx entire in bud, dividing into 2 or 3 sepals when flower opens, persistent. Petals 4 (or occasionally 5 or more), white or pale yellowish or pink, symmetric, entire, deciduous. Stamens numerous, free or united basally in a ring; filaments slender, anthers oblong-elliptic or linear-oblong, without connective gland. Ovary in bisexual flowers either 2-celled with 2 ovules on a basal placenta in each cell or 4-celled with each cell with 1 ovule, in male flowers reduced or absent; style very short; stigma 2–4 lobed, peltate. Fruit a 1–4-celled drupe with edible pulpy mesocarp. Seeds 1–4, large.

A genus of about 50 species, 2 in

America, 2 in Africa, the rest in Madagascar and from Indo-Malaya to Polynesia. The Mamey or Mammee Apple (*M. americana*) is grown as a fruit crop, and the wood of some species is used for construction work.

1. M. americana Linnaeus. Illustration: Martius, Flora Brasiliensis **12**(1): t. 79 (1888); Botanical Magazine, 7562 (1897).

Tree to 20 m, with a very dense columnar crown; latex pale yellow; twigs green to brown. Leaf-blades 12–20 × 5–10 cm, elliptic to obovate, rounded, base broadly wedge-shaped or narrowing. Flowers fragrant, usually 3 together (male) or solitary (bisexual and female), 2–4 cm. Calyx splitting into 2 or 3 sepals, 1–2 cm, obovate. Petals 4–6, white, 1.8–2.4 cm oblong to obovate, spreading. Stamens yellow, *c.* 12 mm, in male flowers shortly united; anthers with connective shortly projecting. Ovary 2–4-celled, *c.* 10 mm (with style); stigmas usually broadly 2-lobed. Drupe 10–15 cm, spherical. 'Seeds' 1–4, endocarp fibrous. *C America, West Indies.* G2. Summer.

4. CLUSIA Linnaeus
N.K.B. Robson

Trees or shrubs, often epiphytic, rarely climbers, evergreen, with milky, white or yellowish latex. Leaves paired, entire, leathery or thick, with numerous parallel lateral veins spreading or ascending, often net-veined towards margin, (lateral and minor venation usually not evident except in dry leaf), sometimes with visible translucent glandular streaks or lines. Flowers usually unisexual, in terminal cymes. Sepals 4–16, free, persistent. Petals 4–19, white, pink or red, deciduous. Male flowers with stamens numerous, free or variously united, sometimes on a column and/or with central staminodes; filaments usually thick; ovary rudiment sometimes present. Female flowers with staminodes 5–numerous, free or variously united; ovary of 4–10 carpels basally or wholly united, each carpel with numerous ovules on a lateral placenta; styles united or free or absent; stigmas separate from each other. Fruit a fleshy or leathery, tardily dehiscent capsule. Seeds numerous, large or small, with arils.

A genus of l45 or more species in tropical America. Several species have been grown under glass in temperate regions but only one seems to be at all widely spread in Europe.

1. C. rosea Jacquin. Illustration: Little & Wadsworth, Common Trees of Puerto Rico and the Virgin Islands **1**: 353 (1964).

Shrub or tree to 18 m, with very broad dense spreading crown and usually prop roots, sometimes epiphytic in natural state; latex yellow; twigs green. Leaf-blades 7–18 × 7–11 cm, broadly obovate, rounded to notched, thickly leathery with ascending lateral veins. Male flowers unknown, female flowers 1–3, *c.* 5 cm but perianth enlarging in fruit. Sepals 4–6, in pairs, lower to 10 mm in fruit, the upper to 2.5 cm in fruit, all rounded. Petals 6–8, white turning pink, 3–4 cm, obovate-spathulate, fleshy. Staminodes united to from resinous cup. Ovary with 6–12 cells, *c.* 12 mm, spherical, stigmas 6–12. Capsule yellow-green to brown, 5–8 cm, spherical *USA (Florida), Bahamas, West Indies, C & N South America.* G2.

5. GARCINIA Linnaeus
N.K.B. Robson

Trees or shrubs (rarely dwarf), evergreen or semi-evergreen, with yellow or white latex, mostly hairless. Leaves paired or rarely in whorls of 3, stalked, entire, leathery, with numerous parallel lateral veins spreading or ascending, net-veined between laterals, sometimes with visible translucent glandular streaks or lines, with red to black glandular lines crossing lateral veins. Flowers functionally unisexual in terminal and/or axillary cymes or false racemes or solitary (female only). Sepals 2–5, free or very rarely united in bud, persistent or deciduous. Petals 2–5 (rarely to 8) or rarely absent, green or white to yellow or pink to red or purple, deciduous. Male flowers with stamens in 2–5 bundles opposite petals or united in a ring or mass, sterile bundles sometimes present; filaments usually thick, often united, or absent; anthers 1–4-celled; ovary rudiment sometimes present. Female flowers sometimes with staminodes and sterile stamen-bundles, free or united; ovary 1–13-celled, each cell with one axile ovule; styles absent or very short, united; stigmas variously lobed or subdivided. Fruit a woody, leathery or fleshy berry. Seeds 1–13, relatively large, without arils but sometimes with fleshy seed coat.

A genus of about 250 species, pantropical if, as here, the Madagascan and American *Rheedia* Linnaeus is included.

Literature: Maheshewari, J.K., Taxonomic studies in Indian Guttiferae III. The genus Garcinia Linnaeus. *Bulletin of*

the Botanical Survey of India **6**: 107–135 (1964); Jones, S.W., Morphology and major taxonomy of Garcinia (Guttiferae). Unpublished Ph.D. thesis. University of Leicester (1980).

1a. Sepals 5 (rarely 4 or 6), petals 5 (rarely 4); male flowers with stamens in 4 or 5 bundles; branches 4–6 sided 2
  b. Sepals 2–4, petals 4–8; male flower with stamens free or united in ring or mass; branches usually cylindric 3
2a. Flower-stalks *c.* 2.5 cm; sepal tips hairy; petals expanded; berry *c.* 6.5 cm **1. xanthochymus**
  b. Flower-stalks *c.* 1 cm; sepal tips not hairy; petals almost closed; berry *c.* 2 cm **2. dulcis**
3a. Male flowers with stamens free; female flowers with free staminodes inserted on fleshy fleshy sterile stamen-bundles; leaves with marginal vein close to sometimes scalloped margin 4
  b. Male flowers with stamens united in ring or mass or unknown; female flowers not as above; leaves without marginal vein or with 2 veins away from entire margin 5
4a. Leaves usually in whorls of 3; sepals 4 (rarely 3), petals 5–8 **3. livingstonei**
  b. Leaves paired; sepals 2 (rarely 3 or 4), petals 4 (rarely 6) **4. humilis**
5a. Male flowers (rare) with anthers 2-celled, dehiscing by 2 slits; female flowers with stigma smooth, deeply 5–8-lobed **5. mangostana**
  b. Male flowers with anthers 4-celled or dehiscing by a horizontal ring; female flowers with stigma rough or ribbed, irregularly rayed or completely divided 6
6a. Male flowers with anthers 4-celled; female flowers with stigma completely 6–12-rayed, ridged and with tubercles; fruit ribbed **6. gummi-gutta**
  b. Male flowers with anthers l-celled, peltate, dehiscing by horizontal ring; female flowers with stigma irregularly 4-rayed, with tubercles; fruit smooth **7. morella**

1. G. xanthochymus J.D. Hooker (*Xanthochymus pictorius* Roxburgh; *X. tinctorius* de Candolle; *Garcinia tinctoria* (de Candolle) W.F. Wright; *G. pictoria* (Roxburgh) D'Arcy, invalid). Figure 12(1), p. 77. Illustration: Beddome, The Flora

Sylvatica for southern India **1**: t. 88 (1869); Pierre, Flore forestière de la Cochinchine **1**: t. 71 (1883).

Shrub or small tree to 10 m, branches drooping, 4-angled, wrinkled. Leaves 12–38 × 4–10 cm, oblong to narrowly elliptic, leathery, bright dark green above, lateral veins spreading, S-shaped. Flowers 4–10, in axils of fallen leaves, clustered, *c.* 1.5 cm. Sepals 4 or 5, circular, with hairy tips. Petals 5, greenish white, *c.* 8 mm, circular, spreading. Male flowers: stamens in 5 strap-shaped bundles of 3–5, anthers 2-celled; sterile bundles 5, fleshy. Female flowers: staminodes in 5 bundles alternating with 5 sterile bundles; ovary flask-shaped, weakly 5-lobed, 5-celled; style evident; stigma 5-lobed, smooth. Berry dark yellow, *c.* 6.5 cm. Seeds 1–4. *E Himalaya, peninsular India, Burma to Malay Peninsula.* G2.

2. G. dulcis (Roxburgh) Kurz (*Xanthochymus dulcis* Roxburgh; *Garcinia elliptica* Choisy not Wallich; *X. javanensis* Blume; *Stalagmitis dulcis* (Roxburgh) Cambessèdes). Figure 12(2), p. 77. Illustration: Little, Woodbury & Wadsworth, Trees of Puerto Rico and the Virgin Islands **2**: 565 (1974).

Tree to *c.* 6 m, similar in form and branches to *G. xanthochymus*. Leaves 11–29 × 3–14 cm, elliptic-oblong to ovate, thickly leathery, dull green above, lateral veins spreading, S-shaped. Flowers 5–12, in axils of fallen leaves, clustered, 1–5 cm, rarely opening fully. Sepals 4–6, *c.* 1.5 mm, circular, rounded. Petals 4 or 5, greenish white, 9–10 mm, cupped. Male flowers: stamens in 4 or 5 strap-shaped bundles of 6–10, anthers 2-celled; sterile bundles 4 or 5, fleshy. Female flowers: staminodes in 4 or 5 bundles, alternating with 4 or 5 sterile bundles; ovary ovoid-spherical, 4- or 5-celled; style evident; stigma 5-lobed, smooth. Berry bright yellow, *c.* 3 cm, Seeds 1–5. *Malay Peninsula to the Moluccas, S Andaman Island.* G2.

3. G. livingstonei T. Anderson (*G. angolensis* Vesque; *G. baikieana* Vesque; *G. ferrandii* Chiovenda). Figure 12(3), p. 77. Illustration: Flora of Tropical East Africa, Guttiferae, 17 (1978).

Shrub or small tree, 1–21 m, with dense, mostly rounded crown and long pendent, angled, wrinkled branches. Leaves usually in whorls of 3–4, 4–17 × 1.5–11.5 cm, oblong or lanceolate to obovate or circular, margin sometimes scalloped, thickly leathery, lateral veins spreading. Flowers 5–15 or more, in axils of older or fallen leaves, bundled, 8–20 mm. Sepals 3 or 4, 1–5.5 mm, oblong to circular, rounded. Petals 5–8, green or creamy white to pale yellow, with orange or reddish glandular lines, 3–11 mm, obovate or elliptic to circular, cupped. Male flowers: stamens apparently numerous, free, inserted on fleshy sterile cushion, anthers 2-celled. Female flowers: staminodes fewer, free, inserted below ovary; ovary spherical, with 2–3 cells; stigma stalkless, weakly 2–3-lobed, smooth. Berry yellow to orange or sometimes pink-tinged or red, 1–4 cm. Seeds 1–3. *Tropical Africa south to Natal & Transvaal.* G2.

4. G. humilis (Vahl) Adams (*Mammea humilis* Vesque; *Rheedia lateriflora* Linnaeus not *G. lateriflora* Blume; *R. sessiliflora* Vesque?). Figure 12(4), p. 77. Illustration: Tussac, Flore des Antilles **3**: t. 32 (1824).

Shrub or tree 10–15 m, erect, with branches angled, wrinkled. Leaves 15–28 × 6–12 cm, elliptic to ovate, margin sometimes scalloped, thickly leathery, dull green above, lateral veins spreading. Flowers 5–30, in axils of fallen leaves, bundled, 9–15 mm. Sepals 2, basally united, or rarely 3 or 4, 1.5–2 mm, circular, rounded. Petals 4–6, white or pale yellow, 6–8 mm, circular, cupped. Male flowers: stamens 12–25, apparently free, inserted beneath fleshy cushion, anthers 2-celled. Female flowers: staminodes few, free, inserted below ovary; ovary spherical, with 3 or 4 cells; style very short or absent; stigma with 3 or 4 lobes, smooth. Berry yellow, 3–4 cm, leathery. Seeds 1–3. *Hispaniola to Trinidad, Panama.* G2.

5. G. mangostana Linnaeus. Figure 12(5), p. 77. Illustration: Little, Woodbury & Wadsworth, Trees of Puerto Rico and the Virgin Islands **2**: 567 (1974).

Tree 10–20 m, with compact crown of nearly horizontal or drooping branches, branches slightly 4-angled, ribbed. Leaves 12–20 × 1–9 cm, elliptic to ovate or obovate, acute to acuminate, entire, base wedged-shaped to rounded, leathery, shiny green above, lateral veins spreading to somewhat ascending. Male flowers in terminal bundles of 5–9, smaller than female flowers; stamens numerous, filaments united below, anthers 2-celled. Female flowers terminal, solitary, 3–5 cm. Sepals 4, 1.5–2 cm, concave-elliptic to circular, rounded. Petals 4 (rarely 5), yellowish to white or rose-pink to purple-red, 1.8–3 cm, obovate, concave. Staminodes *c.* 20 in groups of 2 or 3 below ovary opposite septa, each group with filaments partly united. Ovary ovoid to spherical, 5–8-celled; style absent; stigma 5–8-lobed, smooth. Berry deep purple brown, 3.5–7 cm. Seeds 1–8. *Said to have been found in the Moluccas (Indonesia) but not known wild there or anywhere else now.* G2.

This is the famed mangosteen, one of the most delicious tropical fruits. According to Little, Woodbury & Wadsworth (see above), this species is monoecious. Male flowers, in any case, are rare.

6. G. gummi-gutta (Linnaeus) N. Robson (*Cambogia gummi-gutta* Linnaeus; *G. cambogia* Desrousseaux). Figure 12(6), p. 77. Illustration: Roxburgh, Plants of the coast of Coromandel **3**: t. 298 (1820); Beddome, The Flora Sylvatica for southern India **1**: t. 85 (1869).

Tree to 20 m, erect, with branches horizontal or drooping, striped, black. Leaves 8–13 × 2–5 cm, narrowly elliptic to obovate, rather thin, papery, green to greyish-green above, lateral veins ascending. Flowers solitary or 3–4-clustered, in axils of fallen leaves and sometimes also terminal, 5–10 mm. Sepals 4, *c.* 2 mm, ovate to obovate, rounded. Petals 4, cream-yellow, obovate to elliptic, *c.* 5 mm. Male flowers: stamens 20–30, massed on columnar receptacle, anthers 4-celled. Female flowers (often solitary): staminodes *c.* 20, inserted singly below ovary; ovary spherical, 6–12-celled; style absent; stigma 6–12 rayed, ridged and tuberculate. Berry yellow or red, 5–8 cm, leathery. Seeds 6–8. *Peninsular India, Sri Lanka.* G2.

7. G. morella (Gaertner) Desrousseaux (*Mangostana morella* Gaertner; *Stalagmitis cambogioides* Murray; *Hebradendron cambogioides* (Murray) Graham; *G. gutta* Wight). Figure 12(7), p. 77. Illustration: Beddome, The Flora Sylvatica for southern India **1**: t. 86 (1869).

Tree to 17 m, with branches 4-angled, wrinkled, grey-green. Leaves 4–15 × 2–8 cm, obovate to elliptic, acuminate or obtuse to rounded, entire, base wedge-shaped or narrowing, leathery, dull grey-green to red-brown above, lateral veins ascending. Flowers *c.* 3 (male) or solitary (female) in axils of fallen leaves, *c.* 5 mm. Sepals 4, 4–7 mm, circular. Petals 4, yellow-green, ovate, larger than sepals. Male flowers: stamens 20–30 in spherical mass, anthers circular, peltate, 1-celled. Female flowers: staminodes *c.* 12, united in ring; ovary spherical, 4-celled; style absent; stigma 4-lobed, tuberculate. Berry yellow, 2–3 cm, spherical, smooth, walls

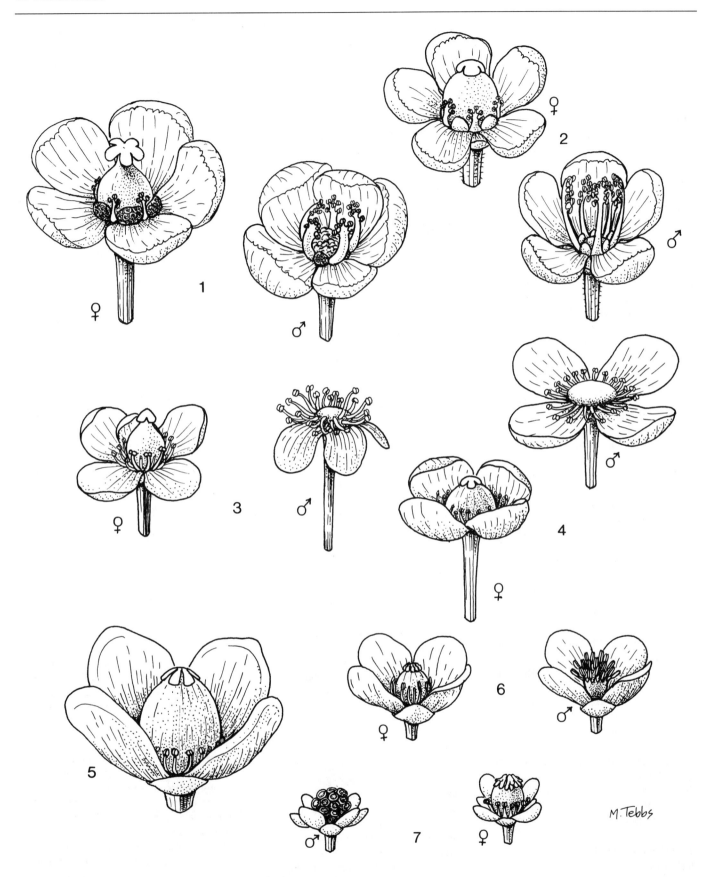

Figure 12. Flowers of *Garcinia* species. 1, *G. xanthochymus*.
2, *G. dulcis*. 3, *G. livingstonei*. 4, *G. humilis*. 5, *G. mangostana*.
6, *G. gummi-gutta*. 7, *G. morella*.

woody. Seeds 4. *Sri Lanka, peninsular India to Thailand.* G2.

XCIII. SARRACENIACEAE

Insectivorous perennial herbs with rhizomes. Leaves modified into flask-like structures (commonly called pitchers) capable of holding fluid and secreting enzymes, and in which insects are trapped; mouth of the pitcher sometimes covered by a hood. Flowers solitary, or few in a raceme, on erect stems, nodding, bisexual; sepals 4 or 5, sometimes petal-like or coloured, persistent, petals 5 and free, or absent. Stamens numerous, free. Ovary with 3–5 cells, style simple, sometimes expanded and umbrella-like. Fruit a capsule; seeds numerous.

An American family of 3 genera which inhabit swamps and bogs and are carnivorous. The species are all avidly collected by carnivorous plant enthusiasts, but relatively few are widely cultivated. These plants thrive only in a moist peaty soil, and relatively humid atmosphere; while insects are desirable for healthy growth, the plants do not normally need artificial 'feeding'. Propagation may be by division or removal of offsets, or by seed which will perhaps only be set following artificial pollination.

Literature: Slack, A., *Carnivorous plants* (1979); Schlauer, J. Nomenclatural synopsis of carnivorous phanerogamous plants. A world carnivorous plant list, *Carnivorous Plant Newsletter* **15**: 59–117 (1986).

1a. Pitchers conspicuously hooded, mouths covered; flowers solitary   2
  b. Pitchers not hooded, mouths open; flowers in racemes
                          2. Heliamphora
2a. Pitchers with forked appendage to hood; petals forming a cone-shaped corolla, shorter than sepals, style with 5 minute apical lobes
                          1. Darlingtonia
  b. Pitchers without forked appendage to hood; petals not shorter than sepals, not forming cone-shaped corolla, style greatly enlarged and umbrella-like   **3. Sarracenia**

1. DARLINGTONIA Torrey
E.C. Nelson
Pitchers to 0.5 m, erect, gradually tapering towards apex, often with slight spiral twist, with swollen hood terminating in forked

appendage which conceals mouth; hood and upper part of pitcher spotted with translucent 'windows'. Flowers solitary; flower-stems usually longer than leaves (to 0.6 m) with several, widely spaced, prominent white-green scales. Sepals 5, green-yellow with purple lines, to 5 × 1 cm, lanceolate with acute apex. Petals 5, red-brown with darker veins, shorter than sepals, margins touching, forming a cone-shaped corolla with 5 pores. Stamens 12–15, anthers with unequal sacs.

A genus of 1 species from North America, which is easy to cultivate and hardy in western Europe. It is propagated by removing plantlets that develop from rhizomes, and thrives in damp (not necessarily sodden) soil. Its English vernacular name, cobra-lily, alludes to the snake-like appearance of the pitchers.

1. D. californica Torrey (*Chrysamphora californica* (Torrey) Greene). Illustration: Botanical Magazine, 5920 (1871); Carnivorous Plant Newsletter **17**: 6, 16, 17, (1988).

W USA (California, Oregon). G1–H5. Late spring–summer.

2. HELIAMPHORA Bentham
E.C. Nelson
Pitchers erect, bulbous about middle, usually with distinct waist at mouth, and a small concave nectar-secreting appendage at tip, with paired wings from mouth to base; hairy inside. Flowers 2–10 in a raceme on erect stem; each flower on stalk with basal bract. Sepals 4, rarely 5 or 6, petal-like, white becoming pink with age, persistent in fruit; petals absent. Stamens 8 or more. Ovary with 3 cells; style simple.

A remarkable genus of 5 species which display extraordinary plasticity, so that many variants are known from its restricted South American habitats. In cultivation, they require warmer, more humid conditions than those of the other genera (10–25 °C – it should not exceed 25 °C for any substantial period). Plants should be liberally watered during the growing season. Plants succeed best in live sphagnum, but a sphagnum/perlite mixture will suffice. None is widely cultivated.

Literature: Steyermark, J.A., Sarraceniaceae, realignment of the genus *Heliamphora*. Annals of Missouri Botanical Garden **71**: 302–312 (1984); Dodd, C. & Powell, C., A practical method for cultivation of *Heliamphora* spp. *Carnivorous Plant Newsletter* **17**(2): 48–50 (1988).

1a. Anthers 15–20, 5–8 mm; lowest floral bract to 20 cm, usually equal to or greatly exceeding lowest flower-stalk; raceme with 2 or 3 flowers (rarely 4); pitchers 15–50 cm   **4. tatei**
  b. Anthers 15 or fewer; lowest floral bract less than 10 cm, usually shorter than lowest flower-stalk; or raceme with 4 or more flowers
                                              2
2a. Racemes with more than 7 flowers; pitchers more than 40 cm, anthers 15, 3.5 mm; lowest floral bract 4–5 cm   **5. ionasii**
  b. Racemes with fewer than 7 flowers; pitchers less than 35 cm; anthers fewer than 15   3
3a. Racemes with 2–5 flowers (very rarely flower solitary), flower-stalks densely downy; anthers 10–14, 3–5 mm; pitchers 5–22 cm   **2. minor**
  b. Flower-stalks hairless   4
4a. Racemes with 4 flowers; anthers 10–14, 3–3.5 mm; floral bract rounded, lowest one 3–3.5 cm
                          3. nutans
  b. Racemes with 2–7 flowers; anthers usually 8–14, 2–7 mm; floral bract with an abrupt point, lowest one to 10 cm   **1. heterodoxa**

1. H. heterodoxa Steyermark. Illustration: Slack, Insect-eating plants, 90 (1986).

Variable. Pitchers green, 12–40 cm, with prominent red appendage usually longer than broad, to 3.5 × 3 cm. Sepals 3–6 × 1–2.5 cm at flowering, increasing to 4.5–8 × 2–3.5 cm in fruit. *Venezuela.* G1. Summer.

2. H. minor Gleason. Illustration: Carnivorous Plant Newsletter **18**(2): 33, 48 (1989).

Pitchers dark red and green, 5–30 cm; appendage 0.3–1.5 × 0.2–1.5 cm. Sepals 2–5 × 0.7–2 cm at flowering, increasing to 4–5.5 × 1–2 cm in fruit. *Venezuela.* G1. Summer.

3. H. nutans Bentham (*Sarracenia nutans* (Bentham) Dietrich). Illustration: Botanical Magazine, 7093 (1890); Slack, Insect-eating plants, 90–91 (1986).

Pitchers to 30 cm, plain green or with red streaks; appendage 2–7 × 5–15 mm, broader than long. Sepals 3.5–4.5 × 1–1.5 cm at flowering, increasing slightly to 4.5 × 2 cm in fruit. *Venezuela.* G1. Summer.

4. H. tatei Gleason (*H. nebulinae* Maguire; *H. macdonaldiae* Gleason; *H. tyleri* Gleason). Illustration: Carnivorous Plant Newsletter

17(2): 35, 64 (1988).

Variable; pitchers to 50 cm, yellow-green or green, with red appendage which is usually longer than broad 1–4 × 0.5–4 cm. Sepals (var. **tatei**) only 0.3 × 0.7–3 cm; (var. **nebulinae** Maguire) 3.5–6 × 1.3–3.5 cm at flowering; increasing in fruit to 4.5–7 × 2–4 cm. *Venezuela*. G1. Summer.

5. H. ionasii Maguire.

Pitchers 40–45 cm, widely flared at the mouth, with red streaks. Racemes with more than 7 flowers; lowest floral bract 4–5 cm; anthers 15, *c*. 3.5 mm. *Venezuela*. G1. Summer.

3. SARRACENIA Linnaeus

E.C. Nelson

Pitchers erect or decumbent, with a conspicuous hood which arches over the mouth; phyllodes produced by some species (these are not flask-like, but have a flat blade). Flowers solitary on erect stems. Sepals 5, usually green, persistent in fruit. Petals 5, longer than sepals, deciduous, distinctively folded (resembling an italic N) to fit over rim of the stigmatic disc; apical portion broadest; stamens numerous. Ovary 5-celled; style with a terminal, expanded, 5-lobed umbrella-shaped disc; stigmas on apices of lobes.

There are 8 species recognised generally in *Sarracenia*, but some American botanists have raised subspecies and varieties to specific rank, and the names used often persist in horticulture. The genus is generally confined to the eastern USA and Canada. However, *S. purpurea* extends into British Columbia and is also naturalised in Europe (Ireland, Switzerland). Some natural hybrids and a swarm of complicated artificial hybrids and selected cultivars are now available in commerce, but the application and validity of their names frequently is questionable; none is described here.

Literature: McDaniel, S., The genus Sarracenia (Sarraceniaceae), *Bulletin of Tall Timbers Research Station* 9 (1971).

1a. Pitchers erect, straight and gradually tapering; petals yellow or red 2
 b. Pitchers decumbent (at least partially), curved and sometimes unequally tapering; petals dark red (very occasionally yellow) 7
2a. Pitchers with translucent or white patches or spots on upper portion 3
 b. Pitchers without translucent or white patches or spots 4
3a. Hood concave, sharply recurved over mouth of pitcher; phyllodes absent; petals yellow **4. minor**
 b. Hood not concave, erect, margins markedly wavy; phyllodes present from late summer through winter into spring; petals red **3. leucophylla**
4a. Petals red; hood erect, margins not wavy and not curved downwards and inwards at base; phyllodes absent **8. rubra**
 b. Petals yellow, or hood margins curved downwards and inwards at base, or phyllodes present from late summer through winter into spring 5
5a. Hood strongly curved downwards and inwards and usually with purple patch at base; petals pure yellow or green-yellow 6
 b. Hood not strongly curved downwards and inwards at base; petals pale cream-white or pale yellow **1. alata**
6a. Phyllodes 5–18 cm, 0.5–3.5 cm, sickle-shaped, more numerous than pitchers; petals yellow-green; flowers sweetly fragrant **5. oreophila**
 b. Phyllodes 12–30 cm, resembling leaves of an iris; petals pure yellow; flowers with faint bitter-sweet fragrance **2. flava**
7a. Mouth of pitcher enclosed and hidden, lateral; pitchers more or less prostrate, hood spherical; petals dark red **6. psittacina**
 b. Mouth of pitcher not lateral and not concealed; hood not spherical; petals pink to dark red, rarely yellow **7. purpurea**

1. S. alata (Wood) Wood. Illustration: Bulletin of Tall Timbers Research Station 9: 16 (1971); Slack, Carnivorous plants, 51 (1979); Carnivorous Plant Newsletter 16: 14 (1987), 18: 81, 96 (1989).

Pitchers to 75 cm, erect, narrowing gradually from mouth, with wing to 1 cm broad, hairless or with short hairs outside, from yellow-green with fine purple veins near mouth to entirely purple; hood broadest towards base, concave, sharply curved over mouth, yellow-green or veined (or suffused) with purple. Flower-stem to 60 cm. Sepals to 6 × 4 cm, yellow green (rarely suffused with red). Petals to 7 × 4 cm, pale cream-white or pale yellow, apical portion ovate. Style-disc yellow. *SE USA (Texas to Alabama)*. G1. Summer.

Much variation is evident in the wild. This species is distinguished easily by the curved hood which 'protects' the pitcher mouth.

2. S. flava Linnaeus. Illustration: Botanical Magazine, 780 (1804); Bulletin of Tall Timbers Research Station 9: 13 (1971); Slack, Carnivorous plants 27: 8–9 (1979); Carnivorous Plant Newsletter 9: 42–43 (1980).

Pitchers to 90 cm, erect, narrowing gradually from mouth, sometimes with a narrow wing, hairless outside, yellow-green (very rarely dark red or suffused or veined with red); hood broader than long, to 10 × 14 cm, tending to be convex, slanting or arched over mouth, from above heart-shaped, yellow-green or with red veins or suffused red, especially at mouth. Flower-stems to 60 cm. Sepals to 5 × 3.5 cm, yellow or green-yellow. Petals to 8.5 × 4 cm, yellow, apical part obovate. Style-disc yellow. *USA (Florida, Alabama to Virginia)*. G1–H5. Spring.

Plants vary considerably ranging from those with pure green pitchers to others with plain dark red pitchers, but none merits recognition above cultivar level.

3. S. leucophylla Rafinesque (*S. drummondii* Croom). Illustration: Bulletin of Tall Timbers Research Station 9: 20 (1971); Slack, Carnivorous plants, 34 (1979); Carnivorous Plant Newsletter 17: 32 (1988), 18: 67, 80 (1989).

Pitchers 0.25–1 m, erect, narrowing gradually from mouth, with wing to 2 cm, hairless outside, green with white, red or purple net-like patterning on hood and upper half of pitcher; hood 2.5–6.5 cm, slanting over mouth, heart-shaped and mucronate with irregular, wavy margins becoming curved downwards and inwards near mouth, hairy inside. Flower-stem to 80 cm. Sepals to 5 × 3 cm, dark red. Petals to 5 cm, dark red, apical part obovate. *S USA (Florida to Mississippi)*. G1. Spring.

4. S. minor Walter (*S. variolaris* Michaux). Illustration: Botanical Magazine, 1710 (1815); Bulletin of Tall Timbers Research Station 9: 18 (1971); Slack, Carnivorous plants, 14 (1979).

Pitchers to 80 cm, erect, gradually narrowing from mouth, hairless or sparsely downy outside, with prominent translucent patches, wing to 3 cm wide. Hood concave, to 6 cm, sharply recurved over mouth, usually marked with purple net-like pattern. Flower-stem to 0.5 m.

Sepals to 3.5 × 3 cm, broadest below middle. Petals to 5 cm, yellow, apical part spathulate; Style disc yellow. *SE USA (North Carolina to Florida)*. G1. Spring.

Most distinctive for the translucent spots on the leaves and the sharply recurved hood.

5. S. oreophila (Kearney) Wherry (*S. flava* var. *oreophila* Kearney). Illustration: Bulletin of Tall Timbers Research Station **9**: 11 (1971), Carnivorous Plant Newsletter **16**: 34 (1987).

Pitchers erect, green or yellow-green, tapering gradually from mouth, to 75 cm, conspicuously winged; hood to 8 cm, kidney-shaped to obovate, curved downwards and inwards and spotted with purple at base, otherwise with purple net-like pattern; with glandular hairs inside. Flowering-stem to 70 cm. Sepals to 5 × 3 cm. Petals to 5.5 × 2 cm, yellow, apical part elliptic to obovate. Style disc yellow. *S USA (Alabama)*. G1. Spring.

One of the rarest species in the wild.

6. S. psittacina Michaux. Illustration: Bulletin of Tall Timbers Research Station **9**: 29 (1971).

Pitchers less than 30 cm; winter leaves less than 7 cm, decumbent, wing triangular, being broadest beyond middle, to 1 cm wide; summer leaves larger, more than 8 cm, ascending, wing to 3 cm wide; margins of hood fused to form the spherical leaf apex, mouth offset to one side, with translucent spots and purple net-like pattern. Flowering-stem to 40 cm. Sepals to 2.5 × 2 cm, dark red shading to green-red below with dark red rim below; petals to 4.5 × 2.5 cm, dark red; apical part obovate. Style disc yellow-green with red tint near stigma. *SE USA (Georgia to Florida & Louisiana)*. G1. Spring.

7. S. purpurea Linnaeus. Illustration: Botanical Magazine, 849 (1805); Bulletin of Tall Timbers Research Station **9**: 26 (1971); Slack, Carnivorous plants, 54–56 (1979); Walsh, Ross & Nelson, An Irish Florilegium, 74 (1983).

Pitchers 5–50 cm, curved and decumbent, hairless or hairy, with wing to 6.5 cm wide, green shading to dark red-purple or veined with red-purple; hood kidney-shaped, to 5 cm, arching over mouth. Flower-stem to 75 cm, dark red. Sepals to 4 × 3.5 cm, dark purple-red; petals to 7 × 2.5 cm, pink to dark red, rarely yellow, apical part oblong or elliptic. Style disc *c.* 5 cm broad, green or yellow-green. *E USA & NE Canada, naturalised in*

Ireland. H3. Late summer.

8. S. rubra Walter (*S. alabamensis* Case & Case; *S. jonesii* Wherry). Illustration: Botanical Magazine, 3515 (1836); Bulletin of Tall Timbers Research Station **9**: 23 (1971); Carnivorous Plant Newsletter **17**: 97, 127 (1988).

Pitchers to 70 cm, erect, green and frequently with red or purple veins; hairless or downy; hood erect but arching slightly over mouth, obovate, with acute tip, green or with red or purple veining. Flowering-stem to 75 cm. Sepals to 3 cm, purple but green below; petals to 4 cm, dark red with grey to dull red undersurface, apical part obovate to 2 cm wide. Style-disc green. *USA (North Carolina to Mississippi)*. G1. Spring–early summer.

Several variants have been segregated into species but these do not warrant segregation; much of the variation is due to the species discontinuous distribution and numerous isolated populations.

XCIV. NEPENTHACEAE

Evergreen shrubs or climbers, dioecious. Stems without exudate or spines. Leaves alternate, usually spiralled, sometimes 2-ranked, without stipules, without gland-dots, with or without a stalk, the blade sometimes extending down the stem (decurrent) as 2 wings. The leaf-blade is lanceolate to oblong or obovate, it has 2 sets of veins, a longitudinal set, which run the length of the blade, and a transverse set of pinnate veins which run from the midrib across the width of the blade to the margin; the midrib is usually extended into a twisting tendril which at the tip is expanded into an elaborate, multicoloured, upright, animal-trapping pitcher containing up to a litre of digestive fluid. The pitcher mouth is encircled by a ribbed or smooth, usually brightly coloured and glistening peristome or lip, cylindric or flattened in section, the inner edge normally terminating in a ring of teeth. A lid, variously appendaged is usually held above the pitcher and prevents the entry of rainwater; a spur projects away from the pitcher at the junction with the lid. Flowers numerous and small in terminal (appearing lateral by growth of a lateral stem shoot after initiation) racemes or panicles. Perianth-segments 3 or 4, usually green or brown, glandular, inconspicuous. Female flowers with a flask-shaped ovary

crowned by a 3 or 4-lobed stigma. Ovary with axile placentation, 3 or 4 locules and numerous ovules. Male flowers with numerous anther thecae crowded in a hemispherical cluster on a short common stalk. Fruit a capsule, splitting along the cells. Seeds numerous, hair-like, dry and very light.

A family of a single genus concentrated in the islands of SE Asia, one species extending as far east as New Caledonia, another south to Queensland, Australia, a third north to India and several species are found far to the west, off the African coast in Madagascar and the Seychelles. The stems are sometimes used locally for fibre, the pitchers as vessels, their contents for medicine. Their greatest and increasing use by man is as ornamental plants for the glasshouse and conservatory in the more affluent, temperate zones of the world.

Literature: Macfarlane, J.M., Nepenthaceae in Engler, A., *Das Pflanzenreich*, **36** (1908); Danser, B.H., The Nepenthaceae of the Netherlands Indies. *Bulletin du Jardin Botanique Buitenzorg* **9**: 249–438 (1928).

1. NEPENTHES

M. Cheek & M. Jepp
Description as for family.

A genus of about 70 species, most of which occur at sea-level at the edge of or in open humid lowland forest or in wetlands and these species ('lowland species') thrive in the hot and steamy conditions normally associated with the genus. However, the majority of the species occur on mountain ridges at altitudes of 1000 m or more and require cooler conditions ('highland species'). *Nepenthes viellardii* is reported from 3520 m in New Guinea, where temperatures reportedly fall below freezing point in the early morning.

In cultivation, an open, well-drained mix of bark, perlite, peat, and charcoal provides best results. Traditionally, orchid baskets are used, but plastic pots are satisfactory. It is important that the compost is never allowed to dry out. *Nepenthes* are less intolerant of fertilisers than most carnivorous plants, but nevertheless, feed should be applied cautiously. About 50% shading should be given in summer to avoid leaf-scorch, but removed in winter. At low light levels, pitcher production ceases. High humidity is important. Species from low altitudes thrive in a temperature range of 15–30 °C, those from higher altitudes (1000 m or more) do better at 10–25 °C, though they can often tolerate temperatures as low as 5 °C.

Propagation is readily achieved from fresh seed, sown on and lightly covered with the sifted compost described above and given bottom heat. Pricking out should be carried out as early as possible to avoid checking. Tip and 3-noded stem cuttings need strong bottom heat to root and must be given a well-aerated (but still highly humid) medium to avoid rotting. Stem layering is viable.

Most species take years to flower in cultivation from seed, and many have yet to do so. For this reason the key below omits floral characters. It is important to note that a single plant will usually produce at least 2 very different sorts of pitchers. 'Lower pitchers' are produced from the basal leaf rosette, before the stem ascends any distance. They are usually broadest at the base and roughly egg-shaped. They face the tendril, which is straight, and have 2 fringed wings down the front. 'Upper pitchers' are produced from higher up the stem; they are usually broadest at the top and roughly funnel-shaped or cylindrical. They face away from the tendril which is usually coiled and often the frontal wings are not fringed.

Until the 1980s, the majority of *Nepenthes* in cultivation, particularly in botanic gardens, were survivors of the numerous 19th century cultivars of complex hybrid origin. They cannot be assigned to a particular species. They fall into 3 main groups, only one of which is now commercially available in Europe; this is included in the key. During the 1980s a very large number of species (the majority of those described below) were reintroduced to cultivation for the first time in the 20th century. All *Nepenthes* appear fully interfertile. Wherever 2 species occur together in the wild, hybrids can usually be found. Some of these have been introduced to cultivation and are listed below under the parent species.

Literature, Kurata, S., *Nepenthes of Mount Kinabalu* (1976).

1a. Leaves thin and membranous, margin often very shallowly toothed, with sparsely scattered long fine hairs   2
  b. Leaves thick and leathery, margin entire, hairless or thickly fringed with short hairs   3
2a. Pitchers green but conspicuously and plentifully splashed with red
                              1. coccinea hybrid group
  b. Pitchers uniformly pale green,

sometimes suffused with pink or red, but never splashed with red
                              2. mirabilis
3a. Lower pitchers about as tall as broad, upper pitchers absent; pitcher lid very narrowly ovate, reflexed, not covering the pitcher
                              3. ampullaria
  b. Lower pitchers usually at least 3 times as tall as broad, upper pitchers present; pitcher lid usually ovate to circular, held over the pitcher   4
4a. Pitcher lid with 2 long, pointed fangs curving from its base into the mouth of the pitcher
                              4. bicalcarata
  b. Pitcher lid without fangs   5
5a. Outer pitcher surface with a conspicuous white band of hairs below the mouth
                              5. albomarginata
  b. Outer pitcher surface without a white band of hairs   6
6a. 'Eye-spots' 2, conspicuous, darker than the background, conspicuous within the top of the pitcher below the attachment of the lid
                              6. reinwardtiana
  b. 'Eye-spots' absent from the inside of the pitcher   7
7a. Leaf-tip deeply notched, appearing 2-lobed   **7. truncata**
  b. Leaf-tip not notched, or if so (very rarely), then very shallowly   8
8a. Leaf-tip constantly and markedly peltate (the pitcher tendril extending not from the tip of the leaf-blade, but from a short distance before the tip of the lower leaf surface; leaves hairy   **8. rajah**
  b. Leaf-tip not peltate, or if so aberrant and leaves hairless   9
9a. Pitchers hour-glass shaped   10
  b. Pitchers, if constricted in the middle, only very slightly so   11
10a. Pitchers white or grey, without wings; pitcher mouth circular
                              9. ventricosa
  b. Pitchers green or red, with 2 shallow wings; pitcher mouth markedly elliptic (upper pitchers)
                              10. lowii
11a. Lower surface of pitcher lid (lower pitchers) thickly covered with long red bristles   12
  b. Lower surface of the pitcher hairless   13
12a. Leaf lacking a distinct stalk
                              11. macfarlanei
  b. Leaf with a distinct stalk   **10. lowii**

13a. Lower surface of lid with one prominent ridge-like protuberance at base near the pitcher and sometimes another at apex   14
  b. Lower surface of lid lacking protuberances   21
14a. Leaves and stems hairless when mature; upper pitcher inflated and broadest at base, narrower and cylindric above, without fringed wings   **12. alata**
  b. Leaves and stems at least partly hairy when mature; upper pitcher more or less funnel-shaped, usually broadest at the top, often with 2 fringed wings   15
15a. Leaves in 2 ranks; rim of pitcher greatly expanded and flattened in frontal view meeting in the centre and extending further each side than the pitcher tube   **13. veitchii**
  b. Leaves usually spiralled; rim of the pitcher not flattened or expanded so dramatically   16
16a. Lower surface of lid with a hair-like appendage from apex
                              14. maxima
  b. Lower surface of lid without a hair-like appendage from apex   17
17a. Upper pitchers more or less cup-shaped, only c. 1.5 times as tall as broad, largely translucent and creamy yellow in colour with a few small red splashes
                              15. burbidgeae
  b. Upper pitchers cylindric, funnel-shaped or ellipsoid, at least 3 times as tall as broad, not translucent
                              18
18a. Upper pitchers ellipsoid, rim of the pitcher broad, ruff-like, more than 1 cm wide   **16. northiana**
  b. Upper pitchers more or less cylindric or funnel-shaped, rim less than 5 mm wide   19
19a. Upper pitchers funnel-shaped, not splashed with yellow   20
  b. Upper pitchers more or less cylindric or tapering slightly to base, splashed with yellow
                              17. stenophylla
20a. Sharply pointed protuberance at base of lower lid surface, hairs on stem c. 5 mm   **18. pilosa**
  b. Blunt, rounded protuberance at base of lower lid surface, hairs on stem to 3 mm   **19. fusca**
21a. Leaves not distinctly stalked   22
  b. Leaves distinctly stalked   27
22a. Leaves decurrent, the stems at least slightly winged   23

b. Leaves not decurrent, stems not
   winged   26
23a. Pitchers 10–15 cm broad; rim 8–15
   cm wide, the ribs ending inside the
   pitcher with teeth twice as long as
   broad   **20. merrilliana**
b. Pitchers 2–5 cm broad; rim 0.5–1.5
   mm wide, the ribs ending inside the
   pitcher with minute teeth about as
   long as broad, or lacking teeth   24
24a. Leaves 20–45 cm; upper pitchers
   10–20 cm   **21. khasiana**
b. Leaves 5–20 cm; upper pitchers
   4–11 cm   25
25a. Leaves often hairy below, inner
   edge of rim lacking teeth
                              22. viellardii
b. Leaves totally lacking hairs, inner
   edge of rim with minute teeth
                                23. gracilis
26a. Upper surface of lid with upright
   hairs *c.* 3 mm   **24. tentaculata**
b. Upper surface of lid lacking hairs
                                           27
27a. Pitcher slender, 6–24 × 1–3.5 cm,
   often purplish black, rim narrow
   and cylindric   **25. gracillima**
b. Pitcher stout, 10–30 × 3–9 cm,
   often green or red, rim often broad
   and flattened, especially in upper
   pitchers   **26. sanguinea**
28a. Pitcher lid with 2 longitudinal
   ribs prominent on lower surface
                              27. rafflesiana
b. Pitcher lid smooth below
                            28. macrovulgaris

1. N. × coccinea Group. Including '*N. ×
coccinea*', '*N. × dormanniana*', '*N. × dominii*',
'*N. × chelsonii*', '*N. × hybrida*', '*N. ×
henryana*', '*N. morganae*', '*N. sanderiana*'
and '*N. × williamsii*'. Illustration: Slack,
Carnivorous plants, 14, 87 (1979).

Climber 2–6 m. Hairs short, white,
appressed, persisting on the stems and
midrib. Stems cylindric, 1–1.5 cm thick.
Leaves not clearly stalked though the
basal part of the blade often conspicuously
narrower, clasping the stem for ½ or ⅔
its circumference in a laterally flattened
sheath, leaf-blade membranous-papery,
narrowly oblong, 18–46 × 5–9 cm, apex
rounded or acute, margin very finely
toothed. Longitudinal veins 3 or 4 on
each side of the midrib, transverse veins
running straight to the edge to oblique and
branching. Lower pitchers flask-shaped, the
lower third swollen, egg-shaped, tapering
gradually to a cylindric upper portion,
8–10 × 3–4 cm, with 2 fringed wings;
mouth ovate, slightly oblique or horizontal

at the front, rising vertically into a neck
at the back; rim 3–7 mm wide, variously
ribbed and toothed. Lid ovate or elliptic,
2–4 × 1.5–3 cm, flat, lacking appendages.
Outer pitcher green or cream, usually
almost obscured by numerous red splashes.
Upper pitchers rarely seen. Inflorescence
25–40 cm with numerous short side-
branches from the upper part each bearing
1 or 2 flowers. *Artificial hybrids*. G2. Late
summer–winter.

These hybrids often have 4 or 5 parents
each, sometimes unknown, but usually
including *N. mirabilis* and *N. rafflesiana*.
Some are known to be raised from the
seeds from the same capsules and although
in the literature these names are cited as
Latin binomials, the majority would be
better treated simply as cultivar names.
Created a century ago, much mislabelling
has since occurred. Several of the hybrids
cannot possibly have been derived from
the parentage originally cited. The group is
characterised by its vigour and resistance
to adverse conditions, prolific production
of small to medium-sized flask-shaped
pitchers heavily splashed with red on thin,
often fimbriate margined leaves.

2. N. mirabilis (Loureiro) Druce.
Illustration: Slack, Carnivorous plants,
80 (1979); Kondo & Kondo, Carnivorous
plants of the world in colour, 112 (1983);
Slack, Insect-eating plants, 144 (1986).

Shrub or climber to 7.5 m. Hairs simple,
long, fine, appressed, very sparse in mature
stems and leaves, but present on younger
growth. Stem cylindric, 5–10 mm thick.
Upper leaves membranous to papery,
stalked, stalk 12–25 cm, not decurrent and
not usually clasping stem, lanceolate to
oblong, 20–40 × 3–8 cm, acute or obtuse,
margin with a line of fine hairs, sometimes
on shallowly pointed teeth. Longitudinal
veins 4–8 on each side of midrib spread
over entire blade. Transverse veins straight
to the margin, forming rectangles with
longitudinal veins. Lower pitchers 12–16
× 2–4 cm, flask-shaped, lower half more
or less egg-shaped, tapering into a broad
cylinder above, with 2 fringed wings;
mouth nearly horizontal, almost circular,
acute near lid; rim usually conspicuously
flattened, 2–8 mm broad, ribs 0.3–0.2
mm apart, not toothed inside. Lid rounded
to egg-shaped, 2.5–3.5 × 1.5–2 cm, flat,
lacking appendages. Outer pitcher surface
green, sometimes suffused with red.
Inflorescence 25–45 cm with numerous
short side-branches each single-flowered.
China, Indo-China, Malay Peninsula,

*Sumatra, Borneo, Celebes, Moluccas,
Philippines, Java, New Guinea, N Australia.*
G2. Late summer & autumn.

3. N. ampullaria Jack. Illustration: Slack,
Carnivorous plants, 83 (1979); Slack,
Insect-eating plants, 134 (1986); Fessler,
Fleischfressende Pflanzen für Haus und
Garten, 45 (1982); Lecoufle, Carnivorous
plants, 131 (1990).

Climber to 6 m. Leaves and stems with
dense, short, scurfy hairs, brown or grey.
Stem 5–8 mm thick, cylindric. Leaves
lacking blades, only with pitchers in the
basal rosettes. Upper leaves leathery, not
clearly stalked, though narrowed towards
the base, not decurrent, but clasping stem
for ½ its circumference or more,
lanceolate to spathulate 12–25 × 3–6 cm,
apex rounded or acute. Longitudinal veins
3–5 on each side of midrib, evenly spaced.
Transverse veins numerous, normal to
midrib. Pitchers present only in basal
rosettes or the lowest of stem leaves, cup-
shaped, 2–10 × 2–10 cm, slightly flattened
laterally, with 2 fringed wings; mouth
horizontal, ovate; rim barely developed on
outside of pitcher, flat, 3–15 mm broad on
the inside, ribs 0.25–0.3 mm apart, teeth
absent. Lid reflexed, not covering pitcher,
very narrowly elliptic, 2–10 × 0.2–0.8
cm, folded, with 2 longitudinal ridges,
appendages absent. Green with vertical red
flecks, rarely completely red. Inflorescence
6–30 cm, with numerous side branches,
lowermost bearing up to 10 flowers. *Malay
Peninsula, Sumatra, Borneo & New Guinea.*
G2. Late summer.

N. × hookeriana Lindley (*N. ampullaria
× N. rafflesiana*). Illustration: Slack, Insect-
eating plants, 147 (1986). Intermediate
between the parents. *Malay Peninsula,
Singapore, Borneo.* G2. Late summer–
winter.

N. × trichocarpa Miquel (*N. ampullaria
× N. gracilis*). Illustration: Slack, Insect-
eating plants, 148 (1986). Intermediate
between the parents. *Malay Peninsula,
Singapore & Sumatra.* G2. Not known to
flower in cultivation.

4. N. bicalcarata J.D. Hooker. Illustration:
Slack, Carnivorous plants, 82 (1979);
Kurata, Nepenthes of Mount Kinabalu,
39 (1976); Kondo & Kondo, Carnivorous
plants of the world in colour, 98 (1983);
Slack, Insect-eating plants, 145, 146
(1986).

Climber 7–15 m. Hairs inconspicuous,
minute, scurfy, stellate, absent from upper
leaf and stem. Stems 1–2 cm thick,
cylindric. Leaf-stalk 3.75–5 cm, sheathing

stem for 3–4 cm, papery, basal rosette leaves narrowly rectangular, 24–28 × 6–9 cm, apex acute. Upper leaves as basal, but to 65 cm, base decurrent for 3–8.5 cm. Longitudinal veins 9–12 on each side of midrib, evenly spaced. Transverse veins numerous, conspicuous. Lower pitchers spherical to egg-shaped, 7–17.5 × 7–15 cm with 2 fringed wings; mouth circular, concave in side view; rim 3–6 mm outside, much deeper inside, topmost teeth below lid elongated into a pair of stout downward pointing spines 1–3 cm. Lid kidney-shaped, 3.5– 6 × 4.7–6 cm. Upper pitchers as the lower, but funnel-shaped and lacking fringed wings, golden orange or greenish in the shade. Inflorescence to 90 cm with numerous side-branches 5.7– 11.5 cm, each bearing 7–15 flowers clustered in a false umbel. *N Borneo*. G2. Not known to flower in cultivation.

5. N. albomarginata Lindley. Illustration: Kondo & Kondo, Carnivorous plants of the world in colour, 96 (1983); Lecoufle, Carnivorous plants, 128 & 129 (1990).

Climber 2 m, rarely to 10 m. Hairs white, inconspicuous, stellate. Stems 3–5 mm, cylindric. Leaves stalkless, clasping the stems but not decurrent, thick, leathery, basal rosette leaves narrowly oblanceolate to spathulate, 20–36 × 2–3.5 cm, apex acute. Upper leaves and pitchers rarely seen in cultivated plants. Longitudinal veins 1 on each side, near the margin, often inconspicuous. Transverse veins numerous but inconspicuous. Lower pitchers flask-shaped, 8.5–15 × 3.5–6 cm wide at base tapering to 2–3 cm wide above, with 2 fringed wings; mouth ovate, oblique; rim more or less cylindric, c. 1 mm wide with fine, closely spaced ridges; outer pitcher green or red, with a band 1–5 mm broad of white shining hairs. Lid ovate, to 4 × 3 cm, without appendages. Inflorescence a raceme 11–45 cm, with 20–80 short side-branches each bearing 1–3 flowers. *Peninsular Malaysia, Sumatra & Borneo*. G2. Not known to flower in cultivation.

6. N. reinwardtiana Miquel. Illustration: Kondo & Kondo, Carnivorous plants of the world in colour, 118 (1983).

Climber to 20 m. Hairs absent from stem and leaves. Stem triangular, rarely rounded, 3–7.5 mm wide, the corners rounded or winged. Leaves stalkless, decurrent for 8.5 cm, clasping the stem for half its circumference, thinly leathery, narrowly elliptic to slightly oblanceolate, basal rosette leaves unknown, upper leaves 11–25 × 1.1–3.1 cm, apex acute. Longitudinal veins 1–3 on each side of the midrib, usually confined to the outer third. Transverse veins inconspicuous, ascending. Lower pitchers flask-shaped, to 11 cm, base c. 4 cm wide, apex c. 3 cm, with 2 fringed wings; mouth ovate, oblique; rim more or less cylindric in section, 0.5–1 mm wide, without ridges, inner edge with a row of holes. Lid ovate to shortly elliptic to 3.9 × 3.7 cm, base truncate, appendages absent. Upper pitchers as the lower, but larger and lacking fringed wings, 9–31 × 3.2–6.8 cm at base, inner pitcher surface often with 2 conspicuous eye-like spots c. 1 cm apart, each 1–5 mm wide. Inflorescence a raceme, 11–50 cm, bearing numerous 2-flowered side branches. *Sumatra & Borneo*. G2. Not known to flower in cultivation.

7. N. truncata Macfarlane. Illustration: Kondo & Kondo, Carnivorous plants of the world in colour, 125 (1983); Slack, Insect-eating plants, 145 (1986).

Shrub, height in wild unknown, to 50 cm in cultivation. Hairs numerous, unbranched, brown, appressed, persisting on leaf-stalks when mature. Stem cylindric, 1.5–2 cm thick. Leaf-stalk 3–24 cm, clasping the stem for half its diameter and forming around it a laterally flattened sheath to 4 cm; leaf-blade leathery, more or less rectangular, 10–45 × 9–22 cm, tapering markedly from apex to base, apex deeply notched for 3–10 cm, base truncate to cordate. Longitudinal veins c. 5, evenly spread. Transverse veins branching near midrib, running straight to the margin. Lower pitchers as upper, but smaller and with 2 fringed wings. Upper pitchers very broadly cylindric, sometimes slightly and gradually waisted in the centre, or flaring slightly to apex, 15–30 × 4–10 cm, fringed wings absent; mouth oblique, transversely elliptic; rim flattened, to 6 cm wide, outer margin wavy, ribs c. 1 mm apart, extending into teeth as long as broad at inner edge. Lid ovate to 8 × 6 cm, flattened, with a crest-like appendage on lower surface near lid. Pitcher red and green. Inflorescence to 50 cm, with numerous short side-shoots towards the apex, bearing 2 flowers each. *Mindanao, Philippines*. G1–2. Not known to flower in cultivation.

8. N. rajah J.D. Hooker. Illustration: Kurata, Nepenthes of Mount Kinabalu, 62–63 (1976); Slack, Carnivorous plants, 85 (1979); Slack, Insect-eating plants, 155 (1986).

Shrub c. 2 m. Hairs long, brown, spreading, sparse in mature leaves and stems. Stems cylindric, 1.5–3 cm thick. Leaf-stalk 5–15 cm, not decurrent, but forming a laterally flattened sheath around the stem for c. ¾ its circumference, blade leathery, oblong, 25–50 × 10–15 cm, apex rounded, with the tendril leaving lower surface before the tip. Longitudinal veins 3–5 on each side of midrib in the outer half. Transverse veins running obliquely to margin. Lower pitcher ellipsoid, 20–35 × 11–18 cm, with 2 fringed wings; mouth elliptic, highly oblique; rim expanded unevenly 1–2 cm wide in front, 2–5 cm broad near the lid, ribs 1–3 mm apart, inner teeth 2–4 times as long as wide. Lid ovate, vaulted, 15–25 × 10–18 cm, with a distinct ridge on the lower surface near the pitcher, entire pitcher scarlet to purple. Upper pitcher as lower, but funnel-shaped, lacking fringed wings. Inflorescence 50–80 cm, with many short side-branches, each with 2 flowers, lacking bracts. *Borneo*. G1. Not known to flower in cultivation.

9. N. ventricosa Blanco. Illustration: Slack, Carnivorous plants, 205 (1979); Kondo & Kondo, Carnivorous plants of the world in colour, 127 (1983); Slack, Insect-eating plants, 158 (1986).

Shrub, sometimes epiphytic, climbing to 1–2 m. Hairs absent in mature plants. Stem cylindric or rounded triangular in section, 7–9 mm thick. Leaves stalkless, base clasping stem for c. ⅔ of its diameter, with prominent wings obliquely decurrent for c. 1 cm; leaf-blade thinly leathery, oblanceolate or narrowly oblong, 15–25 × 2–3 cm, base gradually tapered, apex acute. Longitudinal veins 5 or 6 on each side of midrib, evenly scattered. Transverse veins obscure, oblique. Lower pitchers not seen. Upper pitchers hour-glass-shaped, 10–16 × 3–7 cm, the waist ⅓–½ width of base and apex, fringed wings absent; mouth circular to elliptic, slightly oblique; rim curved, outer margin slightly wavy, 1–2.5 cm wide, ribs 1–2 mm apart. Lid narrowly elliptic, 4–5 × 2–2.5 cm, flat. Outer pitcher white or grey, rim salmon to dark red. Inflorescence 30–60 cm, bearing numerous, single-flowered side-branches in the upper half. *Philippines (Luzon)*. G1–2. Late summer–winter.

10. N. lowii J.D. Hooker. Illustration: Kurata, Nepenthes of Mount Kinabalu, 54–55 (1976); Kondo & Kondo, Carnivorous plants of the world in colour, 106 (1983); Slack, Insect-eating plants, 154 (1986).

Climber to 8 m. Hairs rarely persisting on mature leaves and stems. Stems cylindric, 6–10 mm thick. Leaf-stalk 4– 14 cm, not decurrent but clasping the stem for $\frac{2}{3}$–$\frac{4}{5}$ of its circumference, leaf-blade oblong to lanceolate, 15–30 × 6–9 cm, leathery, rounded at apex, base abruptly contracted into stalk. Longitudinal veins 2–4 on each side of midrib. Transverse veins inconspicuous. Lower pitchers rare, flask-shaped, lower part obliquely ellipsoid, constricted at the middle and then flaring out again to the mouth, to the same breadth as the lower part, with 2 fringed wings; mouth ovate, strongly oblique; rim more or less cylindric, 2–3 mm wide, with well-defined ribs. Lid circular to ovate, with numerous long hair-like growths, to 8 mm on lower surface. Upper pitchers very differently shaped, as the lower but almost woody, 15–28 cm, lower part spherical or obliquely egg-shaped, 5–10 cm wide, then abruptly constricted to a third the width before flaring out to 6–12 cm wide at the mouth, wings reduced to ribs; mouth rotated at 90° from basal part of pitcher through being twisted at waist, transversely elliptic, horizontal; rim reduced, detectable only as a line of corrugations inside the rim, lid elliptic, strongly vaulted and held at 90° to the mouth, without appendages, but with numerous hair-like structures from the lower surface. Pitcher surface dark green outside, dark red inside. Inflorescence 15–30 cm, numerous side-branches with 2 flowers each, lacking bracts. *Borneo*. G1. Not known to flower in cultivation.

11. N. macfarlanei Hemsley. Illustration: Kondo & Kondo, Carnivorous plants of the world in colour, 107 (1983); Slack, Insect-eating plants, 153 (1986).

Climber to 4 m. Hairs absent from mature leaves and stems. Stems more or less cylindric, 4–7 mm thick. Leaves stalkless, not decurrent, base rounded or cordate, clasping for $\frac{1}{2}$ the circumference of the stem, leathery, more or less oblong, 5–18 × 2–5 cm, apex acute to rounded. Longitudinal veins 3–5 on each side of midrib in outer half. Transverse veins indistinct. Lower pitchers more or less ovoid, to 20 × 3–7 cm, conspicuously constricted towards the mouth; mouth horizontal in front, the back rising vertically into a short neck to support the lid; rim flattened, to 8 mm broad, ribs 0.75–1 mm apart, inside with teeth 3–6 times as long as broad. Lid circular, to 7 cm across, lower surface with numerous

pendent bristles to 5 mm, outer surface greenish brown, marbled with dark red, rim bright red. Upper pitchers as lower, abruptly constricted at mouth, but funnel-shaped 8–20 × 3–7 cm. Outer surface creamy yellow, rim banded creamy yellow and bright red. Inflorescences 10–25 cm with numerous 2-flowered side-branches. *Malay Peninsula*. G1. Not known to flower in cultivation.

12. N. alata Blanco. Illustration: Slack, Carnivorous plants, 86 (1979); Kondo & Kondo, Carnivorous plants of the world in colour, 95 (1983); Lecoufle, Carnivorous plants, 118–121 (1990).

Climber to 4 m. Leaves and stems hairless when mature. Stems rounded or roughly triangular, 4–8 mm thick. Leaves not distinctly stalked, not decurrent, but clasping $\frac{4}{5}$ of the stems circumference, thinly leathery, narrowly oblanceolate or spoon-shaped, to 13 × 3.5 cm, apex obtuse to acute. Longitudinal veins up to 5 on each side. Transverse veins running obliquely to margin. Lower pitchers flask-shaped, the lower $\frac{1}{3}$–$\frac{1}{2}$ swollen, tapering to the cylindric apex, to 10 × 2.5 cm, with 2 fringed wings; mouth very oblique; rim cylindric to flattened, 1–3 mm broad, ribs 0.25–0.3 mm apart. Lid circular to elliptic, to 3 × 3 cm, with a low keel below near junction with pitcher. Pitchers bright green, sometimes with a tinge of red. Upper pitchers as lower, but higher and more slender 8–25 × 2.5–6 cm, wings reduced to 2 ribs. Inflorescence 20–60 cm, with numerous side-branches each with 1 or 2 flowers. *Peninsular Malaysia, Sumatra & the Philippines*. G1–2. Late summer.

13. N. veitchii J.D. Hooker. Illustration: Kondo & Kondo, Carnivorous plants of the world in colour, 126 (1983); Slack, Insect-eating plants 153 (1986).

Climber to several metres, by means of the clasping, 2-ranked leaves. Hairs dense, brown and spreading on stem, lower leaf, tendril and pitcher when young, becoming sparser, in places hairless, when mature. Stems elliptic, to 10 mm thick. Leaf-stalk to 5 cm, base forming a laterally flattened sheath around stem, rarely decurrent as 2 shallow wings reaching internode below; leaf-blade leathery, more or less oblong, 16–25 × 4–10 cm, apex acute to shallowly notched. Longitudinal veins 3 or 4 on each side of midrib in outer $\frac{1}{2}$. Transverse veins oblique, petering out before the edge. Lower pitcher broadly cylindric, broadest near middle, 15–28 × 4–10 cm, with 2 fringed wings; mouth horizontal in front,

rising and narrowing at rear into a short neck; rim expanded, 4–15 mm broad in front, 1–6 cm broad near lid, the ribs 0.5–1 mm apart, extending into teeth 3–4 times as broad on the inside. Lid narrowly ovate, 3–9 × 1.75–5 cm, with a laterally flattened appendage on lower surface near pitcher, and a smaller one at apex. Outer pitcher green, fading to grey, rim striped red and green. Upper pitcher as lower, also broadly cylindric, to 17 × 4.5 cm, with 2 fringed wings. Inflorescence 17–40 cm, with numerous side-branches each bearing 2 flowers, lacking bracts. *Borneo*. G1–2. Late summer–winter.

14. N. maxima Nees. Illustration: Kondo & Kondo, Carnivorous plants of the world in colour, 110 (1983); Slack, Insect-eating plants, 153 (1986); Lecoufle, Carnivorous plants, 125 (1990).

Shrub or climber, 0.5–3 m, densely velvety hairy when young on leaves (except upper surface) and stems, hairs largely falling when mature. Stems cylindric or triangular, 3–9 mm thick, 2 of the 3 angles often winged. Leaf-stalk 4.5–6 cm, base completely encircling stem in short side-shoots, but on main stems more usually decurrent sometimes to node below, wings to 8 mm wide; blade thinly leathery, narrowly elliptic to oblong 15–30 × 2.5–7 cm, apex obtuse or acute. Longitudinal veins 2– 4 on each side of midrib in the outer $\frac{1}{3}$. Transverse veins numerous, running straight to margin. Lower pitchers nearly cylindric, lower $\frac{2}{3}$ sometimes slightly inflated, to 20 × 6 cm, with 2 fringed wings; mouth oblique or very oblique, tapering towards lid into a neck 1–2 cm; rim flattened or expanded, 1–5 mm broad in front, 2–20 mm broad at some distance from lid, ribs 0.25–1 mm apart, extending into teeth 3 times as long as broad inside. Lid ovate to triangular, below with 2 appendages, one a curved ridge at the junction with pitcher, the other at the tip, hair-like, *c*. 5 mm. Pitchers light green with brownish or purplish red markings. Upper pitchers tubular to cylindric, lacking fringed wings, 7–30 × 1.8–8 cm; mouth horizontal at front, at rear abruptly vertical and narrowed into a neck to 3 cm. Lid ovate to narrowly triangular, 2–7 mm. Inflorescence 25–50 cm, with numerous side-branches each bearing 2 flowers. *Borneo, Celebes, Moluccas, New Guinea*. G1. Late summer and autumn.

15. N. burbidgeae Burbidge. Illustration: Kurata, Nepenthes of Mount Kinabalu,

42–43 (1976); Slack, Insect-eating plants, 157 (1986).

Climber, 12–15 m. Hairs simple, brown, numerous on stem, lower leaf-blade and tendril when young, falling and sparser when older. Stems cylindric, 12–15 mm thick. Stem leaves long-stalked (lower leaves not known), 7–10 cm, clasping stem for up to half its circumference and decurrent for one internode as 2 wings, leathery, oblong, 20–35 × 6–8 cm, apex abruptly attenuate. Longitudinal veins 3 or 4 on each side of midrib in outer ½ or ⅔. Transverse veins numerous, ascending. Lower pitchers obliquely ellipsoid, 6–18 × 4–12 cm; mouth circular, horizontal; with a thick, curved rim 2–5 mm broad in front, 4–12 mm near lid, 2 fringed wings. Lid circular, slightly vaulted, 3–8 × 3–8 cm, with a blunt appendage below at junction with pitcher, creamy white, with few small red blotches and the translucency of porcelain. Upper pitchers as lower in colour, but lacking fringed wings, funnel-shaped, 5–13 × 3–7 cm; rim thinner, 2–4 mm wide at front, 3–6 mm at lid. Inflorescence 25–35 cm, the short side-branches with 2 flowers. *Borneo.* G1. Not known to flower in cultivation.

16. N. northiana J.D. Hooker. Illustration: Slack, Insect-eating plants, 145 (1986).

Shrub or climber, height unknown, hairs sparse. Stems cylindric, 6–8 mm thick, leaves thinly leathery, not stalked or very shortly stalked, the stalk clasping the stem, decurrent as 2 wings for up to 2 internodes, blade elliptic to oblong, 15–35 × 3–10 cm, apex obtuse or rounded, possibly sometimes peltate. Lower pitchers egg-shaped, with 2 fringed wings; mouth ovate, oblique; rim cylindric to flattened, irregularly expanded, outer margin sinuous, 1–5 cm broad, broadest at sides, finely ribbed, toothed inside. Lid ovate to oblong, 3–14 × 2–10 cm with a blunt, ridge-like appendage at base of lower surface. Upper pitcher curved, narrowly funnel-shaped, to 40 × 12 cm, with 2 fringed wings; mouth, rim and lid as lower pitcher, but narrower. Outer pitcher greyish or greenish white, speckled with a little red, rim orange to dark red with some green streaks. Inflorescence 25–35 cm, with comparatively few flower-stalks, each with 1 or 2 flowers. *Borneo.* G1–2. Not known to flower in cultivation.

17. N. stenophylla Masters. Illustration: Kondo & Kondo, Carnivorous plants of

the world in colour, 122 (1983); Slack, Insect-eating plants, 135, 153 (1986).

Climber to at least 2 m. Hairs numerous, short and brown, c. 1 mm on leaf undersurface, margins and stems, sparser with age. Stem cylindric (but angular if winged), 6–11 mm thick. Leaf-stalk 4–9 cm, forming a laterally flattened sheath at base, or decurrent, with 2 wings 1–2 mm wide descending to internode below, blade leathery oblong to lanceolate, 15–23 × 4–9 cm, apex rounded or acute. Longitudinal veins 2 or 3 on each side of midrib in outer ¼, indistinct. Transverse veins indistinct. Lower pitchers cylindric, slightly waisted c. ⅔ from base, 16–25 × 3.5–5 cm, with 2 unfringed wings; mouth round, oblique, narrowed and upswept at the rear; rim cylindric to flattened, 2–3 × 4–7 mm towards lid, ribs c. 0.3 mm apart, teeth about as long as broad. Lid more or less circular, 3–5.5 cm across, with a blunt, ridge-like appendage on lower surface near lid. Outer pitcher surface largely dark red flecked with creamy-white. Upper pitcher resembling lower, but not waisted, cylindric to slightly funnel-shaped, wings smaller, creamy white with splashes of red. Inflorescence to 30 cm, with numerous short side-shoots each bearing 2 flowers, without a bract. *Borneo.* G1. Not known to flower in cultivation.

18. N. pilosa Danser. Illustration: Phillipps & Lamb, Nature Malaysiana **13**(4): 26–27 (1987).

Climber to at least 3 m. Long, brown hairs covering the stems, lower leaves and tendrils, dense when young, less so in older growth. Stems cylindric, 6–9 mm. Leaf-stalks 2.5–3 cm, not markedly decurrent, but completely clasping stem forming a laterally flattened sheath, leaf-blade thinly leathery, obovate to lanceolate, 10–30 × 4.5–7.5 cm, apex rounded or acuminate. Longitudinal veins 4 or 5 on each side of midrib. Transverse veins indistinct, running straight to margin. Lower pitchers flask-shaped, egg-shaped below, narrowing to a cylinder above, c. 10 × 3–4 cm; mouth oblique, ovate, rising towards lid; rim, flattened, 4–5 mm broad at front, to 7 mm near lid, ribs 0.25–0.3 mm, inner edge toothed. Lid circular, c. 2.5 cm across, flat, lower surface with a keel near lid. Outer surface greenish. Upper pitcher as lower, but cylindric, slightly funnel-shaped and curved at top, 18–28 × 6–7 cm, fringed wings absent; mouth horizontal in front narrowed and curving upwards into a neck 2–3 cm; rim flattened, evenly c. 1.2

cm wide. Lid with a curved, hooked ridge on lower surface near the lid. Inflorescence unknown. *Borneo.* G1. Not known to flower.

19. N. fusca Danser. Illustration: Kurata, Nepenthes of Mount Kinabalu, 50 (1976); Slack, Insect-eating plants, 133 (1986).

Climber to 10 m. Hairs dark-brown, simple or slightly branched, fairly dense on stems and lower leaf surface when young, sparser when mature. Stems cylindric, 4–6 mm thick. Leaf-stalk 2–4 cm, clasping stem for c. ½ its circumference, leathery, narrowly elliptic or oblong, 12–15 × 4–6 cm, apex acute to obtuse. Longitudinal veins 3 or 4 on each side of midrib, in marinal third. Transverse veins branching repeatedly. Lower pitchers cylindric, 10–15 × 2–3 cm, very slightly constricted c. ⅔ of the way from base, with 2 fringed wings; mouth ovate, oblique; rim cylindric, c. 4 mm broad in front, 6–8 mm near lid, ribs 0.2–0.3 mm apart, inner surface finely toothed. Lid oblong, 2–3 × 0.5–1 cm, lower surface with a blunt swelling at base and sometimes another at apex. Outer surface almost completely covered with splashes of blackish purple and sparsely covered with short white hairs. Upper pitchers funnel-shaped, 8–15 × 3–4 cm, with 2 ribs; mouth nearly circular, front horizontal, abruptly vertical at back; rim as lower pitcher. Lid narrowly triangular, 3–4 × 0.5–1 cm, outer surface sparingly splashed with red, or entirely green. Inflorescence a raceme 10–15 cm, short side-branches with 1 or 2 flowers. *Borneo.* G1. Not known to flower in cultivation.

20. N. merrilliana Macfarlane. Illustration: Kondo & Kondo, Carnivorous plants of the world in colour, 111 (1983); Slack, Insect-eating plants, 143 (1986).

Climber or sprawling shrub, height unknown. Leaf and stem stellate-hairy when young, hairless when mature. Stem obtusely rectangular, 7–10 mm thick. Leaves stalkless, narrowing gradually to base, decurrent into 2 wings over ½ –1 internode, thinly leathery, narrowly oblong to lanceolate, 20–60 × 5–7 cm, apex rounded or obtuse. Longitudinal veins 6 or 7 on each side of midrib in outer ⅔. Transverse veins oblique, numerous, branching and inconspicuous near margin. Lower pitchers broadly egg-shaped to 20 × 12 cm, with 2 fringed wings; mouth oblique; rim flattened or expanded, 8–15 mm broad, ribs c. 1 mm apart, extending into teeth as long as broad on interior margin. Lid broadly egg-shaped, c. 10 ×

10 cm. Outer pitcher whitish grey suffused with red. Upper pitcher as lower, but 20–26 × 8–9 cm, lacking fringes to wings. Inflorescence 30–35 cm, with numerous side-branches, those at base bearing 2 flowers, those above single-flowered. *Philippines*. G1–2. Not known to flower in cultivation.

21. N. khasiana J.D. Hooker. Illustration: Kondo & Kondo, Carnivorous plants of the world in colour, 105 (1983); Slack, Insect-eating plants, 138 (1986).

Climber 6–10 m. Hairs absent from mature stems and leaves. Stem cylindric, 8–12 mm. Leaves stalkless, base clasping the stem for *c.* ½ the circumference, decurrent, often to internode below; leaf-blade thinly leathery, narrowly oblanceolate, 20–45 × 5–8 cm, basal ¼ narrow and stalk-like, apex acute. Longitudinal veins 5 or 6 on each side of midrib. Transverse veins oblique, branching. Upper pitchers cylindric to angular, 10–20 × 3.5–4.5 cm, broadest ¼ of the way up from base where ringed by a raised line, then tapering slightly to the upper cylindric part; lacking wings; mouth broadly ovate, slightly oblique; rim cylindric to flattened, 1–2 mm wide, ribs conspicuous, narrow, *c.* 0.5 mm apart, lacking teeth at their inner edge. Lid elliptic, 4–5 × 3–4.5 cm, lacking appendages. Outer surface of pitcher dull or yellowish green often suffused with dark red, particularly near top. Inflorescence 25–60 cm bearing numerous short side-branches, each with 2 flowers. *NE India*. G1. Late summer–winter.

22. N. viellardii J.D. Hooker. Illustration: Lecoufle, Carnivorous plants, 127 (1990).

Shrub, occasionally a climber to 15 m. Hairs variable, in some plants absent, in other densely long-hairy, especially on lower leaf surface. Stem cylindric or slightly angular, 1–6 mm thick. Leaves stalkless, base decurrent into 2 wings sometimes nearly reaching to internode below; blade leathery, lanceolate, 5–20 × 1–4 cm, apex acute. Longitudinal veins 4 or 5 on each side of midrib. Transverse veins variable, straight or oblique, indistinct or distinct. Lower pitchers flask-shaped, 4–11 cm, with 2 fringed wings; mouth oblique, more or less circular; rim more or less cylindric, 0.25–0.6 mm apart. Lid circular, flat. Upper pitcher cylindric, slightly constricted ⅗ from the base or shortly and narrowly funnel-shaped, 4–14 × 1–3 cm, lacking fringed wings, otherwise as lower pitcher. Outer pitcher surface

green, rim red, interior of pitcher pale green, sometimes suffused with red. Inflorescence 7–30 cm, with numerous side-branches each bearing a single flower, lacking bracts. *W New Guinea & New Caledonia*. G1. Not known to flower in cultivation.

23. N. gracilis Korthals. Illustration: Slack, Carnivorous plants, 206 (1979); Kondo & Kondo, Carnivorous plants of the world in colour, 101 (1983); Slack, Insect-eating plants, 144 (1986).

Climber to 2–4 m. Hairs absent from stems and leaves. Stems usually triangular, rarely rounded, 1.5–5 mm. Leaves stalkless, decurrent by 1–4 mm, thinly leathery, margin slightly recurved, narrowly lanceolate, basal leaves to 3 cm, upper leaves 12–19 cm, apex acute, rarely acuminate, almost peltate. Longitudinal veins 4–6 on each side of midrib, usually confined to outer ⅔. Transverse veins numerous, slightly oblique. Lower pitchers flask-shaped 5.5–16.5 × 1.7–3.7 cm wide below, 1.4–2.4 cm broad in the upper half, with 2 fringed wings; mouth ovate, concave; rim more or less cylindric, *c.* 0.5 mm wide, without ridges, inner edge toothed. Lid rounded to egg-shaped, 1–3 × 1–3 cm, without appendages. Upper pitchers as lower, but larger, lacking fringed wings, green, sometimes partly suffused with pink. Inflorescence a raceme 15–22 cm, bearing numerous single-flowered side-branches. *Peninsular Malaysia, Sumatra, Borneo & the Celebes*. G2. Late summer.

24. N. tentaculata J.D. Hooker. Illustration: Kondo & Kondo, Carnivorous plants of the world in colour, 123 (1983); Slack, Insect-eating plants, 153 (1986).

Climber to at least 2–3 m. Leaf and stem hairless when mature, with sparse adpressed stellate hairs when very young. Stems cylindric or triangular, 2–5 mm thick. Leaves stalkless, base clasping stem at an oblique angle, blade leathery, elliptic to lanceolate, 5–20 × 1–8 cm. Longitudinal veins *c.* 4 on each side of midrib in outer ⅔. Transverse veins oblique. Lower pitchers flask-shaped, to 9 × 4 cm, with 2 fringed wings, mouth highly oblique, narrowed and rising at the back to form a neck; rim cylindric, *c.* 1.5 mm wide, ribs 0.5 mm apart or indistinct, lacking teeth. Lid ovate with upright hair-like appendages to 3 mm on upper surface. Outer pitcher purplish. Upper pitchers cylindric, 6–26 × 2–8 cm, with 2 fringed wings, otherwise as lower

pitcher. Inflorescence 7–20 cm, the short side-branches single-flowered, lacking bracts. *Borneo*. G1. Not known to flower in cultivation.

25. N. gracillima Ridley. Illustration: Slack, Insect-eating plants, 152 (1986).

Climber to 5 m. Hairs absent from stem and leaf when mature; pitchers with fine brown stellate hairs, especially near the rim. Stem more or less cylindric, 2.5–4 mm. thick. Leaves stalkless, not decurrent, but rounded at base, clasping stem for about ½ its circumference, leathery, oblong, 4–14 × 0.8–4 cm, apex acute or obtuse. Longitudinal veins often indistinct, to 4 on each side of midrib in outer ¼. Transverse veins obscure. Lower pitchers more or less cylindric, very slightly constricted in the middle 5–15 × 1–3 cm, with 2 narrow fringed wings up to 2 mm wide; mouth ovate, oblique; rim cylindric to flattened, 0.5–1 mm thick, ribs 0.25–0.5 mm apart, inner surface with teeth. Lid circular, 6–28 × 6–28 mm, apex rounded; flattened, without appendages. Outer surface green or white overlain with blackish purple markings or completely black, very pale green inside mouth. Upper pitchers as lower, but taller and more slender, 6–24 × 1–3.5 cm, lacking fringed wings. Inflorescence a raceme 8–20 cm, bearing numerous single-flowered side-branches, each with a single hair-like bract. *Malay Peninsula*. G1. Not known to flower in cultivation.

26. N. sanguinea Lindley. Illustration: Kondo & Kondo, Carnivorous plants of the world in colour, 120 (1983); Slack, Insect-eating plants, 152 (1986).

Climber to 7 m. Hairs absent from stem and leaf, tendril and pitcher sparsely covered with long fine hairs. Stem rounded triangular, 4–9 mm thick. Leaves stalkless, base not decurrent, but clasping stem almost for ½–⅔ its circumference, cordate, not laterally flattened, blade leathery, more or less narrowly oblong, 10–20 × 2–5 cm. Longitudinal veins 3–5 on each side of midrib in outer half. Transverse veins running straight to margin, indistinct. Lower pitchers flask shaped to cylindric, lower half inflated gradually from the base, narrowing to a slight waist *c.* halfway to lid and then cylindric or very slightly flared to mouth, 10–30 × 3–9 cm, with 2 fringed wings; mouth ovate, strongly oblique; rim rounded to flattened, widest at sides, to 2 cm wide. Lid ovate, to 8 cm, lacking appendages. Outer surface greenish, sometimes bronzed, inner pitcher bluish white at mouth, with red mottling. Upper pitcher as lower, but slightly more cylindric and flared at top, but still slightly constricted at middle, fringed wings absent 15–25 × 3–6 cm, rim distinctly flattened and broader. Outer surface sometimes bright red, rim yellow, inner pitcher white at the mouth. Inflorescence 25–60 cm, with many short side-shoots each bearing 2 flowers, at base with a hair-like bract. *Malay Peninsula.* G1. Late summer–early winter.

27. N. rafflesiana Jack. Illustration: Slack, Carnivorous plants, 93 (1979); Fessler, Fleischfressende Pflanzen für Haus und Garten, 50 (1982); Slack, Insect-eating plants, 143, 150 (1986); Lecoufle, Carnivorous plants, 137 (1990).

Climber to 4–15 m. Hairs of 2 sorts, long, fine and spreading and denser, but fine stellate hairs, both are less conspicuous on older growth. Stem cylindric, 4–8 mm thick. Leaf-stalks 2–15 cm, not decurrent but clasping stem and forming a laterally flattened sheath *c.* ½–⅔ its circumference, blade leathery, lanceolate to oblong, 12–30 × 3–10 cm, apex rounded to acuminate. Longitudinal veins 3–5 on each side of midrib in outer ⅔. Transverse veins oblique, not reaching the margin. Lower pitchers egg-shaped, 5–25 × 3–10 cm, with 2 wide, fringed wings; mouth highly oblique, at rear elongated into a neck 2–5 cm; rim flattened, inside mouth vertical and to 15 mm broad, ribs 0.5–1 mm apart, terminating in teeth 3–6 times as long as broad. Lid ovate, 5–10 × 3.5–8 cm with 2 ridges running the length of lower surface, strongly vaulted. Outer surface pale green to creamy white, usually heavily splashed with red, sometimes blackish brown, rim red. Upper pitchers cylindric to funnel-shaped, lacking fringed wings, 10–40 × 3–6.5 cm, mouth, rim and lid as lower pitchers, but front of mouth often raised in a ridge. Outer surface pale green, rim striped red and green. Inflorescence 15–70 cm, with numerous short side-branches each bearing a single flower and lacking bracts. *Malay Peninsula, Sumatra, Borneo.* G2. Late summer–winter.

28. N. macrovulgaris Turnbull & Middleton. Illustration: Nature Malaysiana **13**(4): 17 (1988).

Climber to 6 m. Hairs absent from mature stems and leaves. Stem cylindric 6.5–11 mm thick. Leaf-stalk winged, *c.* 1 cm, clasping the stem for about half its

diameter, rarely decurrent for to ⅓ of an internode, leaf-blade papery to leathery, narrowly elliptic, 7.5–37 × 1.5–6.5 cm, apex obtuse, base gradually tapering. Longitudinal veins 2–3 in outer ⅓ of blade. Transverse veins running obliquely to margin. Lower pitchers flask-shaped, basal ⅖ egg-shaped, upper part cylindric, 7–31 × 3.5–18 cm, with 2 fringed wings; mouth strongly oblique, ovate, acute near lid, neck absent; rim cylindric, 3–10 mm wide at front, often slightly flattened, 4–18 mm wide at side, ribs 0.3–0.5 mm apart, not extending into teeth along inner edge. Lid ovate 2.3–8 × 2–5.3 cm, flat, lacking appendages. Colour of outer pitcher green, mottled and often partly suffused with bright or dark red, rim yellow or cream, sometimes with reddish bands. Inner pitcher whitish green, sometimes flecked with dark red. Upper pitcher more slender than lower, more or less cylindric, slightly and very gradually waisted in central portion, 14–20 × 34–48 cm, lacking fringed wings, otherwise as lower pitcher. Pitcher colour pale or medium green, rim greenish yellow, sometimes with red bands, inner pitcher pale green with red spots. Inflorescence 10.5–18.5 cm, short side-branches each with 2 flowers. *Borneo.* G2. Late summer–winter.

XCV. DROSERACEAE

Herbs or low shrubs. Leaves alternate, or rarely whorled, usually in basal rosettes, usually simple, insectivorous with sticky glandular hairs and sometimes with traps. Flowers regular, bisexual, often in racemes, sometimes solitary. Sepals 4 or 5, fused at the base. Petals 4 or 5, free. Stamens 4–20, free. Ovary superior. Fruit a capsule.

A family of 4 genera found throughout the world but concentrated in Australia and New Zealand. Commonly grown as botanical curiosities on account of their insect-catching abilities. Most species require cool, shaded greenhouse conditions, although a number of species may be grown outside in a moist locality. A growing medium of equal parts peat and sand is suitable. Plants require plenty of water and often thrive best when their pots are stood in water-filled trays. Plants should be watered with soft water and never water from the tap. Exceptional care should be taken with the tuberous species from W Australia in particular, as these are natives of seasonally wet desert

areas and require careful drying off when dormant.

1a. Plant a submerged aquatic, roots absent; leaves crowded in whorls
      1. Aldrovanda
  b. Plant terrestrial, roots present; leaves in a basal rosette   2
2a. Leaf-blade hinged along the middle, margin fringed with bristles; 2 halves of leaf snap together when sensitive hairs on upper surface are stimulated   **2. Dionea**
  b. Leaf-blade not as above   3
3a. Plant herbaceous; leaves coiled upwardly from the tip in bud
      3. Drosera
  b. Plant woody at base; leaves coiled downwardly in bud
      4. Drosophyllum

1. ALDROVANDA Linnaeus
E.H. Hamlet
Submerged aquatic, rootless, perennial herb, generally free-floating. Stems simple or sparsely branched, to 20 × 2–3 cm, leafy and transparent. Leaves crowded in whorls of 6–9, united at their bases; leaf-stalks flattened, 3–6 mm, tapered from apex to base, apex more or less truncate, bearing 4–8 finely toothed bristles; blade 3–6 mm, rounded or kidney-shaped, hinged along the midrib, margin with small teeth and hairs, upper surface with glands. Flowers radially symmetric, bisexual, solitary in the leaf-axils, arising on short stalks, greenish white, to 1 cm across and raised above the water level. Sepals 5, elliptic, 3–4 mm. Petals 5, free, 5–7 mm, later bending over and hood-like. Stamens 5, as long as sepals. Ovary superior, 1-celled. Styles 5 with terminal branched stigmas. Fruit a spherical capsule opening by 5 flaps, borne on a deflexed stalk. Seeds several, glossy black.

A genus of a single species. It grows best in clear, slightly shaded water with a temperature of 21–24 °C and a pH of 4.5–5.5. As it grows, the older parts decay; growth is continuous in warm climates, but in colder areas growth is slower and overwintering turions are produced; these are 6–8 mm across and usually sink to the bottom after breaking free from the decaying parent plant, rising and producing new foliage in the following spring. Flowers are rarely produced in Europe, and are usually cleistogamous.

1. A. vesiculosa Linnaeus. Illustration: Aston, Aquatic plants of Australia, 72 (1973); Sainty & Jacobs, Water plants of New South Wales, 152–153 (1981); Morley & Tölken, Flowering plants in Australia, f. 97 (1983).

Widespread in the Old World tropics & subtropics. H5–G1. Summer.

2. DIONAEA Ellis
E.H. Hamlet

Rhizomatous, perennial. Leaves in a basal rosette, each with a leaf-like, winged stalk and blade modified to trap insects, kidney-shaped to circular, the inner part with reddish glandular dots and variously coloured and patterned in shades of green, yellow and red, hinged along the centre and fringed with marginal bristles to 8 mm; each half of the blade bears 3 sensitive hairs on the upper surface which, when touched in succession by an insect, cause the 2 halves to snap rapidly together, thus trapping the prey. Flowers radially symmetric, borne in a cluster on a leafless stalk 15–45 cm. Sepals 5. Petals 5, white with green veins, spathulate, 1.1–1.3 cm. Stamens 15–20. Ovary superior, 1-celled; style 1. Fruit an ovoid capsule 3–4 mm, which splits irregularly. Seeds numerous, black.

A genus of a single species from the SE USA, cultivated for its interest as an insectivorous plant. It is best grown in a cool greenhouse or on a window sill, and should be allowed to become semi-dormant in winter. During the growing season it requires plenty of light and a humid atmosphere; it grows well in a medium consisting mainly of moss peat, in pots standing in shallow water. Propagation is by seed or by leaf-cuttings.

1. D. muscipula Ellis. Illustration: Rickett, Wild flowers of the United States **2**: 257 (1967); Justice & Bell, Wild flowers of North Carolina, frontispiece & 83 (1968); Schnell, Carnivorous plants of the United States and Canada, 16–19 (1976); Slack, Carnivorous plants, 154, 157, 158 (1979).

SE USA (N & S Carolina). G1. Summer.

3. DROSERA Linnaeus
M. Cheek

Carnivorous perennial, terrestrial herbs, rarely annual, sometimes overwintering as a resting bud or underground tuber. Roots usually fibrous, rarely stout and bootlace-like. Stems very short, sometimes erect and twining. Leaves at first in a basal rosette; some species then producing leafy stems (most tuberous species not producing basal leaves when re-emerging from dormancy). Leaf-shape often changing greatly from basal rosette to aerial stem, and very variable between species, always with numerous stalked sticky glandular hairs on upper surface. Stipules large and chaffy, often joined to basal part of leaf-stalk. Flowers with parts in 4s or 5s, usually lasting a single day. Sepals united at the base, persistent in fruit. Petals free, obovate, white, purple or pink, rarely red or yellow, partly deliquescing then adhering together, forming a cap over the developing fruit. Stamens in a single whorl, anthers sometimes divided by the expanded connective. Ovary with 3 or 5 styles, varying greatly in branching and shape between species. Fruit a dehiscent capsule. Seeds dust-like to 1 mm, shape and sculpture varying greatly between species.

A pantropical genus of about 150 species, concentrated in S Africa and SW Australia but with representatives in cold temperate regions. Confined to continually or seasonally wet areas in the wild, surviving hot and dry or freezing periods as dormant buds or underground tubers.

Small animals, largely insects, which stick to the tips of the longer marginal hairs are pushed by these hairs to the centre of the leaf where shortly stalked glandular hairs secrete digestive juices. In some species (e.g. *D. burmannii*) the marginal hairs move extremely quickly, in others (e.g. *D. capensis*) the whole leaf wraps around the prey.

Literature: Diels, L., Droseraceae in Engler, A. *Das Pflanzenreich* **4**: 112–249 (1906); Schnell, D., *Carnivorous plants of the United States & Canada* (1976); Slack, A., *Carnivorous plants* (1979); Kondo, K. & Kondo, M., *Carnivorous plants of the world in colour* (1983); Lowrie, A. *Carnivorous plants of Australia* **1**. [mostly tuberous *Drosera*] (1987); Lowrie, A. *Carnivorous plants of Australia* **2**. [mostly pygmy *Drosera*] (1989); Bennett, S. & Cheek, M., The cytology & morphology of *Drosera slackii* and its relatives in South Africa. *Kew Bulletin* **45**(2): 375–381 (1990); Cheek, M., A new species of pygmy *Drosera* from Western Australia and a note on the status of section *Bryastrum* and section *Lamprolepis*. *Phytologia* **68**(2): 85–89 (1990); Marchant, N. & Lowrie, A., New names and new combinations in 34 taxa of Western Australian tuberous and pygmy *Drosera*. *Kew Bulletin* **47**(2): 315–328 (1992).

1a. Underground tubers present; stipules absent   2
 b. Underground tubers absent; stipules usually present   45
2a. Tubers more or less round; aerial stems, if present usually with the leaf-stalk distinct, cylindric, much longer than broad, with stalked glandular hairs   5
 b. Tubers long and thin; aerial stems if present with leaf-stalk flattened and glandular, not distinct from the blade   3
3a. Basal rosette leaves of 2 different shapes, the first spathulate, *c.* 7.5 mm, the second thread-like, *c.* 5 cm; aerial stem absent   **3. alba**
 b. Basal rosette leaves of uniform shape, spathulate to narrowly elliptic, never thread-like; aerial stem present   4
4a. Flower borne on leafy aerial stem; petals never black at the base
                                        1. cistiflora
 b. Flower borne on leafless aerial stem; petals black at base
                                        2. pauciflora
5a. Aerial stem entirely absent; leaves uniform in shape, forming a flat basal rosette   6
 b. Aerial stem present; leaves often of 2 shapes, rosette-leaves (usually only seen in seedlings) different from stem-leaves   16
6a. Leaf-stalk clearly distinct from blade
                                                      7
 b. Leaf-stalk not clearly distinct from blade, glandular hairs reaching more or less to leaf-base   11
7a. Leaf-blade kidney- to crescent-shaped, much broader than long
                                        4. zonaria
 b. Leaf-blade circular or obovate   8
8a. Leaf-blade circular or obovate abruptly constricted at leaf-stalk
                                      10. orbiculata
 b. Leaf-blade circular or obovate, gradually tapering into leaf-stalk
                                                      9
9a. Leaves 4–8 per rosette; flowering stems bearing several flowers
                                      11. erythrorhiza
 b. Leaves 9–30; flowering stems with a solitary flower   10
10a. Leaves 9–15, stalks *c.* 4 mm wide; style-branches *c.* 100
                                      12. whittakeri
 b. Leaves 15–30, stalks *c.* 2 mm wide; style-branches *c.* 20   **9. lowriei**
11a. Flowering-stems each bearing several flowers   12
 b. Flowering-stems each bearing a solitary flower   13

12a. Flowering-stem solitary
  11. erythrorhiza
  b. Flowering-stems several
  5. macrophylla
13a. Leaves with glandular trapping hairs only in the apical half, leaf-base shallowly pleated; style branches c. 100 **12. whittakeri**
  b. Leaves with glandular trapping hairs spreading from the tip almost to the base, leaf-base smooth, not pleated; style-branches less than 50 14
14a. Leaves with midrib sunken; sepals with black gland dots **6. rosulata**
  b. Leaves with midrib raised; sepals without black gland dots 15
15a. Leaves yellowish when mature; styles white, hair-like, barely swollen at tips **8. bulbosa**
  b. Leaves coppery red when mature; styles brown-purple, stout and short, inflated like doughnuts at tips **7. tubistylus**
16a. Aerial stems usually single, 15 cm or more when mature (but *D. bulbigena*, *D. salina* & *D. fimbriata* are shorter); leaf-blades peltate; semi-circular or circular (saucer-shaped to conical) 17
  b. Aerial stems usually clustered (except *D. platypoda*), 15 cm or less; leaf-blades not peltate but decurrent, never semi-circular or circular (except rare variants of *D. stolonifera*) 43
17a. Plants free-standing or leaning 18
  b. Plants scrambling or climbing 38
18a. Basal leaf rosette only present in seedlings 22
  b. Basal leaf rosette present in adult plants 19
19a. Leaf-blade of stem leaves kidney-shaped **26. andersoniana**
  b. Leaf-blade of stem leaves semi-circular, with 2 antennae 20
20a. Plants less than 5 cm **29. salina**
  b. Plants 5–37 cm 21
21a. Sepals hairy **27. peltata**
  b. Sepals hairless **28. auriculata**
22a. Petals yellow or dark, purplish red 23
  b. Petals white or pink 24
23a. Petals yellow
  13. neesii subsp. **neesii**
  b. Petals dark, purplish red
  14. microphylla
24a. Petals white 25
  b. Petals pink 33
25a. Leaf-blade semi-circular, with 2 antennae 26

b. Leaf-blade kidney-shaped or circular (saucer-like to conical), without antennae 27
26a. Plant very highly branched, to 1 m, with up to 30 scale-leaves at the base of the stem **19. gigantea**
  b. Plant with 1–3 branches, to 0.2 m, with 2 or 3 scale-leaves at the base of the stem **20. graniticola**
27a. Leaf-blade kidney-shaped 28
  b. Leaf-blade circular (saucer-shaped to conical) 30
28a. Scale-leaves at stem base numerous, 4–10 mm 42
  b. Petals 5; scale-leaves at stem base absent or 1 or 2, to 3 mm 29
29a. Plant 3–6 cm; leaf-stalk attached at margin of blade; leaves not developing axillary tubers after flowering **22. bulbigena**
  b. Plant to 18 cm; leaf-stalk attached near centre of blade; leaves developing axillary tubers after flowering **23. radicans**
30a. Inflorescence borne on a leafless stalk c. 12 cm **15. myriantha**
  b. Inflorescence-stalk less than 6 cm 31
31a. Stem-leaves in clusters of 3 or more **16. macrantha**
  b. Stem-leaves single 32
32a. Stem-leaves conical **17. huegelii**
  b. Stem-leaves saucer-shaped **18. marchantii**
33a. Stem-leaves semi-circular **13. neesii** subsp. **borealis**
  b. Stem-leaves circular (saucer-shaped to conical) 34
34a. Inflorescence borne on a leafless stalk c. 12 cm **15. myriantha**
  b. Inflorescence-stalk less than 6 cm 35
35a. Stem-leaves single **18. marchantii**
  b. Stem-leaves usually in clusters of 3 36
36a. Stem hairless **25. menziesii** subsp. **menziesii**
  b. Stem with stalked glands 37
37a. Lower third of stem hairless **16. macrantha**
  b. Entire stem with stalked glands **24. stricticaulis**
38a. Petals white or pink 39
  b. Petals yellow **32. subhirtella**
39a. Stem-leaves with 2 long antennae **30. modesta**
  b. Stem-leaves circular (saucer to cup-shaped) 40
40a. Upper two-thirds of stem with stalked glands **16. macrantha**

b. Entire stem hairless 41
41a. Stems and leaf-stalks red **25. menziesii**
  b. Stems and leaf-stalks green **31. pallida**
42a. Scale-leaves whorled, fringed **36. fimbriata**
  b. Scale-leaves spiralled, not fringed **21. heterophylla**
43a. Leaves in whorls of 2–4 **33. stolonifera**
  b. Leaves single, alternate 44
44a. Basal rosette usually producing a single, leafy erect stem with flowers at tip **35. platypoda**
  b. Basal rosette usually producing 2–5 stems, the flowers only produced at the tip of the 1–3 short, leafless stems **34. ramellosa**
45a. Leaf-blade slightly cupped, with a distinct outer rim, often peltate or almost so; leaf-stalk constricted at tip; stipules usually forming a large, dense central cluster over apical bud 46
  b. Leaf-blade (if distinct) more or less flat, not peltate, often decurrent, leaf-stalk not constricted at the tip; stipules if present, not as above 78
46a. Flowers mostly with 4 sepals, petals and styles 47
  b. Flowers with 5 sepals and petals, 3 (rarely 5) styles 48
47a. Styles white; all flowers with 4 sepals, petals and styles **37. pygmaea**
  b. Styles purple; flowers mostly as above but always some with 5 sepals, petals and/or styles **54. × badgerupii**
48a. Aerial stem distinct, 1.5–4 cm 49
  b. Aerial stem inconspicuous, less than 1 cm 52
49a. Petals bright orange-red; sepals with red hairs **38. barbigera**
  b. Petals white, rarely pink; sepals without red hairs 50
50a. Flowering-stem thickly covered in non-glandular, white, woolly hairs **39. scorpioides**
  b. Flowering-stems glandular-hairy or more or less hairless 51
51a. Flowering-stems with minute, stalkless glands **41. dichrosepala**
  b. Flowering-stems hairless 72
52a. Leaf-blade deeply cupped; flowering-stems usually with a single flower **42. occidentalis**
  b. Leaf-blade shallowly cupped; flowering-stems always with several flowers 53

53a. Petals yellow and white
    43. citrina
  b. Petals white, pink or orange-red
    54
54a. Petals orange-red   55
  b. Petals white or pink   62
55a. Leaf-stalks to 2 mm broad; styles always 5, tips hooked
    61. pulchella
  b. Leaf-stalks to 1.5 mm broad; styles usually 3, sometimes 4 or 5, not hooked   56
56a. Fruits pendent   **44. miniata**
  b. Fruits erect   57
57a. Style-stigma cylindric or thread-like   60
  b. Style bearing distinctly swollen stigma   58
58a. Styles white   **45. platystigma**
  b. Styles black   59
59a. Styles 3   **46. hyperostigma**
  b. Styles 5   **47. sewelliae**
60a. Stipule-bud *c.* 4 mm   **48. callistos**
  b. Stipule-bud *c.* 7 mm   61
61a. Stipule-bud smooth
    49. leucoblasta
  b. Stipule-bud bristly
    50. echinoblastus
62a. Style cylindric, stigma abruptly swollen   63
  b. Style-stigma cylindric, stigma not swollen   66
63a. Style red-purple   **51. nitidula**
  b. Style white, pale green or yellow   64
64a. Styles 3 or 4; stigma more or less as long as broad; fruit erect   65
  b. Styles 5; stigma slipper-shaped; fruit pendent   **52. mannii**
65a. Stigma trilobed or kidney-shaped
    51. nitidula
  b. Stigma unlobed, bun-shaped
    53. ericksoniae
66a. Stipule-bud 5-angled, acute
    55. androsacea
  b. Stipule-bud not angled   67
67a. Flowering-stem 1–5 cm   68
  b. Flowering-stem 6–9 cm   73
68a. Styles 5, hooked at tip
    61. pulchella
  b. Styles 3 or 4, not hooked   69
69a. Stipule-bud smooth   70
  b. Stipule-bud bristly hairy   71
70a. Flowers up to 20, crowded together; flowering stem densely long glandular-hairy   **56. leioblastus**
  b. Flowers 8–12, distant; flowering stem almost hairless
    57. oreopodion

71a. Flowering-stem densely covered in short white, non-glandular hairs
    59. trichocaulis
  b. Flowering-stem almost hairless
    58. paleacea
72a. Petals white; style-stigma erect
    40. enodes
  b. Petals white with red blotch at base; style-stigma horizontal
    60. parvula
73a. Fruit pendent   74
  b. Fruit erect   75
74a. Petals waisted, constricted at centre
    62. walyunga
  b. Petals obovate-elliptic   **63. spilos**
75a. Stipule-bud more or less globular, glossy, smooth   **64. pycnoblasta**
  b. Stipule-bud ovoid, usually bristly hairy   76
76a. Flower-stalks *c.* 4 mm; petal tip notched   **65. eneabba**
  b. Flower-stalks *c.* 2 mm; petal tip rounded   77
77a. Stipule-bud bristly; style-stigma reddish at base, widest at midpoint
    66. closterostigma
  b. Stipule-bud smooth; style-stigma greenish white, widest at base
    67. helodes
78a. Leaves more than 15 cm   79
  b. Leaves less than 15 cm   81
79a. Leaves thread-like   88
  b. Leaves forked or flattened   80
80a. Leaf-blade forked at the base, sometimes repeatedly, stalk 7–15 cm   **68. binata**
  b. Leaf-blade not forked, tapering to a point from a leaf base *c.* 1 cm wide, stalk indistinct   **69. regia**
81a. Stipules present   83
  b. Stipules absent   82
82a. Stem 15–30 cm, bearing thread-like leaves   **70. indica**
  b. Stem less than 2 cm, leaves 2 or 3, strap-shaped   **71. arcturi**
83a. Petals much shorter or equal to the sepals   84
  b. Petals longer than the sepals   86
84a. Leaf-blade heart-shaped, stalk longer than blade; inflorescence producing plantlets at top
    72. prolifera
  b. Leaf-blade gradually tapering at the base, without a distinct stalk; flowering stem never producing plantlets   85
85a. Leaves narrowly lanceolate, tip acute   **73. adelae**
  b. Leaves obovate, tip notched or truncate   **74. schizandra**

86a. Styles 5, branching only at tip; leaf with longest apical hairs lacking mucilage, sweeping swiftly (*c.* 5 seconds) to centre of leaf when touched   87
  b. Styles 3–5, usually forked at base; leaf hairs usually all with mucilage, sweeping slowly (5 minutes or more) to the centre of the leaf when touched   91
87a. Sepals with colourless hairs; flowering-stems 3–15 cm
    75. burmannii
  b. Sepals with stout red glandular hairs; flowering-stems 15–25 cm
    76. sessilifolia
88a. Stipules ovate, undivided, *c.* 1.5 × 1 cm, brown and chaffy; winter resting bud absent
    77. graminifolia
  b. Stipules divided into numerous teeth, less than 5 mm; resting bud present in winter   89
89a. Leaf-glands lacking red pigment entirely   **80. tracyi**
  b. Leaf-glands with red pigment   90
90a. Leaf-glands markedly red; fruits with seed   **78. filiformis**
  b. Leaf-glands faintly red; fruits without seed   **79. × californica**
91a. Overwintering as a resting bud – all trapping leaves dying back at end of summer   92
  b. Overwintering without a resting bud – trapping leaves always present   95
92a. Leaves horizontal, sometimes blade broader than long
    81. rotundifolia
  b. Leaves erect, blade longer than broad   93
93a. Flowering-stem held at about the same height as the leaves; leaf-blade usually less than 1 cm
    84. intermedia
  b. Flowering-stem more or less twice the height of the leaves; leaf-blade usually more than 1 cm   94
94a. Leaf-blade more or less rectangular to narrowly ellipsoid, 1.5–3 × 0.3–0.4 cm   **83. anglica**
  b. Leaf-blade 0.9–1.7 × 0.6–0.7 cm, obovate   **82. × obovata**
95a. Aerial stem present   96
  b. Aerial stem absent   101
96a. Leaves to 1.7 cm; stem completely erect, completely concealed and supported by a solid cushion of dead leaves   **102. roraimae**
  b. Leaves 2–14.5 cm; stems often prostrate at base, never completely

concealed or supported by dead leaves 97

97a. Leaf-blade obovate, less than 1.5 cm, markedly constricted at junction with leaf-stalk 99

b. Leaf-blade strap-shaped, 2.5–9 cm, not markedly constricted at junction with leaf-stalk 98

98a. Stem and lower leaf more or less hairless **85. capensis**

b. Stem and lower leaf with densely felted hairs **86. hilaris**

99a. Stipules 7.5–10 mm; leaves with adpressed white hairs below **87. glabripes**

b. Stipules to 5 mm; leaves hairless or hairy with red hairs below 100

100a. Lower leaf hairless; flower-stalks 5–10 mm; free part of stipule *c.* 2 mm (rest united to leaf-stalk) **88. collinsiae**

b. Lower leaf hairy with red hairs when young; flower-stalks 1.5–3.5 cm; free part of stipule 2–4.5 mm (only base united to leaf-stalk) **89. madagascariensis**

101a. Leaf-blade broader than long, very markedly constricted at junction with leaf-stalk; leaf-rosette less than 4 cm 102

b. Leaf-blade longer than broad, not always clearly distinct; leaf-rosette often greater than 4 cm 103

102a. Fruits pendent; flowering stem to 7 cm, densely covered in non-glandular hairs; petals orange-red **91. glanduligera**

b. Fruits erect; flowering stem 7–18 cm, mostly hairless, with a few glandular hairs amongst the flowers; petals white to purple **95. burkeana**

103a. Leaf-tip acute **101. villosa**

b. Leaf-tip round to truncate 104

104a. Petals *c.* 1.6 cm; styles 5, unbranched, flattened, united in a column **90. hamiltonii**

b. Petals less than 1.4 cm; styles 3, always branched, never united in a column 105

105a. Flowering-stem covered in long-stalked glands from base to tip; stipules reduced to a few inconspicuous teeth 106

b. Flowering-stem if with long-stalked glands then at tip only; stipules not reduced to a few inconspicuous hairs 107

106a. Leaf-blade oblanceolate; styles forked at base, each branch forking again twice in the upper half **97. trinervia**

b. Leaf-blade more or less spathulate; styles forked at base, each branch unforked, the tip slightly swollen **103. brevifolia**

107a. Lower leaf surface covered with coarse, long, adpressed coppery hairs **102. roraimae**

b. Lower leaf surface hairless or, if with hairs, not coppery and coarse 108

108a. Lower leaf surface with a scattering of short succulent adpressed, bright red hairs **92. slackii**

b. Lower leaf surface hairless or with fine silvery hairs 109

109a. Leaves with a broader blade on a long, narrow, non-glandular stalk 111

b. Leaves without a distinct leaf-stalk, the glandular surface extending almost to the base 110

110a. Leaves with fine white adpressed silky hairs below; styles forked once at base and once to twice at tip **94. aliciae**

b. Leaves mostly hairless below; styles forked once at base, but then unbranched **93. cuneifolia**

111a. Flowering-stem entirely hairless **99. capillaris**

b. Flowering-stem with hairs or glands or both 112

112a. Flowering-stem long-hairy (non-glandular) along its length **100. montana**

b. Flowering-stem with glandular hairs at the tip only 113

113a. Styles 3, forked at the base, the branches dividing again 1 or 2 times in the upper half; leaves membranous **98. natalensis**

b. Styles 3, forked at the base, but then unbranched; leaves almost leathery 114

114a. Style-branches slightly spoon-shaped at the tip; stipules to 2.5 mm, divided into 3 **96. dielsiana**

b. Style-branches not swollen at the tip; stipules 2–4 mm, finely divided **104. spathulata**

1. D. cistiflora Linnaeus. Illustration: Slack, Carnivorous Plants, 136 & 218 (1979); Kondo & Kondo, Carnivorous plants of the world in colour, 26 (1983).

Perennial, dying down in summer. Root-tubers cylindric *c.* 5 × 0.3–0.5 cm, vertical, white and fleshy. Aerial stem erect, 5–30 cm, slightly hairy. Stipules absent. Leaves at first spathulate in a basal rosette, *c.* 2 × 0.7 cm, stem-leaves obovate to oblong-linear or narrowly lanceolate, 2–6 × 0.5–0.7 cm, leaf-stalk absent. Flower usually solitary, terminal. Petals 1.5–3 cm, pink or purple, rarely white or yellow. Styles 3, forked at the base, at the middle and then twice at the flattened tip. *South Africa (Cape Province).* G1. Late summer (rarely flowers in cultivation).

2. D. pauciflora de Candolle. Illustration: Kondo & Kondo, Carnivorous plants of the world in colour, 26 bottom left (1983) – as D. cistiflora.

Perennial, dying down in summer, roots tuberous, cylindric, 4–8 × 0.1–0.2 cm, vertical, white or fleshy. Aerial stem leafless. Stipules absent. Leaves in a basal rosette, spathulate to wedge-shaped, 1–2.5 × 0.5–0.7 cm, stalk not distinct. Flowering-stem 3–30 cm, sparsely glandular-hairy. Flowers 1 or 2, petals 0.8–2.2 cm, pink or purple, rarely white, blackish at the base. Styles 3, forked at the base, then repeatedly at the tip, resembling a shaving-brush. *South Africa (Cape Province).* G1. Flowering unknown in cultivation.

3. D. alba Phillips.

Perennial, dying down in summer. Root tubers cylindric or tapered at ends, *c.* 1.5 × 0.1–0.2 cm, white and fleshy. Aerial stem absent. Stipules absent. First 3–5 leaves produced small and spathulate, horizontal *c.* 0.75 × 0.3 cm, later leaves long and thread-like, held more or less vertical, *c.* 5 × 0.1 cm. Flowering-stem 8–10 cm, hairless, bearing 2 or 3 flowers; petals white, to 6 mm. Styles unknown. *South Africa (Cape Province).* G1. Flowering unknown in cultivation.

4. D. zonaria Planchon. Illustration: Kondo & Kondo, Carnivorous plants of the world in colour, 81 (1983); Lowrie, Carnivorous plants of Australia **1**: 193 & 195 (1987).

Perennial to 7 cm across. Tuber obovoid, *c.* 7 × 10 mm, orange. Aerial stem absent. Stipules absent. Leaves to 30, in a flat, basal rosette; blade broadly rhombic, *c.* 7 × 17 mm, apical half with a crescent-like swathe of glandular hairs; leaf-stalk *c.* 15 × 2–5 mm. Flowering-stem lateral, unfolding before the leaf rosette, spike-like, *c.* 5 cm, hairless, with 1–8 flowers. Flower-stalks 1–4 mm, erect in fruit. Petals to 4 mm, white. Styles *c.* 3, divided from the base into 30–40 radiating, bristle-like segments, each with a slightly club-shaped stigmatic tip. *SW Western Australia.* G1. Winter, very rarely flowering in cultivation.

5. D. macrophylla Lindley. Illustration: Kondo & Kondo, Carnivorous plants of the world in colour, 45 (1983); Lowrie, Carnivorous plants of Australia **1**: 177 & 179 (1987).

Perennial to 12 cm across. Tuber obovoid, *c.* 1.5 × 1 cm, orange-red. Aerial stem absent. Stipules absent. Leaves to 12 in a flat basal rosette, blade obovate, *c.* 5.5 × 3 cm, base wedge-shaped; leaf-stalk absent. Flowering-stems 5–40, produced with the leaf rosette, to 15 cm, hairless, each with 1–4 flowers. Flower-stalks 1–2 cm. Petals *c.* 12 mm, white. Stigmas *c.* 3, divided from the base into *c.* 80 thin, radiating bristles, each with a pointed stigmatic tip, white. *SW Western Australia.* G1. Winter–spring.

6. D. rosulata Lehman. Illustration: Lowrie, Carnivorous plants of Australia **1**: 185 & 187 (1987).

Perennial to 7 cm across. Tuber obovoid, *c.* 8 × 12 mm, orange. Aerial stem absent. Stipules absent. Leaves 8–10, in a flat, basal rosette, blade obovate 3.5 × 1–1.5 cm, base wedge-shaped, midrib sunken; leaf-stalk absent. Flowering stems 1–12, *c.* 2 cm, each single-flowered, produced from the centre of the immature leaf-rosette, hairless, prostrate in fruit. Sepals black-dotted. Petals *c.* 1 cm, notched, white. Styles divided at the base into 20–25 ascending, slender, bristle-like branches, each with a shallowly 2 or 3-lobed stigmatic tip, white. *SW Western Australia.* G1. Autumn.

7. D. tubistylus Marchant & Lowrie. Illustration: Kondo & Kondo, Carnivorous plants of the world in colour, 23 top right (1983) – as D. bulbosa; Lowrie, Carnivorous plants of Australia **1**: 189 & 191 (1987).

Perennial to 3.5 cm across. Tuber spherical, *c.* 0.8 mm, pale orange. Aerial stem absent. Stipules absent. Leaves 5–9 in a flat basal rosette, blade obovate, *c.* 1.5 × 1 cm, wedge-shaped at the base, midrib raised and red in basal half, stalkless. Flowering-stems 1–4, from the centre of the leaf-rosette, single-flowered, *c.* 2.5 cm, prostrate in fruit. Sepals not black-dotted. Petals *c.* 1 cm, notched at the tip, white. Styles divided from the base into about 50 short cylindric segments, thickest towards the tip, the stigmatic tip swollen, like a shallowly-lobed tyre, brown to purple, the whole forming a dome. *SW Western Australia.* G1. Autumn.

8. D. bulbosa J.D. Hooker.

Perennial to 6 cm across. Tuber obovoid, *c.* 10 × 7 mm, red. Aerial stem absent. Stipules absent. Leaves 7–12, golden yellow, in a basal rosette, obovate, *c.* 2.5 × 0.7–1 cm, base wedge-shaped, midrib raised, leaf-stalk indistinct. Flowering-stems 1–8, from the centre of the rosette, single-flowered, 2–3 cm, prostrate in fruit. Sepals not black-dotted. Petals very narrowly obovate, *c.* 1 cm, notched, white. Styles divided at the base into 20–25 radiating hairs, each with a slightly swollen stigmatic tip, white. *SW Western Australia.* G1. Late autumn–early winter.

Subsp. **bulbosa**. Illustration: Lowrie, Carnivorous plants of Australia **1**: 149 & 151 (1987). Description as above.

Subsp. **major** (Diels) Marchant & Lowrie. Illustration: Lowrie, Carnivorous plants of Australia **1**: 153 & 155 (1987). Herb to 11 cm across, brownish-yellow. Tuber *c.* 2 × 1.5 cm, obovoid, red. Leaves *c.* 12, blade *c.* 5.5 × 1.5–2.5 cm. Flowering-stems 12–25, *c.* 6 cm. Sepals black-dotted. *SW Western Australia.* G1. Autumn.

9. D. lowriei Marchant. Illustration: Lowrie, Carnivorous plants of Australia **1**: 173 & 175 (1987).

Perennial *c.* 4 cm across. Tuber ellipsoid, *c.* 10 × 7 mm, pink. Aerial stem absent. Stipules absent. Leaves 15–30, in a flat basal rosette; blade obovate *c.* 1 × 1 cm, stalk *c.* 1 × 0.2 cm, red, midrib not distinct. Flowering-stems 1–9, alternating with the leaves, single-flowered, *c.* 1.5 cm, prostrate in fruit. Sepals not black-dotted. Petals 5 mm, white. Styles divided at the base into *c.* 20 segments, the inner group erect, the outer united in pairs, fully reflexed, clasping the ovary, stigmatic tip slightly swollen, 2-lobed, white. *SW Western Australia.* G1. Winter.

10. D. orbiculata Marchant & Lowrie. Illustration: Kondo & Kondo, Carnivorous plants of the world in colour, 23 top left (1983) – as D. bulbosa; Lowrie, Carnivorous plants of Australia **1**: 181 & 183 (1987).

Perennial to 4 cm across. Tuber spherical, *c.* 0.8 mm, pale orange. Aerial stem absent. Stipules absent. Leaves 4–6 in a flat basal rosette, blade circular, 1.5 cm, midrib indistinct; leaf-stalk 1–1.5 × 0.2–0.3 mm, flattened, red. Flowering-stems 3–8, from the centre of the leaf-rosette, single-flowered, *c.* 1.5 cm, prostrate in fruit. Sepals black-dotted. Petals *c.* 4 mm, white. Styles divided from the base into *c.* 20 bristle-like hairs, the inner ones erect, the outer curving out, then up,

stigmatic tip distinctly captitate. *SW Western Australia.* G1. Autumn.

11. D. erythrorhiza Lindley. Illustration: Kondo & Kondo, Carnivorous plants of the world in colour, 30 (1983); Lowrie, Carnivorous plants of Australia **1**: 157–171 (1987).

Perennial 1.5–6.5 cm across. Tuber spherical, 0.8–1.7 cm, orange red. Aerial stems absent. Stipules absent. Leaves usually 4, in a flat basal rosette; blade rounded-rhombic to broadly ovate, 2.7–3.5 × 1–3 cm, base tapering or wedge-shaped; leaf-stalk indistinct. Flowering-stem single, many-flowered, from the centre of the leaf-rosette, stalk 2–4.5 cm, with 2–10 cymose branches from the tip, branches to 5 cm, each with 3–6 flowers. Flower-stalks 5–10 mm, erect in fruit. Petals *c.* 1 cm, white. Styles 3, distinctly separated at the base, each divided into 10–20 hair-like, erect to ascending branches, swollen at the tip. *SW Western Australia.* G1. Winter.

Subsp. **erythrorhiza** Lindley. Rosette to 6 cm across. Leaves 4, circular in outline. Flowering after leaves develop. Stigmatic tips not lobed.

Subsp. **collina** Marchant & Lowrie. Rosette to 12 cm across. Leaves 8 or 9, obovate. Flowering after leaves develop. Stigmatic tips not lobed.

Subsp. **magna** Marchant & Lowrie. Rosette to 12 cm across. Leaves 4–6, slightly longer than broad, shortly obovate. Flowering after leaves develop. Stigmatic tips slightly 2-lobed.

Subsp. **squamosa** (Bentham) Marchant & Lowrie. Rosette to 6 cm across. Leaves 6–8, obovate, with a thick red band around the tip, horseshoe-shaped. Flowering before leaves develop. Stigmatic tips slightly 2-lobed.

12. D. whittakeri Planchon (*D. praefolia* Tepper; *D. whittakeri* var. *praefolia* (Tepper) J. Black). Illustration: Kondo & Kondo, Carnivorous plants of the world in colour, 80 (1983).

Perennial 2–5.5 cm across. Tuber almost spherical, 0.8–1 cm, orange. Aerial stem absent. Stipules absent. Leaves 9–15 in a flat basal rosette, blade circular to very shortly obovate, 8–12 × 5–10 mm, base decurrent, very broad and flattened, 6–8 × 4–5 mm, tapering towards the base, hairless. Flowering-stems 1–5, rarely to 14, from the centre of the leaf rosette, 2–4 cm, prostrate in fruit. Flowers solitary; petals 0.8–1.2 cm, white. Styles divided at the base into *c.* 100 hair-like brushes each with a blunt swollen stigmatic tip, white. *S Australia.* G1. Winter.

13. D. neesii Lehman. Illustration: Kondo & Kondo, Carnivorous plants of the world in colour, 78 (1983); Lowrie, Carnivorous plants of Australia **1**: 77 & 79 (1987).

Erect or climbing to 60 cm. Tuber hairless, *c.* 1 cm, yellow or yellowish pink. Stem hairless, unbranched. Stipules absent. Basal leaf-rosette absent. Scale-leaves on lower half of stem. Stem-leaves in groups of 3, peltate, semi-circular, *c.* 4 × 4 mm, with 2 antennae 1–2 mm from the corners, more or less horizontal on leaf-stalks 1–1.5 cm, the longest 3–5 cm, horizontal to erect. Flowering-stem terminal, almost hairless, with 5–15 flowers. Flower-stalk 3–10 mm. Sepals with stalked glandular hairs. Petals 8–12 mm, tip rounded, yellowish. Styles 3, each with 10–20 hair-like, flexible branches from near the base, forming a cushion, tip with 3 short, minute, stigmatic prongs, yellowish green. Ovary with glandular hairs. *SW Western Australia.* G1. Spring–early summer.

Subsp. **neesii** (*D. sulphurea* Lehman). Description as above.

Subsp. **borealis** Marchant. Illustration: Kondo & Kondo, Carnivorous plants of the world in colour, 52 (1983); Lowrie, Carnivorous plants of Australia **1**: 81 & 83 (1987). Erect to 25 cm. Tuber almost black, rarely white. Scale-leaves 2 or 3. Stem-leaves reaching almost to the ground. Flowering-stem with 4–8 flowers. Petals *c.* 1 cm, pink. Styles 3, the branches stouter than subsp. *neesii*, and not only branched at the base, slightly swollen with a very shallowly 2-lobed stigmatic tip. *SW Western Australia.* G1. Midwinter–spring.

14. D. microphylla Endlicher. Illustration: Kondo & Kondo, Carnivorous plants of the world in colour, 49 (1983); Lowrie, Carnivorous plants of Australia **1**: 65 & 67 (1987).

Erect perennial to 40 cm. Tuber almost spherical, *c.* 8 mm, warty, red. Stem hairless, unbranched. Stipules absent. Basal leaf-rosette absent. Scale-leaves 4 or 5, minute, hairless. Stem-leaves peltate, single, (rarely in groups of 3 at the top) almost saucer-shaped, but with the top edge straight, 3–3.5 mm across; leaf-stalks to 1.5 cm (where in groups of 3, the shorter pair *c.* 3 mm). Flowering-stem terminal, hairless, with 1–10 flowers. Flower-stalks 5–30 mm. Sepals hairless, larger than the petals. Petals *c.* 8 mm, tip acute, almost rounded, dark red. Stamens 5–6, filaments fleshy, bright red, erect. Styles 3, horizontal, flattened and stout in the lower 1/3–1/2, then divided repeatedly in the horizontal plane, blade broad, rather deeply lobed, webbed, dark red to black. *SW Western Australia.* G1. Winter.

15. D. myriantha Planchon. Illustration: Lowrie, Carnivorous plants of Australia **1**: 73 & 75 (1987).

Erect or slightly climbing to 30 cm. Tuber almost spherical, *c.* 1 cm, white. Stem hairless, with 1–3 horizontal branches to 10 cm. Stipules absent. Basal leaf-rosette absent. Scale-leaves 4 or 5. Stem-leaves single, peltate, saucer-shaped, *c.* 3 mm across, facing horizontally or pendent; leaf-stalks 5–7 mm, almost erect. Flowering-stem terminal, the basal *c.* 12 cm without leaves or bracts, hairless with numerous flowers. Flower-stalks *c.* 3 mm, erect in fruit. Sepals hairless, entire. Petals *c.* 5 mm, white or pink. Styles 3, each divided at the base into 4 unbranched hair-like, straight strands, slightly swollen at the tip. *SW Western Australia.* G1. Early Summer.

16. D. macrantha Endlicher. Illustration: Kondo & Kondo, Carnivorous plants of the world in colour, 44 (1983); Lowrie, Carnivorous plants of Australia **1**: 37 & 39 (1987).

Perennial climbing by sticky leaves, to 1.5 m. Tuber almost spherical, 1.5–2 cm, warty, white. Stem hairless in lower third, glandular hairy above. Stipules absent. Basal leaf-rosette absent. Stem-leaves in clusters of 3, obconical, peltate, *c.* 6 × 6 cm, almost pendent, stalks 1.5–5 cm, the shorter, lateral pair 0.5–1 cm. Flowering-stem terminal, glandular-hairy, with numerous flowers. Flower-stalks 7–9 mm; petals white or pink, tips truncate. Styles 3, erect, the upper 3/4 finely divided into 10–16 thread-like segments, white. *SW Western Australia.* G1. Winter.

A variant with erect stems to 25 cm, with glandular hairs to the base, subsp. **eremeae** Marchant & Lowrie 'Marchant', has been offered for sale and may represent a distinct new species.

17. D. huegelii Endlicher. Illustration: Kondo & Kondo, Carnivorous plants of the world in colour, 39 (1983); Lowrie, Carnivorous plants of Australia **1**: 33 & 35 (1987).

Erect or slightly climbing, perennial 5–45 cm. Tuber spherical, *c.* 1 cm, white. Stem hairless, zig-zagging slightly, unbranched. Stipules absent. Basal leaf-rosette absent. Scale-leaves 1 or 2 at base of stem. Stem-leaves with blades peltate, obconical and pendent 4–6 × 5–8 mm, stalks 1–1.5 cm, erect from the base, the tip narrowing and pendent. Flowering-stem terminal, hairless, with 3–12 flowers. Flower-stalks 4–20 mm, erect. Petals 9–10 mm, white. Styles 3, stout, in the final quarter with numerous branches forming a compact head, like a cauliflower, each with a notched and slightly swollen stigmatic tip. *SW Western Australia.* G1. Winter.

18. D. marchantii de Buhr. Illustration: Kondo & Kondo, Carnivorous plants of the world in colour, 47 (1983); Lowrie, Carnivorous plants of Australia **1**: 45 & 47 (1987).

Erect perennial to 50 cm. Tuber spherical, *c.* 8 cm, white or red. Stem hairless, unbranched. Stipules absent. Basal leaf-rosette absent. Scale-leaves 4 or 5 with sparse, stalked glandular hairs. Stem-leaves single, peltate, saucer-shaped, *c.* 4 mm across, held horizontally on tapering, erect leaf-stalks *c.* 1 cm. Inflorescence a terminal cyme, with 5–10 flowers. Flower-stalks 0.5–2.5 cm, pendent in fruit. Sepals with stalked glandular hairs, longest at the margins, forming a fringe. Petals *c.* 12 mm, tip rounded, pink, opening and closing over several days. Styles 3, short basal third stout and horizontal, each branching into 4 then each of these again into 4, forming a hemispherical cushion, pink, stigmatic tips minutely swollen. *SW Western Australia.* G1. Winter.

Subsp. **prophylla** Marchant & Lowrie is shorter with more glandular and numerous scale leaves and white flowers.

19. D. gigantea Lindley. Illustration: Kondo & Kondo, Carnivorous plants of the world in colour, 33 (1983); Lowrie, Carnivorous plants of Australia **1**: 21 & 23 (1987).

Erect perennial to 1 × 1 m. Tuber rounded, *c.* 1.5 × 3 cm, red. Stem to 1 cm thick at base, hairless, with many scale leaves, particularly at the base, with numerous branches, each branching again. Stipules absent. Basal leaf-rosette absent in adult plants. Stem-leaves single or in clusters of 3, blade peltate, crescent-shaped, *c.* 2.5 × 3.5 mm, the 2 points extended by 1–2 mm; leaf-stalks 8–10 mm. Flowering-stems at the ends of the main and primary branches, with numerous flowers. Flower-stalks 3–8 mm. Petals *c.* 5.5 mm, rounded at tip. Styles 3, broad and stout, orange-brown, divided at the tip into *c.* 4 blunt lobes, each bearing *c.* 8–10 stigmatic lumps. *SW Western Australia.* G1. Spring–summer.

20. D. graniticola Marchant. Illustration: Lowrie, Carnivorous plants of Australia **1**: 25 & 27 (1987).

Erect perennial to 20 cm. Tuber spherical, to 7 mm, white. Stem hairless, often with 1 or 2 branches. Basal leaf-rosette absent. Scale-leaves 1 or 2. Stem-leaves in clusters of 2, later 3, peltate, semicircular to very slightly kidney-shaped, with blunt antennae, c. 2 × 2 mm, leaf-stalks 7–10 mm. Inflorescence terminal, hairless, with 2–10 flowers. Flower-stalks erect in fruit. Petals c. 5 mm, tip truncate, white, not closing at night. Styles 3, stout at the base, divided midway into 6–10 narrow segments, each ending in a slightly swollen stigma, white. *SW Western Australia.* G1. Winter.

21. D. heterophylla Lindley. Illustration: Kondo & Kondo, Carnivorous plants of the world in colour, 37 (1983); Lowrie, Carnivorous plants of Australia **1**: 29 & 31 (1987).

Erect perennial to 30 cm. Tuber spherical, c. 1 cm, white. Stem hairless, unbranched. Stipules absent. Basal leaf-rosette absent. Scale-leaves numerous on the lower stem. Stem-leaves single, peltate, dished, kidney-shaped, c. 3 × 6 mm, lacking antennae; leaf-stalk 5–15 mm. Inflorescence terminal, hairless, with 1–5 flowers. Flower-stalks 2–4 mm, nodding in fruit. Petals 8–12, tip rounded, white, opening and closing over several days, 1–1.8 cm. Styles 3, branching repeatedly, forming a spherical mass of white fibres. *SW Western Australia.* G1. Winter–early spring.

22. D. bulbigena Morrison. Illustration: Kondo & Kondo, Carnivorous plants of the world in colour, 22 (1983); Lowrie, Carnivorous plants of Australia **1**: 17 & 19 (1987).

Erect perennial, 3–6 cm. Tuber spherical, c. 2 mm, red. Stem zig-zagging slightly, hairless, unbranched. Stipules absent. Basal leaf-rosette absent. Stem-leaves almost peltate, inserted near margin, kidney-shaped, c. 2 × 2.5 mm, lacking antennae, stalks c. 2 mm. Inflorescence terminal, hairless, with 1–5 flowers. Flower-stalks 2–4 mm. Petals white, 5–5.5 mm, tip truncate to slightly notched, slightly toothed. Styles 3, with 4 branches from the base, branches entire, or forked midway, stigmatic tips pointed, white. *SW Western Australia.* G1. Winter.

23. D. radicans Marchant. Illustration: Lowrie, Carnivorous plants of Australia **1**: 93 & 95 (1987).

Erect perennial to 18 cm. Tuber spherical, c. 5 mm, white. Stem hairless, usually unbranched. Basal leaf-rosette absent. Scale-leaves absent. Stem-leaves single, with 2 extra leaves developing after flowering from an axillary tuber, peltate, kidney-shaped, 3 × 4 mm, held erect or horizontally. Inflorescence terminal, hairless, with up to 25 flowers. Flower-stalks 4–8 mm, erect in fruit. Sepals very deeply toothed, almost hairless, black-dotted. Petals c. 4 mm, tip truncate, white. Styles c. 3, each divided from the base into 4 ascending, bristle-like segments with a pointed stigmatic tip, white. *SW Western Australia.* G1. Winter.

24. D. stricticaulis (Diels) O. Sargent (*D. macrantha* var. *stricticaulis* Diels). Illustration: Kondo & Kondo, Carnivorous plants of the world in colour, 76 (1983); Lowrie, Carnivorous plants of Australia **1**: 101 & 103 (1987).

Erect, perennial to 30 cm. Tuber spherical, c. 8 mm across, white. Stem erect, with minutely glandular leaves, unbranched. Stipules absent. Basal leaf-rosette absent. Scale-leaves 6–9 at the base of the stem. Stem-leaves in clusters of 3, peltate, the blade saucer-shaped, c. 5 mm across, facing horizontally; leaf-stalk stalk 0.5–1 cm, the longest 1.5–2.5 cm, erect, glandular. Inflorescence terminal, minutely glandular, with 3–12 flowers. Flower-stalks 0.5–1 cm, pendent in fruit. Sepals thickly covered with long-stalked glands. Petals c. 7 mm, tip rounded, pink. Styles c. 3, each with 2 or 3 erect branches from the base, each branch dividing again into 3–5 branches in the upper half, each with 3–6 minute apical stigmatic prongs, white. *SW Western Australia.* G1. Winter–spring.

25. D. menziesii de Candolle. Illustration: Kondo & Kondo, Carnivorous plants of the world in colour, 48 (1983); Lowrie, Carnivorous plants of Australia **1**: 53–63 (1987).

Erect perennial to 25 cm. Tuber spherical, c. 1 cm, red. Stem almost hairless, usually unbranched. Stipules absent. Basal leaf-rosette absent. Scale-leaves 4 or 5, hairless. Stem-leaves in clusters of 3 or 5, blade peltate, saucer-shaped, c. 3.5 mm across, pointing horizontally, stalks erect, 3–5 mm, the longest 1.5–2.5 cm. Flowering-stem with c. 6 flowers. Flower-stalks 5–10 mm. Sepals with stalked glandular hairs, margin with

coarse hairs. Petals c. 14 mm, tip broad and rounded, pink to red. Styles probably 3, much divided at the base (only) into numerous fine hair-like strands forming a cushion, stigmatic apices pointed, whitish. *SW Western Australia* G1. Winter.

Subsp. **penicillaris** (Bentham) Marchant & Lowrie. (*D. penicillaris* Bentham). Stem climbing to 1.15 m, base very spindly, not supporting the upper part which climbs by the sucker-like, glandular leaves. Flowering-stem with up to 18 flowers. *SW Western Australia.* G1. Winter.

Subsp. **thysanosepala** (Diels) Marchant (*D. thysanosepala* Diels). Stems climbing to 40 cm, extremely spindly at the base, supported by the sucker-like glandular leaves. Stem-leaves in clusters of 3. Sepals hairless, margin fringed. Petals white or pink. *SW Western Australia.* G1. Winter–spring.

26. D. andersoniana Ewart & White. Illustration: Kondo & Kondo, Carnivorous plants of the world in colour, 13 (1983); Lowrie, Carnivorous plants of Australia **1**: 13 & 15 (1987).

Erect perennial to 25 cm. Tuber top-shaped, c. 7 × 7 mm, yellow. Stem slightly hairy, 15–25 cm. Basal leaf-rosette present in adult plants, 2–3 cm across, blade rounded, broader than long, c. 6 × 10 mm, stalk c. 12 × 1 mm, hairless. Scale-leaves absent. Stem-leaves in groups of 3, peltate, slightly kidney-shaped, cupped, c. 4 × 5 mm, stalks 1–2 cm. Inflorescence terminal, hairy, with 3–15 flowers. Flower-stalks 4–15 mm, pendent in fruit. Petals white, c. 7 mm, tip rounded or truncate. Styles 3, each branching 5–7 times from the base to the rounded tips. *SW Western Australia* G1. Winter.

27. D. peltata Thunberg. Illustration: Slack, Carnivorous Plants, 144 & 145 (1979); Kondo & Kondo, Carnivorous plants of the world in colour, 57 (right) (1983); Lowrie, Carnivorous plants of Australia **1**: 89 & 91 (1987).

Erect perennial to 20 cm. Tuber spherical, c. 8 mm, red. Stem hairless, unbranched. Stipules absent. Basal leaf-rosette present at maturity, blade rounded, c. 4 × 8 mm, leaf-stalk flattened, c. 12 × 1 mm. Scale-leaves absent. Stem-leaves single, peltate, semi-circular, cupped, c. 3 × 4 mm, with a long antenna 1–2 mm at each corner, pointing horizontally, stalk c. 1 cm, horizontal. Inflorescence terminal, hairless, with 5–10 flowers. Flower-stalks 8–10 mm, erect in fruit. Sepals completely covered outside with long, black-tipped hairs. Petals *c.* 4 mm, tip truncate, white. Styles 3, short, stout, erect, divided midway into *c.* 4 cylindric segments, stigmatic tip shallowly bilobed, orange. *India to Japan through SE Asia to E Australia, also in SW Western Australia.* G1. Summer.

28. D. auriculata Planchon. Illustration: Kondo & Kondo, Carnivorous plants of the world in colour, 18 (1983).

Erect perennial, 5–37 cm. Tuber spherical, 0.5–0.8 mm, pink. Stem hairless, often with 1 or 2 branches, rarely many. Stipules absent. Basal leaf-rosette usually present, sometimes replaced with scales, leaves 4–6 (rarely to 15), circular, sometimes broader than long, 2–3.5 × 3–6 mm, leaf-stalk 4–15 mm. Stem-leaves single, rarely in groups of 3; blade held horizontally, peltate, crescent-shaped, 3–5 × 3–4 mm with two antennae *c.* 1 mm; leaf-stalk 0.6–1.2 cm, erect. Inflorescence terminal, stalk 4–5 cm, bearing 3–15 flowers. Flower-stalks 4–11 mm, erect in fruit. Sepals hairless. Petals 7–8 mm, pink, but sometimes white. Styles as *D. peltata. E Australia & New Zealand.* G1. Summer.

29. D. salina Marchant & Lowrie. (*D.* 'Salt Lake'). Illustration: Lowrie, Carnivorous plants of Australia **1**: 97 & 99 (1987). Erect perennial to 5 cm. Tuber spherical, 4–5 mm across, white. Stem erect, hairless, unbranched. Stipules absent. Basal rosette of 8–10 leaves, blade rounded, broader than long, *c.* 1.5 × 2.5 mm, leaf-stalk *c.* 5 × 0.3 mm. Scale-leaves absent. Stem-leaves single, peltate, pointing horizontally to erect, semi-circular, *c.* 2 × 2.5 mm, with 2 erect, short antennae 0.5–1 mm, stalks erect, 7–8 mm. Inflorescence terminal, of 1 or 2 flowers. Flower-stalks 1–1.8 cm, erect. Sepals *c.* 3 mm, irregularly toothed, black dotted. Petals *c.* 5 mm, rounded, white. Styles 3, the basal half erect and stout with 1 long then 2 or 3 progressively shorter lateral branches, the tip with 3 shorter segments, yellowish. *SW Western Australia.* G1. Winter.

30. D. modesta Diels. Illustration: Lowrie, Carnivorous plants of Australia **1**: 69 & 71 (1987).

Climbing perennial, 40–80 cm. Tuber almost spherical, *c.* 6 mm, red. Stem hairless in the lower third, glandular and hairy above, unbranched. Stipules absent. Basal leaf-rosette present in immature plants. Scale-leaves 4 or 5, barely detectable, hairless, minute. Stem-leaves solitary in lower third, in 3s above, peltate, rounded, *c.* 2.5 mm across, with 2 long,

erect antennae, each *c.* 3 mm; leaf-stalk 0.4–0.6 cm, the longer 1–4 cm. Inflorescence terminal, glandular, with 10–25 flowers. Flower-stalks 5–15 mm. Sepals with stalked glandular hairs, margin with teeth each with several long-stalked glands. Petals 6–6.5 mm, tip truncate to rounded, white. Styles probably 3, divided at the base into numerous hair-like strands, each sometimes divided again to form a diffuse white cushion, tips pointed. *SW Western Australia.* G1. Winter.

31. D. pallida Lindley. Illustration: Kondo & Kondo, Carnivorous plants of the world in colour, 56 (1983); Lowrie, Carnivorous plants of Australia **1**: 85 & 87 (1987).

Climbing perennial, to 1.5 m. Tuber almost spherical *c.* 1.5 × 2 cm, white. Stem hairless, unbranched. Stipules absent. Basal leaf-rosette absent. Scale-leaves 4 or 5 at the base. Stem-leaves in clusters of 3, peltate, shallowly saucer-shaped, *c.* 3 mm, stalks 1.5–4 cm, held at various angles. Flowering-stem hairless, terminal with 20–25 flowers. Flower-stalks 7–10 mm, pendent in fruit. Sepals hairless, black-dotted. Petals 9–10 mm, truncate, white. Styles *c.* 3, divided into numerous hair-like strands from the base, equalling or longer than the stamens, forming a compact cushion. Stigmatic tip pointed, white. *SW Western Australia.* G1. Winter–late spring.

32. D. subhirtella Planchon. Illustration: Kondo & Kondo, Carnivorous plants of the world in colour, 77 (1983); Lowrie, Carnivorous plants of Australia **1**: 105–111 (1987).

Climbing perennial to 40 cm. Tuber almost spherical, *c.* 8 mm, yellow. Stem with up to 3 lateral branches, glandular. Stipules absent. Basal leaf-rosette absent. Scale-leaves 2 or 3, at the base of the stem, inconspicuous. Stem-leaves in clusters of 3, sometimes 5 at the base, blade peltate, circular to shallowly saucer-shaped, *c.* 3 mm, lower surface with a few glandular hairs, stalks 0.4–0.5 cm, the longest 2–4 cm. Inflorescence terminal at the end of the branches, glandular, with up to 20 flowers. Flower-stalks 5–15 mm, pendent in fruit. Sepals thickly covered in stalked glands. Petals 9–10 mm, tip broadly rounded to truncate, yellow. Styles 3, the lower fifth stout and erect, then branching into 6 long, hair-like segments, each with 5 or 6 minute, pointed stigmas, spike-like at the very tip. *SW Western Australia.* G1. Late winter–early spring.

Subsp. **moorei** (Diels) Marchant (*D. moorei* Diels). Climbing perennial to 20

cm. Tuber spherical, *c.* 8 mm, white. Stem hairless, branching unknown. Basal leaf-rosette absent. Scale-leaves 2 or 3, inconspicuous, scattered at the base of the stem. Stem-leaves in groups of 3, blade peltate, circular to slightly kidney-shaped, *c.* 2 × 2.5 mm; stalk 1–1.5 cm, the central stalk only slightly longer than the laterals. Inflorescence terminal, sparsely branched, hairless, with 2–20 flowers. Flower-stalks 1.5–2 cm, hairless. Sepals with deep forward pointing teeth, hairless, finely black-dotted. Petals *c.* 8 mm, bright yellow. *SW Western Australia* G1. Spring.

33. D. stolonifera Endlicher. Illustration: Kondo & Kondo, Carnivorous plants of the world in colour, 75 (1983); Lowrie, Carnivorous plants of Australia **1**: 125–147 (1987).

Erect or prostrate perennial to 15 cm. Tuber spherical or broader than long, to 1.5 × 1 cm, orange or red. Stems 2–5 (rarely 1–7) from the basal rosette, lateral to the central flowering-stem, usually erect, hairless, often flowering. Stipules absent. Basal rosette with 4–7 leaves, blades broadly ellipsoid to circular, *c.* 5 × 8 mm. usually decurrent at the base, blades flattened, 2–3 mm across. Scale-leaves absent. Stem-leaves in whorls of 3, folded along their length, erect; blades kidney-shaped, *c.* 3 × 5 mm; stalks *c.* 5 × 1 mm, grooved on the upper surface. Flowering-stem with an erect, naked stalk 10–15 cm from the basal leaf-rosette, bearing *c.* 6 branches, each with up to 10 flowers; also from the tip of the vegetative stems. Flower-stalks *c.* 5 mm, erect in fruit. Petals 9–10 mm, white. Styles divided from the base into 20 or more tapering branches, the innermost of the branches more or less erect, the outermost more or less horizontal except the erect apices, white. *SW Western Australia.* G1. Winter–spring.

Subsp. **compacta** Marchant. Plants compact, lateral stems 5–12 cm. Leaf-stalks very long, 1–3 cm. Stem-leaves usually inserted at the base of the lateral stems, within the basal rosette. *SW Western Australia.* G1. Winter–spring.

Subsp. **rupicola** Marchant. Lateral stems semi-prostrate, 5–15 cm. Stem-leaves circular to transversely elliptic, *c.* 10 × 15 cm, folded along their length. Flowering-stems 1–4 from the basal rosette, to 10 cm. *SW Western Australia.* G1. Winter.

Subsp. **humilis** (Planchon) Marchant (*D. humilis* Planchon; *D. stolonifera* var. *humilis*

(Planchon) Diels). Plants compact. Lateral stems 3–5, to 15 cm. Basal rosette of very small leaves, obovate, blades *c.* 1.5 × 2 mm, stalks 1.5–2 × 2–3 mm. Stem-leaves kidney-shaped, sinus angular; leaf-stalk cylindric, not grooved. Flowers opening when lateral stems only 1–2 cm.

Subsp. **porrecta** (Lehmann) Marchant & Lowrie (*D. porrecta* Lehmann). Tall plants to 45 cm with a single stem unless in flower. Basal leaf rosette in 2 (rarely 3) distinct whorls, one above the other, leaf-blade circular, *c.* 5 × 6 mm, stalks *c.* 5 × 2–3 mm. Stem-leaves in whorls of 4 or 5, blade kidney-shaped, sinus angular, stalk cylindric, grooved. *SW Western Australia.* G1. Late winter.

Subsp. **prostrata** Marchant & Lowrie. Stems 4 or 5, prostrate, to 5 cm. Basal leaf rosette absent. Stem-leaves in whorls of 4, blade circular to obovate or transversely elliptic, *c.* 3 × 3 mm, stalks twisted to hold blades horizontal above the sand, slightly flattened and grooved. *SW Western Australia.* G1. Winter.

34. D. ramellosa Lehman. Illustration: Kondo & Kondo, Carnivorous plants of the world in colour, 66 (young plant) (1983); Lowrie, Carnivorous plants of Australia **1:** 121 & 123 (1987).

Erect perennial to 12 cm. Tuber obovoid, *c.* 10 × 7 mm, orange. Stem hairless, usually with 2 branches from the basal rosette. Stipules absent. Basal rosette with 6–12 leaves, blade rounded, broader than long, *c.* 5 × 8 mm, base decurrent, stalk broad, 5–6 × 2–3 mm, hairless. Scale-leaves absent. Stem-leaves single, as the basal leaves, but erect, folded along their length and almost adpressed to the stem. Flowering-stems 1–3 from the basal rosette, not from the erect stems, 2–3 cm, hairless, each with 1–3 flowers. Flower-stalks 5–7 mm, pendent in fruit. Sepals toothed at tip, hairless. Petals 5 mm, white or pink. Styles 3, basal part stout, almost horizontal, with 3–7 slender, vertical branches nearly as long as the basal part, the stigmatic apices pointed, white. *SW Western Australia.* G1. Winter–spring.

35. D. platypoda Turczaninow. Illustration: Kondo & Kondo, Carnivorous plants of the world in colour, 60 (1983); Lowrie, Carnivorous plants of Australia **1:** 117 & 119 (1987).

Erect perennial to 20 cm. Tuber obovoid, *c.* 10 × 7 mm, orange. Stem hairless, rarely branching from the basal rosette. Stipules absent. Basal rosette of 3–5 leaves, blade transversely elliptic, *c.* 5 × 9 mm,

stalk broad, 5–6 × 2–3 mm, hairless. Scale-leaves absent. Stem-leaves like basal, but held erect, very close to the stem and folded slightly along their length, the lateral margins of the blade inrolled. Flowering-stem with up to 20 flowers, 2–4 open at a time. Flower-stalks 5–10 mm. Sepals hairless. Petals 7–8 mm, white. Styles divided from the base into *c.* 25 radiating branches, each with a spherical stigmatic tip slightly thicker than the style, white. *SW Western Australia* G1. Winter–early summer.

36. D. fimbriata de Buhr. Illustration: Lowrie, Carnivorous plants of Australia **1:** 113 & 115 (1987).

Erect perennial to 10–15 cm. Tuber elliptic, *c.* 1 × 0.8 cm, orange. Stem hairless, unbranched. Stipules absent. Basal leaf-rosette absent. Scale-leaves at stem base in 1–3 whorls of 4 or 5, spreading, 4–6 mm with a highly fringed, glistening margin, the whole arrangement resembling a spider's web around the stem. Stem-leaves in whorls of 3–5, blade peltate, circular to faintly heart- or kidney-shaped, very slightly cupped, *c.* 4 mm, stalks 5–18 mm, erect, all those in one whorl more or less equal. Inflorescence terminal, hairless, with 5–15 flowers. Flower-stalks 3–5 mm, hairless. Sepals hairless. Petals 5, 7–8 mm, white. Styles divided into *c.* 12 short stout horizontal branches from the very base, gradually curving upwards and tapered to the spike-like stigmatic apices, white. *SW Western Australia.* G1. Spring.

37. D. pygmaea de Candolle. Illustration: Kondo & Kondo, Carnivorous plants of the world in colour, 65 (1983); Lowrie, Carnivorous plants of Australia **2:** 151 & 153 (1989).

Short-lived perennial. Stem inconspicuous. Leaf-rosette *c.* 1.5 cm across. Stipule-bud loose and poorly developed, ovoid, 3.5 × 3 mm. Leaf-blade *c.* 1.5 mm, circular, stalk 5–6 × 0.2–0.5 mm. Flowering-stems 1–4, to 2.5 cm, almost hairless with only a few minute, stalkless glands with a single flower. Flower-stalk erect in fruit. Petals 4, white, 3–3.3 mm. Styles 4, thread-like below, the stigmatic tip slightly swollen and club-like. *SW Western Australia, E Australia & New Zealand.* G1. Summer.

38. D. barbigera Planchon. Illustration: Kondo & Kondo, Carnivorous plants of the world in colour, 29 (1983) – as D. drummondii; Lowrie, Carnivorous plants of Australia **2:** 27 & 29 (1989).

Short-lived perennial. Stem at length aerial, 2–3 cm, clothed with old leaves. Leaf-rosette 3–5 cm across. Stipule-bud ovoid, *c.* 8 × 4 mm. Leaf-blade very narrowly elliptic, *c.* 5 × 1.5 mm, stalk *c.* 12 × 0.5–1 mm. Inflorescence terminal, single, to 9 cm, woolly with glandular hairs, flowers 5–8. Flower-stalks to 1 mm, erect in fruit. Sepals thickly covered in red, usually glandular hairs. Petals bright orange-red, black at base, 9–10 mm. Styles 3, thread-like, black. *SW Western Australia.* G1. Summer.

39. D. scorpioides Planchon (*D.* 'Gidgeganup'). Illustration: Kondo & Kondo, Carnivorous plants of the world in colour, 71 (1983); Lowrie, Carnivorous plants of Australia **2:** 163 & 165 (1989).

Short-lived perennial. Stem conspicuous, 2–3 cm. Leaf-rosette very loose and open, *c.* 3.5 cm high and across. Stipule-bud loose and poorly developed, ovoid, *c.* 10 × 6 mm, hairy at the base. Leaf-blade narrowly elliptic, *c.* 6 × 1.6 mm, stalk *c.* 10 × 0.5–1 mm, covered with long-stalked glandular hairs. Flowering-stem solitary, to 5 cm, the whole densely covered in white woolly hairs, with 20–30 flowers. Flower-stalks *c.* 1 mm, erect in fruit. Petals white to pink, *c.* 5 mm. Styles 3, thread-like, erect and strongly incurved.

40. D. enodes Marchant & Lowrie. (*D. omissa* Marchant not Diels). Illustration: Lowrie, Carnivorous plants of Australia **2:** 55 & 57 (1989).

Short-lived perennial. Stem at length aerial, *c.* 1.5 cm, covered with old leaves. Leaves forming a rather open, cushion-like rosette *c.* 2.5 cm across. Stipule-bud ovoid, *c.* 7 × 5 mm, shaggy. Leaf-blade elliptic 2–3 × 1–1.5 mm, stalk *c.* 8 × 0.8 mm. Flowering-stem to 7.5 cm, hairless with 3–20 sweetly scented flowers. Flower-stalks *c.* 4.5 mm, erect in fruit. Petals white, not closing at night but remaining open for several days, *c.* 3.5 mm. Styles 3, thread-like, white, each with a curved, inverted, club-like stigma. *SW Western Australia* G1. Summer.

41. D. dichrosepala Turczaninow (*D. scorpioides* misapplied not Planchon). Illustration: Slack, Carnivorous Plants, 138 (1979); Kondo & Kondo, Carnivorous plants of the world in colour, 28 (1983); Lowrie, Carnivorous plants of Australia **2:** 43 & 45 (1989).

Short-lived perennial. Aerial stem 2–4 cm, completely covered with old leaves. Leaf-rosette 1.5–2 cm across. Stipule-bud ovoid, *c.* 4 × 2 mm, shaggy. Leaf-blade elliptic, *c.* 3 × 1.5 mm, stalk *c.* 6 × 0.6 mm. Flowering-stem up to 4 cm, almost hairless with 3–12 scented flowers. Flower-stalks *c.* 2.5 mm, erect in fruit. Petals not closing at night, lasting more than one day, white, concave *c.* 3 × 1 mm. Styles 3, thread-like, white. *SW Western Australia.* G1. Early Summer.

42. D. occidentalis Morrison. Illustration: Lowrie, Carnivorous plants of Australia **2:** 115, 117, 119 & 121 (1989).

Short-lived perennial. Stem inconspicuous. Leaf-rosette often sparse, 1.5–2 cm across. Stipule-bud broadly ovoid 2.5–3 × 2.5–3 mm. Leaf-blade circular 1–1.5 mm. Leaf-stalk *c.* 5 × 0.2–0.4 mm. Flowering-stems 1–4, sparsely and minutely glandular, 1–2.5 cm, almost hairless, with 1–8 flowers. Flower-stalks erect. Petals 2–2.5 mm, white–pink. Styles 5, whitish or nearly white, thread-like, incurved, each terminating in a club-shaped stigma. *SW Western Australia.* G1. Summer.

Larger, more robust plants with 4–8 flowers and flowering stems 1.5–2.5 cm can be distinguished as subsp. **australis** Lowrie.

43. D. citrina Lowrie & Carlquist (*D. rechingeri* Strid; *D. chrysochila* Schlaver; *D.* 'Regan's Ford'). Illustration: Kondo & Kondo, Carnivorous plants of the world in colour, 64 (1983) – as 'D. pycnoblasta'; Lowrie, Carnivorous plants of Australia **2:** 155 & 157 (1989).

Short-lived perennial. Stem inconspicuous. Leaf-rosette to 1.5 cm across. Stipule-bud almost spherical, *c.* 5 × 5 mm, with a short apical tuft of hairs *c.* 0.5 mm, slightly shaggy. Leaf-blade shortly elliptic, *c.* 1.5 × 1.2 mm. Leaf-stalk *c.* 5 × 0.3–1 mm, margin with stalked glandular hairs. Flowering-stems 1 or 2, almost hairless, minutely glandular-hairy, densely so amongst the flowers, to 4 cm, with 12 flowers. Flower-stalks to 3 mm, erect in fruit. Sepals with sparse long-stalked glands. Petals 4–5 mm, bright lemon-yellow, basal half white. Styles 3, thread-like, white. *Western Australia.* G1. Summer.

44. D. miniata Diels. Illustration: Lowrie, Carnivorous plants of Australia **2:** 83 & 85 (1989).

Short-lived perennial. Stem inconspicuous. Leaf-rosette to 2 cm across. Stipule-bud barely ovoid, *c.* 5 × 3 mm, acute, silvery. Leaf-blade almost circular

to very broadly elliptic *c.* 1.7 × 1.5 mm, stalk *c.* 3.5 × 0.4 mm. Flowering-stem 4–5 cm, mostly glandular, with 5–12 flowers. Flower-stalk pendent in fruit, to 2 cm. Petals 9–10 mm, orange-red, black at base. Styles 3, thread-like, black. *SW Western Australia.* G1. Summer.

45. D. platystigma Lehmann. Illustration: Kondo & Kondo, Carnivorous plants of the world in colour, 61 (1983); Lowrie, Carnivorous plants of Australia **2:** 139 & 141 (1989).

Short-lived perennial. Stem inconspicuous. Leaf-rosette to 2 cm across. Stipule-bud ovoid, *c.* 5 × 3.5 mm. Leaf-blade shortly elliptic, *c.* 3.5 × 2.5 mm, stalk *c.* 6 × 0.8–1 mm. Flowering-stems 1–2, almost hairless, with a few minute stalkless glandular hairs below, denser amongst the flowers, to 6 cm, with 3–6 flowers. Flower-stalk to 2.5 cm, erect in fruit. Sepals with long-stalked, glandular hairs. Petals 7–8 mm, orange-red, dark red at the base. Styles 3–5, white, thread-like; stigmas ovoid, peltate, black. *SW Western Australia.* G1. Summer.

46. D. hyperostigma Marchant & Lowrie. (*D. platystigma* O'Brien). Illustration: Lowrie, Carnivorous plants of Australia **2:** 67 & 69 (1989).

Short-lived perennial. Stem inconspicuous, to 5 mm. Leaf-rosette to 2.5 cm. Stipule-bud poorly defined and loose, roughly ovoid, *c.* 5 × 5 mm, shaggy. Leaf-blade almost circular to shortly elliptic, *c.* 2.5 × 2 mm, stalk *c.* 4 × 0.8 mm. Flowering-stem to 5 mm, with a few short stout glandular hairs, almost hairless, with 6–8 flowers. Flower-stalks 1–1.5 mm, erect in fruit. Petals 6–7 mm, orange-red with a black basal spot. Styles 3, thread-like, each with an obovoid stigma, black. *SW Western Australia.* G1. Summer.

47. D. sewelliae Diels. Illustration: Kondo & Kondo, Carnivorous plants of the world in colour, 72 (1983); Lowrie, Carnivorous plants of Australia **2:** 167 & 169 (1989).

Short-lived perennial. Stem inconspicuous. Leaf-rosette to 2.5 cm. Stipule-bud ovoid, *c.* 5 × 4 mm, shaggy. Leaf-blade shortly elliptic, *c.* 2.2 × 2 mm, stalk 7–8 × 0.7–0.9 mm, with minute glands. Flowering-stem to 6 cm, minutely glandular, with long-stalked glands amongst the flowers, with *c.* 6 flowers. Flower-stalks *c.* 2 mm, erect in fruit. Sepals with long-stalked glandular hairs. Petals 8–9 mm, orange-red, dark red and black at the base. Styles 5, thread-like, terete, black;

stigmas spherical, black. *SW Western Australia.* G1. Summer.

48. D. callistos Marchant & Lowrie (*D.* 'Brookton'). Illustration: Lowrie, Carnivorous plants of Australia **2:** 31 & 33 (1989).

Short-lived perennial. Stem inconspicuous or to 0.5 cm. Leaf-rosette 1–3 cm across. Stipule-bud broadly ovoid, shaggy, *c.* 4 × 4 mm. Leaf-blade broadly elliptic, *c.* 2.5 × 2 mm, stalk 5–6 × 0.6–1 mm. Inflorescence terminal, to 7 cm, glandular, with 6–12 flowers. Flower-stalks 1–2 mm, erect in fruit. Sepals with stalked glands; petals 8–9 mm, orange, black at base. Styles 3, thread-like, black. *SW Western Australia.* G1. Summer.

49. D. leucoblasta Bentham. Illustration: Kondo & Kondo, Carnivorous plants of the world in colour, 50 (1983) – as 'D. miniata'; Lowrie, Carnivorous plants of Australia **2:** 75 & 77 (1989).

Short-lived perennial. Stem inconspicuous. Leaf-rosette *c.* 2 cm across. Stipule-bud ovoid *c.* 7 × 4 mm, smooth, silvery. Leaf-blade circular, *c.* 2.2 mm, stalk *c.* 5 × 0.7 mm. Flowering-stem to 12 cm, with a few short-stalked or stalkless glands most dense at the top, with 6–9 flowers. Flower-stalks *c.* 1.5 mm, erect in fruit. Petals 9–10 mm, orange, dark red (but not black) at base. Styles 3, more or less thread-like, but broadest at centre, reddish with yellowish tips. *SW Western Australia.* G1. Summer.

50. D. echinoblastus Marchant & Lowrie. (*D.* 'Camallo') Illustration: Lowrie, Carnivorous plants of Australia **2:** 47 & 49 (1989).

Short-lived perennial. Stem inconspicuous, to 7 mm. Leaves forming a rosette to 1.5 cm across. Stipule-bud narrowly ovoid, *c.* 7 × 3 mm, bristly. Leaf-blade circular, 2–2.5 mm, stalk *c.* 5 × 0.8 mm. Flowering-stem to 12 cm, sparsely glandular with 9–12 flowers. Flower-stalks *c.* 2 mm, erect in fruit. Petals 8–9 mm, orange-red, without a darker or black spot at the base. Styles 3, thread-like, yellowish green. *SW Western Australia.* G1. Summer.

51. D. nitidula Planchon (*D. omissa* Diels). Illustration: Slack, Carnivorous Plants, 139 (1979); Kondo & Kondo, Carnivorous plants of the world in colour, 54 (1983); Lowrie, Carnivorous plants of Australia **2:** 87 & 89 (1989).

Short-lived perennial. Stem inconspicuous. Leaf-rosette 1.5–2.5 cm across. Stipule-bud loose, poorly defined, ovoid 3–5 × 3–5 mm, very hairy. Leaf-blade almost circular to obovate, 1.5–3 × 1.5–2 mm, base decurrent, glands extending to leaf-stalk which is 5.5–8 × 0.5–1 mm. Flowering-stems 1–2, 2.5–8 cm, covered with minute glands, with 4–20 flowers. Flower-stalks erect in fruit, sometimes pendent, 3.5–6 mm, rarely 10 mm. Petals 2.5–3.5 mm, white, sometimes shallowly notched at the tip. Styles 3 or 4, white, thread-like, each with a purple-red (rarely white) peltate, kidney-shaped (sometimes 3-lobed) stigma. *SW Western Australia.* G1. Summer.

52. D. mannii Cheek (*D.* 'Bannister'; *D. manniana* Marchant, invalid). Illustration: Lowrie, Carnivorous plants of Australia **2**: 79 & 81 (1989); Phytologia **68**(2): 87 (1990).

Short-lived perennial. Stem inconspicuous. Leaf-rosette to 1.5 cm across. Stipule-bud usually poorly defined and loose, ovoid *c.* 3 × 3 mm. Leaf-blade shortly elliptic *c.* 2.5 × 2 mm, stalk 3.5–5 × 0.5–1 mm. Flowering-stem 6–18.5 cm, hairless except for few glandular hairs where flowers are borne, flowers 10–24. Flower-stalks to 2.5 mm, pendent in fruit, glandular. Petals pale pink, 5–8 mm, slightly notched at tip. Styles 5, thread-like, white, each with a translucent peltate slipper-shaped stigma. *SW Western Australia.* G1. Summer.

53. D. ericksoniae Marchant & Lowrie (*D. omissa* Erickson not Diels). Illustration: Lowrie, Carnivorous plants of Australia **2**: 59 & 61 (1989).

Short-lived perennial. Stems inconspicuous. Leaves forming a rosette to 3 cm across. Stipule-bud loose, not well defined, ovoid, *c.* 3 × 3.5 mm. Leaf-blade obovate *c.* 4 × 3 mm, stalk *c.* 6 × 1 mm. Flowering stem 1–4.5 cm, with short-stalked glands, with up to 7 flowers. Flower-stalks *c.* 3 mm, erect in fruit. Petals 4 or 5, light to dark pink, palest at the base, tip truncate. Styles 3 or 4, white, thread-like, each stigma bun-shaped, peltate, white. *SW Western Australia.* G1. Summer.

54. D. × badgerupii Cheek (*D.* 'Lake Badgerup'). Illustration: Lowrie, Carnivorous plants of Australia **2**: 107 & 109 (1989).

Short-lived perennial. Stem inconspicuous. Leaf-rosette *c.* 1.5 cm across. Stipule-bud broadly ovoid *c.* 3 × 4 mm, shaggy. Leaf-blade circular *c.* 1.3 × 1.7 mm, stalk 4–5 × 0.3–0.4 mm. Flowering-stems 1 or 2, minutely glandular to hairless, *c.* 2 cm, usually with a single flower. Flower-stalk erect in fruit. Petals white, 2.8–3 mm. Styles 4, purple, each thread-like, gradually swelling into 4 club-like stigmas. *SW Western Australia.* G1. Summer.

55. D. androsacea Diels. Illustration: Kondo & Kondo, Carnivorous plants of the world in colour, 14 (1983); Lowrie, Carnivorous plants of Australia **2**: 23 & 25 (1989).

Short-lived perennial. Stem inconspicuous. Leaves forming a rosette to 2 cm across. Stipule-bud distinctively 5-angled, acute, cone-shaped *c.* 5 × 3 mm, greyish green. Leaf-blade almost circular, *c.* 2.5 × 2.7 mm, stalk 4–4.5 × 0.7–0.9 mm. Flowering-stem single, 4–5 cm, sparsely glandular, flowers 4–7. Flower-stalks 2.5–3 mm, pendent in fruit. Petals 4.5–5 mm, pink or white. Styles 5, thread-like, white. *SW Western Australia.* G1. Summer.

56. D. leioblastus Marchant & Lowrie (*D.* 'Steve's paleacea'). Illustration: Lowrie, Carnivorous plants of Australia **2**: 71 & 73 (1989).

Short-lived perennial. Stem inconspicuous. Leaf-rosette to 1.5 cm across. Stipule-bud ovate, *c.* 6 × 4 mm, tip acute, smooth. Leaf-blade circular, 1.5–1.7 mm, stalk *c.* 4.5 × 0.6 mm. Flowering-stem to 2 cm, densely covered in long-stalked glands, with up to 20 crowded flowers. Flower-stalks 7–8 mm, erect in fruit. Petals white, 1.5–2 mm. Styles 3, incurved, thread-like but slightly broader at the middle, white. *SW Western Australia.* G1. Summer.

57. D. oreopodion Marchant & Lowrie (*D.* 'Armadale'). Illustration: Lowrie, Carnivorous plants of Australia **2**: 123 & 125 (1989).

Short-lived perennial. Stem inconspicuous. Leaf-rosette *c.* 1.5 cm across. Stipule-bud rather narrowly ovoid, *c.* 6 × 3 mm, tip acute. Leaf-blade almost circular 1.5–1.7 mm, stalk 3.5–4 × 0.4–0.8 mm. Flowering-stem to 3.5 cm, sparsely covered with minute glands (almost hairless), with 8–12 flowers. Flower-stalk to 2 mm, erect in fruit. Petals 3.5–4 mm, white with red veins. Styles 3, thread-like, slightly broader at centre, white. *SW Western Australia* G1. Summer.

58. D. paleacea de Candolle. Illustration: Kondo & Kondo, Carnivorous plants of the world in colour, 55 (1983); Lowrie, Carnivorous plants of Australia **2**: 127, 129 (1989).

Short-lived perennial. Stem inconspicuous, always less than 1 cm and densely clothed in leaves. Leaf-rosette to 1.5 cm across. Stipule-bud ovoid 4–5 × 3–3.5 mm, densely bristly hairy. Leaf-blade 1.5–2 mm, almost circular, stalk 3–5 × 0.3–0.5 mm. Flowering-stem 1–4, to 3 cm, almost hairless, with 30 or more flowers densely crowded on the upper part. Flower-stalk to 1.5 mm, erect in fruit. Sepals hairless. Petals 2–2.5 mm, white. Styles 3, greenish white, tightly incurved, thread-like, very short, *c.* 0.5 mm. *SW Western Australia.* G1. Summer.

59. D. trichocaulis Diels. Illustration: Lowrie, Carnivorous plants of Australia **2**: 131, 133 (1989) – as D. paleacea subsp. trichocaulis.

Short-lived perennial. Stem inconspicuous. Leaf-rosette to 1.3 cm across. Stipule-bud ovoid *c.* 4 × 3.5 mm with a 1 mm apical brush. Leaf-blade almost circular, to 1.5 mm, stalk *c.* 3.5 × 0.35–0.5 mm. Flowering-stem to 3.5 cm, densely white-woolly, with 10–20 flowers crowded in upper part. Flower-stalks to 1 mm, erect in fruit. Sepals with a mixture of long-stalked glandular and white woolly hairs. Petals to 3 mm, white. Styles 3, thread-like, incurved, greenish white. *SW Western Australia.* G1. Summer.

60. D. parvula Planchon. Illustration: Lowrie, Carnivorous plants of Australia **2**: 135 & 137 (1989).

Short-lived perennial. Stem conspicuous to 1.5 cm. Leaf-rosette to 1.5 cm across. Stipule-bud ovoid *c.* 5 × 3 mm, with an apical narrow tuft of hairs *c.* 4 mm. Leaf-blade elliptic *c.* 1.2 × 0.8 mm, stalk *c.* 4 × 0.7 mm. Flowering-stems 2–4, to 5 cm, more or less hairless, with 10–12 flowers. Flower-stalks to 1 mm, erect in fruit. Sepals hairless except for a few stalked glandular hairs. Petals 3–3.2 mm, white, the basal tip yellow with a red spot above. Styles 3 or 4, rarely 5, thread-like, the tips gradually & slightly swollen, white. *SW Western Australia.* G1. Summer.

61. D. pulchella Lehmann. Illustration: Slack, Carnivorous Plants, 140 (1979); Kondo & Kondo, Carnivorous plants of the world in colour, 63 (1983); Lowrie, Carnivorous plants of Australia **2**: 143 & 145 (1989).

Short-lived perennial. Stem inconspicuous. Leaf-rosette up to 3 cm across. Stipule-bud poorly developed, loose and open, narrowly ovoid, *c.* 6 × 2.5 cm. Leaf-blade more or less circular, 4–4.5

mm, stalk 8.5–9 × 1–2 mm. Flowering-stems 1–4, glandular hairy, 3–4 cm, with 3–5 flowers. Flower-stalks *c.* 2.5 mm, erect in fruit. Sepals with long-stalked glandular hairs. Petals usually pink, sometimes white, white with a red base or orange-red with a dark red or black base. Styles 5, white, thread-like, the tips conspicuously hooked sideways. *SW Western Australia.* G1. Summer.

62. D. walyunga Marchant & Lowrie. Illustration: Lowrie, Carnivorous plants of Australia **2**: 175 & 177 (1989).

Short-lived perennial. Stem inconspicuous. Leaf-rosette 1.5–2 cm across. Stipule-bud compact, ovoid *c.* 5 × 4 mm, tip acute, more or less smooth. Leaf-blade shortly elliptic, *c.* 3.2 × 2.5 mm, stalk 4–5 × 0.5–0.9 mm. Flowering-stems 1 or 2, densely covered in minute stalkless glands, to 9 cm, with about 12 flowers. Flower-stalks *c.* 2.5 mm, pendent in fruit. Petals 8–9 mm, white to pink, waisted. Styles 3, thread-like, horizontal at the base, the tip vertical, as the horns of a bull, white. *SW Western Australia.* G1. Summer.

63. D. spilos Marchant & Lowrie (*D.* 'Muchea'). Illustration: Lowrie, Carnivorous plants of Australia **2**: 171 & 173 (1989).

Short-lived perennial. Stem inconspicuous. Leaf-rosette *c.* 1.5 cm across. Stipule-bud 5 × 3 mm, conical to narrowly ovoid, with long stiff hairs. Leaf-blade shortly elliptic, *c.* 2 × 1.5 mm, stalk 5.5–6 × 0.5–0.6 mm. Flowering-stems 1 or 2, minutely glandular or almost hairless, to 7 cm, with up to 15 flowers. Flower-stalks to 2 mm, pendent in fruit. Petals *c.* 7 mm, white with a pink basal spot. Styles 3, thread-like, straight, horizontal, white. *SW Western Australia.* G1. Summer.

64. D. pycnoblasta Diels. Illustration: Lowrie, Carnivorous plants of Australia **2**: 147 & 149 (1989).

Short-lived perennial. Stem inconspicuous. Leaf-rosette to 1.5 cm across. Stipule-bud very broadly ovoid or globular, *c.* 4 × 5 mm, tip flattened, completely hairless, glossy. Leaf-blade circular, *c.* 2 mm, stalk 5–6 × 0.5–1.3 mm. Flowering-stems 1 or 2, sparsely glandular below, densely so above, to 9 cm, with *c.* 8 flowers. Flower-stalks *c.* 2.5 mm, more or less erect in fruit. Petals 4–5 mm, white (rarely pink) with a pink blotch at the base. Styles 3, thread-like, white. *SW Western Australia.* G1. Summer.

65. D. eneabba Marchant & Lowrie. Illustration: Lowrie, Carnivorous plants of Australia **2**: 51 & 53 (1989).

Short-lived perennial. Stem inconspicuous. Leaves forming a rosette up to 2 cm across. Stipule-bud ovoid, 5 × 3 mm, shaggy. Leaf-blade shortly elliptic, *c.* 2.3 × *c.* 1.8 mm, stalk 4.5 × 0.5 mm. Flowering-stem to 8 cm, very sparsely glandular to hairless, with 10–15 closely spaced flowers. Flower-stalks *c.* 4 mm, mostly erect in fruit. Petals 2–3 mm, pure white to pink, but always with a red spot at the base, tip deeply notched. Styles 3, thread-like, white. *SW Western Australia.* G1. Summer.

66. D. closterostigma Marchant & Lowrie. (*D.* 'Cataby'). Illustration: Lowrie, Carnivorous plants of Australia **2**: 35 & 37 (1989).

Short-lived perennial. Stem inconspicuous, or to 0.5 cm. Leaves forming a rosette to 1.8 cm across. Stipule-bud *c.* 5 × 4 mm, ovoid, bristly. Leaf-blade *c.* 2 × 1.7 mm, almost circular to broadly elliptic, stalk *c.* 4.5 × 0.8 mm. Flowering-stems 1–2, to 6 cm, sparsely glandular, with 6–12 flowers. Flower-stalks *c.* 2 mm, erect in fruit. Petals 5–6 mm, white or light pink with a red spot at the base. Styles thread-like, reddish at the base. *SW Western Australia.* G1. Summer.

67. D. helodes Marchant & Lowrie (*D.* 'Bullsbrook'). Illustration: Lowrie, Carnivorous plants of Australia **2**: 63 & 65 (1989).

Short-lived perennial. Stem inconspicuous. Leaves forming a rosette to 1.3 cm across. Stipule-bud ovoid *c.* 7 × 4 mm, smooth. Leaf-blade circular, 1.6–1.7 mm, stalk *c.* 4 × 0.5 mm. Flowering-stem to 8 cm, with 6–14 flowers, almost hairless, except for a few short-stalked glands. Flower-stalks to 2 mm, erect in fruit. Petals 6–7 mm, white to pink with a red spot at the base. Styles 3, greenish white, thread-like, widest at the base. *SW Western Australia.* G1. Summer.

68. D. binata Labillardière (*D. dichotoma* Smith; *D. pedata* Persoon). Illustration: Slack, Carnivorous Plants, 123, 149–153 (1979); Kondo & Kondo, Carnivorous plants of the world in colour, 20 (1983).

Perennial 8–60 cm. Roots resemble brittle black bootlaces, *c.* 3 mm thick. Leaves produced all year round or dying down to a resting bud in winter. Stem very short, barely detectable. Leaves 3–9, erect, glossy, 5–32 cm, blade flattened,

linear, 2–3 mm across, 5–17 cm, forked dichotomously 1–5 times (see cultivars) from the stalk, producing 2–32 points, stalks flattened, 7–15 cm × 1–3 mm, hairless. Flowering-stem taller than the leaves, to 60 cm, hairless, upper quarter bearing 2 or 3 spike-like branches with 15–30 flowers. Flower-stalks 5–10 mm, erect in fruit. Petals 8–10 mm, white. Styles 3, forked at the base, then once to twice again above, swollen at the tip. *Eastern Australia & New Zealand.* G1. Summer–early autumn.

'Dichotoma Giant' has leaves dying down in winter, 30–40 cm, erect, stout, yellowish green, forking twice. 'Multifida' has leaves not dying down in winter, to 20 cm, drooping, purplish red, forking 3 times. 'Multifida Pink', is as 'Multifida' but with pink-tinged flowers. 'Multifida Extrema', is as 'Multifida' but leaves forked 4 or 5 times.

69. D. regia Stephens. Illustration: Kondo & Kondo, Carnivorous plants of the world in colour, 67 (1983).

Perennial, lacking a resting bud or tubers. Roots black and brittle, 3 mm or more thick. Stem woody, rhizomatous, creeping, *c.* 5 mm thick. Leaves linear, 12–40 × 0.8–1.2 cm, clasping the stem, then tapering gradually to the pointed tip, midrib broad, occupying a third of the leaf width, deeply sunken above, prominent below, upper surface with stalked glandular hairs, except the midrib and the basal 1–2 cm, lower surface hairless, stalk indistinct. Stipules absent. Flowering-stem to 40 cm, hairless. Petals *c.* 2.5 cm, rose-coloured. Styles undivided. Seeds fusiform. *South Africa (Cape Province).* G1. Flowering unknown in cultivation.

70. D. indica Linnaeus. Illustration: Kondo & Kondo, Carnivorous plants of the world in colour, 40 (1983).

Annual. Stem to 30 cm, slender, erect, at length prostrate, never rooting. Leaves borne on stem at internodes of 1–2 cm, thread-like, 2–10 × 0.05–0.3 cm, covered in colourless stalked glandular hairs apart from the base, stalk indistinct. Stipules absent. Flowering-stems axillary, spike-like, with 3–20 flowers on the upper side. Petals 6–8 mm, pink or mauve, but sometimes white or red. Styles divided in two from the base, dilated at tip. *Africa, Asia & Australia.* G2. Late summer.

71. D. arcturi J.D. Hooker. Illustration: Kondo & Kondo, Carnivorous plants of the world in colour, 16 (1983).

Perennial *c.* 7 mm. Stem subterranean, stout, clothed with dead leaves, to *c.* 2 × 0.2–0.3 cm. Leaves 2 or 3, rarely more, almost in 2 ranks, ascending, strap-shaped, 3–4.5 × 0.3–0.6 cm, the lower third without glands, slightly narrower. Flowering-stem 2–7 cm, hairless; flower solitary. Petals 4–7 mm, white. Styles 3, unbranched, stout. Stigmas swollen. *SE Australia & New Zealand*. H1. Flowering unknown in Europe.

72. D. prolifera C. White. Illustration: Kondo & Kondo, *Carnivorous plants of the world in colour*, 62 (1983).

Perennial. Stem 1–3 cm. Roots rarely producing suckers. Leaves in basal rosettes, membranous, rigid. Leaf-blade rounded, 1–2 × 1–2 cm, tip rounded, base cordate, red glandular hairy above, hairless below, stalk 2–5 × 0.2–0.4 cm, slightly winged, hairless. Stipules inconspicuous. Flowering-stem drooping to prostrate, 7–12 cm, hairless, with 3–5 flowers. Petals 2–3 mm, red. Styles 3, forked once. Plantlets 1 or 2, produced from the axils of the inflorescence bracts. *Australia (Queensland)*. G2. Flowering all year round.

73. D. adelae F. von Mueller. Illustration: Kondo & Kondo, *Carnivorous plants of the world in colour*, 12 (1983).

Perennial. Stem 1–3 cm. Roots producing suckers at side of pots. Leaves 5–10 × 0.8–2 cm, in a dense basal rosette, narrowly lanceolate, membranous, limp, tip pointed, base wedge-shaped, tapering into leaf-stalk, upper surface with red-stalked glandular hairs, lower surface with fine white hairs, particularly noticeable when in bud; leaf-stalk 0.5–1.5 cm. Stipules 4–6 mm, much divided. Flowering-stem 8–20 cm, hairy, spirally coiled, single-sided, bearing 8–35 flowers. Flower-stalks 5–10 mm. Petals *c.* 3 mm, bright red or greenish. Styles forked once. Seeds almost spherical. *Australia (Queensland)*. G2. Flowering all year round.

74. D. schizandra Diels. Illustration: Slack, *Carnivorous Plants*, 147 (1979); Kondo & Kondo, *Carnivorous plants of the world in colour*, 70 (1983).

Perennial. Stem 1–3 cm. Roots producing suckers at side of pots. Leaves in dense basal rosette, membranous, rigid, obovate, 7–10 × 4–5 cm, tip notched or truncate, base wedge-shaped, stalked, red glandular hairs above, more or less hairless below apart from the sparsely hairy thickened midrib and veins, stalk not distinct. Stipules inconspicuous. Flowering-stem 8–10 cm, hairy, single-sided, bearing 10–25 flowers. Flowers with petals *c.* 6 mm, red. Styles 3, forked twice in the upper half. *Australia (Queensland)*. G2. Flowering all year round.

75. D. burmannii Vahl. Illustration: Kondo & Kondo, *Carnivorous plants of the world in colour*, 24 (1983).

Annual or short-lived perennial. Stem very short. Leaves very pale green, with reddish flush, in a dense basal rosette, narrowly obovate, 1–1.5 × 0.6–0.8 cm, tip truncate, base wedge-shaped, centre conspicuously concave above, colourless glandular hairs above, margin with non-mucilaginous, fast-moving glandular hairs above, hairless below, stalk undifferentiated. Stipules large, whitish, 4–5 mm, margin fringed. Flowering-stem 3–15 cm, nearly hairless. Flowers 3–15. Petals 2–3 mm, pink. Styles 5, flattened at the tip with 4 or 5 finger-like projections. *India to Japan & N Australia*. G2. Summer–early autumn.

76. D. sessilifolia Saint-Hilaire.

Annual or short-lived perennial. Stem very short. Leaves in dense basal rosette, obovate to oblong, 1–2 × 0.5–0.6 cm, tip truncate to rounded with long non-mucilaginous hairs, base only very slightly tapered, slightly wedge-shaped, upper surface with red glandular hairs, lower surface hairless, stalk not distinct. Stipules 2–3 mm. Flowering-stem 15–25 cm, hairless, the upper ⅕–¹⁄₁₀ fertile, with 4–12 flowers. Sepals with stout, straight, red glandular hairs. Petals 5–7 mm, pink to purple. Styles 5, each forked twice in the upper ¹⁄₁₀. *S America (Venezuela to Brazil)*. G2. Flowering unknown in cultivation.

77. D. graminifolia Saint-Hilaire.

Perennial, possibly with a resting bud. Stem very short. Leaves 4–12, thread-like and erect, 10–21 × 0.1–0.2 cm, the lower 1–2 cm lacking glands, glandular hairs red. Stipules and leaf-bases not hairy, stipules ovate, undivided, *c.* 1.5 × 1 cm, chaffy and brown. Flowering-stem 15–25 cm, densely hairy, the upper ⅙ fertile. Flowers 5–12. Petals purple or rose, 1–1.5 cm. Styles 3, forked once at the base. *E Brazil*. G1. Spring–autumn.

78. D. filiformis Rafinesque (*D. filiformis* var. *typica* Wayne). Illustration: Slack, *Carnivorous Plants*, 125, 130 (1979); Kondo & Kondo, *Carnivorous plants of the world in colour*, 32 (right) (1983).

Perennial, wintering as a resting bud. Stem very short. Leaves 4–12, thread-like, erect, 12–21 × 0.5–0.6 cm, the lower 1–2 cm lacking glands, glandular hairs red. Stipules and leaf-bases concealed by brown hairs. Flowering-stem 16–35 cm, hairless, the upper ⅙ fertile. Flowers 4–14, scentless. Petals *c.* 7 mm, pink. Styles 3, forked once at the base. Seeds obovoid. *E USA*. H4–5. Summer.

79. D. × californica Cheek (*D.* 'California Sunset').

Intermediate between parents in dimensions. Leaf-glands faintly red. More vigorous than both parents. *Artificial hybrid*. H4–5. Summer.

More common in cultivation than either parent. A hybrid between *D. tracyi* and *D. filiformis*.

80. D. tracyi Macfarlane (*D. filiformis* var. *tracyi* Diels).

Perennial, wintering as a resting bud. Stem very short. Leaves 6–12, thread-like and erect, 24–45 × 0.1–0.15 cm, the lower 2–4 cm lacking glands, glandular hairs entirely lacking red pigment. Stipules and leaf-bases densely brown hairy. Flowering-stem 33–60 cm, hairless, the upper quarter fertile. Flowers 12–23, scentless. Petals pink, *c.* 1.5 cm. Styles as *D. filiformis*. Seeds obovoid-ellipsoid. *SE USA*. G1–H5. Summer.

81. D. rotundifolia Linnaeus. Illustration: Slack, *Carnivorous Plants*, 120, 126, 127 (1979); Kondo & Kondo, *Carnivorous plants of the world in colour*, 69 (1983).

Perennial to 7 cm across, overwintering as a resting bud. Aerial stem absent. Leaves 4–7, held horizontally, forming a basal rosette, blade circular, broader than long, 2–6 × 3.5–7 mm, stalk slightly flattened, 0.6–2.5 × 0.1–0.2 cm. Flowering-stem a single erect spike 6–12 cm, bearing 3–7 flowers. Flower-stalks *c.* 3 mm, erect in fruit. Petals 2–3 mm, white. Styles 3, forked at the base, the stigmatic tips slightly club-shaped. *N temperate zone, extending S to New Guinea*. H1. Late summer.

82. D. × obovata Mertens & Koch.

As *D. anglica*, but leaf-blade obovate, 0.9–1.7 × 0.6 cm. *N. temperate Eurasia & America*. H1. Summer.

A hybrid between *D. rotundifolia* and *D. anglica*.

83. D. anglica Hudson. Illustration: Slack, *Carnivorous Plants*, 128, 129, 217 (1979); Kondo & Kondo, *Carnivorous plants of the world in colour*, 15 (1983).

Perennial 4–8 cm across, overwintering

as a resting bud. Aerial stem almost absent. Stipules almost united with leaf-bases for 3–4 mm, with numerous reddish chaffy teeth 3–4 mm from the margin to the tip. Leaves 6–15, erect, blade rectangular to narrowly ellipsoid, 1.5–3 × 0.2–0.3 cm, base sometimes tapering slightly, stalk 30–60 × 0.05–0.1 mm, hairless. Flowering-stem held above the leaves, 8–18 cm, with 1–10 flowers. Flower-stalks 1–2 mm, erect in fruit. Petals 2–3 mm, white. Styles 3, forked at the base, the stigmatic apices slightly club-shaped. *N temperate zone of Eurasia & America*. H1. Summer.

84. D. intermedia Hayne (*D. longifolia* Linnaeus, in part). Illustration: Kondo & Kondo, Carnivorous plants of the world in colour, 41 (1983).

Perennial 3–5 cm across, overwintering as a resting bud. Aerial stem almost absent. Stipules reduced to 3–7 hairs, each 1–1.5 mm, emerging 2 mm from the base of the leaf-stalk. Leaves 5–15, ascending, blade obovoid, 4–6 × 2–3 mm, stalk 2–3 × 0.5–0.6 mm. Flowering-stem with flowers held at leaf-level or just above, 2–8 cm, with 3–10 flowers. Flower-stalks 0.5–1 mm, erect in fruit. Petals 2–3 mm, white. Styles 3, forked at the base, the stigmatic apices slightly club-shaped. *N temperate Eurasia & America*. H1. Summer.

85. D. capensis Linnaeus. Illustration: Slack, Carnivorous Plants, 136, 137 (1979).

Perennial, 10–17 × 7–15 cm. Stem slowly formed, erect, to 6 cm, robust, reddish, hairless. Leaves linear, 7–14 × 0.5–0.6 cm, the lower 2–5 cm lacking glands. Stipules brown, chaffy, rectangular, *c.* 10 × 5 cm. Flowering-stem 15–25 cm, hairy, with 6–15 flowers on the upper ⅓. Petals *c.* 9 × 5 mm. Styles 3, forked at the base, the stigmatic apices with a spherical swelling. *South Africa*. G1. Autumn–winter.

Variants with narrow (2–3 mm) and broad (6–8 mm) leaf-stalks have been offered for sale respectively as 'narrow-leaf' and 'broad-leaf' cultivars. A cultivar entirely lacking red pigment in petals and glands has been discovered recently and has been distributed as 'albino form' or 'alba', neither of which are valid cultivar names.

86. D. hilaris Chamisso & Schlechtendal. Illustration: Kondo & Kondo, Carnivorous plants of the world in colour, 38 (1983).

Perennial 8–15 × 8–9 cm. Stem slowly formed, erect to 15 cm, robust, with densely felted hairs. Leaves strap-shaped, 4–7 × 0.5–0.9 cm, the lower 1.7–3.2 cm lacking glands, stalked. Non-glandular parts covered with thick matted hairs. Stipules brown and chaffy, highly divided into numerous hair-like teeth, 3–6 mm. Flowering-stems almost terminal, 14–30 cm, softly hairy, with 8–10 flowers from the upper ¼. Flower-stalks 4–7 mm, erect in fruit. Petals pale pink. Styles 3, spreading, forked at the base then unbranched and swollen at the stigmatic apices or terminating in 2 short branches *c.* 0.5 mm. *South Africa (Cape Province)*. G1. Summer.

Mature plants are very rarely seen in Europe, though seed is widely available.

87. D. glabripes (Planchon) Stein (*D. ramentacea* de Candolle var. *glabripes* Planchon; *D. glabripes* (Planchon) Salter). Illustration: Kondo & Kondo, Carnivorous plants of the world in colour, 31 (1983).

Perennial 5–35 cm, decumbent, only the upper part erect and aerial. Stipules reddish, 7.5–10 mm, divided to within 1.5 mm of the base into 5–7 hair-like segments. Leaves 5–10, at first erect, slowly reflexing with age, blade narrowly obovoid, *c.* 10 × 4.5 mm, tip rounded, base tapering into the stalk, with adpressed, slightly wavy white hairs *c.* 1 mm beneath, stalk cylindric, 9–25 × 0.5–0.6 mm, hairless or almost downy below. Flowering-stem lateral, 6–19 cm, with glandular hairs above, becoming almost hairless at the base, with 4–10 flowers. Flower-stalks to 7 mm, erect in fruit. Petals *c.* 1 cm, reddish purple. Styles 3, forked at the base, stigmatic apices shortly fan-shaped, multifid. *South Africa (Cape Province)*. G1. Summer.

88. D. collinsiae Burtt Davy.
Perennial 4.5–9 cm across. Aerial stem usually absent, at least in young specimens, rarely 1–2 cm. Stipules united at the base to the leaf-stalk, above divided into numerous hairs *c.* 2 mm. Leaves *c.* 10, more or less erect, blade obovate 0.7–1.2 × 0.45–0.9 cm, tip rounded, base tapering into the stalk, hairless below, stalk 2.5–4.2 × 0.1–0.2 cm, hairless. Flowering-stems 1–2, with a few glandular hairs amongst the flowers, hairless below, 6–13 cm, with 3–6 flowers. Flower-stalks 5–10 mm, erect in fruit. Petals *c.* 7 mm, purple. Styles 3, forked at the base, stigmatic tip spoon-shaped. *South Africa (Transvaal)*. G1. Summer.

Very rarely cultivated, though listed in catalogues. Exell & Laundon, Bulletin Societé Broteriana séries 2, **30**: 219 (1956)

believe that this species may well be a hybrid between *D. burkeana* & *D. madagascariensis*. The type however bears abundant seed and interspecific hybrids in this genus are usually sterile. Much South African material distributed as *D. madagascariensis* is *D. collinsiae*.

89. D. madagascariensis de Candolle. Illustration: Kondo & Kondo, Carnivorous plants of the world in colour, 46 (1983).

Perennial 5–10 × 3.75–5 cm. Stem to 20 cm, decumbent, only the upper part erect and aerial. Stipules 2–4.5 × 1.5–1.6 mm, coppery red, divided by slightly more than half into *c.* 10 hair like segments. Leaves 8–10, radiating, blade obovoid, 7–10 × 3–4 mm, tip rounded, base tapering into the stalk, with scattered reddish hairs when young, stalk 1.4–2.3 cm, cylindric, slightly hairy below when young. Flowering-stem lateral, 11.5–38 cm, with a few scattered stalked glandular hairs above, otherwise hairless, with 2–10 flowers on the upper ¹⁄10. Flower-stalks 1.5–3.5 mm, erect in fruit. Petals 5–8 mm, pink or purple. Styles 3, forked at the base, branches often conspicuously forked at the tip. *Madagascar, subtropical & tropical Africa*. G1. Late summer.

Rarely seen in cultivation.

90. D. hamiltonii C.R.P. Andrews. Illustration: Slack, Carnivorous Plants, 142 (1979); Kondo & Kondo, Carnivorous plants of the world in colour, 36 (1983); Lowrie, Carnivorous plants of Australia **2**: 183 & 185 (1989).

Perennial 3–6 cm across. Roots, *c.* 1 mm thick, brittle and black, readily producing new plants. Aerial stem absent. Stipules 2–3 mm, inconspicuous. Leaves 10–20, held horizontally in a dense rosette; blade 2–3 × 0.5–1 cm, olive-green, oblanceolate to strap-shaped, tip rounded, base wedge-shaped, stalk not distinct. Flowering-stem single, lateral, 30–40 cm, densely covered with red stalkless glands, with 20–30 flowers at the tip. Flower-stalks 3–5 mm, erect in fruit. Petals *c.* 1.6 × 2.2 cm, dark pink. Styles 5, united along their length into an erect cylinder, 2–3 mm. Stigmas erect, ovate *c.* 0.8 mm, with forward pointing teeth, purple. *SW Western Australia* G1. Late summer, but rarely flowers in cultivation.

91. D. glanduligera Lehmann. Illustration: Kondo & Kondo, Carnivorous plants of the world in colour, 35 (1983); Lowrie, Carnivorous plants of Australia **2**: 179 & 181 (1989).

Annual 2–4 cm across, yellowish in full sun. Aerial stem absent. Stipules *c.* 2 mm, divided into *c.* 8 teeth almost to the base, inconspicuous. Leaves 5–9 in a flat basal rosette, blade very shortly obovate to more or less circular, broader than long, *c.* 5 × 7 mm, base decurrent, centre sunken, margin with long hairs lacking mucilage, stalk *c.* 8 × 1.5 mm, flattened, thickly covered with long, non-glandular hairs. Flowering-stems 2, rarely 1–4, from the centre of the leaf rosette, to 7 cm, densely covered with long, non-glandular, colourless hairs, with 5–15 flowers. Flower-stalks *c.* 1 cm, pendent in fruit. Sepals densely glandular-hairy. Petals 4–5 mm, orange-red. Styles 3, ascending, branching in the upper half into 4–7 hair-like branches of various lengths. *SW Western Australia & SE Australia.* G1. Summer.

92. D. slackii Cheek. Illustration: Kew Bulletin **45**(2): 377 (1990).

Perennial 4–7 cm across. Roots 1–2 mm thick, black and brittle. Aerial stem almost absent. Stipules, 5–8 mm, red, divided halfway into 6 or 7 teeth. Leaves 10–20, in a flat basal rosette, blade poorly distinguished from stalk, rounded to rectangular, *c.* 1.1–1.5 × 0.9–1.2 mm, stalk *c.* 1.5–2.1 × 0.8–1 cm, markedly constricted just below the blade, the remainder strongly winged, lower surface of the leaf with short, thick, red, adpressed non-glandular hairs. Flowering-stem solitary, lateral, 12–25 cm, densely covered with red stalkless glands, with 6–15 flowers. Flower-stalks 5.5–7.5 mm, erect in fruit. Petals 0.8–1.1 cm, pink. Styles 3, spreading, forked at the base, each arm with *c.* 3 rounded stigmatic lobes, *c.* 0.5 mm, at the tip. *South Africa (Cape Province).* G1. Summer.

93. D. cuneifolia Linnaeus filius. Illustration: Kew Bulletin **45**(2): 377 (1990).

Perennial 4–6 cm across. Aerial stem absent. Stipules colourless, 2–4 mm, divided by up to ¼ into 6–10 teeth. Leaves 10–20, in a flat basal rosette; blade barely distinct from stalk, almost rectangular, *c.* 1.5 × 0.8 cm, tapering gradually into the leaf-stalk which is *c.* 0.8 × 0.4 cm, narrowing gradually from blade to base, the underside of the leaf hairless or with a few fine silvery hairs. Flowering-stem 11–35 cm, lateral, densely covered with red stalkless glands, with 6–15 flowers. Flower-stalks 5–6 mm, erect in fruit. Petals *c.* 12 mm, pink. Styles 3,

spreading, forked at the base, each arm undivided, the stigmatic tips slightly swollen. *South Africa (Cape Province).* G1. Summer.

94. D. aliciae Hamet (*D. curviscapa* Salter). Illustration: Slack, Carnivorous Plants, 124, 134, 135 (1979); Kew Bulletin **45**(2): 377 (1990).

Perennial 2–5 cm across. Aerial stem almost absent. Stipules colourless, or pinkish, 2–7 mm, divided by about ¾ the length into *c.* 3 teeth. Leaves 10–20 in a flat basal rosette, blade barely distinguishable from stalk, rectangular to strap-shaped, *c.* 0.8 × 0.4 mm, narrowing gradually from blade to base, the underside of the leaf usually densely covered in fine, adpressed, silvery hairs. Flowering-stem solitary, lateral, to 35 cm, densely covered with red stalkless glands, with 6–15 flowers. Flower-stalks 2–5 mm, erect in fruit. Petals 0.7–1 cm, pink. Styles 3, spreading, forked at the base, each usually divided once or twice again in the upper ¼. *South Africa (Cape Province).* G1. Summer.

95. D. burkeana Planchon.

Annual or short-lived perennial 1.2–4.25 cm across. Aerial stem absent. Stipules *c.* 3.5 cm across, divided almost to the base into hair-like segments, reddish. Leaves 5–10 in a basal rosette, blade broadening abruptly from the stalk, almost circular, usually broader than long, 4–10 × 3.5–8 mm, tip rounded, base conspicuously concave to wedge-shaped, lower surface hairless, stalk of uniform width, 6–15 × 0.3–1 mm, hairless or very sparingly and finely long-hairy. Flowering-stem from the centre of the basal rosette, 7–18 cm, with stalked glandular hairs in the upper half, sometimes extending to the lower half, with 4–11 flowers in the upper ¼. Flower-stalks 1–2 mm, erect in fruit. Petals 5–7 mm, white, pink or purple. Styles 3, forked at the base, stigmatic apices spoon-shaped. *S & equatorial Africa, Madagascar.* G1. Summer.

Most material grown under this name is *D. spathulata* or other species.

96. D. dielsiana Exell & Laundon.

Annual or short-lived perennial 1.5–5 cm across. Aerial stem absent. Stipules to 2.5 mm, deeply toothed, 3-lobed almost to the base, red. Leaves 5–10 in a basal rosette, blade obovate-spathulate, 3–15 × 3–7 mm, tip rounded, base contracting gradually, stalk 5–35 × 1–1.5 mm, hairless to conspicuously hairy below.

Flowering-stem 5–12 cm, with scattered, stalked, glandular hairs on the upper half, with 3–9 flowers. Flower-stalks *c.* 2 mm, erect in fruit. Petals *c.* 7 mm. Styles 3, forked at the base, unbranched, but slightly spoon-shaped at the tip. *S Africa (Zimbabwe, Malawi, to Natal).* G1. Summer.

97. D. trinervia Sprengel.

Annual or short-lived perennial 1.8–4.5 cm across. Aerial stem absent. Stipules largely united, reduced to 2 lateral teeth to the base of the leaf-stalk. Leaves 5–15 in a basal rosette, blade not distinct from stalk, oblanceolate (rarely slightly spathulate), 5–17 × 2–6 mm, tip truncate to rounded, base tapering to 0.5–1 mm at the base, undersurface hairless to almost so, sometimes conspicuously 3-nerved. Flowering stems 1–4, from the centre of the rosette, 3–18 cm, thickly covered with stalked glandular hairs from base to tip, rarely becoming hairless, with 4–9 flowers. Flower-stalks 2–5 mm, erect in fruit. Petals white. Styles 3, forked at the base, each branch conspicuously forked in the upper half, and then again just below the stigmatic apices. *South Africa (Cape Province).* G1. Summer.

98. D. natalensis Diels. Illustration: Kondo & Kondo, Carnivorous plants of the world in colour, 51 (1983).

Perennial, 1.8–6 cm across. Aerial stem absent. Stipules 1.5–2.5 mm, divided into numerous bristle-like teeth, reddish. Leaves 5–15 in a basal rosette, spathulate, rarely wedge-shaped, blade usually distinct from the stalk, obovate to broadly elliptic, 3–10 × 3–8.5 mm, tip rounded, base gradually contracting slightly, stalk tapering gradually from the tip (*c.* 2 mm wide) to the base (0.5–1 mm wide); hairless to almost so, rarely hairy below, membranous. Flowering-stem 10.5–30 cm, the upper half with stalked glandular hairs, these rarely extending to the base, flowers 2–6, usually confined to the upper ¹⁄₁₀ of the stalk. Flower-stalk 4–6 mm, erect in fruit. Petals pink to purple, sometimes white. Styles 3, forked at the base, the branches each dividing again 1 or 2 times in the upper ½. *South Africa (Eastern Cape to Transvaal), Madagascar* G1. Summer.

99. D. capillaris Poiret. Illustration: Schnell, Carnivorous Plants of the United States & Canada, 68 (1976); Kondo & Kondo, Carnivorous plants of the world in colour, 25 (1983).

Perennial 2–8 cm across, lacking a winter resting bud. Aerial stem absent.

Stipules 2–4 mm, stiff, divided into numerous teeth, red. Leaves 5–20 in a basal rosette, blade obovate, 4–10 × 3–5 mm, tip rounded, base wedge-shaped, stalk slightly flattened, not tapered, 4–38 × 0.5–1 mm, hairless or sparsely hairy. Flowering-stems 1–4 from the centre of the rosette, 5–20 cm, entirely hairless, bearing 1–10 flowers. Flower-stalks 1–2 mm, or absent, erect in fruit. Petals pink or white, 6–7 mm. Styles 3, forked at the base, the apices slightly club-shaped. *Coastal SE USA, C & S America*. G1. Summer.

100. D. montana Saint-Hilaire.

Perennial or annual, lacking a resting bud. Stem very short. Stipules 1.5–3 mm, divided into numerous teeth. Leaves 8–15, held horizontally in a flat basal rosette, narrowly spathulate, 1–1.5 × 0.25–0.4 cm, tip rounded, hairless below, stalk 3–6 × 0.25–0.5 mm, hairless. Flowering-stem 15–20 cm, slightly to densely long-hairy, the upper tenth fertile. Flowers 3–7. Petals *c.* 8 mm, pink. *Brazil*. G1. Summer.

101. D. villosa Saint-Hilaire. Illustration: Kondo & Kondo, Carnivorous plants of the world in colour, 79 (1983).

Perennial, lacking a resting bud. Stem very short. Leaves 9–20, held horizontally in a crowded basal rosette, strap-shaped, 2–4 × 0.25–0.7 cm, acute at tip, densely white-hairy, with shaggy hairs below, stalk *c.* 1.5 × 0.2 cm, hairless, the upper ⅓–⅐ fertile. Flowers 5–20. Petals 6–7 mm, pink. *E Brazil*. G1.

102. D. roraimae (Diels) Maguire & Laundon (*D. montana* Saint-Hilaire var. *roraimae* Diels). Illustration: Kondo & Kondo, Carnivorous plants of the world in colour, 68 (1983).

Perennial 2–3 cm across, lacking a winter resting bud, the growing point ascending vertically, supported by a dense accumulation of dead leaves, forming a column 2–10.5 cm. Stipules *c.* 5 mm, coppery, deeply divided into numerous teeth. Leaves 5–14, initially erect, at length reflexed, forming a raised rosette, blade distinct from stalk, obovate to oblanceolate, 5–7 × 2.5–3 mm, tip rounded, base wedge-shaped, stalk 8–10 × 0.5 mm, hairless above, the lower surface covered with coarse, long, adpressed coppery hairs, extending to the blade. Flowering-stem axillary, 20–32 cm, hairless or rarely with scattered, spreading non-glandular hairs at the base, with 8–30 flowers. Flower-stalks to 2 mm, erect in fruit. Petals white, 6–8 mm. Styles 3,

forked at the base and again at the tip of each branch. *Guyana & Venezuela*. G1. Summer.

103. D. brevifolia Pursh. Illustration: Schnell, Carnivorous Plants of the United States & Canada, 69 (1976); Kondo & Kondo, Carnivorous plants of the world in colour, 21 (1983).

Annual or short-lived perennial lacking winter resting bud, 1.25–3.5 cm across. Aerial stem absent. Stipules reduced to a few inconspicuous teeth. Leaves 10–15 in a flat basal rosette, blade and stalk not clearly distinct, almost spathulate, 1.2–2 cm, the tip very shallowly rounded, 3.5–5 mm across, tapering abruptly then gradually to the base, glandular trapping hairs often extending almost to the base, lower surface hairless, rarely with a few long hairs. Flowering-stems 1–3 from the centre of the rosette, 2–12 cm, thickly covered with stalked glandular hairs from base to tip, with 1–5 flowers. Flower-stalks 1–3 mm, erect in fruit. Petals 4–10 mm, white or pink. Styles 3, forked at the base, the stigmatic apices slightly swollen. *SE USA & Brazil*. G1. Summer.

104. D. spathulata Labillardière. Illustration: Slack, Carnivorous Plants, 132 (1979); Kondo & Kondo, Carnivorous plants of the world in colour, 73 (1983).

Perennial lacking winter resting bud, 1.3–5.5 cm across. Aerial stem absent. Stipules reddish, 2–4 mm, deeply divided. Leaves 10–20 in a flat basal rosette, spathulate, blade obovate, 3–6 mm across, tip rounded, base abruptly constricted and distinct from the leaf-stalk, to gradually tapering and indistinct, trapping hairs occupying the apical half of the leaf only, to extending almost to the base, lower surface hairless, rarely with a few scattered hairs. Flowering-stems 1–7, from the centre of the rosette, 6–19 cm, hairy and with scattered glandular hairs, with 3–25 flowers. Flower-stalks 2–3 mm, erect in fruit. Petals white or pink, 3.5–6 mm. Styles 3, forked at the base, the branches abruptly curved upwards at the tip, not dilated. *Borneo to Japan, Australia & New Zealand*. G1. Summer.

A variable species, easily the most widespread and soon becoming a nuisance in cultivation. Much seed distributed under the name of other species proves to be *D. spathulata*. Several strains have been offered for cultivation over the last 10 years. From Japan, 'Kansai' and 'Kanto', both producing large plants, and the smaller and weediest, 'spathulate-

rotundate'. The white-flowered variant from New Zealand is less usually seen.

D. × nagamotoi Cheek. A hybrid with *D. anglica* forms a very open resting bud which does not always die down in winter; leaves held horizontally; flowers pink.

4. DROSOPHYLLUM Link
S.G. Knees

Woody-based perennial to 30 cm, stems sometimes branching, older dead leaves often persisting; all parts covered with sticky glandular hairs. Leaves coiled downwardly circinate in bud, crowded 15–20 cm, narrowly linear with thread-like tips. Flowers *c.* 4 cm across, borne in loose branched cymes, sepals 5, petals 5, sulphur-yellow; stamens 10–20; styles several; ovary superior, placentation parietal. Fruit a capsule; seeds many.

A genus of a single species from the southwesten Iberian Peninsula and North Africa. Usually propagated from seed, it can be grown in sphagnum or a light sandy loam and requires sunshine and warmth.

1. D. lusitanicum (Linnaeus) Link. Illustration: Polunin & Smythies, Flowers of south-west Europe, pl. 13 (1973); Heywood, Flowering plants of the world, 139 (1978).

SW Spain, Portugal & N Morocco. G1. Spring–summer.

XCVI. PAPAVERACEAE

Annual or perennial herbs, more rarely herbaceous or woody climbers or shrubs; sap milky or coloured with latex, or clear. Leaves usually alternate, rarely opposite or all basal, usually divided, often in complex ways; veins pinnate or palmate. Flowers solitary or in cymes, racemes, umbels or panicles, varying from small to large, radially symmetric (with 2 obvious planes of symmetry) or bilaterally symmetric. Sepals 2 or 3, usually falling early, sometimes united. Petals 4 or 6, rarely many or absent, usually borne in 2 distinct whorls, either of simple form or variously lobed, free or variably united. Stamens 8–many or 4, when 4 either simple or united in 2 bundles, each bundle consisting of a whole stamen united to 2 half-stamens. Sepals, petals and stamens rarely borne on a perigynous zone. Ovary superior with 2 or more carpels which are usually united; style present or absent, stigmas 2 or more; ovules usually numerous, parietal, more rarely 1 and almost basal. Fruit a capsule or an indehiscent nutlet, or breaking into 1-seeded segments. Seeds with oily endosperm, each often with an appendage (aril, caruncle, arilloid, elaiosome).

A family of 41 genera from northern temperate regions, South America, South Africa and Australia, of which a considerable number is grown. The family as recognised here in its broadest sense, is divided into 7 subfamilies (in a narrower sense it is divided into 2 or 3 families). Of these, 4 (Papaveroideae, Platystemonoideae, Chelidonioideae and Eschscholtzioideae, genera **1–19**) make up the Papaveraceae in the strict sense: plants with milky or coloured latex, radially symmetric flowers with no nectar, many stamens and the carpels 2–many. A fifth subfamily (Fumarioideae, often recognised as a separate family, Fumariaceae, genera **22–27**) consists of a group of genera looking superficially very unlike those of the strict Papaveraceae, having watery sap, flowers usually bilaterally symmetric, nectar usually secreted by the bases of the stamens (often prolonged into a nectar-secreting spur or peg), 4 stamens which are often united as described above, and the carpels always 2. These 2 groups are, however, linked by the final two subfamilies (Hypecoideae, sometimes recognised as a separate family, Hypecoaceae, and Pteridophylloideae, consisting only of the 2 genera *Hypecoum* and *Pteridophyllum*, genera **20 & 21**). Some species of the genus *Fumaria* Linnaeus may, from time to time, occur in gardens, though they are scarcely actively cultivated; except perhaps in wild gardens following introduction in wild flower seed mixus.

Literature: Fedde, F., Papaveraceae (Papaveroideae, Hypecoideae), *Das Pflanzenreich* **40** (1909) – excludes Fumarioideae; Ernst, W.R., A comparative morphology of the Papaveraceae (1962) – excludes Hypecoideae and Fumarioideae; Lidén, M., Synopsis of Fumarioideae (Papaveraceae) with a monograph of the tribe Fumarieae, *Opera Botanica* **88** (1986) – includes only Fumarioideae.

1a. Carpels 3 or more, usually united, occasionally more or less free   2
 b. Carpels always 2, always united   9
2a. Leaves opposite or whorled   3
 b. Leaves alternate or all basal   4
3a. Carpels several, becoming free from each other as they ripen
         7. Platystemon

 b. Carpels 3 or 4, remaining united
         8. Stylophorum
4a. Softly woody shrub with large white flowers; ovary covered with stiff, hard bristles   **1. Romneya**
 b. Herbs; combination of other characters not as above   5
5a. Stigmas forming radii on a flat, domed or conical disc borne on top of the ovary   **3. Papaver**
 b. Stigmas various, not as above   6
6a. Plants glaucous, with prickles on the stems and leaves or leaf-margins; sepals each with an almost apical, hollow horn
         2. Argemone
 b. Plants not usually glaucous, not prickly but often bristly; sepals without horns   7
7a. Capsule more than 5 mm broad, spherical or cylindric, opening by a series of pores just below the apex
         8
 b. Capsule less than 5 mm broad, linear-cylindric, opening by the walls splitting from the placental framework from the apex for some distance downwards   **6. Roemeria**
8a. Style thread-like, stigma head-like; petals orange with a purplish blotch near the base, the claw below the blotch greenish; annual herb   **5. Stylomecon**
 b. Style and stigmas not as above; petals various, but not coloured as above; perennial or monocarpic herbs   **4. Meconopsis**
9a. Stamens more than 4, free; nectar absent   10
 b. Stamens 4, free or in 2 bundles; nectar usually secreted at the bases of the filaments   20
10a. Sepals, petals and stamens perigynous   11
 b. Sepals, petals and stamens hypogynous   12
11a. Sepals united, falling as a candle-snuffer-shaped unit; cotyledons bifid
         17. Eschscholzia
 b. Sepals free, falling individually; cotyledons not bifid
         18. Hunnemannia
12a. Petals absent   13
 b. Petals present   14
13a. Leaves palmately veined and lobed
         15. Macleaya
 b. Leaves pinnately veined, pinnately lobed or simple and entire
         16. Bocconia
14a. Leaves palmately veined and lobed
         15

 b. Leaves pinnately veined and lobed   16
15a. Petals 8–16 or more; flowers solitary   **14. Sanguinaria**
 b. Petals 4; flowers in cymes
         13. Eomecon
16a. Shrub; leaves evergreen
         19. Dendromecon
 b. Herb; leaves deciduous   17
17a. Leaves pinnate into distinct leaflets; seeds each with an appendage (aril)
         18
 b. Leaves pinnatisect to lobed, but without distinct leaflets; seeds not appendaged   19
18a. Petals 2–2.5 cm; flowers solitary, without bracteoles
         10. Hylomecon
 b. Petals 7–15 mm; flowers in umbels, each with 2 bracteoles
         9. Chelidonium
19a. Seeds embedded in a spongy false septum   **11. Glaucium**
 b. Seeds without a false septum
         12. Dicranostigma
20a. Stamens 4, free; no petals spurred
         21
 b. Stamens 4 in 2 bundles of a whole stamen united to 2 half-stamens; upper petal (at least) usually spurred   22
21a. All 4 petals more or less similar; leaves all basal, pinnate into oblong leaflets; fruit a capsule
         20. Pteridophyllum
 b. The 2 outer petals entire to weakly 3-lobed, the 2 inner petals distinctly divided into 3 segments of which the central is very different from the laterals; leaves not as above; fruit breaking into 1-seeded segments (ours)   **21. Hypecoum**
22a. Flowers with 2 distinct planes of symmetry; outer petals swollen or with spurs at the base   23
 b. Flowers with only a single plane of symmetry; upper petal distinctly spurred, the lower sometimes swollen at the base but not spurred
         24
23a. Outer petals rounded or slightly and equally swollen at the base but not spurred; perianth persistent on the fruit   **23. Adlumia**
 b. Outer petals both distinctly and equally spurred; perianth not persistent   **22. Dicentra**
24a. Fruit a capsule containing 2 or more seeds   **24. Corydalis**
 b. Fruit a nutlet containing 1 or rarely 2 seeds   25

25a. Stigma crested or with 2 papillae
27. Fumaria
  b. Stigma not as above   26
26a. Inner petals without yellow lateral
wings; nutlet not ribbed
26. Rupicapnos
  b. Inner petals with yellow lateral
wings; nutlet ribbed
25. Sarcocapnos

1. ROMNEYA Harvey
J. Cullen
Large, rather woody, glaucous perennials,
with creeping, underground rootstocks.
Leaves alternate, pinnatifid to pinnatisect.
Flowers solitary, terminal, large and
showy. Sepals 3. Petals 6, white. Stamens
numerous, filaments thread-like, yellow.
Ovary of 7–12 united carpels, placentation
parietal, ovules borne on inwardly directed
plates which almost meet at the centre.
Stigmas stalkless on top of the ovary. Fruit
a capsule opening at the top.

A genus of 2 species of very handsome,
shrubby perennials which require a warm
situation in full sun, and, in more
northerly parts, the protection of a wall; in
severe winters the aerial parts may be cut
back to ground level, but the underground
stock will generally continue to grow and
produce new shoots in the spring.
Propagation is by seed, when available, or
by division of the rootstock.

1a. Sepals with adpressed bristles
2. trichocalyx
  b. Sepals hairless   **1. coulteri**

1. R. coulteri Harvey. Illustration: Morley
& Everard, Wild flowers of the world, pl.
154 (1970); Rickett, Wild flowers of the
United States **4**: pl. 51 (1970); Botanical
Magazine, n.s., 678 (1975).

Shrub to 2.5 m. Leaves shortly stalked,
ovate, with 3–5 main segments which are
1–2 cm wide, the segments occasionally
further lobed or divided, all sparsely hairy.
Flower-stalks hairless. Flowers to 20 cm
across, buds hairless. Capsule 3–4 cm. *USA
(California)*. H5. Summer.

2. R. trichocalyx Eastwood (*R. coulteri*
var. *trichocalyx* (Eastwood) Jepson).
Illustration: Botanical Magazine, 8002
(1905).

Similar to *R. coulteri* but not as large;
leaf-segments narrower; flower-stalks
bristly below the flower; sepals with
adpressed bristles; flowers to 16 cm across.
USA (California). H5. Summer.

The hybrid between the 2 species, which
has no valid name (though the names
R. × hybrida invalid and *R. × vandedenii*

invalid are found in the literature) is also
cultivated; it is very like *R. trichocalyx* but
reputedly more vigorous and floriferous.

2. ARGEMONE Linnaeus
J. Cullen
Annual to short-lived perennial herbs.
Latex usually yellowish. Stems glaucous,
often prickly. Leaves alternate, glaucous,
lobed and with prickle-tipped teeth,
surfaces prickly or not, often pale or bluish
over the veins, the uppermost leaves
usually stem-clasping. Sepals usually 3,
often prickly, each with an almost apical
horn which is hollow at the base and
prickle-like or flattened towards the apex.
Petals usually 6, orange, yellow, white or
rarely mauve. Stamens numerous,
filaments mostly thread-like. Ovary of
3–5 carpels, 1-celled, placentas scarcely
projecting inwards. Style short or absent,
stigmas usually velvety, purplish. Capsule
usually prickly, opening at the apex. Seeds
numerous.

A genus of about 30 species from N & S
America (1 species widely introduced and
naturalised in all warm parts of the world).
They have spectacular flowers and spiny,
glaucous vegetative parts and can make
interesting plants for borders. Their latex is
very poisonous. Propagation is by seed.

Literature: Ownbey, G.B., Monograph
of the genus Argemone for North America
and the West Indies, *Memoirs of the Torrey
Botanical Club* **21**(1) (1958).

1a. Petals yellow or yellow-orange
1. mexicana
  b. Petals white, rarely white with a
yellowish flush or mauve   2
2a. Horns of sepals flattened; flowers
10–15 cm across   **2. platyceras**
  b. Horns of sepals not flattened,
ending in a prickle; flowers to 10
cm across   3
3a. Buds more or less spherical;
uppermost leaves not conspicuously
stem-clasping   **3. grandiflora**
  b. Buds ellipsoid-oblong; uppermost
leaves conspicuously stem-clasping
4. polyanthemos

1. A. mexicana Linnaeus. Illustration:
Memoirs of the Torrey Botanical Club
20(1): 32 (1958); Rickett, Wild flowers of
the United States **2**: pl. 64 (1967); Polunin,
Flowers of Europe, pl. 29 (1969); Morley &
Everard, Wild flowers of the world, f. 154
(1970).

Annual herb with bright yellow latex.
Stems to 2.5 m (more in the wild, usually
less in cultivation), often prickly. Leaves

glaucous with bluish markings over the
veins, the upper clearly stem-clasping,
all rather deeply lobed, margins prickly
toothed; main veins with prickles beneath.
Flowers 4–6 cm across. Petals bright to
pale yellow or yellow-orange. Capsule
usually spiny. *C America & S USA
(Florida), widely introduced and naturalised
elsewhere*. H4. Summer.

2. A. platyceras Link & Otto. Illustration:
Botanical Magazine, 6402 (1878) – as
A. hispida; Addisonia **20**: t. 665 (1939);
Memoirs of the Torrey Botanical Club
21(1): 106 (1958).

Annual or sometimes short-lived
perennial herb. Stems to 75 cm, with few
prickles. Leaves bluish, lobed, the lobes
toothed with prickle-tipped teeth, prickly or
not on the surfaces, the upper leaves stem-
clasping. Flowers 10–15 cm across. Sepal-
horns flattened. Petals white, white faintly
flushed with yellow or rarely mauve.
Capsule usually densely spiny. *Mexico*. H5.
Spring.

3. A. grandiflora Sweet. Illustration:
Botanical Magazine, 3073 (1831);
Memoirs of the Torrey Botanical Club
21(1): 126 (1958).

Annual or short-lived perennial with
yellow latex. Stem to 1.2 m, usually
prickly. Leaves glaucous, lobed, the lobes
with prickle-tipped teeth, usually prickly
on the veins beneath, the uppermost leaves
scarcely stem-clasping. Buds more or less
spherical. Flowers 6–10 cm across. Petals
white. Capsule with scattered spines.
Mexico. H5. Summer.

4. A. polyanthemos (Fedde) Ownbey (*A.
alba* James). Illustration: Memoirs of the
Torrey Botanical Club **21**(1): 130 (1958);
Rickett, Wild flowers of the United States
4: pl. 52 (1970), **6**: pl. 58 (1973).

Annual herb with yellow latex. Stem to
80 cm, sparsely prickly. Leaves glaucous,
lobed and with prickle-tipped teeth, with
recurved prickles on the main veins
beneath, the uppermost leaves
conspicuously stem-clasping. Buds oblong-
ellipsoid. Flowers 7–10 cm across, petals
white or rarely mauve. Capsule with stout
spines. *USA (Wyoming to Texas & New
Mexico)*. H4. Spring–summer.

3. PAPAVER Linnaeus
J. Cullen
Perennial, biennial or annual herbs, often
bristly or hairy. Latex usually whitish,
sometimes orange or reddish. Leaves
alternate or all basal, usually divided and
toothed, sometimes stem-clasping at the

base. Flowers in raceme- or spike-like inflorescences or solitary, often large and showy, with or without bracts. Sepals usually 2, rarely 3, buds usually pendent until just before the flower opens, sepals falling as the flower opens. Petals usually 4, rarely 6, crumpled in bud, red, pink, orange, yellow or white, often blotched at the base with a contrasting colour. Stamens numerous, filaments thread-like or club-shaped. Ovary of 3 or more united carpels, the ovules borne on placentas which project into the ovary as plates, almost meeting in the centre. Styles borne on a flat, convex or pyramidal disc on the top of the ovary, as velvety rays. Fruit a capsule opening by pores just beneath the stigmatic disc. Seeds numerous, without arils.

A genus of about 70 species, widespread in northern temperate regions (including the Arctic), and in the southern hemisphere in South Africa and Australia. Many of the annual species are weeds, and all are easily grown, even in poor soils. The perennials may be propagated by seed or by division of the roots, the annuals and biennials by seed.

1a. Perennials   2
  b. Biennials or annuals   19
2a. Leaves all basal, plants scapose; capsules usually bristly, if hairless then petals not orange   3
  b. Stems leafy, or if plant almost scapose then capsules not bristly and petals orange   11
3a. Stamens fewer than 60   4
  b. Stamens more than 60   5
4a. Plant strongly cushion-forming; petals falling quickly after flowering; capsule more or less spherical   **13. fauriei**
  b. Plant not or scarcely cushion-forming; petals persisting for some time after flowering; capsule barrel-shaped   **12. radicatum**
5a. Scapes mostly more than 30 cm   6
  b. Scapes mostly less than 30 cm, usually much less   7
6a. Capsule without bristles; petals white   **10. amurense**
  b. Capsule bristly; petals yellow, orange or reddish, rarely white   **9. croceum**
7a. Scapes and buds with dense, brown hairs; leaves densely white hairy, pinnatisect   **11. canescens**
  b. Scapes, buds and leaves variously hairy, but not as above; leaves mostly 1–3 times pinnate   8
8a. Petals yellow or reddish   9
  b. Petals white   10

9a. Ultimate leaf-segments rather broad, obtuse, forwardly pointing, asymmetric   **14. rhaeticum**
  b. Ultimate leaf-segments narrow, spreading, acute, usually symmetric   **17. kerneri**
10a. Ultimate leaf-segments forwardly pointing, asymmetric   **15. sendtneri**
  b. Ultimate leaf-segments spreading, usually symmetric   **16. burseri**
11a. Filaments club-shaped; petals dark red to pink or pink-orange, often black-blotched   12
  b. Filaments thread-like; petals pale orange, not blotched   15
12a. Flowers with 1–8 bracts borne immediately below the sepals and somewhat overlapping them   13
  b. Flowers without any bracts   14
13a. Petals dark red, each with an oblong black blotch running from the base almost to the middle; bracts 3–8; bristles on the sepals broadly triangular at base   **1. bracteatum**
  b. Petals orange or orange-red, each with a rectangular (broader than long) black blotch above the base; bracts 1–4; bristles on sepals slender throughout   **2. pseudo-orientale**
14a. Stem-leaves not extending to upper ⅓ of stem; buds drooping; petals without black blotches   **3. orientale**
  b. Stem-leaves extending to upper ⅓ of stem; buds usually erect; petals each with a rectangular black blotch above the base   **2. pseudo-orientale**
15a. Flowers 3 – many in spike- or raceme-like inflorescences   **4. pilosum** group
  b. Flowers solitary (or rarely 2 together) on long stalks which make up at least ½ the height of the plant   16
16a. Leaves all basal, plant scapose   **6. monanthum**
  b. Leaves on the lower parts of the stems   17
17a. Leaves oblong, extending to half-way up the stem; capsule dark greyish   **5. lateritium**
  b. Leaves obovate, in the lower ⅓ of the stem only; capsule green or brownish   18
18a. Leaves hairy only at the margins and on the veins beneath; sepals hairless   **7. rupifragum**

  b. Leaves hairy all over the surface beneath; sepals hairy   **8. atlanticum**
19a. Plants biennial with greyish or glaucous leaves   **18. fugax**
  b. Plants annual (if biennial in cultivation then leaves not greyish or glaucous)   20
20a. Capsule bristly, at least near the apex   **23. pavoninum**
  b. Capsule not bristly   21
21a. Stem-leaves stalkless, clasping the stem   22
  b. Stem-leaves stalkless or shortly stalked, not clasping the stem   23
22a. Leaves pinnatisect; capsule to 2 cm; buds ovoid, 2–3 cm   **25. glaucum**
  b. Leaves irregularly toothed; capsule 5 cm or more; buds ovoid-oblong, to 2 cm   **24. somniferum**
23a. Stems and leaves with long, rigid, prickle-like bristles   **19. aculeatum**
  b. Stems and leaves hairy or bristly but bristles not rigid and prickle-like   24
24a. Sinuses between the rays of the stigmatic disc obvious, extending inwards for more than ⅓ of the radius of the disc, stigmatic rays keeled   **22. macrostomum**
  b. Sinuses between the rays of the stigmatic disc small, extending inwards for much less than ⅓ of the radius of the disc, stigmatic rays not keeled   25
25a. Plant bristly-hairy; petals usually deep red with black blotches, or pink or white and unblotched   **20. rhoeas**
  b. Plant hairless; petals pale red or pale orange, each with a greenish or pale blotch at the base   **21. californicum**

Section **Macrantha** Elkan (Section *Oxytona* Bernhardi).

Perennial herbs, often bristly. Leafy stems present. Flowers apparently solitary, actually in cymes, the terminal flower of each cyme usually with 3 sepals and 6 petals, the others with 2 sepals and 4 petals; petals red to deep red or paler and somewhat orange, usually dark-blotched. Filaments club-shaped, usually black.

Literature: Goldblatt, P., Biosystematic studies on Papaver section Oxytona, *Annals of the Missouri Botanical Garden* **61**: 264–296 (1974).

1. P. bracteatum Lindley (*P. orientale* var. *bracteatum* (Lindley) Ledebour). Illustration:

Perry, Flowers of the world, 221 (1970); Bloom, Perennials for your garden, 106 (1986).

Stems to 1 m or more. Leaves pinnatisect, margins usually toothed, all bristly: stem-leaves extending into upper third of stem. Bracts 3–8, irregularly toothed or entire. Buds erect, ovoid, becoming ovoid-oblong before opening. Flowers to 16 cm across. Sepals 3, with broadly based, adpressed bristles, petals 6 in most flowers. Petals dark red, each with an oblong (longer than broad) black blotch running from the base to near the middle. Anthers 2.8–5.5 mm. Stigmatic disc flat or convex in flower, often pyramidal in fruit; rays 12–24. Capsule to 4 × 3 cm. *Iran, former USSR (Caucasus).* H2. Summer.

See comment under *P. orientale.*

2. P. pseudo-orientale (Fedde) Medwedew (*P. bracteatum* var. *pseudo-orientale* Fedde). Illustration: Botanical Magazine, 57 (1788); The Garden **110**: 274 (1985).

Like *P. bracteatum* but less robust; stems to 80 cm, bracts absent or 1–4 per flower, small; flowers smaller; sepals 2 or 3, with slender bristles, petals 4 or 6, orange-red, usually with a rectangular black blotch (broader than long) near, but not extending to the base; anthers 1.7–3.5 mm; disc with 9–19 rays, flat or convex in flower and fruit. *E Turkey, adjacent former USSR (Caucasus).* H2. Summer.

See comment under *P. orientale.*

3. P. orientale Linnaeus.

Like both *P. bracteatum* and *P. pseudo-orientale* but less robust; leaves less deeply divided, stem-leaves not extending to the upper third of the stem; bracts absent; flowers smaller, buds drooping during development; sepals with fine bristles, 2 or 3, petals 4 or 6, orange, occasionally marked with a bluish or whitish mark above the base, not black-blotched; anthers 1.8–3 mm; disc with 8–15 rays, flat or convex in flower and fruit. *E Turkey, N Iran, former USSR (Caucasus).* H2. Summer.

Plants belonging to all 3 of these species are widely grown as spectacular, summer-flowering perennials, usually under the name *P. orientale*, though this species is, in fact, the least often cultivated. In the late 19th and early 20th centuries numerous cultivars were developed, some of them very large (and with a tendency for the stems to arch and buckle, making them untidy subjects for a border), with very large flowers, and others with double or pale or white flowers. Most of these are unavailable today; those that remain are difficult to relate directly to the species, though the largest of them belong to *P. bracteatum*. Hybrids between *P. pseudo-orientale* and *P. orientale* occur in the wild, and possibly all 3 species have been hybridised in gardens.

Section **Pilosa** Prantl.

Perennial herbs, stems leafy throughout or only in the lower half or rarely all leaves basal, plant scapose, usually rather densely bristly-hairy. Sepals 2, petals 4, orange. Filaments thread-like.

4. P. pilosum group.

Perennial herbs to 80 cm, generally hairy with bristle-like hairs, sometimes very densely so. Basal leaves stalked, toothed or lobed; stem-leaves numerous to rather few, stalkless and stem-clasping, toothed or lobed. Flowers numerous or as few as 3 per stem, in spike-or raceme-like inflorescences. Petals orange. Capsules mainly narrowly club-shaped, occasionally with rounded bases, usually hairless, occasionally hairy, stalkless or with short to very long stalks. *Turkey.* H4. Spring–summer.

A very complex group in which identification and nomenclature are very difficult. A satisfactory classification has not yet been devised, and the important units are briefly described below; the names attached to them are those that are currently available and will certainly change as research progresses. It is not certain how many of them are genuinely in cultivation.

P. spicatum Boissier & Balansa. Illustration: Das Pflanzenreich **40**: 357 (1909). Plants very densely hairy the hairs obscuring the surface and giving the whole plant a silvery yellow appearance. Stems leafy to the top. Fruits club-shaped, stalkless or very shortly stalked (stalks always shorter than the capsules). *W Turkey (Izmir & Manisa).*

P. pilosum Sibthorp & Smith. Like *P. spicatum* but fruits on long stalks, the capsules with rather rounded bases. *NW Turkey (Bursa, Mt Olympus).*

P. heldreichii Boissier. Illustration: Botanical Magazine, n.s., 331 (1959). Like *P. pilosum* but fruits spindle-shaped, fairly evenly tapered at both ends (more abruptly so to the apex), hairless or hairy. *SW Turkey (Antalya).*

Plants of this species with hairy fruits have been named *P. spicatum* var. *luschanii* Fedde.

P. heldreichii Boissier var. **sparsipilosum** Boissier. Plants much less hairy than those above, the surface clearly visible, hairs whitish or greyish. Stems leafy throughout, all the leaves with rather short, blunt teeth; flowers more than 5 per stem. Capsules tapered gradually to the base, slightly so to the apex, borne on very long, erect stalks. *SW Turkey (Muğla, Burdur & Antalya).* This unit has no available name at species level.

P. apokrinomenon Fedde. Like *P. heldreichii* var. *sparsipilosum* but with fewer stem-leaves, which have acute teeth; fruit not or scarcely tapered to the apex. *Central part of W Turkey (Isparta & Konya).*

P. strictum Boissier & Balansa. Like *P. heldreichii* var. *sparsipilosum* and *P. apokrinomenon* but leaves lobed with narrowly triangular lobes. *N & C parts of W Turkey (Balikeşir, Kütahya).*

P. pseudostrictum Fedde. Like the 3 species above, but stem-leaves abruptly reduced above, and flowers per stem few (3–5). *NW Turkey (Zonguldak, Kastamonu & Ankara).* In its few-flowered inflorescences and reduction of stem-leaves, this species is rather similar to *P. lateritium.*

5. P. lateritium Koch. Illustration: Das Pflanzenreich **40**: 361 (1909); The Garden **110**: 275 (1985).

Perennial, stems to 60 cm. Leaves borne only in the lower half of the stem, at least half the height of the plant made up by the very long flower- and fruit-stalks. Leaves oblong, regularly but unequally toothed (occasionally 1 or 2 of the lower teeth more deeply cut and forming lobes), densely hairy, especially along the veins beneath, the hairs of different sizes, some conspicuously shorter and thinner than the others. Flowers solitary (rarely 2 together). Buds spherical, sepals densely hairy. Petals orange. Capsules club-shaped, gradually tapered to the base, not tapered towards the apex, greyish. *NE Turkey and adjacent former USSR (Caucasus).* H4. Summer.

P. 'Nanum Flore Pleno' is a hybrid between this and *P. pseudo-orientale* (see Mathew, B., The Garden **110**: 271–274, 1985, with illustration on p. 273).

6. P. monanthum Trautvetter.

Similar to *P. lateritium* but leaves obovate, irregularly pinnately lobed, the lobes toothed, all basal. Scapes to 50 cm. *NE Turkey, adjacent former USSR (Caucasus).*

7. P. rupifragum Boissier & Reuter. Illustration: Das Pflanzenreich **40**: 361

(1909); Castroviejo et al., Flora Iberica **1**: 412 (1986).

Perennial to 60 cm or more. Leaves in the lower third of the stem only, obovate, stalked, lobed or rather bluntly toothed, hairy on the margins and on the veins beneath, all the hairs on the leaves of more or less the same size. Buds obovoid, sepals hairless. Petals orange. Capsule club-shaped, tapered gradually to the base, greenish, eventually brownish. *Spain, Morocco*. H3. Summer.

Can become an extremely invasive weed in some situations.

8. P. atlanticum Cosson (*P. rupifragum* subsp. *atlanticum* (Cosson) Maire). Illustration: Das Pflanzenreich **40**: 361 (1909); Maire, Flore d'Afrique du Nord **11**: 317 (1964).

Very like *P. rupifragum* but leaves hairy all over the surface beneath, sepals hairy. *Morocco*. H4. Summer.

Section **Meconella** Spach (Section *Scapiflora* Reichenbach; Section *Lasiotrachyphylla* Bernhardi).

Plants scapose, leaves all basal, the scapes with solitary flowers arising from a rosette of leaves. Filaments thread-like.

A very complex group of an uncertain number of species which are very difficult to identify. The species fall into 3 geographical groups which are not easy to discriminate in terms of their structure. (1) Species from the European mountains (Pyrenees, Alps, Carpathians), generally referred to under the group name *P. alpinum* or *P. pyrenaicum*; both of these names are, however, of uncertain application and are not used below. These plants are all diploid. (2) Species from the Arctic regions of the northern hemisphere, extending south to Japan and the Rocky mountains, known collectively as *P. radicatum*; these species are all polyploid. (3) Species from Asia (former USSR, Himalaya, China) known collectively as *P. nudicaule*; this group includes both diploids and tetraploids.

With cultivated material no use can be made of geographical origin for identification; the differences given in the key above and in the descriptions below should allow for the identification of most garden plants, but some cultivars may be difficult to place under the individual species.

Literature: Randel, U., Beiträge zur Kentniss der Sippenstruktur der Gattung Papaver Sectio Scapiflora Reichenbach, *Feddes Repertorium* **84**: 655–732 (1974).

9. P. croceum Ledebour (*P. nudicaule* of many authors and of gardens, not Linnaeus). Illustration: Botanical Magazine, 1633 (1814), 2344 (1822); Busch, Flora Sibiricae et Orientis extremi, t. 24a & b (1913); Tosco, Mountain flowers, 105 (1973).

Plants robust, scapes usually more than 30 cm, rather thin, hairless or sparsely to densely hairy with hairs which are often dark brown. Leaves pinnatifid to pinnatisect, the lobes themselves sometimes further toothed or divided, sparsely bristly or more rarely densely hairy. Buds ovoid or almost spherical, hairy. Flowers 2–6 cm across. Petals pale yellow to yellow, orange or reddish (rarely white in some cultivars). Stamens 70–100. Capsule narrowly club-shaped, usually covered with whitish or pale bristles. *E former USSR, Mongolia*. H2. Spring–summer.

Almost always found under the name *P. nudicaule* in gardens but the cultivated material should be referred to as *P. croceum* and genuine *P. nudicaule* is not cultivated. Cultivars (generally available as mixed seed) are often larger than the wild plants.

10. P. amurense Karrer (*P. nudicaule* Linnaeus subsp. *amurense* (Karrer) Busch; *P. anomalum* misapplied).

Like *P. croceum* but more robust, leaves and scapes more densely hairy, flowers larger (4–8 cm across), petals usually white, capsule hairless. *Former USSR, China*. H2. Spring–summer.

11. P. canescens Tolmachev (*P. pseudocanescens* misapplied).

Plant cushion-forming, scapes to 15 cm, densely hairy with brown hairs. Leaves usually densely hairy with long, white hairs, pinnatisect with narrowly ovate to ovate lobes. Buds covered with brown hairs. Flowers 2–5 cm across. Petals pale yellow to orange. Stamens 70–100. Capsule narrowly cylindric, bristly. *Former USSR, Korea, China?* H2. Spring–summer.

12. P. radicatum Rottboell. Illustration: Grey-Wilson & Blamey, Alpine flowers of Britain and Europe, 59 (1979).

Tufted perennial, scarcely cushion-forming. Scapes to 25 cm (rarely more), sparsely to densely bristly. Leaves pinnatifid to pinnatisect, variably hairy or bristly. Buds ovoid to almost spherical, hairy. Flowers 2–5 cm across. Petals pale yellow or white, not falling quickly after flowering, drying metallic blue-green. Stamens 20–30. Capsule barrel-shaped,

bristly. *Arctic regions*. H1. Spring–summer.

A very variable species with numerous subspecies (recognised as distinct species by some authors); it is not known which of these are genuinely in cultivation.

P. macounii Greene. Similar, but with the capsule conspicuously narrowed towards the base. *USA (Alaska) & adjacent former USSR*. H1. Spring–summer.

13. P. fauriei Fedde. Illustration: Takeda, Alpine flora of Japan **2**: pl. 115 (1977).

Perennial, cushion-forming herb. Scapes to 10 cm, weakly to densely adpressed-bristly. Leaves pinnate with distant segments which are themselves usually deeply lobed, sparsely to densely hairy. Buds ovoid, brown-hairy. Flowers 2–3 cm across, petals pale yellow or greenish yellow, soon falling. Stamens to 60. Capsule more or less spherical. *N Japan, Kurile Islands*. H2? Spring–summer.

The name **P. miyabeanum** Tatewaki is applied to a very similar (or possibly the same) species from the Kurile Islands. Material in cultivation under this name is generally *P. fauriei*. The name *P. miyabeanum* has been grossly misspelled and garbled in some horticultural literature (see Brough, M.A., Bulletin of the Alpine Garden Society **36**: 186, 1968).

14. P. rhaeticum Leresche (*P. alpinum* Linnaeus subsp. *rhaeticum* (Leresche) Hayek; *P. pyrenaicum* (Linnaeus) Kerner subsp. *rhaeticum* (Leresche) Fedde). Illustration: Polunin, Flowers of Europe, pl. 28 (1969); Hess, Landolt & Hirzel, Flora der Schweiz **2**: 110 (1970) - as P. aurantiacum; Rasetti, I fiori delle Alpi, pl. 54, f. 214 (1980); Guittoneau & Huon, Connaître et reconnaître la flore et la vegetation méditerranéennes, 99 (1983).

Perennial, scapes to 20 cm, with adpressed bristles. Leaf-bases forming a compact tunic around the lower part of the plant. Leaves pinnate, the segments themselves pinnately lobed, ultimate segments 1–6 mm wide, ovate to ovate-lanceolate or obovate, obtuse, forwardly pointing. Flowers 4–5 cm across, petals golden yellow, very rarely reddish or pale. Capsule oblong-ellipsoid, to 1.5 cm, bristly, stigmatic disc usually flat with 5–7 rays. *Pyrenees, Alps*. H2. Spring–summer.

15. P. sendtneri Hayek (*P. alpinum* Linnaeus subsp. *sendtneri* (Hayek) Schinz & Keller; *P. pyrenaicum* (Linnaeus) Kerner subsp. *sendtneri* (Hayek) Fedde). Illustration: Rasetti, I fiori delle Alpi, pl. 54, f. 213 (1980); Hegi, Illustrierte Flora von

Mitteleuropa, edn 2, **4**(1): 37 & t. 123
(1986); Huxley, Mountain flowers of
Europe, f. 185 (1986).

Like *P. rhaeticum* but ultimate leaf-
segments narrower, usually acute,
spreading, scape to 15 cm, with adpressed
or spreading bristles, flowers 3–4 cm
across (rarely more), petals usually white;
stigmatic rays usually 5. *Alps*. H2. Spring–
summer.

16. P. burseri Crantz (*P. alpinum*
Linnaeus subsp. *burseri* (Crantz) Fedde).
Illustration: Bonnier, Flore complète **1**:
pl. 25 (1911); Grey-Wilson & Blamey,
Alpine flowers of Britain and Europe, 59
(1979); Huxley, Mountain flowers of
Europe, f. 187 (1986).

Perennial herb with leaf-bases forming
a loose tunic around the base of the plant,
the whole plant more or less hairless.
Leaves 2–3 times pinnate, ultimate
segments narrow, acute, spreading. Scape
to 20 cm, sometimes with a few adpressed
bristles. Flowers 3–4 cm across, petals
white. Capsule club-shaped, 1–1.2 cm,
stigmatic disc pyramidal with usually 4
rays. *Alps, Carpathians*. H2. Spring–
summer.

17. P. kerneri Hayek (*P. alpinum* subsp.
kerneri (Hayek) Fedde). Illustration:
Botanical Magazine, n.s., 432 (1963);
Rasetti, I fiori delle Alpi, pl. 53, f. 212
(1980); Hegi, Illustrierte Flora von
Mitteleuropa, edn 2, **4**(i): t. 37 & t. 123
(1986); Huxley, Mountain flowers of
Europe, f. 188 (1986).

Very like *P. burseri* but somewhat
hairier, petals yellow, capsule to 1 cm,
broadly club-shaped, stigmatic rays usually
5. *Alps (Austria, Italy, former Yugoslavia)*.
H2. Spring–summer.

Section **Meconidium** Spach (Section
Miltantha Bernhardi).

Glaucous, biennial herbs. Leaves 1–4
times pinnate or pinnatisect into narrow
ultimate segments. Flowers in complex
panicles, flowering branches arising from
almost every leaf-axil. Petals reddish to
orange. Filaments thread-like. Capsule
cylindric-club-shaped or cylindric-ellipsoid,
usually tapered at both ends, stigmatic
disc usually narrower than the capsule,
pyramidal.

18. P. fugax Poiret (*P. caucasicum*
Bieberstein). Illustration: Botanical
Magazine, 1675 (1814); Townsend &
Guest, Flora of Iraq **4**(2): pl. 140 (1980).

Stems stiffly erect, to 60 cm, hairless
or sparsely bristly. Leaves pinnatisect, the

segments toothed or lobed, 5 mm or more
broad, narrowly to broadly triangular.
Buds hairless. Flowers to 5 cm across.
Petals orange. Capsule gradually tapered to
the base, abruptly so to the apex, 1.4–1.8
× 0.4–0.8 cm, usually hairless,
occasionally sparsely bristly. Disc depressed
conical to almost flat. *Turkey, Iran, N Iraq,
former USSR (Caucasus)*. H3. Spring–
summer.

P. triniifolium Boissier. Illustration: Das
Pflanzenreich **40**: 351 (1909); Botanical
Magazine, 9292 (1932). Similar but very
glaucous; leaves 2–4 × pinnatisect,
ultimate segments linear, 1–2 mm broad;
petals orange-pink; capsules 6–14 × 3–4
mm, disc conical. *E & S Turkey*. H3.
Summer.

Both of these species are occasionally
grown from seed, but tend not to persist in
cultivation.

Section **Horrida** Elkan.

Annual (or, in cultivation, biennial or
short-lived perennial) herbs. Stems and
leaves with rigid, prickle-like, tuberculate-
based bristles.

19. P. aculeatum Thunberg. Illustration:
Botanical Magazine, 3623 (1837);
Eliovson, South African wild flowers for the
garden, edn 4, 138A (1965); Gibson, Wild
flowers of Natal (inland region), pl. 49
(1978).

Herb to 60 cm, almost all parts with
stiff, yellowish, spreading bristles, which,
on the stem at least, are tuberculate-based
and prickle-like. Rosette-leaves stalked,
pinnatisect, the segments ovate, oblong
or triangular, angular-toothed, the teeth
ending in bristles. Flower-stalks long. Buds
ovoid, bristly. Flowers to 5 cm across.
Petals orange. Capsule ellipsoid-club-
shaped, gradually tapered or somewhat
rounded to the base, slightly tapered to
the apex, not bristly. *South Africa (Cape
Province, Natal); Australia*. H5–G1. Spring.

Grown occasionally as a curiosity, but
not usually persisting for long.

Section **Rhoeadium** Spach (Section
Orthorrhoeades Fedde).

Annual herbs. Stem-leaves not stem-
clasping. Filaments thread-like. Capsules
hairless; sinuses between the lobes of the
stigmatic disc obscure, not extending for
more than one-third of the radius of the disc.

Literature: Kadereit, J., A revision of
Papaver section Rhoeadium, *Notes from
the Royal Botanic Garden Edinburgh* **45**:
225–286 (1988).

20. P. rhoeas Linnaeus (*P. strigosum*

Boenninghausen). Illustration: Bonnier,
Flore complète **1**: pl. 24 (1911); Polunin
& Huxley, Flowers of the Mediterranean,
pl. 34 (1965); Polunin, Flowers of Europe,
pl. 29 (1969); Hegi, Illustrierte Flora von
Mitteleuropa, edn 2, **4**(1): t. 123 (1986).

Stem to 1 m, bristly. Leaves variable,
usually pinnatifid or pinnatisect with
toothed segments, the terminal segment
usually longer than the others, usually
lanceolate. Flowers on long stalks with
dense, usually spreading, reddish or
brownish bristles. Flowers to 7 cm across.
Petals deep red to crimson, each usually
with a black blotch at the base (petals
pink or white and unblotched in some
cultivars). Stamens black (yellowish in
some cultivars). Capsule rounded to the
base, half-spherical; disc flat. *Eurasia, N
Africa, widely introduced elsewhere are a crop
weed*. H1. Spring–summer.

A variable species, with several
cultivated variants, including some with
double flowers. The most widely cultivated
variant is 'Shirley', in which the flowers
have lost all black coloration, the petals
being pale pink or white (sometimes white
with pink margins), not black-blotched,
and the stamens yellowish; some of these
variants have somewhat doubled flowers.
The seeds of this species can persist for
many years in the ground, plants
appearing when the soil is turned over.
The name *P. strigosum* is sometimes
applied to variants in which the hairs on
the flower-stalks are adpressed.

P. commutatum Fischer & Meyer.
Similar but smaller in all parts, stems
with soft greyish hairs, petals each with a
rectangular black blotch near the middle.
*Crete, Turkey, former USSR (Caucasus), N
Iran*. H3. Summer.

P. dubium Linnaeus. Similar but
smaller, with pinnatisect leaves without
long terminal segments, stems with sparse
adpressed bristles, flowers pale red, petals
unblotched, fruit narrow, gradually
tapered to the base. *Eurasia, widely
introduced elsewhere*. H1. Spring–summer.
This may occur as a weed in neglected
areas, but is generally not actively
cultivated.

21. P. californicum Gray. Illustration:
Rickett, Wild flowers of the United States
4: pl. 51 (1970).

Slender hairless herb to 60 cm. Leaves
pinnatisect. Petals pale red to orange, each
with a greenish or pale blotch at the base.
Capsule broadly club-shaped, tapered to
the base. *USA (California)*. H5. Spring.

This species is not in general cultivation in Europe, though the name is sometimes found in catalogues. Most material grown under this name is *P. rupifragum* (p. 000).

Section **Carinatae** Fedde.

Like Section *Rhoeadium* but sinuses of the stigmatic disc deep, conspicuous, extending for more than one-third of the radius of the disc: stigmatic rays keeled.

22. P. macrostomum Boissier & Huet. Illustration: Botanical Magazine, 9226 (1931); Townsend & Guest, Flora of Iraq **4**(2): pl. 140 (1980).

Stems to 50 cm with spreading bristles below, adpressed bristles above. Leaves pinnatisect, the segments entire or toothed, oblong-or linear-lanceolate. Flower-stalks with adpressed bristles. Flowers to 5 cm (rarely more) across. Petals broader than long, crimson to red-purple, sometimes black-blotched at the base. Capsule oblong-ellipsoid. *Turkey, former USSR (SE European part, Caucasus), Syria, Iran.* H3. Summer.

Section **Argemonidium** Spach (Section *Argemonorrhoeades* Fedde).

Annual herbs. Stem-leaves not stem-clasping. Filaments club-shaped. Capsule variously shaped, usually with bristles, lobes of the stigmatic disc obscure, extending for less than one-third of the radius of the disc.

Literature: Kadereit, J., A revision of Papaver section Argemonidium, *Notes from the Royal Botanic Garden Edinburgh* **44**: 24–43 (1986).

23. P. pavoninum Fischer & Meyer.

Herb to 55 cm, stems with spreading bristles below. Leaves pinnatisect or pinnatifid, segments coarsely toothed or lobed. Buds ellipsoid or ovoid, each sepal with an apical, projecting process. Flowers to 6 cm across. Petals orange-red, each with a blackish blotch at the base. Capsule ellipsoid to almost spherical, with few to many bristles. Stigmatic disc much narrower than the widest part of the capsule. *Iran, former USSR (Turkmeniya, Kazakhstan, Uzbekistan, Tadzhikistan), Afghanistan, Pakistan.* H3. Spring–summer.

Other species of this section, notably **P. argemone** Linnaeus (smaller plants with small flowers, sepals without processes, fruit cylindric-club-shaped with few bristles, these mostly near the top) and **P. hybridum** Linnaeus (similar to *P. argemone* but capsule almost spherical, very bristly) may occur as weeds, but are not actively cultivated.

Section **Papaver** (Section *Mecones* Bernhardi).

Annual herbs. Stem-leaves clasping the stems. Filaments club-shaped. Capsules hairless. Lobes of stigmatic disc conspicuous but sinuses between them not extending to one-third of the radius of the disc.

24. P. somniferum Linnaeus. Illustration: Bonnier, Flore complète **1**: pl. 24 (1911); Polunin, Flowers of Europe, pl. 28 (1969); Hegi, Illustrierte Flora von Mitteleuropa, edn 2, **4**(1): pl. 123 (1986).

Glaucous herb to 1 m or more. Lower leaves shortly stalked; upper leaves stalkless and stem-clasping, variously toothed, occasionally deeply so. Buds ovoid-oblong, 1.5–2 cm. Flowers to 8 cm across. Petals large, white to purplish, red or pink. Capsules hemispherical, rounded at the base to a short but usually distinct stalk, 5–7 × 4–5 cm. Stigmatic rays usually 8–12, rarely fewer or more. *Known only from cultivation, naturalised or escaped in many parts of the world.* H3. Spring–summer.

Cultivated for ornament, for its seeds (used to flavour bread and pastries) and for the narcotic latex (opium) produced from the almost ripe capsules. Legitimate cultivation of the narcotic-yielding varieties is strictly controlled, and the juices are used for the production of morphine and codeine for medical and veterinary purposes. Illegitimate cultivation of these varieties, for the production of hallucinogenic drugs (opium, heroin), is illegal in most countries, though the cash value of the products is so high that such cultivation is actively pursued in some areas. The varieties cultivated for ornament on the whole do not yield narcotics; they are very variable, with pink, white, lilac or red petals (sometimes double, or all the stamens converted to petals).

For further information, see Hussain, A. & Sharma, J. K. (eds), *The opium poppy*, Medicinal and aromatic plants series No. 1 (1983).

25. P. glaucum Boissier & Haussknecht. Illustration: Townsend & Guest, Flora of Iraq **4**(2): pl. 140 (1980).

Like *P. somniferum* but leaves pinnatisect with triangular-oblong to linear-oblong segments, buds ovoid, 2–3 cm, petals red, each with a black blotch at the base, capsule 1.5–2 cm. *Turkey, Syria, N Iraq, W Iran.* H3. Spring–summer.

4. MECONOPSIS Viguier
J.A. Ratter & G. Taylor

Biennial or perennial herbs, some flowering only once and then dying (monocarpic), others flowering for several years before dying (polycarpic). Leaves simple or variously divided, often forming a dense basal rosette. Flowers stalked, borne singly or in a branched, often compound inflorescence, which opens from above downwards. Sepals 2, rarely 3 or 4, falling early. Petals 4–10. Stamens many. Ovary more or less spherical, ovoid or obovoid to cylindric, usually with a distinct style bearing a number of stigmatic lobes (style rarely expanded to form a disc on top of the ovary, or absent). Fruit a capsule (opening above by flaps).

A genus of about 50 species; one confined to W Europe, the others native to SC temperate Asia. At least 27 species have been recorded in cultivation but many of these have been lost. At least six of the species included here are extremely rare.

Literature: Taylor, G., *An Account of the Genus Meconopsis* (1934); Cobb, J.L.S., Cultivation of the genus Meconopsis, Part I, *The Rock Garden* **19**: 343–353 (1986), Part II, *The Rock Garden* **20**: 150–172 (1987); Cobb, J.L.S., *Meconopsis* (1989).

1a. Style expanded at the base to form a disc on top of the ovary **19. discigera**
  b. Style tapered or of almost even thickness throughout 2
2a. Petals blue or lilac to mauve-purple 3
  b. Petal colour not as above 11
3a. Flowers borne singly on leafless stalks arising between the basal leaves 4
  b. Inflorescence not as above, usually a panicle 6
4a. Leaves deeply dissected; flower-stalks *c*. 10 cm **18. bella**
  b. Leaves simple; flower-stalks taller 5
5a. Scapes, slender, weak, recurved above; flowers pendent; petals 2.5–4 cm **10. quintuplinervia**
  b. Scapes, more or less erect; flowers semi-pendent; petals usually 4–5 cm **9. simplicifolia**
6a. Leaf-stalks, inflorescences, and usually other parts covered with small simple spines (prickly to the touch); upper part of the inflorescence lacking bracts 7
  b. Not as above 9

7a. Leaves pinnate pinnatisect to deeply
    pinnatifid   **17. aculeata**
 b. Leaves toothed to more shallowly
    pinnatifid   8
8a. Leaf-blade more or less oblong to
    lanceolate, toothed; shortly stalked
                        15. horridula
 b. Leaf-blade broadly oblong to ovate,
    deeply scalloped; long stalked
                        16. latifolia
9a. Leaves of basal rosette pinnatisect
    to pinnatifid; plant monocarpic
                      8. napaulensis
 b. Basal leaves shallowly toothed or
    scalloped; plant usually polycarpic
                                      10
10a. Basal leaves elliptic-lanceolate,
    blade wedge-shaped at base
                        14. grandis
 b. Basal leaves more or less oblong-
    cordate, blade square cut to cordate
    at base   **13. betonicifolia**
11a. Flowers yellow or yellow-cream 12
 b. Flowers reddish or white   18
12a. Leaves spathulate to almost linear,
    entire   **11. integrifolia**
 b. Leaves variously toothed or
    dissected   13
13a. Sepals, and usually ovary and
    capsule, hairless; plants polycarpic
                                      14
 b. Sepals, ovary and capsule hairy or
    bristly; plants monocarpic   16
14a. Ovary and capsule cylindric with
    stalkless stigmas   **3. villosa**
 b. Ovary ovoid to ellipsoid with a
    distinct style   15
15a. Tall, slender plant, to 1.5 m; stem
    leaves with 3 leaflets which are
    deeply lobed; panicle diffuse and
    slender with 3-flowered cymes
                    2. chelidonifolia
 b. More compact plant, to 75 cm;
    leaves pinnatisect; flowers on long
    stalks arising singly in the axils of
    the upper stem leaves
                        1. cambrica
16a. Leaves of basal rosette narrowly
    elliptic, with forward pointing teeth,
    densely covered with silky hairs
                          5. regia
 b. Basal leaves deeply dissected   17
17a. Plant very robust, often exceeding
    1 m; inflorescence axis with a dense
    covering of very short stellate hairs
    giving a mealy effect (as well as
    longer hairs)   **7. paniculata**
 b. Somewhat smaller plant, to 60
    cm; leaves and other parts of the
    plant bearing dark-basal bristles
                        6. dhwojii

18a. Flowers white   19
 b. Flowers reddish   20
19a. Blades of basal leaves more or less
    elliptic-oblong, gradually tapering
    into stalk; stems with long semi-
    adpressed simple hairs and a mass
    of very short stellate hairs; bracts
    of uppermost flowers vestigial;
    monocarpic   **4. superba**
 b. Blades of basal leaves more or
    less oblong-cordate, square-cut
    to cordate at base; stems almost
    hairless, waxy; uppermost group of
    flowers arising from a well-
    developed whorl of bracts; usually
    polycarpic   **13. betonicifolia**
20a. Dwarf plant with single-flowered
    scapes, c. 10 cm; leaves deeply
    dissected   **18. bella**
 b. Tall herbs with paniculate
    inflorescences, or if smaller and
    bearing single-flowered scapes then
    leaves not deeply dissected   21
21a. Basal leaves deeply dissected:
    pinnatisect or pinnatifid
                      8. napaulensis
 b. Basal leaves more or less elliptic,
    entire or with forward pointing
    teeth at margins   22
22a. A small plant, rarely 30 cm, with
    single-flowered inflorescences
                        12. sherriffii
 b. A large robust plant, to often nearly
    2 m, with paniculate inflorescence
                          5. regia

1. M. cambrica (Linnaeus) Viguier.
Illustration: Taylor, An Account of the
Genus Meconopsis, pl. 1 (1934); Ross-
Craig, Drawings of British Plants **2**: pl. 9
(1948); Keble Martin, The Concise British
Flora, pl. 5 (1965); Polunin, Flowers of
Europe, pl. 28 (1969).

Polycarpic, hairless to sparsely hairy
herbs. Stems erect, leafy, to 75 cm. Basal
leaves to 25 cm, long-stalked, pinnate to
pinnatisect; stem-leaves reduced forms of
the basal leaves. Flowers arising singly in
the axils of upper leaves; stalks to 35 cm.
Petals usually 4, sometimes more, c. 3.5
× 4 cm, bright yellow or orange. Ovary
ellipsoid or ovoid, hairless, with short
distinct styles. Capsule ellipsoid-oblong.
*Wales, W Ireland, SW England, W France to
N Spain, now extensively naturalised.* Late
spring–early summer. H2.

Double-flowered strains are also
cultivated.

2. M. chelidonifolia Bureau & Franchet.
Illustration: Taylor, An Account of the
Genus Meconopsis, pl. 2 (1934); Journal of

the Royal Horticultural Society **9**: fig. 180
(1974).

Polycarpic. Stems erect, leafy, branched,
to 1.5 m, dark-coloured with adpressed
rust-coloured bristles towards base and
whitish adpressed hairs above. Basal leaves
pinnatisect to pinnatifid, not present at
time of flowering; stem-leaves usually with
3 leaflets; leaves sparsely hairy.
Inflorescences of 2- to 3-flowered cymes
(or sometimes compound cymes) borne
in the axils of upper stem leaves. Flower
stalks slender, hairless, to 8 cm. Flower
buds hairless, obcordate. Petals 4, c. 2 × 2
cm, pale yellow. Ovary ellipsoid, hairless
or with short adpressed hairs, with short
distinct style. Capsule ellipsoid. *China
(Sichuan).* Early–mid summer.

3. M. villosa (J.D. Hooker) G. Taylor
(*Cathcartia villosa* Hooker). Illustration:
Botanical Magazine, 4596 (1851); Flore
des Serres **7**: t. 686 (1851) – same plate as
Botanical Magazine; Taylor, An Account of
the Genus Meconopsis, pl. 3 (1934).

Polycarpic. Stems erect, leafy,
unbranched, to 70 cm, densely covered
with spreading rust-coloured hairs below.
Basal leaves palmatifid (blade 5–11 × 5–11
cm), long-stalked, with indumentum as on
stem; upper stem-leaves stalkless. Flowers
borne singly or in pairs in the axils of upper
stem-leaves. Petals 4, c. 3.5 × 3 cm, pale
yellow. Ovary hairless, cylindric with 4–9
stalkless radiating stigmas. Capsule cylindric,
to 9 cm, often with a wrinkled surface. *E
Nepal to Bhutan.* Late spring–early summer.

4. M. superba Prain. Illustration: Journal
of the Royal Horticultural Society **54**: fig.
96 & 97 (1929); Gardener's Chronicle **87**:
fig. 214 & 215 (1930); Taylor, An Account
of the Genus Meconopsis, pl. 4 (1934);
Botanical Magazine, 9513 (1935).

Monocarpic. Stem robust, erect, leafy,
to 2 m, covered in dense, usually silvery,
semi-adpressed and tiny stellate hairs.
Basal leaves forming a dense overwintering
rosette; blade with forward pointing teeth,
oblanceolate, lanceolate or elliptic, to 40
cm with stalk to 12 cm; indumentum
similar to stem. Flowers borne terminally
on stem, and singly in the axils of the
upper stem-leaves; stalks stout, to 20 cm.
Flowers large, to 11 cm across. Petals
4–6, c. 6 × 4.5 cm, white. Ovary almost
spherical to ellipsoid-oblong with dense
adpressed hairs; style long (to over 1 cm)
bearing prominent, dark red-purple
stigmatic lobes. Capsule ellipsoid-oblong.
China (Xizang) & W Bhutan. Late spring &
early summer.

5. M. regia G. Taylor. Illustration: Taylor, An Account of the Genus Meconopsis, pl. 5 & 29 (1934); Botanical Magazine, 9348 (1934); Journal of the Royal Horticultural Society **59**: fig. 171 (1934), **99**: fig. 181 (1974).

Monocarpic. Like *M. superba* but with basal rosettes of narrowly elliptic leaves with forward pointing teeth, densely covered in a silvery, or sometimes gold, silky felt of long adpressed hairs; inflorescences of small (often 4-flowered) cymes of pale yellow or deep red flowers in the axils of the stem-leaves (becoming single-flowered at the top of the stem). *C Nepal.* Early summer.

A very handsome plant now apparently lost from cultivation, at least in pure unhybridised form. It seems to hybridise readily with *M. paniculata* and *M. napaulensis* and many cultivated stocks appear to be hybrids involving these three, and perhaps other, robust monocarpic species.

6. M. dhwojii Hay. Illustration: Gardener's Chronicle **92**: 405 & 407 (1932); Journal of the Royal Horticultural Society **59**: fig. 172 (1932); Taylor, An Account of the Genus Meconopsis, pl. 6 (1934); Botanical Magazine, n.s., 396 (1962).

Monocarpic with dense overwintering leaf rosette. Stem to 60 cm, stout, leafy, often mauve-purple, with erect bristly hairs. Basal and lower stem-leaves pinnate, to 30 cm, often rather dark glaucous green, covered by sparse but conspicuous bristles with dark swollen bases. Inflorescence with a terminal group of flowers, and below these flowers usually occurring singly in the leaf axils but occasionally, particularly near the base, as 3–5-flowered cymes. Flowers 3–6 cm across. Petals 4 or 5, *c.* 2 × 3 cm, lemon-yellow. Ovary spherical to ellipsoid, covered in pale adpressed hairs with well-developed style. Capsule ellipsoid-oblong. *E Nepal.* Late spring.

Cobb (*Quarterly Bulletin of the Alpine Garden Society* **52**: 63–73, 1984) comments that many of the coarser *M. dhwojii* plants which are now so prevalent in cultivation are probably hybridised with *M. napaulensis*.

7. M. paniculata (D. Don) Prain. Illustration: Botanical Magazine, 5585 (1866) – as M. nipalensis; Taylor, An Account of the Genus Meconopsis, pl. 7 (1934); Journal of the Royal Horticultural Society **98**: fig. 53 (1973); Growing from Seed **3**(2): 25 (1989).

Monocarpic. Stem robust, leafy, to 2 m. Almost the whole plant densely covered with long simple hairs and abundant tiny stellate hairs producing a mealy covering. Basal leaves pinnatisect to pinnatifid, to 40 cm, forming a dense overwintering rosette. Flowering-axis very floriferous and most of the flowers borne in 2–6-flowered axillary cymes. Flowers large, to 9 cm across, pendent. Petals 4 (rarely 5), *c.* 4 × 3 cm, pale lemon-yellow. Ovary spherical or almost spherical, densely covered with adpressed golden yellow hairs; style often nearly as long as ovary. Capsule ellipsoid. *E Nepal to India (NE Assam).* Late spring–early summer.

8. M. napaulensis D.Don. (*M. wallichii* Hooker). Illustration: Botanical Magazine, 4668 (1852); Taylor, An Account of the Genus Meconopsis, pl. 9 (1934); L'ami des jardins et de la maison **56** (no. 726): 53 (1986); Growing from Seed **3**(2): 25 (1989).

Similar to *M. paniculata* but with a generally rather coarse indumentum of rust coloured or pale hairs (often with some tiny almost stellate hairs also present); petals red to purple, violet or blue (very rarely white). *C Nepal to China (W Sichuan).* Summer.

A very variable species; some of the variants in cultivation are hybridised with *M. paniculata* and *M. regia*. The blue-flowered, rust coloured hairy form originally described as *M. wallichii* Hooker is very constant and distinct. It flowers distinctly later than the other large monocarpic species, and this seasonal isolation has probably kept it genetically pure.

9. M. simplicifolia (D. Don) Walpers. Illustration: Botanical Magazine, 8364 (1911); Taylor, An Account of the Genus Meconopsis, pl. 10 (1934); Journal of the Royal Horticultural Society **72**: fig. 56 & 62 (1947); RHS dictionary of gardening **3**: 1272 (1981).

Mono- or polycarpic. Leaves all basal, blade more or less narrowly lanceolate, 7–20 cm with entire irregular forward-pointing teeth or shallowly lobed margins, narrowing at base into a stalk more or less equalling the blade, covered in long straw-coloured hairs. Flowers borne on single-flowered, more or less bristly scapes to 85 cm of which 1–6 produced per plant. Flowers almost pendent. Petals 5–10, 4–5 cm, purple to sky-blue. Filaments, mauve, purple or dark blue. Ovary ellipsoid, hairless to bristly with well-developed style. Capsule oblong-ellipsoid, hairless to bristly. *C Nepal to China (SE Xizang).* Late spring.

Some cultivated variants of *M. grandis* with single-flowered scapes will also key out here, as will **M. delavayi** (Franchet) Prain. The former are generally larger-flowered plants than *M. simplicifolia* and can readily be distinguished by the colour of the filaments (whitish or mauve to dark blue) and ovary and capsule shape. *M. delavayi* is difficult to cultivate and is extremely rare. It is a much smaller plant than *M. simplicifolia* and has very characteristic, usually hairless, and often glaucous, long-stalked leaves with elliptic to narrowly rhomboid entire-marginal blades. Good illustrations of this species are provided in: Taylor, An Account of the Genus Meconopsis, pl. 17 (1934); Botanical Magazine, 294 (1957); and Journal of the Royal Horticultural Society **86**: fig. 86 (1961), **92**: fig. 163 (1967).

10. M. quintuplinervia Regel. Illustration: Taylor, An Account of the Genus Meconopsis, pl. 12 (1934); Meikle, Garden flowers, 82 (1963); Hay, Reader's Digest encyclopaedia of garden plants and flowers, 438 (1985).

Polycarpic. Leaves all basal, to 25 cm (including stalk), lanceolate to obovate, longitudinally 3–5-nerved, narrowing into a long stalk, more or less densely covered in straw-coloured to rust-coloured bristles. Flowers pendent borne singly on slender leafless bristly hairy scapes to 45 cm; 1 (rarely to 3) scapes per leaf-rosette. Petals 4 (rarely to 6), 2.5–4 cm, pale milky blue to purplish. Ovary ellipsoid to almost spherical, densely covered in adpressed bristles; style short. Capsule ellipsoid. *China (NE Xizang, S Gansu, NW Sichuan to C Shaanxi).* Spring.

M. punicea Maximowicz. Illustration: Botanical Magazine, 8119 (1907); Taylor, An Account of the Genus Meconopsis, pl. 13 (1934); Journal of the Royal Horticultural Society **84**: fig. 151 (1959). Similar but with larger flowers with spectacular intensely red petals. It has been reintroduced to cultivation recently but is extremely rare.

The hybrid **M. × cookei** G. Taylor (*M. punicea × M. quintuplinervia*) is an attractive plant combining characters of both parents but has probably been lost from cultivation.

11. M. integrifolia (Maximowicz) Franchet. Illustration: Botanical Magazine, 8027 (1905); Taylor, An Account of the

Genus Meconopsis, pl. 14 (1934); Encke, Parey's Blumengärtnerei **1**: 681 (1958); RHS dictionary of gardening **3**: 1272 (1981).

Monocarpic herb, covered throughout by a sparse to dense covering of white or golden yellow to rust-coloured hairs. Basal leaves to 30 cm, spathulate to obovate, entire, with 3 or more prominent longitudinal veins narrowing into the stalk. Flowers borne on single-flowered leafless scapes and/or stout leafy axes to 40 cm terminating in an apical group of long-stalked flowers. Flower buds spherical. Flowers not, or only partially deflexed, large. Petals 6–9, *c.* 5 cm, lemon-yellow. Ovary ellipsoid-ovoid, usually surmounted by a bundle of 4–9 stigmatic rays forming a club-shaped structure *c.* 5 mm. Capsule broadly ellipsoid oblong. *China (W Gansu, Sichuan, Xizang, NW Yunnan) & NE Upper Burma*. Late spring.

The hybrids of *M. integrifolia* with *M. betonicifolia* (*M. × sarsonsii*) and *M. grandis* (*M. × beamishii*) are described under these last two species.

12. M. sherriffii G. Taylor. Illustration: Taylor, New Flora and Silva **9**: opposite p. 160 (1937); Journal of the Royal Horticultural Society **52**: fig. 65 (1947), **76**: fig.146 (1951); Cobb, The rock garden **20**: 211, fig. 46 (1987).

Polycarpic. Leaves densely aggregated in a basal rosette, blade to 6 cm, more or less narrowly elliptic, tapered below into a stalk of about the same length, densely silky felted to hairy. Flowering-stems to 70 cm, solitary from each rosette, bearing a few leaves or sometimes leafless, densely silky felted to hairy, single-flowered. Petals 5–8, to 5 cm, pinkish pale wine-red. Ovary more or less broadly ellipsoid, densely silky felted with style and stigmas usually like *M. integrifolia* (sometimes a slender style present). Capsule ellipsoid-oblong. *China (Xizang), Bhutan*.

An extremely rare and difficult species in cultivation.

13. M. betonicifolia Franchet (*M. baileyi* Prain). Illustration: Botanical Magazine, 9185 (1930); Taylor, An Account of the Genus Meconopsis, pl. 15 (1934); RHS dictionary of gardening **3**: 1271 (1981).

Polycarpic (rarely monocarpic). Basal leaves with stalk 5–30 cm; blade to 30 cm, oblong to elongate-oblong, square-cut or slightly cordate at base, more rarely tapering gradually on to stalk, margin scalloped-toothed, with straw-coloured or brown hairs along margin and sometimes over leaf surface. Stem to 1.5 m, hairless or with scattered rust-coloured hairs, leafy and with a terminal whorl of 4–7 small bracts subtending the apical group of flowers, lower flowers borne singly in the axils of leaves below the apical bract-whorl. Flowers large, semi-pendent, 7–13 cm across. Petals 4–7, 3.5–6 cm, pale blue, more rarely purple-mauve or white. Ovary ellipsoid-oblong, hairless or covered in adpressed hairs. Capsule ellipsoid-oblong. *China (NW Yunnan, SE Xizang) & N Upper Burma*. Mid-summer.

A hybrid with *M. integrifolia*, **M. × sarsonii** Sarsons, is cultivated. Illustration: Journal of the Royal Horticultural Society **84**: fig. 152 (1959). It resembles *M. betonicifolia*, the female parent, much more closely than *M. integrifolia*, but has cream-yellow flowers.

14. M. grandis Prain. Illustration: Botanical Magazine, 9304 (1933); Taylor, An account of the genus Meconopsis, pl. 16 (1934); The Garden **102**: 54 (1977); Hay, Reader's Digest encyclopaedia of garden plants and flowers, 439 (1985); Growing from Seed **3**(2): 24 (1989).

Polycarpic. Basal leaves with stalk to 16 cm; blade to 35 cm, narrowly lanceolate to elliptic, tapering into the stalk at the base, acute or almost acute at apex, shallowly toothed or scalloped, with rust-coloured bristles scattered on both surfaces. Flowering-axis usually a leafy stem, often with rust-coloured hairs to 1.2 cm, with a whorl of 2 or 3 bracts subtending a group of *c.* 3 flowers, or, more rarely, a single-flowered leafless scape. Flowers large, semi-pendent, to 13 cm across. Petals 4–9, to 7 cm, deep sky-blue to mauve-purple. Filaments whitish. Ovary narrowly ovoid to oblong, hairless to more or less densely hairy. Capsule narrowly ellipsoid-oblong. *E Nepal, W China, Sikkim, Bhutan*. Late spring.

Variants of this species with single-flowered scapes may be confused with *M. simplicifolia* – see note under *M. simplicifolia* for the characters to distinguish the two species.

Various strains of the hybrid of this species with *M. betonicifolia*, **M. × sheldonii** G. Taylor, are cultivated. Illustration: Journal of the Royal Horticultural Society **84**: fig. 153 (1959); Growing from Seed **3**(2): 24 (1989). They are vigorous, extremely handsome plants with the larger flowers of *M. grandis* but showing *M. betonicifolia* characteristics in the inflorescence, in that flowers are often borne in the axils of one or two of the

stem-leaves below the uppermost whorl of bracts.

The hybrid of this species with *M. integrifolia*, **M. × beamishii** Prain, is quite common in cultivation. It has yellow petals which are sometimes blotched with purple at the back towards the base. It is very similar to *M. × sarsonii* (*M. betonicifolia × integrifolia*) with which it is frequently confused.

15. M. horridula J.D. Hooker & Thomson. Illustration: Botanical Magazine, 8568 (1914) – as M. rudis, 8619 (1915) – as M. prattii; Taylor, An Account of the Genus Meconopsis, pl. 24 (1934); Quarterly Bulletin of the Alpine Garden Society **52**: 289 (1984).

Monocarpic. Whole plant covered in semi-erect spines. Basal leaves to 25 cm, shortly stalked, oblong-lanceolate, toothed or sinuous, forming a rosette. Flowering-stem to 75 cm, uppermost flowers (to *c.* 12 in number) without bracts, lower flowers borne singly in the axils of bracts usually right to the base of the stem; rarely flowers borne on leafless scapes. Petals 4–8, to 4 cm, pale or deep royal blue to blue-violet. Ovary and capsule ellipsoid to almost spherical, spiny, with long style and lobed green stigma. *China (Gansu, NW Yunnan, SE Xizang), Upper Burma, Bhutan, Sikkim, C Nepal*. Early summer.

16. M. latifolia (Prain) Prain. Illustration: Botanical Magazine, 8223 (1908) – as M. sinuata var. latifolia; Taylor, An Account of the Genus Meconopsis, pl. 25 (1934); Journal of the Royal Horticultural Society **84**: fig. 158 (1959); Growing from Seed **3**(2): 24 (1989).

Similar to *M. horridula* but blade broadly oblong, ovate or lanceolate, coarsely scalloped or with deep forward-pointing teeth, abruptly narrowed to a long stalk (blade to 15 cm, stalk to 13 cm); petals 4, milky blue; stigma purple. *India (N Kashmir)*. Early summer.

17. M. aculeata Royle. Illustration: Botanical Magazine, 5456 (1864); Sayers, Journal of the Royal Horticultural Society **93**: fig. 130 (1968).

Similar to *M. grandis* and *M. horridula* but leaves pinnate, pinnatisect or pinnatifid. *W Himalaya*. Late spring–early summer.

In cultivation, populations of *M. horridula*, *M. latifolia* and *M. aculeata* show great variability, including the occurrence of many 'runts', indicating the presence of much heterozygosity. Cobb (Meconopsis hybrids, *Quarterly Bulletin of the Alpine Garden Society* **52**: 63–73, 1984) indicates that hybridisation between these species is prevalent.

18. M. bella Prain. Illustration: Botanical Magazine, 8130 (1907); Taylor, Journal of the Royal Horticultural Society **52**: fig. 63 (1947); RHS dictionary of gardening **3**: 1271 (1981).

Polycarpic. Leaves all basal and crowded; stalk to 10 cm, blade to 10 × 2.5 cm, more or less irregularly pinnately or bipinnately lobed with the ultimate segments usually 3-fid (rarely simple and rhomboid), hairless or very sparsely bristly. Flowers solitary on scapes to *c.* 10 cm. Petals 4 or 5, *c.* 3.5 cm, pink, pale blue or purple. Ovary ellipsoid to almost spherical, hairless or somewhat bristly below; style distinct becoming much swollen at the base in fruit. Capsule more or less pear-shaped or obovoid. *C Nepal to Bhutan.*

Dissection of the leaves can be very variable: simply pinnate leaves often look rather like those of a scabious, while simple rhomboid leaves resemble those of *M. delavayi* – in fact herbarium specimens of *M. bella* and *M. delavayi* can look very alike. The species is difficult to grow and extremely rare in cultivation.

19. M. discigera Prain. Illustration: Taylor, An Account of the Genus Meconopsis, pl. 28 (1934); Journal of the Royal Horticultural Society **84**: fig. 157 (1959).

Monocarpic. Basal leaves in a dense rosette covered in golden brown bristles (particularly on the stalk), to 14 cm including the long linear stalk, blade narrowly oblanceolate tapering into the stalk and usually trilobed at the apex. Flowering-stem to 40 cm, bristly or hairy, bearing 13–20 flowers, without bracts above but flowers subtended by stalkless bracts below. Petals 4, to 10 cm, dark crimson, red, purple or pale blue. Ovary shortly oblong, covered with golden yellow bristles; style distinct and slender, to 7 mm, expanded at the base into a broad hairless disc extending over the top of the ovary and fringed at the margin. Capsule oblong. *C Nepal & Bhutan to W China.*

A very rare species in cultivation.

5. STYLOMECON Taylor
J. Cullen
Annual herb with yellow sap. Leaves alternate, pinnatifid to pinnatisect, segments entire or themselves further divided. Flowers in the upper leaf-axils on long stalks. Sepals 2, free, falling individually. Petals 4. Stamens numerous. Ovary 1-celled with 4 or more placentas. Style slender, short, with a head-like stigma. Capsule narrowly top-shaped, opening by pore-like flaps towards the apex. Seeds kidney-shaped.

A genus of a single species from California, originally included in *Meconopsis*. It is easily grown in a sunny site. Propagate by seed.

1. S. heterophylla (Bentham) Taylor (*Meconopsis heterophylla* Bentham). Illustration: Botanical Magazine, 7636 (1899); Morley & Everard, Wild flowers of the world, pl. 154 (1970); Rickett, Wild flowers of the United States **4**: pl. 51 (1970).

Annual herb to 60 cm, hairless or stems hairy below. Leaves variable. Sepals 4–10 mm. Petals 1–2 cm, orange-red with a purplish spot near the base, the claw below the spot greenish. Capsule to 1.5 cm. *USA (California).* H3. Spring.

6. ROEMERIA Medikus
J. Cullen
Slender annual herbs with yellowish latex. Leaves alternate, 1–3 times pinnatisect. Flowers solitary. Sepals 2; petals 4. Stamens numerous, filaments club-shaped or thread-like. Ovary of 3 or 4 united carpels; placentation parietal. Fruit a linear-cylindric capsule opening towards the apex, bristly all over or at least in part, or with 3 or 4 long bristles at the apex which overtop the stigmas.

A genus of a few species from the Mediterranean area and the Middle East. They are easily grown and propagated by seed.

1. R. refracta de Candolle. Illustration: Das Pflanzenreich **40**: 240 (1909); Morley & Everard, Wild flowers of the world, pl. 46 (1970).

Stem to 60 cm. Leaves 2–3 times pinnatisect with narrow segments. Flowers to 6 cm across. Petals deep red, each with a black blotch at the base. Filaments club-shaped. Capsule hairless except for 3 or 4 bristles at the apex. *Middle East (Turkey to Afghanistan & Pakistan).* H3. Spring–summer.

Rather like *Papaver rhoeas* (p. 109) in general appearance, but with very different fruit.

R. hybrida (Linnaeus) de Candolle, which has smaller, violet petals, thread-like filaments and a capsule which is bristly, at least in part, though without 3 or 4 bristles at the apex, may occur as a weed in gardens in S Europe.

7. PLATYSTEMON Bentham
J. Cullen
Small, hairy, annual herbs. Leaves opposite or whorled, entire, mostly borne in the lower parts of the stems, more or less stalkless, linear-oblong to lanceolate. Flowers solitary on long stalks. Sepals 3, rather slow to fall as the flower opens. Petals 6, white, yellowish or cream, 8–20 mm. Stamens numerous. Ovary of 6–25 carpels, each with many parietal ovules, at first united, ultimately separating and splitting transversely into 1-seeded segments.

A genus of a single species, easily grown from seed.

1. P. californicus Bentham. Illustration: Botanical Magazine, 3579 (1837), 3750 (1840); Morley & Everard, Wild flowers of the world, pl. 154 (1970); Rickett, Wild flowers of the United States **6**: pl. 57 (1973).

SW USA (California, Arizona, Utah), Mexico (Baja California). H3. Summer.

A very variable species divided into numerous varieties (treated as species by some earlier authors); it is uncertain which of these are in cultivation.

8. STYLOPHORUM Nuttall
J. Cullen
Perennial herbs with yellow-orange or reddish latex. Leaves mostly basal and long-stalked, generally with 2 opposite leaves on the stem between which is the inflorescence; blades pinnatisect (or the lowermost lobes distinct and separate) into 5–7 lobes which are themselves irregularly bluntly lobed or toothed, all green and sparsely hairy with curled hairs. Flowers few in an umbel-like cyme borne between the upper leaves; flower-stalks long, with small bracteoles at the base. Sepals 2, free, falling early. Petals 4, yellow. Stamens yellow, numerous. Ovary of 3 or 4 united carpels; style short, stigmas 3 or 4. Fruit a capsule covered with bristles.

A genus of 3 species, 2 from E Asia, the other from eastern N America; of these, only the American is cultivated. It generally has an ovary of 3 or 4 (rarely 2) carpels, while the Asiatic species have ovaries of 2 carpels. They are woodland plants which require a moist, rich soil and partial shade. Propagate by seed.

1. S. diphyllum (Michaux) Nuttall. Illustration: Botanical Magazine, 4867 (1855); Das Pflanzenreich **40**: 205 (1909);

Rickett, Wild flowers of the United States **1**: pl. 36 (1969).

Flowering stem to 50 cm. Petals elliptic to broadly obovate, to 2.5 cm. Fruits broadly ellipsoid, tapered to both ends, to 2.5 cm. *E USA*. H3. Summer.

9. CHELIDONIUM Linnaeus
J. Cullen

Perennial, somewhat glaucous, rhizomatous herbs with yellow, irritant latex. Leaves deciduous, alternate, long-stalked, pinnate with oblong-ovate or obovate leaflets which are lobed or scalloped. Flowers long-stalked in leaf-opposed umbels; flower-stalks with small bracteoles at the base. Buds pear-shaped. Sepals 2, falling quickly as the flower opens, somewhat yellowish. Petals 4, obovate, yellow, 7–15 mm. Stamens numerous, yellow. Ovary of 2 united carpels; style short; stigma 2-lobed. Fruit a linear-cylindric, wrinkled capsule to 6 cm. Seeds numerous, shiny, each with a small, whitish aril.

A genus of a single species from Eurasia, introduced into eastern N America. It is easily grown; propagate by seed.

1. C. majus Linnaeus. Illustration: Bonnier, Flore complète **1**: pl. 26 (1911); Hess, Landolt & Hirzel, Flora der Schweiz **2**: 116 (1970); Ceballos et al., Plantas silvestres de la peninsula Iberica, 108 (1980); Garrard & Streeter, The wild flowers of the British Isles, pl. 4 (1983).

Eurasia, introduced in N America. H1. Spring–autumn.

The wild species is generally regarded as a weed, but double-flowered ('Flore-Pleno') and cut-leaved ('Laciniatus') variants are occasionally grown as curiosities.

10. HYLOMECON Maximowicz
J. Cullen

Perennial herbs with short, oblique rhizomes and yellowish latex. Leaves soft, bright green, mostly basal and long-stalked, 1 or 2 alternate on the flowering-stems, rather shortly stalked; blades pinnate with usually 5 oblong-obovate or obovate, sharply and irregularly toothed leaflets. Flower solitary, terminal, stalked, without bracteoles. Sepals 2, falling as the flower opens. Buds ovoid, pointed. Petals 4, yellow, elliptic to almost circular, 2–2.5 cm. Stamens numerous, yellow. Ovary of 2 united carpels, styles short, stigmas 2. Fruit a linear-cylindric capsule.

A genus of a single species of woodland plants from E Asia, easily cultivated in moist places, in some shade. Propagate by seed.

1. H. japonica (Thunberg) Prantl & Kundig (*Chelidonium japonicum* Thunberg; *H. vernalis* Maximowicz). Illustration: Botanical Magazine, 5830 (1870) – as Stylophorum japonicum; Das Pflanzenreich **40**: 205 (1909); Gartenpraxis for 1984: 23 (1984).

E Asia (former USSR, China, Japan, Korea). H2. Spring.

A variable species; variants with very dissected leaves (var. **dissecta** (Franchet & Savatier) Fedde and with narrow leaflets (var. **lanceolata** Makino) are known in the wild and may be found in gardens.

11. GLAUCIUM Miller
J. Cullen

Annual, biennial or short-lived perennial herbs. Leaves often glaucous and somewhat fleshy, pinnatifid, pinnatisect, lobed or almost entire, alternate. Flowers solitary or in cymes. Sepals 2, falling as the flower opens. Petals 4, yellow, orange, red or mauve. Stamens numerous. Ovary of 2 united carpels containing many parietal ovules; style absent or very short, stigma mitre-like, conspicuous. Capsule linear-cylindric, usually more than 10 cm, often curved, sometimes contorted, opening from the base upwards or from the apex downwards. Seeds embedded in a spongy false septum.

A genus of about 20 species from Eurasia. Few are cultivated; they are easily grown and propagation is by seed.

Literature: Mory, B., Beitrage zur Kentniss der Sippenstruktur der Gattung Glaucium, *Feddes Repertorium* **89**: 499–594 (1979).

1a. Fruit opening from the apex downwards, covered with small prickles   **4. squamigerum**
  b. Fruit opening from the base upwards, smooth, warty or hairy, not prickly   **2**
2a. Petals yellow or rarely yellow-orange   **3. flavum**
  b. Petals red or red-orange   **3**
3a. Annual; flower- and fruit-stalk shorter than the leaf that subtends it   **1. corniculatum**
  b. Perennial; flower- and fruit-stalk longer than the leaf that subtends it   **2. grandiflorum**

1. G. corniculatum (Linnaeus) Rudolph. Illustration: Bonnier, Flore complète **1**: pl. 25 (1911); Huxley & Polunin, Flowers of the Mediterranean, pl. 35 (1965); Polunin, Flowers of Europe, pl. 29 (1969); Garrard & Streeter, Wild flowers of the British Isles, pl. 4 (1983).

Annual herb to 40 cm, stems ascending to erect. Leaves pinnatifid or pinnatisect, the segments toothed, hairy. Buds ovoid, acuminate. Sepals 1–2.5 cm, rough-hairy. Petals 1.5–2.8 cm, usually reddish or reddish orange. Capsule opening from the base upwards, hairy, 10–25 cm, borne on a stalk shorter than the leaf which subtends it. *Mediterranean area*. H4. Spring–summer.

Variable in flower colour.

2. G. grandiflorum Boissier & Huet.
Perennial herb with a single main stem to 50 cm, roughly hairy. Leaves pinnatifid to pinnatisect, with toothed segments. Buds to 4.5 cm, narrowly ovoid, long-acuminate. Sepals 3–5 cm, roughly hairy. Petals 3–6 cm, dark red or dark orange, each with a darker spot at the base. Capsule opening from the base upwards, hairy, 10–15 cm, borne on a stalk longer than the subtending leaf. *Greece, SW Asia*. H3. Summer.

3. G. flavum Crantz. Illustration: Bonnier, Flore complète **1**: pl. 25 (1911); Polunin, Flowers of Europe, pl. 29 (1969); Parrish, Flowers in the wild, 24 (1983); Garrard & Streeter, Wild flowers of the British Isles, pl. 4 (1983).

Sprawling biennial or perennial herb. Stems to 60 cm or more, hairless or hairy. Leaves pinnatifid to pinnatisect, with curled hairs. Buds 2–3.5 cm, ovoid. Sepals with curled hairs. Petals yellow, occasionally somewhat reddish, sometimes with a purple blotch at the base, to 5 cm. Capsules opening from the base upwards, warty over most of their length, 15–25 cm. *Coasts of Europe, N Africa, SW Asia*. H3. Summer.

Variable in flower colour. Var. **fulvum** (Smith) Fedde has orange-yellow petals, while var. **serpieri** (Heldreich) Halacsy has pale yellow petals, each with a purple spot at the base; var. **flavum** has pure yellow petals.

4. G. squamigerum Karelin & Kiriloff.
Annual. Stems to 40 cm, with scattered, whitish hairs. Leaves usually 2, pinnatisect. Flower solitary. Sepals with thick, white hairs. Petals yellow, each with a darker or orange spot at the base. Capsule opening from the apex downwards, covered with small, whitish prickles. *C Asia (former USSR)*. H3. Summer.

Very doubtfully cultivated, though the name appears in some catalogues.

12. DICRANOSTIGMA Hooker & Thomson
J. Cullen

Like *Glaucium* but always perennial, fruits not longer than 10 cm, the seeds not embedded in a spongy false septum.

A genus of 3 species from the Himalaya and China. They are not often cultivated, though easily grown.

1a. Buds without long-acuminate tips; capsule with a dense covering of short hairs **3. lactucoides**
  b. Buds with long-acuminate tips; capsules not hairy, sometimes warty **2**
2a. Flowers 2–3 cm across; fruits linear-cylindric, not tapering towards the apex **2. leptopodum**
  b. Flowers 4–5 cm across; fruits oblong-cylindric, tapering towards the apex **1. franchetianum**

1. D. franchetianum (Prain) Fedde (*Glaucium vitellinum* misapplied). Illustration: Botanical Magazine, 9404 (1935).

Plant erect, to 1.3 m. Leaves mostly basal, pinnatisect, hairy. Buds with long-acuminate tips. Flowers somewhat cup-shaped, 4–5 cm across. Petals yellow. Capsule oblong-cylindric, tapering to the apex, often somewhat warty. *C & W China.* H3. Summer.

Material offered as *Glaucium vitellinum* (a distinct species of *Glaucium* recognised in Mory's revision) is usually this species.

2. D. leptopodum (Maximowicz) Fedde. Illustration: Baileya **15**: 112 (1967).

Like *D. franchetianum* but much smaller, flowers widely open, 2–3 cm across, fruits linear-cylindric, not tapering above. *C China.* H3. Summer.

Material grown as *D. franchetianum* is often this species.

3. D. lactucoides Hooker & Thomson.

Plant to 60 cm. Leaves pinnatisect, hairy. Flowers to 5 cm across. Buds without long-acuminate tips. Petals yellow-orange. Fruits cylindric, softly hairy. *Himalaya to W China.* H5. Summer.

This species has been introduced several times to cultivation from the Himalaya, but does not persist for long.

13. EOMECON Hance
J. Cullen

Herbaceous perennial to 50 cm. Leaves all basal, long-stalked; blades ovate, main veins palmate, margins rather wavy or scalloped with blunt scallops, base deeply cordate. Inflorescence a loose cyme with several flowers; flower-stalks with small bracteoles at the base. Buds almost spherical but tapering to an acuminate apex. Sepals 2, completely united, falling as a whole as the flower opens. Petals 4, to 1.5 cm, white, obovate. Stamens numerous, yellow. Ovary of 2 united carpels; style present, lengthening in fruit, stigma bilobed. Fruit an ellipsoid capsule tapering into the long, persistent style.

A genus of a single species from China, very beautiful but not widely grown, as it does not persist for long. It requires woodland conditions with partial shade and a rich, moist soil. Propagate by seed, when available.

1. E. chionantha Hance. Illustration: Botanical Magazine, 6871 (1886); Das Pflanzenreich **40**: 205 (1909); The Garden **110**(3): front cover (1977).

S, W & C China. H5. Spring–autumn.

14. SANGUINARIA Linnaeus
J. Cullen

Herbaceous perennials with creeping rhizomes; latex red. Leaf 1 to each flowering-shoot, basal, long-stalked and with small scale-leaves at the base, blade to 30 cm across, circular or somewhat broader than long, main veins palmate, palmately lobed to about half the radius or less, especially towards the apex, the lobes entire or rather bluntly toothed, base cordate, all greyish green. Flowers solitary, borne on a scape with a scale-leaf at the base, appearing just before the leaves. Sepals 2, free, soon falling. Petals 8–16, to 3 cm, obovate, white or white flushed with pink. Stamens many. Ovary of 2 united carpels; style present, stigma small. Fruit an ellipsoid capsule, tapering at both ends.

A genus of a single species from eastern N America. It is a woodland plant and thrives best in partial shade in rather moist, rich soil. Propagate by seed or division of the rhizomes.

1. S. canadensis Linnaeus. Illustration: Das Pflanzenreich **40**: 205 (1909); Justice & Bell, Wild flowers of North Carolina, 73 (1968); Rickett, Wild flowers of the United States **1**: pl. 36 (1969); Quarterly Bulletin of the Alpine Garden Society **54**: 286, 287 (1986).

Eastern N America, from Quebec & Manitoba to Texas & Florida. H3. Spring.

'Multiplex' ('Flore Pleno'), which has all the stamens converted to petals, and the ovary absent, is widely grown.

15. MACLEAYA R. Brown
J. Cullen

Tall, rhizomatous perennial herbs with stems woody below but dying back every year; latex orange. Leaves alternate, basal and on the stems, the lower long-stalked, all with the main veins palmate, blade oblong to circular, cordate at the base, rather deeply lobed, the lobes themselves lobed or bluntly toothed, all glaucous, hairy on the lower surface, at least when young. Flowers in clusters in large, upright, plume-like panicles. Buds narrowly obovoid, whitish or purplish. Sepals 2, falling as the flowers open. Petals absent. Stamens 8–30, with long or short filaments, anthers mucronate. Ovary of 2 united carpels; style short, stigma bilobed. Ovules few and parietal or 1 and basal. Fruit a small, flattened capsule containing 1–few seeds. Seeds each with a small, lateral aril.

A genus of 2 species from Japan and China, once confused with *Bocconia* (p. 117). They form striking large herbs with plumes of small flowers; the inflorescences are sometimes dried for interior decoration. The plants are easily grown and can become invasive in suitable places. Propagate by seed or by division of the rhizome.

Literature: Hutchinson, J., *Bocconia and Macleaya, Kew Bulletin* for 1920: 275–282 (1920).

1a. Stamens 25–30; fruits oblanceolate to obovate, with 4–6 seeds **1. cordata**
  b. Stamens 8–12; fruits almost circular, though pointed, with a single seed **2. microcarpa**

1. M. cordata (Willdenow) R. Brown (*Bocconia cordata* Willdenow). Illustration: Das Pflanzenreich **40**: 216 (1909); The Garden **101**: 150 (1976).

Stems to 2 m, glaucous. Hairs soon falling from leaf undersurfaces. Buds to 1 cm. Stamens 25–30, filaments as long as anthers. Capsule 1.5–2.2 cm, containing a few parietally attached seeds. *China, Japan.* H3. Summer.

2. M. microcarpa (Maximowicz) Fedde (*Bocconia microcarpa* Maximowicz). Illustration: Das Pflanzenreich **40**: 216 (1909); Perry, Flowers of the world, 223 (1972).

Like *M. cordata*, but rather smaller,

leaves persistently hairy beneath, buds to 5 mm, stamens 8–12, filaments shorter than anthers, capsule more or less circular though pointed at each end, to 5 mm, containing a single, basally attached seed. *C China*. II3. Summer.

The hybrid between these 2 species is **M. × kewensis** Turrill. Illustration: Botanical Magazine, n.s., 321 (1958). Probably more widely cultivated than either. It is intermediate between its parents in most characters, and is sterile.

16. BOCCONIA Linnaeus
J. Cullen
Shrubs, trees or climbers with persistent woody shoots, latex yellowish. Leaves alternate, pinnately veined, pinnately lobed or toothed, stalked. Flowers in erect, much branched, rather tight terminal panicles. Buds ovoid to spherical. Sepals 2, falling as the flowers open, ovate or almost circular. Petals absent. Stamens 8–24, with filaments shorter than anthers. Ovary of 2 united carpels, long-tapered to a stalk below, containing a single, basally attached ovule; style present; stigma bilobed. Fruit flattened, elliptic, pointed at both ends, containing a single seed. Seed with its base enclosed in a cup-like aril.

A genus of about 9 species from C & S America, resembling *Macleaya* in being wind-pollinated, the flowers without petals, but otherwise unlike it. Plants are not particularly attractive, and only a few are grown for their botanical interest. Little is known of their precise cultivation requirements.

Literature: Hutchinson, J., *Bocconia* and *Macleaya*, *Kew Bulletin* for 1920: 275–282 (1920).

1a. Leaves toothed or scalloped but not lobed　　**3. integrifolia**
　b. Leaves pinnately lobed　　　　2
2a. Leaves long-tapered at the base, appearing to have a winged stalk; stamens *c.* 12　　**1. arborea**
　b. Leaves rounded or abruptly tapered at the base, stalk not appearing winged; stamens *c.* 16　　　　**2. frutescens**

1. B. arborea Watson. Illustration: Kew Bulletin for 1920: 276 (1920).

Ultimately a tree to 8 m. Leaves pinnately lobed with triangular, acute lobes, the base long-tapered so that the leaf appears to have a winged stalk. Panicle *c.* 20 × 12 cm. Stamens *c.* 12. *C America*. G1.

2. B. frutescens Linnaeus. Illustration: Graf, Exotica, International series 4, **2**: 1921 (1982).

Shrub with hairy branches. Leaves pinnately lobed with toothed lobes, the base abruptly tapered or rounded, the stalk not appearing winged. Panicle to 40 cm. Stamens *c.* 16. *C America*. G1.

3. B. integrifolia (Humboldt & Bonpland) de Candolle.

Very branched shrub, ultimately to 5 m. Leaves toothed or scalloped, not lobed. Panicle to 35 cm. Stamens *c.* 10. *S America (Colombia, Peru, Bolivia)*. G2.

17. ESCHSCHOLZIA Chamisso
J. Cullen
Annual herbs (sometimes occasionally perennial in the wild) with clear sap. Leaves alternate, usually hairless, divided into 3 leaflets which are themselves further divided into linear-oblong segments, green or glaucous. Flowers numerous, cream, yellow, reddish or orange; sepals, petals and stamens borne on a cup- or funnel-shaped perigynous zone. Sepals 2, united, falling as a candle-snuffer-like unit. Petals large, spreading, sometimes crisped at the margins, usually 4 (rarely more). Stamens 16-many. Ovary cylindric, 1-celled, with 2 placentas, 10-veined. Styles 2, each divided into 2 or 3 thread-like segments. Capsule linear-cylindric, opening from the base upwards.

A genus of about 10 species from western N America, formerly divided into a very large number. Only 1 is of importance in horticulture in Europe. It is generally cultivated as an annual, and is easily grown in a sunny position. Propagate by seed.

1. E. californica Chamisso (*E. crocea* Bentham). Illustration: Botanical Magazine, 2887 (1827) & 3495 (1836); Morley & Everard, Wild flowers of the world, pl. 154 (1970); Rickett, Wild flowers of the United States **4**: pl. 50 (1970); Wright, The complete handbook of garden plants, 513 (1984).

Flopping annual herb to 60 cm, almost entirely hairless. Cotyledons bifid. Leaves basal and on the stems, usually glaucous. Perigynous zone with 2 rims, the outer green and spreading, the inner translucent and erect. Calyx 1–4 cm. Petals 2–6 cm. Capsule to 10 cm. Seeds grey-brown. *USA (California)*. H3. Spring–summer.

Very variable, both in the wild and in gardens. Numerous colour-variants

have been selected, and variants with semi-double flowers also occur.

18. HUNNEMANNIA Sweet
J. Cullen
Very like *Eschscholzia*, but perennial, upright and very glaucous; sepals 2, free, falling individually.

A genus of a single species from Mexico, occasionally grown (usually as an annual). Propagate by seed.

1. H. fumariifolia Sweet. Illustration: Sweet, British flower garden **3**: t. 276 (1828); Botanical Magazine, 3061 (1831); Wright, Complete handbook of garden plants, 513 (1984).

Upright perennial herb to 60 cm. Sepals 1.4–1.6 cm. Petals 2.5–3 cm, clear yellow. Capsule to 10 cm, linear-cylindric. *NW Mexico*. H5. Summer.

19. DENDROMECON Bentham
J. Cullen
Evergreen shrub with clear sap. Leaves alternate, leathery, upper surface with a network of raised veins, yellowish or greyish green, entire or very finely toothed (a lens is necessary to see the teeth). Flowers solitary, terminal on short branches. Sepals 2, free, falling individually as the flower opens. Petals 4 (rarely 6), yellow. Stamens numerous, with short filaments. Ovary 1-celled with 2 placentas, styles 2. Capsule cylindric, opening from the base upwards. Seeds finely pitted, appendaged.

A genus of 2 species from California and western Mexico. Both are easily grown in a sunny site in well-drained soil. In much of Europe, glasshouse protection is required, and in much of western Europe, where the plants can be grown out of doors in sheltered sites, severe frosts can damage them, killing them at least to the root. Propagation is usually by seed.

1. D. rigida Bentham. Illustration: Botanical Magazine, 5134 (1859); Spellenberg, Audubon Society field guide to North American wild flowers, western region, t. 346 (1979); Wright, Complete handbook of garden plants, 225 (1984); Brickell, RHS gardener's encyclopaedia of plants and flowers, 115 (1989).

Upright shrub to 6 m in the wild, usually much less in cultivation. Leaves 2.5–10 cm × 7–25 mm, linear- to oblong-lanceolate, held almost vertically on twisted stalks, margins very finely toothed. Flower-stalks exceeding the subtending leaves. Sepals 8–10 mm. Petals 2–3 cm, spreading. Seeds 2–2.5 mm, brownish. *USA (mainland California), Mexico (Baja California)*. H5. Spring–summer.

A variable species. Closely related (and perhaps not distinct) is **D. harfordii** Kellogg (*D. rigida* var. *harfordii* (Kellogg) Brandegee), which has broader leaves with entire margins, flower-stalks not or scarcely exceeding the subtending leaves and seeds 2.5–3 mm. *USA (islands off the coast of California)*. H5. Spring–summer.

20. PTERIDOPHYLLUM Siebold & Zuccarini
J. Cullen

Perennial herb with rhizomes. Leaves all basal, dark green above, paler beneath, to 15 cm, oblong-obovate to narrowly elliptic, pinnately divided into numerous oblong leaflets, blunt and faintly toothed at the tip, truncate at the base, each with a tooth on the upper margin near the base; all with sparse adpressed bristles above and beneath. Scape to 30 cm, bearing a long raceme containing many flowers, the flower-stalks often arising in 2s or 3s at the lower nodes; bracts very small. Sepals 2, very small and very quickly deciduous. Petals 4, to 8 mm, white, spreading, all more or less similar or 1 pair narrower than the other pair. Stamens 4. Ovary of 2 carpels, 1-celled but with a style and 2 short stigmas and 2–4 ovules. Fruit a capsule opening by the sides falling from the central placental framework; seeds usually 2.

A genus of a single species from Japan. It is not easy to grow, requiring a cool, somewhat shaded site with rich, peaty, water-retentive soil. Propagate by seed or occasionally by division of the rhizome.

1. P. racemosum Siebold & Zuccarini. Illustration: Pflanzenreich **40**: 84 (1909); Botanical Magazine, 8743 (1918); Kitamura & Murata, Coloured illustrations of herbaceous plants of Japan, pl. 45 (1961); Takeda, Alpine flora of Japan, pl. 51 (1963).

Japan. H5. Spring–summer.

The foliage is remarkably similar to that of some ferns (species of *Blechnum*, see vol. **1**: 61).

21. HYPECOUM Linnaeus
J. Cullen

Annual herbs, often glaucous. Leaves in a basal rosette, deeply 2–4 times pinnate into narrow segments. Flowers in cymes, bracts similar to the leaves. Sepals 2, soon falling.

Petals 4 in 2 opposite pairs, the outer pair entire to 3-lobed, broadly diamond-shaped, the inner pair each deeply 3-fid, the lateral lobes linear, spreading, the central lobe stalked, the blade folded, those of the pair enclosing the anthers and stigmas. Stamens 4, filaments winged towards the base. Ovary cylindric, of 2 carpels, 1-celled, containing several ovules, and with 2 styles. Fruit linear-cylindric, breaking into 1-seeded joints (a capsule in 1 non-cultivated species).

A genus of perhaps 10 species from the Mediterranean area across Asia to China. A few species may occasionally be grown as curiosities. They grow easily in almost any dry soil and are propagated by seed.

Literature: Dahl, A.E., Taxonomic and morphological studies in *Hypecoum* section *Hypecoum*, *Plant Systematics and Evolution* **163**: 227–279 (1989).

1a. Petals white with pinkish mauve or bluish tips   **3. leptocarpum**
  b. Petals yellow or orange-yellow   2
2a. Outer petals weakly 3-lobed, the lateral lobes smaller than the central lobe; fruit-stalks not thickened   **1. procumbens**
  b. Outer petals strongly 3-lobed, the lateral lobes as large as the central lobe; fruit-stalks usually thickened   **2. imberbe**

1. H. procumbens Linnaeus. Illustration: Bonnier, Flore complète **1**: pl. 26 (1911); Polunin & Huxley, Flowers of the Mediterranean, t. 36 (1965).

Prostrate or ascending, to 30 cm. Leaves glaucous. Petals yellow, those of the outer pair 5–10 mm, weakly 3-lobed with lateral lobes smaller than the central lobe. Fruit-stalks usually not thickened. *Mediterranean area*. H2. Spring–summer.

2. H. imberbe Sibthorp & Smith (*H. grandiflorum* Bentham). Illustration: Bonnier, Flore complète **1**: pl. 26 (1911); Polunin, Flowers of Europe, pl. 30 (1969).

Like *H. procumbens* but more erect, leaves with narrower segments, petals orange-yellow, those of the outer pair strongly 3-lobed with lateral lobes as large as the central lobe; fruit-stalks usually thickened. *W & S Europe, N Africa, SW Asia*. H2. Spring–summer.

3. H. leptocarpum Hooker & Thomson. Sprawling herb, to 60 cm, very glaucous. Petals white with pinkish mauve to bluish tips, the outer pair distinctly 3-lobed. Fruit-stalks not thickened.

Himalaya, SW China. H5. Spring–autumn.

22. DICENTRA Bernhardi
J. Cullen

Hairless perennial herbs with rhizomes, tubers or bulblets, sometimes climbing by means of tendrils which replace some leaflets. Leaves all basal or borne on the stem, usually 1–3 times pinnate or 2–3 times divided into 3, the ultimate segments often lobed or toothed. Flowers solitary or in axillary or terminal racemes or panicles, usually pendent, subtended by a bract and, often, 2 bracteoles. Flowers compressed, with 2 distinct planes of symmetry. Sepals 2, usually deciduous in flower. Petals 4, in 2 pairs, all variably united; outer pair large, enclosing the inner except at the apex, pouched or swollen at the base with free tips which are usually spreading or reflexed; inner pair clawed, crested towards the apex. Stamens 4, in 2 bundles of a central stamen attached to 2 half-stamens; filaments united to varying extents. Nectar secreted by the bases of the filaments or by a spur projecting backwards into the pouched part of the outer petal. Ovary of 2 carpels, 1-celled, with several ovules. Fruit a capsule, with persistent style. Seeds usually with an oily appendage.

A genus of 19 species from N America and E Asia. Several are grown for their ornamental foliage and flowers, the latter sometimes rather bizarre. They are generally easily grown, though they require a cool, damp site protected from wind and too much sun. They are propagated by seed, by bulblets, if present, or by division of the rhizome.

The flowers of this genus are of a remarkable form; being compressed, they present recognisable shapes when seen in profile; these are not easy to describe. In the descriptions, measurements of the flowers are length, from the base of the swelling of the outer petal to the tips of the inner petals, and breadth across the broadest part even when this is at the extreme base of the swollen petals. The stigmas are also of great diagnostic value, but as they are difficult to see and interpret, they are not mentioned here; reference should be made to the monograph cited below for further information.

Literature: Stern, K.R., Revision of *Dicentra* (Fumariaceae), *Brittonia* **8**: 1–57 (1960).

Habit. Climbing: **10**.

Rootstock. A horizontal rhizome:
2–5,7,10?; a short, erect rhizome: **6**;
tuberous: **7**; with bulblets: **7,8,9**; a
tap-root: **1**.
Leaves. All basal: **4–9**; borne on the stems:
1–3,10.
Flowers. In panicles: **1,4,5**; in racemes
of usually 4 or more: **2,3,6,8–10**; in
racemes of 1–3: **6,7**. Basic colour
(excluding free tips of outer petals which
are often contrasting in colour) white:
2,4,6–9; basic colour reddish, pink
or purple: **2–7**; basic colour yellow:
1,4,10.
Seeds. Without an oily appendage: **1** and
the species mentioned under **7**; with an
oily appendage: **2–10**.

1a. Plant climbing by means of tendrils
      10. macrocapnos
  b. Plant not climbing, tendrils absent
      2
2a. Leaves borne on the stems   3
  b. Leaves all basal   5
3a. Inflorescences terminal; flowers
    yellow   **1. chrysantha**
  b. Inflorescences axillary, or terminal
    and flowers pink, white or purplish
     4
4a. Flowers 2–3 cm, cordate at base
      2. spectabilis
  b. Flowers 3.5–5 cm, tapering to the
    base   **3. macrantha**
5a. Plants with rhizomes but without
    bulblets or tubers   6
  b. Plants with tubers or bulblets   8
6a. Rhizomes short, upright; flowers in
    racemes   **6. peregrina**
  b. Rhizomes long, horizontal; flowers
    in panicles   7
7a. Reflexed free tips of outer petals 5–8
    mm   **5. eximia**
  b. Reflexed free tips of outer petals 3–5
    mm   **4. formosa**
8a. Tuber present; nectar secreted by
    the base of the filament; flowers
    1–3   **7. pauciflora**
  b. Bulblets present; nectar secreted
    by a spur-like projection from the
    base of the filament; flowers usually
    more than 3 in each raceme   9
9a. Bulblets white to pink, ovoid; nectar
    secreting spur 2–4 mm; flowers not
    fragrant;   **9. cucullaria**
  b. Bulblets yellow, pea-shaped; nectar
    secreting spur to 1 mm; flowers
    fragrant   **8. canadensis**

1. D. chrysantha (Hooker & Arnott)
Walpers. Illustration: Botanical Magazine,
7954 (1904); Brittonia **13**: 23 (1960);
Rickett, Wild flowers of the United States

4: pl. 68 (1970); The Garden **101**: 433
(1976).

Perennial herb with a stout tap-root;
stems to 1 m or more, glaucous. Leaves
borne on the stem, 2–3 times pinnate,
the ultimate segments variably lobed and
toothed, glaucous. Inflorescence a long,
narrow, terminal panicle with numerous
flowers. Flowers golden yellow with an
unpleasant smell, erect. Corolla more or
less oblong, cordate at base, 1–1.6 ×
0.4–0.5 cm, the outer petals with
spreading or reflexed tips 4–6 mm. Capsule
spindle-shaped to *c.* 3 × 0.7 mm. Seeds
dull, without oily appendages. *USA
(California).* H5. Summer.

D. ochroleuca Engelmann. Illustration:
Brittonia **13**: 18 (1960); Rickett, Wild
flowers of the United States **4**: pl. 68
(1970). Broadly similar, but flowers very
pale yellow, more crowded, 2–2.6 cm, the
tips of the outer petals more or less erect.
USA (California). H5. Spring–summer.

2. D. spectabilis (Linnaeus) Lemaire.
Illustration: Botanical Magazine, 4458
(1847) – as Dielytra spectabilis; Brittonia
13: 25 (1960); Perry, Flowers of the
world, 119 (1970); The Garden **101**: 378
(1976).

Large herb to 1.5 m, with a rhizome.
Leaves borne on the stem, 1–2 times
pinnate, the ultimate segments more or
less obovate, deeply lobed with broad,
acute lobes. Flowers numerous, pendent
in terminal and axillary racemes which
arch to the horizontal or beyond. Corolla
2–3 × 1.5–1.8 cm, ovate, deeply cordate at
the base with rounded lobes. Outer petals
rose-purple to pink or rarely whitish, tips
spreading or reflexed, to 1 cm; inner petals
whitish, blotched purple internally at the
tip. Seeds with oily appendages. *E former
USSR, Korea, N China.* H3. Spring–summer.

The evenly spaced, large flowers hanging
from the long, arching racemes give this
plant a very distinctive appearance.

3. D. macrantha Oliver. Illustration:
Hooker's Icones Plantarum **20**: t. 1937
(1890); Brittonia **13**: 27 (1960).

Herb to 1.2 m, with a rhizome. Leaves
borne on the stem, 3 times divided into 3
leaflets, the leaflets ovate to elliptic with
rather evenly toothed margins. Flowers
3–14, pendent, in erect, axillary racemes.
Corolla 3.5–4.8 × 0.8–1 cm, oblong
though slightly waisted in the middle,
truncate or rounded at the base, pale pink
to dull purplish, tips of outer petals erect.
Capsule 3–4 × 0.5–0.7 cm. Seeds black
and shining, each with a minute oily

appendage. *W China & adjacent Burma.* H5.
Spring–summer.

4. D. formosa (Haworth) Walpers.
Illustration: Botanical Magazine, 1335
(1811) – as Fumaria formosa; Brittonia
13: 28 (1960); Rickett, Wild flowers of the
United States **5**: pl. 65 (1971).

Perennial herb with a stout, horizontal
rhizome. Leaves all basal, 1-pinnate, the
segments pinnatisect, the ultimate
segments bluntly or sharply toothed,
glaucous at least beneath. Scapes to 60 cm
bearing a panicle of 5–15 pendent flowers.
Corolla ovate-cylindric, cordate with a
narrow sinus and rounded lobes at the
base, 1.2–2 × 0.6–1 cm, rose-purple, pink,
creamy white ('Alba') or rarely yellow;
tips of the outer petals usually spreading
to some extent, 3–5 mm, usually darker
in colour than the rest. Central filament
of each bundle curving inwards with the
others at the base. Capsule 1.5–2.5 ×
0.4–0.5 cm. Seeds black and shiny, each
with an oily appendage. *Western N
America, from British Columbia to California.*
H4. Spring–summer.

A variable species. Subsp. **oregana**
(Eastwood) Munz is smaller, has leaves
very bluish glaucous on both surfaces and
yellowish flowers with the tips of the outer
petals reddish. It occurs in Oregon and
California and is sometimes cultivated.

D. nevadensis Eastwood. Illustration:
Brittonia **13**: 28 (1960). Similar but
generally smaller and the central filament
of each bundle separate from the others
and looped outwards towards the base.
USA (California). H5. Summer.

5. D. eximia (Ker) Torrey. Illustration:
Brittonia **13**: 33 (1960); Rickett, Wild
flowers of the United States **1**: pl. 52
(1967); Justice & Bell, Wild flowers of
North Carolina, 75 (1968).

Very similar to *D. formosa* but often
more robust, flowers generally larger,
1.5–3 × *c.* 0.8 cm, less narrowly cordate
at the base, the free, spreading tips of the
outer petals 5–8 mm, generally rather
narrow. *E USA.* H5. Spring–summer.

6. D. peregrina (Rudolph) Makino (*D.
pusilla* Siebold & Zuccarini; *D. peregrina*
var. *pusilla* (Siebold & Zuccarini) Makino).
Illustration: Brittonia **13**: 35 (1960);
Takeda, Alpine flora of Japan, pl. 51
(1963); Botanical Magazine, n.s., 461
(1964); Quarterly Bulletin of the Alpine
Garden Society **54**: 255 (1986).

Perennial herb with a short, upright
rhizome. Leaves all basal, glaucous,

pinnately divided 2 or more times into narrow, crowded segments. Scape erect, 5–20 cm, bearing a raceme of 3–8 flowers which are erect to pendent. Flowers long heart-shaped, shortly cordate with rounded lobes at the base, 2–2.5 × 0.8–1.2 cm. Outer petals white to purple with darker, reflexed free tips to 1.2 cm. Central filament of each bundle free and looping outwards at the base. Nectary-spur absent. Capsule 10–12 × 3–4 mm. Seeds black and shining, each with an oily appendage. *E former USSR, Japan.* Summer.

7. D. pauciflora Watson. Illustration: Brittonia **13**: 37 (1960); Rickett, Wild flowers of the United States **5**: pl. 66 (1971).

Perennial herb with a horizontal rhizome and/or spindle-shaped, tuberous roots which sometimes bear minute bulblets at their tips. Leaves few (1 or 2), basal, pinnately divided 2 or more times into narrow segments, not glaucous. Scapes scarcely exceeding the leaves, bearing 1–3 erect to pendent flowers. Corolla ovate-oblong, deeply cordate at the base with a narrow sinus and rounded lobes, 1.8–2.5 × 0.7–0.9 cm. Outer petals white or pink, the free tips spreading or reflexed, darker, 7–11 mm. Central filament of each bundle free and looping outwards towards the base. Capsule 10–15 × 4–6 mm. Seeds black and shining, each with an oily appendage. *USA (California, Oregon).* H5. Spring–summer.

D. uniflora Kellogg. Illustration: Brittonia **13**: 37 (1960); Rickett, Wild flowers of the United States **5**: pl. 66 (1971). Very similar, but always with a cluster of spindle-shaped tubers; scapes always 1-flowered. the free part of the outer petals 4–6 mm; seeds without oily appendages. *USA (California).* H5. Spring–summer.

8. D. canadensis (Goldie) Walpers. Illustration: Botanical Magazine, 3031 (1830) – as Dielytra canadensis; Brittonia **13**: 40 (1960); Rickett, Wild flowers of the United States **1**: pl. 52 (1967); Justice & Bell, Wild flowers of North Carolina, 75 (1968).

Perennial herb arising from a short stock which bears yellow, pea-shaped bulblets. Leaves all basal, 2–3 times pinnate with oblong, rounded ultimate segments. Scapes to 35 cm, bearing erect racemes of 5–12 pendent, very fragrant flowers. Corolla ovate-triangular, cordate at the base with somewhat diverging, rounded lobes, 1.2–1.8 cm. Outer petals whitish, their tips

broad, spreading almost horizontally, 3–5 mm, purplish. Filaments of each bundle free almost to the base, that of the central looping outwards towards the base. Nectar-secreting spur to 1 mm. Capsule ovoid, 9–17 × 3–6 mm. Seeds black, shining, each with an oily appendage. *E North America.* H2. Spring.

9. D. cucullaria (Linnaeus) Bernhardi. Illustration: Botanical Magazine, 1127 (1808) – as Fumaria cucullaria; Brittonia **13**: 42 (1960); Rickett, Wild flowers of the United States **1**: pl. 52 (1967); The Garden **101**: 433 (1976).

Perennial herb arising from a short stock which bears pink or white ovoid bulblets. Leaves all basal, glaucous, 2–3 times pinnately divided with narrow, acute ultimate segments. Scapes exceeding the leaves, bearing erect racemes of 5–14 pendent flowers. Flowers not fragrant, curiously shaped, deeply cordate at the base with diverging, tapering, ultimately rounded lobes, the whole flower widest at the base of the lobes, tapering smoothly from there to the divergence of the free outer petal tips; 1.3–1.7 × 1.3–1.6 cm. Outer petals white or suffused with pink, the free tips spreading, yellow or orange-yellow, 2–5 mm. Nectar-secreting spur 2–4 mm. Capsule ovoid, 9–15 × 3–5 mm. Seeds black and shining, each with an oily appendage. *E North America.* H2. Spring.

Variable but usually recognisable by the distinctive, almost moth-like shape of the flowers.

10. D. macrocapnos Prain. Illustration: Brittonia **13**: 51 (1960).

Perennial climbing to 3 m or more. Leaves borne on the stem, twice divided into 3, the leaflets ovate-oblong, more or less entire, some of them replaced by fine tendrils. Racemes axillary, dense, spreading, containing up to 15 pendent flowers. Corolla yellow with the tips of the petals brownish, 1.5–2 × 0.6–0.9 cm, ovate-triangular in outline (broadest almost at the extreme base), base cordate, lobes rounded. Outer petals with erect or somewhat spreading tips 2–5 mm. Capsule spindle-shaped, 2–3 × 0.5–0.6 cm. Seeds black, shining, with an oily appendage. *Nepal, NW India.* H5. Spring–summer.

Other climbing species (all belonging to subgenus *Dactylicapnos* (Wallich) Stern, recently treated by Lidén, *Opera Botanica* **88**: 20 (1986) as a distinct genus, *Dactylicapnos* Wallich) have been in cultivation from time to time, but do not seem to persist for long.

23. ADLUMIA de Candolle
J. Cullen

Annual or biennial herbs, scrambling and climbing among other vegetation; stems brittle. Leaves 2–3-pinnate, the ultimate segments ovate or elliptic, entire or variously lobed and toothed; leaflet-stalks tendril-like, coiling around other vegetation. Flowers numerous, pendent, each subtended by a small bract, in axillary panicles whose stalks are united at the base for a short distance with those of the subtending leaves. Flowers with 2 distinct planes of symmetry. Sepals 2, toothed. Petals 4, united for most of their length to form a long urn-shaped corolla, the outer pair enclosing the inner pair except for their tips, swollen but not spurred at the base. Stamens 4 in 2 bundles of a whole stamen united to 2 half-stamens, each bundle united to the corolla for most of its length. Ovary with few ovules; stigma 4-lobed. Fruit a few-seeded capsule surrounded by the detached but persistent corolla. Seeds black, each with an appendage.

A genus of 2 species, 1 from eastern N America, the other from E Asia. Only the American species is cultivated: it requires a cool, damp site protected from both sun and wind. Propagation is by seed, which often germinates and grows *in situ* around the site of the parent plant.

1. A. fungosa (Aiton) Britton, Sterns & Poggenburg. Illustration: Britton & Brown, Illustrated Flora of the northern United States and Canada **2**: 105 (1897) & edn 2, **2**: 143 (1913); Rickett, Wild flowers of the United States **1**: pl. 52 (1967); The Green Scene **5**(2): 33 (1976).

Delicate annual reaching to 3 m in suitable locations. Leaves very thin, long-stalked. Flowers 1.2–1.8 cm, white, cream or pink. *E North America.* H3. Summer.

24. CORYDALIS Ventenat
J. Cullen

Annual to perennial, usually hairless herbs, rarely climbing; rootstock a tap-root, a cluster of fleshy roots, a rhizome or a tuber. Stems leafy or not. Leaves basal, alternate or opposite, usually deeply divided (pinnate or successively divided into 3s) 2–4 times, usually thin and delicate, the terminal leaflet rarely modified into a branched tendril. Flowers in racemes (rarely aggregated into panicles), bracts entire or lobed towards the apex. Sepals 2, persistent in flower, usually somewhat peltate and toothed. Petals 4 in 2 pairs, variously united, the outer pair more or less enclosing the inner pair except at the tips; upper petal with a long or short spur, lower petal sometimes swollen at the base. Stamens in 2 groups of a whole stamen united to 2 half-stamens, the upper group with a nectary at the base which projects backwards into the petal-spur. Ovary of 2 carpels, with 2–many ovules. Fruit a capsule containing 2–several seeds, opening by the sides separating from the persistent placental framework; style usually persistent. Seeds usually with a whitish or brownish appendage (caruncle, arilloid).

A genus of about 300 species from northern temperate areas. About 25 are commonly grown, but several others (especially tuber-bearing species) have been sporadically in cultivation in the past and may appear from time to time.

A broad, traditional concept of the genus is used here. Recently, however, Lidén (reference below) has studied the classification of the whole of the subfamily Fumarioideae and has concluded that some species should be split off from the traditional *Corydalis* and placed in segregate genera (*Pseudofumaria*, *Capnoides*, and *Ceratocapnos*), making particular use of the character of whether or not the style is persistent on the fruit. This arrangement has much to recommend it, but it is not adopted here as most gardeners will not be familiar with it. The names used by Lidén are, however, cited in the appropriate places. Lidén recognises 19 sections in the (reduced) genus, but these are not used here either, though the order of the species follows his arrangement.

The stigmas in this genus are highly variable and are important in the classification at the sectional level; they are not described here, however, as they are difficult to observe and interpret. *Corydalis* is a variable genus, and the cultivated species are spread throughout its various sections, so its cultivation is also variable. The taller species (generally those with tap-roots or rhizomes) can be grown in borders, where they require a cool, damp, partially shaded site with protection from wind (the leaves are delicate and the stems brittle). The smaller species, particularly those with tubers are most often grown in pots, generally in an alpine house or frames; they are treated like bulbs, with a dry resting period in summer. Propagation is by seed, or division of the rootstock.

Literature: Ryberg, M., A taxonomical survey of the genus *Corydalis* Ventenat with reference to the cultivated species *Acta Horti Bergiani* **17**: 115–175 (1955); Lidén, M., Synopsis of the Fumarioideae (Papaveraceae) with a monograph of the tribe Fumarieae, *Opera Botanica* **88** (1986), 17–32; Lidén, M. & Zetterlund, H., Notes on the genus Corydalis, *Bulletin of the Alpine Garden Society* **56**: 146–169 (1988).

1a. Style deciduous after flowering, absent from the fruit 2
 b. Style persistent on the fruit 4
2a. Plant annual, climbing or scrambling by means of tendrils **26. claviculata**
 b. Plant perennial, not climbing, without tendrils 3
3a. Flowers yellow **25. lutea**
 b. Flowers white or cream with yellow tips **24. ochroleuca**
4a. Flowers entirely or predominantly yellow or cream, the tips sometimes of a contrasting, darker colour 5
 b. Flowers not yellow, usually blue to purplish or reddish, rarely white, tips often contrasting in colour and sometimes yellow 16
5a. Plants shortly but distinctly white-hairy on leaves and stems **23. tomentella**
 b. Plants completely hairless 6
6a. Flowering-stem with a scale-leaf at the base; rootstock a rounded tuber 7
 b. Flowering-stem without a scale-leaf at the base; rootstock various, not a tuber 8
7a. Flowering-stem to 40 cm; corolla 2–2.5 cm **5. bracteata**
 b. Flowering-stem to 20 cm; corolla 3.5–4.5 cm **6. schanginii**
8a. Leaves all basal or borne only on the lowermost part of the flowering stem; plants rarely more than 10 cm 9
 b. Leaves clearly borne on the flowering stem above the base; plants more than 10 cm 11
9a. Leaves fern-like, green; primary leaflets more than 10 **18. cheilanthifolia**
 b. Leaves not fern-like, glaucous; primary leaflets fewer than 10 10
10a. Capsule linear; lobes of the leaves rounded **22. wilsonii**
 b. Capsule broadly ovoid; leaf-lobes lanceolate, obtuse **17. rupestris**
11a. Plant annual or biennial 12
 b. Plant perennial 13

12a. Flowers 1–1.5 cm, spur 3–5 mm; capsule 1.5–3 × 0.3–0.5 cm **20. aurea**
 b. Flowers 7–9 mm, spur to 3 mm; capsule 5–15 × 2–3 mm **15. sibirica**
13a. Racemes dense, head-like **14. nobilis**
 b. Racemes not as above 14
14a. Racemes much-branched; spur 5–10 mm; capsule narrowly obovate to obovate **13. chaerophylla**
 b. Racemes simple; spur to 5 mm; capsule linear 15
15a. Racemes long and dense; flowers white or cream, *c.* 1 cm **19. ophiocarpa**
 b. Racemes loose; flowers yellow, to 2.5 cm **15. saxicola**
16a. Stem-leaves 2, opposite or almost so 17
 b. Stem-leaves 1, 2 or more, alternate 20
17a. Leaves deeply dissected into linear lobes less than 2 mm wide **11. verticillaris**
 b. Leaves less deeply divided, the ultimate segments elliptic, ovate or obovate, more than 2 mm wide 18
18a. Flowers to 4.5 cm, spur to 2 cm **8. popovii**
 b. Flowers at most 2.5 cm, spur to 1.5 cm 19
19a. Leaves clearly stalked, divided into 3 at the top of the stalk **9. diphylla**
 b. Leaves not or scarcely stalked, divided into 3 almost at the junction with the stem **10. rutifolia**
20a. Corolla predominantly pink 21
 b. Corolla not pink 23
21a. Inner petal spurs *c.* 2.5 cm **6. schanginii**
 b. Inner petal spurs less than 2.5 cm 22
22a. Capsules linear **1. sempervirens**
 b. Capsules obovoid or club-shaped **16. scouleri**
23a. Flowers vivid blue or purple-blue in condensed, umbel-like racemes; roots several, fleshy, in a cluster, tuber absent **12. cashmeriana**
 b. Flowers usually whitish, pink or mauve pink, rarely blue; rootstock a tuber 24
24a. Flowering-stem without a scale-leaf at the base; tuber in mature plants to 5 cm or more across, hollow **7. cava**

b. Flowering-stem often with a scale-leaf at the base; tuber smaller, solid   25

25a. Flowers blue or purplish blue, inner petals whitish; leaves twice divided into 3, the segments entire, obovate; bracts all entire
4. ambigua

b. Flowers purple, reddish or whitish; leaves twice divided into 3, the segments deeply lobed into narrowly oblong or obovate lobes; at least the lower bracts lobed   26

26a. Lower bracts 3-lobed; corolla whitish or cream, sometimes streaked with purple; capsule linear-cylindric   **3. angustifolia**

b. Lower bracts with more than 3 lobes; corolla purple or reddish, rarely whitish; capsule oblong-lanceolate or linear-cyclindric
2. solida

1. C. sempervirens (Linnaeus) Pursh (*C. glauca* (Curtis) Pursh; *Capnoides sempervirens* (Linnaeus) Borckhausen). Figure 13(1), p. 123. Illustration: Botanical Magazine, 179 (1796); Rickett, Wild flowers of the United States **1**: pl. 52 (1967); Justice & Bell, Wild flowers of North Carolina, 76 (1968); Opera Botanica **88**: 20 (1986).

Annual to biennial herb to 1 m, becoming somewhat woody at the base. Leaves very glaucous, basal and alternate on the stem, the lower stalked, all twice divided into 3, each leaflet deeply lobed into 3 narrow, oblong to elliptic or obovate, blunt segments; upper leaves less divided, shortly stalked. Flowers few to many in terminal and axillary racemes. Corolla pink (rarely whitish) with yellow tip, 1–1.6 cm, the spur 2–4 mm wide, bluntly rounded; base of the lower petal somewhat swollen. Capsule linear, 3–5 × 0.2–0.3 cm, somewhat constricted between the seeds; style persistent. Seeds black, shining, each with a whitish appendage. *Most of N America.* H1. Spring–summer.

Treated by Lidén (reference above) as the sole member of the genus *Capnoides* Miller.

2. C. solida (Linnaeus) Clairville (*C. bulbosa* (Linnaeus) de Candolle; *C. densiflora* Presl & Presl). Figure 13(2), p. 123. Illustration: Bonnier, Flore complète **1**: pl. 27 (1911); Polunin, Flowers of Europe, pl. 30 (1969); Everard & Morley, Flowers of the world, pl. 11c (1970); Huxley & Taylor, Flowers of Greece and the Aegean, pl. 59 (1977).

Tuber ovoid-spherical, solid, 1–1.5 cm

across. Flowering-stem to 20 cm, with a single, often yellowish scale-leaf at the base. Leaves few on the flowering stem, alternate, twice divided into 3, each leaflet deeply lobed with narrow, oblong or narrowly obovate, blunt lobes; all variably glaucous. Flowers in a rather dense terminal raceme; lower bracts with 4 or more lobes at the apex. Corolla pink or mauve-pink, rarely white, 1.5–2 cm, spur broad, slightly tapering, blunt, straight or slightly curved upwards or downwards, 1–1.5 cm. Capsule oblong-lanceolate or linear-cylindric, 1–2 × 0.3–0.4 cm, tipped with a long, persistent style. Seeds black, shining, each with a whitish appendage. *Europe, N Africa, SW Asia.* H1. Late winter–spring.

An early-flowering species which is variable in the degree of toothing of the bracts and the size and precise colour of the flowers. 'Alba' has whitish corollas contrasting with brownish green bracts.

3. C. angustifolia (Biberstein) de Candolle. Figure 13(3), p. 123. Illustration: Gartenflora **9**: t. 304 (1860).

Like *C. solida* but leaves with narrower segments, lower bracts 3-lobed, corolla yellowish white, capsule linear-cylindric, to 20 × 1.5 mm. *E Turkey, N Iran, former USSR (Crimea, Caucasus).* H2. Late winter–spring.

C. paczoskii Busch. Illustration: Quarterly Bulletin of the Alpine Garden Society **56**: 165 (1988). Very like *C. angustifolia* but flowers mauvish pink to purple, inner petals with dark tips. *Former USSR (Crimea).* H2. Spring.

4. C. ambigua Chamisso & Schlechtendahl (*C. jezoensis* Miquel; *C. ambigua* var. *jezoensis* (Miquel) Anon.) Figure 13(4), p. 123. Illustration: Botanical Magazine, n.s., 758 (1978); Quarterly Bulletin of the Alpine Garden Society **56**: 159 (1988).

Tuber solid, spherical, 1–2 cm across. Flowering-stem to 15 cm with a single scale-leaf at the base (sometimes below soil-level) and usually 2 alternate stem-leaves. Leaves somewhat irregularly twice divided into 3, the ultimate segments oblong to obovate, blunt. Flowers 3–12 in a terminal raceme; bracts entire, ovate. Corolla 2–3 cm, blue or sometimes suffused with pink, the inner petals whitish; spur 1–1.5 cm, broad, slightly tapering, blunt. Capsule linear to oblong, 15–25 × 2–2.5 mm, style persistent. Seeds black, each with a small appendage. *Japan, E former USSR, N China; ?Korea.* H3. Spring.

Very variable in leaflet shape; some variants are finely papillose.

5. C. bracteata (Stephan) Persoon. Figure 13(5), p. 123. Illustration: Botanical Magazine, 3242 (1833); Edwards's Botanical Register **19**: t. 1644 (1834); Gartenflora **14**: t. 476 (1865); Mathew, Dwarf bulbs, pl. 23 (1973).

Tuber solid, spherical, 1–1.5 cm across. Flowering-stem to 40 cm, with a conspicuous scale-leaf at the base; stem-leaves 1 or 2, alternate, 1–2 times divided into 3, the segments further lobed. Raceme rather dense, with several flowers; bracts with several lobes towards the apex. Corolla yellow, 2–2.5 cm, spur tapered to a blunt apex, curved upwards, to 1.5 cm. Capsule linear-oblong, to 2 × 0.4–0.5 cm, tapered into the persistent style. Seeds black, shining, each with a whitish appendage. *Former USSR (C Asia, Siberia).* H4. Spring–summer.

C. × allenii Irving (not *C. allenii* Fedde, which is a synonym of *C. scouleri*) is the name applied to a cultivated plant rather like *C. bracteata* but flowering earlier and with white flowers streaked with purple. It is completely sterile and it has been suggested that its origin is *C. bracteata* × *C. solida*, but this is not certain; see The Garden **72**: 290 (1908).

6. C. schanginii (Pallas) Fedtschenko (*C. longiflora* (Willdenow) Persoon). Figure 13(6), p. 123. Illustration: Botanical Magazine, 3230 (1833); Quarterly Bulletin of the Alpine Garden Society **56**: 165 (1988).

Tuber solid, more or less spherical. Flowering-stem to 20 cm, with a scale-leaf at the base and 1 or usually 2 alternate stem-leaves. Leaves glaucous, twice divided into 3, the ultimate segments elliptic or obovate, obtuse but pointed at the apex. Flowers 10–25 (rarely fewer) in a long terminal raceme; bracts entire. Corolla 3.5–4.5 cm, pale rose pink (rarely yellow and white), the inner petals darker-tipped, narrow, tapering into the slender, tapered, downwardly curved spur which is *c.* 2.5 cm. Capsule to 2 × 0.2 cm, linear, tapering into the persistent style. Seeds black, shining, each with a conspicuous appendage. *Former USSR (C Asia).* H3.

Remarkable for its long, tapering, narrow spur.

7. C. cava (Linnaeus) Schweigger & Koerte (*C. bulbosa* misapplied; *C. tuberosa* de Candolle). Figure 13(7), p. 123. Illustration: Bonnier, Flore complète **1**: pl. 27 (1911); Flora SSSR **7**: t. 43 (1937); Acta Horti Bergiani **17**: 125 & pl. 2 (1955).

Figure 13. Leaves of *Corydalis* species (× 2). 1, *C. sempervirens.*
2, *C. solida.* 3, *C. angustifolia.* 4, *C. ambigua.* 5, *C. bracteata.*
6, *C. shanginii.* 7, *C. cava.* 8, *C. popovii.* 9, *C. diphylla.*
10, *C. verticillaris.* 11, *C. cashmeriana.* 12, *C. chaerophylla.*
13, *C. nobilis.* 14, *C. sibirica.*

Tuber hollow, 5 cm or more across. Stems often several, erect, to 20 cm, without scale-leaves at the base, each stem bearing 2 or more alternate leaves which are once or twice divided into 3, the leaflets deeply and irregularly lobed, the ultimate segments oblong-obovate, blunt. Racemes terminal, loose or dense, with 5--many flowers; bracts large, often brownish, entire. Corolla violet or white, 2–3 cm, the spur oblong, slightly tapered, c. 1 cm, the apex somewhat swollen and curved downwards. Capsule ovate-elliptic, to 2.5 cm, tapering into the persistent style. Seeds black, shining, each with a conspicuous whitish appendage. *Most of Europe.* H1. Late winter–spring.

A variable species; a white-flowered variant, with dark bracts is grown as 'Albiflora' (var. *albiflora* Anon.); similar variants occur in the wild.

C. marschalliana (Pallas) Persoon (*C. bulbosa* (Linnaeus) de Candolle subsp. *marschalliana* (Pallas) Chater). Illustration: Gartenflora **15**: t. 511 (1866); Flora SSSR **7**: t. 43 (1937); Quarterly Bulletin of the Alpine Garden Society **56**: 158 (1988). Very similar to *C. cava* and perhaps intergrading with it in E Europe; leaf-segments broader and less divided, corolla yellowish white or pink. *E Europe, SW Asia.* H3. Spring.

8. C. popovii Popov. Figure 13(8), p. 123. Illustration: Flora SSSR **7**: t. 43 (1937).

Tuber solid. Flowering-stem to 15 cm, bearing a pair of opposite leaves; scale-leaf absent. Leaves glaucous, divided twice into 3, ultimate segments more than 2 mm broad, obovate, acute, the terminal sometimes deeply 3-lobed. Raceme distinctly overtopping the leaves, terminal, with 3–6 flowers; bracts entire, ovate. Corolla 3.5–4.5 cm, purplish pink, deep purple at the tip, tapered, spur 2–2.5 cm, somewhat tapered, curved downwards at the blunt tip; inner petals notched at the tips. Capsule broadly ovate, 10–13 × 5–6 mm, narrowed into the persistent style. Seeds black, shining, appendaged. *Former USSR (C Asia).* H3. Spring.

9. C. diphylla Wallich. Figure 13(9), p. 123. Illustration: Quarterly Bulletin of the Alpine Garden Society **56**: 140 (1988).

Tuber spherical, solid. Flowering-stem to 20 cm, without a scale-leaf at the base, bearing a pair of opposite (sometimes not quite opposite) leaves. Leaves distinctly stalked, 2 or 3 times divided into 3, the ultimate segments more than 2 mm broad, elliptic, tapered to the ultimately obtuse

apex, all glaucous. Raceme terminal with 5 or more flowers; bracts entire, rather small. Corolla white or cream, widening to a broad, purple tip 1.5–2 cm, tapered into the usually upwardly directed spur, which is 7–10 mm, its blunt tip curved downwards or upwards. Capsule elliptic, to 15 × 5 mm, tapering into the persistent style. Seeds black, shining, appendaged. *W Himalaya.* H3. Spring.

10. C. rutifolia (Sibthorp & Smith) de Candolle. Illustration: Matthews, Lilies of the field, opposite p. 40 (1968); Everard & Morley, Flowers of the world, pl. 91 (1970).

Tuberous. Flowering-stem 3–12 cm, without a scale-leaf at the base and with a pair of opposite, stalkless stem-leaves which are 2 or 3 times divided into 3, the ultimate segments more than 2 mm broad, obovate to narrowly elliptic. Raceme terminal, rather dense, with up to 7 flowers. Corolla 1–2 cm, dark red or purplish or sometimes pink or whitish, tips of petals darker, tapering into the cylindric spur to 1.5 cm, usually curved downwards at its blunt tip. Capsule ovate. Seeds black, shining, appendaged. *E Mediterranean area.* H5. Spring.

Plants of this species are often very small and inconspicuous, so are not often grown.

11. C. verticillaris de Candolle. Figure 13(10), p. 123. Illustration: Botanical Magazine, 9486 (1937).

Like *C. rutifolia* but leaves deeply dissected into narrow segments which are less than 2 mm wide; flowers pinkish purple to pink, with darker tips, 2.5–3.5 cm, spur to 2 cm, often curved upwards. *Iran.* H4. Spring.

12. C. cashmeriana Royle. Figure 13(11), p. 123. Illustration: Botanical Magazine, n.s., 522 (1968); Hay & Synge, Dictionary of garden plants in colour, pl. 40 (1969); Perry, Flowers of the world, 119 (1970); Gartenpraxis for June, 1985: 17 (1985).

Herb arising from a cluster of fleshy roots, without a tuber. Stems to 20 cm (often much less), erect, bearing 1 or 2 alternate leaves. Stem-leaves stalkless, divided into 3 stalked leaflets which are themselves deeply lobed almost to the base into narrow linear or narrowly obovate, obtuse segments; basal leaves similar but stalked. Flowers few in condensed, umbel-like racemes; bracts similar to the stem-leaves but smaller. Corolla bright blue or purplish blue, inner petals whitish below the tip, 1.4–2.8 cm, spur tapered, narrow,

straight or curved downwards at the blunt apex, 6–12 mm. Capsule pendent from the apex of the erect fruit-stalk, c. 10 × 1 mm, abruptly tapered into the persistent style. Seeds shining, appendaged. *W Himalaya eastwards to N India (Sikkim).* H4. Spring–summer.

A variable species, highly prized for its bright blue flowers. It is difficult to grow well, however, requiring a cool acid soil with plenty of peat or leaf-mould in a well-drained but always moist, partially shaded site; it can be grown in pots, which should be plunged during the summer. The roots should not be planted too deeply after division.

13. C. chaerophylla de Candolle. Figure 13(12), p. 123. Illustration: Acta Horti Bergiani **17**: 144 & pl. 7 & 8 (1955).

Tall perennial herb to 1 m from a stout rhizome. Leaves basal and on the stem, the basal long-stalked, those on the stem shortly stalked or stalkless, all 2–3 times pinnate, the segments lobed or pinnatifid. Flowers in a terminal panicle or raceme, very numerous and dense; bracts entire, equalling the short flower-stalks. Corolla bright yellow, 1.3–2 cm, spur tapered, straight or curved downwards to the blunt apex. Capsules narrowly obovate to obovate, 1–1.5 cm, tapering into the persistent style. Seeds few, black, shining, each with a whitish appendage. *Himalaya.* H3. Summer.

14. C. nobilis (Linnaeus) Persoon. Figure 13(13), p. 123. Illustration: Botanical Magazine, 1953 (1818); Edwards's Botanical Register **5**: t. 395 (1819); Acta Horti Bergiani **17**: 125 & pl. 3 (1955); Gartenpraxis for June, 1985, 17.

Perennial herb with a fleshy, branched tap-root, later with a rhizome. Stems erect, somewhat winged, to 1 m. Leaves basal and on the stem, pinnate, the leaflets deeply pinnatisect with deeply lobed segments, ultimate segments oblong or obovate. Flowers many in a very dense, head-like, terminal raceme; bracts entire or the lowermost toothed at the apex. Corolla 1.5–2.5 cm, basically pale yellow, the tips of the petals usually brownish purple; spur 5–10 mm, straight or curved downwards. Capsule ovate-elliptic, c. 15 × 5 mm, long-tapered into the persistent style. Seeds black, shining, each with a conspicuous, whitish appendage. *Former USSR (C Asia).* Spring–summer.

Probably the most frequently grown of the taller species, spectacular when in flower.

15. C. sibirica (Linnaeus filius) Persoon. Figure 13(14), p. 123. Illustration: Acta Horti Bergiani **17**: 152 (1955).

Annual or biennial herb. Stems branched, erect, very slender, to 70 cm, grooved. Leaves basal and on the stem, 2–3 times pinnate, the ultimate segments narrow, linear. Flowers in terminal racemes; bracts weakly toothed towards the apex, longer than the flower-stalks. Corolla yellow, sometimes violet at the tip, 7–9 mm; spur *c.* 3 mm, wide, tapering, blunt; lower petal somewhat swollen at the base. Capsule 5–15 × 2–3 mm, obovoid, tapering to the base, rather abruptly rounded into the persistent style. Seeds black, shining, each with a minute whitish appendage. *Former USSR (Siberia).* H1. Summer.

It is doubtful if this species is still in general cultivation.

16. C. scouleri J.D. Hooker (*C. allenii* Fedde). Figure 14(1), p. 126. Illustration: Acta Horti Bergiani **17**: 144 & pl. 5 (1955); Rickett, Wild flowers of the United States **5**: pl. 68 (1971); Clark, Wild flowers of British Columbia, 182 (1973).

Perennial with a thick, horizontal rhizome. Stems to 1 m. Leaves basal and on the stem, long-stalked, 3–4 times pinnate, the ultimate segments often deeply lobed, green above, somewhat greyish beneath; stem-leaves with whitish stipules. Racemes with up to 25 flowers, dense, the flowers borne almost vertically with the spur pointing upwards; bracts entire, narrow. Corolla 2.5–3 cm, pink or lilac-pink (rarely white), the spur tapered, straight, obtuse at the apex, to 2 cm. Capsule obovoid or club-shaped; style persistent. Seeds black, shining, each with a small, white appendage. *W North America (British Columbia, Oregon, Washington).* H3. Late spring.

The curious posture of the flowers renders this species immediately recognisable.

17. C. rupestris Boissier. Figure 14(2), p. 126. Illustration: Quarterly Bulletin of the Alpine Garden Society **56**: 159 (1988).

Small, tufted perennial herb with a thick tap-root. Leaves all basal, twice pinnate with 4–10 leaflets, the ultimate segments narrow, longer than broad; old leaf-stalks dry and persisting around the tuft. Flowering-stem to 10 cm, bearing a few, bract-like scale-leaves. Raceme terminal with up to 12 flowers; bracts lanceolate, shorter than the flower-stalks. Corolla *c.* 2 cm, yellow; spur broad, blunt, *c.* 2 mm.

Capsule ovoid, tapering into the persistent style. Seeds black, each with a small white appendage. *Iran.* H5. Late spring.

Not common in cultivation.

18. C. cheilanthifolia Hemsley. Figure 14(3), p. 126. Illustration: The Garden **84**: 234 (1920); Acta Horti Bergiani **17**: 158 & pl. 6 & 8 (1955); Quarterly Bulletin of the Alpine Garden Society **56**: 160 (1988).

Perennial herb with clustered roots. Stems to 20 cm bearing leaves only at its extreme base, the upper part of the stem with simple, bract-like scale-leaves. Foliage leaves 2–3 times pinnate, with 10 or more primary leaflets, the ultimate segments very fine, all green to dark green and very fern-like. Racemes terminal with many flowers; bracts lanceolate, entire, about as long as the flower-stalks. Corolla 1.2–1.5 cm, bright yellow, the tips of the petals sometimes greenish, spur curved upwards, tapered, blunt, 6–7 mm. Capsule linear-cylindric, curved, spreading, slightly constricted between the seeds; style persistent. Seeds black, shining, each with a small, white appendage. *C & W China.* H3. Spring–autumn.

Flowers produced late in the year are sometimes without spurs.

19. C. ophiocarpa Hooker & Thomson. Figure 14(4), p. 126. Illustration: Acta Horti Bergiani **17**: 158 & pl. 6 & 7 (1955); Opera Botanica **88**: 28 (1986).

Perennial herb with a branched tap-root. Stems to 75 cm, winged and grooved, leafy. Leaves greyish green, 2–3 times pinnate, the segments ovate-oblong with shallow, rounded lobes. Racemes terminal with many flowers; bracts narrowly lanceolate, shorter than the flower-stalks. Corolla 1–1.5 cm, pale yellow or cream tipped with purplish or dark red; spur to 3 mm, scarcely tapered, blunt. Capsule linear, somewhat constricted between the seeds, irregularly contorted, to 3 × 0.2 cm; style persistent. Seeds black, shining, each with a large, brownish appendage. *Himalaya, China, Japan, Taiwan.* H3. Spring–summer.

Not particularly ornamental, but sometimes grown as a curiosity.

20. C. aurea Willdenow. Figure 14(5), p. 126. Illustration: Rickett, Wild flowers of the United States **1**: pl. 53 (1967); Clark, Wild flowers of British Columbia, 175 (1973); Porsild, Rocky mountain wild flowers, 197 (1974).

Annual or biennial herb with a stout tap-root. Stems erect or ascending, usually

10–30 cm, leafy throughout, branched. Leaves somewhat glaucous, finely divided (1–3 times pinnate or pinnatisect), the ultimate segments fine, narrow, acute. Flowers numerous in dense terminal racemes, to 1.5 cm, yellow; spur rounded, somewhat swollen, to 5 mm. Capsule linear, constricted between the seeds, often strongly curved, 1.5–3 × 0.3–0.5 cm; style persistent. Seeds black, shining, each with a small, brownish appendage. *N America, S Mexico.* H1. Spring–summer.

C. vaginans Royle, which is taller, has leaves glaucous beneath and flowers with longer spurs, may occasionally be grown. *W Himalaya.*

21. C. saxicola Bunting (*C. thalictrifolia* Franchet). Figure 14(6), p. 126. Illustration: Botanical Magazine, 7830 (1902); Acta Horti Bergiani **17**: 158 & pl. 8 (1955); Opera Botanica **88**: 29 (1986).

Tufted perennial herb with a long, woody rhizome. Stems to 30 cm, leafy. Leaves 2–3 times pinnate, the ultimate segments broad, rather deeply divided into rounded or somewhat acute lobes. Racemes terminal (appearing leaf-opposed) with many flowers; bracts ovate, acuminate, entire, shorter than the flower-stalks. Corolla 1.5–2.5 cm, yellow, the spur more or less parallel-sided, curved downwards towards the blunt apex, *c.* 5 mm. Capsule linear, *c.* 30 × 1.5–2 mm, somewhat constricted between the seeds; style persistent. Seeds black, shining, each with a brownish appendage. *C & W China.* H3. Spring.

22. C. wilsonii N.E. Brown. Figure 14(7), p. 126. Illustration: Botanical Magazine, 7939 (1904); Acta Horti Bergiani **17**: pl. 6 (1955); Amateur Gardening **103** (December): 20 (1987).

Tufted perennial herb with a woody tap-root. Stems to 10 cm (rarely more), bearing leaves only near or at the base in a rosette. Leaves usually very glaucous, pinnate, the leaflets themselves pinnate or pinnatisect with broad, rounded ultimate segments which are shallowly lobed. Racemes terminal with few to many flowers; bracts narrowly lanceolate, shorter than the flower-stalks. Corolla 1.6–2 cm, bright yellow, often greenish at the tip; spur rounded, blunt, 3–6 mm. Capsule oblong-lanceolate, to 20 × 1.5–2 mm, long-tapering into the persistent style. Seeds black, shining, each with a small, brownish appendage. *C China.* H3. Spring–summer.

One of the most attractive species which

HSM

Figure 14. Leaves of *Corydalis* species (× 2). 1a, Apical leaflets of *C. scouleri* (from 1b). 1b, Diagram of *C. scouleri*. 2, *C. rupestris*. 3, *C. cheilanthifolia*. 4, *C. ophiocarpa*. 5, *C. aurea*. 6, *C. saxicola*. 7, *C. wilsonii*. 8, *C. tomentella*. 9, *C. lutea*. 10, *C. claviculata*.

can ultimately build up large clumps in good growing conditions.

23. C. tomentella Franchet (*C. tomentosa* N.E. Brown). Figure 14(8), p. 126.

Similar to *C. wilsonii* but leaves and stems covered with dense, whitish hairs; leaves more deeply divided, with longer, more acute segments; capsules linear, *c.* 30 × 1.5–2 mm. *W China.* H5. Spring–summer.

24. C. ochroleuca Koch (*Pseudofumaria alba* (Miller) Lidén). Illustration: Acta Horti Bergiani **17**: 130 & pl. 5 (1955); Polunin, Flowers of Europe, pl. 30 (1969); Pignatti, Flora d'Italia **1**: 360 (1982); Opera Botanica **88**:31 (1986).

Perennial herb with a rhizome. Stems to 50 cm, leafy, winged and grooved. Leaves 2–3 times pinnate, the ultimate segments lobed or toothed, broad, glaucous on both surfaces; stalks widened towards the base and running into the wings on the stem. Racemes dense with up to 20 flowers; bracts narrow, entire or minutely toothed at the apex, shorter than the flower-stalks. Corolla 1.3–1.7 cm, cream, yellowish at the tip; spur rounded, *c.* 3 mm. Capsule erect or spreading, *c.* 13 × 5 mm, somewhat constricted between the seeds, acute at the apex, the style early deciduous. Seeds black, appendage dissected. *SE Europe, naturalised in W & C Europe.* H2. Spring–summer.

This and the next species comprise the genus *Pseudofumaria* Medikus.

25. C. lutea (Linnaeus) de Candolle (*Pseudofumaria lutea* (Linnaeus) Borckhausen). Figure 14(9), p. 126. Illustration: Bonnier, Flore complète **1**: pl. 27 (1911); Ary & Gregory, The Oxford book of wild flowers, 30 (1962); Polunin, Flowers of Europe, pl. 30 (1969); Garrard & Streeter, Wild flowers of the British Isles, pl. 6 (1983).

Like *C. ochroleuca* but leaves green above, glaucous beneath, stalk not widened below and joining wings on the stem, corolla golden yellow with darker tip. *S Europe, widely naturalised elsewhere.* H2. Spring–summer.

A handsome species which colonises damp ground and, particularly, old walls.

26. C. claviculata (Linnaeus) de Candolle (*Ceratocapnos claviculata* (Linnaeus) Lidén). Figure 14(10), p. 126. Illustration: Bonnier, Flore complète **1**: pl. 27 (1911); Ary & Gregory, The Oxford book of wild flowers, 82 (1962); Polunin, Flowers of Europe, pl. 30 (1969).

Slender, much-branched, climbing annual to 1 m. Leaves twice pinnate, ultimate segments shortly stalked, elliptic-obovate, the terminal leaflet replaced by a branched tendril. Flowers 6–10 in rather dense, terminal racemes (appearing leaf-opposed); bracts ovate, broad-based, entire, shorter than to as long as the flower-stalks. Corolla white or pinkish, often darker tipped, 5–6 mm; spur *c.* 1.5 mm, rounded. Capsule oblong, sometimes hairy, tapering to the apex, style early deciduous. *W Europe from Spain and Portugal to Norway and Denmark.* H3. Spring–summer.

Not often grown, as the flowers are very small, but sometimes allowed to naturalise in shady corners among shrubs. One of the three species placed in the genus *Ceratocapnos* Lidén.

Another plant of similar, climbing habit, which may occasionally be grown is **Cysticapnos vesicarius** (Linnaeus) Fedde. This is more robust and has larger leaf-segments, the flowers are larger and usually pink, and the fruit is inflated and bladdery, becoming hardened and fibrous as it matures; the style is persistent and the mature fruit disintegrates to release the numerous seeds. *South Africa.* H5–G1. Summer.

25. SARCOCAPNOS de Candolle
J. Cullen

Cushion- or clump-forming annual to perennial herbs, leaves and stems fleshy and brittle, often glaucous. Leaves long-stalked, undivided or 1–3 times divided into 3. Flowers in corymb-like racemes with small bracts. Sepals 2, small. Petals 4, white or pink, the upper petal with a backwardly projecting spur and the lower with a long claw and notched blade; lateral petals with yellow lateral wings, blotched internally with blackish red at the apex. Stamens 4 in 2 bundles (upper and lower) of 1 stamen plus 2 half-stamens, the upper with a nectary projecting back into the petal-spur. Ovary with 2 ovules; stigma expanded with receptive tissue along the margins. Fruit a 1–2-seeded nutlet with an apical appendage and ribbed faces, the fruit-stalks lengthening and bent downwards or inwards.

A genus of 3 species from Spain and N Africa. They grow in the wild on cliffs, often forming large clumps or cushions. They may be grown in well-drained soil in a rockery or in pots in an alpine house (when they require careful watering during the growing season). Propagate by seed or cuttings.

Literature: Heywood, V.H., Notulae criticae ad floram hispaniae pertinentes 1, *Bulletin of the British Museum (Botany)* **1**(4): 90–92 (1954); Lidén, M., Synopsis of Fumarioideae (Papaveraceae) with a monograph of the tribe Fumarieae, *Opera Botanica* **88**: 33–38 (1986).

1a. Leaflets 7–18; stigma forming a trapezoidal, membranous flat plate   **1. enneaphylla**
  b. Leaflets 1–7 (rarely to 8); stigma forming a low crest   **2. crassifolia**

1. S. enneaphylla (Linnaeus) de Candolle. Illustration: Botanical Magazine, 9525 (1938); Polunin & Smythies, Flowers of southwest Europe, pl. 8 (1973); Castroviejo et al., Flora Iberica **1**: 434 (1986).

Annual or short-lived perennial, without the remnants of old leaves at the base, hairy with spreading hairs, or hairless. Leaves 2–3 times divided into 3, with 7–18 leaflets which are ovate to elliptic, usually cordate at the base when broad, acute at the apex. Racemes with 5–15 white or pink flowers. Flowers 1.2–1.7 cm (including the spur which is 4–6 mm). Stigma a membranous, trapezoidal flat plate. Fruit 4–5 mm, with rounded appendage and ribs on the faces narrower than those on the margins. *E Spain, Morocco (1 locality).* H5. Spring–summer.

2. S. crassifolia (Desfontaines) de Candolle.

Perennial with the remnants of old leaves at the base, not hairy. Leaves with 1–7 (rarely more) leaflets; leaflets very fleshy, ovate to almost circular, rounded to cordate at the base, obtuse or acute at the apex. Racemes with up to 35 flowers. Flowers fragrant, white or pink, 0.8–2.1 cm (including the spur which is 1–4 mm). Stigma a low crest. Fruit 5–6 mm, with a pointed appendage and the ribs on the faces not or scarcely narrower than those on the margins. *S Spain, Algeria, Morocco.* H5. Spring–summer.

A variable species; most commonly grown is subsp. **speciosa** (Boissier) Rouy (*S. speciosa* Boissier). Illustration: Quarterly Bulletin of the Alpine Garden Society **40**: 287 (1972); Polunin & Smythies, Flowers of southwest Europe, pl. 8 (1973); Castroviejo et al., Flora Iberica **1**: 438 (1986). Large plant with an open growth habit and flowers 1.6–2.1 cm, usually rose pink. *S Spain.* Subsp. **crassifolia**. Flowers 1–1.7 cm, white. *Algeria and Morocco.* Subsp. **atlantis** (Emberger & Maire) Lidén. Flowers smaller. *Morocco.*

The third species of the genus, **S. baetica** (Boissier & Reuter) Nyman. Illustration: Castroviejo et al., Flora Iberica **1**: 440 (1986). Similar to *S. crassifolia*, but has even smaller flowers (5–6 mm). *S Spain*. Occasionally grown.

26. RUPICAPNOS Pomel
J. Cullen

Annual or perennial (often short-lived) herbs forming clumps or loose cushions, stems and leaves very brittle. Leaves long-stalked, pinnate, segments deeply lobed or toothed. Flowers in flat-topped, shortly stalked racemes, each flower subtended by a small bract. Sepals 2, often lobed or toothed. Petals 4, white or pink, the upper with a small, purple-marked apex (ours) and a short, blunt, downwardly curved spur, the lower oblong-linear, greenish at the acute apex, the laterals winged, with purple apices. Stamens in 2 bundles (upper and lower) of a central stamen united to the anther with 2 half-stamens, the upper with a nectary projecting back into the petal-spur. Ovary with 1 ovule; stigma a flattened dome with lobed margins. Fruit a nutlet, the fruit-stalks lengthening and growing downwards or inwards, burying the fruit.

A genus of 7 species (following Lidén, see below; Pugsley recognised 34 species) of cliff-dwelling plants from S Spain and N Africa. Only 1 species seems to be in general cultivation, grown in rockeries or in pots in alpine houses. Propagate by seed, division or cuttings.

Literature: Pugsley, H.W., A revision of the genera *Fumaria* and *Rupicapnos, Journal of the Linnean Society* **49**: 93–113 (1932); Lidén, M., Synopsis of Fumarioideae (Papaveraceae) with a monograph of the tribe Fumarieae, *Opera Botanica* **88**: 92–104 (1986).

1. R. africanus (Lamarck) Pomel. Illustration: Maire, Flore d'Afrique du Nord **12**: 37, 39 (1965); Opera Botanica **88**: 94 (1986); Castroviejo et al., Flora Iberica **1**: 468 (1986).

Perennial to 10 cm, widely spreading, very glaucous. Sepals conspicuously toothed or lobed. Upper petal 1.2–1.8 cm, including the spur which is 2–3 mm. Fruit somewhat compressed, warty, beaked. *S Spain, Morocco, Algeria*. H5. Spring and sporadically to late autumn.

A very variable species which Lidén divides into 6 subspecies. Which of these is cultivated is uncertain, though subsp. **africanus** with lanceolate leaf-segments

and deeply toothed sepals *c*. 2 × 2 mm, and subsp. **decipiens** (Pugsley) Maire with broader leaf-segments and larger (3–4.5 × 2–3 mm), less deeply toothed sepals have both been available.

27. FUMARIA Linnaeus
S.G. Knees

Annual herbs with long, erect, spreading or climbing stems. Leaves 2–4 times pinnately divided; leaf segments flat or channelled, oblong-lanceolate to broadly ovate. Flowers in racemes subtended by bracts, or without bracts; sepals 2; corolla bilaterally symmetric, consisting of 1 spurred upper petal, 1 lower petal and 2 inner petals; ovary with 2 ovules, stigma crested or with 2 papillae. Fruit an indehiscent capsule with 2 apical pits, seeds 1–13 (in the species covered here), light brown or black.

A genus of about 55 species from Europe to central Asia and the Himalaya, south to the tropical east African highlands. Only one is commonly cultivated and many are rather weedy. Propagate from seed.

Literature: Lidén, M. Synopsis of Fumarioideae (Papaveraceae) with a monograph of the tribe Fumarieae. *Opera Botanica* **88**: 1–133 (1986).

1. F. officinalis Linnaeus. Illustration: Sowerby & Smith, English Botany **9**: 589 (1799); Opera Botanica **88**: 82 (1986); Blamey & Grey-Wilson, Illustrated flora of Britain & northern Europe, 130, 131 (1989).

Annual herb, with erect stems to 45 cm. Leaf segments flat, bright green. Flowering raceme longer than flower-stalk, with up to 80 flowers; bracts half as long to as long as fruiting stalks; sepals 1.5–3.5 × 1–1.5 mm, irregularly toothed; corolla 7–9 mm, purplish pink, wings blackish red. Fruit *c*. 2 × 2.5–3 mm; seed solitary, light brown. *Europe*. H2. Spring.

XCVII. CAPPARIDACEAE

Deciduous trees, shrubs or annual herbs. Leaves alternate, simple, with 3 leaflets or palmate. Stipules present or absent. Flowers bisexual, somewhat bilaterally symmetric, solitary or in terminal racemes. Sepals 4, free or united in lower part. Petals 4, usually stalked. Stamens *c*. 6 or numerous. Ovary superior, stalked, the stalk lengthening in fruit. Fruit 1-celled, of 2 united carpels. Seeds few to many.

A mainly tropical family of 45 genera

and about 700 species. Few species are cultivated. In addition to the 3 genera treated below, **Boscia albitrunca** (Burchell) Gilg & Benedict is a tropical African tree grown for its roots, which are a coffee substitute, but is rarely cultivated in Europe. **Cladostemon kirkii** A. Braun & Vatke from tropical Africa, with gourd-like fruits and **Koeberlinia spinosa** Zuccarini from Mexico and adjacent USA, with spiny leafless branches, are occasionally cultivated as curiosities.

1a. Leaves simple **1. Capparis**
 b. Leaves compound **2**
2a. Leaves with 3 leaflets; stipules present **2. Crateva**
 b. Leaves with 3–7 leaflets; stipules absent **3. Cleome**

1. CAPPARIS Linnaeus
J.R. Akeroyd & S.G. Knees

Shrubs with trailing or ascending branched stems to 2 m. Leaves 2–6 × 1–5 cm, ovate to almost circular, entire, rounded or cordate at base, obtuse to acuminate, mucronate, leathery or somewhat succulent, with or without hairs; leaf-stalk 5–10 mm. Stipules rigid or bristle-like, spiny, often recurved. Flowers solitary in leaf axils; flower-stalks stout, longer than leaves. Sepals 4, broadly ovate, the outer pair longer, concave. Petals 1.5–3 cm, broadly obovate, obtuse, white, often tinged pink. Stamens numerous, longer than petals, purplish, especially towards the anthers; anthers purple. Stigma stalkless. Ovary stalked, the stalk lengthening in fruit. Fruit 2.5–4 × 1.5–2.5 cm, ovoid or ellipsoid, ribbed, leathery. Seeds *c*. 3 × 2.5 mm, kidney-shaped, smooth, brown.

A genus of 250 mainly tropical species, only one of which is native and cultivated in Europe. Propagation is by cuttings or from seed.

Literature: Higton, R.N. & Akeroyd, J.R. *Botanical Journal of the Linnean Society* **106**(2): 104–112 (1991).

1. C. spinosa Linnaeus (*C. ovata* Desfontaines). Illustration: Botanical Magazine, 291 (1795); Polunin, Flowers of Europe, pl. 30 (1969); Huxley, Flowers of Greece and the Aegean, t. 65 (1977); Krüssmann, Manual of cultivated broad-leaved trees & shrubs **1**: 272 (1984).

Mediterranean area & Middle East. H5–G1. Spring–summer.

Subsp. **spinosa** has leaves with a distinct mucro and conspicuous, spiny stipules; the flower-buds are the capers of commerce.

Subsp. **rupestris** (Smith) Nyman (*C. rupestris* Smith; *C. inermis* Turra). Has somewhat broader leaves with a minute mucro and inconspicuous, bristle-like stipules.

2. CRATEVA Linnaeus
S.G. Knees

Deciduous small trees or shrubs with pale brown bark; branches with conspicuous white lenticels. Leaves alternate, with basal stipules, hairless, palmately compound, crowded towards ends of branches; leaf-stalks 7–17 cm; leaflets 3, ovate-lanceolate, 3–10 × 3–4 cm, apex long acuminate, base wedge-shaped. Flowers bisexual, apparently bilaterally symmetric, up to 20, in terminal corymbs. Sepals 4, *c.* 4 mm, pale green; petals 4, clawed, obovate-spathulate, 1.25–2 cm, white or creamy yellow, tipped lilac. Disc curved. Stamens numerous, 2–2.5 cm, pale lilac. Ovary 1-celled. Fruit a round berry, 2.2–2.6 cm, greenish, spotted white, pulp fleshy, yellow; seeds *c.* 6 mm, numerous.

A genus of 6 species occurring in the Old World tropics. Only one is cultivated in Europe. Propagation is from cuttings rooted in sand with bottom heat.

1. C. religiosa Forster filius (*C. adansonia* de Candolle). Illustration: Engler, Die natürlichen Pflanzenfamilien **3**(2): 227 (1891); Li, Flora of Taiwan **2**: 672 (1976) – as C. adansonii subsp. formosensis; American Horticulturalist **63**(2): 5 (1984); Blundell, Collins guide to the wild flowers of east Africa, 225 (1987).

Africa, India, S China, Polynesia. G2.

3. CLEOME Linnaeus
J.R. Akeroyd & S.G. Knees

Erect annual or shrubby perennials, often with glandular hairs. Leaves compound. Stipules absent but leaf-stalks sometimes with a solitary spine at base. Flowers in terminal, bracteate racemes. Sepals 4; petals 4; stamens usually 6; ovary stalked, the stalk lengthening considerably in fruit. Fruit a 2-celled, many seeded capsule. Seeds kidney-shaped, smooth, brown.

A genus of some 150 species, mainly tropical, although only three are commonly cultivated in Europe, where they are treated as half-hardy annuals or as glasshouse subjects. Propagation is usually from seed.

1a. Petals yellow, stalkless; leaves with 3 leaflets **3. isomeris**
  b. Petals white, pink or violet, distinctly stalked; leaves usually 5–7-palmate 2
2a. Petals 2.5–4 cm; leaf-stalk often with a spine at base **1. hasslerana**
  b. Petals *c.* 1 cm; leaf-stalk without a spine at base **2. gynandra**

1. C. hasslerana Chodat (*C. spinosa* misapplied; *C. arborea* misapplied; *C. gigantea* misapplied; *C. pungens* misapplied). Illustration: Botanical Magazine, 1640 (1814); RHS Dictionary of gardening **1**: 504 (1951); Graf, Exotica 3, 507 (1963); Hay & Synge, Dictionary of garden plants, 34 (1969).

Erect, glandular-hairy, aromatic annual, 1–2 m. Leaves palmate with 5–7 leaflets; leaflets 3–7 × 0.8–1.5 cm, narrowly elliptic, entire, acuminate, tapered to a short stalk; leaf-stalk about as long as leaflets, often with a spine at base. Flowers in long racemes, each subtended by a heart-shaped bract. Sepals free, lanceolate. Petals 2.5–4 cm, elliptic, distinctly stalked, deep pink or violet (sometimes white), soon fading and falling. Stamens usually 6, to 6 cm. Fruit 5–15 × 0.3 cm, narrowly cylindrical and often curved, hairless. *E South America (SE Brazil to Argentina); widely naturalised in the tropical and warm temperate Americas.* H5–G1. Summer–early autumn.

Many cultivars have been selected, notably 'Rose Queen', 'Pink Queen' and 'Rosea' with pink flowers; 'Helen Ballard' and 'Alba' with white flowers. Often used in summer bedding.

2. C. gynandra Linnaeus (*C. pentaphylla* Linnaeus; *Gynandropsis gynandra* (Linnaeus) Briquet; *G. pentaphylla* (Linnaeus) de Candolle). Illustration: Botanical Magazine, 1681 (1814); Täckholm, Student's Flora of Egypt pl. 50 (1974); Blundell, Collins guide to the wild flowers of East Africa, 199 (1987).

Similar to *C. hasslerana* but smaller, not exceeding 1 m; leaflets broader, to 2 cm wide, the leaf-stalk usually longer than leaflets, without spine at base; bracts more leaf-like; petals *c.* 1 cm; stamens borne on stalk of ovary; fruit glandular-hairy when young. *Old world tropics; widely naturalised as a weed throughout the tropics.* H5–G1. Summer–early autumn.

Sometimes placed in a separate genus on account of the position of the stamens.

3. C. isomeris Greene (*Isomeris arborea* Nuttall). Illustration: Botanical Magazine,

3842 (1840); Munz, A California Flora, 206 (1969); Coyle & Roberts, A field guide to the common and interesting plants of Baja California, 77 (1975).

Shrub 1–3 m, softly and finely hairy. Leaves with 3 leaflets, foetid when bruised; leaflets 2.5–3.5 × 0.8–1 cm, narrowly elliptic, wedge-shaped at base, almost obtuse, slightly mucronate, tapered to a short stalk; leaf-stalk slightly shorter than leaflets, white-hairy, especially above. Flowers in compact racemes, each subtended by a simple, leaf-like bract. Sepals 4, united for the lower third, broadly triangular, awned. Petals 1.2–1.6 cm, elliptic, stalkless, golden yellow. Stamens 6, to 2.5 cm, yellow; anthers yellow. Capsule 2.5–5 × 1.5–2 cm, obconical, inflated, sharply pointed, hairless. Seeds 6–7 mm, brown. *USA (California), Mexico (Baja California).* H5. Summer–early autumn.

XCVIII. CRUCIFERAE

Annual to perennial herbs, rarely small shrubs. Leaves alternate, rarely opposite, without stipules. Flowers usually bisexual, sepals, petals and stamens hypogynous. Sepals 4, free, in two opposite pairs. Petals 4, free, usually clawed, alternating with the sepals. Stamens usually 6, rarely 4, 2 or absent, outer pair with short filaments, two inner pairs with long filaments; filaments sometimes winged or with tooth-like appendage. Nectaries variable and variously arranged around base of stamens and ovary. Ovary of two fused carpels, usually with 2 parietal placentas and 2-celled by the development of a membranous false septum formed from outgrowths of the placentas. Stigma capitate to 2-lobed. Fruit usually a dehiscent capsule opening by 2 valves from below; sometimes 1 or more seeds develop in an indehiscent beak at the base of the style. Fruit called a siliqua when at least 3 times as long as wide or a silicula if less than 3 times as long as wide; sometimes indehiscent or breaking transversely into 1-seeded portions. Seeds often mucilaginous when wet.

A family containing about 3000 species in 390 genera, cosmopolitan but chiefly in north temperate regions.

Identification of genera requires knowledge of fruit characters, petal colour and hairs type. A hand lens (× 10) is necessary to see the hairs, and both leaves and stem should be examined: not only may hairs be commoner on leaves than stems, but different types may also occur on the different organs. Petal colour often changes on drying, not only fading but also sometimes white turning yellow. *Armoracia* rarely produces ripe fruits and is difficult to key out, but the genus can easily be characterised: plant robust with strong tap-root; basal leaves 30–50 cm, ovate or oblong, toothed, long-stalked; lower stem-leaves often pinnatifid; flowers small, white, in large panicles; fruits rarely produced.

Habit. Spiny shrub: **1–7,26,51**; stemless herbs, fruit lengthening and curving down burying fruits: **53,55**.

Smell. Fresh plants smelling of garlic when crushed: **5,41**.

Flowers. Two petals much larger: **43**; petals sometimes yellow: **2,4,9,10,12, 14,18,23–27,32,34,35,38,42–44, 49–56**; petals pinnately lobed: **1**; filaments joined in pairs: **51**.

Fruits. Pendent: **8,22**; flattened and 2-lobed: **44,45**; siliqua: **1–12,14,15, 17–20,22,32,48–50,53–56**; silicula: **13,16,20,21,23–47,51,52,56**; dehiscing explosively: **18**.

1a. Fruit less than 3 times as long as wide   2
  b. Fruit at least 3 times as long as wide   62
2a. Fruit with 2 segments, the upper flat, leaf-like, or with a conspicuous transverse appendage below tip   3
  b. Fruit without a terminal flat, leaf-like segment nor a conspicuous transverse appendage below tip   4
3a. Small perennial shrubs; hairs unbranched   **51. Vella**
  b. Annuals, though becoming woody; hairs stellate   **13. Anastatica**
4a. Fruit inflated, bladder-like   5
  b. Fruit not inflated and bladder-like   11
5a. Petals white, pink or purple; hairs 2-fid   **29. Lobularia**
  b. Petals yellow; hairs stellate or absent   6
6a. Petals short-clawed   **36. Cochlearia**
  b. Petals entire   7
7a. Leaves hairless   **38. Coluteocarpus**
  b. Leaves with stellate hairs   8
8a. Fruits less than 8 mm   9
  b. Fruits at least 10 mm   10
9a. Siliculas 5–20 mm, usually 2-lobed or at least with a prominent apical sinus 2–4 mm deep   **35. Physaria**

  b. Siliculas less than 7 mm, never 2-lobed, usually without an apical sinus, or the sinus not more than 1 mm deep   **34. Lesquerella**
10a. Petals 1.4–1.5 mm; seeds 2 in each cell   **24. Degenia**
  b. Petals 1.5–2 cm; seeds 4–8 in each cell   **23. Alyssoides**
11a. Plant hairless, or with hairs only on leaf veins beneath   12
  b. Plant hairy   31
12a. Petals yellow or creamy yellow   13
  b. Petals white, pink, purple or mauve   16
13a. Fruits 2-lobed   **44. Biscutella**
  b. Fruits not 2-lobed   14
14a. Fruits flattened and winged   **42. Aethionema**
  b. Fruits flattened or not, wings absent   15
15a. Fruits dehiscent, seeds few to many   **32. Draba**
  b. Fruits indehiscent, 1-seeded   **52. Crambe**
16a. At least some leaves pinnatifid, or pinnate   17
  b. All leaves entire or with shallow lobes   22
17a. Plant annual; sepals pouched at base   **56. Heliophila**
  b. Plant perennial; sepals not pouched at base   18
18a. Fruits uncompressed, appearing ovoid to spherical   19
  b. Fruits strongly compressed at right angles to septum   **39. Pritzelago**
19a. Herbaceous perennials, robust, to 1.5 m   20
  b. Annuals, perennial herbs or small shrubs to 0.7 m, usually much smaller   21
20a. Styles short; seeds in two rows in each cell, though rarely produced   **16. Armoracia**
  b. Styles absent; fruits 1-seeded   **52. Crambe**
21a. Style distinct; stigma capitate   **43. Iberis**
  b. Style inconspicuous   **46. Lepidium**
22a. Straggling climber to 5 m   **56. Heliophila**
  b. Plant not a climber   23
23a. Fruits compressed   24
  b. Fruits not compressed   29
24a. Some sepals pouched   25
  b. Sepals not pouched   26
25a. Outer sepals pouched   **42. Aethionema**
  b. Inner sepals pouched   **32. Draba**
26a. Leaves smelling of garlic when bruised   **41. Pachyphragma**

  b. Leaves not smelling of garlic when bruised   27
27a. Stamen filaments expanded at the base   **32. Draba**
  b. Stamen filaments not expanded, with or without appendages   28
28a. Stamen filaments with a tooth-like appendage at the base; seeds 1 or 2 in each cell   **28. Bornmuellera**
  b. Stamen filaments without an appendage; seeds 2–8, rarely only 1 in each cell   **40. Thlaspi**
29a. Style absent; seeds in 1 row in each cell   **52. Crambe**
  b. Style short; seeds in 2 rows in each cell   30
30a. Basal leaves 30–50 cm, ovate or ovate oblong   **16. Armoracia**
  b. Basal leaves variable, but usually 5–20 cm   **36. Cochlearia**
31a. Hairs simple   32
  b. Some hairs forked, branched or stellate   45
32a. Fruits compressed, with the septum across the widest diameter   33
  b. Fruits not compressed with the septum across the narrowest diameter   38
33a. Leaves palmately 3–5-lobed   **33. Petrocallis**
  b. Leaves simple to deeply toothed, or lyrate to pinnate   34
34a. Petals 1–2.5 cm   **21. Lunaria**
  b. Petals 1–8 mm, rarely to 14 mm   35
35a. Petals usually clawed, often with basal appendages   **56. Heliophila**
  b. Petals not clawed, without basal appendages   36
36a. Seeds with a membranous border   **37. Kernera**
  b. Seeds not bordered   37
37a. Basal leaves simple, entire or toothed   **32. Draba**
  b. Basal leaves long-stalked, lyrate, lobes toothed   **46. Lepidium**
38a. Fruits deeply notched, 2-lobed   39
  b. Fruits not deeply notched or 2-lobed   40
39a. Style long   **44. Biscutella**
  b. Style absent, stigma stalkless   **45. Megacarpaea**
40a. Outer 2 petals much larger than inner   **43. Iberis**
  b. Petals equal   41
41a. Leaves lobed or pinnate; seeds one in each cell   42
  b. Leaves simple; seeds 2–numerous, rarely only one in each cell   43
42a. Petals more than 3 mm, with a short claw   **52. Crambe**

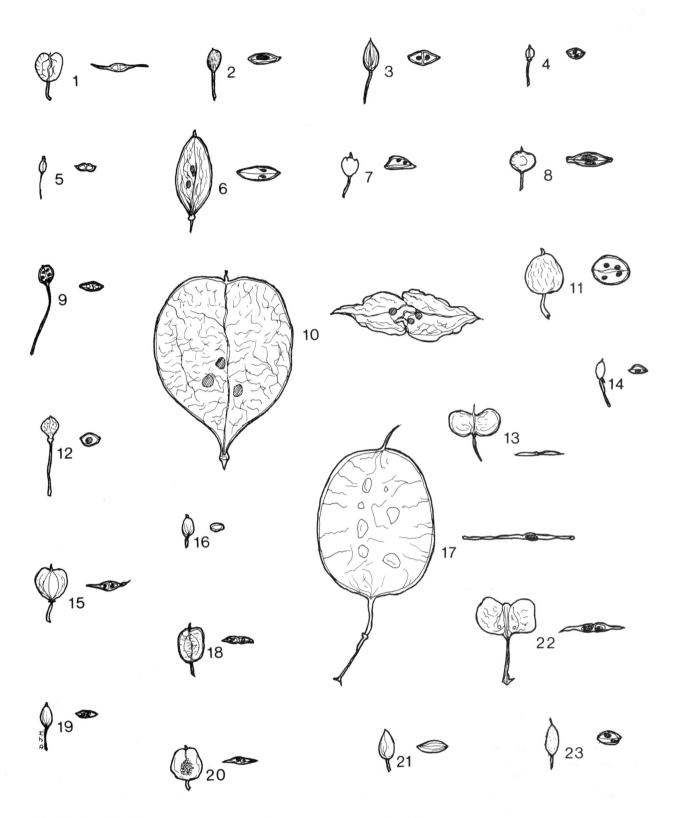

Figure 15. Siliculas of *Cruciferae* species (× 2 except where noted). 1, *Aethionema saxatile*. 2, *Bornmuellera cappadocica*. 3, *Cochlearia officinalis*. 4, *Kernera saxatilis*. 5, *Armoracia rusticana*. 6, *Coluteocarpus vesicaria* subsp. *vesicaria* (× 1). 7, *Alyssum spinosum*. 8, *Aurinia saxatilis*. 9, *Ionopsidum acaule*. 10, *Coluteocarpus vesicaria* subsp. *boissieri* (× 1).

11, *Alyssoides cretica* (× 1). 12, *Crambe cordifolia* (× 1). 13, *Biscutella laevigata* (× 1). 14, *Lobularia maritima*. 15, *Thlaspi bulbosum* (× 1). 16, *Petrocallis pyrenaica* (× 1). 17, *Lunaria annua* (× 1). 18, *Iberis sempervirens* (× 1). 19, *Schivereckia podolica*. 20, *Nothothlaspi rosulatum* (× 1). 21, *Draba aizoides*. 22, *Pachyphragma macrophyllum* (× 1). 23, *Berteroa incana*.

b. Petals to 3 mm, without a short
   claw   **46. Lepidium**
43a. Fruit not strongly compressed or
   winged   **36. Cochlearia**
b. Fruit strongly compressed and
   winged   44
44a. Leaves fleshy; seeds numerous in
   each cell   **47. Notothlaspi**
b. Leaves not fleshy; seeds 2–8, rarely
   only one in each cell   **40. Thlaspi**
45a. Fruit almost spherical to obovoid or
   ellipsoid, not compressed   46
b. Fruits compressed   47
46a. Hairs stellate   **34. Lesquerella**
b. Hairs 2-fid   **29. Lobularia**
47a. Fruits with septum across the
   narrowest diameter   48
b. Fruits with septum across the
   widest diameter   50
48a. Fruits deeply notched, 2-lobed
                                    35. Physaria
b. Fruits not notched, or 2-lobed   49
49a. Leaves simple   **20. Aubrieta**
b. Leaves pinnatisect   **39. Pritzelago**
50a. Petals yellow   51
b. Petals white, pink or purple   54
51a. Inner sepals strongly pouched
                                    25. Fibigia
b. Inner sepals not or only very
   slightly pouched   52
52a. Lower leaves 4–10 cm, rarely only
   2 cm, with persistent and swollen
   bases   **27. Aurinia**
b. Lower leaves 0.2–1.5 cm, rarely
   to 2 cm, never with persistent and
   swollen bases   53
53a. Dwarf perennials with a distinct
   scape   **32. Draba**
b. Annuals to perennials, without a
   scape   **26. Alyssum**
54a. Stamen filaments not pouched,
   toothed or winged at base   55
b. Stamen filaments pouched or
   winged at base   56
55a. Seed 1 in each cell   **29. Lobularia**
b. Seed in 2 rows in each cell
                                    32. Draba
56a. Plant with some stellate hairs 57
b. Plant without stellate hairs (though
   forked hairs may occur)   58
57a. Sepals erect to spreading, not
   pouched at base; petals often
   notched   **26. Alyssum**
b. Sepals spreading, slightly pouched
   at base; petals never notched
                                    31. Schivereckia
58a. Leaf-stalks of basal leaves with
   persistent and swollen bases
                                    27. Aurinia
b. Leaf-stalks of basal leaves never with
   persistent or swollen bases   59

59a. Petals deeply 2-fid   **30. Berteroa**
b. Petals entire or slightly notched
                                    60
60a. Plant herbaceous, mat-forming
                                    20. Aubrieta
b. Plant tufted, or a herb shrubby at
   base   61
61a. Seeds 1 or 2 in each cell
                                    28. Bornmuellera
b. Seeds 4–7 in each cell
                                    31. Schivereckia
62a. Plants stemless; fruits stalkless,
   lengthening and curving down to
   bury the fruits in the soil   63
b. Plants with obvious stem; fruits not
   lengthening and curving down to
   bury the fruits in the soil   64
63a. Leaves pinnatisect with 1–4 pairs of
   lobes   **55. Raffenaldia**
b. Leaves pinnatisect with 10 or more
   pairs of lobes   **53. Morisia**
64a. Fruits pendent   65
b. Fruits erect or spreading   66
65a. Petals yellow   **8. Isatis**
b. Petals lilac   **22. Ricotia**
66a. Hairless or with unbranched hairs
   only   67
b. At least some hairs forked,
   branched or stellate   85
67a. Valves of fruit dehiscing explosively
   from the base   **18. Cardamine**
b. Fruits not dehiscing explosively
                                    68
68a. Plant with stout rhizomes, 1–3 cm
   thick   **6. Wasabia**
b. Rhizomes absent   69
69a. Fruit beaked with an upper seedless
   or seeded part and a lower
   (sometimes very short) segment
                                    70
b. Fruit not beaked, not divided into
   two parts, though sometimes
   constricted between the seeds   73
70a. Fruit breaking transversely into two
   or more segments, not dehiscing by
   valves   **54. Raphanus**
b. Fruit not breaking transversely into
   segments, dehiscing by valves   71
71a. Stigma head-like, inconspicuously
   2-lobed   72
b. Stigma with decurrent lobes joining
   the fruit, prominently 2-lobed
                                    50. Eruca
72a. Plant aquatic; stigma small
                                    17. Nasturtium
b. Plant terrestrial; stigma large
                                    49. Brassica
73a. Petals yellow   74
b. Petals white, pink, lilac, violet or
   blue   77

74a. Stem-leaves auriculate and clasping
   the stem   **14. Barbarea**
b. Stem-leaves not auriculate or
   clasping the stem   75
75a. Stigma with decurrent lobes
                                    10. Hesperis
b. Stigma more or less head-like; style
   without wings   76
76a. Lower leaves pinnate
                                    2. Sisymbrium
b. Leaves all simple   **32. Draba**
77a. Basal leaves simple   78
b. Basal leaves with 3 leaflets,
   pinnatifid, pinnate or bipinnate 84
78a. Petals white   79
b. Petals coloured   81
79a. Plant smelling of garlic when
   crushed   **5. Alliaria**
b. Plant not smelling of garlic when
   crushed   80
80a. Fruits less than 2 cm   **32. Draba**
b. Fruits 2.5 cm or more
                                    56. Heliophila
81a. Petals to 1.3 cm   82
b. Petals 1.4 cm or more   83
82a. Fruits less than 2 cm   **32. Draba**
b. Fruits 2.5 cm or more
                                    56. Heliophila
83a. Leaves very glaucous and
   somewhat fleshy   **48. Moricandia**
b. Leaves not glaucous or fleshy
                                    10. Hesperis
84a. Perennial semi-aquatic herbs, often
   rooting at lower submerged nodes
                                    17. Nasturtium
b. Not aquatic, not rooting at nodes
                                    56. Heliophila
85a. Stigma deeply 2-lobed, the lobes
   sometimes erect and joined to form
   a beak on the fruit   86
b. Stigma capitate, retuse or slightly
   2-lobed   88
86a. Stigma-lobes with a dorsal swelling
   or horn   **12. Matthiola**
b. Stigma-lobes without a dorsal
   swelling or horn   87
87a. Style short, stigma-lobes free
                                    10. Hesperis
b. Style absent, stigma-lobes erect and
   joined   **11. Malcolmia**
88a. Leaves 2-pinnatisect
                                    4. Hueguerina
b. Leaves entire to 1-pinnatisect   89
89a. Petals yellow   90
b. Petals white, pink, purple, orange,
   red, cream or violet   91
90a. Tufted perennials, rarely annuals,
   often forming tight cushions, seeds
   in 2 rows   **32. Draba**

b. Erect, herbaceous annual to perennial herbs, sometimes slightly woody; seeds in 1, rarely 2 rows **9. Erysimum**

91a. Plant covered with medifixed hairs only **9. Erysimum**

b. Hairs various, but not all medifixed 92

92a. At least some leaves pinnatisect or pinnately divided 93

b. Leaves simple, entire, or with a few teeth 94

93a. Petals pinnately lobed **1. Schizopetalon**

b. Petals notched, but not pinnately lobed **3. Murbeckiella**

94a. Seeds in 1 row in each cell 95

b. Seeds in 2 rows in each cell 96

95a. Fruit constricted between the seeds **7. Braya**

b. Fruit not constricted between the seeds **19. Arabis**

96a. Inner sepals distinctly pouched 97

b. Inner sepals not pouched 98

97a. Stem-leaves auriculate, clasping the stem, rarely all leaves basal **19. Arabis**

b. Stem-leaves not auriculate, or clasping the stem, leaves never all basal **20. Aubrieta**

98a. Stem-leaves auriculate, clasping the stem **19. Arabis**

b. Stem-leaves not auriculate, or clasping the stem 99

99a. Stigma 2-lobed **7. Braya**

b. Stigma head-like 100

100a. Fruits less than 2 cm **32. Draba**

b. Fruits 2–8 cm **15. Phoenicaulis**

1. SCHIZOPETALON Sims
S.L. Jury

Annual herbs, downy with stellate hairs. Stems erect. Leaves alternate, simple to pinnately divided; margin wavy, toothed. Flowers in terminal racemes. Sepals erect, green; petals clawed, pinnately lobed, white to purple. Fruit a siliqua.

A genus of 5 species in Chile, only one of which is cultivated. They are best grown as a half-hardy annual, with seed sown where the plants are to flower, preferably in a sunny situation, in well-drained, fertile soil. The plants flower until first frost.

1. S. walkeri Sims. Figure 16(13), p. 134. Illustration: Botanical Magazine, 2379 (1823); Die Natürlichen Pflanzenfamilien **17b**: 403 (1936).

Plants to 45 cm. Leaves 10–14 cm, sinuous pinnate; downy with a grey appearance. Bracts linear. Petals pure white, markedly pinnately lobed so as to appear fringed. Flowers strongly almond scented. *Chile.* H5–G1. Summer–early autumn.

2. SISYMBRIUM Linnaeus
S.L. Jury

Annual to perennial herbs; hairless or with simple hairs. Leaves entire to pinnate. Sepals not pouched at base. Petals yellow, rarely white, entire. Fruit a siliqua; valves usually 3-veined; stigma more or less 2-lobed. Seeds usually less than 2.5 mm.

A genus of 80 species, mainly Eurasian, but also in the New World, North and southern Africa. Many species are weedy and widely naturalised: *S. irio* Linnaeus (London Rocket) from the Mediterranean appeared in profusion in London after the Great Fire of 1666; *S. orientale* Linnaeus (Eastern Rocket), also from the Mediterranean was similarly conspicuous after the Blitz of World War II.

Literature: Schulz, O.E. in Engler, Das Pflanzenreich **86**: 46–157 (1924).

1. S. luteum (Maximowicz) O.E. Schulz (*Hesperis lutea* Maximowicz). Illustration: Das Pflanzenreich **86**: 70 (1924).

Coarsely hairy perennial, 80–120 cm. Leaves stalked, lower pinnate with 1–3 pairs lateral segments, terminal narrowly triangular-ovate, 8–12 × 3–5 cm, acute, short-toothed. Racemes lengthening after flowering. Flower-stalks 1.2–1.5 cm, ascending. Sepals coarsely hairy, 8–9 mm. Petals yellow, 1.2–1.3 cm. Pods narrowly linear, 8–10 cm; stigma shallowly bilobed. *Japan, Korea, Manchuria.* H3. Spring.

3. MURBECKIELLA Rothmaler
S.L. Jury

Perennial, hairless herbs with stellate, or long, unbranched hairs. Leaves entire to pinnatisect. Sepals unequal, the inner more or less pouched at base. Petals white, notched. Fruit a siliqua; valves with a distinct central vein; style very short; stigma slightly 2-lobed. Seed not more than 1.5 mm, often winged at apex.

A genus of 4 species from SW Europe, NW Africa, Turkey and the Caucasus.

1. M. pinnatifida (Lamarck) Rothmaler (*Braya pinnatifida* (Lamarck) Koch; *Phryne pinnatifida* (Lamarck) Bubani; *Sisymbrium pinnatifidum* (Lamarck) de Candolle). Illustration: Reichenbach, Icones Florae Germanicae et Helveticae **2**: t. 73 (1837–1838); Coste, Flore de France **1**: 91 (1937); Hegi, Illustrierte Flora von Mitteleuropa, **4**: 113 (1986).

Plant to 20 cm, sparsely to densely covered in stellate hairs; stems more or less densely leafy throughout. Basal leaves 2.5–5 cm, entire to toothed, stem-leaves 6–9, pinnatifid, 4–6 pairs of lobes. Flower-stalks 3–4 mm in fruit. Sepals 1.5–2.5 mm. Petals 3.5–4 mm, somewhat notched. Siliqua 1.5–2.5 × 0.1–0.2 cm. Seeds *c.* 1 × 0.5 mm, wingless or winged only at apex. *C Pyrenees E to WC Alps.* H2. Spring–early summer.

4. HUGUENINIA Reichenbach
S.L. Jury

Stout hairless to densely grey-hairy herbaceous perennial, 30–70 cm. Leaves 2- pinnatisect; lower to 30 cm, long-stalked; leaf-segments in 8–10 pairs, broadly linear to lanceolate, toothed to pinnatifid, ultimate lobes 4–10 mm wide. Flowers in crowded terminal racemes, lengthening in fruit. Petals *c.* 4 mm, yellow, exceeding sepals. Fruit a siliqua, 6–15 × 1.2–1.5 mm, erect to spreading; valves strongly 1-veined; style very short; stigma slightly 2-lobed. Seeds *c.* 2 mm, unwinged, slimy when wet.

A single species fairly easily propagated by seed. It will grow in any not-too-heavy garden soil in a sunny situation. Suitable for the large rock garden.

1. H. tanacetifolia (Linnaeus) Reichenbach (*Sisymbrium tanacetifolium* Linnaeus). Figure 16(12), p. 134. Illustration: Reichenbach, Icones Florae Germanicae et Helveticae **2**: t. 81 (1837–1838); Bonnier, Flore complète **1**: t. 39 (1911).

SW Alps, Pyrenees & mts of N Spain. H2. Late spring–early summer.

Subsp. **tanacetifolia** is found in the Alps, while the more attractive, densely woolly-haired and grey-leaved subsp. **suffruticosa** (Coste & Soulié) P.W. Ball is found in N Spain and the Pyrenees. Subsp. *suffruticosa* is also characterised by having 1–4, rarely 5, pairs of teeth on the segments of the lower leaves and fruit-stalks 7–11 mm in fruit, as opposed to 4–8, rarely to 10, pairs of teeth and fruit-stalks 5–8 mm.

5. ALLIARIA Scopoli
S.L. Jury

Annual or biennial herbs with simple hairs. Basal leaves simple. Sepals not pouched at base; petals with a short claw. Fruit a siliqua, usually 4-angled, unbeaked, lobes 3-veined, dehiscent; septum thin; style distinct; stigma shallowly 2-lobed; seeds large, *c.* 3 mm, winged, not mucilaginous.

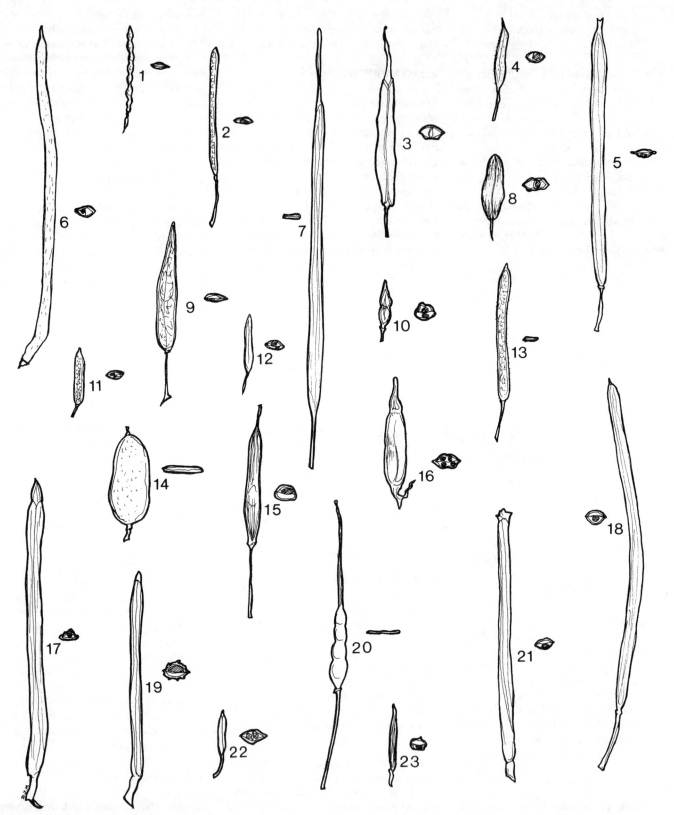

Figure 16. Siliquas of *Cruciferae* species (× 2 except where noted). 1, *Heliophila longifolia* (× 4). 2, *Arabis alpina* (× 4). 3, *Brassica oleracea*. 4, *Aubrieta deltoidea*. 5, *Erysimum cheiri*. 6, *Malcolmia maritima* (× 4). 7, *Orychophragmus violaceaus* (× 1). 8, *Isatis tinctoria*. 9, *Phoenicaulis cheiranthoides* (× 1). 10, *Vella spinosa*. 11, *Nasturtium officinale*. 12, *Hugueninia tanacetifolia* (× 4). 13, *Schizopetalon walkeri* (× 1). 14, *Fibigia clypeata* (× 1). 15, *Cardamine heptaphylla*. 16, *Eruca vesicaria*. 17, *Moricandia arvensis* (× 4). 18, *Hesperis matronalis*. 19, *Alliaria petiolata* (× 4). 20, *Raphanus sativus*. 21, *Matthiola fruticulosa*. 22, *Dieliocharis kotschyi* (× 4). 23, *Barbarea vulgaris* (× 4).

A genus of about 5 species from Europe, N Africa and Asia eastwards to Japan; propagated from seed. Widely naturalised and frequently seen in wild gardens.

1. A. petiolata (Bieberstein) Cavara & Grande. Figure 16(19), p. 134. Illustration: Ross-Craig, Drawings of British plants **3**: pl. 43 (1949); Butcher, A new illustrated British Flora **1**: 345 (1961); Blamey & Grey-Wilson, Illustrated Flora of Britain and Northern Europe, 133 (1989).

To 1.2 m, smelling of garlic when crushed. Leaves pale green, kidney-shaped to triangular-ovate, cordate at base, margins sinuous or distantly toothed; basal leaves long-stalked. Inflorescence short. Petals 4–7 (rarely to 9) mm, white. Fruit 2–7 × 0.1–0.3 cm, erect or erect-spreading. *Temperate Eurasia, N Africa.* H1. Spring–early summer.

Formerly used as a flavouring.

6. WASABIA Matsumura
L.S. Springate
Perennial herbs with rhizomes; stems leafy, simple, decumbent or ascending. Basal leaves more or less circular, cordate with scalloped margins and long stalks with a broad sheathing base; stem-leaves much smaller with a short stalk and fewer, deeper teeth or lobes. Flowers small, white in simple racemes, at least the lowest with bracts. Corolla obovate to oblong with a short claw. Fruit cylindric on a slender spreading or recurved stalk; stigma simple; seeds up to 8 in one row.

A genus of 4 rather similar species from E Asia. The only cultivated species requires a moist, cool, shaded position and is propagated by division.

Literature: Hodge, W.H. Wasabi – native condiment plants of Japan, *Economic Botany* **28**(2): 118–129 (1974).

1. W. japonica (Miquel) Matsumura (*Lunaria japonica* Miquel; *Eutrema wasabi* (Siebold) Maximowicz). Illustration: Das Pflanzenreich **86**: 30 (1924); Economic Botany **28**: 118, 120–126 (1974); Satake et al., Wild Flowers of Japan, pl. 129 (1985).

Evergreen perennial, hairless or with a few crisped hairs above; rhizomes horizontal, stout, 1–3 cm thick; stems ascending, 20–75 cm, thick, hollow. Basal leaves kidney-shaped, apex rounded or short acuminate, scalloped to 14 × 14 cm with thick stalks to 30 cm. Stem-leaves broadly ovate to narrowly triangular, shallowly cordate with larger, rounded teeth, rarely almost entire, much smaller.

Racemes with 30–50 flowers, lengthening in fruit; fruit-stalk 1–3 cm. Calyx 2–4 mm; corolla 5–7 mm. Fruit 1–1.6 cm, rarely formed, with a distinct stalk above the receptacle and a distinct narrow style at least 2 mm long. *Japan, former USSR (S Sakhalin).* H3. Late spring.

Grown as a condiment. Clones have been selected in Japan for drier conditions.

7. BRAYA von Sternberg & Hoppe
S.L. Jury
Perennial, tufted herbs with both branched and unbranched hairs. Leaves undivided. Flowers small. Sepals not pouched at base; petals white or purplish, truncate. Fruit a siliqua or a silicula; valves 1-veined; style short; stigma slightly 2-lobed. Seeds in 1 or 2 rows in each cell, *c.* 1 mm.

A genus of 20 species with a N circumpolar distribution, also from the Alps and Himalaya to C Asia.

1. B. alpina von Sternberg & Hoppe. Illustrations: Das Pflanzenreich **86**: 228 (1924); Engler, Die Natürlichen Pflanzenfamilien **176**: 629 (1936); Hegi, Illustrierte Flora von Mitteleuropa, **4**: 274, 275 (1986).

Loosely tufted herb, flowering stems to 10 cm. Basal leaves lanceolate, entire or toothed; stem leaves 1–several, stalkless. Inflorescence 10–20 flowered. Petals 3–4 mm, white becoming rose-violet. Fruit 5–11 × 1–2 mm, 5–7 times as long as wide. *E Alps.* H1.

8. ISATIS Linnaeus
J.R. Akeroyd
Annuals, biennials or perennials; hairless or with simple hairs. Stems erect, branched above, leafy. Sepals not pouch-like. Petals with a short claw, yellow. Fruit an indehiscent siliqua, flattened, winged, 1-seeded.

A genus of about 40 species in the Middle East and C Asia, extending to the Mediterranean region. Propagation is by seed.

Literature: Davis, P.H., *Notes from the Royal Botanic Garden, Edinburgh* **26**: 11–25 (1964).

1a. Fruit ovate, less than 3 times as long as wide · · · · **3. allionii**
  b. Fruit obovate, 3–5 times as long as wide · · · · · · · · · · 2
2a. Stem-leaves clasping, not leathery · · · · · · · · **1. tinctoria**
  b. Stem-leaves not clasping, leathery · · · · · · · · · **2. glauca**

1. I. tinctoria Linnaeus (*I. canescens* de Candolle). Figure 16(8), p. 134. Illustration: Ross-Craig, Drawings of British Plants **3**: pl. 72 (1949); Keble Martin, Concise British flora, pl. 10 (1965); Rich, Crucifers of Great Britian and Ireland, 245 (1991).

Biennial or short-lived perennial; more or less hairless to hairy, greyish; stems 40–120 cm. Lower leaves to 10 cm, oblong to lanceolate, stalked, entire to toothed. Stem-leaves stalkless, clasping, arrow-shaped, usually entire. Petals 2.5–4 mm. Siliqua 13–22 × 4–8 mm, 3–5 times as long as wide, more or less obovate, hairless or shortly hairy. *SW Asia, but possibly native in parts of SE Europe.* HI. Early summer.

A very variable species, particularly in the size, shape and hairiness of the siliqua, that has been divided into a number of subspecies and varieties. It has been cultivated since ancient times as a source of a blue dye (woad), since replaced by indigo and aniline dyes, and is now mostly grown as an ornamental.

2. I. glauca Boissier. Illustration: Davis, Flora of Turkey and the East Aegean Islands **1**: 293 (1965).

Similar to *I. tinctoria* but always perennial. Stems not more than 1 m; stem-leaves not clasping, oblong, entire, almost hairless, leathery, grey; fruit broader. *SW Asia.* HI. Summer.

A variable species infrequent in cultivation.

3. I. allionii P.W. Ball (*I. alpina* misapplied). Illustration: Pignatti, Flora d'Italia **1**: 380 (1982).

Similar to *I. tinctoria* but always hairless. Stems 10–30 cm; petals 4–5 mm; fruits to 20 × 9 mm, less than 3 times as wide, ovate. *Mts of C & NW Italy.* H1. Summer.

9. ERYSIMUM Linnaeus
J.R. Akeroyd
Annuals, biennials or perennials, sometimes small shrubs, with forked or branched hairs. Leaves narrow, entire or toothed. Sepals erect, the inner usually pouched at base. Petals usually yellow. Fruit a siliqua. Seeds in 1, rarely 2, rows in each cell.

A genus of some 80 species, mostly in Eurasia. The taxonomy is both critical and unresolved, even in Europe, with classification of the cultivated species further complicated by hybridisation. The following account is therefore very provisional. Propagation is by seed or by cuttings in summer. Most species require well-drained soils.

Literature: Gairdner, A.E. The inheritance of factors in Cheiranthus cheiri, *Journal of Genetics* **32**: 479–486 (1936).

1a. Flowers orange, red, purple, purplish-pink or violet   2
 b. Flowers yellow or cream   4
2a. Petals at least 1.5 cm wide, variously coloured; fruit erect   **3. cheiri**
 b. Petals usually less than 1.5 cm wide, orange, purple or violet; fruit usually somewhat spreading   3
3a. Leaves linear to oblong-lanceolate toothed; petals orange   **6. allionii**
 b. Leaves usually linear, entire; petals purple or violet   **8. linifolium**
4a. At least the lower leaves with backwardly directed teeth; stems usually unbranched   5
 b. Leaves entire or, if toothed, the teeth forwardly directed; stems usually branched   6
5a. Petals 7–12 mm wide, yellow; fruit ascending to erect   **1. arkansanum**
 b. Petals 4–6 mm wide, often pale yellow or cream; fruit spreading   **2. capitatum**
6a. Leaves needle-like; stems not more than 10 cm   **10. kotschyanum**
 b. Leaves sometimes linear but not needle-like; stems usually at least 10 cm   7
7a. Petals hairy on lower surface; fruit 0.5–3 cm   8
 b. Petals hairless; fruit usually at least 3 cm   9
8a. Petals 7–11 mm, the lower surface brownish, reddish or purplish; fruit cylindric   **7. mutabilis**
 b. Petals 1.1–1.6 cm, the lower surface yellow; fruit 4-angled   **9. pulchellum**
9a. Petals usually at least 1.2 cm wide; sepals glandular   **3. cheiri**
 b. Petals not more than 1.2 cm wide; sepals without glands   10
10a. Fruit flattened   **4. senoreri**
 b. Fruit cylindric   11
11a. Leaves 1–5 cm; fruit linear, erect   **5. jugicola**
 b. Leaves 5–8 cm; fruit linear-oblong ascending   **6. allionii**

1. E. arkansanum Nuttall. Illustration: Gray & Sprague, Genera florae Americae boreali-orientalis illustrata, t. 63 (1848).

Biennial; stems 30–120 cm, unbranched, slender. Leaves *c.* 5 × 0.5–2 cm, oblong-lanceolate, remotely and backwardly toothed, or entire, green. Petals 1.5–2.5 × 0.7–1.2 cm, yellow. Fruit ascending to erect; flower-stalks ascending or spreading in fruit. *SC USA (Arkansas, Texas)*. H1. Summer.

'Golden Gem' has deep yellow petals.

2. E. capitatum (Douglas) Greene (*E. asperum* de Candolle). Illustration: Britton & Brown, Illustrated flora of northern United States & Canada **2**: 152 (1897); Bailey, Standard cyclopedia of horticulture, 1140 (1914); Rickett, Wild flowers of the United States **4**(1): pl. 70 (1970).

Biennial to perennial; stems 10–60 cm, usually unbranched. Leaves linear to oblong-lanceolate, the lower backwardly or sinuously toothed, rough, greyish with appressed hairs. Petals 1.2–2 × 0.4–0.6 cm, yellow, pale yellow or cream. Fruit 6–10 cm, 4-angled, spreading; flower-stalk ascending or spreading in fruit. *W & C USA*. H1. Summer.

One of the putative parents of *E. allionii*.

3. E. cheiri (Linnaeus) Crantz (*E. helveticum* (Jacquin) de Candolle; *E. murale* Desfontaines; *E. suffruticosum* Sprengel; *Cheiranthus cheiri* Linnaeus). Figure 16(5), p. 134. Illustration: Journal of Genetics **32**: 486 (1936); Opera botanica **13**: 65 (1967); Hay & Synge, RHS Dictionary of flowering plants in colour, 33 (1969).

Biennial to perennial; stems 20–80 cm, branched, the lower part woody and covered with leaf-scars in older plants. Leaves 5–20 × 0.5–2 cm, with a short stalk, lanceolate to narrowly elliptic or obovate, entire or sometimes obscurely toothed, sparsely hairy. Sepals glandular. Petals 1.5–3 × 1–2 cm, yellow, orange, red, purple or pinkish-purple, often streaked. Siliqua 4–10 cm, flattened, almost erect; style stout. *Known only from cultivation, but widely naturalised in Europe.* H1. Spring–summer.

Probably derived from *E. senoreri* and related Aegean taxa by hybridisation or selection. Several other species are involved and the taxonomy is both complex and unresolved. On walls, and as a naturalised plant, *E. cheiri* has a shrubby habit and usually yellow flowers, but is most frequently grown as a biennial. Dwarf plants have been called *E. alpinum* misapplied. 'Bowles Mauve' ('E.A. Bowles'), shrubby, with violet flowers, is of hybrid origin, perhaps *E. cheiri × E. linifolium*. Other variants of *E. cheiri* commonly grown are 'Harpur-Crewe' (flowers double, bright yellow) and 'Bloody Warrior' (flowers double, red). 'Wenlock Beauty' and numerous other cultivars are also of uncertain parentage.

4. E. senoreri (Heldreich & Sartori) Wettstein (*Cheiranthus senoreri* Heldreich & Sartori). Illustration: Opera botanica **13**: 53 (1967).

Similar to *E. cheiri* but smaller; stems little branched, without conspicuous leaf-scars; leaves finely and remotely toothed, leaf-stalk distinct, to 1/5 as long as blade; sepals without glands; petals 1.2–2.5 × 0.3–1.2 cm, yellow; fruit 4–6 cm; fruit and style more slender. *Aegean region (Cyclades)*. H3. Spring.

Infrequent in cultivation.

5. E. jugicola Jordan (*E. pumilium* misapplied).

Tufted perennial, with many basal rosettes of leaves; stems 5–30 cm, unbranched. Leaves 1–5 cm, linear to lanceolate, toothed, stalked. Petals 1.4–2 × 0.5–1 cm, pale yellow. Fruit 2.5–5.5 cm, linear, erect; flower-stalk spreading to erect in fruit. *SW Alps*. H1. Summer.

Extensively hybridised with *E. cheiri* in cultivation. The closely related **E. rhaeticum** (Hornemann) de Candolle (*E. helveticum* misapplied), is a taller and more robust plant from the C & W Alps and may also have been used as a hybrid parent.

6. E. × allionii invalid (*E. perofskianum* Fischer & Meyer; *Cheiranthus allionii* misapplied). Illustration: Addisonia **16**: 508 (1931); Everard & Morley, Wild flowers of the world, pl. 11A (1970).

Biennial or perennial; rather tufted; stems 20–50 cm, unbranched or little branched. Leaves mostly basal, 5–8 cm, linear to oblong-lanceolate, rather coarsely toothed. Petals 1.5–2.5 × 0.5–1.2 cm, usually orange. Fruit 2–4 cm, linear-oblong, more or less cylindric, ascending; somewhat spreading. *Known only from cultivation*. H1. Spring–summer.

A hybrid of garden origin, perhaps derived from the N American *E. capitatum* and the European **E. hieraciifolium** Linnaeus, the latter not now in cultivation. The commonly grown variants of *E. allionii* have orange petals but cultivars such as 'Lemon Delight' (Illustration: Hay & Synge, RHS Dictionary of garden flowers, 33 t. 261, 1969) have pale yellow petals.

7. E. mutabile Boissier & Heldreich (*Cheiranthus mutabilis* misapplied).

Loosely tufted perennial; stems 5–20 cm, procumbent, slender. Lower leaves to 2 cm, oblanceolate; upper leaves linear.

Petals 7–11 × 1.5–3.5 mm, yellow, on upper surface, the lower surface hairy, brownish, reddish or purplish. Fruit 0.5–2 cm, cylindric somewhat spreading. *Crete.* H3.

8. E. linifolium (Persoon) Gray (*Cheiranthus linifolius* Persoon). Illustration: Gardners' Chronicle **68**: 69 (1920).

Tufted, somewhat shrubby perennial, woody below, with branched stock; stems 30–70 cm, single or branched. Leaves 2–9 cm, linear to lanceolate, usually sinuously toothed, green or greyish. Petals 1.2–2 × 0.5–0.6 cm, violet. Fruit 3–7.5 cm, linear, slender, almost erect to spreading. *N Portugal & C Spain.*

The plants in cultivation belong to subsp. **linifolium**. 'Variegatum' has variegated leaves, and 'Glaucum' has grey leaves.

9. E. pulchellum (Willdenow) Gray (*E. rupestre* (Smith) de Candolle). Illustration: Revue Horticole, 421 (1880), 161 (1908); Robinson, English flower garden, 109 (1887).

Tufted perennial, with branched stock and numerous non-flowering basal rosettes of leaves; stems 5–25 cm. Leaves oblong-lanceolate or spathulate, entire or sinuously toothed. Petals 1.1–1.6 × 0.3–0.5 cm, yellow, hairy on the lower surface. Fruit 2–3 cm, linear, slender, 4-angled, sparsely to densely hairy, almost erect. *Balkan Peninsula to Caucasus.* H1. Summer.

10. E. kotschyanum Gray.

Small tufted perennial with non-flowering basal rosettes of leaves; stems to 10 cm. Leaves linear, needle-like, greyish-hairy. Petals 1–1.2 cm, yellow. Fruit to 5 cm, flattened, hairy. *Turkey.* H1. Summer.

10. HESPERIS Linnaeus
L.S. Springate

Biennial or perennial herbs with erect leafy stems; hairless or with simple, branched or glandular hairs. Leaves ovate to spathulate, entire to pinnatifid. Flowers usually with bracts, in racemes or panicles, sometimes strongly fragrant. Sepals erect, the inner pair with a pouch at the base. Petals with a long claw, white to violet, yellowish, greenish or brown. Fruit a slender siliqua, valves 3-nerved, style short; stigma with 2 erect lobes; seeds in one row in each cell.

A genus of about 30 species from W Europe to W China, best grown in a well-drained soil in sun, with adequate moisture for *H. matronalis* and *H. dinarica.*

Literature: Hodgkin, E., The Double Rockets, *Journal of the Royal Horticultural Society* **96**: 188–189 (1971); Lord, T., Double Sweet rocket saved from extinction, *The Garden* **114**(3): 124–126 (1989).

1a. Flowers yellowish green to brown
            1. tristis
  b. Flowers white or pink to violet   2
2a. All hairs branched, dense, matted
            4. bicuspidata
  b. Hairs sparse, not all branched   3
3a. Middle and upper stem-leaves with a short stalk, not clasping the stem; flowers white or pink to violet
    2. matronalis subsp. **matronalis**
  b. Middle and upper stem-leaves without a stalk, clasping the stem; flowers white   **3. dinarica**

1. H. tristis Linnaeus. Illustration: Botanical Magazine, 730 (1804); Hegi, Illustrierte Flora von Mitteleuropa, **4**: 155 (1958).

Stout biennial or perennial herb 25–70 cm, hairs very sparse to numerous, sometimes tawny or reddish, simple, branched or glandular. Leaves ovate-lanceolate to lanceolate, entire or slightly toothed, blade to 17 cm, basal leaves long-stalked; upper leaves tapered to slightly cordate at base. Inflorescence sometimes a leafless raceme, usually an open, corymb-like panicle, sometimes leafy below. Flowers very strongly scented at night. Sepals 0.9–1.5 cm. Petals 1.8–3.2 cm, yellowish green to brownish with purple veins. Stalks of fruit thickened and up to 9 cm. Fruit 4–9 × 0.3–0.5 cm, usually hairless, spreading. *C & E Europe.* H3. Early summer.

2. H. matronalis Linnaeus subsp. **matronalis**. Figure 16(18), p. 134. Illustration: Bonnier, Flore complète **1**: pl. 33 (1912); Polunin, Flowers of Europe, pl. 31 (1969); Pignatti, Flora d'Italia 1: 390 (1982); Rich, Crucifers of Great Britain and Ireland, 191 (1991).

Perennial herb 40–120 cm, hairy, usually without glandular hairs. Leaves ovate-acuminate to elliptic or oblong, toothed; basal leaves with long stalks, blade to 20 cm; middle and upper leaves with short stalks, not clasping the stem, to 15 cm. Inflorescence a simple raceme or panicle with up to 15 branches and leafy bracts below the flowers. Flower-stalk without glandular hairs. Flowers strongly fragrant in the evening. Sepals 5–10 mm. Petals 1.4–2.5 cm, white, lilac, purple or

violet. Fruit 2.5–10 × 0.1–0.3 cm, valves rounded, hairless. *Italy, S France, widely naturalised.* H3. Summer.

There are several double-flowered cultivars.

3. H. dinarica G. Beck.

Like *H. matronalis* but many glandular hairs present. Upper leaves stalkless, base clasping the stem. Flowers always white. *Balkans.* H3. Summer.

4. H. bicuspidata (Willdenow) Poiret (*H. violacea* Boissier).

Perennial herb with few stems, often with a tuft of dense non-flowering shoots, 20–70 cm, covered in soft, branched, matted, silvery or sometimes tawny hairs. Basal leaves obovate to oblong, entire to pinnatifid, apex obtuse, sometimes tapered at base, stalked, blade to 8.5 cm. Uppermost leaves oblong, lanceolate or linear, entire to sharp-toothed, apex acute, base almost clasping the stem or rarely tapered. Inflorescence of 1–12 branches from open, elongated to dense, pyramidal to corymb-like. Sepals 7–11 mm, sometimes coloured. Petals 1.4–2.5 cm, pale lilac to deep violet often with darker veins. Fruit slender, erect, 4.5–7.5 × 0.1–0.2 cm, hairless, with sparse glandular hairs or rarely dense short hairs. *Turkey, Armenia & W Syria.* H3. Summer.

11. MALCOLMIA R. Brown
J.R. Akeroyd

Annuals, biennials or perennials, hairs 2–4 branched. Leaves simple. Sepals erect, the 2 inner usually pouch-like. Petals with a long claw, pink, purple or violet. Fruit a long, slender siliqua; valves 3-veined; stigma deeply 2-lobed. Seeds in 1 row in each cell.

A genus of about 30 species from the Mediterranean area to C Asia. Propagation is by seed.

Literature: Stork, A.L., *Opera Botanica* **33**: 1–118 (1972).

1a. Perennial, woody at the base and with non-flowering-stems; densely white-hairy   **1. littorea**
  b. Annual, not woody and without non-flowering-stems; green or greyish green   2
2a. Petals usually more than 1 cm; sepals 6–10 mm; stigma 2–5 mm
            2. maritima
  b. Petals usually not more than 1 cm; sepals 3–6 mm; stigma 0.5–2 mm
            3
3a. Petals pink with a yellow base; fruit 1–4 cm   **3. graeca**

b. Petals entirely pink or violet; fruit
2.5–7 cm   **4. chia**

1. M. littorea (Linnaeus) R. Brown.
Illustration: Botanical Magazine, 4672
(1852).

Perennial, woody at the base, densely
covered with white felted hairs. Stems 10–40
cm. Leaves entire or sinuous-toothed.
Flowers in loose clusters. Petals 1.4–2.2 cm,
purple. Fruits 3–6 cm; stigma 2–6 mm. *W
Mediterranean area*. H3. Summer.

2. M. maritima (Linnaeus) R. Brown
(*Cheiranthus maritimus* Linnaeus). Figure
16(6), p. 134. Illustration: Botanical
Magazine, 166 (1791); Hay & Synge,
Dictionary of garden plants, 42 (1969);
Polunin, Flowers of Europe, pl. 32 (1969);
Rich, Crucifers of Great Britain and Ireland,
189 (1991).

Annual 5–30 cm. Leaves oblong to
elliptic or obovate, entire or toothed,
obtuse. Flowers fragrant. Sepals 6–10 mm.
Petals 1.2–2.5 cm, pink, lilac, violet, red
or white. Fruits 3–8 cm; stigma 2–5 mm.
Greece & S Albania. H1. Spring–autumn.

One of the most easily and widely grown
of garden plants ('Virginia Stock'),
flowering 4–6 weeks after sowing, and
thus a suitable plant to introduce children
to horticulture.

3. M. graeca Boissier & Spruner.
Illustration: Opera Botanica **33**: 36 (1972).

Annual 5–30 cm; hairs mostly forked.
Leaves ovate to lanceolate-oblong, entire
to minutely toothed. Flowers fragrant,
in small dense clusters. Sepals 3–6 mm.
Petals 4–12 mm, pink or violet, yellow at
the base. Fruits 1–4 cm; stigma 0.5–2 mm.
Greece & S Albania. H1. Summer.

The cultivated plant is subsp. **bicolor**
(Boissier & Heldreich) A.L. Stork (*M. bicolor*
Boissier & Heldreich).

4. M. chia (Linnaeus) de Candolle.
Illustration: Opera Botanica **33**: 19 (1972);
Huxley & Taylor, Wild flowers of Greece
and the Aegean, pl. 68 (1977); Collenette,
An illustrated guide to the flowers of Saudi
Arabia, 202 (1985).

Annual 5–20 cm. Leaves ovate-oblong
to narrowly elliptic, entire to slightly
toothed. Sepals 2.5–5 mm. Petals 6–10
mm, pale pink to violet. Fruits 3–7 cm, on
slightly thickened stalks; stigma 0.5–1.5
mm. *E Mediterranean*. H1. Summer.

12. MATTHIOLA R. Brown
J.R. Akeroyd
Annuals, biennials or perennials,
sometimes almost shrubs, with branched

or stellate hairs. Leaves oblong to linear,
entire, sinuous or pinnatifid. Flowers
conspicuous, fragrant. Sepals erect, the
2 inner pouch-like. Petals with a long
claw. Fruit a long siliqua, usually hairy,
the valves usually 1-veined; stigma-lobes
persistent, often horn-like. Seeds many in
each cell, broadly winged.

A genus of about 40 species, mainly in
the Mediterranean and adjacent regions.
Propagation is by seed.

1a. Plant annual; fruit with 2 or 3
conspicuous horns usually more
than 3 mm   2
 b. Plant usually biennial or perennial
(rarely annual); fruit hornless or
with 2 inconspicuous horns not
more than 3 mm   3
2a. Fruit with 2 equal horns and a
smaller protuberance
                              5. longipetala
 b. Fruit with 3 equal horns
                              6. tricuspidata
3a. Fruit glandular   4
 b. Fruit without glands or glandular
hairs   5
4a. Biennial or annual; fruit
compressed   **1. sinuata**
 b. Perennial (with vegetative rosettes);
fruit more or less cylindric
                              4. fruticulosa
5a. Fruit more or less cylindric
                              4. fruticulosa
 b. Fruit compressed   6
6a. Leaves usually entire; petals violet,
purple, reddish or white
                                   2. incana
 b. Leaves mostly toothed or pinnatifid;
petals cream becoming purplish
                              3. odoratissima

1. M. sinuata (Linnaeus) R. Brown (*M.
oyensis* Menier & Viaud). Illustration:
Botanical Magazine, 7703 (1900); Ross-
Craig, Drawings of British plants **3**: pl. 1
(1949); Polunin, Flowers of Europe, pl. 33
(1969); Rich, Crucifers of Great Britain and
Ireland, 185 (1991).

Somewhat woody biennial, rarely
annual. Stems 20–60 cm, erect. Leaves
densely felted with white and glandular
hairs; lower leaves oblanceolate, sinuous
to pinnatifid, the upper narrowly elliptic,
entire. Sepals 0.8–1.2 cm. Petals 2–2.5
cm, pale purple. Fruits 8–15 cm, erect,
compressed, with conspicuous yellow or
black glands; stigma without conspicuous
horns. Seeds 3–4 mm, elliptic, brown.
Mediterranean & W Europe. H2. Summer.

2. M. incana (Linnaeus) R. Brown
(*Cheiranthus incanus* Linnaeus; *M.
fenestralis* (Linnaeus) R. Brown).
Illustration: Ross-Craig, Drawings of British
plants **3**: pl. 2 (1949); Hay & Synge,
Dictionary of garden plants, 43 (1969);
Polunin, Flowers of Europe, pl. 32 (1969);
Rich, Crucifers of Great Britain and Ireland,
183 (1991).

Often rather shrubby biennials, short-
lived perennial or annual, almost hairless
to densely hairy. Stems 20–80 cm, erect,
often branched. Leaves to 10×2 cm,
linear to oblong-lanceolate or obovate,
entire or sometimes slightly sinuous-
pinnatifid, obtuse or almost acute. Sepals
0.9–1.3 cm. Petals 2–3 cm, violet, purple,
red, pink or white; flowers single or double.
Fruits 4–16 cm, erect to somewhat
spreading, compressed, without glands;
stigma without conspicuous horns.
Mediterranean & W Europe. H2. Summer.

Two principal groups of variants of this
species are cultivated in Europe: Annual
plants, var. **annua** (Linnaeus) Voss
(*Cheiranthus annus* Linnaeus), sown in
spring and flowering 10–12 weeks after
sowing ('10-week' Stock); and biennial
plants, sown in summer and flowering
during the following year ('Brompton' and
'East Lothian' Stock).

3. M. odoratissima (Bieberstein)
R. Brown. Illustration: Botanical Magazine,
1711 (1815).

Densely white-hairy perennial or almost
a shrub. Stems 20–75 cm, erect, branched.
Leaves toothed to pinnatifid. Flowers
scented at night. Sepals 0.6–1.6 cm. Petals
2–3 cm, cream becoming brownish or
yellowish purple. Fruits 8–18 cm, erect to
somewhat spreading, compressed; stigma
without conspicuous horns. *Caucasus &
Iran, extending to Crimea & NE Bulgaria*.
H2. Summer.

4. M. fruticulosa (Linnaeus) Maire (*M.
tristis* R. Brown; *M. varia* (Sibthorp &
Smith) de Candolle; *Cheiranthus fruticulosus*
Linnaeus). Figure 16(21), p. 134.
Illustration: Rich, Crucifers of Great Britain
and Ireland, 187 (1991).

Perennial, woody at the base, densely
to sparsely white-hairy. Stems 10–60
cm. Leaves linear, entire to toothed or
scalloped. Sepals 0.6–1.4 cm. Petals
1.5–2.5 cm, yellowish to greenish brown
or purplish. Fruits 3–12 cm, more or less
cylindric; without glandular hairs, the
valves 1–3 veined; stigma without or with
2 short horns. *S Europe*. H2. Summer.

Subsp. **valesiaca** (Boissier) P.W. Ball

(*M. tristis* subsp. *valesiaca* (Boissier) Rouy & Foucaud. Illustration: Pignatti, Flora d'Italia **1**: 395 (1982). Plant more densely tufted; petals pale pinkish purple; fruit more or less erect, with glandular hairs. *S Alps, Balkans, Pyrenees to N & E Spain.*

Subsp. **perennis** (P. Conti) P.W. Ball (*M. perennis* P. Conti). Illustration: Quarterly Bulletin of the Alpine Garden Society **31**: 325 (1963). Plant tufted, rather woody below; petals purplish pink; fruits to 3 mm wide, fruit valves 3–5 veined. *N Spain (Picos de Europa).*

5. M. longipetala (Ventenat) de Candolle. Illustration: Rich, Crucifers of Great Britain and Ireland, 187 (1991).

Straggling annual 20–50 cm. Lower leaves sinuous-toothed to pinnatifid. Flowers scented at night. Sepals 0.8–1.2 cm. Petals 1.5–2.5 cm, lilac to purplish. Fruits 5–15 cm, cylindric; stigma with 2 conspicuous, upwardly curved horns 2–10 mm. *Greece to SW Asia.* H2. Summer.

The cultivated plant is subsp. **bicornis** (Sibthorp & Smith) P.W. Ball.

6. M. tricuspidata (Linnaeus) R. Brown. Illustration: Polunin, Flowers of Greece and the Balkans, 250 (1980); Pignatti, Flora d'Italia **1**: 395 (1982).

Annual, 10–40 cm. Leaves sinuous, scalloped or pinnatifid, with rounded lobes. Sepals 0.7–1.1 cm. Petals 1.5–2.2 cm, purple. Fruits 3–10 cm, spreading or deflexed, cylindric; stigma with 3 conspicuous horns 2–6 mm. *Mediterranean area.* H2. Summer.

13. ANASTATICA Linnaeus
S.L. Jury
Much-branched annual becoming woody, 3–15 cm. Stems densely covered in stellate hairs. Leaves oblong to diamond-shaped, simple, toothed, 1–4 cm. Flowers numerous, almost stalkless, in short dense stalkless clusters in the angles of the branches. Petals white. Ovary stalkless, green, covered in stellate hairs, 4-seeded; style shorter than ovary, stigma large, bilobed. Fruit a hard silicula, ovoid to almost spherical, beaked; lobes strongly convex with a transverse appendage below the tip.

A genus of 1 species from countries bordering the Mediterranean. Easy to grow in warm situations.

1. A. hierochuntica Linnaeus.
Illustration: Maire, Flore de l'Afrique du Nord **13**: 193 (1967); Danin, Desert vegetation of Israel and Sinai, 112 & 113 (1983); Collenette, an illustrated guide to

the flowers of Saudi Arabia, 192 (1985).
Morocco to Egypt, Arabia, Palestine & Iran. H5. Spring–summer.

The dead plants remain anchored by the tap-root for dozens to hundreds of years. Popularly known as Rose of Jericho, a resurrection plant, as the stems close up into a tight ball on drying and re-open on moistening, releasing the seeds.

14. BARBAREA R. Brown
L.S. Springate
Biennial or perennial erect branching herbs, hairless or with unbranched hairs. Basal leaves stalked, usually lyrate, rarely entire, forming a rosette. Upper leaves stalkless. Flowers in a dense raceme, much elongated in fruit. Petals small, yellow with a short claw. Filaments without appendages. Fruit a siliqua, 4-angled, round or flattened; style distinct, stigma small; seeds in one row in each cell.

A genus of about 12 species, widespread in the northern temperate zone, often weeds. Best grown in a rich soil, in sun or shade but not very dry. Propagated by seed. It is grown as a winter salad plant.

Literature: Rich, T.C.G., The genus *Barbarea* R. Brown in Britain and Ireland, *Watsonia* **16** : 389–396 (1987).

1a. Uppermost stem-leaves/ inflorescence bracts simple, toothed or shallowly lobed, sometimes with 1 or 2 pairs of lateral lobes, the terminal lobe broad, ovate to obovate; fruit 1.5–4 cm **1. vulgaris**
  b. Uppermost stem-leaves/ inflorescence bracts pinnatifid with 2 pairs of lateral lobes, the terminal lobe narrow, oblanceolate to oblong; fruit 2.8–7 cm **2. verna**

1. B. vulgaris R. Brown. Figure 16(23), p. 134. Illustration: Ross-Craig, Drawings of British plants **3**: pl. 8 (1949); Hegi, Illustrierte Flora von Mitteleuropa **4**: 167, f. 90e, (1958); Hess et al., Flora der Schweiz **2**: 216 (1970); Rich, Crucifers of Great Britain and Ireland, 169 (1991).

Biennial or perennial, usually hairless, to 1 m. Basal leaves usually with 2–5 pairs of lateral lobes, rarely entire, usually 10–20 cm. Uppermost stem-leaves or inflorescence bracts simple, toothed or shallowly lobed, sometimes with 1, rarely 2, pairs of lateral lobes; the terminal lobe broad, ovate to obovate. Flower-stalk slender, 3–6 mm, erect or spreading in fruit. Petals 5–8 mm, twice length of sepals. Fruit 1.5–4 ×

0.1–0.2 cm, straight or curved upwards; style 1.5–4 mm, slender. *Eurasia, N W Africa, naturalised elsewhere.* H2. Summer.

'Variegata' has leaves mottled with yellow, and comes true from seed. Illustration: Brickell, RHS gardeners' encyclopedia of plants and flowers, 246 (1989).

2. B. verna (Miller) Ascherson, (*B. praecox* (Smith) R. Brown). Illustration: Ross-Craig, Drawings of British Plants **3**: pl. 11 (1949); Hegi, Illustrierte Flora von Mitteleuropa **4**: 170 (1958); Hess et al., Flora der Schweiz **2**: 216 (1970); Rich, Crucifers of Great Britain and Ireland, 173 (1991).

Hairless or sparsely downy biennial to 75 cm. Basal leaves usually with 4–10 pairs of lateral lobes, usually to 10–20 cm. Uppermost stem-leaves or inflorescence bracts pinnatifid with 2–4 pairs of lateral lobes; the terminal lobe narrow, oblong to oblanceolate. Flower-stalk stout, 3–8 mm, spreading in fruit. Petals 5.5–8.5 mm, twice length of sepals. Fruit 2.8–7 × 0.1–0.2 cm, curved upwards; style 0.5–2 mm, stout. *W Mediterranean & Macaronesia, naturalised elsewhere.* H3. Spring–Summer.

15. PHOENICAULIS Nuttall
S.L. Jury
Perennial tufted herbs. Stems to 20 cm, covered in the remains of old leaf-stalk bases. Leaves in dense basal rosettes, 3–10 cm, grey with stellate hairs, with an entire margin. Inflorescence a showy raceme of small flowers. Sepals *c.* 4 mm. Petals long clawed, 8–10 mm, pinkish, purple or rarely white. Fruit a siliqua, 2–8 × 0.2–0.6 cm, compressed, oblong-lanceolate, often more or less sickle-shaped, hairless.

The single species can be cultivated in a dry, sunny situation with sandy, humus-rich loam on the rock garden. It can be killed by winter wet and thus is more frequently encountered in the alpine house. Propagation is by seed.

1. P. cheiranthoides Nuttall (*Parrya menziesii* (Hooker) Greene). Figure 16(9), p. 134. Illustration: Hitchcock, Vascular plants of the Pacific Northwest **2**: 531 (1964); Rickett, Wild flowers of the United States **5**(1): pl. 61 (1971).

W North America. H1. Spring.

16. ARMORACIA Gaertner, Meyer & Scherbius
L.S. Springate
Hairless perennial herbs with erect leafy branching stems. Rootstock stout with

fleshy, cylindric roots. Leaves simple to pinnatifid; basal leaves very large with long stalks; stem-leaves much smaller, with short stalks or stalkless. Sepals without basal pouches. Petals small, white with a short claw. Filaments broadened but without appendages. Fruit a spherical to ellipsoid silicula; seeds in 2 rows in each cell; style short; stigma large, scarcely lobed.

A genus of 3 similar species from the Balkans, SW and NE Asia. One aquatic species from E USA, with leaves divided into many slender segments may be included in this genus. The cultivated species is grown in moist deep soils and propagated by root-cuttings, for use as a condiment.

Literature: Courter, J.W. & Rhodes, A.M., Historical Notes on Horseradish, *Economic Botany* **23**(2): 156–164 (1969); Al-Shehbaz, I.A., The genera of Arabideae in the Southeastern United States, *Journal of the Arnold Arboretum* **69**(2): 85–166 (1988).

1. A. rusticana Gaertner, Meyer & Scherbius (*A. lapathifolium* misapplied). Figure 15(5), p. 131. Illustration: Ross-Craig, Drawings of British plants **3**: pl. 36 (1949); Hegi, Illustrierte Flora von Mitteleuropa **4**: 184 (1958); Pignatti, Flora d'Italia **1**: 401 (1982); Rich, Crucifers of Great Britain and Ireland, 247 (1991).

Stems to 1.5 m. Basal leaves with blade lanceolate to oblong, with scalloped margin, base truncate or cordate 30–50 cm. Lower stem-leaves often pinnate. Upper stem-leaves linear-lanceolate, margins finely scalloped or entire, with short stalks or stalkless, not clasping the stem. Flower-stalk slender 8–11 mm, erect. Petals 3–6 mm, twice length of sepals. Fruits rarely produced, with 4–6 ovules in each cell. *SW former USSR, widely naturalised.* H3. Summer.

17. NASTURTIUM R. Brown

J.B. Martínez-Laborde
Perennial, aquatic, almost completely hairless herbs. Stem often rooting at lower, submerged nodes. Leaves mostly pinnate. Petals white, longer than sepals. Median nectaries absent. Fruit a siliqua; midrib of valves somewhat inconspicuous; style short, beak seedless. Seeds brown, with net-veined patterns on surface, arranged in 1–2 rows in each cell.

A small genus of only two species which some authors include in *Rorippa* Scopoli. *Nasturtium* is native to Eurasia, but *N.*

officinale (watercress) is cultivated and naturalised throughout the world. The two species are very similar and difficult to differentiate at vegetative or flowering stages: fruits and seeds are needed for proper determination. Vegetative propagation occurs in nature and can easily be carried out, although propagation by seeds is preferable because it prevents virus spreading.

Literature: Al-Shehbaz, I.A. & Rollins, R.C., *Journal of the Arnold Arboretum* **69**: 65–71 (1988).

1a. Fruits 2–3 mm wide; seeds in 2 rows in each cell, surface coarsely net-veined, with 25–50 areoles on each side   **1. officinale**
 b. Fruits more slender, 1–2 mm wide; seeds mostly in one row in each cell, surface finely net-veined, with 100–150 areoles on each side   **2. microphyllum**

1. N. officinale R. Brown (*Rorippa nasturtium-aquaticum* (Linnaeus) Hayek; *Sisymbrium nasturtium-aquaticum* Linnaeus). Figure 16(11), p. 134. Illustration: Ross-Craig, Drawings of British plants **3**: pl. 3 (1949); Collenette, Illustrated guide to the wild flowers of Saudi Arabia, 204 (1985); Hegi, Illustrierte Flora von Mitteleuropa **4**: 187 (1986); Rich, Crucifers of Great Britain and Ireland, 151, 153 (1991).

Stems up to 60 cm or more, hollow, procumbent-ascending, sometimes floating. Leaves with ear-like appendages at base, blade to 18 cm, mostly pinnate; leaflets 5–11, ovate, elliptic or roundish, margin shallowly scalloped or wavy; terminal leaflet usually larger than laterals; lower leaves with 1–3 leaflets. Flower-stalks shortly hairy on innermost side. Sepals hairless or with a few hairs. Petals 3–5 mm. Nectaries 4, 2 at the base of each lateral stamen. Ovules 20–30 per cell. Fruiting flower-stalks 6–20 mm, spreading to deflexed. Fruit 1–2 × 0.2–0.3 cm, frequently curved towards the stem apex; valves somewhat warty. Seeds in 2 rows in each cell, 1–1.2 mm, with 25–50 polygonal depressions on each side. *Almost cosmopolitan.* H2–3. Spring–summer.

2. N. microphyllum Reichenbach (*N. uniseriatum* Howard & Manton; *Rorippa microphylla* (Reichenbach) Hylander). Illustration: Reichenbach, Icones florae Germanicae et Helveticae **2**: t. 50 (1837–1838); Ross-Craig, Drawings of British plants **3**: pl. 4 (1949).

Very similar to *N. officinale*, but fruit more slender, 1–2 mm wide, seeds mostly in 1 row in each cell, more finely net-veined, with 100–150 areoles on each side. *Eurasia, N America, Australia.* H2–3. Spring–summer.

A hybrid (**Rorippa × sterilis** Airy Shaw) betwen *N. officinale* and *N. microphyllum* occurs in nature and has been cultivated (brown cress). It has sterile fruits.

18. CARDAMINE Linnaeus

L.S. Springate
Herbaceous perennials, usually with rhizomes and erect, leafy stems; hairless or with simple hairs; evergreen or deciduous. Leaves simple, palmate or pinnate. Flowers in a compact terminal raceme or leafy panicle, lengthening in fruit. Petals white, pink or purple, rarely pale yellow. Fruit a linear or lanceolate flattened siliqua, the valves dehiscing elastically from the base to eject the seed.

More than 200 species worldwide, most numerous in north temperate regions: the description above only applies to the cultivated species. Inconspicuous hairs occur on some specimens of most cultivated species and are disregarded here. Best grown in moist, rich soil in light shade, some are dormant from late summer to spring. *C. asarifolia* and *C. trifolia* are used as ground cover plants. Propagation is by seed or division, also from adventitious plantlets or bulbils.

Literature: Detling, L.E., The genus Dentaria in the Pacific States, *American Journal of Botany* **23**: 570–576 (1936).

Rootstock. Cylindric rhizomes with scales: **1–5,7**; ovoid rhizomes: **6**; with many slender stolons: **8,9**.
Leaves. Simple: **6,8**; palmately divided with 3 leaflets: **5–7**; palmately divided with 5 leaflets: **3**, pinnately divided: **2,4,9–11**; evergreen: **7**; rhizomal leaves produced in autumn: **6**.
Bulbils. Present in upper leaf axils: **1**.
Flowers. Yellow: **4–5**; white: **1–3,5–9,11**; pink **1–3,6,7,9,11**; purple or violet: **1–3,6,9,10**.

1a. Leaves all similar in size and shape, more or less in one whorl of 2–5 below the flower cluster   2
 b. Leaves dissimilar in size and shape, not in one whorl below the flower cluster   3
2a. Leaves pinnately divided   **4. kitaibelii**

b. Leaves palmately divided
  5. enneaphyllos

3a. Rhizomes ovoid   **6. californica**

b. Rhizomes cylindric or absent   4

4a. Basal leaves simple   **8. asarifolia**

b. Basal leaves compound or absent at flowering   5

5a. Axils of upper leaves with brownish purple bulbils   **1. bulbifera**

b. Axils of upper leaves without bulbils   6

6a. Leaves all palmate   7

b. At least some leaves pinnate   8

7a. Leaflets with sharp teeth, petals more than 1.3 cm, fruit more than 2.5 mm wide   **3. pentaphyllos**

b. Leaflets with obscure teeth, petals less than 1.3 cm, fruit less than 2.5 mm wide   **7. trifolia**

8a. Rhizome 4–10 mm thick, petals 1.4–2 cm, fruit more than 3 mm wide   **2. heptaphylla**

b. Rhizome less than 4 mm thick or absent, petals 0.5–1.2 cm, fruit less than 3 mm wide   9

9a. Rhizome with slender stolons, anthers blackish violet (rarely yellowish white)
  9. amara subsp. **amara**

b. Rhizome without slender stolons or absent, anthers yellow   10

10a. Leaves dissimilar, lower with ovate to kidney-shaped leaflets, upper with oblong or linear-lanceolate leaflets   **11. pratensis**

b. Leaves all similar, all leaflets circular to kidney-shaped
  10. raphanifolia subsp. **raphanifolia**

1. C. bulbifera (Linnaeus) Crantz (*Dentaria bulbifera* Linnaeus). Illustration: Ross-Craig, Drawings of British plants **3**: pl. 24 (1949); Hegi, Illustrierte Flora von Mitteleuropa **4**: 216, t. 132 (1958); Hess et al., Flora der Schweiz **2**: 205 (1970); Rich, Crucifers of Great Britain and Ireland, 149 (1991).

Perennial herb 20–45 cm; rhizomes subterranean, 2–3 mm thick with fleshy triangular scales. Leaves alternate, 6 or more; lower leaves pinnate with 1–3 pairs of lanceolate leaflets to 8 cm with forward-pointing teeth; uppermost leaves simple, smaller. Small brownish purple bulbils present in upper leaf axils. Petals 1.2–2 cm, pale purple, rarely white or violet. Fruit 1.5–3.5 × 0.2–0.3 cm, often absent. *Europe to Iran.* H2. Spring.

2. C. heptaphylla (Villars) Schulz (*Dentaria pinnata* Lamarck). Figure 16(15), p. 134. Illustration: Bonnier, Flore complète **1**: pl.45 (1912); Barneby, European alpine flowers in colour, pl.29 (1967); Hess et al., Flora der Schweiz **2**: 206 (1970).

Perennial herb 30–60 cm; rhizomes subterranean 4–10 mm thick with crescent-shaped scales 1–2 mm. Leaves alternate, usually 3, pinnate; the lower leaves with 3–5 pairs of leaflets, the uppermost with 2 or 3 pairs; leaflets to 15 cm, usually lanceolate, with teeth pointing forwards. Petals 1.4–2 cm, white, pink or purplish. Fruit 4–8 × 0.3–0.5 cm. *W Europe.* H2. Spring.

3. C. pentaphyllos (Linnaeus) Crantz (*Dentaria digitata* Lamarck). Illustration: Botanical Magazine, 2202 (1820); Polunin, Flowers of Europe, pl.34 (1969); Hess et al., Flora der Schweiz **2**: 207 (1970); Brickell, RHS gardeners' encyclopedia of plants and flowers, 227 (1989).

Like *C. heptaphylla* except: rhizomes 1.5–2.5 mm thick, with triangular, 3-lobed, concave scales 6–10 mm. Leaves palmately divided, usually with 5 leaflets, rarely 3 or 7. Fruit 2.5–4 mm wide. *W & C Europe.* H2. Spring.

4. C. kitaibelii Becherer (*Dentaria polyphylla* Waldstein & Kitaibel). Illustration: Bonnier, Flore complète **1**: pl. 45 (1912); Hegi, Illustrierte Flora von Mitteleuropa **4**: 220 (1958); Hess et al., Flora der Schweiz **2**: 206 (1970).

Perennial herb 20–50 cm; rhizomes subterranean, 3–6 mm thick with concave scales 5–8 mm. Leaves usually 3, rarely 4 or 5, in a loose whorl, pinnate with 2–6 pairs of lanceolate leaflets to 12 cm with forward-pointing teeth. Petals 1.5–2.2 cm, pale yellow, twice as long as the stamens. Fruit 4–6.5 × 0.2–0.3 cm. *S Europe (Alps, Apennines, NW former Yugoslavia).* H3. Spring.

5. C. enneaphyllos (Linnaeus) Crantz (*Dentaria enneaphyllos* Linnaeus). Illustration: Hegi, Illustrierte Flora von Mitteleuropa **4**: t. 132 (1958); Polunin, Flowers of Europe, pl. 33 (1969); Brickell, RHS gardeners' encyclopedia of plants and flowers, 229 (1989).

Perennial herb 20–40 cm; rhizomes subterranean, to 6 mm thick, with inconspicuous scales. Leaves usually 3, rarely 2 or 4, in a loose whorl, palmate, usually with 3 ovate-lanceolate leaflets to 15 cm with forward pointing teeth. Flowers somewhat pendent. Petals 1.2–1.6 cm, white or pale yellow, scarcely exceeding the stamens. Fruit 4–7.5 ×

0.3–0.4 cm. *S Europe (E Alps to W Carpathians south to Italy & S former Yugoslavia).* H3. Spring.

6. C. californica (Nuttall) Greene (*Dentaria californica* Nuttall). Illustration: Niehaus, Sierra Wildflowers, 52 (1974); Niehaus & Ripper, Field guide to Pacific States wildflowers, 31 (1976); Wiggins, Flora of Baja California, 572 (1980).

Perennial herb to 50 cm, hairless; rhizome ovoid, 3–8 mm across, deep-buried. Leaves simple to pinnate; leaf-blade or leaflets circular to oblanceolate, entire to deeply lobed, cordate to tapered at base, 1.5–11.5 cm across. Rhizomal leaves produced in autumn with long, slender stalks; stem-leaves 2–5, often differing in shape from rhizomal leaves, the upper ones usually stalkless. Petals 0.7–1.4 cm, white, pink or purple. Fruit 2.5–5 × 0.1–0.3 cm. *W North America (Oregon to Baja California).* H3. Spring.

7. C. trifolia Linnaeus. Illustration: Botanical Magazine, 452 (1799); Hegi, Illustrierte Flora von Mitteleuropa **4**: 212, t. 130 (1958); Hess et al., Flora der Schweiz **2**: 204 (1970); Brickell, RHS gardeners' encyclopedia of plants and flowers: 302 (1989).

Evergreen perennial 20–30 cm; rhizomes 2–4 mm thick, branched, creeping at soil surface, with prominent nodes and remote scales. Leaves on rhizome with 3 leaflets and long stalks; leaflets broad-rhombic to circular with shallow teeth and sparse hairs above, purplish beneath. Stem-leaves absent or 1–3, similar, though smaller, stalkless and often simple. Petals 0.5–1.1 cm, white or pink. Anthers yellow. Fruit 2–2.5 × 0.2 cm. *C Europe.* H2. Spring.

8. C. asarifolia Linnaeus. Illustration: Botanical Magazine, 1735 (1815); Hegi, Flora von Mitteleuropa **4**: 203 (1958); Hess et al., Flora der Schweiz **2**: 197 (1970).

Perennial herb 20–45 cm; rhizomes at soil surface, short, stout, horizontal. Lower leaves simple, kidney-shaped, scalloped, to 10 cm across, with long stalks; upper leaves similar, smaller, with short stalks, rarely with 3 rounded leaflets. Flowering-stems often branched. Petals 0.4–1 cm, white. Anthers violet. Fruit 2–3 × 0.1–0.2 cm. *Pyrenees, Alps, Apennines.* H3. Summer.

9. C. amara Linnaeus subsp. **amara**. Illustration: Bonnier, Flore complète **1**: pl. 44 (1912); Ross-Craig, Drawings of British plants **3**, pl. 19 (1949); Roles, Flora of the British Isles, Illustrations **1**, f. 206 (1957); Hegi, Illustrierte Flora von Mitteleuropa **4**: t. 130 (1958).

Perennial herb 20–60 cm; stems angular; rhizome at soil surface, short, horizontal with many slender stolons. Leaves 6 or more per stem, thin, pinnate, with the terminal leaflet a little larger than the lateral leaflets; lower leaves not in a rosette, with circular to ovate, scalloped leaflets and long stalks; upper leaves with ovate to lanceolate, angular leaflets and short stalks. Flowering-stems often branched. Petals 0.5–1 cm, white to reddish purple. Anthers blackish violet, rarely yellowish white. Fruit 2–4 × 0.1–0.2 cm. *Europe to W Siberia*. H2. Late spring–early summer.

10. C. raphanifolia Pourret subsp. **raphanifolia** (*C. latifolia* Vahl). Illustration: Botanical Magazine, 7628 (1898); Bonnier, Flore complète **1**: pl. 43 (1912); Pignatti, Flora d'Italia **1**: 405 (1982); Rich, Crucifers of Great Britain and Ireland, 147 (1991).

Perennial herb 30–80 cm; stem stout; rhizome at soil surface, long, horizontal, without stolons. Leaves large, thick, dark green, lyrate-pinnate, to 25 cm, with long stalks; leaflets circular to kidney-shaped; basal leaves not in a rosette. Flowering-stems sometimes branched. Petals 0.8–1.3 cm, reddish violet. Anthers yellow. Fruiting heads remain compact. Fruit 1.5–4 × 0.1–0.2 cm. *N Spain to S France* H3. Late spring–early summer.

11. C. pratensis Linnaeus subsp. **pratensis**. Illustration: Bonnier, Flore compléte **1**: pl. 44 (1912); Ross-Craig, Drawings of British plants **3**: 20 (1949); Roles, Flora of the British Isles, Illustrations **1**, 205 (1957); Hegi, Illustrierte Flora von Mitteleuropa **4**: t. 130 (1958).

Perennial herb 20–45 cm; rhizome at soil surface, short, oblique, without stolons. Leaves pinnate, thin. Basal leaves in a rosette, to 15 cm, with 1–7 pairs of ovate to kidney-shaped leaflets and long stalks; terminal leaflets sometimes twice as large as lateral leaflets. Stem-leaves 2–5, with oblong or linear-lanceolate leaflets and short stalks or stalkless; terminal leaflets not much larger than lateral leaflets. Flower-stems sometimes branched. Petals 0.8–1.3 cm, white, pink or lilac. Anthers yellow. Fruit 2.5–4 × 0.1–0.2 cm. *Europe, Asia, N America*. H2. Late spring–early summer.

'Flore Plena' has flowers double, lilac.

19. ARABIS Linnaeus
J.R. Akeroyd
Annuals, biennials or perennials; hairs forked, branched or stellate, rarely the plant hairless. Leaves simple. Inner sepals often pouched at base. Petals usually white or pink. Fruit a slender siliqua; valves with or without a median vein. Seeds in 1 row, rarely 2, in each cell, usually winged.

A genus of about 120 species, mostly in Europe, SW Asia and the Mediterranean region. The majority of the cultivated species are cushion-forming montane plants and grow best in the alpine glasshouse. Several artificial hybrids are cultivated. Propagation is mostly by seed.

Plant. Annual: **17**; forming cushions: **6–11**; with long stolons: **1–2,5**.
Leaves. Hairless (sometimes with hairy margins): **5–6,12,14**; entire: **3,5–10,15**.
Stems. Leafless: **8**.
Petals. Purplish pink: **3–4,11,(13),16**; pale blue: **12**; violet-purple, **17**.

1a. Plant annual; petals violet-purple with yellowish claw **7. verna**
  b. Plant usually perennial; petals usually white, pink or pinkish purple 2
2a. Plant distinctly stoloniferous 3
  b. Plant tufted or forming a mat 5
3a. Plant hairless or sparsely hairy; fruit not more than 3.5 cm **5. procurrens**
  b. Plant more or less densely hairy; fruit usually more than 3.5 cm 4
4a. Basal leaves with few teeth; fruit erect; style 0.5–1 mm **1. stelleri**
  b. Basal leaves coarsely toothed; fruit spreading to ascending; style 1.5–2 mm **2. serrata**
5a. Petals 4–5 mm, pale blue **12. caerulea**
  b. Petals usually more than 5 mm, white, pink or purplish 6
6a. Plant usually compact, densely tufted; fruit not more than 2 cm 7
  b. Plant usually loosely tufted or forming a mat; fruit at least 2 cm 12
7a. Petals purplish-pink; leaves deeply and sharply toothed **11. aubretioides**
  b. Petals white; leaves entire 8
8a. Flowering-stems leafless **8. bryoides**
  b. Flowering-stems with at least a few leaves 9
9a. Petals 5–10 mm 10

  b. Petals 1–1.5 cm 11
10a. Basal leaves almost hairless; petals 5–7 mm **6. vochinensis**
  b. Basal leaves hairy; petals 8–10 mm **7. ferdinandi-coburgii**
11a. Basal leaves linear, with marginal hairs **9. carduchorum**
  b. Basal leaves obovate, silky-hairy **10. androsacea**
12a. Plant hairless or almost hairless **14. soyeri**
  b. Plant hairy 13
13a. Petals 1–2 cm; seeds usually unwinged 14
  b. Petals 0.6–1 cm; seeds usually winged 15
14a. Petals usually white; fruit 4–7 × 0.1–0.3 cm **13. alpina**
  b. Petals usually purplish-pink; fruit 2–4 × 0.2–0.3 cm **16. cypria**
15a. Petals white; fruit 2–4 cm 16
  b. Petals usually purplish pink; fruit 3–9 cm 17
16a. Basal leaves toothed; stem leaves clasping **13. alpina**
  b. Basal leaves entire or with 2 teeth; stem leaves not clasping **15. pumila**
17a. Stems simple; basal leaves more or less entire; fruit spreading **3. blepharophylla**
  b. Stems usually branched; basal leaves deeply toothed; fruit erect **4. collina**

1. A. stelleri de Candolle (*A. japonica* (A. Gray) A. Gray; *A. stelleri* var. *japonica* (A. Gray) F. Schmidt).

Stoloniferous perennial with dense simple hairs; stems 20–40 cm, erect, robust. Basal leaves 2–7 cm, oblong to oblanceolate, stalked, few-toothed; stem-leaves arrow-shaped, clasping, deeply toothed. Flowers many, in often branched racemes. Petals 5–10 mm, white. Fruit 3–6 × 0.1–0.2 cm, erect; style 0.5–1 mm. Seeds 1.5 mm, narrowly winged. *Japan, Korea & Kamtschatka*. H2. Summer.

2. A. serrata Franchet & Savatier.
Similar to *A. stelleri* but stems to 30 cm, with few leaves; basal leaves long-stalked, coarsely toothed; fruit spreading to ascending; style 1.5–2 mm. *Japan, Korea*. H2. Summer.

3. A. blepharophylla Hooker & Arnott. Illustration: Rickett, Wild flowers of the United States **5**(1): pl. 63 (1971).

Tufted perennial, woody at base; stems 5–20 cm, simple, densely and stiffly hairy below. Basal leaves 1–3 cm, broadly spathulate, short-stalked, obtuse, more or less entire; stem-leaves oblong, stalkless, clasping. Flowers in compact racemes, fragrant. Petals 6–9 mm, almost circular, purplish pink or white. Fruit 3–7 × 0.2 cm, spreading. Seeds 1–1.5 mm, narrowly winged. *USA (California, S Oregon)*. H3. Spring.

4. A. collina Tenore (*A. muralis* Bertoloni; *A. rosea* de Candolle). Illustration: Botanical Magazine, 3246 (1833); Flora Republicae Popularis Bulgaricae **4**: 459 (1970).

Perennial; stems 10–30 cm, usually branched, often hairy. Basal leaves obovate, deeply toothed, obtuse, the blade gradually narrowing into stalk, somewhat hairy; stem-leaves oblong to ovate, somewhat heart-shaped and clasping at base. Petals 6–10 × 2–4 mm, purplish pink. Fruit 3–9 × 0.1–0.2 mm, erect. *S & SE Europe*. H1. Summer.

A very variable species the classification of which is unresolved.

5. A. procurrens Waldstein & Kitaibel (*A. mollis* misapplied; *A. praecox* misapplied). Illustration: Flora Republicae Popularis Bulgaricae **4**: 469 (1970).

Perennial, with long leafy stolons; stems 15–30 cm, hairless or sparsely hairy. Basal leaves 2–3 cm, obovate to oblanceolate, acuminate, entire, almost hairless except on the margins; stem-leaves ovate. Petals 8–10 mm, white. Fruit 1.5–3.5 × 0.1–0.2 cm, spreading; flower-stalk more than 1 cm. *Carpathian & Balkan mts*. H1. Spring–early summer.

6. A. vochinensis Sprengel. Illustration: Pignatti, Flora d'Italia **1**: 418 (1982).

Similar to *A. procurrens* but without stolons; stems hairy with forked hairs; basal leaves not more than 1.5 cm, obtuse; stem-leaves oblong; petals 5–7 mm; fruit 8–15 mm; flower-stalk not more than 1 cm. *SE Alps*. H1. Spring–early summer.

7. A. ferdinandi-coburgii J. Kellerer & Sündermann. Illustration: Quarterly Bulletin of the Alpine Garden Society **43**: 121 (1975) – both typical and variegated variants; Brickell, RHS gardeners' encyclopedia of plants and flowers, 328 (1989).

Similar to *A. procurrens* but without stolons, forming a greyish green cushion; basal leaves long-stalked, narrowly oblong to lanceolate, hairy on both surfaces; stem-leaves lanceolate, also hairy; fruit 1.5–2 × *c*. 0.1 cm. *SW Bulgaria (Pirin Mts)*. H1. Spring.

The specific epithet commemorates 'Foxy Ferdinand', Czar of Bulgaria during World War I, and also a keen plantsman. A variegated variant of *A. ferdinandi-coburgii* is widely cultivated. Two hybrids are grown in gardens.

A. × kelleri Sündermann (*A. ferdinandi-coburgii* × *A. bryoides*) forms dense grey cushions with lanceolate leaves, hairy on both surfaces and white petals.

A. × suendermannii Sündermann (*A. ferdinandi-coburgii* × *A. procurrens*) has oblong-lanceolate leaves, smooth above, hairy beneath and on margins, and white petals.

8. A. bryoides Boissier. Illustration: Strid, Wild flowers of Mount Olympus, pl. 108 (1980); Polunin, Flowers of Greece and the Balkans, pl. 10 (1980); Quarterly Bulletin of Alpine Garden Society **57**: 132 (1989).

Compact, densely tufted perennial, with rather woody, branched stock; stems 2–8 cm, slender, leafless. Basal leaves in numerous dense rosettes, overlapping, to 8 mm, broadly oblanceolate, obtuse, entire grey-hairy with dense felted hairs. Flowers 2–6. Petals 6–9 mm, white. Fruit 1–2 × 0.1–0.2 cm, hairless; style to 1 mm. *Mts of S part of Balkan Peninsula*. H1. Summer.

9. A. carduchorum Boissier.

Similar to *A. bryoides* but stems leafy; leaves with marginal hairs, the basal 1–1.5 cm, linear, rigid; stem-leaves lanceolate; fruit 1–1.5 × 0.1–0.2 cm. *Mts of SE Turkey*. H1. Summer.

The hybrid **A. × wilkzekii** Sündermann, between *A. bryoides* and *A. carduchorum* has a similar habit to *A. carduchorum*, with ovate-lanceolate, hairy leaves.

10. A. androsacea Fenzl.

Similar to *A. bryoides* but stems leafy, with simple hairs; leaves obovate, with silvery, silky hairs; fruit *c*. 1.2 cm, the valves somewhat keeled. *S Turkey (Taurus Mts)*. H1. Summer.

11. A. aubretioides Boissier.

Densely tufted perennial; stems 7–15 cm, erect, slender. Leaves deeply and sharply toothed, greenish; basal leaves obovate, obtuse; stem-leaves clasping, acute. Flowers few. Petals 1.2–1.6 cm, purplish-pink. Fruit erect to somewhat spreading, curved. *S Turkey (Taurus Mts)*. H1. Summer.

12. A. caerulea (Allioni) Haenke. Illustration: Hess et al., Flora der Schweiz **2**: 236 (1970); Pignatti, Flora d'Italia **1**: 419 (1982).

Tufted perennial, usually more or less hairless; stems 5–15 cm. Basal leaves obovate, the blade gradually running into stalk, with 2–5 prominent teeth near apex; stem-leaves 1–3, oblong flowers in small nodding clusters. Petals 4–5 mm, pale blue. Fruit 1–3 × 0.2–0.3 mm, erect. *Alps*. H1. Summer.

13. A. alpina Linnaeus (*A. flavescens* Weltot). Figure 16(2), p. 134.

Perennial, forming a mat; flowering stems; stems 5–30 cm, ascending to erect. Basal leaves oblong or obovate, toothed, obtuse, the blade gradually narrowed into stalk; stem-leaves ovate to lanceolate, clasping. Flowers in a loose raceme, fragrant petals usually white. Fruit 2–7 × 0.1–0.3 cm, spreading. *Arctic, mts of Europe, Mediterranean area & W Asia*. H1. Spring–early summer.

Subsp. **alpina**. Illustration: Botanical Magazine, 225 (1793); Caryologia **15**: 258 (1962); Polunin, Flowers of Europe, pl. 34 (1969); Flora Republicae Popularis Bulgaricae **4**: 469 (1970) – all as C. alpina. Compact plant with few leaf rosettes. Leaves small green, sparsely hairy with stellate hairs; stem-leaves more or less heart-shaped at base. Sepals 3–5 mm. Petals 6–10 mm. Fruit 2–4 cm; valves with an indistinct mid-vein. Seeds winged. *Arctic and higher mts*.

Subsp. **caucasica** (Willdenow) Briquet (*A. caucasica* Willdenow; *A. albida* Steven; *A. billardieri* de Candolle; *A. lucida* misapplied). Illustration: Botanical Magazine, 2046 (1819).

Spreading plant forming a loose mat with many leaf rosettes. Leaves grey to greyish-green, softly and densely hairy with stellate hairs; stem-leaves arrow-shaped at base. Sepals 5–8 mm. Petals 1–2 cm. Fruit 4–7 cm; valves with a distinct median vein. Seeds usually unwinged. *S Europe to Iran*. The common plant of cultivation.

Subsp. **brevifolia** (de Candolle) Greuter & Burdet (*A. brevifolia* de Candolle). Similar to subsp. *caucasica* but leaf-stalks distinct from blade, the blade with few or no teeth on margins; flower-stalks curved (straight in the other two subspecies). *S Turkey & Syria*.

A very variable species, with many intermediates between the subspecies, especially in the E Mediterranean area. Subsp. *caucasica* is one of the most ubiquitous of garden plants, mainly in the rock garden and on walls. There are a number of widely grown cultivars, some derived by hybridisation with other species:

'Coccinea' crimson petals; 'Flore plena', illustration: Hay & Synge, RHS Dictionary of garden plants in colour, **2** (1969), somewhat less vigorous; flowers double; petals white or pink; 'Grandiflora superba', larger clusters of flowers; 'Rosabella', 'Rosea' (*A. arendsii* misapplied), pink petals; 'Variegata' and 'Aureovariegata', variegated leaves.

14. A. soyeri Reuter & Huet. Illustration: Pignatti, Flora d'Italia **1**: 419 (1982).

Almost hairless perennial; stems 15–50 cm, sometimes sparsely hairy. Basal leaves oblong-spathulate to obovate, finely toothed, dark green, usually glossy and hairless except on the margin; stem-leaves 4–10 ovate to oblong, usually entire, clasping or not. Petals 5–7 mm, white. Raceme of 10–20 flowers, compact in fruit. Fruit 2.5–5 × 0.1–0.2 cm. Seeds narrowly winged. *Pyrenees.* H1. Summer.

Subsp. **jacquinii** (G. Beck) B.M.G. Jones (*A. jacquinii* G. Beck; *A. bellidifolia* Jacquin). Illustration: Hess et al., Flora der Schweiz **2**: 236 (1970). Plant hairless. Stem-leaves not clasping. *Alps & W Carpathians.*

15. A. pumila Jacquin. Illustration: Hess et al., Flora der Schweiz **2**: 236 (1970); Pignatti, Flora d'Italia **1**: 419 (1982).

Similar to *A. soyeri* but usually somewhat hairy; stems 5–20 cm; basal leaves with 0–2 teeth; stem-leaves 1–4, not clasping; flowers 3–10 in raceme; fruit 2–4 cm. *Alps & Appennines.* H1. Summer.

16. A. cypria Holmboe. Illustration: Quarterly Bulletin of the Alpine Garden Society **45**: 324 (1977).

Shortly creeping perennial, rather woody at base; stems 10–25 cm, erect. Basal leaves in a rosette, 2–8 × 0.1–0.2 cm, obovate to spathulate, softly and shortly hairy; stem-leaves few, small. Petals 1–1.8 cm, purplish pink or white. Fruit 2–4 × 0.2–0.3 cm, spreading. Seeds unwinged. *Cyprus.* H3. Spring.

17. A. verna (Linnaeus) R. Brown. Illustration: Botanical Magazine, 3331 (1834); Polunin, Flowers of Europe, pl. 35 (1969).

Annual, rather roughly hairy, with several stems 5–30 cm, arising from a basal rosette of leaves. Basal leaves stalked, obovate or ovate, with forward-pointing teeth; stem-leaves few, ovate, stalked. Raceme to 10-flowered, loose in fruit. Petals 5–9 mm, violet-purple with a yellowish claw. Fruit 3–7 × 0.1–0.2 mm, somewhat spreading; fruit-stalks thickened in fruit. Seeds narrowly winged.

Mediterranean area. H1. Spring–early summer.

20. AUBRIETA Adanson
J.R. Akeroyd
Perennials, usually forming mats; hairs stellate, forked or simple; stems short, weakly ascending. Leaves simple, entire or with a few teeth. Inner sepals pouch-like. Petals with a long claw, pink, purple or violet. Filaments somewhat winged, the outer ones toothed. Fruit a siliqua or silicula. Seeds in 2 rows in each cell.

A taxonomically difficult genus of about 15 species from Sicily to Iran. Several have been cultivated, but most cultivated plants are variants or hybrids of *A. deltoidea*. Propagation is by seed or cuttings. Plants grow best in well-drained soils containing lime.

Literature: Mattfeld, J., *Quarterly Bulletin of the Alpine Garden Society* **7**: 157–181, 217–227 (1939); Phitos, D., *Candollea* **25**: 69–87 (1971); *The Garden* **112**: 384–385, (1987).

1a. Fruit with both stellate and distinctly longer simple or forked hairs **1. deltoidea**
  b. Fruit with stellate hairs only, or with forked or simple hairs of about the same length **2**
2a. Fruit a siliqua or silicula, not more than 1.2 cm **3**
  b. Fruit a siliqua, more than 1.2 cm **5**
3a. Sepals at least 6 mm; petals purple or violet **2. columnae**
  b. Sepals less than 6 mm; petals white, pink or lilac **4**
4a. Leaves toothed; fruit a silicula, 4–7 mm **5. parviflora**
  b. Leaves entire or with a single pair of teeth; fruit a siliqua, 7–12 mm **6. erubescens**
5a. Fruit 4.5–5.5 mm wide, flattened, hairless or sparsely hairy **3. olympica**
  b. Fruit usually less than 4.5 mm wide, not flattened, hairy (rarely hairless or sparsely hairy) **6**
6a. Leaves narrowly oblanceolate, coarsely toothed **4. pinardii**
  b. Leaves linear, oblong, obovate or spathulate, entire or with 1–3 teeth on each margin **7**
7a. Valves of fruit without or with obscure net-like venation; petals lilac **8. thessala**
  b. Valves of fruit with distinct net-like venation; petals purple or violet **8**

8a. Leaves entire or with 1 pair of teeth; fruit not more than 1.6 cm, straight **2. columnae**
  b. Leaves with 1–3 pairs of teeth; fruit usually at least 1.6 cm, somewhat curved **7. gracilis**

1. A. deltoidea (Linnaeus) de Candolle (*A. cultorum* misapplied). Figure 16(4), p. 134. Illustration: Botanical Magazine, **126** (1790); Polunin, Flowers of Greece and the Balkans, pl. 71 (1980); Huxley & Taylor, Flowers of Greece and the Aegean, pl. 74 (1977); The Garden **112**: 384, 385 (1987).

Plant forming spreading mat; flowering stems 5–15 cm. Leaves obovate to diamond-shaped, with 1–3 teeth on each margin, greyish hairy. Sepals 6–10 mm. Petals 1.5–2 cm, rounded, usually violet-purple, but also in many shades of pink or purple. Fruit 8–15 × 2.5–4 mm, cylindric, with both stellate and simple or forked hairs; valves not or scarcely reticulately veined. *Italy (Sicily) to SW Asia.* H1. Spring.

One of the commonest and best-loved of all garden flowers (misleadingly called Aubretia), especially on walls and in rock gardens. It is a very variable species both in the wild and in cultivation. Much of the variation in gardens has been derived by hybridisation with other species, notably *A. erubescens*. Among the many variants in cultivation are:

Var. **graeca** (Grisebach) Regel (*A. graeca* Grisebach), a tall, robust plant with large flowers. *C Greece (Attica).*

Var. **intermedia** (Boissier) Baldacci (*A. intermedia* Boissier), with a longer and rather narrow fruit 1.4–2 cm. *C Greece.*

Numerous cultivars include 'Eyrei' with deep violet-purple flowers, 'Dr Mules' with purple flowers, and 'Rosea' and 'Leichtlinii' with pink flowers. Two variegated cultivars, 'Aurea', the leaves edged with gold, and 'Variegata' ('Argentea'), with leaves edged with white, are also widely grown; 'Nana Variegata' is a similar but dwarf, tufted variant. New cultivars are still being bred, especially F1 hybrids such as 'Red Cascade' and purple-flowered 'Monarch'.

A. hybrida Haussknecht is a hybrid between *A. deltoidea* and *A. gracilis*.

2. A. columnae Gussone. Illustration: Pignatti, Flora d'Italia **1**: 421 (1982).

Similar to *A. deltoidea* but petals purple or violet; siliqua 5–16 × 2–4.5 mm, usually relatively shorter and wider, with sparse to dense stellate hairs only and valves usually reticulately veined. *S Europe, eastwards from Italy.* H1. Spring.

Subsp. **columnae**. Leaves oblong to spathulate, usually entire. Fruit 5–12 mm; style 7–10 mm. *Italy (C & S Appennines).*

Subsp. **italica** (Boissier) Mattfeld. Leaves broadly obovate, with a tooth on each margin near the apex, sometimes entire. Fruit 8–11 mm; style 4–6 mm. *S Italy (Monte Gargano).*

Subsp. **croatica** (Schott, Nyman & Kotschy) Mattfeld. Leaves broadly obovate to diamond-shaped, with a tooth on each margin near the apex. Fruit 0.7–1.6 cm; style 3–7 mm. *Balkan Peninsula.*

Perhaps not distinct from **A. canescens** (Boissier) Bornmüller subsp. **cilicica** (Boissier) Cullen from SW Turkey, also said to be in cultivation in Europe. Another subspecies of *A. canescens* may also be cultivated, subsp. **macrostyla** Cullen & Huber-Morath (*A. libanotica* Boissier; *A. antilibani* misapplied), distinguished by the more tufted habit and inflated fruit. *S Turkey & Lebanon.*

3. A. olympica Boissier.

Similar to *A deltoidea* but leaves oblanceolate, toothed; fruit 1.5–1.8 × 0.5–0.6 cm, flattened, hairless or with sparse stellate hairs. *NW Turkey (Ulu Dağ).* H1. Early summer.

4. A. pinardii Boissier. Illustration: Davis, Flora of Turkey and the East Aegean Islands **1**: 261 (1965).

Similar to *A. deltoidea* but leaves narrowly oblanceolate, coarsely toothed; flowers many in a raceme; petals *c.* 1.8 cm, lilac or purple; fruit 2–2.5 × 0.2–0.3 cm, hairy with stellate hairs. *C Turkey.* H1. Early summer.

5. A. parviflora Boissier (*A. kotschyi* Boissier).

Similar to *A. deltoidea* but plant white-hairy; leaves spathulate, toothed; petals 7–8 mm, white or pale pink; fruit 4–7 × 3–4 mm, ellipsoid, with long and short stellate hairs. *SW Asia.* H1. Early summer.

6. A. erubescens Grisebach. Illustration: Strid, Mountain flora of Greece **1**: 273 (1986).

Plant loosely tufted, with a taller, looser habit than *A. deltoidea.* Leaves oblong to somewhat spoon-or diamond-shaped, entire or with 1 tooth on each margin, sparsely hairy. Sepals 4–5.5 mm. Petals 0.6–1.1 cm, white, becoming pink or lilac. Fruit 0.7–1.2 × 0.2–0.4 cm, ovoid, flattened, with stellate hairs. *N Greece (Athos).* H1. Spring.

Frequently hybridised in cultivation with *A. deltoidea.*

7. A. gracilis Boissier. Illustration: Polunin, Flowers of Greece and the Balkans, pl. 11 (1980).

Plant more or less tufted, usually hairy. Leaves linear-lanceolate to oblong-obovate, entire or with a few teeth. Sepals 5–7.5 mm. Petals 1.2–1.8 cm. Fruit 1.3–4 cm, usually 5–15 times as long as wide, somewhat curved, hairy with stellate hairs; valves with net-like venation. *S part of Balkan Peninsula.* H1. Spring.

A variable species. Three subspecies can be distinguished in Europe, all of which are apparently in cultivation.

Subsp. **gracilis**. Leaves broadly lanceolate to oblong-obovate, usually 2–3 times as long as wide. Petals purple. *C Greece.*

Subsp. **scardica** (Wettstein) Phitos. Illustration: Strid, Wild flowers of Mount Olympus, pl. 100 (1980) – as A. gracilis. Leaves linear to lanceolate, normally at least 3 times as long as wide. Petals purple. *Widespread.*

Subsp. **glabrescens** (Turrill) Akeroyd. Leaves lanceolate to ovate, hairless or with a few hairs along the margins. Petals violet. *NW Greece (Smolikas).*

8. A. thessala Boissier. Illustration: Strid, Wild flowers of Mount Olympus, pl. 100 (1980); Quarterly Bulletin of the Alpine Garden Society **57**: 126 (1989).

Plant tufted, fanning small, loose mats. Leaves obovate to broadly spathulate, with 1–3 pairs of teeth. Sepals 7–14 mm. Petals 1.4–2 cm, lilac. Fruits 14–20 × *c.* 3 mm, usually with stellate hairs only (rarely with a few short forked hairs; valves without or with obscure net-like venation. *NC Greece (Olympus).* H1. Spring.

21. LUNARIA Linnaeus
T. Upson
Erect branching annual to perennial herbs. Leaves ovate or ovate-triangular, toothed. Flowers violet to purple sometimes white in terminal racemes. Petals long-clawed, sepals erect. Fruit with a persistent round, flat, papery septum.

A genus of 3 species native to C and SE Europe, often naturalised elsewhere. Plant for the mixed border or naturalising (*L. annua*). The silvery papery seed pod septa are used in dried flower arrangements. Easily cultivated in most soils, but best when sandy and in sun but tolerates light shade. Propagation by seed or division in *L. rediviva.*

1a. Upper leaves stalkless or with small stalks **1. annua**
b. Upper leaves distinctly stalked **2. rediviva**

1. L. annua Linnaeus (*L. biennis* Moench). Figure 15(17), p. 131. Illustration: Brickell, RHS gardeners encyclopedia of plants and flowers, 267 (1989) – 'Variegata'; Polunin, Flowers of Europe, 135 (1969); Hay & Beckett, Reader's Digest encyclopaedia of plants and flowers, 420 (1971).

Annual or biennial, occasionally a short-lived perennial up to 1 m. Leaves ovate to lanceolate, acuminate, coarsely and irregularly toothed. Upper leaves stalkless or almost so. Petals 1.5–2.5 cm, lilac, purple or occasionally white. Flowers fragrant. Fruit flat oblong to circular, rounded at base and apex, 2–7 × 1–3.5 cm. *SE Europe, naturalised elsewhere.* H2. Mid–late spring.

Cultivars include 'Alba', 'Variegata', 'Munstead Purple' (dark purple) and 'Atrococcinea' (reddish purple).

2. L. rediviva Linnaeus. Illustration: Thomas, Perennial garden plants, pl. 18D (1990); Hay & Beckett, Reader's Digest encyclopaedia of plants and flowers, 420 (1971).

Perennial to 1.5 m. Leaves ovate, acuminate, sharply toothed. Upper leaves distinctly stalked. Petals lilac to violet 1–2 cm. Fruit 3.5–9 × 1.5–3.5 cm elliptic, almost acute at base and apex. *Most of Europe except the extreme N & S.* H2. Mid–late spring.

22. RICOTIA Linnaeus
J.R. Akeroyd
Annuals or perennials, branched, grey, usually hairless. Leaves with 3 leaflets or pinnatifid. Sepals pouch-like. Petals lilac. Fruit a compressed siliqua. Seeds in 1 or 2 rows in each cell.

A genus of 9 species in the E Mediterranean and SW Asia. Propagation is by seed.

Literature: Burtt, B.L., *Kew Bulletin for 1951:* 123–132 (1952).

1a. Annual; leaves 2–3 pinnate **2. lunaria**
b. Perennial; leaves with 3 leaflets **1. davisiana**

1. R. davisiana B.L. Burtt. Illustration: Quarterly Bulletin of the Alpine Garden Society **43**: 133 (1975).

Perennial, shortly and densely hairy. Leaves with 3 leaflets, the segments ovate to almost circular, fleshy, greyish. Fruit linear-oblong, with 6–12 seeds. *S Turkey (Tahtali Dağ).* H1. Spring.

2. R. lunaria (Linnaeus) de Candolle (*R. aegyptiaca* Linnaeus; *Cardamine lunaria* Linnaeus). Illustration: Edwards's Botanical Register, 49 (1815); Post, Flora of Syria, Palestine and Sinai, edn 2, **1**: 80 (1932).

Annual, 20–40 cm. Leaves 2–3-pinnate, the segments oblong or ovate, sinuate, not fleshy. Fruit 2–3 cm, ovate-lanceolate, with 2–7 seeds. *Lebanon, Syria & Israel.* H3. Summer.

23. ALYSSOIDES Miller
J.R. Akeroyd
Perennials, woody at the base; hairs branched to stellate. Leaves crowded, mostly in non-flowering rosettes, entire. Sepals erect to somewhat spreading, the inner pouch-like at the base. Petals with long claws. Fruit a strongly inflated silicula, the valves without a conspicuous vein; style long. Seeds 4–8 in each cell, winged.

A genus of 2 species in S and SE Europe and Turkey. Propagation is from seed or cuttings. The plants require a well-drained soil and grow best in rocks and walls.

1a. Leaves green; fruit hairless
1. utriculata
  b. Leaves grey or whitish; fruit hairy
2. cretica

1. A. utriculata (Linnaeus) Medicus (*A. graeca* (Reuter) Jávorka; *Vesicaria graeca* Reuter). Illustration: Botanical Magazine, 130 (1790); Quarterly Bulletin of the Alpine Garden Society **43**: 134 (1975). Polunin, Flowers of Greece and the Balkans, 250 (1980).

Leaves dark green; rosette-leaves 2–5 cm. oblong to spathulate, stalked; stem-leaves oblanceolate, stalkless, often more or less hairless. Flowering-stems 10–40 cm, unbranched. Sepals 0.8–1.2 cm. Petals 1.5–2 cm, yellow, with an almost circular blade. Fruit 1–1.5 cm, ovoid to almost spherical, hairless; style up to 1 cm. *S Alps to C Greece, Romania & N Turkey.* H2. Late spring–summer.

A variable species, especially in hairiness.

2. A. cretica (Linnaeus) Medicus (*Alyssum creticum* Linnaeus). Figure 15(11), p. 131. Illustration: Huxley & Taylor, Flowers of Greece, t. 75 (1977); Polunin, Flowers of Greece and the Balkans, pl. 10 (1980).

Like *A. utriculata* but woodier below and more shrubby and branched. Leaves oblanceolate to obovate, grey or whitish,

densely hairy; flowering-stems to 25 cm; blade of petals elliptic; fruit more or less spherical, hairy. *Crete & S Aegean area* H3. Late spring–summer.

24. DEGENIA Hayek
S.L. Jury
Perennial herbs to 10 cm. Leaves linear-lanceolate in rosettes, some rosettes non-flowering. Flowers 1–1.2 cm. Petals yellow, 1.4–1.5 cm, long clawed. Style long. Fruit a silicula, 1–1.4 × 0.7–0.8 cm, ovate, inflated, densely covered with stellate hairs. Seeds 2 in each cell, broadly winged, 5 mm across.

A monotypic genus endemic to NW Yugoslavia.

1. D. velebitica (Degen) Hayek. Illustration: Polunin, Flowers of Greece and the Balkans, 190 (1980).

NW former Yugoslavia. H4. Spring.

25. FIBIGIA Medicus
T. Upson
Erect perennial herbs and small shrubs. Leaves alternate, linear to spathulate. Plant covered with stellate hairs (occasionally a few simple). Sepals erect. Petals yellow, shortly clawed. Fruit a strongly compressed silicula. Seeds 2–8 in each cell, winged.

A genus of 14 species native from the Mediterranean area to Afghanistan. Requires a very free-draining soil in full sun; ideal for a hot dry situation. Fruits used in dried flower arrangements.

1. F. clypeata (Linnaeus) Medicus (*Alyssum clypeata* Linnaeus; *Farsetia clypeata* (Linnaeus) R. Brown). Figure 16(14), p. 134. Illustration: Polunin, Flowers of Europe, pl. 35 (1969).

Perennial to 30–75 cm. Lower leaves oblong or oblanceolate, entire or toothed; linear on stem; covered in stellate hairs, giving a grey felted appearance. Inflorescence a raceme, 10–20 cm, flowers 4–6, petals yellow, 0.8–1.3 cm. Fruits elliptic, 1.4–2.8 × 0.9–1.3 cm, also covered with stellate hairs. *C Europe, Balkans, W Syria, N Iraq, C & W Iran, Transcaucasia, Crimea, ?Egypt.* H5. Late spring–early summer.

26. ALYSSUM Linnaeus
J.R. Akeroyd
Annuals, biennials or herbaceous to shrubby perennials; hairs simple, branched or stellate, sometimes scaly. Basal and stem leaves usually similar. Flower buds ovoid or ellipsoid. Sepals erect to spreading, not pouch-like. Petals yellow, white or rarely

purple, entire to notched or bifid. Filaments of long stamens usually winged, those of short stamens usually with appendages. Fruit a siliqua; valves with short, inconspicuous mid-veins; style often short. Seeds 1 or 2, rarely up to 6, in each cell, usually distinctly winged.

A genus of about 150 species from Europe, Asia and the Mediterranean area. In the wild, many are plants of sandy or rocky ground and cliffs, and in cultivation grow best in well-drained, calcium-rich soils. Propagation is by seed or cuttings.

The genus *Ptilotrichum* is here included within *Alyssum*, although some other species formerly included in *Alyssum* in horticultural texts have been removed to *Alyssoides, Aurinia, Lobularia* and *Schivereckia.* The account of the yellow-flowered species must be regarded as provisional. The wild plants require elucidation, and it is not certain which species are in cultivation.

1a. Petals white, pinkish or purple   2
  b. Petals yellow   8
2a. Petals purple; stems 2–5 cm
7. purpureum
  b. Petals white or sometimes pinkish purple; stems usually more than 10 cm   3
3a. Branches becoming spiny; petals white or pinkish purple
6. spinosum
  b. Branches spineless; petals white 4
4a. Petals 5–6 mm, broadly obovate to almost circular; fruit 6–8 mm   5
  b. Petals 3–4 mm, obovate; fruit 4–6 mm   6
5a. Fruit hairy; leaves obovate-lanceolate   **1. pyrenaicum**
  b. Fruit hairless; leaves spathulate
2. reverchonii
6a. Loosely tufted, more or less prostrate perennial; stem-leaves few, linear-lanceolate
5. longicaule
  b. Shrubby, erect perennial; stem-leaves many, oblong to oblong-lanceolate   7
7a. Fruits in loose, elongate racemes; seeds narrowly winged
3. lapeyrousianum
  b. Fruits in dense corymb-like clusters; seeds broadly winged
4. ligusticum
8a. Seeds narrowly winged or unwinged   9
  b. Seeds distinctly winged   11
9a. Leaves of non-flowering shoots

pleated; petals usually less than
3 mm **20. serpyllifolium**
  b. Leaves of non-flowering shoots not
     pleated; petals at least 3 mm 10
10a. Petals 5–6.5 mm, slightly notched
     or bilobed **16. stribrnyii**
  b. Petals 3–3.5 mm, entire
     19. alpestre
11a. Hairs scale-like 12
  b. Hairs stellate, branched or simple
     but not scale-like 13
12a. Stems usually more than 5 cm;
     petals notched
     15. moellendorfianum
  b. Stems 2–5 mm; petals entire
     17. idaeum
13a. Valves of fruit flat **18. murale**
  b. Valves of fruit inflated (usually with
     flat margin) 14
14a. Non-flowering shoots few or absent
     11. wierzbeckii
  b. Many non-flowering stems or leaf-
     rosettes present 15
15a. Petals entire 16
  b. Petals notched, slightly notched or
     bilobed 18
16a. Plant with both adpressed and
     longer spreading hairs; style 3–5
     mm **12. pulvinare**
  b. Plant with adpressed hairs only;
     style 2–3 mm 17
17a. Stem-leaves larger than basal;
     petals hairless **9. wulfenianum**
  b. Stem- and basal leaves of similar
     size; petals hairy beneath
     10. ovirense
18a. Flowering-stems usually more than
     20 cm; plant with both adpressed
     and longer spreading hairs
     8. repens
  b. Flowering-stems not more than 20
     cm; plant with adpressed hairs only
     19
19a. Racemes long and loose in fruit;
     petals not more than 6 mm
     13. montanum
  b. Racemes short and dense in fruit;
     petals to 8 mm **14. cuneifolium**

1. A. pyrenaicum Picot de Lapeyrouse
(*Ptilotrichum pyrenaicum* (Picot de
Lapeyrouse) Boissier).

Dwarf shrub to 30 cm with woody,
branched rootstock. Leaves obovate-
lanceolate, leathery, covered with stellate,
silvery hairs. Flowers in a dense corymb.
Petals 5–6 mm, white, broadly obovate to
almost circular. Fruits 6–8 mm, obovoid,
compressed, hairy; styles about as long as
fruits. Seeds winged. *SW France (E
Pyrenees)*. H1. Summer.

2. A. reverchonii (Degen & Hervier)
Greuter & Burdet (*Ptilotrichum reverchonii*
Degen & Hervier).

Similar to *A. pyrenaicum* but leaves
broadly spathulate; fruits hairless; styles
2–4 mm. *SE Spain (Sierra de Cazorla)*. H1.
Summer.

3. A. lapeyrousianum Jordan
(*Ptilotrichum lapeyrousianum* (Jordan)
Jordan).

Similar to *A. pyrenaicum* but leaves
oblong to oblong-lanceolate; flowers in
a raceme, becoming loose in fruit; petals
3–4 mm, obovate; fruits 4–6 mm, obovoid,
hairless; styles to 1.5 mm; seeds narrowly
winged. *E Spain & SW France (Pyrenees)*.
H1. Summer.

4. A. ligusticum Breistroffer (*Alyssum
halimifolium* misapplied; *Ptilotrichum
halimifolium* Boissier). Illustration:
Botanical Magazine, 101 (1789).

Similar to *A. pyrenaicum* but leaves
oblong to oblong-lanceolate; petals 3–4
mm, obovate; fruits 4–6 mm, obovoid,
hairless; styles 1–2 mm; seeds broadly
winged. *SE France, NW Italy*. H1. Summer.

5. A. longicaule Boissier (*Ptilotrichum
longicaule* (Boissier) Boissier).

Loosely tufted perennial, woody at base.
Stems 20–50 cm, prostrate. Leaves covered
with adpressed, stellate, silvery hairs. Basal
leaves in rosettes, obovate to spathulate,
acute; stem-leaves few, linear-lanceolate.
Petals 2–4 mm, white. Fruits 4–6 mm,
obovoid, hairless; styles to 0.5 mm. *S
Spain*. H1. Spring.

6. A. spinosum Linnaeus (*Ptilotrichum
spinosum* (Linnaeus) Boissier). Figure 15(7),
p. 131. Illustration: Polunin & Smythies,
Flowers of south-west Europe, pl. 9 (1973);
Quarterly Bulletin of the Alpine Garden
Society **53**: 24 (1985).

Dwarf shrub to 30 cm, sometimes larger,
much-branched with spiny branches
interlocking to form a compact dome.
Leaves of non-flowering rosettes obovate
to spathulate. Leaves of flowering branches
linear-lanceolate, acute, with adpressed,
stellate, silvery hairs. Flowers in corymbs.
Petals *c.* 3 mm, white or pinkish purple.
Fruits 4–6 mm, obovoid, hairless; styles
to 1.5 mm. *S & E Spain, S France*. H1.
Summer.

7. A. purpureum Lagasca & Rodriguez
(*A. lagascae* misapplied; *Ptilotrichum
purpureum* (Lagasca & Rodriguez) Boissier).
Illustration: Polunin & Smythies, Flowers of
south-west Europe, pl. 9 (1973).

Densely tufted perennial; stems 2–5 cm,
branched, leafy. Leaves oblong-elliptic to
linear, with stellate, white or greyish hairs.
Flowers in dense corymbs. Petals 3–4 mm,
purple. Fruits 4–5 mm, ellipsoid, hairy;
styles 1–2 mm. *S & E Spain*. H1. Summer.

8. A. repens Baumgarten (*A.
trichostachyum* Ruprecht; *A. transilvanicum*
misapplied).

Perennial, somewhat woody below,
greyish green; hairs both adpressed and
spreading; stems 20–60 cm, erect or
ascending, hairy; non-flowering stems
terminated by rosettes of leaves. Basal
leaves oblong-obovate to stellate; stem-
leaves linear-oblong, acute. Flower-stalks
spreading. Petals 5–7 mm, notched,
golden-yellow. Fruits 3–6 mm, obovate to
more or less circular; valves inflated with
narrow flat margin; style 1.5–3.5 mm. *SE
Europe to the Caucasus*. H1. Summer.

9. A. wulfenianum Bernhardi.

Perennial; stems to 20 cm, prostrate to
ascending; plant greyish green to whitish;
non-flowering stems with rosettes of leaves.
Basal leaves oblong-obovate, obtuse; stem-
leaves larger, oblanceolate. Flower-stalks
spreading. Petals 5.5–6.5 mm, entire,
hairless. Fruits 6–8 mm, elliptic, obtuse,
slightly hairy; valves inflated; styles 2–3
mm. *SE Alps*. H1. Summer.

10. A. ovirense Kerner. Illustration:
Pignatti, Flora d'Italia **1**: 426 (1982).

Similar to *A. wulfenianum* but stems to
12 cm, prostrate; plant green to greyish
green; basal leaves about same size as
stem-leaves, obovate to almost circular;
stem-leaves oblong-lanceolate; petals hairy
beneath; fruits 6.5–8 mm. *SE Alps & W
former Yugoslavia*. H1. Summer.

11. A. wierzbeckii Heuffel.

Robust biennial or short-lived perennial;
stems 30–60 cm, erect, stiff, little-
branched; non-flowering stems few or
absent. Leaves *c.* 5 cm, ovate-elliptic to
lanceolate, acute, green. Flowers in dense
corymbs 3–4 cm in across. Flower-stalks
spreading to erect. Petals 5–7 mm, entire,
yellow. Fruits 4–6 mm, more or less
circular, notched, hairy; valves inflated
with flat margin; styles 3–4.5 mm. *Balkans*.
H1. Spring.

12. A. pulvinare Velenovsky.

Densely tufted, grey perennial with
many short non-flowering stems and
rosettes of leaves; hairs both adpressed and
spreading; stems 5–15 cm. Basal leaves
oblanceolate or spathulate; stem-leaves
lanceolate to spathulate. Petals 5–6 mm,
entire, yellow. Fruit 4–5.5 cm, elliptic to

almost circular, hairy; valves inflated with flat margin; styles 3–5 mm. Seeds broadly winged. *Balkans*. H1. Summer.

13. A. montanum Linnaeus (*A. atlanticum* Desfontaines; *A. speciosum* Pomel; *A. rostratum* misapplied). Illustration: Botanical Magazine, 419 (1798).

Tufted green to almost white perennial with many non-flowering rosettes of leaves; stems 5–20 cm, usually prostrate. Basal leaves oblong to obovate; stem-leaves linear to narrowly spathulate. Flowers in rather loose racemes, fragrant. Flower-stalks more or less spreading. Petals 4–6 mm, notched, yellow. Fruits 3–6 mm, circular; valves inflated with flat margin; styles 1–3 mm. *C & S Europe & Mediterranean area*. H1. Summer.

14. A. cuneifolium Tenore.

Tufted grey to whitish perennial with numerous non-flowering rosettes of leaves; stems 5–15 cm. Basal leaves oblong-obovate; stem-leaves narrower. Racemes short and dense in fruit. Petals 5–8 mm, notched, yellow. Fruits 5–7 mm, almost circular, grey-hairy; valves inflated with flat margin; style 2–4 mm. *S Europe*. H1. Summer.

15. A. moellendorfianum G. Beck.

Tufted silvery-scaly perennial with non-flowering short stems and rosettes of leaves; stems 5–20 cm. Leaves almost circular to spathulate, stalked. Flowers fragrant in long, loose racemes. Flower-stalks spreading. Petals 5–6 mm, notched, yellow. Fruit 4–5 mm, circular, densely scaly; valves inflated with flat margin; styles 2–2.5 mm. *W former Yugoslavia*. H1. Summer.

16. A. stribrnyi Velenovsky. Illustration: Flora Republicae Popularis Bulgaricae **4**: 501 (1970).

Small shrubby perennial with long non-flowering stems; stems 5–20 cm. Basal leaves obovate to spathulate; basal leaves and leaves of non-flowering stems with dense, silvery grey hairs; stem-leaves oblanceolate; acute, shortly hairy above. Racemes long and loose in fruit. Petals 5–6.5 mm, slightly notched or bilobed, yellow. Flower-stalks strongly spreading. Fruits 4–5 mm, almost circular, with dense grey stellate hairs; valves inflated with flat margin; styles 2–3.5 mm. Seeds not or scarcely winged. *Balkans to Mesopotamia*. H1. Summer.

17. A. idaeum Boissier & Heldreich.

Procumbent, greyish to white perennial; stems 2–5 cm, simple. Leaves oblong to

almost circular, covered with scaly hairs. Racemes short, dense. Flower-stalks spreading, short. Petals 4–6 mm, entire, yellow. Fruits 5–7 mm, circular or almost circular, hairless or somewhat scaly; valves inflated; styles *c.* 2 mm. *Crete (Mt Ida)*. H1. Summer.

18. A. murale Waldstein & Kitaibel (*A. argenteum* misapplied).

Tufted perennial with long non-flowering-stems and rosettes of leaves; plant with stellate hairs (rarely with simple); stems 30–60 cm. Leaves of non-flowering stems oblanceolate to obovate; stem-leaves larger, lanceolate or oblanceolate, greyish above, whitish beneath, acute. Flowers in corymbs. Petals 2–4 mm, usually entire, yellow. Flower-stalks spreading or ascending. Fruit 2–5 mm, ovate to circular, hairy; valves flat; styles 0.5–2 mm. *SE Europe & SW Asia*. H1. Early Summer.

A very variable species that requires taxonomic study.

19. A. alpestre Linnaeus.

Rather woody, tufted perennial with numerous non-flowering rosettes of leaves; stems 5–15 cm, usually prostrate. Leaves oblong-obovate to oblanceolate, white to greyish silver. Petals 3–3.5 mm, entire, pale yellow. Fruits 2–4.5 mm, elliptic, white-hairy; valves asymmetrically inflated; styles 1–1.5 mm. Seeds not or scarcely winged. *Alps*. H1. Summer.

20. A. serpyllifolium Desfontaines.

Like *A. alpestre* but usually taller and more erect to 30 cm, with short non-flowering stems; leaves grey or white beneath, grey or greyish green above, those of the non-flowering stems pleated; petals 2–3 mm; fruit broadly elliptic to obovate. *SW Europe*. H3. Summer.

27. AURINIA Desvaux
J.R. Akeroyd
Perennial herbs, rather woody at the base; stems somewhat woody, spreading or ascending, branched, forming a mat. Hairs simple and branched. Basal leaves forming loose rosettes, distinctly larger than the stem-leaves; stalks with persistent swollen bases. Flowers in corymb-like clusters; buds spherical. Sepals not pouched at base. Petals usually yellow. Fruit a flattened, almost circular silicula; style short. Seeds 2–8 in each cell, usually winged.

A genus of about 12 species from S Europe and SW Asia, included by many in *Alyssum*, but differing in several features. Propagation is by seed or cuttings.

Literature: Dudley, T.R., *Journal of the Arnold Arboretum* **45**: 390–400 (1964).

1a. Petals white, entire   **5. rupestris**
  b. Petals yellow, bifid or deeply notched   2
2a. Valves of fruit almost flat or if inflated with a flat margin; seeds 2 in each cell   3
  b. Valves of fruit inflated; seeds usually more than 2 in each cell   4
3a. Valves of fruit inflated, with flat margin; flowers in raceme
                                        2. petraea
  b. Valves of fruit flat; flowers in corymb   **3. saxatilis**
4a. Petals *c.* 4 mm; seeds 2–4 in each cell   **1. corymbosa**
  b. Petals 5–8 mm; seeds 4–8 in each cell   **4. sinuata**

1. A. corymbosa Grisebach (*Alyssum corymbosum* (Grisebach) Boissier).

Stems 20–50 cm; plant with branched hairs. Basal leaves lanceolate or obovate to spathulate, entire or sinuate; stem-leaves lanceolate, entire. Flowers in dense corymbs. Sepals 1.5–2.5 mm. Petals *c.* 4 mm, bifid, yellow. Fruit 3.5–5.5 mm, almost spherical, the valves inflated, hairless; styles 1–2 mm. Seeds 2–4 in each cell. *W & S Balkan Peninsula*. H2. Early summer.

2. A. petraea (Arduino) Schur (*Alyssum petraeum* Arduino; *A. gemonense* Linnaeus; *A. edentulum* Waldstein & Kitaibel).

Stems 15–40 cm; hairs branched to stellate. Basal leaves oblong to obovate, sinuate to pinnatifid. Flowers in raceme. Sepals *c.* 2 mm. Petals *c.* 4 mm, bifid, yellow. Fruits 3–5 mm, elliptic to obovate, hairless, the valves inflated but with a flat margin; styles 1–1.5 mm. Seeds 2 in each cell. *SE Europe from Austria to N Greece*. H2. Early summer.

3. A. saxatilis (Linnaeus) Desvaux (*Alyssum saxatile* Linnaeus; *A. arduini* Fritsch). Figure 15(8), p. 131. Illustration: Botanical Magazine, 159 (1791); Hay & Synge, Dictionary of garden plants, 1 (1969) – as 'Citrinum'; Everett, New York Botanical Garden encyclopedia of horticulture **1**: 306–7 (1980); Brickell, RHS gardeners' encyclopedia of plants & flowers, 289, 290 (1989).

Stems 10–40 cm; hairs stellate. Basal leaves 5–15 cm, obovate to oblanceolate, entire to somewhat toothed. Sepals 2–3, rarely 4 mm. Petals 3–6 mm, bifid, yellow. Fruits 3.5–5 mm, slightly longer than wide, rounded at apex; style 0.3–0.8 mm.

Seeds 2 in each cell. *C & SE Europe.* H1. Spring–early summer.

Subsp. **orientalis** (Arduino) T.R. Dudley (*Alyssum orientale* Arduino). Basal leaves toothed to sinuate-pinnatifid, sometimes entire. Fruits wider than or about as wide as long, truncate or notched at apex; styles up to 1.5 mm. *S Italy, S Balkan Peninsula, W Turkey.* H1. Early summer.

A. saxatilis is one of the most commonly grown rock-garden plants, especially the typical variant with golden yellow petals; 'Citrina' ('Lutea', 'Sulphurea'), with sulphur-yellow petals, and the double-flowered 'Plena'. Other cultivars include 'Variegata', with variegated leaves, 'Dudley Neville' with brownish yellow petals, and several dwarf variants, notably 'Compacta'.

4. A. sinuata (Linnaeus) Grisebach (*Alyssum sinuatum* Linnaeus).

Stems 15–40 cm; plant grey-hairy. Basal leaves lanceolate to oblanceolate to spathulate, sinuously toothed; stem-leaves lanceolate, entire. Sepals 3–4 mm. Petals 5–8 mm, notched, pale yellow. Fruits 7–11 mm, ellipsoid to spherical, the valves inflated; styles 2–4 mm. Seeds 4–8 in each cell. *S Italy & NW Balkan Peninsula.* H1. Early summer.

5. A. rupestris (Heynhold) Cullen & T.R. Dudley (*Alyssum rupestre* Tenore; *Ptilotrichum rupestre* (Heynhold) Boissier).

Densely tufted; stems 5–20 cm, unbranched; plant silvery-hairy. Basal leaves lanceolate; stem-leaves linear, few. Petals 5–6 mm, entire, white. Fruit obovoid-spherical, scaly or hairless (var. *scardicum* misapplied); style *c.* 0.5 mm. *C Italy & W Balkan Peninsula.* H1. Early summer.

28. BORNMUELLERA Haussknecht
J.R. Akeroyd
Small shrubby perennials; hairs forked, rarely 4–6-fid or the plant hairless. Sepals erect to spreading, not pouch-like at base. Petals entire, white. Filaments with a tooth-like appendage at base. Fruit a flattened silicula; style short. Seeds 1 or 2 in each cell.

A genus of 6 species in the Balkans and Anatolia. Propagation is from seed and cuttings.

1. B. cappadocica (de Candolle) Cullen & T.R. Dudley (*Iberis cappadocica* de Candolle; *Ptilotrichum cappadocica* (de Candolle) Boissier). Figure 15(2), p. 131.

Stems 5–10 cm, erect or ascending, leafy. Leaves mostly basal, linear to elliptic-oblanceolate, greyish when young. Petals

4–6 mm. Fruits *c.* 3 mm, almost circular. *Turkey.* H1. Late spring.

29. LOBULARIA Desvaux
J.R. Akeroyd
Annuals or sometimes short-lived perennials; rather hairy with 2-fid hairs, greyish. Stems ascending, much-branched near the base. Leaves linear-lanceolate, acute. Flowers fragrant, in dense, rounded clusters lengthening in fruit. Sepals spreading, not pouched at the base. Petals entire. Filaments of stamens not winged and without appendages. Fruit an obovate to almost circular silicula; valves slightly inflated, with more or less distinct mid-vein; style distinct; stigma capitate. Seed 1 in each cell.

A genus of 5 species from the Mediterranean and Macaronesia. The one species grown in gardens is propagated from seed, and once established will come up spontaneously each year. Literature: Borgen, L., *Opera Botanica* **91** (1987).

1. L. maritima (Linnaeus) Desvaux (*Alyssum maritimum* Linnaeus). Figure 15(14), p. 131. Illustration: Hay & Synge, Dictionary of garden plants in colour, pl. 29 (1969); Polunin, Flowers of Europe, pl. 37 (1969).

Stems 5–30 cm. Leaves 1–3 cm. Petals 2.5–4 mm, white, or sometimes purple or pink. Fruits 2–4 mm, hairy to nearly hairless. *Mediterranean area.* H1. Spring–autumn (rarely to winter).

There are many cultivars of this popular garden plant, varying in habit and petal-colour. The wild plant is usually perennial and has white petals.

30. BERTEROA de Candolle
J.R. Akeroyd
Annuals to perennials; hairs simple, forked or stellate. Leaves more or less entire. Sepals not pouch-like. Petals deeply 2-lobed, white. Outer filaments toothed at base. Fruit a silicula; valves without conspicuous veins. Seeds 2–6 in each cell.

A genus of 8 species in Europe, the Mediterranean area and SW Asia. Propagation is by seed.

1a. Flowers in dense racemes; fruits somewhat inflated, hairy; stalks of fruit erect **1. B. incana**
b. Flowers in loose racemes; fruits flat, hairless; stalks of fruit somewhat spreading **2. B. mutabilis**

1. B. incana (Linnaeus) de Candolle. Figure 15(23), p. 131. Illustration:

Bonnier, Flore Complète **1**: 47 (1912). Hitchcock, Vascular plants of the Pacific Northwest **2**: 461 (1964); Hegi, Illustrierte Flora von Mitteleuropa **4**: 293, (1986); Pignatti, Flora d'Italia **1**: 430 (1982).

Annual to perennial; stems 30–70 cm. Leaves obovate or lanceolate, entire or remotely toothed, greyish. Flowers in dense racemes. Petals 4.5–7.5 mm. Silicula 4–10 mm, ovoid, somewhat inflated, shortly hairy; style 2–4 mm. Fruiting flower-stalks erect. Seeds not winged. *C & E Europe.* H1. Summer.

2. B. mutabilis (Ventenat) de Candolle. Illustration: Pignatti, Flora d'Italia **1**: 430 (1982).

Biennial or perennial; stems 10–50 cm. Leaves lanceolate or obovate-elliptic, entire. Flowers in loose racemes. Petals 4–6 mm. Fruit 6–12 mm, oval, more or less flat, hairless; style 1–2 mm. Stalks of fruit somewhat spreading. Seeds broadly winged. *SE Europe & SW Asia.* H1. Summer.

31. SCHIVERECKIA de Candolle
J.R. Akeroyd
Perennials with minute stellate and forked hairs. Leaves silvery green; basal leaves in rosettes. Sepals 2–2.5 mm, spreading, the inner slightly pouched. Filaments toothed, the 2 inner winged. Fruit a convex, oblong to ovoid silicula; stigma head-like. Seeds *c.* 1 mm, 4–7 in each cell.

A genus of 2 species in E Europe and W Turkey. Propagation is by seed. The plants require very good drainage and grow well in tufa.

Literature: Kasakova, M.V., *Biologiceskie Nauki* **4**: 57–62 (1984).

1a. Leaves with 1–5 teeth on each margin; stem leaves ovate-oblong, somewhat clasping **1. podolica**
b. Leaves entire or with 1, rarely 2, teeth on each margin; stem leaves lanceolate, not clasping **2. doerfleri**

1. S. podolica (Besser) de Candolle (*Alyssum podolicum* Besser). Figure 15(19), p. 131. Illustration: Sweet, The British flower garden **1**: t. 77 (1824).

Densely tufted. Leaves with 1–5 teeth on each margin. Basal leaves oblanceolate-oblong to spathulate; stem-leaves ovate-oblong, somewhat clasping. Flowering-stems to 25 cm; racemes with up to 30 flowers. Petals 3–5 mm, white. Fruit 2–6 × 2–3 mm; style 1–2 mm. *Ukraine & NE Romania.* H1. Spring.

2. S. doerfleri (Wettstein) Bornmüller (*Draba doerfleri* Wettstein; *S. bornmuelleri* Prantl).

Loosely tufted. Leaves entire, rarely with 1–2 teeth on each margin. Basal leaves oblanceolate; stem-leaves linear-lanceolate, not clasping. Flowering-stems 5–15 cm; racemes with up to 15 flowers. Petals 4–6 cm, white. Fruit 4–6 × 2–4 mm; style 0.5–1 mm. *Balkans to Turkey.* H1. Spring.

32. DRABA Linnaeus
R. Hyam
Perennial herbs, occasionally annual, usually tuft-or cushion-forming, sometimes straggling. Rootstock woody. Leaves simple, frequently hairy, entire or toothed, often forming basal rosettes. Flowering-stems naked or leafy, erect or ascending. Inflorescence a terminal raceme, often corymb-like at first, lengthening in fruit. Flowers usually small; sepals 4, erect or spreading, sometimes slightly pouched; petals yellow or white, longer than sepals, tip rounded, sometimes notched; stamens 6, (4 long, 2 short), filaments simple, somewhat expanded at base; ovary with few to many ovules. Fruit a silicula or occasionally a siliqua. Seeds in 2 rows, not winged.

Around 300 species from north temperate and subarctic regions and South American mountains; commonly grown as rock garden plants. Propagation is by division or from seed. In the wild flowering time varies between spring and summer, depending on altitude; in cultivation plants are mostly spring-flowering.

Literature: Schulz, O.E., Cruciferae – Draba and Erophila in Engler, A., Das Pflanzenreich **89** (1927).

1a. Flowers yellow   9
  b. Flowers white, or cream to white
                                                2
2a. Dense perennial; flowering-stems leafless; petals distinctly notched   3
  b. Flowering-stems leafy   4
3a. Sepals with translucent margin, 3–5 mm; petals 6–9 mm; style more than 2 mm   **23. oreadum**
  b. Sepals 2–3 mm, lacking translucent margin; petals 4–6 mm; style less than 2 mm   **10. dedeana**
4a. Usually over 10 cm in fruit with more than 4 stem-leaves; fruit often twisted   5
  b. Usually less than 10 cm in fruit with fewer than 4 stem-leaves; fruit usually straight   6

5a. Leaves densely white-hairy, hairs simple or branched, rarely stellate   **29. incana**
  b. Leaves sparsely stellate-hairy   **28. arabisans**
6a. Leaves and stems densely stellate-hairy   7
  b. Leaves and stems with simple, branched and occasional stellate hairs   **26. norvegica**
7a. Style more than 1 mm; fruits hairless   8
  b. Style less than 1 mm; fruits hairy   **27. tomentosa**
8a. Leaves spathulate to oblong, usually over 8 mm; raceme usually with more than 10 flowers   **24. ussuriensis**
  b. Leaves oblanceolate, usually less than 8 mm; raceme usually with less than 10 flowers   **25. stellata**
9a. Plants with numerous prostrate stems; leaf-hairs attached at their mid-points   **14. sibirica**
  b. Plants more or less tufted; leaf-hairs not attached at their mid-points   10
10a. Style usually less than 1 mm in fruit   19
  b. Style usually more than 1 mm   11
11a. Flowering-stems hairless   15
  b. Flowering-stems hairy   12
12a. Leaves coated with stellate hairs   **19. cappadocica**
  b. Leaves with simple or forked hairs, or more or less hairless   13
13a. Leaves hairless except for marginal hairs; fruit-hairs simple   **3. cuspidata**
  b. Leaves with simple and forked hairs; fruit-hairs simple and branched   14
14a. Leaves acute, *c.* 1.5 mm wide; flowering-stem hairs less than 0.5 mm   **9. hispanica**
  b. Leaves blunt, *c.* 2.5 mm wide; flowering-stem hairs *c.* 1 mm   **7. parnassica**
15a. Stamens as long as petals   16
  b. Stamens shorter than petals   17
16a. Leaves to 1.5 mm across; inflorescence with 4–20 flowers; fruits compressed; style 1–5 mm   **1. aizoides**
  b. Leaves to 1 mm across; inflorescence with 4–5 flowers; fruits inflated; style 3–7 mm   **2. aspera**
17a. Fruits inflated at base, compressed above, densely hairy all over   **8. haynaldii**

  b. Fruits compressed, hairless at maturity or only hairy on margins   18
18a. Leaves to 2 cm, hairless except on margins; inflorescence remaining fairly dense, even in fruit   **5. lasiocarpa**
  b. Leaves to 7 mm, with simple hairs; inflorescence becoming loose in fruit   **13. rigida**
19a. Leaves rigid, narrow, more or less linear   20
  b. Leaves soft, rarely rigid, more or less spathulate, obovate or oblong   24
20a. Fruit compressed   21
  b. Fruit not compressed, flaps convex or inflated at base   23
21a. Flowering-stems hairless   22
  b. Flowering-stems long-haired   **11. × salamonii**
22a. Petals pale yellow, *c.* 3 mm; fruit sometimes hairy on margins   **6. compacta**
  b. Petals yellow, 4–6 mm; fruit hairless   **4. sauteri**
23a. Leaves 4–6 mm; flowering-stems sometimes hairless; fruits 3.5–6 mm   **12. bruniifolia**
  b. Leaves 6–10 mm; flowering-stems always hairy; fruits 6–7.5 mm   **7. parnassica**
24a. Fruits inflated   25
  b. Fruits compressed   26
25a. Leaves 4–5 mm; flowering-stems 1–4 cm; inflorescence with 4–10 flowers   **18. polytricha**
  b. Leaves 1.5–2 cm; flowering-stems 8–10 cm; inflorescence with 8–20 flowers   **20. rosularis**
26a. Petals usually less than 3 mm, slightly longer than sepals; fruits hairless   **30. crassifolia**
  b. Petals more than 3 mm, clearly longer than sepals; fruits usually hairy   27
27a. Leaves usually less than 6 mm; fruits hairless   28
  b. Leaves usually more than 6 mm; fruits usually hairy   29
28a. Petals to 7 mm; style 0.5–1 mm   **22. longisiliqua**
  b. Petals to 5 mm; style less than 0.5 mm   **21. mollissima**
29a. Leaves usually more than 1.5 cm; flowering-stems to 25 cm   **15. glacialis**
  b. Leaves usually less than 1.5 cm; flowering-stem to 10 cm   30
30a. Leaves to 1.1 cm; fruits 3–7 mm, hairs short   **16. oligosperma**

b. Leaves to 1.5 cm; fruits 7–9 mm,
   often hairless   **17. incerta**

1. D. aizoides Linnaeus. Figure 15(21),
p. 131. Illustration: Marret, Icones Florae
Alpinae Plantarum, **5**: t. 75 (1911);
Engler, Das Pflanzenreich, **89**: 23 (1927);
Keble Martin, The concise British flora, pl.
11 (1974).

Tuft-forming perennial. Leaves borne in
rosettes, hairless, linear-oblong, tapered
at tip and base, 0.5–1.5 mm wide, bristle-
tipped, with marginal hairs. Flowering-
stems 5–15 cm, leafless, hairless.
Inflorescence dense, becoming loose.
Flowers 4–20 per raceme, 8–9 mm across;
sepals tinged yellow, not usually pouched;
petals yellow, 4–6 mm, triangular to
obovate; stamens as long as petals. Fruits
elliptic, 6–12 × 2.5–4 mm, compressed,
hairless; style 1–5 mm. *C & SE Europe,
British Isles*. H3–4.

2. D. aspera Bertoloni (*D. bertolonii*
Nyman; *D. longirostra* Schott, Nyman &
Kotschy; *D. aramata* Schott, Nyman &
Kotschy). Illustration: Botanisk Tidsskrift,
21: 296 (1897–98).

Closely resembles *D. aizoides* except for
narrower leaves, to 1 mm across;
inflorescence with 4–5 flowers; fruits
inflated; style 3–7 mm. *S Europe (Pyrenees
to Balkans)*. H3.

3. D. cuspidata Bieberstein. Illustration:
Flora of the USSR, **8**: 383 (1939).

Dwarf, tuft-forming perennial. Leaves
0.6–1.5 cm, narrow linear, blunt or
somewhat acute, blade hairless; margin
entire, hairy near the tip. Flowering-stems
leafless, densely long-hairy, 2–7 cm.
Inflorescence loose, with 4–14 flowers,
lengthening in fruit. Flowers large; sepals
finely hairy, ovate-oblong, green tinged
with blue, 3–5 mm; petals 6.5–8 mm,
obovate-oblong, yellow, apex rounded,
sometimes shallowly notched; stamens
almost as long as petals. Fruits 6–9 mm,
ovate-oblong, inflated, with dense, mainly
simple hairs; style to 7 mm. *Former USSR
(Crimea)*. H3–2.

4. D. sauteri Hoppe. Illustration: Marret,
Icones Florae Alpinae Plantarum, **5**: t. 84
(1911); Engler, Das Pflanzenreich, **89**: 58
(1927); Grey-Wilson & Blamey, The alpine
flowers of Britain and Europe, 67 (1979).

Loosely tufted perennial. Stems
somewhat decumbent, branching,
spreading, lower parts covered by the
scale-like remains of old leaf-stalks. Leaves
rigid, linear to spathulate, blunt, 5–8 ×
1.5–2 mm, in loose rosettes at stem tips.

Flowering-stems to 3 cm, leafless, hairless,
filamentous. Flowers few; sepals *c.* 2 mm,
ovate; petals 4–6 mm, ovate, yellow,
notched at tip; stamens shorter than petals.
Fruits broadly ovate, frequently somewhat
asymmetric, 3–7 mm, compressed, hairless;
style *c.* 0.5 mm. *NE Alps*. H2–3.

5. D. lasiocarpa Rochel (*D. aizoon*
Wahlenberg; *D. elongata* Host). Illustration:
Grey-Wilson & Blamey, The alpine flowers
of Britain and Europe, 67 (1979): Strid,
Mountain flora of Greece **1**: 313 (1986).

Densely tufted perennial. Leaves broadly
linear, to somewhat oblong, 5–20 × 2–3
mm, hairless, except for marginal hairs.
Flowering-stems to 15 cm, leafless,
hairless. Inflorescence dense, not becoming
loose in fruit, few-flowered. Petals 3–8 mm,
deep yellow, ovate, longer than stamens.
Fruits elliptic, compressed, 5–11 × 2–3.5
mm, hairless or with bristle-like hairs on
margins; style 1–1.5 mm. *Balkan Peninsula,
Carpathians, Alps*. H3.

6. D. compacta Schott, Nyman &
Kotschy. Illustration: Oesterreichische
Botanische Zeitschrift, **11**: t. 1 (1861).

Closely resembles *D. lasiocarpa* and is
sometimes included in it. Very dwarf, tufts
2–3 mm across. Leaves 4–10 × 1.5–2 mm.
Inflorescence with 5–20 flowers. Petals
pale yellow, *c.* 3 mm. Fruit 5–6 × 2.5–3
mm; style *c.* 0.5 mm. *Balkan Peninsula, E
Carpathians*. H3.

7. D. parnassica Boissier & Heldreich.
Illustration: Marret, Icones Florae Alpinae
Plantarum, **5**: t. 83 (1911); Polunin,
Flowers of Greece and the Balkans, pl. 11
(1980); Strid, Mountain flora of Greece **1**:
313 (1986).

Dwarf, tuft-forming perennial. Leaves
broadly linear, blunt, 6–10 × *c.* 2.5 mm,
with simple, branched hairs, veins
prominent below; margin with coarse
hairs. Flowering-stems leafless, ascending
or erect, to 7 cm, long-haired. Raceme
with 6–22 flowers. Sepals *c.* 3.5 mm,
elliptic, tip rounded, hairy on back; petals
yellow, *c.* 5.5 mm, narrowly obovate-
triangular, tip notched; stamens shorter
than petals. Fruits ovoid, inflated at base,
covered in simple, branched hairs, 6–7.5 ×
c. 3 mm; style 0.5–1.5 mm. *C Greece*. H4.

8. D. haynaldii Stur. Illustration: Marret,
Icones Florae Alpinae Plantarum, Fascicle
5: t. 83 (1911); Oesterreichische
Botanische Zeitschrift, **11**: t. 2 (1861).

Forming tight cushions, 2–5 cm across.
Leaves linear, 5–7 × *c.* 1 mm, pointed.
Flowering-stems leafless, to 6 cm, hairless.

Inflorescence with 3–8 flowers; lengthening
greatly in fruit. Sepals *c.* 3 mm, blunt,
margin transparent; petals 4–4.5 mm,
triangular to obovate, yellow, longer than
stamens. Fruits ovoid, inflated below,
compressed above, densely covered with
coarse hairs, 5–7 × 2–3 mm; style 1–2
mm. *S Carpathians*. H3–4.

9. D. hispanica Boissier (*D. atlantica*
Pomel). Illustration: Marret, Icones Florae
Alpinae Plantarum, **5**: t. 79a (1911).

Perennial forming small cushions.
Leaves linear, acute, keeled, 6–8 × *c.* 1.5
mm, coarsely hairy with simple and forked
hairs; margin with long hairs. Flowering-
stems 1–6 cm, hairy. Inflorescence
congested, corymb-like lengthening at
flowering. Sepals 3.5–4.5 mm, oblong-
elliptic; petals yellow, 5–9 mm, longer than
stamens, triangular-obovate, notched at
tip. Fruits with coarse branched, simple
hairs, 6–8 × 2.5–3 mm, somewhat inflated
at base; style 2–4 mm. *E & S Spain*. H3–4.

10. D. dedeana Boissier & Reuter.
Illustration: Marret, Icones Florae Alpinae
Plantarum, **5**: t. 79b (1911); Farrer, The
English rock garden, pl. 32 (1925).

Very dense, cushion-forming perennial.
Stock woody. Leaves broad linear, blunt,
base tapered, hairy, 3–6 × 1–1.5 mm,
with coarse marginal hairs. Flowering-
stems leafless, long-haired. Inflorescence
lengthening in fruit, with 3–10 flowers.
Sepals 2–3 mm, ovate, blunt, often purple-
tipped; petals broadly ovate, 4–6 mm,
white, purple near base, tip notched;
stamens much shorter than petals. Fruits
ovoid, 5–9 × 2.5–3.5 mm, compressed; style
c. 0.5 mm. *Mountains of N & E Spain*. H4.

11. D. × salamonii Sündermann. (*D.
bruniifolia × D. dedeana*).

Resembles *D. dedeana* very closely except
flowers yellow, sterile. *Garden origin*. H4.

12. D. bruniifolia Steven. Illustration:
Farrer, The English rock garden, pl. 32
(1925); Flora of the USSR, **8**: 383 (1939).

Tufted perennial. Leaves 4–6 × *c.* 1
mm, rigid, linear, blunt, hairless or finely
hairy, margins hairy. Flowering-stems
leafless, 1–10 cm, hairless to densely long-
haired. Inflorescence loose, with 8–16
flowers, somewhat corymb-like. Sepals 3
mm, ovate to obovate, tip rounded, hairy;
margin transparent; petals 5–6 mm, bright
yellow, ovate, shallowly notched at tip or
truncate; stamens 2.5–3 mm. Fruits ovoid,
hairy, 3.5–6 × 2.5–3.5 mm, flap somewhat
convex; style *c.* 0.5 mm. *Turkey, Caucasus,
N Iran*. H3–4.

A variable species often split into a number of subspecies.

13. D. rigida Willdenow. Illustration: Quarterly Bulletin of the Alpine Garden Society **35**: 225 (1967).

Tufted perennial. Leaves linear to obovate, with simple hairs, blunt, 3–7 × 1.5–2 mm, spreading, margin broadly hairy. Flowering-stems 5–10 cm, numerous, hairless, erect, smooth. Inflorescence corymb-like, with 5–20 flowers, dense at first, becoming loose. Sepals 2–3.5 mm, ovate, rounded at tip; petals 4.5–6 mm, yellow, obovate-triangular, rounded at tip, sometimes notched. Fruits 4–9 × 2–2.5 mm, ellipsoid, compressed, becoming hairless; style 1–1.5 mm. *Turkey*. H3–4.

Var. **bryoides** (de Candolle) Boissier (*D. bryoides* de Candolle) has leaves curved inwards, to 2 mm or less. *Turkey*. H3–4.

14. D. sibirica (Pallas) Thellung.

Prostrate perennial. Stems slender, creeping, branching. Leaves 0.7–2.5 × 0.2–1 cm, oblong to lanceolate or obovate, pointed, tapering to base, sparsely hairy, hairs attached at their mid-points. Flowering-stems lateral, 2.5–20 mm, leafless, ascending, lengthening in fruit. Racemes with 8–20 flowers, compact at first. Sepals ovate, sparsely to densely hairy; petals 3.5–6.5 mm, oblong-obovate, notched at tip, dark yellow. Fruits oblong-elliptic, hairless, 4–7 × 1–2 mm; style 0.5–1 mm. *C & E former USSR, Siberia & Greenland*. H1–2.

15. D. glacialis Adams. Illustration: Botanisk Tidsskrift, **21**: 295 (1897–98).

Loosely tufted perennial. Leaves 7–30 × 2–7 mm, oblong-obovate, downy, hairs stellate, branched; margin hairy, entire. Flowering-stems leafless, 5–25 cm, hairs stellate, branched with occasional simple hairs. Inflorescence compact or drawn out, with 4–15 flowers; sepals oblong-ovate, with simple hairs; petals obovate-oblong, sulphur-yellow. Fruits 5–9 mm, compressed, hairless or slightly hairy; style c. 0.5 mm. *Arctic Russia*. H1.

16. D. oligosperma J.D. Hooker. Illustration: Botanisk Tidsskrift, **21**: 296 (1897–98).

Mat-forming perennial. Leaves lanceolate or oblanceolate, 3–11 × 1–2 mm, white with branched, stellate hairs; margin hairy. Flowering-stems hairless, except at base, 1–10 cm. Racemes with 3–15 flowers, loose. Sepals 2–2.5 mm; petals 3–4.5 mm, yellow, obovate, truncate or notched.

Fruits 3–7 × 2–4 mm, ovate, compressed, shortly hairy; style less than 1 mm. *W North America (Sierra Nevada to British Columbia)*. H2–3.

17. D. incerta Payson (*D. oligosperma* var. *pilosa* (Regal) Schulz).

Closely resembles *D. oligosperma*. Loosely tufted perennial. Leaves oblanceolate, 7–15 × c. 3.5 mm; with stellate hairs; margin hairy towards base. Flowering-stems with stellate, branched hairs. Inflorescence loose, with 3–14 flowers. Petals yellow, c. 5 mm. Fruits broadly lanceolate, 7–9 × 2–3.5 mm, hairless or with a few simple, branched hairs, compressed, beaked; style 0.5–1 mm. *W North America*. H1–2.

18. D. polytricha Ledebour (*D. reuteri* Boissier & Huet). Illustration: Grey-Wilson, A manual of alpine and rock garden plants, pl. 10 (1989).

Cushion-forming perennial. Leaves 4–5 mm, narrowly spathulate to oblong, blunt, white-hairy; marginal hairs to 2 mm. Flowering-stems numerous, leafless, 1–4 cm, long-haired, occasionally hairless. Inflorescence corymb-like, with 4–10 flowers. Sepals 3–5 mm, membranous, oblong, not pouched, tip rounded; petals obovate, yellow, 3.5–5 mm; stamens c. 2.5 mm. Fruits 2.5–4.5 × c. 2 mm, ovoid, inflated; style c. 0.5 mm. *Turkey & Armenia*. H2–3.

19. D. cappadocica Bossier & Balansa. Illustration: Annalen des K.K. Naturhistorischen Hofmuseums Wien **20**: t. 15 (1905).

Round tuft-forming perennial. Stock woody, leafy near to base. Leaves 3–6 mm, linear to obovate, soft, overlapping, with dense stellate hairs. Flowering-stems leafless, to 1–5 cm with long hairs. Raceme with 5–8 flowers. Flowers small; sepals c. 2 mm, ovate, persistent; petals c. 4 mm, yellow, obovate, tip truncate; stamens c. 3 mm. Fruits short, ovoid, 2–4 × 1.5–2.5 mm; style 1–2 mm, with dense stellate hairs. *C Turkey*. H3–4.

20. D. rosularis Boissier.

Closely resembles *D. cappadocica*. Tuft-forming perennial. Leaves 1.5–2 × 0.3–0.4 cm, soft, elliptic-oblong, blunt, with dense, white, stellate hairs. Flowering-stems 8–10 cm, leafless, ascending or erect, with tufted hairs. Raceme loose, with 8–20 flowers. Sepals c. 2.5 mm, ovate, round-tipped; petals 4–5 mm, yellow, obovate, shallowly notched at tip. Fruits elliptic-ovoid, inflated, with stellate hairs, c. 4 × 2.5 mm; style c. 0.3 mm. *Turkey*. H3–4.

21. D. mollissima Steven. Illustration: Botanisk Tidsskrift, **21**: 297 (1897–98); Flora of the USSR, **8**: 397 (1939); The Rock Garden **21**: Fig. 12 (1988).

Compactly tufted perennial. Leaves 3–6 × 1–2 mm, oblong, blunt, base tapered, densely grey-hairy, hairs stellate, branched, midrib prominent below. Flowering-stems leafless, 1–8 cm, hairless or hairy below. Racemes compact, becoming loose in fruit, with 6–8 flowers. Sepals c. 2 mm, densely hairy, ovate; petals 4–5 mm oblong-ovate, notched at tip, yellow; stamens 2.5–3 mm. Fruits 6–10 × 1.5–2.5 mm, elliptic-oblong, hairless, compressed; style less than 0.5 mm. *Former USSR (Caucasus)*. H2–3.

22. D. longisiliqua Schmalhausen. Illustration: Berichte der Deutschen Botanischen Gesellschaft **10**: t. 16 (1892–1893).

Tightly tufted perennial. Leaves obovate, base tapered, 3–6 × 1.5–2 mm, densely grey-hairy, hairs branched, stellate. Dead leaves remaining attached for some time. Flowering-stems 3–10 cm, leafless, short-branched, hairy or hairless. Raceme loose, with 3–14 flowers; sepals c. 2.5 mm, ovate, hairy; petals 4.5–7 mm, obovate-triangular, bright yellow, notched at tip. Fruits 5–9 × 1.5–2.5 mm, hairless, ovate-elliptic, strongly compressed; style 0.5–1 mm. *Former USSR (Caucasus)*. H2–3.

23. D. oreadum Maire. Illustration: Engler, Das Pflanzenreich **89**: 144 (1927).

Perennial, forming tufts. Stock branching, densely leafy. Leaves 5–14 × 2–3.5 mm, oblong to obovate, tapering to a broad stalk, blunt, hairless to coarsely hairy; margin entire, with hairs to 1 mm. Flowering-stems naked, 2–7 cm. Racemes somewhat corymb-like at first, with 6–10 flowers, lengthening in fruit. Sepals 3–5 mm, oblong-ovate, hairy, margin transparent; petals 6–9 mm, oblong-elliptic, notched at tip, white. Fruits compressed, contorted, lanceolate, 5–12 × 3–4.5 mm; style 3–5 mm. *Morocco (Grand Atlas)*. H3–4.

24. D. ussuriensis Pohle.

Loosely tufted perennial. Leaves 7–14 × 2–4 mm, tapered to the stalk, spathulate to oblong, blunt, densely grey-hairy, hairs stellate. Flowering-stems 2–14 cm. Stem-leaves 1 or 2 or absent, ovate, hairy, few-toothed. Raceme loose, with 7–17 flowers. Sepals c. 2 mm, with short, simple hairs, ovate; petals 4–6 mm, obovate, rounded, shallowly notched, creamy white. Fruits

$5-10 \times 2-2.5$ mm, oblong-ovate, slightly pointed, compressed, hairless; style c. 1 mm. *Japan & China*. H3–4.

25. D. stellata Jacquin. Illustration: Grey-Wilson & Blamey, The alpine flowers of Britain and Europe, 67 (1979).

Tufted perennial. Leaves oblanceolate, 4–8 mm, with dense white stellate hairs; margin hairy. Flowering-stems to 10 cm, erect, stellate-hairy near base. Stem-leaves 1–3 or absent, often toothed. Racemes with 3–12 flowers. Sepals 2–2.5 mm, creamy white with a translucent margin; petals 4.5–8 mm, creamy white, broadly ovate; stamens exceeded by petals, filaments somewhat dilated at base. Fruits elliptic-lanceolate, 6–10 mm, hairless; style 1–2 mm. *E Alps*. H3–4.

26. D. norvegica Gunnerus (*D. rupestris* R. Brown). Illustration: Rhodora **26**: pl. 301, 302 (1934); Keble Martin, The concise British flora, pl. 8 (1974).

Tufted perennial. Leaves $5-20 \times 1.5-4$ mm, spathulate to oblong, blunt to somewhat acute, base tapered. Hairs branched, simple, occasionally stellate; margin toothed or entire, hairy. Flowering-stems 5–25 cm. Stem-leaves 2 or 3 or absent, scalloped, ovate, hairy. Inflorescence dense, with 5–15 flowers, loose in fruit. Flowers small; sepals elliptic, hairy; petals obovate, 3–4 mm, shallowly notched, white. Fruits $5-8 \times 1.5-2.5$ mm, ovate-elliptic or oblong, stellate hairy or hairless; style less than 0.5 mm. *Scandinavia, Siberia, N America*. H2–3.

27. D. tomentosa Clairville. Illustration: Das Pflanzenreich 89: 245 (1927).

Tufted biennial or perennial. Densely covered throughout in stellate hairs. Leaves variable, c. 1 cm, entire, narrowly obovate-elliptic, tip rounded. Flowering-stems erect, to c. 10 cm. Stem-leaves 1–3 or absent, broadly ovate, stalkless. Inflorescence with 3–12 flowers; sepals c. 2 mm, broadly ovate, short-haired; petals white or pale cream, 4–4.5 mm. Fruits $6-10 \times 2.5-4$ mm, broadly elliptic, hairy, somewhat inflated when ripe; style less than 0.5 mm. *C & S Europe*. H3–4.

28. D. arabisans Michaux. Illustration: Rhodora **26**: pl. 314, 315 (1934).

Tufted perennial. Stock much-branched. Leaves 1–7 cm, spathulate to oblong, thin, sparsely stellate hairy. Flowering-stems hairless or hairy, 15–45 cm, occasionally branched. Stem-leaves scattered, toothed. Racemes loose. Sepals 2–2.5 mm; petals white, 3.5–4.5 mm. Fruits $9-15 \times 1.5-2$ mm, lanceolate-elliptic, twisted; style c. 1 mm. *E North America*. H2–3.

29. D. incana Linnaeus. Illustration: Engler, Das Pflanzenreich, **89**: 105 (1927); Grey-Wilson & Blamey, The alpine flowers of Britain and Europe, 67 (1979); Keble Martin, The concise British flora, pl. 8 (1974).

Biennial or occasionally perennial. Leaves to 2.5 cm, lanceolate, blunt. Hairs dense, white, simple, branched or rarely stellate. Flowering stems hairy, 10–35 cm. Stem-leaves numerous, oblong-ovate, somewhat pointed. Inflorescence with 10–40 flowers, dense. Sepals 2–2.5 mm, oblong elliptic, blunt; petals white, 4–5 mm, triangular-obovate, shallowly notched or truncate; filaments sometimes expanded at base. Fruits lanceolate to oblong, hairless, often twisted, $6-12 \times 2-2.5$ mm. *W & C Europe*. H3–4.

30. D. crassifolia R.C. Graham. Illustration: Botanisk Tidsskrift, **21**: 299 (1897–98); Morley & Everard, Wild flowers of the world, pl. 1 (1970).

Tufted perennial. Leaves oblong to spathulate, dense, thick, congested, blunt, with sparse simple hairs; margin remotely fine-toothed. Flowering-stems ascending, hairless. Stem-leaves oblong, hairy. Inflorescence with 3–9 flowers. Sepals c. 2 mm, oblong, blunt, hairless; petals narrow, pale yellow, c. 2.5 mm, shallowly notched; stamens c. 2 mm. Fruits hairless, $4-7 \times 1.5-2.5$ mm, compressed; style very short or lacking. *Scandinavian Arctic*. H4–5.

33. PETROCALLIS R. Brown
S.L. Jury
Tufted perennial herbs resembling many species of *Draba* or *Saxifraga* in habit. Leaves in compact rosettes, palmately 3–5 lobed with simple unbranched hairs. Flowers few on leafless stalks. Fruit a silicula, obovate to elliptic, hairless with 1 seed, rarely 2, in each cell.

A genus of 2 species, 1 in S European mountains, 1 in N Iraq, but only the former is cultivated. It is best suited to rock crevices and screes with abundant limestone.

1. P. pyrenaica (Linnaeus) R. Brown (*Draba pyrenaica* Linnaeus). Figure 15(16), p. 131. Illustration: Jelitto & Schacht, Hardy herbaceous perennials, edn 3, **2**: 477 (1990).

Leaves stiff and greyish, 4–6 mm. Scapes short, 2–3 cm hairy. All hairs unbranched. Flowers fragrant. Petals 4–5 mm, pale lilac, pink or white. *Pyrenees, Alps, Carpathians*. H3.

The white flowered 'Albiflora' is more easily cultivated than the typical wild species.

34. LESQUERELLA S. Watson
S.L. Jury
Low annual to perennial herbs, densely covered with stellate hairs. Basal rosette present; leaves simple to deeply lobed. Inflorescence a raceme, lengthening in fruit. Flowers small, numerous. Fruit a silicula, often inflated, almost spherical to obovoid or ellipsoid; valves without veins, hairless or with stellate hairs.

A genus of 40 species from N America and a few from temperate S America. They need a sunny location in very gritty soil or in a scree or moraine. Propagation is easy by seed, but the plants are only a modest addition to the garden.

Literature: Rollins, R.C. & Shaw E.A. *The genus Lesqueralla (Cruciferae) in North America* (1973).

1a. Stems decumbent, 1–15 cm; fruits with stellate hairs   **1. kingii**
  b. Stems decumbent or erect, 3–40 cm; fruits hairless   **2. fendleri**

1. L. kingii (S. Watson) S. Watson (*Vesicaria kingii* S. Watson). Illustration: Hitchcock et al., Vascular plants of the Pacific Northwest **2**: 527 (1964); Rickett, Wild flowers of the United States **4**(1): pl. 73 (1970).

Plants decumbent, stems 5–15 cm, silvery white with stellate hairs. Basal leaves stalked, ovate to circular, entire, obtuse 2–5 cm. Stem-leaves oblanceolate 0.5–1.5 cm. Fruits almost spherical, 3–5 mm across, hairy with stellate hairs; styles 3–5 mm; seeds c. 2 mm. *W North America*. H2. Spring.

2. L. fendleri (Gray) S. Watson (*Vesicaria fendleri* Gray; *L. foliacea* Greene; *L. stenophylla* (Gray) Rydberg). Illustration: Rickett, Wild flowers of the United States **3**(1): pl. 48 (1969).

Similar to *L. kingii* but often stems more erect, 3–40 cm, fruits 4.5–8 mm, hairless; seeds c. 1.5 mm. *SW USA to N Mexico*. H2. Spring.

35. PHYSARIA (Nuttall) A. Gray
S.L. Jury
Perennial, tufted, herbs, silvery with stellate hairs. Stems simple, arising laterally from a somewhat elongated stock. Basal leaves numerous, stalked, oblanceolate to obovate or almost round, entire, toothed or divided into segments.

Stem-leaves few, entire to toothed. Flowers in racemes; petals yellow, rarely purplish, spathulate. Fruit a silicula, often broader than long, more or less heart-shaped and notched at apex, often inflated. Seeds 2–6 in each segment, brown, wingless.

A genus of about 14 species from western North America, closely related to *Lesquerella*. Propagation is usually by seed, though also by cuttings and division. Suitable for rock gardens with thoroughly well-drained soil in full sun.

Literature: Rollins, R.C. The Cruciferous genus *Physaria*, Rhodora **41**: 392–415 (1939).

1a.	Style 1–2 mm	**1. oregona**
b.	Style more than 4 mm	2
2a.	Apical notch of fruit shallow (less than 1 mm)	3
b.	Apical notch of fruit deep (more than 2 mm)	4
3a.	Fruit slightly inflated, heart-shaped in outline, less than 1 cm wide	**2. geyeri**
b.	Fruit strongly inflated, circular in outline, more than 1 cm wide	**3. alpestris**
4a.	Septum of fruit obovate; ovules 4 in each segment	**4. didymocarpa**
b.	Septum of fruit oblong to linear-oblong; ovules 2–6 in each segment	5
5a.	Basal leaves rounded at apex; apical notch of fruit deep (equalling septum in width and depth)	**5. vitulifera**
b.	Basal leaves acute at apex; apical notch of fruit shallow	6
6a.	Basal leaves entire, or with a single broad tooth on each side, linear-oblanceolate, less than 4 cm; septum narrowly obovate	**6. acutifolia**
b.	Basal leaves divided, broadly oblanceolate, less than 4 cm; septum linear, constricted	**7. floribunda**

1. P. oregona Watson. Illustration: Rhodora **41**: pl. 556 (1939); Hitchcock et al., Vascular plants of the Pacific Northwest **2**: 534 (1964).

Stems 1-many, 10–35 cm, erect or somewhat decumbent. Basal leaves obovate, 4–6 × 0.8–1.5 cm, slender stalked, usually divided or with broad teeth along stalk. Sepals 5–7 × 1–2 mm. Petals lemon-yellow, spathulate, 9–12 × 2–3 mm. Fruits flattened laterally, 1–1.2 × 1.8–2.5 cm, 2-lobed, with stellate hairs, somewhat inflated, apical notch broad and open, basal notch absent. Septum broadly lanceolate, acute at apex, 6–8 × 2–3 mm. Style 1–2 mm. Seeds 2–4 in each segment, brown, 2–3 mm, wingless. *W North America (Idaho & Oregon)*. H1–2. Spring.

2. P. geyeri (Hooker) A. Gray. Illustration: Rhodora **41**: pl. 556 (1939); Hitchcock et al., Vascular plants of the Pacific Northwest **2**: 534 (1964).

Like *P. oregona* but petals yellow, 8–12 × 3–4 mm; fruits much smaller 5–7 × 6–9 mm; style 5–7 mm; seeds 1 or 2 in each segment. *W North America (Montana & Idaho)*. H1–2. Spring.

Variants with purple petals are referred to as var. **purpurea** Rollins.

3. P. alpestris Suksdorf. Illustration: Rhodora **41**: pl. 556 (1939); Hitchcock et al., Vascular plants of the Pacific Northwest **2**: 531 (1964); Rickett, Wild flowers of the United States **6**(1): pl. 83 (1973); The Rock Garden **21**: fig. 20 (1989).

Stems several, 5–15 cm, erect or somewhat decumbent. Basal leaves obovate, rarely somewhat acute, 3–5 × 1–2 cm, entire, tapering abruptly to a slender stalk. Sepals 8–10 × 1.5–2 mm. Petals yellow, spathulate, 1.2–1.4 × 0.2–0.3 cm. Fruits highly inflated, 0.7–1 × 1.4–2 cm, shallow, open notch at apex and slight notch below, lobes somewhat kidney-shaped. Septum lanceolate, acute at apex, 7–10 × 1.5–2.5 mm. Style 5–7 mm. Seeds 1–5 in each segment, brown, 2–3 mm. *W North America (Washington)*. H1–2. Spring.

4. P. didymocarpa (Hooker) A. Gray. Illustration: Engler, Die natürlichen Pflanzenfamilien, **17b**: 430 (1936); Rhodora **41**: pl. 556 (1939); Hitchcock et al., Vascular plants of the Pacific Northwest **2**: 534 (1964), Rickett, Wild flowers of the United States **6**(1): pl. 82 (1973).

Stems numerous, *c.* 10 cm, decumbent. Basal leaves obovate, toothed rarely entire, 1.5–4 × 0.8–1.6 cm, long stalked. Sepals 6–8 × 1.5–2 mm, often keeled. Petals yellow, spathulate 1–1.2 × 0.3–0.4 cm. Fruits inflated, 0.8–1.2 × 1.2–1.6 cm, apical and basal notches deep, narrow. Septum obovate to broadly oblong, not constricted, obtuse at apex, 3–4 × 2–3 mm. Style 7–9 mm. Seeds 2–4 in each segment, brown, *c.* 2.5 mm. *W North America, (Alberta, Montana, Wyoming)*. H1–2 Spring.

5. P. vitulifera Rydberg. Illustration: Rhodora **41**: pl. 556 (1939).

Stems numerous, 10–20 cm, usually decumbent. Basal leaves fiddle-shaped or obovate, obtuse margins deeply cut to rarely almost entire, 3–6 × 1–2 cm. Sepals oblong, 6–8 × 1.5–2 mm. Petals yellow, spathulate 9–10 × 3–4 mm. Fruits inflated, 5–6 × 6–8 mm, somewhat angular, obtuse below, apical notch, broad, open and deep. Septum oblong, often constricted 2–3 × 0.5–1 mm. Style 5–7 mm. Seeds 1 or 2 in each segment, brown, *c.* 2.5 mm, wingless. *W North America (Colorado)*.

6. P. acutifolia Rydberg. Illustration: Rhodora **41**: pl. 556 (1939); Rickett, Wild flowers of the United States **6**(1): pl. 82 (1973).

Stems several to numerous, 5–10 cm, decumbent. Basal leaves oblanceolate or broader, acute, entire or with 1 or 2 broad teeth, 2–3.5 × 0.5–1 cm. Sepals linear, 5–7 × 1–2 mm. Petals yellow, spathulate, often truncate at apex 8–10 × 2–3 mm. Fruits inflated, 6–8 × 8–10 mm, apical notch broad and deep, somewhat heart-shaped at base, segments almost circular. Septum obovate, obtuse at apex, 3 × 1.5 mm. Style 5–7 mm. Seeds 1 or 2 in each segment, brown, 2 mm across. *W North America (Colorado)*. H1–2. Spring.

7. P. floribunda Rydberg. Illustration: Rhodora **41**: pl. 55 (1939); Rickett, Wild flowers of the United States **4**(1): pl. 74 (1970).

Stems numerous, 10–20 cm, decumbent or erect. Basal leaves broadly oblanceolate, pinnatifid or merely toothed, rarely almost entire, 4–8 × 1–2 cm, terminal lobe acute or obtuse, but not rounded, leaf-stalk winged. Sepals linear-oblong, 5–7 × 1–2 mm. Petals yellow, spathulate 9–11 × 2–3 mm. Fruit somewhat inflated, 4–6 × 6–10 mm, deeply and broadly notched above obtuse or slightly heart-shaped at base. Septum linear-oblong, constricted, obtuse at apex, 2.5–4 × 0.5–1 mm. Style 5–8 mm. Seeds 1 or 2 in each each segment, brown, 2 mm, circular, marginless. *W North America (Colorado, New Mexico)*. H1–2. Spring.

36. COCHLEARIA Linnaeus
S.L. Jury
Annual, biennial or perennial herbs. Leaves variable, usually simple, stalked stem-leaves often arrow-shaped with auricles. Flowers small in racemes; sepals erect to spreading, petals short-clawed, white, tinged yellow or purple. Fruit a

silicula, inflated with convex valves. Seeds numerous.

A genus of 25 species from N temperate regions.

1. C. officinalis Linnaeus. Figure 15(3), p. 131. Illustrations: Ross-Craig, Drawings of British plants **3**: pl. 32 (1949); Hitchcock, Vascular plants of the Pacific Northwest **2**: 481 (1964); Polunin, Wild flowers of Britain and Northern Europe, pl. 424 (1988); Rich, Crucifers of Great Britain and Ireland, 267 (1991).

Annual, biennial or perennial to 50 cm. Leaves kidney- to heart-shaped, basal long-stalked, stem-leaves stalkless, toothed. Petals 3–7 mm, white. Raceme lengthening and loose in fruit. Fruit ovoid to spherical, 4–7 mm, shorter than flower-stalk. *NW Europe, Alps*. H3. Spring.

A concentrated source of vitamin C, and the source of a popular breakfast drink in the 17th century.

37. KERNERA Medicus
J.R. Akeroyd

Perennial, sometimes biennial; stems 10–30 cm, usually branched. Basal leaves in a rosette, stalked, lanceolate to obovate or spathulate, entire to deeply toothed, usually hairy; stem-leaves lanceolate or ovate, often clasping and arrow-shaped at base. Racemes many-flowered; bracts few. Sepals erect to spreading. Petals 2–4 mm, with a short claw, white. Filaments curved. Fruit a swollen siliqua, 2–4.5 mm, ovoid to almost spherical, stalked or not; valves convex, rigid, smooth with a more or less prominent mid-vein in lower part.

A genus of a single variable species in the mountains of C & S Europe. Propagation is by seed.

1. K. saxatilis (Linnaeus) Reichenbach (*K. auriculata* (Lamarck) Reichenbach; *K. decipiens* (Willkomm) Nyman). Figure 15(4), p. 131. Illustration: Strid, Wild flowers of Mount Olympus, pl. 82 (1980).

C & S Europe. H1. Summer.

38. COLUTEOCARPUS Boissier
S.L. Jury

Tufted perennial herbs, 4–20 cm. Rosette leaves simple, oblong-lanceolate, hairless; margins with 3–5 sharp teeth; stem-leaves stalkless, entire. Flowers numerous in flat-topped cymes; petals golden-yellow, 6–11 × 2.5–4 mm, clawed. Inflorescence lengthening after flowering; fruit-stalks to 9 mm. Fruit an inflated papery silicula, 1.4–3.3 × 1.6–2.1 cm; style persistent; 1–4 mm. Seeds 2–10.

A genus of a single species, allied to *Alyssum*.

1. C. vesicaria (Linnaeus) Holmboe (*Alyssum vesicaria* Linnaeus; *C. reticulatus* Boissier). Figure 15(6, 10), p. 131.

E Mediterranean, Turkey. H4. Spring.

39. PRITZELAGO O. Kuntze
S.L. Jury

Small, branched perennial herbs to 15 cm; hairless or small, with both branched and unbranched hairs. Basal leaves in a rosette, pinnatisect with 3–9, divisions 1–5 × 1–3 mm. Sepals small to 2 mm; petals 3–5 mm, white, clawed. Fruit a silicula, elliptic to lanceolate, with a narrow septum; seeds 2–4.

A genus of 1 species from the mountains of C and S Europe, southwards to N Spain, C Italy and former Yugoslavia (Macedonia). Formerly known as *Hutchinsia*.

1. P. alpina (Linnaeus) O. Kuntze (*Hutchinsia alpina* (Linnaeus) R. Brown; *Noccaea alpina* (Linnaeus) Reichenbach).

1a. Flowering-stems flexuous, leafy
　　subsp. **auerswaldii**
　b. Flowering-stems more or less
　　leafless, straight　　　　　　2
2a. Petals 3 mm wide, style 1 mm
　　subsp. **alpina**
　b. Petals 1–2 mm wide, stigma
　　stalkless　　subsp. **brevicaulis**

Subsp. **alpina**. Illustration: Fiori, Flora Italiana Illustrata, 3rd edn, fig. 1472 (1933); Pignatti, Flora d'Italia **1**: 442 (1982); Demiri, Flora Ekskursioniste e Shqiperise, 660 (1981). Flowering-stems 5–12 cm, sparsely hairy. Basal leaves with 5–9 ovate-lanceolate divisions; stem-leaves absent. Petals 4–5 × 2–3 mm. Fruit 4–6 × 1.5–2 mm, ovate, acute. *Alps & Jura, Pyrenees, N C Italy*. H1. Spring.

Subsp. **auerswaldii** (Willkomm) Greuter & Burdet. Illustration: Heydenreich, Gartenflora **78**: 140 (1929). Like subsp. *alpina*, but flowering-stems to 15 cm, flexuous and leafy. Stem-leaves with lanceolate to linear divisions. *N Spain*. H1. Spring.

Subsp. **brevicaulis** (Hoppe) Greuter & Burdet. Illustration: Hegi, Illustrierte Flora von Mitteleuropa **4**: 359, t. 134 (1919); Marret, Icones florae alpinae plantarum, ser. 3, 116, (1924); Jávorka & Csapody, A magyar flóra képekten, 201 (1930). Flowering-stems 2–5 cm, hairless, more or less leafless. Basal leaves with 3–7 ovate-lanceolate divisions. Petals 3–4 × 1–2 mm.

Fruit 3.5–4 × 1–2 mm, obtuse. *Alps & Balkan Peninsula*. H1. Spring.

40. THLASPI Linnaeus
J.R. Akeroyd

Annuals, biennials or perennials, more or less hairless. Leaves simple; basal leaves usually in a rosette, more or less spathulate; stem-leaves stalkless, usually clasping. Flowers in bractless racemes, lengthening in fruit. Sepals not pouch-like. Petals white, lilac or purple, with a short claw. Filaments without an appendage. Fruit a compressed silicula, usually notched at apex; valves winged or keeled. Seeds 2–8, rarely only 1 in each cell.

A genus of about 60 species, mostly from Eurasia. The classification is critical and further studies are needed to elucidate species' limits. Propagation is by seed.

Literature: Ingrouille, M.J. & Smirnoff, N., *New Phytologist* **102**: 219–233 (1986).

1a. Annual, without non-flowering
　　leaf-rosettes; petals usually not
　　more than 3 mm　**8. violascens**
　b. Perennial, usually with non-
　　flowering leaf-rosettes; petals
　　usually at least 4 mm　　　　2
2a. Fruits strongly keeled but not
　　winged, notch absent or narrow　3
　b. Fruits winged, notch broad
　　(sometimes shallow)　　　　　4
3a. Plant with long stolons; basal
　　leaves elliptic to almost circular
　　　　　　　6. cepaeifolium
　b. Plant usually without stolons; basal
　　leaves oblong　**7. bellidifolium**
4a. Roots tuberous; petals deep violet
　　　　　　　　3. bulbosum
　b. Roots not tuberous; petals white,
　　lilac or purplish　　　　　　5
5a. Anthers violet or reddish　　　6
　b. Anthers yellow　　　　　　　7
6a. Stems more than 5 cm; petals less
　　than 3 mm, white
　　　　　　　1. caerulescens
　b. Stems not more than 5 cm; petals
　　3–5 mm, purplish　**2. stylosum**
7a. Fruit obcordate, broadly winged,
　　with a distinct notch at apex
　　　　　　　　4. montanum
　b. Fruit narrowly triangular-
　　obcordate, narrowly winged, with a
　　shallow notch at apex
　　　　　　　　5. alpinum

1. T. caerulescens J. & C. Presl (*T. alpestre* misapplied; *T. sylvestre* Jordan). Illustration: Ross-Craig, Drawings of British plants **3**: pl. 68 (1949); Polunin, Collins photoguide to wild flowers, p. 89 (1988).

Biennial or perennial 10–50 cm, hairless, greyish. Leaves entire or toothed, the basal elliptic to obovate; stem-leaves narrowly ovate. Petals 1.5–2.5 mm, white or lilac. Anthers reddish or violet. Fruit 5–8 mm, obovate, winged, distinctly notched. *Europe*. H1. Late spring–early summer.

This species has been divided into a number of taxa by different authors, although probably best treated as a single variable species.

2. T. stylosum (Tenore) Mutel. Illustration: Pignatti, Flora d'Italia **1**: 450 (1982).

Tufted perennial, 2–5 cm. Leaves mostly basal, 5–10 mm, ovate. Petals 3–5 mm, purplish. Anthers violet. Fruit broadly winged, with small notch. *Italy (S & C Apennines)*. H1. Summer.

3. T. bulbosum Spruner. Figure 15(15), p. 131. Illustration: Kew Bulletin **1**: 102 (1954).

Tufted perennial 5–10 cm, hairless, greyish; woody below, with tuberous roots. Stems many, ascending. Basal leaves ovate; stem-leaves ovate-oblong. Petals 6–8 mm, deep violet. Anthers violet. Fruit 8–10 mm, obcordate, broadly winged, broadly and shallowly notched. *C Greece & Aegean Islands*. H1. Early summer.

4. T. montanum Linnaeus. Illustration: Hess et al., Flora der Schweiz **1**: 143 (1970).

Perennial, forming a mat, hairless, greyish; stems 10–30 cm, erect. Leaves entire or finely and sinuously toothed. Basal leaves long-stalked, ovate to circular; stem-leaves ovate-oblong. Petals 5–7 mm, white. Anthers pale yellow. Fruit 6–8 mm, obcordate, broadly winged, with wide notch. *C & S Europe*. H1. Early summer.

5. T. alpinum Crantz. Illustration: Hess et al., Flora der Schweiz **1**: 143 (1970).

Tufted perennial, often with stolons; stem 5–15 cm. Basal leaves oblong to ovate or almost circular; stem-leaves oblong to ovate, obtuse. Petals 6–7 mm, white. Anthers yellow. Fruit narrowly triangular-obcordate, narrowly winged, with a very shallow notch. *Alps*. H1. Early summer.

The closely related **T. alpestre** Jacquin (*T. kerneri* Huter), from the SE Alps and former Yugoslavia, with acute stem-leaves, the raceme more contracted in fruit, and smaller flowers and fruits, has also been cultivated; the two species are easily confused.

6. T. cepaeifolium (Wulfen) Koch (*T. rotundifolium* misapplied; *T. rotundifolium* var. *limosellifolium* Burnet). Illustration: Botanical Magazine, 5749 (1869) – as Iberidella rotundifolia; Hay & Synge, RHS dictionary of garden plants in colour, 27 (1969); Hess et al., Flora der Schweiz **1**: 145 (1970).

Perennial, with long stolons, greyish; stems 5–15 cm. Basal leaves elliptic to almost circular; stem-leaves few, distant, more or less entire. Flowers fragrant. Petals lilac. Fruit oblong-obovate, keeled but not winged, with a narrow notch. *S Alps*. H1. Late spring–early summer.

The widespread plant in Europe, both in the wild and cultivation, is subsp. **rotundifolium** (Linnaeus) Greuter & Burdet.

Subsp. **cepaeifolium** from the SE part of the Alps, with basal leaves not arranged in a distinct rosette, stem-leaves crowded and only slightly clasping, broader and sinuously toothed, is also cultivated.

7. T. bellidifolium Grisebach. Illustration: Polunin, Flowers of Greece and the Balkans, 255 (1980).

Plant densely tufted, sometimes with short stolons; stems 1–3 cm. Basal leaves oblong, toothed. Petals deep purple. Anthers yellow. Fruit oblong to reversed heart-shaped, truncate, keeled but not winged, with a narrow notch. *N Balkan Peninsula*. H1. Early summer.

8. T. violascens Boissier.

Annual, with simple stems 10–25 cm, often purplish below. Basal leaves oblong, entire or scalloped; stem-leaves ovate-oblong, entire or toothed. Petals *c*. 3 mm. Fruit 7–10 mm, triangular, with obtuse wings above the middle narrow notch. *Turkey*. H1. Early summer.

A similar species, also from Turkey, **T. densiflorum** Boissier & Kotschy, a biennial with ovate-oblong basal leaves, slightly larger flowers and fruits, the fruits with acute wings, may also be in cultivation. It is perhaps not specifically distinct from *T. violascens*.

41. PACHYPHRAGMA Busch
J.R. Akeroyd

Rhizomatous perennial herb, smelling of garlic when bruised. Basal leaves with stem to 25 cm, the base dilated, persistent; blade 1.5–1.8 mm, ovate to almost circular, hairless except for short hairs on veins beneath. Flowering-stems 15–40 cm, erect or almost so; stem-leaves few. Flowers slightly scented, in domed terminal clusters which lengthen in fruit. Petals 8–9 mm, white with fine green veins. Fruit a flattened silicula 8–16 mm, broadly winged, deeply notched at apex, with thick septum (fruit rarely produced in cultivation).

A genus of a single species from deciduous forests in the Caucasus region. It is closely related to *Thlaspi*, from which it differs mainly by its rhizomatous habit. Propagation is by division of the rhizome. The plant forms extensive clumps and can be grown in shade.

Literature: Davie, J.H. & Akeroyd, J.R. *Botanical Journal of the Linnaean Society* **87**: 77–82 (1983).

1. P. macrophyllum (Hoffmann) Busch (*Thlaspi macrophyllum* Hoffmann; *T. latifolium* Bieberstein). Figure 15(22), p. 131. Illustration: Flora URSS **8**: t. 32, 4 (1939); Grossheim, Flora Kavkasa **4**: t. 6, 5 (1950); Botanical Journal of the Linnean Society **87**: 77–82 (1983).

Caucasus. H3. Spring–early summer.

42. AETHIONEMA R. Brown
J.R. Akeroyd

Usually woody-based perennials, rarely annuals or biennials. Leaves small, stalkless or nearly so, simple, entire, often leathery or somewhat fleshy. Sepals erect, the 2 outer pouch-like at base. Petals entire, white or pink, rarely yellow. Filaments of the 4 inner stamens winged, the wing sometimes toothed above. Silicula flattened and winged, either with a single indehiscent, 1-seeded cell, or 2-celled – each with 1–4 seeds.

A genus of about 70 species from the Mediterranean area and W Asia. Propagation is by cuttings or seed. Plants require light, well-drained, calcium-rich soil and a sunny aspect, and are usually grown in the rock garden or alpine glasshouse.

Habit. Woody-based: **1–3,5,7**.
Leaves. Linear or linear-oblong: **8,10–12**.
Fruit. Wings less than 0.5 mm, or absent: **3,4,7**.
Wings. Irregularly toothed: **1,2,10,11**.

1a. Wings of fruit irregularly toothed　　2
 b. Wings or fruit entire or scalloped, sometimes absent　　5
2a. Leaves ovate to oblong-lanceolate; petals pink, lilac, white or yellow　　3
 b. Leaves linear-oblong; petals pink　　4

3a. Lower leaves obtuse, clasping; fruit-stalk erect **1. cordatum**
 b. Lower leaves acute, not clasping; recurved **2. stylosum**
4a. Fruit 8–12 mm; petals at least 5 mm **10. grandiflorum**
 b. Fruit 6–7 mm; petals less than 5 mm **11. diastrophis**
5a. Flowering-stems densely covered with linear leaves; petal blade almost circular **8. schistosum**
 b. Flowering-stems with loose, covered with leaves; leaves rarely linear; petal-blade not circular 6
6a. Fruit wings absent or less than 0.5 mm wide; petals usually white 7
 b. Fruit wings distinct, usually at least 1 mm wide; petals usually pink or lilac 9
7a. Petals pink **4. oppositifolium**
 b. Petals white 8
8a. Leaves alternate; style 2–3 mm **3. trinervium**
 b. Leaves opposite; style 0.5–0.7 mm **7. iberideum**
9a. Leaves oblong to ovate or almost circular; often both indehiscent 10
 b. Leaves linear to oblong; mostly dehiscent 11
10a. Fruit in loose racemes, dehiscent or indehiscent **5. saxatile**
 b. Fruit in dense racemes, mostly indehiscent **6. thomasinianum**
11a. Fruit 8–12 mm, with a 3–4 mm apical notch; petals usually at least 6 mm **10. grandiflorum**
 b. Fruit not more than 7 mm; notch not more than 2 mm; petals 3–6 mm 12
12a. Leaves evenly distributed on both flowering- and non-flowering stems; petals 1.5 mm wide **9. armenum**
 b. Leaves mostly on non-flowering stems; petals 2.5–3.5 mm wide **12. cordifolium**

1. A. cordatum (Desfontaines) Boissier (*A. cardiophyllum* Boissier & Heldreich; *A. koenigii* Woronow). Illustration: Flora SSSR **8**: t. 30 (1939).

Woody-based hairless perennial; stems 5–20 cm, erect, simple or branched. Lower leaves opposite, ovate, clasping, obtuse; upper leaves alternate, acute. Sepals 4 mm. Petals 5–9 mm, pink, white or yellow. Fruit *c.* 7 mm, obovate, truncate, irregularly toothed; style 2–4 mm. *SW Asia to C & S Greece*. H1. Summer.

2. A. stylosum de Candolle (*A. lacerum* Boissier & Balansa).

Similar to *A. cordatum*, but leaves ovate to oblong-lanceolate, acute, not clasping; petals white or lilac; ovate, stalk recurved. *S Turkey & Syria (Taurus Mts)*. H1. Summer.

3. A. trinervium (de Candolle) Boissier. Illustration: Flora SSSR **8**: t. 31 (1939).

Woody-based hairless perennial; stems to 20 cm, branched. Leaves oblong or oblong-lanceolate, stalkless or clasping, acute. Petals 6–10 mm, white. Fruit 5–9 mm, oblong or heart-shaped, entire; wings very narrow or absent, style 2–3 mm. *Armenia to Afghanistan*. H1. Summer.

4. A. oppositifolium (Persoon) Hedge (*Eunomia oppositifolia* (Persoon) de Candolle; *A. bourgaei* Boissier; *A. chlorifolium* (Sibthorp & Smith) de Candolle; *A. rotundifolia* misapplied).

Densely tufted perennial; stems 2–5 cm, unbranched. Leaves obovate or almost circular, stalkless, papillose, rather fleshy. Petals 6–8 mm, pink. Racemes compact. Fruit 6–8 mm, ovate; wings absent or very narrow; style 0.5 mm. *SW Asia*. H1. Summer.

5. A. saxatile (Linneaus) R. Brown. Figure 15(1), p. 131. Illustration: Polunin, Flowers of Europe, pl. 36 (1969); Huxley & Taylor, Flowers of Greece and the Aegean, pl. 77 (1977).

Rather woody perennial; stems 5–35 mm, ascending to erect. Leaves alternate, or the lower opposite, 5–15 mm, oblong, elliptic, ovate or almost circular; upper leaves narrower and more acute. Petals 2–8 mm, white or lilac. Fruit 3–8 (rarely to 10) mm, obovate to almost circular, shallowly notched, with 1 or 2 cells, style to 2 mm. *Mediterranean area, extending to the Alps*. H1. Spring.

A very variable species. Several variants are sometimes regarded as species. Four subspecies are reported from European gardens:

Subsp. **saxatile** (*A. gracile* de Candolle). Illustration: Willdenowia **13**: 8 (1983). Stems to 35 cm. Leaves at least 3.5 times as long as wide. Petals 2–4 mm. Most fruits 2-celled; style 0.2–0.4 mm. *Mainly W Mediterranean*.

Subsp. **ovalifolium** (de Candolle) Nyman (*A. ovalifolium* (de Candolle) Boissier). Illustration: Willdenowia **13**: 12 (1983). Stems to 35 cm. Leaves 1.5–3 times as long as wide. Petals 3–5 mm. Most fruit 2-celled; style 0.3–0.7 mm. *Spain & SW France*.

Subsp. **graecum** (Boissier & Spruner) Hayek (*A. graecum* Boissier & Spruner).

Illustration: Huxley & Taylor, Flowers of Greece and the Aegean, pl. 77 (1977) – as *A. saxatile*; Willdenowia **13**: 17 (1983). Stems not more than 12 cm, rather woody below. Leaves 2–4 times as long as wide. Petals 3–7 mm. Most fruit 2-celled; style 0.3–0.9 mm. *Balkan Peninsula & Aegean area*.

Subsp. **creticum** (Boissier & Heldreich) I.A. Anderson et al. (*A. creticum* Boissier & Heldreich; *A. ovalifolium* misapplied). Stems not more than 10 cm, very woody below. Leaves 1.5–2.5 times as long as wide, broadly elliptic to almost circular. Petals 2.5–5 mm; style 0.3–0.9 mm. About half the fruit 1-celled. *S Greece & the Aegean area*.

6. A. thomasinianum Gray. Illustration: Willdenowia **13**: 6 (1983).

Similar to *A. saxatile* subsp. *saxatile*, but stems not more than 15 cm; raceme very dense; most fruits 1-celled, very broadly winged; siliculae 2-celled, 10–12 mm. *SW Alps (Val d'Aosta)*. H1. Spring–early summer.

7. A. iberideum (Boissier) Boissier (*Eunomia iberidea* Boissier; *A. theodorum* misapplied).

A shrubby perennial; stems 8–20 cm, erect or ascending, branched, densely leggy below. Leaves oblong to elliptic or ovate, opposite, acute, greyish. Racemes compact. Petals 5–7 mm, white. Fruit ovate to heart-shaped, notched, without wings; style 0.5–0.7 mm. *SW Turkey & E Greece (Evvoia)*. H1. Early summer.

8. A. schistosum Boissier & Kotschy.

Perennial to 10 cm, with many ascending stems densely covered with linear leaves. Racemes compact. Petals 5–7 mm, pink, almost circular. Fruit 7–9 mm, ovate, with broad entire wings and a deep notch. *S Turkey (Taurus Mts)*. H1. Early summer.

9. A. armenum Boissier (*A. recurvum* Haussknecht & Bornmuller).

Compact perennial, usually hairless; stems many, 10–15 cm, ascending, more or less unbranched. Leaves linear to oblong, acute, grey. Petals 3–6 × 1–2 mm, pink. Fruit 5–7 mm, ovate to obovate, cordate at base, with rather narrow, entire or scalloped wings; style *c.* 0.5 mm. *SW Asia*. H1. Summer.

'Warley Rose' Illustration: Hay & Synge, RHS dictionary of garden plants in colour, 1 (1969). Flowers deep purplish pink in large racemes. Probably of hybrid origin; propagated by cuttings.

10. A. grandiflorum Boissier & Hohenacker (*A. pulchellum* Boissier & Huet; *A. kotschyi* misapplied).

Perennial; stems many, 15–25 cm, ascending to erect. Leaves linear-oblong, greyish. Petals 5–10 × 3–4 mm, deep pink. Fruit 8–12 mm, ovate to almost circular, with broad, entire or irregularly and shallowly toothed wings and a deep apical notch; style very short. *SW Asia*. H1. Early summer.

A variable species, especially in the shape and size of the petals.

11. A. diastrophis Bunge.

Similar to *A. grandiflorum*, but petals *c.* 4 × 2 mm, pale pink; silicula 6–7 mm, with deeply toothed wings and a shallow apical notch; style *c.* 0.5 mm. *Armenia & Caucasus*. H1. Early summer.

12. A. coridifolium de Candolle (*A. jucundum* misapplied). Illustration: Botanical Magazine, 5952 (1872) – as Iberis jucunda.

Similar to *A. grandiflorum*, but stems usually not more than 20 cm, less erect and less branched; leaves mostly on non-flowering-stems, narrower; petals 4–5.5 × 2.5–3 mm; Fruit 6–8 mm with narrower, entire or scalloped wings. *S Turkey & Lebanon*. H1. Spring–summer.

43. IBERIS Linnaeus

T. Upson

Annuals, perennials or almost shrubs. Leaves alternate, linear or obovate, entire or pinnately cut, hairless or with unbranched hairs. Inflorescence a corymb or raceme; petals white, pink or purple, the two outer much larger than the two inner. Fruit a flat silicula, rounded at base, entire or notched at apex. Seeds solitary in each cell, often winged.

A genus of 40 species native to S Europe and W Asia. Usually found in calcareous habitats, frequently in rock crevices, but will grow in most soil types, although best in light, sandy soils, which must be well-drained and in full sun. The small shrubs particularly hate the wet, so are best in rock gardens or even dry stone walling, although suitable for the front of a border. The annual species are propagated by seed sown *in situ* during the spring for summer flowering or late summer to flower the following spring. Biennials are also raised from seed. The perennials can be raised from seed or by division, the small shrubs by seed or from semi-ripe cuttings in summer.

Species under the names of *I. arborea* and *I. correvonicina* are sometimes offered, but these names are invalid. It is not clear to which species they apply.

1a. Fruits much wider than long; flowering in winter **6. semperflorens**
  b. Fruits longer than wide or almost circular; flowering spring or summer **2**
2a. Perennials or small shrubs with woody or herbaceous stems; usually with non-flowering rosettes **3**
  b. Annual; most rosettes flowering **6**
3a. Fruits in racemes **4**
  b. Fruits in corymbs **5**
4a. Leaves 2.5–5 mm wide, obtuse; flowering-stems lateral **5. sempervirens**
  b. Leaves 1–2 mm wide, acute; flowering-stems terminal **4. saxatilis**
5a. Plants procumbent **3. pruitii**
  b. Plants not procumbent **2. gibraltarica**
6a. Fruits in racemes **1. amara**
  b. Fruits in corymbs **7**
7a. Plants covered with stiff bristly hairs **8**
  b. Plants not covered with stiff bristly hairs **9**
8a. Inflorescence surrounded by upper whorl of leaves **7. odorata**
  b. Inflorescence without whorl of leaves **8. crenata**
9a. Leaves entire **10. umbellata**
  b. Leaves pinnatifid **9. pinnata**

1. I. amara Linnaeus. Illustration: Keble Martin, Concise British flora in colour, pl. 10 (1969); Polunin, Flowers of Europe, pl. 36 (1969); Hay & Beckett, Reader's Digest encyclopaedia of plants and flowers, 353 (1971); Brickell, RHS gardeners' encyclopaedia of plants and flowers, 262 (1989).

Erect branching annual 10–40 cm. Leaves spathulate, distinctly toothed towards apex, although sometimes entire. More or less hairy on margins and below. Inflorescence a corymb lengthening in fruit. Petals white or purplish. Fruit almost circular 3–5 mm. *W Europe including Britain*. H2. Summer.

2. I. gibraltarica Linnaeus. Illustration: RHS dictionary of gardening 2: 1041 (1951).

A hairless evergreen tufted perennial with a woody rootstock from which arise leaf rosettes and many flowering-stems, 15–30 cm. Leaves fleshy, entire or with 1–4 teeth on either side near the apex. Lower leaves broadly spathulate *c.* 2.5 × 1.2 cm the upper smaller. Inflorescence a corymb, 4–5 cm across, flat in flower but contracted in fruit. Petals white to lilac, the outer 1.5–1.8 cm. Fruit ovate, 7–8 mm broadly winged from base. *Gibraltar & Morocco*. H5. Late spring–summer.

3. I. pruitii Tineo (*I. candolleana* Jordan; *I. jordanii* Boissier; *I. jucunda* Schott & Kotschy).

A tufted, procumbent to ascending perennial, 3–15 cm. Leaves fleshy, entire or with a few teeth near obtuse apex; lower obovate-spathulate, upper narrower, hairless. Flowers in compact corymbs, petals white occasionally lilac. Fruit rectangular-elliptic, broadly winged, 6–8 mm. *Mediterranean area*. H4. Late spring–early summer.

Extremely variable which in the past has led to a number of forms being recognised as separate species.

4. I. saxatilis Linnaeus. Illustration: Brickell, RHS gardeners' encyclopedia of plants & flowers, 314 (1989).

A small evergreen, spreading shrub, 8–15 cm. Leaves semi-cylindric on non-flowering shoots, flat on the flowering stems, up to 2 cm; entire, linear and rather acute; margins hairy at first but soon hairless. Inflorescence a corymb borne on terminal branches, lengthening in fruit. Petals white, often with a purple tinge when fading. Fruit obovate, broadly winged from base, 5–8 × 4.5–6 mm. *S Europe*. H4. Late spring–summer.

Var. **corifolia** Sims is totally hairless with white petals.

5. I. sempervirens Linnaeus (*I. commutata* Schott & Kotschy). Figure 15(18), p. 131. Illustration: Hay & Beckett, Reader's Digest encyclopaedia of plants and flowers, 354 (1971); Polunin, Flowers of Greece and the Balkans, pl. 11 (1980); Brickell, RHS gardeners' encyclopaedia of plants & flowers, 286 (1989) – as 'Little Gem'.

A spreading, evergreen, small shrub forming a hairless, leafy mat, 10–25 cm. Leaves 2.5–5 mm wide, oblong spathulate, obtuse, entire and flat. Inflorescence raceme-like, lengthening in fruit. Petals white. Fruit 6–7 mm, circular to ovate, broadly winged at base. *Mediterranean area, often naturalised elsewhere*. H3. Spring & summer.

Variants include, 'Little Gem' and 'Snowflake'. Var. **garrexiana** Allioni with narrower leaves and the dwarf var. **nana**

Allioni (*I. pigmaea* invalid) are cultivated.

6. I. semperflorens Linnaeus.

A small spreading evergreen shrub to 80 cm. Leaves thick, entire the lower broadly spathulate, 3–7 × 0.7–1.8 cm the upper oblong-spathulate and much smaller. Inflorescence a corymb, lengthening in fruit. Petals white. Fruit ovate to rhombic, narrowly winged 0.5–0.8 × 1–1.4 cm. *W Italy & Sicily.* H5. Winter.

7. I. odorata Linnaeus (*I. acutiloba* Bertoloni).

A short stiff-haired annual. Numerous leafy stems arise from the base to 15–30 cm. Leaves linear spathulate, pinnatifid with 1–2 pairs of segments near the apex. Inflorescence a flat, dense corymb, shortly stalked and surrounded by upper leaves. Flowers fragrant. Petals white. Fruit ovate, hairless. *Turkey, Greece & Crete.* H4. Summer.

8. I. crenata Linnaeus.

Erect, stiff-haired annual, 15–30 cm. Leaves linear shaped, toothed. Inflorescence a dense convex corymb. Petals white. Fruits almost square, 5–6 mm. *C & S Spain.* H4. Summer.

9. I. pinnata Linnaeus.

Erect annual, 10–30 cm. Leaves obovate-oblong, pinnately cut into linear, blunt lobes. Inflorescence a short, dense, convex corymb. Flowers fragrant. Petals white to lilac. Fruit almost square, hairless, 5–6 mm. *C & S Europe.* H3. Summer.

10. I. umbellata Linnaeus. Illustration: Polunin, Flowers of Greece and the Balkans, pl. 12 (1980); Brickell, RHS gardeners' encyclopedia of plants & flowers, 266 (1989).

Erect hairless annual, 20–70 cm. Leaves linear-lanceolate, acuminate, lower toothed, upper entire. Inflorescence a dense terminal umbel. Petals pink to purplish, highly variable. Fruit ovate, to 1 cm, broadly winged at base. *S Europe.* H4. Spring–summer.

Many varieties are commercially available.

44. BISCUTELLA Linnaeus

T. Upson

Annual to perennial herbs or small shrubs. Leaves entire to pinnatifid. Petals yellow and usually clawed. Fruit a flat, indehiscent silicula. Seeds unwinged.

A genus of 40 species native to southern and central Europe, but few are cultivated. They need full sun and any well-drained

soil, but grow best on limestone. Propagation is by seed. The relative uniformity of the floral and fruiting characters makes the classification of this group difficult. Vegetative features are used in most cases. The generic name refers to the fruit, coming from the Greek *bis* (double) and *scutella* (little shield).

1a. Whole plant covered with dense, white-felted hairs   **1. frutescens**
   b. Plant hairless or with few hairs   **2. laevigata**

1. B. frutescens Cosson. Illustration: Barneby, European alpine flowers in colour, pl. 29 (1967).

Rhizomatous perennial, woody at base. Whole plant covered with dense, white-felted hairs. Stems simple or branched to 50 cm. Basal leaves *c.* 20 × 6 cm, ovate, sinuous toothed or lyrate. Inflorescence much branched. Petals yellow, 3–4 mm. Fruits 3.5–4 × 6–7 mm, with oblique valves. Margin with small swelling. *SW Spain & Balearic Islands.* H4. Late spring–summer.

2. B. laevigata Linnaeus. Figure 15(13) p. 131. Illustration: Everett, The New York Botanical Garden illustrated encyclopedia of horticulture 2: 426 (1981).

Perennial 10–70 cm, simple or branched. Basal leaves 1–15 × 0.3–2 cm, usually in a rosette; linear to ovate-lanceolate, ovate or rarely spathulate, entire, sinous or finely toothed, hairless or hairy. Stem-leaves 2–10, resembling basal leaves or small, linear and entire. Flowers in loose or dense racemes. Petals 3–4 mm, clawed. Highly variable. *C & S Europe, extending N to Belgium.* H2. Late spring–summer.

A plant for the mixed flower border or naturalistic plantings. Seed fruits useful for dried flower arrangements.

45. MEGACARPAEA de Candolle

S.L. Jury

Herbaceous perennials with unbranched hairs. Leaves pinnatisect. Sepals not pouched. Petals white, cream or violet. Fruit a silicula, with a narrow septum, deeply 2-lobed; lobes flat, broadly winged; seed solitary.

A genus of 7 species from Europe to C Asia, Himalaya and China. Best grown in a light, sandy soil. Propagation by seed.

1a. Leaf segments almost entire; petals yellow   **2. bifida**
   b. Leaf segments with many irregularly and acutely toothed

lobes; petals white, cream, yellow or pinkish violet   2
2a. Plants 1–2 m; basal leaves 15–30 cm; fruits 3–3.5 × 3.5–5 cm
     1. polyandra
   b. Plants 20–40 cm; basal leaves up to 15 cm; fruits 1.5–2 × 2.2–3.2 cm
     3. megalocarpa

1. M. polyandra Bentham. Illustration: Botanical Magazine, 8734 (1917); Engler, Die natürlichen Pflanzenfamilien, edn 2, **17b**: 434 (1936); Polunin & Stainton, Flowers of the Himalaya, pl. 15 (1984).

Robust herb 1–2 m, main stem 6–12 cm thick at base. Hairless or somewhat hairy. Basal leaves 15–60 cm, pinnatisect with 7–9 pairs of oblong-lanceolate lobes, 10–20 × 1–5 cm, toothed. Inflorescences large, dense, many-flowered, to 20 cm in fruit. Flowers *c.* 1 cm across. Petals white, cream or yellow. Stamens 8–16. Fruits almost circular, 3–3.5 × 3.5–5 cm, often unequally bilobed, deeply notched at apex. *India (Kashmir) to C Nepal.* H1. Spring–summer.

The young leaves are sometimes eaten.

2. M. bifida Bentham.

Like *M. polyandra*, but smaller and leaf segments more or less entire. Flowers *c.* 5 mm across. Petals yellow. Stamens 7–11. Fruits 4–5 × 6–7 cm, very deeply notched at apex. *India (Kashmir).* H1. Summer.

3. M. megalocarpa (de Candolle) B. Fedtschenko (*Biscutella megalocarpa* de Candolle; *B. laciniata* de Candolle).

Herb 20–40 cm; roots thick; stem solitary, branched above; hairs stiff or crisped. Basal leaves to 15 cm, pinnatisect with many irregular lobes, toothed, densely white-hairy. Flowers irregularly unisexual, those in the upper part of each branch male, in the lower female. Female flowers without sepals and petals. Intermediate flowers sometimes present with smaller, white petals and variously developed stamens and ovaries. Fruit 1.8–2 × 2.2–3.2 cm, bilobed, triangularly notched above and below. *SE Russia, W Kazakhstan, Mongolia.* H1. Spring–summer.

The thick roots are often collected and eaten, being rich in starch.

46. LEPIDIUM Linnaeus

J.R. Akeroyd

Annuals to perennials, hairless or with small, simple hairs. Sepals not pouch-like. Petals usually small and inconspicuous, white, sometimes absent. Fruit a silicula, strongly compressed, keeled, often winged;

style short or stigma stalkless. Seeds usually 1 in each cell.

A cosmopolitan genus of about 140 species, many of them weedy. Propagation of the two cultivated species is by seed.

1. L. sativum Linnaeus. Illustration: Harrison, Masefield & Wallis, Oxford book of food plants, pl. 153 (1969); Collenette, An illustrated guide to the flowers of Saudi Arabia, 201 (1985); Rich, Crucifers of Great Britain and Ireland, 230 (1991).

Annual, 20–60 cm, hairless or somewhat hairy, greyish green. Basal leaves lyrate to pinnate, the lobes toothed, withered by flowering time; stem-leaves 1–2 pinnate, the uppermost lobed or linear and entire; rarely all leaves more or less entire. Petals 2–3 mm, white or lilac. Fruit 5–6 mm, elliptic to almost circular, strongly notched. *Middle East*. H1. Summer.

Cultivated as a salad (cress), usually in containers indoors; eaten at the seedling stage.

47. NOTOTHLASPI J.D. Hooker
J.R. Akeroyd
Compact perennials or biennials with long tap-roots. Leaves spathulate, fleshy. Flowers crowded. Sepals erect. Fruit a reversed heart-shaped to obovate siliqua, strongly compressed, winged. Seeds *c.* 1 mm, many in each cell.

A genus of 2 species endemic to New Zealand. Propagation is by seed. Plants require deep, well-drained soil and grow best under glass in an alpine house.

1. N. rosulatum J.D. Hooker. Figure 15(20), p. 131. Illustration: Botanical Magazine, 130 (1790); Laing & Blackwell, Plants of New Zealand, title page (1927); Moore & Irvin, Oxford book of New Zealand plants, pl. 40 (1978).

Basal leaves up to *c.* 4 cm, stalked, closely overlapping, scalloped or toothed, white-hairy when young. Flowering stems 8–25 cm, erect, unbranched, stout, with few leaves. Flowers crowded in a conical or pyramidal raceme, fragrant. Sepals 4–7 mm. Petals 8–12 mm, obovate-spathulate, white. Fruit 1–2.5 cm; style *c.* 1 mm. *New Zealand (South Island)*. H4. Spring.

Usually behaves as a biennial in cultivation.

48. MORICANDIA de Candolle
J.R. Akeroyd
Hairless annuals to perennials with simple, greyish fleshy leaves. Inner sepals pouch-like at base. Petals violet-purple. Fruit

a linear, compressed siliqua; style short. Seeds many.

A genus of 7 species in the Mediterranean and SW Asia. Propagation is by seed. Plants grow best in warm, sunny sites on light, calcium-rich soil. Although somewhat perennial in the wild, in cultivation they are best treated as annuals or biennials.

Literature: de Bolós, A., *Anales del Instituto Botánico A.J. Cavanilles* **6**: 451–61 (1946).

1a. Racemes usually with not more than 20 flowers; siliqua 3–8 cm; seeds in 1 row. **1. arvensis**
  b. Racemes with at least 20 flowers; siliqua 8–13 cm; seeds in 2 rows **2. moricandioides**

1. M. arvensis (Linnaeus) de Candolle. Figure 16(17), p. 134. Illustration: Botanical Magazine, 3007 (1830); Polunin, Flowers of Europe, pl. 37 (1969).

Annual to short-lived perennial; stems 25–50 cm, branched. Lower leaves obovate, sinuously toothed, obtuse; upper leaves cordate, clasping at base, entire, more or less acute. Raceme loose, with 10–20 (rarely to 25) flowers. Petals *c.* 2 cm. Fruit 3–8 × 0.2–0.3 cm, 4-angled; seeds *c.* 1 mm, in 2 rows. *Mediterranean area, mostly in the west*. H1. Summer.

2. M. moricandioides (Boissier) Heywood (*M. ramburii* Webb). Illustration: Botanical Magazine, 4947 (1856); Polunin & Smythies, Flowers of south-west Europe, 68 (1973).

Similar to *M. arvensis* but longer living; racemes with 20–40 flowers; fruit 8–13 cm, more or less cylindric; seeds 2–2.5 mm, in 1 row. *S Spain*. H1. Summer.

49. BRASSICA Linnaeus
J.R. Akeroyd
Usually perennials, hairless or with simple hairs. Leaves entire to pinnatifid. Sepals erect or spreading. Petals yellow or white, with a claw. Fruit a siliqua with a distinct beak; valves convex, prominently 1-veined. Seeds in 1 row (rarely 2 rows) in each cell, spherical.

A genus of about 35 species, mostly of the Mediterranean area.

Literature: Mitchell, N.D. & Richards, A.J. *Journal of Ecology* **67**: 1087–1096 (1979).

1a. Fruits usually less than 3 cm, beaks not more than 3 mm, seedless **7. nigra**

  b. Fruits usually at least 3 cm, beaks more than 3 cm, often containing seeds 2
2a. Stem-leaves clasping 3
  b. Stem-leaves not clasping (sometimes stalkless or almost so) 6
3a. Open flowers overtopping unopened buds **4. rapa**
  b. Open flowers not overtopping unopened buds 4
4a. Annual to biennial; petals 1–1.5 cm, yellow. **3. napus**
  b. Biennial to perennial; petals 1.5–3 cm, pale yellow or white 5
5a. Basal leaves greyish, green or variously coloured, ovate; petals pale yellow **1. oleracea**
  b. Basal leaves greyish, oblong-ovate; petals white to pale yellow **2. cretica**
6a. Lower leaves lyrate-pinnatisect **5. juncea**
  b. Lower leaves obovate, not or scarcely lobed **6. integrifolia**

1. B. oleracea Linnaeus (*B. sylvestris* (Linnaeus) Miller). Figure 16(3), p. 134. Illustration: Masefield et al., Oxford book of food plants, 157, 159 (1969).

Biennial or perennial; stem 0.8–2.5 m, usually hairless, woody and covered by leaf-scars below, branched above. Basal leaves 1.5–3 cm, ovate, unlobed to lyrate-pinnatifid with a large terminal lobe, stalked; stem-leaves ovate-lanceolate or oblong, usually stalkless. Leaves not overtopping young flowers. Flowers in a raceme with 20–40 flowers that lengthens considerably. Petals 1.5–2 cm, pale yellow. Fruits 5–10 × 0.2–0.5 cm, cylindric; beak 5–10 mm, as wide as valves, conical, with 1 or 2 (rarely absent) seeds. *Mediterranean area & W Europe*. H1. Summer.

A very variable species, long cultivated as a vegetable; more recently plants with leaves variegated, red, purple and yellow, have been cultivated as ornamentals. Many familiar cultivars have been treated as varieties or subspecies by different authors.

Var. **capitata** Linnaeus. Perennial; stem short. Leaves in dense compact giant bud. Cultivated either as a vegetable (cabbage), both green and purple ('red') variants, or plants with variegated red, purple, green, bluish, yellow or white leaves are grown as ornamentals. Similar variants (kales) are biennial, and less woody, with compact heads.

Var. **sabauda** Linnaeus (Savoy

cabbage), differs by the usually dark green, blistered or puckered leaves.

Var. **gemmifera** Linnaeus ('Brussels sprouts'). Stems long, covered with compact buds.

Var. **gongylodes** Linnaeus ('Kohl rabi'). Base of stem and bases of stems of lower leaves grossly swollen.

Chinese cabbage or Pe Tsai, a familiar and attractive vegetable or salad plant, is referrable to **B. pekinensis** Ruprecht.

2. B. cretica Lamarck subsp. **botrytis** (Linnaeus) O. Schwartz (*B. oleracea* var. *botrytis* Linnaeus). Illustration: Masefield et al., Oxford book of food plants, 159 (1969).

Similar to *B. oleracea* but smaller to 1.5 m; leaves greyish; basal leaves oblong-ovate, lyrate; leaves overlapping young flowers; petals white to pale yellow. Fruits 3–90 × 0.3–0.5 cm. *Aegean area.* H1. Summer.

Cultivated plants ('cauliflowers') are grown for the dense, compact young racemes. Similar variants of *B. oleracea* are cultivated ('broccoli').

3. B. napus Linnaeus (*B. chinensis* Linnaeus). Illustration: Masefield et al., Oxford book of food plants, 25 (1969).

More or less hairless, greyish annual or biennial to 1.5 m. Basal leaves lyrate, stalked; stem-leaves lanceolate, stalkless, clasping. Flowers in loose corymb-like racemes, the open flowers not overlapping unopened buds. Petals 1–1.5 cm, yellow. Fruits 3.5–10 × 0.3–0.5 cm, spreading to more or less erect; beak to 2.5 cm, slender. *Only known from cultivation.* H1. Summer.

The plant usually grown in gardens is var. **napobrassica** (Linnaeus) Reichenbach ('swede'), which is biennial, the base of the stem and root are grossly enlarged, spherical and fleshy. Illustration: Masefield et al., Oxford book of food plants, 173 (1969).

B. napus is probably derived by hybridisation of *B. oleracea* and *B. rapa*.

4. B. rapa Linnaeus (*B. campestris* Linnaeus). Illustration: Masefield et al., Oxford book of food plants, 173 (1969).

Similar to *B. napus* but basal leaves green, with bristly hairs; open flowers overtopping unopened buds; petals 0.5–1.2 cm, yellow. *Europe & W Asia, widespread as a weed.* H1. Summer.

The cultivated plant (subsp. **rapa**) is distinguished by its biennial habit and grossly enlarged, spherical and fleshy root ('turnip'). Subsp. **sylvestris** (Linnaeus) Janchen occurs as a weed of cultivation.

5. B. juncea (Linnaeus) Czernajew.

Almost hairless annual, 30–150 cm. Basal leaves to 20 cm, stalked, lyrate-pinnatisect with 1–3 pairs of lateral lobes and a larger, bristly terminal lobe; upper leaves narrowly oblong, almost entire, shortly stalked, hairless. Open flowers borne at about the same level as unopened buds. Petals 0.6–1 cm, pale yellow. Fruits 2–6 × 0.2–0.5 cm, narrowed into a seedless beak, 5–10 mm. *C & E Asia.* H1. Summer.

6. B. integrifolia (West) O.E. Schulz (*B. cernua* Forbes & Hemsley).

Similar to *B. juncea* but basal leaves broadly obovate, scarcely or not lobed, sinuous, variably toothed to entire; beak 4–6 mm. *E Asia.* H1. Summer.

7. B. nigra (Linnaeus) Koch.

Slender annual to 2 m; stems with many long ascending branches. Leaves to 15 cm, all stalked, green, bristly; basal leaves lyrate-pinnatisect, with 1–3 pairs of lateral lobes and a much larger terminal lobe; stem-leaves sinuously lobed, upper leaves lanceolate to linear, entire, hairless. Flowers in a corymb-like raceme. Petals 0.7–1 cm, yellow. Fruits 1–3 × 0.1–0.4 cm, often narrow, seedless beak to 3 mm. *W & S Europe, W Asia & N Africa, but widespread as a weed.* H1. Summer.

Usually grown as a field crop for its edible seeds, used as a condiment (black mustard).

50. ERUCA Miller
J.R. Akeroyd

Annuals, usually hairy. Leaves pinnatifid. Sepals erect, the inner somewhat pouched. Petals with a long claw. Fruit a siliqua, the valves 1-veined; beak long, flat. Seeds in 2 rows in each cell.

A genus of 5 species in the Mediterranean area. Propagation of the single cultivated species is by seed. Plants require warmth but the soil should not be allowed to dry out.

1. E. vesicaria (Linnaeus) Cavanilles. Figure 16(16), p. 134. Illustration: Masefield et al., Oxford book of food plants, pl. 152 (1969); Wild flowers of the world, pl. 30A (1970); Huxley & Taylor, Wild flowers of Greece, 86 (1977); Polunin, Flowers of Greece and the Balkans, 250 (1980).

Plant 20–60 cm, rather bristly, with a pungent smell on bruising. Leaves lyrate-pinnatifid, with a prominent terminal lobe, the upper more or less stalkless. Sepals oblong, reddish. Petals 1.5–2 cm, obovate, whitish or yellowish, veins brown or violet. Siliqua 1.2–3 × 0.3–0.6 cm, hairless or with a few stiff hairs; fruiting flower-stalks thick. *Mediterranean area.* H1. Summer.

Cultivated as a salad (rocket, roquette), and for the oil extracted from the seeds; the cultivated plant is subsp. **sativa** (Miller) Thellung (*E. sativa* Miller).

51. VELLA Linnaeus
S.L. Jury

Small much-branched shrubs, sometimes spiny; leaves entire; hairs unbranched. Flowers shortly stalked in loose, spike-like racemes. Sepals erect. Petals long-clawed, yellow, sometimes violet-veined. Filaments of inner stamens joined in pairs. Fruit transversely articulated, lower segment dehiscent by two lobes, ellipsoid; lobes convex, 3-veined; the upper segment sterile, strongly compressed in the form of a leafy, narrowly oblong, acute, 5-veined, hairless beak.

A genus of 4 species from Spain, Morocco and Algeria. Rare in cultivation. They need a well-drained, sunny, sheltered position, as they are not reliably hardy. Propagation is by seed or cuttings.

Literature: Gomez-Campo, C., Taxonomic and evolutionary relationships in the genus *Vella* L. (Cruciferae), *Botanical Journal of the Linnean Society* **82**: 165–179 (1981); Gardner, M.F., Plants in peril, 8, *The Kew Magazine* **3**: 140–142 (1986).

1a. Spiny bush with short few-flowered inflorescences — 2
  b. Non-spiny bush with elongated many-flowered inflorescences — 3
2a. Leaves entire, linear; fruit hairless — **3. spinosa**
  b. Leaves pinnatisect; fruit hairy — **4. mairei**
3a. Beak of fruit acute; leaves with appressed hairs — **2. anremerica**
  b. Beak of fruit roundish, leaves with erect hairs — **1. pseudocytisus**

1. V. pseudocytisus Linnaeus.

Spineless much-branched shrub, woody at base, to 1 m. Leaves entire, apex rounded, with cavities which tend to orientate themselves into a vertical plane. Fruit with a rounded beak.

1a. Leaves sparsely hairy; fruit-valves hairless — 2
  b. Leaves and fruit-valves densely hairy — subsp. **pseudocytisus**
2a. Leaves with hairs only at the edge; inflorescence-stalk hairless — subsp. **paui**

b. Leaves with hairs on both faces;
inflorescence-stalk hairy
subsp. **glabrescens**

Subsp. **pseudocytisus** (*V. pseudocytisus*
Linnaeus subsp. *iberica* Litardière & Maire;
V. monosperma Menéndez-Amor).
Illustration: The Kew Magazine **3**: 141
(1986). Inflorescence-stalk hairy. Leaves
hairy on both surfaces. Fruit-lobes densely
shortly bristled. *C & SE Spain.* H1. Spring.

Subsp. **paui** Gómez-Campo (*V.
pseudocytisus* Linnaeus var. *glabrescens*
Willkomm; *Pseudocytisus integrifolius*
Rehder var. *badalii* (Pau) Heywood; *V.
badalii* Pau). Inflorescence-stalks hairless.
Leaves with hairs only at the edge. Fruit-
lobes hairless. *E Spain.* H1. Spring.

Subsp. **glabrescens** (Cosson) Litardière
& Maire (*V. glabrescens* Cosson). Illustration:
Cosson, Illustrationes florae Atlanticae **1**:
t 48 (1884). Inflorescence-stalks hairy.
Leaves sparsely hairy on both surfaces.
Fruit-lobes hairless. *Morocco & Algeria.* H1.
Spring.

2. V. anremerica (Litardière & Maire)
Gómez-Campo (*V. pseudocytisus* Linnaeus
subsp. *anremerica* Litardière & Maire).

Spineless much-branched shrub, to 50
cm; with appressed hairs. Leaves entire,
apex acute. Lobes of fruit sparsely hairy;
beak acute, spade-shaped. *Morocco (High
Atlas Mts).* H1. Spring.

3. V. spinosa Boissier (*Pseudocytisus
spinosus* (Boissier) Rehder). Figure 16(10),
p. 134. Illustration: Das Pflanzenreich
105: 43 (1923).

Hemispherical spiny shrub to 1 m.
Leaves linear. Petals less than 2 cm. Lobes
of fruit hairless; beak acute. *S & E Spain.*
H1. Spring.

4. V. mairei Humbert. (*Pseudocytisus
mairei* (Humbert) Maire.)

Hemispherical spiny shrub to 1 m.
Leaves pinnatisect. Petals more than 2 cm.
Lobes of fruit hairy; beak acute. *Morocco
(High Atlas mts).* H1. Spring.

52. CRAMBE Linnaeus
J.R. Akeroyd
Robust perennials, with large rootstock,
hairless or with simple hairs. Stems erect,
branched. Sepals erect to spreading. Petals
with a short claw, white to cream.
Filaments of the inner stamens usually
with a tooth-like appendage. Fruit a 2-
segmented silicula; lower segment short,
sterile; upper segment ovoid to almost
spherical, indehiscent, with 1 seed; stigma
stalkless.

A genus of 25 species, mostly in W &
C Asia. Propagation is by seed or root
cuttings; plants require a rich, well-drained
soil.

1a. Plant hairless; upper segment of
silicula 7–14 mm   **1. maritima**
b. Plant hairy, at least below; upper
segment of silicula less than 7 mm
2
2a. Leaves pinnatifid or pinnatisect   3
b. Leaves entire or lobed   4
3a. Upper segment of silicula 4–7 mm;
lower segment stout   **2. tataria**
b. Upper segment of silicula 3–4 mm;
lower segment slender
4. orientalis
4a. Stems hairy below; leaf-base not
cordate, oblong   **3. koktebelica**
b. Stems hairless; leaf-base cordate,
ovate   **5. cordifolia**

1. C. maritima Linnaeus. Illustration:
Ross-Craig, Drawings of British plants **3**:
pl. 74 (1949); Polunin, Flowers of Europe,
pl. 37 (1969); Rich, Crucifers of Great
Britain and Ireland, 71 (1991).
Rootstock branched; plant hairless,
bluish grey; stems 25–70 cm. Leaves
rather fleshy, tough; lower leaves to 30
cm, long-stalked, ovate, irregularly
pinnatifid; stem-leaves narrower, sinuously
toothed to entire. Flowers in compact
corymbs. Petals 0.6–1 cm. Lower segment
of fruit 1–4 mm, stout; upper segment
7–14 mm, almost spherical. Flower-stalks
usually at least 1.2 cm. Seeds 4–5 mm.
*Atlantic coasts of Europe & coasts of Black
Sea.* H1. Early summer.
The blanched shoots are used as a
vegetable. Plants benefit from occasional
administrations of sodium chloride.

2. C. tataria Sebeók (*C. tatarica* Pallas).
Illustration: Polunin, Flowers of Greece and
the Balkans, pl. 12 (1969).
Rootstock spindle-shaped, apex
branched; plant greyish green, usually
densely and stiffly hairy; stems to 1 m.
Lower leaves pinnatifid to pinnatisect;
primary segments ovate to triangular.
Flowers in rather compact corymbs. Petals
3–6 mm. Lower segment of fruit *c.* 1 mm,
stout; upper segment 4–7 mm, often 4-
angled. Flower-stalks usually not more
than 1.2 cm, stout. *E Europe & W Asia.*
H1. Summer.

3. C. koktebelica (Junge) N. Busch.
Illustration: Flora SSSR **8**: 481 (1939).
Rootstock spindle-shaped. Stems 1–2 m,
bristly below. Lower leaves oblong, almost
entire to sinuous, lobes toothed, hairless or

somewhat hairy on lower surface. Flowers
in spreading corymbs. Petals 4–6.5 mm.
Lower segment of fruit 0.5–1 mm, slender;
upper segment 4–4.5 mm, almost
spherical, somewhat angled. Flower-stalks
to 1 cm, slender. *Caucasus & Crimea.* H1.
Summer.

4. C. orientalis Linnaeus (*C. juncea*
Bieberstein).
Similar to *C. koktebelica* but stems to
1.2 m; lower leaves pinnatisect, with a
prominent terminal lobe, irregularly
toothed, hairy at least on lower surface;
upper segment of fruit usually 3–4 mm.
SW Asia. H1. Summer.

5. C. cordifolia Steven (*C. kotschyana*
Boissier). Figure 15(12), p. 131.
Illustration: Hay & Synge, RHS dictionary
of flowering plants in colour, 133, (1969).
Stems 0.8–2 m, hairless. Leaves cordate,
fleshy, hairy; lower leaves 30–100 ×
20–40 cm, long-stalked, broadly ovate,
unequally lobed or toothed, apex rounded;
upper leaves ovate, toothed. Flowers in
corymb-like clusters in a much-branched,
spreading panicle. Petals 0.5–1 cm. Upper
segment of fruit *c.* 5 mm, spherical.
Caucasus to Himalaya. H1. Summer.

53. MORISIA Gay
T. Upson
A stemless perennial with a rosette of
deeply cut, oblong lanceolate, downy
leaves, Flowers large, petals 0.9–1.2 cm,
produced singly in axils of leaves. Flower-
stalks erect when flowering, 0.5–2.5 cm,
later arching and lengthening to 6 cm,
causing the fruit to lie on the ground or
be buried. Fruit in two segments, lower
almost spherical, 2-celled dehiscing by 2
lobes with 3–5 (rarely to 12) seeds; upper
ovoid, conical, indehiscent with 1 or 2
seeds and smaller than lower segment.
A genus of a single species growing in
sandy places, requiring a poor, sandy soil.
Propagation by seed or from root cuttings.
A plant for either the rock garden or alpine
house.

1. M. monanthos Viviani. (*M. hypoqaea*
Gay). Illustration: Hay & Beckett, Reader's
Digest encyclopaedia of plants and flowers,
449 (1971).
Corsica & Sardinia. H5. Spring.

54. RAPHANUS Linnaeus
J.R. Akeroyd
Bristly annuals, biennials or perennials.
Leaves mostly lyrate-pinnatifid. Sepals
erect, the 2 inner slightly pouched. Petals

with a long claw, prominently veined. Fruits with 2 segments, the lower short, seedless, the upper usually constricted between the seeds; beak narrow, seedless.

A genus of about 6 species in Eurasia and the Mediterranean area. Propagation of the single cultivated species is by seed. Plants grow best in a light, fertile soil that should not be allowed to become too dry.

Literature: Rich, T., Recognizing radishes. *BSBI News* **51**: 13–15 (1989).

1. R. sativus Linnaeus. Figure 16(20), p. 134. Illustration: Masefield et al., Oxford book of food plants, 171 (1969); The Garden **106**: 26 (1981).

Branched annual or biennial, 10–60 cm; root swollen, spherical to cylindric. Leaves stalked; basal leaves lyrate. Sepals 0.5–1 cm, narrowly elliptic. Petals 1.4–1.8 cm, white, yellow or pink. Fruits 2–6 cm, somewhat inflated, with little constriction between the 5–12 seeds. *Garden origin*. H1. Summer.

A variable species with many cultivars. The origin is unknown but it is probably derived from **R. raphanistrum** Linnaeus, a widespread weed of cultivated ground with related variants on seashores. The cultivated plant (radish) was grown in ancient Egypt. Two series of cultivars are grown today: plants grown in spring and summer for salad, notably the round-rooted 'Scarlet Globe' and the cylindric-rooted 'French Breakfast', and plants grown in summer and autumn for winter use cooked or for salad, for example, 'Black Spanish' and 'China Rose'.

55. RAFFENALDIA Godron
S.L. Jury

Stemless perennial herbs with all leaves in a basal rosette. Leaves pinnatisect with 1–4 pairs of lobes, the terminal largest, ovate to oblong, stalked. Flowers long-stalked, solitary amongst the leaves. Fruit-stalks becoming strongly reflexed. Sepals pouched at base. Fruit an indehiscent siliqua, squarish or compressed in cross-section, linear oblong without valves, dehiscing into 3–10 single-seeded segments, usually with thickenings between the segments.

A genus of two species from N Africa; best grown in gritty compost in the alpine house.

Literature: Maire, R., *Flore de l'Afrique du Nord* **12**: 315–321 (1965).

1a. Fruit not falling early, clearly square in cross-section, beak short, 4.5–7 mm **1. primuloides**
  b. Fruit falling early, strongly compressed, beak long, up to 1.2 cm **2. platycarpa**

1. R. primuloides Godron (*Cossonia africana* Durieu). Illustration: Maire, Flore de l'Afrique du Nord **12**: 317 (1965).

Petals 1–2.3 cm, lilac or yellow, rarely white. Fruit hidden in the leaf-rosette or more or less forced into the ground, not separating from their stalks, 2–4.5 × 0.3–0.8 cm, all alike, almost square in cross-section, linear-lanceolate, beak 4.5–7 mm, segments 4–10. *Algeria & Morocco*. H1. Spring.

2. R. platycarpa (Cosson) Stapf (*Cossonia platycarpa* Cosson). Illustration; Maire, Flore de l'Afrique du Nord **12**: 320 (1965).

Petals 1.5–2 cm, pale yellow. Fruit hidden in the leaf-rosette, separating from their stalks at maturity, 6–37 × 5–9 mm, heteromorphic, some linear-oblong, others oblong or ovate, all very compressed, beak to 1.2 cm, segments 1–7. *Morocco*. H1. Spring.

56. HELIOPHILA Linnaeus
S.L. Jury

Annual to perennial herbs or shrubs, to 3 m or more. Hairless or with simple hairs. Stems erect, decumbent or climbing. Leaves entire, lobed, pinnate or bipinnate. Stipules sometimes present, minute. Flowers usually in a raceme, sometimes in the leaf angles, or a few together on short side-branches, or between stem nodes. Sepals sometimes pouched at the base. Petals usually clawed, lanceolate to round, often with appendages at base, white, blue-purple or pink. Fruit a siliqua or a silicula, rounded to linear, sometimes contracted between the seeds, round or compressed in cross-section, dehiscent, hairy or not. Seeds generally in 1 row, few to many, compressed, sometimes winged.

A genus with over 70 species, endemic to South Africa and Namibia. The annual species may be sown in spring in the open border or raised under glass and planted out; *H. scandens* requires greenhouse cultivation.

Literature: Marais, W., Heliophila in Codd L.E. et al., *Flora of Southern Africa* **13**: 17–77 (1970).

1a. Perennials 2
  b. Annuals 3
2a. Herb somewhat woody at base or small shrub to 90 cm with stems erect or creeping **5. linearis**
  b. Straggling woody climber to 3 m **6. scandens**
3a. Leaves thread-like **2. crithmifolia**
  b. Leaves linear-lanceolate to oblong, sometimes pinnate but never thread-like 4
4a. Fruit not constricted between the seed **4. africana**
  b. Fruit constricted between the seeds 5
5a. Flowers white to purple; fruit 2.5–3.5 × 1.5–2.5 cm **1. amplexicaulis**
  b. Flowers blue with a yellow-green centre; fruit 3–9 × 0.1–0.2 mm **3. coronopifolia**

1. H. amplexicaulis Linnaeus filius. Illustration: Flora of Southern Africa **13**: 27, (1970); Roux & Schelpe, South African wild flower guide **1**: 85 (1981).

Annual herb; stems 15–45 cm, glaucous. Leaves entire, 1.5–6 × 0.1–1.6 cm, broadly linear to elliptic heart-shaped, apex blunt to acuminate, usually heart-shaped at base. Racemes terminal, few to many-flowered. Sepals 2–4.3 × 0.9–1.8 mm, inner slightly pouched at base, with broad membranous margins. Petals 3.5–8.5 × 2.5–5.5 mm, white, mauve or pink. Style 1.3–4 mm. Ovules 6–16. Fruits 2.5–3.5 × 0.1–0.3 cm, linear, strongly contracted between the seeds. *South Africa (Cape Province)*. H5. Summer.

2. H. crithmifolia Willdenow. Illustration: Flora of Southern Africa **13**: 35 (1970).

Annual herb, 8–15 cm, hairless to densely hairy in lower half. Minute stipules present. Leaves, 3–12 cm, pinnate with 1–6 pairs of more or less fleshy, thread-like lobes to 5 cm. Racemes loose, with few–many flowers. Sepals 3–4.5 × 1–1.5 mm, inner pouched, with a broad membranous margin. Petals 4–10 × 1.5–6 mm, violet to pink or white, sometimes changing colour. Style short and stout. Ovules 20–40. Fruit 1.5–6 × 0.2–0.5 cm, linear or linear-oblong, margins straight. *Namibia & South Africa*. H5. Summer.

3. H. coronopifolia Linnaeus. Illustration: Flora of Southern Africa **13**: 53 (1970); Mason, Western Cape and veld flowers, pl. 47 (1972); Eliovson, Wild flowers of Southern Africa, edn 6, pl. 176 (1980); Roux & Schelpe, South African wild flower guide **1**: 83 (1981).

Annual herb; stems 10–60 cm, hairy, rarely hairless. Leaves 6–15 cm, simple or pinnate with 3–13 thread-like lobes. Sepals 4.5–7.5 × 1–3 mm, shortly horn-tipped, inner pouched and with a broad membranous margin. Petals 6.5–13 ×

3–8 mm, pale to bright blue with a white or pale greenish yellow base. Style 1–5 mm, stout. Ovules 16–50. Fruit 3–9 × 0.1–0.2 cm, linear with deep or shallow constrictions between the seeds. *South Africa (Cape Province)*. H5. Summer.

4. H. africana (Linnaeus) Marais (*Cheiranthus africanus* Linnaeus; *H. integrifolia* Linnaeus; *Pachystylum glabrum* Ecklon & Zeyher). Illustration: Mason, Western Cape and veld flowers, pl. 47 (1972).

Annual herb; stems 60–70 cm, rarely to 135 cm, erect or decumbent, hairless or not. Leaves 3–13 × 0.3–1.2 cm, linear lanceolate to broadly oblanceolate, entire or with irregular lobes, hairy or not. Sepals 4.8–6.5 mm, oblong, inner pouched. Petals 6.5–11 × 3–7 mm, blue, mauve or purple. Style 2–6, rarely to 1.7 cm. Ovules 30–50. Fruits 3–10 × 0.1–0.3 cm, linear with straight margins. *South Africa (Cape Province)*. H5. Summer.

5. H. linearis (Thunberg) de Candolle (*Cheiranthus linearis* Thunberg; *H. reticulata* Ecklon & Zeyher; *H. linearifolia* de Candolle). Illustration: Batten & Bokelmann, Wild flowers of the Eastern Cape Province, pl. 61 (1966); Flora of Southern Africa **13**: 60 (1970); Burman & Bean, South African wild flower guide **5**: 97 (1985).

Stout erect or decumbent perennial herbs, sometimes woody at base, to 90 cm, sometimes glaucous, hairless or hairy. Leaves 2–12 × 0.8–1.7 cm, entire or few-lobed, thread-like, often fleshy. Sepals 5–9.5 × 1.3–2.6 mm, inner pouched. Petals 7–12 × 3.3–6 mm, white to dark blue, violet or mauve. Style 1.5–8 mm. Ovules 20–50. Fruits 2.5–11 × 0.2–0.4 cm, linear with straight margins or irregularly wavy. Seeds flat, winged. *South Africa (Cape Province)*. H5–G1. Spring–summer.

6. H. scandens Harvey. Illustration: Botanical Magazine, 7668 (1899); Pole Evans, Flowering plants of South Africa **2**: pl. 48 (1922); Flora of Southern Africa **13**: 67 (1970); Gibson, Wild flowers of Natal, pl. 36 (1975).

Straggling climber to 5 m or more, hairless. Leaves entire, 4–8 × 0.7–3 cm, lanceolate to broadly elliptic, apex acuminate, acute or obtuse, base narrowed into stalk, fleshy. Racemes terminal. Flowers scented. Sepals 4.8–6.5 × 1.5–3 mm, inner with broad membranous margins. Petals 8.5–14 × 3–8 mm, white.

Style 2.5–6 mm. Ovules 2. Fruit 2.5–4 × 0.8–1.3 cm, elliptic-lanceolate to elliptic. Seeds narrowly winged. *South Africa (Natal)*. G1. Winter.

XCIX. RESEDACEAE

Annual to perennial herbs. Leaves alternate, simple, or pinnately divided. Flowers in bracteate racemes, bisexual, bilaterally symmetric. Sepals 4–7; petals 2–8, fringed, shortly stalked. Stamens 3–40, filaments free or united at base, not covered by the petals in bud; anthers 2-celled. Ovary of 2–6 carpels. Fruit an open capsule. Seeds kidney-shaped.

A family of 6 genera and some 75 species from the northern hemisphere, especially the Old World.

Literature: Abdallah, M.S., The Resedaceae, a taxonomical revision of the family. *Mededelingen Landbouwhogeschool Wageningen, Nederland 67* **8**: 1–98 (1967); Abdallah, M.S. & de Wit, C.H.D., The Resedaceae. *Mededelingen Landbouw-hogeschool Wageningen, Nederland 78* **14**: 99–416 (1978). Both parts were published together in *Belmontia New Series* **8**: 1–416 (1978), and this reference is cited for the illustrations.

1. RESEDA Linnaeus
J.R. Akeroyd & S.G. Knees
Description as family.

A genus of 55 species, with a similar distribution as the family, cultivated largely for their economic uses in perfumery or as dyes. Propagation is usually from seed. Seeds should be sown in their final position, as plants do not transplant readily.

1a. Leaves mostly pinnatifid; capsule distinctly longer than wide   2
  b. Leaves mostly simple, the upper sometimes with 2 or 3 lobes; capsule about as long as wide   3
2a. Petals white; anthers flesh-coloured   **2. alba**
  b. Petals and stamens yellow   **4. lutea**
3a. Petals 4, yellow; anthers yellow   **1. luteola**
  b. Petals 6, yellowish white or greenish; anthers orange–deep red   **3. odorata**

1. R. luteola Linnaeus. Illustration: Ross-Craig, Drawings of British plants **4**: pl. 2

(1950); Phillips, Wild flowers of Britain, 75 (1977); Belmontia new series **8**: f. 60 (1978); Polunin, Collins guide to wild flowers of Britain and Europe, 196 (1988).

Erect biennial, 50–150 cm; stems little-branched. Leaves 3–12 × 0.4–1.5 cm, strap-shaped to narrowly oblanceolate, entire, tapered towards base, very shortly stalked. Flowers numerous, in dense racemes to 35 cm. Sepals 4; petals 4, 2–4 mm, the upper 4–8-lobed, shortly stalked, the remainder stalked or not, entire or 4-lobed, yellow. Stamens 20–30; anthers yellow. Capsule 3–4 × 5–6 mm, almost spherical, with 3 or 4 terminal teeth. Seeds *c*. 1 mm, rounded to kidney-shaped, blackish. *Mediterranean area & SW Europe*. H1. Summer.

Cultivated both for ornament and as a source of a yellow dye.

2. R. alba Linnaeus. Illustration: Zohary, Flora Palaestina **1**, Plates: 483 (1966); Polunin, Flowers of Europe, pl. 38 (1969); Phillips, Wild flowers of Britain, 97 (1977); Belmontia new series **8**: f. 18–20 (1978).

Erect annual to perennial, 30–100 cm, branched above. Leaves 5–10 cm, 1–2-pinnatifid with 5–15 pairs of entire lobes. Flowers in dense conical racemes. Sepals 5 or 6, 3–4 mm. Petals 5 or 6, 3.5–6 mm, triangular, deeply 3-lobed, narrowed into a long stalk, white. Stamens 10–12; anthers flesh-coloured. Capsule 0.8–1.6 cm, narrowly obovate or ellipsoid, narrowed at apex, with usually 4 terminal teeth. Seeds *c*. 1 mm, almost kidney-shaped, brown. *Mediterranean area to Middle East*. H1. Summer.

3. R. odorata Linnaeus. Illustration: Botanical Magazine, 29 (1787); Hay & Synge, RHS Dictionary of garden plants, 46 (1969); Maire, Flore de l'Afrique du Nord **14**: 193 (1976); Belmontia new series **8**: f. 67 (1978).

Annual to perennial, 10–50 cm; stems branched. Leaves 5–10 × 0.5–1.5 cm, oblanceolate to obovate, usually with 1 or 2 lateral lobes. Flowers in loose racemes, fragrant. Sepals 6; petals 6, 4–4.5 mm, yellowish white or greenish, 9–15-lobed. Stamens 20–25, anthers orange to deep red. Capsule 9–10 × 7–11 mm, nodding, almost spherical. Seeds *c*. 1.5 mm, rounded, ridged, greenish yellow. *North Africa, but naturalised in Mediterranean area*. H5. Summer.

Cultivated for ornament and for use in perfumery. 'Goliath' has compact racemes with conspicuous rusty orange anthers; in 'Red Monarch' they are deep red.

4. R. lutea Linnaeus. Illustration: Ross-Craig, Drawings of British plants **4**: pl. 1 (1950); Polunin, Flowers of Europe, pl. 38 (1969); Phillips, Wild flowers of Britain, 37 (1977); Belmontia new series **8**: f. 57 (1978).

Somewhat untidy annual to perennial, 30–80 cm; stems branched. Leaves 5–10 cm, 1–2-pinnatifid, with 1–4 pairs of entire lobes. Racemes up to 20 cm, dense; bracts falling early. Sepals 6; petals 6, 3–4 mm, entire or 2- or 3-lobed, yellow. Stamens 12–20; anthers yellow. Capsule 6–12 × 4.5–5.5 mm, sometimes nodding, ovoid-cylindric. Seeds 1.4–1.8 mm, obovoid, blackish. *Mediterranean area & S Europe.* H1. Summer.

C. MORINGACEAE

Deciduous trees, bark light grey, trunk often swollen near base. Leaves spirally arranged, 2–3-pinnate, with opposite leaflets. Stipules absent or reduced to glands at bases of leaf stalks and leaflets. Flowers bisexual, radially or bilaterally symmetric, in axillary panicles. Sepals 5, spreading or reflexed, unequal, overlapping. Petals 4 or 5, outermost larger than inner, overlapping. Stamens 5, staminodes 5, both around margin of disc; anthers splitting lengthwise. Carpels 3, ovary 1-celled, superior, on short stalk, style hollow. Fruit a woody pod, opening by 3 slits. Seeds many, 3-winged or wingless.

A family of single genus, often cultivated for ornament. The seeds are a useful source of oil.

1. MORINGA Adanson
S.G. Knees
Description as for family.

A genus of 12 species from the drier part of Africa, Arabia and India. Although easily propagated from seed, the plants die back soon after producing a short stem. The underground storage organ develops quickly and a period of drought should follow. Once established the plant can be treated in a similar way to other caudiciform succulents.

1a. Leaflets 1.5–2.5 cm wide
  1. ovalifolia
  b. Leaflets 0.5–1.2 cm wide   2
2a. Leaflets obtuse, notched
  3. oleifera
  b. Leaflets acute, mucronate
  2. drouhardii

1. M. ovalifolia Dinter & A. Berger (*M. ovalifoliolata* Dinter & A. Berger). Illustration: Marloth, Flora of South Africa **1**: 245 (1935); Jacobsen, Handbook of succulent plants **2**: 695 (1960); Flora of Southern Africa **13**: 185 (1970); Rowley, Caudiciform and pachycaul succulents, 201–202 (1987).

Tree with few branches, 2–6 m, with trunk to 1 m across. Leaves 50–80 × 40–60 cm, twice pinnate; leaflets 2–4 × 1.5–2.5 cm, ovate-narrowly elliptic, mucronate. Flowers radially symmetric; petals 5, white. Fruits 15–30 cm, c. 2 cm thick. Seeds 2.5–3.5 × 0.8–1.5 cm, with 3 membranous wings. *SW Africa.* G1.

2. M. drouhardii Jummelle. Illustration: Koechlin et al., Flore et vegetation de Madagascar, 264 (1974); Flore de Madagascar et Comores. Famille **85**: 35, 39 (1982).

Tree 5–10 m, trunk very swollen towards base. Leaves 20–30 cm, twice or three times pinnate; leaflets 1.5–3 × 0.5–1.2 cm, oblong-ovate, obtuse, mucronate. Flowers radially symmetric; petals 5, yellowish white. Fruits 30–50 cm, constricted; seed 2–2,5 × 1.8–2 cm, ovoid to triangular, without membranous wings, but with 3 ridges. *SW Madagascar.* G1.

3. M. oleifera Lamarck (*M. pterygosperma* Gaertner). Illustration: Engler, Die natürlichen Pflanzenfamilien **3**(2): 243 (1891); Meninger, Flowering trees of the world for tropics and warm climates, pl. 281 (1962); Hortus III, 742 (1976); Flore de Madagascar et Comores. Famille **85**: 35, 39 (1982).

Small trees, 3–10 m. Leaves 30–60 cm, pinnate; leaflets 0.9–2 × 0.5–1.2 cm, ovate, obtuse, notched. Flowers bilaterally symmetric; petals 4, white. Fruits 20–50 cm, 3-sided; seed 2.4–2.6 × 0.4–0.7 cm, including membranous wings, 3-sided to spherical. *NW India, but widely naturalised in tropics and subtropics, often planted as street trees.* G1.

CI. PLATANACEAE

Deciduous, bisexual trees to 50 m, with bark peeling away in thin flakes leaving a smooth, pale surface. Leaves alternate, simple, more or less palmately lobed and veined (rarely unlobed and pinnately veined) with the base of the leaf-stalk covering the axillary bud. Stipules usually prominent, often leaf-like, united at the base into a short tube around the stem. Young growth, especially the underside of leaves, often covered with a dense felt of star-shaped and simple hairs, sometimes persistent. Inflorescence of up to 12 dense, spherical, stalkless or stalked heads of flowers in hanging strings, each inflorescence either male or female. Wind pollinated. Sepals 3–8, free; petals 3–8, spathulate; both sometimes considered to be bracts rather than true perianth. Male flowers with 3–4 almost stalkless anthers and occasionally 3–4 pistillodes; female flowers with 6–9 very short-stalked free pistils, each with a long style and hooked stigma; 3–4 staminodes are sometimes present. Female flowers develop into achenes with often persistent styles, forming characteristic prickly balls. Individual achenes are subtended by a tuft of bristly hairs derived from the perianth or scales.

A family of a single genus occurring throughout the north temperate zone.

1. PLATANUS Linnaeus
A.C. Whiteley
Description as for family.

A genus of about 8 species from North and Central America, south eastern Europe to Iran and Indo-China. Several species are much planted in cities as shade and street trees, being tolerant of pruning and pollution. All species are sun-loving and do best in warmer areas, preferring a deep loamy soil. Propagation is by seed, hardwood cuttings or layering.

1a. Leaves 3–5-lobed; middle lobe as long as or shorter than wide   2
  b. Leaves 5–7-lobed; middle lobe longer than wide   3
2a. Leaves usually 3-lobed; middle lobe shorter than wide; lobes shallowly toothed; fruiting heads solitary
  2. occidentalis
  b. Leaves 3–5-lobed; the middle lobe about as long as wide, set at right angles to the lateral lobes, coarsely toothed at the base; fruiting heads usually 2–4   **3. × acerifolia**
3a. Leaves 5–7-lobed; lobes about half as long as blade, coarsely toothed
  1. orientalis
  b. Leaves 5-lobed, the lobes at least half as long as blade, narrow, entire or slightly toothed   4
4a. Leaves usually entire; fruiting heads 2–4, stalked, not bristly
  5. wrightii

b. Leaves usually slightly toothed; fruiting heads 2–7, stalkless, bristly **4. racemosa**

1. P. orientalis Linnaeus. Illustration: Hora, The Oxford encyclopedia of trees of the world, 121 (1981); Mitchell, Field guide to the trees of Britain & northern Europe, pl. 24 (1974, 1986); Krüssmann, Manual of cultivated broad-leaved trees & shrubs, **2**: 417 (1986).

Tree to 30 m, often branching low and forming a spreading crown, the bark peeling in large plates before becoming rugged with age. Leaves 10–25 cm wide, palmate, slightly shorter than wide, base usually wedge-shaped, occasionally truncate, with usually 5 narrow large lobes and two smaller ones at the base, the larger at least half the length of the blade and with 1–3 large teeth or small lobes each side; leaf-stalks 4–7 cm. Leaves and young shoots at first densely felted with star-shaped hairs, soon falling but persistent on the veins beneath. Fruiting heads 2–6, close together on the stalk, each c. 2.5 cm. Achenes conical and often with a hairy surface with persistent styles. *SE Europe*. H3. Spring.

2. P. occidentalis Linnaeus. Illustration: Sargent, Silva of North America **7**: pl. 326, 327 (1895); Hora, The Oxford encyclopedia of trees of the world, 121 (1981); Krüssmann, Manual of cultivated broad-leaved trees & shrubs **2**: 417 (1986).

Tree to 40 m in the wild, with a short trunk and ascending branches, the bark peeling in small plates. Leaves shallowly 3-lobed, 10–18 cm, the lobes acuminate and shallowly toothed, the base truncate or slightly cordate or wedge-shaped, densely hairy at first, persisting only on the veins beneath; leaf-stalks 7–13 cm. Fruiting heads usually solitary, c. 2.5 cm, on stalks 7–15 cm. Achenes rounded or truncate, tipped with the short persistent base of the style, hairless on the surface. *E & S North America*. H5. Spring.

Not widely grown, being severely affected by plane anthracnose, *Gnomonia platani*, a fungal disease which kills the young growth.

3. P. × acerifolia (Aiton) Willdenow (*P. orientalis × P. occidentalis*; *P. × hispanica* Muenchhausen; *P. × hybrida* Brotero). Illustration: Phillips, Trees in Britain, Europe & North America, 46, 165 (1978); Mitchell, Field guide to the trees of Britain and northern Europe, pl. 24 (1986); Krüssmann, Manual of cultivated broad-leaved trees & shrubs, **2**: 416 (1986).

Tree to 35 m with a smooth, tall trunk and rounded head of often rather contorted branches, the bark peeling in large plates. Terminal shoots pendent on older trees. Leaves 10–20 × 12–25 cm, commonly palmate, 3–5-lobed, the central lobe broadly triangular, set at right angles to the lateral lobes, all lobes toothed at the base, sometimes sparsely throughout; base truncate or cordate, occasionally wedge-shaped; leaf-stalks 5–8 cm. Many variations can occur on a single tree. Shoots and leaves covered by a pale brown felt when young, falling quite quickly. Fruiting heads 2–4, c. 3 cm, usually separated along the stalk, bristly at first, becoming smoother as persistent styles break off. Achenes conical, with hairs on the surface. *Garden origin*. H3. Spring.

The most widely grown species, frequent in towns and cities, valued for its tolerance of pollution. A number of cultivars is attributable to this hybrid.

'Augustine Henry' has pendent lower branches, a less contorted crown, very freely flaking bark and generally larger leaves, truncate at the base, deeply lobed, regularly toothed, the undersides remaining felted along the veins. Fruiting heads 1–3, not freely produced. Achenes with few surface hairs.

'Pyramidalis' has horizontal lower branches (often removed), ascending upper branches and a loose, broad crown. The bark does not flake freely and soon becomes rugged. Leaves relatively small, mostly 3-lobed, sparsely toothed and rather glossy. Fruiting heads 1 or 2, to 5 cm. A widely planted clone, often mixed with *P. × acerifolia*.

'Mirkovec' is slow-growing and bushy, the young foliage has a bronze tinge.

'Suttneri' has the leaves splashed and streaked with white variegation.

4. P. racemosa Nuttall (*P. californica* Bentham). Illustration: Sargent, Silva of North America, **7**: pl. 328 (1895); Krüssmann, Manual of cultivated broad-leaved trees & shrubs, **2**: 418 (1986).

Tree 30–40 m in the wild, often branching low down, forming a rounded crown, the bark dark and rugged low down, flaking on the branches. Leaves 3–5-lobed, 17–32 × 15–30 cm, cordate to somewhat wedge-shaped at the base, the lobes at least half the length of the blade, narrowly tapered, sparsely toothed. Young shoots and the undersides of leaves densely downy, the down persisting on veins and leaf-stalks. Fruiting heads 2–7, c. 2 cm. Achenes densely hairy, at least when young, acute or rounded; styles persistent. *SW USA*. H5. Spring.

5. P. wrightii S. Watson (*P. racemosa* var. *wrightii* (S. Watson) Benson). Illustration: Sargent, Silva of North America, **7**: pl. 329 (1895).

Tree to 25 m, often branching low down into 2 or 3 trunks, lower branches horizontal, upper erect, bark rugged at base of trunk, becoming scaly upwards, finally smooth. Leaves 15–18 cm, 3–5-lobed, the lobes more than half the length of the blade, narrowly tapered, mostly entire; base cordate to wedge-shaped; leaf-stalks 4–8 cm. Young shoots and leaves downy, the down persisting on veins and leaf-stalks. Fruiting heads 2–4, c. 2 cm, individually stalked, not bristly. Achenes truncate, hairless, without a persistent style. *SE USA*. H5. Spring.

CII. HAMAMELIDACEAE

Shrubs or trees, often with stellate hairs, bisexual or unisexual. Leaves usually alternate, simple or palmate; stipules usually present. Flowers in spikes, clusters or pairs, bisexual or unisexual; petals usually 4 or 5, sometimes absent (rarely 2 or 3 or 6–10); sepals 4 or 5, rarely absent; anthers 4 or 5 (rarely to 25); carpels 2, inferior to superior, usually joined, with distinct styles; ovules 1–many, often aborting to 1; fruit a woody capsule, splitting along or between the radial walls.

A family of 28 genera and 90 species, widespread but mainly in the subtropics and tropics of E Asia.

Literature: Weaver, R.E., The Witch Hazel Family (Hamamelidaceae), *Arnoldia* **36**: 69–109 (1976); Wright, D., Hamamelidaceae, a survey of the genera in cultivation, *The Plantsman* **4**: 29–53 (1982).

1a. Leaves deeply palmately divided 2
  b. Leaves simple 3
2a. Leaves deciduous **8. Liquidambar**
  b. Leaves evergreen **4. Exbucklandia**
3a. Leaves at least partly evergreen 4
  b. Leaves quite deciduous 8
4a. Petals absent 5
  b. Petals 4, white **9. Loropetalum**
5a. Flowers in heads 6
  b. Flowers in racemes **3. Distylium**
6a. Flower-heads stalkless 7

b. Flower-heads with long stalks **12. Rhodoleia**

7a. Leaves evergreen, entire; bracts 5 mm **15. Sycopsis**

b. Leaves semi-deciduous, shallowly toothed; bracts 10 mm or more **14. × Sycoparrotia**

8a. Bark flaking **10. Parrotia**

b. Bark not flaking 9

9a. All flowers bisexual 10

b. Some flowers unisexual 14

10a. Flower-clusters with conspicuous white bracts at base **11. Parrotiopsis**

b. Flower-clusters or racemes without conspicuous bracts 11

11a. Leaves palmately veined, base cordate **2. Disanthus**

b. Leaves pinnately veined 12

12a. Flowers in clusters or racemes; petals present 13

b. Flowers in erect white spikes; petals lacking **6. Fothergilla**

13a. Flowers in hanging yellow racemes; petals ovate **1. Corylopsis**

b. Flowers in clusters; petals strap-shaped **7. Hamamelis**

14a. Racemes hanging, unisexual **13. Sinowilsonia**

b. Racemes erect, bisexual in spring, male in autumn **5. Fortunearia**

1. CORYLOPSIS Siebold & Zuccarini
D.O. Wijnands

Deciduous shrubs. Leaves elliptic-ovate to circular, green to glaucous, sharply and finely toothed, veins impressed. Flowers in hanging racemes, bell- to funnel-shaped, yellow, scented.

A genus of 7 species from E Asia.

Literature: Morley, B. & Chao, Jew-Ming, *Journal of the Arnold Arboretum* **58**: 382–415 (1977).

1a. Racemes with 1–5 flowers; leaves to 5 cm **3. pauciflora**

b. Racemes with more than 5 flowers; leaves more than 6 cm 2

2a. Sepals apparently absent; nectaries truncate **2. multiflora**

b. Sepals present; nectaries 2-fid 3

3a. Sepals thin, lanceolate; flowers often fewer than 10 **5. spicata**

b. Sepals fleshy, shortly ovate; flowers often more than 10 4

4a. Sepals bluntly triangular, *c.* 1 mm; leaves to 8 cm **1. glabrescens**

b. Sepals ovate to triangular, *c.* 2 mm; leaves to 12 cm **4. sinensis**

1. C. glabrescens Franchet & Savatier
(*C. coreana* Uyeki; *C. gotoana* Makino; *C.*

platypetala misapplied). Illustration: Taylor's guide to shrubs, 167 (1987).

Shrub to 6 m. Leaf-stalks 1.5–3 cm, almost hairless; blades 5–8 × 3–7 cm, ovate to almost circular, acuminate. Racemes with 10–20 flowers evenly spaced along the hairless axis; petals to 8 mm; anthers yellow or purplish; nectaries bifid, longer than the sepals. *Japan, Korea*. H3. Spring.

2. C. multiflora Hance (*C. wilsonii* Hemsley). Illustration: Hooker's Icones Plantarum **29**: 2819 & 2820 (1909).

Shrub or small tree. Leaf-stalks 1–2.5 cm, hairy, blades 7–15 cm, ovate to elliptic, acute or tailed at tip. Racemes with 10–20 flowers on a hairy axis; petals 5–6 mm, linear-spathulate; nectaries entire and truncate. *C China*. H3. Spring.

3. C. pauciflora Siebold & Zuccarini.
Illustration: Botanical Magazine, 7736 (1900); Hay & Synge, Dictionary of garden plants in colour, 192 (1969); Hillier colour dictionary of trees & shrubs, 81 (1981); Taylor's guide to trees and shrubs, 166 & 167 (1987).

Spreading small shrub to *c.* 3 m. Leaves ovate, 3–5 cm. Racemes few-flowered, 2–3 cm, pale yellow; petals to 8 mm, oblong-ovate; anthers yellow; nectaries entire and truncate. *Korea, Japan, Taiwan*. H3. Spring. Dislikes lime and dry sunny positions.

4. C. sinensis Hemsley var. sinensis (*C. willmottiae* Rehder & Wilson). Illustration: Hay & Synge, Dictionary of garden plants in colour, 192 (1969); Everard & Morley, Wild flowers of the world, pl. 94 (1970); The Garden **102**: 250 (1977)

Shrub to 5 m. Leaf-stalks 5–15 mm, usually silky-hairy; blades 5–12 cm, obovate to oblong-elliptic, acuminate, rounded or obliquely cordate at base, hairy between the veins. Racemes with 10–30 flowers, densely and evenly spaced along the hairy axis; petals 7–8 mm, almost circular; anthers yellow; nectaries 2-fid, longer than the sepals. *China*. H3. Spring.

'Spring Purple' has purple unfolding leaves.

Var. **calvescens** Rehder & Wilson (*C. platypetala* Rehder & Wilson; *C. glaucescens* Handel-Mazzetti; *C. hypoglauca* Cheng). Leaf-stalks mostly hairless, occasionally glandular; blades hairless between the veins. Petals to 6 mm, circular to kidney-shaped. *China, Tibet*. H3. Spring.

Forma **veitchiana** (Bean) Morley & Chao (*C. veitchiana* Bean). Illustration: Botanical Magazine, 8349 (1910). Anthers

red; leaves hairless beneath. *China (W Hubei)*. H3. Spring.

5. C. spicata Siebold & Zuccarini.
Illustration: Botanical Magazine, 5458 (1864); Hay & Synge, Dictionary of garden plants in colour, 192 (1969); Gardening from 'Which', March 1987, 70; Hellyer, Shrubs in colour, 35 (1982).

Shrub to 3 m. Leaf-stalks 1–2.5 cm, hairy, blades 5–10 cm, ovate to obovate, shortly acuminate. Racemes with 5–12 flowers clustered toward the end of the hairy axis; petals pale yellow, 7–9 mm; anthers reddish purple; nectaries 2-fid, shorter than the sepals. *Japan*. H3. Spring.

2. DISANTHUS Maximowicz
D.O. Wijnands

Deciduous shrub to 8 m. Leaves alternate, palmately veined, long-stalked, circular to ovate, to 10 cm across, entire, hairless, turning deep red and orange in autumn. Flowers axillary in short-stalked pairs, dark purple, 1.5 cm across, faintly scented, bisexual; petals, sepals, stamens and staminodes 5, calyx softly hairy with recurved lobes. Capsule 2-celled with several seeds in each cell.

Grown mainly for its autumn colour. Performs best in shade in moist acid soil. A single species native to Japan and China.

1. D. cercidifolius Maximowicz.
Illustration: Botanical Magazine, 8716 (1917); Hay & Synge, Dictionary of garden plants in colour, 192 (1969); Hillier colour dictionary of trees & shrubs, 95 (1981); Gartenpraxis **4**: 43 (1990).

Japan & China. H4. Autumn.

3. DISTYLIUM Siebold & Zuccarini
D.O. Wijnands

Evergreen trees or shrubs. Leaves ovate to lanceolate, entire or remotely toothed above the middle.

A genus of 12 species from E Asia and C America.

Literature: Walker, E.H., A revision of Distylium and Sycopsis. *Journal Arnold Arboretum* **25**: 319–341 (1944).

1. D. racemosum von Siebold & Zuccarini
(*D. myricoides* misapplied; *Sycopsis tutcheri* misapplied). Illustration: Botanical Magazine, 9501 (1937); Krüssmann, Handbuch der Laubgehölze **1**: 480 (1976); Gartenpraxis **4**: 42 (1990).

Tree to 25 m, in cultivation usually a shrub; leaves elliptic, acute, 3–7 cm; racemes to 4 cm, longer in fruit, with stellate hairs; anthers red. *Japan, Korea, SE*

& C China, Ryukyu Islands. H4. Spring–summer.

Earlier-flowering plants with narrower, more pointed leaves and longer racemes are often grown under the name *D. myricoides* Hemsley, a hardier Chinese species. Illustration: Iconographia Cormophytorum Sinicorum **2**: 167 (1987). Many of these are likely to be *D. racemosum*.

4. EXBUCKLANDIA R.W. Brown
D.O. Wijnands

Evergreen trees with alternate leaves; stipules prominent, paired, enclosing the axillary buds and inflorescence. Flowers in heads, unisexual, in groups of 4 sunk in the floral axis.

A genus of 2 timber-bearing species related to *Liquidambar*, distributed from E Himalaya through S China to Sumatra.

1. E. populnea (W. Griffith) R.W. Brown (*Bucklandia populnea* W. Griffith; *Symingtonia populnea* (W. Griffith) van Steenis). Illustration: Botanical Magazine, 6507 (1880); Nakao, Living Himalayan flowers, t. 194 (1964); Iconographia Cormophytorum Sinicorum **2**: 157 (1987).

Leaves ovate, cordate, 10–15 cm, palmately veined, leathery, glossy green above, veins and lower surface reddish; leaf-stalks 6 cm; stipules obovate, *c.* 3 cm. *Himalaya (Nepal to Bhutan), S China, Malaya, Indonesia (Sumatra)*.

5. FORTUNEARIA Rehder & Wilson
D.O. Wijnands

Large deciduous shrub to 8 m. Leaves obovate, acuminate, toothed. Flowers greenish, in erect racemes; male flowers *c.* 2.5 cm formed in autumn, bisexual flowers *c.* 5 cm appearing with the leaves; petals 5, strap-shaped.

A single species from China related to *Hamamelis*. Moderately lime-tolerant; not an attractive garden plant, rare in cultivation.

1. F. sinensis Rehder & Wilson. Illustration: Iconographia Cormophytorum Sinicorum **2**: 166 (1987).

W & C China. H3. Autumn & spring.

6. FOTHERGILLA Linnaeus
D.O. Wijnands

Deciduous shrubs with alternate, coarsely toothed leaves. Flowers cream, without petals, in terminal spikes or heads like bottle-brushes, slightly before or with the leaves; flower-tube bell-shaped, with 4–7 lobes; stamens 15–25, white, thick near the apex; ovary 2-celled; fruit a 2-seeded capsule.

A genus of 2 species from southeastern North America. Propagation is by seed or cuttings, which take long to strike. Requires neutral or lime-free soil.

Literature: Weaver, R.E., The Fothergillas. *Arnoldia* **31**: 89–96 (1971).

1a. Leaves 4–6 cm; flowers appearing before the leaves **1. gardenii**
  b. Leaves 6–15 cm; flowers appearing with the leaves **2. major**

1. F. gardenii Linnaeus (*F. alnifolia* Linnaeus filius; *F. carolina* Britton). Illustration: Botanical Magazine, 1341 (1811); Hay & Synge, Dictionary of garden plants in colour, 202 (1969); Taylor's guide to shrubs, 198 & 199 (1987).

Shrub to *c.* 75 cm. Leaves 4–6 cm, obovate, dull glaucous-green, turning golden yellow in autumn. Flowers cream, before the leaves, in 2–3 cm spikes. *USA (Virginia to Georgia & Alabama)*. G1. Spring–summer.

2. F. major (Sims) Loddiges (*F. monticola* Ashe; *F. gardenii* misapplied). Illustration: Botanical Magazine, 1342 (1811); The Plantsman **4**: 37 (1982); Taylor's guide to shrubs, 198 & 199 (1987); Philips and Rix, Shrubs, 151 & 268 (1989).

Shrub to 3 m. Leaves 6–15 cm with stellate hairs, broadly obovate, glaucous green, turning yellow, orange or scarlet in autumn. Flowers with the leaves, in 3–6 cm heads or spikes. *USA (N Carolina to Alabama)*. H3. Spring–summer.

Very variable in habit, size of leaves and autumn colour. At least 6 unnamed clones are in cultivation; those with small leaves tend to show the best autumn colour.

7. HAMAMELIS Linnaeus
M. de Ridder

Deciduous shrubs or small trees with stellate hairs. Leaves alternate, pinnately veined, short-stalked, unequal at base, ovate to obovate; margin irregularly toothed, sometimes wavy towards the tip. Flowers bisexual, fragrant, in many, very condensed inflorescences in clusters along the branches. Inflorescence usually with 3 flowers; bracts 1; bracteoles 2, more or less fused. Flowers short-stalked, parts in 4s; calyx persistent, cup-shaped and hairy outside; petals strap-shaped, wrinkled and crisped, 5–25 × 1–2 mm, rolled in bud. Fruit a woody capsule splitting with great power, hurling away the black, shiny seeds.

A genus of 4 species, 2 from E USA and 2 from E Asia, and one hybrid of garden origin (*H. × intermedia* Rehder) between the 2 asiatic species. Although this hybrid was first observed and noted in the Arnold Arboretum (USA), most cultivars have been selected and marketed in Western Europe. These are usually grafted onto rootstocks of *H. virginiana*, which is grown from seed. Propagation by cuttings is also possible. *Hamamelis* needs light shade from the midday sun in dryer and warmer conditions and deep, moist, rich, but not heavy soil.

Literature: de Ridder, M. & Wijnands, D.O., De systematiek van Hamamelis, *Dendroflora* **17**: 6–8 (1981); Grootendorst, H.J., Hamamelis, *Dendroflora* **17**: 9–17 (1981); Lamb, J.G.D., & Nutty, F., The propagation of Hamamelis, *The Plantsman* **6**: 45–48 (1984).

1a. Bracteoles not joined 2
  b. Bracteoles joined or mostly so 3
2a. Petals 5–8 mm, yellow to orange; calyx violet inside; leaves obliquely wedge-shaped at base **4. vernalis**
  b. Petals 10–12 mm, lemon-yellow; calyx yellow-green inside, leaves obliquely cordate at base **5. virginiana**
3a. Bracteoles totally joined; leaves hairless above, with stellate hairs on veins and vein-axils beneath; petals yellow; calyx grey-purple inside **2. japonica**
  b. Bracteoles more or less joined; leaves softly hairy, less so during the growing-season 4
4a. Leaves with stellate hairs on both sides; petals flat, bright yellow; bracteoles joined only at base **3. mollis**
  b. Leaves hairless to sparsely hairy; petals wrinkled and crisped, red to orange or yellow; bracteoles more or less joined **1. × intermedia**

1. H. × intermedia Rehder. Illustration: Amateur Gardening 5335, 21 (1987); Phillips & Rix, Shrubs, 14 (1989).

Variable shrub to 4 m, generally intermediate between the parents (*H. mollis* and *H. japonica*) in shape and size of leaf, amount of hair, size of flower etc. *Garden origin.* Winter–spring.

Several clones have arisen, among which are:

'Jelena', a shrub of vigorous, spreading habit with large, broad, softly hairy leaves; flowers in dense clusters; petals *c.* 20 × 2 mm, yellow suffused with rich coppery red, margin yellow. Midwinter.

'Feuer-zauber', vigorous shrub with strong ascending branches; flowers coppery-red; petals *c.* 16 × 1.5 mm. Winter–spring.

'Orange Beauty', flowers deep yellow to orange-yellow; petals *c.* 16 × 1.5 mm. Spring.

2. H. japonica Siebold & Zuccarini (*H. bitchuensis* Makino; *H. megalophylla* Koidzumi). Illustration: Huxley, Deciduous garden trees and shrubs, f. 86 (1973); Journal of the Royal Horticultural Society **99**: 22 (1974); Phillips & Rix, Shrubs, 14 (1989).

Variable; commonly a large spreading shrub to 3 m. Leaves obovate, smaller than those of *H. mollis*, becoming hairless and shiny. Flowers small to medium-sized; petals 1–1.5 cm, very narrow, strap-shaped, much twisted and crumpled, yellow with a brown spot at the base, slightly scented. *Japan & Korea.* G3. Spring.

Var. **arborea** (Ottolander) Gumbleton (*H. arborea* Ottolander) is a wide-spreading shrub, flowers like those of the species.

'Zuccariniana' is a large, distinctly erect shrub, flattening out when older; flowers small, pale sulphur-yellow, with a greenish brown calyx. One of the latest-flowering, usually early spring.

Var. **flavopurpurascens** (Makino) Rehder differs from *H. japonica* in its brownish-red petals, orange-yellow or sulphur yellow towards the tip. *Japan.* Spring.

3. H. mollis Oliver. Illustration: Botanical Magazine, 7884 (1811); Hay & Synge, Dictionary of garden plants in colour, 192 (1969); Everard & Morley, Wild flowers of the world, pl. 94 (1970); Phillips & Rix, Shrubs, 14 (1989).

Large shrub to 5 m; young branches very downy. Leaves softly hairy, rounded or very broadly obovate, toothed, shortly and abruptly pointed, unequally cordate; leaf-stalk short and downy. Flowers rich golden yellow, very fragrant; petals strap-shaped, *c.* 17 × 2 mm, not wavy. *China.* Winter–spring.

'Pallida': leaves unlike those of *H. mollis*, becoming hairless when older; flowers large, strongly scented, sulphur yellow, borne in densely crowded clusters along the naked stems. Winter–early spring.

'Brevipetala': petals only *c.* 1 cm, orange-yellow.

4. H. vernalis Sargent. Illustration: Botanical Magazine, 8573 (1914); Everard & Morley, Wild flowers of the world, pl.

155 (1970); Taylor's guide to shrubs, 171 (1987).

Upright suckering shrub to 3 m. Leaves obovate to oblong, obtuse, broadly wedge-shaped to truncate at base, 6–12 cm, slightly glaucous beneath, hairless. Flowers very small, 5–8 mm, pale yellow-red. *USA (Missouri, Oklahoma, Louisiana, Alabama).* G1. Winter–spring.

5. H. virginiana Linnaeus (*H. macrophylla* Pursh). Illustration: Botanical Magazine, 6684 (1883); Journal of the Royal Horticultural Society **94**: f. 41 (1969); Phillips & Rix, Shrubs, 256 & 269 (1989).

Large shrub or occasionally a small, broad-crowned tree. Leaves 8–15 cm, turning yellow in autumn. Flowers small to medium-sized, yellow. *E USA.* G1. Autumn.

Often used as a rootstock for the larger-flowered species.

8. LIQUIDAMBAR Linnaeus.
D.O. Wijnands

Deciduous unisexual trees with balsamic resin; twigs often with corky ridges. Leaves alternate (opposite in *Acer* which may look similar), long-stalked, palmately lobed. Flowers without petals, numerous, in spherical heads, male heads in terminal catkins, female heads solitary. Fruiting-heads spherical, woody, with conspicuous persistent styles; capsule splitting along the radial wall. Seed flattened, narrowly winged.

Four widely dispersed species from eastern North America, Turkey and China. They need deep, lime-free soil. *Liquidambar* is sometimes placed with *Altingia* in the separate family Altingiaceae.

Literature: Samorodova-Bianki, G., De Genere Liquidambar L., *Notulae Systematicae (Leningrad)* **18**: 77–89 (1957); Thomas, J.L., Liquidambar, *Arnoldia* **21**: 59–66 (1961); Santamour, F.S., Interspecific hybridization in Liquidambar, *Forest Science* **18**: 23–26 (1972).

1a. Small tree or shrub with rounded crown; leaf-lobes themselves lobed **1. orientalis**
  b. Large tree; leaf-lobes entire 2
2a. Leaves with 3 (rarely 5) lobes **3. formosana**
  b. Leaves with 5 (rarely 7) lobes **2. styraciflua**

1. L. orientalis Miller. Illustration: Hora, Oxford encyclopaedia of trees of the world, 122 (1981).

Slow-growing tree to 10 m with a

rounded crown, often a shrub in cultivation. Leaves 5–10 × 6–13 cm, 5-lobed, the lobes themselves lobed. Fruiting-heads 2.5–3 cm, hanging. *W & SW Turkey, Greece (Rhodes).* H4. Spring.

The scented medicinal balsam, liquid storax, is produced by wounding the bark.

2. L. styraciflua Linnaeus. Illustration: Hay & Synge, Dictionary of garden plants in colour, 210 (1969); Perry, Flowers of the world, 136 (1972); Hora, Oxford encyclopaedia of trees of the world, 122 (1981).

Tree to 30 m with a pyramidal crown. Leaves 12–15 cm, with 5 (rarely 7) lobes, toothed, shiny green, hairless above and with tufts of hairs in the axils beneath. *E North America (Connecticut to Florida & Texas), C America.* H3. Spring.

3. L. formosana Hance (*L. acerifolia* Maximowicz; *L. maximowiczii* Miquel). Illustration: Hora, Oxford encyclopaedia of trees of the world, 122 (1981); Iconographia Cormophytorum Sinicorum **2**: 159 (1987).

Tree to 10 m. Leaves 8–15 cm wide, with 3 (rarely 5) lobes, toothed, sparsely hairy, matt green. Fruit-clusters *c.* 3 cm across. *S China & Taiwan.* H5. Spring.

Var. **monticola** Rehder & Wilson does not differ significantly. The Chinese plants known by this name have 3-lobed, almost hairless leaves and are hardier. H4. Spring.

9. LOROPETALUM Reichenbach
D.O. Wijnands

Evergreen shrubs with ovate, slightly asymmetric leaves. Flowers 6–8 in a terminal cluster, petals 4, strap-shaped, anthers 4-chambered.

A genus of one or two species from the Himalaya, China and Japan.

1. L. chinense (R. Brown) Oliver. Illustration: de Noailles & Lancaster, Mediterranean plants and gardens, 96 (1977); Krüssmann, Manual of cultivated broad-leaved trees & shrubs **2**: 257 (1986); Taylor's guide to shrubs, 174 & 175 (1987); Phillips & Rix, Shrubs, 15 (1989).

Evergreen spreading shrub to 2 m, with interlacing spreading branches. Leaves broadly ovate to heart-shaped, 2–4 × 1–2 cm, finely toothed, veins deeply impressed. Flowers creamy white, green at the base, similar to *Hamamelis*. Petals narrow, 20 × 2.5 mm. *Japan, China, Assam.* H5. Late winter–spring.

10. PARROTIA C.A. Meyer
D.O. Wijnands

Deciduous tree or shrub to 12 m, with flaking bark; branches often hanging. Leaves 6–10 × 4–8 cm, oblong-elliptic to narrowly ovate, wavy and shallowly toothed, glossy green turning orange, yellow, crimson and purple in autumn, short-stalked, with large early-falling stipules. Flowers bisexual, *c.* 1.2 cm across, in dense heads surrounded by large bracts; petals absent; anthers 5–7, red, conspicuous.

A genus of a single species from the SW Caspian area; needs a sunny, dry position; tolerant of lime. Propagation is by greenwood cuttings.

1. P. persica (de Candolle) C.A. Meyer. Illustration: Botanical Magazine, 5744 (1868); Hay & Synge, Dictionary of garden plants in colour, 216 (1969); Perry, Flowers of the world, 136 (1972); Hora, Oxford encyclopaedia of trees of the world, 123 (1981).

N Iran, former USSR (Azerbaijan), along the Caspian Sea. H3. Spring.

'Vanessa' is a tree-forming clone, suitable for street planting.

11. PARROTIOPSIS (Niedenzu) Schneider
D.O. Wijnands

Deciduous tree to 6 m, or more usually an *Alnus*-like shrub. Leaves 3–5 cm, circular, slightly truncate, toothed, turning yellow in autumn. Flowers bisexual, in clusters *c.* 1 cm across, surrounded by 4–6 white bracts (as in *Cornus florida*); petals absent; anthers 15–24.

A genus of a single species from the W Himalaya.

1. P. jacquemontiana (Decaisne) Rehder. Illustration: Botanical Magazine, 7501 (1896); The Plantsman **4**: 49 (1982); Gartenpraxis **4**: 42 (1990).

W Himalaya. H3. Spring & autumn.

12. RHODOLEIA Hooker
D.O. Wijnands

Large evergreen shrub or tree, to 25 m in the wild. Leaves 5–13 × 2–6 cm, lanceolate to broadly ovate, entire, glaucous beneath; stalks 2–4.5 cm. Flowers joined, 5–10 together in hanging, stalked heads *c.* 2 cm across, surrounded by 12–20 reddish brown bracts, petals absent or 1–4, red, anthers 7–11, black. Fruits woody, 2-celled; seeds numerous.

A genus of a single very variable species, distributed from S China to Malaysia; performs best in sheltered woodland.

Literature: Vink, W. *Flora Malesiana* **5**: 371–374 (1957).

1. R. championii Hooker. Illustration: Botanical Magazine, 4509 (1850); Flora Malesiana **5**: 373 (1957); Menninger, Flowering trees of the world, pl. 173 (1962); Walden & Hu, Wild flowers of Hong Kong, pl. 6 (1977).

China (Yunnan, Hong Kong), Burma, Malaya, Indonesia (Sumatra). G5. Summer–spring.

13. SINOWILSONIA Hemsley
D.O. Wijnands

Deciduous bisexual tree to 8 m with stellate hairs. Leaves alternate, elliptic to broadly ovate, to 15 cm, entire. Flowers unisexual, without petals; racemes hanging, male to 6 cm, female to 3 cm. Flower-tube urn-shaped; styles projecting.

A genus of one species from China.

1. S. henryi Hemsley. Illustration: Hooker's Icones Plantarum, pl. 2817 (1909); Iconographia Cormophytorum Sinicorum **2**: 166 (1987).

C & W China. H3.

14. × SYCOPARROTIA Endress & Anliker
D.O. Wijnands

Semi-deciduous shrub to 4 m. Similar to *Parrotia* but smaller; stems not flaking. Leaves oblong-elliptic, shallowly spiny-toothed, 5–8 × 2–4 cm, often persistent. Flowers as those of *Parrotia*.

Literature:. Endress, P.K. & Anliker, J., *Mitteilungen Botanisches Museum Universität Zürich* **244** (1968).

1. S. semidecidua Endress & Anliker. Illustration: Philips & Rix, Shrubs, 47 (1989).

Garden origin. H3. Spring.

A hybrid between *Parrotia persica* and *Sycopsis sinensis*, originating in Switzerland around 1950.

15. SYCOPSIS Oliver
D.O. Wijnands

Evergreen shrubs or trees. Leaves entire, with pinnate veins, short-stalked. Flowers without petals, male or bisexual, in short racemes or heads, surrounded by softly hairy bracts; male flowers with 8–10 stamens and minute sepals; female flowers with an urn-shaped, 5-lobed flower-tube. Fruit with 2 shining brown seeds.

A genus of about 7 species distributed from NE India to China, Malaysia and New Guinea. Propagation is by cuttings of fairly ripened wood placed under heat.

1. S. sinensis Oliver. Illustration: Botanical Magazine, 655 (1973); Bean, Trees & shrubs 4, pl. 67 (1980); Hillier colour dictionary of trees & shrubs, 245 (1981); The Plantsman **4**: 51 (1982).

Evergreen shrub or small tree to 7 m. Leaves leathery, strongly veined, entire or slightly toothed towards the apex, elliptic to lanceolate. Male flowers more showy than the females; stamens 10, filaments yellow, anthers red. *C & W China.* H3. Spring.

Plants grown under the name *S. tutcheri* Hemsley, a Chinese species not in cultivation, are usually *Distylium racemosum*.

CIII. CRASSULACEAE

Succulent shrubs and herbs, sometimes small trees. Leaves spiral to opposite or whorled, simple, usually entire; stipules absent. Flowers in cymes, spikes, racemes or panicles, or occasionally solitary, usually with 5 parts, sometimes with 3, 4 or more parts; sepals almost free, persistent; petals usually as many; stamens usually in 2 whorls, twice or the same number as petals, rarely basally united; ovary superior, carpels as many as petals or sepals, usually free, with nectar gland at base, placentation almost marginal. Fruit usually a group of follicles, rarely a capsule.

A family of about 30 genera with an almost cosmopolitan distribution. Most are cultivated.

1a.	Stamens as many as petals	2
b.	Stamens twice as many as petals	6
2a.	Leaves alternate; flowering stems arising from a terminal rosette of leaves	**2. Sinocrassula**
b.	Leaves alternate, opposite or whorled; flowering stems not arising from a terminal rosette of leaves	3
3a.	Leaves opposite; flowers borne in axils of most leaves	**1. Crassula**
b.	Leaves opposite, alternate or whorled; flowers in terminal cymes or corymbs or in axils of upper leaves only	**6. Sedum**
4a.	Leaves opposite	**1. Crassula**
b.	Leaves alternate, in rosettes	5
5a.	Petals with rows of spots	**17. Graptopetalum**
b.	Petals without rows of spots	**2. Sinocrassula**

6a. Flower-parts in 4s, with petals united into a tube **3. Kalanchoe**

b. Flower-parts in 5s or more; if 4 (*Rhodiola rosea*) then petals free 7

7a. Flower-parts 4–5 or, if 6–7, then leaves not in a rosette 8

b. Flower-parts in 6s (rarely 5s) or more; leaves in rosettes (some also scattered in annual and biennial plants) 37

8a. Inflorescence terminal; leaves usually not forming a rosette 9

b. Inflorescence lateral; leaves commonly in a rosette 21

9a. Petals free or nearly so 10

b. Petals more or less united 14

10a. Inflorescence equilateral, with branches in all planes 11

b. Inflorescence a one-sided cyme with flowers all in one plane 13

11a. Flowers cup-shaped, with erect petals spreading only at the tips 12

b. Flowers rotate, with petals spreading from low down **4. Orostachys**

12a. Leaves on non-flowering shoots opposite **5. Lenophyllum**

b. Leaves on non-flowering shoots not opposite **6. Sedum**

13a. Perennial stem base covered in appressed brown scale leaves **7. Rhodiola**

b. No leaf-scales at bases of stems **6. Sedum** (including **8. × Sedadia,** **9. × Sedeveria** in part)

14a. Plants annual **10. Pistorinia**

b. Plants perennial 15

15a. Sepals small, shorter than the corolla tube 16

b. Sepals conspicuous, as long as or longer than the corolla tube, or if shorter then leaf-like **11. Villadia**

16a. Leaves alternate 17

b. Leaves opposite 20

17a. Leaves deciduous, gradually replaced by bracts 19

b. Leaves persistent, abruptly replaced by bracts 18

18a. Plant a small shrub **12. Adromischus**

b. Plant a rosette-forming herb **19. Rosularia**

19a. Plant dying back to a subterranean tuber **13. Umbilicus**

b. Plant with persistent, thick aerial stems **14. Tylecodon**

20a. Flowers erect, yellow **15. Chiastophyllum**

b. Flowers nodding, orange, reddish or yellow **16. Cotyledon**

21a. Petals united into a more or less long bell-shaped tube 22

b. Petals free or only shortly united 24

22a. Flowers vivid magenta **17. Graptopetalum**

b. Flowers white, yellow or pinkish 23

23a. Stem-forming, with loose rosettes 10 cm or more across; leaves *c.* 1 cm thick **18. Cremnophila**

b. Stemless or nearly so, with compact rosettes 2–6 cm (rarely to 11 cm) across; leaves to 5 mm thick **19. Rosularia**

24a. Petals spreading radially from midway 25

b. Petals erect or only slightly spreading above 28

25a. Inflorescence a narrow, equilateral panicle or thyrse with many short few-flowered branches **20. Thompsonella**

b. Inflorescence not as above 26

26a. Petals spotted **17. Graptopetalum**

b. Petals not spotted 27

27a. Stamens erect **21. Dudleya**

b. Stamens spreading **23. × Pachyveria**

28a. Petals with a pair of internal basal appendages **22. Pachyphytum** (including × **Pachyveria** in part)

b. Petals not appendaged 29

29a. Leaves awn-tipped, or scape with large leafy bracts, or petals spotted **24. × Graptoveria**

b. Leaves not awned, scapes with small scattered bracts only, and petals not spotted 30

30a. Petals uniformly bright yellow **9. × Sedeveria**

b. Petals not uniformly bright yellow 31

31a. Corolla showy, red, part yellow or rarely green or white, not rolled up in the bud, usually strongly pentagonal; petals thick and fleshy, sharply keeled; sepals more or less spreading; leaves narrow-based, readily detachable **25. Echeveria**

b. Corolla pallid, rolled up in bud, slightly angled only; petals thin, scarcely keeled; sepals erect or adpressed; leaves broad-based, not readily detachable **21. Dudleya**

32a. Flower-parts 5–16 33

b. Flower-parts 17–32 **26. Greenovia**

33a. Scales at base of carpels large and petal-like **27. Monanthes**

b. Scales at base of carpels not petal-like 34

34a. Petals with rows of spots **17. Graptopetalum**

b. Petals not spotted, or if spotted then spots not in rows 35

35a. Rosettes stemless, tufted (hardy plants) 36

b. Stem-forming plants (tender) 38

36a. Petals united into a short, wide, 6–8-angled tube **19. Rosularia**

b. Petals free or only shortly united 37

37a. Flower-parts 6–7; flower bell-shaped **28. Jovibarba**

b. Flower-parts 8–16; flower star-shaped **29. Sempervivum**

38a. Perennial with terminal leaf rosettes; nectar glands more or less 4-sided **30. Aeonium**

b. Annual or biennial (rarely perennial) with some scattered leaves; nectar glands 1- or 2-horned **31. Aichryson**

1. CRASSULA Linnaeus

S.G. Knees, H.S. Maxwell, R. Hyam & G.D. Rowley

Succulent annual or perennial herbs, shrublets or shrubs with cartilaginous or soft-wooded branches; rarely perennating in tubers. Leaves opposite, more or less united, membranous to thickly fleshy, persistent or deciduous. Inflorescence with 1–several dichasia, very rarely a single terminal flower; stem with leaf-like bracts. Flowers spreading or erect. Calyx of 4–7 or 12 sepals; corolla of 4–7 or 12 petals, sometimes with dorsal projection fused at the base into a short tube or open and star-shaped; stamens 4, 5 or 12 in one whorl; anthers included or exserted, nectary scales 4, 5 or 12, free; carpels 4, 5 or 12 free and constricted into styles; stigma terminal. Seeds ellipsoid, smooth or covered with tubercles.

A genus of about 150 perennial species mainly from southern Africa and a few annuals in Europe, the Americas, Australasia and Africa. Most species and cultivars can be readily propagated from cuttings.

Literature: Higgins, V., *Crassulas in cultivation* (1949); Tölken, H.R., Crassulaceae in Leistner, O.A. ed. *Flora of southern Africa*. **14** (1985).

1a. Aquatic annuals or perennials, stems trailing, rooting at the nodes 2

b. Succulent annual or perennial herbs or shrubs 3

2a. Stems 2–5 cm, solitary or little branched; flowers usually stalkless **53. aquatica**

b. Stems 2–30 cm, branched; flowers stalked **54. helmsii**

3a. Leaves 2–8 mm, scale-like, closely adpressed to form cord-like, erect stems 4

b. Combination of characters not as above 5

4a. Stems often leafless below; flowers in terminal heads; sepals *c.* 4 mm; anthers brown. **14. ericoides**

b. Stems covered with living or dead leaves; flowers borne in leaf axils; sepals *c.* 1 mm; anthers yellow **1. muscosa**

5a. Almost stemless perennial, rosettes with only the tips of leaves emerging above soil-level; anthers black; stigmas red **42. susannae**

b. Combination of characters not as above 6

6a. Leaves in almost stemless basal rosettes 7

b. Stems distinct, leafy throughout 21

7a. Leaves at least twice as long as wide, narrowly triangular to lanceolate 8

b. Leaves about as long as wide, if greater, then broadly ovate to rounded 11

8a. Stems to 25 cm, sometimes producing runners; leaves 2–5 cm 9

b. Stems to 50 cm, without runners; leaves 5–25 cm 10

9a. Flower-stalks hairless, flowers in conical panicles **19. orbicularis**

b. Flower-stalks hairy, leafy; flowers in flat-topped clusters **24. setulosa**

10a. Corolla white, pink-tinged or red; anthers dark brown **26. alba**

b. Corolla yellowish white; anthers yellow **25. vaginata**

11a. Flowers borne in stalkless clusters among the leaves **43. mesembrianthemopsis**

b. Flowers borne on clear stems above the leaves 12

12a. Leaves bluish grey, covered with hard whitish, irregularly shaped papillae **40. tecta**

b. Leaves green or reddish, or if bluish grey, then lacking papillae 13

13a. Leaves crowded together in basal apparently 2-ranked clusters **39. alstonii**

b. Leaves overlapping or free in apparently 4-ranked clusters 14

14a. Leaves broadly ovate to rounded, tapered and almost free at base 15

b. Leaves ovate to diamond-shaped, acute, closely overlapping at base 16

15a. Leaves covered with velvety down, light greyish green with a reddish tinge **51. cotyledonis**

b. Leaves covered with short white hairs, bright green and almost glossy, reddish tinged **28. capitella**

16a. Leaves distinctly grey-green, margins silky white-hairy margin, very succulent; flowers in leafy, spike-like inflorescences **29. barbata**

b. Leaves bright green to yellowish green, often tinged red, margins without silky white hairs; flower-stalks with 1 or 2 pairs of leaf-like bracts 17

17a. Calyx-lobes 0.5–1 mm **27. hemisphaerica**

b. Calyx-lobes 1–3 mm 18

18a. Calyx-lobes 1–1.5 mm 19

b. Calyx-lobes 2–3 mm 20

19a. Leaves 4–7 × 4–6 mm, ovate-elliptic **22. socialis**

b. Leaves 1–3 × 1–3 cm, obovate to rounded **21. intermedia**

20a. Stamens with black anthers **20. montana** subsp. **quadrangularis**

b. Stamens with yellow anthers **23. exilis**

21a. Leaves at least 4 times as long as wide 22

b. Leaves not more than 2.5 times as long as wide 26

22a. Leaves 1.2–3.5 cm wide **47. perfoliata**

b. Leaves 0.1–0.8 cm wide 23

23a. Leaves 0.8–3 cm; sepals 1–2.5 mm; petals 3–5 mm 24

b. Leaves 1.8–10 cm; sepals 0.5–2 mm; petals 1–4 mm 25

24a. Stems hairy when young; leaves 1–3 cm; sepals bluntly acute **15. sarcocaulis**

b. Stems hairless, even when young; leaves 0.8-1.5 cm; sepals rounded **50. subaphylla**

25a. Leaves 0.1–0.4 cm wide; sepals 1–2 mm; petals cream **16. tetragona**

b. Leaves 0.3–1.5 cm wide; sepals 0.5–1 mm; petals white tinged pink **33. macowaniana**

26a. Leaves opposite perfoliate, stem clearly visible between each pair of leaves 27

b. Leaves free, or if united then stem not visible between the leaves 32

27a. Leaves 2–8 cm **44. grisea**

b. Leaves 0.3–2 cm 28

28a. Stems less than 10 cm **4. deltoidea**

b. Stems 10–60 cm 29

29a. Corolla star-shaped; anthers white or purple **5. pellucida**

b. Corolla tubular; anthers brown, black or yellow 30

30a. Leaves 1.5–2 cm; anthers black or yellow **32. brevifolia**

b. Leaves 0.3–2 cm; anthers brown 31

31a. Petals 2–2.5 mm, cream or pale yellow **30. perforata**

b. Petals 3–4 mm, white, pink or red **31. rupestris**

32a. Stout-stemmed, shrubs with rounded to ovate leaves and stalked terminal inflorescences 33

b. Dwarf succulents with stems covered by leaves, or if shrubby then stems slender; flowers stalked or stalkless 38

33a. Leaves 1.0–2.5 cm; anthers yellow **8. cordata**

b. Leaves 2–10 cm; anthers purple or black 34

34a. Corolla tubular; anthers black **49. cultrata**

b. Corolla star-shaped; anthers purple 35

35a. Leaf-stalks 0.5–2 cm, blades with slightly recurved margins **9. multicava**

b. Leaves often stalkless, or if stalked then stalk not more than 5 mm 36

36a. Leaves bright glossy green; petals 7–10 mm **11. ovata**

b. Leaves dull greyish green or whitish green with distinct red margins, not glossy or shiny 37

37a. Leaves stalkless; petals 4–7 mm **10. lactea**

b. Leaf-stalks to 5 mm, occasionally absent; petals 7–10 mm **12. arborescens**

38a. Stems obscured by leaves, forming tightly packed, column-like growths 39

b. Stems clearly visible between the leaves 45

39a. Leaves 0.4–1.2 cm, ovate to diamond-shaped, closely adpressed to stem, forming a tight square-edged column **37. pyramidalis**

b. Leaves not as above 40

40a. Leaves at least 3 times as long as wide, narrowly elliptic to triangular 41

b. Leaves as wide as or wider than long, broadly ovate, deeply keeled below 42

41a. Cushion-forming perennial, with leaves clustered at the ends of the branches; petals 5–7 mm **41. ausensis**

b. Erect perennial or biennial; petals 8–12 mm **36. alpestris**

42a. Leaves 3–4 × 1–15 mm, adpressed to stem, forming a smooth column to 1.5 cm across **35. barklyi**

b. Leaves 0.5–1.5 × 0.6–2.5 mm, closely adpressed, but not forming smooth columns 43

43a. Flowers borne in tight rounded heads, with stalks obscured by leaves **34. columnaris**

b. Flowers borne on slender stalks 2–8 cm above the leaves 44

44a. Unbranched perennial, occasionally branched from the base; leaves dark greyish green **45. plegmatoides**

b. Well-branched perennial; leaves light blue grey-green, conspicuously dotted **46. deceptor**

45a. Leaf-stalks 5–15 mm **13. nemorosa**

b. Leaves stalkless or very shortly stalked 46

46a. Usually tuberous rooted perennial to 8 cm, with thick woody stems at base; leaves 3–5 mm; flowers concealed among the leaves **3. corallina**

b. Annual or perennial herbs, stems thin, or if woody then flowers stalked and held clear of the leaves 47

47a. Leaves 2–12 mm, stems very slender 48

b. Leaves 1.2–8 cm, stems variable in width 49

48a. Erect annual herb; stems and leaves hairless **2. dichotoma**

b. Creeping perennial; stems rooting at the nodes, stems and leaves very hairy **38. lanuginosa**

49a. Leaves with conspicuous marginal hairs **17. ciliata**

b. Leaf margins hairless but sometimes dotted or toothed 50

50a. Shrubby perennials with stout, well-branched, more or less erect stems 51

b. Slightly woody perennials with weak, often wiry, prostrate or scrambling stems 53

51a. Sepals 1.5–3 mm; anthers yellow **52. nudicaulis**

b. Sepals 3–22 mm; anthers black 52

52a. Sepals 3–3.5 mm; corolla white or cream **48. mesembrianthoides**

b. Sepals 1.5–2.2 cm; corolla bright red or white tinged red **18. coccinea**

53a. Stems usually 4-sided; leaves often cordate at base **6. spathulata**

b. Stems rounded, arising from irregularly shaped tubers; leaves elliptic-ovate, constricted at base **7. sarmentosa**

1. C. muscosa Linnaeus (*C. lycopodioides* Lamarck; *C. ericoides* misapplied). Illustration: Flora of southern Africa **14**: 127 (1985); Graf, Exotica, series 4, **1**: 873 (1985).

Scrambling perennial, 10–40 cm, branches woody, hidden by dead and living leaves, not more than 5 mm across. Leaves 2–8 × 1–4 mm, conspicuously 4-ranked, ovate to triangular, flat, pointed or blunt, hairless, leathery, green tinged yellow, grey or brown; leaf sheath 1.5 cm. Flowers solitary or densely clustered in groups of 2–8, malodorous. Sepals *c.* 1 mm, triangular; petals *c.* 2 mm, erect, triangular, keeled, yellow-green to brown; stamens 0.5–1 mm, anthers yellow; ovaries tapered, style less than ½ ovary length. *Southern Africa.* G1. Summer.

2. C. dichotoma Linnaeus (*Grammanthes gentianoides* (Lamarck) de Candolle; *Crassula gentianoides* Lamarck; *Vauanthes dichotoma* (Linnaeus) O. Kuntze). Illustration: Botanical Magazine, 4607 (1851); 6401 (1878); Flora of southern Africa **14**: 127 (1985).

Thin-stemmed annual, 5–25 cm. Leaves 5–12 × 1–6 mm, elliptic-linear, pointed or blunt, hairless; black hydathodes often present. Flowers tubular, subtended by bracts, borne in terminal cymes. Sepals 4–6 mm, lanceolate-elliptic, blunt; petals 0.7–2 cm, elliptic-oblanceolate, blunt or pointed, orange-yellow; stamens 0.5–1.2 cm, anthers yellow; style ½ ovary length. *South Africa (Cape Province).* G1. Spring–summer.

3. C. corallina Thunberg. Illustration: Higgins, Crassulas in cultivation, 36 (1949); Jacobsen, Lexicon of succulent plants, pl. 43 (1974).

Perennial herbs to 8 cm, often with tuberous tap-root; stems sometimes woody below. Leaves 3–5 × 2–5 mm, obovate, stalkless, tapering at tip and base, convex above and below, warty, with a flaking waxy surface. Flowers solitary or clustered, partially obscured by the upper leaves; sepals 1–2 mm, triangular, blunt, grey; petals 2–3.5 mm, obovate-oblong, rounded at tip, pouched below, reflexed above, cream; stamens 1.5–2 mm, anthers yellow; style short or absent. *South Africa, Namibia.* G1.

4. C. deltoidea Thunberg (*C. rhomboidea* N.E. Brown). Illustration: Higgins, Crassulas in cultivation, 40 (1949); Graf, Exotica, series 4, **1**: 881 (1985).

Perennial shrub to 10 cm, with fleshy branching stems. Leaves 1–2 × 0.4–1.5 cm, oblanceolate-ovate or diamond-shaped, fused at base, blunt or pointed, flat above, convex below, grey-green; surface covered in flaking wax. Flowers in rounded terminal clusters; sepals 0.5–2 mm, rounded, hairless; petals 3.5–5 mm, oblanceolate-elliptic, creamy white; stamens 3–4.5 mm, anthers black; styles ½ carpel length. *SE Africa.* G1.

5. C. pellucida Linnaeus (*C. centauroides* Aiton). Illustration: Botanical Magazine, 1765 (1815); Higgins, Crassulas in cultivation, 58 (1949).

Perennials, or rarely annuals, with prostrate stems to 60 cm. Leaves 1–2 × 0.5–1.2 cm, ovate-elliptic, green, sometimes with brown stripes; margin colourless or red. Flowers solitary or bundled, often hidden by leaves. Sepals 2.5–5 mm, acute, green or colourless; petals 3–5 mm, elliptic, acute, white tinged pink; stamens with white or purple anthers. *South Africa.* G1. Spring–summer.

6. C. spathulata Thunberg (*C. cyclophylla* Schönland & E.G. Baker; *C. latispathulata* Schönland & E.G. Baker). Illustration: Higgins, Crassulas in cultivation, 66 (1949).

Prostrate perennial, sparsely branching, rooting, horizontal stems to 20 cm. Leaves 2–3 × 1.5–2.5 cm, ovate, blunt, base cordate; margin with forward-pointing teeth, often tinged red; leaf-stalk short. Sepals 1–2 mm, triangular-linear, blunt, hairless, tinged red; petals 3.5–5 mm, linear-lanceolate, hooded, white tinged pink; stamens 3–4 mm, anthers purple-pink; styles as long as ovaries. *South Africa (Cape Province).* G1. Autumn.

7. C. sarmentosa Harvey (*C. ovata* E. Meyer). Illustration: Jacobsen, Lexicon of succulent plants, pl. 48 (1974); Graf, Exotica, series 4, **1**: 884 (1985).

Tuberous rooted, scrambling perennial with sparsely branched, climbing or hanging stems to 1 m. Leaves 2–6 × 1–3 cm, ovate-elliptic, pointed, flat, green; hydathodes spread below; margins tinged red, entire or with forward-pointing teeth; leaf-stalk short or absent. Inflorescence terminal, flat or rounded. Sepals 1–3 mm, triangular-linear, pointed; petals 4–8 mm, linear-lanceolate, cream-white, tinged pink, ridged with several projections; stamens 4–6 mm, anthers white, tinged red; styles longer than ovaries. *South Africa (Natal to Cape Province)*. G1. Winter.

8. C. cordata Thunberg. Illustration: Higgins, Crassulas in cultivation, 37 (1949); Graf, Exotica, series 4, **1**: 884 (1985).

Perennial to 25 cm, with erect or horizontal, sparsely branched, woody stems. Leaves shortly stalked, 1–2.5 × 0.8–2 cm, ovate, blunt, cordate, grey-green, with red hydathodes scattered above; margin entire, often tinged red. Sepals 1–2 mm, triangular, pointed or blunt; petals 4–5 mm, lanceolate, with a drawn-out point, ridged, light yellowish cream, tinged red; stamens 2–3.5 mm, anthers yellow; style as long as ovaries. *South Africa (Cape Province)*. G1. Early spring.

9. C. multicava Lemaire. Illustration: Higgins, Crassulas in cultivation, 55 (1949); Everett, New York Botanic Garden illustrated encyclopedia of horticulture **3**: 904 (1981); Graf, Exotica, series 4, **1**: 881 (1985).

Prostrate or erect perennial to 40 cm; stems sparsely branched, swollen and rooting at the nodes, woody below. Leaves 2–6.5 × 1.5–4 cm, oblong-ovate or elliptic, tip blunt or notched, hydathodes conspicuous above; margin entire, curved under; leaf-stalk 0.5–2 cm, sheathed at base. Sepals 4 or 5, triangular, 1–2 mm, ridged; petals 4 or 5, lanceolate, pointed, ridged, 3–6 mm, white or cream, red at tip; stamens 3–5 mm, anthers purple; style as long as ovaries. *South Africa (Cape Province)*. G1. Autumn.

10. C. lactea Solander. Illustration: Higgins, Crassulas in cultivation, 48 (1949); RHS dictionary of gardening **2**: 567 (1974); Jacobsen, Lexicon of succulent plants, pl. 44 (1974); Everett, New York Botanic Garden illustrated encyclopedia of horticulture **3**: 904 (1981).

Perennial to 20 cm, with occasionally scrambling, thick horizontal stems to 40

cm. Leaves 2.5–7 × 1–3 cm, stalkless, oblanceolate, tapering towards base, convex above and below, dull green; margin entire, horny, white dotted. Sepals 1.5–3 mm, linear, pointed, very fleshy; petals 4–7 mm, lanceolate, yellow-white, pink above, with a projection; stamens 4–5.5 mm, anthers purple; style equal to or exceeding ovary length. *South Africa (Cape Province)*. G1. Summer.

11. C. ovata (Miller) Druce (*Cotyledon ovata* Miller; *Crassula argentea* Thunberg; *C. portulacea* Lamarck; *C. obliqua* Solander; *C. arborescens* misapplied). Illustration: Higgins, Crassulas in cultivation, 50 (1949); Jacobsen, Lexicon of succulent plants, pl. 45, 46 (1974); Everett, New York Botanic Garden illustrated encyclopedia of horticulture **3**: 903 (1981); Graf, Tropica, edn 3, 364, 366 (1986).

Perennial shrub 1–2 m, with much branched stems to 20 cm across. Leaves 2–3 × 1–1.8 cm, elliptic, shiny; margin often red, horny; sometimes stalked, to 4 mm. Flowers in rounded clusters on stems 1–3 cm. Sepals 1–2 mm, triangular; petals 7–10 mm, elliptic-lanceolate, pointed, hooded; stamens with purple anthers. *South Africa (Natal to Cape Province)*. G1. Winter.

Very popular as a house plant grown under the name 'Jade Plant'. 'Hummels Sunset' has leaves that turn red in sunlight.

12. C. arborescens (Miller) Willldenow (*Cotyledon arborescens* Miller). Illustration: Botanical Magazine, 384 (1897); Higgins, Crassulas in cultivation, 26 (1949); Jacobsen, Lexicon of succulent plants, pl. 42 (1974); Graf, Exotica, series 4, **1**: 878, 879 (1985).

Shrubby perennial 1–2 m, to 6 cm across at base, with freely branched stems; bark peeling. Leaves 3–7 × 2–4 cm, obovate to round, tip obtuse, base tapered; surface grey bloomed; margin entire, horny, often tinged purple; leaf-stalk absent or to 5 mm. Flowers in rounded heads, with several branches. Sepals 5–7, 1–1.5 mm, triangular, ridged, pointed; petals, 5–7, 7–10 mm, lanceolate, hooded, cream tinged red at tip; stamens 4, 5–6 mm, anthers purple; style ½ length of ovary. *South Africa (Cape Province)*. G1. Summer.

13. C. nemorosa (Ecklon & Zeyher) Endlicher (*Petrogeton nemorosum* Ecklon & Zeyher; *C. confusa* Schönland & E.G. Baker; *C. coerulescens* Schönland). Illustration:

Cactus and succulent journal of Great Britain **40**: 53 (1978).

Tuberous perennial with prostrate or erect stems 4–15 cm. Leaf blade 4–13 mm across, nearly round or ovate, apex blunt, grey or brown-green; margin entire; leaf-stalk 3–15 mm. Sepals ovate-triangular, pointed or blunt; petals 2–3.5 mm, lanceolate with a long thin point, slightly ridged below, green-yellow; stamens 1.5–2 mm, anthers yellow; styles ½ ovary length. *South Africa (Western Cape Province)*. G1.

14. C. ericoides Haworth. Illustration: Jacobsen, Lexicon of succulent plants, pl. 44 (1974).

Erect perennial to 12 cm with woody branches. Leaves in 4 series, 3–7 × 1–3 mm, ovate-lanceolate, stalkless, pointed; margins entire. Flowers partly concealed by upper leaves; corolla tubular. Sepals 2–4 mm, linear; petals 3–5 mm, elliptic, pointed, white, ridged on back; stamens 3–4 mm, anthers brown; styles as long as ovaries. *South Africa (Natal & Cape Province)*. G1. Summer.

15. C. sarcocaulis Ecklon & Zeyher. Illustration: Higgins, Crassulas in cultivation, 64 (1949); Graf, Exotica, series 4, **1**: 884 (1985).

Shrubby perennial with many branched, erect, hairy stems 20–60 cm. Leaves 6–30 × 1–8 mm, lanceolate-elliptic, pointed, stalkless, compressed or almost round in cross-section; margin entire, occasionally hairy. Inflorescence terminal, dense; sepals 1–2 mm, triangular, ridged; petals 3–5 mm, oblanceolate-oblong, ridged but without a projection, white; stamens 2.5–3 mm, anthers brown; styles ½ ovary length, ovaries kidney-shaped. *Southern Africa*. G1. Late summer.

16. C. tetragona Linnaeus. Illustration: Higgins, Crassulas in cultivation, 70 (1949); Graf, Exotica, series 4, **1**: 881 (1985); Graf, Tropica, edn 3, 364 (1986).

Perennial with erect, branched, hairless stems to 1 m. Leaves 1.8–5 cm × 1–4 mm, lanceolate, pointed, stalkless. Sepals 1–2 mm, triangular, blunt; petals 1–3 mm, oblanceolate, elliptic, rounded at tip, ridged on back, but without a projection, cream; stamens 1–2 mm, anthers brown; styles less than ⅓ ovary length. *South Africa (Cape Province)*. G1. Summer.

17. C. ciliata Linnaeus. Illustration: Higgins, Crassulas in cultivation, 30 (1949).

Perennial shrublets to 20 cm, well

branched, with persistent old leaves. Leaves 1.5–3 × 0.5–1.2 mm, rounded, oblong to elliptic; margins with green to yellowish hairs. Inflorescence domed on stems 15–25 cm. Sepals 2–2.5 mm, triangular, green-yellow; petals 3.5–4.5 mm, elliptic, cream to pale yellow; stamens with yellow anthers. *South Africa (Cape Province)*. G1. Summer.

18. C. coccinea (Linnaeus) Linnaeus (*Rochea coccinea* Linnaeus). Illustration: Botanical Magazine, 495 (1799); Jacobsen, Lexicon of succulent plants, pl. 122 (1974); Graf, Exotica, series 4, **1**: 919, 920 (1985).

Shrubby perennial to 60 cm, with few branched stems, well covered with leaves. Leaves 1.2–2.5 × 0.4–1.5 cm, elliptic to ovate, pointed, green often tinged red; margin with hairs, curved upwards. Flowers stalkless; sepals 1.5–2.2 cm, lanceolate, unequal, pointed; corolla tubular, 3.5–4.5 cm, petals spathulate, recurved, red or white tinged red with a dorsal appendage; stamens 2–3 cm, anthers black; style 1 or 2 times length of ovary. *South Africa (Cape Province)*. G1. Summer.

19. C. orbicularis Linnaeus (*C. rosularis* Haworth). Illustration: Higgins, Crassulas in cultivation, 49 (1949); Jacobsen, Lexicon of succulent plants, pl. 45 (1974); Graf, Exotica, series 4, **1**: 880, 883 (1985).

Perennial, 5–25 cm, stems sometimes producing runners. Leaves packed in basal rosettes, 1.5–5 × 0.5–2 mm, elliptic to oblanceolate, pointed; hydathodes often visible on upper surface and margins, but not on lower surface; margin with hairs, sometimes tinged red. Inflorescence with scale-like leaves, flowers stalked. Sepals 1–3 mm, lanceolate-triangular, blunt; petals 2–5 mm, oblanceolate, blunt or pointed, pale yellow-white, tinged pink, often with a dorsal appendage; stamens 2.5–3.5 mm, anthers yellow; styles ½ ovary length. *South Africa (Natal)*. G1. Summer.

20. C. montana Thunberg subsp. **quadrangularis** (Schönland) Tölken (*C. quadrangularis* Schönland). Illustration: Higgins, Crassulas in cultivation, 49 (1949).

Cushion-forming, rosetted perennial reaching 4–10 cm. Leaves 0.4–2 × 0.4–1.7 cm, in 4 ranks, obovate to ovate, pointed, adpressed or spreading, greenish brown, dark-spotted; hydathodes red, on margins and over upper surface; margins with

hairs. Flowering stems to 10 cm, flowers stalkless, in flat-topped axillary clusters. Sepals 2–3 mm, triangular; petals 3–6 mm, oblong, blunt, reflexed, white, tinged pink, with dorsal projection; stamens 2.5–3 mm, anthers black; style less than ¼ of ovary length. *South Africa (Cape Province)*.

21. C. intermedia Schönland. Illustration: Higgins, Crassulas in cultivation, 31 (1949).

Perennial to 20 cm, like *C. montana* subsp. *quadrangularis* except for flowers with yellow anthers and short reflexed styles. *South Africa (Cape Province)*. G1. Summer.

22. C. socialis Schönland. Illustration: Higgins, Crassulas in cultivation, 68 (1949); Jacobsen, Lexicon of succulent plants, pl. 48 (1974); Everett, New York Botanic Garden illustrated encyclopedia of horticulture **3**: 906 (1981); Graf, Exotica, series 4, **1**: 882 (1985).

Perennial to 6 cm, like *C. montana* subsp. *quadrangularis* but lower growing, and having pure white flowers with yellow anthers and short, reflexed styles. *South Africa (Cape Province)*. G1. Summer.

23. C. exilis Harvey (*C. bolusii* J.D. Hooker). Illustration: Higgins, Crassulas in cultivation, 35 (1949).

Annual or cushion forming perennial to 10 cm. Leaves 4–4.5 × 1–10 mm, in 4 ranks; blade flat or rounded in cross-section, tip pointed; margin with hairs. Flowers stalked in terminal clusters; sepals 2–2.5 mm, triangular, pointed, margins hairy; petals 4–4.5 mm, oblong-obovate, with a projection, white, tinged pink; stamens 3–3.5 mm, anthers yellow. Fruit spreading at right angles to flower-stalk. *South Africa (Cape Province)*. G1.

Subsp. **cooperi** (Regel) Tölken (*C. cooperi* Regel; *C. picturata* Boom), has dense spreading cushions and oblanceolate, less fleshy leaves.

A complex group of garden hybrids including *C. justi-corderoyi* misapplied and *C. picturata* hybrids, would also key out here.

24. C. setulosa Harvey (*C. curta* N.E. Brown; *C. milfordiae* Byles). Illustration: Higgins, Crassulas in cultivation, 54 (1949).

Woody-stemmed perennial to 25 cm. Leaves 2–3.5 × 0.1–1 cm, pointed, flat, green tinged red, 4-ranked or densely packed in basal rosettes; hydathodes present on margins and upper surface. Stem-leaves reduced towards top.

Inflorescence hairy, bracts triangular. Sepals 1–3 mm, lanceolate-triangular; petals 2.5–4 mm, oblong, white tinged red, with a projection; stamens 2–3 mm, anthers yellow-brown; style ⅓–⅔ ovary length. *Widespread in southern Africa*. G1. Summer.

25. C. vaginata Ecklon & Zeyher (*C. drakensbergensis* Schönland; *Sedum crassifolium* O. Kuntze). Illustration: Flora of southern Africa **14**: 175 (1985).

Tuberous rooted perennial to 50 cm. Leaves 5–25 × 0.3–3.5 cm, linear-lanceolate, fused at base, margin with hairs, in basal rosettes. Flowers numerous; sepals 1.5–3.5 mm, triangular, fine pointed, sometimes toothed; petals 2.5–5 mm, oblong, blunt, hooded, yellowish white, appendage very small; stamens 2.5–3.5 mm, anthers yellow; style ⅓ ovary length. *South Africa to Arabia*. G1. Summer.

26. C. alba Forskål (*C. recurva* N.E. Brown). Illustration: Flora of southern Africa **14**: 175 (1985).

Biennial or perennial to 50 cm. Leaf pairs spirally arranged, leaves mainly in basal rosettes, 6–17 × 0.5–1.5 cm, linear-lanceolate, stalkless, acute, dark geenish yellow, tinged purple; margin with fine recurved hairs. Flowers numerous; sepals 1–4.5 mm, triangular; petals 3–6 mm, oblong, pointed, hooded and recurved at tip, white or red; stamens 3–5 mm, anthers dark brown. *Southern Africa*. H5.

27. C. hemisphaerica Thunberg (*Purgosea hemisphaerica* (Thunberg) G. Don). Illustration: Higgins, Crassulas in cultivation, 31 (1949); Jacobsen, Lexicon of succulent plants, pl. 43 (1974); Everett, New York Botanic Garden illustrated encyclopedia of horticulture **3**: 905 (1981); Graf, Tropica, edn 3, 367 (1986).

Perennial, 5–12 cm, with leaves 4-ranked in few rounded rosettes. Leaves 1–5 × 0.8–2.5 cm, obovate, bristle-tipped, stalkless; hydathodes on margins and upper surface; margins with hairs. Inflorescence long and tapering, with many small stalkless flowers. Sepals 0.5–1 mm, triangular, hairless; petals 2–3 mm, oblong-lanceolate, pointed or rounded with a dorsal ridge and projection, creamy white; stamens 1.5–2.5 mm, anthers black; style short. *South Africa (Cape Province)*. G1. Spring.

28. C. capitella Thunberg. Illustration: Graf, Exotica, series 4, **1**: 868, 869 (1985).

Perennial or biennial to 40 cm, stems usually woody. Leaves 1–12 × 0.3–2 cm, the pairs spirally arranged in basal rosettes, usually smallest near the centre of the rosette, ovate to linear-lanceolate, pointed, stalkless, spreading, hairless or hairy, grooved, often red spotted; hydathodes red, scattered over upper surface; margin with fine recurved hairs or papillae. Inflorescence hairless or hairy, spike-like. Sepals 1–4 mm, triangular-lanceolate, pointed, with marginal hairs; petals 2–5 mm, oblong-lanceolate, blunt, hooded, white tinged pink, with a projection on the back; stamens 2–4.5 mm, anthers dark brown; style very short or absent. *South Africa (Cape Province, Lesotho, Orange Free State & Transvaal)*. G1.

A complex species with two subspecies in cultivation. Subsp. **nodulosa** (Schönland) Tölken (*C. nodulosa* Schönland; *C. elata* N.E. Brown) has tuberous bases and reflexed leaves, a hairy inflorescence with hairy sepals and short or absent style. Subsp. **thyrsiflora** (Thunberg) Tölken (*C. thyrsiflora* Thunberg; *C. turrita* Thunberg) has hairless leaves, except for the margins, hairless sepals which are occasionally toothed, petals with rounded appendages and style ⅓–½ of ovary length.

29. C. barbata Thunberg. Illustration: Higgins, Crassulas in cultivation, 13 (1949); Graf, Exotica, series 4, **1**: 880, 883 (1985).

Annual or biennial to 30 cm. Leaves spreading with erect tips in 4 series and basal rosettes; blade 1–4 × 1–3.5 cm, obovate to nearly round, greyish green; margin with long, tufted hairs. Flowers stalkless in spiked racemes. Sepals 2.5–3 mm, oblong-triangular, blunt; petals 4.5–6 mm, oblong, rounded or pointed with a dorsal projection, white tinged pink; stamens 3–4.5 mm, anthers black, style very short, broad. *South Africa (Cape Province)*. H5–G1. Spring.

30. C. perforata Thunberg. Illustration: Higgins, Crassulas in cultivation, 59 (1949); Jacobsen, Lexicon of succulent plants, pl. 46 (1974); Everett, New York Botanic Garden illustrated encyclopedia of horticulture **3**: 906 (1981); Graf, Exotica series 4, **1**: 873 (1985).

Shrubby or scrambling perennials with little-branched stems, partially covered in old leaves. Leaves 4–20 × 3–15 mm, ovate, stalkless, united in pairs, pointed or blunt, convex below; hydathodes on upper surface only; margins hairless or with hairs, reddish yellow, horny. Inflorescence

with adpressed bracts, flowers stalkless. Sepals 0.5–1 mm, triangular; petals 2–2.5 mm, oblong-elliptic, with small projection; stamens 2 mm, anthers brown; styles ⅓ ovary length. *South Africa (Natal & Cape Province)*. G1.

31. C. rupestris Thunberg. Illustration: Higgins, Crassulas in cultivation, 67 (1949); Jacobsen, Lexicon of succulent plants, pl.47 (1974); Graf, Exotica, series 4, **1**: 873, 880 (1985); Graf, Tropica, edn 3, 366 (1986).

Shrubby perennial to 50 cm, with erect, well-branched stems and flaking bark. Leaves 3–15 × 2–13 mm, ovate-lanceolate, stalkless, united in pairs, hairless, brownish red; hydathodes red, spread over upper surface; margins horny, yellow-red. Inflorescence domed; sepals to 1 mm, triangular; petals 3–4 mm, oblong-elliptic, tip rounded, ridged, projection very small, white tinged red; stamens 2–3.5 mm, anthers brown; style ½ ovary length. *South Africa (Cape Province)*. G1.

32. C. brevifolia Harvey. Illustration: Higgins, Crassulas in cultivation, 67 (1949); Graf, Exotica, series 4, **1**: 871 (1985).

Shrubby perennial to 50 cm, stems branched. Leaves 1.5–5 × 0.2–0.6 cm, oblong elliptic, sharply tapered to a blunt tip, spreading and upturned, convex above, convex or keeled below, very fleshy, greyish green, hydathodes on upper surface and margins; margins horny at edge. Inflorescence domed; sepals triangular, blunt; petals 3–5 mm, oblong-elliptic, rounded, ridged with a projection; stamens 2.5–4 mm, anthers black to yellow; styles ⅓ length of ovary. *Namibia & South Africa (Cape Province)*. G1.

33. C. macowaniana Schönland & E.G. Baker. Illustration: Flora of southern Africa **14**: 194 (1985).

Perennial shrubs with branched, sometimes prostrate stems and flaking bark. Leaves 2.4–8 × 0.3–1.5 cm, linear-lanceolate, stalkless, pointed, flat above, convex below, hairless, brownish green. Flowers in domed clusters; sepals 0.5–1 mm, oblong, rounded; petals 2.5–4 mm, oblanceolate, rounded, ridged, white tinged pink, projection very small; stamens 2.5–3 mm, anthers black; style ½ ovary length; ovaries tapered above. *Namibia, South Africa (Cape Province)*. G1. Summer.

34. C. columnaris Thunberg. Illustration: Higgins, Crassulas in cultivation, 13 (1949); Marshall Cavendish encyclopedia

of gardening **2**: 382 (1968); Flora of southern Africa **14**: 194 (1985); Graf, Exotica, series 4, **1**: 871, 875 (1985).

Biennial or perennial, 3–10 cm, often monocarpic. Leaves 0.3–1.2 × 1–2.5 cm, transversely ovate, blunt, stalkless, grey or brownish green, clasping and dish-shaped with erect hairy margin, densely packed into an erect column 2–3 cm across. Flowers partially hidden by upper leaves, nearly stalkless; sepals 3–4 mm, oblong-elliptic, rounded; petals 7–13 mm, oblong rounded with a small dorsal projection, white or yellowish, tinged red. Stamens 4.5–6 mm, anthers brown to yellow; style very short. *South Africa & Namibia*.

35. C. barklyi N.E. Brown (*C. teres* Marloth). Illustration: Botanical Magazine, 8421 (1921); Higgins, Crassulas in cultivation, 68 (1949); Everett, New York Botanic Garden illustrated encyclopedia of horticulture **3**: 906 (1981); Graf, Tropica, edn 3, 362 (1986).

Perennial 5–9 cm with horizontal or erect stems branching at base. Leaves 3–4 × 1–15 mm, adpressed to stem to form a smooth column up to 1.5 cm across; stalkless, transversely ovate, convex below, concave above, dish-shaped; margin membranous with erect dense hairs. Flowering-stems almost stalkless, hidden by upper leaves. Sepals 4–5 mm, oblanceolate to oblong, rounded; petals hooded, with a projection, cream; stamens 4.5–5 mm, anthers yellow, style short, wider ovary conical. *South Africa (Cape Province)*. H5–G1. Winter.

36. C. alpestris Thunberg. Illustration: Higgins, Crassulas in cultivation, 25 (1949).

Perennial, or rarely biennial, erect and sometimes branched. Leaves 10–20 × 5–8 mm, triangular, sharply acute, leathery, green-brown. Sepals 2.5–3 mm, rounded with spreading marginal hairs, green-brown; petals 8–12 mm, fused at base, elliptic-oblong, white or cream, often tinged red; stamens with brown anthers. *South Africa (Cape Province)*. G1. Winter-spring.

37. C. pyramidalis Thunberg. Illustration: Higgins, Crassulas in cultivation, 50 (1949); Jacobsen, Lexicon of succulent plants, pl. 47 (1974); Flora of southern Africa **14**: 194 (1985); Everett, New York Botanic Garden illustrated encyclopedia of horticulture **3**: 906 (1981).

Perennial with simple or branched, erect or horizontal stems 3–25 cm. Leaves

4–12 × 4–8 mm, ovate, adpressed to stem forming a more or less square column; margins hairless, horny. Inflorescence terminal, with numerous stalkless flowers. Sepals 3–4 mm, lanceolate to spathulate; petals white-cream, 7–11 mm, oblong, hooded, projection transparent; stamens 3.5–4.5 mm, anthers yellow; styles indistinct. *South Africa (Cape Province)*. G1. Spring.

38. C. lanuginosa Harvey. Illustration: Higgins, Crassulas in cultivation, 51 (1949).

Perennial to 15 cm, with horizontal or drooping stems, occasionally rooting at the nodes. Leaves 2–35 × 2–8 mm, obovate to linear-lanceolate, pointed or blunt, convex above and below, hairy; margin with recurved hairs. Flowers terminal, very shortly stalked, clustered; sepals 1.5–2 mm; anthers black; ovaries tapered to lateral stigmas. *South Africa (Cape Province)*. G1. Summer.

39. C. alstonii Marloth. Illustration: Higgins, Crassulas in cultivation, 13 (1949); Jacobsen, Lexicon of succulent plants, pl. 41 (1974); Flora of southern Africa **14**: 203 (1985); Graf, Exotica, series 4, **1**: 872 (1985).

Perennial herb with leaves in 2 rows, forming dense round rosettes 2–5 cm across. Leaf-blade 0.6–1 × 1.2–2 cm, rounded or obovate, tip erect, hairy beneath. Flowers stalkless; sepals triangular, pointed, hairy, margins with hairs; petals 2.5–3 mm, lanceolate-oblong, yellowish cream, pointed; stamens 1.5–2 mm; brownish yellow; stigmas red; ovaries kidney-shaped. *South Africa (Cape Province)*. H3. Autumn.

40. C. tecta Thunberg. Illustration: Higgins, Crassulas in cultivation, 68 (1949); Jacobsen, Lexicon of succulent plants, pl. 48 (1974); Graf, Exotica, series 4, **1**: 885 (1985).

Perennial to 20 cm, often branched to form tufts. Leaves 2–3.5 × 0.5–1.5 cm, oblong-lanceolate, rounded, stalkless, conspicuously white, with papillae, hairy below. Flowers numerous, almost stalkless, in rounded, terminal clusters. Sepals 1.5–2 mm, triangular, hairy; petals 3–4 mm, oblong-lanceolate, blunt with a projection, white-cream; stamens 1.5–2 mm, anthers yellow; stigmas lateral, style very short or absent. *South Africa (Cape Province)*. G1. Summer.

41. C. ausensis P.C. Hutchison (*C. hofmeyeriana* Dinter). Illustration: Higgins, Crassulas in cultivation, 27 (1949).

Perennial with short branches, and tightly grouped, persistent leaves forming dense cushions. Leaves 1.2–3 × 0.4–1 cm, lanceolate to elliptic, sharply acute, flat above, convex below, covered with short hairs, green or brownish. Inflorescence umbel-like, with flowers on stalks 2–4.5 cm. Sepals 2–3 mm, triangular, green, fleshy; petals 5–7 mm, fused at base for *c.* 1 mm, oblong, acute, white-cream, dorsal appendage indistinct; stamens with brown anthers. *Namibia*. G1. Autumn.

42. C. susannae Rauh & Friedrich. Illustration: Jacobsen, Lexicon of succulent plants, pl.48 (1949); Graf, Exotica, series 4, **1**: 871 (1985).

Perennial 1–4 cm. Leaves 6–10 × 4–8 mm, stalkless, oblong, fleshy, channelled, in 4-ranked rosettes. Flowers few, in terminal racemes. Sepals 1.5–2 mm, triangular-oblong, bluntly pointed, hairy; petals 2.5–3.5 mm, sharply pointed, oblong, ridged, white, projection very small; stamens 2–2.5 mm, anthers black; styles short, wide, stigmas lateral, red. *South Africa (Cape Province)*. G1. Autumn.

43. C. mesembrianthemopsis Dinter. Illustration: Higgins, Crassulas in cultivation, 32 (1949); Jacobsen, Lexicon of succulent plants, pl. 45 (1974); Graf, Exotica, series 4, **1**: 871 (1985); Graf, Tropica, edn 3, 367 (1986).

Similar to *C. susannae*, except: leaves wedge-shaped, triangular in cross-section, not grooved; flowers short-stalked, partially hidden by leaves; sepals to 3 mm, petals 5–7 mm, stamens 3–4 mm, anthers yellow. *Namibia & South Africa (Cape Province)*. G1. Autumn.

44. C. grisea Schönland. Illustration: Higgins, Crassulas in cultivation, 45 (1949); Jacobsen, Lexicon of succulent plants, pl. 43 (1974); Flora of southern Africa **14**: 210 (1985).

Perennial with few erect stems to 20 cm. Leaves persistent, scarcely shrivelling, 2–8 × 0.3–0.6 cm, linear-lanceolate, grey-green, hairless, hairy or covered with rounded papillae. Flowers in terminal clusters on stalks 2–10 cm. Sepals 1–1.5 mm, oblong-triangular; petals 2–3.5 mm, obtuse, cream fading to brown; stamens with brown anthers. *South Africa (Cape Province) & Namibia*. G1. Summer–autumn.

45. C. plegmatoides Friedrich (*C. deltoidea* Schönland & Baker; *C. arta* in the sense of Higgins & Jacobsen, but not Schönland).

Illustration: Higgins, Crassulas in cultivation, 26 (1949); Jacobsen, Lexicon of succulent plants, pl. 43 (1974); Flora of southern Africa **14**: 210 (1985); Graf, Exotica, series 4, **1**: 871 (1985).

Perennial to 15 cm, only branched from base or unbranched. Leaves 5–9 × 7–13 mm, ovate, adpressed to stem to form a 4-angled column. Flower-stalk 3–6 cm; sepals 1.5–2 mm, triangular, fleshy, grey-green; petals 2–3 mm, fused at base, cream fading to brown; stamens with brown anthers. *Namibia & South Africa (Cape Province)*. G1. Late summer–early autumn.

46. C. deceptor Schönland & E.G. Baker (*C. deceptrix* Schönland; *C. arta* Schönland). Illustration: Flora of southern Africa **14**: 210 (1985); Graf, Exotica, series 4, **1**: 881, 883 (1985).

Perennial to 15 cm, with numerous branches, usually covered with old leaf bases. Leaves 0.6–1.5 × 0.6–1.5 cm, ovate, stalkless, in 4 ranks, adpressed to stem, forming a square column to 2.5 cm across, hydathodes prominent on both surfaces. Flowers stalkless, spreading; sepals *c.* 1.5 mm, triangular-oblong, blunt-pointed, hairy; petals 2–2.5 mm, elliptic-oblong, pointed or blunt with a dorsal projection, creamish brown; stamens 1–2 mm, anthers brown, stigmas stalkless, red. *South Africa (Cape Province)*. G1. Summer.

47. C. perfoliata (Linnaeus) de Candolle (*Rochea falcata* (Wendland) de Candolle var. *acuminata* Ecklon & Zeyher; *R. perfoliata* Linnaeus, in part). Illustration: Jacobsen, Lexicon of succulent plants, pl. 46 (1974); Everett, New York Botanic Garden illustrated encyclopedia of horticulture **3**: 903 (1981); Graf, Exotica, series 4, **1**: 873 (1985); Graf, Tropica, edn 3, 366 (1986).

Perennial with erect branches to 1.5 m, covered with coarse papillae. Leaves persistent, 4–12 × 1.2–3.5 cm, lanceolate-triangular, acute, densely covered with papillae, green or greyish, sometimes with purple spots. Inflorescence rounded or flat-topped on stems 3–10 cm. Sepals 1–3 mm, triangular, hairy, fleshy and green; petals 4–7.5 mm, fused at base for *c.* 1 mm, oblong-lanceolate, white, pink or scarlet; stamens with black anthers. *South Africa (Cape Province)*. G1.

48. C. mesembryanthoides (Haworth) Dietrich (*Globulea mesembryanthoides* Haworth). Illustration: Higgins, Crassulas in cultivation, 32 (1949).

Perennial, forming shrublet to 40 cm, much branched with spreading woody branches, old leaves deciduous. Leaves 1–5 × 0.2–0.3 cm, linear-triangular to linear-elliptic, green-brown, covered with hairs to 1 mm. Flower-stalks 10–30 cm; sepals 3–3.5 mm, linear-triangular, hairy; petals 4.5–6 mm, fused at base for *c.* 1 mm, with elliptic dorsal appendage, white to cream; stamens with black anthers. *South Africa (Cape Province).* G1. Autumn.

49. C. cultrata Linnaeus. Illustration: Botanical Magazine, 1940 (1817); Jacobsen, Lexicon of succulent plants, pl.43 (1974); Graf, Exotica, series 4, **1**: 879 (1985).

Erect perennial shrub to 80 cm; stems branched, older bark flaking towards base. Leaves 2.5–10 × 1–3 cm, lanceolate, rounded or blunt at tip, hairless or velvety; margin horny, red. Flowers numerous in loose groups, almost stalkless. Sepals 2–3 mm, triangular-oblong, rounded, hairy; petals 3.5–4.5 mm, fiddle-shaped, with apical projection, cream. Stamens 2.5–3.5 mm; anthers black; style very short or absent. *South Africa (Cape & Natal).*

50. C. subaphylla (Ecklon & Zeyher) Harvey (*Sphaeritis subaphylla* Ecklon & Zeyher). Illustration: Higgins, Crassulas in cultivation, 69 (1949); Flora of southern Africa **14**: 217 (1985).

Perennial shrublet to 80 cm, with wiry woody branches. Leaves 8–15 × 2–3 mm, linear-elliptic to lanceolate, green, grey-green or brown, sometimes with short hairs. Flower-stalks 3–15 cm; sepals 2–2.5, triangular, fleshy, grey-green; petals 3–5 mm, with terminal dorsal appendage; stamens with yellow to brown anthers. *South Africa (Cape Province), Namibia.* G1.

51. C. cotyledonis Thunberg. Illustration: Jacobsen, Lexicon of succulent plants, pl. 42 (1974).

Perennial with basal rosettes; stems woody, to 20 cm. Leaves persistent, 3–6 × 1–2.5 cm, oblong-oblanceolate, covered in coarse recurved hairs and marginal hairs, grey-green to yellowish green. Inflorescences 15–30 cm; flowers many; sepals 2–3 mm, oblong triangular; petals 3–5 mm, cream to pale yellow; stamens with yellow anthers. *South Africa (Cape Province) & Namibia.* G1. Summer.

52. C. nudicaulis Linnaeus.
Woody-stemmed perennial with hairy or hairless branches. Leaves in numerous rosettes, 2–9 × 0.4–2.5 cm, oblong-elliptic, sometimes linear or round, convex or flat above, convex below, hairy or hairless. Flowers in a loose cyme-like inflorescence,

sepals 1.5–3 mm, triangular-oblong, hairy or hairless; petals 3–5 mm, fiddle-shaped with a projection, cream; stamens 2.5–3.5 mm, anthers yellow; style sharp; stigmas lateral. *Southern Africa.* G1.

53. C. aquatica (Linnaeus) Schönland (*Tillaea aquatica* Linnaeus). Illustration: Blamey & Grey-Wilson, Illustrated flora of Britain and northern Europe, 164 (1989); Stace, New Flora of the British Isles, 379 (1991).

Tiny annual plant of moist habitats. Stems solitary or little branched, horizontal. Leaves 3–6 mm, linear, fleshy, tips pointed. Flowers borne in leaf axils, solitary, almost stalkless. Sepals 4, triangular, blunt; petals 4, longer than sepals, *c.* 1 mm, white or pale pink, ovate, pointed. *N & C Europe, N Asia & N America.* H1. Summer–autumn.

54. C. helmsii (T. Kirk) Cockayne (*C. recurva* (J.D. Hooker) Ostenfeld; *Tillaea recurva* (J.D. Hooker) Hooker). Illustration: Stace, New Flora of the British Isles, 379 (1991).

Perennial aquatic herb with simple or branched erect stems. Leaves 4–20 mm, linear to lanceolate, pointed, joined at the base. Flowers solitary, stalked; sepals 4, oblong, pointed; petals 4, ovate, 1.5–2 mm, white to pink, blunt. *Australia & New Zealand, naturalised in N Europe.* H5. Summer.

Often cultivated in ponds. It should not be introduced into the wild as it often becomes a noxious weed.

2. SINOCRASSULA Berger
S.G. Knees
Succulent perennials, hairless or with minute hairs. Leaves in rosettes, thickened, obtuse or tapering, hair-tipped, lined, blotched or variously marked with reddish brown. Flowering shoots erect, bracts loose; panicles cyme-like or simple with shortly stalked, erect flowers crowded towards apex. Sepals 5; petals 5, whitish with vivid red tips, joined to form spherical or urn-shaped corolla. Stamens 5, in a single whorl.

A genus of 2 species, formerly included in *Crassula* and *Sedum.* Occurring in the Himalaya and western China these plants are almost hardy, but grow best if kept in a cold frame during winter months. Propagation is by leaf cuttings, seed or division.

1a. Leaves 3.5–6 cm, linear-spathulate **1. indica**

b. Leaves 1.2–2.5 cm, lanceolate **2. yunnanensis**

1. S. indica (Decaisne) Berger (*Crassula indica* Decaisne; *Sedum indicum* (Decaisne) Raymond-Hamet). Illustration: Jacquemont, Voyages dans l'Indie **4**: 73 (1835); Hooker, Flora of British India **2**: 413 (1878).

Stems 10–30 cm, smooth, hairless, creeping and branching. Leaves linear-spathulate, long-tapered, 3.5–6 × 1–1.5 cm, underside convex, greyish green, spotted and lined with reddish brown markings. Flower-stalk 15–28 cm; calyx reddish; petals greenish crimson, *c.* 3 × 1 mm; stamens 2.5–3 mm. *India, Bhutan, Sikkim.* G1. Summer.

Sometimes cultivated under the name *Echeveria maculata.*

2. S. yunnanensis (Franchet) Berger (*Crassula yunnanensis* Franchet; *Sedum indicum* var. *yunnanensis* Raymond-Hamet). Illustration: Jacobsen, A handbook of succulent plants **2**: 846 (1960); Haage, Cacti & succulents, a practical handbook, 77 (1963); Jacobsen, Lexicon of succulent plants, pl. 133 (1977); Lamb, Popular exotic cacti in colour, 159 (1975).

Root-crown short and thick bearing annual stems 3–9 cm. Basal leaves kidney-shaped, 50–70 in dense rosettes; stem-leaves in 3s, lanceolate, tapered, mucronate; upper side roundish, underside very rounded, 12–25 × 4.5–6 mm, dark bluish green, covered with minute, white and glandular hairs. Inflorescence minutely hairy. *W China (Yunnan).* H5. Summer.

'Cristata', a monstrous fasciated form is often cultivated as a curiosity.

3. KALANCHÖE Adanson
L.S. Springate
Annuals to small trees, mostly perennials, erect to prostrate, herbaceous to succulent. Hairs absent or simple, often glandular or branched, sometimes scale-like. Leaves opposite, very rarely alternate or whorled, entire to twice pinnatisect, usually flattened. Inflorescence terminal, very rarely lateral. Flower-parts in 4s. Calyx-tube very short to exceeding lobes. Corolla-tube usually much exceeding lobes. Stamens in 2 whorls, partly fused to corolla-tube, often projecting from tube (position of upper whorl of anthers noted in the following descriptions). Plantlets sometimes produced on leaves, leaf-stalks or on inflorescence.

A genus of about 130 species with greatest diversity in Madagascar, also occurring in Africa and southern Asia to the Ryuku Islands and Indonesia and widely naturalised in the tropics. The species flower in winter naturally. Cultivation in low winter light and temperatures in Europe can result in delayed flowering, abnormal flower-head and flower formation and uncharacteristic regrowth. Grown under glass in most porous composts with restricted nutrients. Control of day-length is needed for continuous pot-plant production. Propagation from seed (except for some cultivars), cuttings or plantlets.

Literature: Raymond-Hamet, Monographie de Genre Kalanchoë, *Bulletin de l'Herbier Boissier, série 2*, **7**: 869–900 (1907), **8**: 17–48 (1908); Boiteau, P. & Mannoni, O, Les plantes grasses de Madagascar. Les Kalanchoë, *Cactus*, **12**: 6–10 (1947); **13**: 7–10 (1948); **14**: 23–28 (1948); **15–16**: 37–42 (1949); **17–18**: 57–58 (1949); **19**: 9–14 (1949); **20**: 43–46 (1949); **21**: 69–76 (1949); **22**: 113–114 (1949) (1947–9); Raymond-Hamet & Marnier-Lapostolle, J, Le Genre Kalanchoë au Jardin Botanique 'Les Cèdres' (1964); Cufodontis, G., The species of Kalanchoë in Ethiopia and Somalia Republic, *Webbia* **19**: 711–744 (1965); Raadts, E., The genus Kalanchoë (Crassulaceae) in tropical East Africa, *Willdenowia* **8**: 101–157 (1977); van Voorst, A., & Arends, J.C., The origin and chromosome numbers of cultivars of Kalanchoë blossfeldiana von Poellnitz: Their history and evolution, *Euphytica* **31**: 573–584 (1982).

1a. Leaf-hairs 3- or 4-branched, non-glandular, sometimes compressed and scale-like   2
 b. Leaf-hairs simple, often glandular, or absent   8
2a. Leaves alternate   3
 b. Leaves opposite   4
3a. Leaves stalked   **25. rhombopilosa**
 b. Leaves stalkless   **23. tomentosa**
4a. Leaves entire   5
 b. Leaves toothed   7
5a. Leaves with long, soft hairs   **24. eriophylla**
 b. Leaves with dense, scale-like hairs   6
6a. Leaves circular to obovate, very obtuse; corolla-tube less than 6.5 mm   **21. hildebrandtii**
 b. Leaves ovate to lanceolate, more-or-less acute; corolla-tube more than 6.5 mm   **19. orygalis**

7a. Leaves triangular to hastate, with irregular teeth and lobes   **20. beharensis**
 b. Leaves ovate to obovate, with regular, shallow teeth only   **22. millotii**
8a. Stems climbing, supported by the leaves   9
 b. Stems erect, free-standing or prostrate   10
9a. Adult leaves with irregular pinnate divisions   **9. schizophylla**
 b. All leaves entire to lobed, never with pinnate divisions   **10. beauverdii**
10a. Plant rather glaucous, rarely hairless; leaves triangular, more or less hastate, stalked, non-peltate; margins sinuous with uneven teeth and lobes   **20. beharensis**
 b. Plant clearly different in at least one feature   11
11a. Ovaries and styles divergent   12
 b. Ovaries and at least lower part of styles adpressed   13
12a. Leaves peltate   **1. peltata**
 b. Leaves not peltate   **3. gracilipes × manginii**
13a. Free part of filament longer than part fused to corolla   14
 b. Free part of filament shorter than part fused to corolla   28
14a. Calyx-segments at least as long as tube   15
 b. Calyx-segments shorter than tube   24
15a. No spurs bearing plantlets on leaf margins   16
 b. Spurs bearing plantlets on leaf margins   22
16a. Stems thick (more than 4 mm at base), more or less erect   17
 b. Stems thin, creeping   19
17a. Mature plants more than 40 cm, or flower-heads with more than 30 flowers   **2. miniata**
 b. Mature plants less than 40 cm, flower-heads with 3–30 flowers   18
18a. Flower-heads usually 3-branched from base, with 15–30 flowers   **6. 'Wendy'**
 b. Flower-heads with single stem, branched at apex, with less than 15 flowers   **7. porphyrocalyx**
19a. Plant glaucous   **26. pumila**
 b. Plant green   20
20a. Flowers funnel-shaped, yellow   **5. jongmansii**
 b. Flowers urn-shaped, red to purple   21

21a. Leaves obovate to spathulate or oblong-obovate; calyx-segments more than 5 mm; corolla-tube more than 2 cm   **4. manginii**
 b. Leaves more or less circular; calyx-segments less than 5 mm; corolla-tube less than 2 cm   **8. uniflora**
22a. Leaves nearly cylindric, entire except for 3–9 teeth at apex   **11. delagoensis**
 b. Leaves flat, whole margin toothed   23
23a. Leaves with less than 20 irregular teeth and few spurs bearing plantlets   **12. aff. 'Hybrida'**
 b. Leaves with more than 20 regular teeth, alternating with plantlet-bearing spurs   **13. daigremontiana**
24a. Leaves simple   25
 b. Leaves pinnately divided   27
25a. Leaf-blade more than 7.5 cm, with more than 30 teeth   **16. gastonis-bonnieri**
 b. Leaf-blade less than 7.5 cm with less than 30 teeth   26
26a. Leaves more or less tapered at base, without auricles, crowded at base of stem   **14. fedtschenkoi**
 b. Leaves with very small to prominent auricles, well-spaced at base of stem   **15. laxiflora**
27a. Leaves pinnatifid to pinnatisect; corolla less than 3 cm   **17. prolifera**
 b. Leaves divided into 3–5 leaflets; corolla more than 3 cm   **18. pinnata**
28a. Leaves at middle of stem half-cylindric or triangular in section, longer than 5 cm   29
 b. All leaves flattened, or shorter than 5 cm   30
29a. Flowers white; leaves entire   **43. bentii**
 b. Flowers pink; leaves toothed or lobed   **42. × kewensis**
30a. Corolla-tube more than 2.7 cm   31
 b. Corolla-tube less than 2.7 cm   32
31a. Leaves stalked   **40. quartiniana**
 b. Leaves stalkless   **41. marmorata**
32a. Inflorescence lateral; stolons long, with a single-node, from leaf axils   **27. synsepala**
 b. Inflorescence terminal; stolons absent   33
33a. Plant with some hairs   34
 b. Plant entirely hairless   38
34a. Some anthers protruding from throat of corolla; flowers pink   **38. petitiana**

b. All anthers inside corolla-tube or flowers red, orange, yellow or whitish 35

35a. Stems sprawling, slender; corolla funnel-shaped **5. jongmansii**

b. Stems erect, more than 4 mm wide at base; corolla-tube cylindric 36

36a. Some leaves deeply lobed to twice pinnatisect **32. laciniata**

b. All leaves scalloped to almost entire 37

37a. Plant hairy throughout **31. lateritia**

b. At least lower part of plant hairless **30. crenata**

38a. All anthers inside corolla-tube 39

b. Some anthers projecting from throat of corolla 44

39a. Leaves more or less green, stalks slightly to distinctly broadened at base, not easily detached 40

b. Leaves glaucous or stalks not broadened at base, easily detached 42

40a. Corolla-lobes 1–2.5 mm wide, filaments free for less than 1.5 mm; flowers red, orange, yellow or rose **28. blossfeldiana**

b. Corolla-lobes 2.5–8 mm wide or filaments free for more than 1.5 mm or flowers purplish 41

41a. Leaves evenly scalloped; flowers yellow to red only; plants at least 30 cm **30. crenata**

b. Leaves with irregular or angular teeth or flowers with pink or purple hue or plants less than 30 cm **29. blossfeldiana** hybrids

42a. Calyx-lobes to 2.5 mm; corolla twisted around carpels in fruit **33. rotundifolia**

b. Calyx-lobes usually longer than 2.5 mm; corolla not twisted around carpels in fruit 43

43a. Calyx more or less persistent; leaves usually stalked **34. glaucescens**

b. Calyx-lobes dropping early and separately; leaves more or less stalkless **35. flammea**

44a. Corolla-lobes more or less as long as tube 45

b. Corolla-lobes clearly shorter than tube 47

45a. Leaves green **29. blossfeldiana** hybrids

b. Leaves glaucous 46

46a. Flowers pink **26. pumila**

b. Flowers yellow **44. grandiflora**

47a. Calyx-lobes to 2.5 mm **45. farinacea**

b. Calyx-lobes longer than 2.5 mm 48

48a. Leaves peltate **39. nyikae**

b. Leaves not peltate 49

49a. Both whorls of anthers projecting from throat of corolla; flowers pink **38. petitiana**

b. One whorl of anthers projecting from throat of corolla or flowers yellow, orange or red 50

50a. Leaves green **30. crenata**

b. Leaves glaucous 51

51a. Flowers pink, orange or red; stems not square in section **35. flammea**

b. Flowers greenish yellow to orange; stems square in section 52

52a. Leaves mostly basal, entire **37. thyrsiflora**

b. Leaves evenly distributed along stem, toothed **36. longiflora**

1. K. peltata (Baker) Baillon. Illustration: Jacobsen, Handbook of succulent plants, 657 (1960); Raymond-Hamet & Marnier-Lapostolle, Le Genre Kalanchoë au Jardin Botanique 'Les Cèdres', f. D, 6, 7 (1964); Maire, Flore de l'Afrique du Nord **14**: 262 (1976).

Perennial to 2 m, hairless or sparsely covered with simple hairs; stems decumbent. Leaves ovate, unevenly scalloped, peltate, often with reddish spots, 3–12.5 × 2.5–6 cm; stalk thin, 2–10 cm. Inflorescence corymb-like; flowers many, pendent. Corolla rose to red; tube narrow 2.1–2.7 cm, lobes ovate, partly spreading, 3–6 × 4.5–9.5 mm. Anthers and styles projecting from throat of corolla, filaments free up to 6 mm; styles divergent, pressed to side of corolla-tube. *Madagascar*. G2.

2. K. miniata Hilsembach & Bojer (*Bryophyllum miniatum* (Hilsembach & Bojer) Berger). Illustration: Jacobsen, Handbook of succulent plants, 655 (1960), Raymond-Hamet & Marnier-Lapostolle, Le Genre Kalanchoë au Jardin Botanique 'Les Cèdres', f. L. 52, 53 (1964).

Perennial to 80 cm, hairless or inflorescence with simple, glandular hairs. Stems erect or decumbent. Leaves ovate to obovate, usually simple, rarely 3-lobed, usually finely scalloped, rarely entire, 2–13 × 1.2–7.5 cm; base tapered to auriculate; stalk often winged, 0.7–4 cm, rarely absent. Inflorescence a panicle of several corymb-like clusters; flowers pendent, often partly replaced by bulbils. Corolla bell-shaped, yellow, pink or red; tube 2.3–3.1 cm; lobes recurved, ovate, sometimes purple, 4–6 × 5–7 mm. Anthers and

stigmas at throat of corolla; filaments *c.* 5 mm, fused to corolla. *Madagascar*. G2.

3. K. gracilipes (Baker) Baillon × **K. manginii** Raymond-Hamet & Perrier de la Bâthie. Illustration: Brickell, RHS gardeners' encyclopedia of plants & flowers, 385 (1989).

Similar to *K. manginii*, but differs in all leaves ovate-oblong, with 2–6 distinct teeth; stalk distinct, 4–13 mm. Inflorescence without bulbils. Flower colour similar in bright light, but very pale in shade. Styles divergent. G1–2.

The main cultivar, 'Tessa' differs from its parents in often having 6 or more flowers per cluster.

K. gracilipes is similar, but differs in: hairs entirely absent, calyx-tube more or less as long as lobes, corolla flesh or buff.

4. K. manginii Raymond-Hamet & Perrier de la Bâthie (*Bryophyllum manginii* (Raymond-Hamet & Perrier de la Bâthie) Northdurft). Illustration: Jacobsen, Handbook of succulent plants, 653 (1960); Raymond-Hamet & Marnier-Lapostolle, Le Genre Kalanchoë au Jardin Botanique 'Les Cèdres', f. 124, 125 (1964); Jacobsen, Das Sukkulenten Lexikon, pl. 41 (1981).

Creeping perennial with sparse, simple, glandular hairs; stems slender; flowering branches 10–40 cm. Leaves obovate to spathulate, entire or rarely 1–3 weak notches near apex, thick, 1.2–3 × 0.6–1.4 cm; stalk absent or indistinct. Inflorescence open; flowers pendent, usually 1–5, partly replaced by bulbils. Corolla narrowly urceolate, red; tube 2.1–2.8 cm; lobes ovate, 3–7 × 4.5–5.5 mm. Anthers and styles projecting from corolla-tube; filaments fused to corolla for 5–9 mm. *Madagascar*. G1–2.

5. K. jongmansii Raymond-Hamet & Perrier de la Bâthie. Illustration: Botanical Magazine, n.s. 388 (1962); Raymond-Hamet & Marnier-Lapostolle, Le Genre Kalanchoë au Jardin Botanique 'Les Cèdres'. f. Q, 43, 44 (1964); Cactus & Succulent Journal (U.S.) **55**: 210 (1983).

Sprawling perennial with few simple glandular hairs mainly on the flowering shoots; stems slender; flowering branches ascending, to 30 cm. Leaves linear-elliptic, entire, stalkless, fleshy, 7.5–42 × 2.5–9.5 mm. Flowers few, at end of branches and from upper leaf-axils, more or less erect. Corolla gradually spreading from base, golden yellow; tube 0.7–2.2 cm; lobes ovate, 6–9 × 4–6 mm. Anthers and stigmas usually inside corolla-tube;

filaments free for up to 6 mm. *Madagascar.* G1–2.

6. K. 'Wendy' (*K. miniata* × *K. porphyrocalyx*). Illustration: Brickell, RHS gardeners' encyclopedia of plants & flowers, 384 (1989).

Shrublet to 40 cm, semi-erect, with very few fine simple hairs only in the inflorescences. Stems soon thickening and epidermis separating into plates. Leaves ovate to oblong-lanceolate with uneven, shallow, rounded teeth, 4–7 × 1–2.5 cm, tapered at base; stalk indistinct. Stem of inflorescence usually 3-branched from base with 15–30 pendent flowers. Corolla urn-shaped; tube rose-purple, *c.* 2.3 × 1.3 cm; lobes ovate, yellow, 5 × 6 mm. Anthers and styles just projecting from corolla-tube; filaments fused to corolla for *c.* 5 mm. G2.

Similar to *K. porphyrocalyx*, but coarser and more floriferous when grown in similar conditions. Sometimes self-fertilised, seedlings very variable.

7. K. porphyrocalyx (Baker) Baillon (*Bryophyllum porphyrocalyx* (Baker) Berger). Illustration: Cactus (France), **17–18:** 57 (1948); Raymond-Hamet & Marnier-Lapostolle, Le Genre Kalanchoë au Jardin Botanique 'Les Cèdres', f. N, 117, 118 (1964).

Shrublet to 30 cm, semi-erect, with fine glandular hairs in the inflorescence. Stems soon thickening and epidermis separating into plates. Leaves oblong to obovate, with uneven, shallow or deep, rounded teeth, 2.3–5.5 × 0.5–3 cm, tapered at base; stalk indistinct. Flowering-stem single, only branched at apex, usually with 5–15 pendent flowers. Corolla urn-shaped; tube rose to red, 2–3.1 × 0.6–1.2 cm; lobes ovate, yellow or orange, 3–5.5 × 3–7.5 mm. Anthers and stalks usually just projecting from corolla-tube. *Madagascar.* G2.

8. K. uniflora (Stapf) Raymond-Hamet (*Bryophyllum uniflorum* (Stapf) Berger. Illustration: Botanical Magazine, 8286 (1909); Raymond-Hamet & Marnier-Lapostolle, Le Genre Kalanchoë au Jardin Botanique 'Les Cèdres'. f. G, H, 119 (1964); Hay et al., The dictionary of indoor plants in colour, pl. 78 (1974); Cactus & Succulent Journal (U.S.) **55:** 201 (1983).

Prostrate perennial, hairless or with fine glandular hairs in the inflorescence; stems slender, rooting. Leaves circular, with few uneven, rounded teeth, both faces convex, 4–15 mm; stalk 1–3 mm. Flowering-stem spreading or pendent, with few pendent flowers. Corolla urn-shaped, red to purple; tube 1–1.9 cm × 6–9 mm across; lobes ovate 3.5–4.5 × 3–6 mm. Anthers and styles just projecting from throat of corolla; filaments fused to corolla to 4 mm. *Madagascar.* G2.

9. K. schizophylla (Baker) Baillon (*Bryophyllum schizophylla* (Baker) Berger). Illustration: Raymond-Hamet & Marnier-Lapostolle, Le Genre Kalanchoë au Jardin Botanique 'Les Cèdres', f. 65–67 (1964).

Woody climber, supported by hooked leaves, hairless; stem to 8 m. Young leaves ovate, toothed; mature leaves pinnatisect, 13–15 × 4–5 cm; segments more or less linear, with teeth and sometimes 1–3 hooked lobes. Inflorescence large, with few flowers, pinnatisect bracts and many bulbils; flowers dull violet, pendent. Corolla narrowly bell-shaped; tube 1.3–1.7 cm; lobes ovate, 2–3.5 × 2–4 mm. Filaments and styles projecting beyond corolla; filaments fused to corolla for *c.* 5 mm. *Madagascar.* G2.

10. K. beauverdii Raymond-Hamet (*Bryophyllum beauverdii* (Raymond-Hamet) Berger; *K. costantinii* Raymond-Hamet; *K. juelii* Raymond-Hamet). Illustration: Cactus (France) **19:** 12, (1949); Raymond-Hamet & Marnier-Lapostolle, Le Genre Kalanchoë au Jardin Botanique 'Les Cèdres', f. 34, 44, 45 (1964); Jacobsen, Das Sukkulenten Lexikon, t. 103 (1981).

Woody climber to 6 m, supported by reflexed leaves, often brownish. Leaves with teeth and plantlets only at apex, stalkless; lower leaves almost cylindric; upper leaves ovate to linear; base rounded to cordate or long-tapered, blade hastate, 2–10 × 0.5–2.5 cm. Inflorescence a loose panicle of small flower-clusters; flowers usually pendent. Corolla funnel-shaped, greenish to purplish; tube 1.1–1.5 cm; lobes rounded, 1.2–1.7 × 1–1.9 cm. Anthers and styles projecting from corolla-tube; filaments fused to corolla for up to 4 mm. *Madagascar.* G2.

11. K. delagoensis Ecklon & Zeyher (*Bryophyllum delagoense* (Ecklon & Zeyher) Schinz; *B. tubiflorum* Harvey; *K. tubiflora* (Harvey) Raymond-Hamet). Illustration: Botanical Magazine, 9251 (1929); The Garden **108:** 238 (1983); Leistner: Flora of Southern Africa **14:** 74 (1985); Brickell, RHS gardeners' encyclopedia of plants & flowers, 388 (1989).

Perennial to 1.2 m, hairless, grey-green; stems erect. Leaves opposite on young shoots, in whorls of 3 or sometimes alternate on older shoots, cylindric, stalkless, 1.5–13 × 0.2–0.6 cm; apex with 3–9 teeth with spurs in between, bearing plantlets. Inflorescence dense, large, corymb-like, flowers pendent. Calyx tube 2–6 mm, lobes 5.5–8 × 3.5–5.5 mm. Corolla pale orange to magenta; tube narrow, 2–2.5 cm; lobes obovate, spreading, 7–12 × 7–10 mm. Anthers and styles just projecting from corolla-tube; filaments fused to corolla for up to 6.5 mm. *Madagascar.* G2.

12. K. aff. **'Hybrida'** (*K. hybrida* invalid; *K. serrata* misapplied).

Fertile plants intermediate between *K. delagoensis* and *K. daigremontiana.* Leaves usually opposite, rarely a few alternate, oblong to lanceolate, with sharp teeth and some bulbils, 3–5.5 cm; stalks 1.5–3 cm. Calyx-lobes about as long as tube.

A race of cultivated plants apparently without a valid name: *K.* 'Hybrida' (Illustration: Jacobsen, Handbook of succulent plants, f. 860, (1960) was a similar clone but was usually sterile. G2.

13. K. daigremontiana Raymond-Hamet & Perrier de la Bâthie (*Bryophyllum daigremontianum* (Raymond-Hamet & Perrier de la Bâthie) Berger). Illustration: Raymond-Hamet & Marnier-Lapostolle, Le Genre Kalanchoë au Jardin Botanique 'Les Cèdres', f. 35–38 (1964); Jacobsen, Das Sukkulenten Lexikon, t. 103 (1981); Brickell, RHS gardeners' encyclopedia of plants & flowers, 383 (1989).

Erect perennial 30–100 cm, hairless, brownish green. Leaves lanceolate, peltate at least at base of plant, with regular, small, sharp teeth alternate with spurs bearing plantlets, marbled beneath with brown-purple, 6–15 × 0.9–5 cm; stalk 1–3.5 cm. Inflorescence an open panicle with several dense long-stemmed clusters of pendent flowers. Calyx-tube 3–4.5 mm; lobes 3–5.5 × 2–3.5 mm. Corolla greyish violet; tube 1.6–1.9 cm; lobes obovate, partly spreading, 7–8 × 3–4.5 mm. Anthers and styles projecting from corolla-tube; filaments fused to corolla for up to 8 mm. *Madagascar.* G2.

14. K. fedtschenkoi Raymond-Hamet & Perrier de la Bâthie (*Bryophyllum fedtschenkoi* (Raymond-Hamet & Perrier de la Bâthie) Lauzac-Marchal). Illustration: Jacobsen, Handbook of succulent plants, 647 (1960); Raymond-Hamet & Marnier-Lapostolle, Le Genre Kalanchoë au Jardin Botanique 'Les Cèdres', f. 39, 40 (1964); Hortus Third, 329 (1976).

Perennial to 50 cm, hairless, blue-glaucous; stem decumbent with leaves more crowded at base. Leaves obovate to oblong with 2–8 prominent teeth in apical half, 1.2–6 × 0.8–4 cm; base tapered; stalk 1–6 mm. Young flowering-stem bent back at apex, resembling a crosier. Inflorescence a small loose corymb; flowers few, pendent. Calyx tube 1.2–1.3 cm, lobes 6–6.5 × 4.5–5 mm. Corolla bell-shaped, dull red to purple; tube 1.5–2 cm; lobes obovate, c. 6.5 × 4.5 mm. Anthers and styles projecting from corolla-tube; filaments fused for 6–9.5 mm. *Madagascar.* G2.

'Variegata'. Illustration: Brickell, RHS gardeners' encyclopedia of plants & flowers, 383 (1989). Leaves with narrow cream margin.

15. K. laxiflora Baker (*Bryophyllum crenatum* Baker). Illustration: Botanical Magazine, 7856 (1902); Raymond-Hamet & Marnier-Lapostolle, Le Genre Kalanchoë au Jardin Botanique 'Les Cèdres', f. 47–49 (1964).

Similar to *K. fedtschenkoi*, but differs in: plant often pale green. Leaves more or less evenly spaced on stem; blade circular or ovate to oblong; teeth blunter; base truncate with very small to large, upturned auricles; stalks 1.2–3.7 cm. *Madagascar.* G2.

16. K. gastonis-bonnieri Raymond-Hamet & Perrier de la Bâthie (*Bryophyllum gastonis-bonnieri* (Raymond-Hamet & Perrier de la Bâthie) Lauzac-Marchal). Illustration: Raymond-Hamet & Marnier-Lapostolle, Le Genre Kalanchoë au Jardin Botanique 'Les Cèdres', f. 41, 42 (1964).

Mealy perennial to 75 cm; fine glandular hairs only on corolla; sterile shoots forming loose rosettes; flowering stems with about 4 pairs of leaves. Leaves lanceolate, scalloped, with transverse bands of spots, 9–16.5 × 3.7–5.5 cm; stalk broad, indistinct, 3.5–6.3 × 1.2–1.5 cm. Inflorescence corymb-like or paniculate; flowers c. 2–4 cm, pendent. Calyx inflated. Corolla pale, yellowish or reddish; tube narrow, 2.9–3 cm; lobes ovate, partly recurved, 9–11 × 5.5–7.5 mm. Anthers and styles projecting from corolla-tube; filaments fused to corolla for 7–10 mm. *Madagascar.* G2.

17. K. prolifera (Bowie) Raymond-Hamet (*Bryophyllum proliferum* Bowie). Illustration: Botanical Magazine, 5147 (1859); Jacobsen, Handbook of succulent plants, 660 (1960); Raymond-Hamet & Marnier-Lapostolle, Le genre Kalanchoë

au Jardin Botanique 'Les Cèdres', f. 60–62 (1964).

Perennial to 1.5 m, hairless; stems erect. Leaves pinnatifid to pinnatisect, to 45 cm; segments lanceolate to oblong, oblique, scalloped to deeply toothed, 7–15 × 1.5–5 cm; stalks 6–16 cm. Inflorescence an open panicle of small clusters of flowers on long stalks with numerous bulbils; flowers pendent. Calyx inflated. Corolla-tube greenish yellow, 1.8–2.4 cm; lobes ovate, recurved, often pink, 2.5–3.5 × 3–4 mm. Anthers and styles projecting from corolla-tube; filaments fused to corolla for 4.5–8 mm; styles recurved at tip. *Madagascar.* G2.

18. K. pinnata (Lamarck) Persoon (*Bryophyllum pinnatum* (Lamarck) Oken; *B. calycinum* Salisbury). Illustration: Botanical Magazine, 1409 (1811); Raymond-Hamet & Marnier-Lapostolle, Le Genre Kalanchoë au Jardin Botanique 'Les Cèdres', f. 57–9 (1964); Flora of East Tropical Africa: Crassulaceae, 29 (1977).

Perennial to 2 m, with sparse, fine, glandular hairs only on the corolla; stems erect. Lower leaves simple, upper leaves usually pinnate with 3–5 leaflets and stalk 2.5–7.5 cm; leaflets oblong, scalloped, 5–20 × 3–12 cm. Inflorescence a loose panicle; flowers pendent. Calyx inflated, green or reddish. Corolla-tube 3–4 cm; lobes ovate, partly recurved, reddish. Anthers and styles projecting from throat of corolla; filaments fused for 9–12 mm. *Widespread in tropics.* G2.

19. K. orygalis Baker. Illustration: Jacobsen, Handbook of succulent plants, 657 (1960); Raymond-Hamet & Marnier-Lapostolle, Le Genre Kalanchoë au Jardin Botanique 'Les Cèdres', f. 80, 81 (1964); Cactus and Succulent Journal (U.S.), **55:** 204–205 (1983).

Shrublet to 1.5 m, covered in fine, scale-like, star-like, golden hairs; stems erect. Leaves opposite, ovate to lanceolate, entire, 7.5–12 × 5.5–10.5 cm; stalk stout, 5–15 mm. Inflorescence a long, open, panicle, with distant, dense, small flower clusters; flowers more or less erect. Corolla urn-shaped, yellow, tube 8–10 mm, lobes ovate, 3–4 × 3–6 mm. Anthers and stigmas at throat of corolla, filaments free for 1–6 mm. *Madagascar.* G1–2.

20. K. beharensis Drake del Castillo. Illustration: Everard, Wild flowers of the world, pl. 70 (1970); Cactus & Succulent Journal (U.S.), **49:** 269 (1977); Jacobsen, Das Sukkulenten Lexikon, t. 103 (1981); Brickell, RHS gardeners' encyclopedia of

plants & flowers, 38 (1989).

Much-branched tree eventually to 6 m, often with a single non-flowering stem in cultivation, covered in dense, short, branched, silver or golden hairs, rarely hairless and glaucous. Leaves triangular to hastate, 4.5–35 × 3.5–25 cm; margins wavy, with uneven teeth and lobes; stalks thick, 1.5–13 cm. Inflorescence mostly lateral with an open panicle of small, dense flower clusters; flowers more or less erect. Corolla urn-shaped, yellow-green, tube 6.5–10 mm, lobes obovate, violet-lined, about as long as tube. Anthers and styles projecting from throat of corolla, filaments free for 4–9.5 mm. *Madagascar.* G1–2.

21. K. hildebrandtii Baillon. Illustration: Jacobsen, Handbook of succulent plants, 650 (1960); Raymond-Hamet and Marnier-Lapostolle, Le Genre Kalanchoë au Jardin Botanique 'Les Cèdres', f. 7, 8 (1964).

Much-branched shrub to 5 m, covered in dense, fine, scale-like, silver, star-like hairs. Leaves circular to obovate, entire, 1.6–4 × 1.3–3.5 cm; stalk 3–7 mm. Inflorescence an open panicle of small, dense flower clusters; flowers more or less erect. Corolla bell-shaped, white or pale green; tube 3.5–5 mm; lobes oblong, 2–3.5 × 1.5–2.0 mm. Anthers and styles projecting from throat of corolla; filaments free for 1.5–3.5 mm. *Madagascar.* G1–2.

22. K. millotii Raymond-Hamet & Perrier de la Bâthie. Illustration: Raymond-Hamet & Marnier-Lapostolle, Le Genre Kalanchoë au Jardin Botanique 'Les Cèdres', f. 26–28 (1964); Hay et al., Dictionary of indoor plants in colour, pl. 309 (1974); Jacobsen, Das Sukkulenten Lexikon, t. 105 (1981).

Much-branched shrublet to 35 cm, covered in dense, short, branched, white hairs. Leaves ovate to obovate, with even, shallow teeth, 3–6 × 2.5–5.5 cm; stalk thick, 5–18 mm. Inflorescence small, dense, corymb-like; flowers erect. Calyx also with simple glandular hairs. Corolla narrowly bell-shaped; tube yellow-green, 9–10.5 mm; lobes oblong, reddish, 3–3.5 mm. Anthers and styles just projecting from throat of corolla; filaments free for 2.5–3 mm. *Madagascar.* G1–2.

23. K. tomentosa Baker. Illustration: Everard, Wild flowers of the world, pl. 70 (1970); Cactus & Succulent Journal (U.S.) **55:** 202 (1983); Brickell, RHS gardeners' encyclopedia of plants & flowers, 387 (1989).

Shrub to 1 m, erect, sparsely branched,

covered in stiff, matted, long, silver, branched hairs. Leaves alternate, oblong, stalkless, thick, 2–8 × 1–3 cm; margin entire or with a few teeth towards apex, often brown. Inflorescence a narrow panicle of small, dense flower-clusters; flowers spreading. Corolla bell-shaped, with simple, reddish, glandular hairs and usually branched, non-glandular hairs; tube greenish yellow, 10–12 mm; lobes ovate, usually purplish, 2.5–3.5 × 4–5.5 mm. Anthers and stigmas usually inside corolla-tube; filaments free for 3–6 mm. *Madagascar.* G1.

Several cultivars based on leaf variation are grown.

24. K. eriophylla Hilsembach & Bojer. Illustration: Raymond-Hamet & Marnier-Lapostolle, Le Genre Kalanchoë au Jardin Botanique 'Les Cèdres', f. M, 24, 25 (1964); Cactus & Succulent Journal (U.S.), **55**: 204 (1983).

Perennial to 20 cm, covered in soft, matted, long, white branched hairs. Leaves oblong, entire, stalkless, very thick, 1.6–3 × 0.8–1.5 cm. Inflorescence a tight cluster of 2–7 more or less erect flowers on a slender stem. Corolla bell-shaped, blue-violet; tube 5–6.5 mm; lobes obovate, 4–6 × 3–5.5 mm. Anthers and stigmas more or less at throat of corolla, filaments free for 2.5–3.5 mm. *Madagascar.* G1–2.

25. K. rhombopilosa Mannoni & Boiteau. Illustration: Raymond-Hamet & Marnier-Lapostolle, Le Genre Kalanchoë au Jardin Botanique 'Les Cèdres', f. 126–8 (1964); Graham, Growing succulent plants, 81 (1987).

Slow-growing shrublet to 50 cm; stems and leaves covered in dense, greyish, scale-like, 4-branched hairs; stems erect. Leaves stalked, alternate, obovate to fan-shaped, grey-green usually with red-brown streaks, 2–3 × 1.5–2.5 cm; the apical margin with prominent teeth; sometimes the hairs confined to the apical margin and the leaves copper-coloured; stalk 2–5 mm. Inflorescence a small, open panicle; flowers erect. Corolla urn-shaped, yellow-green, red-lined; tube c. 4 mm; lobes ovate, c. 2.5 mm. Anthers and styles projecting from throat of corolla; filaments free for 2–3.5 mm. *Madagascar.* G1–2.

26. K. pumila Baker. Illustration: Hay et al., Dictionary of indoor plants in colour, pl. 310 (1974); Cactus & Succulent Journal (U.S.) **55**: 204 (1983); Blundell, Wild flowers of East Africa, 338 (1987).

Sprawling shrublet to 30 cm, covered in white bloom. Leaves glaucous, obovate, toothed towards apex, 2–3.5 × 1.2–2 cm; base tapered; stalk usually indistinct. Inflorescence a tight cluster of about 5 erect flowers; stem slender, to 6 cm. Corolla narrowly bell-shaped, pink with purple lines; tube 4.5–8.5 mm; lobes obovate, 7.5–10 mm. Anthers and styles projecting from throat of corolla; filaments free for 3–6 mm. *Madagascar.* G1–2.

27. K. synsepala Baker. Illustration: Raymond-Hamet & Marnier-Lapostolle, Le Genre Kalanchoë au Jardin Botanique 'Les Cèdres', f. 13–17 (1964); Cactus & Succulent Journal (U.S.) **55**: 206–207 (1983).

Perennial to 30 cm, hairless or with simple glandular hairs in the inflorescence, bearing long, slender stolons with one internode. Leaves lanceolate to oblanceolate, finely or coarsely toothed or pinnatifid to 15 × 7 cm, tapered; stalk indistinct. Inflorescence on slender, erect, lateral stem; flowers erect in dense clusters. Corolla narrowly bell-shaped; tube 7–12 mm; lobes elliptic, 4.5–7 × 3–4 mm. Anthers and styles projecting from throat of corolla; filaments free for 1–2 mm. *Madagascar.* G2.

28. K. blossfeldiana von Poellnitz (*K. globulifera* var. *coccinea* Perrier de la Bâthie). Illustration: Botanical Magazine, 9440 (1936); Euphytica 31, pl. 1 (1982).

Perennial to 30 cm, much-branched, hairless. Leaves ovate to oblong, scalloped, bright green, 2–7.5 × 1.6–4 cm; base rounded; stalk broad, indistinct to 10 mm. Inflorescence of several dense clusters of erect flowers. Calyx-lobes linear, 3–5 × 1–1.5 mm. Corolla-tube cylindric, reddish above, 6–9 × c. 1.5 mm; lobes elliptic, bright red, 4–5.5 × 1.5–2.5 mm. Anthers inside corolla-tube; filaments free for up to 1 mm. *Madagascar.* G2.

Cultivars with compact growth and yellow, orange or rose flowers have been selected. Illustration: Hay et al., Dictionary of indoor plants in colour, pl. 308 (1974); Euphytica 31, pl. 1 (1982), – as 'Vulcan geel'. Scarcely cultivated. Plants not closely matching the description above are referred to *K. blossfeldiana* hybrids.

29. K. blossfeldiana hybrids (*K. blossfeldiana* misapplied).

Perennial to 30 cm, much-branched, hairless. Leaves ovate, deep green, shallowly scalloped, to 12 × 6.5 cm; base rounded or truncate; stalk broad, to 23 × 8 mm. Inflorescence often one broad, dense, corymb-like cluster of erect flowers. Calyx-lobes lanceolate, 4–5.5 × 2.5 mm. Corolla-tube greenish, 6–7 × 4.5 mm; lobes oblong to 8 × 6.5 mm, in many clear colours from pale yellow and pink to brick-red and purple. Anthers and stigmas at throat of corolla; filaments free for up to 2.5 mm. G2.

Current pot-plant cultivars are described above. They are complex hybrids of inadequately documented origin. Earlier hybrids persist in plant collections and are often confused with other species. Some are intermediate between *K. blossfeldiana* hybrids and various other species, particularly *K. pumila* and *K. flammea*; others more closely resemble *K. blossfeldiana* hybrids and include plants with variegated leaves. Illustration: Hay et al., Dictionary of indoor plants in colour, pl. 306–307 (1974); Euphytica 31, pl. 1 (1982) – as 'Chérie'; Brickell, RHS gardeners' encyclopedia of plants & flowers, 385 (1989).

30. K. crenata (Andrews) Haworth (*K. laciniata* misapplied; *K. brasiliensis* Cambessèdes; *K. coccinea* Britten; *K. integra* (Medicus) O. Kuntze, ambiguous). Illustration: Hutchison & Dalziel, Flora of West Tropical Africa, edn 2, **I**: 117 (1954); Heywood, Flowering plants of the world, 145 (1978).

Erect perennial to 2 m, hairless or with simple glandular hairs in the flower-head. Leaves ovate to spathulate, usually scalloped, 4–25 × 1.2–15 cm; stalks flattened, to 4 cm. Inflorescence an open panicle; flowers erect. Corolla red to yellow, tube narrow, 8–16 mm; lobes elliptic, spreading, 4–10 x 2.5–5 mm. Anthers and stigmas inside corolla-tube, filaments free for 0.5–3.5 mm. *Natal to Arabia, naturalised in tropical America.* G2.

Asian plants, from Kashmir to the Ryuku Islands and Java, are sometimes distinguished as *K. spathulata* de Candolle, but cannot be readily separated in gardens. The leaves are narrow and sometimes divided into 3 leaflets; the leaf-stalks are sometimes indistinct. Illustration: Li et al., Flora of Taiwan **3**: 14 (1977).

31. K. lateritia Engler (*K. coccinea* Britten var. *subsessilis* Britten; *K. integra* (Medicus) Kuntze var. *subsessilis* (Britten) Cufodontis; *K. velutina* misapplied; *K. zimbabwensis* Rendle). Illustration: Botanical Magazine, 787 (1902); Raymond-Hamet & Marnier-Lapostolle, Le Genre Kalanchoë au Jardin Botanique 'Les Cèdres', f. I, 89–91 (1964); Everard, Wild flowers of the world, pl. 70 (1970).

Erect perennial to 1.5 m, with brownish, simple, usually dense, glandular hairs. Leaves ovate to obovate, more or less scalloped, 4.5–16 × 3–8 cm; base tapered; stalk 0.5–5 cm, sometimes indistinct. Inflorescence rather dense, paniculate or corymb-like; flowers erect. Corolla red to pale yellow or pink; tube narrow, 8–11 mm; lobes ovate, spreading, 4.5–8 × 1.5–6 mm. Anthers and stigmas inside corolla-tube, filaments free for 1–2 mm. *Kenya to Zimbabwe.* G2.

32. K. laciniata (Linnaeus) de Candolle (*K. schweinfurthii* Penzig).

Erect perennial to 1.2 m, sometimes monocarpic; inflorescence and sometimes upper leaf surface with colourless or brownish, simple, glandular hairs. Leaves simple or pinnate, to 20 cm, usually with 3 or 5 ovate to elliptic, entire or scalloped to pinnatisect leaflets; leaf-stalk 2–5 cm. Inflorescence paniculate or corymb-like, very dense; flowers erect. Corolla greenish white to pale orange; tube narrow, 8–16 mm; lobes ovate, spreading, 3.5–7 × 2–5.5 mm. Anthers and stigmas inside corolla-tube, filaments free for *c.* 1 mm. *Namibia to Ethiopia, S India, Thailand.* G2.

33. K. rotundifolia (Haworth) Haworth (*K. guillauminii* Raymond-Hamet). Illustration: Raymond-Hamet & Marnier-Lapostolle, Le Genre Kalanchoë au Jardin 'Les Cèdres'. f. 108, 109 (1964).

Erect or decumbent perennial, 20–200 cm, hairless, usually glaucous. Leaves ovate to spathulate, circular to linear, entire or scalloped, sometimes deeply 3-lobed, 1–8.5 × 0.5–5.5 cm; stalk to 2.5 cm. Inflorescence open, paniculate or corymb-like; flowers erect. Calyx-lobes persistent. Corolla pink or orange to deep red; tube narrow, 6–10 mm; lobes elliptic, spreading, 2.5–5 × 1.2 mm, twisted around carpels in fruit. Anthers and stigmas inside corolla-tube, filaments free for *c.* 1 mm. *South Africa to Zimbabwe, Socotra.* G2.

34. K. glaucescens Britten (*K. laciniata* misapplied; *K. beniensis* de Wildeman). Illustration: Andrews, Flowering plants of the Anglo-Egyptian Sudan **1**: 77 (1950); Toussaint, Flore du Congo Belge et Ruanda-Urundi **2**: 567 (1951); Blundell, Wild flowers of East Africa, 276 (1987).

Erect or decumbent perennial, 30–120 cm, hairless, glaucous. Leaves ovate to spathulate, with few rounded teeth and sometimes reddish margins and marks on lower surface, 3–10 × 1.2–7 cm; base

tapered; stalk 0.5–2.5 cm. Inflorescences loose or dense, paniculate or corymb-like; flowers erect. Calyx-lobes persistent. Corolla yellow to red or pink; tube narrow, 5–15 mm; lobes elliptic, spreading, 3–7.5 × 1–2.5 mm. Anthers and stigmas inside corolla-tube, filaments free for *c.* 1 mm. *Sudan to Zaire.* G2.

35. K. flammea Stapf. Illustration: Botanical Magazine, 7595 (1898).

Perennial to 40 cm, hairless, glaucous; branches short, thick, erect above. Leaves usually obovate to spathulate, entire, 2.5–9 × 1.5–4 cm, rarely rhombic or scalloped; stalks usually indistinct. Inflorescence dense, corymb-like; flowers erect. Calyx-lobes falling early and separately. Corolla-tube narrow, 8–13 mm; lobes ovate to obovate, spreading, orange or red, rarely rose, never yellow, 4–8 × 4–5 mm. Anthers and stigmas inside corolla-tube, filaments free for *c.* 1 mm. *Somalia.* G2.

36. K. longiflora Schlechter (*K. petitiana* misapplied). Illustration: Raymond-Hamet & Marnier-Lapostolle, Le Genre Kalanchoë au Jardin Botanique 'Les Cèdres', f. 92, 93 (1964); Maire, Flore de l'Afrique du Nord **14**: 256 (1976).

Decumbent perennial to 60 cm, hairless, glaucous, sometimes deep red in sun; branches and flower-stems square in section. Leaves obovate or fan-shaped, somewhat concave with few teeth towards apex, 4–8 × 3–8 cm; stalk indistinct, to 1.5 cm. Calyx persistent. Corolla greenish yellow to orange; tube narrow, 1.1–1.7 cm; lobes obovate, spreading, 2–5 × 3–5 mm. Anthers and stigmas just projecting from throat of corolla, filaments free for 0.5–1.5 mm. *Natal.* G2.

37. K. thyrsiflora Harvey. Illustration: Botanical Magazine, 7678 (1899); Flowering Plants of Africa, 341 (1929); Leistner, Flora of Southern Africa **14**: 70 (1985).

Monocarpic or perennial with basal offsets, to 1.3 m, covered in white bloom; stems square in section. Leaves in a loose basal rosette, oblanceolate, entire, stalkless, 6–14 × 2.5–9 cm, opposite pairs united at base. Inflorescence a long, spike-like panicle to 30 × 8 cm; flowers erect or spreading, strongly scented. Corolla deep yellow, tube narrow, 1.1–2 cm, lobes rounded recurved, 2–5 mm. Anthers and stigmas just projecting from throat of corolla, filaments free for 1–2 mm. *Botswana, Lesotho, South Africa.* G1–2.

38. K. petitiana Richard.

Decumbent perennial to 1.5 m, hairless or with sparse, simple, glandular hairs. Leaves ovate, twice scalloped, 4–16 × 3–11 cm; base tapered to almost cordate with upturned lobes; stalk slender, 1.8–4 cm. Inflorescence an open panicle, flower-clusters dense; flowers erect. Whole calyx shed in fruit. Corolla pale to deep pink, tube narrow, 1.4–2.4 cm, lobes oblong, recurved, 6–7 × 3.5–4.5 mm. Anthers and styles projecting from throat of corolla. *Ethiopia.* G2.

Very scarce in cultivation, the name is usually applied to *K. longiflora*.

39. K. nyikae Engler (*K. hemsleyana* Cufodontis). Illustration: Raymond-Hamet & Marnier-Lapostolle, Le Genre Kalanchoë au Jardin Botanique 'Les Cèdres', f. 102, 103 (1964).

Erect or decumbent perennial, 40–200 cm, hairless, glaucous. Leaves ovate, more or less entire, peltate with base upturned, 6.5–18 × 5–16 cm; stalk stout, 3–10 cm. Inflorescence dense to open, corymb-like or paniculate; flowers erect. Corolla-tube narrow, 1.7–2.1 cm, lobes ovate or lanceolate, spreading, cream, yellow or pinkish, 8–11 × 3–5.5 mm. Anthers at throat of corolla, filaments free for 0.5–1.5 mm. *Uganda, Kenya, Tanzania.* G2.

40. K. quartiniana Richard.

Perennial to 1 m, hairless, more or less glaucous; stems robust, erect. Leaves obovate to oblong, scalloped, unmarked, 10–18 × 6–10 cm; stalk broad, 3–6 cm. Inflorescence paniculate; flowers erect. Corolla white, tube very narrow, 3–5 cm, lobes obovate, spreading, 1–2 × 0.6–0.7 cm. Anthers at throat of corolla, filaments free for 0.5–2 mm. *Ethiopia.* G2.

41. K. marmorata Baker (*K. macrantha* Baker; *K. somaliensis* Baker). Illustration: Botanical Magazine, 7333 (1894), 7831 (1902); Flowering plants of Africa, 1049 (1947); Jacobsen, Handbook of succulent plants, 654 (1960).

Erect or decumbent perennial to 1.3 m, hairless, glaucous. Leaves obovate, entire, scalloped or sharp-toothed, often with purplish marks on both faces, 6–20 × 3.5–13 cm; base tapered, stalkless. Inflorescence paniculate; flowers erect. Corolla white, rarely with pink or yellow tinge; tube very narrow, 4.5–12 cm; lobes lanceolate to obovate, spreading, 1–2.5 × 0.6–1.2 cm. Anthers at throat of corolla, filaments free for 0.5–2 mm. *Sudan to Zaire.* G2.

42. K. × kewensis Thiselton-Dyer (*K. flammea × bentii* subsp. *bentii*). Illustration: Annals of Botany 17, pl. 21–23 (1903); van Laren, Vetplanten, 77 (1932).

Erect or decumbent perennial to 1.2 m, hairless, bronze-green, slightly glaucous. Leaves cylindric, grooved above, often with several lateral lobes or teeth, 10–30 cm, stalkless; sometimes basal leaves flat, elliptic, toothed. Inflorescence dense, paniculate or corymb-like; flowers erect. Calyx more or less erect. Corolla-tube narrow, *c.* 1.45 cm; lobes ovate, spreading, deep pink, *c.* 8 × 6 mm. Anthers at throat of corolla, filaments free for 0.5–4 mm, degree of fusion irregular. G2.

43. K. bentii J.D. Hooker (*K. teretifolia* misapplied).

Erect perennial to 1.5 m, olive-green, glaucous. Leaves cylindric, thickest in middle, grooved on upper surface, entire, stalkless, 7.5–40 × 0.6–1.2 cm. Inflorescence paniculate with flat-topped clusters of erect, scented flowers. Calyx spreading. Corolla white; tube narrow, 3–3.5 cm; lobes lanceolate, recurved, 8–16 × 3.5–6.5 mm. Anthers at throat of corolla; filaments free for *c.* 0.5 mm. G2.

Subsp. **bentii**. Illustration: Botanical Magazine, 7765 (1901). Hairless. *Yemen*.

Subsp. **somaliensis** Cufodontis. Calyx inside and corolla outside with simple glandular hairs. *Somalia*.

44. K. grandiflora Wight & Arnott. Illustration: Botanical Magazine, 5460 (1864); Addisonia, 725 (1945).

Erect perennial to 80 cm, hairless, glaucous. Leaves ovate to obovate, weakly scalloped, 4–10 × 3.5–7.5 cm; base tapered; stalk indistinct. Inflorescence paniculate, compact; flowers more or less erect. Calyx spreading. Corolla bright yellow, tube *c.* 1.2 cm × 4–6 mm, lobes obovate, spreading, 1.3–1.7 cm × 5–7 mm. Anthers and styles just projecting from throat of corolla, filaments free for 0.5–2 mm. *S India*. G2.

45. K. farinacea Balfour (*K. scapigera* misapplied). Illustration: Botanical Magazine, 7769 (1901); Flowering plants of Africa, 1329 (1960); Jacobsen, Das Sukkulenten Lexikon, pl. 42 (1981); British Cactus and Succulent Journal **4**: 6 (1986).

Perennial to 30 cm, hairless; stems short, erect above. Leaves crowded, obovate, entire with thin grey wax, 2–6 × 1.5–5 cm; base tapering; stalk indistinct. Inflorescence rounded, fairly dense, almost stemless; flowers erect. Corolla red, tube

narrow, sometimes yellow towards base, 8–12 mm, lobes ovate, ascending, 4–5.5 × 2–2.5 mm. Anthers and stigmas at throat of corolla, filaments free for 1–2 mm. *Socotra*. G2.

4. OROSTACHYS Fischer
S.G. Knees

Monocarpic perennial herbs with leaves in rosettes. Leaves alternate, linear-elliptic, often with a cartilaginous bristly tip. Flowers in tall, dense spike-like, terminal racemes. Sepals 5, almost equal; petals 5, often joined at the base, spreading, white, yellowish or reddish, often spotted red; stamens 10; styles straight, slender, carpels 5, ovaries stalked. Fruits erect, free.

A genus of about 10 mainly herbaceous species from Europe and temperate Asia, at one time included in *Sedum* but differing in the callus-tipped leaves in dense rosettes and flowers in dense panicles. Plants flower in the second or third year and propagation is usually from seed or offsets. These plants can be grown in the rock garden in a very freely drained soil.

Literature: Ohba, H., Notes towards a monograph of the genus Orostachys (Crassulaceae) (1), *Journal of Japanese Botany* **65**(7): 193–203 (1990).

Leaves with obvious spiny tip: **1,2,3**; lacking spiny tip: **4,5,6**.
Flowers white or pinkish white: **1,3,5,6**; greenish white: **4**; yellowish green: **2**.

1a. Leaves with an obvious, hard spiny tip   2
  b. Leaves with hard margins but lacking a spiny tip   4
2a. Flowering-spike 4–10 × 1.5–2 cm   **1. erubescens**
  b. Flowering-spike 10–30 × 5–8 cm   3
3a. Petals 3–5 mm, greenish yellow   **2. spinosa**
  b. Petals 6–8 mm, white, tipped red   **3. chanetii**
4a. Leaves distinctly bluish grey; sometimes with cream margins   **5. iwarengis**
  b. Leaves green or yellowish green, never bluish   5
5a. Leaves 2–4 × 1–2 cm   **6. aggregata**
  b. Leaves 1–2 × 0.5–1 cm   **4. furusei**

1. O. erubescens (Maximowicz) Ohwi (*O. japonica* Maximowicz). Illustration: Hayashi, Azegami & Hishiyama, Wild flowers of Japan, 439 (1983); Graf, Tropica, 374 (1978).

Leaves 13–30 × 4–6 mm, narrowly spathulate, fleshy; margin and apex hard, with several teeth, spine 1–3.5 mm. Flowering-stem 6–15 cm, with lanceolate leaves; flowering-spike 4–10 × 1.5–2 cm, stalked; petals 5–7 mm, whitish pink; anthers red at first, darkening to purple. *China, Korea & Japan*. H5–G1. Summer–autumn.

Three variants are recognised and occasionally seen in cultivation: var. **erubescens**, var. **japonica** Maximowicz and var. **polycephala** (Makino) Hara.

2. O. spinosa (Linnaeus) Berger. Illustration: Jacobsen, A handbook of succulent plants **2**: 702 (1960); Graf, Exotica **4**(1): 876 (1985); Gartenpraxis **8**: 21 (1987).

Tufted perennial with leaves of 2 lengths, in basal rosettes, the centre of which folds up in winter. Leaves 15–25 × 3–5 mm, oblong with spiny white tips 2–4 mm. Flowering-stems 10–30 cm, leafy, with flowers borne in spike-like panicles 5–20 cm. Petal-stalks 1–2 mm or absent; petals 3–5 mm, lanceolate, greenish yellow; anthers yellow. Fruits 5–6 mm. *NE & C Asia (Xizang to Mongolia)*. H5–G1. Summer.

3. O. chanetii (A. Léveillé) Berger. Illustration: Natürlichen Pflanzenfamilien 18a: 464 (1930).

Leaves of 2 lengths, to 2.5 × 0.5 cm, linear, spine-tipped, grey-green. Flowers borne in branched pyramidal panicles 15–30 × 5–8 cm; petals 6–8 mm, white, tipped reddish. *China*. H5–G1. Summer.

4. O. furusei Ohwi.

Sterile stems ascending to 5 cm, with leaves clustered towards apex. Leaves 1–2 × 0.5–1 cm, obovate, wedge-shaped at base. Flowering-stems 5–10 cm, leafy, flowers many. Sepals *c.* 3.5 mm, erect; petals 4.5–5 mm, ovate, greenish white; styles *c.* 1 mm. *Japan*. H5–G1. Summer–autumn.

5. O. iwarengis (Makino) Hara. Illustration: Jacobsen, Lexicon of succulent plants, pl. 112 (1974); Hayashi, Azegami & Hishiyama, Wild flowers of Japan, 439 (1983); Gartenpraxis **8**: 8 (1987); Kakteen und andere Sukkulenten **41**(10): 222–224 (1990).

Stems erect, to 5 cm, producing many stolons; rosettes persisting over winter. Leaves 3–7 × 0.7–2.8 cm, oblong to spathulate, fleshy, striking bluish grey. Flowering-stems 5–20 cm, forming narrow cones. Flowers solitary, shortly stalked;

sepals 2–3 mm; petals 5–7 mm, white. *Japan, China*. H5–G1. Summer–autumn.

Propagate by decapitating young rosettes and rooting the offsets.

'Fuji', a cultivar with broad cream leaf-margins, is sometimes grown. Illustration: Graf, Tropica, 374 (1978).

6. O. aggregata (Makino) Hara (*Cotyledon aggregata* Makino; *O. malacophylla* (Pallas) Fischer, invalid). Illustration: Botanical Magazine, 4098 (1844); Jacobsen, A handbook of succulent plants **2**: 701 (1960); Journal of Japanese Botany **65**(7): 197 (1990).

Like *O. iwarengis*, but leaves smaller, 2–4 × 1–2 cm, rounded, blunt, green. *Japan*. H5–G1. Summer–autumn.

5. LENOPHYLLUM Rose
S.G. Knees
Hairless, succulent, perennial herbs. Roots pale brown; stock whitish. Stems erect or spreading, pinkish. Leaves opposite on non-flowering shoots, often crowded together, usually boat-shaped. Flowering-stems terminal with scattered alternate leaves, often branched towards apex. Sepals 5, often very fleshy; petals 5, more or less equalling sepals, erect, yellow, tips recurved; stamens 10, 5 arising directly from petals, 5 from between petals; nectaries oblong, orange or deep yellow; carpels erect, joined at base, green; stigmas white, follicles spreading, brown. Seeds oblong-elliptic, brown.

A genus of 5 or 6 species from northern Mexico and adjacent Texas, closely related to *Echeveria* and with similar cultivation requirements. All can be readily rooted from leaf cuttings or seed.

1a. Leaves obtuse or rounded at apex 2
 b. Leaves acute 3
2a. Leaves broadest at base **3. guttatum**
 b. Leaves narrow at base **2. weinbergii**
3a. Stems 3–5 cm; flowers solitary **4. pusillum**
 b. Stems 10 cm or more; flowers clustered 4
4a. Leaf pairs crowded together; flowers yellow with reddish tips **1. texanum**
 b. Leaf pairs distant; flowers greenish yellow **5. acutifolium**

1. L. texanum (J.C. Smith) Rose (*Sedum texanum* J.C. Smith; *Villadia texana* (J.C. Smith) Rose). Illustration: Report of the Missouri Botanic Garden **6**: t. 50 (1895); Addisonia **8**: t. 267 (1923); Clausen, Sedum of North America north of the Mexican plateau, 586 (1975).

Tufted perennial herb, stems to 10 cm. Leaves 1.5–3.5 × 0.8–1.8 cm, ovate, elliptic-lanceolate, acute; boat-shaped above, rounded beneath, green when young, turning lavender-green or reddish with age. Flowering-stems erect or spreading, 10–35 cm, pinkish. Sepals acute, 3 mm, oblanceolate-ovate. Petals oblanceolate, oblong, acute, pale primrose-yellow with dark red blotch at tip. Stamens yellow; follicles spreading. All floral parts persistent. *SE USA (Texas), NE Mexico*. G1. Late summer–early winter.

A rather weedy species propagating very easily from leaves, which may fall off at the slightest touch.

2. L. weinbergii Britton. Illustration: Smithsonian Miscellaneous Collections **47**: 161 (1904); Botanical Magazine n.s., 750 (1978).

Perennial herb. Leaves 2–4 × 1.8–3 cm, obovate to wedge-shaped, trough-shaped, truncate, more or less concave, obtuse at apex. Flowering-stems to 15 cm, few-flowered with 2 pairs of leaf-like bracts to 1.5 × 1 cm; branches 5, with 1–9 flowers. Sepals obtuse, 4 mm, club-shaped, dull purplish grey or light green, speckled purple. Petals *c*. 7 mm, erect with spreading or recurved tips. Fruits to 7 mm. *NE Mexico (Coahuila)*. G1. Summer.

3. L. guttatum Rose. Illustration: Smithsonian Miscellaneous Collections **47**: t. 20 (1904).

Perennial herb. Leaves 1.8–3.5 cm, elliptic-ovate to rhombic, deeply but shallowly grooved above, apex obtuse or rounded, greyish pink or light green, flecked with blackish purple. Flowering-stems to 20 cm; branches 2–6 with 5–12 flowers; flowers almost stalkless. Sepals obtuse, *c*. 6 mm, club-shaped; petals *c*. 5 mm, obtuse, yellowish, eventually drying reddish. *NE Mexico*. Late summer–autumn. G1.

4. L. pusillum Rose. Illustration: Jacobsen, Handbook of succulent plants **2**: 672 (1960).

Perennial herb with matted stems 1–3 cm. Leaves fleshy, 0.8–1.5 cm, narrow, acute, upper surface furrowed, lower surface keeled, dirty reddish green; readily falling and rooting. Flowering-stems to 5 cm, with solitary, terminal flowers. Sepals acute; petals 6–7 mm, lemon yellow. *Mexico*. G1. Summer–autumn.

5. L. acutifolium Rose. Illustration: Smithsonian Miscellaneous Collections **47**: t. 19 (1904).

Perennial herb, well branched at the base with stems to 10 cm. Leaves arranged in 6–8 distant pairs, lanceolate, acute, upper surface grooved. Flowers numerous, in interrupted spikes, stalkless or very shortly stalked; sepals thick, almost equal, greenish yellow; petals erect below, spreading above, greenish yellow. *Mexico (Nuevo Leon)*.

6. SEDUM Linnaeus.
N. Groendijk-Wilders & L. Springate
Plants hairless, downy or glandular hairy, fleshy, erect or decumbent, sometimes tufted or moss-like. Leaves very variable, opposite, alternate or whorled, entire or rarely toothed. Inflorescence usually terminal, rarely lateral, mostly cyme-like or flowers solitary, white, yellow or rose, rarely red or blue, hermaphrodite, floral parts usually in 5s, rarely in 4s, 6s, 7s, 8s or 9s; petals usually free, sometimes fused for almost a third, stamens usually twice as many as petals, rarely equal in number to petals.

A genus of about 280 species, mostly natives of the temperate and colder regions of the northern hemisphere.

Literature: Praeger, R.L., An account of the genus Sedum as found in cultivation. *Journal of the Royal Horticultural Society* **46** (1920–1921); Clausen, R.T., *Sedum of the trans-Mexican volcanic belt* (1959); Jacobsen, H., *A handbook of succulent plants* (1960); Clausen, R.T., *Sedum of North America: North of the Mexican plateau* (1975); Köhlein, F., *Freilandsukkulenten* (1977); 't Hart, H., *Biosystematic studies in the Acre-group and the series Rupestria Berger of the genus Sedum L. (Crassulaceae)* (1978); Fessler, A., *Die Freilandschmuckstauden*, edn 3 (1985).

1a. Plant with somewhat woody stems 2
 b. Plant herbaceous 28
2a. Leaves opposite, ovate to spherical; flowers yellow **1. stahlii**
 b. Leaves alternate 3
3a. Leaves crowded, sometimes seemingly in whorls 4
 b. Leaves not crowded 9
4a. Flowers light pink to deep scarlet-purple; leaves oblong-lanceolate, thick-fleshy, bluish bloomed 5

b. Flowers yellow or greenish yellow
6

5a. Shoots wholly pendent; leaves to 2 cm, slightly bloomed
2. morganianum

b. Shoot-ends curving slightly outwards; leaves to 2.5 cm, densely bloomed
× Sedeveria 'Harry Butterfield'

6a. Flowers orange-yellow; leaves obovate to spathulate, glaucous
3. palmeri

b. Flowers pale yellow or greenish yellow
7

7a. Plant 15–20 cm; leaves flat, fleshy green
8

b. Plant to 30 cm; leaves cylindric, very fleshy, grey-green
4. pachyphyllum

8a. Stems erect; leaves without spur beneath
11. fusiforme

b. Stems prostrate; leaves with spur beneath
× Sedadia amecamecana

9a. Flowers greenish yellow or bright orange
10

b. Flowers white, pink, red or white with red
17

10a. Stem at first fleshy, later somewhat woody; leaves rounded at apex, pale green, often suffused with red
5. rubrotinctum

b. Stem always woody
11

11a. Leaves with waxy white, later reddish margin, spathulate, broadly rounded
6. dendroideum

b. Leaves without waxy margin
12

12a. Leaves glossy or pale green, flat
13

b. Leaves bluish or grey-green or semi-terete
16

13a. Flower to 1 cm across; leaves obovate-oblong
12. nudum

b. Flowers more than 1 cm across; leaves obovate to spathulate or oblanceolate to spathulate
14

14a. Leaves 5–7 cm, oblanceolate to spathulate
7. praealtum

b. Leaves 1.5–4 cm, obovate to spathulate
15

15a. Leaves 1.5–2 × 1.1–1.3 cm, pale green; sepals 2–3 mm
8. decumbens

b. Leaves 2–4 × 1.2–2 cm, mid-green; sepals 1–1.5 mm
9. confusum

16a. Leaves blue-green, densely grey bloomed
10. treleasei

b. Leaves grey-green, not bloomed
3. palmeri

17a. Leaves reddish blue, oblong-elliptic with rounded apex; flowers white, bell-shaped
13. craigii

b. Leaves green, yellowish or grey
18

18a. Leaves less than 8 mm
19

b. Leaves more than 8 mm
20

19a. Leaves obovate or spathulate; nectaries large, red, exposed
14. longipes

b. Leaves ovate to lanceolate; ectaries small, white or pale yellow, concealed
15. moranense

20a. Leaf-margin toothed or notched
21

b. Leaf-margin entire
22

21a. Leaves ovate, cordate at base (poplar-like), margin coarsely and irregularly toothed
16. populifolium

b. Leaves obovate-oblong, spathulate, margin deeply notched at tip
17. retusum

22a. Inflorescence borne at top of main shoots
23

b. Inflorescence borne on short or long lateral branches
26

23a. Leaves more than 5 mm wide
24

b. Leaves less than 5 mm wide
25

24a. Shrub 50–90 cm; flowers pinkish to dull red, *c.* 1.3 cm across
18. oxypetalum

b. Shrub to 40 cm; flowers greenish white, 1.6–1.8 cm across
19. allantoides

25a. Leaves green, average width 1.5–2.1 mm, average thickness 0.6–1.2 mm; nectaries purple, oblong or spathulate
20. bourgaei

b. Leaves grey or green, average width 2.3–2.7 mm, average thickness 1.4–2.0 mm; nectaries white or yellow, kidney-shaped
21. griseum

26a. Leaves dark lustrous green; flower-stalks 1–7 mm
22. lucidum

b. Leaves dull or yellowish green; flower-stalks 8–20 mm
27

27a. Shrub 7.5–12.5 cm; shoots ascending; inflorescence a panicle
23. adolphii

b. Shrub to 7 cm; shoots prostrate; inflorescence a corymb
24. nussbaumerianum

28a. Perennial with thickened rootstock and slender roots; stems mostly annual with flat leaves; flowers yellow to orange
29

b. Rootstock, roots, stems and leaves formed differently
33

29a. Plant 30–80 cm
30

b. Plant 10–30 cm
31

30a. Plant 30–45 cm, densely hairy; leaves greyish downy, 4–5 × 1–1.2 cm; flowers 8–10 mm across
25. selskianum

b. Plant to 80 cm, hairless; leaves 5–8 × 1.9–2.1 cm; flowers *c.* 1.2 cm across
26. aizoon

31a. Stems persistent, creeping, with leaves at tip in winter
27. hybridum

b. Stems annual, not creeping
32

32a. Leaves bright or dark green; flowers 1.5–2 cm across; petals mucronate at top; plant usually with one terminal, leafy or leafless inflorescence
28. kamtschaticum

b. Leaves dark green, crowded; flowers 1.2–1.5 cm across; petals acute at top; plant produces many short axillary leafy floriferous branches
29. floriferum

33a. Annual or biennial with flat broad leaves; basal leaves forming a rosette
34

b. Perennial or, when annual or biennial, leaves linear or not forming a rosette
36

34a. Plant hairy, with linear-ovate leaves, green dotted with red; flowers white, often with a few pinkish spots
30. cepaea

b. Plant downy, flowers rose to crimson red
35

35a. Plant glandular-downy, 5–10 cm; leaves to 1.8 cm, linear to spathulate, obtuse; flowers rose-red, *c.* 9 mm across
31. pilosum

b. Plant downy, 10–30 cm; leaves 2.5–3 cm, ovate, bluntly acuminate; flowers carmine-red, *c.* 1.3 cm across
32. sempervivoides

36a. Plant annual or biennial (perennial in some forms of *S. hispanicum*)
37

b. Plant perennial
40

37a. Annual with reddish stems, 15–25 cm; leaves flat, margin entire; inflorescence very large, of many 2–3 forked branches with flowers in the forks; flowers pale yellow
33. formosanum

b. Plant 5–15 cm; leaves almost or partly terete; inflorescence not as above; flowers white, pink, rose or blue
38

38a. Flowers many, pale or sometimes sky-blue, 5–6 mm across
34. caeruleum

b. Flowers white to pink, never blue, 6–12 mm across 39

39a. Leaves glandular hairy, 6–12 mm, green; flowers pink, dull rose or white, 6 mm across **35. villosum**

b. Leaves hairless, 1.2–2.5 cm, grey-green, often reddish; flowers pink to white, *c.* 1.2 cm across **36. hispanicum**

40a. Plant with short rootstocks and thick roots 41

b. Plant rootstocks not as above or absent 56

41a. Stems to 30 cm 42

b. Stems more than 30 cm 50

42a. Leaves opposite or in whorls of 3 43

b. Leaves alternate 46

43a. Leaves opposite 44

b. Leaves in whorls of 3 **40. sieboldii**

44a. Plant 5–10 cm; leaves oblong-lanceolate **41. cyaneum**

b. Plant 10–25 cm; leaves circular to obovate 45

45a. Leaves not red-dotted, margin usually entire; flowers pink or pale violet **38. ewersii**

b. Leaves minutely red-dotted, margin with a few blunt teeth; flowers rose-purple, becoming carmine-red **39. cauticola**

46a. Flowers to 1 cm across 47

b. Flowers more than 1 cm across 49

47a. Flowers white to pink **45. telephioides**

b. Flowers purplish red to purplish pink 48

48a. Leaves 1.2–2.5 cm, grey-green **37. anacampseros**

b. Leaves 5–7.5 cm, blue-green **46. telephium**

49a. Leaves oblanceolate, entire, grey-green; flowers purplish carmine-rose **41. cyaneum**

b. Leaves linear-lanceolate, toothed towards apex, green; flowers pale pink **42. tatarinowii**

50a. Leaves alternate **46. telephium**

b. Leaves opposite or in whorls 51

51a. Flowers 1–1.3 cm across 52

b. Flowers to 6 mm across 54

52a. Flowers pink; leaves rounded or tapered at base, stalkless, whole plant never purplish **48. spectabile**

b. Flowers white to greenish white or rose-white or leaves with distinct short stalks or whole plant purplish 53

53a. Plant 30–60 cm; leaves usually

opposite, grey-green, 5–7.5 cm, petals rose-white or greenish white **49. erythrostrictum**

b. Plant 50–100 cm; leaves opposite or ternate, dark-green, 5–12.5 cm, flowers yellowish or greenish white or whole plant purplish **50. maximum**

54a. Leaves red-dotted; flowers pale-green **43. verticillatum**

b. Leaves green or green with a red edge; flowers cream to pink 55

55a. Leaves green, base rounded, about as long as internodes **47. pseudospectabile**

b. Leaves glaucous, base wedge-shaped, twice as long as internodes **44. 'Autumn Joy'**

56a. Root stocks thickening horizontally or contracted; leaves evergreen; flowers mostly white, rarely red or yellow 57

b. Rootstocks absent; plant creeping or erect, flowering shoots monocarpic or sterile 65

57a. Stems 15–30 cm, arising from perennial leafy rosette 58

b. Stems 2–15 cm, annual or lasting up to 18 months; rosette absent, or small and loose 61

58a. Plant hairless 59

b. Plant hairy 60

59a. Leaves oblong-ovate; flowers white with purple midrib and markings towards base, petals ovate **51. glabrum**

b. Leaves spathulate; flowers white, petals lanceolate **56. bellum**

60a. Plant hairy; stem-leaves linear or lanceolate, usually less than 10.5 mm across **52. hemsleyanum**

b. Plant minutely hairy; stem-leaves ovate, very fleshy, usually more than 10.5 mm across **53. ebracteatum**

61a. Flowers white or tinged red, petals opaque 62

b. Flowers pink, reddish or yellow or petals translucent 63

62a. Flowers bell-shaped, white; petals oblong-ovate, erect below, wide-spreading above **54. wrightii**

b. Flowers star-like; petals lanceolate, spreading **55. cockerellii**

63a. Leaves *c.* 2.5 cm, green, red at margins and tip **57. versadense**

b. Leaves *c.* 0.6 cm, slightly glaucous 64

64a. Plant hairless, fresh green; leaves slightly glaucous; flowers sulphur yellow **58. greggii**

b. Plant glaucous with grey-green leaves; petals whitish or pinkish, translucent **59. alamosanum**

65a. Flowers yellow 66

b. Flowers pink, purple, red, white or cream 91

66a. All leaves opposite or whorled 67

b. Leaves alternate (at least on flowering shoots) 71

67a. Leaves opposite, thick, obovate to spathulate, green or reddish green; flowers *c.* 1.8 cm across **60. divergens**

b. Plant with 3 or more leaves at one level 68

68a. Plant with light green shining tint; leaves 2–3 mm across, narrowly linear, obtuse, in whorls of 3–5, or rarely solitary **61. mexicanum**

b. Plant without green shining tint; leaves mostly ternate 69

69a. Leaves light green, broadest near the apex, obtuse, 5 mm broad; stems, leaves and sepals with sparse fine papillae **62. chauveaudii**

b. Leaves pale green, linear or lanceolate, acute or nearly acute at the top; plant without papillae 70

70a. Leaves in whorls of 3, sometimes also opposite, broadly lanceolate, 6 mm broad; flowering-stems short **63. sarmentosum**

b. Leaves linear or linear-lanceolate, to 3 mm across; flowering-stems tall **64. lineare**

71a. Leaves spathulate, flat 72

b. Leaves not as above 75

72a. Leaves green, to 1.8 cm; leaves of erect stems elliptic; flowers bright yellow, petals oblong to lanceolate **65. purdyi**

b. Leaves tinged red or orange-red when older and 1.2–3 cm; petals lanceolate 73

73a. Leaves shining green, often reddened, very fleshy, 1.2–1.8 cm; petals long acuminate, ascending; flower-buds *c.* 1.5 cm **66. oreganum**

b. Leaves greyish or bluish green, to 3 cm; petals acute 74

74a. Leaves greyish green or white mealy, often tinged red or purple, white on back; flowers to 1.5 cm across; petals wide-spreading **67. spathulifolium**

b. Leaves bluish green, in age orange-red; flowers 6–10 mm across; petals erect below, somewhat spreading above **68. obtusatum** subsp. **obtusatum**

75a. Leaves with marginal hairs
        69. humifusum
  b. Leaves hairless   76
76a. Plant with cypress-like branches;
    leaves ovate, closely overlapping, *c.*
    1.5 cm   **70. muscoideum**
  b. Branches not as above   77
77a. Leaves shorter than 8 mm   78
  b. Leaves 10 mm or longer   84
78a. Leaves broadest at the base or at
    the top   79
  b. Leaves linear   81
79a. Leaves ovate-triangular to oblong
    or ovate-rounded, densely
    overlapping throughout   **71. acre**
  b. Leaves oblong-elliptic to obovate,
    overlapping only at shoot tips   80
80a. Stems minutely roughened, red
    brown, 2 mm thick near tip;
    follicles erect   **72. oaxacanum**
  b. Stems smooth, greenish, 1 mm
    thick near tip; follicles spreading
        73. alpestre
81a. Leaves arranged in 6 spiral rows,
    bright green, 4–7 mm; flowers
    golden yellow, 5–9 mm across.
        74. sexangulare
  b. Leaves not as above; flowers *c.* 1.2
    cm across   82
82a. Flowers bright yellow, to 1.6 cm
    across   **83. urvillei**
  b. Flowers dull yellow, to 1.3 cm
    across   83
83a. Plant bushy with shrubby growth;
    leaves green, with papillae, all
    winter deciduous except the
    youngest   **75. multiceps**
  b. Plant with ascending branches;
    leaves variable   **76. japonicum**
84a. Vegetative rosettes in leaf axils of
    floral stems   **77. stenopetalum**
  b. Plant not as above   85
85a. Sterile shoot tips forming
    propagules covered in dead leaf
    bases in summer
        78. amplexicaule
  b. Fresh leaves persisting through
    summer at shoot-tips   86
86a. Inflorescence dense, drooping in
    bud; petals 6–7   87
  b. Inflorescence not drooping in bud,
    loose or flowers with 5 petals   88
87a. Leaves in dense rosette-like cones at
    tips of sterile shoots
        79. forsterianum
  b. Leaves evenly spread along sterile
    shoots, not in cones   **80. rupestre**
88a. Plant to 15 cm; leaves to 1.8 cm
        89
  b. Plant to 60 cm; leaves often greater
    than 1.8 cm   **93. sediforme**

89a. Long prostrate sterile shoots with
    tufts of leaves at tip; flowers pale
    yellow, sepals and petals 6
        81. pruinatum
  b. Compact sterile shoots, less than 6
    cm; flowers usually bright yellow,
    sepals and petals 5   90
90a. Leaves oblong-lanceolate, with
    blunt mucros; flowering-shoot
    with long stem dividing near apex
        82. lanceolatum
  b. Leaves ovate to elliptic, blunt
    without mucros; flowering-shoot
    with short stem, soon dividing
        83. urvillei
91a. Flowers red, purple, pink or green
    and red   92
  b. Flowers white or white with tinges
    of pink, red or green, or greenish
    red   97
92a. Flowers rose-red outside, whitish
    inside, *c.* 5 mm across; leaves
    obovate to spathulate, obtuse, 5–6
    mm, spotted with red
        84. stevenianum
  b. Flowers white, white tinged red or
    rose-purple   93
93a. Leaves opposite   94
  b. Leaves alternate or ternate   96
94a. Leaves ovate or obovate, obtuse,
    roundish towards the base,
    narrowed into a very short leaf-
    stalk; petals slender, pointed
        85. obtusifolium
  b. Leaves not as above; petals acute
        95
95a. Leaves obovate, dark green; flowers
    pink to purple, *c.* 1.3 cm across
    with semi-erect petals; inflorescence
    dense and flat   **86. spurium**
  b. Leaves diamond-shaped to
    spathulate, bright green; flowers
    pink with wide-spreading petals,
    1.3–1.8 cm across; inflorescence
    loose and small
        87. stoloniferum
96a. Stems triangular; leaves in whorls
    of 3, uppermost often alternate,
    light green; flowers greenish red,
    1.3 cm across; petals wide-
    spreading, later sharply reflexed
        88. rhodocarpum
  b. All leaves alternate, green, narrow-
    linear; flowers rosy purple, 5–14
    mm across, 4-parted
        89. pulchellum
97a. Leaves opposite, toothed in upper
    half   **86. spurium**
  b. Leaves alternate, in whorls of 3 or
    4 or entire   98

98a. Leaves flat, most in whorls of 3 or
    4   99
  b. Leaves thickened or both alternate
    and opposite on one plant   100
99a. Leaves in whorls of 3 or opposite,
    on the flowering-stems alternate,
    1.2–2.5 cm, pale green, obovate;
    flowers *c.* 1.3 cm across
        90. ternatum
  b. Leaves in whorls of 4, 3–10 mm;
    olive-green, oblong to spathulate;
    flowers 6–10 mm across
        91. monregalense
100a. Inflorescence a raceme; leaves light
    to bright green   **92. magellense**
  b. Inflorescense cyme-like; leaves pale
    green, blue-green, grey-green or
    dark green   101
101a. Plant with thorn-like point on lower
    leaves   **93. sediforme**
  b. Plant lacking thorn-like point on
    lower leaves   102
102a. Plant partly hairy   103
  b. Plant hairless   108
103a. Leaves ovate or diamond-shaped to
    linear to oblong elliptic or slightly
    spathulate, base not or scarcely
    tapered   104
  b. Leaves obovate, diamond-shaped to
    spathulate, distinctly long-tapered
    towards base   107
104a. Leaves glaucous, pale grey-green or
    blue-green, ovate, elliptic or oblong
    to linear, hairy or hairless   105
  b. Leaves not glaucous, dull green,
    ovate to diamond-shaped, elliptic or
    oblong, downy or hairy   106
105a. Leaves linear or linear-oblong
        94. bithynicum
  b. Leaves ovate to elliptic-oblong
        95. dasyphyllum
106a. Petals *c.* 3 mm; much branched
    corymb-like inflorescence of more
    than 25 flowers   **96. gypsicola**
  b. Petals 5–8.5 mm; open, paniculate
    inflorescence of less than 25 flowers
        97. hirsutum
107a. Petals fused for 1–1.5 mm; flowers
    sweetly scented; leaves diamond-
    shaped to spathulate
        98. fragrans
  b. Petals fused for 1.5–2 mm; flowers
    not scented; leaves oblong to
    spathulate
        Rosularia adenotricha
108a. Low plant to 2.5 cm with rosette-
    like, obovate, grey-green leaves;
    flowers white, intensely scented
        99. compactum
  b. Plants more than 2.5 cm, with
    non-scented flowers   109

109a. Leaves of sterile shoots ovoid   110
   b. Leaves long and thin or semi-terete
      to flat   112
110a. Sepals spurred at base
                     110. tenellum
   b. Sepals not spurred at base   111
111a. Leaves alternate, 6–10 mm, with
      uneven covering of waxy scales
                 100. furfuraceum
   b. Leaves in rows of 4 or 5, 2–6 mm,
      mealy or smooth and lustrous
                 101. brevifolium
112a. Leaves green with red or pink tips
      or dots   113
   b. Leaves green or pale brownish or
      bluish green, lacking dots   121
113a. Leaves obovate to spathulate; petals
      fused for 1–3 mm
                  102. oregonense
   b. Leaves ovate to oblong-elliptic or
      linear; petals free or fused up to
      1 mm   114
114a. Leaves glaucous   115
   b. Leaves not glaucous   116
115a. Sterile shoots to 5 cm; leaves ovate
      to oblong; petals 3–5.5 mm
                   95. dasyphyllum
   b. Sterile shoots to 15 cm; leaves
      oblong to linear; petals 4–7 mm
                     103. diffusum
116a. Leaves ovate or ovate-elliptic;
      inflorescence with 1–3 branches
                                      117
   b. Leaves more or less linear or
      cylindric to ovate, or inflorescence
      corymb-like with more than 3
      branches   118
117a. Plant 5–15 cm; leaves 5 mm or
      more, ovate-oblong, green, tipped
      red; old leaves persistent with loose
      silvery bases, mucronate at the
      top; flowers white; petals acute
                 104. liebmannianum
   b. Plant 7–8 cm; leaves 4–5 mm,
      ovate-elliptic, thick, green, often
      red; flowers white-pink; petals
      mucronate   **105. anglicum**
118a. Plant 3–5 cm   119
   b. Plant 5–20 cm   120
119a. Leaves to 6 mm, linear-oblong,
      nearly cylindric, green, dotted red;
      inflorescence branches semi-erect,
      forked, without flowers between
      the primary and secondary forks;
      flowers 9–12 mm across, white,
      with red spots on back
                      106. albertii
   b. Leaves 6–8 mm, narrowly linear to
      cylindric, green or red; inflorescence
      dense, flat; flowers c. 6 mm across,
      white   **107. lydium**

120a. Plants 5–10 cm; leaves c. 8 mm,
      linear-oblong, bright green or
      reddish green; inflorescence
      branches almost horizontal; flowers
      white, dotted red on back; follicles
      spreading   **108. gracile**
   b. Plants 10–20 cm; leaves 6–12 mm,
      cylindric to oblong-ovate, dark
      green, often reddish; inflorescence
      branches semi-erect; flowers white;
      follicles erect   **109. album**
121a. Leaves 12 mm or longer   122
   b. Leaves of sterile stems c. 3 mm
                                      123
122a. Leaves 1.2–1.8 cm, pale brownish
      green   **111. nevii**
   b. Leaves c. 2.5 cm, pale or blue-green
               112. glaucophyllum
123a. Leaves to 8 mm; sepals spurred at
      base; follicles erect
                     110. tenellum
   b. Leaves to 6 mm; sepals not spurred;
      follicles spreading   **108. gracile**

1. S. stahlii Solms. Illustration: Botanical
Magazine, 7908 (1903); Journal of the
Royal Horticultural Society **46:** 222
(1920–1921); Engler, Die natürlichen
Pflanzenfamilien **18a:** 441 (1930);
Clausen, Sedum of the trans-Mexican
volcanic belt, f. 49 (1959).

Evergreen, finely downy, prostrate
shrub, 10–20 cm. Leaves opposite, 7–14 ×
4–8 mm, ovoid to spherical, apex rounded,
dark green to brown. Flowers c. 1.3 cm,
yellow; petals elliptic-lanceolate, short
acuminate. *Mexico.* G1. Late summer.

2. S. morganianum Walther. Illustration:
Jacobsen, Handbook of succulent plants,
789 (1960); Boom, Flora Cultuurgewassen
3, kamer-en kasplanten: pl. 130 (1968);
Hay & Beckett, Reader's Digest
encyclopaedia of garden plants and
flowers, 658 (1978).

Evergreen shrublet with prostrate or
creeping branches, to 7.5 cm. Leaves
alternate, numerous, oblong-lanceolate,
nearly cylindric, acute, c. 20 × 8 mm,
thick-fleshy, pale green, bluish bloomed.
Flowers pale pink to deep scarlet purple, c.
10 mm across. *Mexico.* G1. Spring.

3. S. palmeri S. Watson (*S. compressum*
Rose). Illustration: Journal of the Royal
Horticultural Society **46:** 233, 235
(1920–1921).

Evergreen hairless shrub, to 22 cm,
with ascending branches. Leaves in loose
rosettes, oblong to obovate to spathulate,
obtuse, sometimes mucronate, 2.5–5 ×
1.5–2 cm, usually grey-green. Flowers

usually orange-yellow, 0.7–1.5 cm across.
Petals narrowly lanceolate, acute,
spreading. *Mexico.* G1. Winter–early
summer.

4. S. pachyphyllum Rose. Illustration:
Contributions from the U.S. National
Herbarium **13:** pl. 58 (1911); Journal of
the Royal Horticultural Society **46:** 215
(1920–1921); Gartenschönheit **8:** 289
(1927).

Evergreen, hairless, erect shrub to 30
cm. Stems branched, somewhat prostrate.
Leaves in distinct spiral rows, oblong-
lanceolate, cylindric, bluntly rounded,
1.5–4 × 0.8–1 cm, very fleshy, grey-green
mostly with a reddish tip. Flowers pale
yellow, c. 1.5 cm across. Petals oblong-
lanceolate. *Mexico.* G1. Winter–spring.

5. S. rubrotinctum R.T. Clausen (*S.
guatemalense* misapplied). Illustration:
Evans, Handbook of cultivated Sedums,
pl. 8 (1983).

Small hairless shrublet, 15–30 cm.
Branches at first fleshy, later somewhat
woody. Leaves alternate, stalkless,
cylindric, rounded at apex, 1.3–2 × c. 0.5
cm, pale green, often suffused with red.
Flowers yellow, 1.3 cm across. *Mexico.* G1.
Winter.

'Aurora'. Illustration: Hay & Beckett,
Reader's Digest encyclopaedia of garden
plants and flowers: 659 (1978); Evans,
Handbook of cultivated Sedums, pl. 8
(1983). Has grey-green leaves, tinged
rose-red.

6. S. dendroideum de Candolle.
Illustration: Journal of the Royal
Horticultural Society **46:** 208
(1920–1921), Sedum Society Newsletter
16: 5 (1991).

Shrub to 60 cm. Leaves alternate,
spathulate to circular, 4.5 × 2 cm, bright
green, margin waxy white, later reddish,
with a row of red or dark green glandular
dots. Flowers yellow, c. 1.5 cm across.
Petals lanceolate, acute or obtuse. *Mexico,
Guatemala.* G1. Spring–summer.

7. S. praealtum de Candolle (*S.
dendroideum* de Candolle subsp. *praealtum*
(de Candolle) R.T. Clausen). Illustration:
Journal of the Royal Horticultural Society
46: 210 (1920–1921).

Evergreen hairless shrub, 30–150 cm.
Leaves alternate, oblanceolate to
spathulate, bluntly pointed or rounded,
5–7 × 1.5–2 cm, light green throughout,
margin without glandular dots. Flowers
pale-bright yellow, 1.5–1.8 cm across.
Petals narrowly lanceolate, very acute,

wide-spreading. *Mexico*. H5 or G1. Spring–summer.

8. S. decumbens R.T. Clausen (*S. confusum* misapplied). Illustration: Sedum Society Newsletter **16**: 1, 3 (1991).

Similar to *S. confusum* but differing in terms of its shorter height, smaller leaves, 1.5–2 × 1.1–1.3 cm, pale green; inflorescence branched at apex only, flat-topped, usually 3-branched, flowers *c.* 15; sepals 2–3 mm. *Origin unknown*. G1.

9. S. confusum Hemsley (*S. aoikon* Ulbrich). Illustration: Journal of the Royal Horticultural Society **46**: 212 (1920–1921).

Shrub to 30 cm, branches ascending. Leaves alternate, obovate to spathulate, obtuse, thickish, 2–4 × 1.2–2 cm, glossy green throughout, margin with glandular dots. Inflorescence rounded, with 3–5 terminal and lateral branches and *c.* 25 flowers; flowers yellow, 1.2–1.5 cm across. Petals ovate-lanceolate; sepals 1–1.5 mm. *Mexico*. G1. Winter–spring.

10. S. treleasei Rose. Illustration: Contributions from the U.S. National Herbarium **13**: 300 (1911); Journal of the Royal Horticultural Society **46**: 217 (1920–1921); Gartenschönheit **8**: 290 (1927).

Evergreen shrub to 45 cm, with erect or prostrate stems. Leaves alternate, oblong-obovate, obtuse, very fleshy, upper face more or less flattened, 2–4.5 × 0.9–1.5 cm, blue-green, densely grey-bloomed. Flowers pale-bright yellow, *c.* 1.3 cm across. Petals ovate-lanceolate, acute. *Mexico*. G1. Spring.

11. S. fusiforme Lowe. Illustration: Evans, Handbook of Cultivated Sedums, 154 (1983).

Small shrub to *c.* 12 cm. Leaves alternate, lanceolate to oblong or elliptic, terete, crowded at tips of branches, to 1.6 cm, grey-green. Flowers greenish yellow, red at the centre, *c.* 1.6 cm across. *Madeira*. G1. Spring & autumn.

12. S. nudum Aiton. Illustration: Journal of the Royal Horticultural Society **46**: 252 (1920–1921); Evans, Handbook of cultivated Sedums, 274 (1983).

Evergreen shrub with hairless shoots. Leaves alternate, stalkless, obovate-oblong, obtuse, *c.* 10 × 5 mm, thick, pale green. Flowers greenish yellow, to *c.* 1 cm across. Petals linear-lanceolate, acute, wide-spreading. *Madeira*. G1. Spring–summer.

13. S. craigii R.T. Clausen (*Graptopetalum craigii* (Clausen) Clausen). Illustration: Evans, Handbook of cultivated Sedums, 146 (1983).

Small shrub with fleshy stems. Leaves alternate, stalkless, oblong-elliptic, apex rounded, 2–5 × 0.9–2.2 cm, reddish blue. Flowers white, *c.* 6 mm across. Petals erect, narrowed below, apex recurved. *Mexico*. G1. Autumn.

14. S. longipes Rose. Illustration: Journal of the Royal Horticultural Society **46**: 203 (1920–1921).

Hairless matted perennial with long arching stems developing from short rosettes. Leaves alternate, obovate or spathulate, fleshy, entire, *c.* 8 × 5 mm, bright green. Flowers reddish, purple or in part whitish, *c.* 7.5 mm across. Petals ovate, obtuse, almost spreading, red in upper part, becoming silvery white near base. Nectaries large, red, prominent. *Mexico*. G1. Winter.

15. S. moranense Humboldt, Bonpland & Kunth. Illustration: Journal of the Royal Horticultural Society **46**: 173 (1920–1921); Clausen, Sedum of the trans-Mexican volcanic belt, 260 (1959).

Evergreen hairless small shrub, to 15 cm, much-branched, prostrate, green, later brown and dark red stems, woody below. Leaves alternate, ovate to ovate-lanceolate, obtuse or acute, 3–5 × 2–3 mm, green. Flowers stalkless, white, sometimes pinkish, *c.* 1 cm across. Petals lanceolate, acute or obtuse. *Mexico*. H5–G1. Summer.

Var. **arboreum** (Masters) Praeger is an erect, much-branched, 15–30 cm bush with leaves occasionally in 5 spiral rows.

16. S. populifolium Pallas. Illustration: Journal of the Royal Horticultural Society **46**: 148 (1920–1921); Hegi, Flora von Mitteleuropa **4**: 527 (1921); Botanical Magazine, 211 (1952); The Plantsman **8**: 9 (1986).

Deciduous hairless perennial, woody at base, 20–40 cm, stems greenish or purplish. Leaves alternate, ovate, acute, cordate at base, 1.3–3.5 × 1–2.5 cm, green, coarsely and irregularly toothed. Flowers pink or white, scented, *c.* 1 cm across. Petals lanceolate, acute, spreading. *Former USSR (Siberia)*. H1. Summer.

17. S. retusum Hemsley. Illustration: Journal of the Royal Horticultural Society **46**: 149 (1920–1921).

Evergreen hairless shrub to 30 cm, bark grey, non-peeling. Leaves alternate, obovate-oblong, spathulate, 1.5–2.5 × 0.6–1 cm, deeply notched at apex. Flowers white with red centre, *c.* 1 cm across.

Petals oblong-lanceolate, obtuse, shortly mucronate, erect below, spreading above. *Mexico*. G1. Summer–autumn.

18. S. oxypetalum Humboldt, Bonpland & Kunth. Illustration: Journal of the Royal Horticultural Society **46**: 193 (1920–1921).

Hairless shrub, 50–90 cm, bark peeling in papery layers. Leaves alternate, ovate-lanceolate or spathulate, 1.5–5 cm, finely papillose, green, deciduous after flowering. Flowers stellate, pinkish or dull red especially at centre, *c.* 1.3 cm across, scented. Petals lanceolate, acute, wide-spreading from near base. *Mexico*. G1. Summer.

19. S. allantoides Rose. Illustration: Contributions from the U.S. National Herbarium **12**: pl. 79 (1909); Journal of the Royal Horticultural Society **46**: 154 (1920–1921).

Evergreen, glaucous shrub to 40 cm, branches erect. Leaves alternate, club-shaped, 2–3 × 0.6–2.5 cm, grey-green, white-grey bloomed. Flowers greenish white, 1.6–1.8 cm across. Petals lanceolate, acute, wide-spreading, often red-spotted near tip. *Mexico*. G1. Summer.

20. S. bourgaei Hemsley. Illustration: Journal of the Royal Horticultural Society **46**: 155 (1920–1921); Clausen, Sedum of trans-Mexican volcanic belt, 148 (1959).

Evergreen small shrub, 15–30 cm. Leaves alternate, linear, obtuse, 1–2 cm × *c.* 2 mm, green. Flowers white, usually tipped red, *c.* 1.2 cm across. *C Mexico*. H5–G1. Late summer.

21. S. griseum Praeger (*S. farinosum* misapplied). Illustration: Journal of the Royal Horticultural Society **46**: 158 (1920–1921); Clausen, Sedum of the trans-Mexican volcanic belt, 160 (1959).

Small evergreen shrub 15–20 cm, often white-grey bloomed. Leaves alternate, linear or lanceolate-linear, 1–2 cm × *c.* 2.5 mm, green or glaucous-green. Flowers white, 1.2–1.5 cm across. Petals lanceolate. *Mexico*. G1. Winter.

22. S. lucidum R.T. Clausen. Illustration: Clausen, Sedum of the trans-Mexican volcanic belt, 94 (1959).

Hairless shrub with prostrate stems, to 45 cm. Leaves alternate, elliptic to oblanceolate or spathulate, stalkless, acute to obtuse, 2.5–5 × 1–1.9 cm, thick, dark green, shining (especially young leaves). Flowers white with musky scent, 0.9–1.1 cm across. Petals lanceolate or elliptic-oblong, obtuse or acute. *Mexico*. G1. Late autumn–early spring.

23. S. adolphii Raymond-Hamet.

Evergreen, hairless shrub, 7.5–12.5 cm, branches ascending. Leaves alternate, upper rather crowded, very fleshy, oblong-oblanceolate, *c.* 3.5 × 1.5 cm, yellowish green, sometimes with faint reddish margins. Flowers white, sometimes pinkish on back, *c.* 1.8 cm across. Petals ovate-lanceolate, acuminate, wide-spreading. *Mexico.* G1. Spring.

24. S. nussbaumerianum Bitter (*S. adolphii* misapplied). Illustration: Clausen, Sedum of the trans-Mexican volcanic belt, 108 (1959); Boom, Flora Cultuurgewassen **3**, kamer-en kasplanten, pl. 132 (1968).

Hairless shrub with stout spreading branches, *c.* 7 cm, stems reddish brown. Leaves alternate, oblanceolate-elliptic, acute at apex, 1.2–5 × 0.5–0.8 cm, yellow-green, edged with yellow, orange or red. Flowers white or pinkish, 1.4–1.6 cm across. Petals linear-lanceolate, acute. *Mexico.* G1. Winter.

25. S. selskianum Regel & Maack (*S. aizoon* Linnaeus subsp. *selskianum* (Regel & Maack) Fröderström). Illustration: Regel & Maack, Tentamen Flora Ussuriensis, **66**: t. 6 (1861); Gartenflora **11**: t. 361 (1862); Journal of the Royal Horticultural Society **46**: 114 (1920–1921).

Densely hairy perennial herb with annual stems, erect, 30–45 cm. Leaves stalkless, alternate, linear-oblong, crowded, 4–5 × 1–1.2 cm, dark green, greyish downy, upper half toothed. Flowers yellow, 0.8–1 cm across. Petals broadly lanceolate, acuminate or mucronate, wide-spreading. *China (Manchuria), former USSR (Amur).* H1. Summer.

26. S. aizoon Linnaeus (*S. maximowiczii* Regel). Illustration: Journal of the Royal Horticultural Society **46**: 109 (1920–1921); Hay & Synge, The dictionary of garden plants in colour; pl. 1369 (1969).

Perennial herb, hairless, deciduous, shoots erect, to 80 cm. Leaves alternate, stalkless, ovate-lanceolate to linear-lanceolate, 5–8 × *c.* 2 cm, sharply toothed from below the middle. Flowers numerous, yellow to orange, *c.* 1.2 cm across. Petals linear-lanceolate, mucronate, wide-spreading. *Japan, former USSR (Siberia), China.* H1. Summer–autumn.

'Aurantiacum' has a dark red stem, dark green leaves, dark yellow to deep orange flowers and red fruits.

27. S. hybridum Linnaeus. Illustration: Reichenbach, Icones Florae Germanicae

23: f. 64 (1897); Journal of the Royal Horticultural Society **46**: 126 (1920–1921).

Evergreen perennial herb, 10–30 cm, branches persistent, ascending. Leaves alternate, oblanceolate to spathulate, upper part coarsely toothed, lower part entire, 2.5 × 1.2 cm, bright green, in winter reddish green. Flowers yellow, 1.2–1.4 cm across. Petals mucronate. *Former USSR (Siberia), Mongolia.* H1. Summer.

28. S. kamtschaticum von Fischer & C.A. Meyer (*S. aizoon* Linnaeus subsp. *kamtschaticum* (von Fischer & C.A. Meyer) Fröderström). Figure 17(2), p. 193. Illustration: Gartenflora **32**: 250 (1883); Journal pf the Royal Horticultural Society **46**: 121 (1920–1921).

Perennial herb, 15–30 cm, branches annual, ascending. Leaves opposite or alternate, stalkless, oblanceolate to spathulate, toothed towards apex, 3–5 × *c.* 1.2 cm, dark green. Flowers orange-yellow, 1.5–2 cm across. Petals lanceolate, mucronate. *N & C China, Japan, former USSR (Kamchatka).* H1. Late summer.

'Variegatum'. Illustration: Köhlein, Freilandsukkulenten; 161 (1977). Leaves with a creamish white margin.

Var. **ellacombianum** (Praeger) R.T. Clausen (*S. ellacombianum* Praeger). Illustration: Journal of the Royal Horticultural Society **46**: 118 (1920–1921); Köhlein, Freilandsukkulenten, 143 (1977). Deciduous perennial, 15–25 cm. Leaves opposite, obovate to spathulate, 3.5–4 × *c.* 1.8 cm, bright green. Flowers pure yellow, *c.* 1.5 cm across. *Japan.* H3. Summer.

Var. **middendorffianum** (Maximowicz) R.T. Clausen (*S. middendorffianum* Maximowicz; *S. aizoon* Linnaeus subsp. *middendorffianum* (Maximowicz) R.T. Clausen). Illustration: Journal of the Royal Horticultural Society **46**: 116 (1920–1921). Plant to 20 cm. Leaves alternate, linear-oblanceolate, *c.* 35 × 3 mm, fresh bright green, toothed near apex. Flowers pale yellow, *c.* 1.5 cm across.

Var. **middendorffianum** forma **diffusum** Praeger (*S. middendorffianum* var. *diffusum* Praeger). Leaves larger, lanceolate to narrowly spathulate, 2.5–5 × *c.* 0.6 cm, sharply toothed in upper part. *N & C Asia, former USSR (E Siberia), China (Manchuria), Mongolia.* H1. Summer.

29. S. floriferum Praeger (*S. kamtschaticum* var. *floriferum* Praeger). Illustration: Journal of the Royal Horticultural Society **46**: 123 (1920–1921).

Hairless perennial herb with annual red stems, *c.* 15 cm. Leaves alternate, stalkless, dark green, crowded, obtuse. Flowers yellow, 1.2–1.5 cm across. Petals lanceolate, acute, wide-spreading. *NE China.* H1. Summer.

'Weihenstephaner Gold'. Illustration: Fessler, Die Freilandschmuckstauden, 586 (1985); The Plantsman **8**: 11 (1986). Forms very compact mats, flowers bright yellow, 1.5–1.8 cm across.

30. S. cepaea Linnaeus. Illustration: Edwards's Botanical Register, t. 1391 (1830–1831); Hegi, Flora von Mitteleuropa **4**: 520 (1921).

Annual to biennial herb, hairy, to 30 cm. Leaves alternate, opposite or in whorls of 3–4, linear-ovate, 1.2–2.5 × 0.6–1.2 cm, dotted with red. Flowers white, often with a few pinkish spots, *c.* 9 mm across. Petals lanceolate, acuminate. *W, C & S Europe, W Turkey.* H2. Early summer.

31. S. pilosum Bieberstein. Illustration: Botanical Magazine, 8503 (1913); Journal of the Royal Horticultural Society **46**: 284 (1920–1921); Gartenschönheit **11**: 199 (1930); Köhlein, Freilandsukkulenten, 143 (1977).

Biennial herb, glandular-downy, 5–10 cm, producing a dense rosette of very numerous leaves in the first year. Leaves linear to spathulate, obtuse, to *c.* 1.8 cm × 1.2 mm on flowering stems, dark green. Flowers rose-red, *c.* 9 mm across. Petals elliptic-lanceolate, acute, erect below, divergent above. *SW Asia, Caucasus, Iran.* H3. Spring–summer.

32. S. sempervivoides Bieberstein (*S. sempervivum* Sprengel). Illustration: Botanical Magazine, 2474 (1824); Gartenflora **16**: t. 551 (1876), **33**: t. 1155 (1884); Journal of the Royal Horticultural Society **46**: 282 (1920–1921); Köhlein, Freilandsukkulenten, 143 (1977).

Biennial, 10–30 cm. Leaves in rosettes, alternate on the flowering-stems. Leaves stalkless, obovate to ovate, bluntly acuminate, 2.5–3 × 1.3–1.8 cm, downy, reddened. Flowers carmine-red, *c.* 1.3 cm across. Petals lanceolate, acute, bright crimson. *SW Asia, Caucasus, N Iran, former USSR (Georgia).* H3. Summer.

33. S. formosanum N.E. Brown. Illustration: Journal of the Royal Horticultural Society **46**: 296 (1920–1921).

Annual herb with succulent, reddish stems, 15–25 cm. Leaves alternate, spathulate, obtuse, recurved, *c.* 2.5 × 1.2

Figure 17. Diagnostic details of *Sedum* species. 1, *S. forsterianum* a, flowering-stem; b, habit; c, flower from above; d, fruit. 2, *S. kamtschaticum* var. *ellacombianum* a, flowering-stem; b, habit; c, flower from above; d, fruit. 3, *S. sexangulare*, a, b, habit; c, flower from above; d, fruit. 4, *S. album* 'Murale' a, b, habit; c, flower from side and above; d, fruit.

cm, entire. Flowers yellow, *c.* 1.2 cm across. Petals lanceolate, spreading. *S Japan to Taiwan.* H3–4. Summer.

34. S. caeruleum Linnaeus (*S. azureum* Desfontaines). Illustration: Edwards's Botanical Register, t. 520 (1820–1821); Botanical Magazine, 2224 (1821); Journal of the Royal Horticultural Society **46:** 303 (1920–1921); Hay & Synge, The dictionary of garden plants, pl. 374 (1969).

Annual hairless herb, 5–10 cm. Leaves alternate, ovate to long-ovate, 0.9–1.8 cm, pale green. Flowers pale blue, usually 7-parted, *c.* 6 mm across. Petals oblong, obtuse. *Corsica, Sardinia, Sicily, Algeria.* H5. Summer–autumn.

Often misspelled as *S. coeruleum.*

35. S. villosum Linnaeus. Illustration: Reichenbach, Icones Florae Germanicae **23:** t. 52 (1897); Journal of the Royal Horticultural Society **46:** 302 (1920–1921); Hegi, Flora von Mitteleuropa **4:** pl. 140 (1921); Engler, Die natürlichen Pflanzenfamilien **18a:** 441 (1930).

Biennial or sometimes perennial herb with red stems, 7.5–10 cm. Leaves alternate, linear-oblong, obtuse, 6–12 mm, fleshy, glandular hairy, green. Flowers pink, dull rose or white, *c.* 6 mm across. Petals ovate, rather acute. *Scandinavia across Europe to N Africa, Greenland, Iceland.* H1. Early–summer.

36. S. hispanicum Linnaeus (*S. glaucum* Waldstein & Kitaibel). Illustration: Reichenbach, Iconographia botanica **9:** 1136 (1831–1833); Journal of the Royal Horticultural Society **46:** f. 178 excluding b (1920–1921); Hegi, Flora von Mitteleuropa **4:** 528 (1921).

Annual, biennial or perennial herb, 5–15 cm, hairless or flower-stem with shaggy glandular hairs. Leaves alternate, linear to long-lanceolate, 1.2–2.5 cm, grey-green, often reddish. Flowers pink to white, *c.* 1.2 cm across. Petals lanceolate, acuminate, white often with purplish mid-vein. *Alps, S & SE Europe, Turkey, N Iran.* H3. Summer.

37. S. anacampseros Linnaeus. Illustration: Botanical Magazine, 118 (1790); Journal of the Royal Horticultural Society **46:** 105 (1920–1921); Hegi, Flora von Mitteleuropa **4:** 520 (1921).

Evergreen perennial, 10–25 cm. Leaves alternate, stalkless, ovate to obovate or elliptic, slightly acuminate, 1.2–2.5 × 1.2–1.8 cm, grey-green. Leaf-margins entire, reddish. Flowers purple, *c.* 6 mm

across. Petals oblong-lanceolate, obtuse. *Europe.* H2. Summer.

38. S. ewersii von Ledebour. Illustration: Gartenflora **9:** t. 295 (1860); Journal of the Royal Horticultural Society **46:** 97 (1920–1921); Polunin & Stainton, Flowers of the Himalaya, pl. 45 (1984).

Deciduous perennial herb, stem ascending, 10–20 cm. Leaves opposite, stalkless, circular to broad-ovate, with cordate base, *c.* 1.8 × 1.8 cm, bluish green, usually entire. Flowers pink or pale violet, 1–1.2 cm across. Petals oblong-lanceolate, pinkish with darker spots. *W Himalayas, Mongolia.* H1. Summer.

Var. **homophyllum** Praeger (*S. pluricaule* misapplied; *S. cyaneum* misapplied). Illustration: Journal of the Royal Horticultural Society **46:** 98 (1920–1921). All parts smaller; leaves obovate, 1.2–1.5 × 0.6–0.9 cm, more glaucous than the type. *Only known in cultivation.*

39. S. cauticola Praeger. Illustration: Journal of the Royal Horticultural Society **46:** 100 (1920–1921); Botanical Magazine, 401 (1962–1963); Hay & Synge, The dictionary of garden plants, pl. 1370 (1969).

Sprawling perennial herb, shoots dark purple, 10–20 cm. Leaves opposite, roundish to spathulate, obovate, *c.* 2.5 × 1.8 cm, glaucous, minutely dotted with red. Flowers rose-purple, older ones carmine-red, *c.* 1.5 cm across. Petals lanceolate, acute. *Japan.* H3–4. Summer–autumn.

'Ruby Glow' (*S. 'Robustum'*). Illustration: Hay & Synge, The dictionary of garden plants in colour, pl. 1374 (1969) – as *S. spectabile* hybrid 'Ruby Glow'. Perennial herb with ascending stems, 15–25 cm. Leaves as *S. cauticola*, glaucous blue and purplish red. Flowers pinkish red, *c.* 7 mm across. *Garden origin.* H2. Summer–autumn.

A hybrid between *S. cauticola* × *telephium.*

40. S. sieboldii Sweet. Illustration: Botanical Magazine, 5358 (1863); Journal of the Royal Horticultural Society **46:** 102 (1920–1921).

Deciduous, hairless perennial herb with sprawling purplish stems, 15–20 cm. Leaves 3 whorled, stalkless, almost circular, 1.3–2.5 × *c.* 1.9 cm, upper part toothed, blue-green with red edge. Flowers pink, *c.* 1.3 cm across. Petals broadly lanceolate, acute, spreading. *Japan.* H3. Autumn.

'Mediovariegatum' Illustration: Köhlein, Freilandsukkulenten, 161 (1977). Leaves with large pale yellow central patch. Requires winter protection.

41. S. cyaneum Rudolph. llustration: Gartenflora **27:** t. 972 (1879); Journal of the Royal Horticultural Society **46:** 106 (1920–1921).

Deciduous creeping hairless perennial herb, 5–10 cm. Leaves alternate or opposite, oblong-lanceolate, obtuse, entire, 1–2 × 0.5–0.6 cm, grey-green. Flowers purplish carmine rose, to 1.3 cm across. Petals ovate-oblong. *Former USSR (E Siberia & Sakhalin).* H1. Summer.

42. S. tatarinowii Maximowicz. Illustration: Journal of the Royal Horticultural Society **46:** 103 (1920–1921).

Hairless perennial herb, 10–15 cm, stems almost erect. Leaves alternate, linear-lanceolate, nearly obtuse, 1.3–2.5 cm, fleshy, coarsely toothed towards apex. Flowers pinkish white, *c.* 1.2 cm across. Petals ovate-lanceolate, acute, wide-spreading. *N China, Mongolia.* H1. Summer–autumn.

43. S. verticillatum Linnaeus. Illustration: Journal of the Royal Horticultural Society **46:** 95 (1920–1921).

Hairless perennial herb with erect stems, 30–60 cm. Leaves whorled, lower ones often opposite or in 3s, upper in whorls of 4 or 5, oblong-lanceolate, 5–7.5 × 2–2.5 cm, green, red dotted. Flowers pale green, *c.* 6 mm across. Petals ovate-lanceolate, acute, wide-spreading. Numerous bulbils produced in the axils of old leaves and bracts. *Japan, former USSR (Kamchatka).* H1. Autumn.

Var. **nipponicum** Praeger. Slender, dwarf plant with opposite leaves.

44. S. spectabile Boreau × **telephium** Linnaeus **'Autumn Joy'** (*'Herbstfreude'*). Illustration: Hay & Synge, The dictionary of garden plants in colour, pl. 1372 (1969).

Hairless perennial herb with dark glaucous green leaves, 40–60 cm. Flowers cream to light pink, *c.* 5 mm across. Petals semi-erect. Stamens absent. Carpels dark pink, very fleshy, longer than petals, hence dominating the flower colour. *Garden origin.* H1. Summer–autumn.

45. S. telephioides Michaux. Illustration: Britton & Brown, An illustrated flora of the Northern United States **2:** 208 (1897); Clausen, Sedum of North America, 71, 88 (1975).

Hairless perennial, 15–30 cm. Leaves mostly alternate, elliptic to spathulate, obtuse, 1–4.5 × 0.6–3 cm, wedge-shaped throughout, entire or remotely toothed. Flowers white to pink, *c.* 1 cm across. Petals lanceolate. *E North America.* H1. Summer–early autumn.

46. S. telephium Linnaeus. Illustration: Reichenbach, Iconographia Botanica **8**: 968 (1830); Engler, Die natürlichen Pflanzenfamilien **18a**: 441 (1930).

Hairless, erect perennial herb, 20–70 cm. Leaves usually alternate, oblong to ovate-oblong, obtuse, 5–7.5 × 2–4 cm, irregularly toothed, bluish green. Flowers mostly red-purple, 7–10 mm across. Petals lanceolate, acute, spreading, somewhat recurved. *Europe to Siberia.* H1. Summer.

Subsp. **telephium** (*S. purpureum* (Linnaeus) Schultes; *S. purpurascens* Koch). Illustration: Journal of the Royal Horticultural Society **46**: 83 (1920–1921). Leaves alternate, lower leaves wedge-shaped at base, upper leaves truncate at base, fleshy, *c.* 7.5 × 3.6 cm, dark green. Flowers purple, rarely whitish, *c.* 1 cm across. *Europe, Asia.*

Var. **borderi** Rouy & Camus. Illustration: The Plantsman **8**: 6 (1986). Plant with red stems, 25–40 cm. Leaves deeply toothed. Flowers purplish pink, *c.* 7 mm across. *France.*

Subsp. **fabaria** (Koch) Kirschleger. (*S. fabaria* Koch; *S. vulgare* (Haworth) Link). Illustration: Journal of the Royal Horticultural Society **46**: 84 (1920–1921). Plant 20–40 cm. Leaves sometimes stalked, lanceolate, dark green. Flowers smaller than in the type. *W & C Europe.*

47. S. pseudospectabile Praeger. Illustration: Journal of the Royal Horticultural Society **46**: 91 (1920–1921).

Perennial hairless herb, 30–60 cm. Leaves in 3s or opposite, broadly ovate to obovate, stalkless, base rounded, 3.5–5 × 2.5–3.5 cm, irregularly toothed, green, often with a red edge. Flowers pink, *c.* 6 mm across. Petals ovate-lanceolate, acute, spreading but not at right angles. *China.* H1. Summer–autumn.

48. S. spectabile Boreau. Illustration: Saunders, Refugium Botanicum, f. 32 (1869); Gartenflora **21**: t. 709 (1872); Journal of the Royal Horticultural Society **46**: 93 (1920–1921); Hay & Beckett, Reader's Digest encyclopaedia of garden plants and flowers, 658 (1978).

Deciduous hairless perennial herb, 30–50 cm. Leaves opposite or 3-whorled, ovate to spathulate, *c.* 7.5 × 5 cm, base entire, toothed above, fleshy, bluish white. Flowers pink, 1–1.3 cm across. Petals lanceolate, acute, spreading. *Korea, China (Manchuria), Japan.* H1. Summer–autumn.

The following cultivars are commonly grown, 'Brilliant': petals bright pink with darker carpels and anthers; 'Carmen': petals darker pink; 'Humile': petals pink as in the species; 'Rosenteller': petals bright pink; 'Septemberglut' (illustration: Köhlein, Freilandsukkulenten, 144, 1977): petals deep pink, the darkest cultivar.

49. S. erythrostictum Miquel (*S. alboroseum* Baker; *S. japonicum* misapplied). Illustration: Saunders, Refugium Botanicum f. 33 (1869); Gartenflora **21**: t. 709 (1872); Journal of the Royal Horticultural Society **46**: 89 (1920–1921).

Glaucous, erect, perennial herb with thickened roots, stem 30–60 cm. Leaves usually opposite, often narrowed towards the base, ovate, 5–7.5 × 3.5–4 cm, grey-green. Flowers greenish white or rose-white, carpels pink. Petals oblong-lanceolate, acute, wide-spreading. *E Asia.* H1. Summer–early autumn.

'Mediovariegatum': illustration: Köhlein, Freilandsukkulenten, 1143 (1977). Leaves yellowish white, margin green; flowers rose.

50. S. maximum Suter (*S. telephium* Linnaeus subsp. *maximum* (Linnaeus) Krocker). Illustration: Reichenbach, Iconographia botanica **8**: f. 969 (1830); Journal of the Royal Horticultural Society **46**: 80 (1920–1921); Hegi, Flora von Mitteleuropa **4**: pl. 140 (1921).

Hairless, herbaceous perennial with green or red erect stems, 50–100 cm. Leaves opposite or sometimes 3-whorled, stalkless, broadly ovate with a cordate base, obtuse, 5–12.5 × 3–5 cm, irregularly toothed, dark green. Flowers greenish or greenish white, more rarely pale pink, *c.* 1 cm across. Petals ovate-lanceolate, rather acute. *Europe, Caucasus.* H1. Late Summer.

'Atropurpureum': illustration: Botanical Magazine, 429 (1962–1963); Hay & Synge, The dictionary of garden plants in colour, pl. 1371 (1969). Leaves and stems deep purple. This cultivar is very variable.

51. S. glabrum (Rose) Praeger. Illustration: Journal of the Royal Horticultural Society **46**: 128 (1920–1921).

Evergreen hairless perennial herb to 20 cm. Basal leaves forming a rosette, stem-leaves alternate, oblong-ovate, 3 × 1.2 cm, pale green. Flowers white, *c.* 1 cm across. Petals ovate, white with purple midribs and markings towards base, recurving apically. *Mexico.* G1. Summer–autumn.

52. S. hemsleyanum Rose. Illustration: Clausen, Sedum of the trans-Mexican volcanic belt, 228 (1959).

Evergreen herb to 30 cm, hairy. Basal leaves forming a rosette, circular; stem-leaves alternate linear or lanceolate, 2–3.5 × 0.6–1 cm. Flowers white, 0.8–1.1 cm across. Petals ovate to lanceolate, recurving apically. *Mexico.* G1. Autumn–early winter.

53. S. ebracteatum de Candolle. Illustration: Journal of the Royal Horticultural Society **46**: 131 (1920–1921); Engler, Die natürlichen Pflanzenfamilien **18a**: 441 (1930); Clausen, Sedum of the trans-Mexican volcanic belt, 237 (1959).

Evergreen perennial herb to 30 cm, minutely hairy. Basal leaves forming rosettes, ovate; stem-leaves ovate, 2–2.5 × 1–1.6 cm. Flowers white, 1–1.3 cm across. Petals ovate, acute, recurved apically. *Mexico.* G1. Autumn.

54. S. wrightii A. Gray. Illustration: Journal of the Royal Horticultural Society **46**: 138 (1920–1921); Clausen, Sedum of North America, 203 (1975).

Evergreen hairless perennial herb, 2–15 cm. Leaves crowded into small rosettes, elliptic to ovate or obovate, extremely fleshy, *c.* 10 × 3 mm, pale green with shining papillae. Flowers bell-shaped, white, *c.* 1 cm across. Petals oblong-obovate, obtuse, mucronate, erect below, wide-spreading above. *SW USA.* H4. Summer–autumn.

55. S. cockerellii Britton. Illustration: Clausen, Sedum of North America, 189 (1975).

Evergreen perennial herb to 18 cm, with hairless or papillose shoots. Leaves alternate, on sterile shoots crowded, basal leaves spathulate, acute, *c.* 12 × 6 mm, stem-leaves linear-lanceolate, rather acute, 1.2–2.5 × 0.3–0.6 cm, green, partly becoming red. Flowers white, *c.* 1.2 cm across. Petals narrowly lanceolate, acute. *USA (California to New Mexico & Texas).* H3. Late summer.

56. S. bellum Rose (*S. aleurodes* Bitter; *S. farinosum* Rose). Illustration: Contributions from the U.S. National Herbarium **13**: pl. 54 (1911); Journal of the Royal Horticultural Society **46**: 142 (1920–1921).

Tufted, perennial herb, 7–15 cm. Leaves alternate, in dense rosettes, spathulate, broadest just below the rounded tip, 2–3.5 × 0.8–1 cm, light green, mealy white bloomed. Flowers white, c. 1.2 cm across. Petals lanceolate, spreading. *Mexico*. G1. Spring.

57. S. versadense Thompson (*S. chontalense* Alexander). Illustration: Journal of the Royal Horticultural Society **46**: 143 (1920–1921); Clausen, Sedum of the trans-Mexican volcanic belt, 222 (1959).

Evergreen, tufted perennial with densely hairy stems, 7–15 cm. Leaves alternate, spathulate or obovate, c. 2.5 × 1.2 cm, very fleshy, downy, green, red at tip and edges. Upper leaves in flowering-shoots smaller, more distant, almost acute, hairless. Flowers pale pink, 1.2–1.8 cm across. Petals elliptic, mucronate. *Mexico*. G1. Spring–late summer.

58. S. greggii Hemsley (*S. diversifolium* Rose). Illustration: Journal of the Royal Horticultural Society **46**: 145 (1920–1921); Clausen, Sedum of the trans-Mexican volcanic belt, 252 (1959); Gartenschönheit **8**: 290 (1927).

Perennial herb, 7–15 cm. Basal leaves in rosettes, papillose. Leaves alternate, elliptic or oblanceolate, c. 6 mm or more. Flowers sulphur yellow, c. 1 cm across. Petals elliptic or ovate, acute. *Mexico*. G1. Winter–spring.

59. S. alamosanum S. Watson. Illustration: Journal of the Royal Horticultural Society **46**: 134 (1920–1921).

Low perennial herb, glaucous, partly papillose; sterile stems c. 1 cm; flowering-stems 7–12 cm, dying after flowering, replaced by branches from base. Leaves alternate, densely adpressed, linear-oblong, c. 6 mm, grey-green. Flowers few, reddish white, c. 1 cm across. Petals broadly lanceolate or elliptic, acute, wide-spreading, whitish or pinkish translucent. *NW Mexico*. G1. Winter–spring.

60. S. divergens S. Watson. Illustration: Journal of the Royal Horticultural Society **46**: 220 (1920–1921); Clausen, Sedum of North America, 303 (1975).

Perennial hairless herb, 7–10 cm. Leaves opposite, stalkless, thick, almost spherical, obovate to spathulate, rounded at apex, c. 6 × 5 mm, green or reddish green. Flowers yellow, c. 1.8 cm across. Petals elliptic-lanceolate, acute or obtuse. *W North America*. H4. Summer.

61. S. mexicanum Britton. Illustration: Journal of the Royal Horticultural Society **46**: 231 (1920–1921).

Evergreen hairless perennial herb, stems erect, ascending, to 15 cm. Leaves usually in whorls of 3–5, rarely alternate except at top of flowering-stem. Leaves narrowly linear, obtuse, 6–15 × 2–3 mm, with a light green shining tint. Flowers nearly stalkless, golden yellow, 1–1.3 cm across. Petals elliptic-lanceolate, acute. *Mexico*. G1. Spring–summer.

62. S. chauveaudii Raymond-Hamet. Illustration: Journal of the Royal Horticultural Society **46**: 225 (1920–1921).

Evergreen creeping perennial herb, stems red, leaves and sepals sparsely fine papillose. Leaves oblanceolate, obtuse, 1.2–2 × c. 0.5 cm, light green. Flowers yellow, streaked red on back, c. 1.5 cm across. *W China (Yunnan)*. H4. Summer–autumn.

63. S. sarmentosum Bunge. Illustration: Journal of the Royal Horticultural Society **46**: 226 (1920–1921).

Hairless perennial herb with long creeping stems, shoots dying in winter. Leaves usually in 3s, stalkless, elliptic, usually acute, flat, 1.3–2.5 × 0.4–0.6 cm, pale green. Flowers stalkless, bright yellow, 0.6–1.5 cm across. Petals linear-lanceolate, narrowly acute, wide-spreading. *Japan, Korea, N China*. H3. Summer.

64. S. lineare Murray. Illustration: Journal of the Royal Horticultural Society **46**: 228 (1920–1921).

Evergreen hairless perennial herb with reddish stems, erect or creeping. Leaves in whorls of 3, linear or linear-lanceolate, nearly acute, rounded on back, c. 2.5 × 0.3 cm, pale green. Flowers bright yellow, stalkless, stellate, c. 1.6 cm across. Petals narrowly lanceolate, very acute. *Japan*. H5. Spring–summer.

'Robustum': illustration: Journal of the Royal Horticultural Society **46**: 230 (1920–1921). Plant grey-green, stouter and more branched with paler flowers and broader petals.

'Variegatum'. Leaves with a broad whitish margin.

65. S. purdyi Jepson (*S. spathulifolium* Hooker subsp. *purdyi* Jepson). Illustration: Clausen, Sedum of North America, 469 (1975).

Perennial herb with prostrate or creeping stems, 7–10 cm. Leaves in compact flat rosettes, spathulate, to 2

cm, papillose on margins. Leaves of erect stems elliptic, obtuse, green. Flowers bright yellow to white, c. 1.5 cm across. Petals oblong to lanceolate, wide-spreading. *USA (S Oregon, N California)*. H5. Spring–early summer.

66. S. oreganum Torrey & A. Gray (*S. obtusatum* misapplied). Illustration: Journal of the Royal Horticultural Society **46**: 242 (1920–1921); Clausen, Sedum of North America, 346 (1975).

Evergreen creeping hairless perennial, 5–15 cm. Leaves alternate, wedge-shaped to spathulate, obtuse, 1.2–1.8 × c. 1 cm, very fleshy, shining green, often reddened. Flowers yellow, 1.2–1.5 cm across. Petals narrowly lanceolate, long acuminate, ascending. *USA (Alaska to N California)*. H3. Summer.

67. S. spathulifolium J.D. Hooker. Illustration: Journal of the Royal Horticultural Society **46**: 239 (1920–1921); Clausen, Sedum of North America, 440 (1975); The Plantsman **8**: opposite page 1 (1986).

Evergreen hairless perennial herb, 5–15 cm. Leaves in dense flat rosettes, obovate to spathulate, obtuse, c. 2.5 × 1 cm, fleshy, greyish green, often tinged red, white on back. Leaves of flowering-shoots alternate, stalkless, oblong, very fleshy. Flowers yellow, to 1.5 cm across. Petals lanceolate, acute, wide-spreading. *N America (British Columbia to California)*. H1. Summer.

Var. **majus** Praeger. Illustration: Journal of the Royal Horticultural Society **46**: 239 (1920–1921). Rosettes twice as large; leaves longer and broader, more mucronate, green, scarcely glaucous, not suffused with red. The name is misapplied to large glaucous or mealy plants.

'Purpureum': illustration: Fessler, Die Freilandschmuckstauden, 587 (1985), has large rosettes, 3.5–5 cm across, deep purple leaves except when young, when they are white and mealy.

'Cape Blanco': illustration: Hay & Synge, The dictionary of garden plants in colour, pl. 203 (1969) – as 'Capa Blanca' has rosettes with silver-white mealy leaves.

68. S. obtusatum A. Gray subsp. **obtusatum** (*S. rubroglaucum* Praeger). Illustration: Clausen, Sedum of North America, 359 (1975).

Evergreen hairless prostrate perennial, 6–15 cm. Leaves on the sterile stems in loose rosettes, opposite, spathulate to obovate or wedge-shaped, obtuse or sometimes notched, 1.2–2.5 × 0.6–1.2

cm, entire, bluish green, in age orange-red, leaves on flowering shoots alternate. Flowers bright or pale yellow, 0.6–1 cm across. Petals lanceolate, erect and partly fused below, somewhat spreading above, less than 3 times sepals. *USA (California)*. H3. Summer.

69. S. humifusum Rose. Illustration: Contributions from the U.S. National Herbarium **13**: pl. 55 (1911); Journal of the Royal Horticultural Society **46**: 244 (1920–1921); Gartenschönheit **11**: 199 (1930); Engler, Die natürlichen Pflanzenfamilien **18a**: 441 (1930).

Creeping, evergreen perennial herb, stems to 2.5 cm. Leaves obovate, densely overlapping, 2–3 mm, light green, later reddish, with marginal hairs. Flowers solitary, bright yellow. *Mexico*. G1. Spring–summer.

70. S. muscoideum Rose (*S. cupressoides* misapplied). Illustration: Journal of the Royal Horticultural Society **46**: 245 (1920–1921).

Evergreen perennial herb, branches like a cypress. Leaves alternate, ovate, *c*. 1.5 mm. Flowers bright yellow. *Mexico*. G1. Summer.

71. S. acre Linnaeus (*S. ukrainae* invalid). Illustration: Journal of the Royal Horticultural Society **46**: 246 (1920–1921); Hegi, Flora von Mitteleuropa **4**: pl. 140 (1921); Köhlein, Freilandsukkulenten, 144 (1977).

Low, creeping, hairless, evergreen, perennial herb, with sterile shoots erect, 2–10 cm. Leaves alternate, crowded, ovoid-triangular to oblong, fleshy, 2–8 mm. Flowers bright yellow, 1–1.5 cm across. Petals lanceolate, acute, wide-spreading. *Europe, W & N Asia, N Africa*. H1. Summer.

'Elegans' has pale silvery shoot-tips and young leaves in spring.

Var. **majus** Masters. Illustration: Journal of the Royal Horticultural Society **46**: 247 (1920–1921). Leaves overlapping in 7 rows, pale green. Flowers *c*. 1.5 cm across. Larger and more robust than the type. *Morocco*.

'Aureum'. Illustration: Hay & Beckett, Reader's Digest encyclopaedia of garden plants & flowers, 657 (1978). Leaves and shoot-tips bright golden yellow in spring; flowers somewhat smaller than in the type.

Var. **krajinae** Domin (*S. krajinae* Domin). Leaves broad triangular-ovate to ovate-roundish, obtuse, 4.5–5 × 3–4 mm, very thick, pale or fresh green. Flowers

yellow, 1.2–1.5 cm across. *Slovakia*. H2. Summer.

72. S. oaxacanum Rose. Illustration: Journal of the Royal Horticultural Society **46**: 250 (1920–1921); Jacobsen, A handbook of succulent plants, 790 (1960).

Evergreen, creeping, small hairless perennial with reddish stems. Leaves alternate, obovate to oblong-elliptic, obtuse, *c*. 6 × 3 mm, thick, green, often glaucous. Flowers stalkless, yellow, *c*. 7.5 mm across. Petals nearly oblong, very broad below tip. *Mexico*. G1. Spring.

73. S. alpestre Villars. Illustration: Hegi, Flora von Mitteleuropa **4**: 536, 539 (1921); 't Hart, Biosystematic studies in the Acre-group and the series Rupestria Berger of the genus Sedum L., 36 (1978).

Dwarf tufted perennial, hairless. Leaves crowded in a rosette, obovate to narrowly oblong, *c*. 6 mm, bright green. Flowers green-yellow, often tinged with red, *c*. 4 mm across, petals ascending. *Mountains C & S Europe, Turkey*. H3. Summer.

74. S. sexangulare Linnaeus (*S. mite* Gilibert; *S. boloniense* Loiseleur). Figure 17(3), p. 193. Illustration: Journal of the Royal Horticultural Society **46**: 265 (1920–1921); Reichenbach, Icones Florae Germanicae **23**: t. 57 (1897).

Evergreen hairless perennial herb with creeping stems, 5–15 cm. Leaves usually arranged in 6 rows, terete, linear-oblong, cylindric, obtuse, 4–7 mm, bright green. Flowers yellow, 5–9 mm across. Petals narrowly lanceolate, acute, wide-spreading. *C Europe to Finland, W France to Romania & Ukraine*. H1. Summer.

75. S. multiceps Cosson & Durieu. Illustration: Journal of the Royal Horticultural Society **46**: 264 (1920–1921); Jacobsen, A handbook of succulent plants, f. 1027, (1960); Gardener's Chronicle **6**: 204 (1876), **10**: 717 (1878).

Partly deciduous hairless bushy perennial, 5–15 cm. Leaves alternate, very densely arranged, linear-oblong, obtuse, glaucous, papillose, 6 mm with a flat upper surface. Flowers stalkless, yellow, *c*. 1.3 cm across. Petals oblong-lanceolate, mucronate, wide-spreading. *Algeria*. H5–G1. Summer.

76. S. japonicum von Siebold. Illustration: Gartenflora **15**: t. 513 (1866).

Evergreen perennial herb to 15 cm, with ascending branches. Leaves very variable, linear-oblong, *c*. 6 mm. Flowers yellow, *c*.

1.2 cm across. Petals oblong-lanceolate, acuminate. *Japan, China*. H3. Summer.

77. S. stenopetalum Pursh (*S. douglasii* J.D. Hooker; *S. himalense* misapplied). Illustration: Gartenflora **21**: t. 741 (1872); Journal of the Royal Horticultural Society **46**: 258 (1920–1921); Clausen, Sedum of North America, 265 (1975).

Evergreen hairless perennial herb, stems 7.5–15 cm. Leaves alternate, stalkless, fleshy, linear-lanceolate, acute, 1.2–1.5 cm, bright green often flushed red. Flowers stalkless, pale-bright yellow, 1.2–1.8 cm across. Petals linear-lanceolate, acute. This plant produces vegetative rosettes in the axils of the leaves of the floral stems. *C & W North America*. H3. Summer.

78. S. amplexicaule de Candolle (*S. tenuifolium* (Sibthorp & Smith) Strobl). Illustration: Journal of the Royal Horticultural Society **46**: 280 (1920–1921); 't Hart, Biosystematic studies in the Acre-group and the series Rupestria Berger of the genus Sedum L., 88 (1978).

Hairless perennial herb, 3–15 cm; leaves withering in summer, plant persisting by shoot tips enclosed in the swollen leaf bases, green for the rest of the year. Branches 2.5–10 cm. Leaves alternate, linear, mucronate, membranous at base, 1–2 cm, bluish green. Flowers golden yellow, 1.5–1.8 cm across, floral parts in 5s–8s. Petals oblong-lanceolate. *S Europe, Algeria, Bulgaria, Turkey*. H3. Late spring–early summer.

79. S. forsterianum Smith (*S. elegans* Lejeune; *S. rupestre* misapplied). Figure 17(1), p. 193. Illustration: Reichenbach, Iconographia botanica **9**: 1135 (1831–1833); Journal of the Royal Horticultural Society **46**: 266 (1920–1921); 't Hart, Biosystematic studies in the Acre-group and the series Rupestria Berger of the genus Sedum L., 83 (1978).

Perennial herb. Leaves in dense rosette-like cones at tips of the shoots, linear-elliptic, 1–1.5 cm, green or glaucous. Flowers yellow, *c*. 1.6 cm across. Petals lanceolate-elliptic. *W Europe, Morocco*. H3. Summer.

This species is closely related to *S. rupestre*. Variants with glaucous leaves are sometimes distinguished as subsp. **elegans** (Lejeune) E.F. Warburg.

80. S. rupestre Linnaeus (*S. rupestre* subsp. *reflexum* (Linnaeus) Hegi & Schmid; *S. reflexum* Linnaeus). Illustration: Journal of the Royal Horticultural Society **46**: 266, 269 (1920–1921); Hegi, Flora von

Mitteleuropa **4**: pl. 140 (1921); 't Hart, Biosystematic studies in the Acre-group and the series Rupestria Berger of the genus Sedum L., 100 (1978); The Plantsman **8**: 17 (1986).

Evergreen hairless creeping herb, 15–30 cm. Leaves alternate, linear, cylindric, acute, 1.2–1.8 cm, green or grey-green. Flowers pale yellow, 1.3–1.8 cm across. Petals lanceolate-elliptic. Very variable species. *W & C Europe to S Norway & S Sweden, former USSR (W Ukraine), Sicilia.* H1. Summer.

'Cristatum'. Illustration: Journal of the Royal Horticultural Society **46**: 271 (1920–1921) – as S. reflexum var. cristatum. A fasciated variant with flattened stems often *c.* 5 cm broad. In this condition it never flowers.

81. S. pruinatum Brotero. Illustration: Journal of the Royal Horticultural Society **46**: 278 (1920–1921); 't Hart, Biosystematic studies in the Acre-group and the series Rupestria Berger of the genus Sedum L., 93 (1978).

Evergreen perennial herb, strongly bloomed, to 10 cm. Leaves alternate, stalkless, linear, acute, to 2 cm, grey-green, upper surface flattened. Flowers pale yellow to straw yellow, 1.5–2 cm across. Petals linear, acute, wide-spreading. *N & C Portugal.* H4. Summer.

82. S. lanceolatum Torrey (*S. stenopetalum* misapplied). Illustration: Gartenflora **12**: t. 403 (1862) – as S. rhodiola var. lanceolatum; Journal of the Royal Horticultural Society **46**: 276 (1920–1921); Clausen, Sedum of North America, 212 (1975).

Small, evergreen, hairless perennial. Leaves alternate, stalkless, linear-lanceolate, obtuse or obtusely mucronate, minutely papillose, 1–1.8 × 0.2–0.3 cm, bluish green or somewhat purplish. Flowers bright yellow, 1.5–2 cm across. Petals lanceolate, acute or acuminate, widely spreading. *C & W North America.* H3. Early summer.

83. S. urvillei de Candolle (*S. hillebrandtii* Fenzl; *S. sartorianum* Boissier; *S. stribrnyi* Velenovsky). Illustration: Journal of the Royal Horticultural Society **46**: 249 (1920–1921); 't Hart, Biosystematic studies in the Acre-group and the series Rupestria Berger of the genus Sedum L., 53 (1978).

Hairless perennial herb to 15 cm. Leaves narrow ovate to linear-oblong, semi-terete or sometimes terete, usually obtuse, grey-green, 4–15 mm. Flowers bright yellow, 0.8–1.6 cm across. Petals lanceolate. *E & C Balkans south of Carpathians, SW Asia, SW former USSR, Israel, Lebanon, Syria.* H4. Summer.

84. S. stevenianum Rouy & Camus (*S. roseum* von Steven). Illustration: Journal of the Royal Horticultural Society **46**: 200 (1920–1921).

Tufted hairless perennial herb, 2–5 cm. Leaves opposite, obovate to spathulate, very obtuse, 5–6 × *c.* 1.5 mm, entire, spotted with red. Flowers rose-red outside, white inside, *c.* 6 mm across. Petals ovate-lanceolate, blunt, not widely spreading. *SW Asia, Caucasus.* H3. Summer.

85. S. obtusifolium C.A. Meyer.

Hairless perennial, 15–25 cm, often glandular hairy towards the apex. Leaves opposite, ovate or obovate, obtuse, roundish towards the base, narrowed into a very short leaf-stalk, 0.6–3 × 0.4–2 cm. Flowers pink, reddish or white, *c.* 1 cm across. Petals lanceolate, slender-pointed. *SW Asia, Caucasus, Iran.* H2. Late spring–summer.

86. S. spurium Bieberstein (*S. oppositifolium* Sims; *S. stoloniferum* misapplied). Illustration: Botanical Magazine, 2370 (1823); Journal of the Royal Horticultural Society **46**: 195 (1920–1921); Hegi, Flora von Mitteleuropa **4**: 527 (1921), Hay & Beckett, Reader's Digest encyclopaedia of garden plants & flowers, 658 (1978).

Evergreen, somewhat papillose perennial herb with creeping stems, 10–20 cm. Leaves opposite, obovate, 2.5–3 × *c.* 1.8 cm, with teeth in upper half, dark green, with marginal hairs. Flowers pink to purple, white or cream, *c.* 1.3 mm across. Petals lanceolate, acute, almost erect. *N Iran, Caucasus, Armenia.*

The most common cultivars are: 'Album Superbum': with white flowers, flowers rarely; 'Fuldaglut': leaves dark purplish red, flowers carmine pink; 'Purpurteppich': leaves and flowers dark purplish red; 'Roseum Superbum'. Illustration: Köhlein, Freilandsukkulenten, 143 (1977). Flowers pink; 'Schorbuser Blut'. Illustration: Hay & Synge, The dictionary of garden plants, pl. 204 (1969). Leaves with red margin or dark brown leaves, flowers dark purplish pink; 'Variegatum': leaves with creamish pink margin, flowers pink.

87. S. stoloniferum Gmelin (*S. ibericum* von Steven; *S. hybridum* misapplied). Illustration: Journal of the Royal Horticultural Society **46**: 197 (1920–1921).

Tufted, hairless or somewhat papillose perennial herb with creeping habit, to 15 cm. Leaves opposite, diamond-shaped to spathulate, obtuse, 1–3 × 0.8–1.8 cm, bright green, scalloped in upper half, mature leaves without marginal hairs. Flowers pink, 1.3–1.8 cm across. Petals lanceolate, acute, wide-spreading. *N Iran, SW Asia, Caucasus.* H3. Summer.

88. S. rhodocarpum Rose. Illustration: Contributions from the U.S. National Herbarium **13**: pl. 59 (1911); Journal of the Royal Horticultural Society **46**: 201 (1920–1921).

Evergreen, perennial herb with 3-angled stems. Leaves in 3s, uppermost leaves alternate, spathulate to circular, *c.* 2 cm, light green, entire. Flowers greenish red, *c.* 1.3 mm across. Petals oblong, acute, wide-spreading, later sharply reflexed. *Mexico.* G1. Winter.

89. S. pulchellum Michaux. Illustration: Botanical Magazine, 6223 (1876); Journal of the Royal Horticultural Society **46**: 205 (1920–1921); Clausen, Sedum of North America, 143, 153 (1975); Köhlein, Freilandsukkulenten, 143 (1977).

Annual or short-lived perennial, 10–15 cm, stems ascending. Leaves alternate, narrow-linear, cylindric, obtuse, 1.3–2.5 × 0.2–0.5 cm, green. Flowers stalkless, rosy purple, 0.5–1.4 cm across. Petals lanceolate, acute. *USA (Virginia to Georgia, Indiana, Missouri, Texas).* H3. Summer.

90. S. ternatum Michaux. Illustration: Edwards's Botanical Register, t. 142 (1816–1817); Botanical Magazine, 1977 (1818); Journal of the Royal Horticultural Society **46**: 160 (1920–1921); Clausen, Sedum of North America, 93 (1975).

Evergreen, smooth or finely papillose perennial herb, 7–15 cm. Leaves of sterile shoots usually in whorls of 3, obovate to spathulate, apex obtuse usually rounded, 1.2–2.5 × 0.6–1.2 cm, pale green. Flowers stalkless, white, *c.* 1.3 cm across, parts in 4s. Petals oblong-narrowly lanceolate, acute. *USA (New York & New Jersey to Georgia, Indiana, Michigan, Tennessee).* H3. Summer.

91. S. monregalense Balbis. Illustration: Reichenbach, Iconographia Botanica **3**: 438 (1825); Reichenbach, Icones Florae Germanicae **23**: t. 64a (1897); Journal of the Royal Horticultural Society **46**: 172 (1920–1921).

Tufted perennial herb, 7–15 cm,

minutely hairy. Leaves in whorls of 4, opposite below; oblong to spathulate, obtuse, fleshy, papillose at the apex, 3–10 mm, olive-green. Flowers white, 0.6–1 cm across, glandular hairy. Petals triangular, spreading, white on upper surface, pinkish brown beneath. *SW Alps, Italy, Corsica*. H3. Summer.

92. S. magellense Tenore. Illustration: Tenore, Flora Napolitana **1**: 139, f. 1 (1811–1815); Journal of the Royal Horticultural Society **46**: 167 (1920–1921).

Evergreen perennial herb, 7.5–10 cm, tufted with hairless, light green rosettes. Leaves alternate or opposite, 6–10 mm, obovate to elliptic-oblong, obtuse, flat, fleshy, light green. Inflorescence raceme-like, 2.5–5 cm, flowers white or whitish, 6–9 mm across. Petals lanceolate, acute, mucronate, wide-spreading. *Italy, Greece, SW Turkey, NW Africa*. H3. Late spring–early summer.

93. S. sediforme (Jacquin) Pau (*S. altissimum* Poiret; *S. nicaeense* Allioni). Illustration: Journal of the Royal Horticultural Society **46**: 272 (1920–1921); Jacobsen, A handbook of succulent plants, 798 (1960); Köhlein, Freilandsukkulenten, 161 (1977); 't Hart, Biosystematic studies in the Acre-group and the series Rupestria Berger of the genus Sedum L., 95 (1978).

Evergreen, hairless perennial herb, 15–60 cm. Leaves alternate, oblong-elliptic to linear-elliptic, acuminate, 1–3 × 0.3–0.5 cm, grey-green. Flowers greenish white, rarely yellow, *c.* 1.3 cm across. Petals lanceolate-elliptic. *Mediterranean area, C France, Portugal*. H3. Spring–summer.

94. S. bithynicum Boissier (*S. hispanicum* Linnaeus var. *minus* Praeger; *S. lydium* invalid; *S. lydium aureum* invalid; *S. glaucum* invalid). Illustration: Journal of the Royal Horticultural Society **46**: f. 178b (1920–1921).

Perennial herb, hairless or flower-stems with shaggy glandular hairs, 2–7.5 cm. Leaves alternate, linear to oblong, nearly cylindric, 4–7 mm, pale green or grey-green. Flowers 0.9–1.2 cm across. Petals ovate, acuminate, white often with purplish mid-vein and keel. *Balkans, Turkey*. H3. Summer.

Closely related to the annual *S. hispanicum* Linnaeus from S Europe.

95. S. dasyphyllum Linnaeus (*S. corsicum* Duby). Illustration: Journal of the Royal Horticultural Society **46**: 177

(1920–1921); Hegi, Flora von Mitteleuropa **4**: pl. 140 (1921); Fessler, Die Freilandschmuckstauden, 528, 585 (1985).

Evergreen, glandular hairy herb, sometimes almost hairless, 2–10 cm. Leaves opposite or alternate, stalkless, ovate to elliptic-oblong, 3–12 mm, blue-green. Flowers white, pinkish on back, *c.* 0.9 cm across. *Europe, N Africa*. H4. Summer.

Var. **suendermannii** Praeger. Illustration: Journal of the Royal Horticultural Society **46**: 178 (1920–1921). Larger and with densely overlapping, obovate, densely glandular hairy leaves. Flowers somewhat larger and appearing some weeks later.

96. S. gypsicola Boissier & Reuter. Illustration: Journal of the Royal Horticultural Society **46**: 187 (1920–1921).

Creeping, evergreen perennial herb, with fine dense grey hairs, 5–10 cm. Leaves overlapping at the apices of the stems in 5 spiral rows, ovate to diamond-shaped, obtuse, 6–8 mm, dull grey-green, often tinged red. Flowers white, *c.* 6 mm across. *Spain, Portugal, Morocco*. H4. Early summer.

97. S. hirsutum Allioni. Illustration: Journal of the Royal Horticultural Society **46**: 188 (1920–1921); Valdes et al., Flora Vascular de Andalucía Occidental **2**: 14 (1987).

Evergreen, perennial herb, 5–10 cm, glandular hairy. Leaves alternate in dense rosettes at ends of shoots, oblong, elliptic or slightly spathulate, obtuse, 6–10 × 2–5 mm. Flowers white to pink. *SW Europe: S France, N Italy, Morocco*. H2. Summer.

Subsp. **baeticum** Rouy (*S. winkleri* (Willkomm) Wolley-Dod. Illustration: Evans, Handbook of cultivated Sedums, 313, pl. 15 (1983). Taller and stouter, with larger flowers. *S Spain*.

98. S. fragrans 't Hart (*S. alsinefolium* Allioni misapplied). Illustration: Journal of the Royal Horticultural Society **46**: 170 (1920–1921); Grey-Wilson & Blamey, The alpine flowers of Britain & Europe, 77 (1979).

Perennial hairy herb. Leaves of sterile shoots in a crowded rosette, diamond-shaped to spathulate, obtuse, 6–15 × *c.* 6 mm, dark green. Flowers white, *c.* 6 mm across. Petals oblong-obovate, abruptly acuminate, erect in lower half, spreading above. *S France, N Italy*. H4. Summer.

99. S. compactum Rose. Illustration: Contributions from the U.S. National Herbarium **13**: pl. 53 (1911); Journal of the Royal Horticultural Society **46**: 176 (1920–1921); Gartenschönheit **8**: 290 (1927).

Low perennial herb. Leaves in rosettes, obovate, obtuse, *c.* 3 mm, grey-green. Flowers white, small, urn-shaped, strongly scented. Petals ovate, mucronate. *Mexico*. G1. Summer.

100. S. furfuraceum Moran. Illustration: Evans, Handbook of cultivated Sedums, 154 (1983).

Creeping hairless herb. Leaves alternate, crowded, ovoid, obtuse, 0.6–1 cm, dark green to purplish, with uneven grey scaly covering. Flowers white or pinkish, *c.* 1 cm across. *Mexico*. G1. Early spring.

101. S. brevifolium de Candolle. Illustration: Journal of the Royal Horticultural Society **46**: 179 (1920–1921); Valdes et al., Flora Vascular de Andalucia Occidental **2**: 13 (1987).

Evergreen, hairless tufted herb. Leaves opposite, crowded in 4 rows, ovoid or spherical, 3–4 mm, white-bloomed, flushed with red, sometimes with bloom, purplish or yellow-green; alternate on the flowering-stems. Flowers white, 6–8 mm across. Petals ovate, acute, white with reddish midrib. *SW Europe, Morocco*. H4–5. Summer.

Var. **quinquefarium** Praeger has shoots twice as thick as long, leaves in 5 rows, 5–8 mm, white-bloomed.

102. S. oregonense (Watson) Peck (*S. watsoni* (Britton) Tidestrom; *S. rubroglaucum* misapplied). Illustration: Clausen, Sedum of North America, 411, 413 (1975).

Similar to *S. obtusatum* subsp. *obtusatum*, but differs in terms of its larger size, white or creamish flowers, petals more than 3 times length of sepals. *W USA*. H3. Summer.

103. S. diffusum S. Watson (*S. potosinum* Rose). Illustration: Journal of the Royal Horticultural Society **46**: 139 (1920–1921); Sedum Society Newsletter **7**: 1, 5 (1988).

Evergreen, hairless perennial, 7–25 cm with prostrate stems. Leaves of sterile shoots alternate, dense, oblong to linear, obtuse, almost terete, 3–12 × 1–3 mm. Flowers *c.* 1.5 cm across. Petals white, sometimes flushed red on back. *Mexico*. H5–G1. Spring–autumn.

104. S. liebmannianum Hemsley.
Illustration: Rose, Contributions from the
U.S. National Herbarium **13**: f. 56 (1911);
Journal of the Royal Horticultural Society
46: 175 (1920–1921).

Perennial hairless herb, 5–15 cm, stems
ascending, nearly deciduous. Leaves
alternate, ovate-oblong, obtuse,
overlapping, *c.* 5 mm, green, tipped red,
forming a silvery persistent covering when
dead. Flowers white, stalkless, *c.* 1 cm
across. Petals lanceolate, rather acute.
Mexico & Texas. G1. Summer.

105. S. anglicum Hudson. Illustration:
Journal of the Royal Horticultural Society
46: 181 (1920–1921).

Evergreen, hairless perennial herb, 7–8
cm. Leaves alternate, ovate-elliptic, very
thick, 4–5 mm, green, often tinged red.
Flowers white-pink, *c.* 1 cm across. Petals
lanceolate. *Europe.* H2. Summer.

106. S. albertii Regel. Illustration:
Gartenflora **19**: t. 1019 (1870).

Small, hairless, evergreen perennial
herb, 3–5 cm. Roots thickened, cord-like.
Leaves alternate, crowded, linear-oblong,
nearly cylindric, obtuse, covered with
rough papillae, *c.* 6 mm. Flowers white,
0.9–1.2 cm across, stalks absent or *c.*
1 mm. Petals oblong-ovate, mucronate,
fused at base. Anthers violet. Follicles erect.
Former USSR (W Siberia, Turkestan). H1.
Late spring–early summer.

107. S. lydium Boissier. Illustration:
Journal of the Royal Horticultural Society
46: 189 (1920–1921).

Evergreen, hairless perennial herb, 3–5
cm. Leaves alternate, crowded, narrowly
linear to cylindric, obtuse, 4–8 × *c.* 1 mm,
green or red, minutely papillose at tip.
Flowers white, *c.* 6 mm across. Petals
lanceolate, acute. Follicles erect. *W Turkey.*
H3. Summer.

108. S. gracile C.A. Meyer. Illustration:
Journal of the Royal Horticultural Society
46: 190, 192 – as S. albertii (1920–1921).

Evergreen, perennial herb, 5–10 cm.
Leaves alternate, crowded linear-oblong,
nearly cylindric, *c.* 8 mm, bright green,
sometimes reddish green. Flowers white,
often dotted red on back, 7–8 mm across.
Follicles spreading. *N Iran, Caucasus, E
Turkey.* H3. Early summer.

109. S. album Linnaeus. Figure 17(4),
p. 193. Illustration: Oeder, Flora Danica
t. 66 (1763); Reichenbach, Icones Florae
Germanicae **23**: t. 55 (1897); Journal of
the Royal Horticultural Society **46**: 183

(1920–1921); Hegi, Flora von Mitteleuropa
4: pl. 140 (1921).

Evergreen, creeping, almost hairless,
perennial herb, branches erect, with some
glandular hairs, 10–20 cm. Leaves
alternate, cylindric to ovate, 6–12 mm,
obtuse, dark green, often reddish. Flowers
white, 0.7–1.2 cm across. Petals lanceolate
to ovate, obtuse. *Europe, N Africa, C Asia.*
H3. Summer.

Var. **micranthum** (de Candolle) de
Candolle. Illustration: Hegi, Flora von
Mitteleuropa **4**: 536 (1921). Inflorescence
has many small flowers.

The following cultivars are good garden
plants: 'Coral Carpet', as ground cover
with, in winter greenish and in summer
bronze-red leaves; 'Laconicum', with partly
green and partly red-brown leaves, petals
somewhat broader than in the species;
'Micranthum Chloroticum', is *c.* 5 cm high
with bright green leaves, flowers greenish
white; 'Murale', has brownish red leaves
and pale rose flowers.

110. S. tenellum Bieberstein.
Low, hairless perennial herb. Leaves on
sterile stems alternate, crowded, oblong
to nearly circular, *c.* 3 mm. Leaves on
flowering-stems linear-oblong, cylindric,
obtuse, *c.* 6 mm. Flowers white, tinged red,
6–8 mm across. Sepals spurred at base.
Follicles erect. *Armenia, N Iran, Caucasus.*
H3. Summer.

111. S. nevii A. Gray. Illustration:
Clausen, Sedum of North America: 110
(1975).

Evergreen, tufted, hairless herb, 3–10
cm. Sterile shoots forming rather dense
rosettes, *c.* 1.8 cm across. Leaves of
flowering-shoots alternate, oblanceolate or
linear-lanceolate, mostly nearly cylindric,
1.2–1.8 × *c.* 0.4 cm, entire, pale brownish
green, ratio of width to thickness less than
1.7, glaucous. Flowers white, to *c.* 1.3 cm
across. Petals linear-lanceolate, acute. *USA
(Virginia to Alabama).* H1. Summer.

112. S. glaucophyllum R.T. Clausen (*S.
nevii* misapplied). Illustration: Clausen,
Sedum of North America, 123, 133 (1975).

Perennial herb, 2–10 cm. Leaves of
sterile stems in dense rosettes, obovate to
spathulate, obtuse or nearly acute, *c.* 1.5
× 1 cm; flowering-stems *c.* 2 cm, leaves
oblanceolate, *c.* 2.5 × 0.6 cm, pale or blue-
green, ratio of width to thickness more
than 2. Flowers stalkless, white, to *c.* 1.2
cm across. Petals lanceolate, acuminate,
widely spreading. *E North America.* H3.
Late spring–summer.

Some plants with narrow green leaves
closely resemble *S. nevii* and are best
distinguished by the leaf width to thickness
ratio of flowering-shoots. The correct
application of *S. beyrichianum* Masters and
S. nevii var. *beyrichianum* (Masters) Praeger
is uncertain; the names are applied to both
species in gardens.

7. RHODIOLA Linnaeus
L.S. Springate
Perennial herbs usually with a stout, very
short, fleshy, perennial stem, sometimes
branched above, often partly exposed,
covered in brown triangular scales;
branches simple, annual, leafy with
terminal flower heads developing laterally
below each terminal bud on the perennial
stem. Sometimes perennial stem
lengthening, slender, much-branched
forming tufts or shrublets, branches
crowned with leaves distinct from those
on the annual stems and scales present
or absent. Floral parts usually in 4s or
5s. Flowers solitary or few to many in
spherical, corymb-like, raceme-like or
paniculate heads, bisexual or functionally
unisexual, the plants then usually
dioecious. Stamens twice as long as petals,
absent in female flowers. Carpels present,
opposite the petals in female flowers,
reduced and opposite the sepals in male
flowers.

A genus of 50 species centred on the
mountains of E Asia, extending to N
America and Europe. The species listed
can be grown in pots if protected from
winter rain or in raised beds; the coarser
species can also be grown in open borders.
Flowering may be poor in zones H4 and
H5 at low latitudes.

Literature: Praeger, R.L., *An account
of the genus Sedum as found in cultivation*
(1921); Clausen, R.T., *Sedum of North
America north of the Mexican Plateau*
(1975); Ohba, H. in Ohashi, H., *Flora of
the Eastern Himalaya, 3rd Report,* 283–362
(1975); Ohba, H., A revision of the Asiatic
Species of Sedoideae (Crassulaceae), *Journal
of the Faculty of Science of the University of
Tokyo, Section 3.* **12**: 337–405, (1980); **13**:
65–119, 121–169 (1981–1982); Evans,
R.L., *Handbook of Cultivated Sedums* (1983).

Perennial stems. Long and slender: **2,3**;
stout, but lengthening: **13,14**; compact,
stout: **1,4–12**.
Dead annual stems. Persistent, dense, in
twiggy tufts: **13,14**.
Flowers. Bisexual: **1–7**; unisexual: **8–14**;
bell-shaped: **5**.

Petals. White or cream to pale green: **2,3,4,5,6,7,10**; yellow to yellowish green: **8,10,12,14**; pink or reddish to reddish green or brown: **1,6,7,9,10,12**; dark red or purplish: **11,13(14)**; erect or ascending: **1,2,4,6,7,13,14**; and female flowers of **10,12**; spreading: **3,4,6,8–12**.

1a. Perennial stems long, slender, much-branched, forming tufts, or shrublets   2
   b. Perennial stem solitary, stout, short or if long then not slender   3
2a. Leaves of perennial stems with distinct stalks more than 4 mm   **2. primuloides**
   b. Leaves of perennial stems with indistinct stalks to 3 mm   **3. pachyclados**
3a. Perennial stem (caudex) with stalked green leaves; filaments fixed to back of anthers   **1. hobsonii**
   b. Perennial stem (caudex) without green leaves (triangular scales sometimes green at first); filaments fixed to base of anthers   4
4a. Inflorescence raceme-like   5
   b. Inflorescence corymb-like or head-like   6
5a. Leaves linear, to 7 cm; petals 7–8.5 mm   **6. semonovii**
   b. Leaves oblong, to 2.5 cm; petals 8–13 mm   **7. rhodantha**
6a. Petals erect, more than 7.5 mm   7
   b. Petals spreading or less than 7.5 mm   8
7a. Leaves linear, less than 3 mm wide   **5. dumulosa**
   b. Leaves oblong, more than 3 mm wide   **7. rhodantha**
8a. Inflorescence diffuse; leaf length less than twice width   **11. bupleuroides**
   b. Inflorescence compact, all flowers touching, or leaf length at least twice width   9
9a. Petals 5–11 mm; flowers bisexual   **4. wallichiana**
   b. Petals to 5 mm; flowers unisexual   10
10a. Leaves linear-lanceolate or linear   11
   b. Leaves broader   12
11a. Leaves to 1.2 cm   **14. quadrifida**
   b. Leaves more than 2.5 cm   **12. kirilowii**
12a. Leaves with papillae above   **13. himalensis**
   b. Leaves smooth   13

13a. Inflorescence very condensed; flower-stalks partly fused, to 1 mm   **10. heterodonta**
   b. Inflorescence corymb-like; flower-stalks entirely separate, 1–4 mm   14
14a. Petals red; leaves hardly glaucous   **9. integrifolium**
   b. Petals yellowish; leaves usually glaucous   **8. rosea**

1. R. hobsonii (Raymond-Hamet) Fu (*Sedum hobsonii* Raymond-Hamet). Illustration: Praeger, Account of the genus Sedum as found in cultivation, 70 (1921); Evans, Handbook of cultivated sedums, 75, pl. 1 (1983).

Plant to 12 cm, hairless with two distinct types of leaf. Perennial stem massive, short, crowned by erect, elliptic, sometimes persistent, green leaves 7.5–10 × 3–4 mm, with stalks 7–10 mm. Annual stems spreading, 5–15 cm, with many lanceolate to spathulate, stalkless green leaves, 6–15 × 2–4.5 mm and a cluster of 3–10 flowers. Flowers bisexual, 7–9 mm across, floral parts usually in 5s. Petals elliptic, erect, hooded, rose, 5.5–7.5 × 2–3 mm. Filaments attached to back of anthers. *Bhutan & China (Xizang)*. H3. Late summer.

2. R. primuloides (Franchet) Fu (*Sedum primuloides* Franchet). Illustration: Praeger, Account of the genus Sedum as found in cultivation, 70 (1921); Evans, Handbook of cultivated sedums, 75, pl. 1 (1983).

Plant to 10 cm, hairless. Perennial stem usually long, much-branched, occasionally short, compact; branches sometimes rooting, crowned by rosettes of leaves; leaves bright green, sometimes persistent, circular to oblong to 1 × 0.9 cm, leaf-stalks to 9 mm, clasping stem at base. Annual stems to 5 cm, with 1–4 flowers and leaves not clasping. Flowers bisexual, 6–9 mm across; floral parts in 5s. Petals erect, elliptic, white, 5–6.5 × 2.5–3.5 mm. *SW China*. H3. Late summer.

3. R. pachyclados (Aitchison & Hemsley) Ohba (*Sedum pachyclados* Aitchison & Hemsley). Illustration: Sedum Society Newsletter **6**: 1 (1988).

Plant to 10 cm, hairless with 2 distinct types of leaf. Perennial stem much-branched, rhizomatous below, forming dense tufts; branches crowned by rosettes of persistent leaves; leaves ovate to obovate with some wavy teeth near the apex, 4–9 × 3–6 mm. Annual stems to 4 cm with 1–10 flowers and leaves obovate, entire,

4–5.5 × 1.5–2 mm. Flowers bisexual; floral parts in 5s. Petals spreading, elliptic, concave, white, 5.5–7 × 2–3 mm. *Afghanistan & Pakistan*. H3. Early summer.

4. R. wallichiana (Hooker) Fu (*Sedum wallichianum* Hooker; *R. asiatica* invalid; *S. crassipes* Hooker & Thomson; *S. crassipes* var. *cholaense* Praeger). Illustration: Praeger, Account of the genus Sedum as found in cultivation, 59 (1921).

Plant to 30 cm, hairless. Perennial stem massive, short, sometimes with underground stolons. Annual stems erect, with many leaves; leaves linear-lanceolate, upper half with distant teeth, green, 1.2–3 cm × 1–6 mm. Inflorescence dense, corymb-like. Flowers bisexual; floral parts usually in 5s. Petals usually spreading, elliptic, concave, yellowish white to pale greenish, 5–11 × 1.5–2.5 mm. *Kashmir to Sikkim, China (Xizang)*. H3. Summer.

Cultivated plants with a few characteristics found in *R. cretinii* (Raymond-Hamet) Ohba but not in *R. wallichiana*, e.g. calyx-lobes less than 5 mm, petals less than 6.5 mm and stoloniferous stems, are included in *R. wallichiana* here. Illustration: Praeger, Account of the genus Sedum as found in cultivation, 56, 57 (1921); Evans, Handbook of cultivated sedums, 66, pl. 1 (1983).

5. R. dumulosa (Franchet) Fu (*Sedum dumulosum* Franchet). Illustration: Praeger, Account of the genus Sedum as found in cultivation, 62; Evans, Handbook of cultivated sedums, 67, pl. 1 (1983).

Plant to 18 cm, hairless. Perennial stem massive, short. Annual stems erect with very many leaves; leaves linear with lower surface convex, green, 1–2.5 cm × 1.5–2.5 mm. Inflorescence compact, corymb-like with 4–20 flowers. Flowers bisexual, bell-shaped; floral parts usually in 5s. Petals oblong, greenish white to cream, 7.5–12 × 2–2.5 mm, with mucro 0.5–1.5 mm. *Bhutan to Manchuria & N Korea*. H3. Summer.

6. R. semonovii (Regel & Herder) Borissova (*Sedum semonovii* (Regel & Herder) Masters; *Clementsia semonovii* (Regel & Herder) Borissova). Illustration: Praeger, Account of the genus Sedum as found in cultivation, 66 (1921); Evans, Handbook of cultivated sedums, 82, pl. 2 (1983).

Plant to 60 cm, hairless, perennial stem massive, short. Annual stems erect with very many leaves; leaves linear, entire or with remote teeth, green; the middle leaves largest, to 7 cm × 3 mm. Inflorescence raceme-like, dense. Flowers bisexual; floral parts usually in 5s. Calyx-lobes 4–5.5 mm. Petals lanceolate, partly spreading, greenish white to pink, 7–8.5 mm. Carpels as long as petals at flowering. *Central Asia*. H3. Summer.

7. R. rhodantha (Gray) Jacobsen (*S. rhodanthum* Gray; *Clementsia rhodantha* (Gray) Rose). Illustration: Praeger, Account of the genus Sedum as found in cultivation, 68 (1921); Clausen, Sedum of North America, 476 (1975); Evans, Handbook of cultivated sedums, 78, pl. 1 (1983).

Plant to 40 cm, hairless, perennial stem massive, short. Annual stems erect with many leaves; leaves oblong, entire or with obscure teeth near apex, green, 1.5–2.5 × 0.5–0.6 cm. Inflorescence raceme-like, sometimes condensed and almost capitate. Flowers bisexual, floral parts in 5s. Calyx-lobes 5–8 mm. Petals erect, elliptic, rose to almost white, 8–13 mm. Carpels usually distinctly shorter than petals at flowering. *USA (Montana to Arizona)*. H3. Summer.

8. R. rosea Linnaeus (*Sedum rosea* (Linnaeus) Scopoli; *S. rhodiola* de Candolle; *R. elongata* (Ledebour) Fischer & Meyer; *R. roanensis* (Britton) Britton). Illustration: Praeger, Account of the genus Sedum as found in cultivation, 29 (1921); Clausen, Sedum of North America, 518 (1975); Evans, Handbook of cultivated sedums, 79, pl. 1 (1983).

Plant 5–30 cm, hairless, dioecious. Perennial stem massive, short. Annual stems few, stout with many leaves, leaves broad-ovate to narrow-oblanceolate, entire or with irregular teeth towards apex, glaucous, 0.7–4 × 0.5–1.7 cm. Inflorescence compact, convex, with many flowers; flower-stalks 1–4 mm; floral parts most often in 4s. All parts yellow or greenish yellow in male flowers, the petals narrow-oblong, concave, spreading, 2.5–3.5 × 0.7–1.2 mm. Petals greenish yellow, linear-subulate, 2–2.5 mm in female flowers, the carpels green. *Europe, N Asia, N America, Greenland*. H3. Early summer.

Widespread and very variable; dwarf cultivated plants (illustration: Evans, Handbook of cultivated sedums, pl. 2 (1983)) are sometimes incorrectly referred to a race from the Palaeo-arctic islands as *R. arctica* Borissova; *R. rosea* subsp. *arctica* (Borissova) Löve; *Sedum arcticum* (Borissova) Ronning or *S. rosea* 'Arcticum' invalid.

9. R. integrifolia Rafinesque (*Sedum integrifolium* (Rafinesque) Nelson; *R. atropurpurea* (Turczaninow) Trautvetter & Meyer; *S. rosea* var. *atropurpureum* (Turczaninow) Praeger). Illustration: Praeger, Account of the genus Sedum as found in cultivation, f. 5C (1921); Clausen, Sedum of North America, 488, 507 (1975); Evans, Handbook of cultivated sedums, 73, pl. 1 (1983).

Very similar to *R. rosea* in appearance, but differs in having petals usually red, in male flowers spathulate, 2.5–5 × 1.1–1.7 mm, leaves green, not or hardly glaucous and floral parts more often in 5s than in 4s. *Arctic Siberia to Sakhalin & N America*. H4. Early summer.

10. R. heterodonta (Hooker & Thomson) Borissova (*Sedum heterodontum* Hooker & Thomson). Illustration: Praeger, Account of the genus Sedum as found in cultivation, 35 (1921); Evans, Handbook of cultivated sedums, 69, pl. 1 (1983).

Plant to 40 cm, dioecious, hairless. Perennial stem massive, short. Annual stems few, stout, erect, growth continuing after flowering; leaves distant; leaves ovate, often clasping the stem, nearly entire or with few sharp teeth, green or glaucous, 1.2–2.5 × 1–1.5 cm. Flowers numerous, densely packed in a hemispheric head; stalks absent or to 1 mm; floral parts usually in 4s. Male flowers with petals narrowly elliptic, 3.5–4.5 mm; filaments 5.5–9 mm, very prominent. Female flowers with petals linear, 2–2.5 mm. Petals and filaments reddish, yellowish or greenish; anthers red. *Afghanistan & Pamirs to Nepal*. H3. Late spring.

Male clones with whole flower-head appearing brick-red or purplish are most often grown.

11. R. bupleuroides (Hooker & Thomson) Fu (*Sedum bupleuroides* Hooker & Thomson; *S. elongatum* Hooker & Thomson; *S. bhutanense* Praeger). Illustration: Praeger, Account of the genus Sedum as found in cultivation, 42, 44, 45, 48 (1921); Evans, Handbook of cultivated sedums, 65, pl. 1 (1983).

Plant 7–90 cm, dioecious, hairless. Perennial stem massive, short. Annual stems few, slender, erect or ascending, with distant leaves; leaves broadly ovate to narrowly elliptic or obovate, cordate to tapered at base, entire or with few shallow teeth, green or somewhat glaucous, 0.3–9.5 × 0.4–4.5 cm. Flowers usually numerous in loose, open heads; floral parts usually in 5s, dark purplish. Male

flowers with petals oblanceolate, concave, widespread, 3–4 mm; stamens 3–4 mm. Female flowers with petals linear, 1.5–3 mm. *Nepal to China (Yunnan)*. H3. Early summer.

12. R. kirilowii (Regel) Maximowicz (*Sedum kirilowii* Regel; *S. longicaule* Praeger; *S. kirilowii* var. *rubrum* Praeger). Illustration: Praeger, Account of the genus Sedum as found in cultivation, 36, 38, 40 (1921); Evans, Handbook of cultivated sedums, 74, pl. 1 (1983).

Plant to 90 cm, dioecious, hairless. Perennial stem massive, short. Annual stems few, stout, erect, growth continuing after flowering, with very many leaves; leaves linear-lanceolate or linear, entire to sharply toothed throughout, green, 4–9 cm × 3–6 mm. Flowers numerous in dense, corymb-like heads; flower-stalks 3–4.5 mm; floral parts in 5s, yellowish green or brownish red. Male flowers with petals oblanceolate, 3–5 mm, stamens 4–6 mm. Female flowers with linear petals 2.5–3 mm. *Central Asia, N China*. H3. Late spring.

13. R. himalensis (Don) Fu (*Sedum himalense* Don). Illustration: Journal of the Faculty of Science, University of Tokyo, Section 3, **13**: 134, 136 (1982).

Plant to 50 cm, dioecious, hairless. Perennial stem massive, ascending, to 40 cm exposed. Annual stems many, erect, reddish, with papillae, persisting when dead, to 30 cm with many leaves; leaves elliptic to oblanceolate, entire or with shallow teeth towards apex, green, upper surface with papillae, 6–20 × 2.5–7 mm. Inflorescence corymb-like, to 50 flowers; male head dense, female loose; the floral parts usually in 5s. Petals ovate-oblong, deep red, 2.5–4.5 × 1.5–2.5 mm. Stamens 0.8–3 mm. *Nepal to China*. H3. Early summer.

Cultivated plants (illustration: Evans, Handbook of cultivated sedums, 70, pl. 1 (1983)) differ in some features.

14. R. quadrifida (Pallas) Fischer & Meyer (*Sedum quadrifidum* Pallas).

Plants 3–15 cm, dioecious, hairless. Perennial stem massive, eventually well exposed and branched above. Annual stems many, erect with very many leaves; leaves linear, lower surface convex, entire, 5–12 × 1–1.5 mm. Inflorescence corymb-like with 6–16 flowers, floral parts usually in 4s, mainly yellow. Petals oblong 2.5–4 × 1.5–2 mm. Stamens as long as petals. Nectaries and fruits deep red. *Arctic Siberia to C Asia & Mongolia*. H3. Summer.

Cultivated plants and those from Pakistan differ in having all flower parts deep red, stems or leaf margins sometimes with papillae and stamens 0.8–2.8 mm and may be an undescribed species. Cultivated plants incorrectly named *R. fastigiata* (Hooker & Thomson) Fu or *Sedum fastigiatum* Hooker & Thomson (illustration: Evans, Handbook of cultivated sedums, 68, pl. 1, 1983) are rather similar.

8. × SEDADIA Moran
G.D. Rowley
Hairless, perennial, evergreen shrublet with spirally arranged, entire, succulent leaves crowded towards the branch tips. Pale yellow flowers are borne in crowded corymbs at the ends of the shoots.

An intergeneric hybrid between species of *Sedum* and *Villadia*. The single hybrid in cultivation attributable to this ancestry has long passed as a species of *Sedum*. For cultivation and other information, see under *Sedum*.

1. × S. amecamecana (Praeger) Moran (*Sedum amecamecanum* Praeger). Illustration: Journal of the Royal Horticultural Society **46**: 214 (1921); Clausen, Sedum of the trans-Mexican volcanic belt 79–80 (1959); Evans, Handbook of cultivated Sedums, 136 (1983).

Stems prostrate or creeping, with branches 15–20 cm. Leaves narrow elliptic to oblanceolate, with a short spur below, 1.5–2 cm × 5 mm, acute, somewhat recurved, with some scattered along the shoots. Inflorescence terminal, bearing starry, 5-petaled yellow flowers 1.5 cm across. Sepals unequal, green, spreading. *Mexico.* G1–H5. Summer.

(*Sedum praealtum* × *Villadia batesii*).

9. × SEDEVERIA Walther
G.D. Rowley
Shrubby or stemless perennial, evergreen leaf-succulents. The many, small flowers are borne on lateral inflorescences, and yellow is the dominant colour. They are broadly funnel-shaped with the 5 petals shortly united at the base.

Their diverse habit reflects their origin as hybrids between different species of the large genera *Echeveria* and *Sedum*. Only the 2 best known are covered here.

1a. Stems pendent, clothed all along with leaves; flowers inverted at the shoot tips
     1. × S. 'Harry Butterfield'
  b. Stems prostrate, leafy only towards

the ends; flowers borne on erect scapes   **2. × S. hummelii**

1. × S. 'Harry Butterfield' (*Echeveria ? derenbergii* × *Sedum morganianum*; 'Super Burro's Tail'). Illustration: British Cactus and Succulent Journal **5**: 11 (1987).

Stems *c.* 5 mm thick, long, pendent, branching from near the base, densely packed throughout with pale green, almost terete, pointed, glaucous, very succulent leaves 2.5 cm, of a pale whitish green colour. The flowers resemble those of *Sedum morganianum*. G1.

Basket plant with the habit of the 'Burro's Tail', *Sedum morganianum*, p. 190.

2. × S. hummelii Walther (*Echeveria derenbergii* × *Sedum pachyphyllum*; "*S. hummellii*" was a misspelling). Illustration: Cactus and Succulent Journal (U.S.) **25**: 20–21 (1953); Jacobsen, Lexicon of succulent plants, t. 124/4 (1974).

Shrublet to 15 cm, recalling *Sedum pachyphyllum*, with prostrate stems 7–8 mm across and densely leafy towards the tips. Leaves narrow elliptic, tapering at both ends, biconvex to almost terete, glaucous pearly green with a sharp red tip, 3.5–5 × 0.8–1.1 cm. Inflorescence to 15 cm, slender, with scattered bracts, bearing one-sided cymes of *c.* 12 showy bright yellow flowers *c.* 1.5 cm across. Sepals 5, unequal, green, adpressed, 7–8 mm. Petals 1.1–1.2 cm, acuminate. Spring.

A vigorous, almost hardy, free-flowering hybrid, welcomed for its early blooming. × *S.* 'Alidaea' is apparently the same cultivar. G1–H5.

10. PISTORINIA de Candolle
S.G. Knees
Erect annuals with alternate leaves. Stems hairless below, glandular hairy above. Leaves linear, succulent, stalkless and soon falling. Flowers shortly stalked, numerous in dense corymbs. Sepals 5, petals 5, united into a funnel-shaped corolla, tube long and narrow, lobes spreading or erect. Stamens 10, 5 long, 5 short. Styles slender. Carpels 5. Fruit a many seeded follicle.

A genus of 2 species occurring in the western Mediterranean countries of North Africa and the Iberian Peninsula. Propagation is from seed.

1a. Flowers pink, swelling towards base; stems slender   **1. hispanica**
  b. Flowers yellow, narrowing towards base; stems robust   **2. brevifolia**

1. P. hispanica (Linnaeus) de Candolle (*Cotyledon hispanica* Linnaeus). Illustration:

Polunin & Smythies, Flowers of South-West Europe, pl. 13 (1973); Valdes et al., Flora vascular de Andalucía Occidental **2**: 8 (1987).

Stems erect, sometimes branched, 4–15 cm. Leaves 1.2–2.5 cm, oblong, obtuse, covered with glandular hairs. Sepals 1.6–2 mm, narrowly triangular, joined towards the base. Corolla 2–2.5 cm, pinkish purple, swelling towards base. *S Spain, Portugal, NW Africa (Morocco).* H5. Summer.

2. P. breviflora Boissier. Illustration: Jacobsen, Lexicon of succulent plants, pl. 119 (1977); Valdes et al., Flora vascular de Andalucía Occidental **2**: 8 (1987).

Stems erect, rarely branched, 5–12 cm. Leaves almost cylindric, 5–10 mm, acute. Sepals 2–3 mm, linear-lanceolate, free, or joined only at the base. Corolla 1.8–2 cm, yellow, narrowly cone-shaped. *S Spain, N W Africa (Morocco).* H5. Summer.

11. VILLADIA Rose
S.G. Knees
Succulent, annual or perennial herbs, sometimes woody at base. Stems solitary or branched, trailing or erect. Leaves alternate, cylindric or nearly so. Flowering-stems leafy, terminal. Flowers in racemes or spikes (Section Villadia); or in flat cymes of several branches (Section Altamiranoa). Sepals 5 equal, erect; petals 5, more or less united; stamens 10; carpels 5, united only at base; styles short and erect.

A genus of 25–30 species of ornamental succulent plants, native of the Americas, from Texas, through Mexico to Peru. Most can be grown out of doors in summer, in much of Europe, although they will need glasshouse protection in winter.

1a. Plant forming short-stemmed mats   2
  b. Plant with long, trailing or erect stems   3
2a. Leaves overlapping, closely adpressed to stem; flowers white   **1. imbricata**
  b. Leaves spreading; flowers yellow   **2. parva**
3a. Stems not more than 25 cm; leaves 10–13 mm; flowers white or pinkish   **3. batesii**
  b. Stems usually 35–50 cm; flowers orange, red, yellow, purple, or if white, leaves only 5–6 mm   4
4a. Leaves 5–6 mm, spreading; flowers white or purplish on stems with 2–5 branches   **4. jurgensii**

b. Leaves 6–25 mm; flowers orange-red or yellowish green in narrow spikes   5

5a. Stems spotted, to 40 cm; flowers orange or reddish, petals hooded
   5. cucullata

b. Stems unspotted, to 50 cm; flowers yellow or greenish yellow, petals not hooded   **6. guatemalensis**

1. V. imbricata Rose (*Altamiranoa ericoides* (Rose) Jacobs). Illustration: National Cactus & Succulent Journal **13**: 76 (1958); Jacobsen, A handbook of succulent plants **2**: 904 (1960).

Mat-forming perennial with few branched stems 2–6 cm. Leaves ovate, overlapping, closely adpressed to stem, 5–6 mm, acute, with fine tubercles beneath. Flowers erect, in short spikes; sepals leaf-like; corolla 3–5 mm, white. *Mexico (Oaxaca)*. G1. Summer.

2. V. parva (Hemsley) Jacobsen (*Sedum parvum* Hemsley; *Altamiranoa parva* (Hemsley) Rose).

Freely branched, mat-forming perennial. Leaves 3–6 mm, linear, spreading around stem. Flowers yellow. *Mexico (San Luis Potosi)*. G1.

3. V. batesii (Hemsley) Baehni & Macbride (*Altamiranoa batesii* Hemsley). Illustration: Hemsley, Biologia Centrali-Americana **5**: pl. 19 (1880); Sanchez Sanchez, La Flore del valle de Mexico, pl. 132 (1968).

Hairless perennial with thick roots and several erect or prostrate stems 15–25 cm, arising from base. Leaves 10–13 × 2–4 mm, linear-lanceolate. Flowering-stems with 2–5 branches. Corolla 2–4 mm wide, white or pinkish. *Mexico*. H5–G1. Summer–autumn.

4. V. jurgensii (Hemsley) Jacobsen (*Cotyledon jurgensii* Hemsley).

Erect or trailing perennial, branching from near the base. Stems woody at base, to 45 cm, with very fine downy hairs. Leaves 5–6 mm, spreading, acute. Flowering-stems with 2–5 branches, each with 2 or 3 flowers. Corolla 5–6 mm wide, white or purplish. *E Mexico*. G1. Summer.

5. V. cucullata Rose.
Tuberous-rooted perennial. Stems spotted, hairless, to 40 cm. Leaves to 2.5 cm, acuminate. Flowers in a narrow spike to 20 cm. Sepals *c.* 2 mm, distinct, green; corolla 2–4 mm, orange or reddish, with finely toothed, hooded petals. *NE Mexico (Coahuila)*. H5–G1. Autumn.

6. V. guatemalensis Rose (*V. laevis* Rose; *Altamiranoa guatemalensis* (Rose) Walther). Illustration: Contributions to the U.S. National Herbarium **20**: pl. 81 (1909); National Cactus & Succulent Journal **13**: 77 (1958); Jacobsen, A handbook of succulent plants **2**: 904 (1960); Jacobsen, Lexicon of succulent plants, pl. 141 (1977).

Hairless perennial, with freely branching stems to 50 cm. Leaves linear, 6–25 mm, acute. Flowers in a narrow spike; corolla yellow or greenish yellow, *c.* 6 mm wide. *S Mexico, Guatemala*. G1. Autumn.

12. ADROMISCHUS Lemaire
F.T. de Vries
Perennial shrublet with usually prostrate, rarely erect, fleshy or somewhat woody branches. Leaves fleshy, spirally arranged, usually clustered. Inflorescence spike-like, usually *c.* 35 cm, sometimes *c.* 45 cm, with clusters of 1–3 erect or spreading flowers. Calyx with 5 triangular, acute lobes; corolla fused into a tube at the base, with 5 spreading lobes. Stamens 10 in 2 whorls, filaments fused to the lower third of the corolla-tube, anthers either protruding or not. Nectary scales 5. Carpels 5, usually free, narrowed into a short style. Seeds ellipsoid with a constriction at the end, and vertical ridges.

A genus of about 30 species from South Africa and Namibia, closely related to *Cotyledon*. The species with purple and dark green leaf-markings are especially popular, but unfortunately the spots are often lost in cultivation. They grow well in loose, peaty, slightly moist soil, preferably in full sun to ensure a compact plant with highly coloured leaves. If kept warm enough (minimum 15 °C) they can flower nearly all year round, though they can tolerate lower temperatures. Propagation is not difficult, and leaves root easily, forming small, soon very attractive plantlets. Seed and cuttings are also used.

The measurements given are those of the plant without inflorescence.

Literature: Tölken, H.R., *Flora of Southern Africa* **14**: 37–60 (1985).

1a. Anthers protruding from corolla-tube; corolla-lobes at least as broad as long   2

b. Anthers not protruding from corolla-tube; corolla-lobes up to 3 times longer than broad   7

2a. Corolla-lobes abruptly narrowed at tip, with wavy, ruffled margins   3

b. Corolla-lobes gradually tapered at tip, with wavy margins   6

3a. Leaves covered with flaking wax; corolla-lobes 2–2.5 mm, sometimes with a few club-shaped hairs in the throat of the tube
   2. hemisphaericus

b. Leaves smooth, or with a very slight bloom; corolla-lobes 1–2 mm, without club-shaped hairs   4

4a. Leaves oblong-oblanceolate, more or less flattened
   1. filicaulis subsp. **filicaulis**

b. Leaves linear-elliptic, rarely linear-lanceolate, terete or almost so   5

5a. Stems erect or prostrate and rarely with fibrous adventitious roots
   1. filicaulis subsp. **filicaulis**

b. Stems prostrate or decumbent and with stilt roots
   1. filicaulis subsp. **marlothii**

6a. Leaves elliptic to rounded with a marked marginal ridge extending to the base   **3. trigynus**

b. Leaves oblanceolate to obovate, with marginal ridge rarely extending beyond the middle or if so then narrow and not horny
   4. umbraticola

7a. Buds cylindric or slightly angular and spreading; club-shaped hairs present on lower parts of lobes and in throat of corolla-tube   8

b. Buds distinctly grooved between petals and until flowering adpressed to central axis; club-shaped hairs found mainly in throat of corolla-tube   11

8a. Leaves terete or almost so, apex gradually tapered to both ends, base abruptly wedge-shaped; margin only horny at tip
   5. mammillaris

b. Leaves flattened, oblanceolate to obovate; horny margin extending right around leaf to the base   9

9a. Leaves evenly and gradually narrowed towards base, oblanceolate, never with spots   10

b. Leaves abruptly narrowed towards the base   **7. maculatus**

10a. Corolla-tube 8–9 mm   **6. maximus**

b. Corolla-tube 10–12 mm
   8. sphenophyllus

11a. Leaves flattened, at least 3 times broader than thick at middle of leaf; leaves with a more or less thick white bloom   12

b. Leaves distinctly concave on both surfaces, about as broad, rarely up to twice as broad as thick at middle of leaf; leaves smooth or with a slight white bloom   13

12a. Corolla-lobes 3–5 mm; leaves oblanceolate to round or almost so; leaves with a thick white bloom
9. leucophyllus

b. Corolla-lobes 1.5–3 mm; leaves obovate, abruptly narrowed to the base; leaves with a thin white bloom   **12. marianae** var. **hallii**

13a. Leaves usually acute, concave towards tip on both surfaces or with a longitudinal groove above; apical gland on each anther raised above pollen-sacs   14

b. Leaves obtuse, truncate and/or flattened towards tip; apical gland on each anther stalkless   17

14a. Roots tuberous   15

b. Roots fibrous   16

15a. Leaves smooth, with marginal ridge horny and usually wavy
12. marianae var. **hallii**

b. Leaves warty (sometimes only visible under × 10 lens), with marginal ridge slightly raised but indistinct, not horny and wavy
12. marianae var. **immaculatus**

16a. Leaves concave above, marginal ridge horny and raised
12. marianae var. **marianae**

b. Leaves convex above (rarely somewhat concave towards tip), marginal ridge never horny and scarcely raised
12. marianae var. **kubusensis**

17a. Club-shaped hairs in throat and on corolla-lobes; stem hairless, occasionally with aerial roots; leaves hairless   **11. cooperi**

b. Club-shaped hairs usually only in corolla-throat; stem densely covered with brown aerial roots or glandular hairs; leaves usually covered with glandular hairs   18

18a. Stems 4–8 cm, without aerial roots, covered with glandular hairs
10. cristatus var. **zeyheri**

b. Stems 2–4 cm, covered with aerial roots   19

19a. Ridge at tip of leaves narrower than broadest point on leaf; inflorescence with glandular hairs
10. cristatus var. **schonlandii**

b. Ridge at tip of leaf constitutes broadest point of leaf; inflorescence hairless, rarely with a few hairs when young   20

20a. Leaf-blade 1–1.5 times longer than breadth of apical ridge, leaves reversed-triangular, usually with distinct leaf-stalk
10. cristatus var. **cristatus**

b. Leaves 2–5 times longer than breadth of apical ridge, reversed-triangular to club-shaped, hairless or nearly so
10. cristatus var. **clavifolius**

1. A. filicaulis (Ecklon & Zeyher) C.A. Smith (*A. kleinioides* C.A. Smith; *Cotyledon mammillaris* J.D. Hooker).

Erect to prostrate, *c.* 35 cm; roots fibrous, often with stiff, adventitious stilt roots. Leaves 2–8 × 0.5–1.5 cm, lanceolate, elliptic to oblong or oblanceolate, usually terete, sometimes flattened, grey-green to greyish brown, sometimes with darker spots; tip acute or obtuse; base abruptly wedge-shaped. Margin horny, slightly expanded. Buds terete, abruptly narrowed towards tip, spreading. Flowers 1–1.3 cm, yellowish green; corolla-lobes 1–2 mm, pale, often mauve-red on mucro, rough, without club-shaped hairs. Anthers protruding from corolla-tube. *South Africa.* G2.

Subsp. **filicaulis** has branches 7–12 mm across, with peeling bark, leaves elliptic to oblong-oblanceolate, rarely lanceolate, often distinctly flattened above; tip obtuse, rarely acute.

Subsp. **marlothii** (Schönland) Tölken (*A. tricolor* C.A. Smith). Branches 3–8 mm across, with flaking bark, leaves lanceolate, rarely elliptic, usually more or less terete; tip acute.

2. A. hemisphaericus (Linnaeus) Lemaire (*A. rotundifolius* (Haworth) C.A. Smith; *A. bolusii* (Schönland) Berger; *Cotyledon crassifolius* Salisbury). Illustration: Kakteenkunde, 17 (1940); Rice & Compton, Wild Flowers of the Cape of Good Hope, pl. 49 (1951); Cactus & Succulent Journal, 51 (1953).

Much branched, low-growing *c.* 30 cm, with fibrous roots. Leaves 1–4.5 × 1–3 cm, oblanceolate to obovate, rarely circular, flattened but often convex on both surfaces, especially beneath, grey-green with often darker spots, and flaking wax; tip obtuse or rounded; base abruptly wedge-shaped. Margin indistinctly horny in upper half, sometimes wavy. Buds terete, abruptly narrowed towards tip, spreading. Corolla-lobes 2–2.5 mm, white or tinged pink, darker around throat and on mucro. Anthers protruding from corolla-tube. *South Africa.* G2.

3. A. trigynus (Burchell) von Poellnitz (*A. rupicolus* C.A. Smith; *A. maculatus* misapplied). Illustration: Bothalia **3**: 645 (1939); National Cactus and Succulent Journal, 35 (1951); Jacobsen, Handbook of succulent plants **1**: 52 (1960); Parey's Blumengärtnerei **1**: 726 (1958).

Short, erect, much branched plant, with fibrous roots. Leaves 1.5–4 × 0.8–3 cm, elliptic to rounded, rarely oblanceolate, flattened, usually concave above, more or less convex beneath, grey-green, usually with darker spots; tip rounded; base abruptly narrowed. Horny margin extending right around leaf. Buds terete, slightly grooved, gradually tapered towards tip, adpressed at first, later erect. Flowers pale yellowish green, thickly bloomed; corolla-lobes 1.5–2.5 mm, acuminate, off-white to tinged pink, rough and with some club-shaped hairs mainly in throat. Anthers protruding from corolla-tube. *South Africa.* G2.

4. A. umbraticola C.A. Smith. Illustration: Bothalia **3**: 646 (1938); Letty, Wildflowers of the Transvaal, 74 (1962); Barkhuizen, Succulents of South Africa, 69 (1978).

More or less branched *c.* 10 cm, with fibrous roots. Leaves 1.5–6.5 cm, oblanceolate, rarely linear-oblanceolate or obovate, flattened, usually convex on both sides, green to grey-green; tip obtuse or rounded; base usually gradually constricted. Margin horny, well developed in the upper half to absent. Buds with longitudinal grooves, gradually tapered towards tip, adpressed at first, later erect. Flowers 1–1.3 cm, pale green, usually tinged pink, bloomed; corolla-lobes 2–2.5 mm, acuminate, off-white to tinged pink, rough and with some club-shaped hairs mainly in throat. Anthers protruding from corolla-tube. *South Africa.* G2.

5. A. mammillaris (Linnaeus filius) Lemaire. Illustration: Rooksby, Desert Plant life, 112 (1938); National Cactus and Succulent Journal, 61 (1948); Jacobsen, Handbook of succulent plants **1**: 48 (1960).

Sparsely branched, creeping, *c.* 15 cm, stems rooting. Leaves 2–5 × 0.5–1.1 cm, linear-lanceolate, slightly flattened to almost terete, grey-green, sometimes tinged brown; tip tapered gradually; base abruptly wedge-shaped, shortly stalked. Horny margin restricted to tip. Buds terete, erect to spreading or curved towards tip. Flowers 1.1–1.3 cm, grey-green; corolla-lobes 4–5 mm, acute, white often tinged pink, mauve along margins, spreading to recurved, rough; lower lobes and throat with club-shaped hairs. Anthers not protruding from corolla-tube. *South Africa.* G2.

6. A. maximus P.C. Hutchison. Illustration: Cactus and Succulent Journal (U.S.) **31**: 131, 132 (1959).

Erect to decumbent, *c.* 20 cm, with fibrous roots. Leaves oblanceolate 6–16 × 2.5–6 cm, flattened, grey-green, spotless; tip usually rounded, rarely obtuse; base wedge-shaped. Margin horny, at least in upper half, sometimes right around leaf. Buds terete, slightly grooved; corolla lobes 2.5–4 mm, acute, white or cream, pale pink tinged, rough; lower lobes and throat with club-shaped hairs. Anthers not protruding. *South Africa.* G2.

7. A. maculatus (Salm-Dyck) Lemaire (*A. mucronatus* Lemaire; *Cotyledon hemisphaerica* Harvey; *A. trigynus* misapplied). Illustration: Bothalia **3**: 621 (1938); Barkhuizen, Succulents of South Africa, 68 (1978); Flowering Plants of Africa **45**: t. 1776 (1978).

Sparsely branched, prostrate, *c.* 15 cm, somewhat woody, with fibrous roots. Leaves 2.5–10 × 1.5–4 cm, obovate or spathulate to oblanceolate, flattened, green to greyish green, with or without purple spots; tip obtuse, often with a mucro or notched; base abruptly wedge-shaped, margin horny all round the leaf. Buds terete, gradually tapered towards tip, spreading. Flowers 8–11 mm, pale yellowish green; corolla-lobes 2.5–5 mm, acute, white or tinged pale pink, mauve along margins, spreading to recurved, rough; lower lobes and throat with club-shaped hairs. Anthers not protruding from corolla-tube. *South Africa.* G2.

8. A. sphenophyllus C.A. Smith (*A. rhombifolius* misapplied). Illustration: National Cactus and Succulent Journal (U.S.) **17**: 150, 151 (1945).

Little branched, to 20 cm, with fibrous roots. Leaves 2.5–6.5 × 1–3.5 cm, oblanceolate, gradually tapered towards base, flattened, grey-green without spots; tip rounded to obtuse, mucronate; base wedge-shaped, margin horny, extending right around the leaf, forming a straight line from broadest point to base. Buds terete, gradually tapered towards tip, spreading. Flowers 9–12 mm, pale green, sometimes tinged red; corolla-lobes 2.5–4 mm, acute, white often tinged pink, with mauve margin,

rough; lower lobes and throat with club-shaped hairs. Anthers not protruding from corolla-tube. *South Africa.* G2.

9. A. leucophyllus Uitewaal. Illustration: Cactus and Succulent Journal (U.S.) **32**: 136 (1960); National Cactus and Succulent journal **9**: 58–59 (1954).

Much branched, somewhat erect to decumbent *c.* 10 cm, with fibrous roots. Leaves 1.5–4 × 1.5–3 cm, oblanceolate to obovate or circular, flattened, whitish green, with a thick white bloom; tip usually rounded; base abruptly wedge-shaped. Margin distinctly horny, extending right around the leaf, sometimes darker. Buds terete, slightly grooved, gradually tapered towards tip, erect at first, later spreading. Flowers 1.1–1.3 cm; corolla-lobes 3–5 mm, triangular, white with a pink median band, rough and with club-shaped hairs mainly in throat. Anthers not protruding from corolla-tube. *South Africa.* G2.

10. A. cristatus (Haworth) Lemaire. Illustration: Roeder, Sukkulenten, edn 2, **9**: 3 (1931); Jacobsen, Handbook of succulent plants **1**: 48 (1960); Rauh, Die grossartige Welt der Sukkulenten, 69 (1967).

Erect *c.* 8 cm, with fibrous roots at base, and glandular hairs on stem. Leaves 1.5–5 × 0.5–2 cm, reversed-triangular to oblong-elliptic, terete to somewhat flattened, green to grey-green; tip truncate or rounded to more or less broadened and crisped; base wedge-shaped, sometimes stalked. Margin in upper half of the leaf horny, wavy, often darker. Buds terete, slightly grooved, gradually tapered towards tip, erect at first, later spreading. Flowers 1–1.2 cm, grey-green; corolla-lobes 2--3.5 mm, ovate-triangular, white tinged pink, with darker margin, spreading or recurved, rough and with club-shaped hairs mainly in throat. Anthers not protruding from corolla-tube. *South Africa.* G2.

Var. **cristatus** is much branched, *c.* 4 cm, has red aerial roots and red hairs on the stem, and reversed-triangular leaves which have much broadened margins at the tip.

Var. **clavifolius** (Haworth) Tölken (*A. clavifolius* Haworth; *A. poellnitzianus* Werdermann; *Cotyledon cristata* Harvey). Illustration: National Cactus & Succulent Journal, 33 (1952); Andersohn, Cacti and succulents, 263 (1988). Much branched, *c.* 4 cm, covered with aerial roots, and has reversed-triangular to club-shaped leaves, broadened at the tip, narrowed towards base, with a wavy margin about as broad

as leaf. Inflorescence hairless with a waxy bloom.

Var. **schonlandii** (Phillips) Tölken. A brittle, much branched plant, *c.* 4 cm, with aerial roots and narrowly triangular to oblong-lanceolate leaves with a horny, wavy margin, narrower than the leaf. Inflorescence with glandular hairs.

Var. **zeyheri** (Harvey) Tölken is little branched, *c.* 8 cm, has no aerial roots, but glandular hairs on the stem, and broadly triangular leaves with much broadened, horny, wavy margins. Inflorescence with glandular hairs.

11. A. cooperi (Baker) Berger (*A. festivus* C.A. Smith; *A. pachylophus* C.A. Smith). Illustration: Bothalia **3**: 634 (1938); Barkhuizen, Succulents of Southern Africa, 66 (1978); Flowering plants of Africa **45**: t. 1849 (1982).

Much branched, erect *c.* 10 cm, with thick fibrous roots and sometimes aerial roots. Leaves 2–9 × 1–3 cm, oblanceolate-oblong to spathulate, terete to somewhat flattened, green to grey-green, with or without purple spots; tip obtuse to truncate; base wedge-shaped to stalked. Margin ridged at the tip, wavy and wider than remainder of leaf. Buds terete, slightly grooved, gradually tapered towards tip, erect at first, later spreading. Flowers 9–11 mm, bluish green, glaucous; corolla-lobes 3–4.5 mm, sharply acute, spreading or recurved, pale pink with a thick bloom, wine-red towards margins, rough and with club-shaped hairs mainly in throat. Anthers not protruding from corolla-tube. *South Africa.* G2.

12. A. marianae (Marloth) Berger. Illustration: Rauh, Die grossartige Welt der Sukkulenten, pl. 69 (1967); Barkhuizen, Succulents of Southern Africa, 69 (1978); Court, Succulent Flora of southern Africa, 65 (1981).

Erect, rarely decumbent *c.* 8 cm, with often tuberous roots. Leaves 2–12 × 0.4–2.5 cm, linear-lanceolate to elliptic, rarely obovate, almost terete with a more or less pronounced central groove above, grey-green to greyish brown, rarely with bloom or spots; tip acute or truncate; base wedge-shaped. Margin at least in places horny. Buds terete, slightly grooved, gradually tapered towards tip, erect at first, later spreading. Flowers 1–1.2 cm, pale pink to white, with a thick bloom; corolla-lobes 2–3 mm, acute, spreading or recurved, white with purple margins, rough and with club-shaped hairs mainly in throat. Anthers not protruding from corolla-tube. *South Africa, SW Namibia*. G2.

Var. **marianae**. To 10 cm, sparsely branched, clump-forming, with thick old fibrous roots; leaves fairly narrow, oblanceolate to elliptic, rarely lanceolate, concave above, often more or less spotted, with a horny margin, usually distinct from tip to middle of leaf.

Var. **hallii** (Hutchison) Tölken (*A. casmithianus* von Poellnitz). Illustration: Cactus and Succulent Journal (U.S.) **28**: 144–147 (1956). Much branched with thick stems and thick tuberous roots; leaves 1.5–2.5 cm, obovate to almost circular, rarely oblanceolate, flattened, slightly concave above, rarely faint reddish-spotted, usually bloomed, with a raised horny, often brown and wavy margin at the tip.

Var. **immaculatus** Uitewaal (*Cotyledon herrei* Barker; *A. antidorcadum* von Poellnitz; *A. alveolatus* Hutchison). Illustration: Rauh, Die grossartige Welt der Sukkulenten, pl. 69 (1967); Barkhuizen, Succulents of Southern Africa, 69 (1978); Court, Succulent Flora of southern Africa, 94 (1981). Much branched, small, with thick stems, often constricted at base; leaves oblanceolate to elliptic, rarely obovate or slightly concave above and below, more or less warty, without purple spots, with a brown or white, raised but not horny margin.

Var. **kubusensis** (Uitewaal) Tölken (*A. blosianus* P.C. Hutchison; *A. geyeri* P.C. Hutchison). Illustration: Cactus and Succulent Journal (U.S.) **29**: 36, 37 (1957); **32**: 89, 91 (1960). Much branched, with thick stems, rarely continued in thick fibrous roots; leaves 3–9 cm, oblanceolate, rarely elliptic, somewhat concave on both sides to terete, smooth and rarely purple-spotted, with an often brown, scarcely raised and not horny margin.

13. UMBILICUS de Candolle

S.G. Knees

Perennial, succulent herbs with new shoots arising annually from scaly rhizomes or tubers. Leaves alternate, hairless; stalked at base, becoming stalkless. Flowering-stems terminal, usually solitary, but occasionally branched above into several racemes. Sepals 5, free, about half as long as corolla. Petals 5, united into a bell-shaped corolla; lobes more or less erect. Stamens 10, rarely 5; styles short or absent. Fruit a slender follicle; seeds numerous.

A genus of about 18 species, chiefly in Europe, W Asia, N Africa and the Atlantic

Islands. Propagation is usually from seed. Plants are hardy but are often best kept in a cool greenhouse or cold-frame during winter months.

1a. Racemes occupying more than half the total stem; flowers nodding   **2**
  b. Racemes occupying less than half the total stem; flowers erect or horizontal   **3**
2a. Racemes 15–30 cm; flowers greenish yellow, spotted red   **1. rupestris**
  b. Racemes 10–16 cm; flowers yellowish white   **2. intermedius**
3a. Flowers horizontal, 4.5–6 mm   **3. horizontalis**
  b. Flowers erect, 9–14 mm   **4. erectus**

1. U. rupestris (Salisbury) Dandy (*Cotyledon rupestris* Salisbury; *C. tuberosa* (Linnaeus) Halaçsy; *Umbilicus pendulinus* de Candolle). Illustration: Polunin, Flowers of Europe, pl. 39 (1969); Phillips, Wild flowers of Britain, 73 (1977); Pignatti, Flora d'Italia **1**: 489 (1982); Blamey & Grey-Wilson, Illustrated flora of Britain and northern Europe, 165 (1989).

Erect, tuberous perennial; tuber rounded, 1–3 cm across, usually covered with a web of fine roots. Stem 20–50 cm. Basal leaves almost circular, peltate, blade shallowly funnel-shaped, 3–10 cm across; margin coarsely scalloped; stalk to 15 cm. Upper stem-leaves becoming progressively smaller and very shortly stalked, not peltate. Flowers pendent, numerous in bracteate, long, cylindric racemes, 15–30 cm. Corolla 5–8 mm, greenish yellow, spotted or streaked with red. Fruit 5–6 mm, a narrowly boat-shaped, membranous follicle; seeds dark brown *c.* 0.5 mm. *Europe, W Asia, N Africa, Azores, Madeira.* Spring–early summer. H5.

2. U. intermedius Boissier. Illustration: Bouloumoy, Flore du Liban & Syrie. Atlas, t. 64 (1930).

Like *U. rupestris* but lowest leaves not very succulent. Flowers 6–8 mm, yellowish white, in dense racemes 10–16 cm. *Bulgaria, Cyprus, Turkey, Israel, Egypt*. G1. Spring–summer.

3. U. horizontalis (Gussone) de Candolle (*Cotyledon horizontalis* Gussone). Illustration: Pignatti, Flora d'Italia **1**: 489 (1982); Valdes et al., Flora vascular de Andalucía occidental **2**: 7 (1987).

Like *U. rupestris*, but leaves diminishing abruptly; upper stem covered with scale-like, linear-triangular bracts, *c.* 2 × 0.5

cm or less. Flowers usually horizontal in very congested, unbranched racemes, 10–25 cm. Corolla 4.5–6 mm, yellowish brown. Fruit dark reddish, *c.* 4.5 mm. *Mediterranean area, Azores*. H5–G1. Spring–summer.

4. U. erectus de Candolle (*Cotyledon umbilicus-veneris* Linnaeus; *Cotyledon lutea* Hudson). Illustration: Bouloumoy, Flore Liban & Syrie. Atlas, t. 64 (1930); Pignatti, Flora d'Italia **1**: 489 (1982).

Stout erect perennial with a tuberous root; stems simple, 30–70 cm. Leaves peltate, 3–7 cm across, stem-leaves ovate-triangular to linear, margins toothed. Flowers more or less erect in simple or, more rarely, branched racemes, 8–25 cm. Corolla tubular 9–14 mm, yellowish. Fruit 5–6 mm. *S Europe, Middle East, NW Africa.* G1. Spring–summer.

14. TYLECODON Tölken

G.D. Rowley

Small to large perennial shrublets with greatly swollen, soft, succulent stem bases and few proportionally thick, sometimes tuberculate branches. Leaves spirally arranged near the stem tips, simple, entire, stalkless, soft and fleshy, early deciduous. Inflorescence a tall, erect scape more or less branched into a loose head of usually erect tubular flowers. Bracts few, narrow, scattered. Sepals 5, green. Petals 5, united into a long tube with recurving lobes. Stamens 10 in 2 whorls, arising from the corolla-tube and usually longer than it. Nectar scales 5, free. Seeds ellipsoid, ridged, constricted at one end. The soft, fleshy stems are poisonous to cattle.

A segregate genus from *Cotyledon*, for which over 30 species have been described, but these variable in both vegetative and floral characters and the number could almost certainly be reduced by a more conservative treatment. The spiralled, deciduous leaves and mostly erect flowers most readily distinguish the genus from *Cotyledon*. They grow in the Cape Province of South Africa, extending into Namibia. The miniatures typically inhabit rock crevices and are difficult to see when dormant and leafless. The taller species (notably *T. paniculatus*) are a conspicuous feature of many landscapes, the massive conical trunk with peeling papery bark resembling that of *Cyphostemma* and the Mexican tree sedums. All species appeal to devotees of succulents. As extreme xerophytes they require a dry rest when leafless, a porous, well-drained soil and rather sparing watering at any time. Full sun is preferred. In winter the temperature should not be allowed to fall below 4 °C for any length of time. Propagation is mainly by seed, although the thinner-stemmed cuttings of some can be rooted.

Literature: Tölken, H.R., *Flora of Southern Africa* **14**: 19–35 (1985).

1a. Plant covered in the dry, stiff remains of persistent inflorescences like weak thorns   **5. reticulatus**
  b. Plant not covered in persistent thorny inflorescences   2
2a. Leaves flat, obovate-spathulate
                                2. paniculatus
  b. Leaves terete or semi-terete   3
3a. Stem covered in persistent leaf-bases forming hard tubercles 6 mm or more   4
  b. Stem without elevated tubercles   6
4a. Flowers tubular, bright yellow, erect, to 2.5 cm
          3. papillaris subsp. **papillaris**
  b. Flowers more or less bell-shaped, not bright yellow, pendent or almost erect, to 1.5 cm   5
5a. Inflorescence glandular
          3. papillaris subsp. **wallichii**
  b. Inflorescence not glandular
          3. papillaris subsp. **ecklonianus**
6a. Leaves rarely produced and then short-lived, not in rosettes and confined to a few branches; other branches with minute scales only
                              1. buchholzianus
  b. All branches at some time bearing leaves in small rosettes
                                4. pearsonii

1. T. buchholzianus (Schuldt & Stephan) Tölken (*Cotyledon buchholziana* Schuldt & Stephan). Illustration: Lamb, Illustrated Reference on cacti and other succulents **3**: 758 (1963); Tölken, Flowering plants of Africa, t. 1774 (1978); Rauh, Wonderful world of succulents, t. 59/1 (1984); Rowley, Caudiciform and pachycaul succulents, 177 (1987).

Compact stem-succulent shrublet made up of numerous, closely packed, erect, fleshy, cylindric, brittle shoots, 6–10 mm across, grey and slightly rough from numerous small, spiralled black leaf-bases. Leaves few and erratically appearing, often at long intervals, scattered, terete, short-lived, greyish green, fusiform, 8–15 × 2–4 mm, ascending. Scape thin and wiry, to 10 cm, more or less persistent, with 1 to few erect red flowers 1–1.2 cm. Sepals 1.5–2.5 mm. Corolla-tube cylindric. *S Africa (NW Cape Province to Namibia)*. G1. Winter.

This curious plant can grow for years without producing leaves, the shoots elongating and bearing minute scales only.

2. T. paniculatus (Linnaeus filius) Tölken (*Cotyledon paniculata* Linnaeus filius; *Cotyledon fascicularis* Aiton). Illustration: Botanical Magazine, 5602 (1866); Graf, Tropica, 366 (1978); Rauh, Wonderful world of succulents, t. 59/3 & 4 (1984); Rowley, Caudiciform and pachycaul succulents, 19, 176, 235 (1987).

Stem succulent to 1.6 m, with a swollen soft trunk to 60 cm across at the base, branching above into a few ascending shoots, 1.5–2 cm across, covered in yellowish peeling papery bark. Leaves obovate-spathulate, 6–12 × 3–6 cm, tapered to the base, rounded at the apex, fleshy and brittle, bright green, more or less glandular downy. Scape terminal, in cultivation to 60 cm, branched, with monochasia of scattered nodding flowers, 1.2–1.6 cm. Sepals 4–6 mm. Petals dark red with yellowish stripes. *South Africa (W Cape Province), Namibia*. Spring.

3. T. papillaris (Linnaeus) Rowley (*T. cacalioides* (Linnaeus filius) Tölken; *T. wallichii* (Harvey) Tölken; *Cacalia papillaris* Linnaeus; *Cotyledon eckloniana* Harvey). Illustration: van Laren, Succulents other than cacti, 72 (1934); Graf, Tropica, 365 (1978); Rauh, Wonderful world of succulents, t. 59/5, 7 (1984); Rowley, Caudiciform and pachycaul succulents, 178, 179 (1987).

Shrub to 80 cm, with thick succulent branches covered in spirally arranged tubercles formed from the persistent leaf-bases 6–13 × 3–4 mm, with an obliquely truncate sharp apex. Bark silvery grey. Leaves 5–10 × 0.2–0.4 cm, linear, terete, hairless, short-lived, pale green, dying back from the tip. Scape terminal, to 60 cm, green parts all covered in glands (glands absent in subsp. **ecklonianus** (Harvey) Rowley), branching near the top, with numerous pendent bright yellow flowers 1.7–2.5 cm (dull greenish yellow and not above 1.3 cm in subsp. **wallichii** (Harvey) Rowley). Sepals 4–12 mm. Corolla-tube cylindric, urn-shaped or cupped. *South Africa (W Cape, Karoo)*. G1. Winter.

The natural hybrid with *T. paniculatus*, with tapering stems covered in low tubercles, **T. × fergusoniae** (L. Bolus) Rowley, is also in cultivation.

4. T. pearsonii (Schönland) Tölken (*Cotyledon luteosquamata* von Poellnitz). Illustration: Jacobsen, Handbook of

succulent plants **1**: t. 286 (1960); Lamb, Illustrated reference on cacti and other succulents **4**: 1069 (1966); Rowley, Caudiciform and pachycaul succulents, 177, 261 (1987).

Small, much branched, stem-succulent shrublet with finger-like shoots covered in leaf-scars arising from a much swollen base. Leaves terete with a groove down the upperside, 2–4 cm × 2–4 mm, soft and pliable, smooth or finely downy, grey-green. Scape 6–10 cm, the main axis more or less persistent, sparingly branched at the top, bearing a few whitish, red-striped flowers which are pendent at first but become erect as the buds open. Sepals glandular, 5–6 mm. Corolla-tube urn-shaped, 1.2–1.4 cm. *South Africa (NW Cape) to Namibia*. G1. Winter.

T. pygmaeus (Barker) Tölken is much reduced, with gnarled stems only a few cm high and *c*. 5 mm across, with papillose leaves 2.5 × 0.6–1.2 cm.

T. schaeferianus (Dinter) Tölken (*Cotyledon sinus-alexandri* von Poellnitz). Illustration: Rowley, Caudiciform and pachycaul succulents, 177 (1987). Another curious miniature, much in demand with collectors, having smooth club-shaped leaves and relatively large erect glandless pinkish white flowers.

5. T. reticulatus (Linnaeus filius) Tölken (*Cotyledon reticulata* Linnaeus filius). Illustration: van Laren, Succulents other than cacti, 63 (1934); Jacobsen, Lexicon of succulent plants, t. 40/2 (1974); Rauh, Wonderful world of succulents, t. 59/2 (1984); Rowley, Caudiciform and pachycaul succulents, 176, 262 (1987).

Compact highly stem-succulent shrublet with many stout branches forming a gnarled, roughly hemispherical dome *c*. 30 cm across. Bark greyish, papery, peeling. Leaves almost cylindric, flattened or furrowed above, 2–4.5 cm × *c*. 6 mm, soft and flexible, green downy or smooth. Scape slender, wiry, with a much branched dichasium, the whole persisting to form a dense tangle of weak, silvery thorns enveloping the plant. Flowers erect, 7–15 mm, yellowish green. Corolla-tube cylindric or urn-shaped. *South Africa (W Cape, Karoo) Namibia*. G1. Winter.

15. CHIASTOPHYLLUM Berger
P.G. Barnes
Fleshy hairless evergreen perennial with prostrate or ascending stems often rooting from the lower nodes. Leaves opposite, broadly elliptic or rounded, scalloped,

short-stalked. Inflorescence a loose panicle of arching spike-like branches with small linear bracts and almost stalkless yellow flowers. Calyx-lobes 5, narrowly oblong; petals 5, elliptic, erect, rather fleshy. Stamens 10, filaments attached at base of corolla-lobes. Carpels 5, each with a small scale at the base. Fruiting-head consisting of 5 erect oblong follicles.

A genus of one species. A hardy plant for the rock garden, requiring a moist but well-drained soil and a sunny position. Propagation by cuttings or by division.

1. C. oppositifolium (Ledebour) Berger (*Cotyledon oppositifolia* Ledebour; *Cotyledon simplicifolia* invalid). Illustration: Botanical Magazine, 8822 (1919); RHS dictionary of gardening **1**: 459 (1956); Hay & Synge, Dictionary of garden plants in colour, f. 41 (1969); Graf, Exotica series 4, edn 12, **1**: 905 (1985).

Stems to 25 cm. Leaves 2.5–4 × 1.5–4 cm, stalk 5–10 mm. Flowers 5–7 mm. *Caucasus*. H4. Summer.

A cultivar with variegated foliage has recently been introduced as 'Jim's Pride'.

16. COTYLEDON Linnaeus
G.D. Rowley

Perennial, evergreen, leaf-succulent shrublets with fleshy or later soft woody branches. Leaves simple, entire or scalloped at the apex, thick and soft, opposite in crossed pairs, free, stalkless or nearly so. Inflorescence terminal, an umbrella-like cluster of usually pendent tubular flowers at the top of a long, almost bractless, erect scape; reduced to a few or even a solitary flower in the smallest species. Sepals 5, green. Petals 5 (as distinct from 4 in the otherwise similar *Kalanchoe*), united into a long tube with recurving lobes. Stamens 10 in two whorls, arising from the corolla-tube, and projecting beyond it. Nectary more or less cup-like. Seeds ellipsoid, ridged, constricted at one end.

Linnaeus's genus contained species now classified as *Adromischus*, *Echeveria*, *Tylecodon*, *Umbilicus* and other segregate genera, many of whose species have synonyms under *Cotyledon*. As delimited here, the genus has about 9 species from S Africa extending to the tropics, Ethiopia and Arabia. Most are popular with collectors of succulents and some, notably the different clones of the extremely variable *C. orbiculata*, have long been favoured as conservatory and house plants. They thrive in a porous but

nutritious compost with sharp drainage and ample water in summer but only enough in winter to keep the soil from becoming dust dry. In the dormant state some leaf-drop is inevitable. A winter minimum of 4 °C suits them, and most will not suffer from an occasional drop to zero. Propagation is easy from cuttings, and single fallen leaves often root and grow plantlets.

Literature: Tölken, H.R., *Flora of Southern Africa* **14**: 3–17 (1985).

1a. Corolla-lobes bright yellow, spreading, at least twice as long as the tube   **2. campanulata**
　b. Corolla-lobes not bright yellow, recurved, not longer than tube   2
2a. Leaves hairy, scalloped with small round apical teeth   **4. tomentosa**
　b. Leaves hairless, or if hairy then not toothed   3
3a. Corolla-tube bulging into pouches between sepals   **1. barbeyi**
　b. Corolla-tube cylindric or only gradually widened below   4
4a. Corolla-tube 8–16 mm, or if longer, then bracts several   **3. orbiculata**
　b. Corolla-tube 2–2.5 cm; bracts 1 or 2 pairs
　　　　3. orbiculata var. **oblonga**

1. C. barbeyi Baker (*C. wickensii* Schönland). Illustration: Jacobsen, Handbook of succulent plants **1**: t. 273 (1960); National Cactus and Succulent Journal **35**: 88 (1980); Court, Succulent flora of South Africa, 95 (1981); Tölken, Flora of Southern Africa **14**: f. 1A (1985).

Stiffly erect shrubs to 2 m, with numerous thick fleshy branches covered in semi-circular leaf-scars. Leaves very variable in size, shape and thickness from flat and oblanceolate to nearly terete, 6–12 × 2–4.5 cm, smooth or slightly glandular, grey-green. Scape 20–60 cm with a head of more or less pendent flowers 1–3 cm. Sepals 8–10 mm, glandular. Petals orange to deep red, united into a tube which bulges out into nectar-pouches between the 5 sepals; lobes recurved. *Transvaal through E Africa to Arabia*. G1. Spring.

2. C. campanulata Marloth (*C. teretifolia* Thunberg). Illustration: Botanical Magazine, 6235 (1876); van Laren, Succulents other than cacti, 72 (1934); Jacobsen, Lexicon of succulent plants, pl. 41 (1974).

Sprawling downy shrublets with thick, fleshy, decumbent stems eventually becoming 20 cm or more, branching

basally. Leaves nearly terete, slightly wider midway and usually channelled on the upper side, 4–12 × 0.8–1.5 cm, soft and very succulent, yellowish green, covered in glandular hairs, with a distinctly flattened, obtuse reddish apex. Scape 20–40 cm, with a cluster of more or less pendent, showy, bright yellow flowers about 4 cm across. Sepals 3–5 mm, glandular. Petals united for about a third of their length. *South Africa (E Cape Province)*. G1. Early summer.

3. C. orbiculata Linnaeus (*C. ausana* Dinter; *C. decussata* Sims). Illustration: Jacobsen, Lexicon of succulent plants, pl. 39 (1974); Graf, Tropica, 361, 363–365 (1978); Rowley, Illustrated encyclopedia of succulents, 123 (1978); Rauh, Wonderful world of succulents, pl. 70 (1984).

Shrub to 1 m, with thick, brittle branches, exceedingly variable in stature, leaf size and shape. Leaves typically longer than broad, 3.5–8 cm or longer, to 4.5 cm across; margin entire, notched or even incised, sometimes wavy; white- to grey-bloomed, usually hairless, often with a red margin. Scape 25–40 cm, with few, pendent, orange to pinkish red long-tubed flowers. Corolla-tube more or less cylindric, 8–16 mm, with shorter recurved lobes. *South Africa (Cape Province), Namibia & Angola*. G1. Spring.

The commonest species in cultivation since its introduction in 1690, and the most variable, with 54 synonyms according to Tölken (1985). The most popular variants with growers are the miniatures, especially those with scalloped wavy-edged leaves or egg-shaped leaves (var. **oophylla** Dinter and var. **dinteri** Jacobsen).

Var. **oblonga** (Haworth) de Candolle (*C. undulata* Haworth; *C. coruscans* Haworth; *C. whitei* Schönland & E.G. Baker). Illustration: van Laren, Succulents other than cacti, 67 (1934); Jacobsen, Lexicon of succulent plants, pl. 41 (1974); Rauh, Wonderful world of succulents, pl. 70 (1984). Vigorous, showy-flowered, decumbent shrublet with branches 1–1.8 cm across, bearing narrow obovate to oblanceolate closely packed leaves. Scape with 1 or less frequently 2 pairs of bracts. Corolla-tube 2–2.5 cm.

C. papillaris Linnaeus filius (not to be confused with *Tylecodon papillaris*). Dwarf, clump-forming, with narrow oblong biconvex acuminate leaves 2–3 × 0.6–1 cm, glaucous grey-green, flushed and margined with red. Scape 10 cm or more, with few, pendent, greenish to yellowish red flowers.

C. woodii Schönland & E.G. Baker. An attractive, free-flowering, small-growing species with flat, red-edged, obovate, acuminate leaves and scape bearing 1–3 pendent orange to red flowers.

4. C. tomentosa Harvey (*C. ladismithensis* von Poellnitz). Illustration: Jacobsen, Lexicon of succulent plants, pl. 39, 41 (1974); Graf, Tropica, 363 (1978); Court, Succulent flora of South Africa, 94 (1981); Rauh, Wonderful world of succulents, pl. 70 (1984).

Compact shrublet to 30 cm or more, with many thin branches. Leaves obovate oblong, tapered to a short stalk at the base, very plump and soft, 1.5–5.5 × 0.8–1.5 cm, with a more or less scalloped apex with up to 9 teeth, densely felted with hairs. Scape 10–20 cm with around 10 nodding red flowers. Sepals densely felted. Corolla-tube 1.2–1.6 cm with recurved lobes half as long. *South Africa (Cape Province: Ladismith area)*. G1. Spring.

17. GRAPTOPETALUM Rose
S.G. Knees

Evergreen perennial herbs and shrubs. Leaves alternate, mostly in rosettes. Flowers shortly stalked in 1–several alternating cymes, borne in axillary clusters. Sepals 5, almost equal, closely adpressed to corolla; petals 5–7, united below, overlapping in bud, stamens 5–10, outcurved; carpels erect, mostly short-styled.

About 12 species of succulent perennials from the southern United States and Mexico. Closely allied to *Echeveria*, but petals spreading from the middle, with red dots or blotches, often in more or less transverse rows.

1a. Leaves glossy, dark green   2
  b. Leaves covered with a matt greyish or slight purplish bloom   3
2a. Leaves with a bristle-tip to 12 mm; margins with a white wing
      1. filiferum
  b. Leaves mucronate, margins reddish
      2. bellum
3a. Leaves distinctly club-shaped
      3. pachyphyllum
  b. Leaves linear to spathulate with a distinct point   4
4a. Plants shrubby with distinct stems
      5
  b. Plants herbaceous, often in stemless clusters   7
5a. Shrublet *c.* 15 cm; leaves with distinct purplish bloom, in rosettes of 12–15   **4. amethystinum**

  b. Shrubs 0.4–2 m; leaves with a greyish bloom, in rosettes of 15–30
      6
6a. Leaves 2–4 cm, with abruptly pointed tips, pale green or reddish
      5. fruticosum
  b. Leaves 5–9 cm, rounded at tips, yellowish   **6. grande**
7a. Leaves greyish red; flowers dark red with greenish bands   **7. rusbyi**
  b. Leaves pale lavender, blue-grey or yellowish; flowers white or yellowish with red markings   8
8a. Erect or trailing herb with stems to 30 cm; leaves pale lavender grey
      8. paraguayense
  b. Almost stemless herb, producing many stolons; leaves greyish blue
      9. macdougallii

1. G. filiferum (S. Watson) Whitehead (*Sedum filiferum* S. Watson). Illustration: Rauh, Schöne Kakteen und andere sukkulenten, 223 (1978); Riha & Subik, Illustrated encyclopedia of cacti and other succulents, 276 (1981).

Stemless perennial with rosettes of 50–300 leaves, to 6 cm across. Leaves 1.2–5 × 0.8–1.2 cm, spoon-shaped, shiny green with white-winged edges and a bristle-tip to 12 mm. The flowers are borne on stems 5–10 cm, with 2–5 branches, each with 2–5 flowers. Petals 4–7 mm, white, spotted red. *NW Mexico*. G1. Spring–early summer.

2. G. bellum (Moran & Meyran) D.R. Hunt (*Tacitus bellus* Moran & Meyran). Illustration: Rowley, Illustrated encyclopedia of succulents, 128 (1978); Pacific horticulture **39**(4): 36 (1978); Botanical Magazine n.s., 781 (1979); Brickell, RHS gardeners' encyclopedia of plants and flowers, 1398 (1989).

Hairless, succulent perennial herb, with 25–50 leaves in basal rosettes, 3–8 cm across. Leaves 2–3.5 × 1.5–2.8 cm, obtuse, mucronate, dark grey-green, margin reddish brown. Flower-stalks axillary, with 1–3 branches terminating in large, odourless, open flowers to 3.5 cm across. Flower-stalk 1–2 cm, sepals equal, reflexed; petals 6–10 mm across, overlapping in bud, ovate to elliptic, bright crimson-red; stamens 10, filaments red, anthers bright yellow; carpels 5, free to base, 1–1.7 cm; style *c.* 4 mm. *Mexico (Chihuahua)*. G1. Late spring–early summer.

3. G. pachyphyllum Rose. Illustration: Addisonia **7**: 247 (1922); Lamb, Pocket encyclopedia of cacti and succulents in

colour, 261 (1974); Everett, New York Botanic Garden encyclopedia of horticulture **5**: 1532 (1981).

Perennial herb with trailing stems to 20 cm, branched at the base. Leaves in rosettes of 20–50, club-shaped, 1.2–1.8 cm, bluish green, tipped red, covered with a greyish bloom. Flowering-stems 2.5–10 cm, with 1–4 branches, each with 2–5 flowers. Petals *c.* 11 mm, creamy yellow with few red spots. *Mexico*. G1. Spring–summer.

4. G. amethystinum (Rose) Walther (*Pachyphytum amethystinum* Rose). Illustration: Everett, New York Botanic Garden encyclopedia of horticulture **5**: 1532 (1981); Riha & Subik, Illustrated encyclopedia of cacti and other succulents, 274 (1981).

Shrublet or perennial herb to *c.* 15 cm. Leaves in loose rosettes of 12–15, 5–7.5 × 2–3.5 cm, ovate, to 2 cm thick, bluish grey to green and covered with a purplish bloom. Flowering-stems 15–17 cm, with 3–10 branches, each with 3–6 flowers. Petals *c.* 11 mm, greenish yellow with bands of red dots. *W Mexico*. G1. Spring–early summer.

5. G. fruticosum Moran. Illustration: Cactus & Succulent Journal (U.S.) **40**(4): 152–154 (1968).

Shrub to 40 cm, branching from the base, with smooth silvery bark. Leaves 20–30, borne in loose rosettes, 3–7 cm across; spathulate to rhombic, 2–4 × 0.8–2 cm, pale green or reddish, with a slight grey bloom, abruptly pointed at tip. Flowering-stems 12–30 cm, with 5–11, zig-zagged branches, each with 12–50 flowers; petals 7–9 mm, pale yellow with red bands towards tip. *Mexico (Jalisco)*. G1. Spring–summer.

6. G. grande Alexander. Illustration: Cactus & Succulent Journal (U.S.) **28**: 174–175 (1956).

Large shrub to 2 m, bark grey-green, eventually fissured. Leaves 15–25, rounded, 5–9 × 2–4 cm, and *c.* 5 mm thick, wedge-shaped to spathulate, yellowish with a bluish grey bloom. Flowering-stems to 60 cm, with 8–15 zig-zagged or coiled branches. Flowers to 2.5 cm across; petals *c.* 11 mm, yellow, with red dots; stamens greenish white, anthers red-brown; carpels *c.* 6 mm. *S Mexico*. G1. Winter–spring.

7. G. rusbyi (Greene) Rose (*Cotyledon rusbyi* Greene). Illustration: Addisonia

9: 304 (1924); Lamb, Colourful cacti of the American deserts, 101 (1974); Japan Succulent Society, Colour encyclopaedia of succulents, 60 (1981).

Stemless perennial with rosettes of 10–35 leaves, 1.5–5 × 0.3–1.2 cm, linear to spathulate, the surface covered in rough papillae, greyish red or green. Flowering stems 7–12 cm, with 2 or 3 branches, each with 3–7 flowers. Petals 5–7, *c.* 8 mm, dark red with greenish bands. *N Mexico, USA (Arizona)*. G1. Spring–summer.

8. G. paraguayense (N. E. Brown) Walther. Illustration: Graf, Tropica, 374 (1978); Everett, New York Botanic Garden encyclopedia of horticulture **5**: 1532 (1981); Cactus & Succulent Journal (U.S.) **58**: 48, 49, 53 (1986); Brickell, RHS gardeners' encyclopedia of plants and flowers, 1398 (1989).

Hairless perennial herb to 30 cm. Leaves 2–8 × 1.5–2.5 cm, in loose rosettes of 15–25, flat or hollowed above, keeled beneath, pale lavender-grey, spathulate. Flowers borne on stems to 15 cm with 2–6 branches, each with 3–14 flowers; petals *c.* 1 cm, white, dotted with red. *W Mexico*. G1. Late winter–early spring.

Var. **bernalense** Kimnach & Moran. Illustration: Cactus & Succulent Journal (U.S.) **58**(2): 54 (1986), with yellowish leaves not more than 4 cm. Occasionally offered for sale.

9. G. macdougallii Alexander. Illustration: Everett, New York Botanic Garden encyclopedia of horticulture **5**: 1533 (1981); Riha & Subik, Illustrated encyclopedia of cacti and other succulents, 275 (1981).

Almost stemless perennial, producing many stolons with short tight rosettes of very blue-grey. Leaves spathulate, 2–4 × 1.5–2 cm, apex abruptly pointed. Flowering-stems to 15 cm, weak, with 1–3 branches, each with 2–5 flowers; petals *c.* 1.3 cm, white to yellowish green with red banding towards the tip. *S Mexico*. G1. Winter–early spring.

18. CREMNOPHILA Rose
S.G. Knees
Hairless, succulent perennial herbs. Stems spreading, trailing or erect. Leaves alternate, fleshy, turgid, arranged in basal rosettes or scattered along stems. Flowering-stems leafy, axillary, deciduous, paniculate, each branch with a solitary flower. Sepals 5, unequal, closely adpressed to corolla; petals 5, united; stamens 10; carpels 5, erect.

A genus of 2 species, both from Mexico, closely allied to *Echeveria*, and sometimes hybridising with it.

Literature: Moran, R., Resurrection of Cremnophila. *Cactus & Succulent Journal (U.S.)* **L**. (3): 139–146 (1978).

1a. Leaves grey; petals yellow, spreading   **1. nutans**
  b. Leaves green; petals greenish white, erect   **2. linguifolia**

1. C. nutans Rose (*Sedum cremnophila* R.T. Clausen; *S. nutans* Rose not Haworth). Illustration: Addisonia **1**: t. 25 (1916); Cactus & Succulent Journal (U.S.) **50**(3): 139–141 (1978); Graf, Exotica, 4 **1**: 898 (1985); British Cactus & Succulent Journal **8**(3): 82–83 (1990).

Stems 7–12 cm. Leaves 20–30, crowded, greyish, rhombic, 2.5–7.5 × 3–3.5 cm, and 1–1.2 cm thick, densely clustered, ascending or decumbent. Panicles *c.* 10 cm, nodding. Petals *c.* 6 mm, bright yellow, spreading, overlapping in bud. *Mexico (Morelos)*. H5–G1. Winter.

2. C. linguifolia (Lemaire) Moran (*Echeveria linguifolia* Lemaire). Illustration: Cactus & Succulent Journal (U.S.) **50**(3): 142–144 (1978); Everett, New York Botanic Garden illustrated encyclopaedia of horticulture **3**: 913 (1981).

Stems to 30 cm. Leaves 15–40, crowded or separate, obovate to oblong, spathulate, 2.5–10 × 1.8–5 cm, and 1–1.2 cm thick, obtuse to rounded, green. Flowering-stems 30–45, loose, nodding. Petals *c.* 9 mm, greenish white. *Mexico (Mexico State)*. H5–G1. Winter–spring.

19. ROSULARIA (de Candolle) Stapf
U. Eggli
Herbaceous evergreen rosette-forming perennials with or without rootstock and/or long tap-root; rosettes single or highly tufted and mat-forming; leaves succulent-juicy, broadly spathulate to oblong, hairless or glandular-hairy; margin entire, glandular-hairy or minutely toothed, some with waxy bloom; inflorescence lateral or terminal, upright or decumbent, often drooping in bud, paniculate with sometimes curled lateral branches; perianth segments 5–9, united for ¹⁄₁₀–³⁄₄ of their length, tips more or less reflexed; stamens twice as many as petals, carpels free or basally united, upright. Seeds oblong-ellipsoid, 0.5–1.3 mm, dark brownish to ochre, longitudinally striped.

A genus of 25 species including some formerly placed in *Afrovivella* Berger and *Sempervivella* Stapf, mainly from Turkey and adjacent E countries, Inner Asia, and the Himalaya region. Some (mostly not in cultivation) also occur in Crete, W Mediterranean countries, N Africa and Ethiopia. Although winter-hardy species can be grown in rockeries, most benefit from some protection against excessive rain in winter and should be grown in fast-draining, open compost in shallow pots or pans. Species from the E Mediterranean start their growth cycle in early winter. In species from Turkey, the outer leaves of rosettes dry up during spring and summer, leaving a tightly closed 'bud'. Propagation is usually by offsets or seed. Common pests are aphids on developing inflorescences, and larvae of wine weevil feeding inside enlarged rootstocks or tap-roots. All species of the genus show tremendous variability, which is only partly due to genetic differences. Almost all characters are greatly influenced by the environment, mostly by the availability of water and fertiliser. All subsequent descriptions as well as the key apply to plants grown under 'hard' conditions in order to keep them compact.

Literature: Eggli, U., A monographic study of the genus Rosularia. *Bradleya* **6**: Supplement, 1988.

1a. Rootstock absent, tap-root weakly developed, flowers whitish or yellowish, with or without reddish spots and streaks   2
  b. Rootstock present, or mature plants with somewhat to much thickened tap-root   6
2a. Inflorescence lateral   **1. aizoon**
  b. Inflorescence terminal   3
3a. Leaves hairless, flowers normally with purplish streaks
     2. serpentinica var. **serpentinica**
  b. Leaves glandular-hairy, at least along margin; flowers normally without purplish streaks   4
4a. Outer leaves of rosette flushed bright red   **3. muratdaghensis**
  b. All leaves green throughout   5
5a. Inflorescence 10–20 cm
     4. chrysantha
  b. Inflorescence 5–9 cm
     5. rechingeri
6a. Inflorescence terminal, plants normally dying after flowering
     6. globulariifolia
  b. Inflorescence normally lateral, and plants not dying after flowering   7

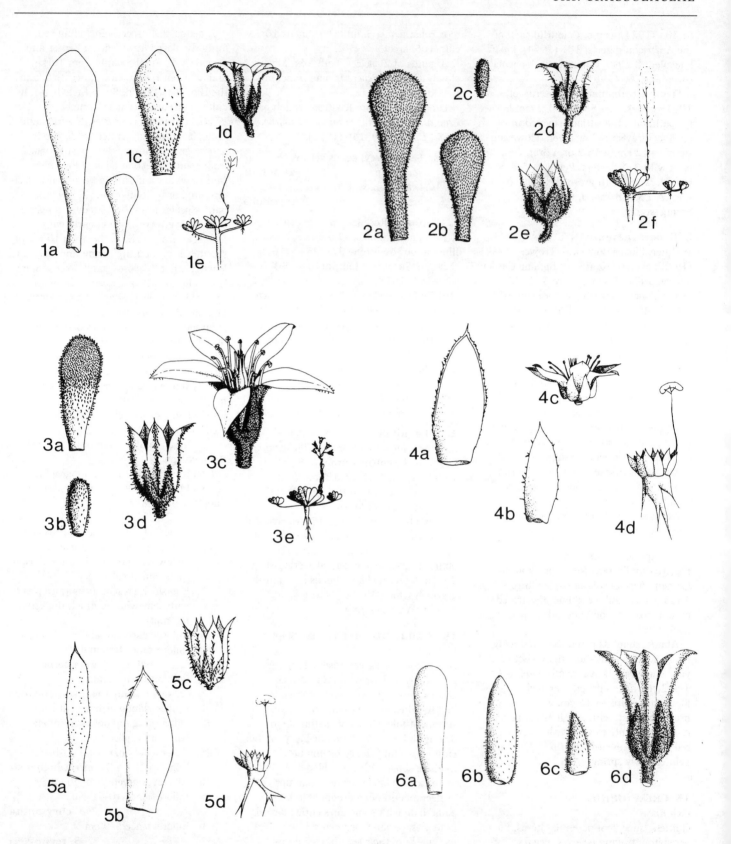

Figure 18. Diagnostic details of *Rosularia* species.
1, *R. adenotricha* subsp. *adenotricha* a, b, rosette leaf;
c, inflorescence leaf; d, flower; e, habit. 2, *R. adenotricha* subsp.
viguieri a, b, rosette leaf; c, inflorescence leaf; d, e, flower;
f, habit. 3, *R. aizoon* a, rosette leaf; b, inflorescence leaf;
c, d, flower; e, habit. 4, *R. alpestris* subsp. *alpestris* a, rosette
leaf; b, inflorescence leaf; c, flower; d, habit. 5, *R. alpestris* subsp.
marnieri a, b, rosette leaf; c, flower; d, habit. 6, *R. serpentinica*
var. *serpentinica* a, rosette leaf; b, c, inflorescence leaves;
d, flower.

7a. Inflorescence flowering for more than ½ of its length, leaves with copious whitish bloom **7. serrata**
  b. Inflorescence flowering for upper ⅓ or less 8
8a. Flowers congested into a head-like inflorescence (which may lengthen after flowering) **8. alpestris**
  b. Flowers not congested into a head-like inflorescence 9
9a. Flowers white, saucer-shaped and widely opening, *c.* 1.5 cm across perianth segments 6–8; cultivated plants normally with long thread-like runners **9. sedoides**
  b. Flowers white or various shades of pink, narrowly funnel-shaped, perianth segments 5; cultivated plants never with thread-like runners 10
10a. Leaves strikingly blue-green or grey-green, sometimes tinged purplish 11
  b. Leaves green 12
11a. Leaves blue-green, flat and slightly succulent, margin toothed **10. sempervivum** subsp. **glaucophylla**
  b. Leaves grey-green, thickened and very succulent, margin more or less entire **11. adenotricha** subsp. **adenotricha**
12a. Leaves hairless, margin toothed; inflorescence hairless or glandular-hairy **10. sempervivum** subsp. **persica**
  b. Leaves glandular-hairy, margin slightly toothed or entire; inflorescence normally glandular-hairy 13
13a. Flowers broadly bell-shaped, 8–12 mm **12. lineata**
  b. Flowers narrowly bell-shaped or tubular, 6–9 mm 14
14a. Rosettes 1.5–3 cm across; leaves 8–18 mm **13. haussknechtii**
  b. Rosettes 3–6 cm across; leaves 1.6–3.2 cm 15
15a. Petals 7–9 mm, pale pink with darker veins **10. sempervivum** subsp. **libanotica**
  b. Petals 6–7 mm, pale pink or white **10. sempervivum** subsp. **pestalozzae**

1. R. aizoon (Fenzl) Berger (*R. pallida* (Schott & Kotschy) Stapf; *R. chrysantha* misapplied). Figure 18(3), p. 212. Illustration: Takhtajan, Flora Armenii, 3 t. 107 (1958); Bradleya **6** suppl.: 42, 49, 64 (1988).

Rosettes globular to slightly flattened-globular, 1.5–3 cm across, pale to fresh green, dry leaves papery and ochre-coloured; leaves glandular-hairy. Inflorescence lateral, 5–9 cm, leafy. Flowers narrowly funnel-shaped (sometimes opening more widely), pale yellowish, rarely with purplish venation. *S, SE, E & NE Turkey, adjacent Russian Armenia.* H3–5.

2. R. serpentinica (Werdermann) Muirhead var. **serpentinica**. Figure 18(6), p. 212. Illustration: Gartenpraxis **8**: 8–11 (1987); Bradleya **6** suppl.: 49, 100 (1988).

Rosettes globular to flattened-globular, 1–3 cm across, offsetting and forming compact cushions; leaves hairless (innermost leaves glandular-hairy before a rosette prepares to flower), bluish green, outer leaves attractively flushed purplish, dry leaves ochre and papery. Inflorescence terminal, leafy, glandular-hairy, flowers funnel-shaped, erect, 1.1–1.3 cm, petals white to cream with purplish longitudinal streaks. *SW Turkey.* H4–G1.

A very attractive species, easily grown in porous soil.

Var. **gigantea** Eggli. Figure 19(1), p. 214. Differs in having much larger rosettes (to 11 cm across) and longer inflorescences. The rosettes do not normally produce offsets and the plant has to be propagated by seeds. *SW Turkey.* G1.

3. R. muratdaghensis Kit Tan (*R. platyphylla* misapplied (not *R. platyphylla* (Schrenk) Berger, which is not in cultivation). Figure 19(2), p. 214. Illustration: Bradleya **6** suppl.: 49 (1988) – as sp. A.

A recently named plant with very attractive, highly tufted fresh green rosettes, outer leaves flushed bright reddish. Flowering very rarely, flowers as in *R. chrysantha* to which is seems to be related. *W Turkey.* H5–G1.

4. R. chrysantha (Boissier) Takhtajan (*R. pallida* misapplied). Figure 19(3), p. 214. Illustration: Jelitto et al. Die Freiland Schmuckstauden, 545 (1985); Gartenpraxis **8**: 8 (1987); Bradleya **6** suppl.: 49, 52 (1988).

Rosettes globular to flattened-globular, 1.5–3 cm across, strongly offsetting, pale green, dry leaves ochre and papery, leaves glandular-hairy. Inflorescence terminal, 10–20 cm, leafy throughout, strongly glandular-hairy, flowers narrowly funnel-shaped, pale greenish cream to pale yellow, with or without purplish venation, or pale dirty olive. *W & SW Turkey.* H3–4.

5. R. rechingeri Jansson (*R. turkestanica* misapplied). Figure 19(4), p. 214. Illustration: Acta Horti Gotoburgensis **28**: 185 (1966); Flora Iranica **72**: t.6 (1970); Bradleya **6** suppl.: 49, 79 (1988).

Very similar to *R. aizoon*, differing in having nearly hairless to slightly glandular leaves, terminal inflorescences and longer flowers. *N Iraq, Iran, Turkey.* H3. Summer.

Cultivated material under this name is often wrongly identified and represents *R. serpentinica*, but note that *R. turkestanica* (Regel & Winkler) Berger is a synonym of the uncultivated *R. platyphylla*.

6. R. globulariifolia (Fenzl) Berger (*Umbilicus cyprius* Holmboe; *Sedum globulariaefolium* (Fenzl) Raymond-Hamet; *Rosularia cypria* (Holmboe) Meikle). Figure 19(5), p. 214. Illustration: Bergens Museums Skrifter **1**(2): (1914); Raymond-Hamet, Crassulacearum Icones Selectae **2**: t. 38 (1956); Flora of Cyprus **1**: 646 (1977); Bradleya **6** suppl.: 57, 61 (1988).

Rosettes flattish, open, 4–11 cm across, single or rarely offsetting; leaves flattish, medium green, glandular-hairy and sticky with resinous smell; inflorescence in cultivation nearly always terminal, narrowly pyramidal, 15–40 cm; flowers 5–8 mm, urn-shaped, with large green sepals, petals white or pale pink, more or less erect or slightly reflexed. Spontaneously sets seeds by self-pollination. *E Mediterranean.* H3. Early summer.

In habitat, this species flowers laterally for several years, before it finally produces a terminal inflorescence and then dies. In cultivation, mostly terminal inflorescences are formed, and the plant has to be propagated by seeds.

7. R. serrata (Linnaeus) Berger (*Cotyledon serrata* Linnaeus). Figure 19(6), p. 214. Illustration: Dillenius, Hortus Elthamensis, t. 95 (1732); Bradleya **6** suppl.: 64, 106 (1988).

Rosettes globular or open with more or less erect leaves, 2.5–6 cm; leaves oblong to spathulate with an obscure tip, flattish, grey-green, with copious whitish bloom, glabrous, margins toothed. Inflorescence lateral (very leafy at first and easily mistaken for a rosette), narrowly pyramidal. Flowers small, 6–8 mm, urn-shaped, petals whitish or pale to brownish pink. *SE Aegean Islands, SW coastal Turkey.* G1.

8. R. alpestris (Karelin & Kiriloff) A. Borissova subsp. **alpestris** (*Sempervivella acuminata* (Decaisne) Berger; *Sempervivella*

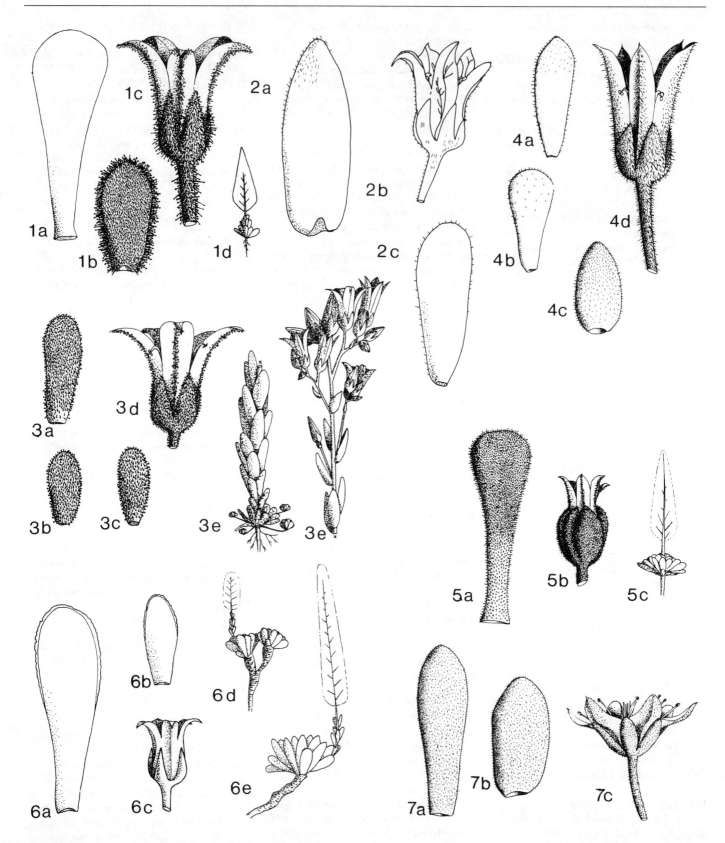

Figure 19. Diagnostic details of *Rosularia* species.
1, *R. serpentinica* var. *gigantea* a, rosette leaf; b, inflorescence
leaf; c, flower; d, habit. 2, *R. muratdaghensis* a, inflorescence
leaf; b, flower, c, rosette leaf. 3, *R. chrysantha* a, rosette leaf;
b, inflorescence leaf; c, rosette leaf; d, flower; e, habit.

4, *R. rechingeri* a, b, rosette leaf; c, inflorescence leaf; d, flower.
5, *R. globulariifolia* a, rosette leaf; b, flower; c, habit. 6, *R. serrata*
a, rosette leaf; b, inflorescence leaf; c, flower; d, e, habit.
7, *R. sedoides* a, rosette leaf; b, inflorescence leaf; c, flower.

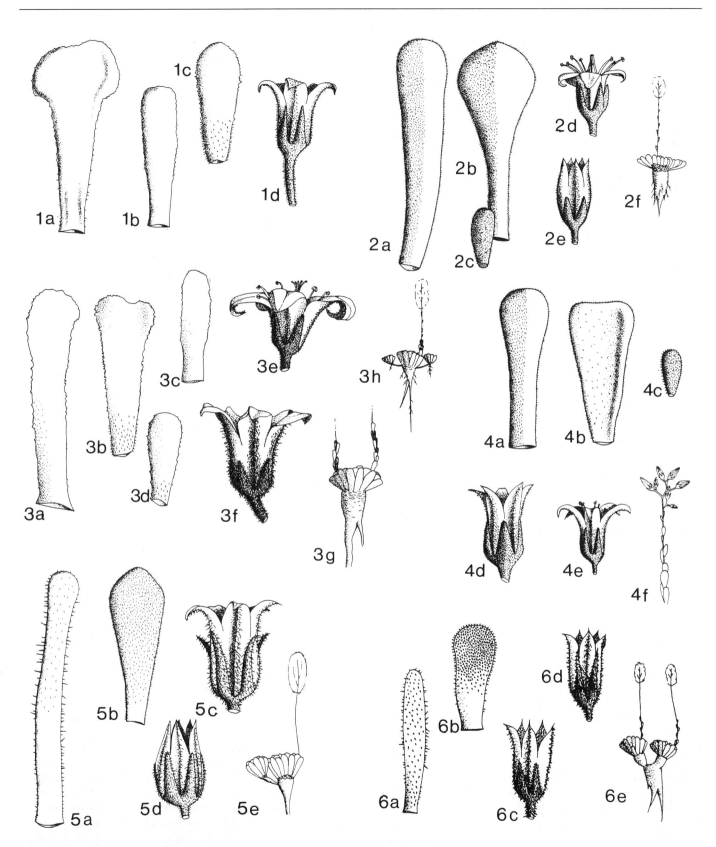

Figure 20. Diagnostic details of *Rosularia* species.
1, *R. sempervivum* subsp. *glaucophylla* a, b, rosette leaf;
c, inflorescence leaf; d, flower. 2, *R. sempervivum* subsp. *libanotica*
a, b, rosette leaf; c, inflorescence leaf; d, e, flower; f, habit. 3, *R. sempervivum* subsp. *persica* a, b, rosette leaf; c, d, inflorescence leaf; e, f, flower; g, h, habit. 4, *R. sempervivum* subsp. *pestalozzae* a, b, rosette leaf; c, inflorescence leaf; d, e, flower; f, habit.
5, *R. lineata* a, b, rosette leaf; c, d, flower; e, habit.
6, *R. haussknechtii* a, b, rosette leaf; c, d, flower; e, habit.

mucronata (Edgeworth) Berger & var. *glabra* Kitamura). Figure 18(4), p. 212. Illustration: Bradleya **6** suppl.: 45, 53 (1988).

Rosettes open-globular, 1.5–6 cm, leaves fresh green, hairless, very lush and succulent, often hard-tipped. Inflorescence 10–20 cm, flowers densely packed at flowering time, later well spaced, opening widely, saucer-shaped, petals white, with or without greenish or purplish venation. *Himalaya, Pamir, China (Xizang), adjacent former USSR.* H3–5.

Subsp. **marnieri** (H. Ohba) Eggli. Figure 18(5), p. 212. Illustration: Bradleya **6** suppl.: 48, 53 (1988). Differs in having very long narrow leaves and smaller, often more greenish flowers, and petals with an irregularly cut margin.

9. R. sedoides (Decaisne) H. Ohba (*Sempervivella sedoides* (Decaisne) Stapf; *Sempervivella alba* (Edgeworth) Stapf; *Rosularia sedoides* var. *alba* (Edgeworth) P. Mitchell). Figure 19(7), p. 214. Illustration: Decaisne, Voyages Inde 4 (Bot) 74 (1844); Jelitto, Die Freiland Schmuckstauden, 590 (1985); Succulenta **66**(1): 20 (1987) – as Sempervivella mucronata; Bradleya **6** suppl.: 53, 83 (1988).

Rosettes flattish to flattened-globular, 2–3.5 cm, freely offsetting, with offsets on long, thread-like runners; leaves fresh green, spathulate, glandular hairy. Inflorescence lateral, upright or decumbent, leafy, flowers 1.2–1.8 cm, saucer-shaped, petals white with greenish venation. *N India (Himalaya).* H2–4.

A commonly grown rock garden plant, easily propagated by the copious offsets formed.

10. R. sempervivum (Bieberstein) Berger.

This group of the genus shows a tremendous variability which makes classification very difficult. Several recognised as species in the past have been subsumed under this species in a recent revision. Intermediates are very common; garden material may moreover be of hybrid origin. Coming from regions with very cold but dry winters, all members of this group are best cultivated in a frost-free greenhouse (G1), some grow adequately under H4–5 conditions. The following subspecies are met with in cultivation:

Subsp. **glaucophylla** Eggli (*R. libanotica* 'Group C' Chamberlain & Muirhead in Flora of Turkey **4**: 219; *R. spathulata* invalid; *Cotyledon globulariifolia* Baker). Figure 20(1), p. 215. Illustration:

Refugium Botanicum **3**: 201 (1871); Bradleya **6** suppl.: 60, 90 (1988). Striking blue-green flattish rosettes; leaves with toothed margin. Inflorescence hairless or glandular-hairy. Flowers broadly funnel-shaped, 6–9 mm, whitish to pale pink. *S Turkey (Taurus).*

Subsp. **libanotica** (Labillardière) Eggli (*R. libanotica* 'Group E' Chamberlain & Muirhead in Flora of Turkey **4**: 219), Figure 20(2) p. 215. Illustration: Raymond-Hamet, Crassulacearum Icones Selectae **3**: 43–44 (1958) – as Sedum sempervivum; Bradleya **6** suppl.: 94 (1988). Rosettes medium green, leaves glandular-hairy, flowers broadly funnel-shaped with conspicuously reflexed petal tips, 8–10 mm, pale pink with darker venation. *S Turkey, W Syria, Lebanon, W Jordan, Israel.*

Subsp. **persica** (Boissier) Eggli (*R. persica* (Boissier) Muirhead; *R. radiciflora* Borissova invalid; subsp. *glabra* (Bossier) Muirhead). Figure 20(3), p. 215. Illustration: Bradleya **6** suppl.: 60, 64, 96, (1988). Rosettes flattish, dark green, leaves mostly hairless, sometimes loosely glandular-hairy, margin normally irregularly toothed, often whitish horny. Inflorescence hairless or strongly glandular-hairy, flowers narrowly funnel-shaped, 8–12 mm, clear to dark pink, rarely nearly white. *SW Syria (Hermon), Lebanon, Israel, E Turkey, N & W Iran, NE Iraq, Transcaucasia.*

Very widespread and extremely variable.

Subsp. **pestalozzae** (Boissier) Eggli (*R. pestalozzae* (Boissier) Samuelsson & Fröderström; *R. sempervivum* var. *pestalozzae* (Boissier) Thiébaut, incl. *R. libanotica* 'Group B' Chamberlain & Muirhead in Flora of Turkey **4**: 219). Figure 20(4), p. 215. Illustration: Bradleya **6** suppl.: 60, 99 (1988). Rosettes flattish to flattened-globular, leaves medium green, finely and densely glandular-hairy. Inflorescence glandular-hairy, flowers narrowly funnel-shaped, 6–8 mm, whitish to pale pink. *S Turkey & adjacent Syria.*

11. R. adenotricha (Edgeworth) Jansson (*Sedum adenotrichum* Edgeworth; *Cotyledon tenuicaulis* Aitchison). Figure 18(1), p. 212. Illustration: Refugium Botanicum **3**: 296 (1871); Journal of the Linnean Society **18**: 101 (1882); Journal of the Royal Horticultural Society **46**: 164 (1921); Bradleya **6** suppl.: 38, 53 (1988).

Rosettes open, 1.5–2.5 cm, leaves more or less erect, exposed stems slightly thickened, brownish, more or less glossy,

often with leaf-scars. Inflorescence 15–20 cm, hairless or glandular-hairy; flowers 6–10 mm, petals white or pale pink with darker venation. *S Himalaya & Pamir.* H5–G1.

Easily propagated from detached leaves.

Subsp. **viguieri** (Raymond-Hamet) Jansson (*Sedum viguieri* (Raymond-Hamet) Fröderström). Figure 18(2), p. 212. Differs in having medium green, glandular hairy leaves and pure white flowers.

12. R. lineata (Boissier) Berger (*R. setosa* Bywater). Figure 20(5), p. 215. Illustration: Nouvelle Flore du Liban et de la Syrie **2**: 67 (1970); Kew Bulletin **34**(2): 403 (1979); Bradleya **6** suppl.: 57, 69 (1988).

Rosettes flattish, 2–7 cm; leaves oblong to spathulate, with rounded or truncate apex, 2–4 cm, deep green, shortly glandular-hairy, margin glandular-hairy with long bristles. Inflorescence lateral, erect, flowers 8–12 mm, broadly bell-shaped, pale pink with dark pink venation, often foetid. *Lebanon, Syria, Jordan, Israel.* G1.

13. R. haussknechtii (Boissier & Reuter) Berger. Figure 20(6), p. 215. Illustration: Bradleya **6** suppl.: 63, 64 (1988).

Rosettes flattish, 1–3 cm; leaves dull or fresh green, narrowly spathulate, basally narrowed into a long claw, margin entire but with some obvious bristles. Inflorescence lateral, 4–8 cm, flowers narrowly funnel-shaped, bright pink. *SE Turkey.* H4–G1.

Probably not in cultivation; this name has been widely misapplied to many different Turkish species.

20. THOMPSONELLA Britton & Rose
G.D. Rowley

Echeveria-like rosette succulents with a tendency to shed their leaves in winter; stem undeveloped or very short, unbranched. Leaves spreading, thick, soft fleshy, hairless except for marginal hairs, concave above, convex below. Inflorescence lateral, a long dense spike or slender thyrse. Flowers almost stalkless, small, with 5 erect sepals, 5 thin spreading petals with recurved tips, 10 erect stamens and 5 free carpels with slender beak-like tips and narrow bases. Nectar scales thin.

A genus of 2 species in Mexico. Rot-sensitive and difficult to keep in cultivation: water sparingly and keep quite dry in winter when the leaves wither. Propagation by seed; adventitious buds sometimes appear in the inflorescence and may be treated as cuttings.

Literature: Moran, R., *Cactus & Succulent Journal (U.S.)* 41: 173–175 (1969).

1. T. minutiflora (Rose) Britton & Rose (*Echeveria minutiflora* Rose). Illustrations: Contributions from the U.S. National Herbarium **12**: t. 44 (1909); Cactus & Succulent Journal (U.S.) **8**: 100–102 (1937); **41**: 173–175 (1969); Martin & Chapman, Succulents and their cultivation, 160–161 (1977).

Roots thickened; rosette to 17 cm, flat; leaves 10–20, oblanceolate, acute, to 10 × 1.5–2.5 cm, 3–7 mm thick, dark green, sometimes flushed with purple, margin upcurved, more or less curly. Inflorescence to 30 cm, unbranched, with a diminishing series of leaf-like bracts below. Flowers 1–7 in tight cymes, to 1 cm across, dull yellow with red markings recalling those of *Graptopetalum*. *Mexico*. G1. Late summer–autumn.

21. DUDLEYA Britton & Rose
E.H. Hamlet
Perennial herbs with simple or branched rootstocks or small spherical to oblong corms; leaves mainly in basal rosettes, fleshy, flattened to nearly terete, more or less ovate to linear, commonly white, glaucous. Flowering-stems axillary, with stem-leaves much reduced. Flowers in terminal paniculate or cyme-like clusters. Calyx deeply divided into 5 erect, lanceolate-linear to ovate segments. Corolla white to yellow or red, cylindric or bell-shaped; petals united near base, erect or spreading from middle or near tips of petals. Stamens 10, borne on the corolla-tube. Seeds numerous, narrowly ovoid, brown or reddish brown.

A genus of 40 species native to SW USA & NW Mexico. They prefer a well-drained fertile soil in a semi-shaded position and little water in winter. Propagation by division or seed.

Flowering stems. Mealy: **5,9**. Twisted: **5**.
Petals. White or cream: **4,6,8**. Yellow: **1–3,6,7,9,10**. Pink or red: **2,3,5,7,8,10**.

1a. Leaves linear or narrowly
lanceolate   2
b. Leaves not as above   4
2a. Leaves narrowly lanceolate; flowers yellow or red   **10. saxosa**
b. Leaves linear; flowers not yellow
   3
3a. Petals narrowly ovate; leaves acute
   8. densiflora

b. Petals oblong-lanceolate; leaves acute to almost acuminate
   4. edulis
4a. Rosette leaves 30 or more   5
b. Rosette leaves fewer than 30   6
5a. Calyx segments triangular; leaves triangular-ovate to oblong-lanceolate   **1. candida**
b. Calyx segments lanceolate; leaves obovate to spathulate
   5. pulverulenta
6a. Flowering-stems mostly white-mealy   **9. farinosa**
b. Flowering-stems not white-mealy
   7
7a. At least inner leaves glaucous   9
b. Leaves not glaucous   8
8a. Leaves oblanceolate to spathulate, 1.5–6 cm   **7. cymosa**
b. Leaves oblong, 5–13 cm
   2. cultrata
9a. Stem to 20 cm   10
b. Stem rarely more than 5 cm
   7. cymosa
10a. Flowers yellow to red; leaves 5–20 cm   **6. caespitosa**
b. Flowers yellow with red markings; leaves 5–8 cm   **3. rigida**

1. D. candida Britton. Illustration: Jacobsen, Lexicon of Succulent Plants, pl. 54 (1974); Cactus & Succulent Journal (U.S.) **46**: 71 (1974); Lamb, The illustrated reference on cacti and other succulents **3**: 765 (1977).

Stems 2–6 cm across, branching to form clumps to 80 cm across. Rosettes 7–21 cm across. Leaves 30–70, glaucous or green, 5–11 × 1–3 cm, triangular-ovate to oblong-oblanceolate. Inflorescence to 50 cm, with 3–5 branches. Petals 8.5–13 × 2–3.5 mm, oblong-lanceolate, acute, pale yellow. *Mexico (Baja California)*. Spring.

2. D. cultrata Rose. Illustration: Cactus & Succulent Journal (U.S.) **40**(4): 170 (1968).

Stems 2–4 cm across, to 20 cm, branching. Rosettes 3–8 cm across. Leaves 20–30, green, 5–13 × 1–1.5 cm, 3–5 mm thick, oblong, pointed. Inflorescence to 40 cm, with about 3 branches. Stem-leaves triangular, acute. Calyx 4.5–6 × 5–6 mm; segments triangular, acute. Petals 10–13 × 3 mm, elliptic, acute, pale yellow. *Mexico (Baja California)*. Spring.

3. D. rigida Rose. Illustration: Jacobsen, Lexicon of Succulent Plants, pl. 55 (1974); Cactus & Succulent Journal (U.S.) **59**(5): 188–189 (1987).

Stems erect, to 10 cm or more, 1–3.5

cm across, branching to form clumps to 25 cm across. Rosettes 6–15 cm across. Leaves 10–25, usually slightly glaucous, 5–8 × 2.5–4 cm, 6–10 mm thick, oblong to triangular-ovate, tapering, short-acuminate. Flowering-stems to 50 cm, with 2 or 3 branches. Petals yellow with red markings. *Mexico (Baja California)*. Spring.

4. D. edulis Moran. Illustration: Lamb, The illustrated reference on cacti and other succulents **5**: 1430 (1978).

Stems 1.5–4.5 cm across, to 15 cm, branching to form clumps to 40 cm across. Rosettes 5–10 cm across; leaves 15–25, green, 5–20 × 0.4–1 cm, linear, slightly glaucous, acute to almost acuminate. Flowering-stems 15–50 cm, stem-leaves 1–5 cm, turgid, triangular-lanceolate, acute. Inflorescence long, rather open. Flower-stalks to 2 mm; calyx-lobes 2.5–4.5 mm, oblong, acute; petals 7–10 mm, joined for 1–1.5 mm, oblong-lanceolate, acutish, creamy white. *USA (California), Mexico (Baja California)*. Early summer.

5. D. pulverulenta Britton & Rose. Illustration: Cactus & Succulent Journal (U.S.) **48**: 4 (1976); Riha & Subik, The illustrated encyclopedia of cacti and other succulents, 254 (1981).

Plant mealy-glaucous; stems to 40 cm, 4–9 cm thick. Rosette leaves 30–80, obovate to spathulate, 8–25 × 4–10 cm. Flowering-stems stout, 40–80 cm; branches twisted so that the flowers are more or less on the underside; stem-leaves many, 1–4 cm, broadly ovate, heart-shaped, clasping at base, acute. Flower-stalks slender, 5–30 mm, spreading. Calyx-segments *c.* 4–8 mm, lanceolate, acute, red to glaucous; petals 1.2–1.8 cm, yellow to mostly deep red, joined nearly to the middle. Seeds brown. *USA (California), Mexico (Baja California)*. Late spring–summer.

6. D. caespitosa Britton & Rose. Illustration: Cactus & Succulent Journal (U.S.) **26**: 10 (1954).

Stems 1.5–4 cm across, to 20 cm, branching. Rosettes 5–15 cm across; leaves 15–30, 5–20 × 1–5 cm, oblong-oblanceolate, almost shining, only the inner glaucous. Inflorescence to 60 cm, with 3–14 flowers. Calyx 4–6 × 4–8 mm, nearly truncate to tapering at base. Petals 0.8–1.6 cm, white to bright yellow, rarely red. *USA (California)*. Late spring.

7. D. cymosa Britton & Rose (*Cotyledon cymosa* Baker). Illustration: Saunders, Refugium Botanicum **1**: f. 68 (1869).

Stems 1–3.5 cm, simple or with few branches. Rosettes usually 6–15 cm across. Leaves 16–25, 1.5–6 × 1–3 cm, mostly evergreen, green to glaucous, usually oblanceolate to spathulate, pointed. Inflorescence with 2 or 3 branches. Calyx 3–6 mm, triangular-ovate, acute. Petals 0.7–1.4 cm, yellow to red. *USA (California)*. Late spring–summer.

8. D. densiflora Moran. Illustration: Jacobsen, Lexicon of succulent plants, pl. 55 (1974).

Stems 1–2.5 cm across, to 10 cm, branching. Rosettes 7–25 cm across; leaves 20–40, linear, 6–15 × 0.6–1.2 cm, 5–8 mm thick, acute, terete above, persistently glaucous. Flowering-stems 15–30 cm, with 3 or more branches; inflorescence dense, more or less rounded. Stem-leaves 1–4 cm, turgid, acute. Flower-stalks 2–5 mm; calyx-lobes 1.5–2.5 mm, triangular-ovate, acute; petals 5–10 mm, narrowly ovate, acute, whitish to pink. *USA (California)*. Summer.

9. D. farinosa Britton & Rose. Illustration: Lamb, The illustrated reference on cacti and other succulents, 470 (1973); Jacobsen, Lexicon of succulent plants, pl. 54 (1974); Riha & Subik, The illustrated encyclopedia of cacti and other succulents, 253 (1981).

Stems stout, 1–3 cm thick, often tapered gradually, usually with several rosettes, each with 15–30 leaves. Rosettes 4–10 cm across; leaves 2.5–6 × 1–2.5 cm, 3–6 mm thick, densely white-mealy to green, ovate-oblong, acute, flat on upper surface, lightly rounded beneath. Flowering-stems 10–35 cm, with 3–5 branches, stout, mostly white-mealy; stem-leaves many, 1–2.5 cm, triangular-ovate, concave. Flower-stalks 1–5 mm, stout; calyx 5–8 mm, mostly 5–6 mm wide, rounded at base; lobes deltoid-ovate. Petals 1–1.4 cm, lemon-yellow, oblong, acute. *USA (California, Oregon)*. Summer.

10. D. saxosa Britton & Rose. Illustration: Munz, A flora of Southern California, 380 (1974).

Plant pale green or almost glaucous; stems 1–3 cm across. Rosettes 3–10 cm across. Rosette leaves 10–25, narrowly lanceolate, 3–15 × 0.5–2.5 cm, 1.5–6 mm thick, almost rounded. Flowering-stems 5–40 cm, more or less reddish, with 2 or 3 branches; leaves ovate-lanceolate, slightly clasping. Flower-stalks 1–2 cm; calyx-lobes *c.* 5 mm, lanceolate-ovate, red; petals 0.9–2 cm, oblong-lanceolate, acute, yellow, more or less reddish in age. *USA (California)*. H2–3. Summer.

22. PACHYPHYTUM Link, Klotzsch & Otto

G.D. Rowley

Small, evergreen, hairless, perennial shrublets with thick, fleshy, sparingly branched stems covered in leaf-scars. Leaves simple, entire, stalkless, smooth, thick and highly succulent, spirally arranged in rosettes, mostly white- to bluish-bloomed, often flushed purple. Scape lateral, erect, with small scattered bracts, ending in 1–few pendent or scrolled one-sided inflorescences, with small tubular flowers in a double row often half-hidden by overlapping bracts, pendent but often becoming erect by subsequent growth of the inflorescence. Sepals 5, equal or unequal, adpressed. Petals 5, shortly united at base, erect with diverging tips, the lower margins inrolled to form 2 scale-like appendages at the base about half as long as the petal. Stamens 10 in 2 whorls. Carpels 5, free.

A genus of around 13 species from Mexico, beloved of succulent collectors for the sculptural effect of the few, thick, chunky leaves with a characteristic waxy bloom, which in *P. oviferum* resemble sugared almonds. They should not be handled, however, as they mark easily. The normal porous but nutritious soil used for most succulents suits *Pachyphytum*, with good drainage, water (preferably from below to avoid stains on the foliage) freely in summer and sparingly in winter. A minimum of 4 °C is adequate, and an occasional drop below zero should do no harm if the plant is dry and resting. Flowering tends to be spread over a long season rather than concentrated into a single month. Most species are adaptable for use as house plants, and add appeal to bowl gardens and floral wreaths. Propagation is by cuttings or single leaves, which throw up one or more plantlets and roots from the leaf-base.

For hybrids with *Echeveria* see under ×*Pachyveria*.

Literature: Poellnitz, K.V., *Cactus Journal (Great Britain)* **5:** 72–75 (1937); Walther, E., *Cactus and Succulent Journal (U.S.)* **3:** 9–13 (1931). (Both with keys to 8–10 species).

1a. Petals without basal appendages
    ×**Pachyveria scheideckeri**
  b. Petals with 2 scale-like basal
    appendages   2
2a. Leaves yellowish green, never
    glaucous   **6. viride**
  b. Leaves grey, pinkish, dark or bluish
    green, glaucous   3
3a. Leaves flattened; sepals large, at
    least as long as petals   4
  b. Leaves more or less terete; sepals
    smaller, shorter than petals   9
4a. Sepals obtuse, widest above the
    middle   5
  b. Sepals acute, widest at or below the
    middle   6
5a. Leaves less than twice as long as
    broad, upperside convex; flowers
    7–15   **5. oviferum**
  b. Leaves more than twice as long as
    broad, upperside flat or concave;
    flowers 12–25   **1. bracteosum**
6a. Inflorescence a recumbent panicle
    ×**Pachyveria sodalis**
  b. Inflorescence erect, raceme-like   7
7a. Flowers on short stalks, crowded   8
  b. Flowers on long stalks, scattered
    ×**Pachyveria mirabilis**
8a. Leaves grey, broadest midway;
    raceme forked
    ×**Pachyveria clavata**
  b. Leaves flushed purple, broadest
    above the middle; racemes mostly
    with 3 branches
    ×**Pachyveria pachyphytoides**
9a. Stems sticky near the top; leaves
    broad, 2.5 cm wide or more
    3. glutinicaule
  b. Stems not sticky; leaves narrower,
    almost as thick as wide   10
10a. Leaves densely packed and faceted
    from mutual pressure   11
  b. Leaves scattered with visible
    internodes, scarcely faceted   12
11a. Leaves 5–6 cm, upperside slightly
    flattened; petal tips recurved
    ×**Pachyveria glauca**
  b. Leaves 2–4 cm, angular terete;
    petals erect   **2. compactum**
12a. Leaves 2–3 times as long as broad;
    petals 1.2–1.5 cm   **4. hookeri**
  b. Leaves to 5 times as long as broad;
    petals 1–1.2 cm
    ×**Pachyveria sobrina**

1. P. bracteosum Link, Klotzsch & Otto. Illustration: Botanical Magazine, 4951 (1856); van Laren, Succulents other than cacti, 80 (1934); Succulents **38:** 20 (1959); Rauh, Wonderful world of succulents, t. 77/3 (1984).

Stem 10–30 cm, 1.2–2.5 cm thick, with few thick branches. Leaves obovate, tapering towards the base, apex obtuse or rounded, often mucronate, 7–10 × 2.5 cm, to 1.2 cm thick, bloomed with sometimes a mauve flush. Scape 15–50 cm, unbranched, with oblong or ovate bracts to 2.5 cm. Flowers 12–25; sepals 1.4–2.2 cm; petals deep red, 9 mm. *Mexico (Hidalgo)*. G1. Summer–autumn.

2. P. compactum Rose. Illustration: Contributions to the U.S. National Herbarium **13**: t. 61 (1911); van Laren, Succulents other than cacti, 79 (1934); Succulenta **38**: 7, 26, 154 (1959); Jacobsen, Lexicon of succulent plants, t. 114/5 (1974).

Compact, with thick stems rarely above 10 cm bearing dense rosettes of 30–60 terete, lanceolate, dark to reddish green, bloomed leaves 2–4 × 0.5–1 cm across. Upper leaf surface somewhat flattened or faceted with net-like, paler marginal lines; apex tapered, acute. Scape to 40 cm with about 10 pendent flowers *c.* 8 mm, on stalks to 1.5 cm. Sepals nearly equal, smaller than the petals. Petals orange-red with a bluish green tip. *Mexico (Hidalgo)*. G1. Spring.

A cristate clone is commonly grown.

3. P. glutinicaule Moran (*P. brevifolium* misapplied). Illustration: Cactus and Succulent Journal (U.S.) **3**: 152 (1932), **4**: 237–238 (1932); **35**: 37–39 (1963); Succulenta **38**: 102–103, 154 (1959); Jacobsen, Lexicon of succulent plants, t. 114/2 (1974).

Stems to 30 cm, decumbent, 0.7–1.5 cm thick, sticky when young. Leaves in loose rosettes 7–12 cm across, bluish glaucous, obovate to oblong, mucronate, 3–6 × 2–3.5 cm, and 5–15 mm thick. Scape 15–20 cm, with 6–23 flowers that are pendent at first, erect later. Flower-stalks 4–15 mm; sepals 1–1.6 cm; petals red, 1.2–1.7 cm. *Mexico (Queretaro, Hidalgo)*. G1. Flowers over a long season, mostly winter–spring.

4. P. hookeri (Salm-Dyck) A. Berger (*Cotyledon adunca* Baker; *Pachyphytum uniflorum* Rose). Illustration: Saunders Refugium Botanicum, t. 60 (1869); Cactus and Succulent Journal (U.S.) **16**: 159 (1944); Succulenta **38**: 132 (1959).

Stems ascending or decumbent when old, little branched, 0.9–1.8 cm thick. Leaves narrow oblong or lanceolate, almost terete, 3–5 × 0.7–1.7 cm, 0.5–1.1 cm thick, obtuse with a minute mucro, green, glaucous. Scape 10–25 cm, with scattered bracts. Flowers 5–18, 1.2–1.5 cm; sepals equal, 5–8 mm; petals *c.* 1.5 cm, yellowish, flushed purple towards the tip. *Mexico (San Luis Potosi)*. G1. Mainly spring.

5. P. oviferum Purpus. Illustration: van Laren, Succulents other than cacti, 80 (1934); Succulenta **38**: 44 (1959); Rowley, Illustrated encyclopedia of succulents, 124 (1978); Rauh, Wonderful world of succulents, t. 77/2 (1984).

Stems short, tending to become prostrate, to 1.3 cm across. Leaves in loose rosettes, obovate, 2.5–5 × 1.8–3 cm, 1–1.6 cm thick, rounded at the edges and somewhat channelled on the upperside, bloomed with often a lavender flush recalling sugared almonds. Scape 8–10 cm, with 7–15 greenish or reddish white flowers. Sepals unequal, 1.5–1.8 cm. Petals *c.* 1 cm. *Mexico (San Luis Potosi)*. G1. Winter–spring.

6. P. viride Walther. Illustration: Cactus and Succulent Journal (U.S.) **8**: 210–211 (1937); Lamb, Illustrated Reference on Cacti and other Succulents **2**: 472 (1959); Succulenta **38**: 51 (1959); Jacobsen, Lexicon of succulent plants, t. 114/3 (1974).

Stem squat, 2–3 cm across, heavily leaf-scarred. Leaves crowded at the stem apex, finger-like, spreading and inclined, almost terete, 8–15 × 1.3–2 cm, 1–1.8 cm thick, blunt, yellowish green, never glaucous. Scape 20–30 cm. Flowers 10–22, nodding at first, erect later. Sepals 1.2–2.5 cm; petals shorter, hidden within the calyx, pinkish red with paler margins. *Mexico (Queretaro)*. G1. Autumn & spring.

P. fittkaui Moran has shorter, broader, narrow obovate to oblanceolate leaves. *Mexico.*

23. × PACHYVERIA Haage & Schmidt
G.D. Rowley
More or less shrubby evergreen perennials with very succulent, spirally arranged entire leaves in terminal rosettes as well as more or less scattered along the branches. Leaves oblong to spathulate to nearly cylindrical, upperside sometimes concave, more or less glaucous, bluish green, often flushed purple. Flowering-stems lateral, ending in a one-sided cyme with few or no branches, with scattered bracts like reduced foliage leaves. Sepals adpressed. Corolla bell-shaped, pinkish to orange-red, often darker spotted or blotched. Petals with or without (× *P. scheideckeri*) the scale-like appendages of *Pachyphytum* at the base.

Intergeneric hybrids between species of *Echeveria* and *Pachyphytum*. For key to hybrids and hybrid cyltivars, see also under *Pachyphytum.*

Literature: Walther, E., *Cactus and Succulent Journal (U.S.)* **6**: 53–56 (1934); Keppel, J.C.V., *Succulenta* **41**: 6–10, 30–31, 45–46, 88–89 (1962); **42**: 86–87, 115–117, 133–135 (1963).

1a. Petals with a pair of basal appendages   2
  b. Petals without basal appendages   **5. scheideckeri**
2a. Leaves flattened; sepals large, as long as petals or longer   3
  b. Leaves almost terete; sepals small, shorter than petals   6
3a. Inflorescence recumbent, paniculate   **7. sodalis**
  b. Inflorescence erect, raceme-like   4
4a. Flowers crowded, on short stalks   5
  b. Flowers not crowded; stalks to 6 mm   **3. mirabilis**
5a. Leaves grey, broadest midway; raceme forked   **1. clavata**
  b. Leaves flushed purple, broadest above the middle; raceme with usually 3 branches
                                          4. pachyphytoides
6a. Leaves crowded and faceted from mutual pressure   **2. glauca**
  b. Leaves loose, not or scarcely faceted   **6. sobrina**

1. × P. clavata Walther (*Echeveria* sp. × *Pachyphytum bracteosum*; *E. clavifolia* misapplied). Illustration: Succulenta **41**: 88–89 (1962); Jacobsen, Lexicon of succulent plants, t. 116/1 (1974).

Small shrublets to 40 cm, with few, stout, erect branches with leaves scattered along the upper parts and crowded in a rosette at the ends. Leaves flat, oblong with an obtuse apex, *c.* 10 × 3 cm, grey-glaucous. Inflorescences erect, with 2 scrolled branches packed with numerous small flowers hanging when young, later becoming erect, on short stalks. Sepals broad, as long as corolla. Petals reddish, appendaged. *Garden origin*. G1. Summer.

A cristate variant is also in cultivation.

2. × P. glauca Walther (*Echeveria* sp. × *Pachyphytum compactum*). Illustration: Succulenta **41**: 45–46 (1962); Jacobsen, Lexicon of succulent plants, t. 116/3 (1974).

Stem undeveloped. Leaves about 40, crowded into a dense rosette, almost terete, slightly flattened on the upperside, 5–6 × 1–1.5 cm, angled from mutual pressure when young, acute, bluish glaucous. Inflorescences 30 cm or more, with scattered small bracts, bearing *c.* 8 pendent flowers. Sepals equal, *c.* 6 mm; petals yellow, with recurving red tips, *c.* 1.2 cm, appendaged at base. *Garden origin.* G1.

'Glossoides' is classified here and develops decumbent stems covered in leaves.

3. × P. mirabilis (Deleuil) Walther (*Echeveria scheeri × Pachyphytum bracteosum*). Illustration: Cactus and Succulent Journal (U.S.) **6**: 54 (1934).

Stem undeveloped. Leaves flat, opaline rose, glaucous, *c.* 60 to a rosette, oblanceolate, acute, *c.* 7 × 2 cm, channelled. Inflorescence simple, erect, to 20 cm, with a single long loose hanging one-sided series of flowers on flower-stalks to 6 mm. Sepals acute, as long as petals. Petals appendaged at the base. *Garden origin.* G1.

4. × P. pachyphytoides (Morren) Walther (*Echeveria gibbiflora × Pachyphytum bracteosum*). Illustration: Succulenta **42**: 86 (1963).

Stem thick, stiffly erect, usually unbranched, 20–40 cm, with spiralled leaf scars, topped by a rosette of flat, spreading, oblanceolate, obtuse or rounded leaves to 12 × 5 cm, bluish glaucous with a purple flush. Inflorescences to 40 cm, with usually 3 scrolled branches bearing numerous, small, crowded, short-stalked flowers. Sepals longer than corolla. Petals pinkish red, appendaged. *Garden origin.* G1.

Similar to *Pachyphytum bracteosum*, but distinguished by the forked inflorescence, more pointed bracts, sepal shape and pale tips to the petals.

5. × P. scheideckeri Walther (*Echeveria secunda × Pachyphytum bracteosum*). Illustration: van Laren, Succulents other than cacti, 75 (1934); Succulenta **41**: 8, 10 (1962); Jacobsen, Lexicon of succulent plants, t. 116/2 (1974).

Short-stemmed, with *Echeveria*-like rosettes of *c.* 50 oblong or tapered flat leaves 5–7 × 1–2 cm, *c.* 6 mm thick, acute, bluish glaucous and somewhat striped. Inflorescences *c.* 13 cm, with a long pendent tip bearing about 10 starry orange-red flowers. Sepals broad, nearly as long as corolla. Petals *c.* 1.3 cm, without appendages. *Garden origin.* G1.

A fasciated 'Cristata' and variegated 'Albocarinata' are also in cultivation.

6. × P. sobrina (Berger) Walther (*Echeveria* sp. × *Pachyphytum hookeri*). Illustration: Cactus and Succulent Journal (U.S.) **6**: 54 (1934).

Stem 10–15 cm, to 2 cm thick, with closely packed leaf-rosettes. Leaves *c.* 20,

narrow, to 5 times as long as broad, *c.* 5 × 0.9 cm, almost terete, oblanceolate, acute, flattened above, bright green, glaucous. Inflorescences to 25 cm with small scattered bracts, forked at the top. Flowers 6–15, on a pendent branch; flower-stalks 1–1.2 cm. Bracts lanceolate, acute. Sepals shorter than the petals, nearly equal, triangular lanceolate, 6–7 mm. Petals 1–1.2 cm, red, keeled, appendaged at the base. *Garden origin.* G1.

7. × P. sodalis (Berger) Walther (*Echeveria* sp. × *Pachyphytum bracteosum*). Illustration: Cactus and Succulent Journal (U.S.) **6**: 54 (1934).

Stems short, thick, little branched, to 3 cm across. Leaves 15–25 in a rosette, flattened, wedge-shaped with a broad blunt tip with a mucro, to 9 × 3.5 cm, ascending, glaucous, bright green flushed purple. Inflorescence recumbent, 40–50 cm, with scattered lanceolate bracts. Branches about 3, pendent, with overlapping bracts. Flower-stalks 6–7 mm. Sepals narrow, acute, unequal, linear-lanceolate, 1–1.5 cm. Petals lanceolate, keeled, finely striped with red, appendaged at base. *Garden origin.* G1.

24. × GRAPTOVERIA Gossot
G.D. Rowley
Shrubby or stemless evergreen perennials with compact or loose rosettes. Leaves entire, succulent, lanceolate to obovate or spathulate, tapered below, acute or mucronate, bluish green, sometimes flushed or blotched purple and more or less glaucous. Inflorescences borne laterally, with usually much branched cymes and a series of bracts transitional from foliage leaves. Flower-stalks 1–3 cm. Sepals 5, short, adpressed to the corolla tube. Corolla more or less urn-shaped, yellow to pinkish, often red-spotted. Stamens 10, 5 of which recurve between the petals.

Intergeneric hybrids between species of *Echeveria* and *Graptopetalum*. In addition to the hybrids described here, several more have been recently offered in the USA. All are notable for the leaf textures and intense colourings.

Literature: Keppel, J.C.V., *National Cactus and Succulent Journal* **35**: 28–31 (1980); **36**: 13–17 (1981).

1a. Leaves papillose, at least when
　　　young　　　　**2. 'Michael Roan'**
　b. Leaves all smooth　　　　　　2
2a. Leaves to 5 cm, awned, with a
　　　purplish bristle tip; stems
　　　undeveloped　　　**3. 'Silver Star'**

　b. Leaves 10 cm or more, not awned;
　　　rosettes on stems above ground
　　　　　　　　　　　1. 'Fred Ives'

1. × G. 'Fred Ives' (*Echeveria gibbiflora × Graptopetalum paraguayense*). Illustration: National Cactus and Succulent Journal **36**: 14 (1981).

Vigorous, stem to 3 cm thick, leaf-rosettes 30–40 cm across when the plant is well fed. Leaves hairless, obovate-oblanceolate, *c.* 15 × 6 cm, with the mucro often oblique, bluish green with a purple flush when exposed to sun. Flowering-stem copiously leafy, branching at the top with groups of 5–15 flowers with large persistent bracts. Sepals long, unequal. Petals yellowish. *Garden origin.* G1. Spring.

Cristate and variegated clones are also grown, and being smaller are preferred when space is limited.

2. × G. 'Michael Roan' (*Echeveria setosa × Graptopetalum paraguayense*). Illustration: Rowley, Illustrated encyclopedia of succulents, 120, (1978).

Stem very short, *c.* 8 mm thick, branching to form a cluster of compact rosettes *c.* 9 cm across. Leaves blue-green, at first papillose, oblong with an acute apex, convex and keeled below, *c.* 5 × 1 cm. Flowering-stem to 15 cm, almost bractless, branching all round, with numerous small cup-shaped flowers with yellow petals spotted with red. Sepals adpressed. *Garden origin.* G1. Spring.

3. × G. 'Silver Star' (*Echeveria agavoides × Graptopetalum filiferum*). Illustration: Cactus and Succulent Journal (US) **46**: 135 (1974); **55**: 261 (1983); Lamb, Neale's Photographic Reference Plates, No. 1787, (undated).

Densely clump-forming with compact almost stemless rosettes 6–10 cm across, made up of numerous, ovate, very fleshy, silvery green leaves, each tapering to a long purplish bristle tip, like a larger version of *Graptopetalum filiferum*. Leaves to 4 × 1.3 cm, lanceolate with the awn 9–11 mm. Flowers on slender lateral inflorescences, intermediate between those of the parents. *Garden origin.* G1. Spring.

25. ECHEVERIA de Candolle
S.G. Knees, H.S. Maxwell, R. Hyam, G.D. Rowley
Succulent, evergreen or occasionally somewhat deciduous, perennial herbs and shrublets. Roots usually fibrous, sometimes thick. Stems usually short, simple, occasionally long or branched. Leaves alternate or spiralled, scattered or in dense

rosettes, usually entire and stalkless, fleshy or thin, pointed, not clasping at base, hairless or hairy, glaucous or shiny. Inflorescence lateral, on an erect stem with numerous bracts; flowers in racemes or panicles; sepals 5, more or less equal, fused at base or free, spreading, erect or adpressed; petals 5, red, orange, yellow, white or greenish, fused into a 5-angled tube at base, segments erect or reflexed, thickly keeled; stamens 10, unequal, carpels 5.

A genus of about 150 species from USA, Mexico, Central and South America. Most are easily propagated from leaf or stem cuttings or offsets.

Literature: Ginns, R., Echeverias, *The National Succulent Society Handbooks, No 1* (1968); Walther, E., *Echeveria* (1972); Carruthers, L. & Ginns, R. *Echeveria: A guide to cultivation and identification* (1973).

1a. Leaves conspicuously hairy or with very fine hairs or papillae   2
 b. Leaves hairless, but often covered with a waxy bloom   13
2a. Leaf-hairs very fine (hand lens required), or leaf surface covered with minute papillae   3
 b. Leaf-hairs obvious without the use of a lens   5
3a. Leaves 10–14 cm, with very fine hairs, or sometimes hairless
      1. semivestita
 b. Leaves 3–7 cm, with fine hairs or pointed papillae   4
4a. Leaves 4–7 cm, surface roughened with minute pointed papillae; corolla 2–2.4 cm   **2. spectabilis**
 b. Leaves not more than 5 cm, surface with fine hairs but not roughened; corolla *c.* 1.6 cm   **3. nodulosa**
5a. Leaf-rosettes stemless, flowers borne in simple or 2-branched, 1-sided racemes   6
 b. Shrublets, often with scattered leaves, or rosettes borne on clear stems; flowers borne all around the stem   7
6a. Leaves hairy over entire surface
      4. setosa
 b. Leaves hairy only along margins and on keel beneath   **5. ciliata**
7a. Stems often branched, 30–60 cm, rusty brown towards the top, especially when young; leaves in loose terminal clusters
      6. coccinea
 b. Stems very short or rarely exceeding 30 cm, solitary or branched   8

8a. Leaves thick, turgid; sepals closely adpressed to corolla   9
 b. Leaves usually thin; sepals spreading   10
9a. Leaf-hairs often reddish; sepals clearly joined at the base, less than half as long as corolla
      7. pulvinata
 b. Leaf-hairs white; sepals almost free at the base, more than half as long as corolla   **8. leucotricha**
10a. Well-branched shrub to 30 cm or occasionally more; flowers few, corolla 3 cm or more   **9. harmsii**
 b. Shrublet 10–20 cm; flowers in racemes, corolla 1.2–2.6 cm   11
11a. Stems to 20 cm; corolla 2.4–2.6 cm
      10. amphoralis
 b. Stems rarely exceeding 10 cm; corolla 1.2–1.5 cm   12
12a. Leaves to 7 cm, oblanceolate-elliptic; flowers borne in compound panicles   **11. pilosa**
 b. Leaves to 4 cm, obovate-lanceolate; flowers borne in simple racemes
      12. pringlei
13a. Leaves 15–40 cm   14
 b. Leaves 2–15 cm   19
14a. Leaves 12–25 cm wide   15
 b. Leaves 5–10 cm wide   16
15a. Bracts to 10 cm, bristle-tipped
      14. gibbiflora
 b. Bracts not more than 4 cm, tapered to a fine point   **15. grandiflora**
16a. Leaves 4 times as long as wide
      16. acutifolia
 b. Leaves not more than 2–3 times as long as wide   17
17a. Leaves rounded and notched at apex; flowering-stem to 2 m
      17. gigantea
 b. Leaves rounded or pointed with a bristle tip; flowering-stem not more than 1 m   18
18a. Bracts spurred; corolla 1.2–1.5 cm, buff within   **18. fimbriata**
 b. Bracts not spurred; corolla 2–2.5 cm, yellow-red within
      19. subrigida
19a. Leaves less than 1 cm wide   20
 b. Leaves usually 1.5 cm or more across   24
20a. Plants stemless, or very short, forming crowded rosettes   21
 b. Stems 10–12 cm   22
21a. Sepals to 3 mm, unequal, almost free   **20. amoena**
 b. Sepals *c.* 6 mm, equal   **21. bella**
22a. Corolla 1.7–1.9 cm
      23. macdougallii
 b. Corolla 9–11 mm   23

23a. Leaves 2.5–3 cm   **13. gracilis**
 b. Leaves 3–3.5 cm   **22. johnsonii**
24a. Leaves at least 4 times as long as wide   25
 b. Leaves 1–3 times as long as wide   34
25a. Leaves not more than 5 cm
      24. whitei
 b. Leaves mostly more than 5 cm (4–12 cm)   26
26a. Corolla creamy white
      25. vanvleitii
 b. Corolla orange-red, pinkish green or yellow   27
27a. Sepals 3–10 mm   28
 b. Sepals 1.1–1.6 cm   32
28a. Stems to 5 cm, branched
      26. rauschii
 b. Plants stemless or very short stemmed, sometimes unbranched   29
29a. Corolla 8–10 mm   30
 b. Corolla 1.2–1.4 cm   31
30a. Leaves 5–7 × 1–1.5 cm; corolla pinkish orange outside, yellow inside   **27. carnicolor**
 b. Leaves 9–10 × 2–2.5 cm; corolla green-yellow   **28. megacalyx**
31a. Stems branched; corolla to 1.4 cm
      29. sessiliflora
 b. Stems unbranched; corolla not more than 1.2 cm   **30. bifida**
32a. Stems tufted; corolla not more than 10 mm, pinkish green
      31. sanchez-mejoradae
 b. Stems horizontal or absent; corolla 1.3–1.5 cm, red, orange or yellow   33
33a. Stems horizontal, thin; sepals to 1.6 cm, linear, pinkish   **32. rosea**
 b. Stems short or absent; sepals to 1.4 cm, triangular to oblong, bluish green or red   **33. strictiflora**
34a. Leaves 10–15 cm   35
 b. Leaves 2–10 cm   41
35a. Stems absent or very short   36
 b. Stems 10–60 cm   38
36a. Stems absent; corolla not more than 1.2 cm, red at base
      34. paniculata
 b. Stems very short; corolla 1.5–1.6 cm, yellow throughout   37
37a. Sepals not more than 8 mm, unequal   **35. maculata**
 b. Sepals *c.* 1.3 cm, almost equal
      36. lutea
38a. Stems 30–60 cm; leaves green with red margins; corolla red at base, green above   **37. nuda**
 b. Stems 10–30 cm; leaves green or purplish, often glaucous   39

39a. Leaves and stems deep purplish green; corolla red **38. atropurpurea**

b. Leaves greenish; corolla pink-yellow or reddish orange 40

40a. Leaf-margins usually wavy; corolla pink outside, yellow inside **39. crenulata**

b. Leaf-margins wavy or flat; corolla red outside, orange inside **40. fulgens**

41a. Plants with obvious stems 2–90 cm 42

b. Plants stemless or with very short stems, or tufted 54

42a. Stems 2–10 cm 43

b. Stems 20–90 cm 48

43a. Stems not more than 2–3 cm **41. moranii**

b. Stems mostly 5–10 cm 44

44a. Stems unbranched 45

b. Stems branched 46

45a. Leaves 2–5 cm; corolla red **42. affinis**

b. Leaves 5–9 cm; corolla red outside, yellow inside **43. laui**

46a. Stems not more than 6 cm; leaves 3–3.5 cm; corolla red throughout **44. pulchella**

b. Stems 6–10 cm; leaves 4–10 cm; corolla red outside, pink or yellow inside 47

47a. Bracts to 4 cm; corolla pink inside **45. pittieri**

b. Bracts not more than 2.5 cm; corolla yellow inside **46. stolonifera**

48a. Stems 20–30 cm 49

b. Stems 30–90 cm 50

49a. Leaves 2–7 × 1.5–2 cm; corolla red outside, pink inside **47. australis**

b. Leaves 8–15 × 4–7 cm; corolla red outside, orange inside **40. fulgens**

50a. Corolla 1.2–1.5 cm, yellow outside, orange-yellow inside **48. bicolor**

b. Corolla 8–11 mm, red, at least in part 51

51a. Sepals not more than 6 mm; corolla red throughout, or yellow inside 52

b. Sepals mostly 8–9 mm; corolla red and white or green 53

52a. Shrublet with many branched stems; leaf-margin and tips red; bracts to 2.5 cm; corolla red, orange-yellow on margin and inside **49. multicaulis**

b. Horizontal or erect herb, stems little branched; bracts to 4 cm; corolla red throughout **50. maxonii**

53a. Leaves 6–13 cm; corolla red at base, green above **37. nuda**

b. Leaves 2.5–7.5 cm; corolla white at base, red above **51. waltheri**

54a. Leaves 3 times as long as wide 55

b. Leaves 1–2.5 times as long as wide 58

55a. Leaves 3–6 × 1–2 cm 56

b. Leaves 6–10 × 2–3 cm 57

56a. Bracts 1–1.2 cm; corolla 1.2–1.5 cm, pink outside, yellow-orange inside **52. elegans**

b. Bracts c. 1.5 cm; corolla to 1 cm, lemon-yellow **53. pulidonis**

57a. Corolla c. 1.2 cm, yellow, red at base **34. paniculata**

b. Corolla 1.4–1.7 cm, red outside, orange inside **54. schaffneri**

58a. Sepals 1.3–1.6 cm 59

b. Sepals 2–10 mm 61

59a. Corolla c. 2 cm **55. runyonii**

b. Corolla 1.2–1.5 cm 60

60a. Leaves 2–4 × 2–2.5 cm; corolla yellow, red on keel and tip **56. derenbergii**

b. Leaves 6 cm or more; corolla pink outside, orange inside **57. cuspidata**

61a. Bracts with 2 or 3 spurs at base 62

b. Bracts entire 63

62a. Bracts with 2 spurs; corolla c. 11 mm, red-pink **58. peacockii**

b. Bracts with 3 spurs; corolla 2.5–3 cm, greenish yellow **59. longissima**

63a. Sepals 3–6 mm 64

b. Sepals 8–10 mm 71

64a. Leaves 1–1.3 cm across; corolla c. 1.6 cm **60. halbingeri**

b. Leaves 1.4 cm or more across; corolla 7–12 mm 65

65a. Plants tufted, with numerous offsetting rosettes 66

b. Rosettes solitary or with few branches 67

66a. Leaves 1.2–3 cm wide; flowering-stems solitary, to 30 cm **61. secunda**

b. Leaves 3–4 cm wide; flowering-stems several, to 50 cm **62. lindsayana**

67a. Bracts 5–10 mm; flowering-stem to 50 cm **63. racemosa**

b. Bracts 1.5–2 cm; flowering-stems 20–40 cm 68

68a. Flowering-stem solitary 69

b. Flowering-stems several 70

69a. Leaves spotted with red and brown, in a dense solitary rosette **64. purpusorum**

b. Leaves not spotted, rosettes usually tufted, rarely solitary **65. agavoides**

70a. Sepals broadly triangular, to 5 mm; corolla pink outside yellow-orange inside, occasionally white **66. potosina**

b. Sepals lanceolate, acute, to 6 mm; corolla orange **67. obtusifolia**

71a. Bracts 8–15 mm 72

b. Bracts 2–3 cm 73

72a. Corolla 1–1.3 cm; pink **68. shaviana**

b. Corolla 1.5–1.8 cm; pink outside, yellow inside **69. albicans**

73a. Stems very short, branched; bracts 2 cm; corolla orange outside, orange-yellow inside **70. humilis**

b. Plant stemless; bracts 3 cm; corolla red outside, orange inside **71. chihuahuaensis**

1. E. semivestita Moran. Illustration: Walther, Echeveria, 167–169 (1972); Graf, Exotica, series 4, **1**: 886 (1985).

Stem short, unbranched. Leaves 10–14 × 1.5–3 cm, lanceolate, covered with very fine hairs, very rarely hairless, margins curved upwards and tinged red, base nearly round in cross-section, borne in a rosette. Flowering-stem to 55 cm, a many-flowered panicle; bracts leaf-like, reduced in upper part; sepals to 1.5 cm, unequal, green-purple, glaucous; corolla to 1.3 cm, pink-red outside, yellow-red within. *Mexico*. G1.

2. E. spectabilis Alexander. Illustration: Walther, Echeveria, 318 (1972).

Shrublet 10–60 cm with much branched stems. Leaves 4–7 × 2.5–3 cm, spathulate to obovate, bristle-tipped, flat, covered with minute, pointed papillae, green with red margins, with leaf-stalks, borne in loose rosettes or scattered. Flowering-stems to 70 cm, several, with c. 10-flowered racemes; bracts c. 5 mm, oblong, stalkless; sepals to 1.8 cm, nearly equal, lanceolate, pointed, ascending or somewhat spreading; corolla 2–2.4 cm, red-orange outside, yellow on margins and within, 5-sided, tapering to mouth, segments keeled, pouched at base. *Mexico*. G1. Summer.

3. E. nodulosa (Baker) Otto (*Cotyledon nodulosa* Baker; *E. discolor* Baker). Illustration: Walther, Echeveria, 229, 315, 316 (1972); Graf, Exotica, series 4, **1**: 895 (1985).

Stems to c. 20 cm, branching. Leaves 4–5 × 1–1.5 cm, wedge-shaped to obovate, thick, concave above, light green, deep red on margins and keel, surface covered with fine hairs, but not roughened; borne in loose rosettes or scattered. Flowers 8–12, horizontal, in racemes, on several stems to 30 cm; bracts *c.* 3 cm; sepals to 1.5 cm, nearly equal, linear, pointed, thick, spreading; corolla to 1.6 cm, red, yellow within and on margins, hardly tapering to mouth, segments sharply keeled, thick, fine-pointed. *Mexico.* G1. Autumn.

4. E. setosa Rose & Purpus. Illustration: Walther, Echeveria, 18, 240, 399, 400 (1972); Reader's Digest encyclopaedia of garden plants and flowers, 240 (1972); Graf, Exotica, series 4, **1**: 892 (1985); Graf, Tropica, edn 3, 370 (1986).

Leaf-rosettes stemless, or very short-stemmed. Leaves 4–5 × 1.8–2 cm, numerous, oblanceolate, flat above, pointed, bristle- tipped, covered with long hairs throughout. Flowering-stems to 30 cm, several, with racemes of *c.* 10 flowers; bracts *c.* 2.5 cm, oblong, pointed, thick, ascending; sepals to 10 mm, nearly equal, triangular to oblong, thick; corolla 1–1.2 cm, yellow with red markings, yellow within, urn-shaped, 5-sided. *Mexico.* G1. Summer.

5. E. ciliata Moran. Illustration: Walther, Echeveria, 240, 401, 402 (1972); Graf, Tropica, edn 3, 386 (1986).

Low-growing, hairy perennial with leaves clustered in basal rosettes; stem to 7 cm. Leaves 3–5 cm, wedge-shaped to obovate, bristle-tipped, hairy only along the margins and on keel beneath, 30–70 to a rosette, leaf-stalk thick. Flowers 4–7 in simple or one-sided racemes; sepals *c.* 4 mm, nearly equal, spreading, lanceolate to triangular, pointed; corolla *c.* 1 cm, yellow-red, tapering to mouth. *Mexico.* G1.

6. E. coccinea (Cavanilles) de Candolle (*Cotyledon coccinea* Cavanilles; *E. pubescens* Schlechtendal). Illustration: Botanical Magazine, 2572 (1825); Walther, Echeveria, 389 (1972); Graf, Exotica, series 4, **1**: 887 (1985).

Densely hairy throughout with branching stems 30–60 cm, deep rusty brown when young, later greyish. Leaves 6–8 × 2 cm, oblanceolate, concave above, convex below, clustered towards ends of branches, stalk-like at base. Flowers *c.* 25, borne in spikes to 30 cm; bracts *c.* 3 cm, leaf-like, soon falling; sepals to 1.3 cm, unequal, lanceolate, round in cross-section, pointed; corolla 1–1.2 cm, red outside, orange-yellow within, 5-sided, cylindric. *Mexico.* G1. Autumn–winter.

7. E. pulvinata Rose. Illustration: Botanical Magazine, 7918 (1903); Walther, Echeveria, 237, 393 (1972); Carruthers & Ginns, Echeverias: A guide to cultivation & identification, pl. 1 (1973); The RHS gardeners' encyclopedia of plants & flowers, 385 (1989); Graf, Tropica, edn 3, 368 (1986).

Densely hairy throughout, with stems to 20 cm. Leaves 2.5–6.5 × 1.8–3.5 cm, spathulate to obovate, fine-pointed, thick, turgid, borne in very loose rosettes, hairs often reddish. Flowering-stems 20–30 cm, with racemes of *c.* 15 flowers; bracts round in cross-section, pointed; sepals 5–9 mm, nearly equal, clearly joined at base, adpressed; corolla 1.2–1.9 cm, yellow, keel red, urn-shaped, 5-angled. *S Mexico.* G1. Winter–spring.

8. E. leucotricha J.A. Purpus. Illustration: Walther, Echeveria, 375 (1972); Graf, Exotica, series 4, **1**: 891 (1985); Graf, Tropica, edn 3, 370 (1986).

Hairy shrublet, to 15 cm, with red branching stems. Leaves 6–8 × 2–2.5 cm, oblong-lanceolate, blunt, bristle-tipped, upcurved and red, borne in loose rosettes, hairs whitish. Flowering-stems to 40 cm, with 12–15 flowers, spike-like or paniculate; bracts *c.* 3 cm, obovate-oblong; sepals 1–1.2 cm, nearly equal, free to base, adpressed, triangular-linear, pointed; corolla to 1.8 cm, orange, red on keel, 5-angled, tapering to mouth, segments spreading at tips. *Mexico.* G1. Spring.

9. E. harmsii J.F. Macbride (*Oliverella elegans* Rose; *Oliveranthus elegans* (Rose) Rose; *Echeveria elegans* (Rose) Berger). Illustration: Walther, Echeveria, 224, 410 (1972); Graf, Exotica, series 4, **1**: 894, 898 (1985); Graf, Tropica, edn 3, 374 (1986).

Shrub, shortly hairy throughout, with branching stems to 30 cm, occasionally more. Leaves 2–5 × 1–1.5 cm, oblanceolate, pointed, light green, margins and tip red, clustered on ends of branches. Flowering-stems many, to 20 cm, few-flowered; bracts *c.* 2 cm, few, pointed; sepals to 1.8 cm, spreading. Corolla 3–3.4 cm, red, yellow on margins and inside, urn-shaped, 5-angled, segments fine-pointed. *Mexico.* Summer.

Many of the hybrids common in cultivation have this as one parent, including 'Set Oliver', 'Victor' and 'Pulv Oliver'. Most would key out here.

10. E. amphoralis Walther. Illustration: Cactus & Succulent Journal of America 30: 149–150 (1958); Walther, Echeveria, 241, 407–408 (1972).

Shrublet, hairy throughout with branched stems to 20 cm. Leaves many, 3.5–4 × 2–2.5 cm, wedge-shaped to obovate, bristle tipped, borne in loose, ill-defined rosettes, leaf-stalk thick. Flowering-stems to 20 cm, several, with few-flowered racemes; bracts 2–3 cm, bristle-tipped, spreading; sepals to 1.4 cm, nearly equal, round in cross-section, pointed, spreading; corolla 2.2–2.6 cm, red at base, yellow above and within. *Mexico.* G1. Summer.

11. E. pilosa J.A. Purpus. Illustration: Jacobsen, A handbook of succulent plants **1**: 382 (1960); Walther, Echeveria, 297 (1972). Stem short, not exceeding 10 cm, unbranched, with long hairs throughout.

Leaves 6–7 × 1.5–2 cm, oblanceolate to elliptic, tapering to a point, concave above, convex below, *c.* 40 in loose rosettes. Flowers borne in a raceme, paniculate below, stems several, to 30 cm; bracts to 3 cm, many, oblong, pointed; sepals to 1.5 cm, nearly equal, almost round in cross-section, pointed; corolla *c.* 1.2 cm, orange, keels red, yellow at tip and within, 5-sided, tapering to the mouth. *Mexico.* G1. Spring–summer.

12. E. pringlei (S. Watson) Rose (*Cotyledon pringlei* S. Watson). Illustration: Walther, Echeveria, 241, 403 (1972).

Densely hairy throughout, stem to 10 cm and branched. Leaves 3.5–4 × 2–2.5 cm, obovate to lanceolate, pointed, borne in loose rosettes at end of branches or scattered. Cymes 15–25 cm, simple, with 5–18 flowers, becoming horizontal; bracts to 3 cm, leaf-like, reduced in upper part; sepals to 1.4 cm, nearly equal, lanceolate, pointed, spreading; corolla to 1.5 cm, red outside, buff within, urn-shaped, 5-angled. *Mexico.* G1.

13. E. gracilis Walther. Illustration: Walther, Echeveria, 300 (1972).

Shrubby at base, stems to 10 cm, horizontal or erect. Leaves 2.5–3 × 0.9–1 cm, *c.* 5 mm thick, oblong to club-shaped, curving upwards, scattered or in loose rosettes. Racemes several with 10 or more flowers; bracts *c.* 2 cm, many, pointed; sepals *c.* 10 mm, nearly equal, adpressed or widely spreading; corolla *c.* 10 mm, red outside, orange within, 5-angled, tapering slightly, segments thick, pointed. *Mexico.* G1.

14. E. gibbiflora de Candolle (*Cotyledon gibbiflora* (de Candolle) Baker). Illustration: Walther, Echeveria, 223, 226, 227 (1972); Graf, Exotica, series 4, **1**: 893 (1985); Graf,

Tropica, edn 3, 369 (1986). Stem to 30 cm, unbranched, erect, about 5 cm thick.

Leaves 35–38 × 10–25 cm, tinged purple, obovate-spathulate, pointed, margins wavy, 12–25 to a rosette. Panicles to 1 m, many-branched with 12 solitary flowers to each branch; bracts 8–10 × 4–5 cm, obovate, bristle-tipped; sepals lavender, to 11 mm, unequal, spreading, lanceolate, pointed; corolla c. 1.6 cm, red outside, buff within, cylinder-shaped to bell-shaped, segments slightly spreading at tips. *Mexico.* G1. Autumn–winter.

15. E. grandiflora Haworth. Illustration: Walther, Echeveria, 230, 231 (1972).

Similar to *E. gibbiflora* except; leaves c. 30 × 12 cm, oblanceolate, margins curved upwards, pointed and bristle-tipped; bracts not more than 4 cm, narrower and tapered to a fine point. *Mexico.* G1. Autumn–winter.

16. E. acutifolia Lindley. (*E. holwayi* Rose; *Cotyledon acutifolia* (Lindley) Baker; *Cotyledon devensis* N.E. Brown). Illustration: Walther, Echeveria, 214, 215 (1972).

Stem to 30 cm, unbranched. Leaves 28–32 × 7–9 cm, obovate to oblong, bristle-tipped or pointed, green tinged red, leaf-stalk long, channelled. Flower-stalks c. 1 cm, with 3 or 4 flowers to each panicle; sepals c. 6 mm, unequal, ascending or erect; petals 11–12 mm, red, mouth of tube contracted. *S Mexico.* G1. Winter.

17. E. gigantea Rose & Purpus. Illustration: Walther, Echeveria, 198–202 (1972); Graf, Exotica, series 4, **1**: 892 (1985).

Stem to 50 cm, unbranched. Leaves 15–20 × 8–10 cm, grape- green, margins purple, spathulate to obovate, rounded and notched at apex, base of leaf stalk-like, few borne in a loose rosette, Flowers solitary in panicles to 2 m, thick at base; bracts green, margins green, spathulate to oblong, pointed or blunt, numerous, c. 1.5 cm; sepals unequal, triangular to oblong-lanceolate, tapering to the tip, spreading; corolla 1.2–1.7 cm, rose-red, not yellow within, 5-sided, not constricted at mouth. *Mexico.* G1. Winter.

18. E. fimbriata C. Thompson. Illustration: Walther, Echeveria, 19, 187 (1972).

Stem to 50 cm. Leaves 15–20 × 6–7 cm, almost stalked at base, rounded with a bristle tip, somewhat hairy at first, few, borne in a loose rosette. Flowering-stems 2 or 3 branched, flowers usually solitary; bracts c. 4 × 1.8 cm, few, pointed, spurred;

sepals light blue to grey-green, to 9 mm, unequal, spreading to erect, pointed. Corolla 1.2–1.5 cm, red-pink outside, buff within, narrowing slightly at mouth, segments thick, spreading at tips. *Mexico.* G1. Winter.

19. E. subrigida (Robinson & Seaton) Rose (*Cotyledon subrigida* Robinson & Seaton; *E. angusta* von Poellnitz; *E. palmeri* Rose; *E. rosei* Nelson & Macbride). Illustration: Jacobsen, A handbook of succulent plants **1**: 386 (1960); Walther, Echeveria, 177, 178 (1972); Graf, Exotica, series 4, **1**: 893 (1985).

Stem to 10 cm, thick. Leaves 15–25 × 5–10 cm, obovate to oblanceolate, pointed, white glaucous like a *Dudleya*, margins red, upturned and with finely rounded teeth, borne in dense rosettes. Flowering-stems 60–90 cm, 1 or 2 with 6–15 branches, each with panicles of c. 15 flowers; bracts few, 3–5 cm; sepals to 2.5 cm, just equal, triangular to lanceolate, ascending, grey-purple; corolla 2–2.5 cm, red, bloomed white outside, yellow-red within, 5-sided, not very constricted at mouth. *Mexico.* G1. Summer.

20. E. amoena L. de Smet (*E. microcalyx* Britton & Rose; *E. pusilla* Berger).

Mat-forming. Leaves 2–3 × 0.6–0.8 cm, spathulate to oblanceolate, pointed, round in cross-section, somewhat club-shaped, grey-green with maroon tips, borne in dense rosettes. Flowering-stalks 10–20 cm, erect, with 6–12 nodding flowers; bracts numerous; sepals to 3 mm, unequal almost free, blunt; corolla 8–9 mm, pinkish rose to red, margins yellow, thin, exceeding stamens. *Mexico.* G1. Spring.

21. E. bella Alexander. Illustration: Walther, Echeveria, 357, 358 (1972).

Stem very short and branched. Leaves 2–2.8 × 0.3–0.5 cm, narrowly oblong, pointed, convex above and below, in dense rosettes of 20–30. Flowering-stems c. 25 cm, red, with racemes of c. 12 flowers; bracts c. 2.3 × 0.5 cm, oblong, pointed and spurred; sepals c. 6 mm, nearly equal, spreading, almost round, pointed; corolla to 10 mm, tapering to mouth, segments red to yellow spreading at tips and bristled below apex. *S Mexico.* G1. Spring.

22. E. johnsonii Walther. Illustration: Walther, Echeveria, 293 (1972); Carruthers & Ginns, Echeverias: A guide to cultivation & identification, pl. 25 (1973).

Stems to c. 10 cm, branched. Leaves 3–3.5 × 0.7–0.9 cm, crowded near end of branches, club-shaped, or oblanceolate,

nearly round in cross-section, faintly purple on edges. Flowers c. 10, borne in spike-like heads, lateral, to 10 cm; bracts c. 2 cm, oblong, narrow, round in cross-section, pointed; corolla 9–11 mm, buff, red on keel, orange-yellow within, 5-sided, straight, segments just spreading at tips. *Ecuador.* G1.

23. E. macdougallii Walther. Illustration: Walther, Echeveria, 303, 304 (1972); Carruthers & Ginns, Echeverias: A guide to cultivation & identification, pl. 32 (1973).

Shrublet with stem to 12 cm. Leaves 2–3 × 0.5–1 cm, glaucous, green with red markings, very thick, obovate to club-shaped, nearly round in cross-section, borne in loose rosettes near end of the stems. Flowers few, borne in racemes on stems 10–25 cm; bracts 1.8–2.5 cm, leaf-like; sepals c. 10 mm, nearly equal, oblong to elliptic, round in cross-section, blunt; corolla 1.7–1.9 cm, red, yellow on upper edges and within, to 5-sided, straight, tapering to mouth, segments slightly spreading at tips. *Mexico.* G1.

24. E. whitei Rose (*E. buchtienii* von Poellnitz; *E. chilonensis* Walther). Illustration: Walther, Echeveria, 354 (1972).

Stem stout, branching. Leaves 3–5 × 0.8–2 cm, spathulate, bristle-tipped, flat above, rounded below, glaucous green-brown, c. 35 borne to each rosette. Flowering-stalks to 30 cm, sometimes one-sided, branched, with racemes of c. 10 flowers; bracts to 1 cm, few, triangular to ovate, pointed; sepals to 6 mm, ascending to erect, linear to lanceolate, pointed; corolla 1–1.5 cm, 5-sided, urn-shaped, red. *Bolivia.* G1. Winter–spring.

25. E. vanvlietii von Keppel.

Stem to 5 cm, 1–2 cm thick. Leaves 4–8 × 1–2 cm, oblanceolate to oblong, with a small point, convex below, grey-green, tinged bronze, 20–25 to each rosette. Flowering-stems 20–50 cm, with 15–40 flowers in each raceme; bracts to 2.5 × 1 cm, ovate; sepals 5–10 mm, unequal, spreading, ascending, triangular to ovate; corolla to 12 mm, creamy white, 5-sided, segments spreading at tips. *Bolivia.* G1. Summer.

26. E. rauschii von Keppel.

Stem to 5 cm, branched. Leaves 4–7 × 1.3–1.5 cm, triangular to oblanceolate, pointed, green, margins and tip dark red, rosettes with 10–15 leaves. Flowering-stems 10–25 cm, tinged red, with racemes of 7–20 flowers; bracts few, to 4 cm, oblong; sepals 3–10 mm, unequal, spreading, oblong to linear; corolla straight, orange-red outside, orange-yellow within. *Bolivia*. G1. Early autumn.

27. E. carnicolor (Baker) Morren (*Cotyledon carnicolor* Baker). Illustration: Jacobsen, A handbook of succulent plants **1**: 370 (1960); Walther, Echeveria, 346, 347 (1972); Graf, Exotica, series 4, **1**: 894 (1985).

Stem short or absent. Leaves 5–7 × 1.5–1.6 cm, in rosettes of about 20, oblanceolate to spathulate, blunt, thick, concave above, convex below, flesh-coloured, papillose. Flowers 6–20, borne in several racemes to 25 cm; bracts *c.* 2 cm, numerous, easily detached, pointed. Sepals 5–6 mm, lanceolate, nearly equal, almost round in cross-section, pointed; corolla 9–10 mm, pink-orange outside, buff-yellow inside, 5-angled, straight, segments thick, keeled. *E Mexico*. G1. Winter.

28. E. megacalyx Walther. Illustration: Walther, Echeveria, 365, 366 (1972).

Stem short, branching. Leaves 9–10 × 2–2.5 cm, thin, oblong to spathulate, many borne in dense rosettes, margin thin, transparent. Flowers solitary, nodding, *c.* 30, in racemes on stems to 45 cm; bracts leaf-like, reduced higher up; sepals 8–10 mm, leaf-like, spreading, nearly equal, pointed; corolla *c.* 8 mm, green-yellow, urn-shaped, tips of segments reflexed. *S Mexico*. G1. Summer–autumn.

29. E. sessiliflora Rose (*E. corallina* Alexander).

Stem 2–3 cm, branched. Leaves 6–8 × 1.5–1.6 cm, oblanceolate, pointed, glaucous-green, borne in terminal rosettes, margins brown-red. Flowering-stems 40–50 cm, pink, with 20–25 flowers; bracts leaf-like, numerous, reduced in upper part; sepals to 7 mm, unequal, blue-green; corolla to 1.4 cm, conical, segments spreading at tips. *Mexico*. G1.

30. E. bifida Schlechtendal (*E. teretifolia* Kunze). Illustration: Walther, Echeveria, 217, 243, 246 (1972).

Stem short, not branched. Leaves tinged red, 3.5–10 × 1.2–2.5 cm, oblanceolate to diamond-shaped, pointed, bristle-tipped, glaucous green. Flowers 20–30, borne in a 2-branched raceme, 25–60 cm; bracts *c.* 3 cm, many, round in cross-section, blunt; sepals to 1 cm, unequal, rigid, rounded, blunt, spreading. Corolla *c.* 1.2 cm, pink-orange outside, yellow within, keeled, petals joined for much of their length, urn-shaped. *Mexico*. G1. Summer.

31. E. sanchez-mejoradae Walther. Illustration: Walther, Echeveria, 109, 110, 213 (1972).

Stems tufted. Leaves *c.* 6 × 1.5 cm, oblanceolate to obovate or linear, tapering towards the base, tip rounded with a small point, borne in a number of rosettes. Flowering-stems to 50 cm, several, unbranched; flowers 10, one-sided; bracts sparse; sepals *c.* 11 mm, triangular to lanceolate, pointed, usually spreading, unequal; corolla 9–10 mm, pink-green, pitcher-shaped. *Mexico*. G1. Spring–summer.

32. E. rosea Lindley (*Cotyledon roseata* Baker). Illustration: Walther, Echeveria, 229, 326 (1972).

Often epiphytic shrublet, stems little branched, thin, horizontal. Leaves 5–9 × 1.5–2 cm, oblanceolate to oblong, up-turned, more or less flat, pointed, leaf-stalk 3-angled, borne in indistinct rosettes. Flowering-stems to 20 cm, flowers borne in spikes or racemes; bracts *c.* 3 mm, oblanceolate and stalked; sepals *c.* 1.6 cm, linear, narrow, spreading and pink-purple; corolla 1.2–1.3 cm, yellow with red tips, straight, segments thin, keeled. *Mexico*. G1. Winter–spring.

33. E. strictiflora A. Gray (*Cotyledon strictiflora* (A. Gray) Baker). Illustration: Walther, Echeveria, 217, 250, 251 (1972).

Stem short or absent. Leaves 7–9 × 1.5–2 cm, spreading and ascending, obovate to diamond-shaped, tapering to base and pointed at tip, thin, glaucous, few borne in a loose rosette, margins curled upwards. Flowering-stem to 20 cm, solitary to several, racemes one-sided, occasionally branched with 10–15 flowers; bracts 2–3 cm, many, keeled, pointed; sepals *c.* 1.4 cm, unequal, oblong to triangular, ascending, pointed; corolla 1.4–1.5 cm, red outside, orange-red within, bell or urn-shaped, 5-sided, lobes sharply keeled. *USA (Texas) & Mexico*. G1. Summer.

34. E. paniculata A. Gray (*Cotyledon grayii* Baker; *E. grayii* (Baker) Morren). Illustration: Walther, Echeveria, 371 (1972).

Stem absent. Leaves 10–12 × 3–5 cm, *c.* 40 per rosette, spreading, oblanceolate, fine-pointed. Flowers solitary, 2–6 to to each branched panicle, stems to 50 cm; bracts few, lanceolate to ovate; sepals to *c.* 10 cm, unequal, triangular to oblong, pointed, spreading; corolla 1–1.2 cm, yellow, red at base, conical, 5-sided, tapering to mouth. *Mexico*. G1. Summer.

35. E. maculata Rose. Illustration: Walther, Echeveria, 373 (1972).

Roots thick, stem short or absent. Leaves 9–15 × 2–3 cm, oblanceolate to obovate, fine-pointed, glaucous, green, up-curved. Flowers solitary, borne in spikes or panicles on stems 30–90 cm; bracts to *c.* 5 cm, nearly round in cross-section, many; sepals to *c.* 8 mm, round in cross-section, pointed, spreading, unequal; corolla 1.5–1.6 cm, narrowly urn-shaped, segments spreading above, yellow. *Mexico*. G1. Spring–summer.

36. E. lutea Rose. Illustration: Walther, Echeveria, 257, 258 (1972).

Stems very short. Leaves 11–13 × 3–4 cm, linear to oblanceolate, pointed, margins and tip curved, somewhat keeled, green, tinged purple, borne in crowded rosettes of *c.* 25. Flowering-stems simple or 2-branched with 12–30 solitary flowers; bracts 3–6.5 cm, numerous, round to triangular in cross-section, lanceolate to linear, bristle tip clear; sepals *c.* 1.3 cm, almost equal; corolla to 1.7 cm, yellow, lobes somewhat spreading, 5-sided. *Mexico*. G1. Summer.

37. E. nuda Lindley (*Cotyledon nuda* (Lindley) Baker; *E. navicularis* L. de Smet). Illustration: Walther, Echeveria, 221, 279, 280 (1972); Graf, Exotica, series 4, **1**: 894 (1985).

Stem 30–60 cm, branched. Leaves 6–13 × 2–5.5 cm, spathulate to obovate, blunt with a bristle tip or pointed, thin, tapering to the base, green, bloomed at first, borne in loose rosettes at ends of branches, margins red. Flowering-stems to 40 cm, with *c.* 20 to each spike; bracts to 2.5 cm, obovate, pointed, thin; sepals to 9 mm, nearly equal, linear to lanceolate, fine-pointed, somewhat ascending; corolla 1–1.1 cm, 5-angled, segments red at base, green above, tip erect or spreading. *Mexico*. G1. Summer.

38. E. atropurpurea (Baker) Morren (*Cotyledon atropurpurea* Baker; *Echeveria sanguinea* misapplied). Illustration: Walther, Echeveria, 232, 329 (1972).

Stem 10–15 × 2.5 cm. Leaves 10–12 × 3–5 cm, oblong or spathulate to obovate, dark purple-green, glaucous, forming dense rosettes of *c.* 20 leaves. Flowering-stems 30–60 cm, with 20–25 flowers to each raceme; bracts leaf-like but smaller; sepals *c.* 4 mm, almost equal, lanceolate, spreading to swept back; corolla to 1.2 cm, red, 5-angled, segments pointed. *Mexico*. G1.

39. E. crenulata Rose. Illustration: Graf, Exotica, series 4, **1**: 893 (1985); Graf, Tropica, edn 3, 369, 370 (1986).

Stem *c.* 10 cm, branching at base. Leaves 10–12 × 7–9 cm, obovate to diamond-shaped, pointed, with or without a bristle tip, few in a loose rosette, margins wavy or flat, brown. Flowers borne in several panicles to 50 cm; bracts many, *c.* 5 mm, reduced higher up, pointed, sepals to 1 cm, unequal, triangular-oblanceolate, pointed, spreading or swept back; corolla to 1.8 cm, pink outside, yellow within, 5-angled, straight. *Mexico.* G1. Winter.

40. E. fulgens Lamarck (*E. retusa* Lamarck). Illustration: Walther, Echeveria, 160, 161, 162 (1972); Graf, Exotica, series 4, **1**: 893 (1985).

Stems to 30 cm, occasionally branched. Leaves 8–15 × 4–7 cm, spathulate to obovate, glaucous green, blunt, bristle-tipped, few, borne in rosettes, base stalk-like, margin wavy or flat. Flowering-stems to 90 cm, several, few branched, flowers 20–30; bracts numerous, to 3.5 cm; sepals to 10 mm, unequal, lanceolate to triangular, pointed, green; corolla to 1.5 cm, red outside, orange within, segments slightly spreading at tips. *Mexico.* G1. Winter.

41. E. moranii Walther. (*E. proxima* Walther). Illustration: Walther, Echeveria, 348, 349 (1972).

Stems 2–3 × 0.6–0.8 cm thick. Leaves 4–6 × 2–3 cm, wedge-shaped to obovate, 1 cm thick at centre, concave above, convex below, bristle tipped, papillose when young, green with maroon spots, margins and tip, *c.* 25 to each rosette. Flowering-stems several, 20–50 cm, with *c.* 15 flowers to each raceme; bracts green to blue-grey, *c.* 3 × 1 cm, oblong to elliptic, with transparent spur; sepals *c.* 4 mm, nearly equal, triangular to ovate, pointed, thick; corolla *c.* 1.3 cm, red outside, orange-buff within, 5-sided, conical, segments slightly spreading at tip. *Mexico.* G1. Summer.

42. E. affinis Walther. Illustration: Walther, Echeveria, 73, 74 (1972).

Stem to 10 cm, unbranched. Leaves 4.5–5 × 1.8–2.2 cm, oblanceolate to oblong, short-pointed, nearly flat above, convex below, 6–9 mm thick, bright green. Flowering-stems *c.* 30 cm, with 2–5 branches, flowers 15–30; bracts few; sepals erect, nearly equal, triangular ovate, pointed; corolla 9–10 mm, red, bell-shaped, just exceeding stamens. *W Mexico.* G1. Summer.

43. E. laui Moran & Meyran. Illustration: Graf, Tropica, edn 3, 361 (1986).
Glaucous perennial, stem to 10 cm, simple.

Leaves white, tinged red, 5–9 × 3–4 cm, 6–8 mm thick, flat to concave above, convex below, 30–50 borne in a dense rosette. Flowering-stems to 10 cm, with 9–17 flowers; bracts 1.1–2 × 0.7–1.4 cm, rounded to obovate, tinged red; sepals to 1.8 cm, ascending, unequal, ovate-elliptic; corolla 1.3–1.6 cm, tapering towards the mouth, red, segments pointed, yellow inside. *Mexico.* G1. Spring.

44. E. pulchella Berger. Illustration: Graf, Exotica, series 4, **1**: 896 (1985).

Stem to 6 cm. Leaves 3–3.5 × 1–1.5 cm, spathulate to oblong, numerous, ascending, thick, concave above, bristle-tipped, green, suffused red in the sun. Flower-stalks 2 or 3 branched, to 20 cm, flowers 10–18; sepals light brown-olive or red, to 4 mm, triangular to ovate, fused for half their length; corolla to 6 mm, red, thin. *Mexico.* G1. Spring–summer.

45. E. pittieri Rose. Illustration: Walther, Echeveria, 321 (1972); Graf, Exotica, series 4, **1**: 896 (1985).

Shrublet, stem branching, to *c.* 10 cm. Leaves 4–10 × 2–3 cm, oblanceolate to elliptic, pointed, stalked, slightly concave above, green-brown at tip, borne in loose rosettes. Flowers in dense spikes with thick erect stems *c.* 20 cm; bracts green-purple, leaf-like, *c.* 4 cm, numerous; sepals to 9 mm thick, pointed; corolla 1.2–1.3 cm, red outside, pink within, bell-shaped, 5-angled, segments spreading at tips. *Guatemala, Nicaragua & Costa Rica.* G1. Winter.

46. E. stolonifera (Baker) Otto (*Cotyledon stolonifera* Baker). Illustration: Walther, Echeveria, 153 (1972).

Tuft forming; stems 5–10 cm, branching, stoloniferous. Leaves 5–8 × 2–4 cm, obovate to spathulate, bristle-tipped, concave above, keeled, glaucous green when young. Flowering-stems 6–10 cm, several; bracts to 2.5 cm, leaf-like, numerous; sepals to 1 cm, spreading, nearly round in cross-section, pointed; corolla *c.* 1.4 cm, red outside, yellow within, segment tips pointed, spreading. *Mexico (possibly a garden hybrid).* G1. Summer.

47. E. australis Rose. Illustration: Walther, Echeveria, 225, 298 (1972).

Stems 20–30 cm, little branched with leaves crowded at their ends, round and smooth. Leaves 2–7 × 1.5–2 cm, wedge-shaped, obovate, rounded, bristle-tipped,

thin, keeled, lime-green tinged purple. Flowering-stems to 25 cm, with dense, many-flowered panicles or racemes; bracts *c.* 2.5 cm, many, spreading, oblong-rounded; sepals 8–12 mm, nearly equal, spreading to ascending; corolla 1.1–1.4 cm, 5-angled, segments thick, red outside, pink within. *Costa Rica, Honduras.* G1. Spring–summer.

48. E. bicolor (Humbolt, Bonpland & Kunth) Walther (*Sedum bicolor* Humboldt, Bonpland & Kunth; *E. bracteolata* Link, Klotzsch & Otto; *Cotyledon bracteolata* (Link, Klotzsch & Otto) Baker; *E. subspicata* (Baker) Berger). Illustration: Walther, Echeveria, 336 (1972).

Stem to 60 × 2 cm, upright and branched from the base. Leaves 9–10 × 3–3.5 cm, wedge-shaped to ovate, rounded, bristle-tipped, green, stalks angled, borne in ill-defined rosettes. Flowering-stems sturdy, many, with 25 flowers to each raceme; bracts *c.* 5 × 3–5 cm, leaf-like, numerous, becoming reduced towards apex; sepals green, *c.* 1.4 cm, nearly equal, lanceolate and fine-pointed, somewhat keeled; corolla 1.2–1.5 cm, yellow outside, orange-yellow within, 5-sided, tapering to mouth, segments oblong to oval, fine pointed, keeled. *Venezuela & Colombia.* G1. Winter.

49. E. multicaulis Rose. Illustration: Walther, Echeveria, 228, 312 (1972); Graf, Exotica, series 4, **1**: 892 (1985).

Shrublet, stems 30–60 cm, much branched and covered with old leaf-scars. Leaves 3–5 × 2–4 cm, wedge-shaped to obovate, blunt, with a bristle tip, flat, green, appearing in terminal rosettes, margins and tip red. Flowers 6–15, borne in racemes on several stems to 25 cm; bracts to 2.5 cm, uppermost orange, numerous, rounded to obovate, pointed; sepals *c.* 6 mm, red, almost equal, ascending, pointed; corolla *c.* 1 cm, scarlet orange-yellow on margins and within, bell-shaped, twice as wide at mouth as at base, segments keeled. *Mexico.* G1. Winter–spring.

50. E. maxonii Rose (*Cotyledon acutifolia* Hemsley). Illustration: Walther, Echeveria, 295, 296 (1972).

Stems 30–80 cm, horizontal or erect and little branched. Leaves 3–10 × 3–4 cm, spathulate to oblanceolate, blunt, bristle-tipped or pointed, concave above, convex below, papillose, scattered or in loose rosettes. Flowers solitary, *c.* 25 to a raceme, on stems to 60 cm; bracts *c.*

4 cm, many, ovate, pointed; sepals *c.* 5 mm, nearly equal, thick, semicircular in cross-section, elliptic to oblong; corolla *c.* 1 cm, red, somewhat tapering, segments pointed, keeled. *Guatemala.* G1. Winter.

51. E. waltheri Moran & Meyran.

Stems to 90 cm, horizontal to erect. Leaves 2.5–7.5 × 1.2–2.5 cm, ovate to spathulate, abruptly pointed, margins red. Flowering-stems 30–60 cm, few with many-flowered spikes; sepals to 9 mm, nearly equal, erect; corolla to 9 mm, white below, red above. *Mexico.* G1. Autumn–winter.

52. E. elegans Rose (*E. perelegans* Berger). Illustration: Botanical Magazine, 7993 (1905); Walther, Echeveria, 53, 98 (1972); Graf, Tropica, edn 3, 369 (1986); Brickell, The RHS gardeners' encyclopedia of plants & flowers, 391 (1989).

Stem short. Leaves 3–6 × 1–2 cm, thick, thinner near the tip, spathulate to oblong, flat above, concave below, bristle-tipped, glaucous white, borne in spherical rosettes, margins translucent. Flowers 5–10, borne in several unbranched stems, 10–15 cm; bracts 1–1.2 cm; sepals *c.* 5 mm, triangular to lanceolate, pointed, unequal, spreading; corolla 1.2–1.5 cm, pink outside, yellow-orange within, spreading at apex. *Mexico.* G1. Spring–summer.

Var. **simulans** (Rose) von Poellnitz (*E. simulans* Rose). Illustration: Walther, Echeveria, 103 (1972). Rosettes flattened, leaves to 7 × 4 cm, white. Inflorescence sometimes 2-branched; sepals to 6 mm; corolla deep pink. *Mexico.* G1. Winter–spring.

53. E. pulidonis Walther. Illustration: Walther, Echeveria, 123 (1972); Carruthers & Ginns, Echeverias: A guide to cultivation & identification, pl. 15 (1973).

Leaves borne in a basal rosette, numerous, *c.* 5 × 1.5 cm, ascending, oblong to ovate, tapering to the base, slightly concave above, convex below, bristle-tipped, light green, margins and tips red. Flowers nodding, 10 or more to each one-sided raceme, flowering-stalk to 18 cm; bracts 1.4–1.6 cm; sepals to 6 mm, triangular to ovate, pointed, spreading; corolla 9–10 mm, lemon-yellow, 5-sided, narrowing slightly at mouth. *Mexico.* G1. Spring–summer.

54. E. schaffneri (S. Watson) Rose (*Cotyledon schaffneri* S. Watson).

Stems short. Leaves 6–10 × *c.* 2 cm, bright green with red margins,

oblanceolate to oblong, spreading, with a small abrupt point, borne in dense rosettes. Flowering-stems solitary, 2-branched, to 20 cm, with 8–24 flowers; bracts *c.* 2.5 × 0.7 cm, green, numerous, oblanceolate; sepals to 8 mm, nearly equal, lanceolate, tapering to the tip, spreading; corolla 1.4–1.7 cm, red outside, orange within, 5-angled, not contracted at mouth, segments sharply keeled. *Mexico.* G1. Winter–spring.

55. E. runyonii Walther. Illustration: Graf, Exotica, series 4, **1**: 896, 898 (1985); Graf, Tropica, edn 3, 372 (1986).

Stems short or absent. Leaves 6–8 × 3–4 cm, wedge-shaped to spathulate, blunt, sometimes shallowly notched, glaucous, borne in open rosettes. Flowering-stems several, 2-branched, racemes nodding at first; bracts 2–4 cm, many, adpressed, oblanceolate to linear; sepals 1.3–1.5 cm, unequal, spreading; corolla 1.7–2 cm, red-pink, 5-angled, erect. *Mexico.* G1.

Var. **macabeana** Walther differs in having leaves pointed or bristle-tipped.

56. E. derenbergii J.A. Purpus. Illustration: Walther, Echeveria, 273 (1972); Hay & Beckett, Reader's Digest encyclopaedia of garden plants and flowers, 240 (1972); Graf, Exotica, series 4, **1**: 892, 894 (1985); Graf, Tropica, edn 3, 370 (1986).

Stem short, much branched. Leaves 2–4 × 2–2.5 cm, wedge-shaped to obovate, thick, bristle-tipped, pale green, margin and tip red, numerous, borne in dense tuft-forming rosettes. Flowers few, borne in several racemes to 10 cm; bracts *c.* 1.5 cm, pointed, keeled; sepals 7–10 mm, light green, margins red, nearly equal, oblanceolate, fine-pointed, ascending; corolla 1.2–1.5 cm, segments yellow, keel and tip red, somewhat bell-shaped, erect. *Mexico.* G1. Spring–summer.

57. E. cuspidata Rose (*E. parrasensis* Walther).

Leaves grey-green somewhat glaucous, red-tipped, 6–8 × 3.5–4 cm, oblong-obovate, flat, rather thin, blunt with a small point, numerous, borne in a dense basal rosette. Flowers *c.* 15, borne in several one-sided racemes; bracts to 1.6 cm, fine-pointed; sepals *c.* 8 mm, unequal, lanceolate-oblong, petals *c.* 1.4 cm, tube pink outside, orange within, conical, segment tips just spreading. *Mexico.* G1. Spring–summer.

58. E. peacockii Croucher. Illustration: Graf, Exotica, series 4, **1**: 886 (1985); Graf, Tropica, edn 3, 368 (1986).

Stem short. Leaves 5–6 × 2–3 cm, obovate-oblanceolate, pointed or bristle-tipped, convex below, flat above, numerous, usually crowded in solitary rosettes. Racemes several, *c.* 30 cm, one-sided; flowers solitary; bracts with 2 spurs at base; corolla *c.* 11 mm, not constricted at mouth, segments keeled, slightly spreading at tip, red-pink. *Mexico.* G1. Summer.

59. E. longissima Walther. Illustration: Walther, Echeveria, 244, 414–416 (1972).

Stem short, unbranched. Leaves 6 × 3 cm, wedge-shaped to ovate, bristle-tipped, thick, 20 to each rosette, margins red. Flowering-stems to 20 cm, one-sided, sometimes branched, racemes with 4–12 flowers; bracts *c.* 2 cm, 3-spurred; sepals to 8 mm, urn-shaped, narrow; corolla 2.5–3 cm, green-yellow below, green above. *Mexico.* G1. Summer.

60. E. halbingeri Walther. Illustration: Walther, Echeveria, 121, 122, 231 (1972).

Leaves glaucous green, 2–2.5 × 1–1.3 cm, obovate, blunt, with a small point, triangular in cross-section at tip, borne in dense, basal rosettes. Flowering-stems simple, to 12 cm, with 6–9 cm racemes; bracts to 1 cm, triangular in cross-section; sepals to 6 mm, spreading unequal; corolla *c.* 1.6 cm, orange outside, light orange within, urn-shaped, segment tips recurved. *Mexico.* G1. Summer.

61. E. secunda W. Booth (*Cotyledon secunda* Baker; *E. spilota* Kunze). Illustration: Botanical Magazine, 8748 (1918); Walther, Echeveria, 130 (1972); Graf, Tropica, edn 3, 368 (1986); Brickell, The RHS gardeners' encyclopedia of plants & flowers, 393 (1989).

Stems short and forming numerous offsets. Leaves 2.5–7.5 × 1.2–3 cm, spathulate to wedge-shaped, blunt, bristle-tipped, keeled, glaucous green, margin and tip often red, numerous borne in dense, basal rosettes. Flowering-stem to 30 cm, with 5–15 flowers borne in a simple raceme; sepals spreading, lanceolate, pointed; corolla 7–12 mm, red outside, yellow within, tube narrowing at mouth, lobes swept back. *Mexico.* G1.

Var. **glauca** (Baker) Otto (*E. glauca* Baker). Leaves thin, scarcely keeled, almost flat, to 2 cm across, truncate to rounded.

Var. **pumila** (Schönland) Otto (*E. pumila* Schönland). Leaves to 1.5 cm across, thin and scarcely keeled, almost flat, acute.

62. E. lindsayana Walther. Illustration: Graf, Exotica, series 4, **1**: 898 (1985); Graf,

Tropica, edn 3, 370 (1986).

Leaves 5–9 × 3–4 cm, oblong to obovate, blunt or bristle-tipped, slightly keeled at tip, light green, tip red, numerous, in dense rosettes at first, later becoming tufted. Flowering-stems several, to 50 cm, tinged pink, 2-branched, 14-flowered; sepals *c.* 3 mm triangular to ovate, pointed; corolla *c.* 1 cm, pink, yellow to orange at tips and within. *Mexico*. G1. Spring–summer.

63. E. racemosa Schöenland & Chamisso (*E. lurida* Haworth). Illustration: Botanical Magazine, 7713 (1900); Walther, Echeveria, 233, 341 (1972).

Stem very short or absent. Leaves 5–8 × 3–3.5 cm, oblanceolate to diamond-shaped, pointed, slightly concave above, borne in dense rosettes of *c.* 15, margins often irregularly cut. Flowering-stems 1 or 2, erect, to 50 cm, thick, racemes with 20–40 flowers; bracts 5–10 mm, obovate, pointed, easily detached; sepals to 6 mm, nearly equal, triangular to ovate, pointed, nearly round in cross-section, spreading; corolla 1–1.2 cm, red-orange outside, orange-buff within, conical or urn-shaped, segments fine-pointed and spreading at tips. *Mexico*. G1. Autumn.

64. E. purpusorum Berger (*Urbinia purpusii* Rose). Illustration: Walther, Echeveria, 124–127 (1972); Carruthers & Ginns, Echeverias: A guide to cultivation & identification, pl. 17 (1973); Graf, Exotica, series 4, **1**: 898 (1985).

Rosette solitary. Leaves 3–4 × 1.5–2 cm, to 1 cm thick, ovate, tapering to a sharply pointed tip, flat above, convex below, spinach-green, spotted red and brown, borne in dense, solitary rosettes. Flowering-stems *c.* 20 cm, simple with racemes of 6–9 flowers; bracts *c.* 1.5 cm, ovate, pointed; sepals 2–3 mm, pointed, adpressed to petals; corolla *c.* 1.2 cm, pink-red outside, yellow within, fused near base, thick. *Mexico*. G1. Early summer.

65. E. agavoides Lemaire (*Cotyledon agavoides* Baker; *E. obscura* (Rose) von Poellnitz; *Urbinia obscura* Rose; *U. agavoides* (Lemaire) Rose). Illustration: Walther, Echeveria, 82, 83 (1972); Graf, Exotica, series 4, **1**: 889 (1985); Graf, Tropica, edn 3, 371 (1986); Brickell, The RHS gardeners' encyclopedia of plants & flowers, 394 (1989).

Rosettes tufted or solitary, stem very short. Leaves few, 3–8 × 2.5–3 cm, ovate to triangular, thick, sharply pointed, deep green, waxy, margins transparent. Flowers

10–16, borne on a 2-branched flowering-stalk; bracts few; sepals olive, to 5 mm, spreading; corolla 1–1.2 cm, thin, spreading at tip, pink-orange outside, yellow within. *Mexico*. G1. Spring–summer.

Var. **prolifera** Walther. Leaves numerous, 10–12 × 3–4 cm; petals to 1.6 cm.

Var. **corderoyi** (Baker) von Poellnitz (*Cotyledon corderoyi* Baker; *E. corderoyi* (Baker) Morren; *Urbinia corderoyi* (Baker) Rose). Leaves numerous, ovate, *c.* 6.5 × 3.5 cm. Flowering-stalks 3-branched; sepals free; petals to 9 mm.

E. × gilva Walther (*E. agavoides* × *E. elegans*). Illustration: Walther, Echeveria, 116 (1972). Stems short, branching. Leaves 5–8 × 2–2.5 cm, oblong, short-pointed, surface crystalline, margin translucent, up to 30 borne in dense rosettes. Flowers borne in several simple racemes to 25 cm; sepals *c.* 5 mm, unequal; corolla *c.* 9 mm, pink below, yellow above. *Garden origin*. G1. Spring.

66. E. potosina Walther. Illustration: Walther, Echeveria, 105, 106, 107 (1972); Graf, Exotica, series 4, **1**: 886 (1985); Graf, Tropica, edn 3, 370 (1986).

Similar to *E. elegans*, differing by leaves *c.* 3 cm across, thickest at apex and often tinged purple in full sun; sepals to 5 mm, broadly triangular; corolla rarely white. *Mexico*. G1. Spring–early summer.

67. E. obtusifolia Rose. Illustration: Walther, Echeveria, 165, 216 (1972).

Stem short. Leaves 4–8 × 2.5–3.5 cm, spathulate to oblanceolate, rounded, bristle-tipped, green with red margins, borne in a rosette. Flowers 9–11, borne in several red, 2-branched stems; bracts numerous, 1.5–2 cm, oblong, pointed; sepals 5–6 mm, unequal, spreading; corolla 1–1.2 cm, orange. *Mexico*. G1. Winter.

68. E. shaviana Walther. Illustration: Walther, Echeveria, 221, 271, 272 (1972); Graf, Exotica, series 4, **1**: 889 (1985); Graf, Tropica, edn 3, 369 (1986).

Stem short or absent. Leaves *c.* 5 × 1.5–2.5 cm, obovate, tapering to a stalk-like base with a small point, somewhat spathulate, glaucous green, flushed pink, margin wavy, toothed near tip, many borne in crowded rosettes. Flowering-stems to 12 cm, solitary to several, simple, one-sided, nodding with 12–15 flowers; bracts 1–1.5 cm, linear, pointed, spurred; sepals to 9 mm, unequal, linear to lanceolate or triangular, pointed, ascending; corolla

1–1.3 cm, 5-sided, pink, segments keeled, tips slender and spreading. *Mexico*. G1. Summer.

69. E. albicans Walther (*E. elegans* var. *kesselringiana* von Poellnitz). Illustration: Walther, Echeveria, 111–113 (1972); Graf, Tropica, edn 3, 371 (1986).

Stem very short or absent, leaves 3–5 × 1.5–5 cm, overlapping, oblong to ovate, thickest near tip, small-pointed, white, borne in solitary rosettes which become tufted. Flowering-stems *c.* 20 cm, unbranched, one-sided; bracts numerous; sepals to 1 cm, fused below, unequal, triangular to oblong, pointed, almost erect; corolla pink outside, yellow inside, 1.5–1.8 cm, more or less obconical, segments slightly spreading at tips. *Mexico*. G1. Summer.

70. E. humilis Rose.

Stem short and branching. Leaves 4–7 × 2.5 cm, lanceolate to ovate, sharply bristle-tipped, convex below, green-brown, borne in dense, long rosettes, Flowering-stems to 20 cm, simple, occasionally branched; flowers in pendent racemes; bracts *c.* 2 cm, pointed, nearly round in cross-section; sepals to 9 mm, unequal, thick, pointed, ascending; corolla to 1.3 cm, orange outside, orange-yellow within, bell to urn-shaped, segments spreading at tips. *Mexico*. G1. Summer.

71. E. chihuahuaensis von Poellnitz. Illustration: Graf, Exotica, series 4, **1**: 887 (1985); Graf, Tropica, edn 3, 361 (1986).

Stem absent. Leaves 4–6 × 3–4 cm, numerous, obovate to oblong, blunt, short-pointed, thin, white-glaucous green, tip reddish purple. Flowering-stems *c.* 20 cm, simple or branched; bracts 2–3 cm; sepals *c.* 8 mm, unequal, oblong-lanceolate to triangular; corolla *c.* 1.4 cm, red outside, orange within, narrowing at mouth, segments spreading at tips. *Mexico*. G1. Spring–summer.

26. GREENOVIA Webb
J.Y. Clark

Dwarf evergreen perennial succulent herbs. Stem small. Leaves fleshy, spoon-shaped, alternate, stalkless, mostly hairless and glaucous, occasionally glandular-hairy, margins usually entire, occasionally hairy, forming a dense rosette which usually dies after flowering. Flowers yellow, bisexual, borne in a terminal glandular cyme; calyx 16–35-merous; petals and carpels narrow, 16–32, half-fused to the hypogynous disc; stamens twice the number of petals.

A genus of 4 species endemic to the Canary Islands and similar to *Aeonium*, there being a number of putative hybrids between the two genera. *G. aurea* is the most commonly cultivated. The main growing season is from October to March. Watering should only be light in the resting season, perhaps restricted to mist-spraying in most species, although *G. aizoon* prefers to be kept generally moist all year round. Propagation is generally easy from seed and cuttings, although leaves are sometimes difficult to root.

Literature: Praeger, R.L., *An account of the Sempervivum group* (1932); Bramwell, D. & Bramwell Z., *Wild flowers of the Canary Islands* (1974).

1a. Leaves densely glandular-hairy, green, remaining as more or less open rosettes in the resting season **1. aizoon**
  b. Leaf surface hairless, glaucous; leaves forming tightly packed rosettes in the resting season 2
2a. Offsets absent; leaf-margin often hairy, especially in small rosettes **2. diplocycla**
  b. Offsets present; leaf margin usually lacking hairs 3
3a. Rosettes less than 8 cm across, rarely widely open, with numerous, often long-stemmed offsets; petals and carpels 16–22 **3. dodrentalis**
  b. Rosettes 8–20 cm across, opening fairly widely in the growing season, with a few usually short-stemmed offsets; petals and carpels 20–35 **4. aurea**

1. G. aizoon Bolle. Illustration: Praeger, An account of the Sempervivum group, 215 (1932); Lamb, The illustrated reference on cacti and other succulents **3**: 840 (1963); Lamb, Reference plates, 2585a; Cactus and Succulent Journal (U.S.) **39**: 94, 134 (1967).

Much branched herb forming clumps of green rosettes, each 4–7 cm across. Rosettes more or less open during the summer drought. Stems usually relatively short, but branches to 15 cm sometimes formed. Leaves oblong to spathulate, densely glandular-hairy all over, 3–4 × 1–1.5 cm, apex more or less truncate, mucronate. Flower-stem 10–13 cm, leafy and glandular-hairy; flower-head 2–3 × 4–8 cm, with 10–40 flowers; flowers *c.* 1.2 cm across; sepals *c.* 4 mm, 2-fid; petals yellow, 6–7 mm; stamens 5–6.5 mm; carpels hairy, green, 4–5 mm. *Tenerife*. G1. Late spring.

2. G. diplocycla Webb. Illustration: Praeger, An account of the Sempervivum group, 217 (1932); Lamb, Reference plates, 2586 (1), 2586 (2).

Herb, forming a usually solitary rosette, 8–25 cm across, tightly closed during the summer drought. Stem short. Leaves spathulate, glaucous, hairless on both surfaces, margins often sparsely to densely hairy especially when young, 5–8 × 4–6 cm, apex more or less rounded to somewhat depressed, sometimes mucronate. Flower-stem 10–20 cm, leafy and glandular hairy; flower-head 13–16 × 15–20 cm; flowers *c.* 1.5 cm across; sepals *c.* 1.7 mm, more or less hairy; petals 15–25, yellow, 5–6 mm; stamens 5–5.5 mm; carpels 15–25, hairy, greenish yellow, *c.* 4 mm. *Palma, Gomera, Hierro.* G1. Late winter–late spring.

3. G. dodrentalis (Willdenow) Webb (*G. gracilis* Bolle). Illustration: Praeger, An account of the Sempervivum group, 222 (1932); Cactus and Succulent Journal (U.S.) **5**: 427 (1933); Jacobsen, Lexicon of succulent plants, pl. 89.4 (1974).

Much branched herb forming clumps of glaucous rosettes, each 3–8 cm across, tightly closed during the summer drought and seldom widely open even during the growing period. Main stem short, offsets long-stemmed. Leaves spathulate, glaucous, hairless on both surfaces, margins finely glandular-hairy only when young, 2–5 × 1–2 cm, apex more or less pointed, rounded or somewhat depressed, sometimes mucronate. Flower-stem 15–25 cm, leafy and glandular-hairy; flowers *c.* 1.5 cm across; sepals hairy; petals 16–22, yellow, *c.* 7 mm; stamens *c.* 5 mm; carpels 16–22, greenish yellow, *c.* 4 mm. *Tenerife.* G1. Late winter–late spring.

4. G. aurea (C. Smith) Webb & Berthelot. Illustration: Marshall Cavendish encyclopedia of gardening, 806 (1968); Bramwell & Bramwell, Wild flowers of the Canary Islands, pl. 161 (1974); Jacobsen, Lexicon of succulent plants, pl. 89.5 (1974); Graf, Tropica, 378 (1984).

Usually branched herb, forming clumps of glaucous rosettes, each 8–25 cm across, tightly closed during the summer drought, fairly open during the growing season. Main stem short, offsets usually also short-stemmed but longer than main stem. Leaves spathulate, glaucous, hairless on both surfaces, margins sometimes finely glandular-hairy only when young, 5–10 × 3–6 cm, apex more or less pointed, rounded or somewhat depressed, sometimes mucronate. Flower-stem 30–45 cm, leafy and glandular-hairy; flowers 2–2.5 cm across; sepals hairy; petals 20–35, yellow, 7–8 mm; stamens *c.* 5 mm; carpels 20–35, greenish yellow, *c.* 5 mm. *Tenerife, Gran Canaria, Gomera, Hierro.* G1. Spring.

27. MONANTHES Haworth
E.H. Hamlet

Small perennial herbs or shrublets. Stems simple or branched, erect or spreading. Leaves alternate, rarely opposite, ovate, diamond or club-shaped, entire. Inflorescence raceme or cyme-like. Flowers greenish, purplish, or yellowish; calyx saucer-shaped, with 6–8 parts. Petals linear or lanceolate, same number as calyx segments. Stamens twice as many as petals. Nectar-scales at base of carpels large, petal-like.

A genus of 12 species from Canary Islands, Salvage Islands and Morocco. In frost-free climates they can be grown outdoors. They prefer well-drained sandy soil. When grown indoors, cool conditions in winter are necessary. Night temperatures of 5–10 °C and 10–15 °C by day are sufficient. Little water is required in winter and only moderate amounts from spring to autumn. Propagation is by seed, stem-and leaf-cuttings.

Literature: Nyffeler, R., A taxonomic revision of the genus Monanthes Haworth (Crassulaceae) *Bradleya* **10**: 49–82 (1992).

1a. Stems short, thick, unbranched 2
  b. Stems slender, branched 3
2a. Rosettes loose **2. brachycaulon**
  b. Rosettes dense **3. pallens**
3a. Leaves opposite **1. laxiflora**
  b. Leaves alternate 4
4a. Shrubby, stem erect **6. anagensis**
  b. Not shrubby, stem prostrate in lower part 5
5a. Rosettes more or less spherical; flowering-stems with long whitish hairs **4. polyphylla**
  b. Rosettes narrower; flowering-stems finely downy **5. subcrassicaulis**

1. M. laxiflora Bolle. Illustration: Praeger, Sempervivums, 234, 236 (1932); Kunkel, Flora de Gran Canaria **3**: 69 (1978); Bradleya, **10**: 61, 64 (1992).

Perennial with grey stems; branches rather twisted, at first erect then sprawling. Leaves 8–10 × 6–8 mm, 3–5 mm thick, opposite, elliptic to oblanceolate or obovate, grooved and slightly flattened on face, dark green, sometimes with silvery mottling. Flowering-shoots erect, 2–8 cm, with opposite leaves below; racemes almost bare with 6–10 flowers. Flowers with 5–8 parts, purplish or yellowish. Calyx green, *c.* 3.5 mm with ovate segments. Petals *c.* 4 mm, linear-lanceolate, acute, mostly greenish with red lines giving a purple effect. Stamens purplish; anthers pale red. *Canary Islands.* Spring.

2. M. brachycaulos (Webb & Berthelot) Lowe (*M. brachycaulon* Lowe). Illustration: Praeger, Sempervivums, 228 (1932); Kunkel, Flora de Gran Canaria **3**: 67 (1978); Bradleya, **10**: 61, 64 (1992).

Minute perennial; stems erect, short, thick, bulb-like or cylindric. Leaves forming a loose rosette, oblanceolate to spathulate, tapering gradually at base, very fleshy, flat or concave on face, very convex on back, 1.5–2 × 0.5 cm, 2–2.5 mm thick, green, mottled with purple, with purple midrib and stain at base. Flowering-branches lateral, from lower leaf-axils, 4–7 cm, erect, leafy in the middle with small loosely rosulate or crowded hairy leaves, raceme-like in the upper part. Racemes with 5–10 flowers, 2–5 cm, glandular-hairy. Flower parts in 6s or 7s; flowers 8–10 mm across, purplish green or greenish purple. Calyx *c.* 4 mm, green, glandular-hairy with ovate-lanceolate segments. Petals lanceolate or oblanceolate, *c.* 4 × 1 mm. Stamens greenish or purplish; anthers red. *Canary Islands.* Spring.

3. M. pallens Christ. Illustration: Praeger, Sempervivums, 231 (1932); Bradleya, **10**: 61 (1992).

Perennial with erect, fleshy stems 1–3 cm, cylindric and bare in old plants; leaf-rosettes dense, 1.5–3.5 cm across, slightly concave, borne at ends of stems. Leaves 10–18 × 3–3.5 mm, 2 mm thick, diamond-shaped to spathulate, thick above, tapering below, green or glaucous, flattish on face, very convex on back, lower part pale, base purple. Flowering-shoots lateral, from outer leaf-axils *c.* 3 cm. Flowers with 6 or 7 parts; calyx *c.* 4 mm, glandular-hairy, segments oblong-lanceolate, *c.* 3 mm, with red mottling and green papillae. Petals 3.5–4 × 0.75–1 mm, linear-lanceolate, acute, yellowish with red lines. Stamens *c.* 4 mm; filaments reddish, anthers red. *Canary Islands.* Spring.

4. M. polyphylla Haworth. Illustration: Praeger, Sempervivums, 241 (1932); Edgar & Lamb, The illustrated reference on cacti and other succulents **4**: 1150 (1963); Bramwell, Wild flowers of the Canary Islands, pl. 143 (1974); Gartenpraxis 1985 (2): 52 (1985).

Minute creeping perennial with prostrate, slender stems forming a mat or dense cushion. Rosettes 1–1.5 cm, bluntly conical or ovoid, very dense. Leaves 6–8 × 2–2.5 mm, closely overlapping near the rounded apex, narrowly wedge-shaped to spathulate, fleshy, convex on face and back. Flower-stem arising from centre of rosette, 1–1.8 cm, with 4–8 flowers. Flowers 1–1.2 cm, with 6–8 parts, green or brownish purple; calyx *c.* 4 mm, hairy; petals 4.5–5 × 0.5–0.8 mm, almost linear, acuminate, hairy on face, back and edges. *Canary Islands.* Spring.

5. M. subcrassicaulis Praeger.

Tufted perennial, not shrubby. Leaf-rosettes dense, *c.* 1 cm across, forming close tufts; leaves overlapping, *c.* 7–10 × 2.5–3 mm, *c.* 1.5 mm thick, wedge- to club-shaped, shining, dark green, base purple, lower leaves often purplish. Flowering-shoots mostly from the centre of the rosettes, one or several to each rosette, leafless, 3–4 cm, with 1–5 flowers. Flowers with 7 or 8 parts, purplish. Calyx hairy, striped red, *c.* 3 mm, with lanceolate segments. Petals *c.* 3.5 mm, linear-deltoid, acuminate, hairy. Stamens purplish; anthers purple. *Canary Islands.*

6. M. anagensis Praeger. Illustration: Praeger, Sempervivums, 238 (1932); Bradleya, **10**: 61 (1992).

Perennial herb or shrublet to 15 cm. Branches grey, twisted; leaves *c.* 2.5 × 0.4 cm, *c.* 3 mm thick, alternate, stalkless, linear, upper surface grooved. Inflorescence raceme-like with 2–6 flowers. Flower parts in 7s, *c.* 1 cm across, greenish yellow. Calyx with triangular segments; petals *c.* 4 mm, triangular-lanceolate, acute, greenish yellow with a reddish vein. Stamens almost equal in length to the petals; filaments reddish; anthers yellow. *Canary Islands.* Spring.

28. JOVIBARBA Opiz
Z.R. Gowler
Like *Sempervivum* but petals 6 or 7 in number, pale yellow, keeled outside, fringed with glandular hairs forming a bell-shaped flower.

A genus of 6 European species in which no natural hybrids occur.

Literature: Mitchell, P.J. *The Sempervivum and Jovibarba handbook* (1979); Praeger, R.L., *An account of the Sempervivum group* (1932).

1a. Plant without stolons   **1. heuffelii**
  b. Plant with stolons   2
2a. Rosette-leaves finely hairy on the surface   **2. allionii**
  b. Rosette-leaves hairless on the surface but with hairs along the margins   3
3a. Rosette-leaves broadest near the middle   **3. arenaria**
  b. Rosette-leaves not broadest near the middle   4
4a. Rosette-leaves broadest ²/₃ of the way up   **4. sobolifera**
  b. Rosette-leaves broadest ¹/₃ of the way up   5
5a. Rosette-leaves an intensive bluish green colour   **5. preissiana**
  b. Rosette-leaves wholly green or yellow-green with brown tips   **6. hirta**

1. J. heuffelii (Schott) A. & D. Löve (*Sempervivum heuffelii* Schott). Illustration: Praeger, An account of the Sempervivum Group, 94 (1932); Quarterly bulletin of the Alpine Garden Society, **6**: 103–104 (1938). Hay & Beckett, Reader's Digest encyclopaedia of garden plants and flowers, 659 (1971); Payne, Plant jewels of the high country, 42 (1972).

Very variable; leaves green, grey, brown, red or purple; rosette-leaves oblong-obovate, gradually narrowing towards the base; margins with stiff white hairs, apex very acute; flowering-stems 8–20 cm, with broad clasping lanceolate leaves; inflorescence dense and flattish; petals 6–7, yellow to white, hairy. *E Carpathians to the Balkan Peninsula.*

This is the only species that does not produce offsets or stolons but increases by the parent rosette splitting into 2 or more equal rosettes. The plant does not die after flowering.

Var. **heuffelii** develops medium-sized rosettes with distinctly hairy leaves.

Var. **glabra** Beck & Szyszylowicz. Rosette-leaves hairless on both sides, obovate and very much contracted to the apex. Petals truncate.

Var. **kopaonikense** (Pančič) P.J. Mitchell (*Sempervivum kopaonikense* Pančič). Hypogynous scales spreading, not erect, causing a bulge at base of calyx and corolla giving the flower a swollen appearance.

Var. **patens** (Grisebach & Schrenk) P.J. Mitchell (*Sempervivum patens* Grisebach & Schrenk). Rosettes minutely hairy.

Cultivars include: 'Aquarius', 'Bermuda', 'Bronze Ingot', 'Cameo', 'Chocoleto', 'Giuseppi Spiny', 'Greenstone', 'Henry Correvon', 'Hystyle', 'Miller's Violet', 'Minuta', 'Pallasii', 'Purple Haze', 'Sundancer', 'Tan', 'Tancredi', 'Torrid Zone', 'Xanthoheuff'.

2. J. allionii (Jordan & Fourreau) D.A. Webb (*Sempervivum allionii* Jordan & Fourreau). Illustration: Praeger, An account of the Sempervivum Group, 96 (1932); Pitschman et al, Bilder-Flora der Sudalpen, 105 (1959); Huxley, Mountain flowers, pl. 243 (1967); Payne, Plant jewels of the high country, 20 (1972).

Rosettes 2–3 cm across, almost globular due to incurving of the leaves. Rosette-leaves lanceolate, acuminate, minutely hairy on both surfaces, with longer hairs on margins, yellowish green with a reddish flush on the apex when fully exposed. Numerous offsets produced on short thin stolons, often stalkless, forming dense matted clumps. Flower-stems slender, 10–15 cm, inflorescence with 2–3, few-flowered, short, branches. Sepals minutely hairy; petals greenish white, obviously fringed. *Europe (S Alps, France to the Tyrol).* H4. Early summer.

A compact and easily grown species which is sometimes slow and reluctant to flower under cultivation.

J. allionii × J. hirta. Illustration: Houselekes **16**(1): 19 (1985). Intermediate with incurving rosettes, producing offsets. Rosette-leaves bright green with very few hairs, outer surface of older leaves and tips of new leaves flushed red. Offsets almost globular, on very thin stolons.

J. allionii × J. sobolifera Leaves deep green with a bright red flush on the leaf backs.

J. × kwediana Mitchell (*J. allionii × J. heuffelii*). Illustration: Houselekes **13**(2): 39, 40 (1982).

Rosettes intermediate between the parents, 3.5–4 cm across. Rosette-leaves ovate, concave, mucronate, with scattered hairs on both surfaces, very succulent, 0.5 cm thick, with marginal hairs. Offsets on very short stolons. Flowering-stem *c.* 12 cm, glandular-hairy. Stem-leaves ovate, *c.* 2 × 1 cm. Inflorescence compact; flowers 1.5–1.75 cm; petals 5–6 erect, fringed, pale yellow.

'Cabaret' (*J. allionii × J. heuffelii* 'Giuseppi Spiny'). Upper half of the incurved rosette leaves flushed red, lower half bright green, rosettes squat and slow-growing.

'Pickwick' has larger and more robust rosettes, longer and more erect, lightish green rosette leaves, flushed red on the upper and outer surfaces. Flowering-stem more than 12 cm.

3. J. arenaria (Koch) Opiz (*Sempervivum arenarium* Koch). Illustration: Reichenbach, Icones Florae Germanicae et Helveticae **23**: t. 74 (1899); Hegi, Illustrierte Flora von

Mitteleuropa **4**: 557, 559 (1922); Praeger, An account of the Sempervivum Group, 62 (1932); Huxley, Mountain flowers, pl. 244 (1967).

Rosettes 0.5–2 cm across, globular, many-leaved. Rosette leaves incurved, lanceolate, broadest near the middle, bright green and flushed with red-brown at apex, hairless except on the leaf-edges. Minute offsets borne on very short (a few mm), slender horizontal stolons from the base of the parent rosette. Flowering-stems 7–12 cm, with erect, lanceolate, acuminate leaves. Inflorescence large for the size of the plant, with 3 forked branches. Petals 6, greenish white, hairy, especially at the tips *c.* 3 times as long as calyx. Flowers are produced infrequently and are seldom seen. *Europe (Alps, S Tirol, Styria & Carinthia).* H4. Summer.

Requires a lime-free soil.

J. × mitchellii Zonneveld (*J. arenaria × J. heuffelii*). Illustration: Houselekes **13**(2): 36 (1982). Rosette 2.5–4 cm across. Leaves mucronate, hairless except on margins. Offsets between the lower leaves on short stolons. Flowering-stem *c.* 8 cm, covered in leaves that are lightly hairy on the outside. Flowers bell-shaped, sterile, variously coloured; petals with 2 or 3 long fringes at tip, but nearly absent on keel.

4. J. sobolifera (Sims) Opiz (*Sempervivum soboliferum* Sims). Illustration: Botanical Magazine, 1457 (1812); Reichenbach, Icones Flora Germanicae et Helveticae **23**: t. 66 (1899); Praeger, An account of the Sempervivum Group, 101 (1932); Grey-Wilson & Blamey, Alpine flowers of Britain and Europe, 76 (1973).

Rosettes 2–3 cm across, usually half-closed with incurved leaves. Rosette leaves oblanceolate, *c.* 10 × 5 mm, very fleshy, bright green often with a coppery red flush on the back near the apex, hairless; margins with stiff glandular hairs. Offsets numerous, spherical, borne among the outer and middle leaves on very slender stolons. Flowering-stems 10–20 cm, glandular-hairy, clothed with many, almost-erect leaves. Inflorescence 5–7 cm across, dense and flattish; flowers seldom produced, petals 6, fringed, greenish yellow. *C & E Europe (C Germany to the E Carpathians, N & C Russia).*

'Green Globe'. Rosettes bright green without the red flush.

J. × nixonii Zonneveld. Illustration: Houselekes **13**(2): 32, 35 (1982). Rosettes 3–5 cm across. Rosette-leaves abruptly mucronate, hairless except along the

margins. Offsets between the lower leaves on very short stolons. Flowering-stem *c.* 8 cm, covered in lightly hairy leaves; flowers sterile. Petals fringed near the top, but not on keel. The plant dies after flowering.

'Jowan' (*J. sobolifera × J. heuffelii*). Rosettes dark almost brownish red, the colour is retained for a considerable proportion of the year. Offsets on very short stolons around the base of the parent rosette, soon forms a compact cluster.

5. J. preissiana (Domin) Omel'chuk-Myakushko & Chopik (*Sempervivum preissianum* Domin). Illustration: Botanicheskii Zhurnal, **60**(8): 1184 (1975).

Rosettes 4–5 cm across, stellate; rosette leaves intense bluish green, 1.5–2.5 cm, ovate-lanceolate to long-lanceolate, broadest ⅓ of the way up from the base, gradually pointed at the tip, surface hairless except along margin. Flowering-stems robust, 20–30 cm, softly hairy, with numerous stem-leaves, ovate-lanceolate, gradually pointed, to 8–10 mm across, margins hairy. Inflorescence dense, large and many-flowered. Flower-stalk fleshy and glandular-hairy; flowers almost bell-shaped; petals 6, pale or whitish yellow. *Carpathians.* H3. Summer.

6. J. hirta (Linnaeus) Opiz (*Sempervivum hirtum* Linnaeus). Illustration: Reichenbach, Icones Flora Germanicae et Helveticae **23**: t. 73 (1899); Jávorka & Csapody, Iconographia Florae Hungaricae, 223 (1930); Praeger, An account of the Sempervivum Group, 98–99 (1932); Huxley, Mountain flowers, pl. 241 (1967).

Rosettes 2.5–7 cm across, open and stellate; rosette-leaves broadly lanceolate, broadest ⅓ of the way up, narrowing to an acute apex. Leaves hairless except along margins, wholly green or yellow-green with brown tips, some forms are a bright red-brown when in full exposure; offsets small, very numerous, arising on very thin stolons from the axils of the middle and lower leaves of the parent rosette; flowering-stems 10–20 cm, stout, glandular-hairy; stem-leaves numerous, erect, broadly lanceolate, hairless. Inflorescence dense, flattish or convex, 5–8 cm across with 3 forked branches, each with *c.* 5 flowers; petals 6, erect, pale yellow to greenish white, hairy along the edges. *E Alps, Carpathians, Hungary, NW Balkan Peninsula.* H4. Summer.

Subsp. **hirta.** Rosettes 2.5–5 cm across; leaves 5–6 mm across at or below the middle, without a red apex; margins with glandular hairs. Stem-leaves and sepals hairy on the lower surface, at least on the midribs.

Subsp. **borealis** (Huber) R. Sóo. Rosettes 1–3 cm across; leaves broadest above the middle, often with a red apex; margins with glandular hairs. Surface of sepals hairless.

Subsp. **glabrescens** (Sabransky) R. Sóo & Jávorka. Stem-leaves and sepals hairless except along margins.

Var. **neilreichii** (Schott, Nyman & Kotschy) Konop & Bendak. Rosette leaves 2–3 mm across, broadest in upper 1/3; stem-leaves hairless.

Forma **hildebrandtii** (Schott) Konop & Bendak. Rosette-leaves 8–12 mm across, rosettes grey-green; stem-leaves hairless.

29. SEMPERVIVUM Linnaeus
Z.R. Gowler & M.C. Tebbitt

Succulent perennial herbs with pointed basal leaves in a rosette. Rosettes 1–15 cm across, monocarpic, plant increases by producing young offsets from the leaf axils, usually on runners (stolons). Flower-stems terminal, erect, usually with a covering of leaves. Flowers in a terminal cyme, with 8–16 parts in a star shape, twice as many stamens as petals, equal number of carpels to petals. Petals yellowish, pink, purple or red.

A genus of about 50 species chiefly from Europe but also Turkey, Iran, the Caucasus and Morocco. Sempervivums hybridise easily both in the wild and in cultivation.

Literature: Mitchell, P.J., *The Sempervivum & Jovibarba Handbook* (1973); Praeger, R.L. *An account of the Sempervivum group* (1932).

1a. Petals and stamens lacking
　　　　　　2. octopodes var. **apetalum**
　b. Petals and stamens present　　　2
2a. Petals yellow or yellow-green　　3
　b. Petals not yellow　　　　　　19
3a. Petals red, pink or purple at the base　　　　　　　　　　　4
　b. Petals not coloured at the base　12
4a. Petals with a green central stripe
　　　　　　　　　　　1. sosnowskii
　b. Petals without a green central stripe　　　　　　　　　　　5
5a. Rosettes not more than 2 cm across
　　　　　　　　　　　　2. octopodes
　b. Rosettes more than 2 cm across　6
6a. Leaves give off a strong 'goaty' smell　　　　　　**3. grandiflorum**
　b. Leaves not smelly　　　　　　7
7a. Leaf-tips not coloured
　　　　　　　　　4. transcaucasicum

　b. Leaf-tips coloured　　　　　　8
8a. Hairs present on the surfaces of mature rosette-leaves　　　9
　b. Hairs absent from the surfaces of mature rosette-leaves　　　11
9a. Rosettes 3–4 cm across; stamen filaments dark crimson
　　　　　　　　　　　　6. zeleborii
　b. Rosettes 4–7 cm across; stamen filaments purple　　　　　　10
10a. Leaf-margins with long glandular hairs; flower-stems very leafy
　　　　　　　　　　　5. kindingeri
　b. Leaf-margins with medium glandular hairs; flower-stems not very leafy　　**4. transcaucasicum**
11a. Juvenile rosettes glandular-hairy
　　　　　　　　　　　　7. armenum
　b. Juvenile rosettes hairless
　　　　　　　　　　　　8. wulfenii
12a. Petals flushed red near the tip
　　× **praegeri** (see **11. ciliosum**)
　b. Petals not flushed red near the tip　　　　　　　　　　　13
13a. Rosettes more than 5 cm across
　　　　　　　　　　　9. ruthenicum
　b. Rosettes 5 cm or less across　14
14a. Rosettes half-open　　　　　　15
　b. Rosettes open　　　　　　　17
15a. Rosettes to 2.5 cm across
　　　　　　　　　　　　10. minus
　b. Rosettes 3 cm or more across　16
16a. Hairs on leaf-margin 2–4 mm
　　　　　　　　　　　11. ciliosum
　b. Hairs on leaf-margin less than 2 mm　　　　　　**12. davisii**
17a. Leaves incurved　　　　　　18
　b. Leaves not incurved
　　　　　　　　　　14. glabrifolium
18a. Hairs on leaf-margin 2–4 mm
　　　　　　　　　　　11. ciliosum
　b. Hairs on leaf-margin less than 2 mm　　　　　　**13. pittonii**
19a. Petals white or greenish white　20
　b. Petals pink, purple or red　21
20a. Petals white with rose-red stripes
　　× **degenianum** (see **9. ruthenicum**)
　b. Petals greenish white with no stripes　　　　　**15. leucanthum**
21a. Petals with white lines or white margins　　　　　　　　22
　b. Petals without white lines or white margins　　　　　　34
22a. Petals with narrow white lines
　　　　　　　　　　16. erythraeum
　b. Petals with white margins　23
23a. Rosette-leaves completely hairless on the surfaces　**17. marmoreum**
　b. Rosette-leaves with some hairs on the surfaces　　　　24

24a. Rosette-leaves with only a few hairs on the backs　　**18. iranicum**
　b. Rosette-leaves with hairs on both sides　　　　　　25
25a. Filaments bright lilac, white striped on the base
　　17. marmoreum var. **angustifolia**
　b. Filaments not as above　26
26a. Leaf-tips coloured　　　　27
　b. Leaf-tips not coloured　　31
27a. Leaf-tips brown　　　　28
　b. Leaf-tips red-purple　　30
28a. Stolons less than 4 cm
　　　　　　　　　　19. giuseppii
　b. Stolons more than 4 cm　29
29a. Rosettes flat, many-leaved; stem-leaves ovate to oblong
　　　　　　　　　　20. ingwersenii
　b. Rosettes not flat, few leaved; stem-leaves triangular-lanceolate
　　　　　　　　　　21. ossetiense
30a. Stolons long, to 12 cm
　　　　　　　　　　22. kosaninii
　b. Stolons very short
　　　　　　　　23. dzhavachischvilii
31a. Stolons slender　　　　32
　b. Stolons stout
　　　　　　　24. reginae-amaliae
32a. Flower-stems covered in shaggy white hairs　**25. dolomiticum**
　b. Flower-stems not covered in shaggy white hairs　　　　33
33a. Anthers yellow
　　　　　　　　26. thompsonianum
　b. Anthers purple　**27. pumilum**
34a. Surfaces of rosette-leaves hairless　　　　　　　　35
　b. Surfaces of rosette-leaves hairy　43
35a. Rosettes semi-open　　　36
　b. Rosettes open　　　　38
36a. Rosette-leaves with brown tips　37
　b. Rosette-leaf-tips not coloured
　　　　　　　　　　28. ballsii
37a. Petals with toothed edges
　　　　　　　　　29. borissovae
　b. Petals with smooth edges
　　　　　　　　　30. andreanum
38a. Leaf-margin hairless, leaves usually grey-green with purple tips
　　　　　　　　　31. calcareum
　b. Leaf-margin with at least a few hairs, leaves not as above　39
39a. Leaf-tips coloured　　　40
　b. Leaf-tips not coloured　41
40a. Flower-stems 5–12 cm
　　　　　　　　　32. balcanicum
　b. Flower-stems 20–40 cm
　　　　　　　　　33. tectorum
41a. Flower-stem-leaves with longitudinal spots　**34. vincentei**

b. Flower-stem-leaves without longitudinal spots   42
42a. Rosette 5–10 cm across; leaves blue-green
    33. tectorum var. **glaucum**
b. Rosette 2.5–3.5 cm across; leaves green   **35. nevadense**
43a. Leaves covered in long white woolly hairs   **36. arachnoideum**
b. Leaves not covered in long white woolly hairs   44
44a. Marginal hairs of unequal lengths   45
b. Marginal hairs all the same length or absent   47
45a. Rosettes 8 cm or more across
    37. charadzeae
b. Rosettes less than 6 cm across   46
46a. Flower-stem *c.* 20 cm   × **funckii**
    (see **36. arachnoideum**)
b. Flower-stem *c.* 30 cm
    38. italicum
47a. Leaf-tips coloured   48
b. Leaf-tips not coloured   51
48a. Leaf-tips dark brown
    39. caucasicum
b. Leaf-tips red or purple   49
49a. Rosettes 5–6 cm across
    40. cantabricum
b. Rosettes 4 cm or less across   50
50a. Flower-stems short
    41. macedonicum
b. Flower-stems tall (*c.* 33 cm)
    42. altum
51a. Rosettes to 2 cm across   52
b. Rosettes 4–8 cm across
    43. atlanticum
52a. Leaf-tips with long woolly hairs
    × **barbulatum**
    (see **36. arachnoideum**)
b. Leaf-tips lacking woolly hairs
    44. montanum

1. S. sosnowskii Ter-Chatschatorova. Illustration: Journal of the Sempervivum Society **10**(1): 26 (1979).

Similar to *S. armenum*. Rosettes 10–12 cm across. Leaves oblong to spathulate, slightly broader in the upper $^1/_3$, hairless except for short hairs along the margins, flushed red at the tips. Offsets numerous, on stout stolons 2–3 cm. Flowering-stem 30–40 cm, covered in short scattered hairs and hairless strap-shaped leaves 6–9 cm. Flowers 3–3.5 cm across. Petals 14–16, yellowish green with a green central stripe and lilac tinge at the base. Petals glandular-hairy; filaments lilac-tinted, glandular-hairy. *Georgia.*

2. S. octopodes Turrill. Illustration: Quarterly Bulletin of the Alpine Garden Society **6**: 99–100 (1938); **8**: 215 (1940); **17**: 233 (1949); Journal of the Sempervivum Society **9**(3): 85 (1978), **10**(3): 81 (1979); Payne, Plant jewels of the high country, 50 (1972).

Rosettes 1–2 cm across, incurved or semi-open. Leaves oblanceolate or obovate, slightly obtuse, shortly mucronate, fleshy, densely hairy on both surfaces, glandular marginal hairs getting longer towards the red-brown tips. Offsets on slender, brown stolons to 7 cm. Flower-stem *c.* 9 cm, slender with a compact inflorescence of few flowers. Petals yellow with a pale red spot at the base. Filaments reddish purple, anthers yellow. *S former Yugoslavia (N Macedonia).* Summer.

Similar to *S. ciliosum* and *S. thompsonianum*. Requires lime-free soil. It is difficult to grow as it dislikes excessive wet or dry at any time of year and is readily attacked by greenfly.

Var. **apetalum** Turrill. Rosettes 2.5–3 cm across with more leaves. Leaves less fleshy and lighter green with very small brown markings on the tips; stolons to 9 cm. Flowers have no petals or stamens but numerous sepals. Easy to grow, withstands winter damp well.

3. S. grandiflorum Haworth (*S. globiferum* Curtis; *S. globiferum* Gaudin; *S. gaudini* Christ; *S. braunii* Arcangeli; *S. wulfenii* Hoppe subsp. *gaudini* Christ). Illustration: Praeger, An account of the Sempervivum Group, 78 (1932); Quarterly Bulletin of the Alpine Garden Society **3**: 273 (1935), **37**: 96 (1969); Nicholson et al., The Oxford book of garden flowers, 91 (1963); Huxley, Mountain flowers, pl. 233 (1967).

Rosettes 2–10 cm across, flat and rather loose. Leaves oblanceolate to wedge-shaped, or almost strap-shaped, green often with a brown tip, densely hairy and quite sticky, giving off a strong resinous, 'goaty' smell. Offsets nearly spherical on long leafy stolons. Flowers large; petals yellow or greenish yellow tinged purple at the base. *S Switzerland & N Italy.* Summer.

Requires lime-free soil. 'Fasciatum' is a curiously congested form; 'Keston' has yellow-green leaves in spring.

S. × christii Wolf (*S. grandiflorum* × *S. montanum*; *S. rupicolum* Chenevard & Schmidely). Varies between the parents. Leaves with a small purple tip or wholly green, lightly or densely hairy. Petal colour varies from yellow to purple but is always yellowish towards the tip. *Switzerland to N Italy.*

S. × hayekii Rowley (*S. grandiflorum*

× *S. tectorum*). Rosettes large, sparingly hairy, with marginal hairs on the leaves. Flowers pale yellowish purple, smaller than in *S. grandiflorum* but larger than *S. tectorum*.

4. S. transcaucasicum Muirhead.

Rosettes 4–7 cm across, semi-open. Leaves obovate or oblanceolate, shortly mucronate, densely and finely glandular-hairy on both surfaces, with many marginal hairs, green or yellow-green, upper half tinged pink when exposed. Few offsets produced on stout stolons *c.* 2 cm. Flower-stem 15–18 cm, glandular-hairy, covered in many overlapping glandular hairy leaves, flushed pink on the outer surfaces. Flowers to 2.5 cm across, consisting of 12–14 petals. Petals greenish yellow, purple at the base, twisted at the tip, densely glandular hairy on the lower surface. Filaments pale purple, the lower half densely hairy, anthers yellow. *Caucasus, Transcaucasus.*

Quite easy to grow but needs some protection from winter damp; seldom flowers. Can be confused with *S. ruthenicum* Schnittspahn & Lehmann and *S. zeleborii* Schott. *S. globiferum* misapplied is a name used to describe all yellow species from the Caucasus and Turkey.

5. S. kindingeri Adamovic. Illustration: Praeger, An account of the Sempervivum Group, 83 (1932); Payne, Plant jewels of the high country, 44 (1972); Quarterly Bulletin of the Alpine Garden Society **3**: 280 (1935); Houslekes, **16**(1): 35 (1985).

Rosettes 4–6 cm across, flat, open. Leaves wedge-shaped to oblong, glandular-hairy, with long marginal hairs, pale yellowish green with a purplish flush at the tips. Offsets on medium-long stolons. Flower-stem to 20 cm, very leafy. Petals 12–14, pale yellow with a pink-red base. Filaments purple, anthers yellow. *Former Yugoslavia (Macedonia), N Greece.*

A rare plant which needs protection in the winter, flowers poorly and develops few offsets.

6. S. zeleborii Schott (*S. ruthenicum* Koch). Illustration: Journal of the Sempervivum Society, **7**(3): 66 (1976).

Rosettes 3–4 cm across, spherical and compact. Leaves oblong-obovate, shortly mucronate, densely hairy, pale or grey-green with outer leaves flushed pink on exposure, sometimes with a small dark tip. Few offsets produced on short stolons. Flower-stem 10–15 cm. Flowers *c.* 2.5 cm across. Petals 12–14, yellow with a crimson base. Filaments dark crimson, anthers yellow. *SE Bulgaria & S Romania.*

7. S. armenum Boissier & Huet. Illustration: Quarterly Bulletin of the Alpine Garden Society **10**: 236 (1942); Notes from the Royal Botanic Garden Edinburgh **29**: 16, pl. 1A (1969).

Rosettes 4–6 cm across, glaucous. Rosette leaves hairless except along the edges, mucronate, dark purple tips. Few stolons; juvenile rosettes glandular-hairy, hairless on maturity. Flower-stem 10–15 cm, upper leaves hairy. Inflorescence a dense cyme 4–5 cm across. Flowers with 12–14 pale yellow or greenish petals, purple tinged at the base. Filaments purple, anthers yellow. *Armenia, Turkey.*

A rare species often confused with *S. globiferum* and some forms of *S. tectorum*, considered to be closely related to *S. sosnowskii*. Slow-growing, apt to damp-off in the winter wet.

Var. **insigne** Muirhead. Rosettes 2–3 cm, highly coloured with many offsets. Stem-leaves overlapping. Petals flushed rose-purple violet at base. Filaments violet.

8. S. wulfenii Merten & Koch (*S. globiferum* Wulfen). Illustration: Reichenbach, Icones Florae Germanicae et Helveticae, **23**: t. 69 (1896–1899); Praeger, An account of the Sempervivum Group, 91 (1932); Huxley, Mountain flowers, pl. 232 (1967); Grey-Wilson & Blamey, Alpine flowers of Britain and Europe, 75 (1979).

Rosettes 5–9 cm across, open. Leaves oblong to spathulate, hairless except for the marginal hairs, grey-green with a purple base and darker tips. Few offsets produced on long stolons to 10 cm. Flower-stem 15–25 cm, covered with hairy leaves which are slightly recurved at the tips. Flowers *c.* 2.5 cm across, funnel-shaped to stellate. Petals 12–15, lemon-yellow with a purple base. Filaments purple, anthers yellow. *C & E Alps.* Summer.

Like *S. tectorum* var. *glaucum* but central leaves closed to form a central bud. Slow-growing, difficult in cultivation, needs winter protection.

9. S. ruthenicum Schnittspahn & Lehmann (*S. globiferum* (Linnaeus) Koch; *S. globiferum* Linnaeus; *S. ruthenicum* Koch). Illustration: Quarterly Bulletin of the Alpine Garden Society **3**: 274, 279; Jelitto & Schacht, Hardy herbaceous perennials **2**: 613 (1990).

Rosettes 5–8 cm across, incurved but more open in summer. Leaves club-shaped, narrowing at the base, shortly glandular-hairy with dense marginal hairs, dark green. Stolons 3–5 cm, stout. Flower-stem 20–30 cm, covered in weakly pointed, oblong leaves with reddish brown tips. Petals yellow with small rounded scales which stand out clearly. Filaments green, anthers yellow. *Romania, Ukraine.*

Quite easy to grow and sometimes confused with *S. zeleborii* Schott (*S. ruthenicum* Koch) and *S. armenum* Boissier & Huet.

S. × degenianum Domokos (*S. banaticum × S. ruthenicum*). Rosettes 8–10 cm across, open. Leaves linear, contracting abruptly towards the tip, broadest in the upper ⅓, dense white hairs on both surfaces, marginal hairs, brown-red. Flower-stem to 30 cm, stem-leaves covered in dense white hairs. Flowers to 2.5 cm across, with 13–14 parts. Petals white with rose-red stripes, glandular-hairy below, hairless above. Filaments densely hairy at the base, anthers pale pink or pale yellow. Summer. *E Europe (Banatu to near the lower Danube).*

10. S. minus Turrill. Illustration: Quarterly Bulletin of the Alpine Garden Society **10**: 235–237 (1942).

Rosettes 1–2.5 cm across, very compact, central leaves closed, outer leaves more open. Leaves oblong-oblanceolate, acute, very short hairs on both surfaces with slightly longer hairs on the margins, dull olive-green, purple at the base, bronze on the outer leaves when exposed. Offsets on very short stolons, almost stemless. Flower-stem 2.5–6.5 cm, slender and with elliptic fleshy leaves. Inflorescence compact with few, large flowers. Petals pale yellow, anthers yellow. *N Turkey.*

The smallest species of *Sempervivum*. Rarely seen in cultivation as it increases slowly and flowers often. May sometimes resemble *S. pumilum.*

11. S. ciliosum Craib (*S. wulfenii* Velenovsky; *S. wulfenii* Hoppe; *S. ciliosum* Pančič; *S. ciliatum* Craib; *S. borisii* Degen & Urumov). Illustration: Praeger, An account of the Sempervivum Group, 89 (1932); Payne, Plant jewels of the high country, 32 (1972); Quarterly Bulletin of the Alpine Garden Society **8**: 207 (1940); Journal of the Sempervivum Society **7**(1): 19 (1976), **9**(3): 77–78 (1978), **10**(3): 75, 77 (1979).

Rosettes 3.5–5 cm across, grey-green, flattened spherical, wholly or half-closed, outer leaves tinged red. Leaves oblong-oblanceolate, acute, strongly incurved,

hairy, marginal hairs very long, giving the leaves their greyness. Many offsets produced on strong hairy stolons. Yellow flowers on short stems with overlapping leaves. Petals 10–12. *Bulgaria, former Yugoslavia, NW Greece.* Late spring.

Not difficult to grow but prone to winter damp.

Var. **borisii** (Degen & Urumov) P. Mitchell, extra densely haired giving a white appearance, stolons short. *Bulgaria.*

Var. **galicicum** A.C. Smith, hairy rosettes becoming plum-red in the summer; stolons extra long, hairy. Flowers small, yellow. *Former Yugoslavia (SW Macedonia).*

S. ciliosum × S. marmoreum. Small green rosettes heavily fringed with hair.

S. × praegeri Rowley (*S. ciliosum × S. erythraeum*). Illustration: Journal of the Sempervivum Society **7**(2): 28 (1976). Some consider this hybrid to be *S. erythraeum × S. leucanthum*, often sold under the name *S. × praegeri*. Intermediate between the parents. Rosettes like *S. erythraeum*, small and open, but the leaves densely glandular-hairy with both long and short hairs. Leaves in outline like *S. erythraeum*, broader than *S. ciliosum*. Flowers intermediate, petals greenish yellow, flushed with red near the tip, hairy on back and edges. Anthers buff flushed purple. Filaments purplish above, whitish below. *Bulgaria.*

12. S. davisii Muirhead. Illustration: Notes from the Royal Botanic Garden Edinburgh **29**: pl. 3B, 26 (1969).

Rosettes 3–4 cm across, half-open. Leaves oblanceolate or obovate, abruptly mucronate, covered in dense glandular hairs, grey-green sometimes with a brown-red tip. Few offsets produced on stolons 2–3 cm. Flower-stem 10–12 cm, erect, covered in short dense glandular hairs. Numerous stem-leaves, short glandular hairs on both sides, grey-green, with a brown-red apex. Inflorescence compact or expanded, with 20–40 flowers, glandular. Flowers *c.* 2 cm across, 12–14 pale yellow petals. Filaments white, anthers yellow. *Turkey.*

13. S. pittonii Schott, Nyman & Kotschy (*S. braunii* Maly). Illustration: Seboth & Bennett, Alpine plants **2**: t. 34 (1880); Hegi, Illustrietre Flora von Mitteleuropa, **4**: 552, 557 (1922); Praeger, An account of the Sempervivum Group, 85 (1932); Huxley, Mountain flowers, pl. 234 (1967).

Rosettes 1.5–3 cm across, dense, flattish, many-leaved. Leaves linear-oblanceolate, acute, incurved, glandular-hairy on both sides with glandular marginal hairs, grey-green with a very small purple tip. Stolons 2–3 cm. Flower-stem 2–3 cm, slender with narrow, overlapping, purple-tipped leaves. Inflorescence comparatively small with large flowers 2–2.5 cm across. Petals 9–12, yellow, filaments greenish yellow, anthers yellow. *E Alps*.

Similar to *S. leucanthum*; not difficult to grow but prone to winter damp.

14. S. glabrifolium Borissova. Illustration: Notes from the Royal Botanic Garden Edinburgh **29**: pl. 1B, 16 (1969).

Rosettes, *c.* 2.5 cm across. Leaves with marginal hairs, olive-green with heavy purple markings on the upper half, especially towards the outside of the rosette. Numerous swollen offsets produced on short stolons. Petals pale greenish yellow; filaments white, anthers yellow. *E Turkey*.

Very similar to *S. armenum* and *S. sosnowskii*, all have hairless rosette-leaves when mature but are glandular when young. Difficult to grow, will damp-off in summer and winter.

15. S. leucanthum Pančić. Illustration: Praeger, An account of the Sempervivum Group, 87 (1932); Quarterly Bulletin of the Alpine Garden Society **3**: 280 (1935); Journal of the Sempervivum Society **6**(1): 8, 12 (1975).

Rosettes 2.5–5 cm across, flattish with the inner leaves closed, outer leaves erect. Leaves narrowly wedge-shaped, widest near the tip and finely hairy on both sides, marginal hairs of unequal lengths, pale yellow-green becoming dark green with a dark red-purple tip. Stolons stout, 5–8 cm. Flower-stem tall, slender, bearing a small inflorescence. Petals 11–13 greenish white; filaments white to purple, anthers yellow. *Bulgaria (Rila Mts)*. Early summer.

Easy to cultivate but increases slowly and soon dies out as all the rosettes often flower at once. Similar to *S. kindingeri* but rosettes less open with many more leaves, stolons much longer.

16. S. erythraeum Velenovsky (*S. montanum* Velenovsky; *S. cinerascens* Adamovic; *S. leucanthum* Stojanoff & Stojanoff; *S. ballsii* misapplied). Illustration: Praeger, An account of the Sempervivum Group, 56 (1932); Quarterly Bulletin of the Alpine Garden Society **3**: 25 (1935), **8**: 201 (1940); Payne, Plant jewels of the high country, 38 (1972).

Rosettes 2–5 cm across, flat and open. Leaves obovate, spathulate, mucronate, covered in very short white hairs that can often only be seen under a magnifying glass, marginal hairs of unequal length, grey-green, often tinged with purple. Offsets always on short stolons. Flower-stem very hairy, to 20 cm. Flowers *c.* 2 cm wide; petals 11 or 12, purplish red with narrow white lines, with fine white hairs on the lower side. *Bulgaria*. Summer.

17. S. marmoreum Grisebach (*S. montanum* Sibthorp & Smith; *S. schlehanii* Schott; *S. assimile* Schott; *S. blandum* Schott; *S. rubicundum* Schur; *S. tectorum* Boissier; *S. reginae-amaliae* Boissier; *S. montanum* var. *assimile* Stojanoff & Stefanoff; *S. blandum* var. *assimile* Stojanoff & Stefanoff; *S. montanum* var. *blandum* (Schott) misapplied; *S. ornatum* misapplied). Illustration: Quarterly Bulletin of the Alpine Garden Society **6**: 100 (1938), **8**: 302 (1940); Huxley, Garden perennials and water plants, pl. 251 (1971); Journal of the Sempervivum Society, **6**(3): 11 (1975).

Rosettes *c.* 6 cm across, flat and open. Leaves obovate to spathulate, broader in the upper part, abruptly mucronate, hairless at maturity except for stout deflexed marginal hairs, green sometimes with darker tips or red on back and base. Offsets on thick stolons *c.* 2 cm. Flower-stem stout, 10–15 cm. Flowers *c.* 2.5 cm wide. Petals 12–13, purple-pink with white margins. *E Europe, Balkans*. Summer.

Var. **angustissimum** Priszter. Rosettes 3–4 cm across. Leaves long and narrow, hairy on both surfaces, top ⅓ flushed deep red. Offsets on short stolons. Flower-stem 25–30 cm, finely hairy with few leaves, flushed purple. Inflorescence 7–10 cm across with over 30 flowers. Flowers 1.6–2 cm across. Petals 11–13, bright rose, hairy on margins and lower surface. Filaments bright lilac, white and striped on the base, lower ⅓ glandular hairy. *NE Hungary*.

Var. **dinaricum** Becker. Rosettes 1–2.5 cm across. Leaves sharply pointed, dark red-brown at the tips. Inflorescence smaller with narrow petals. *Bulgaria, former Yugoslavia (Macedonia)*.

'Brunneifolium': rosettes compact, uniform brown, becoming red in winter, mature leaves hairless; 'Chocolate' ('Brunneifolium Dark Form'): rosettes slightly less compact than 'Brunneifolium', dark chocolate-coloured; 'Rubrifolium': leaves deep red, tips and margins green; 'Ornatum': rosettes large, ruby-red, tips apple-green, colour only lasts from spring to summer.

S. × versicolor Velenovsky (*S. marmoreum × S. zeleborii*). Illustration: Quarterly Bulletin of the Alpine Garden Society **3**: 273 (1935). Rosettes large. Flowers pale yellow to pale lilac with age, seldom produced. On roofs in *Bulgaria* but not found wild.

18. S. iranicum Bornmuller & Gauba. Rosettes 3–5 cm across, open. Leaves narrowly obovate, mucronate, hairs along the margins and few on the backs. Offsets on very short stolons. Petals rose-pink with narrow white margins; filaments rose, anthers brownish pink. *N Iran*.

Not difficult to grow but prone to winter damp.

19. S. giuseppii Wale. Illustration: Quarterly Bulletin of the Alpine Garden Society **9**: 111, 114 (1941); Journal of the Sempervivum Society, **6**(1): 8 (1975).

Rosettes 2.5–3.5 cm across, compact in dense clumps. Leaves ovate, mucronate, pale green with a small brown patch at the apex, densely hairy with long stiff marginal hairs. Numerous offsets on short stolons. Flowers rose-red with narrow white margins. *NW Spain*.

20. S. ingwersenii Wale. Illustration: Quarterly Bulletin of the Alpine Garden Society **10**: 88, 91, 93 (1942).

Rosettes 3–4 cm across, very compact and flattish. Leaves densely but finely hairy and with a small brown apex. Numerous offsets produced on long brownish red stolons. Flowers rarely produced, petals red with narrow white margins. *Caucasus*.

Vigorous and easy to cultivate. Like *S. altum* and *S. ossetiense* in habit but with pea-green rosettes and smaller flowers in a more compact inflorescence.

21. S. ossetiense Wale. Illustration: Quarterly Bulletin of the Alpine Garden Society **10**: 100, 102 (1942).

Rosettes *c.* 3 cm across, dense, few-leaved. Leaves oblanceolate to oblong-oblanceolate, very fleshy, covered in short hairs, marginal hairs slightly longer than surface hairs, pea-green with a small brown tip. Offsets on stout stolons 5–8 cm. Flower-stem 8–10 cm. Inflorescence small, few-flowered but individual flowers large. Petals purple with broad white margins; filaments purple, anthers deep red. *Caucasus*.

Difficult to grow, liable to winter damp, produces few offsets or flowers. Similar to *S. altum* and *S. ingwersenii*.

22. S. kosaninii Praeger. Illustration: Bulletin Institut & Jardin Botaniques Belgrade **1**: 211–212 (1930); Praeger, An account of the Sempervivum Group, 54 (1932); Quarterly Bulletin of the Alpine Garden Society **3**: 268 (1935); Journal of the Sempervivum Society **7**(2): 43 (1976).

Rosettes 4–8 cm across, open, dense, flattish. Leaves oblanceolate, shortly acuminate, glandular-hairy on both sides, marginal hairs twice as long as other hairs, dark green with a red-purple tip. Offsets on strong leafy stolons to 12 cm. Flower-stem stout, covered in loosely overlapping leaves. Petals reddish purple with white margins, greenish on the back. Filaments purple, anthers light red. *SW former Yugoslavia (Macedonia)*.

23. S. dzhavachischvilii Gurgenidze. Illustration: Houslekes **11**: 79, 81 (1980).

Rosettes 3–5 cm. Leaves lanceolate, acuminate, dark green, outside flushed purple, dark purple tips often bent inwards, glandular-hairy on both sides. Offsets with outer leaves longer and increasingly cylindric, on very short stolons. Flower-stem 7–10 cm, stem-leaves covered in unequal length glandular hairs. Flower 2–2.3 cm across. Petals purple-pink with white margins. Filaments dark purple. Anthers pale purple with traces of orange. Rosette-leaves persistent at flowering. *E Caucasus*.

Closely related to *S. pumilum*.

24. S. reginae-amaliae Haláscy. Illustration: Gartenflora **83**: 159 (1934); Quarterly Bulletin of the Alpine Garden Society **8**: 62, 201 (1940); Payne, Plant jewels of the high country, 54 (1972); Houslekes, **14**(2): 55–58 (1983).

Rosettes 2–4 cm across, usually dense-leaved and compact. Leaves spathulate to obovate, acute, thick, evenly hairy on both surfaces with longer marginal hairs, soft green, outer leaves red. Few offsets produced on short stout stolons. Flower-stem 9-12 cm, wider at the base, covered in overlapping ovate leaves, usually flushed rose to red. Petals 10–12, crimson with clearly defined white margins. Filaments crimson, anthers buff sometimes violet. *Greece, S Albania*. Summer.

This name is often wrongly applied to a dark-tipped form of *Jovibarba heuffelii*. Praeger included it under *S. schlehanii* Schott. Very variable in colour, size and compactness of rosette, but floral characters constant. Easy to grow but slow to increase.

25. S. dolomiticum Facchini. (*S. tectorum* var. *angustifolium* Seybold; *S. lehmanni* Schnittspahn; *S. oligotrichum* Dalla Torre). Illustration: Jávorka & Csapody, Iconographia Florae Hungaricae, 222 (1930); Praeger, An account of the Sempervivum Group, 51 (1932); Huxley, Mountain flowers, pl. 236 (1967); Grey-Wilson & Blamey, Alpine flowers of Britain and Europe, 75 (1979).

Rosettes 2–4 cm, erect, semi-open, nearly spherical. Leaves covered in glandular hairs and with long white hairs along the margins, bright green with the outside of older leaves tinged scarlet. Numerous offsets on slender stolons. Flower-stem to 10 cm, thin, covered in shaggy white hairs and purple-tipped leaves. Petals 10–14, rose-red with a dark central stripe and white marginal flecks. *E Alps*. Summer.

Similar to some forms of *S. montanum* but upright leaves denser and more pointed. Not very easy to grow, prone to winter damp, flowers only occasionally.

S. dolomiticum × *S. montanum*. A compact growing plant with small deep green rosettes.

26. S. thompsonianum Wale. Illustration: Quarterly Bulletin of the Alpine Garden Society **8**: 208, 212 (1940); Journal of the Sempervivum Society **7**(2): 41 (1976).

Rosettes 1.5–2 cm across, many-leaved, nearly spherical, outer leaves erect. Leaves ovate-lanceolate, acute, hairy, marginal hairs of unequal length, *c.* 3 times as long as the surface hairs, yellowish green, outer leaves flushed red. Stolons 5–8 cm, slender and brown. Flower-stems *c.* 8 cm, very slender and covered with few, narrow, lanceolate leaves. Inflorescence few-flowered and compact. Petals deep pink with white margins and yellow tips. Filaments purple, anthers yellow. *S former Yugoslavia (Macedonia)*.

27. S. pumilum Bieberstein (*S. montanum* Eichwald; *S. braunii* Ledebour). Illustration: Quarterly Bulletin of the Alpine Garden Society **3**: 260 (1935), **10**: 81, 82, 87 (1942); Praeger, An account of the Sempervivum Group, 58 (1932); Houslekes **11**(2): 41 (1980).

Rosettes 1–2 cm across, spherical. Leaves lanceolate or oblong-lanceolate, acute or shortly acuminate, wholly green, glandular-hairy on both sides, marginal hairs twice as long as surface hairs. Numerous offsets on very slender stolons *c.* 1 cm. Flower-stems 5–8 cm. Inflorescence

small with 4–8 relatively large flowers. Petals rosy purple with white margins; filaments purple, anthers red-purple. *Caucasus*.

28. S. ballsii Wale. Illustration: Kew Bulletin, 1940 (4): pl. 5 (1940); Quarterly Bulletin of the Alpine Garden Society **8**: 205–206 (1940); Journal of the Sempervivum Society **9**(1): 22 (1978).

Rosettes neat, nearly spherical, densely leaved, *c.* 3 cm across, inner leaves closed with the outer leaves more open and erect. Rosette-leaves obovate, abruptly mucronate, hairless at maturity except for a few marginal hairs on the lower $2/3$, uniform green with a bronze to red tinge on the outermost leaves. Offsets 3–4, produced in spring on stout, basal stolons *c.* 1.5 cm. Flowering-stem to 10 cm, with erect, only slightly overlapping leaves. Inflorescence compact, small, *c.* 4 cm across. Flowers 1.8–2 cm across; petals 12, dull pink; filaments crimson. *NW Greece*. Summer.

29. S. borissovae Wale. Illustration: Quarterly Bulletin of the Alpine Garden Society **10**: 94, 97, 99 (1942); Payne, Plant jewels of the high country, 28 (1972).

Rosettes flat, semi-open, *c.* 3 cm across, in large dense clumps. Rosette-leaves with many marginal hairs but otherwise hairless, green at the base but red-brown in the upper half. Petals rose-red with slightly irregular toothed edges. *Caucasus*.

30. S. andreanum Wale. Illustration: Quarterly Bulletin of the Alpine Garden Society **9**: 105, 117 (1941).

Rosettes 1.5–4 cm across, bright green with brown markings. Outer rosette-leaves erect, tipped with small brown markings. Central leaves in a tight conical bud (like *S. wulfenii*). Offsets produced on short stolons causing the rosettes to grow in tight clumps. Petals pale to bright pink, light red at the base. Anthers yellow. *Spain*. Summer.

31. S. calcareum Jordan (*S. tectorum* var. *calcareum* (Jordan) Cariot & Saint-Lager; *S. racemosum* Jordan & Fourreau; *S. columnare* Jordan & Fourreau; *S. californicum* misapplied, Baker; *S. greenii* Baker). Illustration: Quarterly Bulletin of the Alpine Garden Society **12**: 11 (1944), **30**: 79 (1962); Grey-Wilson & Blamey, Alpine flowers of Britain and Europe, 75 (1979); Houslekes **13**(1): 22 (1982); Jelitto & Schacht, Hardy herbaceous perennials **2**: 611 (1990).

Rosettes large, grey-green, 6 cm or more across. Leaves hairless, tipped with purple on both sides. Flowers pale pink but seldom produced. *French Alps.*

Several cultivars are grown including: 'Benz', a large variant with more than the usual number of leaves; 'Greenii', with smaller neater rosettes; 'Mrs Giuseppi', more compact, leaf-tips deep red; 'Sir William Lawrence', similar to 'Mrs Giuseppi' but rosettes larger and more spherical, leaf-tips red; 'Griggs Surprise', ('Monstrosum') a curious congested form with abnormal growth development that occasionally occurs in gardens, leaves grey-green, round and hollow without an opening at the tip, each leaf ends in a red thorn-like beak which is often double.

32. S. balcanicum Stojanov.

Rosettes 1.5–2.5 cm across. Leaves oblong-lanceolate, gradually mucronate, green with red tips, hairless except along the margins. Flower-stem 5–12 cm, densely hairy. Stem-leaves oblong to oblong-linear. Inflorescence with few branches. Petals pale lilac, filaments dirty purple, blackish towards apex. Anthers rounded, sulphur-yellow, margins black. *C Balkans. Summer.*

Closely related to *S. erythraeum.*

33. S. tectorum Linnaeus (*S. arvernense* Lecocq & Lomotee; *S. cantalicum* Jordan & Fourreau; *S. lamottei* Boreau). Illustration: Reichenbach, Icones Florae Germanicae et Helveticae, **23**: t. 67 (1896–99); Praeger, An account of the Sempervivum Group, 5, 66, 71 (1932); Perry, The good gardener's guide, 465 (1974); Quarterly Bulletin of the Alpine Garden Society **47**: 77–78 (1979), **51**: 70 (1983).

Rosettes 2–8 cm (rarely to 18 cm) across, open. Leaves oblong-lanceolate to obovate, sharply mucronate, hairless except for obvious white marginal hairs, very fleshy, dark green sometimes red-brown or purple, red or white at the base and frequently tipped with a darker colour. Stolons to 4 cm, stout, reddish. Flower-stem 20–40 cm, stout, covered in many white hairs, only the upper stem-leaves hairy on the surfaces. Inflorescence large and flat, flowers 40–100; each *c.* 2.5 cm across. Petals 12–16, dull pink or purple; anthers orange brown. *C Europe to Balkans. Summer.*

The best-known cultivated *Sempervivum*, very variable and will hybridise with many other species.

Var. **tectorum** Linnaeus (*S. murale* Boreau). Rosettes to 18 cm across, flat and open. Leaves obovate-lanceolate, hairless, bright green, whitish at the base; tips well marked with purple-brown. Flower-stem 30–50 cm. Inner whorl of stamens mostly sterile and often replaced by carpels.

Var. **alpinum** Praeger (*S. alpinum* Grisebach & Schenk; *S. boutignyanum* Billot & Grenier; *S. arvernense* forma *boutignyanum* Rouy & Camus; *S. fuscum* Lehmann & Schnittspahn; *S. tectorum* subsp. *alpinum* Wettstein & Hayek). Illustration: Blandford, Garden flowers in colour, 110 (1955). Rosettes 2–6 cm across. Leaves green, always red at the base, sometimes with purple-brown tips. Flower-stem 10–30 cm; inflorescence compact and flattish. *Pyrenees, Alps.*

Var. **glaucum** Praeger (*S. glaucum* Tenore; *S. spectabile* Lehmann & Schnittspahn; *S. tectorum* subsp. *schottii* Hayek; *S. acuminatum* Schott; *S. schottii* Baker). Rosettes 5–10 cm across. Leaves blue-green, white at the base and no markings on the tips. Flower-stem to 60 cm. Variable. *S & E Alps.*

Cultivars include: 'Atropurpureum', 'Boissieri', 'Nigrum', 'Red Flush', 'Royanum', 'Sunset', 'Triste', 'Atroviolaceum'.

S. × calcaratum Baker (*S. tectorum* × *S. calcaratum* misapplied; *S. comollii* misapplied). Rosettes to 15 cm across. Leaves spathulate to oblanceolate, noticeably mucronate, hairless, green-blue to purple with a crimson base. Stolons to 5 cm. Flower-stem *c.* 30 cm. Flowers dull red-purple. Easy to grow and increases quickly.

S. × widderi Lehmann & Schnittspahn (*S. tectorum* × *S. wulfenii*; *S. albidum* Lehmann & Schnittspahn). Intermediate between the parents but many variations. Medium-sized blue-green rosettes. Leaves hairless except for marginal hairs, usually red at the base. Petals yellow, towards the base red or streaked red and yellow. *Switzerland.*

34. S. vincentei Pau.

Rosettes *c.* 2 cm across. Leaves oblong to spathulate, mucronate, hairless except for marginal hairs. Juvenile offsets hairy. Flower-stem to 12 cm, stem-leaves oblong-lanceolate with longitudinal spots. Petals pale red with a purplish base. Filaments purple, hairy. *Spain.*

35. S. nevadense Wale. Illustration: Quarterly Bulletin of the Alpine Garden Society **9**: 107–108 (1941); Journal of the Sempervivum Society, **8**(3): 57 (1977).

Rosettes 2.5–3.5 cm across, flattish, compact, many-leaved. Leaves obovate, mucronate, fleshy, strongly incurved, hairless except for the short, stout, curved, marginal hairs, green, outer leaves scarlet in winter and pinkish bronze at flowering time. Offsets on short stolons. Flower-stems covered in many fleshy, overlapping leaves. Inflorescence compact. Petals crimson, filaments dark red, hairless. *Spain.*

36. S. arachnoideum Linnaeus. (*S. sanguineum* Timbal-Lagrave; *Sedum arachnoideum* Krause). Illustration: Botanical Magazine, 68 (1942); Praeger, An account of the Sempervivum Group, 23, 36 (1932); Reichenbach, Icones Florae Germanicae et Helveticae **23**: t. 72 (1899); Grey-Wilson & Blamey, Alpine flowers of Britain and Europe, 75 (1979).

Rosettes small, ball-shaped, densely covered with white woolly hairs. Rosette-leaves deep red or shades of green. Offsets stalkless, crowded. Flower-stem 7–12.5 cm. Flowers usually rose-red with 9–12 petals. *Pyrenees, Alps, Apennines, Carpathians.*

A species very variable in rosette size and quantity of hairs. All forms are easy to grow, increase freely and hybridise easily.

Var. **arachnoideum**. Rosettes to 2 cm across, ovoid or spherical, with variable amounts of arachnoid hairs.

'Fasciatum' A curiosity that is fasciated. The fasciation is not very stable.

Var. **glabrescens** Wilkommen (*S. doellianum* Lehmann; *S. heterotrichum* Schott; *S. moggridgei* J.D. Hooker; *S. arachnoideum* var. *doellianum* Jaccard). Rosettes less than 1.5 cm across, flattish with only a small amount of cobweb. Flowers white and produced so freely that it is difficult to culture as it flowers itself to death.

Var. **tomentosum** (Lehmann & Schnittspahn) Hayek (*S. tomentosum* Lehmann & Schnittspahn; *S. webbianum* Lehmann & Schnittspahn; *S. laggeri* Hallier). Rosettes to 4 cm across, large and flattish with a very dense cobweb and an undercolour of dark red in spring.

Common cultivars include: 'Kappa', 'Robin', 'Sultan', 'Rubrum', 'Stansfieldii'.

S. arachnoideum × *S. nevadense*. Intermediate between the parents. Rosettes green with a flush of red and covered in woolly hair. *Spain (Sierra Nevada).*

S. × barbulatum Schott (*S. arachnoideum* × *S. montanum*; *S. fimbriatum* Schnittspahn & Lehmann; *S. elegans* Lagger; *S. barbatulum* Baker; *S.hausmanii* Nyman; *S. oligotrichum* Baker; *S. × hybridum* Brugger; *S. dolomiticum* Koch; *S. hookeri* misapplied). Illustration: Chanousia

1: 117–118 (1928). Rosettes *c.* 1 cm across. Leaves elliptic-oblong, blunt, finely and densely glandular-hairy on both sides, marginal hairs glandular, longer woolly hairs on the tip. Very small offsets produced on short, slender, bare stolons. Flower-stem to 5 cm, glandular-hairy. Inflorescence with *c.* 6 flowers, glandular-hairy. Flowers with 10 petals, *c.* 2 cm across. Petals purplish rose, glandular-hairy on the back, edges and upper part of face. *European Alps, N Spain.* Summer.

S. × faunconettii Reuter (*S. arachnoideum × S. tectorum; S. flavipilum* Sauter; *S. schnittspahnii* misapplied, not Lagger; *S. villosum* Aiton; *S. × thomayeri* Correvon; *S. thompsonii* Lindsay). Illustration: Bibleotheca Botanica **11**(58): t. 6, fig 171 (1902); Payne, Plant jewels of the high country, 38 (1972). Rosettes small and green with a tuft of straggling white hairs at the leaf tips, and on the edges near the tips. Inflorescence similar to *S. arachnoideum*, but flowers not so bright, dull purple-pink. *Europe.*

S. × funckii Koch (*S. arachnoideum × S. montanum) × S. tectorum; S. montanum* Mertens & Koch). Illustration: Hegi, Illustrietre Flora von Mitteleuropa, **4**: 552 (1922); Huxley, Mountain flowers, pl. 237 (1967). Rosettes 2.5–4 cm across, compact, flattish, open. Leaves obovate-lanceolate, shortly mucronate, shortly but finely hairy on both surfaces, strong marginal hairs of unequal length, bright green, white at the base, older leaves sometimes with a purplish tinge on the tips. Offsets very numerous on short stems. Flower-stem *c.* 20 cm, with glandular-hairy, erect oblong-ovate leaves. Inflorescence compact and flattish, 6–8 cm across. Flowers dull purple-red. *Alps, Switzerland to Syria.*

S. × morelianum Viviand-Morel (*S. arachnoideum × S. calcareum*). Medium-sized rosettes with tufts of woolly hair on the leaf tips.

S. × roseum Hunter & Nyman (*S. arachnoideum × S. ciliosum; S. fimbriatum* Hegi). Rosettes neat grey-green, leaves hairless, sparingly tipped with woolly hairs.

S. fimbriatum Schott. Rosettes small, compact, spherical. Leaves tinted red and fringed with hairs. Petals yellow with red lines or red with a yellow edge or an intermediate shade. *Austria, Switzerland.*

S. × vaccarii Wilczeck (*S. arachnoideum × S. grandiflorum*). Variable. Rosettes much resembling *S. grandiflorum* but smaller and leaf-tips slightly white-hairy.

37. S. charadzeae Gurgenidze. Illustration: Houslekes **15**(2): 37 (1984).

Rosettes 8–12 cm. Leaves spathulate with a sharply narrowing tip, hairs of unequal length on surfaces and margins. Flower-stem 40–50 cm. Inflorescence corymb-like to umbel-like, 3–5 branches, with 60–80 flowers. Flowers dark purple with alternate petals hairless on the inside of the bases and the outside ⅓ hairy. Opposite petals are ¾ covered with short glandular hairs on both sides. Filaments rose; Anthers pale yellow. *E Georgia.* Summer.

38. S. italicum Ricci.

Rosettes to 5 cm across. Leaves linear, sharply mucronate, covered in short glandular hairs on both surfaces, marginal hairs show an almost regular alternation between long and short. Flower-stem to 30 cm. Petals 11 or 12, purple, glandular-hairy. *C Italy (Apennines).*

Sometimes confused with *S. montanum* Linnaeus.

39. S. caucasicum Boissier (*S. tectorum* Bieberstein; *S. montanum* Meyer; *S. tectorum* subsp. *caucasicum* Berger). Illustration: Quarterly Bulletin of the Alpine Garden Society **10**: 103 (1942).

Rosettes made up of few fleshy leaves, dense, 2–5 cm across. Leaves spathulate, abruptly contracted, mucronate with a dark brown apex, shortly hairy along the edges and sparingly hairy on both sides. Offsets 6 or 7, few-leaved and spherical. Stolons 6–8 cm, thick and reddish nearer the rosette, narrower and greenish further from the rosette, *c.* 8 leaves. Flower-stem 12–20 cm. Petals 14, rose-red with a central stripe. *Caucasus.*

40. S. cantabricum Huber & Sündermann. Illustration: Feddes, Repertorium **33**: t. 140 (1934); Polunin & Smythies, Flowers of South-West Europe, pl. 9 (1973); Journal of the Sempervivum Society **8**(2): 38 (1977); Houslekes **11**(1): 7–14 (1980); **15**(2): 33, 35, 36, (1984).

Rosettes 5–6 cm, half-open. Leaves hairy on both sides, deep green with dark purple-red tips. Few offsets produced on stout, stiff, leafy stems, stolons to 5 cm. Flower-stems stout, 25–30 cm. Flowers deep crimson with purple filaments. *N Spain.*

Relatively easy to grow but suffers from winter wet.

Subsp. **guadarramense** M.C. Smith. Small rosettes with incurving leaves and purple tips.

Subsp. **urbionense** M.C. Smith. Rosettes light green, hairy, incurving, with pointed leaves tipped red.

S. cantabricum × S. montanum subsp. *stiriacum* Large rosettes strikingly tipped with purple.

41. S. macedonicum Praeger. Illustration: Praeger, in Bulletin de Institut & Jardin Botaniques Belgrade **1**: 213 (1930); Praeger, An account of the Sempervivum Group, 53 (1932); Quarterly Bulletin of the Alpine Garden Society **8**: 216 (1940).

Rosettes small, similar to *S. montanum* but flatter and more open, often with a reddish tinge, 2–4 cm across, central leaves closed. Leaves broadly oblanceolate, shortly acuminate, densely but minutely hairy, marginal hairs dull green often red near the tip but not purple-tipped. Offsets on long stolons. Flower-stem short and very leafy, bearing a compact inflorescence. Petals dull red-purple; filaments lilac. *SW former Yugoslavia (Macedonia).*

Easy to grow, increases quickly but produces few flowers.

42. S. altum Turrill. Illustration: Quarterly Bulletin of the Alpine Garden Society **10**: 103 (1942).

Rosettes rather loose, 2.5–4 cm across. Leaves oblanceolate, abruptly mucronate-acuminate, light green, glandular-hairy on both surfaces with hairs along the edges. Offsets are produced on stolons 8–12 cm long. Flower-stem *c.* 33 cm, strong, densely glandular-hairy, covered with many linear-lanceolate stem-leaves *c.* 4.5 cm, with glandular hairs around the edges. Inflorescence branching, many-flowered, almost dense, densely glandular-hairy. Flowers 2.6–3 cm across with 12 or 13 pale, red-purple, glandular-hairy petals. Stamens purple, anthers yellow. *Caucasus.* Summer.

On exposure to the sun the tips of the outer leaves may turn red; similar to *S. kosaninii* but flower-stems taller; dislikes winter damp.

43. S. atlanticum Ball (*S. tectorum* var. *atlanticum* J.D. Hooker). Illustration: Botanical Magazine, 6055 (1873); Praeger, An account of the Sempervivum Group, 62 (1932); Quarterly Bulletin of the Alpine Garden Society, **3**: 26 (1935).

Rosettes 4–8 cm across, pale green, flushed red when fully exposed, nearly erect, often asymmetric. Leaves hairy on both sides and abruptly mucronate. Numerous offsets produced on very short stolons forming a regular hump of rosettes. Flowers pink but rarely seen in cultivation. *Morocco (Atlas Mts)*.

'Edward Balls' ('Ball's Form'): larger more symmetrical rosettes, gaining much colour in the summer.

44. S. montanum Linnaeus (*S. flagelliforme* Link). Illustration: Reichenbach, Icones florae Germanicae et Helveticae **23**: t. 68 (1899); Praeger, An account of the Sempervivum Group, 23, 44 (1932); Huxley, Mountain flowers, pl. 238 (1967); Quarterly Bulletin of the Alpine Garden Society **3**: 286 (1935), **49**: 77, 79 (1981), **53**: 143 (1985).

Rosettes less than 2 cm across, open, forming a close mat. Leaves oblanceolate, rather acute, finely and densely sticky-hairy on both surfaces, marginal hairs slightly longer, wholly dull green. Many offsets on slender leafy stolons 1–3 cm. Flower-stem 5–8 cm, leafy, few relatively large flowers. Petals 10–15, soft violet-purple. *Pyrenees, Alps, Carpathians, Apeninnes, Corsica*.

A very variable species with a wide distribution. Hybridises easily in nature with *S. arachnoideum*, *S. grandiflorum*, *S. nevadense*, *S. tectorum* and *S. wulfenii*, and with many other species in cultivation.

Var. **braunii** Praeger (*S. wulfenii* Bertolini; *S. stiriacum* var. *braunii* Hayek; *S. montanum* subsp. *stiriacum* var. *braunii* Hayek). Illustration: Quarterly Bulletin of the Alpine Garden Society **3**: 259 (1935). Rosettes c. 2.5 cm across, compact, green with nut-brown tips, very finely hairy on both sides. Flowers rose.

Var. **burnatii** Hayek. Illustration: Quarterly Bulletin of the Alpine Garden Society **6**: 259 (1938). Rosettes to 8 cm across, open, light green. Leaves obovate to wedge-shaped. Offsets on long strong stolons. Flower-stem twice as tall as var. *montanum*, flowers lighter purple. *SW Alps & Pyrenees*. Similar to *S. grandiflorum*.

Var. **stiriacum** Hayek (*S. braunii* Koch; *S. funckii* Maly; *S. stiriacum* Hayek). Illustration: Quarterly Bulletin of the Alpine Garden Society **3**: 260 (1935); **45**: 59 (1977). Rosettes 2–4.5 cm across, more open than var. *montanum*. Leaves oblanceolate with dark red-brown tips in the summer, marginal hairs distinctly longer than the surface hairs. Flowers larger than the type and often darker. *E Alps*.

'Rubrum' Medium-sized rosettes, mahogany-red in summer.

S. × densum Lehmann & Schnittspahn (probably *S. montanum × S. tectorum*). Small rosettes, leaves finely hairy on both sides, marginal hairs and tips well marked in dark red. Flowers white, anthers rose.

S. × schottii Lehmann & Schnittspahn (*S. montanum × S. tectorum*; *S. verlotti* Lamotte; *S. monticolum* Jordan & Fourreau; *S. modestum* Jordan & Fourreau; *S. parvulum* Jordan & Fourreau; *S. funckii* Jordan & Fourreau; *S. rhaeticum* Brügger; *S. rupestre × candollei* Rouy & Camus). Very variable, generally resembles a small *S. tectorum* but the rosettes are denser and the leaves are glandular-hairy without a purple tip.

30. AEONIUM Webb & Berthelot
S.G. Knees

Succulent, evergreen herbs and shrublets; usually terrestrial, but occasionally epiphytic, monocarpic or perennial. Stems erect or spreading, often woody at base, branched or unbranched; bark fissured or covered with remains of old leaf-bases. Leaves in dense concave or flattish rosettes, alternate, simple, stalked or stalkless, fleshy and succulent, 1–12 mm thick, convex below, flattened or concave above, spathulate, usually hairless, glaucous, sparsely glandular-hairy; margins entire or shallowly toothed, usually hairy. Leaves of vegetative- and flowering-shoots often dissimilar. Inflorescence terminal, subtended by bracts, flowers hermaphrodite; calyx bell-shaped, yellow, green or variegated with red or pink, with 6–16 triangular to lanceolate sepals; corolla like calyx, but always larger and with petals equalling number of sepals. Stamens twice as many as petals, filaments sometimes widened at base, anthers basifixed, white, yellow or brown. Nectar-producing glands present at base of ovary. Carpels usually equalling sepals in number, rarely fewer, free or loosely attached at their bases; styles erect or spreading, stigma apical; ovaries 1-celled, ovules numerous. Fruit a follicle, seeds numerous, ovate to pear-shaped, ridged, yellowish brown or red-brown.

A genus of 31 species, formerly included in *Sempervivum*, and sometimes found in gardens under that name. Largely endemic to the Canary Islands, Cape Verde Islands and southern Morocco, but with isolated species in eastern Africa and Yemen. Many different growth forms from large shrubby perennials to small stemless biennials are found in the genus and have always attracted horticultural interest.

Propagation is by cuttings or from seed, though the species hybridise freely and consideration must be given to the source of the seed. There are more published names of hybrids than species.

Literature: Praeger, R.L., *An account of the Sempervivum Group* (1932); Liu, H.-Y., Systematics of Aeonium (Crassulaceae). *Serial Publications Number 3, National Museum of Natural Science (Taiwan)* (1989).

1a.	Plants stemless or nearly so	2
b.	Plants with distinct stems at least 25 cm	5
2a.	Rosettes usually 4–10 cm across	**4. simsii**
b.	Rosettes usually 10–50 cm across	3
3a.	Leaves glaucous; stems often with stolons	**6. cuneatum**
b.	Leaves green; stolons absent	4
4a.	Leaf-margins smooth with hairs 1–2 mm	**8. tabuliforme**
b.	Leaf-margins sometimes wavy with hairs 0.6–1 mm	**7. canariense**
5a.	Stems unbranched	6
b.	Stems branched	9
6a.	Leaves yellowish or dark green, sometimes with brownish margins	7
b.	Leaves glaucous, sometimes with reddish margins	8
7a.	Bundle-scars visible; leaves oblanceolate to spathulate or oblong; stems to 2.5 m	**11. undulatum**
b.	Bundle-scars indistinct; leaves obovate; stems to 0.6 m	**12. nobile**
8a.	Plant 1–2 m; leaves glaucous green, leaving scars 5–8 mm wide	**17. urbicum**
b.	Plant 2.5–5 m; leaves very glaucous, often with a pink or purplish tinge and reddish margins, leaving scars 6–18 mm wide	**21. hierrense**
9a.	Stems with 1–10 lateral branches	10
b.	Stems with 30–100 lateral branches	17
10a.	Stems with adventitious roots	11
b.	Stems without adventitious roots	12
11a.	Bark with tubercles; leaf-scars slightly raised	**18. ciliatum**
b.	Bark with raised netted lines; leaf-scars depressed	**15. lancerottense**
12a.	Whole plant smelling strongly of balsam	**9. balsamiferum**
b.	Plants generally odourless	13

13a. Stem and leaves very sticky, especially when young
    13. glutinosum

   b. Stems and leaves not sticky   14

14a. Leaves dark green with reddish margin, but without any brownish stripes   **19. percarneum**

   b. Leaves sometimes with reddish or purple margins, but usually with brown stripes   15

15a. Leaf rosettes with raised centre, the young leaves arising erectly
    14. gorgoneum

   b. Leaf rosettes with flattened centres, the young leaves tightly adpressed to the older ones   16

16a. Leaf-apex acuminate; marginal hairs 0.3–1 mm   **10. arboreum**

   b. Leaf-apex acute, mucronate; marginal hairs 0.5–2 mm
    11. undulatum

17a. Adventitious roots common on lower stems   18

   b. Adventitious roots absent   20

18a. Leaf-scars not more than 1 mm wide; bark fissured into plates
    22. castello-paivae

   b. Leaf-scars 2–4 mm wide; bark only slightly fissured, but not into plates   19

19a. Stem surface covered in netted lines; bark grey or brown; leaf scars c. 2 mm wide   **16. haworthii**

   b. Stem surface unlined; bark pale reddish green or whitish; leaf-scars 3–4 mm wide   **20. decorum**

20a. Leaf-scars not more than 0.5 mm wide   **2. goochiae**

   b. Leaf-scars 1–1.5 mm wide   21

21a. Leaf-scars 1.1–1.5 mm wide; plant smelling of balsam   **1. lindleyi**

   b. Leaf-scars c. 1 mm wide; plant usually odourless   22

22a. Leaf-margin with bead-shaped hairs
    5. spathulatum

   b. Leaf-margin without bead-shaped hairs   **3. sedifolium**

1. A lindleyi Webb & Berthelot (*Sempervivum lindleyi* (Webb & Berthelot) Christ; *S. tortuosum* Aiton var. *lindleyi*). Illustration: Praeger, An account of the Sempervivum Group, 205 (1932); Graf, Exotica, 3, 652 (1963).

Perennial, terrestrial shrublet with c. 50 lateral branches; smelling strongly of balsam. Stems to c. 50 × 3–15 cm, young branches softly hairy, green to brownish, sticky, bark grey or whitish, slightly fissured; leaf-scars 1.1–1.5 mm wide. Leaf-rosettes 4–9 cm across. Leaves

2–4.5 × 0.6–1.6 cm, obovate to spathulate, yellowish to dark green, sticky. Inflorescence a cyme-like panicle, 2–7 × 3–9 cm, with 15–85 flowers. Sepals 7–9; petals 5–7 × 1.5–2 mm, narrowly elliptic-lanceolate; anthers yellow, nectar-glands yellow; ovaries 2–3 mm, styles 2–3 mm. *Canary Islands (Tenerife)*. H5–G1. Summer–autumn.

2. A. goochiae Webb & Berthelot (*Sempervivum goochiae* Webb & Berthelot). Illustration: Praeger, An account of the Sempervivum Group, 207 (1932).

Terrestrial shrublet with c. 30 branched stems to 40 cm; leaf-scars c. c. 0.5 mm wide, bark pale brown, slightly fissured. Rosettes 3–12 cm across. Leaves 2–6.5 × 1.5–2.5 cm, elliptic to spathulate, narrowing into a stalk-like part c. 0.25 mm wide, pale green or yellowish green, occasionally variegated with red, shortly hairy; margin with short hairs. Inflorescence 2–5 × 3–11 cm, paniculate with 10–45 flowers. Sepals 7 or 8, narrowly triangular, 3–3.5 × 0.7–1 mm, apex acute; petals 5–7 × 1–2 mm, very pale yellow or whitish, tinged pink towards centre; stamens 5.5–7 mm; nectar-glands c. 0.6 mm wide, yellowish; ovaries 2–3 mm, styles 2.5–3 mm. *Canary Islands (La Palma)*. H5–G1. Late winter–summer.

3. A. sedifolium (Bolle) Pitard & Proust (*Aichryson sedifolium* Bolle; *Greenovia sedifolium* (Bolle) Christ; *Sempervivum sedifolium* (Bolle) Christ). Illustration: Praeger, An account of the Sempervivum Group, 211 (1932); Graf, Exotica, 3, 652 (1963); Bramwell & Bramwell, Wild flowers of the Canary Islands, pl. 153 (1974); Jacobsen, Lexicon of succulent plants, pl. 5 (1977).

Perennial, terrestrial shrublet with up to 100 branches. Stems to 40 cm × 1–5 mm, young branches sparsely covered with fine down, dark brown, lustrous, sticky; bark grey or greyish brown, fissured; leaf scars c. 1 mm wide. Leaf-rosettes 1.4–3 cm across, overlapping, becoming almost spherical in very dry conditions. Leaves stalkless, ovate to obovate, green–yellowish green, with pale reddish markings on midrib and towards the apex. Inflorescence 2–7 × 2–5 cm, with 6–15 flowers. Sepals 9–11, elliptic, 2.5–3 mm, variegated with reddish lines; petals 5–7 × 2–2.5 mm, obovate-oblanceolate, yellow, apex often reflexed, toothed, hairless. Stamens 4–5.5 mm; nectar-glands absent; ovaries c. 1.5–2 × 1 mm, styles 2.5–3 mm. *Canary Islands (La Palma)*. G1. Spring–summer.

4. A. simsii (Sweet) Stearn (*A. caespitosum* (Smith) Webb & Berthelot; *Sempervivum ciliatum* Sims; *S. simsii* Sweet; *S. caespitosum* Smith; *S. ciliare* Haworth; *S. barbatum* Hornemann). Illustration: Botanical Magazine, 1978 (1818); Praeger, An account of the Sempervivum Group, 194 (1932); Graf, Exotica, 3, 654 (1963); Bramwell & Bramwell, Wild flowers of the Canary Islands, pl. 158 (1974).

Perennial terrestrial herb; stems tufted, with c. 8 branches, c. 15 cm, forming low hummocks. Rosettes 4–10 cm across, withered leaf remains persisting beneath new growth. Leaves 2–6 × 0.6–2 cm, lanceolate, with red multicellular hairs along margin. Flowers borne on stems 5–30 cm, arising from leaf axils. Sepals 7–9, elliptic; petals 5–6 mm, oblanceolate, yellow; stamens 5–6 mm, anthers yellow. Nectar-glands 0.3–0.4 mm; carpels 1.5–2 mm. *Canary Islands (Gran Canaria)*. G1. Spring–late summer.

Hybridises with *A. canariense*, *A. arboreum*, *A. percarneum*, *A. spathulatum* and *A. undulatum*. Of these the most widely cultivated is **A. × velutinum** Hill (*A. simsii × A. canariense*).

5. A. spathulatum (Hornemann) Praeger (*Sempervivum spathulatum* Hornemann; *S. lineolare* Haworth; *S. villosum* Lindley; *A. cruentum* Webb & Berthelot; *A. strepsicladum* Webb & Berthelot; *Aichryson pulchellum* C.A. Meyer). Illustration: Edwards's Botanical Register, t. 1553 (1832); Praeger, An account of the Sempervivum Group, 202 (1932); Bramwell & Bramwell, Wild flowers of the Canary Islands, pl. 157 (1974).

Perennial, terrestrial shrublet. Stems to 60 cm, ascending or almost erect, branches c. 80, green or greyish brown, covered with fine hairs, bark slightly fissured; leaf-scars c. 1 mm wide. Rosettes 1–5 cm across. Leaves 0.5–2.5 × 0.3–0.9 cm, ovate to spathulate, margin with multicellular hairs to 0.06 mm, and bead-shaped hairs to 1 mm. Inflorescence 3–10 × 3–15 cm, stalks 5–20 cm. Sepals 8–10; petals 8–12 with red stripes; stamens 4–5.5 mm, anthers yellow; nectar-glands absent. Ovaries 2–2.5 cm, styles 1.5–2 mm. *Canary Islands*. G1. Spring–early summer.

6. A. cuneatum Webb & Berthelot (*Sempervivum cuneatum* Webb & Berthelot). Illustration: Praeger, An account of the Sempervivum Group, 142 (1932); Bramwell & Bramwell, Wild flowers of the Canary Islands, pl. 147 (1974).

Perennial terrestrial or epiphytic herb. Stems 1 or 2 erect, very short, often with brown stolons, to 25 cm. Rosettes 15–50 cm across. Leaves 10–25 × 5–8 cm, obovate to oblanceolate, glaucous; margins hairy, with unicellular hairs to 0.4 mm. Inflorescence 18–60 × 12–30 cm. Sepals triangular, 3–4 mm; petals 6.5–7.5 mm, oblanceolate, yellow; stamens 5–6 mm, anthers yellow; nectar-glands obovate c. 0.7 mm, greenish. Carpels 3–3.5 mm, styles 3.5–4 mm. *Canary Islands (Tenerife).* G1. Spring–early summer.

Very similar to *A. canariense*, with which it sometimes hybridises.

7. A. canariense (Linnaeus) Webb & Berthelot (*Sempervivum canariense* Linnaeus; *S. latifolium* Salisbury; *A. giganteum* Christ). Illustration: Praeger, An account of the Sempervivum Group, 133–135 (1932); Graf, Exotica, 3, 655 (1963); Graf, Tropica, **1**: 362 (1978); Everett, New York Botanical Garden illustrated encyclopedia of horticulture **1**: 66 (1980).

Perennial terrestrial herb. Stems stout and unbranched to c. 5 cm. Rosettes 10–40 cm across. Leaves obovate to oblanceolate, 6–20 × 3–8 cm, green, velvety; margins sometimes wavy with hairs 0.6–1 mm. Inflorescence 15–60 × 12–30 cm. Sepals 6–12; petals 7–10 mm, elliptic to lanceolate; stamens 6–9 mm; styles 2.5–3.5 mm. *Canary Islands.* G1. Spring–summer.

Var. **canariense** from Tenerife, and var. **virgineum** (Christ) H.-Y. Liu (*A. virgineum* Christ) from Gran Canaria are cultivated. Var. *virgineum* is early flowering and has more cup-shaped rosettes, rarely exceeding 25 cm across, and red-tinged leaves.

8. A. tabuliforme (Haworth) Webb & Berthelot (*Sempervivum tabuliforme* Haworth; *S. complanatum* A. de Candolle; *A. berthelotianum* Bolle; *A. macrolepum* Christ). Illustration: Praeger, An account of the Sempervivum Group, 146 (1932); Graf, Exotica, 3, 652 (1963); Bramwell & Bramwell, Wild flowers of the Canary Islands, pl. 151 (1974); Jacobsen, Lexicon of succulent plants, pl. 5 (1977).

Biennial or perennial, terrestrial herb; stems solitary, short, erect, occasionally tufted, 0.5–15 mm across. Rosettes 10–40 cm across, with closely overlapping leaves 4–20 × 2–4 cm, blade obovate to oblanceolate, narrowed towards the base. Flowering-stems 15–30 cm. Sepals 7–9, elliptic, 3–4 × 1.5–2 mm; petals narrowly elliptic 6–7 × 1.5–2 mm, pale yellow; stamens 5–6.5 mm, anthers yellow; nectar-glands 1–1.5 × 0.3–0.5 mm, oblong, whitish. Ovaries 2.5–3.5 × 1.5–2 mm; styles 2.5–3 mm. *Canary Islands (Tenerife).* H5–G1. Spring–summer.

A fasciated variant 'Cristata' is common in cultivation. It hybridises with *A. lindleyi*.

9. A. balsamiferum Webb & Berthelot (*Sempervivum balsamiferum* (Webb & Berthelot) Christ). Illustration: Praeger, An account of the Sempervivum Group, 157 (1932).

Perennial terrestrial shrublet; stems with c. 8 erect or ascending branches to 1.5 m; rhombic leaf scars 3–9 mm wide. Rosettes 7–18 mm across, generally flattened in the centre. Leaves 3–7 × 1.5–3.5 cm, spathulate, recurved, greyish green, hairless, acute to acuminate. Sepals 7–8; petals 6–8 × 1.2–1.5 mm, lanceolate, yellow, nectar-glands 0.5 mm, wedge-shaped, yellow; ovaries c. 3 × 1 mm, hairless; styles c. 3.5 mm. *Canary Islands (Lanzarote, Fuerteventura).* G1. Late spring–early summer.

10. A. arboreum (Linnaeus) Webb & Berthelot (*Sempervivum arboreum* Linnaeus; *Sedum arboreum* Linnaeus; *A. manriqueorum* Bolle). Illustration: Praeger, An account of the Sempervivum Group, 158, 161 (1932); Graf, Exotica, 3, 656 (1963); Bramwell & Bramwell, Wild flowers of the Canary Islands, pl. 152 (1974); Everett, New York Botanical Garden encyclopedia of horticulture **1**: 64 (1980).

Perennial, terrestrial shrublet, occasionally epiphytic, with erect branched stems to 2 m × 1–3 cm. Branches c. 8; bark light brown or greyish brown; leaf-scars distinct, 3–8 mm wide. Rosettes 10–25 cm across, shrinking to c. 5 cm when dry, flattened in centre. Leaves 5–15 × 1.4–5 cm, obovate to oblanceolate, apex acuminate, straight or slightly recurved, occasionally with wavy margins, green or purplish green, or with purple or whitish lines near margin and midrib; margin with curved hairs 0.3–1 mm. Inflorescence dense, conical, ovoid or hemispherical, 7–30 × 7–15 cm. Sepals 9–10, triangular, 2–3.5 × 1–1.4 mm, acuminate; stamens 5–6.5 mm, anthers yellow, nectar glands 0.5–1.2 mm wide. Ovaries 2–3.5 mm, styles 1.5–2.5 mm. Seeds c. 0.7 × 0.2 mm. *Canary Islands (Gran Canaria), widely naturalised in the Mediterranean, C & S America and New Zealand.* H5–G1. Flowering for most of the year.

'Albovariegatum': leaf-margins yellowish white. 'Zwartkop': leaves intense blackish purple: unique and highly regarded.

11. A undulatum Webb & Berthelot (*Sempervivum undulatum* Webb & Berthelot). Illustration: Praeger, An account of the Sempervivum Group, 155 (1932); Bramwell & Bramwell, Wild flowers of the Canary Islands, pl. 149 (1974); Jacobsen, Lexicon of succulent plants, pl. 5 (1977); Everett, New York Botanical Garden encyclopedia of horticulture **1**: 64 (1980).

Perennial shrublet, terrestrial or epiphytic; stems branched or unbranched to 2.5 m, hairless, smooth; bark green, pale brown or grey. Leaf-scars distinct, narrowly transversely rhombic, 3–9 mm wide. Rosettes 10–30 cm across, with flattened centres. Leaves 6–18 × 3–5 cm, oblanceolate or spathulate to oblong, dark green, often with brownish wavy margins; marginal hairs curved, 0.5–2 mm. Inflorescence 12–50 × 12–40 cm, sepals 9–12, obovate, 1.5–1.8 × 0.8–1 mm, hairless, apex slightly notched; petals oblong-lanceolate 6–8 × 1.2–1.5 mm, yellow, hairless, slightly notched. Stamens 5.5–7 mm, anthers yellow; nectar-glands 0.6–0.8 mm wide. Ovaries c. 3 mm, hairless; styles 2 mm. *Canary Islands (Gran Canaria).* G1. Spring.

12. A. nobile (Praeger) Praeger (*Sempervivum nobile* Praeger). Illustration: Praeger, An account of the Sempervivum Group, 150, 151 (1932); Bramwell & Bramwell, Wild flowers of the Canary Islands, pl. 159 (1974); Jacobsen, Lexicon of succulent plants, pl. 5 (1977).

Terrestrial monocarpic shrublet with simple, erect stems to 60 cm; bark brown or greyish brown, hairless, rough; leaf-scars narrowly elliptic, 5–15 mm wide. Rosettes 15–60 cm across. Leaves 7–30 × 4–20 cm, thick and leathery, obovate, yellowish green, occasionally with irregular brownish lines, especially on or near midrib, acute; margin with straight lines c. 0.5 mm, sometimes hairless when mature. Inflorescence 20–40 × 30–60 cm, broadly dome-shaped; sepals 7–9, 2–3 mm, green, often striped with reddish lines; petals 3–5 mm, lanceolate, purplish red; stamens 3–6 mm, hairless, whitish, tinged red, anthers yellow, glands 4-sided, c. 1 mm wide; ovaries 2–3 mm, styles 2–3 mm. *Canary Islands (La Palma).* G1. Spring–summer.

13. A. glutinosum (Aiton) Webb & Berthelot (*Sempervivum glutinosum* Aiton). Illustration: Praeger, An account of the Sempervivum Group, 144 (1932); Botanical Magazine, 1963 (1818).

Terrestrial shrublet to 1.5 m, with *c.* 5 erect or ascending branches, overtopping central or oldest stem; bark green, brown or greyish brown; sticky, hairless; leaf-scars distinct, 6–12 mm wide. Rosettes 12–22 cm across, glutinous towards centre. Leaves 7–12 × 3.5–5 cm, obovate, spathulate, slightly folded, dull pale green, usually with brown stripes near midrib and apex; apex acute, margin with scattered hairs *c.* 0.5 mm. Inflorescence loose, 15–40 × 2.5–3.5 mm, marked with reddish stripes. Petals 5–7 × 2–3 mm, oblong-obovate to oblong-lanceolate, yellow with red lines; stamens 4 or 5, hairless, anthers yellow; nectar-glands 0.5–0.8 mm wide, yellow; ovaries 2–2.5 mm, styles 1–3 mm. *Madeira.* G1. Spring–autumn.

14. A. gorgoneum J.A. Schmidt (*Sempervivum gorgoneum* J.A Schmidt). Illustration: Praeger, An account of the Sempervivum Group, 165 (1932).

Terrestrial shrublet, stems with *c.* 7 branches to 2 m; bark green, yellowish orange or greyish brown, irregularly fissured; leaf-scars 4–9 mm wide. Rosettes 9–20 cm across. Leaves 5–10 × 1.5–3 cm, oblanceolate to spathulate, glaucous green, often red towards midrib and margin; marginal hairs *c.* 0.05 mm, straight or curved. Inflorescence 5–8 × 7–10 cm, pyramidal. Sepals 8–10, elliptic, 2–3 mm, often tinged red, acuminate; petals 5–6 × 1–1.5 mm, oblong-lanceolate, yellowish, with red markings, acuminate and sometimes recurved. Stamens 5–6.5 mm, anthers yellow; nectar-glands 4-sided, 0.3–0.5 mm wide. Ovaries to 3 mm; styles *c.* 2 mm. Seeds 0.5 × 0.2 mm. *Cape Verde Islands.* G1. Autumn–winter.

15. A. lancerottense (Praeger) Praeger (*Sempervivum lancerottense* Praeger). Illustration: Praeger, An account of the Sempervivum Group, 191, 192 (1932); Bramwell & Bramwell, Wild flowers of the Canary Islands, pl. 155 (1974).

Terrestrial shrublet with *c.* 10 ascending branches to 60 cm, often with adventitious roots; bark pale brown or silver-grey; leaf-scars 3–8 mm wide. Rosettes 10–18 cm across. Leaves 5–9 × 1.5–4 cm, obovate to oblanceolate or spathulate, concave, green to yellowish green, often tinged reddish, glaucous; margin weakly toothed with distant hairs. Inflorescence 8–30 × 8–25 cm, dome-shaped. Sepals 7 or 8, triangular, 1.7–3 mm, acuminate, occasionally pinkish; petals 6–9 × 1–1.5 mm, linear-lanceolate, whitish, tinged pink,

hairless, acuminate. Stamens 5–8 mm, whitish, anthers yellow or pinkish; glands 4-sided, 0.5 mm wide. Ovaries 2.5–3 mm; styles 3–4 mm, occasionally marked with pink lines. Seeds *c.* 0.5 mm. *Canary Islands (Lanzarote).* H5–G1. Spring–summer.

16. A. haworthii Webb & Berthelot. Illustration: Praeger, An account of the Sempervivum Group, 175 (1932); Graf, Exotica, 3, 652 (1963).

Terrestrial shrublet, stems with *c.* 40 branches to 60 cm; bark grey or brown, smooth or fissured, surface covered with netted lines, adventitious roots often present; leaf-scars *c.* 2 mm wide. Rosettes 6–11 cm across. Leaves 3.5–5 × 1.5–3 cm, obovate, sometimes folded towards the apex, green, often with glaucous bloom, apex acute or heart-shaped; margin with curved hairs 0.4–0.8 mm. Inflorescence loose, hemispherical, 6–16 cm across; sepals 7–9, lanceolate; petals pale yellow or whitish, often with pinkish variegation, margin finely toothed. Stamens 5.5–8 mm, filaments white or pinkish, anthers pale yellow; nectar-glands 4-sided, *c.* 0.8 mm wide; ovaries 3–5 mm; styles *c.* 3.5 mm. Seeds *c.* 0.4 × 0.2 mm. *Canary Islands (Tenerife).* G1. Spring–summer.

17. A. urbicum (Hornemann) Webb & Berthelot (*Sempervivum urbicum* Hornemann). Illustration: Praeger, An account of the Sempervivum Group, 168 (1932); Engler & Prantl, Natürlichen Pflanzenfamilien **18a:** 431 (1930); Graf, Exotica, 3, 655 (1963); Bramwell & Bramwell, Wild flowers of the Canary Islands, pl. 54 (1974).

Terrestrial shrublet usually with unbranched stems to 2 m. Bark pale brown or greyish, hairless, smooth or slightly fissured; leaf-scars 5–8 mm wide, 1–2 mm tall. Rosettes 15–32 cm across. Leaves 8–25 × 3–5.5 cm, obovate-oblanceolate, glaucous green, margin with straight hairs 0.5–1 mm. Inflorescence 15–75 × 10–45 cm, dome-shaped. Sepals 8–10, triangular, 2–3 × 1–1.5 mm, sometimes reddish variegated, acuminate. Petals 7–10 × 1.2–2 mm, lanceolate, whitish except for pink midrib, hairless, acuminate; stamens 6–10 mm, whitish, anthers pale yellow or whitish; nectar-glands 4-sided, *c.* 1 cm wide, whitish. Ovaries 3.5–4.5 mm, hairless; styles 4–6 mm. Seeds *c.* 0.5 mm. *Canary Islands (Tenerife, Gran Canaria).* H5–G1. Spring–autumn.

18. A. ciliatum Webb & Berthelot (*Sempervivum ciliatum* Willldenow). Illustration: Praeger, An account of the Sempervivum Group, 182, 183 (1932); Santos, A., Vegetacíon y flora de La Palma, 125 (1983).

Terrestrial shrublet; stems with *c.* 6 branches to 1 m. Bark pale brown or greyish, hairless, rough and with many adventitious roots; leaf-scars 3–5 mm wide, slightly raised. Rosettes 8–20 cm across. Leaves 4–12 × 2–5 cm, obovate to oblanceolate or spathulate, often slightly folded near apex, dark green or yellowish green, glaucous, apex acute, often recurved; margin with hairs 0.4–0.8 mm. Inflorescence 15–40 × 10–35 cm. Sepals 7–9, triangular, 2.5–3 mm, acuminate; petals 7–10 × 1.2–2 mm, lanceolate, slightly keeled, whitish; ovaries 2.5–4 mm; styles 2.5–5 mm. Seeds *c.* 0.5 × 0.2 mm. *Canary Islands (Tenerife).* H5–G1. Spring–summer.

19. A. percarneum (R.P. Murray) Pitard & Proust (*Sempervivum percarneum* R.P. Murray). Illustration: Praeger, An account of the Sempervivum Group, 188, 189 (1932).

Terrestrial shrublet; unbranched or occasionally with up to 8 branches to 1.5 m; bark light brown or greyish, hairless, smooth or slightly fissured; leaf-scars 3–13 mm wide. Rosettes 8–20 cm across. Leaves 4.5–10 × 2–4 cm, obovate to oblanceolate or spathulate, concave, dark green, glaucous, hairless; margin reddish, entire or weakly toothed, with curved hairs 0.5–1 mm. Inflorescence 10–30 × 10–25 cm, dome-shaped. Sepals 8–10, triangular, 2–3 mm, green, acuminate; petals 7–8 mm, lanceolate, whitish with pink markings near midrib, acuminate; stamens 5–7 mm, filaments white or pinkish, anthers yellow or whitish; glands 0.5–0.8 mm wide, 4-sided, greeni.h; ovaries 2–3 mm, pinkish; styles 3–4 mm; seeds *c.* 0.6 × 0.2 mm. *Canary Islands (Gran Canaria).* H5–G1. Spring–summer.

20. A. decorum Bolle (*Sempervivum decorum* Bolle). Illustration: Praeger, An account of the Sempervivum group, 180 (1932); Graf, Exotica, 3, 654 (1963); Everett, New York Botanical Garden encyclopedia of horticulture **1**: 64 (1980);

Terrestrial shrublet, stems with *c.* 50 slender branches to 60 cm; bark slightly fissured, pale reddish green or whitish, hairless, rough with abundant adventitious roots; leaf-scars 3–4 mm wide. Rosettes 5–10 cm across. Leaves 2.5–5 × 1–1.5

cm, obovate-oblanceolate, often slightly folded and recurved near apex, acuminate, glaucous dark green; margin reddish with very short hairs to c. 0.5 mm. Inflorescence loose, cylindric, 8–30 × 8–20 cm. Sepals 6–8, triangular, 3–4 mm, green with reddish tinge; petals 7–8 × 2–2.5 mm, lanceolate, whitish with pink coloration towards midrib; stamens 5–7 mm, filaments whitish, anthers pale yellow or whitish; nectar glands c. 1 mm wide, whitish; ovaries 2–3.5 mm; styles 3–4 mm. Seeds c. 0.5 × 0.2 mm. *Canary Islands (La Gomera)*. G1. Spring–summer.

21. A. hierrense (R.P. Murray) Pitard & Proust (*Sempervivum hierrense* R.P. Murray). Illustration: Praeger, An account of the Sempervivum Group, 170 (1932).

Terrestrial shrublet, with erect stems to 1.2 m, usually unbranched; bark greyish, hairless, often slightly fissured; leaf-scars 6–18 mm wide, 2.5–5 mm tall. Rosettes 15–60 cm across. Leaves 10–30 × 3–8 cm, obovate to oblanceolate, glaucous green with reddish margin and often pink or purple tinged; margin with hairs 1–2 mm. Inflorescence 15–30 × 12–50 cm, dome-shaped. Sepals 6–9, triangular, 2.5–3 mm; petals 7–9 × 1.5–2 mm, lanceolate, acuminate, whitish; anthers pale yellow; nectar-glands 0.6 mm wide, whitish; ovaries c. 3.5 mm; styles c. 3.5 mm; seeds c. 0.5 × 0.2 mm. *Canary Islands (Hierro, La Palma)*. H5–G1. Spring.

22. A. castello-paivae Bolle (*Sempervivum castello-paivae* (Bolle) Christ; *S. paivae* R. Lowe; *A. paivae* (R. Lowe) Lamarck). Illustration: Botanical Magazine, 5593 (1866); Praeger, An account of the Sempervivum Group, 178 (1932).

Terrestrial, perennial shrublet often with c. 60 twisted branched stems to 70 cm; bark pale brown or greyish, fissured into plates; adventitious roots common; leaf-scars c. 1 mm wide. Rosettes 3–7 cm across. Leaves 1.5–3.5 × 0.8–2 cm, obovate to spathulate, pale glaucous green or yellowish green, sometimes with reddish lines, hairless; margin with hairs to 0.2 mm. Inflorescence loose, hemispherical, 6–20 × 6–20 cm. Sepals 7–9, triangular, 2.5–3.5 mm, green, acuminate; petals 8–10 × 1–1.5 mm, greenish white. Stamens 5–7 mm, filaments white, anthers white or pale yellow; nectar-glands 0.8–1.5 mm wide, pinkish; ovaries c. 3 mm; styles c. 3 mm; seeds c. 0.5 × 0.2 mm. *Canary Islands (La Gomera)*. H5–G1. Spring.

31. AICHRYSON Webb & Berthelot
A.C. Whiteley

Annual to perennial herbs, occasionally shrubs, some species monocarpic. Branching often appearing dichotomous, including that of the inflorescence. Leaves alternate, simple, mostly entire, usually hairy, fleshy. Flowers in terminal panicles, yellow. Calyx cup-shaped with 5–12 fleshy lobes. Petals as many as calyx lobes, almost free. Stamens twice as many as calyx lobes. Carpels as many as calyx lobes, their bases fused with the receptacle. Nectar-glands 2-horned or with several teeth.

A genus of 15 species from Macaronesia and Morocco. They are adapted to survival on rocks and walls. In cultivation they tend to grow much larger than in nature. Several species make interesting pot-plants suitable for a sunny windowsill, requiring the minimum of attention. Propagation by seed or cuttings.

Literature: Praeger, R.L., *An Account of the Sempervivum Group* (1932).

1a. Plants densely clothed with long
      hairs   2
  b. Plants downy or sparsely hairy   3
2a. Leaves 2.5–4 cm, the blade longer
      than the stalk   **4. villosum**
  b. Leaves 3–8 cm, the blade about as
      long as the stalk   **3. laxum**
3a. Congested shrub to 10 cm; leaves to
      1.2 × 0.6 cm   **1. tortuosum**
  b. Loose shrub to 30 cm; leaves to 4 ×
      1 cm   **2. × domesticum**

1. A. tortuosum (Aiton) Praeger (*Sempervivum tortuosum* Aiton; *Aeonium tortuosum* Berger). Illustration: Praeger, An account of the Sempervivum Group, 105 (1932); Botanical Magazine, 296 (1795).

A compact, downy shrub to 10 cm. Stems woody at the base, much branched, widely diverging, downy at first, becoming hairless. Leaves crowded at the tips of non-flowering-shoots, stalkless, obovate to spathulate, fleshy, to 1.2 × 0.6 cm, and 2–4 mm thick, sticky, sometimes reddish. Flowering-shoots short with a few widely spaced leaves. Panicles with 2 or 3 branches to 3 cm, few-flowered, stickily hairy. Flowers c. 1 cm across, with parts in 8s, on stalks to 7 mm. Calyx c. 4 mm, with acute lobes to 3 mm. Petals lanceolate, c. 6 mm, the margins finely hairy. Stamens 3–4 mm. Nectar-glands orange, to 1 × 0.5 mm, 2-horned, each horn 2-fid. Carpels 3–4 mm. Styles short. *Canary Islands*. G1. Early summer.

2. A. × domesticum Praeger (*Aeonium domesticum* Berger; *Sempervivum tortuosum* de Candolle). Illustration: Praeger, An account of the Sempervivum Group, 107 (1932); Beckett, The Royal Horticultural Society encyclopaedia of houseplants, 67 (1987).

A sparsely hairy, glandular shrub to 30 cm. Stems woody at base, much branched, widely diverging. Leaves crowded at the tips of non-flowering-shoots, spathulate to obovate, obtuse, narrowed gradually to the base, to 4 × 1 cm, and 1.5 mm thick, sticky and with a resinous scent. Flowering-branches 10–20 cm, branched up to 3 times, with widely spaced leaves at the base. Flowers 1.3–1.5 cm across, with parts in 7s or 8s; stalks 5–10 mm. Calyx 3–5 mm with acute lobes 2–3 mm. Petals lanceolate, acuminate, 6–7 mm, the tips reflexed. Stamens 4–4.5 mm. Nectar-glands orange or reddish, 0.5–0.75 mm, 2-fid or toothed. Carpels 3–4 mm. Styles recurved. *Garden origin*. G1. Summer.

Probably a hybrid between *A. tortuosum* (Aiton) Praeger and *A. punctatum* (C.A. Smith) Webb & Berthelot. Widely cultivated, especially as the cultivar 'Variegatum', which has leaves margined with creamy white.

3. A. laxum (Haworth) Bramwell (*A. dichotomum* (de Candolle) Webb & Berthelot). Illustration: Praeger, An account of the Sempervivum Group, 111 (1932); Bramwell, Wild flowers of the Canary Islands, 58 (1974); Kunkel, Flora de Gran Canaria 3: 57 (1978); Graf, Exotica, series 4, 860 (1982).

An erect, softly hairy annual or biennial, 10–100 cm. Stems strong, sparingly dichotomously branched, densely covered with hairs to 4 mm. Leaves densely hairy, 3–8 cm, the blade rounded to rhombic, about as long as the stalk, the tip occasionally notched. Inflorescence terminal, large and much branched. Flowers to 1.5 cm across, with parts in 9s to 12s; stalks 6–12 mm. Calyx shallow, 4–5 mm, with narrow lobes to 3.5 mm. Petals linear to lanceolate, acuminate, 6–7 mm. Stamens c. 5 mm. Nectar-glands irregularly 2-horned, orange or yellow, c. 0.5 mm. Carpels 3.5–4 mm. Styles recurved. *Canary Islands*. G1. Early summer.

4. A. villosum (Aiton) Webb & Berthelot. Illustration: Botanical Magazine, 1809 (1819); Praeger, An account of the Sempervivum Group, 117 (1932); Graf, Exotica, series 4, 860 (1982).

A glandular annual to 20 × 30 cm, completely clothed with long hairs. Stem stout at the base, soon branching into long dichotomous shoots. Leaves sometimes crowded, spathulate to diamond-shaped, blade 1.5–2 × 1.5–2 cm, stalk 1–2 cm. Flowers 1.2–1.5 cm across, with parts in 8s (rarely 6s, 7s or 9s); stalks to 1 cm. Calyx 4–5 mm with lanceolate lobes to 3 mm. Petals lanceolate to oval, acute, 5–6 mm. Stamens *c.* 4 mm. Nectar-glands orange, *c.* 0.5 × 0.75 mm, with 4–6 teeth. Carpels 3.5–4.5 mm. Styles recurved. *Azores.* G1. Early summer.

CIV. CEPHALOTACEAE

Perennial, insectivorous herbs with short, branching rhizomes. Leaves 5–7 cm, in a basal rosette, some with flattened, spathulate blades, others adapted for trapping insects, with coarsely hairy, flask-like lidded pitchers to 5 × 2.5 cm, marked with red. Inflorescences borne on erect, leafless stems to 60 cm; flowers numerous. Sepals 6, white; petals absent. Stamens 12, in 2 whorls. Ovary superior with 16 cells; ovules 1 or 2 per follicle. Fruits surrounded by persistent calyx.

A family of a single Australian species superficially resembling *Nepenthes*, although not closely allied. In cultivation it requires a deep pot and a well-drained but moist mixture of peat, leaf mould and sand; overwatering in winter can damage it. Propagation is by root or leaf cuttings or division.

1. CEPHALOTUS Labillardière
E.C. Nelson
Description as for family.

1. C. follicularis Labillardière.
Illustration: Botanical Magazine, 3118–3119 (1831); Slack, Carnivorous plants, 88, 206 (1979); Morley & Toelken, Flowering plants in Australia, 141 (1983).
Western Australia. H5–G1. Summer.
A protected plant confined to seasonal swamps.

CV. SAXIFRAGACEAE

Low to medium-sized herbs and shrubs, usually perennial. Leaves usually simple and alternate, without stipules, with or without stalks. Flowers with regular or irregular symmetry, in cymes, racemes, or rarely solitary. Sepals 4 or 5, occasionaly more showy than petals. Petals 4 or 5, rarely absent. Stamens usually twice as many as petals, occasionally fewer or more. Ovary superior or inferior, usually with 2 carpels, styles as many as carpels. Fruit usually a capsule, rarely a berry.

A difficult family to delimit due to its somewhat intermediate position in the classification system. Here the family is given a wide interpretation, but in the past many genera, and in particular the woody ones, have been placed in their own families.

In this wide interpretation the family contains about 80 genera and is almost cosmopolitan. The majority of species however occur in E Asia, the Himalayas and in North America.

1a.	Plant herbaceous	2
b.	Plant a woody shrub, tree or climber	43
2a.	Flowering-stems with at least one leaf	3
b.	Flowering-stems lacking leaves	33
3a.	Carpels 4 or 5	4
b.	Carpels 2 or 3	5
4a.	Staminodes present, carpels 4 **30. Parnassia**	
b.	Staminodes absent, carpels 5 **1. Penthorum**	
5a.	Carpels 3	6
b.	Carpels 2	7
6a.	Stamens 10, rhizome bearing bulb-like tubers **15. Lithophragma**	
b.	Stamens 15, rhizome not as above **34. Kirengeshoma**	
7a.	Leaves peltate	8
b.	Leaves not as above	9
8a.	Leaves palmately lobed; stamens 10 **26. Peltoboykinia**	
b.	Leaves almost circular; stamens 6–8 **4. Astilboides**	
9a.	Flowers in bracted dichotomous cymes on erect shoots, bracts leaf-like, greenish to bright yellow **27. Chrysosplenium**	
b.	Flowers not as above	10
10a.	Basal leaves stalkless or with very short winged stalks	11
b.	Basal leaves stalked, often long-stalked	13
11a.	Basal leaves with very short winged stalks; styles usually indistinct **5. Leptarrhena**	
b.	Basal leaves stalkless; styles usually distinct	12
12a.	Leaves lanceolate with a long gradually diminishing point at both ends **1. Penthorum**	

b.	Leaves not as above	**8. Saxifraga**
13a.	Flowering-stems with scale-like leaves at base and 2–4 large opposite leaves near the top **25. Deinanthe**	
b.	Flowering-stems not as above	14
14a.	Flowers with a variable number of stamens present within a single inflorescence	**18. × Heucherella**
b.	Flowers with a constant number of stamens present within a single inflorescence	15
15a.	Stamens 10, rarely 8 or 5, if 5, leaves 2 or 3 pinnate and petals absent	16
b.	Stamens 3 or 5, if 5, leaves not 2 or 3 pinnate and petals present	25
16a.	Stamens 5	17
b.	Stamens 10, rarely 8	18
17a.	Leaflets leathery, wrinkled **3. Rodgersia**	
b.	Leaflets not as above	**2. Astilbe**
18a.	Fruit a follicle	**11. Jepsonia**
b.	Fruit a capsule	19
19a.	Fruit a few-seeded capsule with 2 unequal parts	**16. Tiarella**
b.	Fruit a many-seeded capsule with 2 equal parts	20
20a.	Petals deeply divided, often comb-like	21
b.	Petals entire or rarely shallowly toothed	23
21a.	Calyx usually 1–3 mm; styles less than 1 mm	**20. Mitella**
b.	Calyx 4–8 mm; styles over 1 mm	22
22a.	Stamens 10	**21. Tellima**
b.	Stamens 5	**14. Elmera**
23a.	Leaves compound, or rarely simple and then leaves ovate, coarsely double-toothed and often somewhat lobed	**2. Astilbe**
b.	Leaves not as above	24
24a.	Styles partially fused	**10. Telesonix**
b.	Styles free above the ovule-bearing portion of the ovary	**8. Saxifraga**
25a.	Stamens 3	**22. Tolmiea**
b.	Stamens 5	26
26a.	Fruit a pair of follicles	**11. Jepsonia**
b.	Fruit a many-seeded capsule	27
27a.	Petals deeply divided	28
b.	Petals not as above	29
28a.	Calyx usually 1–3 mm; styles less than 1 mm	**20. Mitella**
b.	Calyx 6–8 mm; styles over 1 mm	**14. Elmera**
29a.	Rootstock bearing bulbils	**19. Bolandra**
b.	Rootstock lacking bulbils	30

30a. Plant with slender horizontal
    stolons   **9. Sullivantia**
  b. Plant lacking stolons   31
31a. Leaves deeply palmately lobed;
    petals absent   **3. Rodgersia**
  b. Leaves kidney-shaped or shallowly
    palmately lobed; petals usually
    present   32
32a. Ovary with 1 cell   **17. Heuchera**
  b. Ovary with 2 cells   **13. Boykinia**
33a. Carpels 4, staminodes present
    29. Francoa
  b. Carpels 2, staminodes absent   34
34a. Leaves all peltate   **7. Darmera**
  b. Leaves not peltate   35
35a. Fruit capsule with 2 unequal parts
    16. Tiarella
  b. Fruit capsule with 2 equal parts
    36
36a. Petals deeply divided, often comb-
    like   **20. Mitella**
  b. Petals not deeply divided, or
    sometimes absent   37
37a. Stamens 5 or 6   38
  b. Stamens 10, rarely 8   40
38a. Petals fused into a short tube at
    base or absent   **17. Heuchera**
  b. Petals only joined at base   39
39a. Flowers on stems to 25 cm
    23. Bensoniella
  b. Flowers almost stalkless
    24. Mukdenia
40a. Leaves compound, or rarely simple
    and then ovate, coarsely double-
    toothed and often somewhat lobed
    2. Astilbe
  b. Leaves not as above   41
41a. Petals absent; flowers inconspicuous
    12. Tanakaea
  b. Petals present; flowers often
    conspicuous   42
42a. Carpels almost free from each other
    and from the perigynous zone
    6. Bergenia
  b. Carpels fused for most of their
    length   **8. Saxifraga**
43a. Leaves alternate   44
  b. Leaves opposite or whorled   53
44a. Leaves pinnately divided
    52. Davidsonia
  b. Leaves not divided   45
45a. Ovary superior   46
  b. Ovary inferior or half-inferior   49
46a. Ovary with 1 or 2 cells; flowers in
    unbranched racemes   47
  b. Ovary with 5 cells; flowers in
    umbels or branched panicles   48
47a. Leaves with large, rounded teeth;
    flower parts in 6s or 9s; ovary
    1-celled   **51. Anopterus**

  b. Leaves spiny or finely toothed;
    flower parts in 5s; ovary 2-celled
    39. Itea
48a. Flowers *c.* 8 mm across; in panicles;
    fruit a berry, *c.* 8 mm
    41. Abrophyllum
  b. Flowers *c.* 3.5 cm wide, in umbels;
    fruit woody, *c.* 5 cm   **40. Brexia**
49a. Ovary 1-celled; fruit a berry
    28. Ribes
  b. Ovary with 2–5 cells; fruit a capsule
    50
50a. Outer flowers of cluster sterile;
    stamens numerous
    42. Cardiandra
  b. All flowers fertile; stamens 5 or 6
    51
51a. Evergreen or deciduous shrub;
    leaf-stalks to 1 cm   **49. Escallonia**
  b. Evergreen tree, if a shrub then
    leaf-stalks *c.* 2 cm   52
52a. Leaves to 6 cm; margins not
    glandular-hairy   **48. Carpodetus**
  b. Leaves 6 cm or more; margins
    glandular-hairy   **35. Quintinia**
53a. Deciduous or evergreen climber
    54
  b. Deciduous or evergreen shrub   57
54a. Outer flowers of cluster sterile, or
    if all flowers fertile, then corymbs
    at first enclosed by 4 papery bracts
    55
  b. All flowers similar, fertile   56
55a. Sterile flowers reduced to a single
    bract 2.5 or more long
    37. Schizophragma
  b. Sterile flowers not as above or
    absent and then corymbs at first
    enclosed by 4 papery bracts
    36. Hydrangea
56a. Petals 7–10; stamens 20–30; ovary
    with 7–10 cells   **38. Decumaria**
  b. Petals 4–5; stamens 8–10; ovary
    with 4 or 5 cells   **47. Pileostegia**
57a. Stamens numerous   58
  b. Stamens 10 or fewer   61
58a. Evergreen shrub; ovary superior or
    mostly so   59
  b. Deciduous shrub; ovary inferior
    60
59a. Leaves divided; flowers solitary,
    axillary; styles 2   **50. Bauera**
  b. Leaves simple; flowers 3–7 per
    cluster; style 1   **43. Carpenteria**
60a. Outer flowers of cluster sterile, with
    petal-like calyx; styles 2
    46. Platycrater
  b. All flowers similar, fertile
    33. Philadelphus
61a. Outer flowers of cluster larger,
    sterile   **36. Hydrangea**

  b. All flowers similar, fertile   62
62a. Sepals and petals 4; stamens 8
    45. Fendlera
  b. Sepals and petals almost always 5;
    stamens 10   63
63a. Ovary more or less superior; leaves
    with thick down beneath
    31. Jamesia
  b. Ovary inferior; leaves without thick
    down beneath   64
64a. Deciduous shrub, rarely evergreen,
    usually with stellate hairs; filaments
    winged   **32. Deutzia**
  b. Evergreen shrub without stellate
    hairs; filaments not winged
    44. Dichroa

1. PENTHORUM Linnaeus

M.F. Gardner

Erect perennials with stolons. Leaves
alternate, lanceolate with a long gradually
diminishing point at both ends, stalkless,
with minute teeth. Flowers in terminal,
spirally coiled clusters, yellow-green; sepals
5; petals 5, or absent; stamens 10, carpels
5, united in the lower half; capsules
flattened, 5-beaked, upper portion
deciduous.

A genus of 1–3 species native to East
and South-East Asia and North America.
Suitable for planting in boggy waterside
margins or even in shallow water.

1. P. sedoides Linnaeus. Illustration:
Rickett, Wild flowers of the United States,
1: 201 (1966); Correll & Correll, Aquatic
& wetland plants of southwestern United
States **2**: 1000 (1975); Voss, Michigan
flora, pl. 4 (1985).

Stems hairless to 60 cm, branched
above. Leaves not fleshy, pale green
becoming bright orange with age.
Flowering-stems glandular; petals
sometimes absent. Seeds pink with warts,
elliptic to ovoid, *c.* 7 mm. *E North America.*
H4. Summer.

2. ASTILBE D. Don

P.G. Barnes

Herbaceous perennials with stout rhizomes,
sometimes stoloniferous, generally with
conspicuous brown hair-like scales. Leaves
alternate, both basal and on the stem,
pinnately compound or rarely simple;
leaflets lanceolate to broadly ovate,
strongly toothed, teeth forward-pointing;
stipules scarious. Inflorescence a terminal
panicle, the branches usually spike-like.
Flowers small, white, pink or purplish,
sometimes unisexual. Calyx-tube short, 5-
lobed; petals 5, linear, narrowly spathulate
or rarely absent; stamens 5 or 10; carpels

2, more or less joined at the base, superior; capsule 2-celled, the cells dehiscing inwardly.

A genus of about 14 species, in north-east Asia, the Himalaya and North America. The astilbes are popular plants best suited by moist acid or neutral soils, and a position in full sun. The smaller species and hybrids may be grown in the rock garden or on peat-banks. Propagation is by division in autumn or early spring. Many cultivars are grown, mostly of complex hybrid origin. These show a wide range of flower colour, from whitish pink to lilac and deep red and may have extra floral parts. The name *A. astilboides* Lemoine is of uncertain application: some plants under this name probably belong to *A. japonica*.

Literature: Ievina, S.O. & Lusinya, M.A., *Astilby: introdutsiya v Latviiskoi SSR* (1975). Descriptions of numerous cultivars of hybrid origin may be found in G.S. Thomas, *Perennial Garden Plants*, edn 2 (1982) and in Jelitto, L., Schacht, W. & Fessler, A., *Die Freiland Schmuckstauden* (1985).

1a. Petals absent; stamens 5
                   5. rivularis
  b. Petals present; stamens 10   2
2a. Leaves simple   **1. simplicifolia**
  b. Leaves compound   3
3a. Petals pink, red, purple or white; inflorescence often with long curled hairs   4
  b. Petals white; inflorescence often with short glandular hairs   6
4a. Leaflets wedge-shaped at base; sepals red   **6. rubra**
  b. Leaflets rounded or cordate at base
                          5
5a. Inflorescence with erect branches; petals pink   **2. chinensis**
  b. Inflorescence with spreading branches; petals white
                8. koreana
6a. Leaflets wedge-shaped at base, sharply double-toothed
                3. japonica
  b. Leaflets rounded to cordate at base
                     7
7a. Plant to 80 cm; leaflets strongly double-toothed; stamens shorter than petals   **4. thunbergii**
  b. Plant to 1.5 m; leaflets simply toothed; stamens longer than petals
                7. grandis

1. A. simplicifolia Makino. Figure 21(1), p. 247. Illustration: Gardeners' Chronicle, 101 (1912); Terasaki, Nippon Shokubutsu Zufu f. 697 (1933); Kitamura & Murata, Colored illustrations of herbaceous plants of Japan, **2**: pl. 35 (1986); Hayashi et al., Wild flowers of Japan, 424 (1988).

Compact plant with a short rhizome. Stem to 30 cm. Leaves 3–10 × 2–6 cm, simple, ovate, coarsely double-toothed and often somewhat lobed, rather glossy. Inflorescence 10–15 cm, a narrow loose panicle with spreading branches, usually leafless, somewhat glandular-hairy. Flowers short-stalked; petals 2–3 mm, linear, white. Stamens 10, equalling petals; anthers pale yellow. *Japan*. H4. Early summer.

The species is sometimes grown on rock-gardens but many plants under the name are hybrids with species with compound leaves.

2. A. chinensis (Maximowicz) Franchet & Savatier. Figure 21(2), p. 247. Illustration: Iconographia cormophytorum Sinicorum **2**: 122 (1972).

Stems to 80 cm, rather hairy above; basal leaves 2–3 times compound or pinnately divided, leaflets 3–8 × 2–4 cm, ovate, acuminate, coarsely toothed. Inflorescence rather narrow with ascending spike-like branches. Flowers almost stalkless, very crowded; petals *c*. 5 mm, linear, rosy-pink; stamens 10, shorter than petals. *NE Asia*. H4. Summer.

More often cultivated are var. **davidii** Franchet: to 1 m, with purplish pink petals; a parent of many garden hybrids, especially those included in **A. × arendsii** Arends; and 'Pumila' (var. *pumila* misapplied), a dwarf variant smaller in all parts. It has dense narrow inflorescences usually less than 30 cm; petals purplish pink. The name var. *taquetii* Vilmorin is of uncertain significance but it is said to be a parent of many hybrids.

3. A. japonica (Morren & Decaisne) A. Gray. Figure 21(3a, c), p. 247. Illustration: Step, Favourite flowers of garden and greenhouse **2**: pl. 92 (1897); Makino, Illustrated flora of Nippon, 496 (1942); Satake et al., Wild flowers of Japan **2**: pl. 156 (1985); Kitamura & Murata, Colored illustrations of herbaceous plants of Japan **2**: pl. 36 (1986).

Stems 50–90 cm. Leaves compound, leaflets 3–7 × 1–2 cm, lanceolate, acute, sharply double-toothed, base wedge-shaped. Inflorescence with ascending glandular-hairy branches. Petals 3–4 mm, narrowly spathulate, white. Stamens 10, shorter than petals. *Japan*. H4. Summer.

Subsp. **glaberrima** (Nakai) Kitamura (*A. japonica* var. *terrestris* (Nakai) Murata; *A. glaberrima* Nakai var. *saxatilis* (Nakai) Ohba). figure 21(3b), p. 247. Illustration: Quarterly Bulletin of the Alpine Garden Society **19**: 386 (1951); Satake et al., Wild Flowers of Japan **2**: pl. 157 (1985). A dwarf variant, 20–40 cm. Leaflets relatively broader, glossy and incised-toothed. Calyx pink. *Japan (Yakushima)*.

4. A. thunbergii (Siebold & Zuccarini) Miquel. Figure 21(4), p. 247. Illustration: Makino, Illustrated Flora of Nippon, 496 (1942); Satake et al., Wild flowers of Japan **2**: pl. 156 (1985); Kitamura & Murata, Colored illustrations of herbaceous plants of Japan **2**: pl. 36 (1986).

Stem to 80 cm. Basal leaves compound, leaflets 4–12 × 2–5 cm, ovate, long-acuminate, sharply double-toothed, base rounded or cordate. Inflorescence a loose panicle to 30 cm, with dense, fine glandular hairs. Petals 3–4 mm, linear, white. Stamens 10, shorter than petals. *Japan*. H4. Summer.

Var. **formosa** (Nakai) Ohwi has leaflets acute or short acuminate; petals 5–7 mm.

A. × lemoinei Lemoine includes hybrids with **A. astilboides** Lemoine.

5. A. rivularis Hamilton. Figure 22(1), p. 249. Illustration: Collett, Flora Simlensis, 174 (1902); Polunin & Stainton, Flowers of the Himalaya, 466 (1985).

Stems to 1.5 m. Leaves 2 or 3 pinnate, leaflets 5–12 × 3–7 cm, ovate, acuminate, scalloped toothed, base cordate. Inflorescence a large panicle to 50 cm; calyx-lobes spreading, white or pale pink, petals absent. Stamens 5, *c*. 3 mm. *Himalaya*. H4. Summer.

6. A. rubra Hooker & Thomson. Figure 22(2), p. 249. Illustration: Botanical Magazine, 4959 (1857); Grierson & Long, Flora of Bhutan **1**(3): f. 35 (1987).

Stems to 1 m, often with abundant long brown hairs in inflorescence. Leaves bipinnate or compound, leaflets 4–7 × 2–4 cm, ovate, toothed, acute or obtuse, base wedge-shaped. Petals 4–5 mm, linear, red to purple. Stamens 10, shorter than petals. *Himalaya*. H4. Summer.

7. A. grandis Wilson. Figure 22(3), p. 249. Illustration: Gardeners' Chronicle 426 (1905); Journal of Horticulture **59**: 331 (1909).

Stems to 1.5 m, with soft hairs and often short glandular ones in the inflorescence. Leaves compound, leaflets 4–12 × 3–7 cm, ovate, acute or shortly acuminate, teeth pointing forwards, cordate or rounded at

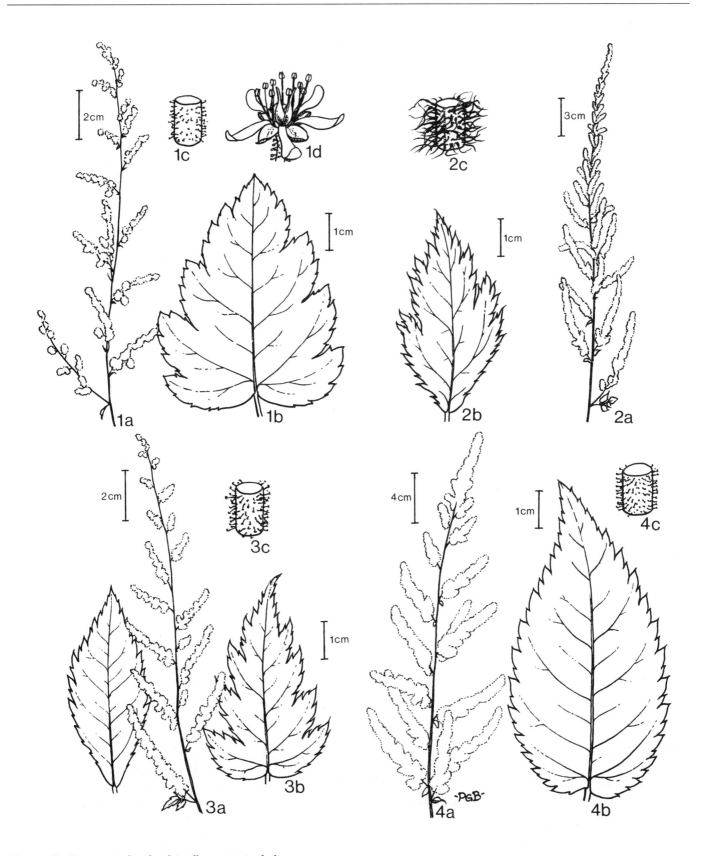

Figure 21. Diagnostic details of *Astilbe* species including inflorescences, leaves, flowers and rachis sections. 1a, b, c, d, *A. simplicifolia*. 2a, b, c, *A. chinensis*. 3a, c, *A. japonica*, 3b, *A. japonica* subsp. *glaberrima*. 4a, b, c, *A. thunbergii*.

base. Flowers in a loose panicle of spike-like branches. Petals 4–5 mm, linear, white. Stamens 10, longer than petals. *W China*. H4. Summer.

8. A. koreana Nakai. Figure 22(4), p. 249. Illustration: Hay, Plants for connoisseurs, 18 (1938).

Stems 30–60 cm. Leaves 2 or 3-pinnate; leaflets broadly ovate, teeth pointing forwards, acute or shortly acuminate, base cordate or rounded. Inflorescence a large panicle with spreading hairy branches; flowers pink in bud, petals linear, white. Stamens 10, shorter than petals. *Korea & N China*. H4. Summer.

3. RODGERSIA A. Gray
J. Cullen
Perennial, rhizomatous herbs. Flowering-stems with reduced leaves, most of the foliage arising directly from the rhizome. Leaves long-stalked, pinnate, apparently pinnate or palmate, leaflets toothed, generally hairy or downy beneath. Inflorescence a large, corymb-like, flat-topped cyme with many flowers. Flowers shortly stalked, each with a bract. Sepals 5, greenish, whitish or red, united below. Petals absent. Stamens 5, spreading. Ovary of 2 carpels which are united in their lower part. Fruit a capsule. Seeds numerous (often not ripened in cultivation).

A genus of 6 species from E Asia, grown for their bold foliage and large inflorescences. The leaves are pinnate, palmate or apparently pinnate, the last a term referring to leaves which are divided into a variable number of leaflets staggered along a long or short axis; the lowermost set of leaflets is most commonly 3, and these are borne in a plane opposite to that of all the rest (of which there may be 3, 5 or 7). They are generally easily cultivated in moist, rich soil and are propagated by division of the rhizome or by seed when available.

Literature: Cullen, J., Taxonomic notes on the genus Rodgersia, *Notes from the Royal Botanic Garden Edinburgh* **34**: 113–123 (1975).

1a. At least some of the basal and
    lower stem-leaves pinnate or
    apparently pinnate   2
  b. All leaves strictly palmate   3
2a. Leaflets with adpressed bristles
    above   **2. sambucifolia**
  b. Leaflets hairless above   **1. pinnata**
3a. Leaflets with 3–5 lobes at the tip;
    sepals acuminate with straight tips
                                   3. podophylla

  b. Leaflets not lobed at the tips; sepals
     ovate, obtuse to rounded, with
     usually reflexed tips   4
4a. Leaf-teeth hairy on under surface;
    sepals conspicuously enlarging after
    flowering   **4. aesculifolia**
  b. Leaf-teeth hairless on under surface;
     sepals not enlarging after flowering
                                   5. henrici

1. R. pinnata Franchet. Illustration: Botanical Magazine, 7892 (1903); Notes from the Royal Botanic Garden Edinburgh **34**: 114 (1975).

Stems to 1 m. Leaves mostly pinnate or apparently pinnate, leaflets leathery, wrinkled, hairless above, erect to spreading and channelled. Sepals white or pink, enlarging or not after flowering, almost as long as to longer than the stamens. Capsule hard, 5–6 mm. *W China (Yunnan, Sichuan)*. H4. Summer.

A variable species of which 3 cultivars are available: 'Alba', with white sepals, 'Elegans' with pink sepals, and 'Superba' a very robust plant with red-suffused leaves and pink sepals, which may perhaps be a hybrid between *R. pinnata* and *R. aesculifolia*. A further presumed hybrid between these species is sometimes grown as *R. purdomii* invalid, a name of no botanical standing.

2. R. sambucifolia Hemsley. Illustration: Parey's Blumengärtnerei **1**: 754 (1958); Notes from the Royal Botanic Garden Edinburgh **34**: 114 (1975).

Similar, but smaller in stature, leaves always strictly pinnate, the leaflets with adpressed bristles above. *W China (Yunnan, Sichuan)*. H5. Summer.

3. R. podophylla Gray (*R. japonica* Regel). Illustration: Botanical Magazine, 6691 (1883); Journal of the Royal Horticultural Society **96**: September, front cover (1971); Notes from the Royal Botanic Garden Edinburgh **34**: 114 (1975).

Stems to 1 m. Leaves strictly palmate, the leaflets drooping, flat, with 3–5 teeth at the apex. Sepals acuminate at their tips, not enlarging in fruit, considerably exceeded by the stamens. Capsule hard, 4–5 mm. *Japan, Korea*. H3. Summer.

4. R. aesculifolia Batalin. Illustration: Parey's Blumengärtnerei **1**: 753 (1958); Notes from the Royal Botanic Garden Edinburgh **34**: 114 (1975).

Stems to 1 m. Leaves strictly palmate, leaflets erect to spreading, channelled, not divided at the apex, hairy on the veins and teeth beneath. Sepals ovate with reflexed

tips, white or pink, conspicuously enlarging after flowering. Stamens pink. Capsule hard, 5–7 mm. *N China*. H3. Summer.

5. R. henrici (Franchet) Franchet. Illustration: Notes from the Royal Botanic Garden Edinburgh **34**: 114 (1975).

Very similar to *R. aesculifolia*, but leaf-teeth not hairy beneath, sepals not enlarging in fruit. *W China, Burma*. H5. Summer.

4. ASTILBOIDES Engler
J. Cullen
Perennial, brown-hairy rhizomatous herbs to 1.5 m. Leaves mostly basal, long-stalked, blade peltate, almost circular, to 90 cm across, lobed and irregularly toothed. Flowers to 8 mm across, creamy white, in spike-like, terminal cymes to 5 cm; stem-leaves very small. Sepals 4 or 5, triangular, blunt or notched. Petals 4 or 5, oblong or oblong-lanceolate, somewhat unequal. Stamens 6–8, about as long as the petals. Ovary usually 2-celled (rarely 4-celled), each cell with *c*. 8 ovules; styles 2 (rarely 4), long, with small, capitate stigmas. Capsule usually 2-celled, many-seeded. Seeds pointed at each end.

A genus of a single species from E Asia, easily grown in semi-shade in moist, rich soil. Propagation is by division of the rhizome or by seed.

1. A. tabularis (Hemsley) Engler (*Saxifraga tabularis* Hemsley; *Rodgersia tabularis* (Hemsley) Komarov). Illustration: Hay & Synge, Dictionary of garden plants in colour, pl. 1351 (1969); Huxley, Garden perennials and water plants, pl. 228 (1970).

N China, Mongolia, North Korea. H2. Summer.

5. LEPTARRHENA R. Brown
D.M. Miller
Herbs with horizontal, spreading rootstocks. Leaves in basal rosettes, 5–15 cm, oblong to obovate, narrowed to the short, winged stalk, toothed, leathery, dark glossy green above, paler beneath. Inflorescence-stalk 5–30 cm, with 1–3 small leaves clasping the stem. Flowers bisexual in dense, terminal inflorescences. Sepals 5, minute. Petals 5, 2–3 mm, spathulate, persistent, white or tinged with pink. Stamens 10, longer than petals. Stigmas borne on top of the carpels, more or less without styles; carpels 2. Fruit a group of 2 follicles *c*. 8 mm, bright red.

A genus of a single species from North

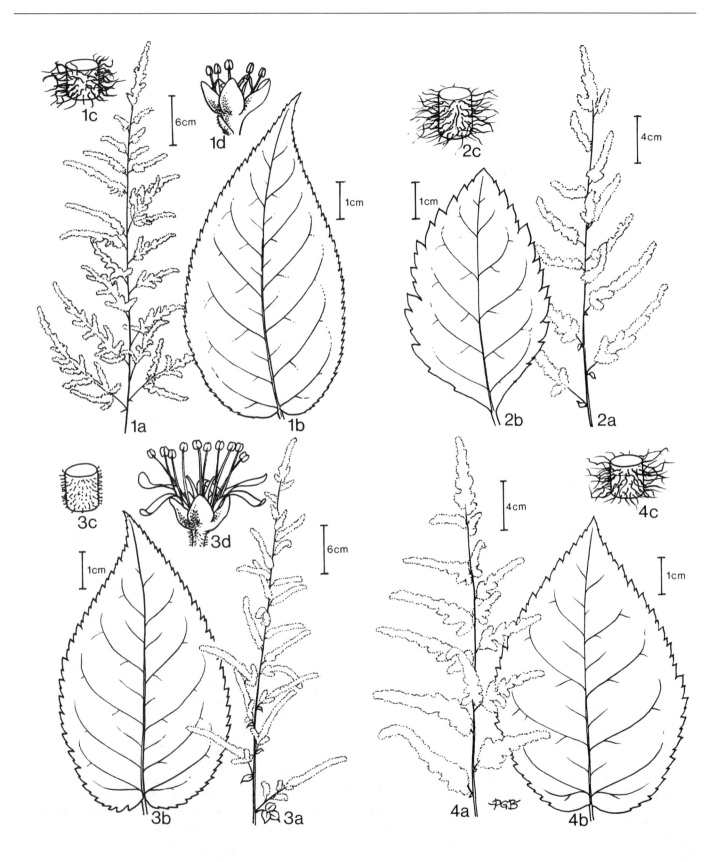

Figure 22. Diagnostic details of *Astilbe* species including
inflorescences, leaves, flowers and rachis sections.
1a, b, c, d, *A. rivularis* 2a, b, c, *A. rubra*. 3a, b, c, d, *A. grandis*.
4a, b, c, *A. koreana*.

America which can be grown in semi-shade or sun in a moist but not wet soil. Propagation is by cuttings or division, or by seed in spring.

1. L. amplexifolia (Sternberg) Seringe (*L. pyrolifolia* (D. Don) Seringe). Illustration: Hitchcock et al., Vascular plants of the Pacific Northwest **3**: 11 (1971); Rickett, Wild flowers of the United States **5**: pl. 80 (1971); Clark, Wild flowers of British Columbia, 199 (1973).

W North America (Alaska to Oregon). H1. Summer.

6. BERGENIA Moench
P.F. Yeo

Perennial herbs, with stout prostrate rooting stems terminating in leaf-rosettes. Leaves 6–25 × 5–17 cm, usually evergreen; stalk stout, sheathing at base; blade simple, shallowly toothed or nearly entire, leathery. Flowers numerous, in panicles supported on stout scapes, differing from those of Saxifraga in being perigynous and having the carpels nearly free from each other and from the perigynous zone.

A genus of 6–8 species from cool-temperate Asia; 5–6 species are widely cultivated, being used for borders, large rock gardens and wild gardens. Their large tough leaves give ground-cover and winter colour and they produce masses of long-lasting but somewhat frost-sensitive flowers in spring. The species are interfertile and individuals are self-incompatible; garden seed, therefore, usually produces hybrids. Indeed, a great many hybrids are now in circulation in gardens. This should be borne in mind when attempting identifications as only a few of them can be described here. The plants are easy to grow and extremely resistant to neglect; however, for good effect propagation every 5 years is recommended. This is done by breaking up the rhizomes and selecting terminal pieces for replanting.

Literature: Borissova, A., De speciebus generis Bergenia Moench Asiae Mediae, *Botanicheskie Materialy* **16**: 97–103 (1954); Pan, J.T., A conspectus of the genus Bergenia Moench, Acta Phytotaxonomica Sinica **26**: 120–129 (1988), in Chinese with English summary; Yeo, P.F., Two Bergenia hybrids, *Baileya* **9**: 20–28 (1961); Yeo, P.F., A revision of the genus Bergenia: *Kew Bulletin* **20**: 113–148 (1966); Yeo, P.F., Further observations on Bergenia in cultivation: *Kew Bulletin* **26**: 47–56 (1971); Yeo, P.F., Cultivars of Bergenia in the British Isles, Baileya **18**: 96–112 (1972).

1a. Leaf-blade hairless or slightly hairy on margins at base 2
 b. Leaf-blade densely hairy on margins 4
2a. Petals 1.1–2.5 cm; some glands on the flowering branches stalked **3. purpurascens**
 b. Petals 0.8–1.3 cm; glands on the flowering branches not stalked 3
3a. Leaves ovate, base wedge-shaped or rounded; flowers more or less reflexed; petals elliptic to broadly ovate, tapering into the claw **1. crassifolia**
 b. Leaves circular, base rounded or cordate, often blistered; flowers mostly erect or upwardly inclined; petals circular or broadly ovate, abruptly contracted into the claw **2. cordifolia**
4a. Petals at first pink or purple, never white 5
 b. Petals normally white at first, sometimes reddish in age 6
5a. Flowers reflexed, usually purple, late **3. purpurascens**
 b. Flowers not reflexed, pink, redder in age, early **4. schmidtii**
6a. Leaves circular or broadly ovate, usually deciduous; flowers erect or ascending; petals nearly circular **5. ciliata**
 b. Leaves obovate, base wedge-shaped base; petals obovate **6. stracheyi**

1. B. crassifolia (Linnaeus) Fritsch (*Megasea crassifolia* (Linnaeus) Haworth; *B. pacifica* Komarov). Illustration: Botanical Magazine, 196 (1792); Step, Favourite flowers of garden and greenhouse **2**: t. 89 (1897); Bloom, Flowers for your garden, 37 (1975) – as B. cordifolia.

Plant to 45 cm. Leaf-blade oblong, obovate or broadly ovate, base shallowly cordate, rounded or wedge-shaped, apex rounded or truncate, hairless. Flowering-stem branched near the top with the branches turned to one side, reddish, with stalkless glands. Flowers more or less nodding, bell-shaped. Perigynous zone and calyx together 5–9 mm. Petals 8–15 × 5.5–8 mm, elliptic or obovate, purplish pink. *Former USSR (Siberia & Far Eastern Region).* H1. Spring.

2. B. cordifolia (Haworth) Sternberg (*Megasea cordifolia* Sternberg). Illustration: Perry, Collins guide to border plants, t. 214 (1957); Kew Bulletin **20**: 142 (1966).

Plant to 40 cm. Leaf-blade circular, base rounded or cordate, sometimes blistered, hairless. Flowering-stem branched near the top; panicle spherical, becoming umbrella-shaped. Flowering-stem, branches, perigynous zone and sepals flushed with red, with stalkless glands. Flowers very numerous, broadly bell-shaped, becoming erect. Perigynous zone and calyx together 6–8 mm. Petals to 1.2 cm, circular or broadly ovate, with distinct claw, pink. *Former USSR (Siberia).* H1. Spring.

Not regarded as distinct from B. crassifolia by Soviet botanists. 'Purpurea' (illustration: Brickell, RHS gardeners' encyclopedia, 226, 1989), leaves thicker, redder, petals deep purple, more widely overlapping and spreading. Frequently grown.

3. B. purpurascens (J.D. Hooker & Thomson) Engler (*B. delavayi* (Franchet) Engler; *B. beesiana* invalid; *B. yunnanensis* invalid). Illustration: Botanical Magazine, 5066 (1858), 117 (1950); Hay & Beckett, Readers' Digest encyclopaedia of garden plants and flowers, 81 (1978); Stainton, Flowers of the Himalaya, supplement, t. 37 (1988).

Plant to 40 cm. Leaf-blade elliptic or ovate-elliptic, base rounded or acute, usually convex and glossy above, reddish, hairless or with a few marginal hairs at base. Panicle compact, with most branches near the top and turned to one side. Flowering-stem, branches, perigynous zone and sepals deep purplish or brownish red with numerous glands, some of them stalked. Flowers few to several, nodding, bell-shaped. Perigynous zone and calyx together 0.8–1.4 cm. Petals 1.5–2.5 × 0.7–0.9 cm, broadly spoon-shaped or obovate, deep purplish red or bright pink. *E Himalaya, N Burma, W China.* H2. Late spring.

'Ballawley' ('Delbees'). Illustration: Bloom, Perennials for your garden 37 (1975). Leaves large, circular, doubly toothed or scalloped, base cordate, petals to 1.7–1.2 cm, purple.

4. B. × schmidtii (Regel) Silva-Tarouca (*B. ligulata* Engler var. *speciosa* invalid; *B. ornata* Guillaumin; *B. leichtlinii* misapplied; *B. stracheyi* misapplied; *Saxifraga ligulata* var. *speciosa* Verlot). Illustration: Revue Horticole **40**: 261 (1868); Botanical Magazine, 5967 (1872); Gartenflora **27**: t. 946 (1878); Addisonia **6**: 29 (1921).

Plant to 40 cm. Leaf-blade usually to 25 cm, obovate to obovate-elliptic, base rounded or shallowly cordate, usually bright green, often drooping, margin hairy. Panicle about as long as its stalk, more or less spherical at first, later becoming hemispherical and rather loose with arching branches. Flowering-stem, branches, perigynous zone and sepals flushed with bright red, with sparse short-stalked or stalkless glands. Flowers at first nodding, soon horizontal to erect, broadly bell-shaped. Perigynous zone and calyx together 6–9 mm. Petals 1.3–1.7 × 0.8–1.3 cm, oblong or broadly ovate, the claw lengthening with age, pink, darker at the claw. *Garden origin.* H2. Winter–spring.

This is the most commonly grown *Bergenia* in Atlantic Europe.

5. B. ciliata (Haworth) Sternberg (*B. ligulata* Engler). Illustration: Botanical Magazine, 3406 (1835), 4915 (1856); Edwards's Botanical Register **32**: t. 33 (1846); Polunin & Stainton, Flowers of the Himalayas, t. 40 (1985).

Plant to 30 cm. Leaves usually deciduous except for a few especially small ones produced in autumn. Stalks of larger leaves often with the sheath not reaching to within 1 cm of base of blade. Leaf-blade circular to broadly ovate, with rounded or cordate base, margin hairy, surface hairy (forma **ciliata**) or hairless (forma **ligulata** Yeo). Panicle as long as or longer than the stalk, loose, sparingly branched, without stalked glands; branches green to pink. Flowers erect or upwardly inclined, broadly bell-shaped or cup-shaped. Perigynous zone and calyx together 7–12 mm, more or less pink. Petals 1.1–1.8 × 0.7–1.8 cm, broadly obovate or circular, white, becoming red in age. *Himalaya, Assam.* H2. Spring.

6. B. stracheyi (Hooker & Thomson) Engler. Illustration: Edwards's Botanical Register **29**: t. 65 (1843); Blatter, Beautiful flowers of Kashmir **1**: 126 (1927); Rau, Illustrations of West Himalayan wild flowering plants, 15 (1964); Kew Bulletin **20**: 143 (1966); Polunin & Stainton, Flowers of the Himalayas, t. 40 (1985).

Plant to 25 cm. Leaf-stalk with sheath often reaching on to blade and always to within 1 cm of it. Leaf-blade 6–20 × 3–10 cm, obovate, base more or less wedge-shaped, rarely as much as ⅘ as wide as long, margin hairy, convex above. Panicle to *c.* ½ as long as stalk, compact, usually with stalked glands; branches usually green. Flowers nodding, bell-shaped. Perigynous zone and calyx together 0.8–1.3 cm, green. Petals 1–1.5 × 0.6–0.8 cm, obovate or spoon-shaped, white,

becoming pale pink. *E Afghanistan, former USSR (Tadzhikistan), W Himalaya.* H2. Spring.

B. × spathulata Guillaumin (*B. ciliata × B. stracheyi*) is evergreen but has some leaves with the sheath not reaching to within 1 cm of the base of the blade. It is most commonly represented by 'Gambol', dwarf, with smooth, neat, obovate, sometimes glossy foliage; inflorescence softly hairy.

7. DARMERA Voss
M.F. Gardner
Perennial herb to 65 cm, rhizomes stout, tips clothed by broad stipular leaf-sheaths. Leaves basal, almost circular, peltate, 5–40 cm across, with 7–15 shallow lobes, each lobe cut and sharply toothed, leaf-stalks 10–150 cm. Flowering-stems 1–20 cm, with long hairs and shorter glandular hairs; flowers borne in many-flowered corymbs, calyx 5-lobed, 2.5–3.5 mm; petals white or pale pink, oblong-elliptic to obovate, 4.5–7 mm. Follicles purplish, 6–10 mm, joined with calyx for 1–2 mm.

A genus of a single species native to W North America. A majestic plant for a cool, moist site in dappled shade.

1. D. peltata (Torrey) Voss (*Peltiphyllum peltatum* (Torrey) Engler). Illustration: Botanical Magazine, 6074 (1874); RHS dictionary of gardening, 1517 (1951); Hitchcock et al., Flora of the Pacific Northwest, 192 (1987); Brickell, RHS gardeners' encyclopedia of plants and flowers, 197 (1989).

British Columbia to N California. H2. Spring.

8. SAXIFRAGA Linnaeus
R.J. Gornall
Perennial, biennial or annual herbs, erect or forming mats or cushions. Leaves alternate (rarely opposite). Flowers solitary or in branched, usually cyme-like inflorescences. Sepals 5. Petals 5. Stamens usually 10, rarely 8, inserted at the junction of the ovary wall and floral-tube. Ovary usually of 2 carpels, with separate styles.

A genus of about 440 species, mostly from N temperate and arctic areas, with a few species as far south as Thailand, Ethiopia and the Andes south to Tierra del Fuego. The richest areas are the mountains of Europe, western North America, the Himalayan-Tibetan area and eastern Asia. Most species require shade from the midday sun, good drainage, a high

atmospheric humidity and ample watering in the growing season but less so at other times. Species of woodland or other shady habitats are usually best grown in sandy loam with a mulch of organic material, and a more even supply of water. Species vary as to their preferred pH; this is indicated in the text where appropriate. In cultivation, most lime-loving species grow well in standard 'alpine' compost. Propagation may be readily effected by seed, or, in the case of cushion or mat-forming species, by cuttings consisting of the apical centimetre or two of a leafy shoot. Full cultivation details may be found in Köhlein (1984) and Harding (1992).

The arrangement followed here is based on Gornall (1987), which recognises 15 sections. By far the most horticulturally important are sections *Porphyrion* (the 'Kabschia' and 'Engleria' saxifrages), *Ligulatae* (the 'silver' or 'encrusted' saxifrages) and *Saxifraga* (which includes the 'mossy' saxifrages). Accounts of the first and last of these, which are by far the best represented in gardens, in addition make use of the ranks of subsection and series where appropriate in an attempt to indicate groups of apparently closely related species. The accounts of the European species are based on the work of Professor D.A. Webb, of Trinity College, Dublin.

In gardens hybrids are commoner than species but only brief accounts of these are given here. Details of the many hybrids and other cultivars in section *Porphyrion* are given in the comprehensive account by Horny et al. (1986); descriptions of many of these hybrids and also of those from sections *Ligulatae* and *Saxifraga* can also be found in Köhlein (1984) and Harding (1992).

Literature: Engler, A. & Irmscher, E., Saxifragaceae–Saxifraga, *Das Pflanzenreich*, **67, 69** (IV. 117): 1–448 (1916), 449–709 (1919); Harding, W., *Saxifrages. The genus Saxifraga in the wild and in cultivation* (1970); Köhlein, F., *Saxifrages and related genera* (1984); Horny, R. et al., *Porophyllum saxifrages* (1986); Gornall, R.J., An outline of a revised classification of Saxifraga L., *Botanical Journal of the Linnean Society* **95**: 273–292 (1987); Webb, D.A. & Gornall, R.J., *Saxifrages of Europe* (1989); Harding, W. *Saxifrages. A gardeners' guide to the genus* (1992).

KEY TO GROUPS
1a. Leaves hard or stiff and usually encrusted with at least some lime,

without a distinct stalk, margins apparently entire (though sometimes fringed with tooth-like hairs or minute serrations)   **Group A**

b. Leaves soft, fleshy or herbaceous, not leathery, hardly ever encrusted with lime, often with a distinct stalk and a lobed blade   2

2a. Ovary at least ⅓ inferior   **Group B**

b. Ovary superior to less than ⅓ inferior   3

3a. Petals yellow or orange, rarely purple   **Group C**

b. Petals whitish or pink, sometimes with red, green or yellowish spots   **Group D**

Group A

1a. Petals bright yellow or orange   2

b. Petals white, pink, purple, red, pale creamy yellow or greenish   13

2a. Calyx bell- or urn-shaped, enclosing and hiding the petals   **79. corymbosa**

b. Calyx cup- or saucer-shaped, not hiding the petals   3

3a. Larger leaves at least 7 mm across   **98. mutata**

b. Larger leaves to 6 mm across   4

4a. Leaves fleshy, rather soft, usually with only one hydathode   **100. aizoides**

b. Leaves usually stiff and hard, with 3–9 hydathodes   5

5a. Stamens at least as long as petals; apex of leaves with a short point   6

b. Stamens shorter than petals; apex of leaves with a short point or rounded   12

6a. Leafy shoots forming secondary rosettes of leaves on the stem above the primary basal rosettes   7

b. Leafy shoots forming primary basal rosettes only   8

7a. Flowering-stem, stalks and perigynous zone hairless   **51. subverticillata**

b. Flowering-stem, stalks and perigynous zone glandular-hairy   **52. scleropoda**

8a. Lowest flower-stalks 6–10 mm; leaves soft, recurved at the tip   **49. kotschyi**

b. Lowest flower-stalks absent or to 6 mm; leaves hard, straight   9

9a. Leaves oblong, mucronate; inflorescence narrowly oblong   **50. pseudolaevis**

b. Leaves linear to lanceolate, apiculate; inflorescence spherical to broadly oblong   10

10a. Flower-stalks absent   **48. desoulavyi**

b. Flower-stalks present but short   11

11a. Flowering-stem, flower-stalks and perigynous zone glandular-hairy   **46. juniperifolia**

b. Flowering-stem, flower-stalks and perigynous zone hairless   **47. sancta**

12a. Inflorescence usually with not more than 6 flowers; most leaves with a short, straight point; petals notched   **45. aretioides**

b. Inflorescence with more than 6 flowers; leaves obtuse or with an incurved point; petals entire   **44. ferdinandi-coburgi**

13a. Leaves mostly opposite   14

b. Leaves alternate   20

14a. Margins of the leaf-pairs confluent, base of each leaf without tooth-like hairs   15

b. Margins of the leaf-pairs meeting at an acute angle, base of each leaf with tooth-like hairs   18

15a. Sepals without marginal hairs   **84. retusa**

b. Sepals with marginal hairs, at least at the base   16

16a. Leaves nearly circular, rather soft and fleshy   **85. biflora**

b. Leaves obovate to oblong, hard and rigid   17

17a. Petals usually pink to purple; leaves at least 5 mm   **83. oppositifolia**

b. Petals white; leaves to 5 mm   **86. vacillans**

18a. Leaves more than 3 mm   **88. georgei**

b. Leaves to 2.5 mm   19

19a. Hairs on flowering-stem with pale red glands   **89. alpigena**

b. Hairs on flowering-stem with white glands   **87. quadrifaria**

20a. Flowering-stems obsolete or nearly so (less than 1 cm)   21

b. Flowering-stems longer than 1 cm   27

21a. Leaves not markedly thickened or truncate at the tip   **76. lowndesii**

b. Leaves thickened and truncate at the tip   22

22a. Leaves with 3–5 equal-sized chalk-glands set in the margin of thickened tip; flowers often more than 1   **61. clivorum**

b. Leaves with 1 chalk-gland (sometimes with a much smaller one on each side of it) set in the margin of thickened tip; flowers usually solitary   23

23a. Tip of leaf fringed with tooth-like hairs   **75. hypostoma**

b. Tip of leaf without hairs   24

24a. Sepals broader than long; petals broadly obovate to nearly circular, overlapping   25

b. Sepals at least as long as broad; petals obovate, barely overlapping   26

25a. Leafy shoots forming a loose cushion; petals a bright rose-lilac, entire   **76. lowndesii**

b. Leafy shoots forming a dense cushion; petals creamy white, sometimes lobed at the tip   **73. lolaensis**

26a. Sepals with 1 chalk-gland, slightly recurved; petals 3.5–4.5 mm across   **74. matta-florida**

b. Sepals without chalk-glands, straight; petals 2–3 mm across   **72. pulvinaria**

27a. Calyx bell- or urn-shaped, enclosing and hiding the petals   28

b. Calyx cup- or saucer-shaped, not hiding the petals   33

28a. Petals yellow   **79. corymbosa**

b. Petals pink or purple   29

29a. Most flowers without stalks; inflorescence not branched   30

b. Most flowers clearly with stalks; inflorescence often branched   32

30a. Leaves linear to linear-oblong, not widened at the tip   **82. sempervivum**

b. Leaves obovate to spathulate, widened near the tip   31

31a. Leaves 4–10 × 2–3 mm; inflorescence dull pink, sometimes tinged with green, usually bearing 4–7 flowers   **80. porophylla**

b. Leaves 12–35 × 3–8 mm; inflorescence bright red to dark crimson, bearing 10–20 flowers   **81. federici-augusti**

32a. Inflorescence sparingly branched, with 1 or 2 flowers on each primary branch   **77. media**

b. Inflorescence freely branched, with 3 or 4 flowers on each primary branch   **78. stribrnyi**

33a. All leaves less than 1.5 cm   34

b. Larger leaves at least 1.5 cm   61

34a. Leaves recurved in the upper half   35

b. Leaves straight   51

35a. Flowering-stems usually with a solitary flower   36

b. Flowering-stems usually with more than one flower, at least in bud **38**

36a. Leaf-tips rounded or almost acute **83. oppositifolia**

b. Leaf-tips mucronate (shortly pointed) **37**

37a. Leafy shoots more than 10 cm, forming a loose cushion; leaves at least 5 mm **54. poluniniana**

b. Leafy shoots less than 10 cm, forming a dense cushion; leaves mostly less than 5 mm **53. lilacina**

38a. Leaves glandular-hairy over the whole lower surface and at least half the upper **59. spruneri**

b. Leaves usually hairless on both surfaces, but if hairs are present then they are non-glandular **39**

39a. Leaves linear or oblong, 4–5 times as long as broad, acutely pointed **40**

b. Leaves obovate or spathulate, rarely linear, less than 4 times as long as broad, mucronate to obtuse **43**

40a. Leaves 2–4 mm across **41**

b. Leaves to 2 mm across **42**

41a. Leaves tapered gradually to a point; petals white or pale pink, rarely crimson **56. scardica**

b. Leaves tapered more abruptly to a point; petals deep rose **63. rhodopetala**

42a. Leaves 10–12 mm; petals 8–10 mm **65. cinerea**

b. Leaves *c.* 4 mm; petals 4–6 mm **66. caesia**

43a. Leaves with thickened tips in which are set the chalk-glands **44**

b. Leaf-tips not thickened **45**

44a. Leaves heavily encrusted with lime; petals 5–9 × 3–7 mm **64. stolitzkae**

b. Leaves with only traces of lime; petals 3–6.5 × 1.5–4 mm **62. andersonii**

45a. Petals at least 5 mm **46**

b. Petals to 6 mm **50**

46a. Leaves usually 5–15 mm, if shorter then petals without a notch at the tip **47**

b. Leaves 3–4 mm; petals often with a notch at the tip **49**

47a. Flowering-stems at least 2.5 cm **48**

b. Flowering-stems less than 2.5 cm **60. alberti**

48a. Leaf-margins fringed with hairs only in the basal half **55. marginata**

b. Leaf-margins fringed with hairs from base to mid-way or more **56. scardica**

49a. Petals obovate, usually with a notch at the tip, fading to deep purple at the base **57. iranica**

b. Petals broadly obovate, without a notch at the tip, fading to pink **58. wendelboi**

50a. Leaves curved outwards from near the base; upper part of flowering stem hairier than lower part **66. caesia**

b. Leaves curved outwards only from near the tip; upper part of flowering-stem less hairy than lower part **67. squarrosa**

51a. Tips of leaves tapered to a sharp point **52**

b. Tips of leaves obtuse **57**

52a. Leaves to 5 mm **53**

b. Leaves at least 6 mm **54**

53a. Leaf-points sharply incurved **69. tombeanensis**

b. Leaf-points straight **68. diapensioides**

54a. Flowers solitary, rarely 2 per stem **70. burseriana**

b. Flowers 3–14 per stem **55**

55a. Leaves deep green, not clearly encrusted with lime, lanceolate **71. vandellii**

b. Leaves glaucous, usually clearly lime-encrusted, oblong to obovate-elliptic **56**

56a. Middle part of leaf-margin entire **55. marginata**

b. Middle part of leaf-margin finely toothed **56. scardica**

57a. Petals 7–12 mm **58**

b. Petals 3–6 mm **60**

58a. Flower-stalks hairless, or very nearly so **92. cochlearis**

b. Flower-stalks densely glandular-hairy **59**

59a. Leaves usually more than 1.5 mm across, broadest above the middle **55. marginata**

b. Leaves usually less than 1.5 mm across, broadest at or below the middle **68. diapensioides**

60a. Leaves finely toothed in the upper half; chalk-glands confined to near the margins of the leaves **95. paniculata**

b. Leaves entire in the upper half; chalk-glands scattered over the upper surface of the leaf **97. valdensis**

61a. Leaf-rosette solitary, without offsets; plant dying after flowering **62**

b. Leaf-rosettes producing offsets which persist after the flowering rosette dies, thereby forming a clump **63**

62a. Leaves glaucous, lime-encrusted, tip almost acute; petals white **90. longifolia**

b. Leaves dull green, not obviously lime-encrusted, mucronate; petals dull pink **99. florulenta**

63a. Flowering-stem to 10 cm, bearing up to 8 flowers **55. marginata**

b. Flowering-stem taller than 10 cm, bearing more than 10 flowers **64**

64a. Leaves entire to scalloped; chalk-glands embedded in leaf-margin itself **65**

b. Leaves toothed; chalk-glands next to teeth, on upper leaf-surface **67**

65a. Leaves spathulate, obtuse **92. cochlearis**

b. Leaves linear to oblong, if a little expanded near the tip, then almost acute **66**

66a. Leaves 2.5–7 mm across; lower branches of inflorescence bearing at least 5 flowers **91. callosa**

b. Leaves 2–3 mm across; lower branches of inflorescence bearing not more than 4 flowers **93. crustata**

67a. Panicle occupying at least half the flowering-stem; lower primary branches bearing at least 13 evenly spaced flowers **96. cotyledon**

b. Panicle occupying less than half the flowering-stem; lower primary branches bearing up to 12 flowers crowded at the tips **68**

68a. Basal leaves curving downwards near the tip, to make a flat or convex leaf-rosette; primary branches of the inflorescence usually with 4–12 flowers **94. hostii**

b. Basal leaves tending to curve upwards, to make a concave or almost hemispherical leaf-rosette; primary branches of the inflorescence usually with 1–3 flowers **95. paniculata**

Group B

1a. Bulbils present, at or below ground level, in the axils of the basal leaves **2**

b. Bulbils absent from axils of the basal leaves **11**

2a. Petals glandular-hairy on the upper suface, at least near the base **3**

b. Petals hairless **6**

3a. Bulbils present in the axils of the stem-leaves and bracts
120. bulbifera

b. Bulbils present only in the axils of the lower leaves **4**

4a. Flowering-stem usually branched from the middle or below, bearing a diffuse inflorescence
117. haenseleri

b. Flowering-stem usually branched only in the top ¼, bearing a compact inflorescence **5**

5a. Basal leaves deeply divided into linear-oblong or oblanceolate lobes
118. dichotoma

b. Basal leaves scalloped or somewhat pinnately lobed, but not deeply so
119. carpetana

6a. Basal leaves deeply divided into 3 primary lobes, which are narrowed or stalked at the base **7**

b. Basal leaves divided up to ⅔ of the way to the base; lobes not narrowed or stalked at the base **8**

7a. Petals 12–20 mm **121. biternata**

b. Petals 5–8 mm **122. bourgaeana**

8a. Petals 2–6 mm **9**

b. Petals 7–16 mm **10**

9a. Basal leaves semi-circular or kidney-shaped, 9 mm or more across; ovary not fully inferior
114. rivularis

b. Basal leaves wedge-shaped, 4–9 mm across; ovary fully inferior
117. haenseleri

10a. Leaves 3-lobed; flowering-stem branched usually from near the base **116. corsica**

b. Leaves scalloped rather than lobed; flowering-stem usually branched from above the middle
115. granulata

11a. Carpels united to above the middle at fruiting stage; flowering stems usually with at least one leaf **12**

b. Carpels separated nearly to the base at fruiting stage; flowering-stems usually leafless **76**

12a. Leaves linear-oblong, fleshy; flowers yellow or dark red **100. aizoides**

b. Leaves not fleshy; flowers white, cream or greenish yellow **13**

13a. Leaves all entire **14**

b. Some leaves toothed, scalloped or lobed **25**

14a. Petals conspicuous, longer than sepals, usually pure white **15**

b. Petals inconspicuous, barely longer than sepals, dull in colour **24**

15a. Leafy shoots with dormant axillary buds conspicuous at flowering time **16**

b. Leafy shoots without conspicuous, summer-dormant, axillary buds **17**

16a. Leaves obovate, obtuse; petals turning pink after pollination
155. erioblasta

b. Leaves linear-oblong, apiculate; petals staying white after pollination **153. conifera**

17a. Leaves hairless except for a few on the margin **18**

b. Leaves more or less hairy on the surface **20**

18a. Leaves 3–6 mm across; flowering stems with 1 or 2 leaves and 1–3 flowers **159. androsacea**

b. Leaves 1–2 mm across; flowering-stems with 4–7 leaves and 3–9 flowers **19**

19a. Leaves and sepals apiculate
164. tenella

b. Leaves and sepals obtuse
163. glabella

20a. Petals 2–3 mm; plant annual
165. tridactylites

b. Petals 4–7 mm; plant perennial **21**

21a. Leaves usually lime-encrusted, firm, with a translucent margin; inflorescence with 5–12 flowers
59. spruneri

b. Leaves not lime-encrusted, soft, without a translucent margin; inflorescence with 1–4 flowers **22**

22a. Leafy shoots to 10 cm, crowded into a compact cushion; leaves mostly narrower than 2.5 mm; petals touching **161. muscoides**

b. Leafy shoots shorter and fewer, forming a low mat; leaves mostly wider than 2.5 mm; petals not touching **23**

23a. Leaves 3–6 mm across; flowering-stems with 1–3 flowers
159. androsacea

b. Leaves 6–9 mm across; flowering-stems with 3–7 flowers
160. depressa

24a. Petals gradually tapered to a narrow base, tip notched
162. presolanensis

b. Petals not tapered to a narrow base, tip rounded **149. exarata**

25a. Petals greenish, cream or dull yellow **26**

b. Petals white, pink or bright red **30**

26a. Larger leaves more than 6 mm across, usually wider than long **27**

b. Larger leaves to 6 mm across, usually longer than wide **28**

27a. Blade of leaves covered in cobweb-like hairs **128. arachnoidea**

b. Blade of leaves hairless, or nearly so
129. paradoxa

28a. Leaf-segments furrowed on the upper surface **29**

b. Leaf-segments flat on the upper surface **149. exarata**

29a. Leaf-segments mucronate, more or less hairless **150. hariotii**

b. Leaf-segments obtuse or almost acute, glandular-hairy
149. exarata

30a. Mature leaves hairless, though often with stalkless glands **31**

b. Mature leaves hairy, at least on the margin or leaf-stalk (hairs may be visible only with a lens) **38**

31a. Flowering-stem axillary **32**

b. Flowering-stem terminal **35**

32a. Ultimate leaf-segments not more than 9 **33**

b. Ultimate leaf-segments 9–38 **34**

33a. Petals 10–11 mm
143. portosanctana

b. Petals 5–8 mm **141. cuneata**

34a. Leaves usually deeply lobed, antler-like **139. trifurcata**

b. Leaves usually shallowly lobed, not antler-like **142. maderensis**

35a. All leaf-segments obtuse or almost acute **36**

b. Some leaf-segments mucronate **37**

36a. Leaf-segments usually not more than 2 mm across, sides parallel, upper surface furrowed; petals 3.5–5 mm **136. pentadactylis**

b. Leaf-segments usually more than 2 mm across, sides often curved, upper surface flat; petals 7–14 mm
137. fragilis

37a. Leaf-segments oblong to elliptic, 1.5–2.5 times as long as wide
138. camposii

b. Leaf-segments linear, 3–6 times as long as wide **140. canaliculata**

38a. Leafy shoots without conspicuous, dormant axillary buds at flowering time **39**

b. Leafy shoots with conspicuous, dormant axillary buds at flowering time **68**

39a. Petals prominently notched at the tip, often unequal **40**

b. Petals barely notched, if at all, equal **41**

40a. Basal leaves divided almost to the base; petals 8–11 mm, touching; biennial **126. petraea**

b. Basal leaves divided not more than halfway to the base; petals 4–8 mm, not touching; perennial **127. berica**

41a. Perennial 42
  b. Annual or biennial 65
42a. Petals at least 7 mm 43
  b. Petals usually not more than 7 mm 48
43a. Ultimate leaf-segments more than 11 44
  b. Ultimate leaf-segments less than 11 46
44a. Flowering-stems 2–5 mm across; aquatic **123. aquatica**
  b. Flowering-stems less than 2 mm across; not aquatic 45
45a. Stems forming a compact tuft, herbaceous at the base **124. irrigua**
  b. Stems spreading to form a loose cushion, woody at the base **132. geranioides**
46a. Plant with prostrate, non-flowering-shoots bearing at least some entire leaves **151. hypnoides**
  b. Plant without prostrate, non-flowering-shoots 47
47a. Petals more than 1 cm **131. pedemontana**
  b. Petals not more than 1 cm **144. rosacea**
48a. Leaf-segments furrowed on the upper surface 49
  b. Leaf-segments flat on the upper surface 54
49a. Hairs on leaves and calyx with stalks barely longer than the diameter of the gland, giving a finely warty appearance 50
  b. Hairs on leaves and calyx with stalks much longer than the diameter of the gland, giving a downy appearance 52
50a. Larger leaves 2–3 cm, in cushions to 30 cm across **133. moncayensis**
  b. Larger leaves not more than 15 mm, in cushions rarely more than 10 cm across 51
51a. Stems woody below; leaves dark green; forming a loose cushion; petals 4–5 mm across **135. intricata**
  b. Stems scarcely woody below; leaves fresh green; forming a compact cushion; petals to 4 mm across **149. exarata**
52a. Leaves with short, glandular hairs, sometimes sparse; petals about twice as long as wide, not touching **149. exarata**

b. Leaves with a dense cover of long, glandular hairs; petals about 1.5 times as long as wide, usually touching 53
53a. Leaves dark green; rosettes often 2–2.5 cm across; stem-leaves usually lobed **146. pubescens**
  b. Leaves fresh green; rosettes not more than 1.5 cm across; stem-leaves usually entire **148. cebennensis**
54a. Stems and leaf-stalks covered with sticky, cobweb-like hairs 5–10 mm **128. arachnoidea**
  b. Stems and leaf-stalks with much shorter hairs 55
55a. Leaves with broadly triangular or ovate lobes; leafy shoots short, forming a tuft or mat 56
  b. Leaves with lanceolate, oblong or elliptical lobes; leafy shoots rather long, forming a cushion or rarely a mat 58
56a. Leaves at least 9 mm across with a distinct leaf-stalk **114. rivularis**
  b. Leaves to 9 mm across without a distinct leaf-stalk 57
57a. Leaves 3–6 mm across; flowering-stems with 1–3 flowers **159. androsacea**
  b. Leaves 6–9 mm across; flowering-stems with 3–7 flowers **160. depressa**
58a. Petals more than 5 mm 59
  b. Petals to 5 mm 63
59a. Hairs on the leaves much less than 0.5 mm; foliage with a strong spicy scent **134. vayredana**
  b. Hairs on the leaves mostly more than 0.5 mm; foliage scarcely scented 60
60a. Leaves with hairs mostly or entirely non-glandular **144. rosacea**
  b. Leaves with numerous glandular hairs 61
61a. Flowers not more than 10 mm across; sepals and perigynous zone crimson; petals often with red veins **147. nevadensis**
  b. Flowers at least 10 mm across; sepals and perigynous zone mainly green; petals without red veins 62
62a. Petals 5.5–6.5 mm, usually tinged with green or yellow; leaves mostly 3-lobed **145. cespitosa**
  b. Petals usually more than 6.5 mm, pure white; some leaves usually with 5 or more lobes **144. rosacea**
63a. Leafy shoots erect and sometimes very short 64

b. Leafy shoots prostrate or ascending, usually long **130. praetermissa**
64a. Sepals and perigynous zone crimson; petals often with red veins; anthers red or orange before shedding pollen **147. nevadensis**
  b. Sepals and perigynous zone greenish; petals without red veins; anthers yellow before shedding pollen **149. exarata**
65a. Basal leaves kidney-shaped, with a long, well-defined leaf-stalk **125. latepetiolata**
  b. Basal leaves oblanceolate, with a short, ill-defined leaf-stalk 66
66a. Flower-stalks to 1 cm in fruit **166. adscendens**
  b. Flower-stalks 1–2 cm in fruit 67
67a. Petals 2.5–3 mm; annual **165. tridactylites**
  b. Petals c. 5 mm; biennial **167. blavii**
68a. Petals 1.1–2 cm 69
  b. Petals 0.4–1 cm 72
69a. Leaves deeply divided, with 3 primary lobes narrowed to stalk-like bases **121. biternata**
  b. Leaves palmately lobed, the lobes not narrowed at the base 70
70a. Outer leaves of axillary buds lobed, leaf-like in colour and texture, without cobweb-like hairs **131. pedemontana**
  b. Outer leaves of axillary buds entire, more or less papery, fringed with cobweb-like hairs 71
71a. Leaves deeply divided into linear-oblong lobes **154. rigoi**
  b. Leaves divided ½–¾ of the way to the base into lobes which are wedge-shaped at the base **158. maweana**
72a. Leaf-segments mucronate or apiculate 73
  b. Leaf-segments obtuse to acute 74
73a. Dormant axillary buds without stalks, their outer leaves partly green **151. hypnoides**
  b. Dormant axillary buds on stalks, their outer leaves papery and translucent **152. continentalis**
74a. Dormant axillary buds clothed in white, woolly hairs **155. erioblasta**
  b. Dormant axillary buds hairy, but the hairs not white and woolly 75
75a. Sepals c. 4 × 3 mm; petal-margins recurved; flowering-stem with 1–3 flowers **156. reuteriana**
  b. Sepals c. 2 × 1.5 mm; petals flat; flowering-stem with at least 4 flowers **157. globulifera**

76a. Petals purplish crimson at the sides, green elsewhere   **33. hieracifolia**

b. Petals cream or white   77

77a. Leaves sparingly toothed to almost entire; reddish brown hairs absent from base of stem and leaf-stalks   **32. pensylvanica**

b. Leaves scalloped to sharply toothed; reddish brown hairs often present at the base of the stem or leaf-stalks   78

78a. Petals with 2 yellow spots near the base, equalling or shorter than the sepals   **29. reflexa**

b. Petals unspotted, equalling or longer than the sepals   79

79a. Reddish brown hairs present at least in the leaf axils and at the base of the leaf-stalk   80

b. Reddish brown hairs absent, replaced by chaffy white hairs   82

80a. Leaves evenly and sharply toothed, often densely covered with reddish brown hairs beneath; inflorescence more or less flat-topped   **31. rufidula**

b. Leaves usually scalloped or coarsely and unevenly toothed; reddish brown hairs mostly confined to the axil and base of the leaf-stalk, sometimes sparingly so; inflorescence pyramidal or or rounded   81

81a. Filaments linear   **26. nivalis**

b. Filaments club-shaped   **30. occidentalis**

82a. Petals 2–3 mm, less than twice as long as the sepals   **27. tenuis**

b. Petals 3–6 mm, 2–3 times as long as the sepals   **28. virginiensis**

Group C

1a. Leaves lobed or toothed   2

b. Leaves entire (but margins may be fringed with stout hairs)   5

2a. Leafy summer-dormant buds in the axils of the stem-leaves   **7. strigosa**

b. Stem-leaves without leafy summer-dormant buds   3

3a. Sepals reflexed in fruit   **17. sibthorpii**

b. Sepals upright or spreading in fruit   4

4a. Petals not more than 3 mm, white or pale yellow   **18. hederacea**

b. Petals more than 4 mm, bright yellow   **16. cymbalaria**

5a. Slender runners produced from the axils of the leaves   6

b. Runners absent   9

6a. Petals not overlapping or touching   **13. neopropagulifera**

b. Petals overlapping or touching   7

7a. Leaves shiny; sepals less than ⅓ length of petals   **14. brunonis**

b. Leaves matt; sepals more than ⅓ length of petals   8

8a. Runners at most only sparingly glandular-hairy   **11. flagellaris**

b. Runners densely glandular-hairy   **12. mucronulata**

9a. All non-glandular hairs white or cream, not reddish brown   10

b. Reddish brown non-glandular hairs present on leaves, leaf-stalks, leaf-axils or stems   14

10a. Leaf-margins fringed with stout hairs or bristles   11

b. Leaf-margins without stout hairs or bristles   13

11a. Flowering-stems to 1 cm   **15. eschscholtzii**

b. Flowering-stems more than 2 cm   12

12a. Flowers stellate, petals widely separated, golden-yellow   **13. neopropagulifera**

b. Flowers cup-shaped, petals touching or overlapping, pale yellow   **8. punctulata**

13a. Sepals ascending   **9. serpyllifolia**

b. Sepals reflexed   **10. chrysantha**

14a. Basal leaves soon deciduous; stem-leaves numerous, most more or less equal in size   **6. cardiophylla**

b. Basal leaves persistent; stem-leaves progressively reduced in size upwards   15

15a. Blade of basal leaves tapering or rounded at the base, the stalk ill-defined, if present at all   16

b. Blade of basal leaves cordate at the base, with a well-defined stalk   18

16a. Stems and leafy shoots forming a dense cushion   **3. montana**

b. Stems and leafy shoots solitary or in a loose mat   17

17a. Veins in the sepals remaining separate   **1. hirculus**

b. Veins in the sepals converging to a point near the tip   **2. hookeri**

18a. Petals yellow   **4. diversifolia**

b. Petals orange   **5. pardanthina**

Group D

1a. Flowering-stems leafy   2

b. Flowering-stems leafless   18

2a. Bulbils present in axils of basal leaves   3

b. Bulbils absent from axils of basal leaves   7

3a. Bulbils replacing at least some of the flowers   4

b. Bulbils not replacing any flowers   5

4a. Filaments awl-shaped   **113. cernua**

b. Filaments club-shaped   **43. mertensiana**

5a. Petals more than 7 mm, white   **112. sibirica**

b. Petals to 5 mm, off-white or pink   6

6a. Filaments club-shaped   **43. mertensiana**

b. Filaments awl-shaped   **114. rivularis**

7a. Blade of basal leaves heart-shaped or truncate at the base, with a distinct leaf-stalk   8

b. Blade of basal leaves tapering or parallel-sided at the base, without a well-defined leaf-stalk   11

8a. Bulbils often replacing some flowers; sepals reflexed in fruit; filaments club-shaped   **43. mertensiana**

b. Bulbils absent; sepals erect or spreading in fruit; filaments awl-shaped   9

9a. Largest leaves at least 2.5 cm across, with at least 15 scalloped or triangular teeth   **110. rotundifolia**

b. Largest leaves to 2.5 cm across, with 3–11 scalloped or triangular teeth   10

10a. Plant annual; petals white, without spots   **18. hederacea**

b. Plant perennial; petals white, with purplish-red spots   **11. taygetea**

11a. Capsule streaked or spotted with purple   12

b. Capsule green to brown   13

12a. Leaves of basal rosettes more than 1 cm; filaments linear   **19. merkii**

b. Leaves of basal rosettes to 1 cm; filaments club-shaped   **20. tolmiei**

13a. Axillary buds on leafy shoots conspicuous at flowering time   14

b. Axillary buds on leafy shoots not conspicuous at flowering time   15

14a. Leaves of non-flowering-shoots straight, much longer than subtended axillary bud; leaves of flowering-stem c. 1 cm, spreading   **101. aspera**

b. Leaves of non-flowering-shoots incurved, barely longer than subtended axillary bud; leaves of flowering-stem not more than 5 mm, almost erect   **102. bryoides**

15a. Leaves usually with 3 teeth at the tip **105. tricuspidata**

b. Leaves usually entire 16

16a. Flowers cup-shaped **8. punctulata**

b. Flowers saucer-shaped 17

17a. Leaves narrowed gradually to a spine-like tip **103. bronchialis**

b. Leaves mucronate at the tip **104. cherlerioides**

18a. Flowers with irregular or bilateral symmetry 19

b. Flowers with regular symmetry 25

19a. Blade of basal leaves cordate or truncate at the base, with a slender leaf-stalk 20

b. Blade of basal leaves tapering or parallel-sided at the base, with a broad, often indistinct leaf-stalk 24

20a. Slender runners produced from the axils of the leaves 21

b. Runners usually absent 22

21a. Leaves green on upper surface **42. veitchiana**

b. Leaves marked with grey veins on upper surface **41. stolonifera**

22a. Upper petals linear-lanceolate, without a claw, unspotted; seeds smooth **38. fortunei**

b. Upper petals broadly ovate, clawed at the base, spotted yellow or red; seeds warty 23

23a. Leaves palmately divided **39. cortusifolia**

b. Leaves scalloped or shallowly lobed **40. nipponica**

24a. Lower bracts leaf-like; most basal leaves with 5–10 coarse teeth on each side; flowers not replaced by bulbils **21. clusii**

b. Lower bracts small; most basal leaves with shallow teeth on each side; at least some flowers often replaced by bulbils **22. ferruginea**

25a. Leaf-margin with a narrow, translucent border 26

b. Leaf-margin without a translucent border 29

26a. Petals rarely with red spots; ovary almost white; leaves usually entire in basal $\frac{1}{3}$ of blade **106. cuneifolia**

b. Petals usually with red spots; ovary pink or green; leaves toothed or scalloped almost to the base of the blade 27

27a. Leaf-stalk slender, more or less cylindric; leaves usually hairy on both surfaces **109. hirsuta**

b. Leaf-stalk broad, flattened; leaves hairless at least beneath 28

28a. Leaf-stalk with numerous marginal hairs, usually shorter than the scalloped blade **107. umbrosa**

b. Leaf-stalk with few marginal hairs, mostly near the base, often at least as long as the boldly toothed blade **108. spathularis**

29a. Blade of basal leaves circular to kidney-shaped, base heart-shaped; leaf-stalk well defined, slender 30

b. Blade of basal leaves lanceolate to obovate, base tapered or truncate; leaf-stalk ill defined or, if present, then broad 33

30a. At least some flowers replaced by bulbils; leaf-stalks with long white hairs **43. mertensiana**

b. Inflorescence without bulbils; leaf-stalks usually only sparingly hairy 31

31a. Petals more than 5 mm, shorter than stamens in mature flowers **37. manschuriensis**

b. Petals less than 5 mm, longer than stamens 32

32a. Petals broadly elliptic to circular, with 2 green or yellow spots **35. odontoloma**

b. Petals ovate to oblong, unspotted or with orange spots **34. nelsoniana**

33a. Leaves densely overlapping on leafy shoots 34

b. Leaves forming a basal rosette 35

34a. Petals *c.* 3 mm across **8. punctulata**

b. Petals less than 3 mm across **15. eschscholtzii**

35a. Filaments club-shaped 36

b. Filaments linear 39

36a. Blade of basal leaves more than 10 cm **25. micranthidifolia**

b. Blade of basal leaves less than 10 cm 37

37a. Axils of the leaves hairless or with white hairs **36. lyallii**

b. Axils of the leaves with at least a few reddish brown hairs 38

38a. Petals shorter than or equalling sepals, with 2 yellow spots near the base **29. reflexa**

b. Petals longer than sepals, unspotted **30. occidentalis**

39a. Carpels a dark, blackish crimson **24. melanocentra**

b. Carpels pinkish, green or brown 40

40a. Petals less than 4 mm; leaves with numerous, sharp, even teeth, often covered with reddish brown hairs **31. rufidula**

b. Petals 3–8 mm; leaf-margins scalloped or with 3 or 4 blunt teeth on each side, with whitish hairs at least at the base 41

41a. Leaf-margins scalloped or with numerous teeth; ovary at least partly inferior **28. virginiensis**

b. Leaves with 3 or 4 blunt teeth on each side; ovary superior **23. stellaris**

Section **Ciliatae** Haworth (section *Hirculus* Tausch). Perennials, usually evergreen, some monocarpic; stems solitary or clumped into mats or cushions. Leaves herbaceous, sometimes stiff or leathery, with or without a distinct stalk, margin usually entire. Chalk-glands absent. Flowering-stems usually leafy, bearing 1–several flowers. Petals usually yellow, with swellings near the base. Ovary superior to $\frac{1}{2}$ inferior; capsule splitting above the middle.

Although several species belonging to this section have been introduced to cultivation, it seems that many have now disappeared from gardens or are at best extremely rare. The following, however, have been more persistent, or have been reintroduced.

1. S. hirculus Linnaeus. Illustration: Ross-Craig, Drawings of British plants, **10**: pl. 3 (1957); Keble Martin, The new concise British flora, pl. 32 (1982); Garrard & Streeter, The wild flowers of the British Isles, 91 (1983).

Plant with loose rosettes of basal leaves forming a loose mat. Basal leaves 1–3 × 0.3–0.6 cm, narrowly lanceolate, entire without a distinct leaf-stalk. Reddish brown, non-glandular hairs present on leaf-bases and leaf-axils. Flowering-stem to 35 cm, leafy, bearing 1–4 flowers in a loose corymb. Petals 0.9–1.6 × 0.4–0.6 cm, elliptic to obovate-oblong, bright yellow. Ovary superior. *Circumboreal, Colorado, Romania & Caucasus.* H2. Late summer–autumn.

Several subspecies are recognised but none is common in cultivation. Difficult to grow; needs well-irrigated conditions, such as in a pan of moss.

2. S. hookeri Engler & Irmscher. Like *S. hirculus* but the veins in the sepals converging to a point near the tip, rather than remaining separate. *E Himalaya, Bhutan, Sikkim, China (S Xizang).* H2. Late summer–autumn.

3. S. montana H. Smith.

Plant forming a cushion. Basal leaf blades 5–15 × 2–4.5 mm, elliptic-lanceolate, tapering to a stalk. Flowering-stems 3–25 cm, leafy, reddish brown, non-glandular hairs densely present especially on upper part. Flowers 1-few. Petals 6–10 × 3–4 mm, elliptic or obovate, yellow. Ovary superior. *China (Yunnan, Sichuan), Bhutan, Sikkim.* H2? Summer.

Probably best in a well-drained peaty soil.

4. S. diversifolia Seringe. Illustration: Köhlein, Saxifrages and related genera, 157 (1984).

Plant upright, forming a clump, with a cluster of long-stalked basal leaves with reddish brown, non-glandular hairs; blade 1.5–3 cm, almost as wide, heart-shaped, margin entire. Stem-leaves diminishing in size and with progressively shorter leaf-stalks up the flowering-stem. Flowering-stem 20–30 cm, branched above the middle to form a panicle of numerous dull or golden yellow flowers, each to 2 cm across. Ovary superior. *W China, Kashmir to Bhutan.* H3. Late summer–autumn.

Prefers an acid soil rich in humus in partial shade.

5. S. pardanthina Handel-Mazzetti.

Like *S. diversifolia* but flowering-stems to 30 cm. Petals deep orange, marked with red. *SW China.* H3? Late summer–autumn.

6. S. cardiophylla Franchet.

Plant upright, forming a clump. Basal leaves soon deciduous; stem-leaves 2–2.5 × 1–1.5 cm, broadly ovate, base cordate. Reddish brown, non-glandular hairs at base of leaf-stalks. Inflorescence more or less flat-topped, sparingly branched with only 4–8 orange-yellow flowers. Ovary superior. *Sikkim, China (Sichuan, Yunnan).* H3? Late summer–autumn.

Prefers an acid soil rich in humus in partial shade.

7. S. strigosa Wallich.

Plant upright. Basal leaves soon falling; stem-leaves forming a rosette *c.* ⅓ of the way up the stem; leafy summer dormant buds in the axils of the leaves; leaves toothed at the tip, with long, white, non-glandular hairs appressed on both surfaces. Flowering-stem bracteate, bearing a solitary (rarely more) flower. Petals 5–6 × 2–3 mm, buttercup-yellow, sometimes white marked with yellow or red. Sepals reflexed. *N India, W China to upper Burma.* H3? Summer–autumn.

Prefers a moist acid soil rich in humus, in partial shade.

8. S. punctulata Engler. Illustration: Harding, Saxifrages, 11 (1992).

Plant forming a dense cushion. Basal leaves 4–6 × 2–3 mm, spathulate, without a distinct stalk. Flowering-stems 2–5 cm, leafy, glandular-hairy, bearing 1–6 flowers. Petals 6–8 × 3–4 mm, white with 2 yellow blotches or pale yellow, and with several red (or black?) spots. *China (Xizang), Nepal, Sikkim.* H2? Late summer-early autumn.

9. S. serpyllifolia Pursh. Illustration: Harding, Saxifrages, 17 (1992).

Plant forming a mat, with non-flowering, procumbent leafy shoots produced from axils of basal or lower stem-leaves. Basal leaves 3–9 mm, spathulate, entire, without a distinct leaf-stalk, hairless except for margins with white, non-glandular hairs. Flowering-stems 2–8 cm, leafy, bearing a solitary flower. Sepals ascending. Petals 4–7 mm, obovate with a short claw, yellow or rarely purple. *Arctic Siberia, Japan, W North America.* H2. Summer.

10. S. chrysantha A. Gray.

Like *S. serpyllifolia* but leaves almost hairless; sepals slightly more oblong and reflexed and petals golden. *USA (Rocky Mts).* H2. Summer.

11. S. flagellaris Willdenow. Illustration: Botanical Magazine, 4261 (1851); Evans, Alpines '81, 36 (1981); Webb & Gornall, Saxifrages of Europe, pl. 1 (1989).

Evergreen perennial, with prominent rosettes of basal leaves, from the axils of which emerge thread-like runners that terminate in a leafy bud. Rosette-leaves 0.7–1.6 × 0.2–0.5 cm, narrowly lanceolate to elliptic-obovate, margin fringed with stout hairs. Flowering-stems *c.* 6 cm, leafy, glandular-hairy, bearing a solitary flower, or sometimes a loose cyme of 1–4 flowers. Petals 6–10 mm, obovate, bright yellow. Ovary superior to almost superior. *Circumpolar Arctic, W North America (Rocky Mts), Caucasus & Himalaya to Baikal.* H1. Summer.

Eight to 10 subspecies can be recognised, of which the following may be seen in cultivation.

Subsp. **flagellaris**. Runners without glandular hairs. Stem-leaves longer than the internodes. Sepals narrowly triangular. Ovary superior, with a shallow perigynous zone. *Caucasus.*

Subsp. **platysepala** (Trautvetter) Porsild. Runners sparingly glandular-hairy. Stem-leaves longer than the internodes. Sepals ovate-triangular. Ovary almost superior, with a bowl-shaped perigynous zone. *Circumpolar Arctic.*

Subsp. **setigera** (Pursh) Tolmatchev. Runners with glandular hairs. Stem-leaves shorter than the internodes. Sepals narrowly triangular. Ovary superior, with a shallow perigynous zone. *C & E Asia, W North America.*

Subsp. **crandallii** (Gandoger) Hultén. Like subsp. *platysepala* but with narrower, erect petals, and a smaller, funnel-shaped perigynous zone with white, glandular hairs in addition to the usual black ones. *W North America.*

Best grown in a cool glasshouse in well-drained medium, dryish in winter but wetter in the spring/summer growing season.

12. S. mucronulata Royle. Illustration: Harding, Saxifrages, 17 (1992) – as S. flagellaris subsp. sikkimensis.

Like *S. flagellaris* but runners densely glandular-hairy; flowers 3–10, funnel-shaped, petals only twice as long as the sepals. Most plants in cultivation belong to subsp. **sikkimensis** (Hultén) Hara. *E Himalaya, China (S Xizang).* H2? Summer.

Cultivation as for *S. flagellaris.*

13. S. neopropagulifera Hara. Illustration: Harding, Saxifrages, 18 (1992).

Like *S. flagellaris*, but flowering-stems 8–12 cm, bearing solitary, stellate flowers in which the golden yellow petals do not overlap. Runners produced in the wild but not so far seen in cultivation. *Nepal.* H2? Late spring.

Cultivation as for *S. flagellaris.*

14. S. brunonis Seringe (*S. brunoniana* Wallich, invalid). Illustration: Botanical Magazine, 8189 (1908); Harding, Saxifrages, 126 (1970).

Like *S. flagellaris* but stem nearly hairless, leaves shiny, runners hairless, flowers more numerous, sepals shorter, 2–2.5 mm, and only ¼–⅓ the length of the petals. *Kashmir to Bhutan & China (S Xizang).* H3. Summer.

Cultivation as for *S. flagellaris.*

15. S. eschscholtzii Sternberg. Illustration: Hultén, Flora of Alaska and neighbouring territories, 566 (1968).

Plant forming a compact cushion of leafy shoots. Leaves 1–3 mm, oblong to obovate, without a distinct leaf-stalk, densely overlapping, the margin fringed with stout bristles. Flowering-stems to 1 cm, leafless, bearing a solitary flower. Petals yellow, white or pink, soon falling.

Dioecious. *E Siberia to Yukon*. H1. Summer.

Difficult to grow; cultivation in a cool glasshouse is probably essential.

Section **Cymbalaria** Linnaeus. Annuals or biennials. Leaves soft or slightly fleshy, with a distinct stalk, margin lobed, toothed or nearly entire. Chalk-glands absent. Flowering-stems leafy, producing flowers on long axillary flower-stalks. Petals usually yellow or orange, rarely white, often with swellings near the base. Ovary superior, or nearly so; capsule splitting above the middle.

16. S. cymbalaria Linnaeus. Illustration: Harding, Saxifrages, 98 (1970); Köhlein, Saxifrages and related genera, t. 11 (1984); Webb & Gornall, Saxifrages of Europe, pl. 2 (1989).

Annual (sometimes biennial?), with ascending, diffusely branched, leafy stems 10–25 cm. Leaf-stalk to 35 cm, blade usually *c.* 1 × 1.3 cm but sometimes larger, kidney-shaped to circular or ovate, more or less 5–9 lobed, the lobes acute to rounded. Flowers 2–6 in a loose cyme. Petals 4.5–6 mm, elliptic-oblong with a short claw, bright yellow towards the tip, duller orange-yellow at the base, with 2 small swellings located above the short claw. *Turkey & Caucasus, Romania, Algeria*. H3. Spring–early autumn.

Three varieties are recognised, the one usually seen in gardens, as a handsome weed, is var. **huetiana** (Boissier) Engler & Irmscher. This is distinguished by the leaves that are sometimes opposite and have rounded lobes and truncate or wedge-shaped bases. Prefers a shady place in the garden.

17. S. sibthorpii Boissier. Illustration: Webb & Gornall, Saxifrages of Europe, pl. 3 (1989).

Like *S. cymbalaria* but with reflexed, rather than spreading, sepals, and solitary flowers on leafless stalks. Doubtfully in cultivation, most plants grown under this name are *S. cymbalaria*. *Greece & SW Turkey*. H4. Summer.

18. S. hederacea Linnaeus.

Like *S. cymbalaria*, but petals 2–3 mm, obovate-elliptic, very pale yellow or white. *E Mediterranean*. H4. Spring–early summer.

Section **Merkianae** (Engler & Irmscher) Gornall. Evergreen perennials, somewhat woody below, forming mats. Leaves without a well-defined leaf-stalk; margins entire or 3-lobed at the tip. Chalk-glands absent. Flowering-stems more or less leafy, bearing 1–4 flowers. Sepals reflexed in

fruit. Petals white. Ovary almost superior; capsule splitting above the middle, streaked or spotted with purple.

19. S. merkii Fischer.

Leaves of basal rosettes 1.2–1.5 × 0.3–0.4 cm, obovate or spathulate to lanceolate, acute at the tip, margin usually entire but fringed with hairs; leaf-stalk indistinct. Flowering-stem 3–6 cm, sparingly leafy, bearing 1–3 flowers at the top. Petals 6–7 mm, ovate, tapered at the base to a claw, white. Filaments linear. *Japan*. H4. Summer.

20. S. tolmiei Torrey & A. Gray. Illustration: Hitchcock et al., Vascular plants of the Pacific Northwest, **3**: 57 (1961); Harding, Saxifrages, 16 (1992).

Leaves of basal rosettes 3–10 mm, linear, spathulate or oblanceolate, tapered to an indistinct leaf-stalk, base usually with long hairs; margin entire, cylindric or curved slightly downwards and inwards. Flowering-stems 3–12 cm, leafy, bearing 1–4 flowers in a loose cyme. Petals 3–6 mm, spathulate to broadly oblanceolate, white. Filaments club-shaped. *W North America*. H4. Summer.

Section **Micranthes** (Haworth) D. Don (section *Boraphila* Engler). Perennials, usually evergreen, occasionally deciduous. Stems solitary or few. Leaves usually all basal, herbaceous, or rather fleshy or leathery, with or without a distinct stalk; margin entire, toothed or variously lobed. Chalk-glands absent. Flowering-stems leafless, though sometimes with large, leaf-like bracts, terminating in a loose or dense panicle. Petals usually white, variously spotted or flushed with red or dark purple-black, occasionally greenish. Ovary superior to more than ½ inferior; capsule splitting to below the middle.

21. S. clusii Gouan. Illustration: Webb & Gornall, Saxifrages of Europe, pl. 5 (1989).

Blades of basal leaves 4–12 × 1.2–4.5 cm, oblanceolate to elliptic-oblong, with 4–10 coarse teeth on each side, tapered to an ill-defined leaf-stalk. Flowering-stems to 40 cm, leafless, branches many, forming a broad, diffuse, bracteate panicle, the terminal flower of which is regular, but the others irregular. Petals 4–7 × 2–3 mm, lanceolate, acute, with a prominent claw, white, unequal, the 3 upper ones longer and with 2 mustard spots at the base. Filaments linear. Ovary superior. *SW Europe*. H4. Summer.

Subsp. **clusii** has sparse, short hairs on the flowering-stem, and no leafy buds

replacing the flowers. *Pyrenees, SW France*.

Subsp. **lepismigena** (Planellas) D.A. Webb has numerous long hairs on the flowering-stem, and at least some flowers replaced by leafy buds. *N Spain, N Portugal*. Probably best grown in a cool glasshouse, in acid soil.

Also keying out with this species is **S. michauxii** Britton, which is disconcertingly like *S. clusii*, but probably can be distinguished by the larger, coarser teeth of its leaves. *E North America*. H4? Summer–early autumn.

22. S. ferruginea Graham. Illustration: Hitchcock et al., Vascular plants of the Pacific Northwest, **3**: 43 (1961).

Blade of basal leaves usually 1.5–5 cm, occasionally more, spathulate to oblanceolate, tapered to a broad, indistinct leaf-stalk, margin sharply but irregularly toothed. Flowering-stems 10–30 cm, panicle diffuse; flowers sometimes replaced by bulbils. Petals 3–6 mm, upper 3 clawed, white with 2 yellow spots, lower 2 elliptic to spathulate, not clawed or spotted. Filaments linear. Ovary superior. *W North America*. H4. Summer.

Best grown in a cool glasshouse.

23. S. stellaris Linnaeus. Illustration: Ross-Craig, Drawings of British plants **10**: pl. 2 (1957). Phillips, Wild flowers of Britain, 189 (1977); Keble Martin, The new concise British flora, pl. 32 (1982); Garrard & Streeter, The wild flowers of the British Isles, 90 (1983).

Leaf-rosettes sometimes forming carpets. Blades of basal leaves 1.2–7 × 0.8–2 cm, oblanceolate to broadly elliptic, with 3–5 blunt, forward-pointing teeth on each margin, tapered at the base to a more or less distinct leaf-stalk. Flowering-stems 5–20 cm, leafless, bearing a loose but rather narrow panicle, bracts smaller than in *S. clusii*. Flowers usually regular. Petals 3–8 × 2–3 mm, lanceolate, with a prominent claw, white with 2 yellow spots near the base. Filaments linear. Ovary superior. *Europe, Greenland & NE Canada*. H2. Summer.

Prefers a moist, acidic soil in a cool glasshouse.

24. S. melanocentra Franchet. Illustration: Harding, Saxifrages, 21 (1992).

Blades of leaves 2–3 × 0.5–2 cm, ovate-oblong with a scalloped or coarsely toothed margin, narrowed to a distinct leaf-stalk. Flowering stems 5–10 cm, bearing 1–5 flowers in a loose cyme. Petals 6–8 × 4–5 mm, obovate, upper surface white with 2 yellow or orange spots at the base, lower surface sometimes becoming purplish. Filaments linear. Ovary superior, carpels a dark, blackish crimson. *SW China, Nepal.* H3? Summer.

Grows best in a cool alpine house.

25. S. micranthidifolia (Haworth) Britton (*S. erosa* Pursh). Illustration: Gleason, The new illustrated flora of the northeastern United States & adjacent Canada **2**: 263 (1952).

Like *S. clusii* in general appearance but usually larger. Leaf-blades to 35 cm, lanceolate to oblanceolate, margin coarsely to rather finely, but irregularly, toothed, tapered to an indistinct leaf-stalk. Axils of basal leaves often with reddish brown, non-glandular hairs. Flowering-stems 30–75 cm, panicle loose. Petals 1.5–3.5 mm (shorter than the stamens), equal, elliptic to spathulate, narrowed at the base to a short claw, with a yellow spot. Filaments club-shaped. Ovary superior. *E North America.* H4? Late spring–summer.

26. S. nivalis Linnaeus. Illustration: Ross-Craig, Drawings of British plants, pl. 1 (1957); Keble Martin, The new concise British flora, pl. 32 (1982); Garrard & Streeter, The wild flowers of the British Isles, 90 (1983).

Blades of basal leaves 1–4 cm, ovate, diamond-shaped or almost circular, tapered to a broad leaf-stalk, usually with some reddish brown, non-glandular hairs on the margin or lower surface; margin scalloped or toothed. Flowering-stem 5–20 cm, more or less leafless, simple or branched near the top in a congested or, occasionally, loose cyme, with conspicuous narrow bracts. Petals 2–3 mm, slightly longer than the sepals, oblong to obovate, white, usually with pink tips. Filaments linear. Ovary half-inferior. *Circumpolar Arctic, southwards in W North America; C Europe & the Altai.* H1. Summer.

Best kept cold and dry in winter (cold-frame) and brought into an alpine house in spring and watered through the summer.

27. S. tenuis (Wahlenberg) Lindman (*S. stricta* Hornemann). Illustration: Ronning, Svalbards Flora, f. 28 (1964).

Like *S. nivalis*, but without the reddish brown, non-glandular hairs. It is also usually smaller in all its parts and sometimes has a more open inflorescence. *Distribution uncertain, but probably more strictly arctic than* S. nivalis. H1. Summer.

Cultivation conditions as for *S. nivalis*.

28. S. virginiensis Michaux. Illustration: Botanical Magazine, 1664 (1814); Gleason, The new illustrated flora of the northeastern United States & adjacent Canada **2**: 264 (1952).

Leaf-blades to 9 cm, ovate to elliptic, tapered to a short, broad leaf-stalk; margin scalloped to sharply toothed; reddish brown, non-glandular hairs present in the axils. Flowering-stems 6–50 cm, bearing a congested, conical inflorescence. Petals 3–6 mm, mostly 2–3 times as long as sepals, elliptic to spathulate, unspotted. Filaments linear. Ovary less than ⅓ inferior. *E North America.* H3. Spring.

Probably best in a humus-rich soil in partial shade.

29. S. reflexa W.J. Hooker. Illustration: Hultén, Flora of Alaska and neighbouring territories, 580 (1968).

Leaf-blades 1–2 cm, circular to ovate, oblanceolate or spathulate, tapered (sometimes abruptly) to a broad leaf-stalk; margin coarsely toothed; reddish brown, non-glandular hairs often present in the axils, short greyish hairs on the upper leaf-surface, lower surface purplish. Flowering-stems 9–60 cm, bearing a loose, rather flat-topped cyme. Petals 2–3 mm, shorter than or equalling the sepals, elliptic-oblong, without a claw, white with 2 yellow spots. Filaments club-shaped. Ovary less than ⅓ inferior. *NW Canada to USA (Alaska).* H3? Summer.

30. S. occidentalis S. Watson. Illustration: Hitchcock et al., Vascular plants of the Pacific Northwest **3**: 52 (1961).

Like *S. reflexa* but differing chiefly in its unspotted petals (sometimes narrowed to a claw), which are longer than the sepals, and in its linear to only slightly club-shaped filaments. *W North America.* H4. Spring–summer.

31. S. rufidula (Small) Macoun. Illustration: Hitchcock et al., Vascular plants of the Pacific Northwest **3**: 52 (1961).

Like *S. reflexa* but basal leaves sharply and evenly toothed, often densely covered with reddish brown, non-glandular hairs beneath. Flowering-stems to 15 cm, bracts and calyces with reddish brown, non-glandular hairs. Filaments linear. *W North America.* H4? Spring–summer.

Probably best grown in a cool glasshouse or in a shaded, moist spot in the rock garden.

32. S. pensylvanica Linnaeus. Illustration: Gleason, The new illustrated flora of the northeastern United States & adjacent Canada, **2**: 265 (1952).

Leaf-blades 10–25 cm, linear to oblanceolate, tapered to a broad, indistinct leaf-stalk; margin sparingly toothed to nearly entire; reddish brown, non-glandular hairs absent. Flowering-stems 25–60 cm, bearing a loose, cylindric inflorescence of flowers clustered at the ends of ascending branches. Petals 2.5–4 mm, linear to elliptic or narrowly obovate. Filaments linear. Ovary less than ½ inferior, becoming much less so in fruit. *E North America.* H4? Spring–summer.

A distinctive variant, with carpels markedly divergent when ripe, is known as var. **forbesii** (Vasey) Engler & Irmscher. Probably best in a humus-rich soil in partial shade.

Also keying out here is **S. oregana** Howell, a species very much like *S. pensylvanica* and from which there seem to be no consistent differences, apart from in chromosome number and geographical distribution! *W North America.* H4? Spring–summer.

33. S. hieracifolia Willdenow. Illustration: Rasetti, I fiori delle Alpi, f. 243 (1980); Webb & Gornall, Saxifrages of Europe, pl. 4 (1989).

Deciduous perennial. Leaf-blades 2–6 × 1–3 cm, ovate-elliptic, tapering to a short, winged leaf-stalk, margin entire or bluntly toothed, reddish brown, non-glandular hairs absent. Flowering-stem to 50 cm, more or less leafless, terminating in a narrow, dense, spike-like panicle, bracts often very conspicuous. Petals 1.5–3 mm, roughly equalling the sepals, but narrower, green with a purplish crimson margin. Filaments linear. Ovary half-inferior, less so in fruit. *Circumpolar Arctic, Carpathians, the Altai.* H1. Summer.

Best grown under glass in shade, cold and dry in the winter but with a good supply of water in the summer.

34. S. nelsoniana D. Don (*S. punctata* misapplied). Illustration: Hultén, Flora of Alaska and neighbouring territories, 572–574 (1968). Hitchcock et al., Vascular plants of the Pacific Northwest **3**: 57 (1961).

Leaves in a loose, basal rosette; blade 2–8 × 3–9 cm, kidney-shaped to nearly circular, base cordate, margin regularly scalloped; leaf-stalk to 4 times as long as blade, with non-glandular hairs (if present) whitish. Flowering-stem 10–35 cm, bearing a loose to compact panicle. Petals 2.5–4.5 mm, ovate to oblong, narrowed at the base to a short claw, white or pale pink, occasionally with orange spots. Filaments club-shaped. Ovary superior. *Arctic & subarctic regions from the Urals eastwards to NW North America.* H2. Summer.

Rarely cultivated in Europe; best grown on the edge of a bog-garden.

35. S. odontoloma Piper (*S. arguta* misapplied; *S. odontophylla* invalid). Illustration: Hitchcock et al., Vascular plants of the Pacific Northwest **3**: 37 (1961).

Like *S. nelsoniana* and often confused with it, differing in its slightly irregular flowers, the larger petals of which are 2–4 mm, broadly elliptic to circular, contracted sharply at the base to a slender claw, with 2 yellow or green spots. *W North America.* H4? Summer.

36. S. lyallii Engler. Illustration: Hitchcock et al., Vascular plants of the Pacific Northwest **3**: 48 (1961); Hultén, Flora of Alaska and neighbouring territories, 578 (1968).

Basal leaf-blades to 8 cm, spathulate to fan-shaped, base contracted to a more or less well-defined leaf-stalk; margin regularly toothed in the upper part. Flowering-stems usually 7–30 cm, bearing a loose, slender, cylindric, few-flowered cyme. Petals 2.5–5 mm, broadly elliptic to almost circular, with a short claw, white, cream or reddish, usually with 2 yellow or green spots. Filaments club-shaped. Ovary superior. *NW North America.* H3? Summer–early autumn.

37. S. manschuriensis (Engler) Komarov. Illustration: Botanical Magazine, 8707 (1917); Harding, Saxifrages, 97 (1970).

Leaf-blades 4–7 × 6–8 cm, kidney-shaped to nearly circular, base deeply cordate, margin regularly and coarsely toothed or scalloped; leaf-stalk 1.5–2.5 times longer than the blade. Flowering-stem 15–35 cm, branched at the top to form a fairly compact panicle. Flowers with 6–8 petals. Petals 5–6 mm, narrowly oblong-elliptic. Stamens becoming longer than the petals with age, the filaments club-shaped. Ovary superior. *N China (Manchuria), Korea.* H5. Summer.

Best grown in a cool glasshouse in a well-drained soil but with plenty of water in the growing season.

Section **Irregulares** Haworth (section *Diptera* (Borkhausen) Sternberg).

Perennials, deciduous outdoors, evergreen under glass. Leaves usually all basal, herbaceous, or rather fleshy or leathery, margin entire, toothed or variously lobed; leaf-stalks well-defined. Chalk-glands absent. Flowering-stems leafless, though sometimes with large, leaf-like bracts, terminating in a loose panicle. Flowers with a bilateral or irregular symmetry, the upper 3 or 4 petals smaller than the lower 2 or 1. Petals usually white, variously spotted. Ovary superior to almost superior; capsule splitting only in the upper half.

All the cultivated species of this section thrive best in light woodland, in acidic soils with plenty of humus and a degree of shelter. They do not tolerate much frost and can only be reliably grown outside in H5 areas.

38. S. fortunei J.D. Hooker (*S. cortusifolia* var. *fortunei* (J.D. Hooker) Maximowicz). Illustration: Botanical Magazine, 5377 (1863); Blume, Alpines for your garden, 94 (1980).

Leaf-blades 4–6 × 6–10 cm, kidney-shaped to nearly circular, base deeply cordate, margin scalloped; leaf-stalk to 1.5 times as long as the blade, expanded at the base. Flowering-stems to 50 cm, bearing a loose panicle of numerous, irregular flowers. Petals unequal, linear-lanceolate, unspotted, lower 1 or 2 to 2.5 cm, occasionally toothed in the upper part, upper 3 or 4 to 1 cm, entire. Filaments linear, scarcely broadened above. *Japan.* H5. Early autumn.

A variable species, with varieties distinguished mainly on leaf-shape. The cultivars 'Rubrifolia' and 'Wada' are suffused with red throughout, the latter more deeply so.

39. S. cortusifolia Siebold & Zuccarini. Illustration: Botanical Magazine, 6680 (1883).

Like *S. fortunei* but flowers smaller, upper, 'short' petals *c.* 4 mm, broadly ovate, spotted with yellow or red, contracted at the base to a claw. *Japan, Korea.* H5. Summer.

40. S. nipponica Makino.

Like *S. fortunei* but with smaller leaves, and flowers usually with 2 long petals rather than 1. *Japan.* H7. Summer.

Dislikes growing in a pot.

41. S. stolonifera Curtis (*S. sarmentosa* Linnaeus filius; *S. cuscutiformis* Loddiges). Illustration: Botanical Magazine, 92 (1789).

Leaf-blades 4–9 cm across, kidney-shaped to circular, dark green, hairy and

marked with grey veins on the upper surface, but reddish and hairless beneath. Thread-like runners emmanate from the axils, root at intervals and generate new plants. Flowering-stems 20–50 cm, bearing a loose panicle of numerous, irregular flowers. Petals unequal: the upper 3 are 3–4 mm, ovate, and spotted red and yellow near the base; the lower 2 are 1–2 cm, and narrowly elliptic. Filaments club-shaped. *SE China to Japan.* H5–G1. Summer.

There are a few variants, some of which are grown: var. **tricolor** (Lemaire) Maximowicz is characterised by leaves variegated with red and grey. G1. A dwarf variant, often seen under the name *S. cuscutiformis* Loddiges, is hardier than most. H5.

42. S. veitchiana I.B. Balfour.

Like *S. stolonifera* but smaller in stature and leaves not veined with grey. *W China.* H5. Summer.

Section **Heterisia** (Small) A.M. Johnson. Evergreen perennial, with basal leaves in loose rosettes. Leaf-blades with the margins scalloped or toothed; leaf-stalks well-defined. Chalk-glands absent. Flowering-stems more or less leafy, terminating in a loose panicle of regular flowers. Petals white. Ovary superior; capsule splitting to just above the middle.

43. S. mertensiana Bongard. Illustration: Hitchcock et al., Vascular plants of the Pacific Northwest **3**: 48 (1961).

Leaf-blades 2–8 cm, kidney-shaped to circular; margin shallowly scalloped, lobes toothed or scalloped; leaf-stalk to 5 times as long as blade, often with long hairs. Flowering-stems 15–40 cm, bearing a loose panicle of numerous flowers, some of which are usually replaced by bulbils. Sepals reflexed in fruit. Petals 4–5 mm, oblong-elliptic to obovate, base truncate or with a short claw, white. *W North America.* H4? Spring–summer.

Best cultivated in woodland conditions.

Section **Porphyrion** Tausch. Evergreen perennials, forming cushions or mats. Stems somewhat woody below, producing cylindric leafy shoots or rosettes. Leaves usually hard or leathery, without a distinct stalk, margin entire or finely toothed. Chalk-glands located in pits, usually evident by the conspicuous calcareous encrustation on the leaves. Flowering-stems leafy, produced from persistent leaf-rosettes, bearing flowers singly or in small cymes or racemes. Petals white, pink, purple or yellow. Ovary almost superior to nearly inferior; capsule splitting above the middle.

Recent exploration of the Himalayan region has led to the introduction of several new saxifrages belonging to this section. In many cases the identity of these plants is not yet known for certain and, regrettably, they have therefore been omitted from the following account.

Hybrids: Since hybrids between species of this section (**44–89**) are more widely grown than the parents, it is appropriate to detail them in a check-list, together with the names of some of the more common cultivars. Most have been made in cultivation. In general they are intermediate between the parents in their appearance, but so many selections of each hybrid exist that any general description is not diagnostic and thus of only limited use for purposes of identification. Nevertheless an indication of at least flower colour is given in the notes below. A key to the hybrids is given in Horny et al. (1986), where descriptions of individual cultivars may also be found.

S. × abingdonensis Arundel & Gornall (*S. burseriana × S. poluniniana*). Flowers like the first parent and foliage like the second. 'Judith Shackleton'.

S. × anglica Horny, Soják & Webr (*S. aretioides × S. lilacina × S. media*). Inflorescence of 1–3 flowers with pink to rose-purple petals. 'Beatrix Stanley', 'Cranbourne', 'Winifred'.

S. × anormalis Horny, Soják & Webr (*S. pseudolaevis × S. stribrnyi*). Petals carmine-red initially, becoming a drab orange-yellow with reddish streaks later.

S. × apiculata Engler (*S. marginata × S. sancta; S. × pungens* Sündermann; *S. × malyi* invalid). A vigorous hybrid with pale yellow petals in most of its cultivars. 'Alba' (white petals!), 'Gregor Mendel'.

S. × arco-valleyi Sündermann (*S. lilacina × S. marginata*). A slow-growing plant bearing usually solitary flowers with very pale lilac-pink petals in most cultivars. 'Arco'.

S. × baccii Young & Gornall (*S. aretioides × S. lilacina × S. media × S. stolitzkae*). Petals pink-lilac. 'Irene Bacci'.

S. × bertolonii Sündermann (*S. sempervivum × S. stribrnyi; S. × amabilis* Stapf). A vigorous hybrid, more or less intermediate between the parents, but with the inflorescence often more resembling that of *S. sempervivum*. 'Amabilis'.

S. × biasolettoi Sündermann (*S. federici-angusti* subsp. *grisebachii × S. sempervivum*). Similar to subsp. *grisebachii*; inflorescence purple-red to purple-violet. 'Crystalie', 'Feuerkopf', 'Phoenix'.

S. × bilekii Sündermann (*S. ferdinandi-coburgi × S. tombeanensis*). Petals a pale yellow. 'Castor', often referred to incorrectly in catalogues as *S. diapensioides lutea* (see under *S. × malbyana*).

S. × boeckeleri Sündermann (*S. ferdinandi-coburgi × S. stribrnyi*). Petals orange-yellow. 'Armida'.

S. × borisii Sündermann (*S. ferdinandi-coburgi × S. marginata*). Petals usually yellow to pale yellow. 'Kyrilli', 'Margarete', 'Pseudo-borisii', 'Vesna'.

S. × boydii Dewar (*S. aretioides × S. burseriana; S. aretiastrum* Engler & Irmscher). Flowers 1–3, like those of *S. burseriana*, but with yellow petals. 'Aretiastrum', 'Cherrytrees', 'Pilatus', 'Sulphurea'.

S. × boydilacina Horny, Soják & Webr (*S. aretioides × S. burseriana × S. lilacina*). Flowers solitary with pink to pale creamy yellow petals. 'Moonbeam', 'Penelope'.

S. × bursiculata Jenkins (*S. burseriana × S. marginata × S. sancta*). Flowers circular in outline with white petals. 'King Lear'.

S. × byam-groundsii Horny, Soják & Webr (*S. aretioides × S. burseriana × S. marginata*). Flowers usually solitary with pale yellow petals. 'Lenka'.

S. × caroliquarti Lang (*S. albertii × S. lowndesii*). Petals violet-pink. 'Ivana'.

S. × clarkei Sündermann (*S. media × S. vandellii*). Flowers in a small panicle, with soft pink petals. 'Sidonia'.

S. × doerfleri Sündermann (*S. federici-angusti* subsp. *grisebachii × S. stribrnyi*). A vigorous hybrid with inflorescences intermediate; flowers pink or purple-red to purple-violet. 'Ignaz Doerfler'.

S. × edithae Sündermann (*S. marginata × S. stribrnyi*). Petals pink or pale pink. 'Bridget'.

S. × elisabethae Sündermann (*S. burseriana × S. sancta; S. × godseffiana* invalid). A very vigorous hybrid bearing flowers usually with yellow petals. 'Boston Spa', 'Carmen', 'Mrs Leng', 'Ochroleuca'.

S. × eudoxiana Sündermann (*S. ferdinandi-coburgi × S. sancta*). Flowers in compact cymes. Petals yellow. 'Gold Dust', 'Haagii'.

S. × fallsvillagensis Horny, Soják & Webr (*S. burseriana × S. marginata × S. tombeanensis*). Petals white. 'Swan'.

S. × finnisiae Horny, Soják & Webr (*S. aizoides × S. aretioides × S. lilacina × S. media*). Petals pale yellow, tinged with orange. 'Parcevalis'.

S. × fleischeri Sündermann (*S. federici-angusti* subsp. *grisebachii × S. corymbosa*). Petals yellowish to red. 'Mephisto'.

S. × fontanae Sündermann (*S. diapensioides × S. ferdinandi-coburgi*). Slow-growing, with yellow petals. 'Amalie'.

S. × geuderi Sündermann (*S. ferdinandi-coburgi × S. aretioides × S. burseriana*). Flowers supposedly solitary, with deep yellow petals. Plants in cultivation under this hybrid binomial do not correspond to the original description, being multiflorous. See Horny et al. (1986) for details. 'Eulenspiegel'.

S. × gloriana Horny, Soják & Webr (*S. lilacina × S. obtusa*). Flowers solitary with pale pinkish lilac petals. The cultivar 'Godiva' is often sold as 'Gloriana'. The whole history of this hybrid is discussed by Horny et al. (1986).

S. × goringana Young & Gornall (*S. aretioides × S. cinerea × S. lilacina × S. media*). Petals deep cerise-pink fading to pale lilac-pink. 'Nancye'.

S. × grata Engler & Irmscher (*S. aretioides × S. ferdinandi-coburgi*). Petals deep yellow. Plants of this hybrid may not be distinguishable from those going under the name *S. × guederi*; see Horny et al. (1986) for details. 'Loeflingii'.

S. × gusmusii Irving & Malby (*S. corymbosa × S. sempervivum*). Flowering-stems drab purple or yellowish, branched or not. Petals yellowish or drab orange. 'Subluteiviridis'.

S. × hardingii Horny, Soják & Webr (*S. aretioides × S. burseriana × S. media*). Petals creamy yellow, apricot, orange or pink. 'C.M. Prichard', 'Iris Prichard'.

S. × heinrichii Sündermann (*S. aretioides × S. stribrnyi*). Flowers in small cymes, with yellow to orange petals. 'Ernst Heinrich'.

S. × hoerhammeri Engler & Irmscher (*S. federici-angusti* subsp. *grisebachii × S. marginata*). Panicles irregular; petals bright rose. 'Lohengrin'.

S. × hofmannii Sündermann (*S. burseriana × S. sempervivum*). Flowering-stems more like those of *S. sempervivum*, with erect, pink to drab pink petals. 'Bodensee', 'Ferdinand'.

S. × hornibrookii Horny, Soják & Webr (*S. lilacina × S. stribrnyi*). Forms distinctly domed cushions. Flowers 1–5 per stem, according to cultivar and influence of *S. strbrnyi*. Petals spreading, deep wine-red to violet-purple. 'Riverslea'.

S. × ingwersenii Horny, Soják & Webr (*S. lilacina × S. tombeanensis*). Flowering stem with 1–3 flowers with pale pink petals. 'Simplicity'.

S. × irvingii May et al. (*S. burseriana × S. lilacina*). Flowers like those of *S. burseriana*, but with pinkish petals. 'His Majesty', Walter Irving', 'Jenkinsiae'.

S. × kayei Horny, Soják & Webr (*S. aretioides × S. burseriana × S. ferdinandi-coburgi × S. sancta*). Petals a very deep yellow. 'Buttercup'.

S. × kellereri Sündermann (*S. burseriana × S. stribrnyi; S. × suendermannii* Sündermann; *S. × hedwigii* invalid). Flowering-stems with few flowers. Petals spreading, pale to purplish rose. 'Johann Kellerer', 'Kewensis', 'Sundermannii Major'. Plants grown in the British Isles as *S. pseudo-kellereri* invalid are not separable from 'Johann Kellerer'.

S. × laeviformis Horny, Soják & Webr (*S. marginata × S. pseudolaevis*). Flowers in small, compact cymes, with muddy yellow petals. 'Egmont'.

S. × landaueri Horny, Soják & Webr (*S. burseriana × S. marginata × S. stribrnyi*). Petals pale rose. 'Leonore'.

S. × leyboldii Sündermann (*S. marginata × S. vandellii*). Flowers in small cymes, with white petals. 'August Hayek'.

S. × lincolni-fosteri Horny, Soják & Webr (*S. aretioides × S. burseriana × S. diapensioides*). Flowers 1–3, with large, pale yellow petals. 'Salome'.

S. × lismorensis Young & Gornall (*S. aretioides × S. georgei × S. lilacina × S. media*). Petals pink to red. 'Lismore Carmine', 'Lismore Pink'.

S. × luteopurpurea Lapeyrouse (*S. aretioides × S. media; S. lapeyrousii* D. Don; *S. × ambigua* de Candolle; *S. × benthamii* Engler & Irmscher). A natural hybrid found in the Pyrenees. It is at least partly fertile and hence very variable. Leaves and inflorescence are more or less intermediate between those of the parents. Petals yellow to pinkish purple. 'Godroniana'.

S. × malbyana Horny, Soják & Webr (*S. aretioides × S. diapensioides; S. diapensioides lutea* invalid). A slow-growing, dense cushion, like *S. diapensioides* in appearance. Flowers in a loose cyme, with pale yellow to cream petals. 'Primulina'.

S. × margoxiana Horny, Soják & Webr (*S. ferdinandi-coburgi × S. marginata × S. sancta*). Flowers stellate with pale yellow petals. 'Parsee'.

S. × mariae-theresiae Sündermann (*S. burseriana × S. federici-angusti* subsp. *grisebachii; S. × marie-meresiae* invalid). Petals pink. 'Gaertneri', 'Theresia'.

S. × megaseaeflora May et al. (*S. aretioides × S. burseriana × S. lilacina × S. media*). Flowers usually solitary with petals varying from pale purple to pink or yellow. 'Robin Hood', 'Jupiter'.

S. × millstreamiana Horny, Soják & Webr (*S. burseriana × S. ferdinandi-coburgi × S. tombeanensis*). Petals pale yellow. 'Eliot Hodgkin'.

S. × paulinae Sündermann (*S. burseriana × S. ferdinandi-coburgi; S. kolbyi* invalid). Flowers in loose cymes, with yellow petals. 'Franzii', 'Kolbiana'.

S. × petraschii Irving (*S. burseriana × S. tombeanensis; S. × assimilis* invalid). Flowers 1–5, with white petals. 'Funkii', 'Hansii', 'Kaspar Maria Sternberg', 'Schelleri'.

S. × poluanglica Bürgel (*S. aretioides × S. lilacina × S. media × S. poluniniana*). Petals rich rose-pink, deepening toward base. 'Redpoll'.

S. × polulacina Bürgel (*S. lilacina × S. poluniniana*). Close to latter parent. Petals clear pink, fading to pale pink with age. 'Kathleen'.

S. × poluteo-purpurea Bürgel (*S. aretioides × S. media × S. poluniniana*). Petals light pink, nectary red. 'Sásava'.

S. × pragensis Horny, Soják & Webr (*S. ferdinandi-coburgi × S. marginata × S. stribrnyi*). Inflorescence racemose. Petals initially yellow, later turning orangeish or salmon-coloured. 'Golden Prague'.

S. × prossenii (Sündermann) Ingwersen (*S. sancta × S. stribrnyi*). Flowering-stems with small cymes of stellate flowers, with bronze to rosy orange petals. 'Prometheus', 'Regina'.

S. × pseudo-kotschyi Sündermann (*S. kotschyi × S. marginata*). Flowering-stems with compact, flat-topped cymes of stellate flowers with yellow petals. 'Denisa'.

S. × rosinae Horny, Soják & Webr (*S. ?diapensioides × S. marginata*). Forms a cushion of nearly spherical rosettes. Flowering-stems with flat-topped cymes. Petals off-white. 'Rosina Sundermann'.

S. × saleixiana Gaussen & Le Brun (*S. aretioides × S. caesia*). A natural hybrid possibly of this parentage is reported fom the Pyrenees.

S. × salmonica Jenkins (*S. burseriana × S. marginata; S. × salomonii* Sündermann, invalid). Flowering-stems with few-flowered cymes. Petals white. 'Friesei', 'Kestoniensis', 'Maria Luisa', 'Obristii', 'Pichleri', 'Pseudo-salomonii', 'Salomonii', 'Schreineri'.

S. × schottii Irving & Malby (*S. corymbosa × S. stribrnyi*). Petals drab yellow or vermillion red. 'Sub-stribrnyi'.

S. × semmleri Horny, Soják & Webr (*S. ferdinandi-coburgi × S. pseudolaevis ×* *S. sancta*). Inflorescence compact and flat-topped with yellow, stellate flowers. 'Martha'.

S. × smithii Horny, Soják & Webr (*S. marginata × S. tombeanensis*). Flowering-stems short, bearing flat-topped cymes of large flowers with white petals. 'Vahlii'.

S. × steinii Sündermann (*S. aretioides × S. tombeanensis*). Flowers several. Petals creamy white. 'Agnes'.

S. × stormonthii Horny, Soják & Webr (*S. ?desoulavyi × S. sancta*). Flowers in compact cymes, with yellow petals. 'Stella'.

S. × stuartii Sündermann (*S. aretioides × S. media × S. stribrnyi*). Paniculate cyme. Flowers with pale yellow to drab rose-pink petals. 'Lutea', 'Rosea'.

S. × thomasiana Sündermann (*S. stribrnyi × S. tombeanensis*). Flowering-stems tall (6–10 cm) bearing a loose cyme of small flowers with pale pink petals. 'Magdalena'.

S. × tiroliensis Kerner (*S. caesia × S. squarrosa*) occurs as a natural hybrid in the Dolomites and Julian Alps, associated with the parents.

S. × urumoffii Horny, Soják & Webr (*S. corymbosa × S. ferdinandi-coburgi*). Flowering-stems tall (6–14 cm), bearing a short cyme of small flowers with yellow petals. 'Ivan Urumov'.

S. × webrii Horny (*S. sancta × S. scardica; S. sartorii* misapplied). Flowers in flat-topped cymes. Petals yellow. 'Pygmalion'.

S. × wehrhahnii Horny, Soják & Webr (*S. marginata × S. scardica*). Flowers in flat-topped cymes. Petals white. 'Pseudo-scardica'.

S. × wendelacina Horny & Webr (*S. lilacina × S. wendelboi*). Flowering-stem bearing 1 or 2 flowers. Petals pale pink to lilac. 'Wendrush'.

S. × youngiana Harding & Gornall (*S. lilacina × S. marginata × S. stribrnyi*). Petals lilac. 'Lilac Time'.

Subsection **Kabschia** (Engler) Rouy & Camus. Leafy shoots aggregated into cushions or mats. Leaves alternate, the margins with a translucent border. Flowering-stems distinct or virtually lacking. Sepals spreading or semi-erect, shorter than the petals and not concealing them. Petals white, pink, purple or yellow, hairless. Ovary almost superior to inferior.

Series **Aretioideae** (Engler & Irmscher) Gornall. Leaves usually with 3–13 chalk-glands; tips erect or patent. Flowering-stems distinct, bearing 3–15 flowers. Petals longer than the stamens, yellow. Ovary almost superior to ¾ inferior.

44. S. ferdinandi-coburgi Kellerer &
Sündermann. Illustration: Horny et al.,
Porophyllum saxifrages, f. 20–22 (1986).

Leafy shoots forming an untidy cushion.
Leaves 4–8 × 1–2 mm, oblong-lanceolate,
with a short point usually bent upwards;
margin entire, except for a few hair-like
teeth near the base; chalk-glands
producing a conspicuous calcareous
encrustation and making the leaves look
grey. Flowering stems 3–12 cm, reddish,
bearing 3–15 flowers in a flat-topped or a
rather long cyme. Petals 5–8 mm, broadly
to narrowly obovate, scarcely touching,
bright yellow. Ovary 1/2–3/4 inferior. *E
Bulgaria, N Greece.* H2. Spring.

Var. **radoslavoffii** Stojanov (var.
pravislavii misspelling) and var. **macedonica**
Drenov are based on variation in the height
of the flowering-stem, in the number of
flowers and size of petals; these features,
however, are diagnostic of individuals rather
than populations, so cultivar status is more
appropriate. Plants distributed under the
name of the former variety, however, often
turn out to be var. **rhodopea** Kellerer &
Stojanov, which differs consistently from the
type only in its longer rosette-leaves, 6–13
mm.

Prefers an alkaline soil.

45. S. aretioides Lapeyrouse. Illustration:
Horny et al., Porophyllum saxifrages, f. 23
(1986); Webb & Gornall, Saxifrages of
Europe, pl. 13 (1989).

Similar to *S. ferdinandi-coburgi*, but leaves
shorter, wider, apex obtuse, or if with a
short point, then this held straight and not
angled upwards and inwards. Flowering-
stems usually with up to 5 flowers, these
packed more densely into a more
consistently flat-topped cyme. Petals
sometimes narrower, tip often notched.
Pyrenees, NW Spain. H2. Spring.

Prefers an alkaline soil, does best in a
cool glasshouse.

Series **Juniperifoliae** (Engler &
Irmscher) Gornall. Leaves usually with 3–9
chalk-glands; tips erect or patent, barely if
at all recurved. Flowering-stems distinct,
bearing 3–19 flowers. Petals shorter than
or barely equalling the stamens, not
touching, yellow (very rarely white). Ovary
3/4 inferior.

46. S. juniperifolia Adams (*S. juniperina*
Bieberstein; *S. pseudosancta* Janka; *S.
macedonica* Degen). Illustration: Botanical
Magazine, n.s. 137 (1951); Köhlein,
Saxifrages and related genera, t. 23
(1984); Horny et al., Porophyllum

saxifrages, f. 64–66 (1986).

Leafy shoots woody below, columnar,
forming a fairly dense cushion. Leaves
9–13 × 1–2.5 mm, linear to linear-
lanceolate, tapered to a stiff point; margin
minutely toothed towards the base;
producing little (if any) calcareous
encrustation. Flowering-stems 3–6 cm,
usually with many shaggy hairs, bearing
3–8 flowers in a dense, oblong cyme.
Petals 5–6 mm, obovate, sometimes
narrowly so, yellow. *Bulgaria, NE Turkey,
Caucasus.* H2. Spring.

May be more vigorous in partial shade.

47. S. sancta Grisebach (*S. juniperifolia*
subsp. *sancta* (Grisebach) D.A. Webb).
Illustration: Botanical Magazine, n.s. 137
(1951); Köhlein, Saxifrages and related
genera, t. 23 (1984); Horny et al.,
Porophyllum saxifrages, f. 63, 65 (1986).

Like *S. juniperifolia* but leaves usually less
than 1 cm, margin with long tooth-like
hairs up to near the apex. Flowering-
stems hairless; inflorescence more nearly
spherical or flat-topped. Stamens barely, if
at all, exceeding the petals. *NE Greece, NW
Turkey (Mt Ida).* H2. Spring.

Prefers an alkaline soil.

48. S. desoulavyi Oettingen (*S. caucasica*
var. *desoulavyi* (Oettingen) Engler &
Irmscher. Illustration: Horny et al.,
Porophyllum saxifrages, f. 67 (1986).

Like *S. sancta* but flowering-stems 1.5–2
cm, sparingly glandular-hairy, producing
a spherical head of 3–7 flowers. *Caucasus.*
H2. Early spring.

Prefers an alkaline soil and some shade.

49. S. kotschyi Boissier. Illustration:
Botanical Magazine, 6065 (1873); Horny
et al., Porophyllum saxifrages, f. 71
(1986).

Leafy shoots forming a deep cushion.
Leaves 4–12 × 1–4 mm, oblong to
spathulate, tip mucronate, basal margin
with tooth-like hairs; producing an obvious
calcareous encrustation. Flowering-stems
1–8 cm, glandular-hairy, bearing a fairly
loose, somewhat flat-topped panicle of
4–13 flowers. Petals 3–6 mm, obovate to
spathulate, yellow. *C & E Turkey to NW
Iran.* H2?. Early spring.

Probably prefers an alkaline soil.

50. S. pseudolaevis Oettingen (*S. laevis*
misapplied). Illustration: Horny et al.,
Porophyllum saxifrages, f. 68 (1986).

Leafy shoots forming a mat. Leaves 4–8
× 2.5–3 mm, oblong-obovate to
spathulate, shortly pointed at the tip,
shiny, with tooth-like hairs on the margins

towards the base. Flowering-stems 2–5
cm, variably glandular-hairy to hairless,
bearing a small, oblong cyme of 5–7
flowers. Petals 2.5–4 mm, obovate, tip
irregularly lobed, yellow. Styles markedly
spreading. *Caucasus.* H2. Early spring.

Prefers an alkaline soil and partial
shade.

51. S. subverticillata Boissier.
Illustration: Horny et al., Porophyllum
saxifrages, f. 73 (1986).

Leafy shoots forming a loose cushion,
with secondary leaf-rosettes terminating
extended shoots. Leaves 12–20 × 1–2.5
mm, oblong to linear-oblong, sharply
pointed; tooth-like hairs on margin at
base. Flowering-stems 2.5–4 cm, hairless,
bearing a loose panicle of 6–9 flowers.
Petals 3–4 mm, narrowly obovate, well
separated one from another, pale yellow
to cream. Styles unusually long, to 7 mm.
Caucasus. H2. Early spring.

Var. **colchica** (Albow) Horny & Webr
differs in its shorter leaves and slightly
longer petals.

Prefers partial shade.

52. S. scleropoda Sommier & Levier.
Illustration: Horny et al., Porophyllum
saxifrages, f. 72 (1986).

Leafy shoots columnar, woody below,
forming a cushion with secondary leaf-
rosettes terminating extended shoots. Leaves
3.5–10 × 1–2 mm, lanceolate, oblong or
elliptic, tip mucronate, the lower 2/3 with
tooth-like hairs on margin, tips recurved.
Flowering-stems 2–4 cm, glandular-hairy,
bearing a spherical cyme of 12–19 flowers.
Petals 2.5–5 mm, narrowly obovate to
linear-oblong, well separated, sulphur
yellow. *Caucasus.* H2. Spring.

Series **Lilacinae** Gornall. Leaves with
5–9 chalk-glands, leaf-tips usually strongly
recurved but not prominently thickened.
Flowering-stems usually distinct, bearing
solitary flowers. Petals longer than the
stamens, usually more or less touching or
overlapping, white, pink or purple. Ovary
1/4–3/4 inferior.

53. S. lilacina Duthie. Illustration: Horny
et al., Porophyllum saxifrages, f. 113
(1986); Quarterly Bulletin of the Alpine
Garden Society, **58**(3): 272 (1990).

Leafy shoots forming a compact cushion.
Leaves 3.5–5.5 × 1–1.5 mm, oblong to
spathulate, tip obtuse or shortly pointed,
margin recurved, with tooth-like hairs in
the lower half. Flowering-stem 1–2 cm.
Petals 8–11 mm, obovate, often with wavy

margins, tips recurved, deep violet at first but fading later to reveal dark veins. Ovary ¾ inferior. *W Pakistan, Kashmir, India*. H2. Spring.

This species is a key parent of many Kabschia cultivars. Prefers an acid soil; does best in a cool glasshouse.

54. S. poluniniana H. Smith. Illustration: Horny et al., Porophyllum saxifrages, f. 107 (1986); Harding, Saxifrages, 54 (1992).

Leafy shoots forming a loose cushion. Leaves 5–6 × 1–2 mm, oblong-spathulate to linear, tip almost acute but with a short point, recurved, moderately thickened, base with tooth-like hairs on margin, calcareous encrustation prominent. Flowering-stems 1.5–2 cm. Petals *c.* 1 cm, obovate, tapering to a slender claw, overlapping, white becoming pink with age (always pink in the sun). Ovary inferior. *Nepal*. H2. Spring.

Prefers an alkaline soil in partial shade.

Series **Marginatae** (Engler & Irmscher) Gornall. Leaves usually with 5–18 chalk-glands, tips usually strongly recurved and often prominently thickened. Flowering-stems distinct, bearing mostly 2–12 flowers. Petals longer than the stamens, usually more or less touching, white, pink or purple. Ovary ½–¾ inferior.

55. S. marginata Sternberg. Illustration: Botanical Magazine, 6702 (1883); Köhlein, Saxifrages and related genera, t. 22 (1984); Horny et al., Porophyllum saxifrages, f. 26–38 (1986); Webb & Gornall, Saxifrages of Europe, f. 16 (1989); Harding, Saxifrages, 71 (1992).

Leaves obovate to narrowly elliptic, 3–13 × 1–5 mm, tip obtuse or mucronate, sometimes recurved; margin entire, but fringed with hairs at least at the base; calcareous encrustation usually obvious. Flowering-stems 3–12 cm, terminating in a small, compact panicle usually of 2–8 flowers. Petals 5–15 mm, obovate, scarcely touching, white but sometimes turning pink with age. *S Italy, Romania, Balkans*. H2. Spring.

The variation in the species is great but difficult to treat owing to its more or less continuous nature. It seems, however, that it is possible to recognise at least two subspecies.

Subsp. **marginata** (*S. rocheliana* Sternberg; *S. coriophylla* Grisebach; *S. boryi* Boissier & Heldreich). Distinguished by its leaves, in which the tips are rounded or obtuse and the marginal hairs are confined to the basal region. A diminutive variant from the Balkans with columnar shoots of small leaves is known as var. **coriophylla** (Grisebach) Engler, and a similar one but with hairless flowering-stems is known as subsp. **bubakii** (Rohlena) Chrtek & Soják.

Subsp. **karadzicensis** (Degen & Kosanin) Chrtek & Soják has small leaves (3–4.5 × 1–2 mm), in which the tips are shortly pointed, bear only one chalk-gland, and the margins are fringed with long hairs from the base almost to the tip; non-glandular hairs also sometimes occur on the leaf surface. *Former Yugoslavia (Karadzice)*. Many plants under this name in cultivation do not appear to match those found in the wild.

It is possible that the enigmatic **S. obtusa** (Sprague) Horny & Webr (*S. scardica* Grisebach var. *obtusa* Sprague; *S. dalmatica* invalid) should be included within *S. marginata* or one of its hybrids. It is possibly sterile and not known for certain in the wild; a full discussion of it and its alleged hybrids is presented by Horny et al. (1986).

Prefers an alkaline soil.

56. S. scardica Grisebach (*S. sartorii* Boissier). Illustration: Botanical Magazine, 8243 (1909); Horny et al., Porophyllum saxifrages, f. 45–47 (1986); Webb & Gornall, Saxifrages of Europe, f. 17 (1989).

Leafy shoots crowded (but remaining distinct) into a hard, fairly dense cushion. Leaves 5–15 × 2–4 mm, oblong, acute or with a short point, rather glaucous; margin entire at tip only, middle minutely toothed; calcareous encrustation evident. Flowering-stems 4–12 cm, with a compact cyme of 5–12 flowers. Petals 7–12 mm, obovate, nearly touching, white but sometimes becoming pink or purple with age. *Balkan Peninsula*. H2. Spring.

There is some variation in leaf-size, and those plants with the smallest leaves (*c.* 6 mm) are known as var. **pseudocoriophylla** Engler & Irmscher. Prefers an alkaline soil in partial shade; probably best in a cool glasshouse.

57. S. iranica Bornmüller. Illustration: Horny et al., Porophyllum saxifrages, f. 74 (1986).

Leafy shoots forming a very compact, hard cushion. Leaves 3–4 × 2.5–4 mm, oblong to broadly obovate, tip obtuse. Flowering-stems 2.5–5 cm, bearing a small cyme of 3–6 flowers. Petals 5–12 mm, obovate, tip usually slightly notched, touching or overlapping, recurved, white becoming deep purple with age mainly at the base. *N Iran (Elburz Mts)*. H2. Spring.

Prefers an alkaline soil.

58. S. wendelboi Schönbeck-Temesy. Illustration: Horny et al., Porophyllum saxifrages, f. 76 (1986).

Like *S. iranica*, of which it is probably only a variant, but flowering-stem 3–6 cm; petals more broadly obovate, less recurved, ageing to a pink colour. *Iran*. H2. Spring.

Prefers an alkaline soil.

59. S. spruneri Boissier. Illustration: Horny et al., Porophyllum saxifrages, f. 42–43 (1986); Webb & Gornall, Saxifrages of Europe, f. 18 (1989).

Leafy shoots mostly with dead leaves, but terminating in living rosettes aggregated into a deep cushion. Leaves 4–8 × 2–4 mm, oblong-spathulate to obovate-oblanceolate, obtuse or with a short point, uniquely (in the section) the whole lower surface and at least half the upper glandular-hairy; margin entire. Flowering-stems 3–8 cm, bearing 4–15 flowers in a flat-topped cyme. Petals 4–6 mm, obovate, more or less touching, white. *Balkan Peninsula*. H2. Spring.

Prefers an alkaline soil; best grown in a cool glasshouse.

60. S. alberti Regel & Schmalhausen. Illustration: Horny et al., Porophyllum saxifrages, f. 103 (1986).

Leafy shoots forming a dense, compact cushion. Leaves 5.5–9 × 1–2 mm, linear to oblong, obtuse, margins fringed with tooth-like hairs in the lower half, dull green. Flowering-stems 1.5–2.5 cm, bearing a compact, flat-topped cyme of 4–8 flowers. Petals 6–7 mm, obovate, touching, white. *C Asia (Pamirs, Tien Shan)*. H2? Spring.

Probably prefers an acid soil in partial shade.

61. S. clivorum H. Smith. Illustration: Horny et al., Porophyllum saxifrages, f. 119 (1986).

Leafy shoots forming a dense cushion. Leaves 5–6 × 2–2.5 mm, linear-oblong, tip truncate with chalk-glands embedded in margin, margin fringed with a few thick, glandular hairs in the lower half. Flowering-stems *c.* 8 mm, bearing 1–3 flowers. Petals 3–4 mm, obovate, often broadly so, touching, white. *Bhutan, Nepal?* Spring.

62. S. andersonii Engler. Illustration: Horny et al., Porophyllum saxifrages, f. 122 (1986).

Leafy shoots forming a compact (loose in shade) cushion. Leaves 4–10 × 1–4 mm, linear-obovate to obovate, tip obtuse or shortly pointed, thickened into a horseshoe-shaped rim; margin fringed with hairs from the base to about the middle. Flowering-stems 1.5–3.5 cm, bearing 1–5 flowers. Petals 3–6 × 1.5–4 mm, obovate, white ageing to pink, or sometimes pinkish to start with. *Nepal, China (Xizang), Bhutan.* H2? Late spring–summer.

Prefers a well-watered, but well-drained neutral compost.

S. afghanica Aitchison & Hemsley appears to be similar but is doubtfully in cultivation. *Afghanistan, W Pakistan, Nepal, China (SE Xizang).* H2? Spring.

63. S. rhodopetala H. Smith. Illustration: Horny et al., Porophyllum saxifrages, f. 118 (1986).

Like *S. andersonii* but distinguished by its usually 5–9 flowers with petals a deep rose-pink. *Nepal.* H2? Late spring–summer.

64. S. stolitzkae Engler & Irmscher. Illustration: Evans, Alpines '81, 26 (1981); Horny et al., Porophyllum saxifrages, f. 124 (1986); Elliot & Bird, Quarterly Bulletin of the Alpine Garden Society **56**(1): 55 (1988); Harding, Saxifrages, 71 (1992).

Leafy shoots forming a dense cushion. Leaves 7–9 × 2–3 mm, oblong to slightly spathulate, tip shortly pointed, thickened into a horseshoe-shaped rim, margin fringed with tooth-like hairs in the lower ⅓, glaucous, with a prominent calcareous encrustation. Flowering-stems 3–8 cm, bearing 1–6 flowers. Petals 5–9 mm, obovate to broadly so, tapered to a short claw, margins often wavy, overlapping, white becoming pink with age, especially at the base. *India, Bhutan, Nepal.* H2? Spring.

Prefers an alkaline soil; best in a cool glasshouse.

65. S. cinerea H. Smith. Illustration: Horny et al., Porophyllum saxifrages, f. 123 (1986); Harding, Saxifrages, 56 (1992).

Leafy shoots forming a low, pad-like cushion. Leaves 10–12 × 1.5–2 mm, linear-oblong, tip acute, base expanded and fringed with a few tooth-like hairs, glaucous-green. Flowering-stems *c.* 8 cm, red, bearing 2–6 flowers in a compact cyme. Petals 8–10 mm, obovate, touching, white. *Nepal.* H2? Spring.

Series **Squarrosae** (Engler & Irmscher) Gornall.

Leaves with 5–7 chalk-glands, tips recurved. Flowering-stems distinct, bearing 2–8 flowers. Stamens united at the base

by a ribbon of tissue. Petals longer than the stamens, white. Ovary more than ¾ inferior.

66. S. caesia Linnaeus (*S. recurvifolia* Lapeyrouse). Illustration: Webb & Gornall, Saxifrages of Europe, f. 15 (1989); Quarterly Bulletin of the Alpine Garden Society **58**(3): 273 (1990); Harding, Saxifrages, 36 (1992).

Leafy shoots forming a moderately dense cushion. Leaves *c.* 4 × 1–1.5 mm, oblong to spathulate, upper half curving outwards and expanded near the tip, glaucous, with conspicuous calcareous encrustation; margins entire but fringed with hairs at the base. Flowering-stems 4–12 cm, branched near the top to form a small panicle of 2–8 flowers. Petals 4–6 mm, white. *Pyrenees, Alps, Tatra & W Carpathians, W former Yugoslavia.* H2. Late spring.

An alleged hybrid with *S. mutata*, called **S. × forsteri** Stein, is so much like *S. caesia* that the parentage must be doubted. For a hybrid with *S. aizoides* see under that species. Prefers an alkaline soil; best grown in tufa under glass.

67. S. squarrosa Sieber. Illustration: Finkenzeller & Grau, Alpenblumen, 107 (1985).

Like *S. caesia* but with narrower, denser leafy shoots aggregated to form a taller, more compact cushion. Leaves oblong (not expanded towards the tip), only the tip curving outwards, less glaucous, with fewer chalk-glands (usually 3 versus 7). Flowering-stem hairier at the base than at the top (the reverse is true of *S. caesia*). *SE Alps.* H2. Spring.

Prefers an alkaline soil; best grown in tufa under glass.

Series **Rigidae** (Engler & Irmscher) Gornall. Leaves with 3–7 chalk-glands, the tips erect or patent, barely, if at all, recurved. Flowering-stems distinct, bearing 1–6 flowers. Petals longer than the stamens, white. Ovary ¾ inferior.

68. S. diapensioides Bellardi (*S. glauca* Clairville). Illustration: Quarterly Bulletin of the Alpine Garden Society **25**(2): 132 (1957); Finkenzeller & Grau, Alpenblumen, 105 (1985); Horny et al., Porophyllum saxifrages, f. 39 (1986).

Leafy shoots forming dense, hard cushions, but remaining more or less distinct, not merging into the mass of foliage. Leaves 3–5 × 1–1.5 mm, oblong, obtuse or with a very short point, glaucous; margin entire; calcareous

encrustation prominent. Flowering-stems 3–8 cm, terminating in a small flat-topped cyme of 3–5 flowers. Petals 7–9 mm, white. *W Alps.* H2. Spring.

Prefers an alkaline soil in partial shade.

69. S. tombeanensis Engler. Illustration: Horny et al., Porophyllum saxifrages, f. 41 (1986); Webb & Gornall, Saxifrages of Europe, pl. 10, 11 (1989).

Leafy shoots very tightly packed to form a dense, moderately hard cushion. Leaves 2–4.5 × 1–1.5 mm, narrowly oblong to elliptic-lanceolate, with a short point usually bent upwards, darkish green and only slightly glaucous; margin entire; calcareous encrustation slight or not apparent. Flowering-stems 3–7 cm, bearing a small, compact cyme of 1–4 flowers. Petals 8–12 mm, broadly obovate, touching, white. *SE Alps.* H2. Spring.

Prefers an alkaline soil in partial shade.

70. S. burseriana Linnaeus. Illustration: Botanical Magazine, 747 (1977); Horny et al., Porophyllum saxifrages, f. 1–19 (1986); Webb & Gornall, Saxifrages of Europe, pl. 12 (1989).

Leafy shoots forming a dense mat or cushion. Leaves 6–12 × 1.7–2 mm, mostly pointing upwards, narrowly lanceolate, apex tapered, acuminate, distinctly glaucous, with 5–7 chalk-glands but little evident calcareous encrustation; margin entire, except for a few tooth-like hairs at the base. Flowering-stems 2.5–5 cm, crimson, usually with a solitary flower. Petals 7–12 mm (sometimes more in cultivars), white. *E Alps.* H2. Spring.

Variation in the size of petals and height of flowering-stems is continuous and difficult to treat. Epithets such as *crenata, major, minor* or *tridentina*, used by gardeners to describe particular clones, are best regarded as cultivars. Prefers an alkaline soil.

71. S. vandellii Sternberg. Illustration: Finkenzeller & Grau, Alpenblumen, 105 (1985); Horny et al., Porophyllum saxifrages, f. 40 (1986).

Leafy shoots densely crowded, forming a deep, hard cushion. Leaves 7–9 × 2–3 mm, dark green, narrowly triangular-lanceolate, tapered to a sharp point; margin entire except for a few tooth-like hairs at the base; chalk-glands 5–7, generating only a slight calcareous encrustation which is soon gone. Flowering-stems 4–6 cm, bearing 3–6 flowers in a compact cyme. Petals 7–9 mm, obovate, touching, white. *Italian Alps.* H2. Spring.

Prefers an alkaline soil in partial shade; does best in a vertical crevice in tufa.

Series **Subsessiliflorae** Gornall. Leaves with a solitary chalk-gland in the margin of the usually thickened, truncated tip which is not recurved. Flowers stemless or very nearly so, solitary. Petals longer than the stamens, touching or overlapping, white or cream. Ovary half-inferior.

72. S. pulvinaria H. Smith (*S. imbricata* Royle, not Lamarck). Illustration: Horny et al., Porophyllum saxifrages, f. 92 (1986).

Leafy shoots forming a low cushion or mat. Leaves 2–4 × 0.5–1.5 mm, oblong, margin fringed with tooth-like hairs from near the tip to the base, lower surface with a prominent keel, green. Petals 4–5 mm, obovate, not recurved, more or less touching, white. *Sino-Himalayan area.* H2? Late spring.

Prefers an acid soil; best in a cool glasshouse.

73. S. lolaensis H. Smith. Illustration: Horny et al., Porophyllum saxifrages, f. 93 (1986).

Leafy shoots forming a dense cushion. Leaves *c.* 3 × 1.5 cm, oblong to elliptic, margin fringed with tooth-like hairs from near the tip to the base, or hairless. Petals *c.* 3.5 mm, broadly obovate to nearly circular, base contracted to a short claw, sometimes irregularly and shallowly lobed, recurved, overlapping, creamy-white. *China (SE Xizang), Nepal?* H2? Late winter–early spring.

Prefers an acid soil; best grown in a cool glasshouse.

74. S. matta-florida H. Smith. Illustration: Horny et al., Porophyllum saxifrages, f. 89 (1986).

Leafy shoots forming a dense, compact cushion. Leaves 4–6 × 1.5–2 mm, linear to oblong, margin more or less entire except possibly for a few minute teeth. Petals 5–6 mm, broadly obovate, base tapered to a short claw, recurved, overlapping, white. *Bhutan, China (SE Xizang).* H2? Spring.

Prefers an acid soil; best grown in a cool glasshouse.

75. S. hypostoma H. Smith. Illustration: Evans, Alpines '81, 27 (1981); Horny et al., Porophyllum saxifrages, f. 88 (1986).

Leafy shoots forming a dense, compact cushion. Leaves *c.* 4 × 1.5 mm, linear to linear-obovate, tip margin fringed with tooth-like hairs, base entire, green. Petals *c.* 4 mm, broadly obovate, tapered to a short claw, margin wavy, recurved, overlapping, white, set round a glistening green ovary. *Nepal.* H2? Spring.

Prefers an acid soil; best grown in a cool glasshouse.

76. S. lowndesii H. Smith. Illustration: Horny et al., Porophyllum saxifrages, f. 95 (1986); Quarterly Bulletin of the Alpine Garden Society **58**(1): 39 (1990); Harding, Saxifrages, 54 (1992).

Leafy shoots forming a loose cushion.

Leaves 5–7 × 2–2.5 mm, linear-obovate to spathulate, soft, obtuse but not markedly thickened at the tip, margin sparingly fringed, base with a few short, glandular-hairs, shiny green. Petals *c.* 7 mm, nearly circular, base tapered to a short claw, overlapping, bright rose-lilac. Ventral walls of the carpels united only at the base (rather than to at least half-way). *Nepal.* H2? Spring.

Subsection **Engleria** (Sündermann) Gornall. Leaves alternate; margin with a translucent border. Flowering-stems distinct, bearing an inflorescence with coloured bracts and sepals. Flowers broadly ellipsoid, with erect sepals which usually exceed and largely hide the petals. Petals pink to purple, white or yellow, the margins fringed with hairs toward base. Ovary ⅔ or more inferior.

77. S. media Gouan (*S. calyciflora* Lapeyrouse). Illustration: Botanical Magazine, 7315 (1893); Quarterly Bulletin of the Alpine Garden Society **23**(1): 39 (1955); Horny et al., Porophyllum saxifrages, f. 48 (1986); Webb & Gornall, Saxifrages of Europe, pl. 15 (1989).

Leafy shoots terminating in prominent rosettes, arranged into a cushion. Leaves 6–17 × 2–4 mm, linear-oblong to oblanceolate, acute, slightly glaucous; margin entire; chalk-glands 7–17, producing a calcareous encrustation of variable density. Flowering-stems 3–12 cm, reddish pink, sometimes with a green tinge. Flowers 2–12, all stalked, in a raceme-like or sparingly branched panicle. Petals 3.5–4 mm, obovate to nearly circular, bright pinkish purple. *Pyrenees.* H2. Late spring.

Prefers an alkaline soil; best grown in a cool glasshouse.

78. S. stribrnyi (Velenovsky) Podpera. Illustration: Botanical Magazine, 8946 (1913); Horny et al., Porophyllum saxifrages, f. 55–58 (1986); Webb & Gornall, Saxifrages of Europe, pl. 16 (1989).

Like *S. media* but with larger, more spathulate leaves, and a more freely branched panicle (with 3 or 4 flowers on each primary branch, rather than the 1 or 2 in *S. media*). *Balkans.* H2. Spring.

Prefers an alkaline soil; best grown in a cool glasshouse.

79. S. corymbosa Boissier (*S. luteoviridis* Schott & Kotschy). Illustration: Horny et al., Porophyllum saxifrages, f. 53–54 (1986); Webb & Gornall, Saxifrages of Europe, pl. 18 (1989).

Like *S. media* but flowers in a crowded cyme; petals pale greenish yellow. *SE Europe, Balkans, Turkey.* H2. Spring.

Prefers an alkaline soil.

80. S. porophylla Bertoloni. Illustration: Horny et al., Porophyllum saxifrages, f. 52 (1986); Webb & Gornall, Saxifrages of Europe, f. 20 (1989); Harding, Saxifrages, 35 (1992).

Leafy shoots short, forming distinct rosettes aggregated into a cushion. Leaves 4–10 × 2–3 mm, obovate-spathulate to oblong-oblanceolate, tip obtuse or with a short point, glaucous; margin entire; chalk-glands 5–11, producing a calcareous encrustation of variable density. Flowering-stems 3–8 cm, pink to bright rose-pink. Flowers 4–7, rarely to 12, at least some without stalks, in a slender raceme. Petals *c.* 1.5 mm, obovate, white, sometimes with a pink stripe, becoming pink with age. *Italy (Apennines).* H2. Late spring.

Prefers an alkaline soil.

81. S. federici-augusti Biasoletto. Illustration: Botanical Magazine, 8308 (1910); Horny et al., Porophyllum saxifrages, f. 49–51 (1986); Webb & Gornall, Saxifrages of Europe, pl. 14 (1989); Harding, Saxifrages, 18, 35 (1992).

Like *S. porophylla* but leaves larger, 1.2–3.5 × 0.3–0.8 cm; inflorescence cherry-red to dark purple-crimson, bearing 10–20 flowers with purplish pink petals. *Balkan Peninsula.* H2. Spring.

Subsp. **federici-augusti** (*S. montenegrina* Engler & Irmscher). Basal-leaves to 1.9 cm, obovate-lanceolate; inflorescence dark purple-crimson, with up to 16 flowers.

Subsp. **grisebachii** (Degen & Dörfler) D.A. Webb (*S. grisebachii* Degen & Dörfler). Basal-leaves to 3.5 cm, usually spathulate; inflorescence crimson to cherry-red, with up to 20 flowers.

Prefers an alkaline soil.

82. S. sempervivum C. Koch. Illustration: Köhlein, Saxifrages and related genera, t. 23 (1984); Horny et al., Porophyllum saxifrages, f. 59–62 (1986).

Like *S. porophylla* but leaves, 5–20 × 1–2 mm, narrower, linear to linear-oblong, apex never widened; flowering-stems 6–20 cm, with 10–20 flowers on well-grown plants; petals reddish purple. *Balkan Peninsula, Turkey*. H2. Spring.

Very narrow-leaved variants have been distinguished as *S. thessalica* Schott or, in gardens, as forma **stenophylla** Boissier. All intermediates occur, however. An albino variant with a pale green inflorescence is in cultivation under the name 'Zita'. Prefers an alkaline soil; best grown in a cool glasshouse.

Subsection **Oppositifoliae** Hayek. Leaves usually opposite, margin without a translucent border. Flowering-stems very short or obsolete. Flowers in small cymes of 1–3, the sepals suberect or spreading, shorter than and not concealing the petals. Petals deep purple, pink or white. Ovary inferior or nearly so.

Series **Oppositifoliae** (Hayek) Gornall. Leafy stem not perfoliate, the leaf-pairs with margins meeting at an acute angle, with tooth-like hairs at the base. Chalk-glands usually 3–7 (rarely solitary). Flowers 1–3 per stem.

83. S. oppositifolia Linnaeus. Illustration: Ross-Craig, Drawings of British plants, **10**: pl. 15 (1957); Köhlein, Saxifrages and related genera, t. 8 (1984); Webb & Gornall, Saxifrages of Europe, pl. 17, f. 21 (1989); Harding, Saxifrages, cover, 99 (1992).

Stems branched, forming a mat or loose cushion of leafy shoots. Leaves mostly 2–5 × 1.5–2 mm, elliptic or obovate, tip almost acute to rounded, margins entire but with bristly hairs at least near the base; chalk-glands 1–5, producing a calcareous encrustation variable in its intensity. Flowering-stems 1–2 cm, with a solitary flower. Petals mostly 5–12 × 2–7 mm, obovate to elliptic-oblong, pink to deep purple (rarely white), fading to violet. *Circumboreal, extending to the Altai & Himalaya*. H1. Spring–summer.

The species is very variable and has been split into a large number of somewhat ill-defined subspecies and varieties. The following subspecies are the chief sources of garden plants.

Subsp. **oppositifolia**. Leafy shoots forming a mat. Leaves usually oblong to narrowly obovate, hairy up to the tip. Flowering-stems well-developed. Hairs on the sepals usually non-glandular. Petals usually 7–12 mm. *Circumpolar*.

Plants with glandular-hairy sepals have been distinguished as *S. latina* Terraciano (with densely overlapping leaves) or *S. murithiana* Tissière, but intermediates to subsp. *oppositifolia* always occur.

Subsp. **rudolphiana** (Koch) Engler & Irmscher. Leafy shoots short, dense, forming a tight, flattish cushion. Leaves less than 2 mm long, oblong-obovate, hairy in the basal half only. Flowering-stems very short, nearly obsolete. Sepals with glandular hairs. Petals 5–7 mm. *Alps*.

Subsp. **blepharophylla** (Hayek) Engler & Irmscher. Leafy shoots columnar, forming a compact cushion. Leaves 3–5 mm, broadly obovate, hairy up to the tip. Flowering-stems *c*. 5 mm. Sepals with long, non-glandular hairs. Petals 5–8 mm. *Austrian Alps*.

Subsp. **speciosa** (Dörfler & Hayek) Engler & Irmscher. Leafy shoots forming a moderately compact cushion. Leaves 4–5 mm, broadly obovate to nearly circular, hairy in the basal half. Flowering-stems less than 5 mm. Sepals with non-glandular hairs. Petals 8–12 mm. *Italy (C Apennines)*.

Usually prefers a slightly alkaline soil in partial shade.

84. S. retusa Gouan (*S. purpurea* Allioni). Illustration: Webb & Gornall, Saxifrages of Europe, pl. 21 (1989).

Like *S. oppositifolia* but with shiny leaves, often more than one flower per stem, sepals lacking marginal hairs. *Pyrenees, Alps, Tatra, Carpathians & Rila mountains*. H2. Late spring–summer.

Subsp. **retusa**. Flowering-stems to 1.5 cm, bearing 1–3 flowers with hairless perigynous zones.

Subsp. **augustana** (Vaccari) Fournier. Flowering-stems to 5 cm, bearing 3–5 flowers with hairy perigynous zones.

The former subspecies prefers an acid soil, whereas the latter prefers an alkaline one; both do better under glass.

85. S. biflora Allioni (*S. kochii* Bluff, Nees & Schauer; *S. macropetala* A. Kerner). Illustration: Webb & Gornall, Saxifrages of Europe, pl. 20 (1989); Harding, Saxifrages, 100 (1992).

Like *S. oppositifolia* but with soft, fleshy, nearly circular leaves; petals widely separated, dull purple or white; prominent yellow nectary disc in the centre of flower. *Alps, NW Greece*. H2. Summer.

Subsp. **biflora**. Leaves with 1 chalk-gland and negligible calcareous encrustation. Flowers usually at least 2 per stem, barely raised above the leaves. Sepals hairy beneath.

Subsp. **epirotica** D.A. Webb. Leaves with 3–5 chalk-glands and prominent calcareous encrustation. Flowers solitary, well above the leaves. Sepals hairless beneath.

Best grown in a cool glasshouse.

86. S. vacillans H. Smith. Illustration: Horny et al., Porophyllum saxifrages, f. 84 (1986).

Leafy shoots forming a fairly compact cushion. Leaves 3.5–5 × 1.5–2 mm, linear, acute, margin fringed with small tooth-like hairs at the base, mostly opposite but alternate on vigorous shoots, with 3–7 chalk-glands. Flowering-stems 1.5–2 cm, bearing 1–2 flowers. Petals 5–6 mm, obovate, margin with irregular, shallow lobes, white. *Bhutan*. Late spring.

Doubtful whether in cultivation.

Series **Tetrameridium** (Engler) Gornall. Leafy stem perfoliate, with margins of the leaf-pairs completely confluent, entire, lacking basal hairs, usually with a solitary chalk-gland. Flowers solitary.

87. S. quadrifaria Engler & Irmscher. Illustration: Horny et al., Porophyllum saxifrages, f. 79 (1986).

Leafy shoots forming a dense, compact cushion. Leaves 2–2.5 × 1–1.5 mm, ovate, thickened at the somewhat truncated tip, green. Flowers almost stemless. Petals 3–4.5 mm, obovate, more or less touching, white. *China (SE Xizang), Nepal*. H2? Spring.

Prefers an acid soil in partial shade; best grown in a cool glasshouse.

88. S. georgei Anthony. Illustration: Horny et al., Porophyllum saxifrages, f. 81 (1986); Harding, Saxifrages, 100 (1992).

Leafy shoots forming a loose cushion. Leaves 3–3.5 × 2–3 mm, oblong-obovate, tip obtuse, light green. Flowers stemless or nearly so, 4–5-parted. Petals *c*. 5 mm, broadly obovate, base tapered to a short claw, barely touching, white or pale pink. *Nepal, Bhutan, China (Sichuan, NW Yunnan, SE Xizang)*. H2? Spring.

Best grown in a cool glasshouse.

89. S. alpigena H. Smith. Illustration: Horny et al., Porophyllum saxifrages, f. 82 (1986).

Like *S. georgei* but leaves 1.5–2.5 × 1–2 mm, ovate to obovate. Flowers 5-parted. Petals *c*. 6 mm, obovate, base tapered to a long claw, white. *Nepal*. H2? Spring.

Prefers an acid soil; best grown in a cool glasshouse.

Section **Ligulatae** Haworth. Evergreen perennials. Stems somewhat woody at the base, producing leaf-rosettes that aggregate to form cushions or mats. Leaves alternate, somewhat fleshy, hard or leathery, usually glaucous, without a distinct stalk, margins entire or finely toothed, often with a translucent border. Chalk-glands located in pits, nearly always evident by the conspicuous calcareous encrustation on the leaves. Flowering-stems leafy, produced from monocarpic leaf-rosettes (sometimes the entire plant is monocarpic), bearing an often large and many-flowered panicle. Petals usually white, often with red spots, rarely pink, yellow or orange. Ovary at least half-inferior; capsule splitting above the middle. The rosette bearing the flowering-stem dies after flowering but, in those species which are not monocarpic, offsets arise from the base of the flowering rosette. The species and their hybrids are well known to gardeners as the 'silver' or 'encrusted' saxifrages; hybrids and cultivars are treated in more detail by Köhlein (1984) and Harding (1992).

90. S. longifolia Lapeyrouse. Illustration: Schacht, Rock gardens and their plants, 125 (1960); Köhlein, Saxifrages and related genera, t. 18, 19 (1984); Webb & Gornall, Saxifrages of Europe, f. 22, 33 (1989).

Stem normally unbranched, producing a single leaf-rosette to 15 cm across, with up to 200 living leaves. Leaves 6–11 × 0.4–0.7 cm, linear, sometimes slightly expanded below the acute tip, margin entire with the numerous chalk-glands set into it and not on the upper surface. Flowering-stem to 60 cm, branched from near the base to form a cylindric to conical panicle with up to 800 flowers, the primary branches bearing 4–10 flowers. Petals *c.* 7 mm, obovate, white, sometimes with crimson spots. Ovary inferior. The species is strictly monocarpic; during flowering the leaves die, and by the time seed is ripe, the plant is dead. *Pyrenees, locally in E Spain, Morocco (High Atlas).* H2. Summer.

Hybridises in the wild with *S. cotyledon*: **S. × superba** Rouy & Camus, (*S. × imperialis* invalid; *S. × splendida* invalid) and *S. paniculata*: **S. × lhommei** Coste & Soulié. Garden hybrids with *S. callosa* (*S. × calabrica* invalid) and *S. cochlearis* are grown. That with the former is known as 'Tumbling Waters', that with the latter may include 'Dr Ramsey' and 'Francis Cade', but there is little firm evidence.

Prefers an alkaline soil in partial shade; grow in a vertical crevice for full effect.

91. S. callosa Smith (*S. lingulata* Bellardi; *S. lantoscana* Boissier & Reuter). Illustration: Botanical Magazine, 8434 (1912); Köhlein, Saxifrages and related genera, t. 14 (1984); Webb & Gornall, Saxifrages of Europe, f. 24 (1989).

Leaf-rosettes formed in clumps, offsets usually present, the larger rosettes to 16 cm across. Leaves 4–9 × 0.25–0.7 cm, linear, sometimes expanded near the tip, hard; margin entire except for a few non-glandular hairs near base, numerous chalk-glands set into it and not on the upper surface, calcareous encrustation variable. Flowering-stem 15–40 cm, bearing a many-flowered, narrow panicle which occupies 40–60% of the stem, primary branches bearing 3–7 flowers. Petals 6–12 mm, obovate to oblanceolate, sometimes with a long claw, white, sometimes with crimson spots near the base. Ovary about ¾ inferior. *NE Spain, SW Alps, Apennines to S Italy, Sicily & Sardinia.* H2. Late spring–summer.

Subsp. **callosa**. Inflorescence hairless or sparingly glandular-hairy. Plants with leaves linear, scarcely expanded near the tip are known as var. **callosa**. *SW Alps, N Italy.*

Those with leaves oblanceolate to linear, with a more or less diamond-shaped tip are called var. **australis** (Moricand) D.A. Webb. *SW Alps, C & S Italy, Sicily, Sardinia.*

Plants from the western Maritime Alps are commonly found in gardens under the name var. *lantoscana*, but they scarcely differ from var. *australis*.

Subsp. **catalaunica** (Boissier & Reuter) D.A. Webb. Inflorescence fairly densely glandular-hairy. Leaves oblanceolate, fairly short. *NE Spain.*

Hybrids with *S. cotyledon*: **S. × macnabiana** Lindsay (*S. × lindsayana* Engler & Irmscher); *S. cochlearis* (*S. × farreri* invalid) and *S. hostii* (*S. × florairensis* invalid?) have been recorded. The artificial hybrid *S. callosa* × *S. paniculata* probably includes the commonly-grown cultivar 'Kathleen Pinsent', and see *S. longifolia* for a hybrid with that species. Prefers an alkaline soil.

92. S. cochlearis Reichenbach. Illustration: Botanical Magazine, 6688 (1883); Köhlein, Saxifrages and related genera, t. 15 (1984); Finkenzeller & Grau, Alpenblumen, 101 (1985); Webb & Gornall, Saxifrages of Europe, f. 25 (1989).

Leaf-rosettes produced in a dense, irregular cushion. Leaves to 45 × 7 mm, usually much less, spathulate to oblanceolate, leathery and fleshy, base often tinged red; margin entire, numerous chalk-glands set into it and not on the upper surface, base with a few hairs. Flowering-stems 5–30 cm, reddish, branched in the upper ⅓ to form an open, densely glandular-hairy (at least in the lower ½) panicle of 15–25 flowers, rarely to 60. Petals 7–11 mm, oblong-obovate, white, sometimes with crimson spots near the base. Ovary ¾ inferior. *Maritime Alps, Italy (Portofino Peninsula).* H2. Late spring–summer.

A commonly grown dwarf variant is known by the cultivar name 'Minor'. The cultivar 'Major', also known by the invalid name *S. cochleata*, may be of hybrid origin.

S. × burnatii Sündermann (*S. cochlearis* × *S. paniculata*) has been recorded once from the Maritime Alps; 'Esther' and 'Whitehill' are commonly grown cultivars probably of this parentage. For hybrids with *S. callosa* and *S. longifolia* see under those species. Prefers an alkaline soil.

93. S. crustata Vest (*S. incrustata* invalid; *S. vochinensis* invalid). Illustration: Finkenzeller & Grau, Alpenblumen, 103 (1985); Webb & Gornall, Saxifrages of Europe, f. 26 (1989).

Leaf-rosettes 2.5–8 cm across, crowded to form a thick cushion. Leaves usually 10–25 × 2–3 mm, linear, tip obtuse and scarcely expanded, margin with a few long hairs at the base, entire or very slightly scalloped; chalk-glands numerous, sunk into the margin. Flowering-stems 12–35 cm, glandular-hairy, branched from the middle or above to form a panicle of up to 35 flowers, each primary branch usually with 1–3 flowers aggregated near the tip. Petals 5–6 mm, obovate, white, rarely with red spots. Ovary inferior. *E Alps, N & C former Yugoslavia.* H2. Summer.

Natural hybrids with *S. hostii* (**S. × engleri** Huter; *S. × paradoxa*, invalid) and *S. paniculata* (**S. × pectinata** Schott, Nyman & Kotschy; *S. × fritschiana* invalid; *S. × portae* invalid) are known. Prefers an alkaline soil.

94. S. hostii Tausch (*S. elatior* Mertens & Koch). Illustration: Webb & Gornall, Saxifrages of Europe, f. 27, 28 (1989).

Leaf-rosettes 6–18 cm across, forming a loose mat. Leaves 3–10 × 0.4–1 cm, oblong to broadly linear, tip scarcely expanded, more or less glaucous; margins finely toothed, with the chalk-glands situated on the upper surface of the minute marginal teeth, fringed with a few long hairs at the base. Flowering-stems 25–50 cm, glandular-hairy above, branched in the upper ½ to form a panicle, the primary branches of which usually bear 5–12 flowers. Petals 4–8 mm, elliptic-obovate, white often with purple-red spots. Ovary inferior. *E Alps*. H2. Late spring–summer.

Subsp. **hostii** (*S. hostii* var. *altissima* (Kerner) Engler & Irmscher). Basal leaves to 10 cm, at least 6 mm across, tip obtuse, marginal teeth distinct. Primary branches of panicle with at least 5 flowers. *NE Italy to SE Austria & NW Slovenia*.

Subsp. **rhaetica** (Kerner) Braun-Blanquet. Basal leaves to 5 × 0.7 cm, tapered to an acute tip, marginal teeth obscure. Primary branches of panicle with 3–5 flowers. *N Italy*.

Natural hybrids with *S. paniculata* (**S. × churchillii** Huter) occur; for those with *S.callosa* and *S. crustata* see under those species. Prefers an alkaline soil.

95. S. paniculata Miller (*S. aizoon* Jacquin; *S. recta* Lapeyrouse; *S. zelebori* invalid?). Illustration: Finkenzeller & Grau, Alpenblumen, 103 (1985); Webb & Gornall, Saxifrages of Europe, f. 29 (1989).

Leaf-rosettes 1–9 cm across, aggregated into a loose cushion. Leaves mostly 8–35 × 4–5 mm, obovate-oblong to broadly linear, tip obtuse to acute or acuminate, stiff and rather fleshy; margin with forward-pointing teeth, these grading into hairs at the base, chalk-glands situated on the upper surface at the base of each tooth. Leaves tending to curve upwards making the rosette hemispherical rather than flat as in other species. Flowering-stems 6–40 cm, variably glandular-hairy, branched in the upper half to form a narrow panicle, the primary branches of which usually bear 1–3 flowers. Petals 3–6 mm, elliptic-obovate, white, rarely pink or pale yellow, often with reddish purple spots. Ovary ¾ inferior. *E North America, Greenland, Iceland, S Norway, N Spain through C & S Europe to Caucasus*. H2. Late spring–summer.

A very variable species but one in which most of the variation is continuous. Populations from the Caucasus and neighbouring regions, however, have acuminate leaf-tips and sometimes pink or crimson petals, and these are designated as subsp. **cartilaginea** (Willdenow) D.A. Webb (*S. kolenatiana* Regel; *S. sendtneri* invalid). Populations from the Balkans

with a rather similar leaf-shape have been assigned to *S. aizoon* var. *orientalis* Engler. Plants from Monte Baldo, Italy, with neat, crowded rosettes of small leaves (*c.* 5 × 1.5 mm) and reddish flowering-stems are known to gardeners as *S. aizoon* var. *baldensis* Farrer.

There is a host of other variants in gardens to which many names have been given; taken in the context of the variation pattern in the wild, however, it is not practical to give any of them recognition, except possibly as cultivars. Accounts of a representative selection are given by Harding (1992). For hybrids with *S. callosa, S. cochlearis, S. crustata, S. cotyledon, S. hostii, S. longifolia, S. aizoides* and *S. cuneifolia* see under those species. **S. × andrewsii** Harvey (possibly *S. paniculata* × *S. spathularis*; *S. × guthrieana* invalid) looks very like a hybrid between sections *Ligulatae* and *Gymnopera*. Its origin is obscure but was probably made accidentally in an Irish garden. It is more or less intermediate between the parents, although the leaves are longer and narrower than might be expected. The species is variable in its preference for acidic or alkaline soils.

96. S. cotyledon Linnaeus (*S. pyramidalis* Lapeyrouse; *S. montavoniensis* Kolb; *S. nepalensis* misapplied; *S. linguaeformis* invalid). Illustration: Quarterly Bulletin of the Alpine Garden Society **22**(1): 50 (1954); Köhlein, Saxifrages and related genera, t. 16 (1984); Webb & Gornall, Saxifrages of Europe, f. 30, 31 (1989); Harding, Saxifrages, 117 (1992).

Principal leaf-rosette 7–12 cm across, usually accompanied by smaller rosettes arising from short axillary runners which soon die, so that the plant never forms large clumps or cushions. Leaves 2–8 × 0.6–2 cm, oblong to oblanceolate or spathulate, margin finely and regularly toothed except at the base where it is fringed with non-glandular hairs, slightly fleshy, not very glaucous; chalk-glands situated in the middle of each marginal tooth on the upper surface, bearing a small amount of calcareous encrustation. Flowering-stems to 70 cm, branched from near the base, or at least from below the middle, to form a pyramidal panicle; primary branch with 8–40 flowers. Petals 7–10 mm, oblanceolate, narrowed at the base to a claw, white, sometimes with red spots or veins. Ovary inferior. *Scandinavia & Iceland, Pyrenees, Alps*. H2. Summer.

Attempts to give taxonomic recognition

to populations from the 3 areas listed above cannot be justified, although cultivar status may be appropriate if desired, e.g. 'Norvegica', 'Icelandica' and 'Caterhamensis'.

A natural hybrid with *S. paniculata* (**S. × gaudinii** Brügger; *S. × timbalii* Rouy & Camus; *S. canis-dalmatica* invalid; *S. × speciosa* invalid) is known. For hybrids with *S. callosa* and *S. longifolia* see under those species. The cultivar 'Southside Seedling', whose petals are blotched with crimson, is possibly a hybrid of *S. cotyledon*. Prefers acidic soils, but will grow in moderately calcareous conditions; probably better outside than under glass.

97. S. valdensis de Candolle (*S. compacta* Sternberg; *S. rupestris* Seringe). Illustration: Webb & Gornall, Saxifrages of Europe, f. 32 (1989).

Leaf-rosettes 1–3 cm across, aggregated into a dense, hard, mounded cushion. Leaves 3–8 × 2–3 mm, obovate to oblanceolate or spathulate, tip obtuse and somewhat curved downwards, margin entire except for a few hairs fringing the base and lacking a distinct translucent border, markedly glaucous, some chalk-glands set in the margin and others scattered on the upper surface. Flowering-stems 3–11 cm, branched above the middle to form a panicle of 6–12 flowers. Petals 4–5 mm, obovate, white, without spots. Ovary inferior. *SW Alps*. H2. Summer.

Usually prefers alkaline soils.

98. S. mutata Linnaeus. Illustration: Webb & Gornall, Saxifrages of Europe, f. 33 (1989); Harding, Saxifrages, 93 (1992).

Leaf-rosettes 5–15 cm across, often solitary, but sometimes with a few offsets. Leaves 2.5–7 × 0.7–1.5 cm, oblong–oblanceolate, tip obtuse, margin irregularly toothed in the middle section, with a prominent translucent border, base fringed with hairs; somewhat fleshy; chalk-glands situated on the upper surface inside the translucent border, but producing little, if any, calcareous encrustation. Flowering-stem 10–50 cm, bearing a narrow panicle with a very variable number of flowers. Petals 6–8 mm, linear, acute at the tip. Ovary ¾ inferior. *Alps, S Carpathians, Low Tatra Mts*. H2. Summer–early autumn.

Subsp. **mutata** has the flowering-stem branched from the middle or above. *Alps, Low Tatra Mts*.

Subsp. **demissa** (Schott & Kotschy) D.A. Webb has the flowering-stem branched from or near the base. *S Carpathians*.

For hybrids with *S. caesia* and *S. aizoides* see under those species. Prefers an alkaline soil with a good supply of water.

99. S. florulenta Moretti. Illustration: Botanical Magazine, 6102 (1874); Rasetti, I fiori delle Alpi, pl. 263–264 (1980); Heath, Collectors' alpines, their cultivation in frames and alpine houses, 441 (1981); Harding, Saxifrages, 118 (1992).

Monocarpic perennial. Leaf-rosette 5–15 cm across, nearly always solitary. Leaves 3–6 × 0.4–0.7 cm, narrowly spathulate, with a short, spine-like tip; margin with a conspicuous translucent border, entire near the tip, becoming irregularly scalloped below, basal half fringed with tooth-like hairs; chalk-glands near the margin on the upper surface, producing little, if any, calcareous encrustation. Flowering-stem 10–25 cm, branched from near the base to form a dense cylindric panicle of numerous flowers. Petals 5–7 mm, oblanceolate, flesh-pink. Ovary inferior; carpels 3, rather than the 2 found in nearly all other species. *Maritime Alps*. H2. Summer.

Prefers a very well-drained acid soil in partial shade; difficult to grow.

Section **Xanthizoon** Grisebach. Evergreen perennial, forming a mat or cushion. Leaves fleshy, narrow, without a distinct stalk, margin more or less entire. Chalk-glands flush with the surface, not in pits, often without any apparent calcareous encrustation. Flowering-stems leafy, bearing flowers in a loose cyme. Petals yellow, orange or reddish. Ovary ½ inferior; capsule splitting above the middle.

100. S. aizoides Linnaeus (*S. autumnalis* Jacquin; *S. atrorubens* Bertoloni; *S. crocea* Gaudin). Illustration: Ross-Craig, Drawings of British Plants, **10**: pl. 14 (1957); Webb & Gornall, Saxifrages of Europe, pl. 22, 24 (1989).

Leafy shoots forming a thick mat or loose cushion. Leaves 4–22 × 1.5–4 mm, linear to oblong, tips obtuse, acute or with a short point; medium to darkish green, not glaucous; margin entire or occasionally with 2 short teeth near the tip, usually fringed with forward-pointing tooth-like hairs; chalk-glands usually solitary, near the tip. Flowers solitary or in short, leafy cymes of 2–15. Petals 3–7 mm, usually yellow, often with orange spots, sometimes orange or brick-red. Ovary ½ inferior, covered on top by a prominent nectary disc. *Arctic-alpine in Europe & North America*. H1. Summer–early autumn.

Var. **atrorubens** (Bertoloni) Sternberg with brick-red petals and deep crimson nectary disc is commonly grown.

The cultivar 'Primulaize' is traditionally supposed to be *S. aizoides* × *S. umbrosa* 'Primuloides'; **S. × larsenii** Sündermann (*S. aizoides* × *S. paniculata*) has been made artificially. Natural hybrids with *S. mutata* (**S. × hausmannii** Kerner; **S. × regelii** Kerner), *S. caesia* (**S. × patens** Gaudin) and *S. squarrosa* are also recorded. The species prefers neutral to alkaline, moist but well-drained soils.

Section **Trachyphyllum** (Gaudin) W.D.J. Koch. Evergreen perennials, forming loose cushions or mats. Leaves narrow, usually lanceolate, without a distinct stalk, stiff, margin entire or 3-lobed at tip, the latter with a short point, fringed with stout, often hooked, hairs. Chalk-glands absent. Axillary buds often prominent but not summer-dormant. Flowering-stems leafy, bearing 1–several flowers in a cyme. Petals white to pale yellow, sometimes with reddish spots or a yellow patch at the base. Ovary superior or nearly so; capsule splitting above the middle.

101. S. aspera Linnaeus (*S. hugueninii* Brügger; *S. etrusca* Pignatti). Illustration: Heathcote, Plants of the Engadine, pl. 91 (1891).

Leafy shoots forming an irregular mat, with nearly spherical leaf-rosettes. Leaves of prostrate shoots 5–8 mm, pressed close to the stem at first but spreading later; axils with conspicuous leafy buds at the time of flowering, which are shorter than the subtending leaf. Flowering-stems 7–22 cm, usually bearing 2–7 flowers in an open cyme. Petals 5–7 mm, oblong, base narrowed to a very short claw, white or pale cream, often with a deep yellow patch at the base and reddish spots near the middle. *Pyrenees, Alps, N Apennines*. H2. Summer.

Probably best grown under glass to avoid damage by birds.

102. S. bryoides Linnaeus. Illustration: Webb & Gornall, Saxifrages of Europe, pl. 23, f. 34 (1989).

Leafy shoots forming a dense mat or low cushion, with nearly spherical leaf-rosettes. Leaves of prostrate shoots 3.5–8 × 1– 1.5 mm, oblong-lanceolate, incurved and remaining so, tip with a short point, surface shiny; axils with leafy buds at time of flowering, which equal or exceed the subtending leaf. Flowering-stems 2–5 cm, bearing a solitary flower. Petals 5–7 mm,

elliptic-oblong to obovate, white with a large patch of deep yellow at the base and some reddish spots near the middle. *High mountains of Europe*. H2. Summer–early autumn.

103. S. bronchialis Linnaeus (*S. spinulosa* Adams). Illustration: Hultén, Flora of Alaska and neighbouring territories, 570 (1968).

Leafy shoots forming a thick mat or low cushion. Leaves usually 5–12 × 1–2.5 mm, narrowly oblong to oblong-lanceolate, tip tapered to a white spine; axils with leafy buds, inconspicuous at flowering. Flowering-stems 5–20 cm, bearing 2–12 flowers in a fairly compact cyme. Petals 3.5–8 mm, oblong-elliptic, yellowish white, variously spotted with yellow or crimson. *Arctic-alpine areas from the Urals eastwards to the Rocky Mts*. H1. Summer.

104. S. cherlerioides D. Don (*S. stelleriana* Merk). Illustration: Hultén, Flora of Alaska and neighbouring territories, 570 (1968).

Like *S. bronchialis* but shoots columnar, leafy; leaves incurved, overlapping, tip mucronate, marginal hairs without glands; flowering-stems 2–6.5 cm. *Aleutian Islands, Kamchatka, E Siberia*. H2. Summer.

105. S. tricuspidata Rottböll. Illustration: Hultén, Flora of Alaska and neighbouring territories, 571 (1968).

Leafy shoots forming a mat or low cushion. Leaves 6–15 × 1.5–6.5 mm, oblanceolate to obovate, tip usually with 3 teeth; axillary buds inconspicuous. Flowering-stems 4–24 cm, bearing an open cyme. Petals 4–7 mm, elliptic, base truncate, white or cream, with yellow, orange and red spots in sequence from the base upwards. *Arctic-alpine North America, Greenland*. H2. Summer.

Section **Gymnopera** D. Don. Evergreen perennials with rather succulent, leathery, stalked leaves in basal rosettes, margins toothed or scalloped. Chalk-glands absent. Flowering-stem leafless, terminating in a many-flowered panicle. Sepals reflexed. Petals white, usually with yellow and reddish pink spots. Ovary superior; capsule splitting above the middle.

106. S. cuneifolia Linnaeus. Illustration: Köhlein, Saxifrages and related genera, t. 9 (1984); Webb & Gornall, Saxifrages of Europe, f. 9, 10 (1989).

Stems prostrate, runner-like, producing leaf-rosettes, ultimately forming a mat. Leaf-blade 0.8–2.5 × 0.7–2.2 cm, usually wedge-shaped, broadly ovate or nearly circular, apex often truncate; hairless; margin toothed, scalloped or entire, with a translucent border. Leaf-stalk flat, 0.4–1.3 times as long as the blade into which it grades, with a very few hairs at the extreme base. Flowering-stem 10–25 cm. Petals 3–5.5 mm, oblong, white, often with a yellow patch at the base and rarely with reddish spots near the middle. *Pyrenees, Alps, Carpathians.* H2. Late spring–summer.

Subsp. **cuneifolia** (*S. cuneifolia* var. *capillipes* Reichenbach; *S. capillaris* invalid) has stems prostrate, rosettes spaced 3–6 cm apart, leaves small (the largest to 2.5 cm, including the stalk), margin almost entire. Flowering-stem slender, nearly hairless, usually bearing up to 10 flowers. *S France (Maritime Alps), N Italy (Tuscany).*

Subsp. **robusta** D.A. Webb usually has leaf-rosettes less than 2 cm apart on the prostrate stems. Largest leaves more than 2.5 cm (including stalk), margin distinctly scalloped-toothed. Flowering-stem robust, clearly glandular-hairy, usually bearing more than 10 flowers.

Variegated plants of the species are occasionally found in gardens and are best treated as cultivars, e.g. 'Aureo-maculata' and 'Variegata'.

107. S. umbrosa Linnaeus. Illustration: Keble Martin, The new concise British Flora, pl. 32 (1982); Garrard & Streeter, The wild flowers of the British Isles, 91 (1983); Webb & Gornall, Saxifrages of Europe, f. 12 (1989).

Evergreen perennial producing leaf-rosettes from prostrate stems, ultimately forming a low cushion. Leaf-blade 1.5–3 × 1–2 cm, oblong-elliptic, sometimes nearly obovate, usually hairless; margin with 5–10 shallow, forward-pointing scalloped teeth on either side; translucent border 0.2–0.3 mm wide. Leaf-stalk usually 1/3 –1/2 as long as the blade, flat, margin densely glandular hairy. Flowering-stem to 35 cm. Petals *c.* 4 mm, broadly elliptic, tapered to a short claw, white with crimson spots in the middle and 2 yellow ones near the base. *Pyrenees.* H2. Summer.

Var. **hirta** D.A. Webb. Plant dwarf; leaf-blades usually with some hairs. Very similar to those known by the invalid names of *S. umbrosa* var. *primuloides* or *S. umbrosa* var. *minor*, which are sometimes assigned to *S.* × *urbium* (see next entry); of these 'Clarence Elliott' is popular. Best grown in a somewhat sheltered, shady spot in the garden.

108. S. spathularis Brotero (*S. hibernica* Sternberg; *S. serrata* Sternberg). Illustration: Ross-Craig, Drawings of British plants **10**: pl. 5 (1957); Garrard & Streeter, The wild flowers of the British Isles, 91 (1983); Webb & Gornall, pl. 6, f. 13 (1989).

Evergreen perennial producing leaf-rosettes from prostrate, runner-like stems, forming a loosely tufted mat. Leaf-blade 1.5–5 × 1.2–3 cm, circular to oblong-elliptic, hairless; margin with 5–11 teeth on each side (3 or 4 in dwarf plants), teeth usually sharply, but occasionally bluntly, triangular; translucent border *c.* 0.1 mm wide. Leaf-stalk of at least some leaves longer than the blade, flat, with only a few marginal hairs. Flowering-stem to 50 cm. Petals *c.* 5 mm, elliptic, white with several crimson spots near the middle and 2 yellow spots near the base. *SW Europe, Ireland.* H3. Late spring–summer.

S. × **urbium** D.A. Webb (*S. umbrosa* misapplied; *S. spathularis* × *S. umbrosa*). Illustration: Phillips, Wild flowers of Britain, 67 (1977); Stace, New Flora of the British Isles, 384 (1991). This garden hybrid is best distinguished from its parents by the leaf-stalk which is barely longer than the blade, with at least a few hairs on its margins, marginal teeth of the blade projecting more nearly at right angles, and the translucent border of the blade measuring 0.2–0.25 mm in width.

One of the commonest Saxifrages in gardens, often called 'London Pride'. Variegated plants are grown under the names 'Aureo-variegata', 'Aureo-punctata' and 'Variegata', are commonly naturalised or persisting in Britain.

S. × **andrewsii** Harvey, a putative hybrid with *S. paniculata* is grown and is more or less intermediate between the parents. For a hybrid with *S. hirsuta* see under that species.

109. S. hirsuta Linnaeus (*S. geum* misapplied). Illustration: Ross-Craig, Drawings of British plants **10**: pl. 4 (1957); Garrard & Streeter, The wild flowers of the British Isles, 91 (1983). Webb & Gornall, Saxifrages of Europe, f. 14 (1989); Stace, New Flora of the British Isles, 384 (1991).

Evergreen perennial producing sprawling, loose rosettes of leaves from prostrate, rhizome-like stems. Leaf-blade 1.5–4 × 1–5 cm, kidney-shaped to broadly elliptic or circular, hairs at least on the upper surface; margin usually 2–13 scalloped or toothed on either side,

translucent border inconspicuous. Leaf-stalk usually 2–3 times as long as the blade, nearly cylindric, hairy all round. Flowering-stem 12–40 cm. Petals 3.5–4 mm, oblong, white but usually with a yellow patch at the base and a few pink spots near the middle. *SW Europe, SW Ireland.* H3. Late spring–summer.

Subsp. **hirsuta** has a leaf-stalk at least twice as long as the blade, which is kidney-shaped to circular, 0.8–1.2 times as long as wide, and whose margin bears at least 6 crenations or teeth on each side. *Range of species.*

Subsp. **paucicrenata** (Gillot) D.A. Webb has a leaf-stalk only to 1.5 times as long as the blade, which is ovate or elliptic-oblong to nearly circular, 1.1–1.8 times as long as wide, and whose margin bears 2–6 crenations on each side. *Pyrenees.*

S. × **geum** Linnaeus (*S. hirsuta* × *S. umbrosa*). Illustration: Stace, New Flora of the British Isles, 384 (1991). Distinguished from its parents by the leaf-stalk longer than the oblong blade, with hairs only on the margins of the stalk and only very sparsely scattered over the surface of the blade, margin scalloped. *Pyrenees.*

S. × **polita** (Haworth) Link (*S. hirsuta* × *S. spathularis*). Illustration: Stace, New Flora of the British Isles, 384 (1991). Distinguished from its parents by the leaf-stalk longer than the nearly circular blade, hairs scattered sparingly all over the stalk as well as on the surface of the blade, teeth numerous, acutely pointed; translucent border 0.1–0.15 mm. *Ireland, Spain.*

Section **Cotylea** Tausch. Evergreen perennials. Leaves round or kidney-shaped, somewhat fleshy, with distinct stalks, margins scalloped, toothed or slightly lobed. Chalk-glands absent. Flowering-stem leafy, terminating in a many-flowered panicle. Sepals erect or spreading. Petals white, usually with red and yellow spots. Ovary superior; capsule splitting above the middle.

110. S. rotundifolia Linnaeus (*S. repanda* Willdenow; *S. lasiophylla* Schott; *S. olympica* Boissier). Illustration: Botanical Magazine, 424 (1798); Webb & Gornall, Saxifrages of Europe, pl. 7, f. 8 (1989); Harding, Saxifrages, 105 (1992).

Rhizomatous, producing clumps of leafy shoots with leaves in loose rosettes. Leaf-blade 1.7–4.5 × 3–8.5 cm, kidney-shaped to circular, base cordate, margin scalloped, toothed or palmately lobed. Leaf-stalk 4–18 cm. Flowering-stem 15–100 cm. Petals 6–11 × 2.5–5 mm, narrowly oblong to broadly elliptic, contracted to a short claw, white, usually with crimson-purple spots in the middle, orange to yellow at the base. *C & S Europe, SW Asia.* H2. Late spring–summer.

Subsp. **rotundifolia** has a uniformly narrow leaf-stalk and blade with a narrow, translucent border, usually without fringing hairs. *C & S Europe, SW Asia.*

Var. **rotundifolia** has stellate flowers with narrow, spreading petals to 9 mm, lightly spotted.

Var. **heucherifolia** (Grisebach & Schenk) Engler has cup-shaped flowers with wide, more or less erect petals to 11 mm, heavily spotted.

Var. **apennina** D.A. Webb has large basal leaves, more than 7 cm wide; petals 10.5–11.5 mm, spreading, narrow, heavily spotted.

Subsp. **chrysosplenifolia** (Boissier) D.A. Webb (*S. chrysosplenifolia* Boissier) has a leaf-stalk that widens at the top to grade into the blade, which is nevertheless cordate but lacks a translucent border and is usually fringed with hairs. *Aegean & Balkan Peninsula.*

Var. **chrysosplenifolia** has scalloped leaf-margins and its petals lack red spots.

Var. **rhodopea** (Velenovsky) D.A. Webb has finely toothed leaf-margins and petals with red spots.

Best naturalised in woodland, except in very cold climates.

111. S. taygetea Boissier & Heldreich.

Like *S. rotundifolia* but of smaller stature, with smaller leaves, 0.5–1.3 × 0.8–2.3 cm, margin with up to 9 scalloped teeth versus at least 13 in *S. rotundifolia. Balkan Peninsula.* H2. Summer.

Section **Mesogyne** Sternberg. Usually winter-dormant perennials, with bulbils in the axils of the basal leaves, and sometimes also the stem-leaves, or replacing flowers. Basal leaves thin, with a slender leaf-stalk and a semicircular to kidney-shaped blade, palmately divided into 5–11 lobes. Chalk-glands absent. Flowering-stems leafy; flowers solitary or in a small cyme. Petals white or pink. Ovary superior to 1/3 inferior or more; capsule splitting above the middle. All the species are best kept cold and dry in winter (cold-frame) and brought into an alpine house and watered in spring and summer.

112. S. sibirica Linnaeus. Illustration: Köhlein, Saxifrages and related genera, t. 11 (1984); Webb & Gornall, Saxifrages of Europe, pl. 25 (1989); Harding, Saxifrages, 106 (1992).

Stems solitary or in tufts. Basal leaf-blades 0.5–2 × 0.8–3 cm, kidney-shaped, palmately divided into 5–7 broadly ovate, obtuse lobes; stalk long, thin. Flowering-stems 5–18 cm, bearing 2–7 flowers in a compact cyme; stalks to 2.5 cm. Petals 7–14 cm, narrowly obovate, white. Ovary almost superior. *SE Europe, Turkey, Caucasus, C Asia, W Himalaya, S Siberia, China.* H2. Spring–summer.

113. S. cernua Linnaeus. Illustration: Webb & Gornall, Saxifrages of Europe, pl. 27 (1989).

Stems solitary or in tufts. Basal leaf-blades 0.5–1.8 × 0.9–2.5 cm, semicircular to kidney-shaped, divided into 3–7, ovate to oblong, almost acute lobes; stalk 2–3 times as long as blade, base sheathing. Flowering-stem 3–30 cm, usually simple but sometimes branched from near the middle, bearing red to blackish purple bulbils in the axils of the stem-leaves; flowers usually solitary at the ends of each stem, or aborted. Petals 7–12 mm, obovate, white. Ovary almost superior. *Circumboreal arctic-alpine.* H1. Summer.

114. S. rivularis Linnaeus. Illustration: Ross-Craig, Drawings of British Plants 10: pl. 9 (1957); Keble Martin, The new concise British flora, pl. 32 (1982); Garrard & Streeter, The wild flowers of the British Isles, 92 (1983); Webb & Gornall, Saxifrages of Europe, pl. 26 (1989).

Stems in tufts or cushions. Basal axillary bulbils germinating before flowering to form slender runners, which produce new plants at the tips. Basal leaf-blades usually 0.5–1.2 × 0.9–1.7 cm, semicircular or kidney-shaped, divided into 3–7 broadly ovate, obtuse lobes; stalk 2–6 times as long as blade, base sheathing. Flowering-stems 3–15 cm, usually bearing a solitary flower, but sometimes branched from near the middle and then bearing 2–5 flowers on long stalks. Petals 4–5 mm, obovate, white, sometimes tinged with pink. Ovary at least 1/3 inferior. *Circumpolar Arctic, W North America.* H1. Late spring–summer.

Section **Saxifraga**. Usually perennial, occasionally annual or biennial; perennials mostly evergreen, but some summer-dormant perennating by bulbils. Habit varied but often with leafy shoots forming a cushion or mat. Leaves usually rather soft, often lobed or scalloped, usually with a distinct leaf-stalk. Chalk-glands absent. Flowering-stems usually leafy, bearing 1–several flowers in a cyme. Petals usually white, rarely yellowish, pink or red. Ovary 1/2 to fully inferior; capsule splitting above the middle.

Subsection **Saxifraga**. Perennials, more or less summer-dormant, forming loose rosettes of basal leaves, with axillary bulbils, or these in the axils of scales on the short stock; bulbils consisting of an outer series of loosely imbricate, papery scales with marginal or surface hairs, surrounding an inner series of fleshy scales. Leaves ovate to kidney-shaped, variously lobed. Flowering-stems terminal. Petals white, rarely tinged or veined with pink. Ovary 1/2 or more inferior.

115. S. granulata Linnaeus. Illustration: Ross-Craig, Drawings of British Plants 10: pl. 7 (1957); Phillips, Wild flowers of Britain, 31 (1977); Keble Martin, The new concise British flora, pl. 32 (1982); Garrard & Streeter, The wild flowers of the British Isles, 91 (1983).

Plant forming loose rosettes of stalked basal leaves; bulbils clustered. Basal leaf-blades 0.6–3 × 0.8–5 cm, kidney-shaped, base cordate, margin scalloped or toothed, with 5–13 rounded or flat-topped lobes. Leaf-stalk to 5 cm, usually 2–5 times as long as the blade. Flowering-stem usually 10–30 cm, variably branched to form a panicle of 4–30 flowers. Petals 0.7–1.6 cm, obovate to broadly oblanceolate, white, rarely with red veins, hairless. Ovary 3/4 inferior. *Europe (except the SE), N Africa.* H3. Spring–summer.

A variable species, one variant of which, with doubled, sterile flowers, has been grown in gardens since the 17th century and is known as 'Flore Pleno'. For details of the 'mossy' saxifrages, a complex of hybrids between *S. granulata, S. exarata, S. rosacea*, and *S. hypnoides*, see under *S. exarata*. The species is best grown in partial shade in deep soil on the rock garden, or in a meadow.

116. S. corsica (Seringe) Grenier & Godron.

Like a small *S. granulata* but at least some basal leaves deeply 3-lobed. Flowering-stems usually branched, widely spreading, at or from near the base; flower-stalks thread-like, longer than the capsule (unlike *S. granulata* where they mostly equal the capsules). *E Spain, Balearic Islands, Corsica, Sardinia.* H5? Summer.

Subsp. **corsica**. Basal leaves divided to half-way; inflorescence branched at the base. *Corsica, Sardinia.*

Subsp. **cossoniana** (Boissier & Reuter) D.A. Webb. Basal leaves divided to more than half-way; inflorescence branched from above the base. *E Spain.*

Plants from the Balearic Islands (Formentera) are intermediate between the 2 subspecies, but slightly closer to the first.

117. S. haenseleri Boissier & Reuter. Illustration: Webb & Gornall, Saxifrages of Europe, pl. 30 (1989).

Basal leaf-blades 6–10 × 4–9 mm, somewhat wedge-shaped, rather deeply divided into 3–7 oblong, obtuse lobes, tapered to a leaf-stalk to 1.2 cm. Flowering-stem 8–30 cm, branched from the middle or the base, bearing a diffuse panicle of usually 6–12 flowers, but sometimes as many as 40. Petals 5–6 mm, narrowly obovate, white, sometimes with glandular hairs on the upper surface at the base. Ovary inferior. *S Spain.* H5? Spring–early summer.

118. S. dichotoma Willdenow (*S. hervieri* invalid; *S. albarracinensis* Pau). Illustration: Maire, Flore de l'Afrique du Nord **15**: f. 5 (1980).

Basal leaf-blades 0.5–1.8 × 0.7–3 cm, semi-circular to fan-shaped, base truncate to wedge-shaped, divided about ⅔ of the way into 3–7 obtuse lobes, which may themselves be lobed, thus generating 5–15 ultimate segments. Leaf-stalk 1.5–4 times as long as the blade. Flowering-stem 6–25 cm, branched above the middle, bearing a narrow cyme of 2–7 flowers. Petals 5–10 mm, narrowly obovate, glandular-hairy, white, often veined or tinged pink. Ovary ½–¾ inferior. *Spain, Morocco, Algeria.* H5? Spring.

Cultivation difficult.

119. S. carpetana Boissier & Reuter (*S. atlantica* Boissier & Reuter; *S. veronicifolia* Dufour; *S. graeca* Boissier & Heldreich). Illustration: Webb & Gornall, Saxifrages of Europe, pl. 31 (1989).

Basal leaf-blades circular to kidney-shaped, cordate, scalloped, with a long leaf-stalk. Flowering-stems 12–25 cm, branched above the middle to form a compact cyme of 4–13 flowers. Petals 8–12 mm, narrowly obovate, pure white, upper surface glandular-hairy. Ovary ¾ inferior. *Mediterranean.* H5. Late spring–summer.

Subsp. **carpetana.** Later-formed basal leaf-blades ovate, base wedge-shaped or rounded, with a short leaf-stalk. Sepals *c.* 2 mm. *Iberian Peninsula, N Africa.*

Subsp. **graeca** (Boissier & Heldreich) D.A. Webb. All basal leaf-blades like first-formed. Sepals 2.5–3 mm. *Balkans, Italy, Algeria.*

Subsp. *graeca* can be grown in a pan under glass; subsp. *carpetana* is more difficult.

120. S. bulbifera Linnaeus.

Basal leaves like those of *S. granulata* or *S. carpetana* but distingushed by the numerous axillary bulbils present on the flowering-stem, and by the compact inflorescence of usually 3–8 flowers. Petals 6–10 mm, obovate-oblong, upper surface glandular-hairy, white. *C & S Europe.* H4. Late spring–summer.

Best grown in a pan, outside and fairly dry in summer but under glass and rather wetter during autumn and winter.

121. S. biternata Boissier. Illustration: Harding, Saxifrages, 20 (1970); Botanical Magazine, n.s., 670 (1974).

Stems sprawling, woody below, often with persistent leaf-stalks of dead leaves. Leafy shoots densely glandular-hairy, arranged in a diffuse cushion-like habit. Blade of largest leaves *c.* 4 × 4.5 cm, deeply divided into 3 primary lobes, each narrowed to a stalk-like base (the leaf thus appearing compound), primary lobes themselves 3-lobed, with the secondary lobes narrowed at the base to a short stalk, scalloped or divided again into segments at the tip. In a poorly grown plant the blade may consist only of 3 stalked, scalloped lobes. Leaf-stalks with a sheathing base. Flowering-stems to 10 cm, bearing a loose cyme of 2–6 (occasionally to 15) flowers. Petals 1.2–2 cm, broadly oblanceolate to obovate, white with green veins. Ovary inferior. *S Spain.* H5. Late spring–summer.

Best grown under glass in a poor, alkaline soil.

122. S. bourgaeana Boissier & Reuter.

Like *S. biternata* but with bulbils restricted to lowest leaf-axils, secondary lobes of leaf-blade not narrowed to a stalk, and shorter petals, 5–8 mm. *S Spain.* H5.

Subsection **Triplinervium** (Gaudin) Gornall. Biennials or evergreen perennials, often forming mats or cushions. Leaves obovate to kidney-shaped with the margins variously lobed, or linear to lanceolate with the margins entire. Bulbils absent but summer-dormant leafy buds sometimes present in leaf-axils. Flowering-stems terminal or axillary. Petals white to greenish yellow, sometimes tinged or veined with red. Ovary ⅔ or more inferior.

Series **Aquaticae** (Engler) Pawlowska. Biennials or evergreen perennials, forming

mats or clumps. Leaves longer than 4 cm (including the leaf-stalk), with the leaf-stalk longer than the many-lobed blade, robust. Hairs, at least on the leaf-stalks, long and wavy. Petals white. Flowering-stems terminal or axillary. Ovary ⅔ or more inferior.

123. S. aquatica Lapeyrouse (*S. petraea* misapplied). Illustration: Webb & Gornall, Saxifrages of Europe, pl. 51 (1989).

Perennial, forming dense mats to 2 m across. Blade of basal leaves to 2.5 × 3.5 cm, more or less semi-circular, divided almost to the base into 3 primary lobes, which are themselves further divided to give 15–27 ovate to narrowly triangular ultimate segments, the latter often overlapping one another. Flowering-stems 25–60 cm, robust, axillary, branched above the middle to form a narrow panicle. Petals 7–9 mm, narrowly obovate, usually white. Ovary ¾ inferior. *Pyrenees.* H4. Summer.

Requires flowing, neutral or acidic water; probably best by a suitable stream.

124. S. irrigua Bieberstein (*S. ranunculoides* Haworth). Illustration: Botanical Magazine, 2207 (1821).

Perennial, forming a tuft of leafy shoots. Blade of basal leaves usually 2.5–3 × 3.5–4 cm, kidney-shaped or semi-circular, divided nearly to the base into 3 primary lobes, which are further divided to give 11–35 oblong-lanceolate, almost acute to apiculate, ultimate segments. Flowering-stems 10–20 cm, robust, terminal, bearing in the upper part a flat-topped cyme of 5–12 flowers. Petals 1.2–1.6 cm, oblanceolate, tip almost acute, white. Ovary ⅔ inferior. *Crimea.* H4. Late spring–summer.

Grows well as a large clump in a rock garden.

125. S. latepetiolata Willkomm. Illustration: Botanical Magazine, 7056 (1889); Webb & Gornall, Saxifrages of Europe, pl. 56, f. 57 (1989); Harding, Saxifrages, 135 (1992).

Biennial, densely covered with long, sticky, glandular hairs. Basal leaves arranged in a domed 'rosette' to 6 cm high; blade 0.8–1.5 × 1–2.7 cm, kidney-shaped to semi-circular, base more or less cordate, divided to half-way into 5–7 obovate to wedge-shaped, truncate lobes, these further divided in larger leaves into 3 segments; leaf-stalk brittle. Flowering-stem 15–25 cm, robust, terminal, branched from near the base to give a narrow, pyramidal panicle of numerous flowers. Petals 7–10 mm, narrowly obovate, white. Ovary inferior. *E Spain*. H5. Late spring–summer.

Probably best grown in a pan in a cool glasshouse.

Series **Arachnoideae** (Engler & Irmscher) Gornall. Biennials or evergreen perennials, with straggling or ascending leafy stems. Leaves usually longer than 4 cm (including the leaf-stalk), usually with leaf-stalk longer than scalloped or lobed blade, thin. Hairs on leaf-stalks long. Flowering-stems terminal. Petals white or pale greenish yellow. Ovary inferior.

126. S. petraea Linnaeus. Illustration: Köhlein, Saxifrages and related genera, t. 11 (1984); Finkenzeller & Grau, Alpenblumen, 115 (1985); Webb & Gornall, Saxifrages of Europe, pl. 52 (1989).

Biennial, covered with long, soft, sticky glandular hairs. Blade of basal leaves 1.2–3 × 2.2–3.5 cm, semi-circular to diamond-shaped, deeply divided into a narrow central lobe, with 3–5 teeth, and 2 broader lateral lobes, variously toothed or lobed, giving a total of 19–23 ultimate segments. Flowering-stems to 35 cm, brittle, freely branched, bearing small, loose, leafy cymes. Flowers usually slightly irregular, petals to 1.1 cm, tip notched, pure white. *E Alps*. H5. Spring–summer.

Best grown in a pan in alkaline soil in a cool glasshouse.

127. S. berica (Béguinot) D.A. Webb. Illustration: Webb & Gornall, Saxifrages of Europe, f. 58 (1989).

Like *S. petraea* but perennial, leaves with a brownish tint; flowers smaller, more asymmetrical; hairs shorter (*c.* 0.5 mm on the flower of *S. berica* versus *c.* 1.5 mm on *S. petraea*). *NE Italy (Colli Berici)*. H5. Spring–early summer.

Appears to grow best in crevices in tufa under glass.

128. S. arachnoidea Sternberg. Illustration: Finkenzeller & Grau, Alpenblumen, 115 (1985); Webb & Gornall, Saxifrages of Europe, pl. 53 (1989).

Perennial, densely covered with wavy, sticky, glandular hairs. Blade of basal leaves usually *c.* 1.2 × 1.4 cm, occasionally to 2 × 3 cm, fan-shaped, circular or elliptic, divided near the tip into 3–5 broadly ovate, obtuse lobes. Flowering-stems 10–20 cm, ascending, brittle, bearing up to 5 flowers in a loose

cyme. Petals 2.5–3 mm, oblong, not touching, off-white or cream. Ovary inferior, surmounted by a prominent nectary disc. *N Italy (Giudicarian Alps)*. H4? Summer.

Cultivate in a cool glasshouse in a fine, alkaline soil, in shade, with a fairly high humidity.

129. S. paradoxa Sternberg. Illustration: Webb & Gornall, Saxifrages of Europe, f. 59 (1989).

Perennial, probably short-lived, almost hairless except for a few long wavy hairs on leaf-stalks and base of the stem. Blade of basal leaves 1.5–2 × 2.5–4 cm, kidney-shaped, base cordate, thin, shiny, with 5–9 shallow, obtuse lobes usually broader than long. Flowering-stems to 20 cm, brittle, barely projecting above the basal leaves, bearing few-flowered cymes. Petals *c.* 1.5 mm, linear, greenish yellow. Ovary inferior, surmounted by a conspicuous nectary-disc. *SE Alps*. H4? Summer.

Difficult to grow; requires deep shade and a moist soil in the growing season.

Series **Axilliflorae** (Willkomm) Pawlowska. Evergreen perennials with prostrate leafy shoots, forming loose mats. Leaves usually less than 1.5 cm long (including the leaf-stalk), divided into 3–5 primary lobes, with the leaf-stalk more or less the same length as the blade. Hairs short. Flowering-stems terminal or axillary. Petals white. Ovary inferior. Capsule narrowly cylindric rather than ovoid or ellipsoid.

130. S. praetermissa D.A. Webb (*S. ajugifolia* misapplied). Illustration: Webb & Gornall, Saxifrages of Europe, f. 46 (1989).

Flowering-stems 10–15 cm, arising from the axils of leaves on prostrate shoots some distance from the upturned tip. *Pyrenees, NW Spain*. H4. Summer.

Sporadically in cultivation.

Series **Ceratophyllae** (Haworth) Pawlowska. Evergreen perennials. Leafy shoots more or less woody below, forming cushions. Leaves usually less than 7 cm (including the leaf-stalk), lobed, often stiff or leathery, the leaf-stalk longer than or equalling the blade; usually with very short, often stalkless, glandular hairs. Flowering-stems terminal or axillary. Petals white. Ovary inferior or nearly so.

131. S. pedemontana Allioni (*S. allionii* Terraciano; *S. pedatifida* misapplied). Illustration: Botanical Magazine, n.s., 687 (1975); Finkenzeller & Grau, Alpenblumen,

113 (1975); Webb & Gornall, Saxifrages of Europe, pl. 43–45, f. 48–50 (1989); Harding, Saxifrages, 136 (1992).

Leafy shoots forming a loose cushion. Blade of rosette leaves 0.8–1.5 × 0.9–2 cm, palmately divided into 3–9 elliptic to linear-oblong segments, not furrowed on the upper surface, covered with very short glandular hairs. Flowering-stems 5–18 cm, terminal, branched in the upper half to form a narrow panicle of 2–12 flowers. Petals 0.9–2.1 cm, oblanceolate, basal part erect, upper bent outwards, pure white, rarely tinged or veined with red. *Caucasus, Carpathians, Balkan Peninsula, SW Alps, Corsica, Sardinia, Cevennes, Morocco*. H3. Summer–early autumn.

Five subspecies are currently recognised, and a recently discovered dwarf variant from the Caucasus may represent a sixth.

Subsp. **cymosa** Engler. Leaf-blade tapered gradually into a wide leaf-stalk; segments ovate-oblong, obtuse, short, forward-pointing. Flowering-stem to 8 cm. Petals 9–15 × 3–5 mm. *Carpathians, Balkan Peninsula*.

Subsp. **pedemontana**. Leaf-blade tapered gradually to a narrow leaf-stalk; segments oblong, almost acute, short. Flowering-stem to 18 cm. Petals 15–21 × 6–8 mm, sometimes tinged pink at the base. *SW Alps*.

Subsp. **cervicornis** (Viviani) Engler. Young leaves on non-flowering rosettes incurved; blade contracted to a narrow leaf-stalk; segments narrowly oblong, obtuse, acute or with a short point, long, forward-pointing or divergent. Flowering-stem to 15 cm. Petals 10–13 × 4–5 mm. *Corsica, Sardinia*.

Subsp. **demnatensis** (Battandier) Maire. Like subsp. *cervicornis* but with larger, more leathery leaves and oblong sepals which are equal to or are longer than the perigynous zone. *Morocco*.

Subsp. **prostii** (Sternberg) D.A. Webb. Leaf-blade contracted to a fairly narrow leaf-stalk; segments broader than in subsp. *cervicornis*, but narrower than in the first 2 subspecies, acute or with a short point. Flowering-stem to 18 cm. Petals 9–12 × 2.5–4 mm. *France (Cevennes)*.

Cultivate in acidic soils, either in partial shade on the rock garden or in a cool glasshouse. The last subspecies is best outside.

132. S. geranioides Linnaeus. Illustration: Webb & Gornall, Saxifrages of Europe, f. 51 (1989).

Leafy shoots forming a loose cushion.

Leaf-blade *c.* 1.5 × 2.5 cm, semi-circular, deeply divided into 3 primary lobes, these further lobed or toothed, giving a total of 17–25 acute segments, not furrowed on the upper surface, covered with very short glandular hairs. Flowering-stems 15–25 cm, terminal, usually branched from near the middle to give a loose cyme of up to 20 flowers. Petals *c.* 1.2 cm, oblanceolate, white. *Pyrenees.* H4. Summer.

Prefers a moist, acidic soil in partial shade on the rock garden.

133. S. moncayensis D.A. Webb. Illustration: Webb & Gornall, Saxifrages of Europe, pl. 48, f. 52 (1989).

Leafy shoots forming a fairly dense, deep cushion. Leaf-blade 0.8–1.1 × 0.9–1.5 cm, deeply divided into 3 primary lobes, which are narrowly oblong to oblanceolate, obtuse, furrowed on the upper surface, outer lobes sometimes with a secondary lobe, covered with very short glandular hairs. Flowering-stems 5–10 cm, terminal, branched from the middle to give a loose panicle of 9–35 flowers. Petals 6–7 mm, broadly oblong, narrowed at the erect base, bent outwards in the upper part, not touching. *NE Spain (Sierra de Moncayo).* H4. Late spring–summer.

Prefers partial shade on the rock garden or in a pan in a cool glasshouse.

134. S. vayredana Luizet.

Leafy shoots forming a large, fairly dense cushion. Leaf-blade 5–9 × 8–13 mm, deeply divided into 3 primary lobes, these often themselves shortly lobed to give a total of 5–9 acute, spreading segments, not furrowed on the upper surface, covered in very short glandular hairs and smelling very strongly of spice when crushed. Flowering-stems 7–12 cm, terminal, bearing 3–9 flowers in a compact, narrow panicle. Petals 6–7 mm, obovate, pure white, nearly touching. *NE Spain (Sierra de Montseny).* H4. Summer.

Grows well in partial shade on the rock garden or in a pan under glass.

135. S. intricata Lapeyrouse (*S. nervosa* invalid). Illustration: Webb & Gornall, Saxifrages of Europe, pl. 47 (1989).

Like *S. vayredana* but leaf-segments obtuse, forward-pointing, furrowed on the upper surface. Petals 5–6 mm, broadly elliptic to nearly circular, overlapping, white, spreading horizontally. *Pyrenees.* H4. Summer.

Best grown in an acidic soil in a cool glasshouse.

136. S. pentadactylis Lapeyrouse.

Leafy shoots forming an open, rather brittle cushion. Leaf-blade divided nearly to the base into 3 primary lobes, the lateral ones further divided into 2 segments, the central one sometimes into 3; segments linear to oblong, obtuse, furrowed on the upper surface; covered with stalkless glands. Flowering-stems 7–17 cm, terminal, bearing 5–50 flowers. Petals 3.5–5 mm, obovate to oblong, usually white. *Pyrenees, N & C Spain.* H4. Summer.

A variable species, rarely found in cultivation despite the frequency with which the name is encountered. Four subspecies are currently recognised, of which subsp. **willkommiana** (Willkomm) Rivas Martinez seems to be the easiest to grow. It is recognised by its leaf-stalk which equals or is slightly shorter than the blade and is as narrow as the broadest segments, which are widely divergent.

The subspecies differ with respect to their preference for soil pH. The easiest one, subsp. *willkommiana*, likes an acidic soil in a pan in a cool glasshouse.

137. S. fragilis Schrank (*S. corbariensis* Timbal-Lagrave). Illustration: Botanical Magazine, n.s., 701 (1975); Webb & Gornall, Saxifrages of Europe, pl. 49, f. 53 (1989).

Leafy shoots forming a loose cushion. Leaf-blade 1–1.7 × 1–3 cm, usually semi-circular, divided almost to the base into 3 primary lobes, the lateral ones usually (and the central one always) divided into 2 or 3 segments, usually giving a total of 5–11 obtuse to almost acute, narrowly oblong segments, not furrowed on the upper surface, covered with stalkless glands. Flowering-stems 10–22 cm, terminal, branched in the upper half to form a loose cyme of 5–20 flowers. Petals 0.7–1.4 cm, obovate to oblanceolate, tip sometimes notched, pure white. *E Spain, S France.* H4. Spring–summer.

Subsp. **fragilis**, from the northern part of the range, can be distinguished from the more southerly subsp. *valentina* (Willkomm) D.A. Webb by its longer petals (at least 10 mm) and longer stamens (exceeding sepals by 3 mm or more). Subsp. *fragilis* is the more garden-worthy plant. Best grown in a pan of alkaline soil in a cool glasshouse.

138. S. camposii Boissier & Reuter. Illustration: The plant figured under this name in Botanical Magazine, 6640 (1882) is probably *S. corsica* subsp. *cossoniana* or one of its hybrids.

Like *S. fragilis* subsp. *valentina* but with stiffer leaves, at least some leaf-segments mucronate. *S & SE Spain.* H5. Late spring–summer.

139. S. trifurcata Schrader (*S. ceratophylla* Dryander). Illustration: Botanical Magazine, 1651 (1814); Webb & Gornall, Saxifrages of Europe, f. 54 (1989).

Leafy shoots forming a fairly open cushion of dark, shiny-green leaves. Leaf-blade to 2 × 3 cm, more or less semi-circular, divided ¾ of the way to the base into 3 primary lobes, these divided further to give 9–17 triangular apiculate segments, not furrowed on the upper surface, many overlapping, the lateral ones strongly recurved, covered with stalkless glands. Flowering-stems to 30 cm, axillary, bearing a loose cyme of 5–15 flowers. Petals 8–11 mm, elliptic-oblong, white. *N Spain.* H3. Late spring–summer.

S. × schraderi Sternberg, a garden hybrid between *S. continentalis* and *S. trifurcata*, approaches *S. trifurcata* in appearance, and is frequent in cultivation under the name *S. trifurcata*.

Grows well in alkaline soil in partial shade on the rock garden.

140. S. canaliculata Engler. Illustration: Webb & Gornall, Saxifrages of Europe, pl. 46, f. 55 (1989).

Leafy shoots forming a loose cushion. Leaf-blade to 1.2 × 1.8 cm, semi-circular, deeply divided into 3–11 linear, deeply grooved segments, covered with stalkless glands. Flowering-stems 8–15 cm, terminal, bearing a compact cyme of 5–12 flowers. Petals 8–10 mm, broadly obovate, overlapping, tip often recurving with age, white. *N Spain.* H3. Late spring–summer.

Cultivation requirements as for the previous species.

141. S. cuneata Willdenow. Illustration: Webb & Gornall, Saxifrages of Europe, pl. 50 (1989).

Leafy shoots in loose tufts. Leaf-blade to 2.2 × 2.6 cm, diamond- to fan-shaped, divided to half-way into 3, broadly triangular to ovate, mucronate, primary lobes, which are secondarily 3-lobed on larger leaves, covered with stalkless glands. Flowering-stems 7–30 cm, axillary, branched above the middle to form a narrow panicle of 7–15 flowers. Petals 5–8 mm, obovate, more or less touching, pure white. *N Spain, SW France.* H4. Late spring–summer.

Grows best in a pan of alkaline soil in a cool glasshouse.

142. S. maderensis D. Don. Illustration: Bocagiana **33**: 3 (1973).

Leafy shoots forming a loose cushion. Leaf-blade semi-circular, kidney-, diamond- or fan-shaped, base cordate or tapered to the leaf-stalk, divided up to half-way into 9–38 obtuse, acute or apiculate segments, covered with stalkless glands, though sometimes with some very short-stalked glandular hairs on the leaf-stalk. Flowering-stems to 10 cm, axillary, bearing up to 13 flowers in a compact to loose cyme. Petals 4–12 mm, white. *Madeira*. H4. Summer.

Var. **maderensis**. Leaf-blades kidney-shaped to semi-circular, base cordate to somewhat tapered; segments obtuse to acute. Flowering-stems with up to 13 flowers. Petals 4–10 mm. *Madeira*.

Var. **pickeringii** (C. Simon) D.A. Webb & Press. Leaf-blades diamond- to fan-shaped, base strongly tapered; segments acute to apiculate. Flowering-stems with up to 6 flowers. Petals 8–12 mm. *Madeira*. The latter variety performs well in partial shade in alkaline soil on the rock garden; the former is rather better under glass.

143. S. portosanctana Boissier. Illustration: Bocagiana **33**: 3 (1973).

Like *S. maderensis* but leaf-blade much longer than broad, base strongly tapered, divided to half-way or more into up to 9 obtuse segments. Petals 1–1.1 cm, *c.* 3 times as long as broad (versus only 2 times in *S. maderensis*). *Porto Santo*. H4. Summer.

Grows well in alkaline soil on the rock garden.

Series **Cespitosae** (Reichenbach) Pawlowska. Evergreen perennials, with leafy shoots forming cushions. Leaves less than 2.5 cm (including the stalk), the blade divided in to lobes and usually longer than or equalling the leaf-stalk. Hairs on leaf-stalk not long and wavy. Flowering-stems terminal. Petals white to dull yellow, sometimes tinged or veined with red. Ovary ⅔ or more inferior.

144. S. rosacea Moench (*S. decipiens* Sternberg; *S. palmata* Smith; *S. sternbergii* Willdenow; *S. hirta* misapplied; *S. hibernica* Haworth; *S. quinquifida* Haworth; *S. affinis* D. Don). Illustration: Ross-Craig, Drawings of British plants **10**: pl. 12 (1957); Garrard & Streeter, The wild flowers of the British Isles, 92 (1983); Webb & Gornall, Saxifrages of Europe, f. 44 (1989).

Leafy shoots forming a dense or loose cushion or thin mat. Rosette-leaves 0.6–2.5 cm, including the fairly distinct leaf-stalk which may be longer or shorter than blade; blade divided to half-way or more into 3–5 primary lobes, these sometimes further divided to give a total of up to 11 broadly elliptic to linear-oblong segments which are obtuse, acute or shortly pointed, but not furrowed on the upper surface. Flowering-stems 4–25 cm, bearing 2–6 flowers in an open cyme. Petals 6–10 mm, obovate, pure white. Ovary ⅔ inferior. *C & NW Europe*. H4. Late spring–summer.

Subsp. **rosacea**. Leaf-segments variable but not shortly pointed, usually broad. Leaf-hairs mostly non-glandular. *C & NW Europe*.

Subsp. **sponhemica** (Gmelin) D.A. Webb. Leaf-segments shortly pointed, narrowly oblong. Leaf-hairs mostly non-glandular. *C Europe*.

Subsp. **hartii** (D.A. Webb) D.A. Webb. Leaf-segments almost acute, broad. Leaf-hairs mostly glandular. *Ireland (Arranmore Island)*.

For a commonly grown complex of hybrids with *S. exarata*, *S. hypnoides* and *S. granulata* see under *S. exarata*. Grows well on the rock garden and under glass.

145. S. cespitosa Linnaeus (*S. groenlandica* Linnaeus). Illustration: Keble Martin, The new concise British flora, pl. 32 (1982); Garrard & Streeter, The wild flowers of the British Isles, 92 (1983).

Leafy shoots packed into a dense cushion, rarely a loose mat. Leaves 0.4–1.5 cm, including the indistinct leaf-stalk, most divided into 3 oblong-elliptic, obtuse lobes, not furrowed on the upper surface; margin densely glandular-hairy. Flowering-stems 2–10 cm, bearing 1–5 flowers in an open cyme. Petals 5.5–6.5 mm (in European plants), obovate, usually off-white. Ovary ⅔ inferior. *Arctic-circumpolar, W North America*. H1. Summer.

Requires cold and dry winter conditions (cold-frame) and standard alpine house treatment during spring and summer.

146. S. pubescens Pourret (*S. mixta* Lapeyrouse). Illustration: Webb & Gornall, Saxifrages of Europe, pl. 36 (1989).

Leafy shoots forming a loose to compact cushion with rather flat rosettes. Leaves to 1.8 cm, including the broad, indistinct leaf-stalk, but usually much shorter, divided into 5–9 oblong, obtuse lobes, 2–3 times as long as broad, most furrowed on the upper surface, dark green, densely glandular-hairy all over. Flowering-stems 3–10 cm, apparently axillary but actually terminal, bearing a flat-topped cyme of 5–9 flowers. Petals 4–6 mm, broadly obovate to nearly circular, overlapping, white. Ovary nearly inferior. *Pyrenees*. H4. Summer.

Subsp. **pubescens**. Leafy shoots short, loose, with dead leaves soon dropping. Leaf-segments about 3 times as long as broad, rather spreading. Petals pure white. Anthers yellow. Ovary green.

Subsp. **iratiana** (Schultz) Engler & Irmscher. Leafy shoots long, columnar, in a tight cushion, with persistent dead leaves. Leaf-segments about 2 times as long as broad, nearly parallel. Petals sometimes veined with red. Anthers and ovary reddish. The former subspecies grows well in acidic soil under glass; the latter is more difficult.

147. S. nevadensis Boissier. Illustration: Webb & Gornall, Saxifrages of Europe, pl. 40 (1989).

Like *S. pubescens* but leaves with a more distinct leaf-stalk of 3–4 mm; blade 4–7 × 5–8 mm, divided to half-way into 3–5 oblong, obtuse segments, not furrowed on the upper surface. Petals more consistently red-veined. *S Spain (Sierra Nevada)*. H3. Late spring–summer.

Best kept outside in a cold-frame in autumn and winter (protected from the wet) but brought under glass and watered in spring and early summer.

148. S. cebennensis Rouy & Camus (*S. prostiana* (Seringe) Luizet). Illustration: Köhlein, Saxifrages and related genera, t. 5 (1984).

Leafy shoots forming a soft, domed cushion of light green foliage. Leaf-blade *c.* 6 × 5 mm, tapered to a distinct leaf-stalk of *c.* 6 mm, usually divided to half-way into 3–5 obtuse lobes (though some leaves are entire), furrowed on the upper surface, densely glandular-hairy all over. Flowering-stems 5–8 cm, bearing 2 or 3 flowers. Petals 6–8 mm, broadly obovate, overlapping, creamy white. Ovary inferior. *S France*. H4. Late spring–summer.

Best grown under glass in alkaline soil.

149. S. exarata Villars. Illustration: Finkenzeller & Grau, Alpenblumen, 107 (1985); Webb & Gornall, Saxifrages of Europe, pl. 41, 42 (1989); Quarterly Bulletin of the Alpine Garden Society **58**(1): 39 (1990).

Leafy shoots forming a compact cushion. Leaves 0.4–2 cm, including the fairly distinct leaf-stalk, divided to at least half-way, usually into 3 oblong, obtuse lobes (sometimes entire or with 5–7 lobes), furrowed or flat on the upper surface, variably hairy. Flowering-stems 3–10 cm, bearing a flat-topped cyme of 1–5 flowers. Petals 2.5–6 mm, oblong or narrowly elliptic to broadly obovate, never properly touching, white to greenish yellow, rarely tinged with red. Ovary inferior. *C & S Europe eastwards to the Caucasus*. H3. Late spring–early autumn.

Subsp. **exarata**. Leaf lobes usually 3–5, spreading, furrowed on the upper surface. Petals 4–6 mm, about 1.5 times as long as broad, broadly obovate, white or pale cream. *Alps, Balkan Peninsula, Turkey, Caucasus.*

Subsp. **pseudoexarata** (Braun-Blanquet) Webb. Leaf lobes mostly 3, slightly spreading, furrowed on the upper surface. Petals 3.5–4.5 mm, twice as long as broad, oblong to narrowly elliptic, well separated, pale dull yellow. *Alps, Apennines, Balkan Peninsula.*

Subsp. **moschata** (Wulfen) Cavillier (*S. moschata* Wulfen; *S. pyrenaica* Villars; *S. pygmaea* Haworth; *S. rhei* Schott; *S. varians* Sieber). Leaves entire or 3-lobed, not furrowed on the upper surface. Petals 3–4 mm, twice as long as broad, oblong to narrowly elliptic, well separated, yellowish (sometimes tinged with red). *C & S Europe eastwards to the Caucasus, except Apennines & S Balkan Peninsula.*

Subsp. **lamottei** (Luizet) Webb (*S. lamottei* Luizet). Leaves entire to 5-lobed, segments not furrowed on the upper surface, sparingly hairy. Petals 4–5 mm, usually less than twice as long as broad, obovate, nearly touching. *SC France.*

Subsp. **ampullacea** (Tenore) Webb (*S. ampullacea* Tenore). Like subsp. *lamottei* but leaves entire or with 3 very short, parallel lobes; leaf-stalk indistinct. Flowers 1–3. *C Apennines.*

This species is a parent (along with *S. rosacea, S. hyphenoides* and others) of the invalidly named *S. × arendsii* hybrid complex. Also known as the 'mossy' saxifrages, their precise parentage is unknown, but descriptions of many of the cultivars may be found in Harding (1970, 1992) and Köhlein (1984).

150. S. hariotii Luizet & Soulié. Illustration: Webb & Gornall, Saxifrages of Europe, pl. 37 (1989).

Leafy shoots forming a dense to fairly loose cushion. Leaves 5–9 mm, including the indistinct leaf-stalk, mostly 3-lobed but some entire; lobes oblong and with a short point, furrowed on the upper surface, sparingly glandular-hairy. Flowering-stems 3–7 cm, bearing 3–12 flowers. Petals

3.5–4 mm, oblong, not touching, dull creamy-white, usually with reddish veins. Ovary inferior. *W Pyrenees*. H4. Summer.

Best grown in an alkaline soil under glass.

Series **Gemmiferae** (Willkomm) Pawlowska. Evergreen perennials with leafy shoots usually forming mats or cushions. Summer-dormant, leafy buds, with long wavy hairs, present in the leaf-axils of at least some shoots. Leaves lobed or entire, usually less than 4 cm, including leaf-stalk, with the latter longer or shorter than the blade. Flowering stems terminal. Petals white, rarely tinged with pink. Ovary inferior or nearly so.

151. S. hypnoides Linnaeus (*S. leptophylla* D. Don; *S. rupestris* Salisbury). Illustration: Ross-Craig, Drawings of British plants **10**: pl. 10 (1957); Phillips, Wild flowers of Britain, 44 (1977); Keble Martin, The new concise British flora, pl. 32 (1982); Garrard & Streeter, The wild flowers of the British Isles, 92 (1983).

Leafy shoots usually forming a loose mat. Leaves mostly entire and linear-lanceolate on the prostrate shoots, but 3–7-lobed and fan-shaped, with a broad leaf-stalk, in the terminal rosettes; lobes with bristle-like tips, 0.5–0.75 mm; hairs mostly confined to the leaf-stalk. Summer-dormant buds 5–10 × 2–4 mm, sometimes present in the leaf axils of the prostrate shoots; outer bud-leaves with broad, membranous margins and a green centre. Flowering-stems 5–20 cm, branched above to form a loose panicle of 2–7 flowers, the buds of which are nodding. Petals 7–12 mm, elliptic, white, not touching. Ovary ¾ inferior. *NW Europe*. H3. Spring–summer.

For a complex hybrid with *S. exarata* and others, see under that species. Grows well in partial shade on the rock garden, or under glass.

152. S. continentalis (Engler & Irmscher) D.A. Webb.

Leafy shoots prostrate, forming a compact mat. Stem-leaves entire, linear at top of shoot, to 1–2 cm; shorter and 3-lobed in middle, shortest and 5-lobed at base of shoot. All lobes with a short, bristle-like point. Winter rosette-leaves divided into 3 primary lobes, which are further divided to give a total of 5-13 elliptic to oblong-lanceolate segments. Hairs mostly confined to the leaf-stalk. Summer-dormant buds 7–11 mm, narrowly ellipsoid, on short stalks, always

present in at least some leaf axils; outer bud-leaves lanceolate, membranous and translucent except for the midrib. Flowering-stems 8–25 cm, bearing 4–11 flowers. Petals 4–8 mm, elliptic, white, not touching. Ovary ¾ inferior. *SW Europe*. H5. Spring–summer.

Var. **continentalis**. Leaf-segments very narrow, outer ones usually strongly recurved. *France, NE Spain.*

Var. **cantabrica** (Engler) D.A. Webb. Leaf-segments broad, mostly forward-pointing. *Portugal, W Spain.*

For a garden hybrid with *S. trifurcata* see under that species. Probably usually best in an alkaline soil, either on a rock garden or under glass.

153. S. conifera Cosson & Durieu. Illustration: Webb & Gornall, Saxifrages of Europe, pl. 35 (1989).

Leafy shoots forming a dense mat. Leaves 3–10 mm, linear-lanceolate, entire, shortly pointed, shiny, silvery green; margin fringed with hairs, these long, wavy and non-glandular on the outer leaves of the summer-dormant buds, but short and mainly glandular nearer the top of the shoot. Summer-dormant buds 9–12 × 3–4 mm, spindle-shaped to obconical, on stalks 4–10 mm. Flowering-stems 4–8 cm, bearing 3–7 flowers in a compact cyme. Petals 3–4 mm, narrowly obovate, white. Ovary inferior. *N Spain*. H5. Late spring–summer.

Fairly difficult to grow; an alkaline soil is preferred.

154. S. rigoi Porta. Illustration: Webb & Gornall, Saxifrages of Europe, pl. 34 (1989).

Leafy shoots forming a loose or fairly dense cushion. Leaves to 2.5 cm, including the variable leaf-stalk; blade to 8 × 12 mm, semi-circular, deeply 3-lobed, the lateral pair (and sometimes the central one) divided to give 5–7 linear-oblong, almost acute segments; covered in short glandular hairs. Summer-dormant buds 5–8 × 2.5–4 mm, ovoid, oblong or obovoid, on stalks; outer bud-leaves linear, herbaceous. Flowering-stems 7–13 cm, bearing a fairly compact cyme of 2–5 flowers. Petals 1.2–2 cm, oblanceolate, more or less erect. Ovary inferior. *SE Spain, Morocco (Rif Mts)*. H5. Late spring–summer.

Probably best grown in an alkaline soil in a cool glasshouse.

155. S. erioblasta Boissier & Reuter. Illustration: Webb & Gornall, Saxifrages of Europe, pl. 39, f. 42 (1989).

Leafy shoots forming a small, compact cushion. Leaves 4–7 mm, oblong-oblanceolate to spathulate, entire or with 3 short, obtuse lobes, leaf-stalk indistinct; covered in short, glandular hairs. Summer-dormant buds 1.5–3 mm across, obovoid, on short stalks; outer bud-leaves similar to the smaller, entire leaves on the leafy shoots, but the inner bud-leaves broader, mostly membranous and translucent. Flowering-stems 4–7 cm, bearing 1–4 flowers in a small cyme. Petals 3.5–5 mm, obovate, nearly touching, white but becoming cherry-pink with age. Ovary inferior. *S Spain*. H5? Late spring–summer.

Probably best grown in an alkaline soil in a cool glasshouse.

156. S. reuteriana Boissier. Illustration: Webb & Gornall, Saxifrages of Europe, pl. 32, f. 40, 41 (1989).

Leafy shoots crowded into a compact, hemispherical cushion. Blade of leaves usually 0.6–1 × 1–1.5 cm, semi-circular but divided to about half-way into 3 lobes, the lateral pair usually further divided into 2–4 segments, the central more rarely into 3, to give a total of 5–11 almost acute segments; leaf-stalk much longer than blade, grooved; leaf glandular-hairy. Summer-dormant buds 4–6 mm across, globular to obovoid; outer bud-leaves herbaceous, entire. Flowering-stems 2–5 cm, bearing 1–4 flowers. Petals 8–10 mm, elliptic, margins reflexed near the tip, slightly greenish white. Ovary inferior. *S Spain*. H5? Late spring–summer.

Probably best grown in an alkaline soil in a cool glasshouse.

157. S. globulifera Desfontaines. Illustration: Webb & Gornall, Saxifrages of Europe, pl. 38 (1989).

Leafy shoots forming a cushion. At least some basal leaves spathulate, without a distinct stalk. Blade of most other leaves to 0.8 × 1.7 cm, semi-circular but divided to about half-way or more into 5 elliptic, acute lobes; leaf-stalk 2–3 times as long as the blade; leaf covered with glandular hairs. Summer-dormant buds 2.5–3.5 mm across, obovoid to nearly spherical, stalked. Flowering-stems 5–15 cm, bearing usually 2–7 flowers. Petals 5–7 mm, narrowly obovate, white. Ovary inferior. *S Spain, Gibraltar, N Africa*. H5? Spring–summer.

A variable species, especially so in N Africa, where several varieties have been described. Probably best grown in an alkaline soil in a cool glasshouse.

158. S. maweana Baker. Illustration: Botanical Magazine, 6384 (1878).

Similar to *S. globulifera*. Leafy shoots forming fairly long rosettes of kidney-shaped basal leaves which are divided into segments. Summer-dormant buds narrowly oblong. Petals 1.2–1.4 cm. H5? *Morocco (Rif Mts)*.

Doubtfully in cultivation.

Subsection **Holophyllae** Engler & Irmscher. Evergreen perennials forming cushions or mats. Leaves entire or with 3 short lobes at the tip. Bulbils and summer-dormant leafy buds absent. Flowering-stems terminal. Petals white or pale greenish yellow. Ovary inferior or nearly so. Most of the species, coming as they do from snow-lie habitats, are very difficult to cultivate and do not persist for long in gardens. Most, consequently, are seen only very infrequently.

159. S. androsacea Linnaeus. Illustration: Rasetti, I fiori delle Alpi, pl. 247 (1980).

Leafy shoots axillary, short, erect, forming rosettes of apparently basal leaves which cluster into tufts or mats. Dead leaves brownish in colour. Leaves 7–30 × 3–6 mm, including the indistinct stalk; blade linear-oblong to narrowly obovate, usually entire but sometimes with 3 almost acute teeth, tip obtuse or almost acute; margin with glandular hairs, some of them to 2 mm. Flowering-stems 2–8 cm, bearing 1–3 flowers. Petals 4–7 mm, oblong or narrowly obovate, not touching, white. *Mountains of Europe, Siberia*. H3. Late spring–summer.

160. S. depressa Sternberg. Illustration: Rasetti, I fiori delle Alpi, pl. 248 (1980); Finkenzeller & Grau, Alpenblumen, 11 (1985); Webb & Gornall, Saxifrages of Europe, pl. 54 (1989).

Like *S. androsacea* but leaves 6–9 mm across, shortly 3-lobed, covered with short glandular hairs. Flowering-stems with 3–7 pure white flowers. *Dolomites*. H3. Summer.

161. S. muscoides Allioni (*S. planifolia* Sternberg). Illustration: Webb & Gornall, Saxifrages of Europe, pl. 57 (1989).

Leafy shoots axillary, long, columnar, erect, forming soft, deep cushions. Old leaves a greyish colour in the upper half. Leaves usually *c.* 5 × 1.5 mm, including the indistinct stalk, oblong to narrowly elliptic, entire, tip rounded; margin with glandular hairs. Newly dead leaves silvery grey, especially near the tip. Flowering-stems usually 1–5 cm, with 1 or 2 flowers.

Petals 3.5–5 mm, obovate, tip obtuse or slightly notched, overlapping, white, cream or pale yellow. *Alps*. H3. Summer.

162. S. presolanensis Engler. Illustration: Finkenzeller & Grau, Alpenblumen, 109 (1985); Webb & Gornall, Saxifrages of Europe, pl. 58 (1989).

Like *S. muscoides* but flowering-stems 6–10 cm, bearing 2–8 flowers, in a loose, flat-topped cyme. Petals 3–4 mm, oblong to wedge-shaped, tip usually deeply notched, well-separated, translucent, off-white, variably tinged with pale greenish yellow. Capsule nearly spherical. *Italian Alps*. H3. Summer.

163. S. glabella Bertoloni. Illustration: Strid, Wild-flowers of Mount Olympus, pl. 104 (1980).

Leafy shoots short, ascending or straggling, forming a loose cushion or mat. Dead leaves brownish in colour. Leaves mostly 5–8 × 1–1.5 mm, narrowly oblanceolate, obtuse, more or less hairless. Flowering-stems 4–10 cm, leafy, bearing 3–8 flowers in a fairly compact cyme. Petals 2–2.5 mm, broadly obovate, overlapping, white. *C Apennines, Balkan Peninsula*. H4. Summer.

164. S. tenella Wulfen. Illustration: Rasetti, I fiori delle Alpi, pl. 258 (1980); Webb & Gornall, Saxifrages of Europe, pl. 59 (1989).

Leafy shoots short, prostrate to ascending, forming a fairly dense mat. Dead leaves a shiny straw or silvery grey colour. Leaves 8–11 × 1–2 mm, linear, without a distinct stalk, tip long, pointed, keeled beneath, shiny, straw-coloured or silvery; margin with a narrow, translucent border. Flowering-stem to 15 cm, bearing 3–9 flowers in loose cyme. Petals *c.* 3 mm, obovate, creamy white. *E Alps*. H3. Summer.

Relatively easy to grow either in a shady part of the rock garden or in a cool glasshouse.

Subsection **Tridactylites** (Haworth) Gornall. Annuals or biennials. Leaves entire or with 3–5 short lobes. Bulbils and summer-dormant leafy buds absent. Flowering-stems terminal, bearing a loose, leafy panicle of flowers on long stalks. Petals usually white and notched at the tip. Ovary inferior or nearly so.

165. S. tridactylites Linnaeus. Illustration: Ross-Craig, Drawings of British plants **10:** pl. 6 (1957). Phillips, Wild flowers of Britain, 30 (1977); Keble Martin,

The new concise British flora, pl. 32 (1982); Garrard & Streeter, The wild flowers of the British Isles, 91 (1983).

Winter-annual, very variable according to growing conditions, ranging from 20 cm, with a large panicle of more than 50 flowers, to less than 3 cm with a solitary flower. First-formed basal leaves spathulate, later ones oblong, diamond-shaped or semi-circular, usually with 3–5 spreading lobes, rarely entire, with or without a stalk, very variable in size but usually *c.* 10 × 4 mm, including the stalk, not forming a well-defined rosette. Flower-stalks usually 1–2 cm in fruit. Petals 2.5–3 mm, narrowly obovate, tip entire or slightly notched, white. Capsule almost spherical, rounded at the base. *Europe, N Africa, SW Asia, Caucasus.* H3. Spring–summer.

Best grown in a pan in a cool glasshouse; water freely until the first flowers appear, then cut of the water supply and keep in the sun. The leaves should turn red, setting off the small white flowers.

166. S. adscendens Linnaeus. Illustration: Harding, Saxifrages, 123 (1970).

Biennial or winter-annual, differing from *S. tridactylites* in its well-developed, persistent rosette of basal leaves; flower-stalks shorter, to 8 mm in fruit; petals longer, 3–5 mm; fruit usually an obovoid capsule tapered at the base. *Scandinavia, mountains of C & S Europe, Turkey, Caucasus, W North America.* H3. Summer.

A variable species, very difficult to cultivate; it seems to need cold winters.

167. S. blavii (Engler) G. Beck. Illustration: Webb & Gornall, Saxifrages of Europe, pl. 60 (1989).

Like *S. adscendens* but with a basal rosette of small leaves that wither at flowering. Leaf-lobes short, forward-pointing. Flower-stalks usually 1–2 cm in fruit. Perigynous zone much smaller than in *S. adscendens*. Petals *c.* 5 mm. Capsule almost spherical, base rounded, broadly ellipsoid. *Balkan Peninsula.* H4? Summer.

Very attractive; probably worth trying in a pan of alkaline soil in a cool glasshouse.

9. SULLIVANTIA Torrey & Gray
D.M. Miller
Herbaceous perennials with slender horizontal stolons. Leaves basal and on the stem, alternate, kidney-shaped or rounded, toothed and lobed. Flowers bisexual, radially symmetric in branched inflorescences. Perigynous zone bell-shaped. Sepals 5, erect. Petals 5, clawed. Stamens 5. Carpels 2, ovary 2-celled, stigmas without obvious styles.

A genus of 6 species from C & W USA, represented in cultivation by a single species which thrives in moist woodland conditions in partial shade. Propagate by division or seed in spring.

1. S. oregana Watson. Illustration: Abrams & Ferris, Illustrated flora of the Pacific States **2**: f. 2227 (1944); Hitchcock et al., Vascular plants of the Pacific Northwest **3**: 61 (1971).

Slender plant to 20 cm. Leaves 2–7 cm across, kidney-shaped, incised to halfway into 7–9 lobes, coarsely toothed, hairless, yellowish green; stalk to 10 cm. Inflorescence to 10 cm, flowers erect. Sepals to 1 mm. Petals to 2 mm, white. Stamens not projecting from the corolla. Flower-stalks reflexed in fruit. *USA (Oregon).* H4. Late spring.

10. TELESONIX Rafinesque
D.M. Miller
Low-growing, herbaceous perennials, with short rhizomes, glandular-hairy. Basal leaves 2–5 cm wide, kidney-shaped, shallowly lobed and scalloped, stalks to 12 cm; stem-leaves smaller. Flowers bisexual, radially symmetric, *c.* 2 cm across, in clusters of 5–25 on reddish stems to 20 cm. Calyx 5-lobed, lobes *c.* 4 mm. Petals 5, to 5 mm, ovate to rounded, reddish purple. Stamens 10. Styles 2, carpels 2. Fruit a many-seeded capsule.

A genus of a single species for the alpine house or rock garden which is not always easy to establish in cultivation. It grows best in full sunlight, except in the hottest season, in a very well-drained, gritty soil which, however, should not be allowed to dry out completely. Propagate by seed or cuttings.

1. T. jamesii (Torrey) Rafinesque (*Saxifraga jamesii* Torrey; *Boykinia jamesii* (Torrey) Engelmann). Illustration: Hitchcock et al., Vascular plants of the Pacific Northwest **3**: 61 (1971); Rickett, Wild flowers of the United States **6**: pl. 98 (1973).

C to W USA. H2. Summer.

11. JEPSONIA Torrey & Gray
M.F. Gardner
Herbaceous perennial; rootstock corm-like. Leaves mostly basal, 2–6 cm across, rounded to heart-shaped, shallowly lobed, hairy. Flowers borne in terminal cymes, stalks 10–30 cm, slender, glandular-hairy becoming hairless later, flower-tube bell-shaped, usually purple-veined; sepals 5; petals 5, white, spathulate; stamens 10, shorter than sepals; carpels 2, ovary 2-celled. Fruit a pair of follicles, beaked; seeds 4-ridged.

A genus of 1 or 2 species native to N America. Best cultivated in the alpine house, where they require a freely drained compost and semi-shade. Plants should be liberally watered in the autumn and winter, when there growth is active; allow to dry out after flowering.

1. J. parryi (Torrey) Small (*Saxifraga parryi* Torrey). Illustration: Torrey Bulletin **23**: pl. 256 (1896); Abrams, Illustrated flora of Pacific States **2**: 353 (1944); Quarterly Bulletin of the Alpine Garden Society **37**: 276 (1969).

California. G1. Autumn–winter.

12. TANAKAEA Franchet & Savatier
P.G. Barnes
Rhizomatous dioecious perennials with creeping rhizomes and slender stolons. Leaves all basal, evergreen, ovate-oblong, acute, base cordate to rounded, with forward-pointing teeth, leathery, stalked. Scapes erect, with many small creamy white flowers in a narrow or broad panicle, with small bracts. Sepals 5, lanceolate; petals absent; stamens 10, longer than sepals. Carpels 2, joined for most of their length, styles free.

A genus of 1 or 2 species in NE Asia. Hardy and easily cultivated in a moist, well-drained soil or in light shade. Propagate by division or by detaching rooted stolons.

1. T. radicans Franchet & Savatier. Illustration: Botanical Magazine, 7943 (1904); Quarterly Bulletin of the Alpine Garden Society **30**: 154 (1962); Jelitto, Schacht & Fessler, Die Freiland Schmuckstauden, 619 (1985); Kitamura & Murata, Colored illustrations of herbaceous plants of Japan **2**: pl. 35 (1986).

Stems to 30 cm. Leaves 3–8 × 2–5 cm, mostly longer than the stalks. Flowers 2–3 mm. *Japan, China.* H4. Spring.

The Chinese plants are sometimes separated as **T. omeiensis** Nakai: this appears to be smaller (to 10 cm), with leaves 1–4 cm on stalks to 5 cm.

13. BOYKINIA Nuttall
D.M. Miller
Herbaceous perennials with short rootstocks, glandular-hairy. Leaves mostly

basal (stem-leaves small), kidney-shaped, lobed and toothed; stipules present. Flowers bisexual in loose panicles. Perigynous zone bell-shaped to obconical. Sepals 5. Petals 5, often deciduous. Stamens 5. Carpels 2; styles 2; ovary 2-celled. Fruit a many-seeded capsule.

A genus of about 8 species from N America and E Asia, of which about 3 are in cultivation. They are easily grown in woodland conditions in light shade and fairly moist, humus-rich, preferably lime-free soil. Propagate by seed or by division.

1a. Petals more or less equal to sepals
     1. rotundifolia
  b. Petals distinctly longer than sepals
     2
2a. Leaf-lobes with sharp bristle-pointed teeth **2. elata**
  b. Leaf-lobes without bristle-pointed teeth **3. aconitifolia**

1. B. rotundifolia Parry. Illustration: Abrams & Ferris, Illustrated Flora of the Pacific States **2**: f. 2230 (1944).

Plant 45–90 cm. Basal leaves 7–15 cm across, rounded, with shallowly rounded, toothed lobes; stipules very small; stalk to 20 cm. Flowers shortly stalked; perigynous zone bell-shaped. Sepals ovate, *c.* 2 mm. Petals obovate, *c.* 2 mm, sometimes unequal, white. Fruits drooping. *W USA (California)*. H4. Early summer.

2. B. elata (Nuttall) Greene (*B. occidentalis* Torrey & Gray). Illustration: Abrams & Ferris, Illustrated Flora of the Pacific States **2**: f. 2229 (1944); Hitchcock et al., Vascular plants of the Pacific Northwest **3**: 6 (1971); Clark, Wild flowers of the Pacific Northwest, 202 (1976).

Plant 30–60 cm with brownish, gland-tipped hairs on the stem. Basal leaves 2–8 cm across, rounded, base cordate, fairly deeply 5–7-lobed, with sharp, bristle-pointed teeth; stalk to 15 cm; stipules brown and scarious or reduced to bristles. Inflorescence with leaf-like bracts. Perigynous zone obconical, especially in fruit. Sepals *c.* 2 mm. Petals 5–6 mm, oblanceolate, white, sometimes tinged with pink. *W North America*. H4. Early summer.

B. major Gray is very similar.

3. B. aconitifolia Nuttall. Illustration: Justice & Bell, Wild flowers of North Carolina, 85 (1968); Gleason, Illustrated flora of the northeastern States and adjacent Canada **2**: 267 (1974); Jelitto, Schacht & Fessler, Die Freiland Schmuckstauden, 102 (1985).

Plant 30–80 cm. Basal leaves 5–12 cm across, rounded to kidney-shaped, divided to about halfway into 5–9 sharply toothed lobes; stalk to 10 cm (stem-leaves shortly stalked). Inflorescence with leaf-like bracts. Perigynous zone obconical, sticky. Sepals small, lanceolate. Petals 3–5 mm, obovate, white. *E USA*. H4. Summer.

14. ELMERA Rydberg
D.M. Miller
Low-growing herbaceous perennials with slender, horizontal rhizomes. Basal leaves *c.* 2 × 3–5 cm, kidney-shaped, lobes rounded, toothed, somewhat hairy; stalk to 7 cm, glandular-hairy with large, membranous stipules; leaves on flowering-stems 1–4, alternate, smaller. Flowers bisexual, 10–30 in a raceme to 25 cm. Sepals 5, triangular, erect, greenish yellow, *c.* 4 mm. Petals 5, erect, deeply 3–5-fid, yellowish white, 4–6 mm. Stamens 5, shorter than the sepals. Styles 2, thick, carpels 2. Fruit a many-seeded capsule.

A genus of a single species which thrives in a well-drained, fertile soil in full sun to semi-shade. Propagate by division in spring or autumn, or by seed.

1. E. racemosa (Watson) Rydberg (*Heuchera racemosa* Watson; *Tellima racemosa* (Watson) Greene). Illustration: Hitchcock et al., Vascular plants of the Pacific Northwest **3**: 11 (1971); Rickett, Wild flowers of the United States **5**: pl. 77 (1971).

NW North America (British Columbia to Washington). H2. Summer.

15. LITHOPHRAGMA (Nuttall) Torrey & Gray
D.M. Miller
Herbaceous perennial with rhizomes bearing bulb-like tubers. Leaves mostly basal, kidney-shaped or rounded, deeply or shallowly lobed, with long stalks. Flowers bisexual, slightly bilaterally symmetric, in few-flowered racemes. Calyx 5-lobed. Petals 5, clawed, often deeply divided. Stamens 10. Carpels and styles 3, ovary 1-celled. Fruit a many-seeded capsule.

A genus of 9 species from W North America represented by a single species in cultivation. It is best grown in a fertile soil in a rock garden or woodland, and should be kept relatively dry in summer when it is dormant. Propagate by division, bulbils, or seed.

1. L. parviflora (Hooker) Nuttall (*Tellima parviflora* Hooker). Illustration: Hitchcock et al., Vascular plants of the Pacific

Northwest **3**: 23 (1971); Rickett, Wild flowers of the United States **5**: pl. 79 (1971).

Plant with glandular hairs in all parts. Basal leaves 1–3 × 1–2.5 cm, divided nearly to the base into 3–5 segments, each deeply 3–5-lobed; stalk to more than 6 cm; stem-leaves 2 or 3, divided into narrower, slightly smaller segments. Raceme with 10 or more flowers, 25–40 cm, often purplish; flower-stalks 2–5 mm, erect. Flowers 1.8–2 cm across, slightly irregular. Sepals 2 mm, acute, green tinged with brown. Petals 8–10 × 5–6 mm, deeply 3–5-lobed, white, veins pink, or pale pink. *W North America*. H3. Spring.

16. TIARELLA Linnaeus
D.M. Miller
Herbaceous perennial with rhizomes. Leaves mostly basal, simple, lobed or made up of 3 leaflets; stipules small; stalks long. Inflorescence a raceme or panicle. Flowers bisexual, small, white, radially symmetric. Sepals 5, coloured. Petals 5, clawed. Stamens 10, protruding, sometimes of unequal size. Styles 2. Fruit a few-seeded capsule with 2 unequal flaps.

A genus of about 7 species of herbaceous plants, 1 from E Asia, the rest from N America. They are easily grown in shaded woodland conditions in a moist, humus-rich soil. Propagate by division in spring or autumn, or by seed sown in a peaty compost.

Literature: Lakela, O., A Monograph of the genus *Tiarella* Linnaeus in North America, *American Journal of Botany* **24**: 344–51 (1937).

1a. Inflorescence panicle-like 2
  b. Inflorescence raceme-like 4
2a. Leaves simple **6. unifoliata**
  b. Leaves compound 3
3a. Leaflets shallowly 3-lobed
     3. trifoliata
  b. Leaflets deeply divided into narrow segments **4. laciniata**
4a. Petals subulate to linear; inflorescence open **5. polyphylla**
  b. Petals narrow but not subulate; inflorescence fairly dense, expanding as fruit develops 5
5a. Plants producing stolons after one year; petals elliptic to lanceolate
     1. cordifolia
  b. Plants never producing stolons; petals narrowly lanceolate
     2. wherryi

1. T. cordifolia Linnaeus. Illustration: Botanical Magazine, 1589 (1813); Justice

& Bell, Wild flowers of North Carolina, 85 (1968).

Plant with slender, rooting stolons. Basal leaves 5–10 × 3–8 cm, broadly ovate to almost circular, base cordate, 3–5-lobed, unequally toothed, hairy, often marbled bronze; stalks 5–9 cm. Inflorescence 10–30 cm, raceme-like, stalk sometimes with 1 or 2 small stem-leaves. Sepals 2–4 mm, white. Petals 4–8 mm, elliptic to lanceolate, white. Stamens 2–7 mm, all equal; anthers orange. Capsule 4–10 mm. *E North America (Ontario to Alabama)*. H1. Late spring–early summer.

Some plants varying in flower or leaf colour have been given cultivar names.

2. T. wherryi Lakela (*T. cordifolia* var. *collina* Wherry).

Plant without stolons. Basal leaves 7–14 × 6–9 cm, broadly ovate, 3-lobed, unevenly toothed, base cordate, apex acute, hairy, often becoming reddish in autumn; stalks 10–20 cm. Inflorescence raceme-like, 15–35 cm, occasionally with 1–3 small leaves on the stalk. Sepals 1.5–2 mm, white to purple-tinged. Petals 3–5 mm, narrowly lanceolate, white. Stamens 3–5 mm, equal in length, anthers orange. Capsule 5–10 mm. *SE USA*. H3. Late spring–early summer.

3. T. trifoliata Linnaeus. Illustration: Hitchcock et al., Vascular plants of the Pacific Northwest **3**: 68 (1971); Rickett, Wild flowers of the United States **6**: pl. 98 (1973).

Plant without stolons. Basal leaves to 9 × 12 cm, made up of 3 leaflets, each leaflet 3–8 × 2–5 cm, diamond-shaped, 3-lobed, unevenly toothed, hairy; leaf-stalk 5–17 cm. Inflorescence narrow panicle-like, open, 15–50 cm, with 2 or 3 leaves. Sepals 1–2 mm, white or sometimes pink-tinged. Petals 2–5 mm, subulate, with twisted tips, white. Stamens 3–5 mm, unequal, anthers cream. Capsule 3–7 mm. *NW North America*. H3. Late spring–early summer.

4. T. laciniata J.D. Hooker.

Very similar to *T. trifoliata*, but has a smaller leaf with leaflets deeply divided into narrow segments. *W North America (Alaska to Oregon)*. H3. Late spring–early summer.

5. T. polyphylla D. Don. Illustration: Ito, Alpine plants in Hokkaido, pl. 397 (1981).

Plant with underground stolons. Basal leaves 2–7 cm, almost circular, cordate, shallowly 5-lobed with unequal teeth, hairy; stalk 2–10 cm. Inflorescence raceme-like, 10–40 cm, open, with 2 or

3 shortly stalked leaves. Sepals 1–2 mm, white. Petals 2–3 mm, subulate to linear, white. Stamens 3–5 mm, slightly unequal, anthers cream. Capsule 7–12 mm. *Japan, China*. H3. Late spring–early summer.

6. T. unifoliata J.D. Hooker.

Like *T. polyphylla*, but differing in its paniculate inflorescence. *W North America (Alaska to Oregon)*. H3. Late spring–early summer.

May be in cultivation.

17. HEUCHERA Linnaeus
D.M. Miller

Herbaceous perennials with semi-woody, scaly, branched rootstocks. Leaves mostly basal, simple, palmately lobed, usually with cordate bases and long stalks. Inflorescences paniculate, sometimes with leafy bracts. Flowers bisexual, usually radially symmetric, occasionally bilaterally symmetric where the perigynous zone is longer on 1 side. Sepals 5. Petals 5 or rarely absent. Stamens 5. Carpels 2, ovary 1-celled, styles 2. Fruit a many-seeded capsule.

A genus of about 55 species from N America, some of which are grown for ground-cover, or for their flowers in rock gardens and borders. They are best grown in a well-drained, fertile soil in full sun or partial shade. Propagate by division in spring or autumn, or by seed (except for the named cultivars). It is advisable to divide and replant large clumps.

Literature: Rosendahl, O.C., Butters, F.K. & Lakela, O., A Monograph of the genus *Heuchera*, *Minnesota Studies in Plant Science* **2**: 1–180 (1936).

Inflorescence. Diffuse and very open: **4,5**; spike-like: **1,2**; compact but not spike-like: **3,6–10**.
Flowers. Red or red-tinged: **3,4,6,9,10**; white **2,5,6**.
Petals. Exceeding calyx: **2,4–8,10**; included in calyx or absent: **1,3**.

1a. Petals bright red **3. sanguinea**
 b. Petals white, pink, greenish or tinged with red, or absent 2
2a. Stamens shorter than sepals 3
 b. Stamens equalling or exceeding sepals 4
3a. Petals shorter than sepals or absent, greenish; leaves usually longer than broad **1. cylindrica**
 b. Petals equalling or exceeding sepals, white; leaves usually broader than long **2. grossulariifolia**
4a. Inflorescence wide, open, diffuse 5

 b. Inflorescence narrow to spike-like 6
5a. Leaves broader than long, lobes acute; leaf-stalk usually hairless **5. glabra**
 b. Leaves longer than broad, lobes obtuse; leaf-stalk usually with long hairs **4. micrantha**
6a. Flowers bilaterally symmetric 7
 b. Flowers radially symmetric 9
7a. Flowers narrow, pinkish red, or white tinged with pink **6. rubescens**
 b. Flowers almost spherical, green, pink or white 8
8a. Inflorescence-stalk usually leafless, with stiff hairs **7. richardsonii**
 b. Inflorescence-stalk with 1–3 small leaves, hairs not stiff **8. pubescens**
9a. Petals more or less equal to sepals; panicle often leafy **9. americana**
 b. Petals distinctly longer than sepals; panicle usually leafless 10
10a. Plant very hairy; stamens more or less equalling or just exceeding sepals; leaves 3–9 cm across **10. pilosissima**
 b. Plant slightly hairy; stamens obviously exceeding sepals; leaves 1–5 cm across **6. rubescens**

1. H. cylindrica Douglas. Illustration: Edwards's Botanical Register, t. 1924 (1837); Rickett, Wild flowers of the United States **6**: pl. 99 (1973); Clark, Wild flowers of British Columbia, 199 (1973).

Leaves basal, usually 2–7 × 2–6 cm, broadly ovate to almost circular, deeply lobed, lobes rounded, margin scalloped, often hairy; stalk 2–15 cm. Inflorescence-stalk to 90 cm, leafless, with glandular hairs. Panicle 3–15 cm, narrow, spike-like. Flowers 4–7 mm. Sepals 2–4 mm, cream or tinged with green. Petals 1–2 mm or absent. Stamens and styles shorter than sepals. *W North America*. H2. Summer.

2. H. grossulariifolia Rydberg. Illustration: Hitchcock et al., Vascular plants of the Pacific Northwest **3**: 15 (1971); Quarterly Bulletin of the Alpine Garden Society **50**: 343 (1982).

Leaves all basal, 1–6 cm across, circular to kidney-shaped, broader than long, deeply 3–5-lobed, coarsely toothed, sometimes hairless; stalk 1–6 cm. Inflorescence-stalk to 40 cm, leafless, sparsely glandular-hairy. Panicle 1–6 cm, narrow, spike-like. Flowers 3–4 mm. Sepals 1–1.5 mm. Petals 1.5–3 mm, oblanceolate, white. Stamens and styles shorter than sepals. *C USA*. H2. Summer.

3. H. sanguinea Engelmann. Illustration: Botanical Magazine, 6929 (1887); Rickett, Wild flowers of the United States **4:** pl. 89 (1970); Hay & Beckett, Readers' Digest encyclopaedia of garden plants and flowers, 340 (1987).

Basal leaves 2–6 cm across, kidney-shaped to almost circular, with 5–7 rounded lobes, sharply toothed, glandular-hairy; stalk 4–12 cm. Flowering-stem to 50 cm, often with 2 or 3 very small leaves. Panicle 5–15 cm, open; flowers 4–6 in pendent clusters. Flowers 6–10 mm. Sepals *c.* 4 mm, red. Petals, stamens and styles shorter than sepals. *SW USA.* H3. Summer.

Many named cultivars have been raised from this species, including one with white leaves spotted with green, becoming pink in winter, known as 'Taff's Joy'.

H. × brizoides Anon. is presumed to be a hybrid between *H. sanguinea* and *H. micrantha.* It is very similar to, and often confused with, × *Heucherella tiarelloides* but differs in the absence of stolons and the constant presence of 5 stamens. Many named cultivars are grown.

4. H. micrantha Lindley. Illustration: Edwards's Botanical Register, t. 1302 (1829); Hitchcock et al., Vascular plants of the Pacific Northwest **3:** 15 (1971); Clark, Wild flowers of British Columbia, 206 (1973).

Plants usually with long hairs. Basal leaves 2–8 × 2–7 cm, kidney-shaped to oblong, shallowly 5–7-lobed, lobes rounded with broad teeth; stalk 5–15 cm. Flowering-stem to 1 m, with or without small leaves, hairy. Panicle 30 cm or more, very loose. Flowers 1–3 mm, greenish white tinged with red. Sepals less than 1 mm. Petals 1–2 mm, very narrow, white. Stamens and styles projecting. *W North America.* H2. Summer.

A plant with reddish purple foliage, grown as 'Palace Purple' is usually assigned to this species.

5. H. glabra Willdenow. Illustration: Hitchcock et al., Vascular plants of the Pacific Northwest **3:** 15 (1971); Rickett, Wild flowers of the United States **5:** pl. 78 (1971); Clark, Wild flowers of British Columbia, 207 (1973).

Plant with few hairs. Basal leaves 2–8 × 3–9 cm, rounded, deeply 5–7-lobed, lobes acute and with acute teeth; stalk 5–20 cm. Flowering-stem to 70 cm, usually with 1–3 small leaves; panicle to 25 cm, open, with long internodes. Flowers 2–4 mm, greenish white. Sepals *c.* 1 mm. Petals 2–4 mm, very narrow, white. Stamens and styles

projecting. *NW North America (Oregon to Alaska).* H1. Summer.

H. villosa Michaux is very similar, differing mainly in the more hairy leaf- and flower-stalks.

6. H. rubescens Torrey. Illustration: Rickett, Wild flowers of the United States **6:** pl. 100 (1973).

Leaves all basal, 1–5 cm across, broadly ovate to rounded, deeply 3–7-lobed, sharply toothed; stalk 2–10 cm. Inflorescence stalk to 35 cm, leafless; panicle *c.* 15 cm, narrow. Flowers 3–5 mm, slightly irregular, narrow, pink or white tinged with pink. Sepals *c.* 2 mm. Petals 3–4 mm, very narrow, almost white. Stamens and styles usually exceeding petals. *W USA.* H2. Summer.

H. versicolor Greene is very similar, differing in the position of the stamens, and is considered by some authors to belong to this species.

7. H. richardsonii Britton. Illustration: Hitchcock et al., Vascular plants of the Pacific Northwest **3:** 19 (1971); Rickett, Wild flowers of the United States **6:** pl. 99 (1973).

Leaves all basal, 4–13 × 4–11 cm, rounded, shallowly lobed and toothed, hairy beneath; stalk 6–20 cm. Inflorescence-stalk to 80 cm or more, leafless; panicle to 40 cm, narrowly cylindric. Flowers 5–10 mm, irregular, greenish. Sepals 2–3 mm. Petals as long as or slightly longer than sepals. Stamens exceeding sepals, styles about equal to them. *C North America.* H1. Summer.

8. H. pubescens Pursh. Illustration: Gleason, Illustrated flora of the northeastern States and adjacent Canada **2:** 271 (1952).

Plant with glandular hairs. Basal leaves 4–11 × 3–10 cm, almost circular, with 5–7 triangular lobes, toothed; stalk 10–20 cm. Flowering-stem to 75 cm, with 1–3 small leaves; panicle *c.* 25 cm, conical. Flowers 5–10 mm, irregular, greenish tinged with purple. Sepals 2–3 mm. Petals exceeding sepals, reddish purple. Stamens and styles mostly equalling petals, sometimes longer. *E USA.* H2. Summer.

9. H. americana Linnaeus (*H. glauca* Rafinesque). Illustration: Gleason, Illustrated flora of the northeastern United States and adjacent Canada **2:** 269 (1952).

Basal leaves 5–14 × 4–12 cm, rounded, 5–9-lobed, toothed, with stiff hairs; stalk 8–25 cm. Flowering-stem to 1 m, often with 1–3 small leaves; panicle 30 cm or

more, narrowly cylindric. Flowers 2–5 mm, regular, green tinged with red. Sepals 1–2 mm. Petals about as long as sepals. Stamens and styles much longer than petals. *E North America.* H2. Summer.

10. H. pilosissima Fischer & Meyer. Illustration: Rickett, Wild flowers of the United States **5:** pl. 78 (1971).

All parts very hairy. Basal leaves 5–9 × 3–9 cm, with 5–7 rounded, toothed lobes; stalk 7–20 cm, with brown hairs. Flowering-stems 20–50 cm, usually leafless but sometimes with 1–3 very small leaves; panicle narrow but loose. Flowers 2–5 mm, spherical, very hairy, tinged red. Sepals 1–2 mm. Petals longer than sepals, oblanceolate, pinkish white. Stamens and styles scarcely projecting. *W USA (California).* H3. Summer.

18. × HEUCHERELLA Wehrhahn
D.M. Miller

Herbaceous perennials. Leaves mainly basal, ovate or rounded, long-stalked. Flowers small, numerous, in panicles. Sepals 5. Petals 5. Stamens 5–10, often 7–8. Carpels 2, slightly unequal.

Hybrids between species of *Heuchera* and *Tellima* which can be distinguished from the parents by the variable number of stamens within one inflorescence (5–10 but usually 7 or 8). The plants are easily grown in any good garden soil in sun or partial shade but may take 2 years or more to flower freely. They are sterile, but may be propagated by division or cuttings.

1a. Plant spreading by stolons
1. tiarelloides
 b. Plant forming a compact clump
2. alba

1. × H. tiarelloides (Lemoine) Wehrhahn (*Heuchera tiarelloides* Lemoine). Illustration: Botanical Magazine, n.s., 31 (1948).

A hybrid between a pink- or red-flowered hybrid *Heuchera* and *Tiarella cordifolia.* Plant spreading by thin, rooting stolons. Basal leaves *c.* 9 cm across, mostly more or less rounded, with 7 shallow lobes with rounded teeth, light green blotched with brownish red when young; leaf-stalk 7–15 cm, hairy, reddish. Inflorescence-stalk to 40 cm, brownish red. Panicle narrow, open. Sepals 2 mm, ovate, spreading, pinkish red. Petals 3 mm, erect, pale pink. Stamens *c.* 3 mm, anthers yellow. *Garden origin.* H3. Late spring–early summer.

2. × H. alba (Lemoine) Stearn (*Heuchera tiarelloides* Lemoine var. *alba* Lemoine).

A hybrid between a white-flowered

Heuchera and *Tiarella wherryi*. Clump-forming, without stolons. Basal leaves *c*. 10 × 9 cm, broadly ovate, 7–9-lobed, sharply toothed, green marked with grey blotches. Inflorescence-stalk to 60 cm, green. Panicle narrow. Sepals *c*. 3 mm, ovate, white. Petals narrow, 3–4 mm, white. Stamens as long as or shorter than petals, anthers brown. *Garden origin*. H3. Late spring–early autumn.

'Bridget Bloom', with light pink flowers, is the variant found most frequently in cultivation.

19. BOLANDRA A. Gray
D.M. Miller

Herbaceous perennials with short rootstocks bearing bulbils. Leaves alternate with large stipules; lower leaves kidney-shaped, palmately veined and long-stalked, the upper leaves smaller, stalkless and with leaf-like stipules. Flowers bisexual, few, in loose panicles with conspicuous leafy bracts. Sepals 5, spreading, linear-lanceolate. Petals 5, linear, erect. Stamens 5, shorter than the petals. Ovary 2-celled; carpels 2. Fruit a capsule containing many seeds.

A genus of 2 species native to western North America, useful for partially shaded areas of rock or wild gardens in moist but not waterlogged soil. Propagate by seed or by division.

1. B. oregana Watson. Illustration: Abrams & Ferris, Illustrated Flora of the Pacific States **2**: f. 2224 (1944); Hitchcock et al., Vascular plants of the Pacific Northwest **3**: 6 (1971).

Plant to 60 cm. Basal and lower stem-leaves to 7 cm across, shallowly lobed into 9–12 irregularly and sharply toothed segments; stalk to 15 cm. Panicle 3–7-flowered. Sepals 5–10 mm, purplish. Petals purple, equalling sepals. Stamens with reddish purple filaments. *NW North America*. H4. Late spring–early summer.

20. MITELLA Linnaeus
D.M. Miller

Low-growing herbaceous perennials. Leaves mostly basal, simple, lobed, with cordate bases and long stalks. Inflorescences simple, often 1-sided, occasionally leafy racemes. Flowers numerous, small. Perigynous zone bell-shaped; sepals 5, small. Petals 5, usually deeply cut, green or white. Stamens 5 or 10, alternate or opposite to the petals. Styles 2, short, carpels 2, ovary 1-celled. Capsule containing numerous glossy, black seeds.

A genus of about 20 species native to

N America and E Asia, useful as ground cover. All thrive in partial shade in woodland conditions in moist soils, especially if these are rich in organic matter. Propagate by division in spring or autumn, or by seed.

Inflorescence-stalk. Leafy: **2,4**; leafless: **1,3,5–8**.
Petals. Comb-like: **1–4,7,8**; divided into 3: **5,6**.
Stamens. 5: **3–8**; 10: **1,2**.

1a.	Stamens 10	2
b.	Stamens 5	3
2a.	Inflorescence-stalk usually leafless; petals greenish yellow	**1. nuda**
b.	Inflorescence-stalk with 2 opposite leaves; petals white	**2. diphylla**
3a.	Stamens on the same radii as the petals	**3. pentandra**
b.	Stamens on the same radii as the sepals	4
4a.	Inflorescence-stalk with 1–3 leaves	**4. caulescens**
b.	Inflorescence-stalk without leaves or occasionally with 1 very small leaf	5
5a.	Petals never divided into more than 3 divisions	6
b.	Petals comb-like, usually with 5 or more divisions	7
6a.	Petals spreading, segments thread-like	**5. stauropetala**
b.	Petals erect, segments narrow but not thread-like	**6. trifida**
7a.	Leaves 4–8 cm wide, broader than long, slightly hairy	**7. breweri**
b.	Leaves 1–4 cm wide, longer than broad, coarsely hairy	**8. ovalis**

1. M. nuda Linnaeus. Illustration: Hitchcock et al., Vascular plants of the Pacific Northwest **3**: 25 (1971); Rickett, Wild flowers of the United States **6**: pl. 98 (1973).

Plant with stolons. Basal leaves 1–3 cm, rounded to kidney-shaped, margins scalloped, sparsely hairy. Inflorescence-stalk to 20 cm, rarely with a small leaf at the base; inflorescence 3–12-flowered, flower-stalks 2–6 mm. Sepals 1–2 mm, ovate. Petals to 4 mm, comb-like, greenish yellow. Stamens 10. *N America to E Asia*. H1. Summer.

2. M. diphylla Linnaeus. Illustration: Edwards's Botanical Register, t. 166 (1816).

Basal leaves 3–6 cm, broadly ovate, 3–5-lobed, toothed, apex acute to acuminate, hairy. Inflorescence-stalk to 45 cm, with 2 opposite, almost stalkless

leaves similar to but smaller than the basal leaves; inflorescence 5–20 flowered, flower-stalks *c*. 2 mm. Sepals 1–2 mm, Petals 2–3 mm, comb-like, white. Stamens 10. *E North America*. H1. Spring–early summer.

3. M. pentandra W.J. Hooker. Illustration: Botanical Magazine, 2933 (1829); Hitchcock et al, Vascular plants of the Pacific Northwest **3**: 25 (1971); Rickett, Wild flowers of the United States **6**: pl. 99 (1973).

Plant sometimes with stolons. Basal leaves 3–8 × 2–5 cm, ovate, shallowly 5–9-lobed, margins scalloped, sometimes hairy. Inflorescence-stalk to 30 cm; inflorescence with 5–25 flowers, flower-stalks 2–7 mm. Sepals *c*. 2 mm, spreading. Petals 2–3 mm, comb-like, greenish. Stamens 5, on the same radii as the petals. Stigma 2-lobed. *C to W North America*. H2. Spring–early summer.

4. M. caulescens Nuttall. Illustration: Hitchcock et al., Vascular plants of the Pacific Northwest **3**: 23 (1971); Rickett, Wild flowers of the United States **6**: pl. 99 (1973).

Plant usually with stolons. Basal leaves 3–7 cm across, rounded, mostly 5-lobed, toothed, sparsely hairy; stem-leaves 1–3, alternate, smaller than basal leaves. Inflorescence-stalk to 40 cm, with up to 25 flowers, opening from top downwards; flower-stalks 2–8 mm. Sepals 1–2 mm, spreading. Petals 3–4 mm, comb-like, greenish purple. Stamens 5, on the same radii as the sepals. Stigma capitate. *C to W North America*. H2. Late spring–early summer.

5. M. stauropetala Piper. Illustration: Hitchcock et al., Vascular plants of the Pacific Northwest **3**: 30 (1971).

Basal leaves 2–8 cm across, rounded to broadly ovate, indistinctly lobed, margins scalloped. Inflorescence-stalk to 50 cm, leafless, with 10–30 flowers on one side; flower-stalks *c*. 1 mm. Sepals 2–3 mm, recurved. Petals 2–4 mm, spreading, divided into 3 thread-like segments, greenish white to purple. Stamens 5, on the same radii as the sepals. Stigmas flattened. *C to W North America*. H1. Late spring–early summer.

6. M. trifida Graham. Illustration: Hitchcock et al., Vascular plants of the Pacific Northwest **3**: 30 (1971).

Basal leaves 2–6 cm across, rounded to broadly ovate, indistinctly lobed, margin scalloped. Flower-stalk thickened. Sepals 2–3 mm, erect. Petals 2–4 mm, erect, 3-fid, white to purple-tinged. Stamens 5, on the same radii as the sepals. Stigmas flattened. *W North America*. H2. Late spring–early summer.

7. M. breweri A. Gray. Illustration: Hitchcock et al., Vascular plants of the Pacific Northwest **3**: 23 (1971); Rickett, Wild flowers of the United States **6**: pl. 99 (1973); Clark, Wild flowers of the Pacific Northwest, 215 (1976).

Plants with slender rhizomes. Basal leaves 4–8 cm across, rounded, indistinctly lobed and scalloped, sparsely hairy. Inflorescence-stalk to 30 cm, leafless, inflorescence with 20–40 flowers, flower-stalks 1–2 mm. Sepals very small. Petals 1–2 mm, comb-like, greenish yellow. Stamens 5, on the same radii as the sepals. Stigmas bilobed. *C to W North America*. H1. Late spring–early summer.

8. M. ovalis Greene. Illustration: Hitchcock et al., Vascular plants of the Pacific Northwest **3**: 25 (1971).

Basal leaves 4–6 × 2–4 cm, ovate, indistinctly lobed, margin scalloped or toothed, coarsely hairy. Inflorescence-stalk to 30 cm, leafless; inflorescence with 20–40 flowers, flower-stalks 1–2 mm, stout. Sepals very small. Petals *c.* 2 mm, dissected into 3–5 thread-like segments, greenish yellow. Stamens 5, on the same radii as the sepals. Stigmas bilobed. *W North America*. H2. Late spring–early summer.

21. TELLIMA R. Brown
D.M. Miller

Herbaceous perennials with thick, short rootstocks. Basal leaves to 10 cm across, kidney-shaped, stiffly hairy, 5–7-lobed, coarsely toothed, light green; stalks 15–25 cm, with long hairs; stem-leaves 2 or 3, smaller, almost stalkless. Flowers bisexual, radially symmetric, 15–30, more or less drooping in a raceme to 30 cm, stalk to 75 cm; flower-stalks to 4 mm. Perigynous zone 8–10 mm, inflated, glandular-hairy. Sepals 5, 3 mm, ovate, pale green. Petals 5, reflexed, 4–6 mm, lanceolate, fringed into linear segments, greenish white turning pinkish red. Stamens 10, filaments very short. Styles 2, divided to half-way, carpels 2. Fruit a many-seeded capsule.

A genus of a single species which grows well in rock gardens or woodland conditions often naturalising in moist soil rich in organic material. Propagate by seed or division in spring or autumn.

1. T. grandiflora (Pursh) Douglas (*T. odorata* Howell). Illustration: Edwards's Botanical Register, t. 1178 (1828); Rickett, Wild flowers of the United States **5**: pl. 82, 83 (1971).
W North America (Alaska to California). H3. Late spring–early summer.

The names 'Rubra' and 'Purpurea' have been used for cultivated variants with leaves which become reddish purple, especially in winter.

22. TOLMIEA Torrey & Gray
D.M. Miller

Herbaceous perennials with creeping rhizomes. Basal leaves 5–12 × 4–10 cm, heart-shaped, palmately veined, apex acute, shallowly lobed and toothed, hairy; stalks 5–20 cm, hairy; stem-leaves smaller with shorter stalks. Many leaves produce plantlets at the junction of blade and stalk. Inflorescence a narrow raceme to 60 cm, with 20–50 bisexual, shortly stalked, bilaterally symmetric flowers. Perigynous zone cylindric to funnel-shaped. Sepals 5, 3–4 mm, 2 smaller, ovate, greenish purple. Petals 4, *c.* 6 mm, thread-like, purplish brown. Stamens usually 3, shorter than petals, unequal. Styles 2. Fruit a narrow, many-seeded capsule.

A genus of a single species which thrives in moist woodland conditions where it often naturalises. It is best grown in a humus-rich soil, and is useful as ground cover; it is also grown as an unusual pot-plant. Propagate by division or planting leaves bearing plantlets in a sandy, peaty soil.

1. T. menziesii (Pursh) Torrey & Gray. Illustration: Everett, The New York Botanical Garden illustrated encyclopedia of horticulture **10**: 3359 (1982); Jelitto, Schacht & Fessler, Die Freiland Schmuckstauden, 630 (1985).
W North America (S Alaska to California). H4. Late spring–early summer.

Plants with foliage streaked and splashed with yellow or white are grown as 'Taff's Gold' or 'Maculata'.

23. BENSONIELLA Morton
D.M. Miller

Herbaceous perennials with slender, scaly, branching rhizomes. Leaves all basal, 4–8 cm wide, shallowly 5–7-lobed, scalloped, base cordate, slightly hairy beneath; stalk to more than 7 cm. Flowers bisexual in rather narrow, dense racemes on stems to 25 cm. Perigynous zone saucer-shaped, creamy white. Sepals 5, *c.* 2 mm, creamy white. Petals 5, 2–3 mm, linear, white. Stamens 5, anthers pink. Styles 2. Capsules with many seeds.

A genus of a single species (formerly known as *Bensonia* Abrams & Bacigalupi), suitable for growing as ground cover in moist woodland conditions in semi-shade. Propagate by division in autumn.

1. B. oregona (Abrams & Bacigalupi) Morton (*Bensonia oregona* Abrams & Bacigalupi). Illustration: Abrams & Ferris, Illustrated flora of the Pacific States **2**: f. 2280 (1944).
NW USA (Oregon). H3. Early summer.

24. MUKDENIA Koidzumi
P.G. Barnes

Herbaceous perennials with short scaly rhizomes. Leaves 1 or 2, palmately lobed, with forward-pointing teeth, hairless and slightly fleshy. Scape leafless. Inflorescence a compact bractless panicle of small, almost stalkless white flowers. Sepals 5 or 6, oblong, white; petals 5–6. Stamens 5–6, about equalling the petals; carpels 2, free to about the middle. Fruit a capsule of 2 locules.

A genus of 2 species in China and Korea. Easily cultivated in a humus-rich acid soil that does not dry out. Propagate by seed or division in early spring.

1. M. rossii (Engler) Koidzumi (*Aceriphyllum rossii* Engler). Illustration: Quarterly Bulletin of the Alpine Garden Society **2**: 286 (1933) – as Aceriphyllum borisii; Iconographia cormophytorum Sinicorum **2**: 137 (1972); Jelitto, Schacht & Fessler, Die Freiland Schmuckstauden, 10 (1985); The Garden **113**(12): 569 (1988).

Leaves 4–7 × 5–8 cm, circular, palmately lobed to about half-way, lobes ovate, acute, with forward-pointing teeth; leaf-stalks 7–9 cm. Scape 20–40 cm, flowers *c.* 6 mm. *N China*. H4. Summer.

25. DEINANTHE Maximowicz
P.G. Barnes

Herbaceous perennials with woody, creeping rhizomes. Stems erect, with scale-like leaves at the base and 2–4 large leaves near the top. Leaves opposite, rather large, broadly ovate or elliptic, stalked, with forward-pointing teeth, coarsely hairy. Inflorescence a hairless corymb, often with a few small sterile flowers consisting of 3 or 4 sepals. Fertile flowers nodding, with an inferior, conical ovary and 5 rounded sepals. Petals 5 or more, rounded, deciduous; stamens numerous, style 1, rather prominent with 5 small lobes. Ovary 5-celled, seeds numerous, with a short tail at each end.

A genus of 2 species from NE Asia. They

are fairly hardy but require cool, moist and somewhat shady conditions and an acid soil. Propagate by division or seed.

1a. Leaves notched; flowers white, flat
1. bifida
  b. Leaves mostly acuminate; flowers blue, cupped   **2. caerulea**

1. D. bifida Maximowicz. Illustration: Quarterly Bulletin of the Alpine Garden Society **50**: 223 (1982); Satake et al., Wild flowers of Japan **2**: pl. 146 (1985); Kitamura & Murata, Coloured illustrations of herbaceous plants of Japan **2**: f. 278 (1986).

Stems 20–50 cm. Upper leaves 10–20 × 6–10 cm, elliptic, with forward-pointing teeth, apex deeply notched. Flowers *c.* 2 cm across, white. Petals *c.* 1 cm, rounded, spreading widely; anthers and filaments yellow. *Japan.* H4. Summer.

2. D. caerulea Stapf. Illustration: Botanical Magazine, 8373 (1911); Quarterly Bulletin of the Alpine Garden Society **50**: 214 (1982); Thomas, Perennial garden plants, f. 9 (1982); Jelitto, Schacht & Fessler, Die Freiland Schmuckstauden, 188 (1985).

Stems to 35 cm. Leaves 10–20 × 5–15 cm, broadly ovate, apex acuminate or occasionally deeply notched. Flowers 2–3 cm across, pale violet-blue. Petals 5–8, rounded, 1–1.5 cm, concave. Filaments and anthers blue. *China.* H4. Summer.

26. PELTOBOYKINIA (Engelmann) Hara
D.M. Miller
Large herbaceous perennials with short, thick rhizomes. Basal leaves few, 10–25 cm, rounded, peltate on long stalks, palmate with 7–13 shallowly toothed lobes, almost hairless, glossy green; stem-leaves 2 or 3, smaller, almost stalkless. Flowers bisexual, radially symmetric, in terminal cymes on shoots to 60 cm. Calyx with 5 erect lobes, each 4–5 mm. Petals 5, 1–1.2 cm × 5 mm, oblanceolate, toothed at the tip, erect, pale yellow. Stamens 10, anthers dark brown. Styles 2, *c.* 3 mm. Capsules 1–1.3 cm.

A genus of 1 or 2 species from Japan which are easily grown in the wild garden in partial shade in humus-rich, moist soils. Propagate by seed or division in spring.

1. P. tellimoides (Maximowicz) Hara (*Saxifraga tellimoides* Maximowicz; *Boykinia tellimoides* (Maximowicz) Engelmann). Illustration: Botanical Magazine, 9002 (1924); Wehrhahn, Die Gartenstauden, 555 (1930).

Japan. H4. Spring–early summer.

27. CHRYSOSPLENIUM Linnaeus
P.G. Barnes
Herbaceous perennials with underground rhizomes or rooting prostrate stems. Leaves stalked, alternate or opposite, slightly fleshy and often hairy. Erect stems with leaves often clustered towards the tips, the lower leaves much smaller. Flowers in bracted dichotomous cymes on erect shoots, the bracts leaf-like, sometimes coloured. Sepals 4 or 5, petals absent. Stamens 8 or 10, surrounded at the base by a fleshy, 8- or 10-lobed disc. Styles 2, free, carpels 2, united, more-or-less inferior, opening along the inner edge. Seeds numerous, black.

About 60 species in Europe, NE Asia, N Africa, N America and temperate S America. Hardy perennials, mostly easily grown in a moist soil and a sunny position. A few species are occasionally grown on peat walls or in damp rock-gardens or stream-sides. Propagate by seed or more usually by division.

1a. Leaves alternate, bracts bright yellow   **1. davidianum**
  b. Leaves opposite, bracts greenish yellow   **2. oppositifolium**

1. C. davidianum Maximowicz. Illustration: Iconographia cormophytorum Sinicorum **2**: 143 (1972).

Mat-forming perennial with stems 5–12 cm. Stem-leaves alternate, the upper with blades 1.5–5 cm, broadly ovate or rounded, with sparse coarse hairs. Bracts 5–10 mm, obovate, slightly scalloped, hairless, bright yellow. Flowers 3–4 mm across, yellow. *W China.* H4. Spring.

2. C. oppositifolium Linnaeus. Illustration: Ross-Craig, Drawings of British plants, **10**: pl. 16 (1957); Keble Martin, Concise British flora in colour, pl. 32 (1965); Phillips, Wild flowers of Britain, 30 (1977).

Leafy stems prostrate and rooting; flowering-stems erect, to 15 cm. Lower leaves with blades 1–2 cm, circular, scalloped or entire, mostly longer than the stalk. Flower-stems with 1–3 pairs of smaller leaves; bracts greenish yellow. Flowers 3–4 mm across. *W, C & S Europe.* H4. Spring.

28. RIBES Linnaeus
M.C. Tebbitt
Deciduous or occasionally evergreen, low to medium shrubs; branches with or without bristles and or thorns. Leaves alternate, simple, usually lobed and toothed or scalloped, stalked. Flowers usually bisexual, sometimes unisexual

and plants dioecious, usually 5-, rarely 4-parted, in few–many-flowered racemes. Ovary inferior, styles 2. Fruit a berry, crowned by remains of calyx.

A genus of about 150 species from cool and temperate regions of the northern hemisphere and S America. The majority of species are grown for their flowers or occasionally their attractive foliage; a few species are grown for their edible fruit. All are easily grown in a fertile, open soil in full sun, either in the open or trained against a wall. Plants should be pruned immediately after flowering if straggly, but care should be taken not to cut back shoots too severely as the current year's flowers are borne on the previous year's growth. Propagate by nodal hardwood cuttings in autumn to winter, or in the case of *R. laurifolium* by semi-ripe cuttings during the summer.

The most frequently encountered species are *R. sanguineum*, *R. odoratum* and *R. uva-crispa*.

1a. Branches lacking thorns and bristles   2
  b. Branches with thorns and or bristles   53
2a. Plant evergreen   3
  b. Plant deciduous   6
3a. Leaves lobed   **56. gayanum**
  b. Leaves not lobed   4
4a. Leaves 2–4 cm, margin entire or with a few small teeth
        22. viburnifolium
  b. Leaves 5–10 cm, margin scalloped   5
5a. Shoots hairy when mature   **70. henryi**
  b. Shoots hairless when mature   **71. laurifolium**
6a. Flowers consistently in clusters or racemes of 1–3   7
  b. Flowers usually in clusters or racemes of more than 3   11
7a. Calyx tubular   8
  b. Calyx bell-shaped   9
8a. Flowers white, greenish to yellowish; styles with long hairs   **18. cereum**
  b. Flowers usually pink; styles usually hairless   **19. inebrians**
9a. Fruit green   **11 ambiguum**
  b. Fruit red to black   10
10a. Young branches grey; racemes compact   **45. hirtellum**
  b. Young branches reddish; racemes elongated   **44. grossularioides**
11a. Plant with flowers only of one sex   12

b. Plant with bisexual flowers   19
12a. New leaves appearing late spring; fruit black   13
b. New leaves appearing early spring; fruit red, yellow or green   14
13a. Leaves to 2 cm   **65. vilmorinii**
b. Leaves to 6 cm   **68. luridum**
14a. Flowers purple-brown or brownish red   15
b. Flowers yellow-green to green-red   16
15a. Flowers purple-brown   **67. glaciale**
b. Flowers brownish red   **66. tenue**
16a. Shoots glandular hairy   17
b. Shoots hairless   18
17a. Leaves usually constantly 3-lobed; racemes 1–5 cm   **62. orientale**
b. Leaves 3–5-lobed; racemes 3–20 cm   **69. maximowiczii**
18a. Leaf-lobes obtuse   **63. alpinum**
b. Leaf-lobes acute   **64. distans**
19a. Plant prostrate   20
b. Plant erect, occasionally low-growing   24
20a. Fruit with glandular hairs or bristles   21
b. Fruit smooth, lacking hairs and bristles   23
21a. Leaves 5–7-lobed; leaves unpleasent smelling   **57. glandulosum**
b. Leaves usually 5-lobed; leaves not unpleasant smelling   22
22a. Leaf-lobes triangular; racemes compact   **13. coloradensis**
b. Leaf-lobes ovate; racemes loose   **12. looseiflorum**
23a. Leaves mostly 3-lobed; fruit red   **5. triste**
b. Leaves equally 3- and 5-lobed; fruit brownish   **23. procumbens**
24a. Fruit red   25
b. Fruit black, blue, purple, brown, yellowish white, green   33
25a. Flowers with a ring around style-base   26
b. Flowers without a ring around style-base   27
26a. Leaf-base deeply cordate with a narrow sinus   **4. rubrum**
b. Leaf-base shallowly to mediumly cordate, with a wide sinus   **6. warszewiczii**
27a. Calyx with 5 'warts' inside   **2. mandshuricum**
b. Calyx lacking 5 'warts' inside   28
28a. Flowers yellow   29
b. Flowers greenish, brown, reddish   30
29a. Branches hairy when young   **1. multiflorum**

b. Branches hairless when young   **55. fasciculatum**
30a. Young shoots hairy   **1. multiflorum**
b. Young shoots hairless   31
31a. Leaf-base truncate to shallowly cordate, sinus broad; flowers large   **3. spicatum**
b. Leaf-base usually cordate, sinus medium; flowers small to medium   32
32a. Sepals short round, with hairs on margin   **7. petraeum**
b. Sepals broadly ovate, without hairs on margins   **8. emodense**
33a. Leaf-lobes 5–7   **20. bracteosum**
b. Leaf-lobes 3–5   34
34a. Flowers tubular   35
b. Flowers cup- to bell-shaped   41
35a. Racemes to 30 cm   **9. longiracemosum**
b. Racemes to 15 cm   36
36a. Leaves with white or grey matted hairs below   37
b. Leaves hairless or hairy below   38
37a. Leaves 2–5 cm across, 3-lobed, roughly hairy above   **16. malvaceum**
b. Leaves 5–10 cm across, 3–5-lobed, softly hairy above   **15. sanguineum**
38a. Leaves with yellow gland spots on both surfaces   **30. americanum**
b. Leaves without yellow gland spots on both surfaces   39
39a. Leaf- and flower-stalks with black-stalked glands   **17. ciliatum**
b. Leaf- and flower-stalks without black-stalked glands   40
40a. Young shoots hairless or sparsely hairy   **24. aureum**
b. Young shoots hairy   **25. odoratum**
41a. Racemes to 30 cm   **9. longeracemosum**
b. Racemes to 15 cm   42
42a. Sepals white on outer surface   43
b. Sepals yellow, green, brown, pink, or red on outer surface   44
43a. Leaves kidney-shaped, wider than long   **26. hudsonianum**
b. Leaves circular   47
44a. Branches with prominent yellow glands; plant very aromatic   **27. nigrum**
b. Branches without prominent yellow glands; plant not particularly aromatic   45
45a. Calyx-cup interior with a slightly recognizable ring   **6. warszewiczii**

b. Calyx-cup interior without a slightly recognizable ring   46
46a. Leaf-lobes acute, sharply toothed   **28. petiolare**
b. Leaf-lobes obtuse, scalloped   **14. nevadense**
47a. Flowers stalkless   **10. moupinense**
b. Flowers stalked   48
48a. Plant lacking stolons   49
b. Plant stoloniferous   **29. ussuriense**
49a. Leaves to 15 cm across; racemes erect   **21. japonicum**
b. Leaves to 10 cm across; racemes pendent   50
50a. Leaves both 3- and 5-lobed on the same plant   **6. warszewiczii**
b. Leaves mostly 3-lobed   51
51a. Leaf-lobes obtuse   **58. magellanicum**
b. Leaf-lobes acute   52
52a. Sepals round, with hairs on margin   **7. petraeum**
b. Sepals broadly ovate, without hairs on margin   **8. emodense**
53a. Flowers in racemes of 5–20   54
b. Flowers in racemes of 1–5   57
54a. Fruit black   **31. lacustre**
b. Fruit red, yellow or green   55
55a. Branches thornless   **69. maximowiczii**
b. Branches with at least a few thorns   56
56a. Leaves 3-lobed; racemes erect   **60. pulchellum**
b. Leaves 5-lobed; racemes pendent   **32. montigenum**
57a. Plant evergreen; flowers *Fuchsia*-like   **33. speciosum**
b. Plant deciduous; flowers not *Fuchsia*-like   58
58a. Fruit smooth, hairless   72
b. Fruit glandular-hairy to bristly   59
59a. Plant with flowers of one sex only   **61. giraldii**
b. Plant with bisexual flowers   60
60a. Petals rolled inwards   61
b. Petals not rolled inwards   63
61a. Young shoots densely bristly   **35. menziesii**
b. Young shoots hairy, but not bristly   62
62a. Sepals often only 4; petals pink   **34. lobbii**
b. Sepals constantly 5; petals white   **36. roezlii**
63a. Calyx tubular   **47. setosum**
b. Calyx bell- to urn-shaped   64
64a. Petals strap-shaped   65
b. Petals fan-shaped, obovate   66

65a. Branches wavy
42. stenocarpum
b. Branches straight 68
66a. Plant with prickles, hairless or hairy but not bristly 67
b. Plant bristly, without prickles
44. grossularioides
67a. Flowers greenish **48. uva-crispa**
b. Flowers orange-red
38. pinetorum
68a. Leaves hairless to sparsely hairy 69
b. Leaves hairy 70
69a. Ovaries glandular bristly
43. alpestre
b. Ovaries bristly, but only the very shortest occasionally glandular
37. californicum
70a. Fruit green **40. burejense**
b. Fruit red or black 71
71a. Sepals white **47. setosum**
b. Sepals brownish green
50. cynosbati
72a. Branches grey and thorns large, hooked **51. divaricatum**
b. Branches and thorns not as above 73
73a. Sepals white or yellowish 74
b. Sepals green to red 79
74a. Sepal-lobes yellow
39. quercetorum
b. Sepal-lobes white 75
75a. Thorns constantly single, thin, bristles absent 76
b. Thorns 1–3 on a single plant, often stout, bristles often present 77
76a. Leaf-margin toothed, sparsely hairy
54. curvatum
b. Leaf-margin scalloped, usually hairless **53. niveum**
77a. Leaves hairless
49. oxyacanthoides
b. Leaves finely hairy 78
78a. Prickles usually less than 1 cm, awl-shaped **47. setosum**
b. Prickles usually c. 1 cm, not awl-shaped **46. leptanthum**
79a. Branches occasionally with 5–7 prickles at node **41. aciculare**
b. Branches only with 1–3 prickles at node, never 5–7 80
80a. Branches with stout thorns 81
b. Branches without stout thorns, bristly or prickly only 83
81a. Branches wavy
42. stenocarpum
b. Branches straight 82
82a. Leaves slightly 3-lobed
52. rotundifolium
b. Leaves deeply 3–5-lobed
49. oxyacanthoides

83a. Sepal-lobes twice as long as calyx-cup **59. diacanthum**
b. Sepal-lobes same length as calyx-cup **45. hirtellum**

1. R. multiflorum Kitaibel. Illustration: Janczewski, Monographie des Groseilliers, 274 (1907); Krüssmann, Manual of cultivated broad-leaved trees & shrubs **3**: 215 (1986).

Erect shrub to 2 m; branches thornless, ash-grey, hairy when young; buds large. Leaves c. 10 cm, circular, 3–5-lobed, toothed, grey-white beneath. Flowers bell-shaped, to 50, in racemes of c. 12 cm. Calyx golden yellow-green. Stamens wide-spreading. Fruit red. *SE Europe, Balkan Peninsula.* H2. Late spring–early summer.

R. × urceolatum Tausch (*R. multiflorum × R. petraeum*). Illustration: Janczewski, Monographie des Groseilliers, 487, (1907). Similar to *R. multiflorum*, but flowers c. 25 in loose racemes, to 12 cm. Flowers brownish. Fruit red. *Garden origin.* H1–2. Early summer.

R. × koehneanum Janczewski (*R. multiflorum × R. rubrum*). Illustration: Janczewski, Monographie des Groseilliers, 486 (1907). Dome-shaped shrub. Leaves 6.5 cm across, 3–5-lobed. Flowers brownish, up to 35 in racemes, to 10 cm. Stamens pink; stamens and style same length. *Garden origin.* H1–2. Late spring.

The commonly grown garden currant.

2. R. mandshuricum (Maximowicz) Komarov. Illustration: Janczewski, Monographie des Groseilliers, 275 (1907); Krüssmann, Manual of cultivated broad-leaved trees & shrubs **3**: 205 (1986).

Shrub to 2 m, bark usually almost black, branches spineless. Leaves to 9 × 11 cm, broadly ovate, 3-lobed, lobes usually acute, coarsely toothed. Flowers in many-flowered pendent racemes. Calyx cup-shaped, the inside with 5 'warts', not interconnected by a raised ring. Fruit red. *NE Asia.* Early summer.

3. R. spicatum Robson (*R. rubrum* Linnaeus in part; *R. schlechtendahlii* Lange). Illustration: Krüssmann, Manual of cultivated broad-leaved trees & shrubs **3**: 204 (1986).

Erect shrub to 2 m; branches hairless. Leaves to 10 cm across, circular, 3–5-lobed, base truncate to shallowly cordate, sinus broad. Flowers light green, tinged brown-red, usually in erect racemes; calyx cup-shaped, lacking a ring around style-base. Fruit red. *N Europe, N Asia.* H1. Early summer.

4. R. rubrum Linnaeus (*R. silvestre* (Lamark) Mertens; *R. sativum* (Reichenbach) Symes; *R. vulgare* Lamark). Illustration: Krüssmann, Manual of cultivated broad-leaved trees & shrubs **3**: 204 (1986).

Erect, broad shrub; young shoots slightly hairy, glandular. Leaves to 6 cm across, circular, 3–5-lobed, lobes acute, base deeply cordate, sinus narrow. Flowers greenish to reddish, in pendent to spreading racemes; calyx cup-shaped, style-base with a pentagonal ring. Fruit red. *W Europe.* H1–2. Early summer.

R. × houghtonianum Janczewski (*R.* 'Houghton Castle'). Illustration: Janczewski, Monographie des Groseilliers, 479 (1907). Intermediate between the parents. Vigorous shrub; new growth appearing late. Leaves c. 6 cm across, hairy, base slightly cordate. Flowers green, tinged brown, 8–18 in racemes, to 5 cm. Fruit red, edible. *Garden origin.* H1. Early summer.

A hybrid between *R. silvestre × R. spicatum.*

5. R. triste Pallas (*R. albinervium* Michaux). Illustration: Janczewski, Monographie des Groseilliers, 283 (1907); Gleason, New illustrated flora of the north-eastern United States and adjacent Canada **2**: 279 (1952); Krüssmann, Manual of cultivated broad-leaved trees & shrubs **3**: 206 (1986)

Shrub to 50 cm, stems decumbent, spineless. Leaves 6–10 cm across, 3-lobed, lobes toothed, base usually cordate. Flowers in racemes, the later glandular, mostly shorter than the leaves. Flowers reddish. Calyx broadly bell-shaped, sepal-lobes spreading. Fruit c. 6 mm across, smooth, red. *N North America.* H1. Early summer.

6. R. warszewiczii Janczewski. Illustration: Janczewski, Monographie des Groseilliers, 285 (1907); Krüssmann, Manual of cultivated broad-leaved trees & shrubs **3**: 204 (1986)

Similar to *R. spicatum.* Erect shrub to 1.5 m. Leaves to 9 cm, rounded, 3–5-lobed, base cordate. Flowers green tinged with red in pendent racemes, the later 5–7 cm. Calyx-cup interior with a slightly recognisable ring around the style-base. Fruits red-black. *E Siberia.* H1. Early summer.

7. R. petraeum Wulfen (*R. petraeum* var. *bullatum* (Otto & Dietrich) Schneider). Illustration: Janczewski, Monographie des Groseilliers, 291 (1907); Huxley, Mountain

flowers of Europe, 1040 (1986);
Krüssmann, Manual of cultivated broad-
leaved trees & shrubs **3**: 205 (1986).

Erect shrub to 1.5 m or more; branches
grey-brown, hairless. Leaves 7–10 cm,
circular, usually 3-lobed, lobes acute,
toothed, hairy beneath, base cordate to
truncate. Flowers small, bell-shaped, green
to reddish, in dense many-flowered
racemes. Sepals short, round, hairy on
margins. Petals ½ as long as sepals. Style
cone-shaped. Fruit red to purple. *W & C
Europe to Siberia*. H1. Early summer.

Var. **altissimum** (Turczaninow)
Janczewski. Shrub to 3 m. Leaves to 15 cm
across. Flowers pale red, *c.* 20, in racemes,
5–7 cm. *Siberia*.

Var. **atropurpureum** (C.A. Meyer)
Schneider. Leaves to 15 cm across, 3-
lobed. Flowers purple outside, lighter
inside, *c.* 15 in racemes, 2–4 cm. *Siberia*.

Var. **biebersteinii** (Berland) Schneider
(*R. caucasicum* Bieberstein; *R. biebersteinii*
Berland). Leaves *c.* 12 cm across, usually
5-lobed, hairy beneath. Flowers reddish,
racemes to 10 cm. *Caucasus*.

R. × gondouinii Janczewski (*R.
petraeum × R. silvestre*). Illustration:
Janczewski, Monographie des Groseilliers,
484 (1907). Intermediate between parents.
Young shoots red, hairless, appearing
early. Flowers bell-shaped, in almost
horizontal racemes. Fruit red. *Garden
origin*. H1–2. Early summer.

8. R. emodense Rehder. (*R. himalayense*
Decaisne not Royle; *R. meyeri* Schneider
not Maximowicz). Illustration: Krüssmann,
Manual of cultivated broad-leaved trees &
shrubs **3**: 205 (1986).

Similar to *R. petraeum*, but sepals
broadly ovate, lacking hairs on margin;
petals wedge-shaped, erect. Fruit large, red
to black. *Himalayas, China (Yunnan)*. H1.
Early summer.

9. R. longeracemosum Franchet.
Illustration: Janczewski, Monographie
des Groseilliers, 301 (1907); Krüssmann,
Manual of cultivated broad-leaved trees &
shrubs **3**: 205 (1986).

Shrub to 3 m; branches thornless,
hairless. Leaves to 14 cm, circular, 3–5-
lobed, lobes acute, hairless. Flowers tubular
to bell-shaped, reddish to greenish, to 15 in
loose, pendent racemes to 30 cm; sepals and
petals erect. Stamens and style protruding.
Fruit black, glossy. *W China (Sichuan, Hubei,
Xizang)*. H1. Early summer.

10. R. moupinense Franchet. Illustration:
Janczewski, Monographie des Groseilliers,

299 (1907); Krüssmann, Manual of
cultivated broad-leaved trees & shrubs **3**:
205 (1986).

Erect shrub, 1–2 m; branches thornless,
hairless. Leaves variable, to 16 cm across,
3–5-lobed, with scattered stalkless glands
on both surfaces. Flowers red or green-red,
in racemes 4–12 cm. Petals and sepals
erect. Stamens included. Fruit black,
glossy. *China*. H1. Early summer.

11. R. ambiguum Maximowicz.
Illustration: Janczewski, Monographie des
Groseilliers, 304 (1907).

Shrub to 60 cm; branches thornless.
Leaves 2–5 cm across, kidney-shaped,
3–5-lobed, lobes obtuse, stocky glandular
beneath. Flowers 1–2, stalks *c.* 1 cm.
Sepals elliptic, greenish. Fruit green,
glandular-hairy, translucent. *Japan, China
(Sichuan)*. H1–2. Early summer.

Commonly found growing on old trees
in the wild.

12. R. laxiflorum Pursh (*R. affine*
Douglas). Illustration: Janczewski,
Monographie des Groseilliers, 307 (1907);
Krüssmann, Manual of cultivated broad-
leaved trees & shrubs **3**: 204 (1986).

Prostrate shrub, branches thornless.
Leaves 6–8 cm across, circular, deeply
5-lobed, base cordate, sharply toothed,
hairless above, hairy beneath. Flowers
6–12, in loose erect racemes to 8 cm;
sepals reddish; petals fan-shaped. *N
America (Alaska to N Carolina)*. H1. Early
summer.

13. R. coloradensis Coville. Illustration:
Janczewski, Monographie des Groseilliers,
310 (1907); Krüssmann, Manual of
cultivated broad-leaved trees & shrubs **3**:
204 (1986).

Low-growing shrub to 50 cm; young
shoots hairy. Leaves 5–8 cm; broadly
ovate, usually 5-lobed, lobes triangular,
hairless above, hairy on veins beneath.
Flowers 6–12, in erect, glandular-hairy
racemes, to 5 cm. Sepals greenish to
reddish; petals fan-shaped, purple. Fruit
black. *USA*. H1. Summer.

14. R. nevadense Kellogg. Illustration:
Janczewski, Monographie des Groseilliers,
316 (1907); Krüssmann, Manual of
cultivated broad-leaved trees & shrubs **3**:
203 (1986).

Erect shrub to 1.5 m; branches
thornless, usually hairless. Leaves 3–6 cm
across, circular, usually 3-lobed, leaves
rounded, scalloped. Flowers *c.* 20, in
pendent racemes. Sepals pink; petals white,
rounded-oblong. Fruit blue, glandular.

USA (California, Sierra Nevada). H2–3. Early
summer.

15. R. sanguineum Pursh. Illustration:
Janczewski, Monographie des Groseilliers,
320 (1907); Hay & Beckett, Reader's
Digest encyclopedia of garden plants and
flowers, 592 (1978); Krüssmann, Manual
of cultivated broad-leaved trees & shrubs
3: 203 (1986); Rix & Phillips, Shrubs, 40
(1989).

Shrub to 4 m; branches spineless,
aromatic, glandular. Leaves 5–10 cm
across, circular, 3–5-lobed, base cordate,
dark green, hairy above, white felted
beneath, stalk hairy. Flowers in many-
flowered, glandular-hairy, erect or pendent
racemes, to 8 cm. Calyx reddish purple,
lobes longer than tube. Petals white to
reddish purple, ½ as long as sepals. Fruit
to 1 cm across, black, blue-white bloomed.
W America. H1. Early Summer.

The most frequently grown of all the
ornamental currents. Many cultivars exist
and include: 'Albescens', flowers whitish;
'Atrorubens', compact habit, flowers deep
red, small; 'Brocklebankii', slow-growing,
flowers yellow; 'Grandiflorum', flowers red,
in large racemes; 'Koja', flowers dark red;
'Plenum', slow-growing, flowers double,
red; 'Splendens', flowers dark red and
'Pulborough Scarlet', flowers deep red with
a white centre, in large racemes, one of the
best cultivars available today.

R. × fontenayense Janczewski (*R.
sanguineum × R. uva-crispa*). Illustration:
Janczewski, Monographie des Groseilliers,
492 (1907). Intermediate between parents.
Shrub to 1 m; branches thornless. Leaves
6–8 cm, circular, 3–5-lobed, hairy
beneath. Flowers red, 3–6 in somewhat
pendent racemes. Calyx tube short, broader
than long. Free-flowering. Fruit purple-
black, seldom produced. *Garden origin*. H1.
Early summer.

16. R. malvaceum Smith. Illustration:
Krüssmann, Manual of cultivated broad-
leaved trees & shrubs **3**: 203 (1986).

Very similar to *R. sanguineum* but leaves
roughly hairy above and grey felted
beneath. Ovary white hairy. *USA
(California)*. H2. Early Summer.

R. × bethmontii Janczewski (*R.
malvaceum × R. sanguineum*). Habit like *R.
sanguineum*; leaves roughly hairy, like
R. malvaceum. Flowers bright pink; calyx
urceolate, style almost twice as long as
tube. Pollen sterile. Fruit black; sets fruit
only after pollination by another species.
France. H1–2. Early summer.

17. R. ciliatum Roemer & Schultes. Illustration: Janczewski, Monographie des Groseilliers, 329 (1907).

Dome-shaped shrub to 2 m; much branched, branches glandular-hairy when young; leaf- and flower-stalks with black-stalked glands. Leaves 3–5 cm across, 3–5-lobed, coarsely double-toothed, with flat bristles above, glandular-hairy beneath. Flowers green, outer surface hairy, 6–10 in a pendent raceme. Fruit black, glossy. *Mexico*. H5. Early summer.

18. R. cereum Douglas. Illustration: Janczewski, Monographie des Groseilliers, 337 (1907); Krüssmann, Manual of cultivated broad-leaved trees & shrubs **3**: 209 (1986)

Shrub to 1 m, much branched, without spines. Leaves 1–4 cm across, kidney-shaped, 3–5-lobed, toothed. Flowers white, greenish to yellowish, in short, few-flowered racemes. Calyx tubular, sepals longer than petals. Styles with long hairs. Fruit red. *W North America*. H1. Early Summer.

19. R. inebrians Lindley (*R. pumilum* Nuttall). Illustration: Janczewski, Monographie des Groseilliers, 335 (1907); Krüssmann, Manual of cultivated broad-leaved trees & shrubs **3**: 209 (1986).

Very similar to *R. cereum*, but flowers usually pink, in few-flowered, pendent racemes; ovary usually hairless. Fruit red, glandular. *W North America*. H1. Early Summer.

20. R. bracteosum Douglas. Illustration: Janczewski, Monographie des Groseilliers, 339 (1907); Krüssmann, Manual of cultivated broad-leaved trees & shrubs **3**: 206 (1986).

Erect shrub to 3 m; young shoots sparsely glandular hairy. Leaves 5–20 cm across, circular, 5–7-lobed, lobes lanceolate, double-toothed. Flowers in erect racemes, to 20 cm, much longer than leaves. Calyx green, tinged purplish red. Petals larger than sepals, white. Fruit spherical, black, white bloomed. *W North America (Alaska to N California)*. H1. Early summer.

R. × fuscescens Janczewski (*R. bracteosum × nigrum*). Habit similar to *R. bracteosum*, but calyx reddish brown, bracts small, linear. Fruit larger. *Garden origin*. H2. Early summer.

21. R. japonicum Maximowicz. Illustration: Janczewski, Monographie des Groseilliers, 340 (1907).

Shrub to 2 m; branches thornless. Leaves *c.* 15 cm across, circular, 5-lobed, lobes acute, toothed. Flowers bell-shaped, green or brown, in erect racemes. Fruit black, hairless. *Japan*. H1–2. Early summer.

22. R. viburnifolium A. Gray. Illustration: Janczewski, Monographie des Groseilliers, 341 (1907); Krüssmann, Manual of cultivated broad-leaved trees & shrubs **3**: 211 (1986).

Evergreen shrub to 1.5 m; branches thornless. Leaves 2–4 cm, broadly ovate to elliptic, base round, margin entire or with a few- small teeth. Flowers pink, in erect racemes, to 2.5 cm. Fruit red. *USA (California)*. H4–5. Early summer.

23. R. procumbens Pallas. Illustration: Janczewski, Monographie des Groseilliers, 342 (1907); Krüssmann, Manual of cultivated broad-leaved trees & shrubs **3**: 215 (1986).

Low shrub to 70 cm. Leaves to 8 cm, kidney-shaped, 3–5-lobed. Flowers reddish, in racemes to 4 cm, bracts absent. Fruit brownish, smooth. *E Siberia*. H1. Early summer.

24. R. aureum Pursh (*R. tenuiflorum* Lindley). Illustration: Janczewski, Monographie des Groseilliers, 334 (1907); Hay & Beckett, Reader's Digest encyclopedia of garden plants and flowers, 592 (1978); Krüssmann, Manual of cultivated broad-leaved trees & shrubs **3**: 206 (1986).

Erect shrub to 2 m, branches spineless, bark brown. Leaves 3–5 cm, broadly elliptic, 3–5-lobed, coarsely toothed, hairless. Flowers yellow, fragrant, in few- to many-flowered, pendent racemes. Calyx tubular, sepals spreading, but inclining after flowering. Petals commonly becomming reddish. Fruit purple-brown to black. *USA (California), Mexico*. H1. Early summer.

25. R. odoratum Wendland (*R. aureum* misapplied; *R. fragrans* Loddiges). Illustration: Gleason, New illustrated flora of the north-eastern Unites States and adjacent Canada **2**: 279 (1952); Rix & Phillips, Shrubs, 41 (1989).

Similar to and often confused with *R. aureum*; differs in its hairier young shoots. Branches spineless; leaves 3–8 cm across, maple-like, 3–5-lobed, lobes coarsely toothed. Flowers yellow, 5–10, in racemes, cupule 6–10 × *c.* 1.5 mm, sepals spreading, lobes to half as long as tube. Fruit *c.* 1 cm across, spherical, black. *C North America*. H1. Early summer.

R. × gordonianum Beaton (*R. petraeum × R. sanguineum*). Illustration: Rix & Phillips, Shrubs, 40 (1989). Shrub to 2.5 m; branches spineless, hairless. Leaves *c.* 4 cm, circular, 3–5-lobed, shallowly to coarsely toothed, glandular-hairy on both sides. Flowers in long, many-flowered racemes, more erect than in parents; calyx tube *c.* 3 mm, calyx reddish yellow, or tube red and yellow and lobes red outside, yellow inside. *Garden origin*. H1. Early summer.

26. R. hudsonianum Richards. Illustration: Janczewski, Monographie des Groseilliers, 346 (1907); Gleason, New illustrated flora of the north-eastern United States and adjacent Canada **2**: 278 (1952); Krüssmann, Manual of cultivated broad-leaved trees & shrubs **3**: 203 (1986).

Erect shrub to 1.5 m; branches thornless. Leaves to 10 cm across, broadly ovate, more or less hairy, with resin glands, 3–5-lobed, lobes ovate, obtuse to acute, coarsely toothed. Flowers white, in loose erect racemes to 6 cm; calyx cup-shaped, sepals ovate, spreading. Ovary with resin glands. Fruit black, hairless. *N North America*. H1. Early summer.

27. R. nigrum Linnaeus. Illustration: Janczewski, Monographie des Groseilliers, 347 (1907); Gleason, New illustrated flora of the north-eastern United States and adjacent Canada **2**: 278 (1952); Krüssmann, Manual of cultivated broad-leaved trees & shrubs **3**: 204 (1986).

Rounded shrub to 2 m; branches with yellowish glands, hairy to hairless when young, yellow, aromatic. Leaves 5–10 cm across, circular, 3–5-lobed, hairy beneath. Flowers bell-shaped, 4–10 cm, in pendent racemes. Calyx greenish outside, reddish white inside. Fruit black, edible. *Europe to C Asia & Himalaya*. H1. Early summer.

'Apiifolium', leaves 3-lobed, very deeply incised and toothed; 'Chlorocarpum', fruit green; 'Coloratum', leaves variegated white; 'Heterophyllum', leaves deeply cleft; 'Marmoratum', leaves deeply lobed, marbled cream; 'Xanthocarpum', fruit yellow to white.

R. × culverwellii MacFarlane (*R. schneideri* Maurer). Habit like *R. nigrum*, leaves like *R. uva-crispa*. Fruit sterile. Resembling a gooseberry without the bristles. *Garden origin*. H2. Early summer. A hybrid between *R. nigrum* and *R. uva-crispa*.

28. R. petiolare Fischer. Illustration: Krüssmann, Manual of cultivated broad-leaved trees & shrubs **3**: 203 (1986).

Hairless shrub to 1.5 m. Leaves 10–15 cm across, with resin glands beneath. Flowers white, in erect racemes, to 12 cm. Fruit black, not bloomed. *E Siberia, Manchuria, W North America*. H1. Early summer.

29. R. ussuriense Janczewski. Illustration: Janczewski, Monographie des Groseilliers, 349 (1907).

Stoloniferous shrub to 1 m; young shoots hairy, with yellow resin-glands, camphor-scented. Flowers yellowish green, 5–9 in racemes. Fruit bluish black. *Manchuria to Korea*. H1. Early summer.

30. R. americanum Miller (*R. floridum* Miller). Illustration: Gleason, New illustrated flora of the north-eastern United States and adjacent Canada **2**: 278 (1952); Krüssmann, Manual of cultivated broad-leaved trees & shrubs **3**: 203 (1986).

Shrub to 1.5 m; branches spineless, grey-brown. Leaves 5–8 cm, circular, usually 3-lobed, lobes acute, sharply toothed, with yellow gland spots on both surfaces (under magnification). Flowers *c.* 10 in racemes, to slightly longer than leaves, pendent. Calyx tube greenish yellow. Petals small. Fruit *c.* 6 mm across, black. *N North America*. H1. Summer.

31. R. lacustre (Persoon) Pourret (*R. grossularioides* Michaux; *R. echinatum* Lindley). Illustration: Janczewski, Monographie des Groseilliers, 352 (1907); Gleason, New illustrated flora of the north-eastern United States and adjacent Canada **2**: 277 (1952); Krüssmann, Manual of cultivated broad-leaved trees & shrubs **3**: 207 (1986).

Shrub to 1 m; branches slightly pendent, bristly and slightly prickly. Leaves 3–6 cm across, circular, deeply lobed. Flowers greenish red, 4–10 in loose, pendent racemes, to 9 cm. Fruit small, black, densely bristly. *N North America*. H1. Early summer.

Often wrongly named *R. grossularioides* in gardens; the true **R. grossularioides** Maximowicz is a little known, medium-height, bristly to smooth shrub closely related to *R. alpestre* and distinguished from this species by its anthers which lack glands at their tips.

32. R. montigenum McClatchie (*R. lacustre* var. *molle* A. Gray). Illustration: Janczewski, Monographie des Groseilliers, 355 (1907).

Loosely branched shrub, *c.* 75 cm; branches bristly, with a few thorns. Leaves 1–4 cm across, kidney-shaped, 5-lobed, lobes acute, hairy. Flowers shortly tubular, in short, few-flowered, pendent racemes; calyx brownish green, tube glandular-bristly. Fruit dark red, glandular-bristly. *W North America*. H1. Early summer.

33. R. speciosum Pursh. Illustration: Janczewski, Monographie des Groseilliers, 357 (1907); Hay & Beckett, Reader's Digest encyclopedia of garden plants and flowers, 592 (1978); Krüssmann, Manual of cultivated broad-leaved trees & shrubs **3**: 205 (1986).

Evergreen shrub to 4 m; branches grey-brown at first, densely spiny; thorns to 1 cm, in groups of 2 or 3, at nodes. Leaves 1–4 cm, broadly ovate, 3–5-lobed, hairless. Flowers in groups of 3–5, 4-parted, calyx bell-shaped, sepals narrow, red-purple; stamens long protruding. Fruit densely glandular-bristly, red. *USA (California)*. H2–3. Early summer.

One of the most attractive species.

34. R. lobbii Gray (*R. subvestitum* Hooker & Arnott). Illustration: Janczewski, Monographie des Groseilliers, 359 (1907); Krüssmann, Manual of cultivated broad-leaved trees & shrubs **3**: 210 (1986).

Shrub to 2 m; branches thorny, hairy when young, thorns in groups of 3, 1–2 cm. Leaves 2–3.5 cm across, circular to heart-shaped, 3–5-lobed, finely hairy above. Flowers large, single or in pairs; sepals deep purple, often only 4. Anthers black, filaments pink. Fruit purple, densely glandular-hairy. *N America (British Columbia to California)*. H2–3. Late spring.

35. R. menziesii Pursh (*R. subvestitum* Hooker & Arnott). Illustration: Janczewski, Monographie des Groseilliers, 362 (1907); Krüssmann, Manual of cultivated broad-leaved trees & shrubs **3**: 207 (1986).

Shrub to 2 m. Young shoots densely bristly, with thin spines in groups of 3, 1–2 cm. Leaves 2–4 cm across, 3–5-lobed, hairless to sparsely hairy above, velvety hairy and glandular beneath. Flowers 1 or 2, calyx tubular, lobes 3 times as long as tube, purple. Petals whitish. Stamens long protruding. Fruit densely bristly black. *USA (Oregon to California)*. H5–G1. Early summer.

R. × darwinii F. Koch (*R. menziesii × R. niveum*). Illustration: Krüssmann, Manual of cultivated broad-leaved trees & shrubs **3**: pl. 70 (1986). Vigorous, erect shrub to 3 m. Intermediate between parents. Leaves similar to those of *R. niveum*. Flowers 1–4 in pendent racemes; sepals red, apex whitish, reflexed; petals white. Style to 1.5 cm, reddish. Fruit black, hairless. *Garden origin*. H1–2. Early summer.

36. R. roezlii Regel (*R. amictum* Greene). Illustration: Krüssmann, Manual of cultivated broad-leaved trees & shrubs **3**: 210 (1986).

Erect shrub to 1.5 m; branches with thin spines in groups of 3 at nodes, each to 1.5 cm, young branches hairy but not bristly. Leaves 1.5–2.5 cm, circular, 3–5-lobed, toothed, base cuneate to almost cordate, leaf-stalk short. Flowers in clusters of 1–3; calyx bell-shaped, lobes purple; petals 1/2 as long as sepals, white; filaments included, very bristly. Fruit 1–1.5 cm, purple, very bristly. *USA (California)*. H3. Early summer.

37. R. californicum Hooker & Arnott. Illustration: Krüssmann, Manual of cultivated broad-leaved trees & shrubs **3**: 210 (1986).

Similar to *R. menziesii* but branches lacking bristles; leaves hairless or almost so; flowers green or reddish. *USA (California)*. H2. Early summer.

38. R. pinetorum Greene. Illustration: Janczewski, Monographie des Groseilliers, 370 (1907); Krüssmann, Manual of cultivated broad-leaved trees & shrubs **3**: 205 (1986).

Shrub to 2 m; branches thorny, but lacking bristles. Leaves 2–3 cm, heart-shaped, deeply 3–5-lobed, hairless above, hairy below. Flowers large, usually solitary, occasionally paired; calyx bell-shaped, orange-red, sepals twice as long as calyx-tube; style hairless. Fruit purple. *USA*. H1–2. Early summer.

39. R. quercetorum Greene.
Erect shrub to 1.5 m; branches pendent, thorny, new growth appearing very early. Leaves 1–2 cm, circular, deeply 3–5-lobed. Flowers shortly tubular, yellowish or whitish; sepals spreading; petals short. Ovary and style hairless. Fruit black, hairless. *USA (California)*. H3–4. Early summer.

40. R. burejense F. Schmidt. Illustration: Janczewski, Monographie des Groseilliers, 371 (1907).

Shrub to 1 m; branches usually very bristly, with thorns at the nodes, 1 cm. Leaves 2–6 cm across, circular, deeply 3–5-lobed, lobes obtuse, toothed, hairy on both surfaces. Flowers single or in pairs, stalks 3–6 mm; calyx broadly bell-shaped. Stamens protruding, styles hairless. Fruit green, very bristly, edible. *NE Asia, Alaska, N California*. H1. Early summer.

41. R. aciculare Smith. Illustration: Janczewski, Monographie des Groseilliers, 373 (1907); Krüssmann, Manual of cultivated broad-leaved trees & shrubs **3**: 205 (1986).

Very similar to *R. burejense* but branches short prickled and bristly, occasionaly with 5–7 prickles at the nodes. Flowers pink or light green. Fruit red, green or yellow. *Siberia*. H1. Early summer.

42. R. stenocarpum Maximowicz. Illustration: Janczewski, Monographie des Groseilliers, 375 (1907); Krüssmann, Manual of cultivated broad-leaved trees & shrubs **3**: 205 (1986).

Shrub to 2 m; branches very prickly and bristly, thorny. Leaves *c.* 3 cm across, deeply 3–5-lobed. Flowers in groups of 1–3; calyx bell-shaped, reddish; petals white, 3/5 as long as sepals. Fruit greenish to reddish. *NW China*. H1. Early summer.

43. R. alpestre Wallich. Illustration: Janczewski, Monographie des Groseilliers, 376 (1907); Polunin & Stainton, Flowers of the Himalayas, 468 (1984); Krüssmann, Manual of cultivated broad-leaved trees & shrubs **3**: 205 (1986).

Vigorous, erect shrub to 3 m; branches very prickly and bristly, reddish when young. Leaves 2–5 cm across, circular, 3–5-lobed, toothed, hairless to slightly hairy. Flowers 1 or 2 on short stalks, small, greenish red and white. Ovary glandular-bristly. Fruit purple-red, glandular-bristly. *Himalaya, W China*. H1. Early summer.

Var. **giganteum** Janczewski. Shrub to 5 m; thorns to 3 cm. Fruit green, lacking bristles. *W China*. H1. Early summer.

44. R. grossularioides Maximowicz. Illustration: Janczewski, Monographie des Groseilliers, 377 (1907); Krüssmann, Manual of cultivated broad-leaved trees & shrubs **3**: 205 (1986).

Similar to *R. alpestre*, but distinguished by its anthers which lack glands at their tips. *Japan*. Early summer.

Uncommon in cultivation.

45. R. hirtellum Michaux (*R. oxyacanthoides* J.D. Hooker not Linnaeus; *R. gracile* of Janczewski). Illustration: Gleason, New illustrated flora of north-eastern United States and adjacent Canada **2**: 277 (1952).

Shrub to 1 m; branches thin, grey when young, later dark brown, thornless or with small prickles. Leaves ovate to circular, 3–5-lobed. Flowers 1–3 in short racemes, narrowly bell-shaped; calyx-lobes erect or spreading, greenish or reddish. Fruit purple to black, mostly hairless. *N USA*. Summer.

46. R. leptanthum A. Gray. Illustration: Janczewski, Monographie des Groseilliers, 379 (1907); Krüssmann, Manual of cultivated broad-leaved trees & shrubs **3**: 205 (1986).

Erect, dense, spreading shrub usually to 1 m; branches stoutly thorny. Leaves *c.* 2 cm across, circular, 3–5-lobed, lobes obtuse, finely hairy. Flowers numerous, single or in pairs; calyx-base greenish, sepals greenish white, hairy; petals spathulate, white or light reddish. Fruit black. *USA*. H1. Summer.

R. × lydia F. Koch (*R. leptanthum × R. quercetorum*). New growth appearing early. Leaves like *R. leptanthum*; flowers in pairs, light yellow. Fruit small, black. *Garden origin*. H1–2. Late spring.

R. × magdalenae F. Koch (*R. leptanthum × R. uva-crispa*). Very similar to *R. leptanthum*, but flowers pink and white. Fruit black, mostly sterile. *Garden origin*. H1–2. Early summer.

47. R. setosum Lindley (*R. saximontanum* E. Nelson). Illustration: Janczewski, Monographie des Groseilliers, 382 (1907); Gleason, New illustrated flora of the north-eastern United States and adjacent Canada **2**: 277 (1952).

Shrub to 1 m; branches with awl-shaped prickles, to 1 cm. Leaves 1–4 cm across, circular, 3–5-lobed, finely hairy. Flowers in groups of 1–3; calyx shortly tubular to bell-shaped, white, sepals 1/2 as long as tube. Fruit red to black, slightly bristly to smooth. *NW USA*. H1. Early summer.

48. R. uva-crispa Linnaeus (*R. grossularia* var. *uva-crispa* Smith; *R. grossularia* var. *pubescens* Koch). Illustration: Polunin, Flowers of Europe, pl. 43 (1969); Krüssmann, Manual of cultivated broad-leaved trees & shrubs **3**: 205 (1986).

Broad shrub to 1 m; shoots hairy, prickly. Leaves 2–6 cm across, circular, 3–5-lobed, scalloped, softly hairy beneath. Flowers greenish, 1–3 in racemes; calyx cup-shaped. Ovary downy. Fruit small, yellow or green, hairy. *NE & C Europe*. H1. Early summer.

49. R. oxyacanthoides Linnaeus. Illustration: Janczewski, Monographie des Groseilliers, 387 (1907); Krüssmann,

Manual of cultivated broad-leaved trees & shrubs **3**: 205 (1986).

Shrub *c.* 1 m; branches bristly, with 1–3 thorns at nodes, each *c.* 1 cm. Leaves 2–4 cm, circular, 3–5-lobed, toothed, hairless. Flowers 1–2, on short stalks. Calyx shortly tubular, lobes longer than tube, greenish white. Fruit purple-red, hairless. *N America*. H1. Early summer.

50. R. cynosbati Linnaeus. Illustration: Janczewski, Monographie des Groseilliers, 383 (1907); Gleason, New illustrated flora of the north-eastern United States and adjacent Canada **2**: 276 (1952); Krüssmann, Manual of cultivated broad-leaved trees & shrubs **3**: 207 (1986).

Shrub to 1.5 m; branches pendent, with spines simple or in groups of 3 at nodes, or occasionally thornless. Leaves 3–5 cm across, circular, 3–5-lobed, lobes coarsely toothed. Flowers in groups of 2 or 3; calyx bell-shaped, sepals and petals brownish green, sepals longer than petals. Stamens slightly protruding from calyx. Fruit black, top half with bristles. *E USA*. H1. Early summer.

51. R. divaricatum Douglas. Illustration: Janczewski, Monographie des Groseilliers, 390 (1907); Krüssmann, Manual of cultivated broad-leaved trees & shrubs **3**: 211 (1986).

Shrub to 3 m; branches grey-brown, with stout, hooked thorns to 2 cm. Leaves 2–6 cm across, ovate to circular, 5-lobed, coarsely toothed. Flowers greenish purple, in clusters of 2–4; calyx bell-shaped, sepals slightly longer than petals. Anthers protruding. Fruit dark red to black, white bloomed. *W North America*. H1. Early summer.

R. × succirubrum Zabel (*R. divaricatum × R. niveum*). Illustration: Janczewski, Monographie des Groseilliers, 500 (1907). Vigorous, erect shrub; branches hairless, with prickles at nodes, to 2 cm. Leaves 3–5 cm across. Flowers bright pink, 2–4 in pendent racemes. Stamens widely spreading. Fruit black, bloomed. *Garden origin*.

52. R. rotundifolium Michaux (*R. triflorum* Willdenow; *R. gracile* of Pursh not Michaux). Illustration: Janczewski, Monographie des Groseilliers, 392 (1907); Gleason, New illustrated flora of the north-eastern United States and adjacent Canada **2**: 276 (1952); Krüssmann, Manual of cultivated Broad-leaved trees & shrubs **3**: 205 (1986).

Shrub to 1 m; branches sparsely thorny.

Leaves 2–5 cm across, more or less broadly heart-shaped, usually 3-lobed, lobes obtuse, finely hairy. Flowers greenish red, in clusters of 1–3; calyx bell-shaped, sepals twice as long as tube, stamen and style long protruding. Fruit purple. *E & C USA*. H1. Early summer.

Often confused with *R. divaricatum* in cultivation.

53. R. niveum Lindley. Illustration: Janczewski, Monographie des Groseilliers, 394 (1907); Krüssmann, Manual of cultivated broad-leaved trees & shrubs **3**: 207 (1986).

Shrub to 3 m; branches with thin solitary thorns, hairless; reddish brown; young growth early. Leaves to 3 cm across, circular, 3–5-lobed, lobes usually rounded, few-toothed, usually hairless. Flowers 1–4, nodding, on thin stalks. Cupule bell-shaped; sepals *c.* 1 cm, white; petals very short, white. Fruit blue-black, hairless. *NW North America*. H1–2. Early summer.

R. × kochii Krüssmann (*R. niveum × R. speciosum*). Slow-growing shrub to 1 m; branches thorny. Flowers 4 or 5 parted; sepals red; petals white to reddish. Fruit brownish, bristly, rarely produced. *Garden origin*. H2–3. Early summer.

R. × robustum Janczewski. Thorns small. Flowers white or pinkish white. Fruit black. *Garden origin*. H1–2. Early summer.

54. R. curvatum Small. Illustration: Janczewski, Monographie des Groseilliers, 395 (1907).

Shrub to 1 m, very similar to *R. niveum*. Thorns *c.* 5 mm. Leaves to 3 cm across, usually slightly hairy. Flowers bell-shaped, usually solitary or in pairs; sepals spreading; petals small, white. Ovary glandular. Fruit purple, hairless. *SE North America*. H2–3. Early summer.

55. R. fasciculatum Siebold & Zuccarini. Illustration: Janczewski, Monographie des Groseilliers, 396 (1907); Krüssmann, Manual of cultivated broad-leaved trees & shrubs **3**: 209 (1986).

Shrub to 1.5 m, branches spineless, usually hairless when young. Leaves appearing very early and persisting a long time, 4–7 cm, circular to broadly ovate, 3–5-lobed. Flowers yellow, fragrant, in racemes of 2–9; calyx cup-shaped; petals longer than sepals. Fruit red. *Japan, Korea*. H1. Early summer.

56. R. gayanum (Spach) Steudel (*R. villosum* C. Gray not Nuttall). Illustration:

Janczewski, Monographie des Groseilliers, 427 (1907); Krüssmann, Manual of cultivated broad-leaved trees & shrubs **3**: 202 (1986).

Evergreen shrub, to 1 m; young branches downy. Leaves 3–6 cm, circular, shallowly 3–5-lobed, lobes obtuse, toothed, hairy on both surfaces. Flowers bell-shaped, yellow, fragrant, hairy, in many flowered racemes, to 6 cm. Fruit black, edible. *Chile*. H4. Early–mid summer.

57. R. glandulosum Weber (*R. prostratum* L'Héretier). Illustration: Janczewski, Monographie des Groseilliers, 431 (1907); Gleason, New illustrated flora of the north-eastern United States and adjacent Canada **2**: 278 (1952); Krüssmann, Manual of cultivated broad-leaved trees & shrubs **3**: 204 (1986).

Vigorous prostrate shrub to 40 cm, wide-spreading; young branches sparsely glandular-hairy. Leaves appearing early, 3–8 cm across, circular, 5–7-lobed, unpleasant smelling, hairless above, hairy on veins beneath. Flowers reddish white, 8–12 in ascending racemes, to as long as leaves. Fruit red, glandular-bristly. *N America*. H1. Early summer.

58. R. magellanicum Poiret. Illustration: Janczewski, Monographie des Groseilliers, 443 (1907); Moore, Flora of Tierra del Fuego, pl. 6 (1983).

Erect shrub to 2 m; branches spineless. Leaves *c.* 4 cm, circular, usually 3-lobed, sharply toothed, hairless. Flowers in many-flowered racemes; calyx cup-shaped, whitish green; corolla lobes orange-red. Fruit black, edible. *Chile, Argentina*. H4. Summer.

59. R. diacanthum Douglas (*R. saxatile* Pallas). Illustration: Janczewski, Monographie des Groseilliers, 390 (1907); Krüssmann, Manual of cultivated broad-leaved trees & shrubs **3**: 207 (1986).

Erect shrub to 2 m; branches with a few short prickles below the nodes. Leaves 2.5–3.5 × 1.5–3 cm, ovate, 3-lobed, lobes toothed. Flowers small, 1–6 in racemes, shorter than the leaves. Calyx greenish yellow, tinged red. Petals red. Fruit small, spherical, red, hairless. *N Asia*. H1. Late spring.

60. R. pulchellum Turczaninow. Illustration: Janczewski, Monographie des Groseilliers, 453 (1907).

Prickly shrub to 2 m. Similar to *R. diacanthum*, but leaves to 5 cm, more deeply 3-lobed, base wedge-shaped to slightly cordate. Flowers reddish, in

racemes to 6 cm. Fruit red. *N China*. H1. Early summer.

Often confused with *R. orientale* var. *heterotrichum*.

61. R. giraldii Janczewski. Illustration: Janczewski, Monographie des Groseilliers, 455 (1907); Krüssmann, Manual of cultivated broad-leaved trees & shrubs **3**: 202 (1986).

Spreading shrub to 1 m; branches hairy, bristly and thorny. Leaves *c.* 3.5 cm, circular. Flowers greenish brown, in non-glandular racemes to 7 cm. Fruit red, glandular hairy. *N China (Shaansi)*. H1–2. Early summer.

62. R. orientale Desfontaines (*R. villosum* Wallich; *R. resinosum* Pursh; *R. punctatum* Lindley). Illustration: Janczewski, Monographie des Groseilliers, 457 (1907); Polunin & Stainton, Flowers of the Himalaya, pl. 44 (1984); Krüssmann, Manual of cultivated broad-leaved trees & shrubs **3**: 202 (1986); Stainton, Flowers of the Himalaya, supplement, pl. 35 (1988).

Shrub to 2 m; branches thornless, red-brown, glandular glutinous, new growth appearing very early. Leaves *c.* 4.5 × 5.5 cm, broadly ovate, usually 3-lobed, hairy beneath, fragrant when crushed. Flowers 5–20, in 1–5 cm, erect racemes. Calyx bell-shaped, green, turning reddish. Fruit scarlet red, glandular-hairy. *Greece to Himalaya & Siberia*. H1. Early summer.

Var. **heterotrichum** (C.A. Meyer) Janczewski. Young branches reddish; flowers reddish; fruit not glandular. *Siberia*.

63. R. alpinum Linnaeus. Illustration: Janczewski, Monographie des Groseilliers, 461 (1907); Polunin, Flowers of Europe, pl. 43 (1969); Rix & Phillips, Shrubs, 74 (1989).

Erect shrub to 2 m; much branched, branches spineless, hairless, light grey. Leaves to 5 cm, circular, 3–5-lobed, lobes rounded, toothed, appearing early. Flowers dioecious, in small, erect racemes; calyx cup-shaped, yellowish green; petals longer than sepals; male racemes many-flowered; female racemes few-flowered. Fruit dark red. *Europe to Siberia*. H1. Early summer.

64. R. distans Janczewski (*R. alpinum* var. *mandshuricum* Maximowicz; *R. maximowiczianum* Komarov). Illustration: Krüssmann, Manual of cultivated broad-leaved trees & shrubs **3**: 202 (1986).

Similar to *R. alpinum*; leaves *c.* 5 × 6 cm, base cordate; sepals green. Fruit red, stalk almost absent. *China (Manchuria)*. H1. Early summer.

65. R. vilmorinii Janczewski. Illustration: Janczewski, Monographie des Groseilliers, 462 (1907); Krüssmann, Manual of cultivated broad-leaved trees & shrubs **3**: 202 (1986).

Dioecious shrub; similar to *R. alpinum*; differing in young shoots red, appearing very late. Flowers greenish, 6–12 in pendent racemes, to 6 cm. Fruit black-red. *W China*. H4. Late winter–early spring.

66. R. tenue Janczewski.

Dioecious shrub to 2 m; branches thornless. Flowers brownish red, in racemes. Fruit red. *W Asia*. H4–5. Early Summer.

67. R. glaciale Wallich. Illustration: Krüssmann, Manual of cultivated broad-leaved trees & shrubs **3**: 202 (1986); Stainton, Flowers of the Himalaya supplement, pl. 35 (1988).

Dioecious shrub; similar to *R. alpinum*; differing in leaf-lobes acute; flowers appearing soon after leaves; petals purple-brown. Fruit scarlet-red. *China (Yunnan, Hubei, Xizang)*. H1. Late spring.

68. R. luridum J.D. Hooker & Thomson. Illustration: Krüssmann, Manual of cultivated broad-leaved trees & shrubs **3**: 202 (1986).

Dioecious shrub; similar to *R. glaciale*. Branches smooth; new growth appearing later. Leaves to 6 cm, rounded ovate to kidney-shaped, 3–5-lobed. Fruit black. *Himalaya, Sikkim, Nepal, W China (Xizang)*. H1. Early summer.

69. R. maximowiczii Batalin. Illustration: Krüssmann, Manual of cultivated broad-leaved trees & shrubs **3**: 202 (1986).

Dioecious shrub; similar to *R. luridum* but shoots glandular-bristly. Leaves ovate, entire or shallowly 3–5-lobed, coarsely toothed. Flowers red-green, in erect, many-flowered racemes, 3–20 cm. Fruit glandular-hairy or finely bristly, red, yellow or green. *C China (Gansu)*. H1. Early summer.

70. R. henryi Franchet.

Evergreen shrub to 1.2 m, similar to *R. laurifolium* but shoots hairy; leaves to 10 cm, thinner. *C China*. H3–4. Early spring.

71. R. laurifolium Janczewski. Illustration: Rix & Phillips, Shrubs, 41 (1989); Brickell, The RHS gardeners' encyclopedia of plants and flowers, 143 (1986).

Evergreen shrub to 1.5 m; young shoots glandular, later hairless. Leaves 5–10 cm, ovate-oblong, scalloped, entire, apex acute,

base round, leathery. Flowers greenish yellow, 6–12 in pendent racemes, to 6 cm. Fruit black-red. *W China*. H4. Late winter–early spring.

29. FRANCOA Cavanilles
A.C. Whiteley

More or less evergreen herb. Leaves in a basal rosette, lyrate, often with winged leaf-stalk, softly hairy, to 30 × 10 cm. Flowers white or pink, in dense terminal racemes on stems to 90 cm, occasionally branched. Sepals 4, acute, *c.* 5 mm. Petals 4, oblong, white or pink, with or without darker central markings, *c.* 10 × 3 mm. Stamens 8. Staminodes simple. Ovary superior, cylindric. Fruit a cylindric capsule. Seeds numerous, winged.

A genus of a single, very variable species from Chile, often split into 4 or 5 species. Easily grown in most soils in a sunny position. Also good in pots. Propagate by seed or occasionally division.

1. F. sonchifolia Cavanilles (*F. appendiculata* Cavanilles; *F. glabrata* de Candolle; *F. ramosa* D. Don). Illustration: Botanical Magazine, 3178 (1832), 3309 (1834), 3824 (1840); Revue Horticole, 428–429 (1906).

Chile. H5. Summer.

White-flowered plants are sometimes separated as *F. ramosa* or cultivar 'Alba'; 'Rogerson's Form' has deep pink flowers.

30. PARNASSIA Linnaeus
P.G. Barnes

Herbaceous perennials with short rootstocks and alternate, mostly basal leaves. Flowering-stems bearing 1–6 leaves, flowers solitary. Sepals and petals 5, petals white. Stamens 5, alternating with the petals. Staminodes 5, yellowish, nectar-bearing, often fringed, opposite the petals. Ovary superior, ovoid, usually of 4 united carpels, style 1, short, stigmas 4. Fruit an ovoid or obovoid capsule, dehiscing along the middle of each cell.

Estimates of the number of species vary from 15 to 50, in the northern temperate areas. Most are typical of wet acid grassy places and require moist conditions in cultivation. Propagate by division or seed.

1a.	Stem-leaves 2–8; petals fringed all round	**2. foliosa**
b.	Stem-leaf solitary	2
2a.	Petals not fringed	**1. palustris**
b.	Petals fringed in lower half	3
3a.	Leaves ovate; petals 1.2–1.5 cm	**4. nubicola**

b.	Leaves rounded or kidney-shaped; petals *c.* 1 cm	**3. fimbriata**

1. P. palustris Linnaeus. Figure 23(1), p. 295. Illustration: Ross-Craig, Drawings of British plants **10**: 18 (1957); Keble Martin, Concise British flora in colour, pl. 32 (1969); Kitamura & Murata, Coloured illustrations of herbaceous plants of Japan **2**: pl. 35 (1986).

Plant 10–30 cm. Basal leaves 1–5 cm, ovate, base cordate, long-stalked. Stem with a single stalkless leaf. Flowers 1.5–3 cm across; petals 0.8–1.2 cm, broadly elliptic. Staminodes 3–4 mm, spathulate, divided at apex into 7–13 filaments, each tipped with a yellow gland. *Europe & N Asia*. H4. Summer.

2. P. foliosa Hooker & Thomson. Figure 23(4), p. 295. Illustration: Terasaki, Nippon Shokubutsu Zufu, t. 1440 (1932); Kitamura & Murata, Coloured illustrations of herbaceous plants of Japan **2**: pl. 35 (1986); Inami, Illustrations of selected plants from Hiroshima prefecture **4**: 35 (1988).

Plant 15–30 cm. Basal leaves 2–4 × 2–4 cm, circular, base cordate, long-stalked. Stems bearing 2–8 stalkless leaves, 1–2 cm. Petals 1–1.5 cm, ovate, fringed all round the margin. Staminodes divided into 3 gland-tipped filaments. *Japan*. H4. Summer.

The above description is of subsp. **nummularia** (Maximowicz) Kitamura & Murata, which appears to be the one that is generally cultivated. Subspecies **foliosa** from China, is larger in all its parts.

3. P. fimbriata K. König. Figure 23(2), p. 295. Illustration: Quarterly Bulletin of the Alpine Garden Society **6**: 118 (1938); Hitchcock et al., Vascular plants of the Pacific Northwest **3**: 30 (1971); Clark, Wild flowers of the Pacific Northwest, 206, 219 (1976).

Plant 15–30 cm. Basal leaves 2–5 cm, rounded or kidney-shaped, stalk 3–10 cm. Stem-leaf solitary, stalkless. Flower 2–2.5 cm across; petals obovate, *c.* 1 cm, the margins conspicuously fringed near the base. Staminodes obovate, bluntly lobed or fringed with gland-tipped filaments. Capsule *c.* 1 cm. *W North America*. H4. Summer.

Most cultivated plants appear to be var. **hoodiana** Hitchcock, with staminodes bearing 6–10 gland-tipped filaments.

4. P. nubicola Royle. Figure 23(3), p. 295. Illustration: Grierson & Long, Flora of Bhutan **1**(3): f. 36 e–h (1987).

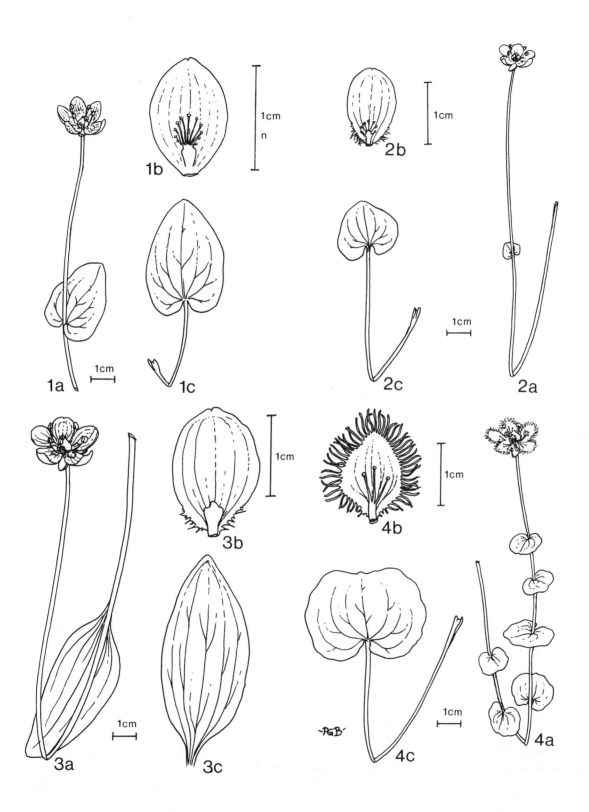

Figure 23. Diagnostic details of *Parnassia* species.
1a, b, c, *P. palustris*. 2a, b, c, *P. fimbriata*. 3a, b, c, *P. nubicola*.
4a, b, c, *P. foliosa*.

Plant 20–30 cm. Basal leaves 3–10 × 2–5 cm, ovate, acute, base cordate, rounded or broadly tapered into a stalk to 12 cm. Stem-leaf solitary, stalkless, the margins hairy towards the base. Sepals 8–10 mm, ovate. Petals obovate, 1.2–1.5 × 0.7–1 cm, usually fringed towards the base. Staminodes oblong, shortly 3-lobed at apex. Capsule 1–1.5 cm. *Himalaya.* H4–5. Summer.

31. JAMESIA Torrey & Gray
D.M. Miller

Deciduous shrubs to 1 m or sometimes more, with peeling, brown, papery bark. Branches with solid pith, downy when young. Leaves 2–7 × 1.5–5 cm, opposite, simple, ovate, acute, coarsely toothed, wrinkled, dull green above, with thick, grey-white down beneath; stalk downy, 2–15 mm. Leaves on flowering-shoots similar but smaller. Flowers *c.* 1.5 cm across, bisexual, slightly fragrant, in erect, terminal, many-flowered panicles to 5 cm. Sepals 5, *c.* 3 × 2 mm, ovate to lanceolate, downy. Petals 5, 8–10 × 5–6 mm, oblong to obovate, white or pinkish. Stamens 10. Styles 3–5, united at base. Ovary more or less superior. Fruit a many-seeded capsule to 4 mm.

A genus of a single species from N America, with orange-red autumn colour. It grows best in a sunny position in well-drained, fertile soil. Propagate by seed or cuttings.

1. J. americana Torrey & Gray. Illustration: Botanical Magazine, 6142 (1875); Krüssmann, Manual of cultivated broad-leaved trees & shrubs **2**: pl. 71 (1986).

E North America. H3. Summer.

32. DEUTZIA Thunberg
D.R. McKean

Deciduous or rarely evergreen shrubs with pith-filled branches and peeling bark. Hairs often stellate with a varying number of rays; sometimes a long ray is borne erect, at right angles to the others. Leaves opposite, usually shortly stalked, sometimes toothed, without stipules. Flowers in racemes, cymes, panicles or corymbs, or solitary on terminal or axillary shoots. Calyx-teeth 5. Petals 5, edge-to-edge or overlapping in bud. Stamens 10, in 2 series of 5, those of the inner series smaller; filaments mostly broadly winged, with 2 teeth at the top on either side of the anther; anthers sometimes shortly stalked above the filament. Ovary 3- or 4-celled,

inferior. Styles 3 or 4, thickened at apex. Fruit a capsule.

A genus of about 60 species, of which some 50 have at one time or another been in cultivation (some of them perhaps only in America). Those included here are all Asiatic, ranging from the Himalaya to Japan and the Philippines. They are notoriously difficult plants to identify because of the small differences which separate several of the species and the great variability of other species. They have long been favourite shrubs in gardens and many cultivars have been raised and distributed (especially by Lemoine's nursery in the late 19th and early 20th centuries). A large number of these cultivars and hybrids still exists; many of them are of doubtful origin and are particularly difficult to identify. Cultivation as for *Philadelphus* (p. 303).

Literature: Zaikonnikova, I.T., A key to the species of the genus Deutzia, *Baileya* **19**: 133–144 (1975).

1a.	Petals all edge-to-edge in bud 2
b.	Petals overlapping in bud (or partly edge-to-edge in the flowers of some hybrids) 27
2a.	Flowers in panicles or racemes 3
b.	Flowers solitary or in corymbs or cymes 14
3a.	Filaments of outer stamens tapered to apex **11. scabra**
b.	Filaments of all stamens broad, toothed at apex 4
4a.	Leaves hairless or almost so beneath or with 4–6-rayed hairs 5
b.	Leaves with 8–16-rayed hairs beneath 8
5a.	Inflorescence a long panicle or raceme, or with very few white flowers 6
b.	Inflorescence a broad, loose panicle; flowers pinkish or rarely white 7
6a.	Leaves almost hairless beneath **1. gracilis**
b.	Leaves with a moderate to dense covering of 4- or 5-rayed hairs beneath **7. taiwanensis**
7a.	Calyx-teeth longer than tube **3. × rosea**
b.	Calyx-teeth shorter than tube **2. × carnea**
8a.	Leaves to 2.5 cm wide and obscurely toothed 9
b.	Leaves more than 2.5 cm wide and distinctly toothed 10
9a.	Panicles dense, 20–60-flowered **4. ningpoensis**
b.	Panicles loose, 12–20-flowered **6. maximowicziana**
10a.	Inflorescence a raceme, all stamens distinctly toothed **8. crenata**
b.	Inflorescence a panicle, not all stamens distinctly toothed 11
11a.	Calyx-teeth about as long as ovary **9. × magnifica**
b.	Calyx-teeth shorter than ovary 12
12a.	Leaves leathery, with dense 14–16-rayed hairs beneath; styles 5 **5. pulchra**
b.	Leaves thinner, less densely hairy with hairs with fewer rays; styles less than 5 13
13a.	Leaves obscurely and finely scalloped; panicle narrow **8. crenata**
b.	Leaves with fine forward-pointing teeth; panicle broad and loose **10. schneideriana**
14a.	Inflorescence with 1–3 flowers 15
b.	Inflorescence with more than 3 flowers 16
15a.	Leaves white beneath **17. grandiflora**
b.	Leaves pale green beneath **16. coreana**
16a.	Calyx-teeth shorter than ovary, broadly ovate or triangular; leaves not whitish beneath **13. setchuenensis**
b.	Calyx-teeth as long as or longer than ovary; leaves sometimes whitish beneath 17
17a.	Mature leaves on flowering-shoots *c.* 9 × 3 cm **19. longifolia**
b.	Mature leaves on flowering-shoots smaller 18
18a.	Petals with reflexed margins **20. reflexa**
b.	Petals without reflexed margins 19
19a.	Leaves sharply toothed; calyx reflexed; flowers white **18. vilmoriniae**
b.	Leaves finely toothed; calyx not reflexed; flowers pink or white 20
20a.	Inflorescence a dense cyme borne on a very short shoot **21. calycosa**
b.	Inflorescene not as above 21
21a.	Leaf underside with stellate hairs, not spreading hairs 22
b.	Leaf underside with spreading hairs 23
22a.	Leaves pale green beneath; flowers *c.* 0.7 cm **22. rehderiana**
b.	Leaves white beneath; flowers *c.* 1 cm **14. monbiegii**

Figure 24. Leaf hairs of *Deutzia* species; top, upper surface; bottom, lower surface (× 30). 1, *D. corymbosa*. 2, *D. purpurascens*. 3, *D. schneideriana*. 4, *D. ningpoensis*. 5, *D. glomeruliflora*. 6, *D. crenata*. 7, *D. scabra*. 8, *D. discolor*. 9, *D. pulchra*. 10, *D. rubens*. 11, *D. setchuenensis*. 12, *D. gracilis*. 13, *D. monbeigii*. 14, *D. mollis*. 15, *D. hookeriana*. 16, *D. grandiflora*. 17, *D. longifolia*. 18, *D. staminea*. 19, *D. parviflora*. 20, *D. calycosa*.

23a. Leaves on flowering shoots *c.* 3 × 1.5 cm, with 4- or 5-rayed stellate hairs beneath   **23. glomeruliflora**

   b. Leaves on flowering-shoots usually larger, stellate hairs with more rays   24

24a. Leaves moderately 5–7-rayed stellate hairs beneath (hairs not overlapping)   **26. × elegantissima**

   b. Leaves with densely (overlapping) 7–12-rayed hairs   25

25a. Leaf-underside with 7–9-rayed stellate hairs   **25. purpurascens**

   b. Leaf-underside with 9–17-rayed stellate hairs   **24. discolor**

26a. Leaves with very dense 10–13-rayed hairs beneath   **15. staminea**

   b. Leaves densely shaggy-hairy, moderately stellate-hairy or almost hairless beneath   27

27a. Leaves densely shaggy-hairy beneath   28

   b. Leaves moderately stellate-hairy or almost hairless beneath   29

28a. Petals all overlapping in bud; leaves with only spreading hairs beneath   **35. mollis**

   b. Some petals not overlapping in bud, others overlapping; leaves with both stellate and spreading hairs   **36. × wilsonii**

29a. Inflorescence a panicle; some petals overlapping, others edge-to-edge in bud; filaments toothed   30

   b. Inflorescence a corymb; all petals overlapping in bud; filaments toothed or not   33

30a. Flowers white   31

   b. Flowers pinkish outside   32

31a. Long filaments, some tapered at apex   **12. × candida**

   b. Long filaments, all toothed   **29. × lemoinei**

32a. Calyx teeth-longer than ovary; flowers *c.* 1.5 cm across   **28. × maliflora**

   b. Calyx-teeth as long as ovary; flowers *c.* 2 cm across   **27. × kalmiiflora**

33a. Longer stamens, all tapered to the apex, not toothed   **30. parviflora**

   b. Longer stamens, all broad, distinctly toothed at apex   34

34a. Longer stamens, some toothed, others tapered   **34. compacta**

   b. Longer stamens, all toothed   35

35a. Lower leaf-surface densely hairy with 5–9-rayed stellate hairs, the surface usually not visible   **31. hookeriana**

   b. Stellate hairs less dense on lower leaf-surface, the surface visible through the hairs   36

36a. Longer stamens truncate or shallowly lobed at apex   **32. rubens**

   b. Longer stamens distinctly toothed at apex, the anthers borne on short stalks   **33. corymbosa**

1. D. gracilis Siebold & Zuccarini. Figure 24 (12), p. 297; 25(17), p. 299. Illustration: Hay & Beckett, Reader's Digest encyclopaedia of garden plants and flowers, 221 (1972); Davis, The gardener's illustrated encyclopaedia of trees & shrubs, 116 (1987).

Erect, deciduous shrub, 1–2 m. Leaves 3.5–6 (rarely to 10) × 2–4 cm, lanceolate to ovate, with minute, forward-pointing teeth, very thin, bright green, with scattered hairs, 3- or 4-rayed on upper surface and 4- or 5-rayed on the lower. Flowers in racemes or narrow panicles, 40–80 cm; flower-stalks long and slender. Flowers 1–2 cm across, pure white, strongly scented. Calyx-teeth *c.* 1.5 mm, triangular, usually greenish. Ovary *c.* 2 mm. Anthers borne on short stalks above the filament teeth. Styles 5, about the same length as stamens. Capsules *c.* 4 mm across. *Japan.* H3. Spring–early summer.

Plants of this species may need protection from late spring frosts.

Var. **nagurae** (Makino) Makino has smaller flowers (*c.* 5 mm across).

D. × candelabrum (Lemoine) Rehder (*D. gracilis* × *D. scabra*). Illustration: Revue Horticole, 175 (1908). Like *D. gracilis* but with larger, many-flowered panicles, coarser leaves and more spreading growth. *Garden origin.* H5. Early summer.

2. D. × carnea (Lemoine) Rehder (*D. × rosea* 'Grandiflora' × *D. scabra*).

Differs from *D. gracilis* by its broad, loose panicles of pink flowers with purplish sepals. *Garden origin.* H3. Spring–early summer.

3. D. × rosea (Lemoine) Rehder (*D.gracilis* × *purpurascens*). Illustration: Kaier, Garden trees and shrubs, pl. 60 (1963); Seabrook, Shrubs for your garden, 63 (1974); Hessayon, Tree and shrub expert, 23 (1983); Krüssmann, Manual of cultivated broad-leaved trees & shrubs **1**: pl. 170, 171 (1986).

Dwarf arching shrub to 1 m. Leaves ovate or oblong to lanceolate, margins with forward-pointing teeth, and with scattered, 4–6-rayed hairs. Flowers in

short, broad panicles, carmine outside, white inside. Calyx-teeth lanceolate, longer than the ovary. Styles usually longer than the stamens. *Garden origin.*

Several cultivars are grown, including 'Campanulata', with dense panicles of white flowers; 'Carminea' with reddish pink flowers and 'Eximia' with pale pink flowers, white insides.

4. D. ningpoensis Rehder (*D. chunii* Hu). Figure 24(4), p. 297; 25(16), p. 299. Illustration: Hu & Chun, Icones Plantarum Sinicarum **5**: f. 222 (1937); Krüssmann, Manual of cultivated broad-leaved trees & shrubs **1**: pl. 168, 172 & (1986).

Shrub to 2 m. Leaves 3.5–7 × *c.* 2.5 cm, ovate, generally entire, with 5- or 6-rayed hairs above and very dense 12–14-rayed hairs beneath. Inflorescence a narrow panicle to 10 cm. Flowers 5–10 mm across, white or pink, densely crowded. Calyx-teeth *c.* 1 mm, shorter than the ovary (*c.* 2 mm). Stamens indistinctly toothed, the anthers borne on stalks *c.* 0.75 mm above the filaments. *E China (Zhejiang, Anhui).* H4. Summer.

5. D. pulchra Vidal. Figure 24(9), p. 297; 25(19), p. 299. Illustration: Botanical Magazine, 8962 (1923); Li, Woody flora of Taiwan, f. 89 (1963); Li et al., Flora of Taiwan **3**: f. 468 (1977).

Shrub 2–4 m. Leaves 5–10 × 2–5 cm, lanceolate to narrowly ovate, base tapered or rounded, entire or toothed, thick and leathery, with moderately dense 5–8-rayed (rarely to 13-rayed) hairs above and very dense 14–16-rayed hairs beneath. Flowering-shoot terminal or almost so, bearing a few-flowered panicle to 12 × 7 cm. Flowers upright to reflexed, white. Calyx-teeth *c.* 1 × 2 mm, broadly triangular, shorter than the ovary (to 3.2 mm). Stamens *c.* 8 mm, narrow, toothed. Styles 5, as long as stamens. *Philippines (Luzon) & Taiwan.* H4. Spring.

6. D. maximowicziana Makino (*D. hypoleuca* Maximowicz). Illustration: Kitamura & Okamoto, Coloured illustrations of trees & shrubs of Japan, pl. 202 (1977).

Deciduous shrub to 1.5 m; bark brown, current growth stellate downy. Leaves 3–8 × 1.2–1.8 cm, narrowly oblanceolate, long-pointed, white beneath with dense stellate hairs; moderately stellate-hairy above. Flowers white, in loose, erect 12–20-flowered panicles 5–7 cm. Petals *c.* 7 mm. Longest stamens *c.* 8 mm, slightly toothed, shorter stamens untoothed. Styles

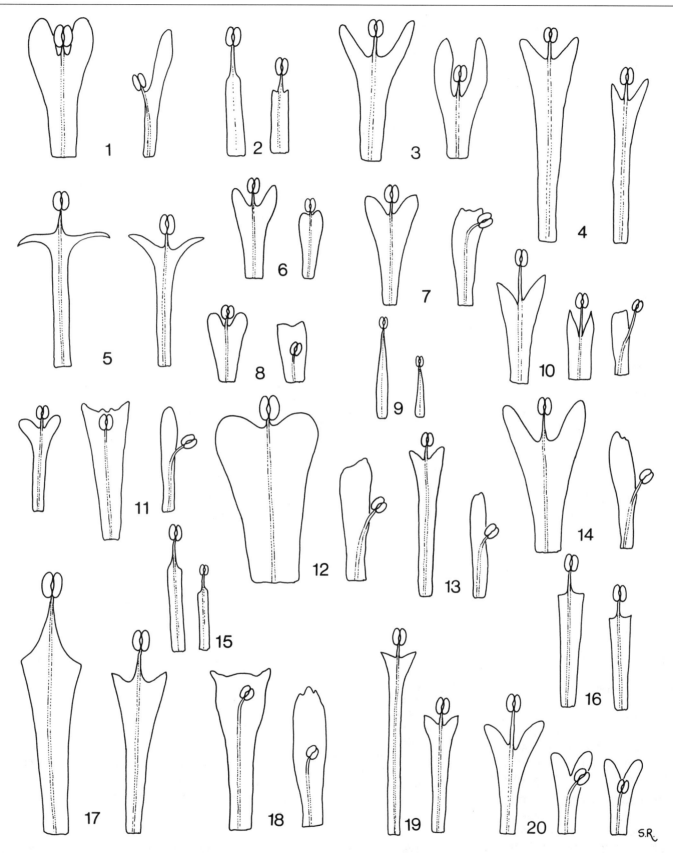

Figure 25. Stamens of *Deutzia* species (× 6.6). 1, *D. calycosa*.
2, *D. parviflora*. 3, *D. longifolia*. 4, *D. crenata*. 5, *D. grandiflora*.
6, *D. staminea*. 7, *D. monbeigii*. 8, *D. hookeriana*. 9, *D. mollis*.
10, *D. corymbosa*. 11, *D. setchuenensis*. 12, *D. glomeruliflora*.
13, *D. schneideriana*. 14, *D. purpurascens*. 15, *D. scabra*.
16, *D. ningpoensis*. 17, *D. gracilis*. 18, *D. rubens*. 19, *D. pulchra*.
20, *D. discolor*.

3 or 4, *c.* 1 cm. *Japan.* H4. Spring–early summer.

7. D. taiwanensis (Maximowicz) Schneider (*D. crenata* var. *taiwanensis*). Illustration: Ito, Taiwan shabubutu dyusetu (Illustration Formosan Plants) t. 620 (1927).

Deciduous shrub to 2 m; current growth thinly stellate-hairy. Leaves 5–8 × 2–3 cm, ovate lanceolate, rough with very short stiff hairs, with fine forward-pointing teeth, moderately 3- or 4-rayed stellate hairy above, more densely 4- or 5-rayed hairy beneath. Flowers few in terminal panicles to 6 cm. Petals *c.* 8 × 4 mm, oblong. Calyx-teeth *c.* 1 × 1 mm. Stamens *c.* 5–7 mm indistinctly toothed. Capsule almost spherical, *c.* 3.5 mm across. *Taiwan.* H4. Spring.

Probably rare in cultivation and of doubtful horticultural value because of the small number of insignificant flowers.

8. D. crenata Siebold & Zuccarini. Figure 24(6), p. 297; 25(4), p. 299. Illustration: Siebold & Zuccarini, Flora of Japan **1**: t. 6 (1835); Botanical Magazine, 3838 (1841) – as D. scabra; Hu & Chun, Icones Plantarum Sinicorum **5**: t. 220 (1937) – as D. scabra; Makino, New illustrated flora of Japan, f. 950 (1963) – as D. sieboldiana.

Shrub to 2.5 m, with erect branches. Leaves 3–6 × 1.5–3 cm on fertile shoots, much larger on non-flowering-shoots, ovate to ovate-lanceolate, obscurely scalloped, stalked; hairs on upper surface with 4–6 spreading rays and a longer, erect, central ray, those on the lower surface 6–13-rayed. Inflorescence mainly a raceme, more rarely a narrow panicle, 10–15 cm. Flowers 1.5–2 cm across. Calyx-teeth *c.* 1.8 mm, shorter than the ovary (*c.* 2.3 mm), covered with apparently simple as well as stellate hairs. Petals long and narrow, 1–1.5 cm. All stamens toothed. Styles 3 or 4. Capsules *c.* 5 mm across. *Japan, ?SE China.* H3. Spring.

A species with several cultivars and a parent of several hybrids. It has been much confused with *D. scabra*, from which it differs in its stalked leaves subtending the inflorescences and its toothed filaments.

9. D. × magnifica (Lemoine) Rehder (*D. crenata* var. *magnifica* Lemoine). Illustration: Seabrook, Shrubs for your garden, 62 (1974); Hofman, Ornamental shrubs, 77 (1979).

Shrub to 2 m; similar to *D. crenata*, with strong, upright growth and stout branches. Leaves 4–6 cm, ovate-oblong, sharply but finely toothed, rough above and with dense 10–15-rayed hairs beneath. Flowers white, single or double, in short, dense panicles 4–6 cm. Calyx-teeth as long as ovary. Filaments with large teeth. *Garden origin.* H5. Early summer.

Recent authors have indicated the parentage of this hybrid as *D. scabra* × *D. vilmoriniae*. The original author considered it to be a variety of *D. crenata*, however the parentage given below seems more likely. Several cultivars are available, including 'Eburnea' with single, white, bell-shaped flowers in loose panicles; 'Latiflora' with single, white flowers to 3 cm across in erect panicles; 'Longiflora', with single flowers with long, narrow petals; and 'Macrothyrsa', a tall plant with many umbel-like panicles along each branch.

D. crenata? × *D. vilmoriniae.*

10. D. schneideriana Rehder. Figure 24(3), p. 297; 25(13), p. 299. Illustration: Hu & Chun, Icones Plantarum Sinicarum **5**: f. 221 (1937); Iconographia cormophytorum Sinicorum **2**: t. 1931 (1972).

Shrub 1–2 m. Leaves 9–11 × 3–4 cm, ovate or lanceolate-ovate, margin with fine, forward-pointing teeth, hairs on upper surface with 4 or 5 spreading rays and a longer, erect, central ray, those on lower surface dense, 9–13-rayed, some occasionally with a longer central ray. Inflorescence corymb-like, 9–11 × 4–6 cm. Calyx-teeth *c.* 1 mm, broadly triangular, ovary *c.* 2 mm. Petals narrowly lanceolate, *c.* 1 cm; longer stamens toothed, shorter stamens not toothed, anther borne just below the apex. Styles 3. *China (W Hubei).* H4. Summer.

Var. **laxiflora** Rehder has looser and broader inflorescences and larger petals (1.2–1.4 cm), and is more often grown.

11. D. scabra Thunberg (*D. sieboldiana* Maximowicz). Figure 24(7), p. 297; 25(15), p. 299. Illustration: Makino, New illustrated Flora of Japan, f. 951 (1963); Bean, Trees & shrubs hardy in the British Isles, edn 8, **2**: 49 (1973); Hofman, Ornamental shrubs, 78 (1978).

Shrub to 2.5 m. Leaves 3–8 × 2–4 cm, broadly ovate, stalked except for those subtending the inflorescences, with coarse, forward-pointing teeth; hairs on the upper surface 3 or 4-rayed, those on the lower 4–6-rayed, some on the veins occasionally with a longer central, erect ray. Panicles broadly pyramidal, loose. Flowers 1–1.5 cm across, white, honey-scented. Calyx-teeth *c.* 1 mm, ovary *c.* 2 mm. Filaments not toothed. Styles usually 3. *Japan.* H4. Summer.

Sometimes confused with *D. crenata.* Several widely cultivated hybrids and cultivars formerly attributed to this species should be referred to *D. crenata*. For *D. × elegantissima* see *D. glomeruliflora.*

12. D. × candida (Lemoine) Rehder (*D. discolor* var. *candida* Lemoine).

Growth upright. Leaves 3.5–5 cm, ovate, with forward-pointing teeth, slightly rough above with short, stiff hairs, with 5–7-rayed hairs beneath. Flowers numerous in panicles, *c.* 2 cm across, white. Calyx-teeth oblong-ovate, about as long as ovary. Petals partly overlapping and partly edge-to-edge in bud. Filaments toothed. Styles 3, shorter than the longer stamens. *Garden origin.* H3. Summer.

A hybrid between *D. × lemoinei* and *D. scabra.*

Var. **compacta** Lemoine is more compact and has smaller flowers. 'Boule de Niege' has a compact habit and denser inflorescences of larger flowers.

13. D. setchuenensis Franchet. Figure 24(11), p. 297; 25(11), p. 299. Illustration: Iconographia cormophytorum Sinicorum **2**: t. 1926 (1972); Bean, Trees & shrubs hardy in the British Isles edn 8, **2**: f. 9 (1973); Krüssmann, Manual of cultivated broad-leaved trees & shrubs **1**: f. 316, 317 (1986).

Shrub 1.5–2 m. Leaves *c.* 6 × 2 cm, ovate, usually long acuminate, margin with fine, forward-pointing teeth, upper surface with 4–6-rayed hairs, lower with 3–5-rayed-hairs, some of which have longer central rays. Inflorescnce composed of loose corymbs. Flowers to 1 cm across, white, on long and slender stalks. Calyx-teeth *c.* 0.8 mm. Longer stamens toothed, shorter not toothed, with anthers attached below the apex. Ovary *c.* 1 mm. Styles 3, short. *W China.* H5. Summer.

Var. **corymbiflora** (Lemoine) Rehder. Illustration: Botanical Magazine, 8255 (1909); Krüssmann, Manual of cultivated broad-leaved trees & shrubs **1**: pl. 169, 170 (1986). Leaves 3–11 × 1.5–3 cm; flowers *c.* 1.5 cm across in larger corymbs. This is more commonly found in cultivation than var. **setchuenensis**, and prefers a limey soil.

14. D. monbeigii W.W. Smith. Figure 24(13), p. 297; 25(7), p. 299. Illustration: Botanical Magazine, n.s., 123 (1950).

Shrub 1–1.5 m, with slender branches. Leaves 1.5–3 × 0.5–1.2 cm, ovate-lanceolate, margin with forward-pointing teeth, white beneath with very dense 12–15-rayed hairs, the hairs above less dense, 6–9-rayed. Inflorescence corymb-like or cyme-like, c. 6 × 4 cm. Calyx-teeth c. 1.5 mm, broad, about the same length as the ovary. Petals c. 7 × 3 mm, ovate. Longer stamens c. 5 mm, toothed, smaller stamens c. 3 mm, with anthers attached below the apex, minutely toothed. Styles 5. *SW China (Yunnan)*. H4. Spring–summer.

15. D. staminea Wallich. Figure 24(18), p. 297; 25(6), p. 299. Illustration: Edwards's Botanical Register **33**: pl. 13 & 265 (1847) – as D. corymbosa; Schneider, Illustriertes Handbuch der Laubholzkunde **1**: f. 244 (1906).

Small shrub, 1 m (rarely to 2 m). Leaves ovate, apex acuminate, base rounded or wedge-shaped, margin with fine, forward-pointing teeth; upper surface hairy, hairs 9–11- (rarely as few as 7-) rayed, lower surface with dense 10–13- (rarely to 17-) rayed hairs. Inflorescence a panicle or rarely corymb-like, c. 5 cm wide. Flowers c. 1.5 cm across, white or pink. Calyx-teeth narrowly triangular, 1.5–2 mm, ovary 2–3.5 mm, densely stellate-hairy. Petals boat-shaped, oblong-elliptic, 7–10 × 3–5 mm. Stamens 6–7.5 mm, outer toothed, inner with minute teeth, anthers almost stalkless, attached below the apex. Styles 4, equal in length to the outer stamens. *Himalaya*. H5. Early summer.

Var. **brunoniana** Hooker & Thomson is less hairy and larger flowered.

16. D. coreana H. Léveillé. Illustration: Nakai, Flora sylvatica Koreana **15**: t. 16 (1926); Krüssmann, Manual of cultivated broad-leaved trees & shrubs **1**: f. 316 (1986).

Shrub to 2 m; current growth rough, warty. Leaves ovate to ovate-elliptic, paler beneath, margin finely double-toothed, with teeth pointing forwards, both surfaces moderately covered with 4–6-rayed hairs. Flowers on previous years growth, solitary, white, in leaf axils. Calyx-tube c. 2.5 mm, stellate, teeth c. 2 mm, broadly triangular. Both whorls of stamens toothed. Styles 3–4, c. 7 mm. *Korea*. H4. Spring.

Probably rare in cultivation.

17. D. grandiflora Bunge. Figure 24(16), p. 297; 25(5), p. 299. Illustration: Hu & Chun, Icones Plantarum Sinicarum **5**: f. 228 (1937); Iconographia cormophytorum Sinicorum **2**: t. 1935 (1972).

Shrub 1.5–2 m, young shoots grey at first with stellate hairs. Leaves on flowering-shoots 2.5–3 × 1–1.5 cm, increasing later to 5 × 2 cm, margins with fine forward-pointing teeth, white beneath with dense, 9- or 10-rayed hairs, above with 5-rayed hairs which have additional central rays. Flowers 1–3 on short leafy shoots, 2.5–3 cm across, white. Calyx-teeth c. 3.8 mm, linear-lanceolate. Both sets of filaments toothed. Ovary c. 3 mm. Styles 3, longer than stamens. *N China*. H3. Early spring.

This is the earliest flowering species, with the largest flowers, the 1–3-flowered inflorescence is distinctive.

18. D. vilmoriniae Lemoine. Illustration: Krüssmann, Manual of cultivated broad-leaved trees & shrubs **1**: pl. 170 (1986).

Vigorous deciduous shrub to 2 m; current growth rough with stellate hairs. Leaves ovate to oblong-lanceolate, slender pointed, 3–8 × 2–5 cm, sharply toothed, dull green above, grey beneath with stellate hairs; hairs on the midrib appearing simple but with a minute stellate base. Flowers white, c. 2.5 cm across; panicles c. 8 cm. Calyx-teeth linear lanceolate, reflexed. Filaments toothed. *SW China*. H4. Summer.

19. D. longifolia Franchet. Figure 24(17), p. 297; 25(3), p. 299. Illustration: Botanical Magazine, 8493 (1913); Hu & Chun, Icones Plantarum Sinicarum **5**: f. 227 (1937); Iconographia cormophytorum Sinicorum **2**: t. 1933 (1972); Krüssmann, Manual of cultivated broad-leaved trees & shrubs **1**: pl. 168 & f. 316–7 (1986).

Shrub 1.5–2 m, young shoots sparsely hairy with stellate hairs at first, later hairless. Leaves 5–7 × 1–2.5 cm on fertile shoots, lanceolate, base rounded or tapered, apex acuminate, thick and with prominent veins, margins with forward pointing, fine teeth, upper surface dull green, with 5 or 6-rayed hairs, lower surface pale green, with dense 8–12-rayed hairs, hairs on the veins simple. Cymes broad, loose or compact, c. 8 × 6 cm. Flowers 2–2.5 cm across; calyx-teeth c. 3 mm, narrowly triangular; petals broadly ovate-oblong, white with a purplish pink stripe outside. Outer stamens toothed, inner with a single large tooth and 1 or 2 variable, minute teeth. Ovary c. 2.8 mm. *China (Sichuan, Yunnan)*. H4. Early summer.

Var. **macropetala** Zaikonnikova. Leaves on fertile shoots 6–11 × 2.5–3 cm; flowers 2.5–3 cm across.

Several cultivars of D. longifolia are known, including 'Veitchii' with larger, darker purple flowers and purple flowering-stems, possibly the most handsome Deutzia, and 'Elegans' with slender, drooping branches and flowers to 2 cm wide.

D. × hybrida Lemoine (*D. longifolia × D. discolor*). Illustration: National Horticultural Magazine, **30**: 87 (1951); Hay & Beckett, Reader's Digest encyclopaedia of garden plants and flowers, 221 (1972).

Like D. longifolia but leaves 6–10 cm; flowers larger and wider, petals pink with wavy edges; anthers bright yellow. *Garden origin*. H4. Early summer. Usually found as cultivar 'Magician'.

D. × excellens (Lemoine) Rehder (*D. × rosea* 'Grandiflora' × *D. longifolia*). Like D. longifolia but leaves 3–6 cm, ovate-oblong, margins with fine, forward-pointing teeth, upper surface rough with short, stiff hairs, lower surface greyish white with 8–12-rayed hairs and simple spreading hairs on the main veins. Flowers in broad, loose corymbs 4–6 cm across, white. *Garden origin*. H4. Summer.

20. D. reflexa Duthie.

Deciduous shrub c. 1 m, current growth hairless. Leaves 5–10 × 1–2.5 cm, broadly oblong-lanceolate, long-pointed, faintly toothed, thinly stellate downy above, more densely hairy beneath. Hairs on midrib appearing simple, but with a minute stellate base and a long central arm. Flowers in dense panicles, petals c. 8 mm, white, margin reflexed. Stamens toothed. Calyx-teeth narrowly oblong. *C China*. H4. Summer.

21. D. calycosa Rehder. Figure 24(20), p. 297; 25(1), p. 299.

Shrub to 2 m. Leaves 2–7 × 1–2 cm, ovate, the upper surface with hairs which have 5 or 6 rays and a central ray, lower surface moderately to densely hairy with hairs 9- or 10- (rarely to 15-) rayed. Flowering-shoots with 2 or 3 (rarely to 6) pairs of leaves. Inflorescence a dense cyme, 4–7 cm (rarely to 10 cm), borne on a very short shoot. Flowers c. 2 cm across, white with purplish exterior. Calyx-teeth c. 3 mm. Ovary c. 2 mm. Petals c. 9 × 7 mm, broadly ovate. Inner and outer stamens c. 6 mm, outer toothed, inner not toothed, anthers attached just below the apex. *SW China (Yunnan)*. H4. Spring.

A recent introduction rare, as yet, in cultivation.

22. D. rehderiana Schneider. Illustration: Krüssmann, Manual of cultivated broad-leaved trees & shrubs **1**: pl. 168, 316 (1986).

Deciduous, dense shrub to 1.5 m; current growth rough with stellate hairs. Bark red-brown, peeling. Leaves 1.5–3 × 1–1.5 cm, ovate, with fine forward-pointing teeth, rough and densely stellate-hairy above, with 4–6-rayed stellate-hairs, paler beneath with 4–8-rayed stellate hairs. Flowers in 3–8-flowered cymes in leaf axils, *c.* 7 mm across, white. Calyx *c.* 4–5 mm, densely stellate, teeth 1.5–2 mm. Styles free *c.* 3 mm. Capsule *c.* 2–3 mm. *W China.* H4. Spring.

23. D. glomeruliflora Franchet. Figure 24(5), p. 297; 25(12), p. 299. Illustration: Hu & Chun, Icones Plantarum Sinicarum **5**: f. 224 (1937); Iconographia cormophytorum Sinicorum **2**: t. 1934 (1972); Krüssmann, Manual of cultivated broad-leaved trees & shrubs **1**: pl. 170 (1986).

Shrub 2–3 m, with arching branches. Leaves to 4 × 1.5 cm on flowering-shoots, lanceolate to ovate, acuminate, margin with fine forward-pointing teeth, upper surface with hairs with 3 or 4 rays and a central ray, lower surface densely hairy with 4- or 5-rayed hairs. Cymes numerous, few-flowered, borne on shoots 2.5–5 cm. Flowers white, *c.* 2 cm wide. Calyx-teeth *c.* 4 mm, narrowly lanceolate. Petals 1.3 × 1 cm. Inner and outer stamens *c.* 6 mm, the outer with broad teeth, inner untoothed and with the anthers attached about half-way down each filament. Styles 3–5, *c.* 6 mm. Ovary *c.* 2.5 mm. Capsule *c.* 3.5 × 4 mm. *W China (Sichuan).* H4. Spring.

Possibly no longer commercially available.

24. D. discolor Hemsley. Figure 24(8), p. 297; 25(20), p. 299. Illustration: Hu & Chun, Icones Plantarum Sinicarum **5**: f. 226 (1937); Iconographia cormophytorum Sinicorum **2**: f. 1932 (1972); Krüssmann, Manual of cultivated broad-leaved trees & shrubs **1**: pl. 168, 170 (1986).

Arching shrub 1–2 m. Leaves 4–11 × 1.5–3 cm, narrowly ovate-oblong, thin, margin with fine, forward-pointing teeth; upper surface with 4- or 5-rayed hairs, a few of them with an additional erect ray, lower surface with dense, 9–12-rayed hairs, some of them, especially on the veins, with an additional central ray, rendering the veins shaggy. Cymes loose. Flowers 1.3–2.5 cm across, white to pink.

Calyx-teeth 3–4 mm. Longest stamens *c.* 6 mm, distinctly toothed, inner stamens *c.* 4 mm, less distinctly toothed, anthers borne on stalks above the filaments. Ovary 2–3.5 mm. *China (W Hubei).* H4. Spring.

Several cultivars of this species are available. **D. globosa** Duthie is similar to *D. discolor* but with smaller leaves and cream-coloured flowers. *C China (W Hubei).*

25. D. purpurascens (Henry) Rehder (*D. discolor* Hemsley var. *purpurascens* Henry). Figure 24(2), p. 297; 25(14), p. 299. Illustration: Botanical Magazine, 7708 (1900); New Flora and Silva **4**: pl. 103 (1930) & **6**: pl. 106 (1932); Krüssmann, Manual of cultivated broad-leaved trees & shrubs **1**: f. 316 (1986).

Slender arching shrub to 1.5 m. Leaves 4–7 × 1.2–3 cm, ovate to ovate-lanceolate, margin with forward-pointing teeth, upper surface with mainly 5-(rarely 3- or 4-) rayed hairs, lower surface with 7- or 8-rayed hairs, some hairs on both surfaces with erect, central rays. Cymes on short shoots, 4–6 cm. Flowers *c.* 2 cm across, white inside, purplish outside. Calyx-teeth *c.* 4.5 mm, lanceolate. Stamens broadly winged, outer series with broad teeth, inner with minute teeth, anthers attached below apex. Ovary *c.* 2.5 mm. Styles 4. *W China.* H4. Late spring–early summer.

26. D. × elegantissima (Lemoine) Rehder (*D. discolor elegantissima*). Illustration: Hay & Beckett, Reader's Digest encyclopaedia of garden plants and flowers, 221 (1972); Krüssmann, Manual of cultivated broad-leaved trees & shrubs **1**: pl. 172 (1986).

Upright shrub to 1.5 m. Leaves ovate to oblong-ovate, irregularly and sharply toothed, with rather sparse 4–6-rayed hairs beneath. Cymes loose. Flowers *c.* 2 cm across, pink. Inner filaments toothed. *Garden origin.* H4. Early summer.

A hybrid between *D. scabra* × ?*D. crenata*. Distinguished from *D. purpurascens* by its broader, more abruptly acuminate leaves, shorter leaf-stalks, and larger stamens, the filaments of the inner series toothed. Due to the confusion between *D. scabra* and *D. crenata*, *D. scabra* is often recorded as one of the parents of this hybrid. However, the toothing of the inner filaments suggests that *D. crenata* is involved.

27. D. × kalmiiflora Lemoine (*D. purpurascens* × *parviflora*). Illustration: Davis, The gardener's illustrated encyclopaedia of trees and shrubs, 117 (1987).

Like *D. purpurascens* but an arching shrub to 1.5 m, less densely hairy.

Flowering-shoots 10–30 cm; leaves 3–6 cm, finely toothed. Flowers in upright, umbel-like panicles, each to 2 cm across, deep pink outside, white inside. *Garden origin.* H5. Early summer.

28. D. × maliflora Rehder (*D. × lemoinei* × *D. purpurascens*).

Like *D. purpurascens* but an upright shrub to 2 m; leaves 2.5–4 cm, ovate-oblong, acuminate, margin with fine, forward-pointing teeth, covered with scattered 5–8-rayed hairs, flowers reddish outside, *c.* 1.5 cm across, in corymbs 3–6 cm broad.

'Boule Rose' is a very floriferous cultivar, a rounded shrub to 1 m with white petals edged with pink.

29. D. × lemoinei Lemoine (*D. gracilis* × *D. parviflora*). Illustration: Garden and Forest **9**: 285 (1896).

Upright shrub to 2 m. Leaves 3–6 cm, elliptic-lanceolate to lanceolate, long-pointed, with sharp, forward-pointing teeth, and 5–8-rayed hairs beneath. Flowers numerous, in panicles or pyramidal corymbs. Calyx-teeth triangular, shorter than ovary. Petals partly overlapping, some edge-to-edge in bud. Filaments toothed. *Garden origin.* H2. Early summer.

Many cultivars have been raised, including 'Avalanche', with dense corymbs on drooping branchlets and 'Boule de Niege', a compact dwarf shrub with large, white flowers.

30. D. parviflora Bunge. Figure 24(19), p. 297; 25(2), p. 299. Illustration: Schneider, Illustriertes Handbuch der Laubholzkunde **1**: 379, 381 (1905); Csapody & Tóth, Colour atlas of flowering trees and shrubs, 79 (1982).

Shrub to 2 m. Leaves 3–11 × 2–3 cm, ovate, ovate-lanceolate or elliptic, with coarse, forward-pointing teeth, hairs moderately dense above, 5–8-rayed, sparse beneath, 10–12-rayed; simple hairs present along the main veins. Inflorescence broadly corymb-like. Flowers *c.* 1–1.5 cm across, white with orange disc. Calyx-teeth *c.* 1 mm. Stamens tapered at apex or obscurely toothed. Ovary *c.* 2 mm. Styles 3, shorter than the longer stamens. *N China.* H5. Early summer.

Rare in cultivation.

Var. **amurensis** Regel (*D. amurensis* (Regel) Airy-Shaw) is more commonly grown: it is as above but without the simple hairs on the leaf-veins beneath. *N China, Korea.*

D. × myriantha Lemoine (*D. parviflora × setchuenensis*). Illustration: Gardeners' Chronicle **52**: 45 (1912). Upright shrub to 1 m; leaves 4–6 cm, oblong-lanceolate, finely toothed, rough on both surfaces, with 5- or 6-rayed hairs beneath. Flowers *c.* 2 cm across, white. Filaments strongly toothed. *Garden origin*. H5. Early summer.

31. D. hookeriana (Schneider) Airy-Shaw (*D. corymbosa* Brown var. *hookeriana* Schneider). Figure 24(15), p. 297; 25(8), p. 299.

Shrub, roughly 2 m; like *D. corymbosa* except for the narrower leaves with 3–5-rayed hairs above and dense 5–9-rayed hairs beneath; anthers borne on the side of the filament; flowers sometimes tinged pink. *Himalaya to W China (Yunnan)*. H5. Uncommon in cultivation.

32. D. rubens Rehder (*D. hypoglauca* Rehder). Figure 24(10), p. 297; 25(18), p. 299. Illustration: Botanical Magazine, 9362 (1934); Krüssmann, Manual of cultivated broad-leaved trees & shrubs **1**: pl. 170, 317 (1986).

Erect shrub to 2 m; young shoots stellate-hairy, becoming hairless. Leaves 4–7 × 1.5–3 cm, oblong to ovate-oblong, apex acuminate, base narrowed, or rarely rounded, minutely toothed, thin, with sparse 4-rayed hairs above, with sparse to dense 4–7-rayed hairs beneath. Inflorescence cyme-like. Flowers 8–25 mm across, pink. Calyx-teeth 1–2 mm. Anthers of larger stamens attached at apex of filament or just below it, smaller stamens untoothed but with a small lobe at the apex. Ovary *c.* 2 mm. Styles 3, equalling or shorter than the stamens. *C China (Sichuan, Hubei, Shaanxi)*. H4. Early summer.

33. D. corymbosa R. Brown. Figure 24(1), p. 297; 25(10), p. 299. Illustration: Edwards's Botanical Register **26**: t. 5 (1840); Schneider, Illustriertes Handbuch der Laubholzkunde **1**: 381 (1906).

Shrub to 2 m. Leaves 6–10 × 2.5–5 cm, ovate, apex acuminate, finely to coarsely toothed; hairs not dense, but often denser above than beneath, those above with 6 or 7 short, fat rays, those beneath with 12–14 long, thin rays. Inflorescence corymb-like, borne on a shoot 10–18 cm. Flowers 1–1.5 cm across, white, hawthorn-scented. Calyx-teeth, *c.* 1 × 1.8 mm, Petals broadly obovate. Larger stamens with shallow teeth, smaller stamens mostly lacking teeth, anthers on short stalks 1–2 mm above the filament. Ovary *c.* 2 mm.

Styles 3–4, longer than stamens. *W Himalaya*. H5. Early summer.

A frost-tender species. Var. **staurothrix** (Airy-Shaw) Zaikonnikova differs from the above in its large, almost uniformly 4-rayed, cross-shaped hairs.

34. D. compacta Craib. Illustration: Botanical Magazine, 8795 (1919); Krüssmann, Manual of cultivated broad-leaved trees & shrubs **1**: f. 316 (1986).

Deciduous shrub to 1.5 m. Young growth stellate-downy at first. Leaves 5–6 × 2–2.5 cm, lanceolate to oblanceolate, apex long-pointed, dull green above, paler beneath. Flowers numerous, in broad corymb-like panicles, white. Petals *c.* 4 × 4, almost circular. *Only known from cultivation, supposed origin China*. H4. Midsummer.

Doubtfully distinct from *D. corymbosa*. 'Lavender Time' with flowers pale purple at first is a most attractive cultivar.

35. D. mollis Duthie. Figure 24(14), p. 297; 25(9), p. 299. Illustration: Botanical Magazine, 8559 (1914); Bean, Trees & shrubs hardy in the British Isles edn 8 **2**: 42 (1973); Krüssmann, Manual of cultivated broad-leaved trees & shrubs **1**: pl. 168 & f. 316, 317 (1986).

Vigorous shrub to 1.5–2 m. Leaves large, to 9 × 4 cm, margin coarsely doubly toothed, densely shaggy-hairy beneath, with 4-rayed stellate hairs above which have conspicuous central rays. Inflorescence paniculate or corymb-like, to 12 × 13 cm, borne on shoots 9–20 cm. Flowers white, *c.* 1 cm across. Calyx-teeth *c.* 1.5 mm. Stamens long, narrow and without teeth. Ovary *c.* 2 mm, densely hairy. *China (W Hubei)*. H5. Summer.

36. D. × wilsonii Duthie. Illustration: Botanical Magazine, 8083 (1906); Schneider, Illustriertes Handbuch der Laubholzkunde **2**: 933 (1912); Krüssmann, Manual of cultivated broad-leaved trees & shrubs **2**: pl. 168 (1986).

Vigorous shrub to 2 m; leaves 7–11 cm, apex acuminate, margin with forward-pointing teeth, densely covered with 5–10-rayed stellate hairs, and with spreading hairs on the veins also beneath. Flowers in loose broad corymbs, each *c.* 2 cm across, white. Some stamens toothed, others tapered at the apex. *Garden origin*. H4. Early summer.

33. PHILADELPHUS Linnaeus
D.R. McKean
Shrubs with mainly peeling bark; leaves usually deciduous, opposite, simple.

Flowers in racemes, panicles or cymes, or solitary, often strongly scented. Sepals 4. Petals 4. Stamens numerous. Ovary inferior, surmounted by a nectar-secreting disc; carpels 4, united; styles 4, partially or wholly united. Fruit a many-seeded capsule. Seeds usually with tails.

A genus of about 40 species, mainly from E Asia and the Himalaya, N America, S Europe and the Caucasus. They prefer a loamy soil in full sun; pruning should consist of thinning some of the older wood only, as flowers are produced on the previous year's growth. Vigorous shoots should be left unpruned. Propagate by softwood cuttings.

Literature: Koehne, E., *Gartenflora* **45**: 450–1, 486–8, 500–8, 541–2, 561–3, 596–7, 618–9, 651–2 (1896); Schneider, C.K., *Illustriertes Handbuch der Laubholzkunde* **1**: 362–374 (1906); Hu, S.Y., The genus Philadelphus, *Journal of the Arnold Arboretum* **35**: 275–333 (1954); **36**: 52–109, 325–368 (1955); **37**: 15–90 (1956).

1a. Epigynous zone hairless outside 2
 b. Epigynous zone hairy outside 15
2a. Flowers solitary, in 3s or in few-flowered panicles 3
 b. Flowers in distinct, many-flowered racemes or panicles 5
3a. Flowers mainly semi-double or double; bark not peeling; stamens absent or fewer than 20
  16. × cymosus
 b. Flowers single (except in some cultivars of *P.* × *lemoinei*); bark peeling; stamens more than 20 4
4a. Petals elliptic, acute; stamens 20–35 **17. × falconeri**
 b. Petals oblong, obtuse; stamens 60–90 **3. inodorus**
5a. Calyx greenish purple; flowers bell-shaped; flower-buds tinged pink
  15. purpurascens
 b. Calyx green; flowers mainly disc- or cross-shaped; flower-buds white 6
6a. Leaves uniformly long-hairy above and on the veins beneath; stamens 30–40 **14. delavayi**
 b. Leaves with hairs only on the veins and/or vein-angles or hairless or almost so; stamens mainly 25–35 7
7a. Flowers in racemose panicles 8
 b. Flowers in simple racemes 9
8a. Upper leaf-surface hairless or becoming so, lower surface moderately hairy **13. × lemoinei**
 b. Leaves hairy on veins and/or vein-angles only **30. californicus**

9a. Leaves hairless or almost so (rarely slightly hairy beneath in *P.* × *purpureo-maculatus*) 10

b. Leaves hairy on the veins on either or both surfaces 11

10a. Sepals each with a small tail at the apex; petals *c.* 1.5 × 1.2 cm; bark grey; stamens 38 **8. intectus**

b. Sepals not tailed; petals 0.9–1.1 × 0.8 cm; bark dark brown; stamens 25 **10. pekinensis**

11a. Leaves usually shaggy-hairy beneath, becoming hairless above, long-acuminate **9. tomentosus**

b. Leaves hairy only on veins and/or vein-angles beneath 12

12a. Current growth shaggy at first; leaves sparsely stiff-hairy above; stamens 30–40 **11. brachybotrys**

b. Current growth hairless or sparsely downy; leaves hairy only on veins and/or vein-angles; stamens to 30 13

13a. Current growth sparsely downy; leaves softly hairy on veins and vein-angles; flowers fragrant **12. coronarius**

b. Current growth hairless; leaf-veins stiffly hairy; flowers not fragrant 14

14a. Leaves 6–11 cm; flowers disc-shaped **7. × splendens**

b. Leaves 4–5 cm; flowers mainly cross-shaped (except for some varieties) **6. lewisii**

15a. Epigynous zone hairy outside, the surface visible through the hairs 16

b. Epigynous zone very densely hairy outside, the surface not visible through the hairs 28

16a. Leaves 1–1.5 cm **27. microphyllus**

b. Leaves more than 2.5 cm 17

17a. Flowers 3–5, in cymes or corymbs 18

b. Flowers more numerous in racemose panicles or in distinct racemes 23

18a. Flowers with a pink or purplish centre 19

b. Flowers without a pink or purplish centre 20

19a. Petals obovate-oblong, *c.* 2 × 1 cm, hairy inside **2. × burkwoodii**

b. Petals ovate, *c.* 1.1 × 1 cm, not hairy **5. × purpureo-maculatus**

20a. Leaves densely hairy beneath; winter-buds exposed **32. hirsutus**

b. Leaves moderately hairy beneath; winter-buds hidden beneath the bark 21

21a. Leaves evergreen, with long curved hairs above and beneath **1. mexicanus**

b. Leaves deciduous, upper surface hairless or soon becoming so 22

22a. Stamens 60–90 **4. floridus**

b. Stamens 30 or fewer **28. × polyanthus**

23a. Flowers in panicles **31. insignis**

b. Flowers in racemes 24

24a. Leaves hairy on veins and in vein-angles only **26. satsumi**

b. Leaves hairy on the surfaces as well as on veins or not hairy 25

25a. Leaves densely long, shaggy-hairy beneath **9. tomentosus**

b. Leaves hairy only on the veins beneath 26

26a. Hairs on leaf veins sparse, each with a swollen base **24. kansuensis**

b. Hairs on leaf veins dense, bases of hairs not swollen 27

27a. Disc and style hairless **22. sericanthus**

b. Disc and style hairy **23. subcanus**

28a. Leaves uniformly long-hairy above and on the veins beneath **22. sericanthus**

b. Leaves not as above 29

29a. Flowers solitary **29. argyrocalyx**

b. Flowers in racemes 30

30a. Leaves densely hairy beneath 31

b. Leaves not densely hairy beneath 32

31a. New growth hairy **21. incanus**

b. New growth hairless **18. pubescens**

32a. Current year's growth hairless; flowers usually double **19. × nivalis**

b. Current year's growth with long shaggy hairs; flowers single or double 33

33a. Flowers single; styles hairy **25. schrenkii**

b. Flowers usually double; styles hairless **20. × virginalis**

1. P. mexicanus Schlechtendahl. Illustration: Botanical Magazine, 7600 (1898).

Tender, climbing, evergreen shrub to 5 m; branches long, drooping; bark dark brown, wrinkled, current growth long-bristly, winter-buds prominent. Leaves 5–11.5 × 2–5 cm, ovate, adpressed-bristly on both surfaces, apex long-acuminate, base rounded to cordate, more or less entire or with a few tiny teeth. Flowers 3–4 cm across, solitary or in 3s, yellowish white, rose-scented. *Mexico, Guatemala.* G1. Summer.

2. P. × burkwoodii Burkwood & Skipwith.

Dwarf shrub with prominent winter-buds, bark very dark brown in the second year and eventually peeling; current growth adpressed stiffly hairy. Leaves 3.5–6.6 × 1.5–3 cm, ovate-elliptic, hairless above, sparsely hairy beneath. Flowers 5–6 cm across, 1–5, in panicles, cross-shaped, white with purplish inside, fragrant. Sepals short-hairy or partially so; disc and style hairless. *Garden origin.* H4. Summer.

A hybrid between *P. mexicanus* and another, unknown species.

3. P. inodorus Linnaeus. Illustration: Journal of the Arnold Arboretum 35: pl. 4 (1954); Krüssmann, Manual of cultivated broad-leaved trees & shrubs, f. 273 (1986).

Arching shrub to 2–3 m, bark of the second year chestnut brown, peeling; current year's growth hairless. Leaves 5–9 × 2–3.5 cm, ovate-elliptic or elliptic, more or less entire or faintly toothed, sparsely adpressed hairy or almost hairless above, hairy on main veins and vein-angles beneath. Flowers 4–5 cm across, in cymes of 1, 3 or rarely 9; flower-stalk, epigynous zone and calyx all hairless. Stamens 60–90. Style equal to the longest stamens, hairless, stigmas swollen. Seeds long-tailed. *SE USA.* H3. Summer.

4. P. floridus Beadle. Illustration: Schneider, Illustriertes Handbuch der Laubholzkunde **1**: 366 (1905).

Shrub to 3 m; current year's growth hairless, brown, second year's growth chestnut. Leaves 4–10 × 2–6 cm, mostly ovate-elliptic, apex sharply acuminate, base rounded or obtuse, almost entire or inconspicuously and remotely toothed, evenly adpressed bristly beneath, hairless above except for a few adpressed bristles on the veins. Flowers in 3s, rarely solitary or in racemes, epigynous zone and calyx shaggy-hairy. Flower 4–5 cm across, disc-shaped, petals almost circular, *c.* 2.5 cm, pure white. Stamens 80–90, to 1.5 cm. Stigma oar-shaped. Seeds long-tailed. *USA (Georgia).* H3. Early summer.

Var. **faxonii** Rehder has smaller flowers in the shape of a cross, and is sometimes regarded as a cultivar. *Origin unknown.* H3. Early summer.

5. P. × purpureo-maculatus Lemoine (*P.* × *lemoinei* × *P. coulteri*). Illustration: Botanical Magazine, 8193 (1908).

Shrub to 1.5 m; bark blackish brown, eventually peeling in the second year; current growth hairy. Leaves 1–3.5 × 0.6–2.5 cm, broadly ovate, apex acute, base rounded, entire or almost so, with a few scattered hairs beneath. Flowers solitary or in 3s or 5s, fragrant, epigynous zone with short, rough hairs. Corolla 2.5–3 cm across, disc-shaped, petals almost circular, white, purplish at base. Stamens c. 30. *Garden origin*. H3. Summer.

Several cultivars are known, all with the characteristic pink- or purplish-centred flowers, flowering late and with a long flowering season; they include 'Belle Etoile', 'Bicolore', 'Etoile Rose', 'Fantaisie', 'Galathee', 'Nuage Rose', 'Sybille'.

6. P. lewisii Pursh. Illustration: Schneider, Illustriertes Handbuch der Laubholzkunde **1**: 368 (1905); Krüssmann, Manual of cultivated broad-leaved trees & shrubs, 380 (1986).

Erect shrub to 3 m; second year's growth yellowish to chestnut brown, bark not peeling but with transverse cracks, current year's growth hairless except at the nodes. Leaves 4–5.5 × 2–3.5 cm, apex acute, base rounded, obtuse or shortly acuminate, more or less entire or inconspicuously finely toothed, very sparsely covered with long, rough hairs on the veins above and with tufts of hair in the vein-angles beneath, margin hairy. Flowers 3–4.5 cm across, in racemes of 5–11, cross-shaped. Sepals 5–6 × 3 mm, ovate, wide at the base. Stamens 28–35, the longest being half the length of the petals. Anthers and disc hairless. Style shorter than the longest stamens, hairless, undivided or slightly divided above. Seeds long-tailed. *W North America (British Columbia to California)*. H3. Early summer.

Var. **gordonianus** (Lindley) Koehne. Illustration: McMinn, Manual of Californian shrubs, 139 (1939); Krüssmann, Manual of cultivated broad-leaved trees & shrubs **3**: 381 (1986). Leaves more densely hairy and more strongly toothed, flowers disc-shaped. *W USA (Washington to California)*. H3. Summer.

7. P. × splendens Rehder. Illustration: Journal of the Arnold Arboretum **35**: pl. 3 (1954); Krüssmann, Manual of cultivated broad-leaved trees & shrubs, f. 271 (1986).

Upright shrub; bark of the second year dark brown, peeling; current growth hairless. Leaves 6–11.5 × 2.5–5 cm, oblong-elliptic, apex acuminate, base rounded, inconspicuously and finely toothed or almost entire, hairless or rough, shaggy hairy on the veins and vein-angles beneath. Flowers in crowded racemes of 5–9, disc-shaped, c. 4 cm across, slightly scented. Petals rounded, pure white. Epigynous zone, calyx, disc and style hairless. Stamens 30. *Garden origin*. H3. Summer.

Thought to be *P. inodorus* var. *grandiflorus* × *P. lewisii* var. *gordonianus*.

8. P. intectus Beadle (*P. pubescens* Loiseleur var. *intectus* (Beadle) Moore).

Erect shrub to 5 m; bark silvery not peeling; current growth hairless. Leaves on non-flowering-shoots 6–10 × 4–6 cm, ovate to oblong-elliptic; on flowering-shoots 3–6 × 1.5–3.5 cm; apex acuminate, base rounded to obtuse, hairless or rarely hairy beneath, margin with a few forward-pointing teeth. Flowers c. 3 cm across, in racemes of 5–9, disc-shaped. Epigynous zone hairless. Sepals ovate, tailed. Stamens 38. Disc and style hairless. Seeds long-tailed. *SE USA*. H3. Summer.

9. P. tomentosus Royle (*P. coronarius* Linnaeus var. *tomentosus* (Royle) Hooker & Thomson). Illustration: Royle, Illustrated botany of the Himalayas **2**: pl. 46 (1839); Schneider, Illustriertes Handbuch der Laubholzkunde **1**: 371 (1905); Krüssmann, Manual of cultivated, broad-leaved trees & shrubs **3**: 383 (1986).

Shrub to 3 m, similar to *P. coronarius* but for the downy undersides of the leaves; second year's bark cinnamon, eventually peeling off; current growth hairless or becoming so. Leaves 4–10 × 2–5 cm, ovate or rarely lanceolate, apex acuminate (often strikingly so), base rounded or obtuse, becoming hairless above, uniformly shaggy hairy beneath or rarely almost hairless. Flowers c. 3 cm across, 5–7 in a raceme, cross-shaped, fragrant. Petals obovate-oblong, cream. Disc and styles hairless, stigmas club-shaped. *N India, Himalaya*. H3. Early summer.

10. P. pekinensis Ruprecht. Illustration: Schneider, illustriertes Handbuch der Laubholzkunde **1**: 237, 238 (1905); Journal of the Arnold Arboretum **35**: pl. 3, 4 (1954) & **37**: pl. 5 (1956); Krüssmann, Manual of cultivated broad-leaved trees & shrubs **3**: f. 276 (1986).

Low compact shrub to 2 m; bark of the second year dark brown, peeling; current year's growth hairless. Leaves on non-flowering shoots 6–9 × 2.5–4.6 cm, ovate, those of flowering-shoots 3–7 × 1.5–2.5 cm, apex long-pointed, base rounded or obtuse, toothed, hairless on both surfaces or sometimes with tufts of hairs in the vein-angles nearest the stalk beneath. Racemes with 3–9 yellowish white, fragrant, disc-shaped flowers, 2–3 cm across. Calyx and disc hairless. Sepals ovate, c. 4 mm. Stamens 25. Seeds with short tails. *N & W China*. H3. Early summer.

11. P. brachybotrys (Koehne) Koehne (*P. pekinensis* Ruprecht var. *brachybotrys* Koehne).

Shrub to 3 m; bark of the second year brownish grey, not peeling; current growth shaggy hairy, becoming hairless. Leaves 2–6 × 1–3 cm, ovate, apex shortly acuminate, base rounded, finely toothed or almost entire, sparsely adpressed bristly above and on the veins beneath. Flowers c. 3 cm across, 5–7 in short racemes, cream, disc-shaped. Epigynous zone, calyx, disc and style hairless. Stamens 30–40. Style about as long as stamens, stigma spathulate, inner surface 1 mm, outer 2 mm. Seeds short-tailed. *SE China*. H5. Summer.

There is some doubt as to whether the plants in cultivation under this name are genuine or of hybrid origin.

12. P. coronarius Linnaeus. Illustration: Journal of the Arnold Arboretum **35**: pl. 1, 3, 4 (1954); Krüssmann, Manual of cultivated broad-leaved trees & shrubs **3**: f. 274 (1986).

Shrub to 3 m; bark dark brown, slowly peeling in the second year; current growth sparsely downy, becoming hairless. Leaves 4.5–9 × 2–4.5 cm, ovate, almost hairless but downy on the the major veins and in the vein-angles beneath, margins irregularly and shallowly toothed, apex acuminate, base obtuse or acute. Flowers 2.5–3 cm across, 5–9 in short terminal racemes, creamy white, strongly fragrant. Sepals triangular, acute, hairless. Stamens c. 25. Disc and style hairless. Seeds with long tails. *S Europe (Austria, Italy, Romania), former USSR (Caucasus)*. H4. Early summer.

Several cultivars (formerly recognised as varieties or forms) are known, e.g. 'Aureum' (var. *aureus* Anon.) with the young leaves yellowish; 'Deutziflorus' with the petals pointed at the apex; 'Duplex' with double flowers; and 'Variegatus' with white margins to the leaves.

13. P. × lemoinei Lemoine (*P. coronarius* × *P. microphyllus*). Illustration: Journal of the Arnold Arboretum **35**: pl. 3 (1954);

Krüssmann, Manual of cultivated broad-leaved trees & shrubs **3**: f. 279 (1986).

A low compact shrub as wide as high, with peeling bark. Leaves 1.5–2.5 × 0.7–1.2 cm, ovate, hairless above, sparsely bristly beneath, apex acuminate, base rounded or obtuse, with *c.* 6 teeth. Flowers usually in 3s, more rarely solitary or in 5s, cross-shaped, *c.* 3 cm across. Sepals ovate, hairless. Petals notched. Stamens *c.* 5. *Garden origin.* H4. Summer.

Several cultivars have been raised, e.g. 'Avalanche', 'Candelabre', 'Coup d'Argent', etc.

14. P. delavayi Henry. Illustration: Schneider, Illustriertes Handbuch der Laubholzkunde **1**: 370 (1905); Botanical Magazine, 9022 (1924); Bean, Trees & shrubs hardy in the British Isles, edn 8, **3**: 130 (1976).

Shrub to 4 m; bark of second year grey-brown, grey or chestnut brown, not peeling; current growth hairless, glaucous. Leaves 2–8 × 2–5 cm on flowering-shoots, much larger on non-flowering-shoots, ovate-lanceolate or ovate-oblong, apex acuminate, base rounded, usually with forward-pointing teeth but sometimes entire; all sparsely bristly above, densely adpressed shaggy-hairy beneath. Flowers 2.5–3.5 cm across, in racemes of 5–9 (rarely more), disc-shaped, pure white, fragrant. Calyx hairless, glaucous, tinged with purple. Stamens *c.* 35. Disc and style hairless. *SW China.* H4. Early summer.

15. P. purpurascens (Koehne) Rehder (*P. brachybotrys* var. *purpurascens* Koehne). Illustration: Botanical Magazine, 8324 (1910) – as P. delavayi; Krüssmann, Manual of cultivated broad-leaved trees & shrubs **3**: pl. 154 (1986).

Shrub to 4 m; bark of second year brown or grey, smooth; current growth hairless. Leaves 1.5–6 × 0.5–3 cm on flowering-shoots, much longer on non-flowering-shoots, ovate to ovate-lanceolate, usually uniformly adpressed bristly above and on the veins beneath, finely toothed. Racemes usually with 5–9 bell-shaped flowers 3–4 cm wide and very fragrant. Calyx green tinged with purple, glaucous. Stamens 25–30. Style and disc hairless. *S China.* H4. Summer.

Var. **venustus** (Koehne) S.Y. Hu (*P. venustus* Koehne) has shaggy hairs on the young growth. *SW China.* H4. Summer.

16. P. × cymosus Rehder (*P. floribundus* Schrader).

Erect shrub to 2.5 m; bark brown, not

peeling. Leaves ovate, sparsely toothed, hairy beneath especially on the veins. Flowers in cymes of 1–5. Sepals and disc hairless. Some stamens may be petal-like. *Garden origin.*

Parentage unknown. Several cultivars are known, e.g. 'Amalthée', 'Bonniere' (semi-double), 'Bouquet Blanc' (double), 'Conquete' and 'Dresden'.

17. P. × falconeri Nicholson. Illustration: Journal of the Arnold Arboretum **35**: pl. 1, 2 (1954); Krüssmann, Manual of cultivated broad-leaved trees & shrubs **3**: pl. 155 (1986).

Shrub to 3 m; bark brown, peeling in the second year; branches slender and pendent, current growth hairless. Leaves 3–6.5 × 1–2.5 cm, ovate or ovate-elliptic, base rounded or obtuse, faintly toothed, veins with adpressed bristles beneath. Flowers to 3 cm across, 3–5 in cymes, abundant, pure white, stellate, petals elliptic, pointed. Styles much longer than stamens, sterile. *Garden origin.* H3. Summer.

A hybrid of unknown parentage which is often shy to flower.

18. P. pubescens Loiseleur. Illustration: Schneider, Illustriertes Handbuch der Laubholzkunde **1**: 369 (1905); Krüssmann, Manual of cultivated broad-leaved trees & shrubs **3**: f. 272 (1986).

Shrub to 5 m; second year's bark grey, first year's bark not peeling, current growth hairless. Leaves 4–8 × 3–5.5 cm on flowering-shoots, ovate, apex abruptly acuminate, base rounded, remotely toothed or entire, hairless above except for rough, short, stiff hairs on the veins, shaggy bristly hairy beneath. Flowers *c.* 3.5 cm across, 5–11 in racemes, white, scentless. Stamens *c.* 35. Disc and style hairless. Seeds large, with short tails. *SE USA.* H3. Early summer.

P. × monstrosus (Späth) Schelle (*P. inodorus* × *P. lewisii* var. *gordonianus*) differs only in its less hairy epigynous zone. A hybrid between *P. pubescens* and perhaps *P. inodorus* is also rarely cultivated (*P. × pendulifolius* Carrière).

19. P. × nivalis Jacques.

Arching shrub to 2.5 m; bark dark brown, peeling; current growth hairless. Leaves 5–10 × 2.5–6 cm, ovate or ovate-elliptic, apex acuminate, base rounded or obtuse, faintly toothed, hairless above, uniformly shaggy-hairy beneath. Racemes with 5–7 double flowers. Sepals with dense, long, rough hairs. Corolla 2.5–3.5

cm across, disc-shaped. Disc and style hairless. *Garden origin.* H3. Early summer.

Probably *P. coronarius* × *P. pubescens.*

20. P. × virginalis Rehder. Illustration: Krüssmann, Manual of cultivated broad-leaved trees & shrubs **3**: pl. 156 (1986).

Stiffly upright shrub to 2.5 m, second year's bark grey, peeling only when old, current growth with shaggy hairs. Leaves 4–7 × 2.5–4.5 cm, ovate, apex shortly acuminate, base rounded, becoming hairless above, uniformly tough-shaggy-hairy beneath. Flowers 4–5 cm across, in racemes, usually double, pure white, very fragrant. Calyx densely hairy. Style hairless. *Garden origin.* H4. Summer–late summer.

Of doubtful origin but with strong characteristics of *P. pubescens.* Several cultivars are grown, including 'Argentine', 'Boule d'Argent', 'Enchantment', 'Fleur de Neige'.

21. P. incanus Koehne. Illustration: Journal of the Arnold Arboretum **35**: pl. 3 (1954).

Erect shrub to 3.5 m; bark grey and smooth to the second year, later peeling; current growth hairy. Leaves 4–8.5 × 2–4 cm on flowering-shoots, to 10 × 6 cm on non-flowering-shoots, ovate-elliptic, apex slender-pointed, base tapered to rounded, sparsely bristly above, adpressed-bristly beneath. Flowers *c.* 2.5 cm, 7–11 in racemes, white across. Calyx and flower-stalk adpressed bristly. Stamens *c.* 34. Seeds very short-tailed. *China (Hubei, Shaanxi).* H4. Late summer.

22. P. sericanthus Koehne. Illustration: Botanical Magazine, 8941 (1922); New Flora and Silva, pl. 85 (1934); Csapody & Tóth, Colour atlas of flowering trees and shrubs, 79 (1982); Krüssmann, Manual of cultivated broad-leaved trees & shrubs **3**: pl. 154 (1986).

Shrub to 3 m; current growth hairless or soon becoming so, second year's bark grey or grey-brown, slowly peeling. Leaves 4–11 × 1.5–5 cm, ovate-elliptic or elliptic-lanceolate, apex acuminate, base obtuse or rounded, usually coarsely toothed, sparsely adpressed-bristly above and on the veins beneath. Flowers *c.* 2.5 cm across, 7–15 in racemes, pure white, unscented. Calyx and flower-stalk densely adpressed bristly. Disc and style hairless. Seeds short-tailed. *China (Sichuan & Hubei).* H4. Summer.

23. P. subcanus Koehne. Illustration: New Flora and Silva, pl. 85 (1934); Csapody & Tóth, Colour atlas of flowering trees and shrubs, 79 (1982).

Erect shrub to 6 m; current growth brown, hairless or soon becoming so, older bark grey-brown, smooth, peeling late. Leaves 4–14 × 1.5–7.5 cm, ovate or ovate-lanceolate, apex acuminate, base rounded or obtuse, obscurely finely toothed on flowering-shoots but with forward-pointing teeth on non-flowering-shoots, all sparsely covered with upright hairs above, shaggy-hairy on the veins beneath. Flowers 5–29 in racemes which are 2.5–22 cm; flowers 2.5–3 cm across, disc-shaped, pure white, slightly fragrant. Calyx curly-hairy. Petals circular to obovate, long curly-hairy at base. Stamens *c.* 30. Disc downy, lower style hairy. Seeds with very short tails. *W China*. H4. Early summer.

Rare in cultivation; most commonly seen is var. **magdalenae** (Koehne) S.Y. Hu, which is somewhat smaller in all its parts and has the flower-stalks and calyx only slightly downy with curly hairs. *SW China*. H4. Early summer.

24. P. kansuensis (Rehder) S.Y. Hu (*P. pekinensis* Ruprecht var. *kansuensis* Rehder). Illustration: Journal of the Arnold Arboretum **35**: pl. 4 (1954).

Upright shrub to 7 m; current year's growth with curly hairs, eventually becoming hairless, second year's bark grey-brown, peeling. Leaves to 11 × 6.5 cm on non-flowering-shoots, 3–5 × 1–2 cm on flowering-shoots, ovate or ovate-lanceolate, more or less entire or faintly toothed, apex pointed, base obtuse or rounded, uniformly bristly-hairy above, hairs on the veins beneath with swollen bases. Flowers *c.* 2.5 cm across, 5–7 in racemes, disc-shaped, flower-stalks bristly-hairy. Petals oblong-rounded. Stamens *c.* 30, disc bristly at its rim. Style hairless. Seeds short-tailed. *NW China*. H5. Summer.

25. P. schrenkii Ruprecht. Illustration: Schneider, Illustriertes Handbuch der Laubgehölze **1**: 237 (1905); Krüssmann, Manual of cultivated broad-leaved trees & shrubs **3**: f. 278 (1986).

Upright shrub to 4 m, second year's bark grey or rarely brown, with transverse cracks, rough hairy at first. Leaves 7–13 × 4–7 cm on non-flowering-shoots, 4.5–7.5 × 1.5–4 cm on flowering-shoots, ovate, occasionally ovate-elliptic, apex acuminate, base acute or obtuse, remotely finely toothed or almost entire, sparsely shaggy-hairy on the main veins beneath, usually hairless above. Flowers 2.5–3.5 cm across, 3–7 in racemes, cross-shaped, very fragrant. Sepals 3–7 mm, ovate. Stamens 25–30. Disc hairless. Style hairy. Seeds short-tailed. *Korea to former USSR (E Siberia)*. H3. Summer.

Var. **jackii** Koehne. Illustration: Krüssmann, Manual of cultivated broad-leaved trees & shrubs **3**: f. 278 (1986). Leaves more obviously toothed, the veins hairy beneath. *N China, Korea*. H3. Summer.

26. P. satsumi (Siebold) S.Y. Hu. Illustration: Schneider, Illustriertes Handbuch der Laubholzkunde **1**: 237 (1905); Makino, Illustrated flora of Japan, 1461 (1956); Krüssmann, Manual of cultivated broad-leaved trees & shrubs **3**: f. 276 (1986).

Upright shrub to 3 m, second-year twigs brown, bark eventually peeling; current growth becoming hairless. Leaves on non-flowering-shoots 6–9 × 3–5 cm, ovate or broadly elliptic, with coarse, forward-pointing teeth, apex long-acuminate, base obtuse or rounded, those on flowering-shoots 4.5–7 × 1.5–4.5 cm, ovate or ovate-lanceolate, tapered, apex long-acuminate, base obtuse or sometimes rounded; all with sparse bristles or hairless above, with stiff hairs on the veins beneath and in the vein-angles. Flowers *c.* 3 cm across, 5–7 in a raceme, cross-shaped, slightly fragrant. Petals oblong-obovate. Stamens *c.* 30. Style hairless, shortly divided at the apex. Seeds with medium tails. *Japan*. H5. Summer.

P. satsumanus Miquel. Similar but with uniformly downy leaves. *Japan*. H5. Summer.

Seldom found in cultivation.

27. P. microphyllus Gray. Illustration: Schneider, Illustriertes Handbuch der Laubholzkunde **1**: 234 (1905); Journal of the Arnold Arboretum **35**: pl. 1, 2, 4 (1954); Csapody & Tóth, Colour atlas of flowering trees and shrubs, 79 (1982).

Low, erect, graceful shrub to 1 m; current growth adpressed downy, second year's bark chestnut brown, shiny, soon flaking off. Leaves 1–1.5 × 0.5–0.7 cm, ovate-elliptic or sometimes lanceolate, entire and with marginal hairs, hairless or becoming so above, softly shaggy hairy beneath, base obtuse, apex acute or obtuse. Flowering-shoots 1.5–4 cm. Flowers *c.* 3 cm across, 1–2, pure white, very fragrant, cross-shaped. Sepals lanceolate. Stamens *c.* 32. Seeds with very short tails. *SW USA*. H5. Early summer.

28. P. × polyanthus Rehder.
Erect shrub, bark dark brown, eventually peeling; current year's growth with a few shaggy hairs. Leaves 3.5–5 × 1.5–2.5 cm, ovate, apex acuminate, base rounded or obtuse, entire or with a few sharp teeth, hairless above, sparsely hairy with short, adpressed bristles beneath. Flowers *c.* 3 cm across, 3–5 in cymes or corymbs, cross-shaped. Epigynous zone and calyx downy, sepals tailed at apex. Stamens *c.* 30. *Garden origin*. H3. Summer.

Thought to be *P. insignis* × *P. × lemoinei*. Various cultivars are grown, including 'Etoile Rose', 'Fantaisie', 'Galathée', 'Nuage Rose', 'Sybille'.

29. P. argyrocalyx Wooton. Illustration: Journal of the Arnold Arboretum **35**: pl. 14 (1954); Krüssmann, Manual of cultivated broad leaved trees & shrubs **3**: f. 273 (1986).

Erect shrub to 2 m; twigs grey-brown, the second-year bark usually intact, current growth with rusty, shaggy hairs. Leaves 1–3.5 × 0.4–1.5 cm, ovate, ovate-lanceolate or elliptic, apex acute or obtuse, base obtuse, dark green, hairless or becoming so above, sparsely shaggy-bristly and paler green beneath. Flowers solitary to 3.5 cm across, on short stalks (1–2 mm), cross-shaped, white and slightly fragrant. Calyx densely white woolly. Seeds long-tailed. *USA (New Mexico)*. Summer–late summer.

30. P. californicus Bentham (*P. lewisii* var. *californicus* (Bentham) Torrey). Illustration: Schneider, Illustriertes Handbuch der Laubholzkunde **1**: 234 (1905); Dippel, Handbuch der Laubholzkunde **1**: 181 (1889); Journal of the Arnold Arboretum **35**: pl. 2 (1954); Krüssmann, Manual of cultivated broad-leaved trees & shrubs **3**: f. 272 (1986).

Erect shrubs to 3 m; second year's bark dark brown, current year's growth soon becoming hairless. Leaves on non-flowering-shoots 4.5–8 × 3–5 cm, those on flowering-shoots 3–5 × 2–3 cm (rarely to 8 cm), ovate or ovate-elliptic, hairless except for tufts in the vein-angles beneath, apex acute, base acute, obtuse or sometimes rounded, entire or obscurely toothed. Flowers to 2.5 cm across, 3–5 in a panicle, cross-shaped, fragrant. Sepals ovate, hairless. Stamens 25–37. Disc and style hairless. Seed short-tailed. *USA (California)*. H5. Summer.

31. P. insignis Carrière (*P. × insignis* Carrière). Illustration: Schneider, Illustriertes Handbuch der Laubholzkunde **1**: 236 (1905); Krüssmann, Manual of cultivated broad-leaved trees & shrubs **3**: f. 277 (1986).

Erect shrub to 4 m; second year's bark grey, (rarely brown) smooth. Leaves 3.5–8 × 1.5–6 cm, ovate or ovate-elliptic, obtuse, apex acute or strongly acuminate, base acute or rounded, more or less entire or faintly toothed, coarsely adpressed-hairy beneath. Flowers 2.5–3.5 cm across, in 3s in panicles. Sepals ovate, usually adpressed hairy. Stamens *c.* 30. Disc and style hairless. Seeds short-tailed. *W USA (California, Oregon).* H5. Summer.

Considered by some authors to be a hybrid (*P. californicus* × *P. pubescens*).

32. P. hirsutus Nuttall. Illustration: Edwards's Botanical Register **24**: t. 14 (1839); Botanical Magazine, 5334 (1862); Schneider, Illustriertes Handbuch der Laubholzkunde **1**: 234 (1905).

Low, spreading shrub with slender, slightly twisted arching branches, to 2.5 m; shoots widely divergent, second year's bark dark brown, peeling, current year's growth with shaggy hairs. Leaves 2.5–7 × 1–5 cm, ovate-elliptic or ovate-lanceolate, apex acuminate, base rounded, sharply toothed, uniformly covered with hairs with swollen bases above and densely shaggy beneath. Flowers *c.* 2.5 cm across, 1–5 on very short shoots with 1 or 2 pairs of leaves, disc-shaped. Sepals broadly triangular, shaggy hairy. Style and disc hairless. Seeds without tails. *SE USA (North Carolina to Georgia).* H5. Early summer.

34. KIRENGESHOMA Yatabe
P.G. Barnes
Erect herbaceous perennial, with short, thick rhizomes. Leaves opposite, the lower long-stalked, the upper stalkless, all palmately lobed and coarsely sinuous-toothed. Flowers in terminal and axillary cymes, somewhat nodding on long stalks. Sepals 5, small, petals 5, narrowly ovate, rather thick, pale yellow, not opening widely. Stamens 15, styles usually 3; ovary inferior, 3-celled. Capsule ovoid, 3-celled, seeds flat, with an irregular wing.

A genus of 2 species from NE Asia. Easily cultivated hardy perennials, requiring a moisture retentive but well-drained soil and a position in full sun or part shade. Propagate by division in autumn or early spring. Although seeds are often produced quite freely, germination may be slow and erratic.

1. K. palmata Yatabe. Illustration: Botanical Magazine, 7944 (1904); Thomas, Perennial garden plants, f. 13 (1982); Jelitto, Schacht & Fessler, Die

Freiland Schmuckstauden, 338 (1985); Kitamura & Murata, Coloured illustrations of herbaceous plants of Japan **2**: pl. 33 (1986).

Stem to 1 m or more, often purplish above. Leaves 10–20 cm, slightly hairy beneath. Flowers 2.5–3.5 cm. *Japan.* H4. Summer.

Plants from Korea have been distinguished as *K. koreana* Nakai. This is said to be taller, with more erect flowers.

35. QUINTINIA A. de Candolle
M.F. Gardner
Evergreen trees or shrubs. Leaves leathery, alternate, entire or with obscurely forward-pointing teeth with marginal glandular hairs. Flowers borne in many-flowered terminal and axillary racemes or many-flowered panicles; petals 5, white or pale lilac, oblong, spreading or turned-back, overlapping; stamens 5; ovary inferior, 3–5 celled, stigma head-like, 3–5 lobed. Capsules 3–5 ribbed, opening at the top; seeds winged on each side.

A genus of 4 species from C Malaysia, New Guinea, Australia and New Zealand. Successful cultivation is achieved in the cool glasshouse in a loam-enriched soil. Propagate by seed or cuttings.

1a. Petals white; leaf-margins wavy, entire   **2. acutifolia**
  b. Petals pale lilac; leaf-margins wavy, toothed   **1. serrata**

1. Q. serrata A. Cunningham. Illustration: Hooker's Icones Plantarum, **6**: 558 (1843); Moore & Irwin, The Oxford book of New Zealand plants, 79 (1978); Salmon, The native trees of New Zealand, 196 (1980); Salmon, A field guide to the native trees of New Zealand, 104 (1986).

Tree to *c.* 7 m, shoots clammy. Leaves often blotched, 6–12.5 × 1–2.5 cm, oblong, with coarsely forward-pointing teeth and wavy. Racemes 6–8 cm, flower-stalks *c.* 4 mm; petals 2.5–3 cm, pale lilac. Capsules 4–5 mm, obovoid. *New Zealand.* G1. Summer.

2. Q. acutifolia Kirk. Illustration: Salmon, The native trees of New Zealand 194–195 (1980); Salmon, A field guide to the native trees of New Zealand, 104 (1986).

Tree or shrub to 8 m. Leaves yellowish with green veins, 6–16 × 3–5 cm, broadly elliptic-obovate to wedge-shaped, margins wavy, leaf-stalks to 2 cm. Racemes 4–7 cm, flower-stalks *c.* 3 mm; petals 3–3.5 mm, white. Capsules 4–6 mm, obovoid or oblong. *New Zealand.* G1. Summer.

36. HYDRANGEA Linnaeus
S.T. Buczacki
Deciduous or evergreen shrubs, small trees or climbers. Leaves usually rounded-ovate and toothed, opposite or in whorls of 3. Bark often flaking when mature. Fertile flowers bisexual (rarely unisexual), radially symmetric, in panicles or corymbs. Sepals 4 or 5, small, inconspicuous. Petals 4 or 5, white, blue or pink. Stamens 8 or 10 (rarely more). Ovary inferior, 2–5-celled, containing many ovules. Fruit a 2–5-celled, many-seeded capsule. Many species also bear larger, sterile flowers borne at the outside of the corymb like inflorescences.

A genus of up to 100 species from China, Japan, the Himalaya, the Philippines, Indonesia, N and S America. Many variants have been bred and selected, those with sterile flowers having particular appeal. Many of the popular types of mop-head garden hydrangea are particularly favoured for seaside planting and adopt differing flower colours depending on the relative availability of aluminium ions in the soil. In alkaline soils where aluminium is unavailable, the natural flower colour is pink but a change to blue can be encouraged by supplying the plants with aluminium sulphate, known commercially as blueing powder. Many species are prone to damage from late frosts. Propagate by cuttings or layers.

Literature: McClintock, E., Monograph of the genus Hydrangea *Proceedings of the Californian Academy of Science* (1957); Haworth-Booth, M., *The Hydrangeas*, edn 5 (1984).

1a. Evergreen, clinging climber   2
  b. Deciduous shrub, tree or climber
                                          3
2a. Flowers in a series of clusters one above the other   **16. serratifolia**
  b. Flowers in a single, terminal cluster
                                17. seemanni
3a. Clinging climber   4
  b. Shrub or small tree   5
4a. Leaves coarsely toothed; sterile flowers few; stamens usually fewer than 15   **8. anomala**
  b. Leaves finely toothed; sterile flowers numerous; stamens usually 15 or more   **9. petiolaris**
5a. Leaves deeply 5–7-lobed, like those of an oak   **2. quercifolia**
  b. Leaves usually toothed, but not deeply lobed   6
6a. Leaves small, very coarsely toothed, like those of a nettle   **11. hirta**

b. Leaves fairly finely toothed 7
7a. Flowers in panicles
12. paniculata
b. Flowers in corymbs 8
8a. Corymbs enclosed by c. 6 persistent bracts **3. involucrata**
b. Corymbs without persistent bracts 9
9a. Leaves hairless on both surfaces, often more or less shiny above 10
b. Leaves at least slightly downy, hairy or bristly beneath 11
10a. Sterile flowers white, each 1–1.8 cm across **1. arborescens**
b. Sterile flowers pink, each 3–5 cm across **14. macrophylla**
11a. Leave with hairs confined to the veins beneath 12
b. Leaves with hairs or bristles not confined to the veins beneath 15
12a. Corymbs with up to 12 pink or blue sterile flowers **15. serrata**
b. Corymbs with white, white-blue or cream sterile flowers 12
13a. Sepals toothed **10. scandens**
b. Sepals entire 14
14a. Sterile flowers more than 2 cm across **13. heteromalla**
b. Sterile flowers less than 2 cm across **1. arborescens**
15a. Leaves densely softly downy beneath 16
b. Leaves hairy, bristly or downy beneath but not with a dense covering 17
16a. Fertile flowers dull white **1. arborescens**
b. Fertile flowers white-purple or pink **4. aspera**
17a. Shoots densely covered with hairs and stiff bristles **7. sargentiana**
b. Shoots at most finely downy 18
18a. Corymbs to 30 cm across **5. longipes**
b. Corymbs to 15 cm across **6. robusta**
19a. Flower-stalks bristly; leaf-stalks 4–17 cm **5. longipes**
b. Flower-stalks downy; leaf-stalks 1.5–3 cm **13. heteromalla**

1. H. arborescens Linnaeus. Illustration: Botanical Magazine, 437 (1799); Haworth-Booth, The Hydrangeas, edn 5, 35 (1984); Davis, The gardener's illustrated encyclopaedia of trees and shrubs, 142 (1987); Taylor's guide to shrubs, 140 (1987).

Fairly loose, open, deciduous shrub, 1–3.5 m. Shoots at first downy, then hairless. Leaves 7.5–17.5 × 5–15 cm, broadly ovate, acuminate, with coarse, forward-pointing teeth, rather shiny, dark green above, paler beneath, hairless or slightly downy beneath on veins and in vein axils;, stalks 2.5–7.5 cm. Corymbs fairly flat, much branched, 5–15 cm across; sterile flowers absent or 1–8; sterile flowers long-stalked, creamy white, each 1–1.8 cm across; fertile flowers numerous, small, dull white; flower-stalks downy. Capsule 8–10-ribbed. *E USA*. H2. Summer.

A variable species; the commonest cultivated variant is subsp. **arborescens** 'Grandiflora'; it was originally found wild in Ohio and has a large, cushion-like, heavy head of white sterile flowers only.

Subsp. **discolor** Seringe (*H. cinerea* Small) has tiny warts on the leaves, with sparse downy hairs beneath, and is usually seen as the cultivar 'Sterilis' in which most of the flowers are sterile.

Subsp. **radiata** (Walter) McClintock (*H. radiata* Walter) is a striking plant with darker green leaves and a unique indumentum of thick, downy, white hairs on the leaves beneath.

2. H. quercifolia Bartram (*H. platanifolia* invalid). Illustration: Botanical Magazine, 8447 (1912); American Horticulturist **54**: 15 (1975); Practical gardening for August 1987, 37; Davis, The gardener's illustrated encyclopaedia of trees and shrubs, 145 (1987).

Fairly loose, rounded, deciduous shrub 1–2.5 m. Shoots thick, stout, at first finely reddish downy, then hairless and flaky. Leaves 7.5–20 × 5–17 cm, broadly ovate-rounded and deeply 5–7-lobed (like those of the oak, *Quercus rubra* – see volume **3**: 79 & figure 19), minutely toothed; stalks 2.5–6 cm. Panicles more or less erect, long-pyramidal, 10–25 cm, with numerous long-stalked, white sterile flowers which gradually turn purplish, each 2.5–3.5 cm across; fertile flowers numerous, small, white. Flower-stalks loosely hairy. *SE USA*. H2. Summer.

3. H. involucrata Siebold. Illustration: Kitamura & Okamoto, Coloured illustrations of trees and shrubs of Japan, 197 (1977); Davis, The gardener's illustrated encyclopaedia of trees and shrubs, 143 (1987).

Fairly loose, open, deciduous shrub 1–1.2 m (to 2 m in milder areas). Shoots at first bristly-downy, later hairless. Leaves 7.5–15 × 2.5–6 cm, broadly ovate-oblong, acuminate, finely toothed, bristly, especially above; stalks 0.6–2.5 cm. Corymbs irregular, 7.5–12.5 cm across, at first enclosed by c. 6 broadly ovate bracts which later open outwards and persist, covered with flattened, whitish down. Sterile flowers few, long-stalked, pale blue or faintly pink, each 1.8–2.5 cm across; fertile flowers numerous, small, blue. Flower-stalks slightly downy. *Japan, Taiwan*. H4. Late summer.

Most frequently seen in cultivation as the cultivar 'Hortensis' with more numerous, double, pink-white sterile flowers; it is of Japanese garden origin.

4. H. aspera Don (*H. fulvescens* Rehder; *H. kawakamii* Hayata; *H. rehderiana* Schneider; *H. villosa* Rehder). Illustration: Millar Gault, The dictionary of trees and shrubs in colour, pl. 237 (1976); Davis, The gardener's illustrated encyclopaedia of trees and shrubs, 142, 146 (1987); Amateur gardening for November 28, 1987, 11.

Spreading deciduous shrub or small tree to 4 m. Shoots at first with flattened or spreading hairs, later hairless and often peeling. Leaves 10–25 × 2.5–10 cm, mostly lanceolate to narrowly ovate, acute or acuminate, base rounded or tapered, with spreading or forward-pointing marginal teeth, densely covered with soft down beneath, sparsely hairy above; stalks 2.5–10 cm. Corymbs fairly flattened, to 25 cm across, with few to many white to pale pink or purple, darker-veined, sterile flowers, each with 4 rounded, toothed or entire sepals, each to 2.5 cm across. Fertile flowers small, numerous, white-purple or pink, each with 5 petals, falling early. Flower-stalks hairy. Capsule 2.5–3 mm, hemispherical. *Himalaya, W & C China, Taiwan, Indonesia (Java, Sumatra)*. H5. Summer.

A variable species in respect of overall habit (including some very robust, more or less climbing or scrambling variants), Most familiar in cultivation is the variant previously known as *H. villosa*.

Subsp. **strigosa** (Rehder) McClintock (*H. aspera* var. *macrophylla* Hemsley; *H. strigosa* Rehder; *H. strigosa* var. *macrophylla* (Hemsley) Rehder). Illustration: Botanical Magazine, 9324 (1933). As above, but leaves with short, stiff hairs beneath. *China*. H5. Summer–autumn.

5. H. longipes Franchet (*H. aspera* subsp. *robusta* misapplied). Illustration: Schneider, Illustriertes Handbuch der Laubholzkunde **2**: 939 (1912).

Loose, spreading, deciduous shrub, 2–2.5 m. Shoots at first loosely downy, later hairless. Leaves 7.5–17.5 × 3–9 cm, rounded-ovate, abruptly acuminate, base rounded-cordate, sharply toothed, bristly, especially beneath; stalks 4–17 cm. Corymbs fairly flat, 10–15 cm across. Sterile flowers 8–9, each 1.9–2.4 cm across, white or faintly purple. Fertile flowers numerous, small, white. Flower-stalks bristly. Capsule rounded, hairless, with the calyx at the top. *C & W China.* H5. Summer–autumn.

6. H. robusta Hooker & Thomson (*H. aspera* subsp. *robusta* (Hooker & Thomson) McClintock; *H. rosthornii* Diels). Illustration: Schneider, Illustriertes Handbuch der Laubholzkunde **2**: 939 (1912).

Loose, spreading, deciduous shrub to 4 m, very like *H. longipes* but leaves larger, thicker, densely bristly beneath; corymbs to 30 cm across, with 20 or more large white sterile flowers and blue fertile flowers. *China.* H5. Summer–autumn.

7. H. sargentiana Rehder (*H. aspera* subsp. *sargentiana* (Rehder) McClintock). Illustration: Botanical Magazine, 8447 (1933); Bean, Trees & shrubs hardy in the British Isles, edn 8, **2**: 630 (1970); The Garden **100**: 599 (1975); Davis, The gardener's illustrated encyclopaedia of trees and shrubs, 145 (1987).

Loose, spreading, deciduous shrub to 3 m. Shoots densely covered with small, erect hairs and stiff, translucent bristles. Leaves 10–25 × 5–18 cm, broadly ovate, base rounded, densely covered with short, velvety hairs above, densely bristly beneath; stalks 2.5–11 cm, bristly. Corymbs fairly flattened, 12.5–22.5 cm across, with a few, pink-white sterile flowers confined to the periphery, each to 3 cm across and with 4–5 entire, irregular sepals; fertile flowers numerous, small, pale purple. Flower-stalks hairy. *China (W Hubei).* H5. Summer.

8. H. anomala D. Don (*H. altissima* Wallich).

Clinging, deciduous climber to 12 m. Shoots hairless or hairy, becoming very rough and peeling. Leaves 7.5–13 × 4–10 cm, ovate, shortly acuminate, base cordate, regularly toothed, hairless except for downy tufts in the vein-axils beneath; stalks 1.5–7.5 cm. Corymbs fairly flat, 15–20 cm across, with few, white, peripheral sterile flowers each 1.5–3.7 cm across and numerous, small, cream fertile flowers. *Himalaya, China.* H4. Early summer.

9. H. petiolaris Siebold & Zuccarini (*H. anomala* subsp. *petiolaris* (Siebold & Zuccarini) McClintock; *H. scandens* Maximowicz). Illustration: Botanical Magazine, 6788 (1884); Bean, Trees & shrubs hardy in the British Isles, edn 8, **2**: 628 (1970); Practical gardening for November 1986, 16; Davis, The gardener's illustrated encyclopaedia of trees and shrubs, 145 (1987).

Clinging, deciduous climber to 20 m. Shoots at first finely hairy or hairless, later rough and peeling. Leaves 3.5–11 × 2.5–8 cm, ovate-rounded, shortly acuminate, base more or less cordate, finely toothed, hairless above, sometimes downy beneath, especially on the veins; stalks 5–40 mm. Corymbs flat, 15–25 cm across, with up to 12 white, peripheral sterile flowers each 2.5–4.5 cm across. Fertile flowers small, numerous, off-white. *Japan, former USSR (Sakhalin), Korea, Taiwan.* H3. Summer.

10. H. scandens (Linnaeus) de Candolle (*H. scandens* Maximowicz; *Viburnum scandens* Linnaeus; *V. virens* Thunberg). Illustration: Haworth-Booth, The Hydrangeas, edn 5, 49 (1984).

Spreading or almost pendent shrub to 1 m. Shoots hairless or very finely downy. Leaves 5–9 × 2–4 cm, lanceolate or oblong-ovate, shortly toothed, hairless above, usually finely downy on the veins beneath; stalks *c.* 5 mm. Corymbs fairly flattened, to 7.5 cm across, often abundant, with few white-blue sterile flowers each 1.7–3.8 cm across, with toothed sepals. Fertile flowers numerous, white, with small, clawed petals. *S Japan.* H5. Summer.

Subsp. **chinensis** (Maximowicz) McClintock (*H. chinensis* Maximowicz; *H. davidii* Franchet; *H. umbellata* Rehder) is an altogether larger plant from the E Asian mainland and other areas outside Japan, with tough, woody twigs and more leathery leaves. The name *H. chinensis* covers a number of different variants, including forma **lobbii** Maximowicz from the Philippines with large, white sterile flowers and forma **macrosepala** Hayata from Japan.

11. H. hirta Siebold. Illustration: Schneider, Illustriertes Handbuch der Laubholzkunde **1**: 385, 386 (1905).

Very similar to *H. scandens* but plant with nettle-like leaves and smaller flowers. *Japan.* H5. Summer.

Occasionally grown but of little merit as a garden plant.

12. H. paniculata Siebold. Illustration: Botanical Magazine, n.s., 301 (1957);

Huxley, Deciduous garden trees and shrubs in colour, pl. 92 (1979); Davis, The gardener's illustrated encyclopaedia of trees and shrubs, 145 (1987).

Large deciduous shrub or small tree to 4 m (or, in very favourable conditions, 6 m). Shoots at first downy, then hairless. Leaves 7.5–15 × 3.7–7.5 cm, ovate, acuminate, rounded or sometimes tapered at the base, toothed, sparsely bristly above and on the veins beneath; stalks 1.2–2.5 cm. Panicles pyramidal, 15–20 × 10–13 cm wide at the base, with few white-pink sterile flowers each 1.7–3 cm across; fertile flowers numerous, yellow-white. Flower-stalks downy. *E & S China, Japan, former USSR (Sakhalin).* H3. Summer–autumn.

A variable species including some notable cultivated variants, especially the cultivar 'Grandiflora', in which almost all of the white-pink flowers are sterile, and which, with pruning, can produce panicles to 45 cm. Other widely grown cultivars are 'Floribunda' with more sterile flowers than the species but fewer than 'Grandiflora'; 'Praecox' an earlier-flowering, more upright plant; and 'Tardiva' a poorer but later-flowering variant.

13. H. heteromalla D. Don (*H. dumicola* W.W. Smith; *H. hypoglauca* Rehder; *H. khasiana* Hooker & Thomson; *H. mandarinorum* Diels; *H. vestita* Wallich; *H. xanthoneura* Diels). Illustration: Schneider, Illustriertes Handbuch der Laubholzkunde **1**: 385, 389 (1905); Millar Gault, Dictionary of shrubs in colour, pl. 238 (1976); Haworth-Booth, The Hydrangeas, edn 5, 57 (1984).

Deciduous shrub to 3 m. Shoots at first with short hairs, later hairless and smooth. Leaves 8.7–20 × 3–14 cm, variable, mostly narrowly ovate, base rounded, wedge-shaped or sometimes cordate, toothed and bristly at the margins, hairless above, downy beneath, at least on the veins; stalks 1.5–3 cm. Corymbs flattened, 15 cm across, with few white sterile flowers, each 2.5–5 cm across. Fertile flowers numerous, small, white. Flower-stalks downy. *Himalayas, W & N China.* H4. Summer.

Variable, especially in the size of the corymbs and colour of the flowers, some being markedly yellow. The cultivar 'Bretschneideri' (*H. pekinensis* invalid) is distinguished by its peeling bark.

14. H. macrophylla (Thunberg) Seringe (*H. hortensis* Siebold; *H. opuloides* (Lamarck) Anon.). Illustration: Siebold & Zuccarini, Flora Japonica, t. 52 (1845); Millar Gault, Dictionary of shrubs in colour, pl. 239–246 (1976); Davis, The gardener's illustrated encyclopaedia of trees and shrubs, 143, 144 (1987).

Fairly spreading deciduous shrub to 3 m. Shoots more or less hairless. Leaves broadly ovate, acute or acuminate, very coarsely toothed, shining and almost greasy above, smooth beneath; blade 10–20 × 6.5–14 cm, stalks 1.7–5 cm. Corymbs flattened, much branched, with few pink sterile flowers each 3–5 cm across and numerous small blue or pink fertile flowers; flower-stalks hairless. Capsule yellow-brown, erect, 6–8 × 1–3 mm, with 3 apical, diverging, woody styles 1–3 mm. *Japan*. H3. Summer.

The above description is of a maritime plant sometimes called *H. macrophylla* var. *normalis* Wilson (*H. maritima* Haworth-Booth) with the corymb-type called 'Lace-cap'. It is considered to be the ancestor of many 'Hortensia' cultivars of both mop-head and lace-cap type. The mop-heads, such as the very common 'Generale Vicomtesse de Vibraye', have spherical corymbs almost wholly composed of sterile flowers. The plant described above is believed to be a true wild species, but it was its variety, grown in Japanese gardens and later called 'Sir Joseph Banks', together with certain old Japanese cultivars of hybrid origin involving *H. macrophylla* and other species, especially 'Otaksa', that were the plants first introduced to the West and from which many modern cultivars derive. 'Sea Foam', which arose as a branch-sport on 'Sir Joseph Banks' is the modern cultivar closest to the wild plant. In most garden variants the flower colour is influenced by the presence and availability of aluminium ions in the soil (see above). Some cultivars have variegated leaves.

Some cultivated lace-caps, such as the familiar 'Bluebird' are, however, derived from *H. serrata* (see below), considered by some authorities as a subspecies of *H. macrophylla*. It is possible that several others, both lace-cap and mop-head are actually hybrids between these 2 species.

15. H. serrata (Thunberg) Seringe (*H. japonica* Sielbold, in part; *H. macrophylla* var. *acuminata* (Siebold & Zuccarini) Makino; *H. macrophylla* subsp. *serrata* (Thunberg) Makino; *H. serrata* forma *acuminata* (Siebold & Zuccarini) Wilson). Illustration: Millar Gault, Dictionary of shrubs in colour, pl. 247–249 (1976); Hillier's Manual of trees & shrubs, edn 5, 152 (1981); Davis, The gardener's

illustrated encyclopaedia of trees and shrubs, 146 (1987); Practical gardening, August, 37 (1987).

Spreading deciduous shrub to 2 m. Shoots at first finely downy, later hairless. Leaves 5–15 × 2.5–7 cm, lanceolate, acuminate, hairless above, veins beneath with short hairs. Corymbs flattened, 5–10 cm across, with up to 12 pink or blue sterile flowers, each 1–1.5 cm across, with 3 entire or variously toothed, variously shaped sepals. Fertile flowers numerous, small, pink or blue. Capsules 1–3 × 0.5–1 cm, erect, yellow-brown, with 3 tiny, apical, diverging woody styles. *Japan & Korea (Quelpaert Island)*. H2. Summer.

Lace-cap cultivars derived from *H. serrata* include 'Bluebird' and 'Grayswood' and tend to be hardier than those of *H. macrophylla* origin. Some cultivars never have blue flowers, even in very acid soils. A smaller, compact variant, known as var. **thunbergii** Siebold (sometimes treated as a species, *H. thunbergii* Siebold) is occasionally seen, and differs in its very dark stems and leaves toothed only towards the apex, slightly hairy above.

16. H. serratifolia (Hooker & Arnott) Philippi (*H. integerrima* (Hooker & Arnott) Engler). Illustration: Botanical Magazine, n.s., 153 (1951).

Dioecious evergreen, clinging climber, to 30 m in favourable situations. Shoots at first with fine down, later hairless, and bearing aerial roots. Leaves 5–15 × 2.5–7.5 cm, elliptic, acuminate, base tapered, often cordate, almost always entire, leathery, hairless; stalks 5–45 mm. Inflorescences terminal and axillary, to 15 × 9 cm, composed of numerous small corymbs, each at first enclosed by 4 papery bracts. Flowers normally all small, fertile, white, although some variants exist with 1 or few white sterile flowers. *Chile, Argentina*. H4. Summer.

17. H. seemanni Riley.

Similar to *H. serratifolia* but with flowers in a single terminal cluster, sterile flowers sometimes present. *Mexico*. H5–G1.

Occasionally seen in cultivation.

37. SCHIZOPHRAGMA Siebold & Zuccarini
A.C. Whiteley

Deciduous climbing shrubs clinging by aerial roots. Leaves opposite, on long stalks. Flowers white in large terminal cymes, the central flowers small and hermaphrodite, outer ones sterile, reduced to a single large bract on a slender stalk.

Fertile flowers with 4 or 5 sepals and petals. Stamens 10. Ovary inferior, united with the calyx tube, 10-celled. Stigma lobed. Fruit a ribbed capsule, dehiscing between the ribs.

A genus of 2 species from E Asia. They do well in most soils, climbing up walls or trees. Propagate by cuttings.

1a. Leaves coarsely toothed; inflorescence 20–25 cm across, bracts 2.5–4 cm **1. hydrangeoides**
 b. Leaves sparingly toothed; inflorescence to 30 cm across, bracts to 9 cm **2. integrifolium**

1. S. hydrangeoides Siebold & Zuccarini. Illustration: Botanical Magazine, 8520 (1913); Everett, New York Botanical Garden illustrated encyclopedia of horticulture, **9**: 3081 (1982); Krüssmann, Manual of cultivated broad-leaved trees & shrubs, **3**: pl. 106 (1986); Phillips & Rix, Shrubs, 237 (1987).

Stems climbing to 12 m. Leaves broadly ovate, base rounded, heart-shaped or tapering, 10–15 × 6–10 cm, with deep veins and coarse teeth, underside slightly glaucous, with silky hairs. Inflorescence 20–25 cm across, somewhat downy. Fertile flowers slightly scented. Sterile bracts terminal on main branches of cyme, ovate or heart-shaped, 2.5–4 cm, yellowish white. *Japan & Korea*. H4. Summer.

'Roseum' has pinkish bracts.

2. S. integrifolium (Franchet) Oliver (*S. hydrangeoides* var. *integrifolium* Franchet). Illustration: Botanical Magazine, 8991 (1923); Hillier colour dictionary of trees & shrubs, 278 (1981); Krüssmann, Manual of cultivated broad-leaved trees & shrubs, **3**: pl. 106 (1986).

Stems climbing to 10 m. Leaves ovate, acuminate, base heart-shaped or rounded, 7–17 × 4–11 cm, entire or with a few small teeth, hairy on the veins beneath. Inflorescence to 30 cm across. Fertile flowers *c.* 5 mm across. Sterile bracts terminal on main branches of cyme, narrowly ovate, to 9 × 4 cm, white with darker veins. *S & C China & Taiwan*. H4. Summer.

38. DECUMARIA Linnaeus
A.C. Whiteley

Deciduous or evergreen climbing shrubs clinging by aerial roots. Young shoots and buds downy. Leaves opposite, ovate, with or without shallow teeth. Flowers in terminal corymbs or panicles. Petals

7–10, oblong, white. Calyx-teeth 7–10, alternating with petals. Stamens 20–30. Ovary inferior with 7–10 cells. Stigma head-like. Fruit a ribbed capsule dehiscing between the ribs.

A genus of 2 species from SE United States and C China. They do well in most soils in sheltered positions. Propagate by cuttings or occasionally seed.

1a. Deciduous; leaves 7–12 cm
1. barbara
  b. Evergreen; leaves 3–8 cm
2. sinensis

1. D. barbara Linnaeus. Illustration: Gleason, New illustrated flora of the north-eastern United States and adjacent Canada, **2**: 272 (1952); Krüssmann, Manual of cultivated broad-leaved trees & shrubs **1**: 426 (1984); Phillips & Rix, Shrubs, 237 (1989).

Deciduous to 10 m. Leaves 7–12 × 3–8 cm, slightly downy beneath when young, entire or shallowly toothed towards the tip, leaf-stalk 2.5–5 cm. Flowers *c.* 6 mm across in corymbs 5–8 cm long and wide. Petals 7–10, narrowly oblong, white. Fruit *c.* 8 mm. *SE United States.* H5. Summer.

2. D. sinensis Oliver. Illustration: Botanical Magazine, 9429 (1936).

Evergreen to 4 m. Leaves 3–8 × 1–4 cm, hairless, glossy, leaf-stalk 0.6–2 cm. Flowers *c.* 5 mm across in panicles 4–9 cm long and wide. Petals 7–10, oblong, yellowish white. Stamens prominent. *C China.* H5. Summer.

39. ITEA Linnaeus
D.M. Miller
Deciduous or evergreen trees or shrubs. Branches with chambered pith. Leaves alternate, entire or toothed. Flowers small, bisexual, radially symmetric, in terminal or axillary racemes or spikes. Sepals 5, persistent. Petals 5, linear. Stamens 5. Styles 2, united; ovary superior, 2-celled. Fruit a many-seeded capsule.

A genus of about 10 species from N America to E Asia. They grow in most garden soils, unless very dry, in sun or partial shade. Pruning is not essential, except for the removal of dead or crowded branches. Propagate by cuttings taken from ripened wood, or seed if available.

1a. Deciduous shrub   **1. virginica**
  b. Evergreen shrub   2
2a. Leaves with sharp, spiny, holly-like teeth, ovate to rounded; flowers greenish white in mid- to late summer
2. ilicifolia

  b. Leaves with fewer, fine spines, ovate to lanceolate; flowers dull white in early to mid summer
3. yunnanensis

1. I. virginica Linnaeus. Illustration: Botanical Magazine, 2409 (1823); Bean, Trees & shrubs hardy in the British Isles, edn 8, **2**: pl. 59 (1973); Krüssmann, Manual of cultivated broad-leaved trees & shrubs **2**: pl. 70 (1986).

Deciduous shrub 1–2 m with rather upright habit. Leaves to 10 × 3 cm, narrow-ovate to ovate, somewhat downy beneath, margins finely toothed; stalk to 1 cm. Flowers fragrant, in dense, cylindric, erect racemes 6–15 cm. Sepals to 2 mm, lanceolate, downy. Petals to 5 mm, narrow, white. Stamens as long as petals. Seeds to 5 mm or more, borne in narrow, downy capsules. *E USA.* H2. Early summer.

2. I. ilicifolia Oliver. Illustration: Botanical Magazine, 9090 (1925); Bean, Trees & shrubs hardy in the British Isles, edn 8, **2**: pl. 57 (1973).

Evergreen shrub to 4 m or more. Leaves to 9 × 7 cm, broadly ovate to rounded, apex usually rounded with a short point, margins with stiff spines, dark glossy green above, with small tufts of hairs in the vein-axils beneath; stalk to 1 cm. Flowers in very narrow, pendent racemes to 30 cm. Sepals minute. Petals to 3 mm, narrow, greenish white. *C China.* H5. Mid–late summer.

3. I. yunnanensis Franchet. Illustration: Iconographia cormophytorum Sinicorum **2**: 119 (1972).

Evergreen shrub to 3 m. Leaves to 11 × 4 cm, ovate to lanceolate, apex acute, spine-tipped, margins with fine, spiny teeth, dark glossy green above, more or less hairless on both surfaces; stalk to 1.5 cm. Flowers in narrow arching racemes to 18 cm. Petals to 3 mm, very narrow, dull white. Stamens and styles shorter than petals. *W China (Yunnan).* H5. Early–midsummer.

40. BREXIA Thouars
A.C. Whiteley
Evergreen shrub or small tree to 6 m with grey bark. Leaves alternate, oblanceolate to oblong, leathery, obtuse, entire or with a few teeth (especially on young plants), 10–15 × 3–4 cm, with short, stout leaf-stalks. Flowers *c.* 3.5 cm across in umbels of 6–20 on axillary stalks. Calyx of 5 united, rounded lobes. Petals 5, oblong, *c.* 1.2 cm, greenish white. Stamens 5, with

fleshy filaments nearly as long as petals, alternating with teeth of the prominent epignyous disc. Ovary superior, conical, 5-celled. Stigma head-like with 5 lobes. Fruit oblong, *c.* 5 cm, woody, indehiscent. Seeds numerous, black.

A genus of a single, somewhat variable species needing a heated greenhouse, but seedlings sometimes used for tropical effect in summer bedding. Propagate by seed or cuttings.

1. B. madagascariensis (Lamarck) Ker-Gawler. Illustration: Edwards's Botanical Register, 730 (1823); Nicholson, Illustrated dictionary of gardening **1**: 211, (1884); Nicholson & Mottet, Dictionaire practique d'horticulture et de jardinage, **1**: 413 (1892).

E tropical Africa, Madagascar & Seychelles. G2. Spring–summer.

41. ABROPHYLLUM J.D. Hooker
M.F. Gardner
Shrub or tree. Leaves alternate, ovate to lanceolate, pointed, with forward-pointing teeth. Flowers borne in terminal or axillary panicles, sepals 5 or 6, deciduous; petals 5 or 6, spreading; stamens 5 or 6; ovary superior, 5-celled, stigma stalkless, 5-lobed. Fruit a berry, crowned by a stigma, many seeded.

A genus of 2 species from E Australia. The cultivated species is an attractive plant suited to a cool, moist, woodland site, in a well-drained soil. Propagate from freshly sown seed or cuttings. Not very widely cultivated.

1. A. ornans J.D. Hooker. Illustration: Hooker's Icones Plantarum pl. 1323 (1880); Jones, Ornamental rainforest plants in Australia, 196 (1986).

Shrub or tree, 3–6 × 2–5 m, young shoots hairy. Leaves 10–22 × 4–10 cm, dark green, toothed in upper part. Flowers white to yellowish, fragrant, each flower *c.* 8 mm across; filaments very short, anthers broadly oblong; Fruit a berry, purplish black, *c.* 8 mm across, spherical, many seeded. *Australia (New South Wales & Queensland).* G1. Summer.

42. CARDIANDRA Siebold & Zuccarini
M.F. Gardner
Deciduous small shrubs. Leaves alternate, lanceolate, with coarse forward-pointing teeth; stipules absent. Flowers numerous, borne in loose corymbs, outer flowers sterile with 3 petal-like calyx lobes; inner flowers fertile, sepals 4 or 5, triangular-ovate; petals 5, overlapping; stamens

numerous; ovary inferior, 3-celled; styles 3. Capsules egg-shaped, crowned by calyx-limb and styles, opening between the styles, many-seeded, seeds flattened, winged on each side.

A genus of 5 species from China, Japan, and Taiwan.

1. C. alternifolia Siebold & Zuccarini. Illustration: Schneider, Handbuch der Laubholzkunde I: 383, 385 (1906); Satake et al., Wild flowers of Japan **2**: pl. 146 (1981); Hayashi, Azegami & Hishiyama, Wild flowers of Japan, 432 (1983).

Stems semi-woody, 45–75 cm, downy towards the ends. Leaves 5–15 × 2–2.5 cm, slender-pointed at both ends, with forward-pointing teeth. Flowers borne in terminal, long-stalked corymbs, 7.5–10 cm across, sepals triangular, blunt; petals *c.* 3 mm, white becoming rose-lilac; capsules broadly egg-shaped, *c.* 4 mm. *Japan.* G1. Summer.

43. CARPENTERIA Torrey
D.M. Miller

Evergreen bushy shrubs to 6 m, stems angled, branches pithy. Leaves opposite, to 11 × 2.5 cm, simple, lanceolate, entire, acute, base narrowed, hairless and bright green above, glaucous with short white hairs beneath; stalk to 5 mm. Flowers 5–7 cm across, bisexual, fragrant, in terminal clusters of 3–7. Sepals usually 5, to 10 × 6 mm, ovate, downy. Petals 5, to 2.5 × 2.5 cm, circular, white. Stamens numerous with conspicuous yellow anthers. Style 1, 5–7-lobed. Ovary superior. Fruit a conical capsule containing numerous seeds.

A genus of a single species from western USA which grows best in a well-drained, sandy soil in a sunny, sheltered position. Pruning is not necessary, but dead or weak shoots should be cut out. It may be grown from seed, but some seedlings may eventually have inferior flowers. Variants with larger flowers, such as 'Ladham's Variety' should be propagated by layering or by cuttings rooted in mist.

1. C. californica Torrey. Illustration: Botanical Magazine, 6911 (1886); Gardeners' Chronicle **40**: f. 5 (1906); Bean, Trees & shrubs hardy in the British Isles, edn 8, **1**: pl. 28 (1970).

W USA (California). H4. Early summer.

44. DICHROA Loureiro
A.C. Whiteley

Evergreen shrubs. Leaves opposite, simple, usually coarsely toothed. Flowers in terminal panicles, white, blue or pink. Calyx-lobes 5, rarely 4 or 6. Petals 5,

sometimes 6, usually more brightly coloured on inner surfaces. Stamens 10. Ovary inferior, 4-celled. Styles 4. Fruit a small berry. Seeds numerous.

A genus of around 13 species in SE Asia and Indonesia. One species is cultivated. It requires a frost-free climate and will succeed in most soils given ample moisture and slight shade. Propagate by seed or cuttings.

1. D. febrifuga Loureiro (*D. cyanitis* Miquel; *Adamia cyanea* Wallich). Illustration: Botanical Magazine, 3046 (1841); Everett, The New York Botanical Gardens illustrated encyclopedia of horticulture **4**: 1061 (1981).

Shrub to 2.5 m. Leaves elliptic, acuminate, tapering gradually to the base, coarsely toothed, 10–20 × 4–9 cm, the veins beneath often reddish. Leaf-stalks 1–2 cm, often reddish. Flowers in panicles to 15 × 20 cm. Petals white to blue or pink, somewhat fleshy, *c.* 5 × 2 mm, boat-shaped. Calyx-lobes triangular, *c.* 1 mm. Stamens blue or violet, *c.* 5 mm. Receptacle green to pink. Fruit spherical, blue, to 5 mm wide. *SE Asia.* H5. Spring–autumn.

45. FENDLERA Engelmann & Gray
D.M. Miller

Deciduous bushy shrubs with ribbed branches. Leaves opposite, ovate to lanceolate, entire, more or less stalkless, 1–3-veined. Flowers bisexual, 1–3, stalked, on short, lateral branches. Sepals 4, small. Petals 4, clawed, ovate with toothed margin. Stamens 8. Styles 4, completely free. Ovary half-inferior. Fruit a many-seeded capsule opening by 4 flaps, surrounded by the persistent calyx.

A genus of 2 or 3 species from SW USA and Mexico. They should be grown in a hot, sunny location in well-drained soil; moist or cool conditions should be avoided. Old flowering branches and weak or crowded shoots should be thinned after flowering. Propagate from seed or by half-ripe cuttings under mist with gentle heat.

1a. Leaves green, more or less hairless or bristly beneath **1. rupicola**
  b. Leaves with thick, grey-white, densely felted hairs beneath **2. wrightii**

1. F. rupicola Gray. Illustration: Bailey, Standard cyclopedia of horticulture **2**: f. 1480 (1915).

Shrub to 2 m with downy, grey-yellow young bark. Leaves 2–3 × *c.* 1 cm on vegetative shoots, lanceolate to narrowly oblong, 3-veined, stalkless, rough above, but smaller, linear and clustered on

flowering-shoots. Flowers usually solitary, to 3 cm across, fragrant, on short side-branches. Sepals very small. Petals *c.* 1 cm, spreading, white or pink-tinged. Stamens ½ as long as petals. Capsule more than 1 cm. *SW USA, N Mexico.* H3. Late spring–early summer.

2. F. wrightii (Gray) Heller (*F. rupicola* var. *wrightii* Gray). Illustration: Botanical Magazine, 7924 (1903).

Very similar to *F. rupicola*, but leaves and flowers smaller, generally more hairy. *SW USA (Texas to New Mexico), NW Mexico.* H3. Late spring–early summer.

46. PLATYCRATER Siebold & Zuccarini
M.F. Gardner

Deciduous shrub, sometimes prostrate with papery bark. Leaves opposite, 10–15 × 3–7 cm, oblong to broadly lanceolate, sparsely hairy above, hairy below, with forward-pointing teeth, with slender points; leaf-stalk 5–30 cm. Flowers with long slender stalks; of 2 types; sterile ornamental, at the top of cyme branches, 1–3 cm across; calyx large, shallowly 3- or 4-lobed; fertile less showy flowers lower down, 2–3 cm across; calyx smaller, deeply 4-lobed to near the base, sepals ovate-lanceolate; petals 4, white, very thick; stamens numerous; styles 2, persistent; ovary inferior; capsules cone-shaped.

A genus of a single species native to Japan, which prefers a cool, moist site in partial shade.

1. P. arguta Siebold & Zuccarini (*P. serrata* Makino). Illustration: Siebold, Flora Japonica **1**: 27 (1837); Schneider, Handbuch der Laubholzkunde **1**: 383–385 (1906); Journal of Japanese Botany, **61**: 70 (1986); Hayashi, Woody plants of Japan, 228 (1988).

Japan (Honshu, Shikoku, Kyushu). H5. Summer.

47. PILEOSTEGIA Hooker & Thomson
A.C. Whiteley

Evergreen climbing or prostrate shrubs, clinging by aerial roots. Leaves opposite, simple, usually entire, leathery. Flowers small, white, in terminal panicles. Calyx cup-shaped with 4 or 5 short lobes. Petals 4 or 5, coherent, falling quickly. Stamens 8–10, prominent. Ovary inferior, 4- or 5-celled. Fruit a spherical ribbed capsule, dehiscing between the ribs.

A genus of 3 species from E Asia. One species is cultivated. It does well in most soils, climbing up walls or trees. Propagate by cuttings.

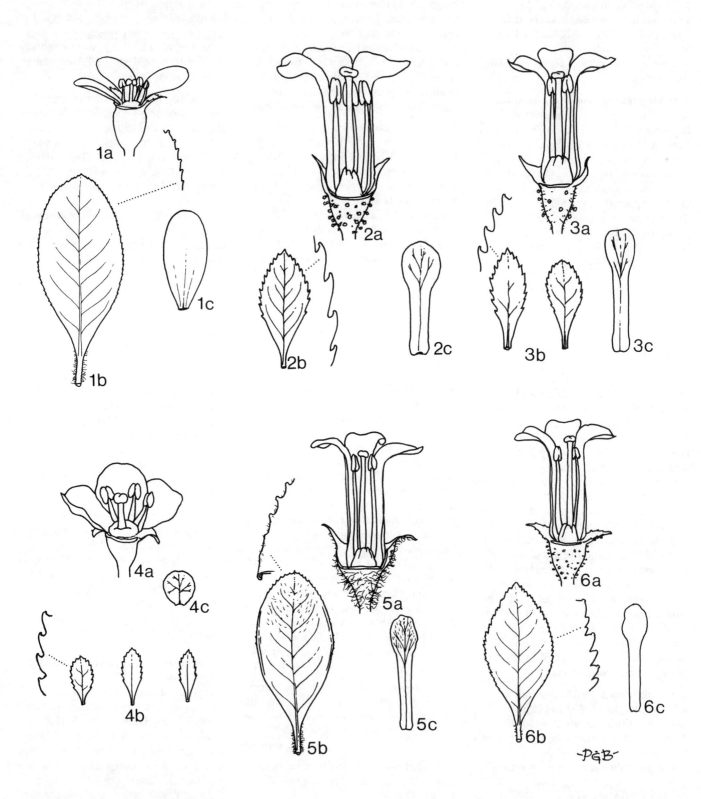

Figure 26. Diagnostic details of *Escallonia* species (a, flower; b,
leaf; c, petal). 1a, b, c, *E. pulverulenta*. 2a, b, c, *E. rosea*.
3a, b, c, *E. alpina*. 4a, b, c, *E. virgata*. 5a, b, c, *E. revoluta*.
6a, b, c, *E. rubra*.

1. P. viburnoides Hooker & Thomson
(*Schizophragma viburnoides* (Hooker &
Thomson) Stapf). Illustration: Botanical
Magazine, 9262 (1929); Bean, Trees &
shrubs hardy in the British Isles, **3**: pl. 31
(1976); Miller Gault, Dictionary of shrubs
in colour, fig. 322 (1976); Phillips & Rix,
Shrubs, 232 (1989).

Stems climbing to 6 m. Leaves entire,
soon hairless, narrowly oblong to ovate,
acute, 6–15 × 2–6 cm, leaf-stalk 5–25
mm. Flowers creamy white, *c.* 9 mm
across in panicles 10–15 × 10–15 cm
with a central axis and several pairs of
dichotomous branches. Stamens 5 mm.
Himalaya, China & Taiwan. H4. Autumn.

48. CARPODETUS J.R. Forster &
G. Forster
A.C. Whiteley
Shrubs or small trees. Leaves alternate,
without stipules. Flowers small, in few-
flowered panicles. Calyx-tube united with
ovary, bearing 5 or 6 deciduous lobes.
Petals and stamens 5 or 6, alternating,
inserted at the margin of an epigynous
disc. Ovary half-inferior. Stigma capitate.
Fruit a spherical, indehiscent capsule with
3–5 cells. Seeds numerous.

A genus of 10 species, 9 in New Guinea
and 1 endemic to New Zealand which
is occasionally cultivated. It requires a
moist, acid to neutral soil rich in humus.
Propagate by seed or cuttings.

1. C. serratus J.R. Forster & G. Forster.
Illustration: Laing & Blackwell, Plants of
New Zealand, 187 (1907); Cheeseman,
Illustrations of the New Zealand flora **1**: pl.
41 (1914); Salmon, New Zealand flowers
& plants in colour, 62 (1970); Irwin, The
Oxford book of New Zealand plants, 79
(1978).

Evergreen tree to 10 m. Stems with
prominent lenticels, downy when young.
Adult leaves elliptic, somewhat leathery,
acute or obtuse with a few small teeth, 4–6
× 2–3 cm. Young leaves 1–3 × 1–2 cm, on
zig-zag shoots. Leaf-stalks to 1 cm, downy.
Flowers 5–6 mm across, in panicles to 5
× 5 cm. Calyx-lobes *c.* 1 mm. Petals ovate,
3–4 mm, white. Capsule 4–6 × 4–6 mm.
New Zealand. H5. Summer.

49. ESCALLONIA Mutis
P.G. Barnes & A.C. Whiteley
Evergreen or rarely deciduous shrubs,
occasionally small trees; branches, leaves
and inflorescences hairy, hairless or with
stalked glands. Leaves alternate, usually
somewhat leathery, toothed. Flowers in
short racemes, panicles or solitary in axils

of upper leaves. Calyx-tube short, lobes
5, erect or spreading; petals 5, obovate or
circular and spreading or with an erect
long narrow claw and spreading circular
apical portion, the latter type of flower
appearing tubular. Stamens 5, alternating
with and about as long as petals. Ovary
inferior, 2- or 3-celled; style 1 with a
capitate or 2-fid stigma, or rarely 2-fid at
apex with 2 distinct stigmas; disk conical
or cushion-shaped and surrounding base of
style, or flat. Fruit a capsule with
numerous seeds.

A genus of 39 species, native to S
America, easily cultivated in most soils, in
a sunny situation. Many are not very hardy
in temperate areas. Resistant to maritime
conditions and useful for hedges or as
specimen shrubs. Propagate by cuttings.
Literature: Sleumer, H. Die Gattung
Escallonia (Saxifragaceae), *Verhandelingen
der Koninklijke nederlandsche Akademie van
Wetenschappen, afdeeling natuurkunde.* **2**(58)
Nr. 2. (1968).

1a. Flowers open; petals spreading, not
clawed   2
  b. Flowers appearing tubular; petals
erect with a long narrow claw,
stamens included   5
2a. Leaves 4–10 cm   3
  b. Leaves less than 4 cm   4
3a. Flowers in slender racemes
  1. pulverulenta
  b. Flowers in broad panicles
  11. bifida
4a. Leaves deciduous   **4. virgata**
  b. Leaves evergreen   **7. leucantha**
5a. Petals pink–red   6
  b. Petals white rarely pale pink   8
6a. Leaves less than twice as long as
wide; disc taller than wide
  6. rubra
  b. Leaves more than 2 times longer
than wide   7
7a. Leaves 3–7.5 × 1–2.5 cm; disk
almost flat   **9. laevis**
  b. Leaves 1.5–3 cm; disk as tall as
wide   **3. alpina**
8a. Leaves 1–2.5 cm, branchlets
winged   **2. rosea**
  b. Leaves usually more than 3 cm,
branchlets smooth or angled   9
9a. Leaves clearly hairy, at least
beneath   **5. revoluta**
  b. Leaves hairless or almost so   10
10a. Leaves sticky when young, strongly
scented   **8. illinita**
  b. Leaves not sticky when young, not
strongly scented
  10. tucumanensis

1. E. pulverulenta (Ruiz & Pavon)
Persoon. Figure 26(1), p. 314. Illustration:
Sweet, British flower garden **7**: pl. 310
(1835).

Evergreen shrub to 4 m or more.
Branches slightly angled, hairy, sticky
when young. Leaves 5–10 × 2–4 cm,
oblong or elliptic, apex rounded, base
abruptly narrowed to a short stalk. Young
leaves with dense, soft hairs, persisting
beneath, margin with fine sharp teeth.
Inflorescence a narrow terminal raceme,
10–20 × 2–3 cm. Flower-stalks 2–4 mm,
hairy. Receptacle 1–2 mm. Calyx-lobes
1–1.5 mm, narrowly trianglular. Petals
3–5 × *c.* 2 mm, obovate, erect or
spreading, white. Stamens and style 2–2.5
mm. Disk flat, the margin 5-lobed. Capsule
4–6 mm, obovoid. *Chile*. H5. Summer.

2. E. rosea Grisebach (*E. pterocladon*
Hooker; *E. montana* Philippi). Figure 26(2),
p. 314. Illustration: Botanical Magazine,
4827 (1855); Phillips & Rix, Shrubs, 222
(1989); Krüssman, Manual of cultivated
broad-leaved trees & shrubs **2**: f. 26c
(1986).

Evergreen shrub to 3 m. Branchlets
downy, reddish, distinctly angled with
sinuous wings. Leaves 8–25 × 4–8 mm,
narrowly obovate, acute, tapering to a
short stalk, hairless except for the midrib
above, glossy green, margin toothed,
somewhat glandular. Inflorescence a stiff
axillary raceme of 10–15 fragrant flowers,
each borne in the axil of a leaf, or, towards
the tip, a leafy bract. Flower-stalks short,
reddish, with 2 narrow glandular-toothed
bracteoles. Receptacle 1.5 mm. Calyx-tube
1–2 mm, lobes 1.5 mm, triangular,
glandular-toothed. Petals 2.5–3 × *c.* 3
mm, erect, claw 5–6 × 1–1.5 mm, ovate,
white, sometimes tinged with red, reflexed.
Stamens 6–7 mm. Disk cylindric, slightly
lobed, 2–2.5 mm. Style *c.* 8 mm, stigma
capitate to peltate, 1–1.5 mm across. *S
Chile*. H4. Summer.

3. E. alpina de Candolle (*E. fonkii*
Philippi). Figure 26(3), p. 314. Illustration:
Botanical Magazine, n.s., 642 (1973).

Evergreen shrub 1–3 m, hairless or
clothed with short, erect hairs. Shoots
reddish at first, sparsely glandular,
becoming grey, bark flaking. Leaves to
2.7 × 1.2 cm, spathulate to obovate, tip
rounded with a sharp point, base wedge-
shaped, stalkless, glossy green above with
minute down on the midrib towards the
base, paler beneath, sharply toothed
towards the tip, each tooth gland-tipped.
Inflorescence an axillary raceme of 4–15

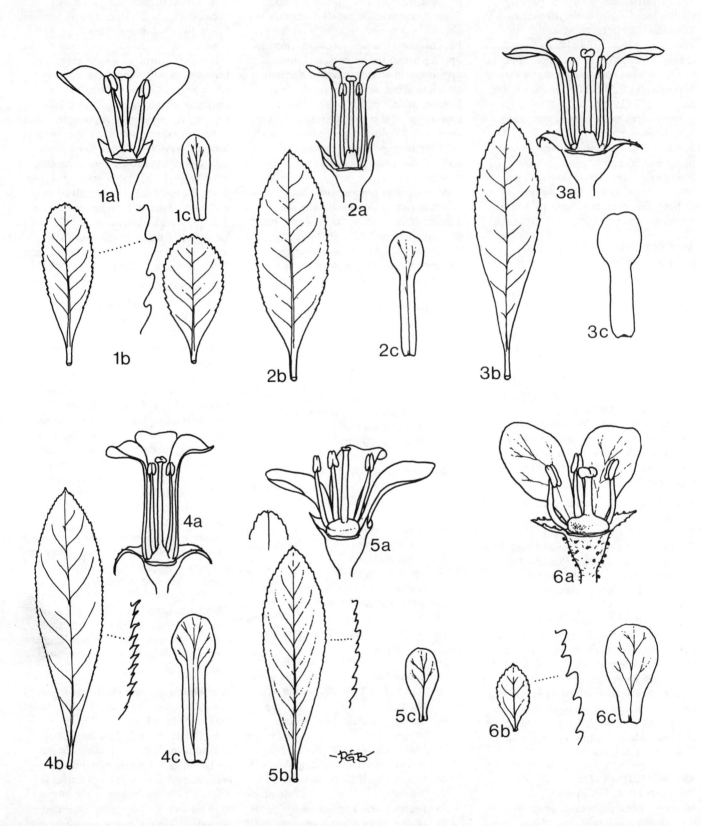

Figure 27. Diagnostic details of *Escallonia* species (a, flower; b, leaf; c, petal). 1a, b, c, *E. leucantha*. 2a, b, c, *E. illinita*. 3a, b, c, *E. laevis*. 4a, b, c, *E. tucumanensis*. 5a, b, c, *E. bifida*. 6a, b, c, *E. × langleyensis*.

flowers, leafy at the base and with leaf-like bracts towards the tip. Receptacle c. 2 mm. Calyx-tube c. 1.5 mm, lobes 2–3 mm, triangular, somewhat hairy and glandular. Petals erect, red, the claw 6–7 × 1.5 mm, reflexed. Stamens c. 8 mm. Disk conical, slightly lobed. Style 7–8 mm, stigma capitate, c. 1.5 mm across. Capsule conical to spherical, 4–5 mm. *Chile, Argentina.* H4. Summer.

4. E. virgata (Ruiz and Pavon) Persoon (*E. philippiana* (A. Engler) Masters. Figure 26(4), p. 314. Illustration: Bean, Trees & shrubs hardy in the British Isles, edn 8, **2**: 125 (1973); Krüssmann, Manual of cultivated broad-leaved trees & shrubs **2**: f. 26a & e (1986); Phillips & Rix, Shrubs, 223 (1989).

Deciduous shrub to 2 m. Branches erect, lateral shoots short, angled, with very short hairs. Leaves 1–2 × 0.4–0.6 cm, obovate, hairless, apex acute or obtuse, almost stalkless, margin somewhat toothed. Flowers solitary or few in the upper leaf axils and terminal on short shoots, forming a leafy raceme. Receptacle 1.5–2.5 mm. Calyx-lobes c. 2 mm, triangular-ovate, with glandular teeth. Petals 3–6 × 2–3 mm, rounded, spreading, white or pale pink. Stamens and style 1.5–2 mm. Capsule obovoid, 4–6 mm. Disk flat. *Chile & S Argentina.* H4. Summer.

Hybrids with *E. rubra* are named **E. × langleyensis** Veitch. Figure 27(6), p. 316. The numerous cultivars vary considerably between the parental extremes. All are evergreen. The cultivar 'Apple Blossom' is typical. Illustration: Hay & Beckett, Reader's Digest encyclopedia of gardening, edn 4, 260 (1987); Phillips & Rix, Shrubs, 222 (1989).

5. E. revoluta (Ruiz & Pavon) Persoon. Figure 26(5), p. 314. Illustration: Botanical Magazine, 6949 (1887).

Evergreen shrub to 6 m. Branchlets downy to densely hairy. Leaves obovate, 2–5 × 1.2–3 cm, acute to obtuse, narrowed to a short stalk, more or less downy on both surfaces, margin sharply toothed towards tip, somewhat curved under. Inflorescence a panicle or raceme, 4–10 cm, terminating a side-branch. Flower-stalks short, each with a small bracteole at the base. Receptacle 2–3 mm. Calyx 2–3 mm, lobes c. 2 mm, hairy, narrow, acute. Petals white, erect, claw 1.1–1.2 × 0.2–0.4 mm, tip ovate, c. 3 × 2 mm. Stamens c. 9 mm. Disk 1–2 mm, conical. Style c. 1.2 cm, stigma peltate to capitate. Capsule 4–5 mm, conical, hairy. *Chile.* H4. Late summer–autumn.

6. E. rubra (Ruiz & Pavon) Persoon (*E. punctata* de Candolle; *E. rubra* var. *punctata* de Candolle) Hooker. Figure 26(6), p. 314. Illustration: Botanical Magazine, 6599 (1881); Krüssmann, Manual of cultivated broad-leaved trees & shrubs **2**: f. 26d (1986); Phillips & Rix, Shrubs, 222 (1989).

Evergreen shrub 2–4 m. Branches slightly hairy, with conspicuous stalked glands. Leaves 2–8 × 1.5–4 cm, ovate to obovate, blunt or short acuminate, narrowed into a short stalk, glossy above glandular-spotted beneath, margin coarsely toothed. Flowers in loose terminal and lateral panicles. Flower-stalks slender, 5–8 mm Receptacle c. 4 mm. Calyx-lobes 2–3 mm, with stalked glands. Petals 1–1.2 cm, including a long claw c. 2 mm wide, erect except for the broader spreading tip, deep pink to deep red. Stamens and style c. 9 mm, disk conical, 5-lobed at tip. Capsule 7–8 mm, obovoid. *Argentina, Chile.* H4–5. Summer.

Var. **macrantha** (Hooker & Arnott) Reiche. Leaves 3–8 × 2–4 cm, narrowly to broadly elliptic, deeply toothed. Petals 1.5–1.8 cm, claw c. 4 mm wide. Stamens and style 1.2–1.4 mm, capsule 6–7 mm. *Chile.* H5. Summer–autumn.

7. E. leucantha Rémy (*E. bellidifolia* Philippi). Figure 27(1), p. 316.

Evergreen shrub to 5 m. Shoots downy. Leaves 1.2–2.5 × 0.8–1.2 cm, obovate to oblanceolate, acute or obtuse, tapered to a short stalk, hairless except on the midrib beneath, margin finely toothed. Inflorescence a large leafy panicle to 30 × 15 cm, composed of short terminal racemes on side-shoots. Flower-stalks short, downy. Receptacle c. 1 mm. Calyx-lobes short, triangular, finely hairy. Petals c. 7 × 3 mm, spathulate, spreading, white. Stamens c. 5 mm. Style 6–7 mm, stigma peltate. Disk flat, slightly lobed. Capsule 3–4 mm, conical to bell-shaped. *Chile.* H4. Summer.

8. E. illinita Presl (*E. grahamiana* Hooker & Arnott). Figure 27(2), p. 316. Illustration: Edwards's Botanical Register **22**: pl. 1099 (1836).

Evergreen shrub to 3 m or more. Young branches densely glandular. Leaves 2–6 × 1–3 cm, obovate or elliptic, glossy above, sticky when young and strongly scented, narrowed to a long stalk 3–6 mm. Inflorescence a panicle, 1–1.2 cm, branches slightly hairy and glandular, bracts leaf-like. Receptacle c. 3 mm. Calyx-lobes 2.5 mm, awl-shaped. Petal white, erect, claw 9–11 × c. 2 mm, tip c. 3 mm wide, spreading. *Chile.* H4–5. Summer.

9. E. laevis (Vellozo) Sleumer (*E. organensis* Gardner). Figure 27(3), p. 316. Illustration: Botanical Magazine, 4274 (1846); Graf, Exotica, series 4, edn 12, 2042 (1985).

Evergreen shrub to 2 m. Branchlets stout, angled, glandular. Leaves to 7.5 × 2.5 cm, obovate to oblong, obtuse, tapering to a short reddish stalk, hairless, margin toothed towards the tip, often reddish. Inflorescence a short, dense terminal panicle. Receptacle 2–3 mm. Calyx-tube c. 1 mm, lobes c. 3 mm, narrowly triangular, hairless, sometimes glandular. Petals erect, red, claw c. 10 × 2 mm, tip c. 5 × 4 mm, ovate to rounded, reflexed at right angles to the claw. Stamens c. 8 mm. Style 9–10 mm, stigma capitate, somewhat bilobed. Disk flat. Capsule 4–5 mm, spherical. *Brazil.* H4–5. Autumn.

10. E. tucumanensis Hosseus. Figure 27(4), p. 316. Illustration: Botanical Magazine, n.s., 565 (1970); Phillips & Rix, Shrubs, 223 (1989).

Evergreen shrub to 6 m. Young shoots reddish, hairless or slightly downy. Leaves 5–8 × 1.5–2 cm, oblong to elliptic, thin, shortly acuminate, base wedge-shaped, stalk short. Lower surface of leaf dotted with blackish glands, midrib sometimes downy, upper surface glossy, margin finely toothed. Inflorescence an axillary panicle of 4–15 flowers, leafy at the base and with leafy bracts to 15 × 4 mm subtending the flowers on slender stalks to 1.2 cm. Receptacle c. 3 mm. Calyx-tube 1–2 mm, lobes 3–5 × 1–1.5 mm, linear, toothed and glandular. Petals white, erect, claw 10–14 × 2–3 mm, tip 3–5 × 3–4 mm, reflexed. Stamens c. 1 cm. Disk thin, papillose, slightly conical. Style 9–11 mm, stigma capitate, c. 2 mm across. Capsule 5–6 mm, approximately spherical. *Argentina.* H4. Summer.

11. E. bifida Link & Otto (*E. montivedensis* (Chamisso & Schlechtendal) de Candolle; *E. floribunda* misapplied). Figure 27(5), p. 316. Illustration: Botanical Magazine, 6404 (1879); Journal of the Royal Horticultural Society **86**: f. 34 (1961); Krüssmann, Manual of cultivated broad-leaved trees & shrubs **2**: f. 26b (1986).

Evergreen shrub to 3 m or more. Young shoots bluntly angled, hairy, becoming smooth. Leaves 3–8 × 1–1.5 cm, oblanceolate, blunt or rounded, base tapered to a short stalk. Margin finely toothed. Inflorescence a broad terminal panicle to 8 cm, becoming loose with age. Flower-stalks to 7 mm; receptacle c. 2

mm. Calyx-lobes 1.5–2 mm, triangular, with minute glandular teeth. Petals c. 7 × 3 mm, obovate, white, somewhat spreading. Stamens and style 6–7 mm, stigma capitate; disk low, flat-topped. Capsule 3–4 mm, broadly obovate to spherical. *E South America*. H4–5. Autumn.

50. BAUERA Andrews
P.G. Barnes
Low evergreen shrubs, leaves opposite, with 3 leaflets, stalkless, appearing whorled. Flowers regular, solitary in upper leaf axils, long-stalked or nearly stalkless. Sepals 4–10; petals 4–10, pink or white. Stamens numerous, on a conspicuous nectar-secreting disk. Ovary almost superior, 2-celled, styles 2, recurved. Fruit a 2-celled capsule.

A genus of 3 species in E Australia. Slightly tender shrubs requiring a sheltered situation in full sun. In temperate regions some winter protection is required although plants originating at high altitudes may be hardier. The smaller variants are well-suited to cultivation in pans in an alpine house. Propagate by cuttings in late summer or seed.

1a. Flowers stalked; leaflets less than 1.5 cm **1. rubioides**
  b. Flowers stalkless; leaflets usually more than 1.5 cm **2. sessiliflora**

1. B. rubioides Andrews. Illustration: Botanical Magazine, 715 (1804); Curtis, Student's Flora of Tasmania **1**: f. 42 (1956); Galbraith, Wild flowers of South-east Australia, pl. 14, (1977).

Shrub of variable habit, prostrate, scrambling or erect, to 1 m. Leaflets 5–15 mm, lanceolate, slightly toothed, glossy dark green. Flowers 5–18 mm across, on stalks usually longer than the leaves. Petals obovate, white, pale pink or occasionally bicoloured. *Australia*. H5. Summer.

Var. **microphylla** (Seringe) Bentham leaflets 5–7 mm; flowers 7–8 mm wide, mostly with 6 petals, white, on stalks mostly shorter than the leaves.

2. B. sessiliflora von Mueller. Illustration: Galbraith, Wild Flowers of South-east Australia, pl. 14 (1977).

Erect shrub to 2 m. Leaflets 1.2–3.5 cm, entire, hairy. Flowers stalkless, pink. *Australia*. H5–G1. Summer.

51. ANOPTERUS Labillardière
A.C. Whiteley
Evergreen shrubs or small trees. Leaves alternate, simple, leathery, obovate to oblanceolate, tapered to both ends and with large rounded teeth, each gland-tipped. Flowers in upright terminal racemes, cup-shaped, white or pink-tinged, petals 6–9 concave, obovate; calyx small, with 6–9 toothed, triangular lobes. Stamens as many as petals, with flattened tapering filaments. Ovary superior, 1-celled with numerous ovules. Fruit a cylindric capsule with 2 recurving valves. Seeds winged.

A genus of 2 species from Australia. They require an acid or neutral soil, rich in humus and prefer some shade. Propagate by seed or cuttings.

1a. Leaves to 12 cm; flowers with 6 calyx-lobes, petals and stamens **1. glandulosus**
  b. Leaves to 30 cm; flowers with up to 9 calyx-lobes, petals and stamens **2. macleayanus**

1. A. glandulosus Labillardière. Illustration: Botanical Magazine, 4377 (1848); Revue Horticole, 310 (1868); Galbraith, Wild flowers of South-east Australia, pl. 27 (1977); Krüssmann, Manual of cultivated broad-leaved trees & shrubs **1**: 159 (1984).

Tall shrub or tree to 10 m, hairless. Leaves 5–12 × 4–5 cm, dark glossy green, leaf-stalk 6–20 mm. Flowers c. 1.6 cm across. Calyx-lobes, petals and stamens 6. Capsule c. 1.2 cm. *Australia (Tasmania)*. H5. Spring.

2. A. macleayanus F.J. Mueller.
Similar but has leaves to 30 cm and flowers with up to 9 calyx-lobes, petals and stamens. *SE Australia*. H5.

52. DAVIDSONIA F.J. Mueller
M.F. Gardner
Evergreen shrubs or trees, branches densely hairy. Leaves spirally arranged, pinnate, winged between the leaflets, margins irregularly toothed, covered with bright red irritant hairs when young; stipules conspicuous, palmately veined. Flowers borne in drooping racemes; bracts with glandular teeth; calyx 4- or 5-lobed; petals absent; stamens 8–10; ovary with 2 cells. Fruit a drupe-like berry; seeds flattened, fibrous.

A genus of 1 or 2 species native to Australia. Best cultivated in dappled shade in a well-drained, acid soil in mild areas, possibly under glass or as a house plant elsewhere. Propagate from fresh seed.

1. D. pruriens F.J. Mueller. Illustration: Gardener's Chronicle, 819 (1877); Elliot & Jones, Encyclopaedia of Australian plants suitable for culivation **3**: 199 (1984); Jones, Ornamental rainforest plants in Australia, 216 (1986).

Slender, unbranched shrub or tree, 4–8 × 1–2.5 m; bark brown, corky; Leaves dark green, 30–80 cm, leaflets 5–25 × 3–9 cm, ovate-lanceolate, opposite, densely hairy. Flowers brown or reddish brown, c. 6 mm across. Berries purple to blue-black, 3–5 cm, hairy, ovoid; seeds flat with lacerated margins. *Australia (Queensland & New South Wales)*. G1. Summer.

CVI. CUNONIACEAE

Evergreen trees or shrubs, rarely climbers. Leaves leathery, often glandular, opposite, rarely whorled, occasionally simple but more often compound, with 3 leaflets or pinnate. Stipules united in pairs, often conspicuous. Flowers unisexual or bisexual, in solitary or branched racemes or compact heads. Petals 4 or 5, free or united, sometimes absent, sepals larger than petals, 3–6, free or united. Stamens numerous, sometimes 4 or 5 (often alternating with petals), or 8–10. Ovary superior, with 2–5 free or united carpels; ovules in two ranks. Styles 2. Fruit a capsule, rarely a drupe or nut. Seed small, winged or hairy.

A southern hemisphere family of about 24 genera and 340 species, mostly from Australasia and the Pacific. **Schizomeria ovata** D. Don, **Aphanopetalum clematideum** (Drummond & Harvey) C. Gardner and **A. resinosum** Endlicher were briefly cultivated after their original introduction from Australia but are apparently no longer widely grown.

1a. Leaves in whorls of 3 **4. Calycomis**
  b. Leaves in opposite pairs 2
2a. Stipules persistent, with forward pointing teeth **3. Caldcluvia**
  b. Stipules falling early 3
3a. Leaves winged between the leaflets **2. Weinmannia**
  b. Leaves not winged between the leaflets 4
4a. Leaves with 3 leaflets or pinnately divided, with glandular forward-pointing teeth **1. Cunonia**
  b. Leaves simple or with 3 leaflets, not glandular 5
5a. Leaves simple; flowers in spherical heads **7. Callicoma**
  b. Leaves simple or with 3 leaflets; flowers in panicles or racemes 6
6a. Flowers in loose panicles; sepals

lengthening, becoming reddish, rarely white after flowering **6. Ceratopetalum**

b. Flowers in dense racemes; sepals inconspicuous **5. Geissois**

1. CUNONIA Linnaeus
M.F. Gardner

Trees or shrubs. Leaves opposite, with 3 leaflets or pinnate; leaflets with glandular forward-pointing teeth; stipules large, falling early. Flowers white or cream, bisexual, small, borne in dense spike-like axillary racemes, sepals and petals 5, overlapping; stamens 10, flattened at the base. Ovary superior, 2-celled; styles 2, persistent. Capsules 2-valved, seeds numerous.

A genus of 17 species, 1 from South Africa and the rest from New Caledonia. Easily grown in a well-drained, peat-enriched soil, outdoors in mild areas or a cool glasshouse elsewhere. Propagate from half-ripened cuttings taken in August and rooted under glass with bottom heat.

1. C. capensis Linnaeus. Illustration: Loddiges' Botanical Cabinet, 826 (1824); Botanical Magazine, 8504 (1913); Palgrave, Trees of Southern Africa, 205 (1977); Everett, New York Botanical Garden illustrated encyclopedia of horticulture **2**: 942 (1981).

A large shrub or small tree to 15 m with pale flaky bark when young. Leaves with 3–5 pairs and one terminal leaflet; leaflets 5–10 cm, lanceolate to ovate; margins with forward-pointing teeth. Racemes to 14 cm, opposite; flowers white or cream. Fruit a 2-horned capsule, brownish. *South Africa (Cape Province & Natal).* H5–G1. Summer.

2. WEINMANNIA Linnaeus
M.F. Gardner

Trees or shrubs. Leaves opposite, simple, with 3 leaflets or pinnate with wings between the leaflets; stipules falling early. Flowers small, clustered in terminal or axillary racemes; sepals and petals 4 or 5; stamens 8 or 10; ovary superior, 2-celled; styles 2. Fruit a dry capsule; seeds often hairy, rarely slightly winged.

A genus of about 190 species widely distributed throughout Central and South America, Pacific islands, New Zealand, Malaysia and Madagascar. Easily grown in a well-drained, loamy soil, in the open in milder areas or in a cool glasshouse elsewhere. Propagated by seed sown freshly under glass or autumn cuttings rooted under a mist unit with bottom heat.

1a. Leaves pinnately divided **1. trichosperma**

b. Leaves simple **2. racemosa**

1. W. trichosperma Linnaeus. Illustration: Dimitri, La region de los bosques Andino-Patagonicos **2**: 35 (1982); Hoffmann, Flora silvestre de Chile zona austral, 101 (1982); Rodriguez et al., Flora arborea de Chile, 337, 338 (1982); Krüssmann, Manual of cultivated broad-leaved trees & shrubs, **3**: 457 (1986).

Tree or shrub to 20 m. Leaves pinnate, 8–10 cm with triangular wings between the leaflets; leaflets 11–13, broadly elliptic, deep glossy green, with 3 coarse teeth on each side. Flowers white, crowded in erect 3–5 cm racemes, fragrant. Fruit glossy red when young. *Chile.* H5–G1. Spring.

2. W. racemosa Linnaeus filius. Illustration: Salmon, The native trees of New Zealand, 186, 187 (1980); Wilson, Stewart Island plants, 73 (1982); Salmon, A field guide to the native trees of New Zealand, 100 (1986); Evans, New Zealand in flower, 64 (1987).

Tree to 20 m. Leaves *c.* 10 cm, simple, oblong-lanceolate to elliptic, coarsely toothed, leaves of younger plants varying from simple to 3–5-parted, or 3-lobed. Flowers white to light pink borne in narrow, 10–12 cm racemes. Fruit reddish brown. *New Zealand.* H5–G1. Summer.

3. CALDCLUVIA D. Don
M.F. Gardner

Trees with opposite branchlets. Leaves opposite, simple, or pinnate, with glandular forward-pointing teeth, finely nerved; stipules leaf-like, sickle-shaped, with forward-pointing teeth, persistent. Flowers borne in axillary panicles; calyx 4 or 5, sepals deciduous; petals 4 or 5; stamens 8–10; styles 2. Capsules 2-beaked, 2-celled, many-seeded.

A genus of 11 species from southern Chile, New Zealand, tropical Australia, Malesia and New Guinea. Cultivation as for *Weinmannia.*

Literature: *Blumea* **25**: 481–505 (1979).

1a. Leaves simple, to 12 cm **1. paniculata**

b. Leaves pinnately lobed, to 25 cm **2. rosifolia**

1. C. paniculata (Cavanilles) D. Don. Illustration: Dimitri, La region de los bosques Andino-Patagonicos **2**: 35 (1982); Hoffmann, Flora silvestre de Chile zona austral, 99 (1982); Rodriguez et al., Flora

arborea de Chile, 96, 97 (1983); Flora Patagonica **4b**: 43 (1984).

Tree to 6 m. Leaves simple, 5–12 × 4–5 cm, oblong-lanceolate, shiny above, downy beneath, with glandular forward-pointing teeth, shortly stalked. Stipules 1.2–1.7 cm, toothed. Flowers 5–6 cm across, cream, borne in axillary panicles; sepals 2.5–3 cm, brownish, petals 4–4.5 cm; stamens free, 4–5.5 cm. Fruit a many-seeded leathery capsule, softly hairy, 7.5–8 cm. *S Chile.* H5–G1. Spring–summer.

2. C. rosifolia (A. Cunningham) Hoogland (*Ackama rosaefolia* A. Cunningham, *Weinmannia rosaefolia* (A. Cunningham) A. Gray. Illustration: Kirk, Forest flora of New Zealand, 63 (1889); Cheeseman, Illustrations of the New Zealand flora pl. 42 (1914); Eagle's 100 trees of New Zealand, pl. 54 (1978); Salmon, The native trees of New Zealand, 188–189 (1980).

Tree to 12 m, covered with short brownish hairs. Leaves pinnately lobed with a terminal leaflet, to 25 cm; lobes 6–20, unequal, with sharp forward-pointing teeth, almost stalkless; stipules leaf-like. Flowers *c.* 3 mm across, borne in many-flowered branched panicles, to 15 cm; sepals ovate, *c.* 1 mm, persistent. Styles persistent. Capsules ovoid, 3–4 mm, softly hairy. *New Zealand (North Island).* G1. Spring–summer.

4. CALYCOMIS D. Don
M.F. Gardner

Shrub, 1–2 × 1–1.5 m. leaves dark shiny green, 4–10 × 2–4.5 cm, in whorls of 3, ovate, stalkless, pointed at apex, rounded at base, with regular forward-pointing teeth, prominently veined beneath. Flowers white tinged pink, 5–6 mm across, borne at the ends of the branches in dense axillary clusters, to 8 cm. Capsules 2-celled, seeds covered with glands.

A genus of a single species from Australia. This is a very decorative plant grown for both its showy flowers and handsome foliage. Easily cultivated in a humus-rich, well-drained soil in a cool shady or semi-shady position. Propagate from seed or cuttings.

1. C. australis (A. Cunningham) Hoogland (*Acrophyllum venosum* A. Cunningham). Illustration: Maund, The Botanist **2**: pl. 95 (1839); Botanical Magazine, 4050 (1844) – as Acrophyllum verticillatum.

Australia (New South Wales). H5–G1. Summer.

5. GEISSOIS Labillardière
M.F. Gardner

Trees or shrubs. Leaves opposite, simple or with 3 leaflets (in ours), with forward-pointing teeth; stipules falling early, basal stipules conspicuous. Flowers bisexual, borne in racemes, axillary or clustered near ends of branches; calyx deeply divided into 4 or 5 lobes; petals absent, stamens 10, exserted; ovary 2-celled; styles 2. Fruit a slender capsule, seeds numerous, flattened, winged in upper part.

A genus of 18 species from Australia, New Caledonia, Vanuatu and Fiji. Although **G. racemosa** Labillardière was introduced into cultivation in 1851, it is now rarely seen in gardens. These handsome plants are grown for their colourful young foliage and decorative leaves. They are easily cultivated in a deep organically rich, loamy soil which is moisture retentive.

1. G. benthamii F. Mueller. Illustration: Maiden, The forest flora of New South Wales **4**: pl. 221 (1916); Wrigley & Fagg, Australian native plants, edn 3, 512 (1979); Elliot & Jones, Encyclopaedia of Australian plants **4**: 352 (1986).

Tree, 10–18 × 5–12 m; trunk with large buttresses; bark grey and wrinkled. Leaves 20–30 cm, leaflets 3, leaf-stalks 2–7 cm, leaflets 8–18 × 3–5 cm, leathery, dark green and shiny, ovate to elliptic, veins prominent, coarsely toothed. Flowers cream to yellowish, borne in dense, axillary racemes, 10–15 cm; capsules 1.5–1.8 cm, cylindric, hairy. *Australia (New South Wales)*. G1. Summer.

6. CERATOPETALUM Smith
M.F. Gardner

Trees or shrubs. Leaves opposite, simple or with 3 leaflets, dark green; stipules falling early. Flowers borne in branched axillary or terminal panicles; sepals 5, spreading after flowering; petals 5 or absent, linear, 3-lobed; stamens 10. Ovary semi-superior to almost superior, 2-celled, styles 2. Fruit 1-seeded.

A genus of 5 species from New Guinea and eastern Australia. Although the flowers are not very showy, the sepals enlarge after fertilisation and become very colourful. All species are easily cultivated in a well-drained soil in a sunny or semi-shady site. Propagate by seed or cuttings from half-ripened wood.

1a. Leaves simple   **1. apetalum**
  b. Leaves with 3 leaflets
                                     2. gummiferum

1. C. apetalum D. Don. Illustration: Maiden, The forest flora of New South Wales, pl. 21 (1904); Maiden, Some principal common trees of New South Wales, 209 (1917); Holliday & Hill, Field guide to Australian trees, 75 (1969); Elliot & Jones, Encyclopaedia of Australian plants, **3**: 10 (1989).

Tree 10–20 × 5–8 m; young shoots with 4 conspicuous ridges. Leaves simple, 3–25 × 1–7 cm, ovate to oblong-lanceolate, tapered at both ends, margins with shallow forward-pointing teeth; leaf-stalks 6–12 mm. Flowers borne in axillary and terminal panicles 7–10 cm across. Sepals *c*. 6 mm, white, lengthening to 1.8 cm and becoming crimson or purple; petals usually absent. *Australia (New South Wales)*. G1. Spring.

2. C. gummiferum J. Smith. Illustration: Maiden, Flowering plants and ferns of New South Wales **7**: pl. 25 (1898); Botanical Magazine, n.s., 312 (1958); Hortus III, 344 (1976); Elliot & Jones, Encyclopaedia of Australian plants, **3**: 10, 11 (1989).

Shrub or tree 3–10 × 2–6 m; young shoots rounded. Leaves with 3 leaflets; leaflets 3–7 × 0.6–1.4 cm, narrowly oblong, blunt at the apex, dark green above, paler beneath; margins with shallow forward-pointing teeth; leaf-stalks 1.2–3.6 cm. Flowers *c*. 6 mm across, white, borne in terminal panicles. Sepals 3–4 mm lengthening to 1.2 cm and becoming bright red, rarely white; petals 2–3 mm. *Australia (New South Wales)*. G1. Spring.

7. CALLICOMA Andrews
M.F. Gardner

Tree or shrub, 3–10 × 4–6 m, with long and slender branches. Leaves opposite, simple, 5–12 × 2–4 cm, elliptic to lanceolate, apex pointed, tapered or rounded at base, shiny above, whitish or rust-downy beneath, coarsely toothed; stipules falling early. Flowers stalkless, borne in many-flowered spherical heads, 1–2 cm across, petals absent; stamens 8–10, long-exserted, anthers pale yellow; ovary 2- or 3-celled, styles 2 or 3. Fruit a capsule, enclosed by calyx-lobes.

A genus of a single species native to Australia. Successful cultivation depends on a cool protected position in a moisture-retentive soil. Propagate from seed or cuttings.

1. C. serratifolia Andrews. Illustration: Maiden & Campbell, Flowering plants and ferns of New South Wales (2) pl. 6

(1895); Blombery, A guide to native plants Australian plants, 107 (1967); Wrigley & Fagg, Australian native plants, edn 3, 227 (1979); Elliot and Jones, Encyclopaedia of Australian plants **2**: 409 (1982).

Australia (New South Wales). H5–G1. Spring–summer.

CVII. PITTOSPORACEAE

Evergreen trees, shrubs or vines. Leaves alternate or whorled, usually entire. Inflorescences axillary or terminal; flowers bisexual (rarely unisexual) solitary or in clusters. Sepals 5. Petals 5, longer than sepals, hypogynous. Stamens 5, alternating with petals, hypogynous. Ovary with 2–5 carpels, ovules numerous; style simple. Fruit a capsule or berry.

A family of about nine genera and 300 species, concentrated in the southern hemisphere and Pacific region.

Literature: Bennett, E.M., New taxa and new combinations in Australian Pittosporaceae. *Nuytsia* **2**(4): 184–199 (1978); Bennett, E.M., Pittosporaceae. *Australian Plants* **15**: 20–36 (1988) [all Australian species except Billardiera].

1a. Robust shrubs; anthers ovate; fruit
    dehiscent   2
  b. Twining plants with weak stems,
    woody at base; anthers linear or
    ovate; fruit indehiscent berries   4
2a. Branches becoming spiny; flowers
    numerous; capsule not woody,
    stalked; seeds flattened   **2. Bursaria**
  b. Branches without spines, or flowers
    more or less solitary; capsule
    leathery or woody   3
3a. Petals less than 2.5 cm; seeds not
    winged   **4. Pittosporum**
  b. Petals more than 3 cm; seeds
    winged   **3. Hymenosporum**
4a. Flowers blue; anthers linear, fused
    around style   **5. Sollya**
  b. Flowers not blue; anthers ovate,
    free   **1. Billardiera**

1. BILLARDIERA Smith
E.C. Nelson

Vines or shrubs. Leaves entire. Flower solitary, or a few in clusters, pendent, bell-shaped, axillary or terminal. Sepals 5, petals 5, anthers 5, free. Fruit a berry.

A genus of about 30 species from Australia.

1a. Young shoots hairy; flowers yellow;
    fruit yellow-green   **2. scandens**

b. Young shoots hairless; flowers green-yellow; fruit white, red, blue or purple (never green) **1. longiflora**

1. B. longiflora Labillardière. Illustration: Botanical Magazine, 1507 (1812); Morley & Toelken, Flowering plants in Australia, 139 (1983); Brickell, The RHS gardeners' encyclopedia of plants & flowers, 176 (1989); Macoboy, What shrub is that? 57 (1989).

Vine, young shoots hairless. Leaves 2–5 × 0.4–0.6 cm, linear-lanceolate, hairless, dark green, sometimes tinged with purple. Sepals 0.2–0.7 cm. Petals 2–3 cm, green-yellow, solitary, terminal on short branchlets, urceolate (broadest towards mouth). Fruit a berry, *c.* 2 × 1 cm, shining purple. *Australia (New South Wales, Victoria, Tasmania).* G1–H5. Summer.

Variants with red, blue or white berries are cultivated, and the best of these may be perpetuated by cuttings, as seedlings will not always be true.

2. B. scandens Smith. Illustration: Botanical Magazine, 801 (1805), 1313 (1810); Cochrane et al., Flowers and plants of Victoria, t. 368 (1968).

Weak vine (rarely a shrub), young stems hairy. Leaves 1–5 × 0.4–0.6 cm, ovate lanceolate, margin wavy, usually hairy on lower surface. Sepals 0.6–0.9 cm; petals 1.5–2.5 cm, pale yellow. Berry 2–3 cm, elliptic, green-yellow, often downy. *E Australia.* G1–H5. Summer (most of year).

2. BURSARIA Cavanilles
E.C. Nelson

Trees or shrubs; branches often with spines. Leaves entire, almost stalkless. Inflorescence terminal, pyramidal. Capsule flattened, walls thin and parchment-like, with 2 carpels, each with 1 or 2 seeds.

A genus of 6 species, endemic to Australia. Only one species is currently listed from European gardens, but given the polymorphic nature of the species, more than one may be present.

Literature: Bennett, E.M., New taxa and new combinations in Australian Pittosporaceae. *Nuytsia* **2**(4): 184–199 (1978); Elliott, W.R. & Jones, D., *Encyclopaedia of Australian plants suitable for cultivation* **2**: 392–394 (1982).

1a. Leaves hairless; sepals persistent **1. spinosa**
  b. Leaves hairy on lower surface; sepals deciduous **2. lasiophylla**

1. B. spinosa Cavanilles. Illustration: Botanical Magazine, 1767 (1815); Macoboy, What shrub is that? 66 (1989).

Shrub or rarely a small tree to 10 m; branches hairless, spiny. Leaves 1–5 × 0.5–0.7 cm, obovate, shining green above, hairless. Panicles large and showy; sepals 0.1–0.2 cm; corolla to 1 cm across, cream to white, fragrant. Capsule heart-shaped. *Australia (all states except Northern Territory).* G1–H5. Summer.

An extremely variable species sometimes split into 6 varieties.

2. B. lasiophylla E.M. Bennett.
Similar to *B. spinosa* but with leaves hairy on lower surface and sepals deciduous. *Australia (New South Wales, Victoria, South Australia).* G1–H5. Spring–summer.

3. HYMENOSPORUM Mueller
E.C. Nelson

Evergreen tree or shrub, to 20 m. Leaves alternate, dark glossy green. Flowers in terminal clusters, to 5 cm, yellow turning pale cream or white, fragrant. Petals united into tube, silky. Capsule with 2 cells, seeds numerous, with a membranous wing.

A genus of 1 species, confined to rainforests in Australasia.

1. H. flavum (R. Brown) Mueller. Illustration: Botanical Magazine, 4799 (1854); Morley & Toelken, Flowering plants in Australia, 139 (1983).
Australia (Queensland, New South Wales), New Guinea. G1. Summer.

4. PITTOSPORUM Gaertner
E.C. Nelson

Trees or shrubs, usually evergreen. Leaves usually simple, often clustered towards tips of shoots. Flowers terminal or axillary, usually solitary. Sepals free or occasionally fused at base.

A genus of about 200 species from the Pacific region and eastern Asia. Those best known in European gardens are from Australasia. They are cultivated principally as foliage plants. Propagation by cuttings is easy, but species may also be raised from seeds; cultivars must be perpetuated by vegetative means.

Literature: Allan, H.H., *Flora of New Zealand*, **1**: 305–318 (1961); Metcalf, L.J., *The cultivation of New Zealand trees and shrubs* (Revised edn) (1987).

1a. Branching widely spreading; leaves less than 1 × 0.2 cm; flowers *c.* 0.4 cm across **1. anomalum**

  b. Plant not as above 2
2a. Leaves entirely hairless 3
  b. Leaves at least felted beneath 7
3a. Leaf-margins toothed (rarely entire) **4. dallii**
  b. Leaf-margins entire 4
4a. Flowers dark red **7. tenuifolium**
  b. Flowers cream to yellow-green 5
5a. Leaf-margins wavy, not curved downwards and inwards; flowers to 1 cm across 6
  b. Leaf-margins not wavy, often curved downwards and inwards; flowers large, to *c.* 2.5 cm across **8. tobira**
6a. Leaves lemon-scented when crushed **5. eugenioides**
  b. Leaves not as above **9. undulatum**
7a. Leaves to 1 cm across, dark green, margins recurved; flowers dark red and cream **2. bicolor**
  b. Leaves more than 2 cm across; flowers uniformly dark maroon 8
8a. Leaves elliptic, base rounded, not tapering; flowers bisexual **6. ralphii**
  b. Leaves elliptic to obovate, base tapering; flowers functionally unisexual **13. crassifolium**

1. P. anomalum Laing & Gourlay. Illustration: Metcalf, The cultivation of New Zealand trees and shrubs, 260 (1987).

Shrub to about 1 m; branches widely diverging and so intricately interlaced, densely felted with grey-white hairs when young, soon becoming black. Foliage sparse. Adult leaves to 1 × 0.2 cm, linear-oblong, pinnatifid or prominently lobed. Flowers *c.* 0.4 cm across, pale cream, fragrant, almost stalkless, terminal or occasionally in leaf-axils, petals to 3 mm. *New Zealand.* H5. Spring–early summer.

An astonishing shrub, with pale grey interlaced branchlets. It is a veritable botanical curiosity and is hardly recognisable as a species of *Pittosporum*.

2. P. bicolor Hooker. Illustration: Cochrane et al., Flowers and plants of Victoria, 445 (1968).

Shrub to 5 m. Leaves to 3 × 0.5 cm, oblong, leathery, margins often recurved, dull dark green above, hairy and paler beneath. Flowers on long, slender stalks, petals recurved, red shading to almost translucent cream. Ovary hairy. Fruits more than 1 cm across, valves wrinkled on inside. *Australia (New South Wales, Tasmania, Victoria).* H5. Spring.

3. P. crassifolium A. Cunningham. Illustration: Salmon, The native trees of New Zealand, 144–145 (1980); Brickell, The RHS gardeners' encyclopedia of plants & flowers, 71 (1989).

Tree or shrub to 9 m. Leaves 5–10 × 2.4–2.6 cm, elliptic to obovate, margins recurved slightly, upper surface glossy, crackled. Flowers functionally unisexual; male flowers in umbels of 5–10; female flowers in pairs or solitary. Fruits to 3 cm, sparsely hairy; seeds black in golden yellow glutin. *New Zealand (North & South Islands).* H5. Summer.

'Variegatum' is often grown.

4. P. dallii Cheeseman. Illustration: Salmon, The native trees of New Zealand, 135 (1980); Brickell, The RHS gardeners' encyclopedia of plants & flowers, 72 (1989).

Tree or shrub to 6 m, spreading. Leaves 5–10 × 1–4 cm, alternate or almost whorled, crowded at tips of branches; blades thick, leathery, dark green and somewhat glossy above, paler beneath, margins regular toothed (rarely entire); leaf-stalks to 2 cm. Flowers to 1.5 cm across, in dense, terminal compound umbels, yellow-green (or white), fragrant. Sepals to 1 cm. Petals to 2 cm. *New Zealand (South Island).* H5. Summer.

A rare species both in the wild and in cultivation, but reputed to be the most hardy under European conditions.

5. P. eugenioides A. Cunningham. Illustration: Salmon, The native trees of New Zealand, 132 (1980); Brickell, The RHS gardeners' encyclopedia of plants & flowers, 71 (1989).

Tree or small shrub to 12 m, shoots hairless. Leaves 5–12.5 × 2.5–4 cm, alternate, elliptic, margin wavy, glossy with pallid undersurface, aromatic (lemon-scented when crushed); leaf-stalks slender to 2.5 cm long. Flowers green-yellow, fragrant, sometimes unisexual (shrubs dioecious), to 1 cm diameter, in compound, terminal umbels, flower-stalks silky. Sepals to c. 2 mm. Petals less than 7 mm, spreading or recurved. Fruits black, wrinkled outside. *New Zealand.* H5. Summer.

Variegated plants are most commonly found in cultivation.

6. P. ralphii Kirk. Illustration: Salmon, The native trees of New Zealand, 142 (1980).

Tree or shrub to 4 m; shoots, flowers and leaf-stalks densely felted. Leaves 7–12.5 × 2–2.5 cm, alternate, elliptic, margin sometimes wavy, glossy green above, densely felted below. Flowers dark maroon, in terminal clusters. Fruits hairy outside; seeds black. *New Zealand (North Island).* H5. Summer.

7. P. tenuifolium Gaertner (*P. colensoi* Hooker; *P. nigricans* invalid; *P. mayi* invalid). Illustration: Salmon, The native trees of New Zealand, 140–141, 146–147 (1980); Walsh, An Irish Florilegium, 149–150 (1984); Brickell, The RHS gardeners' encyclopedia of plants & flowers, 95, 145 (1989).

Monoecious or dioecious tree or shrub, sometimes reaching over 10 m; bark usually very dark, almost black. Leaves to 6 × 2.5 cm, alternate elliptic, green and shining above, paler beneath, margin entire, sinuously wavy. Flowers c. 1 cm across, unisexual or rarely bisexual, solitary or in clusters, axillary, dark crimson, fragrant, almost stalkless. Sepals silky when young. Petals c. 1.2 cm, recurved, paler towards the base. Fruit grey-black at maturity; seeds black. *New Zealand.* H5. Spring.

P. tenuifolium is now most frequently seen in the less specialised gardens in its purple-leaved (e.g. 'Purpureum' and 'Tom Thumb') and variegated variants (particularly as 'Silver Queen' with white margins and grey-green blade). A similar plant, with grey green leaves, irregularly variegated with creamy white and invariably spotted with crimson is 'Garnettii' ('Sandersii'), a hybrid of uncertain parentage.

8. P. tobira Aiton. Illustration: Woody plants of Japan, 234 (1985); Macoboy, What shrub is that? 274 (1989).

Shrub to 10 m; branches spreading. Leaves 3–10 × 2–4 cm, obovate, leathery, dark green shining above, ribs paler. Flowers to 2.5 cm across, in terminal cluster, cream-white becoming yellow when mature. Fruit yellow-green; seeds red. *Japan, China, Korea.* G1–H5. Spring.

9. P. undulatum Ventenat. Illustration: Cochrane et al., Flowers and plants of Victoria, 567 & 568 (1968).

Tree or shrub to 15 m. Leaves 7–15 × 2.5–5.0 cm, obovate, margin wavy, leathery, dark green and shining above, pale beneath, hairless. Flowers in terminal umbels, creamy white, fragrant. Fruits to 1 cm across, smooth; seeds red-brown. *E Australia.* H5. Spring–early summer.

5. SOLLYA Lindley
E.C. Nelson

Subshrubs with thin twining stems, usually hairless. Leaves stalkless, entire. Flowers in pendent clusters, or solitary, on slender stalks, blue. Anthers longer than filaments, joined to form cone around style; pollen shed inwards. Ovary with 2 cells. Fruit a juicy berry; seeds embedded in pulp.

A genus of 2 species native to SW Western Australia, only one of which is widely cultivated in Europe. It is tolerant of a considerable range of conditions but seems to prefer a moist humus-rich soil. Propagation is by seed.

Literature: Burtt, B.L., The correct names of the Australian bluebell creepers, *Kew Bulletin* for 1948, 74–76.

1. S. heterophylla Lindley (*Billardiera fusiformis* Labillardière; *Sollya fusiformis* (Labillardière) Briquet). Illustration: Edwards's Botanical Register, 1466 (1832); Morley & Toelken, Flowering plants in Australia 139 (1983); Bennett & Dundas, The bushland plants of Kings Park, Western Australia, 58 (1988); Phillips & Rix, Shrubs, 221 (1989).

Weak-stemmed vine. Leaves 3–5 × 0.5–2 cm, oblong. Sepals less than 0.3 cm, acute, slightly pouched at base. Petals c. 1 cm, ovate, acute, overlapping to form an elegant bell. Filaments enlarged at base, anthers yellow. Ovary downy. Fruit a blue berry to 2.5 cm, cylindric, succulent. *SW Western Australia (naturalised in S Australia, Tasmania).* G1–H5. Summer–autumn.

The fruits are edible when blue. White- and pink-flowered cultivars are reported in Australia but are very rarely seen in Europe.

CVIII. BYBLIDACEAE

Insectivorous herbs. Leaves simple, alternate, linear, spirally coiled when immature, covered with stalked glandular hairs and stalkless glands. Flowers solitary, bisexual, axillary on long stalks. Sepals 5, fused at base, persistent in fruit. Petals 5, fused at base. Stamens 5, alternating with petals, markedly turned to one side. Ovary superior, carpels 2. Style simple, persistent. Fruit a capsule.

A single genus from Australia, superficially resembling the Droseraceae, although not related. It is closely allied to the Roridulaceae, with which it is sometimes combined.

1. BYBLIS Salisbury

E.C. Nelson

Description as for family.

A genus of 2 species, cultivated by enthusiastic collectors of insectivorous plants. They do not release pollen unless stamens are vibrated; in natural habitats, insects' wing beats effect release. Propagation is from seed or (*B. gigantea*) by root cuttings in a peat : sand : perlite or peat : sand mixture, which is kept continually moist but not sodden, in a well-lit, heated glasshouse.

1a. Leaves to 30 cm; flowers 2–4 cm
     across   **1. gigantea**
  b. Leaves to 5 cm; flowers *c*. 1 cm
     across   **2. linifolia**

1. B. gigantea Lindley. Illustration: Botanical Magazine, 7846 (1902); Erickson et al., Flowers and plants of Western Australia, 33 (1973); Morley and Toelken, Flowering plants in Australia, 141 (1983).

Perennial herb or subshrub with woody rootstock, to 60 cm. Leaves to 30 cm. Flowers 2–4 cm across, iridescent; sepals 0.8–2 cm, narrow, ovate; petals lilac-pink, blue or purple. Seeds numerous. *SW Western Australia*. G1. Summer.

2. B. linifolia Salisbury. Illustration: Jessop, Flora of Central Australia 103 (1981).

Annual herb (rarely perennial), to 30 cm. Leaves to 5 cm. Flowers *c*. 1 cm across; petals pink to pale mauve. *New Guinea, N tropical Australia (Western Australia, Northern Territory, Queensland)*. G2. Summer.

CIX. RORIDULACEAE

Dwarf to medium shrubs. Leaves clustered at tips of branches, stalkless, simple, alternate, with long gland-tipped hairs. Flowers solitary or clustered in terminal racemes. Sepals 5, persistent, with glandular hairs. Petals 5. Stamens 5; filaments erect; anthers with 2 partly fused cells, opening by apical pores and with a swelling near the base. Ovary with 3 carpels, each with 1–4 seeds. Style simple.

This distinctive family contains a single genus, sometimes included in the Byblidaceae. The glandular hairs suggest that they are insectivorous though this has not been proved. They require well-ventilated conditions in a well-drained but moist soil. Propagation is from seed.

Literature: Obermeyer, A.A., Roridulaceae, in *Flora of southern Africa* **13**: 201–204 (1970).

1. RORIDULA Linnaeus

E.C. Nelson

Leaves yellow-green, with gland-tipped hairs ('tentacles') and non-glandular hairs on the upper surface; lower side hairless or nearly so. Flowers pink to magenta. Sepals with glandular hairs, more or less equalling petals.

A genus of 2 species from the Cape Province of South Africa. They have explosive anthers which when touched suddenly reverse their position showering a pollinating insect with pollen.

1a. Leaves with prominent linear-
     lanceolate lobes   **1. dentata**
  b. Leaves entire   **2. gorgonias**

1. R. dentata Linnaeus (*R. muscicapa* Gaertner). Illustration: Gardener's Chronicle 10: 367 (1891); Flora of southern Africa **13**: 203 (1970); Kondo & Kondo, Carnivorous plants of the world in colour, 133 (1983).

Shrub to 2 m. Leaves to 5 × 0.3 cm, linear-lanceolate with opposite, linear-lanceolate marginal lobes. Petals persistent, *c*. 1.2 cm. Ovary hairy; style with scarcely broadened stigma; carpels with 1 ovule per cell. *South Africa (SW Cape Province)*. G1–H5. Late winter–spring.

2. R. gorgonias Planchon (*R. crinita* Gandoger). Illustration: Flora of southern Africa **13**: 203 (1970); Kondo & Kondo, Carnivorous plants of the world in colour, 134 (1983).

Low shrub to 0.5 m. Leaves entire, linear-lanceolate, to 12 × 0.5 cm. Petals deciduous, *c*. 1.5 cm. Ovary hairless; style with broad stigma; carpels with 2–4 ovules per cell. *South Africa (SW Cape Province)*. G1–H5. Spring.

CX. BRUNIACEAE

Shrubs or small trees, usually ericoid (heather-like), with long unicellular hairs. Leaves alternate, small, simple, often overlapping with centric structures; stipules absent or vestigial. Flowers stalkless in spikes or heads, involucrate, resembling Compositae, rarely solitary, usually small, bisexual, regular, usually epigynous; sepals 4 or 5, united or free; petals 4 or 5; stamens 4 or 5, alternate with petals, intrastaminal disc sometimes present, anthers opening by longitudinal slits; ovary inferior, of 2 (rarely 3) fused carpels. Fruit dry, often with persistent calyx, achene like, 1-seeded or, 1- or 2-seeded with carpels separating and opening along the ventral suture.

A family of 11 genera and 69 species from southern Africa, a few of which are sometimes grown for the cut-flower trade.

CXI. ROSACEAE

Annual or perennial herbs, shrubs and trees. Branches often thorny. Leaves deciduous or evergreen, alternate or borne in apparently whorled clusters; entire, palmately compound or pinnately divided; stipules usually present, and attached to leaf-stalk, sometimes falling early. Flowers mostly bisexual, radially symmetric, perigynous to epigynous, solitary or in cymes. Sepals 5, rarely 3–10, often occurring as lobes on the perigynous zone; petals 5, rarely absent or 3–10, variously coloured, but mostly white, pink, red or yellow; stamens usually numerous, occasionally 1 or 5, free or attached to the perigynous zone; ovary superior or inferior, 1–many, when solitary carpels 1–5; ovules 1–several; placentation basal, axile or parietal; styles as many as carpels. Fruit various, sometimes an achene, follicle, hip, pome or drupe; (the arrangement of floral parts and different fruit types in selected genera and species of the family are illustrated in figure 28, p. 326).

A family of about 3200 species in 115 genera, widely distributed throughout the world, but most abundant in temperate regions. In several genera (e.g. *Alchemilla*, *Rubus*) numerous apomictic microspecies occur; these are not included in the figure for species given above. Various attempts to segregate the family have been made in the past and although some parts of the family are easily distinguished by fruit characters, they are united by similar floral characters and are best treated as tribes or subfamilies rather than separate families.

1a. Leaves simple or shallowly lobed;
     if deeply lobed then at least part of
     the blade visible on either side of
     midrib   2
  b. Leaves pinnately or palmately
     divided to mibrib; leaflets stalked or
     stalkless   62
2a. Leaves evergreen, needle-like;
     solitary or clustered
                                33. Adenostoma

b. Leaves deciduous, or if evergreen then not needle-like   3

3a. Deciduous perennial herbs, or plants with woody stock and annual leafy and flowering shoots   4

b. Evergreen or deciduous trees and shrubs   5

4a. Leaves palmately divided, sepals yellowish green; petals absent
47. Alchemilla

b. Leaves simple or 3–5-lobed; sepals reddish green; petals white
23. Rubus

5a. Dwarf or prostrate shrublets; non-flowering stems not exceeding 10 cm   6

b. Shrubs or trees; or leafy shoots at least 20 cm   9

6a. Leaves deeply 3–5-lobed in upper part, resembling a mossy saxifrage
16. Luetkea

b. Leaves entire or margins scalloped or toothed   7

7a. Dense, cushion-forming tufted shrublet; leaves 2.5–4 mm; flowers almost stalkless, pink to purple
15. Kelseya

b. Loose shrublets, stems not tufted; leaves 5 mm or more; flowers borne on long stalks, creamy white, white or rarely yellow   8

8a. Leaves stalkless, light green above and beneath; flowers borne in spike-like racemes   **14. Petrophytum**

b. Leaves stalked; margins scalloped or toothed, dark green above, white felted beneath; flowers solitary on long stalks   **39. Dryas**

9a. Leaves evergreen   10

b. Leaves deciduous   27

10a. Leaves often deeply lobed to more than half-way towards midrib   11

b. Leaves entire or shallowly toothed   13

11a. Leaves 0.6–1.5 cm, deeply divided into 3–9 very narrow lobes
37. Cowania

b. Leaves 2 cm or more, lobes broad   12

12a. Flowers 2–5, in umbels; stamens 30–50   **50. Docynia**

b. Flowers many, in corymbs or panicles, very rarely solitary; stamens 5–25   **71. Crataegus**

13a. Ovaries or ovary superior   14

b. Ovary inferior   19

14a. Petals absent; style very long, conspicuously plumose; fruit a 1-seeded achene   **34. Cercocarpus**

b. Petals present; other characters not as above   15

15a. Stamens numerous   16

b. Stamens 10–20   17

16a. Fruit a hip   **24. Rosa**

b. Fruit a drupe   **76. Prunus**

17a. Stamens 10   **1. Quillaja**

b. Stamens 15–20   18

18a. Male flowers in racemes; female flowers solitary; borne on separate plants; fruit of 5 follicles united at base   **2. Kageneckia**

b. Male and female flowers borne on the same plant; fruit a dry woody capsule   **4. Lindleya**

19a. Leaf-margin entire   20

b. Leaf-margin toothed   24

20a. Flowers borne in erect racemes or panicles; flower-stalks fleshy
63. Rhaphiolepis

b. Flowers solitary or in cymes at the end of lateral spurs   21

21a. Sepals with 2 bracteoles at base; fruit a dry capsule surrounded by calyx-lobes   **68. Dichotomanthes**

b. Sepals without bracteoles; fruit fleshy, crowned by persistent calyx-lobes   22

22a. Stamens 10   **66. Heteromeles**

b. Stamens c. 20   23

23a. Branches conspicuously thorny
69. Pyracantha

b. Branches thornless
67. Cotoneaster

24a. Fruits 1–4 cm, usually pear-shaped, yellow, fragrant   **59. Eriobotrya**

b. Fruits 1 cm or less, orange, red, dark red or bluish black, rarely yellow   25

25a. Fruits bluish black
63. Rhaphiolepis

b. Fruits orange, red, dark red or yellow   26

26a. Branches conspicuously thorny; fruits usually compressed laterally
69. Pyracantha

b. Branches usually thornless; fruits rounded or egg-shaped
65. Photinia

27a. Ovary or ovaries superior   28

b. Ovaries inferior   44

28a. Leaves opposite; sepals and petals 4
20. Rhodotypos

b. Leaves alternate; sepals and petals 5 or more   29

29a. Stipules absent   30

b. Stipules present, but sometimes falling early   31

30a. Leaves lobed and toothed, widest below the middle; fruit of 5 hairy achenes enclosed in a persistent calyx   **18. Holodiscus**

b. Leaves entire, widest above the middle; fruit of 5, 2-seeded follicles joined at the base   **13. Sibiraea**

31a. Petals absent   **22. Neviusia**

b. Petals present   32

32a. Petals yellow   33

b. Petals white, pink, purple or red   35

33a. Branches thornless; fruit a group of 5–8 achenes surrounded by a persistent calyx   **21. Kerria**

b. Branches thorny or with prickles; fruit a drupe or a hip   34

34a. Pith of young branches channelled, bark flaking; fruit a 1-seeded drupe
77. Prinsepia

b. Pith of young branches entire, bark smooth; fruit a hip   **4. Rosa**

35a. Fruit a 1-seeded drupe   36

b. Fruit not as above   38

36a. Sepals 10, petals reddish brown
78. Maddenia

b. Sepals 5, petals white, pink or purplish   37

37a. Pith of young branches channelled, each flower-stalk with 2 bracts
79. Oemleria

b. Pith of young branches entire, flower-stalks without bracts
76. Prunus

38a. Leaves entire or lobed, or if divided then not more than half-way towards midrib   39

b. Leaves divided almost to midrib   42

39a. Leaf-margins entire or toothed (rarely lobed)   40

b. Leaf-margins lobed and toothed   41

40a. Leaf-margins entire; flowers at least 2 cm across; fruit a woody capsule
3. Exochorda

b. Leaf-margins toothed or lobed, or if entire then flowers not more than 8 mm across; fruit a dehiscent follicle
12. Spiraea

41a. Flowers in rounded, umbel-like racemes; partial fruit bladder-like, opening by two slits
9. Physocarpus

b. Flowers in long conical racemes or panicles; fruit not bladder-like
10. Niellia

42a. Leaves hairless or with few hairs, widest below the middle, ovate, doubly toothed and lobed; flowers in panicles; fruit an irregularly shaped follicle   **11. Stephanandra**

b. Leaves densely hairy especially beneath, widest above the middle, obovate, deeply divided into 3–7

lobes; flowers solitary; fruit an
achene   43
43a. Flowers stalkless, *c.* 1 cm across;
fruit an achene with persistent style
  36. Purshia
  b. Flowers borne on long stalks, 2–3
cm across; fruit a head of hairy
achenes   **38. Fallugia**
44a. Leaves palmately or pinnately lobed
  45
  b. Leaves entire but often toothed   51
45a. Leaves rounded and entire in the
upper half, partly pinnate or lobed
below the middle
  62. × Amelosorbus &
  58. × Sorbaronia
  b. Leaves lobed above the middle,
palmately lobed or even lobed
throughout   46
46a. Leaves lobed on upper half only
  72. × Crataemespilus
  & 73. + Crataegomespilus
  b. Leaves palmately or evenly lobed
throughout   47
47a. Fruit with 1–5 bony nutlets
  71. Crataegus
  b. Fruit with 1–5 cells, each
containing 1 or more seeds   48
48a. Styles free   **53. Pyrus**
  b. Styles joined at the base   49
49a. Flowers not more than 1 cm across
  58. × Sorbaronia
  b. Flowers at least 1.5 cm across   50
50a. Leaves palmately lobed
  56. Eriolobus
  b. Leaves never palmately lobed
  55. Malus
51a. Stamens not more than 25   52
  b. Stamens numerous   54
52a. Stipules persistent, kidney-shaped
  49. Chaenomeles
  b. Stipules deciduous, triangular   53
53a. Petals not more than 1 cm; fruit
not more than 1 cm, red to black
  64. Aronia
  b. Petals 1.5–2 cm; fruit 2.5 cm
across, green at first becoming
brown   **70. Mespilus**
54a. Stipules persistent, conspicuous,
large and kidney-shaped
  49. Chaenomeles
  b. Stipules deciduous, usually
inconspicuous   55
55a. Petals 2–3 cm   **48. Cydonia**
  b. Petals 0.6–1.8 cm   56
56a. Fruit with 1–5 bony nutlets
  67. Cotoneaster
  b. Fruit with 1–5 cells, each
containing 1 or more seeds   57
57a. Flowers in racemes
  61. Amelanchier

  b. Flowers in umbels, corymbs,
panicles in solitary   58
58a. Fruits 6–8 cm   **51. × Pyronia &**
  52. + Pyrocydonia
  b. Fruit 0.4–4 cm   59
59a. Leaves narrowly oblong, stalkless
or very shortly stalked, clustered at
the ends of short shoots
  60. Peraphyllum
  b. Leaves not as above   60
60a. Calyx-lobes deciduous   **57. Sorbus**
  & 54. × Sorbopyrus
  b. Calyx-lobes persistent on fruit   61
61a. Fruit 5–12 mm   **65. Photinia**
  b. Fruit 2.5–4 cm
  54. × Sorbopyrus
62a. Deciduous perennial herbs or plants
with woody stock and annual
flowering-shoots   63
  b. Evergreen or deciduous trees or
shrubs   79
63a. Leaves with 3 leaflets   64
  b. Leaves pinnately divided or
palmately divided with more than 3
leaflets   71
64a. Petals sometimes absent
  44. Sibbaldia
  b. Petals present, usually 5–7   65
65a. Plants with spreading stolons or
rhizomes   66
  b. Plants without stolons or rhizomes
  68
66a. Flowers white or pink; fruiting
receptacle juicy   **45. Fragaria**
  b. Flowers yellow or creamy-yellow;
fruiting receptacle dry   67
67a. Flowers solitary   **46. Duchesnia**
  b. Flowers in corymbs of 3–8
  41. Waldsteinia
68a. Styles lengthening in fruit, often
with feathery hairs, or jointed
  40. Geum
  b. Styles not lengthening in fruit   69
69a. Calyx bell-shaped with 5 teeth
  5. Gillenia
  b. Calyx 5-lobed, with 5 epicalyx
segments between the lobes   70
70a. Stamens 4 or 5   **44. Sibbaldia**
  b. Stamens 10–30   **43. Potentilla**
71a. Petals absent   72
  b. Petals present   74
72a. Leaves palmately divided or lobed;
sepals 4 with 4 alternating epicalyx
segments   **47. Alchemilla**
  b. Leaves pinnately divided; sepals 3–6
or if 4, epicalyx absent   73
73a. Stipules forming a sheath around
leaf-stalk, with a leafy lobed apex
on each side; receptacle usually
with barbed spines   **31. Acaena**
  b. Stipules leaf-like, crescent-shaped,

joined to leaf-stalk; receptacle
without barbed spines
  27. Sanguisorba
74a. Epicalyx segments present   75
  b. Epicalyx absent   77
75a. Leaves pinnate, with terminal leaflet
much larger than lateral leaflets
  76
  b. Leaves palmate or if pinnate then
all leaflets similar in size
  43. Potentilla
76a. Styles persistent as awns on the
fruit   **40. Geum**
  b. Styles deciduous   **42. Coluria**
77a. Flowers yellow   **25. Agrimonia**
  b. Flowers cream, white, pink or
purplish   78
78a. Flowers unisexual in spike-like
racemes; fruit a group of follicles
  17. Aruncus
  b. Flowers bisexual in large panicles;
fruit a group of achenes
  19. Filipendula
79a. Leaves evergreen   80
  b. Leaves deciduous   89
80a. Petals absent   81
  b. Petals present   82
81a. Fruit a white berry
  30. Margyricarpus
  b. Fruit an achene; receptacle with
barbed spines   **31. Acaena**
82a. Epicalyx segments present between
calyx lobes   **43. Potentilla**
  b. Epicalyx segments absent   83
83a. Stamens numerous   84
  b. Stamens 1–20   85
84a. Fruit a hip   **24. Rosa**
  b. Fruit a 1-seeded achene terminated
by a feathery style   **37. Cowania**
85a. Leaves 11–40 cm   86
  b. Leaves not more than 7 cm   87
86a. Leaves 11–17 cm, pinnately divided
into 3–9 leaflets, each further
divided to midrib
  8. Lyonothamnus
  b. Leaves 20–40 cm with 10–16
leaflets   **26. Hagenia**
87a. Leaves to 2 cm, deeply 3-lobed and
further divided   **16. Luetkea**
  b. Leaves 2.5–7 cm   88
88a. Leaves 2.5–3.5 cm; stamens 12–15
  75. × Pyracomeles
  b. Leaves 3–7 cm; stamens 15–20
  74. Osteomeles
89a. Ovary inferior; fruit a pome   90
  b. Ovary or ovaries superior; fruit not
a pome   92
90a. Leaves only pinnate in lower part
  62. × Amelasorbus
  & 58. × Sorbaronia
  b. Leaves pinnate throughout   91

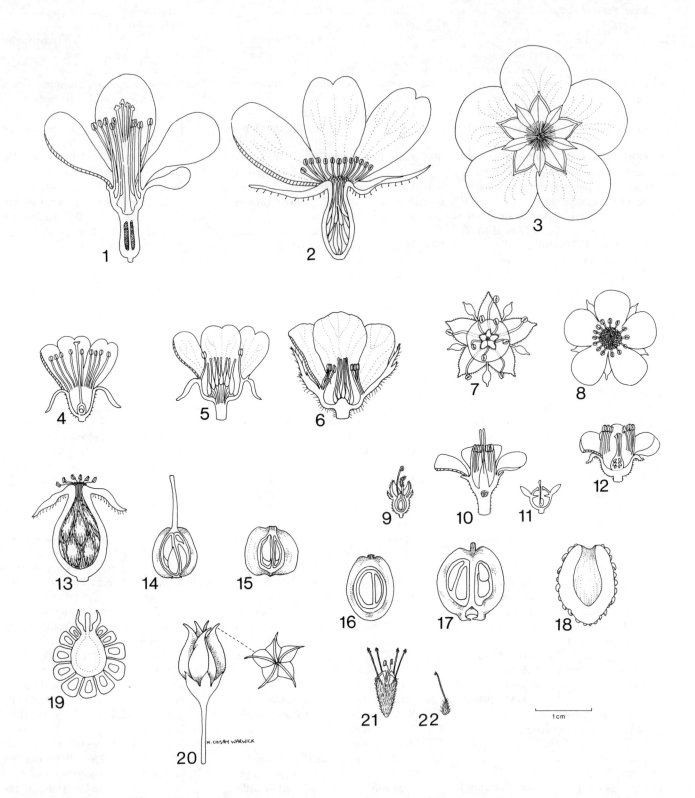

Figure 28. Diagnostic details of Rosaceae genera. 1, *Chaenomeles* half-flower. 2, *Rosa* half-flower. 3, *Potentilla fruticosa* flower from below, showing calyx and epicalyx. 4, *Prunus* half-flower. 5, *Geum urbanum* half-flower. 6, *Rubus tricolor* half-flower. 7, *Quillaia saponaria* flower from above. 8, *Fragaria* from above. 9, *Acaena caesiiglauca* half-flower. 10, *Raphiolepis umbellata* half-flower. 11, *Alchemilla* half-flower. 12, *Sorbaria* half-flower.

13, *Rosa* half-fruit. 14, *Malus* half-fruit. 15, *Cotoneaster roseus* half-fruit. 16, *Prunus domestica* half-fruit. 17, *Sorbus cashmiriana* half-fruit. 18, *Fragaria alpina* half-fruit. 19, *Rubus* half-fruit. 20, *Physocarpus amurensis* whole fruit from side and above. 21, *Acaena caesiiglauca* whole fruit from side. 22, *Geum urbanum* whole achene from side.

91a. Leaflets entire, 0.5–1.5 cm
74. Osteomeles
  b. Leaflets toothed, 1.5–5 cm
57. Sorbus
92a. Petals absent   93
  b. Petals present   96
93a. Stamens 2–7   **31. Acaena**
  b. Stamens 25 or more   94
94a. Leaves with 5–7 leaflets
32. Polylepis
  b. Leaves with 7–25 leaflets   95
95a. Leaflets 1–8 mm
28. Sarcopoterium
  b. Leaflets 2–2.5 cm   **29. Marcetella**
96a. Leaves 3-pinnate
35. Chamaebatia
  b. Leaves 1- or 2-pinnate or with 3
     leaflets   97
97a. Leaves 2-pinnate   98
  b. Leaves pinnate or with 3 leaflets   99
98a. Leaves fern-like, primary leaflets not
     more than 6 mm; fruit of 5 follicles
7. Chamaebatiaria
  b. Leaves not as above, primary
     leaflets 4–20 cm; fruit a berry of
     fleshy druplets   **23. Rubus**
99a. Fruit a hip   **24. Rosa**
  b. Fruit not as above   100
100a. Fruit a berry of fleshy druplets
23. Rubus
  b. Fruit not as above   101
101a. Leaves with 15–25 leaflets; fruit a
     capsule of united carpels
6. Sorbaria
  b. Leaves usually with 3–9 leaflets;
     fruit a head of achenes   102
102a. Style persistent, feathery
38. Fallugia
  b. Style deciduous, simple
43. Potentilla

1. QUILLAJA Molina
J. Cullen

Evergreen trees or shrubs. Leaves alternate,
leathery, toothed, wavy or almost entire,
shortly stalked; stipules small, soon falling.
Flowers few, in short racemes, often
functionally unisexual. Sepals 5, edge-to-
edge in bud. Petals 5, narrowly obovate,
greenish white. Stamens 10, 5 borne at
the apices of the lobes of a 5-lobed nectar-
secreting disc (the lobes on the same radii
as the sepals) and 5 borne in the centre of
the flower on the same radii as the petals.
Ovary of 5 carpels, free or almost so. Fruit
a group of 5 follicles, each many-seeded
and opening widely, united at their bases.

A small genus from temperate and
subtropical S America, with one
occasionally grown species. Propagation is
usually by seed.

1. Q. saponaria Molina. Illustration:
Botanical Magazine, 7568 (1897);
Rodriguez, Matthei & Quezada, Flora
arborea de Chile, pl. 73 (1983).

Shrub to 15 m. Young shoots densely
downy. Leaves elliptic to oblong, obtuse,
usually toothed, to 6 cm or more. Flowers
inconspicuous, to 1 cm across. Sepals
densely downy outside and inside. Petals
much narrower than sepals. Disc greenish
yellow. Fruit to 2 cm across, velvety when
young. *Chile*. H5–G1. Summer.

The bark contains saponins and extracts
from it have been used as soap substitutes.

2. KAGENECKIA Ruiz & Pavon
M.F. Gardner

Evergreen trees or shrubs. Leaves alternate,
margins with forward-pointing teeth;
stipules very small, falling early. Flowers
unisexual, terminal, borne separately on
different plants; male flowers in racemes
or corymbs, female flowers solitary. Calyx-
tube bell- or cone-shaped, sepals 5,
overlapping, petals 5; stamens 16–20,
inserted at the mouth of the calyx; carpels
5, free, pouch-like at base, styles 5. Ovary
superior. Fruit with 5 follicles, united at
the base, covered with spreading stellate
hairs; seeds numerous, broadly winged in
the upper part.

A genus of 3 species from Chile which
prefer a well-drained soil and a very
protected, sunny position. Propagated
under glass from seed sown in the spring.

1. K. oblonga Ruiz & Pavon. Illustration:
Edwards's Botanical Register, 1836 (1836)
– as K. crataegoides Don; Rodriguez, Flora
arborea de Chile, 178 (1983); Hoffmann,
Flora silvestre de Chile zona austral, 175
(1982).

Tree or shrub, 4–8 m, shoots hairless.
Leaves 3–10 × 1–5 cm, leathery, ovate-
lanceolate, tapering at the base, rounded
or pointed at the tip. Female flowers *c.* 2
cm across, petals pure white, rounded, in
corymbs of 6–9; male flowers with 16–20
stamens. Fruit 2–3 cm across, light brown.
C Chile. H5–G1. Summer.

3. EXOCHORDA Lindley
E.C. Nelson

Deciduous shrubs. Leaves alternate, simple,
stalked. Flowers frequently unisexual,
in terminal racemes on previous year's
shoots. Sepals 5. Petals 5, white.
Hypanthium broadly top-shaped with a
large disc. Stamens 15–30, inserted on
the margin of the disc. Styles 5, fused
towards base. Ovary superior. Fruit with 5

wings or 5-angled, at maturity separating
into 5 hard capsules; 1 or 2 seeds in each
capsule, flattened, winged.

A genus of about 4 poorly differentiated
species from E Asia. For best results
selectively prune weakest young shoots
immediately after flowering. Propagation by
cutting for selected cultivars, or from seed.

1a. Petal margins not overlapping
1. giraldii
  b. Petal margins overlapping   2
2a. Petals broadly obovate; stamens in
    groups of 3–5   **2. × macrantha**
  b. Petals round; stamens in groups of
     3 or 5;   **3. racemosa**

1. E. giraldii Hesse var. **wilsonii** Rehder.
Illustration: Kew Magazine 3: 156 (1986);
Phillips & Rix, Shrubs, 86 (1989).

Shrub to 4 m; branches erect. Leaves
ovate to elliptic, margins usually entire,
leaf-stalk 1–2.5 cm, slender, green. Flowers
to 5 cm across, 6–8 in racemes, uppermost
stalkless. Petals elliptic, long-clawed,
margins not overlapping, occasionally with
2–4 apical teeth. Stamens 20–25 in weakly
defined clusters. Fruit 1–1.5 cm. *China*. H3.
Spring.

E. serratifolia Moore differs in having
sharp, forward-pointing teeth. Rare in
gardens.

2. E. × macrantha (Lemoine) Schneider.
Illustration: Phillips & Rix, Shrubs, 82
(1989).

Shrub to 3 m. Flowers to 3 cm across;
racemes terminal on leafy axillary shoots,
to 10 cm, with 6–10 flowers. Petals
broadly obovate, short-clawed, margins
overlapping. Stamens 15–25, in groups of
3–5. H3. *Garden origin*. Spring.

A hybrid between *E. korolkowii* × *E.
racemosa*. It differs from the parents in
having stamens in unequal clusters. The
commonest cultivar is 'The Bride'.

3. E. racemosa (Lindley) Rehder (*E.
grandiflora* (Hooker) Lindley; *Spiraea
grandiflora* Hooker). Illustration: Botanical
Magazine, 4795 (1854).

Shrub to 3 m. Leaves elliptic to obovate,
apex obtuse or with a short point; leaves to
7.5 × 2.5 cm, on flowering-shoots usually
entire, those on sterile shoots weakly
toothed towards apex; leaf-stalk *c.* 1.2 cm.
Flowers 3–5 cm across; racemes erect, to
10 cm, with 6–10 flowers. Petals round,
abruptly forming short basal claw, margins
overlapping. Stamens 15–25, in clusters
of 3 or 5. Fruit *c.* 1 cm. *N China*. H3. Early
summer.

E. korolkowii Lavallée (*E. albertii* Regel).

Flowers smaller; stamens 5 in clusters of 5. *Former USSR (Turkestan)*. H3. Spring.

4. LINDLEYA Humbolt, Bonpland & Kunth
M.F. Gardner
Evergreen trees with leaves scattered; stipules present. Flowers bisexual, solitary, borne in leaf-axils towards the end of branches; flower-stalks with 2 bracts; calyx-tube persistent, sepals 5, overlapping; petals 5, spherical, stalkless; stamens 15–20, inserted in mouth of calyx, filaments unequal, flattened at base, free, anthers recurved; ovary superior, 5-celled; styles 5, free, terminal, erect. Fruit an oblong, 5-sided, woody capsule; seeds *c.* 10, thinly winged.

A genus of 2 species native to Mexico.

1. L. mespiloides (Humbolt, Bonpland & Kunth) Rydberg. Illustration: Revue Horticole, 81 (1854); Schneider, Handbuch der Laubholzkunde **1**: 496 (1906).

Tree to 3 m, shoots grey, hairless and warty. Leaves 1.2–1.8 cm, oblong to lance-shaped, pointed, with minutely glandular teeth. Flowers terminal, mostly solitary, *c.* 1.8 cm across; calyx tubular, sepals 5; petals 5, white; stamens 20. Fruit roundish to ovate, *c.* 8 mm, 5-angled. *Mexico*. H5–G1. Summer.

5. GILLENIA Moench
S.G. Knees
Perennial herbs with erect branches to 1.2 m. Leaves with 3 lanceolate leaflets, margins irregularly and sharply toothed. Stipules paired, small, inconspicuous or larger and leaf-like. Flowers hermaphrodite in long-stalked terminal panicles; petals 5, oblong-ovate, white or pinkish. Calyx bell-shaped with 5 teeth; stamens 10–20; styles 5; carpels with 2–4 ovules. Fruit of 5 leathery follicles, each with 1–4 seeds, testa leathery, endosperm striped.

A genus of just 2 species native to E North America. Both are suitable for a shady position in the larger rock garden or herbaceous border. Propagation is either by division of the rootstock in spring or by seed.

1a. Stipules ovate-lanceolate, coarsely toothed, persistent; petals mainly white **1. stipulata**
  b. Stipules narrowly linear, inconspicuous, margin smooth or slightly toothed, falling early; petals pink **2. trifoliata**

1. G. stipulata (Mühlenberg) Baillon (*Spiraea stipulata* Mühlenberg). Illustration:

Steyermark, Flora of Missouri, pl. 94 (1963); Strausbaugh & Core, Flora of West Virginia, edn 2, 467 (1978).

Stems to 1.2 m, sparsely hairy. Leaves 5–8 cm, those near the base often with deeply divided leaflets. Stipules 4–6 cm, ovate, leaf-like, persistent. Leaves and stipules gland-dotted beneath. Petals 1–1.3 cm, narrowly oblanceolate, white (rarely pink-tinged). Stamens 10–20. Fruit 6–7 mm. *E North America (New York to Illinois S to Georgia, Kansas & Texas)*. H2. Early–mid summer.

2. G. trifoliata (Linnaeus) Moench (*Spiraea trifoliata* Linnaeus). Illustration: Baillon, Histoire de plantes **1**: f. 442, 389 (1869); Strausbaugh & Core, Flora of West Virginia, edn 2, 467 (1978).

Stems 1–1.2 m, sparsely hairy. Leaves 5–7 cm, leaflets oblong-ovate, margins toothed, lower leaves not markedly different from upper leaves, leaf-stalks *c.* 8 mm. Stipules inconspicuous, narrowly linear, falling early. Lower surface of leaves and stipules without glands. Petals 1.2–2.2 cm, narrowly lanceolate, pink. Stamens 10–20. Fruit 5–6 mm. *E North America (New York, Ontario, Michigan S to Georgia & Alabama)*.

6. SORBARIA (de Candolle) Braun
J. Cullen
Deciduous shrubs, frequently with stellate hairs. Leaves large, pinnate with up to 25 toothed leaflets, alternate, with stipules. Inflorescence a large, terminal panicle. Flowers with sepals, petals and stamens perigynous, the perigynous zone cup-shaped, lined by a nectar-secreting disc. Sepals 5. Petals 5. Stamens 20–50. Ovary of 5 (rarely 4) carpels which are united towards the base, each with several ovules. Fruit a 'capsule' formed from the more or less united carpels, the follicles opening lengthwise along their outer margins. Seeds long, pale brown.

A genus of about 10 species, mostly from Asia, grown for their handsome pinnate leaves and large panicles of white or cream flowers produced rather late in the year. Propagation is by cuttings.

1a. Panicle-branches spreading very widely, borne at more or less 90° to the main axis 2
  b. Panicle-branches erect or curving upwards, borne at an acute angle to the main axis 4
2a. Leaflets less than 1.5 cm wide, simply toothed **3. aitchisonii**
  b. Leaflets more than 1.5 cm wide, doubly toothed 3

3a. Young leaves usually with some stellate hairs beneath; panicle and often the lower leaf surface with prominent stalked glands; stamens exceeding petals **1. arborea**
  b. Young leaves with simple or clustered hairs beneath; stalked glands absent or very inconspicuous; stamens about as long as petals **2. tomentosa**
4a. Stamens *c.* 20; leaflets with *c.* 25 pairs of veins **6. assurgens**
  b. Stamens 40–50; leaflets with up to 20 pairs of veins 5
5a. Panicle lacking long glandular hairs **4. sorbifolia**
  b. Panicle with long glandular hairs **5. sograndiflora**

1. S. arborea Schneider (*Spiraea arborea* (Schneider) Bean). Illustration: Horticulture **18**: 489 (1913), **19**: 393 (1919); Huxley, Deciduous garden trees and shrubs, pl. 174 (1979).

Shrub 1.3–8 m. Young shoots thinly downy, sometimes with a few stellate hairs as well, more rarely hairless. Leaves with 13–17 (rarely as few as 9) leaflets which are lanceolate, acuminate, doubly toothed and more than 1.5 cm wide; usually with stellate hairs beneath at least when young, often with stalked brown glands as well. Panicle-branches widely spreading; axis densely downy or with stellate hairs and with dark, stalked glands. Sepals 2.5–3 mm, reflexed. Petals white or cream, almost circular, shortly clawed, 2.5–3 mm. Stamens clearly exceeding petals. Capsule hairless, 3–4 mm. *W China*. H4. Late summer–autumn.

The type and density of hairs is very variable in this species. The 3 varieties formerly recognised: var. *arborea*, leaves with thin stellate hairs beneath; var. *glabrata* Rehder, with the leaves more or less hairless beneath; and var. *subtomentosa* Rehder, leaves with dense stellate hairs beneath, are merely different selections from the range of variation available in the wild.

2. S. tomentosa (Lindley) Rehder (*S. lindleyana* Maximowicz; *Spiraea lindleyana* Loudon).

Very similar to *S. arborea* but leaflets to 23, very long acuminate and with simple or somewhat clustered hairs beneath and stamens about as long as the petals. *Himalaya (E Afghanistan to W Nepal)*. H4. Late summer–autumn.

3. S. aitchisonii (Hemsley) Rehder (*Spiraea aitchisonii* Hemsley).

Shrub to 3 m. Young shoots hairless, usually reddish. Leaflets 15–21, narrowly lanceolate, acuminate, less than 1.5 cm wide, simply toothed, hairless. Panicle-branches widely spreading, axis and branches hairless. Sepals *c.* 2 mm. Petals white, *c.* 3.5 mm, circular, shortly clawed. Stamens exceeding petals. Capsule hairless. *W Himalaya.* H5. Late summer–autumn.

4. S. sorbifolia (Linnaeus) Braun (*Spiraea sorbifolia* Linnaeus). Illustration: Kitamura & Okamoto, Coloured illustrations of trees and shrubs of Japan, f. 214 (1977).

Shrub to 2 m. Young shoots hairless or downy. Leaves with 13–23 very narrowly elliptic to oblong-lanceolate, acuminate, doubly toothed leaflets, each with *c.* 20 pairs of veins, hairless or with stellate hairs beneath. Panicle-branches erect, curving upwards, downy and with dark, stalked glands. Sepals *c.* 2 mm; petals white, *c.* 3.5 mm. Stamens 40–50, considerably exceeding petals. Ovary hairless or with sparse stellate hairs. Capsule to *c.* 5 mm, hairless or sparsely hairy. *N Asia, including Japan.* H3. Late summer–autumn.

A variant with stellate hairs on leaves and carpels has been called var. **stellipila** Maximowicz.

5. S. grandiflora (Sweet) Maximowicz (*S. pallasii* (G. Don) Pojárkova; *Spiraea grandiflora* Sweet).

Similar to *S. sorbifolia* but petals 4–5 mm, panicle narrow, more or less corymb-like, with long, glandular bristles. *E former USSR.* H3. Late summer–autumn.

6. S. assurgens Vilmorin & Bois (*Spiraea assurgens* Vilmorin & Bois.

Very similar to *S. sorbifolia* but leaflets with 25 or more pairs of veins, stamens 20, scarcely exceeding the petals. *W China?* H5. Late summer–autumn.

The origin of this species is uncertain. It was named from cultivated plants thought to have come from China.

7. CHAMAEBATIARIA (Brewer & Watson) Maximowicz
J. Cullen
Aromatic, deciduous, upright shrub to 1.5 m, with dense stellate hairs and stalked glands, at least on the young growth. Leaves alternate (sometimes appearing almost whorled on condensed lateral shoots), to 3 cm, ovate-lanceolate, bipinnate with very numerous small leaflets; stipules small, linear-lanceolate. Flowers to 1 cm across, in dense terminal panicles. Sepals 5, densely hairy, often

greyish or whitish. Petals 5, circular or almost so, spreading, white. Stamens numerous, borne on the margins of the entire nectar-secreting disc. Carpels 5, hairy, united only at their extreme bases. Fruit a group of 5 follicles, each few-seeded, united only at their extreme bases.

A genus of a single species, a handsome small shrub with fern-like foliage. It requires a sunny, well-drained position and is propagated by seed or cuttings.

1. C. millefolium (Torrey) Maximowicz (*Spiraea millefolia* Torrey). Illustration: Journal of the Royal Horticultural Society **95**: f. 125 (1970); Lenz & Dourley, California native trees and shrubs, 74 (1981).

W USA. H5. Spring–summer.

8. LYONOTHAMNUS A. Gray
J. Cullen
Slender trees to 16 m; bark reddish, scaling in narrow strips. Leaves evergreen, opposite, simple and entire to bipinnate (only variants with bipinnate leaves cultivated); leaflets 3–9, divided into triangular segments with a curved lower side and more or less straight upper side, close packed, not exactly opposite each other; all dark green above, whitish, greyish or greenish beneath, with soft hairs. Inflorescence a many-flowered panicle borne on a lateral shoot. Flowers bisexual, radially symmetric. Sepals 5, densely white, cotton-hairy outside. Petals 5, white, spreading, 2–3 mm. Stamens *c.* 15. Carpels 2, free, glandular; style solitary. Fruit of 2 glandular follicles borne in the small, persistent perigynous zone; seeds 4 in each follicle.

A genus of a single species, from islands off the coast of California. It requires a sunny site, and is propagated by seed, cuttings or root suckers.

1. L. floribundus Gray subsp. **asplenifolius** (Greene) Raven (*L. asplenifolius* Greene). Illustration: Abrams & Ferris, Illustrated flora of the Pacific States **2**: 407 (1944).

W USA (islands off the coast of California). H5–G1.

Subsp. *asplenifolius* has bipinnate leaves and rarely flowers in cultivation.

9. PHYSOCARPUS Maximowicz
J. Cullen & H.S. Maxwell
Deciduous shrubs with bark peeling in thin strips; hairs, when present, mostly stellate. Leaves alternate, with small, often toothed stipules, stalked, usually with 3–5

palmately arranged main veins (other venation pinnate), margins toothed or scalloped and often 3- or 5-lobed; buds in the leaf-axils several, one above the other. Flowers in corymb- or umbel-like racemes; bracts usually small and falling early. Sepals, petals and stamens perigynous, perigynous zone cup-shaped. Sepals 5, generally hairy on both surfaces. Petals 5, usually almost circular, sometimes irregularly notched or toothed. Stamens 20–40, anthers usually reddish. Ovary superior of 2–5 (rarely 1) carpels, united at least at the base, each containing 2–5 ovules. Fruit a group of usually bladder-like follicles opening along both sutures. Seeds usually 2 per follicle, yellowish and shining, each with an appendage (caruncle).

A genus of about 10 rather similar species, 2 from E Asia, the rest from N America. They are easily grown, succeeding best in moist soils. Propagation is by seed or by cuttings.

1a. Carpels 3–5, united only towards the base 2
 b. Carpels 2 (rarely 1), united for at least half their length 6
2a. Carpels and follicles completely hairless 3
 b. Carpels densely hairy; follicles hairy (sometimes only along the sutures) 5
3a. Sepals hairy only around their margins; flower-stalks hairless or almost so **3. ribesifolius**
 b. Sepals hairy on both surfaces; flower-stalks usually hairy 4
4a. Follicles about twice as long as sepals; leaves mostly longer than broad; racemes rather open **1. opulifolius**
 b. Follicles scarcely exceeding sepals; leaves generally as broad as long; racemes condensed **2. capitatus**
5a. Largest leaves 6 cm or more, leaves hairy beneath **4. amurensis**
 b. Largest leaves to 6 cm, leaves hairless beneath, or with a few hairs along the veins only **5. intermedius**
6a. Follicles bladdery, not flattened; styles more or less spreading **6. monogynus**
 b. Follicles flattened and keeled towards the apex; styles erect **7. malvaceus**

1. P. opulifolius (Linnaeus) Maximowicz (*Spiraea opulifolia* Linnaeus). Illustration: Schneider, Illustriertes Handbuch der

Laubholzkunde **2**: 443, 445, 447 (1906); Gleason, Illustrated Flora of the northeastern United States and adjacent Canada **2**: 284 (1952); Botanical Magazine, n.s., 459 (1964); Justice & Bell, Wild flowers of North Carolina, 94 (1968).

Much-branched shrub to 3 m; branches hairless or almost so. Leaves mostly ovate, longer than broad, 6–10 cm, usually 3-lobed (occasionally 5-lobed), irregularly toothed, mostly hairless, sometimes with a few hairs on the veins beneath. Racemes on short leafy shoots, many-flowered, rather open; flower-stalks usually with stellate hairs. Flowers 1–1.3 cm across. Sepals usually hairy on both surfaces. Petals white. Carpels usually 5, hairless or almost so. Follicles united only towards the base, hairless, shining, about twice as long as the persistent sepals. *E & C North America*. H1. Summer.

'Luteus' (var. *luteus* Zabel): young leaves bright yellow.

2. P. capitatus (Pursh) Kuntze (*Spiraea capitata* Pursh). Illustration: Schneider, Illustriertes Handbuch der Laubholzkunde **2**: 445, 447 (1906); Clark, Wild flowers of British Columbia, 246 (1973); Krüssmann, Manual of cultivated broad-leaved trees and shrubs **2**: 403 (1986).

Very similar to *P. opulifolius* but racemes many-flowered, condensed and rather head-like, follicles scarcely exceeding the sepals. *W North America*. H4. Summer.

3. P. ribesifolius Komarov (*P. amurensis* misapplied).

Shrub to 3 m; branches hairless. Leaves mostly broadly ovate, usually 3-lobed, the largest 6–10 cm or more, irregularly doubly toothed, hairless beneath or with a few hairs along the veins. Racemes on short leafy shoots, many-flowered, rather open; flower-stalks hairless. Flowers to 1 cm across. Sepals hairy only on margins. Petals white. Carpels 5, hairless. Follicles united only towards the base, hairless, somewhat longer than the persistent sepals. *E former USSR*. H1. Summer.

Much confused in gardens and in the literature with *P. amurensis*.

4. P. amurensis (Maximowicz) Maximowicz (*Spiraea amurensis* Maximowicz). Illustration: Schneider, Illustriertes Handbuch der Laubholzkunde **2**: 443, 445, 447 (1906); Krüssmann, Manual of cultivated broad-leaved trees and shrubs **2**: 403 (1986).

Shrub to 3 m; branches hairless or somewhat hairy when young. Leaves broadly ovate, 3-lobed, the largest more than 6 cm, irregularly double-toothed, pale and densely hairy beneath with stellate hairs. Racemes on short leafy shoots, few- to many-flowered, open; flower-stalks usually densely stellate-hairy. Flowers to 1.5 cm across. Sepals hairy on both surfaces. Petals white. Carpels usually 5, with dense stellate hairs. Follicles united only towards the base, rather sparsely hairy, sometimes with hairs only near the sutures, somewhat longer than the sepals. *E Asia (former USSR, China, Korea)*. H1. Summer.

5. P. intermedius Schneider.

Usually a small shrub to 1.5 m; branches usually erect, crowded, hairless or almost so. Leaves circular to broadly ovate, simple or shallowly 3-lobed, irregularly double-toothed or scalloped, to 6 cm, hairless or sparsely hairy on the veins beneath. Racemes on short leafy shoots, many-flowered, rather dense; flower-stalks hairy. Flowers to 1.5 cm across. Sepals hairy on both surfaces. Petals white. Carpels 3–4, densely hairy. Follicles united only towards the base, hairy, about twice as long as the sepals. *E North America*. H2. Summer.

6. P. monogynus (Torrey) Coulter (*Spiraea monogyna* Torrey). Illustration: Schneider, Illustriertes Handbuch der Laubholzkunde **2**: 445, 447 (1906); Krüssmann, Manual of cultivated broad-leaved trees and shrubs **2**: 403 (1986).

Small shrub to 1 m; branches hairless or with a few stellate hairs. Leaves rounded to almost kidney-shaped, rarely longer than broad, 1–3 cm, usually 3- or 5-lobed (rarely entire), coarsely and irregularly toothed, hairless or with a few hairs on the veins beneath. Racemes on short leafy shoots, many-flowered, rather dense; flower-stalks hairless or with a few stellate hairs. Flowers 0.8–1.3 cm across. Sepals hairy on both surfaces. Petals white or pinkish. Carpels usually 2 (rarely 1), densely hairy. Follicles united for up to half their length, not flattened, densely hairy; styles spreading. *C USA*. H3. Summer.

7. P. malvaceus (Greene) Nelson (*Neillia torreyi* of Botanical Magazine). Illustration: Botanical Magazine, 7758 (1901).

Shrub to 2 m; branches often with stellate hairs. Leaves ovate, 3- or 5-lobed, doubly toothed or scalloped, with stellate hairs beneath, 2–6 cm. Racemes on short leafy shoots, many-flowered, dense; flower-stalks with stellate hairs. Flowers to 1.5

cm across. Sepals densely hairy on both surfaces. Petals white. Carpels 2, with dense stellate hairs. Follicles united for about half their length, flattened and keeled towards the apex, densely hairy; styles erect. *W North America*. H3. Summer.

10. NEILLIA D. Don
J. Cullen

Arching deciduous shrubs to 3 m with brown, shredding bark. Leaves with variably deciduous stipules, ovate to lanceolate, toothed, variably divided into 1–5 (usually 3) lobes, usually hairy, at least on the veins beneath. Axillary buds multiple and one above the other on the vegetative shoots, sometimes also on flowering-shoots. Inflorescence a terminal panicle or racemes borne on short, leafy, axillary shoots of the previous year. Flowers with bracts, shortly stalked. Sepals, petals and stamens perigynous, the perigynous zone bell-shaped or tubular, persisting and enlarging in fruit and often becoming bristly-glandular. Calyx-lobes 5, acute, downy within. Petals 5, white, pink or red. Stamens 15–30 (rarely as few as 10) in 2 or 3 whorls, inflexed in bud. Ovary superior. Carpels free, usually 1, rarely 2–5, each with 2–10 ovules. Fruit a follicle or group of follicles borne within the persistent, enlarged perigynous zone.

A genus of about 10 species from the Himalaya, China and E Asia. A few are easily grown as ornamental shrubs. They are propagated by seed or by cuttings.

Literature: Vidal, J., Le genre Neillia (Rosaceae), *Adansonia* **3**: 142–166 (1963); Cullen, J., The genus Neillia in mainland Asia and in cultivation, *Journal of the Arnold Arboretum* **52**: 137–158 (1971).

1a. Leaves on flowering-shoots with multiple buds one above the other in their axils; perigynous zone with adpressed bristles in flower; inflorescence a terminal panicle
   1. thyrsiflora
 b. Leaves on flowering-(but not vegetative) shoots with single buds in their axils; perigynous zone hairless, downy or with spreading bristles in flower; inflorescence a raceme borne on a short, leafy, axillary shoot   2
2a. Perigynous zone bell-shaped, as broad as or broader than long
   2. affinis
 b. Perigynous zone cylindric or tubular, longer than broad   3

3a. Axis of inflorescence with tubercule-based, stellate hairs; ovules 2
3. uekii

b. Axis of inflorescence hairless or hairy with simple hairs; ovules 5–9
4

4a. Perigynous zone hairless; racemes with up to 17 (rarely to 21) flowers
4. sinensis

b. Perigynous zone with fine adpressed hairs; racemes with 23–60 flowers (rarely as few as 19) **5. thibetica**

1. N. thyrsiflora D. Don. Illustration: Schneider, Illustriertes Handbuch der Laubholzkunde **1**: 445, 447 (1905); Journal of the Arnold Arboretum **52**: 148 (1971).

Shrub to 2 m. All leaves (on vegetative shoots and beneath the inflorescence) with multiple buds, one above the other, in their axils. Inflorescence a terminal panicle. Perigynous zone with adpressed bristles in flower. Petals usually white. *Himalaya, N Burma, China (Yunnan, Kwangsi), Vietnam & Indonesia (Java, Sumatra).* H5. Summer.

2. N. affinis Hemsley. Illustration: Journal of the Arnold Arboretum **52**: 148 (1971).

Shrub to 2 m. Only leaves on vegetative shoots with multiple buds, one above the other, in their axils. Racemes borne on short, leafy, axillary shoots. Petals pink. Carpels often more than 1. Perigynous zone hairless, downy or with spreading bristles in flower, bell-shaped, shaped, as broad as or broader than long. *W China (Yunnan, Sichuan) & N Burma.* H5. Summer.

Plants cultivated as *N. ribesioides* (see below under *N. sinensis*) are often this species. Two varieties have occurred in cultivation: var. **affinis** has racemes with 10 or more flowers which are spaced along the axis, and var. **pauciflora** (Rehder) Vidal, has racemes with fewer flowers clustered towards the ends of the flowering shoots.

3. N. uekii Nakai. Illustration: Nakai, Flora sylvatica Koreana **4**: t. 13 (1916).

Shrub to 2 m. Only leaves on vegetative shoots with multiple buds one above the other, in their axils. Racemes borne on short leafy axillary shoots, the axis with tuberculately based stellate hairs (which are ultimately deciduous). Petals usually white. Perigynous zone cylindric or tubular, longer than broad; ovules 2. *Korea.* H3. Summer.

4. N. sinensis Oliver. Illustration: Schneider, Illustriertes Handbuch der

Laubholzkunde **1**: 445, 447 (1905); Journal of the Arnold Arboretum **52**: 148 (1971); Everett, New York Botanical Garden illustrated encyclopedia of horticulture **7**: 1270 (1981).

Shrub to 3 m. Leaves variable, ovate to oblong, often lobed. Only leaves on vegetative shoots with multiple buds, one above the other, in their axils. Racemes with 9–17 (rarely to 21) flowers, borne on short, leafy, axillary shoots. Flower-stalks 2–7 mm. Perigynous zone hairless (in cultivated varieties), cylindric or tubular, longer than broad, whitish or pink, 5–11 mm. Petals white or pink. Carpel 1. *China.* H5. Summer.

A variable species with *c.* 6 varieties in the wild; only 2 are known in cultivation: var. **sinensis** with perigynous zone 6.5–11 mm and var. **ribesioides** (Rehder) Vidal (*N. ribesioides* Rehder) with shorter perigynous zone. The name *N. ribesioides* is often misapplied in gardens (see *N. affinis* above).

5. N. thibetica Bureau & Franchet (*N. longeracemosa* Hemsley). Illustration: Botanical Magazine, n.s., **3** (1968); Journal of the Arnold Arboretum **52**: 148 (1971).

Shrub to 3 m. Leaves ovate or ovate-oblong, usually downy beneath, often lobed. Only leaves on vegetative shoots with multiple buds, one above the other, in their axils. Racemes borne on short, leafy, axillary shoots, with 23–60 (rarely as few as 19) flowers. Flower-stalks 0.5–2 mm. Perigynous zone cylindric or tubular, longer than broad, pink, finely adpressed-downy and with a few bristles developing near the base after fertilisation. Petals usually pink. *W China (Sichuan).* H4. Summer.

The most widely grown of the species, usually found under the name *N. longeracemosa.*

11. STEPHANANDRA Siebold & Zuccarini
J. Cullen

Arching shrubs to 2 m or more. Leaves deciduous, ovate to narrowly ovate, lobed and doubly toothed, with stipules and multiple buds, one above the other in their axils. Flowers in a panicle borne terminally on long or short shoots, sepals, petals and stamens perigynous. Perigynous zone hemispherical, lined with a nectar-secreting disc. Sepals 5. Petals 5, white. Stamens 10–20. Ovary superior, hairy, of a single carpel which contains a single ovule. Fruit an irregularly shaped follicle. Seed 1, pale to dark brown.

A genus of 3 or 4 species from Japan and China. They are easily grown for their small but numerous white flowers and handsome leaves. Propagation is by seed or division or by cuttings.

1a. Stamens 10; leaves (at least near the base) lobed for more than ⅓ of the distance from margin to midrib; sepals without a conspicuous apical tooth **2. incisa**

b. Stamens 15–20; leaves lobed at most to ⅓ of the distance from margin to midrib; sepals each with a conspicuous apical tooth
1. tanakae

1. S. tanakae (Franchet & Savatier) Franchet & Savatier. Illustration: Kitamura & Okamoto, Coloured illustrations of trees and shrubs of Japan, f. 213 (1971).

Shrub to 2 m. Leaves broadly ovate, weakly 3–5-lobed, the lobes extending for at most ⅓ of the distance from margin to midrib, hairless above, downy along the main veins beneath. Stipules large, lanceolate, toothed. Bracts conspicuous, entire or with a few gland-tipped teeth. Perigynous zone hairless. Sepals 2.5–3 mm, each with a conspicuous apical tooth. Petals rounded to the base, *c.* 1.5 mm. Follicle hairy. *Japan.* H4. Summer.

2. S. incisa (Thunberg) Zabel (*Spiraea incisa* Thunberg; *Stephanandra flexuosa* Siebold & Zuccarini). Illustration: Schneider, Illustriertes Handbuch der Laubholzkunde **1**: 445, 447, 448 (1905); Gartenpraxis for 1987: 8 (1987).

Shrub to 2 m. Leaves ovate to narrowly ovate, with 3–7 or more deep lobes, with adpressed bristles above and beneath along the veins and stalk. Stipules long-lanceolate, more or less entire. Bracts inconspicuous. Perigynous zone with sparse adpressed bristles. Sepals 3–3.5 mm. Petals *c.* 1.5 mm, distinctly clawed. Follicle spreading-hairy. *China (Shandong), Japan & Korea.* H3. Summer.

12. SPIRAEA Linnaeus
H.S. Maxwell & S.G. Knees

Deciduous shrubs, buds small with 2–8 exposed scales. Leaves alternate, simple, lobed, toothed or occasionally entire; usually with short stalks, stipules usually absent. Flowers bisexual or unisexual in umbel-like racemes, corymbs or panicles; receptical cup-shaped; sepals 5, petals 5, stamens 15–60, inserted between the disc and the sepals. Ovary superior, carpels 5, styles 5. Fruit a group of dehiscent follicles, seeds 2–10, oblong.

A genus of 80–100 species from temperate Asia, Europe and North America as far south as Mexico. Some of the species can be propagated from suckers or cuttings of well-ripened wood. As fertile seed is freely produced and cross fertilisation is very common seed-raised plants cannot be guaranteed to come true. However, some of the most widely grown spiraeas are hybrids.

Literature: Silverside, A.J. The nomenclature of some hybrids of the *Spiraea salicifolia* group naturalised in Britain, *Watsonia* **18**: 147–151 (1990).

1a. Flowers in terminal conical panicles which are longer than wide   2
b. Flowers in lateral corymbs, or if terminal then not longer than wide   12
2a. Leaves toothed for ⅔ of their length or more   3
b. Leaves toothed only at apex, or not more than ⅓ of their length, occasionally entire   8
3a. Leaves felted beneath   4
b. Leaves hairless beneath or with few scattered hairs   5
4a. Lower leaf surface completely covered with greyish brown felted hairs; margins with round teeth, occasionally doubly toothed
  1. tomentosa
b. Lower leaf surface partially covered with whitish felted hairs; margins with acute teeth, usually doubly toothed   **2. × sanssouciana**
5a. Leaves 2.5–4 cm wide
  3. menziesii
b. Leaves 1.5–2.5 cm wide   6
6a. Flowering-stems totally hairless, leaf-margins with regular rounded teeth   **4. latifolia**
b. Flowering-stems hairy; leaf-margins irregular and doubly toothed   7
7a. Stems light yellowish brown, rounded, softly downy
  5. × semperflorens
b. Stems dark reddish brown, grooved, almost hairless   **6. salicifolia**
8a. Flowers in narrow conical panicles not more than 2.5 cm wide   9
b. Flowers in broad conical panicles 3–7 cm wide   10
9a. Leaves often entire towards inflorescence; leaf midrib and veins beneath light greyish green
  7. douglasii
b. Leaves usually toothed, even towards inflorescence; leaf midrib and veins beneath light brown
  8. × pseudosalicifolia

10a. Leaves 1–3.5 cm; densely hairy beneath   **9. × brachybotrys**
b. Leaves 3.5–7 cm; with few hairs or hairless   11
11a. Flowers in loose panicles 7–17 cm at widest point   **10. alba**
b. Flowers in tight panicles not more than 5 cm at widest point
  11. × pyramidata
12a. Flowers in terminal panicles or corymbs   13
b. Flowers in lateral corymbs   24
13a. Flowers in broadly cone-shaped panicles   14
b. Flowers in flat topped or spreading panicles or corymbs   17
14a. Leaves toothed in upper ½ or ⅓   15
b. Leaves toothed for ⅔ or more   16
15a. Medium shrub 1.5–1.8 m; leaves 5–8 cm   **12. × watsoniana**
b. Small shrub 0.6–1.2 m; leaves 1.5–4.5 cm   **13. densiflora**
16a. Petals half as long as stamens
  14. × conspicua
b. Petals as long as stamens
  15. × fontenaysii
17a. Dwarf straggly branched shrub not more than 30 cm; leaves widest above the middle   **16. decumbens**
b. Erect shrubs usually 50 cm or more; leaves widest at or below middle   18
18a. Leaves widest at the middle   19
b. Leaves widest below the middle
  21
19a. Dwarf shrub 20–50 cm; leaves 1.7–2.7 cm
  17. betulifolia var. **aemiliana**
b. Erect shrub to 1 m; leaves 2–8 cm
  20
20a. Leaf-margins scalloped
  17. betulifolia
b. Leaf-margins doubly toothed
  18. × foxii
21a. Dwarf shrub to 40 cm; leaves 1–3 cm, thick, conspicuously puckered, leathery   **19. bullata**
b. Shrub usually 1–2 m; leaves 5–15 cm, thin   22
22a. Fruit hairless   **20. japonica**
b. Fruit hairy   23
23a. Corymbs to 13 cm across, yellow, downy; flowers 5–7 mm
  21. × revirescens
b. Corymbs to 15 cm across, covered with brown felted hairs; flowers 4–5 mm   **22. micrantha**
24a. Leaves mostly 3 cm or less   25
b. Leaves mostly 3 cm or more   38
25a. Leaves 1.5–3 cm   26

b. Leaves commonly less than 1.5 cm
  34
26a. Leaves conspicuously 3–5-lobed and toothed towards the apex
  23. trilobata
b. Leaves entire, with 1–3 shallow notches towards apex or shallowly toothed   27
27a. Leaves broadly ovate-rounded; less than twice as long as wide   28
b. Leaves narrowly ovate-elliptic; more than twice as long as wide   29
28a. Leaves often entire, rounded; petals longer than the stamens
  24. nipponica var. **rotundifolia**
b. Leaves ovate-rounded, petals as long as stamens   **25. blumei**
29a. Leaves ovate-elliptic, mostly widest at or below the middle   30
b. Leaves linear-lanceolate or obovate to ovate, mostly widest above the middle   32
30a. Petals shorter than stamens
  26. canescens
b. Petals as long as or longer than stamens   31
31a. Petals rounded, equalling stamens
  27. cana
b. Petals oblong, longer than stamens
  28. × cinerea
32a. Leaves linear to lanceolate, at least 4 times longer than wide
  29. thunbergii
b. Leaves obovate to ovate, mostly less than twice as long as wide   33
33a. Fruit almost hairless; petals longer than or as long as stamens
  30. hypericifolia
b. Fruit covered with fine down; petals shorter than stamens
  31. crenata
34a. At least some leaves 3-lobed or conspicuously toothed at apex   35
b. Leaves mostly entire or with 1–3 small teeth or indentations at apex
  36
35a. Leaves downy above, grey or white felted beneath; fruit downy
  32. yunnanensis
b. Leaves hairless, bright green above and below; fruit hairless
  33. gemmata
36a. Leaves mostly 4–7 mm; flowers in groups of 6–8   **34. calcicola**
b. Leaves 8–15 mm; flowers in groups of 10 or more   37
37a. Corymbs hairless but with a bluish grey bloom   **35. baldschuanica**
b. Corymbs downy, not bloomed
  36. arcuata

38a. Leaves toothed at apex or only for
     upper ⅓   39
  b. Leaves toothed for most of their
     length or at least ⅔   51
39a. Fruit hairless   40
  b. Fruit downy or with longer hairs
                                                46
40a. Leaves hairless   41
  b. Leaves hairy   43
41a. Petals longer than stamens
                            37. × **schinabeckii**
  b. Petals equalling or shorter than
     stamens   42
42a. Shrub to about 4 m, branches
     downy   **38. veitchii**
  b. Shrub to about 1 m, branches
     hairless   **39. virginiana**
43a. Leaves deeply toothed to 3-lobed
                              40. pubescens
  b. Leaves finely toothed or entire   44
44a. Leaves entire   **38. veitchii**
  b. Leaves finely toothed   45
45a. Leaves 0.4–1.5 cm across, widest at
     or above the middle
                            41. sargentiana
  b. Leaves 1.5 cm or more across,
     widest at or below the middle
                              39. virginiana
46a. Leaves hairless   **42. trichocarpa**
  b. Leaves hairy   47
47a. Leaves 2–3 cm wide   48
  b. Leaves 1.5 cm wide or less   49
48a. Petals shorter than stamens
                        43. media subsp. **polonica**
  b. Petals longer than stamens
                                44. henryi
49a. Leaves almost entire or with a few
     very shallow teeth   **45. wilsonii**
  b. Leaves conspicuously toothed   50
50a. Leaves obovate; margins with
     conspicuous white hairs   **43. media**
  b. Leaves ovate-elliptic, margins
     without conspicuous hairs
                              46. mollifolia
51a. Leaf-margin with fine even teeth
                                                52
  b. Leaf-margin irregularly and often
     doubly toothed or lobed   59
52a. Leaves narrowly ovate-elliptic to
     lanceolate or if broadly ovate then
     very finely toothed   53
  b. Leaves broadly ovate, distinctly
     wider at the base, teeth conspicuous
                                                55
53a. Flowers borne in leafless, almost
     stalkless umbels; leaves ovate with
     very short even teeth
                              47. prunifolia
  b. Flowers borne in leafy, stalked
     umbels; leaves narrowly ovate to
     elliptic-lanceolate   54

54a. Flowers borne in downy clusters
                                48. bella
  b. Flowers borne in hairless clusters
                          49. cantoniensis
55a. Fruit almost hairless
                          50. longigemmis
  b. Fruit hairy, often densely so   56
56a. Flowers borne in hairless clusters
                                                57
  b. Flowers borne in hairy clusters   58
57a. Flowers perfect; leaves wedge-
     shaped at base
                   51. miyabei var. **glabrata**
  b. Male and female flowers borne
     separate; leaves truncate at base
                                52. amoena
58a. Flowers c. 8 mm, in corymbs 3–6
     cm   **51. miyabei**
  b. Flowers c. 6 mm, in corymbs 5–9
     cm   **53. rosthornii**
59a. Petals twice as long as stamens
                                                60
  b. Petals half as long or as long as
     stamens   61
60a. Leaves densely felted beneath
                                54. × blanda
  b. Leaves hairless beneath
                          55. × vanhouttei
61a. Underside of leaves and young
     shoots covered with yellowish
     brown, felted hairs   **56. chinensis**
  b. Underside of leaves hairless or with
     few scattered hairs
                          57. chamaedryfolia

1. S. tomentosa Linnaeus. Illustration:
Gleason, The Illustrated flora of the north-
eastern United States and adjacent Canada
2: 286 (1952); House, Wild flowers, pl.
94A (1961); Justice & Bell, Wild flowers
of North Carolina, 93 (1968); Krüssmann,
Manual of cultivated broad-leaved trees &
shrubs **3**: 347 (1986).

Upright shrub, to 1.2 m, suckering,
strong-growing, shoots with dense brown
felted hairs when young, erect, angled.
Leaves 4–8 × 2–4 cm, oblong-ovate, acute,
coarsely and often doubly toothed, almost
to base, wrinkled, dark green and almost
hairless above, densely felted greyish
brown hairs below, stalks 1–4 mm.
Flowers purple-pink, in narrow, erect,
dense brown felted cone-shaped panicles,
8–20 × 4–6.5 cm. Petals obovate, shorter
than the stamens. Fruit downy; sepals
reflexed. *E North America*. H2. Late
summer–autumn.

Var. **alba** Weston. Differs in its white
flowers.

2. S. × sanssouciana Koch (*S. nobleana*
Hooker; *S. japonica* 'Paniculata').

Illustration: Botanical Magazine, 5169
(1860); Dippel, Handbuch der
Laubholzkunde **3**: 496, (1893).

Shrub to 1.5 m, shoots with fine downy
greyish hairs when young, brownish,
angular, erect. Leaves 5–10 × 2–3 cm,
oblong-lanceolate, doubly toothed on the
apical ½–⅔, light green, densely whitish
felted below. Flowers pink to bright rose in
terminal panicles, 10–17 × 3–7 cm, made
up of several multi-topped cymes; petals
half as long as stamens. Fruit hairless,
slender, longer than the reflexed calyx;
style spreading; sepals recurved. *Garden
origin*. H2. Summer.

A hybrid beteen *S. douglasii* and *S.
japonica*.

3. S. menziesii Hooker (*S. douglasii* var.
menziesii (Hooker) Presl). Illustration:
Schneider, Illustriertes Handbuch der
Laubholzkunde **1**: 474, 482 (1905).

Shrub 1–1.5 m, young shoots finely
downy, striped. Leaves 5–11 × 2.5–4 cm,
oblong-lanceolate, usually hairless but
sometimes with hairs on the veins, paler
green below, stalks 3–5 mm. Flowers pink,
in terminal conical panicles which are
longer than wide; stamens more than twice
as long as the petals. *W North America
(Alaska to Oregon)*. H2. Summer.

4. S. latifolia (Aiton) Borkhausen (*S.
bethlehemensis* misapplied; *S. carpinifolia*
Willdenow; *S. salicifolia* Linnaeus var.
latifolia Aiton; *S. canadensis* misapplied).
Illustration: Bailey, The standard
cyclopedia of horticulture **3**: 3213 (1935);
Gleason, The Illustrated flora of the north-
eastern United States and adjacent Canada
2: 286 (1952); House, Wild flowers, pl.
93A (1961); Krüssmann, Manual of
cultivated broad-leaved trees & shrubs **3**:
347 (1986).

Shrub 1.5–1.8 m, young shoots hairless,
brownish red, erect, angular. Leaves 3–7
cm, broadly elliptic to obovate or oblong,
acute at both ends, bright green above,
bluish green below; margins with regular
rounded teeth. Flowers white to pale pink,
in hairless, broad conical shaped panicles,
longer than 20 cm, often with branches
spreading horizontally at the end of the
current years growth. Petals shorter than
the stamens, disc usually pink. Fruit
hairless; style spreading. *N America*. H2.
Summer–early autumn.

5. S. × semperflorens Zabel (*S. spicata*
Dippel).

Shrub to 1.5 m, shoots downy, yellowish
brown, round, finely striped. Leaves 6–10

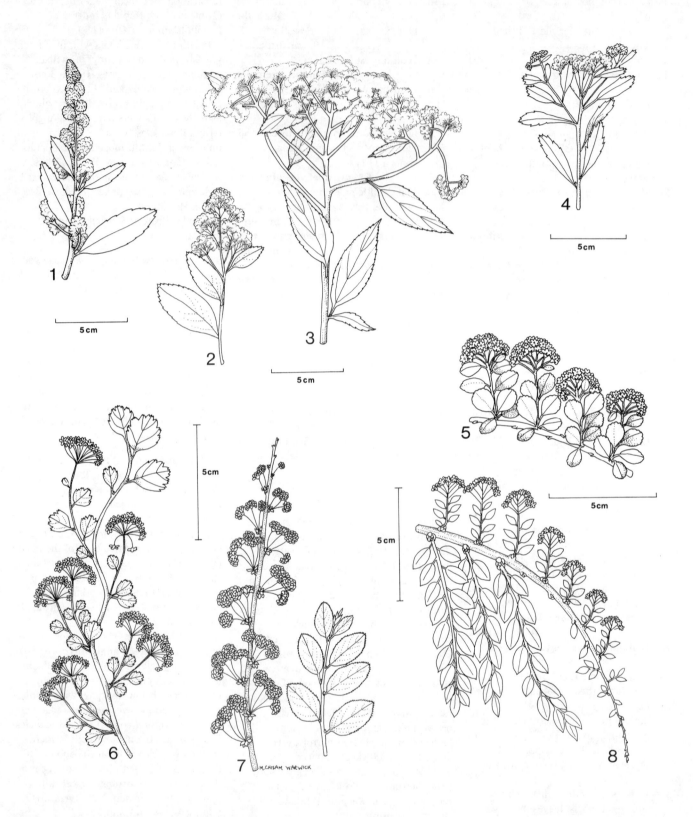

Figure 29. Inflorescences of *Spiraea* species. 1, *S. douglasii*.
2, *S.* × *watsoniana*. 3, *S. micrantha*. 4, *S. decumbens*. 5, *S. blumei*.
6, *S. trilobata*. 7, *S. prunifolia*. 8, *S. calcicola*.

cm, oblong-lanceolate, warty and sharply, doubly toothed almost to base, acuminate, glaucous and almost hairless below. Flowers rose-pink, large, in well-branched, finely downy, narrowly cone-shaped panicles 17–20 × 3–5 cm. Petals half as long as the stamens. Fruit hairless. *Garden origin*. H2. Summer–early autumn.

A hybrid between *S. japonica* and *S. salicifolia*. 'Syringaeflora' is a spreading shrub 0.9–1.2 m, branches slightly angled. Leaves *c.* 7.5 × 2 cm, lanceolate to oblong, acuminate, almost hairless beneath, margins toothed above the middle. Flowers pink in densely felted conical panicles; stamens longer than petals. H2. Summer–early autumn.

6. S. salicifolia Linnaeus. Illustration: Reichenbach, Icones Florae Germanicae et Helveticae **24**: pl. 152 (1909); Csapody & Tóth, A colour atlas of flowering trees and shrubs, 96 (1982); Pignatti, Flora d' Italia **1**: 539 (1982); Daykin, Pictorial guide to shrubs and climbing plants, 96 (1986).

Narrowly erect shrub, to 2 m, suckering and strong growing, shoots finely downy when young, reddish brown, slightly angular. Leaves 4–8 × 1.5–2.5 cm, elliptic to oblong-lanceolate, acute at both ends, scabrous, sometimes doubly toothed, hairless on both sides, darker green above, short-stalked. Flowers pink to whitish, in slim, downy, erect pyramid-shaped terminal panicles *c.* 20 × 8 cm; petals half as long as stamens; calyx downy. Fruit hairy, style recurved. *SE Europe to NE Asia & Japan (naturalised in many parts of Europe)*. H2. Summer.

A hybrid **S. × rosalba** Dippel (*S. × rubella* Dippel; *S. salicifolia × S. alba*) differs from *S. salicifolia* in its narrowly cone-shaped panicle and very pale pink petals.

7. S. douglasii Hooker. Figure 29(1), p. 334. Illustration: Botanical Magazine, 5151 (1859); Schneider, Illustriertes Handbuch der Laubholzkunde **1**: 474, 482 (1905); Clark, Wild flowers of British Columbia, 282 (1973); Krüssmann, Manual of cultivated broad-leaved trees & shrubs **3**: 347 (1986).

Erect shrub 1.5–2.5 m, shoots producing suckering stolons, brown to reddish, finely downy when young, slender. Leaves 4–10 × 1.5–2.5 cm, oblong, rounded at both ends, toothed towards the apex, leaves below inflorescence usually entire, dark green above, white dense felted hairs below, stalks 2–4 mm. Flowers purple-pink, in erect terminal panicles, to 20 cm,

narrowly cone-shaped. Axis of flower with white densely felted hairs, stalk and calyx downy grey; stamens pink, longer than the petals. Fruit shiny, hairless, with erect or spreading styles; sepals recurved. *W North America (British Columbia to N California, naturalised in parts of N & C Europe)*. HΣ. Summer.

S. × billardii Hérincq (*S. alba × S. douglasii*), intermediate between the parents in leaf shape, hairiness, panicle shape and flower colour but usually closer to *S. douglasii*.

8. S. × pseudosalicifolia Silverside (*S. × billardii* misapplied not Herincq). Illustration: Krüssmann, Manual of cultivated broad-leaved trees & shrubs **3**: 347 (1986).

Shrub to 2 m, shoots brown, finely downy. Leaves 5–8 cm, oblong-lanceolate, acute at both ends, coarsely toothed on the apical ²⁄₃, greyish green with densely felted light brown hairs on midrib and veins below when young. Flowers pink, *c.* 5 mm, in narrow densely felted 10–20 cm panicles. Stamens twice as long as petals. Fruit hairless; styles spreading. *Garden origin*. H2. Summer.

A hybrid between *S. douglasii* and *S. salicifolia*.

9. S. × brachybotrys Lange (*S. pruinosa* Zabel; *S. luxuriosa* Zabel).

Shrub, to 2.5 m, shoots finely downy, striped, angular, more or less erect. Leaves 1–3.5 × 1–2.5 cm, narrowly elliptic-oblong, toothed towards the apex, or sometimes with only a few teeth at the tip, dull dark green, slightly downy above, paler with grey felted hairs below. Flowers in dense, usually leafy terminal panicles 3–10 cm, covered with felt-like hairs and with short side-shoots. Flower-stalks and calyx hairy; petals pale pink; sepals reflexed. *Garden origin*. H2. Summer.

A hybrid between *S. canescens* and *S. douglasii*. Similar to *S. fontenaysii* but leaves with densely felted hairs below and reflexed sepals.

10. S. alba Duroi (*S. lanceolata* Borkhausen; *S. salicifolia* Linnaeus var. *paniculata* Aiton). Illustration: Gardeners' Chronicle **11**: 753 (1879); Schneider, Illustriertes Handbuch der Laubholzkunde **1**: 482 (1905); Gleason, The Illustrated flora of the north-eastern United States and adjacent Canada **2**: 286 (1952); Stace, New flora of the British Isles, 398 (1991).

Suckering shrub, similar to *S. salicifolia*, to 2 m, branches slightly downy when

young, outspread, reddish brown, angled. Leaves oblong-oblancelate, 3.5–5 × 1.3–3 cm, acute, sharply toothed, veins hairless or downy beneath, stalk *c.* 5 mm. Flowers white, in finely hairy, broadly conical panicles 15–30.5 cm. Sepals erect; stamens white, as long or longer than petals. Fruit hairless. *E USA to Canada*. H2. Summer.

11. S. × pyramidata Greene. Illustration: Clark, Wild flowers of British Columbia, 271 (1973).

Shrub to 1 m, spreading by suckers, branches hairless, reddish brown, round, upright. Leaves 3.5–7 cm, elliptic-oblong, obtuse to acute, coarsely and occasionally doubly toothed towards the apex, hairless below, stalks 2–5 mm. Flowers many, white to pinkish, in pyramidal or rounded, downy or hairless panicles. *NW USA*. H2. Summer.

A hybrid between *S. betulifolia* var. *lucida* and *S. menziesii*, which does sometimes occur in the wild.

12. S. × watsoniana Zabel (*S. nobleana* Zabel not J.D. Hooker; *S. subvillosa* Rydberg). Figure 29(2), p. 334.

Shrub, 1.5–1.8 m, erect, shoots shoots. Leaves 5–8 cm, elliptic-oblong to almost obtuse, toothed only towards the apex, base usually rounded, with densely grey felted hairs below. Flowers pink, many, in finely downy pyramidal terminal panicles. Fruit small. *Has occurred in cultivation and in the wild with parents. WN USA (Oregon)*. H2. Summer.

A hybrid between *S. douglasii* and *S. densiflora* subsp. *splendens*. Similar to *S. × sanssouciana*.

13. S. densiflora Rydberg. Illustration: Munz, California mountain wildflowers, pl. 16 (1963); Fries, Wild flowers of Mount Rainier and the Cascades, 130 (1970).

Shrub to 1.2 m, young shoots round, hairless, reddish brown. Leaves 1.5–4.5 cm, elliptic, rounded at both ends, toothed or scalloped towards the apex, hairless on both surfaces, deep green above, paler below, stalks 2 mm or less. Flowers pink, in dense 1.5–4 cm, hairless, dome-shaped erect panicles on current year's growth; petals shorter than stamens. Fruit hairless; sepals obtuse, ovate, erect. *W North America (British Columbia to Oregon)*. H2. Summer.

Subsp. **splendens** (K. Koch) Abrams (*S. splendens* K. Koch; *S. arbuscula* Greene). Illustration: Schneider, Illustriertes Handbuch der Laubholzkunde **1**: 468, 474 (1904). Shrub, to 1.2 m, young shoots finely downy. Leaves ovate to elliptic-lanceolate, sometimes acute, toothed or doubly toothed towards the apex, with acuminate teeth. Flowers in finely downy panicles; sepals acute, triangular. *USA (Oregon to California)*. H2. Summer.

'Nobleana': leaves mostly tapered, sometimes rounded at base, downy on the veins above, covered with a dull greyish down below. Flowers many, in broad corymbose panicles 7.5–25.5 cm, at the end of the new growth. Flower-stalks grey-felted. H2. Summer.

14. S. × conspicua Zabel.

Upright shrub, *c.* 1 m, young shoots downy, dark brown, slightly angled. Leaves 3–6 cm, elliptic-oblong, acute at both ends, toothed or doubly toothed, almost hairless. Flowers pale pink to white, in finely downy, broad pyramidal panicles on long erect shoots; petals half as long as the stamens. *Garden origin*. H2. Summer–early autumn.

A hybrid between *S. japonica* 'Albiflora' and *S. alba*.

15. S. × fontenaysii Lebas (*S. fontenaysii* forma *alba* Zabel; *S. fontenaysiensis* Dippel).

Erect slender branched shrub to 2 m, shoots downy when young, angular. Leaves 2–5 cm, elliptic-oblong, obtuse at both ends, scallop-toothed towards the apex, blue-green, almost hairless below, short-stalked. Flowers white, in downy 3–8 × 3–8 cm, pyramid-shaped panicles on short side-shoots at the end of last year's growth; petals as long as the stamens. Fruit almost hairless; styles spreading. *Garden origin*. H2. Summer.

A hybrid between *S. canescens* and *S. salicifolia*.

16. S. decumbens W. Koch. Figure 29(4), p. 334. Illustration: Schneider, Illustriertes Handbuch der Laubholzkunde **1**: 459, 474 (1905); Reichenbach, Icones Florae Germanicae et Helveticae **24**: pl. 151 (1909); Pignatti, Flora d' Italia **1**: 540 (1982); Krüssmann, Manual of cultivated broad-leaved trees & shrubs **3**: 352 (1986).

Dwarf shrub, often prostrate, *c.* 25 cm, shoots slender, erect, buds obtuse, small with few scales. Leaves 1.3–3.8 × 0.6–1.3 cm, elliptic-oblong, widest above the middle, acute at both ends, hairless on both surfaces, toothed or doubly toothed, stalks *c.* 3 mm. Flowers white, *c.* 6 mm, in small 3–5 cm terminal flat-topped panicles or corymbs on the current year's growth. Flowers not quite dioecious; petals as long as stamens. Fruit hairless, style erect;

sepals spreading or recurved. *SE Europe (S Germany to S Tyrol)*. H2. Summer.

Subsp. **tomentosa** Poech (*S. hacquetii* Fenzl & Koch; *S. lancifolia* Hoffmannsegg). Illustration: Dippel, Handbuch der Laubholzk **3**: 476, (1893); Schneider, Illustriertes Handbuch der Laubholzkunde **1**: 459, 474 (1905); Reichenbach, Icones Florae Germanicae et Helveticae **24**: pl. 151 (1909). Dwarf shrub, often prostrate, 20–30 cm, shoots grey-downy. Leaves elliptic, 1.5–2.5 cm, toothed at apex, grey dense felted hairs below. Flowers white, in 2.5–3.5 cm flat-topped panicles, not projecting above the apical leaves. Sepals half spreading in fruit. *Austria, Italy (Tyrol)*. H2. Early summer.

17. S. betulifolia Pallas. Illustration: Schneider, Illustriertes Handbuch der Laubholzkunde **1**: 468, 474 (1905); Gleason, The Illustrated flora of the north-eastern United States and adjacent Canada **2**: 285 (1952); Clark, Wild flowers of British Columbia, 279 (1973); Krüssmann, Manual of cultivated broad-leaved trees & shrubs **3**: 352 (1986).

Densely bushy shrub of rounded habit, 0.5–1 m. Branches reddish brown, striped, hairless, slightly angled. Leaves 2–4 cm, broadly ovate to elliptic, usually rounded at apex, rounded to tapered at base, doubly or simply scalloped, sometimes only in the upper half, dark green above, grey-green with net-like venation below, usually hairless, sometimes downy below, stalks hairless, 1–6 mm. Flowers white or very occasionally pinkish, hairless, in dense terminal corymbs 2.5–9 cm; stamens twice as long as petals; sepals reflexed in fruit. Fruit hairless; style erect. *NE Asia to C Japan*. H2. Summer.

Var. **corymbosa** (Rafinesque-Schmaltz) Maximowicz (*S. corymbosa* Rafinesque-Schmaltz not Muhlenberg; *S. betulifolia* misapplied, not Pallas). Illustration: Schneider, Illustriertes Handbuch der Laubholzkunde **1**: 474 (1905). Leaves 3–8 cm, ovate to roundish, broadly tapered at base, scalloped or coarsely and doubly toothed towards the apex, glaucous beneath with a few hairs, stalks 3–8 mm. Flowers white, to 5 mm, in rounded densely packed 5–10 cm corymbs; stalks downy 3–8 mm. Sepals erect in fruit. Fruit erect, hairless, glossy. *E USA*. H2. Summer.

Var. **lucida** (Douglas) C.L. Hutchinson (*S. lucida* J.D. Hooker; *S. corymbosa* var. *lucida* (Douglas) Zabel). Illustration: Schneider, Illustriertes Handbuch der Laubholzkunde **1**: 468, 474 (1905).

Shoots brown to yellowish. Leaves 2–6 cm, broadly ovate to oblong, acute or sometimes rounded at the apex, rounded or broadly tapered at the base, scalloped or doubly and deeply toothed on the apical ½–⅔, shiny above, paler below, stalks 3–6 mm. Flowers white, in densely packed hairless 3–10 cm corymbs. *NW USA*. H2. Summer.

Var. **aemiliana** (Schneider) Koidzume (*S. beauverdiana* Schneider). Dwarf shrub; 20–50 cm, leaves 1.7–2.7 cm, broadly rounded, scalloped, with distinct net-like venation. Flowers 4–5 mm, on a inflorescence 2–2.5 cm. *Japan*. H2. Summer.

18. S. × foxii (Vos) Zabel.

Compact shrub 0.5–1 m, shoots brown, almost hairless, flexuous. Leaves 5–8 cm, elliptic, doubly toothed on the upper 2/3 only, dull green above often with brown markings, light green beneath, hairless on both sides. Flowers whitish, sometimes tinged bluish pink, in finely downy, widely branched, flattish corymbs 10–20 cm across. Fruit obtuse at apex; styles spreading. *Garden origin*. H2. Summer.

A hybrid between *S. betulifolia* var. *corymbosa* × *S. japonica*, similar to *S. × margaritae*.

19. S. bullata Maximowicz (*S. crispifolia* misapplied). Illustration: Schneider, Illustriertes Handbuch der Laubholzkunde **1**: 468, 474 (1905); Garden Answers, 55 (1987).

Dwarf shrub to 40 cm, branches with long white hairs when young, eventually brown and hairless. Leaves 1–3 cm, rounded, ovate, conspicuously puckered, leathery, deeply toothed, greyish green beneath. Flowers deep rosy pink, in small dense corymbs forming a terminal panicle 4–8 cm across. Stamens reddish, slightly longer than petals. Fruit with style remnants stunted. *Japan, only known in cultivation*. H2. Summer.

20. S. japonica Linnaeus filius (*S. × bumalda* Burvénich; *S. callosa* Thunberg; *S. pumila* Zabel). Illustration: Gleason, The Illustrated flora of the north-eastern United States and adjacent Canada **2**: 286 (1952); Csapody & Tóth, A colour atlas of flowering trees and shrubs, pl. 176 (1982); Pignatti, Flora d' Italia **1**: 540 (1982); Krüssmann, Manual of cultivated broad-leaved trees & shrubs **3**: 352, pl. 123 (1986).

Shrub to 2 m, usually only slightly branched, shoots erect, downy when young, striped or nearly round. Leaves 5–12 cm, short-stalked, ovate-oblong to narrowly elliptic, acute, deeply doubly toothed, hairless on both surfaces or grey-green and downy on the veins below. Flowers light to dark pink-red, occasionally white, *c.* 6 mm, in flat terminal downy 15–20 cm corymbs, leafy at the base. Petals shorter than the stamens; calyx-lobes erect at flowering, later spreading. Fruit hairless. *Himalaya to China, Korea & Japan.* H1. Summer.

Var. **acuminata** Franchet. Leaves oblong-ovate to lanceolate, acuminate, green, downy on the venation. Flowers pink, in 10–14 cm corymbs. *C & W China.* H2. Summer.

Var. **fortunei** (Planchon) Rehder (*S. fortunei* Planchon; *S. callosa* Lindley not Thunberg). Taller than 1.5 m, shoots downy when young, rounded. Leaves 5–10 cm, oblong-lanceolate, sharply and doubly toothed with hardened tips, wrinkled above, hairless and glaucous below. Inflorescence downy, many branched; flowers pink. *E & C China.* H2. Summer.

'Albiflora' ('Alba'; *S. albiflora* (Miquel) Zabel). Illustration: Bailey, The standard cyclopedia of horticulture **3**: 3212 (1935); Krüssmann, Manual of cultivated broad-leaved trees & shrubs **3**: 348 (1986). Dwarf shrub 30–60 cm.

'Anthony Waterer' (*S.* × *bumalda* 'Anthony Waterer'). Illustration: The Hillier colour dictionary of trees & shrubs, 242 (1981); Csapody & Tóth, A colour atlas of flowering trees and shrubs, pl. 177 (1982); Amateur Gardening **105**: 5443 (1989). Compact shrub, 1.2–1.5 m, with narrow often variegated leaves and bright crimson flowers.

'Little Princess'. Illustration: Gardening from 'Which' May **4**: 149 (1988). Dwarf shrub to 50 cm, wider than taller, erect, low, well branched. Leaves ovate, *c.* 2.5 × 1.2 cm, dull green above, often wrinkled, hairless below. Flowers many, pale lilac-pink, in *c.* 4 cm downy cymes, short-stalked.

S. × **margaritae** Zabel. Illustration Krüssmann: Manual of cultivated broad-leaved trees & shrubs **3**: 355 (1986). Shrub to 1.5 m, branches dark reddish brown, finely downy, round, finely striped. Leaves 5–8 × 3–4 cm, ovate-elliptic, acute to wedge shaped, coarsely to doubly toothed, dark green above, paler below with a few hairs on the veins, short-stalked. Flowers rose-pink becoming lighter, 7–8 mm, in downy, loose, flat corymbs, to 15 cm. Petals half as long as

the stamens; calyx downy. Fruit small, hairless; styles usually erect. *Garden origin.* H2. Summer–autumn.

A hybrid between *S. japonica* and *S. superba* (*S. albiflora* × *S. corymbosa*). Foliage brightly coloured in autumn.

21. S. × **revirescens** Zabel.
Shrub, to 1.2 m, branches brown, downy when young, slightly angled. Leaves 5–10 cm, ovate-oblong, acuminate, tapered at base, deeply or doubly toothed, light green, blue-green below, veins downy yellow. Inflorescence downy, corymbs 10–13 cm, terminal; flowers rose-pink, 5–7 mm; petals shorter than the stamens. Fruit downy. *Garden origin.* H2. Summer–autumn.

A hybrid between *S. amoena* and *S. japonica.*

22. S. micrantha J.D. Hooker (*S. japonica* Linnaeus var. *himalaica* Kitamura). Figure 29(3), p. 334.
Small to medium-sized shrub, *c.* 1.5 m, branches hairy. Leaves 7.5–15 cm, ovate-lanceolate, toothed to doubly toothed, warty, tapering to a long point at the apex, almost rounded at the base, downy below. Flowers mostly bisexual, 4–5 mm, pale pink, in loose leafy corymbs to 15 cm, with dense brown felted hairs. Fruit densely hairy. *E Himalaya.* H2. Summer.

23. S. trilobata Linnaeus (*S. aquilegiifolia* misapplied; *S. rotundifolia* misapplied; *S. grossulariifolia vera* misapplied). Figure 29(6), p. 334. Illustration: Schneider, Illustriertes Handbuch der Laubholzkunde **1**: 455, 459 (1905); Bailey, The standard cyclopedia of horticulture **3**: 3210 (1935); Krüssmann, Manual of cultivated broad-leaved trees & shrubs **3**: 351 (1986).
Shrub, 1–1.5 m, dense, compact, broad habit, shoots hairless, slender, spreading, often flexuous. Leaves 1.5–3 cm rarely to 4 cm, rounded, scallop-toothed, usually 3–5 lobed, base rounded to broadly tapering, bluish green above, stronger in colour below, stalks 5–8 mm. Flowers white, small, in many-flowered corymbs, 2–4 cm on the end of short leafy twigs on last year's growth, flower-stalks 0.6–1.9 cm, hairless, slender. Petals longer than the stamens. Fruit somewhat spreading; sepals erect. *N Asia (China to Siberia & Turkestan).* H2. Summer.

Similar to *S. vanhouttei* but smaller in all parts.

24. S. nipponica Maximowicz var. **rotundifolia** (Nicholson) Makino (*S. bracteata* Zabel not Rafinesque). Illustration: Botanical Magazine, 7429 (1895); Bean,

Trees & shrubs hardy in the British Isles **4**: 489 (1980); Krüssmann, Manual of cultivated broad-leaved trees & shrubs **3**: 351 (1986); Daykin, Pictorial guide to shrubs and climbing plants, 96 (1986).
Upright bushy shrub, branches hairless, reddish brown, rounded, angular, long, arching. Leaves 1.5–3 cm, ovate to rounded, scalloped, apex rounded, usually notched, sometimes entire, broadly tapering at base, hairless, dark green above, blue-green below, stalks to 4 mm. Flowers *c.* 8 mm, white, many, in semi-circular corymbs; stalks leafy. Petals roundish, longer than the stamens. Fruit erect, slightly hairy; styles usually spreading. *Japan.* H2. Summer.

Var. **tosaensis** (Yatabe) Makino. Differs in having smaller flowers in dense umbel-like corymbs and is probably one of the commonest spiraeas in cultivation.

25. S. blumei D. Don (*S. chamaedryfolia* Blume not Linnaeus; *S. obtusa* Nakai). Figure 29(5), p. 334. Illustration: Schneider, Illustriertes Handbuch der Laubholzkunde **1**: 455, 459 (1905); Krüssmann, Manual of cultivated broad-leaved trees & shrubs **3**: 351 (1986).
Shrub to 1.5 m, branches spreading, arching, rounded, hairless. Leaves 2–3.5 cm, ovate-rounded, obtuse, sharply scalloped-toothed, sometimes almost 3–5-lobed, distinctly veined below, bluish green; stalks 6–8 mm. Flowers white, both uni- and bisexual, in many small corymbs. Petals rounded-obovate, equal in length to the stamens. Styles spreading, longer or as long as the stamens. *Japan, Korea.* H2. Summer.

A variable species often confused with *S. trilobata.* The main difference is the shape of the leaves, i.e. more diamond-shaped and pinnately veined in *S. blumei.*

26. S. canescens D. Don (*S. vaccinifolia* misapplied, not Don; *S. flagelliformis* misapplied; *S. laxiflora* Lindley). Illustration: Schneider, Illustriertes Handbuch der Laubholzkunde **1**: 468, 474 (1905); Krüssmann, Manual of cultivated broad-leaved trees & shrubs **3**: 352 (1986).
Dense shrub to 2–3.5 m, branches downy, angular, ribbed, arching. Leaves 1–2.5 cm, elliptic to obovate, toothed at the apex, tapering to a short stalk, downy on the margins above, greyish green hairs below. Flowers many, white, in semi-circular 3–5 cm corymbs on the upper side of the branches; petals shorter than the stamens. Flower-stalk and calyx with greyish felted hairs. Fruit with long hairs; styles angled. *Himalaya.* H2. Summer.

27. S. cana Waldstein & Kitaibel. Illustration: Schneider, Illustriertes Handbuch der Laubholzkunde **1**: 451, 455 (1905); Reichenbach, Icones Florae Germanicae et Helveticae **24**: pl. 148 (1909).

Dense shrub, to 1 m, rarely 2.5 m, branches downy, slender, round, buds small. Leaves 1–2.5 cm, elliptic-oblong, acute, mucronate, entire or sometimes toothed at the apex, downy grey on both sides, thicker below, silky when young. Flowers white, *c.* 6 mm, in downy, dense umbel-like corymbs on leafy stalks. Petals round, as long as the stamens. Fruit downy; styles spreading; sepals reflexed. *SE Europe*. H2. Early summer.

28. S. × cinerea Zabel.

Densely branched shrub, 1.5 m, sometimes more, downy, brown, striped, angular. Leaves 2.5–3.5 cm, oblong, acuminate, entire or with 1 or 2 teeth at the apex with short recurved leathery tips, grey-downy when young becoming grey-green above, paler below. Flowers white, *c.* 6 mm, in many-flowered small, stalkless corymbs at the apex, or on leafy stalks to *c.* 2 cm towards the base of the shoots; petals oblong rounded, longer than the stamens; styles slanting to erect, usually falling before the fruit ripens. *Garden origin.* H2. Early summer.

A hybrid between *S. cana* and *S. hypericifolia*.

29. S. thunbergii Siebold. Illustration: Gleason, Illustrated flora of the north-eastern United States and adjacent Canada **2**: 284 (1952); Csapody & Tóth, A colour atlas of flowering trees and shrubs, 96 (1982); Krüssmann, Manual of cultivated broad-leaved trees & shrubs **3**: 351 (1986); Gardening from 'Which' March, 67 (1987) – as S. thunbergii 'Kew'.

Shrub, 1–1.5 m, young shoots downy at first, eventually hairless, bright green. Leaves 2–3.5 × 0.3–0.6 cm, linear-lanceolate, acute, warty and finely toothed towards the apex, hairless, often yellow, orange-red in the autumn. Flowers 6–8 mm, white, in 3–5-flowered stalkless corymbs. Calyx shallow, smooth; petals obovate, longer than the stamens; flower-stalks *c.* 8 mm. Styles spreading. *China, Japan.* H2. Spring–early summer.

'Arguta' (*S. × arguta* Zabel). Illustration: Schneider, Illustriertes Handbuch der Laubholzkunde **1**: 450 (1905); Huxley, Deciduous garden trees and shrubs, pl. 180 (1979); Hellyer, The Collingridge illustrated encyclopedia of gardening, pl.

533 (1983); Gardening Now, August, 12 (1987). Shrub, 1–2.5 m, branches slender, nodding, downy. Leaves 2–4 × 0.5–1.5 cm, bright green, oblong-obovate or lanceolate, often doubly toothed, downy at first, becoming hairless. Flowers *c.* 8 mm, white, many, in corymbs. H2. Spring.

A hybrid between *S. thunbergii* and *S. multiflora*. Similar in habit to *S. thunbergii* but taller and more vigorous.

30. S. hypericifolia Linnaeus. Illustration: Schneider, Illustriertes Handbuch der Laubholzkunde **1**: 450, 451 (1905); Pignatti, Flora d' Italia **1**: 540 (1982); Krüssmann, Manual of cultivated broad-leaved trees & shrubs **3**: 351 (1986).

Shrub, 1–1.8 m, shoots finely downy, brown, rounded, arching. Leaves 1.5–3.5 × 1–1.4 cm, obtuse, obovate to lanceolate, entire or slightly scalloped towards the apex, grey-green above, lighter below, 3-veined at the base, slightly downy to hairless, almost stalkless. Flowers white, 3–6, in almost stalkless corymbs, if stalked then downy; petals round, longer than the stamens. Fruit nearly hairless; styles erect-recurved. *SE Europe to Siberia & C Asia.* H2. Late spring.

Subsp. **obovata** (Waldstein & Kitaibel) Huber. (*S. obovata* Waldstein & Kitaibel; *S. hypericifolia* var. *obovata* (Waldstein & Kitaibel) Maximowicz). Illustration: Schneider, Illustriertes Handbuch der Laubholzkunde **1**: 451 (1905). Leaves narrowly obovate, rounded, entire or with 3–5-scalloped teeth towards the apex. Flowers *c.* 5 mm; sepals as long as calyx cup; petals *c.* 3 mm, as long as the stamens. *SE Europe.* H2. Spring.

S. × multiflora Zabel. Shrub, *c.* 1.5 m, branches finely downy when young, brownish, slender. Leaves 2–3 × 1–1.5 cm, obovate, greyish green, paler and downy below when young, entire towards the long tapering base, veins 3–5. Flowers white, many, in usually stalkless corymbs, the lower ones usually on short stalks, leafy at the base.

A hybrid between *S. crenata* × *S. hypericifolia*.

31. S. crenata Linnaeus (*S. crenifolia* C.A. Meyer; *S. vaccinifolia* misapplied). Illustration: Schneider, Illustriertes Handbuch der Laubholzkunde **1**: 451, 455 (1905); Reichenbach, Icones Florae Germanicae et Helveticae **24**: pl. 147 (1909); Krüssmann, Manual of cultivated broad-leaved trees & shrubs **3**: 351 (1986).

Densely bushy shrub to 1.5 m, shoots downy at first becoming hairless, reddish

brown, striped, slender, round, slightly angled. Leaves 2–3.5 × 0.5–3 cm, obovate-oblong, acute or rounded, tapered at base, entire or with a few scalloped teeth towards the apex, 3-veined, greyish green, downy at first. Flowers white, *c.* 5 mm, in dense, semi-circular umbel-like corymbs to 2.5 cm, on small leafy stalks. Petals round, obovate, usually shorter than the stamens. Fruit finely downy, styles erect, enclosed by sepals. *SE Europe to Caucasus & Altai Mts.* H2. Summer.

Similar to *S. hypericifolia* which has 3-veined leaves from base but differs in its almost stalkless inflorescence.

32. S. yunnanensis Franchet (*S. sinobrahuica* W.W. Smith).

Shrub *c.* 2 m, shoots yellowish brown with dense felted hairs when young. Leaves 0.8–2 cm, ovate-rounded to obovate, doubly toothed to shallowly lobed or sometimes entire, rounded at apex, tapered towards the base, dull green and downy above, grey or white felted hairs below, stalks 2–5 mm. Flowers 1–2 cm, white, 10–20 in densely hairy corymbs on short leafy stalks. Calyx and flower-stalk downy. Petals roundish, nearly twice as long as the 20 stamens. Fruit downy. *W China (Yunnan).* H2. Summer.

33. S. gemmata Zabel (*S. mongolica* Koehne not Maximowicz). Illustration: Schneider, Illustriertes Handbuch der Laubholzkunde **1**: 451, 455 (1905); Krüssmann, Manual of cultivated broad-leaved trees & shrubs **3**: 351 (1986).

Shrub 2–3 m, almost hairless, shoots reddish, slender, angled; winter buds slender, to 5 mm, pointed, longer than the leaf-stalk. Leaves 1–2 × 0.3–0.8 cm, narrowly elliptic to oblong, entire, sometimes with 3 teeth at the apex, hairless, bright green above and below, stalk short to 2.5 mm. Flowers 6–8 mm, white, 2–6 in very short or stalkless corymbs *c.* 2.5 cm; petals almost round, longer than the stamens. Fruit hairless; sepals spreading; style erect or spreading. *NW China, Mongolia.* H1. Summer.

34. S. calcicola W.W. Smith. Figure 29(8), p. 334.

Shrub 0.7–1.5 m, shoots reddish, hairless, slender, arching. Leaves 4–7 × 3–4 mm, ovate to elliptic, obtuse, entire, rounded at apex, tapered at base, hairless (leaves of strong shoots to 1.3 cm, semicircular, deeply 3-lobed, lobes coarsely and doubly scallop-toothed, truncate at base). Weaker shoots obovate-tapered, deeply toothed at apex. Flowers white tinged reddish on the outside, 6–8 in corymbs, stalks 2–5 mm, forming an inflorescence of 10–13 cm. Calyx-lobes hairless. Fruit hairless with erect styles. *W China (Yunnan)*. H2. Summer.

35. S. baldschuanica B. Fedtschenko.

Dwarf shrub, rounded compact habit, densely branched, thin, hairless. Leaves 0.8–1.5 cm, obovate, bluish green, toothed at the apex. Flowers white in small erect hairless corymbs; flower-stalks often with a bluish grey bloom. *SE former USSR*. H3. Summer.

36. S. arcuata J.D. Hooker. Illustration: Schneider, Illustriertes Handbuch der Laubholzkunde **1**: 462 (1905); Krüssmann, Manual of cultivated broad-leaved trees & shrubs **3**: 351 (1986).

Shrub with short arching branches; stems reddish brown, strongly ribbed, downy when young, eventually almost hairless. Leaves 0.8–1.5 cm, oblong-elliptic, slightly downy, entire or occasionally with a few small teeth towards apex. Flowers *c.* 6 mm, white occasionally pink, 10–15 in downy 2–4 cm umbel-like corymbs, at the end of short leafy stalks; stamens 18–25, slightly longer than petals. Fruit hairless. *Himalaya (Nepal) to SW China*. H2. Early summer.

37. S. × schinabeckii Zabel.

Shrub 1.2–1.8 m, shoots with scattered hairs, brownish yellow, striped towards the apex, flexuous, base 5-sided. Leaves 4.5–5 cm, oblong to ovate, hairless, doubly to deeply toothed, dark green above, bluish below. Flowers white, large, in corymbs; petals rounded, longer than the stamens. Styles almost curving outwards. *Garden origin*. H2. Summer.

A hybrid between *S. chamaedryfolia × S. trilobata*. Similar to *S. chamaedryfolia* but flowers larger.

38. S. veitchii Hemsley. Illustration: Botanical Magazine, 8383 (1911); Schneider, Illustriertes Handbuch der Laubholzkunde **2**: 960 (1912).

Shrub, to 4 m, strong-growing, branches slightly downy, striped, reddish, 0.5–1 m, arching. Leaves 2–5 × 0.8–2 cm, elliptic to oblong, occasionally obovate, obtuse, entire, base broadly tapered, hairless, or downy and glaucous below, stalks *c.* 2 mm. Flowers 4–5 mm, white, in finely downy, densely packed corymbs, 3–6 cm; petals shorter than the stamens; calyx and stalks finely downy. Fruit hairless. *C & W China*. H2. Summer.

39. S. virginiana Britton. Illustration: Gleason, Illustrated flora of the northeastern United States and adjacent Canada **2**: 286 (1952).

Shrub, 1 m or taller, hairless, with many branches. Leaves 2–5 × 1.5–1.8 cm, oblong-lanceolate, entire or with a few teeth towards the apex, acute, rounded to tapered at base, paler to glaucous below, hairless or finely hairy on veins beneath. Flowers white, many in slightly downy or hairless corymbs to 5 cm. Fruit hairless. *E USA*. H2. Summer.

Similar to *S. betulifolia* var. *corymbosa* but with leaves entire or with a few teeth and glaucous below.

40. S. pubescens Turczaninow. Illustration: Schneider, Illustriertes Handbuch der Laubholzkunde **1**: 455, 459 (1905).

Shrub, 1–2 m, branches densely felted, downy hairy when young, arching, slender, rounded. Leaves 3–4 cm, ovate to diamond-shaped to elliptic, downy above with grey felted hairs below, deeply toothed to almost 3-lobed, stalks 2–3 mm. Flowers 6–8 mm, white, in hairless almost semi-circular corymbs. Petals as long as the stamens. Fruit hairless, styles spreading. *N China*. H2. Summer.

41. S. sargentiana Rehder.

Shrub, to 2 m, shoots downy when young, arching, round, slender, spreading, buds ovoid to obtuse with several scales. Leaves 2–4 × 0.4–1.5 cm, narrowly elliptic to obovate, finely toothed at the apex, entire at base, dull green and finely downy above, paler with shaggy hairs below, stalks 1–3 mm. Flowers *c.* 6 mm, creamy white, in dense, shaggy numerous corymbs 2.5–4 cm; petals as long as stamens. Fruit almost hairless; styles spreading. *W China*. H2. Summer.

42. S. trichocarpa Nakai. Illustration: Nakai, Flora sylvatica Koreana **4**: pl. 12 (1916); Bean, Trees & shrubs hardy in the British Isles **3**: 464 (1933).

Shrub, 1–2 m, shoots hairless, slender, rigid, angular, spreading. Leaves 2.5–6 × 1–2.5 cm, oblong to lanceolate, hairless or with scattered hairs, green, almost acute, entire or toothed near the apex, tapered at the base, stalk to 6 mm. Flowers *c.* 8 mm, white, in downy corymbs, 2.5–5 cm, at the end of short leafy twigs on the previous year's growth, forming an inflorescence over 30 cm, with the lower branches having 2–7 flowers, and often a leafy bract, stalks downy. Petals rounded,

notched. Fruit downy. *Korea*. H2. Summer.

43. S. media Schmidt (*S. confusa* Regel & Koernicke). Illustration: Bean, Trees & shrubs hardy in the British Isles **4**: 487 (1980); RHS dictionary of gardening **4**: 2002 (1981); Krüssmann, Manual of cultivated broad-leaved trees & shrubs **3**: 351 (1986).

Upright shrub, 1–1.5 m, smooth, round branches, downy when young, sometimes hairless, yellow to brownish. Leaves 3–5 × 0.8–2 cm, ovate-oblong, base wedge-shaped, deeply toothed towards the apex, occasionally entire, bright green on both surfaces, more or less downy below, margins with long white hairs, stalks to 4 mm. Flowers *c.* 8 mm, white, many, in hairless or almost hairless racemes at the ends of small leafy side-shoots or sometimes in groups of small panicles. Petals round, shorter than the stamens. Fruit downy; sepals reflexed; styles spreading or reflexed. *E Europe to NE Asia*. H2. Spring–early summer.

Subsp. **polonica** (Blocki) Pawloski (*S. polonica* Blocki). Leaves 5–6 × 2–3 cm; inflorescence covered with soft down, petals yellowish, margins fringed. *Poland, former Czechoslovakia*. H2. Summer.

S. × pikoviensis Besser (*S. nicoudiertii* misapplied). Illustration: Reichenbach, Icones Florae Germanicae et Helveticae **24**: pl. 149 (1909). Closely related to *S. crenata* but shoots hairless, yellowish brown, round. Leaves 2.5–5 cm, oblong, entire or with a few teeth at the apex, very slightly downy below. Flowers white, many, in almost hairless corymbs, stalks 1.5–3 cm. Petals round, shorter than the stamens. Fruit downy; styles usually straight, erect. *W Ukraine, former USSR*. H2. Late spring.

A hybrid between *S. crenata* and *S. media*.

44. S. henryi Hemsley. Illustration: Botanical Magazine, 8270 (1909); RHS dictionary of gardening **4**: 2002 (1956); Huxley, Deciduous garden trees and shrubs pl. 181 (1979); Bean, Trees & shrubs hardy in the British Isles **4**: 480 (1980).

Shrub, 2–3 m, open spreading habit, shoots roundish, reddish brown, downy when young. Leaves 3–8 × 2–3 cm, oblanceolate, acute or rounded, coarsely toothed at apex, entire on the smaller leaves, hairless or almost downy above, densely hairy below, stalks 4–6 mm. Flowers white, *c.* 6 mm, in rounded corymbs *c.* 5 cm on the end of short leafy shoots; petals round, longer than the stamens; stalk and calyx downy. Fruit hairy, slightly spreading. *C & W China*. H2. Summer.

45. S. wilsonii Duthie. Illustration: Botanical Magazine, 8399 (1911); Schneider, Illustriertes Handbuch der Laubholzkunde **2**: 960 (1912).

Shrub 2–2.5 m, shoots downy when young, reddish purple, arching. Leaves 2–5.5 cm, elliptic-obovate, entire or with a few coarse teeth at the apex, base tapered, dull green and downy above, greyish green with longer and denser hairs below especially on the veins, very short stalked. Flowers *c.* 6 mm, white, many, in hairless, semi-circular, terminal corymbs 3–5 cm, on short leafy side-shoots, stalks hairless. Petals as long as stamens; calyx hairless. Fruit spreading, downy. *C & W China*. H2. Summer.

Differs from *S. henryi* in having hairless ovaries and hairless or slightly silky flower-stalks. Leaves of flowering-shoots entire, downy above and duller green.

46. S. mollifolia Rehder. Illustration: Krüssmann, Manual of cultivated broad-leaved trees & shrubs **3**: 356 (1986).

Shrub 1.5–2 m, branches reddish purple, hairy when young, arching. Leaves 2–3 × 0.5–1 cm, elliptic-obovate, tapered at both ends usually more so at the apex, usually entire but sometimes with a few teeth towards the apex, with a few grey silky hairs on both sides, light greyish green beneath, short-stalked. Flowers *c.* 8 mm, white, in downy corymbs to 2.5 cm, corymbs with short leafy stalks; stamens 20. Fruit downy; sepals spreading; styles erect or spreading. *W China*. H2. Summer.

47. S. prunifolia Siebold & Zuccarini (*S. prunifolia* var. *plena* Schneider). Figure 29(7), p. 334. Illustration: Schneider, Illustriertes Handbuch der Laubholzkunde **1**: 450, 451 (1905); Gleason, Illustrated flora of the north-eastern United States and adjacent Canada **2**: 284 (1952); Krüssmann, Manual of cultivated broad-leaved trees & shrubs **3**: 356 (1986).

Upright dense shrub, to 2–3 m, branches downy at first, arching, nodding, slender. Leaves 2.5–4.5 × 1–2 cm, ovate to oblong-elliptic, acute at both ends, finely toothed, bright green, shiny above, soft downy grey below especially when young, stalk to 5 mm, orange to reddish brown in the autumn. Flower-stalks 1.5–2 cm, slender, hairless, flowers *c.* 1 cm, white, double with numerous broadly-ovate petals, 3–6 in stalkless corymbs. Petals longer than

stamens. Fruit hairless, spreading. *Japan, China in cultivation but also found wild in Hubei*. H2. Spring.

Only the double variant is found in cultivation. The single-flowered variant is *S. prunifolia* forma *simpliciflora* Nakai. *Korea, China, Taiwan*.

48. S. bella Sims (*S. expansa* Wallich; *S. fastigiata* Wallich). Illustration: Botanical Magazine, 2426 (1823); Schneider, Illustriertes Handbuch der Laubholzkunde **1**: 468, 474 (1905); Bean, Trees & shrubs hardy in the British Isles **4**: 471 (1980); Krüssmann, Manual of cultivated broad-leaved trees & shrubs **3**: 348 (1986).

Shrub to 1 m, occasionally to 1.8 m, branches slender, spreading, angular, downy when young. Leaves 2.5–5.5 cm, ovate-elliptic to lanceolate, acute, rounded to broadly tapered at base, doubly toothed towards the apex, hairless on both surfaces, or sometimes downy on the veins below, glaucous beneath, stalks to 0.6 cm. Flowers dioecious, pink to white, *c.* 6 mm, in downy, loose, terminal 2–4 cm corymbs on previous year's growth. In male flowers stamens longer than the petals, in female flowers stamens shorter. Sepals reflexed; styles spreading. *Himalaya*. H2. Summer.

49. S. cantoniensis Loureiro (*S. reevesiana* Lindley). Illustration: Schneider, Illustriertes Handbuch der Laubholzkunde **1**: 459, 464 (1905); Csapody & Tóth, A colour atlas of flowering trees and shrubs, 96 (1982) – as S. cantoniensis 'Lanceata'.

Shrub to 1.8 m, hairless with slender, arching, round branches. Leaves 2.5–5.5 × 1.3–1.8 cm, diamond-shaped to lanceolate, 3-lobed or finely toothed, dark green above, glaucous below with distinct net-like venation, stalks 6–8 mm. Flowers white, *c.* 1 cm, many in semi-circular 2.5–5 cm corymbs on leafy stalks; petals rounded to elliptic, longer than the stamens. Fruit with sepals erect; styles spreading. *China, Japan*. H2. Summer.

50. S. longigemmis Maximowicz. Illustration: Bailey, The standard cyclopedia of horticulture **3**: 3212 (1935); Krüssmann, Manual of cultivated broad-leaved trees & shrubs **3**: 352 (1986).

Shrub 1–1.5 m, branches hairless when young, thin, spreading, angular. Leaves 3–6 cm, broadly ovate, distinctly wider at the base, acute, toothed to doubly toothed, glandular-tipped, bright green, hairless, sometimes hairy on the veins below. Flowers *c.* 6 mm, white, in loose, downy 5–7 cm corymbs; stalk and calyx hairy;

petals shorter than the prominent stamens. Fruit almost hairless; style spreading. *NW China*. H2. Summer.

51. S. miyabei Koidzumi (*S. silvestris* Nakai). Illustration: Nakai, Flora sylvatica Koreana **4**: t. 5 (1916).

Upright shrub, 1–1.2 m, erect branches, downy when young, slightly angular; buds ovoid, small with several scales. Leaves 3–6 cm, ovate to elliptic, acute, rounded or tapering to rounded at base, doubly toothed, green and hairless or almost so on both sides, stalks 2–5 mm. Flowers *c.* 8 mm, white, in downy flat corymbs, 3–6 cm across. Petals almost round; stamens 2–3 times longer than the petals. Fruit with dense felted hairs; styles spreading. *Japan*. H2. Summer.

Var. **glabrata** Rehder. Differs in its larger leaves which are wedge-shaped at the base and flowers borne in hairless clusters. *C China*. H2. Summer.

52. S. amoena Spae (*S. expansa* K. Koch). Illustration: Schneider, Illustriertes Handbuch der Laubholzkunde **1**: 474 (1905).

Upright shrub to 2 m, stems downy, rounded, slender, few branches; buds hairy. Leaves to 10 × 3 cm, broadly-ovate, truncate at base, acute to acuminate, rounded to tapered, with fine even teeth, both simply and doubly toothed, usually only on the upper half, dark green above, sometimes hairless, glaucous, downy and hairy on the veins beneath, stalks 0.5–0.8 cm. Flowers dioecious, white to slightly pinkish red, in flat compound corymbs 5–20.5 cm borne on the current year's growth. Fruit hairy. *NW Himalaya*. H2. Summer.

53. S. rosthornii Pritzel.

Shrub, to 2 m, shoots downy at first, spreading. Leaves 3–8 × 2–4 cm, broadly-ovate, acuminate, tapered or almost rounded at base, deeply and doubly toothed to slightly lobed, bright green above, often with soft downy hairs on both sides, especially on the veins below, stalks 5–8 mm. Inflorescence downy, flowers *c.* 6 mm, white, on long-stalked corymbs 5–9 cm across, at the end of leafy shoots, petals shorter than the stamens. Fruit downy; styles spreading. *W China*. H2. Summer.

54. S. × blanda Zabel (*S. reevesiana* misapplied). Illustration: Schneider, Illustriertes Handbuch der Laubholzkunde **1**: 464 (1905).

Shrub, 1.5–2 m, branches finely downy, brown, striped, angular, almost straight.

Leaves 2.5–6 cm, ovate-oblong, acute, scalloped to sharply toothed, hairless above, grey dense felted hairs below. Flowers white, in downy corymbs; petals round, longer than the stamens. Fruit hairless; styles reflexed. *Garden origin*. H2. Spring–summer.

A hybrid between *S. cantoniensis* × *S. chinensis*.

55. S. × vanhouttei (Briot) Zabel (*S. aquilegiifolia* Briot var. *vanhouttei*). Illustration: Schneider, Illustriertes Handbuch der Laubholzkunde **1**: 464 (1905); Gleason, Illustrated flora of the north-eastern United States and adjacent Canada **2**: 285 (1952); Hellyer, The Collingridge illustrated encyclopedia of gardening pl. 535 (1983); Gartenpraxis **8**: 8 (1989).

Shrub, to 2 m, branches, hairless, rod-like, long, arching. Leaves 1.5–4 × 1.5–3 cm, diamond-shaped to obovate, dark green above, bluish below, hairless, scalloped or coarsely toothed and occasionally 3–5-lobed at the apex. Flowers *c.* 8 mm, white, many, in flat corymbs 2.5–5 cm across; petals round, twice as long as the partly sterile stamens. Fruit spreading; styles half-erect; sepals spreading. *Garden origin*. H2. Summer.

A hybrid between *S. cantoniensis* × *S. trilobata*.

56. S. chinensis Maximowicz (*S. pubescens* Lindley not Turczaninow). Illustration: Schneider, Illustriertes Handbuch der Laubholzkunde **1**: 455, 459 (1905).

Dense shrub, to 1.5 m, shoots with yellowish felted hairs when young, stems nodding, angular. Leaves 2.5–5 × 1.5–3.5 cm, ovate-rhombic to obovate, acute to rounded, rounded or broadly tapered at base, deeply toothed, glandular-tipped, occasionally 3-lobed, dark green and finely downy above, yellowish densely felted hairs below; stalks 2–8 mm. Flowers white, *c.* 1 cm, in densely hairy, many-flowered, 2.5–5 cm corymbs. Sepals ovate-lanceolate, spreading at first, becoming erect. Stamens shorter or as long as petals. Fruit downy; styles spreading. *NE China*. H2. Summer.

57. S. chamaedryfolia Linnaeus (*S. flexuosa* Fischer). Illustration: Schneider, Illustriertes Handbuch der Laubholzkunde **1**: 455, 459 (1905); Gleason, Illustrated flora of the north-eastern United States and adjacent Canada **2**: 285 (1952); Pignatti, Flora d' Italia **1**: 540 (1982); Krüssmann,

Manual of cultivated broad-leaved trees & shrubs **3**: 351, 357 (1986).

Shrub to 2 m, with stolons; branches hairless, yellowish, angled, flexuous. Leaves 2–7 × 2–3 cm, ovate-elliptic to oblong-lanceolate, acute, tapered, deeply toothed, bright green, usually hairless, occasionally with few scattered hairs below, stalk 5–10 mm. Flowers white, *c.* 1 cm, in corymbs of *c.* 4 cm, borne at the ends of short leafy shoots, occasionally stalkless on the upper shoots. Sepals reflexed, petals round, shorter than the stamens. Fruit slightly downy, outspread, styles spreading. *NE Asia*. H2. Summer.

Var. **ulmifolia** (Scopoli) Maximowicz (*S. ulmifolia* Scopoli). Illustration: Schneider, Illustriertes Handbuch der Laubholzkunde **1**: 455 (1905); Reichenbach, Icones Florae Germanicae et Helveticae **24**: pl. 150 (1909). Taller shrub, shoots less spreading, not as stiff. Leaves ovate, usually rounded at base, doubly toothed. Many-flowered semi-circular inflorescence *c.* 8 cm with longer stalks. Fruit with upright styles. *SE Europe to NE Asia & Japan*. H2. Summer.

Var. **flexuosa** (Fischer) Maximimowicz (*S. flexuosa* Fischer). Illustration: Schneider, Illustriertes Handbuch der Laubholzkunde **1**: 459 (1905). Smaller shrub, stems more angled or with wings; leaves smaller and narrower, simply toothed towards the apex or sometimes entire. Fewer flowers. *Siberia*. H2. Summer.

S. × gieseleriana Zabel. Illustration: Krüssmann, Manual of cultivated broad-leaved trees & shrubs **3**: pl. 129 (1986). Shrub, 1.5–3 m, downy grey, shoots obtuse, angular, somewhat flexuous. Leaves 3–4 × 1–2 cm, ovate, acute, downy, sharply toothed on the upper ⅔, sometimes entire. Flowers *c.* 0.8 mm; petals as long as or shorter than the stamens. H2. Summer.

A hybrid between *S. cana* and *S. chamaedryfolia*.

S. × nudiflora Zabel (*S. hookeri* Zabel). Differs in having inflorescences to 8 cm with flowers in compound or sometimes simple, downy corymbs and fruits with reflexed sepals.

A hybrid between *S. chamaedryfolia* var. *ulmifolia* and *S. bella*.

13. SIBIRAEA Maximowicz

J. Cullen

Deciduous, prostrate to erect shrubs to 1 m; young shoots dark reddish, hairless. Leaves alternate (sometimes appearing whorled on condensed lateral shoots), linear-oblong to narrowly obovate, shortly

stalked, obtuse but slightly mucronate, 3–10 × 0.6–2 cm, slightly hairy beneath and on the margins. Stipules absent. Flowers in racemes clustered at the tips of the shoots, to 6 mm across, mostly functionally unisexual (plants effectively dioecious). Sepals 5, to 1 mm, triangular, erect. Petals white to greenish, 2–2.5 mm, almost circular but shortly clawed. Stamens numerous, borne on the margins of a cup-shaped, indented nectar-secreting disc. Ovary superior. Carpels 5, joined only at their extreme bases. Fruit a group of 5, 2-seeded follicles.

A genus of a single species with a remarkable distribution, centred in E Asia but with disjunct localities in SE Europe. The plant is easily grown and is propagated by seed or cuttings.

1. S. laevigata (Linnaeus) Maximowicz (*S. altaiensis* (Laxmann) Schneider; *Spiraea laevigata* Linnaeus; *S. altaiensis* Laxmann). Illustration: Csapody & Tóth, A colour atlas of flowering trees and shrubs, 99 (1982).

Former USSR (Siberia), China, former Yugoslavia. H3. Summer.

The correct name of the species is uncertain, both names cited above having been originally published within a very short time of each other. Variants with narrow, oblong-linear leaves have been described as var. **angustata** Rehder (E Asia). The European populations have been treated as var. **croatica** (Degen) Schneider, but are not in any way distinctive.

14. PETROPHYTUM (Torrey & Gray) Rydberg

J. Cullen

Small, low, hummock-forming, shrubs with dense mats of evergreen basal leaves. Stems bearing reduced, bract-like leaves. Inflorescence a very dense, spike-like raceme. Flowers with bracts, sepals, petals and stamens perigynous, perigynous zone top-shaped to hemi-spherical, lined with a nectar-secreting disc. Sepals 5. Petals 5, creamy white. Stamens 20–40. Ovary superior of usually 5 (more rarely 3 or 7) free carpels, each containing 2–4 ovules, style long. Fruit a group of few-seeded follicles.

A genus of 3 species from N America. They are small shrubs which grow best in crevices of rock or walls, as they are prone to injury from too much water. Propagation is by seed.

1a. Leaves 1-veined   **1. caespitosum**

b. Leaves 3-veined   2
2a. Leaves ash-grey; follicles *c.* 3 mm
3. cinerascens
b. Leaves green; follicles to 2 mm
2. hendersonii

1. P. caespitosum (Nuttall) Rydberg
(*Spiraea caespitosa* Nuttall). Illustration:
Schneider, Illustriertes Handbuch der
Laubholzkunde **1**: 487 (1905); Rickett,
Wild flowers of the United States **5**: pl. 47
(1973).

Basal leaves oblanceolate, tapering
slightly towards the base, 7–15 × 1.5–4
mm, with a single vein visible, densely
silvery silky. Stems to 20 cm; stem-leaves
narrowly oblanceolate. Racemes 1.5–4
cm, dense. Sepals erect, densely hairy, *c.*
2 mm. Petals oblong-oblanceolate, 2–2.5
mm. Stamens 20, exceeding the petals.
Carpels densely hairy. Follicles *c.* 2 mm. *W
USA, extending E to Arizona and Texas.* H4.
Summer.

2. P. hendersonii (Canby) Rydberg.
Illustration: The Rock Garden, **20**(79): 190
(1987).

Basal leaves oblanceolate to obovate,
1–2.5 × 0.2–0.6 cm, with 3 veins visible
beneath, hairy or hairless. Stems to 20
cm, stem-leaves narrowly obovate. Spikes
dense, 2–4 cm. Sepals with marginal hairs,
reflexed. Petals oblong to oblong-obovate,
2–2.5 mm. Stamens 35–40, as long as or
slightly longer than petals. Carpels hairy.
Follicles to 2 mm. *NW USA (Washington,
Olympic Mts).* H5. Summer.

3. P. cinerascens (Piper) Rydberg.
Dense shrublet; shoots short, stout.
Leaves to 2.5 cm, oblanceolate, obtuse or
almost acute, thick, leathery, ash-grey with
3 veins, loosely hairy. Inflorescence-stem
to 15 cm. Flower-stalks to 4 mm. Sepals
2 mm, lanceolate, ash-grey. Petals *c.* 2
mm, spathulate or oblanceolate, obtuse.
Follicles *c.* 3 mm, sparsely hairy. *NW USA
(Washington).* H5. Summer–autumn.

15. KELSEYA (Watson) Rydberg
M.F. Gardner
Evergreen, cushion-forming, tufted
shrublets, 5–8 × 7–8 cm. Leaves entire,
2.5–4 mm, leathery, densely overlapping,
covered with fine silky hairs. Flowers
solitary, pinkish to purplish, fading to
brown, nearly stalkless; sepals 5; petals
oblong-elliptic, 2–3 mm; stamens 10,
longer than petals, reddish purple; styles 5.
Fruit a follicle, *c.* 3 mm, splitting
completely at 2 seams.

A genus of a single species from W
North America, best grown in the alpine

house, but also thriving out of doors in a
trough or sink. It thrives if grown jammed
between two pieces of rock in a well-
drained gritty soil. Propagation is usually
from seed.

1. K. uniflora (Watson) Rydberg.
Illustration: Quarterly Bulletin of the
Alpine Garden Society **31**: 358 (1963);
Everett, New York Botanical Garden
illustrated encyclopedia of horticulture **6**:
1893 (1981); Hitchcock & Cronquist, Flora
of the Pacific Northwest, 214 (1987).
W North America. H4. Spring–summer.

16. LUETKEA Bongard
J. Cullen
Low, evergreen shrubs with creeping
woody stems and erect, herbaceous,
sparsely hairy flowering-shoots, to 20 cm.
Leaves to 2 cm, mostly towards the bases
of the flowering-stems, alternate, stalked,
usually divided into 3 narrow segments
which are themselves deeply 3-lobed, the
ultimate segments more or less linear;
leaves on the flowering stems smaller and
less divided. Flowers in terminal racemes,
5–8 mm across; stalks short. Sepals 5,
triangular, acute. Petals white, *c.* 4 mm,
exceeding the sepals. Stamens 20,
somewhat united towards their bases,
attached to the margins of a 10-lobed,
fleshy nectar-secreting disc. Ovary
superior, of 5 free carpels. Fruit a group of
5 follicles, each containing several seeds.

A genus of a single species from W
North America. It may be grown in a
sunny, well-drained position and is
propagated by seed.

1. L. pectinata (Torrey & Gray) Kunze
(*Spiraea pectinata* Torrey & Gray).
Illustration: Rickett, Wild flowers of the
United States **5**: pl. 47 (1971).
W North America. H4. Spring–summer.

17. ARUNCUS Linnaeus
J. Cullen
Large herbs with rather thick, woody
rhizomes, dioecious or almost so. Leaves
alternate, arising from swollen nodes,
divided into 3 (rarely 4 or 5) leaflets which
themselves are similarly divided or more
commonly pinnate, the ultimate segments
irregularly doubly toothed. Flowers
unisexual, in spike-like racemes which
form a panicle (rarely a simple raceme).
Sepals 5, small. Petals white or cream,
slightly exceeding the sepals. Stamens
numerous, small and vestigial in female
flowers, attached at the margins of a cup-
shaped, entire nectar-secreting disc. Carpels

3 (rarely more), free, absent in male
flowers. Fruit a group of follicles.

A genus of a few species from the
northern hemisphere. They are similar in
general appearance to species of *Astilbe*
(Saxifragaceae, p. 244) and the two genera
are sometimes confused; they can be most
easily distinguished by the flowers of
Aruncus having many stamens, those
of *Astilbe* 4–10. They are easily grown,
requiring rather damp, rich soil to do well,
and are propagated by seed or by division
of the rhizome.

1a. Flowers with conspicuous bracts
3. aethusifolia
b. Flowers without bracts   2
2a. Plant to 2 m or more; racemes to 8
cm   **1. dioicus**
b. Plant to 30 cm; racemes to 3 cm
2. parvulus

1. A. dioicus (Walter) Fernald (*A.
sylvestris* Kosteletzky; *A. vulgaris*
Rafinesque; *Spiraea aruncus* Linnaeus).
Illustration: Justice & Bell, Wild flowers
of North Carolina, 93 (1968); Polunin,
Flowers of Europe, pl. 44 (1969); Hess,
Landolt & Hirzel, Flora der Schweiz **2**: 313
(1970).

Stems erect, unbranched, to 2 m or
more, hairless. Leaves long-stalked, divided
into 3 leaflets which are themselves
pinnate with 5 or more segments (the
lowermost of which may also be somewhat
pinnately divided); ultimate segments
ovate-oblong, acuminate, irregularly
doubly toothed, bright green, sparsely
hairy on the veins beneath. Racemes
spreading more or less at right angles to
the panicle-axis, to 8 cm, axis with dense,
crisped hairs. Flowers without bracts, to
5 mm across. Carpels usually 3. *Northern
hemisphere.* H1. Summer.

A very variable species which has been
divided into several varieties (sometimes
recognised as species); the most commonly
seen, apart from the species as described
above is var. **triternata** (Maximowicz)
Hara, with the leaflets divided into 3
segments (not pinnate) and 'Kneiffii' which
has the ultimate segments deeply cut.

2. A. parvulus Komarov.
Similar to *A. dioicus*, but smaller, 15–30
cm; with shorter racemes to 3 cm. *E former
USSR.* H1. Summer.

3. A. aethsusifolius (Léveillé) Nakai.
Smaller, leaf-segments themselves deeply
lobed, dark green above, racemes upright,
flowers with conspicuous bracts. *Korea.* H3.
Summer.

Very *Astilbe*-like, and doubtfully in cultivation.

18. HOLODISCUS Maximowicz
J. Cullen

Shrubs or small trees to 8 m. Leaves alternate, deciduous, simple, toothed and/or lobed, usually whitish or greyish hairy at least beneath. Stipules absent. Inflorescence a large terminal panicle with many flowers, lateral branches spreading or drooping; axis, branches and calyces densely hairy. Flowers small, shortly stalked, each subtended by a bract. Sepals 5, triangular. Petals white, almost circular, shortly clawed, spreading. Stamens 20, borne on the margins of an entire, cup-shaped disc. Ovary superior, of 5 free carpels, usually hairy, each containing 2 ovules. Fruit a group of 5 indehiscent, 1-seeded achenes.

A genus of 8 species from W America, from Canada to Mexico. They grow well in sunny, open positions and may be propagated by seed, semi-ripe hardwood cuttings or by layering.

Literature: Ley, A., A taxonomic revision of the genus Holodiscus (Rosaceae), *Bulletin of the Torrey Botanical Club* **70**: 275–288 (1943).

1a. Leaves with deep major teeth or even lobes which are themselves toothed; blade of leaf truncate, rounded or somewhat tapered to base, distinct from the obvious stalk which is 0.5–2.5 cm   **1. discolor**
   b. Leaves simply toothed; blade very gradually tapered to base, stalk indistinct, less than 2 mm   2
2a. Leaves toothed only at the rounded apex   **3. microphyllus**
   b. Leaves toothed almost to the base   **2. dumosus**

1. H. discolor (Pursh) Maximowicz (*Spiraea discolor* Pursh). Illustration: Bulletin of the Torrey Botanical Club **70**: 287 (1943); Muller, Wild flowers of Santa Barbara, unnumbered pages (1958); Thomas, Flora of the Santa Cruz mountains, 198 (1961); Csapody & Tóth, A colour atlas of flowering trees and shrubs, 99 (1982).

Arching or spreading shrub to 6 m; bark reddish brown to grey or almost black, peeling; young shoots downy. Leaves ovate to elliptic, deeply toothed or even lobed, the lobes or major teeth themselves with teeth; base rounded, truncate or somewhat tapered, but clearly distinct from the stalk; blades 3–9 × 2–8 cm, sparsely hairy to

hairless above, whitish hairy beneath; stalk 0.5–2.5 cm. Flowers to 5 mm across; petals white, to 2 mm, hairy towards the base outside. *W North America, from British Columbia to Baja California.* H3. Summer.

2. H. dumosus (Nuttall) Heller (*H. discolor* var. *dumosus* (Nuttall) Dippel). Illustration: Bulletin of the Torrey Botanical Club **70**: 287 (1943).

Widely spreading shrub to 3 m; bark of older shoots reddish, becoming grey and peeling; young shoots pale, very downy. Leaves 1–3 × 0.5–2 cm, ovate to spathulate, densely whitish hairy beneath, toothed to beyond half their length from the apex, teeth rarely themselves toothed, hairy above, base tapered very gradually; stalk absent or inconspicuous. Flowers to 5 mm across. Petals white, hairy towards the base outside. *W USA & NW Mexico.* H5. Summer.

Variable and not always clearly distinguishable from *H. discolor*.

3. H. microphyllus Rydberg. Illustration: Bulletin of the Torrey Botanical Club **70**: 287 (1943).

Generally a small spreading shrub, occasionally to 2 m; bark of older shoots reddish becoming grey or black and peeling; young shoots pale, hairy, sometimes glandular. Leaves obovate to spathulate, 0.5–1.8 × 0.3–1 cm, densely greyish downy on both surfaces, toothed only near the rounded apex, long-tapered at the base, without a distinct stalk. Petals 1.5–2 mm, white with a few hairs towards the base outside. *W USA, Mexico (Baja California).* H5. Summer.

19. FILIPENDULA Miller
J. Cullen

Perennial herbs (often large) with tuberous roots or with rhizomes. Basal and lower stem-leaves pinnate with 3–5 or more major leaflets, often with smaller leaflets or lobes between them; terminal leaflet usually larger than the laterals, 3–5-lobed, the laterals lobed or not, all toothed. Inflorescence a large panicle with many flowers, branches arching erect or spreading. Flower-buds more or less spherical; flowers without bracts. Sepals 5 or 6, small, erect in flower, persistent and reflexed in fruit. Petals 5 or 6, white, cream, pink or purplish red, spreading. Stamens 20–40, attached at the margin of a small, cup-shaped nectar-secreting disc. Carpels 5–10, free, sometimes compressed, 2-seeded, usually attached by their bases but in 1 species attached on their inner

faces, with part of the carpel projecting below the point of attachment. Fruit a group of achenes, erect and separate, or (in the species with laterally attached carpels) those of each flower spirally coiling together.

A genus of about 10 species of herbs, some of them very large, from north temperate regions, though with a concentration of species in E Asia. The name 'Spiraea palmata' has been used by different authors for several of the species, and material under this name (or under *F. palmata*) generally requires re-identification. The species are generally easily grown, requiring moist, rich soil, and are propagated by seed or by division.

Roots. Tuberous: **7**.
Basal and lower stem-leaves. Pinnate with many, more or less equal, pinnately lobed or toothed leaflets: **7**; pinnate with the terminal leaflet larger than the 3–5-lobed lateral leaflets: **5,6**; pinnate with the terminal leaflet larger than the unlobed lateral leaflets: **1–4**. Usually white- or greyish hairy beneath: **1,2,4, 5**.
Sepals. With a small patch of sparse hairs inside: **1,3,6**.
Petals. White or cream: **1,2,3** (an uncommon cultivar), **4,5,7**; pink or purplish red: **3,6**. Blade tapering gradually into an indistinct claw: **1,2,7**; blade abruptly tapered or truncate into a distinct claw: **4–6**; blade deeply truncate or even auriculate at junction of blade and distinct claw: **3**.
Achenes. Flattened, hairy on the keels: **1–3**; not flattened, hairy all over: **7**; neither flattened nor hairy: **4–6**. Attached above the base, all those of each flower coiling spirally as they mature: **4**.

1a. Leaves with numerous leaflets all more or less similar in size; carpels hairy all over the surface, not flattened; roots tuberous   **7. vulgaris**
   b. Leaves with rarely more than 5 leaflets, the terminal larger than the laterals; carpels compressed, with long hairs on the keels only, or neither compressed nor hairy   2
2a. Lateral leaflets of the basal and lower stem-leaves 3–5-lobed   3
   b. Lateral leaflets of the basal and lower stem-leaves not lobed   4
3a. Petals pink or purplish red; leaves green beneath, hairs only on the veins   **6. rubra**

b. Petals white or cream; leaves densely white-hairy over the whole surface beneath **5. palmata**

4a. Carpels hairless, not compressed, attached above the base, achenes of each flower coiling spirally as they mature **4. ulmaria**

b. Carpels compressed with hairs along the keels, attached at their bases, not coiling spirally **5**

5a. Stems hairless; petals pink or pinkish red (white in an uncommon cultivar), the blade deeply and abruptly truncate or even auriculate at the junction with the distinct claw **3. purpurea**

b. Stems variably hairy; petals white or cream, the blade gradually tapered into the indistinct claw **6**

6a. Plant 1.5–3.5 m; sepals with a small patch of sparse hairs inside **1. kamtschatica**

b. Plant to 80 cm; sepals entirely hairless **2. vestita**

1. F. kamtschatica (Pallas) Maximowicz (*Spiraea kamtschatica* Pallas; *S. palmata* Miquel). Illustration: Flora SSSR **10**: 287 (1941).

Very large rhizomatous perennial herb, stems 1.5–3.5 m, shortly hairy. Basal and lower stem-leaves pinnate with 3–5 leaflets, the terminal leaflet deeply 3–5-lobed, cordate at base, much larger than the unlobed lateral leaflets, all dark green and hairless above, densely or sparsely whitish hairy beneath, all irregularly doubly toothed. Stipules large, narrowly oblong, toothed. Inflorescence rather spreading, branches hairy. Sepals usually 5, hairless outside, with a small patch of sparse hairs inside. Petals usually 5, white, 3–4 mm, the blade gradually tapered into an indistinct claw. Carpels usually 5, attached at their bases, compressed with long hairs on the keels. Achenes 1–5, erect, with long hairs on the keels. *Japan, N China, E former USSR.* H3. Summer.

The largest and most spectacular of the species.

2. F. vestita (G. Don) Maximowicz.

Perennial herb to 80 cm. Stem with rather fine adpressed hairs of variable density. Basal and lower stem-leaves pinnate with 3–5 leaflets, terminal leaflet 3–5-lobed, cordate at base, much larger than the unlobed lateral leaflets, all leaflets irregularly doubly toothed, dark green and hairless above, densely grey or white-hairy beneath. Stipules broadly oblong, toothed. Inflorescence-branches erect arching, finely hairy. Sepals usually 5, hairless. Petals usually 5, *c.* 3 mm, white or very pale yellowish cream, gradually tapered to an indistinct claw. Carpels 5–10, attached at their bases, compressed, the keels with long hairs. *Himalaya from Afghanistan to Nepal; W China.* H4. Summer.

3. F. purpurea Maximowicz (*Spiraea palmata* Thunberg). Illustration: Botanical Magazine, 5726 (1868); Kitamura & Murata, Coloured illustrations of herbaceous plants of Japan, pl. 29 (1978).

Perennial rhizomatous herb to 1.3 m. Stems hairless. Basal and lower stem-leaves pinnate with 3–5 leaflets, terminal leaflet deeply 3-lobed, truncate at base, much larger than the unlobed lateral leaflets, all dark green and hairless above, with short hairs on the veins beneath, all irregularly doubly toothed. Stipules rather narrow, lanceolate, little-toothed. Inflorescence spreading, branches hairless or hairy. Sepals usually 5, pinkish, hairless outside, with a small patch of sparse hairs inside. Petals usually 5, 2–3.5 mm, pink or purplish red, the blade abruptly truncate or even auriculate where it joins the short but distinct claw. Carpels usually 5, attached by their bases, compressed, with long hairs along the keels. Follicles very compressed, with long hairs along the keels. *Originated in gardens in Japan, not known for certain in the wild.* H4. Summer.

Some plants in cultivation as *F. purpurea* 'Alba' are similar but have white petals; some of these, at least, have hairless carpels (and follicles?) and may be of hybrid origin.

4. F. ulmaria (Linnaeus) Maximowicz (*Spiraea ulmaria* Linnaeus). Illustration: Gleason, Illustrated flora of the north-eastern United States and adjacent Canada **2**: 300 (1952); Ary & Gregory, The Oxford book of wild flowers, 80 (1962); Polunin, Flowers of Europe, pl. 44 (1969); Garrard & Streeter, Wild flowers of the British Isles, pl. 35 (1983).

Upright perennial herb to 1.2 m, with pinkish rhizomes. Basal and lower stem-leaves pinnate with 2–5 pairs of leaflets, terminal leaflet larger than lateral leaflets, deeply divided into 3 lobes, dark green and hairless above, usually densely white-downy beneath, irregularly doubly toothed. Stipules rounded, toothed. Inflorescence-branches usually arching, with dense or sparse short hairs. Flowers fragrant. Sepals usually 5, yellowish, shortly hairy outside, hairless inside. Petals usually 5, cream-white, 3–5 mm, rather abruptly tapered into a distinct claw. Carpels 6–10, hairless, attached by part of their inner angle, projecting a little below the point of attachment. Achenes of each flower coiling spirally together as they mature. *Europe, W Asia, naturalised in E North America.* H1. Summer.

A double-flowered variant is known as 'Plena' and a variant with the leaves variegated is known as 'Aurea' or 'Aureo-variegata' (var. *aureovariegata* Voss).

5. F. palmata (Pallas) Maximowicz (*Spiraea palmata* Pallas not other authors; *S. digitata* Willdenow). Illustration: Flora SSSR **10**: 287 (1941).

Rhizomatous perennial herb to 1 m. Basal and lower stem-leaves pinnate with 3–5 leaflets, all leaflets deeply 3–5-lobed, the terminal leaflet considerably larger than the laterals, all doubly toothed, dark green and hairless above, densely white-downy beneath. Stipules oblong-rounded, toothed. Inflorescence rather spreading, branches hairless or with short hairs. Sepals usually 5, hairless. Petals usually 5, white, 2–3 mm, rather abruptly tapered into a distinct claw. Carpels 5–8, attached at their bases, hairless. Achenes erect. *E former USSR.* H1. Summer.

Uncommon in cultivation, often confused with other species (especially *F. purpurea*).

6. F. rubra (Hill) Robinson (*Spiraea lobata* Gronovius; *S. palmata* Murray). Illustration: Gleason, Illustrated flora of the north-eastern United States and adjacent Canada **2**: 300 (1952); Rickett, Wild flowers of the United States **1**: pl. 41 (1965).

Upright rhizomatous perennial herb to 3 m. Basal and lower stem-leaves pinnate, all leaflets 3-lobed, the terminal leaflet larger than the laterals, all deeply doubly toothed, dark green and hairless above, green and hairy only on the veins beneath. Stipules oblong, toothed. Inflorescence-branches arching upright, branches hairless. Sepals usually 5, hairless outside, with a small patch of sparse hairs inside. Petals usually 5, pink or (in 'Venusta') purplish red, rather abruptly tapered or truncate into a distinct claw. Carpels 6–10 attached at their bases, hairless. Achenes 6–10, erect, hairless. *E & C USA.* H3. Summer.

7. F. vulgaris Moench (*Spiraea filipendula* Linnaeus; *Ulmaria filipendula* (Linnaeus) Hill; *F. hexapetala* Gilibert). Illustration: Bonnier, Flore complète **3**: pl. 169 (1914); Polunin, Flowers of Europe, pl. 44 (1969); Garrard & Streeter, Wild flowers of the

British Isles, pl. 35 (1983).

Perennial herb to 80 cm, with swollen, tuberous roots; vegetative parts hairless or almost so. Leaves pinnate with numerous leaflets all more or less the same size (diminishing slightly downwards), pinnately lobed or toothed. Stipules large, ovate-oblong, toothed. Inflorescence branches erect arching or spreading. Flowers fragrant. Sepals usually 6, whitish, hairless. Petals usually 6, creamy white, obovate, 6–7 mm, tapering gradually into an indistinct claw. Carpels 6–12, attached by their bases, hairy all over, not compressed. Achenes erect, hairy all over. *Europe, N Africa, W Asia.* H1. Summer.

'Flore Pleno' has double flowers and 'Grandiflora' has slightly larger cream-coloured flowers.

20. RHODOTYPOS Siebold & Zuccarini
J. Cullen

Deciduous shrub to 2 m or more; branches greyish, hairless. Leaves opposite, shortly stalked, 4–10 cm ovate to ovate-oblong, apex acuminate, base rounded, irregularly doubly toothed, with long whitish hairs on both surfaces. Stipules linear or thread-like. Flowers terminating short, leafy, lateral shoots, 2.5–4 cm across. Sepals 4, toothed, alternating with 4 additional small lobes. Petals 4, white, almost circular. Stamens numerous. Ovary superior. Carpels 4, free. Fruit a group of up to 4 shining, black, almost dry drupes surrounded by the persistent and enlarged calyx.

A genus of a single species from China, Japan and Korea, grown for its rose-like flowers. It is easily grown and propagated by softwood cuttings taken in summer.

1. R. scandens (Thunberg) Makino (*R. kerrioides* Siebold & Zuccarini). Illustration: Botanical Magazine, 5805 (1869); Addisonia **23**: pl. 759 (1954–59); Bean, Trees & shrubs hardy in the British Isles, edn 8, **3**: 942 (1976); Csapody & Tóth, Atlas of flowering trees and shrubs, pl. 52 (1982).

China, Japan, Korea. H1. Late spring–summer.

21. KERRIA de Candolle
J. Cullen

Deciduous shrub to 3 m; branches green, hairless. Leaves simple, alternate, stalks short, 2.5–11 cm, lanceolate to ovate, apex acuminate, base cordate to truncate, doubly toothed, hairless or with adpressed bristles above, sparsely white-hairy beneath. Stipules narrow, falling early.

Flowers solitary, terminating short, leafy lateral shoots, 2–5 cm across. Sepals 5. Petals 5 (many in 'Pleniflora', see below), yellow or orange-yellow. Stamens numerous. Ovary superior; carpels 5–8, free. Fruit a group of up to 8 achenes surrounded by the persistent calyx.

A genus of a single species probably originally from China, but widely cultivated there and elsewhere in the Far East for a long period, often as the double-flowered variant. It is easily grown and can be propagated by softwood cuttings taken in summer.

1. K. japonica (Linnaeus) de Candolle. Illustration: Perry, Flowers of the world, 262 (1972); Kitamura & Okamoto, Coloured illustrations of trees and shrubs of Japan, 216 (1977); Everett, New York Botanical Gardens illustrated encyclopedia of horticulture, 1895 (1981); Il Giardino Fiorito **54**: 31 (1988).

China, Japan & E Asia. H1. Late spring–summer

The double-flowered 'Pleniflora' is most commonly grown. Variants with variegated leaves ('Variegata', 'Aureo-variegata') and with branches striped green and yellow ('Aureo-vittata') are also grown.

22. NEVIUSIA A. Gray
J. Cullen

Deciduous shrub, 1–2 m. Stems brownish, hairless. Leaves alternate, shortly stalked, ovate to ovate-oblong, 3–7 cm, acute or acuminate at the apex, rounded to slightly cordate at the base, irregularly doubly toothed, with adpressed, white bristles on both surfaces. Stipules very small, thread-like. Flowers in cymes or solitary, terminating short, leafy, lateral shoots, to 1.5 cm across. Sepals 5, spreading, toothed. Petals absent. Stamens numerous, yellowish white, exceeding the sepals. Carpels 2–4, free. Fruit a group of 2–4 achenes surrounded by the persistent calyx.

A genus of a single species from the SE USA, grown occasionally for its attractive stamens. It is easily grown and propagated by softwood cuttings in summer.

1. N. alabamensis Gray. Illustration: Botanical Magazine, 6806 (1885); Dean, Mason & Thomas, Wild flowers of Alabama, 81 (1973); Arnoldia **36**: 59 (1976); Csapody & Tóth, Colour atlas of flowering trees and shrubs, pl. 52 (1982).

SE USA (Alabama). H3. Summer.

23. RUBUS Linnaeus
J. Cullen

Erect, scrambling, trailing or prostrate shrubs or low shrubs, shoots rarely herbaceous and dying down in winter, often with prickles on the stems, leaf- and leaflet-stalks and inflorescences. Leaves deciduous or evergreen, entire, lobed or divided into 3–many leaflets which are usually toothed and may themselves be lobed; stipules always present, often conspicuous, sometimes falling early. Flowers in clusters, racemes or panicles, sometimes solitary, usually bisexual. Sepals 4 or 5 or rarely more, spreading, erect or reflexed. Petals 4 or 5 or rarely more, very variable in size, white, pink, red or purple, spreading or erect and adpressed to the stamens. Stamens numerous; anthers sometimes hairy. Ovary superior; carpels 5-many, free, borne on a usually cylindric or conical receptacle. Fruit a 'berry' composed of 5–many variably coherent fleshy drupelets, which may separate from the receptacle as a (hollow) unit, or may fall from the plant by abscission of the flower-stalk while still attached to the receptacle.

A large genus of an uncertain number of species from most parts of the world. Many of the species produce edible fruit, and some are grown commercially for this purpose; the fruit of other wild species is also collected and eaten.

The identification of the species is in general reasonably straightforward except for those of Subgenus *Rubus* (see p. 356). In many of the species the shoots (stems, canes) are biennial, bearing leaves during their first year and shoots bearing leaves and inflorescences arising from the axils of the first-year leaves in the second. The leaves of the first and second year may differ considerably in degree of division and/or lobing. In this account it is the leaves of the flowering-shoots that are described unless otherwise indicated. In some plants, shoots borne low down on the second-year stems do not flower, but grow rapidly and mimic first-year shoots; the leaves on these shoots are often aberrant.

Different species of the genus show a remarkable range of hair types on stems, leaf-stalks and leaves. Many species have long, soft, sometimes felted hairs which may be mixed with stalked or stalkless glands, bristles (long, straight, parallel-sided hairs) or prickles (hard, straight or variously hooked small spines).

In a few species the flowers are

functionally unisexual; stamens and carpels are both present, but the members of one or other set are infertile. This characteristic is sometimes difficult to judge.

Literature: Focke, W.O., Species Ruborum, *Bibliotheca Botanica* **72** (1910–11); Bailey, L.H., Certain cultivated Rubi, *Gentes Herbarum* **1**: 139–200 (1925); Newton A. & Edees, E.S. *Brambles of the British Isles* (1988).

1a. Plants herbaceous, stems dying down each winter   2
  b. Plants with persistent, aerial, usually woody stems   3
2a. Leaves divided into 3 distinct leaflets; flowers bisexual
                                      1. arcticus
  b. Leaves simple, sometimes 3-lobed; flowers unisexual
                                2. chamaemorus
3a. Aerial woody stems thin, creeping close to the soil surface, giving rise to short, erect, herbaceous flowering-shoots   4
  b. Aerial woody stems stout, prostrate, trailing, arching, scrambling or erect, not as above   6
4a. Stipules deeply divided into narrow segments; anthers with sparse, long hairs; leaves persistent, leathery and conspicuously wrinkled above
                                    3. pentalobus
  b. Stipules entire or toothed; anthers hairless; leaves not persistent, leathery and wrinkled as above   5
5a. Leaves simple, 3-lobed
                                      4. calycinus
  b. Leaves divided into 3 leaflets
                                    5. nepalensis
6a. Stipules broad, attached directly to shoot, free from leaf-stalks, and/or sometimes falling early   7
  b. Stipules narrow, lanceolate or linear, attached directly to the leaf-stalks   16
7a. Stems without prickles; stipules persistent; leaves always simple
                                        6. tricolor
  b. Stems with at least some prickles; stipules usually falling early; leaves simple or compound   8
8a. Leaves divided into leaflets, or lobed for more than 1/3 of the distance from margin to stalk-attachment
                                                              9
  b. Leaves simple, not lobed or more shallowly lobed   11
9a. Leaves divided into distinct, shortly-stalked leaflets which have many,

conspicuous, parallel, lateral veins which end in many, even, hair-tipped teeth   **8. lineatus**
  b. Leaves lobed, but not to the base, venation not as above   10
10a. Leaves densely white- or pale brown-hairy beneath; flowers in racemes longer than broad, borne in leaf-axils   **7. henryi**
  b. Leaves densely brown-hairy beneath; flowers in more or less stalkless clusters in leaf-axils
                                      12. reflexus
11a. Inflorescence a large, widely spreading terminal panicle   12
  b. Inflorescences forming condensed clusters in leaf-axils   14
12a. Leaves rather leathery, lanceolate (rarely lanceolate-ovate), usually hairless or very sparsely hairy beneath; margins with small, distant teeth   **9. ichangensis**
  b. Leaves rather thin, ovate-lanceolate, usually hairy, at least on the veins beneath; margins with regular, close, conspicuous teeth
                                                              13
13a. Flowering-shoots densely hairy; branches of inflorescence with conspicuous stalked glands
                                        10. parkeri
  b. Flowering-shoots hairless; branches of inflorescence without glands
                                    11. lambertianus
14a. Leaves as broad as long or broader, rounded to the short, abruptly acute apex   **15. irenaeus**
  b. Leaves longer than broad, tapering gradually to the pointed apex   15
15a. Leaves not, or only extremely shallowly lobed   **13. flagelliflorus**
  b. Leaves distinctly 3-, 5- or 7-lobed
                                  14. kumaonensis
16a. Flowers all unisexual, in large, usually widely spreading panicles; leaflets usually short compared to their very long, prickly stalks   17
  b. Flowers usually bisexual in inflorescences of various kinds, not as above; at least the lateral leaflets longer than their stalks   19
17a. Leaves simple (reduced to a terminal leaflet)   **20. parvus**
  b. Leaves with 3–5 leaflets (the blades sometimes very reduced, but leaflet-stalks evident)   18
18a. Leaflets ovate or ovate-oblong, acutely toothed; fruits yellowish
                                      21. australis
  b. Leaflets lanceolate or oblong-lanceolate, with few, rather deep,

triangular lobes; fruits red to orange   **22. squarrosus**
19a. Stems without prickles, upright, with flaky bark; leaves simple, lobed   20
  b. Stems with prickles, upright, arching, scrambling, trailing or prostrate, bark not as above; leaves usually compound   23
20a. Flowers solitary, 5 cm or more across   21
  b. Flowers 2 or more together in racemes or panicles, less than 5 cm across   22
21a. Sepals in fruit brownish outside, reddish inside; leaf-lobes acute
                                      17. trilobus
  b. Sepals in fruit greenish or brownish inside and out; leaf-lobes rather rounded   **16. deliciosus**
22a. Petals purple; flowers many, in panicles   **18. odoratus**
  b. Petals white; flowers few, in racemes or rarely solitary
                                      19. parviflorus
23a. Fruit separating from cylindric or conical receptacle, therefore hollow
                                                              24
  b. Fruit not separating from receptacle, falling from parent plant by abscission of fruit-stalks   43
24a. Leaves on flowering-shoots simple, though often lobed   25
  b. Leaves on flowering-shoots with 3 or more distinct leaflets   26
25a. Leaves to 3.5 cm; flowers solitary
                                  23. microphyllus
  b. Leaves 4 cm or more; flowers in racemes or clusters
                                  24. crataegifolius
26a. Most flowers on a plant solitary, occasionally 2 or 3 together   27
  b. Most flowers on a plant in racemes, panicles or clusters, occasionally a few solitary   30
27a. Leaflets mostly 3; petals pink
                                    25. spectabilis
  b. Leaflets always 5 or more; petals white   28
28a. Leaves, at least beneath, with scattered, stalkless, glistening yellow glands   **27. rosifolius**
  b. Leaves without stalkless, glistening yellow glands   29
29a. Lateral leaflets 1–3 cm; sepals usually hairy and glandular outside, sparingly hairy inside towards the apex   **28. amabilis**
  b. Lateral leaflets 3.5–9 cm; sepals usually hairless outside, densely hairy inside towards the apex
                                  26. illecebrosus

30a. Leaflets on leaves of flowering-
shoots mostly 5 or more, if 3, then
terminal leaflet tending to be deeply
3-lobed 31
  b. Leaflets on leaves of flowering-
shoots always 3, the terminal not
deeply 3-lobed 34
31a. Terminal leaflet very broadly obovate
to diamond-shaped, broadest near
the tip **33. coreanus**
  b. Terminal leaflet ovate, lanceolate or
elliptic, broadest near the middle or
below 32
32a. Leaflets 7–15, the terminal
lanceolate, pinnately lobed
  31. thibetanus
  b. Leaflets 5 or 7 (rarely 3), the
terminal not as above 33
33a. Inflorescence raceme-like, long,
not flat-topped; leaflets scarcely
lobed; stems conspicuously bloomed
  32. cockburnianus
  b. Inflorescence corymb-like, flat-
topped; leaflets conspicuously lobed;
stems not bloomed
  40. mesogaeus
34a. Plant with numerous bristles as
well as soft hairs and prickles 35
  b. Plant without bristles 37
35a. Leaflets broadly elliptic, rounded
at the tip, very evenly toothed
  35. ellipticus
  b. Leaflets ovate to narrowly elliptic,
acute, irregularly toothed or
shallowly lobed 36
36a. Terminal leaflet much larger than
the lateral leaflets; flowers 1.5–2.5
cm across; bristles reddish or
purplish **34. phoenicolasius**
  b. Terminal leaflet little larger than
the laterals; flowers 1–1.8 cm
across; bristles brownish or pale
  41. idaeus
37a. Calyx hairless outside, though
occasionally with small prickles
near the base 38
  b. Calyx variously hairy outside 39
38a. Carpels and styles hairy, the fruit
very densely hairy; flowers 2.5–3.5
cm across **30. lasiostylus**
  b. Carpels, styles and fruits not as
above; flowers 1–2 cm across
  29. biflorus
39a. Inflorescence an open raceme,
not flat-topped; prickles mostly
straight or nearly so; fruit usually
red **41. idaeus**
  b. Inflorescence a dense, flat-topped
cluster; prickles often hooked; fruit
usually purplish or black 40

40a. Plant with gland-tipped hairs at
least on flower-stalks and/or calyx
  39. glaucifolius
  b. Plant entirely without gland-tipped
hairs 41
41a. Leaves rather finely and regularly
toothed **38. glaucus**
  b. Leaves coarsely and irregularly
toothed 42
42a. Mature fruit black; prickles on
flower-stalk thin, not conspicuously
broadened at base, little hooked
  36. occidentalis
  b. Mature fruit purplish; prickles on
flower-stalks conspicuously
broadened at base, hooked
  37. leucodermis
43a. Stems prostrate or trailing on the
ground, the flowering-shoots erect,
arising from them 44
  b. Stems upright, arching or
scrambling, sometimes arching over
and rooting at the tip 50
44a. Stems conspicuously bloomed 45
  b. Stems not conspicuously bloomed
  48
45a. Flowers bisexual; prickles broadly
based **55. caesius**
  b. Flowers unisexual; prickles
narrowly based, needle-like 46
46a. Most parts of the plant with
conspicuous, stalked glands
  58. macropetalus
  b. Stalked glands absent or few,
inconspicuous 47
47a. Leaves bright green, thinly hairy to
hairless beneath **56. vitifolius**
  b. Leaves dull or grey-green, densely
hairy beneath **57. ursinus**
48a. At least some of the leaves of the
first-year stems leathery and
persistent; bristles usually present
  52. trivialis
  b. Leaves all deciduous; bristles
usually absent 49
49a. Flowers 2–3 cm across
  53. flagellaris
  b. Flowers 3–4 cm across
  54. roribaccus
50a. All leaves with a whitish
undersurface produced by dense
hairs beneath 51
  b. All leaves more or less green
beneath, though often hairy 56
51a. Inflorescence a few-flowered cluster,
not obviously longer than broad,
without a distinct, long axis
  51. cuneifolius
  b. Inflorescence generally many-
flowered, raceme-like, longer than
broad, with a distinct, long axis 52

52a. Stems conspicuously bloomed
  47. ulmifolius
  b. Stems usually not bloomed 53
53a. Each leaf with the lowermost
leaflets unstalked **49. linkianus**
  b. Each leaf with the lowermost
leaflets with short, but distinct
stalks 54
54a. Prickles in the inflorescence mostly
straight 55
  b. Prickles in the inflorescence hooked
  50. procerus
55a. Petals white or pale pink
  48. bifrons
  b. Petals deep pink
  43. elegantispinosus
56a. Inflorescence, leaf-stalks and new
shoots with conspicuous, stalked
glands **44. allegheniensis**
  b. Stalked glands absent or few and
inconspicuous 57
57a. Leaflets deeply lobed
  42. laciniatus
  b. Leaflets toothed but not deeply
lobed 58
58a. Flower-clusters raceme-like, much
longer than broad **46. bellobatus**
  b. Flower-clusters corymb-like, little
longer than broad **45. frondosus**

Subgenus **Cylactis** (Rafinesque) Focke.
Plants herbaceous, stems dying down
each winter. Leaves divided into 3 distinct
leaflets; stipules broad or narrow, generally
free from the leaf-stalk. Flowers bisexual.
Fruit consisting of few, large, loosely
coherent drupelets.

1. R. arcticus Linnaeus. Figure 30(1),
p. 348. Illustration: Botanical Magazine,
132 (1797); Polunin, Flowers of Europe,
pl. 43 (1969).

Stems to 30 cm, borne on a persistent
rhizome, dying down annually, softly hairy
or hairless, without prickles. Leaves long-
stalked, divided into 3 distinct leaflets (very
rarely deeply 3-lobed only); leaflets broadly
elliptic to almost diamond-shaped, 1.2–4.5
× 1.5–3.5 cm, shortly stalked, margins
unevenly toothed, lateral leaflets often
oblique at the base, usually softly hairy
on both surfaces. Flowers 1–3 together,
long-stalked, 1.5–2.5 cm across. Sepals
and petals 5–7 or more, petals longer than
sepals, pink. Carpels downy. Fruit made
up of about 20 drupelets, red. *N Europe, N
Asia, N America.* H1. Summer.

Subsp. **acaulis** (Michaux) Focke (*R.
acaulis* Michaux) is smaller, has strictly
1-flowered stems, and the flowers are
somewhat larger (to 3 cm across). *N North
America.*

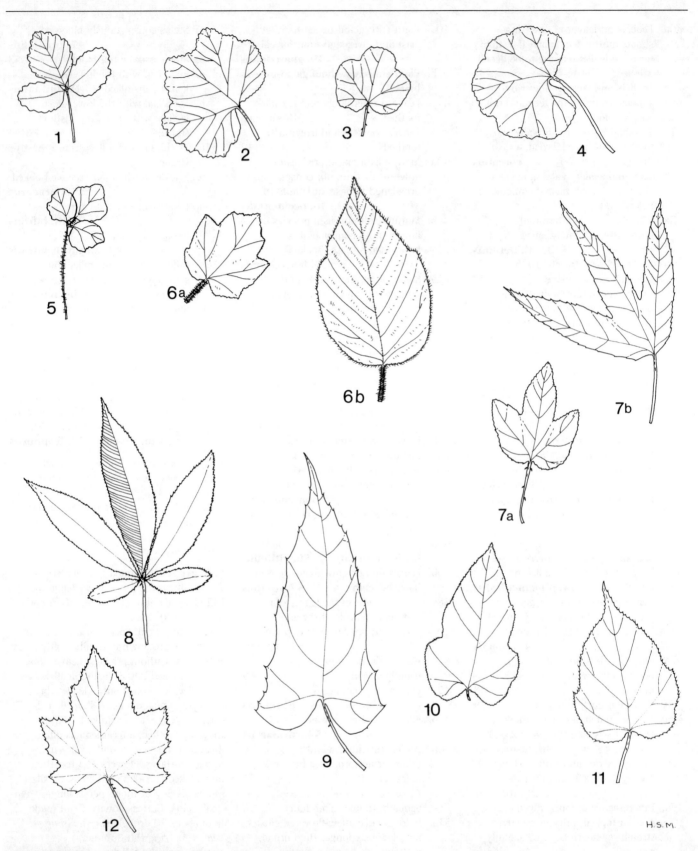

H.S.M.

Figure 30. Leaves of *Rubus* species (× 0.5). 1, *R. articus*.
2, *R. chamaemorus*. 3, *R. pentalobus*. 4, *R. calycinus*.
5, *R. nepalensis*. 6a, b, *R. tricolor*. 7a, b, *R. henryi*. 8, *R. lineatus*.
9, *R. ichangensis*. 10, *R. parkeri*. 11, *R. lambertianus*.
12, *R. reflexus*.

Subgenus **Chamaemorus** Focke. Plants herbaceous, stems dying down each winter, without prickles. Leaves simple or 3-lobed; stipules broad, leaf-like. Flowers solitary, terminal, unisexual. Fruits of rather few coherent drupelets.

2. R. chamaemorus Linnaeus. Figure 30(2), p. 348. Illustration: Ary & Gregory, The Oxford book of wild flowers, 80 (1962); Rickett, Wild flowers of the United States **1**: pl. 141 (1967); Polunin, Flowers of Europe, pl. 44 (1969); Keble Martin, The concise British flora in colour, edn 2, pl. 29 (1969).

Stems 5–20 cm, borne on a persistent rhizome, dying down each winter, with downwardly directed, soft hairs, some stalked glands and persistent stipules. Leaves 2–6.5 × 3–10.5 cm, almost circular to kidney-shaped, cordate at base, simple, usually shallowly 3–5-lobed, margins irregularly toothed, upper surface rough, lower surface with soft, white hairs and sometimes stalked glands. Flower solitary, 1.5–2.5 cm across, terminal; flower-stalks softly hairy and glandular. Sepals erect-spreading, ovate, reddish. Petals 5 or more, white, exceeding the sepals. Fruit golden orange when ripe. *N Europe, N Asia, N America.* H1. Summer.

Subgenus **Chamaebatus** Focke. Plants with woody, persistent aerial stems lying close to the soil surface and giving rise to erect, leafy flowering shoots; small prickles present. Leaves simple, sometimes 3-lobed; stipules broad, ovate, sometimes divided, free from the leaf-stalks. Flowers solitary or 1–3 together, bisexual. Fruit with a few rather large drupelets.

3. R. pentalobus Hayata (*R. calycinoides* Hayata not Kuntze; *R. calycinoides* var. *macrophyllus* Li; *R. fockeanus* misapplied). Figure 30(3), p. 348. Illustration: Botanical Magazine, 9644 (1944); Li et al., Flora of Taiwan **3**: 110 (1977).

Prostrate small shrub, sometimes with arching branches. Stems creeping and rooting, hairy and with a few small prickles. Flowering-shoots hairy and with a few prickles. Leaves leathery, persistent, conspicuously wrinkled above, simple, rounded at the apex, base cordate, shallowly 3–5-lobed, margins unevenly and coarsely toothed, hairy above when young, persistently brownish hairy beneath. Stipules 7–10 mm, deeply divided into narrow, tapering segments. Flowers solitary, shortly stalked, to 2 cm across. Sepals ovate, irregularly toothed at the apex or entire. Petals white, shorter than to almost as long as sepals. Anthers with a few hairs. Fruit scarlet. *Taiwan.* H4. Summer.

This species was placed in Subgenus *Chamaebatus* by its original author. However, Sealy in the notes to the Botanical Magazine plate cited above, suggested it should be placed in Subgenus *Malachobatus*. It is anomalous in both subgenera, but apparently less so in *Chamaebatus*.

Its naming is very complex. It was first described as *R. calycinoides* by Hayata (a name under which it is frequently found in gardens at present), who unfortunately overlooked an earlier use of this name, by Kuntze, for a Himalayan species of Subgenus *Malachobatus* which is perhaps not cultivated (see below). This oversight was missed by Sealy and by the authors of the *Flora of Taiwan*, who use the name *calycinoides*; in the Flora, a variety, *macrophyllus* Li was recognised, under which *R. pentalobus* Hayata was cited as a synonym. The difference between the varieties was based mainly on size, and seems insignificant; the name *R. pentalobus* is therefore the earliest that can legitimately be used for the species.

The plant was introduced in the 1930s under the name *R. fockeanus*; genuine *R. fockeanus* Kurz is a species from Subgenus *Dalibarda* Focke (which has no other cultivated representatives), differing from *R. pentalobus* in its deciduous leaves with 3 distinct leaflets, slightly toothed stipules and lack of prickles; it occurs in the Himalayas and SW China.

4. R. calycinus Wallich. Figure 30(4), p. 348.

Aerial stems woody, thin, creeping, rooting, giving rise to short, leafy flowering-stems which bear backwardly directed prickles and some long hairs. Leaves simple, slightly 3-lobed, margins finely toothed, 2–4.5 × 3.5–6.5 cm, almost circular to kidney-shaped, base cordate, with long, fine hairs above and prickles on the veins beneath; stalks long and prickly; stipules entire and toothed. Flowers terminal, solitary or 2 together, 1.5–3 cm across. Sepals ovate, toothed at the apex. Petals white, about as long as sepals. Anthers hairless. Fruit reddish, of numerous drupelets. *Himalaya.* H3. Spring.

Subgenus **Dalibardastrum** Focke. Plants with trailing or scrambling woody aerial stems, without prickles. Leaves simple or divided into 3 leaflets; stipules attached to the stem, free from the leaf-stalks, large and broad, persistent in our species. Flowers bisexual.

5. R. nepalensis (J.D. Hooker) Kuntze (*R. nutans* G. Don). Figure 30(5), p. 348. Illustration: Botanical Magazine, 5023 (1857).

Prostrate, small shrub; stems rooting, thin, creeping, covered with soft, curled hairs and irregularly spreading bristles. Leaves divided into 3 distinct leaflets; leaflets ovate-oblong to diamond-shaped, evergreen, shortly stalked, margins irregularly and finely toothed, softly hairy above and beneath; leaf-stalks hairy like the stems. Stipules ovate, entire or sparsely toothed, hairy. Flowers solitary, terminal, 2–3 cm across. Sepals narrowly triangular, acute or acuminate, entire or somewhat toothed at the tip; perigynous zone densely bristly. Petals as long as to a little longer than the sepals, white. Fruit red, somewhat concealed by the persistent sepals. *W Himalaya, from NW India to Nepal and perhaps Sikkim.* H3. Summer.

R. fockeanus Kurz is very similar (see under *R. pentalobus*).

6. R. tricolor Focke. Figure 30(6), p. 348. Illustration: Botanical Magazine, 9534 (1938); Bean, Trees & shrubs hardy in the British Isles, edn 8, **4**: 237 (1980).

Large, deciduous, trailing or scrambling shrub to 2 m high or wide, with short stems covered with dense, spreading, brownish bristles. Flowering-shoots arching or prostrate, many-leaved, with similar bristles. Leaves simple, scarcely lobed or with 3–5 or more small lobes, ovate or broadly ovate, base cordate, apex acute or acuminate, 4.5–11.5 × 4–9 cm, margins finely toothed, dark green and hairless or with stiff, adpressed bristles along the veins above, densely white-hairy and with dense bristles along the veins beneath. Stipules large, persistent, ovate, toothed. Flowers few in terminal racemes, sometimes with solitary flowers in a few leaf-axils below the raceme, to 3 cm across. Sepals white-hairy and bristly outside, long-tapered, toothed. Petals white, somewhat shorter than the sepals. Anthers hairless. Fruit bright red. *SW China (Yunnan, Sichuan).* H5. Summer.

Subgenus **Malachobatus** Focke. Mostly shrubs with woody, upright, arching, scrambling or somewhat prostrate aerial branches, with prickles. Leaves simple or divided into several leaflets, when simple, often lobed; stipules large, attached directly to the stems, often early deciduous.

7. R. henryi Hemsley & Kuntze. Figure 30(7), p. 348. Illustration: Botanical Magazine, n.s., 33 (1948).

Evergreen, scrambling shrub to 6 m. Stems white-hairy, at least when young, with few prickles. Leaves rather deeply 3- or 5-lobed, the lobes extending to more (usually much more) than one-third of the distance from margin to the point of stalk-attachment, rarely a few leaves almost unlobed; lobes elliptic to narrowly elliptic, to 12 cm, long-tapered at the apex, toothed, dark green above, densely white- or pale brown-hairy beneath. Stipules narrowly elliptic, entire, deciduous. Flowers borne in long racemes in the leaf-axils, 1–2 cm across. Sepals triangular, densely hairy, reflexed. Petals pink, more or less erect, somewhat shorter than the sepals, soon falling. Anthers with sparse, long hairs. Fruit black and shining, to 1.5 cm across. *C & W China*. H3. Spring–summer.

Var. **bambusarum** (Focke) Rehder (*R. bambusarum* Focke) has narrow leaf-lobes, and is the most commonly cultivated variant.

8. R. lineatus Blume. Figure 30(8), p. 348. Illustration: Bean, Trees & shrubs hardy in the British Isles, edn 8, **4**: 228 (1980).

Deciduous or evergreen, scrambling shrub to 3 m; stems with close, parallel, adpressed white hairs at least when young, with a few short prickles or prickles virtually absent. Leaves divided into 3 or 5 shortly stalked leaflets; leaflets elliptic or elliptic-oblanceolate, 9–15 × 1.5–4 cm, tapering to base and apex, dark green above, greenish or silvery with dense, parallel, adpressed hairs beneath, lateral veins very conspicuous, numerous, parallel, ending in close, even, hair-tipped teeth at the margins. Stipules ovate, entire. Flowers few in short axillary clusters. Sepals ovate to ovate-lanceolate, acuminate, silky hairy outside, to 1 cm. Petals greenish white, 4–5 mm. Fruit red. *Himalaya & W China to Indonesia (E to Borneo)*. H5. Spring–summer.

The leaflets of this species are very striking with numerous, parallel lateral veins, numerous regular teeth and, usually, shining, silvery undersurfaces.

R. splendidissimus Hara. Similar, but leaves usually with 3 leaflets, the hair on all parts more woolly, less regularly arranged and including some gland-tipped bristles. *Nepal, Bhutan*. H5. Spring–summer.

9. R. ichangensis Henry & Kuntze. Figure 30(9), p. 348.

Deciduous, scrambling or more or less prostrate shrub; stems with stalked glands and a few hooked prickles. Leaves thinly leathery, lanceolate or more rarely ovate-lanceolate, margins sometimes somewhat shallowly and sinuously lobed, with small, distant teeth, hairless or very sparsely hairy beneath and on the main veins above; stalks prickly. Stipules narrowly lanceolate, toothed, quickly deciduous. Flowers in a large, spreading, terminal panicle to 25 cm, 7–10 mm across. Sepals erect, ovate-triangular, acute, with dense short hairs inside and out, outside also with adpressed bristles and spreading, gland-tipped bristles. Petals white, about as long as sepals. Fruit of 12–20 drupelets, red. *C & W China*. H3. Summer.

10. R. parkeri Hance. Figure 30(10), p. 348.

Like *R. ichangensis* but shoots densely hairy and prickly, leaves thinner, broader and rather more conspicuously but still shallowly lobed, rather densely hairy; panicle conspicuously glandular; fruit of fewer drupelets, black. *C China*. H3. Early summer.

11. R. lambertianus Seringe. Figure 30(11), p. 348.

Like *R. ichangensis* but leaves broadly ovate, thinner but half-evergreen, conspicuously toothed, abruptly acuminate, stipules divided into narrow segments; petals shorter than sepals; fruit of 15–20 drupelets, red. *C & S China, Japan, Taiwan*. H3. Summer.

12. R. reflexus Ker Gawler. Figure 30(12), p. 348. Illustration: Botanical Magazine, 7716 (1900); Gartenpraxis for January 1981, 28.

Deciduous scrambling shrub; stems with very dense and close brown hairs and sparse, minute prickles. Leaves 6–13.5 × 6–15 cm, rather deeply 3- or 5-lobed, the lobes triangular or narrowly triangular, acute, finely to coarsely toothed, with brown hairs along the veins above and densely and closely covering the whole surface beneath. Stipules large, persistent, deeply divided into narrow, brown-hairy lobes. Flowers to 1.5 cm across, almost stalkless in condensed axillary clusters. Sepals ovate, densely hairy within, very densely brown-hairy outside. Petals about as long as sepals. Anthers hairless or with a few long hairs. Fruit spherical, purple-black. *S China*. H5. Early summer.

Plants cultivated as *R. moluccanus* are often this species. The name *R. moluccanus* Linnaeus applies to a species from Indonesia not in general cultivation.

13. R. flagelliflorus Focke.

Evergreen scrambling or trailing shrub. Stems hairy, at least when young, and with small prickles. Leaves ovate to ovate-lanceolate, not lobed, acute, cordate at base, 8–16 cm, margins toothed, hairy above when young, densely hairy with yellowish hairs beneath. Stipules deeply divided. Flowers in condensed axillary clusters. Sepals ovate, acute, densely hairy outside, hairless and purple inside, reflexed. Petals white, about as long as sepals, quickly deciduous. Fruit hemispherical, purple-black. *C & W China*. H4. Summer.

14. R. kumaonensis Balakrishnan (*R. reticulatus* J.D. Hooker).

Large scrambling shrub. Stems whitish hairy, with a few short, straight or weakly curved prickles. Leaves ovate to almost circular, with 3-, 5- or 7 conspicuous but shallow, acute lobes, 10–20 × 10–20 cm, margins finely toothed, base cordate, upper surface with a conspicuous network of veins and with long spreading hairs, lower surface whitish with short, curled hairs and longer, somewhat spreading hairs. Stipules to 1 cm, lanceolate, deeply divided into narrow segments. Flowers to 1.5 cm across, in axillary clusters. Sepals ovate-acuminate, whitish hairy. Petals white, shorter than sepals. *N India*. H3. Summer.

Part of a complex of rather similar Himalayan species, including **R. rugosus** Smith with smaller leaves with rounded lobes and dense flower-clusters, and the genuine *R. calycinoides* Kuntze (see p. 349) which is larger with prickles on the main veins and stalks of the leaves and larger stipules. The name *R. moluccanus* Linnaeus has been applied to all of these species from time to time (see above).

15. R. irenaeus Focke. Figure 31(1), p. 351.

Trailing or almost prostrate shrub; stems densely hairy with short, curled hairs and scattered, longer, adpressed hairs, sometimes with a few prickles. Leaves very broadly ovate to kidney-shaped, 7–10 × 9.5–12 cm, very shallowly 3-lobed, margins regularly toothed, cordate at base, rounded to the abruptly acute apex, dark green above, whitish or pale brown beneath with dense, curled, whitish or brownish hairs and longer, spreading hairs. Stipules ovate-oblong, deeply

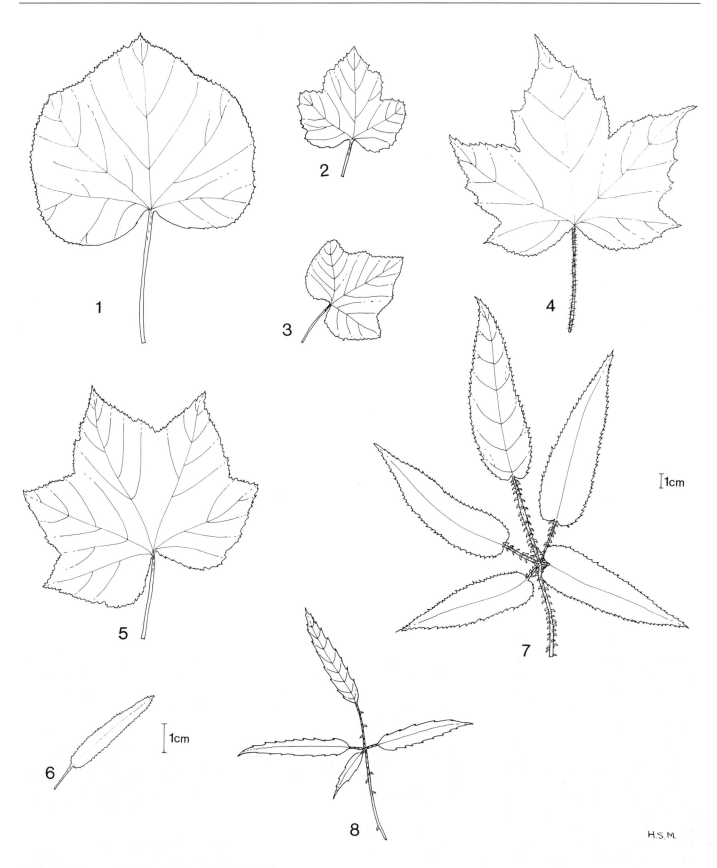

Figure 31. Leaves of *Rubus* species (× 0.5). 1, *R. irenaeus.*
2, *R. delicious.* 3, *R. trilobus.* 4, *R. odoratus.* 5, *R. parviflorus.*
6, *R. parvus.* 7, *R. australis.* 8, *R. squarrosus.*

toothed, to 3 cm. Flowers 1.5–2 cm across, in short, condensed clusters in the leaf axils. Sepals ovate, the apex acuminate-tailed, densely hairy. Petals white, slightly exceeding the sepals. Anthers with long hairs. Fruit red. *C China*. H3. Summer.

Subgenus **Anoplobatus** Focke. Arching or upright shrubs without prickles; leaves simple, lobed; stipules, narrow, attached directly to the leaf-stalks. Flowers bisexual, usually large.

16. R. deliciosus Torrey. Figure 31(2), p. 351. Illustration: Botanical Magazine, 6062 (1873); Perry, Flowers of the world, 264 (1970); Bean, Trees & shrubs hardy in the British Isles, edn 8, **4**: 220 (1980).

Deciduous shrub to 3 m, with arching and spreading branches; bark brownish, peeling, without prickles, but with soft, white hairs while young. Flowering-shoots softly hairy. Leaves 2.5–6.5 × 2–8.5 cm, semi-circular to ovate in outline, with 3 or 5 rather rounded, coarsely and irregularly toothed lobes, base truncate to cordate, softly hairy above and beneath. Stipules oblong, entire. Flower solitary, 5–6.5 cm across. Sepals ovate, tailed, hairy and with scattered glands outside, remaining greenish or becoming completely brown in fruit. Petals exceeding the sepals. Fruit hemispherical, dark purple. *W USA*. H5. Summer.

17. R. trilobus Seringe. Figure 31(3), p. 351. Illustration: Botanical Magazine, n.s., 452 (1964).

Very similar to *R. deliciosus* but leaves 4–7.5 × 4–7 cm, darker green and with more acute lobes, sepals less glandular outside, becoming brown outside and red inside in fruit. *S Mexico*. H5–G1. Spring–summer.

Very difficult to distinguish from *R. deliciosus*, though from a different area. In gardens, the hybrid between them, 'Benenden', often known as *R.* × 'Tridel', makes identification even more difficult.

18. R. odoratus Linnaeus. Figure 31(4), p. 351. Illustration: Botanical Magazine, 323 (1801); House, Wild flowers, pl. 101 (1961); Justice & Bell, Wild flowers of North Carolina, pl. 89 (1968).

Upright or scrambling deciduous shrub to 3 m; bark peeling, pale brown; shoots without prickles, hairy when young, flowering-shoots with soft hairs and conspicuous stalked glands. Leaves broadly triangular, 7–17 × 8–24 cm, cordate at base, with 3 or 5 triangular, acute, toothed lobes, sparsely hairy above, hairy along

the veins beneath. Stipules lanceolate. Flowers 3.3–5 cm across, usually numerous in spreading panicles, fragrant; branches of the panicle conspicuously glandular to the naked eye. Sepals ovate, tailed, glandular-hairy. Petals exceeding the sepals, bright purple. Fruit red. *N America, N Mexico*. H2. Late spring–summer.

19. R. parviflorus Nuttall. Figure 31(5), p. 351. Illustration: The Garden **102**: 224 (1977); Stace, New flora of the British Isles, fig. 405 (1991).

Upright or rather low and tangled deciduous shrub to 2 m; bark peeling, shoots without prickles but with soft hairs and shortly stalked glands when young, flowering-shoots with soft hairs and similar glands. Leaves almost circular in outline, 4–19 × 5–22 cm, with 3 or 5 (rarely 7) triangular, acute, coarsely toothed lobes, cordate at base, with soft spreading hairs and shortly stalked glands above and beneath. Stipules lanceolate. Flowers 2–10 in corymb-like racemes (rarely solitary), 2.5–4 cm across; branches of racemes softly hairy and with shortly stalked glands (not conspicuous to the naked eye). Sepals ovate, tailed, hairy and glandular. Petals white, exceeding the sepals. Fruit broadly convex, red. *N America, N Mexico*. H2. Summer.

Subgenus **Lampobatus** Focke. Scrambling evergreen shrubs with leathery leaves, usually compound, the leaflets often reduced and much shorter than their prickly stalks. Flowers small, unisexual, in usually widely spreading panicles.

20. R. parvus Buchanan. Figure 31(6), p. 351. Illustration: Salmon, Field guide to the alpine plants of New Zealand, pl. 204, 205 (1968).

Low-growing shrub with the main stems creeping and rooting, without prickles when mature. Leaves simple (reduced to a single, terminal leaflet), leathery, linear to linear-oblong, acute, shallowly cordate or rounded at base, margins regularly and finely toothed, borne on long, prickly stalks. Flowers in panicles to 2 cm across. Sepals ovate-acuminate, hairy. Petals white. Fruit red. *New Zealand (South Island)*. H5–G1.

21. R. australis Forster. Figure 31(7), p. 351.

Young plants creeping and rooting. Shoots slender, hairy, usually with numerous slender prickles. Leaves evergreen, divided into 3–5 leaflets with

long, prickly stalks; leaflets thin, oblong, toothed, acute, 1–3 × 1–2 cm. Adult plants forming mounds to 10 m or more. Adult leaves similar to the juvenile but relatively broader, leathery, rounded to the base, 3–5 × 1–3.5 cm; stalks and leaflet-stalks very prickly. Flowers in panicles 6–12 cm across. Sepals ovate, obtuse, hairy. Petals white, somewhat exceeding sepals. Fruit yellowish. *New Zealand*. H5–G1.

Normally seen only in the juvenile phase in cultivation.

22. R. squarrosus Fritsch. Figure 31(8), p. 351.

Similar to *R. australis* but with yellowish prickles, leaflets lanceolate or oblong-lanceolate, with few, rather triangular large teeth or lobes, to 7 × 3 cm, rounded at the base; the blades of the leaflets are often very reduced or absent; petals yellowish; fruit orange-red. *New Zealand*. H5–G1.

The most frequently cultivated of the New Zealand species (often called *R. australis* in gardens). When the leaflet-blades are reduced, as is often the case, the plant resembles a mass of barbed wire, consisting mostly of the prickly stems, leaf- and leaflet-stalks.

Subgenus **Idaeobatus** Focke. Most plants with biennial stems, with or rarely without prickles. Leaves simple or mostly divided into several leaflets. Stipules narrow, attached to the leaf-stalks. Flowers bisexual. Fruit separating from the cylindric or conical receptacle when ripe, and therefore hollow.

23. R. microphyllus Linnaeus filius (*R. incisus* Thunberg). Figure 32(1), p. 353.

Low, trailing shrub; stems reddish, white-bloomed, with a few prickles. Leaves simple, ovate or broadly ovate in outline, 1.5–3.5 × 1.5–3.2 cm, entire or shallowly 3-lobed, margins finely toothed, hairless above or sparsely hairy along the veins towards the base, hairy along the veins beneath, undersurface glaucous; base truncate or cordate, apex acute. Flowers 1.5–2.2 cm across, erect, solitary in the leaf-axils. Sepals lanceolate, hairless outside, densely hairy inside. Petals white or white flushed with pink, considerably longer than the sepals. *Japan, China*. H3. Spring.

24. R. crataegifolius Bunge. Figure 32(2), p. 353.

Upright shrub to 3 m; stems prickly, reddish, not bloomed. Leaves simple, palmately 3- or 5-lobed, 4–16 × 3.5–19

Figure 32. Leaves of *Rubus* species (× 0.5). 1, *R. microphyllus*.
2, *R. crataegifolius*. 3, *R. spectabilis*. 4, *R. illecebrosus*.
5, *R. rosifolius*. 6, *R. amabilis*. 7, *R. biflorus*. 8, *R. lasiostylus*.
9, *R. thibetanus*. 10, *R. cockburnianus*.

cm, broadly ovate in outline, the middle lobe often narrowed to its base, base truncate or cordate, apex acute, margins irregularly doubly toothed; hairless or sparsely hairy on the veins above, more densely so on the veins beneath. Flowers 1.5–2 cm across, few in racemes or clusters. Sepals lanceolate, hairless or sparsely hairy outside, densely hairy inside. Petals white, exceeding the sepals. Fruit red. *Japan, China, Korea*. H3. Summer.

25. R. spectabilis Pursh. Figure 32(3), p. 353. Illustration: Edwards's Botanical Register **17**: pl. 1425 (1831).

Upright shrub to 2 m, spreading rapidly by rhizomes; stems with numerous fine prickles below. Leaves divided into usually 3 leaflets, hairless or with long hairs along the veins above, with rather dense long hairs along the veins beneath; leaflets deeply toothed or somewhat lobed, terminal leaflet ovate or broadly ovate, 4.5–9 × 2.5–7 cm, lateral leaflets 2.5–7 × 1.5–5 cm. Flowers solitary (occasionally 2 or 3 together) in the leaf-axils, long-stalked, drooping, fragrant, 2–3.5 cm across. Sepals lanceolate, acuminate, hairy inside and out. Petals pink or pinkish purple, considerably exceeding the sepals. Fruit orange. *W North America*. H3. Late spring.

This can become a very invasive weed in some situations; once established, it is very difficult to eradicate. A double-flowered variant ('Flore Pleno') is sometimes grown.

26. R. illecebrosus Focke. Figure 32(4), p. 353. Illustration: Botanical Magazine, 8704 (1917).

Small shrub with creeping stems and annual, upright flowering-shoots to 1 m; stems prickly. Leaves mostly with 5–7 lanceolate, doubly toothed leaflets with conspicuous pinnate veins, hairless or sparsely bristly above, hairless beneath; terminal leaflet 4–10.5 × 1.5–2.5 cm, lateral leaflets 3.5–9 × 1–2.5 cm. Flowers mostly solitary in the leaf axils, or 2–3 together forming a flat-topped cluster, 2.5–4.5 cm wide. Sepals usually hairless outside, very densely hairy towards the apex inside. Petals white, exceeding the sepals. Fruit large, scarlet. *Japan*. H3. Late summer.

27. R. rosifolius Smith. Figure 32(5), p. 353. Illustration: Botanical Magazine, 6970 (1887).

Erect or sometimes scrambling shrub to 2.5 m (though often less), stems softly hairy and prickly. Leaves with 5–7 lanceolate or more rarely ovate, doubly toothed leaflets, sparsely hairy above with adpressed bristles, beneath with similar bristles along the veins and with glistening, yellow stalkless glands on the surface; terminal leaflet 3–8 × 1.5–2.5 cm, lateral leaflets 2–7 × 1–2 cm. Flowers 2–3.5 cm across, solitary. Sepals lanceolate, long-acuminate, with fine spreading hairs and yellow glands outside, and dense, velvety hairs inside. Petals as long as to slightly longer than the sepals, white. Fruit red. *Himalaya to Japan & Indonesia (Sumatra)*. H5. Summer.

A double-flowered variant, var. **coronarius** Sweet is sometimes grown.

28. R. amabilis Focke. Figure 32(6), p. 353.

Upright shrub to 2 m or more, stems with small prickles. Leaves with 7–11 ovate to lanceolate, conspicuously doubly toothed leaflets, hairless above, with long fine hairs and a few small prickles along the veins beneath; terminal leaflet 2.5–6 × 1.5–3.5 cm, lateral leaflets 1–3 × 0.5–2 cm. Flowers 3–5 cm across, solitary. Sepals lanceolate, hairy and glandular outside, sparingly hairy inside towards the apex. Petals white, longer than the sepals. Fruit red. *W China*. H5. Summer.

29. R. biflorus Smith. Figure 32(7), p. 353. Illustration: Botanical Magazine, 4678 (1852).

Upright shrub to 3 m; stems very glaucous or whitish bloomed, with numerous prickles. Leaves with mostly 3 leaflets, sparsely hairy above, densely white-hairy beneath, margin with irregular, rounded teeth, sometimes somewhat lobed. Flowers 1–3 together in the leaf-axils, to 2 cm across. Sepals ovate, hairless outside, hairy inside towards the acuminate tip. Petals white, as long as or slightly longer than the sepals. Fruit yellow. *Himalaya*. H3. Spring–summer.

Var. **quinqueflorus** Focke, with conspicuously white stems and more flowers in the cluster, is usually grown.

30. R. lasiostylus Focke. Figure 32(8), p. 353. Illustration: Botanical Magazine, 7426 (1895).

Upright shrub to 2 m; stems glaucous or whitish bloomed, with numerous prickles. Leaves with mostly 3 (rarely 5) leaflets, sparsely bristly above, densely white-hairy beneath; terminal leaflet broadly ovate to almost circular, base somewhat cordate, 7.5–11 × 6.5–9 cm, considerably larger than the lateral leaflets (which are 6–9 × 3.5–6 cm), all irregularly coarsely doubly toothed or somewhat lobed. Flowers 1–few, drooping, in axillary clusters, 2.6–3.4 cm across. Sepals ovate-lanceolate, acuminate, hairless outside, densely white-hairy within. Petals reddish, shorter than the sepals. Styles hairy. Fruit red but densely covered in long, whitish hairs which obscure the fruit colour. *W China*. H5. Summer.

31. R. thibetanus Franchet. Figure 32(9), p. 353.

Shrub to 2 m with very bloomed, prickly stems. Leaves divided into 7 or more leaflets, lanceolate overall, tapering to the apex, rather densely greyish hairy above, very densely white-hairy beneath, toothed; terminal leaflet lanceolate, pinnately lobed towards its base, toothed towards its apex, 3–4.5 × 1–2 cm, lateral leaflets 0.9–2 × 0.4–1 cm. Flowers in racemes terminating short leafy shoots (sometimes some solitary in axils below the raceme), 7–9 mm across. Sepals densely hairy inside and out. Petals red-purple, hairy outside especially towards the base. Fruit black, bloomed. *W China*. H5. Spring–summer.

32. R. cockburnianus Hemsley (*R. giraldianus* Focke). Figure 32(10), p. 353. Illustration: Journal of the Royal Horticultural Society **94**: 514 (1969).

Upright shrub to 3 m; stems whitish bloomed, with scattered prickles. Leaves with 5–7 (rarely to 9) leaflets, dark green and sparsely hairy at least along the veins above, with whitish hairs beneath giving a greenish white colour to the surface, all irregularly doubly toothed; terminal leaflet ovate, sometimes lobed, to 10 × 5 cm; lateral leaflets narrower, 6–7.5 × 2.5–3 cm. Flowers numerous, in terminal racemes. Petals pinkish purple. Fruit black. *China*. H3. Spring–summer.

33. R. coreanus Miquel. Figure 33(1), p. 355.

Upright or arching shrub to 3 m; stems with conspicuous white bloom and numerous prickles. Leaves with usually 5 (rarely 7) leaflets, hairless above except along the veins, usually densely white-hairy beneath, margins irregularly toothed; terminal leaflet usually wedge-shaped or obovate, broadest towards the top, 2–5.5 × 1.5–4.5 cm; lateral leaflets ovate, 1.5–5 × 1–3.5 cm. Flowers 1–1.7 cm across, in flat-topped terminal racemes. Sepals lanceolate, acuminate, rather loosely hairy outside, more densely so inside. Petals pink, shorter than the sepals. Fruit red or blackish.

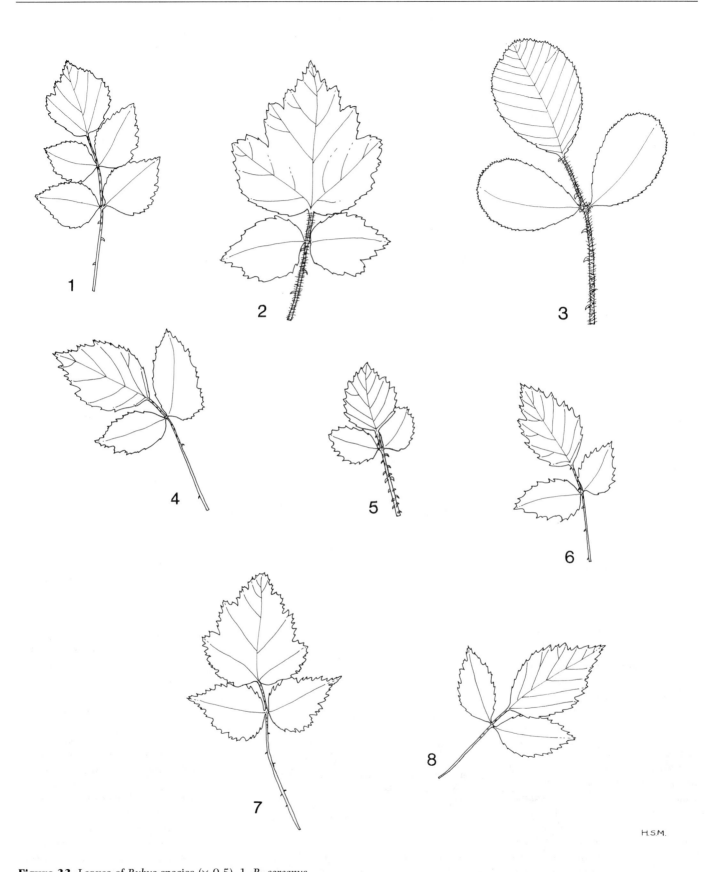

Figure 33. Leaves of *Rubus* species (× 0.5). 1, *R. coreanus*.
2, *R. phoenicolasius*. 3, *R. ellipticus*. 4, *R. occidentalis*.
5, *R. leucodermis*. 6, *R. glaucifolius*. 7, *R. mesogaeus*. 8, *R. idaeus*.

China, Japan, Korea. H3. Spring–summer.

34. R. phoenicolasius Maximowicz. Figure 33(2), p. 355. Illustration: Botanical Magazine, 6479 (1889); Gardening from 'Which', February, 28 (1988).

Upright shrub to 3 m or more; stems hairy and with numerous reddish purple bristles and few fine, almost straight prickles. Leaves with 3 broadly ovate to ovate, acute, shallowly lobed and toothed leaflets, hairless or sparsely hairy above, densely white-hairy beneath, often also with a few bristles or prickles along the veins; terminal leaflet 4.5–11 × 2–11 cm, lateral leaflets 3.5–8.5 × 1.5–6 cm. Flowers 1.5–2.5 cm across, few, in short, rather congested racemes, stalks very bristly. Sepals lanceolate, acuminate, very bristly. Petals pink, shorter than the sepals. Fruit bright red. *Japan, Korea, N China.* H3. Summer.

35. R. ellipticus Smith. Figure 33(3), p. 355.

Erect shrub to 3 m; stems softly hairy and with numerous bristles and scattered, downwardly hooked prickles. Leaves with 3 distinct broadly elliptic to almost circular, rounded, evenly toothed leaflets, dark green and hairless above, densely white-hairy beneath, with bristles and a few prickles on the main veins; terminal leaflet 4–10 × 2.5–7.5 cm, lateral leaflets 2.5–6.5 × 1.5–4.5 cm. Flowers 8–14 mm across, numerous in rather dense panicles, the panicle- and flower-stalks bristly. Sepals oblong, acute, softly hairy on the lobes, softly hairy and bristly on the perigynous zone. Petals white, about as long as the sepals. Fruit yellow. *Himalaya to SW China.* H5. Spring.

Very distinctive with its bristly stems and broad leaflets with rounded tips.

36. R. occidentalis Linnaeus. Figure 33(4), p. 355. Illustration: Gentes Herbarum **5**: 881 (1945).

Upright shrub to 2.5 m, the first-year stems ultimately arching and rooting at their tips, usually strongly white-bloomed. Leaves with 3 leaflets which are coarsely and irregularly toothed, dark green above, densely white-hairy beneath; terminal leaflet 3–12 × 1.5–8 cm, long-stalked, often somewhat cordate at the base, lateral leaflets 2.5–10 × 1–6 cm. Flowers 1.2–1.7 cm across, mostly in a close, flat-topped cluster; flower-stalks with soft white hairs and thin, scarcely hooked, rather narrowly based prickles. Sepals triangular, acuminate, downy. Petals white, almost as long as sepals. Fruit black, bloomed. *E North America, extending west to Colorado.* H3. Summer.

37. R. leucodermis Douglas. Figure 33(5), p. 355. Illustration: Gentes Herbarum **5**: 885 (1945).

Very similar to *R. occidentalis* but leaflets smaller (terminal 2–7.5 × 1.5–4.5 cm, laterals 1–6 × 1–3.5 cm), prickles on the flower-stalks usually broadly based and hooked; fruit purplish. *W North America.* H4. Summer.

38. R. glaucus Bentham.

Similar to *R. occidentalis* but leaflets finely toothed, hairless above, fruit reddish black. *C & S America.* G1.

Doubtfully in cultivation in Europe.

39. R. glaucifolius Kellogg. Figure 33(6), p. 355.

Upright arching or trailing shrub, the main stems not rooting at the tip. Leaves with 3 leaflets which are coarsely and irregularly toothed, finely downy to hairless above, densely white-hairy beneath; terminal leaflet 3–5 × 2–3.5 cm, long-stalked, often more or less 3-lobed, lateral leaflets 2–4 × 1–2.5 cm. Flowers 1–1.5 cm across, mostly in close, flat-topped clusters; flower-stalks with stalked glands and dense hairs, prickles few to many, rather broadly based, mostly straight. Sepals upright, triangular, downy and usually with stalked glands. Petals white, somewhat shorter to slightly longer than the sepals. Fruit red to purple. *Western USA (California).* H5. Summer.

40. R. mesogaeus Focke. Figure 33(7), p. 355.

Stems scrambling to 3 m or more, hairy and with few broadly based prickles. Leaves with 5 leaflets, or if with 3 then the terminal leaflet tending to be 3-lobed; leaflets ovate, acute, lobed and toothed, dark green and sparsely hairy above, whitish or greyish with dense hairs beneath and with a few small prickles along the midrib; terminal leaflet 3–12 × 2–8 cm, lateral leaflets 1.5–9.5 × 1–5 cm. Flowers 0.9–1.6 cm across in short, flat-topped, corymbose racemes. Sepals very narrowly lanceolate, acuminate, densely white-hairy. Petals pink or white, shorter than sepals. Fruits black. *C & W China.* H3. Summer.

41. R. idaeus Linnaeus. Figure 33(8), p. 355. Illustration: Ary & Gregory, The Oxford book of wild flowers, 78 (1962); Keble Martin, The concise British flora in colour, edn 2, pl. 28 (1969).

Very variable, widely cultivated and often escaping. Stems to 2.5 m, first-year stems often glaucous or bloomed, with or without prickles. Leaves with 3 leaflets (up to 7 leaflets on the leaves of first-year stems), hairy or hairless above, densely white-hairy beneath, margins toothed; leaflets very variable in shape and size, the terminal stalked, 3–13 × 1.5–12.5 cm, laterals 2.5–10.5 × 1–6.5 cm. Flowers mostly in racemes which are not close and flat-topped, 1–1.7 cm across, stalks with or without prickles. Sepals triangular, acuminate, usually downy. Petals white, shorter than the sepals. Fruit red (rarely yellowish). *Europe, N Asia, introduced elsewhere.* H1. Early summer.

The European Raspberry, widely cultivated for fruit, less commonly for ornament, but spreading rapidly in gardens and orchards by both suckers and seed.

R. strigosus Michaux (*R. idaeus* var. *strigosus* (Michaux) Maximowicz. Illustration: Gentes Herbarum **5**: 869 (1945). Very like *R. idaeus* but often with glandular hairs and numerous bristles on stems, leaf- and flower-stalks. *N America.* H4. Summer.

Subgenus **Rubus**. Like Subgenus *Idaeobatus* but fruit not separating from the receptacle, dispersed with it by abscission of the fruit-stalk.

The species of this Subgenus are particularly difficult to identify. Because of their tendency to produce seed without fertilisation (apomixis), their ability, when fertile, to hybridise easily, (the resultant hybrids often persisting by apomixis), and the occurrence of polyploidy, a vast complex of variants is found in Europe and N America. Many hundreds of species (microspecies) have been named, and their identification in the wild is difficult, even for experts. Though only a few species are reputedly grown for ornament, their relationship to the wild species is uncertain and identification is thus even more difficult. Fortunately, the plants are not particularly ornamental and very few are deliberately grown for this purpose, though wild plants (from seed distributed by birds) may well appear and persist in wild gardens or neglected areas.

The account below covers those named species that are reported to have occurred in cultivation (whether they actually have been or still are cultivated remains a matter of considerable doubt). It is intended for guidance only; for further information the copious literature on the group should be studied.

Elaborate hierarchies of sections, subsections and series have been devised to group these microspecies; unfortunately, work in Europe and North America has not been well coordinated, and it is not possible to reconcile the various systems here, as only a small number of the microspecies are included. The literature listed below should be consulted for details.

Literature: Sudre, H., *Rubi Europae* (1908–13); Bailey, L.H., *Rubus in North America*, *Gentes Herbarum* **5** (1941–45); Watson, W.C.R., *Handbook of Rubi of Great Britain and Ireland* (1958); Weber, H.E., *Die Gattung Rubus L. (Rosaceae) in nordwestlichen Europa* (1972); Weber, H.E., *Rubi Westfalici* (1985); Edees, E.S. & Newton, A., *Brambles of the British Isles* (1988).

42. R. laciniatus Willdenow.

Robust plant, stems armed with numerous, equal, hooked prickles. Leaves with 5 leaflets which are deeply lobed, greenish and hairy or not beneath. Inflorescence widely spreading, with numerous, hooked prickles. Sepals deflexed. Petals white or pink, toothed at the apex. *Origin unknown, widely cultivated and naturalised.* H3. Summer.

A prickleless variant is also cultivated.

43. R. elegantispinosus (Schumann) Weber.

Like *R. laciniatus* but stems sometimes bloomed, prickles straight or deflexed, usually purple; leaflets 5, not lobed, with felted white hairs beneath; petals deep pink. *W Europe.* H3. Summer.

44. R. allegheniensis Bailey.

Robust, upright shrub to 3 m, yellowish green overall; stems with strong, scattered prickles. Leaves of 3 or 5 leaflets, toothed, blunt, the terminal cordate, greenish and hairy beneath, usually with stalked glands. Inflorescence raceme-like, with conspicuous stalked glands. Petals white, narrow. *E North America.* H3. Summer.

R. rosa Bailey. Illustration: Gentes Herbarum **5**: 541, 543 (1944). Similar, but inflorescence shorter and more compact. *E USA.*

45. R. frondosus Bigelow. Illustration: Gentes Herbarum **5**: 733 (1945).

Upright, to 2.5 m, branches sometimes arching and rooting at the tip, with few, straight or hooked prickles. Leaves usually with 3 toothed leaflets, green and hairy beneath. Flower-clusters corymb-like, flat, without an long axis. Petals white, rather narrow. *E USA.* H3. Summer.

46. R. bellobatus Bailey. Illustration: Gentes Herbarum **5**: 665, 667, 668 (1945).

Like *R. frondosus* but prickles straighter, flower-clusters raceme-like, longer than broad, with an long axis, and larger fruits. *E USA.* H3. Summer.

47. R. ulmifolius Schott. Illustration: Keble Martin, The concise British flora in colour, edn 2, pl. 28 (1969).

Stems robust, arching, conspicuously bloomed, with broadly based, straight or hooked prickles. Leaves divided into 3–5 leaflets which are densely white-hairy beneath. Inflorescence raceme-like. Sepals deflexed. Petals pink or white, crumpled. *S, W & C Europe.* H3. Summer.

A variant with double flowers ('Bellidiflorus') is sometimes grown, as is one with variegated leaves ('Variegatus').

48. R. bifrons Trattinick.

Similar to *R. ulmifolius* but stems not bloomed, lowermost leaflets with short but distinct stalks, prickles in the inflorescence mostly straight; petals white or pale pink. *W & C Europe.* H3. Summer.

49. R. linkianus Seringe. Illustration: Addisonia **22**: pl. 722 (1943–46).

Similar to *R. ulmifolius* but stems not bloomed; lowermost leaflets without stalks; prickles in the inflorescence hooked; flowers always double. *Origin unknown.* H3. Summer.

50. R. procerus Boulay (*R. discolor* misapplied; *R. armeniacus* Focke).

Similar to *R. ulmifolius* but stem not bloomed, lowermost leaflets with short, distinct stalks, prickles in the inflorescence mostly hooked, petals white or pale pink; anthers hairy. *S, W & C Europe.* H3. Summer.

'Himalayan Giant' ('Theodore Reimers') has larger leaves and flowers and is generally grown for its fruit; it is naturalised in various parts of Europe (see Kent, D.H., Rubus procerus 'Himalayan Giant', *Kew Magazine* **5**: 32–36, 1988).

51. R. cuneifolius Pursh. Illustration: Gentes Herbarum **5**: 424, 429 (1943).

Small, stiffly erect, ash-grey plant, stems with numerous hooked prickles. Leaves usually with 3 leaflets, these rather obovate and obtuse, whitish beneath with dense, felted hairs. Flower-clusters with 3–5 flowers, short and not raceme-like, without a distinct axis. Petals white, rather broad. *E USA.* H3. Summer.

52. R. trivialis Michaux. Illustration: Gentes Herbarum **5**: 200, 204 (1941).

Variable trailing plant, the tips of the branches rooting; stem bearing numerous, hooked prickles and bristles. Leaves of first-year shoots leathery, persisting until flowering, usually with 5 leaflets; those on the flowering shoots similar but smaller and often with 3 leaflets. Flowers solitary or 2 or 3 together, on long, prickly and glandular stalks. Petals white to pinkish, rather broad, often overlappping. *Widely distributed in S & E USA.* H3. Summer.

53. R. flagellaris Willdenow. Illustration: Gentes Herbarum **5**: 245, 248 (1943).

Trailing plant, rooting at the tips of the branches and sometimes also at the nodes; stems with sparse, hooked prickles. Leaves all deciduous, those on the flowering-shoots generally with 3 leaflets which are usually obovate and then abruptly tapered, sometimes shouldered. Flowers in clusters of 1–3 on long stalks, 2–3 cm across. Petals white, obtuse, not overlapping. *E North America.* H3. Summer.

54. R. roribaccus Rydberg. Illustration: Gentes Herbarum **5**: 255 (1943).

Similar to *R. flagellaris* but larger and more vigorous; leaflets ovate, gradually tapered to the apex; flowers 3–4 cm across. *SE USA.* H3. Summer.

Two other generally very similar species may also be found: **R. velox** Bailey with more oblong leaflets and **R. almus** (Bailey) Bailey with elliptic leaflets softly hairy beneath, both from Texas.

55. R. caesius Linnaeus. Illustration: Polunin, Flowers of Europe, pl. 44 (1969).

Stem trailing, hairless, bloomed, with weak, broadly based, straight or hooked prickles. Leaves with 3 leaflets, downy beneath, the lateral leaflets often somewhat lobed. Inflorescence corymb-like with 2–5 flowers. Flowers 2–2.5 cm across, bisexual. Sepals erect. Petals white, broad. Stamens greenish. Fruit black, bloomed, the individual drupelets easily separable. *Most of Europe, introduced & naturalised elsewhere.* H1. Summer.

56. R. vitifolius Chamisso & Schlechtendahl. Illustration: Gentes Herbarum **5**: 49 (1941).

Plant trailing; stems with straight, narrowly based prickles (almost narrow enough to be considered bristles). Leaves on first-year shoots with 3 leaflets, or deeply 3-lobed, those on flowering-shoots

3-lobed or scarcely lobed at all, ovate, bright green, thinly hairy when young. Flowers few to numerous, unisexual (stamens or carpels sterile), in large or small, raceme-like clusters. Petals white, oblong or ovate, not overlapping. Fruit black, oblong, hairy. *W USA (California)*. H5. Summer.

57. R. ursinus Chamisso & Schlechtendahl. Illustration: Gentes Herbarum **5**: 55 (1941).

Like *R. vitifolius* but all parts, especially the leaves, more densely greyish hairy, the leaves dull green; leaves on first-year shoots with 3 or 5 leaflets, those on flowering-shoots with 3 leaflets or 3-lobed. Fruit oblong to conical. *W USA (Oregon, California)*. H5. Summer.

R. loganobaccus L.H. Bailey. Like *R. ursinus*, but with stems to 4 m. Leaves with dense down beneath. Fruit to 4 cm, blackish, glossy. *California*. The loganberry of commerce.

58. R. macropetalus Douglas. Illustration: Gentes Herbarum **5**: 63 (1941).

Like *R. vitifolius* and *R. ursinus*, but plant with numerous, conspicuous, stalked glands; all leaves with 3 or 5 leaflets; petals white, very variable in size. Fruit oblong, not hairy. *W USA*. H5. Summer.

24. ROSA Linnaeus
V.A. Mathews

Deciduous, or sometimes evergreen, shrubs with erect, arching or scrambling, occasionally trailing stems, usually armed with prickles and/or bristles. Stipules usually present, persistent, usually joined to leaf-stalk for most of their length. Leaves alternate, usually odd-pinnate, rarely with 3 leaflets or simple; leaflets toothed. Perigynous zone spherical to urn-shaped. Flowers solitary or in corymbs, usually borne at the end of short branches, single to double. Sepals 5 (rarely 4), entire or the 2 outer and half the third one with lateral lobes, and the inner 2 and half the third one entire, the tips acute to attenuate or broadened and leafy. Petals 5 (rarely 4) in single flowers, usually obovate, tip often notched, white, cream, pink, red, purplish, orange or yellow. Stamens (in single flowers) 30–200, in several whorls. Carpels many, each with 1 ovule, free, carried on the base and/or sides of the enclosing receptacle (perigynous zone). Styles free or united into a column, protruding from mouth of receptacle or not. Fruit (hip or hep) containing many

achenes, usually red or orange, sometimes blackish or green, enclosed by the fleshy receptacle. Achenes l-seeded, bony, often hairy.

A genus of over 100 species (possibly 150), distributed through temperate and subtropical zones of the northern hemisphere. Problems of classification and naming are rife in this genus, many species of which have been cultivated and hybridised for centuries, and some of which are very polymorphic. The last complete account of the genus was by Lindley (*Rosarum Monographia*) in 1820: a new revision is urgently needed which will incorporate investigations into the complex cytology and the problems produced by hybridisation. Thus, it must be stressed that this account is somewhat provisional.

Literature: Krüssmann, G. *The complete book of roses* (1981) (English translation of Rosen, Rosen, Rosen, 1974); *Modern Roses 9*. The international checklist of Roses (1987); Thomas, G.S. *The old shrub roses* (1955, revised edn. 1983); Thomas, G.S. *Shrub roses of today* (1962); Thomas, G.S. *Climbing roses old and new* (1983); Willmott, E. *The genus Rosa*. 2 vols. (1910–1914).

1a. Leaves simple, lacking stipules; if leaves pinnate and/or stipules present, then petals with a basal crimson blotch   2
  b. Leaves pinnate with 3–13 leaflets, stipules present   3
2a. Leaves simple, grey-green; fruit very bristly   **1. persica**
  b. Leaves simple or with up to 7 leaflets, dark green; fruit with a few bristles   **2. × hardii**
3a. Flowers yellow   4
  b. Flowers white, cream, green, orange, pink, red or purplish   13
4a. Margin of leaflets with simple teeth   5
  b. Margin of leaflets with compound teeth   10
5a. Leaflets 11–19   **6. xanthina**
  b. Leaflets 3–9   6
6a. Flowers 2–3 cm across   7
  b. Flowers 3.8–8 cm across   8
7a. Flowers single or double, borne in many-flowered clusters; sepals reflexed and falling after flowering   **104. banksiae**
  b. Flowers single, solitary; sepals erect and persistent after flowering   **7. ecae**
8a. Sepals reflexed and falling after flowering   **103. × odorata**

  b. Sepals erect and persistent after flowering   9
9a. Flowers double; prickles decurved   **12. hemisphaerica**
  b. Flowers single or semi-double; prickles straight or slightly curved, flattened, often reddish   **6. xanthina**
10a. Flowers 2.5–4.5 cm across   **9. primula**
  b. Flowers 5–7.5 cm across   11
11a. Flowers double   **11. × harisonii**
  b. Flowers single   12
12a. Leaflets 0.8–1 cm; tips of sepals not expanded; flowers scentless   **8. kokanica**
  b. Leaflets 1.5–4 cm; tips of sepals expanded; flowers smelling unpleasant   **10. foetida**
13a. Bracts absent   14
  b. Bracts present, sometimes very small and falling early   52
14a. Flowers coppery red inside, yellow-buff outside; margin of leaflets with double teeth   **10. foetida**
  b. Flowers not as above; margin of leaflets with simple teeth; if teeth double then flowers white, ageing to pinkish   15
15a. Flowers solitary or 2–4 together   16
  b. Flowers in clusters of 5–100 or more   24
16a. Leaves evergreen   17
  b. Leaves deciduous   19
17a. Petals white or creamy white   18
  b. Petals pale pink with deeper veins   **107. × anemonoides**
18a. Flowers in clusters of 3–20   **98. luciae**
  b. Flowers solitary   **106. laevigata**
19a. Receptacle and fruit prickly   **111. stellata**
  b. Receptacle and fruit lacking prickles   20
20a. Fruit dark brown, becoming blackish   **3. pimpinellifolia**
  b. Fruit yellow, orange, red, deep crimson or purplish red   21
21a. Fruit purplish red, drooping   **4. × reversa**
  b. Fruit neither purplish red nor drooping   22
22a. Sepals falling after flowering   **100. arvensis**
  b. Sepals persistent after flowering   23
23a. Sepals glandular on the back; petals white tinged with pink   **5. koreana**

b. Sepals hairless or silky on the back; petals white, cream or rose-pink
13. sericea

24a. Flowers double 25

b. Flowers single 27

25a. Stigmas hairless; leaflets 7–9 (rarely 5–11) **82. multiflora**

b. Stigmas hairy; leaflets 3–5
85. × beanii

26a. Flowers pale to deep pink 27

b. Flowers white or cream 30

27a. Sepals entire; leaflets linear-lanceolate **83. watsoniana**

b. Sepals with lateral lobes; leaflets elliptic to oblong, ovate or roundish 28

28a. Flowers 5–7.5 cm across
84. setigera

b. Flowers 1.5–5 cm across 29

29a. Prickles more or less curved; sepals smooth or sparsely glandular on the back; fruit 6–25 mm
100. arvensis

b. Prickles curved downwards; sepals glandular-bristly on the back; fruit 6–7 mm **82. multiflora**

30a. Leaflets linear-lanceolate
83. watsoniana

b. Leaflets narrowly elliptic to almost round 31

31a. Sepals persistent after flowering
90. moschata

b. Sepals falling after flowering 32

32a. Petals silky or slightly downy on the back 33

b. Petals hairless on the back 34

33a. Flowers *c.* 5 cm across, up to 15 in a cluster **91. longicuspis**

b. Flowers 2–2.5 cm across, up to 100 or more in a cluster **94. filipes**

34a. Flowers solitary, or up to 8 in 5–5.5 cm across on stalks 2.5–3.5 cm **88. mulliganii**

b. Flowers many in a cluster 39

35a. Leaflets obtuse **93. wichuraiana**

b. Leaflets acute to acuminate 35

36a. Leaflets downy beneath 36

b. Leaflets more or less hairless or hairy only the veins beneath 37

37a. Flowers 2.5–3.8 cm across on stalks 1–2.5 cm **87. rubus**

b. Flowers 4.5–5.5 cm across on stalks 2.5–3.5 cm **88. mulliganii**

38a. Stigmas hairless; leaves deciduous
100. arvensis

b. Stigmas woolly; leaves evergreen or semi-evergreen **98. luciae**

39a. Leaflets usually 7 or 9, rarely 5 or 11 40

b. Leaflets usually 5, rarely 3 or 7 46

40a. Flowers to 100 or more in a cluster
94. filipes

b. Flowers to 50 in a cluster 41

41a. Leaves evergreen or semi-evergreen
97. wichuraiana

b. Leaves deciduous 42

42a. Stigmas hairless **82. multiflora**

b. Stigmas hairy, sometimes sparsely so 43

43a. Stems 3–4 m **96. soulieana**

b. Stems 5–12 m 44

44a. Leaflets bluish green or grey-green and usually somewhat hairy above
89. brunonii

b. Leaflets green and hairless above 45

45a. Flowers 2–4 cm across
86. helenae

b. Flowers 4.5–5.5 cm across
88. mulliganii

46a. Leaves evergreen or semi-evergreen 47

b. Leaves deciduous 48

47a. Leaflets obtuse, very shiny; fruit orange to dark red, 1–1.5 cm
97. wichuraiana

b. Leaflets acute to acuminate, less glossy, thinner; fruit red to purplish, *c.* 0.7 cm **98. luciae**

48a. Stems ultimately 5–12 m 49

b. Stems ultimately to 5 m 51

49a. Leaflets green, hairless above 50

b. Leaflets bluish green, somewhat hairy above **89. brunonii**

50a. Flowers 2–2.5 cm across; fruit orange, becoming crimson-scarlet, smooth **94. filipes**

b. Flowers 4.5–5.5 cm across; fruit dark red, bristly **88. mulliganii**

51a. Leaflets hairless beneath or more or less so **92. cerasocarpa**

b. Leaflets downy beneath, at least on the veins **87. rubus**

52a. Leaves evergreen or semi-evergreen 53

b. Leaves deciduous 60

53a. Flowers *c.* 1.5 cm across
105. cymosa

b. Flowers at least 2 cm across 54

54a. Flowers 10–15 cm across
102. gigantea

b. Flowers 2–10 cm across 55

55a. Bracts very small, falling quickly 56

b. Bracts larger, persistent 57

56a. Styles not protruding; fruit *c.* 0.7 cm **104. banksiae**

b. Styles protruding; fruit 1–1.6 cm
99. sempervirens

57a. Petals pale pink to scarlet or crimson, or green, or yellow with an orange back 58

b. Petals white; leaflets 5–11, obtuse 59

58a. Fruit 1.5–2 cm; leaflets 3–5, downy on midrib beneath
101. chinensis

b. Fruit 2–3 cm; leaflets 5–7, completely hairless beneath
103. × odorata

59a. Leaflets obtuse; bracts large, deeply toothed; sepals brown hairy on the back **108. bracteata**

b. Leaflets acuminate; bracts neither large nor deeply toothed; sepals smooth on the back
103. × odorata

60a. Sepals entire 61

b. At least some sepals with lateral lobes 149

61a. Flowers white 62

b. Flowers pale to deep pink, crimson, mauve-pink or purplish pink 72

62a. Flowers 6–7.5 (rarely to 9) cm across; leaflets conspicuously wrinkled above **45. rugosa**

b. Flowers 1.2–6.5 cm across; leaflets not wrinkled above 63

63a. Flowers 8 or more in a cluster; fruit red, becoming purplish
81. beggeriana

b. Flowers solitary or up to 7 in a cluster; fruit red to orange-red 64

64a. Sepals hairy inside 65

b. Sepals not hairy inside 66

65a. Sepals glandular-bristly outside; leaflets broadly elliptic to obovate
74. fedtschenkoana

b. Sepals smooth outside; leaflets narrowly ovate to elliptic
76. elegantula

66a. Margin of leaflets with compound teeth 67

b. Margin of leaflets with simple teeth 69

67a. Receptacle and fruit glandular-bristly **67. wardii**

b. Receptacle and fruit smooth 68

68a. Stipules with glandular teeth
58. nutkana

b. Stipules lacking glands **54. laxa**

69a. Leaflets 9–15 **68. murieliae**

b. Leaflets 5–9 70

70a. Margin of leaflets toothed in upper ⅔ only **71. webbiana**

b. Margin of leaflets toothed for whole length 71

71a. Flower-stalks smooth, hairless
56. woodsii

b. Flower-stalks glandular **54. laxa**

72a. Sepals falling after flowering 73

b. Sepals persisting after flowering 80

73a. Leaflets acute   74
  b. Leaflets obtuse   78
74a. Mature leaflets green   75
  b. Mature leaflets grey-green or tinged
     with brown or purple, covered with
     a dull waxy bloom   **39. glauca**
75a. Leaflets dull above   **40. palustris**
  b. Leaflets shiny above   76
76a. Sepals with expanded leafy tips;
     fruit 1–1.5 mm wide
                           41. virginiana
  b. Tips of sepals not leafy and
     expanded; fruit 7–10 mm wide   77
77a. Stems with sparse prickles and
     dense bristles; flower-stalks 1–3 cm
                               43. nitida
  b. Stems with sparse prickles, bristles
     rarely present; flower-stalks 0.5–1
     cm   **44. foliolosa**
78a. Flowers 5–7 cm across; receptacle
     glandular-bristly
                    16. × francofurtana
  b. Flowers 2.5–3.8 cm across;
     receptacle smooth   79
79a. Flowers rose-pink; stems with
     prickles intermixed with bristles
                          80. gymnocarpa
  b. Flowers purplish pink; stems with
     prickles only   **79. willmottiae**
80a. Margin of leaflets with compound
     teeth   81
  b. Margin of leaflets with simple teeth
                                            102
81a. Styles protruding   82
  b. Styles not protruding   83
82a. Leaflets 3–5, acute; flowers 1.9–2.5
     cm across   **59. corymbulosa**
  b. Leaflets 7–9 (rarely 5), obtuse;
     flowers 2.5–3.8 cm across
                        77. multibracteata
83a. Receptacle and fruit smooth   84
  b. Receptacle and fruit glandular-
     bristly   96
84a. Leaflets acute to acuminate   85
  b. Leaflets obtuse   91
85a. Sepals smooth on the back   86
  b. Sepals glandular on the back   88
86a. Stems usually unarmed,
     occasionally with slender prickles
     and bristles   **50. pendulina**
  b. Stems with large, straight prickles
     in pairs at the nodes   87
87a. Stems 1–1.5 m; fruit 1–1.3 cm
                            53. davurica
  b. Stems 1.5–3 m; fruit 1.5–2 cm
                            58. nutkana
88a. Fruit usually drooping   89
  b. Fruit erect   90
89a. Stem to 2 m; receptacle smooth or
     glandular-bristly   **50. pendulina**

  b. Stem 2–4 m; receptacle smooth
                            51. oxyodon
90a. Flowers 5–6.5 (rarely 4) cm across;
     fruit 1.5–2 cm   **58. nutkana**
  b. Flowers 1.2–2.5 cm across; fruit
     rarely more than 1 cm
                            73. giraldii
91a. Flowers 1.2–2.5 cm across
                            73. giraldii
  b. Flowers 3.7–6.5 cm across   92
92a. Petals pale pink   **54. laxa**
  b. Petals deep pink to purplish pink or
     red   93
93a. Fruit 7–15 mm   **57. californica**
  b. Fruit 1.5–3 cm   94
94a. Fruit drooping, usually with a neck
     at the top   95
  b. Fruit erect, lacking a neck at the
     top   **58. nutkana**
95a. Stems up to 2 m; receptacle smooth
     or glandular-bristly
                           50. pendulina
  b. Stems 2–4 m; receptacle smooth
                            51. oxyodon
96a. Flowers solitary or in clusters of
     2–6; petals hairless on the back
                                             97
  b. Flowers in clusters of 5–20; petals
     slightly downy on the back
                            64. setipoda
97a. Fruit 0.8–2 cm, if more, then
     drooping   98
  b. Fruit 2.5–7.5 cm, erect   100
98a. Fruit drooping   **50. pendulina**
  b. Fruit erect   99
99a. Flowers 1.2–2.5 cm across; fruit
     rarely more than 1 cm
                            73. giraldii
  b. Flowers (4–)5–6.5 cm across; fruit
     1.5–2 cm   **58. nutkana**
100a. Sepals hairy inside   **69. moyesii**
  b. Sepals not hairy inside   101
101a. Flowers 3.5–5 cm across; fruit *c.*
     2.5 cm   **66. sweginzowii**
  b. Flowers 5–7.5 cm across; fruit
     2.5–4 (rarely to 7.5) cm
                          63. macrophylla
102a. Styles protruding   103
  b. Styles not protruding   104
103a. Leaflets acute; flowers 3.8–5 cm
     across; fruit 1.5–2.5 cm
                             60. davidii
  b. Leaflets obtuse; flowers 2–3.8 cm
     across; fruit to 1.3 cm
                           78. forrestiana
104a. Leaflets acute to acuminate   105
  b. Leaflets obtuse   129
105a. Flowers 1.5–3.8 cm across   106
  b. Flowers 3–9 cm across   111
106a. Sepals smooth and hairless on the
     back, or hairy on the margin   107

  b. Sepals hairy or glandular-bristly on
     the back   109
107a. Fruit 7–15 mm   108
  b. Fruit 2.5–5 cm   **62. banksiopsis**
108a. Sepals hairy inside; leaflets glaucous
     green above   **76. elegantula**
  b. Sepals hairless inside; leaflets shiny
     green above   **47. arkansana**
109a. Leaflets usually 5–9, downy
     beneath   **55. pisocarpa**
  b. Leaflets usually 9–15, hairless
     beneath or downy only on the veins
                                        110
110a. Stems 50–120 cm; leaflets 2–6 cm;
     flowers 2.5–4 cm across; sepals
     smooth or glandular-bristly on the
     back; fruit 1 cm or more
                           47. arkansana
  b. Stems 1.2–2.5 m; leaflets 0.6–2
     cm; flowers 2–2.5 cm across; sepals
     downy on both sides; fruit 1 cm or
     less   **75. prattii**
111a. Fruit smooth   112
  b. Fruit glandular-bristly, sometimes
     only sparsely so   119
112a. Fruit 0.7–1.5 cm   113
  b. Fruit 1.5–2.5 cm   117
113a. Leaflets usually 5–7 (rarely 9)   114
  b. Leaflets usually 7–11   116
114a. Stems with pairs of hooked prickles
     at the nodes   **52. majalis**
  b. Stems unarmed, or with straight or
     slightly curved prickles   115
115a. Flowers 3–4.5 cm across; sepals
     hairless or downy on the back
                             56. woodsii
  b. Flowers 4.5–6.5 cm; sepals downy
     and glandular on the back
                             49. blanda
116a. Leaflets shiny above; sepals hairless
     inside   **47. arkansana**
  b. Leaflets glaucous on both sides;
     sepals hairy inside
                           76. elegantula
117a. Leaflets conspicuously wrinkled
     above, margin toothed in the upper
     ⅔ only   **45. rugosa**
  b. Leaflets not wrinkled above, margin
     toothed for the whole length   118
118a. Fruit dark red; leaflets 0.6–2.5 cm
                             72. sertata
  b. Fruit scarlet or mid-red; leaflets
     1.5–6 cm   **46. acicularis**
119a. Petals pale pink or lilac-pink; sepals
     hairy inside   120
  b. Petals mid-pink to deep purplish
     pink or red; sepals hairy or not
     inside   121
120a. Leaflets elliptic to narrowly ovate;
     flowers 2–3.8 cm across
                           76. elegantula

b. Leaflets broadly elliptic to obovate; flowers *c.* 5 cm across
    74. fedtschenkoana

121a. Fruit 1–2.5 cm   122

b. Fruit 2.5–7.5 cm   127

122a. Flowers 2–4 cm across; fruit 1–1.5 cm   123

b. Flowers 3.5–6.5 cm across; fruit 1.5–2.5 cm   124

123a. Leaflets shiny above; sepals hairless inside   **47. arkansana**

b. Leaflets glaucous on both sides; sepals hairy inside
    76. elegantula

124a. Leaflets with short stalks
    70. bella

b. Leaflets stalkless   125

125a. Fruit lacking a neck at the top
    58. nutkana

b. Fruit with a neck at the top   126

126a. Fruit dark red   **72. sertata**

b. Fruit orange-red   **61. caudata**

127a. Flowers solitary or in clusters of 2–5; petals hairless on the back
    128

b. Flowers in clusters of 5–20; petals downy on the back   **64. setipoda**

128a. Leaflets elliptic to narrowly ovate; sepals hairless inside
    63. macrophylla

b. Leaflets broadly elliptic to ovate; sepals downy inside   **69. moyesii**

129a. Leaflets usually 5–7, rarely 9   130

b. Leaflets usually 7–13, rarely less
    141

130a. Sepals smooth or downy on the back   131

b. Sepals with glands on the back, hairy or not   136

131a. Flower-stalks smooth, hairless
    132

b. Flower-stalks glandular   134

132a. Flowers 3–4.5 cm across; stipules narrow, flat   133

b. Flowers *c.* 5 cm across; stipules broad, those on strong shoots usually inrolled   **52. majalis**

133a. Prickles straight or slightly curved; flowers mid-pink   **56. woodsii**

b. Prickles recurved; flowers deep pink to crimson   **57. californica**

134a. Flowers pale pink   **54. laxa**

b. Flowers bright or deep pink to crimson or purple-pink   135

135a. Flowers 3.7–4 cm across; prickles recurved; fruit with a neck at the top   **57. californica**

b. Flowers (4–)5–6.5 cm across; prickles usually straight; fruit lacking a neck at the top
    58. nutkana

136a. Fruit 0.8–1.3 cm   137

b. Fruit 1.5–2.5 cm   138

137a. Flowers 2.5–3 cm across
    55. pisocarpa

b. Flowers 4.5–6.5 cm across
    49. blanda

138a. Flowers pale pink or lilac-pink
    139

b. Flowers deep pink to purplish pink or bright red   140

139a. Prickles recurved, hooked or straight, intermixed with bristles
    54. laxa

b. Prickles straight, bristles absent
    71. webbiana

140a. Leaflets 0.5–2.5 cm; stems lacking bristles   **72. sertata**

b. Leaflets 1.9–5 cm; young stems with bristles   **58. nutkana**

141a. Fruit to 2.5 cm   142

b. Fruit 2.5–6 cm   148

142a. Leaflets dull blue-green or somewhat glaucous   143

b. Leaflets bright green, sometimes shiny above   146

143a. Sepals hairy inside and on the margin, not glandular on the back; fruit 0.7–1.3 cm   **76. elegantula**

b. Sepals not hairy inside, glandular on the back; fruit 1.5–2.5 cm   144

144a. Stems with dense bristles
    46. acicularis

b. Stems lacking dense bristles   145

145a. Flowers pale pink or lilac-pink, borne on flower-stalks 1–1.3 cm
    71. webbiana

b. Flowers deep pink to purplish pink, borne on flower-stalks 1.5–3 cm
    72. sertata

146a. Flowers pale pink, 4–5 cm across
    54. laxa

b. Flowers mid-pink to deep pink, 2.5–4 cm across   147

147a. Stems to 50 cm; leaflets downy beneath   **48. suffulta**

b. Stems 50–120 cm; leaflets usually hairless beneath, sometimes downy on the veins   **47. arkansana**

148a. Flower-stalks densely glandular-bristly; fruit dark red; flowers smelling of unripe apples; petals downy on the back   **64. setipoda**

b. Flower-stalks moderately glandular-bristly or smooth; fruit orange-red; flowers scentless; petals hairless on the back   **69. moyesii**

149a. Sepals persistent after flowering, spreading to erect   150

b. Sepals falling after flowering, usually reflexed, sometimes spreading   168

150a. Flowers double   151

b. Flowers single or semi-double   152

151a. Leaflets 5–7, broadly ovate to more or less round; fruit red, not prickly
    17. × centifolia

b. Leaflets usually 9–15, narrowly ovate to obovate; fruit yellow-green, prickly   **109. roxburghii**

152a. Flowers white, pale pink or pale lilac-pink   153

b. Flowers mid-pink to deep pink or purple-pink   156

153a. Receptacle prickly or glandular-bristly   154

b. Receptacle smooth   **34. coriifolia**

154a. Margin of leaflets with compound teeth, glandular and downy beneath, aromatic when crushed; flowers 2.5–5 cm across
    26. rubiginosa

b. Margin of leaflets with simple teeth, downy or hairless beneath, not aromatic; flowers 5–10 (rarely 3) cm across   155

155a. Leaflets 1–2.5 cm; fruit prickly
    110. hirtula

b. Leaflets 2.5–3.5 cm; fruit slightly glandular-bristly   **36. britzensis**

156a. Fruit yellow-green, prickly
    109. roxburghii

b. Fruit orange-red to dark red or blackish, smooth or glandular-bristly   157

157a. Fruit with a neck at the top   158

b. Fruit lacking a neck at the top
    159

158a. Stems to 2 m   **65. helmsleyana**

b. Stems 2.4–3 m   **64. setipoda**

159a. Prickles straight or slightly curved
    160

b. Prickles hooked   164

160a. Margin of leaflets with simple teeth
    161

b. Margin of leaflets with compound teeth   162

161a. Stems 50–120 cm; leaflets usually hairless beneath, sometimes downy on the veins   **47. arkansana**

b. Stems to 50 cm; leaflets downy beneath   **48. suffulta**

162a. Stems with prickles intermixed with bristles   163

b. Stems with prickles only   164

163a. Margin of leaflets with 3–10 teeth on each side; flower-stalks 2–5 mm, usually smooth   **27. sicula**

b. Margin of leaflets with 8–20 teeth on each side; flower-stalks 2–15 (rarely to 25) mm, usually glandular-bristly
    28. pulverulenta

164a. Young stems not bloomed
    22. villosa
  b. Young stems bloomed   165
165a. Leaflets more or less round
    23. mollis
  b. Leaflets elliptic to broadly ovate
    24. sherardii
166a. Stems 30–70 cm; leaflets 0.7–1.5
  cm   **28. pulverulenta**
  b. Stems 1–3 m; leaflets 1–3.5 cm
    167
167a. Leaflets dark green above,
  glandular and downy beneath
    26. rubiginosa
  b. Leaflets often bloomed above,
  hairless or downy beneath, but not
  glandular   **37. dumalis**
168a. Styles united   169
  b. Styles free   171
169a. Flowers 5–7.5 cm across; stigmas
  woolly; fruit 1.5–2.5 cm
    38. jundzillii
  b. Flowers 1.5–5 cm across; stigmas
  hairless or sparsely hairy; fruit to
  1.2 cm   170
170a. Bracts usually absent; flowers fruit
  scented   171
  b. Bracts present; flowers not scented
    172
171a. Stigmas hairless; fruit red, 0.6–0.7
  cm   **82. multiflora**
  b. Stigmas sparsely hairy; fruit pale to
  dark orange, c. 1 cm
    96. soulieana
172a. Young stems with bristles as well
  as prickles; leaflets 7–9, hairless on
  both sides   **93. maximowiczii**
  b. Young stems bearing a few prickles,
  but no bristles; leaflets 5 (rarely 3–7),
  more or less downy on both sides,
  especially beneath   **95. phoenicia**
173a. Flowers semi-double or double
    174
  b. Flowers single   187
174a. Prickles hooked   175
  b. Prickles straight or curved   178
175a. Margin of leaflets with compound
  teeth   **14. gallica**
  b. Margin of leaflets with simple teeth
    176
176a. Flowers c. 5 cm across; fruit
  orange-red   **21. × collina**
  b. Flowers 6–8 cm across; fruit red
    177
177a. Styles protruding   **20. × alba**
  b. Styles not protruding
    17. × centifolia
178a. Margin of leaflets with compound
  teeth   179
  b. Margin of leaflets with simple teeth
    180

179a. Flowers pale pink, fading to
  whitish; sepals hairy inside and on
  margin   **15. × macrantha**
  b. Flowers rose-pink or crimson, or
  striped white, pink and red; sepals
  neither hairy inside nor on margin
    14. gallica
180a. Leaflets obtuse   181
  b. Leaflets acute to acuminate   184
181a. Styles protruding; flowers semi-
  double   **19. × damascena**
  b. Styles not protruding; flowers
  double   182
182a. Bracts narrow   183
  b. Bracts very large
    16. × francofurtana
183a. Flowers 6–8 cm across; stems to 2
  m   **17. × centifolia**
  b. Flowers 3.8–5 cm across; stems to
  50 cm   **42. carolina**
184a. Flowers double   185
  b. Flowers semi-double   186
185a. Flowers 6–8 cm across; stems to
  2 m   **17. × centifolia**
  b. Flowers 3.8–5 cm across; stems to
  50 cm   **42. carolina**
186a. Styles protruding; leaflets downy
  beneath   **19. × damascena**
  b. Styles not or only slightly
  protruding; leaflets hairless beneath
  but with glands and tiny prickles
  on the midrib   **15. × macrantha**
187a. Margin of leaflets with simple teeth
    188
  b. Margin of leaflets with compound
  teeth   197
188a. Sepals hairy inside and on the
  margin   **15. × macrantha**
  b. Sepals hairless inside   189
189a. Petals deep pink, white at the base;
  mature leaflets grey-green, often
  tinged with brown or purple
    39. glauca
  b. Petals white or pale pink to mid-
  pink, if deeper then not white at the
  base; mature leaflets green   190
190a. Leaflets acute to acuminate   191
  b. Leaflets obtuse   195
191a. Flowers 2.5–5 cm across   192
  b. Flowers 5–7.5 cm across   194
192a. Prickles straight; receptacle
  glandular-bristly   **42. carolina**
  b. Prickles hooked; receptacle smooth
    193
193a. Styles protruding   **35. canina**
  b. Styles not protruding or only
  slightly so   **32. stylosa**
194a. Leaflets 3–5; styles protruding
    18. × richardii
  b. Leaflets 5–9; styles not protruding
    41. virginiana

195a. Stems to 1.5 m bearing prickles
  intermixed with bristles
    42. carolina
  b. Stems 1.5–5 m bearing prickles but
  no bristles   196
196a. Leaflets narrowly elliptic to ovate,
  usually hairless on both sides,
  although sometimes downy on the
  veins beneath   **35. canina**
  b. Leaflets broadly ovate to roundish,
  downy on both sides
    33. corymbifera
197a. Flowers rose-pink or crimson, or
  striped white, pink and red
    14. gallica
  b. Flowers white to pale pink   198
198a. Receptacle smooth   199
  b. Receptacle glandular-bristly   202
199a. Leaflets with glands beneath   200
  b. Leaflets lacking glands beneath
    35. canina
200a. Prickles unequal   **31. serafinii**
  b. Prickles equal   201
201a. Sepals glandular on the back
    29. micrantha
  b. Sepals not glandular on the back
    35. canina
202a. Leaflets 3–5 (rarely to 7); flowers c.
  7.5 cm across   **15. × macrantha**
  b. Leaflets 5–7; flowers 2.5–5.5 cm
  across   203
203a. Prickles equal; styles more or less
  protruding   204
  b. Prickles unequal; styles not
  protruding   **30. horrida**
204a. Flowers 2.5–3 cm across
    29. micrantha
  b. Flowers 3.8–5.5 cm across
    25. tomentosa

1. R. persica A.L. Jussieu (*Hulthemia persica* (A.L. Jussieu) Bornmüller; *H. berberifolia* (Pallas) Dumortier; *Rosa berberifolia* Pallas; *R. simplicifolia* Salisbury). Illustration: Harkness, Roses, pl. 1 (1978); Thomas, A garden of roses, 127 (1987); Austin, The heritage of the rose, 352 (1988); Phillips & Rix, Roses, 19 (1988).

Straggling, often decumbent, suckering shrub. Stems yellowish brown, 40–90 cm, hairless or downy, bearing straight or curved, slender, yellow or reddish prickles. Stipules absent. Leaves deciduous, simple, stalkless, grey-green, 1.3–3.2 cm, broadly elliptic to obovate, acute, toothed towards apex, usually downy. Receptacle bristly. Flowers solitary, single, 2.5–3 cm across. Sepals lanceolate, downy and more or less prickly on the back. Petals yellow with a basal crimson spot. Fruit spherical, sometimes flattened, blackish, very prickly.

Iran, Afghanistan, former USSR (Central Asia, SW Siberia). H5–G1. Early summer.

2. R. × hardii Cels (× *Hulthemosa hardii* (Cels) Rowley). Illustration: Paxton's Magazine of Botany **10**: 195 (1843); Harkness, Roses, pl. 2 (1978); Thomas, A garden of roses, 93 (1987); Phillips & Rix, Roses, 18 (1988).

A hybrid between *R. persica* and *R. clinophylla*. Stems to 2 m. Stipules present. Leaves simple or with up to 7, dark green, hairless leaflets. Flowers *c.* 5 cm across. Sepals with some lateral lobes. Petals deep yellow with a basal crimson spot. Fruit yellowish orange, shortly hairy with a few bristles. *Garden origin.* H5–G1. Early summer.

3. R. pimpinellifolia Linnaeus (*R. spinosissima* Linnaeus (1771) not Linnaeus (1753). Illustration: Ross-Craig, Drawings of British Plants, **9**: pl. 15 (1956); Phillips & Rix, Roses, 20 (1988); Blamey & Grey-Wilson, The illustrated flora of Britain and northern Europe, 179 (1989).

Much-branched, suckering shrub. Stems purple-brown, 90–200 cm, bearing dense straight, slender prickles and stiff bristles, especially dense at the stem-base. Stipules narrow. Leaves deciduous; leaflets 7–9 (rarely 7–11), broadly elliptic to broadly obovate or more or less circular, 0.6–2 cm, obtuse, hairless although mid-vein sometimes downy beneath, margin with simple glandular teeth. Bracts absent. Receptacle usually hairless. Flowers solitary, single or double, 3.8–6 cm across. Sepals entire, narrowly lanceolate, acuminate, much shorter than petals, hairless on the back but with woolly margin, erect and persistent in fruit. Petals creamy white. Fruit spherical or more or less so, dark brown becoming blackish, 0.7–1.5 cm, smooth and hairless. *W & S Europe, SW & C Asia east to China and Korea.* H2. Summer.

'Grandiflora' ('Altaica'; *R. pimpinellifolia* var. *altaica* (Willdenow) Thory; *R. pimpinellifolia* var. *grandiflora* Ledebour). Illustration: Beales, Classic roses, 139 (1985); Thomas, A garden of roses, 131 (1987); Phillips & Rix, Roses, 20 (1988). Stems less bristly than the species and flowers 5–7.5 cm across. 'Hispida' (*R. pimpinellifolia* var. *hispida* (Sims) Boom; *R. spinosissima* var. *hispida* (Sims) Koehne). Illustration: Harkness, The rose, pl. 103 (1979). Leaflets 1.9–3.2 cm; flowers 5–7.5 cm across which open pale yellow and fade to cream. 'Andrewsii' (*R. spinosissima* var. *andrewsii* Willmott). Flowers double,

rose-pink. 'Nana' (*R. spinosissima* var. *nana* Andrews). Flowers semi-double, white.

R. × hibernica Templeton is a hybrid between *R. pimpinellifolia* and *R. canina* which occurred naturally in Ireland although it is now thought to exist only in gardens. Illustration: Gault & Synge, The dictionary of roses in colour, pl. 22 (1971); Thomas, A garden of roses, 97 (1987); Walsh & Nelson, An Irish florilegium **2**: 19 (1987); Austin, The heritage of the rose, 367 (1988). Stems 4 m or more. Leaflets 5–7, sparsely hairy beneath, especially on the veins. Flowers single. Sepals with lateral lobes and the apex expanded. Petals pale pink, paler towards the base. Fruit dark red, obovoid.

4. R. × reversa Waldstein & Kitaibel. Illustration: Phillips & Rix, Roses, 16 (1988).

A hybrid between *R. pimpinellifolia* and *R. pendulina*. Leaflets densely glandular. Petals carmine. Fruit spherical-ovoid, deep purplish red, drooping, *c.* 2 cm. *S Europe north to Switzerland and S France.* H4. Summer.

5. R. koreana Komarov.

Dwarf, densely branched shrub. Stems dark red, to 1 m, bearing dense bristles. Stipules narrow. Leaves deciduous; leaflets 7–11 (rarely to 15), elliptic to elliptic-obovate, 1–2 cm, more or less obtuse, hairless, sometimes slightly downy beneath, margins with simple glandular teeth. Bracts absent. Flowers solitary, single, 4–6 cm across, scented. Sepals entire, lanceolate, glandular on the back, erect and persistent in fruit. Petals white, tinged with pink. Fruit ovoid to oblong, orange-red, 1–1.5 cm. *N Korea.* H4. Summer.

6. R. xanthina Lindley.

Erect shrub. Stems brown when young, ageing to grey-brown, 1.5–3.5 m, bearing straight or slightly curved, broad-based prickles which are sometimes much flattened on non-flowering shoots. Stipules narrow, with margins curved downwards and inwards. Leaves deciduous; leaflets 7–13, broadly elliptic to obovate or spherical, 0.8–2 cm, obtuse, usually hairless above and hairy beneath. margins with simple teeth. Bracts absent. Receptacle smooth. Flowers solitary, rarely 2, semi-double, 3.8–5 cm across. Sepals entire, lanceolate, acuminate, leafy and toothed at tip, hairless or sparsely hairy, erect and persistent in fruit. Petals bright yellow. Fruit spherical or broadly ellipsoid, brown-red or maroon, 1.2–1.5 cm, smooth and

hairless. *N China, Korea.* H4. Early summer.

Forma **hugonis** (Hemsley) Roberts (*R. hugonis* Hemsley). Illustration: Bois & Trechslin, Roses, 15 (1962); Gibson, Shrub roses for every Garden, pl. 2 (1973); Thomas, A garden of roses, 99 (1987); Phillips & Rix, Roses, 17 (1988). Differs in having leaflets elliptic to obovate and hairless flower-stalks bearing single flowers 4–6 cm across. *C China.* H4.

Forma **spontanea** Rehder (*R. xanthina* forma *normalis* Rehder & Wilson, in part). Illustration: Austin, The heritage of the rose, 353 (1988); Phillips & Rix, Roses, 20 (1988). Differs from *R. xanthina* itself in having single flowers 5–6 cm across. It is one of the parents of the splendid 'Canary Bird', the other parent being forma *hugonis*.

7. R. ecae Aitchison (*R. xanthina* var. *ecae* (Aitchison) Boulenger). Illustration: Gault & Synge, The dictionary of roses in colour, pl. 7 (1971); Harkness, The rose, pl. 101 (1979); Thomas, A garden of roses, 69 (1987); Phillips & Rix, Roses, 16 (1988).

Much-branched, erect, suckering shrub. Stems to 1.5 m, bearing dense, straight, flattened, reddish prickles. Stipules very narrow. Leaves deciduous, aromatic; leaflets 5–9, broadly elliptic to obovate or spherical, 4–8 mm, obtuse, glandular beneath, margins with simple, often glandular teeth. Bracts absent. Flowers solitary, single, 2–3 cm across. Sepals entire, hairless, spreading or deflexed in fruit, persistent. Petals deep yellow. Fruit spherical, shiny red-brown, 0.5–1 cm, smooth and hairless. *NE Afghanistan, NW Pakistan and adjacent former USSR, N China.* H4. Summer.

A parent (with 'Canary Bird') of 'Golden Chersonese' (Illustration: Phillips & Rix, Roses, 15, 1988) which grows to 2 m and has inherited the aromatic leaves and deep yellow flowers of *R. ecae*. 'Helen Knight' (Illustration: Phillips & Rix, Roses, 16, 1988) is a hybrid between *R. ecae* and probably *R. pimpinellifolia* 'Grandiflora' and has large yellow flowers and red-brown stems which will reach 3 m against a wall.

8. R. kokanica (Regel) Juzepczuk (*R. xanthina* var. *kokanica* (Juzepczuk) Boulenger). Illustration: Phillips & Rix, Roses, 14 (1988).

Erect suckering shrub. Stems to 2 m, reddish brown when young. Stipules narrow. Leaves deciduous, aromatic; leaflets 5–7, broadly elliptic to obovate or more or less circular, glandular beneath, margins with compound glandular teeth. Bracts absent. Flowers solitary, single, *c.* 5 cm across. Sepals entire, hairless, erect and persistent in fruit. Petals bright yellow. Fruit spherical, brownish. *Former USSR (Central Asia) possibly extending into China.* H4. Early summer.

9. R. primula Boulenger (*R. ecae* subsp. *primula* (Boulenger) Roberts). Illustration: Gault & Synge, The dictionary of roses in colour, pl. 39 (1971); Gibson, The book of the rose, pl. 15 (1980); Beales, Classic roses, 89 (1985); Phillips & Rix, Roses, 15 (1988).

Erect shrub. Stems slender, to 3 m, reddish brown when young, bearing stout, straight, somewhat compressed, broad-based prickles. Stipules narrow. Leaves deciduous, very aromatic; leaflets 9 (rarely 7–13), elliptic to obovate or oblanceolate, 0.6–2 cm, acute or obtuse, hairless above, beneath with large glands, margins with compound glandular teeth. Bracts absent. Receptacle smooth. Flowers solitary, single, 2.5–4.5 cm across. Sepals entire, hairless, erect and persistent in fruit. Petals primrose-yellow. Fruit spherical to obconical, brownish red to maroon, 1–1.5 cm, smooth. *Former USSR (Central Asia) to N China.* H4. Early summer.

10. R. foetida Herrmann (*R. lutea* Miller; *R. eglanteria* Linnaeus, in part). Illustration: Harkness, The rose, pl. 100 (1979); Thomas, A garden of roses, 75 (1987); Phillips & Rix, Roses, 15 (1988); Blamey & Grey-Wilson, The illustrated flora of Britain and northern Europe, 179 (1989).

Erect shrub. Stems erect or arching, 1–3 m, at first dark brown, later greyish, bearing few, straight or curved prickles of unequal length, abruptly broadened at the base, usually mixed with bristles. Stipules very narrow. Leaves deciduous; leaflets 5–9, elliptic to obovate, 1.5–4 cm, obtuse to almost acute, bright green and more or less hairless above, dull green, downy and more or less glandular beneath, margins with rather few, compound glandular teeth. Bracts absent. Receptacle smooth or bristly. Flowers solitary or sometimes 2–4, single, 5–7.5 cm across, with an unpleasant smell. Sepals entire or with a few lateral lobes, lanceolate, apex expanded, hairless or glandular-bristly, erect and persistent in fruit. Petals deep yellow. Stigmas hairy. Fruit spherical, red, 0.8–1 cm, smooth or bristly. *SW & WC Asia.* H4. Summer.

'Bicolor' (*R. bicolor* Jacquin; *R. foetida* var. *bicolor* (Jacquin) Willmott). Illustration:

Thomas, A garden of roses, 77 (1987); Austin, The heritage of the rose, 356 (1988); Phillips & Rix, Roses, 15 (1988). Possibly a very old hybrid between *R. kokanica* and *R. hemisphaerica*. Petals coppery red inside and yellow-buff on the back. Branches occasionally revert to the yellow-flowered form. 'Persiana' (*R. foetida* var. *persiana* (Lemaire) Rehder). Illustration: Austin, The heritage of the rose, 372 (1988); Phillips & Rix, Roses, 15 (1988); Brickell, The RHS gardeners' encyclopedia of plants and flowers, 152 (1989). Flowers very double, yellow, freely produced, smaller than those of *R. foetida* itself.

11. R. × harisonii Rivers. Illustration: Gault & Synge, The dictionary of roses in colour, pl. 20 (1971); Gibson, The book of the rose, pl. I (1980); Beales, Classic roses, 140 (1985).

A hybrid between *R. pimpinellifolia* and *R. foetida*. Erect shrub with occasional suckers. Stems 50–200 cm. Leaves deciduous; leaflets 5–9, elliptic, glandular beneath, margins with compound, somewhat glandular teeth. Bracts absent. Receptacle bristly. Flowers solitary, loosely double, 5–6 cm across, with an unpleasant smell. Sepals lobed at apex, slightly hairy on the back, margin glandular, erect and persistent in fruit. Petals sulphur-yellow. Fruit almost black, vertically compressed, bristly. *Garden origin.* H4. Summer.

'Lutea maxima' (*R. spinosissima* Linnaeus var. *lutea* Bean) has the same parentage as *R. × harisonii* but shows more influence of *R. foetida* in that it has buttercup-yellow flowers, the sepals sometimes with lateral lobes, and a smooth fruit.

12. R. hemisphaerica Herrmann (*R. rapinii* Boissier & Balansa). Illustration: Anderson, The complete book of 169 Redouté roses, 29 (1979); Thomas, A garden of roses, 95 (1987); Phillips & Rix, Roses, 14 (1988).

Stiffly erect, much-branched shrub. Stems to 2 m, bearing scattered, slender decurved prickles with broad bases, on the young shoots mixed with bristles. Stipules narrow. Leaves deciduous, slightly aromatic; leaflets 5–9, obovate, obtuse to slightly notched, grey-green and hairless above, glaucous and downy on the veins beneath, margins with simple teeth in upper ⅔. Bracts absent. Receptacle sometimes glandular. Flowers solitary, double, 4–5 cm across, often sweetly scented. Sepals broadly lanceolate, toothed at the apex, erect and persistent in fruit.

Petals sulphur-yellow. Stigmas woolly. Fruit spherical, dark red, 1.2–1.5 cm, smooth. *Gardens of SW Asia.* H4. Summer.

Uncommon in cultivation. The double-flowered plant, 'Flore Pleno' (*R. sulphurea* Aiton), is more usually cultivated. Illustration: Phillips & Rix, Roses, 15 (1988).

13. R. sericea Lindley. Illustration: Phillips & Rix, Roses, 16 (1988) – as subsp. omeiensis.

Upright or somewhat spreading shrub. Stems 2–4 m, grey or brown, bearing straight or curved, often upward-pointing, reddish prickles with broad bases, often mixed with slender bristles. Stipules narrow. Leaves deciduous; leaflets 7–11, elliptic, oblong or obovate, 0.6–3 cm, obtuse or more or less acute, hairy or not beneath, margins with simple teeth often confined to upper half. Bracts absent. Receptacle hairless. Flowers solitary, on short lateral shoots, single, 2.5–6 cm across. Sepals entire, hairless or silky on the back, erect or spreading and persistent in fruit. Petals usually 5, white or cream, notched at the apex. Fruit spherical to pear-shaped with a narrow stalk, dark crimson, scarlet, orange or yellow, 0.8–1.5 cm, smooth. *Himalaya, NE India, N Bhutan, SW & C China.* H4. Early summer.

Subsp. **omeiensis** (Rolfe) Roberts. (*R. omeiensis* Rolfe; *R. sericea* var. *omeiensis* (Rolfe) Rowley). Illustration: Thomas, A garden of roses, 134 (1987); Phillips & Rix, Roses, 17 (1988) – as R. sericea. Leaflets 11–19, silky beneath, petals usually 4, white. Fruit with a fleshy stalk. *China (Sichuan, Hubei, Yunnan).*

Subsp. **omeiensis** forma **pteracantha** Franchet (*R. sericea* var. *pteracantha* (Franchet) Boulenger; *R. omeiensis* forma *pteracantha* (Franchet) Rehder & Wilson). Illustration: Gibson, The book of the rose, pl. 1 (1980); Austin, The heritage of the rose, 353 (1988) – as R. sericea var. pteracantha; Welch, Roses, 62 (1988); Phillips & Rix, Roses, 17 (1988). The stems bear flattened prickles which can be up to 3.8 cm wide at the base and 1.2–2 cm deep, abruptly and sharply pointed, forming interrupted wings down the stems; prickles on young stems are red, turning grey and woody in the second year. *W China.*

14. R. gallica Linnaeus (*R. provincialis* Herrmann; *R. rubra* Lamarck). Illustration: Phillips & Rix, Roses, 28 (1988).

Low, erect shrub, suckering. Stems 50–200 cm, green to dull red, bearing slender, unequal, curved or sometimes hooked prickles, mixed with glandular bristles. Stipules narrow with spreading pointed tips. Leaves deciduous; leaflets 3–5 (rarely to 7), leathery, broadly elliptic to almost circular, acute or obtuse, dark green and hairless above, hairy and glandular beneath, at least on the veins, margins with compound, usually glandular teeth. Bracts present. Receptacle with stalked or stalkless glands. Flowers solitary, or sometimes 2–4, single or semi-double, fragrant, 4–8 cm across. Sepals with some lateral lobes, glandular on the back, reflexed and falling after flowering. Petals rose-pink or crimson. Styles free, not protruding; stigmas woolly. Fruit spherical to ellipsoid, brick-red, *c.* 1.3 cm, densely glandular-bristly. *S & C Europe, E to Ukraine, Turkey, Iraq and the Caucasus (naturalised in E North America).* H4. Summer.

'Officinalis' (*R. gallica* var. *officinalis* Thory). Illustration: Anderson, The complete book of 169 Redouté roses, 43 (1979); Welch, Roses, 106 (1988); Phillips & Rix, Roses, 49 (1988). Shrub rarely more than 70 cm, flowers semi-double, crimson, very fragrant. 'Versicolor' (*R. gallica* var. *versicolor* Linnaeus; *'Rosa Mundi'*). Illustration: Gibson, The rose gardens of England, 22 (1988); Le Rougetel, A heritage of roses, 11 (1988); Phillips & Rix, Roses, 40 (1988). A sport of 'Officinalis' with stems 1–2 m and semi-double flowers 7–9 cm across and striped white, pink and red. It often reverts to the non-striped type. It is frequently confused with *R. damascena* 'Versicolor'.

15. R. × macrantha misapplied, not Desportes. (*R.* 'Macrantha'). Illustration: Edwards, Wild and old garden roses, 34 (1975); Beales, Classic roses, 195 (1985); Austin, The heritage of the rose, 352 (1988); Phillips & Rix, Roses, 109 (1988).

Vigorous shrub. Stems 1.5–2 m, arching or spreading, green, bearing sparse, scattered, straight or slightly curved prickles, mixed with small straight bristles and stalked glands. Stipules narrow. Leaves deciduous; leaflets 3–5 (rarely to 7), dull green, ovate to oblong-ovate, acute to acuminate, hairless on both sides but glandular beneath and with tiny prickles on the midrib, margins with simple or compound teeth. Bracts present. Receptacle slightly glandular below. Flowers 2–5 in clusters, single to semi-double, fragrant, *c.* 7.5 cm across. Sepals with many lateral lobes, hairy inside and on margin, slightly

glandular on the back, reflexed and falling after flowering. Petals pale pink, fading to whitish. Styles free, not or only slightly protruding. Fruit almost spherical, red, to 1.5 cm wide. *Garden origin.* H4. Summer.

A hybrid between *R. gallica* and possibly *R. canina.*

16. R. × francofurtana Muenchhausen (*R. turbinata* Aiton).

Shrub with erect stems to 2 m, grey-green, the flowering-stems unarmed or bearing a few straight or curved prickles and bristles. Stipules broad. Leaves deciduous; leaflets 5–7, grey-green, broadly ovate to more or less round, obtuse, downy on the veins beneath, margins with coarse simple teeth. Bracts very large. Receptacle glandular-bristly. Flowers solitary or 2–6, usually double, slightly fragrant, 5–7 cm across. Sepals entire or with a few lateral lobes, hairy and glandular on margin and back, erect or reflexed, falling after flowering. Petals purplish pink with darker veins. Styles free, not protruding. Fruit obconical, red. *Garden origin.* H4. Summer.

A hybrid, probably between *R. gallica* and *R. majalis.*

17. R. × centifolia Linnaeus (*R. provincialis* Miller in part (1788) not Herrmann, 1762). Illustration: Phillips & Rix, Roses, 60 (1988).

Loosely branched shrub producing a few suckers. Stems to 2 m, bearing many small, almost straight, scattered prickles and larger hooked ones. Stipules narrow. Leaves deciduous; leaflets 5–7, dull green, broadly ovate to more or less round, acute to obtuse, usually hairless above and downy beneath, margins with simple glandular teeth. Bracts narrow. Receptacle bearing glands. Flowers 1–few, double, more or less spherical, very fragrant, 6–8 cm across. Sepals with lateral lobes, glandular on the back, spreading, somewhat persistent in fruit. Petals usually pink, more rarely white or dark red. Styles free, not protruding. Fruit ellipsoidal or spherical, red. *Garden origin.* H4. Summer.

A complex hybrid involving *R. gallica, R. moschata, R. canina* and *R. damascena.* There are several cultivars: 'Bullata'. Illustration: Anderson, The complete book of 169 Redouté roses, 51 (1979); Thomas, A garden of roses, 51 (1987); Phillips & Rix, Roses, 59 (1988). Leaflets crinkled, brownish above when young; flowers pink. 'Cristata' (*R. centifolia* var. *cristata* Prévost).

Illustration: Thomas, A garden of roses, 53 (1987); Phillips & Rix, Roses, 60, 61 (1988). Lower-growing; sepals with many marginal segments which provide each pink flower with a green fringe; flowers less globular. 'Muscosa' (*R. muscosa* Miller; *R. centifolia* var. *muscosa* (Miller) Seringe). Illustration: Bois & Trechslin, Roses, 13 (1962); Thomas, A garden of roses, 55 (1987); Phillips & Rix, Roses, 64, 65 (1988). Shrub *c.* 1 m. Calyx and flower-stalks bearing much-branched scented glands which form the so-called 'moss', hence the common name of 'moss rose'. 'Parvifolia' (*R. parvifolia* Ehrhart (1791) not Pallas (1788); *R. centifolia* var. *parvifolia* (Ehrhart) Rehder). Illustration: Phillips & Rix, Roses, 58 (1988). Stems to 1 m; leaflets small, dark green; flowers fragrant, deep pink suffused with purple and with a paler centre. It is possible that it is not in fact related to *R. centifolia* at all.

18. R. × richardii Rehder (*R. sancta* Richard, not Andrews; *R. centifolia* var. *sancta* (Richard) Zabel). Illustration: Beales, Classic roses, 196 (1985); Austin, The heritage of the rose, 350 (1988); Phillips & Rix, Roses, 57 (1988).

Low, spreading shrub. Stems to 1.3 m, bearing small, scattered, hooked prickles of unequal size. Stipules broad. Leaves deciduous; leaflets 3–5, ovate to narrowly elliptic, acute, wrinkled above, downy beneath, margins with simple glandular teeth. Bracts narrow. Receptacle smooth. Flowers several in loose clusters, single, 5–7.5 cm across. Sepals with lateral lobes, the apex leafy, downy and glandular on the back, reflexed and falling after flowering. Petals pale pink. Styles free, protruding; stigmas woolly. *Garden origin.* H4. Summer.

A hybrid of *R. gallica* and possibly *R. phoenicia* or *R. arvensis.*

19. R. × damascena Miller. Illustration: Bois & Trechslin, Roses, 11 (1962).

Shrub with stems to 2.2 m, densely armed with stout, curved, equally sized prickles and stiff bristles. Stipules often with toothed margins. Leaves deciduous; leaflets 5 (rarely to 7), grey-green, ovate to elliptic, acute to obtuse, hairless above, downy beneath, margins with simple teeth. Bracts narrow. Receptacle glandular-bristly. Flowers in clusters of up to 12, semi-double, fragrant. Sepals with lateral lobes and slender tips, glandular and hairy on the back, reflexed and falling after flowering. Petals pink. Styles free, protruding. Fruit obconical, red, to 2.5 cm, bristly. *Turkey*. H4. Summer.

A hybrid between *R. gallica* and *R. moschata*. *R. × damascena* is the summer damask rose. The autumn damask rose is var. **semperflorens** (Loiseleur & Michel) Rowley (*R. bifera* (Poiret) Persoon). Illustration: Krüssmann, The complete book of roses, 259 (1981); Phillips & Rix, Roses, 54 (1988) – as 'Quatre Saisons'. It cannot be distinguished from *R. × damascena* except by the production of autumn flowers. 'Versicolor' (*R. damascena* var. *versicolor* Weston). Illustration: Anderson, The complete book of 169 Redouté roses, 115 (1979); Phillips & Rix, Roses, 56 (1988). Known as the York and Lancaster rose, it has loosely double flowers with deep pink or very pale pink petals, or petals which are partly deep pink and partly pale pink. 'Trigintipetala' (*R. × damascena* var. *trigintipetala* (Dieck) Keller). Flowers semi-double, *c.* 8 cm across, with *c.* 30 red petals. 'Portlandica' has bright red semi-double flowers, faintly scented and produced from midsummer to autumn.

20. R. × alba Linnaeus.

Spreading shrub with stout, arching stems 1.8–2.5 m, bearing scattered, hooked, unequal prickles, often mixed with bristles. Stipules broad. Leaves deciduous; leaflets 5 (rarely to 7), dull green, ovate to more or less round, shortly acuminate to obtuse, hairless above and downy beneath especially on the veins, margins with simple teeth. Bracts narrow. Receptacle slightly glandular-bristly, at least below. Flowers 1–3, semi-double or double, 6–8 cm across, fragrant. Sepals with lateral lobes and leafy tips, glandular-bristly on the back, reflexed and falling after flowering. Petals white to pale pink. Styles free, protruding. Fruit more or less spherical, red, 2–2.5 cm. *Garden origin*. H4. Summer.

Parentage of ths hybrid is disputed. It may be *R. gallica* × *R. arvensis* or *R. corymbifera* or even *R. canina* × *R. × damascena*.

'Maxima' (*R.* 'Alba Maxima'). Illustration: Phillips & Rix, Roses, 40 (1988). Double flowers open pink but fade to creamy white. 'Semiplena'. Illustration: Gibson, The book of the rose, pl. xxii (1980); Phillips & Rix, Roses, 41 (1988). Flowers semi-double, white. Is a sport of 'Maxima'. 'Incarnata' (*R. incarnata* Miller; *R. × alba* var. *incarnata* (Miller) Weston). Illustration: Phillips & Rix, Roses, 41 (1988). Stems with very few prickles but densely bristly on the flowering branches below the bracts. Leaflets usually 7. Flowers double, pale pink.

21. R. × collina Jacquin.

Erect shrub with stems 1.5–2 m, bearing strong, hooked prickles which are red when young. Leaves deciduous; leaflets 5 (rarely to 7), hairless above, downy on the veins beneath, margins with simple teeth. Bracts present. Receptacle often glandular. Flowers 1–3, more or less double, fragrant, *c.* 5 cm across. Sepals with lateral lobes and leafy tips, reflexed and falling after flowering. Petals pink. Fruit ovoid, orange-red, sometimes glandular. *Garden origin.* H4. Summer.

Possibly a hybrid between *R. canina* and *R. gallica*.

22. R. villosa Linnaeus (*R. pomifera* Herrmann, invalid). Illustration: Blamey & Grey-Wilson, The illustrated flora of Britain and northern Europe, 181 (1989).

Densely branched shrub, often suckering. Branches stiff, straight, 1–2.4 m, reddish when young, bearing scattered, slender, straight or slightly curved prickles of unequal size. Stipules broad. Leaves deciduous; leaflets 5–9, bluish green, elliptic, 3.2–6.5 cm, acute or obtuse, hairy on both sides, often with dense, apple-scented glands beneath, margins with compound glandular teeth. Bracts broad, often concealing receptacle. Receptacle densely glandular-bristly. Flowers 1–3 or more, single, slightly fragrant, 2.5–6.5 cm across. Sepals with a few lateral lobes, acuminate, glandular on the back, persistent in fruit. Petals deep pink. Styles free, not protruding; stigmas woolly. Fruit spherical to pear-shaped, dark red, 1–3 cm, glandular-bristly. *C & S Europe, Turkey, former USSR (Caucasus)*. H4. Summer.

'Duplex' (*R. pomifera* forma *duplex* (Weston) Rehder). Illustration: Gault & Synge, The dictionary of roses in colour, pl. 38 (1971); Beales, Classic roses, 6 (1985). Leaflets grey-green; flowers semi-double, clear pink, *c.* 6.5 cm across; more free-flowering. Possibly a hybrid between *R. villosa* and a tetraploid rose of garden origin.

23. R. mollis Smith (*R. villosa* misapplied not Linnaeus). Illustration: Phillips & Rix, Roses, 29, 215 (1988); Blamey & Grey-Wilson, The illustrated flora of Britain and northern Europe, 181 (1989).

Similar to *R. villosa* and thought by some to be synonymous. It differs in the young stems being bloomed, the smaller more or less round leaflets, 1.2–3.5 cm, the less bristly receptacle and a smaller fruit, 1–1.5 cm. *Europe, W Asia*. H3. Summer.

24. R. sherardii Davies. Illustration: Ross-Craig, Drawings of British Plants, **9**: pl. 22 (1956); Blamey & Grey-Wilson, The illustrated flora of Britain and northern Europe, 181 (1989).

Shrub with dense branches to 2 m and with a whitish waxy bloom when young, bearing straight or curved prickles. Leaves deciduous; leaflets 5–7 (rarely 3–7), bluish green, elliptic to broadly ovate, hairy on both surfaces, margins with compound teeth. Bracts present. Receptacle glandular-bristly. Flowers several in a cluster, single, 3–5 cm across. Sepals with lateral lobes, glandular-hairy on the back, erect and persistent in fruit. Petals deep pink. Styles free, not protruding; stigmas woolly. Fruit ovoid to obconical, red, 1.2–2 cm wide, glandular-bristly. *N & C Europe*. H3. Summer.

25. R. tomentosa Smith. Illustration: Ross-Craig, Drawings of British Plants **9**: pl. 23 (1956); Blamey & Grey-Wilson, The illustrated flora of Britain and northern Europe, 181 (1989).

Shrub with arching, often zig-zag branches to 3 m, green when young and often bloomed, bearing sparse, straight or slightly curved prickles of equal size and often in pairs. Stipules with short free tips. Leaves deciduous, smelling resinous when crushed; leaflets 5–7, light- to grey-green, elliptic to ovate, obtuse, acute or acuminate, usually hairy on both sides and glandular beneath, margins with compound glandular teeth. Bracts present. Receptacle glandular-bristly. Flowers solitary or few, single, fragrant, 3.8–5.5 cm across. Sepals with lateral lobes and expanded tips, glandular-bristly on the back, reflexed and falling after flowering. Petals pale pink or white. Styles free, somewhat protruding; stigmas woolly or not. Fruit ovoid to spherical, orange-red, 1.8–2.5 cm with stalked glands. *Europe, Turkey, former USSR (Caucasus) (occasionally naturalised in USA)*. H4. Summer.

26. R. rubiginosa Linnaeus (*R. eglanteria* Linnaeus, in part). Illustration: Krüssmann, The complete book of roses, 263 (1981); Phillips & Rix, Roses, 18, 106 (1988); Blamey & Grey-Wilson, The illustrated flora of Britain and northern Europe, 181 (1989); Brickell, The RHS gardeners' encyclopedia of plants and flowers, 149 (1989).

Dense, much-branched shrub with arching branches 2–3 m, bearing many, scattered, stout, hooked, unequal prickles and often with stiff bristles on the flowering branches. Stipules broad. Leaves deciduous, aromatic especially in damp weather; leaflets 5–9, dark green, ovate to roundish, 1–3 cm, obtuse to acute, usually hairless above, glandular and downy beneath, margins with compound teeth. Bracts present. Receptacle glandular-bristly. Flowers 1–7, occasionally more, single, fragrant, 2.5–5 cm across. Sepals with lateral lobes, glandular-bristly on the back, erect and persistent in fruit. Petals pale to deep pink. Styles free, not protruding; stigmas woolly. Fruit ovoid to nearly spherical, red or orange, 1–2.5 cm, smooth or glandular-bristly. *Europe, N Africa, Asia (naturalised in N America)*. H4. Summer.

27. R. sicula Trattinick.

Densely branched, suckering shrub. Stems 50–150 cm, red when young, bearing sparse, slender, straight or slightly curved prickles of equal length, intermixed with glandular bristles. Leaves deciduous, slightly aromatic; leaflets 5–9, broadly ovate to roundish, 0.6–2 cm, hairless above, glandular and sometimes downy beneath, margins with glandular compound teeth. Bracts present. Receptacle with sparse glandular bristles. Flowers usually solitary, sometimes 2 or 3, single, 2.5–3.3 cm across. Sepals with a few lateral lobes, glandular on the back, margin hairy, erect or spreading and persistent in fruit. Petals bright pink. Styles free, not protruding; stigmas woolly. Fruit ovoid to nearly spherical, red, eventually blackish, 1–1.3 cm, sparsely glandular-bristly. *Mediterranean area*. H4. Summer.

28. R. pulverulenta Bieberstein (*R. glutinosa* Smith; *R. dalmatica* Kerner; *R. glutinosa* var. *dalmatica* (Kerner) Keller).

Compact shrub with stems 30–70 cm, bearing numerous stiff, straight or decurved, whitish prickles intermixed with small glandular bristles. Stipules broad. Leaves deciduous; leaflets 5–7 (rarely 3 to 9), elliptic to obovate or roundish, 0.7–1.5 cm, both sides hairless or somewhat downy, glandular, margins with glandular compound teeth. Bracts present. Receptacle smooth or glandular-bristly. Flowers 1 or 2, single, 2.5–3.8 cm across. Sepals with a few gland-edged lateral lobes, the tips slightly expanded, erect and persistent

in fruit. Petals rose-pink. Styles free, not protruding; stigmas woolly. Fruit ellipsoidal or nearly spherical, dark red, to 2.5 cm, smooth or glandular-bristly. *S Europe to Turkey, Lebanon, Iran, Afghanistan, former USSR (Caucasus)*. H4. Summer.

29. R. micrantha Smith. Illustration: Ross-Craig, Drawings of British Plants **9**: pl. 25 (1956); Krüssmann, The complete book of roses, 258 (1981).

Much-branched shrub with arching stems to 3.5 m, bearing curved or hooked prickles of equal size; bristles usually absent. Stipules narrow. Leaves deciduous; leaflets 5–7, elliptic to broadly ovate, 1.5–4 cm, acute, hairless or hairy above, densely hairy and glandular beneath, margins with glandular compound teeth. Bracts present. Receptacle smooth or sparsely glandular-bristly. Flowers 1–4 (rarely to 8), single, 2.5–3 cm across. Sepals with lateral lobes and an expanded tip, glandular on the back, spreading or reflexed and falling after flowering. Petals pale pink or white. Styles free, more or less protruding; stigmas hairless. Fruit ovoid to nearly spherical, red, 1.2–1.8 cm, smooth or sparsely glandular-bristly. *Europe, except extreme north, NW Africa, SW Asia (naturalised in N America)*. H3. Summer.

30. R. horrida Fischer (*R. biebersteinii* Lindley).

Low, stiffly branched shrub. Stems bearing short, stout, hooked prickles of unequal length, interspersed with glandular bristles. Leaves deciduous; leaflets 5–7, broadly elliptic to roundish, 0.8–2.2 cm, more or less hairless, glandular beneath, margins with compound teeth. Bracts present. Receptacle glandular-bristly. Flowers solitary, single, 2.5–3 cm across. Sepals with lateral lobes, glandular on the back, reflexed and falling after flowering. Petals white. Styles free, not protruding; stigmas not woolly. Fruit ovoid to spherical, red, 1–1.6 cm, glandular-bristly. *SE Europe to Turkey and former USSR (Caucasus)*. H4. Summer.

31. R. serafinii Viviani. Illustration: Phillips & Rix, Roses, 18 (1988).

Densely branched shrub with stems to 1.2 m, bearing stout, curved or hooked prickles of unequal length, often intermixed with bristles. Stipules narrow. Leaves deciduous, aromatic; leaflets 5–7 (rarely to 11), ovate to roundish, 0.5–1.2 cm, obtuse, glossy and hairless above, glandular beneath, margins with glandular compound teeth. Bracts present. Receptacle

smooth. Flowers 1 (rarely to 3), single, 2.5–5 cm across. Sepals with lateral lobes, often glandular on the back, reflexed and falling after flowering. Petals pale pink. Styles free, not protruding; stigmas not woolly. Fruit obovoid to spherical, red, 0.8–1.2 cm wide, smooth. *Italy, Sicily, Corsica, Sardinia, Bulgaria & S former Yugoslavia*. H4. Early summer.

32. R. stylosa Desvaux.

Shrub with arching stems to 3 m, bearing stout, hooked, wide-based prickles. Stipules narrow. Leaves deciduous; leaflets 5–7, dark green, narrowly elliptic to ovate, 1.5–5 cm, acute or acuminate, glossy and hairless above, downy beneath, lacking glands, margins with simple teeth. Bracts narrow. Receptacle smooth. Flowers 1–8 or more, single, 3–5 cm across. Sepals with lateral lobes, often slightly glandular on the back, reflexed and falling after flowering. Petals usually white, sometimes pale pink. Styles free, not or slightly protruding; stigmas not woolly. Fruit ovoid to spherical, red, 1–1.5 cm, smooth. *W Europe, Bulgaria, W Asia*. H4. Summer.

33. R. corymbifera Borkhausen (*R. dumetorum* Thuillier).

Shrub with spreading or erect branches 1.5–3 m, bearing stout, hooked prickles. Leaves deciduous; leaflets 5–9, broadly ovate to roundish, 2.5–6 cm, obtuse, downy on both sides, margins with simple teeth. Bracts present. Flowers in clusters, single, 4–5 cm across. Sepals with some lateral lobes, usually hairless, sometimes slightly glandular, reflexed and falling after flowering. Petals white to pale pink. Styles free, not or slightly protruding. Fruit ovoid to nearly spherical, orange-red, 1.2–2 cm. *Cooler parts of Europe and SW Asia, N Africa*. H3. Summer.

34. R. coriifolia Fries.

Densely branched shrub with erect or arching stems *c.* 2 m, bearing curved prickles of equal length. Leaves deciduous; leaflets 5–7, oblong to broadly elliptic, obtuse, hairless above, downy beneath at least on the veins, margins with simple or compound teeth. Bracts large. Receptacle hairless. Flowers 1–4, single. Sepals with lateral lobes, hairy or not on the back, erect to spreading and persistent in fruit. Petals white or pink. Styles free; stigmas woolly. Fruit ovoid to spherical, red, to 2.5 cm, smooth. *Europe, former USSR (Caucasus)*. H3. Summer.

Very similar to *R. dumalis* and considered by some botanists to be the same. Alternatively, other botanists place it as a variety of *R. dumalis* subsp. *boissieri*, and due to the nomenclatural principle of priority its name would then become *R. dumalis* Bechstein subsp. *boissieri* (Crépin) Nilsson var. *boissieri*.

35. R. canina Linnaeus. Illustration: Bois & Trechslin, Roses, 9 (1962); Gibson, The book of the rose, pl. 2 (1980); Phillips & Rix, Roses, 28 (1988); Blamey & Grey-Wilson, The illustrated flora of Britain and northern Europe, 179 (1989).

Vigorous shrub with arching, sometimes climbing stems 1.5–5.5 m, bearing scattered, strong, hooked, more or less equal prickles; bristles absent. Stipules narrow to broad. Leaves deciduous; leaflets 5–7, narrowly elliptic to ovate, 1.5–4 cm, acute or obtuse, usually hairless on both sides, sometimes glandular beneath or downy on veins, margins with simple or compound teeth. Bracts often broad. Receptacle smooth. Flowers solitary or 2–5, single, fragrant, 2.5–5 cm across. Sepals with lateral lobes, usually hairless on the back, reflexed and falling after flowering. Petals white or pale pink. Styles free, protruding; stigmas woolly or not. Fruit ovoid to nearly spherical, red to orange, 1–3 cm, smooth. *Most of Europe & SW Asia, NW Africa (naturalised in North America)*. H3. Summer.

36. R. britzensis Koehne. Illustration: Krüssmann, The complete book of roses, 267 (1981).

Erect shrub with stems 2–3 m, bearing sparse, small, scattered, slender prickles on the flowering-branches. Leaves deciduous; leaflets 7–11, grey-green, elliptic to ovate, 2.5–3.5 cm, hairless above, beneath with sparse glands on midrib, margins with simple teeth. Bracts present. Receptacle glandular-bristly. Flowers 1 or 2, single, 7–10 cm (rarely 3–10 cm) across. Sepals with lateral lobes, glandular-bristly on the back, erect and persistent in fruit. Petals pale pink fading to white, notched. Styles free; stigmas woolly. Fruit ovoid, dark red or brownish, 2.5–3 cm, slightly glandular-bristly. *S Turkey, N Iraq*. H4. Early summer.

37. R. dumalis Bechstein (*R. glauca* Loiseleur, not Pourret).

Shrub with more or less erect, often bloomed stems 1–2 m, bearing hooked, broad-based prickles. Stipules broad. Leaves deciduous; leaflets 5–7, broadly ovate to roundish, 1.2–3.5 cm, acute or obtuse, hairless and often bloomed above,

hairless or downy beneath, margins with simple or compound teeth. Bracts broad. Receptacle smooth to glandular-bristly. Flowers 1–many, single. Sepals with lateral lobes, hairless or slightly glandular on the back, margin downy, erect and persistent in fruit. Petals pink. Styles free, scarcely protruding; stigmas woolly. Fruit ovoid to spherical, red, 1.5–2.2 cm, smooth or glandular-bristly. *Europe, Turkey*. H3. Summer.

38. R. jundzillii Besser (*R. marginata* misapplied, not Wallroth).

Erect or trailing, suckering shrub with stems 1–2.4 m, bearing a few slender, scattered, straight or decurved prickles, sometimes intermixed with bristles. Stipules large and broad. Leaves deciduous; leaflets 5–7, obovate to broadly elliptic, 2.5–4.5 cm, acute or acuminate, hairless above, more or less glandular and sometimes downy beneath, margins with compound glandular teeth. Bracts present. Receptacle often glandular-bristly. Flowers usually solitary, sometimes 2–8, single, slightly fragrant, 5–7.5 cm across. Sepals with lateral lobes, glandular on the back, spreading to reflexed and falling after flowering. Petals pale to rosy pink, fading with age. Styles united; stigmas woolly. Fruit ovoid to spherical, red, 1.5–2.5 cm, smooth or glandular-bristly. *W Europe to S former USSR (including Caucasus), Turkey*. H3. Summer.

39. R. glauca Pourret (*R. rubrifolia* Villars). Illustration: Gault & Synge, The dictionary of roses in colour, pl. 47, 48 (1971); Thomas, A garden of roses, 91 (1987); Phillips & Rix, Roses, 18 (1988); Kew Magazine **6:** pl. 115 (1989).

Erect, sparsely branched shrub with arching stems 1.5–3 m, dark red and bloomy when young, bearing sparse, straight or decurved, broad-based prickles, the strong shoots also with bristles. Stipules entire or with glandular teeth. Leaves deciduous; leaflets 5–9, grey-green, often tinged with brown or purple and covered with a dull waxy bloom, ovate to narrowly elliptic, 2–4.5 cm, acute, hairless on both sides, margins with simple teeth. Bracts present. Receptacle often with sparse glandular bristles. Flowers solitary or 2–12, single, 2.5–4 cm across. Sepals entire or with a few lateral lobes, smooth or glandular-bristly on the back, spreading and falling after flowering. Petals deep pink, white at the base. Styles free, protruding. Fruit ovoid to nearly spherical, brownish red, 1.3–1.5 cm, smooth or with

sparse glandular bristles. *Mts of C & S Europe*. H3. Summer.

40. R. palustris Marshall (*R. corymbosa* Ehrhart). Illustration: Krüssmann, The complete book of roses, 268 (1981); Phillips & Rix, Roses, 21 (1988).

Erect, rather broadly spreading, suckering shrub with slender reddish or purplish brown stems to 2 m, bearing stout, more or less curved, broad-based prickles in pairs at the nodes, Stipules narrow, the halves of each pair rolled inwards to form a tube, Leaves deciduous; leaflets 5–7 (rarely to 9), dull green, oblong to elliptic, 2–6 cm, acute, hairless above, downy beneath at least on veins, margins with simple teeth. Bracts broad. Receptacle spherical, glandular-bristly. Flowers usually in clusters, rarely solitary, single, 4–5.5 cm across. Sepals entire, expanded at tips, glandular-bristly on the back, spreading and falling after flowering. Petals pink. Styles free, not protruding. Fruit spherical, red, *c.* 8 mm, smooth or glandular-bristly. *E North America*. H3. Summer.

41. R. virginiana Herrmann (*R. lucida* Ehrhart). Illustration: Niering & Olmstead, The Audubon Society field guide to North American wildflowers: eastern region, pl. 581 (1979); Gibson, The book of the rose, pl. 15 (1980); Thomas, A garden of roses, 149 (1987); Phillips & Rix, Roses, 27 (1988).

Erect, often suckering shrub, with brownish red stems to 1.5 m, unarmed or bearing straight or recurved prickles in pairs at the nodes, and also bristles on the young stems. Stipules widening towards tip, glandular-toothed. Leaves deciduous; leaflets 5–9, obovate to oblong-elliptic, 2–6 cm, acute, glossy green and hairless above, beneath hairless, or downy on midrib, margins with coarse simple teeth except at base. Bracts present. Receptacle smooth or glandular-bristly. Flowers 1–8, single, 5–6.5 cm across, fragrant. Sepals entire or with a few lateral lobes, tips leafy, glandular and hairy on the back, spreading or reflexed and falling after flowering. Petals pale pink to bright pink. Styles free, not protruding; stigmas woolly. Fruit nearly spherical, red, 1–1.5 cm wide, smooth or glandular-bristly. *E North America*. H3. Summer.

R. × mariae-graebneriae Ascherson & Graebner is a hybrid between *R. virginiana* and probably *R. palustris*. An erect shrub with a spherical shape and stems to 1.6 m, which bear slightly curved prickles and sometimes scattered bristles. Leaves deciduous; leaflets glossy green, margins with simple teeth. Bracts present. Petals rose-pink. Fruit nearly spherical, red. *Garden origin.* H3. Summer–early autumn.

R. × suionum Almquist is a hybrid (probably complex) whose origin is uncertain, but recent opinion suggests that *R. virginiana* may be involved. Leaves distinct, with 5 or 7 leaflets, and with a much smaller pair of leaflets below the terminal one. Flowers double, clear pink. *Garden origin.* H2. Summer.

42. R. carolina Linnaeus (*R. humilis* Marshall). Illustration: Peterson & McKenny, A field guide to wildflowers, 219, 257 (1968); Thomas, A garden of roses, 49 (1987); Phillips & Rix, Roses, 77 (1988).

Suckering shrub with slender stems 1–1.5 m, bearing scattered, straight, slender, unequal prickles in pairs at the nodes, and dense bristles especially on young stems. Stipules narrow, the halves of each pair often rolled inward to form a tube, entire to glandular-toothed. Leaves deciduous; leaflets 5–9, lanceolate to narrowly ovate or almost round, acute to obtuse, hairless on both sides or downy beneath, margins with simple teeth. Bracts present. Receptacle glandular-bristly. Flowers solitary or few in a cluster, single, 3.8–5 cm across. Sepals with lateral lobes, glandular-bristly on the back, spreading and falling after flowering. Petals pale to mid-pink. Styles free, not protruding. Fruit nearly spherical, red, 0.7–0.9 cm, glandular-bristly. *E & C North America.* H3. Summer.

'Alba' (*R. virginiana* Herrmann var. *alba* Rafinesque). Illustration: Thomas, A garden of roses, 151 (1987). Leaflets always hairy beneath; flowers white.

'Plena'. Illustration: Thomas, Shrub roses of today, pl. 2 (1974). Dwarf shrub to 50 cm; flowers double, outer petals ageing almost to white.

43. R. nitida Willdenow. Illustration: Krüssmann, The complete book of roses, 268 (1981); Thomas, A garden of roses, 121 (1987); Phillips & Rix, Roses, 21, 214 (1988).

Erect, suckering shrub with often reddish stems 50–100 cm, bearing sparse, straight prickles and dense purplish brown or reddish bristles. Stipules broad, with glandular teeth. Leaves deciduous; leaflets 5–9, elliptic, 1–3 cm, acute, shiny green and hairless above, hairless or sparsely hairy beneath, margins with fine, simple teeth. Bracts present. Receptacle glandular-bristly. Flowers 1 (rarely to 3), single,

fragrant, 4.5–6.5 cm across. Sepals entire, bristly or glandular on the back, spreading and falling after flowering. Petals deep pink. Styles free, not protruding. Fruit spherical or more or less so, dark scarlet, 0.8–1 cm, glandular-bristly. *E North America.* H3. Summer.

44. R. foliolosa Torrey & Gray. Illustration: Beales, Classic roses, 20, 212 (1985); Phillips & Rix, Roses, 27 (1988).

Shrub, suckering, with rather weak, reddish stems 50–100 cm, unarmed or bearing sparse slender, straight prickles, and only rarely with bristles. Stipules narrow. Leaves deciduous; leaflets 7–9 (rarely to 11), narrowly oblong, 1–5 cm, acute, glossy and hairless above, downy on midrib beneath, margins with fine simple teeth. Bracts present. Receptacle glandular-bristly. Flowers 1–5, single, fragrant, 5–6.5 cm across. Sepals entire, glandular-bristly on the back, spreading and falling after flowering. Petals white to rose-pink. Styles free, not protruding. Fruit nearly spherical, red, *c.* 0.8 cm wide, glandular-bristly. *SE USA.* H4. Summer–early autumn.

45. R. rugosa Thunberg. Illustration: Niering & Olmstead, The Audubon Society field guide to North American wildflowers: eastern region, pl. 582 (1979); Phillips & Rix, Roses, 100 (1988).

Erect shrub with stout stems 1.2–2.5 m, downy when young, bearing very dense prickles of unequal size and dense bristles. Stipules large. Leaves deciduous; leaflets 5–9, dark green, oblong to elliptic, 2.5–5 cm, usually acute, hairless and conspicuously wrinkled above, downy beneath with conspicuous veins, margins with shallow teeth except at base. Bracts large, enfolding the flower-stalks. Receptacle smooth. Flowers solitary or few, single, fragrant, 6–7.5 (rarely to 9) cm across. Sepals entire, expanded at tip, downy, erect and persistent after flowering. Petals purplish pink. Styles free, not protruding. Fruit nearly spherical, with a neck at the top, red to orange-red, 2–2.5 cm, smooth. *Eastern former USSR, Korea, Japan, N China (naturalised in Britain & NE USA).* H2. Early summer–autumn.

Var. **rosea** Rehder has single, rose-pink flowers. Var. **albo-plena** Rehder has double, white flowers. 'Alba'. Illustration: Beales, Classic roses, 239 (1985); Austin, The heritage of the rose, 238 (1988); Phillips & Rix, Roses, 101 (1988)). Flowers solitary, white, opening from pale pink buds; very free fruiting.

R. × kamtchatica Ventenat is a hybrid

involving *R. rugosa* and either *R. davurica* Pallas or *R. amblyotis* C.A. Meyer. Some authorities consider it to be a variety of *R. rugosa*, in which case its correct name would be var. *ventenatiana* C.A. Meyer (*R. rugosa* var. *kamtchatica* (Ventenat) Regel). It differs from *R. rugosa* in the more slender stems which bear fewer prickles, fewer wrinkled leaflets, and smaller flowers and fruit. *Former USSR (E Siberia, Kamtchatka).* H2. Summer.

R. rugosa has been crossed with *R. arvensis* to produce **R. × paulii** Rehder ('Paulii'). Illustration: Gault & Synge, The dictionary of roses in colour, pl. 34 (1971); Gibson, The book of the rose, pl. 36 (1980); Austin, The heritage of the rose, 356 (1988). Stems vigorous, trailing to 4 m, flowers in clusters, single, white, fragrant, to 6 cm across, with entire glandular sepals. 'Rosea'. Illustration: Gault & Synge, The dictionary of roses in colour, pl. 35, 1971; Austin, The heritage of the rose, 349 (1988); Phillips & Rix, Roses, 101 (1988). Flowers pink with paler centres.

R. × microgosa Henkel (*R. vilmorinii* Bean). Illustration: Gault & Synge, The dictionary of roses in colour, pl. 28 (1971). A hybrid between *R. rugosa* and *R. roxburghii* which is intermediate between the parents. Flowers pale pink, 10–12.5 (rarely 8–12.5) cm across. Fruit orange-red, nearly spherical, with prickles. 'Alba'. Illustration: Beales, Classic roses, 229 (1985) has white fragrant flowers produced over a period, and more erect growth.

46. R. acicularis Lindley. Illustration: Krüssmann, The complete book of roses, 272 (1981).

Loose shrub with stems 1 (rarely to 2.5) m, bearing straight or slightly curved, weak, slender, unequal prickles intermixed with dense, slender bristles. Stipules narrow. Leaves deciduous; leaflets 5–7 (rarely 3–9), ovate to elliptic, 1.5–6 cm, acute, blue-green and usually hairless above, greyish and downy beneath, margins with simple teeth. Bracts narrow, more or less equalling flower-stalks. Receptacle smooth. Flowers 1 (rarely to 3), single, fragrant, 3.8–6.2 cm across. Sepals entire, glandular on the back, erect and persistent after flowering. Petals rose-pink to purplish pink. Styles free, not protruding; stigmas woolly. Fruit ellipsoidal or spherical to pear-shaped, with a neck at the top, red, 1.5–2.5 cm, smooth. *Circumpolar: N America, N Europe, former*

USSR (Siberia), south to N China & Japan. H1. Early summer.

Var. **nipponensis** (Crépin) Koehne. Illustration: Phillips & Rix, Roses, 24 (1988). Leaflets 7–9, oblong, obtuse; flowers bright pink. *C & S Japan.*

47. R. arkansana Porter. Illustration: Krüssmann, The complete book of roses, 273 (1981); Phillips & Rix, Roses, 20 (1988).

Erect, suckering shrub with stems 50–120 cm, bearing usually dense, straight, unequal prickles, and bristles. Stipules narrow. Leaves deciduous; leaflets 9–11 (rarely 3–7), obovate to elliptic, 2–6 cm, acute or obtuse, shiny and hairless above, sometimes downy on the veins beneath, margins with simple teeth. Bracts present. Receptacle smooth or slightly glandular. Flowers few to many in lateral clusters, single, 2.5–4 cm across. Sepals entire or with lateral lobes, smooth or glandular-bristly on the back, spreading to erect, and persistent after flowering. Petals deep pink. Styles free, not protruding. Fruit pear-shaped to nearly spherical, red, 1–1.5 cm, smooth or slightly glandular. *C USA.* H3. Summer.

48. R. suffulta Greene (*R. arkansana* var. *suffulta* (Greene) Cockerell). Illustration: Beales, Classic roses, 249 (1985); Phillips & Rix, Roses, 2 (1988).

Low shrub with stems to 50 cm, bearing dense, straight prickles and bristles. Stipules broad. Leaves deciduous; leaflets 7–11, ovate-oblong to broadly elliptic, 1.5–4 cm, obtuse, bright green and hairless or downy above, downy beneath, margins with simple teeth. Bracts present. Receptacle smooth. Flowers in clusters, single, *c.* 3 cm across. Sepals entire or sometimes with lateral lobes, spreading to erect, and persistent after flowering. Petals mid-pink. Styles free, not protruding. Fruit spherical, red, *c.* 1 cm, smooth. *E & C North America.* H3. Summer.

Very similar to *R. arkansana* and considered to be conspecific by some authorities.

49. R. blanda Aiton. Illustration: Thomas, A garden of roses, 39 (1987).

Erect shrub 1–2 m, stems brown, unarmed or sometimes bearing a few, scattered, slender, straight prickles near the base. Stipules broadening towards top. Leaves deciduous; leaflets 5–7 (rarely to 9), elliptic to oblong-obovate, 2–6 cm, acute to obtuse, dull green and hairless above, usually downy beneath, margins with

simple teeth. Bracts large, enfolding the flower-stalks. Receptacle smooth. Flowers solitary or 3–7, single, 4.5–6.5 cm across, fragrant. Sepals entire, downy and glandular on the back, erect and persistent in fruit. Petals rosy pink. Styles free, not protruding. Fruit ovoid to pear-shaped, red, *c.* 1 cm, smooth. *E & C North America.* H3. Early summer.

50. R. pendulina Linnaeus (*R. alpina* Linnaeus; *R. cinnamomea* Linnaeus (1753) not (1759); *R. pyrenaica* Gouan). Illustration: Gibson, The book of the rose, pl. 34 (1980); Rosetti, I fiori delle Alpi, pl. 267 (1980); Phillips & Rix, Roses, 24, 25 (1988).

Suckering shrub of variable habit, with green or red-brown stems 60–200 cm, usually unarmed but sometimes bearing slender prickles and bristles. Stipules broadening towards top. Leaves deciduous; leaflets 5–9 (rarely to 11), elliptic to broadly so, 2–6 cm, acute or obtuse, hairless or downy above, usually downy beneath and sometimes glandular, margins with compound glandular teeth. Bracts enfolding and equalling the flower-stalks, soon falling. Receptacle smooth or glandular-bristly. Flowers 1 (rarely to 5), single, 3.8–6.5 cm across. Sepals entire with an expanded tip, usually smooth, sometimes glandular on the back, erect and persistent in fruit. Petals deep pink or purplish pink. Styles free, not protruding; stigmas hairy. Fruit often pendent, nearly spherical to ovoid, usually with a neck at the top, red, 1.5–3 cm, smooth or glandular-bristly. *Mts of S & C Europe.* H3. Early summer.

'Nana' is a dwarf, freely suckering selection to 30 cm.

R. × lheriteriana Thory. Illustration: Thomas, A garden of roses, 107, (1987). Thought to be a hybrid between *R. pendulina* and *R. chinensis*. Shrub with reddish stems climbing to 4 m, unarmed or with a few prickles. Leaves deciduous; leaflets 3–7, ovate-oblong, with simple teeth, hairless on both sides. Bracts present. Receptacle smooth. Flowers many, in clusters, semi-double. Sepals entire. Petals purplish red, whitish at the base. Styles free, not protruding. Fruit nearly spherical, smooth. *Garden origin.* H4. Summer.

51. R. oxyodon Boissier (*R. alpina* var. *oxyodon* (Boissier) Boulenger; *R. pendulina* var. *oxyodon* (Boissier) Rehder).

Very similar to *R. pendulina* but stems 2–4 m; receptacle smooth, sepals usually

glandular on the back; fruit smooth. *Former USSR (E Caucasus).* H4. Summer.

52. R. majalis Herrmann (*R. cinnamomea* Linnaeus (1759) not (1753); *R. cinnamomea* var. *plena* Weston; *R. foecundissima* Muenchhausen; *R. majalis* 'Foecundissima'). Illustration: Thomas, A garden of roses, 115 (1987); Phillips & Rix, Roses, 26 (1988); Blamey & Grey-Wilson, The illustrated flora of Britain and northern Europe, 179 (1989).

Erect, suckering shrub with slender, red-brown stems 1.5–3 m, much-branched towards the top, bearing slender, hooked prickles in pairs at the nodes as well as scattered prickles on lower parts of stem, intermixed with bristles. Stipules broad, those on strong shoots often rolled inwards. Leaves deciduous; leaflets 5–7, elliptic to obovate, 1.5–5 cm, obtuse to acuminate, sometimes downy above, usually downy beneath, margins with simple teeth except at the base. Bracts large, equal to or longer than flower-stalks. Receptacle smooth. Flowers few, single or (in gardens) more usually double, *c.* 5 cm across. Sepals entire, with woolly margin and often hairy on the back, erect and persistent in fruit. Petals mid-pink to purplish pink. Styles free, not protruding; stigmas woolly. Fruit spherical or slightly lengthening, dark red, 1–1.5 cm, smooth. *N & C Europe, former USSR (Siberia).* H2. Summer.

Herrmann based the species on the double-flowered form, which has been cultivated since the 16th century.

53. R. davurica Pallas.

Shrub with stems 1–1.5 m, bearing large, more or less straight, prickles in pairs at the nodes. Stipules narrow. Leaves deciduous; leaflets *c.* 7, oblong-lanceolate, 2.5–3.5 cm, acute, hairy and glandular beneath, margins with compound teeth. Bracts present. Receptacle smooth. Flowers 1–3, single. Sepals entire, margins hairy, erect and persistent in fruit. Petals pink. Styles free, not protruding. Fruit ovoid, red, 1–1.3 cm, smooth. *NE Asia & N China.* H2. Summer.

Very similar to *R. majalis.*

54. R. laxa Retzius. Illustration: Phillips & Rix, Roses, 21 (1988).

Shrub with slender, arching stems to 2.5 m, bearing few, large, recurved, hooked or straight prickles, intermixed with bristles. Stipules broad, not glandular. Leaves deciduous; leaflets 5–9, ovate to elliptic or oblong, 1.5–4.5 cm, obtuse, hairless above, usually hairy beneath, margins with simple or compound teeth. Bracts present. Receptacle smooth. Flowers 1–6, single, 4–5 cm across. Sepals entire, usually glandular on the back, erect and persistent after flowering. Petals white to pale pink, notched. Styles free, not protruding. Fruit ellipsoidal to spherical, red, *c.* 1.5 cm, smooth. *Former USSR (Siberia south to Tienshan & Pamir Alai), NW China.* H3. Summer.

55. R. pisocarpa A. Gray. Illustration: Krüssmann, The complete book of roses, 276 (1981); Phillips & Rix, Roses, 26 (1988).

Erect shrub with slender, arching stems 90–250 cm, unarmed or bearing few, straight, weak prickles in pairs at the nodes, and some bristles towards base. Stipules broad. Leaves deciduous; leaflets 5–7 (rarely to 9), elliptic to ovate, 1.3–4 cm, obtuse or somewhat acute, hairless above, downy beneath, margins with simple teeth. Bracts broad. Receptacle smooth. Flowers 4 or 5 (rarely 1–5), single, 2.5–3 cm across. Sepals entire, glandular-bristly on the back, erect and persistent in fruit. Petals rosy pink to purplish pink. Styles free, not protruding. Fruit ellipsoidal to spherical, purplish red to red or orange, to 1.3 cm, smooth. *W North America from British Columbia south to N California.* H3. Summer.

56. R. woodsii Lindley (*R. macounii* Greene).

Stiffly branched shrub with stems to 2 m, purplish or reddish brown, becoming grey, bearing many, straight or slightly curved prickles, sometimes intermixed with bristles. Stipules narrow. Leaves deciduous; leaflets 5–7 (rarely to 9), obovate to elliptic, 1–3 cm, acute to obtuse with simple teeth, hairless above, hairless or downy beneath. Bracts present. Receptacle smooth. Flowers 1–3 (rarely to 5), single, 3–4.5 cm across. Sepals entire, hairless or downy on the back, erect to spreading and persistent in fruit. Petals mid-pink, occasionally white. Styles free, not protruding. Fruit ovoid to spherical, usually with a neck at the top, 0.5–1.5 cm, smooth. *W & C North America.* H3. Summer.

Var. **fendleri** (Crépin) Rydberg (*R. fendleri* Crépin). Illustration: Phillips & Rix, Roses, 26 (1988). Differs in having glandular-bristly stipules and flower-stalks, leaflets which usually have compound teeth, smaller lilac-pink flowers and smaller fruit. *W North America to N Mexico.*

57. R. californica Chamisso & Schlechtendahl.

Erect shrub with stems 1.5–3 m, greenish when young, red-brown when older, bearing stout, broad-based, recurved prickles in pairs at the nodes, and often with bristles on the young shoots. Stipules narrow. Leaves deciduous; leaflets 5–7, ovate to broadly elliptic, 1–3.5 cm, obtuse, hairless or downy above, downy and often glandular beneath, margins usually with simple teeth. Bracts broad. Receptacle smooth or hairy when young. Flowers in clusters, single, fragrant, 3.7–4 cm across. Sepals entire, hairy on the back, erect and persistent in fruit. Petals deep pink to bright crimson. Styles free, not protruding. Fruit spherical or slightly lengthened, with a neck at the top, 0.7–1.5 cm, smooth. *W USA south to Mexico (Baja California).* H4. Summer.

'Plena' (forma *plena* Rehder). Illustration: Thomas, A garden of roses, 45 (1987); Austin, The heritage of the rose, 355 (1988); Phillips & Rix, Roses, 27 (1988). Stems to 2 m, flowers semi-double. The rose which is usually cultivated under this name is in fact *R. nutkana*.

58. R. nutkana C. Presl. Illustration: Beales, Classic roses, 237 (1985).

Robust shrub with stout, purplish brown stems 1.5–3 m, bearing stout, usually straight, broad-based prickles in pairs at the nodes and slender bristles on the young stems. Leaves deciduous; leaflets 5–7 (rarely to 9), dark green, ovate to elliptic, 1.9–5 cm, acute or obtuse, hairless above, glandular and sometimes downy beneath, margins with compound glandular teeth. Bracts present. Receptacle smooth. Flowers 1 (rarely to 3), single, 5–6.5 (rarely 4–6.5) cm across, with glandular-bristly stalks. Sepals entire, glandular-bristly or not, and more or less downy on the back, erect and persistent in fruit. Petals bright red to purple-pink, occasionally white. Styles free, not protruding. Fruit spherical to more or less so, red, 1.5–2 cm, smooth. *W North America from Alaska to N California.* H3. Summer.

Frequently cultivated under the name *R. californica* 'Plena'.

Var. **hispida** Fernald (*R. spaldingii* Crépin). Illustration: Phillips & Rix, Roses, 20 (1988). Differs in having leaflets downy beneath and simple non-glandular teeth, hairless flower-stalks and glandular-bristly receptacles and fruit. *W North America from British Columbia to Utah* (more easterly than *R. nutkana* itself).

59. R. corymbulosa Rolfe. Illustration: Botanical Magazine, 8566 (1914).

Shrub with erect or sometimes prostrate or climbing stems to 2 m, unarmed or bearing a few straight, slender prickles. Leaves deciduous, turning purple in autumn; leaflets 3–5, elliptic to ovate-oblong, 1.3–5 cm, acute, dark green and sparsely downy above, glaucous and downy beneath, margins with compound teeth. Bracts present. Receptacle obovoid, glandular-bristly. Flowers in clusters of up to 12, single, 1.9–2.5 cm across. Sepals entire, downy and glandular-bristly on the back, erect and persistent in fruit. Petals deep pink, paler at the base. Styles free, slightly protruding; stigmas downy. Fruit ovoid-spherical to spherical, coral-red, 1–1.3 cm, glandular-bristly. *China (Hubei, Shensi).* H4. Summer.

60. R. davidii Crépin. Illustration: Phillips & Rix, Roses, 24, 217 (1988).

Shrub with erect or arching stems, 1.8–4 m, bearing scattered, stout, straight or slightly curved, reddish, broad-based prickles. Leaves deciduous; leaflets 7–9 (rarely 5–11), dark green, ovate to elliptic, 0.6–5 cm, acute, wrinkled and hairless above, downy beneath especially on the veins, margins with simple, sometimes glandular teeth. Bracts present. Receptacle glandular-bristly, more or less downy. Flowers 3–12 or more, single, fragrant, 3.8–5 cm across. Sepals entire, often downy and glandular, erect and persistent in fruit. Petals rosy pink. Styles free, protruding; stigmas woolly. Fruit pendent, ovoid with a slender neck at the top, scarlet, 1.5–2 cm. *W & C China including SE Xizang.* H4. Summer.

Var. **elongata** Rehder & Wilson which has leaflets 5–7.5 cm which may be hairless beneath, fewer (3–7) flowers, and more elongated fruit to 2.5 cm, is now considered by many to represent part of the variation of *R. davidii*.

61. R. caudata Baker.

Erect shrub with reddish stems 1–4 m, bearing few, stout, scattered, broad-based prickles. Stipules broad. Leaves deciduous; leaflets 7–9, elliptic to ovate, 2.5–5 cm, acute, hairless on both sides except for midrib beneath, margins with simple teeth. Bracts present. Receptacle glandular-bristly. Flowers few in a tight cluster, single, 3.5–5 cm across. Sepals entire, glandular-bristly on the back, expanded at the tip, erect and persistent in fruit. Petals deep pink. Styles free, not protruding. Fruit ovoid-oblong with a neck at the top, orange-red, 2–2.5 cm, glandular-bristly. *W China*. H4. Summer.

Very similar to and possibly not distinct from *R. setipoda*.

62. R. banksiopsis Baker.

Similar to *R. caudata*: differs in the receptacle and sepals being smooth and the smaller flowers (2–3 cm across). *W China*. H4. Summer.

63. R. macrophylla Lindley. Illustration: Thomas, A garden of roses, 111 (1987); Phillips & Rix, Roses, 24, 25, 216 (1988).

Shrub with erect and arching, dark red to purple stems 2.5–4 (rarely to 5) m, unarmed or bearing a few stout, upward-pointing, straight prickles which are often paired at the nodes. Stipules usually broad. Leaves deciduous; leaflets 9–11 (rarely 7–11), elliptic to narrowly ovate, 2.5–6.5 cm, acute or acuminate, hairless above, downy and sometimes glandular beneath, margins with simple or compound teeth. Bracts broad. Receptacle glandular-bristly. Flowers 1–5, single, 5–7.5 cm across. Sepals entire, bristly and more or less glandular on the back, expanded at the tip, erect and persistent in fruit. Petals mid-pink or deep pink to mauvish pink. Styles free, not protruding; stigmas woolly. Fruit nearly spherical to bottle-shaped with a neck at the top, red, 2.5–4 (rarely to 7.5) cm, glandular-bristly. *Himalaya from Pakistan east to W China*. H4. Summer.

64. R. setipoda Hemsley & Wilson. Illustration: Gault & Synge, The dictionary of roses in colour, pl. 51 (1971); Phillips & Rix, Roses, 22 (1988).

Shrub with stout, reddish, erect or arching stems 2.4–3 m, bearing sparse, short, straight, broad-based prickles intermixed with bristles. Stipules large. Leaves deciduous, slightly aromatic; leaflets 7–9, elliptic to elliptic-ovate, 3–6 cm, acute to obtuse, mid-green and hairless above, greyish, glandular and downy on the veins beneath, margins with simple or compound teeth. Bracts large. Receptacle narrowly ellipsoidal, glandular-bristly, purplish. Flowers in loose clusters of up to 20, single, smelling of unripe apples, 3–5 cm across. Sepals usually entire, expanded at the tip, margin glandular, erect and persistent in fruit. Petals mid-pink to deep purplish pink, shading to white at the base, slightly downy on the back. Styles free, not protruding. Fruit bottle-shaped with a neck at the top, dark red, 2.5–5 cm, glandular-bristly. *W China (Hubei, Sichuan)*. H4. Summer.

65. R. hemsleyana Täckholm.

Very similar to, and possibly better regarded as a form of *R. setipoda*. *R. hemsleyana* differs in its smaller size (to 2 m) and sepals with lateral lobes. *N & C China*. H4. Summer.

66. R. sweginzowii Koehne (*R. moyesii* misapplied, not Hemsley & Wilson). Illustration: Gault & Synge, The dictionary of roses in colour, pl. 58, 59 (1971); Phillips & Rix, Roses, 24 (1988).

Shrub with reddish, spreading stems to 3.5 (rarely to 5) m, sometimes unarmed or bearing dense, 3-angled prickles, intermixed with bristles. Stipules broad. Leaves deciduous; leaflets 7–11, elliptic to broadly so, 2.5–5 cm, acute, hairless above, downy on veins beneath, margins with compound teeth. Bracts broad. Receptacle glandular-bristly. Flowers 1–3 (rarely to 6), single, 3.5–5 cm across. Sepals entire, glandular-bristly on the back, erect and persistent in fruit. Petals bright pink. Styles free, not protruding; stigmas woolly. Fruit bottle-shaped, glossy or orange-red, *c.* 2.5 cm, glandular-bristly. *N & W China*. H4. Summer.

67. R. wardii Mulligan. Illustration: Phillips & Rix, Roses, 24 (1988).

Similar to *R. sweginzowii*, differing in generally lacking prickles and bristles, in the smaller leaflets (1.3–1.9 cm), and in the white flowers. *China (SE Xizang)*. H4. Summer.

68. R. murieliae Rehder & Wilson.

Shrub with reddish, erect or arching stems 1.5–3 m, unarmed or bearing a few slender, straight prickles intermixed with pinkish bristles, dense on young shoots. Leaves deciduous; leaflets 9–15, elliptic to oblong, 1–4 cm, more or less acute, hairless on both sides, but downy on midrib beneath, margins with simple glandular teeth. Bracts present. Receptacle smooth. Flowers 3–7, single, 2–2.5 cm across. Sepals entire, expanded at the tip, hairy or not on the back, erect and persistent in fruit. Petals white. Styles free, not protruding. Fruit bottle-shaped, orange-red, 1–2 cm, smooth. *W China*. H4. Summer.

69. R. moyesii Hemsley & Wilson. Illustration: Thomas, A garden of roses, 117 (1987); Austin, The heritage of the rose, 368 (1988); Le Rougetel, A heritage of roses, 39 (1988); Phillips & Rix, Roses, 22, 217 (1988).

Erect shrub with stout, red-brown stems 2–3.5 m, bearing pale, scattered, stout, broad-based prickles especially on non-flowering-shoots, the lower parts of the stems also with bristles. Leaves deciduous; leaflets 7–13, broadly elliptic to ovate, 1–4 cm, acute, dark green above, somewhat glaucous beneath, both sides hairless except for midrib beneath which is downy and sometimes prickly, margins usually with simple teeth. Bracts with glandular margins. Receptacle glandular-bristly. Flowers 1–2 (rarely to 4), single, 4–6.5 cm across. Sepals entire, expanded at the tip, downy inside, hairy and sparsely glandular on the back, erect and persistent in fruit. Petals pink to blood-red. Styles free, not protruding. Fruit bottle-shaped, with a neck at the top, orange-red, 3.8–6 cm, glandular-bristly at least towards the base. *W China (Sichuan, Yunnan)*. H4. Summer.

'Fargesii' (*R. moyesii* var. *fargesii* Rolfe; *R. fargesii* misapplied, not Boulenger). Illustration: Beales, Classic roses, 232 (1985) – as R. fargesii; Phillips & Rix, Roses, 22 (1988). Flowers pink to rose-red; leaflets shorter (1–2 cm) and wider, obtuse. 'Geranium' Illustration: Phillips & Rix, Roses, 22, (1988); Brickell, The RHS gardeners' encyclopedia of plants and flowers, 151 (1989). Compact, growing only to 2.5 m; leaflets paler green; flowers scarlet; fruit broader, crimson, with a shorter neck. It is the cultivar most often grown. 'Sealing Wax'. Illustration: Phillips & Rix, Roses, 216 (1988). Fruits large, bright red; flowers pink.

Forma **rosea** Rehder & Wilson, with pink flowers and coarsely toothed leaflets, may not in fact be related to *R. moyesii*, but appears to be closer to *R. davidii*, which comes from the same area. However, some botanists give it as a synonym of *R. holodonta*. There is confusion about what rose this name should be attached to, and investigation is needed.

R. holodonta Stapf is closely related to *R. moyesii* but has leaflets to 5 cm. *W China*. It may not be worthy of recognition at species level.

A hybrid between *R. moyesii* and *R. setipoda* is called **R. × wintoniensis** Hillier. Differs from *R. moyesii* in having clusters of 7–10 flowers with crimson petals which are white at the base, and aromatic leaves.

R. × highdownensis Hillier. Illustration: Gault & Synge, The dictionary of roses in colour, pl. 23 (1971); Beales, Classic roses, 232 (1985); Austin, The heritage of the rose, 371 (1988). A hybrid of *R. moyesii* with clusters of mid-red to deep pink flowers.

70. R. bella Rehder & Wilson. Illustration: Phillips & Rix, Roses, 2 (1988).

Shrub with spreading, purplish branches to 3 m, bearing a few slender, straight prickles, intermixed with bristles on the lower part of the stems. Stipules broad. Leaves deciduous; leaflets 7–9, shortly stalked, elliptic to broadly ovate, 1–2.5 cm, more or less acute, hairless on both sides, sometimes glandular on the veins beneath, margins with simple teeth. Bracts glandular. Receptacle glandular-bristly. Flowers 3, single, 4–5 cm across. Sepals entire, expanded at the tip, glandular on the back, erect and persistent in fruit. Petals bright pink. Styles free, not protruding. Fruit ovoid or ellipsoid, orange-red, 1.5–2.5 cm, glandular-bristly. *NW China*. H4. Summer.

71. R. webbiana Royle. Illustration: Edwards, Wild and old garden roses, 33 (1975); Thomas, A garden of roses, 153 (1987); Phillips & Rix, Roses, 22 (1988).

Shrub with slender stems 1–2 m, purplish brown, often bloomed when young, bearing few straight, slender, yellowish, broad-based prickles. Leaves deciduous; leaflets 5–9, obovate to broadly elliptic or roundish, 0.6–2.5 cm, obtuse, somewhat glaucous and hairless above, often slightly downy beneath, margins with simple teeth except in the lower third. Bracts present. Receptacle smooth or glandular-bristly. Flowers 1 (rarely to 3), single, faintly fragrant, 3.8–5 cm across. Sepals entire, glandular and often hairy on the back, erect and persistent in fruit. Petals white to pale pink or lilac pink, sometimes pink with a whitish base. Styles free, not protruding. Fruit spherical to broadly bottle-shaped, with a neck at the top, shiny red, 1.5–2.5 cm, smooth or glandular-bristly. *W & C Himalaya, China (Xizang), Afghanistan, former USSR (Central Asia)*. H4. Early Summer.

Var. **microphylla** Crépin (*R. nanothamnus* Boulenger). Illustration: Phillips & Rix, Roses, 23 (1988) – as R. nanothamnus. Stems only to 50 cm, leaflets smaller (to 1.6 cm), and flowers smaller (2.5–3.8 cm across). *Former USSR (Central Asia), Afghanistan, Kashmir*. More common in cultivation than the species.

72. R. sertata Rolfe.

Similar to *R. webbiana* but leaflets obtuse to acute, flowers deep pink to purplish pink on longer stalks (1.5–3 cm rather than 1–1.3 cm), and fruit dark red. *W & C China*. H4. Summer.

73. R. giraldii Crépin. Illustration: Phillips & Rix, Roses, 18 (1988).

Differs from *R. webbiana* in having acute or obtuse leaflets with compound teeth, very short flower-stalks often concealed by large bracts, smaller pink flowers 1.2–2.5 cm across with a white centre, and a smaller fruit, rarely exceeding 1 cm. *China*. H4. Summer.

74. R. fedtschenkoana Regel. Illustration: Gault & Synge, The dictionary of roses in colour, pl. 15 (1971); Beales, Classic roses, 222 (1985); Thomas, A garden of roses, 71 (1987); Phillips & Rix, Roses, 22, 214 (1988).

Suckering shrub with vigorous erect stems 1–2.5 (rarely to 3) m, bearing slender, straight or curved prickles in pairs at the nodes, pinkish when young, sometimes reduced to bristles. Stipules acuminate. Leaves deciduous; leaflets 5–9, glaucous green, broadly elliptic to obovate, acute, hairless above, downy beneath, margins with simple teeth. Bracts present. Receptacle glandular-bristly. Flowers 1–4, single, smelling slightly unpleasant, *c.* 5 cm across. Sepals entire, glandular-bristly on the back, margin and inner surface downy, erect and persistent in fruit. Petals usually white, rarely pink. Styles free, not protruding. Fruit pear-shaped, red to orange-red, 1.5–2.5 cm, glandular-bristly. *Former USSR (Central Asia), NW China*. H4. Summer–yearly autumn.

75. R. prattii Hemsley. Illustration: Phillips & Rix, Roses, 18 (1988).

Shrub with erect purplish or reddish stems 1.2–2.5 m, unarmed or bearing a few, straight, yellow or pale brown prickles. Stipules entire, acute. Leaves deciduous: leaflets usually 11–15 (rarely 7–9), obovate to narrowly ovate or elliptic, 0.6–2 cm, acute, hairless above, downy on the veins beneath, margins with obscure teeth. Bracts present. Receptacle glandular-bristly. Flowers 3–7 (rarely 1–7), single, 2–2.5 cm across. Sepals entire, downy on both sides, erect and persistent in fruit. Petals pale to more usually deep pink. Styles free, not protruding. Fruit ovoid to bottle-shaped, scarlet to orange-red, 6–10 cm, glandular-bristly. *China (Sichuan)*. H5. Summer.

76. R. elegantula Rolfe (*R. farreri* Stearn).

Shrub producing densely prickly suckers, and stems 1–2 m, unarmed or bearing a few prickles and dense, red bristles, particularly in the lower part. Stipules narrow. Leaves deciduous, glaucous green,

turning purple to crimson in the autumn; leaflets 7–11, narrowly ovate to elliptic, 1–2.5 cm, acuminate or obtuse, downy on midrib beneath, margins with simple teeth. Bracts present. Receptacle smooth or glandular. Flowers solitary to few, single, 2–3.8 cm across. Sepals entire, abruptly acuminate, hairy inside and on margin but hairless on the back, spreading and persistent in fruit. Petals white to pale pink or rose-pink. Styles free, not protruding. Fruit ovoid to top-shaped, red, 1.2–1.3 cm, smooth or glandular. *NW & W Central China*. H4. Early summer.

'Persetosa' (*R. elegantula* forma *persetosa* Stapf; *R. farreri* 'Persetosa'). Illustration: Harkness, The rose, pl. 98 (1979); Gibson, The book of the rose, pl. 1 (1980). More prickles on the stems and smaller leaflets (to 1.3 cm), smaller whitish to salmon-pink flowers to 1.2–2 cm across, opening from coral-pink buds, and orange-red fruit *c.* 0.7 cm. *NW China*.

77. R. multibracteata Hemsley & Wilson (*R. reducta* Baker). Illustration: Phillips & Rix, Roses, 25, 214 (1988).

Shrub with vigorous, arching stems 2–4 m, green when young, turning red-brown, bearing slender, straight prickles usually in pairs at the nodes. Stipules glandular. Leaves deciduous; leaflets 7–9 (rarely 5–9), obovate to elliptic or more or less round, 0.6–1.5 cm, obtuse, dark green and hairless above, greyish green beneath and downy on the midrib, margins with compound teeth. Bracts broad. Receptacle glandular-bristly. Flowers solitary to many in terminal clusters, single, smelling slightly unpleasant, 2.5–3.8 cm across. Sepals entire, downy inside, glandular on the back, erect and persistent in fruit. Petals bright pink to lilac pink. Styles free, protruding; stigmas woolly. Fruit spherical to ovoid or bottle-shaped, orange-red, 1–1.5 cm, with a few glandular bristles. *W China*. H4. Summer.

78. R. forrestiana Boulenger. Illustration: Gibson, Shrub roses for every garden, pl. 1 (1973); Austin, The heritage of the rose, 350 (1988); Phillips & Rix, Roses, 23, 214 (1988).

Shrub with vigorous, erect or spreading stems 2 m, bearing straight or forward-pointing, brown prickles in pairs at the nodes. Stipules broadly ovate. Leaves deciduous; leaflets 5–7 (rarely to 9), elliptic to round, 1–1.3 cm, obtuse, hairless above, hairless or hairy to glandular on veins beneath, margins with simple teeth. Bracts very broad. Receptacle smooth or glandular-bristly. Flowers 1–5, single, fragrant, 2–3.8 cm across. Sepals entire, often glandular on the back, erect and persistent in fruit. Petals pale to bright pink. Styles free, protruding. Fruit ovoid with a distinct neck, red, to 1.3 cm, smooth or glandular-bristly. *W China (Yunnan)*. H5. Summer.

R. latibracteata Boulenger. Similar but has larger leaflets to 2.5 cm; styles not protruding. *W China (Yunnan)*. H5. Summer.

79. R. willmottiae Hemsley. Illustration: Thomas, A garden of roses, 157 (1987).

Densely branched shrub with erect or arching stems to 3 m, glaucous-bloomed when young, becoming red-brown, bearing slender, straight, prickles mostly in pairs at the nodes, bristles absent. Stipules glandular-hairy on the margins. Leaves deciduous; leaflets 7–9, obovate to oblong or almost round, 0.6–1.5 cm, obtuse, hairless on both sides, margins with simple or compound teeth. Bracts present. Receptacle smooth. Flowers usually solitary, single, slightly fragrant, 2.5–3.8 cm across. Sepals entire, hairless on the back, falling after flowering. Petals purplish pink. Styles free, not protruding. Fruit ovoid, pear-shaped or more or less spherical, orange-red, 1–1.8 cm. *W & NW China*. H4. Early summer.

'Wisley'. Illustration: Phillips & Rix, Roses, 214 (1988). Flowers deeper pink with narrower petals and sepals persistent in fruit.

R. × pruhoniciana Schneider (Illustration: Gault & Synge, The dictionary of roses in colour, pl. 40, 1971) is a hybrid between *R. moyesii* and *R. willmottiae* or *R. multibracteata*. It is very similar to *R. willmottiae*, differing in having maroon-crimson flowers, and fruit which persist after the leaves have fallen. *Garden origin*.

80. R. gymnocarpa Torrey & Gray. Illustration: Phillips & Rix, Roses, 26 (1988).

Shrub with erect, slender stems 1–3 m, more or less unarmed or bearing slender, straight prickles, intermixed with bristles. Stipules narrow. Leaves deciduous; leaflets 5–9, elliptic to ovate or roundish, 1–4 cm, obtuse, usually hairless on both sides, sometimes with glands beneath, margins with compound, often glandular teeth. Bracts present. Receptacle smooth. Flowers 1–4, single, 2.5–3.8 cm across. Sepals entire, hairless and usually without glands, falling after flowering. Petals rose-pink. Styles free, not protruding. Fruit spherical,

ellipsoid or pear-shaped, red, 0.6–1 cm, smooth. *W North America*. H3. Summer.

81. R. beggeriana Fischer & Meyer. Illustration: Phillips & Rix, Roses, 29 (1988).

Shrub with erect, sometimes climbing stems 1.8–3 m, reddish when young, bearing pale, more or less flattened, hooked prickles usually in pairs at the nodes. Stipules narrow. Leaves deciduous, aromatic; leaflets 5–9, narrowly elliptic to obovate, 0.7–3 cm, obtuse, grey-green and hairless above, glandular and sometimes downy beneath, margins with simple teeth. Bracts narrowly ovate. Receptacle smooth. Flowers in clusters of 8 or more at the ends of new shoots, single, smelling slightly unpleasant, 2–3.8 cm across. Sepals entire, hairy or not on the back, glandular or not, erect after flowering but eventually falling. Petals white. Styles free, not protruding. Fruit spherical, red, turning purplish, 0.6–1 cm, smooth. *SW & C Asia*. H4. Summer.

82. R. multiflora Murray. Illustration: Niering & Olmstead, The Audubon Society field guide to North American wildflowers: eastern region, pl. 211 (1979); Beales, Classic roses, 66 (1985); Thomas, A garden of roses, 119 (1987); Phillips & Rix, Roses, 38 (1988).

Strong-growing shrub with arching, trailing or sometimes climbing stems 3–5 m, bearing many small, stout, decurved prickles. Stipules with laciniate margins, usually glandular-bristly. Leaves deciduous; leaflets 7–9 (rarely 5–11), obovate or elliptic, 1.5–5 cm, acute, acuminate or obtuse, hairless above, hairless or downy beneath, margins with simple teeth. Bracts usually absent. Receptacle hairy. Flowers few to many in branched clusters, single, fruit-scented, 1.5–3 cm across. Sepals with lateral lobes, glandular-bristly on the back, shorter than petals even in bud, reflexed and falling after flowering. Petals cream, fading to white, occasionally pink. Styles united, protruding; stigmas hairless. Fruit ovoid to spherical, red, 0.6–0.7 mm. *Japan, Korea (naturalised in USA)*. H3. Summer.

The lower branches root where they touch the soil. Much used as a stock for grafting, especially for ramblers.

Var. **cathayensis** Rehder & Wilson (*R. cathayensis* (Rehder & Wilson) Bailey; *R. gentiliana* Léveillé & Vaniot). Illustration: Phillips & Rix, Roses, 35 (1988). Single, rosy pink flowers to 4 cm across, in flattish clusters. *China (Hubei, Gansu, Sichuan, Yunnan)*.

Var. **nana** misapplied is a dwarf variety with pale pink or cream single, fragrant flowers. 'Grevillei' (*R. multiflora* var. *platyphylla* Thory; 'Platyphylla'). Illustration: Anderson, The complete book of 169 Redouté roses, 59 (1979). Leaflets large, wrinkled; flowers in clusters of 25–30 (rarely to 50) usually double flowers, deep pinkish purple fading to whitish. 'Carnea' (*R. multiflora* var. *carnea* Thory) has double, flesh-pink flowers. 'Wilsonii' has single white flowers *c*. 5 cm across.

R. × iwara Regel is a hybrid between *R. multiflora* and *R. rugosa*. It differs from *R. multiflora* in the stems bearing large, hooked prickles, and the leaflets being grey-downy beneath. The flowers are white and single. *Japan*.

83. R. watsoniana Crépin (*R. multiflora* var. *watsoniana* (Crépin) Matsumura).

Shrub with arching, trailing or climbing stems to 1 m, bearing small, scattered prickles. Stipules very narrow, entire. Leaves deciduous; leaflets 3–5, linear-lanceolate, 2.5–6.5 cm, hairless above and mottled with yellow or grey near the midrib, downy beneath, margins wavy and bearing simple teeth. Bracts absent. Flowers in clusters, single, 1–1.7 cm across. Sepals entire, hairy on the back, reflexed after flowering, finally falling. Petals pale pink, occasionally white. Styles united, protruding. Fruit spherical, red, 0.6–0.7 cm. *Japan, garden origin*. H3. Summer.

The fruits are usually sterile. Any seedlings which are produced are normal *R. multiflora* suggesting that *R. watsoniana* may be a mutant.

84. R. setigera Michaux. Illustration: Thomas, A garden of roses, 141 (1987); Phillips & Rix, Roses, 34 (1988).

Shrub with slender, spreading, trailing or rambling stems 2–5 m, bearing stout, more or less straight, broad-based, scattered prickles. Stipules narrow. Leaves deciduous; leaflets 3–5, ovate to ovate-oblong, 3–8 cm, acute to acuminate, deep green and hairless above, pale green and downy on the veins beneath, margins with coarse, simple teeth. Bracts absent. Receptacle glandular-bristly. Flowers 5–15 in loose clusters, single, sometimes fragrant, 5–7.5 cm across. Sepals with lateral lobes, downy and glandular-bristly on the back, reflexed after flowering, then falling. Petals deep pink, fading to pale pink or nearly white. Styles united, protruding; stigmas hairless. Fruit spherical, red to greenish brown, *c.* 0.8 cm, glandular-bristly. *E & C North America.* H3. Summer.

85. R. × beanii Heath (*R. anemoniflora* Lindley; *R. triphylla* Roxburgh). Illustration: Thomas, A garden of roses, 31 (1987).

Shrub with spreading stems, bearing few, small, scattered slender, hooked prickles. Leaves deciduous; leaflets 3–5, ovate to narrowly ovate, 3.8–7.5 cm, acute or acuminate, hairless on both sides, margins with simple teeth. Bracts absent. Flowers in loose clusters, double, 2.5–4.5 cm across. Sepals with lateral lobes. Petals pale pink, the inner ones narrow and rather ragged. Styles united, protruding; stigmas hairy. *E China, garden origin.* H4. Summer.

There is enormous confusion about the origin of this rose: some botanists think it is a hybrid between *R. multiflora* and either *R. laevigata* or *R. banksiae*, others that *R. banksiae* and *R. moschata* are the parents.

86. R. helenae Rehder & Wilson. Illustration: Le Rougetel, A heritage of roses, 146 (1988); Phillips & Rix, Roses, 36 (1988).

Shrub with rambling stems 5–6 m, purplish brown when young, bearing short, stout, hooked prickles. Leaves deciduous; leaflets 7–9 (rarely 5–9), ovate to elliptic or obovate, 1.9–6 cm, acute, bright green and hairless above, greyish beneath and downy on the veins, margins with simple teeth. Bracts absent. Receptacle densely glandular. Flowers many in flattish clusters, single, fragrant, 2–4 cm across. Sepals with lateral lobes, glandular on the back, reflexed after flowering, eventually falling. Petals white. Styles united into a hairy column, protruding; stigmas hairy. Fruit ovoid, ellipsoid or pear-shaped, scarlet or orange-red, 1–1.5 cm, glandular. *C China (Shaanxi, Hubei, Sichuan).* H4. Summer.

87. R. rubus Léveillé & Vaniot (*R. ernestii* Bean; *R. ernestii* forma *nudescens* Stapf and forma *velutescens* Stapf).

Vigorous shrub with spreading or semi-climbing, often purplish stems 2.4–5 m, bearing few, short, hooked prickles (young shoots hairy or more or less hairless). Stipules narrow. Leaves deciduous; leaflets 5 (rarely 3 to 5), elliptic ovate to oblong-obovate, 3–9 cm, acute, glossy and hairless above, greyish and usually downy beneath, often purple-tinged when young, margins with simple teeth. Bracts absent.

Receptacle glandular-bristly. Flowers 1 or 2 in tight clusters, single, fragrant, 2.5–3.8 cm across, borne on stalks 1–2.5 cm. Sepals entire or with lateral lobes, downy and glandular on the back, reflexed after flowering and falling. Petals white, yellowish at the base. Styles united into a shortly protruding downy column; stigmas woolly. Fruit spherical or ovoid, dark red, 0.9–1.5 cm, glandular-bristly. *W & C China.* H4. Late summer.

88. R. mulliganii Boulenger. Illustration: Austin, The heritage of the rose, 370 (1988); Phillips & Rix, Roses, 38 (1988).

Similar to *R. rubus*, but stems 6 m or more, leaflets 5–7, flowers 4.5–5.5 cm across on stalks 2.5–3.5 cm, carried in looser clusters; the sepals always have lateral lobes. *W China (Yunnan).* H5. Summer.

89. R. brunonii Lindley (*R. moschata* Herrmann var. *napaulensis* Lindley). Illustration: Thomas, A garden of roses, 43 (1987); Phillips & Rix, Roses, 36, 39 (1988).

Vigorous shrub with arching or climbing stems 5–12 m, bearing short, stout, hooked prickles. Stipules narrow with spreading tips. Leaves deciduous, drooping, 17–21 cm; leaflets 5–7 (rarely to 9), narrowly ovate to elliptic or oblong-elliptic, 3–6 cm, acute or acuminate, grey-green or bluish green and somewhat hairy above, downy beneath at least on veins, sometimes glandular, margins with simple teeth. Bracts absent. Receptacle hairy and usually glandular. Flowers in clusters, often with several clusters combined into a large compound inflorescence, single, fragrant, 2.5–5 cm across. Sepals with lateral lobes, hairy and slightly glandular on the back, reflexed at flowering time, falling after flowering. Petals white. Styles united, protruding; stigmas downy. Fruit nearly spherical to obovoid, reddish brown, 0.7–1.8 cm, smooth. *Himalaya, from Afghanistan to SW China (Yunnan, Sichuan).* H4. Summer.

Many of the plants grown in gardens as *R. moschata* are in fact *R. brunonii*, having been distributed under the wrong name by nurseries.

'La Mortola'. Illustration: Thomas, Climbing roses old and new, pl. 1, 1983, is a very strong-growing selection.

90. R. moschata Herrmann. Illustration: Thomas, Climbing roses old and new, pl. 3 (1983); Beales, Classic roses, 259 (1985); Phillips & Rix, Roses, 34 (1988).

Robust shrub with arching or semi-climbing, purplish or reddish stems 3–10 m, bearing few, scattered, straight or slightly curved prickles. Stipules very narrow. Leaves deciduous; leaflets 5–7, broadly ovate to broadly elliptic, 3–7 cm, acute or acuminate, shiny and hairless above, downy or not on veins beneath, margins with simple teeth. Bracts absent. Receptacle finely adpressed-hairy, rarely slightly glandular. Flowers in few-flowered loose clusters, single, with a musky scent, 3–5.5 cm across. Sepals entire or with lateral lobes, hairy on the back, reflexed and persistent after flowering. Petals white or cream, reflexing with age. Styles united, protruding; stigmas woolly. Fruit nearly spherical to ovoid, orange-red, 1–1.5 cm, usually downy and sometimes glandular. *Unknown in the wild, cultivated in S Europe. Mediterranean & SW Asia (naturalised in N America).* H4. Late summer–autumn.

Var. **nastarana** Christ (Illustration: Wilson & Bell, The fragrant year, 156, 1971) has smaller leaflets always hairless beneath, and larger, more numerous, pink-tinged flowers.

'Dupontii' (*R. × dupontii* Déséglise, in part; *R. nivea* Lindley, not de Candolle). Illustration: Gault & Synge, The dictionary of roses in colour, pl. 6 (1971); Beales, Classic roses, 255 (1985); Thomas, A garden of roses, 67 (1987); Austin, The heritage of the rose, 351 (1988). Was thought for some time to be a hybrid between *R. moschata* and *R. gallica.* but this seems to be unlikely. It differs from *R. moschata* in its smaller size (2–3 m), leaflets with compound teeth, single creamy pink flowers 6–7.5 cm across, sepals glandular on the back, and free styles. *Garden origin.* H3. Summer.

91. R. longicuspis Bertoloni (*R. lucens* Rolfe; *R. yunnanensis* (Crépin) Boulenger). Illustration: Thomas, Climbing roses old and new, pl. 2 (1983); Phillips & Rix, Roses, 39 (1988).

Vigorous shrub with scrambling and climbing stems to 6 m or more, reddish when young, bearing few, short, curved or hooked prickles which may be absent on flowering-branches. Stipules narrow. Leaves evergreen or semi-evergreen, to 20 cm, reddish when young; leaflets 5–7 (rarely 3–7), narrowly ovate to elliptic, leathery, 5–10 cm, acuminate, hairless on both sides although occasionally downy on midrib beneath, margins with simple teeth. Bracts absent. Receptacle hairy and usually glandular. Flowers to 15 in a loose cluster, single, smelling of bananas, *c.* 5 cm across, opening from narrowly ovoid buds. Sepals with lateral lobes, hairy and glandular on the back, reflexed after flowering and eventually falling. Petals white, silky on the back. Styles united, protruding. Fruit broadly ellipsoid to spherical, red to orange, 1.5–2 cm, often hairy and glandular. *NE India, W China, ?Burma.* H4. Early–midsummer.

Var. **sinowilsonii** (Hemsley) Yu & Ku. (*R. sinowilsonii* Hemsley). Illustration: Phillips & Rix, Roses, 39 (1988). Leaves to 30 cm, leaflets slightly downy beneath, flower-buds broadly ovoid, sepals hairless or nearly so on the back. *SW China.* H4. Summer.

92. R. cerasocarpa Rolfe (*R. gentiliana* of Rehder & Wilson in part, not Léveillé & Vaniot).

Shrub with climbing or semi-climbing stems to 4.5 m, bearing few, stout, scattered, recurved prickles. Stipules narrow. Leaves deciduous, 17–20 cm; leaflets 5 (rarely 3–5), narrowly ovate to elliptic, leathery, 5–10 cm, acute to acuminate, hairless or more or less so on both sides, margin with simple teeth. Bracts absent. Receptacle downy and glandular. Flowers many in clusters, single, fragrant, 2.5–3.5 cm across, opening from abruptly pointed buds. Sepals usually with lateral lobes, downy and glandular on the back, reflexed and falling after flowering. Petals white. Styles united. Fruit spherical, deep red, 1–1.3 cm, downy. *W & C China.* H4. Summer.

93. R. maximowicziana Regel.

Shrub with arching and climbing stems bearing few, small, scattered, straight and hooked prickles, with bristles on the young branches. Stipules very narrow. Leaves deciduous; leaflets 7–9, ovate-elliptic to oblong, 2.5–5 cm, acute to acuminate, hairless on both sides, margins with simple teeth. Bracts present. Receptacle smooth. Flowers many in small clusters, single, 2.5–3.5 cm across. Sepals with lateral lobes, reflexed after flowering and finally falling. Petals white. Styles united; stigmas hairless. Fruit ovoid, red, 1–1.2 cm, smooth. *Manchuria, Korea.* H3. Summer.

94. R. filipes Rehder & Wilson. Illustration: Phillips & Rix, Roses, 38 (1988).

Shrub with arching and climbing stems to 9 m, purple when young, bearing few, small, hooked prickles. Stipules narrow. Leaves deciduous, coppery when young;

leaflets 5–7, narrowly ovate to narrowly elliptic, 3.5–8 cm, acuminate, hairless above and beneath or sometimes downy on veins beneath, margins with simple teeth. Bracts absent. Flowers to 100 or more in large clusters to 30 cm or more wide, single, fragrant, 2–2.5 cm across. Sepals with a few lateral lobes, glandular and slightly downy or hairless on the back, reflexed and falling after flowering. Petals creamy white, sometimes downy on the back. Styles united, protruding; stigmas woolly. Fruit spherical to broadly ellipsoid, orange becoming crimson-scarlet, 0.8–1.5 cm. *W China.* H4. Summer.

In cultivation it is most commonly found as the cultivar 'Kiftsgate' (Illustration: Phillips & Rix, Roses, 36, 1988; Brickell, The RHS gardeners' encyclopedia of plants and flowers, 169, 1989) in which the inflorescence can be to 45 cm across.

95. R. phoenicia Boissier.

Shrub with slender, climbing stems 3–5 m, bearing few, short, curved or hooked, broad-based, more or less equal prickles. Stipules glandular-toothed. Leaves deciduous; leaflets 5 (rarely 3–7), elliptic to roundish, 2–4.5 cm, acute to obtuse, more or less downy on both sides, more so beneath, margins with simple or compound teeth. Bracts usually hairy. Receptacle hairless or slightly hairy. Flowers 10–14 in clusters, single, 4–5 cm across, opening from ovoid, rounded buds. Sepals with lateral lobes, often downy on the back, reflexed and falling after flowering. Petals white. Styles united, protruding; stigmas hairless. Fruit ovoid, red, 1–1.2 cm, smooth. *NE Greece, Cyprus, Turkey, Syria, Lebanon.* H5. Summer.

96. R. soulieana Crépin. Illustration: Thomas, A garden of roses, 143 (1987).

Robust shrub with erect, spreading or semi-climbing stems 3–4 m, bearing scattered, stout, decurved, compressed, broad-based prickles. Stipules narrow. Leaves deciduous, 6.5–10 cm; leaflets 7–9 (rarely 5–9), grey-green, obovate to elliptic, 1–3 cm, acute to obtuse, hairless on both sides although more or less downy on midrib beneath, margins with simple teeth. Bracts absent. Receptacle glandular. Flowers in many-flowered branched clusters 10–15 cm across; flowers single, with a fruity scent, 2.5–3.8 cm across, opening from yellow buds. Sepals entire or with a few lateral lobes, hairless or downy and often glandular on the back, reflexed and falling after flowering. Petals white. Styles united, protruding; stigmas sparsely

hairy. Fruit ovoid to nearly spherical, pale to dark orange, *c.* 1 cm, glandular. *W China (Sichuan).* H5. Summer.

97. R. wichuraiana Crépin. Illustration: Thomas, Climbing roses old and new, pl. 5 (1983); Thomas, A garden of roses, 155 (1987); Phillips & Rix, Roses, 35 (1988).

Shrub with prostrate, trailing or climbing stems 3–6 m, bearing strong, curved prickles. Stipules broad, toothed. Leaves evergreen or semi-evergreen; leaflets 5–9, dark green, elliptic to broadly ovate or roundish, obtuse, hairless on both sides except for midrib beneath, margins with simple teeth. Bracts absent. Receptacle sometimes glandular. Flowers 6–10 in loose clusters, single, fragrant, 2.5–5 cm across. Sepals entire or with a few lateral lobes, often downy or slightly glandular on the back, reflexed and falling after flowering. Petals white. Styles united, protruding; stigmas woolly. Fruit ovoid or spherical, orange-red to dark red, 1–1.5 cm. *Japan, Korea, E China, Taiwan (naturalised in N America).* H4. Summer–early autumn.

'Variegata' has leaflets which are cream with pink tips when young, turning green, marked with cream.

R. × jacksonii Baker (Illustration: Thomas, A garden of roses, 101, 1987) is a hybrid between *R. wichuraiana* and *R. rugosa.* It differs from *R. wichuraiana* by its larger, wrinkled leaflets and bright crimson flowers. *Garden origin.*

R. × kordesii Wulff was produced by Wilhelm Kordes who selfed 'Max Graf'. Like *R. × jacksonii* it is a hybrid between *R. wichuraiana* and *R. rugosa. R. × kordesii* is tetraploid and has been used as a parent for a number of repeat-flowering climbers. It differs from *R. wichuraiana* in having deep pink to red, recurrent flowers which are single to semi-double and 7–8 cm across. *Garden origin.*

98. R. luciae Crépin.

Shrub with prostrate or climbing stems to 3.5 m, bearing small, scattered pale brown, rather flattened, slightly hooked prickles. Stipules very thin. Leaves evergreen or semi-evergreen; leaflets 5 (rarely to 7), rather thin, 2–4.5 cm, the terminal one longer than the others, acute to acuminate, both sides hairless or more or less so, margins with simple teeth. Bracts absent. Receptacle smooth or glandular. Flowers 3–20 in clusters, single, fragrant, 2–3 cm across, opening from short, rounded buds. Sepals reflexed, falling after flowering. Petals white. Styles united, protruding; stigmas woolly. Fruit ovoid to spherical, red to purplish, *c.* 0.7 cm, smooth or glandular. *E China, Japan, Korea.* H4. Early summer and again in late summer.

Very similar to *R. wichuraiana* and considered by some botanists to be the same: if so then the name *R. luciae* has priority.

99. R. sempervirens Linnaeus. Illustration: Thomas, A garden of roses, 137 (1987); Phillips & Rix, Roses, 34 (1988); Blamey & Grey-Wilson, The illustrated flora of Britain and northern Europe, 179 (1989).

Shrub with prostrate, trailing or scrambling stems 6–10 m, unarmed or bearing straight or hooked, broad-based prickles. Stipules narrow. Leaves evergreen or semi-evergreen; leaflets 5–7 (rarely 3–7), narrowly ovate to elliptic, 1.5–6 cm, the terminal usually larger than the upper laterals, acuminate, hairless on both sides except for midrib which may be downy beneath, margins with simple teeth. Bracts short, entire, more or less hairless, falling early. Receptacle often glandular-bristly. Flowers 3–10 (rarely 1–10) in clusters, single, slightly fragrant, 2.5–4.5 cm across, opening from blunt, ovoid buds. Sepals entire, glandular-bristly on the back, reflexed to spreading and falling after flowering. Petals white. Styles united, protruding; stigmas usually woolly. Fruit ovoid or spherical, orange-red, 1–1.6 cm, often glandular-bristly. *S Europe, NW Africa, Turkey.* H5. Summer.

100. R. arvensis Hudson. Illustration: Ross-Craig, Drawings of British plants **9**: pl. 13 (1956); Beales, Classic Roses, 252 (1985); Thomas, A garden of roses, 33 (1987); Phillips & Rix, Roses, 35 (1988).

Shrub with trailing or climbing stems 1–2 m, bearing sparse, scattered, stout, more or less curved, equal prickles. Stipules narrow. Leaves deciduous; leaflets 5–7 (rarely 3–7), deep green, elliptic to broadly ovate or roundish, 1–4 cm, more or less acute, hairless on both sides, although sometimes downy on veins beneath, margins with simple teeth. Bracts absent. Receptacle smooth or somewhat glandular. Flowers 1–8, single, usually fragrant, 2.5–5 cm across. Sepals with lateral lobes, smooth or sparsely glandular on the back, reflexed and falling after flowering. Petals white to pink. Styles united, protruding; stigmas hairless. Fruit spherical to ovoid, red, 0.6–2.5 cm, usually smooth. *S, W & C Europe, S Turkey.* H2. Summer.

R. × polliniana Sprengel is a hybrid between *R. arvensis* and *R. gallica* which differs from *R. arvensis* in its rather leathery leaflets which are slightly downy beneath, and flowers 5–6 cm across. *N Italy.*

101. R. chinensis Jacquin (*R. indica* Loureiro, not Linnaeus).

Shrub, varying in habit from dwarf to semi-climbing, to 6 m. Stems more or less unarmed or bearing scattered, more or less hooked, somewhat flattened prickles. Stipules narrow. Leaves evergreen; leaflets 3–5, lanceolate to broadly ovate, 2.5–6 cm, acuminate, glossy and hairless above, hairless beneath except for downy midrib, margins with simple teeth. Bracts narrow. Receptacle smooth or glandular. Flowers solitary or in clusters, single or semi-double, *c.* 5 cm across, often fragrant. Sepals entire or with a few lateral lobes, smooth or glandular on the back, reflexed and falling after flowering. Petals pale pink to scarlet or crimson. Styles free, somewhat protruding. Fruit ovoid to pearshaped, greenish brown to scarlet, 1.5–2 cm. *China, garden origin.* H5. Summer–early autumn.

A repeat-flowering rose which contributed this feature to European roses when it was introduced around 1800.

Var. **spontanea** (Rehder & Wilson) Yu & Ku (*R. chinensis* forma *spontanea* Rehder & Wilson). Illustration: Phillips & Rix, Roses, 32 (1988). Climber or bush, usually 1–2.5 m. Leaflets lanceolate. Flowers 1–3, single, pink, turning red, 5–6 cm across. Sepals entire. Fruit orange. *W China.* It is the wild type of cultivated *R. chinensis*.

'Minima' (*R. chinensis* var. *minima* (Sims) Voss). Stems 20–50 cm; flowers rose-red, single or double, usually solitary, *c.* 3 cm across, with pointed petals. 'Rouletii' (*R. roulettii* Correvon) which has stems 10–25 cm, and double, rosy pink flowers 1.9–2.5 cm across, is thought by some botanists to be synonymous with 'Minima'. 'Mutabilis' (*R. mutabilis* Correvon; *R. chinensis* forma *mutabilis* (Correvon) Rehder; 'Tipo Ideale'). Illustration: Thomas, Shrub roses of today, pl. VI (1974); Phillips & Rix, Roses, 68, 69 (1988). Stems 1–1.7 m, leaflets purplish or coppery when young; flowers single, fragrant, 4.5–6 cm across, petals yellow with an orange back, turning coppery salmon-pink and eventually deep pink. 'Pallida' has stems to 1 m or more, and clusters of semi-double, fragrant, blush-pink flowers. 'Semperflorens' (*R. chinensis* var. *semperflorens* (Curtis) Koehne) has stems 1–1.5 m, and semi-double, deep

pink or crimson-scarlet, delicately scented flowers. 'Viridiflora' (*R. chinensis* var. *viridiflora* (Lavallée) Dippel). Illustration: Phillips & Rix, Roses, 69 (1988); Welch, Roses, 114 (1988). A monstrous variant probably derived from 'Pallida' in which the petals are green streaked with crimson or purplish and the stamens and pistils become leafy, narrow, with toothed segments.

102. R. gigantea Crépin (*R. × odorata* var. *gigantea* (Crépin) Rehder & Wilson; *R. odorata* 'Gigantea'). Illustration: Thomas, A garden of roses, 89 (1987); Phillips & Rix, Roses, 33 (1988).

Shrub with climbing stems 8–12 (rarely to 30) m, bearing stout, scattered, uniform, hooked prickles. Stipules narrow. Leaves evergreen or semi-evergreen; leaflets 5–7, elliptic to ovate, 3.8–9 cm, acuminate, glossy above, hairless on both sides, margins with simple, often glandular teeth. Bracts present. Receptacle smooth. Flowers solitary (rarely to 3), single, fragrant, 10–15 cm across, opening from slender, pale yellow buds. Sepals entire, smooth on the back, reflexed and falling after flowering. Petals white or cream. Styles free; stigmas downy. Fruit spherical or pear-shaped, red, or yellow flushed with red, 2–3 cm. *NE India, Burma, China (Yunnan).* H5. Early summer.

103. R. × odorata (Andrews) Sweet.

Thought to be a hybrid between *R. chinensis* and *R. gigantea*. It differs from *R. gigantea* in having single or double flowers 5–8 cm across, with white, pale pink or yellowish petals. *China, garden origin.* H5. Mid–late summer.

104. R. banksiae Aiton filius.

Shrub with strong climbing stems to 12 m, unarmed or bearing very sparse, hooked prickles. Stipules very narrow, soon falling. Leaves evergreen; leaflets 3–7, oblong-lanceolate to elliptic-ovate, 2–6.5 cm, acute or obtuse, glossy and hairless above, sometimes downy beneath on the midrib, margin wavy, with simple teeth. Bracts very small, soon falling. Flowers many in umbels, single or double, fragrant, 2.5–3 cm across. Sepals entire, reflexed and falling after flowering. Petals white or yellow. Styles free, not protruding. Fruit spherical, dull red, *c.* 0.7 cm. *W & C China.* H4. Early summer.

Var. **banksiae** (*R. banksiae* var. *alboplena* Rehder; *R. banksiae* 'Banksiae'; *R. banksiae* 'Albo-Plena' or 'Alba Plena'). Illustration: Beales, Classic roses, 399 (1985); Phillips

& Rix, Roses, 30 (1988). Flowers double, white, violet-scented.

Var. **normalis** Regel. Illustration: Phillips & Rix, Roses, 31 (1988). Flowers single, white, very fragrant, and stems usually bearing hooked prickles.

'Lutea' (*R. banksiae* var. *lutea* Lindley). Illustration: Harkness, Roses, pl. 4 (1978); Harkness, The rose, pl. 1 (1979); Thomas, A garden of roses, 35 (1987); Phillips & Rix, Roses, 30 (1988) – as 'Lutescens'. Flowers double, yellow, slightly fragrant; leaflets usually 5; stems generally unarmed. The hardiest taxon and the most floriferous. 'Lutescens' (*R. banksiae* forma *lutescens* Voss). Illustration: Austin, The heritage of the rose, 299, 330 (1988); Phillips & Rix, Roses, 30 (1988) as 'Lutea'. Has single, yellow, highly scented flowers.

R. × fortuneana Lindley. Illustration: Beales, Classic roses, 400, (1985); Thomas, A garden of roses, 79, (1987). Thought to be a hybrid between *R. banksiae* and *R. laevigata*. It differs from *R. banksiae* in having 3–5, rather thin leaflets, and solitary, double, creamy white flowers 5–10 cm across. *China, garden origin*. H5. Summer.

105. R. cymosa Trattinick (*R. microcarpa* Lindley, not Besser nor Retzius). Illustration: Phillips & Rix, Roses, 31 (1988).

Differs from *R. banksiae* in having more prickly stems, larger, branched inflorescences, and smaller, always single, white flowers (c. 1.5 cm across) with sepals bearing lateral lobes. *C & S China*. H4. Early summer.

Specimens have been found which are intermediate between *R. cymosa* and *R. banksiae*: it is likely that future work will prove that the two are the same.

106. R. laevigata Michaux. Illustration: Anderson, The complete book of 169 Redouté roses, 89 (1979); Krüssmann, The complete book of roses, 293 (1981); Thomas, A garden of roses, 105 (1987); Phillips & Rix, Roses, 33 (1988).

Vigorous shrub with climbing, green stems to 10 m or more, bearing scattered, stout, red-brown, hooked prickles. Stipules united at the base, soon falling. Leaves evergreen; leaflets 3 (rarely to 5), lanceolate to elliptic or ovate, rather leathery, 3–6 (rarely to 9) cm, acute or acuminate, glossy above, hairless on both sides, midrib sometimes prickly beneath, margins with simple teeth. Bracts absent. Receptacle very bristly. Flowers solitary,

single, fragrant, 5–10 cm across. Sepals entire, bristly on the back, erect and persistent in fruit. Petals white or creamy white. Styles free, not protruding. Fruit pear-shaped, orange-red to red, 3.5–4 cm, bristly. *S China, Taiwan, Vietnam, Laos, Cambodia, Burma (naturalised in S USA)*. H5. Early summer.

'Cooperi'. Illustration: Gault & Synge, The dictionary of roses in colour, pl. 5 (1971); Phillips & Rix, Roses, 32 (1988). Some leaves with 5 or 7 leaflets and red, rather than green, stems. As they age the petals become pink-spotted. It is possible that it is a hybrid between *R. laevigata* and *R. gigantea*.

107. R. × anemonoides Rehder (*R.* 'Anemone'). Illustration: Thomas, A garden of roses, 29 (1987); Phillips & Rix, Roses, 32 (1988).

Is thought to be a hybrid between *R. laevigata* and *R. × odorata*. Less vigorous than *R. laevigata* from which it differs in the stipules being more united and the petals pale pink with deeper veins. *Garden origin*. H5. Early summer.

108. R. bracteata Wendland. Illustration: Thomas, A garden of roses, 41 (1987); Phillips & Rix, Roses, 32 (1988).

Shrub with prostrate or climbing, brown-downy stems 3–6 m, bearing stout, broad-based, hooked prickles in pairs at the nodes, and numerous glandular bristles. Stipules united at the base, fringed. Leaves evergreen; leaflets 5–11, dark green, obovate to elliptic or oblong, 1.5–5 cm, obtuse, glossy and hairless above, downy beneath at least on midrib, margins with simple teeth. Bracts large, deeply toothed, downy. Receptacle hairy. Flowers usually solitary, single, smelling of fruit, 5–8 cm across. Sepals entire, brown-hairy on the back, reflexed and falling after flowering. Petals white. Styles free, not protruding. Fruit spherical, orange-red, 2.5–3.8 cm, hairy. *SE China, Taiwan (naturalised in USA from Virginia to Texas & Florida)*. H5. Summer–late autumn.

109. R. roxburghii Trattinick (*R.* 'Roxburghii'; *R. roxburghii* 'Plena'). Illustration: Thomas, A garden of roses, 135 (1987); Phillips & Rix, Roses, 19 (1988).

Flowers double, pink, darker in the centre. It was originally introduced from Chinese gardens and is rarely grown.

Forma **normalis** Rehder & Wilson (*R. microphylla* Lindley). Illustration: Gault & Synge, The dictionary of roses in colour,

pl. 45, 46 (1971); Harkness, Roses, pl. 3 (1978); Gibson, The book of the rose, pl. 34 (1980); Phillips & Rix, Roses, 18, 19 (1988). Spreading shrub with rather stiff stems to 5 m or more, with peeling grey or pale brown bark, and bearing few, straight, hooked prickles in pairs at the nodes. Stipules narrow, united. Leaves deciduous, 5–10 cm; leaflets usually 9–15 (rarely 7 or 17–19), narrowly ovate to obovate, 1–2.5 cm, acute or obtuse, hairless on both sides, margins with simple teeth. Bracts falling early. Receptacle prickly. Flowers usually solitary, single, fragrant, 5–7.5 cm across. Sepals with lateral lobes, downy and prickly on the back, erect and persistent in fruit. Petals midpink to deep pink. Styles free, not protruding. Fruit flattened-spherical, yellow-green, 3–4 cm, prickly. *W China*. H4. Summer.

R. × coryana Hurst. Illustration: Phillips & Rix, Roses, 19, 1988. Like *R. roxburghii* but stems shorter, less prickly (to 2 m); flowers bright carmine. *Garden origin*.

A hybrid between *R. roxburghii* and probably *R. macrophylla*.

110. R. hirtula (Regel) Nakai (*R. roxburghii* var. *hirtula* (Regel) Rehder & Wilson). Illustration: Phillips & Rix, Roses, 18 (1988).

Like *R. roxburghii* but leaflets elliptic to oblong-elliptic, downy beneath; flowers single, pale pink or lilac-pink. *Japan*.

Similar to *R. roxburghii*: some botanists agree with Rehder & Wilson that it would be better treated as a variety.

111. R. stellata Wooton (*Hesperhodos stellatus* (Wooton) Boulenger). Illustration: Thomas, A garden of roses, 147 (1987).

Shrub of loose habit with erect, slender, stellate-hairy stems 60–120 cm, bearing dense, straight, slender, pale yellow, often paired prickles intermixed with glandular bristles. Stipules united. Leaves deciduous; leaflets 3–5, wedge-shaped to obovate, 0.5–1.2 cm, stellate-hairy on both sides or only beneath, margins with simple teeth except towards the base. Bracts absent. Receptacle with pale spines. Flowers solitary, single, 3.5–6 cm across. Sepals with lateral lobes, glandular and spiny on the back, margin woolly, erect and persistent in fruit. Petals soft pink to bright rose or dark purplish red. Stamens to 160 or more. Styles free, not protruding; stigmas woolly. Fruit hemi-spherical, flat-topped, not fleshy, dull red to brown-red, to 2 cm, prickly. *SW USA (New Mexico, Texas, Arizona)*. H4. Summer.

Var. **mirifica** (Greene) Cockerell (*R. mirifica* Greene; *R. stellata* subsp. *mirifica* (Greene) Lewis; *Hesperhodos mirificus* (Greene) Boulenger). Illustration: Thomas, Shrub roses of today, pl. 4 (1974); Phillips & Rix, Roses, 18 (1988). Differs in lacking stellate hairs and having flowers with fewer (to 150) stamens. *SW USA (New Mexico, S Texas)*. H4. Summer.

25. AGRIMONIA Linnaeus
J. Cullen

Erect perennial herbs with rhizomes. Stems with short glandular hairs borne almost on the surface and longer, spreading or deflexed, simple hairs. Leaves sometimes forming a rosette at the base of the stem, pinnate with 7–13 large leaflets alternating in opposite pairs with much smaller leaflets; all deeply toothed. Stipules large, stem-clasping, deeply toothed. Flowers many in a raceme, each subtended by a bract and 2 bracteoles (all of which may be 3-lobed or entire), shortly stalked. Perigynous zone cylindric, bell-shaped or top-shaped, its upper part with numerous spreading or deflexed, hard, hooked bristles outside and below the calyx. Sepals 5, spreading in flower, persistent and more or less erect in fruit. Petals 5, yellow, golden yellow or orange-yellow, rarely whitish. Stamens 10–20, borne on the edge of a yellowish disc which roofs the perigynous zone. Carpels 2 (rarely more), free, their styles projecting through a small central hole in the roof of the perigynous zone. Fruit a group of 1 or 2 (rarely more) achenes enclosed in the hardened, often grooved perigynous zone surmounted by the bristles and the persistent calyx.

A genus of a small number of species, all rather difficult to distinguish, mainly from the north temperate areas, but 1 species in South Africa and C Africa. They have a remarkable floral structure and their fruits are dispersed by attachment to animal fur by means of the hooked bristles. They are easily grown, responding best to fairly moist conditions, and are propagated by seed or by division.

Literature: Skalicky, V., Ein Beitrag zur Erkenntniss der Europaischen Arten der Gattung Agrimonia L., *Acta Horti Botanici Pragensis*, 87–108 (1962).

1a. Non-glandular hairs on the stem of 2 sizes, longer and spreading, shorter and finer, usually somewhat to very deflexed; glands on the lower leaf surface not easily visible, obscured by hairs   **1. eupatoria**

  b. Non-glandular hairs on the stem all more or less similar, spreading; glands on the lower leaf surface easily visible   2

2a. Lowermost bristles on the perigynous zone sharply deflexed, even in flower, clearly so in fruit; fruit-stalks 4–10 mm   **3. repens**

  b. Lowermost bristles on the perigynous zone spreading in flower, sometimes somewhat deflexed in fruit; fruit-stalks usually 1–3 mm   **2. procera**

1. A. eupatoria Linnaeus. Illustration: Coste, Flore de la France **2**: 58 (1903); Bonnier, Flore complète **4**: pl. 183 (1914); Ary & Gregory, The Oxford book of wild flowers, 16 (1962); Pignatti, Flora d'Italia **1**: 566 (1982).

Stems to 1.5 m with, spreading non-glandular hairs and shorter and finer, somewhat to very deflexed non-glandular hairs as well as almost stalkless glands. Glands on the lower leaf surface not easily visible, obscured by hairs. Bracts 3-lobed. Flower-stalks very short, extending to 1–3 mm in fruit. Petals 3–6 mm, yellow, not notched. Perigynous zone tapered to the base, deeply grooved for most of its length, the lowermost bristles spreading. *Europe, N Africa, W & C Asia*. H1. Summer.

Very variable especially in size and density of hairs. 'Alba' has whitish petals.

2. A. procera Wallroth (*A. odorata* misapplied). Illustration: Coste, Flore de la France **2**: 58 (1903); Pignatti, Flora d'Italia **1**: 566 (1982).

Like *A. eupatoria* but non-glandular hairs on the stem all more or less the same in length and thickness, spreading; glands on the lower leaf surface easily visible; petals golden yellow, 5–8 mm, often notched; perigynous zone rather rounded to the base, scarcely grooved or with shallow grooves for up to half its length, outermost bristles of perigynous zone ultimately slightly deflexed. *Most of Europe, N Africa, SW Asia*. H1. Summer.

3. A. repens Linnaeus (*A. odorata* Miller). Like *A. eupatoria* but non-glandular hairs on the stem all more or less the same in length and thickness, spreading; glands on the lower leaf surface easily visible; petals golden yellow, 5–7 mm, not notched; perigynous zone rounded to the base, deeply grooved over most of its length, the lowermost bristles conspicuously deflexed, even in flower. *Turkey, N Iraq, naturalised in parts of Europe*. H3. Summer.

26. HAGENIA Gmelin
J. Cullen

Dioecious trees to 18 m (in the wild) with open, umbrella-shaped crowns. Branches with dense, golden hairs. Leaves evergreen, 10–40 cm, pinnate with 5–8 pairs of leaflets, silvery silky beneath. Stipules large, sheathing the stems. Flowers numerous in panicles, those of the female flowers broader and more substantial than the male; each flower subtended by 2 bracts. Sepals 4 or 5 alternating with 4 or 5 epicalyx segments. Male flowers with 4 or 5 white or orange petals (rarely absent), 10–20 stamens with sparsely hairy anthers, borne on the edge of a nectar-secreting disc, and a pistillode. Female flowers reddish; petals not known; staminodes to 20, carpels 1 or 2, free, surrounded by the perigynous zone. Fruit an achene or 2 achenes surrounded by the persistent perigynous zone surmounted by the calyx and epicalyx.

A genus of a single species from tropical E Africa. Though spectacular, it is rarely grown. It is propagated by seed, which is relatively short-lived.

1. H. abyssinica (Bruce) Gmelin. Illustration: Menninger, Flowering trees of the world, pl. 326 (1962); Flora Zambesiaca **4**: 21 (1978).

Tropical E Africa from Ethiopia to Zambia. G2.

27. SANGUISORBA Linnaeus
J. Cullen

Herbs, often large, with thick, woody stocks. Leaves pinnate, often with numerous, shortly stalked, toothed leaflets. Stipules leaf-like, often crescent-shaped. Flowers in dense heads (spikes) which are spherical to narrowly cylindric, flowers female or bisexual, opening from the top of the head downwards or the bottom upwards; each flower with a bract and 2 bracteoles which are often hairy. Perianth-segments 4, greenish, whitish or reddish, spreading. Stamens 4–30, filaments thread-like or dilated and flattened above, often projecting well beyond the perianth-segments. Carpels 1–3, free, borne within the rounded or 4-sided perigynous zone, the style(s) long or short, stigma a mop-like group of papillae. Fruit an achene or a group of achenes surrounded by the persistent perigynous zone, which is often winged or ornamented, surmounted by the remains of the perianth.

A genus of about 10 species from north temperate areas, including species formerly placed in *Poterium*. There are some woody species in the Canary Islands but these are not in general cultivation. The plants are easily grown and propagated by seed.

Literature: Nordborg, G., Sanguisorba L., Sarcopoterium Spach and Bencomia Webb et Berth., *Opera Botanica* **11**(2): (1966); The genus Sanguisorba section Poterium, *Opera Botanica* **16**: (1967).

1a. Head with female flowers at the top and bisexual flowers lower down; stamens in the lower bisexual flowers 20–30   **4. minor**
  b. Head containing only bisexual flowers; stamens in all flowers usually 4, rarely up to 12   2
2a. Stamens and styles scarcely projecting beyond the perianth-segments, filaments not dilated and flattened above; heads dark red, little longer than broad   **3. officinalis**
  b. Stamens and styles conspicuously projecting beyond the perianth-segments, filaments dilated and flattened above; heads whitish, pinkish or magenta, much longer than broad   3
3a. Flowers opening from the bottom of the head upwards; filaments white   **1. canadensis**
  b. Flowers opening from the top of the head downwards; filaments usually pink or pale magenta   **2. obtusa**

1. S. canadensis Linnaeus. Illustration: Rickett, Wild flowers of the United States **1**: pl. 37 (1965); Justice & Bell, Wild flowers of North Carolina, 91 (1968).

Stems to 2 m, hairy or hairless below. Leaflets *c.* 13, oblong to ovate-oblong, base cordate, apex rounded, regularly and rather deeply toothed, midrib often with sparse hairs beneath. Inflorescences cylindric, 6–14 × 0.5–0.7 cm (excluding the stamens). Flowers white, opening from the bottom of the head upwards. Perianth-segments 2–3 mm. Stamens 4, projecting well beyond the perianth-segments, filaments white, dilated and flattened above. Carpel usually 1; style projecting beyond the perianth. *N America & E Asia.* H1. Summer.

Plants with particularly broad leaves are sometimes separated as **S. sitchensis** Meyer (illustration: Rickett, Wild flowers of the United States **5**: pl. 48 (1971)).

2. S. obtusa Maximowicz (*S. magnifica* Schischkin). Illustration: Botanical Magazine, 8690 (1916); Huxley, Garden perennials and water plants in colour, pl. 25 (1970); Kitamura & Murata, Coloured illustrations of herbaceous plants of Japan, pl. 30 (1977).

Large herb to 2.5 m or more. Stems usually hairy below. Leaflets *c.* 13, rather close and sometimes overlapping, ovate-oblong, base cordate, apex very broadly rounded, coarsely toothed, usually with some hairs on the midrib and veins beneath. Inflorescences cylindric, 5–9 × *c.* 1 cm (excluding the stamens), flowers opening from the top downwards, white or pink. Perianth-segments *c.* 3 mm, greenish or pink. Stamens 4, projecting well beyond the perianth-segments, filaments dilated and flattened above, pink or white. Carpel usually 1; style projecting well beyond the perianth. *Japan, E former USSR.* H2. Summer.

White-flowered variants have been called var. **albiflora** Makino.

S. hokusanensis Makino. Illustration: Takeda, Alpine flora of Japan, pl. 41 (1963); Kitamura & Murata, Coloured illustrations of herbaceous plants of Japan, pl. 30 (1977). Very similar, but stamens variable in number (to 12), filaments magenta. *Japan.* H5. Summer.

3. S. officinalis Linnaeus. Illustration: Bonnier, Flore complète **4**: pl. 183 (1914); Ary & Gregory, The Oxford book of wild flowers, pl. 4 (1962); Polunin, Flowers of Europe, pl. 45 (1969); Garrard & Streeter, Wild flowers of the British Isles, pl. 39 (1983).

Herb to 1 m or more, hairless or rarely sparsely hairy below. Leaflets 7–15, very variable in shape, ovate, ovate-oblong to linear-oblong, base usually truncate or somewhat cordate (abruptly tapered when leaflets long and narrow), apex rounded to bluntly acute, coarsely toothed. Inflorescences ovoid to broadly cylindric, 1.5–5 × 0.5–1.2 cm, opening from the top downwards, dark red. Perianth-segments *c.* 3 mm. Stamens 4, not or scarcely projecting beyond the perianth, filaments thread-like. Carpel usually 1; style scarcely projecting beyond the perianth. *N temperate areas.* H1. Summer.

4. S. minor Scopoli (*Poterium sanguisorba* Linnaeus). Illustration: Bonnier, Flore complète **4**: pl. 183 (1914); Garrard & Streeter, Wild flowers of the British Isles, pl. 39 (1983).

Herb smelling of cucumber when crushed. Stems to 50 cm or more, with long, wavy hairs below. Lower leaves with up to 25 ovate to almost circular leaflets, shortly stalked, base and apex rounded, deeply toothed, the terminal tooth smaller than those adjacent; upper leaves with fewer, narrower leaflets. Inflorescences oblong to spherical, greenish or tinged with purple, female flowers above, bisexual flowers below. Perianth-segments *c.* 3 mm. Stamens 20–30 in the lower bisexual flowers, fewer in those above, projecting beyond the perianth. Carpels usually 2, styles projecting beyond the perianth or almost so. *Europe, N Africa, W & C Asia.* H1. Summer.

28. SARCOPOTERIUM Spach
J. Cullen

Mound-forming shrub to 75 cm, the ends of the branches persisting as spines, the outer bark often silvery, peeling in strips to reveal the brown under-bark. Leaves pinnate with 9–25 very small, oblong to ovate, entire to 3–5-toothed leaflets 1–8 mm, downy above, whitish hairy beneath, margins turned under. Flowers in leaf-opposed or terminal spikes, unisexual, spikes often with male flowers above, female below. Perianth-segments 4, greenish and often white-margined, *c.* 2 mm. Stamens numerous. Carpels 2, free, enclosed in the perigynous zone. Fruit berry-like, the perigynous zone becoming fleshy, 3–5 mm across, red to yellowish brown.

A genus of a single species from the Mediterranean area where it is often a dominant shrub in certain communities (phrygana). It is not very frequently grown and is propagated by seed.

1. S. spinosum (Linnaeus) Spach (*Poterium spinosum* Linnaeus). Illustration: Polunin & Huxley, Flowers of the Mediterranean, pl. 48 (1965); Polunin, Plants of Europe, pl. 46 (1969); Pignatti, Flora d'Italia **1**: 568 (1982); Il Giardino Fiorito, June 1987: 19 (1987).

Mediterranean area (Sardinia & Tunisia eastwards). H5. Spring.

29. MARCETELLA Sventenius
S.G. Knees

Dioecious trees or shrubs, with more or less erect branches. Leaves alternate, pinnate with a terminal leaflet; leaflets toothed. Flowers very small in tight racemes; sepals 5; petals absent; stamens 25–30, on male flowers; carpels 5. Fruit a dry samara.

A genus of 2 species from Macronesia, only one is sufficiently ornamental to be cultivated regularly.

1. M. moquiniana (Webb & Berthelot) Sventenius (*Bencomia moquiniana* Webb & Berthelot). Illustration: Webb & Berthelot, Histoire naturelle des Iles Canaries **3**: pl. 39 (1836–1850); Bramwell & Bramwell, Wild flowers of the Canary Islands, pl. 163 (1974); Kunkel, Flora de Gran Canaria **1**: pl. 16 (1974); Kunkel, Die Kanarischen Inseln und ihre Pflanzenwelt, 170 (1987).

Small tree or shrub to 4 m. Leaves 15–20 cm, leaflets 2–2.5 cm, toothed, glaucous green above, pale beneath; new leaves reddish, glossy. Sepals 5, concave, 1–2 mm; female flower spikes reddish purple; male flower spikes yellowish green. Fruit a dry samara; wings unequal. *Canary Islands (Gran Canaria, Tenerife)*. H5–G1. Summer–autumn.

30. MARGYRICARPUS Ruiz & Pavon
S.G. Knees
Creeping or prostrate, evergreen dwarf shrub; branches ascending towards apex. Stems straw-coloured, internodes covered for most of their distance by clasping, papery stipules. Margins of stipules with white silky hairs. Leaves 1.4–1.8 × 0.8–1.2 cm, broadly ovate, pinnate with 3–5 pairs of glossy, linear lobes; margins strongly curved downwards and inwards, apex of lobes shortly pointed. Flowers solitary, stalkless, borne in leaf axils, but not on current year's growth. Sepals 5, persistent; corolla absent; stamens 1–3; carpel 1. Fruit 5–6 × 7–9 mm, an almost spherical, soft white berry, often tinged pale pink. Seeds ovate, c. 4 mm, reddish brown.

A genus of just one species, from the Andes. The evergreen foliage and pearly berries are the most attractive features of this plant, which can be grown in poor or sandy soils, provided drainage is good. Plants are easily raised from seed, or from cuttings rooted in late summer. Alternatively, plants can be raised from layers.

1. M. pinnatus (Lamarck) O. Kuntze (*M. setosus* Ruiz & Pavon; *Empetrum pinnatum* Lamarck). Illustration: Ruiz & Pavon, Flora Peruviana, t. 8 (1798–1802); Marshall Cavendish encyclopaedia of gardening, 1198 (1968).

South America (Colombia to S Chile, S Brazil, Uruguay to Argentina). H2. Summer.

31. ACAENA Linnaeus
P.F. Yeo
Perennial evergreen herbs or undershrubs. Stems usually prostrate, often rooting at nodes. Leaves alternate, pinnate with a terminal leaflet; stipules forming a sheath, often with an entire or divided leafy lobe on either side; leaflets usually toothed, asymmetric at the base. Plants bisexual or female and bisexual. Flowers borne in stalked spikes or dense ovoid or spherical heads. Cleistogamous flowers sometimes present in axils at base of flowering-stem. Receptacle hollow, almost closed at the mouth, usually bearing barbed spines. Sepals 3–6, usually 4. Petals absent. Stamens 2–7, with dorsifixed anthers. Carpels 1 or 2, occasionally to 5, concealed in the receptacle except for the curved feathery stigmas. Fruit an achene, dispersed within the receptacle.

A genus of about 100 species in the south temperate area, California and the Hawaiian Islands. The natural variation pattern presents great difficulty for classification, especially in South America. The treatment here might not be sustainable if many more introductions from that region were made. The species are useful for ground cover and present a diverse array of leaf-form and leaf-colour; the flowers are inconspicuous but sometimes the spines are decorative. Propagation is by cuttings, division or seed. Hybridisation is frequent in gardens.

Literature: Bitter, G., Die Gattung Acaena, *Bibliotheca Botanica* **17**(74): (1910–1911); Bitter, G., Weitere Untersuchungen Über die Gattung Acaena, *Feddes Repertorium* **10**: 489–501 (1912); Dawson, J.W., Natural Acaena hybrids in the vicinity of Wellington, *Transactions of the Royal Society of New Zealand* **88**: 13–27 (1960); Grondona, E., Las especies argentinas del genero "Acaena" ("Rosaceae"), *Darwiniana* **13**: 308–342 (1964); Walton, D.W. & Greene, S.W., The South Georgian species of Acaena and their probable hybrid, *British Antarctic Survey Bulletin* **25**: 29–44 (1971); Yeo, P.F., Acaena, 193–221; The species of Acaena with spherical heads cultivated and naturalized in the British Isles, 51–55, in Green, P.S., ed., *Plants: Wild and Cultivated*, (1973); Yeo, P.F., Acaenas, *The Garden* **107**: 326–328 (1982).

Leaves. Feathery: **1**. Bluish glaucous and densely hairy: **12**.
Leaflet-colour. Silvery on both sides: **3**. Silvery beneath and green above: **2,9**. Green above, tinged with brown: **7,11,13**. Bright green above, not silvery beneath: **1,4,8,10**. Bluish or greenish glaucous above: **4–6,12,14,15**.
Leaflet-length. In most leaves not more than 11 mm: **4,6,7,11,13–15**. In some leaves more than 11 mm: **1,3,4,5,8–10,12**.
Inflorescences. Hidden beneath leaves: **15**.
Spines on fruiting receptacle. All over surface: **1–3**. At apex only: **4–6,8–15**. Absent: **7,14**. Not rigid: **13,14**.

1a. Fruiting receptacle with spines not confined to apex   2
  b. Fruiting receptacle with spines confined to apex, occasionally absent   4
2a. Leaves feathery   **1. myriophylla**
  b. Leaves not feathery   3
3a. Leaflets silvery silky on both sides, with 8–24 teeth   **3. splendens**
  b. Leaflets silvery silky beneath, green above though with some silky hairs, with 2–7 teeth   **2. sericea**
4a. Leaflets near tip of leaf at least 1.3 times as long as broad, often 1.5–2.5 times as long as broad   5
  b. Leaflets near tip of leaf at most 1.5 times as long as broad   8
5a. Leaflets near tip of leaf usually with 17 or more teeth   6
  b. Leaflets near tip of leaf with not more than 13 teeth   7
6a. Leaflets not leathery, green beneath; spines on fruiting receptacle 2   **8. ovalifolia**
  b. Leaflets leathery, silvery silky beneath; spines on fruiting receptacle   **49. argentea**
7a. Leaflets bright glossy green, those near tip of leaf 2–2.5 times as long as broad   **10. novaezelandiae**
  b. Leaflets near tip of leaf light matt green, edged and veined with brown, those near base and stems, brown; leaflets near tip of leaf not more than twice as long as broad   **11. anserinifolia**
8a. Leaves not glaucous above   9
  b. Leaves glaucous above   11
9a. Largest leaves 3 cm; spines of fruiting receptacle 6–13 mm, soft, unbarbed, bright red   **13. microphylla**
  b. Largest leaves 5–9.5 cm; spines of fruiting receptacle not as above, sometimes absent   10
10a. Leaflets cut no more than ⅓ of the way to the midrib into 7–11 teeth; spines of fruiting receptacle barbed at tip   **4. magellanica** subsp. **laevigata**
  b. Leaflets cut ½ way to the midrib into 5–8 teeth; spines of fruiting receptacle not barbed, sometimes absent   **7. glabra**

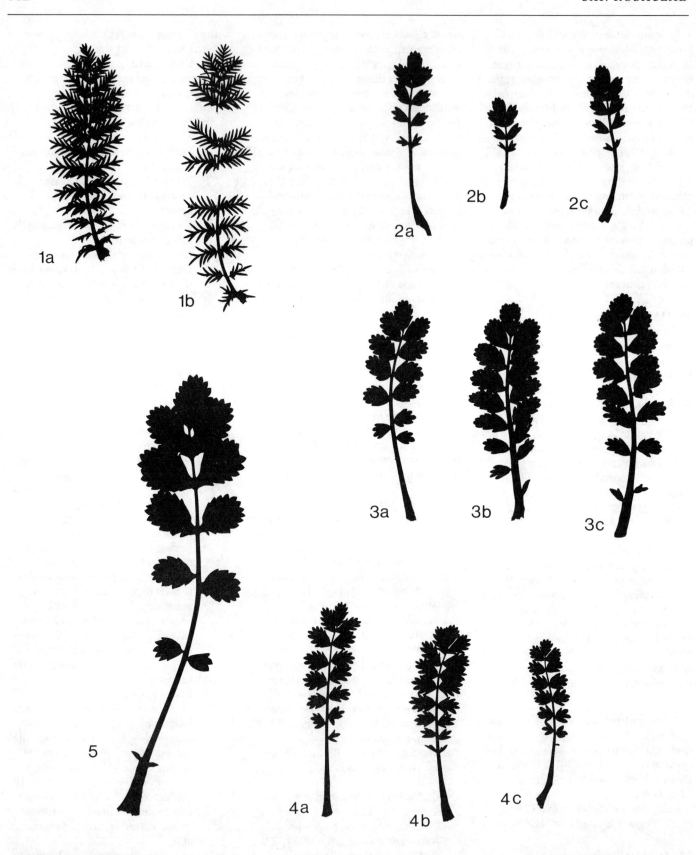

Figure 34. Leaf silhouettes of *Acaena* species (× 0.75).
1a, b, *A. myriophylla*. 2, *A. sericea* (a, b, upper; c, under).
3a, b, c, *A. magellanica* subsp. *laevigata*. 4a, b, c, *A. magellanica*
subsp. *magellanica*. 5, *A. affinis*.

11a. Leaflets 7–9   **12. caesiiglauca**
  b. Leaflets 11 or more in the majority
     of leaves   12
12a. Leaflets near tip of leaf mostly more
     than 10 mm   **5. affinis**
  b. Leaflets near tip of leaf mostly less
     than 10 mm   13
13a. Leaf-sheaths with an entire leafy
     lobe on each side or none
                          7. saccaticupula
  b. Leafy lobes of at least some of the
     leaf-sheaths with 2–6 teeth   14
14a. Inflorescences hidden beneath
     leaves   **15. buchananii**
  b. Inflorescences not hidden beneath
     leaves   15
15a. Leaflets near tip of leaf longer than
     wide; inflorescences 7–11 mm
           4. magellanica subsp. **magellanica**
  b. Leaflets near tip of leaf as broad as
     or broader than long; inflorscences
     less than 7 mm   **14. inermis**

1. A. myriophylla Lindley (*A. hieronymi*
Kuntze). Figure 34(1), p. 382. Illustration:
Bibliotheca Botanica **17**(74): 74, 77, t.
5, 7 (1910); Darwiniana **13**: 255–257
(1964); Cabrera, Flora de la Provincia de
Buenos Aires **3**: 392 (1967).

Perennial silky-hairy herb with trailing
leafy stems to 75 cm. Leaves to 12 × 3
cm, stalk very short, with a short sheath,
its lobes either absent or looking like the
leaflets; leaflets 13–23, sometimes more,
pinnatisect into linear-lanceolate acute
segments. Flowering-stems 20–30 cm,
leafy. Flowers in dense cylindric spikes,
becoming interrupted below in fruit.
Stamens longer than sepals; filaments
white; anthers dark red. Stigmas green,
longer than sepals. Fruiting receptacle with
a dense covering of slender straw-yellow
spines. *C & S Argentina*. H4. Summer.

2. A. sericea Jacquin filius. Figure 34(2)
p. 382. Illustration: Darwiniana **13**:
333–335 (1964); Moore, Flora of Tierra
del Fuego, 134 (1983).

Dwarf shrub with ascending stems,
terminated by clusters of erect leaves, not
rooting. Leaves to 11 × 2 cm, covered in
silky hairs; stalk 0.7–1 times as long as
blade, with a long sheath; leaflets 7–11,
bright green above, whitish beneath, those
near tip of leaf mostly 6–15 × 2.5–6 mm,
1⅓–3 times as long as wide, obovate to
wedge-shaped, with 2–7 teeth. Flowering-
stems to 30 cm, usually with one leaf near
the middle subtending a few flowers. Most
flowers in 1–3 dense ovoid or spherical
heads. Stamens dark red, shorter than
sepals. Stigmas flesh-pink, becoming dark

red, shorter than sepals. Fruiting receptacle
with short, conical spines throughout.
Patagonia. H1. Summer.

3. A. splendens Hooker & Arnott.
Illustration: Darwiniana **13**: 338 (1964).

Dwarf shrub with ascending stems,
terminated by clusters of erect leaves, not
rooting, forming a cushion. Leaves 4–22
cm, covered in silky hairs; stalk ⅓–⅔ as
long as blade, with a long sheath which
usually lacks leafy lobes; leaflets usually
7–11, silvery white on both surfaces, those
near tip of leaf 1.5–3.5 cm, 2¼–4 times
as long as broad, elliptic or obovate, with
8–24 shallow teeth. Flowering-stems *c.*
40 cm. Flowers in a spherical head, or
solitary, or clustered along the stalk.
Stamens shorter than sepals; anthers
purple or blackish. Fruiting receptacle with
barbed spines throughout. *Andes of Chile &
Argentina*. H3. Summer.

4. A. magellanica (Lamarck) Vahl.

Dwarf shrub with short or long trailing
stems; rosettes in leaf-axils. Leaves to 9.5
cm, stalk short, sheath with or without
an entire or 2-toothed leafy lobe. Leaflets
9–15, with 3–11 teeth. Flowering-stems *c.*
17 cm in fruit. Flowers in dense spherical
or slightly lengthening heads; much
smaller subsidiary heads sometimes
present. Stamens and stigma dark red,
as long as or slightly longer than sepals.
Fruiting receptacle with 2–4 barbed spines
at the apex; spines 2–6.5 mm.

Subsp. **magellanica** (*A. magellanica*
subsp. *venulosa* (Grisebach) Bitter; *A.
glaucophylla* Bitter). Figure 34(4), p. 382.
Illustration: Green, Plants: wild and
cultivated, 197 (1973); Bibliotheca
Botanica **17**(74): 155, 156, t. 16a (1910).
Leaves 2–4.5 cm. Leaflets hairless, light
glaucous grey-green above, paler beneath,
those near tip of leaf mostly 3.5–8 mm,
fan-shaped, cut ⅓–½ way towards
midrib into 3–8 blunt teeth. Flower-stems
grey-green, flushed with purplish red.
Main inflorescences 1.8–2.4 cm in fruit. *S
Patagonia (Chile, Argentina)*. H1. Summer.

Subsp. **laevigata** (Aiton filius) Bitter
(*A. adscendens* Vahl; *A. laevigata* Aiton
filius). Figure 34(3), p. 382. Illustration:
Bibliotheca Botanica **17**(74): 171 (1910);
Green, Plants: wild and cultivated, 199
(1973). Leaves mostly 3–9.5 cm, very
sparingly glandular-hairy. Leaflets thickish,
upper surface dark green, with a slight
removable bloom, hairless; lower surface
glaucous green with conspicuously netted
veins, hairless or nearly so, those near
tip of leaf 5.5–16 mm, oblong or nearly

square, cut ¼–⅓ way towards midrib
into 7–9, sometimes 11 teeth. Flower-stems
green, sometimes flushed with red. Main
inflorescences 2.1–2.4 cm in fruit. *Falkland
Islands*. H1. Summer.

5. A. affinis J.D. Hooker (*A. distichophylla*
Bitter; *A. adscendens* misapplied). Figure
34(5), p. 382. Illustration: Hooker, Flora
Antarctica **2**: t. 96 (1845); Bibliotheca
Botanica **17**(74): 178, 209, t. 20 (1910);
Vallentin & Cotton, Illustrations of the
flowering plants and ferns of the Falkland
Islands, t. 17 (1921); Green, Plants: wild
and cultivated, 201 (1973).

Stems trailing *c.* 1 m, herbaceous parts
bright pink. Leaves mostly 6–12 cm,
sometimes with subsidiary leaflets; stalk
short; sheath *c.* 1.3 cm, with 2 entire or
toothed leafy lobes. Leaflets 9–13, light
bluish or greyish glaucous green and
hairless above, or hairy on margins, grey-
green with conspicuous veins and finely
hairy beneath, those near tip of leaf mostly
0.9–1.8 cm, 1–1½ times as long as broad,
oblong or nearly circular, with the lower
edge running down the midrib for up to
4 mm, with 9–14 teeth. Flowering-stems
red, 11–21 cm in fruit. Inflorescences
spherical, 2–2.5 cm in fruit. Stamens and
stigmas dark red, longer than the sepals.
Spines 4, to 8 mm, barbed; additional
minute spines sometimes present. *Magellan
area of S America, Antarctic Islands*. H3–4.
Summer.

6. A. saccaticupula Bitter. Figure 35(1),
p. 384. Illustration: Green, Plants: wild and
cultivated, 206 (1973); The Garden **107**:
327 (1982).

Stems trailing to 50 cm, coppery red.
Leaves 2–7 cm; stalk *c.* ¼ as long as
blade, sheath to 6 mm, with 1–6 leafy
lobes on each side. Leaflets 11–13, upper
surface glaucous grey or grey-blue, tinged
with purple, red or brown, margins finely
hairy, lower surface more strongly
glaucous, sometimes with red veins; leaflets
near leaf-tip 4.5–11 × 3.5–9.5 mm, fan-
shaped or broadly oblong, with 5–9 teeth.
Flowering-stems brownish red or deep
red, 13–19 cm in fruit. Inflorescences
spherical, 1.5–1.8 cm in fruit. Stamens
slightly longer than sepals, filaments white,
anthers red. Stigmas about as long as
sepals, blackish red. Spines 4, 3–4 mm,
barbed, thickened at base, pinkish red. *New
Zealand*. H3. Summer.

This plant remained unidentified for
some time after its introduction and was
then named 'Blue Haze'.

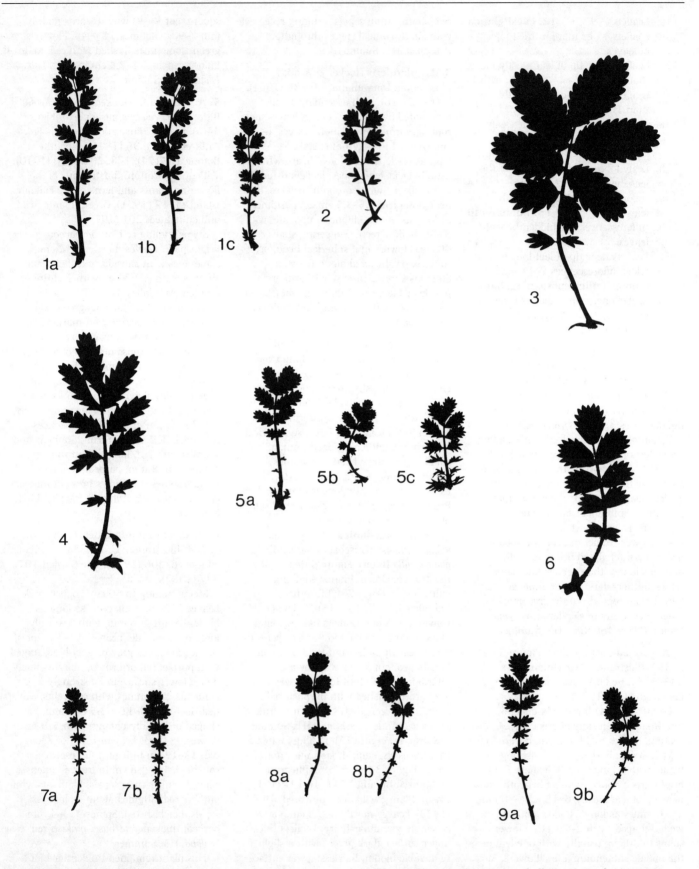

Figure 35. Leaf silhouettes of *Acaena* species (× 0.75 except where noted). 1a, b, c, *A. saccaticupula*. 2, *A. glabra*. 3, *A. ovalifolia*. 4, *A. novaezelandiae*. 5a, b, c, *A. anserinifolia*. 6, *A. caesiiglauca*. 7a, b, *A. microphylla* (× 1). 8a, b, *A. inermis*. 9a, b, *A. buchananii*.

7. A. glabra Buchanan. Figure 35(2), p. 384. Illustration: Bibliotheca Botanica **17**(74): 279, t. 34 (1911); Salmon, Field guide to alpine plants of New Zealand, 127 (1968); Green, Plants: wild and cultivated, 206 (1973).

Stems trailing to 30 cm; greenish straw-coloured. Leaves 3–5 cm; stalk not longer than interval between first two pairs of leaflets; sheath to 6 mm, usually with two entire leafy lobes. Leaflets 9–11, shortly stalked, rather thick and firm, hairless, upper surface slightly glossy, yellowish green, brown-tinged at edges, lower surface glaucous grey-green; leaflets near tip of leaf 5.5–11 mm, 1⅓–1½ times as long as broad, broadly obovate with wedge-shaped base, with 5–8 teeth. Flowering-stems brownish, 6–10 cm in fruit. Inflorescences spherical, *c.* 11 mm in fruit. Stamens white, just exceeding the sepals. Stigma white, shorter than sepals. Spines 4, to 2 mm or sometimes vestigial, not barbed. *New Zealand (Southern Alps).* H3. Summer.

8. A. ovalifolia Ruiz & Pavon. Figure 35(3), p. 384. Illustration: Darwiniana **13**: 230–233 (1964); Cabrera, Flora de la Provincia de Buenos Aires **3**: 392 (1967); Green, Plants: wild and cultivated, 209 (1973); The Garden **107**: 327 (1982).

Stems trailing to 1 m; green, sometimes red-flushed. Leaves 5–12 mm; stalk to half as long as blade; sheath to 5 mm, leafy lobes with 2–4 teeth. Leaflets 7–9, bright green, finely wrinkled and hairless above, slightly glaucous and finely hairy beneath, silky beneath when young, at least on veins, those near tip of leaf 1.5–3 cm, 1¾–2 times as long as broad, elliptic or oblong, with 12–23 teeth. Flowering-stems green, 6–12 cm in fruit. Inflorescences spherical, 1.8–3 cm in fruit. Stamens and stigmas white, slightly longer than sepals. Spines 2, 8–10 mm, barbed, red. *S America (Magellan area to Colombian Andes).* H3. Summer.

9. A. argentea Ruiz & Pavon. Illustration: Darwiniana **13**: 236–238 (1964).

Like *A. ovalifolia* but with leaf-sheaths to 1.8 cm, their leafy lobes sometimes entire, leaflets leathery, green above, white with silky hairs beneath, anthers purple, spines of fruit 4, often unequal. *S Argentina.* H3. Summer.

10. A. novaezelandiae Kirk (*A. sanguisorbae* misapplied; *A. anserinifolia* misapplied). Figure 35(4), p. 384. Illustration: Cheeseman, Illustrations of the New Zealand Flora **1**: t. 39 (1914); Salmon, Field guide to alpine plants of New Zealand, 85 (1968); Cochrane et al., Flowers and plants in Victoria, 120 (1968); Green, Plants: wild and cultivated, 211 (1973).

Stems trailing to 1 m; flushed with pinkish red. Leaves 3–10 cm; stalk not longer than interval between first 2 pairs of leaflets; sheath to 5 mm, with leafy-lobes, teeth 1–4 or absent; leaflets usually 9–13, bright glossy green, finely wrinkled, hairless or nearly so above, glaucous green and sparsely to densely silky-hairy beneath, those near tip of leaf 0.7–2.2 cm, 2–2½ times as long as broad, oblong or obovate-lanceolate, with 8–15 teeth. Flowering stem pale green, sometimes flushed with red, to 11 cm, sometimes to 25 cm in fruit. Inflorescences spherical, 1.5–3 cm in fruit. Stamens white, longer than sepals. Stigmas white with a purplish tip, longer than sepals. Spines 4, 6–9.5 mm, barbed, red. *New Zealand, Australia, Tasmania.* H3–4. Summer.

11. A. anserinifolia (J.R. & G. Forster) Druce (*A. sanguisorbae* invalid). Figure 35(5), p. 384. Illustration: Transactions of the Royal Society of New Zealand **88**: 14, 15 (1960); Green, Plants: wild and cultivated, 212 (1973); The Garden **107**: 328 (1982).

Stems trailing to 30 cm, brown, finely hairy. Leaves 2–5 cm; stalk shorter than interval between first 2 pairs of leaflets, sheath to 3 mm with a pinnatisect leafy lobe on each side; leaflets 9–13; upper surface light matt green, more or less flushed with brown, sparsely silky-hairy, lower surface slightly glaucous green, sometimes tinged with purple, sparsely to densely silky-hairy; leaflets near tip of leaf 3–8 mm, mostly 1⅓–2 times as long as broad, oblong or oblong-obovate, the lower edge often running down midrib for *c.* 1 mm, with 10–13 brush-tipped teeth. Flowering-stems brown, silky-hairy, 3–7.5 cm in fruit. Inflorescences spherical, 1.2–1.6 cm in fruit. Stamens and stigmas white, longer than sepals. Spines 4, 3.5–6 mm, barbed, red. *New Zealand.* H4. Summer.

Hybrids between this species, *A. novaezelandiae* and *A. inermis* are probably the most common hybrids in gardens.

12. A. caesiiglauca (Bitter) Bergmans. Figure 35(6), p. 384. Illustration: Bibliotheca Botanica **17**(74): t. 27 (1911); Green, Plants: wild and cultivated, 215 (1973); The Garden **107**: 327 (1982).

Stems trailing to 60 cm, pale brown, densely woolly. Leaves mostly 4–8 cm; stalk 1–2 times as long as interval between first 2 pairs of leaflets; sheath to 3 mm, lobes with 1–3 teeth or entire; leaflets 7–9; upper surface blue-grey glaucous, with sparse fine adpressed hairs; lower surface slightly paler and more glaucous, silky-hairy; both surfaces slightly purple-tinged in age; leaflets near tip of leaf 6–14 mm, 1⅓–1½ times as long as broad, oblong or broadly obovate, with 6–10 brush-tipped teeth. Flowering-stems pale brown, woolly, 10–14 cm in fruit. Flower-heads spherical, to 2.3 cm in fruit. Stamens white, much longer than sepals. Stigmas white, sometimes tinged with pink. Spines 4, 4.5–7 mm, barbed, olive-brown. *New Zealand.* H3. Summer.

13. A. microphylla J.D. Hooker. Figure 35(7), p. 384. Illustration: Wooster, Alpine plants, 29 (1872); Salmon, New Zealand flowers and plants in colour, edn 2, 128 (1967); Salmon, Field guide to alpine plants of New Zealand, 85 (1968); Green, Plants: wild and cultivated, 216 (1973); The Garden **107**: 328 (1982).

Stems trailing to 30 cm; light brown or brownish green, quickly rooting. Leaves 1–3 cm; stalk very short; sheath to 2 mm, with 2 entire leafy lobes. Leaflets 11–13, yellowish green above, strongly flushed with brown at edges, becoming entirely dull purplish brown, hairless, similar beneath but slightly glaucous and with fine adpressed hairs on the veins, those at tip of leaf 2–4.5 mm, with 3–7 brush-tipped teeth. Flowering stems brown, 1–4 cm in fruit. Inflorescences spherical, to 3 cm in fruit. Stamens and stigmas longer than sepals; stamens white; stigmas white with pink midline. Spines 4, to 1.3 cm, soft, thick, not barbed, bright red. *New Zealand (North Island).* H3. Summer.

14. A. inermis J.D. Hooker (*A. microphylla* misapplied). Figure 35(8), p. 384. Illustration: Salmon, Field guide to alpine plants of New Zealand, 84 (1968); Green, Plants: wild and cultivated, 217 (1973); The Garden **107**: 328 (1982).

Stems trailing to 30 cm; flesh-coloured to brownish glaucous, quickly rooting. Leaves 2–4 cm, sometimes to 6 cm; stalk ⅛–⅕ as long as leaf; sheath to 2.5 mm, with an entire leafy lobe on each side. Leaflets 11–13; upper surface dull bluish grey or brownish grey, slightly tinged with green (appearing marbled under a lens), becoming orange to straw-coloured in age, hairless; lower surface pale grey-green glaucous to purplish glaucous, with fine adpressed hairs on veins; leaflets at tip of leaf usually 2–5 mm, fan-shaped, square or transversely oblong, with 5–10 brush-tipped teeth. Flowering-stems brownish, 1–6 cm in fruit. Inflorescences spherical, to 1.6 cm in fruit if spiny. Stamens white, much longer than sepals. Stigmas white, about as long as sepals. Spines absent or to 6 mm, often unequal, soft, thick, not barbed, bright red. *New Zealand (South Island)*. H3. Summer.

15. A. buchananii J.D. Hooker. Figure 35(9), p. 384. Illustration: Bibliotheca Botanica **17**(74): 290, t. 27, 29 (1911); Green, Plants: wild and cultivated, 219 (1973).

Stems trailing to 30 cm; pale green to pale brown. Leaves 1.5–5.5 cm; stalk ⅕–¼ as long as blade; sheath 3–8 mm, with 2 entire leafy lobes. Leaflets 11–17, pale glaucous green above, hairless or with hairs near edge, about the same colour and hairy beneath, those at tip of leaf 3–9 mm, 1¼–1½ times as long as broad, oblong or broadly ovate, with 6–12 brush-tipped teeth. Flowering-stems pale green, 3–11 mm. Inflorescences hidden beneath the leaves, shorter than broad, *c.* 3 cm wide in fruit, with conspicuous bracts. Stamens 6–7 mm, white. Stigmas much longer than sepals, white, becoming pink. Spines 4, 7–13 mm, greenish yellow, with a tuft of reflexed hairs at the tip. *New Zealand (South Island)*. H3. Summer.

32. POLYLEPIS Ruiz & Pavon
M.F. Gardner
Small trees or shrubs; branches twisted, covered with scars of fallen leaves. Leaves alternate, with 3 leaflets or pinnate; leaflets in few pairs, leathery; leaf-stalks broadly membranous, sheathing at base. Racemes slender, flowers mainly pendent, bracts present, sepals 3–5, persistent, forming a calyx with a constricted throat, usually with 3 or 4 wings; petals absent; stamens numerous, inserted in calyx throat. Fruit a leathery achene enclosed in a hardened angular spiny or winged calyx tube.

A genus of 15 species from the Andes. Only occasionally cultivated but will thrive in a well-drained soil in a sunny position. Propagate from seed.

1. P. australis Bitter. Illustration: Botanische Jahrbücher, **14**: 620 (1911); Krüssmann, Manual of cultivated broad-leaved trees & shrubs **2**: 423 (1986).

A broad, deciduous shrub to 3 m; branches contorted in a zig-zag manner, with long lengths of brown bark flaking away, shoots with long internodes. Leaves unevenly pinnate, the first leaf has only 1 leaflet, the second 5 leaflets and the third 5–7, with only the latter 2 having axillary shoots; leaflets bluish green, 2–4 × 1 cm, elliptic, apex notched, base rounded, margin scalloped. Fruit 3-winged. *Argentina*. H4.

33. ADENOSTOMA Hooker & Arnott
J. Cullen
Evergreen, resinous shrubs. Leaves needle-like, borne in bundles or singly and alternately; stipules present or absent. Flowers small, bisexual, numerous, in erect or spreading terminal panicles. Sepals 5, translucent, erect. Petals 5, white, longer than the sepals. Stamens 10–15 borne at the mouth of the perigynous zone above some nectar-secreting glands. Perigynous zone cylindric, somewhat tapering to the base, 10-grooved. Carpel 1, containing a single ovule and with an oblique or lateral style. Fruit an achene surrounded by the persistent perigynous zone which hardens and becomes contracted towards the apex, the whole ellipsoid, black and hard.

A genus of 2 species from USA (California) and adjacent Mexico. Uncommon in gardens; propagated by seed or cuttings taken in spring.

1a. Leaves mostly borne in bundles, each bundle borne on an expanded base which bears 2 small stipules; flower subtended by a bract and 2 bracteoles, each of which is green and 3-lobed with a narrowly triangular central lobe and much smaller lateral lobes
      1. fasciculatum
  b. Leaves all borne singly, without stipules; flower subtended by a single green, simple bract with, above it, several broader, mostly translucent bracts
      2. sparsifolium

1. A. fasciculatum Hooker & Arnott. Illustration: Baillon, Natural history of plants (English translation) **1**: 373 (1871); Ornduff, Introduction to California plant life, pl. 4 (1974); Munz, Flora of southern California, 790 (1974); Wiggins, Flora of Baja California, 790 (1980).

Shrub to 3.5 m (often with a swollen, tuber-like base in the wild). Leaves mostly borne in bundles, each bundle borne on a hairy, swollen base which bears 2 hairy, narrowly triangular stipules. Leaves needle-like, somewhat swollen above, 4–12

mm. Flowers subtended by a bract and 2 bracteoles, each 3-lobed with a narrowly triangular terminal lobe and shorter lateral lobes. Petals *c.* 1.5 mm. Perigynous zone hairy or hairless. *USA (California), Mexico (Baja California)*. H5–G1. Summer.

Occurs in the wild in areas subject to fires, which it survives by means of its woody tuber.

2. A. sparsifolium Torrey.
Erect, tree-like shrub to 6 m. Shoots and leaves with scattered, scale-like, brown, resin-secreting glands. Leaves borne singly, alternate, 6–15 mm, narrowly linear to thread-like, without stipules. Flowers subtended by a single, simple, narrowly triangular, green bract and several broader, mostly translucent bracts borne above it. Petals 2 mm or more. Perigynous zone with hairs in the grooves. *USA (S California), Mexico (Baja California)*. H5–G1. Summer.

34. CERCOCARPUS Kunth
J. Cullen
Shrubs or small trees. Leaves more or less evergreen, alternate or clustered on short shoots, simple, entire or toothed, shortly stalked. Stipules rather small. Flowers solitary or in groups of 3 (rarely these groups somewhat aggregated), axillary or terminal on short lateral shoots. Perigynous zone hairy, narrow and tubular, broadening abruptly above into a 5-lobed perianth which ultimately opens very widely, the lobes and upper part of the widened tube reflexed. Petals absent. Stamens 10–25 (in our species), inserted in 2 or more series on the upper (reflexed) part of the expanded perianth; anthers hairy or hairless. Ovary superior, carpel 1, borne in the scarcely swollen base of the perigynous zone; ovule solitary; style terminal. Fruit an achene surmounted by the long, persistent, plumed-hairy style, surrounded at the base by the base of the perigynous zone which becomes dry and brownish.

A genus of about 6 species from the south and west USA and adjacent Mexico. They are infrequently grown but can be quite ornamental in flower and fruit. They require a sunny, well-drained position and can be propagated by seed or by cuttings. Literature: Martin, F.L., A revision of Cercocarpus, *Brittonia* **7**: 91–111 (1950).

1a. Anthers hairless; leaves with entire, downwardly rolled margins, acute at the apex
      2. ledifolius

b. Anthers hairy; leaves usually toothed, at least towards the rounded apex, margins not rolled downwards **1. montanus**

1. C. montanus Rafinesque.

Shrub or small tree to 8 m; bark grey or brown, fissured on old trunks. Leaves firm, dark green above, paler beneath, variable in shape, mostly oblanceolate or elliptic, usually conspicuously tapered to the base, usually with at least a few teeth towards the rounded apex, 1–7 × 0.3–3.5 cm, veins conspicuous beneath, forming a pattern of near-squares which are more densely hairy between than on the veins. Stipules to 8 mm. Flowers 3–6 mm wide. Perigynous zone 3–9 mm, densely hairy. Stamens 20–25, anthers hairy. Style 2–9 cm in fruit, densely hairy. *W USA, adjacent Mexico (Baja California).* H3. Spring–summer.

Var. **montanus**. Generally a shrub to 4 m; leaves with rather coarse, ovate teeth which are not apiculate. *Rocky Mts.*

Var. **glaber** (Watson) Martin (*C. betuloides* Torrey & Gray; *C. betulaefolius* Nuttall). Illustration: Hooker's Icones Plantarum **4**: t. 322 (1841); Elias, The complete trees of North America, 559 (1980). Shrub or small tree to 8 m; leaves with few, small, apiculate teeth, or entire. *USA (Oregon, California, Arizona), Mexico (Baja California).*

Var. **blancheae** (Schneider) Martin. Leaves very broadly elliptic, very leathery, coarsely toothed. *Islands off California.*

The naming of this species is in some doubt, and it is usually found in the literature as *C. betuloides*; however, Martin's revision has been followed here as the most general review of the genus.

2. C. ledifolius Torrey & Gray.

Illustration: Hooker's Icones Plantarum **4**: t. 324 (1841); Elias, The complete trees of North America, 556 (1980); Krüssmann, Manual of cultivated broad-leaved trees and shrubs **1**: 314 (1986).

Shrub or small tree to 8 m; bark red-brown, furrowed on old trunks, the younger branches whitish. Leaves leathery, narrowly elliptic to linear, 1–4 cm, tapered to the base and the acute apex, margins entire, rolled downwards, ultimately hairless and shining dark green above, shortly brown-hairy beneath. Stipules to 2 mm. Flowers solitary or in pairs, 2–6 mm wide. Perigynous zone 3–6 mm, densely hairy. Stamens 10–25, anthers hairless. Style in fruit to 7 cm, often twisted or curved, densely hairy. *W USA, adjacent*

Mexico (Baja California). H5. Spring–summer.

35. CHAMAEBATIA Bentham

J. Cullen

Small, aromatic shrubs, most parts with dense, short, somewhat crisped, simple, whitish hairs and much longer, stalked glands. Leaves partially evergreen, 3-pinnate with very small ultimate segments. Stipules small, lanceolate or narrowly triangular. Flowers rather few in a flat-topped terminal panicle, each subtended by a bract and 2 alternate bracteoles. Perigynous zone bell-shaped, densely hairy and glandular outside, densely hairy inside towards the base. Sepals 5, spreading, ultimately somewhat reflexed, white-hairy inside, white-hairy and glandular outside. Petals 5, white, spreading. Stamens numerous. Carpel 1, hairy, with a short style. Fruit an achene enclosed in the persistent perigynous zone.

A genus of 2 species from western USA and adjacent Mexico, superficially very similar to *Chamaebatiaria millefolium* (p. 329), differing in its lack of stellate hairs, 3-pinnate leaves, more open, fewer-flowered panicle, somewhat larger flowers and single carpel forming a 1-seeded achene. It requires a sunny position and well-drained soil and can be propagated by seed or semi-ripe hardwood cuttings in summer.

1. C. foliolosa Bentham. Illustration: Botanical Magazine, 5171 (1860); Jepson, Flora of California **2**: 213 (1936).

Shrub to 1 m. Bark of older shoots dark, often bloomed. Leaves more or less stalkless. Flowers 1.5–2 cm across. *W USA (California).* H5–G1. Summer.

36. PURSHIA de Candolle

J. Cullen

Shrubs to 3 m; shoots hairy when young. Leaves deciduous alternate, mostly borne in clusters on short lateral shoots, obovate, long-tapered to the base, rounded to the apex which is divided into 3 oblong, blunt lobes, slightly hairy above, densely white-hairy beneath, margins rolled under. Stipules small, narrowly triangular. Flowers bisexual, solitary at the ends of the short lateral shoots, stalkless or almost so. Perigynous zone more or less funnel-shaped, white-hairy and glandular. Sepals 5, oblong, blunt, reflexed in flower, yellowish and hairless inside. Petals creamy white, spoon-shaped. Stamens *c.* 25. Ovary superior. Carpels 1 (rarely 2),

style short. Achene(s) ovoid, hairy, *c.* 1 cm, projecting from the perigynous zone beyond the persistent sepals, tipped by the short, persistent style.

A genus of 2 species from western North America. The single cultivated species requires a well-drained soil in a sunny position, and is propagated by layering.

1. P. tridentata (Pursh) de Candolle. Illustration: Edwards's Botanical Register **17**: t. 1446 (1831); Abrams & Ferris, Illustrated Flora of the Pacific States **2**: 452 (1944); Clark, Wild flowers of the Pacific Northwest, 254 (1976).

Sprawling shrub to 3 m. Leaves 0.5–3 cm, deciduous or mostly so. Flowers to 1 cm across. *W USA (Oregon, California to Wyoming & New Mexico).* H5. Summer.

37. COWANIA D.Don

M.F. Gardner

Evergreen shrubs or small trees. Leaves alternate, lobed or pinnatifid, leathery, dotted with glands on upper surface, densely felted with white hairs beneath, margins rolled inwards; stipules joined to the leaf stalk. Flowers white, pale yellow or rich rose, solitary, very shortly stalked, terminal, on short leafy twigs; sepals 5, overlapping; petals 5, obovate, spreading; stamens numerous, inserted at the mouth of the calyx in 2 rows; styles 1–12, stalkless, covered with long hairs, lengthening in fruit; ovary superior. Fruit a one-seeded achene, terminating with a long feathery style.

A genus of 5 species from SW North America to central Mexico, occasionally seen in cultivation. These plants prefer a well-drained, limey soil in a warm sunny, protected position where low winter temperatures occur infrequently. Propagate from seed sown under glass in spring.

1a. Leaves with 5–9 lobes; flowers rich rose **1. plicata**
  b. Leaves with 3–5 lobes, flowers white to pale yellow **2. stansburiana**

1. C. plicata D. Don. Illustration: Botanical Magazine, 8889 (1921); The New Flora and Silva **5**(1): fig. 21 (1932); The RHS dictionary of gardening, **2**: 564 (1956).

A rigid, well-branched shrub to 1.5 m, with peeling bark; young shoots reddish, very glandular, covered with white woolly hairs which soon fall away. Leaves 8–15 mm, obovate, pinnatisect, with 5–9 lobes. Flowers rich rose, *c.* 3 cm wide, sepals swept back, glandular; petals rounded obovate; anthers yellow; fruiting style *c.* 3 cm. *N Mexico* H4–5. Summer.

Usually considered to be the most garden worthy species of the genus.

2. C. stansburiana Torrey. Illustration: McMinn, Manual of the Californian shrubs 226–227 (1951); Everett, New York Botanical Garden illustrated encyclopedia of horticulture **2**: 901 (1981); Martin & Hutchins, Summer wildflowers of New Mexico, 117 (1986).

A freely branched, aromatic shrub to 3 m, bark dark and shredding, twigs reddish brown, glandular. Leaves 6–15 mm, obovate, pinnatisect, with 3–5 lobes; leafstalks 3–8 mm. Flowers tube-shaped, fragrant; flower-stalks glandular, *c.* 5 mm; sepals broadly obovate, 4–6 mm; forming a funnel-shaped calyx covered with glands; petals white or yellow; styles 5–10, 3–5 cm in fruit. *SW North America to N Mexico.* H4–5. Summer.

Some authorities consider this species to be a variety of *C. mexicana* D. Don.

38. FALLUGIA Endlicher

J. Cullen

Deciduous, upright shrub to 1.5 m; bark ultimately greyish or whitish. Leaves alternate, 0.8–1.5 cm, often with short leafy shoots in their axils, pinnate, lobes 3–7, oblong-linear, blunt, variably hairy above, densely white- or brown-hairy beneath between the raised main veins and the downwardly rolled margins. Flowers solitary or in few-flowered racemes terminating the branches, each with a bract and 2 alternate bracteoles, bisexual or female, 2–3 cm across. Perigynous zone shallowly cup-shaped, hairy outside, very densely white-hairy inside. Sepals 5, ovate, with 5 narrow epicalyx-segments alternating with them. Petals 5, white, obovate to almost circular. Stamens numerous. Carpels numerous, hairy, borne on a conical elevation within the perigynous zone, each with a single ovule and a long style. Fruit a head of hairy achenes, each surmounted by the long, persistent, reddish, feathery-hairy style.

A genus of a single species from southern USA and adjacent Mexico. It has attractive flowers and fruits and requires a well-drained soil in a very sunny position. It is propagated by seed.

1. F. paradoxa (D. Don) Endlicher. Illustration: Botanical Magazine, 6660 (1882); Abrams & Ferris, Illustrated Flora of the Pacific States **2**: 452 (1944); Lenz & Dourley, California native trees and shrubs, 99 (1981); Martin & Hutchins, Summer wildflowers of New Mexico, 115 & unnumbered colour plate (1986).

S USA & adjacent Mexico. H5–G1. Summer.

39. DRYAS Linnaeus

J. Cullen

Low shrubs with creeping, branched, woody stems giving rise to tufts or rosettes of leaves. Leaves evergreen, shortly stalked, toothed or scalloped, margins usually rolled under to some extent, usually dark green and conspicuously wrinkled above, densely white hairy beneath with cobwebby hairs, often also with dark purplish or blackish stalked glands and/or long, bristle-like processes which themselves bear white hairs, sometimes also with stalkless, resin-secreting glands. Stipules lanceolate. Flowers solitary on long stalks which rise above the leaves, sometimes functionally unisexual. Perigynous zone short, cup-like; sepals 7–10, often densely covered with dark stalked glands as well as cobwebby hairs. Petals 7–20, often 8, yellowish or white, forwardly directed or spreading. Stamens numerous, filaments sometimes hairy, deciduous. Ovary superior, carpels numerous. Fruit a head of achenes each with a long, persistent, plumed (silkyhairy) style, surrounded at the base by the persistent perigynous zone and sepals; the styles often twist together spirally while the fruits are ripening.

A genus whose classification is very confused; many authors recognise only a single species (*D. octopetala*) which may be divided into subspecies, varieties, etc., whereas others recognise up to 20 or more species. In this account the treatment proposed by Hultén (reference below) is followed; Hultén recognises 4 species (3 of them cultivated), some of which are further divided.

The plants are easily grown in a sunny position and may be propagated by seed, by cuttings taken in early autumn, or by layering.

Literature: Hultén, E., Studies in the genus Dryas, *Svensk Botanisk Tidskrift* **53**: 507–542 (1959).

1a. Flowers pendent or horizontal at flowering, not fully opening, the yellow petals directed forwards; filaments with sparse long hairs **1. drummondii**
  b. Flowers erect, opening widely, the white, cream or faintly yellowish petals spreading; filaments hairless 2
2a. Midrib of the leaves beneath with hairy processes and/or stalked glands **2. octopetala**
  b. Midrib of the leaves beneath with white hairs only (or rarely almost hairless) **3. integrifolia**

1. D. drummondii Richardson (*D. octopetala* Linnaeus var. *drummondii* (Richardson) Watson). Illustration: Botanical Magazine, 2972 (1830); Porsild, Rocky Mountain wild flowers, 233 (1974); Clark, Wild flowers of the Pacific Northwest, 250 (1976).

Leaves elliptic to narrowly obovate, 1–4 × 0.5–1.5 cm, distinctly toothed, densely white-hairy beneath with hairy processes on the midribs. Flowering-stems 5–25 cm, flowers pendent or horizontal. Sepals covered with dark purple stalked glands. Petals yellow, to 1.5 cm, not opening widely, forwardly directed. Fruit erect. *N America.* H1. Late spring–summer.

2. D. octopetala Linnaeus. Illustration: Hay & Synge, Dictionary of garden plants in colour, pl. 60 (1969); Everard & Morley, Wild flowers of the world, pl. 20 (1970); Rickett, Wild flowers of the United States **6**: pl. 65 (1971); Garrard & Streeter, Wild flowers of the British Isles, pl. 35 (1983).

Leaves oblong to ovate or somewhat obovate, 0.5–5 × 0.2–2.5 cm, clearly toothed or scalloped, densely white-hairy beneath and with either or both stalked glands and hairy processes on the midrib. Flowering-stems to 50 cm. Sepals with dark purple stalked glands. Flowers erect. Petals white, 0.7–1.7 cm, spreading. *N temperate regions.* H1. Late spring–summer.

A variable species. Subsp. **octopetala** has hairy processes on the midribs of the leaves beneath and occurs throughout the range of the species; minor variants of it include forma **argentea** (Blytt) Hultén (*D. lanata* Stein; *D. octopetala* var. *lanata* of catalogues), which has the leaves densely hairy above; a small, low-growing variant has been called var. **minor** Hooker; var. **pilosa** Babington (*D. babingtoniana* Porsild) has leaves some of which lack hairy processes on the midrib beneath.

Subsp. **hookeriana** (Juzepczuk) Hultén (*D. hookeriana* Juzepczuk). Illustration: Porsild, Rocky Mountain wild flowers, 235 (1974). This has only stalked glands on the midribs of the leaves beneath, and is from western North America.

The hybrid with *D. drummondii* has

been named as **D. × suendermannii**
Sündermann. It is very like subsp.
octopetala, with ovate leaves, erect flowers
with spreading petals and hairless
filaments, but the petals are yellowish
in bud, fading gradually to white as the
flower matures. It is widely grown.

3. D. integrifolia Vahl (*D. tenella* Pursh;
D. octopetala var. *integrifolia* (Vahl) Hooker).
Illustration: Rickett, Wild flowers of the
United States **6**: pl. 65 (1971).

Like *D. octopetala* but smaller and more
compact. Leaves often almost entire or
with the margins folded under so that the
teeth (which are only towards the base
of the leaf) are not visible; midribs of the
leaves with only white cobwebby hairs
(sometimes completely without hairs).
*Greenland, N America, extreme eastern part
of former USSR.* H1. Summer.

Variable and divided into 2 subspecies
by Hultén; it is uncertain which of these
is cultivated. Hybrids with *D. octopetala*
are known throughout the range of the
species, and may well also be cultivated.

40. GEUM Linnaeus
D.A.H. Rae
Perennial herbs with 2 types of leaves.
Basal leaves usually unequally pinnate
with the terminal leaflet often distinctly
larger than the laterals. Stem-leaves much
smaller. Flowers solitary or in corymbs
on simple or branched stems, bisexual,
yellow, orange, white or red. Sepals 5
with 5 smaller lobes between. Petals 5
in a saucer or bell-shaped arrangement.
Stamens many. Carpels many on a conical
or cylindric receptacle. Fruit a group of
achenes with long persistent styles; often
feathery or jointed.

A genus of about 40 species with a wide
distribution, mostly in temperate and cold
regions. They are easily grown in rock
gardens or borders in any reasonable soil.
Propagated by seed and division, they
hybridise easily in the garden giving rise to
much confusion in their identification.

1a. Terminal leaflet of basal leaves less
than twice size of lateral leaflets   2
  b. Terminal leaflet of basal leaves at
least twice size of lateral leaflets   5
2a. Plants with creeping non-flowering-
stems (stolons); flowers solitary
1. reptans
  b. Plants without creeping non-
flowering-stems; flowers solitary or
to 7 per stem   3
3a. Leaflets lobed to at least half their
length   4

  b. Leaflets shallowly lobed
2. elatum
4a. Flowers yellow; calyx green
3. rossii
  b. Flowers cream to purple, never
bright yellow; calyx often purple
4. triflorum
5a. Plants with creeping non-flowering-
stems; flowers always solitary
6. uniflorum
  b. Plants without creeping non-
flowering-stems; flowers solitary or
to 7 per stem   6
6a. Petals dark orange-red to buff-pink;
flowers nodding   **5. rivale**
  b. Petals yellow, white, cream, bright
orange or scarlet; flowers erect   7
7a. Lateral leaflets of basal leaves
usually less than 5 mm
7. parviflorum
  b. Lateral leaflets of basal leaves more
than 5 mm   8
8a. Stems to 30 cm   9
  b. Stems more than 30 cm   10
9a. Basal leaves to 10 cm; terminal
leaflet *c.* 2.5 cm across
8. montanum
  b. Basal leaves to 20 cm; terminal
leaflet to 7.5 cm across
12. coccineum
10a. Leaf-stalks shaggy, with hairs to 5
mm   **10. bulgaricum**
  b. Leaves with short hairs or hairless
11
11a. Petals yellow   **9. urbanum**
  b. Petals scarlet, sometimes copper
coloured   **11. chiloense**

1. G. reptans Linnaeus. Illustration:
Huxley, Mountain flowers of Europe, pl.
294 (1986); Parish, Flowers in the wild,
129 (1983).

Stems to 20 cm, also with creeping
non-flowering-stems to 40 cm. Basal leaves
pinnate to 15 cm. Terminal leaflet only
slightly larger than laterals, 2.5 cm across,
heart to kidney-shaped, deeply irregularly
lobed. Lateral leaflets in 4–7 pairs, usually
with 3–5 deeply cut wedge-shaped lobes.
Stem-leaves few, slender to 2 cm, 3–5
deeply divided lobes. Flowers erect,
solitary, *c.* 3 cm across. Petals yellow,
scarcely longer than sepals. *European Alps.*
H3. Summer.

G. × rhaeticum Brugger (*G. montanum*
× *G. reptans*), stems to 20 cm, solitary to
numerous. Basal leaves pinnate to 25 cm.
Terminal leaflet only slightly larger than
laterals, heart to wedge-shaped, unevenly
lobed. Lateral leaflets in 3–7 pairs, entire
or shallowly 3-lobed. Stems-leaves few,

small, slender, shallowly lobed. Flowers
erect, *c.* 2.5 cm across. Petals golden
yellow. *European Alps.* H3. Summer.

2. G. elatum Wallich. Illustration:
Botanical Magazine, 6568 (1881); Polunin
& Stainton, Flowers of the Himalaya, pl.
36, 38 (1984).

Stems slender and branched, to 35 cm.
Basal leaves pinnate to 20 cm. Terminal
leaflet usually smaller than largest lateral
and similarly shaped, roundish and
shallowly lobed. Lateral leaflets in 8–12
pairs, often with small and large pairs
arranged alternately. Stem-leaves small
with stipules. Flowers erect, 1–3 per stem,
to 3 cm across. Petals golden yellow, often
2-lobed. Inflorescence stalks long and
slender. *Himalaya.* H3. Summer.

3. G. rossii (R. Brown) Seringe.
Illustration: Steere, Wild flowers of the
United States, 175 (1966).

Stems to 15 cm, slightly downy. Basal
leaves pinnate to 10 cm, minutely downy.
All leaflets, including terminal, to 1.5 cm,
in 6–10 unequally shaped pairs, linear
to wedge-shaped, deeply divided into 3 or
more lobes. Stems-leaves similar to leaflets
but longer, usually 3–5. Flowers erect, 3–5
per stem, to 2 cm across. Petals round,
yellow. Sepals to same length as petals.
Alaska. H3. Summer.

4. G. triflorum Pursh. Illustration:
Botanical Magazine, 2858 (1828); Steere,
Wild flowers of the United States, **1**: 135
(1966).

Stems erect to 45 cm, hairy and with
fine glandular hairs. Basal leaves
irregularly pinnate to 15 cm with 7–15
principal pairs of leaflets each to 2.5 cm.
Terminal leaflet not significantly different
from laterals. Terminal and lateral leaflets
narrowly wedge-shaped, margins very
irregular; leaflets deeply divided, almost to
base, glandular. Stem-leaves few, similar to
basal leaflets, slender, to 2.5 cm. Flowers
on long stems to about twice the length
of the leaves, up to 3 flowers per stem,
2.5 cm across. Petals cream to purple,
sometimes with a purplish red margin,
oblong, not spreading, about as long as
calyx. Calyx often tinged purplish to dark
red. *N America (east of the Cascade Mts).*
H3. Summer.

5. G. rivale Linnaeus. Illustration:
Bonnier, Flore complète **3**: pl. 170 (1914);
Huxley, Mountain flowers of Europe, pl.
302 (1987); Steere, Wild flowers of the
United States **1**: 135 (1966)
Stems 20–80 cm, glandular-downy.

Basal leaves pinnate to 15 cm. Terminal and (often) top 2 laterals to 5 cm across, obovate or wedge-shaped, coarsely double-toothed. Other lateral leaflets in 1–5 pairs, to 1.5 cm, with deep forward-pointing teeth. Stem-leaves few, often 3-lobed and slender to 3 cm. Flowers nodding, 3–5 per stem, *c.* 2.5 cm across. Petals dark orange-red to buff-pinkish, scarcely longer than calyx lobes. *Europe, Asia & N America.* H3. Early summer.

G. × tirolense Kerner (*G. montanum × G. rivale*) usually almost identical to *G. rivale*.

6. G. uniflorum Buchanan. Illustration: Evans, New Zealand in flower, 107 (1987); Mark & Adams, New Zealand alpine plants, pl. 25 (1973); Salmon, Alpine plants of New Zealand, 249 (1968).

Stems to 1.5 cm, creeping rootstock, forming broad patches. Basal leaves pinnate to 7.5 cm, hairy, especially on margin. Terminal leaflet *c.* 2.5 cm wide, heart to broadly kidney-shaped, slightly lobed to deeply scalloped. Lateral leaflets in 1 or 2 pairs, very small (to 3 cm), deeply lobed. Stem-leaves narrow and slender, to 5 mm. Flowers solitary on 10–15 cm slender stems, 2–3 cm across. Petals white or cream, roundish. *New Zealand (South Island).* H4–5. Summer.

7. G. parviflorum Smith. Illustration: Mark & Adams, New Zealand alpine plants, pl. 25 (1973); Moore & Irwin, The Oxford book of New Zealand plants, 31 (1978).

Stems erect 10–50 cm across, hairy. Basal leaves pinnate, to 15 cm. Terminal leaflet to 5 cm across, kidney-shaped, faintly 3–5 lobed, scalloped, hairy on both sides. Lateral leaflets in 4–8 pairs, small, to 5 mm, margins deeply cut or with forward-pointing teeth. Stem leaves unevenly lobed, slender, to 1 cm, hairy. Flowers *c.* 1.2 cm across, on loose, few-flowered panicles, with long slender flower-stalks. Petals white or cream. *New Zealand & S America.* H4–5. Summer.

8. G. montanum Linnaeus. Illustration: Huxley, Mountain flowers of Europe, pl. 295 (1987); Parish, Flowers in the wild, 129 (1983).

Stems to 30 cm (usually less than 10 cm in the wild), hairy. Basal leaves pinnate, to 10 cm. Terminal leaflet *c.* 2.5 cm across, heart-shaped to kidney-shaped or rounded, irregularly and deeply toothed. Lateral leaflets in 3–6 pairs, to 1 cm, toothed. Stem-leaves *c.* 1.5 cm, deeply cut. Flowers

erect, usually solitary, 2–3 cm across. Petals golden yellow, almost round. Sepals almost to length of petals. *S Europe.* H3. Summer.

9. G. urbanum Linnaeus. Illustration: Bonnier, Flore complète **3**: pl. 170 (1914); Polunin, Flowers of Europe, pl. 45 (1969).

Stems erect to 60 cm, slightly downy. Basal leaves pinnate, to 1.5 cm, hairy. Terminal leaflet to 8 cm across, roundish, scalloped. Lateral leaflets in 2 or 3 pairs, often unequal in size, usually 5–10 mm. Upper pair of lateral leaflets often similar in size to terminal leaflet. Stem-leaves to 2 cm, wedge-shaped to linear, unevenly lobed and with forward-pointing teeth. Stipules 1–3 cm, leaf-like. Flowers to 1.5 cm, erect, 1–3 per stem, in open cymes to 2 cm across. Petals yellow, spreading, about same size as sepals, obovate or oblong. *N Europe.* H3. Summer.

G. × intermedium Ehrhart (*G. urbanum × G. rivale*), variable in all aspects between the 2 parents.

10. G. bulgaricum Pančič. Illustration: Polunin, Flowers of Greece and the Balkans, pl. 16 (1980).

Stems to 60 cm with soft and glandular hairs. Basal leaves pinnate, to 30 cm, with dense soft hairs to 5 mm. Terminal leaflet to 10 cm across, heart to kidney-shaped with uneven jagged and rounded forward-pointing teeth; lateral leaflets variable in size and number, but usually in 5–7 pairs, *c.* 1.5 cm. Stem-leaves clasping stem, to 2.5 cm, slender, deeply lobed. Flowers 2.5–3.5 cm across, 3–7. Petals bright yellow or orange, triangular. Styles with fine hairs, *c.* 1.2 cm. *Bulgaria.* H3. Summer.

G. × borisii Keller (*G. bulgaricum × G. reptans*). Illustration: Brickell, RHS gardeners' encyclopedia, 249 (1989). Variable, but usually more similar to *G. bulgaricum.*

11. G. chiloense Balbis (*G. quellyon* Sweet). Illustration: Brickell, RHS gardeners' encyclopedia, 247 (1989).

Stems to 60 cm, hairy and glandular. Basal leaves pinnate, 10–30 cm, hairy. Terminal leaflet to 75 cm across, heart to kidney-shaped with uneven jagged and rounded forward-pointing teeth; lateral leaflets *c.* 1.5 cm, variable in size and number, usually in 3 pairs. Stem-leaves clasping stem, to 2.5 cm, unevenly and deeply lobed. Flowers erect, 1–5, either solitary or in a corymb, *c.* 2.5 cm across. Petals scarlet, sometimes copper-coloured.

Filaments red. *Chile.* H3. Summer.

Most of the frequently grown cultivars are derived from *G. chiloense*, e.g. 'Mrs Bradshaw', red, double; 'Lady Stratheden', yellow, double; 'Red Wings', red, semi-double.

12. G. coccineum Sibthorp & Smith.

Stems to 30 cm with soft dense hairs. Basal leaves pinnate, to 20 cm, with soft hairs. Terminal leaflet 3–4 times larger than lateral leaflets, to 7.5 cm across, rounded or heart-shaped, deeply lobed (sometimes 3-lobed) or unevenly sharply toothed. Lateral leaflets to 2 cm, in 1–3 pairs, deeply lobed, variable. Stem-leaves few, to 2.5 cm, unequally lobed. Flowers *c.* 2.5 mm across, erect, 1–3 per stem. Petals yellow or orange, almost rounded. Epicalyx (lobe between sepals) distinctly different to sepals, *c.* 1 × 2 mm. *S Europe, Greece.* H3–4. Summer–autumn.

G. × jankae G. Beck (*G. coccineum × G. rivale*), similar to *G. coccineum* but leaves more deeply toothed and less hairy.

41. WALDSTEINIA Willdenow

S.G. Knees

Rhizomatous perennial herbs, often with stolons. Leaves mostly basal, ternate or lobed; lobes 3–7. Flowers terminal on slender branched stems. Bracteoles 5; sepals 5, narrowly lanceolate, apex acute; petals 5, rounded, yellow; stamens numerous; style lengthening, soon deciduous. Ovary. Fruits of 2–6 achenes.

These low spreading perennials are valued as good ground cover plants with their ability to quickly colonise otherwise difficult areas, such as shady corners, or north facing borders. Propagation is usually by division, although plants may be grown from seed.

1a. Flowers 1–1.5 cm across; leaves
  with 5–7 lobes   **1. geoides**
  b. Flowers 1.5–2 cm across; leaves
   usually ternate   2
2a. Leaflets 1.2–3 cm   **2. ternata**
  b. Leaflets 3.5–5 cm   **3. fragarioides**

1. W. geoides Willdenow. Illustration: Botanical Magazine, 2595 (1825); Reichenbach, Icones florae Germanicae et Helveticae **25**: t. 65 (1912).

Rhizome erect or shortly creeping. Leaves broadly cordate to kidney-shaped, lobes 5–7, coarsely toothed. Flowering-stems 15–25 cm with 3–7 flowers, bracts leafy, flowers 1–1.5 cm across, petals acuminate at base. *EC Europe (Bulgaria to Ukraine).* H1. Spring

2. W. ternata (Stephan) Fritsch (*W. trifolia* Rochel; *W. sibirica* Trattinick). Illustration: Reichenbach, Icones florae Germanicae et Helveticae **25**: t. 66 (1912).

Plants with creeping, branched rhizomes and rooting stolons. Leaves ternate, leaflets 1.2–3 cm, stalkless, cuneate at base, shallowly lobed, margins with forward-pointing teeth. Flowering stems 10–15 cm with 3–7 flowers, bracts inconspicuous; flowers 1.5–2 cm across, petals rounded. *EC & S Europe, E Siberia, Japan.* H1. Spring.

3. W. fragarioides (Michaux) Trattinick. Illustration: Revue Horticole, 510 (1890); Gleason, Illustrated flora of north-eastern United States and adjacent Canada **2**: 291 (1952).

Plants with rhizomes and stolons. Leaves usually ternate, 9–18 cm, leaflets 3.5–5 cm, cuneate at base, toothed towards apex. Flowering-stems 8–20 cm with 3–8 flowers; flowers 1.5–2 cm across; calyx 3–5 mm, silky, bracts 5–10 mm. Petals obovate-broadly elliptic, 8–10 mm, greatly exceeding sepals. *E North America.* H1. Spring.

42. COLURIA R. Brown
J. Cullen

Herbs with rhizomes. Stems erect. Leaves mostly basal, pinnate with 7 or more large leaflets with much smaller leaflets in between, terminal leaflet largest, often lobed, all toothed; stem-leaves few, reduced; stipules lanceolate. Flowers solitary or few in racemes. Perigynous zone cylindric-bell-shaped to funnel-shaped, conspicuously 10-veined. Sepals 5, alternating with 5 smaller, persistent epicalyx segments. Petals 5–7, almost circular, very shortly clawed. Stamens numerous, filaments hardened and persistent in fruit. Carpels rather few. Styles long, hairy towards the base and constricted there, deciduous. Fruit a group of achenes in the persistent perigynous zone surmounted by the persistent filaments.

A genus of about 5 species from eastern Asia, very rarely cultivated. Little is known of the requirements of the single species treated below, but it is propagated by seed and perhaps by division of the rhizome.

Literature: Evans, W.E., The genus Coluria, *Notes from the Royal Botanic Garden Edinburgh* **15**: 48–54 (1925).

1. C. geoides (Pallas) Ledebour. Illustration: Flora USSR **10**: 245 (1941).

Stems to 35 cm, densely hairy like the leaves. Basal leaves in outline gradually tapered to the base, terminal leaflet truncate at its base. Stem-leaves alternate, small, borne all along the stem. Flowers *c.* 2 cm across, bright yellow. Achenes covered in translucent papillae. *E Former USSR (Siberia), Mongolia.* H1. Spring.

43. POTENTILLA Linnaeus
A.C. Leslie & S.M. Walters

Perennial, rarely annual or biennial herbs or small shrubs. Leaves pinnate, palmate, with narrow leaflets or divided into 3 leaflets. Flowers solitary or in cymes, sepals and petals usually 5; epicalyx present. Perigynous zone more or less flat, receptacle dry or spongy. Petals yellow, white, pink, red, orange or bicoloured. Stamens 10–30. Carpels usually 10–80, rarely as few as 4. Fruit a head of achenes. Style usually deciduous.

A genus of about 500 species, chiefly in the northern temperate and arctic regions. The flower is superficially like that of *Ranunculus* (volume III, p. 364), and most easily distinguished by the presence of the epicalyx. Most species are easily grown in a well-drained soil in a sunny position, though a few may require alpine house treatment. Herbaceous species may be propagated by seed, division or cuttings, woody species by seed or cuttings.

1a.	Shrub	2
b.	Perennial herb, more or less woody at the base	6
2a.	Flowering shoots annual, herbaceous; leaves toothed	**5. salesoviana**
b.	Flowers borne on woody perennial twigs; leaves not toothed	3
3a.	Leaves large, to 2 cm wide, leaflets broadly elliptic; stipules ovate, brownish, conspicuous on the twigs	**4. arbuscula**
b.	Leaves smaller, leaflets oblong, elliptic or linear; stipules lanceolate, pale, not conspicuous on the twigs	4
4a.	Petals white; epicalyx-segments broad, obtuse	**3. davurica**
b.	Petals yellow; epicalyx-segments narrow, acute	5
5a.	Flowers not more than 1.5 cm, borne on long, slender stalks; leaflets 5–9, usually 7, linear, small	**2. parvifolia**
b.	Flowers usually more than 1.5 cm, on relatively short stalks; leaflets 5 or 7, oblong-lanceolate to elliptic	**1. fruticosa**
6a.	Leaves pinnate	7
b.	Leaves palmate with narrow leaflets, or made up of 3 leaflets	13
7a.	Petals purple	**6. palustris**
b.	Petals yellow or white	8
8a.	Petals white; leaflets ovate to almost circular	**11. rupestris**
b.	Petals yellow; leaflets various	9
9a.	Leaflets deeply pinnatisect	**12. multifida**
b.	Leaflets entire or toothed	10
10a.	Leaflets entire or toothed only at the apex	**10. bifurca**
b.	Leaflets toothed along the margins	11
11a.	Flowers solitary	**7. anserina**
b.	Flowers in cymes	12
12a.	Plant with stems producing stolons	**8. lineatus**
b.	Plant producing rhizomes	**9. peduncularis**
13a.	Leaflets 3–7	14
b.	Leaflets 3	32
14a.	Petals yellow (sometimes with an orange spot at the base)	15
b.	Petals white, pink, red or bicoloured yellow and pink/red	26
15a.	Plant stoloniferous, freely rooting at the nodes	**32. reptans**
b.	Plant not stoloniferous (though sometimes rooting from more or less prostrate woody stems)	16
16a.	Leaves densely white-hairy beneath	17
b.	Leaves not densely white-hairy beneath	20
17a.	Stems prostrate or ascending	18
b.	Stems usually erect; petals usually more than 1 cm	**20. gracilis**
18a.	Stems always prostrate; inflorescence densely white-hairy	**14. calabra**
b.	Stems sometimes ascending; inflorescence loosely covered with white hairs	19
19a.	Petals 4–5 mm; styles conical	**13. argentea**
b.	Petals 3–4 mm; styles club-shaped and often distorted	**15. collina**
20a.	Flowering-stems erect, robust, to 70 cm	**16. recta**
b.	Flowering-stems usually ascending, less than 40 cm	21
21a.	Petals at least 1 cm	22
b.	Petals less than 1 cm	23
22a.	Flowering-stems to 40 cm; terminal tooth of leaflets as long as laterals	**24. pyrenaica**
b.	Flowering-stems less than 25 cm; terminal tooth of leaflet shorter than laterals	**28. aurea**

23a. Leaves densely stellate-hairy
beneath **31. cinerea**
b. Leaves not densely stellate-hairy
beneath 24
24a. Plant mat-forming, with more or
less prostrate stems rooting at the
nodes **30. neumanniana**
b. Plant clump-forming, not rooting at
the nodes 25
25a. Leaves more or less silvery-hairy
beneath; petals without an orange
basal spot **19. nevadensis**
b. Leaves not silvery-hairy; petals
usually with an orange basal spot
29. crantzii
26a. Petals usually white (rarely pale
pink) 27
b. Petals deep pink, red or bicoloured
30
27a. Filaments hairy, at least below
37. caulescens
b. Filaments hairless 28
28a. Stems usually shorter than leaves
42. alba
b. Stems usually exceeding leaves 29
29a. Stems 5–10 cm; leaflets often 5;
flowers solitary or in groups of up
to 4 **38. clusiana**
b. Stems 10–30 cm; leaflets often 7;
flowers in groups of more than 4
40. alchemilloides
30a. Gland-tipped hairs present on the
stem **21. thurberi**
b. Gland-tipped hairs absent 31
31a. Stems erect **23. nepalensis**
b. Stems prostrate or ascending
33. × tonguei
32a. Leaves stellate-hairy beneath
31. cinerea
b. Leaves not stellate-hairy beneath
33
33a. Leaves densely white hairy beneath
34
b. Leaves not densely white hairy
beneath 35
34a. Leaflets obovate to circular,
coarsely toothed **17. villosa**
b. Leaflets elliptic to obovate, finely
toothed **22. atrosanguinea**
35a. Petals more than 9 mm 36
b. Petals to 9 mm 41
36a. Flowers solitary (rarely paired) 37
b. Flowers 3–7 per inflorescence
(rarely paired) 39
37a. Petals pink, rarely white; sepals
purplish **39. nitida**
b. Petals yellow; sepals not purplish
38
38a. Stock with persistent, hairy leaf-
bases **36. eriocarpa**

b. Stock lacking persistent, hairy
leaf-bases **35. cuneata**
39a. Epicalyx-segments obtuse
34. megalantha
b. Epicalyx-segments acute 40
40a. Stem with patent or somewhat
adpressed hairs; petals yellow
18. grandiflora
b. Stem with adpressed hairs; petals
deep orange
28. aurea subsp. **chrysocraspeda**
41a. Petals white, cream, pink or orange
42
b. Petals yellow 46
42a. Stems 30–60 cm
22. atrosanguinea
b. Stems to 30 cm 43
43a. Leaves usually shiny green and
hairless above **45. tridentata**
b. Leaves not shiny green and hairy
above 44
44a. Petals 6–9 mm 45
b. Petals *c.* 5 mm **44. sterilis**
45a. Clump forming herbaceous
perennial to 10 cm (rarely more)
41. speciosa
b. Rhizomatous or stoloniferous
herbaceous perennial to 25 cm
43. montana
46a. Stems 30–60 cm
22. atrosanguinea
b. Stems to 30 cm 47
47a. Petals not notched at apex
41. speciosa
b. Petals notched at apex 48
48a. Plant lacking glands; leaves hairless
above **25. brauniana**
b. Plant more or less glandular hairy;
leaves at least sparsely hairy above
49
49a. Petals usually 1.5 times sepals;
leaves densely hairy above; teeth
on leaflets of basal leaves elliptic or
oblanceolate, very obtuse
26. fridgida
b. Petals at least 1.5 times sepals;
leaves usually only sparsely hairy
above; teeth on leaflets of basal
leaves triangular-oblong, acute, or
almost obtuse **27. hyparctica**

1–4. Potentilla fruticosa Group.

Deciduous shrubs to *c.* 1.5 m; much-
branched; twigs thin. Leaves pinnate with
5–9 linear-lanceolate to broadly ovate-
elliptic, entire leaflets (more rarely with 3
leaflets), more or less silky-hairy (rarely
hairless). Stipules persistent, entire, acute
or obtuse. Calyx with 5, more or less
ovate, acute lobes. Epicalyx lobes 5, of
very varying size and shape, from linear

to ovate and sometimes 2-fid or 3-fid.
Corolla of 5 (rarely more in semi-double
variants) large, obovate to circular, free
petals, commonly yellow or white. Stamens
numerous. Carpels few, free, each with a
club-shaped style attached near the base.
Fruit a head of small achenes, each with
a ring of basal hairs. Plants sometimes
dioecious; male flowers (often larger) with
numerous well-developed stamens, carpels
reduced to a clump of hairs; female flowers
with sterile staminodes (often well-
developed), carpels well developed. *N
temperate to arctic areas of Asia & N
America, extending to the Caucasus,
Himalaya & China; rare and local in N
Europe and mountains of S Europe.* H1. Late
spring–summer.

The classification of the group is
complex, and there is yet little agreement
on an appropriate treatment; the group
includes the first four species listed below.
As a whole they are tolerant of lime,
preferring a well-drained soil, and are
intolerant of shade. They are easily
propagated by cuttings, but open-
pollinated seed germinates readily and,
since cross-pollination is frequent in the
bisexual taxa and obligatory in the
dioecious ones, new variants continually
arise, especially those with unusual petal-
colour; good examples are the red-flowered
cultivars 'Red Ace' and 'Royal Flush', the
pink-flowered semi-double 'Princess' and
the orange-flowered 'Tangerine'.

Literature: Bean, W.J., *Trees and shrubs
hardy in the British Isles,* edn 8, **3**:
328–343 (1967); Klackenberg, J., The
holarctic complex *Potentilla fruticosa*
(Rosaceae), *Nordic Journal of Botany* **3**:
181–191 (1983); Brearley, C., The shrubby
Potentillas, *Plantsman* **9**: 90–109 (1987).

1. P. fruticosa Linnaeus.

Illustration:
Raven & Walters, Mountain flowers, pl. 10
(1956); Rickett, Wild flowers of the United
States **1**: 129 (1967); Garrard & Streeter,
Wild flowers of the British Isles, 87 (1983).

Habit variable, usually more or less erect
and loosely branched. Leaflets 5 or 7, 1–2
cm, oblong-lanceolate to elliptic, more or
less hairy with long, white hairs; stipules
lanceolate, acute, pale. Flowers usually
in cyme-like groups, more rarely solitary.
Petals 6–16 mm across, yellow, obovate to
circular. *Widespread.*

May be divided into 2 subspecies,
differing mainly in chromosome number
and sexuality.

Subsp. **fruticosa**, dioecious, a rare plant
of open calcareous rocks and riverside

habitats. *N Europe.*

Subsp. **floribunda** (Pursh) Elkington (*P. floribunda* Pursh) is bisexual. *N America and very locally in the Pyrenees, W Alps and Bulgaria. Its distribution in Asia is uncertain.*

The European plant was cultivated early, both in Britain and Scandinavia, but is rarely seen in modern gardens, having largely been replaced by cultivars. Two modern cultivars, 'Goldfinger' and 'Jackman's Variety', are, however, thought to be selections from *P. fruticosa*, differing most obviously in their larger flowers.

2. P. parvifolia Lehmann (*P. fruticosa* 'Farreri'). Illustration: Krüssmann, Handbuch der Laubgeholze **2**: 462 (1977) – as P. fruticosa 'Parvifolia'.

Like *P. fruticosa*, but leaves usually not exceeding 2 cm, with 5–9, usually 7, linear or linear-lanceolate leaflets, often densely hairy, grey-green above and white beneath, and small, bisexual, yellow flowers on long, slender stalks. *Former USSR (Siberia, C Asia), Mongolia, ?Himalaya.*

Represented in gardens by several cultivars, e.g. 'Buttercup', 'Gold Drop' and 'Klondyke', which have compact habit and long flowering period. The commonly grown, pale yellow-flowered 'Katherine Dykes' is probably a hybrid of *P. parvifolia*.

3. P. davurica Nestler (*P. glabra* Loddiges; *P. mandshurica* Maximowicz; *P. veitchii* Wilson). Illustration: Krüssmann, Handbuch der Laubgeholze **2**: 462 (1977) – as P. fruticosa var. davurica; Brickell, RHS gardeners' encyclopedia of plants and flowers, 126 (1989).

Like *P. fruticosa* but more compact, hairless; flowers white; epicalyx segments broad and obtuse. *Former USSR (E Siberia to Pacific), Japan, China (N & W Xizang).*

Var. **mandshurica** (Maximowicz) Wolf) and 'Manchu' differ in having a thick, adpressed silky hair covering. Represented in gardens mainly by several white-flowered cultivars, especially 'Abbotswood', with dark green foliage and profuse flowering over a long period, and 'Farrer's White', a very vigorous shrub with larger flowers than 'Abbotswood'. Some yellow-flowered cultivars (e.g. 'Elizabeth') may be hybrids involving *P. davurica*.

4. P. arbuscula D. Don (*P. rigida* Lehmann). Illustration: Plantsman **9**: 93 (1987).

A compact shrub to 60 cm with stouter twigs than *P. fruticosa*, large, ovate leaflets to 2 cm across and broadly ovate, brownish stipules, conspicuous on the twigs. Flowers large (usually 2–3 cm across); petals usually yellow; epicalyx-segments often 2-fid or 3-fid. *Himalaya, China (W & N Xizang).*

Typically the leaves have 5 leaflets, but plants with leaves with 3 leaflets, to which the name *P. rigida* is usually applied, are also in cultivation.

This is represented in gardens by several variants, e.g. 'Beesii' (*P. fruticosa* 'Nana Argentea'), a dense, leafy shrub with silvery-hairy leaflets and large, golden-yellow, male flowers. There is much confusion as to the specific limits of *P. arbuscula*, partly because flower-colour is not preserved in herbarium material and not always stated by the original collector.

5. P. salesoviana Stephan. Illustration: Botanical Magazine, 7258 (1892); Krüssmann, Manual of cultivated broad-leaved trees & shrubs **2**: 439 (1986).

Upright shrub 30–100 cm, woody below, with annual, erect, flowering-shoots. Leaves pinnate; leaflets 7–13, sharply toothed, the terminal 2–4 cm, oblong, dark green and hairless above, white-hairy beneath. Flowers 3–7 together, 3–3.5 cm across. Petals obovate, apex not notched, white, sometimes tinged pink. Receptacle almost spherical, densely hairy. Achenes densely hairy. *Former USSR (Siberia, C Asia), Himalaya, China (Xizang). H1. Summer.*

6. P. palustris (Linnaeus) Scopoli (*Comarum palustre* Linnaeus). Illustration: Ross-Craig, Drawings of British plants **8**: t. 39 (1955); Polunin, Flowers of Europe, t. 46 (1969); Garrard & Streeter, Wild flowers of the British Isles, 82 (1983).

Herbaceous perennial with long, creeping, woody rhizomes. Stems decumbent, 15–45 cm, sparsely hairy. Leaves pinnate; leaflets 5–7, 2.6–6 × 1–2 cm, oblong, coarsely toothed, slightly greyish green and hairless above, greyish green and almost hairless beneath. Flowers several, in loose cymes, to 3 cm across. Petals 5–8 mm, ovate-lanceolate, acuminate, deep purple, persistent, half as long as the purplish sepals which enlarge in fruit. Receptacle spongy. *Europe, Asia, N America. H1. Summer.*

Suitable for peaty places in a water garden.

7. P. anserina Linnaeus. Illustration: Ross-Craig, Drawings of British plants **8**: t. 35 (1955); Pignatti, Flora d'Italia **1**: 575 (1982); Garrard & Streeter, Wild flowers of the British Isles, 36 (1983).

Patch-forming herbaceous perennial. Stock short, ending in a rosette of leaves and producing prostrate, stoloniferous stems to 80 cm. Leaves 5–40 × 0.5–1.5 cm, pinnate, with 3–12 pairs of large leaflets alternating with smaller ones; large leaflets 1–6 cm, ovate or oblong, toothed, silvery silky-hairy above and beneath or only beneath, rarely green and sparsely hairy or hairless on both surfaces. Flowers solitary, 1–2 cm across. Petals 7–10 cm, obovate, apex not notched, yellow, much longer than the sepals. *Most of Europe, N & C Asia, N & S America, Australasia. H1. Summer.*

Attractive in its silver-leaved variants, but generally too invasive for all but the wildest parts of the garden.

8. P. lineatus Treviranus (*P. fulgens* Hooker) Illustration: Botanical Magazine, 2700 (1826) – as P. splendens; Iconographia cormophytorum Sinicorum **2**: 290 (1972).

Similar, but with a leafy stem and cymose inflorescence, and producing stolons. *Himalaya, China. H1. Summer.*

9. P. peduncularis D. Don. Illustration: Polunin & Stainton, Flowers of the Himalaya, t. 37 (1984).

Also similar, but rhizomatous and larger in all its parts and lacking the alternating, smaller leaflets; habit much more stiffly erect, the flowers 2–5, in cymes. *W Nepal to SW China. H1. Summer.*

10. P. bifurca Linnaeus. Illustration: Flora SSSR **10**: t. 7 (1941).

Rhizomatous perennial. Stems 10–30 cm, woody at the base, silky-hairy to almost hairless. Leaves pinnate; leaflets 5–15, 8–20 × 3–8 mm, oblong-ovate, entire or with 2 or 3 apical teeth. Flowers in a loose cyme. Petals 4–8 mm, yellow, longer than the sepals. Achenes hairy at the base when young. *SE Europe, Asia. H1. Summer.*

11. P. rupestris Linnaeus. Illustration: Ross-Craig, Drawings of British plants **8**: t. 36 (1955); Grey-Wilson & Blamey, Alpine flowers of Britain and Europe, 99 (1979); Pignatti, Flora d'Italia **1**: 575 (1982).

Clump-forming herbaceous perennial. Stems 20–60 cm, erect, with gland-tipped hairs above, often reddish at the base. Leaves 7–15 cm, pinnate; leaflets in 2–4 distant pairs, decreasing in size below, 1–6 × 0.5–3.5 cm, ovate to almost circular, coarsely toothed, green above and beneath, hairy on both surfaces. Flowers in a loose cyme, 1.5–2.5 cm wide. Petals obovate, entire, 8–14 mm, white, longer than the sepals. *W & C Europe, N Africa, W & C Asia, N America.* H1. Spring–summer.

Dwarf variants from Corsica and Sardinia, with stems only 4–10 cm, fewer leaflets and 1 or 2 smaller flowers have been distinguished as var. **pygmaea** Duby ('Nana').

12. P. multifida Linnaeus. Illustration: Huxley, Mountain flowers, 54 (1967); Grey-Wilson & Blamey, Alpine flowers of Britain and Europe, 99 (1979); Pignatti, Flora d'Italia **1**: 575 (1982).

Clump-forming herbaceous perennial. Stems 5–40 cm, erect or ascending, sparsely to densely hairy. Leaves pinnate, leaflets 5–9, often crowded so appearing almost palmate, 0.5–4 × 0.3–2 cm, deeply pinnatisect with up to 5 linear ultimate segments, green above, silvery silky-hairy beneath. Flowers few to many, 1–1.5 cm across. Petals obovate, 5–7 mm, apex notched, yellow, equalling or slightly exceeding sepals. *Europe to Asia.* H1. Summer.

13. P. argentea Linnaeus. Illustration: Ross-Craig, Drawings of British plants **8**: t. 37 (1955); Keble Martin, The concise British Flora in colour, t. 27 (1965); Garrard & Streeter, Wild flowers of the British Isles, pl. 82 (1983).

Mat-forming herbaceous perennial. Stems 15–50 cm, prostrate or ascending, usually downy. Leaves palmate with narrow leaflets; leaflets 5, 1–3 cm × 0.5–1.5 cm, wedge-shaped to obovate, with 2–7 obtuse teeth or lobes, green and hairless above, densely white-hairy beneath. Flowers numerous, 1–1.5 cm across. Petals 4–5 mm, obovate, yellow, only slightly longer than the sepals. Styles conical, tapering to the apex. *Throughout Europe, W & C Asia, N America.* H1. Summer.

14. P. calabra Tenore.

Similar to *P. argentea*, but always with prostrate stems. Leaves smaller, greyish green to white-hairy above and the inflorescence more densely white-hairy. *C & S Italy, Sicily, western part of Balkan Peninsula.* H1. Summer.

15. P. collina Wibel (*P. alpicola* Fauconnet).

Similar to *P. argentea*, but with club-shaped styles which are often distorted. *C Europe.* H1. Summer.

16. P. recta Linnaeus. Illustration: Butcher, A new illustrated British Flora **1**: 667 (1961); Pignatti, Flora d'Italia **1**: 578 (1982); Brickell, RHS gardeners' encyclopedia of plants and flowers, 246 (1989).

Clump-forming herbaceous perennial. Stems 10–70 cm, erect, with both short, gland-tipped hairs and longer, white, glandless hairs. Leaves palmate; leaflets 5–7, 5–10 × 0.5–3.5 cm, oblong to obovate, coarsely toothed or shallowly lobed, green or greyish above, hairy above and beneath. Flowers numerous in loose cymes, 2–2.5 cm across. Petals 6–12 mm, obovate, apex deeply notched, yellow, longer than sepals. *C, E & S Europe, N Africa, W & C Asia, naturalised elsewhere.* H1. Summer

A variable species. Commonly grown variants include 'Sulphurea' ('Pallida') with pale yellow flowers and 'Warrenii' ('Macrantha') with larger, bright yellow flowers.

17. P. villosa Pursh. Illustration: Abrams & Ferris, Illustrated Flora of the Pacific states **2**: 436 (1944); Clark, Wild flowers of British Columbia, 259 (1973); Clark, Wild flowers of the Pacific northwest, 262 (1976).

Cushion-forming herbaceous perennial. Stems ascending, 10–30 cm, densely white or yellowish hairy. Leaves with 3 leaflets; leaflets 1.5–4 cm, obovate to almost circular, thick, coarsely toothed around the margins, green and slightly hairy above, densely white-hairy beneath, with impressed veins. Flowers few, 2–3 cm across. Petals 6–12 mm, broadly obovate, deeply notched at apex, golden yellow, much longer than sepals. *W North America, NE Asia.* H1. Early summer.

18. P. grandiflora Linnaeus. Illustration: Huxley, Mountain flowers, 55 (1967); Grey-Wilson & Blamey, Alpine flowers of Britain and Europe, 101 (1979); Pignatti, Flora d'Italia **1**: 579 (1982).

Clump-forming herbaceous perennial. Stems 10–40 cm, ascending, with dense, spreading hairs. Leaves with 3 leaflets; leaflets 1.5–4 × 1–3 cm, obovate to almost circular, coarsely and bluntly toothed, green above and beneath, sparsely hairy above, more densely so beneath. Flowers 2–5, 1.5–3.5 cm across. Epicalyx-segments acute. Petals broadly obovate, 1–1.5 × 0.9 cm, apex notched, yellow, much longer than sepals. *C & E Pyrenees, Alps.* H1. Late summer.

19. P. nevadensis Boissier. Illustration: Boissier, Voyage botanique dans le midi d'Espagne **1**: t. 59 (1838–45).

Clump-forming herbaceous perennial. Stems 15–30 cm with long, spreading hairs. Leaves palmate; leaflets 5, 0.4–2 × 0.3–1.5 cm, obovate or oblanceolate, scalloped-toothed, green above, silvery-hairy beneath. Flowers 1–4 together, 1–1.5 cm across. Petals 4–7 mm, obovate, apex notched, yellow, longer than the sepals. *S Spain (Sierra Nevada).* H1. Spring–summer.

Var. **condensata** Boissier is smaller, more densely tufted and has the leaves silvery-silky on both surfaces.

20. P. gracilis J.D. Hooker. Illustration: Jepson, Manual of the plants of California, 489 (1925); Abrams & Ferris, Illustrated Flora of the Pacific states **2**: 433 (1944); Clark, Wild flowers of the Pacific Northwest, 254 (1976).

Clump-forming herbaceous perennial. Stems 40–70 cm, usually erect, with dense, spreading hairs. Leaves palmate: leaflets 5–7 (or more), 2–6 cm, obovate or oblanceolate, conspicuously toothed along their whole length, green but with some silky hairs above, densely white-hairy beneath. Flowers numerous in loose cymes, 1–2 cm across. Petals obovate, usually more than 1 cm, apex notched, yellow, longer than sepals. *W North America.* H1. Summer.

Very variable. The description above is of var. **gracilis**. Plants with the leaves more sparsely hairy and often glandular beneath are distinguished as subsp. **nuttallii** (Lehmann) Keck (*P. nuttallii* Lehmann).

21. P. thurberi Lehmann. Illustration: Rickett, Wild flowers of the United States **4**: t. 53 (1970).

Clump-forming herbaceous perennial. Stems 30–60 cm, erect, with both gland-tipped and glandless hairs. Leaves palmate; leaflets 5–7, 2.5–5 cm, broadly oblanceolate, coarsely toothed, green above, hairy beneath. Flowers in loose cymes, 1.5–2 cm across. Petals dark brownish red, slightly longer than sepals. *SW USA.* H1. Summer.

Similar to the Himalayan *P. nepalensis* (see below), but differing in having gland-tipped hairs.

22. P. atrosanguinea D. Don. Illustration: Coventry, Wild flowers of Kashmir **1**: t. 18 (1923); Blatter, Beautiful flowers of Kashmir **1**: t. 21 (1928); Stainton, Flowers of the Himalaya, a supplement, t. 137 (1988).

Clump-forming herbaceous perennial. Stems 30–60 cm, hairy. Leaves usually of 3 leaflets; leaflets 2–5 × 1–3 cm, elliptic, ovate or obovate, sharply toothed, dark green and hairless or grey silky-hairy above, densely white-hairy beneath. Flowers in loose cymes, 2–3 cm across. Petals 9–11 cm, obovate, apex notched, yellow, orange or red, longer than sepals. *W & C Himalaya*. H1. Summer–autumn.

Var. **atrosanguinea** has red flowers; yellow-flowered plants have been distinguished as var. **argyrophylla** (Lehmann) Grierson & Long (*P. argyrophylla* Lehmann). Var. **cataclina** (Lehmann) Wolf and *P. argyrophylla* var. *leucochroa* J.D. Hooker represent smaller variants of these two varieties.

Garden hybrids and selections of these varieties (and hybrids with *P. nepalensis*) are widely grown. Many are larger plants with more than 3 leaflets, and often have larger, double flowers; several are bicoloured, e.g. 'Yellow Queen', single to double yellow; 'Gibson's Scarlet', single, bright red; and 'Monsieur Rouillard', double, mahogany red with yellow blotches.

23. P. nepalensis J.D. Hooker. Illustration: Botanical Magazine, 9182 (1929); Perry, Collins' guide to border plants, edn 2, t. 35 (1966); Brickell, RHS gardeners' encyclopedia of plants and flowers, 237 (1989).

Clump-forming herbaceous perennial. Stems 30–75 cm, hairy, often tinged red. Leaves palmate; leaflets 5, 2–8 × 1–3 cm, narrowly elliptic or oblong-ovate, coarsely toothed, sparsely hairy, green on both surfaces. Flowers in loose cymes, 1.3–3 cm across. Petals 7–14 mm, obovate, apex notched, typically pinkish red with darker bases, longer than the sepals. *Pakistan to C Nepal*. H1. Summer–autumn.

Variable in flower-colour: 'Miss Wilmott' is a crushed strawberry pink with a darker eye; 'Roxana' is yellowish pink with a darker eye and almost yellow rim. Both come largely true from seed.

P. × hopwoodiana Sweet (*P. nepalensis × P. recta*) is similar but the basal leaves often have 6 leaflets and the petals are pale yellow in the lower half (with a deep rose basal spot) and bright rose in the upper part, though with a paler margin.

24. P. pyrenaica de Candolle. Illustration: Coste, Flore de la France **2**: 22 (1903).

Clump-forming herbaceous perennial. Stems to 50 cm, ascending from a curving base, sparsely to densely hairy with somewhat adpressed hairs. Leaves with long stalks, palmate with 5 oblong, adpressed-hairy leaflets, toothed in the upper two-thirds, with terminal tooth equalling the others; stipules short, blunt. Flowers few, in condensed cymes. Petals large, yellow, obovate, apex notched, usually twice as long as the sepals. Epicalyx-segments shorter or narrower than the sepals. Style thread-like at apex. *Pyrenees, mts of N & C Spain*. H1. Summer.

25. P. brauniana Hoppe (*P. minima* Haller). Illustration: Coste, Flore de la France **2**: 25 (1903).

Dwarf plant, woody at base, with sparsely hairy stems not exceeding 5 cm, and small leaves with 3 obovate leaflets, hairless on the upper surface, toothed in the upper half. Flowers small, solitary. Petals yellow, obovate, apex notched, hardly exceeding sepals. Epicalyx-segments obtuse, more or less equalling sepals. *S Europe (Pyrenees to E Alps)*. H1. Late spring–summer.

26. P. frigida Villars. Illustration: Coste, Flore de la France **2**: 24 (1903); Fenaroli, Flora della Alpi, 175 (1955).

Like *P. brauniana*, but stems to 10 cm and densely hairy throughout, with some glandular hairs mixed with long, spreading, glandless hairs. *S Europe, (Pyrenees to E Alps)*. H1. Late spring–summer.

27. P. hyparctica Malte. Illustration: Polunin, Arctic flora, 271 (1959). Like *P. brauniana* but often strongly glandular-hairy, with larger flowers with petals about 1.5 times as long as sepals. *Circumpolar, Arctic areas*. H1. Summer.

These last 3 species are best cultivated in a sink garden. *P. frigida* is said by some alpine gardeners to be the most attractive.

28. P. aurea Linnaeus. Illustration: Coste, Flore de la France **2**: 25 (1903); Fenaroli, Flora della Alpi, opposite 96 (1955); Grey-Wilson & Blamey, Illustrated Flora of Britain and northern Europe, 187 (1989).

Mat-forming perennial, woody at the base, with silkily hairy stems to 25 cm. Leaves palmate with 3 or 5 leaflets, silkily adpressed hairy on the edge and the veins beneath, toothed in the upper half with the terminal tooth obviously shorter than the others. Stipules narrow, pointed. Cyme with few relatively large flowers. Petals orange-yellow, darker at base, broadly obovate, apex shallowly notched, longer than the sepals. Epicalyx-segments narrow, pointed, more or less equalling sepals. *S & C Europe*. H1. Late spring–summer.

'Rathboneana' has semi-double flowers.

Subsp. **chrysocraspeda** (Lehmann) Nyman, with 3 leaflets only, occurs in the eastern part of the range. It is sometimes grown as the variant 'Aurantiaca' with deep orange flowers.

29. P. crantzii (Crantz) Fritsch (*P. alpestris* Haller; *P. salisburgensis* Haenke). Illustration: Garrard & Streeter, Wild flowers of the British Isles, 83 (1983); Grey-Wilson & Blamey, Illustrated Flora of Britain and northern Europe, 187 (1989).

Clump-forming herbaceous perennial, woody at base, with hairy, ascending flowering-stems to 30 cm. Leaves palmate with 5 (rarely 3) obovate leaflets with few broad, blunt teeth, the terminal almost equalling the others, with spreading hairs beneath and on the margins; stipules rather broad, more or less acute. Flowers 1–10, in a loose cyme. Petals deep yellow, often with an orange basal spot, obovate, apex rather deeply notched, longer than sepals. *N, C & S Europe*. H1. Late spring–summer.

A variable plant in the wild. Variants with well-developed basal spots on the petals make very attractive rock-garden plants.

30. P. neumanniana Reichenbach (*P. verna* Linnaeus; *P. tabernaemontani* Ascherson). Illustration: Garrard & Streeter, Wild flowers of the British Isles, 83 (1983); Grey-Wilson & Blamey, Illustrated flora of Britain and northern Europe, 189 (1989).

Like *P. crantzii*, but mat-forming with more or less prostrate, somewhat woody stems rooting at the nodes, leaves often with 7 leaflets, and narrow, pointed stipules. The flowers are smaller and the petals lack the orange basal spot. *N, W & C Europe*. H1. Late spring–summer.

A very variable plant of limited value in gardens. 'Nana' is a very dwarf variant with flowering-stems less than 5 cm.

31. P. cinerea Villars (*P. arenaria* Borckhausen; *P. tommasiniana* Schulz). Illustration: Coste, Flore de la France **2**: 26 (1903); Grey-Wilson & Blamey, Illustrated flora of Britain and northern Europe, 189 (1989).

Plant with the habit of *P. verna*, but easily recognised by its grey colour, due to a dense covering of stellate hairs mixed with long, simple hairs and sometimes also glandular hairs. Leaves with 3 or 5 obovate leaflets, toothed in the upper half, with the terminal tooth shorter than the others. The flowers resemble those of *P. verna*. *C, S & E Europe*. H1. Late spring–summer.

A very variable plant, prettier than *P. verna* in rock-gardens and occasionally grown.

32. P. reptans Linnaeus. Illustration: Ross-Craig, Drawings of British plants **8**: t. 34 (1955); Pignatti, Flora d'Italia **2**: 582 (1982); Garrard & Streeter, Wild flowers of the British Isles, 83 (1983).

Patch-forming herbaceous perennial. Stock short, ending in a rosette of leaves and producing prostrate, stoloniferous stems 20–100 cm, rooting at the nodes. Leaves palmate; leaflets usually 5, 0.5–7 × 0.3–2.5 cm, obovate or oblong-obovate, toothed, green above and beneath, hairless or sparsely hairy. Flowers solitary, 1.7–2.5 cm across, on long stalks. Petals 7–12 mm, obovate, apex notched, much longer than sepals. *Widespread in Europe, N Africa & Asia, naturalised elsewhere*. H1. Summer.

Generally regarded as a weed, but a double-flowered variant, forma **pleniflora** Bergmans is sometimes cultivated.

33. P. × tonguei Baxter (*P. tormentilla-formosa* Maund, invalid). Illustration: Maund, Botanic Garden **7**: t. 585 (1837); Gentes Herbarum **4**: 313 (1941).

Clump-forming herbaceous perennial. Stems prostrate or ascending, to 35 cm, hairy, not rooting at the nodes. Leaflets 3–5, 0.5–2.5 × 0.3–1.5 cm, narrowly obovate to obovate, coarsely toothed, often tinged bronze, hairy above and beneath. Flowers solitary or in few-flowered cymes, *c*. 1.5 cm across. Petals broadly obovate, apex notched, yellow with a red base, longer than sepals. *Garden origin*. H1. Summer–autumn.

Originally described as being "*Tormentilla reptans × Potentilla formosa*" (i.e. *P. reptans × P. nepalensis*) but later authors have given *P. anglica* or *P. aurea* as the first parent.

34. P. megalantha Takeda (*P. fragariformis* Schlechtendahl). Illustration: Kew Bulletin for 1911, 252; Brickell, RHS gardeners' encyclopedia of plants and flowers, 247 (1989).

Clump-forming, softly hairy herbaceous perennial. Stems to 30 cm, ascending. Leaves with 3 leaflets; leaflets 3–8 × 3–8 cm, thick, broadly elliptic to obovate, with few obtuse teeth, densely hairy but green above and beneath. Flowers 3–7, 3–4 cm

across. Epicalyx-segments obtuse. Petals obovate, apex slightly notched, golden yellow, longer than sepals. *NE Asia*. H1. Summer.

A variant with double flowers, 'Perfecta Plena' is occasionally offered.

35. P. cuneata Lehmann (*P. ambigua* Cambessedes). Illustration: Polunin & Stainton, Flowers of the Himalaya, t. 39 (1984).

Mat-forming herbaceous perennial, the woody stock without persistent leaf-bases. Stems 2–10 cm, erect. Leaves all basal, of 3 leaflets; leaflets 0.6–1.3 × 0.4–1 cm, obovate, shallowly 3-toothed or lobed at the apex, green and hairy or hairless above, adpressed-hairy beneath. Flowers solitary, 1.3–2.5 cm across. Petals broadly obovate, 9–10 × 7–8 mm, yellow, longer than the sepals. Achenes covered with long silky hairs. *India (Kashmir) to SW China*. H1. Summer–autumn.

36. P. eriocarpa Lehmann. Illustration: Coventry, Wild flowers of Kashmir **3**: t. 26 (1930); Polunin & Stainton, Flowers of the Himalaya, t. 38 (1984).

Mat- or cushion-forming herbaceous perennial, the woody stock with persistent, hairy leaf-bases. Stems 2–18 cm. Leaves of 3 leaflets; leaflets 1.2–3 × 0.5–1.5 cm, wedge-shaped to obovate, long-stalked, deeply toothed or 3–5-lobed in the upper half, bright green above, adpressed-hairy beneath. Flowers usually solitary, 2–3 cm across. Petals broadly obovate, apex deeply notched, yellow, longer than sepals. Achenes hairy. *Pakistan to China*. H1. Summer.

37. P. caulescens Linnaeus. Illustration: Huxley, Mountain flowers, 149 (1967); Grey-Wilson & Blamey, Alpine flowers of Britain and Europe, 101 (1979); Pignatti, Flora d'Italia **1**: 583 (1982).

Clump-forming herbaceous perennial. Stems 5–30 cm, ascending, with adpressed or slightly spreading hairs. Leaves palmate; leaflets 5–7, 1–3 × 0.5–0.8 cm, oblong or oblong-obovate, with a few teeth at the apex, green above, silvery silky-hairy beneath. Flowers numerous in loose cymes, to 2 cm across. Petals obovate, 6–10 × 2.5–5 mm, apex sometimes notched, white (rarely pinkish), slightly longer than the sepals. Filaments hairy at base. *S Europe*. H1. Summer.

38. P. clusiana Jacquin. Illustration: Hegi, Illustrierte Flora von Mitteleuropa **4**: 820 (1923); Grey-Wilson & Blamey, Alpine flowers of Britain and Europe, 101 (1979);

Pignatti, Flora d'Italia **1**: 583 (1982).

Clump-forming herbaceous perennial. Stems 5–10 cm, with hairs which are more or less adpressed. Leaves palmate; leaflets usually 5, 7–13 × 3–5 mm, obovate, 3–5-toothed at the apex, silky-hairy beneath. Flowers solitary or in few-flowered cymes. Petals broadly obovate, 9–10 × 6–8 mm, apex notched, white, much longer than sepals. Filaments hairless. *E Alps, W former Yugoslavia, Albania*. H1. Summer.

39. P. nitida Linnaeus. Illustration: Polunin, Flowers of Europe, t. 46 (1969); Grey-Wilson & Blamey, Alpine flowers of Britain and Europe, 103 (1979); Pignatti, Flora d'Italia **1**: 583 (1982).

Densely tufted perennial, with a woody stock. Stems 2–10 cm, densely silvery silky-hairy. Leaflets usually 3 (rarely 4 or 5), 5–10 mm, obovate or oblanceolate, entire or with a few teeth at the apex, silvery silky-hairy above and beneath. Flowers 1–2, *c*. 2.5 cm across. Petals 1–1.2 × 0.7–1 cm, broadly obovate, apex notched, pink, rarely white (forma **albiflora** Sauter), longer than the purplish sepals. *SW & SE Alps, N Apennines*. H1. Summer.

Variable in petal colour. 'Rubra' and 'Lissadell' are selections with darker pink flowers.

40. P. alchemilloides Lapeyrouse. Illustration: Coste, Flore de la France **2**: 18 (1903); Huxley, Mountain flowers, 149 (1967).

Clump-forming herbaceous perennial. Stems 10–30 cm, silky-hairy, exceeding the leaves. Leaves palmate; leaflets 5–7, 1–2.5 cm, oblong-elliptic, usually with 3 small teeth at the apex, green and hairless above, silvery silky-hairy beneath. Petals 8–10 mm, obovate, apex notched, white, longer than the sepals. Achenes hairy. *Pyrenees*. H1. Summer.

41. P. speciosa Willdenow. Illustration: Wehrhahn, Die Gartenstauden **2**: 625 (1930); Polunin, Flowers of Greece and the Balkans, 114 (1980).

Clump-forming herbaceous perennial. Stems 2–10 cm (rarely more), densely grey- or white-hairy. Leaves of 3 leaflets; leaflets 1.5–3 × 1–2 cm, broadly obovate to broadly ovate, thick, scalloped or scalloped-toothed at least in the upper two-thirds, green and almost hairless or densely hairy above, white-hairy beneath. Flowers several, 1–2 cm across. Petals 6–10 mm, obovate, white, rarely pale yellow, slightly longer than the sepals. *W*

& S parts of Balkan Peninsula, Crete, Syria, Iraq. H2. Summer.

42. P. alba Linnaeus. Illustration: Polunin, Flowers of Europe, 161 (1969); Pignatti, Flora d'Italia **1**: 584 (1982); Brickell, RHS gardeners' encyclopedia of plants and flowers, 313 (1989).

Clump-forming herbaceous perennial with adpressed or slightly spreading hairs. Stems 5–20 cm, decumbent or ascending, usually shorter than the leaves. Leaves palmate; leaflets 5, 2–4 cm, oblong to obovate-lanceolate, with a few teeth near the tip, hairless and green above, silvery silky-hairy beneath. Flowers 1–5, 1.5–2.5 cm wide. Petals 7–10 mm, obovate, white, longer than sepals. Achenes smooth, with a few long hairs at the base. *C & E Europe.* H1. Spring–summer.

43. P. montana Brotero (*P. splendens* de Candolle). Illustration: Coste, Flore de la France **2**: 19 (1903); Bonnier, Flore complète **3**: t. 172 (1914).

Rhizomatous or stoloniferous herbaceous perennial. Stems 5–25 cm, downy. Leaflets usually 3, rarely 4 or 5, 1–3 cm, oblong to obovate, bluntly toothed in the upper half or only at the tip, hairy and green above, grey silky-hairy beneath. Flowers 1–4, 1.5–2.5 cm across. Petals 6–9 mm, obovate, tip notched, white, much longer than sepals. Achenes smooth. *N Iberian Peninsula, W & C France.* H1. Spring–summer.

44. P. sterilis (Linnaeus) Garcke (*P. fragariastrum* Persoon; *Fragaria sterilis* Linnaeus). Illustration: Ross-Craig, Drawings of British plants **8**: t. 29 (1955); Pignatti, Flora d'Italia **1**: 585 (1982); Garrard & Streeter, Wild flowers of the British Isles, 82 (1983).

Mat-forming herbaceous perennial. Stems decumbent, 5–15 cm, with spreading hairs; stock usually also producing stolons. Leaves of 3 leaflets; leaflets 0.8–2.5 × 0.8–2 cm, broadly obovate, scalloped-toothed, sparsely hairy and slightly bluish green above, with more dense, spreading, grey hairs beneath. Flowers 1–5, 1–1.5 cm across. Petals *c.* 5 mm, obovate, apex notched, white or very pale pink, equalling or slightly longer than the sepals, widely separated. *W, C & S Europe.* H1. Spring.

45. P. tridentata Solander (*Sibbaldiopsis tridentata* (Solander) Rydberg). Illustration: Britton & Brown, Illustrated flora of the northern United States and Canada **2**: 262 (1913); Fernald, Gray's manual of botany, 807 (1950).

Mat-forming perennial, with extensively creeping rhizomes and prostrate, woody bases. Flowering-stems ascending, 2.5–30 cm, adpressed hairy. Leaves of 3 leaflets; leaflets oblong to wedge-shaped, with usually 3 teeth at the apex, usually shiny green and hairless above, adpressed hairy beneath, leathery, evergreen. Flowers 1–6, 6–15 mm across. Petals ovate or obovate, apex entire, white or pale pink (forma **aurora** Graustein), longer than the sepals. *NE North America, Greenland.* H1. Summer.

44. SIBBALDIA Linnaeus
H.S. Maxwell

Tufted perennial herbs, often woody at base. Leaves deciduous, alternate, with 3 leaflets, hairy; stipules papery, attached at the base of leaf-stalks. Flowers hermaphrodite, in cymes; sepals 5, epicalyx present; petals 5, sometimes absent; stamens 5, rarely 4 or 10; carpels 5–12. Fruit a cluster of 5–10 achenes, styles deciduous in fruit.

A genus of about 8 species from the colder parts of the northern hemisphere, only one of which is widely grown. **S. parviflora** Willdenow (*S. cuneata* Hornemann) is occasionally listed in specialist catalogues but only appears to differ from the species described below in being more hairy. *Sibbaldia* is very closely related to *Potentilla*. Propagation is usually from seed or by division of the rootstock.

1. S. procumbens Linnaeus (*Potentilla sibbaldi* Haller filius). Illustration: Takeda, Alpine flora of Japan, pl. 42 (1963); Fitter, Wild flowers of Britain & Northern Europe, 113 (1974); The Living Countryside **8**: 1893 (1988); Blamey & Grey-Wilson, Illustrated flora of Britain & Northern Europe, 193 (1989).

Prostrate or tufted herb to 10 cm. Leaflets 0.5–2.5 cm, obovate to narrowly wedge-shaped, with 3–6 teeth or small lobes at apex. Flowering-stems 1–30 cm, with 3–7 flowers in dense terminal clusters. Sepals 1–4 mm; petals 1.5–4 mm, rarely absent, narrowly obovate, yellow; style lateral. Fruit 1–3 mm, shiny. *N America, Eurasia.* H1. Summer.

45. FRAGARIA Linnaeus
A.C. Leslie

Perennial herbs with stolons. Leaves all basal, divided into 3 leaflets. Flowers usually in cymes (rarely solitary) on axillary scapes. Perianth parts usually in 5s, additional petals sometimes present. Epicalyx present. Petals usually white,

rarely partly or entirely pink. Stamens and carpels numerous, but one or other reduced or poorly formed in functionally unisexual flowers. Receptacle becoming fleshy and usually brightly coloured in fruit, either bearing achenes on its surface or sunk in pits.

A genus of 12–15 species from north temperate and subtropical areas and South America. They grow best in rich loamy soils, but are tolerant of a wide variety of conditions, providing the soil is not waterlogged. They are all tolerant of shade but flower and fruit best in open, sunny sites. *F. vesca* is especially tolerant of chalky soils. Propagation of the species is generally by seed, but that of named cultivars must be by division or from stolons.

Literature: Staudt, G., Taxonomic studies in the genus Fragaria, *Canadian Journal of Botany* **40**: 869–886 (1962); Darrow, G.M., The Strawberry (1966).

1a. Calyx-lobes adpressed to the young fruit   2
   b. Calyx-lobes spreading or reflexed in young fruit   5
2a. Achenes sunk in deep pits in the receptacle   **3. virginiana**
   b. Achenes in shallow pits or almost superficial   3
3a. Flowers 1–2, *c.* 1.5 cm wide   **6. daltoniana**
   b. Flowers more than 2, more than 2 cm wide   4
4a. Leaflets thick, dark green, with a conspicuous network of veins, silky-hairy beneath; flowers usually functionally unisexual   **4. chiloensis**
   b. Leaflets not as above; flowers usually functionally bisexual   **5. × ananassa**
5a. Achenes sunk in deep pits in the receptacle   **3. virginiana**
   b. Achenes sunk in shallow pits or superficial   6
6a. Uppermost flower-stalks with adpressed hairs   7
   b. Uppermost flower-stalks with spreading or reflexed hairs   8
7a. Flowers less than 2 cm wide; leaflets hairy above; lateral leaflets stalkless or very shortly stalked   **1. vesca**
   b. Flowers more than 2 cm wide; leaflets hairless or very sparsely hairy above; lateral leaflets distinctly stalked   **5. × ananassa**

8a. Terminal leaflets pale or yellowish green, often more or less diamond-shaped with rounded angles; flowers often functionally unisexual   **2. moschata**

b. Terminal leaflets usually dark or bluish green, often obovate or obovate to diamond-shaped; flowers usually bisexual   **5. × ananassa**

1. F. vesca Linnaeus. Illustration: Ross-Craig, Drawings of British plants **8**: t. 28 (1955); Keble Martin, The concise British Flora in colour, 27 (1965); Grey-Wilson & Blamey, The alpine flowers of Britain and Europe, 103 (1979).

Scapes 5–30 cm. Leaflets 1–10 × 1.2–3.8 cm, ovate, obovate or diamond-shaped; lateral leaflets stalkless or very shortly stalked; all bright green and hairy above, with adpressed silky hairs beneath. At least the upper flower-stalks with adpressed hairs. Flowers 1.2–1.8 cm wide, usually bisexual. Calyx-lobes spreading or reflexed in fruit. Fruiting receptacle 1–2 cm, ovoid or almost spherical, rarely ovoid-conical, usually red; achenes projecting from the surface, sometimes lacking at the base of the fruit. *Most of Europe, eastern N America; naturalised elsewhere.* H1. Spring–autumn.

Variants that, under suitable conditions, flower and fruit throughout the year have long been cultivated as Alpine Strawberries (forma **semperflorens** (Duchesne) Staudt). Other variants sometimes encountered are forma **roseiflora** (Boulay) Staudt, with pinkish petals; forma **alba** (Duchesne) Staudt, with whitish fruits; forma **eflagellis** (Duchesne) Staudt, which lacks stolons (a character now incorporated in some Alpine Strawberry cultivars), 'Monophylla' has leaves of a single leaflet; 'Multiplex' ('Flore Pleno', 'Bowles Double') has double flowers and 'Muricata', (Plymouth Strawberry) has leaf-like petals and stamens and enlarged achenes, each with a long, spine-like tip.

2. F. moschata Duchesne (*F. elatior* Ehrhart). Illustration: Coste, Flore de la France **2**: 27 (1903); Roles, Flora of the British Isles, illustrations **2**: t. 578 (1960); Pignatti, Flora d'Italia **1**: 586 (1982).

Scapes 10–50 cm. Leaflets 1.6–12.5 × 2.2–9.2 cm, the terminal often diamond-shaped with rounded angles, all distinctly stalked, pale or yellowish green and hairy above, with loosely adpressed hairs beneath. All flower-stalks with spreading or reflexed hairs. Flowers 1.5–3 cm wide, often functionally unisexual, the male flowers larger than the female. Calyx-lobes spreading or reflexed in fruit. Fruiting receptacle c. 2 cm, ovate, obovate or almost spherical, dark red, with a musky flavour; achenes projecting from the surface but absent from the contracted basal neck. *C Europe, naturalised elsewhere.* H1. Spring–summer.

3. F. virginiana Duchesne. Illustration: Bailey, The evolution of our native fruits, 427–429 (1898); Gleason, Illustrated flora of the north-eastern United States and adjacent Canada **2**: 289 (1952); Darrow, The strawberry, pl. 8–1 (1966).

Scapes 5–25 cm. Leaflets 1.5–7.5 × 0.9–4.5 cm, broadly ovate to obovate, all distinctly stalked, pale to mid green and sparsely hairy above, thinly to densely hairy beneath. Upper flower-stalks usually with adpressed, rarely spreading hairs. Flowers 0.6–2.5 cm wide, usually functionally unisexual, the male flowers larger than the female. Calyx-lobes adpressed to the young fruit. Fruiting receptacles 0.5–2 cm wide, spherical or ovoid, red. Achenes sunk in deep pits. *E North America, naturalised elsewhere.* H1. Spring–summer.

4. F. chiloensis Duchesne. Illustration: Nicholson, Dictionary of gardening **3**: 21 (1885); Abrams & Ferris, Illustrated Flora of the Pacific States **2**: 444 (1944); Darrow, The strawberry, pl. 5-2 & 9-1 (1966).

Scapes 3–25 cm. Leaflets 1.8–5 × 1.2–4 cm, obovate to broadly obovate, all distinctly stalked, dark glossy green and almost hairless above, with a strong network of veins and densely adpressed hairy beneath. Flower-stalks with adpressed or spreading hairs. Flowers 2–5.2 cm wide, usually functionally unisexual. Calyx-lobes adpressed to the young fruit. Fruiting receptacle 1.5–2 cm wide, spherical or ovoid, red. Achenes more or less superficial or in shallow pits. *West coast of N America (Alaska to California), west coast of S America (Peru to Argentina), Hawaii; naturalised elsewhere.* H1. Spring–summer.

5. F. × ananassa Duchesne. Illustration: Roles, Flora of the British Isles, illustrations **2**: t. 579 (1960); Darrow, The strawberry, pl. 6-1 to 6-4 (1966).

Scapes 15–50 cm. Leaflets very variable in size and shape, all distinctly stalked, rather bluish green and almost hairless above, with adpressed hairs beneath. Flower-stalks usually with adpressed hairs, rarely with spreading hairs in a few cultivars. Flowers 2–3.5 cm wide, usually bisexual. Calyx-lobes reflexed or adpressed to the young fruit. Fruiting receptacle very variable in size and shape, to 5 cm wide, red. Achenes more or less superficial or in shallow pits. *Garden origin.* H1. Spring–summer.

A hybrid between *F. chiloensis* and *F. virginiana* and now the most commonly cultivated strawberry. Very variable in all its characters, some variants closely resembling one or other parent. There are many named cultivars, two of which are of ornamental value: 'Variegata', with leaves boldly blotched creamy white, and 'Serenata' which has pink flowers. The latter derives from a cross with *Potentilla palustris* (p. 393).

6. F. daltoniana Gay. Illustration: Polunin & Stainton, Flowers of the Himalaya, t. 38 (1985).

Scapes 1.5–5 cm, often with a solitary flower. Leaflets 0.7–2 × 0.5–1.5 cm, elliptic to obovate, all shortly stalked, dark glossy green and sparsely hairy above, with adpressed hairs beneath. Flower-stalks with adpressed hairs. Flowers to 1.5 cm wide. Calyx-lobes adpressed to young fruit; epicalyx lobes toothed. Fruiting receptacle 1.2–2.5 × 1–1.3 cm, long-ovoid, red to white. Achenes sunk in shallow pits. *Himalaya (Uttar Pradesh to Sikkim).* H1. Spring–summer.

46. DUCHESNEA Smith
P.S. Green
Like *Fragaria*, but the flowers yellow and the fruit not juicy.

A genus of 6 species from India and SE Asia.

1. D. indica (Andrews) Focke (*Fragaria indica* Andrews). Illustration: Edwards's Botanical Register **1**: t. 61 (1815); Hegi, Illustrierte flora von Mitteleuropa, **4**: 241 f. 188 (1964); Everett, New York Botanical Garden illustrated encyclopedia of horticulture **4**: 1148 (1981).

Perennial herb, stems short, leaves in rosettes and on long runners which root at the nodes and tips to form new rosettes. Leaves with 3 leaflets, hairy, leaflets obovate to angular-ovate, margins coarsely toothed, terminal leaflet 1–3 × 0.7–2.5 cm, lateral leaflets slightly smaller. Flower-stalks 2.5–10 cm, slightly longer than the subtending leaves. Epicalyx and sepals hairy, persistent in fruit. Petals bright yellow. Fruit with achenes spread over a bright red, swollen, insipid receptacle to *c*.

2 cm. *China & India, widely introduced into warm climates.* H5–G1. Summer.

47. ALCHEMILLA Linnaeus
S.M. Walters

Deciduous perennials with a more or less developed woody stock and annual flowering-shoots. Basal leaves palmately 5–9 (rarely to 11) lobed, sometimes compound with separate leaflets. Stem-leaves with fewer lobes and relatively large toothed or lobed stipules. Inflorescence cymose, much-branched, with numerous small flowers on short flower-stalks. Perianth greenish, of 4 outer epicalyx segments and 4 alternating sepals. Petals absent. Stamens 4, often with poorly developed anthers. Carpel 1, with basal style and head-like stigma, more or less protruding from a cup-like perigynous zone (receptacular cup, hypanthium), surrounded by a nectar-secreting disk (see figure 28, p. 326). Fruit an achene enclosed in the dry perigynous zone.

Some 300 species have been described, mostly from the mountains of Europe and the Caucasus; many of these are known to be apomictic and high polyploids. Widespread in Europe and Asia, particularly in the north and on mountains; rare (and mostly introduced) in N America; also on mountains in Africa and the Andes. Most species are easily identified on well-grown material, but late-season second growth can be misleading, and should not be used for identification. The cultivated species of *Alchemilla* may be placed into three groups based upon their growth habit. These are: medium or tall, often robust plants, shorter, softly hairy plants and low growing alpines. Several species, especially *A. mollis*, reproduce freely from seed, and all can be propagated vegetatively.

In descriptions of the leaves, the depth of lobing may be indicated by a fraction (e.g. 1/3) which represents the depth of the lobe as a proportion of the distance to the centre of the leaf. If the number of leaf-teeth is given, it refers to one side of the middle lobe.

1a. Low-growing with creeping, sometimes prostrate, annual stems freely rooting at the nodes 2
  b. Habit various, rhizomatous, not freely rooting 4
2a. Plant hairless or nearly so; epicalyx-segments represented by very small teeth **23. pentaphyllea**
  b. Plant hairy; epicalyx-segments well developed, at least half as long as sepals 3
3a. Mat-forming; leaves 5-lobed (rarely 3-lobed), with a few small, blunt teeth **24. ellenbeckii**
  b. Loosely spreading; leaves 7-lobed (rarely 5-lobed), with long, acute teeth **25. abyssinica**
4a. Flowering-stems less than 2 cm **5. faeroensis** var. **pumila**
  b. Flowering-stems at least 5 cm 5
5a. Leaves compound, or lobed half-way to centre or more 6
  b. Leaves lobed less than half-way to centre 10
6a. Leaves with 5–7 free leaflets (sometimes outermost joined basally) 7
  b. Leaves with 7–9 (rarely 5) very deep lobes (sometimes the middle one free) 8
7a. Leaf hairless above; teeth of leaflets *c.* 1 mm **1. alpina**
  b. Upper leaf surface thinly adpressed-hairy; teeth of leaflets at least 2 mm **2. sericea**
8a. Middle leaf-lobe free to base or nearly so **3. plicatula**
  b. All leaf-lobes obviously joined at base 9
9a. Leaves circular in outline, with basal lobes touching or overlapping; leaf-teeth not conspicuous **4. conjuncta**
  b. Leaves kidney-shaped in outline, with widely spaced basal lobes; leaf-teeth conspicuous **5. faeroensis**
10a. Hairs on flowering-stems and leaf-stalks, if present, more or less adpressed, often silky 11
  b. Hairs on flowering-stems and leaf-stalks spreading, at an angle of 45° or more 17
11a. Plant hairless except for lower leaf surface and lowest part of stem **18. glabra**
  b. Plant with hairs on stems up to the inflorescence branches 12
12a. Upper surface of leaves hairless or slightly hairy along the folds only 13
  b. Upper surface of leaves evenly hairy 15
13a. Epicalyx segments equalling sepals **21. venosa**
  b. Epicalyx segments shorter than sepals 14
14a. Upper surface of leaves hairless; leaf-outline kidney-shaped **5. faeroensis**
  b. Upper surface of leaves usually slightly hairy along folds; leaf-outline circular **13. vetteri**
15a. Robust plant with flowering-stems to 30 cm and leaves thick in texture **6. fulgens**
  b. Small plants with slender flowering-stems to 20 cm and leaves thin in texture 16
16a. Basal lobes of mature summer leaves overlapping; leaf-teeth 5 or 6, acute **10. sericata**
  b. Basal lobes of leaves not overlapping; leaf-teeth 4 or 5, often obtuse **11. rigida**
17a. Robust plants with flowering-stems to 60 cm; epicalyx-segments at least as long as sepals 18
  b. Small or medium-sized plants with flowering-stems rarely more than 50 cm; epicalyx-segments shorter than sepals 20
18a. Upper leaf surface hairless **20. epipsila**
  b. Upper leaf surface hairy 19
19a. Leaf-lobes shallow, not more than 1/5; flower-stalks hairless **19. mollis**
  b. Leaf-lobes relatively deep, to 2/5; flower-stalks hairy **22. speciosa**
20a. Upper leaf surface hairless **17. xanthochlora**
  b. Upper leaf surface hairy 21
21a. Whole plant, including individual flower-stalks, hairy 22
  b. At least some individual flower-stalks hairless 24
22a. Hairs on stems and leaf-stalks ascending **9. elisabethae**
  b. Hairs on stems and leaf-stalks spreading or slightly deflexed 23
23a. Leaf-lobes truncate, separated by deep, toothless incisions **8. erythropoda**
  b. Leaf-lobes rounded, separated by shallow incisions **7. glaucescens**
24a. Flowers less than 2.5 mm across; lower part of stem and leaf-stalks densely hairy with some downwardly directed hairs **16. tytthantha**
  b. Flowers at least 2.5 mm across; lower part of stem and leaf-stalks hairy but without downwardly directed hairs 25
25a. Hairs on stem and leaf-stalks ascending, sometimes almost silky **12. lapeyrousii**
  b. Hairs on stems and leaf-stalks spreading at right angles 26

Figure 36. Leaf silhouettes of *Alchemilla* species (× 0.5). 1, *A. alpina*. 2, *A. sericea*. 3, *A. plicatula*. 4, *A. conjuncta*. 5a, *A. faeroensis*, b, *A. faeroensis* var. *pumila*. 6, *A. fulgens*. 7, *A, glaucescens*. 8, *A. erythropoda*. 9, *A. elisabethae*. 10, *A. sericata*. 11, *A. rigida*. 12, *A. lapeyrousii*. 13, *A. vetteri*. 14, *A. monticola*. 15, *A. acutiloba*. 16, *A. tytthantha*. 17, *A. xanthochlora*. 18, *A. glabra*. 19, *A. mollis*. 20, *A. epipsila*. 21, *A. venosa*. 22, *A. speciosa*. 23, *A. pentaphylla*. 24, *A. ellenbeckii*. 25, *A. abyssinica*.

26a. Leaf-lobes of summer leaves almost
triangular **15. acutiloba**

b. Leaf-lobes rounded
14. monticola

Section **Alpinae** Camus. Leaves palmate
or deeply palmately lobed, more or less
silvery-hairy on the lower surface;
epicalyx-segments less than half as long as
sepals.

1. A. alpina Linnaeus. Figure 36(1),
p. 400. Illustration: Coste, Flore de la
France **2**: 63 (1903); Ross-Craig, Drawings
of British plants **9**: pl. 16 (1956); Garrard
& Streeter, Wild flowers of the British Isles,
pl. 37 (1983).

Mat-forming, with creeping woody
rootstock and slender, erect, densely
adpressed-hairy flowering stems to 15 cm.
Basal leaves compound with 5 (rarely 7)
lanceolate-obovate leaflets with acute teeth
towards the apex, dark green and hairless
above, silvery-hairy beneath with closely
adpressed hairs. *W & WC Europe.* H1.
Summer.

Prefers non-calcareous soil. Most plants
sold under this name are either *A. plicatula*
or *A. conjuncta,* which are both more
attractive and easier to cultivate.

2. A. sericea Willdenow. Figure 36(2),
p. 400. Illustration: Flora USSR **10**: pl. 22
(1941).

Like *A. alpina* but with much more
conspicuous and irregular teeth in the
upper half of the leaflets, and with thin,
adpressed hairs on the upper surface of
the leaves. *Former USSR (Caucasus),
Turkey.* H1. Summer.

Rarely cultivated and apparently rather
difficult to grow.

3. A. plicatula Gandoger (*A. asterophylla*
Buser; *A. hoppeana* misapplied). Figure
36(3), p. 400; 37(1), p. 402. Illustration:
Coste, Flore de France **2**: 63 (1903).

Like *A. alpina* but more robust, with
stems to 20 cm; leaves with 7 (rarely 9)
segments, the middle one free to the base,
the others joined near the base, and leaf
outline circular, with somewhat
overlapping basal segments. *Europe
(Pyrenees to Balkan Peninsula), Turkey.* H1.
Summer.

4. A. conjuncta Babington (*A. alpina*
misapplied). Figure 36(4), p. 400.
Illustration: Coste, Flore de France **2**: 63
(1903); Ross-Craig, Illustrations of British
plants **9**: pl. 5 (1956).

Robust plant with flowering-stems to
30 cm, silky-hairy throughout. Leaves

relatively thick, hairless above and densely
silky-hairy beneath, more or less circular
in outline, with 7 (rarely 9) broadly
lanceolate to elliptic segments all distinctly
joined at the base for up to 1/5 of their
length and with indistinct teeth almost
hidden by the silky hair-covering. *Jura
& SW Alps; naturalised in Scotland.* H1.
Summer.

Often sold as *A. alpina.*

Section **Alchemilla.** Leaves palmately
lobed, rarely to more than halfway;
epicalyx-segments more than half as long
as sepals.

5. A. faeroensis (Lange) Buser. Figure
36(5), p. 400.

Like *A. conjuncta,* but leaves kidney-
shaped in outline, usually 7-lobed, with
deep incisions to about halfway to the
centre, and with rather large, acute teeth
around the upper edge of the lobes. *E
Iceland, Faeroe Islands.* H1. Summer.

Both the typical plant and the dwarf var.
pumila (Rostrup) Simmons, which does
not exceed 5 cm, are easily cultivated.

6. A. fulgens Buser (*A. splendens*
misapplied). Figure 36(6), p. 400.
Illustration: Coste, Flore de la France **2**: 64
(1903).

Like *A. conjuncta,* but leaf-lobes broader
and shorter, with acute teeth, separated by
toothless, V-shaped incisions to less than
half-way; upper surface of leaf more or less
adpressed hairy. *Pyrenees.* H1. Summer.

7. A. glaucescens Wallroth (*A. minor*
misapplied; *A. hybrida* misapplied; *A.
pubescens* misapplied). Figure 36(7), p. 400.
Illustration: Coste, Flore de la France **2**:
64 (1903); Ross-Craig, Drawings of British
plants **9**: pl. 1 (1956); Garrard & Streeter,
Wild flowers of the British Isles, pl. 37
(1983).

Plant to 20 cm, covered with densely
spreading hairs throughout. Leaves
circular in outline, 7–9-lobed, basal lobes
usually overlapping; all lobes shallow,
rounded, with 4–6 teeth: incision between
lobes shallow or absent. Inflorescence with
rather dense clusters of flowers. *Europe.*
H1. Summer.

Neat and easily grown, but infrequently
cultivated.

8. A. erythropoda Juzepczuk (*A.
erythropodioides* Pawlowski). Figure 36(8),
p. 400. Illustration: Beckett, Concise
encyclopaedia of garden plants, 17 (1983).

Like *A. glaucescens* but grey or blue-
green with deep incisions to 2/5, between

the truncate leaf-lobes; dense hairs on
leaf-stalks and lower part of flowering-
stems slightly downwardly directed. Often
developing a reddish colour on stems in
sun. *Balkan Peninsula, W Carpathians,
Turkey, Former USSR (Caucasus).* H1.
Summer.

Plants referable to the closely allied
A. caucasica Buser, differing mainly in
lacking downwardly directed hairs, are
apparently also in cultivation under the
name *A. erythropoda.*

9. A. elisabethae Juzepczuk. Figure 36(9),
p. 400.

Dwarf, very blue-green plant with
flowering stems to 15 cm, thickly clothed
with ascending hairs throughout. Leaves
7-lobed, with deep incisions (to 2/5)
between the flattened lobes; teeth on lobes
3 or 4, broad and blunt. *Former USSR
(Caucasus).* H1. Summer.

A very attractive rock-plant, beginning
to be popular.

10. A. sericata Reichenbach. Figure
36(10), p. 400. Illustration: Reichenbach,
Iconographia Botanica **1**: t. 4 (1823).

Small, slender plant with thin flowering-
stems to 20 cm, thinly silky-hairy
throughout. Leaves 7-lobed, circular in
outline, with overlapping basal lobes,
with deep incisions to halfway between
the semicircular lobes; teeth on lobes 5 or
6, acute; upper leaf surface light green,
sparsely adpressed hairy, lower surface
pale green with a denser, rather silky hair-
covering. Flowering-stems and leaf-stalks
quickly turning reddish in sun. *Former
USSR (Caucasus).* H1. Summer.

11. A. rigida Buser. Figure 36(11),
p. 400.

Like *A. sericata* but hair-covering less
closely adpressed (a difference especially
obvious on the underside of the leaves,
which have a less silky appearance), and
leaves a deeper green, kidney-shaped in
outline, with basal lobes not or scarcely
overlapping, with fewer (4 or 5) broader
teeth on the lobes. *Turkey, Former USSR
(Caucasus).* H1. Summer.

An attractive rock-plant.

12. A. lapeyrousii Buser (*A. hybrida*
(Linnaeus) Linnaeus). Figure 36(12),
p. 400. Illustration: Watsonia **5**: 259 & pl.
12b (1963).

Medium-sized, rather robust plant with
flowering-stem to 25 cm, clothed with
forwardly directed hairs throughout except
the flower-stalks, which are often hairless

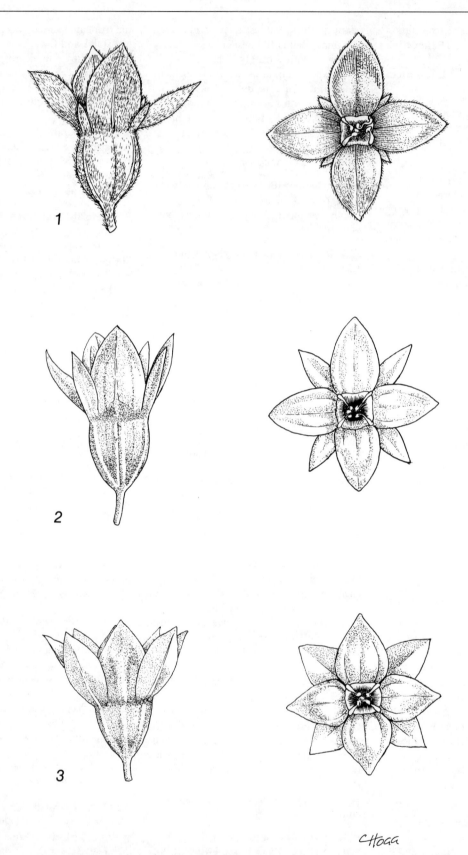

Figure 37. Flowers of *Alchemilla* species, side and front view.
1, *A. plicatula* (Section *Alpinae*) (× 20). 2, *A. tytthantha* (Section *Alchemilla*) (× 20). 3, *A. mollis* (Section *Alchemilla*, Series *Elatae*) (× 10).

(a lens is needed to see this). Leaves 7–9-lobed, ascending-hairy on both surfaces and on leaf-stalks, kidney-shaped in outline, with more or less triangular lobes, with short incisions (to ⅓) between lobes; teeth on lobes 5 or 6, acute. *Pyrenees & SC France*. H1. Summer.

13. A. vetteri Buser. Figure 36(13), p. 400.

Small plant with flowering-stems to 20 cm, clothed throughout with more or less adpressed, silky hairs. Leaves hairless or only thinly hairy along the folds above, silky-hairy beneath, circular in outline, with 7–9 rather truncate lobes separated by long incisions to ⅖; teeth on lobes 5 or 6, acute. *SW Europe, Spain to France (Maritime Alps)*. H1. Summer.

An attractive rock plant, rarely seen in cultivation.

Species **14–18**, which could be grouped together under the aggregate name *A. vulgaris*, have a limited value as garden plants, mainly because their flowers are relatively small (usually less than 3 mm across) and are greenish rather than yellowish.

14. A. monticola Opiz (*A. pastoralis* Buser). Figure 36(14), p. 400. Illustration: Garrard & Streeter, Wild flowers of the British Isles, pl. 37 (1983).

Medium-sized plant with flowering-stems to 50 cm, mature summer leaves to 8 cm across, densely spreading-hairy throughout except for the flower-stalks, which are usually hairless (use a lens). Leaves more or less circular in outline, 9–11-lobed; lobes rounded, separated by short incisions; leaf-teeth 7–9, rather regular, acute. *Europe*. H1. Summer.

One of the commonest 'Vulgaris' Alchemillas throughout much of Europe. Occasionally brought into cultivation from the wild.

15. A. acutiloba Opiz (*A. acutangula* Buser). Figure 36(15), p. 400. Illustration: Garrard & Streeter, Wild flowers of the British Isles, pl. 37 (1983).

Robust plant to 65 cm, with unmistakable mature summer leaves to 10 cm across, with 9–13 almost triangular lobes with straight sides, and very unequal, acute teeth, the largest in the middle of each side, spreading hairy except for the inflorescence which is almost hairless and the upper leaf surface with is only sparsely hairy. *Europe*. H1. Summer.

16. A. tytthantha Juzepczuk (*A. multiflora* Rothmaler). Figures 36(16), p. 400, 37(2), p. 402. Illustration: Watsonia **4**: 282 & pl. 17 (1961).

Medium-sized plant with relatively slender flowering-stems to 50 cm and almost circular leaves with dense spreading hairs on both surfaces, with 9 (rarely to 11) shallow lobes and 6–8 small, acute teeth. Leaf-stalks and lower part of flowering-stems densely hairy with some downwardly pointing hairs. Flowers hairless, small (less than 2.5 mm). *Former USSR (Crimea)*. H1. Summer.

Naturalised in several places in S Scotland, presumably from Botanic Gardens. Tolerates light shade.

17. A. xanthochlora Rothmaler (*A. pratensis* misapplied). Figure 36(17), p. 400. Illustration: Ross-Craig, Drawings of British plants **9**: pl. 3 (1956); Garrard & Streeter, Wild flowers of the British Isles, pl. 37 (1983).

Medium-sized plant, often yellow-green, with robust flowering-stems to 50 cm and more or less kidney-shaped leaves, hairless or nearly so on the upper surface, with dense spreading hairs below. Leaf-lobes 9–11, rounded, with 7–9 wide, acute, almost equal teeth. Lower part of flowering-stems and leaf-stalks with spreading hairs, often somewhat forwardly directed; upper part of inflorescence, including flower-stalks and flowers, usually hairless. *N, W & C Europe, extending to Sweden & Greece*. Summer. H1.

One of the commonest 'Vulgaris' Alchemillas in the wild. Not uncommonly brought into cultivation from the wild, but not recommended for gardens.

18. A. glabra Neygenfind (*A. alpestris* misapplied). Figure 36(18), p. 400. Illustration: Ross-Craig, Drawings of British plants **9**: pl. 4 (1956); Garrard & Streeter, Wild flowers of the British Isles, pl. 37 (1983).

Medium-sized plant with flowering-stems to 60 cm, easily distinguished from all other species likely to be seen in gardens by being almost hairless. Leaves more or less kidney-shaped, with 9–11 triangular-ovate lobes and no incisions between; leaf-teeth 7–9, rather wide but acute, and very unequal. Adpressed hairs are present on lowest stem internodes and on the leaf-stalks and on the outer half of veins on the lower leaf surface. *N & C Europe, extending to the Pyrenees & Balkans*. H1. Summer.

Sometimes cultivated from the wild.

Species **19–22** belong to the Series **Elatae** Rothmaler, and have larger flowers and a generally more decorative, yellower inflorescence than the other species.

19. A. mollis (Buser) Rothmaler (*A. grandiflora* Anon.). Figure 36(19), p. 400, 37(3), p. 402. Illustration: Journal of the Royal Horticultural Society **73**: 309, f. 106 (1948); Beckett, Concise encyclopaedia of garden plants, 17 (1983).

Large, robust plant with flowering-stems to 80 cm, densely spreading-hairy throughout except for the flower-stalks (use a lens). Leaves to 15 cm, circular in outline with overlapping basal lobes; lobes 9–11, very shallow, semicircular, with 7–9 wide, ovate, unequal teeth. Inflorescence showy, yellowish; flowers relatively large, to 5 mm across, with epicalyx-segments at least as long as sepals, and giving the appearance of an 8-pointed star. *E Carpathians, Turkey, former USSR (Caucasus)*. H1. Summer.

In recent years this species has become by far the commonest Alchemilla in cultivation in the British Isles. It flowers relatively late (summer rather than late spring), and is much used in flower-arrangements. 'Variegata', with variably-developed yellowish leaf-markings, is sometimes grown. Plants grown as 'Robusta', and said to be more vigorous than the type, have proved on comparative cultivation to be indistinguishable. 'Mr Poland's Variety' is *A. venosa* (see below).

20. A. epipsila Juzepczuk (*A. indivisa* Rothmaler). Figure 36(20), p. 400.

Like *A. mollis* in stature and showy, yellowish inflorescence, but easily distinguished by the hairless upper leaf surface. The leaf-lobes are less shallow than in *A. mollis*. *Balkan Peninsula*. H1. Summer.

In cultivation in some gardens in Britain, and probably more widespread than presently recognised.

21. A. venosa Juzepczuk. Figure 36(21), p. 400.

Medium-sized plant, with rather slender flowering-stems to 40 cm, clothed up to the main inflorescence branches with more or less adpressed hairs. Inflorescence and flowers resembling *A. mollis*. Leaves somewhat thin and papery, circular in outline, with overlapping basal lobes; lobes 9–11, semi-circular, with clear incisions to ⅓; leaf-teeth 6–8, unequal, acute. Upper leaf surface hairless, lower with more or less adpressed hairs; leaf-stalks more or less adpressed-hairy. *Former USSR (Caucasus)*. H1. Late summer.

Widespread in gardens in Britain, but not usually distinguished from *A. mollis*, which it resembles in flower. Distinguished by its more delicate stature, more or less adpressed hairs and earlier flowering time.

22. A. speciosa Buser. Figure 36(22), p. 400.

Rather large plant, with the general appearance and floral characters of *A. mollis*, but differing most obviously in its deeply lobed leaves (to ⅖) with narrower, acute teeth, its ascending hairs and hairy flower-stalks. *Former USSR (Caucasus)*. H1. Summer.

Rare in cultivation.

Section **Pentaphyllea** Camus. Stoloniferous, rooting at the nodes. Leaves palmate. Epicalyx-segments reduced to 4 small teeth.

23. A. pentaphyllea Linnaeus. Figure 36(23) p. 400. Illustration: Coste, Flore de la France **2**: 62 (1903); Guinochet & Vilmorin, Flore de France **5**: t. 1674 (1984).

Dwarf, often prostrate plant, hairless or nearly so, with short and slender flowering-shoots rooting at the nodes. Leaves to 2 cm, palmate with 5 leaflets deeply cut into long, narrow segments in the upper half. Flowers few; epicalyx-segments represented by very small teeth, much shorter than the ovate, blunt sepals. *Alps*. H1. Summer.

Very different from other European species and reproducing sexually. Rare in cultivation. It needs careful alpine house or trough cultivation.

Section **Longicaules** Rothmaler (including Section *Parvifoliae* Rothmaler). Wide-creeping plants freely rooting at the nodes. Epicalyx-segments at least half as long as sepals. The classification of this and other African groups is as yet relatively little understood, and there is some evidence that they are at least partly sexual, with consequent free hybridisation. For further information, see Hedberg, O., Afro-alpine vascular plants, *Symbolae Botanicae Upsalienses* **15**(1): 102–117 & 281–292 (1957).

24. A. ellenbeckii Engler. Figure 36(24), p. 400.

A small, prostrate, sparsely hairy, wide-creeping plant freely rooting at the nodes of slender stolons, and producing short, erect, few-flowered lateral shoots. Leaves not more than 2 cm across, deeply 5-lobed; lobes

wedge-shaped, with few small blunt teeth. *Mts, from Ethiopia to Kenya*. H4. Summer.

Very variable, but the clones now popularly on sale in garden nurseries are very uniform dwarf carpeting plants suitable for damp ground, and easily reproduced vegetatively. It will not, however, survive more than a little frost and needs protection in winter.

25. A. abyssinica Fresenius. Figure 36(25), p. 400. Illustration: Senckenbergia Biologia (Frankfurt) **61**: 130, 131 (1980).

Wide-creeping plant with prostrate stems rooting at the nodes, and short, erect, few-flowered branches. Leaves to 4 cm, kidney-shaped in outline, mostly 7-lobed; lobes shortly wedge-shaped with incisions to almost half-way, with 3–5 long, acute, curved teeth. Leaf surfaces sparsely hairy; leaf-stalks and stems obviously hairy with long, spreading hairs. *Ethiopia to Kenya*. H4. Summer.

Another African species, **A. pedata** A. Richard, is in cultivation in Britain; it may prove to be hardier than the others, but is not yet widely available.

48. CYDONIA Miller
E.C. Nelson

Deciduous shrubs or trees, without thorns, to 8 m; young shoots with grey-white felt. Leaves entire, ovate to elliptic, to 10 × 5 cm, densely felted below; stipules falling early, hairy, glandular. Flowers solitary, white to pink. Sepals 5, persistent, hairy outside, toothed. Petals 5. Stamens 15–25. Carpels 5, walls cartilaginous in fruit; ovules numerous. Styles 5, free. Fruit a pome, closed at top with persistent sepals, pear-shaped, aromatic, golden yellow, covered with felt of short hair.

A genus of 1 species, the quince, has been cultivated for many centuries and was employed as an aphrodisiac in classical times. Plants are used as stock for grafting pears. It can be propagated by cuttings and from seed.

1. C. oblonga Miller (*Pyrus cydonia* Linnaeus; *C. vulgaris* Persoon; *C. maliformis* Miller). Illustration: Bianchini et al., Fruits of the earth, 135 (1975); Graf, Tropica, 849 (1978).

Perhaps W Asia, naturalised in S Europe. H3. Spring.

As it has been so long in cultivation, numerous minor variants are reported including 'Lusitanica' (*Cydonia lusitanica* Miller) which is more vigorous, produces larger, pink flowers, but is not as hardy as the normal variants.

49. CHAENOMELES Lindley
E.C. Nelson

Deciduous shrubs or trees, sometimes with thorns. Leaves simple, toothed; stipules persistent, large, kidney-shaped. Flowers in axillary clusters. Sepals 5, hairless, deciduous, erect, entire. Petals 5 (numerous in some 'double-flowered' cultivars). Stamens more than 20. Carpels 5, ovules numerous; styles 5, free. Fruit a pome, closed at top, stalkless.

Cultivars must be propagated vegetatively, by cuttings (half-ripened wood, in June and July); species may be raised from seed, but variation should be expected and hybrids may be formed when other species or cultivars are grown nearby. Commonly treated as wall-plants; prune after flowering, and in late summer train and remove surplus shoots.

Literature: Weber, C., Cultivars in the genus *Chaenomeles*, *Arnoldia* **23**: 17–75 (1963); Weber, C., The genus *Chaenomeles* (Rosaceae), *Journal of the Arnold Arboretum* **45**: 302–345 (1964).

1a. Tree or tall shrub at least 2 m; shoots stout, erect; young leaves with dense adpressed hairs beneath
    1. cathayensis
  b. Shrub with spreading shoots, rarely over 1 m (except when artificially trained); leaves hairless or at most sparsely hairy beneath   **2**
2a. Leaves coarsely toothed; young shoots felted becoming hairless, distinctly warty; petals orange-red to pink   **2. japonica**
  b. Leaves finely toothed; young shoots hairless, never warty; petals pink rarely white   **3. speciosa**

1. C. cathayensis (Hemsley) Schneider (*Cydonia cathayensis* Hemsley; *C. mallardii* invalid; *Chaenomeles lagenaria* var. *cathayensis* (Hemsley) Rehder; *C. lagenaria* var. *wilsonii* Rehder). Illustration: Hooker's Icones plantarum, 2657 & 2658 (1901).

Shrub to small bushy tree, to 3 m or more; branches few, stout, stiffly erect, with numerous short, blunt spurs; shoots hairy when young, becoming hairless. Leaves elliptic to lanceolate, 5–8 × 2–3.5 cm, apex acute, margins toothed, teeth forward-pointing, with awn-like tips, felted underneath when young; stipules pointing forwards, to 1.5 × 2 cm. Flowers white to pink. Fruits to 15 cm, occasionally to 20 cm, apple-like, green, yellow or red. *China, Burma, Bhutan*. H4. Spring.

This produces the largest quinces which have long been used medicinally in China.

C. × vilmoriniana Weber. A hybrid between *C. cathayensis* and *C. speciosa*. Similar to the former in general habit and leaf-shape, young leaves felted beneath; flowers white, flushed pink. Differs in its shoots which are more slender, spurred and sharply spined; fruits few, *c.* 8 cm. *Garden origin*. H4. Spring.

Selections include 'Mount Everest', flowers large, white changing to pink with yellow and lavender; 'Vedrariensis' (*C. hybrida* var. *vedrariensis* invalid), flower white flushed pink.

C. × californica Weber. A hybrid between *C. cathayensis* and *C. × superba*. Similar to the former in having stiff, erect branches, differs in being more slender and sharply spurred; young shoots hairy. Leaves lanceolate, felted beneath when young. Flowers large, white (Vilminiana group), pink to red or bicoloured. Fruits apple-shaped. *Garden origin*. H4. Spring.

C. × clarkeana Weber. A hybrid between *C. cathayensis* and *C. japonica*. Low, spreading shrub, spines numerous; leaves intermediate in shape and character between the 2 parents; flowers large, pink to red.

2. C. japonica (Thunberg) Spach (*Pyrus japonica* Thunberg; *P. maulei* Masters; *Cydonia japonica* (Thunberg) Persoon).

Shrub to 1 m; branches loose, spreading, with short felt when young, becoming hairless and distinctly warty. Leaves ovate to spathulate, apex often notched, margins coarsely toothed, to 5 × 3 cm, hairless even when young; stipules toothed, to 1 × 2 cm. Flowers orange-red to pink. Stamens 40–60. Fruits *c.* 4 cm, aromatic. *Japan*. H2. Spring.

Often used to flavour confectionery and jams. A cultivar, 'Maulei', with single, orange to salmon-pink flowers is also grown.

Var. **pygmaea** Maximowicz has subterranean branches and lacks spines.

Var. **alpina** Maximowicz is smaller in all parts, and more branched than var. **japonica**.

3. C. speciosa (Sweet) Nakai (*Cydonia speciosa* Sweet; *C. japonica* (Loiseleur) Persoon; *Chaenomeles lagenaria* (Loiseleur) Koidzumi). Illustration: Botanical Magazine, 692 (1803).

Shrub 2–4 m; shoots erect or spreading, spined, hairless or occasionally hairy when young. Leaves to 10 × 4 cm, ovate to oblong, hairless, rarely sparsely hairy on veins below when young, sharply toothed; stipules pointing forward, to 1 × 2 cm.

Flowers red, rarely white to pink. Fruit variable in shape and size. *China, Burma*. H3. Spring.

C. × superba (Frahm) Rehder (*Cydonia maulei* T. Moore var. *superba* Frahm). A hybrid between *C. japonica* and *C. speciosa*. Shrub to 1.7 m, branches spreading, spined; young shoots with short, rough hairs, becoming warty. Leaves resembling those of *C. japonica*. Flowers white to orange to red. Fruit apple-shaped. *Garden origin*. H3. Spring.

Numerous cultivars have been raised which belong to this hybrid, among the best is 'Rowallane Seedling' with single, red flowers.

50. DOCYNIA Decaisne
M.F. Gardner
Evergreen or semi-evergreen, trees or shrubs. Leaves ovate-elliptic to lance-shaped, entire, lobed, or margins with forward-pointing teeth; stipules present. Flowers, 2–5 borne in stalkless umbels; calyx densely felted, sepal lobes lance-shaped; petals 5, broadly ovate, narrowed towards the base; stamens 30–50; ovary inferior, 5-celled with 3–10 ovules, styles 5, joined, with long hairs at base. Fruit fleshy, egg-shaped to pear-shaped, calyx persistent.

A genus of 5 species native to China, the East Indies and Vietnam. The name *Docynia* is an anagram of *Cydonia*. Plants of *D. rufifolia* (Léveillé) Rehder may be confused with *D. indica* and *D. delavayi*, particularly the latter.

1a. Leaves entire   **1. delavayi**
  b. Leaves divided or lobed, sometimes entire, but with finely toothed margins   **2. indica**

1. D. delavayi (Franchet) Schneider (*Pyrus delavayi* Franchet). Illustration: Revue Horticole, 45–47 (1918); Flora Republicae Popularis Sinicae **36**: 349 (1974); Wu, Wild flowers of Yunnan, **2**: 109 (1986).

Evergreen spreading shrub or tree to 7 m; young shoots hairy, becoming chocolate-brown and hairless, later black, sometimes developing spine-tipped short shoots. Leaves 3–7 × 2.3–2.8 cm, ovate to lance-shaped, pointed, entire, felted beneath; leaf-stalks, 1–1.5 cm, downy. Flowers pink in bud, very fragrant; calyx felted, petals white. Fruit yellow, *c.* 3 cm across, egg-shaped, downy. *China (Yunnan)*. H4. Spring.

2. D. indica Decaisne. Illustration: Brandis, Indian Trees, fig. 124 (1906); Schneider, Handbuch der Laubholzkunde **1**: 726, 727 (1906).

Evergreen tree, 3–5 m, branches short, with thorns; young branches densely white woolly, later almost hairless. Leaves 5–7 × 2.5–2.8 cm, ovate to oblong-ovate, sharply pointed or tapering to a point, rounded towards the base, finely toothed, sometimes entire, woolly hairy beneath, later hairless; leaf-stalks 1–1.5 cm; juvenile leaves lobed. Flowers borne in axillary umbels; calyx grey–yellow, hairy; petals white. Fruit yellow, 4–5 cm, egg-shaped to rounded. *India (Assam), Upper Burma, China (Yunnan)*. H4. Spring.

51. × PYRONIA Veitch
S.G. Knees & P.G. Barnes
Shrub or small tree to 8 m, resembling *Cydonia* in habit. Shoots brown, densely dotted with lenticels. Leaves deciduous, 5–8 × 2–4 cm, elliptic, downy beneath; margins entire or finely scalloped; leaf-stalk 1.5–2 cm. Flowers borne in groups of 3 at the shoot tips; sepals triangular, hairy on both surfaces, with glandular teeth; corolla *c.* 1.7 × 1.5 cm, rounded with notch at apex, white with pink-tinged edges, especially in bud; stamens 20, anthers deep pink–violet; styles more or less free at base, very hairy. Fruit 7–8 cm, egg-shaped, greenish yellow, spotted red; flesh white, lacking stone or grit cells.

A genus of sexual hybrids between *Cydonia* and *Pyrus* originally developed around 1895 in the Veitch nursery.

1. × P. veitchii (Trabut) Guillaumin (*Cydonia oblonga* × *Pyrus communis*). Illustration: Journal of Heredity **7**: 416 (1916).

Originated in cultivation. H4. Spring–early summer.

'Luxemburgiana' is more like *Pyrus* in habit with greenish yellow, pear-shaped fruits. Illustration: Bulletin de la Société Dendrologique de France **56**: 68 (1925).

52. + PYROCYDONIA Daniel
S.G. Knees & P.G. Barnes
Deciduous tree to 5 m. Leaves 3.5–7.5 × 2.5–4 cm, ovate to elliptic, rounded at base, acute at apex, more or less downy; margin toothed; leaf-stalk to 6 mm, downy. Sepals triangular, hairless except for inner-surface, margin toothed with glandular hairs; petals *c.* 1 × 1.2 cm, broadly ovate, white; stamens 20; styles free, hairless at base. Fruits 6–8 cm,

brownish with paler spotting.

This is a result of a graft chimaera, forming a tree which resembles quince.

1. + P. daniellii Daniel (*Cydonia oblonga × Pyrus communis*). Illustration: Revue Horticole, 28–29 (1914).

'Winkleri' has elliptic to ovate leaves to 4.5 cm, which are more downy.

53. PYRUS Linnaeus
S.G. Knees

Deciduous trees, sometimes with thorny branches. Buds with overlapping scales. Leaves alternate, inrolled in bud, stalked, entire or, very rarely lobed; margins often with forward-pointing teeth. Flowers in umbel-like racemes, appearing with or before the leaves. Petals 5, white, or rarely pinkish, clawed, rounded to broadly oblong; stamens 18–30, anthers usually reddish; styles 2–5, free, closely constricted at base by nectariferous disk; ovules 2 per cell. Fruit a spherical or pear-shaped pome; flesh with grit cells; seeds black or brownish black.

A genus of about 25 species, from Europe to E Asia, south to N Africa. Ornamental trees with conspicuous white flowers and foliage which often turns a good reddish colour in autumn. Propagation of the species is from seed or by grafting on to stock of *P. communis*, while most of the cultivars are grafted. A sunny position and not too moist soil will suit most, and several of the Mediterranean species can tolerate long dry periods.

1a. At least some branches terminating in thorns or spines   2
  b. Branches without thorns   14
2a. Leaves remaining hairy or downy, even when mature
         15. eleagrifolia
  b. Leaves hairless when young, or becoming so with maturity   3
3a. Leaf-blades hairless when young, very glossy   4
  b. Leaf-blades downy or hairy when young, eventually hairless   8
4a. Leaf-stalks 2–5 cm   7
  b. Leaf-stalks 5–6 cm   5
5a. Leaves 2.5–5 cm   **11. fauriei**
  b. Leaves at least 7 cm   6
6a. Leaf-margin scalloped
         9. calleryana
  b. Leaf-margin deeply 3–5 lobed
         10. dimorphophylla
7a. Leaf-blades ovate, wedge-shaped, rounded, or cordate at base; stalk often longer than blade
         2. cordata

  b. Leaf-blades ovate to oblong-lanceolate; stalk as long or shorter than blade   **4. syriaca**
8a. At least some leaves pinnately divided or lobed   7
  b. All leaves entire   8
9a. Petals 0.6–1.2 cm   10
  b. Petals 1.5–1.7 cm   **1. communis**
10a. Petals circular   **7. pashia**
   b. Petals ovate to oblong   **19. regelii**
11a. Fruit with deciduous sepals
         3. cossonii
   b. Fruit with persistent sepals   11
12a. Leaf-blades 5–9 mm wide   13
   b. Leaf-blades 1–2 cm wide
         12. amygdaliformis
13a. Branches pendent   **5. salicifolia**
   b. Branches spreading   **20. salvifolia**
14a. Leaf-blades with acute tips   15
   b. Leaf-blades acuminate or obtuse
         16
15a. Leaves 3–6 cm, leaf-stalk hairy
     15. elaegrifolia var. **kotschyana**
   b. Leaves 5–8 cm, leaf-stalk hairless
         14. nivalis
16a. Styles 2 or 3   17
   b. Styles 4 or 5   18
17a. Leaf-margin with broad, spreading teeth; leaf-stalk 2.5–6 cm
         17. phaeocarpa
   b. Leaf-margin with saw-like teeth; leaf-stalk 2–3.5 cm   **8. betulifolia**
18a. Young shoots hairless or nearly so
         19
   b. Young shoots covered with felted hairs   20
19a. Shoots of second year purplish brown; leaf-blade long acuminate
         13. bretschneideri
   b. Shoots of second year yellowish brown; leaf-blade shortly acuminate
         6. ussuriensis
20a. Petals 7–9 mm   **16. × michauxii**
   b. Petals 1.5–1.7 cm   21
21a. Leaf base truncate   **6. ussuriensis**
   b. Leaf base rounded, cordate or broadly wedge-shaped
         18. pyrifolia

1. P. communis Linnaeus. Illustration: Ross-Craig, Drawings of British plants **9**: pl. 27 (1956); Polunin & Everard, Trees and bushes of Europe, 75 (1976); Phillips, Trees in Britain and North America, 179, 219 (1978); Krüssmann, Manual of cultivated broad-leaved trees & shrubs **3**: 77 (1986).

Tree to 15 m, with broad crown and occasionally thorny branches. Leaf-blade 4–5 × 2.5–3.5 cm, ovate to elliptic, almost hairless, acuminate, margins with forward-pointing teeth; leaf-stalks reddish, *c.* 1.5

cm. Flowers in groups of 7–9, often tinged pink in bud. Calyx-lobes 7–9 mm, lanceolate, covered with light brown hairs. Petals 1.5–1.7 × 1–1.2 cm, ovate-oblong, free. Stamens 18–20, anthers deep pinkish red. Styles 3–5. Fruit pear-shaped to spherical, 2–5 cm, yellowish green. *S Europe, SW Asia.* H1. Spring.

Numerous cultivated variants exist.

2. P. cordata Desvaux. Illustration: Decaisne, Jardin fruitier du Muséum **1**: pl. 3 (1858); Ross-Craig, Drawings of British plants **9**: pl. 28 (1956).

Like *P. communis*, but smaller in all its parts, shrubby, rarely exceeding 4 m. Leaves hairless; blades 1–4 × 2–3 cm, ovate, or rounded, often cordate or wedge-shaped at base; leaf-stalks 3–5 cm. Petals 8–10 × 5–9 mm. Fruit spherical to obovoid, 1–1.8 cm across, not tapering towards the stalk, brown with white spots, eventually turning reddish. *SW England, SW Europe (France, Iberian Peninsula)*. H3. Spring.

3. P. cossonii Rehder (*P. longipes* Cosson & Durieu; *P. communis* Linnaeus var. *longipes* (Cosson & Durieu) Henry). Illustration: Decaisne, Jardin fruitier du Muséum **1**: pl. 4 (1858).

Closely related to *P. cordata* but differing in its more elliptic leaves. Leaf-blades 2.5–5 × 1–3 cm, white woolly beneath, eventually hairless; leaf-stalks 2.5–5 cm, reddish. Petals 1.2–1.5 × 0.9–1.1 cm. Fruits spherical, *c.* 1.5 cm across, brown; calyx-lobes deciduous. *N Africa (Algeria)*. H3. Spring.

4. P. syriaca Boissier. Illustration: Decaisne, Jardin fruitier du Muséum **1**: pl. 9 (1858) Meikle, Flora of Cyprus **1**: 632 (1977); Krüssmann, Manual of cultivated broad-leaved trees & shrubs **3**: 77, pl. 26 (1986).

Similar to *P. cossonii*. Tree to 6 m, with erect branches and spreading twigs. Leaves hairless; blades 2–5 cm, oblong-lanceolate; margin finely scalloped; leaf-stalk 2–5 cm, green. Flowers in groups of 6–10 in dense hemispherical corymbs. Petals 7–9 × 5–6 mm, oblong-ovate. Fruits nearly spherical, 2–2.5 cm across, greenish yellow, on stalks to 5 cm. *Cyprus, SW Asia.* H4. Spring.

5. P. salicifolia Pallas. Illustration: Decaisne, Jardin fruitier du Muséum **1**: pl. 12 (1858); Phillips, Trees in Britain and North America, 180 (1978); Krüssmann, Manual of cultivated broad-leaved trees & shrubs, **3**: pl. 27, 28, (1986).

Tree 5–8 m, branches usually pendent. Young shoots very grey, densely covered in forward-pointing silky hairs; short shoots ending in a spine. Leaves 3–5 × 0.7–0.9 cm, lanceolate to narrowly elliptic, wedge-shaped at base, pale greyish green above, whitish green beneath, darkening and becoming hairless with age, stalkless. Flowers in groups of 5–7. Calyx-lobes *c.* 3 mm, broadly triangular, obtuse. Petals 9–12 × 8–9 mm, broadly ovate, free. Stamens 20–22, anthers deep pinkish red. Styles 5. Fruit 2–3 cm, greenish, on short stalks. *W Asia (Caucasus to Turkey)*. H1. Spring.

Most commonly grown as the cultivar 'Pendula' which is said to be a more pendent selection than the wild species.

6. P. ussuriensis Maximowicz. Illustration: Decaisne, Jardin fruitier du Muséum **1**: pl. 5 (1858); Krüssmann, Manual of cultivated broad-leaved trees & shrubs, **3**: 74, 77, pl. 27 (1986).

Tree to 15 m. Young twigs yellowish brown, hairless. Leaves eventually hairless, very glossy on both surfaces; stalks 2.5–4.5 cm, green; blade 5.5–8 × 3.5–4 cm, broadly ovate to elliptic, acuminate, base truncate, margin with forward-pointing teeth. Flowers in groups of 5–8. Calyx-lobes 4–6 mm, ovate, margin with red teeth. Petals 1.5–1.9 × 1.3–1.7 cm, sometimes 3-lobed towards apex, overlapping. Stamens *c.* 20, anthers deep reddish pink; styles 5. Fruit nearly spherical, greenish yellow, 3–4 cm across, on short thick stalks. *E Asia (E former USSR, Korea to Japan)*. H1. Spring.

Var. **hondoensis** (Kikuchi & Nakai) Rehder, differs in its rusty brown, hairy young leaves which are slightly narrower than in var. **ussuriensis**. *C Japan*.

Var. **ovoidea** (Rehder) Rehder. Illustration: Krüssmann, Manual of cultivated broad-leaved trees & shrubs **3**: 77, pl. 27 (1986). Differs in its more ovate leaves and distinctly ovoid fruits with yellow flesh. *N China, Korea*.

7. P. pashia D. Don. Illustration: Decaisne, Jardin fruitier du Muséum **1**: pl. 7 (1858); Polunin, Flowers of the Himalaya, pl. 35 (1984); Krüssmann, Manual of cultivated broad-leaved trees & shrubs, **3**: 73, pl. 27 (1986).

Tree 10–12 m, usually with thorns. Branches hairy when young, soon becoming hairless. Leaves hairy when young, stalks reddish, 3.5–4 cm; blade 7–9 × 3.7–4.5 cm, broadly ovate, apex acuminate, base rounded, margin with forward-pointing teeth. Flowers 6–8, flower-stalks hairy. Calyx-lobes *c.* 3 mm, rounded, concave. Petals 1–1.2 × 0.8–1 cm, circular. Stamens *c.* 20, anthers pale reddish pink. Styles 5. Fruit almost spherical, *c.* 2 cm across, brown, warty, on stalks 2–3 cm. *Himalaya to W China*. H1. Spring.

8. P. betulifolia Bunge. Illustration: Decaisne, Jardin fruitier du Muséum **1**: pl. 20 (1858); Bailey, Standard cyclopedia of horticulture **3**: 2869 (1927); Duke & Ayensu, Medicinal plants of China **1**: 552 (1985); Krüssmann, Manual of cultivated broad-leaved trees & shrubs, **3**: 73, 74, pl. 27 (1986).

Slender tree, 5–10 m; young branches thickly covered with persistent greyish felt, hairless in second year. Leaves hairy when young, with down persisting on veins beneath throughout the year. Blades ovate to oblong or diamond-shaped, acuminate, tapered or rounded at base, margin with coarse forward-pointing teeth, glossy dark green above, stalks 2–3.5 cm. Flowers in groups of 8–12; calyx-teeth 2–3 mm, triangular; petals 7–9 mm, ovate to oblong; styles 2 or 3. Fruit nearly spherical, 1–1.5 cm across, brown with whitish lenticels. *N China*. H2. Spring.

9. P. calleryana Decaisne. Illustration: Krüssmann, Manual of cultivated broad-leaved trees & shrubs, **3**: 71, 73, pl. 27 (1986).

Tree to 16 m, branches hairless, often very thorny. Leaf-blades 7–10 × 4.5–5.5 cm, ovate to broadly ovate, shortly acuminate, hairless; margin scalloped, stalks 5–6 cm. Flowers in groups of 6–12; calyx-lobes with shaggy hairs, 4–5 mm; petals 1.1–1.4 × 0.9–1.1 cm, oblong to oblong to ovate, rounded, often notched at apex; styles 2 or 3. Fruit spherical, *c.* 1 cm across, brownish; calyx-lobes deciduous. *C & S China*. H3. Early spring.

Two selections commonly grown are 'Bradford', a thornless, fast-growing cultivar with good autumn colour and rusty brown fruits; and 'Chanticleer' (illustration: Brickell, RHS gardeners' encyclopedia of plants and flowers, 48, 1989) with a narrowly conical crown and reddish purple leaves in autumn.

10. P. dimorphophylla Makino.
Similar to *P. calleryana* but with deeply 3–5-lobed young leaves. *Japan*. H3. Spring.

11. P. fauriei Schneider.
Similar to *P. calleryana*, but smaller. Leaves 2.5–5 cm, elliptic, narrowly wedge-shaped at base; leaf-stalk 2–2.5 cm, downy. Flowers in clusters of 2–8. Fruit *c.* 1.3 cm. *Korea*.

12. P. amygdaliformis Villmorin. Illustration: Bean, Trees & shrubs hardy in the British Isles, **3**: 446 (1976); Phillips, Trees in Britain and North America, 179 (1978); Polunin, Flowers of Greece and the Balkans, pl. 18 (1980); Krüssmann, Manual of cultivated broad-leaved trees & shrubs **3**: 72, pl. 27 (1986).

Shrub or tree to 16 m, usually with thorny branches. Young shoots with long thin hairs. Leaves variable; blades 2.5–7 × 1–2 cm, ovate, oblong or obovate, acute or obtuse, eventually hairless; leaf-stalk 1–3 cm. Flowers in groups of 8–12; calyx lobes 4–6 mm, lanceolate; petals 1–1.2 × 1–1.1 cm, broadly ovate; stamens *c.* 20, anthers dark red. Fruits spherical, 2–3 cm across, yellowish brown, on stalks 2–3 cm. *S Europe*. H3. Spring.

13. P. bretschneideri Rehder. Illustration: Bailey, Standard cyclopedia of horticulture **3**: 2869 (1927); Krüssmann, Manual of cultivated broad-leaved trees & shrubs, **3**: 76 (1986).

Tree to 15 m, with almost hairless branches, purplish brown in second year. Leaf-blades 5–11 cm, ovate to elliptic, eventually hairless; apex long acuminate, margins with bristly, forward-pointing teeth; leaf-stalks 2.5–7 cm. Flowers in groups of 7–9. Petals 8–10 mm, ovate to oblong. Stamens *c.* 20, styles 4 or 5. Fruit 2.5–3 cm, with white flesh, spherical to ovoid, on stalks 3–4.5 cm. *N China*. H2. Spring.

14. P. nivalis Jacquin. Illustration: Polunin & Everard, Trees and bushes of Europe, 77 (1976); Krüssmann, Manual of cultivated broad-leaved trees & shrubs **3**: 72, pl. 27 (1986).

Small thornless tree to *c.* 13 m. Young shoots thickly covered with white woolly hairs. Leaves downy when young, eventually almost hairless above; blade 5–8 × 2–4 cm, elliptic to obovate, acute, wedge-shaped at base; margin entire or very shallowly scalloped towards apex; leaf-stalk 1–3 cm, whitish green. Flowers in groups of 6–9; calyx-lobes 3–4 mm, triangular, covered with very silky hairs; petals 1–1.2 × 0.7–0.8 cm, ovate; stamens *c.* 20; styles 5. Fruit 3–5 cm across, yellowish green, spherical, on stalks 3–6 cm. *C & SE Europe*. H2. Spring.

15. P. eleagrifolia Pallas. Illustration: Decaisne, Jardin fruitier du Muséum **1**: pl.

17 (1858); Polunin & Everard, Trees and bushes of Europe, 76 (1976); Polunin, Flowers of Greece and the Balkans, pl. 18 (1980); Krüssmann, Manual of cultivated broad-leaved trees & shrubs **3**: 72, pl. 27 (1986).

Small tree or shrub to 7 m, similar to *P. nivalis*, but with thorny branches, and smaller in all its parts. Leaf-blades 4–7 × 1.2–2.5 cm, usually covered on both surfaces with white or greyish woolly hairs. Fruit 2–2.5 cm across, greenish. *Turkey.* H4. Spring.

Var. **kotschyana** (Decaisne) Boissier, illustration: Decaisne, Jardin fruitier du Muséum **1**: pl. 18 (1858), is more shrubby, usually thornless, and has shorter, wider leaves.

16. P. × michauxii Poiret (*P. amygdaliformis × P. nivalis*). Illustration: Decaisne, Jardin fruitier du Muséum **1**: pl. 16 (1858).

Small thornless tree to 9 m, with rounded crown. All parts covered with white woolly hairs when young. Leaf-blade 3–7 × 2–3.5 cm, ovate to oblong-elliptic, obtuse or abruptly acuminate, eventually glossy and hairless above; margin entire; leaf-stalks 2–4 cm. Flowers in groups of 8 or 9. Calyx-lobes 3–4 mm, narrowly triangular; petals 7–9 × 5–6 mm; stamens 18–21, anthers dark red; styles 4 or 5. Fruit *c.* 3 × 2 cm, spherical or top-shaped, yellowish green, spotted brown. *Origin uncertain, possibly Middle East.* H3. Spring.

17. P. phaeocarpa Rehder. Illustration: Moeller, Deutsche Gaertnerzeitung **31**: 112 (1916); Bailey, Standard cyclopedia of horticulture **3**: 2870 (1927).

Tree to 15 m, young branches covered in downy hairs, eventually reddish brown and hairless. Leaf-blades 6–10 cm, ovate-elliptic to oblong-ovate, with long hairs when young, eventually hairless; margin with spreading teeth, apex long acuminate, base broadly wedge-shaped; leaf-stalks 2–6 cm. Flowers in groups of 3 or 4; petals 8–10 mm, ovate-oblong; styles 2 or 3 (rarely 4). Fruit 2–2.5 cm, pear-shaped, brown with paler spots. *N China.* H2. Spring.

18. P. pyrifolia (Burman) Nakai (*P. serotina* Rehder). Illustration: Botanical Magazine, 8226 (1908); RHS Dictionary **4**: 1722 (1974); Krüssmann, Manual of cultivated broad-leaved trees & shrubs **3**: 76, pl. 26, 27 (1986); Hyashi, Woody plants of Japan, 324, 325 (1988).

Tree 5–12 m, young shoots with tufts of woolly hairs, hairless in second year. Leaf-blades 6.5–11 × 3.5–6 cm, oblong-ovate to ovate, apex acuminate, base rounded, cordate, or broadly wedge-shaped; eventually hairless; margin with bristly forward-pointing pointing teeth; leaf-stalk 3–4.5 cm. Flowers in groups of 6–9, calyx-teeth 4–5 mm, lanceolate; petals 1.5–1.7 × 1.1–1.3 cm, ovate-elliptic; stamens 18–21; styles 4 or 5. Fruit 2.5–3.5 cm across, spherical, brownish, dotted with white lenticels. *C & W China.* H3. Spring.

Var. **culta** (Makino) Nakai has longer, wider leaves, to 15 cm and larger, apple-shaped, yellow or brown fruit.

Var. **stapfiana** (Rehder) Rehder has longer pear-shaped fruit to 6 cm.

19. P. regelii Rehder (*P. heterophylla* Regel & Schmalhausen. Illustration: Bean, Trees & shrubs hardy in the British Isles **3**: 452 (1976); Krüssmann, Manual of cultivated broad-leaved trees & shrubs, **3**: 72, pl.27 (1986).

Shrub or small tree, 5–9 m, often with thorny twigs. Leaves very variable, covered with soft downy hairs when young; blade 2–6 cm, entire or with 1–7 shallow or deep lobes; margins coarsely and irregularly toothed; leaf-stalks 2–6 cm. Flowers in groups of 5–7. Petals 6–9 mm, ovate to oblong. Fruit 2–3 cm across, pear-shaped or spherical. *Turkestan.* H3. Spring.

20. P. salvifolia de Candolle. Illustration: Edwards's Botanical Register, t. 1482 (1832); Decaisne, Jardin fruitier du Muséum **1**: pl. 21 (1858); Schneider, Illustriertes Handbuch der Laubholzkunde **1**: f. 361 (1906).

Very thorny tree to 15 m, closely allied to *P. nivalis* and believed to be a hybrid between it and *P. communis*. Leaf-blades *c.* 5 cm, leaf-stalks 4–6 cm. Fruit *c.* 2.5 cm, pear-shaped, on stalks 2–3 cm. *S Europe (France eastward to Hungary).* H3. Spring.

54. × SORBOPYRUS Schneider
S.G. Knees & P.G. Barnes
Tree to 15 m; branches and buds covered in short whitish hairs. Leaves 6–10 cm, broadly elliptic, base rounded, apex acute, irregularly and coarsely toothed, hairy beneath. Flowers in 5-many-flowered corymbs, on slender stalks. Sepals 5, *c.* 5 mm, triangular, downy; petals 5, *c.* 9 × 6 mm, creamy white; stamens *c.* 20, anthers pale pink; styles 2–5. Fruit *c.* 2.5 cm across, pear-shaped, downy, green or reddish yellow.

1. × S. auricularis (Knoop) Schneider (*Pyrus communis × Sorbus aria*). Illustration: Edward's Botanical Register, t. 1437 (1831).

Garden origin. H4. Spring–summer.

Var. **bulbiformis** (Tatar) Schneider is closer to *Pyrus*, but has puckered leaves with coarse forward-pointing teeth; sepals irregularly recurved, to 6 mm; petals *c.* 1.3 × 1.1 cm; *c.* 25 deep pink anthers and fruits to 4 cm across.

55. MALUS Miller
H.S. Maxwell, S.G. Knees, & M.F. Gardner
Deciduous trees or shrubs, occasionally semi-evergreen; lateral shoots often thorned. Leaves alternate, folded or rolled in bud, toothed or lobed, green to reddish purple. Flowers bisexual, stalked, borne in umbel-like clusters; calyx of 5 persistent lobes; petals 5, semi-circular to broadly ovate, white, rose or crimson; stamens 15–50, anthers rounded, ovary inferior, 3–5-celled, styles 2–5, always united at base, basal part with long shaggy hairs; fruit fleshy, with or without grit cells, almost spherical, calyx mostly persistent; seeds 1 or 2 per cell, brownish or black.

A genus of 25–30 species occurring throughout temperate parts of the northern hemisphere. As all the species freely hybridise with each other seed cannot be relied on as an effective means of propagation. Although some can be rooted from cuttings taken from leafless shoots during the winter, most are increased by grafting.

1a. Leaf-margins irregularly and deeply toothed or often lobed   2
  b. Leaf-margins shallowly or evenly toothed but never lobed   15
2a. Flowers borne in umbel-like clusters; leaves usually with 2 lobes towards the base   3
  b. Flowers in corymbs; leaves variously lobed   5
3a. Fruit 1.6–3 cm, yellow   **1. spectabilis**
  b. Fruit 0.6–1.5 cm, yellowish red to brown   4
4a. Umbels almost stalkless; fruits 1–1.5 cm, yellowish red   **2. toringoides**
  b. Umbels stalked; fruits 0.6–0.8 cm, red or yellowish brown   **3. sieboldii**
5a. Leaves 2–3 cm, deeply cut almost to midrib   **4. transitoria**
  b. Leaves 3 cm or more, leaves lobed but not deeply cut   6

6a. Flowers 1–2 cm across   7
  b. Flowers 2.5–4 cm across   11
7a. Leaves with conspicuous densely felted hairs beneath, margins lobed for most of their length; styles 5   8
  b. Leaves almost hairless or with scattered hairs below; margins irregularly lobed; styles 3 or 4   9
8a. Leaves 5–7 cm; lobes pointed, often ½ distance to midrib or more   **5. florentina**
  b. Leaves 6–12 cm; lobes rounded, not more than ⅙ distance to midrib or less   **6. yunnanensis**
9a. Leaves narrowly ovate-elliptic or lanceolate; flower-stalks and calyx covered with dense hairs   **7. fusca**
  b. Leaves broadly ovate to rounded; flower-stalks and calyx hairless   10
10a. Leaves with 1–2 pairs of lobes, usually towards apices; flowers *c.* 1.5 cm across; fruit yellow or purple   **8. kansuensis**
  b. Leaves with 2 or more pairs of lobes; flowers *c.* 2 cm across; fruit yellow or green   **9. honanensis**
11a. Flowers 2.5–3 cm across; fruit 1.5–2.5 cm   12
  b. Flowers 3–4 cm across; fruit 3–4 cm   13
12a. Leaves 3–7 cm, almost evergreen, hairless when mature   **10. angustifolia**
  b. Leaves 7–12 cm, deciduous, remaining hairy, especially below   **11. tschonoskii**
13a. Flower-stalks and outer calyx-lobes covered with matted hairs   **12. ioensis**
  b. Flower-stalks and outer calyx-lobes hairless or with few scattered hairs   14
14a. Leaves on flowering-shoots distinctly lobed; fruit yellow, fragrant   **13. glaucescens**
  b. Leaves on flowering-shoots unlobed or only slightly lobed; fruit green ribbed at apex   **14. coronaria**
15a. Leaves 3–4 times as long as wide; blade narrowly ovate-elliptic to oblong   16
  b. Leaves usually not more than twice as long as wide; blade broadly ovate to rounded   22
16a. Fruits 3–4 cm, green, even when mature   17
  b. Fruits 0.6–1.5 cm, red or brownish   18
17a. Flower-stalks and calyx-lobes hairy   **12. ioensis**
  b. Flower-stalks and calyx-lobes hairless   **14. coronaria**
18a. Flower-stalks 0.5–1.5 cm   **16. brevipes**
  b. Flower-stalks 2–5 cm   19
19a. Flowers 2.5–3 cm across   20
  b. Flowers 3–4 cm across   21
20a. Leaf-margin with jagged teeth; fruit 0.8–1 cm   **15. floribunda**
  b. Leaf-margin smooth or with fine teeth; fruit 1.3–1.5 cm   **17. sikkimensis**
21a. Flowers usually deep pink; fruit 0.6–0.8 cm, red or brownish   **18. halliana**
  b. Flowers white, rarely pink tinged; fruit 0.8–1 cm, red or yellow   **19. baccata**
22a. Fruit 1–1.5 cm   23
  b. Fruit 2–6 cm   24
23a. Flowers *c.* 2 cm across, white   **20. prattii**
  b. Flowers 3.5–4 cm across, pink at first, later white   **21. hupehensis**
24a. Flowers *c.* 3 cm across   25
  b. Flowers 4–5 cm across   26
25a. Leaves densely covered with matted hairs beneath; fruit *c.* 2.5 2.5 cm, brownish yellow, flushed purple   **11. tschonoskii**
  b. Leaves hairless or with few scattered hairs beneath; fruit *c.* 2 cm, yellowish green or red   **22. prunifolia**
26a. Leaves always green; young branches usually thorny; outer calyx lobes hairless; petals white, often tinged pink   **23. sylvestris**
  b. Leaves often bronze or purplish; young branches thornless; outer calyx lobes hairy; petals white or deep pink   **24. pumila**

1. M. spectabilis (Aiton) Borkhausen (*M. spectabilis* var. *plena* Bean; *Pyrus spectabilis* Aiton). Illustration: Botanical Magazine, 267 (1794); Bailey, Standard cyclopedia of horticulture, 3292 (1950) Journal of the Royal Horticultural Society **86**: 42 (1961); Phillips, Trees in Britain, Europe and North America, 139 (1978).

Tall shrub or small tree to 8 m; young branches downy, eventually reddish brown. Leaves 5–8 cm, elliptic to oblong, dark green above, shiny, paler below, downy; apex shortly acuminate, margins with adpressed teeth and occasionally with 2 lobes towards the base. Flowers borne in umbel-like clusters, dark pink in bud, eventually whitish; stalks 2–3 cm; calyx hairless to almost downy, lobes triangular to ovate; corolla 4–5 cm across, occasionally semi-double. Fruit 1.6–3 cm, yellow. *Believed to be from China but not known in the wild.* H2. Spring.

2. M. toringoides (Rehder) Hughes (*M. transitoria* var. *toringoides* Rehder; *Pyrus transitoria* var. *toringoides* Bailey; *P. toringoides* Osborne, *Sinomalus toringoides* Koidzumi). Illustration: Botanical Magazine, 8948 (1922); Krüssmann, Manual of cultivated broad-leaved trees & shrubs **2**: pl. 121 (1986).

Shrub or tree to 8 m; branches hairy at first eventually hairless. Leaves 3–8 cm, ovate, usually with 2 scalloped lobes on either side, rarely entire, eventually hairless except on the veins beneath. Flowers 3–6, in almost stalkless umbel like clusters; calyx with densely felted hairs, corolla *c.* 2 cm across, white, slightly fragrant. Fruit 1–1.5 cm, spherical to almost pear-shaped, yellow or red, persistent. *W China.* H2. Late spring.

3. M. sieboldii (Regel) Rehder (*M. toringo* Nakai; *Pyrus sieboldii* Regel; *P. toringo* Siebold). Illustration: Bailey, Standard cyclopedia of horticulture **2**: 3295 (1950); Makino, Illustrated flora of Japan, 1407 (1956); Massachusetts Horticultural Society, Horticulture, **34**(10): 506 (1956); Krüssmann, Manual of cultivated broad-leaved trees & shrubs **2**: pl. 121 (1986).

Low shrub or small tree to 4 m; branches blackish brown, nodding. Leaves 3–6 cm, ovate-elliptic, coarsely toothed, with 3–5 lobes, dark green above paler below, downy on both sides. Flowers in umbel-like clusters of 3–6, stalks 2–2.5 cm, corolla *c.* 2 cm, pale pink fading to white. Fruit 6–8 mm, spherical, red to yellowish brown, persistent. *Japan.* H2. Late spring.

Var. **sargentii** Rehder. Low-growing shrub to 2 m, with very thorny dense, spreading branches and larger flowers to 2.5 cm across, fruit dark red. Sometimes treated as a separate species.

M. × sublobata (Dippel) Rehder (*M. ringo* forma *sublobata* Dippel). Illustration: Massachusetts Horticultural Society, Horticulture **44**(30): 20 (1966). Similar but with leaves 4–8 cm and 1 or 2 lobes. Flowers to 4 cm across; fruits to 1.5 cm yellow. *Only known in cultivation.* H2. Late spring.

A hybrid between *M. prunifolia* and *M. sieboldii.*

M. × zumi (Matsumura) Rehder (*Pyrus zumi* Matsumura). Illustration: Krüssmann, Manual of cultivated broad-leaved trees & shrubs **2**: 287 (1986). Pyramidal tree;

leaves 5–9 cm; flowers *c.* 3 cm across; fruits *c.* 1 cm, red.

A hybrid between *M. sieboldii* and *M. baccata* var. *mandshurica*.

4. M. transitoria (Batalin) Schneider (*Pyrus transitoria* Batalin; *Sinomalus transitoria* Koidzumi). Illustration: Icones cormophytorum Sinicorum, 2207 (1972); Krüssmann, Manual of cultivated broad-leaved trees & shrubs **2**: pl. 121 (1986).

Very closely related to *M. toringoides* but smaller and branches densely felted when young. Leaves 2–3 cm, with deeper narrower lobes, very rarely entire. Flowers borne in corymbs; fruit usually less than 1 cm, bright red. *NW China*. H2. Late spring.

5. M. florentina (Zuccagni) Schneider (*Pyrus crataegifolia* Savi; *Crateagus florentina* Zuccagni; *Eriolobus florentina* (Zuccagni) Stapf). Illustration: Botanical Magazine, 7423 (1895); Brickell, RHS dictionary of gardening **3**: 1239 (1956); Phillips, Trees in Britain, Europe and North America, 137 (1978); Krüssmann, Manual of cultivated broad-leaved trees & shrubs **2**: pl. 121 (1986).

Deciduous tree *c.* 8 × 6 m, pyramidal when young, becoming rounded; branches slender, dark brown, twigs with long shaggy hairs when young. Leaves 3–7 cm, broadly ovate, irregularly sharply toothed, wedge-shaped or heart-shaped at base, dull green above, grey-yellow felted hairs beneath, leaf-stalk 0.5–2 cm, downy, reddish. Flowers 1.5–2 cm across, in loose clusters of 2–6, flower-stalks slender, *c.* 2.5 cm, downy, pinkish; calyx very woolly, lobes narrow, pointed, falling early; petals white. Fruit *c.* 1 cm, round-ovate, stone-cells present, yellowish turning red. *Italy, former Yugoslavia & Greece*. H2. Summer.

Considered by some authorities to be × *Malosorbus florentina* (Zuccagni) Browicz, a bigeneric hybrid of *Malus sylvestris* × *Sorbus torminalis*.

6. M. yunnanesis (Franchet) Schneider (*Pyrus yunnanensis* Franchet). Illustration: Botanical Magazine, 8629 (1915); Icones cormophytorum Sinicorum, 2210 (1972); Krüssmann, Manual of cultivated broad-leaved trees & shrubs **2**: pl. 119 (1986).

Tree to 10 m, with narrow upright habit; shoots with densely felted hairs when young. Leaves 6–12 cm, broadly ovate, base rounded to cordate, margin double-toothed, sometimes with 3–5 pairs of lobes, densely felted beneath. Flowers borne in corymbs, 4–5 cm across; calyx with shaggy hairs; corolla *c.* 1.5 cm

across, white, styles 5. Fruit 1–1.5 cm, deep red, often dotted. *W China*. H2. Late spring.

Var. **veitchii** Rehder. Differs in having distinctly cordate leaf-bases and flowers less than 1.2 cm across. Fruits to 1.3 cm with white dots. *C China*. H2. Late spring.

7. M. fusca (Rafinesque) Schneider (*M. rivularis* (Hooker) Roemer; *Pyrus fusca* Rafinesque; *P. rivularis* Hooker). Illustration: Sargent, The silva of North America **4**: 170 (1892); Botanical Magazine, 8798 (1919); Elias, The complete trees of North America, 604 (1980); Krüssmann, Manual of cultivated broad-leaved trees & shrubs **2**: pl. 117 (1986).

Deciduous shrub or tree, 6–8 m; crown rounded, spreading; bark with large flattened scales, reddish brown; shoots hairy when young. Leaves 3–10 × 1.5–4 cm, lanceolate, sharply toothed, 3-lobed on long shoots, leaf-stalks hairy. Flowers 1.5–2.2 cm across, in clusters of 6–12, flower-stalks hairy; calyx hairy, falling early; petals white to pinkish white; styles 3 or 4. Fruit 1.5 cm, broadest near the tip, yellow to red. *W North America*. H2. Spring–summer.

8. M. kansuensis (Batalin) Schneider (*Pyrus kansuensis* Batalin; *Eriolobus kansuensis* (Batalin) Schneider. Illustration: Icones cormophytorum Sinicorum, 2206 (1972); Krüssmann, Manual of cultivated broad-leaved trees & shrubs **2**: pl. 121 (1986).

Shrub or small tree to 5 m; young branches downy at first eventually hairless and reddish brown. Leaves 5–8 cm, broadly ovate with 3–5 lobes, usually towards the apex, margin finely toothed, dark green above, paler and hairy below. Flowers in corymbs of 4–10; calyx with shaggy hairs, corolla *c.* 1.5 cm across, creamy-white, styles 3. Fruit *c.* 1 cm, yellow to purple with paler dots. *NW China*. H2. Late spring.

9. M. honanensis Rehder (*Sinomalus honanensis* (Rehder) Koidzumi). Illustration: Icones cormophytorum Sinicorum, 2209 (1972).

Very like *M. kansuensis* but slower growing and with thinner twigs. Leaves 6–8 cm, broadly ovate, rarely ovate-oblong, with 2–5 pairs of ovate toothed lobes, downy beneath. Flowers to 10, in hairless corymbs; corolla *c.* 2 cm across, white; styles 3 or 4. Fruit *c.* 1 cm, yellow to green. *NE China*. H2. Late spring.

10. M. angustifolia (Aiton) Michaux (*Pyrus angustifolia* Aiton). Illustration: Sargent, The silva of North America **4**: 169 (1892); Britton & Brown, Illustrated flora of the Northern States and Canada **2**: 234 (1897); Elias, The complete trees of North America, 603 (1980); Krüssmann, Manual of cultivated broad-leaved trees & shrubs **2**: pl. 121 (1986).

Semi-evergreen tree or shrub, 5–7 m, broad, rounded; bark reddish brown, scaly, narrowly ridged, shoots hairy when young. Leaves 3–7 × 1.5–4 cm, lanceolate-oblong to ovate-oblong, tip rounded or pointed, tapering towards base, entire to coarsely toothed especially towards tip, sometimes lobed, shining green above, paler beneath, hairless when mature. Flowers 2–3 cm across, borne in few-flowered corymbs on short lateral shoots; calyx teeth white, wooly inside; petals rose or nearly white, fragrant; style 1. Fruit 1.5–2.5 cm across, yellowish green, fragrant. *SE North America*. H2. Spring–summer.

11. M. tschonoskii (Maximowicz) Schneider (*Pyrus tschonoskii* Maximowicz; *Eriolobus tschonoskii* Rehder). Illustration: Botanical Magazine, 8179 (1908); Bailey, Standard cyclopedia of horticulture **2**: 3294 (1950); Makino, Illustrated flora of Japan 1411 (1956); Krüssmann, Manual of cultivated broad-leaved trees & shrubs **2**: pl. 119 (1986).

Tree to 12 m; branches erect, pyramidal, shoots with densely felted white hairs. Leaves 7–12 cm, ovate-elliptic to oblong, irregularly toothed and often shallowly lobed, with densely felted white hairs especially below. Flowers in corymbs of 2–5, corolla *c.* 3 cm across, white tinged pink. Fruit 2–3 cm, yellowish green with a red cheek. *Japan*. H2. Late spring.

Foliage especially good for autumn colour, turning red, yellow or purple.

12. M. ioensis (Wood) Britton (*Pyrus coronaria* var. *ioensis* Wood; *P. ioensis* (Wood) Bailey). Illustration: Britton & Brown, Illustrated flora of the Northern States and Canada **2**: 235 (1897); Botanical Magazine, 8488 (1913); Sargent, Manual of the trees of North America, 278 (1933); Krüssmann, Manual of cultivated broad-leaved trees & shrubs **2**: pl. 121 (1986).

Small tree with open habit, developing ornamental peeling bark with age; branches with densely felted hairs at first, becoming reddish brown, hairless. Leaves 5–10 cm, oblong-ovate, coarsely to sharply toothed, often shallowly lobed, dark green above, yellowish green below with densely felted hairs. Flowers in corymbs of 4–6, calyx-lobes covered with matted hairs; corolla *c.* 4 cm across, white or pale pink, fragrant. Fruit *c.* 3 cm, waxy, green, sometimes angled, calyx persisent. *C USA.* H2. Summer.

Several naturally occuring varieties are listed in the American literature including var. **bushii** Rehder, var. **creniserrata** Rehder, var. **palmeri** Rehder and var. **texana** Rehder; however, these are not thought to be widely available in Europe.

M. × denboerii Krüssmann. Differs from *M. ioensis* in its purple to bronze leaves and reddish pink flowers *c.* 2 cm across; fruits to 2.5 cm, reddish. *Cultivated origin.* H2. Summer.

A hybrid involving *M. ioensis* as the female parent, the other parent is uncertain but may be *M. × purpurea.*

M. × soulardii (Bailey) Britton. Illustration: Britton & Brown, Illustrated flora of the Northern States and Canada **2**: 235 (1897). Very like *M. ioensis* but with broader leaves to 8 cm; fruits to 5 cm, flattened, yellowish green. *USA.* H2. Summer.

13. M. glaucescens Rehder (*P. glaucescens* (Rehder) L.H. Bailey. Illustration: Sargent, Trees & Shrubs **2**: 139 (1913); Sargent, Manual of the trees of North America, 381 (1933); Bailey, Standard cyclopedia of horticulture, 3297–3299 (1950).

Shrub or small tree to 8 m; branches hairless, with few sharp thorns. Leaves 5–8 cm, ovate to triangular, with short triangular lobes, dark green above, bluish below, densely hairy at first. Flowers in corymbs of 5–7, calyx slightly shaggy on the outside, densely felted inside, corolla *c.* 3.5 cm, white to pink, fragrant. Fruit 3–4 cm, flattened, greenish yellow, indented at both ends. *N America.* H2. Late spring.

14. M. coronaria (Linnaeus) Miller (*M. bracteata* Rehder in Sargent; *Pyrus coronaria* Linnaeus). Illustration: Botanical Magazine, 2009 (1818); Britton & Brown, Illustrated flora of the Northern States and Canada **2**: 235 (1897); Elias, The complete trees of North America, 604 (1980); Krüssmann, Manual of cultivated broad-leaved trees & shrubs **2**: pl. 119 (1986).

Similar to *M. angustifolia* except shoots white woolly at first. Leaves 5–10 cm, ovate-elliptic to ovate-lanceolate, rounded to heart-shaped at base, irregularly toothed and lobed, hairy at first beneath. Flowers in corymbs of 4–6, flower-stalks and calyx-lobes hairless, corolla 3–4 cm across, pale pink, fragrant. Fruit 3–4 cm across, slightly flattened, greenish, ribbed toward apex. *E North America.* H2. Summer.

Var. **dasycalyx** Rehder. Differs in its smaller flowers to 3 cm wide, calyx often hairy; fruits yellowish green. *C North America.* H2. Summer.

Var. **lancifolia** (Rehder) Fernald (*Pyrus lancifolia* (Rehder) L.H. Bailey. Has more pointed leaves 3–11 cm, lanceolate. *E North America.* H2. Summer.

'Charlottae' (*Pyrus charlottae* misapplied) and 'Nieuwlandiana' (*P. nieuwlandiana* misapplied) both have larger semi-double flowers.

15. M. floribunda Van Houtte (*M. pulcherrima* (Ascherson & Graebner) K.R. Boynton; *Pyrus floribunda* Kirchner not Lindley). Illustration: Revue Horticole pl. 591 (1871); Flore des Serres **15**: pl. 1585 (1862–65); Phillips, Trees in Britain, Europe and North America, 137 (1978); Krüssmann, Manual of cultivated broad-leaved trees & shrubs **2**: pl. 75, pl. 117 (1986).

Shrub or small tree 4–8 m; branches long arching, shoots downy at first, eventually hairless. Leaves 4–8 cm, ovate-acuminate, sharply toothed, especially on the stronger shoots, dark green and hairless above, paler and downy beneath. Flowers in clusters of 4–7, stalks 2.5–4 cm, corolla 2.5–3 cm across, rose in bud, fading to pink or white; styles 4. Fruit 0.8–1 cm, red or yellow, calyx deciduous. *Japan.* H2. Spring.

16. M. brevipes (Rehder) Rehder (*M. floribunda* var. *brevipes* Rehder; *Pyrus brevipes* (Rehder) L.H. Bailey).

Resembling *M. floribunda* but low densely branched shrub; leaves 5–7 cm, ovate, acuminate, finely toothed. Flower-stalks 0.5–1.5 cm, corolla *c.* 3 cm across, white. Fruit 1.5 cm across, bright red, slightly ribbed. *Origin unknown.* H2. Spring.

17. M. sikkimensis (Wenzig) Schneider (*M. pashia* Wenzig var. *sikkimensis* Wenzig; *Pyrus sikkimensis* J.D. Hooker). Illustration: Bailey, Standard cyclopedia of horticulture, 7430 (1950); Bean, Trees & shrubs hardy in the British Isles **2**: 709 (1973); Krüssmann, Manual of cultivated broad-leaved trees & shrubs **2**: pl. 119 (1986).

Tree 5–7 m; young branches covered in woolly hairs. Leaves 5–7 cm, ovate to oblong, margins entire or fine-toothed. Flowers in clusters of 4–9, calyx woolly, corolla *c.* 2.5 cm across, white. Fruit 1.3–1.5 cm across, almost pear-shaped, red or yellow, dotted. *N India.* H2. Late spring.

18. M. halliana Koehne. Illustration: Bailey, Standard cyclopedia of horticulture 3289 (1950); Makino, Illustrated flora of Japan, 1408 (1956); Csapody & Tóth, A colour atlas of flowering trees and shrubs, pl. 203 (1982); Krüssmann, Manual of cultivated broad-leaved trees & shrubs **2**: 286 (1986).

Shrub 2–4 m, rarely a small tree; young shoots soon becoming hairless. Leaves 4–8 cm, oblong-ovate, leathery, hairless. Flowers in clusters of 4–7, corolla 3–4 cm across, dark pink; styles 4. Fruit 6–8 mm, reddish brown. *Japan, China.* H2. Late spring.

Perhaps best known in cultivation as 'Parkmanii' which has paler pink, double flowers.

Var. **spontanea** (Makino) Koidzumi (*M. floribunda* var. *spontanea* Makino). This differs in its vase-shaped habit, almost white flowers and shorter stalks. Fruit yellowish green.

M. × atrosanguinea (Späth) Schneider (*M. floribunda atrosanguinea* misapplied; *Pyrus atrosanguinea* Späth). Illustration: Massachusetts Horticultural Society, Horticulture, **34**(10): 506 (1956) & **35**(2): 76 (1957); American Horticulturist **56**(2): 23 (1977).

Leaves dark green; flowers simple, buds deep rosy red, not fading. Fruit *c.* 1 cm across, red or yellow.

A hybrid between *M. halliana* and *M. sieboldii.*

19. M. baccata (Linnaeus) Borkhausen (*M. sibirica* Borkhausen). Illustration: Icones cormophytorum Sinicorum, 2198 (1972); Bean, Trees & shrubs hardy in the British Isles **2**: 694 (1973); Philips, Trees in Britain, Europe and North America, 136 (1978); Krüssmann, Manual of cultivated broad-leaved trees & shrub **2**: pl. 121 (1986).

Tree or shrub to 5 m; young shoots hairless; leaves 3–8 cm, ovate acuminate, margin with long fine teeth, glossy above, hairless. Flowers borne in small clusters, calyx hairless, lobes acuminate, corolla 3–3.5 cm across, creamy white, fragrant. Fruit *c.* 1 cm, bright red or yellow. *NE Asia to N China.* H2. Spring.

Var. **mandshurica** (Maximowicz) Schneider (*Pyrus baccata* Linnaeus var. *mandshurica* Maximowicz). Illustration; Botanical Magazine, 6112 (1874). Differs in its leaves which are often entire towards the base, and have small distant teeth towards the apex, downy below when young. Calyx-lobes often downy; corolla *c.* 4 cm across, pure white. Fruit 1–1.2

cm, bright red. *C Japan, C China*. H2. Early spring.

M. × hartwigii Koehne. Illustration: Krüssmann, Manual of cultivated broad-leaved trees & shrub **2**: pl. 120 (1986). Differs in its larger leaves, usually 6–9 cm; flowers slightly double, dark pink at first; fruit pear-shaped, yellowish green.

A hybrid between *M. baccata* and *M. halliana*.

20. M. prattii (Hemsley) Schneider (*Pyrus prattii* Hemsley; *Docynopsis prattii* (Hemsley) Koidzumi). Illustration: Lee, Forest botany of China, 170 (1935); Icones cormophytorum Sinicorum, 2208 (1972); Krüssmann, Manual of cultivated broad-leaved trees & shrub **2**: pl. 119 (1986). Upright shrub or tree to 7 m; young shoots hairy at first, soon becoming hairless. Leaves 6–15 cm, ovate-elliptic to oblong, base rounded, slightly hairy beneath. Flowers in clusters of 7–10, corolla *c.* 2 cm, white; styles 5. Fruit 1–1.5 cm across, red or yellow, dotted, calyx persistent. *China (Hubei, Sichuan)*. H3. Spring.

21. M. hupehensis (Pampanini) Rehder (*M. theifera* Rehder). Illustration: Botanical Magazine, 9667 (1943–48); Icones cormophytorum Sinicorum, 2200 (1972); Phillips, Trees in Britain, Europe and North America, 138 (1978); Krüssmann, Manual of cultivated broad-leaved trees & shrub **2**: pl. 121 (1986). Shrub or tree 5–7 m; twigs stiff, outspread, hairy at first soon becoming hairless. Leaves 5–10 cm, ovate to oblong, sharply toothed, base rounded or heart-shaped. Flowers in clusters of 3–7; calyx-lobes hairy, reddish; corolla 3.5–4 cm across, pink at first eventually white. Fruit *c.* 1 cm, greenish yellow, with reddish cheeks, calyx deciduous. *China, India*. H2. Early spring.

22. M. prunifolia (Willdenow) Borkhausen (*Pyrus prunifolia* Willdenow). Illustration: Botanical Magazine, 6158 (1875); Il Giardino Fiorito **47**: 557 (1981); Icones cormophytorum Sinicorum, 2203 (1972); Krüssmann, Manual of cultivated broad-leaved trees & shrub **2**: 286 (1986). Small tree 5–10 m; twigs softly hairy when young. Leaves 5–10 cm, elliptic or ovate, sparsely hairy beneath, margin toothed. Flowers in clusters of 6–10, buds pink; calyx covered with white matted hairs; corolla *c.* 3 cm across, opening pure white. Fruit *c.* 2 cm across, yellow-green or red, calyx persistent. *Probably NE Asia (origin uncertain)*. H2. Spring.

var. **rinki** (Koidzumi) Rehder (*M. pumila* var. *rinki* Koidzumi; *M. ringo* Siebold; *M. asiatica* Nakai; *Pyrus ringo* Wenzig). Illustration: Botanical Magazine, 8265 (1909). Differs only in being slightly more downy, with shorter flower-stalks and usually pinkish flowers. *Cultivated in China*. H2. Spring.

M. × robusta (Carrière) Rehder (*Pyrus microcarpa* C. Koch var. *robusta* Carrière). Illustration: Hay & Synge, The dictionary of garden plants, in colour with house and greenhouse plants, pl. 1703 (1986). Vigorous shrub or small tree; branches spreading, arching. Leaves 7–10 cm, narrowly elliptic; flowers 2.5–4 cm across, mostly white. Fruit 2–2.5 cm, purplish red or yellow.

A hybrid between *M. baccata* and *M. prunifolia*.

'Persicifolia' (*M. robusta* var. *persicifolia* Rehder). Differs in its oblong-lanceolate, peach-like leaves.

23. M. sylvestris (Linnaeus) Miller (*Pyrus malus* Linnaeus var. *sylvestris* Linnaeus). Illustrations: Reichenbach, Icones Florae Germanicae et Helveticae **25**(2): pl. 111 (1914); Ross-Craig, Drawings of British plants, **9**: pl. 35 (1952); Phillips, Trees in Britain, Europe and North America 139 (1978); Blamey & Grey-Wilson, The illustrated flora of Britain and northern Europe 193 (1989).

Deciduous tree or shrub, *c.* 7 m, shoots more or less spiny, short, bark brown, fissured. Leaves 3–11 × 2.5–5.5 cm, ovate to rounded, margin with fine irregular teeth, apex shortly pointed, base broadly wedge-shaped to rounded; leaf-stalk 1.5–3 cm. Flowers 3–4 cm across, borne on short shoots, flower-stalks hairless; sepals 3–7 mm, densely hairy on inside; petals white flushed with pink; anthers yellow; styles hairless or with a few long hairs at base. Fruit 2.5–4 cm, globe-shaped, yellowish green flushed with red. *Europe*. H1. Spring–summer.

Subsp. **mitis** (Wallroth) Mansfeld (*M. domestica* Borkhausen). Shoots densely hairy, not spiny; leaves 4–13 × 3–7 cm, egg-shaped to elliptic with a rounded base, slightly hairy above, densely hairy beneath. Fruits *c.* 5 cm, variable in colour. *Europe*. H1. Spring–early summer.

This is the cultivated apple which is widely naturalised and the parent of numerous orchard hybrids.

24. M. pumila Miller (*M. communis* Poiret; *M. domestica* Borkhausen; *M. pumila* var. *paradisiaca* (Linnaeus) Schneider; *M.*

sylvestris var. *paradisiaca* (Linnaeus) Bailey; *M. dasyphylla* Borkhausen; *Pyrus pumila* (Miller) Tausch; *P. malus* var. *paradisiaca* Linnaeus; *P. malus* var. *pumila* (Miller) Henry). Illustration: Schneider, Handbuch der Laubholzkunde **1**: 718 (1906); Reichenbach, Icones Florae Germanicae et Helveticae **25**(2): pl. 112 (1914); Edlin & Nimmo, The world of trees, 57 (1974); Krüssmann, Manual of cultivated broad-leaved trees & shrubs **2**: 286 (1986).

Small tree, 5–7 m, habit open, twigs usually thornless, young shoots covered with soft down. Leaves 4–10 cm, broadly ovate to rounded, hairy at first, eventually hairless above. Flowers 4–5 cm across; calyx hairy, lobes longer than the cup; corolla white, turning pink. Fruit 2–6 cm, green, indented at each end. *Europe, SE Asia*. H2. Spring.

M. × adstringens Rehder. Flowers usually pink; fruit shortly stalked, 4–5 cm across, red, yellow or green. This is a source of many garden cultivars. A hybrid between *M. baccata* and *M. pumila*.

M. × astracanica Dumont de Courset. Differs from *M. pumila* in its coarser leaves with saw-like teeth, bright red flowers and bloomed fruits. A hybrid between *M. prunifolia* and *M. pumila*.

M. × heterophylla Spach. Differs in its wider leaves, flowers pink in bud, opening white, *c.* 4 cm. Fruits green *c.* 6 cm across.

'Niedzwetzkyana'. (*Malus niedwetzkyana* Dieck). Illustration: Botanical Magazine, 7975 (1904). Shrub to 4 m, branches upright, spreading, young shoots bright red, wood purplish. Leaves bronze-brown; flowers in clusters of 4–7, corolla *c.* 4.5 cm, dark red. Fruit 5–6 cm across, dark red, pulp red.

Much used in hybridisation and the origin of many of the dark leaved and flowered forms.

M. × purpurea (Barbier) Rehder (*Malus floribunda* var. *purpurea* Barbier; *Pyrus purpurea* of gardens). Illustration: Phillips, Trees in Britain, Europe and North America 139 (1978); Csapody & Tóth, A colour atlas of flowering trees and shrubs, pl. 208 (1982). Wood purplish; leaves 8–9 cm, brownish red at first, eventually dark green; flowers 3–4 cm across, purplish red at first, but soon fading. Fruit 1.5–2.5 cm across, purplish red.

A hybrid between *M. pumila* 'Niedzwetzkyana' and *M. × atrosanguinea*.

M. × moerlandsii Doorenbos. Abundantly flowering; fruit 1–1.5 cm across, purple. A hybrid between *Malus × purpurea* 'Lemoinei' and *M. sieboldii*.

56. ERIOLOBUS (de Candolle) Roemer
S.G. Knees

Deciduous trees. Leaves alternate, palmately lobed with long stalks; stipules falling early. Flowers hermaphrodite in terminal stalkless or shortly stalked umbels. Calyx of 5 reflexed lobes, united at the base; corolla 5, white; stamens *c.* 20; ovary inferior, 5-celled; styles 5, united at the base. Fruit spherical to pear-shaped with grit cells, not sunken at calyx; seed 1 or 2, not compressed.

A genus of about 6 species from SE Europe and SW Asia. Propagation is usually from seed or grafting onto *Sorbus* species.

1. E. trilobatus (Poiret) Roemer (*Crataegus trilobata* Poiret; *Pyrus trilobata* (Poiret) de Candolle; *Sorbus trilobata* (Poiret) Heynold; *Malus trilobata* (Poiret) Schneider). Illustration: Schneider, Illustriertes Handbuch der Laubholzkunde **1**: 726, 727 (1906); Botanical Magazine, 9305 (1933); Davis, Flora of Turkey **4**: 89 (1972).

Tree to 6 m; branches upright, young shoots densely covered with white hairs at first, eventually hairless and brownish red. Leaves 1.5–8 × 2–11 cm, usually 3-lobed, with each lateral lobe further divided, lobes acute, margins with saw-like teeth; leaf-stalks about as long as blade. Flowers in clusters of 3–10; sepals 8–12 × 2–3 mm, triangular; corolla *c.* 4 cm across. Fruit *c.* 3 cm across, yellowish green or reddish, calyx persistent; seeds *c.* 10 × 5 mm, brown. *SE Europe, SW Asia & Middle East.* H3. Summer.

57. SORBUS Linnaeus
H. McAllister & N.P. Taylor

Deciduous small trees to rhizomatous shrubs. Buds ovoid-conic, reddish to black, with only 2 or 3 bud-scales visible. Leaves alternate, simple and toothed or pinnate; stipules deciduous. Flowers usually bisexual in flat or pyramidal clusters, white or more rarely pink; sepals and petals 5; stamens 15–20; carpels 2–5, more or less united, each with 2 ovules; ovaries inferior or semi-inferior; styles free or united at the base. Fruit usually a small pome, spherical or ovoid-spherical, orange-red to crimson to white; cells 2–5, with cartilaginous walls, each with 1 or 2 seeds.

A genus of about 100 species from the northern hemisphere. Most are suitable for gardens or the open landscape and are grown for their flowers, decorative fruits, attractive habit and foliage. Almost all species are small, exposure-tolerant, light-demanding, shallow-rooted mountain trees or shrubs preferring light soils and cool moist summers. Many species will grow well even in dry soils. *S. poteriifolia* is best grown in a peat garden. Apomictic species should be grown from seed and, especially with the smaller shrubby species, allowed to form multi-stemmed bushes. Sexual species may need to be grafted, especially those which are self-incompatible such as *S. scalaris*, *S. esserteauiana*, and *S. sambucifolia*. Plants can also be layered or budded on to *S. americana* or *S. aucuparia*

Measurements are mostly given as an upper limit for a healthy vigorous plant because size is so susceptible to environmental factors. Drought greatly reduces the size of many parts, especially leaflets and fruit. Leaf characters are for leaves on fruiting spur shoots; leaves on long vegetative shoots are often larger with more lanceolate much more deeply toothed leaflets, with the toothing extending much further towards the base of each leaflet. Identification is usually easiest in fruit though the pink flowers are a useful distinguishing character of the *S. microphylla* group. Non-fruiting specimens may often be unidentifiable without experience.

Literature: Kalkman, C., *Blumea* **21**(2): 413 (1973).

1a. Leaves compound with 2 or more pairs of free leaflets and a similar terminal leaflet   **2**
  b. Leaves simple, lobed or with one or more pairs of free leaflets at base, but upper part of leaf-blade lobed or merely toothed and not resembling the lateral leaflets   **35**
2a. Fruit orange to vermilion, rarely yellow   **3**
  b. Fruit crimson to white, rarely yellowish or creamy-white   **15**
3a. Fruit yellow   **4**
  b. Fruit orange to vermilion   **6**
4a. Buds ovoid-conic, reddish with rust-coloured hairs; leaflets without papillae beneath at ×100 magnification   **13. 'Joseph Rock'**
  b. Buds ovoid-spherical, dark or blackish red with predominantly white hairs; leaflets with papillae beneath at ×100 magnification   **5**
5a. Fruit broader than long; stipules large, leafy, persistent, even in inflorescence   **2. esserteauiana**
  b. Fruit about as broad as long; stipules small, often soon falling, especially in inflorescence   **1. aucuparia**
6a. Buds very sticky, ovoid, chestnut-coloured   **5. sargentiana**
  b. Buds sticky or not, if sticky then conic, reddish or blackish   **7**
7a. Buds never sticky, ovoid to spherical with predominantly white hairs; leaflets with many papillae beneath at ×100 magnification   **8**
  b. Buds sticky or not, conic with rust-coloured hairs; leaflets without papillae beneath at ×100 magnification   **10**
8a. Stipules small, often soon falling, especially in inflorescence; leaflets not glossy or leathery   **1. aucuparia**
  b. Stipules large, persistent, leafy, even in inflorescence; leaflets glossy, leathery   **9**
9a. Leaflets fewer than 15, large and broad to 9 × 3 cm; fruit broader than long   **2. esserteauiana**
  b. Up to 29 leaflets, each long and narrow to 4.9 × 1 cm, fruit about as broad as long   **3. scalaris**
10a. Shrub with delicate thin twigs; stipules large, leafy, especially in inflorescence; leaves kite-shaped in outline with leaflet size increasing from base of leaf   **4. gracilis**
  b. Not as above   **11**
11a. Leaflets more than 3 times as long as broad, apex drawn out into a fine point   **12**
  b. Leaflets less than 3 times as long as broad, apex more abruptly acute   **14**
12a. Buds very sticky; leaflets *c.* 15 per leaf; gaps visible between calyx-lobes   **6. americana**
  b. Buds more or less sticky or not sticky; leaflets 11–15 per leaf; calyx-lobes overlapping   **13**
13a. Tree or few-stemmed shrub; leaves deep green   **7. commixta**
  b. Many-stemmed shrub; leaves light green and very glossy   **8. scopulina**
14a. Small tree or large shrub; buds large, conic, blackish, sticky; 12–17 leaflets per leaf   **9. decora**
  b. Shrub; buds ovoid, greenish red, more or less sticky; 7–11 leaflets per leaf   **10. cascadensis**
15a. Usually rhizomatous shrubs less than 1 m; leaves less than 10 cm, bearing fewer than 14 leaflets, which are less than 2 cm   **16**
  b. Larger shrub or tree; leaflets larger and/or more numerous   **17**
16a. Shrub more than 20 cm; flowers

white; fruit crimson, becoming
pinkish white   **11. reducta**

b. Shrub less than 20 cm; flowers
pink; fruit crimson, turning white
with pink flecking only around
calyx   **12. poteriifolia**

17a. Fruit yellow-green, ripening to red
29. domestica

b. Fruit not as above   18

18a. Fruit uniformly crimson at first,
becoming almost white (if flower
pink see lead 29)   19

b. Fruit never uniformly crimson,
white more or less flecked with
crimson   22

19a. Leaflets oblong, more than 5 cm,
with papillae beneath at a
magnification of ×100; fruit
ellipsoid but truncate and flat at
calyx, crimson   **16. 'Ghose'**

b. Leaflets lanceolate, less than 5
cm, without papillae beneath; fruit
apple-shaped, crimson becoming
white   20

20a. Leaves with fewer than 23 leaflets
each over 3 cm; fruit hairless, flushed
pink, styles 3   **13. 'Pink Pearl'**

b. Leaves with 23 leaflets or more;
fruit with scattered rust-coloured
hairs, especially around stalks;
styles 3–5   21

21a. Leaves with more than 25 leaflets
each less than 3 cm long; fruit
flecked with crimson; styles 3 or 4
14. vilmorinii

b. Leaves with usually 23–25 leaflets
c. 3 cm; fruit flushed pink; styles 5
rarely 4   **15. pogonopetala**

22a. Buds conic, pointed, more or less
hairless except at tip and scale
margins; up to 17 (rarely to 21)
leaflets per leaf; inflorescence
pyramidal or corymb-like; styles less
than 2.25 mm, separate at base   23

b. Bud ovoid, hairless or hairy; 17 or
more leaflets per leaf; inflorescence
corymb-like; styles more than 2.25
mm, united at base   30

23a. Leaves with sheathing base;
inflorescence corymb-like; fruit
small to 7 × 6 mm, spindle-shaped
but truncate and flat at calyx, fruit
crimson to creamy-pearly white
24

b. Leaves without sheathing base;
inflorescence corymb-like or
pyramidal; fruit more than 7 × 6
mm, more or less apple-shaped,
mostly porcelain-white sometimes
more or less flushed or flecked pink
26

24a. Leaflets not leathery, margin
toothed, recurved only at extreme
base; buds more or less covered
with rust-coloured hairs; fruit
crimson   **16. 'Ghose'**

b. Leaflets leathery, margin toothed
in upper half or only towards apex,
recurved at least in lower half;
buds more or less hairless; fruit
creamy-pearly white   25

25a. Leaflets 9 or more, less than 10 cm
17. insignis

b. Leaflets less than 9, more than 10
cm   **18. harrowiana**

26a. Inflorescence corymb-like; fruit
initially crimson becoming creamy
to pearly white, soft; styles 3 rarely
to 5, 2.25–2.5 mm; calyx-lobes
fleshy only in fruit   **14. vilmorinii**

b. Inflorescence pyramidal; fruit
becoming more crimson on ageing,
white sometimes flushed or flecked
crimson, hard; styles 4 or 5 rarely 3,
c. 2 mm; calyx-lobes very fleshy   27

27a. Fruit almost pure white except
around calyx   28

b. Fruit white flushed pink distant
from as well as around calyx,
especially where exposed to direct
sunlight   29

28a. Leaflets less than 4 cm; fruit mostly
more than 8 × 8.5 mm; styles 3 or
4, rarely 5   **19. forrestii**

b. Leaflets more than 4 cm; fruit
smaller than 8 × 8.5 mm; styles 4
or 5   **20. glabrescens**

29a. Leaves blue-green, kite-shaped
leaflets decreasing in size towards
base from topmost or second
topmost pair; fruit very hard, to 7 ×
8 mm   **21. hupehensis**

b. Leaves green with most leaflets
more or less equal in size; fruit less
hard, some more than 7 × 8 mm
22. laxiflora

30a. Fruit soft, often more than 1.2 cm
across, with fleshy white calyx-
lobes protruding somewhat at fruit
apex   **23. cashmiriana**

b. Fruit firmer, most less than 1 cm
across, with green to blackish
(depending on maturity of fruit),
largely non-fleshy calyx-lobes in
depression at fruit apex (i.e. fruit
apple-shaped)   31

31a. Flowers pink   **24. microphylla**

b. Flowers white   32

32a. Leaflets with papillae beneath at
×100 magnification, oblong, ovate,
blunt or mucronate   33

b. Leaflets without papillae beneath,
ovate to lanceolate, acute   34

33a. Mostly more than 25 leaflets per
leaf; leaflets ovate, less than 1.8 ×
0.7 cm; fruit less than 2.5 × 9.5
mm; style over 3 mm   **25. prattii**

b. Up to 25 leaflets per leaf; leaflets
oblong, more than 1.8 × 0.7 mm;
fruit often more than 7.5 × 9.5
mm; style *c.* 2.5 mm
26. foliolosa

34a. Multi-stemmed shrub usually less
than 2 m; buds blackish with
whitish hairs   **27. fruticosa**

b. Fewer-stemmed larger shrub; buds
reddish to reddish black with rust-
coloured hairs   **28. koehneana**

35a. Calyx and top of receptacle
eventually falling from fruit as
a unit, leaving fruit apex with a
small, clean, sometimes depressed
scar; leaves not or scarcely lobed
36

b. Calyx and top of receptacle
persisting on fruit or the calyx-
lobes detaching individually; leaves
sometimes with well-developed
lobes or free leaflets at base   39

36a. Leaves with 16–24 pairs of lateral
veins   **30. caloneura**

b. Leaves with less than 16 pairs of
lateral veins   37

37a. Leaves persistently white-felted
beneath and/or fruit red, pinkish or
orange-yellow, at least on one side
31. folgneri

b. Leaves hairless or with rusty hairs
mainly on the veins beneath at
maturity; fruits green or brownish
when ripe   38

38a. Leaves with 10–16 pairs of lateral
veins; fruit spherical, 6–10 mm
across   **32. epidendron**

b. Leaves with 7–10 pairs of lateral
veins; fruit depressed-spherical,
1.3–1.8 cm across   **33. keissleri**

39a. Fruit brown to brownish orange,
orange, dull yellow or greenish,
sometimes flushed reddish on one
side, but not wholly red   40

b. Fruit entirely bright to deep red,
without brownish or yellowish tints
45

40a. Leaves hairless beneath except
for small tufts in the axils of the
veins, finely toothed but never
lobed   **34. megalocarpa**

b. Leaves hairy beneath and/or with 3
to many pairs of lobes   41

41a. Leaves closely and persistently
white-woolly beneath, single to

double-toothed but scarcely lobed, often more than 13 cm or smaller and nearly circular with more than 8 pairs of lateral veins   42

b. Leaves loosely grey-woolly to hairless beneath, usually lobed, or less than 13 cm, with less than 9 pairs of veins   43

42a. Styles mostly 4 or 5   **35. vestita**

b. Styles 1–3   **36. thibetica**

43a. Leaves becoming hairless, with 3 or 4 pairs of large, well-defined lobes; fruit brown   **40. torminalis**

b. Leaves more or less hairy beneath, with more than 5 pairs of lobes or these lacking; fruit not entirely brown, often reddish, orange or yellowish tinged   44

44a. Leaves closely white-woolly beneath at least when young   **37. umbellata**

b. Leaves loosely greyish woolly beneath   **41. latifolia**

45a. Leaves green above and below at maturity; flowers with erect reddish petals   **44. chamaemespilus**

b. Leaves with white or greyish hairs beneath at maturity, not concolorous; petals spreading, whitish   46

46a. Leaves not or only shallowly lobed   47

b. Leaves distinctly lobed or with free leaflets at base   48

47a. Leaves widest above middle, tapered to base; fruit depressed-spherical   **38. graeca**

b. Leaves widest at or below middle, rounded at base; fruit ovoid to spherical   **39. aria**

48a. Leaves more or less deeply lobed but lacking free leaflets at base   **42. intermedia**

b. At least some leaves with 1 or more pairs of free leaflets at base   **43. aucuparia × sp.**

Section **Aucuparia** K. Koch. Contains 40 to 50 described species but there are many undescribed apomictic microspecies, a few of which are in cultivation. Two major subsections can be distinguished. The orange-red fruited subsection has a circumboreal distribution and is largely diploid and sexual (only *S. decora*, *S. cascadensis* and *S. sitchensis* being tetraploid and presumably apomictic). The crimson-white-fruited subsection is found only in N and W China and the Himalayan region; most of the species in cultivation are apomictic tetraploid microspecies,

often with no known closely related diploid sexual species.

1. S. aucuparia Linnaeus (*S. pohuashanensis* (Hance) Hedlund; *S. fastigiata* (Loudon) Hartweg & Rümpler; *S. rossica* Späth; *S. moravica* Dippel). Illustration: Hay & Synge, The dictionary of garden plants in colour, 239 (1969); Brickell, RHS gardeners' encyclopedia of plants & flowers, 54 (1989).

Tree to 20 m; bark smooth, grey with laterally elongated lenticels; buds ovoid, blackish with white, rarely a few rust-coloured, hairs. Leaves with up to 15, rarely to 19 coarsely toothed leaflets 6–9 cm, with papillae beneath. Flowers white; fruit red, rarely yellow, late summer, to 0.9–1.2 × 0.9–1.4 cm, like minute apples in shape, with calyx in depression at fruit apex. Carpels 3 or 4, apices hairy, not fused, forming a conical structure within calyx-lobes. *Eurasia & NW Africa*. H1. Spring–summer.

The populations in several regions have been named as separate species but differ morphologically only in minor characteristics. *S. pohuashanensis* of supposedly N Chinese origin is unlike known-origin plants from that region, having large leaflets and large clusters of flowers and fruit. It may be of SE European origin. Many cultivars of *S. aucuparia* are grown including 'Dickeana' with yellow fruit; 'Beissneri' with orange bark and fastigiate habit; 'Aspleniifolia' with deeply cut leaflets.

2. S. esserteauiana Koehne (*S. conradinae* Koehne). Illustration: Botanical Magazine, 9403 (1935); Krüssmann, Manual of cultivated broad-leaved trees & shrubs **3**: pl. 117 (1986).

Similar to *S. aucuparia* but with thicker twigs, more ovoid-spherical, lighter coloured, redder buds; larger, leathery leaflets which are felty white-hairy beneath; large leafy persistent stipules; to 6.5–7 × 8.5–9.5 mm, hard, late-ripening fruits which are broader than long. *China (W Sichuan)*. H2.

A yellow-fruited variant, 'Flava' occurs. Uncommon in cultivation. Many trees grown under this name are *S. esserteauiana* × *S. aucuparia*, presumably raised from seed of the self-incompatible *S. esserteauiana*.

3. S. scalaris Koehne. Illustration: Hay & Synge, The dictionary of garden plants in colour, 239 (1969).

Very closely related to *S. esserteauiana* but with more numerous narrow leaflets,

to 33; fruits about as broad as long. *China (W Sichuan)*. H2.

4. S. gracilis (Siebold & Zuccarini) K. Koch. Illustration: Krüssmann, Manual of cultivated broad-leaved trees & shrubs **3**: pl. 117 (1986).

A delicate shrubby species with leaves kite-shaped in outline with up to 11 leaflets. Stipules large, leafy, persistent. Flowers and fruit in small clusters; flowers greenish white; fruit red, elongated and somewhat pear-shaped in the most commonly cultivated clone. *C Japan*. H2.

Very uncommon in cultivation. Requires an acid soil.

5. S. sargentiana Koehne. Illustration: Hay & Synge, The dictionary of garden plants in colour, 239 (1969); Krüssmann, Manual of cultivated broad-leaved trees & shrubs **3**: pl. 115 (1986).

A slow-growing small tree or large bush with thick twigs and conspicuous large ovoid, chestnut-coloured, very sticky buds. Stipules large, leafy, persistent. Leaves to 30 cm with up to 13 leaflets each to 13 cm, with veins impressed above. Flowers and fruit in large corymbs, fruit small, usually broader than long, to 8 × 9 mm, with calyx-lobes separated, not overlapping. Styles 3 or 4, to 2 mm, inserted on fused flat carpel tops. *W China*. H2.

6. S. americana Marsh. Illustration: Krüssmann, Manual of cultivated broad-leaved trees & shrubs **3**: pl. 116 (1986).

Small tree or shrub very similar and closely related to *S. commixta* but with stouter twigs, often stickier, darker buds, longer narrower leaflets; fruits with calyx-lobes separated, not overlapping. Fruit to 7.5 × 8 mm, orange-red, hard, glossy, usually borne in large clusters. *C & E USA*.

7. S. commixta Hedlund (*S. serotina* Koehne; *S. matsumurana* misapplied; *S. rufoferruginea* (Schneider) Schneider; *S. wilfordii* Koehne; *S. 'Embley'*; *S. 'Jermyns'*; *S. discolor* (Maximowicz) Hedlund). Illustration: Botanical Magazine n.s., 166 (1951); Krüssmann, Manual of cultivated broad-leaved trees & shrubs **3**: pl. 117 (1986); Brickell, RHS gardeners' encyclopedia of plants & flowers, 54 (1989).

Tree or shrub; buds conic, greenish red to red, to dark red, sticky or not, more or less hairless except at tip and scale margins where there are rust-coloured hairs. Leaves at first covered with rust-coloured hairs, soon more or less hairless except along veins beneath; leaflets to

17, oblong lanceolate, finely and evenly toothed, tapered to a fine point, colouring brilliantly in autumn. Flower white, fruit orange-red to red, often small, hard, shiny, spherical with calyx not in depression at fruit apex. *Japan, Korea*. H1.

The commonest clone in cultivation is a fast-growing, tall, fastigiate tree with rather sparse clusters of small hard fruits, but the species is very variable in the wild with respect to stature, leaflet size and breadth, and fruit size but is constant in the characters mentioned in the description. The very closely related **S. randaiensis** (Hayata) Koidzumi, which has only recently been introduced, differs in having more numerous, more glossy leaflets and fruit with 4 or 5 carpels.

8. S. scopulina Greene.

Shrub to 2 m, but otherwise differing from *S. americana* only in its usually greener buds and lighter green leaf with fewer glossier leaflets. *NW America (Rocky Mts north to Alaska)*. H1.

Very uncommon in cultivation.

9. S. decora (Sargent) Schneider.

Small tree or shrub. Twigs stout bearing large conical blackish, usually very sticky buds. Leaves large with up to 15 leaflets, to 7.5 cm, each less than 2.5 times as long as broad. Fruits to 1.2 × 1.2 cm, in large more or less drooping clusters somewhat hidden by leaves, very attractive to birds. *E North America north to Greenland*. H1.

10. S. cascadensis G.N. Jones (*S. sitchensis* misapplied).

Shrub to 2 m with more or less sticky ovoid buds; flowering buds often incompletely covered by the bud-scales. Leaves with up to 11 leaflets each less than 2.5 times as long as broad. Fruits borne in showy erect corymbs, very attractive to birds. *W North America (Cascade Mts)*. H1.

Uncommon in cultivation.

S. sitchensis Roemer. An apomictic complex, which has only recently been introduced to cultivation in Europe and differs in its non-sticky buds which have a glaucous bloom and bear rust-coloured hairs, bluish green matt leaves and cerise-pink, not orange-red fruits. *W North America (Rocky Mts)*.

11. S. reducta Diels. Illustration: Krüssmann, Manual of cultivated broad-leaved trees & shrubs **3**: pl. 118 (1986); Brickell, RHS gardeners' encyclopedia of plants & flowers, 300 (1989).

Shrub, rhizomatous or not, to 50 cm. Buds reddish with rust-coloured hair.

Leaves to 10 cm with up to 15 ovate, glossy leaflets which colour orange-red in autumn. Flowers white. Fruit to 9.5 × 11 mm, at first dull crimson, becoming more white with age. Style about 2 mm. *N Burma to W China (Yunnan)*. H2.

12. S. poteriifolia Handel-Mazzetti (*S. pygmaea* misapplied).

Strongly rhizomatous low shrub to 10 cm; buds reddish. Leaves glossy to 8 cm long with up to 15 ovate leaflets. Flowers pinkish with each petal having a deep pink zone in the centre. Fruit to 8.25 × 9.5 mm first dull crimson, ripening to almost pure white. Carpel apices fused; styles 3–5, about 2 mm, *N Burma to W China (Yunnan)*. H2.

Uncommon in cultivation.

13. S. 'Joseph Rock' & S. 'Pink Pearl' (*S. pluripinnata* (Schneider) Koehne, misapplied). Illustration: Botanical Magazine, n.s., 554 (1969); Hay & Synge, The dictionary of garden plants in colour, 239 (1969); Brickell, RHS gardeners' encyclopedia of plants & flowers, 55 (1989).

Tall fastigiate (*S.* 'Joseph Rock') or small spreading (*S.* 'Pink Pearl') tree with reddish ovate buds bearing rust-coloured hairs. Leaves with up to 21 neat lanceolate leaflets almost all of much the same size to 4.4 × 1.6 cm. Flowers white. Fruit to 10 × 11 mm, yellow with red around calyx or dull crimson becoming almost white. Styles 2.5 mm, mostly 3, distantly inserted on the more or less fused carpel tops. Seeds 1 to rarely 3 per fruit, to 4 × 2.75 mm. *W China (Yunnan)*. H2.

An unnamed diploid sexual species which exists in cultivation in two variants, fastigiate, yellow fruited and small spreading, crimson-white fruited. Isolated trees of both breed true except that many seedlings of *S.* 'Joseph Rock' have much darker orange fruit. Very similar to *S.* 'Pink Pearl' is a triploid apomict with more numerous smaller leaflets and slightly smaller fruit, somewhat intermediate between *S.* 'Pink Pearl' and *S. vilmorinii*.

14. S. vilmorinii Schneider. Illustration: Botanical Magazine, 8241 (1889); Hay & Synge, The dictionary of garden plants in colour, 239 (1969); Brickell, RHS gardeners' encyclopedia of plants & flowers, 66 (1989).

Shrub with slender twigs bearing ovoid reddish buds more or less covered with rust-coloured hairs. Leaves to 14 cm with up to 29 lanceolate leaflets to 2.3 × 0.7

cm. Flowers white. Fruit *c.* 9.5 x 9 mm, crimson becoming white flecked with crimson; carpel tops fused, more or less hairless, sunk in depression at fruit apex; styles *c.* 2 mm, more or less wedge-shaped at base; seeds 1 or 2 per fruit, dark brown, to 5 × 2.5 mm. *W China*. H2.

15. S. pogonopetala Koehne.

A thick-twigged shrub related to *S. vilmorinii* but thicker, larger and stiffer in all its parts. Leaves with up to 25 leaflets. Fruit at first dull crimson becoming almost white to 10.5 × 12 mm; styles *c.* 2.75 mm. *China (Yunnan, Sichuan)*. H2.

Uncommon in cultivation.

16. S. 'Ghose'.

Tree to 30 m with thick stiff twigs bearing large (to 3 cm) ovoid-conic, dark red buds, more or less covered with rust-coloured hairs which are particularly dense at the apex; leaf bases somewhat sheathing, leaves to 32 cm, with 13–15 leaflets, each to 10 × 2.7 cm, acute and toothed to base, with numerous rust-coloured hairs beneath. Flowers white in large corymbs. Fruit in large corymbs, small, to 7.5 × 7.75 mm, crimson, broadly ellipsoid but truncate at calyx; styles 3 or 4, *c.* 2.25 mm, more or less distantly inserted in the more or less fused hairy carpel apices. Seed to 3.5 × 2.25 mm, pale yellow brown. *NE India (Naga Hills)*. H2.

An as yet undescribed species, uncommon in cultivation.

17. S. insignis J.D. Hooker. Illustration: Krüssmann, Manual of cultivated broad-leaved trees & shrubs **3**: pl. 115 (1986).

Large tree with thick stiff twigs bearing large, ovoid-conic greenish to reddish more or less hairless buds. Leaf base sheathing. Leaves to 33 cm with up to 17 leathery leaflets, each to 9 × 1.9 cm, toothed only in upper part and with margin recurved throughout, more or less hairless beneath. Flowers white in large corymbs. Fruit in large corymbs, pinkish white, ovoid, to 6.75 × 5.5 mm; styles 3 or 4, about 2 mm, more or less distantly inserted on the more or less fused hairy carpel apices. *C & W Himalaya*. H3.

Uncommon in cultivation.

18. S. harrowiana (Balfour & W.W. Smith) Rehder. Illustration: Krüssmann, Manual of cultivated broad-leaved trees & shrubs **3**: pl. 115 (1986).

Very similar to *S. insignis* but with fewer, to 5, even larger, often over 10 cm, leaflets. *N Burma to W China (Yunnan)*. H4.

Uncommon in cultivation.

19. S. forrestii McAllister & Gillham. Illustration: Botanical Magazine n.s., 792 (1981).

Small tree to 5 m. Buds ovoid-conic, acute, reddish brown, more or less hairless except for rust-coloured hairs at scale margins and especially at apex. Leaves to 22 cm with up to 19 distantly inserted, oblong-elliptic leaflets to 4.5 × 1.7 cm, all of much the same size and toothed in the upper 1/2–2/3. Flowers white. Fruit at first green, becoming almost pure white except for reddish colour around calyx, to 9.5 × 9 mm; styles 4–5, c. 2.5 mm, distantly inserted on the fused almost hairless carpel apices. Seed brown. *W China (NW Yunnan)*. H2.

Uncommon in cultivation.

20. S. glabrescens (Cardot) McAllister (*S. hupehensis* Schneider misapplied; *S. wilsoniana* Schneider misapplied). Illustration: Hay & Synge, The dictionary of garden plants in colour, 239 (1969).

Small tree to 8 m. Twigs stiff. Buds ovoid-conic, acute, greenish red, more or less hairless except for whitish hairs at scale margins and especially at apex. Leaves with up to 17 oblong leaflets, each to 6 × 2.2 cm, mucronate, all of much the same size. Flowers in pyramidal panicles; petals white, becoming reflexed. Fruit green, becoming white, hard, to 7.5 × 8 mm, calyx-lobes fleshy; styles 4 or 5, c. 2.25 mm, distantly inserted on fused more or less hairless carpel apices; seed brown. *W China (NW Yunnan)*. H2.

21. S. hupehensis Schneider (*S. hupehensis* var. *rosea* misapplied). Illustration: Botanical Magazine, n.s. 167 (1950); Krüssmann, Manual of cultivated broad-leaved trees & shrubs **3**: 341 (1986); Brickell, RHS gardeners' encyclopedia of plants & flowers, 54 (1989).

Small tree to 8 m. Buds ovoid-conic, acute, reddish, more or less hairless except for whitish hairs at scale margins and especially at apex. Leaf bluish green with up to 15 ovate leaflets to 5 × 2.2 cm but often, especially on the weaker shoots, decreasing in size from the apex towards the base giving the leaves a kite-shaped outline. Flowers in pyramidal panicles, white, petals becoming reflexed. Fruit, green flushed crimson, becoming white flushed crimson, especially around the calyx, very hard, to 7 × 8 mm; calyx-lobes fleshy; styles 5, c. 2.25 mm, distantly inserted on fused more or less hairless carpel apices; seed brown. *China (Hubei)*. H3.

22. S. laxiflora Koehne.

Very closely related to *S. hupehensis* but with leaflets longer, narrower, more distantly inserted, more equal sized and fruits larger, to 8.5 × 9 mm, with 4–5 styles. *China (Sichuan)*. H2.

Very uncommon in cultivation.

23. S. cashmiriana Hedlund. Illustration: Hay & Synge, The dictionary of garden plants in colour, 239 (1969); Brickell, RHS gardeners' encyclopedia of plants & flowers, 66 (1989).

Small tree or large shrub. Buds ovoid-conic, acute, reddish, more or less hairless except for rust-coloured hairs at scale margins and especially at apex. Leaves to 20 cm, with 17–21 lanceolate leaflets each to 3.7 × 1.5 cm. Flowers in corymbs, pink, more than 1 cm across. Fruit to 1.5 × 1.3 cm green, becoming almost pure white with white fleshy calyx-lobes protruding; styles 4–5, c. 3 mm, closely inserted on the fused carpel apices. *W Himalaya*. H2.

Related recent introductions from Sichuan differ in having thinner twigs, smaller ovoid leaflets, usually white flowers which are not so large as in *S. cashmiriana*; but the flowers and fruits are still larger and the protruding fleshy calyx more prominent that in any other white-fruited *Sorbus*.

24. S. microphylla (Wallich) Decaisne. Illustration: Botanical Magazine n.s., 879 (1983).

Small tree or shrub. Buds ovoid, Leaves very variable to 17 cm with up to 33 leaflets. Flowers pale pink to almost crimson. Fruit white, sometimes crimson at first; calyx never fleshy, style bases inserted close together on the fused carpel apices. *Himalayas*. H2.

25. S. prattii Koehne. Illustration: Botanical Magazine, 9460 (1936); Hay & Synge, The dictionary of garden plants in colour, 239 (1969).

Shrub with greyish twigs. Buds ovoid, dark-red brown, covered with rusty-grey coloured hairs. Leaves to 14 cm with up to 31 ovate, blunt leaflets to 3 × 1 cm, with papillae beneath. Flowers white. Fruit at first green, becoming almost pure white, to 7.5 × 9.5 mm; calyx partially fleshy; styles 5, c. 3.5 mm, closely inserted on the more or less hairless fused carpel apices. *China (Sichuan)*. H2.

26. S. foliolosa (Wallich) Spach (*S. ursina* (Wenzig) Decaisne).

Tree or shrub with stiff twigs bearing conic dark red buds with rust-coloured

hairs confined to scale margins and bud apex. Leaves to 21 cm bearing up to 25 oblong, blunt or mucronate leaflets all of much the same size to 4.7 × 1.5 cm and with papillae beneath. Fruit at first green, becoming white, to 1 × 1.1 cm, often broader than long, conspicuously black at calyx; styles 5, c. 2.5 mm, closely inserted on the more or less hairless fused carpel apices. *C Himalaya*. H2.

Uncommon in cultivation.

27. S. fruticosa invalid (*S. koehneana* misapplied).

Shrub to 2 m. Twigs dark brown with ovate, pointed blackish buds, hairless except for whitish hairs at scale margins and apex. Leaves dark green with up to 29 deeply toothed leaflets to 2.2 × 0.8 cm, without papillae beneath. Fruit green becoming pure white, to 9.5 × 12 mm; styles 5, c. 2.75 mm, closely inserted on the fused slightly hairy carpel tops, seed initially reddish becoming dark brown. *China (Qinghai)*. H1.

28. S. koehneana Schneider. Illustration: Krüssmann, Manual of cultivated broad-leaved trees & shrubs **3**: pl. 118 (1986).

Small tree or shrub. Buds ovoid-conic, dark-red, hairless except for rust-coloured hairs at scale margins and apex. Leaves with up to 25 leaflets, without papillae beneath. Flowers white. Fruit green, becoming white, to 7.5 × 10 mm, but most much less broad. Styles 5, c. 2.5 mm, more or less closely inserted on the fused carpel tops. Seed initially reddish, becoming brown. *N & W China*. H1.

Uncommon in cultivation. This diploid sexual species is widespread in China but no introduction to date has proved popular. Likely to prove more popular are very closely related shrubs from Sichuan with very small glossy leaflets which have been named **S. setschwanensis** Koehne and an apomictic tetraploid with up to 33 leaflets.

Section **Cormus** (Spach) Boissier. Now sometimes treated as a separate genus. Leaves pinnate, carpels 5, fully united; ovary inferior. Fruits green to brown.

29. S. domestica Linnaeus. Illustration: The Plantsman **7**(2): frontispiece (1985).

Tree to 20 m; bark deeply furrowed; branches becoming hairless; buds sticky and shiny. Leaflets 11–21, narrowly oblong, 3–8 cm, with coarse, forward-pointing teeth, base symmetric, hairy beneath. Flowers c. 1.5 cm, in conical corymbs to 10 cm. Fruit to 3 cm, apple or pear-shaped, yellow-green, ripening to red. *C & S Europe, N Africa, Turkey*. H2. Spring–early summer.

'Signalman' (*S. domestica* × *S. scopulina*), low narrowly pyramidal tree; leaves small; fruit large, pale orange.

Forma **pomifera** (Hayne) Rehder, fruit apple-shaped, 2–3 cm.

Forma **pyriformis** (Hayne) Rehder, fruit pear-shaped, 3–4 cm.

Section **Micromeles** (Decaisne) Rehder. Distinguished by its fully inferior ovaries, and receptacle apex and calyx usually deciduous as a unit, leaving a clean scar at the apex of the mature fruit.

30. S. caloneura (Stapf) Rehder (*Micromeles caloneura* Stapf; *S. meliosmifolia* Rehder). Illustration: Botanical Magazine, 8335 (1910); Krüssmann, Manual of cultivated broad-leaved trees & shrubs **3**: 333 (1986).

Tree to 10 m. Stems and winter buds hairless. Leaves simple, ovate-elliptic to oblong, mostly to 11 × 5 cm, tapered at both ends, lateral veins in 16–24 pairs, more or less hairless except in the vein axils, margins double-toothed; leaf-stalk 6–12 mm. Flowers white, *c.* 1.2 cm across in dense, downy corymbs 5–10 cm across; anthers pinkish brown. Fruit spherical to pear-shaped, *c.* 1 cm, brown to bronze-coloured, covered with lenticels. *E & SE Asia*. H1.

S. japonica (Decaisne) Hedlund (*Micromeles japonica* (Decaisne) Koehne; *S. japonica* var. *calocarpa* Rehder) is a tree to 20 m, with larger, shallowly lobed leaves and smaller flowers. *Japan & Korea*. H5.

S. alnifolia (Siebold & Zuccarini) K. Koch (*Micromeles alnifolia* (Siebold & Zuccarini) Koehne; *S. alnifolia* var.

submollis Rehder). Illustration: Botanical Magazine, 7773 (1901). Tree to 20 m. Leaves sometimes shallowly lobed, with 6–12 pairs of lateral veins, more or less hairy beneath at first, margins double-toothed. Fruit bright red to deep pink, covered with lenticels. *E Asia*. H1.

32. S. epidendron Handel-Mazzetti.

Tree or shrub to 15 m. Stems and foliage loosely hairy at first, winter buds hairless. Leaves simple, elliptic to obovate, to 15 × 7 cm, but often much smaller, tapered at both ends, lateral veins 10–12 pairs, margins simply toothed; leaf-stalk of variable length. Flowers white, *c.* 6–7 mm across. Fruit spherical, truncate at apex, only 6–8 mm across, covered with lenticels, in dense clusters. *S to C China & adjacent Burma*. H2.

Very similar is **S. rhamnoides** (Decaisne) Rehder, distinguished by its smaller leaves with 12–14 pairs of veins, flowers to *c.* 1 cm across and fruit lacking lenticels. *E Himalaya & Burma*. H5.

33. S. keissleri (Schneider) Rehder (*Micromeles keissleri* Schneider).

Tree to 12 m. Stems densely covered with lenticels, hairy at first. Leaves simple, oval to obovate, leathery, to 7.5 × 6 cm, soon hairless, tapering very gradually at base, abruptly at apex, lateral veins in 7–10 pairs, margins simply toothed; leaf-stalk ill-defined, to 5 mm. Flowers white, to 1 cm across, petals soon falling. Fruit greenish, covered with lenticels, to 1.8 cm across. *C to SW China, N Burma*. H1.

Section **Aria** (Persoon) Beck. Sometimes now treated as a genus including Section *Micromeles*, above. Leaves simple, lobed or toothed. Base of calyx-lobes persisent.

34. S. megalocarpa Rehder. Illustration: Botanical Magazine, n.s. 259 (1955); Krüssmann, Manual of cultivated broad-leaved trees & shrubs **3**: pl. 118 (1986).

Shrub or tree to 8 m, with large, glossy winter buds to 1.8 cm resembling those of a horse-chestnut. Stems stout, to 5 mm thick the first year, hairless, dark but marked with numerous lenticels. Leaves simple, ovate to obovate, rather variable in size but sometimes to 20 × 11 cm, gradually tapered to rounded at base, lateral veins to 20 pairs, more or less hairless on both sides except for small tufts in the vein axils, margins finely simple to double-toothed; leaf-stalks 2–5 cm. Flowers expanding with or before the leaves, white, *c.* 1.8 cm across, in dense corymbs to 15 cm across. Fruit egg-shaped, to 3 × 2 cm,

brown. *S to C China (Sichuan)*. H1.

The more commonly cultivated var. **cuneata** Rehder has shorter leaf-stalks and smaller fruits.

35. S. vestita (G. Don) Loddiges (*S. cuspidata* (Spach) Hedlund). Illustration: Botanical Magazine, 8259 (1909); Krüssmann, Manual of cultivated broad-leaved trees & shrubs **3**: pl. 119 (1986).

Tree to 23 m. Stems white-woolly at first, stout (*c.* 5 mm across at 1 year old), with obtuse, nearly hairless winter buds. Leaves simple, variably oblong-ovate to broadly elliptic, to 22 × 12 cm, obtuse to acuminate at apex, rounded to tapered at base, lateral veins in 8–16 pairs, closely and finely white-felted beneath, margin single or irregularly double-toothed; leaf-stalk 2.5–5 cm. Flowers white, 1.5–2.5 cm across, in very woolly corymbs to 7.5 cm across; styles 4 or 5, rarely 3. Fruit more or less spherical, *c.* 1.5 cm across or larger, yellow-brown sometimes tinged red, covered with lenticels. *Himalaya & N Burma*.

S. hedlundii Schneider differs mainly in its leaves with brownish hairs on the lateral veins beneath. *E Nepal to W Bhutan*. H3.

36. S. thibetica (Cardot) Handel-Mazzetti (*S. mitchellii* invalid; *S. wardii* Merrill?). Illustration: Brickell, RHS gardeners' encyclopedia of plants & flowers, 52 (1989).

Like *S. vestita* but with more slender stems and leaves to only *c.* 13 cm, sometimes almost circular, more loosely and less woolly beneath. Flowers smaller; styles 1–3. Fruit brown, yellow tinged reddish or green-orange. *E Himalaya to SC China*.

S. pallescens Rehder is similar, but has flowers with 2–5 styles and yellow fruit only *c.* 9 mm across. *China*. H1.

37. S. umbellata (Desfontaines) Fritsch. Illustration: Krüssmann, Manual of cultivated broad-leaved trees & shrubs **3**: pl. 120 (1986).

Small tree or shrub to 6 m. Leaves simple, varying from fan-shaped to rounded, 2.5–7 × 3–6.5 cm, tapered at base, lateral veins in 5–8 pairs, brilliant white-felted beneath, shallowly lobed or with large single to double toothing at margin; leaf-stalk 0.6–1.8 cm. Flowers white, 1–1.5 cm across, in woolly corymbs to 7.5 cm across; styles 2. Fruit spherical, *c.* 1.2 cm. *SE Europe and adjacent parts of Asia*. H1.

S. lanata (D. Don) Schauer is larger in

all its parts, with leaves less persistently white-woolly beneath, lateral veins in 9 or more pairs and fruit to 2.5 cm across or larger. *E Himalaya.* H5.

38. S. graeca (Spach) Kotschy.

Like *S. umbellata*, but leaves to 9 cm, toothed rather than lobed, lateral veins 9–11 pairs, margins double-toothed, the teeth spreading and symmetrical. Fruit almost spherical, usually less than 1.2 cm, crimson, with few large lenticels. *Mediterranean area, EC Europe & Iraq.* H3.

S. rupicola (Syme) Hedlund. Illustration: Clapham, Tutin & Warburg, Flora of the British Isles, edn 2, fig. 46E (1962).

Shrub to 2 m. Leaves to 14.5 cm, lateral veins mostly 7–9 pairs, marginal teeth forward-pointing, curved on their outer edges. Fruit larger, carmine, with numerous lenticels. *NW Europe.* H3.

Rarely cultivated is **S. lancastriensis** E.F. Warburg, with leaves and fruits somewhat intermediate between the above. *NW England.*

39. S. aria (Linnaeus) Crantz (*S. majestica* invalid). Illustration: Botanical Magazine, 8184 (1908); Clapham, Tutin & Warburg, Flora of the British Isles, edn 2, fig. 45J (1962); Krüssmann, Manual of cultivated broad-leaved trees & shrubs **3**: pl. 116 (1986); Brickell, RHS gardeners' encyclopedia of plants & flowers, 51 (1989) – as 'Lutescens'.

Tree 10–25 m, rarely a shrub. Leaves simple, variable in shape, 5–12 cm, lateral veins mostly 10–14 pairs, densely white-felted beneath, margins double-toothed; leaf-stalk 0.7–2 cm. Flowers white, *c.* 1.2 cm across, in woolly corymbs to 7.5 cm across. Fruit usually longer than broad, bright red, with lenticels. *W, C & S Europe.* H1.

Section **Torminaria** (de Candolle) Dumortier. Sometimes now treated as a separate genus. Leaves sharply lobed; styles 2 united for more than half their length; carpels 2, surrounded by a dense layer of stone-cells. Fruit brownish, densely covered with lenticels.

40. S. torminalis (Linnaeus) Crantz. Illustration: Bean, Trees & shrubs hardy in the British Isles, edn 8, **4**: pl. 55, dust jacket (1980); Krüssmann, Manual of cultivated broad-leaved trees & shrubs **3**: pl. 115 (1986).

Tree 10–22 m. Young parts loosely woolly at first, eventually hairless. Leaves simple, broadly ovate-triangular, *c.* 13 ×

13 cm, with 3 or 4 pairs of large, acute lobes, which grade into teeth towards leaf apex, margins double-toothed; leaf-stalk 2.5–5 cm. Flowers white, 1–1.5 cm across, in loose, woolly inflorescences; styles 2. Fruit spherical to ovoid, *c.* 1.2 cm, brownish, densely covered with lenticels. *Europe, SW Asia, N Africa.* H2.

41. S. latifolia (Lamarck) Persoon. Illustration: Bean, Trees & shrubs hardy in the British Isles, edn 8, **4**: pl. 55 (1980); Krüssmann, Manual of cultivated broad-leaved trees & shrubs **3**: pl. 114 (1986).

Tree 10–20 m. Stems downy at first. Leaves broadly elliptic, to 10 cm, nearly as wide or narrower, apex pointed, base rounded to broadly tapered, lateral veins and lobes in 7–9 pairs, persistently greyish felted beneath, margins irregularly single- to double-toothed; leaf-stalk 1.2–2.5 cm, hairy. Flowers white, 1.5–2 cm across, in woolly corymbs to 7.5 cm across. Fruit spherical, dull brownish red, with lenticels. *Portugal to SW Germany.* H1.

Believed to be of hybrid origin, involving *S. torminalis* and one of the *S. aria* group. The following are similar (see also *S. intermedia*):

S. croceocarpa P.D. Sell (*S.* 'Theophrasta'; *S. theophrasta* invalid). Illustration: Ross-Craig, Drawings of British plants **9**: 33 (1956); Dendroflora **3**: 62, fig. 1 (1966). Leaves 7.5 cm, with 8–11 pairs of lateral veins, double-toothed, not lobed. *Naturalised in S Britain.*

Rarely cultivated are **S. karpatii** Boros, a shrub with lobed leaves *c.* 7.5 × 6 cm, lateral veins 7–9 pairs, and **S. pseudovertesensis** Boros, with obscurely lobed leaves *c.* 8.5 × 5 cm, lateral veins 9–11 pairs. *Hungary.*

42. S. intermedia (Ehrhart) Persoon (*S. scandica* (Linnaeus) Fries). Illustration: Clapham, Tutin & Warburg, Flora of the British Isles, edn 2, fig. 45G (1962); Krüssmann, Manual of cultivated broad-leaved trees & shrubs **3**: pl. 114 (1986).

Tree to 13 m. Leaves elliptic to oblong-elliptic, 7–12 cm, often rounded at base, lobed, lobe sinuses extending from 1/6–1/3 of the way to the midrib (on sucker shoots sometimes with a free leaflet at base), lateral veins in 7–9 pairs, yellowish grey-woolly beneath, margins single to double-toothed; leaf-stalk 1–1.5 cm. Flowers white, *c.* 1.8 cm across, in densely woolly corymbs to 12.5 cm across; anthers cream; styles 2. Fruit oblong, 1.2–1.5 cm, scarlet, sparsely covered with lenticels. *W Europe (Scandinavia & Baltic region).* H1.

Apomictic and of hybrid origin, like the following, which are similar in most details:

S. anglica Hedlund. Illustration: Clapham, Tutin & Warburg, Flora of the British Isles, edn 2, fig. 45H (1962). Shrub to 2 m. Leaves more or less obovate, tapered at base, whitish grey-woolly beneath. Anthers pink tinged. Fruit depressed-spherical, 0.7–1.2 cm, crimson. *W Britain & Eire.* H3.

S. minima (A. Ley) Hedlund. Illustration: Clapham, Tutin & Warburg, Flora of the British Isles, edn 2, fig. 45F (1962). Shrub to 3 m. Leaves 6–8 cm, narrower, 1.8–2.2 times as long as broad. Flowers to 9 mm across. Fruit depressed-spherical, 6–8 mm. *Wales (Brecon).* H1.

S. arranensis Hedlund. Illustration: Clapham, Tutin & Warburg, Flora of the British Isles, edn 2, fig. 45D (1962). Leaves with lobe sinuses extending to half or more of the way to the midrib. Flowers *c.* 9 mm across; anthers cream or pink. Fruit 8–10 mm. *Scotland (Isle of Arran).* H2.

S. bristoliensis Wilmott. Illustration: Clapham, Tutin & Warburg, Flora of the British Isles, edn 2, fig. 46J (1962). Leaves more or less obovate, very shallowly lobed; leaf-stalk 1.2–2 cm. Flowers *c.* 1.2 cm across; anthers pink. Fruit 9–11 mm, bright reddish orange. *England (Avon Gorge).* H3.

S. bakonyensis (Jávorka) Kárpáti is similar to the preceding. *Hungary.* H3.

S. mougeotii Soyer-Willemet & Godron is somewhat intermediate between *S. intermedia* and *S. anglica*. *W Alps, Pyrenees.* H2.

S. × hostii (Jacquin) K. Koch is a hybrid between *S. mougeotii* and *S. chamaemespilus*, which forms a shrub to 4 m, with lobed leaves and flowers white tinged pink.

43. S. aucuparia × S. sp. (*S. × thuringiaca* (Ilse) Fritsch; *S. thuringiaca*, misapplied in part; *S. quercifolia* invalid; *S. decurrens* invalid; *S. lanuginosa* invalid). Illustration: Clapham, Tutin & Warburg, Flora of the British Isles, edn 2, fig. 45B (1962).

Tree to 15 m, with woolly winter buds. Leaves oblong-lanceolate to narrowly elliptic, 7.5–15 × 3.5–7.5 cm, dull grey-woolly beneath, with 1 or 2 pairs of free leaflets at base on extension growths (to 5 or more pairs in 'Decurrens' (*S. lanuginosa*)), or only lobed to the midrib on old spur shoots, principal lateral veins in 10–12 pairs, margins double-toothed; leaf-stalk 1.5–3 cm. Flowers white, *c.* 1.2

cm across, in corymbs 7.5–12 cm across; styles 2 or 3. Fruit depressed-spherical to ellipsoid, *c.* 9 mm across, bright red. *Europe.*

A hybrid, the unnamed parent being either *S. aria* or *S. intermedia*. Sometimes confused with the following, which is a tetraploid apomict (possibly derived from *S. aucuparia* × *S. rupicola*):

S. hybrida Linnaeus (*S. fennica* (Kalm) Fries; *S. meinichii* misapplied). Smaller. Leaves to 10.5 cm, white-woolly beneath. Flowers *c.* 1.5 cm across; styles 3. Fruit spherical, 1–1.5 cm across. *Scandinavia.* H1.

Of similar origin, but rarely cultivated, are **S. meinichii** (Hartmann) Hedlund and **S. teodorii** Liljefors, notable for their leaves with 5 or 4 pairs of free leaflets at base, respectively.

Section **Chamaemespilus** (Medikus) Schauer. Sometimes now treated as a separate genus. Petals pink, erect; sepal erect.

44. S. chamaemespilus (Linnaeus) Crantz. Illustration: Krüssmann, Manual of cultivated broad-leaved trees & shrubs **3**: pl. 120 (1986).

Dwarf shrub to 2 m, rarely more. Leaves simple, ovate to obovate, 3–7.5 × 1.5–3.7 cm, green and almost hairless above and below, finely toothed; leaf-stalks 3–8 mm. Flowers pinkish red, in compact corymbs composed of small umbel-like clusters, petals remaining erect. Fruit 8–12 mm, scarlet. *C & S Europe.* H2.

58. × SORBARONIA Schneider
S.G. Knees

Deciduous shrubs or small trees, with slender, sometimes pendent branches; young shoots usually covered with white hairs. Leaves simple or lobed, margins toothed. Flowers in small dense clusters. Sepals 5; petals 5, white or pale pink; styles 3 or 4; fruit red or nearly black.

Hybrids between *Sorbus* and *Aronia* and intermediate between the two. Several other hybrids have been named but do not appear to be widely cultivated.

1a. Leaves pinnately divided or lobed, especially towards the base; margins without glandular hairs   **1. hybrida**
 b. Leaves simple; margins with glandular hairs   **2. alpina**

1. × S. hybrida (Moench) Schneider (*Sorbus aucuparia* × *Aronia arbutifolia*). Illustration: Edwards's Botanical Register

14: 1196 (1828).

Leaves 3–8 cm, ovate to oblong, with 2 or 3 pairs of lobes towards the base, scalloped towards the apex, downy beneath, margins with forward-pointing teeth. Flowers in clusters 2–3 cm across, white or pinkish white. Fruit spherical to pear-shaped. 8–10 mm, dark purple. *Garden origin, before 1785.* H3. Spring–summer.

2. × S. alpina (Willdenow) Schneider (*Sorbus aria* × *Aronia arbutifolia*). Illustration: Schneider, Illustriertes Handbuch der Laubgeholzkunde **1**: 699 (1906).

Shrub to 8 m, similar to *Sorbus aria.* Leaves simple, margins glandular with forward-pointing teeth. Flowers in clusters 4–7 cm across, creamy white; styles 3–4. Fruit 7–9 mm across, red or brownish red. *Garden origin, before 1809.* H4. Spring–summer.

59. ERIOBOTRYA Lindley
C.M. Mitchem

Evergreen tree or large shrub. Leaves spirally arranged, stalked or almost stalkless, simple, leathery, with prominent veins extending to leaf-margin; stipules present. Inflorescence a terminal panicle, to 15 cm. Calyx-tube 5-lobed, fused to receptacle. Petals 5, usually white; stamens *c.* 20. Ovary inferior, with 2–5 chambers, each containing 2 ovules; styles 2–5, fused at base. Fruit a fleshy pome containing 1 or 2 large seeds; calyx-lobes usually persistent, except 2.

A genus of 20 species from the Himalayas to Japan and SE Asia, grown for fruit, timber or ornament. Propagation is by seed in spring or autumn; selected cultivars by budding or grafting on quince or seedling loquat. They require a deep, rich, sandy loam, in a sunny position. Fruit only ripens reliably in S Europe.

Literature: Sargent, C.S., *Plantae Wilsonianae* **1**: 194 (1912); Popenoe, W., *Manual of tropical & subtropical fruit*, 20 (1960); Vidal, J.E., Notes sur quelques Rosacéaes Asiatiques (3): Révision du genre Eriobotrya (Pomoideae). *Adansonia* **5**: 537–580 (1965); Kalkman, C., The Malesian species of the subfamily Maloideae (Rosaceae). *Blumea* **21**: 413–442 (1973).

1a. Leaf-stalk less than 1 cm   2
 b. Leaf-stalk greater than 1 cm   3
2a. Underside of leaves covered with densely felted hairs; leaf-stalk 1–5 mm; styles 5   **1. japonica**

 b. Underside of leaves sparingly covered with felted hairs; leaf-stalk 5–10 mm; styles 2   **2. hookeriana**
3a. Underside of leaves covered with densely felted hairs; leaf-stalk 1.5–2 cm   **3. prinoides**
 b. Under-side of leaves almost hairless; leaf-stalk 2–5 cm   **4. deflexa**

1. E. japonica (Thunberg) Lindley (*Crataegus bibas* Loureiro; *Mespilus japonica* Thunberg; *Photinia japonica* Franchet & Savatier). Illustration: Hay & Synge, Colour dictionary of garden plants, pl. 1591 (1969); Masefield & Harrison, The Oxford book of food plants, 105 (1969); Polunin & Everard, Trees & bushes of Europe, 84 (1976).

Large shrub to 6 m. Leaves elliptic, 12–30 × 3.5–10 cm, upper half of leaf-margin smooth or toothed; hairless above, woolly below; 12–22 pairs of veins. Leaf-stalk less than 5 mm; stipules subulate and persistent, to 1.5 cm. Inflorescence to 12 cm, flowers stalkless. Calyx-tube to 5 mm, woolly, lobes to 3 mm. Petals obovate, *c.* 10 × 6 mm, white; stamens *c.* 20, styles 5. Fruit ovoid to spherical, to 3 cm across, yellow. *China & Japan.* H5. Autumn.

2. E. hookeriana Decaisne (*E. dubia* (Lindley) Decaisne).

Slender tree to 10 m. Leaves elliptic to oblanceolate, 15–25 × 4–10 cm, margin with coarse forward-pointing teeth; hairless above, sparsely felty below; 15–30 pairs of veins. Leaf-stalk 5–10 mm; stipules semi-lunar, to 11 mm. Inflorescence a spreading panicle, to 15 cm, flowers almost stalkless. Calyx-tube 2–3 mm, lobes to 3 mm, white or pinkish; styles 2. Fruit an ellipsoid pome, 1.2–1.8 × 0.2–0.8 cm, yellow. *E Himalaya (Sikkim, Bhutan).* H5. Autumn.

3. E. prinoides Rehder & Wilson (*E. bengalensis* (Hooker) Dunn; *E. dubia* (Decaisne) Franchet). Illustration: Adansonia **5**: 564 (1965).

Tree to 12 m. Leaves elliptic, 8–15 × 3–7.5 cm, upper ⅔ of margin toothed; hairless above, densely felted hairs below; 10–12 pairs of veins. Leaf-stalk 1.5–2 cm. Inflorescence to 10 cm; calyx-tube to 3 mm, woolly, lobes to 2 mm, falling early. Petals white, oval, to 5 mm. Styles 2 or 3, woolly at base. Fruit ovoid, to 10 × 7 mm. *China (Sichuan, Yunnan).* H5. Autumn.

Var. **laotica** Vidal has almost circular leaves to 7 × 5 cm. Var. **prinoides** is most commonly cultivated.

4. E. deflexa (Hemsley) Nakai (*Photinia deflexa* Hemsley).

Large tree, to 12 m. Leaves oblanceolate or elliptic, 6–15 × 2.5–5 cm, coarsely lobed, hairless above, with a few hairs below; 10–15 pairs of veins; leaf-stalk 2–5 cm. Inflorescence a spreading panicle, to 15 cm. Calyx-tube to 4 mm, woolly, lobes to 4 mm. Petals white, almost circular; stamens *c.* 20; styles 2 or 3. Fruit ovoid or spherical, to 1.5 cm across, yellow. *Taiwan, S Vietnam*. H5. Autumn.

60. PERAPHYLLUM Torrey & Gray
C.M. Mitchem

Deciduous shrubs to 2 m, stems widely divergent, much-branched, bark grey. Leaves 2.5–4 × 0.4–1 cm, narrowly lanceolate to obovate, mucronate; margin smooth or with fine forward-pointing teeth. Stipules falling early. Inflorescence an erect corymb of 2 or 3 flowers. Flowers white, bisexual, to 2 cm across, subtended by 2 small bracts. Calyx persistent, to 1 cm, covered with silky hairs, fused to bell-shaped receptacle. Petals 8–10 mm, circular. Stamens 20, as long as petals, arranged around rose-coloured dish; anthers yellow. Ovary 2–4-celled; stigma red, capitate; styles 2 or 3, not exceeding stamens. Fruit a yellow, fleshy, spherical drupe to 2 cm, containing 1 seed.

A genus of one species from W North America. Prefers hot, dry summers and sandy or loamy soil. Propagation by seed or layering.

1. P. ramosissimum Torrey & Gray. Illustration: Botanical Magazine, 7420 (1895).

W North America (Oregon to California & Colorado). H1. Late spring–early summer.

61. AMELANCHIER Medikus filius
J.E. Richardson

Deciduous trees or shrubs. Leaves alternate, simple, entire or with forward-pointing teeth. Stipules deciduous. Flowers in terminal racemes, rarely solitary. Calyx-tube bell-shaped, more or less fixed to the ovary, lobes 5, narrow, reflexed, persistent. Petals 5, white, rarely pink. Stamens 10–20, inserted on the throat of the calyx; filaments awl-shaped. Styles 2–5. Ovary inferior, apex usually hairy, cells becoming double the number of styles. Fruit a small berry-like pome with 1–5 cells each containing 1 or more seeds.

A genus of about 30 species from N America, N and C Europe, Japan, China and Korea. Propagate from seed, layering

of branches or by division of the suckers. Slow-growing and sun-loving.

1a. Leaves folded in bud not overlapping; flowers several to many, racemose **2**
 b. Leaves overlapping in bud; flowers 1–3 together **14. bartramiana**
2a. Leaf-margin coarsely toothed (less than 6 teeth per cm) **3**
 b. Leaf-margin finely toothed (more than 6 teeth per cm) **7**
3a. Plant stoloniferous **4**
 b. Plant not stoloniferous **5**
4a. Petals 0.7–1 cm; apex of ovary hairy **4. humilis**
 b. Petals 1–1.5 cm; apex of ovary hairless **6. ovalis**
5a. Leaf apex usually truncate **1. alnifolia**
 b. Leaf apex acute to rounded to obtuse **6**
6a. Shrub or small tree; petals 0.6–0.8 cm **2. utahensis**
 b. Straggling or arching slender shrub; petals 1–1.5 cm **3. sanguinea**
7a. Stoloniferous shrub **8**
 b. Non-stoloniferous tree or shrub **9**
8a. Tree or shrub to 8 m; leaves hairless except for on midrib **7. canadensis**
 b. Shrub to 2 m; leaves densely white-hairy beneath when young **5. stolonifera**
9a. Apex of ovary hairless **10**
 b. Apex of ovary hairy **13**
10a. Racemes more than 8 cm **11**
 b. Racemes less than 8 cm **12**
11a. Leaves hairless when young; fruit to 1.8 cm across **10. laevis**
 b. Leaves densely white-hairy when young; fruit more than 1.8 cm across **9. × grandiflora**
12a. Tree or shrub to 20 m; leaf apex acute; fruit red-purple **8. arborea**
 b. Tree or shrub to 10 m; leaf apex obtuse to nearly acute or bluntly shortly acuminate; fruit purple-black **11. lamarckii**
13a. Shrub or tree to 12 m; racemes pendent **12. asiatica**
 b. Shrub to 4 m; racemes erect **13. spicata**

1. A. alnifolia (Nuttall) Nuttall. (*Aronia alnifolia* Nuttall; *Amelanchier canadensis* var. *alnifolia* Torrey & Gray; *A. canadensis* var. *florida* Schneider; *A. oreophila* A. Nelson in part). Illustration: Britton & Brown, Illustrated flora of the Northern United States and Canada 239 (1897).

Shrub or small tree to 12 m, becoming hairless and somewhat glaucous. Leaves 2–5.5 cm, thick, broadly elliptic or almost circular, apex usually truncate (rarely acutish), base rounded or almost cordate, coarsely toothed above the middle, vein pairs 7–12. Racemes to 4 cm, erect, with 5–15 flowers. Petals 0.8–1.6 cm, oblanceolate, tapered to the base, creamy white. Fruit *c.* 1 cm across, dark blue, juicy and edible. *W & C North America*. H1. Late spring–early summer.

Var. **cusickii** (Fernald) C. Hitchcock. (*A. cusickii* Fernald). Shrub to 3 m; young branches red and glossy, becoming grey; leaves to 5.5 cm, hairless.

Var. **semiintegrifolia** (Hooker) C. Hitchcock. (*A. oxyodon* Koehne; *A. florida* Lindley). Shrub or small tree to 12 m, young branches rusty hairy, becoming hairless; leaves to 4 cm, hairy.

2. A. utahensis Koehne (*A. crenata* Greene; *A. mormonica* Schneider; *A. oreophila* A. Nelson in part; *A. prunifolia* Greene; *A. purpusii* Koehne).

Shrub or small tree to 5 m. Leaves to 3 cm, rounded to ovate, apex rounded, notched, base rounded to wedge-shaped, finely hairy, coarsely toothed, vein pairs 11–13. Racemes to 3 cm, erect or ascending, 3–6-flowered. Petals 6–8 mm, linear. Fruit to 1 cm across, spherical, purple-black. *W North America*. H1. Spring.

3. A. sanguinea (Pursh) de Candolle (*A. amabalis* Wiegand; *A. rotundifolia* Roemer). Illustration: Britton, North American trees, 439 (1908).

Straggling or arching slender shrub to 3 m. Young branches red or grey. Leaves 2.5–7 cm, ovate to nearly rounded, apex acute to obtuse, base rounded to slightly wedge-shaped, densely white-yellow-hairy beneath, becoming hairless, coarsely toothed, almost reaching base, vein pairs 11–13. Racemes to 8 cm, erect to pendent, with 4–10 flowers. Petals 1–1.5 cm, white or light pink. Fruit 6–9 mm across, spherical, purplish black, sweet. *E North America, S Canada*. H1. Spring.

4. A. humilis Wiegand.

Shrub to 1.25 m, stoloniferous, patch forming. Leaves to 5 × 3 cm, elliptic to elliptic oblong, base rounded or almost cordate, with coarse forward-pointing teeth to below the middle, very densely white-hairy beneath. Racemes erect, many-flowered. Petals 0.7–1 cm, oblong obovate. Ovary hairy at apex. Fruit nearly black, sweet. *E North America*. H1. Spring.

5. A. stolonifera Wiegand (*A. spicata* misapplied). Illustration: Taylor, Taylor's guide to shrubs, pl. 86 (1987).

Shrub to 2 m, stoloniferous, patch-forming. Leaves to 5 × 3 cm, oblong to almost circular, base rounded or nearly cordate, with fine forward-pointing teeth in the upper ⅔, densely white-hairy beneath when young, vein pairs 7–10. Racemes erect. Petals to 9.5 mm, obovate oblong. Ovary hairy at apex. Fruit blue-black, juicy. *E North America*. H1. Spring.

6. A. ovalis Medikus filius (*A. vulgaris* Moench; *A. rotundifolia* (Lamarck) Dumont de Courset). Illustration: Botanical Magazine, 2430 (1823).

Erect or spreading stoloniferous shrub to 3 m. Young twigs woolly. Leaves 2.5–5 cm, ovate to obovate, apex rounded or notched and mucronate, with coarse forward-pointing teeth, white-hairy beneath when young, becoming hairless. Racemes to 4 cm, erect, with 3–8 flowers, white hairy. Petals 1–1.5 cm, ovate oblong. Apex of ovary hairless. Styles 5, free. Fruit 6 mm across, spherical, dark blue to black. *C & S Europe*. H1. Spring.

7. A. canadensis (Linnaeus) Medikus filius (*A. botryapium* (Linnaeus filius) Borkhausen; *A. intermedia* Spach; *A. oblongifolia* (Torrey & Gray) M.J. Roemer). Illustration: Britton, North American trees, 237 (1908); Botanical Magazine, 8611 (1915).

Shrub to 8 m, stoloniferous. Leaves to 5 × 2.5 cm, elliptic, apex acute or rounded, base usually rounded, with fine, shallow, sharp teeth, hairless except for midrib and stalk, vein pairs 9–13. Racemes to 6 cm, erect, initially white-hairy. Petals to 9 × 3 mm, ovate to oblanceolate, obtuse, white. Fruit *c.* 1 cm across, spherical, purple-black. *E North America*. H1. Spring.

'Micropetala': tall, erect, leaves richly coloured in autumn; flowers small. 'Prince William': to 3.5 m, fruits purple, edible, abundant. 'Springtyme': to 3.5 m, ovate, erect, compact; leaves yellow-orange in autumn. 'Tradition': to 7.5 m, ovate, erect, early flowering.

8. A. arborea (Michaux filius) Fernald (*A. canadensis* Siebold & Zuccarini). Illustration: Preston, North American trees, 250 (1969); Petrides, A field guide to eastern United States trees, pl. 42 (1988); Foote & Jones, Native shrubs and woody vines of the southeast, pl. 18 (1989).

Tree or shrub to 20 m. Bark light brown, tinged with red. Leaves to 10 cm, with fine forward-pointing teeth, oblong-ovate to ovate or ovate, apex acute, dark green and hairless above and pale below, vein pairs 11–17. Racemes to 7.5 cm, pendent, with 4–10 flowers. Petals 1–1.5 cm, spathulate to strap-shaped, pure white. Fruit *c.* 1 cm across, red-purple, dry and tasteless. *E North America*. H1. Spring.

9. A. × grandiflora Rehder (*A. botryapium* var. *lanceolata* misapplied; *A. canadensis* var. *grandiflora* Zabel). Illustration: Taylor, Taylor's guide to trees, pl. 128, 129 (1988).

Differs from *A. arborea* in the larger flowers, longer, more slender, less hairy racemes and leaves purple with densely matted woolly hairs to densely white-hairy. Differs from *A. laevis* in the densely white-hairy young leaves, more flowers, shorter flower-stalks and larger fruit. *Garden origin*. H1. Spring–summer.

'Autumn Brilliance': bark light grey; leaves brilliant red in autumn. 'Ballerina': flowers pure white in large heads. 'Coles Select': leaves with bright autumn colours. 'Cumulus': erect, leaves leathery, orange-red in autumn; flowers abundant, fruits red, becoming purple. 'Princess Diana': small tree, gracefully spreading, leaves bright red in autumn; fruit purple. 'Robin Hill': narrowly erect; flowers pale pink, fading to white, nodding; fruit small, red, juicy. 'Rubescens': flowers bluish pink. 'Strata': branches horizontally.

10. A. laevis Wiegand. Illustration: Voss, Michigan Flora, fig. 187 (1985); Taylor, Taylor's guide to trees, pl. 128, 129 (1988).

Shrub or tree to 13 m. Leaves 4–6 cm, ovate, apex acute, base rounded, finely toothed nearly to base, hairless, vein pairs 12–17. Racemes to 12 cm, pendent, flowers numerous. Petals 1.2–2.2 cm. Fruit to 1.8 cm across, spherical, purple to nearly black, sweet. *C North America*. H1. Spring.

'Prince Charles': somewhat rounded habit, flowers abundant; leaves orange and red in autumn; fruit blue, edible.

11. A. lamarckii Schröder (*A. botryapium* (Linnaeus) de Candolle; *A. canadensis* K. Koch; *A. canadensis* var. *botryapium* Koehne; *A. confusa* misapplied; *A. × grandiflora* Franco; *A. laevis* Clapham, Tutin & Warburg; *A. laevis* forma *villosa* Pelkwijk). Illustration: Bird, Flowering trees and shrubs, 15 (1989).

Tree or shrub to 10 m. Young branches bristly, becoming hairless. Leaves 8.5 × 5 cm, elliptic, oblong-elliptic or oblong-obovate, apex obtuse to nearly acute or bluntly shortly acuminate, base rounded to wedge-shaped, with fine, sharp forward-pointing teeth, nearly all the way around, copper-red and bristly when flowering, becoming hairless. Racemes to 7.5 cm, loose, with 6–10 flowers. Petals *c.* 14 × 5 mm, oblanceolate to elliptic. Ovary hairless at apex. Fruit *c.* 9.5 mm across, spherical, purple-black. *E Canada*. H1. Spring.

12. A. asiatica (Siebold & Zuccarini) Walpers (*A. canadensis* var. *japonica* Miquel). Illustration: Kurata, Illustrated important trees of Japan **4**: pl. 13 (1973).

Shrub or tree to 12 m, with slender spreading branches. Leaves to 7 cm, ovate to elliptic oblong, apex acute, base rounded or almost cordate, with fine forward-pointing teeth almost to the base, densely white or yellow hairy beneath, becoming hairless. Racemes pendent, densely flowered, woolly. Flowers *c.* 3 cm across. Petals *c.* 1.5 cm, ovate oblong. Ovary woolly at apex. Fruit blue-black. *Japan, China, Korea*. H1. Spring.

13. A. spicata (Lamarck) K. Koch (*A. ovalis* Borkhausen). Illustration: Botanical Magazine, 7619 (1898); Foote & Jones, Native shrubs and vines of the southeast, pl. 18 (1989).

Shrub to 4 m. Leaves to 5 cm, finely toothed, densely white-hairy when young, becoming hairless, vein pairs 7–9. Racemes to 4 cm, erect, with 4–10 flowers, woolly. Petals 4–10 mm, oblanceolate, white or pink. Styles united at base. Ovary apex hairy. Fruit to 8 mm across, spherical, purple-black. *NE North America*. H1. Spring.

14. A. bartramiana (Tausch) M. Roemer (*A. oligocarpa* M. Roemer). Illustration: Botanical Magazine, 8499 (1913).

Shrub to 2.5 m. Leaves 3–5 cm, elliptic to elliptic-oblong, apex acute or rounded, base wedge-shaped, with sharp and fine forward-pointing teeth to below the middle or nearly to the base, hairless when young, overlapping when in bud, vein pairs 10–16. Flowers 1–3, to 2.5 cm across. Petals *c.* 8 mm, obovate. Ovary woolly at apex. Fruit to 1.5 cm across, purplish black. *C to E North America*. H1. Spring.

62. × AMELASORBUS Rehder
J.E. Richardson
Hybrid between *Amelanchier* and *Sorbus*. Differs from *Amelanchier* in the partly pinnate leaves and paniculate

inflorescences, and from *Sorbus* in the partly simple leaves, partly lobed or incompletely pinnate leaves and paniculate flowers with 5 pistils.

Cultivation as for *Amelanchier*.

1. × A. jackii Rehder.

Deciduous shrub to 2 m. Leaves to 10 cm, ovate to elliptic, toothed, margin slightly sinuous, irregularly lobed or entire, hairy becoming hairless. Inflorescence a panicle to 5 cm, petals *c*. 2 cm, oblong, white. Fruit small, almost spherical, dark red, blue bloomy. *NW USA*. H1. Spring–summer.

63. RHAPHIOLEPIS Lindley
C.M. Mitchem

Small evergreen trees or shrubs. Leaves leathery, simple, stalked; young leaves densely covered in felty hairs. Inflorescence a terminal panicle, felted. Floral parts in 5s, subtended by bracts. Stamens 15–20. Ovary inferior, 2-celled; styles 2. Fruit a dry ovoid or spherical berry with a distinct scar at apex, with 1 or 2 large seeds.

The genus is currently being revised; it is estimated that there are 3–5 species all from the subtropical and warm temperate regions of Southeast and East Asia. Two species and one hybrid are in cultivation. They prefer a warm, sunny position in a well-drained fertile soil. Plants are slow-growing and resent disturbance; usually grown as wall plants or as a low hedge. Propagation is by seed, autumn cuttings, layering or grafting.

Literature: Kalkman, C., The Malesian Species of the Subfamily Maloideae (Rosaceae). *Blumea* **21**: 413–442 (1973); Kitamura, S., Short Reports of Japanese Plants. *Acta Phytotaxonomica et Geobotanica.* **26**: 1–2 (1974); Ohashi, H., Rhaphiolepis (Rosaceae) of Japan. *Journal of Japanese Botany* **63**(1): 1–7 (1988).

1a. Leaves thin, to 3 cm across, deeply toothed **1. indica**
 b. Leaves thick and leathery, more than 3 cm across, shallowly toothed **2**
2a. Panicles erect; flowers fragrant **2. umbellata**
 b. Panicles ascending; flowers scentless **3. × delacourii**

1. R. indica (Linnaeus) Lindley (*R. fragrans* Geddes; *R. japonica* Siebold & Zuccarini; *R. major* Cardas; *R. salicifolius* Lindley; *Crateagus indica* Linnaeus). Illustration: Botanical Magazine, 1726 (1815).

Small tree or shrub, to 1 m. Leaves 7–11 × 2–3 cm, thin, dark glossy green above, paler below, narrowly-elliptic to oblanceolate, margin sharply toothed. Inflorescence a loose panicle. Bracts to 5 mm, narrowly lanceolate. Petals to 8 mm, pinkish white. Fruit to 1 cm, spherical, blue-black. *Japan*. H5. Spring–summer.

2. R. umbellata (Thunberg) Makino (*R. japonica* misapplied; *R. japonica* Ziebold & Zuccarini var. *integerrima* Hooker & Arnott; *R. ovata* Briot). Illustration: Botanical Magazine, 5510 (1865); Addisonia **2**: t. 70 (1917).

Sturdy shrub to 2 m. Leaves 3–9 × 3.5–6, thick and leathery, ovate to oblanceolate, dark green and hairless above, paler and covered with felted hairs below, margin with fine forward-pointing teeth in upper half. Inflorescence an erect terminal panicle; flowers fragrant. Bracts to 1 cm, awl-shaped. Petals to 1 cm, white, tinged rose. Fruit a single-seeded dry berry, to 1 cm, blue-black with slight bloom. *Japan & Korea*. H4. Spring–summer.

3. R. × delacourii André. Illustration: Botanical Magazine, n.s. 362 (1960); Miller Gault, The dictionary of shrubs in colour, pl. 341 (1976); Krüssmann, Manual of cultivated broad-leaved trees & shrubs **2**: pl. 5 (1986).

Dome-shaped shrub to 2 m. Leaves 6–9.5 × 3–4.5 cm, leathery, broadly obovate to oblanceolate, margin finely toothed in upper half. Inflorescence an ascending panicle; flowers scentless. Bracts to 7 mm, narrowly lanceolate. Sepals to 5 mm, lanceolate. Petals to 1 cm, rose pink, narrowly obovate. *Garden origin*. H5. Early spring–summer.

A hybrid between *R. indica* and an unknown cultivar of *R. umbellata*, with a long flowering period.

Two cultivars are grown: 'Coates Crimson' and 'Spring Song'.

64. ARONIA Medikus filius
S.G. Knees

Deciduous shrubs; branches with closely adpressed, slender pointed buds. Leaves alternate, stalked, simple with toothed margins and blackish glands on midrib above. Stipules small, falling early. Flowers white or pale pink in small corymbs. Calyx-lobes 5, joined at base. Petals 5, spreading; stamens numerous, anthers purplish pink; ovary 5-celled, with woolly hairs towards the apex; styles 5, joined at their bases, carpels partly free. Fruit apple-like, with persistent remains of calyx.

Three closely related species from North America, cultivated for their attractive flowers and fruit; in addition, the leaves colour well in autumn. The fruits are inedible and probably gave rise to the widely used common name of chokeberry. Aronias have a confused history and have been included in *Pyrus*, *Sorbus* and *Mespilus*, but are probably most closely related to *Photinia*. All species are hardy and easily propagated from seed or by cuttings.

Literature: Hardin, J.W., The enigmatic chokeberries (Aronia, Rosaceae). *Bulletin of the Torrey Botanical Club* **100**: 178–184 (1973); Wiegers, J., Aronia Medik. in the Netherlands 1. Distribution and taxonomy. *Acta Botanica Neerlandica* **32**(5/6): 481–488 (1983).

1a. Leaves and inflorescence usually hairless; fruit purplish black **1. melanocarpa**
 b. Leaves and inflorescence covered with whitish hairs; fruit red or dark brownish purple **2**
2a. Fruit red **2. arbutifolia**
 b. Fruit dark brownish purple **3. × prunifolia**

1. A. melanocarpa (Michaux) Elliott (*Mespilus arbutifolia* Michaux var. *erythrocarpa* Michaux; *Pyrus melanocarpa* (Michaux) Willdenow). Illustration: Britton & Brown, Illustrated Flora of the Northern States & Canada, **2**: 291 (1913); Botanical Magazine, 9052 (1924); Brown & Brown, Woody plants of Maryland, 123 (1972); Krüssmann, Manual of cultivated broad-leaved trees and shrubs **1**: 173 (1984).

Shrub 0.5–3.5 m, with hairless branches. Leaves 2–7 × 1–6 cm, elliptic or obovate to oblong-lanceolate, apex acuminate or almost obtuse; margins finely toothed. Inflorescence a hairless, loosely branched corymb with 4–11 flowers. Petals white, *c*. 7 mm, almost round and abruptly narrowed into a stalk-like base. Fruit 6–8 mm across, almost spherical or slightly flattened, glossy purplish black. *E North America*. H2. Spring.

Three varieties have been described and plants are sometimes grown under these names. Var. **elata** Rehder is the largest with measurements for all its parts at the top of the ranges given above. Var. **grandifolia** (Lindley) Schneider (*Pyrus grandifolius* Lindley) has larger leaves, to 6 cm wide, but is otherwise similar to var. *elata*. A third variant with softly hairy, young leaves is var. **subpubescens** (Lindley) Schneider.

2. A. arbutifolia (Linnaeus) Elliott (*Pyrus arbutifolia* Linnaeus). Illustration: Botanical Magazine, 3668 (1838); Britton & Brown, Illustrated Flora of the Northern States & Canada, **2**: 290 (1913); Brown & Brown, Woody plants of Maryland, 133 (1972); Krüssmann, Manual of cultivated broad-leaved trees and shrubs **1**: 173 (1984).

Shrub usually to 3 m, more rarely reaching 6 m; branches covered with densely felted hairs. Leaves 4–7 × 2–4.5 cm elliptic to oblong or obovate, apex acute or abruptly acuminate, hairless above except for the midrib, greyish with felted hairs beneath. Inflorescence 3–4 cm across, a dense corymb of 9–20 flowers. Calyx with glandular hairs; petals white or reddish pink, 6–8 mm. Fruit spherical or slightly pear-shaped, 5–7 mm across (8–10 mm in some varieties), bright scarlet or very dark red. *E USA*. H2 Spring.

Amongst the varieties sometimes grown are var. **leiocalyx** Rehder, with hairless sepals and flower-stalks; var. **macrophylla** (Hooker) Rehder, which forms a large shrub or small tree to 6 m; var. **macrocarpa** Zabel has fruits 8–10 mm across. Var. **pumila** (Schmidt) Rehder (*Mespilus pumila* Schmidt) is a dwarf shrub with small leaves and dark red fruits. Like *A. melanocarpa*, these varieties are probably just extreme variants within the natural range of the species and may not merit formal recognition. However they are included here as they are sometimes encountered in the trade.

3. A. × prunifolia (Marshall) Rehder (*Pyrus floribunda* Lindley; *A. atropurpurea* Britton). Illustration: Britton & Brown, Illustrated Flora of the Northern States & Canada, **2**: 291 (1913); Brown & Brown, Woody plants of Maryland, 133 (1972);

Shrub to 4 m, very similar to *A. arbutifolia*, but with looser corymbs of 10–20 flowers. Calyx less hairy and usually lacking glandular hairs. Fruit almost spherical, 8–10 mm across, dark brownish purple. *E North America*. H2. Spring–early summer.

Considered by most recent authorities to be a hybrid between the two preceding species. A selection with good autumn colour and dark red fruits is 'Brilliant'.

65. PHOTINIA Lindley
M.F. Gardner & S.G. Knees
Evergreen or deciduous trees and shrubs. Leaves alternate, simple, usually with fine forward-pointing teeth or entire, with short stalks, stipules sometimes almost leaf-like,

free. Flowers, normally white, woolly in bud, borne in terminal or axillary umbel-like panicles; petals 5 with a distinct claw, hairless or hairy at the base, sepals 5, persisting in fruit; stamens 15–25. Ovary semi-inferior, 2–5-celled, styles 5. Fruit more or less fleshy, seeds 1 or 2 to each cell.

A genus of about 40 species from the Himalaya to Japan and Sumatra. Most are easily cultivated in freely drained soils; some evergreen species are noted for their tolerance of chalky soils, but the deciduous ones require a soil which is neutral to acid. Many of the evergreen species require a sheltered position. Most are grown for their outstanding autumn colour and attractive berries which last long into the winter; others are valued for their attractive reddish young leaves. Propagate from seed sown in early spring or by cuttings of half-ripened wood taken in early autumn and rooted under mist with bottom heat.

Literature: Kalkman, C., The Malesian species of the subfamily Maloideae (Rosaceae). *Blumea* **21**: 413–442 (1973).

Leaves. Evergreen: **1–10**; deciduous: **11–14**.
Leaf-margins. Entire: **7,8,10**; spiny: **1**; toothed: **2–6,9,11–14**; gland-tipped: **11, 14**.
Fruits. Globe-shaped: **1–4,6–10**; egg-shaped: **5,11–14**.

1a. Leaf-margins entire   2
  b. Leaf-margins toothed   4
2a. Fruit red to blackish purple, not more than 5 mm across
       7. integrifolia
  b. Fruit bright red, pinkish red or yellow, 8 mm or more across   3
3a. Upper leaf surface shiny
       10. niitakayamensis
  b. Upper leaf surface dull, often reddish   **8. davidiana**
4a. Leaf-margin conspicuously spiny
       1. prionophylla
  b. Leaf-margin shallowly toothed, occasionally gland-tipped   5
5a. Non-flowering or internal shoots usually spiny   **5. davidsoniae**
  b. All shoots without spines   6
6a. Leaves evergreen; fruit globe-shaped   7
  b. Leaves deciduous; fruit egg-shaped   11
7a. Leaves 10–18 cm, very glossy above   **4. serratifolia**
  b. Leaves 3–10 cm, dull or slightly shiny   8

8a. Petals white with a pinkish tinge; fruit black   **2. glabra**
  b. Petals white; fruit yellowish, orange or red   9
9a. Shoots with conspicuous lenticels; fruit yellowish orange-red
       6. lasiogyna
  b. Lenticels inconspicuous; fruit red   10
10a. Flowers 6–8 mm across, in umbels 10–12 cm wide   **3. × fraseri**
  b. Flowers 1.1–1.3 cm across, in umbels 5–10 cm wide   **9. nussia**
11a. Leaf-margin with gland-tipped teeth   12
  b. Marginal teeth without glands, spine-tipped   13
12a. Leaves 12–18 × 5–6 cm, veins hairy beneath   **14. glomerata**
  b. Leaves 5–13 × 2–4 cm, veins hairless   **11. beauverdiana**
13a. Leaves with persistent shaggy hairs beneath; flower-stalks hairy
       12. villosa
  b. Leaves and flower-stalks hairless
       13. parvifolia

1. P. prionophylla (Franchet) Schneider (*Eriobotrya prionophylla* Franchet). Illustration: Botanical Magazine, 9134 (1927); Flora Reipublicae Popularis Sinicae **36**: 235 (1974); Wu, Wild flowers of Yunnan **1**: 133 (1986).

Evergreen shrub to 2 m, growth habit rigid; young shoots covered with a greyish down. Leaves 3.5–9 × 2.5–5 cm, very hard, leathery, obovate to ovate, wedge-shaped at the base, rounded and tipped with a short point, spiny-toothed, persistently downy and strongly veined beneath, finely downy above when young, soon becoming hairless and dark green, leaf-stalks c. 1.2 cm. Flowers c. 8 mm across, white, borne in crowded, flattish umbels 5–7.5 cm across; petals obovate, incurved, stamens 20, yellow; calyx-tube woolly, sepals short, triangular, downy or hairless towards their tips. Fruit crimson, c. 6 mm across, globe-shaped, woolly at the apex. *China (Yunnan)*. H5–G1. Summer.

Requires the protection of a warm wall in cooler districts.

2. P. glabra (Thunberg) Maximowicz (*Crataegus glabra* Thunberg). Illustration: Edwards's Botanical Register 1956 (1837); Makino, Illustrated flora of Japan, 1405 (1956); Kurata, Illustrated important forest trees of Japan **3**: 57 (1971); Hayashi, Woody plants of Japan, 337 (1988).

Evergreen shrub to 4.5 m. Leaves 5–8 × 2–4 cm, bronzy when young, elliptic to oblong–obovate, pointed, tapered at base, with shallow forward-pointing teeth; stalks 1–1.5 cm, hairless. Flowers white tinged pink, hawthorn-scented, borne in loose terminal, many-branched umbels, 5–10 cm across, petals narrowly ovate, hairy at base on the inside. Fruit 4–6 mm wide, red changing to black, globe-shaped. *Japan & China*. H4–5. Summer.

The young growth is often susceptible to late spring frosts. Mainly represented in cultivation by the clones 'Rubens', which has been selected for its bronzy red young leaves and 'Parfait' ('Variegata'), which has creamy white margins.

3. P. × fraseri Dress (*P. glabra* × *P. serratifolia*). Illustration: Brickell, RHS gardeners' encyclopedia of plants & flowers, 85 (1989).

Large evergreen shrub, 3–5 m, intermediate between the parents. Leaves 7–9 cm, glossy green above, lighter beneath, elliptic to more obovate, with fine forward-pointing teeth, abruptly short-pointed, broadly wedge-shaped at base, leaf-stalk 2.8–5.6 cm, hairy above when young. Flowers 6–8 mm across, in umbels 10–12 cm across, petals white, hairy on inside at base. *Garden origin*. H3–4. Summer.

'Birmingham', Red Robin' and 'Robusta' have been selected for their attractive bright coppery or red young leaves which are susceptible to late spring frosts.

4. P. serratifolia (Desfontaines) Kalkman (*Crataegus serratifolia* Desfontaines; *P. serrulata* Lindley). Illustration: Botanical Magazine, 2105 (1819); Icones cormophytorum Sinicorum, 2145 (1972); Guo-zi Hsu, The wild woody plants of Taiwan, 131 (1984); Krüssmann, Manual of cultivated broad-leaved trees & shrubs 2: pl. 159, 397 (1986).

Evergreen shrub or tree 8–12 m, with sturdy branches. Leaves 10–18 × 5–8 cm, leathery, reddish when young, oblong, rounded or tapering at base, dark glossy green above, yellowish beneath, midrib hairless, margins shallowly toothed; leaf-stalks 2.5–4 cm, covered with whitish hairs. Flowers in umbels 10–18 cm across, petals hairless. Fruit 5–6 mm across, globe-shaped, red. *China*. H4–5. Spring–summer.

Sometimes used as a windbreak or hedge in less exposed sites as young growth is often susceptible to late spring frosts. The clone 'Rotundifolia' is lower growing and has leaves which are smaller and rounded.

5. P. davidsoniae Rehder & Wilson. Illustration: Gardener's Chronicle **71**: 199 (1922); Icones cormophytorum Sinicorum, 2146 (1972); Flora of Jiangsu 2: 283 (1982); Krüssmann, Manual of cultivated broad-leaved trees & shrubs 2: pl. 159 (1986).

Evergreen tree or shrub, 6–10 m; inner branches often with conspicuous straight spines *c.* 3 cm; young shoots reddish, downy; buds extremely small. Leaves 5–15 × 2–4.5 cm, dark glossy green above, pale beneath, oblanceolate to narrowly elliptic, tapering at both ends, leaf-stalk 7–14 mm. Flowers many, *c.* 1 cm across, borne in terminal umbels 8–10 cm across, flower-stalks downy, calyx-tube funnel-shaped, lobes triangular, downy, persistent in fruit. Fruit egg-shaped, orange-red, *c.* 8 mm. *China (W Hubei)*. H4–5. Summer.

6. P. lasiogyna (Franchet) Franchet (*Eriobotrya lasiogyna* Franchet). Illustration: Icones cormophytorum Sinicorum, 2149 (1972); Wu, Wild flowers of Yunnan **2**: 115 (1986).

Evergreen shrub or small tree, 3–6 m; branches slender, bark covered with lenticels. Leaves 5–9 × 2.5–4 cm, obovate, leathery, shiny above; apex acute or rounded, margins shallowly toothed; stalk 1.2–1.9 cm. Flowers in corymbs 8–10 cm across, fragrant; calyx-lobes short, broadly triangular; petals 5–7 mm, rounded, creamy white; styles 2–4. Fruit 5–8 mm across, globe-shaped, reddish or orange-yellow, woolly at apex. *China (Yunnan)*. H4–5. Summer.

7. P. integrifolia Lindley. Illustration: Flora Reipublicae Popularis Sinicae **36**: 223 (1974); Polunin & Stainton, Flowers of the Himalaya, 466 (1984); Wu, Wild flowers of Yunnan **2**: 113 (1986); Stainton, Flowers of the Himalaya, a supplement, pl. 29 (1988).

Evergreen tree or shrub, young branches hairless. Leaves 8–10 × 3–4 cm, leathery, oblong or ovate oblong, wedge-shaped or rounded at base, tapered at apex, entire, hairless, glossy above paler beneath; leaf-stalk 1–1.5 cm, stipules 2–4 mm, falling. Flowers borne in terminal, spreading panicles, hairless to slightly hairy, flower-stalks 1–3 mm; sepals hairless. Fruit reddish to black-purple globe-shaped, *c.* 5 mm across. *C & E Himalayas, S China & N Vietnam, Malaysia*. H5. Summer.

8. P. davidiana (Decaisne) Kalkman (*Stranvaesia davidiana* Decaisne). Illustration: Botanical Magazine, 9008

(1923); Krüssmann, Manual of cultivated broad-leaved trees & shrubs 3: pl. 130 (1986); Wu, Wild flowers of Yunnan 1: 143 (1986); Brickell, RHS gardeners' encyclopedia of plants & flowers, 67 (1989).

Evergreen, sparsely branched tree or shrub, 8–9 m; young shoots silky hairy, soon becoming hairless. Leaves 6–12 × 2–3 cm, often reddish, dull, entire, oblong-lanceolate, narrowed at the apex, sharply tipped, hairless, except on veins beneath; leaf-stalks 1–2 cm, often reddish, covered with silky hairs; stipules awl-shaped, soon falling. Flowers arranged in loose, hairy, flat umbels 5–8 cm across, anthers red or pinkish. Fruit bright red, globe-shaped, 7–9 mm wide. *China*. H3. Summer.

'Fructuluteo' has very attractive yellow fruit. 'Prostrata', is often used for ground cover on the account of its prostrate stems, although erect branches often occur.

'Palette' is a slow-growing variant with leaves blotched creamy white, sometimes tinged pink.

'Redstart' (*P. davidiana* 'Fructuluteo' × *P. × fraseri* 'Robusta'). This seedling has red young growths and flowers very freely. Fruit red, tipped with yellow. *Garden origin*. H4. Summer.

'Winchester' is similar to 'Redstart' and was raised from the same cross but has thinner elliptic leaves to 13.5 cm and fruit orange-red, flushed yellow.

9. P. nussia (D. Don) Kalkman (*Stranvaesia nussia* (D. Don) Decaisne; *S. glaucescens* Lindley). Illustration: Edwards's Botanical Register, 1956 (1837); Polunin & Stainton, Flowers of the Himalaya, 466 (1984).

Evergreen shrub to 5 m, young branchlets covered with whitish down, becoming hairless. Leaves 6–10 × 3–5 cm, tough and leathery, dark glossy green above, paler beneath with a downy midrib, lanceolate to obovate, finely toothed. Flowers 1.1–1.3 mm wide, borne in flat almost nodding, terminal umbels, 5–10 cm across, flower-stalks usually woolly, occasionally hairless. Fruit orange, globe-shaped, *c.* 8 mm, hairy when young, later becoming red and hairless. *E Himalayas, S China to Philippines* H5. Summer.

Requires the warmth of a south-facing wall in colder districts.

10. P. niitakayamensis Hayata. Illustration: The Kew Magazine **5**: 148 (1988); Li, Woody flora of Taiwan, 327 (1963); Li, Flora of Taiwan 3: 142 (1977).

Evergreen shrub to 3 m, densely

branched. Leaves 5–10 × 2–3 cm, entire, oblanceolate, tapered at the apex, rounded to wedge-shaped at the base, dark shiny green above, pale green almost dull beneath with up to 15 pairs of veins, mid-vein and margins hairy, leaf-stalk 0.7–1.7 cm, curved, grooved, densely hairy above; stipules *c.* 5 mm, joined to leaf-stalk at base. Flowers, 9–10 mm across, borne in hemispherical, terminal umbels, *c.* 4 × 8 cm, anthers lilac-pink. Fruit *c.* 8 × 9 mm, bright pinkish red, globe-shaped. *Taiwan.* H4–5. Summer.

11. P. beauverdiana Schneider. Illustration: Icones cormophytorum Sinicorum 2153 (1972); Flora of Jiangsu **2**: 283 (1982); Wu, Wild flowers of Yunnan **2**: 111 (1986); Krüssmann, Manual of cultivated broad-leaved trees & shrubs **2**: pl. 159, 397 (1986).

Deciduous shrub or slender tree to 9 m, bark downy when young, soon hairless. Leaves 5–13 × 2–4 cm, leathery, narrowly obovate to lanceolate, narrowly wedged-shaped at base, long, pointed at tip, veins in 8–14 pairs, conspicuous; margin with stiff teeth, each bearing a small black gland, leaf-stalk *c.* 1 cm. Flowers numerous, 6–8 mm wide, borne in large, flattish, terminal or axillary umbels *c.* 5 cm wide. Fruit *c.* 6 mm, deep red, egg-shaped. *China (W Hubei)* H2. Summer.

Var. **notabilis** (Schneider) Rehder & Wilson. Illustration: Liu, Illustrations of native & introduced ligneous plants of Taiwan **1**: 435 (1960); Guo-zi Hsu, The wild and woody plants of Taiwan, 128 (1984); Krüssmann, Manual of cultivated broad-leaved trees & shrubs **2**: pl. 159 (1986). Leaves broad oblong-elliptic, 7–12 cm, with 12 pairs of veins. Inflorescences more loose, 8–10 cm wide. Fruit orange-red. In fruit this variety is considered to be far superior to the species.

12. P. villosa (Thunberg) de Candolle. Illustration: Botanical Magazine, 9275 (1932); Icones cormophytorum Sinicorum, 2159 (1972); Flora of Jiangsu **2**: 284 (1982); Krüssmann, Manual of cultivated broad-leaved trees & shrubs **2**: 397 (1986).

Deciduous shrub or small tree to 5 m, young shoots downy, later hairless. Leaves 3–8 × 1.5–3 cm, obovate or ovate-lanceolate, tapered at the base with a long fine point at the apex, very tough, dark green above, pale yellow-green beneath with shaggy hairs, finely toothed, each tooth spine-tipped. Flowers borne in hairy umbels, *c.* 5 cm wide, stalks warty. Fruit *c.* 8 mm, bright red, egg-shaped. *Japan, China & Korea.* H2. Summer.

Very variable in the amount of down present on the leaves, young shoots and flower-stalks. Much valued for its red fruits and brilliant autumn colour.

Var. **laevis** (Thunberg) Dippel. Illustration: Botanical Magazine, 9275 (1929); Krüssmann, Manual of cultivated broad-leaved trees & shrubs **2**: 397 (1986). Leaves longer, pointed; branchlets and flowers only slightly hairy to hairless; fruit to 1.5 cm.

Forma **maximowicziana** (Léveillé) Rehder. Leaves almost stalkless, distinctly puckered, rounded and abruptly pointed at the tip, base wedge-shaped, veins conspicuously sunken, autumn colour very striking. According to some authorities this variant is better recognised as a seperate species.

Var. **sinica** Rehder & Wilson. Slender tree to 10 m, young shoots downy. Leaves not tough, mostly elliptic, bright green, soon hairless above, paler beneath, especially downy on the midrib and veins, hairless later. Flowers borne in racemes, 2.5–5 cm across. Fruit *c.* 1.2 cm, orange-scarlet, conspicuously warted.

13. P. parvifolia (Pritzel) Schneider. Illustration: Icones cormophytorum Sinicorum 2158 (1972); Flora of Jiangsu **2**: 285 (1982); Guo-zi Hsu, The wild woody plants of Taiwan, 130 (1984); Krüssmann, Manual of cultivated broad-leaved trees & shrubs **2**: 397 (1986).

Deciduous shrub, 2–3 m, young shoots dark red, hairless. Leaves 3–6 cm, ovate to obovate, slenderly pointed, broadly wedge-shaped at the base, sharply toothed, deep green above, lighter beneath. Flowers 5 or 6 in terminal umbels, 5 or 6 cm across; flower-stalks 1.2–2.5 cm, hairless. Fruit dullish to orange-red, egg-shaped, *c.* 8 mm, crowned by persistent sepals. *China (Hubei).* H2. Spring–summer.

14. P. glomerata Rehder & Wilson. Illustration: Icones cormophytorum Sinicorum 2148 (1972); Flora Reipublicae Popularis Sinicae **36**: 223 (1974); Wu, Wild flowers of Yunnan **2**: 112 (1986).

Deciduous tree or shrub, shoots reddish when young, with long shaggy hairs. Leaves 12–18 × 5–6 cm, thin, leathery, narrow oblong to oblanceolate, tapered towards the base, margins with fine, glandular teeth, slightly rolled inwards, yellowish green above, lighter beneath with 6–9 pairs of veins, with long shaggy deciduous hairs. Flowers fragrant, nearly stalkless, in umbels 6–10 cm across, covered with dense shaggy hairs. Fruit

5–7 mm, egg-shaped, red. *China (Yunnan).* H4–5. Summer.

66. HETEROMELES M.J. Roemer
M.F. Gardner
Evergreen shrubs, 2–10 m; bark grey, branchlets slightly hairy. Leaves 5–10 cm, elliptic or lance-shaped to oblong, leathery, sharply toothed, pale beneath; leaf-stalks 1–2 cm, downy. Flowers many, borne in flattish, corymb-like panicles; sepals triangular, 1–1.5 mm; petals white, *c.* 3 mm; stamens 10; styles 2 or 3, separate. Ovary inferior. Fruit bright red, 5–6 mm, persistent throughout the winter.

A genus with a single species from North America, grown for its attractive fruit and handsome foliage. Although easily cultivated in the open in milder areas, in colder districts it is best grown against a wall for protection. Propagation as for *Photinia*.

1. H. arbutifolia M. Roemer. Illustration: Edwards's Botanical Register, 491 (1820) – as Photinia arbutifolia; Sargent, Manual of the trees of North America **1**: 392 (1965); Munz, A flora of Southern California, 744 (1974); Lenz & Dourley, California native trees & shrubs 107 (1981).

USA (California). H4–5. Summer.

67. COTONEASTER Medikus
J. Fryer & B. Hylmö
Evergreen or deciduous shrubs and small trees. Leaves alternate, simple, entire. Flowers in cymes, small clusters, or solitary; stamens 10–20; carpels 1–5, usually with free styles. Fruit (pomes) somewhat succulent, containing 1–5 one-seeded, hard-walled nutlets.

The genus is distributed throughout Eurasia and North Africa. A marked concentration of species is found in the Himalaya and western China.

As with several other genera of Rosaceae, (e.g. *Crataegus, Rubus* and *Sorbus*) species with apomictic breeding systems are common. This means that plants raised from seed are usually genetically identical to their mother. The morphological differences between species with such breeding systems are often small making their identification difficult.

The species show a great variation in growth habit: from tall trees 15–20 m (*C. frigidus*) to prostrate subalpine shrubs (*C. radicans*). Species from colder areas are deciduous, whereas those from warmer zones are mostly evergreen or semi-evergreen. In central and northern Europe, most cultivated species are deciduous. One species (*C. lucidus*), native to the shores of Lake Baikal, Siberia, is one of the most hardy of all garden shrubs, and is frequently planted in Russia and Scandinavia (even north of the Arctic Circle). In those parts of Europe with milder climates, such as the British Isles, evergreen species are more common in cultivation. Many of these are cultivars selected from sexually reproducing species such as *C. salicifolius*, *C. dammeri*, and *C. conspicuus*.

The fruits are much liked by birds, and in many areas several species have become naturalised following their escape from gardens. Bees also visit the shrubs in great numbers.

Shrubs thrive in any adequately drained soil and are easily propagated from cuttings, taken in late summer; or seed, which needs a period of chilling before it will germinate. Apomictic species breed true from seed while outbreeding species will give a degree of variation. Most cultivars should be increased by cuttings only.

Fireblight can sometimes be a problem, especially with some evergreen species; it can be controlled by spraying an antibiotic (streptomycin) or copper salts, and cutting out affected branches 30 cm into the green unaffected wood. It is important to burn removed branches immediately.

Literature: Klotz, G., Uebersicht über die in Kultur befindlichen Cotoneaster, Arten und Formen; *Wissenschaftliche Zeitschrift der Martin-Luther-Universitat Halle-Wittenburg* **6**: 945–982 (1957); Flinck, K.E. & Hylmö, B., A list of series & species in the genus Cotoneaster; *Botaniska Notiser* **119**: 445–463 (1966); Hurusawa, I. & Kawakami, *Informationes Annuales Hortorum Botanicorum Facultatis Scientiarum Universitatis Tokyoensis* (1967); Klotz, G., Synopsis der Gattung Cotoneaster Medic. Phipps, *1-Beiträge zur Phytotaxonomie* **10**: 7–81 (1982); Phipps, J.B. et al. A checklist of the subfamily Maloideae (Rosaceae) *Canadian Journal of Botany* **68**: 2209–2269 (1990).

Habit. Evergreen or semi-evergreen: **1,3, 4,43–46,69,71–99**; deciduous: **2, 5–42, 47–68, 70**; height less than 1 m (without support): **4,6–12,15,59,77, 78,87,90–92,95–99**; 1–3 m: **1–3, 5–6,12–20,22–24,26–30,33,35,36, 39–52,54–55,58–60,62,67–69,79, 86,88,89,93,94**; 3–6 m: **21,25, 31–34,37,38,53,56,57,61,63,66, 70–76,80–85**; over 6 m: **63–65,70**.

Leaf size. Less than 2.5 cm: **1–17,46–48, 59,78,87–93,95–99**. 2.5–8 cm: **18–37,39,40,42–45,49–58,60–62, 67–69,72–74,76,77,79–82,84–86, 94**. Some leaves more than 8 cm: **21,38,41,63–66,70,71,72,75,83**.

Number of flowers in cyme. Solitary: **1–3, 4,5,7,14–16,31,77,78,87–93,95–99**; 1–5: **2,5–17,31–33,48,77,78,87,88, 92,93,94,97**; 3–15: **18–24,27–30, 34–36, 39, 40,42–47,49,50,54–56, 58–66,69,72,79**; 5–25: **25,26,37,39, 51–57,67–69,74,80,82,86**; more than 25: **70–73,75,76,83–85**.

Flowers. Petals erect: **1–4,7–24,29–45, 46–50**; semi-spreading: **25–28,46,51**; spreading: **5,6,52–99**.

Flower colour. Multi-coloured (purple, red, pink or off-white): **1–50**; white (sometimes pink in bud): **51,53–99**; pink: **3–6,52,60**.

Flower. Stamen number 10–15: **3–18**; 15–20: **1–2,19–99**.

Fruit colour. Orange-red: **1,2,10–13,39, 42–45,48,69,80,86**; red: **3–9,14–38, 40,41,46,47,49–68,70–79,81–85, 87–99**; becoming maroon: **27,28,40, 49,55,56,66,68**; becoming black: **16–19,21–26,36,41,61–63,65,66**.

Fruit. Nutlets 1: **67–69**; nutlets 2 (joined together as 1): **53–57**; nutlets 1 or 2: **61,62,66,88**; nutlets 2: **3,4–6,7,9,15, 16,18,19,21–23,29,31,43,45,49–52, 58–60,63,65,70,71,73,74,79–99**; nutlets 2 or 3: **12–14,17,24–28,31,32, 34,46,75,78**; nutlets 3: **1,2,8,10, 19–23,43,44,51,93–95**; nutlets 3,4 or 5: **29,30,33,35–42,44,46–48,71,72, 77,78**.

1a. Flowers in each group opening in sequence (or flowers solitary); petals mostly obovate, usually erect, (rarely spreading), purple, red or pink rarely white; filaments usually pink or red   **2**
  b. Flowers in each group opening more or less simultaneously (or flowers solitary); petals nearly circular, spreading, usually white or cream; filaments usually white   **14**
2a. Stamens 10–15   **3**
  b. Stamens *c.* 20   **6**
3a. Branches warted; leaves to 2 cm, thick, shiny, edges wavy; flowers solitary, pendent; fruit red, nearly spherical; nutlets 2
    (series **Verruculosi**) 24
  b. Branches not warted; leaves to 5 cm, thin to moderately thick, dull or shiny; edges flat or wavy; flowers

usually 2 or 3 per group, (rarely 4–6), if solitary then spreading or erect; fruit red or black, cylindric to obovate or rarely spherical or almost so; nutlets 2 or 3 (rarely 4)   **4**
4a. Petals spreading, plain red, falling early; filaments dark red; fruit red; nutlets 2   (series **Sanguinei**) 25
  b. Petals erect, multicoloured, purple, red, pink, white, not falling early; filaments dark red, red or pale pink; fruit red or black; nutlets 2 or 3 (rarely 4)   **5**
5a. Leaves initially with sparse adpressed bristles beneath, edges flat or wavy; filaments dark red or pink; fruit orange to red   (series **Adpressi**) 26
  b. Leaves initially with downy or shaggy hairs beneath, edges flat; filaments pink (rarely red); fruit black   (series **Nitentes**) 35
6a. Leaves with adpressed bristles beneath   **7**
  b. Leaves with densely felted, shaggy or sparse hairs beneath   **9**
7a. Leaves less than 2 cm; flowers solitary or 2–3 per cyme, pendent; calyx hairless or very sparsely hairy or bristly; fruit orange or red; nutlets 3   (series **Distichi**) 23
  b. Leaves usually more than 2 cm; flowers 1–11 per cyme, nearly erect or if spreading to pendent; calyx with adpressed bristles; fruit red or black; nutlets 2–5   **8**
8a. Flowers very wide or narrowly bell-shaped; fruit obovoid or nearly spherical 8–12 mm, red; nutlets 2–4   (series **Acuminati**) 46
  b. Flowers small, spreading, not bell-shaped; fruit sperical, to 7 mm, red or black; nutlets mostly 5
    (series **Glomerulati**) 49
9a. Surface of leaves blistered or wrinkled; fruit red, rarely orange (or in one species purple-black)   **10**
  b. Surface of leaves flat, with the veins only slightly impressed; fruit black, red or maroon   **11**
10a. Leaves 1–3.5 cm (rarely 6 cm), wrinkled, often dull, leaf-edges sometimes recurved; anthers pale purple or white; fruit red; nutlets 2–4   (series **Franchetioides**) 55
  b. Leaves 1–15 cm, blistered, shiny, leaf-edges not recurved; anthers white; fruit red or purple-black; nutlets 3–5   (series **Bullati**) 50
11a. Lower surface of leaves with short

erect or shaggy hairs often only
sparsely so; fruit black
                    (series **Lucidi**) 37

b. Lower surface of leaves with
   densely felted hairs; fruit red,
   maroon or black   12

12a. Petals erect or semi-spreading; fruit
   maroon or black
                  (series **Melanocarpi**) 41

b. Petals erect; fruit red   13

13a. Nutlets 3–5
                  (series **Cotoneaster**) 45

b. Nutlets 2   (series **Zabelioides**) 60

14a. Anthers white   15

b. Anthers purple or black   18

15a. Nutlets apparently 1 (= 2 nutlets
   joined)   (series **Multiflori**) 62

b. Nutlets 1 (not joined) or 2 (rarely
   3)   16

16a. Fruit black, (in one species red-
   purple); nutlets 1 or 2
                  (series **Insignes**) 68

b. Fruit red; nutlets 2 (rarely 3)   17

17a. Leaves hairy beneath; petals and
   filaments pink, or leaves hairless
   beneath; petals and filaments white;
   nutlets 2 (rarely 3)
                  (series **Megalocarpi**) 61

b. Leaves hairy beneath; petals and
   filaments white; nutlets 2
                  (series **Racemiflori**) 66

18a. Fruit black   (series **Insignes**) 68

b. Fruit red, orange or salmon-pink
                                    19

19a. Branches growing at right angles;
   nutlets 1 (rarely 2) per fruit
                  (series **Hebephylli**) 73

b. Branches growing at acute angles;
   nutlets 2–5   20

20a. Leaf-edges not recurved; flowers
   many per cyme; nutlets 2   21

b. Leaf-edges usually recurved; flowers
   solitary, 2–7 or many per cyme;
   nutlets 2–5   22

21a. Leaves thin, blade 10–15 cm
           (series **Chaenopetalum**) 75

b. Leaves thick, blade usually less
   than 5 cm (rarely to 9 cm)
                  (series **Pannosi**) 82

22a. Leaves to 10 cm, upper surface
   wrinkled or with veins slightly
   impressed; flowers solitary, 2–4
   or many per cyme; nutlets 2–5
                  (series **Salicifolii**) 76

b. Leaves to 3 cm, upper surface flat;
   flowers solitary or 2–7 per cyme;
   nutlets 2   (series **Microphylli**) 89

23a. Flowers solitary; leaves nearly
   circular   **1. nitidus**

b. Flowers solitary or 2 or 3 per cyme;
   leaves elliptic   **2. marquandii**

24a. Young branches densely warted,
   stiffly erect; leaves to 1.5 cm, rarely
   notched at apex   **3. verruculosus**

b. Young branches sparsely warted,
   spreading, erect; leaves to 1 cm,
   notched at apex   **4. cavei**

25a. Leaves with hairs only on the veins
   beneath   **5. sanguineus**

b. Leaves with felted hairs beneath
                           6. rubens

26a. Branching irregular, rooting at
   nodes; leaves thin, usually wavy-
   edged; fruit spherical, succulent
                                    27

b. Branching regular, seldom rooting
   at nodes; leaves moderately thick,
   usually not wavy-edged; fruit
   ellipsoid or cylindric, rarely
   spherical, not succulent   30

27a. Shrub to 30 cm   **7. adpressus**

b. Shrub 0.5–1 m   28

28a. Leaves broadly obovate
                      12. atropurpureus

b. Leaves spherical   29

29a. Leaves very wavy-edged; flower
   9–10 mm; nutlets usually 2
                           9. nanshan

b. Leaves not or only slightly wavy-
   edged; flower 6–7 mm; nutlets
   usually   **38. apiculatus**

30a. Fruit orange-red   31

b. Fruit red to rich dark red   34

31a. Leaves to 2 cm, concave
                      13. hjelmqvistii

b. Leaves less than 1.5 cm, flat, or
   only slightly wavy-edged   32

32a. Leaves broadly obovate, thin to
   moderately thick, slightly wavy-
   edged   **12. atropurpureus**

b. Leaves nearly circular to broadly
   elliptic, thick, flat.   33

33a. Leaves to 1.2 cm; fruit 5–6 mm
                       10. horizontalis

b. Leaves to 0.8 cm; fruit 4–5 mm
                         11. perpusillus

34a. Branching regular; leaves
   moderately thick, to 1.5 cm
                          14. ascendens

b. Branching irregular, straggly;
   leaves thin, to 2.5 cm
                         15. divaricatus

35a. Leaves 2–5 cm; flowers 3–6 per
   cyme   **18. tenuipes**

b. Leaves to 2.5 cm; flowers 1–3 per
   cyme   36

36a. Shrub to 3 m; stamens 10,
   filaments pink   **16. nitens**

b. Shrub to 1.5 m; stamens usually
   12, filaments red
                       17. harrysmithii

37a. Fruit red   **20. acutifolius**

b. Fruit black   38

38a. Leaves falling very early; nutlets
   usually 3   **19. lucidus**

b. Leaves falling medium to late;
   nutlets usually 2   39

39a. Leaves thin, flat; upper surface not
   blistered   **23. laetevirens**

b. Leaves moderately thick; upper
   surface slightly blistered; with
   shaggy hairs   40

40a. Leaves to 8 cm (rarely 10 cm)
                          21. villosulus

b. Leaves to 5 cm   **22. ambiguus**

41a. Petals erect; fruit black or bluish
   black   42

b. Petals semi-spreading; fruit maroon
                                    44

42a. Shrub 1.5–2 m; flowers to 15 per
   cyme   **24. melanocarpus**

b. Shrub 2–4 m; flowers more than 15
   to 40 per cyme   43

43a. Flower-stalks with shaggy hairs;
   late flowering and ripening of fruit
                       25. polyanthemus

b. Flower-stalks with only a few hairs;
   earlier flowering and ripening of
   fruit   **26. laxiflorus**

44a. Leaves to 8 cm; cymes loose
                            27. ignavus

b. Leaves to 3.5 cm; cymes compact
                       28. zeravschanicus

45a. Leaves broadly ovate, apex acute,
   upper surface nearly hairless
                       29. integerrimus

b. Leaves nearly circular to broadly
   elliptic, apex obtuse, upper surface
   persistently hairy
                        30. tomentosus

46a. Leaves thin, somewhat wavy edged,
   elliptic-ovate; nutlets 2
                        31. acuminatus

b. Leaves moderately thick, flat;
   nutlets 2–4   47

47a. Leaves circular-ovate, to 3 cm;
   nutlets 3 or 4   **33. simonsii**

b. Leaves ovate-lanceolate, to 5 cm;
   nutlets 2 or 3   48

48a. Leaves acuminate, upper surface
   and calyx-lobes persistently hairy
                         32. mucronatus

b. Leaves acute, upper surface and
   calyx-lobes becoming
                        34. newryensis

49a. Leaves shiny above; fruit red
                       35. glomerulatus

b. Leaves dull above; fruit black
                        36. foveolatus

50a. Fruit orange-red or red   51

b. Fruit dark maroon or purplish
   black   54

51a. Fruit orange-red   52
  b. Fruit red   53
52a. Leave thin, usually 1.5 times as
     long as wide, lower surface
     becoming thinly hairy; nutlets 3 or
     4 (rarely 5)   **39. boisianus**
  b. Leaves thick or moderately thick,
     usually twice as long as wide, lower
     surface with persistent felted or
     shaggy hairs; nutlets 4 or 5 (rarely
     3)   **42. sikangensis**
53a. Leaves to 7 cm   **37. bullatus**
  b. Leaves to 15 cm   **38. rehderi**
54a. Leaves to 5 cm; fruit maroon
                               40. obscurus
  b. Leaves to 12 cm; fruit purplish
     black   **41. moupinensis**
55a. Leaves evergreen or semi-evergreen,
     lower surface with silvery white
     felted hairs; very late flowering and
     fruiting   56
  b. Leaves deciduous, lower surface
     with greyish felted hairs; medium to
     early flowering and fruiting   59
56a. Shrub 1–1.5 m; leaves to 2 cm;
     fruit spherical, red   **46. amoenus**
  b. Shrub 1.5–3 m; leaves 2–6 cm;
     fruit obovoid or spherical, orange-
     red   57
57a. Leaves to 6 cm; anthers white
                               44. sternianus
  b. Leaves to 3.5 cm; anthers pale
     purple   58
58a. Calyx-lobes triangular-acuminate;
     nutlets 3 (rarely 2)
                               43. franchetii
  b. Calyx-lobes drawn out into a long
     mucro; nutlets 2   **45. wardii**
59a. Shrub to 3 m; fruit red
                               47. dielsianus
  b. Shrub *c.* 1 m (rarely 1.5 m); fruit
     orange-red   **48. splendens**
60a. Leaves to 4 cm, elliptic to ovate,
     apex acute; flowers off-white to
     pale pink; fruit obovoid, dark red to
     nearly maroon   **49. zabelii**
  b. Leaves to 3 cm, broadly ovate, apex
     obtuse; flowers pink and red; fruit
     spherical, red   **50. fangianus**
61a. Leaves with lower surface
     persistently hairy; flowers white
                               51. megalocarpus
  b. Leaves with lower surface hairless;
     flowers pink   **52. roseus**
62a. Fruit carmine or red   63
  b. Fruit maroon   64
63a. Leaves initially sparsely hairy,
     soon hairless; leaf-stalk and calyx
     hairless; flowers to 20 per cyme
                               53. multiflorus
  b. Leaves with lower surface

persistently sparsely hairy; leaf-stalk
and calyx initially hairy; flowers to
10 per cyme   **54. hupehensis**
64a. Leaves thick, dark green, lower
     surface, leaf-stalk and calyx with
     felted hairs; very late flowering and
     leaf fall   **56. veitchii**
  b. Leaves thin, green or light green,
     lower surface, leaf-stalk and calyx
     persistently sparsely hairy or
     hairless; early or medium-early
     flowering and leaf fall   65
65a. Leaf-stalk and calyx hairless;
     flowers and petals flat, all white;
     early flowering   **55. calocarpus**
  b. Leaf-stalk and calyx sparsely hairy;
     flowers wide bell-shaped, petals
     concave, white, bases pale purple;
     medium-late flowering
                               57. przewalskii
66a. Shrub usually less than 1 m; leaves
     circular, to 2 cm (rarely 3 cm).
                               59. nummularius
  b. Shrub to 3.5 m; leaves ovate to
     elliptic, to 5 cm   67
67a. Leaf apex acuminate or acute
                               58. racemiflorus
  b. Leaf apex acute or obtuse
                               60. tauricus
68a. Leaves circular   69
  b. Leaves elliptic or obovate   70
69a. Shrub or small tree 3–6 m; leaves
     to 6 cm   **61. ellipticus**
  b. Shrub 1.5–2 m; leaves to 2.5 cm
                               62. hissaricus
70a. Leaves obovate, hairless beneath,
     except midrib sparsely hairy; nutlets
     spherical 2   **65. bacillaris**
  b. Leaves lanceolate to elliptic, at
     least initially hairy beneath; nutlets
     obovoid, 1 or 2   71
71a. Leaves lanceolate-elliptic, lower
     surface initially sparsely hairy;
     nutlets 1 or 2   **66. transens**
  b. Leaves elliptic, lower surface
     persistently hairy; nutlets 2   72
72a. Cymes compact; fruit black
                               63. affinis
  b. Cymes loose; fruit red to purple
                               64. gamblei
73a. Shrub to 2 m, densely branched;
     leaves elliptic-lanceolate, apex acute
                               69. ludlowii
  b. Shrub to 3 m, branches slender,
     arched; leaves broadly elliptic to
     nearly circular, apex obtuse rarely
     acute.   74
74a. Fruit red   **67. hebephyllus**
  b. Fruit dark maroon
                               68. monopyrenus

75a. Leaves deciduous, thin, surface flat
                               70. frigidus
  b. Leaves evergreen or semi-evergreen,
     moderately thin, surface wrinkled
                               71. × watereri
76a. Prostrate, creeping shrub with
     rooting branches; leaves less than 3
     cm   77
  b. Erect shrub, to 5 m; leaves to 10
     cm   78
77a. Leaves to 3 cm; nutlets usually 5
                               77. dammeri
  b. Leaves less than 1.5 cm; nutlets
     usually 4   **78. radicans**
78a. Leaves strongly wrinkled; nutlets
     2–5   79
  b. Leaves only slightly wrinkled;
     nutlets 2 (rarely 3)   81
79a. Leaves broadly elliptic; nutlets 2
                               74. rugosus
  b. Leaves elliptic-lanceolate; nutlets
     3–5   80
80a. Leaves elliptic-lanceolate, lower
     surface with felted hairs to
     somewhat thinly woolly hairs
                               72. salicifolius
  b. Leaves narrowly lanceolate, lower
     surface extremely woolly
                               76. floccosus
81a. Leaves hairless, lower surface
     glaucous   **73. glabratus**
  b. Leaves with felted hairs beneath
                               75. henryanus
82a. Leaves hairless; calyx-lobes hairless
                                                      83
  b. Leaves persistently hairy beneath;
     calyx-lobes hairy   84
83a. Leaves to 8 cm
                               80. glaucophyllus
  b. Leaves to 4 cm   **86. meiophyllus**
84a. Leaves less than 3 cm; flowers to
     12 per cyme   **79. pannosus**
  b. Leaves more than 3 cm; flowers
     more than 12 per cyme   85
85a. Leaves narrowly elliptic or
     narrowly obovate, more than twice
     as long as wide   **82. harrovianus**
  b. Leaves broadly elliptic or broadly
     obovate, less than twice as long as
     wide   86
86a. Calyx only sparsely hairy with long
     silky hairs   **85. serotinus**
  b. Calyx with felted hairs   87
87a. Leaves thick, obovate or broadly
     elliptic, upper surface with
     impressed and lower surface raised
     veins   **83. lacteus**
  b. Leaves thin or moderately thick,
     elliptic, surfaces flat   88
88a. Flowers to 30 per cyme; calyx-lobes
     woolly   **81. vestitus**

b. Flowers to 75 per cyme; calyx-lobes felted with silky hairs
        84. turbinatus

89a. Flowers usually 2–7 per cyme   90

b. Flowers usually solitary   94

90a. Calyx with dense adpressed bristles
        93. buxifolius

b. Calyx with sparse adpressed bristles   91

91a. Leaves to 3 cm   **94. marginatus**

b. Leaves less than 1.5 cm   92

92a. Leaves narrowly elliptic, sometimes ovate, apex acute to acuminate with, a long mucro
        97. lidjiangensis

b. Leaves elliptic, broadly-elliptic, to broadly obovate, apex obtuse or acute, with a short mucro   93

93a. Leaves 5–8 mm
        87. microphyllus

b. Leaves 8–15 mm, apex acute
        92. 'Ruby'

94a. Upper surface of leaves pale green, dull   **95. congestus**

b. Upper surface of leaves medium to dark green, shiny, (or only slighty dull), hairless, hairy or initially sparsely hairy or bristly   95

95a. Leaf-blade narrow, 2–5 times as long as wide   96

b. Leaf-blade broad, less than twice as long as wide (usually 1.5)   98

96a. Leaves narrowly elliptic, upper surface only slightly shiny, papillose   **88. conspicuus**

b. Leaves oblong to oblanceolate or narrowly obovate, upper surface very shiny, not papillose   97

97a. Leaves to 1.5 cm; fruit 8–10 mm
        98. integrifolius

b. Leaves less than 8 mm; fruit 4–5 mm   **99. linearifolius**

98a. Shrub to 2 m, robust, erect or prostrate; leaves to 2 cm
        89. rotundifolius

b. Shrub prostrate to creeping; leaves less than 1.5 cm   99

99a. Flowers to 1 cm, often pink in bud; fruit 0.8–1 cm   100

b. Flowers to 9 mm, pink or white in bud; fruit 4–6 mm   101

100a. Shrub to 20 cm, branches flat to the ground   **90. cochleatus**

b. Shrub to 50 cm, branches somewhat raised above ground
        92. prostratus

101a. Leaves elliptic, upper surface shiny, midrib initially with long adpressed hairs; flower buds often pinkish   **91. cashmiriensis**

b. Leaves broadly obovate, upper surface somewhat dull, hairless; flower buds white
        96. procumbens

In each series the main species is fully described, comparison is then made with other species in that series.

Series **Distichi** Yü.

1. C. nitidus Jacques (*C. distichus* Lange; *C. rupestris* Charlton). Illustration: Saunders Refugium Botanicum **1**: t. 54 (1869) – as C. rotundifolius).

Shrub, erect, *c.* 3 m. Branches straggly, initially with adpressed bristles, stipules *c.* 4 mm, more or less persistent. Flowers solitary, pendent; stalks *c.* 5 mm. Calyx hairless; lobes obtuse. Petals erect, red and off-white. Stamens 20; anthers white; filaments red. Fruit 9–11 mm, pendent, obovate to nearly spherical, orange, finally turning red, shiny; nutlets 3. Apomictic. *China (Yunnan)*. H3. Spring.

2. C. marquandii Klotz.

Branches fastigiate. Leaves deciduous; blade elliptic to obovate-elliptic, apex acuminate. Flowers solitary, or 2 or 3 per cyme; calyx with sparse adpressed bristles. Fruit 7–8 mm; bright orange. *Himalaya (Burma, Bhutan)*. H3. Spring.

Series **Verruculosi** Klotz.

3. C. verruculosus Diels.

Shrub, erect, 0.6–1.5 m. Branches erect, densely warted, with adpressed bristles. Leaves deciduous or semi-evergreen; blade 8–14 mm, moderately thick, circular to broadly elliptic or broadly obovate, edges slightly wavy; apex obtuse, mucronate, upper surface shiny, mid-green, with a few adpressed bristles, lower surface hairless; stalk 2–5 mm, slender, stipules red, oblong, some with hairy edges, persistent. Flowers solitary, pendent, nearly stalkless. Calyx hairless; lobes usually obtuse. Petals erect, red, pink-edged. Stamens 12–15; anthers white, filaments pale pink. Fruit 8–10 mm, pendent, nearly spherical, red, shiny; nutlets 2. Apomictic. *China (Yunnan)*. H3. Spring.

4. C. cavei Klotz.

Shrub *c.* 80 cm; branches not so noticeably warted. Leaf-blade to 2 cm, apex notched, upper surface dark-green; stalk 1–3 mm. Petals deep-pink to red. Fruit 7–9 mm, spherical. *Himalaya (Nepal, Sikkim)*. H3 Spring.

Series **Sanguinei** Klotz.

5. C. sanguineus Yü. Illustration: Bulletin British Museum Botany **1**: 5, pl. 4 (1954).

Shrub, erect, 1.5–2.5 m. Branches dense, straggly, initially with adpressed yellowish bristles. Leaves deciduous, blade 1–3 cm, moderately thick, broadly ovate, apex acute to obtuse, sometimes mucronate, upper surface shiny, sparsely hairy, veins impressed, lower surface with only a few hairs on the veins; leaf-stalk 3–6 mm with adpressed bristles; stipules to 2 mm, lanceolate, margin with shaggy hairs, not persistent. Flowers *c.* 1 cm across, solitary or in pairs, semi-erect, almost stalkless. Calyx hairless, wide bell-shaped, red, lobes shortly triangular, edges with shaggy hairs. Petals spreading, dark-red, falling early. Stamens 10 (rarely to 15); anthers white, filaments dark red. Fruit 9–11 mm, pendent, broadly obovoid to nearly spherical, light red; nutlets 2. Apomictic. *Himalaya*. H3. Spring.

6. C. rubens W.W. Smith.

Shrub 0.5–1.5 m. Leaf-blade 2–4 cm, circular or wide elliptic, apex obtuse, mucronate, lower surface with felted hairs; leaf-stalk *c.* 1 mm, with felted hairs. Flowers *c.* 8 mm across. Calyx densely downy. *China (Yunnan)*. H3. Spring.

Series **Adpressi** Hurusawa.

7. C. adpressus Bois (*C. horizontalis* var. *adpressus* Schneider). Illustration: Tarouca & Schneider, Freiland-Laubgeholz, 164 (1922).

Shrub, prostrate, to 30 cm. Branches rigid, irregular, initially downy, rooting at nodes, internodes short. Leaves deciduous; blade 5–15 mm, thin, falling early, broadly ovate to obovate, margin wavy, initially hairy, apex obtuse to nearly acute, mucronate, upper surface dull green, hairless, lower surface initially with adpressed hairs; stalk 1–2 mm. Flowers including calyx 6–7 mm, erect, solitary or in pairs, nearly stalkless. Calyx slightly downy, lobes triangular, sometimes mucronate, margin hairy. Petals erect, red and pink. Stamens 10 (rarely to 13); anthers white, filaments pink to red. Fruit 6–7 mm, erect, spherical, bright red, succulent; nutlets 2. Diploid, variable. *W China*. H3. Spring.

'Little Gem'. Smaller, forming a compact cushion, sometimes flowers but rarely sets fruit.

8. C. apiculatus Rehder & Wilson.

Prostrate, ends of branches ascending, 0.5–1 m. Leaf-blade 1–2 cm, somewhat circular, margin usually flat, (rarely slightly wavy), apex terminating abruptly to a small point, upper surface shiny; stalks becoming purple. Fruit 1–1.2 cm, succulent; nutlets usually 3. Apomictic. *China (Sichuan)*. H2. Spring.

9. C. nanshan Mottet (*C. adpressus* var. *praecox* Bois & Berthault).

Prostrate, ends of branches ascending, *c.* 1 m. Leaf-blade 1.2–2.5 cm, circular, apex somewhat acute or acuminate, edge extremely wavy. Flowers (including calyx) 9–10 mm. Petals red and pink, fringed off-white. Fruit 1–1.2 cm, nearly spherical, very succulent; nutlets usually 2. Apomictic. *China (Sichuan)*. H2. Spring.

10. C. horizontalis Decaisne.

Height *c.* 1 m (3 m if supported). Branches regular (herringbone), not rooting at nodes. Leaves not falling early; blade to 1.2 cm across, shiny, nearly circular to broadly elliptic, edges usually flat. Filaments dark red. Fruit 5–6 mm orange-red; nutlets 3. Apomictic. *W China*. H2. Spring.

11. C. perpusillus Flinck & Hylmö (*C. horizontalis* var. *perpusillus* Schneider; *C. horizontalis* 'Saxatilis').

Branches regular (herringbone), not rooting at nodes. Leaves not falling early; blade thick, shiny, edges usually flat. Fruit 4–5 mm, oblong to nearly spherical, orange-red. Apomictic. *China (Hubei)*. H3. Spring.

12. C. atropurpureus Flinck & Hylmö (*C. horizontalis* var. *prostratus* misapplied).

Prostrate or ascending shrub; 0.5–1 m (3 m supported). Branching regular. Leaf-blade thin to moderately thick, broadly obovate, edges slightly wavy; apex obtuse to truncate; upper surface somewhat shiny. Petals red with a black base. Filaments dark red. Fruit orange-red; nutlets 2 or 3. Apomictic. *China (Hubei)*. H2. Spring.

'Variegatus'. Shrub to 50 cm; leaves edged white, tinged pink in autumn, sometimes wavy edged.

13. C. hjelmqvistii Flinck & Hylmö (*C. horizontalis* var. *robustus* Grootendorst).

Prostrate and ascending to 1.5 m (4 m if supported). Branching regular, not rooting at nodes. Leaves with intense autumn colour; blade to 2 cm, moderately thick, ovate to circular, concave, upper surface shiny, light-green. Fruit orange-red. Apomictic. *China (Sichuan)*. H3. Spring.

Sometimes grown as *C. horizontalis* 'Darts Splendid' or 'Coralle'.

14. C. ascendens Flinck & Hylmö (*C. horizontalis* var. *wilsonii* Wilson; *C. horizontalis* var. *fructo-sanguineus* invalid).

Ascending; 1–2 m. Branches regular, not rooting at nodes. Leaf-blade moderately thick, elliptic, edges only slightly wavy, upper surface shiny light green. Flowers solitary or 2 or 3 per cyme. Petals pink and off-white. Fruit ellipsoid to nearly spherical, red, nutlets 2 or 3. Apomictic. *China (Hubei)*. H2. Spring.

15. C. divaricatus Rehder & Wilson.

Erect; 1–2 m. Branches wide-spreading, not rooting at nodes. Leaf-blade 0.8–2.5 cm, thin, ovate to elliptic, edges rarely wavy, upper surface shiny dark green. Flowers solitary or 2–4 per cyme. Stamens 12–15. Fruit 7–9 mm, ellipsoid to cylindric, very dark red. Apomictic. *China (Hubei)*. H2. Spring.

'Gracia'. Prostrate and ascending, to 1 m; branches somewhat regular, wide spreading, not rooting at nodes; leaves with intense autumn colour; blade to 2 cm, thick, apex sometimes obtuse, upper surface with veins impressed; petals pink; Fruit light red, not numerous. H3. Spring.

'Valkenburg'. Prostrate and ascending, to 1.5 m; branching dense, spreading, not rooting at nodes; leaf-blade to 2 cm, elliptic, moderately thick, upper surface with veins slightly impressed; petals pink, rarely sets fruit. H3. Spring.

Series **Nitentes** Flinck & Hylmö.

16. C. nitens Rehder & Wilson. Illustration: Botaniska Notiser **115**: 32 (1962).

Shrub, erect, 2–3 m. Branches arched, initially light red-brown with sparse yellow-grey hairs. Leaves deciduous, blade 1–2.5 cm, thin, circular to ovate, apex obtuse, upper surface hairless, shiny green, lower surface initially sparsely hairy; stalk 2–3 mm, sparsely hairy. Flowers erect, solitary or 2 or 3 per cyme; stalks 2–4 mm, soon hairless. Calyx sparsely hairy, lobes broadly triangular. Petals erect, purple-red and off-white. Stamens 10; anthers white; filaments pink. Fruit 7–8 mm, spreading, obovoid, purple-black; nutlets 2. Apomictic. *China (Sichuan)*. H2. Spring.

17. C. harrysmithii Flinck & Hylmö.

Prostrate, ascending, to 1.5 m. Branches regular, somewhat horizontal. Leaf-blade 1.5–2.5 cm, elliptic to ovate, apex acute to acuminate, sparsely hairy above, the newly unfolded leaves

papillose, sparsely hairy beneath. Flower-stalks 1–3 mm, sparsely hairy. Stamens usually 12, filaments red. Fruit 6–7 mm. *China (Sichuan)*. H2. Spring.

18. C. tenuipes Rehder & Wilson.

Branches slender, graceful, pendent, initially shiny, light purple-brown. Leaf-blade 2–5 cm, narrowly elliptic, lower surface with shaggy hairs; leaf-stalk slender. Flowers 3–6 per cyme. Petals pink and off-white. Fruit 6–10 mm, cylindric. *China (Sichuan)*. H2. Spring.

Series **Lucidi** Pojárkova.

19. C. lucidus Schlechtendal. Illustration: Flora of the USSR **9**: 252 (1939).

Shrub, erect, 1–3 m. Branches fairly dense, initially with adpressed downy hairs. Leaves deciduous, moderately thick, intense autumn colour, falling early, blade 2–7 cm, ovate-elliptic, apex acute, upper surface shiny, flat, soon hairless; lower surface initially with shaggy yellowish hairs, later with sparse adpressed down; leaf-stalk 2–6 mm, hairy. Flowers 5–15 per loose spreading cyme; stalks with yellowish adpressed-downy hairs. Calyx nearly hairless; lobes broadly triangular, edges hairy. Petals erect, pink, green and off-white. Stamens 20; anthers white, filaments pink. Fruit 8–10 mm, pendent, obovoid to spherical, black, shiny; nutlets 3 (rarely 2). Apomictic. *Siberia (Lake Baikal)*. H1. Early spring.

20. C. acutifolius Turczaninov.

Similar to *C. lucidus*, but fruit red. *China (Nei Mongol)*.

Probably not in cultivation; plants grown under this name are usually either *C. lucidus* or *C. villosulus*, both of which have black fruit.

21. C. villosulus Flinck & Hylmö (*C. acutifolius* var. *villosulus* Rehder & Wilson).

Height 2–4 m. Branching vigorous. Leaves falling medium to late; blade 3–10 cm, ovate to oblong-ovate, apex acute to acuminate, upper surface somewhat blistered, lower surface and calyx with dense shaggy hairs. Fruit obovoid slightly downy; nutlets 2 (rarely 3). *China (Hubei)*. H2. Early spring.

22. C. ambiguus Rehder & Wilson.

Leaves falling medium to late; blade 2.5–5 cm, apex finely acuminate, upper surface slightly blistered. Fruit nearly spherical; nutlets 2. *C China*.

Very rarely grown.

23. C. laetevirens Klotz (*C. acutifolius* var. *laetevirens* Rehder & Wilson).

Shrub 2–3 m. Branching open and more graceful. Leaf-edges hairy, upper surface light green, lower surface sparsely hairy; stalk 3–5 mm. Flowers 3–7 per cyme. Calyx-lobes acute or acuminate, mucronate. Fruit oblong to nearly spherical; nutlets usually 2. *China (Sichuan)*. H2. Early spring.

Many Cotoneasters labelled *C. ambiguus* are this species.

Series **Melanocarpi** Pojárkova.

24. C. melanocarpus Loddiges. Illustration: Loddiges' Botanical Cabinet **154**: 1531 (1830).

Shrub, erect, 1.5–2 m. Branches loosely erect, initially shiny red-brown, thickly downy. Leaves deciduous, blade 2–6 cm, somewhat thin, ovate to elliptic, apex acute, mucronate, upper surface dull dark green, initially with sparse hairs, lower surface with greyish felted hairs; stalks 4–7 mm, felted. Flowers 3–15 per cyme, spreading to pendent; flower-stalks sparsely hairy, the axils more so. Calyx nearly hairless, lobes obtuse. Petals erect, pink and off-white. Stamens 20; anthers white, filaments pink. Fruit 7–9 mm, pendent, obovoid to spherical, black, with a waxy bloom; nutlets 2 or 3. Apomictic. *Russia*. H1. Early spring.

25. C. polyanthemus Wolf.

Height to 4 m. Upper surface of leaves shiny. Flowers 15–40 per cyme, flowering much later than *C. melanocarpus*; stalks with shaggy hairs. Calyx sparsely hairy. Fruit obovoid to pear-shaped with only a very weak waxy bloom, ripening irregularly. *C Asia (Tianshan Mts)*. H1. Spring.

26. C. laxiflorus Lindley.

Height 2–2.5 m. Branches initially brownish purple with a grey peeling cuticle. Flowers 20–40 per cyme, loose, pendent, stalks nearly hairless. Fruit nearly spherical. *Origin unknown*. H2. Early spring.

27. C. ignavus Wolf.

Branches stiffly erect. Leaf-blade thin, nearly circular, obovate to broadly elliptic. Flower-stalks noticeably dichotomously branched. Petals semi-spreading. Fruit nearly ellipsoid, maroon. *Turkmenistan*. H2. Early spring.

28. C. zeravschanicus Pojárkova.

Leaf-blade 1.5–3.5 cm, nearly circular, upper surface a medium green. Flowers 4–8 per cyme, compact, erect; stalks usually purple. Petals semi-spreading, very pale pink to off-white. Fruit 5–8 mm, dull dark maroon, with the dried remains of the calyx persisting as a coronet. *Tadzhikstan*. H2. Early spring.

Series **Cotoneaster**.

29. C. integerrimus Medikus. Illustration: Nordens Flora **2**: 331 (1964).

Shrub, prostrate to erect, *c.* 1.5 m. Branches arched, initially with felted hairs. Leaves deciduous, blade 1–5 cm, somewhat thin, usually broadly-ovate, apex acute (rarely obtuse), usually mucronate, upper surface dull, initially very sparsely hairy, lower surface with felted greyish hairs; stalks 4–7 mm with felted hairs. Flowers 3–7 per cyme, spreading to pendent, axils thinly felted; stalks with sparse hairs. Calyx hairless, lobes obtuse, edges sometimes with downy hairs. Petals erect-incurved, short, scarcely longer than calyx-lobes, pale pink to off-white. Stamens 20; anthers white, filaments very pale pink. Fruit spreading and pendent, nearly spherical, 8–11 mm, red; nutlets 3–5, (rarely 2). Apomictic. *Europe*. H1. Early spring.

30. C. tomentosus Lindley (*C. nebrodensis* Koch).

Shrub to 2 m. Leaf-blade 3–6 cm, nearly circular to broadly elliptic, apex obtuse, upper surface persistently hairy, lower surface with dense felted hairs. Flowers 5–12 per cyme, stalks and calyx with felted hairs. Nutlets 4 or 5. *C Europe (Alps)*. H1. Early spring.

Series **Acuminati** Yü.

31. C. acuminatus Lindley. Illustration: Transactions of the Linnean Society of London **101**: 13 (1821) Loddiges' Botanical Cabinet **41**: 919 (1824).

Shrub, erect, 2–4 m. Branches stiffly erect, vigorous, initially dark greenish brown, with adpressed bristles. Leaves deciduous, blade 3–6 cm, thin, elliptic-ovate, margin somewhat wavy, apex acuminate, with a small point, upper surface shiny light green, with adpressed bristles, lower surface thickly covered with adpressed bristles, becoming thinner. Leaf-stalk 2–5 mm with adpressed bristles. Flowers solitary or 2–4 per cyme, spreading to pendent; stalks usually 5–10 mm with adpressed bristles. Calyx wide bell-shaped, with adpressed bristles; lobes broadly triangular, obtuse or acute. Petals

wide-erect, red-pink, green and off-white, stamens 20; anthers white, filaments pale pink. Fruit 8–12 mm, mostly pendent, broadly obovoid, bright red, shiny, hairy near apex; nutlets 2 (rarely 3). Diploid, variable. *Himalaya*. H3. Spring.

32. C. mucronatus Franchet.

Leaf-blade to 5 cm, thicker and flatter than *C. acuminatus*, ovate-lanceolate, apex acute to acuminate, mucronate, upper surface dark green, lower surface with dense adpressed bristles. Calyx bell-shaped, lobes acuminate. Nutlets 2 or 3. Apomictic. *China (Yunnan)*. H3. Spring.

33. C. simonsii Baker.

Height to 4 m. Branches wide-spreading. Leaf-blade 1.5–3 cm, thicker and flatter than *C. acuminatus*, broadly ovate, apex acute to somewhat acuminate, lower surface with adpressed bristles. Calyx-lobes acuminate. Fruit 8–10 mm, obovoid; nutlets 3 or 4. Apomictic. *Himalaya (Sikkim, Bhutan)*. H3. Spring.

Commonly used for hedging.

34. C. newryensis Barbier.

Shrub to 5 m. Branches spreading, initially with downy hairs. Leaf-blade moderately thick, 2–3 cm, elliptic, edges flat, apex acute, upper surface mid-green, becoming hairless, lower surface sparsely hairy; stalk with shaggy hairs. Flowers 5–11 per cyme, spreading. Calyx not bell-shaped, lobes obtuse or acuminate, hairless, margin woolly. Petals pink with very little white. Fruit 8–10 mm spreading, obovoid to nearly spherical, light red; nutlets 2 or 3, usually 3. *China (Sichuan, Yunnan)*. H3. Spring.

Series **Glomerulati** Flinck & Hylmö.

35. C. glomerulatus W.W. Smith (*C. nitidifolius* Marquand). Illustration: Hooker's Icones Plantarum, t. 3145 (1930).

Shrub, erect, very leafy, with intense autumn colour, 1–3 m. Branches spreading, initially golden brown, with sparse yellowish grey felted hairs. Leaves deciduous; blade 4–6 cm, moderately thick, lanceolate-ovate, apex acuminate, upper surface shiny, pale to mid-green, lower surface with sparse adpressed bristles; stalks 2–3 mm, with adpressed bristles. Flowers small, *c.* 4 mm across, 3–9 per cyme, spreading; stalks with adpressed bristles. Calyx with adpressed bristles, lobes triangular, acute or acuminate. Petals erect, pink to off-white. Stamens 20 (rarely

16); anthers white, filaments pink. Fruit *c.* 5 mm, spreading or pendent, nearly spherical, red, shiny; nutlets 3–5, usually 5. Diploid, variable. *China (Yunnan).* H3. Spring.

36. C. foveolatus Rehder & Wilson.

Leaf-blade 3–8 cm, thin, broadly-elliptic to ovate, edges with adpressed bristles, apex acuminate, upper surface dull, mid-green, lower surface with adpressed bristles. Flowers 3–7 per cyme. Fruit black. Apomictic. *China (Hubei).* H2. Spring.

Rare in cultivation.

Series **Bullati** Flinck & Hylmö.

37. C. bullatus Bois (*C. bullatus* var. *floribundus* Rehder & Wilson). Illustration: Botanical Magazine, 8284 (1909) – as *C. moupinensis* forma *floribunda*).

Shrub or small tree, erect, 2.5–4 m. Branches with a spreading open habit, initially light brown, with greyish yellow down, older bark becoming blackish. Leaves deciduous, blade 3–7 cm, moderately thick, ovate to oblong-elliptic, apex acute, upper surface shiny, dark green, blistered, slightly downy, lower surface with greyish green shaggy hairs, densely so on the veins; stalks 3–6 mm. Flowers 12–30 per cyme, spreading; stalks with greyish downy hairs. Calyx with sparse adpressed bristles, lobes short, triangular. Petals erect, red, green and off-white. Stamens 20; anthers white. Fruit 6–8 mm, spreading, obovoid to nearly spherical, red, somewhat shiny; nutlets 5 (rarely 4). Apomictic. *China (Sichuan).* H2. Early summer.

'Firebird': Leaf-blade somewhat convex, lower surface with whitish down. Cymes compact. Fruit 8–10 mm obovoid, orange-red. *China (Yunnan).* H2. Early summer.

38. C. rehderi Pojarkova (*C. bullatus* var. *macrophyllus* Rehder & Wilson).

Shrub 4–5 m. Branches with older bark not so dark as *C. bullatus*. Leaf-blade 5–15 cm, apex acuminate, lower surface with only a few hairs; stalks 1–4 mm. Calyx nearly hairless, lobes with hairy edges. Fruit 8–11 mm, rich red. *China (Sichuan).* H2. Early Summer.

39. C. boisianus Klotz.

Shrub 2.5–3 m. Branching habit more dense. Leaf-blade 3–6 cm, thin, elliptic-ovate, lower surface becoming thinly hairy; stalks 2–3 mm. Flowers 9–18 per cyme.

Calyx with adpressed bristles. Fruit bright orange-red; nutlets 3–4 (rarely 5). *China (Sichuan).* H3. Early summer.

40. C. obscurus Rehder & Wilson.

Shrub 1.5–2 m. Leaf-blade 3–5 cm, ovate, apex acute or acuminate, upper surface dull. Fruit maroon, nutlets 3 or 4. *China (Sichuan).* H2. Early summer.

41. C. moupinensis Franchet.

Shrub to 3 m. Leaf-blade to 12 cm, apex acuminate. Fruit purple-black; nutlets 5. *China (Sichuan).* H2. Early summer.

42. C. sikangensis Flinck & Hylmö.

Height 2–3 m. Branches initially with adpressed bristles, older bark greyish. Leaf-blade 2–5 cm, thick, ovate to elliptic-ovate, apex acuminate or rarely acute, lower surface with shaggy to felted hairs. Flowers 6–15 per cyme. Fruit *c.* 1 cm, obovoid, reddish orange; nutlets 4 or 5 (rarely 3). *China (Sichuan).* H2. Early summer.

Series **Franchetioides** Flinck & Hylmö.

43. C. franchetii Bois. Illustration: Revue Horticole, 379–381 (1902); Botanical Magazine, 8571 (1914).

Shrub, erect, 1.5–3 m. Branches slender, spreading, arched, initially with yellowish brown felted hairs. Leaves semi-evergreen; blade 2–3.5 cm, thick, slightly wrinkled, elliptic to ovate, apex acute or acuminate, upper surface shiny, sparsely hairy, lower surface with silvery white felted hairs; stalks 1–3 mm with felted hairs. Flowers 5–15 per long stalked cyme, erect and spreading; stalks with downy hairs. Calyx with long silky hairs, lobes acuminate, terminating in a long narrow point. Petals erect, red and off-white. Stamens 20; anthers pale purple; filaments white. Fruit 6–9 mm, spreading, obovoid, orange to light red, downy; nutlets 3 (rarely 2). Apomictic. *China (Yunnan).* H3. Summer.

Often confused with *C. sternianus* which now seems to be the more commonly grown of the two species.

44. C. sternianus Boom (*C. franchetii* var. *sternianus* Turrill). Illustration: Stace, New flora of the British Isles, 459 (1991).

Branches stiffly erect. Leaf-blade 2.5–6 cm, very thick. Cymes shorter stalked, more erect. Calyx with woolly hairs, lobes triangular, sometimes mucronate. Petals fringed. Anthers white. Fruit spherical, 8–11 mm, dark orange-red; nutlets 3 (rarely 4). *China (Yunnan).* H3. Summer.

45. C. wardii W.W. Smith (*C.* 'Gloire de Versailles').

Shrub 1.5–2.5 m. Branches thin and very graceful. Leaves thin. Calyx-lobes often drawn out into a long point. Nutlets 2. *China (Xizang).* H3. Summer.

Plants grown as *C. wardii* are usually *C. sternianus*. *C. wardii* is very rare in cultivation.

46. C. amoenus Wilson.

Shrub 1–1.5 m. Branching dense. Leaves evergreen, blade 1–2 cm; stalks 2–5 mm. Flowers 6–10 per cyme, compact, more erect. Petals semi-spreading, white with pink-red. Anthers pink-purple. Fruit 5–6 mm, spherical, red; nutlets 2 or 3, (rarely 4). *China (Yunnan).* H3. Summer.

47. C. dielsianus Pritzel.

Shrub 2–3 m. Branches erect, very long and thin often with fan-like branching at tips. Leaves deciduous, blade 1.5–2.5 cm, ovate to obovate, lower surface with felted yellowish grey hairs. Flowers 3–7 per cyme. Calyx with downy hairs, lobes acuminate with a long mucro. Anthers white. Fruit *c.* 6 mm, nearly spherical, rich red; nutlets 3–4. *China.* H2. Early summer.

'Rubens' (*C. rubens* misapplied, not W.W. Smith). Height 1–1.5 m. Branches erect and spreading. Leaves deciduous; blade elliptic, apex acute to acuminate, lower surface with grey felted hairs. Fruit obovoid, shiny, rich red; nutlets 3. *Origin unknown.* H2. Early summer.

48. C. splendens Flinck & Hylmö (*C.* × 'Sabrina').

Shrub spreading and ascending to 1–2 m. Leaves deciduous; blade 1–2 cm, thin, broadly elliptic to nearly circular, upper surface somewhat shiny, bright green, sparsely hairy, lower surface with yellowish grey hairs. Flowers 3 (rarely 2–7) per cyme. Anthers white. Fruit 9–11 mm, nearly spherical, shiny, bright orange-red; nutlets usually 4. *China (Sichuan).* H2. Early summer.

Series **Zabelioides** Flinck & Hylmö.

49. C. zabelii Schneider. Illustration: Die Gartenwelt **25**: 429 (1931).

Shrub, erect, 1.5–2 m. Branches slender, spreading initially with densely, downy hairs. Leaves deciduous, blade 1.5–4 cm thin, ovate to elliptic, apex acute (sometimes obtuse), upper surface dull light green, sparsely hairy, lower surface with yellowish grey shaggy hairs; stalks to 3 mm, thickly hairy. Flowers 4–12 per cyme, loose, spreading, stalks thickly hairy. Calyx with shaggy hairs, lobes broadly triangular, obtuse. Petals erect, off-white to pale pink. Stamens 20 (rarely 18); anthers white; filaments pale pink. Fruit 7–8 mm pendent, obovoid, with downy hairs, dark red to nearly maroon; nutlets 2. Apomictic. *China*. H2. Early summer.

50. C. fangianus Yü.

Shrub to 3 m. Leaf-blade 1–3 cm, broadly ovate, apex usually obtuse. Flowers 5–15 per cyme, more compact. Petals pink and red. Fruit spherical, rich-red. *China (Hubei)*. H2. Early summer.

Series **Megalocarpa** Pojárkova.

51. C. megalocarpus Popov. Illustration: Flora of Kazakhstana **4**: 399 (1961).

Shrub, erect, 1.5–2.5 m. Branches thin, spreading, initially dark purple-brown, with downy hairs. Leaves deciduous, blade 2–5 cm, thin, ovate, apex acute, mucronate, upper surface shiny light green, initially sparsely hairy, lower surface with somewhat shaggy hairs; stalk 3–5 mm with shaggy hairs. Flowers 3–20 per cyme, erect, stalks with sparse downy hairs. Calyx nearly hairless, lobes acute. Petals semi-spreading, white. Stamens 20; anthers and filaments white. Fruit *c.* 1 cm, erect, spherical, red, shiny; nutlets 2 (rarely 3). Apomictic. *C Asia (Tianshan Mts)*. H2. Early spring.

52. C. roseus Edgeworth.

Leaf-blade ovate-elliptic, both surfaces hairless. Cymes flat-topped. Calyx hairless; lobes acute to obtuse. Petals bright, clear pink. Nutlets 2. *Himalaya (Kashmir, Afghanistan)*. H2. Spring.

Series **Multiflori** Yü.

53. C. multiflorus Bunge. Illustration: Ledebour, Icones Plantarum Novarum Flora Rossicum, t. 274 (1831).

Shrub or small tree, erect, 3–5 m. Branches slender, arched, initially shiny purple-brown, slightly downy. Leaves deciduous; blade 2–5 cm, thin, broadly ovate to nearly circular, apex usually obtuse, upper surface dull, yellowish green, hairless, lower surface initially sparsely hairy; stalk 3–10 mm, hairless. Flowers 8–10 mm across, erect and spreading, 10–20 per cyme; stalks nearly hairless. Calyx hairless, lobes triangular, acute or obtuse. Petals spreading, white. Stamens 20; anthers and filaments white. Fruit 1–1.2 cm, spreading, spherical, carmine, shiny, succulent; nutlets apparently 1 (= 2 nutlets joined). Apomictic. *C Asia (Kazakhstan)*. H2. Spring.

54. C. hupehensis Rehder & Wilson.

Shrub 1.5–3 m. Leaf-blade 15–35 mm, elliptic to ovate, apex usually acute, upper surface green, lower surface sparsely hairy. Flowers *c.* 1.3 cm across, 5–10 per cyme, stalks sparsely hairy. Fruit carmine to red. *China (Hubei)*. H3. Spring.

55. C. calocarpus Flinck & Hylmö (*C. multiflorus* var. *calocarpus* Rehder & Wilson).

Leaf-blade broadly elliptic, apex acute; upper surface initially sparsely hairy, lower surface initially with shaggy hairs becoming sparsely hairy. Flowers to 1.5 cm across; petals flat, sometimes reflexed. Fruit dark red to maroon. *China (Sichuan)*. H2. Early spring.

56. C. veitchii Klotz (*C. racemiflorus* var. *veitchii* Rehder & Wilson).

Shrub 3–4 m. Branches wide-spreading. Leaf-blade thick, broadly ovate, sometimes elliptic, apex acute, upper surface dark green, initially with shaggy hairs, becoming sparsely so, lower surface and leaf-stalks with felted hairs. Flowers 1.2–1.5 cm across, 5–15 per cyme. Calyx with felted hairs. Petals somewhat concave, pink in bud. Fruit 1.2–1.3 cm, dark red to maroon, frequently splitting. *China (Hubei)*. H2. Early summer.

57. C. przewalskii Pojárkova.

Shrub 2–5 m. Leaf-blade to 6 cm, broadly ovate, apex acute to acuminate; stalks sparsely hairy. Flowers *c.* 1.3 cm across. Calyx campanulate, sparsely hairy. Petals concave, bases pale purple. *China (Gansu)*. H2. Spring.

Series **Racemiflori** Klotz.

58. C. racemiflorus Koch.

Shrub, erect, 1.5–3 m. Branches initially shiny dark purple, with grey shaggy hairs. Leaves deciduous, thin, blade 2–5 cm, broadly ovate to elliptic, apex acute or acuminate, mucronate, upper surface dull, hairless, lower surface with greyish felted hairs; stalks 4–8 mm, with shaggy hairs. Flowers *c.* 8 mm across, 6–15 per cyme, compact, erect, flower-stalks and calyx with felted hairs; lobes triangular, acute. Petals spreading, white. Stamens 20; anthers white; filaments white. Fruit 8–10 mm, spreading, nearly spherical, red, dull; nutlets 2. Apomictic. *Caucasus*. H2. Early summer.

59. C. nummularius Fischer & Meyer.

Shrub to 1 m. Leaf-blade 2 cm (rarely to 3 cm), circular, (rarely broadly elliptic), apex mucronate, upper surface sparsely hairy. Flowers 3–7 per cyme, densely compact. Petals red in bud. Fruit 7–8 mm. *W Asia*. H3. Early summer.

60. C. tauricus Pojárkova.

Leaf-blade 1.5–4 cm, ovate or elliptic, apex obtuse or acute, upper surface hairless. Flower-stalks and calyx with sparse hairs. Petals red in bud. Fruit 7–8 mm. *Ukraine (Crimea)*. H2. Early summer.

Series **Insignes** Pojárkova.

61. C. ellipticus Loudon (*C. lindleyi* Schneider; *C. insignis* Pojarkova).

Shrub or small tree, erect, 3–6 m. Branches wide-spreading, initially red-brown to dark brown, with downy hairs. Leaves deciduous, moderately thick, blade 2.5–6 cm, circular (rarely broadly elliptic to obovate), apex mucronate, upper surface hairless, or sparsely hairy on midrib, lower surface with greyish white hairs; leaf-stalk to 8 mm, sparsely hairy. Flowers *c.* 1 cm across, 5–20 per cyme, erect, stalk and calyx with felted hairs, lobes broadly triangular, acute or obtuse. Petals spreading, white, with tufts of hair at base, pink in bud. Stamens 20; anthers and filaments white. Fruit 7–9 mm, spreading, spherical, black with a bluish waxy bloom, apex open; nutlets 1 or 2. Apomictic. *W Himalaya*. H3. Early summer.

62. C. hissaricus Pojárkova.

Shrub 1.5–2 m, densely branched. Leaf-blade 1.5–2.5 cm (rarely longer), circular. Flowers *c.* 8 mm across, 5–12 per cyme. Petals usually white in bud. Fruit 6–7 mm. *C Asia (Tadzhikistan)*. H3. Early summer.

63. C. affinis Lindley.

Shrub or small tree to 8 m. Leaf-blade 4–10 cm, thin, elliptic. Flowers to 15 per cyme, compact. Anthers mauve. Fruit 1–1.2 cm, cylindric, purple becoming black, succulent; nutlets 2, obovate. *Himalaya*. H3. Early summer.

64. C. gamblei Klotz.

Very similar to *C. affinis* but with cymes loose, and the fruit remaining red to purple. *Himalaya (Sikkim, Bhutan)*. H3. Early summer.

65. C. bacillaris Lindley.

Branches suckering, arched and pendent, graceful. Leaf-blade 4.5–10 cm, obovate, lower surface hairless, except for sparsely hairy midrib. Calyx sparsely hairy. Petals with large tufts of hairs at base. Anthers mauve. Fruit 8 mm; nutlets 2, spherical. *Himalaya (Kashmir to Bhutan)*. H3. Early summer.

66. C. transens Klotz.

Small tree, to 5 m. Leaf-blade 5–10 cm, reddish when young, lanceolate to elliptic, lower surface initially sparsely hairy. Anthers dark mauve. Fruit *c.* 1 cm, maroon to purple-black; nutlets 1 or 2. *China (Yunnan).* H3. Early summer.

Probably more closely related to series *Hebephylli.* Frequently grown as *C. affinis* or *C. bacillaris.*

Series **Hebephylli** Klotz.

67. C. hebephyllus Diels (*C. tibeticus* Klotz). Illustration: Wissenschaftliche Zeitschrift der Friedrich-Schiller-Universitat, Jena, 334 (1968).

Shrub, erect, 1.8–3 m. Branches spreading at right angles, graceful, initially shiny purple, with downy hairs. Leaves deciduous; blade 1.5–4 cm, thin, broadly obovate, apex obtuse, shortly mucronate, upper surface dull, initially reddish brown, later light green, hairless, lower surface initially sparsely hairy; stalks to 5 mm, sparsely hairy. Flowers *c.* 7 mm across, 6–20 per cyme, erect; stalks slender, sparsely hairy. Calyx initially sparsely hairy, lobes broadly triangular, acute. Petals spreading, white, sometimes with small tufts of hair at base. Stamens 20; anthers purple to black, filaments white. Fruit 6–7 mm, pendent, depressed spherical, carmine, becoming red, dull; nutlets 1 per fruit. Apomictic. *China (Yunnan, Xizang).* H3. Early summer.

Plants grown as *C. hebephyllus* are usually *C. monopyrenus.*

68. C. monopyrenus Flinck & Hylmö (*C. hebephyllus* var. *monopyrenus* W.W. Smith).

Branches initially dark brown. Leaf-blade moderately thick, 2.5–5 cm, broadly elliptic, apex obtuse, sometimes acute, mucronate, upper surface dark green, reddish at fruiting, lower surface initially felted. Flowers *c.* 1 cm across. Fruit *c.* 1 cm, cylindric, becoming dark maroon. *China (Yunnan).* H3. Early summer.

69. C. ludlowii Klotz.

Shrub to 2 m; densely branched. Leaves semi-evergreen; blade 1.7–3 cm, elliptic to elliptic-lanceolate, apex acute, upper surface rich green; stalks to 4 mm. Flowers *c.* 1 cm across, 4–12 per cyme, stalks thickly hairy. Fruit 8–9 mm, salmon-pink and orange. *Himalaya (Bhutan, Nepal, Xizang).* H3. Early summer.

Series **Chaenopetalum** Koehne.

70. C. frigidus Lindley. Illustration: Edwards's Botanical Register **15**: 1229 (1829).

Large shrub or tree, erect; 5–15 m. Branches spreading, widely arched,

initially dark brown, with downy hairs. Leaves deciduous; blade 6–15 cm, thin, narrowly elliptic to obovate, apex acute, upper surface flat, dull green, hairless except for sparsely hairy midrib, lower surface initially with shaggy hairs; stalks 5–8 mm, downy. Flowers *c.* 7 mm across, 20–60 per cyme, erect and spreading; stalks and calyx with felted hairs, lobes triangular, acute. Petals spreading, white (also white in bud). Stamens 20; anthers purple; filaments white. Fruit *c.* 6 mm, spreading, nearly spherical, red; nutlets 2. Diploid, variable. *Himalaya.* H3. Early summer.

The largest species in the genus. Many varieties cultivated. 'Fructo-Luteo' has creamy-yellow Fruit. 'Anne Cornwallis' has thinner, wavy-edged leaf-blades than *C. frigidus;* fruit pear-shaped, orange.

71. C. × watereri Exell. Illustration: Stace, New flora of the British Isles, 459 (1991).

Shrub or small tree, to 6 m; branches vigorous, pendent, initially dark grey-brown. Leaves evergreen or semi-evergreen; blade 5–9 cm, moderately thick, narrowly elliptic, apex acute-acuminate, upper surface somewhat shiny, dark green, wrinkled, lower surface initially downy; stalks 5–7 mm, densely downy. Flowers 30–50 per cyme; stalks densely downy. Calyx densely downy, lobes broad, triangular. Fruit 8–9 mm; nutlets 2–4 (rarely 5). *Garden origin.* H3. Early summer.

The possible hybrid between *C. frigidus* and *C. salicifolius* to which the type name 'John Waterer' is given.

'Salmon Spray'. Fruit 6–7 mm, pinkish orange.

'Coral Bunch'. Similar to 'Salmon Spray', but with fewer, larger fruit of a lighter shade.

'Cornubia'. Fruit red, very abundant.

'Aldenhamensis'. Shrub, narrowly erect, to 4 m; fruit 5–7 mm, rich-red.

'St Monica'. Shrub, nearly evergreen; branches initially very shiny purple-brown; fruit in large pendent bunches.

'Vicarii'. Leaves to 15 cm; fruit dark-red, 5–7 mm.

72. C. salicifolius Franchet. Illustration: Havens Planteleksikon, 169 (1987).

Shrub, erect, 2–5 m. Branches erect and spreading, initially red purple to greenish, thinly downy. Leaves evergreen; blade 3–10 cm, thick, elliptic-lanceolate, edges recurved, apex acute to acuminate, upper surface shiny, dark green, wrinkled,

hairless, lower surface with felted to thinly woolly hairs, sometimes hairless by autumn. Midrib often reddish; stalks 5–8 mm, often reddish with felted hairs. Flowers *c.* 6 mm across, 30–100 per cyme, erect-spreading, dense; stalks with felted hairs. Calyx with downy hairs, lobes shortly triangular acute. Petals spreading, white. Stamens 20, anthers purple to black, filaments white. Fruit *c.* 5 mm, spreading, nearly spherical, bright-red shiny; nutlets 3–5. Many varieties cultivated. Diploid, variable. *China (Sichuan).* H3. Early summer.

'Exburyensis'. Leaf-blade 8–12 cm, upper surface mid-green; fruit golden-yellow.

'Rothschildianus'. Leaf-blade upper surface pale-green; fruit cream to pale yellow.

'Pink Champagne'. Similar to above two cultivars; fruit yellow, becoming pinkish.

'Repens'. Prostrate to 1.5 m; branches spreading; leaf-blade 2.5–3.5 cm; stalk to 4 mm; flowers 3–10 per cyme.

'Herbstfeuer' ('Autumn Fire'). Prostrate 20–25 cm; branches spreading; leaf-blade 4–6 cm, colouring well in autumn; flowers 5–12 per cyme.

'Gnom'. Prostrate; to 15 cm; branches spreading; leaves evergreen, blade to 2.5 cm, upper surface very dark green; fruit 4–6 mm.

'Parkteppich' ('Park Carpet'). Shrub to 1 m; branches spreading; leaf-blade 2–3 cm; fruit light red.

'Hybridus Pendulus'. Prostrate 20–25 cm; branches spreading; as with previous four cultivars it can be trained (or top grafted onto a *C. × watereri* cultivar) to form a pendent bush or small tree. May also be trained against a wall.

This is probably a hybrid between *C. salicifolius* and *C. dammeri.*

73. C. glabratus Rehder & Wilson.

Branching dense. Leaf-blade 4–7 cm, upper surface hairless, lower surface glaucous, hairs only on midrib; stalks sparsely hairy. Flower-stalks initially hairy. Calyx sparsely hairy; lobes broadly triangular, obtuse. Fruit 4–5 mm, bright orange-red; nutlets 2. *China (Sichuan).* H3. Early summer.

74. C. rugosus Pritzel.

Leaf-blade very thick, broadly elliptic, apex acute, upper surface very strongly wrinkled, lower surface with dense shaggy hairs. Flower-stalk and calyx with shaggy hairs. Nutlets 2. *C China.* H3. Early summer.

Possibly not in cultivation. Plants grown

as *C. rugosus* are usually *C. salicifolius* or other species.

75. C. henryanus Rehder & Wilson. Shrub 2–4 m; robustly branched.

Leaves semi-evergreen, blade 8–12 cm, oblong-elliptic to oblong-lanceolate, upper surface very sparsely hairy, lower surface remaining felted with woolly hairs, leaf and flower-stalks with shaggy hairs. Calyx with dense shaggy hairs, lobes triangular, acute. Fruit 6–7 mm, dark red; nutlets 2 or 3. *China (Hubei)*. H3. Early summer.

Rare in cultivation. Plants grown as this species are usually *C. salicifolius*.

76. C. floccosus Flinck & Hylmö (*C. salicifolius* var. *floccosus* Rehder & Wilson).

Shrub 2–4 m. Leaf-blade narrowly lanceolate, lower surface with a dense covering of woolly hairs which fall away in tufts. Apomictic. *China (Sichuan)*. H3. Early summer.

Uncommon in cultivation. Plants grown as this species are usually *C. salicifolius*.

77. C. dammeri Schneider (*C. humifusus* Veitch).

Prostrate. Branches long, creeping, rooting at nodes. Leaf-blade 1.5–3 cm, broadly obovate to elliptic, apex usually obtuse, upper surface with veins slightly impressed, lower surface sparsely hairy, veins somewhat prominent; stalks sparsely hairy. Flowers *c.* 9 mm across, solitary or 2–4 per cyme; stalks to 1.2 cm, initially sparsely hairy. Calyx initially sparsely hairy. Fruit 5–7 mm, shy-fruiting; nutlets usually 5. *China (Hubei)*. H3. Early summer.

'Major'. More robust than *C. dammeri*. Leaf-blade 2.5–3.5 cm, nearly circular. Often found as *C. dammeri* var. *radicans*.

78. C. radicans Klotz (*C. dammeri* var. *radicans* Schneider).

Prostrate. Branches long, creeping, rooting at nodes. Leaf-blade 1–1.5 cm, obovate to elliptic, upper surface mid-green, initially with a few long hairs, somewhat dull, veins slightly impressed, lower surface initially sparsely hairy; stalks 2–4 mm, initially sparsely hairy. Flowers *c.* 9 mm across, solitary or in pairs; stalks sparsely hairy. Calyx sparsely hairy, lobe apex acute to acuminate. Fruit red; nutlets 3 or 4, usually 4. *China (Sichuan)*. H3. Early summer.

Often grown as *C. dammeri* 'Eichholz' or 'Oakwood'.

C. × suecicus Klotz. Prostrate, 40–60 cm. Branches arched, rooting at nodes. Leaves semi-evergreen; blade 1–2 cm,

obovate to oblong, apex obtuse, upper surface not wrinkled, lower surface somewhat glaucous, initially sparsely hairy; stalk and calyx sparsely hairy. Flowers *c.* 8 mm across, to 6 per cyme. Fruit red, uncommon. H3. Early summer.

This is probably the hybrid between *C. conspicuus* and *C. dammeri* to which the commonly grown 'Skogholm' belongs.

'Coral Beauty'. Similar to 'Skogholm' but more leafy, and with more abundant coral-red fruit.

'Jürgl'. Prostrate, to 50 cm. Branches widely arched. Leaf-blade to 1.5 cm, ovate, upper surface slightly wrinkled. Flowers solitary or in pairs. Fruit to 9 mm, red; nutlets 2 or 3.

Series **Pannosi** Flinck & Hylmö.

79. C. pannosus Franchet. Illustration: Acta Phytotaxonomica et Geobotanica **13**: 225–237 (1967).

Shrub, erect, 2–3 m. Branches long and slender, initially with whitish felted hairs. Leaves semi-evergreen; blade 1–3 cm, thick, elliptic to ovate, apex acute, mucronate, upper surface dull, greyish green, initially sparsely hairy, lower surface with whitish felted hairs; stalks 3–7 mm with felted hairs. Flowers to 8 mm, 6–12 per cyme, erect to spreading; stalk and calyx plus lobes with felted hairs; lobes triangular acuminate, mucronate. Petals spreading, white. Stamens 20; anthers purple-black; filaments white. Fruit 6–8 mm, spreading, nearly spherical, red, shiny, apex persistently hairy; nutlets 2. Apomictic. *China (Yunnan)*. H3. Summer.

Late flowering, winter fruiting.

80. C. glaucophyllus Franchet.

Shrub 2–4 m. Leaf-blade 3–8 cm, ovate or elliptic, apex acute, upper surface green, hairless, lower surface glaucous, hairless; stalks 5–10 mm. Flowers *c.* 6 mm across, 15–40 per cyme; stalks with sparse downy hairs. Calyx lobes hairless. Fruit 6–7 mm, obovoid, orange, hairless. *China (Yunnan)*. H3. Summer.

81. C. vestitus Flinck & Hylmö (*C. glaucophyllus* var. *vestitus* W.W. Smith).

Shrub to 4 m. Leaf-blade to 6 cm, elliptic, upper surface green. Flowers to 30 per cyme. Fruit with woolly hairs. *China (Yunnan)*. H3. Summer.

82. C. harrovianus Wilson.

Shrub to 4 m. Leaf-blade 3–5 cm, elliptic or obovate, apex acute with a long point, upper surface somewhat shiny, lower surface of older leaves finally hairless;

stalks 4–5 mm, with silky hairs. Flowers *c.* 6 mm across, 20 to 60 per cyme; stalk and calyx, with silky hairs. Fruit 4–5 mm, hairy. *China (Yunnan)*. H3. Summer.

83. C. lacteus W.W. Smith.

Shrub or small tree, to 5 m. Branches initially reddish brown, with shaggy to felted yellowish hairs. Leaf-blade 3.5–9 cm, obovate or broadly elliptic, apex acute or obtuse, upper surface somewhat shiny, with veins impressed, lower surface with felted yellowish hairs, becoming thinner with age. Flowers *c.* 100 per cyme. Petals milky white. Fruit to 6 mm, obovoid. *China (Yunnan)*. H3. Summer.

84. C. turbinatus Craib.

Shrub 3–5 m. Leaf-blade 3.5–6 cm, green. Flowers *c.* 6 mm across, to *c.* 75 per cyme, stalk and calyx, including lobes, thickly covered with silky hairs. Fruit 4–5 mm, obconical, with silky hairs. *China (Yunnan)*. H3. Summer.

85. C. serotinus Hutchinson.

Shrub or small tree, 3–5 m. Leaf-blade 4–7 cm, elliptic or obovate-elliptic, apex acute, with a long fine point, upper surface somewhat shiny, green, hairless, lower surface becoming nearly hairless; stalks to 1 cm, with silky hairs. Flowers 5–6 mm, to 40 per cyme, loose; stalk and calyx, including lobes, with silky hairs. Fruit obovoid to nearly spherical, with sparse silky hairs. *China (Yunnan)*. H3. Summer.

86. C. meiophyllus Klotz (*C. glaucophyllus* var. *meiophyllus* W.W. Smith).

Branching dense. Leaves evergreen; blade 2–4 cm, initially reddish, becoming green, broadly elliptic, apex obtuse (rarely acute), both surfaces hairless. Flowers 5–6 mm, 15–40 per cyme. Calyx-lobes hairless. Fruit orange-red, hairless. *China (Yunnan)*. H3. Summer.

87. C. microphyllus Lindley. Illustration: Edwards's Botanical Register **13**: t. 114 (1827).

Shrub, prostrate or ascending, 0.5–1 m. Branches dense, initially black-purple, with adpressed whitish bristles. Leaves evergreen, blade 5–8 × 3–5 mm, thick; broadly elliptic to broadly obovate, edges somewhat recurved, apex obtuse or acute, upper surface shiny, dark green, initially midrib sparsely hairy, lower surface with sparse adpressed bristles, glaucous, with a prominent vein pattern; stalks 1–2 mm, with adpressed bristles. Flowers *c.* 6 mm across, solitary or 2–5 per cyme, erect; stalks 1–2 mm with adpressed bristles.

Calyx with adpressed bristles, lobes triangular, obtuse. Petals spreading, white, often pink in bud. Stamens 20; anthers purple to black; filaments white. Fruit 5–8 mm, depressed spherical, carmine, dull; nutlets 2. *Himalaya*. H3. Spring.

Several other species may be grown as *C. microphyllus*.

'Donard Gem'. Leaf-blade very dark-green, margin hairy; upper surface slightly shiny, sparsely hairy, lower surface with dense adpressed hairs; fruit carmine to red, abundant.

88. C. conspicuus Marquand (*C. conspicuus* var. *decorus* Russell; *C. nanus* Klotz; *C. permutatus* Klotz; *C. pluviflorus* Klotz).

Shrub to 1.5 m. Leaf-blade 5–20 × 2–8 mm, green, narrowly elliptic, upper surface slightly shiny, initially sparsely hairy, papillose, lower surface without vein pattern; stalks 1–1.5 mm. Flowers *c.* 1 cm across, usually solitary, except at ends of branches, and then up to 5 per cyme. Calyx sparsely hairy, lobes acute or acuminate. Fruit 7–9 mm, red, shiny. Diploid, variable. *China (Xizang)*. H3. Spring.

Many varieties are cultivated:

'Red Glory'. Erect; to 2 m; leaf-blade 1–2 cm.

'Highlight'. Fruit light orange; nutlets 1 or 2. This is probably identical to *C. sherriffii* Klotz.

89. C. rotundifolius Lindley (*C. microphyllus* var. *uva-ursi* Lindley).

Shrub 0.5–2 m. Branches more erect. Leaf-blade very thick, 7–20 × 5 × 11 mm. Flowers 1–1.3 cm across, always solitary. Calyx with sparse hairs, lobes acute. Fruit 7–10 mm, nearly spherical. Apomictic. *Himalaya*. H3. Spring.

90. C. cochleatus Klotz (*C. buxifolius* var. *cochleatus* Franchet).

Prostrate, to 20 cm. Branches downward curved, creeping, rooting at nodes. Leaf-blade 5–14 × 3–9 mm, broadly elliptic, apex obtuse. Flowers 8–10 mm across, always pink in bud. Fruit carmine-red. Apomictic. *W China (Yunnan)*. H3. Spring.

Plants grown as *C. cochleatus* in the British Isles are usually *C. cashmiriensis*.

91. C. cashmiriensis Klotz.

Prostrate, to 30 cm. Branches creeping, often rooting at nodes. Leaf-blade 4–11 × 2–6 mm, elliptic, upper surface midrib initially with long adpressed hairs; stalks 1–3 mm. Flowers solitary. Fruit nearly spherical, red. Apomictic. *Himalaya (Kashmir)*. H3. Spring.

92. C. prostratus Baker.

Prostrate, 0.5 m. Branches long and thin, creeping. Leaf-blade 6–13 × 4–8 mm, apex not notched; stalks 2–4 mm. Flowers *c.* 1 cm across, solitary or sometimes in pairs. Fruit 9–10 mm, nearly spherical, red. *Himalaya*. H3. Spring.

'Ruby' (*C. rubens* misapplied not W.W. Smith). Branches stiffly projecting, becoming wide spreading. Leaf-blade 8–15 × 4–7 mm, elliptic; stalks reddish. Flowers *c.* 8 mm across, 3–9 per cyme. Fruit *c.* 8 mm, carmine. Often found as *C. sikkimensis* misapplied.

93. C. buxifolius Lindley.

Shrub to 1.5 m. Leaf-blade 1–1.8 × 5–7 mm, elliptic, upper surface with long adpressed hairs, lower surface with dense adpressed bristles; stalks 2–4 mm. Flowers 3–5 per cyme; calyx felted, lobes triangular, acute. Fruit nearly spherical, hairy. *S India (Nilgiri Hills)*. H3. Spring.

The true species is rare in cultivation.

94. C. marginatus Schlechtendal.

Prostrate to erect, 2–3 m. Leaf-blade 1–2.5 × 0.5–1.1 cm, obovate or elliptic, apex obtuse, sometimes acute, margin with hairs. Stalk 3–5 mm. Flowers 8–10 mm across, 2–7 per cyme, white in bud. Fruit *c.* 8 mm, carmine to red; nutlets 2 (rarely 3). *Himalaya*. H3. Spring.

Often grown as *C. prostratus* var. *lanatus* of gardens.

'Eastleigh'. Shrub to 4 m; branches erect and spreading, robust; leaves to 3 cm; fruit to 1.3 cm, deep red, succulent.

95. C. congestus Baker (*C. pyrenaicus* Chancerel).

Shrub to 70 cm; densely branched, congested, often rooting at nodes. Leaf-blade 5–14 × 3–8 mm, thin, obovate, upper surface dull, pale-green, hairless, lower surface very sparsely hairy; stalks 3–5 mm, thin, often reddish. Flowers 7–9 mm across, solitary; stalk and calyx sparsely hairy. Fruit 6–10 mm, red; nutlets 2 (rarely 3). Diploid, variable. *Himalaya*. H3. Spring.

'Nanus'. Mat-forming, hugs the ground; fruit rarely produced.

96. C. procumbens Klotz.

Prostrate, to 20 cm. Branches often very dense. Leaf-blade 6–12 × 4–11 mm, broadly obovate, upper surface somewhat dull, hairless, purple when very young; stalks 4–6 mm, thin, becoming hairless. Flowers 8–9 mm across, solitary, white in bud. Fruit 5–6 mm, spherical. *China*. H3. Spring.

'Streibs Findling' is probably this species.

97. C. lidjiangensis Klotz (*C. exellens* Marquand, invalid).

Shrub to 1.5 m. Branches slender, spreading and erect, initially with felted hairs. Leaf-blade 6–12 × 3–6 mm, elliptic, sometimes ovate, margin hairy, apex acute or acuminate, not notched, with a long point, upper surface with sparse adpressed bristles, lower surface with whitish densely felted hairs; stalks 2–3 mm, with felted hairs. Flowers 8–9 mm across, solitary or 2–4 (rarely to 8) per cyme; stalks 1–3 mm. Calyx felted, lobes acute or acuminate, often mucronate. Fruit 5–6 mm, obovoid or nearly spherical, carmine to red, shiny; calyx-lobes persisting as a coronet. Apomictic. *China (Yunnan)*. H3. Early summer.

Often found as *C. buxifolius*.

98. C. integrifolius (Roxburgh) Klotz (*C. thymifolius* Baker).

Leaf-blade 7–15 × 3–6 mm, oblanceolate to oblong or obovate, apex notched, upper surface very dark green, initially with sparse adpressed bristles, lower surface with dense adpressed bristles; stalks 2–4 mm. Flowers *c.* 1.1 cm, usually solitary. Calyx-lobes triangular-acute. Fruit 8–10 mm. Diploid, variable. *Himalaya*. H3. Early summer.

99. C. linearifolius Klotz.

Shrub to 60 cm. Branches initially prostrate, dense. Leaf-blade 4–7 × 1–3 mm, oblong to oblanceolate, apex notched, upper surface very dark green, lower surface glaucous, without prominent veins; stalks 0.5–1.5 mm. Flowers solitary, *c.* 5 mm across, pink in bud. Fruit 4–5 mm. Diploid, variable. *Himalaya (Nepal)*. H3. Early summer.

68. DICHOTOMANTHES Kurz
M.F. Gardner

Evergreen trees or shrubs, young branches covered with dense white woolly hairs. Leaves alternate, entire, ovate, pointed, tapered towards the base, 3–10 × 1.5–3 cm, dark green, hairless above, glossy with pale silky hairs beneath; stipules very small, thread-like, falling early. Flowers borne in terminal corymbs *c.* 5 cm across; sepals woolly outside with 2 bracteoles at the base; petals 2–3 mm, white; stamens 15–20; ovary inferior. Fruit a dry, oblong capsule 6 mm, almost entirely surrounded by the calyx which becomes fleshy with maturity.

A genus of a single species closely resembling *Cotoneaster*. Young plants are frost susceptible and are therefore best given the protection of a wall. Propagate by seed or late summer cuttings rooted under mist with gentle bottom heat.

1. D. tristaniicarpa Kurz. Illustration: Journal of Botany **11**: pl. 133 (1873); Hooker's, Icones Plantarum, 2653 (1900); Icones cormophytorum Sinicorum, 2362 (1972); Wu, Wild flowers of Yunnan **2**: 109 (1986).

China (Yunnan). H4–5. Summer.

69. PYRACANTHA M.J. Roemer
S.G. Knees

Evergreen shrubs; branches usually thorny; buds small, softly hairy. Leaves alternate, shortly stalked, margins entire, shallowly scalloped or with 5 forward-pointing teeth; stipules minute, falling early. Flowers hermaphrodite in compound corymbs. Calyx of 5 short triangular teeth; petals white, rounded, often with a notched apex and sometimes narrowed into a short basal stalk; stamens *c.* 20, anthers yellow. Ovary inferior, carpels 5, free on central axis, joined for about half their length to calyx-tube; styles 5. Fruit apple-like with persistent calyx, red, orange or yellowish, often slightly compressed laterally. Pyrenes 5.

A genus of 6 ornamental species from southeastern Europe, Himalaya and China, often grown as wall shrubs. Pruning is not necessary but branches should be tied to supports if grown against walls. Many of the species and cultivars are susceptible to attack by fungal pathogens, particularly scab and fireblight, but resistant clones may become available following research and hybridisation programmes.

Literature: Egolf, D.R. & Drechsler, R.F. Chromosome numbers of Pyracantha (Rosaceae) *Baileya* **15**: 82–88 (1967); Egolf, D.R. Pyracantha 'Mohave', a new cultivar (Rosaceae). *Baileya* **17**: 79–82 (1970).

1a. Leaf-margin scalloped or toothed   2
 b. Leaf-margin not scalloped or toothed   3
2a. Leaves 3–7 cm, elliptic to lanceolate, underside densely covered with hairs   **4. atalantioides**
 b. Leaves 2.5–4.5 cm, oblanceolate to narrowly obovate, underside with few hairs or hairless   **7. koidzumii**
3a. Calyx and flower-stalks hairless   4
 b. Calyx and flower-stalks covered with hairs   5

4a. Young shoots with persistent rusty brown hairs   **2. crenulata**
 b. Young shoots with greyish hairs, eventually hairless and dark reddish brown   **6. rogersiana**
5a. Underside of leaves densely grey hairy   **5. angustifolia**
 b. Underside of leaves hairless or sparsely hairy when young   6
6a. Leaves elliptic to ovate-oblong, widest above the middle, underside hairless   **3. crenatoserrata**
 b. Leaves narrowly ovate-lanceolate, widest below the middle, underside hairy when young   **1. coccinea**

1. P. coccinea M.J. Roemer. Illustration: Hegi, Illustriertes Flora von Mitteleuropa, **4**: 688 (1922); Bean, Trees & shrubs hardy in the British Isles **3**: pl. 55 (1976); Krüssmann, Manual of cultivated broad-leaved trees & shrubs **3**: pl. 26 (1986); Phillips & Rix, Shrubs, 262 (1989).

Erect shrub to 6 m; young shoots grey downy; thorns 1.2–1.8 cm. Leaves 2–4 × 0.6–1.8 cm, narrowly ovate-lanceolate, acute; margin slightly scalloped, hairless except along margin near the base when young. Flowers-stalks slightly downy; flowers *c.* 8 mm across, creamy white; calyx-lobes broadly triangular, slightly downy. Fruit *c.* 1 × 0.8 cm, orange, in dense clusters. *SE Europe, east to the Caucasus*. H2. Summer.

Most widely grown as the selection 'Lalandei' which is slightly larger in all its parts and more colourful in fruit than the species.

2. P. crenulata (D. Don) M.J. Roemer (*Mespilus crenulata* D. Don; *Crataegus crenulata* (D. Don) Lindley). Illustration: Edwards's Botanical Register **30**: 52 (1844); Polunin & Stainton, Flowers of the Himalaya, pl. 34 (1984); Krüssmann, Manual of cultivated broad-leaved trees & shrubs **3**: pl. 26 (1986).

Spiny shrub to 2 m; leaves crowded along the branches, lateral branches often terminating in spines; young shoots with rusty brown hairs. Leaves 1.5–3 cm, narrowly oblong, blunt, shiny, leathery; margin scalloped; leaf-stalk covered with rusty hairs when young. Flowers arranged in axillary clusters along the branches. Calyx with 5 blunt lobes, hairless. Petals 4–5 mm, rounded, white. Fruit spherical, *c.* 6 mm, orange-red with nuts protruding from the persistent calyx. *India, Burma & SW China*. H3. Spring–summer.

3. P. crenatoserrata (Hance) Rehder (*P. yunnanensis* Chittenden; *P. gibbsii* var. *yunnanensis* Osborn; *P. fortuneana* (Maximowicz) Li; *Photinia crenato-serrata* Hance). Illustration: Revue Horticole, pl. 204 (1913); Gardener's Chronicle **65**: 265, 266 (1919); Botanical Magazine, 9099 (1929).

Shrub 4–6 m, similar to *P. atalantioides* and *P. rogersiana*. Young shoots rust brown, hairy; leaves 2.5–7 × 1–2.5 cm, elliptic to ovate-oblong, broadest in upper third of blade, apex rounded, base long tapered; margin coarsely toothed, hairless beneath. Flowering-stems downy; calyx downy; petals 4–5 mm, white. Fruit *c.* 7 mm across, coral-red. *NW China*. H3–4. Summer.

4. P. atalantioides (Hance) Stapf (*P. gibbsii* A.B. Jackson; *Sportella atalantioides* Hance). Illustration: Gardener's Chronicle **60**: 310 (1916); Botanical Magazine, 9099 (1925); Krüssmann, Manual of cultivated broad-leaved trees & shrubs **3**: pl. 26 (1986); Phillips & Rix, Shrubs, 262 (1989).

Shrub to 6 m. Leaves 3–7 × 1.2–3 cm, oblong-elliptic to lanceolate, hairless above, densely covered with greyish hairs beneath, eventually becoming blue-green; margin entire. Flowers 8–9 mm across, white to cream, in dense downy corymbs. Fruit 6–7 mm across, orange-red. *W China*. H3–4. Spring.

5. P. angustifolia (Franchet) Schneider (*Cotoneaster angustifolia* Franchet). Illustration: Botanical Magazine, 8345 (1910); RHS dictionary of gardening **4**: 1719 (1956); Krüssmann, Manual of cultivated broad-leaved trees & shrubs **3**: pl. 25, 26 (1986); Wu, Wild flowers of Yunnan **2**: 116 (1986).

Densely bushy shrub 3–4 m; stems rigid, horizontal, spine-tipped, covered in grey down during the first year. Leaves 1.2–3 × 0.3–1.2 mm, narrowly oblong or slightly obovate, apex rounded, often with a slight notch and a few stiff dark teeth on either side, base rounded or tapered, dark green and hairless above, grey felted beneath. Flowers 2–4 cm across, white, in downy corymbs to 5 cm across. Fruit *c.* 8 mm across, grey downy when young, eventually hairless and brilliant orange-yellow. *W China*. H2. Summer.

6. P. rogersiana (Jackson) Chittenden (*P. crenulata* var. *rogersiana* A.B. Jackson). Illustration: Botanical Magazine n.s., 74 (1949); Phillips & Rix, Shrubs, 173, 262 (1989).

Shrub to 4 m. Leaves 1.5–3.5 × 0.5–1 cm, oblanceolate or narrowly obovate, tapered towards stalk; margin shallowly toothed. Flowers c. 7 mm across, calyx-tube hairless, lobes triangular; petals creamy white; stamens with white filaments. Fruit c. 6 mm across, orange-red to yellow. *W China*. H1. Spring.

'Flava'. Illustration: Phillips & Rix, Shrubs, 262 (1989), a commonly grown, pale yellow fruiting form.

7. P. koidzumii (Hayata) Rehder (*Cotoneaster koidzumii* Hayata). Illustration: Botanical Magazine n.s., 205 (1952); Li, Woody flora of Taiwan, 289 (1963); Li, Flora of Taiwan 3: 91 (1977); Krüssmann, Manual of cultivated broad-leaved trees & shrubs 3: 68 (1986).

Similar to *P. rogersiana*, but leaves usually entire, hairless beneath; calyx-tube and inflorescence with soft hairs. *Taiwan*. H4. Spring–early summer.

The cultivar 'Mohave' is a selection from the hybrid between *P. koidzumii* and *P. coccinea* 'Wyatt', raised in Washington in 1963. It is resistant to fireblight and *Fuscicladium* attack and bears large orange-red fruits from autumn to winter.

70. MESPILUS Linnaeus
S.G. Knees
Shrub or small tree, 2–5 m; branches occasionally spiny, young shoots woolly. Leaves alternate, deciduous, entire; blade 6–12 × 3–6 cm, oblong lanceolate; margin toothed, glandular; stalks 2–5 mm. Stipules deciduous. Flowers solitary, terminal on short shoots. Calyx softly hairy, with 2 bracteoles at the base, lobes 5, up to 4 times as long as tube. Petals 5, rounded, 1.5–2 cm, white; stamens 30–40, anthers almost joined. Carpel solitary at the base of the calyx-tube, 1-celled; style lateral; stigma capitate; ovules 2. Fruit dry, often 1-seeded, 2–5 cm across, slightly exerted from the fleshy calyx which becomes enlarged in fruit, green at first, becoming brown.

A genus of a single species. Several cultivars have been selected for larger or seedless fruits, which are usually eaten once frosted, when they become sweet and palatable. Propagation is from seed, which may be slow to germinate, or by grafting or budding on to pear (*Pyrus*), quince (*Cydonia*) or hawthorn (*Crataegus*). A fertile or clay-rich soil is ideal for the medlar which does not thrive in very dry soils. A position in full sun is preferable to semi-shade.

1. M. germanica Linnaeus. Illustration: Reichenbach, Icones Florae Germanicae et Helveticae 25, t. 100 (1914); Bean, Trees & shrubs hardy in the British Isles 2: pl. 101 (1973); Krüssmann, Manual of cultivated broad-leaved trees & shrubs 2: pl. 124 (1986).

SE Europe east to W Asia & Iran. H1. Spring.

71. CRATAEGUS Linnaeus
S.G. Knees & M.C. Warwick
Deciduous, or rarely evergreen trees or shrubs, often with spiny stems and branches. Leaves alternate with entire or lobed margins, stipules present. Flowers bisexual, solitary or in corymbs. Sepals 5, petals 5, stamens 5–25, ovules solitary or 2–5, joined at base, free at apex. Fruit a drupe with 1–5, hard nutlets, each containing a single seed.

An important ornamental genus in which over 1000 species have been described, though this number has been reduced by recent authors. About 100 species are found in the old world, but the vast majority are from North America. Propagation is normally from seed but hybrids and selected clones are grafted or reproduced by cuttings.

Literature: Sargent, C.S., *Manual of the Trees of North America* (1905).

Fruits. Bluish green, green or yellowish green: **14,31,36**. yellow: **5,7,14,22,31**. orange-yellow: **2,4,16,21**. orange-red: **3,7,12,13,16,33**. reddish brown or brown: **17,23,35,37**. red or pinkish red: **5,8,9,10,11,15,18,19,20,24,25, 26,27,28,29,32,34,35,36,38,39,40, 42,43,44,45,46,47**. bluish red: **11,41, 47**. dark reddish purple to black: **1,6, 23,30,36,46,47** blackish brown: **23**.

1a. Inflorescence and young shoots covered with densely matted white hairs   2
 b. Inflorescence and young shoots hairless, or with long straight hairs   17
2a. Leaves deeply cut or lobed   3
 b. Leaves more or less entire, but margins with forward-pointing teeth or very shallowly lobed   7
3a. Leaf-stalk 1.5 cm or longer; fruit blackish purple   **1. pentagyna**
 b. Leaf-stalk not more than 1.2 cm; fruit yellow, orange or reddish   4
4a. Leaf-surface thickly covered with soft felted hairs   5
 b. Leaf-surface thinly covered with short hairs, but not felty to the touch   6

5a. Flowers 2–2.5 cm across; fruit yellow-orange   **2. tanacetifolia**
 b. Flowers 1.6–1.9 cm across; fruit orange-red   **3. orientalis**
6a. Leaf-tips 3–5- lobed, surface almost hairless; flowers not more than 1.2 cm across   **4. azarolus**
 b. Leaves 7–9-lobed, surface with closely adpressed, white hairs; flowers 1.8 cm or more across   **5. × dippeliana**
7a. Leaves widest at or below the middle, broadly ovate, normally 5–10 cm wide; margins doubly saw-toothed, mostly truncate at base   8
 b. Leaves usually widest at or above the middle, usually not more than 2–5 cm wide, narrowly ovate to oblanceolate   13
8a. Flowers 0.6–1.2 cm across   9
 b. Flowers 2–3 cm across   10
9a. Leaves truncate at base; fruit black   **6. chlorosarca**
 b. Leaves wedge-shaped at base; fruit yellowish orange or reddish   **7. calpodendron**
10a. Stipules persistent after fruiting; anthers rose-pink   **8. ellwangeriana**
 b. Stipules falling early; anthers yellow or white   11
11a. Fruit pear-shaped, c. 1 cm across, light red; seeds 5   **9. submollis**
 b. Fruit spherical, 1.2–1.8 cm across, light bluish red or red   12
12a. Stamens 20; fruit red, seeds 4 or 5   **10. mollis**
 b. Stamens 10; fruit light bluish red, seeds 3 or 4   **11. arnoldiana**
13a. Leaves longer than wide, ovate to lanceolate   14
 b. Leaves more or less as wide as long, ovate to rounded   15
14a. Stamens 15–20; fruit c. 2.5 cm across, yellowish to dull orange-red, with conspicuous lenticels   **12. pubescens**
 b. Stamens usually 5–10; fruit 1.6–1.9 cm across, orange-red   **13. × lavallei**
15a. Leaves not more than 4 cm; thorns very slender; flowers solitary or in groups of 2 or 3   **14. uniflora**
 b. Leaves 4–10 cm; thorns stout or absent; flowers numerous   16
16a. Anthers yellow; fruit not more than 1 cm across, dull red; seeds 5   **15. collina**
 b. Anthers pink, fruit c. 1.8 cm across, yellowish red; seeds 2–5   **16. punctata**

17a. Inflorescence and young shoots hairless　18

b. Inflorescence and young shoots with few straight hairs which sometimes persist　31

18a. Fruit 5–8 mm across　19

b. Fruit at least 10 mm across　22

19a. Leaves often deeply lobed, especially towards the base; fruit light brown　**17. dahurica**

b. Leaves shallowly lobed; fruit red　20

20a. Leaves broadly triangular, truncate at base; fruit *c.* 8 mm　**18. phaenopyrum**

b. Leaves ovate to lance-shaped or oblong, wedge-shaped at base; fruit 5–8 mm　21

21a. Thorns 2.5–3 cm, stout; flowers 1–1.2 cm across　**19. canbyi**

b. Thorns 3–4 cm, slender; flowers *c.* 1.8 cm across　**20. viridis**

22a. Fruit brownish black, orange or yellowish　23

b. Fruit bright red to very dark red　25

23a. Leaves deeply lobed　24

b. Leaves shallowly lobed　**21. wattiana**

24a. Fruit golden yellow　**22. altaica**

b. Fruit brownish black　**23. dsungarica**

25a. Stamens 10　**24. holmesiana**

b. Stamens 15–20　26

26a. Anthers purple　**25. sanguinea**

b. Anthers pink or red　27

27a. Anthers red　28

b. Anthers pink　29

28a. Leaves always lobed; stamens 20; styles and seeds 5　**26. × durobrivensis**

b. Leaves entire and lobed on the same plant; stamens 15–20; style and seed solitary　**27. heterophylla**

29a. Leaves truncate, very rarely wedge-shaped　**28. coccinioides**

b. Leaves wedge-shaped at base　30

30a. Leaves with margins very sharply lobed and doubly toothed; fruit rounded, red　**29. flabellata**

b. Leaves with margins slightly lobed and toothed; fruit egg-shaped, red, eventually blackish　**30. douglasii**

31a. Calyx and inflorescence branches more or less hairless　32

b. Calyx and inflorescence branches covered with long straight hairs　36

32a. Leaves broadly ovate to rounded or diamond-shaped　33

b. Leaves ovate-elliptic to lanceolate　35

33a. Fruit yellowish green　**31. flava**

b. Fruit red or orange-red　34

34a. Leaves 6–8 cm; fruit bright red　**32. wilsonii**

b. Leaves 2–3.5 cm; fruit orange-red　**33. aprica**

35a. Flowers 1.8–2 cm across　**34. nitida**

b. Flowers 1.2–1.5 cm across　**35. crus-galli**

36a. Leaves irregularly and deeply lobed for most of their length with at least some divisions almost to midrib, others dissecting each side of leaf-blade to at least half-way　46

b. Leaves almost entire or with shallow regular lobes, particularly in upper half of leaf　37

37a. Stamens 20, anthers red　**36. pruinosa**

b. Stamens 5–10, or if 20 anthers white, pale yellow or pink　38

38a. Flowers few, in clusters of 3–10　39

b. Flowers many, in clusters of 12 or more　40

39a. Leaves truncate at base; flowers in groups of 3–7; fruit dull reddish brown　**37. intricata**

b. Leaves wedge-shaped at base; flowers in groups of 6–10; fruit red　**38. jackii**

40a. Thorns 7–12 cm　**39. succulenta** var. **macracantha**

b. Thorns less than 7 cm　41

41a. Leaves narrowly ovate-elliptic, almost evergreen　**41. persistens**

b. Leaves broadly ovate to rounded, deciduous　42

42a. Leaf-margin toothed, but rarely lobed　**40. × prunifolia**

b. Leaf-margin lobed and irregularly toothed　43

43a. Stamens 20　**39. succulenta**

b. Stamens 5–10　44

44a. Flowers *c.* 2.5 cm across　**42. jonesiae**

b. Flowers 2 cm or less　45

45a. Stamens 5–10; flowers *c.* 2 cm; fruit spherical, orange-red　**43. chrysocarpa**

b. Stamens 10; flowers 1.5–2 cm across; fruit pear-shaped, glossy red　**44. pedicellata**

46a. Leaves 5–10 cm, broadly angular to ovate; fruit *c.* 1.5 cm　**45. pinnatifida**

b. Leaves less than 6 cm, broadly diamond-shaped, deeply lobed; fruit

0.6–1.3 cm; flowers often double or pink　47

47a. Flowers 1.5–1.8 cm across; fruit 0.6–1.3 cm across; seeds 2 or 3　**46. laevigata**

b. Flowers 0.8–1.5 cm across; fruit 6–8 mm, seed solitary　**47. monogyna**

1. C. pentagyna Willdenow. Figure 38(10), p. 441. Illustration: Revue Horticole, 310 (1901); Reichenbach, Icones Florae Germanicae et Helveticae **25**: pl. 681 (1902); Davis, Flora of Turkey **4**: 143 (1972); Krüssmann, Manual of cultivated broad-leaved trees & shrubs **1**: pl. 153 (1984).

Tree to 5 m, young branches covered with soft hairs and few thorns to 1 cm. Leaves 2–6 cm, diamond-shaped to broadly ovate, with short hairs beneath; lobes 3–7, margins with forward-pointing teeth. Flowers borne in loose shaggy panicles 4–7 cm across; corolla *c.* 1.5 cm across; stamens 20, anthers red. Fruit 1.1–1.3 cm, egg-shaped, blackish purple. *E Europe, Caucasus to SW Asia, Iran.* H5. Late spring–early summer.

2. C. tanacetifolia (Lamarck) Persoon. Illustration: Edwards's Botanical Register, **22**: 1884 (1836); Schneider, Illustriertes Handbuch der Laubholzkunde **1**: 786–787 (1906); Davis, Flora of Turkey **4**: 143 (1972); Krüssmann, Manual of cultivated broad-leaved trees & shrubs **1**: pl. 153 (1984).

Shrub or small tree to 12 m, branches upright, covered with short hairs at first. Leaves *c.* 2.5 cm, ovate to diamond-shaped, both sides with shaggy hairs; margins with forward-pointing teeth, lobes 5–7, narrow. Stipules conspicuous. Flowers in hairy, umbel-like groups of 4–8; corolla 2–2.5 cm across. Fruit 2–2.5 cm across, orange-yellow. *W Asia.* H4. Spring.

3. C. orientalis Pallas (*Mespilus odoratissima* Andrews). Figure 38(11), p. 441. Illustration: Botanical Magazine, 2314 (1822); Phillips, Trees in Britain, Europe and North America 107 (1978); RHS Dictionary of gardening **2**: 570 (1981); Krüssmann, Manual of cultivated broad-leaved trees & shrubs **1**: pl. 154 (1984).

Tree to 7 m, straggly, with short hairy branches, eventually becoming thorny. Leaves 4–5 cm, triangular or diamond-shaped, deeply cut into 5–9 lobes, with short hairs above and dense matted hairs beneath. Flowers in white-hairy clusters of

Figure 38. Leaves of *Crataegus* species. 1, *C. calpodendron*.
2, *C. × prunifolia*. 3, *C. collina*. 4, *C. viridis*. 5, *C. crus-galli*.
6, *C. × lavallei*. 7, *C. dahurica*. 8, *C. douglasii*. 9, *C. pinnatifida*.
10, *C. pentagyna*. 11, *C. orientalis*. 12, *C. azarolus*. 13, *C. laevigata*.
14, *C. phaenopyrum*. 15, *C. wilsonii*. 16, *C. chlorosarca*.
17, *C. wattiana*. 18, *C. pedicellata*. 19, *C. flabellata*.

7–10; corolla 1.6–1.9 cm across, white; stamens 20. Fruit *c.* 1.5 cm across, spherical, orange-red, hairy. *SE Europe, W Asia.* H3. Early summer.

4. C. azarolus Linnaeus. Figure 38(12), p. 441. Illustration: Revue Horticole, 441 (1856); Schneider, Illustriertes Handbuch der Laubholzkunde **1**: 789, 790 (1906); Phillips, Trees in Britain, Europe and North America 106 (1978); Krüssmann, Manual of cultivated broad-leaved trees & shrubs **1**: pl. 152 (1984).

Small tree or shrub, with thornless branches, often covered with soft hairs. Leaves 3–7 cm, ovate to diamond-shaped, with 3–5 deep lobes, both sides hairy at first, eventually glossy green above, greyish beneath. Flowers few, in hairy umbel-like clusters. Fruit *c.* 2 cm across, spherical, yellowish orange. *S Europe, N Africa & W Asia.* H3. Spring.

5. C. × dippeliana Lange. Illustration: Schneider, Illustriertes Handbuch der Laubholzkunde **1**: 797 (1906).

Shrub or small tree resembling *C. tanacetifolia*, but with leaves lobed only to the middle of the blade. Lobes acute, with forward-pointing teeth, hairy on both surfaces. Flowers in dense clusters; calyx and corolla hairy; stamens *c.* 20, anthers red. Fruit *c.* 1.5 cm across, yellow or dull red. *Garden origin. England.* H3. Spring.

Raised in cultivation *c.* 1830 and thought to be a hybrid between *C. punctata* and *C. tanacetifolia.*

6. C. chlorosarca Maximowicz. Figure 38(16), p. 441. Illustration: Dippel, Handbuch der Laubholzkunde **3**: 450 (1893); Schneider, Illustriertes Handbuch der Laubholzkunde **1**: 772, 774 (1906); Krüssmann, Manual of cultivated broad-leaved trees & shrubs **1**: 396, pl. 153 (1984).

Small tree of pyramidal habit; young branches covered with fine hairs when young; thorns 1–1.2 cm, straight. Leaves 5–9 cm, broadly triangular, hairy on both surfaces, eventually hairless above; lobes 3–5 pairs, margins with forward-pointing teeth. Flowers in umbel-like panicles 4–7 cm across; petals *c.* 9 mm. Fruit black, flesh green. *China (Manchuria).* H2. Spring–summer.

7. C. calpodendron (Ehrhart) Medikus (*C. tomentosa* Duroi). Figure 38(1), p. 441. Illustration: Britton & Brown, Illustrated flora of the Northern United States and Canada, 2002 (1898); Bailey, Standard cyclopedia of horticulture, 885 (1935);

Gleason, Illustrated flora of the north-eastern United States and adjacent Canada **2**: 375 (1952); Krüssmann, Manual of cultivated broad-leaved trees & shrubs **1**: 396, pl. 153 (1984).

Small tree to 6 m; branches erect, thornless or with short thorns. Leaves 5–12 cm, broadly ovate, often lobed, hairy beneath, dull green and hairless above, margins with forward-pointing teeth. Flowers in umbel-like panicles 6–12 cm across. Fruit *c.* 1 cm, elliptic, yellowish or orange-red, fleshy. *Canada (Ontario), south to C USA.* H2. Summer.

8. C. ellwangeriana Sargent (*C. coccinea* Linnaeus, in part). Illustration: Botanical Magazine, 3432 (1835); Sargent, The silva of North America **13**: pl. 671 (1902); Hough, Handbook of the trees of the northern states and Canada, 250, 251 (1947); Botanical Magazine, n.s., 105 (1950).

Small tree very like *C. pedicellata* but with leaves and flowers covered in downy hairs. *E USA.* H3. Spring–summer.

9. C. submollis Sargent. Illustration: Sargent, The silva of North America **4**: pl. 182 (1892); Schneider, Illustriertes Handbuch der Laubholzkunde **1**: 797, 799 (1906); Gleason, Illustrated flora of the north-eastern United States and adjacent Canada **2**: 371 (1952).

Large shrub or tree, 8–10 m, branches thorny, softly hairy at first. Leaves 4–8 cm, ovate, abruptly wedge-shaped at base, occasionally rounded or heart-shaped, hairy above and below; margins with forward-pointing teeth, with 4 or 5 lobes on each side. Flowers many in loose hairy corymbs; calyx hairy, sepals with red-stalked glands; corolla *c.* 2 cm across; stamens 10, anthers white to yellow. Fruit *c.* 1 cm across, pear-shaped, light red, seeds 5. *NE North America.* H2. Spring.

10. C. mollis (Torrey & Gray) Scheele. Illustration: Sargent, The silva of North America **13**: pl. 659 (1902); Gleason, Illustrated flora of the north-eastern United States and Canada **2**: 371 (1952); Phillips, Trees in Britain, Europe and North America 108 (1978); Krüssmann, Manual of cultivated broad-leaved trees & shrubs **1**: pl. 153 (1984).

Tree to 11 m, branches with or without thorns. Leaves 6–10 cm, broadly ovate, covered with dense hairs beneath, eventually only on the veins; margins doubly toothed, with 4 or 5 pairs of lobes. Flowers in densely hairy corymbs; corolla

c. 2.5 cm across; stamens 20, anthers pale yellow. Fruit 1.2–1.8 cm across, almost spherical, red, downy; seeds 4–5. *N America.* H2. Spring–early summer.

Plants sometimes offered as **C. arkansana** Sargent (illustration: Sargent, The silva of North America **13**: pl. 660, 1902) with slightly longer fruits are now considered to represent part of the natural range of the species.

11. C. arnoldiana Sargent. Illustration: Sargent, The silva of North America **13**: pl. 668 (1902); Bailey, Standard cyclopedia of horticulture, **1**: 822 (1935); Gleason, Illustrated flora of the north-eastern United States and adjacent Canada **2**: 373 (1952).

Small tree 7–10 m, with stiff thorny branches, covered with long shaggy hairs at first. Leaves 4–5 cm, with 3–5 pointed, saw-toothed lobes on each side, ovate with rounded or truncated base; the leaves on long shoots much larger than those on short shoots. Flowers *c.* 3 cm across, numerous in many loose-flowered corymbs; stamens 10, anthers light yellow. Fruit *c.* 1.5 cm across, light bluish red, rarely red, spherical, with shaggy hairs at each end; seeds 3 or 4. *NE USA.* H4. Spring.

12. C. pubescens (Humbolt, Bonpland & Kunth) Steudel forma **stipulacea** (Loudon) Stapf (*C. stipulacea* Loudon; *C. mexicana* de Candolle). Illustration: Dippel, Handbuch der Laubholzkunde **3**: 427 (1893); Botanical Magazine, 8589 (1914).

Shrub to 3 m, branches with few thorns. Leaves 4–8 cm, wedge-shaped to elliptic or oblong-lanceolate, dark green above, paler with tufted hairs beneath; margins with scalloped teeth. Flowers in clusters of 6–12; corolla *c.* 2 cm wide. Fruit yellowish to dull orange-red. *Mexico.* H5. Early spring.

13. C. × lavallei Hérincq. (*C. pubescens* forma *stipulacea* × *C. crus-galli*). Figure 38(6), p. 441. Illustration: Phillips, Trees in Britain, Europe and North America 108 (1978); RHS dictionary of gardening **2**: 569 (1981).

Tree 5–7 m, young branches downy, thorns to 4 cm, few. Leaves 4–11 cm, oblong to elliptic, acute; margins unevenly toothed. Flowers borne in erect corymbs up to 7 cm across; corolla *c.* 2 cm across; stamens 5–20, anthers pink or yellowish. Fruit 1.6–1.9 cm across, spherical, orange-red, speckled, persisting into winter. *Raised in cultivation, France.* H4. Spring–early summer.

Most commonly grown as 'Carrieri', which has only 5–10 pink anthers.

14. C. uniflora Muenchhausen (*C. champlainensis* Sargent). Illustration: Sargent, The silva of North America **4**: pl. 191 (1892), **13**: 669 (1902); Gleason, Illustrated flora of the north-eastern United States and adjacent Canada **2**: 345 (1952); Krüssmann, Manual of cultivated broad-leaved trees & shrubs **1**: 400 (1984).

Straggly shrub to 3 m, branches with thin thorns to 4 cm. Leaves to 4 cm, ovate, wedge-shaped at base, covered with short hairs at first, eventually hairless; margins finely toothed. Flowers solitary or in clusters of 2–3, almost stalkless; corolla *c.* 1.5 cm across, creamy white; stamens 20, anthers whitish. Fruit *c.* 1.2 cm, rounded to pear-shaped, yellow or greenish. *E North America*. H4. Spring–early summer.

15. C. collina Chapman. Figure 38(3), p. 441. Illustration: Sargent, The silva of North America **13**: 654 (1902); Britton & Brown, Illustrated flora of the northern United States and Canada **2**: 301 (1913).

Shrub or small tree to 8 m; branches spreading, with stout thorns. Leaves broadly ovate to elliptic, about as long as wide, margins doubly toothed; anthers yellow. Fruit *c.* 1 cm, spherical, dull red; pulp yellowish, mealy; seeds 5. *C & E USA*. H2. Spring.

16. C. punctata Jacquin. Illustration: Sargent, The silva of North America **4**: pl. 184 (1892); Hough, Handbook of the trees of the northern states and Canada, 246–247 (1947); Gleason, Illustrated flora of the north-eastern United States and adjacent Canada **2**: 349 (1952); Elias, The complete trees of North America, 615 (1980).

Tree 6–10 m, with a rounded crown and many horizontal branches, usually with stout thorns 5–8 cm, or occasionally thornless. Leaves 5–10 cm, ovate to obovate, with shaggy hairs beneath; margins irregularly toothed. Flowers numerous; corolla 1.5–2 cm across; stamens 20, anthers pink. Fruit rounded *c.* 1.8 cm across, yellowish red, lightly spotted; seeds 2–5. *E North America*. H3. Spring.

17. C. dahurica Koehne. Figure 38(7), p. 441. Illustration: Schneider, Illustriertes Handbuch der Laubholzkunde **1**: 772, 774 (1906); Flora of the USSR **9**: 324 (1971); Krüssmann, Manual of cultivated broad-leaved trees & shrubs **1**: pl. 153 (1984).

Small tree or shrub, branches dark brown, with thorns to 4 cm. Leaves 2–5 cm, ovate to elliptic or diamond-shaped, acutely lobed, dark green, appearing early in spring, turning bright red in autumn. Flowers borne in loose hairless panicles; corolla *c.* 1.5 cm across; stamens 20, anthers purple. Fruit 6–8 mm across, light brown. *SE Siberia*. H1. Spring–early summer.

18. C. phaenopyrum (Linnaeus filius) Medikus. Figure 38(14), p. 441. Illustration: Gleason, Illustrated flora of the north-eastern United States and adjacent Canada **2**: 345 (1952); Elias, The complete trees of North America, 619 (1980); Everett, New York Botanical Garden illustrated encyclopedia of horticulture **3**: 909 (1981); Krüssmann, Manual of cultivated broad-leaved trees & shrubs **1**: pl. 153 (1984).

Tree to 10 m, branches with thorns to 7 cm. Leaves 3–7 cm, almost triangular; lobes 3–5, margins with sharp forward-pointing teeth. Flowers many in umbel-like panicles; corolla *c.* 1.2 cm across; stamens 20, anthers pink. Fruit *c.* 8 mm across, a glossy red, flattened sphere; seeds 2–5. *NE North America (Virginia to Alabama)*. H4. Late spring–early summer.

19. C. canbyi Sargent. Illustration: Sargent, The silva of North America **13**: pl. 638 (1902); Britton, North American trees, 450 (1908); Britton & Brown, Illustrated flora of the northern United States and Canada **2**: 299 (1913).

Large shrub or small tree to 6 m, with a broad open crown, branches with thorns 2.5–3 cm, stout; resembling *C. crus-galli* but with broader leaves to 7 cm, widest near the middle. Flowers in hairless corymbs, corolla 1–1.2 cm across; stamens 10–20, anthers pink. Fruit 5–7 mm across, glossy and bright red with red pulp. *E North America*. H3. Spring.

20. C. viridis Linnaeus. Figure 38(4), p. 441. Illustration: Sargent, The silva of North America **4**: pl. 187 (1892); Britton & Brown, Illustrated flora of the Northern United States and Canada **1**: 1996 (1897); Gleason, Illustrated flora of the north-eastern United States and adjacent Canada **2**: 345 (1952); Elias, The complete trees of North America, 607, 613 (1980).

Tree to 12 m, with spreading branches and thin thorns to 4 cm. Leaves 2–6 cm, ovate to oblong, always wedge-shaped at the base; margins with forward-pointing teeth. Flowers borne in clusters 4–5 cm across; corolla *c.* 1.8 cm, white. Fruit 6–8 mm across, almost spherical, bright red, persistent into winter. *E USA*. H4. Spring–early summer.

21. C. wattiana Hemsley & Lace (*C. korolkowii* misapplied). Figure 38(17), p. 441. Illustration: Revue Horticole, 308 (1901); Botanical Magazine, 8818 (1919).

Small tree resembling *C. sanguinea*; young branches red-brown, thornless or with short thorns. Leaves 5–9 cm, ovate, acute, hairless; lobes 3–5 pairs. Stamens 15–20, anthers whitish. Fruit *c.* 1 cm across, rounded, orange-yellow, fleshy. *C Asia (Altai Mts to SW Pakistan)*. H2. Summer.

22. C. altaica (Loudon) Lange. Illustration: Schneider, Illustriertes Handbuch der Laubholzkunde **1**: 772, 774 (1906); Flora of the USSR **9**: 324 (1971); Icones cormophytorum Sinicorum **1**: 2142 (1972); Krüssmann, Manual of cultivated broad-leaved trees & shrubs **1**: pl. 153 (1984).

Small straggly tree to 4 m. Branches with few thorns 2–3 cm. Leaves ovate, deeply lobed with warty toothed margins, bright green. Flowers borne in loose umbel-like panicles; corolla *c.* 1 cm across. Fruit 1–1.2 cm across, spherical, golden yellow. *C Asia*. H1. Spring.

23. C. dsungarica Zabel. Illustration: Schneider, Illustriertes Handbuch der Laubholzkunde **1**: 772, 774 (1906).

Like *C. altaica* but with leaves 3–8 cm, hairless, ovate to diamond-shaped to broadly ovate, with 2–4 pairs of lobes. Flowers in hairless clusters and fruit blackish brown. *E Siberia, N China*. H1. Spring.

24. C. holmesiana Ashe. Illustration: Sargent, The silva of North America **13**: pl. 676 (1902); Hough, Handbook of the trees of the northern states and Canada, 252, 253 (1947); Gleason, Illustrated flora of the north-eastern United States and adjacent Canada **2**: 367 (1952); Krüssmann, Manual of cultivated broad-leaved trees & shrubs **1**: pl. 153 (1984).

Tree-like shrub occasionally reaching 10 m, with a loose conical crown of ascending branches; thorns slender. Leaves 6–8 cm, ovate to oblong-ovate, with 4–6 pairs of saw-toothed lobes. Flowers many, in hairless corymbs; stamens 10, anthers red. Fruit *c.* 1 cm across, light red; seeds 3, pulp dry or mealy. *E North America*. H4. Spring.

25. C. sanguinea Pallas. Illustration: Nuttall, The North American sylva, pl. 44 (1865); Schneider, Illustriertes Handbuch der Laubholzkunde **1**: 772, 774 (1906).

Tree to 7 m, with glossy, purplish brown branches and few stout thorns to 3 cm. Leaves 5–8 cm, diamond-shaped to broadly ovate, dark green above, with short hairs and paler beneath; margins with 2 or 3 pairs of rough, doubly toothed lobes. Flowers borne in small umbel-like panicles; anthers purple. Fruit 1–1.2 cm across, bright red, almost translucent. *E Siberia*. H1. Spring.

26. C. × durobrivensis Sargent. Illustration: Sargent, Trees & shrubs **1**: pl. 2 (1902).

Shrub 3–5 m, with hairless young shoots and thorns 3–5 cm. Leaves broadly ovate, wedge-shaped at base, hairless, divided into 2–4 triangular lobes. Flowers in hairless clusters; corolla *c.* 2 cm across; stamens 20, anthers red. Fruit *c.* 1.1 cm across, rounded, shiny crimson red. *N America*. H3. Spring.

Often described as one of the most ornamental species with among the largest flowers in the genus. It is thought by some to be a hybrid between *C. pruinosa* and *C. suborbiculata* Sargent.

27. C. heterophylla Flugge. Illustration: Edwards's Botanical Register **14**: pl. 1161 (1828), **22**: pl. 84 (1836); Loudon, Arboretum et fruticens Britanicum **2**: 864, **4**: pl. 31 (1838); Schneider, Illustriertes Handbuch der Laubholzkunde **1**: 780, 782 (1906).

Shrub or small tree, 1.5–3 m; shoots hairless, with thorns 1–1.5 cm. Leaves 3–4.5 cm, shallowly 3–7-lobed or entire, rounded or broadly ovate, wedge-shaped at base. Flowers in groups of 8–15. Fruit 1.5–1.8 cm. *C & SW Asia*. H2. Summer.

28. C. coccinioides Ashe. Illustration: Sargent, The silva of North America **13**: pl. 674 (1902); Bailey, Standard cyclopedia of horticulture, 1099 (1950); Gleason, Illustrated flora of the north-eastern United States and adjacent Canada **2**: 373 (1952); Bean, Trees & shrubs hardy in the British Isles **1**: 768 (1970).

Small tree, thorns 3–5 cm, straight. Leaves dull red on new growth, brownish beneath, truncate at base, very rarely wedge-shaped, orange-red in autumn. Flowers in clusters of 5–7, corolla *c.* 1.2 cm, stamens 20, anthers pink. Fruit rounded, dark reddish pink. *C USA*. H2. Spring.

29. C. flabellata (Bosc) Koch. Figure 38(19), p. 441. Illustration: Gleason, Illustrated flora of the north-eastern United States and adjacent Canada **2**: 361 (1952); Krüssmann, Manual of cultivated broad-leaved trees & shrubs **1**: pl. 153 (1984); Duncan & Duncan, Trees of the southeastern United States, 189 (1988).

Shrub to 6 m, branches with curved thorns, 4–10 cm. Leaves 3–7 cm, broadly ovate, with few shaggy hairs above and on veins beneath at first; margins very sharply lobed and doubly toothed. Flowers borne in downy corymbs; calyx-lobes with forward-pointing teeth; corolla 1.5–2 cm across; stamens 15–20. Fruit 1–1.1 cm across, rounded, red. *E North America*. H3. spring.

30. C. douglasii Lindley (*C. rivularis* Nuttall). Figure 38(8), p. 441. Illustration: Sargent, The silva of North America **4**: 175 (1892); Gleason, Illustrated flora of the north-eastern United States and adjacent Canada **2**: 375 (1952); Phillips, Trees in Britain, Europe and North America 106 (1978); Krüssmann, Manual of cultivated broad-leaved trees & shrubs **1**: 397 (1984).

Tree to 12 m with slender drooping branches, thorns to 3 cm, few. Leaves 3–8 cm, broadly oblong to ovate, hairless, except on midrib beneath; margins toothed and slightly lobed. Flowers in hairless panicles of 10–20. Corolla *c.* 1 cm wide; stamens 20. Fruit *c.* 1.2 cm, egg-shaped, wine-red, eventually reddish black. *W North America*. H2. Spring.

31. C. flava Aiton. Illustration: Sargent, The silva of North America **13**: pl. 693 (1902); Gleason, Illustrated flora of the north-eastern United States and adjacent Canada **2**: 351 (1952); Krüssmann, Manual of cultivated broad-leaved trees & shrubs **1**: 297 (1984); Brickell, RHS gardeners' encyclopedia of plants and flowers, 63 (1989).

Slow-growing, wide-spreading shrub or small tree to 6 m, branches with thorns to 2.5 cm. Leaves *c.* 2.5 cm, broadly ovate to rounded or diamond-shaped; stipules rounded, conspicuous. Flowers borne in small clusters of 3–7; corolla *c.* 1 cm across; stamens 10–20, anthers purple. Fruit *c.* 1 cm across, rounded to pear-shaped, greenish yellow. *E North America*. H4. Spring.

32. C. wilsonii Sargent. Illustration: Icones cormophytorum Sinicorum **1**: 2140 (1972).

Tree to 6 m, resembling *C. calpodendron*, but with leaves broadly ovate to rounded, glossy above, with shaggy hairs beneath. Fruit to 1 cm, elliptic, bright red. *C China*. H4. Spring.

33. C. aprica Beadle. Illustration: Sargent, The silva of North America **13**: pl. 698 (1902); Schneider, Illustriertes Handbuch der Laubholzkunde **1**: 792, 793 (1906).

Shrub or small tree to 6 m with outspread, undulating branches and straight, slender thorns 2–3.5 cm. Leaves 2–3.5 cm, broadly ovate to rounded, apex acute to rounded, hairy when young; margins with saw-like, gland-tipped, teeth and shallow lobes towards the apex. Flowers borne in groups of 3–6, in small corymbs; sepals with gland-tipped, saw-like teeth; stamens 10; anthers yellow. Fruit *c.* 12 mm across, spherical, dull orange-red; seeds 3–5. *SE USA*. Spring. H4–5.

34. C. nitida (Britton & Brown) Sargent (*C. viridis* var. *nitida* Britton & Brown). Illustration: Gleason, Illustrated flora of the north-eastern United States and adjacent Canada **2**: 349 (1952); Sargent, The silva of North America **13**: 703 (1902); Schneider, Illustriertes Handbuch der Laubholzkunde **1**: 797, 799 (1906).

Tree to 7 m, like *C. viridis* but branches covered with short matted hairs at first; eventually hairless or with few long hairs; thornless or with few short thorns. Leaves 2–8 cm, oblong to elliptic, slightly lobed. Flowers 1.8–2 cm across. Fruit *c.* 1.5 cm, egg-shaped to spherical, dull red with a glaucous bloom. *S USA*. H5. Spring–early summer.

35. C. crus-galli Linnaeus. Figure 38(5), p. 441. Illustration: Sargent, The silva of North America **4**: pl. 178 (1892); Gleason, Illustrated flora of the north-eastern United States and adjacent Canada **2**: 345 (1952); Phillips, Trees in Britain, Europe and North America 106 (1978); Krüssmann, Manual of cultivated broad-leaved trees & shrubs **1**: pl. 158 (1984).

Tree to 10 m, with a flat spreading crown and vase-shaped branching pattern; thorns *c.* 8 cm, slender, straight. Leaves 2–8 cm, oblong to ovate, wedge-shaped at base, rounded at tip, leathery; margins smooth. Flowers to 1.5 cm across; petals white; stamens 10, anthers pink. Fruit persistent, dullish red, pulp red. *E USA*.

Often grown as 'Pyracanthifolia', illustration: Sargent, The silva of North America **13**: pl. 637 (1902) with wide-spreading, thorny branches.

C. × grignonensis Mouillefert (*C. crus-galli × C. pubescens*). Illustration: Krüssmann, Manual of cultivated broad-leaved trees & shrubs **1**: 398 (1984). Shrub to *c.* 5 m, branches vase-shaped, almost thornless. Leaves ovate, base wedge-shaped, covered with long and short hairs, persisting for some time; margins scalloped, with 2–4 pairs of lobes. Flowers in small, umbel-like clusters. Fruit to 1.5 cm, brownish red with grey spots.

Raised in Frankfurt *c.* 1873.

36. C. pruinosa (Wendland) K. Koch. Illustration: Sargent, The silva of North America **13**: 648 (1902); Schneider, Illustriertes Handbuch der Laubholzkunde **1**: 797, 799 (1906); Gleason, Illustrated flora of the north-eastern United States and adjacent Canada **2**: 365 (1952); Elias, The complete trees of North America, 618 (1980).

Tree to 6 m, branches hairless, with stout thorns 2.5–4 cm. Leaves 3–5 cm, broadly elliptic, wedge-shaped at base, red at first, eventually blue-green, margins irregularly toothed. Flowers borne in loose corymbs; corolla 2–2.5 cm across; stamens 20, anthers red. Fruit *c.* 1 cm across, rounded, blue-green at first, eventually dark red, pulp yellowish. *NE North America*. H2. Spring.

37. C. intricata Lange. Illustration: Schneider, Illustriertes Handbuch der Laubholzkunde **1**: 800, 801 (1906); Gleason, Illustrated flora of the north-eastern United States and adjacent Canada **2**: 355 (1952); Everett, New York Botanical Garden illustrated encyclopedia of horticulture **3**: 911 (1981); Krüssmann, Manual of cultivated broad-leaved trees & shrubs **1**: pl. 152 (1984).

Shrub 1–3 m, branches with curved thorns 2–4 cm. Leaves ovate-elliptic, acutely lobed, bright green above, hairless; margins doubly toothed. Flowers borne in corymbs of 3–7; corolla 1.2–1.5 cm across; stamens 10, anthers yellow or pale pink. Fruit to 1.3 cm across, rounded, dull reddish brown; seeds 3–5. *E North America*. H3. Spring.

38. C. jackii Sargent. Illustration: Britton & Brown, Illustrated flora of the northern United States and Canada **2**: 306 (1913).

Small tree-like shrub, to 3 m. Leaves 2–3 cm, ovate-elliptic to broadly ovate, broadly wedge-shaped at base, rarely truncate; margins with shallow indistinct lobes and rough teeth. Flowers borne in groups of 6–10; anthers yellow. Fruit *c.* 1.2 cm, red,

egg-shaped; seeds 2 or 3. *Canada (Quebec)*. H2. Summer.

39. C. succulenta Link. Illustration: Sargent, The silva of North America **4**: pl. 181 (1892); Schneider, Illustriertes Handbuch der Laubholzkunde **1**: 775, 776 (1906); Gleason, Illustrated flora of the north-eastern United States and adjacent Canada **2**: 375 (1952); Elias, The complete trees of North America, 611 (1980).

Tree or shrub to 5 m, branches red-brown, straggly, with stout thorns to 7 cm. Leaves 5–8 cm, broadly ovate; margins doubly toothed. Flowers many, in hairy, umbel-like panicles; corolla *c.* 1.8 cm across, white; stamens 20, anthers pink, rarely white. Fruit *c.* 1 cm across, rounded, glossy red. *E North America*. H3. Early summer.

Var. **macracantha** (Loddiges) Eggleston, illustration; Hough, Handbook of the trees of the northern states and Canada, 258, 259 (1947). Differs in its more numerous thorns 7–12 cm and having only 10 stamens with white or pale yellow anthers.

40. C. × prunifolia (Poiret) Persoon (*C. crus-galli × C. succulenta* var. *macracantha* Loudon). Figure 38(2), p. 441. Illustration: Edwards's Botanical Register **22**: pl. 1808 (1836); Phillips, Trees in Britain, Europe and North America 109 (1978); Everett, New York Botanical Garden illustrated encyclopedia of horticulture **3**: 910 (1981); Krüssmann, Manual of cultivated broad-leaved trees & shrubs **1**: pl. 153 (1984).

Shrub or small tree to 6 m, like *C. crus-galli* but branches with curved thorns 4–7 cm. Leaves *c.* 7 cm, broadly ovate to rounded or elliptic; flowers borne in dense, many-flowered corymbs. *Garden origin*. H3. Spring.

Sometimes considered a cultivar of *C. crus-galli* ('Splendens') rather than a hybrid.

41. C. persistens Sargent. Illustration: Sargent, Trees and shrubs, 190 (1913).

Tall shrub or small tree, 3–4 m, with wide-spreading branches, many with thick thorns to 5 cm. Leaves lanceolate to oblong to ovate, coarsely toothed towards apex, almost evergreen, persisting well into the winter. Corolla *c.* 2 cm across; stamens 20; anthers white. Fruit *c.* 1.5 cm across, egg-shaped, dull bluish red, persistent; seeds 2 or 3. *Garden origin*. H3. Spring.

42. C. jonesiae Sargent. Illustration: Sargent, the silva of North America, **13**: pl. 684 (1902); Britton & Brown, Illustrated flora of the northern United States and Canada **2**: 300 (1913).

Tall shrub or small tree to 6 m, young branches with long and short hairs, eventually hairless, glossy brown; thorns 5–7 cm. Leaves 7–10 cm, broadly ovate to rounded, wedge-shaped at base; margins with rough teeth and acute lobes above the middle, glossy dark green above, hairy beneath when young. Corolla *c.* 2.5 cm across; stamens 10, anthers pink. Fruit *c.* 1.5 cm across, bright red; seeds 2 or 3. *NE North America*. H2. Summer.

43. C. chrysocarpa Ashe (*C. rotundifolia* Moench). Illustration: Gleason, Illustrated flora of the north-eastern United States and adjacent Canada **2**: 355 (1952); Canadian Journal of Botany **65**: 2648 (1987); Looman & Best, Budd's flora of the Canadian prairie provinces, 441 (1987).

Tree to 6 m, branches dense, thorns slender. Leaves 3–5 cm, broadly ovate to rounded; lobes 3 or 4 pairs, doubly toothed. Flowers *c.* 2 cm across; sepals with glandular hairs. Fruit almost spherical, red, pulp yellow. *NE North America*. H2. Spring.

44. C. pedicellata Sargent (*C. coccinea* Linnaeus, in part). Figure 38(18), p. 441. Illustration: Sargent, The silva of North America **13**: pl. 677 (1902); Gleason, Illustrated flora of the north-eastern United States and adjacent Canada **2**: 367 (1952); Phillips, Trees in Britain, Europe and North America 108 (1978); Krüssmann, Manual of cultivated broad-leaved trees & shrubs **1**: pl. 152 (1984).

Tree to 7 m, mature branches hairless, with straight or slightly curved thorns to 5 cm. Leaves 7–10 cm, broadly ovate, margins doubly toothed with 4 or 5 pairs of lobes. Flowers borne in loose downy corymbs; corolla 1.5–2 cm across; stamens 10, anthers pink. Fruit 1–1.8 cm, pear-shaped, glossy red. *E USA*. H3. Spring–summer.

45. C. pinnatifida Bunge. Figure 38(9), p. 441. Illustration: Schneider, Illustriertes Handbuch der Laubholzkunde **1**: 769, 770 (1906); Bailey, Standard cyclopedia of horticulture **1**: 888 (1935); Icones cormophytorum Sinicorum **1**: 2137 (1972); Krüssmann, Manual of cultivated broad-leaved trees & shrubs **1**: 400 (1984).

Tree to 6 m, branches hairless, with few thorns to 1 cm. Leaves 5–10 cm, angular to ovate, lobes 5–9, glossy on both sides, soon falling. Flowers few in loose umbel-like panicles; corolla *c.* 1.5 cm; stamens 20. Fruit *c.* 1.5 cm across, red, minutely dotted; seeds 3 or 4. *N China.* H2. Late spring–early summer.

Var. **major** N.E. Brown is the variant most commonly seen in cultivation, it differs in its larger, thicker leaves and fruit to 2.5 cm across.

46. C. laevigata (Poiret) de Candolle (*Mespilus laevigata* Poiret; *C. oxyacantha* misapplied, not Linnaeus; *C. oxyacanthoides* Thuill). Figure 38(13), p. 441. Illustration: Hough, Handbook of the trees of the northern states and Canada, 260–261 (1947); Flora of the USSR **9**: 350 (1971); Phillips, Trees in Britain, Europe and North America 107 (1978); Everett, New York Botanical Garden illustrated encyclopedia of horticulture **3**: 909 (1981).

Shrub or small tree, 2–5 m, branches eventually hairless, with few short thorns. Leaves 4–5 cm, ovate with 3–5 rounded lobes, margins with forward-pointing teeth; stipules acuminate. Flowers in umbel-like panicles of up to 10; corolla 1.5–1.8 cm across, white or pink; stamens 20, anthers red. Fruit 0.6–1.3 cm, rounded to egg-shaped, deep red; seeds 2 or 3. *Europe, N Africa, W Asia.* H2. Spring–early summer.

Among the many cultivars of this species are 'Paul's Scarlet', with bright red flowers; 'Francois Rigaud', with yellow fruits and 'Rubra Plena' with double pink flowers.

C. × mordenensis Boom. Like *C. laevigata*, but with larger leaves and flowers and differing from *C. succulenta* in its shorter thorns and more deeply cut leaves, hairless inflorescence and glandless sepals. A hybrid between *C. laevigata* and *C. succulenta*).

47. C. monogyna Jacquin. Illustration: Schneider, Illustriertes Handbuch der Laubholzkunde **1**: 780, 782 (1906); Davis, Flora of Turkey **4**: 143 (1972); Phillips, Trees in Britain, Europe and North America 108 (1978); Everett, New York Botanical Garden illustrated encyclopedia of horticulture **3**: 909 (1981).

Shrub or tree 2–10 m, branches thorny, hairless or with few short hairs, thorns 2–2.5 cm. Leaves 3–5 cm, with 3–7 deeply cut lobes, whitish green beneath, hairy; stipules entire. Flowers borne in umbellate panicles; calyx-lobes oblong; corolla 0.8–1.5 cm across, white; stamens 20, anthers red. Fruit 6–8 mm across, rounded to egg-shaped, bright red; seed solitary. *Europe, N Africa, Asia.* H2. Late spring.

72. × CRATAEMESPILUS G. Camus
S.G. Knees
Tall shrubs or small trees with densely leaved crowns; branches covered with soft hairs. Leaves deciduous, ovate to obovate, usually lobed; margins smooth or with forward-pointing teeth, hairy beneath. Flowers solitary or in groups of 2 or 3. Sepals and petals 5; stamens 14–28; styles 2 or 3. Fruit with 2 or 3 nutlets, almost spherical, covered with soft hairs.

Hybrids between the genera *Crataegus* and *Mespilus* which originated through normal fruiting rather than by grafting, vegetative means or chimaeras. Both hybrids are of French origin and are very intermediate between their parents.

1a. Leaves lobed towards tip, margin with forward-pointing uneven teeth **1. grandiflora**
  b. Leaves lobed, with smooth margin **2. gillotii**

1. × C. grandiflora G. Camus (*Crataegus laevigata × Mespilus germanica*). Illustration: Botanical Magazine, 3442 (1845) – as Mespilus lobata; Revue Horticole, 80 (1869); Reichenbach, Icones Florae Germanicae et Helveticae **25**: pl. 107 (1914); Krüssmann, Manual of cultivated broad-leaved trees & shrubs **1**: pl. 152 (1986).

Shrub or small tree to 5 m. Leaves 3–7 cm, ovate to obovate, lobed in upper third, margin with uneven forward-pointing teeth; becoming yellow-brown in autumn. Flowers 2–2.5 cm across; sepals hairy; petals white. Fruit 1–1.6 cm across, spherical to egg-shaped, brown. *France.* H4. Early summer.

2. × C. gillotii Beck & Reichenbach (*Crataegus monogyna × Mespilus germanica*). Illustration: Reichenbach, Icones Florae Germanicae et Helveticae **25**: pl. 107 (1914).

Shrub or small tree to 4 m. Leaves 3–6 cm, ovate to obovate, lobed, margin smooth. Flowers *c.* 2 cm across; sepals hairy; petals white; styles 2. Fruit 1–1.5 cm. *France.* H4. Early summer.

73. + CRATAEGOMESPILUS Simon-Louis
S.G. Knees
Deciduous shrub or small tree to *c.* 5 m, with thorny twigs. Leaves 7–15 cm, narrowly oblong to elliptic, dark green above, lighter green and woolly beneath; margins with fine forward-pointing teeth. Flowers in groups of 5–8; sepals 5, erect, united in a woolly calyx; petals 5, white,

5–8 mm, rounded; stamens 15–20; styles 1–3. Fruit 1.5–2 cm across with persistent calyx lobes; seeds 1–3, sterile.

A graft hybrid in which new shoots commonly revert back to the habit and form of their original parents.

1. + C. dardarii Simon-Louis (*Crataegus monogyna + Mespilus germanica*). Illustration: Schneider, Illustriertes Handbuch der Laubholzkunde **1**: 765, 766 (1906); Kew Bulletin of Miscellaneous information: pl. 268 (1911); Reichenbach, Icones Florae Germanicae et Helveticae **25**: 108 (1914); Krüssmann, Manual of cultivated broad-leaved trees & shrubs **1**: 394 (1986).

Garden graft hybrid which originated in France. H4. Spring–summer.

Two cultivars are occasionally grown. 'Jules d'Asniers' ('Asnieresii') is closer to *Crataegus* with flowers in groups of 3–12, and glossy brown fruit. 'Jouinii' is similar but flowers earlier and is totally sterile.

74. OSTEOMELES Lindley
S.G. Knees
Deciduous or semi-evergreen shrubs or small trees. Leaves alternate, pinnately divided; stipules present. Flowers white in terminal nodding corymbs. Calyx-teeth 5, acute; petals 5, ovate to oblong; stamens 15–20; styles 5. Fruit with persistent remains of calyx and 5 seeds.

A genus of about 10 species from eastern Asia, Polynesia and New Zealand. All species can be propagated from semi-hardwood cuttings, or seed, although the latter is not commonly produced in cultivation.

1a. Leaves with 15–31 obovate or oblong to elliptic leaflets; apices straight, acute **1. schweriniae**
  b. Leaves with 9–29, obovate or heart-shaped leaflets; apices very obtuse and usually recurved **2**
2a. Leaflets 4–8 mm, rounded, obovate or heart-shaped **2. subrotunda**
  b. Leaflets 8–15 mm, obovate-oblong **3. anthyllidifolia**

1. O. schweriniae Schneider (*O. anthyllidifolia* Lindley, misapplied). Illustration: Botanical Magazine, 7354 (1894); Krüssmann, Manual of cultivated broad-leaved trees and shrubs **2**: 338 (1986); Brickell, The RHS gardeners's encyclopedia of plants & flowers, 106 (1989); Phillips & Rix, Shrubs, 34 (1989).

Deciduous or semi-evergreen shrub to 3 m. Branches and leaf-stalks with greyish hairs; leaves 3–7 cm, hairy beneath, with

15–31 leaflets, each 4–12 mm, obovate
or oblong to elliptic; apices acute, straight,
ending with a short point. Flowers *c.* 1.5
cm across. Sepals lanceolate-ovate, hairless
on inner face, hairy on outer face. Petals
ovate-oblong, white; stamens 15–18; styles
hairy. Fruit spherical or ovoid, 6–8 mm,
red at first, eventually bluish black,
smooth. *W China.* H5. Spring–summer.

Var. **microphylla** Rehder & Wilson.
Leaves with fewer leaflets, 3–5 mm, often
hairless; inflorescence more compact.
Thought to be more hardy in cultivation.

2. O. subrotunda K. Koch. Illustration:
Hooker, Icones Plantarum **27**: pl. 2644
(1900); Gardeners' Chronicle **103**: 125
(1938); Makino, Illustrated flora of Japan,
1389 (1956).

Slow-growing, rigid, evergreen or semi-
deciduous shrub, rarely exceeding 1 m.
Branches twisted, hairy when young.
Leaflets 9–17, 4–8 mm, rounded or
obovate, sometimes heart-shaped, hairs
beneath sparse and closely adpressed,
apex obtuse, recurved, margin hairy.
Flowers *c.* 1 cm across, in corymbs 2–3
cm wide; styles smooth. *SE China.* H5–G1.
Summer.

Perhaps best grown in containers and
overwintered in a cool glasshouse.

3. O. anthyllidifolia (Smith) Lindley
(*Pyrus anthyllidifolia* Smith). Illustration:
Krüssmann, Manual of cultivated broad-
leaved trees and shrubs **2**: 338, pl. 145
(1986); Wagner et al., Manual of the
flowering plants of Hawaii **2**: 1101 (1990).

Semi-evergreen shrub to 2 m; shoots
densely hairy. Leaves glossy, sparsely hairy
above, densely silky hairy beneath; leaflets
13–29, 8–15 mm, obovate-oblong, apex
obtuse. Flowers similar to *O. subrotunda*,
but with styles shaggy. Fruit spherical, to
8 mm, red and hairy at first, eventually
bluish black and completely smooth.
Hawaii, Polynesia, Bonin Isles. H5–G1.
Summer.

75. × PYRACOMELES Guillaumin
S.G. Knees
Evergreen shrub to 2 m, with slender
young shoots, greyish downy at first, soon
hairless. Leaves 2.5–3.2 cm, pinnate at
base, pinnately lobed towards apex; leaflets
5–9, ovate, rounded and toothed at apex,
hairless, or nearly so. Flowers *c.* 1 cm
across, numerous in terminal corymbs.
Sepals 5; petals white; stamens 12–15.
Fruit *c.* 4 mm across, spherical, coral-red
with 4 or 5 nutlets.

A genus of hybrids between *Pyracantha*

and *Osteomeles*, said to come true from
seed.

1. × P. vilmorinii Rehder (*Pyracantha
atalantioides × Osteomeles subrotunda*).
Illustration: Gardeners' Chronicle **103**:
125 (1938).

Raised in cultivation. H4. Spring.

76. PRUNUS Linnaeus
*J.C.M. Alexander, assisted by J. Cullen, C.J.
King & P.F. Yeo*
Deciduous or evergreen trees or shrubs
of very variable habit, sometimes spiny.
Leaves alternate, usually toothed, generally
with 1 or more conspicuous glands (extra-
floral nectaries) on the stalk; stipules
present, sometimes falling early. Flowers
solitary or borne in racemes, umbels,
corymbs or clusters, with distinct
campanulate or cylindric perigynous zones.
Sepals 5, often toothed, sometimes reddish.
Petals 5 (more in variants with semi-
double or double flowers), white, cream,
or pink to red, usually spreading. Stamens
numerous. Ovary of a single carpel borne
in the base of the perigynous zone; ovules
2. Fruit a drupe, generally fleshy, mostly
1-seeded, sometimes large. Stone pitted,
ridged or smooth, sometimes keeled.

A genus of more than 200 species
widespread in the North Temperate area,
a few on mountains in the tropics (Andes,
SE Asia). Several of the species produce
economically important fruits – cherries,
peaches, plums, damsons, nectarines,
almonds, etc. – and many more are grown
for their flowers which are often very
abundantly produced, frequently before the
leaves have opened. Many are grown as
street-trees and some are used for hedging.

Most are easily grown in good soil and
succeed in full sun. Propagation of most
species is by seed (which requires
stratification), though seed produced in
gardens may well be of hybrid origin.
Many may also be propagated by semi-ripe
hardwood cuttings. Most of the cultivars
require budding or grafting, with species
such as *P. avium*, *P. mahaleb* and *P. padus*
as stocks. Several of the evergreen species
may be propagated by layering. The plants
are susceptible to a wide range of fungal
and insect pests and may require careful
management.

A large number of species is grown for
ornament, and many cultivars have been
selected and raised (both from the species
and from numerous hybrids between
them). The classification and identification
of the plants is therefore somewhat

difficult. There is a hierarchy of subgenera
and sections, based mainly on fruit, seed
and bud characters. As these are not
helpful in identification, they are not
included here. The key below avoids fruit
characters and is based on the flowers and
fully expanded foliage leaves. For those
species which flower well before the leaves
open, specimens of flowers and mature
leaves will be required.

Some authors raise some of the subgenera
to the rank of genus. These are: *Prunus* in
the strict sense (species related to *P. spinosa*),
Cerasus Miller (species related to *P. avium*),
Armeniaca Scopoli (*P. armeniaca*), *Persica*
Miller (*P. persica*), *Amygdalus* Linnaeus
(species related to *P. dulcis*), *Padus* Miller
(species related to *P. padus*) and *Laurocerasus*
Duhamel (species related to *P. laurocerasus*).
Names based on these genera may be found
occasionally in the botanical literature; they
are not cited here as synonyms.

In the descriptions below, the perigynous
zone should be assumed to be campanulate
(that is, widening gradually from the
rounded base for at least half its length)
unless it is stated to be cylindric (that is,
widening abruptly at the base and parallel-
sided for usually more than half its length).
A few species are intermediate and are
described as tubular-campanulate. It
should be noted that the account in the
1991 edition of the *Royal Horticultural
Society's Dictionary of Gardening* uses these
terms in a different sense. It should also be
assumed that the flowers of the deciduous-
leaved species are borne before the leaves,
unless otherwise stated.

Literature: Miyoshi, M., Japanese
mountain cherries, their wild forms and
cultivars, *Journal of the College of Science of
the Imperial University, Tokyo* **24**: 1–175
(1916); Wilson, E.H., The cherries of Japan
(1916); Koehne, E., Die Kirschenarten
Japans, *Mitteilungen der Deutsche
Dendrologische Gesellschaft* for 1917: 1–65
(1917); Ingram, C., Ornamental Cherries
(1948); Sano, T., Flowering cherries (1961);
Chadbund, G., Flowering cherries (1972);
Ohwi, J. & Ohta, Y., Flowering cherries of
Japan (1973); Flower Society of Japan,
Manual of Japanese flowering cherries (1982).

1a. Flowers usually 10 or more in
    narrow, cylindric racemes, if fewer
    then very dense and bark shiny
    brownish yellow   2
  b. Flowers solitary or 2–10 in clusters
    or short, broad open racemes   18
2a. Evergreen trees or shrubs   3
  b. Deciduous trees or shrubs   9

3a. Racemes to 4 cm   4
  b. Racemes more than 5 cm   5
4a. Leaves sparsely toothed
                          88. caroliniana
  b. Leaves spiny   **86. ilicifolia**
5a. Racemes leafy towards base   6
  b. Racemes not leafy towards base   7
6a. Leaves with narrow acute tips
                          74. salicifolia
  b. Leaves with obtuse or rounded tips
                          73. virens
7a. Racemes erect   **85. laurocerasus**
  b. Racemes ascending, spreading or
     hanging   8
8a. Leaves coarsely toothed
                          84. lusitanica
  b. Leaves entire to finely and distantly
     toothed   **87. lyonii**
9a. Racemes not leafy towards base   10
  b. Racemes leafy towards base   11
10a. Leaves gland-dotted beneath; bark
     shiny yellowish brown
                          82. maackii
  b. Leaves not gland-dotted beneath
                          83. buergeriana
11a. Margins of mature leaves scalloped
     or with incurved teeth at least in
     apical part   12
  b. Margins of mature leaves with
     straight projecting teeth   13
12a. Main stalk of inflorescence densely
     hairy; leaf-tips acute-acuminate
                          76. vaniotii
  b. Main stalk of inflorescence hairless
     or sparsely hairy; leaf-tips rounded-
     acuminate   **72. serotina**
13a. Mature leaves with narrow tips 1
     cm long or more   14
  b. Mature leaves with short abrupt
     tips less than 1 cm   15
14a. Leaf-bases rounded; styles
     projecting   **81. grayana**
  b. Leaf-bases cordate; styles not
     projecting   **80. ssiori**
15a. Petals more than 6 mm
                          75. padus
  b. Petals 5 mm or less   16
16a. Main stalk of inflorescence densely
     hairy   **78. napaulensis**
  b. Main stalk of inflorescence hairless
     or sparsely hairy   17
17a. Petals 2–3 mm; leaves shiny above,
     broadest towards tip, mature blades
     usually 10 cm or less
                          79. virginiana
  b. Petals 4–5 mm; leaves dull above,
     broadest near middle, mature
     blades usually more than 10 cm
                          77. cornuta
18a. Flowers stalkless or stalks to 9 mm
                                       19

  b. Flower-stalks more than 9 mm   64
19a. Style or ovary (and fruit) hairy 20
  b. Style, ovary and fruit hairless 39
20a. Leaves folded in bud   21
  b. Leaves rolled in bud   36
21a. Mature leaves to 2 cm wide   22
  b. Most mature leaves 2 cm wide or
     more   28
22a. Leaves densely white-felted beneath,
     of 2 types   23
  b. Leaves not white-felted beneath, all
     similar   24
23a. Flowers hairless or sparsely hairy
     outside, stalkless   **63. incana**
  b. Flowers densely hairy outside,
     stalks c. 3 mm   **65. bifrons**
24a. Young wood with a greyish or
     whitish bloom   25
  b. Young wood greenish, reddish, or
     purplish, not bloomed   27
25a. Often thorny; petals overlapping;
     sepals reddish, not or scarcely
     toothed   26
  b. Not thorny; petals not overlapping;
     sepals greenish, densely toothed
                          33. tenella
26a. Leaves pale greyish green, widest
     near base; stalks reddish; flowers
     white or pale pink   **26. fenzliana**
  b. Leaves dark greyish green, widest
     near middle; stalks green; flowers
     pink or red   **31. tangutica**
27a. Leaves narrowly elliptic, acute,
     entire to finely toothed, hairless
     above   **28. kansuensis**
  b. Leaves broadly elliptic, acuminate,
     coarsely toothed, bristly above
                          37. canescens
28a. Leaves broadly elliptic or ovate,
     sometimes 3-lobed, with large
     broad triangular teeth   29
  b. Leaves narrowly elliptic or
     lanceolate, scalloped or finely
     toothed   32
29a. Stipules narrowly linear, fringed,
     sometimes falling early   30
  b. Stipules broadly elliptic to circular,
     persistent   **37. canescens**
30a. Leaves hairless beneath except on
     midrib and veins; stipules falling by
     midsummer   **42. subhirtella**
  b. Leaves hairy beneath; stipules
     persistent   31
31a. Flowers pink, 1.8–3.5 cm wide,
     often double   **32. triloba**
  b. Flowers white with pinkish centre,
     c. 1.5 cm wide, single
                          62. tomentosa
32a. Leaves with acute, forward-pointing
     teeth, 2 mm or more apart at
     widest part of leaf   **29. davidiana**

  b. Leaves scalloped or more closely
     toothed   33
33a. Leaves tapered at base, widest near
     the middle   34
  b. Leaves rounded at base, widest
     below the middle   35
34a. Sepals hairy; angle between side-
     veins and midrib of leaf obtuse
                          27. persica
  b. Sepals hairless; angle between
     side-veins and midrib acute
                          67. glandulosa
35a. Leaves hairy beside midrib beneath;
     stone smooth   **30. mira**
  b. Leaves hairless; stone pitted
                          25. dulcis
36a. Leaves fringed with dense, long,
     narrow teeth to 2 mm or more
                          20. mandshurica
  b. Leaves scalloped, or with broad
     triangular teeth to 1 mm   37
37a. Leaves wedge-shaped to narrowly
     rounded at base; blade length c. 2
     times the width   **23. mume**
  b. Leaves broadly rounded to truncate
     at base; blade length c. 1.5 times
     the width   38
38a. Most leaves with a gradual point c.
     1.5 cm   **21. sibirica**
  b. Leaves with an abrupt point to 1
     cm   **22. armeniaca**
39a. Leaves rolled in bud   40
  b. Leaves folded in bud   51
40a. Flowers 1.8 cm or more across   41
  b. Flowers 1.5 cm or less across   44
41a. Leaves truncate or slightly cordate
     at base; flowers almost stalkless
                          19. brigantina
  b. Leaves wedge-shaped to rounded at
     base; flower-stalks 5 mm or more
                                     42
42a. Leaves upright, narrow, with few
     veins diverging at a very acute
     angle   **7. simonii**
  b. Leaves horizontal or drooping,
     veins diverging at an angle of c.
     45°   43
43a. First-year twigs shiny, green (purple
     in some cultivars); leaves hairless
     beneath except on midrib and veins
                          3. cerasifera
  b. First-year twigs dull, brown to grey;
     leaves hairy beneath, especially
     when young   **2. domestica**
44a. Mature leaves boat-shaped, often
     curved along midrib   45
  b. Mature leaves not boat-shaped   46
45a. Calyx-lobes hairless, margin
     without glands   **16. angustifolia**
  b. Calyx-lobes hairy inside, margin
     glandular   **17. reverchonii**

46a. Leaves acute or acuminate at tip 47
  b. Leaves rounded or obtuse at tip 49
47a. Flowers becoming pink with age 48
  b. Flowers remaining white **15. munsoniana**
48a. Leaves rounded at base; sepals *c.* 1.6 cm **10. alleghaniensis**
  b. Leaves tapered at base; sepals 2.2–2.5 cm **18. × orthosepala**
49a. Flowers becoming pink with age, stalks hairy **8. subcordata**
  b. Flowers remaining white, stalks hairless 50
50a. Leaf-stalks hairless **4. cocomilia**
  b. Leaf-stalks with short hairs **1. spinosa**
51a. Leaves purplish at first, remaining reddish brown when mature **70. × cistena**
  b. Leaves green when mature 52
52a. Leaves long-acuminate; tips acute or obtuse 53
  b. Leaves not acuminate, sometimes abruptly short-pointed 58
53a. Leaves with acute teeth standing out from margin 54
  b. Leaves scalloped or with obtuse teeth, if teeth acute then not standing out from margin 56
54a. Sepals with gland-tipped marginal teeth **66. japonica**
  b. Sepals without marginal teeth 55
55a. Mature leaves with 8–10 teeth per cm; flowers deep or mid-pink **44. cerasoides**
  b. Mature leaves with 4–6 teeth per cm; flowers white to pale pink **47. hirtipes**
56a. Flowers 2 cm across or more **27. persica** var. **nucipersica**
  b. Flowers less than 2 cm across 57
57a. Sepals hairy; leaf-bases rounded **10. alleghaniensis**
  b. Sepals hairless; leaf-bases tapered **6. consociiflora**
58a. Leaves greyish glaucous beneath, not toothed or scalloped near the base 59
  b. Leaves not glaucous beneath, toothed or scalloped all round 60
59a. Leaves tapered equally at tip and base, ascending **69. pumila**
  b. Leaves tapered narrowly at base, abruptly at tip, spreading **71. besseyi**
60a. Perigynous zone tubular; flowers stalkless or very shortly stalked **64. jacquemontii**

  b. Perigynous zone cup-shaped; flower-stalks 5 mm or more 61
61a. Leaves broadly elliptic, length less than 2 times the width **9. maritima**
  b. Leaves narrowly ovate or elliptic, length *c.* 2½ times the width or more 62
62a. Tree to 8 m; leaves to 10 × 4 cm **15. munsoniana**
  b. Shrub to 2 m; leaves to 7 × 3 cm 63
63a. Leaves to 4 × 2 cm, boat-shaped, often curved along midrib **68. humilis**
  b. Leaves to 7 × 3 cm, flat **67. glandulosa**
64a. Leaves rolled in bud 65
  b. Leaves folded in bud 69
65a. Ovary and base of style (and fruit) hairy **24. × dasycarpa**
  b. Ovary and style hairless 66
66a. Leaf-tips obtuse to rounded **8. subcordata**
  b. Leaf-tips acute or acuminate 67
67a. Leaf-margin scalloped or with rounded teeth **5. salicina**
  b. Leaf-margin with obtuse to acute teeth 68
68a. Flowers usually solitary, to 2.5 cm wide; leaves often purple **3. cerasifera**
  b. Flowers 2–4 together, to 1.2 cm wide **10. alleghaniensis**
69a. Sepals becoming reflexed 70
  b. Sepals not becoming reflexed 88
70a. Flower-clusters lacking a distinct stalk; flower-stalks all arising from the same point in each cluster (stalkless umbels) 71
  b. Flower-clusters clearly stalked; flower-stalks arising at one or more points (stalked umbels or corymbs) 80
71a. Leaves purplish at first, becoming reddish brown when mature **70. × cistena**
  b. Leaves green when mature 72
72a. Leaves with acute or acuminate teeth 73
  b. Leaves with obtuse or rounded teeth 75
73a. Leaves hairy beneath **12. mexicana**
  b. Leaves hairless or almost so beneath 74
74a. Leaf-tips strongly acuminate **11. americana**
  b. Leaf-tips evenly tapered **58. pensylvanica**
75a. Flowers 2 cm wide or more 76

  b. Flowers less than 2 cm wide 78
76a. Flower-clusters with leaf-like bracts **35. cerasus**
  b. Flower-clusters without leaf-like bracts (large bud-scales may be present at base of cluster) 77
77a. Leaves with coarse, narrow, projecting rounded teeth; flower-stalks longer than 2 cm **34. avium**
  b. Leaves scalloped or with short, broad rounded teeth; flower-stalks to 2 cm **13. nigra**
78a. Mature leaves 7 cm or more, rounded or broadly tapered at base **14. hortulana**
  b. Mature leaves less than 6 cm, narrowly tapered at base 79
79a. Twigs grey or black; leaves scalloped, with dark marginal glands **36. fruticosa**
  b. Twigs reddish brown; leaves indistinctly toothed **69. pumila**
80a. Flowers 3 cm or more across 81
  b. Flowers 2.5 cm or less across 82
81a. Leaf-stalks and perigynous zone hairless **56. cyclamina**
  b. Leaf-stalks and perigynous zone hairy **55. dielsiana**
82a. Teeth standing out from leaf-margin, at least 2 mm long 83
  b. Teeth 1 mm or less, or leaves finely scalloped 85
83a. Teeth on leaf-margin with hair-like tips **59. pilosiuscula**
  b. Teeth on leaf-margin rounded or obtuse 84
84a. Teeth on leaf-margin rounded; flowers *c.* 2.5 cm across; stalks arising close together **34. avium**
  b. Teeth on leaf-margin obtuse; flowers *c.* 1.5 cm across, stalks arising from different points **61. maximowiczii**
85a. Mature leaves 6 cm or less, finely scalloped, with short abrupt tips **57. mahaleb**
  b. Mature leaves more than 7 cm, toothed, with drawn out tips 86
86a. Sepals and perigynous zone hairy **54. pseudocerasus**
  b. Sepals and perigynous zone hairless 87
87a. Flowers white; stalks of flower-clusters *c.* 6 mm **60. litigiosa**
  b. Flowers pink; stalks of flower-clusters 2.5 mm or more **43. campanulata**
88a. Stalks of mature leaves at most very sparsely hairy 89

b. Stalks of mature leaves densely downy or hairy   98

89a. Mature leaf-blades 5 cm or less   90

b. Mature leaf-blades usually more than 5 cm   91

90a. Leaf-tips acuminate; midribs ginger hairy beneath   **45. rufa**

b. Leaf-tips round or obtuse, rarely acute; midribs not ginger hairy   **40. mugus**

91a. Teeth on mature leaves very fine and regular, at least 5 major teeth per cm   92

b. Teeth on mature leaves coarse, fewer than 4 per cm   93

92a. Flowers white; bark glossy red, peeling in strips   **46. serrula**

b. Flowers pink; bark not glossy   **44. cerasoides**

93a. Teeth on mature leaves acuminate or drawn out into hair-like tips   94

b. Teeth on mature leaves acute or obtuse, not hair-like   97

94a. Stems and leaves hairless or very slightly hairy   95

b. Stems, undersides of leaves or perigynous zone distinctly hairy   96

95a. Flowers in umbels; leaf-teeth acuminate   **49. sargentii**

b. Flowers in racemes; leaf-teeth drawn out into hair-like tips   **51. serrulata**

96a. Flowers white   **52. speciosa**

b. Flowers pink   **50. juddii**

97a. Flowers produced before the leaves from January–March; bracts falling as flowers open   **47. hirtipes**

b. Flowers produced with the leaves in April and May; bracts persistent   **41. nipponica**

98a. Teeth on mature leaves fine, less than 2 mm   99

b. Teeth on mature leaves coarse, at least 3 mm   100

99a. Flowers 1 cm wide or more; perigynous zone with dense brown hairs   **45. rufa** var. **tricantha**

b. Flowers less than 1 cm wide; perigynous zone hairless or sparsely downy   **10. alleghaniensis**

100a. Mature leaves 6 cm or less   **38. incisa**

b. Mature leaves more than 6 cm   101

101a. Bracts on fully-developed inflorescence 1 cm or more   102

b. Bracts of fully developed inflorescence less than 1 cm   103

102a. Petals falling early; flowers solitary or paired   **39. apetala**

b. Petals not falling early; flowers 5 or 6 in racemes   **48. × yedoensis**

103a. Flowers double or semi-double, usually more than 3 cm wide   **53. sieboldii**

b. Flowers single, usually less than 3 cm wide   **51. serrulata** var. **pubescens**

1. P. spinosa Linnaeus. Illustration: Ross-Craig, Drawings of British plants **8**: t. 1 (1955); Masefield et al., The Oxford book of food plants, 67 (1969); Phillips, Trees in Britain, Europe and North America, 176 (1978); Krüssmann, Manual of cultivated broad-leaved trees and shrubs **3**: pl. 9 & f. 25 (1986).

Deciduous shrub or small tree to 4 m, often suckering; young shoots with thorns, downy when young. Leaves mostly 1–3 cm, rolled in bud, broadest above the middle, obtuse or very shortly acute, wedge-shaped at the base, more or less hairy; stalks 2–10 mm, shortly hairy. Flowers to 1.5 cm across, solitary or 2 together, white, on hairless stalks to 6 mm. Style hairless. Fruit 1–1.5 cm across, spherical, blue-black, bloomed, with bitter green flesh. Stone nearly spherical. *Europe to W Siberia, Mediterranean area.* H1. Early spring.

Used for hedging, and, in double- and pink-flowered, purple-leaved variants ('Plena', 'Purpurea') for ornament. For making sloe gin the fruit is gathered in the wild.

2. P. × domestica Linnaeus.
Deciduous shrub or small tree; young shoots downy or hairless, not shiny; thorns few or none. Leaves 4–10 cm, rolled in bud, broadest above the middle, usually acute, wedge-shaped at the base, hairy beneath; borne horizontally or drooping; veins diverging at an angle of *c.* 45°; stalks 5–25 mm. Flowers white, 1.5–2 cm across, appearing with the leaves, solitary or 2–3 together on downy or hairless stalks 5–20 mm. Fruit 2–8 cm. *Europe, W Asia.* H1. Spring.

Probably a hybrid between *P. spinosa* and *P. cerasifera* var. *divaricata*; we follow Sell (*Nature in Cambridgeshire* **33**: 29–39, 1991 & **34**: 59–60, 1992) in taking a narrow view of *P. spinosa* and *P. cerasifera* and including deviating variants in *P. × domestica*.

Three subspecies are recognised, but there are many intermediates between them.

Subsp. **domestica**. Illustration: Boswell, English Botany, edn 3, **3**: t. 410 (1886); Hegi, Illustrierte Flora von Mitteleuropa **4**: f. 1275k (1923); Masefield et al., The Oxford book of food plants, 69, f. 2–6 (1969). Thornless tree to 12 m, with the leaves hairy on both sides, the fruits 4–8 cm, nearly spherical to oblong-ovoid, sweet, red or purple and the stone strongly flattened, sharply angled and free from the flesh.

The plum, widely cultivated as a fruit crop.

Subsp. **institia** (Linnaeus) Bonnier & Layens (*P. institia* Linnaeus; *P. × damascena* Dierbach). Illustration: Boswell, English Botany, edn 3, **3**: t. 409 (1886); Taylor, Plums of England, 33 (1949); Masefield et al., The Oxford book of food plants, 67, f. 5, 6 (1969). Shrub or tree to 6 m, often thorny, leaves hairy on both sides, the fruits usually 2–3 cm, spherical or shortly ovoid, sweet or acid, yellow, red or purple, the stone not strongly flattened but angled and adhering to the flesh.

The damson or bullace (fruit flesh acid) or mirabelle (fruit flesh sweet).

Subsp. **italica** (Borkhausen) Gams & Hegi (*P. × italica* Borkhausen). Illustration: Hegi, Illustrierte Flora von Mitteleuropa **4**: f. 1275i & l (1923); Taylor, Plums of England, 73, 129 (1949); Masefield et al., The Oxford book of food plants, 69, f. 5, 6 (1969). Small tree, intermediate in size between subsp. *domestica* and subsp. *institia*, with the leaves hairless above and the fruit 3–5 cm, almost spherical, sweet, green, the stone neither strongly flattened nor sharply angled, adhering to the flesh.

The greengage, probably of garden origin.

3. P. cerasifera Ehrhart (*P. myrobalana* Loiseleur). Illustration: Botanical Magazine, 6519 (1880); Taylor, Plums of England, 104 (1949); Masefield et al., The Oxford book of food plants, 67, f. 1 (1969); Phillips, Trees in Britain, Europe and North America, 171 (1978).

Deciduous shrub or small tree to 8 m, without suckers; young shoots hairless, remaining green into the second year (purple in some cultivars), occasionally with spines. Leaves 3–7 cm, rolled in bud, broadest below the middle, acute or acuminate, base rounded or wedge-shaped, hairless above, hairy only on the midrib and veins beneath; borne horizontally or drooping; veins diverging at an angle of *c.* 45°; stalks 5–10 mm, hairy. Flowers 2–2.5 cm across appearing with the leaves, usually solitary, on stalks 0.5–2.5 mm. Fruit spherical, yellow or red, sweet or insipid, spherical, 2–2.5 cm across; stone nearly circular in outline and slightly flattened. *SE Europe, SW Asia*. H2. Spring.

Var. **divaricata** (Ledebour) Bailey (*P. monticola* Koch) is the species as found in the wild; it has a more slender habit, with looser branching and the leaves are rounded at the base; the flowers are small, white and the fruits are yellow, to 2 cm across. It is rarely cultivated.

The cultivated plant is usually found as dark-leaved cultivars grown as small ornamental trees or for hedging. These are: 'Pissardii' (var. *atropurpurea* Jaeger). Illustration: Nicholson & Clapham, The Oxford book of trees, 180 (1975); Mitchell & More, The complete guide to trees of Britain and northern Europe, 79 (1985). This has pink buds, pinkish white petals and blackish purple leaves. 'Nigra'. Illustration: Nicholson & Clapham, The Oxford book of trees, 180 (1975); Mitchell & More, The complete guide to trees of Britain and northern Europe, 79 (1985). Petals pink, leaves dark red.

P. × blireana André. *P. cerasifera* × *P. mume*. Illustration: Revue Horticole for 1905: 392 (1905); Macoboy, What tree is that? 177 (1979). Like *P. cerasifera* but with broader, purple, acuminate leaves, double pink flowers on stalks 1–1.5 cm, and a downy ovary. *Garden origin*.

4. P. cocomilia Tenore (*P. pseudoarmeniaca* Heldreich & Sartorelli). Illustration: Polunin, Flowers of Greece and the Balkans, pl. 19 (1980); Pignatti, Flora d'Italia **1**: 616 (1982).

Like *P. cerasifera* but plant spiny, hairless, with leaves to 4 cm, rounded or obtuse, flowers usually 2 together, to 1 cm across on stalks no longer than the perigynous zones, fruit ⅓ longer than wide, yellow. *S Italy, Sicily, S Balkan Peninsula*. H5. Spring.

5. P. salicina Lindley (*P. triflora* Roxburgh). Illustration: Revue Horticole for 1895, 160 (1895); Krüssmann, Manual of cultivated broad-leaved trees and shrubs **3**: f. 24, 25 (1986).

Deciduous tree to 10 m; young shoots hairless, becoming red-brown and shiny. Leaves 6–10 cm, rolled in bud, oblong-obovate to elliptic-obovate, acute or acuminate, wedge-shaped at the base, glossy, doubly round-toothed; stalks 1–2 cm. Flowers 1.5–2 cm across, solitary to 3 together on hairless stalks 1–1.5 cm. Sepals slightly toothed, hairless. Petals white. Style hairless. Fruit 5–7 cm, almost spherical though with a depression at the attachment, green, yellow or red, with a deep furrow on the side. *China, Korea*. H2. Spring.

The Japanese plum, grown commercially for its fruit outside Europe.

6. P. consociiflora Schneider. Illustration: Botanical Magazine n.s., 805 (1980).

Like *P. salicina* but leaves to 7 cm, folded in bud, with glandular teeth; flowers to 1.2 cm across, 2–5 together. *C China*. H2. Spring.

7. P. simonii Carrière. Illustration: Revue Horticole for 1872, 110 (1872); The Garden **70**: 225 (1906); Krüssmann, Manual of cultivated broad-leaved trees and shrubs **3**: f. 23 (1986).

Small deciduous tree with erect branches; young shoots hairless. Leaves 7–10 cm, rolled in bud, narrowly ovate to oblong-obovate, acuminate, wedge-shaped or rounded at the base, finely bluntly toothed, hairless; borne upright; veins few, diverging at a very acute angle. Flowers 2–2.5 cm across, white, solitary or to 3 together on stalks less than 1 cm. Style hairless. Fruit to 3 × 5 cm, with a deep furrow on the side, dark red with sweet, fragrant, yellow flesh. Stone small, nearly circular in outline, rough, adhering to the flesh. *N China*. H2. Spring.

The apricot plum, cultivated for its fruit in areas free from frost.

8. P. subcordata Bentham. Illustration: Sargent, The silva of North America **4**: 154 (1892); Hitchcock et al., Vascular plants of the Pacific Northwest **3**: 163 (1961); Krüssmann, Manual of cultivated broad-leaved trees and shrubs **3**: f. 11 (1986).

Deciduous shrub or small tree to 8 m, branches spreading; young shoots hairy or hairless, at first bright red, later greyish brown. Leaves 3–7 cm, rolled in bud, broadly ovate to almost circular, sometimes weakly cordate, obtuse, sharply and often doubly toothed, hairy beneath at least when young; stalks 1–2 cm. Flowers to 1.5 cm across, 2–4 together on stalks to 1.2 cm. Sepals downy on both sides. Petals white becoming pink. Style hairless. Fruit 1.5–3 cm, ellipsoid, dark red, bloomed, variably acid. *W USA*. H4. Spring.

The Pacific or Oregon plum, cultivated for its fruit in California.

9. P. maritima Marsh (*P. acuminata* Michaux; *P. pubescens* Pursh). Illustration: Botanical Magazine, 8289 (1909); Gleason, Illustrated Flora of the north-eastern United States and adjacent Canada **2**: 332 (1952).

Straggling shrub to 2 m, lower branches often arching downwards; young shoots hairy. Leaves 4–7.5 cm, folded in bud, ovate, elliptic or obovate, acute, broadly wedge-shaped at base, finely saw-toothed, hairless above, downy beneath; stalks 4–6 mm, downy. Flowers 1.2–1.5 cm across, in umbels of 2–10, on stalks 5–7 mm. Sepals often toothed, downy on both sides. Petals white. Style hairless. Fruit 1.5–2 cm, spherical, red to purple, bloomed, with a deep furrow on the side; stone ovoid in outline, flattened. *E USA*. H2. Spring.

The Beach plum, grown for its fruit in North America and for ornament in Europe.

P. × dunbarii Rehder. *P. americana* × *P. maritima*. Similar, but shoots ultimately hairless, leaves larger, more coarsely toothed, acuminate, less downy beneath; fruits larger, purple, the stone more compressed. *Garden origin*. H2. Spring.

10. P. alleghaniensis Porter. Illustration: Sargent, Silva of North America **3**: t. 153 (1892); Gleason, Illustrated Flora of the north-eastern United States and adjacent Canada **2**: 333 (1952).

Straggling shrub or tree to 6 m; shoots becoming reddish brown, sometimes downy at first, sometimes with thorns. Leaves rolled or folded in bud, 6–9 cm, narrowly elliptic to lanceolate, acuminate, wedge-shaped at the base, finely saw-toothed, hairy at least when young; stalks 7–12 mm, downy, rarely with glands. Flowers 1–1.2 cm across, appearing with the leaves, 2–5 together on stalks 5–16 mm. Sepals more or less hairy. Petals white becoming pinkish. Style hairless. Fruit 8–16 mm across, almost spherical, dark purple, bloomed; stone 6–12 mm, ovoid. *E USA*. H2. Spring.

11. P. americana Marsh (*P. lanata* (Sudworth) Mackenzie & Bush). Illustration: Sargent, The silva of North America **3**: t. 150 (1892); Sargent, Manual of the trees of North America, edn 2, 562, 563 (1922); Gleason, Illustrated Flora of the north-eastern United States and adjacent Canada **2**: 332 (1952).

Deciduous small shrub or tree to 8 m, suckering and forming thickets; trunk divided 1–2 m above the ground, branches pendulous at tips; twigs with thorns. Leaves 5–12 cm, folded in bud, narrowly to broadly ovate or obovate, abruptly or gradually acuminate, broadly tapered or rounded at the base, finely and acutely saw-toothed, hairless or with hairs along the midrib beneath; stalks usually without glands. Flowers 2–3 cm across, 2–5 together on stalks 8–16 mm. Sepals sometimes hairy, often toothed but almost without glands on the teeth, reflexed at the end of flowering. Petals white. Style hairless. Fruit 2–3 cm, spherical or almost so, red or yellowish; stone not or slightly flattened. *E & C USA*. H2. Spring.

12. P. mexicana Watson (*P. arkansana* Sargent). Illustration: Sargent, Manual of the trees of North America, edn 2, 565 (1922); Steyermark, Flora of Missouri, 859 (1963).

Like *P. americana* but a non-suckering tree to 12 m, leaf-stalks with glands, leaves slightly cordate at base, downy beneath when mature, flowers 1.5–2 cm across; fruit purplish. *SC USA & N Mexico*. H3. Spring.

13. P. nigra Aiton (*P. americana* Marsh var. *nigra* Wanghenhein; *P. borealis* Poiret; *P. emarginata* (J.D. Hooker) Eaton). Illustration: Botanical Magazine, 1117 (1808); Sargent, Silva of North America 3: t. 149 (1892); Sargent, Manual of the trees of North America, edn 2, 561 (1922); Rosendahl, Trees and shrubs of the upper Midwest, edn 2, 253 (1955).

Like *P. americana* but narrow-crowned, with upright branches; leaf-stalks 1.2–1.5 cm, with glands near the top, leaf-teeth obtuse and gland-tipped, flower-stalks, perigynous zone and sepals reddish, sepals gland-toothed, petals sometimes becoming pink; fruit more ovoid, stone flattened. *E Canada, E & NC USA*. H3. Spring.

14. P. hortulana Bailey. Illustration: Sargent, The silva of North America 3: t. 151 (1892); Sargent, Manual of the trees of North America, edn 2, 568 (1922); Steyermark, Flora of Missouri, 859 (1963).

Tree to 10 m, not suckering; young shoots hairless. Leaves 7–11 cm, folded in bud, rather thick, ovate-lanceolate to oblong-obovate, gradually acuminate, broadly tapered or rounded at the base, shallowly saw-toothed with incurved, gland-tipped teeth, downy beneath when young; stalks 1.5–3 cm with several glands. Flowers 1.2–1.5 cm across, appearing with the leaves, 2–4 together, on stalks *c*. 1.2 cm. Sepals glandular-hairy, downy inside and sometimes outside, usually finally reflexed. Petals white. Style hairless. Fruit spherical, 1.8–3 cm, red or yellow; stone not much flattened, pointed at both ends. *C USA*. H4. Spring.

15. P. munsoniana Wight & Hedrick. Illustration: Sargent, Manual of the trees

of North America, edn 2, 568 (1922); Gleason, Illustrated Flora of the northeastern United States and adjacent Canada 2: 333 (1952); Braun, The woody plants of Ohio, 218 (1961); Steyermark, Flora of Missouri, 859 (1963).

Like *P. hortulana* but suckering and forming thickets; leaves rolled (or folded?) in bud, leaf-teeth each with a gland on the side, flowers borne on short spurs, sometimes appearing before the leaves, fruit 1.5–2 cm, stone truncate or obliquely truncate at the base. *SC USA*. H2. Spring.

There is some confusion in the literature as to whether the young leaves of this species are rolled or folded in bud.

16. P. angustifolia Marsh. Illustration: Sargent, The silva of North America 3: 152 (1892); Sargent, Manual of the trees of North America, edn 2, 570 (1922); Gleason, Illustrated Flora of the northeastern United states and adjacent Canada 2: 333 (1952).

Deciduous shrub or small tree to 7 m, suckering and forming thickets; twigs flexuous, hairless, reddish, spiny. Leaves 2–5 cm × 5–20 mm, rolled in bud, lanceolate to oblong-lanceolate, acute, tapered to the base, hairless beneath except on the midrib, edges more or less incurved to give a boat-shape; stalk 6–12 mm, red, usually with glands at the top. Flowers 8–9 mm across, 2–4 together on stalks 3–12 mm. Sepals hairy inside at the base, without glands, finally reflexed. Petals white. Style hairless. Fruit to 1.2 cm, almost spherical, red or yellow; stone ovoid. *SE USA*. H4. Spring.

Var. **angustifolia** is described above.

Var. **watsonii** (Sargent) Bailey (illustration: Garden & Forest 7: 135, 1894), grows only to 4 m and has smaller leaves and flowers, finer leaf-teeth and a thicker-skinned fruit. *SC USA*. H4. Spring.

Var. *angustifolia* has not been grown successfully in most of Europe, whereas var. *watsonii* is sometimes seen.

17. P. reverchonii Sargent.

Like *P. angustifolia* but a shrub to 2 m with leaf-stalks to 1.2 cm, leaves bluntly toothed, to 7 cm, sepals glandular-hairy, fruit to 2 cm, and the stone pointed at both ends. *USA (Oklahoma & Texas)*. H2. Spring.

18. P. × orthosepala Koehne. Illustration: Garden & Forest 7: 187 (1894).

Like *P. angustifolia* but leaf-stalks to 1.8 cm, leaves to 7.5 × 3.5 cm, saw-toothed with incurved, thickened but rarely glandular teeth, acuminate, not boat-

shaped; flowers to 1.6 cm across, sepals densely hairy inside, petals becoming pink, fruit to 2.5 cm, dark blue, bloomed, stone flattened. *Garden origin*. H2. Spring.

P. americana × *P. angustifolia* var. *watsonii*.

19. P. brigantina Villars. Illustration: Krüssmann, Manual of cultivated broad-leaved trees and shrubs 3: f. 5 (1986).

Shrub or small tree to 6 m; young shoots hairless. Leaves rolled in bud, ovate to elliptic, shortly acuminate, truncate or slightly cordate at base, doubly toothed, hairy beneath at least on the veins. Flowers almost stalkless. Flowers white or pale pink, to 2 cm across, 2–5 in clusters. Fruit rounded, yellow, smooth. *SE France*. H5. Spring.

Long cultivated for the scented oil obtained from the seeds.

20. P. mandshurica (Maximowicz) Koehne. Illustration: Krüssmann, Manual of cultivated broad-leaved trees and shrubs 3: pl. 6 (1986).

Small wide shrub to 6 m (perhaps a much larger tree in the wild); young shoots dark red-brown. Leaves rolled in bud, broadly elliptic to ovate, abruptly acuminate, fringed with dense, long, narrow teeth to 2 mm or more, hairy in the vein-axils beneath. Flowers solitary, pale pink, to 3 cm across, stalks to 5 mm. Style or ovary (and fruit) hairy. Fruit spherical to wider than long, to 2.5 cm across, yellow, sour; stone small, smooth. *N China, Korea*. H1. Spring.

21. P. sibirica Linnaeus. Illustration: Krüssmann, Manual of cultivated broad-leaved trees and shrubs 3: pl. 6 & f. 5 (1986).

Small upright tree or shrub to 5 m. Leaves rolled in bud, ovate, blade length about 1.5 times the width, long-acuminate, point *c*. 1.5 cm long, broadly rounded to truncate at base, simply and finely toothed, reddish when young, hairless. Flowers solitary, almost stalkless, pink or white, to 3 cm across. Style or ovary (and fruit) hairy. Fruit almost spherical, 1.5–2 cm across, yellow with reddish patches; stone smooth, winged or angular. *E former USSR, N China*. H1. Spring.

Sometimes regarded as a variety of *P. armeniaca*.

22. P. armeniaca Linnaeus. Illustration: Gleason, Illustrated flora of the northeastern United States and adjacent Canada 2: 331 (1952); Krüssmann, Manual of cultivated broad-leaved trees and shrubs 3:

pl. 6 & f. 5, 24 (1986).

Round-crowned tree to 10 m or a smaller shrub; young shoots reddish brown, glossy, hairless. Leaves rolled in bud, broadly ovate or ovate-circular, broadly rounded to truncate at base; blade length about 1.5 times the width, abruptly acuminate, point to 1 cm, closely and obtusely toothed, with teeth to 1 mm. Flowers solitary or 2 together, pink, to 2.5 cm across, stalkless or stalks to 9 mm. Style and ovary hairy. Fruits yellow, often with a red patch, shortly hairy, to 3 cm across; stone large, smooth, with a furrowed margin. *N China.* H3. Spring.

The apricot, long cultivated for its fruit. Variegated ('Variegata') and pendulous-branched ('Pendula') variants are grown for ornament.

23. P. mume Siebold & Zuccarini. Illustration: Revue Horticole for 1885, t. 564 (1885); Krüssmann, Manual of cultivated broad-leaved trees and shrubs **3**: pl. 6 & f. 5, 16, 23 (1986).

Similar to *P. armeniaca*, but young shoots green, leaves ovate to elliptic, long-acuminate, blade length about 2 times the width, wedge-shaped to narrowly rounded at base. Flowers stalkless, white to dark pink, to 3 cm across, fragrant; fruit sour to bitter, not separating from the pitted stone. *S Japan.* H3. Spring.

24. P. × dasycarpa Ehrhart. Illustration: Krüssmann, Manual of cultivated broad-leaved trees and shrubs **3**: f. 5 (1986).

Like *P. armeniaca* but leaves hairy on the veins beneath, flower-stalks more than 9 mm, hairy, ovary and base of style hairy; fruit red or blackish purple, bloomed, sour. *Origin uncertain, long cultivated in W & C Asia.* H4. Spring.

P. armeniaca × *P. cerasifera*.

25. P. dulcis (Miller) D.A. Webb (*P. amygdalus* Batsch; *P. communis* Arcangeli). Illustration: Bonnier, Flore complète **3**: pl. 167 (1914); Lancaster, Trees for your garden, 106 (1974); Phillips, Trees in Britain, Europe and North America, 171 (1978); Edlin, The tree key, 145 (1978).

Upright tree to 10 m; young shoots hairless. Leaves folded in bud, narrowly elliptic or lanceolate, scalloped or finely toothed, rounded at base, widest below the middle, hairless. Flowers solitary or paired, stalkless or almost so, white or pale pink, 3–5 cm across. Style, ovary and fruit hairy. Fruit ovoid, flattened; stone pitted. *SW Asia, N Africa.* H4. Early spring.

The almond, long cultivated for its fruits

(the kernels of the stones). Three varieties occur in the wild and in commercial cultivation, but they are not important in gardens, where several cultivars are grown.

P. × persicoides Ascherson & Graebner (*P. × amygdalopersica* (Weston) Rehder, invalid) is the hybrid between *P. dulcis* and *P. persica*. Illustration: Macoboy, What tree is that? 176 (1979). It has occasional thorns, more finely toothed leaves and flowers 4–5 cm across, pink with a darker centre. Its naming is complicated and the name used above may prove not to be correct.

26. P. fenzliana Fritsch. Illustration: Krüssmann, Manual of cultivated broad-leaved trees and shrubs **3**: pl. 12 & f. 10 (1986).

Like *P. dulcis* but often thorny and with the young wood with a greyish or whitish bloom; leaves widest near the base, greyish or bluish green. *Former USSR (Caucasus).* H1. Spring.

27. P. persica (Linnaeus) Batsch. Illustration: Bonnier, Flore complète **3**: pl. 166 (1914); Edlin, The tree key, 145 (1978); Everett, New York Botanical Garden illustrated encyclopedia of horticulture **8**: 2822 (1981); Macdonald encyclopedia of shrubs and trees, 127 (1988).

Tree to 8 m; young shoots hairless. Leaves folded in bud, narrowly elliptic or lanceolate, finely toothed, tapered at base, widest near the middle; angle between side-veins and midrib of leaf obtuse. Flowers solitary, stalkless or almost so, 2.5–3.5 cm across, pink or red. Sepals, style, ovary and fruit hairy. Fruit almost spherical, to 7 cm across. *N & C China.* H1. Late spring.

The peach, widely cultivated for its fruit since ancient times. Var. **nucipersica** Schneider, the nectarine, also long-cultivated, has hairless fruits; it is unknown in the wild.

28. P. kansuensis Rehder. Illustration: Journal of the Royal Horticultural Society **84**: 165 (1959).

Like *P. persica* but young shoots greenish, yellowish or purplish; flowering earlier. *NW China.* H1. Early spring.

29. P. davidiana (Carrière) Franchet. Illustration: Krüssmann, Manual of cultivated broad-leaved trees and shrubs **3**: pl. 6 & f. 10 (1986); The Garden **112**: 228 (1987).

Tree to 10 m; young shoots hairless. Leaves folded in bud, narrowly elliptic

or lanceolate, finely toothed with acute, forward-pointing teeth 2 mm or more apart. Flowers solitary, stalkless or almost so, pale pink, to 2.5 cm across. Style and ovary hairy. Fruit hairy, yellowish, spherical, to 3 cm across; stone pitted. *NW China.* H1. Spring.

30. P. mira Koehne. Illustration: Botanical Magazine, 9548 (1939).

Tree to 10 m; young shoots hairless. Leaves folded in bud, narrowly elliptic or lanceolate, sparsely scalloped, rounded at base, widest below the middle, hairy beside midrib beneath. Flowers solitary or 2 together, stalkless or almost so, white, 2–2.5 cm across. Style and ovary hairy. Fruit hairy, yellowish, spherical, to 3 cm across; stone smooth. *W China.* H1. Spring.

31. P. tangutica (Batalin) Koehne. Illustration: Botanical Magazine, 9239 (1931).

Dwarf shrub to 4 m; young wood with a greyish or whitish bloom; often thorny. Leaves folded in bud, to 2 cm wide, dark greyish green, widest near middle, not white-felted beneath; stalks green. Flowers pink or red, solitary, stalkless, to 2.5 cm across. Sepals reddish, not or scarcely toothed. Petals overlapping. Style or ovary hairy. Fruit hairy, to 2 cm, across, with thin flesh, keeled. *W China (Sichuan).* H1. Spring.

32. P. triloba Lindley. Illustration: Botanical Magazine, 8061 (1906); Hillier colour dictionary of trees and shrubs, 172 (1981); Everett, New York Botanical Garden illustrated encyclopedia of horticulture **8**: 2821 (1981).

Shrub or small tree to 5 m; young shoots hairless or downy. Leaves folded in bud, broadly elliptic or ovate, sometimes 3-lobed, with large, broad, triangular teeth, hairy beneath. Stipules narrowly linear, fringed, persistent or sometimes falling early. Flowers pink, 1.8–3.5 cm wide, often double, solitary or 2 together, stalkless or stalks to 9 mm. Style, ovary (and fruit) hairy. Fruit almost spherical, to 1.5 cm across, red. *China.* H1. Spring.

Cultivated material generally has double flowers; the single-flowered variant (var. **simplex** (Bunge) Rehder) is rarely seen.

33. P. tenella Batsch. Illustration: Botanical Magazine, 161 (1791); Polunin, Flowers of Greece and the Balkans, pl. 19 (1980); Reader's Digest encyclopaedia of garden plants and flowers, 556 (1985).

Branched, upright shrub to 1.5 m; young wood with a greyish or whitish bloom, not thorny. Leaves folded in bud, obovate to oblanceolate, to 2 cm wide, not white-felted beneath, all similar. Flowers solitary or to 3 together, stalkless, pink to red, to 3 cm wide. Sepals greenish, densely toothed. Petals not overlapping. Style and ovary hairy. Fruit hairy, yellowish, to 2 cm; stone rough. *C Europe to former USSR (Siberia).* H1. Spring.

A few cultivars are available.

34. P. avium Linnaeus. Illustration: Bonnier, Flore complète **3**: pl. 167 (1914); Phillips, Trees in Britain, Europe and North America, 170 (1978); Edlin, The tree key, 147 (1978); Macdonald encyclopedia of shrubs and trees, 126 (1988).

Tree to 20 m (rarely more); young shoots hairless. Leaves folded in bud, ovate-oblong, toothed, with teeth coarse, narrow, rounded or obtuse, standing out from leaf-margin and at least 2 mm long. Flower-clusters clearly stalked or stalkless, without leaf-like bracts (large bud-scales may be present at base of cluster); flower-stalks arising at one or more points (stalked umbels or corymbs), or all arising from the same point in each cluster (stalkless umbels). Flowers 2 cm wide or more, white. Sepals becoming reflexed. Fruit dark red. *Europe, N Africa, SW Asia, former USSR (W Siberia).* H1. Spring.

'Plena' is a widely cultivated variant with double flowers.

P. × schmittii Rehder is the hybrid *P. avium × P. canescens*; it is intermediate between the parents but has glossy, brown bark with conspicuous lenticels.

P. × fontanesiana (Spach) Schneider is *P. avium × P. mahaleb*; similar to *P. avium* but with softly hairy young shoots.

35. P. cerasus Linnaeus. Illustration: Gleason, Illustrated flora of the north-eastern United States and adjacent Canada **2**: 330 (1956); Vedel & Lange, Trees and bushes in wood and hedgerow, 83 (1960); Brosse, Arbres de France et d'Europe occidentale, 161 (1977).

Tree to 10 m, suckering, hairless or with a few hairs on the veins of the leaves beneath. Leaves folded in bud, elliptic to ovate, green when mature, with obtuse or rounded teeth. Flowers in dense clusters borne with the leaves, stalks 1–3.5 cm; clusters lacking a distinct stalk but with leaf-like bracts; flower-stalks all arising from the same point in each cluster (stalkless umbels). Flowers 2 cm wide or more. Sepals becoming reflexed. Fruit depressed-spherical, reddish black. *S Europe to N India.* H3. Spring.

A very variable species, with numerous varieties and cultivars available.

P. × eminens Beck (*P. reflexa* misapplied) is the hybrid *P. cerasus × P. fruticosa*, which is intermediate between its parents.

36. P. fruticosa Pallas. Illustration: Hegi, Illustrierte Flora von Mitteleuropa **4**: t. 157 (1923); Jávorka & Csapody, Iconographia Flora Hungarica, 263 (1931); Krüssmann, Manual of cultivated broad-leaved trees and shrubs **3**: pl. 6 & f. 8 (1986).

Spreading shrub to 1 m; young shoots grey or black, hairless. Leaves folded in bud, scalloped, with obtuse or rounded teeth with dark marginal glands, green when mature, elliptic to obovate, less than 6 cm, narrowly tapered at base. Flowers 2–4 in stalkless umbels, white, to 1.5 cm across; flower-stalks 1.5–2.5 cm. Sepals becoming reflexed. Fruit dark red, more or less spherical. *C Europe to W Siberia.* H1. Spring.

37. P. canescens Bois. Illustration: Krüssmann, Manual of cultivated broad-leaved trees and shrubs **3**: pl. 7 (1986).

Shrub to 2 m, branchlets hairy, young wood greenish, reddish, or purplish, not bloomed. Leaves folded in bud, broadly elliptic or ovate, acuminate, coarsely toothed, with large, broad triangular teeth, sometimes 3-lobed, bristly above, not white-felted beneath, all similar. Stipules broadly elliptic to circular, persistent. Flowers 2–5 in corymbs, pale pink, to 1.2 cm across, petals not opening widely, hairy towards the base; bracts leaf-like; flowers stalkless or stalks to 9 mm. Perigynous zone cylindric. Style, ovary (and fruit) hairy. Fruit spherical, to 1 cm across, red. *W China (Sichuan, Hubei).* H2. Late spring.

P. × dawyckensis Sealy. Illustration: Botanical Magazine, 9519 (1938); Krüssmann, Manual of cultivated broad-leaved trees and shrubs **3**: pl. 7 (1986). Similar to *P. canescens* but much hairier, petals notched, fruits ellipsoid, yellow-red. *China.*

Thought by Rehder and others to be the natural hybrid *P. canescens × P. dielsiana*, but accepted as a species by Krüssmann.

38. P. incisa Thunberg. Illustration: Botanical Magazine, 8954 (1923); Bean, Trees and shrubs hardy in the British Isles, edn 8, **3**: t. 46 (1976); Manual of Japanese flowering cherries, 131, 132 (1982); Brickell, RHS gardeners' encyclopaedia of plants and flowers, 59 (1989).

Usually a rounded shrub, more rarely tree-like and to 10 m. Leaves folded in bud, purplish when young, 6 cm or less, ovate to obovate, acute, teeth coarse, at least 3 mm long; stalks of mature leaves densely downy or hairy. Flowers solitary or to 3 in clusters, white to pale pink, to 2.5 cm across, perigynous zone cylindric; flower-stalks more than 9 mm. Sepals not becoming reflexed. Petals soon falling. Fruit ovoid, black-purple, to 8 mm across. *Japan.* H2. Spring.

P. × hillieri Hillier is the hybrid between *P. incisa* and *P. sargentii*. It is intermediate between the parents, forming a small, densely branched tree which gives good autumn colour.

39. P. apetala Franchet & Savatier. Illustration: Manual of Japanese flowering cherries, 121 (1982); Krüssmann, Manual of cultivated broad-leaved trees and shrubs **3**: pl. 7 & f. 26 (1986).

Like *P. incisa* but mature leaves more than 6 cm long, flowers solitary or in pairs, bracts on fully developed inflorescence 1 cm or more. *Japan.* H2. Spring.

40. P. mugus Handel-Mazzetti. Illustration: Krüssmann, Manual of cultivated broad-leaved trees and shrubs **3**: pl. 7 (1986).

Like *P. incisa* but stalks of mature leaves at most very sparsely hairy; leaf-tips round or obtuse, rarely acute. Perigynous zone tubular-campanulate. *W China (Xizang).* H1. Spring.

41. P. nipponica Matsumura (*P. nikkoensis* Koehne). Illustration: Makino, New illustrated flora of Japan, t. 1160, 1161 (1963); Kurata, Illustrated important forest trees of Japan **2**: 47 (1971); Manual of Japanese flowering cherries, 311 (1982).

Tree to 6 m; young shoots hairless. Leaves ovate or rarely obovate, folded in bud, teeth coarse, fewer than 4 per cm, acute or obtuse, not hair-like; stalks of mature leaves at most very sparsely hairy. Flowers produced with the leaves in April and May; bracts persistent. Flowers solitary or to 3 together, white or pale pink, to 2.5 cm across, perigynous zone tubular-campanulate; flower-stalks more than 9 mm. Sepals not becoming reflexed. Petals falling early. Fruit spherical, black, to 1 cm across. *Japan.* H1. Late spring.

P. kurilensis Miyabe (*P. nipponica* var. *kurilensis* (Miyabe) Wilson). Illustration: Manual of Japanese flowering cherries, 311 (1982). This has somewhat larger leaves and hairy leaf- and flower-stalks. *Japan (Kurile Islands)*.

42. P. subhirtella Miquel. Illustration: Phillips, Trees in Britain, Europe and North America, 175 (1978); Krüssmann, Manual of cultivated broad-leaved trees and shrubs **3**: pl. 8, 17 & f. 26 (1986); Macdonald encyclopedia of shrubs and trees, 129 (1988).

Tree to 10 m; young shoots hairy. Leaves folded in bud, broadly elliptic or ovate, sometimes 3-lobed, with large, broad triangular teeth, hairless beneath except on midrib and veins. Stipules narrowly linear, fringed, falling by midsummer. Flowers 2–5 together in clusters, white or pink, to 2 cm across, perigynous zone cylindric, stalkless or stalks to 9 mm. Style, ovary (and fruit) hairy. Fruit to 1 cm across, black. *Japan*. H1. Autumn–spring.

'Autumnalis' is a selection with widely angled branches and double white flowers, which flowers from early winter to spring.

43. P. campanulata Maximowicz. Illustration: Botanical Magazine, 9575 (1939) – as P. cerasoides var. campanulata; Bean, Trees and shrubs hardy in the British Isles, edn 8, **3**: 357 (1976); Macoboy, What tree is that? 177 (1979); Manual of Japanese flowering cherries, 122 (1982).

Shrub or small tree to 8 m; young shoots hairless. Leaves folded in bud, elliptic-ovate to ovate-oblong, teeth 1 mm or less. Flowers 2–5 in stalked umbels, pink or red, borne before or with the leaves; perigynous zone cylindric; stalks of flower-clusters 2.5 mm or more; stalks of flowers more than 9 mm. Sepals becoming reflexed. Sepals and perigynous zone hairless. Fruit ovoid, red, to 1.5 cm, stone smooth. *Japan, Taiwan*. H4. Spring.

44. P. cerasoides D. Don. Illustration: Polunin & Stainton, Flowers of the Himalaya, pl. 33 (1984); Manual of Japanese flowering cherries, 122 (1982); Krüssmann, Manual of cultivated broad-leaved trees and shrubs **3**: pl. 7 & f. 22 (1986); The Garden **112**: 225 (1987).

Like *P. campanulata* but leaves more leathery, ovate-rounded, more sharply toothed; fruit pointed. *Himalaya*. H4. Spring.

Possibly no longer in cultivation (though the name may be found misapplied in catalogues).

45. P. rufa J.D. Hooker. Illustration: Journal of the Royal Horticultural Society **102**: 354 (1977); Polunin & Stainton, Flowers of the Himalaya, pl. 33 (1984); Krüssmann, Manual of cultivated broad-leaved trees and shrubs **3**: pl. 7 & f. 26 (1986).

Small tree to 7 m; young shoots reddish hairy. Leaves narrowly ovate to obovate-lanceolate, folded in bud, to 5 cm, toothed, acuminate, midribs ginger-hairy beneath, stalks of mature leaves at most very sparsely hairy. Flowers usually solitary, white to pale pink, to 1.5 cm across; stalks more than 9 mm. Sepals not becoming reflexed, perigynous zone tubular, not hairy. Fruit broadly ellipsoid, dark red. *Himalaya*. H4. Spring.

Var. **tricantha** (Koehne) Hara. Inflorescence (flower-stalks, perigynous zone) hairy. *N India*.

46. P. serrula Franchet. Illustration: Bean, Trees and shrubs hardy in the British Isles, edn 8, **3**: t. 47 (1976); Phillips, Trees in Britain, Europe and North America, 174 (1978); Hillier colour dictionary of trees and shrubs, 175 (1981); Krüssmann, Manual of cultivated broad-leaved trees and shrubs **3**: pl. 17 (1986).

Tree to 10 m, bark glossy red, peeling in strips, in old plants scarred with rough, horizontal lenticels; young shoots downy. Leaves folded in bud, lanceolate, acuminate, teeth on mature leaves very fine and regular, at least 5 major teeth per cm; stalks of mature leaves at most very sparsely hairy. Flowers solitary or to 3 together, white, to 2 cm across, borne with the leaves; perigynous zone cylindric; stalks more than 9 mm. Sepals not becoming reflexed. Style hairy below. Fruits ovoid, to 7 mm. *W China*. H2. Late spring.

47. P. hirtipes Hemsley (*P. conradinae* Koehne; *P. helenae* Koehne). Illustration: The Garden **87**: 97 (1923); Krüssmann, Manual of cultivated broad-leaved trees and shrubs **3**: pl. 7 (1986).

Tree to 8 m, or smaller shrub. Leaves folded in bud, ovate to oblong-obovate, long-acuminate, green when mature; teeth 4–6 teeth per cm or fewer than 4 per cm, acute, coarse, standing out from the margin, not hair-like; stalks of mature leaves at most very sparsely hairy. Flowers solitary or to 4 together, white to pale pink, borne before the leaves from January–March; perigynous zone cylindric; bracts falling as flowers open. Sepals not becoming reflexed. Fruits ovoid, red. *C China*. H4. Winter–early spring.

48. P. × yedoensis Matsumura. *?P. subhirtella × P. speciosa*. Illustration:

Phillips, Trees in Britain, Europe and North America, 175 (1978); Everett, New York Botanical Garden illustrated encyclopedia of horticulture **8**: 2823 (1981); Krüssmann, Manual of cultivated broad-leaved trees and shrubs **3**: pl. 1, 6, 21 (1986); Brickell, RHS gardeners' encyclopedia of plants and flowers, 60 (1989).

Tree to 15 m; young shoots downy. Leaves folded in bud, elliptic, acuminate, more than 6 cm, teeth on mature leaves coarse, at least 3 mm long; stalks of mature leaves densely downy or hairy. Flowers 5 or 6 in stalked racemes, pinkish at first, becoming white, to 3.5 cm across; perigynous zone cylindric. Sepals not becoming reflexed. Petals not falling early. Fruits spherical, to 8 mm across, black. *Garden origin*. H2. Spring.

49. P. sargentii Rehder. Illustration: Phillips, Trees in Britain, Europe and North America, 172 (1978); Hillier colour dictionary of trees and shrubs, 175 (1981); Krüssmann, Manual of broad-leaved trees and shrubs **3**: pl. 2, 18 & f. 20 (1986); Brickell, RHS gardeners' encyclopedia of plants and flowers, 39 (1989).

Tree to 20 m; young shoots hairless or sparsely hairy. Leaves folded in bud, oblong-elliptic to obovate-oblong, teeth on mature leaves coarse, fewer than 4 per cm, acuminate or drawn out into hair-like tips; stalks of mature leaves at most very sparsely hairy; stems and leaves hairless or very slightly hairy. Flowers 2–4 in stalkless umbels, pink, to 4 cm across; perigynous zone tubular-campanulate; stalks more than 9 mm. Sepals not becoming reflexed. Petals notched at the apex. Fruits ovoid or oblong-ovoid, dark red. *Japan, former USSR (Sakhalin), Korea*. H1. Spring.

50. P. × juddii Anderson. (*P. sargentii × P. × yedoensis*).

Like *P. sargentii* but young leaves copper-coloured, flowers larger, deeper pink, fragrant. *Garden origin*.

51. P. serrulata Lindley. Illustration: Botanical Magazine, 8012 (1905); Wilson, The cherries of Japan, pl. 6 (1916); Chadbund, Flowering cherries, pl. 2 (1972).

Tree to 10 m, though usually much smaller in cultivation; young shoots hairy or hairless. Leaves folded in bud, narrowly ovate, glossy, long acuminate; teeth on mature leaves coarse, fewer than 4 per cm, acuminate or drawn out into hair-like tips; stalks of mature leaves at most very sparsely hairy. Stems and leaves hairless or very slightly hairy. Flowers 3–5 together in racemes, white or pink, single or double; perigynous zone cylindric; stalks more than 9 mm. Sepals not becoming reflexed. Fruits ovoid, black, shining. *China, Japan, Korea*. H1. Early–late spring.

A highly variable species, generally considered to consist of 3 varieties (though it is uncertain which of these are genuinely in cultivation) and a large group of cultivars, long-cultivated in Japan and now widely grown throughout the temperate world.

Var. **spontanea** (Maximowicz) Wilson is a tall tree with grey-brown bark with large persistent lenticels, the young leaves copper to brown, hairless, flowers white, 2.5–3.5 cm across in shortly stalked corymbose clusters.

Var. **hupehensis** Ingram is similar to var. *spontanea* but the flowers appear earlier than the leaves, the young leaves are bronze and the flowers are in almost stalkless clusters.

Var. **pubescens** Wilson is similar to var. *spontanea* but the leaf-stalks and undersurfaces are hairy, the young leaves are green or slightly bronze and the flowers are in long-stalked corymbs.

The Japanese flowering cherries or Sato-Zakura Group (*P. lannesiana* Wilson (Carrière); *P. donarium* Siebold) are usually considered under this species. The group is of complex hybrid origin and no attempt is made here to list or describe the many cultivars. They vary greatly in habit (strictly upright, vase-shaped, round-headed or weeping), colour of the young leaves (green to dark red), type of inflorescence, degree of doubleness of the flowers (petals 5–50 or more), petal colour (from cream through white, pinkish white, pink and dark pink) and the presence of leaf- or petal-like bodies replacing the ovary.

The following key, translated and adapted from an original by Arie Peeters (Wageningen, The Netherlands), covers 42 of the most commonly encountered cultivars and the varieties described above. If in doubt at any stage, the user is recommended to follow both alternative leads. Other keys can be found in *The RHS dictionary of gardening* **2**: 1086 (edn 1, 1951) and in Krüssmann, *Manual of cultivated broad-leaved trees & shrubs* **3**: 47–48 (1986). References should also be made to Ingram, *Flowering Cherries* (1948) and *Manual of Japanese flowering cherries* (1982).

1a. Petals 20 or fewer   2
b. Petals more than 20   41
2a. Petals 5–10   3
b. Petals 10–20   25
3a. Petals 5 (occasionally 6–8)   4
b. Petals 6–10 (occasionally 5)   20
4a. Flowers pure white when fully open   5
b. Flowers pale pink or darker when fully open   14
5a. Flowers 4.8 cm wide or less   6
b. Flowers more than 4.8 cm wide   13
6a. Flowers less than 4 cm wide   7
b. Flowers 4–4.8 cm wide   10
7a. Leaf-stalks hairy   8
b. Leaf-stalks hairless   9
8a. Flowers 2.5–3 cm wide, opening in May; inflorescence-stalks 1.5 cm or more; young leaves green or bronze-green   var. **pubescens**
b. Flowers 3–3.7 cm wide, opening mid to late April; inflorescence-stalks usually 1 cm or less; young leaves coppery brown   **'Fudanzakura'**
9a. Young leaves coppery red to brown; flower clusters shortly stalked.   var. **spontanea**
b. Young leaves bronze; flower clusters almost stalkless   var. **hupehensis**
10a. Petaloids (small extra deformed petals) present in some flowers   11
b. Petaloids absent   12
11a. Conical tree with flattened top; flowers fragrant; petals 12–13 mm wide   **'Jo-nioi'**
b. Broadly columnar tree; flowers not very fragrant; petals 16–19 mm wide   **'Hatazakura'**
12a. Petals often ragged   **'Washino-o'**
b. Petals entire or slightly notched   **'Taki-nioi'**
13a. Flowers 5–5.5 cm wide, sometimes with more than 5 petals; young leaves bronze-green   **'Ojochin'**
b. Flowers 5.5–6 cm wide, rarely with more than 5 petals; young leaves brown   **'Taihaku'**
14a. Leaf-stalks hairy   15
b. Leaf-stalks hairless   16
15a. Flowers 2.5–3 cm wide, open in May; inflorescence-stalks 1.5 cm or more; young leaves green or bronze-green   var. **pubescens**
b. Flowers 3–3.7 cm wide, open mid to late April; inflorescence-stalks usually 1 cm or less; young leaves coppery brown   **'Fudanzakura'**

16a. Flowers less than 4.7 cm wide   17
b. Flowers 4.7 cm wide or more   19
17a. Flowers sparse, 2 or 3 per cluster, open mid April; petals *c*. 16 × 12 mm; growth rapid   **'Benden'**
b. Flowers abundant, 3–6 per cluster, open late April to early May; petals 18 × 13 mm or more; growth slow to medium   18
18a. Flowers 4–6 per cluster, open late April; petals *c*. 18 × 13 mm; growth slow   **'Taguiarashi'** (**'Ruiran'**)
b. Flowers 3 or 4 per cluster, open late April to early May; petals *c*. 19 × 16 mm; growth medium   **'Hizakura'** (name often misapplied to **'Kanzan'**)
19a. Flowers on short shoots; petals smooth; clusters less than 5 cm   **'Kirigayatsu'** (**'Mikuruma-gaeshi'**)
b. Flowers not on short shoots; petals folded and wavy; clusters often more than 5 cm   **'Ojochin'**
20a. Flowers creamy white to yellowish green   **'Ukon'**
b. Flowers white or pink   21
21a. Columnar tree with more or less erect inflorescences   **'Amanogawa'**
b. Conical or broadly conical tree with spreading or hanging inflorescences   22
22a. Flower-stalks and calyx purplish red; leaf-base wedge-shaped   **'Taoyame'**
b. Not as above   23
23a. Flowers barely fragrant, often with only 5 petals; flower-stalk not firm   **'Ojochin'**
b. Flowers very fragrant, rarely with 5 petals; flower-stalks firm   24
24a. Conical tree; flowers often slightly pink when fully open; style shorter than anthers; calyx-lobes clearly toothed   **'Ariake'**
b. Spreading tree; flowers pure white when fully open; style as long as anthers; calyx-lobes occasionally toothed   **'Shirotae'**
25a. Leaves covered in woolly hairs   **'Takasago'**
b. Leaves not covered in woolly hairs   26
26a. Columnar tree with more or less erect inflorescences   **'Amanogawa'**
b. Not as above   27
27a. Flowers cream-coloured to yellowish green   **'Ukon'**

b. Flowers white or pink   28
28a. Flowers-stalks and calyx purplish to
brownish red   29
b. Not as above   30
29a. Flowers usually with 1 style; petals
5–15, somewhat notched; leaves
broadest towards the tip; growth
rapid   **'Taoyame'**
b. Flowers sometimes with 1–14 extra
styles among the anthers; petals 14
or 15, unnotched; leaves narrow,
broadest towards the base; growth
slow   **'Horinji'**
30a. Flowers almost always with only
one perfect style and ovary   31
b. More than 50% of flowers either
with 2 perfect styles and ovaries,
or with 1 or 2 ovaries which are
leaf-like at base (sample at least 10
flowers)   38
31a. Flowers 5–5.5 cm wide   32
b. Flowers 3.5–4.8 cm wide   34
32a. Flowers with more than 13 petals,
barely fragrant, pale pink when
fully open   **'Sumizome'**
b. Flowers with fewer than 13 petals,
fragrant, white or very slightly pink
when fully open   33
33a. Conical tree; flowers often slightly
pink when fully open; style shorter
than anthers; calyx-lobes clearly
toothed   **'Ariake'**
b. Spreading tree; flowers pure white
when fully open; style as long as
anthers; calyx-lobes occasionally
toothed   **'Shirotae'**
34a. Flowers pure white when fully
open; young leaves green
'Albo Plena'
b. Flowers pale pink or pink when
fully open; young leaves bronze-
green to brownish bronze   35
35a. Petals 18–21 mm wide; calyx-
lobes obtuse, narrowed at base
'Uzuzakura' ('Hokusai')
b. Petals 10–17 mm wide; calyx-lobes
acute, not narrowed at base   36
36a. Petals scalloped, soft purplish pink
(like *P. sargentii*), 10–13 mm wide,
'Yae-marasaki'
b. Petals slightly notched, very pale
pink, 13–17 mm wide   37
37a. Inflorescence-stalks longer than 2.3
mm; calyx-lobes 8–9 mm; leaves
very finely toothed, the teeth hardly
visible in young leaves   **'Shujaku'**
b. Inflorescence-stalks less than 2.3
mm; calyx-lobes 6–7 mm; leaves
more coarsely toothed, the teeth
visible in young leaves
'Edozakura'

38a. More than 50% of flowers with 2
perfect styles and ovaries, if only 1
then never leaf-like at base (sample
at least 10 flowers)   **'Imose'**
b. More than 50% of flowers with 1
or 2 imperfect styles and ovaries
which are leaf-like at base   39
39a. Mature flower-buds conical to
cylindric, lacking stripes on petal
backs; flowers saucer-shaped when
fully open; leaf-like ovary bases not
hidden   **'Ichiyo'**
b. Mature flower-buds ovoid, with
dark pink stripes on petal backs;
flowers cup-shaped when fully
open; leaf-like ovary bases usually
hidden   40
40a. Flowers 4–4.5 cm wide, often with
unopened petals held in calyx-tube
'Itokukuri'
b. Flowers 4.8–5.3 cm wide, rarely
with unopened petals in calyx-tube
'Okikuzakura'
41a. Petals 20–50   42
b. Petals more than 50   54
42a. Leaf-stalks hairy   **'Taizanfukan'**
b. Leaf-stalks hairless   43
43a. Flowers pure white when fully
open, less than 4 cm wide, with 1
perfect style and ovary
'Albo Plena'
b. Not as above   44
44a. More than 50% of flowers with
two perfect styles and ovaries, if
only one then never leaf-like at
base (sample at least 10 flowers)
'Imose'
b. Not as above   45
45a. More than 50% of flowers with
only one style and ovary, the latter
leaf-like only at base (sample at
least 10 flowers)   46
b. More than 50% of flowers with at
least 2 leaf-like ovaries   48
46a. Mature flower-buds conical to
cylindric, lacking stripes on petal
backs; flowers saucer-shaped when
fully open; leaf-like ovary-bases not
hidden   **'Ichiyo'**
b. Mature flower-buds ovoid, with dark
pink stripes on petal backs; flowers
cup-shaped when fully open; leaf-like
ovary-bases usually hidden   47
47a. Flowers 4–4.5 cm wide, often with
unopened petals held in calyx-tube
'Itokukuri'
b. Flowers 4.8–5.3 cm wide, rarely
with unopened petals in calyx-tube
'Okikuzakura'
48a. Flowers less than 4.4 cm wide
'Fugenzo'

b. Flowers more than 4.4 cm wide
49
49a. Flowering middle to end of May
(very late, after 'Kanzan'); flowers
almost white when fully open
(sometimes turning purplish pink
before falling); flower-clusters often
more than 10 cm   50
b. Not as above   51
50a. Young leaves green; flowers
remaining white
'Okumiyaku' ('Shimidsu')
b. Young leaves brown to bronze-
green; flowers often turning
purplish pink before falling
'Shirofugen'
51a. Flowers almost white when fully
open   **'Ichiyo'**
b. Flowers distinctly pink when fully
open (beware of 'Shirofugen, see
50)   52
52a. Flowers with 2 (rarely 3) leaf-like
ovaries (sample at least 10 flowers)
'Sekiyama' ('Kanzan')
b. Some flowers with 4 leaf-like
ovaries   53
53a. Many thin bare branches on crown;
calyx-lobes clearly toothed on
mature buds; anther connectives
elongated   **'Pink Perfection'**
b. No thin bare branches on crown;
calyx-tubes not toothed on mature
buds; anther connectives not
elongated   **'Daikoku'**
54a. Habit weeping   **'Kiku-shidare'**
b. Habit ascending   55
55a. Flowering late April to early May;
flowers less than 4 cm wide;
inflorescence-stalk less than 2 cm
'Geraldinae' ('Asano')
b. Flowering early to mid May; flowers
more than 4 cm wide;
inflorescence-stalk more than 2 cm
'Hiyodora'

52. P. speciosa (Koidzumi) Ingram (*P. lannesiana* (Carrière) Wilson forma *albida* Wilson). Illustration: Krüssmann, Manual of cultivated broad-leaved trees and shrubs **3:** pl. 13 (1986).

Tree to 12 m, bark pale grey-brown. Stems, undersides of leaves or perigynous zones distinctly hairy. Leaves folded in bud, elliptic-obovate to obovate, teeth on mature leaves coarse, fewer than 4 per cm, acuminate or drawn out into hair-like tips; stalks of mature leaves at most very sparsely hairy. Flowers 3 or 4, borne with the leaves in loose, rather long-stalked corymbs, white, to 1.5 cm across; perigynous zone cylindric; stalks more than 9 mm. Sepals not becoming reflexed. Fruit ovoid, black and shining. *Japan*. H2. Spring.

Sometimes considered as part of the Sako-Zakura group of *P. serrulata*.

53. P. × sieboldii (Carrière) Wittmack. (*P. speciosa × P.apetala*). Illustration: Wilson, The cherries of Japan, pl. 8 (1916); Manual of Japanese flowering cherries, 322 (1982);

Tree to 8 m, with smooth, grey bark. Leaves folded in bud, elliptic to ovate, more than 6 cm, teeth coarse, at least 3 mm long; stalks of mature leaves densely downy or hairy. Flowers 3 or 4 in corymbs, double or semi-double, usually more than 3 cm wide; perigynous zone cylindric; stalks more than 9 mm, hairy. Sepals not becoming reflexed. *Garden origin*. H2. Spring.

54. P. pseudocerasus Lindley (*P. involucrata* Koehne). Illustration: Edwards's Botanical Register **10:** t. 800 (1824); Krüssmann, Manual of cultivated broad-leaved trees and shrubs **3:** pl. 7, 15 & f. 22 (1986).

Tree to 8 m; young shoots hairless or slightly hairy. Leaves folded in bud, broadly ovate to oblong-ovate; teeth 1 mm or less. Flowers 3–6 in stalked umbels or corymbs, before or with the leaves, pink in bud, white when open, to 2 cm across; stalks more than 9 mm. Sepals becoming reflexed. Sepals, perigynous zone and style hairy. Fruit ovoid, yellow to red, to 1.5 cm. *China (Hubei)*. H2. Spring.

P. cantabrigiensis Stapf (*P. pseudocerasus* var. *cantabrigiensis* (Stapf) Ingram). Illustration: Botanical Magazine, 9129 (1928). This has longer, more acuminate leaves, a more dense inflorescence and petals deep pink in bud, opening paler. See Yeo, *Baileya* **20:** 11–17 (1976).

55. P. dielsiana Schneider. Illustration: Krüssmann, Manual of cultivated broad-leaved trees and shrubs **3:** pl. 7 (1986).

Shrub or tree to 20 m; young shoots hairless. Leaves folded in bud, oblong-obovate to oblong; stalks hairy. Flowers 3–5 in stalked umbels or corymbs, 3 cm or more across borne before the leaves; stalks more than 9 mm. Perigynous zone hairy. Sepals becoming reflexed. Fruit spherical, red, to 1 cm across. *C China*. H2. Spring.

56. P. cyclamina Koehne. Illustration: Botanical Magazine, n.s., 338 (1959); Krüssmann, Manual of cultivated broad-leaved trees and shrubs **3:** pl. 7 (1986).

Very like *P. dielsiana*, but leaf-stalks and perigynous zone not hairy. *C China*. H2. Spring.

57. P. mahaleb Linnaeus. Illustration: Bonnier, Flore complète **3:** pl. 167 (1914); Phillips, Trees in Britain, Europe and North America, 172 (1978); Krüssmann, Manual of cultivated broad-leaved trees and shrubs **3:** pl. 17 & f. 15 (1986); Brickell, RHS gardeners' encyclopedia of flowers and plants, 59 (1989).

Tree to 10 m; young shoots downy. Leaves folded in bud, almost circular to broadly ovate, downy along the midrib beneath, 6 cm or less, finely scalloped, with short abrupt tips 1 mm or less. Flowers 6–10 in stalked umbels or corymbs, white, fragrant, to 1.5 cm across; stalks more than 9 mm. Sepals becoming reflexed. Fruit *c.* 6 mm across, black. *E Europe to C Asia*. H1. Late spring.

58. P. pensylvanica Linnaeus filius. Illustration: Botanical Magazine, 8486 (1913); Phillips, Trees in Britain, Europe and North America, 172 (1978); Krüssmann, Manual of cultivated broad-leaved trees and shrubs **3:** f. 15, 19 (1986).

Shrub or tree to 10 m; branches hairless. Leaves folded in bud, ovate to oblong-lanceolate, green when mature, with acute or acuminate teeth, hairless or almost so beneath; tips evenly tapered. Flowers 3–6 before or with the leaves in stalkless umbels, white, to 1.5 cm across; stalks more than 9 mm. Sepals becoming reflexed. Fruit red, spherical, to 6 mm across. *N America*. H1. Spring.

59. P. pilosiuscula Koehne. Illustration: Botanical Magazine, 9192 (1930).

Tree to 12 m; young shoots downy or hairless. Leaves folded in bud, oblong to oblong-obovate, teeth standing out from leaf-margin, at least 2 mm long, with hair-like tips. Flowers 2 or 3, borne with the leaves in stalked corymbs, white or pale pink, to 2 cm across; stalks more than 9 mm. Sepals becoming reflexed. Fruit red, ellipsoid, to 1 cm across. *C & W China*. H1. Late spring.

60. P. litigiosa Schneider. Illustration: Krüssmann, Manual of cultivated broad-leaved trees and shrubs **3:** pl. 8 (1986).

Small tree to 6 m; young shoots hairless. Leaves folded in bud, narrowly obovate to oblong-obovate, teeth 1 mm or less. Flowers 2 or 3 in stalked umbels, white, to 2.5 cm across; stalks of flower-clusters *c.* 6 mm; flower-stalks more than 9 mm. Sepals and perigynous zone hairless. Sepals becoming reflexed. Style hairy at the base.

Fruit ellipsoid, red, to 1 cm. *W China (Hubei)*. H2. Spring.

61. P. maximowiczii Ruprecht. Botanical Magazine, 8641 (1915); Krüssmann, Manual of cultivated broad-leaved trees and shrubs **3:** pl. 8 & f. 15 (1986).

Tree to 8 m in cultivation (larger in the wild); young shoots downy. Leaves folded in bud, obovate, teeth rounded or obtuse standing out from leaf-margin, at least 2 mm long. Flowers 5–10, borne after the leaves in stalked corymbs, *c.* 1.5 cm across, cream-white; stalks more than 9 mm. Sepals becoming reflexed. Fruit spherical, black. *E Asia*. H1. Late spring.

62. P. tomentosa Thunberg. Illustration: Botanical Magazine, 8196 (1908); Everett, New York Botanical Garden illustrated encyclopedia of horticulture **8:** 2825 (1981); Krüssmann, Manual of cultivated broad-leaved trees and shrubs **3:** pl. 8, 20 & f. 28 (1986).

Shrub to 3 m, or rarely a small tree; young shoots downy. Leaves folded in bud, broadly elliptic or ovate, sometimes 3-lobed, with large, broad triangular teeth, hairy beneath. Stipules narrowly linear, fringed, sometimes falling early. Flowers usually 2 together, white or pinkish, 1.5–2 cm across, appearing with the leaves; stalkless or stalks to 9 mm. Style and ovary hairy. Fruit spherical, red, ultimately hairless or slightly hairy. *Himalaya to W China, Japan*. H1. Late spring.

63. P. incana (Pallas) Batsch. Illustration: Edwards's Botanical Register **25:** t. 28 (1939); Krüssmann, Manual of cultivated broad-leaved trees and shrubs **3:** f. 14 (1986).

Open shrub to 2 m; young shoots finely downy. Leaves folded in bud, oblong-ovate to obovate, to 2 cm wide, densely white-felted beneath, of two types. Flowers stalkless, solitary or 2 or 3 in stalkless clusters, borne with the leaves, pink, to 1 cm across, hairless or sparsely hairy outside. Perigynous zone cylindric. Style and ovary hairy. Fruit hairy, ovoid, red, 6–8 mm. *SE Europe, Turkey, Caucasus, Iran*. H2. Late spring.

64. P. jacquemontii J.D. Hooker. Illustration: Botanical Magazine, 6976 (1888); Krüssmann, Manual of cultivated broad-leaved trees and shrubs **3:** pl. 8 & f. 14 (1986).

Spreading shrub to 3 m; young shoots hairy or hairless. Leaves folded in bud, elliptic to obovate-oblong, green when mature, not acuminate, sometimes abruptly short-pointed; not glaucous beneath, toothed all round. Flowers solitary or 2 together, pink, to 2 cm across, stalkless or very shortly stalked. Perigynous zone cylindric. Style, ovary and fruit hairless. Fruit spherical, to 1.5 cm across, red. *W Himalaya*. H4. Spring.

65. P. bifrons Fritsch. Illustration: Krüssmann, Manual of cultivated broad-leaved trees and shrubs **3**: f. 14 (1986).

Like *P. jacquemontii* but smaller (to 1.5 m), flowers borne with the leaves, petals often downy towards the base inside, style, ovary and fruit hairy. *W Himalaya*. H2. Spring.

66. P. japonica Thunberg. Illustration: Makino, New illustrated flora of Japan, t. 1145 (1963); Krüssmann, Manual of cultivated broad-leaved trees and shrubs **3**: pl. 8 & f. 13 (1986).

Shrub to 1.5 m; young shoots hairless. Leaves folded in bud, ovate-lanceolate, ovate or broadly ovate, green when mature, long-acuminate, with acute teeth standing out from margin. Flowers borne with the leaves, 2 or 3 in clusters, pink, to 2 cm or more across, shortly stalked. Sepals with gland-tipped marginal teeth. Fruit spherical to ellipsoid, dark red, to 1 cm across. *C China, Korea, introduced early into Japan*. H1. Late spring.

67. P. glandulosa Thunberg (*P. japonica* misapplied). Illustration: Botanical Magazine, 8260 (1909); Hillier colour dictionary of trees and shrubs, 172 (1981); Krüssmann, Manual of cultivated broad-leaved trees and shrubs **3**: f. 13 (1986).

Shrub to 2 m, young shoots usually hairless. Leaves folded in bud, narrowly ovate or elliptic, tapered at base, widest near the middle, length *c*. 2.5 times the width or more, to 7 × 3 cm, flat, green when mature, not acuminate, sometimes abruptly short-pointed, not glaucous beneath. Flowers solitary or 2 together in clusters, white to pale pink, to 1.2 cm across, usually stalked. Sepals hairless. Style and ovary hairy. Fruit hairy, spherical, to 1.2 cm across, red. *C & N China, Japan*.

68. P. humilis Bunge (*P. bungei* Walpers). Illustration: Botanical Magazine, 7335 (1894); Krüssmann, Manual of cultivated broad-leaved trees and shrubs **3**: f. 13 (1986).

Shrub to 2 m, young shoots downy. Leaves folded in bud, to 4 × 2 cm, narrowly ovate or elliptic, length *c*. 2.5 times the width or more, boat-shaped, often curved along midrib, green when mature, not acuminate, sometimes abruptly short-pointed; not glaucous beneath. Flowers borne with the leaves, solitary or 2 together in clusters, pale pink, to 1.5 cm across, stalks 5 mm or more. Fruit almost spherical, bright red, to 1.5 cm across. *N China*. H1. Spring.

69. P. pumila Linnaeus. Illustration: Gleason, Illustrated flora of the north-eastern United States and adjacent Canada **2**: 330 (1952); Krüssmann, Manual of cultivated broad-leaved trees and shrubs **3**: pl. 8, 16 & f. 13, 14 (1986).

Small shrub to 1.5 m; young shoots reddish brown, hairless. Leaves folded in bud, indistinctly toothed, oblanceolate to narrowly obovate, green when mature, less than 6 cm, narrowly tapered at base, not acuminate, sometimes abruptly short-pointed, greyish glaucous beneath; not toothed or scalloped near the base. Flowers 2–4 together, white, to 1.5 cm across, in stalkless umbels. Sepals becoming reflexed. Fruit more or less spherical, shiny purple-black. *E North America*. H1. Late spring.

70. P. × cistena (Hansen) Koehne (*P. cerasifera* 'Atropurpurea' × *P. pumila*). Illustration: Hillier colour dictionary of trees and shrubs, 172 (1981); Everett, New York Botanical Garden illustrated encyclopedia of horticulture **8**: 2825 (1981); Krüssmann, Manual of cultivated broad-leaved trees and shrubs **3**: pl. 2 (1986); Brickell, RHS gardeners' encyclopedia of plants and flowers, 123 (1989).

Shrub to 2 m at most. Leaves folded in bud, lanceolate-obovate, hairy beneath, purplish at first, becoming reddish brown when mature. Flowers solitary or 2 together in stalkless umbels, white. Sepals becoming reflexed. Fruit blackish purple. *Garden origin*. H1. Spring.

71. P. besseyi Bailey. Illustration: Botanical Magazine, 8156 (1907); Krüssmann, Manual of cultivated broad-leaved trees and shrubs **3**: pl. 8 (1986).

Prostrate shrub. Leaves folded in bud, elliptic to elliptic-lanceolate, green when mature, abruptly tapered, sometimes abruptly short-pointed, greyish glaucous beneath; not toothed or scalloped near the narrowly tapered base. Flowers 2–4 in clusters, white, to 1.5 cm across; flowers stalkless or stalks to 9 mm. Fruit spherical, purple-black, to 1.5 cm across. *W USA*. H2. Spring.

72. P. serotina Ehrhart. Illustration: Sargent, The silva of North America **4**: t. 159 (1892); Phillips, Trees in Britain, Europe and North America, 174 (1978); Bärtels, Gartengehölze, 357 (1981).

Deciduous trees to 35 m or shrubs; bark brown, inner aromatic; young shoots hairless. Leaves oblong, acuminate, margins scalloped or with incurved teeth at least in apical part, glossy above, pale beneath and hairy along the veins. Flowers white, 1–1.4 cm across, usually 10 or more in narrow, cylindrical racemes, leafy towards base. Fruit ovoid, 8–10 mm, dark purple. *N America*. H1. Late spring.

73. P. virens (Wooton & Standley) Shreve. Illustration: Sargent, Manual of the trees of North America, edn 2, 578 (1922).

Like *P. serotina*, but smaller, some leaves tending to be evergreen, usually elliptic and acute, more finely toothed, the flowers smaller, in shorter racemes. *USA (Texas to Arizona), Mexico*. H4. Late spring.

74. P. salicifolia Kunth (*P. capuli* Sprengel). Illustration: Revue Horticole for 1893 – as Cerasus capuli, unnumbered plate (1893); Popenoe, Manual of tropical and subtropical fruits, pl. 13 (1974); Krüssmann, Manual of cultivated broad-leaved trees and shrubs **3**: f. 31 (1986).

Evergreen trees or shrubs to 12 m; young shoots hairless. Leaves lanceolate with narrow acute tips, finely toothed, completely hairless or slightly hairy beneath. Flowers white *c*. 1 cm across, usually 10 or more in loose, cylindric racemes more than 5 cm, leafy towards base. Fruit spherical, purple-red, to 1.7 cm across. *Mexico to Peru*. H4. Late spring.

75. P. padus Linnaeus. Illustration: Phillips, Trees in Britain, Europe and North America, 173 (1978); Krüssmann, Manual of cultivated broad-leaved trees and shrubs **3**: f. 18 (1986); Brickell, RHS gardeners' encyclopedia of plants and flowers, 49 (1989).

Deciduous trees or shrubs to 15 m; young twigs at first downy. Leaves elliptic, with short abrupt tips less than 1 cm, margins with straight projecting teeth, hairless, bluish green beneath. Flowers white, fragrant, to 1.5 cm across, 10 or more in narrow, cylindric racemes, leafy towards base. Perigynous zone hairy within. Petals more than 6 mm. Fruit spherical, to 8 mm across, black. *Europe & W Asia to C Japan*. H1. Spring.

76. P. vaniotii Léveillé. Illustration: Krüssmann, Manual of cultivated broad-leaved trees and shrubs **3**: pl. 5 (1986).

Deciduous trees or shrubs to 15 m; young shoots hairless. Leaves oblong to obovate, margins scalloped or with incurved teeth at least in apical part, acute-acuminate. Flowers white, to 8 mm wide, 10 or more in narrow, loose, cylindric racemes, leafy towards base, main stalk of raceme densely hairy. Fruit spherical, orange-red to brown-red, to 8 mm across. *W China to Taiwan.* H2. Spring.

77. P. cornuta (Royle) Steudel. Illustration: Botanical Magazine, 9423 (1935); Polunin & Stainton, Flowers of the Himalaya, pl. 33 (1984); Krüssmann, Manual of cultivated broad-leaved trees and shrubs **3**: pl. 5 & f. 18 (1986).

Deciduous trees or shrubs to 5 m (more in the wild); young shoots hairless. Leaves elliptic to obovate, with short abrupt tips less than 1 cm, margins with straight projecting teeth, dull above, broadest near middle, mature blades usually more than 10 cm. Flowers white, 6–10 mm across, 10 or more in narrow, cylindric racemes, main stalk of raceme hairless or sparsely hairy, leafy towards base. Fruit ovoid, hairy, to 2 cm. *Afghanistan, Himalaya to W China.* H2. Spring.

78. P. napaulensis (Seringe) Steudel. Illustration: Stainton, Flowers of the Himalaya – a supplement, 30 (1988)

Like *P. cornuta* but leaves hairy beneath, stalks of the racemes densely hairy. *Himalaya, W China.* H5. Spring.

Similar to *P. sericea* (Batalin) Koehne, which is rare in cultivation; the two are sometimes regarded as a single species.

79. P. virginiana Linnaeus. Illustration: Sargent, The silva of North America **4**: t. 158 (1892); Phillips, Trees in Britain, Europe and North America, 176 (1978); Krüssmann, Manual of cultivated broad-leaved trees and shrubs **3**: pl. 8, 20 & f. 31 (1986).

Deciduous trees or shrubs, somewhat stoloniferous; young shoots hairless. Leaves broadly obovate to broadly elliptic, with short abrupt tips less than 1 cm, broadest towards tip, shiny above, mature blades usually 10 cm or less; margins with straight projecting teeth. Flowers usually 10 or more in narrow, cylindric racemes; stalk of raceme hairless or sparsely hairy, leafy towards base. Petals 2–3 mm. Fruit spherical, at first red, later black, to 1 cm across. *N America.* H1. Late spring.

80. P. ssiori Schmidt. Illustration: Shirasawa, Iconographie des essences forestières du Japon **2**: t. 28 (1908);

Krüssmann, Manual of cultivated broad-leaved trees and shrubs **3**: pl. 5 & f. 18 (1986).

Deciduous trees or shrubs to 25 m in the wild; young shoots hairless. Leaves obovate-elliptic to oblong, with narrow tips 1 cm long or more, margins with straight projecting teeth, base cordate; dark green and hairless above, hairy in the vein-axils beneath. Flowers white, to 1 cm across, many in narrow, cylindric racemes which are leafy towards the base. Styles not projecting. Fruit more or less spherical, black, to 1 cm across. *Japan.* H2. Spring.

81. P. grayana Maximowicz. Illustration: Shirasawa, Iconographie des essences forestières du Japon **1**: t. 46 (1900); Krüssmann, Manual of cultivated broad-leaved trees and shrubs **3**: pl. 8 & f. 18 (1986).

Small deciduous trees or shrubs, to 7 m; young shoots hairy or hairless. Leaves oblong-ovate, with narrow tips 1 cm long or more, margins with straight projecting teeth, bases rounded. Flowers white, to 1 cm across, many, in narrow, cylindric racemes which are leafy towards the base. Styles projecting. Fruit spherical but pointed, ultimately black, to 8 mm across. *Japan.* H4. Late spring.

82. P. maackii Ruprecht. Illustration: Phillips, Trees in Britain, Europe and North America, 172 (1978).

Deciduous trees or shrubs to 10 m; bark shiny brownish yellow; young shoots downy. Leaves ovate-oblong, finely toothed, with dotted glands (and some hairs on the veins) beneath. Flowers white, to 1 cm across, 6–10 in narrow, irregular, cylindric racemes on the old wood, not leafy towards the base. Styles longer than stamens, hairy below. Fruit spherical, black, to 5 mm across. *N China, Korea.* H1. Spring.

83. P. buergeriana Miquel. Illustration: Shirasawa, Iconographie des essences forestières du Japon **1**: to 46 (1900); Krüssmann, Manual of cultivated broad-leaves trees and shrubs **3**: f. 18 (1986).

Deciduous trees or shrubs to 10 m; young shoots hairy or not. Leaves elliptic or oblong-elliptic, acuminate, toothed, not gland-dotted beneath, but hairy in the vein axils. Flowers white, to 7 mm across, usually many in narrow, cylindric racemes which are not leafy towards the base. Styles very short. Fruit spherical, black, borne on the persistent perigynous zone. *Japan, Korea.* H2. Spring.

84. P. lusitanica Linnaeus. Illustration:

Phillips, Trees in Britain, Europe and North America, 172 (1978); Hillier colour dictionary of trees and shrubs, 173 (1981); Mitchell & More, The complete guide to trees of Britian and northern Europe, 83 (1985); Reader's Digest field guide to the trees and shrubs of Britain, 97 (1988).

Evergreen trees or shrubs to 20 m; young shoots hairless, red. Leaves oblong-ovate, coarsely toothed, dark green and glossy above, paler beneath. Flowers white, to 1 cm across, usually 10 or more in narrow, cylindric racemes, ascending, spreading or hanging, more than 5 cm, not leafy towards base. Fruit ovoid, to 8 mm, dark red. *Spain, Portugal, Azores.* H2. Late spring.

85. P. laurocerasus Linnaeus. Illustration: Nicholson & Clapham, The Oxford book of trees, 140 (1975); Phillips, Trees in Britain, Europe and North America, 172 (1978); Mitchell & More, The complete guide to trees and shrubs of Britain and northern Europe, 83 (1985); Reader's Digest field guide to the trees and shrubs of Britain, 96 (1988).

Evergreen trees or more usually shrubs to 8 m; young shoots hairless, green. Leaves oblong to obovate or elliptic, entire or toothed, dark green above, paler beneath. Flowers white, to 8 mm across, many in narrow, erect, cylindric racemes, more than 5 cm, not leafy towards base. Fruit conical, dark red, to 8 mm. *E Europe, SW Asia.* H2. Spring.

Many selections have been made and treated as cultivars. See Krüssmann, Manual of cultivated broad-leaved trees and shrubs **3**: 36–38 (1986) for descriptions and a key.

86. P. ilicifolia (Nuttall) Walpers. Illustration: Sargent, The silva of North America **4**: t. 162 (1892); Phillips, Trees in Britain, Europe and North America, 171 (1978); Krüssmann, Manual of cultivated broad-leaved trees and shrubs **3**: pl. 5 & f. 12 (1968); Hickman, The Jepson manual, 977 (1993).

Evergreen trees or shrubs to 9 m, holly-like. Leaves spiny, ovate to broadly lanceolate, 5–7 cm, acute, entire to finely and distantly toothed, hairless. Flowers white, to 8 mm across, usually 10 or more in narrow, ascending, spreading or hanging, cylindric racemes, more than 5 cm, not leafy towards base. Fruit almost spherical, purple-black, to 1.5 cm across. *USA (California).* H5. Summer.

87. P. lyonii (Eastwood) Sargent (*P. integrifolia* Sargent not Walpers). Illustration: Abrams & Ferris, Illustrated Flora of the Pacific States **2**: 468 (1944); Krüssmann, Manual of cultivated broadleaves trees and shrubs **3**: pl. 5 (1986); Hickman, The Jepson manual, 977 (1993).

Evergreen trees or shrubs to 14 m. Leaves ovate to ovate-lanceolate, acute, entire or finely and sparsely toothed, hairless. Flowers white, to 8 mm across, usually many in narrow, cylindric racemes, more than 5 cm, not leafy towards base. Fruit spherical, black, 1–2.5 cm across. *USA (California)*. H5. Spring.

88. P. caroliniana (Miller) Aiton. Illustration: Sargent, The silva of North America **4**: t. 160 (1892); Krüssmann, Manual of cultivated broad-leaved trees and shrubs, f. 7 (1986).

Evergreen trees or shrubs to 12 m. Leaves entire or sparsely toothed. Flowers cream-white, to 1 cm across, usually 10 or more in short, dense, cylindric racemes, to 4 cm. *SE USA*. H5. Spring.

77. PRINSEPIA Royle
H.S. Maxwell
Deciduous, arching shrubs with axillary spines and flaking bark; branches with chambered pith; buds small, naked and or covered with a few small hairy scales. Leaves alternate, often clustered, simple, membranous to leathery, entire or toothed. Stipules, if present, small, persistent. Flowers stalked, in axillary bracteate racemes or 1–8 (rarely to 13) in the axils of previous years shoots. Flowers tubular, cup-shaped; calyx 5-lobed, lobes equal or unequal, broad, short. Petals 5, distinct, equal, almost circular, clawed, white or yellow, spreading. Stamens 10 or many. Ovary superior. Fruit an oblique 1-seeded drupe, red or purple, juicy and edible.

A genus of 4 species from China, Taiwan and the Himalaya. Only 3 species are of any horticultural importance. *P. utilis* is cultivated as a hedging plant, *P. uniflora* and *P. sinensis* as ornamentals.

They grow well in ordinary, well-drained soil and are best in full sun, although they can stand slight shade. No special care is required except a little thinning of the branches in late winter or early spring. They generally do not fruit very well in cultivation. Propagation is by seed, cuttings or layering.

1a. Leaves mucronate **2. uniflora**
  b. Leaves long acuminate 2

2a. Flowers white; calyx-lobes irregular; stamens numerous; spines 1–5 cm, with leaves, buds or scales **3. utilis**
  b. Flowers yellow; calyx-lobes equal; stamens 10; spines 3–12 mm, without leaves, buds or scales **1. sinensis**

1. P. sinensis (Oliver) Bean (*Plagiospermum sinense* Oliver). Illustration: Botanical Magazine, 8711 (1917); Bean, Trees and shrubs hardy in the British Isles, edn 8, 344 (1980); Krüssmann, Manual of cultivated broad-leaved trees and shrubs **3**: 442 (1986).

Shrub with rather loose spreading habit, 2–3 m; bark yellowish, hairless or sometimes only with tufts of hair on the nodes and base of stipules. Spines 0.3–1.2 cm, curved, without leaves, buds or scales. Leaves oblong-lanceolate to lanceolate, apex long acuminate, base obtuse to rounded, entire or sparingly toothed, margin hairy. Leaves on fertile branches 2.5–5.6 × 0.6–1.6 cm; stalks 0.3–1.4 cm; sterile branches 3–9 × 0.8–2.4 cm; stalks 0.8–2.4 cm. Flowers yellow, single, in clusters of 1–8, borne in leaf-axils, on very short spurs, 1.3–2 cm across; calyx-lobes equal; flower-stalks hairless, 0.8–2.1 cm. Stamens 10. Fruit red or purple, 1.1–1.3 cm, ripening in August. *E Asia (Manchuria, Ussuri-land & N Korea)*. H3. Spring.

2. P. uniflora Batalin. Illustration: Everett, The New York Botanical Gardens illustrated encyclopedia of horticulture **8**: 2799 (1981); Brickell, The RHS gardeners' encyclopedia of plants and flowers, 105 (1989).

Loose spreading shrub to 1.6 m; bark reddish brown; stipules present; young shoots hairless. Spines 0.3–1.6 cm, curved, without leaves, buds or scales. Leaves membranous, dark glossy green, linear-oblong to narrow-oblong, acute or obtuse, base tapering, apex mucronate, entire or toothed. Leaves on fertile branches 0.8–4.4 × 0.3–0.8 cm; stalks 0.1–0.6 cm; on sterile branches 4.7–8.2 × 0.7–1.4 cm; stalks 0.2–0.4 cm. Flowers white; calyx lobes equal with marginal hairs, 1–8 among the clustered leaves on very short spurs, *c.* 1.4 cm; stalks hairless, 0.3–0.7 cm. Petals *c.* 5 mm, obovate; stamens 10; anthers yellow. Fruit 1–1.5 cm, purple or dark red with a slight bloom. *Inner Mongolia, NW & C China*. H1. Early spring.

3. P. utilis Royle. Illustration: Botanical Magazine, n.s. 194 (1952); Polunin &

Stainton, Flowers of the Himalayas, 31 (1985).

Vigorous spiny shrub, 1.5 m, occasionally to 3.6 m, young shoots green, downy at first, becoming hairless. Spines numerous, produced in every leaf-axil, 1–5 cm, straight, often leafy or with buds or scales. Leaves 1.1–10.5 × 0.5–2 cm, very variable, thin, leathery, more or less persistent, elliptic to elliptic-lanceolate, apex usually long acuminate, base narrowly tapered, usually toothed, marginal teeth glandular, stipules absent, dull green, hairless. Leaf-stalks 4–6 mm, base densely hairy, glandular, especially on young shoots. Inflorescence 1.5–6 cm, with hairless or finely downy stalks to 1.3 cm, flowers in racemes, 1–13 (usually 5–7), 1.3–2 cm across, creamy white, fragrant; bracts occasionally leaf-like; calyx irregular, with 2 small and 3 large calyx-lobes, margin entire or irregularly toothed. Stamens numerous. Fruit oblong-obovoid, 1–1.5 cm, dark purple, bloomed. *W Pakistan, India, Nepal, Bhutan & China*. H3. Late autumn–early winter.

78. MADDENIA J.D. Hooker & Thomson
M.F. Gardner
Deciduous, dioecious trees or shrubs. Leaves alternate, with glandular forward-pointing teeth; stipules large, glandular-toothed. Flowers unisexual, on short stalks, borne in short terminal dense racemes; sepals 10, small, some long and petal-like; stamens 25–40, inserted in calyx mouth in 2 more or less distinct rows. Male flowers with a solitary carpel, female flowers with 2. Ovary superior. Fruit a single-seeded drupe; stone ovoid, 3-keeled on one side.

A genus of 4 species from the Himalaya and China which are easily cultivated in most garden soils.

1. M. hypoleuca Koehne. Illustration: Deutsche Baumschule **11**: 71 (1959); Krüssmann, Manual of cultivated broad-leaved trees & shrubs **2**: 265 (1986); Phillips & Rix, Shrubs, 33 (1989).

Shrub or small tree to 6 m; young twigs dark brown, hairless. Leaves ovate-oblong, 4–7 cm, double-toothed, tapering towards the end, rounded at base, dark green above, blue-white beneath, with 14–18 pairs of veins. Flowers red-brown at first, later green, borne in racemes, 3–5 cm; stamens 20–30, anthers yellow. Fruit *c.* 8 mm, ovoid, black. *C & W China*. H4. Winter–spring.

79. OEMLERIA Reichenbach
C.M. Mitchem

Deciduous, bisexual shrubs to 3 m, which form dense thickets. Bark purple-brown; lenticels prominent on branchlets. Leaves 2.5–9 × 0.8–3 cm, obovate to lanceolate, undersides paler. Inflorescence a raceme of up to 13 flowers each subtended by 2 bracteoles. Flowers to 1 cm across, white. Calyx to 2 cm. Petals ovate and upright. In male flowers: numerous stamens around perigynous zone, to 4 mm, incurved; smaller and non-functional in female flowers. Ovary superior. Carpels 5; stigma bilobed. Fruit a 1-seeded, thinly fleshy drupe, bitter-tasting.

Literature: Allen, G.A., Flowering pattern and fruit production in the dioecious shrub. *Oemleria cerasiformis (Rosaceae)* Canadian Journal of Botany **64**(6): 1216–1220 (1986); Anon., Osmaronia cerasiformis (Torrey & Gray) Greene Davidsonia **5**: 12–15 (1974).

A genus of one species, grown mainly for its early, fragrant flowers and decorative fruits.

1. O. cerasiformis (W.J. Hooker & Arnott) Landon (*Osmaronia cerasiformis* W.J. Hooker & Arnott; *Nuttalia cerasiformis* (W.J. Hooker & Arnott) Greene; *N. davidiana* Baillon; *Exochorda davidiana* Baillon). Illustration: Botanical Magazine, 582 (1970); Krüssmann, Manual of cultivated broad-leaved trees and shrubs **2**: pl. 146 (1986).

Female plants are less commonly planted, as they are coarser and not as free-flowering. Prefers semi-shade and moist, humus-rich soils. Propagation by seed or layering. *W North America (British Columbia to California).* H1. Early spring.

CXII. CHRYSOBALANACEAE

Trees or shrubs. Leaves simple, entire, alternate, often leathery. Inflorescence racemose, paniculate or cymose. Flowers bisexual, rarely unisexual. Calyx-lobes 5, overlapping, often unequal, erect or reflexed. Petals 5, rarely 4 or occasionally absent, commonly unequal, overlapping, usually falling early. Filaments thread-like, free. Style thread-like; stigma 3-lobed. Fruit a dry or fleshy drupe.

A family of 17 genera and about 500 species distributed throughout tropical regions of both hemispheres. Propagation by seed, also cuttings.

Literature: Prance, G.T. & White, F., The Genera of Chrysobalanaceae: A study in practical and theoretical taxonomy and its relevance to evolutionary biology, *Philosophical Transactions of the Royal Society of London* **320**: 1–184 (1988).

1a. Flowers radially symmetric; ovary inserted at or near base of perigynous zone   2
 b. Flowers bilaterally symmetric; ovary inserted at or near mouth of perigynous zone   3
2a. Stamens projecting, joined in groups; filaments hairy   **1. Chrysobalanus**
 b. Stamens included or projecting; filaments smooth, free   **2. Licania**
3a. Stamens included   **3. Parinari**
 b. Stamens projecting   4
4a. Inflorescence of little-branched panicles or racemes   **5. Atuna**
 b. Inflorescence of much branched corymbose panicles   **4. Maranthes**

1. CHRYSOBALANUS Linnaeus
E.H. Hamlet

Small trees and shrubs. Leaves smooth or with adpressed hairs beneath. Bracts and bracteoles small, without glands. Inflorescence terminal or axillary. Flowers bisexual. Calyx-lobes 5, acute; petals 5, longer than calyx-lobes. Stamens 12–26; filaments hairy, united at base. Ovary densely hairy, inserted at base of perigynous zone. Style downy. Fruit a fleshy drupe.

A genus of 2 species from tropical America, tropical Africa & West Indies. They can be grown in ordinary soil and need a frost-free climate. Propagation usually by seed, also by cuttings.

1. C. icaco Linnaeus. Illustration: van Steenis & de Wilde, Flora Malesiana, series 1, **18**(4): 644 (1989).

Small tree or shrub, to 5 m, branches smooth with lenticels. Leaf-stalks 2–5 mm; stipules 1–3 mm, deciduous. Leaves 2–8 × 1.2–6 cm, alternate, leathery, rounded to ovate-elliptic, slightly notched, rounded to obtuse, slightly tapered at base. Inflorescences terminal and axillary panicles or cymes with grey-brown matted hairs. Perigynous zone cup-shaped with densely felted hairs. Calyx-lobes 4 or 5, *c.* 2.5 mm, rounded to acute; petals 4 or 5, *c.* 5 mm, white, smooth. Stamens 12–16, longer than petals; filaments joined in groups, densely hairy. Ovary hairy. Fruit ovate to obovate, 1.5–5 cm. *Tropical W Africa (Guinea to Angola), C America to S Brazil, W Indies.* G2.

2. LICANIA Aublet
E.H. Hamlet

Trees or shrubs. Leaf-stalks smooth or with 2 or more stalkless glands. Leaves entire. Inflorescence a simple or branched racemose panicle, less often a panicle of cymes or spikes. Flowers bisexual. Calyx-lobes 5, acute; petals 4 or 5, or absent. Stamens 3–40, unilateral or forming a circle; filaments rarely joined and usually smooth. Fruit a fleshy drupe.

A genus of about 190 species from Africa, tropical Asia & America.

1a. Stamens 5–7   **3. incana**
 b. Stamens 8–14   2
2a. Inflorescence with grey downy hairs   **2. rigida**
 b. Inflorescence brown to rusty brown   **1. arborea**

1. L. arborea Seemann. Illustration: Contributions from the United States National Herbarium **27**: t. 29 (1928); Pennington & Sarukhan, Arboles Tropicales de Mexico, 161 (1968).

Tree to 60 m. Leaves ovate-rounded to oblong, 5–12 × 2.5–8 cm, on fertile branches, larger on sterile branches. Leaf-stalks 5–12 mm, terete with densely felted hairs when young, later hairless. Inflorescences axillary or terminal racemose panicles, main axis and branches with densely felted brown to rusty brown hairs. Flowers 2.5–3 mm, solitary and densely clustered. Calyx-lobes acute with densely felted hairs; petals 5, oblong, downy. Stamens 8–12, inserted in a circle; filaments as long as calyx-lobes, joined for half their length. Fruit oblong, to 3 cm. *C America & W South America to Peru.* G2.

2. L. rigida Bentham.

Small tree to 15 m. Leaves 6–16 × 2.8–6.5 cm, oblong to elliptic, leathery, rounded to slightly notched at apex, rounded to cordate at base. Inflorescences in racemose panicles; main axis and branches with densely felted grey hairs. Flowers 2.5–3.5 mm, in small groups. Calyx-lobes acute; petals 5, densely downy. Stamens *c.* 14; filaments densely downy. Fruit elliptic, 4–5.5 cm. *NE Brazil.* G2.

3. L. incana Aublet (*L. crassifolia* Bentham). Illustration: Martius, Flora Brasiliensis **14**(2): t. 3 (1867).

Shrub or rarely small tree. Leaves

2.5–8.5 × 1.3–5.5 cm, ovate to oblong, acute to acuminate at apex, with point to 1 cm long, rounded at base, rarely slightly wedge-shaped. Inflorescences terminal and axillary spikes. Flowers c. 2 mm, in small clusters along the stem. Calyx-lobes acute; petals absent. Stamens 5–7, unilateral; filaments shorter than calyx-lobes, smooth. Style equalling filaments. Fruit ovoid to spherical, c. 1.6 cm. *S America (Colombia, Venezuela & Brazil). G2.*

3. PARINARI Aublet
E.H. Hamlet

Trees or shrubs, occasionally woody at base with herbaceous branches. Leaves alternate, entire. Leaf-stalks usually with 2 stalkless glands. Inflorescence in terminal or axillary much-branched panicles. Flowers 4–11 mm, bisexual. Perigynous zone cone to bell-shaped, slightly swollen at one side. Calyx-lobes 5, acute; petals 5. Stamens 6–8; filaments not exceeding calyx-lobes, unilateral with staminodes inserted opposite them. Ovary inserted laterally at the mouth of the receptacle. Style thread-like. Fruit a fleshy drupe.

A genus of about 40 species from tropical Africa, tropical Asia, tropical America and the Pacific Islands.

1a. Stipules persistent, partly clasping
  stem **1. campestris**
  b. Stipules falling early, not clasping
  stem **2. excelsa**

1. P. campestris Aublet. Illustration: Flora of the Guianas **85**: 102 (1986).

Tree to 25 m, young branches downy, becoming hairless and greyish with age. Leaf-stalks 2–7 mm, downy when young, terete; blades 6–13 × 3–6.5 cm, ovate, acuminate at apex, rounded to cordate at base, hairless above, with densely felted hairs beneath. Stipules broad, to 3 cm, acute at apex, persistent, partly clasping the stem. Inflorescences axillary and terminal panicles. Perigynous zone slightly bell-shaped to obconical. Petals 5, white, shorter than calyx-lobes. Stamens 7, fertile, unilateral, with 7 or 8 staminodes opposite them. Ovary and lower part of style densely hairy. Fruit 4–6 × 2–3 cm, oblong. *Tropical America (Venezuela, Trinidad, Guianas to Brazil). G2.*

2. P. excelsa Sabine. Illustration: Hutchison & Dalziel, Flora of West Tropical Africa **1**: 429 (1958).

Tree to 40 m. Leaf-stalks 3–7 mm, downy when young, terete; blades 3–9 × 1.5–5 cm, ovate to oblong-elliptic, rounded to wedge-shaped at base, acuminate at apex. Stipules c. 1 mm, falling early. Inflorescences terminal, in loose panicles. Flower-stalks 1–2 mm. Perigynous zone slightly bell-shaped to obconical. Petals 5, white, ovate, falling early, shorter than calyx-lobes. Stamens 7, fertile, unilateral with 7 or 8 short thread-like staminodes opposite them. Ovary and base of style hairy. Fruit 2.5–4 × 1.8–2.5 cm, ellipsoid. *Tropical America & tropical Africa. G2.*

4. MARANTHES Blume
E.H. Hamlet

Trees with leaf-undersides hairless or with dense woolly hairs. Leaf-stalks usually with 2 glands, rarely without. Bracts and bracteoles without glands, not enclosing the young flowers in groups. Perigynous zone shape various, but narrowed to the base, hairless inside at base. Calyx-lobes rounded. Stamens 18–60, projecting beyond calyx-lobes, often forming a complete circle, or slightly unilateral; filaments hairless. Ovary inserted laterally at the mouth of the receptacle. Fruit a fleshy drupe.

A genus of 11 species from tropical Africa, except 1 from Malaysia to New Guinea, NE Australia and W Polynesia.

1. M. corymbosa Blume (*Parinari corymbosa* Miquel). Illustration: Prance, Flora Neotropica **9**: 203 (1972); Candollea **20**: 143–5 (1965); Van Steenis & de Wilde, Flora Malesiana, series 1, **10**(4): 672 (1989).

Tree to 40 m, young branches shortly downy or hairless. Leaves 6.5–15 × 2–8 cm, oblong-elliptic to lanceolate, leathery, slightly wedge-shaped at base, apex acuminate. Leaf-stalks 4–6 mm, terete with 2 glands near base of blade. Inflorescences in terminal corymb-like panicles. Flowers 5–6 mm, obconical, bell-shaped; flower-stalks 2–3 mm. Calyx-lobes ovate-elliptic; petals 5 with marginal hairs. Stamens inserted in a semi-circle. Fruit oblong to pear-shaped. *SE Asia, Pacific Islands, tropical Australia & New Guinea. G2.*

5. ATUNA Rafinesque
E.H. Hamlet

Trees with complicated widely divergent branching. Leaves almost hairless; stipules large. Inflorescence a raceme or sparsely branched, contracted panicle. Flowers bisexual. Calyx-lobes 5, broadly ovate to lanceolate; petals 5, hairless, exceeding calyx-lobes. Stamens 10–25; filaments free, exserted. Fruit densely warty.

A genus of about 11 species from southern India, Thailand, Malaysia, Indonesia, Pacific Islands and New Guinea.

1. A. racemosa Rafinesque. Illustration: Smith, Flora Vitiensis Nova **3**: 49 (1985); Van Steenis & de Wilde, Flora Malesiana, series 1, **10**(4): 668 (1989).

Tree with young branches hairless or having adpressed bristles. Leaves 4.5–35 × 2–11 cm, ovate, elliptic, oblong or lanceolate, papery to leathery. Flowers in axillary racemes or little branched with up to 3 or more racemose branches on short main inflorescence-stalk. Calyx-lobes 4–7 mm, ovate to ovate-oblong. Petals to 10 mm, ovate or oblong, blue or white. *Thailand, Malaysia, Indonesia, Philippines, Pacific Islands & New Guinea. G2.*

Subsp. **racemosa** (*Parinari scabra* Hasskarl) has leaves 10–35 cm, usually elliptic, oblong or lanceolate but sometimes ovate, papery or thickly leathery, apex long, finely acuminate, 0.6–2.5 cm; leaf-stalks thick; flowers 1–1.7 cm.

Subsp. **excelsa** (Jack) Prance (*Parinari laurina* A. Gray; *P. macrophylla* Teijsmann & Binnendijk) with leaves 4.5–12 cm, usually ovate or oblong-ovate, leathery, apex bluntly acuminate, 3–10 mm; leaf-stalks thin; flowers 8–11 mm.

CXIII. LEGUMINOSAE

Trees, shrubs, herbaceous or woody climbers (sometimes with tendrils) or herbs, often with nodules containing nitrogen-fixing bacteria on the roots. Leaves usually alternate, divided into 3 leaflets, pinnate (sometimes bipinnate or tripinnate) or palmate, rarely completely absent on the mature plant; stipules usually present. Flowers in inflorescences of various kinds, usually racemose or congested into heads, sometimes solitary, radially symmetric or more commonly bilaterally symmetric. Perigynous zone present or absent. Calyx usually of 5 united sepals, 5-toothed or -lobed at the apex, sometimes 2-lipped, sepals rarely free (or almost so). Petals usually 5 (sometimes up to 4 suppressed), often diversified among themselves, free at the base or variously united. Stamens usually 10, sometimes numerous, more rarely fewer, the filaments free or more commonly variously united; anthers opening by longitudinal slits or more rarely by terminal pores. Carpel 1, with 1–many ovules. Fruit a legume containing 1–many seeds, sometimes modified so as to be indehiscent, or breaking transversely into 1-seeded segments.

A very large, diverse and cosmopolitan family (about 600 genera and 13 500 species) especially well developed in the tropics. It contains many important crop plants (peas, beans of various kinds, lentils, etc.), many tropical species are important timber trees and many species are grown as green manures because of their ability (through the bacteria present in the root nodules) to fix atmospheric nitrogen, thus improving the fertility of the soil. Many species are also grown as ornamentals. The family (often known alternatively as **Fabaceae**), is divided into 3 subfamilies, which are often treated as individual families. Though these differ conspicuously among themselves, they are held together by a number of characteristics, of which the possession of a legume as the fruit is the most important. The family is treated in the broad sense here, but, because of its diversity, the 3 subfamilies are briefly described and are keyed separately below.

Literature: Polhill, R.M., Raven, R.H. (eds) *Advances in Legume Systematics* **1** & **2** (1984), **3** (1987).

1a. Corolla radially symmetric, petals edge-to-edge in bud; leaves usually bipinnate or modified into phyllodes; seeds with a U-shaped lateral line   **Mimosoideae**
 b. Corolla bilaterally symmetric (sometimes weakly so), petals variously overlapping in bud; leaves various; seeds usually without a lateral line, rarely with a closed lateral line   2
2a. Uppermost petal overlapped by the laterals (interior); seed usually with a straight radicle
             Caesalpinioideae (p. 470)
 b. Uppermost petal overlapping the laterals (exterior); seed usually with a curved radicle
             Papilionoideae (p. 480)

Mimosoideae (*Mimosaceae*)

Trees, shrubs or herbs. Leaves usually bipinnate or modified into phyllodes (the expanded leaf-axis, without leaflets). Flowers usually small. Corolla radially symmetric, petals edge-to-edge in bud, often united. Stamens 4–many. Seed with a U-shaped lateral line.

1a. Stamens more than 10   2

 b. Stamens 10 or fewer, usually free, sometimes united at the base   7
2a. Stamens free or nearly so   **7. Acacia**
 b. Stamens united into a tube   3
3a. Fruits splitting longitudinally along 1 or 2 margins   4
 b. Fruits not splitting or breaking longitudinally into 1-seeded units
             6
4a. Seeds with arils
             11. Pithecellobium
 b. Seeds without arils   5
5a. Flowers of the same part-inflorescence uniform; valves of pod elastically dehiscing from apex, recurving   **10. Calliandra**
 b. Flowers of the same part-inflorescence heteromorphic; valves of pod not recurving from the apex or, indehiscent   **9. Albizia**
6a. Leaflets 1–many times pinnate
             12. Samanea
 b. Leaflets not pinnate   **8. Inga**
7a. Fruits splitting longitudinally along 1 or 2 margins   8
 b. Fruits not splitting or breaking transversely into individual 1-seeded units   10
8a. Anthers hairy   **5. Leucaena**
 b. Anthers hairless   9
9a. Trees; seeds red, or red and black
             1. Adenanthera
 b. Aquatic or terrestrial herbs; seeds usually brown   **6. Neptunia**
10a. Plants spineless or very rarely sparsely spiny; fruit splitting transversely into 1-seeded units
             11
 b. Plants usually spiny; fruit indehiscent   **3. Prosopis**
11a. Trees, shrubs or herbs; fruits to 3 cm   **4. Mimosa**
 b. Trees, shrubs or woody climbers; fruit usually more than 30 cm
             2. Entada

1. ADENANTHERA Linnaeus
D. Fränz
Evergreen, unarmed trees. Leaves bipinnate, leaflets opposite with several to many pairs of alternate segments. Racemes spike-like, axillary, solitary or sometimes paired, often aggregated at the tips of the shoots. Flowers bisexual, bracts minute. Calyx short, bell-shaped, 5-lobed. Petals 5, united below, not overlapping. Stamens 10, alternately long and short, each anther tipped with a deciduous gland. Ovary stalkless, with many ovules. Style thread-like, stigma small, terminal. Pods linear, curved or spirally twisted, opening

longitudinally into 2 leathery or somewhat leathery valves which do not separate from the sutures nor lose their outer layer. Seeds thick, with a hard, shining, red or red and black seed coat.

A genus of 8 species from tropical Asia, Africa and Australia. They are grown in warm, humid conditions in a compost of peat and loam. Propagation is easy by cuttings taken at a node and placed in sand in a closed frame, or by seeds which require soaking in hot water before sowing.

1. A. pavonina Linnaeus. Illustration: Graf, Exotica, edn 12, 1370 (1985).

Tree 4–20 m; young shoots usually hairless. Leaves to 40 cm, with 3–5 pairs of leaflets each with 5–9 oblong or ovate segments, 1.5–4.5 × 1.2–2.3 cm, rounded at the apex, minutely downy beneath. Racemes 9–26 cm, hairless or slightly downy; flower-stalks 2–3.5 cm. Flowers yellowish or white and yellow in the same racemes. Calyx 0.75–1 mm, usually hairless. Petals 3–4.5 cm. Filaments 2.5–4 mm. Pods 18–22 × 1.3–1.7 cm, brown, much coiled after opening. Seed lens-shaped, red, 8–10 × 7–9 mm. *India, Burma, SE Asia; often cultivated in the tropics elsewhere.* G1. Early summer.

2. ENTADA Adanson
D. Fränz
Trees, shrubs or woody climbers. Leaves bipinnate, each leaflet with 1–many segments. Flowers in spikes or spike-like racemes, solitary or clusters. Flowers white or yellow, sometimes unisexual. Calyx with 5 teeth. Petals 5, free or nearly so, borne on a short perigynous zone. Stamens 10, fertile; anthers each with an early-falling apical gland. Pods straight or curved, flat or rarely spirally twisted, sometimes to 90 cm, woody, made up of many circular, 1-seeded joints whose sides break away from the persistent sutures, and whose outer layer breaks off. Seeds large, circular, compressed, deep brown, smooth.

A genus of about 30 species, widespread in the tropics, whose large seeds are frequently washed up on beaches in western Europe. The 1 species cultivated in Europe requires warm glasshouse treatment.

1. E. phaseoloides (Linnaeus) Merrill (*Lens phaseoloides* Linnaeus; *E. scandens* (Roxburgh) Bentham; *E. gigas* (Linnaeus) Fawcett & Rendle; *Mimosa scandens* Roxburgh; *E. schefferi* Ridley). Illustration: Polunin & Stainton, Flowers of the Himalaya, pl. 24 (1984).

Large woody climber to 25 m, unarmed; young shoots more or less hairless or downy. Leaves with 2 (rarely 1) pairs of leaflets, ending in a forked tendril. Leaflets stalked, segments oblong or obovate, 2.5–5 cm, leathery. Flowers small, cream to greenish or yellowish, on distinct, slender flower-stalks 1–2 mm. Calyx hairless or shortly downy, 1–1.25 mm. Petals 2.5–3 mm. Filaments hairless, 3.5–6 mm. Pods spirally twisted, 60 cm or more × 7.5–10 cm, outer layer falling away to reveal the thick, papery, somewhat flexible inner layer. Seeds hard, dark brown or purplish, 4–5.5 cm across. *West Indies, tropical Africa & Asia, Pacific islands.* G2.

3. PROSOPIS Linnaeus
D. Fränz

Trees or shrubs, with or without axillary, solitary or paired spines, or sometimes the stipules spine-like. Leaves bipinnate with 1 or 2 pairs of leaflets; segments usually numerous, small, entire. Flowers small, greenish, in cylindric or spherical axillary spikes. Calyx bell-shaped, teeth very short and not overlapping. Petals 5, not overlapping, united below the middle or ultimately free, woolly on the inner surface. Stamens 10, free, projecting; each anther with a deciduous gland at the apex. Ovary covered with shaggy hairs; style thread-like. Pod sickle-shaped or spirally coiled, leathery and indehiscent, usually pulpy within, breaking into 1-seeded segments. Seeds many, ovate, compressed.

A genus of 44 species from the tropics, subtropics and W North America, differing from *Acacia* in having fewer stamens and indehiscent pods which break into segments. A few species are grown in cool glasshouse conditions. Propagation is by cuttings of firm young shoots taken close to the stem, rooted in sand under gentle heat. Some species are used as forage crops in desert regions.

1a. Tree with few spines **1. chilensis**
  b. Tree with numerous spines 2
2a. Leaflet-segments close, *c.* 2 cm; pods beaked **2. juliflora**
  b. Leaflet-segments distant, *c.* 1 mm; pods not beaked **3. cineraria**

1. P. chilensis (Molina) Stuntz.

Large trees with few spines. Leaves with 1 or 2 pairs of leaflets, each with 13–20 segments which are 1.5–4 cm, 10 or more times as long as broad. Pods sickle-shaped. *Chile, Argentina.* G1.

Fast-growing in dry conditions.

2. P. juliflora (Swartz) de Candolle (*P. cumanensis* Humboldt Bonpland & Kunth; *P. pallida* Humboldt Bonpland & Kunth).

Tree to 15 m, sometimes to 1 m across, or a shrub, with many spines 1.2–5 cm. Bark grey. Leaflet-segments in 11–19 pairs, very close (*c.* 6 mm apart), oblong or linear-oblong, thin in texture, to 2 cm, less than 5 times as long as broad, hairless or slightly hairy on the margin. Flowers fragrant, in cylindric spikes 5–10 × 1–1.5 cm, yellow (mainly from the projecting stamens). Pods almost straight, beaked, hairless, 10–20 cm. *Coastal Central & N South America, West Indies, Caribbean.* G1.

3. P. cineraria (Linnaeus) Druce (*P. spicigera* Linnaeus).

Small to medium tree with many broadly based spines, and with thick, fibrous grey, rough bark and purplish brown, hard wood. Leaves with distant leaflet-segments *c.* 12.5 mm. Pods cylindric, beaded, pulpy. *Oman & Iran to India & SE Asia.* G1.

4. MIMOSA Linnaeus
D. Fränz

Trees, shrubs or herbs of varying habit (rarely woody climbers), usually with spines. Leaves bipinnate (sometimes superficially palmate because of the short axis), often sensitive to touch, occasionally reduced to phyllodes; leaflets with few to many pairs of segments. Flowers small, stalkless, in spherical heads or cylindric spikes. Calyx minute or rudimentary. Petals 4 or 5, united. Stamens 4–10, projecting; pollen granular. Pods flat, straight to much curved, usually splitting into 1-seeded segments; outer layer not separating; sutures persistent.

A genus of between 400 and 500 species, widely distributed in the tropics and subtropics, but mostly found in S America. They require cool glasshouse treatment in Europe and grow well in a compost of equal parts of loam and peat to which a little sand should be added. They are propagated by seed or by cuttings in a sandy compost with bottom heat.

1a. Annual herb **4. pudica**
  b. Shrubs 2
2a. Leaflet-segments ovate **1. sensitiva**
  b. Leaflet-segments linear-oblong 3
3a. Stems and leaves with recurved spines; flowers in spikes **2. spegazzinii**
  b. Stems and leaves with straight spines; flowers in heads **3. pigra**

1. M. sensitiva Linnaeus (*M. floribunda* Bentham).

Somewhat climbing, prickly evergreen shrub to 1.8 m. Leaves with 2 unequal leaflets; leaflet-segments ovate, acute, with adpressed hairs beneath, hairless above; leaf-stalks prickly. Flowers purple. *Tropical America.* G2.

Less sensitive to touch than *M. pudica.*

2. M. spegazzinii Pirotta. Illustration: Graf, Exotica, edn 12, 1385 (1985).

Spiny, much branched, somewhat climbing shrub, with recurved spines at the bases of the leaf-stalks. Leaves with 2 leaflets 5–7.6 cm; segments very numerous, stalkless and close together, oblong or linear-oblong, acute, 3-nerved. Flowers rose-purple in spherical heads to 3.8 cm across borne in terminal racemes. Filaments rose-purple, anthers yellow. Pods 3- or 4-seeded, prickly, to 2.5 cm, linear. *Argentina, naturalised in USA (California).* G2.

3. M. pigra Linnaeus (*M. asperata* Linnaeus).

Shrub to 4.5 m, sometimes climbing or scrambling. Stems with broadly based spines to 7 mm, usually more or less adpressed-bristly. Leaf-stalks 3–15 mm, axis with erect or forwardly pointing slender prickles; leaflet-segments 6–16, sometimes with prickles between the segment-pairs, linear-oblong, 3.8–12.5 × 0.5–2 mm, margins often minutely bristly. Flowers mauve or pink in almost spherical, stalked heads to 1 cm across, borne 1–3 together in the upper axils. Stamens 8. Pods clustered, densely bristly all over, 3–8 × 0.9–1.4 cm. *Widespread in tropical Africa & America; introduced in tropical Asia.* G2.

4. M. pudica Linnaeus. Illustration: Rucker, Die Pflanzen im Haus, 291 (1982); Graf, Exotica, edn 12, 1387 (1985); Graf, Tropica, edn 3, 546, 562 (1986); Encke, Kalt- und Warmhauspflanzen, 302 (1987).

Grown as an annual, but probably perennial, sometimes woody below, to 50 cm, often prostrate or straggling; stems sparsely spiny, spines 2.5–5 mm, also bristly or almost hairless. Leaves not spiny, stalk 1.5–5.5 cm; axis short, the usually 2 leaflets appearing almost palmate; leaflet-segments 10–26 pairs, linear-oblong, 6–15 × 1.2–3 mm. Flowers lilac or pink, in shortly ovoid, stalked heads 1–1.3 × 0.6–1 cm, borne 1–5 together in the leaf-axils. Stamens 4. Pods clustered, densely prickly on the margins, 1–1.8 cm × 3–5 mm.

Originally from Brazil but now widely naturalised in warm countries. G1.

5. LEUCAENA Bentham
D. Fränz

Trees or shrubs, without spines. Leaves evergreen, bipinnate, with a gland often present on the axis above, between the lowest pair of leaflets, sometimes glands present elsewhere on the axis; leaflets with 1–several pairs of segments. Flowers in rounded, stalked, axillary heads, borne 1–3 together. Flowers bisexual, stalkless. Calyx with 5 teeth. Petals 5, free, downy or hairless outside. Stamens 10, fertile, anthers without glands at their tips. Ovary stalked, downy or hairless. Pods oblong or linear-oblong, compressed, usually thinly leathery, splitting into 2 non-recurving lobes. Seeds lying more or less transversely in the pod, glossy, unwinged.

A genus of about 50 species from the SE USA, C America and S America; some species are grown as fodder and are naturalised in Texas, California and Hawaii. Cultivation as for *Acacia*.

1. L. leucocephala (Lamarck) de Wit
(*Mimosa glauca* Linnaeus; *L. glauca* (Linnaeus) Bentham; *Acacia glauca* (Linnaeus) Moench; *Acacia leucocephala* Link; *Mimosa leucocephala* (Link) Lamarck; *Acacia frondosa* Willdenow). Illustration: Graf, Exotica, edn 12, 1385 (1985); Graf, Tropica, edn 3, 556 (1986).

Shrub or small tree to 9 m; young shoots densely and shortly grey-downy. Leaves with stalk 2.4–7 cm, often with a gland above at the junction of the lowest pair of leaflets, glands usually otherwise absent; leaflets 4–8 pairs, opposite; segments 10–20 pairs, 7–8 × 1.5–5 mm, glaucous beneath. Flower-heads spherical; stalks 2–5 cm. Calyx 2–3.5 mm, pale green, very shortly downy outside. Filaments 6.5–7.5 mm; anthers hairy. Pods 8–18 × 1.8–2.1 cm, on a stalk to 3 cm. Seeds elliptic to obovate, 7.5–9 × 4–5 mm. *Tropical America, extending to USA (Florida, Texas).* G2.

This species is widely naturalised in the Old World from cultivation as a crop.

6. NEPTUNIA Loureiro
D. Fränz

Aquatic or terrestrial, annual or perennial herbs, erect, prostrate or floating, without spines, branches often compressed or angled. Leaves bipinnate; leaflets with 8–15 pairs of segments. Flowers in spherical or ellipsoid heads which are solitary and axillary. Flowers small, stalkless, the upper bisexual, the lower male and the lowest sterile with flattened staminodes. Calyx bell-shaped, small, 5-toothed. Petals 5, free or united at the base, not overlapping. Stamens 5 or 10, free; anthers each with an apical gland. Ovary stalked, with many ovules; style thread-like, stigma minute, terminal, concave. Pod flat, membranous, oblong, not contorted or spiralled, dehiscent. Seeds transverse, compressed, oblong-ellipsoid to obovoid, smooth.

A genus of 12 or more species widely distributed, mainly in the tropics. They are not easy to cultivate. The leaves are sensitive to touch, like those of species of *Mimosa*.

1a. Leaf without a gland on the axis between the lowest pair of leaflets
          1. oleracea
  b. Leaf with a gland on the axis between the lowest pair of leaflets
          2. plena

1. N. oleracea Loureiro (*Mimosa prostrata* Lamarck; *N. prostrata* (Lamarck) Baillon; *Mimosa natans* Roxburgh; *Desmanthus natans* (Roxburgh) Wight & Arnott). Illustration: Graf, Exotica, edn 12, 1385 (1985); Graf, Tropica, edn 3, 560 (1986).

Annual aquatic herb with stems usually floating or creeping, rooting especially at the nodes, hairless or shortly downy when young; branches zig-zag. Leaves with stalk 2.5–9 cm; leaflets in 2–4 pairs, each divided into 7–22 pairs of oblong segments, 5–20 × 1.5–4 mm, hairless or with a few hairs on the margins. Flowers yellow, in heads 1.5–2.5 cm on stalks 6.5–30 cm. Calyx 1–3 mm. Petals 3–4 mm. Stamens 10, anthers completely without glands at their apices; staminodes 1.7–2.1 cm. Pods bent at an angle to the short stalk, shortly oblong, 1.3–3.8 × 1–1.2 cm. Seeds 5–5.5 × 3–3.5 mm. *Tropics.* G2.

2. N. plena (Linnaeus) Bentham (*Mimosa plena* Linnaeus; *Desmanthus plenus* Willdenow). Illustration: Bruggeman, Tropical plants, pl. 176 (1957); Graf, Exotica, edn 12, 1385, 1387 (1985).

Plant perennial, somewhat woody. Leaves with 2–4 pairs of leaflets, each with 10–30 pairs of segments. Flowers mostly pale yellow, male flowers brown. Heads ovoid, *c.* 3.8 × 2.5 cm on stalks to 15 cm. The sterile flowers consist of a mass of reflexed, narrowly lanceolate staminodes. *American & Asian tropics.* G2.

7. ACACIA Miller
E.C. Nelson

Trees and shrubs, evergreen, Leaves compound, bipinnate and persistent in mature plants, or bipinnate leaves entirely absent after the seedling stage and replaced by phyllodes. Inflorescence fluffy, spherical or cylindric, solitary or in clusters, axillary; flowers yellow, hermaphrodite, Sepals 3–5, minute. Petals 3–5, fused into bell-shaped corolla. Stamens numerous, much longer than corolla, thus conspicuous. Ovary stalkless; ovules numerous. Legumes usually dehiscent, with 2 valves; seeds usually with a fleshy aril.

A huge genus of at least 900 (possibly 1200) species concentrated in Australia (Western Australia possesses more than 500 species), but also distributed in tropical and subtropical regions of Africa, Asia and America; especially abundant in arid and semi-arid habitats.

Acacia is undergoing intensive taxonomic studies in Australia at present and numerous new species are likely to be named. There is a proposal to split *Acacia* (cf. Pedley 1986) into three genera - most taxa included here would be assigned to the controversial genus *Racosperma* (cf. Maslin 1989).

Because of these taxonomic problems, and because in European gardens only a few of the hundreds of species are in general cultivation (although botanical gardens may harbour many more species) I have taken a very restricted view of *Acacia* in this treatment. The genus will receive comprehensive treatment in a forthcoming volume of *Flora of Australia*. Several species are sold as "mimosa".

Literature: Pedley, L., Derivation and dispersal of *Acacia* (Leguminosae) with particular reference to Australia, and the recognition of *Senegalia* and *Racosperma*. *Botanical Journal of the Linnean Society* **92**: 219–254 (1986); Maslin, B.R., Wattle become of Acacia. *Australian Systematic Botany Society Newsletter* **58**: 1–13 (1989).

1a. Mature plants with true pinnate or bipinnate leaves   2
  b. Mature plants with phyllodes (not pinnate or bipinnate leaves)   3
2a. Leaves with glaucous bloom; *c.* 5 cm, with 2–4 pairs of pinnae
        1. baileyana
  b. Leaves not glaucous, much more than 5 cm, with more than 4 pairs of pinnae   **2. dealbata**

3a. Phyllodes triangular
                                    5. pravissima
  b. Phyllodes linear or lanceolate, not
     triangular   4
4a. Phyllodes linear, much less than
     0.5 cm across   5
  b. Phyllodes not linear, more than 0.5
     cm across   6
5a. Inflorescence a loose spike, bright
     yellow   **7. riceana**
  b. Inflorescence cylindric, pale yellow
                                 9. verticillata
6a. Inflorescence cylindric
                                   3. longifolia
  b. Inflorescence spherical   7
7a. Phyllodes with several parallel
     veins prominent, to 15 × 3 cm dark
     green to grey green; inflorescence
     pale creamy yellow
                                  4. melanoxylon
  b. Phyllodes with only mid-vein
     prominent; inflorescences bright
     yellow   8
8a. Phyllodes blue-green   **6. retinodes**
  b. Phyllodes green   **8. saligna**

1. A. baileyana Mueller. Illustration:
Botanical Magazine, 9309 (1933); Elliot &
Jones, Encyclopaedia of Australian plants
suitable for cultivation **2**: 20, 21 (1982);
Phillips & Rix, Shrubs, 18 (1989); Brickell,
The RHS gardeners' encyclopedia of plants
and flowers, 69 (1989).

Tree to 10 m. Leaves bipinnate,
glaucous blue-green to grey; pinnules *c.*
0.5 × 0.2 cm, pinnae oblong, 2–4 pairs
per leaf, 2–5 × 0.5–1 cm. Inflorescences
spherical in a compound spike. *Australia (S
New South Wales)*. G1–H5. Summer.

Widely cultivated in southern Europe,
but intolerant of damp winters and thus
not common in northwestern gardens. A
prostrate cultivar, and others with purple
or golden foliage are reported in Australian
gardens.

2. A. dealbata A. Cunningham.
Illustration: Phillips & Rix, Shrubs, 18
(1989); Brickell, The RHS gardeners'
encyclopedia of plants and flowers, 56
(1989).

Tree to 30 m. Leaves bipinnate, usually
green; pinnules linear, 0.2–0.5 cm,
coarsely hairy; pinnae 10–20 pairs per
leaf, *c.* 3 cm. Inflorescences spherical in
spike. *Australia (New South Wales, Victoria,
Tasmania)*. G1–H5. Summer.

Probably the most tolerant of the species
cultivated in Europe, and commonly sold
as "mimosa"; fast-growing in suitable
conditions and not particular about soil
conditions.

3. A. longifolia (Andrews) Willdenow.
Illustration: Botanical Magazine, 1827
(1816), 2166 (1822).

Tree or shrub to 10 m. Phyllodes
curved, lanceolate-elliptic with numerous
veins, 7–15 × 1–2.5 cm, green.
Inflorescences cylindric, to 5 cm. *SE
Australia.*

Exceptionally tolerant of lime.

4. A. melanoxylon R. Brown (*Racosperma
melanoxylon* (R. Brown) L. Pedley).
Illustration: Botanical Magazine, 1659
(1814); Elliot & Jones, Encyclopaedia of
Australian plants suitable for cultivation **2**:
82 (1982).

Tree to 20 m or bushy shrub. Bipinnate
foliage only on juvenile plants, soon
replaced; phyllodes lanceolate to elliptic,
dark green to grey-green, with longitudinal
veins, to 15 × 3 cm. Inflorescences
spherical, pale cream-yellow, in spikes
which are shorter than phyllodes. *E
Australia.* G1–H5. Spring–early summer.

Possibly the most abundant of the
species with phyllodes cultivated in
western Europe. Plants will produce
suckers.

5. A. pravissima Mueller. Illustration:
Elliot & Jones, Encyclopaedia of Australian
plants suitable for cultivation **2**: 99
(1982); Brickell, The RHS gardeners'
encyclopedia of plants and flowers, 69
(1989).

Shrub *c.* 1.5 m, with long, sparsely
branching arching shoots. Phyllodes
approximately triangular, closely set along
stems, 0.5–2 × *c.* 1 cm, with 2 or 3
conspicuous veins and gland on margin
near branch. Inflorescences spherical, in
spikes *c.* 10 cm, bright yellow. *Australia
(New South Wales, Victoria)*. G1–H5.
Spring.

A. cultriformis G. Don. Illustration:
Botanical Magazine, n.s. 322 (1958–59);
Brickell, The RHS gardners' encyclopedia
of plants and flowers, 103 (1989). Similar,
but phyllodes silver-grey.

6. A. retinodes Schlechtendal.
Illustration: Botanical Magazine, 9177
(1929–30); Elliot & Jones, Encyclopaedia of
Australian plants suitable for cultivation **2**:
106 (1982).

Shrub, rarely a small tree to 6 m, with
pendent branches. Phyllodes linear-
lanceolate, often slightly curved, with
prominent mid-vein, 5–20 × 0.5–2 cm,
apex pointed, blue-green. Inflorescences
spherical, in spikes shorter than phyllodes.
Australia. H5–G1. Summer.

Tolerant of salt and thus suitable for
cultivation in coastal gardens.

7. A. riceana Henslow. Illustration:
Botanical Magazine, 5835 (1870); Curtis
& Stones, Endemic flora of Tasmania **2**: 42
(1969).

Shrub to 6 m with pendent branches.
Phyllodes arranged in whorls, linear,
sharply pointed, 1–5 × *c.* 0.2 cm, dark
green. Inflorescence a loose cylindric spike,
lemon-yellow, *c.* 5 cm (longer than
phyllodes).
Australia (Tasmania). G1–H5. Spring.

8. A. saligna (Labillardière) Wendland (*A.
cyanophylla* Lindley). Illustration: Elliot &
Jones, Encyclopaedia of Australian plants
suitable for cultivation **2**: 110 (1982);
Bennett and Dundas, The bushland plants
of Kings Park, Western Australia, 17
(1989).

Shrub, sometimes with pendent
branches. Phyllodes variable, usually
curved slightly, otherwise linear-lanceolate,
8–30 × 1–8 cm, midrib prominent, green.
Inflorescences spherical, in axillary spikes.
Australia (Western Australia). G1–H5.
Spring.

9. A. verticillata (L'Héritier) Willdenow.
Illustration: Botanical Magazine, 110
(1797).

Shrub to 5 m. Phyllodes linear, dark
green, sharply pointed, 1–2 cm.
Inflorescences cylindric, solitary or in
groups of 2–3, to 1.5 cm, pale yellow. *SE
Australia*. G1–H5. Spring.

This is reputed to be tolerant of lime
and wet soils, but it is not common in
European gardens.

8. INGA Miller
D. Fränz
Trees or shrubs without spines. Leaves
pinnate, without a terminal leaflet but with
2–5 pairs of rather large lateral leaflets,
usually with glands on the axis between
the leaflet-pairs. Flowers white or
yellowish, in heads, spikes, racemes or
umbels, borne in the leaf-axils in clusters of
1–5. Stamens many, united below. Corolla
small, tubular or bell-shaped. Pods narrow,
more or less indehiscent, often thickened
at the sutures, often 4-angled, with a white
fleshy pulp around the seeds.

A genus of about 200 species from
tropical America, extending to USA
(Florida, S California). A few are grown
in tropical house conditions, in a peat and
loam compost, with abundant watering
during the summer, but scarcely any in

winter. The genus is similar to both *Calliandra* and *Acacia*, differing from the latter in its united stamens.

1a. Corolla white   **1. feuillei**
 b. Corolla brown, stamens white
                                          2. edulis

1. I. feuillei de Candolle (*I. dulcis* Willdenow; *I. anomala* misapplied).

Tree to 9 m. Leaves simply pinnate, leaflets in 3 or 4 pairs, ovate-oblong, acute at both ends, hairless; leaf-stalks winged. Corolla white. Pods 30–60 cm, linear, flat, hairless, white inside with sweet, edible pulp. *Peru*. G2.

2. I. edulis Martius. Illustration: Graf, Exotica, edn 12, 1384 (1985); Graf, Tropical, edn 3, 563 (1986).

Tree to 15 m, with broad crown and grey bark. Leaves simply pinnate, glossy dark green, the leaflets separated by the winged axis. Corolla brown, hairy, contrasting with the white stamens. Pods with thickened, furrowed margins, 4-angled, containing edible, sweet pulp. *C & S America*. G2.

9. ALBIZIA Durazzini
D. Fränz
Trees or shrubs without spines, rarely climbing. Leaves bipinnate, leaflets each with 1–many pairs of variably sized ultimate segments; stipules usually small but sometimes large and leaf-like. Flowers in spherical heads or spikes or spike-like racemes, stalked, axillary and solitary or clustered. Flowers bisexual or occasionally some male only; 1 or 2 flowers in each head often larger and different in form from the others, apparently male. Calyx with 5 teeth or lobes. Corolla funnel-shaped or bell-shaped with 5 lobes. Stamens 19–50, fertile, their filaments united below into a tube which may project from the corolla. Ovary stalkless or shortly stalked, with many ovules; style thread-like, stigma minute. Pods oblong, straight, flat, usually dehiscent, without cross-walls, papery or leathery but not thickened or fleshy. Seeds ovate or circular, compressed.

A genus of 100–150 species mostly found throughout the tropics, but extending into the subtropics. It is closely related to *Acacia*, but has united filaments. The species require glasshouse conditions in much of Europe, and are usually propagated by seed.

1a. Ultimate segments of leaves fewer
      than 20 pairs   2

 b. Ultimate segments of leaves more
      than 20 pairs   3
2a. Leaflets linear, silky downy beneath
                                          1. lophantha
 b. Leaflets oblong, downy only on the
      midrib, or hairless   **2. julibrissin**
3a. Shrub   **3. basaltica**
 b. Tree   4
4a. Leaflets less than 1 cm; flowers
      cream   **4. chinensis**
 b. Leaflets more than 1 cm; flowers
      yellowish or greenish white   5
5a. Leaflets 0.8–2.4 cm wide, hairless
      beneath   **5. lebbeck**
 b. Leaflets 0.4–1.1 cm wide, very
      downy beneath   **6. adianthifolia**

1. A. lophantha (Willdenow) Bentham (*Acacia lophantha* Willdenow; *Albizia distachya* (Ventenat) McBride). Illustration: Parey's Blumengärtnerei, 854 (1958); Blombery, What flower is that? 42 & pl. 60 (1972); Encke, Kalt- und Warmhauspflanzen, 300 (1987).

Shrub or small tree to 6 m; branches, leaf-stalks and inflorescence-stalks usually velvety-downy. Leaves with 14–24 leaflets, each with 40–60 ultimate segments which are to 1 cm, linear, silky-downy beneath. Flowers stalked in usually paired spikes to 5 cm, yellowish or white. *SW Australia*. H5–G1. Summer.

2. A. julibrissin (Willdenow) Durazzini (*Acacia julibrissin* Willdenow; *Mimosa japonica* Thunberg). Illustration: Everett, New York Botanical Garden illustrated encyclopedia, 23 (1960); Graf, Exotica, edn 12, 1368 (1985).

Tree to 12 m, with broad, spreading crown; young shoots angular, hairless. Leaves bipinnate with 6–12 pairs of leaflets each divided into 20–30 pairs of ultimate segments, the entire leaf 23–46 cm long and half as wide; ultimate segments 8–13 × 2–3 mm, oblong, oblique, sometimes downy on the midrib beneath. Flowers light pink in terminal clusters of dense heads each terminating in a stalk 2.5–5 cm. Stamens numerous, thread-like, 2.5 cm or more. Pods flat, to 15 × 2 cm, constricted between the seeds. *Ethiopia, Iran to Japan, C China*. H4.

Though this is the hardiest of the species, young plants require glasshouse protection during their first year. The leaflets fold up in a sleeping position during the night.

Forma **rosea** (Carrière) Mouillefent (*A. rosea* Carrière; *A. nemu* invalid). Illustration: Noailles & Lancaster, Plantes de jardins mediterranée, 38 (1977).

Dwarfer and more bushy than the genuine species, flowers bright pink.

3. A. basaltica Bentham.
Shrub with cylindric branchlets, minutely glandular rusty-downy. Leaves with 1 or 2 pairs of leaflets, the leaf-stalk at most 1.3 cm; ultimate segments in 5–10 pairs, oblong or ovate, very obtuse, mostly 4–6 mm, leathery, minutely hoary-downy. Flowers in dense spherical heads of about 20–30 in upper axils, inflorescence-stalks scarcely exceeding the subtending leaves. Flowers to 3 mm. Calyx downy, shortly lobed, about ⅔ as long as the corolla. Staminal tube nearly as long as corolla, free parts of the filaments much longer. Pod to 7.5 × 0.8–1 cm, leathery, very flat, with somewhat thickened margins. Seeds flat, circular. *Australia (Queensland)*. G1.

4. A. chinensis (Osbeck) Merrill (*Mimosa chinensis* Osbeck; *A. stipulata* Roxburgh). Illustration: Polunin & Stainton, Flowers of the Himalaya, pl. 24 (1984).

Large deciduous tree to 40 m, trunk to 1.5 m across, with a broad, flat-topped crown and smooth grey bark. Leaves like those of *A. julibrissin* but with conspicuous stipules and ultimate segments scarcely 3 mm wide; midrib of ultimate segments running along or near the upper margin. Flowers mostly cream, though sometimes tinged with purple. Pods 10–18 cm, light brown. *SE Asia, from Pakistan to Indonesia (Java)*. G2.

5. A. lebbeck (Linnaeus) Bentham (*Mimosa lebbeck* Linnaeus; *Acacia lebbeck* (Linnaeus) Willdenow; *Mimosa sirissa* Roxburgh). Illustration: Everett, New York Botanical Garden illustrated encyclopedia of horticulture, 177 (1960); Graf, Exotica, edn 12, 1371 (1985); Graf, Tropica, edn 3, 539, 540 (1986).

Deciduous tree to 15 m, bark grey, young shoots downy. Leaves with 2–4 pairs of leaflets divided into 3–11 pairs of ultimate segments 1.5–4.8 × 0.8–2.4 cm, oblong or elliptic-oblong, somewhat asymmetric with midrib near upper margin, rounded at apex, hairless or rarely thinly downy above. Flowers yellowish white on stalks 1.5–4.5 mm. Calyx 3.5–5 mm, not slit on one side, sometimes shortly downy. Corolla 5.5–9 mm, outside of lobes finely downy. Staminal tube not or scarcely projecting beyond corolla; filaments 1.5–3 cm, pale green or greenish yellow above, white below. Pods oblong, 15–33 × 3–5.5 cm, hairless or almost so, leathery, glossy, more or less veined, straw-coloured. Seeds 7–11.5 × 7–9 mm, flattened. *India to Australia, widely naturalised elsewhere.* G1.

6. A. adianthifolia (Schumacher) Wight (*Mimosa adianthifolia* Schumacher; *A. fastigiata* (Meyer) Oliver).

Tree to 40 m with flattened crown; bark grey to yellowish, rough; young shoots coarsely and persistently rusty- or brown-downy. Leaves with 5–8 pairs of leaflets each divided into 9–17 pairs of obliquely diamond-shaped or oblong ultimate segments mostly 7–20 × 4–11 mm, thinly downy above, very downy beneath. Flowers on stalks 0.5–1 mm. Calyx 2.5–4 mm, downy. Corolla 6–11 mm, white or greenish white, downy. Staminal tube projecting 1.3–2 cm beyond corolla, red to greenish or pink. Pod oblong, flat or slightly transversely folded, densely and persistently downy, not glossy, prominently veined, usually pale brown. Seeds 7–9.5 × 6.5–8.5 mm. *E & S Africa.* G1.

10. CALLIANDRA Bentham
D. Fränz

Shrubs or small trees. Leaves bipinnate. Flowers in large, rounded or spherical heads. Calyx 5-toothed or -lobed. Petals 5, united from the base to about the middle into a 5-lobed, funnel-shaped or bell-shaped corolla. Stamens numerous, long-projecting, their coloured filaments providing most of the colour of the inflorescence. Ovary with many ovules; style thread-like. Pods linear, usually narrowed towards the base, flat, straight or almost so, not pulpy within, opening elastically from the apex, the flaps leathery with raised margins. Seeds circular to obovate, compressed.

A genus of about 200 species from tropical America, Asia and Madagascar. Several species are grown in an open compost of loam and peat. Most require warm greenhouse conditions and abundant sunshine; *C. portoricensis* may survive in the open in very warm areas. In all species, the branches require cutting once a year. Propagation is by seed or by cuttings, which require bottom heat to root successfully.

1a. Leaves with more than 2 pairs of leaflets   2
  b. Leaves with 1 or 2 pairs of leaflets   4
2a. Leaves with 7–12 pairs of leaflets   **1. inermis**
  b. Leaves with 3–7 pairs of leaflets   3

3a. Inflorescence to 35 cm wide   **2. eriophylla**
  b. Inflorescence to 7.5 cm wide   **3. tweedii**
4a. Leaves 1-pinnate   **4. surinamense**
  b. Leaves 2-pinnate   5
5a. Flowers pink   6
  b. Flowers white or yellowish   7
6a. Inflorescence spherical   **5. fulgens**
  b. Inflorescence hemispherical   **6. inaequilatera**
7a. Inflorescence spherical; flowers white   **7. portoricensis**
  b. Inflorescence hemispherical; flowers yellowish   **8. gracilis**

1. C. inermis (Linnaeus) Druce (*Gleditsia inermis* Linnaeus; *Mimosa houstonii* L'Héritier; *Inga houstonii* (L'Héritier) de Candolle; *C. houstonii* (L'Héritier) Bentham).

Slender shrub to 6 m. Leaves with 7–12 pairs of leaflets; ultimate segments in 10–40 pairs, oblong-linear, somwehat curved, to 6 mm, acute. Flower-heads clustered in a terminal, raceme-like inflorescence to 36 cm. Flowers to 5 cm; corolla hairy, anthers purple-red. Pods to 12 cm, with dense, brown, stiff hairs. *C America.* G2.

2. C. eriophylla Bentham. Illustration: Orr, Wildflowers of western America, pl. 181 (1974); Graf, Exotica, edn 12, 1375 (1985).

Low, much-branched shrub to 30 cm, older branches grey, young shoots with broad ridges bearing downwardly pointing hairs. Stipules bristle-like. Leaflets 1–7 pairs; ultimate segments usually 5–8 pairs, oblong, 3–4 mm, obtuse or somewhat acute, bearing more or less flattened bristles. Heads few-flowered, in racemes or axillary. Calyx 1–1.5 mm. Corolla 4–6 mm. Stamens to 2 cm, united at the base into a short tube, reddish purple. Pods 3–6 cm × *c.* 5 mm, tapering below the middle, densely downy with downwardly pointing hairs. *USA (Texas to Arizona & California), Mexico.* G1.

3. C. tweedii Bentham (*Inga pulcherrima* Sweet), Illustration: Graf, Exotica, edn 12, 1374 (1985); Tropica, edn 3, 548 (1986).

Shrub or small tree to 2 m, sparsely hairy. Leaves bipinnate with 3–6 or more pairs of leaflets each 3.8–6.3 cm; ultimate segments very numerous, overlapping, narrowly oblong, 6–8 mm, silky-hairy when young. Flower-heads hemi-spherical, 5–7.5 cm wide, each axillary on a downy stalk to 5 cm. Calyx and corolla yellowish

green, shaggy-hairy, lobes erect. Stamens red, 2.5–3.8 cm, numerous. Pods to 5 cm, shaggy-hairy. *Brazil.* G1–2. Late winter– autumn.

4. C. surinamensis Bentham. Illustration: Milne & Milne, Living plants of the world, 107 (1967); Chin, Malaysian flowers in colour, 51 (1977); Graf, Exotica, edn 12, 1374 (1985); Graf, Tropica, edn 3, 548 (1986).

Small tree, spreading shrub or woody climber. Leaves with 1 pair of leaflets; ultimate segments in 8–12 pairs, oblong-lanceolate, to 1.3 cm. Flowers in axillary heads. Stamens united below into a projecting column, filaments red apically, white in the lower part. *Brazil, Surinam.* G2.

5. C. fulgens J.D. Hooker.

Evergreen shrub or small tree. Leaves with 2 leaflets each divided into 3 pairs of ultimate segments; leaf-stalk slender, downy; ultimate segments narrowly oblong, 2.5–6.5 × 0.8–1.6 cm. Flowers in spherical, shortly stalked heads to 6.5 cm wide. Corolla bright pink, to 1.3 cm. Stamens with scarlet filaments and crimson anthers. *Mexico.* G2. Spring.

6. C. inaequilatera Rusby. Illustration: Graf, Exotica, edn 12, 1374 (1985); Graf, Tropica, edn 3, 547 (1986).

Loose shrub or small tree, to 7.5 m, spreading. Leaves with 2 pairs of leaflets, each divided into 4–10 pairs of ultimate segments; terminal ultimate segments to 9.2 cm, others 2–3.8 cm, obliquely oblong-lanceolate. Flowers in hemi-spherical heads 5–7.5 cm wide, bisexual or functionally male. Corolla small, pink. Stamens *c.* 25, 2.5–3.1 cm, filaments white towards the base, red apically, united at the base into a tube as long as or longer than the corolla, with an irregular internal fringe; anthers black. Pods densely downy. *Bolivia, Brazil.* G2. Late winter.

7. C. portoricensis (Jacquin) Bentham (*Mimosa portoricensis* Jacquin; *Acacia portoricensis* (Jacquin) Willdenow). Illustration: Bruggeman, Tropical plants, pl. 230 (1957); Graf, Exotica, edn 12, 1375 (1985); Graf, Tropica, edn 3, 548 (1986).

Shrub or small tree to 8 m, with grooved, downy young shoots. Leaves bipinnate with 2–7 pairs of leaflets 3.8–7.5 cm, ultimate segments linear to narrowly oblong, overlapping, densely set; stalks and margins downy. Flowers in spherical axillary heads to 5 cm across. Corolla to 6 mm, green. Stamens many, white, *c.*

1.9 cm. Pods to 12.5 cm. *West Indies, C America, W Africa*. G2. Summer.

8. C. gracilis Baker (*C. moritziana* Cardenas).

Evergreen shrub, much branched with hairy shoots. Leaves bipinnate with 1–3 pairs of leaflets, each divided into 4–6 pairs of ultimate segments 1-3–3.2 cm, obliquely ovate-oblong, downy on both surfaces. Flowers in hemi-spherical heads on a stalk 5–10 cm. Corolla yellowish, to 3 mm. Stamens 30–40, 2–2.5 cm, pale cream. *C America*. G2.

11. PITHECELLOBIUM Martius
D. Fränz
Erect, broad trees or shrubs, with or without axillary, stipular spines. Leaves bipinnate, the leaflets either small and themselves pinnately divided into numerous segments, or large and with 1–3 pairs of segments, rarely reduced to a single segment; glands usually present on the leaf-stalk. Flowers in head-like spikes, bisexual or variously unisexual, calyx and corolla 5- or rarely 6-lobed. Calyx bell-shaped, shortly toothed. Corolla tubular or funnel-shaped. Stamens few to many, much projecting, united into a tube, at least at the base. Ovary stalkless or stalked, with many ovules, style thread-like, stigma terminal, small or head-like. Pods compressed, curved, sickle-shaped, twisted or rarely straight, leathery, thick or somewhat fleshy, 2-lobed at the apex, not divided between the seeds. Seeds pulpy, ovate to circular, compressed, often dark-coloured, with variously expanded fleshy arils.

A genus of about 200 species from tropical America. Cultivation as for *Inga*.

1a. Pods twisted   **1. dulce**
  b. Pods curved into a ring
                             2. bigeminum

1. P. dulce (Roxburgh) Bentham (*Mimosa dulcis* Roxburgh; *Inga dulcis* (Roxburgh) Willdenow).

Very spiny tree to 20 m. Leaves bipinnate; ultimate segments obovate to oblong, obtuse, to 2.5 cm. Flowers in shortly stalked heads in a raceme-like panicle, white, downy. Calyx 1–1.5 mm. Corolla 3–4.5 mm. Pods twisted, 12.5–15.5 cm, red. Seeds black, glossy, 9–10 × 7–8 mm; aril pulpy. *C America, N South America; introduced in the Philippine Islands & elsewhere*. G2.

2. P. bigemninum (Willdenow) Martius (*Inga bigemina* Willdenow).

A small tree, like *P. dulce* but with white

flowers and broad, flat pods which are curved into a ring, orange-red inside and containing black seeds. *Himalayas, SE Asia to the Philippine Islands*. G2.

12. SAMANEA (Bentham) Merrill
D. Fränz
Spineless or rarely spiny trees or shrubs. Leaves several to many times pinnate, the leaflets 1–many times pinnate. Flowers in spherical heads, calyx and corolla 5-lobed. Stamens many, united into a tube at the base. Pods straight or somewhat curved, rigid, more or less constricted between the seeds, flat or cylindric, usually indehiscent, with cross-walls between the seeds.

A genus of about 30 species from C & S America, though introduced and cultivated elsewhere.

1. S. saman (Jacquin) Merrill (*Mimosa saman* Jacquin; *Pithecellobium saman* (Jacquin) Bentham; *Enterolobium saman* (Jacquin) Prain). Illustration: Graf, Exotica, edn 12, 1390, 1391 (1985); Graf, Tropica, edn 3, 560, 563, 564 (1986).

Large tree to 35 m, trunk reaching 1 m or more across; branches widely spreading, young shoots velvety-downy. Leaves 2–4-pinnate, shining above, downy beneath. Flowers shortly stalked, in heads which are borne on stalks 10–12.5 cm. Calyx to 7 mm, downy. Corolla to 1.3 cm, yellowish, silky with shaggy hairs. Stamens 20, light crimson, only shortly united at the base. Pods stalkless, straight, thick-margined, leathery-fleshy, hairless, 15–20 × 1.3–2.5 cm, flattened or almost cylindric. *West Indies, C America*. G2.

Fast-growing, widely cultivated in the tropics as an ornamental shade-tree.

Caesalpinioideae (*Caesalpiniaceae*)
Trees, shrubs or herbs. Leaves various, not often bipinnate. Corolla usually bilaterally symmetric, petals overlapping so that the uppermost is innermost, overlapped by the 2 lateral petals. Stamens 10 or fewer. Seeds usually with a straight radicle.

1a. Leaves simple, consisting of a single leaflet (which may be deeply lobed at the apex)   2
  b. Leaves pinnate or bipinnate or consisting of 2 completely distinct leaflets   3
2a. Flowers in clusters borne on the old wood, to 2 cm   **21. Cercis**
  b. Flowers not in clusters on the old wood, much longer than 2 cm
                             22. Bauhinia

3a. Leaves 1-pinnate (or apparently so), sometimes of only 2 leaflets   4
  b. Leaves clearly 2- or 3-pinnate   12
4a. Leaflets 2   **22. Bauhinia**
  b. Leaflets more than 2   5
5a. Petals and sepals similar, both small, or petals absent; flowers in catkin-like inflorescences, usually functionally unisexual   6
  b. Petals and sepals clearly differing; flowers not in catkin-like inflorescences, usually bisexual   7
6a. Petals absent; plants not spiny
                             19. Ceratonia
  b. Petals present; plants usually spiny
                             13. Gleditsia
7a. Fertile stamens 3
                             26. Tamarindus
  b. Fertile stamens 7 or more   8
8a. Flowers bright red or orange-red, borne in large, pendent heads; young leaves limp, brightly coloured, at first enclosed in a tube of scale-leaves   **23. Brownea**
  b. Flowers yellow, or if red then not in large pendent heads; young leaves not as above, though sometimes limp   9
9a. Leaves divided into 2 leaflets at the extreme base, each easily mistaken for a 1-pinnate leaf; leaflet-axes flattened, ultimate segments numerous, small for the size of the leaf   **18. Parkinsonia**
  b. Leaves strictly 1-pinnate; axis not flattened; leaflets to 8 pairs, not conspicuously small for the size of the leaf   10
10a. Anthers opening by terminal or basal pores; petals yellow or orange-yellow, rarely pink or whitish   **20. Cassia**
  b. Anthers opening by slits; flowers red or pink   11
11a. Petals 5, 3 very large (to 5 cm) and 2 very small; flowers in hanging racemes   **25. Amherstia**
  b. Petals 5, more or less equal, or some reduced to threads, the largest to 1.8 cm; flowers in erect panicles   **23. Schotia**
12a. Petals and sepals similar, both small; flowers usually functionally unisexual   13
  b. Petals clearly differing from the sepals, often large; flowers bisexual   14
13a. Some leaves simply pinnate, others 2-pinnate; plant usually spiny
                             13. Gleditsia

b. All leaves 2-pinnate; plant not
spiny   **14. Gymnocladus**

14a. Stigma broad, peltate; pod flattened,
winged, indehiscent, elliptic, 1- or
2-seeded   **15. Peltophorum**

b. Stigma not as above; pod variable,
not as above   15

15a. Leaves with 2 leaflets, each with a
flattened axis and ultimate segments
conspicuously small for the size of
the leaf   **18. Parkinsonia**

b. Leaves with more than 2 leaflets,
not as above   16

16a. Leaflets 15–25 pairs; petals usually
red or scarlet, rarely yellow
                       16. Delonix

b. Leaflets to 12 pairs; petals yellow
or orange-yellow, sometimes red-
spotted   **17. Caesalpinia**

13. GLEDITSIA Linnaeus
C.J. King

Deciduous trees, with trunks and branches
usually armed with simple or branched
spines. Leaves alternate, pinnate or
bipinnate, with numerous, slightly scalloped
leaflets, the terminal often absent. Flowers
small, radially symmetric, greenish,
unisexual and bisexual on the same plant, in
racemes or more rarely panicles. Sepals 3–5.
Petals 3–5, similar to the sepals. Stamens
6–10. Style short with a large terminal
stigma. Fruit a flattened pot, indehiscent or
very late dehiscent. Seeds 1–many, flattened,
ovate to almost circular.

A genus of about 14 species, mainly from
the Caspian region of E Asia, but with 2
or 3 in E North America and one in warm
temperate S America. Three species, one of
these having several cultivars, are generally
cultivated in Europe, usually for their
ornamental foliage and conspicuous spines.
They are rather tender when young,
thriving best in a sunny position in loamy
soils in the warmer parts of Europe.
Propagation is normally by seed, though
cultivars of *G. triacanthos* are increased by
grafting or budding on seedings of the type.

1a. Spines terete; leaves usually simple
pinnate; leaflets rarely more than
14; pods usually less than 25 cm,
almost straight   **1. sinensis**

b. Spines flattened, at least at the base,
rarely absent; leaves pinnate or
bipinnate; leaflets often more than
14; pods twisted   2

2a. Young shoots hairless, dark
purplish brown; leaflets 24 or
fewer; pods usually less than 30 cm
                        2. japonica

b. Young shoots slightly downy at
base, green; leaflets 14–22; pods
usually more than 30 cm
                        3. triacanthos

1. G. sinensis Lamarck (*G. horrida*
Willdenow). Illustration: Krüssmann,
Handbuch der Laubgeholze, edn 2, **2**: t.
42 (1977); Bean, Trees & shrubs hardy
in the British Isles, edn 8, **2**: t. 41 (1978);
Csapody & Tóth, Colour atlas of flowering
trees and shrubs, t. 61 (1982).

Trees to 15 m, with stout, conical, often
branched spines. Young shoots hairless or
soon becoming so. Leaves usually simply
pinnate, 12–18 cm, downy on the axis;
leaflets 8–14 or sometimes more, 3–8 ×
1.5–2.8 cm, ovate or ovate-lanceolate,
obtuse or acute, obliquely tapered at the
base, margins slightly wavy, more or less
downy on midrib and stalk. Flowers in
slender, pendent, downy racemes 6–8 cm.
Pods 12–25 × 2–3 cm, almost straight,
dark purplish brown, the walls dotted with
minute pits. *China*. H3.

2. G. japonica Miquel (*G. horrida*
(Thunberg) Makino; *Fagara horrida*
Thunberg). Illustration: Krüssmann,
Handbuch der Laubgehölze, edn 2, **2**: t. 35
(1977).

Tree to 20 m, with stout, branched
spines, flattened at least at the base. Young
shoots dark purplish brown, hairless and
shining. Leaves pinnate or bipinnate,
20–30 cm, usually downy on the axis;
leaflets 14–24, 2–4 × 0.6–1.2 cm, ovate
to lanceolate, obtuse or acute, sometimes
matched, margins entire or shallowly
scalloped and slightly wavy, more or less
downy on midrib and stalk. Flowers in
slender racemes. Pods 20–30 × 2.5–3.5
cm, curved and twisted. *Japan*. H3.

In small cultivated trees the leaflets are
only 1–1.5 cm, giving the plants a very
different appearance.

3. G. triacanthos Linnaeus. Illustration:
Edlin, The tree key, 153 (1978); Edlin &
Nimmo, Illustrated encyclopaedia of trees
of the world, 183 (1978); Phillips, Trees
in Britain, Europe and North America
121 (1978), Everett, New York Botanical
Garden illustrated encyclopedia of
horticulture **5**: 1493 (1981).

Tree to 25 m in cultivation, with stout,
often 3-branched spines, somewhat
flattened at least at the base. Young shoots
slightly downy at the base. Leaves pinnate
or bipinnate, 10–20 cm, downy on the
axis; leaflets 14–32, 1.5–3.5 × 0.5–1.5
cm, oblong-lanceolate, obtuse or acute,

margins wavy or shallowly scalloped,
downy at first above and beneath. Flowers
in slender, downy racemes 5–7 cm. Pods
30–45 × 2.5–4 cm, more or less curved
and twisted, shining dark brown.
E North America. H3.

There are several different growth forms,
including a pendent variant, 'Bujotii'
(Pendula), 'Elegantissima', a cultivar with
shrubby habit, and a spineless variant,
forma **inermis** de Candolle. The foliage
of this species turns golden yellow in the
autumn, and in the spineless cultivar
'Sunburst' the young growth remains
bright yellow throughout the year
(illustration: Lancaster, Trees for your
garden, edn 2, 71, (1979); The Hillier
colour dictionary of trees and shrubs, 114,
1981).

14. GYMNOCLADUS Lamarck
J. Cullen

Trees, without spines. Leaves deciduous,
bipinnate, stipules present or absent, stipels
present or absent (ours). Inflorescence a
terminal or axillary raceme or panicle.
Flowers radially symmetric, functionally
unisexual, the males with a rudimentary
ovary, the females with abortive or sterile
anthers; both sexes on the same plant.
Calyx tubular, 5-toothed with equal teeth,
the calyx not completely covering the
corolla in bud. Petals 4 or usually 5,
similar to the sepals, hairy, cream or
marked with purple outside. Stamens 10
in 2 whorls, those of 1 whorl longer than
those of the other, filaments free. Ovary
stalkless, with 4 or more ovules. Pod
oblong, often curved, woody, pulpy inside,
containing 1 or more seeds, opening along
the upper suture.

A genus of 4 species from E North
America and E China. Only the American
species is cultivated; it lacks the stipels which
are found in the other 3 species. It requires
a deep, rich soil, but grows slowly and does
not flower regularly; propagation is by seed.

Literature: Lee, Y.T., The genus
Gymnocladus and its tropical affinity,
Journal of the Arnold Arboretum **57**: 91–112
(1976).

1. G. dioicus (Linnaeus) Koch (*G.
canadensis* Lamarck). Illustration:
MacDonald encyclopedia of trees, 123
(1978); Elias, The complete trees of North
America, 652 (1980); Csapody & Tóth, A
colour atlas of flowering trees and shrubs,
139 (1982); Taylor's guide to trees, 156,
157 (1988).

Tree to 10 m, bark dark grey, fissured.

Leaflets ovate, 3–7 × 2–4 cm, abruptly acuminate, broadly rounded to truncate at the base, hairless except for a few scattered hairs on the midrib beneath; in autumn the leaflets fall individually, leaving the bare axes. Racemes or panicles of female flowers longer than those of male flowers, branches of all white-hairy; flowers fragrant. Calyx-tube 0.6–1.2 cm, hairy, teeth 4–6 mm. Petals 5–7 mm, hairy on both surfaces, greenish white. Pod 15–25 cm. *E North America*. H3. Summer.

15. PELTOPHORUM (Vogel) Bentham
J. Cullen

Round-topped deciduous trees. Young shoots, leaf-stalks and axes, and calyces covered in red-brown or greyish hairs. Leaves bipinnate, ultimate segments with short mucros at their tips. Stipules soon falling. Racemes terminal. Calyx with a short tube and 5 more or less equal lobes which overlap and completely enclose the corolla in bud. Corolla bright yellow; petals more or less equal, each with a long claw and broader blade, margins wavy. Stamens 10, filaments free, unequal, hairy at the base. Ovary flattened, shortly hairy; style hairless, twisted near the base and again towards the apex; stigma broad, peltate. Pods pendent, flattened, indehiscent, elliptic, 1- or 2-seeded.

A genus of about 9 species from the tropics. Only one is grown; it will tolerate a small degree of frost, but generally requires glasshouse treatment in most of Europe. Propagation is by seed.

1. P. africanum Sonder. Illustration: Flowering Plants of Africa **36**: t. 1434 (1964); Palgrave, Trees of southern Africa, t. 90 (1977); Palmer, Field guide to the trees of southern Africa, t. 11 (1977).

Tree to 6 m or more. Leaves to 10 cm, densely or sparsely adpressed hairy on both surfaces. Calyx-tube bell-shaped, *c.* 2 mm, lobes longer, green, spreading. Petals to 2 cm, bases of the claws hairy. Pod to 6 × 1.7 cm. *C Tropical Africa, South Africa*. G1.

The leaves are touch-sensitive, the leaflets folding upwards together on stimulation.

16. DELONIX Rafinesque
J. Cullen

Widely branched, rather flat-topped trees, without spines. Leaves bipinnate, with very numerous ultimate segments; stipules absent. Flowers very showy, in terminal or axillary racemes. Calyx of 5 sepals, all equal, free almost to the base, attached to a short perigynous zone, edge-to-edge in bud, usually reddish. Petals 5, red, scarlet or yellow, long-clawed, with abruptly broadened blades which are rounded, often broader than long and wavy on their margins; the uppermost is more gradually tapered than the rest and is sometimes marked with a contrasting colour. Stamens 10, widely spreading, filaments free, anthers dorsifixed. Ovary stalkless with many ovules. Pod woody, compressed, containing many seeds, more or less solid between them.

A genus of about 10 species, 2 from Africa, the rest from Madagascar. One of the Madagascan species (below) is now widely cultivated in the tropics and subtropics world-wide, though it is not easily grown well in Europe, requiring warm glasshouse conditions and considerable space. It is tolerant of most soils and is fast-growing in suitable sites. Propagation is by seed.

1. D. regia (Hooker) Rafinesque (*Poinciana regia* J.D. Hooker). Illustration: Botanical Magazine, 2884 (1829); Pertchik and Pertchik, Flowering trees of the Caribbean, 95 (1951); Menninger, Flowering trees of the world, t. 102, 108 (1962); Leathart, Trees of the world, 167 (1977).

Tree to 15 m. Leaves to 60 cm, with very numerous ultimate segments, bright green above, paler beneath. Flowers 7–10 cm across, red or scarlet (rarely yellow), the upper petal marked with white or yellow; claws of petals hairy outside. Filaments with swollen, hairy bases. Pod to 60 × 5–7 cm, many-seeded. *Originally from Madagascar, now widespread in the tropics & subtropics*. G2. Summer.

The Flamboyant, one of the most spectacular and widely cultivated tropical trees. A variant with completely yellow flowers is sometimes grown.

17. CAESALPINIA Linnaeus
J. Cullen

Trees or shrubs (ours), sometimes climbing or scrambling, often with thorn-like prickles. Leaves bipinnate with up to 12 pairs of leaflets; stipules deciduous or persistent. Flowers in terminal or axillary racemes, panicles or clusters, bracts usually evident. Calyx of 5 almost completely free, overlapping sepals, the lowest outermost, often larger than and partially or completely enclosing the others, sometimes toothed. Petals 5, the uppermost often slightly smaller than the others and more distinctly clawed, all yellow or reddish. Stamens 10, filaments free, shorter than to much longer than the petals. Ovary stalkless, containing few ovules. Pod variably shaped, dehiscent (sometimes very tardily) or indehiscent, woody, few-seeded.

A genus of over 100 species, mainly from the tropics but extending into temperate areas in N & S America and South Africa. They require a dryish, warm soil in a sunny position, and, when established, grow rapidly. Propagation is by seed.

Literature: Isely, D., Leguminosae of the United States: II. Subfamily Caesalpinioideae, *Memoirs of the New York Botanical Garden* **25**(2): 33–51 (1975).

Stems. With prickles: **1–4,6,7**; without prickles: **4,5**. Scrambling or climbing: **3**.
Ultimate segments of leaves. With black dots beneath: **2,5–7**; without black dots beneath: **1,2–4**.
Flower-stalks. To 1 cm: **6,7**; 1–1.5 cm: **2**; more than 1.5 cm: **1,3–5**.
Lowest sepal. Deeply toothed: **7**.
Stamens. 5–8 cm, much longer than petals: **4,5**; to 3 cm, not or little longer than petals: **1–3,6,7**.

1a. Trees; young shoots covered with rusty brown hairs; pod prickly
                        1. echinata
  b. Trees or shrubs, sometimes scrambling; young shoots not as above; pods not prickly 2
2a. Stamens 5–8 cm, considerably exceeding the petals 3
  b. Stamens at most to 3 cm, not or scarcely exceeding the petals 4
3a. Inflorescence glandular, sticky; leaflets 3–8 mm **5. gilliesii**
  b. Inflorescence neither glandular nor sticky; leaflets 1–2.5 cm
                      4. pulcherrima
4a. Lowest sepal deeply toothed, comb-like **7. spinosa**
  b. Lowest sepal not toothed 5
5a. Flower-stalks 1–2 mm; leaflets to 1 cm **6. coriaria**
  b. Flower-stalks 1 cm or more; leaflets 1 cm or more 6
6a. Leaflets asymmetric, oblong to diamond-shaped, attached by 1 corner; flower-stalks 1–1.5 cm; upright shrub **2. sappan**
  b. Leaflets not as above; flower-stalks more than 1.5 cm; plant scrambling or climbing
                      3. decapetala

1. C. echinata Lamarck. Illustration: Die Natürlichen Pflanzenfamilien **3**(3): 175 (1894).

Tree, branches densely prickly, the young shoots covered with rusty brown hairs. Leaves without prickles, with 2–4 pairs of leaflets, each divided into 7–10 pairs of ultimate segments which are oblong to diamond-shaped, rounded at the apex, somewhat tapered to the base, 1.2–2 cm. Racemes axillary and terminal; flower-stalks *c.* 2 cm. Sepals hairy, the lowest not toothed. Petals yellow. Stamens shorter than petals. Pod oblong, to 7 cm, prickly. *Brazil*. G2.

Doubtfully in cultivation in Europe, but reported in some lists.

2. C. sappan Linnaeus. Illustration: Flora of Thailand **4**(1): 66 (1984).

Shrub or tree, with prickles, hairy or not. Leaves with 7–10 pairs of leaflets, these divided into 10–20 pairs of ultimate segments which are asymmetric, oblong to diamond-shaped, attached at one corner, 1.4–2.5 × 0.7–1 cm, apex rounded, with or without black dots beneath. Racemes (rarely panicles) terminal, brown hairy at first; flower-stalks 1–1.5 cm. Sepals unequal, 0.8–1.2 cm, the lowest somewhat hooded in bud. Petals 1–1.4 cm, yellow. Filaments slightly longer than petals, densely hairy. Pod elliptic to oblong, 6–10 × 3–4 cm, eventually dehiscent. *Malaysia, Indonesia, Indo-China, S China*. G2.

3. C. decapetala (Roth) Alston. Illustration: Botanical Magazine, 8207 (1908); Hay & Synge, Dictionary of garden plants in colour, pl. 1471 (1969); Gartenpraxis for September 1976, 422; Noailles & Lancaster, Mediterranean plants and gardens, 50 (1977).

Scrambling or climbing shrub; stems and leaf-stalks with few to many prickles. Leaves with 6–10 pairs of leaflets, these divided into 7–12 pairs of ultimate segments which are obovate to elliptic, rounded at both ends, very shortly stalked, 1–2.5 × 0.5–1.5 cm, veins conspicuous, silky-hairy on both surfaces. Racemes axillary and terminal, many-flowered, flowers often hanging; flower-stalks 1.5–3 cm. Sepals to 1 cm, unequal, the lowest not toothed. Petals pale yellow, sometimes some or all of them red-spotted, 1–1.3 cm. Stamens slightly longer than petals. Pod 7–10 × 2–3 cm, flattened, eventually opening along the upper suture. *Tropical & subtropical Asia, widely introduced elsewhere*.

Var. **decapetala**, from the more tropical parts of Asia has rather distant flowers on stalks 2–3.5 cm. G2.

Var. **japonica** (Siebold & Zuccarini) Isely (*C. japonica* Siebold & Zuccarini; *C.*

sepiaria Roxburgh var. *japonica* (Siebold & Zuccarini) Makino) is from China and Japan, and has the flowers closely packed, on stalks 1.5–2.5 cm. H5–G1.

4. C. pulcherrima (Linnaeus) Swartz (*Poinciana pulcherrima* Linnaeus). Illustration: Botanical Magazine, 995 (1807); Bruggeman, Tropical plants, pl. 227 (1957); Perry, Flowers of the world, 159 (1970); Der Palmengarten, 1987 No. 2, 67.

Shrub or small tree to 5 m or more, with or without prickles, hairless, often somewhat glaucous. Leaves divided into 5–10 pairs of leaflets, these divided into 6–10 pairs of ultimate segments which are elliptic to obovate, 1–2.5 × 0.5–0.8 cm, rounded to the apex, somewhat tapered to the base. Flowers in large, terminal racemes; flower-stalks 4–6 cm. Sepals unequal, orange-red, the lowest broader than the others and enclosing them in bud. Petals 1.5–2 cm, obovate, with wavy margins, orange-yellow to yellow. Stamens with filaments 5 cm or more, much longer than the petals. Pod flattened, 6–12 × 1.5–2 cm, dehiscent. *West Indies, widely introduced in the tropics*. G1.

5. C. gilliesii (Hooker) Dietrich. Illustration: Botanical Magazine, 4006 (1843); American Horticulturist **56**: 23 (1977); Der Palmengarten, 1987 No. 2, 67.

Shrub or tree to 5 m, without prickles, young shoots shortly hairy. Leaves divided into 8–12 leaflets, these divided into 7–11 pairs of ultimate segments which are elliptic to elliptic-oblong, 3–8 × 1–5 mm, rounded to the apex, slightly tapered to the base, with black dots along the margins beneath. Racemes terminal, glandular and sticky, and with hairs without glands; flower-stalks 1.5–3 cm or more. Sepals 1.5–2 cm, the lowest broader than the others. Petals bright yellow with orange markings, 2–3.5 cm. Stamens with red filaments 5–8 cm, much exceeding the petals. Pod flattened, 6–12 × 1.5–2 cm, oblong, often curved, dehiscent, hairy at least when young. *Southern tropical & temperate S America, naturalised in several countries*. G1.

6. C. coriaria (Jacquin) Willdenow.

Small tree or shrub, without prickles. Leaves with 4–8 pairs of leaflets, each divided into 15–25 ultimate segments which are 4–10 × 1–2 mm, stalkless, oblong, hairless, rounded at apex and base, often with dark dots beneath. Flowers in axillary or apparently terminal panicles

or clusters; flower-stalks 1–2 mm. Sepals 3–4 cm, the lowermost not toothed. Petals yellow or pale yellow, 4–6 cm. Stamens scarcely exceeding the petals. Pod 3–5 × 0.8–2 cm,, ovoid to oblong, often becoming twisted, indehiscent. *C & N South America, West Indies*. G2.

7. C. spinosa (Molina) Kuntze (*C. tinctoria* Humboldt, Bonpland & Kunth).

Shrub or small tree; young shoots brown-hairy. Prickles usually present on stems and leaves. Leaves divided into 2–5 leaflets which are divided into 5–7 pairs of ultimate segments which are 1.5–4.5 × 0.8–2 cm, rounded to notched at the apex, slightly tapered towards the base, with conspicuous veins and, usually, dark dots beneath. Racemes clustered, terminal, many-flowered; flower-stalks 0.5–1 cm. Sepals unequal, 5–6 mm, usually deciduous as the flower opens, the lowest deeply toothed, comb-like. Petals yellow and reddish, 6–7 mm. Stamens very unequal, the longest equalling or slightly exceeding the petals. Pod oblong, 6–10 × 1–2.5 cm, indehiscent. *S America*. G2.

18. PARKINSONIA Linnaeus
J. Cullen

Trees or shrubs, usually with a pair of spines at each node. Leaves 2–3-pinnate, the 1–3 leaflets divided into rather small, numerous, ultimate segments. Flowers in axillary or terminal racemes, almost radially symmetric; flower-stalks jointed. Calyx divided almost to the base into 5 segments which are reflexed or deciduous after flowering. Corolla yellow, sometimes red-spotted, petals clawed. Stamens 10, filaments free and usually hairy below, almost equal, about as long as the petals. Pod leathery or woody, indehiscent or very slowly dehiscent. Seeds few.

A genus of about 10 species from southern USA, C America and Africa, some of them (not ours) sometimes separated as the genus *Cercidium* Tulasne. Only 1 is commonly grown, requiring a warm, well-drained site. Propagation is by seed.

1. P. aculeata Linnaeus. Illustration: Menninger, Flowering trees of the world, pl. 320 (1970); Gartenpraxis for January 1978, 47; Everett, New York Botanical Garden illustrated encyclopedia of horticulture, 2494 (1981); Il Giardino Fiorito for April 1987, 4.

Shrub or small tree to 10 m, shoots conspicuously green. Leaves 2-pinnate, stalkless, so that each leaflet appears to be a 1-pinnate leaf; axes of the leaflets flattened, to 30 cm, bearing numerous, shortly stalked ultimate segments which are ovate to oblong and 2–5 mm, often quickly deciduous. Calyx 6–7 mm. Corolla yellow with some red spots, to 2 cm across. Pod 2–10 × 0.5–0.6 cm, mostly rounded in section but some parts flattened. *S USA & adjacent Mexico, widely introduced in other parts of the world.* H5–G1. Spring.

The flattened leaflet-axes bearing small ultimate segments render this species easily identifiable among the cultivated woody legumes.

19. CERATONIA Linnaeus
J. Cullen

Evergreen shrubs or small trees to 10 m. Leaves pinnate, with or without a terminal leaflet; leaflets oblong to almost circular, rounded to the base and the apex, 3–5 × 3–4 cm, finely veined; stipules small, deciduous. Flowers small, numerous, borne in catkin-like racemes on spurs from the old wood, often unisexual, plants monoecious or dioecious or rarely with mixed unisexual and bisexual flowers in each inflorescence. Perigynous zone disc-like. Sepals 5, soon falling. Petals absent. Stamens 5, filaments free, anthers somewhat versatile. Ovary with several ovules; stigma peltate. Pod oblong, 10–20 × 1.5–2 cm, woody, dark brown, indehiscent.

A genus of a single species from the Mediterranean area, remarkable for its very catkin-like inflorescences. It is easily grown in well-drained soil in a sunny site and is usually propagated by seed, though cuttings will root if given bottom heat.

1. C. siliqua Linnaeus. Illustration: Bonnier, Flore complète **3**: pl. 165 (1914); Polunin, Flowers of Europe, pl. 49 (1969); Polunin & Everard, Trees and bushes of Europe, 100 (1976); Edlin, The illustrated encyclopaedia of trees, 183 (1978).

Mediterranean area. H5–G1. Autumn.

The edible fruits yield carob and are processed into various health-food products, and also fed to stock often in cattle cake.

20. CASSIA Linnaeus
J. Cullen

Trees, shrubs, annual or perennial herbs. Leaves pinnate, without a terminal leaflet, often with 1 or more glands on the leaf-stalk or axis. Stipules variable, usually deciduous. Flowers usually in axillary or terminal racemes or panicles, more rarely solitary and axillary. Perigynous zone absent. Calyx of 5 almost completely free sepals, often unequal in size. Corolla usually yellow or orange-yellow, more rarely pink or whitish, the petals somewhat unequal in size, the uppermost usually the smallest. Stamens 10, usually the 1–3 upper small and sterile; of the fertile stamens 2 or 3 are usually longer than the others and may be borne on very long, curved, swollen filaments; anthers usually somewhat arrow head-shaped at the base, opening by terminal, or more rarely basal pores. Pod often large, variable in shape, usually divided into 1-seeded chambers, dehiscent or indehiscent.

As recognised here (including *Senna* Linnaeus and *Chamaecrista* Moench), a very large genus, mostly from the tropics but with extensions into the subtropics and temperate areas in many regions. Only a small number is cultivated for ornament, but many more have been cultivated for their medicinal properties.

Most of the species require glasshouse protection in much of Europe. The woody species may be propagated by seed or by cuttings, and the perennial herbaceous species by division.

Literature: Andrews, S. & Knees, S.G., Confusing Cassias, *The Kew Magazine* **5**(2): 76–81 (1988).

1a. The 3 lower stamens with long, curved filaments much more than twice as long as the anthers, which are short and hairy; large trees   2
  b. The 2 or 3 lower stamens often longer than the others, but filaments at most twice as long as anthers, not as above; anthers hairless; small trees, shrubs or herbs   3
2a. Flowers pink or pinkish orange, whitish outside; leaves with 8–20 pairs of leaflets   **1. grandis**
  b. Flowers yellow; leaves with up to 7 (rarely to 8) pairs of leaflets   **2. fistula**
3a. Herbs, sometimes woody at the base, but without persistent aerial branches   4
  b. Shrubs or small trees, with persistent aerial branches   6
4a. Leaflets obovate, rounded at apex; pods cylindric, very narrow, 10–20 × 3–4 cm   **13. obtusifolia**
  b. Leaflets ovate, ovate-lanceolate or elliptic, acute; pods shorter and broader   5

5a. Leaves with 4 or 5 pairs of leaflets which are acuminate   **12. occidentalis**
  b. Leaves with 6–8 pairs of leaflets, acute but not acuminate   **11. hebecarpa**
6a. Flower-buds concealed in greenish black bracts; anthers bent outwards at the top where the pores are; stipules persistent, conspicuous   **3. didymobotrys**
  b. Flower-buds not as above; anthers not bent at the top; stipules deciduous or persistent but not conspicuous   7
7a. Leaflets narrowly linear, terete, grooved   **10. artemisioides**
  b. Leaflets broader, flat, not grooved   8
8a. Plant completely covered with short, dense hairs   **5. tomentosa**
  b. Plant hairless or sparsely hairy   9
9a. Leaves with a gland between the lowest pair of leaflets or between each pair of leaflets   10
  b. Leaves completely without glands   11
10a. Gland present only between the lowermost pair of leaflets; raceme corymb-like   **4. corymbosa**
  b. Glands present between each pair of leaflets; racemes not corymb-like   **9. laevigata**
11a. Small tree; flower-buds conspicuously spherical, hard; pods linear-oblong, woody, to 30 cm   **6. siamea**
  b. Shrubs; flower-buds not as above; pods papery, much shorter   12
12a. Leaflets oblong-obovate, rounded at the apex   **7. italica**
  b. Leaflets lanceolate or narrowly elliptic, acute   **8. senna**

1. C. grandis Linnaeus filius (*C. renigera* invalid).

Tree to 30 m in the wild, trunk sometimes spiny. Leaves with 8–15 pairs of leaflets which are oblong, rounded and very shortly stalked at the base, bluntly acute at the apex, shortly spreading hairy on both surfaces. Stipules deciduous. Flowers numerous in erect racemes. Sepals more or less equal. Petals pink inside, whitish outside, 1.5–2.5 cm. Fertile stamens 7, the central 4 with short, straight filaments and long anthers opening by apical pores, the lower 3 with very long, curved, swollen filaments and short anthers opening by basal pores; anthers covered with spreading hairs. Pod pendent, woody, 40–100 × 4–6 cm, with thickened sutures. *Tropical America, widely introduced in Asia.* G2.

2. C. fistula Linnaeus. Illustration: Menninger, Flowering trees of the world, pl. 120 (1970); Everard & Morley, Wild flowers of the world, pl. 112 (1970); Perry, Flowers of the world, pl. 161 (1970); Polunin & Stainton, Concise flowers of the Himalaya, pl. 24 (1987).

Tree to 16 m, with widely spreading branches. Leaves with 3–8 pairs of leaflets which are shortly stalked, ovate to lanceolate, 6–16 × 3–6 cm, acute or somewhat rounded at the apex, closely veined, finely downy at least when young. Stipules deciduous. Flowers numerous in long, open, hanging racemes. Sepals unequal. Petals yellow, veined, 2.5–5 cm. Fertile stamens 7, the upper 4 with short, straight filaments, the lower 3 with long, curved, swollen filaments many times longer than the anthers; all anthers finely adpressed hairy, opening by terminal pores. Pod to 50 × 1.7 cm, terete, cylindric, dark brown to black. *India, SE Asia, widely introduced elsewhere.* G2.

3. C. didymobotrys Fresenius (*C. myrsifolia* misapplied; *C. nairobensis* Bailey & Bailey). Illustration: Vieira, Flores de Madeira, t. 86 (1986).

Shrub or small tree to 3 m. Leaves with 7–15 pairs of leaflets which are elliptic or elliptic-obovate, rounded but mucronate at the apex, finely hairy on both surfaces, 1.5–4 × 0.7–1.5 cm; stipules persistent, conspicuous, to 1 cm. Flowers numerous in erect racemes, each concealed by a large, greenish black bract. Sepals more or less equal. Petals golden yellow, veined, 1.5–2.5 cm. Fertile stamens 7, the lower 2 longer, anthers sharply bent at the top where the pores are. Pod stalked, oblong, flat, 8–10 × 1.5–2 cm. *Tropical Africa, introduced elsewhere.* G2.

4. C. corymbosa Lamarck (*Adipera corymbosa* (Lamarck) Britton & Rose; *C. floribunda* invalid). Illustration: Botanical Magazine, 633 (1803); Gartenpraxis, April 1986, 41.

Hairless shrub or small tree to 4 m or more. Leaves with 2 or 3 pairs of leaflets which are oblong-lanceolate, acute; a shortly stalked gland present between the lowest pair of leaflets above. Stipules deciduous. Flowers few in corymbs at the branch apices. Sepals unequal. Petals golden yellow, 1–1.5 cm. Fertile stamens 7, the lower 3 longer; anthers slightly

arrow head-shaped, opening by apical pores. Pod cylindric, finally membranous, 4–10 × 0.7–1 cm. *Temperate S America.* G1.

The plant known as var. **plurijuga** Bentham (*C. floribunda* invalid) has up to 6 pairs of leaflets and is possibly a hybrid between *C. corymbosa* and *C. coluteoides* Colladon (a species not generally cultivated in Europe).

5. C. tomentosa Linnaeus filius (*Adipera tomentosa* (Linnaeus filius) Britton & Rose).

Densely downy shrub or small tree to 6 m. Leaves evergreen with 5–9 pairs of leaflets which are elliptic to oblong or slightly obovate, 1–4 × 0.8–1.2 cm, rounded or acute at the apex, yellowish or brownish hairy when young; inconspicuous and deciduous glands present between some of the lower leaflet-pairs. Stipules deciduous. Flowers in axillary racemes. Sepals unequal, 2 narrower than the others and more hairy. Petals yellow, *c.* 1.5 cm. Fertile stamens 7, 2 longer; anthers arrow head-shaped, opening by terminal pores. Pods oblong, somewhat compressed, 8–12 × 0.8–1 cm, hairy at first. *Tropical America.* G1.

6. C. siamea Lamarck (*C. floribunda* misapplied; *Sciacassia siamea* (Lamarck) Britton & Rose).

Hairless tree to 10 m. Leaves with 6–12 pairs of leathery leaflets which are elliptic-oblong, rounded or slightly notched at the apex, 3–8 × 1–2 cm. Stipules deciduous. Flowers in dense, erect, corymb-like panicles, the buds hard and spherical, the axes robust. Sepals unequal. Petals yellow, 1–1.5 cm. Fertile stamens to 9, sterile 1–3, 2 or 3 of the fertile longer; anthers slightly arrow head-shaped, opening by terminal pores. Pod narrowly oblong, flat, to 30 × 1.5 cm, woody, sutures thickened. *SE Asia, introduced elsewhere.* G2.

7. C. italica (Miller) Steudel (*C. obovata* Colladon).

Perennial herb, somewhat woody at the base. Leaves with 5 or 6 pairs of leaflets which are oblong-obovate, rounded at the apex, 2–3.5 × 1–1.5 cm, somewhat glaucous; stipules deciduous. Racemes axillary. Sepals more or less equal. Petals yellow, veined, *c.* 1.5 cm. Fertile stamens 7, 3 longer; anthers arrowhead-shaped, dark brown with a yellowish stripe on each side, opening by terminal pores. Pods papery, very flattened, usually curved, rounded at both ends, brownish to black, to 5 × 1.5 cm. *Tropical Africa to India.* G1.

8. C. senna Linnaeus (*C. angustifolia* Vahl; *C. acutifolia* Delile; *Senna angustifolia* (Vahl) Batka). Illustration: Bentley & Trimen, Medicinal plants **2**: t. 89–91 (1877); Gartenpraxis, April 1986, 39.

Perennial herb, somewhat woody below, with sparse, adpressed, bristle-like hairs. Leaves with 4–7 (rarely more) pairs of leaflets which are lanceolate or narrowly elliptic, acute at the apex, 2–5 × 0.5–1.2 cm; stipules deciduous. Racemes axillary. Sepals more or less equal. Petals yellow, veined, 1–1.5 cm. Fertile stamens 7, 2 longer; anthers arrowhead-shaped, opening by terminal pores. Pods papery, very flattened, somewhat curved, rounded at both ends, brownish to black, to 5 × 1.5 cm. *N & E Africa to India.* G1.

9. C. laevigata Willdenow (*Adipera laevigata* (Willdenow) Britton & Rose; *C. floribunda* misapplied). Illustration: Gartenflora **3**: t. 77 (1854).

Hairless shrub to 3 m. Leaves with 3–5 pairs of leaflets which are elliptic-ovate to lanceolate, tapered to the acute apex, 3–7 × 1.5–2 cm, with a shortly stalked gland between each leaflet-pair above. Stipules deciduous. Racemes axillary, few-flowered. Sepals unequal. Petals orange-yellow, veined, *c.* 1.5 cm. Fertile stamens 7, 2 or 3 longer; anthers slightly arrowhead-shaped, opening by terminal pores. Pod cylindric-oblong, 5–10 × 1–1.5 cm, sutures thickened. *Tropical America.* G1.

10. C. artemisioides de Candolle (*C. chatelainiana* misapplied). Illustration: Everard & Morley, Wild flowers of the world, pl. 130 (1970); Botanical Magazine, n.s., 599 (1970–72); Rotherham et al., Flowers and plants of New South Wales and southern Queensland, pl. 484 (1975).

Shrub to 2 m, covered in ashy-white hairs. Leaves with 3–8 pairs of leaflets which are linear, terete and grooved, 10–25 × 1–2 mm; stipules deciduous. Racemes axillary, clustered towards the tips of the branches. Sepals more or less equal, the 3 outer hairy. Petals yellow, 0.7–1.6 cm. Stamens 10, the 2 lower a little longer; anthers slightly arrowhead-shaped, opening by terminal pores. Pods oblong, flat, 4–8 × 0.8–1 cm. *E Australia.* G1.

11. C. hebecarpa Fernald (*C. hebecarpa* var. *longipila* Baun; *C. marilandica* misapplied; *Ditremexa marilandica* (Linnaeus) Britton & Rose). Illustration: Rickett, Wild flowers of the United States **1**: pl. 61 (1967).

Perennial herb to 1 m or more, usually conspicuously downy. Leaves with 6–8 pairs of leaflets which are elliptic to elliptic-lanceolate, often with marginal hairs, acute, 2–5 × 0.8–2 cm; leaf-stalk with a conspicuous gland near the base; stipules deciduous. Flowers in axillary and terminal racemes, hanging in bud. Sepals unequal. Petals yellow, veined, c. 1 cm. Fertile stamens 7, 3 longer; anthers slightly arrowhead-shaped, dark, opening by terminal pores. Pod oblong, flattened, 7–10 × 5–7 cm, hairy at least when young, black, distinctly segmented, the segments almost square. *E & C USA*. H3. Summer.

Much confused and intergrading with **C. marilandica** Linnaeus, which is of more southerly distribution, is less hairy, and has narrower pod-segments.

12. C. occidentalis Linnaeus (*Ditremexa occidentalis* (Linnaeus) Britton & Rose). Illustration: Edwards's Botanical Register **1**: t. 83 (1816); Revue Horticole for 1897, 156.

Small or large hairless annual herb. Leaves with 4 or 5 pairs of leaflets which are ovate to ovate-lanceolate, long-tapered to the acuminate apex; leaf-stalks with 1 or rarely 2 glands near the base; stipules deciduous. Flowers few, in axillary racemes. Sepals unequal. Petals yellow, 1.2–1.5 cm. Fertile stamens 7, 2 or 3 longer; anthers slightly arrowhead-shaped, opening by terminal pores. Pods narrowly oblong, flattened, 8–14 × 0.6–0.8 cm, sutures thickened, pale. *Tropical & subtropical America, introduced elsewhere*. G1.

13. C. obtusifolia Linnaeus (*C. tora* misapplied; *Emelista tora* (Linnaeus) Britton & Rose). Illustration: Rickett, Wild flowers of the United States **2**: pl. 116 (1967); Polunin & Stainton, Concise flowers of the Himalaya, pl. 24 (1987).

Small to large annual herb, hairless or downy. Leaves with usually 3 pairs of leaflets which are obovate or broadly obovate, rounded to the ultimately finely pointed apex, 2–8 × 1–2.5 cm; an inconspicuous gland present on the leaf-stalk just below the lowest leaflet-pair; stipules somewhat persistent. Flowers axillary, mostly solitary. Sepals unequal. Petals yellow, veined, c. 1 cm. Fertile stamens 7, 2 or 3 longer; anthers opening by terminal pores. Pod linear-cylindric, 4-angled, curved downwards from near the base, 10–20 × 0.3–0.4 cm. *Tropical & temperate America north to Pennsylvania & Michigan; introduced elsewhere*. H4. Summer.

21. CERCIS Linnaeus
J. Cullen

Shrubs or small trees, flowering before the leaves expand. Leaves simple, palmately veined, ovate, circular or kidney-shaped; stipules deciduous. Flowers usually in umbel-like clusters (strictly short racemes but with very short axes) borne on short shoots on the previous year's growth, more rarely in pendent racemes. Calyx obliquely and widely bell-shaped, shallowly 5-toothed. Corolla usually rose-pink to purple, rarely white; petals clawed, keel longer than standard. Stamens 10, filaments free, curved downwards and included between the keel petals. Ovary several-seeded. Pods oblong, flat, with a narrow wing along the lower suture, several-seeded, eventually dehiscent.

A genus of about 6 species from N America, the Mediterranean area and E Asia. In spite of their wide and scattered distribution, the species (apart from *C. racemosa*) are very similar and difficult to distinguish. This difficulty is compounded by the fact that the plants flower before their leaves are expanded.

They require a rich soil in a sunny site. Propagation is mainly by seed, but some can be grafted on to stocks of *C. siliquastrum* or *C. canadensis*.

Literature: Hopkins, M., *Cercis* in North America, *Rhodora* **44**: 193–211 (1942); Isely, D., Leguminosae of the United States: 2. Subfamily Caesalpinioideae, *Memoirs of the New York Botanical Garden* **25**: 134–150 (1975).

1a. Flowers in distinct, long, pendent racemes　　**1. racemosa**
　b. Flowers in umbel-like clusters　　2
2a. Flowers 1.5 cm or more; all leaves rounded or notched at the apex, 6–12 cm　　**3. siliquastrum**
　b. Flowers to 1.5 cm; usually at least some of the leaves acute or shortly acuminate at the apex, if all rounded or notched then leaves 2.5–7 cm　　3
3a. Buds of next season's inflorescences with black, overlapping scales clearly visible in the leaf-axils by midsummer; flowers 1.1–1.5 cm, stalks 1–2 cm　　**2. chinensis**
　b. Buds of next season's inflorescences small, without obvious black, overlapping scales, not conspicuous; flowers 1–1.5 cm, stalks 0.8–1.5 cm　　**4. canadensis**

1. C. racemosa Oliver. Illustration: Hooker's Icones Plantarum **19**: t. 1894

(1889); Schneider, Illustriertes Handbuch der Laubholzkunde **2**: 5, 6 (1907); Botanical Magazine, 9316 (1933).

Small tree to 10 m, with black bark. Leaves broadly ovate, 6–10 × 6–10 cm, somewhat cordate or truncate at the base, pointed at the apex, pale and densely crisped hairy beneath, especially on or near the veins; stalks 2–4 cm. Racemes pendent, sometimes clustered; axes and flower-stalks densely crisped hairy; flower-stalks 0.8–1.2 cm in flower, 1.5–2 cm in fruit. Calyx 4–5 mm, margins of teeth with sparse fine hairs. Corolla rose-pink, 1–1.3 cm. Pods oblong, tapered at both ends, reddish to dark maroon, 6–11 × 1–1.6 cm, wing 1.5–3 mm. *W China*. H4. Early summer.

2. C. chinensis Bunge. Illustration: Schneider, Illustriertes Handbuch der Laubholzkunde **2**: 5, 6 (1907); Bailey, Standard cyclopedia of horticulture **1**: 721 (1935).

Shrub or small tree to 10 m, bark variable, pale brown to grey or almost black. Leaves ovate to almost circular, with narrow, translucent margins, 7–13 × 8–12 cm, base cordate (sometimes widely so), apex usually shortly acute or acuminate, more rarely some leaves with rounded or notched apices, usually with tufts of hair in the vein-axils beneath; stalks 2.5–4.5 cm. Buds of next year's inflorescences conspicuous in the leaf-axils by midsummer, with overlapping, black scales. Flower-clusters with axis *c*. 2 mm. Calyx 3–4 mm, margins finely hairy; flower-stalks 0.5–1 cm in flower, 1.5–2 cm in fruit. Corolla bright pinkish purple, 1.2–1.5 cm. Pods oblong, tapered at both ends, flat, 8–13 × 1–1.7 cm, reddish when young, wing 2–2.5 mm. *China*. H3. Spring.

3. C. siliquastrum Linnaeus. Illustration: Botanical Magazine, 1138 (1808); Bonnier, Flore complète **3**: pl. 165 (1914); Polunin & Huxley, Flowers of the Mediterranean, pl. 54 (1965); Polunin, Flowers of Europe, pl. 50 (1969).

Small tree or shrub to 8 m; bark greyish, young shoots dark. Leaves broadly ovate to almost circular, 6–10 × 8–14 cm, base cordate with a wide sinus, apex smoothly rounded or notched, hairless beneath even when young; stalks 3–4 cm. Flower clusters each with a hairy axis to 8 mm; flower-stalks 1–2 cm in flower, 1.5–2.5 cm in fruit. Calyx 4–6 mm, each tooth with a patch of short hairs at and just below the apex. Corolla bright purple or purplish red, 1.5–2 cm. Pods oblong, 8–12 × 1.4–2 cm, tapered at both ends, reddish when young, wing 1.5–2 mm. *Mediterranean area*. Spring. H4.

A variant with hairy flower-stalks, calyces and pods, subsp. **hebecarpa** (Bornmuller) Yaltirik, is native in the eastern Mediterranean area and may be grown there. 'Alba' (forma *alba* Rehder) has white flowers.

4. C. canadensis Linnaeus.

Shrub or small tree to 10 m; bark blackish, smooth or rough. Leaves ovate to almost circular or kidney-shaped, 3.5–12 × 5–14 cm, with translucent margin very narrow or absent, cordate to truncate at the base, acute, shortly acuminate or rounded at the apex, hairless or finely hairy along the veins beneath. Flower-clusters without an axis or the axis very short, hairy. Calyx 2.5–3.5 cm, often reddish purple, margins finely hairy. Corolla bright purple-pink or rarely white, 0.9–1.4 cm. Pods oblong, flattened, 4–10 × 0.8–2 cm, tapered at both ends, sometimes glaucous. *N America*.

The above description covers a bewildering array of variants from N America. Following Isely (reference above) there are 2 species, *C. canadensis* in the east, *C. occidentalis* in the west. In intervening areas, especially Texas, various intermediate forms are found, rendering discrimination of the species extremely difficult. Isely remarks that "taxa . . . are most conveniently identifiable on the basis of where they come from". This is of no help to the gardener, who will probably know nothing of the ultimate wild origins of his material.

Isely treats *C. canadensis* as consisting of 3 varieties (2 of them possibly cultivated) and *C. occidentalis* as a simple species. Brief diagnoses are given below, but it must be stressed that the possibility of making accurate identifications is not great.

Var. **canadensis**. Illustration: Gleason, Illustrated Flora of the north-eastern United States and adjacent Canada **2**: 383 (1952); Horticulture **35**: 76 (1957); Justice & Bell, Wild flowers of North Carolina, 98 (1968). Leaves mostly with acute or shortly acuminate apices (if apices rounded then leaves 7 cm or more), thin; flowers 0.9–1.25 cm. *E North America from Canada to Florida*. H3. Spring.

'Alba' (var. *alba* Anon.) has white flowers.

Var. **texensis** (Watson) Hopkins (*C. reniformis* Watson). Leaves leathery, usually rounded at the apex, if acute, to 8 cm at most, flowers 0.9–1.3 cm. *SE USA (Texas, Oklahoma)*. H5. Spring.

A white-flowered variant is known in the USA as 'Texas White' (illustration: American Nurseryman **167**: 31, 1988).

C. occidentalis Gray. Illustration: Abrams & Ferris, Illustrated Flora of the Pacific States **2**: 477 (1944); Munz, California spring wild flowers, 78 & pl. 58 (19). Leaves thick, usually rounded at the apex, flowers 1.2–1.4 cm. *W USA (California, Arizona, Utah)*. H5. Spring.

22. BAUHINIA Linnaeus
J. Cullen

Trees, shrubs or woody climbers, some with tendrils, a few spiny (not ours). Leaves of 2 distinct leaflets or more commonly of a single leaflet which is notched or lobed at the apex; stipules variable, usually deciduous. Flowers in racemes, terminal, axillary or terminating lateral shoots. Calyx either of 5 sepals, edge-to-edge in bud, or united in bud and splitting along 1 or more lines as the flower opens, often spathe-like and reflexed. Petals 5, slightly unequal, spreading, usually conspicuously clawed. Fertile stamens 10, 5, 3 or 1, filaments free, often hairy. Ovary stalked with few to many ovules. Pod dehiscent or not, stalked, leathery or woody.

There are about 250 species from the tropics, of which only a few are in general cultivation. They succeed in a well-drained soil in a sunny position. Propagation is normally by seed, but propagation by suckers (in those species that produce them) is also possible; cuttings are sometimes used, but they are difficult to root.

Literature: Isely, D., Leguminosae of the United States: 2. Caesalpinioideae, *Memoirs of the New York Botanical Garden* **25**(2): 15–31 (1975).

Leaves. Of 2 distinct leaflets: **1**; of a single leaflet variably 2-lobed at the apex: **2–6**. To 6 cm: **1** (leaflets), **2,6**; 7 cm or more: **3–5**.
Tendrils. Present: **1**.
Calyx. Of 5 sepals: **1**; splitting once or more, but not into 5 sepals: **2–6**.
Petals. White or cream: **2–4**; pink, violet, violet-blue or deep violet: **1,4,5**; salmon pink to orange: **6**. Without distinct claws: **2,3**; with distinct claws: **1,4–6**.
Fertile stamens: Ten: **2,3**; five: **4**; three or one: **1,5,6**.

1a. Leaves made up of 2 distinct leaflets; tendrils present
    1. yunnanensis

  b. Leaves of a single leaflet variably notched or lobed at the apex; tendrils absent   **2**
2a. Petals less than 2 cm   **2. racemosa**
  b. Petals more than 2 cm   **3**
3a. Fertile stamens 10; buds narrowly ovoid, very long-acuminate, terminating in 4 or 5 thread-like points   **3. acuminata**
  b. Fertile stamens fewer than 10; buds not as above   **4**
4a. Fertile stamens 5   **4. variegata**
  b. Fertile stamens 1 or 3   **5**
5a. Petals salmon-pink to orange-red, 2.5–4.5 cm, obviously clawed   **6. galpinii**
  b. Petals whitish pink or purple-mauve, 4–5 cm, scarcely clawed   **5. purpurea**

1. B. yunnanensis Franchet. Illustration: Botanical Magazine, 7814 (1902).

Prostrate or erect woody shrub to 3 m or more often a woody climber with paired, flattened, usually tightly coiled tendrils in the leaf-axils. Leaves made up of 2 distinct leaflets with a short projection between them; leaflets asymmetric, ovate, 2.5–5 cm, hairless. Flowers in racemes terminating lateral branches; stalks 2–4 cm. Buds pear-shaped. Sepals 5, becoming free and reflexed or spreading, 5–7 mm. Petals pink or pinkish violet, 1–1.5 cm, obovate, tapered gradually into narrow claws, some hairy outside and on the margins. Fertile stamens 3. Pod stalked, oblong, to 12 × 2 cm, dehiscent. *SW China (Yunnan)*. G1.

2. B. racemosa Lamarck. Illustration: Hooker's Icones Plantarum **2**: t. 141 (1837).

Shrub or small tree to 6 m, branches pendent, without tendrils. Leaves broader than long, 3–6 × 4.5–7 cm, truncate or slightly cordate at the base, bilobed to about one-third, lobes rounded, downy or hairless, greyish beneath. Flowers in long, hairy racemes; stalks to 4 mm. Buds obliquely obovoid, pointed, hairy. Calyx spathe-like. Petals white or pale yellow, *c.* 1 cm, spreading. Fertile stamens 10. Pod stalked, straight or curved, to 20 × 2 cm. *Burma, NE India*. G2.

3. B. acuminata Linnaeus. Illustration: Botanical Magazine, 7866 (1902).

Shrub to 3 m, without tendrils. Leaves ovate to almost circular, 7–15 × 7–15 cm, cordate at the base, lobed to one-third or less at the apex, the lobes usually acute, with short, crisped hairs on the conspicuous veins beneath. Racemes short, few-flowered. Buds narrowly ovoid, very long-acuminate, terminating in 4 or 5 thread-like, erect or spreading points. Calyx splitting once, spathe-like, 2.5–4 cm. Petals white or cream, obovate or elliptic, scarcely clawed, 2.5–4.5 cm. Fertile stamens 10. Pod stalked, oblong, flattened, 5–15 × 1–2 cm. *India, Malaysia, S China.* G2.

4. B. variegata Linnaeus. Illustration: Botanical Magazine, 6818 (1885); Menninger, Flowering trees of the world, pl. 78 (1962); Journal of the Royal Horticultural Society **98**: f. 21 (1973); Gartenpraxis for December 1977, 586.

Shrub or small tree to 10 m, without tendrils. Leaves broadly circular in outline, truncate or cordate at the base, lobed to about one-third at the apex with rounded or obtuse lobes, 6–20 × 7–18 cm, hairless. Flowers few, in short, finely hairy racemes; flower-stalks 2–3 mm, perigynous zone long and stalk-like. Buds appearing stalked, obovoid, blunt at the apex. Calyx 1–2 cm, spathe-like, notched at the apex. Petals white to pinkish, reddish or bluish purple, often mottled and with pronounced, coloured veins, 4–6 cm, obovate, clawed. Fertile stamens 5. Pod stalked, to 30 × 2.5 cm, flattened. *Himalaya, China.* G2.

The white-flowered variant, 'Candida' (var. *candida* Hamilton) is often grown (illustration: Botanical Magazine, 7312, 1893).

B. × blakeana Dunn. Illustration: Menninger, Flowering trees of the world, pl. 79 (1962). Supposedly the hybrid between *B. variegata* and *B. purpurea*. It has more numerous, larger flowers than *B. variegata* and the petals are deep rose-lavender changing to dark red as the flower matures.

5. B. purpurea Linnaeus. Illustration: Menninger, Flowering trees of the world, pl. 80 (1962); Perry, Flowers of the world, 159 (1970); Polunin & Stainton, Concise flowers of the Himalaya, t. 206 (1987).

Tree or shrub to 7 m. Leaves more or less circular or broadly elliptic, truncate or cordate at the base, lobed to one-third or half with rounded or pointed lobes, 7–15 × 7.5–12 cm, hairless or with fine adpressed hairs beneath. Racemes many-flowered, hairy; flower-stalks 2–10 mm. Buds narrowly pear-shaped, pointed, ribbed. Calyx 2–4 cm, splitting on more than 1 line. Petals whitish, pink or purple-mauve, 4–5 cm, distinctly clawed, obovate. Fertile stamens 3. Pod stalked, flattened,

to 25 × 2.5 cm. *Himalaya, Burma, S China.* G2.

6. B. galpinii N.E. Brown. Illustration: Botanical Magazine, 7494 (1896); Palgrave, Trees of Central Africa, 75 (1956); Menninger, Flowering vines of the world, f. 87 (1970).

Shrub to 4 m, usually scrambling or climbing, without tendrils. Leaves broadly circular to kidney-shaped, rounded to slightly cordate at the base, lobed at most to one-third at the apex with rounded lobes, finely hairy beneath, 2.5–5 × 2–5.5 cm. Racemes few-flowered, rusty-brown hairy; flower-stalks 2–3 cm. Buds narrowly obovoid, pointed, hairy. Calyx spathe-like, 2–5 cm. Petals salmon-pink to orange, 2.5–4.5 cm, each with a more or less diamond-shaped, pointed blade narrowing abruptly into the long, conspicuous claw. Fertile stamens 3. Ovary hairy. Pod to 10 × 2.5 cm, flattened, dehiscent. *Zimbabwe, South Africa.* G2.

23. SCHOTIA Jacquin
J. Cullen

Trees or shrubs. Leaves pinnate, without a terminal leaflet, leaflets leathery; stipules small, deciduous. Flowers numerous in panicles borne terminally or sometimes on the older wood. Perigynous zone bell-shaped or funnel-shaped. Calyx of 4 unequal lobes, overlapping in bud. Petals 5, occasionally some or all of them reduced to linear threads. Stamens 10, alternately longer and shorter, filaments entirely free or fused at the base. Ovary containing several ovules, stalked, the stalk united with the perigynous zone. Pod oblong, compressed, woody, tapered to the apex, eventually dehiscent, several-seeded. Seeds with or without arils.

A genus of about 20 species from Africa south of the Zambesi. Only a small number is grown. They require greenhouse protection in most of Europe (though they can be grown out of doors in the Mediterranean area), with a well-drained soil and sunny site. They are propagated by seed.

1a. Stamens free to the base; leaves with 6–18 pairs of leaflets   **1. afra**
 b. Stamens united at the base; leaves with 3–8 pairs of leaflets   2
2a. Petals 5, normally developed; flower-stalks very short, to 2 mm   **2. latifolia**
 b. All or some of the petals reduced to linear threads; flower-stalks at least 5 mm   **3. brachypetala**

1. S. afra (Linnaeus) Thunberg (*S. speciosa* Jacquin). Illustration: Botanical Magazine, 1153 (1809) – as *S. tamarindifolia*; Flowering Plants of Africa 42: p. 1665 (1973); Palgrave, Trees of southern Africa, t. 76 (1977); Palmer, Field guide to the trees of southern Africa, pl. 10 (1977).

Shrub or small tree to 7 m. Leaves with 6–18 pairs of leaflets which are hairless or finely downy, linear to oblong or elliptic, 5–20 × 1–10 mm, obtuse and mucronate at the apex. Flowers in almost spherical panicles; flower-stalks 3–9 mm. Calyx leathery, red. Petals 5, red to pink, 1–1.8 cm, downy within. Stamens exceeding the petals, filaments free to the base. Pods 3–15 × 2–5 cm. Seed without an aril or aril very small. *South Africa (Cape Province), Namibia.* H5–G1.

Variable in the number, size and shape of the leaflets.

2. S. latifolia Jacquin. Illustration: Flora of southern Africa **16**(2): 29 (1977).

Tree to 10 m or more with rounded crown. Leaves with 3–5 pairs of leaflets which are hairless or downy, elliptic-oblong to obovate, 1.5–6.5 × 1–3.5 cm, tapered or rounded at the base, rounded or acute but scarcely mucronate at the apex. Flowers in rather open, terminal panicles; flower-stalks very short (to 2 mm). Calyx reddish brown. Petals 5, pink to pale pink, 0.9–1.1 cm. Stamens slightly exceeding the petals, united at the base with the tube split along 1 side. Pods 5–14 × 3–4.5 cm. Seeds each with a large, yellow aril. *South Africa (Cape Province, Transvaal).* H5–G1.

3. S. brachypetala Sonder. Illustration: Flowering Plants of South Africa **20**: pl. 777 (1940); Flora of southern Africa **16**(2): 29 (1977); Palgrave, Trees of southern Africa, t. 77 (1977); Palmer, Field guide to the trees of southern Africa, pl. 10 (1977).

Tree to 16 m with rounded crown. Leaves with 4–8 pairs of leaflets which are elliptic to oblong or obovate, 2.5–8.5 × 1–4.5 cm, hairless or shortly downy, rounded to the apex. Flowers in congested, more or less spherical panicles; flower-stalks 0.5–1.2 cm. Calyx leathery, red. Petals 5, all or some of them reduced to linear threads, those fully developed red, 1.3–1.8 cm. Stamens with filaments united at the base, tube split at 1 side or not. Pods 5–17 × 3.5–5 cm. Seeds each with a large, yellow aril. *Mozambique, Zimbabwe, South Africa.* H5–G1.

24. BROWNEA Jacquin

J. Cullen

Small trees with spreading crowns, branches ultimately pendent. Leaves evergreen, pinnate with 2–18 pairs of leaflets; young leaves brightly coloured, enclosed at first in a tube of scale-leaves, later limp and hanging for a few days, ultimately expanding; stipules long and thread-like, quickly deciduous. Flowers in pendent heads terminating the branches (rarely borne on the old wood); heads enclosed in large bracts which fall as the flowers open; each flower subtended by a large bract and by 2 bracteoles which are united into a 2-lobed tube around the flower. Sepals 5, red or orange-red. Petals 5, free, more or less equal, clawed, red or orange-red. Stamens 10–11 (rarely more), filaments united at least at the extreme base (sometimes into 2 or 3 bundles). Ovary hairy, with few ovules. Pod oblong, woody, often contorted, few-seeded.

A genus of about 30 species from tropical S America and the Caribbean, whose classification is very confused. They require glasshouse treatment in Europe and are prized not only for their brightly coloured, spectacular inflorescences, but also for the interesting, coloured, limp young leaves. They require a rich soil and high temperatures, and can be damaged by overwatering. Propagation is by hardwood cuttings.

1a. Flower-heads to 5 cm across, funnel-shaped when open; young leaves purplish, becoming pinkish brown; stipules 1–2 cm   **3. coccinea**
  b. Flower-heads more than 5 cm across, spherical when open; young leaves mottled or pinkish buff becoming pinkish brown; stipules more than 10 cm   2
2a. Shoots persistently hairy; stipules to 30 cm; petals exceeding stamens   **1. grandiceps**
  b. Shoots soon hairless, warty; stipules to 12 cm; stamens exceeding petals   **2. ariza**

1. B. grandiceps Jacquin. Illustration: Botanical Magazine, 4839 (1855); Corner, Wayside trees of Malaya, pl. 80 (1940); Everard & Morley, Wild flowers of the world, pl. 170 (1970).

Tree to 5 m or more. Shoots with dense, persistent, pale brown hairs. Leaves with 11 or more pairs of leaflets (3–5 pairs in those near the inflorescences), stalk and axis densely hairy with pale brown hairs; leaflets increasing in size upwards, oblong-obovate or elliptic-oblong, the longest to 15 × 4 cm, unequal and rounded at the base (cordate in the lowermost leaflets), acuminate at the apex, densely and persistently hairy on the midrib beneath; young leaves brownish green mottled with pink and white, later pinkish brown, the stalk and axes whitish. Stipules to 30 cm. Flower-heads spherical, *c.* 15 cm across, many-flowered; bracteoles 2.5–3 cm, tube to 1.5 cm, densely hairy; sepals red, darker than petals, 4–4.5 cm; petals long-clawed, red-scarlet, 7–8 cm. Stamens usually 11, to 5 cm, filaments united at the base for *c.* 8 mm. *N South America.* G2.

2. B. ariza Bentham. Illustration: Botanical Magazine, 6469 (1880).

Similar to *B. grandiceps*, but bark warty, shoots hairless or soon so if slightly hairy when young, leaflets hairless, the young leaves uniformly pinkish buff becoming pinkish brown, then green, stipules to 12 cm; flower-heads somewhat smaller, petals *c.* 6.5 cm, stamens exceeding petals. *N South America (Venezuela only?).* G2.

3. B. coccinea Jacquin. Illustration: Botanical Magazine, 3964 (1843); Corner, Wayside trees of Malaya, pl. 79 (1940).

Tree to 10 m or more. Shoots hairless, shining, warty. Leaflets 3–9 pairs, leathery, lanceolate to obovate, the largest 13–17 × 5–7 cm, unequal and rounded (cordate in lowest leaflets) at the base, acuminate at the apex, greyish green, hairless. Young leaves purplish, becoming pinkish brown. Stipules 1–2 cm. Flower-heads spherical in bud, funnel-shaped when open, *c.* 5 cm across. Bracteoles *c.* 3 cm, tube *c.* 1.5 cm, covered with adpressed, bristle-like hairs. Sepals bright red, *c.* 4 cm. Petals long-clawed, *c.* 5.5 cm, vermilion. Stamens exceeding petals, filaments united into a tube for *c.* 3 cm. *N South America, Jamaica?* G2.

Material grown in the tropics as *B. latifolia* Jacquin and *B. capitella* Jacquin (*B. speciosa* Reichenbach) is very similar; further research is needed on these plants to establish their distinctions (if any) and the correct names.

25. AMHERSTIA Wallich

J. Cullen

Evergreen trees to 20 m or more. Leaves pinnate with 4–7 pairs of leaflets, without a terminal leaflet; leaflets oblong, oblong-lanceolate or obovate, abruptly acuminate, rounded at the base, 10–30 × 2–8 cm, whitish or pale beneath; young leaves bronze, limp and hanging; stipules 2.5–4 cm, soon deciduous. Flowers in hanging racemes to 45 cm; flower-stalks long, each bearing 2 large, persistent bracteoles. Sepals 4, coiling when the flower is open. Petals 5, 3 large and 2 very small, the uppermost to 5 cm long and wide, with a whitish, pink-striped and -spotted claw, the blade deep pink with a yellow blotch at the apex, the other 2 large petals pink with the apices yellow-blotched. Stamens 9 or 10, 5 longer and 4 or 5 shorter, the filaments united into a tube for about half their length. Ovary stalked, with few ovules, yellow, the stalk and style red. Pods to 15 × 4 cm, containing 4–6 seeds.

A genus of a single species from Burma, widely grown as a tropical ornamental. In Europe it requires high temperatures in a glasshouse, rich, well-drained soil and constantly moist air to succeed, but will flower when quite small if well grown. Propagation is by seed or by hardwood cuttings.

1. A. nobilis Wallich. Illustration: Botanical Magazine, 4453 (1849); Corner, Wayside trees of Malaya, pl. 78 (1940); Menninger, Flowering trees of the world, pl. 100 (1962); Everard & Morley, Wild flowers of the world, pl. 113 (1970).

Burma, widely cultivated elsewhere. G2.

26. TAMARINDUS Linnaeus

J. Cullen

Tree to 20 m or more with a rounded crown and ultimately drooping branches. Leaves pinnate without a terminal leaflet; leaflets opposite, in 9–16 pairs, oblong, symmetric, very shortly stalked, rounded at the base, rounded or slightly notched and shortly mucronate at the apex, 8–20 × 4–8 mm, hairless. Flowers 6–10 in short racemes borne on the branches, each flower initially enclosed in a large, quickly deciduous bract. Perigynous zone narrowly funnel-shaped. Sepals 4, 8–9 mm, unequal, reflexed and soon falling after the flower opens. Petals 5, the 2 lower minute, the upper 3 larger, 0.9–1.3 cm, reddish-veined on a pale yellow background, margins wavy. Fertile stamens 3, filaments united below, the top of the fused zone with a number of small teeth. Pod indehiscent, oblong-cylindric though somewhat 4-cornered, often irregular due to abortion of some ovules, 7–12 × 1–3 cm, succulent and pulpy when young, later dry, indehiscent. Seeds few.

A genus of a single species from Africa or India, widely cultivated in tropical regions, especially E Asia and mainly for the sake of the sour pulp of the immature pods, which is used in cookery, for flavouring drinks and in folk-medicine. It will grow well in most soils, preferring moist conditions but surviving some degree of aridity. Propagation is by seed (seedlings very prone to damping off) or by budding.

1. T. indica Linnaeus. Illustration: Bentley & Trimen, Medicinal plants **2**: t. 92 (1876); Corner, Wayside trees of Malaya, pl. 119, 120 (1940); Palgrave, Trees of southern Africa, t. 79 (1977); Matthew, Illustrations on the Flora of the Tamil Nadu Carnatic, 238 (1982).

Widespread in the tropics, origin probably Africa or India. G2.

Papilionoideae (*Papilionaceae* also known as *Fabaceae* in the strict sense).

Trees, shrubs, herbs or climbers (some with tendrils). Leaves various or rarely absent. Corolla bilaterally symmetric so that the uppermost petal is outermost, overlapping the others (petals rarely reduced to 1). Stamens usually 10, filaments free, or all united into a tube or sheath, or 9 united, the uppermost free, at least at the base. Seeds usually with an incurved radicle.

In almost all genera of this subfamily the 5 petals are diversified. The uppermost petal, which overlaps all the others in bud is generally large, with a long claw and the upright or reflexed blade; this is known as the standard. Within this are 2 lateral petals, often rather narrow, known as the wings. Finally, within these are 2 larger petals which are usually united at least along their lower margins; these, which are known as the keel, generally enclose the stamens, ovary and style. In some cases the wings are coupled to the keel by means of a fold or tuck, and their decoupling is sometimes important in pollination. In a few cases (**Amorpha**), the wings and keel are absent and in **Clitoria** and **Centrosema** the flowers are reversed so that they are borne on the plant with the standard lowermost.

The way in which the filaments of the stamens are united is important in the identification of the genera. The filaments may be entirely free (or just minutely joined at the extreme base), or may be all united (monadelphous stamens), either into a tube around the ovary and style, or into a sheath which is open along the top or 9 of them may be fused into a sheath while that of the uppermost is free (diadelphous stamens); in this latter case the uppermost

filament may be entirely free from the rest, or may be free at the base, but united with the others higher up. Also, in the case of a few genera with all 10 filaments united into a tube, the uppermost may progressively become free from the apex of the tube as the flower ages. This character often causes confusion, as it is often difficult to discern in herbarium specimens; however, in living flowers it is generally easy to see and to understand.

1a. Trees or shrubs with flowers of 3 types **119. + Laburnocytisus**
  b. Woody or herbaceous plants with flowers of 1 type 2
2a. Filaments of all 10 stamens completely free for at least 90% of their length (sometimes all slightly fused at the extreme base) 3
  b. Filaments of all or most (usually 9) of the stamens united for much of their length, often the uppermost completely free at the base (sometimes united to the others higher up), rarely the stamens variably united in groups, or rarely the stamens 9 only, filaments all united 25
3a. Most leaves pinnately divided, with more than 3 leaflets 4
  b. Most leaves simple, or with 3 leaflets, or leaves with blades absent 10
4a. Plant sticky with stalkless glands; leaflets fleshy and irregularly toothed; pod glandular, breaking into 1-seeded segments **62. Adesmia**
  b. Plant without the above combination of characters 5
5a. Leaflets narrow, heather-like (ericoid) **100. Burtonia**
  b. Leaflets expanded, not narrow and heather-like 6
6a. Lower 4 petals similar to each other, not differentiated into wings and keel; pod woody, inflated, containing 3–5 chestnut-like seeds **27. Castanospermum**
  b. Lower 4 petals differentiated into wings and keel; pods and seeds not as above 7
7a. Bracts borne on the flower-stalks; corolla dark blue **28. Bolusanthus**
  b. Bracts borne at the bases of the flower-stalks; corolla variously coloured, not dark blue 8

8a. Base of the leaf-stalk swollen, enclosing the axillary bud **30. Cladrastis**
  b. Base of the leaf-stalk not swollen as above, axillary bud not enclosed 9
9a. Pods flattened, linear-oblong, dehiscent; standard reflexed **29. Maackia**
  b. Pods terete, constricted between the seeds, often winged, not dehiscent but breaking into 1-seeded segments; standard reflexed or forwardly-directed **31. Sophora**
10a. Leaf-blades absent 11
  b. Leaf-blades present, simple or of 3 leaflets 12
11a. Mature leaves reduced to thread-like stalks; flowers 1 per bract **102. Viminaria**
  b. Leaves entirely absent from mature plants; flowers 2 per bract **101. Sphaerolobium**
12a. Leaves simple (sometimes formed from a single leaflet) 13
  b. Leaves of 3 leaflets (sometimes these side-by-side, without a common stalk) 20
13a. Calyx umbilicate at base **110. Podalyria**
  b. Calyx not umbilicate at base 14
14a. Leaves opposite or in whorls 15
  b. Leaves alternate 16
15a. Leaves opposite; stipules present; calyx 1–1.5 cm **96. Brachysema**
  b. Leaves in whorls; stipules absent; calyx 6–7 mm **97. Oxylobium**
16a. Ovules 4 or more 17
  b. Ovules 2 18
17a. Leaves ovate, cordate at base, usually with the margins sharply toothed **98. Chorizema**
  b. Leaves linear to linear-lanceolate, not cordate at the base, margins not toothed **97. Oxylobium**
18a. Leaves not heather-like, margins not rolled under, or leaves terete; bracteoles persistent, borne close beneath and overlapping the calyx **104. Pultenaea**
  b. Leaves heather-like, margins rolled under; bracteoles absent or distant from and not overlapping the calyx 19
19a. Pod flattened, triangular in outline; calyx 2.5–3 mm **103. Daviesia**
  b. Pod ovoid or almost spherical, not flattened; calyx 3–6 mm **105. Dillwynia**
20a. Leaflets side-by-side, without a common stalk, heather-like **100. Burtonia**

b. Leaflets not as above, borne on a common leaf-stalk, not heather-like 21

21a. Plant herbaceous, shoots dying back to the root in winter 22

b. Plant woody, aerial shoots persistent through the winter 23

22a. Plants hairless or almost so, often glaucous; corolla violet-blue, yellow or white; pod inflated **116. Baptisia**

b. Plants usually conspicuously hairy, not glaucous; corolla deep purple or yellow; pod flat **115. Thermopsis**

23a. Calyx very deeply 5-lobed, lobes edge-to-edge in bud, tube very short, black outside; keel deep, fringed along the suture; leaves often almost opposite **99. Gompholobium**

b. Combination of characters not as above 24

24a. Standard well developed, about as long as wings; corolla more than 2.5 cm **114. Piptanthus**

b. Standard much shorter than wings; corolla 1.8–2.5 cm **113. Anagyris**

25a. Stamens 10 with all their filaments united for a great proportion of their length, certainly all united in the lower part, forming a tube or sheath 26

b. Stamens 10 with the filament of the uppermost free at the base and for some distance above it, sometimes united with those of the other 9 well above the free base, or filaments variably united in groups or stamens 9, filaments all united 60

26a. Most leaves with more than 3 leaflets, mostly pinnate, occasionally palmate 27

b. All leaves simple or of 3 leaflets only 36

27a. Leaves palmate **117. Lupinus**

b. Leaves pinnate 28

28a. Corolla reduced to the standard only, which envelops the stamens and style **60. Amorpha**

b. Corolla with wings and keel as well as standard 29

29a. Leaves with conspicuous, brown, stalkless glands beneath; flowers c. 3 cm **61. Amicia**

b. Leaves without conspicuous, brown stalkless glands beneath; flowers to 2.5 cm , often much smaller 30

30a. Stipules and stipels persistent; corolla blue to purple **53. Hardenbergia**

b. Stipules present but usually falling early, or absent, stipels absent; corolla variously coloured, not usually blue to purple 31

31a. Evergreen woody climber; pod indehiscent, compressed, hairy, few-seeded, somewhat winged along the sutures **34. Derris**

b. Usually deciduous trees, shrubs, subshrubs or herbs; pod various, not as above 32

32a. Trees or upright shrubs 33

b. Low shrublets or herbs 34

33a. Corolla violet or pink; the 2 upper calyx-teeth reflexed **109. Virgilia**

b. Corolla golden yellow to orange; the 2 upper calyx-teeth forwardly-directed **32. Pterocarpus**

34a. Stems and leaves sticky, glandular-hairy; stipules united to the leaf-stalk **90. Ononis**

b. Stems and leaves not glandular and sticky; stipules absent or free from the leaf-stalk 35

35a. Calyx-teeth about as long as tube; stipules persistent; corolla pale violet to white **73. Galega**

b. Calyx-teeth shorter than the tube; stipules absent; corolla yellow or red, rarely creamy white **79. Anthyllis**

36a. Mature leaves consisting of a single leaflet or mature plants without leaves 37

b. Mature leaves all or mostly with 3 leaflets 48

37a. Plants spiny, either the ends of the branches hardened and spiny, or leaves replaced by spiny phyllodes 38

b. Plants not at all spiny 40

38a. Corolla blue **122. Erinacea**

b. Corolla yellow 39

39a. Leaves absent from mature plants, replaced by spine-tipped phyllodes **130. Ulex**

b. Leaves present on mature plants, spine-tipped phyllodes absent **128. Genista**

40a. Stamens united into a sheath which is open on the upper side 41

b. Stamens united into a tube closed on the upper side 44

41a. Corolla purple **107. Hovea**

b. Corolla yellow or red or yellow and red 42

42a. Stamens all of the same length; standard to 1.5 cm **108. Bossiaea**

b. Stamens alternately longer and shorter; standard more than 1.5 cm 43

43a. Keel beaked; wings puckered or sculptured between the veins outside; corolla yellow **112. Crotalaria**

b. Keel not beaked; wings not puckered or sculptured; corolla red **106. Templetonia**

44a. Calyx divided to the base on the upper side 45

b. Calyx not divided to the base on the upper side 46

45a. Young stems broadly winged **125. Chamaespartium**

b. Young stems not broadly winged **124. Spartium**

46a. Seeds appendaged; upper lip of calyx with 2 very short teeth **121. Cytisus**

b. Seeds not appendaged; upper lip of calyx deeply toothed or notched 47

47a. Pod dehiscent, not inflated **128. Genista**

b. Pod indehiscent, inflated **126. Retama**

48a. Leaflets toothed **90. Ononis**

b. Leaflets not toothed 49

49a. Plant spiny 50

b. Plant not spiny 51

50a. Corolla blue **122. Erinacea**

b. Corolla yellow **128. Genista**

51a. Leaves dotted with dark glands, often with a resinous or bituminous smell 52

b. Leaves not dotted with dark glands, not smelling as above 53

52a. Foetid biennials to perennial herbs; leaves with 3 leaflets **58. Bituminaria**

b. Scented herbs or shrubs; leaves pinnate with a terminal leaflet **59. Psoralea**

53a. Stamens united into a sheath open at the top; pods inflated and hard, the seeds ultimately loose and rattling within them **112. Crotalaria**

b. Stamens united into a tube closed at the top; pods various, not usually as above 54

54a. Seed with an appendage along a long side; corolla purple with a yellow spot at the base of the standard; leaflets usually folded upwards along the midrib **111. Hypocalyptus**

b. Seed without an appendage, or with a small appendage along a short side; corolla yellow, pink or purple; leaflets not usually as above 55

55a. Pod covered with glandular warts **129. Adenocarpus**

b. Pod not covered with glandular warts   56
56a. Small tree with flowers in long, hanging racemes   **118. Laburnum**
b. Shrubs (rarely small trees) with flowers in erect inflorescences   57
57a. Upper lip of calyx with 2 very short teeth; seeds appendaged   **121. Cytisus**
b. Upper lip of calyx deeply 2-toothed; seeds not appendaged   58
58a. Tall shrubs; leaf-stalk 1.5–5 cm; pod 3.5–5 cm   **120. Petteria**
b. Small shrubs; leaf-stalk to 1.5 cm; pod less than 3.5 cm   59
59a. Leaves and branches mostly alternate; calyx not inflated   **128. Genista**
b. Leaves and branches mostly opposite; calyx somewhat inflated   **123. Echinospartium**
60a. Leaves of 3 leaflets, or of 5 leaflets and palmate, rarely simple or absent (sometimes replaced by a tendril)   61
b. Leaves pinnate with more than 3 leaflets   88
61a. Leaves simple or absent from mature plants   62
b. Leaves of 3 or 5 leaflets, present on mature plants   68
62a. Leaves simple   63
b. Leaves absent from mature plants   65
63a. Spineless annual; flowers yellow, several, in heads   **83. Coronilla**
b. Spiny perennial or small shrub; flowers yellow, pink or red, solitary or in pairs   64
64a. Corolla yellow   **79. Anthyllis**
b. Corolla pink or red   **72. Alhagi**
65a. Leaves replaced by a tendril subtended by 2 enlarged leaf-like stipules   **86. Lathyrus**
b. Tendrils absent; stipules minute or absent   66
66a. Pods linear, many-seeded, somewhat compressed between seeds   **127. Notospartium**
b. Pods ovoid to oblong, 1–few-seeded, not compressed between seeds   67
67a. Flower-stalk woolly; raceme with c. 20 flowers; pods indehiscent   **75. Chordospartium**
b. Flower-stalk hairless or finely downy; racemes usually with fewer flowers, if with c. 20, then pods dehiscent   **76. Carmichaelia**
68a. Leaflets with the veins running to the margin, which is usually toothed   69

b. Leaflets with veins joining and looping near the margin, which is not toothed   73
69a. Standard and keel deep blue, wings pink; stipules free from the leaf-stalk   **91. Parochetus**
b. Flowers not of the above colours; stipules joined to leaf-stalk to some extent   70
70a. At least 5 of the filaments dilated below the anthers; pod usually 1-or 2-seeded, usually enclosed in the persistent corolla and calyx   **95. Trifolium**
b. Filaments not dilated below the anthers; pods usually several-seeded, not enclosed as above, corolla usually falling   71
71a. Pods coiled or sickle-shaped, often spiny   **94. Medicago**
b. Pods not as above   72
72a. Pods nutlet-like, with 1–few seeds; plants smelling of new-mown hay (coumarin)   **92. Melilotus**
b. Pods straight or curved, long, many-seeded; plants smelling variously, but not of new-mown hay   **93. Trigonella**
73a. Standard with a projecting spur on the back   **54. Centrosema**
b. Standard without a spur on the back   74
74a. Stipels absent   75
b. Stipels present (rarely soon falling or gland-like)   78
75a. Shrub; flowers in axillary racemes congested into panicles at the tips of the branches   76
b. Herb or small, spiny shrub; flowers solitary or in pairs   77
76a. Keel somewhat curved, acute; bracts each subtending 1 flower; bracteoles quickly deciduous   **43. Campylotropis**
b. Keel straight, blunt; bracts each subtending 2 flowers; bracteoles persistent   **44. Lespedeza**
77a. Small, spiny shrub   **79. Anthyllis**
b. Herb, without spines   **80. Tetragonolobus**
78a. Plant a shrub with spines an the stems and often also on the leaves   **45. Erythrina**
b. Plant a shrub, woody or herbaceous climber or herb, without spines   79
79a. Flowers reversed, with the standard lowermost   **55. Clitoria**
b. Flowers not reversed, standard uppermost   80

80a. Style coiled through 2 or 3 coils   **57. Phaseolus**
b. Style straight or curved but not coiled as above   81
81a. Bracteoles absent   82
b. Bracteoles present   83
82a. Corolla blue-purple or whitish   **53. Hardenbergia**
b. Corolla reddish or yellow and black   **52. Kennedia**
83a. Trees or woody climbers with flowers in pendent racemes   84
b. Combination of characters not as above   85
84a. Pod not flattened, often bristly or velvety (causing irritation)   **46. Mucuna**
b. Pod flattened, wing-like, initially hairy, becoming beige with age   **48. Butea**
85a. Calyx obviously 2-lipped, the upper lip larger, somewhat lobed, the lower consisting of 3 small teeth   **50. Canavalia**
b. Calyx not 2-lipped, 5-toothed, teeth not as above   86
86a. Style flattened and bearded   **56. Dolichos**
b. Style not flattened and bearded   87
87a. Stipules attached to the leaf-base above their bases, part extending below the point of attachment   **51. Pueraria**
b. Stipules attached to the leaf-stalks by their bases, often early deciduous   **42. Desmodium**
88a. Most leaves terminating in a short or minute point, a spine or a tendril, without a terminal leaflet   89
b. Most leaves with a terminal leaflet   99
89a. At least some leaves terminating in tendrils   90
b. Tendrils absent   94
90a. Leaflets with parallel veins and/or stems winged   **86. Lathyrus**
b. Leaflets with pinnate veins; stems not winged   91
91a. Calyx-teeth all equal, all at least twice as long as the tube   **87. Lens**
b. Calyx-teeth unequal, at least 2 of them less than twice as long as the tube   92
92a. Calyx-teeth more or less leaf-like; stipules to 1 cm   **88. Pisum**
b. Calyx-teeth not at all leaf-like; stipules much smaller   93
93a. Style hairy all round, or on the lower side, or entirely hairless   **85. Vicia**

b. Style hairy on the upper side only
86. Lathyrus
94a. Stamens 9, filaments all united
33. Abrus
b. Stamens 10, filaments of 9 united, that of the uppermost free at least at the base 95
95a. Leaves ending in a spine which hardens and persists after the leaflets have fallen 96
b. Leaves ending in a short or minute soft point which does not harden and persist 98
96a. Creeping or mat-forming shrubs or herbs with shoots not clearly differentiated into long and short shoots **70. Astragalus**
b. Mostly upright or spreading shrubs with shoots clearly differentiated into short (flowering and leafing) shoots and long, extension shoots 97
97a. Flowers solitary or clustered on the short shoots on jointed stalks or rarely in a few-flowered umbel; corolla usually yellow, more rarely orange or pink **68. Caragana**
b. Flowers 2–5 in a raceme; corolla purplish pink to magenta
67. Halimodendron
98a. Flowers in axillary racemes; stipules soon falling, never spine-like
40. Sesbania
b. Flowers solitary or clustered on jointed stalks; stipules persistent, often hardened and spine-like
68. Caragana
99a. Stipels present and persistent, though sometimes small 100
b. Stipels entirely absent 103
100a. Flowers reversed, the standard lowermost **55. Clitoria**
b. Flowers not reversed, standard uppermost 101
101a. Large woody climber with numerous flowers in long, pendent racemes **38. Wisteria**
b. Trees or shrubs; inflorescences various, not usually as above 102
102a. Hairs mostly medifixed; stipules not spine-like; shrubs **41. Indigofera**
b. Hairs basifixed or absent; stipules usually persisting as spines; trees or shrubs **39. Robinia**
103a. Veins of leaflets running out to the toothed margin **89. Cicer**
b. Veins of leaflets looping and joining within the entire margin 104
104a. Stipules absent or represented by minute glandular points, the lowermost pair of leaflets sometimes mimicking stipules 105

b. Stipules present, developed, not represented by minute glandular points 108
105a. Leaves more or less stalked, with a variable number of leaflets (from 1 to 27), but generally some with more than 5 leaflets 106
b. Leaves almost stalkless, all with 5 leaflets, the lowermost pair sometimes mimicking stipules 107
106a. At least some flowers in terminal or apparently terminal inflorescences
79. Anthyllis
b. All flowers axillary or in axillary inflorescences **69. Calophaca**
107a. Keel beaked **81. Lotus**
b. Keel not beaked **82. Dorycnium**
108a. Pods inflated, translucent, papery 109
b. Pods not as above 111
109a. Flowers yellow; leaflets 2–6 pairs
66. Colutea
b. Flowers not yellow; leaflets usually more numerous 110
110a. Corolla scarlet; keel not coiled upwards **65. Sutherlandia**
b. Corolla purplish blue, pink, reddish brown or rarely white; keel sometimes coiled upwards
64. Swainsona
111a. Pod indehiscent, breaking transversely into 1-seeded segments 112
b. Pod dehiscent or indehiscent, not breaking transversely into 1-seeded segments, sometimes the pod itself 1-seeded 114
112a. Flowers in racemes, the flower-stalks arising serially from the axis
77. Hedysarum
b. Flowers in umbel-like clusters or heads, all the flower-stalks arising from more or less the same point 113
113a. Segments of the pod crescent-shaped, horseshoe-shaped or rectangular with an arched sinus
84. Hippocrepis
b. Segments of the pod linear to oblong, straight or slightly curved
83. Coronilla
114a. Back of the standard and/or other parts of the corolla hairy 115
b. Corolla hairless 117
115a. Flowers in axillary panicles; pod 4-winged **36. Piscidia**
b. Flowers in terminal or leaf-opposed racemes; pod not 4-winged 116
116a. Corolla bluish purple
35. Mundulea

b. Corolla yellow, pinkish, cream or white **37. Tephrosia**
117a. Twiner, roots bearing strings of tubers; keel coiled upwards within the standard **49. Apios**
b. Combination of characters not as above 118
118a. Shrubs; anthers consisting of a single sac; pod small, opening by the falling of the sides from a persistent framework to which the seeds are attached
76. Carmichaelia
b. Shrubs or herbs; anthers consisting of 2 sacs; pod not as above 119
119a. Standard reflexed, acuminate, keel and wings deflexed, narrow like a parrot's beak, all scarlet or rarely pink or white **63. Clianthus**
b. Flowers not as above 120
120a. Leaves dotted with orange-brown glands **74. Glycyrrhiza**
b. Leaves not dotted with orange-brown glands 121
121a. Pod indehiscent, more or less circular in outline, margins usually toothed, the sides often with toothed veins or pitted
78. Onobrychis
b. Pod not as above 122
122a. Keel toothed on the upper side; leaflets oblique at the base, or if narrow, then curved
71. Oxytropis
b. Keel toothed on the lower side or not toothed; leaflets symmetric at the base **70. Astragalus**

27. CASTANOSPERMUM Cunningham
J. Cullen

Trees to 30 m in the wild, with smooth, grey to brown bark and spreading branches. Leaves odd-pinnate with 8–17 leaflets. Flowers long-stalked in racemes arising on the old (leafless) shoots. Calyx cup-shaped, very weakly and equally 5-toothed. Standard reflexed, the other 4 petals similar to each other. Stamens 10, free; anthers attached at their middles to filaments. Fruit a large pod containing 3–5 large, rounded, chestnut-like seeds.

A genus of a few species from N & E Australia, New Caledonia, the New Hebrides, etc. The one cultivated species is easily grown in rich, moist soils, but good drainage is essential. Propagation is by seed, which germinates rapidly.

1. C. australe Hooker. Illustration: Elliot & Jones, Encyclopaedia of Australian plants **2**: 478 (1982); Morley & Tölken,

Flowering plants in Australia, 154 (1983).

Tree to 8 m in cultivation. Leaves dark green and glossy above. Racemes 5–15 cm, flower-stalks *c.* 2.5 cm. Flowers 3–4 cm, reddish or yellowish (var. **brevivexillum** F.M. Bailey). Pods 10–25 × 4–6 cm, woody, inflated. *Australia (Queensland, northern New South Wales).* G2.

Seedlings are grown as house plants in Australia.

28. BOLUSANTHUS Harms

J. Cullen

Tree to 15 m in the wild, often much smaller in cultivation. Bark grey. Branches arching, branchlets with brownish silky hairs. Leaves deciduous, pinnate with 9–15 leaflets including a terminal leaflet; leaflets slightly unequal-sided and sickle-shaped, shortly stalked, brownish silky hairy above and beneath, especially when young. Inflorescence a many-flowered raceme 14–20 cm; flower-stalks to 2 cm, silky hairy, each bearing a bract towards the base. Calyx cup-shaped, deeply 5-toothed, densely silky-hairy. Corolla dark blue, the standard to 1.5 cm, sometimes with a pale patch towards the base of the blade. Stamens 10, filaments free. Pod linear-oblong, flattened, to 7 × 1.2 cm, indehiscent, containing several seeds.

A genus of a single species from S & E Africa, requiring a warm, sunny site. Propagation is by seed (though this is slow).

1. B. speciosus (Bolus) Harms. Illustration: Flowering plants of South Africa **1**: pl. 23 (1921); Eliovson, South African wild flowers for the garden, edn 4, opposite 195 (1965); Wild flowers of South Africa, 67 (1980); Fabian & Germishuizen, Transvaal wild flowers, pl. 62 (1982).

South Africa (Transvaal, Natal, Swaziland), Botswana, Mozambique, Zimbabwe. G1–2.

29. MAACKIA Ruprecht & Maximowicz

J. Cullen

Deciduous trees. Leaves pinnate with up to 17 opposite leaflets and a terminal leaflet; axillary buds exposed. Flowers numerous, rather small, in terminal racemes or panicles; bracts at the bases of the flower-stalks. Calyx cylindric to bell-shaped, obliquely 5-toothed. Corolla whitish, blade of the standard reflexed upright. Stamens 10, their filaments united at the extreme base only. Pods compressed, linear-oblong, dehiscent, containing 1–5 seeds.

A genus of about 6 species from E Asia.

They are not widely grown, being less ornamental than *Cladrastis*, but are of easy cultivation, requiring good soil and a sunny position. Propagation is normally by seed, but root cuttings have also been used.

Literature: Takeda, H., *Maackia and Cladrastis, Notes from the Royal Botanic Garden Edinburgh* **8**: 95–104 (1913).

1. M. amurensis Ruprecht (*Cladrastis amurensis* (Ruprecht) Koch). Illustration: Botanical Magazine, 6551 (1881); Csapody & Tóth, A colour atlas of flowering trees and shrubs, 131 (1982).

Tree to 20 m in the wild, usually much smaller and shrubby in cultivation; bark peeling; young shoots minutely hairy at first. Leaves to 30 cm, with 7–11 ovate to elliptic leaflets, rounded to the base, tapered, sometimes rather bluntly, to the apex, hairy beneath when young, later often hairless. Flowers in stiffly erect racemes 10–15 cm. Petals whitish; keel and wings to 1.1 cm. Pods 3.5–5 cm, compressed, slightly winged. *E former USSR, N China, Korea, Taiwan.* H1. Summer.

Var. **buergeri** (Maximowicz) Schneider (*Cladrastis amurensis* var. *buergeri* Maximowicz). Illustration: Addisonia **3**: pl. 87 (1918). Plant generally smaller; leaflets persistently and densely hairy beneath. *Japan.* H3? Summer.

M. chinensis Takeda (*M. hupehensis* Takeda). Very similar to *M. amurensis* var. *buergeri* but leaflets 7–17, shiny silvery-hairy beneath when young, hairs persistent, apex rather obtuse, flowers in panicles 15–20 cm. *C China.* H3? Summer.

30. CLADRASTIS Rafinesque

J. Cullen

Trees or shrubs. Leaves pinnate with 7–13 alternate, shortly stalked leaflets including a terminal leaflet, the base of the leaf-stalk swollen and enclosing the axillary bud; stipels sometimes present. Flowers usually many in erect or pendent panicles; bracts at the bases of the flower-stalks. Calyx usually bell-shaped to cylindric with 5 short, equal teeth or lobes. Corolla white or pinkish, sometimes marked with yellow. Stamens 10, filaments almost completely free. Pod membranous, flattened, sometimes winged, often somewhat constricted between the few seeds, indehiscent.

A genus of about 6 species, 1 from N America, the rest from E Asia. They require a rich soil and a sunny position,

and may be propagated by seed or by root cuttings.

Literature: Takeda, H., *Cladrastis and Maackia, Notes from the Royal Botanic Garden Edinburgh* **8**: 95–104 (1913).

1a. Leaflets green beneath, with small, linear stipels; pod conspicuously winged   **3. platycarpa**
  b. Leaflets grey-green beneath, without stipels; pod not conspicuously winged   2
2a. Panicles pendent; leaflets 7–9, 4–6 cm or more wide; calyx 7–8 mm   **1. kentukea**
  b. Panicles upright; leaflets 9–13, to 3.5 cm wide; calyx 4–5 mm   **2. sinensis**

1. C. kentukea (Dumont de Courset) Rudd (*C. lutea* (Michaux) Koch; *C. tinctoria* Rafinesque). Illustration: Botanical Magazine, 7767 (1901); Justice & Bell, Wild flowers of North Carolina, 99 (1968); Bean, Trees & shrubs hardy in the British Isles, edn 8 **1**: 631 (1970); Il Giardino Fiorito **48**: 564 (1982).

Tree to 20 m; bark smooth, wood yellow. Leaves with 7–9 ovate to broadly obovate leaflets 8–15 × 4–6 cm or more, apex rather abruptly tapered or acuminate, base rounded to tapered, dark green above, greyish green beneath, almost hairless. Panicles pendent, branches hairless or with sparse curled hairs. Flowers white, fragrant. Calyx cylindric to cup-shaped, 7–8 mm, with 5 rounded, equal, hairy teeth. Standard 2–2.5 cm. Pod 5–10 × 1–1.5 cm, not winged. *S & C USA.* H4. Early summer.

2. C. sinensis Hemsley. Illustration: Notes from the Royal Botanic Garden Edinburgh **8**: pl. 26 (1913); Botanical Magazine, 9043 (1924).

Tree to 25 m or shrubby, bark brown, smooth. Leaves with 9–13 oblong to oblong-lanceolate leaflets 6–11 × 1.5–3.5 cm, long-tapered to the apex, rounded to the base, dark green and hairless above, greyish green and usually hairy, at least along the midrib, beneath. Panicles erect, branches hairy. Flowers white or pinkish. Calyx bell-shaped, 4–5 mm, equally 5-lobed with obtuse lobes, hairy all over, brownish. Standard 1–1.5 cm. Pod 5–10 × 0.7–1.1 cm, not winged. *W & C China.* H5. Summer.

3. C. platycarpa (Maximowicz) Makino (*Platyosprion platycarpum* (Maximowicz) Maximowicz). Illustration: Kurata, Illustrated important forest trees of Japan

1: pl. 76 (1971); Kitamura & Murata, Coloured illustrations of woody plants of Japan 1: pl. 71 (1977); Krüssmann, Manual of cultivated broad-leaved trees and shrubs 1: 337 (1986).

Tree to 20 m, bark smooth, greyish. Leaves with 7–9 narrowly ovate to ovate-oblong leaflets 7.5–10 × 2.5–4 cm, tapered to acuminate at the apex, tapered to rounded at the base, green above, paler beneath, generally with some crisped hairs on stalks, margins and main veins, each leaflet with 2 linear stipels at the base. Panicles upright, many-flowered, densely hairy. Calyx bell-shaped, c. 6 mm, with 5 equal, obtuse teeth, densely hairy all over, brownish. Petals white, the standard c. 1.5 cm , with a yellow spot at the base. Pod c. 8 × 1.5 cm, conspicuously winged. *Japan.* H5. Summer.

31. SOPHORA Linnaeus
J. Cullen

Trees or shrubs. Leaves pinnate with up to 43 leaflets (including a terminal leaflet) which are generally opposite; stipules small; axillary buds small, exposed. Flowers bisexual, in terminal or axillary racemes or panicles, pea-flower-like (with blade of standard reflexed upright) or with all the petals forwardly directed; bracts borne at the bases of the flower-stalks. Calyx cup-shaped, truncate, usually obliquely so, with 5 small teeth. Stamens 10, filaments slightly united at the base. Pod stalked, terete, constricted between the seeds, sometimes longitudinally winged, often not dehiscing but breaking into 1-seeded segments.

A genus of about 50 species, widespread (though more than half of the cultivated species are from the southern hemisphere). It is very variable and has in the past been divided into several genera. Only a small number generally cultivated. They require warm, sunny sites and rich soil, and may need the protection of a wall or glasshouse in northern parts. Propagation is generally by seed, but some species will root from cuttings of young wood which have a heel of older wood.

Habit. Very large trees: **1**; shrubs with short branches which terminate in spines: **3**; small shrubs with branches diverging at very wide angles: **7**.
Leaflets. More than 21: **6**; 2.5–5 cm: **1,2**; less than 2.5 cm: **3–7**.
Inflorescence. Panicle: **1**; raceme with 4–10 flowers: **2–6**; raceme with 2–3 flowers: **7**

Flowers. Pea-flower-like, with blade of standard reflexed upright: **1–3**; with all petals forwardly directed: **4–7**; white, cream, pinkish or bluish: **1–3**; yellow: **4–7**.
Standard. To 1.5 cm: **1–3**; more than 2.5 cm: **4–7**.
Pods. With up to 6 seeds: **1–4**; with 6 or more seeds: **5–7**; longitudinally winged: **5–7**.

1a. Flowers to 1.5 cm, pea-flower-like, the blade of the standard reflexed upright, white, cream, pinkish or bluish   2
  b. Flowers 3 cm or more, all the petals forwardly directed   4
2a. Plants with spines formed from the ends of short lateral shoots   **3. davidii**
  b. Plants without spines   3
3a. Large trees; inflorescence a panicle   **1. japonica**
  b. Small trees to 6 m; inflorescence a raceme   **2. affinis**
4a. Leaflets 1.8 cm or more; pod not winged   **4. macrocarpa**
  b. Leaflets to 1.8 cm; pod longitudinally winged   5
5a. Small shrub with branches diverging at very wide angles; racemes with 2–3 flowers   **7. prostrata**
  b. Small trees or shrubs with branches diverging at narrow angles; racemes with 4–10 flowers   6
6a. Leaflets 11–21; wings exceeding the standard   **5. tetraptera**
  b. Leaflets 23–43; wings about as long as standard   **6. microphylla**

1. S. japonica Linnaeus. Illustration: Botanical Magazine, 8764 (1918); Polunin & Everard, Trees and bushes of Europe, 105 (1976); Csapody & Tóth, Colour atlas of flowering trees and shrubs, 135 (1982); Everett, New York Botanical Gardens illustrated encyclopedia of horticulture 9: 3183 (1982).

Tree to 30 m with rounded crown and often contorted and twisted main branches; bark greyish and corrugated; young shoots green, downy. Leaves 15–25 cm with 7–17 lanceolate to ovate leaflets 2.5–5 cm, broadly tapered to rounded at the base, acute at apex, shortly stalked, dark green and hairless above, paler or glaucous and downy beneath. Flowers 1–1.5 cm in loose terminal panicles to 30 cm. Calyx 3–4 mm. Petals cream, white or marked with pinkish purple ('Violacea'), the standard with blade reflexed upright. Pods 5–8 cm,

shortly stalked, terete, hairless, constricted between the 1–6 seeds. *China, Korea.* H4. Late summer.

A widely planted tree often used to line avenues in the warmer parts of Europe. It is seldom grown further north, where it rarely flowers. Var. **pubescens** (Tausch) Bosse has narrower leaflets and is more densely hairy. Variants with drooping branches ('Pendula') or variegated leaves ('Variegata') are sometimes grown. A survey of currently available cultivars is provided by Schalk, P.H., *Sophora japonica*, Dendroflora **22**: 69–27 (1985).

Robinia pseudoacacia (p. 489) is often planted with *S. japonica* and is very similar to it when not in flower and fruit; however, the *Robinia* has dark reddish brown young shoots, leaflets with more rounded tips and, usually, paired, short spines at the base of each bud.

2. S. affinis Torrey & Gray. Illustration: Elias, The complete trees of North America, 662 (1980).

Tree to 6 m, with rounded crown. Leaves 10–20 cm with 13–19 elliptic leaflets 2–4 cm, tapered to the base, somewhat rounded at the apex, hairless or slightly hairy beneath. Flowers 1–1.5 cm in axillary racemes to 15 cm. Calyx bell-shaped, downy. Petals white tinged with pink, the standard with blade reflexed upright. Pods 3–8 cm, black, downy, constricted between the few seeds. *S USA (Arkansas, Texas).* H4. Early summer.

3. S. davidii (Franchet) Skeels (*S. viciifolia* Hance). Illustration: Botanical Magazine, 7883 (1903); Csapody & Tóth, Colour atlas of flowering trees and shrubs, 135 (1982).

Rounded shrub to 3 m; branchlets brownish, downy, the ends of many short branches persisting as spines. Leaves 3–9 cm, of 11–17 leaflets; leaflets elliptic-oblong, 0.5–1.2 cm, rounded to the base, rounded or somewhat tapered to the mucronate, often somewhat notched apex, hairy when young. Flowers 1–1.5 cm in racemes terminating short branches. Calyx downy, cylindric to bell-shaped, 4–7 mm, shortly 5-toothed, often violet. Petals bluish white, standard with blade reflexed upright. Pods hairy, 3–6 cm × 3–5 mm, deeply but gradually constricted between the 1–4 distant seeds. *C & W China.* H4. Spring–summer.

4. S. macrocarpa J.E. Smith. Illustration: Botanical Magazine, 8647 (1916); Perry, Flowers of the world, 165 (1972).

Shrub or small tree; young branches

densely downy. Leaves 7–10 cm, divided into 11–17 leaflets; leaflets oblong or lanceolate-oblong 1.8–2.5 (rarely more?) × 0.7–1.2 cm, broadly rounded at the base and apex, finely downy above and beneath. Flowers in short axillary racemes. Calyx downy, cup-shaped, obliquely truncate with 5 very small teeth, 1–1.4 cm. Petals yellow, all forwardly directed, standard 2.8–3.5 cm, wings and keel 2.5–3 cm. Pods downy, 7–12 cm, long-stalked, very deeply constricted between the 1–4 seeds, not winged. *Chile*. H5.

5. S. tetraptera Miller. Illustration: Botanical Magazine, 167 (1796); Morley & Everard, Flowers of the world, pl. 129 (1970); Johnson, The international book of trees, 211 (1973); Salmon, The native trees of New Zealand, 197–199 (1980).

Large shrub or small tree to 12 m; bark greyish, young shoots hairy. Leaves 8–15 cm, with 11–21 leaflets; leaflets ovate to elliptic-oblong, tapered to the base, rounded or notched at the apex, 0.7–1.8 cm, silky hairy on both surfaces. Flowers bright to golden yellow in racemes of 4–10. Calyx bell-shaped, obliquely truncate with 5 small teeth, silky hairy, 1–1.5 cm. Petals all forwardly pointing; standard 3–3.5 cm, wings 3.7–4 cm, keel exceeding wings. Pod to 20 cm, stalked, very deeply constricted between the 6 or more seeds, with 4 wings extending the whole length of the pod. *New Zealand*. H5–G1. Spring.

A variant which flowers as a low shrub is cultivated in New Zealand (and perhaps elsewhere) as 'Gnome'.

6. S. microphylla Aiton. Illustration: Botanical Magazine, 1442 (1812); Journal of the Royal Horticultural Society **96**: f. 247 (1971); Salmon, The native trees of New Zealand, 200, 201 (1980).

Like *S. tetraptera* but a smaller tree with a distinct juvenile phase as a shrub with slender branches which diverge at narrow angles; leaves tending to droop, with 23–43 leaflets which are generally smaller (in var. **longicarinata** (Simpson) Allan the leaflets are numerous, small and broadly elliptic); standard 3–4 cm, wings about as long as the standard, keel longer. *New Zealand*. H5–G1. Spring.

Var. **fulvida** Allan. More densely hairy with brown hairs.

A very closely related plant occurs in temperate South America; it has been named **S. macnabiana** Graham or **S. microphylla** subsp. **macnabiana** (Graham) Yakovlev. Illustration: Botanical Magazine, 3735 (1840). Standard

somewhat shorter than the wings. It may be in cultivation.

7. S. prostrata Buchanan. Illustration: Salmon, The native trees of New Zealand, 201 (1980).

Like *S. tetraptera* and *S. microphylla* but a low shrub to 2 m with branches diverging at very wide angles; leaves to 2.5 cm with *c*. 17 leaflets; flowers in racemes of 2 or 3, the keel to 3 cm, longer than the wings which are themselves longer than the standard. *New Zealand*. H5–G1. Spring.

32. PTEROCARPUS Jacquin
J. Cullen
Evergreen or deciduous trees; sap often red. Leaves pinnate, the leaflets several, opposite or not, terminal leaflets present; stipules often falling quickly, stipels absent. Flowers in racemes or panicles. Calyx funnel- or bell-shaped, shortly 5-lobed, the upper 2 lobes united for much of their length. Corolla yellow, orange, whitish (ours) or violet; wings longer than keel. Stamens 10, filaments all united. Ovary sometimes shortly stalked, with few ovules. Pod compressed, indehiscent, the central part hardened and seed-bearing, winged. Seeds with small arils.

A tropical genus of about 20 species; only 1 is occasionally grown, and little is known of its requirements. Propagation is by seed.

1. P. angolensis de Candolle. Illustration: Palgrave, Trees of Central Africa, 330, 331 (1956); Flora of Tropical East Africa, Leguminosae 3: 90 (1971); Palmer, A field guide to the trees of southern Africa, pl. 13 (1977).

Deciduous tree to 20 m, with open crown (rarely a shrub); bark grey, fissured. Young shoots grey- or silver-hairy. Leaves to 35 cm with 11–19 lanceolate to elliptic, shortly acuminate leaflets, generally persistently hairy beneath. Racemes often appearing before the leaves; flowers scented. Calyx 0.8–1 cm, hairy. Corolla 1.6–2 cm, golden yellow to orange. Fruit almost circular, 1–1.5 cm, stalked, with feathery bristles on the central, seed-bearing part. *Tropical Africa*. G2.

33. ABRUS Adanson
J. Cullen
Woody climbers, scramblers or low shrubs. Leaves pinnate without a terminal leaflet, with numerous, opposite lateral leaflets; stipules small, persistent; stipels small, linear, usually present. Flowers aggregated

on short, wart-like outgrowths in racemes or spikes; bracts and bracteoles small, often falling early. Calyx funnel-shaped, truncate with 5 small teeth. Corolla pale purple to yellowish; standard notched at the apex, keel longer than wings. Stamens 9, the filaments all united and united at the base to the standard. Ovary with many ovules, downy; style curved, persistent. Pods oblong to linear, irregularly swollen or flattened, beaked. Seeds with arils.

A genus of 4 (Breteler) or more than 15 (Verdcourt) species from tropical regions. Only 1 is grown, mainly for the sake of its seeds, which, though extremely poisonous, have often been used as beads. It is easily grown and propagated by seed or by cuttings.

Literature: Breteler, F.J., Revision of *Abrus* Adanson with special reference to Africa, *Blumea* **10**: 607–624 (1960); Verdcourt, B., Studies in the Leguminosae-Papilionoideae for the Flora of East Tropical Africa 2: a reappraisal of the species of the genus *Abrus* Adanson, *Kew Bulletin* **24**: 235–253 (1970).

1. A. precatorius Linnaeus. Illustration: Bentley & Trimen, Medicinal plants **2**: pl. 77 (1877); Marloth, The Flora of South Africa **2**: pl. 29 (1925); Blumea **10**: 618 (1960).

Leaves with 8–17 pairs of leaflets; leaflets oblong to slightly obovate, rounded to the base and the mucronate apex, 6–25 × 3–10 mm, with sparse adpressed whitish hairs. Calyx *c*. 3 mm. Corolla *c*. 1.5 cm, reddish, purplish or yellowish. Pod oblong, compressed, hairy, sometimes warty, 2–5 × 1–1.5 cm. Seeds 5–7 × 4–5 mm, scarlet with a black blotch around the hilum, glossy. *Tropics*. G2.

Verdcourt divides the species into 2, subsp. **precatorius** with hairy pods and subsp. **africanus** Verdcourt, with pods both hairy and warty (mainly African); both may be in cultivation.

34. DERRIS Loureiro
J. Cullen
Evergreen (ours) or deciduous woody climbers. Leaves odd-pinnate with more or less opposite lateral leaflets and a terminal leaflet; stipules falling early, stipels absent. Flowers numerous in racemes or panicles. Calyx broadly bell-shaped, truncate, with 5 short teeth. Corolla white, pink or yellowish. Stamens 10, filaments all united into a tube (in the cultivated species). Ovary hairy, several-seeded. Pod indehiscent, compressed, hairy, few-seeded,

somewhat winged along both sutures.

A genus of about 60 species from the tropics; only a few are grown, mostly as a source of the insecticide 'Derris' (rotenone).

1. D. elliptica (Wallich) Bentham. Illustration: Botanical Magazine, 8530 (1913) – as D. oligosperma.

Large, vigorous, evergreen climber with blackish stems and hairy young shoots. Leaves 20–40 cm, with 9–13 leaflets; leaflets obovate-oblong, acute, persistently brown-hairy beneath. Racemes to 25 cm, many-flowered, axis and branches hairy. Calyx to 8 mm, hairy. Corolla pink or rarely white, to 1.2 cm; standard with 2 thickenings at the base of the blade, hairy on the back; keel hairy towards the tip. Pod hairy with 1–4 seeds. *Burma to Malaysia, cultivated elsewhere.* G2.

D. malaccensis Prain. Illustration: Duke, Handbook of Legumes of world economic importance, 76 (1981). Very similar but leaves and petals hairless. *Malaysia to New Guinea.* G2.

Both species are grown commercially in the tropics as a source of rotenone.

35. MUNDULEA (de Candolle) Bentham
J. Cullen
Small trees or shrubs. Leaves pinnate, leaflets opposite or almost so, terminal leaflet present; stipules small, triangular, stipels absent. Flowers in terminal racemes. Calyx bell-shaped, 5-toothed. Corolla bluish purple, standard silky-hairy outside; wings and keel downy on the margins near the base. Stamens 10, filaments of 9 united into a tube from the base, the uppermost free at the base, united to the others higher up. Ovary with several ovules, hairy. Pod densely hairy, several-seeded, indehiscent, often constricted between the seeds. Seeds without arils.

A genus of about 15 species from Africa to Malaysia, the rest from Madagascar, of which only the non-Madagascan species is grown. Little is known of its requirements in Europe. Propagation is by seed.

1. M. sericea (Willdenow) Chevalier. Illustration: Flora of Tropical East Africa, Leguminosae Part **3**: 156 (1971); Palmer & Pitman, Trees of southern Africa **2**: 920, 921 (1972); Palgrave, Trees of southern Africa, pl. 100 (1977).

Tree or shrub to 7 m; bark smooth or fissured. Leaves to 10 cm, leaflets 13–17, ovate to lanceolate, hairy at least beneath. Flowers in pairs in hairy racemes. Calyx c. 6 mm, the upper 2 teeth united for most of their length. Standard to 2 cm. Pod yellowish brown, velvety. *S & tropical Africa, India, Malaysia.* G2.

36. PISCIDIA Linnaeus
J. Cullen
Deciduous trees. Leaves pinnate with 5–9 opposite leaflets and a terminal leaflet; stipules large, falling early, stipels absent. Flowers in dense, hairy, axillary panicles. Calyx 5-lobed, the upper 2 lobes united for a greater part of their length than the others. Corolla pink, white (sometimes striped with pink or red) or red, standard hairy on the back. Stamens 10, 9 of the filaments united at the base and for most of their length, the uppermost free at the base, united to the others higher up; anthers versatile. Pod 4-winged, indehiscent, containing 3–7 seeds.

A genus of about 10 species in subtropical and tropical America. Only 1 is grown; it can be propagated by seed or by cuttings.

1. P. piscipula (Linnaeus) Sargent. Illustration: Fawcett & Rendle, Flora of Jamaica **4**(2): 83 (1920); Correll & Correll, Flora of the Bahama Archipelago, 676 (1982).

Tree to 10 m. Leaves to 25 cm, with elliptic-ovate, acute or obtuse leaflets, usually hairless above, minutely hairy beneath. Flowers appearing before the leaves. Corolla to 1.2 cm, hairy, standard 1.3–1.5 cm, more or less circular, notched. Pod stalked, 3–8 cm, wings lobed and crisped. *USA (Florida), West Indies.* G2.

37. TEPHROSIA Persoon
J. Cullen
Perennial herbs or shrubs, usually conspicuously hairy. Leaves pinnate with 5–31 leaflets (in the cultivated species) including a terminal leaflet; stipules small, deciduous or persistent, stipels absent. Flowers in terminal or leaf-opposed racemes. Calyx 5-lobed, the upper lobes united for more of their length than the others. Corolla small to large, standard hairy on the back, wings and keel hairy or not. Stamens 10, with 9 of the filaments united at the base and for most of their length, the uppermost filament usually free at the base but united to the others above. Ovary with 4–16 ovules, usually hairy; style hairy or hairless. Pod linear to oblong, usually compressed, hairy at least on the sutures, the walls often cracking obliquely with age. Seeds without arils.

A genus of about 400 species mostly from the tropics, a few extending into temperate areas. Only a few are grown as ornamentals (though others are grown as green manures or for fish poisons in some areas), are of easy culture and propagated by seed.

1a. Style hairy throughout its length; standard 1.5 cm or more   2
  b. Style hairless (rarely hairy just at the extreme base); standard to 1 cm   3
2a. Pods and ovary hairy only along the sutures; standard pink inside, orange outside   **1. grandiflora**
  b. Pods and ovary hairy all over the surface; standard yellow to cream outside, cream to white inside   **2. virginiana**
3a. Leaves white-hairy beneath; pods with short, erect hairs   **3. purpurea**
  b. Leaves hairy but green beneath; pods with adpressed hairs   **4. capensis**

1. T. grandiflora (Aiton) Persoon. Illustration: Batten & Bokelmann, Wild flowers of the Eastern Cape, pl. 70 (1966); Fabian & Germishuizen, Transvaal wild flowers, pl. 61b (1982).

Shrub 50–200 cm; young shoots with adpressed hairs. Leaves with 7–17 leaflets; leaflets narrowly elliptic to narrowly obovate, tapered to the base, abruptly rounded to the blunt or notched, mucronate apex, adpressed hairy on both surfaces, densely so, giving a whitish appearance beneath. Calyx deeply 5-lobed, densely adpressed hairy, tube 3–5 mm, lobes 4–7 mm. Corolla pink, standard orange outside, densely hairy, 1.8–2.5 cm. Style hairy throughout its length. Pod compressed, oblong, hairy only on the sutures. Seeds 9–16. *South Africa, naturalised elsewhere.* H5–G1.

2. T. virginiana (Linnaeus) Persoon. Illustration: Gleason, Illustrated flora of the north-eastern United States and adjacent Canada **2**: 413 (1952); Rickett, Wild flowers of the United States **1**: 281 (1966); Justice & Bell, Wild flowers of North Carolina, 102 (1968).

Low shrub; young shoots with whitish spreading hairs. Leaves with 15–37 leaflets; leaflets narrowly elliptic to obovate, tapered to the base, tapered or somewhat rounded to the mucronate apex, hairless above (except along the margins), with spreading, often somewhat curled hairs beneath. Calyx deeply 5-lobed, densely covered with whitish spreading hairs, tube 2–3 mm, lobes 3–6 mm. Standard 1.5–2 cm, lemon-yellow to cream and densely hairy outside, cream to white inside; wings and keel usually pink, hairy. Style hairy throughout its length. Pods hairy all over the surface. Seeds 6–11. *E & C North America*. H3. Spring–summer.

T. vogelii J.D. Hooker. Illustration: Transactions of the Linnean Society of London **29**: t. 31 (1873). Similar, but leaflets 11–31; upper calyx-lobes united for most of their length; standard *c*. 2.5 cm; pod 10–12 × 1–1.3 cm. *Tropical Africa*. G2.

T. candida de Candolle. Similar to *T. vogelii* but standard, wings and keel white, pod 6–9 × 0.7–0.9 cm. *Tropics*. G2.

3. T. purpurea (Linnaeus) Persoon. Illustration: Flora of East Tropical Africa, Leguminosae, **3**: 187 (1971).

Many-stemmed shrub to 1.3 m; young shoots whitish with silky, adpressed hairs. Leaves with 7–15 leaflets; leaflets obovate, gradually tapered to the base, abruptly rounded to the blunt, truncate or notched, mucronate apex, adpressed silky hairy, densely so and whitish beneath. Calyx deeply 5-lobed, tube 1.5–3 mm, teeth 2–4 mm, adpressed hairy. Corolla purplish red; standard 0.5–1 cm, densely hairy outside, wings and keel densely hairy. Style hairless. Pods somewhat flattened, with erect, short hairs all over the surface. Seeds up to 6. *Tropics*. G2.

A variable species, divided into several subspecies by some authors; it is uncertain which of these are in cultivation.

4. T. capensis Persoon. Illustration: Kidd, Cape Peninsula, 107 (1983); Burman & Bean, Hottentots Holland to Hermanus, 117 (1985).

Many-stemmed slender shrub; young shoots with adpressed hairs. Leaves with 9–15 leaflets; leaflets very narrowly elliptic to narrowly obovate, long-tapered to the base, rather abruptly rounded to the blunt, shortly mucronate apex, hairless above, adpressed hairy beneath. Calyx deeply 5-lobed, tube 2–2.5 mm, lobes 2–2.5 mm, all adpressed hairy. Corolla bright pink or purplish; standard 0.6–1 cm, adpressed hairy outside; wings and keel not hairy. Style hairless except at the extreme base. Pods flattened, with adpressed hairs all over the surface, containing *c*. 6 seeds. *South Africa, Mozambique*. G1.

38. WISTERIA Nuttall

J. Cullen

Large, deciduous woody climbers. Leaves pinnate with up to 19 opposite lateral leaflets and a terminal leaflet; stipules falling early, narrow stipels present; leaflet-stalks somewhat swollen. Flowers numerous in often large, pendent racemes. Calyx bell-shaped or somewhat cylindric, obliquely 5-lobed, the 2 upper lobes often united for most of their length. Corolla blue, purple, lilac, pink or white; standard large, usually with 2 humps at the base of the blade. Stamens 10, 9 of them with their filaments united, that of the uppermost free. Ovary with several ovules. Pod large, compressed, with several seeds.

A genus of 6 species from N America and E Asia, most of them valued as ornamental climbers for growing against walls, on arches or in arbours. Though climbing, some of them can be grown as standards by careful pruning. They are generally easily grown, requiring a sunny site and appropriate support. Propagation by seed is possible but not recommended, as many inferior variants will be produced; selected variants are best propagated by grafting.

Literature: McMillan-Browse, P., Some notes on members of the genus *Wisteria*, The Plantsman **6**: 109–122 (1984), which gives much detail on selected clones and on propagation techniques.

1a. Flower-stalks more than 1 cm, usually flexuous; pods hairy 2
  b. Flower-stalks up to 1 cm, usually stiff, ascending; pods hairless 4
2a. Leaflets persistently silky-hairy on both surfaces **3. venusta**
  b. Leaflets becoming hairless or almost so on both surfaces 3
3a. Stems twining clockwise; leaflets 13–19; flowers opening serially from the base of the raceme, usually when the leaves are fully expanded **1. floribunda**
  b. Stems twining anticlockwise; leaflets 7–13, usually 9–11; flowers in the raceme all opening at more or less the same time, before the leaves are fully expanded **2. sinensis**
4a. Racemes 4–10 cm, without glands **4. frutescens**
  b. Racemes 20–35 cm, with glands on axis and flower-stalks **5. macrostachya**

1. W. floribunda (Willdenow) de Candolle (*W. brachybotrys* Siebold & Zuccarini). Illustration: Botanical Magazine, 7522 (1897); Rose, Climbers and wall plants, t. 47 (1982); The Plantsman **6**: 117 (1984).

Large woody climbers to 8 m or more, twining clockwise. Leaflets 13–19, narrowly ovate or elliptic to lanceolate, rounded to the base, acuminate towards the ultimately blunt apex, hairless above, hairless or very sparsely hairy beneath (except when young); leaves usually fully expanded as flowering begins. Racemes 20–50 cm or more (to 1.5 m in some cultivars), with many fragrant flowers. Flower-stalks 1–3 cm, flexuous and spreading; flowers opening serially from the base of the raceme. Calyx 4–6 mm, silky-hairy. Corolla white, blue, violet or pink, standard 1.4–2 cm, broadly oval. Pod 10–15 cm, narrowed towards the base, finely downy. *Japan*. H3. Late spring–early summer.

Introduced in the form of various clones differing in flower-colour and size of raceme; new clones have been regularly produced since. The most striking are: 'Alba' (flowers white), 'Multijuga' (racemes very long), 'Rosea' (flowers pink), 'Violacea Plena' (flowers violet, double) and 'Issai' (flowers opening before the leaves are fully expanded) the latter perhaps of hybrid origin.

2. W. sinensis (Sims) Sweet. Illustration: Botanical Magazine, 2083 (1819) – as Glycine sinensis; Menninger, Flowering vines of the world, f. 107 (1970); Perry, Flowers of the world, 162 (1970); The Plantsman **6**: 117 (1984).

Very similar to *W. floribunda* but less vigorous, climbing anticlockwise; leaflets 7–13 (usually 9–11), densely hairy at first, later more or less hairless; flowers in the raceme opening more or less simultaneously before the leaves are fully expanded. *China*. H4. Spring.

Variable, with numerous selected clones available. The species name is often misspelled 'chinensis'.

W. × formosa Rehder is the hybrid between *W. sinensis* and *W. floribunda*; it is generally intermediate between its parents, but all the flowers in the raceme open more or less simultaneously.

3. W. venusta Rehder & Wilson. Illustration: The Plantsman **6**: 113 (1984).

Climbing to about 10 m. Leaflets 9–13, ovate to elliptic or oblong to oblong-lanceolate, persistently hairy on both surfaces, the hairs producing a silvery sheen. Racemes downy, 10–15 cm; flower-stalks more than 1 cm, flexuous, spreading, downy. Flowers white or pinkish (purple in 'Violacea'), of similar size to those of *W. floribunda*. Pod 15–20 cm, hairy. *Japan*. H3. Summer.

First described from cultivated, white-flowered plants from Japan; the common wild-type there has purple flowers. A variant with double, white flowers ('Plena') has been grown.

4. W. frutescens (Linnaeus) Poiret. Illustration: Botanical Magazine, 2103 (1819) – as Glycine frutescens; Dean, Mason & Thomas, Wild flowers of Alabama, 89 (1973).

Climbing to 12 m; young shoots hairless or almost so. Leaves with 9–15 leaflets which are narrowly ovate or lanceolate to elliptic, tapered or somewhat rounded at the base, acuminate at the apex, sparsely hairy along the main veins above, hairy and pale beneath, margins more densely hairy. Racemes 4–10 cm, dense, axis and flower-stalks hairy; flower-stalks to 1 cm, stiff, ascending. Calyx cylindric to bell-shaped, densely hairy, 6–9 mm, the upper teeth very short. Corolla lilac-purple, often with a yellowish spot at the base of the standard; standard 1.5–2 cm. Pod hairless, compressed, 5–10 cm. *SE USA*. H4.

A white-flowered variant ('Nivea') has been grown.

5. W. macrostachya (Torrey & Gray) Torrey & Gray. Illustration: Loughmiller & Loughmiller, Texas wild flowers, 140 (1984).

Slender climber to 8 m; young shoots hairless. Leaflets usually 9, ovate to elliptic or lanceolate, hairy beneath when young, rounded to almost cordate at the base, acuminate at the apex. Racemes dense, 20–35 cm, axis and flower-stalks with glandular hairs; flower-stalks to 1 cm, stiff, ascending. Corolla lilac-purple. Pod hairless, somewhat constricted between the seeds. *SE USA*. H4. Summer.

39. ROBINIA Linnaeus
J. Cullen

Trees or shrubs. Leaves pinnate with opposite or somewhat alternate leaflets and a terminal leaflet; stipules often persistent as paired spines; stipels present, small, needle-like. Flowers in erect or pendent racemes; bracts present, soon falling. Calyx bell-shaped, unequally 5-toothed. Corolla white to pink or pale purple; standard broad, reflexed upwards, keel-petals united to each other at the base. Stamens 10, 9 of them with united filaments, that of the uppermost free or partly so. Pod flattened, bristly or not, rarely winged along the upper suture, containing several seeds.

A genus of 4 species (see below) from eastern and southern USA. Traditionally

many more species have been recognised, but recent work (cited below) indicates that only 4 have any real structural basis, and even these hybridise (in the wild) to some extent.

The plants are easy to grow, particularly *R. pseudoacacia*, and will tolerate quite poor soils (*R. pseudoacacia* tends to grow very rapidly and coarsely in good soils). They are propagated by seed, by suckers (root division), or, in the case of cultivars and those species or varieties that rarely form ripe pods, by grafting on to *R. pseudoacacia* stocks.

Literature: Isely D. & Peabody, F.J., *Robinia* (Leguminosae: Papilionoideae), *Castanea* **49**(4): 187–202 (1984).

1a. Large trees; flowers usually white (pink in some cultivars), standard 1.5–2 cm; pods hairless; racemes without glands   **1. pseudoacacia**
  b. Small trees or shrubs; flowers usually pink to purple; pods sparsely to densely bristly; racemes usually with glands on axis and flower-stalks   2
2a. Leaves persistently hairy on both surfaces   **2. neomexicana**
  b. Leaves ultimately hairless at least above   3
3a. Leaflets mostly 13–21, persistently hairy beneath; plant not conspicuously bristly   **4. viscosa**
  b. Leaflets 9–13, becoming hairless beneath; plant often conspicuously bristly   **3. hispida**

1. R. pseudoacacia Linnaeus. Illustration: Justice & Bell, Wild flowers of North Carolina, 101 (1968); Edlin & Nimmo, The world of trees, 60 (1974); Polunin & Everard, Trees and bushes of Europe, pl. 111, 112 (1976); Il Giardino Fiorito **48**: 281 (1982).

Usually large trees with dark brown, fissured bark; young shoots hairless or sparsely hairy. Leaves with 11–23 leaflets (reduced to a large terminal leaflet and 0–2 pairs of small lateral leaflets in 'Unifoliola' also known as 'Monophylla'); leaflets lanceolate, ovate, elliptic or oblong-elliptic, rounded at the apex, tapering to rounded at the base, shortly stalked, hairless or with some fine soft hairs beneath at maturity; stipules persisting as triangular spines (absent in 'Inermis'). Flowers numerous, white, fragrant, in long, pendent racemes. Calyx to 8 mm. Standard 1.5–2 cm, yellow-blotched at the base of the blade. Pod oblong, flattened, hairless, narrowly winged along the upper

suture. *E USA*. H2. Spring.

A very variable tree in which numerous cultivars ('varieties') have been named, based on crown-shape, leaf-coloration, shape and texture and habit. It is very widely planted in C & S Europe, and naturalised there, and is used (often together with the superficially very similar *Sophora japonica*, p. 485) as a street or avenue tree.

'Decaisneana' (*R. pseudoacacia* var. *decaisneana* Carrière). Illustration: Addisonia **19**: pl. 624 (1935–36). Stipule-spines reduced and pink flowers; widely grown, as is 'Bella Rosea' with glandular branches and pink flowers. Hybridises with other species of the genus, and these hybrids are sometimes grown:

R. × slavinii Rehder (*R. margarettae* Ashe). *R. pseudoacacia × R. hispida*. Shrub or small tree with leaflets rather variable in number, bristly inflorescences and large, pink flowers and warty pods. It occurs sporadically in the wild as well as in gardens.

R. × holdtii Beissner. Thought to be *R. pseudoacacia × R. neomexicana* and possibly including back-crosses to *R. pseudoacacia*. Like *R. pseudoacacia* but with keel and wings white, the standard pinkish, pods glandular-bristly.

R. × ambigua Poiret (*R. pseudoacacia × R. viscosa*). Shrub with glandular twigs and whitish or deep pink corolla.

2. R. neomexicana Gray (*R. luxurians* (Dieck) Silva-Taroucana & Schneider). Illustration: Martin & Hutchins, Summer wild flowers of New Mexico, 145 (1986).

Shrub or small tree to 10 m; young shoots downy at first. Leaves with 13–25 leaflets; leaflets lanceolate, narrowly ovate or oblong, rounded at the base and apex, permanently downy on both surfaces. Stipule-spines straight, slender, to 1 cm. Calyx 0.85–1 cm, downy and glandular. Corolla pink, standard 2–2.7 cm. Pod covered with glandular bristles. *S USA (New Mexico, Arizona)*. H4? Summer.

3. R. hispida Linnaeus.

Shrub to 3 m, suckering widely, branches usually with spreading bristles to 1.5 mm, these glandular when young, and fine, soft white hairs. Leaflets 9–13, ovate to broadly elliptic to oblong, rounded to the base, rounded or somewhat acute at the apex, hairless on both surfaces when mature. Racemes rather few-flowered, axis and flower-stalks with glandular bristles and soft, whitish hairs. Calyx 0.9–1.4 cm. Standard 2.8–4 cm, pink to rose purple, as are with wings and keel. Pods rarely formed (see below), densely bristly. *E USA.* H2. Summer.

According to Isely & Peabody, this species consists mainly of triploid, usually sterile clones, which rarely produce fruit, but spread by suckering. Because of this, it is very variable and divided into several varieties, of which all but var. *kelseyi* are known in the wild.

Var. **hispida**. Illustration: Justice & Bell, Wild flowers of North Carolina, 101 (1968); Hay & Synge, Dictionary of garden plants in colour, pl. 1836 (1969); Everard & Morley, Wild flowers of the world, pl. 153 (1970); Gartenpraxis for 1987 **7**: 10. With conspicuous, spreading bristles on young shoots and leaves as well as on the raceme-axis and flower-stalks; fruit very rarely produced.

Var. **nana** (Elliott) de Candolle (*R. elliottii* (Chapman) Small). Low shrubs to 1 m, with spreading bristles only on the raceme-axis and flower-stalks; leaflets usually less than 1.5 cm; stipular spines usually conspicuous; fruit very rarely formed.

Var. **rosea** Pursh (*R. boyntonii* Ashe). Shrub or small tree more than 1 m, with spreading bristles only on raceme-axis and flower-stalks; leaflets usually 1.5–3 cm; stipular spines reduced or often absent; fruit very rarely produced.

Var. **fertilis** (Ashe) Clausen (*R. fertilis* Ashe). Shrub to 3 m with dense, spreading bristles on young shoots and leaves; leaflets mostly 1.2–1.8 times as long as broad; fruit regularly produced.

Var. **kelseyi** (Hutchinson) Isely. Illustration: Botanical Magazine, 8213 (1908); Journal of the Royal Horticultural Society **36**: f. 134 (1911); Addisonia **1**: t. 3 (1916). Shrub to 3 m, with bristles only on raceme-axis and flower-stalks; leaflets mostly 1.8–2.4 times as long as broad; fruit regularly produced.

4. R. viscosa Ventenat.

Shrubs or small trees to 12 m; young shoots with stalked or stalkless glands and soft, white, sometimes stellate hairs. Leaves with 15–21 leaflets; leaflets lanceolate-oblong to ovate-oblong, rounded to the base and apex, persistently hairy beneath. Racemes glandular and sticky, with 10–25 flowers. Calyx 0.7–1 cm; corolla pink to pinkish purple, standard 2–2.5 cm. Pods infrequent, glandular-bristly. *SE USA.* H3. Summer.

Var. **viscosa**. Illustration: Gleason, Illustrated Flora of the north-eastern

United States and adjacent Canada **2**: 414 (1952). Young shoots and racemes with flat, stalkless or almost stalkless glands (stalked glands present in the raceme).

Var. **hartwigii** (Koehne) Ashe (*R. hartwigii* Koehne). Illustration: Gardeners's Chronicle **90**: 389 (1931). Young shoots with conspicuously stalked glands; racemes with stalked glands. Pods more frequently formed than in var. *viscosa.*

40. SESBANIA Scopoli
H.S. Maxwell & J. Cullen
Short-lived shrubs or small trees. Leaves pinnate, terminating in a short extension of the axis (rarely with a terminal leaflet), leaflets often numerous; stipules present, soon falling; stipels small or absent. Flowers usually in axillary racemes. Calyx broadly bell-shaped, truncate or oblique, usually weakly toothed. Petals long-clawed; standard spreading or reflexed, often notched at the apex, often with 2 appendages on the inner surface towards the base. Stamens 10, 9 of them with their filaments united, that of the uppermost free. Pod long, thin, sometimes winged. Seeds numerous, without arils.

A genus of about 50 species from the tropics and subtropics, often found in seasonally damp places. Only a small number are grown. Propagation is by seed.

1a. Calyx 1.7–2.5 cm; standard 7.5–10 cm **1. grandiflorus**
  b. Calyx and standard much smaller 2
2a. Leaflets 14–20; pod 5–10 cm with 4 conspicuous wings **2. puniceus**
  b. Leaflets 20–40; pod 6.5–25 cm, thin, unwinged **3. sesban**

1. S. grandiflorus (Linnaeus) Poiret.
Small, soft-wooded tree to 10 m. Leaflets 30 or more, oblong or slightly obovate-oblong, tapered or rounded to the base, rounded to the sometimes notched apex; axis of leaf hairless, leaflet-stalks and often the undersurface of the leaflets hairy. Flowers few. Calyx broadly bell-shaped, oblique, scarcely toothed, hairless, 1.7–2.5 cm. Corolla white or red; standard reflexed, ovate-oblong, 7.5–10 cm; keel parrot's beak-shaped, 9.5–13 cm. Pod 30 cm or more, flattened, ridged on the corners, very slightly constricted between the numerous seeds. *Areas around the Indian Ocean, from Mauritius to India, Sri Lanka & further east.* G2. Summer.

2. S. puniceus (Cavanilles) Bentham (*S. tripetii* invalid). Illustration: Botanical

Magazine, 7353 (1894); Herter, Flora illustrada del Uruguay, t. 1607 (1954); Menninger, Flowering trees of the world, t. 284, 285 (1962) as – Daubentonia punicea and D. tripetii; Cabrera, Flora de la provincia de Buenos Aires **3**: 499 (1967).

Shrub or small tree. Leaflets 14–20, oblong or slightly obovate-oblong, rounded or tapered to the base, rounded to the mucronate apex, very slightly hairy beneath. Flowers *c.* 10 in axillary racemes. Calyx bell-shaped, truncate, wavy-toothed, hairless, 5–6 mm. Corolla bright reddish purple; standard reflexed, 1.5–2 cm; keel parrot's beak-shaped, 2–2.5 cm. Pod 5–10 cm, conspicuously and broadly 4-winged, slightly constricted betweeen the seeds. *S Brazil, Uruguay, Argentina.* G2.

3. S. sesban (Linnaeus) Merrill (*S. aegyptiaca* Poiret).

Much-branched shrub to 3 m. Leaflets 20–40, oblong, parallel-sided, rounded to the base and apex, sparsely hairy to almost hairless on both surfaces, glaucous. Flowers few to many. Calyx broadly bell-shaped, truncate, weakly 5-toothed, 4–7 mm. Standard 1–1.6 cm, notched at apex, yellow or partly or completely suffused with purple; wings and keel mostly yellow, keel 1.3–1.8 cm. Pod 6.5–25 cm, linear-cylindric, thin, twisted, often somewhat to deeply constricted between the seeds, not winged. *Old world tropics.* G2. Summer.

Very variable in flower colour and pod size and degree of constriction. Many varieties have been named, but it is uncertain which of these have been cultivated.

41. INDIGOFERA Linnaeus
J. Cullen
Shrubs (ours). Hairs usually medifixed (i.e. attached at a point along their length, often at the middle, rather than by one end), mostly adpressed. Leaves with a terminal leaflet and 7–27 opposite lateral leaflets (ours); stipules small to large, persistent or soon falling; stipels present, small. Racemes axillary, few- to many-flowered, usually borne on the current year's growth, more rarely borne on older wood; bracts usually small. Calyx bell-shaped, deeply 5-toothed with the lowermost tooth usually the longest, more rarely truncate and scarcely toothed. Corolla usually pink to purplish, rarely white or yellowish. Standard generally obovate, reflexed upwards, variously hairy outside. Wings borne alongside the keel

or forming a horizontal plate above it, variously hairy towards the base. Keel with a projection or spur on each side, usually hairy, at least along the suture. Stamens 10, 9 of them with filaments united, the uppermost free (ours); anthers appendaged at their tips. Ovary several-seeded. Pod variable, usually linear-cylindric, scarcely compressed, more rarely compressed or constricted between the seeds, always with thin partitions between the individual seeds.

A genus of about 700 species from the tropics and subtropics. Only a small number is grown. The classification of the genus is poorly understood and co-ordinated, and the identification of plants from gardens (which may have originated almost anywhere in the vast geographical range of the genus) is particularly difficult. These difficulties are compounded by the fact that existing descriptions are often incomplete or conflicting, caused, in part, by the complexities of the flower-structure. The pollination mechanism is explosive, involving the decoupling of the wings from 2 backwardly directed outgrowths or spurs of the keel, which causes a discharge of the pollen; following this, the keel tends to reflex downwards and backwards, and then falls, together with the wings, so most herbarium specimens lack these organs. In some species, e.g. *I. heterantha*, the wings form a horizontal plate between the standard and keel, whereas in others, e.g. *I. australis*, the wings are in the more normal position, lying along each side of the keel. For many of the species the exact dispositions of these parts are not known, or not recorded in the literature on classification and identification.

This account attempts to draw together existing knowledge of the cultivated species, but it is clear that much more investigation is necessary before a reliable account can be produced. The species are generally easily grown and valued for their late, pink to purple flowers. They are propagated by seed or by cuttings of semi-ripe wood rooted in a heated case.

1a. Leaves of mature plants reduced to slightly flattened axes, leaflets present only in very young plants **14. filifolia**
  b. Leaves of mature plants with leaflets (but see comment under no. **13**) 2
2a. Leaflets 5–13 (count leaflets on several leaves) 3
  b. Leaflets 13–27 9

3a. Standard 7 mm or more; young shoots hairless or sparsely hairy 4
  b. Standard to 6.5 mm; young shoots densely hairy 6
4a. Leaflets pointed, usually acute **1. decora**
  b. Leaflets very rounded to notched at the (often mucronate) apex 5
5a. Leaflets more densely hairy above than beneath; bracts narrow, not enclosing the flower-buds **3. kirilowii**
  b. Leaflets more densely hairy beneath (sometimes hairless above); bracts boat-shaped, abruptly tapered to a narrow tip, enclosing the flower buds **2. hebepetala**
6a. Racemes with a distinct (flowerless) stalk at the base **6. potaninii**
  b. Racemes with flowers to the base, without a distinct inflorescence-stalk 7
7a. Pods deflexed, slightly curved, slightly constricted between the seeds; standard to 5 mm **7. tinctoria**
  b. Pods horizontal, straight, not constricted between the seeds; standard 5–6.5 mm 8
8a. Pods persistently hairy; leaf-stalk conspicuous, *c.* 3 cm **4. amblyantha**
  b. Pods hairless or almost so, though ovary often hairy; leaf-stalk inconspicuous, much less than 3 cm **5. pseudotinctoria**
9a. At least some racemes borne on old wood, with numerous scales at the base **8. cassioides**
  b. All racemes borne on young shoots, without scales at the base 10
10a. Racemes exceeding the leaves which subtend them 11
  b. Racemes shorter than the leaves which subtend them 12
11a. Racemes 2.5–9 cm, ascending or erect, very dense **10. splendens**
  b. Racemes much longer, arching over or pendent, loose **9. pendula**
12a. Young shoots densely hairy with many spreading hairs **11. dosua**
  b. Young shoots hairless or with closely adpressed hairs 13
13a. Calyx deeply toothed; bracts narrowly triangular, soon falling; ovary and pod hairy **12. heterantha**
  b. Calyx cup-like, scarcely toothed; bracts short, broad, cup-like, persistent; ovary and pod hairless **13. australis**

1. I. decora Lindley. Illustration: Edwards's Botanical Register **32:** t. 22 (1846); Botanical Magazine, 5063 (1858); Bailey, Standard cyclopedia of horticulture **2:** 1645 (1935).

Shrub to 1 m; young shoots reddish brown, hairless. Leaves with 7–13, ovate or narrowly ovate leaflets, 2.5–5 × 1–2.5 cm, tapered from below the middle to an acute, mucronate apex, rounded to the base, pale and sparsely adpressed hairy beneath, hairless above. Racemes to 12 cm, long-stalked. Calyx *c.* 2 mm, sparsely hairy. Standard 1–1.5 cm, purple or white with a pale purple base. Wings and keel persistent, purple. Pod hairless, flattened, 1-seeded or constricted between the few seeds, hairless. *W China.* H5. Late summer.

Not reliably hardy in much of Europe, and often killed to the root in winter.

2. I. hebepetala Baker. Illustration: Botanical Magazine, 8208 (1908).

Shrub to 2 m; young shoots hairless or very sparsely hairy. Leaves with 7–13 elliptic, elliptic-oblong or rarely ovate or somewhat obovate leaflets 2.5–5 × 1.6–3 cm, rounded to the sometimes notched, mucronate apex, broadly rounded to the base, hairy or hairless above, hairy beneath. Racemes to 20 cm, long-stalked; bracts boat-shaped, tapered abruptly to a narrow point, enclosing the flower-buds. Calyx 1.5–2.5 mm, densely hairy. Standard 0.7–1.1 cm, sparsely to densely hairy outside, deep red; wings and keel soon falling, paler. Pod straight, cylindric, not constricted between the seeds, hairless, 3.5–5 cm. *Himalaya.* H3. Late summer.

Plants with the leaflets hairless above have been named var. **glabra** Ali.

3. I. kirilowii Palibin. Illustration: Schneider, Illustriertes Handbuch der Laubholzkunde **2:** 65 (1907); Botanical Magazine, 8580 (1914); Bailey, Standard cyclopedia of horticulture **2:** 1646 (1935).

Shrub to 1 m or more; young shoots hairless or very sparsely hairy. Leaves with 7–11 broadly obovate to almost circular leaflets 2–3.2 × 1.5–2.5 cm, pale beneath, hairy on both surfaces though more densely so above. Racemes to 15 cm, long-stalked; bracts narrow, soon falling, not enclosing the flower-buds. Standard 1.2–1.8 cm, rose-pink; wings and keel soon falling, pink. Pod straight, cylindric though slightly flattened, hairless, 3–5.5 cm. *N China, Japan, Korea.* H3. Summer.

4. I. amblyantha Craib. Illustration: Dendroflora **22:** 67 (1985).

Shrub to 2 m; young shoots hairy. Leaves with 7–11 elliptic to obovate-elliptic leaflets 1–2.5 × 0.6–1.3 cm, rounded to the mucronate, sometimes notched apex, tapered to the base, hairy on both surfaces; leaf-stalk conspicuous, usually 3 cm or more. Racemes to 9 cm, densely covered with flowers from the base upwards, inflorescence-stalk absent or very short. Standard 5–6.5 mm, lilac-purple; wings and keel soon falling. Pod straight, horizontal, slightly flattened, not constricted between the seeds, persistently hairy. *C & W China.* H3. Late summer.

This and the next 3 species are very similar and the precise differences between them require further study.

5. I. pseudotinctoria Matsumura.

Similar to *I. amblyantha*, but leaflets 6–9 mm wide, leaf-stalk inconspicuous, much less than 3 cm, ovary hairy or hairless, pods hairless. *China, Japan.* H3. Late summer.

6. I. potaninii Craib. Illustration: Journal of the Royal Horticultural Society **90**: f. 180 (1965).

Similar to *I. amblyantha* but racemes longer, clearly stalked (without flowers at the base). *N China (Gansu).* H2. Late summer.

Perhaps no longer in cultivation.

7. I. tinctoria Linnaeus. Illustration: Bentley & Trimen, Medicinal plants **2**: t. 72 (1878); Bailey, Standard cyclopedia of horticulture **2**: 1647 (1935).

Like *I. amblyantha* but racemes shorter, standard to 5 mm, pale pink or yellowish, pods deflexed, curved, somewhat constricted between the seeds. *Origin uncertain, formerly widely cultivated.* H4? Late summer.

The major source of the dye indigo, once much-cultivated; it is doubtful if it is grown much nowadays for ornament.

8. I. cassioides de Candolle (*I. pulchella* misapplied). Illustration: Botanical Magazine, 3348 (1834).

Large shrub to 4 m; young shoots hairy. Leaves with 13–21 ovate-elliptic or oblong-elliptic leaflets 1–2.5 × 0.7–1.5 cm, usually rounded to the mucronate, sometimes notched apex, rounded to the base, hairy on both surfaces. Racemes 5–12 cm, at least some of them borne on old shoots and with a cluster of small scales at the base. Standard 1.2–1.5 cm, deep pink; wings and keel persistent. Pods linear-cylindric, straight, not constricted between the seeds, crimson when immature, hairless. *Himalaya.* H5? Late summer.

9. I. pendula Franchet. Illustration: Schneider, Illustriertes Handbuch der Laubholzkunde **2**: 65 (1907); Botanical Magazine, 8745 (1918).

Sprawling shrub to 3 m; young shoots hairy. Leaves with 13–27 narrowly elliptic-oblong leaflets 1.2–2.5 × 0.5–1 cm, pale beneath, sparsely hairy on the lower surface, more or less hairless above. Racemes to 45 cm, conspicuously stalked, many-flowered, exceeding the subtending leaves, arching over or pendent. Standard 0.9–1.2 cm, pale grey and very densely hairy outside, pink or reddish purple inside; wings and keel reddish purple, soon falling. Pods cylindric-linear, straight, not constricted between the seeds, sparsely hairy. *W China.* H5. Summer.

10. I. splendens Ficalho & Hiern. Illustration: Transactions of the Linnean Society of London, series 2, **2**: pl. 3 (1881).

Shrub with flexuous, somewhat hairy branches. Leaves with 13–23 oblong-elliptic leaflets 1–1.5 cm, hairy beneath, hairless above, rounded to the mucronate apex, rounded to the base. Racemes 2.5–9 cm, exceeding the subtending leaves, stalked, erect, very dense. Standard *c.* 1.7 cm, pink(?); wings and keel persistent, *c.* 1.3 cm. Pod hairy. *Angola.* G2.

Though recorded in some lists and catalogues, it is doubtful if this species is in cultivation.

11. I. dosua D. Don.

Shrub 0.3–4 m, young shoots densely hairy with some spreading hairs. Leaves with 13–27 obovate leaflets 9–15 × 3–6 mm, rounded to the sometimes notched apex, rounded or tapered to the base, densely hairy on both surfaces. Racemes 3–10 cm, erect, scarcely stalked (with flowers to the base), shorter than the subtending leaves. Standard 6–9 mm, pink to purple, sparsely hairy outside; wings and keel pink to purple. Pods deflexed, straight, linear-cylindric, persistently hairy, 2–3.5 cm. *Himalaya.* H3. Late summer.

Much confused with the next, and not certainly in cultivation.

12. I. heterantha Brandis (*I. gerardiana* Baker). Illustration: Edwards's Botanical Register **28**: t. 57 (1842) – as I. dosua.

Shrub to 2 m or more, young shoots adpressed hairy. Leaves with 13–27 obovate or oblanceolate leaflets 5–10 × 3–4 mm, rounded to the mucronate apex, tapered to the base, hairy above and beneath. Racemes to 5 cm, with flowers to the base or with a short stalk, shorter than

the subtending leaves; bracts narrow, soon falling. Calyx deeply 5-toothed. Standard 0.5–1 cm, pink to purple; wings and keel soon falling. Pods linear-cylindric, straight, not deflexed, to 3 cm, not constricted between the seeds, persistently hairy. *Himalaya.* H3. Late summer.

13. I. australis Willdenow. Illustration: Edwards's Botanical Register **5**: t. 386 (1819); Cochrane et al., Flowers and plants of Victoria, f. 384 (1968); King & Burns, Wild flowers of Tasmania, 33 (1971); Rotherham et al., Flowers of plants of New South Wales, pl. 244 (1975).

Shrub to 2 m, young shoots hairless or sparsely hairy. Leaves with 13–27 elliptic-oblong leaflets 1–2 × 0.4–0.7 cm, abruptly rounded to the sometimes notched apex, broadly tapered to the base, finely hairy on both surfaces or hairless above; leaflet-stalks conspicuous. Racemes to 9 cm, shorter than the subtending leaves, many-flowered; bracts small, broadly cup-shaped, persistent. Calyx cup-like, truncate, scarcely toothed. Standard to 8 mm, deep pink, wings and keel persistent. Pods linear-cylindric, straight, deflexed, to 5 cm, hairless. *Australia (New South Wales, Victoria, South Australia, Tasmania).* G1.

Var. **signata** Bentham has leaflets very reduced and the leaf-axis flattened and somewhat broadened; this variety, however, does not seem to have been cultivated.

14. I. filifolia Thunberg. Illustration: Botanical Magazine, 2214 (1821) – as Lebeckia nuda; Jackson, Wild flowers of the fairest Cape, 31 (1980); Burman & Bean, Hottentots Holland to Hermanus, 112 (1985).

Shrub to 2 m. Leaves, except in the very young plant, without leaflets, reduced to their axes, which are 7.5–12 cm, hairless and somewhat sharp at the apex. Racemes conspicuously stalked, with several flowers. Standard purple, 0.8–1.1 cm, almost hairless; wings and keel purple. Pods linear-cylindric, 3.5–6 cm, hairless. *South Africa (Cape Province).* G1.

42. DESMODIUM Desvaux
J. Cullen & H.S. Maxwell
Herbs with woody bases, or shrubs. Leaves usually with 3 leaflets, occasionally reduced to the terminal leaflets only; stipules present, attached by their bases, often early deciduous, stipels present, small. Flowers in axillary and/or terminal racemes or panicles with more than 1 flower to each major bract; bracteoles

present. Calyx funnel-shaped, deeply lobed. Standard large, reflexed upwards. Stamens 10, filaments of 9 of them united, that of the upper stamen free or free at the base but united to the others above the base. Pod generally segmented and breaking into 1-seeded segments, more rarely opening along the lower suture. Seeds with or without arils.

A genus of about 300 species (split into several genera by recent authors), mostly from the tropics and subtropics, where several are grown as fodder plants. Only a few are grown for ornament in Europe, where they require a warm, sunny site or the protection of a glasshouse. Propagation is by seed or by division of the roots.

Literature: Ohashi, H., A monograph of the subgenus *Dollinera* of the genus *Desmodium* (Leguminosae), *The University Museum, Tokyo, Bulletin* 2: 259–320 (1971); Ohashi, H., The Asiatic species of *Desmodium* and its allied genera (Leguminosae), *Ginkgoana* No.1 (1973).

1a. Pods not jointed, opening along the lower suture; leaflets usually 3, occasionally 1, the terminal leaflet 4–6 times longer than the lateral leaflets **1. motorium**
  b. Pods jointed, breaking into 1-seeded segments; leaflets 1 only, or, if 3, then the terminal 1–2 times longer than the laterals 2
2a. Leaflet almost always 1, very large and densely white woolly hairy beneath and with a conspicuous network of lateral veins **4. praestans**
  b. Leaflets almost always 3, not as above 3
3a. Filament of upper stamen free to the base; leaflets much longer than broad, usually lanceolate to oblong **2. canadense**
  b. Filament of the upper stamen united to the others for about half its length; leaflets almost as broad as long, broadly ovate to almost circular **3. elegans**

1. D. motorium (Houttuyn) Merrill (*D. gyrans* (Linnaeus filius) de Candolle; *Codariocalyx motorium* (Houttuyn) Ohashi).

Erect, branched shrub to 2 m; young shoots hairless. Leaves mostly with 3 leaflets, occasionally reduced to the terminal leaflet; terminal leaflet ovate-elliptic or narrowly so, 4–6 times as long as the small lateral leaflets; all leaflets hairy above when young, persistently whitish hairy beneath. Racemes or panicles

terminal and axillary, axes and flower-stalks with spreading or hooked hairs. Calyx 2–5 mm. Corolla mauve, lilac or orange, standard 0.7–1.2 cm. Pod curved, flattened, hairy, not segmented but opening along the lower suture, scarcely constricted between the seeds on the upper suture, more deeply so on the lower; seeds 3–11, with arils. *Himalaya & China, SE Asia to N Australia.* G1. Autumn.

Not particularly attractive but of interest because of the rotatory movements performed by the lateral leaflets when exposed to bright sunlight. The species is placed by many authors in the small genus *Codariocalyx* Hasskarl on account of its dehiscent pods containing seeds with conspicuous arils.

2. D. canadense (Linnaeus) de Candolle. Illustration: Gleason, Illustrated Flora of the north-eastern United States and adjacent Canada 2: 429 (1952); Rickett, Wild flowers of the United States 1: pl. 79 (1966); Horticulture 51: 26 (1973).

Large, erect, perennial herb to 2 m; young shoots with sparse, spreading or ascending, somewhat kinked or hooked, bristle-like hairs. Leaves all of 3 leaflets, the terminal 1–1.4 times as long as the laterals; leaflets oblong-lanceolate to oblong, sparsely hairy above with bristle-like hairs, pale and somewhat more densely bristly beneath. Panicles upright, terminal and axillary. Calyx 2–5 mm, densely hairy. Corolla purple, standard 0.9–1.8 cm. Upper stamen with filament free to the base. Pod with 3–5 joints, scarcely notched along the upper suture, deeply so along the lower, covered with short, spreading, hooked, bristle-like hairs. *E North America.* H1. Summer.

3. D. elegans de Candolle subsp. **elegans** (*D. tiliifolium* (D. Don) G. Don). Illustration: Revue Horticole for 1902: 458, 459; Schneider, Illustriertes Handbuch der Laubholzkunde 2: 109 (1907); Polunin & Stainton, Flowers of the Himalaya, pl. 26 (1984).

Shrub to 4 m; young shoots densely adpressed hairy with curled hairs. Leaves with 3 leaflets, the terminal 1.3–1.7 times longer than the laterals; all leaflets broadly ovate to almost circular, adpressed bristly above, whitish or pale and more densely bristly beneath, often markedly silvery hairy when young. Panicles spreading, terminal and axillary, spreading hairy. Calyx 4–6 mm. Corolla pinkish purple, standard 0.9–1.5 cm. Upper stamen with its filament united to the others for about

half its length. Pod with up to 10 joints, notched between the seeds along the upper suture, very deeply notched along the lower, adpressed bristly. *Himalaya, China.* H5. Summer.

D. elegans is a very variable species, divided into several subspecies and varieties by Ohashi (references above); the cultivated variant, commonly known as *D. tiliifolium*, belongs to subsp. *elegans* var. *elegans*.

4. D. praestans Forrest (*D. yunnanense* Franchet subsp. *praestans* (Forrest) Ohashi). Illustration: Journal of the Royal Horticultural Society 68: f. 36 (1943); Botanical Magazine, n.s., 407 (1962).

Shrub to 4 m; young shoots with very densely white-woolly hairs. Leaves usually of a single terminal leaflet (its stalk jointed or with tiny stipels indicating the position of the suppressed lateral leaflets); leaflet mostly 10–25 cm (rarely as little as 5.5 cm) × 5–17.5 cm, broadly ovate to almost circular, greyish above, with dense white-woolly hairs and with a conspicuous network of veins beneath. Panicles terminal, much branched and widely spreading, with white-woolly hairs. Calyx 3–6 mm. Corolla pinkish purple, standard 0.9–1.5 cm. Upper stamen with its filament united to the others for about half its length. Pod with 4–7 (more?) joints, deeply constricted along both sutures, hairy with silky hairs. *W China.* H5. Summer.

43. CAMPYLOTROPIS Bunge
J. Cullen
Deciduous shrubs. Leaves made up of 3 leaflets, the terminal leaflet usually conspicuously stalked; stipules present, stipels absent. Flowers in axillary racemes crowded into panicles at the tips of the branches; bracts each subtending a single flower, bracteoles soon deciduous. Calyx bell-shaped, 5-toothed, the 2 upper teeth united for most of their length; flower-stalk jointed just beneath the calyx. Corolla purple; standard reflexed upright; keel curved, sickle-shaped, acute. Stamens 10, 9 of them with united filaments, that of the uppermost free. Pod short, 1-seeded, ovoid, ellipsoid or almost spherical, compressed, not opening.

A genus of about 65 species from Asia. Only 1 is grown in Europe, where it requires a well-drained soil and a sunny, open position. It is propagated by seed or by division.

1. C. macrocarpa (Bunge) Rehder (*Lespedeza macrocarpa* Bunge). Illustration:

Schneider, Illustriertes Handbuch der Laubholzkunde **2**: 111, 112 (1907).

Shrub to 1 m; young shoots angled, with adpressed or somewhat spreading hairs. Leaflets elliptic to oblong, 2–5 cm, rounded to the base and the conspicuously mucronate (sometimes also notched) apex, hairless above, pale and adpressed hairy beneath; stipules narrowly lanceolate. Racemes dense. Calyx to 4 mm. Corolla purple, standard 1–1.2 cm. Pod ellipsoid, flattened, 1.2–1.5 cm. *N & C China*. H1. Late summer.

44. LESPEDEZA Michaux
J. Cullen

Annual or perennial herbs or shrubs. Leaves borne in spirals or in 2 distinct ranks; leaflets 3, more or less equal; terminal leaflet stalked or not; stipules persistent, stipels absent; winter buds with spiral scales or scales in 2 ranks. Flowers axillary or in axillary racemes, each bract subtending 2 flowers; in several species corolla-less, cleistogamous flowers are produced as well as normal, opening flowers; bracteoles 2, closely overlapping the base of the calyx, persistent. Calyx funnel- or bell-shaped, 5-toothed, the upper teeth often united for much of their length. Standard clawed or not; keel straight, blunt. Stamens 10, filaments of 9 united, that of the uppermost usually free. Pod 1-seeded, not dehiscent.

A genus of about 40 species from temperate and subtropical parts of N America, Asia and Australia. All of the cultivated species are known from Japan, where they have been extensively studied. They are easily grown in a warm, sunny site and are propagated by seed.

Literature: Akiyama, S., A revision of the genus *Lespedeza* section *Microlespedeza* (Leguminosae), *Bulletin of the University Museum, Tokyo* **33** (1988) – contains numerous diagnostic illustrations which are not further cited here.

1a. Plant annual   2
  b. Plant a shrub or long-persisting herb   3
2a. Stems with downwardly pointing hairs; calyx hairy; pod slightly longer than the calyx   **7. striata**
  b. Stems with upwardly pointing hairs; calyx hairless; pod twice as long as the calyx   **8. stipulacea**
3a. Terminal leaflet not or very shortly stalked; cleistogamous flowers present with normal flowers
                                                          6. cuneata
  b. Terminal leaflet conspicuously stalked; cleistogamous flowers absent   4
4a. Leaves and scales of the winter-buds in 2 ranks   5
  b. Leaves and scales of the winter-buds spirally arranged   6
5a. Standard and keel pale yellow
                                                         4. buergeri
  b. Standard and keel reddish purple
                                                       5. maximowiczii
6a. Standard conspicuously and abruptly clawed at the base
                                                          3. formosa
  b. Standard gradually tapered to the base   7
7a. Racemes congested, usually shorter than the subtending leaves; wings longer than keel   **2. cyrtobotrya**
  b. Racemes open, much longer than the subtending leaves; wings shorter than keel   **1. bicolor**

1. L. bicolor Turczaninow. Illustration: Gartenflora **9**: t. 299 (1860); Huxley, Deciduous garden trees and shrubs, pl. 99 (1979).

Shrub to 2 m with persistent aerial branches; young shoots adpressed hairy. Leaves and scales of winter-buds spirally arranged. Terminal leaflet stalked, 2–6 × 1–3.5 cm, elliptic to ovate or obovate, slightly longer than the laterals; all hairless or almost so above, adpressed hairy beneath. Racemes exceeding the subtending leaves, 2–10 cm, with 4–12 loosely arranged flowers; bracteoles narrow, scarcely ribbed, hairy outside. Calyx 3.1–4.7 mm, teeth acute. Standard 0.9–1.2 cm, gradually tapered to the base, red-purple inside; keel shorter than the standard but longer than the wings. Pod 5–7 × 4–6 mm, hairy or not, scarcely stalked. *Japan, Korea, N China, E former USSR*. H3. Autumn.

2. L. cyrtobotrya Miquel. Illustration: Kitamura & Murata, Coloured illustrations of herbaceous plants of Japan, pl. 23 (1981).

Similar to *L. bicolor* but to 2.5 m; racemes congested, 1–2 cm, shorter than the subtending leaves; calyx 4.5–6.2 mm, teeth acuminate; pods slightly smaller, always hairy. *Japan, Korea, N China, E former USSR*. H3. Autumn.

3. L. formosa (Vogel) Koehne (*L. sieboldii* Miquel). Illustration: Thrower, Plants of Hong Kong, 106 (1971).

Shrub to 2 m, many aerial parts dying in winter; young shoots adpressed hairy.

Leaves and scales of winter-buds spirally arranged. Terminal leaflet conspicuously stalked, 2–6 × 1.5–3.5 cm, slightly larger than the laterals, all densely hairy beneath, hairy or not above. Racemes 2–15 cm, exceeding the subtending leaves, open, with 4–14 flowers; bracteoles narrow, scarcely ribbed, hairy. Calyx 3.5–6 mm, teeth acute. Standard 0.95–1.35 cm, abruptly and conspicuously narrowed into a distinct claw, red-purple; keel longer than the red-purple wings. Pod compressed, 7–12 × 4–5 mm, hairy, stalked. *Himalaya to Japan*. H3. Late summer.

A very variable species, particularly in the shape and size of the calyx-teeth (see Akiyama, S. & Ohba, H., Taxonomy of *Lespedeza formosa* (Vogel) Koehne, Bulletin of the University Museum, Tokyo **31**: 217–229, 1988). Related to it are 2 species known with certainty only from cultivation (ancient cultivation in Japan) which have been confused among themselves and with *L. formosa* and are probably of hybrid origin:

L. thunbergii (de Candolle) Nakai (*Desmodium penduliflorum* Oudemans; *L. penduliflora* (Oudemans) Nakai). Illustration: Kitamura & Murata, Coloured illustrations of herbaceous plants of Japan, pl. 23 (1981); Amateur Gardening for September 1987, 21. Similar to the above but leaflets 3.5–9 × 2.5–5 cm, hairless above except along the midrib; racemes 10–15 cm; calyx 5.3–5.5 mm; standard 1.2–1.5 cm, shortly clawed, red-purple; wings deeper red-purple. H3. Summer.

Possibly a hybrid of **L. patens** Nakai, a species very similar to *L. formosa*.

L. japonica Bailey. Illustration: Kitamura & Murata, Coloured illustrations of herbaceous plants of Japan, pl. 23 (1981). Very similar to both *L. formosa* and *L. thunbergii*, but leaflets usually persistently hairy above, and flowers white ('Japonica') or at least partially white (other cultivars). H3. Late summer.

The name *L. japonica* covers a range of variants widely cultivated in Japan and thought to be hybrids of *L. formosa* with various other species.

4. L. buergeri Miquel. Illustration: Schneider, Illustriertes Handbuch der Laubgeholz **2**: 112 (1907); Kitamura & Murata, Coloured illustrations of herbaceous plants of Japan, pl. 23 (1981).

Shrub to 3 m; young shoots adpressed hairy. Leaves and scales of winter-buds arranged in 2 ranks. Terminal leaflet stalked, 1–5 × 0.5–3 cm, narrowly to broadly ovate or elliptic, usually hairless above, adpressed hairy beneath. Racemes 2–7 cm, usually exceeding the subtending leaves, open, with 4–10 flowers; bracteoles brownish, broad, several-ribbed, hairy. Calyx 1.8–3 mm, teeth acute. Standard 7–9 mm, pale yellow with purple patches inside, distinctly clawed; wings about as long as standard, purple or reddish purple; keel pale yellow, exceeding the standard. Pod compressed, 1–1.5 cm × 5–5.5 mm, hairy or not, stalked. *Japan, E China*. H3. Summer.

5. L. maximowiczii Schneider.

Like *L. buergeri* but inflorescences often shorter, bracteoles *c.* 5-ribbed, standard 9–10 mm, red-purple, wings deeper red-purple, shorter than the standard, keel paler than wings, exceeding the standard; pod almost stalkless. *Japan, Korea, S China*. H5. Summer.

6. L. cuneata G. Don (*L. sericea* Miquel). Illustration: Schneider, Illustriertes Handbuch der Laubholzkunde **2**: 112, 114 (1907); Gleason, Illustrated flora of the north-eastern United States and adjacent Canada **2**: 436 (1952); Kitamura & Murata, Coloured illustrations of herbaceous plants of Japan, pl. 23 (1981).

Shrub to 1 m with ascending-upright, densely leafy branches; young shoots with hairs which are spreading at their bases and then curve upwards. Terminal leaflet not or very shortly stalked, 0.7–2 cm, long-tapered to the base, truncate or notched at the apex, hairless above, sparsely to densely hairy beneath. Racemes dense, shorter than the subtending leaves; bracteoles narrowly triangular, hairy. Cleistogamous flowers present as well as normal flowers. Normal flowers with calyx to 4 mm, teeth acuminate, corolla whitish, standard 6–7 mm, keel longer than standard. Pod 1.5–2 mm, not stalked, hairy. *E & SE Asia to N Australia; introduced in E & S USA*. H5. Late summer.

7. L. striata (Murray) Hooker & Arnott (*Kummerowia striata* (Murray) Schindler). Illustration: Gleason, Illustrated flora of the north-eastern United States and adjacent Canada **2**: 436 (1952); Kitamura & Murata, Coloured illustrations of herbaceous plants of Japan, pl. 22 (1981).

Erect annual herb, much branched; stems with downwardly pointing whitish hairs. Stipules broadly lanceolate, brown, obliquely inserted. Terminal leaflet 1–1.5 × 0.5–0.8 cm, not notched, slightly larger than the laterals, margin and main vein beneath hairy. Flowers axillary, normal and cleistogamous. Normal flowers *c.* 5 mm, reddish purple; calyx 3–3.5 mm, hairy. Cleistogamous flowers without corollas. Pod flat, almost circular, abruptly acute, *c.* 3.5 mm, slightly exceeding calyx. *Japan, China, Korea, Taiwan, introduced in E & S USA*. H5. Flowering almost throughout the year.

8. L. stipulacea Maximowicz (*Kummerowia stipulacea* (Maximowicz) Makino). Illustration: Gleason, Illustrated Flora of the north-eastern United States & adjacent Canada **2**: 436 (1952).

Similar to *L. striata* but stems with adpressed, upwardly pointing hairs, leaflets usually notched, calyx hairless, pod about twice as long as the calyx, abruptly rounded at the apex. *Japan, China, Korea, E former USSR, introduced in E & S USA*. H5. Flowering almost throughout the year.

45. ERYTHRINA Linnaeus
U. Oster & J. Cullen
Trees, shrubs or rarely herbs, usually deciduous and spiny. Leaves of 3 leaflets; stipules persistent, glandular stipels usually present at the bases of the stalks of the lateral leaflets. Flowers in terminal racemes, produced with the leaves or when the plant is leafless. Calyx mostly cup-shaped, truncate. Corolla usually red, standard large, folded over and concealing the much shorter wings and keel in bud, remaining so, or spreading and reflexing to reveal the other petals. Stamens 10, filaments of 9 united, that of the uppermost free. Pods woody, flat or cylindric, usually constricted between the seeds. Seed often red to scarlet and with a black blotch.

A genus of more than 100 species of tropical and warm temperate regions of the world, of which only a few are grown in Europe (many more are cultivated in the United States). In most areas they require glasshouse protection. They are propagated by seed or by cuttings.

1a. Standard at last erect and widely spreading at full-flowering, revealing the wings and keel **1. crista-galli**
  b. Standard constantly folded over and concealing the wings and keel **2**
2a. Standard at most to 0.9 cm wide; leaflets mostly shallowly 3-lobed **4. herbacea**
  b. Standard more than 1 cm wide; leaflets not 3-lobed **3**

3a. Leaf-stalks hairless, usually spiny; calyx truncate, without obvious teeth **2. corallodendron**
  b. Leaf-stalks downy, not spiny; calyx with 5 small teeth **3. caffra**

1. E. crista-galli Linnaeus. Illustration: Perry, Flowers of the world, 165 (1970); Noailles & Lancaster, Mediterranean plants and gardens, 78 (1977); Gartenpraxis for July 1977, 324; The Garden **112**: 392 (1987) & **113**: 577 (1988).

Shrub to small or large tree, dependent on growing conditions; stems with strong spines. Leaves with spines on the stalk and midribs; leaflets ovate, ovate-lanceolate or almost circular, rounded to the base, abruptly pointed at the apex; glandular stipels conspicuous. Flowers appearing with the leaves, in terminal racemes. Standard bright red, to 5 cm, at least 1 cm wide, ultimately spreading and revealing the much shorter wings and keel, which are greenish with red tips. Pod woody, to 40 cm. Seeds black with brownish markings. *S Brazil, Uruguay, Paraguay, N Argentina*. G1. Summer.

Copious nectar is produced by the flowers and often drips from them.

2. E. corallodendron Linnaeus.

Tree to 8 m, usually spiny. Leaves spiny or not; leaflets ovate to diamond-shaped, the terminal often larger and more acuminate than the laterals; stalks not hairy; glandular stipels not conspicuous. Flowers few, in racemes, produced when the plant is leafless, deep red-crimson. Standard to 7.5 cm, at least 1 cm wide, constantly folded over the much shorter wings and keel. Pod to 10 cm, beaked. Seeds scarlet with a black spot. *S USA, Mexico*. H5–G1.

3. E. caffra Thunberg. Illustration: de Wit, Plants of the world **2**: pl. 275 (1963–5); Flowering Plants of Africa **43**: pl. 1707 (1974–6).

Spiny tree to 8 m. Leaves without spines, or with very small spines on the downy stalks; leaflets triangular-ovate, rounded at the base, long-acuminate at the apex; glandular stipels conspicuous. Flowers in dense racemes produced when the plant is more or less leafless, brilliant scarlet. Calyx hairy, distinctly 2-lipped, toothed. Standard 3.5–5 cm, at least 1 cm wide, constantly folded over the much shorter wings and keel. Pod to 12.5 cm. Seeds red with a black spot. *E & S Africa*. H5–G1.

4. E. herbacea Linnaeus (*E. arborea* (Chapman) Small). Illustration: Botanical

Magazine, 877 (1805); Dean et al., Wild flowers of Alabama and adjoining states, 87 (1983).

Perennial herb, shrub or small tree, spiny. Leaves with or without spines on the stalk or midrib; leaflets triangular to arrowhead-shaped, usually with 2 rounded lateral lobes and a narrow, acuminate terminal lobe; glandular stipels conspicuous. Flowers scarlet, in terminal racemes, produced with the leaves. Calyx truncate, without obvious teeth. Standard to 5 × 0.9 cm, constantly folded over the much shorter wings and keel. Pods to 20 cm, shortly beaked. Seeds bright scarlet with a black blotch. *S USA, adjacent Mexico*. H5–G1. Spring–summer.

46. MUCUNA Adanson
S.G. Knees

Woody climbers, climbing herbs and erect shrubs. Leaves with 3 stalked leaflets; stipules falling early, stipels often present. Flowers large, in axillary clusters or racemes, often long-stalked and hanging, very showy, purple, red, greenish, yellow or white; bracts and bracteoles deciduous. Calyx-lobes 4 or 5, 2-lipped above; standard rounded, shorter than other petals; keel hardened at apex, sharply beaked; ovaries few–several, ovules *c.* 12; style fine, stigma very small; stamens 10, 9 united, 1 free, anthers in 2 alternate rings of 5, the inner ring shorter. Fruit a pod, often bristly or velvety; seeds oblong to spherical, hilum present or absent.

A genus of about 100 species from tropical Asia, Africa and the Americas. All can be cultivated from seed and will need support from a wall or strong trellis.

Literature: Verdcourt, B., A manual of New Guinea Legumes (1979); Wilmot-Dear, M., A revision of *Mucuna* in China & Japan *Kew Bulletin* **39**(1): 23–65 (1984); Wilmot-Dear, M., A revision of *Mucuna* in the Pacific *Kew Bulletin* **45**(1): 1–35 (1990).

1a. Semi-woody, climbing herb, short-lived perennial or annual to 4 m; leaves conspicuously hairy beneath
                       1. pruriens
  b. Vigorous woody climbers, 12–30 m; leaves usually hairless beneath, or if hairy then only when young
                             2

2a. Evergreen climber to 12 m; flowers blackish purple, unpleasantly scented   **2. sempervirens**
  b. Climber 20–30 m; flowers bright orange-red   3

3a. Leaflets 5–7.5 cm across
                     3. bennettii
  b. Leaflets 8.5–13.5 cm across
               4. novaguineensis

1. M. pruriens (Linnaeus) de Candolle (*Marcanthus cochinchinensis* Loureiro; *Mucuna nivea* (Roxburgh) Wight & Arnott; *M. aterrima* (Piper & Tracy) Holland; *M. deeringiana* (Bort) Merrill). Illustration: Botanical Magazine, 4945 (1856); Kew Bulletin, **39**(1): 32 (1984).

Semi-woody climbing herb, short-lived perennial or annual to 4 m. Stems rough, covered with long bristly hairs when young, eventually hairless. Terminal leaflet 9–16 × 5–10 cm, ovate, obovoid or elliptic, leaf-stalks 2–30 cm. Racemes to 30 cm; calyx covered with pale brown hairs; corolla 2–4 cm, deep blackish purple to lilac or white. Fruit 5–9 × 1–2 cm, oblong, covered with irritant, orange-brown hairs; seeds 1.3–1.7 cm across. *Asia (widely naturalised in tropical Africa & America)*. G2.

Var. **utilis** (Wight) Burck, differs in having stems without long hairs and light brown hairs on the fruit.

2. M. sempervirens Hemsley. Illustration: Botanical Magazine, 7978 (1904); Kew Bulletin **39**(1): 27 (1984); Kew Magazine **3**: 139 (1986); Phillips & Rix, Shrubs, 155 (1989).

Vigorous evergreen climber to 12 m, old stems to 30 cm across. Terminal leaflets 8–16 × 3.5–9 cm, narrowly elliptic to ovate, leaf-stalks 7–16 cm. Flower-clusters 10–36 cm; calyx with red and brown bristles; corolla 2–4 cm, dark purple or reddish. Fruit 30–50 × 3–3.5 cm, covered with red-brown velvety hairs and bristles; seed 2–3 cm across. *India, Sikkim, Bhutan, Burma, W China*. H5–G1.

3. M. bennettii F.J. Mueller. Illustration: Graf, Exotica, series 4, **2**: 1388 (1985); Graf, Tropica, edn 3, 533 (1986); Kew Bulletin **45**(1): 33 (1990).

Woody climber to 20 m; stems rough, mostly hairless. Terminal leaflets 11–15 × 5–7.5 cm, narrowly elliptic, leaf-stalks 10–14 cm. Flower clusters 2–10 cm; calyx to 1.5 cm, with sparse bristles; corolla 3–6 cm; bright red. Fruit unknown. *New Guinea*. G2.

The name of this species is often misapplied to *M. novaguineensis*.

4. M. novaguineensis R. Scheffer (*M. bennettii*, misapplied). Illustration: Verdcourt, A manual of New Guinea Legumes, 449, 456 (1979).

Woody climber to 30 m; stems to 5 cm across, covered with hairs at first, eventually hairless. Terminal leaflets 10–19 × 8.5–13.5 cm, elliptic, apex acute, base rounded, hairless. Flower clusters 7–60 cm; calyx 0.6–1.3 cm; corolla 5–8 cm, brilliant orange-red. Fruit 16–27 × 4–6 cm; seeds *c.* 4 cm across. *New Guinea*. G2.

47. STRONGYLODON Vogel
M.C. Warwick

Evergreen, woody stemmed twining climbers and shrubs. Leaves pinnate, with 3 leaflets; stipules falling early; terminal leaflets lanceolate to circular, apex acuminate, rounded at base; lateral leaflets slightly smaller in size, oblique at base. Flowers axillary in many-flowered drooping racemes. Calyx bell-shaped, 5-lobed. Corolla orange, red, blue or bluish green, standard reflexed, acute at apex with 2 appendages above the claw, wing half as long as standard, attached to the keel, petals at base, keel petals united, as long as standard. Stamens 10, 9 united and 1 free, filaments hairless. Ovary with 1–12 ovules; pods wrinkled; seeds black or brown, smooth or wrinkled.

A genus of about 20 species native to Old World tropics, only one of which is commonly grown. It is frost-tender (min. 18 °C) and requires a humus-rich, moist but well-drained soil and partial shade in summer. After 3 years flowers are produced on new or old wood and stems should be trained along wires to allow flowers to hand down.

Seeds have short viability and must be planted soon after pod opens. Overnight soaking and an incision in the seed coat will assist germination. Cuttings of half-ripened wood can be rooted in a few weeks in a misty, warm greenhouse. Light pruning may be required if plant becomes too rampant.

1. S. macrobotrys Gray. Illustration: Botanical Magazine, n.s., 627 (1972); Perry, Flowers of the world, 164 (1972); Kew Magazine **1**: 189 (1984); Brickell, RHS gardeners' encyclopedia of plants and flowers, 164 (1989).

Woody climber to 20 m, stems to 3 cm thick. Leaves 12–15 × 5–7 cm, pinnate with leaflets oblong to elliptic; stipules 3–4 mm, ovate to triangular. Racemes pendent, to 3 m, bracteoles *c.* 1.5 mm, ovate, with short hairs at margin. Flower-stalk 1.8–4 cm, hairless. Calyx 0.8–1.4 cm, 5-lobed. Flowers 4–6 cm, bluish green, standard 3.7–4.8 cm, ovate, reflexed with claw 3–50 mm; with 2 appendages 0.7–1 cm, above the claw; wings 2–2.4 cm, oblong to elliptic, with claw c. 1.1 cm; keel 1.1–1.3 cm with claw tapered into a gently curved beak. Stamens 4.8–7.2 cm, filaments hairless. Style 3.8–5.5 cm, hairless, stigma terminal; ovary c. 6 mm, covered with adpressed hairs. Pods 8.5–15 × 5–7 cm, elliptic, inflated and wrinkled with 6–12 smooth, black seeds with a long hilum and thin seed coat. *Philippine Islands.* G2. Winter–spring.

S. macrobotrys with its pale flowers in long racemes away from the foliage suggests bat pollination whereas other species are bird pollinated.

48. BUTEA Willdenow
U. Oster & J. Cullen

Trees or woody climbers (ours), rarely herbs. Leaves divided into 3 leaflets; stipules deciduous, stipels present. Flowers in dense terminal racemes or panicles; bracteoles present. Calyx bell-shaped, 5-toothed, the upper 2 teeth partly united and often larger than the others. Corolla large, standard often reflexed. Stamens 10, the filaments of 9 united, that of the uppermost free. Ovary with 1 ovule. Pod flattened, wing-like, the single seed borne at the uppermost end.

A genus of 7 species from tropical Asia, rarely grown and requiring glasshouse protection. Propagation is by seed.

1. B. superba Roxburgh. Illustration: Menninger, Flowering vines of the world, t. 110 (1970).

Large woody climber. Leaflets leathery, broadly ovate, widely tapered to the base, rounded to the apex, to 28 × 20 cm. Racemes long, hanging, axis densely brown hairy. Calyx bell-shaped, 5-toothed, brown-hairy, c. 2 cm. Corolla reddish (drying yellowish), all the petals densely to loosely white hairy outside. Standard reflexed upwards, c. 5 cm, somewhat shorter than the curved, pointed keel. Pods to 15 × 4 cm. *Burma, Thailand.* G2.

Other species of *Butea* are cultivated in the Asian tropics and perhaps in N America, e.g. **B. monosperma** (Lamarck) Taubert (*B. frondosa* Willdenow), which is a tall tree with hairy leaflets and pods. *India, Burma.*

49. APIOS Medikus
U. Oster & J. Cullen

Twining perennial herbs, some (ours) with tuberous roots. Leaves alternate, pinnate with several leaflets, including a terminal leaflet; stipules subulate, persistent; stipels absent. Flowers in short racemes, sometimes leaf-opposed. Calyx cup-like, scarcely toothed. Corolla with a broad standard, the keel much longer but coiled upwards within the standard, blunt and notched at the apex. Stamens 10, 9 with their filaments united, that of the uppermost free. Ovary with coiled style. Pod flattened, several-seeded, the sides spiralling after dehiscence.

A genus of about 10 species from America and Asia. Only 1 is occasionally grown, propagated by seed and by tubers.

1. A. americana Medikus (*A. tuberosa* Moench). Illustration: Botanical Magazine, 1198 (1809); Everard & Morley, Wild flowers of the world, pl. 153 (1970); Dean et al., Wild flowers of Alabama and adjoining states, 89 (1983).

Twiner to 3 m, roots bearing strings of tubers. Leaflets ovate or narrowly ovate, acuminate, rounded to cordate at the base. Racemes dense, generally leaf-opposed; flowers fragrant, brown, calyx c. 5 mm, standard c. 1 cm. Pod to 10 cm. *E North America (New Brunswick to Texas).* H3. Summer.

The tubers are edible.

50. CANAVALIA Adanson
J. Cullen

Climbers, often large, but sometimes cultivated as bushy annuals. Leaves long-stalked, made up of 3 leaflets; stipules soon falling; terminal leaflet with 2 deciduous or occasionally persistent stipels. Flowers in axillary racemes or panicles; bracteoles 2, soon falling. Calyx cylindric, 2-lipped, the upper lip very large, somewhat 2-lobed, the lower lip of 3 small teeth. Corolla showy, usually purplish or violet, more rarely reddish, bluish or almost white; blade of standard reverse heart-shaped. Stamens 10, filaments of 9 united, that of the uppermost free, at least at the base. Pods large, somewhat flattened. Seeds large, each with a long, linear hilum.

A genus of about 50 species from the tropics and subtropics, some grown in these areas for food (young leaves, young pods) or as fodder. They are rarely grown as ornamentals, though 3 species are recorded. They require rich soil in a warm situation and are propagated by seed.

Literature: Sauer, J., Revision of *Canavalia*, Brittonia **16**: 106–181 (1964).

1a. Leaflets acute but not acuminate
   1. ensiformis
   b. Leaflets gradually or abruptly acuminate 2
2a. Leaflets gradually acuminate to an acute tip; calyx at least 1.5 cm; hilum more than 1.2 cm
   2. gladiata
   b. Leaflets abruptly acuminate to an obtuse tip; calyx less than 1.5 cm; hilum less than 1.2 cm **3. virosa**

1. C. ensiformis (Linnaeus) de Candolle. Illustration: Botanical Magazine, 4027 (1843); Masefield et al., The Oxford book of food plants, pl. 41 (1969); Duke, Handbook of legumes of world economic importance, 39 (1981).

Leaflets to 20 cm, ovate or broadly ovate, base rounded to truncate, apex acute but not acuminate, sparsely hairy along the veins above and beneath. Bracteoles obtuse, to 2 mm. Flower-stalk to 2 mm. Calyx to 1.4 cm, sparsely hairy, upper lip as long as tube, its apex abruptly constricted behind the pointed tip; lowest tooth to 2.5 mm, acute, exceeding the acute lateral teeth. Standard to 2.7 cm. Pod to 30 × 3.5 cm, somewhat compressed, the sides rolling spirally on dehiscence. Seeds to 2 cm, oblong, somewhat compressed, cream or white with an inconspicuous mark near the hilum which is c. 0.9 cm. *SW USA, C America, West Indies, S America, introduced elsewhere.* G2.

2. C. gladiata (Jacquin) de Candolle. Illustration: Duke, Handbook of Legumes of world economic importance, 41 (1981).

Similar to *C. ensiformis* but leaflets thinner, conspicuously but gradually acuminate; calyx to 1.6 cm, almost hairless, upper lip constricted just behind the tip which is not pointed; standard c. 3.5 cm; pod to 40 × 5 cm; seeds to 3.5 cm, reddish brown (rarely whitish), hilum c. 2 cm. *Origin uncertain, probably E Asia, naturalised in many parts of the tropics & subtropics.* G2.

3. C. virosa (Roxburgh) Wight & Arnott (*C. polystachya* (Forsskål) Schweinfurth).

Like *C. ensiformis* and *C. gladiata* but leaflets to 15 cm, abruptly acuminate to an obtuse apex, more densely hairy on both surfaces; calyx to 1.2 cm, hairy, upper lip shorter than tube; standard c. 3 cm; pod to 17 × 3 cm; seeds c. 2 cm, brown or reddish brown with black marbling; hilum c. 1 cm. *Tropical & S Africa to India.* G2.

51. PUERARIA de Candolle
J. Cullen

Woody climbers (ours), often with large underground tubers. Leaves of 3 leaflets, the lateral leaflets conspicuously unequal-sided, tending to be lobed; stipules attached near the middle, projecting above and below the point of attachment (ours); stipels narrow. Racemes axillary, dense, with 2 or more flowers at each node. Bracteoles 2, borne at the base of the calyx. Calyx bell-shaped, deeply 5-toothed, the lowest tooth the longest. Corolla white, purplish or bluish. Stamens with the filaments united into a tube, the uppermost free at the base, becoming more so with age. Pod oblong, flattened, more or less straight, containing 5–20 seeds.

A genus of 17 species from tropical and temperate eastern Asia and the Pacific. Only a single species is grown and is propagated by seed; this species is widely grown in Asia and introduced elsewhere for the sake of starch produced from the tuber, as a fodder plant and for erosion control.

1. P. lobata (Willdenow) Ohwi (*P. thunbergiana* (Siebold & Zuccarini) Bentham).

Woody climber (to 30 m in the wild) with large tubers. Terminal leaflet ovate to almost circular, acute or abruptly acuminate, densely hairy above and beneath, 8–20 × 5–22 cm; lateral leaflets smaller. Racemes dense, very hairy. Flowers 1–2.5 cm. Pods 4–13 cm, densely shaggy-hairy. *E Asia, N to Japan, Pacific Islands; introduced elsewhere.* G1.

52. KENNEDIA Ventenat
J. Cullen & H.S. Maxwell

Woody scramblers or prostrate, scrambling herbs. Leaves with 3 (rarely more) leaflets; stipules sometimes falling early; stipels usually present, linear. Flowers in loose or dense axillary racemes (rarely solitary); bracteoles absent. Calyx tubular, deeply 5-toothed. Corolla reddish or black and yellow. Stamens with 9 of the filaments united, that of the uppermost free. Pods flattened, containing several seeds with spongy partitions between them. Seeds appendaged.

A genus of about 11 species from Australia; only a few are in cultivation. Propagation is by seed or cuttings taken in spring or early summer. The name is often spelled 'Kennedya'.

Literature: Silsbury, J.H. & Brittan, N.H., Distribution and ecology of the genus

Kennedya Vent. Western *Australia, Australian Journal of Botany* **3**: 113–135 (1955).

1a. Flowers in pairs in loose racemes
    2
  b. Flowers not in pairs, in dense umbels or racemes or rarely solitary
    3
2a. Flowers black and yellow; leaflets ovate to almost circular
    4. nigricans
  b. Flowers red or reddish purple; leaflets narrower   **5. rubicunda**
3a. Flowers numerous, almost stalkless in dense, umbel-like racemes
    1. coccinea
  b. Flowers few, clearly stalked, in racemes or rarely solitary   **4**
4a. Stipels persistent; bracts persistent, conspicuous   **2. prostrata**
  b. Stipels soon falling; bracts absent or soon falling   **3. eximia**

1. K. coccinea Ventenat. Illustration: Farrall, West Australian native plants in cultivation, 172 (1970).

Prostrate or scrambling, somewhat woody. Leaflets usually 3 (rarely more), leathery, adpressed bristly above and beneath, elliptic or ovate to narrowly lanceolate, rounded or acute at the apex, 3–6 × 0.5–3.2 cm. Flowers almost stalkless, numerous in dense umbel-like racemes which are long-stalked. Calyx densely hairy. Corolla red, 1.3–1.5 cm, keel shorter than wings. Pod compressed, hairy. *Australia (Western Australia).* H5–G1. Spring.

2. K. prostrata R. Brown. Illustration: Morcombe, Australia's western wild flowers, 92 (1968); Cochrane et al., Flowers and plants of Victoria, f. 43 (1968); Galbraith, Collins' field guide to wild flowers of SE Australia, pl. 15 (1977).

Prostrate, scarcely woody. Leaflets 3, lanceolate to almost circular, 1.5–2.5 × 1.2–1.6 cm, rounded though mucronate at the apex, hairless above, adpressed-bristly beneath; stipules large, persistent, stipels linear, persistent. Flowers 1–6 in loose racemes, bracts conspicuous, inflorescence-stalk longer than flower-stalks. Calyx with spreading bristles. Corolla *c.* 2 cm, scarlet-pink. Pod compressed, hairy or not. *Australia.* H5–G1. Spring.

3. K. eximia Lindley. Illustration: Paxton's Magazine of Botany **16**: 36 (1849).

Like *K. prostrata* but stipels soon falling, bracts absent or soon falling. *Australia (Western Australia).* H5–G1. Spring.

4. K. nigricans Lindley. Illustration: Edwards's Botanical Register **20**: t. 1715 (1835); Erickson et al., Flowers and plants of Western Australia, pl. 238 (1973).

Woody climber. Leaflets 3, leathery, 2.5–12 cm, broadly ovate to almost circular, rounded at the apex, hairy on both surfaces, less so above. Flowers in pairs in loose axillary racemes. Calyx densely hairy. Corolla 3–4 cm, black with a patch of yellow on the standard. Pod compressed. *Australia (Western Australia).* H5–G1. Spring.

5. K. rubicunda (Schneevoogt) Ventenat. Illustration: Cochrane et al., Flowers and plants of Victoria, f. 479 (1968); Rotherham et al., Flowers and plants of New South Wales and S Queensland, pl. 237 (1975); Galbraith, Collins' guide to the wild flowers of SE Australia, pl. 15 (1977).

Woody climber. Leaflets 3, leathery, elliptic, narrowly ovate or lanceolate, acute or acuminate at the apex, hairless above, adpressed hairy beneath, 4–9 × 2–4.5 cm. Flowers in pairs in loose axillary racemes. Calyx densely adpressed hairy. Corolla 2.5–3.5 cm, red or reddish purple. Pod compressed, hairy, to 8 cm. *Australia.* H5–G1. Spring.

53. HARDENBERGIA Bentham
J. Cullen & H.S. Maxwell

Shrubs or woody twiners. Leaves of 3–5 leaflets, when the terminal leaflet stalked and the laterals opposite or whorled, or reduced to a single leaflet; stipules and stipels persistent (even when leaf reduced to a single leaflet). Flowers in pairs in axillary racemes. Bracteoles absent. Calyx cylindric to bell-shaped, 5-toothed, the upper teeth united for most of their length. Corolla blue or purple, rarely white. Stamens 10 with all their filaments united into a tube or the uppermost at least partially free. Pod flattened or not, with or without spongy partitions between the seeds.

A genus of a small number (perhaps only 2) of species from Australia. They are easily grown and propagated by seed or by cuttings taken in late summer.

1a. Leaves reduced to a single leaflet
    2. violacea
  b. Leaves of 3–5 leaflets
    1. comptoniana

1. H. comptoniana (Andrews) Bentham. Illustration: Botanical Magazine, 8992 (1924); Gardner, Wild flowers of Western Australia, 73 (1959); Marchant et al., Flora of the Perth region **1**: 267 (1987).

Shrub or woody climber. Leaves of 3–5 leaflets, the 2–4 lateral leaflets opposite or whorled, shortly stalked, the terminal leaflet with a longer stalk; leaflets leathery, 3–14.5 × 1.3–5 cm, lanceolate to ovate, rounded to the base, long-tapered to the ultimately blunt though mucronate apex, hairless or almost so (except for their stalks); stipels small, persistent. Racemes very long, axillary, flower-stalks hairy. Calyx bell-shaped, hairless. Corolla blue to purple, rarely white, 0.7–1.1 cm. Pod cylindric, swollen, 3.5–4 cm. *Australia (Western Australia).* H5. Winter–spring.

2. H. violacea (Schneevoogt) Stearn. Illustration: Rotherham et al., Flowers and plants of New South Wales and Southern Queensland, pl. 160 (1975); Cunningham et al., Plants of western New South Wales, 396 (1981).

Shrub or woody climber. Leaves of a single, leathery, ovate to lanceolate leaflet, 3.5–8 × 0.7–3.7 cm, obtuse at the apex, rounded to truncate at the base, hairless or almost so. Racemes sometimes branched at the base. Calyx bell-shaped, hairless, purple. Corolla 0.7–0.9 cm, deep purple with yellow spots at the base of the standard. Pods flattened, *c.* 3 cm. *Australia.* H5. Winter–spring.

54. CENTROSEMA (de Candolle) Bentham
J. Cullen
Herbaceous climber. Leaves with 3 leaflets, stipules and stipels persistent. Flowers 1–5, in axillary racemes with swollen nodes. Bracteoles 2, conspicuous. Flowers borne upside-down (standard below). Calyx 5-toothed. Standard large, with a small projecting spur towards the base. Stamens 10, filaments of 9 united that of the uppermost free, anthers alternately longer and shorter. Stigma hairy. Pods long-beaked, with thickened sutures.

A genus of about 45 species mostly from the American tropics; only 1 species, from both temperate, subtropical and tropical America is grown. It is propagated by seed.

1. C. virginianum (de Candolle) Bentham. Illustration: Rickett, Wild flowers of the United States **2**: pl. 118 (1966); Dean et al., Wild flowers of Alabama, 91 (1973).

Stems thin from a woody base, twining, finely hairy with hooked hairs. Terminal leaflet stalked, *c.* 4–5 cm, laterals similar in size, very shortly stalked, hairy on both surfaces. Bracteoles large, ribbed, enclosing the calyx in bud. Corolla 2.5–4 cm, bluish violet blotched with yellow or white. Pod

6–12 cm, valves twisting when open. *E USA, N South America.* H5. Late summer.

55. CLITORIA Linnaeus
J. Cullen
Herbaceous or somewhat woody climbers. Leaves pinnate with 3–9 leaflets including a terminal leaflet; stipules and stipels persistent. Flowers large, axillary, solitary or in racemes, borne upside-down (standard below); bracteoles large, leaf-like but much shorter than the calyx. Calyx tubular to funnel-shaped, 5-toothed, the upper teeth more or less united. Standard large, without a spur. Stamens 10, 9 of them with their filaments united, that of the uppermost at least partially free. Style hairy. Pod flattened, containing several seeds.

A genus of about 70 species from the tropics. Only 1 is grown, in rich soil in a warm, sunny, humid situation. It is propagated by seed or by cuttings taken in early spring.

Literature: *Clitoria ternatea*: a select bibliography, *Medicinal and aromatic Plants Abstracts*: **10**(2): 163–166 (1988).

1. C. ternatea Linnaeus. Illustration: Botanical Magazine, 1542 (1813); Reichenbach, Flora exotica **4**: t. 226 (1835); Bruggemann, Tropical plants, t. 10 (1957); Menninger, Flowering vines of the world, pl. 150 (1970).

Climber with annual stems from a more or less persistent stock. Leaves with 5–9 leaflets with smaller and larger, rather sparse, hooked hairs. Flowers solitary. Calyx green, 2–2.5 cm. Standard 4–5 cm, blue with paler markings and white centre, or entirely white ('Album'). Pod flattened, 7–12 cm, hairy with hooked hairs. *Tropics, precise origin unknown.* G2. Summer.

56. DOLICHOS Linnaeus
J. Cullen
Annual to perennial herbs or climbers. Hairs, when present, not hooked. Leaves with 3 leaflets; stipules and stipels persistent. Flowers in groups of 2–4 in racemes; bracteoles present. Calyx broadly bell-shaped, 5-toothed. Corolla large, standard broad and with auricles at the base; wings joined to the keel. Stamens 10, 9 with their filaments united, that of the uppermost free, at least at the base. Style flattened, bearded. Pod broad containing several black or white seeds.

A genus of about 60 species from the Old World tropics, the single cultivated species often separated into the genus *Lablab* Adanson (of which it is the sole

species); it is widely grown in the tropics as a vegetable and many cultivars are known. Only occasionally grown as an ornamental, its propagation is by seed.

1. D. lablab Linnaeus (*Lablab niger* Medikus; *L. purpureus* (Linnaeus) Sweet; *L. vulgaris* Savi). Illustration: Masefield et al., The Oxford book of food plants, pl. 45 (1969); Smitinand, Wild flowers of Thailand, pl. 101 (1975); Gartenpraxis for April 1988, 10.

Perennial but usually grown as an annual, twining or rarely erect, to 10 m, hairy or hairless. Leaflets ovate, 4–15 × 4–10 cm, truncate at base, abruptly acuminate at apex. Flowers purple or white, 1.2–2.5 cm. Pod papery, hairy or not, the upper margin straight, the lower curved. *Origin uncertain, now found throughout the tropics.* H5–G1. Summer.

57. PHASEOLUS Linnaeus
J. Cullen
Like *Dolichos* but plants with hooked hairs, corolla usually smaller, often red, style coiled through a number of revolutions, stigma terete and bearded longitudinally, pods usually narrower and more or less cylindric.

A genus of 60 species from the New World, mainly from the tropics. Many of its species (together with those of the allied genus *Vigna* Savi, some species having names in both genera) are widely cultivated as vegetables (beans of various kinds). Few of the species are cultivated purely as ornamentals; those that are so cultivated are generally grown as annuals and propagated by seed.

Literature: Marechal, R., Mascherpa, J.-M. & Stainer, F., Étude taxonomique des genres *Phaseolus* et *Vigna*, Boissiera **28** (1978).

1. P. coccineus Linnaeus (*P. multiflorus* Lamarck). Illustration: Horticulture **52**: 38 (1974); Gartenpraxis for March 1976, 111 & for April 1988, 12.

Perennial grown as an annual, climbing. Leaflets ovate, to 12 cm. Flowers bright scarlet. Style with 1–2 revolutions. Pods to 30 cm, linear-cylindric, hairy or not. Seeds broadly oblong. *Widely cultivated; origin uncertain, perhaps Mexico.* H2.

The Scarlet runner bean; very variable and with many cultivars of economic importance.

P. caracalla Linnaeus (*Vigna caracalla* (Linnaeus) Verdcourt). Illustration: Horticulture **40**: 203 (1962); Il Giardino Fiorito **54**: 35 (1988, July/August).

Perennial; flowers with reflexed and contorted, pink, white or yellow standard and pink to violet wings; style with 3–5 spirals. *Tropical S America.* G2.

58. BITUMINARIA Fabricius
J.R. Akeroyd

Foetid biennial to perennial herb, somewhat woody at the base, shortly and stiffly hairy; stems 30–120 cm, erect but often rather straggling, branched below. Stipules small, narrow. Leaves with 3 leaflets on long stalks, dotted with glands; leaflets 1–6 cm, narrowly lanceolate to ovate or almost circular, entire. Flowers in dense heads on axillary stalks longer than leaves, subtended by paired bracts. Calyx with 5 unequal teeth. Corolla 1.5–2 cm, bluish violet, sometimes pink or white. Stamens united for much of their length. Fruit indehiscent, ovoid, flattened, 1-seeded, hairy, with long, curved beak; seed 5–6 mm.

A genus of three species from the Mediterranean area, one of which is cultivated. Propagation is by seed.

1. B. bituminosum (Linnaeus) Stirton (*Psoralea bituminosa* Linnaeus; *Aspalthium bituminosum* (Linnaeus) Fourreau). Illustration: Polunin, Flowers of Europe, pl. 54 (1969); Zohary, Flora Palaestina **2**, pl. 66 (1972).

Mediterranean area. H1. Summer.

A variable species. The whole plant has a distinctive smell of pitch, and the sap will blister the skin in sunlight.

59. PSORALEA Linnaeus
S.G. Knees

Scented herbs or shrubs. Leaves alternate, pinnate with a terminal leaflet, rarely simple, with translucent dots. Flowers solitary, in heads, racemes, spikes or sometimes clustered. Calyx-lobes 5, nearly equal; corolla wings about as long as keel; standard ovate or rounded; stamens 10, all united or 1 free. Fruit a short, 1-seeded, indehiscent legume.

1. P. pinnata Linnaeus. Illustration: Andrews, The Botanist's Repository **7**: 474 (1807); Journal of Horticulture, series 3, **7**: 281 (1884); Fryson, Flora of Nilgiri & Pulney hill-tops **2**: 85 (1915); Gibson, Wild flowers of Natal, pl. 43 (1975).

Compact shrub to 4 m. Leaves crowded along erect stems, with 5–11, needle-like leaflets, 2.5–3 cm, linear to linear-lanceolate, acute. Flowers solitary or clustered; calyx *c.* 9 mm; corolla *c.* 1.3 cm, blue, striped white; stamens 9 united,

1 free; style *c.* 1.1 cm, curved. *S Africa, naturalised elsewhere.* H5–G1. Summer–autumn.

60. AMORPHA Linnaeus
J. Cullen

Shrubs. Leaves pinnate with up to 45 leaflets, including a terminal leaflet, leaflets usually with brownish glands on the lower surface; stipules soon falling, stipels thread-like. Flowers very numerous in dense, spike-like racemes which may be clustered to form an apparent panicle, flower-stalks short. Calyx funnel-shaped, 5-toothed, sometimes with glands near the top. Corolla reduced to the standard (wings and keel absent) which is purplish, clawed and envelops the stamens and style. Stamens 10, their filaments all united into a tube below; anthers yellow, brown or purplish. Pod 1-seeded, indehiscent, glandular.

A genus of 15 very similar species from N America and adjacent Mexico. They are best grown in a sunny site and are propagated by division or by cuttings.

Literature: Wilbur, R.L., A revision of the north American genus *Amorpha* (Leguminosae-Psoraleae), *Rhodora* **77**: 337–409 (1975).

1a. Shrub of 1 m or more; leaf-stalks obvious, longer than the width of the lowermost leaflet; pod 5–9 mm **3. fruticosa**
  b. Shrub to 1 m, often much less; leaf-stalks very short or absent; pods 3–5.5 mm **2**
2a. Racemes clustered; plants often ash-grey, hairy; glands on lower leaf surface small and inconspicuous **1. canescens**
  b. Racemes solitary; plants not hairy as above; glands on the lower leaf surface conspicuous **2. nana**

1. A. canescens Pursh. Illustration: Botanical Magazine, 6618 (1882); Rickett, Wild flowers of the United States **1**: pl. 81 (1966); Gartenpraxis for September 1976, 425.

Shrub to 80 cm, rarely more. Leaves spreading, to 12 cm, stalk very short or almost absent, with 27–47 leaflets; leaflets 1–2.5 cm, narrowly elliptic, tapered to base and the acute or somewhat rounded, mucronate apex, shortly stalked, variably (often densely) with grey-white hairs on both surfaces and small, inconspicuous brownish glands beneath. Racemes clustered, to 20 cm or more. Calyx 3–4 mm, densely hairy. Standard 4.5–6 mm, bright violet, abruptly tapered into the

claw. Anthers yellow to brownish. Pod 3–5 mm, glandular and usually hairy. *E North America (Manitoba to Texas).* H3. Summer.

2. A. nana Nuttall (*A. microphylla* Pursh). Illustration: Botanical Magazine, 2112 (1820); Rickett, Wild flowers of the United States **4**: pl. 122 (1970).

Small shrub to 60 cm. Leaves spreading, to 7 cm, stalk very short or almost absent, with 7–22 leaflets; leaflets 0.6–1.2 cm, narrowly elliptic to oblong, broadly tapered to the base, rather rounded to the mucronate apex, sparsely hairy on the margins and main vein beneath or completely hairless, with a few, large, conspicuous brownish glands. Racemes solitary, to 9 cm. Calyx *c.* 3 mm, usually hairless except on the margins of the teeth. Standard 4.5–6 mm, abruptly tapered into the claw, dark purple. Anthers purplish. Pods 4.5–5.5 mm, glandular, not hairy. *C North America (Manitoba & Saskatchewan to New Mexico).* H3. Summer.

3. A. fruticosa Linnaeus. Illustration: Rickett, Wild flowers of the United States **1**: pl. 81 (1966); Justice & Bell, Wild flowers of North Carolina, 100 (1968); Botanical Magazine, n.s., 604 (1970–72); Polunin & Everard, Trees and bushes of Europe, 113 (1976).

Shrub 1–4 m. Leaves spreading, to 25 cm, stalk obvious, at least as long as the width of the lowermost leaflet; leaflets 9–21 (rarely more), 2–5 cm, oblong or elliptic, tapering or rounded to the base and the mucronate apex, variably hairy beneath and with small, inconspicuous glands or glandless. Racemes clustered, to 20 cm. Calyx 3–4 mm, sparsely hairy and with a few glands. Standard 5–6 mm, gradually tapering into the claw, dark reddish purple. Anthers yellow. Pod 5–9 mm, glandular, hairy or not. *N America (Quebec to Florida, California, Wyoming & N Mexico).* H3. Summer.

61. AMICIA Humboldt, Bonpland & Kunth
J. Cullen

Softly wooded shrubs. Stems upright. Leaves pinnate, with 4 or 5 leaflets including a terminal leaflet, all with orange-brown glands on the lower surface; stipules large, almost circular, united at the base, enclosing the young growth, soon falling; stipels represented by dense tufts of hair. Flowers axillary, solitary or in few-flowered racemes; bracts similar to stipules but smaller. Calyx with a large, almost circular (stipule-like) upper tooth, the lateral teeth very small, the lower teeth narrow. Corolla large, yellow, keel blunt, standard glandular outside. Stamens 10, their filaments all united, the tube so formed split above. Pod several-seeded, indehiscent, breaking into 1-seeded segments.

A genus of 7 species in C & S America. The cultivated species requires a sunny site with protection from cold and wind; it may be killed to the root in hard winters. Propagation is by cuttings taken in summer.

1. A. zygomeris de Candolle. Illustration: Botanical Magazine, 4008 (1843); Paxton's Magazine of Botany **13**: 173 (1847); McVaugh, Flora Novo-Galiciana **5**: 277 & frontispiece (1987).

Shrub to 2 m; stems with spreading hairs of varying lengths. Leaflets green above, glaucous beneath, oblong, rounded to the base, deeply and widely notched at the apex, to 6 × 5 cm. Upper calyx-tooth to 1.5 cm. Standard to 3 × 2.5 cm, yellow or orange-yellow, keel yellow with a brownish or purplish blotch towards the apex. *Mexico.* H5.

A handsome flowering shrub and striking foliage plant, whose leaflets show sleep-movements.

62. ADESMIA de Candolle
J. Cullen

Herbs or shrubs, usually with stalkless or stalked glands, often sticky. Leaves pinnate (ours) with many leaflets, including a terminal leaflet; stipules small, persistent, stipels absent. Flowers in axillary racemes (ours). Calyx bell-shaped, 5-toothed. Corolla yellow or reddish. Stamens 10, their filaments free. Pod indehiscent, breaking transversely into 1-seeded segments, often glandular.

A genus of 230 species in S South America (Peru and Brazil to Tierra del Fuego), of which 1 is occasionally grown as a curiosity. Propagation is by seed when available.

Literature: Burkart, A., Sinopsis del genero sudamericano de Leguminosas, *Adesmia* DC., *Darwiniana* **14**: 463–568 (1967).

1. A. boronioides Reiche. Illustration: Botanical Magazine, 7748 (1900).

Low shrub, very glandular and sticky, glands stalkless. Leaves with many, irregularly toothed, notched, fleshy leaflets. Flowers yellow (standard sometimes with red lines), 0.8–1 cm. Pod with few segments, the upper suture straight, the lower deeply indented, the segments

rounded, very glandular. *Chile, Argentina.* H5–G1.

63. CLIANTHUS Lindley
J. Cullen

Herbs or soft-wooded shrubs. Leaves pinnate with numerous lateral leaflets and a terminal leaflet; stipules large, persistent, stipels absent. Flowers very showy in erect or pendent axillary racemes, each with a bract and 2 bracteoles. Calyx bell- or cup-shaped, 5-toothed. Corolla scarlet or scarlet and black, pink or white (a cultivar). Standard reflexed, narrow, acuminate, wings deflexed, shorter than the keel; keel large, acuminate, curved like a parrot's beak. Stamens 10, filaments of 9 of them united, that of the uppermost free. Ovary shortly stalked; style bearded. Pod firm, turgid and beaked, containing several seeds.

A genus of 2 species, 1 from Australia, the other from New Zealand. Both are cultivated and are among the most spectacular of the smaller legumes. Their cultivation is not entirely easy in Europe, though they are more easily grown in the southern hemisphere, where they are often treated as annuals. They require a sunny, open site and well-drained, deep, neutral to alkaline soil. Propagation is by seed (which requires scarification) or cuttings; young plantlets are very susceptible to damping-off. Seedlings of *C. formosus* are sometimes grafted on to seedling stocks of *C. puniceus* or species of *Colutea*.

1a. Densely woolly prostrate or trailing herb; racemes erect; standard with a black or red swelling at the base of the blade   **2. formosus**
  b. Soft-wooded shrub with sparse, adpressed hairs; racemes pendent; standard without a swelling at the base of the blade   **1. puniceus**

1. C. puniceus (G. Don) Lindley. Illustration: Botanical Magazine, 3584 (1837); Journal of the Royal Horticultural Society **84**: f. 38 (1959); Moore & Irwin, The Oxford book of New Zealand plants, 162 (1978); Moggi & Guignolini, Fiori da balcone e da giardino, 86 (1982).

Soft-wooded shrub with branches to 2 m, sparsely adpressed hairy throughout. Leaves with up to 15 pairs of leaflets; leaflets narrowly lanceolate to linear-oblong, rounded to the blunt apex. Racemes with up to 15 flowers, pendent; bracteoles small, distant from the calyx. Calyx broadly bell-shaped, tube 0.7–1 cm, teeth 1.5–3 mm. Corolla scarlet, reddish

or rarely white; standard to 6 cm, erect or somewhat reflexed, without a swelling at the base of the blade; wings much shorter than keel; keel curved forwards, to 8 cm, acuminate and beaked. Pods to 8 cm. *New Zealand (North Island).* H5–G1.

In 'Albus' the flowers are white and in 'Roseus', reddish.

2. C. formosus (G. Don) Ford & Vickery (*C. dampieri* R. Brown). Illustration: Botanical Magazine, 5051 (1858); Erickson et al., Flowers and plants of Western Australia, 143, 147 (1973); Rotherham et al., Flowers and plants of New South Wales and southern Queensland, 162 (1975); Elliot & Jones, Encyclopaedia of Australian plants **3**: 50, 51 (1984).

Sprawling annual or perennial herb, densely woolly throughout. Leaves with up to 10 pairs of leaflets; leaflets elliptic to oblong, acute or rounded at the apex, greyish with dense woolly hairs, at least beneath. Racemes erect, few-flowered; bracteoles close to and overlapping the calyx. Calyx tubular to bell-shaped, tube 6–7 mm, teeth 5–7 mm, all greyish with woolly hairs. Corolla clear scarlet; standard erect, vertical, with a usually black swelling at the base of the blade; wings shorter than keel; keel deflexed downwards vertically but curving outwards, acuminate. Pods to 6 cm, hairy. *Australia (Queensland, New South Wales, Western Australia, South Australia, Northern Territories).* H5–G1.

64. SWAINSONA Salisbury
J. Cullen

Herbs or shrubs, covered generally with short, adpressed hairs. Leaves pinnate, with a terminal leaflet; stipules conspicuous, persistent, stipels absent. Flowers in axillary racemes; bracts present, bracteoles present, rarely very small or absent. Calyx bell-shaped, with 5 nearly equal teeth. Standard upright, blade more or less circular; wings curved or twisted; keel broad, curving or somewhat coiled upwards, obtuse (ours). Stamens 10, 9 of them with filaments united, that of the uppermost free. Ovary stalked or not, with several ovules; style longitudinally bearded. Pod ovoid, membranous, translucent and inflated (ours).

A genus of about 50 species from Australia; only 2 are in general cultivation. They are easily grown in a dry, sunny site, and are propagated by seed or by cuttings.

1. S. galegifolia R. Brown. Illustration: Botanical Magazine, 792 (1805) – as

Colutea galegifolia & 1725 (1815) – as S. coronillifolia; Rotherham et al., Flowers and plants of New South Wales and southern Queensland, 63 (1975); Galbraith, Collins' field guide to the wild flowers of south-east Australia, pl. 15 (1977); Cunningham et al., Plants of western New South Wales, 413 (1981).

Shrub to 1 m or more. Leaves with 5–10 pairs of oblong, obtuse or notched leaflets. Calyx 3–6 mm. Corolla purplish blue, pink, red, reddish brown or rarely white; standard to 1.6 cm wide, with a spot of paler colour at the base, and with 2 oblique plate-like swellings at the base; keel broad, obtuse. Pod inflated, membranous, 2.5–5 cm, on a short stalk. *Australia (Queensland, New South Wales)*. H5–G1.

S. procumbens Mueller. Illustration: Cochrane et al., Flowers and plants of Victoria, 89 (1968); Hodgson & Paine, Field guide to Australian wild flowers **1**: 183 (1979). Similar, but leaflets narrow and distant, corolla mauve-blue, the standard broader than long, deeply notched, keel coiled upwards through somewhat more than 1 revolution. *Australia (South & Western Australia, New South Wales)*. H5–G1. Summer.

65. SUTHERLANDIA R. Brown
H.S. Maxwell

Evergreen erect or prostrate shrubs, 0.6–2 m. Leaves alternate, pinnate with a terminal leaflet. Leaflets green with greyish hairs. Stipules small, usually linear, broader at base. Inflorescence a slender, axillary raceme. Flowers showy, rose to bright red or purple. Calyx bell-shaped, lobes 5, triangular, all segments pointing forwards; standard folded, veined, shorter than the keel. Stamens 10, 9 united and 1 separate. Flowers followed by 'puffy' pale green pods.

A genus of 5 species native to South Africa, only 1 species is in general cultivation. They are easily grown in a greenhouse or in full sun out of doors in a well-drained leaf-rich soil and can be tolerant of a few degrees of frost. Propagation is easy from seed or from cuttings; they should be pruned in late winter.

When the pods of *S. frutescens* float on water they are seen as being like ducks hence the name 'Duck Plant'; it was also at one time thought to be a cancer cure.

1. S. frutescens (Linnaeus) R. Brown (*Colutea frutescens* Linnaeus). Illustration: Everett, The New York Botanical Garden

illustrated encyclopedia of horticulture **10**: 3263 (1982); Phillips & Rix, Shrubs 26 (1989); Brickell, The RHS gardeners' encyclopedia of plants & flowers, 133 (1989); Macoboy, What shrub is that? 322 (1989).

Shrub 0.6–1.6 m. Leaves 4.5–9 cm, with 13–21 narrowly oblong to linear-elliptic leaflets, 0.5–2 cm, sometimes slightly mucronate, usually hairy beneath, hairless above or almost so. Inflorescence a 6–10 flowered raceme, 2.5–8 cm, flower-stalks to 8 mm, with stiff hairs; bracts *c.* 2 mm, ovate. Flowers bright red, 2.5–5 cm; keel 2.5–3.5 cm, claw 0.8–1.5 cm. Ovary stalked. Pods papery, translucent, inflated, broadly elliptic to almost spherical, 3.5–5.5 × 2–4 cm. *S Africa, Namibia*. Late spring–summer.

66. COLUTEA Linnaeus
J. Cullen

Shrubs or rarely small trees. Leaves pinnate, with 2–6 pairs of leaflets and a terminal leaflet, often somewhat glaucous; stipules persistent, stipels absent. Flowers rather few in axillary racemes, with bracts. Calyx usually broadly bell-shaped, with 5 short teeth. Corolla yellow to orange-red, the claws of the petals usually projecting from the calyx. Standard upright, with 2 small swellings above the claw; wings shorter than to longer than the keel, each with a distinct auricle on the upper margin and sometimes a spur on the lower; keel rounded or beaked at the apex. Stamens 10, 9 with their filaments united, that of the uppermost free. Ovary stalked. Pod inflated, papery, and translucent, indehiscent or splitting towards the apex, stalked, the stalk often projecting from the persistent calyx; seeds several.

A genus of 26 rather similar species from Eurasia and E Africa (Ethiopia). The species have been frequently confused, both in the wild and in gardens and the most commonly seen cultivated *Colutea* is the hybrid *C.* × *media*. They are easily grown in a sunny position in well-drained soil and are propagated by seed or by cuttings taken in autumn.

Literature: Browicz, K., [A revision of the genus *Colutea*], *Monographiae Botanicae* **16** (1963).

1a. Ovary and pod hairless   2
  b. Ovary hairy, pod often hairy   4
2a. Keel terminating in a distinct beak; flowers orange-red   **4. orientalis**
  b. Keel blunt, not beaked; flowers yellow   3

3a. Wings shorter than to as long as keel, rarely longer, without a spur on the lower margin   **1. arborescens**
  b. Wings longer than keel, with a distinct spur on the lower margin   **2. cilicica**
4a. Older shoots grey-brown, matt; leaflets usually more than 1.5 cm   **1. arborescens**
  b. Older shoots red- or purple-brown, shining; leaflets to 1.5 cm   5
5a. Keel distinctly beaked; corolla orange-yellow   **5. buhsei**
  b. Keel blunt, not beaked; corolla yellow   **3. gracilis**

1. C. arborescens Linnaeus. Illustration: Polunin & Everard, Trees and bushes of Europe, 112 (1976); Krüssmann, Manual of cultivated broad-leaved trees and shrubs **1**: 361 (1986); Il Giardino Fiorito, June 1987, 55.

Shrub to 5 m; young shoots hairy at first, older shoots grey-brown, peeling, matt. Leaves with 3–6 pairs of leaflets; leaflets ovate to broadly elliptic or rarely obovate, to 3 × 2 cm, rounded to the base and to the obtuse or notched apex, hairless above, sparsely adpressed hairy beneath. Racemes with 3–8 flowers. Calyx 6–8 mm, covered with adpressed hairs which are white, dark brown to black or a mixture of the two. Corolla yellow, standard with reddish veins; wings shorter than to as long as keel, rarely longer, without a spur on the lower margin; keel 1.6–2 cm, blunt at apex. Ovary hairless or hairy. Pod 5–8 cm, usually hairless. *C, S & E Europe*. H2. Summer.

Often recorded from east of Europe, but there mistaken for *C. cilicica*. Two subspecies occur: subsp. **arborescens** from C & E Europe, with ovary hairless, and subsp. **gallica** Browicz, from SW and part of C Europe, with ovary hairy. Both are apparently in cultivation. Browicz mentions 2 cultivars, 'Crispa' with the leaflet-margins wavy, and 'Bullata' with the surface wrinkled.

C. × media Willdenow is the hybrid between this and *C. orientalis*, and is more widely cultivated than either of its parents. It is intermediate between the 2 in most characters, with brownish red or orange flowers with the keel 1.5–1.6 cm, scarcely beaked. Illustration: Seabrook, Shrubs for your garden, 45 (1975); Csapody & Tóth, Color atlas of flowering trees and shrubs, pl. 67 (1982).

2. C. cilicica Boissier & Balansa. Illustration: Flora SSSR **11**: pl. 22 (1945).

Similar to *C. arborescens* but leaflets usually smaller (mostly to 2 × 1.4 cm), calyx broadly bell-shaped, 7–9 mm, wings always longer than the keel and with a spur on the lower margin, keel 2–2.2 cm, ovary always hairless. *Greece, Turkey, Syria, Lebanon, Israel, former USSR (Caucasus, Crimea & adjacent areas).* H4. Summer.

Difficult to distinguish from *C. arborescens*; intermediates occur in E Europe and may also be in cultivation.

3. C. gracilis Freyn & Sintenis. Illustration: Flora SSSR **11**: pl. 22 (1945).

Like *C. arborescens* but bark on older shoots brownish red or brownish violet, shining; leaflets 3–7 × 2–6 mm, rounded or very slightly notched at apex, corolla pale yellow, wings longer than keel, without a spur on the lower margin, keel 1.4–1.6 cm; ovary always hairy, pod 3–4.5 cm, sparsely hairy. *Iran/former USSR border area (Kopet Daĝ).* H3? Summer.

4. C. orientalis Miller. Illustration: Flora SSSR **11**: pl. 22 (1945); Krüssmann, Manual of cultivated broad-leaved trees and shrubs **1**: 361 (1986).

Shrub to 3 m, young shoots hairless, older shoots brownish grey, matt. Leaves with 2–4 pairs of leaflets; leaflets often bluish green, broadly obovate, tapered to the base, rounded or notched at the apex, to 1.8–1.5 cm. Racemes with 3–5 flowers. Calyx 5–6 mm, covered in mostly black hairs. Corolla orange-red, standard with darker veins and a pale yellow spot near the base of the blade; wings shorter than keel; keel 1.1–1.3 cm, beaked. Ovary hairless. Pod to 5 × 2 cm, stalk short or absent. *Former USSR (Caucasus).* H3. Summer.

5. C. buhsei (Boissier) Shaparenko (*C. persica* misapplied var. *buhsei* Boissier). Illustration: Flora SSSR **11**: pl. 22 (1945).

Shrub to 3 m; young shoots hairy at first, older shoots brownish red and shining. Leaves with 3 or 4 pairs of leaflets; leaflets to 1.5 × 1.2 cm, broadly obovate, rounded or somewhat notched at the apex, hairless above, with adpressed hairs beneath. Raceme with 2–5 flowers. Calyx 7–8 mm, sparsely hairy with mixed white and dark brown to black hairs. Corolla orange-yellow, wings longer than keel, keel 2–2.2 cm, beaked. Ovary hairy. Pod 6–7.5 cm, hairy, shortly stalked. *N Iran & adjacent former USSR.* H2. Summer.

67. HALIMODENDRON de Candolle
J. Cullen

Round-headed shrub with pale brown, striped bark, with weakly distinguished long and short shoots. Leaves on the extension shoots persisting as stout spines to 6 cm, each with 2 much smaller stipular spines at the base. Foliage leaves with usually 4 leaflets, pinnately arranged, the axis tipped with a short spine; leaflets glaucous, oblong-obovate, tapered to the base, rounded to the apex, to 4 cm, hairy or hairless; stipules of the short-shoot leaves membranous. Racemes terminating the (axillary) short shoots, with 2–5 flowers each with a bract and 2 bracteoles. Calyx 5–7 mm, bell-shaped, finely hairy, truncate, with 5 small teeth. Corolla 1.5–2 cm, pale purplish pink or magenta, standard erect with its sides reflexed. Stamens 10, filaments of 9 of them united, that of the uppermost free. Ovary shortly stalked, with several ovules. Pod stalked, inflated, obovoid or obovoid-oblong, leathery, beaked, yellowish brown, containing several seeds.

A genus of a single species occurring from E Europe (Russia) to C & E Asia, usually found on saline soils. It is often grown grafted on to stocks of *Caragana arborescens* (p. 504) but can also be grown on its own roots. It requires a sunny position on well-drained soil. Propagation is by seed or by grafting.

1. H. halodendron (Pallas) Voss (*H. argenteum* (Lamarck) de Candolle). Illustration: Flora SSSR **11**: t. 20 (1945); Schischkin, Botanicheskii Atlas, t. 33 (1963).

E Europe to C & E Asia. H3. Late spring.

68. CARAGANA Lamarck
J. Cullen

Shrubs or small trees, often with widely spreading or pendent, little-branched branches. Shoots generally of 2 kinds: long, extension shoots on which the leaves are borne alternately, their axes often persisting as spines, their stipules also often spiny and persistent; and short shoots borne in the axils of the extension-shoot leaves, very long-lived, terminated by usually membranous stipules and bearing clusters of leaves and the flowers. Leaves with 4 or more leaflets, mostly pinnate but some with the leaflets borne in a close cluster (palmate), the terminal leaflet replaced with a spine. Stipules present, joined at the base, divergent at the apex, membranous at first, persistent, some with

thickened midribs which persist as spines as the membranous tissue decays. Flowers solitary or in clusters, each with its own jointed stalk arising from the short shoot, rarely in few-flowered, stalked umbels; bracteoles sometimes persisting at the joint. Calyx bell-shaped or cylindric, truncate or oblique at the mouth, 5-toothed. Corolla usually yellow, more rarely orange or pink, the petals with long claws projecting from the calyx; standard erect, its margins reflexed, wings usually each with a backwardly projecting auricle. Stamens 10, filaments of 9 united, that of the uppermost free. Ovary containing several ovules. Pod several-seeded, hairless or hairy inside.

A genus of perhaps 65 species from Asia (just extending into Europe in the western former USSR). They are often of picturesque form and old plants can make striking specimens. Their shoot architecture, with strongly differentiated long and short shoots is remarkable; it is very difficult to identify material consisting only of extension shoots. The classification of the genus is very confused and in need of considerable revision; the identification of cultivated material is not easy.

They are in general easily grown in a very sunny position in well-drained soil. Propagation is best by seed, though cuttings and layering can also be used.

Literature: Komarov, V.L., *Monografiya roda Caragana, Acta Horti Petropolitani* **29**: 178–362 (1908).

Leaf-spines. Present on long shoots: **3–14**; absent: **1–5**.

Leaves. Mostly pinnate with 4 leaflets in 2 pairs, or more: **1–10**; leaflets 4 in a cluster, not pinnately arranged: **11–14**.

Flowers. In an umbel: **4**. With persistent bracteoles: **4,7**.

Calyx. Truncate at the apex with well-separated teeth: **1**; oblique at the apex with adjacent teeth: **2–14**. Hairy on the surface: **1,2,4,6–9**; hairless on the surface (though hairy on the margins of the teeth): **3,5,10–14**.

Corolla. Orange: **14**; white flushed with pink, or pink: **9**.

1a. Leaves composed of 4 leaflets which form a cluster (i.e. are not pinnately arranged) usually below the tip of the leaf-axis   2
  b. Leaves composed of 4 or more leaflets which are pinnately arranged   5
2a. Leaves of the short shoots stalked (sometimes shortly so), those on

the long shoots also shortly stalked (leaving a single scar on the spine) 3

b. Leaves on the short shoots stalkless (those on the long shoots shortly stalked) 4

3a. Leaflets obovate to reversed-triangular, broadest almost at the apex which is rounded or truncate, completely hairless; flower-stalk jointed above the middle **12. frutex**

b. Leaflets very narrowly obovate or linear-obovate, broadest ¼–⅓ of the length from the apex, which is tapered, with flaky scales beneath; flower-stalk jointed at the middle or below **11. densa**

4a. Calyx to 5 mm; corolla orange or pinkish yellow; ovary hairless **14. aurantiaca**

b. Calyx 7–10 mm; corolla clear yellow; ovary downy **13. pygmaea**

5a. Leaf-stalks and axes of the leaves on the long shoots all deciduous, not persisting as spines 13

b. Leaf-stalks and axes of the leaves on the long shoots persisting as spines, or at least some of them so doing 6

6a. Flowers borne in an umbel of 3–5 on a stalk which arises from the short shoot **4. brevispina**

b. Flowers borne singly or in clusters, each with its own stalk arising from the short shoot 7

7a. Most leaves with 2 pairs of leaflets (occasionally a few with 3 pairs) 8

b. Most leaves with 3 or more pairs of leaflets (occasionally a few with 2 pairs) 11

8a. Surface of calyx hairy, ribbed **6. maximowicziana**

b. Surface of calyx hairless (margins of teeth usually hairy) 9

9a. Calyx 1 cm or more; leaflets broadly obovate-elliptic, 1–1.5 × 0.5–0.7 cm, smoothly rounded to the apex which has a point to 1 mm, hairless **10. sinica**

b. Calyx to 1 cm; leaflets various, not as above, narrower (to 5 mm) or if broader then with the point at the apex at least 2 mm and with sparse silvery hairs on both surfaces 10

10a. Leaves dark green, hairless above, clearly stalked, leaflets oblong-obovate, in 2 very close pairs (some almost with the 4 leaflets clustered); point at the apex to 1 mm **5. spinosa**

b. Leaves pale grey-green, with sparse silky hairs on both surfaces, indistinctly stalked, leaflets obovate, clearly pinnately arranged, point at the apex 1–2.5 mm **3. decorticans**

11a. Corolla 2.5–3 cm, cream flushed with pink or entirely pale to deep pink; spines very numerous, thin, flexible; whole plant covered with long, spreading hairs **9. jubata**

b. Corolla to 2.5 cm, yellow (standard sometimes orange); spines rather few, stout, not flexible; plant variously hairy, not as above 12

12a. Mature leaflets densely silky-hairy on both surfaces; flower-stalks jointed in the lower half, without persistent bracteoles **8. gerardiana**

b. Mature leaflets with only the margins hairy; flower-stalks jointed in the upper half with 3 persistent bracteoles at the joint **7. franchetiana**

13a. Leaflets to 8 mm at most, veins diverging from midrib at a very acute angle; calyx cylindric, mouth oblique, teeth adjacent **2. microphylla**

b. Leaflets more than 8 mm, veins diverging from the midrib at a wide angle; calyx truncate, teeth distant **1. arborescens**

1. C. arborescens Lamarck. Illustration: Huxley, Deciduous garden trees and shrubs in colour, pl. 28 (1979); Csapody & Tóth, Colour atlas of trees and shrubs, 147 (1982); Davis, Gardener's illustrated encyclopaedia of trees and shrubs, 21, 22 (1987); Gartenpraxis for May 1987, 56 and August 1988, 24.

Shrub or small tree to 6 m, bark grey-brown. Short shoots well developed. Leaves of main shoots deciduous, the leaflets falling first, the axes sometimes shortly retained but not persistent and forming spines. Leaves pinnate with 4–6 pairs of leaflets; leaflets oblong or oblong-elliptic, rounded to the base and minutely pointed apex, to 2.5 × 1 cm, veins diverging from the midrib at a wide angle, hairless above, silky-hairy beneath; stipules membranous, not spiny. Flower-stalks to 2.5 cm, hairy, jointed above the middle. Calyx bell-shaped, 6–8 mm, truncate with small, distant teeth, hairy all over the surface. Corolla yellow, to 1.8 cm. Pod 2.5–3 cm, hairless. *E former USSR, Mongolia, NE China.* H2. Summer.

A widespread and variable species; 'Pendula' has pendent branches and 'Nana' is rather low-growing. Several species related to *C. arborescens* are reputedly in cultivation; these are scarcely distinguishable from *arborescens* or from each other.

C. boisii Schneider (*C. arborescens* var. *crasseaculeata* Bois). Smaller, leaves dark but clear green above, paler beneath, stipules of the main shoot leaves spiny, calyx-teeth somewhat larger and more conspicuous. *W China.* H5. Summer.

C. fruticosa Besser. Like *C. boisii* with spiny stipules, but leaves bright green above and beneath. *E former USSR, Korea.* H2. Summer.

C. × sophorifolia Tausch is reputedly the hybrid between *C. arborescens* and *C. microphylla.* It is very like *arborescens*, showing little influence of its other parent, but is lower growing, with smaller leaflets and spiny stipules on the main shoots. *Garden origin.* H2. Summer.

2. C. microphylla Lamarck.

Shrub to 3 m, but often much smaller; bark pale grey-brown. Short shoots well developed, covered in membranous stipules. Leaves of the main shoots not persistent as spines, but stipules forming short spines. Leaves with 4 or more pairs of leaflets; leaflets mostly to 5 × 3 mm (rarely to 8 mm), obovate-elliptic, rounded to the base and to the conspicuously mucronate apex, grey-green, densely silvery-hairy on both surfaces, veins diverging from the midrib at a very acute angle. Flower-stalk to 1.2 cm, densely hairy, jointed at about the middle. Calyx 1 cm or more, hairy all over the surface, cylindric, ribbed, mouth oblique, teeth *c.* 1 mm, triangular, adjacent. Corolla yellow, to 2.5 cm. Pod 2.5–3 cm. *E former USSR, Mongolia.* H2. Summer.

3. C. decorticans Hemsley. Illustration: Acta Horti Petropolitani **29**: t. 14 (1908).

Shrub or small tree, bark shining yellow-brown. Short shoots well developed, covered in membranous stipules. Axes of leaves of long shoots persistent or not as spines; stipules of these leaves spiny, persistent. Leaves with usually 2 pairs (rarely 3 pairs) of leaflets, clearly pinnately arranged; leaflets *c.* 1 × 0.5 cm, grey-green, obovate, tapered to the base, abruptly narrowed to a long point 1–2.5 mm, veins diverging from the midrib at an acute angle, persistently and sparsely silvery-hairy on both surfaces. Flower-stalks to 1.8 cm, shortly hairy, jointed above the middle. Calyx cylindric, to 6 mm, mouth oblique, hairless, teeth *c.* 1.5 mm, triangular, adjacent. Corolla yellow, to 2 cm. Pod to 4 cm, dark brown, hairless. *Afghanistan.* H3. Summer.

4. C. brevispina Royle.

Shrub to 3 m, bark grey-brown, striped. Short shoots small, covered by membranous stipules. Leaves of long shoots persisting as spines to 5 cm, each with 2 much smaller spiny stipules at the base. Leaves pinnate with 5 or more pairs of leaflets; leaflets oblong to elliptic, rounded to the base and to a short point at the apex, hairless, veins conspicuous, spreading at a wide angle from the midrib, dark green above, pale greyish green beneath. Flowers 3–5 in a stalked umbel; flower-stalks hairy; bracteoles persistent. Calyx cylindric to bell-shaped, to 1 cm, finely hairy throughout, mouth oblique, teeth triangular, 2–3 mm, adjacent. Corolla yellow, sometimes flushed with orange, to 2 cm. Pod to 5 cm, hairy outside and inside. *W Himalaya.* H4. Summer.

5. C. spinosa de Candolle (*C. ferox* Lamarck).

Shrub to 2 m with long arching branches; bark grey-brown, striped. Short shoots well developed, covered in membranous stipules. Axes of long shoot leaves persistent as spines to 3 cm; stipules of these leaves membranous, their midribs ultimately becoming spiny. Leaves with 2 pairs of leaflets close but usually pinnately arranged, rarely some with the 4 leaflets clustered; leaflets to 1.8 × 0.4 cm, oblong-obovate, tapered to the base, rounded to the ultimately pointed apex, hairless when mature (silky-hairy when young), veins at an acute angle to the midrib. Flower-stalks hairy, very short, mostly hidden within the short shoot. Calyx cylindric, mouth oblique, hairless except for the margins of the teeth, to 8 mm, teeth to 1.5 mm, adjacent. Corolla yellow to 1.8 cm. Pod to 2 cm, hairless. *E former USSR, N China.* H2. Summer.

6. C. maximowicziana Komarov. Illustration: Acta Horti Petropolitani **29**: t. 11 (1908).

Much-branched, intricate shrub to 2 m; bark brown, shining. Short shoots well developed, covered in spiny stipules and short, somewhat persistent leaf-axes. Leaves on long shoots persistent as spines 1.5–4 cm, with rather broad stipular spines at the base. Leaves of 2 or rarely 3 pairs of pinnately arranged leaflets; leaflets narrowly obovate, gradually tapered to the base, rounded to the ultimately mucronate apex, to 10 × 3 mm, hairy. Flower-stalks short, hairy, mostly hidden within the short shoot. Calyx cylindric, ribbed, to 1 cm, hairy all over the surface, mouth oblique, teeth *c.* 1 mm, adjacent. Corolla yellow, *c.* 2 cm. Pod to 2 cm, hairy. *W China.* H4. Summer.

7. C. franchetiana Komarov. Illustration: Acta Horti Petropolitani **29**: t. 13 (1908).

Spreading or compact shrub to 2 m; bark grey-brown, young shoots hairless. Short shoots well developed, covered with membranous stipules. Leaves of the main shoots persistent as distinct, stout spines to 5 cm; stipules of these leaves membranous, not spiny. Leaves with 4–6 pairs of leaflets, pinnately arranged; leaflets to 1.2 cm × 4 mm, obovate, tapered to the base, rather abruptly tapered to a long fine point, densely silky-hairy when young, only hairy on the margins when mature. Flower-stalks hairy, to 1.5 cm, jointed above the middle and bearing 3 persistent bracteoles at the joint. Calyx cylindric, finely hairy throughout, to 1.5 cm, mouth oblique, teeth *c.* 3 mm, triangular, adjacent. Corolla yellow, the standard sometimes with orange markings, to 2.5 cm. Pod to 5 cm, hairy. *W China.* H4. Summer.

8. C. gerardiana (Graham) Bentham. Illustration: Acta Horti Petropolitani **29**: t. 13 (1908).

Like *C. franchetiana* but old leaf-spines dense, and shoots and leaves densely hairy; flower-stalks very short, without persistent bracteoles at the joint; corolla pale yellow to almost white; pod to 2.5 cm. *W Himalaya.* H3. Summer.

9. C. jubata (Pallas) Poiret. Illustration: Csapody & Tóth, Colour atlas of trees and shrubs, 147 (1982); Bärtels, Gardening with dwarf trees and shrubs, 115 (1986).

Shrub to 5 m, though usually much less; branches covered with spines and dense hair, the bark scarcely visible, but brown and striped. Short shoots well developed, covered with slender spines from old leaves. Leaves on the main shoots persistent as dense spines which are thin, flexible and hairy; stipules membranous. Leaves pinnate with 4–7 pairs of leaflets, the whole densely covered with long, spreading hairs; leaflets oblong, rounded to the base, shortly tapered to an acute but scarcely mucronate apex. Flower-stalks very short, hidden within the short shoot. Calyx reddish, broadly cylindric, to 1.2 cm, mouth oblique, teeth triangular, to 2 mm, adjacent, the whole densely hairy. Corolla cream streaked with pink or entirely pale to deep pink, to 3 cm. Pod to 2 cm, hairy. *C & E former USSR, N China.* H3. Spring–summer.

Old plants have a bizarre appearance due to their long branches densely covered with spines.

10. C. sinica (Buc'hoz) Rehder (*C. chamlagu* Lamarck).

Shrub to 2 m, bark grey-brown. Short shoots small, covered with membranous stipules. Leaves of the main shoots persistent as spines which are rather distant, to 2 cm, each with a pair of shorter spiny stipules at the base. Leaves with 2 pairs of leaflets, pinnately arranged; leaflets to 1.5 × 0.7 cm, broadly obovate-elliptic, tapered to the base, rounded to the apex, which is shortly pointed, hairless. Flower-stalks hairless, to 1.2 cm, jointed above the middle. Calyx broadly cylindric, 1.1–1.4 cm, mouth oblique, teeth triangular, *c.* 2 mm, adjacent, the whole hairless except for the margins of the teeth. Corolla yellow, 2.5–3 cm. Pod to 3.5 cm, hairless. *N & W China.* H3. Summer.

11. C. densa Komarov. Illustration: Acta Horti Petropolitani **29**: t. 7 (1908).

Shrub to 2 m, bark glossy, brown. Short shoots rather small, covered by spiny stipules and leaf-stalks. Leaves of long shoots persisting as distant spines to 1 cm, each with 2 much smaller stipular spines at the base. Leaves with 4 leaflets, clustered (not pinnately arranged) in the upper third of the spine; leaflets glaucous, to 1.2 × 0.3 cm, very narrowly obovate or linear-obovate, broadest 1/4–1/3 below the tapered apex, veins at an acute angle to the midrib, all with flaky scales beneath. Flower-stalks to 1.8 cm, jointed somewhat below the middle, finely crisped hairy. Calyx cylindric, *c.* 6 mm, mouth oblique, teeth triangular, *c.* 2 mm, adjacent, the whole hairless except for the margins of the teeth. Corolla yellow, 1.8–2 cm. Pod to 2 cm, hairless. *W China.* H4. Summer.

12. C. frutex (Linnaeus) Koch (*C. frutescens* de Candolle). Illustration: Acta Horti Petropolitani **29**: t. 5 (1908); Schischkin, Botanicheskii Atlas, t. 35 (1963); Csapody & Tóth, Colour atlas of trees and shrubs, 147 (1982); Gartenpraxis for August 1988, 24.

Shrub to 3 m or more, with long branches; bark glossy brown, striped.

Short shoots well developed, covered in stipules which are mostly membranous. Long shoot leaves persisting as spines to 6 cm, each with 2 small stipular spines at the base. Leaves of 4 clustered leaflets (not pinnately arranged) borne together in the upper third of the spine; leaflets obovate-reversed triangular, to 1.3 cm × 5 mm, tapered to the base from the widest part which is almost at the truncate or broadly rounded, finely pointed apex, all glaucous and hairless, veins at an acute angle to the midrib. Flower-stalks *c.* 1.2 cm, jointed above the middle. Calyx cylindric, *c.* 7 mm, mouth oblique, teeth triangular, *c.* 1.5 mm, adjacent, the whole hairless except for the margins of the teeth. Corolla yellow, *c.* 1.8 cm. Pod to 4 cm, hairless. *European & Asiatic former USSR.* H3. Summer.

A variable species. Plants grown under this name in the early part of the century are not the same as the wild material. They have no obvious spines (though the material available is all of young extension shoots) and the leaves are long-stalked, with 4 long, broad, rounded leaflets borne in a cluster at the top of the stalk; in some leaves there is a small soft, pointed tip between the leaflets, in others this is absent. The calyx is small, rather truncate, with small, distant teeth, and the corolla is larger. These specimens have not been matched with any known wild species, and may perhaps be of hybrid origin (the lack of spines, broad leaflets, truncate calyx and large corolla all suggest *C. arborescens*). It is uncertain whether or not any such material still survives in the living state.

13. C. pygmaea de Candolle. Illustration: Bärtels, Gartengeholze, 122 (1973); Davis, The gardener's encyclopaedia of trees and shrubs, 92 (1987); Gartenpraxis for August 1988, 24.

Upright to prostrate shrub to 1 m, with long, slender, sometimes trailing branches; bark grey-brown, striped. Short shoots rather small, covered in membranous stipules. Leaves of the long shoots persisting as spines to 8 mm, each with 2 somewhat shorter stipular spines at the base. Leaves with 4 leaflets borne in a cluster (not pinnately arranged) almost at the base of the spine; leaflets very narrowly linear-obovate, to 13 × 2 mm, tapered to the base and to the pointed apex, veins diverging at an acute angle from the midrib, all sparsely hairy. Flower-stalks to 1.7 cm, jointed just above the middle, at least the upper part sparsely hairy. Calyx cylindric, 0.7–1 cm, mouth oblique, teeth

triangular, 2–4 mm, adjacent, all hairless except for the teeth margins. Corolla pale yellow, 1.8–2 cm. Ovary downy, pod to 3 cm, hairless when mature. *E former USSR, adjacent China.* H3. Summer.

14. C. aurantiaca Koehne. Illustration: Bean, Manual of trees & shrubs hardy in the British Isles, edn 8, **1**: 493 (1970).

Very similar to *C. pygmaea* but flower-stalks completely hairless, calyx to 5 mm, corolla orange-yellow or pinkish-yellow, 1.5–2 cm, ovary completely hairless. *Afghanistan, adjacent former USSR.* H4. Summer.

69. CALOPHACA de Candolle
S.G. Knees
Deciduous shrubs or low perennial herbs with alternate leaves. Leaves pinnate with 3–13 pairs of leaflets and a terminal leaflet. Flowers solitary or in racemes; calyx tubular with 5 slender teeth; corolla violet or yellow; stamens 10, with 9 united and 1 free. Fruit a cylindric legume.

A genus of about 10 species, all native to Asia, often grown in mixed borders or on rocky slopes. They require good drainage and can be propagated from seed or by grafting onto *Laburnum* species.

1a. Leaves with 5–8 pairs of leaflets; flowers 4–9 in racemes 7–10 cm
    1. wolgarica
 b. Leaves with 8–12 pairs of leaflets; flowers up to 12 in racemes 13–20 cm **2. grandiflora**

1. C. wolgarica (Linnaeus filius) de Candolle. Illustration: Loudon, Arboretum & Fruticeum Britannicum **2**: 635 (1838); Schneider, Illustriertes Handbuch der Laubholzkunde **2**: 104 (1907); Bailey, Standard cyclopedia of horticulture, 636 (1914).

Shrub 90–120 cm. Leaves 5–8 cm, downy beneath; leaflets 0.6–1.2 cm, rounded–ovate, bristle-tipped, in 5–8 pairs. Flowers 4–9 in racemes 7–10 cm; calyx *c.* 8 mm, shaggy; corolla *c.* 2.5 cm, bright yellow. Pod to 3 cm, with 1–2 seeds. *S Former USSR (Turkestan).* H2. Summer.

2. C. grandiflora Regel. Illustration: Gartenflora, pl. 1231 (1886); Dippel, Handbuch der Laubholzkunde **3**: 717 (1893); Schneider, Illustriertes Handbuch der Laubholzkunde **2**: 104 (1907).

Like *C. wolgarica*, but less downy and larger in all its parts. Leaves 4–8 cm; leaflets 2–2.5 × 1.1–1.3 cm, ovate, pointed, in 8–12 pairs. Flowers up to 12 in racemes 13–20 cm; corolla 2.5–2.8 cm,

bright yellow. Pod oblong, with 1–2 seeds. *S Former USSR (Turkestan).* H2. Summer.

70. ASTRAGALUS Linnaeus
J. Cullen
Herbs or small shrubs. Leaves pinnate, either with a terminal leaflet or the axis continuing and persistent as a hardened spine; leaflets symmetric at base; stipules persistent, often conspicuous; stipels absent. Hairs, when present, basifixed or medifixed, grey, white or dark brown to black, colours often mixed. Flowers in racemes or spikes, axillary or arising directly from the basal rosette on a scape or scape absent. Bracts usually present. Calyx bell-shaped to tubular, 5-toothed, sometimes very deeply so. Corolla with wings and keel usually shorter than the variably-shaped standard. Keel sometimes toothed on the lower side. Stamens 10, the filaments of 9 united, that of the uppermost free. Pod variously shaped, usually several-seeded, divided along its length by a septum developed from an infolding of the lower suture.

One of the largest genera of flowering plants, with perhaps 2000 or more species, mainly from W & C Asia, but extending across the northern hemisphere and scattered in the southern hemisphere; in W & C Asia they are characteristic components of steppe vegetation. Fortunately for the present purpose, very few of them are grown as ornamentals. An elaborate hierarchy of subgenera and sections exists, but is of little help with the cultivated species, which are scattered through it; this hierarchy is ignored here, and the cultivated species fall into 3 groups: a) the annuals (no. **1**); b) the perennial, non-spiny species (nos. **2–6, 9 & 11**) and c) the spiny ('tragacanthoid') species (nos. **7, 8, 10, 12 & 13**).

Two characters require a little explanation. The hairs are either basifixed or medifixed, and this difference is important in distinguishing the species. It is sometimes difficult to decide whether hairs are of one type or the other; in general, however, under a magnification of 15 times or more, medifixed hairs can be rotated somewhat on their attachment when one end is gently pushed with a needle. Hairs attached closer to one end than the other occur in the genus as a whole, but are not important in the recognition of the cultivated species. The shape of the standard is sometimes important in the distinction of the spiny species; in most, the blade is truncate or hastate, abruptly contracted into a narrow claw, while in others the claw is broadened upwards, becoming almost as wide as the blade, from which it is demarcated by a quite small constriction; the terms 'stenonychioid' and 'platonychioid' are used to describe these conditions in some of the literature.

All the species are easily grown in a sunny site in well-drained soil, and are propagated by seed.

Literature: There is a very extensive literature on the genus, but this is of little help with the cultivated species. All the species included here are covered in either *Flora Europaea* **2**: 108–124 (1968) or Davis (ed), *Flora of Turkey* **3**: 49–254 (1970), which should be referred to for further information.

Habit. Annual herb: **1**; perennial herb: **2–6,9–11**; small shrub: **7,8,12,13**. Stemless, with scape: **4,11**; stemless without scape: **6**.

Leaves. With a terminal leaflet: **1–6,9,11**; with a terminal spine: **7,8,10,12,13**. Leaflets fewer than 10 pairs: **2–5,7,8, 10–13**; leaflets 10–20 pairs: **1–4,6, 11–13**; leaflets more than 20 pairs: **9**.

Hairs. All basifixed: **1–10**; all medifixed: **11–13**.

Calyx. Deeply divided (almost to the base), the teeth obscured by long, white hairs: **7,8**; with a distinct tube, bell-shaped or tubular, variously hairy but not obscured by long, white hairs: **1–6,9–13**. Usually less than 1 cm: **1–5,7,8,10,12,13**; usually more than 1 cm: **6,9–11**.

Corolla. Yellowish (sometimes buff or flushed with pink): **1,2,5–9,12**; whitish: **1,4,10–13**; bluish or purple: **3,4,10–12**.

1a. Leaves terminating in a leaflet, the axis not continuing as a persistent, hardened spine 2
　b. Leaves terminating in a persistent, hardened spine, without a terminal leaflet 9
2a. Plant annual **1. boeticus**
　b. Plant perennial 3
3a. Most hairs clearly and distinctly medifixed **11. monspessulanus**
　b. Most hairs clearly and distinctly basifixed 4
4a. Stems absent, the inflorescence-stalk either absent (the inflorescence borne between the basal leaves) or present 5
　b. Stems present, the inflorescences borne in the axils of the stem-leaves 6

5a. Leaves (including stalks and axes) with spreading, brown hairs; spikes stalkless or almost so, borne between the rosette leaves; calyx 1.2–1.5 cm; corolla yellow **6. exscapus**
　b. Leaves with adpressed, white hairs; spikes borne on a distinct, though sometimes short stalk; calyx 4–5 mm; corolla white or bluish purple **4. depressus**
6a. Corolla purplish or bluish **3. danicus**
　b. Corolla yellow or whitish 7
7a. Calyx 1.4–2 cm; leaves with 20 or more pairs of leaflets **9. centralalpinus**
　b. Calyx 0.5–1 cm; leaves with up to 15 pairs of leaflets 8
8a. Leaves with 3–7 pairs of leaflets; stipules free **5. glycyphyllos**
　b. Leaves with 8–15 pairs of leaflets; at least the stipules of the upper leaves united at the base **2. cicer**
9a. Most hairs on the plant medifixed 10
　b. Most hairs on the plant basifixed 11
10a. Calyx-teeth ¼–½ as long as tube; corolla pinkish, purple, yellow or white **12. angustifolius**
　b. Calyx-teeth up to ¼ as long as the tube; corolla always whitish **13. massiliense**
11a. Calyx with distinct and visible tube, hairy, but not as below; corolla white to purple, 1–2.2 cm **10. sempervirens**
　b. Calyx with tube very short, the teeth and tube plumed and obscured by dense, long, white hairs; corolla whitish, yellow, buff or pinkish, sometimes with purple veins 12
12a. Leaves densely greyish white-hairy; standard with blade truncate or hastate, abruptly contracted into the narrow claw **8. microcephalus**
　b. Leaves green, sparsely hairy or hairless; standard with claw broadened towards the top, demarcated from the blade by a slight constriction **7. gummifer**

1. A. boeticus Linnaeus. Illustration: Bonnier, Flore complète **3**: pl. 144 (1914).

Annual herb. Stems erect or sprawling, to 60 cm. Leaves with 10–15 pairs of leaflets, with a terminal leaflet; leaflets 0.8–1.8 cm, oblong or oblong-obovate,

tapered to the base, truncate or notched at the apex, rather sparsely adpressed hairy beneath with basifixed hairs. Stipules free, triangular. Flowers 5–15 in dense spikes. Calyx 5–8 mm, tubular, the teeth as long as the tube, covered in dark brown to black and white hairs. Corolla yellow or whitish, 1–1.4 cm. Pod 2–4 cm, oblong, triangular in section, grooved beneath, hooked at apex, shortly hairy. *Canary Islands, Mediterranean area east to Iran.* H1. Spring.

2. A. cicer Linnaeus. Illustration: Bonnier, Flore complète **3**: pl. 144 (1914); Polunin, Flowers of Europe, pl. 54 (1969).

Erect or spreading perennial to 50 cm. Leaves with 8–15 pairs of leaflets, with a terminal leaflet; leaflets 1.5–3.5 cm, lanceolate or ovate-lanceolate to oblong, tapered to base and apex, with short, adpressed, rather sparse basifixed hairs on both surfaces; stipules narrowly lanceolate, united at the base. Flowers 10–25 in dense spikes. Calyx 0.7–1 cm, tubular, teeth about half as long as tube, covered with adpressed, basifixed hairs, some white, most dark brown to black. Corolla yellow, 1.4–1.7 cm. Pods ovoid-spherical to spherical, inflated, hooked at apex, covered with long, somewhat spreading white hairs and shorter, adpressed, dark brown to black hairs. *Most of Europe, Turkey.* H1. Summer.

3. A. danicus Retzius (*A. hypoglottis* misapplied). Illustration: Keble Martin, The concise British flora in colour, pl. 24 (1965); Garrard & Streeter, The wild flowers of the British Isles, pl. 32 (1983).

Perennial herb, stems prostrate to ascending, to 30 cm. Leaves with 6–13 pairs of leaflets, with a terminal leaflet; leaflets 0.5–1.6 cm, oblong-elliptic to ovate-oblong, tapered to base and apex (rarely somewhat notched at apex), with rather sparse, long, white, basifixed hairs on both surfaces; stipules united round the stem for at least one-third of their length. Flowers numerous in dense, oblong to almost spherical spikes. Calyx tubular, 6–8 mm, teeth half or more as long as tube, covered in dense, adpressed, basifixed, dark brown hairs. Corolla purplish or bluish, 1.5–1.8 cm. Pod ovoid, inflated, with long, dense, spreading, somewhat swollen-based, white, basifixed hairs. *Most of Europe, W Asia.* H3. Summer.

4. A. depressus Linnaeus. Illustration: Bonnier, Flore complète **3**: pl. 144 (1914).

More or less stemless perennial with

woody base. Leaves in rosettes, with 6–14 pairs of leaflets, with a terminal leaflet; leaflets 5–9 mm, obovate, tapered to the base, truncate or notched at the apex, with adpressed, basifixed (rarely the hairs attached close to, but not quite at the end) hairs on at least the lower surface; stipules free. Spikes borne on a scape, with 7–30 flowers, cylindric or spherical. Calyx 4–5 mm, tubular, teeth short, with sparse, mostly dark brown hairs. Corolla white or bluish purple, 1–1.4 cm. Pod cylindric, slightly curved, hairless or with sparse white hairs. *C & S Europe, Cyprus, Turkey.* H3. Summer.

5. A. glycyphyllos Linnaeus. Illustration: Bonnier, Flore complète **3**: pl. 143 (1914); Ary & Gregory, The Oxford book of wild flowers, pl. 3 (1962); Keble Martin, The concise British flora in colour, pl. 24 (1969); Polunin, Flowers of Europe, pl. 54 (1969).

Perennial herb with trailing or ascending stems to 2 m. Leaves with 3–7 pairs of leaflets, with a terminal leaflet; leaflets ovate to elliptic or broadly elliptic, 2.5–4.5 cm, rounded to the base and the sometimes shortly mucronate apex, hairless above, sparsely adpressed hairy with white, basifixed hairs beneath; stipules free, triangular. Racemes (flowers slightly stalked) dense, cylindric, with 12–30 or more flowers. Calyx bell-shaped to tubular, 5–7 mm, hairless or with dark brown or black hairs on the teeth. Corolla pale yellow or cream, 1–1.7 cm. Pod cylindric, curved upwards, hairless or with white, adpressed hairs. *Eurasia.* H1. Summer.

A variant with adpressed, dark brown to black hairs on the calyx is known as var. **glycyphylloides** (de Candolle) Matthews (*A. glycyphylloides* de Candolle).

6. A. exscapus Linnaeus. Illustration: Bonnier, Flore complète **3**: pl. 143 (1914).

Stemless perennial with stout, woody stock. Leaves mostly densely hairy (including stalk and axis) with brownish, spreading, basifixed hairs; leaflets in 12–19 pairs, terminal leaflet present, 1–2.5 cm, elliptic-oblong, rounded at base and apex, upper surface sometimes sparsely hairy or almost hairless; stipules free, narrowly triangular. Spikes stalkless or almost so, borne between the rosette leaves, with 3–10 flowers. Calyx tubular, 1.2–1.5 cm, with dense, brown, spreading hairs. Corolla yellow, 2–3 cm. Pod oblong, with dense, brown, spreading hairs. *Spain to C & E Europe.* H3. Late spring–summer.

Part of a complex of species which are difficult to distinguish.

7. A. gummifer Labillardière.
Shrub to 30 cm. Hairs all basifixed. Leaflets 4–7 pairs, the leaf terminating in a hardened, persistent spine; leaflets elliptic, 0.5–1 cm, usually mucronate, sparsely hairy or hairless; stipules triangular ovate. Flowers in groups of 2 or 3 in the leaf-axils, forming ovoid or cylindric inflorescences. Calyx 5–7 mm, divided into teeth almost to the base, the whole covered with long, spreading, white hairs. Corolla buff or yellow (sometimes flushed with pink), 1–1.2 cm, persistent in fruit; standard with the claw broadened above, then slightly constricted into the blade. Pods 1- or 2-seeded. *Turkey, Lebanon.* H5. Summer.

A source of gum tragacanth.

8. A. microcephalus Willdenow.
Like *A. gummifer* but cushion-forming, leaves (including axis and spine) greyish white with dense, spreading, basifixed hairs; stipules narrowly lanceolate; corolla yellow with purple veins, standard with the blade truncate or hastate, abruptly contracted into the narrow claw. *Turkey, former USSR (Caucasus), Iran.* H5. Summer.

Belonging to a different section of the genus from *A. gummifer*, but superficially very similar.

9. A. centralalpinus Braun-Blanquet (*A. alopecuroides* misapplied). Illustration: Botanical Magazine, 3193 (1832); Bonnier, Flore complète **3**: pl. 144 (1914).

Perennial herb with stout, erect stems to 1 m, covered with spreading, basifixed hairs. Leaves with 20–30 pairs of leaflets, with a terminal leaflet; leaflets 1–3 cm, elliptic, ovate or lanceolate, rounded to the base, tapered to the apex, hairless above, densely spreading-hairy with basifixed hairs beneath; stipules conspicuous, lanceolate to narrowly triangular. Flowers numerous in stalkless, ovoid to cylindric spikes. Calyx 1.4–2 cm, cylindric, densely brown-hairy, the teeth as long as or slightly shorter than the tube. Corolla yellow, 1.5–2 cm. Pod ovoid, compressed, with long, spreading, brown hairs. *W Alps (France, Italy), Bulgaria.* H4. Summer.

For many years confused with **A. alopecuroides** Linnaeus from Spain and extreme SW France. Illustration: Bonnier, Flore complète **3**: pl. 144, 1914, – as A. narbonensis). Much less hairy, has spherical, somewhat stalked inflorescences, calyx-teeth longer than the tube and the corolla 2.2–2.7 cm.

10. A. sempervirens Lamarck (*A. aristatus* L'Héritier). Illustration: Loddiges' Botanical Cabinet **13**: t. 1278 (1827); Bonnier, Flore complète **3**: pl. 142 (1914).

Somewhat tufted perennial herb with woody stock; stems to 40 cm, trailing or ascending. Leaves with 4–10 pairs of leaflets, the axis persisting as a weak spine; leaflets narrowly oblong, narrowly oblanceolate or linear, covered, like the axis, in whitish, spreading, basifixed hairs. Stipules joined to the leaf-stalks for about half their length. Spikes with up to 10 flowers. Calyx 0.7–1.5 cm, slightly inflated, covered with dense, greyish, somewhat spreading hairs. Corolla white to purple, 1–2.2 cm. Pod ovoid, densely hairy. *S Europe, N Africa.* H4. Summer.

11. A. monspessulanus Linnaeus. Illustration: Botanical Magazine, 373 (1797); Bonnier, Flore complète **3**: pl. 143 (1914); Polunin & Smythies, Flowers of southwest Europe, pl. 19 (1972).

Erect, stemless perennial herb, scapes to 30 cm. Leaves with 7–20 pairs of leaflets, with a terminal leaflet; leaflets 0.5–1.5 cm, ovate or oblong to almost circular, rounded to the base and apex, hairless above, with sparse, whitish, medifixed hairs beneath. Spikes ovoid or oblong, often rather loose, with up to 30 flowers. Calyx 0.9–1.6 cm, covered with mostly brown, medifixed hairs. Corolla purplish (rarely whitish), 2–3 cm. Pod cylindric, sparsely hairy with medifixed hairs (hairs soon falling), somewhat curved. *S Europe, N Africa.* H5. Summer.

12. A. angustifolius Lamarck.
Dwarf, cushion-forming shrublet. Leaves with 5–12 pairs of leaflets, the axis continuing as a persistent spine which is sparsely hairy at first; leaflets 3–7 mm, narrowly elliptic to obovate, rounded to the sometimes mucronate apex, densely ashy grey-hairy on both surfaces with medifixed hairs; stipules lanceolate, united for about half their length. Flowers 3–14 in racemes. Calyx tubular, 0.6–1 cm, ribbed, with a mixture of white and dark brown to black medifixed hairs; teeth one quarter to half as long as tube. Corolla pinkish, purple, yellow or whitish, 1.5–1.8 cm. Pod oblong-cylindric, usually hairy. *Balkan Peninsula to Turkey.* H5. Summer.

13. A. massiliensis (Miller) Lamarck (*A. tragacantha* Linnaeus, in part). Illustration: Bonnier, Flore complète **3**: pl. 142 (1914); Polunin & Smythies, Flowers of southwest Europe, pl. 19 (1973).

Rather open shrub to 30 cm. Leaves with 6–12 pairs of leaflets, the axis continuing as a stout spine; leaflets 4–6 mm, oblong to elliptic, rounded or truncate to the obtuse apex, densely silvery-silky above and beneath with medifixed hairs. Racemes with 3–8 flowers. Calyx 5–7 mm, tubular, the teeth to one-quarter as long as the tube, covered with dark brown to black and white, medifixed hairs. Corolla whitish, 1.3–1.7 cm. Pod oblong, densely hairy. *SW Europe*. H5. Summer.

71. OXYTROPIS de Candolle
J.R. Akeroyd

Herbaceous perennials or subshrubs. Leaves pinnate; leaflets entire. Flowers in axillary spikes or racemes, similar to those of *Astragalus* but the keel with an acute beak at the apex.

A genus of about 100 species from the arctic and mountainous regions of Eurasia and North America. Propagation is by seed or division in the spring. The species in cultivation in Europe are compact perennials, suitable for the rock garden or alpine glasshouse; most of them prefer calcareous soils.

1a. Plant glandular and foetid
    3. foetida
 b. Plant neither glandular nor fetid   2
1a. Flowering stems leafy   3
 b. Leaves all basal   5
3a. Leaflets obtuse   **5. jacquinii**
 b. Leaflets acute   4
4a. Corolla bluish   **4. lapponica**
 b. Corolla pale yellow   **6. pilosa**
5a. Corolla yellow to cream, sometimes tinged violet   6
 b. Corolla blue, purple, pink or violet   7
6a. Raceme ovoid; calyx-hairs whitish   **1. campestris**
 b. Raceme almost spherical; calyx-hairs blackish   **8. ochroleuca**
7a. Leaflets not distant or in whorls   8
 b. Leaflets distant or in whorls   9
8a. Leaves silky-hairy, usually with fewer than 25 leaflets   **2. halleri**
 b. Leaves downy, with at least 25 leaflets   **7. pyrenaica**
9a. Leaflets not in whorls; corolla *c.* 2 cm, distinctly longer than calyx   **9. lambertii**
 b. Leaflets in whorls of 3–4; corolla 1.2-1.5 cm, slightly longer than calyx   **10. splendens**

1. O. campestris (Linnaeus) de Candolle. Illustration: Keble Martin, Concise British flora in colour, pl. 23 (1967); Barneby, European alpine flowers in colour, pl. 42 (1967); Macdonald encyclopedia of alpine flowers, pl. 173 (1985); Blamey & Grey-Wilson, Illustrated flora of Britain and northern Europe, 207 (1989).

Tufted, downy perennial, with stout rootstock. Leaves all basal; leaflets 17–25, elliptic or lanceolate, acute, hairy. Flowering-stems 8–20 cm, about as long as leaves, hairy. Raceme ovoid, dense. Corolla 1.5–2 cm, pale yellow or cream, the wings and keel often tinged violet. Fruit 1.4–1.8 cm, ovoid-oblong, hairy. *Mts of Europe*. H1. Summer.

2. O. halleri Koch (*O. sericea* (de Candolle) Simonkai; *O. uralensis* misapplied). Illustration: Keble Martin, Concise British flora in colour, pl. 23 (1965); Blamey & Grey-Wilson, Illustrated flora of Britain and northern Europe, 207 (1989).

Similar to *O. campestris*, but silky-hairy; leaflets 11–28; flowering-stems longer than leaves; corolla 1.5–2 cm, blue to purple; fruit 1.5–2 cm, ovoid, more densely hairy. *Mts of Europe*. H1. Summer.

3. O. foetida (Villars) de Candolle.
Similar to *O. campestris*, but glandular-hairy and foetid; leaflets 21–51, linear-lanceolate to narrowly oblong, the margins down-curled; flowering-stems longer than leaves; corolla *c.* 2 cm, pale yellow; fruit narrower. *W Alps*. H1. Summer.

4. O. lapponica (Wahlenberg) Gay. Illustration: Polunin & Stainton, Flowers of the Himalaya, pl. 30 (1984); Polunin, Collins photoguide to wild flowers, 274 (1988); Blamey & Grey-Wilson, Illustrated flora of Britain and northern Europe, 207 (1989).

Loosely tufted perennial. Leaflets 17–29, lanceolate to oblong-lanceolate, acute, adpressed-hairy. Flowering-stems 5–25 cm, decumbent to ascending, leafy, hairy. Racemes almost spherical, rather loose. Corolla 0.8–1.2 cm, bluish violet. Fruit 0.8–1.5 cm, narrowly ellipsoid, pendent, densely appressed-hairy with dark hairs. *Mts of Eurasia, Scandinavia*. H1. Summer.

Grows best on lime-poor soils.

5. O. jacquinii Bunge (*O. montana* misapplied).
Similar to *O. lapponica*, but leaflets 25–41, lanceolate to narrowly ovate, obtuse, sparsely hairy; corolla violet to purple; fruit ovoid, somewhat swollen, not pendent. *Alps, French Jura*. H1. Summer.

6. O. pilosa (Linnaeus) de Candolle (*Astragalus pilosus* Linnaeus). Illustration: Barneby, European alpine flowers in colour, pl. 42 (1967); Blamey & Grey-Wilson, Illustrated flora of Britain and northern Europe, 207 (1989).

Perennial with long silvery hairs. Leaflets 19–27, oblong to linear-oblong, acute. Flowering-stems 15–40 cm, erect, leafy. Racemes ovate-oblong, compact. Corolla 1.2–1.4 cm, pale yellow. Fruit 1.5–2 cm, ovoid to cylindric, densely hairy. *Mts of Europe & W Asia*. H1. Summer.

7. O. pyrenaica Godron & Grenier (*O. montana* subsp. *samnitica* (Arcangeli) Hayek). Illustration: Polunin & Smythies, Flowers of south-west Europe, 148 (1973).

Tufted, downy perennial. Leaflets 25–41, lanceolate to narrowly elliptical or oblong, acute, silky-hairy. Flowering-stems 5–20 cm. Raceme compact, but extending in fruit. Corolla bluish-violet or purplish. Fruit 1.5–2 cm, oblong to narrowly ovoid. *Pyrenees, Alps & mts of N part of Balkan Peninsula*. H2. Summer.

8. O. ochroleuca Bunge.
Leaflets 21–31, oblong-lanceolate or elliptic. Flowering-stems 5–25 cm, ascending, longer than leaves. Raceme almost spherical, dense, many-flowered. Corolla small, pale yellowish. Calyx with blackish hairs. Fruit oblong-ovoid, pendent, finely hairy. *Mts of C Asia*. H1. Summer.

9. O. lambertii Pursh. Illustration: Botanical Magazine, 2147 (1820); Edwards's Botanical Register, 1054 (1827).

Tufted, silky-hairy perennial. Leaflets 11–23, 1–4 cm, rather distant, linear-lanceolate, acute. Flowering-stems 15–30 cm. Racemes to 15 cm. Corolla *c.* 2 cm, pinkish purple. Fruit 2–3 cm, distinctly longer than calyx, cylindric, shortly hairy. *N America*. H1. Summer.

10. O. splendens Douglas.
Tufted, densely hairy perennial. Leaflets in whorls of 3 or 4, lanceolate, acute, silky-hairy. Flowering-stems 10–30 cm, longer than leaves, with spreading hairs. Raceme up to 15 cm, loose. Calyx white-hairy. Corolla 1.2–1.5 cm, blue to purplish. Fruit *c.* 1 cm, slightly longer than calyx, ovoid, hairy. *NW North America*. H1. Summer.

72. ALHAGI Gagnepain
J. Cullen

Bushy, somewhat woody perennial herbs with rigid spines on the lower branches and with the upper, short, lateral flowering branches terminating in spines. Leaves simple, stalkless; stipules minute. Flowers borne singly or in pairs in the axils of minute bracts; flower-stalks very short or absent. Calyx bell-shaped, shallowly 5-toothed. Corolla pink to red, standard and keel longer than the wings, the keel obtuse. Stamens 10, 9 of them with their filaments united, that of the uppermost free. Ovary containing several ovules. Pod indehiscent, constricted between the 1–5 seeds.

A genus of 3 or more species from arid habitats from Europe and North Africa to Central Asia, occasionally grown as curiosities. They require winter-protection in most of Europe, a well-drained soil and a sunny site. Propagation is by seed or cuttings.

Literature: Keller, B.A. & Shaparenko, K.K., Materiali k sistematiko-ekologicheskogo monografii roda Alhagi Tournefort ex Adanson, *Sovetskaya Botanika* **3–4**: 151–185 (1933).

1a. Plant hairless; calyx-teeth with very shallowly U-shaped sinuses between them   **1. pseudalhagi**
 b. Plant hairy (stems, calyx, ovary, pods); calyx-teeth with rather deeper, V-shaped sinuses between them   **2. maurorum**

1. A. pseudalhagi (Bieberstein) Desvaux (*A. camelorum* Fischer). Illustration: Flora SSSR **13**: 371 (1952).

To 1 m, hairless; branches mostly spreading at an acute angle. Leaves 1–2 cm × 3–4 mm, lanceolate to oblanceolate. Calyx *c.* 2 mm, shallowly toothed, the sinuses between the teeth shallowly U-shaped. Corolla pink, 0.7–1 cm. Pod 0.8–3 cm × 2–3 mm, hairless, dark brown, strongly contracted between the seeds. *E Europe to C Asia.* H4. Summer.

2. A. maurorum Medikus (*A. mannifera* Desvaux).

Very similar to *A. pseudalhagi*, but hairy (stems, calyx, ovary, pods), branching at wider angles; calyx-teeth with larger, V-shaped sinuses between them. *E Mediterranean area, Arabia to E Asia.* H4.

73. GALEGA Linnaeus
J.R. Akeroyd
Erect herbaceous perennials with a bushy habit. Leaves irregularly pinnate; leaflets obtuse or acute. Stipules conspicuous. Flowers many in stalked racemes longer than leaves. Stamens 10, 9 of them with filaments united. Fruit cylindric, slender, constricted between seeds, beaked; seeds many.

A genus of 6 species in Europe, SW Asia and East Africa. Propagation is by seed or division of plants during winter. The two species grown in European gardens are also grown as fodder crops and are locally naturalised in W & N Europe.

1a. Stipules arrow-shaped; fruits spreading to somewhat erect
          1. officinalis
 b. Stipules ovate; fruits deflexed
          2. orientalis

1. G. officinalis Linnaeus (*G. bicolor* Regel, *G. persica* Persoon, *G. tricolor* Hooker). Illustration: Hay & Synge, Dictionary of garden plants in colour, t. 1140, 1141 (1969); Polunin, Flowers of Europe, pl. 53 (1969).

Plant 30–160 cm, hairless to sparsely hairy. Leaflets 9–17, 1.5–5 × 0.4–1.5 cm, elliptic to lanceolate, mucronate. Flowers 30–50 in racemes. Calyx-teeth about as long as tube. Corolla 1–1.5 cm, lilac, purple or white. Fruit 2–5 cm, spreading to somewhat erect; seeds 2–10. *C & S Europe to W Pakistan.* H1. Summer–autumn.

2. G. orientalis Lamarck. Illustration: Brickell, RHS gardeners' encyclopedia of plants and flowers, 211 (1989).

Similar to *G. officinalis* but leaflets 3–6 × 1–2.5 cm, acuminate; stipules ovate to broadly ovate; calyx hairy, the teeth shorter than tube; corolla bluish violet; fruit deflexed. *Caucasus.* H1. Summer.

A variable species, especially in flower colour, and numerous varieties and cultivars have been recognised. Several of these are derived from the hybrid

G. × hartlandii Clarke (*G. officinalis* × *G. orientalis*). Illustration: Brickell, RHS gardeners' encyclopedia of plants and flowers, 190 (1989).

74. GLYCYRRHIZA Linnaeus
J. Cullen
Glandular, rhizomatous perennial herbs. Leaves pinnate, with a terminal leaflet, the leaflets dotted with orange-brown glands; stipules lanceolate, minute; stipels absent. Flowers numerous in axillary racemes or spikes which are loose or dense and head-like. Calyx 5-toothed, 2-lipped, the tube narrowly funnel-shaped, the upper teeth short, the lower as long as or longer than the tube. Corolla white, yellow, mauve, violet or bluish; keel obtuse or acute. Stamens 10, 9 of them with their filaments united, that of the uppermost free. Ovary with few ovules. Pod compressed, dehiscent with the sides contorting, seeds 1–several.

A genus of about 20 species from Eurasia, North America and temperate South America. The 2 species generally grown are both European and require a rich, deep soil. Propagation is by division.

1a. Racemes or spikes congested, head-like, 2–5 cm; corolla 5–7 mm; pod covered with spine-like bristles
          2. echinata
 b. Racemes or spikes loose, not head-like, 5 cm or more; corolla 0.9–1.8 cm; pod not covered with bristles
          1. glabra

1. G. glabra Linnaeus. Illustration: Bonnier, Flore complète **3**: t. 148 (1914); Masefield et al., The Oxford book of food plants, 199 (1969).

Perennial to 60 cm; stems sparsely hairy, glandular above. Leaves with 5–9 pairs of leaflets which are ovate to elliptic, rounded to the base, pointed or rounded at the apex. Racemes loose, 5 cm or more; flower-stalks short. Calyx-teeth to 6 mm. Corolla 0.9–1.5 cm, blue or violet. Pod oblong, 1.5–2.5 cm × 4–5 mm, red-brown, with thickened sutures, laterally compressed, with 1–6 seeds, hairless or hairy but without bristles. *S Europe, N Africa, SW Asia.* H3. Summer.

The species is extensively grown as a crop in some areas for its rhizomes which yield liquorice.

2. G. echinata Linnaeus. Illustration: Botanical Magazine, 2154 (1820); Jordanov, Flora Reipublicae popularis Bulgaricae **6**: 183 (1976).

Erect perennial to 1 m or more; stems sparsely hairy above. Leaves with 4–8 pairs of leaflets which are narrowly elliptic, acute at both ends, 1–3 × 0.5–1.5 cm. Racemes or spikes dense, head-like, 2–5 cm; flower-stalks not visible. Calyx *c.* 2 mm. Corolla whitish to violet, 5–7 mm. Pod obovoid, flattened, reddish brown, 1–1.6 × 0.6–0.8 cm, covered in glands and spine-like bristles, 1–3-seeded. *E Mediterranean area, SW Asia.* H5.

75. CHORDOSPARTIUM Cheeseman
J. Cullen
Shrub to 8 m, often with a trunk to 3 cm across in the wild. Branches drooping, leafless except for small, adpressed, triangular, brownish scales at the nodes, terete, grooved. Flowers usually numerous in cylindric racemes, borne singly or in groups of 2–5 at the nodes; inflorescence-stalk hairy, flower-stalks woolly. Flowers to 1 cm, pale lavender or whitish, the standard with purple veins. Calyx 3–4 mm, cup-shaped, hairy, with 5 minute teeth. Stamens 10, 9 of them with their filaments

united, that of the uppermost free. Ovary hairy, containing 1–5 ovules; styles long, incurved, hairy on the upper side towards the apex. Pods hairy, indehiscent, swollen, *c.* 5 mm, usually 1-seeded.

A genus of a single species from New Zealand, notable for its drooping, leafless branches. It is relatively easily grown in a warm site and is propagated by seed.

1. C. stevensonii Cheeseman. Illustration: Botanical Magazine, 9654 (1943); Metcalf, The cultivation of New Zealand trees and shrubs, pl. 4 (1972); Moore & Irwin, The Oxford book of New Zealand plants, 165 (1978); Salmon, The native trees of New Zealand, 202, 203 (1980).

New Zealand (South Island). H5–G1. Early summer.

Very threatened in its native habitat, but established in cultivation.

76. CARMICHAELIA R. Brown
J. Cullen

Shrubs or small trees; habit diverse, ranging from small, compact, patch-forming plants to larger, erect shrubs or trees. Leaves pinnate with 3–7, often notched leaflets, usually absent from mature plants, their axils marked by small notches in the flattened or terete branchlets. Flowers solitary or in racemes borne in the notches, sometimes more than 1 raceme from each notch; flower-stalks hairy or finely downy. Calyx bell-shaped, 5-toothed. Petals distinctly clawed. Stamens 10, the filaments of 9 united, that of the uppermost free; anthers each with only a single pollen-sac. Pod dry with thickened margins which may project into the cell, hard, indehiscent or dehiscent only at the apex or by one or both sides separating from the persistent, thickened margins. Seeds 1–few.

A genus of 39 species from New Zealand and Lord Howe Island. A considerable number of names can be found in seed catalogues and plant lists, but it is uncertain how many are in general cultivation in Europe. Most of the species are leafless when mature and have a broom-like appearance. Their classification is very difficult and is based largely on the pods and the seeds, which may not be produced in cultivation. The account below covers most of the species likely to be found in gardens, and is based largely on Allan's *Flora of New Zealand* **1**: 373–397 (1961). The key given here avoids the use of pod and seed characters, and should therefore be used with caution.

Most of the species are doubtfully hardy in Europe; they are normally propagated by seed.

Literature: Simpson, G., A revision of the genus *Carmichaelia*, *Transactions of the Royal Society of New Zealand* **75**: 231–287 (1945); Slade, B.F., Cladode anatomy and leaf trace systems in New Zealand brooms, ibid., **80**: 81–96 (1952).

1a. Mature plant with leaves present during spring and summer **2**
  b. Mature plant leafless or almost so **3**
2a. Erect or spreading shrub; racemes tight, with 10–40 flowers **1. angustata**
  b. Scrambling shrub; racemes loose, with 2–6 flowers **6. kirkii**
3a. Erect or spreading shrubs or trees, 50 cm or more in height **4**
  b. Prostrate, sprawling or patch-forming plants, rarely higher than 15 cm **7**
4a. Branchlets 1–2 mm wide **5**
  b. Branchlets 3 mm or more wide **6**
5a. Branchlets very compressed; racemes with a conspicuous stalk to 1 cm **4. arborea**
  b. Branchlets scarcely compressed; racemes stalkless or stalk very short **3. flagelliformis**
6a. Branchlets compressed, 0.8–1.2 cm wide; flowers 2–3 cm **2. williamsii**
  b. Branchlets terete, 3–4 mm across; flowers at most 7 mm **5. petriei**
7a. Branchlets terete, not compressed **10. curta**
  b. Branchlets conspicuously compressed **8**
8a. Branches arising from rhizomes; flowers solitary or in pairs at each notch **8. uniflora**
  b. Branches arising from a taproot; flowers mostly in racemes of 3 or more arising from each notch **9**
9a. Branches and calyx hairless **7. enysii**
  b. Branches and calyx hairy **10**
10a. Flowers *c.* 3 × 2 mm **11. nigrans**
  b. Flowers *c.* 10 × 5–7 mm **9. astonii**

1. C. angustata Kirk.

Erect or spreading shrub to 2 m; branches hairless or almost so, slightly compressed, to 2 mm across, grooved, borne at wide angles and drooping. Leaves present during spring and summer, pinnate with 3–7 leaflets. Racemes usually solitary, tight, with 10–40 flowers; inflorescence-

stalk hairy, *c.* 2 cm. Flowers *c.* 4 × 3 mm. Calyx *c.* 2 × 2 mm, hairy, teeth very short. Petals whitish, veins purple, standard with a purple blotch at the base, or all flushed purplish. Pods brownish, 7–8 × 3–4 mm, ovate-oblong, opening at the apex only; beak *c.* 2 mm. Seeds 2–4, pale to dark brown. *New Zealand (South Island).* H5–G1.

C. odorata J.D. Hooker. Illustration: Botanical Magazine, 9479 (1937). Similar, but racemes with not more than 12 flowers, calyx shorter, branches usually hairy, pod broader. *New Zealand (North Island).* G1.

2. C. williamsii Kirk. Illustration: Botanical Magazine, n.s., 70 (1949); Moore & Irwin, The Oxford book of New Zealand plants, 165 (1978); Salmon, The native trees of New Zealand, 206 (1980).

Shrub to 4 m. Branchlets very compressed, 0.8–1.2 cm wide, grooved and hairless. Leaves present only in young plants. Racemes 1 or 2 per notch, each with 1–5 flowers, inflorescence-stalks hairy, *c.* 1 cm. Flowers 2–3 cm. Calyx 5–6 × 4–5 mm, teeth narrowly triangular. Corolla yellowish, veins purple. Pod 2–3 × 0.6–0.7 cm, opening along its whole length, elliptic-oblong, dark brown to black. Seeds 6–15, red or red mottled with black. *New Zealand (North Island).* H5–G1.

C. aligera Simpson (*C. australis* misapplied). Illustration: Allan, Flora of New Zealand **1**: 374 (1961); Salmon, Native trees of New Zealand, 206, 207 (1980). Similar, but racemes with 8–12 flowers; flowers much smaller (*c.* 4 × 3 mm, calyx 1.5–2 × 1.5–2 mm), pods 0.8–1 cm × 4–5 mm, black; seeds orange-red mottled with black. *New Zealand (North Island).* H5–G1.

3. C. flagelliformis J.D. Hooker. Illustration: Allan, Flora of New Zealand **1**: 374 (1961).

Shrub to 3 m. Branchlets somewhat compressed, grooved, *c.* 1 mm across. Leaves borne only on young plants. Racemes 2–4 per notch, each with 3–7 flowers disposed in an almost umbellate manner; inflorescence-stalk very short or almost absent. Flowers *c.* 4 × 3 mm. Calyx *c.* 1 mm, with minute teeth. Corolla whitish, veined and flushed with purple. Pods compressed, ovoid, opening along their whole length, dark brown to black, with straight, stout beaks. Seeds 1 or 2, red mottled with black. *New Zealand (North Island).* H5–G1.

C. arenaria Simpson. Similar, but a shrub to 50 cm, racemes with short stalks, flowers slightly larger, pods longer, with 2–4 seeds. *New Zealand (South Island)*. H5–G1.

4. C. arborea (Forster) Druce (*C. australis* misapplied). Illustration: Salmon, The native trees of New Zealand, 206 (1980).

Shrub or small tree to 5 m. Branchlets compressed, hairless, mostly 1–2 mm wide. Leaves present only in young plants. Racemes 1–3 to each notch, each with 3–5 flowers disposed in an almost umbellate manner; inflorescence-stalk to 1 cm. Flowers *c.* 5 × 4 mm. Calyx *c.* 2 × 2 mm, with minute teeth. Corolla whitish, the standard with purple centre and veins, the keel greenish at the base. Pods oblong, compressed, 0.8–1 cm × 4–5 mm, opening along their whole length, dark brown to black; beak sharp, oblique. Seeds 2–4, pale yellow-green mottled with black. *New Zealand (South Island)*. H5–G1.

C. ovata Simpson. Similar but pods shorter, dark brown, the beak also shorter. *New Zealand (South Island)*. H5–G1.

C. violacea Kirk. Also similar, but a smaller shrub (to 1 m), flowers in racemes of 3–8, pods slightly hairy, smaller. *New Zealand (South Island)*. H5–G1.

5. C. petriei Kirk.

Shrub to 2 m. Branchlets terete, 3–4 mm across. Leaves borne usually only on young plants. Racemes 1–3 per notch, each with 3–8 flowers. Flowers *c.* 6 × 5 mm. Calyx *c.* 3 × 2 mm, with minute, acute teeth. Corolla greenish white flushed with purple and with purple veins. Pods oblong, 0.8–1 cm × *c.* 4 mm, turgid, dark brown. Seeds greenish yellow mottled with black. *New Zealand (South Island)*. H5–G1.

6. C. kirkii J.D. Hooker. Illustration: Allan, Flora of New Zealand **1**: 374 (1961); Moore & Irwin, The Oxford book of New Zealand plants, 165 (1978).

Woody scrambler to 2 m or more. Branchlets more or less terete, striped, silky-hairy or hairless, *c.* 1 mm across. Leaves borne on adult plants during spring and summer, pinnate with 3–5 leaflets; leaflets hairless, stalk and axis silky hairy. Racemes with 2–6 flowers. Flowers 0.6–1 × 0.4–0.8 cm. Calyx *c.* 4 × 2 mm, with narrowly triangular teeth. Corolla white to cream with purple veins and blotches. Pods 0.8–1.5 × 0.4–0.6 cm, turgid, ellipsoid, opening from the base to the apex. Seeds 2–4, white, sometimes tinged with pale blue or mottled with black. *New Zealand (South Island)*. H5–G1.

7. C. enysii Kirk. Illustration: Allan, Flora of New Zealand **1**: 374 (1961); Salmon,

Collins' guide to the alpine plants of New Zealand, 165 (1985).

Dwarf shrub to 5 cm, forming dense patches to 10 cm across, arising from a stout taproot. Adult plants leafless. Branchlets compressed, 1–2 mm wide. Racemes 1 or 2 from each notch, each with 1–3 flowers. Flowers *c.* 5 × 4 mm. Calyx *c.* 1 mm with short acute or blunt teeth. Corolla greenish to purplish, veins dark purple. Pods 6–8 × 4–5 mm, pale, more or less circular in profile, one side falling completely, the other persistent. Seeds 1 or rarely more, dull black. *New Zealand (South Island)*. H5–G1.

C. orbiculata Colenso (*C. enysii* var. *orbiculata* (Colenso) Kirk). Very similar, but branchlets 2–3 mm wide, pods shorter, seeds dull olive-green mottled with black. *New Zealand (North Island)*. H5–G1.

8. C. uniflora Kirk.

Small shrub with underground runners (rhizomes), forming compact or open patches, to 6 cm. Branchlets compressed, 1–1.5 mm across. Leaves absent in mature plants. Flowers 0.6–1 × 0.4–0.6 cm, solitary or in pairs in the notches. Calyx *c.* 3 × 2 mm, with short, acute teeth. Corolla whitish, standard purplish towards the base, and with purple veins; keel purple. Pods elliptic-oblong, *c.* 1 × 0.4 cm, somewhat compressed, dark brown to black. Seeds 4–6, bluish black. *New Zealand (South Island)*. H5–G1.

9. C. astonii Simpson. Illustration: Allan, Flora of New Zealand **1**: 374 (1961).

Dwarf shrub arising from a woody root, forming hard patches to 15 cm, and 20 cm across. Branchlets very compressed, grooved, hairy, 4–8 mm wide. Leaves absent from adult plants. Racemes with 3–7 flowers. Flowers *c.* 1 cm × 5–7 mm. Calyx *c.* 5 × 3 mm, hairy with teeth 2–3 mm. Corolla whitish, veined with purple, standard with a purple basal blotch. Pods oblong, indehiscent, 1.2 × 0.4–0.7 mm, dark brown to black, beak to 3 mm, oblique. Seeds 4–8, pale green with sparse black spots, ultimately brownish. *New Zealand (South Island)*. H5–G1.

C. monroi J.D. Hooker. Illustration: Allan, Flora of New Zealand **1**: 374 (1961); Abrams, New Zealand alpine plants, pl. 26 (1973). Similar but branchlets 2–4 mm wide, pods *c.* 1.5 cm × 4 mm, hooked; seeds 8–12. *New Zealand (South Island)*. H5–G1.

10. C. curta Petrie.

Sprawling shrub, rather little branched,

branches to 1 m, terete, 1–2 mm across, grooved, hairy at least when young. Racemes tight with 8–10 flowers. Flowers *c.* 4 × 3 mm. Calyx *c.* 1 × 1 mm, hairy, with minute teeth. Corolla creamy yellow striped with purple; standard with 2 bands of purple along the blade. Ovary hairy. Pod hairless when mature, 3–5 × 2–3 mm, turgid, brown; beak *c.* 1 mm, curved. Seeds 1–3, usually 2, pale green mottled with black. *New Zealand (South Island)*. H5–G1.

11. C. nigrans Simpson. Illustration: Allan, Flora of New Zealand **1**: 374 (1961).

More or less prostrate shrub, branches to 1 m. Branchlets ascending, compressed, *c.* 1 cm wide, sparsely hairy. Racemes 1–3 at each notch, each with 5–10 flowers. Flowers *c.* 3 × 2 mm. Calyx short, sparsely hairy, with very short teeth. Corolla white veined with purple, standard with a purple blotch at the base. Pods *c.* 4 × 1.5 mm, oblong, somewhat compressed, black; beak curved, oblique. Seeds 2–4, pale brown. *New Zealand (South Island)*. H5–G1.

77. HEDYSARUM Linnaeus
S.G. Knees & M.C. Warwick

Annual or perennial herbs or low shrubs. Stems several, grooved, often covered with adpressed hairs. Leaves alternate, pinnate with a terminal leaflet, the leaflets almost stalkless; stipules united or free, papery. Flowers erect or nodding, borne in stalked axillary racemes; calyx with 5 teeth, corolla with 5 petals, standard longer than keel, pink, purple, yellow or white, ovary with 4–8 ovules. Fruit a stalked segmented pod, with or without wings; each segment containing 1 seed, segments flattened or rounded; seeds 2–6.

A genus of about 100 species from much of the north temperate world. Most species are best grown from seed, but some can be propagated by cuttings. The shrubby species can be cut back to near ground level each spring to encourage fresh vigorous growth.

Literature: Rollins, R., The genus *Hedysarum* in North America, *Rhodora* **42**: 217–239 (1940).

Flowers. Yellowish or creamy white: **4**; red or purple: **1–3,5–8**.
Pods. spiny: **1,6**; smooth **2–5,7–8**.
Stipules. free: **1**; joined **2–8**.

1a.	Stipules free	**1. coronarium**
b.	Stipules joined	2
2a.	Leaflets 21–27	**6. multijugum**
b.	Leaflets usually less than 21	3

3a. Flowers creamy white

  4. boutingyanum

  b. Flowers pink, red or purplish   4

4a. Pods 0.7–1.2 cm wide

  7. occidentale

  b. Pods less than 0.7 cm wide   5

5a. Stems 1–1.5 m   **5. microcalyx**

  b. Stems 10–70 cm   6

6a. Leaflets narrowly elliptic to
    lanceolate, not more than 5 mm
    wide   **8. boreale**

  b. Leaflets broadly ovate, mostly *c.* 1
    cm or more wide   7

7a. Bracts 6–15 mm   **2. hedysaroides**

  b. Bracts 1–4 mm   **3. alpinum**

1. H. coronarium Linnaeus. Illustration:
Flore de Serres series 2 **3**: 1382 (1858);
Robinson, The English flower garden, pl.
128 (1883); Reichenbach, Icones Florae
Germanicae et Helveticae **22**: 195 (1903);
Bonnier, Flore complète **3**: pl. 164 (1914).

Perennial or biennial to 50 cm; stipules
free. Leaflets 7–15, elliptic, 3–4 cm.
Flowers in crowded racemes of 10–35,
fragrant, on long stalks; calyx-teeth as
long as tube, sparsely hairy; corolla *c.* 1.8
cm, deep red. Pod minutely roughened,
becoming spiny; seeds 2–4. *Europe.* H3.
Summer.

'Album'. Illustration: Waldstein &
Kitaibel, Plantarum Rariorum Hungaricae,
2: 3 (1805). Flowers white, sometimes
cultivated.

2. H. hedysaroides (Linnaeus) Schinz &
Thellung (*Astragalus hedysaroides* Linnaeus;
H. obscurum Linnaeus; *H. caucasicum*
Bieberstein). Illustration: Bonnier, Flore
complète **3**: pl. 164 (1914); Jávorka &
Csapody, Iconographia florae Hungaricae,
291 (1932); Hess, Landholt & Hizel, Flora
der Schweiz **2**: 572 (1970); Jelitto &
Schacht, Hardy herbaceous perennials **1**:
277 (1990).

Perennial with mostly unbranched stems
10–60 cm, hairless or with few hairs;
stipules 1–2 cm, joined, brown. Leaflets
7–21, ovate to elliptic, 1–2.5 cm. Racemes
3–8 cm, with 15–20 flowers; bracts
0.6–1.5 cm; calyx-teeth 1.5–3 mm,
triangular; corolla 1.3–2.5 cm, reddish
violet or white. Pod 3–5 mm wide, with
2–5 seeds. *Arctic Russia to SC Europe.* H1.
Summer.

3. H. alpinum Linnaeus (*H. sibiricum*
Ledebour). Illustration: Botanical
Magazine, 3316 (1821); Edwards's
Botanical Register **10**: 808 (1824);
Hitchcock & Cronquist, Flora of the Pacific
Northwest, 261 (1973); Budd, Flora of the

Canadian prairie provinces, 478 (1987).

Multi-stemmed herbaceous perennial,
stems 20–70 cm, grooved; stipules 1.5–2
cm, joined, brown; leaves with 17–21
leaflets, narrowly elliptic, 2.5–3.5 cm.
Bracts 1–4 mm; calyx-teeth almost equal,
narrowly triangular; corolla 1.1–1.5 cm,
red or purplish carmine. Pod 3.5–6 mm
across, wings *c.* 0.5 mm; seeds 1–5.
Circumpolar. H5. Spring–summer.

Var. **grandiflorum** Rollins is very
similar but lower growing with larger
flowers, 1.4–1.8 cm.

4. H. boutignyanum Alleizette.
Illustration: Bulletin de la Société
Botanique de France **75**: 41 (1928).

Hairless perennial, 50–60 cm. Leaflets
7–11, ovate to elliptic, 2–3 cm; stipules
2–3 cm, joined. Racemes 5–20 cm, bracts
1–1.5 cm; calyx-teeth shorter than tube;
corolla 1.2–1.6 cm, cream or white,
sometimes with bluish veins. Pod *c.* 5 mm
across, with 2–5 seeds, margins hairy.
France (SW Alps). H3. Summer

5. H. microcalyx Baker. Illustration:
Botanical Magazine, 6931 (1887).

Almost hairless perennial with slender
stems 1–1.5 m upper leaves whorled;
stipules 1.5–4 cm, elliptic, conspicuous.
Leaflets 3–4 cm, hairless. Corolla 1.6–1.9
cm, bluish purple, wings shorter than keel.
Pod 5–7 mm across, margins narrowly
winged, seeds 2 or 3. *Himalaya (Kashmir to
Uttar Pradesh).* H4. Summer–autumn.

6. H. multijugum Maximowicz.
Illustration: Dippel, Handbuch der
Laubholzkunde **3**: 718 (1893); Schneider,
Illustriertes handbuch der Laubholzkunde
2: 108 (1907); Bean, Trees & shrubs hardy
in the British Isles, edn 7, **2**: 93 (1951);
Hegi, Illustrierte Flora von Mitteleuropa **4**:
1484 (1975).

Woody based perennial with stems to
1.5 m. Leaflets 21–27, ovate, 1–1.2 cm;
stipules 2–3 mm. Calyx 4–6 mm; corolla
1.9–2.3 cm, crimson-purple. Pods *c.* 5 mm
across, with conspicuous spines, seeds 2–4.
Mongolia. H2. Summer–autumn.

Var. **apiculatum** Sprague, illustration:
Botanical Magazine, 8091 (1906), with
loose erect panicles of flowers, is probably
the most commonly cultivated variant of
this species.

7. H. occidentale Greene. Illustration:
Rickett, Wild Flowers of the United States
5(2): 361 (1971); Hitchcock & Cronquist,
Flora of the Pacific Northwest, 261 (1973);
Stewart, Wildflowers of the Olympic
Cascades, pl. 127 (1988).

Stems woody, 40–80 cm, branched
above. Leaves hairy above, with 13–19
leaflets, each 1–3 cm, ovate to lanceolate;
stipules 2–3 cm, joined. Flowers in racemes
of 20–80; calyx-lobes 4–5 mm, corolla
1.6–2.2 cm, reddish purple. Pod 0.7–1.2
cm across, margins winged for 1–2 mm,
seeds 1–4. *W USA & Canada.* H2.
Summer–autumn.

8. H. boreale Nuttall. Illustration:
Hitchcock & Cronquist, Flora of the Pacific
Northwest, 261 (1973); Kuijt, A flora
of Waterton Lakes National Park, 376
(1982); Scotter & Flygare, Wildflowers of
the Canadian Rockies, 139 (1986).

Herbaceous perennial, with many stems
to 45 cm, arising from a woody rootstock.
Leaves 4–8 cm, with 9–13 leaflets, each
2–5 mm wide, hairy; stipules *c.* 1 cm,
joined. Flowers numerous, erect; upper
calyx-lobes slender; corolla 1.2–1.9 cm,
red. Pod 5–7 mm wide, with 2–6 seeds.
*N America (Saskatchewan to Oklahoma and
Arizona).* H3–4. Spring–summer.

Some of the material offered as this
species may be *H. alpinum.*

Var. **mackenzii** (Richardson) C.L.
Hitchcock. Illustration: Botanical Magazine,
6386 (1878). Leaves hairless or very
sparsely hairy above, adpressed hairs
beneath, with 13–17 leaflets and darker
flowers. *North America (Canada to N USA),
Arctic, E Siberia.* H2. Spring.

78. ONOBRYCHIS Miller
J.R. Akeroyd

Herbaceous or subshrubby perennials, or
annuals. Leaves irregularly pinnate; leaflets
entire. Flowers in elongate, stalked axillary
racemes. Calyx bell-shaped, with 5 equal
linear teeth. Stamens united except for the
uppermost which is free. Fruit indehiscent,
flattened, almost circular, usually spiny on
veins and margins; seed solitary.

A genus of about 100 species in Eurasia
and N Africa. Propagation is by seed sown
in the spring.

1a. Annual; flowers 2–8 in racemes

  7. caput-galli

  b. Perennial; flowers more than 8 in
    racemes   2

2a. Standard hairy on back; margin of
    fruit entire or with 2 rows of teeth

                                          3

  b. Standard hairless; margin of fruit
    with 1 row of teeth   4

3a. Leaflets greyish and densely hairy
    beneath   **1. hypargyrea**

  b. Leaflets green and sparsely hairy
    beneath   **2. radiata**

4a. Fruit 4–6 mm, with teeth on sides
and margins   **4. arenaria**
 b. Fruit usually at least 6 mm, with
teeth on margin only   5
5a. Calyx-teeth 1–2 times as long as
tube   **5. montana**
 b. Calyx-teeth 2–4 times as long as
tube   6
6a. Hairs of calyx whitish; margin
of fruit with 6–8 teeth to 1 mm
                            3. viciifolia
 b. Hairs of calyx brownish; margin
of fruit with 3–5 teeth to 1.5 mm
                                 6. alba

1. O. hypargyrea Boissier (*O. tournefortii*
misapplied).

Velvety-hairy perennial; stems erect,
30–100 cm. Leaflets 9–15, 2.5–5 cm,
ovate-oblong, acute, hairless above,
densely greyish-hairy beneath. Calyx with
shaggy hairs, the teeth slightly longer
than tube. Corolla 1.5–2 cm, pale yellow
with pink veins. Fruit 1.4–1.8 cm, with
shaggy hairs, the margin entire or shortly
toothed. *C & W Turkey, former Yugoslavia
(Macedonia), N Greece.* H1. Summer.

2. O. radiata (Desfontaines) Bieberstein.
Illustration: Edwards's Botanical Register
32: 37 (1815).

Softly-hairy perennial; stems 30–60
cm, erect. Leaflets 11–21, 1–3 cm, ovate-
oblong, obtuse or almost acute, hairless
above, green and sparsely hairy beneath.
Calyx with shaggy hairs, the teeth 2–3
times as long as tube. Corolla 1.5–2 cm,
pale yellow with red veins. Fruit 1.1–1.8
cm, shaggy-hairy to hairless, the margin
with small teeth to 1.5 mm. *E Turkey,
Caucasus.* H1. Summer.

3. O. viciifolia Scopoli (*O. sativa*
Lamarck). Illustration: Keble Martin,
Concise British flora in colour, pl. 24
(1965).

Robust, hairy perennial; stems 30–80
cm, usually erect. Leaflets 7–25, 1–3.5
cm, narrowly ovate to oblong, mucronate.
Calyx with rather shaggy hairs, the teeth
2–3 times as long as tube. Corolla 1–1.4
cm, the standard about as long as keel,
bright pink with purplish veins. Fruit 5–8
mm, strongly net-veined, with teeth to 1
mm on lower margin. *C & W Europe.* H1.
Summer.

Mostly grown as a fodder crop and in
amenity seed mixtures.

4. O. arenaria (Kitaibel) de Candolle.
Illustration: Polunin, Flowers of Greece and
the Balkans, 310 (1980).

Similar to *O. viciifolia* but smaller and

less erect; corolla 0.8–1.2 cm; fruit 4–6
mm, softly-hairy, toothed on sides and
margin. *C & SE Europe to C Asia.* H1.
Summer.

5. O. montana de Candolle (*A. viciifolia*
var. *montana* (de Candolle) Lamarck & de
Candolle). Illustration: Barneby, European
alpine flowers in colour, pl. 43 (1967);
Macdonald encyclopedia of alpine flowers,
pl. 172 (1975).

Similar to *O. viciifolia*, but smaller and
more woody at base; stems procumbent
to almost erect; leaflets 11–17, lanceolate
to oblong; raceme denser; corolla purplish
pink, the standard shorter than keel; calyx-
teeth 1–2 times as long as tube; fruit
0.7–1.2 cm, shortly hairy, toothed on
upper margin. *Mountains of C & SE Europe.*
H1. Summer.

6. O. alba (Waldstein & Kitanov) Desvaux.

Perennial, rather woody at base; stems
10–30 cm, procumbent to ascending.
Leaflets 13–19, linear to narrowly elliptic
or oblong, shortly appressed-hairy. Raceme
dense. Calyx hairy, the teeth 2–2.5 times
as long as tube. Corolla 0.7–1.3 cm, pink
to pinkish purple. Fruit 4–7 mm, with
shaggy hairs, net-veined, toothed on sides
and margin, the marginal teeth to 1.5
mm. *Balkan Peninsula, C & S Italy.* H1.
Summer.

A variable species; the plant grown
in gardens is subsp. **laconica** (Boissier)
Hayek (*O. laconica* Boissier).

7. O. caput-galli (Linnaeus) Lamarck.
Illustration: Polunin & Huxley, Flowers of
the Mediterranean, 33 (1965).

Ascending to erect annual, 20–40 cm.
Leaflets 11–15, narrowly oblong to
obovate, greyish-hairy. Flowers 2–8 in
raceme. Calyx hairy, the teeth 1–2 times
as long as tube. Corolla 5–6 mm, pinkish.
Fruits *c.* 8 mm, strongly net-veined,
toothed, especially on margin.

Mediterranean area & S Europe. H2.
Summer.

79. ANTHYLLIS Linnaeus
J. Cullen

Annual or perennial herbs or low shrubs.
Leaves usually pinnate with a terminal
leaflet, rarely reduced to 3 leaflets or to a
solitary (terminal) leaflet. Stipules small,
falling early. Flowers usually in dense
heads, each subtended by 2 usually large
bracts, more rarely in clusters or solitary
in the axils of single bracts. Calyx tubular,
bell-shaped or unequally swollen,
sometimes constricted towards the mouth,

with equal or unequal teeth. Corolla
usually yellow, more rarely cream or red.
Stamens with the filaments all united or
the filament of the uppermost stamen free
for up to half its length. Pod shortly stalked
or stalkless, often indehiscent, usually
enclosed in the persistent, papery calyx,
containing 1–many seeds.

A genus of about 20 species mainly
from the Mediterranean area but extending
north to Finland and Iceland, west to
Madeira, east to the Caucasus and perhaps
beyond, and south to Ethiopia. The plants
are easily grown and will tolerate poor
soils. They are best propagated by seed
which gives faster germination if scarified
before sowing.

Literature: Cullen, J., The *Anthyllis
vulneraria* complex – a resumé, *Notes from
the Royal Botanic Garden Edinburgh* **35**:
1–38 (1976).

Habit. Shrubs: **1,2**; perennial herbs: **3,4**;
annual herbs: **4**: (rarely), **5**.
Leaves. All simple or with 3 leaflets: **1,5**;
at least some with 5 or more leaflets:
2,4,5; most with 17–41 leaflets: **3**.
Flowers. In heads: **2–4**; in clusters or
solitary: **1,5**.
Calyx. Tubular: **1–3**; bell-shaped: **2**:
unequally swollen and constricted
towards the apex: **4,5**. Teeth as long as
tube, fringed with long hairs: **3**.
Standard. Much exceeding the other petals:
3.

1a. Low shrubs with woody branches
                                      2
 b. Herbs, occasionally woody at the
extreme base   3
2a. Flowers in heads; bracts palmately
lobed   **2. barba-jovis**
 b. Flowers in clusters or solitary;
bracts simple   **1. hermanniae**
3a. Calyx tubular, not swollen, not
constricted towards the mouth;
leaflets 17–41   **3. montana**
 b. Calyx unequally swollen,
constricted towards the mouth;
leaflets 1–17, rarely more   4
4a. Calyx-teeth unequal, mouth of
calyx oblique; legume 1-seeded
                            4. vulneraria
 b. Calyx-teeth equal, mouth of calyx
straight; legume 2-seeded,
constricted between the seeds
                             5. tetraphylla

1. A. hermanniae Linnaeus. Illustration:
Botanical Magazine, 2576 (1825); Huxley
& Taylor, Flowers of Greece and the
Aegean, pl. 117 (1977).

Shrub to 50 cm, branches woody and contorted, their tips becoming spine-like. Leaves simple or with 3 leaflets, silky hairy. Flowers axillary, solitary or in clusters; bracts simple. Calyx tubular, 3–5 mm, teeth shorter than tube, mouth straight. Corolla yellow. Pod with 1 seed. *Mediterranean area*. H5. Summer.

2. A. barba-jovis Linnaeus. Illustration: Botanical Magazine, 1927 (1817); Bonnier, Flore complète **3**: pl. 126 (1914).

Shrub to 1 m with woody branches. Leaves with 9–19 more or less equal leaflets, silky hairy above and beneath. Flowers in terminal heads, bracts palmately lobed. Calyx 4–6 mm, tubular to bell-shaped, teeth shorter than tube, mouth straight. Corolla rather pale yellow. Pod with 1 seed. *S Europe*. H5. Summer.

3. A. montana Linnaeus. Illustration: Bonnier, Flore complète **3**: pl. 126 (1914); Botanical Magazine, n.s., 333 (1959); Polunin, Flowers of Europe, pl. 621 (1969); Ceballos et al., Plantas silvestres de la peninsula Iberica, 162 (1980).

Perennial herb, often woody at the base. Leaves mostly near the base of the flowering-stems, with 17–41 more or less equal leaflets, downy on both surfaces. Flowers in heads; bracts deeply palmately lobed. Calyx tubular, teeth as long as tube with long hairs on their margins, mouth of calyx straight. Corolla red to purple, the standard much exceeding the other petals. *Alps & mts of S Europe*. H3. Summer.

4. A. vulneraria Linnaeus. Illustration: Bonnier, Flore complète **3**: pl. 127 (1914); Polunin, Flowers of Europe, pl. 622 (1969).

Perennial (rarely annual) herbs. Leaves with 3–15 (rarely to 19) leaflets or the lowermost reduced to 3 leaflets or a solitary terminal leaflet, when more than 1 leaflet present the terminal leaflet sometimes larger than the others, all variably silky-hairy. Flowers in heads; bracts palmately lobed. Calyx unequally inflated, with 5 short unequal teeth, mouth oblique. Corolla cream, yellow, pink or red. Pod 1-seeded (rarely 2-seeded, when not constricted between the seeds). *Europe, N Africa, SW Asia*. H1. Spring–summer.

A complex species divided into more than 30 subspecies (some treated as separate species by some authors) across its range. It is uncertain which of the subspecies are in cultivation, though the following have been seen:

Subsp. **vulneraria**. Stem-leaves with

more or less equal leaflets; bract-lobes acute; calyx to 5 mm wide; corolla yellow. *NW Europe*. H1. Spring–summer.

Var. **langei** (Sagorski) Jalas is very similar but is more frequently branched, very silver-silky and occurs only in coastal areas.

Subsp. **polyphylla** (de Candolle) Nyman. Similar to subsp. *vulneraria* but leaves with more numerous leaflets and stems shaggy-hairy in the lower part. *C & E Europe*. H1. Summer.

Subsp. **argyrophylla** (Rothmaler) Cullen. Small plants; lower leaves with 1–3 leaflets, the terminal much larger than the laterals, stem-leaves small, leaflets more or less equal, all very densely silver-silky; bract-lobes acute; corolla red. *S Spain*. H5. Summer.

Sometimes grown as *A. webbiana* Hooker, a name of uncertain application.

Subsp. **alpestris** (Schultes) Ascherson & Graebner (*A. alpestris* Schultes). Illustration: Rasetti, I fiori delle Alpi, f. 314 (1980). Small plants; leaves with few leaflets, stem-leaves with the terminal leaflet larger than the laterals; bract-lobes obtuse; calyx more than 5 mm wide, grey-hairy; corolla pale yellow. *European mts, from N Spain to the Balkans*. H3. Summer.

Subsp. **pyrenaica** (Beck) Cullen (*A. pyrenaica* Beck). Illustration: Bonnier, Flore complète **3**: pl. 127 (1914) – as A. dillenii. Like subsp. *alpestris* but calyx white-hairy, corolla pinkish red. *Pyrenees, mts of N Spain*. H3. Summer.

5. A. tetraphylla Linnaeus. Illustration: Botanical Magazine, 108 (1788); Bonnier, Flore complète **3**: pl. 127 (1914); Huxley & Taylor, Flowers of Greece and the Aegean, pl. 115 (1977).

Prostrate annual herb. Leaves with 3–5 leaflets, the terminal much larger than the others, all spreading, hairy. Flowers in axillary clusters. Calyx unequally inflated at flowering, with equal, short teeth, mouth straight. Corolla yellow, keel often red at apex. Pod 2-seeded, constricted between the seeds. *Mediterranean area*. H3. Spring.

80. TETRAGONOLOBUS Scopoli
J. Cullen
Plants similar to Lotus but leaves with 3 leaflets, genuine stipules better-developed and leaf-like. Flowers solitary or 2 together. Calyx-teeth equal. Corolla yellow or dark red. Pod almost square in section, the angles winged.

A small genus, often included in *Lotus*,

from Europe and the Mediterranean area, easily grown and propagated by seed.

1a. Perennial; corolla yellow
1. maritimus
  b. Annual; corolla crimson or dark red
2. purpureus

1. T. maritimus (Linnaeus) Roth (*Lotus siliquosus* Linnaeus; *T. siliquosus* (Linnaeus) Roth). Illustration: Botanical Magazine, 151 (1791); Hegi, Illustrierte Flora von Mitteleuropa **4**: 1375–6 (1923); Polunin & Smythies, Flowers of southwest Europe, pl. 22 (1973).

Perennial to 40 cm, hairy or not. Leaflets oblanceolate to obovate, to 3 cm; stipules ovate. Flowers solitary, borne on stalks considerably longer than the leaves. Corolla pale yellow, 2.5–3 cm. Style with a membranous wing on one side. Pod 3–6 cm, wings to 1 mm wide. *Europe, N Africa, SW Asia*. H3. Summer.

2. T. purpureus Moench (*Lotus tetragonolobus* Linnaeus; *T. edulis* Link). Illustration: Bonnier, Flore complète **3**: pl. 144 (1914); Fenaroli, Flora Mediterranea, 122 (1962); Polunin & Huxley, Flowers of the Mediterranean, pl. 72 (1965); Huxley & Taylor, Flowers of Greece and the Aegean, pl. 144 (1977).

Annual herb to 40 cm, downy. Leaflets to 4 cm, obovate to almost diamond-shaped. Stipules ovate. Flowers solitary, stalks shorter than or equalling leaves. Corolla crimson or dark red. Style without a membranous wing at one side. Pod 3–9 cm, wings 2–4 mm wide. *S Europe, Mediterranean area*. H3. Summer.

81. LOTUS Linnaeus
J. Cullen
Annual or perennial herbs, sometimes woody below. Leaves pinnate with 5 leaflets (ours) including a terminal leaflet, axis often short, the 2 lowermost leaflets often resembling stipules. Genuine stipules very small, glandular. Flowers in heads or solitary. Calyx bell-shaped to almost tubular, with 5 equal or unequal teeth. Corolla scarlet, crimson to dark purplish brown or yellow, the keel with a conspicuous beak. Stamens 10, filaments of 9 united, that of the uppermost free. Pod cylindric, sometimes broadly so, straight or curved. Seeds numerous.

A genus of perhaps 150 species from temperate regions of which only a small number is grown. They will tolerate a wide range of soils and are propagated by seed or cuttings.

Habit. Annual: **4**; perennial: **1–3,5**.

Leaflets. Very narrow, linear to narrowly elliptic: **1,2**; relatively broader: **3–5**. Grey- or whitish-hairy: **1,5**; green, hairy or not: **2–4**.

Flowers. Red or purplish brown: **1,2**; yellow, sometimes tinged with red: **3–5**. Solitary or 2 together: **1,4**; in heads of 3 or more: **2,3,5**. Keel 3–4 cm: **1**. Standard reflexed, with a broad black line down the centre: **1**.

Calyx-teeth. Clearly unequal: **1,5**; equal or almost so: **2–4**.

Pod. 4–8 cm across, grooved above: **4**; narrower, not obviously grooved above: **1–3,5**.

1a. Flowers scarlet, the keel to 4 cm, curved, the standard narrow, reflexed, with a broad black line down its centre **1. berthelotii**
  b. Flowers yellow or brownish, keel much smaller, standard not as above **2**

2a. Leaflets very narrow, linear to very narrowly elliptic, much longer than wide; flowers brownish purple **2. jacobaeus**
  b. Leaflets relatively broader; flowers yellow, sometimes tinged with red **3**

3a. Calyx-teeth unequal, the 2 upper longer, curved upwards; leaves and stems grey- or whitish-hairy **5. creticus**
  b. Calyx-teeth more or less equal, not as above; leaves and stems green, hairy or not **4**

4a. Flowers solitary or 2 together; pod short, curved, 4–8 mm in diameter, grooved above; annual **4. edulis**
  b. Flowers 3 or more together in heads; pod straight, 2–3 mm in diameter, scarcely grooved above; perennial **3. corniculatus**

1. L. berthelotii Masferrer. Illustration: Botanical Magazine, 6733 (1884) – as L. peliorhynchus; Menninger, Flowering vines of the world, pl. 93 (1970); Bramwell & Bramwell, Wild flowers of the Canary Islands, 64 (1974); Gartenpraxis for 1981, 464–465.

Sprawling or scrambling perennial. Leaves and stems conspicuously white- or grey-hairy. Leaves of 5 leaflets all attached at the same point, leaflets to 1.5 cm, linear; axillary shoots usually present in the leaf-axils. Flowers solitary or paired in the leaf-axils. Calyx unequally toothed. Corolla bright red, the standard narrow, reflexed, black-centred, the keel 3–4 cm, curved,

claw-like. Fruit rarely formed. *Canary Islands, Cape Verde Islands.* H5–G1. Summer.

Very spectacular with its red keel looking like a parrot's beak. Now very rare and threatened in the wild, but well established in cultivation, propagated by cuttings.

2. L. jacobaeus Linnaeus. Illustration: Botanical Magazine, 79 (1787); Kew Magazine **4**: 87 (1987).

Perennial. Leaves of 5 leaflets, the upper 3 usually separated from the lower 2 by a stalk; leaflets linear to very narrowly elliptic or very narrowly obovate, to 4 cm. Flowers 3 or more together in axillary heads. Calyx-teeth more or less equal. Corolla purplish brown, keel to 1 cm. Pod linear-cylindric. *Cape Verde Islands.* H5–G1. Summer.

3. L. corniculatus Linnaeus. Illustration: Bonnier, Flore complète **3**: pl. 142 (1914); Keble Martin, The concise British flora in colour, pl. 23 (1965); Hay & Synge, Dictionary of garden plants in colour, 156 (1969).

Sprawling to erect perennial, hairy or not. Leaflets 5, the upper 3 separated from the lower 2 by a short stalk, all obovate to almost circular. Flowers 3 or more together in heads. Calyx-teeth more or less equal. Corolla bright yellow, occasionally tinged with red, keel to 1.5 cm. Pod linear-cylindric, to 3 mm across, not conspicuously grooved above. *Eurasia, N Africa.* H1. Spring–summer.

A very widespread, common and variable plant in Europe. A variant with double flowers ('Pleno') is sometimes grown.

4. L. edulis Linnaeus. Illustration: Bonnier, Flore complète **3**: pl. 142 (1914); Pignatti, Flora d'Italia **1**: 746 (1982).

Annual herb, somewhat downy. Leaves with a short stalk between the 3 upper and 2 lower leaflets; leaflets obovate. Flowers solitary or paired, on a long (common) stalk. Calyx-teeth more or less equal. Corolla yellow, keel to 1.5 cm. Pod curved, widely cylindric, 4–8 mm across, deeply grooved above. *Mediterranean area.* G1–H5. Summer.

5. L. creticus Linnaeus. Illustration: Bonnier, Flore complète **3**: pl. 141 (1914); Polunin & Smythies, Plants of southwest Europe, pl. 22 (1973).

Rather upright perennial herb. Leaves and stems grey- or whitish-hairy. Leaves with 5 leaflets all arising from more or less the same point, obovate. Flowers 3 or more together in heads. Calyx-teeth

unequal, the 2 upper longer than the others and curved upwards. Corolla yellow, keel to 1.5 cm, with a purple beak. Pod linear-cylindric, 2–3 mm across, not conspicuously grooved above. *Mediterranean area.* H5. Spring.

82. DORYCNIUM Miller
J. Cullen

Perennial herbs or small shrubs. Leaves almost stalkless, pinnate with 5 leaflets, the lower pair often appearing superficially like stipules; true stipules minute. Flowers in axillary heads. Calyx bell-shaped with 5 equal or unequal teeth. Corolla whitish, often with red or pink spots or lines; keel obtuse, dark red. Stamens 10, 9 with the filaments united, that of the uppermost free. Pod oblong, ovoid or narrowly cylindric. Seeds 1–many.

A genus of about 8 species, mainly from the Mediterranean area, often included in *Lotus*. They are not especially attractive and are rarely grown. They are, however, easy to cultivate; propagation is by seed.

Literature: Rikli, M., Die Gattung *Dorycnium* Vill., *Botanische Jahrbucher* **31**: 314–404 (1901).

1a. Leaflets all very shortly stalked, without an obvious longer stalk to the upper 3 leaflets **1. hirsutum**
  b. Leaves with a very obvious stalk between the upper 3 and lower 2 leaflets **2. broussonetii**

1. D. hirsutum (Linnaeus) Seringe (*Bonjeania hirsuta* (Linnaeus) Reichenbach). Illustration: Bonnier, Flore complète **3**: pl. 140 (1914); Polunin & Walters, Guide to the vegetation of Britain and Europe, 135 (1985).

Shrub to 50 cm. Leaflets to 1.8 cm, all equally and shortly stalked, elliptic to narrowly obovate, with long, dense, spreading hairs. Flowers in heads of 4–10 subtended closely by bracts which are similar to the leaves. Calyx-teeth slightly unequal. Corolla 1–2 cm, white or pinkish. Pod 0.6–1.2 cm. *Mediterranean area.* H4. Spring–summer.

D. pentaphyllum Scopoli (*D. suffruticosum* Villars). A small shrub less hairy than the above, with narrower leaflets and corollas 3–7 mm. *C & E Europe, Mediterranean area.* H3. Spring–summer.

May occasionally be grown.

2. D. broussonetii (Choisy) Webb & Berthelot. Illustration: Bramwell & Bramwell, Flores silvestres de las Islas

Canarias, 67 (1974).

Shrub to 1.5 m. Leaves with a conspicuous stalk between the lower 2 leaflets, which are borne at the junction with the stem, and the upper 3. Leaflets elliptic to obovate, 2–4 cm. Flowers in heads of 5–8. Calyx-teeth slightly unequal. Corolla white with pink spots or lines. Pod much exceeding the calyx. *Canary Islands.* H5. Spring.

83. CORONILLA Linnaeus
J.M. Lees

Annual or perennial herbs or low shrubs. Leaves pinnate with a terminal leaflet, rarely simple or with only 3 leaflets; stipules variable, free or united to each other. Flowers in heads in the leaf-axils. Calyx bell-shaped, more or less 2-lipped. Petals yellow or pinkish, keel acute. Stamens 10, filaments of 9 united, that of the uppermost free. Pod breaking up into 1-seeded linear to oblong segments, which are not constricted between them, round in section or ridged or angled.

A genus of about 25 species from Europe, western and central Asia and the Canary Islands. They are easily grown in most soils. Propagation is by seed.

Literature: Uhrova, A., Revision der Gattung *Coronilla* L., *Beihefte Botanischer Centralblatt* **53** (13): 1–174 (1935); Jahn, A., Beitrage zur Kentniss der Sippenstruktur einiger Arten der Gattung *Coronilla* L., *Feddes Repertorium* **85**: 455–532 (1974).

1a. Lower leaves simple or with 3 leaflets, the terminal leaflet much larger than the laterals; annual
    7. scorpioides
  b. Lower leaves pinnate, the leaflets more or less equal; perennial or shrubby   2
2a. Corolla white, pink or purple
    6. varia
  b. Corolla yellow   3
3a. Claw of standard 2–3 times as long as calyx   **1. emerus**
  b. Claw of standard equalling or only slightly longer than calyx   4
4a. Corolla 1.2–1.8 cm; leaves glaucous
    5. orientalis
  b. Corolla 0.5–1.2 cm; leaves not glaucous   5
5a. Perennial herb; heads with 12–22 flowers   **4. coronata**
  b. Small low shrub; heads with 4–15 flowers   6
6a. Stipules wedge-shaped to obovate, deciduous, 2–10 mm; leaflets without

scarious margins   **2. valentina**
  b. Stipules united, persistent, *c.* 1 mm; leaflets with scarious margins
    3. minima

1. C. emerus Linnaeus. Illustration: Bonnier, Flore complète **3**: pl. 162 (1914).

Shrub to 2 m. Leaves with 2–4 pairs of leaflets which are 1–2 cm, obovate, mucronate, glaucous. Stipules 1–2 mm, free, membranous. Corolla 1.4–2 cm, pale yellow; claw of standard 2–3 times as long as calyx. Pod 5–11 cm, with 3–12 segments which are 0.8–1 cm. *C & SE Europe, also locally extending to S Norway, the Pyrenees & E Spain.* H2. Summer.

2. C. valentina Linnaeus. Illustration: Bonnier, Flore complète **3**: l. 162 (1914).

Shrub to 1 m. Leaves with 2–6 pairs of leaflets which are obovate and notched. Stipules wedge-shaped to obovate, free, deciduous. Heads with 4–12 flowers. Corolla 0.7–1.2 cm, yellow. Legume 1–5 cm, with 1–10 segments which are 5–7 mm, spindle-shaped, somewhat compressed and with 2 obtuse angles. *Mediterranean area, S Portugal.* H5. Summer.

C. glauca Linnaeus (*C. valentina* subsp. *glauca* (Linnaeus) Battandier). Illustration: Bonnier, Flore complète **3**: pl. 162 (1914); Polunin & Smythies, Flowers of southwest Europe, pl. 23 (1973). Similar but with 2 or 3 pairs of leaflets, stipules ovate or lanceolate, membranous, 2–6 mm; pod with 1–4 (rarely to 10) segments. *Mediterranean area.* H5. Summer.

3. C. minima Linnaeus. Illustration: Bonnier, Flore complète **3**: pl. 161 (1914).

Small shrub to 50 cm. Leaves with 2–6 pairs of leaflets which are elliptic, obovate or almost circular, 2–15 mm, stalkless and with translucent margins. Stipules *c.* 1 mm, united to each other, membranous, persistent. Heads with up to 15 flowers, flower-stalks 2–4 mm. Corolla 0.5–1.2 cm, yellow. Pod 1–3.5 cm, with 1–7 segments which are 4.5–5.5 mm, oblong, 4-angled. *W Europe extending to Switzerland & Italy.* H3. Summer.

4. C. coronata Linnaeus (*C. montana* Scopoli). Illustration: Bonnier, Flore complète **3**: pl. 161 (1914).

Perennial herb, 10–70 cm. Leaves with 3–7 pairs of leaflets which are elliptic to obovate, 1.5–4 cm, shortly stalked and with narrow translucent margins. Stipules 3–5 mm, united to each other, membranous, deciduous. Heads with 12–22 flowers, flower-stalks 4–6 mm. Corolla 0.7–1.1 cm, yellow. Pod 1.5–3 cm,

segments 1–9, 6–7.6 mm, ovoid-oblong, obtusely angled. *C Europe, extending E to the Crimea.* H4. Summer.

5. C. orientalis Miller (*C. iberica* Bieberstein; *C. cappadocica* Willdenow).

More or less prostrate perennial herb with a woody base. Stems to 50 cm. Leaves with 3–5 pairs of leaflets which are broadly wedge-shaped, glaucous and more or less notched. Stipules 3–5 mm, wedge-shaped to obovate. Heads with 3–9 flowers. Flowers 1.2–1.8 cm, yellow, claws of petals about as long as the sepals. Pod 2–4.5 mm, curved, with 2–11 segments. *SW Asia.* H5.

6. C. varia Linnaeus. Illustration: Bonnier, Flore complète **3**: pl. 161 (1914); Huxley & Taylor, Flowers of Greece and the Aegean, t. 113 (1977).

Perennial herb to 1.2 m. Leaves with 5–12 pairs of leaflets which are oblong or elliptic, to 3 cm, with narrow, translucent margins. Stipules 1–6 mm, free, membranous. Heads with 10–20 (rarely as few as 5) flowers. Corolla 0.8–1.5 cm, white, pink or purple. Pod 1.5–8 cm, with 3–12 segments which are 4–6 mm, oblong, 4-angled. *C & S Europe to C Russia.* H2. Summer.

7. C. scorpioides (Linnaeus) Koch.

Annual herb to 40 cm. Leaves simple or with 3 leaflets, the terminal leaflet to 4 cm, elliptic to almost circular, much larger than the kidney-shaped lateral leaflets. Stipules 1–2 mm, united to each other, membranous. Heads with 2–5 flowers. Corolla 4–8 mm, yellow. Pod 2–6 cm, curved, segments 2–11, oblong, obtusely 4-angled. *S Europe.* H5. Summer.

84. HIPPOCREPIS Linnaeus
J.R. Akeroyd

Herbaceous or shrubby perennials (cultivated species). Leaves irregularly pinnate; leaflets entire. Stipules small, linear to lanceolate. Flowers in umbels on long axillary stems. Calyx elongately bell-shaped with 5 teeth. Corolla yellow. Stamens united except for the uppermost which is free. Fruit flattened, indehiscent, of a single segment or segmented and breaking up when ripe, the segments with deep sinuses.

A genus of some 15 species in Europe and the Mediterranean area. Propagation is by seed or division of plants.

1a. Leaflets obtuse; fruit covered with reddish or brown papillae
    1. comosa

b. Leaflets acute; fruit smooth
2. balearica

1. H. comosa Linnaeus. Illustration: Polunin, Collins photoguide to wild flowers, p. 216 (1986); Brickell, RHS gardeners' encyclopedia of plants and flowers, 326 (1989).

Stems 10–40 cm, prostrate to ascending, somewhat woody at base. Leaflets 7–17, 5–15 × 2–4 mm, obovate to linear, obtuse, usually hairy beneath. Flowers 5–12 in umbel. Corolla 0.6–1.2 cm, yellow. Fruit 1.5–3 cm, covered with reddish or brown papillae; sinuses of segments semi-circular to almost circular. *S & W Europe & Mediterranean area.* H1. Summer.

A variable species; a variant with pale yellow corollas is often grown in gardens.

2. H. balearica Jacquin. Illustration: Botanical Magazine, 427 (1798); Polunin & Smythies, Flowers of south-west Europe, pl. 22 (1973).

Subshrub; stems 20–60 cm, erect, woody below. Leaflets 11–21, 4–12 × 1–4 mm, linear or oblong, acute. Flowers 2–10 in umbel. Corolla 1–1.5 cm. Fruit 1.5–4.5 cm, smooth; sinuses of segments semicircular to circular. *Balearic Islands.* H5. Summer.

85. VICIA Linnaeus
F.K. Hibberd

Perennial and annual herbs; stems angular in cross-section but never winged. Leaves usually with many pairs of leaflets, the axis ending in a tendril or short point; leaflets folded in bud (*V. faba* exceptionally sometimes has rolled leaflets). Flowers blue, purple, yellow, orange or whitish. Wing petals attached to keel. Style cylindric or flattened, hairy all round or with a tuft or hairs on the outer face. Pod diamond-shaped or flattened-cylindric. Germination hypogeal.

A genus of about 150 species distributed throughout the temperate northern hemisphere, extending into tropical Africa and south America. Propagation is by seed.

Literature: Kupicha, F.K. The infrageneric structure of *Vicia. Notes from the Royal Botanic Garden Edinburgh* **34**: 287-326 (1976).

Leaves. With a tendril: **3,5**: without a tendril: **1,2,4,6**. Stipules with a nectary (a dark spot, visited by ants): **4–6**; stipules without nectaries: **1–3**.
Flowers. Inflorescence distinctly stalked: **1–3**; flowers more or less stalkless in leaf axils: **4–6**.

1a. Leaf-axis ending in a short point   2
b. Leaf-axis ending in a tendril   5
2a. Racemes distinctly stalked   4
b. Racemes with very short stalk, or flowers stalkless in the leaf axil   3
3a. Annual, with large edible seeds 2 cm or more across   **6. faba**
b. Perennial, with small inedible seeds   **4. oroboides**
4a. Leaves with 3–6 pairs of leaflets; flowers orange-yellow   **1. crocea**
b. Leaves with 1 pair of leaflets, flowers purple   **2. unijuga**
5a. Flowers in a dense stalked raceme   **3. cracca**
b. Flowers 1–2, stalkless in leaf-axils   **5. sativa**

1. V. crocea (Desfontaines) Fedtschenko (*Orobus croceus* Desfontaines).

Softly-hairy perennial with erect stems 40–70 cm, the pair of stipules at each node very unequal in shape and size. Leaf-axis ending in a short point; leaflets 3–6-paired, ovate-acuminate, 3.5–6 × 1.5–3 cm. Racemes many-flowered, arranged in a panicle; flowers 1.5–2 cm, corolla orange-yellow. *Turkey, N Iran, Caucasia.* H2. Early summer.

Very rare in cultivation. *V. crocea* is remarkably similar to *Lathyrus aureus*, but the two can be distinguished as follows: *V. crocea* has folded young leaflets, unequal stipules, the end of the style hairy all round, and a short pod narrowed to a slender stalk, whereas *L. aureus* has rolled leaflets, equal stipules, the style hairy only on the inner face, and a long linear pod.

2. V. unijuga A. Braun (*Orobus lathyroides* Linnaeus). Illustration: Botanical Magazine, 2098 (1819); Phillips & Rix, Perennials **2**: 61 (1990).

Erect perennial 30–40 cm. Leaves with one pair of leaflets, the axis ending in a short point; leaflets ovate-acuminate, with hairy margins. Flowers 3–18 in a loose raceme; corolla purple. *Mongolia, Siberia, China, Korea & Japan.* H1. Early summer.

3. V. cracca Linnaeus. Illustration: Hegi, Illustrierte Flora von Mitteleuropa **4**: t. 170, fig. 1 (1923); Keble Martin, Concise British flora in colour, pl. 24 (1965); Polunin, Flowers of Europe, 55 (1969); Grey-Wilson & Blamey, Alpine flowers of Britain & Europe, 119 (1979).

Downy perennial with stems scrambling or climbing to 2 m. Leaves with many pairs of leaflets and a branched tendril; leaflets linear-lanceolate, 1–2.5 cm. Racemes dense, many-flowered, on stalks

2–10 cm. Flowers 1–1.2 cm, purplish blue. *Widespread in N Europe & Asia.* H1. Summer.

4. V. oroboides Wulfen (*Orobus lathyroides* Sibthorp & Smith, not Linnaeus). Illustration: Hegi, Illustrierte Flora von Mitteleuropa **4**: 1549 (1923).

A hairless or downy perennial with erect stems 5–60 cm. Leaves with 1–3 pairs of leaflets, the leaf axis ending in a short point; leaflets ovate to oblong, acute, up to 8 × 4 cm. Racemes 3–8 flowered, inflorescence stalks very short; flowers *c.* 1.8 cm, whitish or pale yellow. *Italy, Austria, Hungary & former Yugoslavia.* H2. Summer.

Rare in cultivation.

5. V. sativa Linnaeus. Illustration: Keble Martin, Concise British flora in colour, pl. 25 (1965); Polunin, Flowers of Europe, 55 (1969).

Annual, with stems to 1 m. Leaves with 4–8 pairs of leaflets and a tendril; leaflets linear to obovate, notched at apex, 1–2 cm, sparsely hairy. Flowers 1 or 2, stalkless at a node, 1–3 cm, purple. *Europe, Asia, N Africa.* H2. Summer.

6. V. faba Linnaeus. Illustration: Bonnier, Name this flower, pl. 16 (1917); Hegi, Illustrierte Flora von Mitteleuropa **4**: t. 170, fig. 4 & p. 1558 (1923); Masefield et al., Oxford book of food plants, 41 (1969).

Stout erect hairless annual with angular stems from 30–200 depending on the cultivar. Leaves with 2–4 pairs of leaflets, the axis ending in a point; leaflets elliptic, ovate or obovate, 5–10 cm. Flowers 1–several, almost stalkless in leaf-axils, 1.6–3 cm, often white with blackish wings, sometimes red. Pods large, 5–20 × 1–2 cm, containing edible seeds in a "woolly" lining. *Unknown in the wild, but widely cultivated as a crop (broad bean, tic bean or horse bean).* H2. Early summer.

86. LATHYRUS Linnaeus
F.K. Hibberd

Herbaceous perennials and annuals with unbranched climbing or sprawling stems which are often winged. Leaves with 2 or more leaflets, the axis ending in a tendril or a short point; leaflets rolled up in bud. Racemes axillary, 1 to many-flowered. Wing petals attached to keel. Style flattened, with a brush of hairs on the inner face, often twisted through 90 degrees. Germination hypogeal.

A genus of about 150 species distributed throughout the temperate northern hemisphere and extending into tropical

east Africa and South America.
Propagation is by seed or division. Species
with erect stems and leaves with more
than a pair of leaflets are woodland, while
the rest (including *L. japonicus*) grow best
in sun and well-drained soil. Despite
occasional references to hybrids in
horticultural literature, hybridisation
between Eurasian species has been proved
virtually impossible (Davies, 1958).
Lathyrus is a rare example of a genus with
flowers in the three primary colours, red,
blue and yellow.

Literature: Davies, A.J.S., A
cytotaxonomic study in the genus
Lathyrus. Ph.D. Thesis, Manchester
University (1958); Bässler, M., Revision
der eurasiatischen Arten von Lathrys L.
Sect. Orobus (L.) Grenier & Godron. *Feddes
Repertorium* **84**: 329-477 (1973). Kupicha,
F.K., The infrageneric structure of *Lathyrus*.
*Notes from the Royal Botanic Garden
Edinburgh* **41**: 209-244 (1983).

Stems. Clearly winged: **7–13**; angular
   in cross-section but not winged:
   1–6,14,15.
Leaves. With 4 or more leaflets: **1–4**; with
   a pair of leaflets: **5–15**. Tendril present:
   1,4–13,15: absent: **2,3,14**. Leaf-stalk
   absent: **15**.
Flowers. Corolla orange or yellow: **1,2,12**;
   pink or crimson: **3,5–11**; blue or purple:
   3,4,11,13–15. (Several species have
   white-flowered forms.) Flowers sweetly
   scented: **6,11**. Style twisted through
   90°: **5–13**.

1a. Leaves with 4 or more leaflets   2
  b. Leaves with a pair of leaflets   5
2a. Leaf-axis ending in a branched
      tendril   3
  b. Leaf-axis ending in a short point   4
3a. Stems erect; flowers orange
                              1. davidii
  b. Stems prostrate; flowers pale purple
                             4. japonicus
4a. Leaflets elliptic; flowers orange
                               2. aureus
  b. Leaflets ovate-acuminate; flowers
      blue-purple, pink or white
                               3. vernus
5a. Plants hairy   6
  b. Plants hairless   8
6a. Stems winged; tall annuals with
      tendrils   7
  b. Stems unwinged; prostrate
      perennial without tendrils
                             14. laxiflorus
7a. Flowers blue, purple, pink, red or
      white, sweetly scented
                             11. odoratus

  b. Flowers yellow, unscented
                            12. chloranthus
8a. Stems winged   9
  b. Stems not winged   13
9a. Flowers blue; annual, with stems
      to 45 cm; pod oblong, *c.* 3 cm
                              13. sativus
  b. Flowers crimson, pink, purplish
      or white; perennial, with stems
      climbing or sprawling to 2 m or
      more; pod linear, 5 cm or more
                                        10
10a. Racemes with 1 or 2 flowers;
      standing 3.5 cm   **10. tingitanus**
  b. Racemes with 3–11 flowers;
      standing 2 cm or less   11
11a. Leaflets net-veined, 4.5 cm or less
                           7. rotundifolius
  b. Leaflets parallel-veined, 5 cm or
      more   12
12a. Stipules more than half as wide
      as stem; lowest calyx-tooth 8 mm
                             8. latifolius
  b. Stipules less than half as wide as
      stem; lowest calyx-tooth 4 mm
                             9. sylvestris
13a. Leaf-stalk absent, the two leaflets
      and two large stipules forming a
      cluster at each node
                             15. nervosus
  b. Leaf-stalk present, stipules much
      smaller than leaflets   14
14a. Tendrils branched; leaflets elliptic
      (widest at the middle); standard *c.* 4
      cm   **5. grandiflorus**
  b. Tendrils usually unbranched;
      leaflets often broadest above the
      middle; standard *c.* 1 cm
                             6. tuberosus

1. L. davidii Hance. Illustration:
Gartenflora **32**: t. 1127 (1883).

Hairless perennial, with erect
unbranched stems 80 cm; stems not
winged. Leaves ending with a branched
tendril, and having 2–4 pairs of leaflets,
these ovate, 5–9 × 2.5–5.5 cm, yellowish
green, soft-textured; stipules large,
conspicuous. Racemes with 8–20 flowers;
inflorescence-stalk 2–6 cm. Flower-stalk
17–20 cm. Calyx-teeth represented by
mere points around the mouth of the tube.
Corolla orange. *Siberia, China, Korea &
Japan.* H4. Summer.

2. L. aureus (Steven) Brandza (*L. luteus*
misapplied). Illustration: Stojanoff &
Stefanoff, Flora of Bulgaria, edn 2: 655
(1933); Flora Republicii Populare Romine
5: t. 78 (1957).

Perennial with erect unbranched stems
up to 80 cm; stems not winged; stems

and leaf-stalks bearing dark glandular
hairs (hand-lens needed), and plants also
sparsely downy with curly hairs. Leaves
with 3–5 pairs of leaflets, the leaf-axis
ending in a short point 3–6 mm; leaflets
elliptic, acute to slightly acuminate, 3.5–5
× 2–2.5 cm. Racemes dense, many-
flowered, with inflorescence-stalk 7–11 cm.
Flower 1.7–1.9 cm. Calyx-teeth unequal,
the lowest one longest, *c.* 3 mm. Corolla
orange. *Native to countries bordering the
Black Sea.* H2. Early summer.

L. aureus belongs to a complex of closely
related species differentiated by leaflet size
and shape, flower size, presence or absence
of glandular hairs and relative lengths of
calyx-teeth (Bässler, 1973). Other members
of this group, apart from *L. aureus*, may
possibly be in cultivation.

3. L. vernus (Linnaeus) Bernhardi (*Orobus
vernus* Linnaeus; *L. cyaneus* misapplied).
Figure 39(1), p. 520. Illustration: Botanical
Magazine **15**: 521 (1801); Hegi, Illustrierte
Flora von Mitteleuropa **4**: t. 172, fig. 3 &
p. 1576 (1923); Grey-Wilson & Blamey,
Alpine flora of Britain and Europe, 121
(1979); Phillips & Rix, Perennials **1**: 60
(1990).

Hairless perennial with unwinged stems
25 cm. Leaves with 2 or 3 pairs of leaflets,
the leaf-axis ending in a point 0.5–1.3 cm;
leaflets ovate-acuminate, 3–8 × 1.2–4 cm.
Racemes with 3–6 flowers; inflorescence-
stalks 3–5 cm. Flowers 1.3–1.7 cm. Calyx-
teeth unequal, the lowest longest, 2.5–3
mm. Corolla purplish blue, vivid blue, pure
white or with pale pink standard and keel
and white wings. *Continental Europe to E
Siberia.* H1. Spring and early summer.

4. L. japonicus Willdenow (*L. maritimus*
(Linnaeus) Bigelow). Figure 39(2), p. 520.
Illustration: Hegi, Illustrierte Flora von
Mitteleuropa **4**: t. 172 & p. 1586 (1923);
Keble Martin, Concise British flora in
colour, pl. 25 (1965); Polunin, Flowers of
Europe, 56 (1969); Phillips & Rix,
Perennials **2**: 58 (1990).

Hairless or hairy perennial, with
prostrate branches to 1 m; stems not
winged. Leaves ending in a short,
branched tendril and having 2–5 pairs
of leaflets, these elliptic, 1.5–5 × 0.8–2.5
cm, leathery and glaucous; stipules *c.* 2
cm; resembling leaflets in size and texture.
Racemes with 3–10 flowers, inflorescence-
stalks 3.5–5 cm. Flowers 1–2.5 cm. Calyx-
teeth unequal, the lowest tooth 4–5 mm.
Corolla bicoloured: standard pale purple,
wings and keel white tinged with purple.
Circumpolar, on shingle & sandy beaches of

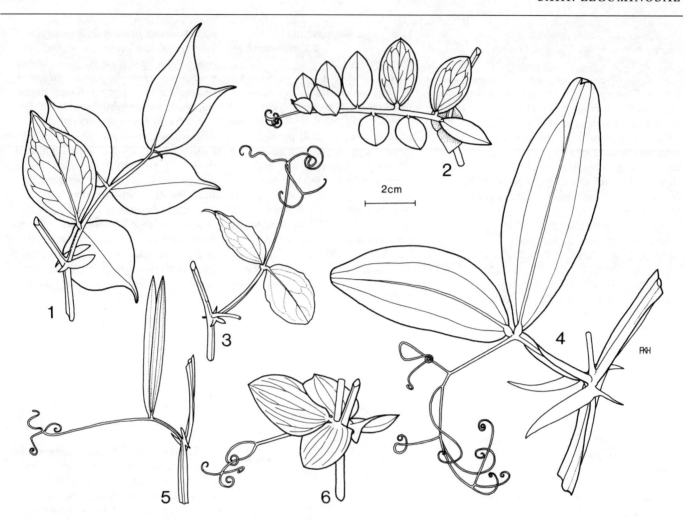

Figure 39. Leaves of *Lathyrus* species. 1, *L. vernus*.
2, *L. japonicus*. 3, *L. grandiflorus*. 4, *L. latifolius*. 5, *L. sativus*.
6, *L. nervosus*.

oceans & lakes. H1. Summer.

Cultivated plants of *L. japonicus* do not have such leathery leaflets as those grown in the wild.

Subsp. **japonicus**. Flowers up to 2.5 cm with pods over 1 cm wide. *E Asia, arctic Europe & North America.*

Subsp. **maritimus** (Linnaeus) Ball. Flowers up to 1.6 cm and pods less than 1 cm wide. *Baltic area, British coasts, Alaska, California & the Great Lakes area of N America.*

5. L. grandiflorus Sibthorp & Smith. Figure 39(3), p. 520. Illustration: Botanical Magazine, 1938 (1817); Polunin, Flowers of Greece & the Balkans, pl. 23 (1980); Huxley & Taylor, Flowers of Greece & the Aegean, fig. 99 (1984); Phillips & Rix, Perennials **2**: 58 (1990).

A far-running hairless perennial, with slender but strongly climbing stems to 2 m or more; stems not winged. Leaves with leaf-stalk 1.5–2.5 cm, one pair of leaflets and a branched tendril; leaflets elliptic, 23–37 × 1.2–2.5 cm, slightly wavy-margined. Flowers 1 or 2, on inflorescence-stalks 3–4 cm. Flowers *c.* 3 cm, standing 4 cm when flattened; calyx-teeth unequal, the lowest tooth 3 mm, standard deep pink, wings crimson, keel pale pink. *Italy, Sicily, former Yugoslavia, Albania & Bulgaria.* H3. Early summer.

6. L. tuberosus Linnaeus. Illustration: Botanical Magazine **4**: 111 (1790); Hegi, Illustrierte Flora von Mitteleuropa **4**: t. 171, fig. 4 (1923); Polunin, Flowers of Europe, 56 (1969); Phillips & Rix, Perennials **2**: 59 (1990).

Hairless perennial with slender rhizomes bearing ovoid tubers 3–4.5 cm, these

edible and nutty-flavoured; stems scrambling to 1 m, unwinged. Leaves with leaf-stalk 5–8 mm, one pair of leaflets and a simple tendril; leaflets elliptic or narrowly obovate, 2–4.3 × 0.7–1.8 cm. Racemes with 3–9 flowers; inflorescence-stalk 4.5–5 cm. Flowers sweetly scented, *c.* 9 mm; calyx-teeth unequal, the lowest 3–5 mm; corolla bright pink. *Most of Europe east to C Asia.* H1. Summer.

7. L. rotundifolius Willdenow (*L. heterophyllus*, misapplied). Illustration: Botanical Magazine, 6522 (1880); Phillips & Rix, Perennials **2**: 59 (1990).

Hairless perennial with stems to 2 m or more; stems winged. Leaves with leaf-stalk 1–2.5 cm, one pair of leaflets and a branched tendril; leaflets elliptic, 3–4.5 × 1.5–3.5 cm. Racemes 4–11-flowered; inflorescence-stalk 4–8 cm. Flowers

1.4–1.7 cm, calyx-teeth unequal, the lowest 3 mm; corolla either bright purplish pink or a soft brownish red. *Crimea, Caucasia, E. Turkey & neighbouring Iraq & Iran*. H2. Summer.

L. undulatus Boissier, a close relative of *L. rotundifolius* distinguished by its wavy-edged leaflets, is probably not in cultivation.

8. L. latifolius Linnaeus. Figure 39(4), p. 520. Illustration: Hegi, Illustrierte Flora von Mitteleuropa **4**: 1597 (1923); Grey-Wilson & Blamey, Alpine flowers of Britain and Europe, 121 (1979); Phillips & Rix, Perennials **2**: 58 (1990).

Hairless perennial with robust, strongly winged stems climbing or sprawling to form dense mounds. Leaves with leaf-stalk 3.5–5 cm, one pair of leaflets and a branched tendril; leaflets oblong-elliptic, 8–11 × 2.5–4 cm. Racemes with 6–11 flowers; inflorescence-stalks 12–14 cm. Flowers *c.* 1.8 cm; calyx-teeth unequal, the lowest *c.* 8 mm. Corolla of uniform coloration, varying from deep pink to white. *S Europe*. H2. Summer.

9. L. sylvestris Linnaeus. Illustration: Ross-Craig, Drawings of British plants **7**: t. 73 (1954); Keble Martin, Concise British flora in colour, pl. 25 (1965); Phillips & Rix, Perennials **2**: 59 (1990).

Hairless perennial with stems to 2 m or more; stems winged. Leaves with leaf-stalk 2.5–4 cm, one pair of leaflets and a branched tendril; leaflets linear-elliptic, 5–15 × 0.5–4 cm. Racemes with 3–8 flowers on inflorescence-stalks 10–20 cm. Flowers *c.* 1.4 cm; calyx-teeth unequal, the lowest *c.* 4 mm; wing petals purplish, standard pink, the colour fading towards the margins and much paler on the reverse. *Europe, NW Africa, Caucasia*. H1. Summer.

10. L. tingitanus Linnaeus. Illustration: Botanical Magazine, 100 (1789); Hegi, Illustrierte Flora von Mitteleuropa **4**: 1565 (1923); Polunin, Flowers of Europe, 56 (1969) – as L. latifolius; Polunin & Smythies, Flowers of south-west Europe, 20 (1973).

Hairless perennial with winged stems climbing strongly to 3 m or more. Leaves with leaf-stalk 2–3 cm, one pair of leaflets and a branched tendril; leaflets linear-elliptic, narrowly elliptic or narrowly oblong-elliptic, 4–7 × 0.5–1 cm. Racemes with 1 or 2 flowers on inflorescence-stalk 4–9 cm. Flowers 1.8–2.4 cm; calyx-teeth subequal, *c.* 3 mm; corolla pale pink or

crimson-magenta, *c.* 3.5 cm when flattened. *Iberian Peninsula, Sardinia, Morocco, Algeria, Canary Islands*. H3. Summer.

11. L. odoratus Linnaeus. Illustration: Botanical Magazine, 60 (1788); Everard & Morley, Wild flowers of the world, pl. 29 (1970); Perry, Flowers of the world, 160 (1972).

Hairy annual with erect, strongly winged stems climbing to 2 m or more. Leaves with leaf-stalk 3–4.5 cm, one pair of leaflets and a branched tendril; leaflets ovate-elliptic, 5–6.5 × 3.5–4 cm. Racemes with 2–4 flowers; inflorescence-stalk *c.* 18 cm. Flowers sweetly scented, 3 cm; calyx-teeth equal, 7–8 mm; standard wine-red, keel and wings purple. *Italy, Sicily*. H3. Summer.

The wild species is described above. Modern "sweet peas" are descended from the cultivars 'Countess Spencer' and 'Gladys Unwin', which both had wavy standards. After 1900 this 'Spencer' type ousted the old-fashioned varieties with stiff upright or hooded standards, and there are now innumerable cultivars in every colour except yellow.

12. L. chloranthus Boissier & Balansa. Illustration: Flora Azerbaijan **5**: t. 49 (1954).

Hairy annual with winged stems climbing to 70 cm. Leaves with 1 pair of leaflets and a branched tendril; leaflets elliptic, 2–6 × 7–20 mm. Flowers 1 or 2 on inflorescence-stalk *c.* 6 cm. Calyx teeth subequal, *c.* 7 mm. Flowers *c.* 2 cm, bright greenish yellow, standard downy on the back. *C & E Turkey, Armenia, N Iraq, Iran*. H3. Summer.

13. L. sativus Linnaeus. Figure 37(5). p. 520. Illustration: Botanical Magazine, 115 (1790); Hegi, Illustrierte Flora von Mitteleuropa **4**: t. 171, fig. 6 (1923); Polunin, Flowers of Europe, 56 (1969).

Hairless annual with winged stems climbing weakly to 45 cm. Leaves with leaf-stalk 1 or 2 cm, one pair of leaflets and a simple or branched tendril; leaflets narrowly elliptic, 4–6 cm × 5–8 mm. Flowers solitary; inflorescence-stalk 1.8–3.2 cm. Flowers *c.* 1.3 cm, calyx-teeth subequal, 3–4 mm; corolla blue, fading towards the base and sometimes with pink veins. *S & C Europe, N Africa & SW Asia*. H3. Early summer.

14. L. laxiflorus (Desfontaines) O. Kuntze (*Orobus hirsutus* Linnaeus). Illustration:

Botanical Magazine, 2345 (1822); Stojanoff & Stefanoff, Flora of Bulgaria, edn 2: 654 (1933); Phillips & Rix, Perennials **2**: 58 (1990).

Softly downy perennial, with unwinged stems arching and spreading, 13–18 cm. Leaves with leaf-stalk 1–1.5 cm and one pair of leaflets; leaf-axis ending in a short point 3–5 mm; leaflets broadly elliptic, 2–3 × 1.5–2 cm. Racemes with 3 or 4 flowers; inflorescence-stalk 2.5–4 cm. Flowers 1.3–1.5 cm; calyx-teeth equal and slender, 6–7 mm; standard purple, wings and keel bright blue. *E Mediterranean area, Caucasia, Iran & Syria*. H3. Early Summer.

15. L. nervosus Lamarck. (*L. magellanicus* misapplied). Figure 39(6). p. 520. Illustration: Botanical Magazine, 3987 (1842); Phillips & Rix, Perennials **1**: 60 (1990).

Hairless perennial with unwinged stems trailing or weakly climbing to 1 m, remaining evergreen in winter. Leaves very thick and leathery, glaucous with a white bloom; leaf-stalk absent, leaf-axis ending in a stiff, branched tendril; stipules as conspicuous as the pair of leaflets; leaflets elliptic, 3.5–4 × 2.5–3 cm. Racemes with *c.* 10 flowers borne in whorl-like clusters of 3 or 4; inflorescence-stalk 7–8.5 cm. Flowers 1.5–1.8 cm; calyx-teeth unequal, the lowest *c.* 5 mm; corolla purplish blue, paler on keel and wings. *S America*. H5. Summer.

87. LENS Miller
F.K. Hibberd

Low, hairy annuals, with stems angular but not winged. Leaves with a few pairs of small leaflets and tendrils; leaflets fold in bud. Racemes few-flowered; flowers small, pale and inconspicuous; calyx-teeth equal, as long as the corolla, wing petals adhering to the keel; style flattened, hairy on the inner surface. Pod stalked, oblong, 1–3-seeded; seeds lens-shaped. Germination hypogeal.

A genus of 5 species, distributed around the Mediterranean, in south-west Asia and tropical Africa. Propagation is by seed.

Literature: Davis, P.H. and Plitmann, U. in Davis, *Flora of Turkey* **3**: 325–328 (1970).

1. L. culinaris Medikus (*L. esculenta* Moench). Illustration: Reichenbach, Icones Florae Germanicae **22**: t. 265 (1903); Hegi, Illustrierte Flora von Mitteleuropa **4**: t. 171, fig. 1 (1923); Masefield, et al., Oxford book of food plants, 43 (1969).

Slender hairy annual to 40 cm. Leaflets

elliptic, 7–15 × 1.5–3.5 mm. Flowers pale mauve or white, *c*. 8 mm. *Unknown in the wild*. Summer.

Widely cultivated in Europe and Asia for its seeds (lentils).

88. PISUM Linnaeus.
F.K. Hibberd
Hairless annuals with stems circular in cross-section. Leaves with large leafy stipules, 1–4 pairs of leaflets and a strong, branched tendril; leaflets fold in bud. Racemes few-flowered, flowers showy, calyx-teeth subequal, wing petals united with the keel; style with retroflexed margins, hairy on the inner face. Pod cylindric. Germination hypogeal.

A genus of 2 species, occurring in the Mediterranean area and south-west Asia. Propagation is by seed.

1. P. sativum Linneaus subsp. **sativum**
Stems 10–200 cm. Inflorescence-stalk up to twice as long as stipules, 1–3-flowered. Flowers 1.6–3 cm.

Var. **sativum**. Illustration: Reichenbach, Icones Florae Germanicae **22**: t. 270 (1903); Bonnier, Name this flower, pl. 17 (1917); Masefield et al., Oxford book of food plants, 43 (1969). Flowers white. Seeds spherical, green, yellow or white, smooth or wrinkled, sweet. *The garden pea is unknown in the wild*. H2. Summer.

Var. *arvense* (Linnaeus) Poiret. Illustration: Hegi, Illustrierte Flora von Mitteleuropa **4**: t. 170, fig. 5 (1923); Polunin, Flowers of Europe, 56 (1969). Flowers bicoloured, with pink standard and purple wings. Seeds dark, angular. *Europe & SW Asia*. H2. Summer.

Var. *arvense*, the field pea, is grown for fodder and silage.

89. CICER Linnaeus
F.K. Hibberd
Perennial and annual herbs, often spiny, conspicuously glandular-hairy. Leaves ending in a leaflet, spine or tendril; leaflets 3–many, toothed, the veins terminating in the teeth. Racemes 1–few-flowered. Wing petals free from the keel. Style hairless. Fruit inflated, seeds beaked. Germination hypogeal.

A genus of 40 species distributed around the Mediterranean and east to the Himalayas and central Asia. Propagation is by seed.

Literature: van der Maesen, L.J.F., *Cicer* L., a monograph of the genus, with special reference to the chickpea (*Cicer arietinum* L.), its ecology and cultivation.

Mededelingen Landbouwhogeschool Wageningen 72, **10** (1972).

1. C. arietinum Linnaeus. Illustration:
Botanical Magazine, 2274 (1821); Hegi, Illustrierte Flora von Mitteleuropa **4**: 1500 (1923); Purseglove, Tropical crops, dicotyledons **1**: t. 37 (1968); Masefield et al., The Oxford book of food plants, 39 (1969).

Sturdy annual, *c*. 30 cm. Leaves with 5–7 pairs plus a terminal leaflet. Flowers solitary, 1–1.2 cm thick, mauve or white. Pod 1.7–3 cm, containing 1–3 seeds. *Unknown in the wild*.

Extensively cultivated in S Europe, N Africa and Asia for the seeds (chick pea).

90. ONONIS Linnaeus.
J.M. Lees
Annual or perennial herbs or dwarf shrubs, usually sticky, glandular-hairy. Leaves usually with 3 leaflets, rarely reduced to a single leaflet or pinnate with a terminal leaflet; leaflets usually toothed. Stipules united to the leaf-stalk. Flowers in spikes, racemes or panicles. Calyx bell-shaped or tubular. Corolla yellow, pink or purple, rarely almost white; keel more or less beaked. Stamens 10, their filaments all united. Pod oblong or ovate. Seeds 1–many.

A genus of about 75 species, mostly from the Mediterranean area, but also occurring in the Canary Islands and east to NW India and Mongolia. They are easily grown. Propagation is usually from seed.

Literature: Sirjaev, G., Generis *Ononis* L., Revisio critica, *Beihefte Botanischer Centralblatt* **49**(2): 381–665 (1932).

1a. Flowers in spikes or racemes, sometimes with several flowers at each node; flower-stalks not jointed, or with a joint not more than 1.5 mm from the base; pod ovate or diamond-shaped   **5. arvensis**
  b. Flowers in panicles, sometimes condensed and with the primary branches 1-flowered; flower-stalks distinctly jointed, the joint more than 1.5 mm from the base; pod linear to oblong   2
2a. Corolla yellow, sometimes with red or violet veins   **4. natrix**
  b. Corolla pink or purple, occasionally white   3
3a. Terminal leaflet with a long stalk   **1. rotundifolia**
  b. Terminal leaflet stalkless or with a very short stalk   4
4a. Stem 25–100 cm, erect, usually

many-flowered; most bracts with a single leaflet   **2. fruticosa**
  b. Stem 5–35 cm, prostrate, 1–6-flowered; bracts with 1 or 3 leaflets   **3. cristata**

1. O. rotundifolia Linnaeus. Illustration:
Bonnier, Flore complète **3**: pl. 124 (1914); Polunin & Smythies, Flowers of south-west Europe, pl. 21 (1973); Ceballos et al., Plantas silvestres de la peninsula Iberica, 169 (1980).

Erect, branched dwarf shrub, 10–50 cm. Stem densely glandular-hairy. Leaves with 3 leaflets which are *c*. 2.5 cm, elliptic to almost circular, obtuse, coarsely toothed, sparingly glandular; terminal leaflet with a long stalk. Primary branches of the inflorescence *c*. 3 cm, increasing to 6 cm in fruit, ending in a spine. Flower-stalks 3–6 (rarely to 20) mm. Corolla 1.6–2 cm, pink or whitish. Pod 2–3 cm. Seeds 10–20, *c*. 3 mm, minutely warty. *SE Spain to E Austria & C Italy*. H4. Summer.

2. O. fruticosa Linnaeus. Illustration:
Bonnier, Flore complète **3**: pl. 124 (1914); Polunin & Smythies, Flowers of south-west Europe, pl. 21 (1973).

Erect dwarf shrub to 1 m. Stem downy towards the apex, the young stems downy throughout. Leaves mostly with 3 leaflets, more or less stalkless; leaflets 0.7–2.5 cm, oblong-oblanceolate, somewhat leathery, hairless, toothed. Primary branches of the inflorescence 1–3 cm, in the axils of translucent bracts. Flower-stalks 2–10 mm. Corolla 1–2 cm, pink. Pod 1.8–3 cm. Seeds *c*. 4, to 2.5 mm, minutely warty. *S Europe*. H4. Summer.

3. O. cristata Miller (*O. cenisia* Linnaeus).
Illustration: Bonnier, Flore complète **3**: pl. 124 (1914); Ceballos et al., Plantas silvestres de la peninsula Iberica, 170 (1980).

Creeping perennial, with rhizomes, 5–35 cm. Stems shortly glandular-hairy. Leaves with 3 leaflets which are 0.5–1 cm, oblong or oblanceolate, somewhat leathery. Primary branches of the inflorescence 0.6–3 cm, shortly pointed. Corolla 1–1.5 cm, pink. Pod 0.9–1.2 cm. Seeds *c*. 5, to 2.5 mm, warty. *Mts of S Europe*. H3. Summer.

4. O. natrix Linnaeus. Illustration:
Bonnier, Flore complète **3**: pl. 126 (1914); Taylor, Wild flowers of Spain and Portugal, 55 (1972); Ceballos et al., Plantas silvestres de la peninsula Iberica, 167 (1980).

Erect, much-branched dwarf shrub,

15–60 cm. Stem densely glandular-hairy. Leaves with 3 leaflets or the lower leaves pinnate with more leaflets; leaflets variable, ovate to linear. Flowers in loose, leafy panicles; primary branches 1-flowered. Corolla 0.6–2 cm, yellow, frequently with red or violet veins. Pod 1–2.5 cm. Seeds 4–10, *c.* 2 mm, smooth or minutely warty. *SW Europe.* H3. Summer.

5. O. arvensis Linnaeus (*O. hircina* Jacquin; *O. altissima* Lamarck; *O. spinosa* Linnaeus subsp. *hircina* (Jacquin) Gams).

Shrubby perennial herb, 30–100 cm. Stems erect, variably hairy. Leaves mostly with 3 leaflets; leaflets 0.6–3 cm, elliptic to ovate. Flowers stalked, borne in pairs at each node in dense spikes at the ends of the branches; bracts with 1–3 leaflets; flower-stalk not jointed or joint not more than 1.5 mm from the base. Corolla 1–2 cm, pink. Pod 5–9 mm, about equalling the calyx. Seeds 1–3, *c.* 2.5 mm, dark brown, warty. *Eurasia.* H1. Summer.

O. spinosa Linnaeus. Similar but with flowers borne singly; stem downy with hairs in a single line, pod equalling or slightly exceeding the calyx. *W, C & S Europe.* H3. Summer.

91. PAROCHETUS D. Don
E.H. Hamlet

Creeping herbaceous perennial. Leaves palmate with 3 heart-shaped, finely toothed leaflets, veins running out to margins. Stipules free from the leaf-stalk. Flowers blue or pale purple, 1–4, stalked, in umbels, those in the lower leaf axils very small and not opening with pods ripening on or below the soil surface. Calyx bell-shaped, deeply cleft; lobes 5, acute, almost equal. Petals free, obovate, short-clawed; standard deep blue (rarely purple) wings pinkish, long-clawed, oblong-obovate; keel shorter than wings, turned inwards. Stamens 10, filaments united except for that of the uppermost, which is free. Style bent; stigma small. Pod linear, acutely beaked; seeds 8–20.

A genus of 1 species from Asia, tropical E and southern Africa. It can be grown outside preferring a well-drained soil and partially shaded position or in a cool greenhouse. Suitable for rock gardens or hanging baskets. Propagation is by cuttings.

1. P. communis D. Don. Illustration: Flora of Tropical East Africa, **4**(2): 1015 (1971); Polunin & Stainton, Flowers of the Himalaya, pl. 28 (1984).

E tropical & S Africa to E & SE Asia.

92. MELILOTUS Miller
E.H. Hamlet

Annual or biennial herbs, smelling of newly mown hay. Leaves with 3 pinnately veined leaflets; leaflets usually linear to elliptic-oblong, short-stalked, veins running out to the margin; stipules lanceolate to awl-shaped, fused to the leaf-stalk to some extent. Flowers small, yellow, white or white tipped with blue, in erect, axillary, spike-like racemes. Calyx shortly bell-shaped; lobes 5, almost equal, awl-shaped to lanceolate, acute to acuminate, shorter than the tube. Petals falling early; standard obovate-oblong, narrow at base, nearly stalkless; wings oblong, auricled at the base. Keel blunt, clawed, obtuse. Stamens 10, filaments of 9 united, that of the uppermost free. Style thread-like; stigma small. Pod nutlet-like, straight, beaked, spherical or obovoid. Seeds 1–few.

A genus of 20 species from Mediterranean Europe, SW Asia and adjacent Africa. They are moderately hardy, drought-resistant and prefer alkaline soils. Propagation is by seed.

1a. Flowers white **1. alba**
  b. Flowers yellow **2**
2a. Standard and wings equal, longer than keel; pod transversely wrinkled **3. officinalis**
  b. Standard, wings and keel equal; pod net-veined **2. altissima**

1. M. alba Medikus. (*M. officinalis* (Linnaeus) Lamarck subsp. *alba* (Medikus) Ohashi & Tateishi). Illustration: Flora Palaestina **2**: pl. 224 (1972); Cunningham et al., Plants of western New South Wales, 405 (1981); Keble Martin, The new concise British Flora, pl. 22 (1982); Pignatti, Flora d'Italia **1**: 706 (1982).

Erect, branched annual or perennial, 30–150 cm. Leaves with 3 pinnately veined leaflets; leaflets 1–3 × 0.5–1.5 cm, narrowly oblong-obovate to almost circular, with teeth pointing forwards. Stipules bristle-like, entire. Racemes loose and slender, many-flowered. Petals 4–5 mm, white; wings and keel nearly equal, shorter than standard. Stamens united for about half their length. Pod 3–5 mm, obovoid, mucronate, net-veined, hairless, greyish brown when ripe. *Eurasia.* H1. Summer-autumn.

2. M. altissimus Thuillier. Illustration: Polunin, Flowers of Europe, pl. 57 (1969); Keble Martin, The new concise British Flora, pl. 22 (1982); Reader's Digest, Field guide to the wild flowers of Britain, 113 (1985).

Erect branched biennial or short-lived perennial, 60–150 cm. Leaflets oblong-ovate or wedge-shaped, obtuse, with teeth pointing forwards. Stipules subulate to bristle-like, entire. Racemes 2–5 cm, many-flowered, lengthening in fruit. Flowers 5–7 mm, yellow; wings, standard and keel equal. Pod 5–6 mm, obovoid, acute, net-veined, downy, black when ripe, usually 2-seeded. Style long and persistent. *Europe & Mediterranean area.* H1. Summer.

3. M. officinalis (Linnaeus) Lamarck. Illustration: Phillips, Wild flowers of Britain, 73 (1977); Chinery, Field guide to the plant life of Britain & Europe, 49 (1987); Jimenez et al., Plantas silvestres de la Peninsula Iberica, 165 (1980).

Trailing or erect, branched biennial, 40–250 cm. Leaflets of lower leaves obovate to ovate, those of the upper ovate-lanceolate, all with teeth pointing forwards. Racemes loose and slender, many-flowered. Flowers 4–7 mm, yellow; wings and standard equal, longer than keel. Pod 3–5 mm, transversely wrinkled, mucronate, hairless, brown when ripe, usually 1-seeded. Style often deciduous. *Eurasia, naturalised in N America.* H1. Summer.

93. TRIGONELLA Linnaeus
E.H. Hamlet

Erect annual herbs, often strongly scented. Leaves with 3 toothed leaflets, veins running out to the leaflet margins; stipules joined to the leaf-stalk. Flowers yellow, blue or white in heads, umbels or short dense racemes in leaf-axils, rarely solitary. Calyx 5-toothed, teeth ovate, acuminate. Petals free, standard obovate or oblong; wings oblong, auricled; keel oblong, shorter than wings, obtuse. Stamens free from petals, filaments of 9 united that of the uppermost free. Fruit variable, oblong or oblong-linear, compressed or terete, with 1-many seeds.

A genus of 80 species from Eurasia, Southern Africa and Australia. They prefer a well-drained soil and sunny position. Propagation is by seed.

1a. Flowers 1.2 cm or more **3. foenum-graecum**
  b. Flowers less than 1 cm **2**
2a. Stems hollow; pod abruptly contracted into a beak **1. caerulea**
  b. Stems solid; pod tapering into a beak **2. procumbens**

1. T. caerulea Seringe. Illustration: Polunin, Flowers of Europe, pl. 57 (1969);

Hegi, Illustrierte Flora von Mitteleuropa, **4**: 1234 (1975); Everett, The New York Botanical Garden illustrated encyclopedia of horticulture, 3401 (1982); Pignatti, Flora d'Italia **1**: 710 (1982).

Stems 20–100 cm, erect, sparsely hairy, hollow. Leaflets 2–5 × 0.5–2 cm, ovate to oblong, notched, finely toothed. Racemes spherical, dense, many-flowered. Inflorescence-stalk 2–5 cm. Calyx *c.* 3 mm. Flowers *c.* 6 mm, blue or white. Pod 4–5 × 3 mm, erect or almost horizontal, diamond-shaped to obovate, abruptly contracted into a beak, *c.* 2 mm. Seeds ovoid, brown, finely warty. *Cultivated for fodder in Europe & widely naturalised as a weed.* Summer.

2. T. procumbens (Besser) Reichenbach. Illustration: Polunin & Walters, A guide to the vegetation of Britain & Europe, 152 (1985).

Stems 20–50 cm, decumbent or erect, solid. Leaflets 1–3 × 0.3–1 cm, oblong to linear-oblong, finely toothed. Inflorescence stalk 2–6 cm. Flowers numerous in dense ovoid heads. Calyx 3–3.5 mm, bell-shaped, teeth lanceolate. Flowers 5.5–7 mm, lilac-blue. Pod 1–3-seeded, oblong, somewhat compressed, tapering into a beak. Seeds oblong, smooth. *Europe & SW Asia.*

3. T. foenum-graecum Linnaeus. Illustration: Polunin, Flowers of Europe, pl. 58 (1969); Zohary, Flora Palaestina **2**: pl. 199 (1972); Hegi, Illustrierte Flora von Mitteleuropa, **4**: 1232 (1975).

Stems 10–50 cm, erect. Leaflets 1–3 × 0.5–1.5 cm, obovate to oblanceolate, toothed or sometimes incised. Flowers 1 or 2 in leaf-axils. Calyx 7–8 mm, tubular; teeth lanceolate-linear, as long as calyx-tube. Flowers 1.2–1.8 cm, yellowish white, sometimes tinged with lilac. Pods terete or somewhat compressed, linear, straight or somewhat curved, net-veined, gradually tapering into a beak. Seeds 10–20. *Mediterranean area, probably as an escape from cultivation. Of doubtful origin.* Summer.

Widely cultivated as a spice plant.

94. MEDICAGO Linnaeus
E.H. Hamlet
Annual or perennial herbs or shrubs. Leaves with 3 pinnate leaflets; leaflets obovate, margins toothed, veins running out to margins. Stipules fused to leaf-stalk to some extent. Flowers small, yellow or violet, rarely variegated, in axillary spikes or small heads. Calyx bell-shaped; lobes 5, almost equal. Petals free from staminal tube, standard obovate or oblong,

narrowed at the base; wings oblong, auricled, clawed, longer than the obtuse keel. Stamens 10, filaments of 9 united that of the uppermost free. Style awl-shaped, smooth; stigma almost capitate, oblique. Pod spirally coiled, or sickle-shaped, often covered with spines, with 1–several seeds.

A genus of about 50 species from temperate Eurasia, the Mediterranean area and Africa. They prefer a well-drained alkaline soil. Propagation is by seed.

Literature: Lesins, K.A. & Lesins, I. *Genus Medicago (Leguminosae), A Taxogenetic Study* (1979).

1a. Flowers yellow 2
 b. Flowers violet, lavender, pink or
    white **2. sativa**
2a. Herb; florets less than 5 mm
    3. lupulina
 b. Shrub; florets more than 5 mm
    1. arborea

1. M. arborea Linnaeus. Illustration: Polunin & Everard, Trees & bushes of Europe, 113 (1976); Polunin, Flowers of Greece & the Balkans, 97 (1980).

Shrub 1–4 m, densely covered with adpressed silky hairs. Stipules triangular, entire. Leaflets variable in size, 1–2 × 0.8–1.8 cm, obovate to ovate, wedge-shaped at base, entire or finely toothed at apex. Inflorescence short, with 4–8 flowers. Calyx-teeth shorter than tube. Petals 1–1.5 cm, yellow, standard elliptic. Pods greyish yellow, 1.2–1.5 cm across, net-veined. *Mediterranean, & Canary Islands, S Europe to SW Asia.* Summer.

2. M. sativa Linnaeus. Illustration: Zohary, Flora Palaestina **2**: pl. 204 (1972); Phillips, Wild flowers of Britain, 84 (1977); Cunningham et al., Plants of western New South Wales, 404 (1981); Reader's Digest, Field guide to the wild flowers of Britain, 110 (1985).

Perennial with stems 30–120 cm, prostrate to erect, arising from the crown. Leaflets 0.8–2.8 × 0.3–1.5 cm, obovate at lower nodes, wedge-shaped or linear-oblanceolate at upper nodes. Inflorescence with 7–35 flowers; petals 0.6–1.2 cm. Calyx half the length of the petal. Corolla violet, lavender, rarely pink or white; standard twice or more as long as wide; wings longer than keel. Pod yellow-brown. *Eurasia, naturalised in N America and probably elsewhere.* Summer–autumn.

3. M. lupulina Linnaeus. Illustration: Zohary, Flora Palaestina **2**: pl. 203 (1972); Phillips, Wild flowers of Britain, 53 (1977);

Reader's Digest, Field guide to the wild flowers of Britain, 112 (1985).

Perennial, biennial or annual herbs; stems decumbent, 20–80 cm, branching from the base. Leaflets 1.1–1.4 × 0.6–1.7 cm, broadly ovate to obovate. Inflorescence with 14–24 flowers, petals 2.5–3.5 mm. Calyx 1.5–2.3 mm. Corolla yellow; standard rounded to broadly ovate; wings shorter than keel. Pod an ash-grey to black nutlet with a coiled tip, 1-seeded. *Eurasia, N Africa, naturalised in N America.* Summer–autumn.

95. TRIFOLIUM Linnaeus
U. Oster
Annual, biennial or perennial herbs, occasionally somewhat woody. Leaves usually with 3 leaflets, rarely palmately divided with 5–8 leaflets, the leaflets usually toothed, the veins running out to the margins; stipules large, persistent, joined to the leaf-stalk to some extent. Flowers in heads or short spikes, rarely solitary. Calyx tubular, 5-toothed, teeth equal or unequal. Petals persistent or deciduous, attached to each other and to the staminal tube. Stamens 10, 9 with their filaments united, that of the uppermost free, with all or 5 of the filaments swollen towards the apex. Pod enclosed in the persistent calyx or shortly protruding, indehiscent or dehiscent by the inner suture or by a hardened lid; seeds 1–4, rarely to 10.

A genus of about 300 species, mostly from north temperate regions, a few from the southern hemisphere. Many are important as fodder or green manure crops, but only a small number is grown for ornament. They are easily grown in good soil and are propagated by seed (which may require scarification) or, in the case of the perennial species, by division.

1a. Flowers solitary **5. uniflorum**
 b. Flowers in heads 2
2a. Leaflets 5 or more **1. lupinaster**
 b. Leaflets 3 3
3a. Calyx with only 5 distinct veins
    4. hybridum
 b. Calyx with 10–20 distinct veins
    4
4a. Calyx hairless 5
 b. Calyx hairy 6
5a. Stipules attached to the leaf-stalk
    for at least $^6/_7$ of their length; seeds
    1 or 2 **2. alpinum**
 b. Stipules attached to the leaf-stalk
    for up to ½ their length; seeds 3 or
    4 **3. repens**

6a. Calyx with 20 distinct veins
　　　　　　　　　　　10. rubens
　b. Calyx with 10 distinct veins　　7
7a. Flowers stalked; fruiting calyx
　　inflated　　　　　　　　　　8
　b. Flowers not stalked; fruiting calyx
　　not inflated　　　　　　　　9
8a. Flowers reversed so that the
　　standard is lowermost
　　　　　　　　　7. resupinatum
　b. Flowers not reversed, standard
　　uppermost　　　　**6. fragiferum**
9a. Inflorescence terminal
　　　　　　　　　11. pannonicum
　b. Inflorescence axillary　　　　10
10a. Inflorescence long egg-shaped to
　　cylindric　　　　　**8. incarnatum**
　b. Inflorescence spherical, egg-shaped
　　or oblong-conical　　　　　11
11a. Lowest tooth of calyx 3-veined, at
　　least at the base
　　　　　　　　　12. alexandrinum
　b. Lowest tooth of the calyx 1-veined
　　　　　　　　　　9. pratense

1. T. lupinaster Linnaeus. Illustration: Botanical Magazine, 879 (1805); Hegi, Illustrierte Flora von Mitteleuropa, **4**: 1313 (1975).

Herbaceous perennial, stems 15–50 cm, erect or ascending, hairless or with a few adpressed hairs. Leaf-stalks to 1 cm, shorter than stipules, and united to them; leaflets 5–8 (only 3 in the first-developed leaves), lanceolate-oblong, toothed. Inflorescence-stalk 10–30 cm, with axillary heads forming an umbel; each head with up to 20 flowers; flower-stalks 1–2 cm, deflexed after flowering. Calyx-teeth about as long as tube. Corolla 1–2 cm, crimson red, or rarely yellowish white. Pod twice as long as the calyx, with 1–9 seeds. *E Europe to E Asia*. H2. Early summer.

2. T. alpinum Linnaeus. Illustration: Polunin, Flowers of Europe, pl. 60 (1969); Hegi, Illustrierte Flora von Mitteleuropa, **4**: 1313 & t. 163 (1975).

Hairless herbaceous perennial to 18 cm, with a tap-root. Leaf-bases covering the stems; leaf-stalks 2–12 cm, stipules 4–9 cm, attached to the leaf-stalks; leaflets 3, 1–5 cm, lanceolate to linear. Inflorescence-stalk 5–15 cm; heads with 3–12 flowers on stalks to 2 mm. Calyx-teeth *c.* 6 mm, the lower to 1 cm. Corolla 1.8–2.5 cm, pink, purple or rarely cream. Pod stalked, obovoid, containing 1 or 2 seeds. *S Europe, from Spain to the Alps*. H3. Summer.

A high-altitude plant whose cultivation in the lowlands is difficult.

3. T. repens Linnaeus. Illustration: Jávorka & Csapody, Iconographia Florae Hungaricae, t. 2010 (1929); Keble Martin, The new concise British Flora in colour, pl. 22 (1969); Hegi, Illustrierte Flora von Mitteleuropa, **4**: t. 164 (1975).

More or less hairless perennial with a tap-root, stems creeping and rooting at the nodes. Leaf-stalks to 2 cm; leaflets 3, 0.5–3 cm, broadly obovate, base tapered, with pale marks above and translucent veins, margins with fine forwardly pointing teeth; stipules membranous, sheathing, with subulate tips, veins green or red. Flowers-heads spherical; flowers stalked, scented. Calyx-tube whitish, teeth lanceolate, green, the 2 upper slightly longer than the rest and separated by a narrow, acute sinus. Corolla 0.8–1.3 cm, cream to pink, deflexed and brown after flowering. Pod linear, flat, constricted between the 3 or 4 seeds. *Europe, W Asia, N Africa, widely introduced elsewhere*. H1. Late spring to autumn.

A variable species, widely cultivated. 'Atropurpureum' (forma *atropurpureum* Anon.), with dark purple-red leaflets, is often cultivated in gardens.

4. T. hybridum Linnaeus. Illustration: Jávorka & Csapody, Iconographia Florae Hungaricae, f. 2009 (1929); Polunin, Flowers of Europe, pl. 59 (1969); Keble Martin, The new concise British Flora in colour, pl. 22 (1969); Hegi, Illustriertes Flora von Mitteleuropa, **4**: t. 164 (1975).

Hairless biennial or perennial herb, usually to 40 cm, sometimes to 90 cm; stems erect or ascending, if prostrate, not rooting at the nodes. Leaves with 3 leaflets, leaf-stalks to 10 cm; leaflets 1–2 × 0.5–1.5 cm, obovate or inverted-heart-shaped, rarely elliptic; stipules greenish, broadly ovate, gradually narrowing to the tip. Flower-heads spherical, apparently terminal; inflorescence-stalk about twice as long as the subtending leaf. Flowers stalked, stalks to twice as long as calyx-tube, deflexed after flowering. Calyx-tube 1–1.5 mm, with 5 distinct veins and 5 more obscure veins; teeth 2–3 mm, the upper 2 slightly longer than the lower 3 and separated by a broad sinus. Corolla 0.5–1 cm, at first whitish, later pink and finally turning brown. Pod ellipsoid, protruding from the calyx, with 2–4 seeds. *Now found in most of Europe & elsewhere, but probably native only to W Europe*. H1. Late spring–summer.

5. T. uniflorum Linnaeus. Illustration: Sweet, British flower garden, ser. 2, **2**: 200

(1838); RHS dictionary of gardening, 2146 (1956); Polunin & Huxley, Flowers of the Mediterranean, pl. 63 (1965).

Perennial with woody tap-root. Stems prostrate, tufted, with very short internodes. Leaf-stalks 1–8 cm, leaves with 3 leaflets; leaflets 0.4–1.5 cm, circular, obovate or diamond-shaped, acute or obtuse, with a short, sharp, flexible point, strongly veined, hairless or with adpressed hairs; stipules broadly triangular, membranous, long-acuminate, overlapping. Flowers solitary, axillary, stalks 1–7 mm, usually shorter than calyx-tube, deflexed and sometimes thickened in fruit. Calyx-tube 5–7 mm, hairless or downy; teeth almost equal, narrowly lanceolate, half as long as tube or less. Corolla 1.2–2.5 cm, cream, purple or parti-coloured. Pod linear, acute, hairy above. *Mediterranean area from Sicily eastwards*. H4. Late spring–summer.

6. T. fragiferum Linnaeus. Illustration: Keble Martin, The new concise British Flora in colour, pl. 23 (1969); Hegi, Illustrierte Flora von Mitteleuropa, **4**: t. 164 (1975).

Perennial (rarely biennial) herb, hairy or hairless. Stems 2–40 cm, usually prostrate and often rooting at the nodes, sometimes tufted. Leaf-stalks to 15 cm; leaflets 3, obovate, elliptic or inverted-heart-shaped, 0.3–2 cm; stipules *c.* 1 cm, membranous, lanceolate. Inflorescence-stalk 5–20 cm, hairless or with scattered hairs; head hemi-spherical and 1–1.5 cm wide in flower, later somewhat ellipsoid. Calyx with shaggy, glandular hairs, at least in the upper part, the upper lip greatly inflated and helmet-like in fruit. Corolla pale pink or rarely white. Pod not projecting from the calyx, with 1 or 2 seeds. *Most of Europe*. H2. Summer.

Tolerates somewhat brackish soils.

7. T. resupinatum Linnaeus. Illustration: Jávorka & Csapody, Iconographia Florae Hungaricae, t. 1996 (1929); Hegi, Illustrierte Flora von Mitteleuropa, **4**: 1320, 1321 (1975).

Hairless annual herb 10–60 cm, stems ascending, prostrate or erect. Lower leaves congested and rosette-like, long-stalked, the upper shorter, almost stalkless. Leaflets 3, 0.5–2.5 cm, obovate, rarely elliptic, tapered to the base; margins with fine, forwardly pointing teeth; stipules lanceolate, membranous. Flower-heads axillary, spherical, 0.8–2 cm. Flowers inverted (standard downwards). Calyx 0.5–1 cm, pear-shaped in fruit, hairy on the upper side, with clearly visible veins, upper 2 teeth divergent. Corolla 2–8 mm, pink to purple. Pod spherical to ovoid, membranous, containing 1 or 2 seeds. *Mediterranean area, introduced elsewhere.* H4. Spring–early summer.

8. T. incarnatum Linnaeus. Illustration: Botanical Magazine, 328 (1796); Jávorka & Csapody, Iconographia Florae Hungaricae, t. 2026 (1929); Polunin, Flowers of Europe, pl. 59 (1969); Hegi, Illustrierte Flora von Mitteleuropa, **4**: 1329 (1975).

Annual herb. Stems 20–50 cm, simple or branched from the base, erect or ascending; stems and leaves with adpressed or spreading hairs. Leaflets 3, obovate to almost circular, 0.8–3.5 cm, finely toothed towards the tip; stipules ovate, blunt, membranous or green, tips obscurely toothed, often reddish. Inflorescence long-stalked, elongate-egg-shaped to cylindric, 10–50 cm, flowers stalkless. Calyx-teeth about as long as the tube, acute, linear, spreading stellately in fruit. Corolla blood-red, pink, cream or white. Pod ovoid. *S Europe.* H4. Late spring–early summer.

Sensitive to frost and to drought during spring.

9. T. pratense Linnaeus. Illustration: Jávorka & Csapody, Iconographia Florae Hungaricae, t. 2021 (1929); Hegi, Illustrierte Flora von Mitteleuropa, **4**: 1331 & t. 162 (1975); The Living Countryside No. **7**: 1523 (1977).

Tufted perennial herbs (often short-lived), more or less adpressed hairy throughout. Stems to 1 m. Leaflets 3, obovate or oblong-lanceolate to almost circular, 1–4 × 0.5–2 cm, often hairy only beneath, often with a lighter green or reddish stripe above; stipules ovate-lanceolate, abruptly contracted into bristle-like tips. Heads axillary, 2–4 cm, spherical to ovoid, enclosed at the base by stipules of reduced leaves; flowers stalkless. Calyx-tube adpressed hairy, 10-veined; teeth subulate, the lowest 1-veined, twice as long as the tube. Corolla 1.2–1.5 cm, reddish purple or pink, rarely white. Pod ovoid, 1-seeded. *Europe, introduced elsewhere.* H1. Late spring–summer.

Widely grown as a fodder crop.

10. T. rubens Linnaeus. Illustration: Jávorka & Csapody, Iconographia Florae Hungaricae, t. 2013 (1929); Polunin, Flowers of Europe, pl. 59 (1969); Hegi, Illustrierte Flora von Mitteleuropa, **4**: 1347 & t. 162 (1975).

Usually hairless perennial with a creeping rhizome. Stems erect, 20–60 cm. Leaflets 3, oblong-lanceolate or rarely elliptic, to 6–7 × 1–1.5 cm, toothed; stipules to 7 cm, lanceolate, attached to the stems for more than half their length. Inflorescence axillary. Heads cylindric, to 8 × 2.5 cm, stalk to 4 cm. Calyx-tube with 20 veins; teeth subulate, hairy, the lowest much longer than the others. Corolla *c.* 1.5 cm, purple or rarely white. Pod ovoid, 1-seeded. *C & S Europe.* H4. Early summer.

11. T. pannonicum Jacquin. Illustration: Jávorka & Csapody, Iconographia Florae Hungaricae, t. 2016 (1929); Hegi, Illustrierte Flora von Mitteleuropa, **4**: 1350 (1975).

Perennial with long tap-root and short rhizomes. Stems erect, 20–50 cm (rarely more), hairy. Leaflets 3, oblong-lanceolate, 3–6 × 0.8–1.8 cm, tips pointed or rounded; free tips of stipules linear, herbaceous, to 3 cm. Inflorescence-stalk to 8 cm. Heads terminal, ovoid or cylindric, 5–8 cm. Calyx-teeth linear to subulate, the lowest twice as long as the others, all hairy. Corolla 2–2.5 cm, yellowish white. Pod 1-seeded. *E, C & S Europe.* H4. Early summer.

12. T. alexandrinum Linnaeus. Illustration: Hegi, Illustrierte Flora von Mitteleuropa, **4**: 1282 (1975).

Annual herb to 70 cm, erect, branched, sparsely hairy. Leaflets 3, elliptic or oblong, 1.3–2.5 × 0.5–1 cm; leaf-stalks 2–5 cm, stipules 0.7–1.4 cm, free part subulate, the upper stipules dilated at the base. Inflorescence-stalk to 3 cm; heads ovoid or oblong-conical, 1.5–2 cm. Calyx hairy, tube tapered towards the base, teeth unequal, spine-like, the lowest 3-veined at least at the base, about as long as the tube, the others shorter and 1-veined. Corolla 0.8–1 cm, cream. Pod slightly protruding from the calyx. *E Mediterranean area.* H4. Summer.

96. BRACHYSEMA R. Brown
J. Cullen
Shrubs, sometimes prostrate, sometimes scrambling or climbing. Leaves opposite (ours) or alternate, simple (reduced to a single leaflet), leathery, occasionally reduced to small scales on flattened stems (not ours). Stipules present (ours) or absent. Flowers axillary, solitary or in clusters. Calyx 5-toothed. Corolla usually red; standard recurved, narrow, shorter than the wings and keel. Stamens 10, filaments free. Ovary with several ovules.

Pods swollen, leathery, containing several seeds with arils.

A genus of 7 species from Western Australia. Only a single species is generally grown, though others have been cultivated in the past. It is propagated by seed or by cuttings.

1. B. celsianum Lemaire (*B. lanceolatum* Meissner). Illustration: Botanical Magazine, 4652 (1852); Erickson et al., Flowers and plants of Western Australia, 127 (1973).

Shrub, sometimes scrambling, to 2 m and 3 m wide; young shoots covered with silver or white hairs. Leaves opposite, ovate, entire, 2–5 cm, dark green and more or less hairless above, densely silvery-hairy beneath; stipules narrow, subulate. Flowers 1–3 in the leaf-axils. Calyx to 1.5 cm, funnel-shaped, deeply 5-toothed, densely covered with silver hairs. Corolla to 2.5 cm, red. Pods hairy, *c.* 1.5 cm. *Western Australia.* H5–G1.

97. OXYLOBIUM Andrews
J. Cullen
Shrubs, sometimes somewhat scrambling. Leaves simple (reduced to a single leaflet), leathery, alternate, opposite or in whorls (these sometimes rather irregular), margins often recurved. Stipules absent or subulate. Flowers in terminal and axillary racemes or panicles which are sometimes dense and head-like, yellow, red or orange. Calyx funnel-shaped, not umbilicate at base, deeply 5-toothed. Standard broad. Stamens 10, filaments free. Ovary stalked or not, ovules 4 or more, usually densely hairy, with several seeds; seeds without appendages.

A genus of about 30 species from Australasia. Several have been cultivated, but only 2 seem to be generally available. They are easily grown in a warm, rather dry site, and may be propagated by cuttings.

1a. Leaves mostly in whorls of 3–4, mostly more than 4 mm wide; stipules apparently absent
　　　　　　　1. ellipticum
　b. Leaves alternate, mostly less than 4 mm wide; stipules present, subulate
　　　　　　　2. linearifolium

1. O. ellipticum R. Brown. Illustration: Botanical Magazine, 3249 (1833).

Erect shrub to 2.5 m; young shoots with dense adpressed or somewhat spreading hairs. Leaves crowded, usually in whorls of 3 or 4 but the whorls sometimes irregular, elliptic to linear, 1–3 × 0.3–1.1 cm (rarely longer and proportionately narrower), apex sharply mucronate, margins recurved, hairless above, densely brown-hairy beneath; stipules apparently absent. Flowers in dense, head-like racemes or panicles at the ends of the branches. Calyx 6–7 mm, with adpressed or somewhat spreading hairs. Corolla yellow or orange-yellow. Standard broad, 1–1.2 cm, longer than wings and keel. Pod shortly cylindric, *c.* 8 mm, densely spreading-hairy. *Australia (Tasmania, Victoria, New South Wales).* H5–G1.

2. O. linearifolium (G. Don) Domin (*O. callistachys* misapplied).

Shrub; young shoots with adpressed hairs. Leaves sparse, alternate, very narrowly linear or linear-lanceolate, to 8 cm × 4 mm, apex sharply mucronate, margins recurved, hairless above, brown-hairy beneath; stipules small, subulate. Flowers sparse in long-stalked racemes. Calyx 0.8–1 cm, densely hairy. Standard yellowish, broad, *c.* 1.5 cm; wings and keel reddish, about as long as the standard. *Western Australia.* H5–G1.

98. CHORIZEMA Labillardière
J. Cullen
Shrubs, sometimes sprawling or scrambling. Leaves alternate, leathery, simple (reduced to a single leaflet), margins usually toothed (ours); stipules small or absent. Flowers in terminal or rarely axillary racemes. Calyx 5-toothed. Corolla red, orange or pinkish purple, standard longer than the wings. Stamens 10, filaments free. Ovary with 8–many ovules. Pod ovoid or compressed, rather soft, several-seeded. Seeds without appendages.

A genus of 18 species from Australia (mostly Western Australia), a few cultivated for their bright flowers and interesting, often holly-like foliage. They require a moist but well-drained soil in a partially shaded but sunny site and are propagated by seed, which requires scarification.

1. C. cordatum Lindley. Illustration: Edwards's Botanical Register **24**: pl. 10 (1838); Botanical Magazine, n.s., 237 (1954); Morcombe, Australia's western wild flowers, 101 (1968); Growing Native Plants **11**: 269 (1981).

Shrub, often sprawling or climbing through other vegetation. Leaves to 5 cm, ovate, acute at apex, cordate at base, margin usually with coarse, broadly triangular, sharply pointed teeth, but occasionally entire or almost so, more or

less hairless above and beneath. Racemes to 15 cm. Standard scarlet or orange, yellow towards the base of the blade, wings and keel pinkish purple. Pods inflated. *Western Australia.* H5–G1.

There are 2 very similar (probably not distinct) species whose names sometimes appear in the literature:

C. varium Lindley. Very similar but young shoots and the undersides of the leaves hairy. *Western Australia.* H5–G1.

C. ilicifolium Labillardière. Illustration: Botanical Magazine, 1029 (1807); Erickson et al., Flowers and plants of Western Australia, 58 (1973). A very small shrub with longer, narrower leaves which are somewhat hairy beneath. *Western Australia.* H5–G1.

99. GOMPHOLOBIUM J.E. Smith
J. Cullen
Shrubs. Leaves alternate to almost opposite, usually compound with 3 (ours) or more leaflets; stipules minute or absent. Flowers axillary, rather large and showy; buds long-stalked. Calyx very deeply 5-lobed, the tube very short, black outside, the lobes edge-to-edge in bud. Petals shortly clawed; standard large, broad, longer than the wings and keel; keel very deep. Stamens 10, their filaments free. Ovary with usually 8 ovules. Pod inflated, ovoid to spherical, several-seeded.

A genus of 26 species from Australasia. They are striking plants which caused great interest when they were introduced to Europe in the 19th century, but only 1 is currently available. It is showy, with narrow, greyish leaflets, black, caper-like flower-buds and yellow flowers. It is not particularly easy to grow, requiring good drainage and a sheltered site. Propagation is by seed or by cuttings taken from current growth when just firm, but young plants are difficult to establish.

1. G. latifolium J.E. Smith. Illustration: Botanical Magazine, 4171 (1845); Rotherham et al., Flowers and plants of New South Wales and southern Queensland, 59 (1975); Elliot & Jones, Encyclopaedia of Australian plants **4**: 380 (1986).

Upright shrub to 2 m; shoots somewhat hairy or rough, at least at the nodes. Leaves alternate or almost opposite, shortly stalked, divided into 3 leaflets; leaflets 3–6 cm × 3–6 mm, linear-obovate, tapered to the base, abruptly acute or truncate (though with a short mucronate point) at the apex, hairless. Calyx 1–1.5 cm, black

outside, the lobes pale inside, sometimes with glands there, margins densely fringed. Corolla bright to dull yellow, standard broad, 2–2.5 cm long; keel deep, fringed along the suture between the petals. Pods ovoid, stalked, to 1.8 cm. *Australia (Victoria, New South Wales, Queensland).* H5–G1.

100. BURTONIA R. Brown
J. Cullen
Small shrubs; young shoots hairy or hairless. Leaves alternate, pinnate or made up of 3 leaflets borne side-by-side, without a common leaf-stalk; leaflets narrow, heather-like, with downwardly curved margins in ours; stipules minute or absent. Flowers solitary in the upper leaf-axils but numerous, forming a leafy terminal raceme. Calyx deeply 5-lobed with very short tube, the lobes edge-to-edge in bud. Corolla purple, red, orange or yellow; standard with broad blade, longer than wings and keel; keel deep, wider than wings. Stamens 10, filaments free. Ovary shortly stalked or stalkless, containing 2 ovules. Pods more or less spherical. Seeds without appendages.

A genus of about 12 species from Australia. Like *Gompholobium*, several species have been introduced, but only 2 are available today. Both of these have narrow, heather-like (ericoid) leaflets. They require moist but well-drained soil, a sheltered site and are very susceptible to frost damage. Propagation is by seed, which requires scarification; young plants are difficult to establish.

1a. Young branches hairy; standard orange-red with a yellow blotch, wings and keel brownish
        2. hendersonii
  b. Young branches hairless; standard pinkish purple, wings and keel reddish   **1. scabra**

1. B. scabra (Smith) R. Brown. Illustration: Botanical Magazine, 4392 (1848) – as B. pulchella, & 5000 (1857); Gardner, Wild flowers of Western Australia, 61 (1959); Erickson et al., Flowers and plants of Western Australia, 73 (1973); Elliot & Jones, Encyclopaedia of Australian plants **2**: 395 (1982).

Shrub with hairless young shoots. Leaves dense, usually longer than the internodes and flower-stalks, to 1.5 cm × 1 mm, greyish, rough. Calyx black, 5–7 mm, the teeth with fringed margins. Standard pinkish purple, 1.5–1.7 cm; wings and keel reddish. *Australia (Western Australia).* H5–G1.

2. B. hendersonii (Paxton) Bentham. Illustration: Erickson et al., Flowers and plants of Western Australia, 119 (1973).

Like *B. scabra* but young shoots hairy, leaves not rough, usually shorter than the internodes and flower-stalks, standard orange-red with a yellow blotch at the base, wings and keel brownish. *Australia (Western Australia)*. H5–G1

101. SPHAEROLOBIUM J.E. Smith
J. Cullen

Hairless perennial herbs or shrubs, leafless (ours). Flowers in a raceme or in lateral clusters, small, 2 per bract. Calyx 5-lobed, the 2 upper lobes broad and with diverging tips, united for most of their length, the 3 lower narrowly triangular to lanceolate, more deeply divided. Standard with a broad blade, longer than wings; keel longer or shorter than wings. Stamens 10, filaments free. Ovary with 2 ovules; style incurved, usually with a membrane or ring of hairs just below the stigma. Pod spherical or compressed, 1–2-seeded. Seeds without appendages.

A genus of about 15 species from Australia. Only 1 is grown for its curious appearance; little is known of its cultivation requirements, but these are presumably similar to those of *Burtonia* (p. 527).

1. S. vimineum J.E. Smith. Illustration: Botanical Magazine, 969 (1807); Rotherham et al., Flowers and plants of New South Wales and southern Queensland, 37 (1975); Galbraith, Collins' field guide to the wild flowers of south-east Australia, pl. 38 (1977).

Perennial herb or low shrub, prostrate, sprawling or erect, to 70 cm; stems wiry, branched, leafless. Flowers small, in pairs (often with a projection between them) in racemes, each pair subtended by a scale-like bract; flower-stalks recurved, jointed just beneath the calyx. Calyx dark or black, 2–3 mm, lobes with fringed margins. Standard 4–6 mm, yellow with a red flush or blotch in the centre; wings yellow, keel greenish. *Australia (New South Wales, Victoria, South Australia, Tasmania)*. H5–G1.

102. VIMINARIA J.E. Smith
J. Cullen

Shrub to 3 m or more with branched shoots, sometimes with a trunk to 1.5 m; young shoots minutely hairy. Leaves alternate, usually reduced to thread-like stalks, leaflets absent; stipules lanceolate,

fringed or minutely toothed. Flowers stalked in long terminal racemes, 1 per bract; bracts similar to the stipules. Calyx green, membranous, cylindric to funnel-shaped, 3–4 mm, with 5 short equal teeth which are sparsely fringed. Standard yellow with a dark blotch, lines or flush towards the base, broad, 0.8–1 cm, longer than wings; wings and keel reddish brown. Stamens 10, filaments free. Ovary short, containing 2 ovules. Pod ovoid-oblong, indehiscent, containing 1 seed which has a small appendage.

A genus of a single species widespread in Australia. Cultivation as for *Sphaerolobium* (p. 528).

1. V. juncea (Schrader) Hoffmannsegg (*V. denudata* J.E. Smith). Illustration: Botanical Magazine, 1190 (1809) – leaves unusual; Cochrane et al., Flowers and plants of Victoria, 25 (1968); Galbraith, Collins' field guide to the wild flowers of south-east Australia, pl. 36 (1977).
   Australia. H5–G1.

103. DAVIESIA J.E. Smith
J. Cullen

Shrubs, sometimes spiny. Leaves alternate, simple, needle- or heather-like, sometimes ending in spines, or absent (see comment under species **2**). Stipules usually absent. Flowers solitary or in clusters in the leaf-axils, usually concentrated towards the tips of the branches. Bracteoles not close beneath and overlapping the calyx. Calyx deeply or shallowly 5-toothed, not umbilicate at base. Corolla yellowish, often marked or blotched with red or other colours; standard with a broad blade, about as long as the wings and keel. Stamens 10, filaments free. Ovules 2. Pod flattened, triangular in outline, containing 2 seeds.

A genus of about 75 species from Australia. Only a few are grown, requiring a well-drained soil and a sunny position. They are propagated by seed, which requires scarification.

1a. Leaves of 2 forms, most of them needle-like but some laterally compressed and with a hump towards the apex on the upper edge; stems hairless   **2. incrassata**
  b. Leaves all of the same form, heather-like; stems with crisped hairs   **1. acicularis**

1. D. acicularis J.E. Smith. Illustration: Botanical Magazine, 2679 (1826); Cunningham et al., Plants of western New

South Wales, 388 (1981).

Shrub to 60 cm or more; shoots rough with short, crisped hairs. Leaves crowded, ascending-erect, heather-like with margins curved downwards and a prominent midrib beneath, not spine-tipped though acute, hairy or not. Flowers solitary or in small clusters in the leaf-axils, much shorter than the subtending leaves. Calyx *c.* 3 mm, deeply 5-toothed. Standard yellow, reddish-blotched towards the base; keel reddish. Pod *c.* 1 cm. *Australia (New South Wales, Queensland, South Australia)*. H5–G1.

2. D. incrassata J.E. Smith. Illustration: Botanical Magazine, 4244 (1846); Gardner, Wild flowers of Western Australia, 66 (1959).

Glaucous shrub to 1.5 m, shoots hairless. Leaves (see comment below) mostly terete, grooved, ending in a sharp spine; some laterally compressed, widening upwards and with a hump on the upper side towards the hard, spiny apex; all leaves distant, spreading, internodes clearly visible. Flowers in axillary clusters of 3–8 in the leaf-axils, exceeding the subtending leaves. Calyx shallowly 5-toothed, *c.* 2.5 mm. Standard reddish orange or yellow, sometimes marked with green or black; keel reddish, sometimes black-tipped. Pod slightly inflated, to 1.4 cm. *Australia (Western Australia)*. H5–G1.

Some authors regard this species as leafless, considering the organs described above as leaves to be modified shoots.

104. PULTENAEA J.E. Smith
J. Cullen

Shrubs. Leaves usually alternate, simple, flat or terete (though grooved above); stipules usually present, small, lanceolate (ours). Flowers in clusters in the leaf-axils towards the tips of the shoots or in terminal heads with individual flowers very shortly stalked. Bracteoles persistent, borne close below and overlapping the calyx. Calyx bell-shaped, not umbilicate at base, shallowly to deeply toothed with 5 more or less equal teeth. Corolla yellow or pink; standard with a broad, notched blade, somewhat longer than the wings and keel. Stamens 10, filaments free. Ovary containing 2 ovules. Pod ovate, compressed, exceeding the persistent calyx. Seeds with arils.

A genus of about 120 species from Australia. Only a small number is available in Europe today, though many more were introduced in the past. They require a well-drained soil in a sunny position and may be propagated by seed or by cuttings.

Literature: Woolcock, D.T. & C.E., *Pultenaea*, with reference to related genera of the pea-flower family, *Australian Plants* **12**: 95–103, 116–121 (1982), 304–309 (1984); **13**: 72–83 (1984), 250–257, 324–329 (1986); **14**: 130–137, 179–185 (1987) and continuing.

1a. Leaves terete, warty; flowers pink **3. subalpina**
  b. Leaves flat, not warty; flowers mostly yellow **2**
2a. Leaves narrowly ovate, 4–7 × 2–3 mm **1. gunnii**
  b. Leaves narrowly obovate or rarely narrowly oblong, 1–2 cm × 2–3 mm **2. flexilis**

1. P. gunnii Bentham. Illustration: Cochrane et al., Flowers and plants of Victoria, 33 (1968); Galbraith, Collins' field guide to the wild flowers of south-east Australia, pl. 39 (1977); Australian Plants **12**: 121 (1982).

Wiry shrub to 1 m; young shoots hairy. Leaves alternate, 4–7 × 2–3 mm, narrowly ovate, convex and hairless above, hairy beneath; stipules pointed. Flowers 3–8 in terminal heads. Calyx *c.* 4 mm, very deeply 5-toothed with lanceolate teeth, hairy. Corolla mainly yellow, keel dark red. Pod *c.* 5 mm, somewhat swollen, hairy. *Australia (Victoria, Tasmania)*. H5–G1.

2. P. flexilis J.E. Smith. Illustration: Australian Plants **12**: 306 (1984).

Erect shrub to 4 m; shoots hairy at least when young. Leaves alternate, 1–2 cm × 2–3 mm, usually narrowly obovate, occasionally narrowly oblong, hairless or very sparsely hairy. Flowers in clusters in the leaf-axils towards the tips of the branches. Calyx *c.* 3 mm, with 5 rather broad, shallow, more or less equal teeth, hairless or the tooth-margins hairy. Corolla yellow. Pod *c.* 6 mm. *Australia (New South Wales, S Queensland)*. H5–G1.

3. P. subalpina (Mueller) Druce (*P. rosea* Mueller). Illustration: Botanical Magazine, 6941 (1887).

Erect shrub; shoots hairy with dark, spreading hairs. Leaves alternate, to 1.5 cm, terete, grooved above, warty, hardened into a short point at the apex. Flowers in stalkless terminal heads. Calyx 4–5 mm, densely hairy, deeply 5-toothed with lanceolate teeth. Corolla pink. Pod 4–5 mm. *Australia (Victoria)*. H5–G1.

105. DILLWYNIA J.E. Smith
J. Cullen

Heath-like shrubs; young shoots usually hairy. Leaves alternate, simple, needle-like or heather-like, often spine-tipped. Flowers axillary, solitary, paired or in small clusters, or in short terminal racemes; bracteoles not close beneath and overlapping the calyx. Calyx 5-toothed, the 2 upper teeth larger and more united than the 3 smaller, lower teeth; not umbilicate at base. Petals yellow, yellow-orange or reddish, standard broader than long, wings narrow, keel straight or incurved. Stamens 10, filaments free. Ovary containing 2 ovules, style hooked near the apex. Pod ovoid to almost spherical, short.

A genus of 22 species from Australia. They require a well-drained, sandy soil and a sunny position and can be propagated by seed or by cuttings.

1a. Leaves warty, blunt at the apex **1. floribunda**
  b. Leaves not warty, spine-tipped **2. juniperina**

1. D. floribunda J.E. Smith. Illustration: Botanical Magazine, 1545 (1813); King & Burns, Wild flowers of Tasmania, 31 (1971); Rotherham et al., Flowers and plants of New South Wales and southern Queensland, 36 (1975); Elliot & Jones, Encyclopaedia of Australian plants **3**: 277 (1984).

Erect shrub, 15–200 cm; shoots hairy, sometimes densely so. Leaves needle-like, more or less terete, 1–1.5 cm, spreading or ascending, warty, blunt at the apex. Flowers in pairs in the leaf-axils, numerous and forming long, many-flowered, leafy racemes, the shoot often continuing vegetative growth above. Calyx top-shaped (narrowed to the base), 5–6 mm, sparsely hairy with long, wavy hairs. Corolla yellow or yellow-orange, petals falling after flowering. Pods to 7 mm, scarcely exceeding calyx. *Australia (New South Wales, Queensland)*. H5–G1.

2. D. juniperina Loddiges. Illustration: Elliot & Jones, Encyclopaedia of Australian plants **3**: 278 (1984).

Erect shrub; shoots densely adpressed hairy. Leaves needle-like, narrow, more or less triangular in section, 1–1.5 cm, widely spreading, not warty, ending in hardened spines. Flowers in short, few-flowered, head-like racemes terminating the shoots. Calyx bell-shaped (rounded at base), 3–4 mm, adpressed hairy. Petals yellow or yellow-orange, persistent. Pod *c.* 6 mm.

Australia (Queensland, New South Wales, Victoria). H5–G1.

106. TEMPLETONIA R. Brown
J. Cullen

Hairless shrubs with angular or grooved branches. Leaves alternate, simple, entire (absent in some non-cultivated species); stipules minute or present as small spines. Flowers axillary, solitary or 2 or 3 together, red in ours. Calyx 5-toothed, the 2 upper teeth almost completely united, the 2 lateral teeth shorter, the lowermost somewhat longer. Standard reflexed. Stamens with filaments all united into a sheath open on upper side; anthers alternately longer and shorter, the longer erect, the shorter versatile. Ovary with several ovules; style incurved, thread-like. Pod very flattened, often oblique, the sides completely separating. Seeds with appendages.

A small genus from Australia. Only 1 species is grown, requiring well-drained soil and a sunny site.

1. T. retusa R. Brown. Illustration: Botanical Magazine, 2088 (1819) – as T. glauca, & 2334 (1822); Journal of the Royal Horticultural Society **91**: f. 211 (1966); Fairall, West Australian native plants in cultivation, 227 (1970); Erickson et al., Flowers and plants of Western Australia, 20 (1973).

Hairless shrub to 2 m; young branches deeply grooved. Leaves thick, leathery, pale green, oblong-obovate, shortly stalked, rounded at the base, rounded, truncate or notched at the apex, but often with a small point. Calyx bell-shaped, 7–9 mm, teeth shallow, margins hairy. Corolla red; standard rather narrow, strongly reflexed, 4–4.5 cm; wings and keel about as long as standard. Pod stalked, flat, oblong, tapering more gradually at base than apex, 3–4.5 × 0.6–1.1 cm, containing several seeds. *Australia (Western Australia, South Australia)*. H5–G1.

107. HOVEA R. Brown
J. Cullen

Shrubs, usually hairy. Leaves alternate, simple, entire or prickly-toothed. Stipules thread-like or absent. Flowers blue or purple in axillary clusters or short racemes, rarely solitary. Calyx with a large upper lip formed from the 2 upper teeth which are almost completely united and truncate; lower 3 teeth much smaller. Petals purple, clawed; standard broad, notched, longer than the wings and keel;

keel short, incurved. Stamens 10, with filaments all united into a sheath split on the upper side, sometimes also on the lower side, some stamens sometimes almost entirely free; anthers alternately longer and shorter, the longer erect, the shorter versatile. Ovary containing 2 or more ovules; style incurved, rather thick. Pod swollen, spherical or ovoid. Seeds with appendages.

A genus of perhaps 20 species from Australia, of which only a few are cultivated. They require a well-drained soil and a sunny site and are propagated mainly by seed.

1a. Leaves heather-like, very narrowly lanceolate, margins curved downwards alongside the prominent midrib, sharply pointed
       3. pungens
   b. Leaves broad, not heather-like, sharply pointed only if toothed   2
2a. Leaves entire   **1. elliptica**
   b. Leaves with toothed margins
       2. chorizemifolia

1. H. elliptica (Smith) de Candolle (*H. celsii* Bonpland). Illustration: Botanical Magazine, 2005 (1818); Edwards's Botanical Register **4:** pl. 280 (1818); Erickson et al., Flowers and plants of Western Australia, 59 (1973).

Shrub to 3 m; young branches slightly grooved, adpressed hairy. Leaves elliptic, entire, to 7 × 3 cm on main shoots, usually smaller on lateral branches, tapered to the base and to the blunt though slightly mucronate apex, hairless above, with brown or blackish hairs beneath. Flowers in clusters. Calyx to 5 mm, densely brown-hairy. Corolla bright purple, standard broad, to 1.2 cm. Pod ovoid, to 1 cm. *Australia (Western Australia).* H5–G1.

2. H. chorizemifolia de Candolle. Illustration: Gardner, Wild flowers of Western Australia, 70 (1959).

Shrub to 2 m, with thickened roots; young shoots slightly grooved, adpressed hairy. Leaves narrowly to broadly oblong-elliptic, 3.5–6 × 1–2.5 cm, with coarse, spine-like teeth on margins, tip spine-like, hairless on both surfaces. Flowers in shortly stalked clusters. Calyx funnel-shaped, 3.5–5 mm, adpressed hairy. Corolla purple, standard to 8 mm. Pods ovoid, often broader than long. *Australia (Western Australia).* H5–G1.

3. H. pungens Bentham. Illustration: Erickson et al., Flowers and plants of Western Australia, 42 (1973).

Shrub to 1 m; young shoots slightly grooved, adpressed hairy. Leaves heather- or needle-like, with margins folded downwards alongside the conspicuous midrib, very narrowly lanceolate, 2.5–3.5 cm × 2–3 mm, hard, widely spreading, sharply pointed, hairless above, with distant, long hairs on the midrib beneath. Flowers solitary or in shortly stalked clusters in the axils. Calyx 5–6 mm, adpressed hairy. Corolla bright purple, standard broad, to 1 cm. Pod very turgid, shiny, broader than long. *Australia (Western Australia).* H5–G1.

108. BOSSIAEA Ventenat
J. Cullen
Shrubs; branches terete, angular, winged or flattened. Leaves alternate (ours) or opposite, simple, entire or rarely toothed; stipules small, brown, lanceolate or thread-like. Flowers axillary or in clusters of 2 or 3 in the leaf-axils, yellow, orange, red or brownish. Calyx 5-toothed with the 2 upper teeth much larger than the lower 3. Petals yellow or red, clawed; standard broad, reflexed; wings narrow, longer than keel. Stamens 10, with filaments united into a sheath open along the upper side; anthers all of the same length, versatile. Ovary stalked or not, with several ovules; style incurved. Pods flat, not winged, the sides completely separating, sutures thickened. Seeds with appendages.

A genus of about 20 species from Australia. Only a few are grown, requiring a well-drained but constantly moist soil and a sunny site. Propagation is by seed or by cuttings of just hardened wood.

1a. Stems conspicuously flattened; leaves linear to elliptic or obovate on the one plant   **4. heterophylla**
   b. Stems terete or somewhat winged; leaves various, but all of similar form on the one plant   2
2a. Leaves not in 2 distinct ranks; young shoots adpressed hairy
       1. cinerea
   b. Leaves in 2 distinct ranks; young shoots with some spreading hairs or hairless or almost so   3
3a. Branches spreading at very wide angles; leaves well spaced; standard 0.9–1.2 cm   **3. disticha**
   b. Branches ascending at narrow angles; leaves crowded; standard to 0.8 cm   **2. linophylla**

1. B. cinerea R. Brown. Illustration: Botanical Magazine, 3895 (1842) – as B. tenuicaule; Cochrane et al., Flowers and

plants of Victoria, 30 (1968); Galbraith, Collins' field guide to wild flowers of south-east Australia, pl. 40 (1977); Elliot & Jones, Encyclopaedia of Australian plants **2:** 358 (1982).

Shrub to 2 m; young shoots terete or angular, adpressed hairy, sometimes whitish. Leaves borne in several ranks, narrowly ovate, to 2 × 0.8 cm, rounded to the base, tapered to the acute, mucronate apex, sparsely hairy above, more densely so beneath with adpressed hairs. Flowers solitary in the leaf-axils, long-stalked. Calyx *c.* 4 mm, with shallow teeth, hairless except for the tooth-margins. Corolla yellow, standard reddish on the back, to 1.5 cm. Pod to 2 × 0.5 cm. *Australia (Victoria, South Australia, New South Wales, Tasmania).* H5–G1.

2. B. linophylla R. Brown. Illustration: Botanical Magazine, 2491 (1824); Elliot & Jones, Encyclopaedia of Australian plants **2:** 359 (1982).

Shrub to 3 m; shoots terete or somewhat winged, hairless or sparsely hairy. Leaves in 2 ranks, crowded, linear, 1.5–2 cm × 1–2 mm, acute, sparsely hairy above and beneath, margins somewhat curved downwards. Flowers in small, stalked racemes in the leaf-axils. Calyx 3–4 mm, sparsely adpressed hairy. Corolla mostly yellowish, keel red; standard to 0.8 cm. Pods stalked, flattened, elliptic to narrowly elliptic, 1–2 × 0.5–0.6 cm, hairless. *Australia (Western Australia).* H5–G1.

3. B. disticha Lindley. Illustration: Edwards's Botanical Register **4:** pl. 55 (1841).

Shrub to 1.5 m; branches terete, long and very widely spreading, rather spreading-hairy. Leaves in 2 ranks, very narrowly ovate to oblong, 6–15 × 2–4 mm, rounded at the base, gradually then ultimately abruptly tapered to the mucronate apex, sparsely adpressed hairy above, more densely so beneath. Flowers long-stalked, solitary in the leaf-axils. Calyx 4–5 mm, teeth rather deep and narrow, all adpressed hairy. Corolla mainly yellow, keel red; standard 0.9–1.2 cm. Pod stalked, oblong, abruptly tapered at each end, *c.* 2 × 0.8 cm, hairy along the upper suture, at least at first. *Australia (Western Australia).* H5–G1.

4. B. heterophylla Ventenat. Illustration: Botanical Magazine, 1144 (1808) – as B. lanceolata; Rotherham et al., Flowers and plants of New South Wales and southern Queensland, 48 (1975).

Shrub to 2 m; shoots conspicuously flattened, hairless, sparsely hairy or with tufts of minute hairs in the leaf-axils. Leaves in 2 ranks, linear to elliptic or obovate, very variable in shape on the one plant, 7–15 × 2–10 mm, tapered to the acute apex, tapered or rounded to the base, hairless. Flowers solitary in the leaf-axils, stalked. Calyx 4–6 mm, hairless except for the fringed tooth-margins. Corolla yellow; standard 1–1.5 cm, reddish on the back. Pod stalked, oblong, flat, abruptly tapered at apex and base, to 3 × 0.6 cm, hairless. *Australia (New South Wales, Victoria, Queensland).* H5–G1.

109. VIRGILIA Poiret
J. Cullen

Shrubs or small trees to 15 m or more in the wild. Leaves pinnate with a terminal leaflet and 6–12 pairs of lateral leaflets (rarely more or fewer); stipules linear, acute; stipels absent. Flowers in racemes or occasionally panicles, axillary and terminal; bracts persistent or falling early. Flowers violet to pink, rarely whitish. Calyx bell-shaped, umbilicate at the base, 2-lipped, upper lip of 2 reflexed teeth, the lower of 3 straight teeth. Standard broad, reflexed, wings curved, keel curved and beaked. Stamens 10, filaments united into a tube; anthers versatile. Ovary with 5–8 ovules; style curved, stigma with a fringe of hairs. Seeds with small appendages.

A genus of 2 species from the Cape Province of South Africa. They require a light, sandy, well-drained soil and a sunny site. Propagation is by seed.

Literature: Van Wyk, B.E., A revision of the genus *Virgilia* (Fabaceae), *South African Journal of Botany* **52**: 347–353 (1986).

1. V. oroboides (Bergius) Salter (*V. capensis* Linnaeus). Illustration: Botanical Magazine, 1590 (1813); Flowering plants of South Africa **8**: pl. 305 (1928); Journal of the Royal Horticultural Society **99**: f. 194 (1974); Palgrave, Trees of southern Africa, t. 96 (1977).

Shrub or small tree; bark smooth or rough. Leaflets with dense adpressed hairs beneath; stipules 3–12 mm. Flowers in a raceme or panicle; bracts persistent, 0.7–1.5 × 0.4–1 cm; bracteoles present but minute. Calyx 0.6–1 cm, densely hairy. Corolla pale pink to violet; standard to 2 cm; beak of keel pink, yellowish green or dark purple. Pods oblong, gradually tapered to the base, abruptly tapered to the apex, to 4 × 1 cm, densely hairy. *South Africa (Cape Province).* H5–G1. Spring–summer.

Van Wyk recognises 2 subspecies: subsp. **oroboides** with pale pink flowers, the keel-beak pink or yellowish green, bark rough; and subsp. **ferruginea** van Wyk, with violet to violet-purple flowers, the keel-beak dark purple, bark usually smooth. Both subspecies are likely to have been introduced into cultivation in Europe.

V. divaricata Adamson. Illustration: Menninger, Flowering trees of the world, pl. 308 (1962); Palmer & Pitman, Trees of southern Africa **2**: 903 (1973). Similar, but leaflets hairless beneath or hairy only on the midrib, bracts 2–5 × 1–3 mm, falling early, bracteoles absent; bark smooth. *South Africa (Cape Province).* H5–G1.

This species has also been introduced to cultivation, but it is uncertain whether or not it is still to be found. Van Wyk mentions that intermediates between the 2 species are cultivated in South Africa, and some of these have certainly been grown in Europe.

110. PODALYRIA Lamarck
J. Cullen

Small to large, usually densely hairy shrubs. Leaves simple, very shortly stalked, somewhat leathery; stipules usually falling early. Flowers solitary in the leaf-axils or in few-flowered axillary racemes which are sometimes concentrated at the shoot tips; bracteoles usually present. Calyx 5-toothed, the 2 upper teeth united further than the 3 lower, the base umbilicate when the flower is open. Standard reflexed, broad, notched, longer than the keel and wings. Stamens 10, filaments free or very slightly united at the extreme base. Ovary with several ovules. Pod hard, woody, containing several seeds.

A genus of 25 species from South Africa, mainly from the Cape Province. They need a well-drained soil and a sunny site; propagation is best by seed, though cuttings of well-ripened wood have also been used.

1a. Standard 2.5 cm wide or more; bracteoles united and forming a hood over the flower-bud, deciduous as the flower opens **3. calyptrata**
　b. Standard to 2 cm wide; bracteoles not as above　**2**
2a. Inflorescence-stalks 2–4-flowered, exceeding the leaves; standard 1.5–2 cm　**1. biflora**
　b. Inflorescence-stalks mostly 1-flowered, not exceeding the leaves; standard to 1 cm **2. sericea**

1. P. biflora (Retzius) Lamarck (*P. argentea* Salisbury). Illustration: Botanical Magazine, 753 (1804); Rice & Compton, Wild flowers of the Cape of Good Hope, t. 34 (1950); Jackson, Wild flowers of Table Mountain, 26 (1977).

Densely branched shrub to 60 cm; young shoots densely hairy. Leaves ovate to elliptic or almost circular, 1–2.5 cm, tapered to the base, rounded or obtusely tapered to the apex which is mucronate and bent downwards, densely silky-hairy on both surfaces and with longer hairs on the margins, at least when young. Flowers in axillary racemes of 2–4; inflorescence-stalk exceeding the subtending leaf; bracteoles small. Calyx 0.7–1 cm, densely hairy, teeth usually longer than the tube. Corolla pink or white, standard 1.5–2 × 1.5–2 cm, longer than the whitish keel and wings. Pod cylindric, to 4 cm, densely hairy. *South Africa (Cape Province).* H5–G1.

2. P. sericea R. Brown. Illustration: Botanical Magazine, 1923 (1817); Batten & Bokelmann, Wild flowers of the eastern Cape Province, pl. 67(5) (1966); Mason, Western Cape sandveld flowers, 129 (1972).

Like *P. biflora* but the dense hairs giving the plant a silvery sheen; flowers solitary in the axils, on short stalks; standard pink with a purple blotch at the base, to 1 cm; pod to 3 cm. *South Africa (Cape Province).* H5–G1.

3. P. calyptrata Willdenow. Illustration: Botanical Magazine, 1580 (1813) – as P. styraciflora; Menninger, Flowering trees of the world, 307 (1962); Palgrave, Trees of southern Africa, t. 98 (1977).

Large shrub to 3 m or more, young shoots hairy. Leaves elliptic to elliptic-obovate, sometimes broadly so, 2–5 cm, tapered or rounded to the base, rounded to the apex where there is a recurved mucro, finely hairy on both surfaces. Flowers few, fragrant, in stalked axillary racemes; bracteoles forming a cup-shaped hood over the flower bud, deciduous as the bud opens. Calyx to 1.3 cm, silky. Corolla pink, standard 2 × 2.5 cm or more. Pod densely shaggy hairy, to 4 cm. *South Africa (Cape Province).* H5–G1.

111. HYPOCALYPTUS Thunberg
J. Cullen

Shrubs; shoots often dark reddish. Leaves made up of 3 leaflets; stipules triangular to linear. Flowers numerous in a terminal raceme; bracts narrow, usually falling early; bracteoles present. Calyx bell-shaped to cylindric, ultimately umbilicate at the base, 5-toothed, the 2 upper teeth united for most of their length, the lower 3 narrowly triangular. Corolla violet; standard with broad, reflexed blade; keel curved, scarcely beaked. Stamens 10, their filaments united into a tube; anthers alternately longer and shorter, the longer erect, the shorter versatile. Ovary stalked, containing 3–30 ovules. Pod stalked, linear-oblong (ours) or inflated, containing several seeds. Seeds more or less oblong, each with an appendage along a long side.

A genus of 3 species from the Cape Province of South Africa. They require a well-drained soil and a sunny site, and are propagated by seed.

Literature: Dahlgren, R., The genus *Hypocalyptus* Thunb. (Fabaceae), *Botaniska Notiser* **125**: 102–125 (1972).

1. H. sophoroides (Bergius) Baillon. Illustration: Botanical Magazine, 3894 (1842); Botaniska Notiser **125**: 107 (1972); Moriarty, Outeniqua, Tsitsikamma and eastern Little Karroo, 119 (1981).

Much branched shrub to 3 m. Leaflets oblanceolate to obovate, 1–3 × 0.6–2 cm, tapered to the base, obtuse to notched, though ultimately pointed, at the apex, hairless above, hairless or very sparsely hairy beneath, usually folded upwards along the midrib so that the upper surface is scarcely exposed; stipules 1–4 mm, often soon falling. Raceme dense, usually with 30 or more flowers; axis hairy. Calyx hairless, 6–8 mm. Corolla purple; standard 1 cm or more, with a yellow spot at the base of the blade. Pod oblong-linear, abruptly tapered at each end, 3.5–6.5 cm × 4–6 mm. *South Africa (Cape Province)*. H5–G1. Summer.

112. CROTALARIA Linnaeus
J. Cullen
Annual or perennial herbs or shrubs. Leaves simple (reduced to a single leaflet) or compound with 3 (ours) or more leaflets; stipules usually falling early; stipels absent. Flowers in racemes which are often long; bracts and bracteoles usually present. Calyx deeply 5-toothed, variable, sometimes 2-lipped. Flowers yellow (ours); standard with a large, reflexed blade with a fold or groove down the middle; wings often conspicuously puckered between the veins outside; keel usually very curved, beaked in cultivated species, often fringed above and on the suture. Stamens 10, filaments all united into a sheath open at the top, anthers alternately longer and

versatile, shorter and erect. Ovary with several ovules; style hairless or with 1 or 2 lines of hairs. Pods turgid, hard, the seeds ultimately loose and rattling within them.

A genus of about 600 species, mainly from tropical regions and the southern hemisphere, only a few cultivated. They are easily grown and propagated by seed or, in the case of the shrubby species, by cuttings.

1a. Leaves stalked, of 3 leaflets 2
 b. Leaves stalkless, of a single leaflet 3
2a. Leaves hairy above; stipules very narrowly lanceolate, inconspicuous **1. micans**
 b. Leaves hairless above; stipules obovate, leaflet-like, conspicuous but often falling **2. capensis**
3a. Shrub; leaves hairy above; calyx-tube less than 3 mm **3. juncea**
 b. Tall annual herb; leaves hairless above; calyx-tube 4–6 mm **4. retusa**

1. C. micans Link (*C. anagyroides* Humboldt, Bonpland & Kunth). Illustration: Bernal, Flora de Colombia **4**: 35 (1986).

Erect or spreading shrub to 1 m; shoots densely hairy. Leaves stalked, of 3 leaflets, the terminal larger than the laterals, all obovate to elliptic, tapered to the base, rounded or tapered to the mucronate apex, hairy above and beneath; stipules very narrowly lanceolate. Racemes rather short. Calyx bell-shaped, tube 3.5–5 mm, lowest tooth 4–6 mm, all sparsely hairy. Corolla yellow; standard 1.5–2 cm; wings as long as or slightly longer than keel; keel densely fringed along its upper margin, bases of the suture also fringed. Pod stalked, broadly cylindric, 2.5–4 × 1–1.5 cm, finely hairy. *N South America*. G1.

2. C. capensis Jacquin. Illustration: Botanical Magazine, 7950 (1904); Flowering Plants of South Africa **10**: 386 (1930); Rice & Compton, Wild flowers of the Cape of Good Hope, t. 41 (1950); Kidd, Cape Peninsula, 99 (1983).

Erect shrub to 2 m; shoots densely hairy. Leaves stalked, of 3 leaflets, the terminal somewhat larger than the laterals, all obovate to elliptic, tapered to the base, usually rounded at the apex, hairless above, adpressed hairy beneath, margins fringed; stipules obovate, leaflet-like, soon falling. Racemes rather short. Calyx bell-shaped, tube 5–6 mm, lowest tooth 6–8 mm, all sparsely hairy. Corolla mostly yellow striped with red brown, fading

orange; standard 2–2.5 cm; wings shorter than keel; keel whitish, fringed along the upper margin and also along the suture. Pod stalked, broadly cylindric, 4–6 × 0.5–1 cm, finely adpressed hairy, bluish green. *South Africa (Cape Province, Natal)*. H5–G1.

3. C. juncea Linnaeus. Illustration: Botanical Magazine, 1933 (1815); Bernal, Flora de Colombia **4**: 61 (1986).

Spreading shrub to 3 m; young shoots ridged, hairy. Leaves reduced to a single leaflet, stalkless, linear-oblong to oblong-elliptic, tapered or rarely rounded to the acute apex, hairy on both surfaces; stipules minute or absent. Racemes to 30 cm, many-flowered. Calyx very obviously 2-lipped, with a very short tube, densely spreading hairy, and with a pronounced bulge at the base beneath; lower teeth 1–1.5 cm. Corolla bright yellow; standard 1.5–2 cm, with brown hairs outside; wings shorter than keel; keel fringed on the upper surface towards the base and along the suture, twisted towards the apex. Pod stalked, ovoid-cylindric, blunt at both ends, to 3.5 × 1.5 cm, hairy, often blackish. *Origin unknown, long cultivated in many areas, especially India*. G1. Late summer.

Cultivated as a fibre plant in many areas.

4. C. retusa Linnaeus. Illustration: Edwards's Botanical Register **3**: t. 253 (1818); Botanical Magazine, 2561 (1825); Bernal, Flora de Colombia **4**: 88 (1986).

Annual herb to 1 m; shoots adpressed hairy. Leaves reduced to a single leaflet, stalkless, obovate to elliptic-obovate, tapered to the base, abruptly rounded to the usually notched apex, hairless above, adpressed hairy beneath, often wavy-margined towards the apex; stipules inconspicuous or absent. Racemes long, many-flowered. Calyx bell-shaped, sparsely hairy or hairless, tube 4–6 mm, upper lip much larger than lower. Corolla yellow; standard 1.5–2 cm; wings somewhat shorter than keel, keel fringed along the suture, twisted towards the apex. Pod stalked, cylindric, hairless, 3–4 × 0.8–1 cm. *Tropical Africa & Asia*. G1. Late summer.

113. ANAGYRIS Linnaeus
J.R. Akeroyd
Deciduous shrub 1–3 m, foetid and poisonous. Twigs green. Leaves with 3 leaflets 3–8 × 1–3 cm, elliptic, more or less obtuse, hairy beneath. Stipules minute, lanceolate. Flowers up to 20 in short axillary racemes on previous year's wood.

Calyx bell-shaped, with 5 triangular teeth. Corolla 1.8–2.5 cm, yellow, the standard about half as long as the other petals, usually with a blackish spot; petals of keel free. Stamens free. Fruit 10–18 cm, shortly stalked, pendent, flat, constricted between seeds, hairless; seeds few, large.

A genus of a single species. Propagation is by seed or cuttings taken in summer.

1. A. foetida Linnaeus (*A. neapolitana* Tenore). Illustration: Sibthorp & Smith, Flora Graeca **4**: t. 366 (1823); Polunin & Everard, Trees and bushes of Europe, 106 (1976).

Mediterranean area & SW Asia. H3. Spring.

114. PIPTANTHUS Sweet
J. Cullen

Shrubs or small trees to 4 m, branches with wide pith. Leaves stalked, of 3 leaflets; stipules conspicuous, fused for two-thirds or more of their length. Flowers in a loose or dense stalked terminal raceme, 3 arising from each bract. Calyx bell-shaped, 5-toothed, the upper 2 teeth fused for most of their length. Corolla yellow, standard reflexed, as long as the wings, with a conspicuous claw, blade notched at the apex. Stamens 10, their filaments free. Ovary hairless or hairy, with 3–10 ovules. Pod oblong, flattened, leathery, with up to 10 seeds.

A genus of 2 species from the Himalaya and SW China. They are relatively easy to grow in most soils and are best propagated by seed.

Literature: Turner, B.L., Revision of the genus *Piptanthus* (Fabaceae: Thermopsideae), *Brittonia* **32**: 281–285 (1980).

1. P. nepalensis (J.D. Hooker) D. Don (*P. bicolor* Craib; *P. concolor* Craib; *P. forrestii* Craib; *P. concolor* subsp. *harrowii* Stapf; *P. concolor* subsp. *yunnanensis* Stapf; *P. laburnifolius* (D. Don) Stapf; *P. laburnifolius* forma *sikkimensis* Stapf and forma *nepalensis* Stapf). Illustration: Sweet, British Flower Garden **3**: t. 264 (1828); Schneider, Illustriertes Handbuch der Laubholzkunde **2**: 20, 22 (1907); Botanical Magazine, 9234 (1931); Polunin & Stainton, Concise flowers of the Himalaya, pl. 26 (1987).

Shrub to 4 m, with brittle stems. Leaflets generally lanceolate, acute, the terminal 2–15 cm, somewhat larger than the laterals; all hairless to finely hairy and bright green above, generally downy and paler beneath, veins conspicuous but not raised.

Calyx 1.3–1.6 cm, downy. Corolla bright yellow, sometimes with brownish markings towards the base of the blade of the standard; keel 2.8–3.2 cm. Pod shortly stalked, hairy or hairless, 3–22 × 0.8–2 cm. *Himalaya, SW China.* H5. Early summer.

The species, subspecies and forms described by Craib (Gardeners' Chronicle **60**: 228, 1916) and Stapf (in the notes to the Botanical Magazine plate cited above) are merely selections from the total range of variability of the species.

P. tomentosus Franchet. Leaves densely hairy on both surfaces, veins beneath prominently raised. Flowers somewhat smaller (keel 1.4–1.6 cm). *SW China.*

115. THERMOPSIS R. Brown
C. Fraile

Tall, usually hairy perennial herbs with woody rhizomes. Leaves with 3 leaflets which vary from linear to elliptic, oblanceolate or obovate; stipules conspicuous, persistent, often leaf-like. Flowers in terminal or axillary, compact or loose racemes. Calyx 5-toothed, sometimes 2-lipped with the upper lip truncate or notched. Corolla deep purple or yellow. Stamens 10, filaments free. Ovary more or less stalkless, containing numerous ovules. Pod flat, straight or recurved, many-seeded.

A genus of about 20 species from N America and Asia. They require deep well-drained soil and are propagated by seed or by division of established plants.

Literature: Larisey, M.M., A revision of the North American species of the genus *Thermopsis*, *Annals of the Missouri Botanical Garden* **27**: 245–258 (1940); St John, H., New and noteworthy Northwestern plants, part 9, notes on North American *Thermopsis*, *Torreya* **41**: 112–115 (1941).

1a.	Corolla deep purple	**1. barbata**
b.	Corolla yellow	2
2a.	Calyx to 7 mm	3
b.	Calyx more than 7 mm	6
3a.	Leaflets to 2 cm wide; fruit strongly recurved	**6. rhombifolia**
b.	Leaflets more than 2 cm wide; fruit spreading or erect	4
4a.	Leaflets hairless above, finely downy beneath	**2. caroliniana**
b.	Leaflets hairy on both surfaces	5
5a.	Stem and leaflets with densely felted hairs throughout; pods erect	**7. macrophylla**
b.	Stem with adpressed bristles, leaflets adpressed hairy; pods spreading	**5. gracilis**

6a.	Leaflets more than 3 cm wide	**8. fabacea**
b.	Leaflets less than 3 cm wide	7
7a.	Leaves adpressed hairy on both surfaces; pods spreading	**4. mollis**
b.	Leaves hairless above, slightly downy beneath; pods erect or recurved	8
8a.	Leaves almost stalkless; pods recurved	**3. lanceolata**
b.	Leaves clearly stalked; pods erect, closely adpressed to the axis	**8. fabacea**

1. T. barbata Royle. Illustration: Botanical Magazine, 4868 (1855); Hara, Photo-album of plants of eastern Himalaya, pl. 213 (1968); The Garden **100**: 353 (1975).

Stems erect, to 45 cm, the whole plant (stems, leaf-stalks, bracts, flower-stalks and calyces) covered with shaggy hairs. Leaves stalkless, opposite, the leaflets also stalkless, forming whorls on the stem, 1–3 × 0.5–1 cm, lanceolate, acuminate or acute, hairy on both surfaces. Flowers in short racemes, opposite. Calyx with 5 free teeth. Corolla deep purple. Pod broadly oblong, 2.5–3.5 cm, contracted to a mucronate apex, shaggy. *Himalaya.* H3. Summer.

2. T. caroliniana Curtis. Illustration: The Green Scene **5**(1): pl. 1 (1971).

Stems erect, sparingly downy to hairless. Leaves alternate, leaflets 5–7.5 × 2.5–4 cm, ovate to obovate, hairless above and finely downy beneath; stipules broadly ovate, leaf-like. Flowers in a compact terminal raceme. Calyx to 7 mm. Corolla yellow. Pod to 4.5 cm, erect, covered with dense, shaggy hairs. *E USA (N Carolina to Georgia).* H5. Summer.

3. T. lanceolata R. Brown.
Stems erect, covered with shaggy hairs. Leaves almost stalkless, leaflets oblong-lanceolate, hairless above, silky-hairy beneath. Flowers apparently whorled in the raceme, yellow. Calyx more than 7 mm. Pod 4–5 cm, recurved, covered with shaggy hairs. *Former USSR (Siberia).* H3. Summer.

4. T. mollis Curtis.
Stems erect, hairy. Leaves shortly stalked; leaflets 2–4 × 1–2 cm, elliptic-ovate, with tawny hairs along the veins both above and beneath. Flowers in compact racemes, yellow. Calyx more than 7 mm. Pod 3–4 cm, spreading, downy. *SE USA.* H5. Summer.

5. T. gracilis Howell. Illustration: Abrams & Ferris, Illustrated flora of the Pacific States **2**: 488 (1944).

Stem erect, covered with adpressed bristles, somewhat angular. Leaflets 3.5–5.5 × 2–3 cm, ovate or elliptic, with adpressed hairs on both surfaces. Flowers in loose racemes, yellow. Calyx to 7 mm, with broadly triangular teeth, covered in shaggy hairs. Pods 4–5 cm, shaggy-hairy, spreading. *W North America*. H4. Summer.

6. T. rhombifolia Richardson. Illustration: Rickett, Wild flowers of the United States **6**: pl. 137 (1973).

Stem erect, adpressed downy throughout. Leaflets 2–3 × 1–2 cm, ovate, tapered to the base, hairless above, adpressed hairy beneath and with a prominent central vein; stipules exceeding the leaf-stalks. Flowers in loose racemes, yellow. Calyx less than 7 mm. Pod strongly recurved. *W North America*. H4. Summer.

7. T. macrophylla Hooker & Arnott. Illustration: Abrams & Ferris, Illustrated Flora of the Pacific States **2**: 488 (1944); Rickett, Wild flowers of the United States **5**: pl. 109 (1971).

Stem erect; stem, leaves and pods woolly with densely felted hairs. Leaflets 4–7 × 3–4 cm, obovate or oblanceolate, tapered to the base; stipules longer than the leaf-stalks, leaf-like. Calyx to 7 mm. Flowers yellow. Pod 6–8 cm, erect. *W USA*. H5. Summer.

8. T. fabacea (Pallas) de Candolle. Illustration: Edwards's Botanical Register **15**: t. 1272 (1829).

Stems erect, shaggy-hairy to hairless. Leaflets 3.5–7.5 × 1.5–3.5 cm, elliptic to obovate, hairless above, slightly downy beneath; stipules longer than the leaf-stalks, leaf-like. Flowers yellow, alternate or in pairs in erect racemes. Calyx 7 mm or more. Pod 4.5–7.5 cm, erect and adpressed to the axis, downy. *W USA*. H5. Summer.

The western North American **T. montana** Torrey & Gray (including *T. ovata* (Robinson) Rydberg) may sometimes be misidentified as *T. fabacea*; it usually has narrower, linear to obovate leaflets.

116. BAPTISIA Ventenat
J. Cullen
Perennial herbs with woody rhizomes, often glaucous, stems somewhat woody towards the base when old. Leaves with 3 leaflets (ours), shortly stalked; stipules minute, thread-like and deciduous or large, leaf-like and persistent. Flowers in racemes. Calyx bell-shaped, 2-lipped, the upper lip of 2 almost completely fused teeth, the lower of 3 distinct teeth. Corolla white,

cream, yellow or blue; standard reflexed. Stamens 10, their filaments free. Ovary with numerous ovules. Pod inflated, variably shaped, many-seeded.

A genus of about 50 species from E & C USA; many hybrids occur in the wild and some may have been introduced. The 3 species generally found in cultivation are easily grown and can be propagated by seed or by division of the rhizome.

Literature: Larisey, M.M., A monograph of the genus *Baptisia*, *Annals of the Missouri Botanical Garden* **27**: 119–244 (1940).

1a. Corolla blue or violet-blue
  3. australis
  b. Corolla white, cream or yellow 2
2a. Corolla white, keel 2–2.5 cm;
   leaflets 2.5–7 cm **1. leucantha**
  b. Corolla cream or yellow, keel 1–1.3 cm; leaflets 1–1.5 cm **2. tinctoria**

1. B. leucantha Torrey & Gray. Illustration: Botanical Magazine, 1177 (1809) – as Podalyria alba; Rickett, Wild flowers of the United States **1**: pl. 76 (1966).

Plant to 2 m, hairless almost throughout, glaucous. Leaflets 2.5–7 × 1.5–3 cm, obtuse, sometimes notched but mucronate at apex; stipules lanceolate, almost as long as to as long as leaf-stalks. Racemes long, flowers numerous, stalks 3–10 mm. Calyx densely white-hairy inside and on the margins. Corolla white, the standard sometimes purple-blotched; wings and keel 2–2.5 cm. Pod black, glaucous, shortly stalked, ovoid to oblong. *E & C USA*. H3. Spring–summer.

2. B. tinctoria (Linnaeus) Ventenat. Illustration: Botanical Magazine, 1099 (1808) – as Podalyria tinctoria; Rickett, Wild flowers of the United States **2**: pl. 121 (1967).

Plant to 1 m, almost completely hairless. Leaves shortly stalked, leaflets 1–1.5 × 0.6–1 cm, obovate, rounded or slightly notched at apex, tapered to the base; stipules minute, thread-like, deciduous. Racemes terminating lateral branches, short; flower-stalks 4–5 mm. Calyx rather sparsely hairy inside and on the margins. Corolla cream or yellow; wings and keel 1–1.3 cm. Pod black, glaucous, wrinkled, ovoid to spherical. *E USA*. H2. Spring–summer.

3. B. australis (Linnaeus) R. Brown (*B. exaltata* Sweet). Illustration: Botanical Magazine, 509 (1801); Rickett, Wild flowers of the United States, **1**: pl. 77 (1966); Hay & Synge, Dictionary of garden

plants in colour, pl. 1001 (1969); Taylor's guide to perennials, 232 (1986).

Plant to 1.5 m, more or less hairless, glaucous. Leaves with stalks to 1 cm, leaflets obovate to oblanceolate, usually rounded at the apex, long-tapered to the base, 4–8 × 1.5–3 cm; stipules ovate-lanceolate to lanceolate, usually persistent, longer than the leaf-stalk. Racemes loose, terminal. Calyx sparsely hairy inside and on the margins. Corolla violet-blue to blue, wings and keel 2–2.7 cm. Pod greyish to brownish black, wrinkled, oblong-ellipsoid. *E USA*. H2. Spring–summer.

117. LUPINUS Linnaeus
D.R. McKean
Annual, biennial or perennial herbs or woody low shrubs. Leaves palmate, stipules attached to the base of the stalk. Flowers in racemes or spikes, in whorls or alternate. Calyx 2-lipped, each lip either deeply cleft to about the mid-point or each with 2 or 3 tiny teeth (almost entire) or entire; bracteoles attached to the calyx. Stamens 10, filaments of all joined together in a tube. Fruit a flat hairy pod.

A genus of about 200 species, widely distributed but absent from Australasia and South Africa. Most are from North, Central and South America. They prefer poor, well-drained, gravelly, acid soils, in full sun. In cultivation they can be forage crops, ornamentals or human food. The annuals are generally disease-free, but the perennials can suffer from root and crown rot. The leaves may be attacked by powdery mildew, or cucumber mosaic virus (mottled leaves). Honey fungus can also kill plants. Particular strains of the edible species are selected to avoid the toxic alkaloids which they all contain. Enthusiasts may be able to obtain some dwarf alpine lupins from America, e.g. *L. alopecuroides, L. confertus, L. lepidus, L. lyallii* and *L. ornatus*, but they are not commonly available and all need the protection of an alpine house.

Literature: Agardh, J. G., *Synopsis Generis Lupini* 1–43 (1835); Gladstones, J.S. Lupines of the Mediterranean Region and North Africa, *Western Australian Department of Agriculture Technical Bulletin* **26**: 1–48 (1974).

Habit. Woody low shrub: **1–3**; perennial: **4–7,9**; annual: **8–18** Shaggy hairy: **4,6,8,13,16**; shortly or silky hairy: **1,2,3,5,7,9–12,14,15,17,18**; hairless or nearly so: **9**.
Flowers. Mainly whorled: **2,5–7,11–13, 16,17**; lower flowers (at least) alternate

1,3,5,7,8,10,13,14,18. Upper calyx-lip. Deeply divided: **2,3,8,10,11,15–18**; entire or almost entire: **1,4–7,9,12–14**. Lower calyx-lip. Deeply divided: **8**; entire or almost entire: **1–7,9–18**.

Corolla. Yellow: **1,11**, (**10,12**, rarely); blue to pink, rarely white: **2–8,10,13,15, 16–18**; white: **7,12,14**.

1a. Woody low-shrub   2
  b. Herbaceous plant   4
2a. Flowers yellow (rarely lilac or blue)
       1. arboreus
  b. Flowers mainly blue   3
3a. Lower leaf-stalks longer than
    leaflets; keel hairy on margins
       2. albifrons
  b. Leaf-stalks about equalling the
    leaflets; keel hairless on margins
       3. chamissonis
4a. Perennial   5
  b. Annual   8
5a. Upper surface of leaves hairy
       4. nootkatensis
  b. Upper surface of leaves hairless   6
6a. Leaflets 9–17, 5–12 cm
       5. polyphyllus
  b. Leaflets usually fewer and shorter   7
7a. Prostrate plant   **6. littoralis**
  b. Erect plant   **7. perennis**
8a. Lower calyx-lip deeply cut
       8. micranthus
  b. Lower calyx-lip entire or with
    shallow notches   9
9a. Stems almost hairless
       9. mutabilis
  b. Stems hairy   10
10a. Plant less than 30 cm; flowers
    yellow   **10. subcarnosus**
  b. Plant more than 30 cm or flowers
    yellow or not   11
11a. Flowers yellow   12
  b. Flowers not yellow   13
12a. Upper leaf surface hairy
       11. luteus
  b. Upper leaf surface hairless
       12. densiflorus
13a. Upper calyx-lip deeply cut to about
    mid-point   17
  b. Upper calyx-lip with a shallow
    notch or entire   14
14a. Stems fairly densely long-hairy;
    flowers dark blue   15
  b. Stems much less hairy; flowers
    white or tinged blue   16
15a. Leaflets oblanceolate; flowers white,
    tinged or veined pink or violet,
    purple or blue   **12. densiflorus**
  b. Leaflets oblong-obtuse; flowers
    blue, standard with a pink centre
       13. mexicanus

16a. Keel hairy on upper margin; pod *c.*
    2 × 1 cm   **12. densiflorus**
  b. Keel hairless on upper margin; pod
    6–10 × 1.1–2 cm   **14. albus**
17a. All flowers alternate   18
  b. At least the upper flowers in whorls
       19
18a. Leaves slightly fleshy and acute;
    lower calyx-lip simple
       15. texensis
  b. Leaves not fleshy, obtuse; lower
    calyx-lip minutely 3-toothed
       10. subcarnosus
19a. Stems densely shaggy-hairy
       16. pilosus
  b. Stems shortly or sparsely hairy   20
20a. Flowers distinctly whorled
       17. nanus
  b. Lower flowers alternate
       18. angustifolius

1. L. arboreus Sims. Illustration: Botanical Magazine, 682 (1803); Dunn & Gillett, The Lupines of Canada and Alaska, f. 53, 55 (1966); Hay & Beckett, Reader's Digest encyclopaedia of garden plants and flowers, 421 (1972); Rickett, Wild flowers of the United States **4**(2): pl. 126 (1970); Rose, The wild flower key, 179 f. 1 (1981).

Much-branched shrub to 3 m. Stems minutely hairy. Leaves with 5–12 leaflets, 2–6 × 0.5–1 cm, obovate-oblong, mucronate, silky-hairy beneath, hairless above; stipules subulate. Raceme-stalk to 10 cm, racemes to 30 cm. Flowers 1.4–1.7 × 1.5 cm, alternate or in whorls, golden yellow, rarely lilac or blue, scented. Upper calyx-lip notched or almost entire; lower entire. *USA (California), but naturalised near the sea in Britain & Ireland.* H5. Summer.

'Golden Spire' is a cultivar with golden yellow flowers.

2. L. albifrons Bentham. Illustration: Edwards's Botanical Register **19**: t. 1642 (1834); Rickett, Wild flowers of the United States **4**(2): pl. 131 (1970).

Shrub 75–150 cm. Leaves with 7–10 leaflets, to 3 × 0.7 cm, spathulate-obovate, densely silvery-silky, leaf-stalk to 10 cm. Raceme-stalk 5–13 cm, racemes 8–30 cm. Flowers *c.* 1.4 cm across, mainly in whorls, petals blue or reddish purple, standard hairy on the back. Calyx upper lip deeply divided, lower lip entire. Pods 3–5 × *c.* 0.8 cm, 5–9 seeded. *USA (California).* H5. Summer.

The name *L. excubitus* Jones sometimes appears in horticulture, but it may not be distinct from *L. albifrons*.

3. L. chamissonis Eschscholtz. Illustration: Botanical Magazine, 8657

(1916); Abrams & Ferris, Illustrated flora of the Pacific States **2**: f. 2606 (1944); Rickett, Wild flowers of the United States **4**(2): pl. 127 (1970).

Low, dense, bushy shrub, 75–150 cm, minutely spreading hairy. Leaves with 6–9 leaflets, to 2.5 × 0.4–0.6 cm , minutely soft-hairy on both sides. Racemes 6–15 cm; flowers alternate or in whorls, blue or lavender, standard with a central yellow spot, slightly hairy on reverse, 1.2–1.6 cm across, keel hairy on upper margin, stalks *c.* 7 mm, spreading-hairy; bracts deciduous. Calyx upper lip deeply cleft, lower lip entire. Pods dull yellow, 3–5 × 0.7–0.8 cm. *USA (California).* H5. Spring–summer.

Succeeds best in well-drained soil by a south-facing wall.

4. L. nootkatensis Don. Illustration: Botanical Magazine, 1311 (1810); Loddiges' Botanical Cabinet **9**: t. 897 (1824); Dunn & Gillett, The Lupines of Canada and Alaska, f. 70, 71 (1966).

Stout perennial to 70 cm, stem and upper side of leaves shaggy-hairy. Leaves with 7 or 8 leaflets, 2–6 × 1–1.5 cm, oblanceolate, mucronate. Calyx upper lip notched, lower lip entire. Racemes *c.* 10 cm; flowers purple-blue, pink, white or parti-coloured, 1.2–1.6 cm across. Bracts deciduous. Pods *c.* 5 cm, brown and shaggy-hairy. *W USA, NE Asia.* H3. Early summer.

Naturalised in shingle and sand in Norway, Ireland and Scotland. Possibly best regarded as a subspecies of *L. perennis*.

5. L. polyphyllus Lindley. Illustration: Edwards's Botanical Register **13**: t. 1096 (1827); Reichenbach, Flora Exotica **3**: t.176 (1835); Dunn & Gillett, The Lupines of Canada and Alaska, f. 73–75 (1966).

Herbaceous, mainly unbranched perennial, adpressed short hairy, 50–150 cm. Leaves with 9–17 leaflets, elliptic-oblanceolate, 5–12 × 1–2 cm, hairless above, sparsely hairy beneath. Raceme stalk 4–14 cm, raceme 18–40 cm. Flowers mainly in whorls, sometimes alternate, blue, purple or reddish. Calyx silvery-hairy, lips more or less equal, each entire or minutely toothed, 3.6–7.5 mm. Pod curved, 2.8–5 × 0.8–1 cm, densely woolly. *W North America (California to Alaska).* H5. Summer–autumn.

The following cultivars are known: 'Moerheimii', with pink and white flowers, 'Albiflorus' and 'Albus' with white flowers, 'Atroviolaceus', with dark purple flowers, 'Caeruleus' with blue, 'Carmineus' with red and 'Roseus' with pink flowers.

6. L. littoralis Douglas. Illustration: Edwards's Botanical Register **14**: t. 1198 (1828); Botanical Magazine, 2952 (1829); Dunn & Gillett, Lupines of Canada and Alaska, f. 57–59 (1966).

Prostrate perennial forming mats, usually shaggy-hairy. Leaves with 6–8 leaflets, linear-oblanceolate, obtuse to rounded, mucronate, hairy, becoming hairless above, 1.3–2.5 cm × 3.5–6 mm, leaf-stalks 2–4 cm, bristly; stipules subulate. Raceme-stalk 3.5–5.5 cm, raceme to 10 cm. Flowers mainly in whorls, *c.* 1.2 cm across, standard pale blue, purplish to white, wings bright blue to pale purple, keel finely hairy on inside margin; bracts long-tapering, *c.* 1.5 cm, flower-stalks *c.* 4 mm. Calyx upper lip notched, lower lip entire, soft silvery-hairy, *c.* 5 mm. Pods 3–4 × 0.6 cm, sparsely rough-hairy. *W North America (California to British Columbia).* H5. Summer.

7. L. perennis Linnaeus. Illustration: Botanical Magazine, 202 (1792); Barton, Flora of North America **2**: t. 38 (1822); Strong, American flora **3**: 128 (1849); Dunn & Gillett, The Lupines of Canada and Alaska, f. 60, 61, (1966).

Stout perennial to 70 cm, stems minutely hairy. Leaves with 7–11 leaflets, 3–3.8 × 0.8–1.2 cm, oblanceolate, hairless above, hairy beneath, leaf-stalks to 9 cm, stipules subulate. Racemes to 20 cm, loose. Flowers alternate or whorled, purple-blue, pink, white, or parti-coloured, stalks *c.* 6 mm. Calyx upper lip notched, 4–6 mm, lower lip almost entire, *c.* 6 mm. Pod 3–5 × 0.8–0.9 cm, hairy. *E North America.* H2. Summer.

8. L. micranthus Gussone (*L. hirsutus* misapplied). Illustration: Fiori et al., Flora analytica d'Italia **2**: f. 1866 (1965); Zohary, Flora Palaestina **2**: pl. 56 (1972).

Annual to 50 cm, branching from base, stems and leaf-stalks coarsely hairy, stipules linear, pointed, attached to the leaf-stalk for about half of their length, leaves with 5–9 leaflets, 1–5 × 0.6–2 cm, obovate, tapered at base, mucronate, coarsely hairy above and beneath. Raceme to 12 cm, raceme-stalk to *c.* 2 cm. Lower flowers alternate, upper ones whorled, mainly 1–1.4 cm, but very variable in size, blue with a white basal spot on standard; bracts subulate, persistent; bracteoles linear. Calyx upper lip deeply 2-toothed, lower lip about twice as long as upper, slightly 3-toothed. Pods 3–5 × 0.9–1.2 cm, coarsely hairy. *Mediterranean area.* H5. Summer.

9. L. mutabilis Sweet (*L. cruckshanksii* Hooker). Illustration: Sweet, British flower garden **2**: pl. 130 (1825); Botanical Magazine, 2682 (1826); Edwards's Botanical Register **18**: t. 1539 (1832).

Annual or short-lived perennial to 1.5 m, almost hairless, glaucous. Leaves with 7–9 leaflets, oblanceolate, spathulate, 6 × 1.2 cm, rarely shortly hairy beneath, leaf stalks 4–8 cm. Raceme-stalk *c.* 10 cm, raceme 10–20 cm, bracts deciduous. Flowers *c.* 2 cm across, stalks 0.5–1.4 cm, hairless or minutely hairy; petals, blue or white, standard yellow centered, wings very broad, keel hairy on the margin. Calyx silvery-hairy, upper lip slightly notched, lower lip entire. Pods *c.* 8 × 1.6 cm, adpressed hairy. *Colombia, Peru.* H5–G1. Summer.

10. L. subcarnosus Hooker. Illustration: Botanical Magazine, 3467 (1836); Rickett, Wild flowers of the United States, **3**(1): pl. 81 and title page (1969).

Annual to 25 cm, silky-hairy. Leaves with 5–7 leaflets *c.* 3.5 × 0.6 cm, obovate-lanceolate, slightly fleshy, becoming hairless above, silky beneath, stipules subulate, bristle-pointed. Raceme-stalk to 10 cm, raceme *c.* 15 cm. Flowers alternate, dark blue, rarely white or yellow, with a white spot in centre of standard. Calyx silky, upper lip shorter than lower, deeply 2-toothed, lower lip almost entire, minutely 3-toothed. Pods *c.* 4 × 0.6 cm, densely long-hairy. *USA (Texas).* H5. Spring–early summer.

11. L. luteus Linnaeus. Illustration: Botanical Magazine, 140 (1791); Reichenbach, Icones Flora Germanicae et Helveticae **22**: t. 6 (1903); Hegi, Illustrierte Flora von Mitteleuropa, **4**: f. 1310 (1975).

Annual to 80 cm, branched from the base, hairy. Leaves with 7–11 leaflets, 3–6 × 0.8–1.5 cm, obovate, oblong, long-hairy above, less densely so beneath. Raceme-stalk to 12 cm, raceme to 25 cm. Flowers golden yellow, sweetly scented, in well-spaced whorls, on stalks *c.* 2 mm. Bracts obovate, soon falling; bracteoles linear. Calyx upper lip deeply 2-toothed, lower lip with 3 shallow teeth. Pod 4–6 × 1–1.4 cm, densely long hairy. *Mediterranean area.* H5. Summer.

12. L. densiflorus Bentham. Illustration: Edwards's Botanical Register **20**: t. 1689 (1834); Abrams & Ferris, Illustrated flora of the Pacific States **2**: f. 2563 (1944).

Erect annual to 50 cm, adpressed hairy, single-stemmed or branched towards the top. Leaves with 7–9 leaflets, oblanceolate, 3–5 cm, long-hairy beneath. Racemes to 25 cm. Flowers white, tinged or veined pink or violet, purple or blue, rarely yellow, 1.4–1.8 cm, in close or distant whorls, standard elliptic, apex rounded, keel hairy on upper margin, stalks *c.* 2 mm, bracts reflexing. Calyx upper lip with 2 short teeth, lower lip with 3 larger teeth. Pod 2 × 1 cm, long spreading hairy. *USA (California).* H4. Summer.

13. L. mexicanus Lagasca. (*L. hartwegii* Lindley; *L. bilineatus* Bentham). Illustration: Edwards's Botanical Register **25**: t. 31 (1839); Sweet, Ornamental flower garden **1**: t. 28 (1854); Hay & Beckett, Reader's Digest encyclopaedia of garden plants and flowers, 421 (1972).

Erect annual to 1 m with long spreading shaggy hairs, and shorter-hairy beneath the long hairs. Leaves with 7–9 leaflets, oblong-obtuse, to 4 × 0.7 cm, long adpressed-hairy beneath, hairless above; stalks to 9 cm; stipules linear. Raceme-stalk *c.* 10 cm, raceme to 20 cm. Flowers in whorls or alternate, 1.4–1.6 cm, blue, standard with a pink centre. Bracts feathery, deciduous. Calyx upper lip slightly notched, lower lip entire, densely hairy; bracteoles present. Pod densely felted, to 3.8 × 0.6 cm, about 9-seeded. *Mexico.* H4. Summer.

14. L. albus Linnaeus. Illustration: Reichenbach, Icones Florae Germanicae et Helveticae **22**: t. 10 (1903); Polunin, Flowers of Europe, pl. 521 (1969).

Annual to 1.2 m, stems and leaf-stalks sparsely silky. Leaves with 5–9 leaflets, 2–6 cm × 1–2 mm, oblong-obovate, finely mucronate, almost hairless above, shaggy beneath, margins hairy; stipules finely pointed. Raceme to 50 cm, raceme-stalk to *c.* 2 cm. Lower flowers alternate, upper ones whorled, *c.* 1.5 cm across, white, sometimes tinged blue or violet with a white central spot on standard, on short stalks to 2 mm. Bracts soon falling; bracteoles tiny or absent. Calyx-lips more or less equal, upper lip entire, lower lip entire or minutely 3-toothed. Pods 7–15 × 1–2 cm, shaggy at first, longitudinally wrinkled on drying. *Greece to W Turkey & Crete.* H5. Summer.

15. L. texensis Hooker. Illustration: Botanical Magazine, 3492 (1836); Rickett, Wild flowers of the United States **3**(1): pl. 79 (1969).

Annual to 25 cm, branching at ground level. Leaves with 5 leaflets, lanceolate, acute, 3 × 0.8 cm, hairless above, silky beneath; stipules subulate. Raceme-stalk to 8 cm, raceme to 8 cm. Flowers alternate, c. 1 cm across, deep blue, standard with a white, yellow or reddish central blotch, stalks c. 5 mm. Calyx silky, upper lip shorter than lower, deeply toothed, lower lip entire. Pod 2.5 cm × 4 mm. *USA (California, Texas)*. H4. Early summer.

This species is very similar to *L. subcarnosus* but with more acute and less fleshy leaves and an entire lower calyx lip.

16. L. pilosus Murray (*L. hirsutus* Linnaeus; *L. varius* Linnaeus). Illustration: Sibthorp & Smith, Flora Graeca **7**: t. 684 (1830).

Erect, little-branched annual to 80 cm, long shaggy-hairy. Leaves with 7–11 leaflets, oblong-obovate, 2.5–6 × 1–1.8 cm, with long soft hairs on both surfaces. Raceme-stalk to 8 cm, raceme to 20 cm. Flowers 1.5–2 cm across, in whorls, deep blue, rarely pink with a white spot in centre of standard, base of keel white, darkening with age; stalks c. 5 mm. Calyx upper lip deeply divided, lower lip entire. Pod 8–2.5 cm, densely woolly. *Greece & Crete to Israel*. H4. Spring.

17. L. nanus Bentham. Illustration: Edwards's Botanical Register **20**: t. 1705 (1835); Sweet, British flower garden **3**: pl. 257 (1835); Abrams & Ferris, Illustrated flora of the Pacific States **2**: t. 2579 (1944).

Erect annual 50 cm, branching at base or simple-stemmed, minutely adpressed- or spreading-hairy. Leaves with 5–7 leaflets, 1.5–3 cm × 4 mm, linear-lanceolate, channelled, adpressed-hairy above; leaf-stalks 4–8 cm. Raceme-stalk to 8 cm, raceme to 20 cm. Flowers in whorls, c. 1.4 cm across, wings azure blue, keel white with dark purple tip, standard azure with yellowish or white spot, fragrant. Calyx upper lip deeply toothed, lower lip with 3 tiny teeth. Pods silky at first, to 3 cm × 4 mm, up to 7-seeded. *USA (California)*. H4. Spring–early summer.

18. L. angustifolius Linnaeus. Illustration: Reichenbach, Icones Flora Germanicae et Helveticae **22**: t. 10 (1903); Polunin, Flowers of Europe, pl. 520 (1969).

Annual to 1.5 m, stems sparsely silky. Leaves with 5–9 leaflets, to 5 × 0.6 cm, linear to linear-spathulate, hairless above, sparsely silky beneath. Flowers alternate, upper ones in whorls, c. 1–1.5 cm across, light to dark blue, tinged purple, rarely pink or white. Bracts oblanceolate-obovate, deciduous; bracteoles short. Calyx upper lip deeply 2-toothed, lower lip entire or with 2 or 3 tiny teeth. Pods 3.5–6 × 0.7–1.5 cm. *W France to Morocco & Mediterranean area*. H5. Summer.

118. LABURNUM Fabricius
H. Ern
Small trees. Leaves made up of 3 leaflets. Flowers in simple, axillary or apparently terminal, leafless racemes, pendent while flowering. Calyx bell-shaped, slightly 2-lipped, lips undivided or shortly toothed. Corolla yellow. Stamens 10, all filaments united into a tube. Pod dehiscent, flattened, slightly constricted between the seeds. Seeds numerous, compressed.

A genus of 2 species native to the mountains of SC and SE Europe, both of which (and their hybrid) are widely used as ornamentals. The pods of all are very poisonous.

1a. Racemes more than 40 cm **3. × watereri**
 b. Racemes at most 40 cm **2**
2a. Twigs adpressed-downy, greyish green; pod adpressed-downy, the hairs more or less persisting till maturity **1. anagyroides**
 b. Twigs almost hairless, green; pod hairless **2. alpinum**

1. L. anagyroides Medikus (*Cytisus laburnum* Linnaeus; *L. vulgare* Berchtold & Presl). Illustration: Polunin & Everard, Trees and bushes of Europe, 106 (1976).

Small tree to 7 m. Twigs greyish green, adpressed-downy. Leaflets 3–8 cm, elliptic to elliptic-obovate, usually obtuse and shortly mucronate, greyish green, adpressed-downy beneath when young. Racemes 10–30 cm, loose. Corolla c. 2 cm, golden yellow. Pod 4–6 cm, adpressed-downy when young, almost hairless when mature but some hairs usually persisting, upper suture unwinged. Seeds black. *SC Europe*. H1. Late spring.

2. L. alpinum (Miller) Berchtold & Presl (*Cytisus alpinus* Miller). Illustration: Botanical Magazine, 176 (1791).

Shrub or tree to 5 m or to 10 m in cultivation. Twigs hairless (sometimes hairy when young), green. Leaflets 3–8 cm, light green beneath. Racemes 15–30 cm, rather dense. Corolla 1.5 cm, yellow. Pod 4–5 cm, hairless, upper suture with wing 1–2 mm wide. Seeds brown. *SC & SE Europe*. H1. Early summer.

3. L. × watereri (Kirchner) Dippel. Illustration: Hora, The Oxford encyclopaedia of trees of the world, 209 (1981).

The hybrid between *L. anagyroides* and *L. alpinum*, intermediate in all characters except for the very long racemes. *SC Europe & of garden origin*.

Very widely cultivated; currently it has almost replaced the wild species in most parks and gardens. 'Vossii' has even longer, very many-flowered racemes.

119. + LABURNOCYTISUS C.K. Schneider
S.G. Knees
Deciduous spreading tree to 8 × 6 m, with greyish bark. Leaves with three elliptic-ovate leaflets, each 2–6 cm, dark green, adpressed-downy, especially beneath. Flowers of three types: yellowish *Laburnum*-like, purplish *Cytisus*-like and yellow and pink flowers in *Laburnum*-like racemes.

A graft chimera, originally raised in 1826. Needs full sun and will grow in most soil types, except those in waterlogged situations. Propagation is by grafting onto *Laburnum anagyroides* in late summer.

1. + L. adamii (Poiteau) Schneider (*Cytisus adamii* Poiteau; *Laburnum adamii* (Poiteau) Kirchner). Illustration: Edwards's Botanical Register 1965 (1837); La Belgique Horticole **21**: 16–18 (1871); Phillips, Trees in Britain, Europe and North America, 50, 128 (1978); Krüssmann, Manual of cultivated broad-leaved trees & shrubs **2**: pl. 38 (1986).

Graft chimera raised in gardens. H3. Spring–early summer.

The graft-hybrid between *Cytisus purpureus* and *Laburnum anagyroides*.

120. PETTERIA C. Presl
H. Ern
Non-spiny shrubs. Leaves made up of 3 leaflets. Flowers in terminal, erect, leafless racemes. Calyx bell-shaped to tubular, two lipped; upper lip divided to about two-thirds, lower 3-toothed. Corolla yellow; Pod linear-oblong, dehiscent, straight or slightly curved, somewhat inflated. Seeds without appendages.

A genus of one species in the west Balkan Peninsula, resembling *Laburnum*, but less attractive.

1. P. ramentacea (Sieber) C. Presl (*Cytisus ramentaceus* Sieber). Illustration: Polunin, Flowers of Greece and the Balkans, 288 (1980).

Erect shrub to 3 m. Branches terete or obscurely angled, hairless; twigs loosely adpressed-hairy when young, later more or less hairless. Leaves with stlks 1.5–5 cm; leaflets 2–7 × 1.2–3 cm, elliptic to obovate, rounded, occasionally slightly notched; dull green on both surfaces; hairless above, with adpressed hairs along the mid-vein and margins beneath. Flowers borne in terminal racemes 4–7 cm of 10–20. Calyx adpressed-hairy. Standard 1.6–2 × 1.4–1.5 cm, pentagonal, notched. Pod 3.5–5 × 0.8–1 cm, beaked, light brown, hairless, margins slightly thickened. Seeds 5–9, orange-brown. *W former Yugoslavia, N Albania*. H2. Late spring.

121. CYTISUS Linnaeus
F.T. de Vries

Shrubs or small trees without spines, from 10 cm to more than 6 m. Leaves with 1 or 3 leaflets, alternate, sometimes crowded, often soon falling, the branches and branchlets performing most of the photosynthesis. Flowers axillary, forming leafy or leafless, terminal or lateral racemes. Calyx 2-lipped, upper lip with 2 usually short teeth (rarely deeply cleft). Corolla white, yellow, purple or dark-brown; keel more or less sickle-shaped. Stamens 10, filaments all united into a tube. Stigma curved upwards, or rarely rolled up. Pods linear or oblong, with numerous seeds which bear appendages (strophioles).

A genus of about 100 species from Europe and N Africa, closely related to *Ulex* and *Genista*, but differing in the appendaged seeds and lateral racemes. It is often divided into several genera, but a wide circumscription is appropriate here. They require full sun and a dry, neutral to acid, poor sandy soil. They are propagated by seed, cuttings or grafting, and should not be moved once established.

Literature: Van de Laar, H.J., *Cytisus en Genista, Dendroflora* 8: 3–18 (1971).

1a. Calyx upper lip with 2 short teeth; keel sickle-shaped   2
  b. Calyx upper lip deeply cleft; keel oblong   35
2a. Calyx bell-shaped; leaves of 3 leaflets, simple, or both on the same plant; style curved or rolled up   3
  b. Calyx tubular; leaves of 3 leaflets; style rolled up   22
3a. Leaves all simple   4
  b. At least some leaves of 3 leaflets   8
4a. Branches grooved   5
  b. Branches angled, sometimes terete when older   6

5a. Leaf-stalks to 1 cm; leaves small, to 1.5 cm; procumbent shrub to *c*. 40 cm; flowers yellow   **9. × beanii**
  b. Leaves stalkless, to 2 cm; shrub to *c*. 3 m; flowers white, yellow, pink or red   **14. × praecox**
6a. Calyx and pod hairless   **20. diffusus**
  b. Calyx and pod downy   7
7a. Shrub to *c*. 30 cm; angles unwinged; flower-stalk 1–2.5 cm; bracteoles 2 or 3   **19. decumbens**
  b. Shrub to *c*. 60 cm; angles winged; flower-stalk to *c*. 1 cm; bracteole 1   **18. procumbens**
8a. Leaves all of 3 leaflets   9
  b. Leaves on lower branches of 3 leaflets, on younger or flowering branches simple, stalkless   15
9a. Inflorescence a leafless raceme   10
  b. Inflorescence a raceme consisting of leafy clusters or forming an umbel-like head   11
10a. All leaves stalked; calyx downy   **2. nigricans**
  b. Leaves on flowering branches stalkless; calyx hairless   **1. sessilifolius**
11a. Flowers white or pinkish-white   12
  b. Flowers yellow, sometimes red-streaked   13
12a. Branches angled; calyx more or less hairless   **6. filipes**
  b. Branches grooved; calyx downy   **5. supranubius**
13a. Inflorescence an umbel-like head, sometimes several combined to a raceme; leaf-stalk to *c*. 4 mm   **4. fontanesii**
  b. Inflorescence a raceme, consisting of leafy clusters; leaf-stalk to *c*. 3 cm   14
14a. Branches grooved; leaves downy on both sides; style rolled up   **7. ardoinii**
  b. Branches angled; leaves hairless above; style curved up   **3. emeriflorus**
15a. Style rolled up; flowers usually solitary or paired in leafy bundles; branches angled; calyx usually hairless   16
  b. Style curved up; flowers to 5 in leafy bundles; branches grooved or angled; calyx usually downy   19
16a. Compound leaves stalkless or nearly so; pod black, white-haired   17
  b. Stalk of compound leaves to 1.5 cm; pod hairless, downy only on the margins   18

17a. Calyx 5–7 mm; compound leaves each with a short stalk to *c*. 5 mm   **17. cantabricus**
  b. Calyx 4–5 mm; compound leaves stalkless   **16. grandiflorus**
18a. Stalk of compound leaves not more than 1.5 cm; calyx hairless   **15. scoparius**
  b. Stalk of compound leaves to *c*. 2.5 cm; calyx somewhat downy   **13. × dallimorei**
19a. Branches grooved   20
  b. Branches angled   21
20a. Procumbent shrub to *c*. 30 cm; stalk of compound leaves 0.5–1 cm   **8. × kewensis**
  b. Upright shrub to *c*. 1 m; compound leaves stalkless   **10. purgans**
21a. Leaves hairless above; flower-stalk 0.7–1 cm   **11. ingramii**
  b. Leaves downy on both sides; flower-stalk to *c*. 0.5 cm sides   **12. multiflorus**
22a. Leaves 4–11 cm; leaf-stalk to 3.5 cm; flowers borne singly in leafless racemes   **32. battandieri**
  b. Leaves to *c*. 5 cm; leaf-stalk to *c*. 2.5 cm; flowers in clusters of more than 1, in leafy bundles or umbel-like heads   23
23a. Flowers 1–7 in several leafy clusters, forming a leafy raceme   24
  b. Flowers many in umbel-like heads, terminal and lateral   31
24a. Leaf-stalk not more than 1.5 cm   25
  b. Leaf-stalk to *c*. 2.5 cm   28
25a. Flowers white or yellowish   26
  b. Flowers yellow with reddish brown markings   27
26a. Leaves softly-downy   **27. palmensis**
  b. Leaves silky-downy   **25. proliferus**
27a. Leaves 1–3.5 mm, downy on both sides; keel yellow   **28. supinus**
  b. Leaves not more than 2 cm, hairless above; keel reddish brown   **26. demissus**
28a. Flowers yellow with reddish brown markings; branches terete, at least when older   29
  b. Flowers purple-pink, or standard white, wings yellow and keel lilac-pink; branches angled   30
29a. Leaves more or less hairless when older; terminal flower with 1 bracteole   **22. glaber**
  b. Leaves on both sides downy; bracteoles absent   **23. hirsutus**

30a. Calyx hairless except on margins;
   flowers purple-pink   **21. purpureus**
   b. Calyx downy; standard white,
   wings yellow, keel lilac-pink   **24. ×**
   versicolor
31a. Flowers white or yellowish   32
   b. Flowers deep yellow; standard
   sometimes with a dark brown spot
   in the centre   33
32a. Calyx dark brown; flower-stalk
   5–15 mm   **25. proliferus**
   b. Calyx whitish; flower-stalk 2–6 mm
   30. albus
33a. Branches long, erect, downy;
   standard with a dark brown spot in
   the centre   **28. supinus**
   b. Branches densely grey-green
   downy; standard yellow   34
34a. Standard narrowly obovate,
   rounded   **31. tommasinii**
   b. Standard broad, slightly notched
   29. austriacus
35a. Leaves of 3 leaflets   **28. supinus**
   b. Leaves simple   36
36a. Standard uniformly downy; leaves
   to *c.* 6 cm, or stalkless
   36. linifolia
   b. Standard hairless or nearly so, or
   downy on the midrib, especially
   near the tip; leaves not more than
   4.5 cm, always stalked   37
37a. Stipules 2.5–6 mm, persistent,
   prominent, giving a scaly
   appearance to the older twigs;
   mucro of leaflet 0.4–0.6 mm
   35. maderensis
   b. Stipules 0.5–2.5 mm, not persisting
   or prominent; mucro of leaflet to
   0.3 mm, or leaf apex obtuse   38
38a. Inflorescences indeterminate,
   flowers borne in congested racemes
   or heads on short axillary branches;
   standard hairless
   37. monspessulanus
   b. Inflorescences determinate, flowers
   borne in short to long terminal
   racemes; standard downy or
   hairless on midrib   39
39a. Leaf-stalk 5–25 mm; most leaflets
   at least 10 mm, narrowly elliptic-
   obovate to oblanceolate
   34. stenopetalus
   b. Leaf-stalk 1–6 mm; most leaflets
   3–8 mm, obovate-oblong to
   oblanceolate   **33. canariensis**

1. C. sessilifolius Linnaeus. Figure 40(4),
p. 540. Illustration: Botanical Magazine,
255 (1794); Vicioso, Genisteas Españolas,
43 (1953); Csapody & Tóth, Colour atlas of
flowering trees and shrubs, 141 (1982).

Upright bushy shrub to *c.* 2 m, entirely
hairless. Branches long, green, angled when
young. Leaves of 3 leaflets, usually reduced
to bracts and stalkless on flowering
branches, otherwise stalk to 2.5 cm. Leaflets
0.6–1.8 × 0.5–1.4 cm, obovate to broadly
elliptic, rounded or mucronate, light green.
Flowers 4–12 in loose, leafless racemes, light
yellow; stalk 4–7 mm, with 2 or 3 persistent
bracteoles. Calyx 2–3 mm, bell-shaped,
hairless. Standard 1–1.2 cm, rounded to
notched; keel beaked. Pods 1.5–4 cm × 4–9
mm, oblong to linear, curved at the base,
smooth; seeds 4–10. *S Europe, N Africa.* H3.
Late spring.

2. C. nigricans Linnaeus. Figure 40(1),
p. 540. Illustration: Botanical Magazine,
8479 (1913); Bean, Trees & shrubs hardy
in the British Isles, edn 8, 822 (1970);
Reader's Digest encyclopaedia of garden
plants and flowers, 205 (1972).

Upright deciduous shrub to *c.* 2 m,
entirely adpressed silky-downy; branches
rod-like. Leaves of 3 leaflets, stalk 2–20
mm. Leaflets 0.6–3 × 0.5–1.6 cm, obovate-
oblong to linear-elliptic, nearly hairless
above, adpressed downy beneath. Racemes
to 20 cm, leafless; flowers many, solitary,
yellow, becoming darker. Flower-stalk
4–8 mm, with a long, persistent bracteole.
Calyx 2–4 mm, bell-shaped, downy.
Standard 0.9–1.2 × 0.6–0.9 cm, obovate-
oblong, slightly notched; wings shorter.
Pods 2–4 cm × 3–7 mm, linear-oblong,
adpressed silky-downy; seeds 2–8. *C & SE
Europe & C Russia.* H3. Early summer.

3. C. emeriflorus Reichenbach (*C.
glabrescens* Sartorius not Schrank).
Illustration: Botanical Magazine, 8201
(1908); Bean, Trees & shrubs hardy in the
British Isles, edn 8, 817 (1970).

Procumbent to erect shrub to *c.* 70 cm,
rounded; branches rigid, angled, shaggy-
hairy when young. Leaves of 3 leaflets;
stalk 1.5–3 cm, slightly downy; leaflets
1–2 cm × 3–4 mm, obovate to elliptic,
hairless above, adpressed downy beneath.
Racemes consisting of leafy bundles, loose;
flowers 1–5 per bundle, golden yellow,
sometimes red-streaked; flower-stalk 1–2.5
cm, with 1 long, persistent bracteole. Calyx
2–3 mm, bell-shaped, downy. Standard
1–1.5 cm, broadly obovate, slightly
notched. Pods 2.5–4 cm × 6–8 mm,
oblong to narrowly obovate, somewhat
bowed, pointed, smooth black; seeds to 6.
SE Europe. H3. Late spring.

4. C. fontanesii Bell (*Spartium biflorum*
Desfontaines; *Chronanthus biflorus*

(Desfontaines) Frodin & Heywood.
Illustration: Vicioso, Genisteas Españolas,
44 (1953–1955).

Upright or also ascending shrub, to
c. 50 cm; branches angled, downy only
when young. Leaves of 3 leaflets; stalk
2–4 mm; leaflets 4–10 × *c.* 1 mm, linear to
lanceolate, with some erect hairs. Flowers
solitary or paired, or to 8 in umbel-like
heads, sometimes combined to leafy
racemes, golden yellow; flower-stalk 2–4
mm, white-downy; bracteole 1. Calyx 3–4
mm, bell-shaped, with bristle-like white
hairs. Standard 1–1.7 cm, notched; keel
about as long. Pods 1–1.5 cm × 6–8 mm,
broadly obovate, pointed, enveloped by the
withered standard, with translucent valves,
brown-black, few-seeded. *E & S Spain,
Balearic Islands.* H4. Late spring.

5. C. supranubius (Linnaeus filius) Kuntze
(*C. fragrans* Lamarck). Illustration: Botanical
Magazine, 8509 (1913); Roles, Flora of the
British Isles, Illustrations, 172 (1957–65).

Upright shrub, to *c.* 3 m; branches thick,
grooved, blue-green bloomed, downy only
when young. Leaves of 3 leaflets, soon
falling; stalk 0.5–1.2 cm, adpressed-downy;
leaflets 4–8 × 1–2 mm, linear to narrowly
lanceolate, adpressed-downy on both sides.
Racemes consisting of leafy clusters;
flowers 1–4 or many per cluster, milk-
white, pink-toned, fragrant. Flower-stalk
2–4 mm, with 2 small bracts. Calyx 2–3
mm, bell-shaped, downy, soon falling;
upper lip reduced, hardly toothed.
Standard 0.9–1.5 cm × 6–8 mm, slightly
notched; keel shorter than wings. Pods
1.5–2.5 cm × 3–5 mm, elongate, tapered
at base and tip, black, hairless; seeds 2–7.
Canary Islands (Tenerife). H5. Spring.

6. C. filipes (Webb & Berthelot) Masferrer
(*Spartocytisus filipes* Webb). Illustration:
Roles, Flora of the British Isles,
Illustrations, 171 (1957–65).

Small shrub, almost entirely hairless;
branches densely arranged, thin, striped
to angled, green. Leaves of 3 leaflets, soon
falling; stalk 3–10 mm, thread-like; leaflets
5–12 × 1–4 mm, oblanceolate to narrowly
obovate, thin, somewhat downy beneath.
Racemes to *c.* 40 cm, consisting of leafy
clusters; flowers 1–4 per cluster, pure
white, fragrant. Flower-stalk 3–6 mm, with
3 bracteoles. Calyx 2–3 mm, bell-shaped,
membranous, soon falling. Standard
0.8–1.2 cm × 4–6 mm, narrowly obovate,
rounded; keel not very curved. Pods 2–3
× 0.5–0.6 cm, oblong, curved at the base,
hairless, black, few-seeded. *Canary Islands.*
H5. Late winter–spring.

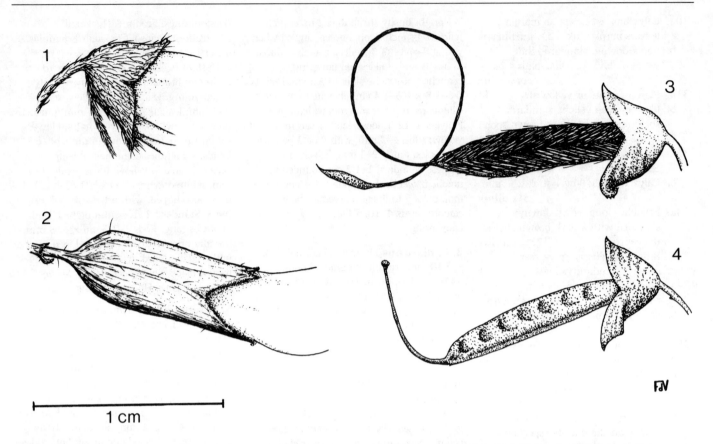

1 cm

Figure 40. Floral characters used in the identification of *Cytisus* species. 1, Adpressed downy bell-shaped calyx, with one linear bracteole (*C. nigricans*). 2, Nearly hairless tubular calyx, with three minute bracteoles of which two have fallen off (*C. purpureus*). 3, Rolled-up style; pod with long hairs on the margin (*C. scoparius*). 4, Curved-up style; pod hairless (*C. sessilifolius*).

7. C. ardoinii Fournier. Illustration: Hay & Synge, The colour dictionary of flowering plants, pl. 45 (1971).

Low shrub to *c.* 60 cm, flat to domed; branches grooved, downy only when young. Leaves of 3 leaflets; stalk 0.6–1.2 cm, with persistent bud-scales; leaflets 4–10 × 1–3 mm, narrowly oblong to obovate-oblong, rounded, both sides long-adpressed-downy. Racemes consisting of leafy clusters; flowers 1–4 per cluster, golden yellow. Flower-stalk 4–7 mm. Calyx 2–4 mm, bell-shaped, long-downy. Standard *c.* 1 cm, rounded to slightly notched; keel hardly beaked; style rolled up. Pods 1–2.5 cm × 4–7 mm, linear-oblong, curved, adpressed-hairy, few-seeded. *S France.* H3. Spring.

8. C. × kewensis Bean. Illustration: Botanical Magazine, n.s., 299 (1956–57); Millar Gault, Dictionary of shrubs in colour, pl. 133 (1976); Gardening from Which, July 1987, 235.

Prostrate shrub, to *c.* 30 cm; branches stiff, grooved, hairless. Leaves usually of 3 leaflets; stalk 0.5–1 cm, adpressed-downy; leaflets 0.7–1.5 cm × 1.5–3 mm, oblong to obovate, adpressed-downy beneath, nearly hairless above. Racemes to *c.* 40 cm, consisting of leafy clusters; flowers 1–3 per cluster, cream-white to sulphur-yellow. Flower-stalk 5–8 mm, downy; bracteoles 3, small, linear; calyx 2–4 mm, bell-shaped, somewhat downy. Standard 1.2–1.5 cm, notched, hairless. Pods 1.5–2 cm, narrowly oblong, straight, densely downy. *Garden origin.* H2. Late spring.

The hybrid between *C. ardoinii* and *C. multiflorus*, developed at Kew Gardens around 1900.

9. C. × beanii Nicholson. Illustration: Hay & Synge, Colour dictionary of flowering plants, pl. 46 (1971); Millar Gault, Dictionary of shrubs in colour, pl. 132 (1976); Gardening from Which, May 1988, 151.

Procumbent shrub, to *c.* 40 cm and to twice as wide; branches grooved, greenish brown, hairless. Leaves simple, 0.8–1.5 cm × 2–5 mm, elliptic to narrowly obovate, both sides slightly downy; stalk to *c.* 1 cm, downy. Racemes leafy; flowers 1–3 per axil, deep yellow. Flower-stalk 3–8 mm, slightly downy, with 3 small bracteoles. Calyx 1–2 mm, bell-shaped, downy. Standard 0.8–1.4 cm, rounded. Pods black with long white hairs when young. *Garden Origin.* H2. Spring.

The hybrid between *C. ardoini* and *C. purgans*, developed at Kew Gardens in 1892.

10. C. purgans (Linnaeus) Boissier. Illustration: Botanical Magazine, 7618 (1898); Vicioso, Genisteas Españolas, 46 (1953–1955); Krüssmann, Handbuch der Laubgehölze, 1: pl. 157 (1976).

Upright shrub, to *c.* 1 m and as wide or wider; branches stiff, grooved to striped, downy when young. Leaves stalkless, on young and flowering twigs simple, otherwise of 3 leaflets, usually falling soon. Leaflets 0.8–1.2 cm × 2–3 mm, linear-lanceolate to narrowly obovate, hairless above, silky-downy beneath. Racemes from compact, head-like to elongate and *c.* 20

cm, consisting of leafy clusters; flowers 1–4 per cluster, fragrant, golden yellow. Flower-stalk 4–9 mm, with 2 small bracteoles. Calyx 2–3 mm, bell-shaped, downy. Standard 1–1.2 × 0.8–1.1 cm, broadly obovate, slightly notched; all petals equal. Pods 1–3 cm × 5–8 mm, obovate-oblong, black; seeds 3–6. *SE Europe & N Africa*. H3. Late spring–early summer.

11. C. ingramii Blakelock. Illustration: Botanical Magazine, n.s., 211 (1952–1953).

Densely leafy upright shrub, to *c.* 2 m; branches angled, hairless. Leaves simple, stalkless on flowering and younger branches, otherwise of 3 leaflets; leaflets 1–3 × 0.6–1.5 cm, elliptic to oblong, silvery-downy beneath. Racemes leafy, few-flowered; flowers 1 or 2 per axil, cream-white to yellow. Flower-stalk 0.7–1 cm, adpressed-downy, with 3 minute bracteoles. Calyx 4–7 mm, bell-shaped, adpressed-downy. Standard 1.5–2.5 cm, slightly notched, cream-white. Pods 3–3.5 × 0.7–1 cm, elliptic-oblong, straight, brownish black, with long hairs; seeds to 8. *N Spain*. H4. Late spring–early summer.

12. C. multiflorus (L'Héritier) Sweet (*C. albus* Link not Hacquet). Illustration: Botanical Magazine, 8693 (1917); Polunin & Smythies, Flowers of southwest Europe, pl. 16 (1973); Polunin & Everard, Trees and bushes of Europe, 108 (1976); Davis, The gardener's illustrated encyclopedia of trees & shrubs, 111 (1987).

Tall shrub to *c.* 3 m; branches rod-like, angled, downy only when young. Leaves simple on flowering and younger branches, otherwise of 3 leaflets, soon falling; stalk to 3 mm; leaflets 0.7–1 cm × 0.5–2 mm, linear-lanceolate to oblong, silky-downy. Racemes long, consisting of many leafy clusters; flowers 1–3 per cluster, pure white to yellowish. Flower-stalk 2–5 mm, silky-downy, with one minute bract. Calyx *c.* 5 mm, bell-shaped, silky-downy. Standard *c.* 1 cm, rounded; keel often shorter than wings. Pods 1.5–2.5 × 0.5–1 cm, oblong, pointed, adpressed-downy; seeds 2–6, with large appendages. *Spain & N Africa*. H3. Late spring.

This species has several cultivars, that are hardier (e.g. 'Durus') or have slightly pinkish-blushed (e.g. 'Incarnatus') or earlier appearing flowers ('Toome's Variety').

13. C. × dallimorei Rolfe. Illustration: Gardeners' Chronicle **51**: 198 (1912); Botanical Magazine, 8482 (1913).

Upright shrub, like *C. scoparius*, to *c.* 2 m; branches angled, hairless. Leaves of 3 leaflets, sometimes also simple on flowering branches; stalk 0.5–2.5 cm, simple leaves stalkless; leaflets like those of *C. scoparius*. Raceme leafy, with solitary or paired, lilac-pink, rosy purple or yellow flowers. Flower-stalk 0.5–1 cm, sometimes downy, with 3 minute bracteoles. Calyx 3–7 mm, bell-shaped, sometimes downy. Standard 1–2 cm, notched; keel long, beaked; wings carmine-red; style rolled up. Pods *c.* 2.5 cm, with long hairs on the margin. *Garden origin*. H3. Late spring.

The hybrid between *C. multiflorus* and *C. scoparius* 'Andreanus', developed at Kew Gardens about 1900.

14. C. × praecox Bean. Illustration: Millar Gault, Dictionary of shrubs in colour, pl. 135 (1976); Gardening from Which, March 1987, 66; Amateur Gardening, September 1987, 19; Taylor's guide to shrubs 176, 177 (1987).

Bushy shrub, to *c.* 3 m; branches long, slightly grooved, grey-green. Leaves usually simple, stalkless, 1–2 cm × *c.* 2 mm, lanceolate to linear-spathulate, silky-downy on both sides, soon falling. Raceme long, leafy, many-flowered; flowers solitary or paired, showy cream-white, yellow, pink or red, with unpleasant fragrance. Flower-stalk 3–8 mm, silky-downy, with 3 bracteoles. calyx *c.* 2 mm, bell-shaped, slightly downy. Standard 1–1.5 cm × 5–9 mm, obovate, rounded. Pods 1.5–2.5 cm × 4–5 mm, ovate, tapered gradually, blackish, patent hairy, few-seeded. *Garden origin*. H3. Spring.

The hybrid between *C. multiflorus* and *C. purgans*, developed in Warminster (England). The species has many cultivars, that have variations in flower-colour, time of flowering or hardiness.

15. C. scoparius (Linnaeus) Link (*Sarothamnus scoparius* (Linnaeus) Wimmer). Figure 40(3), p. 540. Illustration: Polunin & Everard, Trees & bushes of Europe, 108 (1976); Millar Gault, Dictionary of shrubs in colour, pl. 142 (1976); Bärtels, Gardening with dwarf trees and shrubs, 126 (1983).

Erect shrub, to *c.* 2.5 m; branches rod-like, angled, deep green, downy only when young. Leaves simple on flowering and younger branches, otherwise of 3 leaflets; stalk 0.5–1.5 cm, simple leaves stalkless; leaflets 0.5–2 cm × 2–9 mm, elliptic-oblong to obovate. Racemes leafy, many-flowered; flowers 1 or 2 per cluster, golden yellow. Flower-stalk 0.5–2 cm, with 2

minute bracteoles. Calyx 4–7 mm, bell-shaped, hairless. Standard 2–3 × 1–1.6 cm, notched; keel beaked; style rolled up. Pods 2.5–7 × 0.4–1.2 cm, oblong, black, with white or brownish hairs on the margins; seeds 6–15. *W, SE & C Europe, introduced elsewhere*. H2. Spring.

From this species very many varieties and hybrids with many different species have been developed, varying in flower-colour (deep-brown and yellow, e.g. 'Andreanus') and time of flowering (very long, e.g. 'Dukaat').

Subsp. **maritimus** (Rouy) Heywood, from the coasts of NW Europe, has most of its branches prostrate; it is occasionally grown.

16. C. grandiflorus (Brotero) de Candolle (*Sarothamnus grandiflorus* (de Candolle) Webb). Illustration: Vicioso, Genisteas Españolas, 50 (1953–1955); Polunin & Everard, Trees & Bushes of Europe, 108 (1976).

Erect shrub, to *c.* 3 m; branches angled and silvery-downy only when young. Leaves stalkless, simple on flowering and younger branches, otherwise of 3 leaflets; leaflets 0.6–1.2 cm, awl-shaped to elliptic-obovate, slightly silvery-downy, especially on margins. Raceme leafy, many-flowered; flowers 1 or 2 per axil, golden yellow. Flower-stalk 6–9 mm, with 3 minute bracteoles. Calyx *c.* 4 mm, bell-shaped, hairless. Standard 1.5–2.5 × 1.5–2.5 cm, rounded to notched; all petals equal; style rolled up. Pods 2–4.5 × *c.* 1 cm, straight or somewhat bowed, black with long, white hairs; seeds 2–11. *S Spain, S & W Portugal*. H4. Spring.

17. C. cantabricus (Willkomm) Reichenbach (*Sarothamnus cantabricus* Willkomm).

Prostrate, compact shrub to *c.* 2 m; branches angled, hairless. Leaves simple on flowering and younger branches, otherwise of 3 leaflets; stalk 2–5 mm, simple leaves stalkless; leaflets 0.7–1.1 × *c.* 0.5 cm, obovate to lanceolate, somewhat downy above, more densely so beneath. Racemes leafy, with solitary, golden yellow flowers. Flower-stalk *c.* 1 cm, with 2 or 3 small bracteoles. Calyx 5–7 mm, bell-shaped, hairless. Standard 2–2.5 cm, deeply notched; style rolled up. Pods 3–5 cm × 6–8 mm, pointed, black, covered with long white hairs; seeds 3–10. *Spain & SW France*. H5. Late spring–early summer.

18. C. procumbens (Willdenow) Sprengel. Illustration: Edwards's Botanical

Register **47**: 1150 (1817); Hegi, Illustrierte Flora von Mitteleuropa **4**: 1207 (1923); Krüssmann, Handbuch der Laubgehölze, **1**: 157 (1976).

Procumbent, leafy shrub to *c.* 60 cm, entirely adpressed-downy; branches angled when young, terete and striped when older. Leaves simple, stalkless, 1.5–2.5 cm × 2–4 mm, oblong-obovate to lanceolate, less downy above. Raceme to *c.* 15 cm, consisting of leafy clusters; flowers 1–5 per cluster, yellow. Flower-stalk 0.75–1 cm, with one small bracteole. Calyx 2–5 mm, bell-shaped. Standard 1–1.5 cm, rounded; all petals hairless. Pods 3–3.5 × *c.* 0.5 cm, narrowly oblong, adpressed-downy; seeds 5–10. *SE Europe*. H3. Late spring–early summer.

19. C. decumbens (Durande) Spach (*C. prostratus* Simonkai). Illustration: Botanical Magazine, 8230 (1908); Bärtels, Das grosse Buch der Gartengehölze, 142 (1973); Krüssmann, Handbuch der Laubgehölze, **1**: pl. 157 (1976).

Like *C. procumbens*, but to *c.* 30 cm, downy only when young; branches angled. Leaves simple, stalkless, 0.8–2 cm × 3–6 mm, oblong obovate to oblanceolate, adpressed-downy above, more densely so beneath. Racemes consisting of leafy clusters; flowers 1–3 per cluster, golden yellow. Flower-stalk 1–2.5 cm, with 2 or 3 minute bracteoles. Calyx *c.* 5 mm, bell-shaped, with spreading hairs. Standard 1–1.5 cm, rounded; all petals hairless. Pods 2–3.5 cm × *c.* 6 mm, elliptic, black, erect downy, with 3–9 seeds. *S Europe*. H3. Late spring.

20. C. diffusus (Willdenow) Visiani. Illustration: Schlechtendahl et al., Flora von Deutschland, edn 5, 2295 (1880–1887).

Procumbent or ascending shrub, to *c.* 40 cm, almost entirely hairless; branches thin, angled. Leaves simple, stalkless, 1–2 cm, oblong to lanceolate, punctate. Racemes consisting of leafy clusters; flowers 1–3 per cluster, yellow. Flower-stalk 0.6–1 cm, slightly erect-downy, with 2 or 3 minute bracteoles. Calyx *c.* 4 mm, bell-shaped. Standard 1–1.2 cm, rounded; all petals equal. Pods 1.5–2.5 cm × 4–5 mm, elliptic to oblong, curved, black and hairless; seeds 2–6. *W Germany to Austria*. H2. Late spring.

21. C. purpureus Scopoli (*Chamaecytisus purpureus* (Scopoli) Link). Figure 40(2), p. 540. Illustration: Botanical Magazine, 1176 (1809); Bean, Trees & shrubs hardy in the British Isles, edn 8, 825 (1970); Bärtels, Das grosse Buch der Gartengehölze, 141 (1973); Millar Gault, Dictionary of shrubs in colour, pl. 136 (1976).

Procumbent or ascending shrub to *c.* 60 cm, entirely hairless or with occasional long hairs; branches rod-like, slightly angled. Leaves of 3 leaflets; stalk 1–2.5 cm; leaflets 1–2.5 cm × 4–7 mm, obovate-elliptic to oblong, pointed, dark green above. Racemes to *c.* 35 cm, consisting of leafy clusters; flowers 1–3 per cluster, purple-red to rose-pink. Flower-stalk 2–5 mm, with 2 or 3 minute bracteoles. Calyx 7–11 mm, tubular; lip-margins woolly hairy. Standard 1.5–2.5 cm, notched, with a dark patch in the center. Pods 2.5–4 cm × 3–5 mm, narrowly oblong to obovate, curved, black, bloomed; seeds 4–7. *C & SE Europe*. H2. Late spring–early summer.

This species has a few cultivars: with darker purple (e.g. 'Atropurpureus'), lighter red (e.g. 'Albocarneus') to even whitish (e.g. 'Albus') flowers.

22. C. glaber Linnaeus filius (*C. biflorus* L'Héritier; *C. ratisbonensis* Schäffer; *C. elongatus* Waldstein & Kitaibel; *Chamaecytisus glaber* (Linnaeus filius) Rothmaler). Illustration: Edwards's Botanical Register **47**: 1191 (1815); Botanical Magazine, 8661 (1916); Hegi, Illustrierte Flora von Mitteleuropa, **4**: pl. 159 (1923).

Variable, prostrate to erect shrub to *c.* 2 m; branches terete, rough grey or black downy when young. Leaves of 3 leaflets; stalks 0.5–2.5 cm, downy; leaflets 1–2.5 × 0.5–1.2 cm, obovate to oblanceolate, erect downy when young, more or less hairless when older. Racemes consisting of leafy clusters; flowers 2–4 per cluster, yellow with red-brown markings. Flower-stalk 2–7 mm, downy, without bracteoles, or with 1, linear, on terminal flowers. Calyx *c.* 1 cm, tubular, erect downy. Standard 1.5–3 cm, notched; petals sometimes downy. Pods 2–3.5 cm × 4–5 mm, narrowly ovate to oblong, densely downy; seeds 4–8. *C Europe, former USSR (Caucasus to Siberia)*. H2. Late spring.

23. C. hirsutus Linnaeus (*Chamaecytisus hirsutus* (Linnaeus) Link). Illustration: *Botanical Magazine*, 6819 (1885); Hegi, Illustrierte Flora von Mitteleuropa **4**: 1177 (1923); Polunin, Flowers of Europe, 51 (1969).

Variable, prostrate to upright shrub, covered with erect, long hairs; branches terete. Leaves of 3 leaflets; stalk 1–2.5 cm; leaflets 1–3 × 0.4–1.8 cm, obovate to elliptic, downy on both sides, especially on the margins. Raceme consisting of 1–many dense, leafy clusters; flowers 1–4 per cluster, golden yellow. Flower-stalk 1–7 mm, without bracteoles. Calyx 1–1.5 cm, tubular. Standard 2.5–3 cm, rounded or slightly notched, with a red-brown central patch, hairless; keel with a central row of long hairs. Pods 2.5–4 × 0.5–0.8 cm, oblong to obovate-oblong, curved, with rough hairs, especially on the margins. *S & SE Europe to former USSR (Caucasus)*. H3. Late spring–early summer.

24. C. × versicolor (Kirchner) Dippel. Illustration: Krüssmann, Handbuch der Laubgehölze, **1**: pl. 157 (1976).

Broad or upright shrub to *c.* 70 cm; branches rod-like, slightly angled, shaggy downy when young. Leaves of 3 leaflets; stalk 1–2.5 cm, adpressed downy; leaflets 1–3 × 0.4–1 cm, obovate to elliptic, densely downy beneath, nearly hairless above. Raceme to *c.* 35 cm, consisting of leafy clusters; flowers 1–4 per cluster, multicoloured. Flower-stalk 3–7 mm, densely downy, with 2 or 3 minute bracteoles. Calyx 0.7–1,3 cm, tubular, downy. Standard 1.8–2.2 cm, rounded, whitish; wings yellow; keel lilac-pink. Pods downy. *Garden origin*. H3. Late Spring.

A hybrid between *C. hirsutus* and *C. purpureus*.

25. C. proliferus Linnaeus filius. Illustration: Botanical Magazine, 1908 (1817); Roles, Flora of the British Isles illustrations, 28 (1957–1965); Kunkel, Flora de Gran Canaria **1**: 21 (1974).

Evergreen, open shrub to *c.* 5 m, entirely silky-downy; branches flexible, terete, dark brown. Leaves of 3 leaflets, stalk 0.5–1 cm, winged; leaflets 2–3 cm × 3–10 mm, elliptic to elongate-oblong, pointed, shortly stalked. Racemes from compact and umbel-like to *c.* 75 cm, consisting of leafy, shortly stalked clusters; flowers 4–7 per cluster, white. Flower-stalk 0.5–1.5 cm; bracteole 1, long, soon falling. Calyx 1–1.2 cm, tubular, dark brown, downy. Standard 1.5–2.5 cm, notched, downy on the outside, like the keel. Pods 3–5.5 × 0.7–1.1 cm, oblong, pointed, black-brown, downy; seeds 3–9. *Canary Islands (Tenerife, Gomera)*. H5. Spring–summer.

26. C. demissus Boissier (*C. hirsutus* Linnaeus var. *demissus* (Boissier) Halácsy). Illustration: Krüssmann, Handbuch der Laubgehölze, **1**: 126 (1976).

Low, wide shrub, to *c.* 10 cm; branches

thin, grooved, grey-downy. Leaves of 3 leaflets, stalk 4–8 mm, erect downy; leaflets 1–2 cm × 3–6 mm, oblong-obovate to oblanceolate, more or less hairless above, with spreading hairs beneath. Racemes consisting of leafy clusters; flowers yellow, standard and keel reddish brown. Flower-stalk 2–5 mm, dark, erect downy, without bracteoles. Calyx 1.2–1.5 cm, tubular, downy. Standard 2–3 × 1.2–1.5 cm, notched; keel much shorter. *N & E Turkey, Greece*. H3. Late spring–early summer.

27. C. palmensis (Christ) Hutchinson (*C. proliferus* var. *palmensis* Christ).

Evergreen shrub to 2.5 m; branches long, grooved only when young, yellowish downy. Leaves of 3 leaflets; stalk 0.7–1.5 cm, winged, downy; leaflets 1–3.5 × 0.4–1 cm, narrowly elliptic to oblanceolate, pointed, shortly stalked, finely and softly downy. Raceme leafy, consisting of clusters; flowers 2–5 per cluster, white or yellowish. Flower-stalk 0.5–1 cm, densely downy, with 3 linear bracteoles. Calyx 0.7–1 cm, tubular, densely downy, lips deeply cut. Standard 1.5–2 × *c*. 1 cm, rounded; all petals equal. Pods 2.5–4.5 × *c*. 0.5 cm, oblong, curved, brown-yellow-downy; seeds 6–12, glossy black. *Canary Islands (La Palma)*. H5. Late winter–spring.

28. C. supinus Linnaeus (*C. rochelli* Wierzbicki; *Chamaecytisus supinus* (Linnaeus) Link). Illustration: Schlechtendahl et al., Flora von Deutschland, edn 5, 2313 (1880–1897); Hegi, Illustrierte Flora von Mitteleuropa **4**: 1174 (1923); Vicioso, Genisteas Españolas, 42 (1953–5).

Rounded shrub to *c*. 1.2 m and as wide; branches terete, with long erect hairs. Leaves of 3 leaflets, stalk 3–15 mm, adpressed downy; leaflets 1–3.5 × 0.5–1.5 cm, oblong-elliptic to broad obovate, more downy beneath. Flowers 2–10 in umbel-like heads with bracts at the base, sometimes combined to form leafy racemes; flower-stalk 2–5 mm, downy, with 1 bracteole. Calyx 0.8–1.4 cm, tubular, upper lip deeply cut, erect downy. Standard 1.7–2.5 cm, rounded, dull yellow with brown spots in the center, silky outside. Pods 1.5–3.5 cm × 4–6 mm, narrowly obovate to oblong, adpressed or erect downy; seeds 3–6. *C & S Europe*. H3. Summer.

The name *C. rochelii* Wierzbicki is sometimes applied to plants of this species with conspicuously adpressed-downy pods.

29. C. austriacus Linnaeus (*C. canescens* Presley; *Chamaecytisus austriacus* (Linnaeus) Link). Illustration: Schlechtendahl et al., Flora von Deutschland, edn 5, 2312 (1880–87); Hegi, Illustrierte Flora von Mitteleuropa **4**: 1174 (1923).

Upright or procumbent shrub to *c*. 1 m; branches thin, terete, or very slightly grooved, dense grey-downy. Leaves of 3 leaflets; stalk 2–12 mm, downy; leaflets 1–3 cm × 3–10 mm, narrowly elliptic to obovate, downy, especially on margins. Flowers 4 or more, in umbel-like heads, yellow; flower-stalk 2–5 mm, downy, with 3 linear bracteoles. Calyx 1.2–1.5 cm, tubular, downy. Standard 1.5–2 cm, slightly notched, downy outside; keel and wings much shorter. Pods 2–4 cm × 5–7 mm, narrowly oblong to obovate, downy, yellowish when young; seeds to 10. *C & SE Europe*. H2. Summer.

30. C. albus Hacquet (*Chamaecytisus albus* (Hacquet) Rothmaler). Illustration: Botanical Magazine, 1438 (1812).

Upright shrub to *c*. 80 cm; branches slender, terete, downy. Leaves of 3 leaflets; stalk 0.4–1.2 cm, downy; leaflets 1–3 × 0.7–1.3 cm, narrowly elliptic to oblong-obovate, downy on both sides. Flowers 3–many in umbel-like heads, white or yellowish; flower-stalk 2–6 mm, with 2 or 3 bracteoles. Calyx *c*. 1 cm, tubular, whitish downy. Standard 1.6–2 cm, silky outside. Pods 2–3.5 × 0.5–0.6 cm, rough-haired, with numerous seeds. *Hungary & W Germany*. H2. Late spring–early summer.

31. C. tommasinii Visiani (*Chamaecytisus tommasinii* (Visiani) Rothmaler).

Rounded deciduous shrub to *c*. 50 cm, entirely downy; branches terete, greyish green. Leaves of 3 leaflets; stalk 0.5–1.5 cm, winged; leaflets 1.2–2.5 cm × 2–8 mm, elliptic-lanceolate to obovate, both sides adpressed downy. Flowers many in umbel-like heads, yellow; flower-stalk 2–5 mm, downy; bracteoles 2 or 3, linear. Calyx 1–1.3 cm, tubular, deeply cut. Standard 1.5–2.2 cm × *c*. 8 mm, narrowly obovate, rounded. Pods 1.5–2.5 cm × 4–5 mm, oblong, somewhat bowed, yellowish brown, downy; seeds 3–8. *W former Yugoslavia & N Albania*. H3. Late Spring.

32. C. battandieri Maire. Illustration: Botanical Magazine, 9528 (1938–1939); Millar Gault, Dictionary of shrubs in colour, pl. 131 (1976); de Bray, Manual of old-fashioned shrubs, pl. 33 (1986); Gardening from Which, November 1986, 340.

Strong-growing, upright, semi-deciduous shrub to *c*. 5 m and as wide, entirely densely silver-grey-downy. Leaves of 3 leaflets; leaf-stalk 1–3.5 cm; leaflets 4–11 × 2.5–5 cm, oval-elliptic, mucronate, light green. Racemes 2–15 cm, erect, leafless; flowers many, solitary, golden yellow. Flower-stalk 2–3 mm, with 3 deciduous bracteoles. Calyx tubular, with lower teeth deeply cut, soon falling. Standard 1–2 cm, rounded; all petals equal, silvery-grey-downy. Pods 3–5 × 0.6–0.8 cm, narrowly oblong, bowed at the base, yellow-brown, densely silvery-grey-downy; seeds 5–10, bulging. *N Africa*. H4. Early summer.

33. C. canariensis (Linnaeus) Kuntze (*Cytisus candicans* Lamarck; *C. ramosissimus* Poiret; *Spartium albicans* Cavanilles; *Teline canariensis* (Linnaeus) Webb & Berthelot). Illustration: Webb & Berthelot, Phytographia Canariensis **3**(2): t. 41 (1842); Boletim da Sociedade Broteriana **65**: 277 (1971); Kunkel, Flora de Gran Canaria **1**: pl. 23 (1974); Bramwell & Bramwell, Wild flowers of the Canary Islands, pl. 29 & 168 (1974).

Erect or spreading, much-branched shrub, to *c*. 3 m; branches downy when young. Leaf-stalk 1–6 mm; stipules 0.5–2 mm. Leaflets 3–13 × 1.5–4.5 mm, obovate-oblong to oblanceolate, almost hairless above, densely downy beneath; tip mucronate; stalk 0.2–0.6 mm. Racemes 1–6 cm, terminal, with 4–20 flowers; bracts of the lower 3–10 flowers leaf-like with 3 leaflets, the rest simple. Flower-stalk 1–5 mm; calyx 2–6 mm, lower teeth distinct. Standard 1–1.3 cm, notched, with V-shaped downy area on the midrib towards tip. Pod 1.5–3 cm × 3–5 mm, somewhat bowed, shortly downy, with 5–8 seeds. *Canary Islands (Tenerife, Gran Canaria)*. H5. Early Spring.

34. C. stenopetalus (Webb & Berthelot) Christ (*Teline stenopetala* (Webb & Berthelot) Webb & Berthelot). Illustration: Botanical Magazine, n.s., 327 (1958–59); Bramwell & Bramwell, Wild flowers of the Canary Islands, pl. 30 (1974).

Shrub or small tree, to *c*. 6 m; branches downy when young. Leaf-stalk 0.5–2.5 cm; stipules 1–2.5 mm. Leaflets 0.8–4.5 × 0.3–1.8 cm, narrowly elliptic-obovate to oblanceolate, very sparsely to densely downy above, adpressed-downy beneath; stalk 1–2.5 mm; tip acute or mucronate. Racemes 5–13 cm, terminal, with 6–26 flowers; bract of lowest flower sometimes leaf-like with 3 leaflets, but usually simple like the rest. Flower-stalk 2–4 mm; calyx

3–5 mm, upper teeth acute, lower teeth minute. Standard 1–1.6 cm, notched, hairless or downy on midrib. Pods 1.5–4 cm × 3–7 mm, densely shaggy-downy, with 2–5 seeds. *Madeira*. H5. Spring.

Var. **sericeus** Pitard & Proust (*Cytisus stenopetalus* var. *magnofoliosus* Kuntze). Illustration: Guerra, Vegetacion y Flora de la Palma, 202 (1983). Leaves sparsely to densely downy above, leaflets 2–4.5 cm.

Var. **microphyllus** Pitard & Proust (*Cytisus stenopetalus* var. *gomerae* Pitard & Proust). Leaves hairless above, leaflets 0.8–2.5 cm.

35. C. maderensis Masferrer (*Cytisus candicans* Holle; *Teline maderensis* Webb & Berthelot). Illustration: Boletim da Sociedade Broteriana **65**: 292 (1971); Vieria, Flores da Madeira, pl. 4 (1986).

Erect shrub to *c.* 2 m; branches downy when young, stiff. Leaf-stalk 0.7–1.1 cm; stipules 2.5–6 mm. Leaflets 0.6–1.5 cm × 2–10 mm, oblong-obovate to oblanceolate, sparsely downy above, densely so beneath; stalk 0.5–1 mm; tip acute, with a mucro 0.4–0.6 mm. Raceme 1–4 cm, terminal, rather compact, with 5–15 flowers. Bracts of lower 1 or 2 flowers leaf-like with 3 leaflets, those of the rest simple, reduced. Flower-stalk 2.5–4 mm; calyx 5–7 mm, lower teeth distinct. Standard 1.1–1.5 cm, hairless. Pods 2–3.5 cm × 4–8 mm, densely downy. *Madeira*. H5. Late spring–summer.

36. C. linifolius (Linnaeus) Lamarck (*Teline linifolia* (Linnaeus) Webb & Berthelot). Illustration: Bramwell & Bramwell, Wild flowers of the Canary Islands, pl. 169 (1974).

Erect shrub to *c.* 3 m, branches adpressed downy when young. Leaf-stalk absent or to 6.5 mm; stipules 0.2–6 mm. Leaflets 1–6 × 0.2–1 cm, narrowly oblanceolate to elliptic, sparsely white-adpressed-downy above, densely so beneath; tip obtuse to acute, mucronate. Raceme 1–3 cm, terminal, dense with 4–20 flowers; bracts of lower flowers leaf-like with 3 leaflets, those of the rest simple, reduced. Flower-stalk 2.5–7 mm; calyx 0.55–1.5 cm, lower teeth distinct, linear. Standard 0.9–1.8 cm, densely downy. Pods 1.5–3.5 cm × 4–6 mm, adpressed to spreading hairs. *W Mediterranean (excluding Corsica, Sicily, Sardinia), Canary Islands, Moroccan coastlands*. H5. Late spring–early summer.

Subsp. **linifolius**. Illustration: Botanical Magazine, 442 (1799); Boletim da Sociedade Broteriana **65**: 295 (1971). Leaves stalkless, without stipules; leaflets 1.7–2.5 cm × 2–4.5 mm, narrowly oblong, margins markedly curved upwards and inwards; stalk 1–2.5 mm. Flowers in racemes of 6–16; bracteoles 0.6–2.3 mm; calyx 0.6–1 cm; standard 1–1.4 cm.

Subsp. **rosmarinifolia** (Webb & Berthelot) Gibbs & Dingwall. Illustration: Webb & Berthelot, Phytographia canariensis, **3**(2): t. 44 (1836). Leaves stalkless, without stipules; leaflets 0.8–1 cm × *c.* 2 mm, linear-oblong, margins markedly curved downwards and inwards; base stalked *c.* 0.3 mm; tip acute. Raceme 1 cm, with usually 4 flowers; flower-stalk to 2.5 mm; bracteoles *c.* 4 mm; calyx *c.* 5.5 mm; standard *c.* 1 cm. Pod *c.* 1.7 × 0.5 cm, downy.

Subsp. **pallida** (Poiret) Gibbs & Dingwall (*Genista splendens* Webb & Berthelot). Illustration: Webb & Berthelot, Phytographia canariensis **3**(2): t. 43 (1836); Boletim da Sociedade Broteriana **65**: 298 (1971). Leaf-stalk to 2.5 mm and stipules to 6 mm; leaflets 3–6 × 0.4–1.1 cm, narrowly elliptic, margins slightly curved downwards and inwards; base stalked 1.5–2.2 mm; tip acute. Flower-stalk 4–9 mm; bracteoles 0.7–1.3 cm; calyx 0.9–1.5 cm; standard 1.4–1.7 cm, densely downy. Pods *c.* 3–3.5 × 0.4–0.5 cm, densely downy.

37. C. monspessulanus Linnaeus (*Genista candicans* Linnaeus; *Teline medicagoides* Medicus; *Cytisus pubescens* Moench; *T. monspessulana* (Linnaeus) Koch). Illustration: Botanical Magazine, 8685 (1916); Polunin & Smythies, Flowers of southwest Europe, pl. 18 (1973); Polunin & Everard, Trees & bushes of Europe, 109 (1976); Lamarda & Valsecchi, Alberi e Arbusti spontanei delle Sardegna, 280–283 (1982).

Erect, much branched shrub, to 3 m, branches densely downy when young; leaf-stalk 0.8–5 mm; stipules 0.5–1.5 mm. Leaflets 0.5–1.3 × 0.2–0.9 cm, broadly obovate to oblanceolate, sparsely to densely adpressed or spreading-downy on both surfaces; stalk 0.3–2 mm; tip mucronate or obtuse. Raceme short, indeterminate, with dense clusters of 4–7 flowers at the ends of short lateral branches; bracts of the lower 2 or 3 flowers like the leaves, the rest simple, reduced. Flower-stalk 1.5–3 mm; calyx 4–7 mm, lower teeth minute. Standard 1–1.3 cm, hairless. Pods 1.5–2.5 cm × 3–5 mm, densely woolly, with 3–6 seeds. *Mediterranean area*. H5. Late spring.

Nearly all of species **33–37** have spontaneously or in cultivation crossed with nearly all other species to form attractive hybrids. It is often difficult to trace back to the parents of these hybrids.

122. ERINACEA Adanson
H. Ern

Spiny shrubs with opposite or alternate branches. Leaves simple, sometimes made up of 3 leaflets; shortly stalked. Flowers 1–3, in axillary or more or less terminal clusters. Calyx inflated, bell-shaped, 2-lipped; upper lip with 2 teeth; lower with 3 teeth, the teeth one-third as long as the tube. Corolla blue-violet. Pod narrowly oblong, dehiscent. Seeds without appendages.

One species in SW Europe and N Africa. An interesting plant for the sunny rock-garden, but growing very slowly and only flowering for a short period.

1. E. anthyllis Link (*E. pungens* Boissier). Illustration: Botanical Magazine, 676 (1803); Vicioso, Genisteas Españolas **2**: 162 (1955); Polunin & Smythies, Flowers of southwest Europe, pl. 18 (1973).

Hummock-forming shrub, 10–30 cm; branches with stout spines. Leaves *c.* 5 mm, narrowly oblanceolate, falling early. Corolla 1.6–1.8 cm. Pod 1.2–2 cm, glandular and with shaggy hairs. Seeds 4–6. *Mts of Spain, just extending to France in E Pyrenees; NW Africa*. H2. Late spring.

123. ECHINOSPARTIUM (Spach) Rothmaler
H. Ern

Small shrubs with opposite branches. Leaves made up of 3 leaflets, shortly stalked or stalkless. Calyx somewhat inflated, bell-shaped, 2-lipped, upper lip deeply 2-fid; lower with 3 prominent teeth; all teeth as long as or longer than the tube; corolla yellow. Pod dehiscent. Seeds without appendages.

A genus of 3–5 species in the Pyrenees and the Iberian Peninsula. Only 1 species is in cultivation.

1. E. horridum (Vahl) Rothmaler (*Cytisanthus horridus* (Vahl) Gams; *Genista horrida* (Vahl) de Candolle; *Spartium horridum* Vahl). Illustration: Vicioso, Genisteas Españolas **1**: 28 (1953); Polunin & Smythies, Flowers of southwest Europe, 227 (1973).

A spiny shrub up to 40 cm. Leaflets 4–9 mm, narrowly oblanceolate, silky-hairy beneath, nearly hairless above; pulvini prominent. Flowers usually 2 on each branch, opposite. Calyx 0.7–1.2 cm., sparsely silky-hairy; teeth acute. Standard 1.2–1.6 cm, nearly hairless to sparsely silky-hairy. Pod 0.9–1.4 × 0.4–0.5 cm, with shaggy hairs. Seeds 1–3, ovoid, brown. *Pyrenees & SC France.* H2. Early summer.

124. SPARTIUM Linnaeus
J.R. Akeroyd
Deciduous, unarmed shrub 1–4 mm. Branches many, erect, rush-like, grooved, green, hairless. Leaves few, falling early, 1–3 cm, linear-oblong to narrowly elliptic or lanceolate, with appressed-silky hairs beneath. Stipules absent. Flowers in loose terminal racemes, fragrant. Calyx sheath-like, split above, 1-lipped, with 5 small teeth. Corolla 2–3 cm, bright yellow. Stamens united. Fruits 3–8 cm, linear-oblong, flat, silky-hairy when young. Seeds many.

A genus of a single species. Propagation is by seed. Plants require a well-drained, preferably calcareous, soil and are drought-tolerant.

1. S. junceum Linnaeus (*Genista odorata* Moench). Illustration: Botanical Magazine, 85 (1789) – as Genista juncea; Polunin, Hay & Synge, RHS dictionary of garden plants in colour, t. 1914 (1969); Polunin, Flowers of Europe, pl. 52 (1969); Brickell, RHS gardeners' encyclopedia of plants and flowers, 115 (1989).

Mediterranean area, but now naturalised elsewhere. H1. Summer–early autumn.

'Ochroleucum' is a cultivar with pale yellow corollas.

125. CHAMAESPARTIUM Adanson
H. Ern
Dwarf shrubs without spines, the young stems distinctly winged and flattened. Leaves simple or absent. Flowers in dense, terminal racemes. Calyx tubular, 2-lipped; upper lip deeply 2-fid, lower with 3 distinct teeth; corolla yellow, the standard broadly ovate, equalling the wings and keel. Pod dehiscent. Seeds with or without appendages.

A small genus of 2–4 species in C Europe and the Mediterranean area, systematically placed well in between *Cytisus* and *Genista*. Only 1 species is in horticultural use.

1. C. sagittale (Linnaeus) P. Gibbs (*Genista sagittalis* Linnaeus; *Genistella sagittalis* (Linnaeus) Gams; *Pterospartum sagittale* (Linnaeus) Willkomm). Illustration: Vicioso, Genisteas Españolas **1**: 138 (1953); Botanical Magazine, n.s., 332 (1959).

Dwarf shrub with procumbent, woody, mat-forming stems and usually erect,

herbaceous, simple or little-branched flowering-stems 10–50 cm; wings constricted at the nodes, without teeth or lobes. Leaves 0.5–2 × 0.4–0.7 cm, elliptic, hairless or almost so above, downy beneath. Calyx 5–8 mm; corolla 1–1.2 cm, the standard usually hairless. Pod 1.4–2 cm × 4–5 mm, downy. Seeds 2–5, without appendages. *Continental & Mediterranean Europe.* H1. Early Summer.

A nice plant for the sunny rock-garden. The W Mediterranean subspecies **delphinense** (Verlot) Soják from S France and **undulatum** (Ern) Soják from Andalucía may prove to be even more attractive (but less hardy) than the species, due to their silvery indumentum and their prostrate growth.

126. RETAMA Rafinesque
H. Ern
Shrubs without spines. Leaves simple; falling early. Flowers in racemes. Calyx urn-shaped, bell-shaped or obconical, 2-lipped. Corolla white to yellow. Pod ovoid to spherical, indehiscent or finally incompletely dehiscent along ventral suture. Seeds without appendages.

A genus of 4 species in the Canary Islands, Mediterranean area and W Asia. Only 1 species is in cultivation.

1. R. monosperma (Linnaeus) Boissier (*Genista monosperma* Lamarck; *Lygos monosperma* (Linnaeus) Heywood). Illustration: Botanical Magazine, 683 (1803); Polunin & Smythies, Flowers of southwest Europe: 44 (1973); Bramwell & Bramwell, Wild Flowers of the Canary Islands, f. 173 (1974); Maire, Flore de l'Afrique du Nord **16**: 199 (1987).

Erect shrub up to 4 m. Branches pendent, silky-hairy when young. Leaves linear-lanceolate, silky-hairy, falling early. Flowers fragrant, in loose racemes. Calyx *c.* 3.5 mm, without hairs, splitting along a line around the circumference and falling after flowering; upper lip with 2 triangular teeth, lower with 3 linear-subulate teeth, teeth often hairy on the margin. Corolla 1–1.2 cm, white; standard rhombic-ovate, hairy; wings oblong, obtuse, as long as or shorter than the keel. Pod 1.4–1.8 cm, obovoid with a short beak; wrinkled when mature. Seeds 1 or 2, blackish. *S Portugal, SW Spain, Canary Islands, N Africa.* H5. Early spring.

127. NOTOSPARTIUM J.D. Hooker
S.G. Knees
Shrubs or small trees with slender, grooved and drooping, leafless branches. Flowers borne in racemes; calyx bell-shaped;

corolla with shortly reflexed standard, keel obtuse, wings less than keel; stamens 10, 9 united, 1 free; ovary almost stalkless, style curved. Pods flattened, jointed between the seeds; seeds 6–15.

A genus of three species all native to South Island, New Zealand, requiring a sunny site with freely drained, sandy soil. They are easily propagated by seeds sown in autumn or from semi-ripe cuttings taken in summer. Some may need staking if not grown against a wall.

1a. Branchlets not more than 2 mm across; flowers with narrow standard **1. torulosum**
  b. Branchlets 2.5–3.5 mm across; flowers with broad standard 2
2a. Flowers *c.* 0.8 cm, pink-veined, densely packed on racemes; fruits to 2.5 mm across **2. carmichaeliae**
  b. Flowers *c.* 1.2 cm, purple-veined, loosely arranged; fruits to 4 mm across **3. glabrescens**

1. N. torulosum Kirk.
Shrub to 4 m; branchlets *c.* 2 mm across. Racemes slender, to 5 cm; flowers *c.* 0.8 cm, loosely arranged, lilac, flushed purple; standard narrow. Pods 1.5–2.5 cm, not more than 2 mm across; seeds *c.* 1 mm, dark brown, up to 15 per pod. *New Zealand.* H5–G1. Summer.

2. N. carmichaeliae J.D. Hooker. Illustration: Botanical Magazine, 6741 (1884); RHS dictionary of gardening **3**: 1383 (1981); Everett, New York Botanical Gardens encyclopedia of horticulture, 2340 (1981).

Shrub to 4 m; branchlets compressed, to 3.5 mm across. Racemes slender, densely flowered, to 5 cm; flowers *c.* 0.8 cm, pink-flushed and -veined; standard broad. Pods 0.8–1.7 cm, to 2.5 mm across with 6–10 seeds per pod. *New Zealand.* H5–G1. Summer.

3. N. glabrescens Petrie. Illustration: Botanical Magazine, 9530 (1939); Salmon, The native trees of New Zealand, 204, 205 (1980); Everett, New York Botanical Gardens encyclopedia of horticulture, 2340 (1981).

Shrub or small tree to 10 m, with slender drooping branches; branchlets compressed, 2.5–3.5 mm across. Racemes slender, open flowered, to 5 cm; flowers *c.* 1.2 cm, purple flushed and veined; standard broad. Pods 0.8–2.5 cm, up to 4 mm across; seeds *c.* 2.5 mm, reddish yellow, mottled black, about 6 per pod. *New Zealand.* H5–G1. Summer.

128. GENISTA Linnaeus
H. Ern

Spiny or non-spiny shrubs with alternate or opposite branching. Leaves mostly deciduous, sometimes very early so, shortly stalked, simple or made up of 3 leaflets, alternate or opposite. Flowers bisexual, alternate or opposite in racemes or axillary clusters or heads. Calyx tubular, usually with prominent upper and lower lips, the upper lip 2-fid, the lower with 3 distinct teeth. Corolla yellow. Standard broadly ovate or triangular, acute, hairless or downy, as long as or shorter than the keel. Keel narrowly oblong, hairless or downy. Wings as long as standard, hairless. Stamens 10, their filaments all united into a tube. Pod either narrowly oblong and compressed or sickle- or diamond-shaped and more or less inflated, several-seeded, more rarely ovoid acuminate and 1- or 2-seeded, hairless or downy. Seeds without appendages.

A genus of about 100 species in Europe (most numerous in the Iberian Peninsula), SW Asia, N Africa, the Canary Islands and Madeira; introduced elsewhere. Genistas vary from dwarf and prostrate shrubs to more than 6 m. In many species the leaves fall very early. All prefer a sunny position and a well-drained, light, loamy or even stony soil. Whereas some species (e.g. *G. cinerea*, *G. fasselata*, *G. hispanica*, *G. januensis*, *G. pulchella*, *G. radiata* and *G. sylvestris*) tolerate or even prefer calcareous soils, others, like *G. anglica*, *G. berberidea*, *G. falcata* or *G. florida* will not thrive under such conditions since they inhabit acid soils in nature. The smaller species do well in rock-gardens: *G. anglica*, *G. berberidea* and *G. falcata* can give a distinctive note to the heather garden, and the taller species like *G. aetnensis*, *G. florida* and *G. tenera* make fine solitary plants or may be used on sunny edges of tree-plantings. Most are relatively short-lived. All are preferably propagated by seed. There seem not to be any hybrids in cultivation, though there are some sterile clones on the market, some of them with double flowers.

Literature: Balls, E.K., Two months' collecting in Morocco, *Journal of the Royal Horticultural Society* **69**: 357–362 (1944); Vicioso, C., Genisteas Españolas, I: *Genista-Genistella*, 18–135 (1953); Gibbs, P.E., A revision of the genus Genista, *Notes from the Royal Botanic Garden Edinburgh* **27**: 11–99 (1966); Gibbs, P.E. Taxonomic notes on some Canary Island and North African species of *Cytisus* and *Genista*, *Lagascalia* **4**: 33–41 (1974); Canto, P. &

Jesus Sanchez, M., Revision del agregado *Genista cinerea* (Leguminosae), *Candollea* **43**: 73–92 (1988).

Habit. Prostrate or low-growing: **1,7,9,10, 14–16,19**; large shrubs more than 2 m: **11–13**. Spiny shrubs: **1–10**; non-spiny: **1,11–21**.

Leaves. Of 3 leaflets: **3,4,20,21**; simple: **1–3,5–19**.

Flowers. Congested in heads: **7,20**.

Pods. Ovoid-acuminate, 1–2-seeded: **3,4,7,10**.

1a. Plant more or less spiny   2
 b. Plant not spiny   11
2a. Pod narrowly oblong and compressed; seeds 1–12; standard broadly ovate, as long as wings and keel   3
 b. Pod ovoid-acuminate, usually 1-seeded, or diamond- or sickle-shaped and inflated, 2–8-seeded; standard triangular or ovate with a rounded or acute apex, usually shorter than the keel   4
3a. Plant with weak spines; bracteoles absent   **1. pulchella**
 b. Plant with stout spines; bracteoles present   **2. scorpius**
4a. Flowers or flowering branches borne directly on the axillary spines   **3. fasselata**
 b. Flowers or flowering branches not borne directly on the axillary spines   5
5a. Most leaves of 3 leaflets   **4. triacanthos**
 b. All leaves simple   6
6a. Leaves with spine-like stipules   **5. berberidea**
 b. Leaves without spine-like stipules   7
7a. Bracts, at least of the lowermost flowers, 1 mm or less, or absent; bracteoles minute   8
 b. Bracts, at least of the lowermost flowers more than 1 mm; bracteoles usually conspicuous   10
8a. Calyx hairless or very sparsely silky; pod sickle-shaped; seeds up to 18   **6. falcata**
 b. Calyx densely hairy; pod ovoid-acuminate; seeds 1 or 2   9
9a. Flowers congested in heads; standard about as long as the keel   **7. hispanica**
 b. Flowers in loose racemes; standard ½–⅔ as long as the keel   **8. germanica**
10a. Leaves hairless; pod slightly sickle-shaped; seeds 4–6   **9. anglica**

 b. Leaves downy beneath; pod ovoid-acuminate; seeds 1 or 2   **10. sylvestris**
11a. Leaves simple, but sometimes withering early and plant appearing leafless   12
 b. At least the lower leaves of 3 leaflets   21
12a. Large shrubs, 1–4 m or more   13
 b. Prostrate or erect shrubs 10–100 cm   15
13a. Branches practically without leaves when flowering; pod 1–1.5 cm   **11. aetnensis**
 b. Leaves present at flowering and persisting later; pod 1.5–2.5 cm   14
14a. Flowers in loose racemes which are 7–18 cm   **12. florida**
 b. Flowers in rather dense racemes which are 3–5 cm   **13. tenera**
15a. Keel and standard hairless   16
 b. Keel and usually the standard silky-hairy   18
16a. Stems usually 3-winged; leaves with a narrow, translucent, finely toothed margin   **14. januensis**
 b. Stems not winged; leaves without a translucent, toothed margin   17
17a. Leaves on the main stem 9–50 mm, ovate, lanceolate, elliptic, oblong or oblanceolate, downy, or hairless and hairy on margin and midrib beneath   **15. tinctoria**
 b. Leaves on the main stem 3–10 mm, linear-oblanceolate, almost hairless   **16. lydia**
18a. Plant erect with long, flexuous branches   **17. cinerea**
 b. Plant decumbent, much-branched   19
19a. Upper surface of leaves downy   **1. pulchella**
 b. Upper surface of the leaves hairless or nearly so   20
20a. Flowers in dense, terminal racemes; bracteoles present; leaves almost stalkless   **18. sericea**
 b. Flowers usually in long racemes on ascending branches; bracteoles absent; leaves usually shortly stalked   **19. pilosa**
21a. Flowers in 2–12-flowered terminal heads   **20. radiata**
 b. Flowers in usually long and loose racemes   **21. ephedroides**

1. G. pulchella Visiani (*G. humifusa* Villars; *G. villarsii* Clementi). Illustration: Coste, Flore de la France **1**: 301 (1901).

Spreading, non-spiny or weakly spiny

shrub, 10–30 cm. Young branches and surfaces of leaves with dense, long, silky, adpressed or spreading hairs. Leaves 2–9 × 1.5–3 mm, stalkless, narrowly elliptic. Flowers borne singly in the axil of each bract in a congested raceme; bracteoles absent; flower-stalks 2–5 mm. Calyx *c.* 4 mm. Standard 0.7–1 cm, ovate, with dense silky hairs. Pod 1.2–1.5 × 0.5–0.6 cm, felted. Seeds 2–4. *Mts of SE France, Albania, W former Yugoslavia.* H4. Spring.

2. G. scorpius (Linnaeus) de Candolle. Illustration: Bonnier, Flore complète **2**: pl. 119 (1912); Vicioso, Genisteas Españolas **1**: 79 (1953).

Erect, rarely spreading, intricately branched shrub, 50–200 cm, with stout axillary spines. Leaves 3–11 × 1.5–2 mm, simple, sparsely hairy beneath, almost hairless above. Flowers borne on short branches arising from the spines, or directly on the spines. Flower-stalks 2–5 mm. Calyx 3–5 mm, hairless or almost so, lips shorter than tube. Standard 0.7–1.2 cm. Pod 1.5–4 cm, hairless. Seeds 2–7, compressed, ovoid, olive-green. *Spain, S France, Morocco.* H4. Spring.

3. G. fasselata Decaisne (*G. sphacelata* Spach). Illustration: Zohary, Flora Palaestina **2**: pl. 61 (1972).

Spiny shrub, 1–2 m, hairy, with dark scale-like leaf-bases (pulvini). At least some leaves with 3 leaflets; leaflets 3–15 × 1–3 mm, narrowly oblanceolate, silky, falling early. Flowers borne singly or in loose clusters on spines or unarmed branches; bracts leaf-like, bracteoles minute. Calyx 4–5 mm, hairless, lips *c.* one-third as long as tube. Standard 6–7 mm. Pod 0.8–1 × 0.5 cm, ovoid, obliquely beaked, wrinkled. Seeds 2. *S Aegean area, Crete, Cyprus, Israel.* H5. Summer.

4. G. triacanthos Brotero (*G. scorpioides* Spach). Illustration: Vicioso, Genisteas Españolas **1**: 49, 52 (1953).

Erect shrub, 50–100 cm, with axillary spines. Leaves of 3 leaflets; leaflets 3–8 × 1–2 mm, oblanceolate, almost hairless, without spiny stipules. Flowers in loose racemes; lowest bracts *c.* 2 mm, simple, uppermost reduced. Calyx 2.5–4 mm, hairless. Standard *c.* 6 mm, triangular, hairless, shorter than the keel. Pod ovate to diamond-shaped, 0.8–1 cm, blackish, becoming hairless at maturity. Seeds 1–2. *W Iberian Peninsula, Morocco.* H5. Spring.

5. G. berberidea Lange. Illustration: Vicioso, Genisteas Españolas **1**: 77 (1953).

Erect shrub 50–100 cm; young branches

with dense spreading hairs. Leaves lanceolate, *c.* 5 × 3 mm, almost hairless. Stipules spine-like. Flowers 1–4 together at the ends of short branchlets; bracteoles *c.* 1 mm. Calyx 5–6 mm, covered with dense spreading hairs, lower lip twice as long as the tube. Standard 0.8–1 cm. Pod 0.5–1.1 cm, with sparse spreading hairs along the sutures, and with 4–6 dark, olive-green, shining seeds. *NW Spain, N Portugal.* H4. Spring.

6. G. falcata Brotero. Illustration: Vicioso, Genisteas Españolas **1**: 74 (1953).

Erect spiny shrub, 50–100 cm. Young branches and lower surfaces of the leaves sparsely silky. Leaves ovate to oblong-lanceolate, obtuse or acute, 4–12 × 2–6 mm. Flowers in short lateral racemes; bracts and bracteoles minute or absent. Calyx *c.* 5 mm, hairless to sparsely downy, lips as long as the tube. Standard *c.* 9 mm. Pod hairless, 1–2.5 cm, with 10–18 shining yellow seeds. *Western part of the Iberian Peninsula.* H4. Spring.

7. G. hispanica Linnaeus. Illustration: Bonnier, Flore complète **2**: pl. 118 (1912); Vicioso, Genisteas Españolas **1**: 65 (1953); Polunin, Flowers of Europe, pl. 52 (1969).

Decumbent to erect shrub, 10–50 cm, with axillary spines. Leaves simple, stalkless, 0.6–1 cm × 3–5 mm, lanceolate to oblanceolate, lower surface with dense adpressed or spreading hairs. Flowers in dense, terminal, almost head-like racemes; bracts *c.* 1 mm; bracteoles absent. Standard hairless, broadly ovate, slightly notched, about equalling wings and keel. Pod ovate to diamond-shaped, 6–9 mm, becoming hairless at maturity. Seeds 1 or 2, brown, somewhat shining. *N Spain, S France.* H3. Early summer.

Subsp. **hispanica**, from S France westwards to the E Pyrenees and E Spain, has branches and leaves with spreading hairs and standard 6–8 mm, whereas subsp. **occidentalis** Rouy (*G. occidentalis* (Rouy) Coste) occurs in the western Pyrenees and northern Spain and has branches and leaves with adpressed hairs and the standard 0.8–1.1 cm.

8. G. germanica Linnaeus. Illustration: Coste, Flore de la France **1**: 298 (1901); Bonnier, Flore complète **2**: pl. 118 (1912); Polunin & Smythies, Flowers of southwest Europe, pl. 17 (1973); Pignatti, Flora d'Italia **1**: 641 (1982).

Erect shrub 30–60 cm, usually with axillary spines. Leaves simple, 0.8–2 cm × 4–5 mm, elliptic or lanceolate, lower

surface with long, more or less spreading hairs. Flowers in loose racemes; bracts *c.* 1 mm; bracteoles absent. Calyx *c.* 5 mm, silky. Standard *c.* 8 mm, ovate with an acute apex, about two-thirds as long as the keel. Pod 0.8–1 × 0.5–0.6 cm, hairy. Seeds 2–4, ovoid, brown. *Most of Europe except the western fringe.* H1. Early summer.

9. G. anglica Linnaeus. Illustration: Bonnier, Flore complète **2**: pl. 119 (1912); Vicioso, Genisteas Españolas **1**: 71 (1953); Garrard & Streeter, Wild flowers of the British Isles, pl. 29 (1983).

Decumbent to erect shrub, 40–100 cm, with hairless or hairy branches and axillary spines. Leaves 4–10 × 2–3 mm, lanceolate or elliptic, hairless. Stipules not spine-like. Flowers in short racemes; bracts leaf-like, in groups; bracteoles less than 1 mm. Calyx 3–4 mm, hairless, lips longer than tube. Standard 6–8 mm. Pod hairless, 1.4–2 × 0.5–0.6 cm, slightly inflated and sickle-shaped, with 4–6 shining black seeds. *W Europe, extending eastwards to S Sweden, N Germany & SW Italy, Morocco.* H3. Spring.

10. G. sylvestris Scopoli. Illustration: Botanical Magazine, 8075 (1906); Pignatti, Flora d'Italia **1**: 641 (1982).

Decumbent shrub, 20–50 cm, with weak axillary spines; young branches with silky hairs. Leaves 1–2 cm × 1–3 mm, simple, narrowly oblong or elliptic, lower surface with sparse hairs. Flowers in long, loose terminal racemes. Calyx 5–7 mm, with sparse silky hairs, upper lip about as long as lower, the lower teeth equal. Standard 7–8 mm, triangular, heart-shaped, claw less than 2 mm. Pod hairless at maturity, *c.* 8 × 4 mm, ovoid, beaked. Seeds 1 or 2, blackish. *Albania, former Yugoslavia, C & S Italy.* H3. Early summer.

11. G. aetnensis (Bivonia-Bernardi) de Candolle. Illustration: Botanical Magazine, 2674 (1826); Polunin & Everard, Trees and bushes of Europe, 109 (1976); Pignatti, Flora d'Italia **1**: 643 (1982); Gardening from Which, December 1986, 404.

Shrub to 6 m, without spines. Leaves simple, withering early, as do the leaf-like bracts. Flowers in loose racemes. Calyx *c.* 3 mm, almost hairless; upper teeth obtuse and about one-third as long as the tube, lower teeth minute. Standard more or less hairless. Pod 1–1.5 × 0.5 cm, hairless, sickle-shaped. Seeds 2–4. *Italy (Sardinia, Sicily).* H3. Summer.

12. G. florida Linnaeus. Illustration: Vicioso, Genisteas Españolas **1**: 104

(1953); Polunin & Smythies, Flowers of southwest Europe, pl. 17 (1973).

Erect shrub without spines, 1–3 m. Leaves simple, 0.5–2.5 × 0.2–0.5 cm, oblanceolate, sometimes sparingly hairy above, shortly stalked to almost stalkless. Flowers borne singly in the axils of each bract in long, loose racemes; lowermost bracts leaf-like, upper reduced. Calyx 4–6 mm. Standard broadly ovate, almost hairless. Pod flattened, oblong to oblong-lanceolate, pointed, 1.5–2.5 cm, greyish silky. Seeds 2–4, black and shining. *Iberian Peninsula, W Pyrenees, Morocco*. H3. Early summer.

There are several variants (or even species) belonging to this somewhat complex species. Subsp. **leptoclada** (Spach) Coutinho. *N Spain, just extending in to the French Pyrenees*. H3. In Berlin it flowers 4–6 weeks earlier than subsp. **florida**. Plants from Morocco (e.g. those cultivated as Balls 1944) show a beautiful contrast between the silvery leaves and the golden yellow, fragrant flowers (H4).

13. G. tenera (Murray) Kuntze (*G. virgata* (Aiton) de Candolle; *Cytisus tenera* Jacquin; *Spartium virgatum* Aiton). Illustration: Botanical Magazine, 2265 (1821).

Shrub without spines to 2 m or rarely to 4 m. Leaves simple, grey-green, more or less stalkless, *c.* 1.2 × 0.3 cm, silky beneath. Flowers in racemes 3–5 cm, terminating the short shoots of the current year. Calyx with silky-hairs. Flowers bright yellow, standard roundish, *c.* 1.2 cm across. Pods *c.* 2.5 cm, densely silky-hairy. Seeds 3–5. *Madeira*. H4. Summer.

A sterile variant of *G. tenera* with a more graceful habit and producing a profusion of rich golden yellow, fragrant flowers has been known for a long time as 'G. cinerea Hort.' and is now being distributed under the cultivar name 'Golden Shower'.

14. G. januensis Viviani (*G. lydia* Boissier var. *spathulata* (Spach) Hayek; *G. scariosa* Viviani; *G. triangularis* Willdenow; *G. triquetra* Waldstein & Kitaibel). Illustration: Botanical Magazine, 9574 (1939); Pignatti, Flora d'Italia 1: 637 (1982).

Prostrate to erect shrub, 10–50 cm; stems and branches usually 3-winged. Leaves of the flowering branches 0.5–1.2 cm × 2–4 mm, elliptic to obovate; leaves on the non-flowering branches 0.5–4 cm × 3–7 mm, elliptic to lanceolate; all hairless and with narrow, translucent, finely toothed margins. Flowers in short racemes on ascending lateral branches. Calyx 3.5–4 mm, almost hairless, lips shorter than the

tube. Standard 0.9–1 cm, broadly ovate. Pod 2 cm × 4 mm, hairless. Seeds 5–8. *Italy, NW Balkan Peninsula*. H3. Spring.

15. G. tinctoria Linnaeus. Illustration: Vicioso, Genisteas Españolas 1: 115 (1953); Garrard & Streeter, Wild flowers of the British Isles, pl. 29 (1983).

Prostrate to erect shrub, 10–200 cm, without spines, very variable in habit, leaf-shape and degree of hairiness. Leaves 0.9–5 cm × 2.5–15 mm, simple, hairless or densely silky-hairy. Flowers borne singly in the axils of bracts in short racemes towards the ends of the branches or in long simple or compound racemes; bracts leaf-like; bracteoles *c.* 1 mm; flower-stalk 1–2 mm. Calyx 3–7 mm, hairless to densely silky hairy. Corolla hairless; standard 0.8–1.5 cm, broadly ovate. Pod linear-lanceolate, straight or slightly curved, 2–3 × 0.2–0.3 cm, generally becoming hairless at maturity. Seeds 6–12, greenish. *Europe, W Asia*. H1. Summer.

'Plena' is a garden variety which grows as a dwarf, semi-prostrate shrub with double flowers; owing to the more numerous petals it is very brilliant in colour and is regarded as one of the best of all dwarf yellow-flowered shrubs.

16. G. lydia Boissier (*G. antiochia* Boissier; *G. rhodopea* Velenovský; *G. rumelica* Velenovský). Illustration: Botanical Magazine, n.s., 292 (1957).

Prostrate or erect shrub, without spines, 10–200 cm. Stems not winged. Leaves 3–16 × 1–3 mm, simple, linear-oblanceolate or linear-oblong, almost hairless, entire and without a translucent margin. Flowers in short racemes of 2–8 on the lateral branches. Calyx 3.5–5 mm, hairless, lips almost as long as the tube. Standard 1–1.2 cm, broadly ovate. Pod hairless, 2–3 × 0.3–0.5 cm, with 2–6 seeds. *E Mediterranean area*. H3. Late spring.

17. G. cinerea (Villars) de Candolle. Illustration: Botanical Magazine, 8086 (1906); Vicioso, Genisteas Españolas 1: 99 (1953); Ceballos et al., Plantas silvestres de la peninsula Iberica, 157 (1980).

Erect shrub without spines, 50–150 cm. Leaves simple, 0.5–1 × 0.2–0.3 cm, variable in shape, silky-hairy beneath. Flowers mostly paired and borne directly on the main branches; bracts in tufts. Standard hairless or with a central ridge of hairs, rarely uniformly silky hairy. Pod 1.2–1.9 × 0.4–0.5 cm, oblong or lanceolate, straight, shortly pointed,

greyish silky or with shaggy hairs. Seeds 2–5, kidney-shaped to ovoid, brown, shining. *W Mediterranean area*. H3. Summer.

Out of the 5 subspecies recognised by Canto & Jesus Sanchez (reference above), subsp. **cinerea** and subsp. **ausetana** Bolòs & Vigo are hardy in the milder parts of Europe. The more Mediterranean subspecies, namely subsp. **murcica** (Cosson) Cantó & Sanchez, subsp. **speciosa** Rivas Martínez et al., (*G. ramosissima* misapplied) and subsp. **valentina** (Sprengel) Rivas Martínez (*G. valentina* (Sprengel) Steudel) may be recommended for zones H4 and H5.

18. G. sericea Wulfen. Illustration: Pignatti, Flora d'Italia 1: 638 (1982).

Much-branched shrub without spines, 80–120 cm. Leaves 0.5–2.5 × 0.2–0.5 mm, almost stalkless, narrowly elliptic, oblanceolate or obovate, shortly mucronate, silky beneath, hairless or almost so above. Flowers borne singly in the bract-axils, in terminal clusters of 2–5; flowering branches often slender and flexuous. Bracteoles *c.* 1 mm, borne at the middle of the flower-stalk, which is 2–3 mm. Standard 1–1.4 cm, broadly ovate, tapered to the base, silky. Pod 1–1.5 × 0.5–0.6 cm, downy. Seeds 1–4. *Mts of the W part of the Balkan Peninsula, NE Italy*. H3. Early summer.

19. G. pilosa Linnaeus. Illustration: Vicioso, Genisteas Españolas 1: 120 (1953).

Prostrate to more or less erect shrub without spines, 20–50 cm. Leaves 0.5–1.2 cm, simple, usually oblanceolate, shortly stalked to almost stalkless, adpressed silky beneath, hairless above. Flowers borne singly or in pairs in the bract-axils, in loose racemes on ascending branches. Bracteoles absent. Calyx 4–5 mm. Standard 0.8–1 cm, broadly ovate with sparse, adpressed, silky hairs. Pod 1.8–2.8 cm, oblong, compressed, with shaggy hairs. Seeds 3–8, ovoid, greenish, shining. *W & C Europe*. H2. Early summer.

20. G. radiata (Linnaeus) Scopoli (*Cytisanthus radiatus* (Linnaeus) Lang; *Spartium radiatum* Linnaeus). Illustration: Botanical Magazine, 2260 (1821); Coste, Flore de la France 1: 299 (1901); Hegi, Illustrierte Flora von Mitteleuropa, 4: t. 159 (1975); Pignatti, Flora d'Italia 1: 642 (1982).

Erect shrub without spines, 30–100 cm. Branches opposite at almost every node.

Leaves of 3 leaflets, opposite; leaflets 0.5–2 × 0.2–0.4 mm, oblanceolate, silky beneath, almost hairless above. Flowers almost opposite and almost stalkless in terminal clusters of 4–12; lowermost flowers with simple, usually shortly 3-fid, translucent bracts much shorter than the flowers; bracteoles 1–3 mm. Calyx 4–6 mm, lips about as long as the tube. Standard 0.8–1.4 cm, broadly ovate, as long as or slightly longer than the keel, hairless or with a central ridge of silky hairs. Pod c. 5 × 3 mm, with a short curved beak, silky hairy. Seeds 2 or 3. *S Alps to C Italy & W former Yugoslavia, very locally to SW Romania & C Greece.* H3. Summer.

21. G. ephedroides de Candolle (*Cytisanthus ephedroides* (de Candolle) Gams). Illustration: Pignatti, Flora d'Italia **1**: 642 (1982).

Erect shrub without spines, 50–100 cm. Leaves (at least the lower) of 3 leaflets, falling early; leaflets 4–15 × 2–3 mm, silky on both surfaces. Flowers alternate or almost opposite in loose racemes; lowermost bracts of 3 leaflets, the uppermost simple. Calyx 3–6 mm, silky-hairy, lips and upper teeth about as long as tube. Standard 0.7–1 cm, broadly ovate or diamond-shaped, c. two-thirds as long as the keel, sparsely silky. Pod c. 1 × 0.5 cm, with a curved beak, silky-hairy. Seeds 2–3. *S Italy, Sardinia, Sicily.* H5. Early summer.

129. ADENOCARPUS de Candolle
H. Ern

Unarmed shrubs with alternate branches. Leaves divided into 3 leaflets, the leaves sometimes on short shoots. Flowers in terminal racemes or clusters. Calyx tubular, 2-lipped, sometimes with glandular tubercles, the upper and lower lips prominent, the upper deeply 2-fid, the lower with 3 distinct teeth. Corolla orange-yellow. Stamens 10 with their filaments united into a tube. Pod oblong, dehiscent, covered with glandular warts. Seeds numerous, oblong, each with a small appendage along one of the shorter sides.

A genus of about 15 species, most of them from mountain areas of N Africa and southern Spain; however, one species (*A. mannii*) is widely distributed in the high mountains of tropical Africa, and 3 are restricted to the Canary Islands.

Most *Adenocarpus* are very showy and can be highly recommended as ornamental shrubs, both for the deep yellow flowers and for their interesting foliage. In spite of this, only a few species are in horticultural use. They all need well-drained soil and full exposure to sunlight.

Literature: Rivas-Goday, S. & Fernandez-Galiano, E., *Adenocarpus hispanicus* (Lamk.) DC. como planta ornamental, *Anales del Instituto Botanico A.J. Cavanilles* **12** (2): 1–7 (1954); Vicioso, C., *Genisteas Españolas* **2**: 232–252 (1955); Gibbs, P.E., A revision of the genus *Adenocarpus*, *Boletim Sociedade Broteriana* **41**: 67–121 (1967).

1a. Leaflets 1.5–3.5 × 1–1.5 cm, broadly ovate or elliptic **1. anagyrifolius**
  b. Leaflets less than 1.5 cm, or if more then very narrowly elliptic or very narrowly oblanceolate 2
2a. Leaf-stalks 4–15 mm 3
  b. Leaf-stalks 1–4 mm 5
3a. Leaflets 1–2 mm wide, markedly rolled upwards at the margins and appearing almost linear **2. decorticans**
  b. Leaflets 3–7 mm wide, broadly elliptic to oblanceolate, margins only slightly rolled upwards 4
4a. Flower-stalks 1–7 mm; standard 1–1.5 cm **3. complicatus**
  b. Flower-stalks 7–15 mm; standard 1.5–2.5 cm **4. hispanicus**
5a. Calyx with glandular papillae; standard more or less hairless **5. viscosus**
  b. Calyx without glandular papillae; standard covered with silky hairs **6. foliolosus**

1. A. anagyrifolius Cosson & Balansa. Illustration: Boletim Sociedade Broteriana **41**: 114, map 6 (1967).

Erect shrub to 2 m. Leaf-stalks 1–2 cm; leaflets 1.5–3.5 × 1–1.5 cm, broadly ovate or elliptic, apex with a short mucro, lower surface with sparse adpressed hairs, upper surface more or less hairless. Flowers in loose racemes; flower-stalks 2–4 mm. Calyx 5–7 mm, covered with adpressed to somewhat spreading hairs. Standard c. 1.2 cm, broadly ovate, covered with short adpressed hairs. Pods 3–4.5 × 0.4–0.6 cm, covered with dense glandular papillae. *Morocco (Haut & Moyen Atlas).* H5. Early summer.

2. A. decorticans Boissier (*A. boissieri* Webb; *A. speciosus* Pomel). Illustration: Botanical Magazine, n.s., 48 (1949); Boletim Sociedade Broteriana **41**: 107, map 4 (1965).

Erect shrub to 3 m, sometimes becoming tree-like. Leaflets 0.9–1.8 × 0.1–0.2 cm, very narrowly elliptic and with margins markedly rolled upwards, silky-hairy on both surfaces. Flowers in racemes; stalks 4–8 mm. Calyx c. 8 mm, densely silky-hairy, not glandular. Standard c. 1.5 cm, broadly ovate, silky-hairy. Pod 2–6 × 0.8–1 cm, oblong, densely covered with glandular papillae. *S Spain, Morocco, Algeria.* H5. Early summer.

3. A. complicatus (Linnaeus) Grenier & Godron (*A. bivonae* (Presl) Presl; *A. divaricatus* Sweet; *A. graecus* Grisebach; *A. intermedius* de Candolle; *A. parvifolius* (Lamarck) de Candolle). Illustration: Botanical Magazine, 1387 (1811).

Erect shrub to 2 m or more; twigs and leaves sparsely to densely silky-hairy or with spreading hairs. Leaflets 0.5–2.5 × 0.2–0.7 cm, oblanceolate. Flower-stalks 1–7 mm. Calyx 5–7 mm, with or without glandular papillae, more or less hairless or silky-hairy. Standard 1–1.5 cm, silky hairy. Pod 1.5–4.5 × 0.4–0.6 cm, narrowly oblong, covered with glandular papillae. *Mediterranean area, extending to NW France.* H3. Early summer.

A variable and widespread species divided into several subspecies, though it is uncertain which of these is in cultivation. Hardiness depends very much on the origin of the plants or seeds; specimens from higher elevations of C Spain (province of Segovia) are hardy enough to produce seed in Berlin.

4. A. hispanicus (Lamarck) de Candolle. Illustration: Vicioso, Genisteas Españolas **2**: 234 (1955); Boletim Sociedade Broteriana **41**: 107, map 4 (1965).

Erect shrub to 2 m. Leaflets 1.5–3 × 0.3–0.8 cm, oblanceolate, acuminate, silky on both surfaces. Flowers in congested racemes; stalks 0.7–1.5 cm. Calyx 0.8–1.2 cm, with somewhat spreading hairs and glandular papillae, especially on the teeth. Standard 1.5–2.5 cm, broadly ovate, silky-hairy. Pod 2–5 × 0.8–1 cm, oblong, densely covered with glandular papillae. *Western half of the Iberian Peninsula & one small area in the coastal ranges of Morocco.* H4. Summer.

Subsp. **argyrophyllus** (Rivas Goday) Rivas Goday, which is found in the Moroccan locality cited above, has leaves densely silvery white-hairy above, the calyx with few or no glandular papillae and the standard densely downy. This is a highly decorative plant which may prove somewhat more tender than subsp. **hispanicus**.

5. A. viscosus (Willdenow) Webb &
Berthelot. Illustration: Bramwell &
Bramwell, Wild flowers of the Canary
Islands, f. 164 (1974).

Erect, densely leafy and somewhat sticky
shrub. Leaves on short shoots, stalks 2–5
mm, leaflets 3–7 × 1.5–2.5 mm, narrowly
elliptic, margins conspicuously rolled
upwards, both surfaces with spreading
hairs, the lower densely so. Flowers in
terminal racemes. Calyx 7–8 mm, with
long hairs and glandular papillae.
Standard c. 1.2 cm, broadly ovate, more or
less hairless or with sparse hairs towards
the apex. Pod 2–3.5 × 0.5–0.6 cm, with
glandular papillae and sparse hairs. *Canary
Islands (La Palma, Tenerife)*. G1. Spring.

6. A. foliolosus (Aiton) de Candolle.
Illustration: Bramwell & Bramwell, Wild
flowers of the Canary Islands, f. 165
(1974).

Erect, densely leafy shrub. Leaves
usually dense on short shoots, stalks 1–3
mm; leaflets 3–6 × 1.5–2.5 mm, obovate
or narrowly lanceolate, margins somewhat
rolled upwards, lower surface with
adpressed or spreading hairs, upper
hairless. Flowers in terminal racemes;
stalks 4–12 mm. Calyx 7–8 mm, usually
with dense, long hairs. Standard c. 1.2
cm, broadly ovate, with dense silky hairs.
Pod 1.2–4 × 0.4–0.5 cm, narrowly oblong,
with glandular papillae (sometimes very
sparse) and sparse hairs, containing 3–5
seeds. *Canary Islands (Gomera, Gran Canaria
& Tenerife)*. G1. Spring.

130. ULEX Linnaeus
H.S. Maxwell
Spiny, dense shrub. Leaves usually
alternate, to 1.5 cm, ternate, linear,
sharply pointed when young, reduced
to green spines or scales when mature;
stipules absent. Flowers fragrant, 1.5–2.5
cm, solitary or few in small axillary
clusters or racemes, produced in the leaf
axis of the previous years growth, shortly
stalked, golden-yellow; bracteoles 2, small,
below flower. Calyx persistent, 2-lipped,
lower lip with 3 small teeth, upper lip with
2 small teeth. Corolla persistent, standard
ovate, wing and keel obtuse. Stamens 10,
all united, alternating in 2 lengths; style
slightly curved. Fruit a dehiscent, small,
hairy, broadly-ovate to linear-oblong,
explosive pod containing 1–6 seeds.

A genus of about 20 species from Europe
and N Africa but now naturalised in the
mid United States. Only 3 species are in
general cultivation in Europe. The plants

grow best in poor, sandy, acidic soil in full
sun, on grassland or scrub. They are often
used as windbreaks in coastal areas or as
ground cover for dry banks. Propagation
can be from cuttings or seed but they do
tend to hybridise in the wild. They can be
grown in pots but once planted out they
do not transfer well; pruning should be
done after flowering.

1a. Bracteoles 2–5 mm; spines rigid,
deeply grooved; flowers 2–2.5 cm
1. europaeus
 b. Bracteoles 0.5–0.8 mm; spines
softer, slightly grooved; flowers to
1.6 cm   2
2a. Calyx 1–1.5 cm; standard 1.2–2
cm; fruit 0.8–1.5 cm with 1–2 seeds
2. gallii
 b. Calyx 0.5–1 cm; standard 0.7–1.5
cm; fruit 5–8.5 mm with 3–6 seeds
3. minor

1. U. europaeus Linnaeus (*U. hibernicus*
misapplied; *U. strictus* (Mackay) Webb).
Illustration: Ross-Craig, Drawings of British
plants **7**: pl. 5 (1954); Fitter, Fitter &
Blamey, The wild flowers of Britain and
Northern Europe, 121, 1 (1974); Csapody
& Tóth, A colour atlas of flowering trees
and shrubs, pl. 64, 140 (1982); Keble
Martin, The new concise British Flora, pl.
21 (1982).

Densely spiny glaucous shrub to 2.5 m;
stems erect or ascending, sparsely hairy
to densely felted with reddish-brown to
black hairs. Spines to 2.5 cm, rigid, deeply
grooved. Flowers 2–2.5 cm, coconut
smelling, stalks to 6 mm, with adpressed
hairs; bracteoles 2–5 mm; calyx persistent,
1–2 cm, with spreading hairs, shorter than
the corolla; wing petals straight, longer
than keel; standard even longer, 1.2–1.8
mm. Fruit to 1–2 cm, black with dense
shaggy grey or brownish black hairs;
seeds 2 or 3, dehiscing in summer. *W & C
Europe*. H2. Throughout year.

'Flore Pleno' (*U. europaeus* 'Plenus'; *U.
europaeus* var. *flore-pleno* G. Don; *U.
europaeus* forma *plenus* Schneider).
Illustration: Bean, Trees & shrubs hardy
in the British Isles **4**: pl. 92 (1980). Has
a more compact habit and longer lasting
double flowers. Propagation has to be from
cuttings, as no seed is set.

2. U. gallii Planchon. Illustration: Ross-
Craig, Drawings of British plants **7**: pl. 6
(1954); Fitter, Fitter & Blamey, The wild
flowers of Britain and Northern Europe,
121, 1b (1974); Keble Martin, The new
concise British Flora, pl. 21 (1982).

Densely spiny, dark green shrub, 1.5–2
m; stems ascending but often prostrate
in coastal areas, branches hairy. Spines
slightly grooved, to 2.5 cm. Flowers to
1.6 cm; stalks 3–5 mm, with adpressed
hairs, bracteoles 0.5–0.8 mm; calyx finely
downy, 1–1.5 cm, smaller than standard
and keel; wing petals curved, longer than
keel; standard 1.2–2 cm. Fruit 0.8–1.5
cm; seeds 1 or 2, dehiscing in spring. *W
Europe*. H3. Summer–autumn.

3. U. minor Roth (*U. nanus* Symons).
Illustration: Ross-Craig, Drawings of British
plants **7**: pl. 7 (1954); Fitter, Fitter &
Blamey, The wild flowers of Britain and
Northern Europe, 121, 1a (1974); Keble
Martin, The new concise British Flora, pl.
21 (1982).

Dense dwarf shrub 5–150 cm, with
close habit, usually prostrate; stems with
brownish hairs, young twigs and spines
not glaucous. Spines slender, straight or
slightly curved, soft, to 1.5 cm, very
slightly grooved, with shaggy hairs at
base. Flowers to 1.5 cm, paler yellow than
U. europaeus; stalks 3–5 mm, with
adpressed hairs; bracteoles 0.6–0.8 mm;
calyx 0.5–1 cm, slightly downy when
young, becoming hairless; wing petals
straight, as long as or longer than keel;
standard 0.7–1.5 cm, longer than calyx.
Fruit 5–8.5 mm, shaggy haired, seeds 3–6,
dehiscing in spring. *W Europe*. H3. Late
summer–autumn.

CXIV. KRAMERIACEAE

Semi-parasitic trees, shrubs or herbs.
Leaves spirally arranged, simple or rarely
with 3 leaflets, entire; stipules absent.
Flowers bisexual, solitary and axillary or,
in terminal racemes, with stalkless glands;
sepals 4 or 5, overlapping, the 3 outer
often larger than the 3 inner; petals 4 or
5, the 3 upper long-clawed, the 2 lower
smaller; stamens 3 or 4, alternating with
upper petal, filaments thick, sometimes
basally united or fused to corolla, anthers
opening by terminal pores; ovary superior,
with 2 fused carpels, 1 carpel reduced
and empty. Fruit dry, dehiscent, 1-seeded,
usually armed with barbed bristles or
spines.

A family of one genus and 15 species
from southern USA and South America.
Being difficult to grow they are rarely
cultivated but may be found in botanic
gardens or specialist collections.

GLOSSARY

abscission-zone. A predetermined layer at which leaves or other organs break off.

achene. A small, dry, indehiscent, 1-seeded fruit in which the fruit-wall is of membranous consistency and free from the seed.

acuminate. With a long, slender point.

adpressed. Closely applied to a leaf or stem and lying parallel to its surface but not adherent to it.

adventitious. (1) Of roots: arising from a stem or leaf, not from the primary root derived from the radicle of the seedling. (2) Of buds: arising somewhere other than in the axil of a leaf.

aggregate fruit. A collection of small fruits, each derived from a single carpel, closely associated on a common receptacle, but not united. *Ranunculus* and *Rubus* provide familiar examples.

alternate. Arising singly, 1 at each node; not opposite or whorled (figure 41(2), p. 552).

anastomosing. Describes veins of leaves which rejoin after branching from each other or from the main vein or midrib.

anatropous. Describes an ovule which turns through 180° in the course of development, so that the micropyle is near the base of the funicle (figure 44(2), p. 555).

annual. A plant which completes its life-cycle from seed to seed in less than 1 year.

anther. The uppermost part of a stamen, containing the pollen (figure 43(3), p. 554).

apetalous. Describes a flower without a corolla (petals).

apical. Describes the attachment of an ovule to the apex of a 1-celled ovary (figure 44(8), p. 555).

apiculate. With a small point.

apomictic. Reproducing by asexual means, though often by the agency of seeds, which are produced without the usual sexual nuclear fusion.

arachnoid. Describes hairs which are soft, long and entangled, suggestive of cobwebs.

aril. An outgrowth from the region of the hilum, which partly or wholly envelops the seed; it is usually fleshy.

ascending. Prostrate for a short distance at the base but then curving upwards to that the remainder is more or less erect; sometimes used less precisely to mean pointing obliquely upwards.

attenuate. Drawn out to a fine point.

auricle. A lobe, normally 1 of a pair, at the base of the blade of a leaf, bract, sepal or petal.

awn. A slender but stiff bristle on a sepal or fruit.

axil. The upper angle between a leaf-base or leaf-stalk and the stem that bears it (figure 41(1), p. 552).

axile. A form of placentation in which the cavity of the ovary is divided by septa into 2 or more cells, the placentas being situated on the central axis (figure 44(10), p. 555).

axillary. Situated in or arising from an axil (figure 41(1), p. 552).

back-cross. A cross between a hybrid and a plant similar to one of its parents.

basal. (1) Of leaves: arising from the stem at or very close to its base. (2) Of placentation: describes the attachment of an ovule to the base of a 1-celled ovary (figure 44(6,7), p. 555).

basifixed. Attached by its stalk or supporting organ by its base, not by it back (figure 43(3), p. 554).

berry. A fleshy fruit containing 1 or more seeds embedded in pulp, as in the genera *Berberis*, *Ribes* and *Phoenix*. Many fruits (such as those of *Ilex*) which look like berries and are usually so called in popular speech, are, in fact, drupes.

biennial. A plant which completes its life-cycle from seed to seed in a period of more than 1 year but less than 2.

bifid (2-fid). Forked; divided into 2 lobes or points at the tip.

bilaterally symmetric. Capable of division into 2 similar halves along 1 plane and 1 only (figure 43(9), p. 554).

bipinnate. Of a leaf: with the blade divided pinnately into separate leaflets which are themselves pinnately divided (figure 41(18), p. 552).

blade. A broadened part, furthest from the base of a petal or similar organ, which has a relatively narrow basal part – the claw or tube (figure 41(1), p. 552).

bract. A leaf-like or chaffy organ bearing a flower in its axil or forming part of an inflorescence, differing from a foliage-leaf in size, shape, consistency or colour (figure 42(2,3), p. 553).

bracteole. A small, bract-like organ which occurs on the flower-stalk, above the bract, in some plants.

bulbil. A small bulb, especially one borne in a leaf-axil or in an inflorescence.

calyx. The sepal; the outer whorl of a perianth (figure 43(1), p. 554).

campylotropous. Describes an ovule which becomes curved during development and lies with its long axis at right angles to the funicle (figure 44(4), p. 555).

capitate. Compact and approximately spherical, head-like.

capitulum. An inflorescence consisting of small flowers (florets), usually numerous, closely grouped together so as to form a 'head', and often provided with an involucre.

capsule. A dry, dehiscent fruit derived from 2 or more united carpels and usually containing numerous seeds.

carpel. One of the units (sometimes interpreted as modified leaves) situated in the centre of a flower and together constituting the gynaecium or female part of the flower (ovary). If more than 1, they may be free or united. They contain ovules and bear a stigma (figure 43(2), p. 554).

caruncle. A soft, usually oil-rich appendage attached to the seed near the hilum.

catkin. An inflorescence of unisexual flowers, made up of relatively conspicuous, usually overlapping bracts, each of which subtends a small apetalous flower or a group of such flowers; catkins are generally pendent, but some are erect.

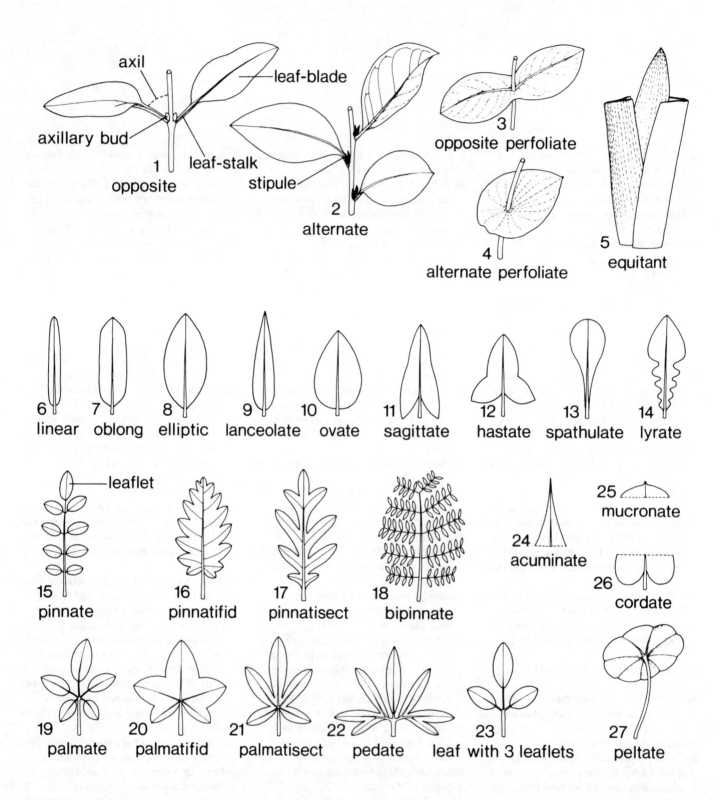

Figure 41. Leaves 1–5, Leaf insertion types. 6–14, leaf-blade outlines. 15–23, Leaf dissection types. 24–26. Leaf apex and base shapes. 27, Attachment of leaf-stalk to leaf-blade.

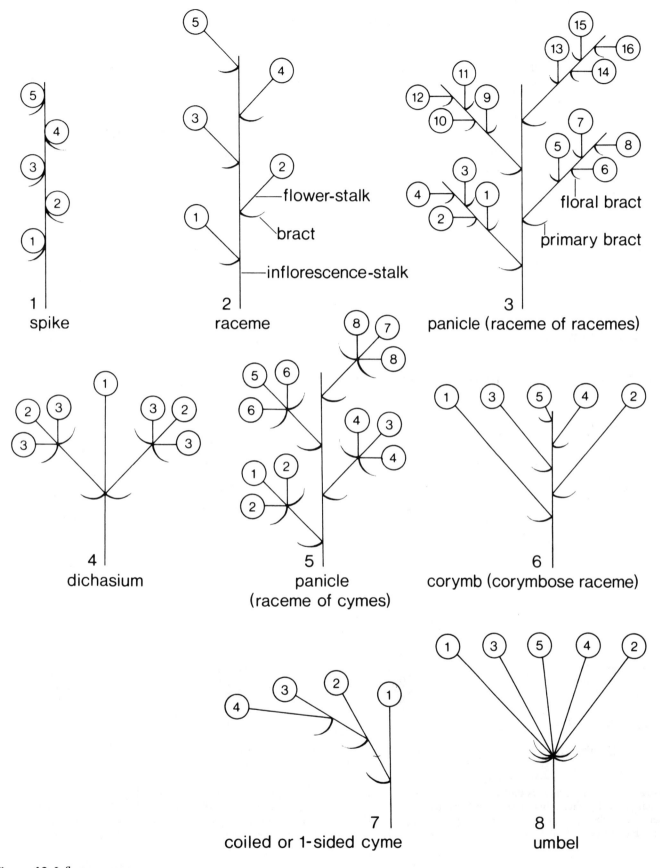

Figure 42. Inflorescences.

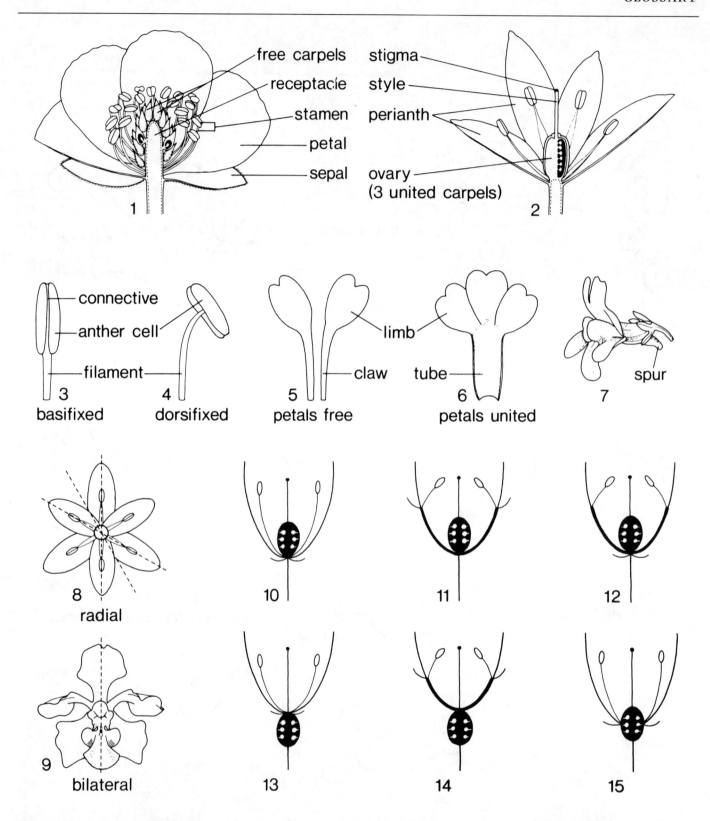

Figure 43. 1, 2, Two flowers illustration floral parts. 3, 4, Two stamens showing alternative types of anther attachment. 5–7, Some terms relating to petals. 8, 9, Floral symmetry, planes of symmetry shown by broken lines. 10–15, Position of ovary.

10–12, Superior ovaries. 13, 14, Inferior ovaries. 15, Half-inferior ovary. 11, Perigynous zone bearing sepals, petals and stamens. 12, Perigynous zone bearing petals and stamens. 14, Epigynous zone bearing sepals, petals and stamens.

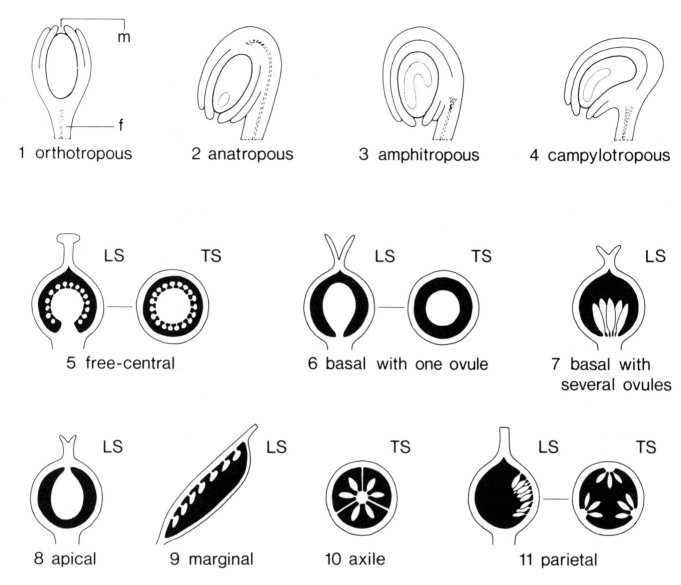

Figure 44. Ovules and placentation. 1–4, Ovule forms (f, funicle; m, micropyle). 5–11, Placentation types (LS. longitudinal section: TS, LS. transverse section).

chromosome. One of the small, thread-like or rod-like bodies consisting of nucleic acid and containing the genes, which become visible in a cell nucleus shortly before cell division.

ciliate. Fringed on the margin with usually fine hairs.

cladode. A branch which takes on the function of a leaf (the leaves being usually vestigial).

claw. The narrow base of a petal or sepal, which widens above into the limb or blade (figure 43(5), p. 554).

cleistogamous. Describes a flower with a reduced corolla which does not open but sets seed by self-pollination.

clone. The sum-total or the plants derived from the vegetative reproduction of an individual, all having the same genetic constitution.

compound. (1) Of a leaf: divided into separate leaflets. (2) Of an inflorescence: bearing secondary inflorescences in place of single flowers. (3) Of a fruit: derived from more than 1 flower.

compressed. Flattened from side to side.

connective. The tissue which separates the 2 lobes of an anther, and to which the filament is attached (figure 43(3), p. 554).

cordate. Describe the base of a leaf-blade which has a rounded lobe on either side of the central sinus (figure 41(26), p. 552).

corolla. The petals: the inner whorl of a perianth (figure 43(1), p. 554).

corymb. A broad, flat-topped inflorescence. In the strict sense the term indicates a raceme in which the lowest flowers have stalks long enough to bring them to the level of the upper ones (figure 42(6), p. 553).

cotyledon. One of the leaves preformed in the seed.

crisped. (1) Of hairs: strongly curved so that the tip lies near the point of attachment. (2) Of leaves, leaflets or petals: finely and complexly wavy.

cristate. With elevated, irregular ridges.

cultivar. A more or less uniform

assemblage of plants (usually selected in cultivation) which is clearly distinguished by 1 or more characters (morphological, physiological, cytological, chemical or others) and which, when reproduced retains its distinguishing characters. This term is derived from *cultivated variety*. Its name can be Latin in form, e.g. 'Alba', but is more usually in a modern language, e.g. 'Madame Lemoine', 'Frühlingsgold', 'Beauty of Bath'.

cupule. A group of bracts, united at least at the base, surrounding the base of a fruit or a group of fruits.

cyme. An inflorescence in which the terminal flower opens first, other flowers being borne on branches which arise below it (figure 42(7), p. 553).

cystolith. A concretion of calcium carbonate found within the cells of the leaf in some plants; they can sometimes be seen when the leaf is viewed against the light, or felt as tiny hard lumps when the leaf is drawn between finger and thumb.

decumbent. More or less horizontal for most of its length but erect or semi-erect near the tip.

decurrent. Continued down the stem below the point of attachment as a ridge or ridges.

dehiscent. Splitting, when ripe, along 1 or more predetermined lines of weakness.

dichasial. Resembling a dichasium.

dichasium. A form of cyme in which each node bears 2 equal lateral branches (figure 42(4), p. 553).

dichotomous. Divided into 2 equal branches; regularly forked.

dioecious. With male and female flowers on separate plants.

diploid. Possessing in its normal vegetative cells 2 similar sets of chromosomes.

disc. A variously contoured, ring-shaped or circular area (sometimes lobed) within a flower, from which nectar is secreted.

dissected. Deeply divided into lobes or segments.

distylic. Having flowers of different plants either with long styles and shorter stamens or with long stamens and shorter styles.

dorsifixed. Attached to its stalk or supporting organ by its back, usually near the middle (figure 43(4), p. 554).

double. Of flowers: with petals much more numerous than in the normal wild state.

drupe. An indehiscent fruit in which the outer part of the wall is soft and usually fleshy but the inner part is stony. A

drupe may be 1-seeded as in *Prunus* or *Juglans*, or may contain several seeds, as in *Ilex*. In the latter case, each seed is enclosed in a separate stony endocarp and constitutes a *pyrene*.

drupelet. A miniature drupe forming part of an aggregate fruit.

ellipsoid. As elliptic, but applied to a solid body.

elliptic. About twice as long as broad, tapering equally both to the tip and the base (figure 41(8), p. 552).

embryo. The part of a seed from which the new plant develops; it is distinct from the endosperm and seed-coat.

endocarp. The inner, often stony layer of a fruit-wall in those fruits in which the wall is distinctly 3-layered.

endosperm. A food-storage tissue found in many seeds, but not in all, distinct from the embryo and serving to nourish it and the young seedling during germination and establishment.

entire. With a smooth, uninterrupted margin; not lobed or toothed.

epicalyx. A group of bracts attached to the flower-stalk immediately below the calyx and sometimes partly united with it.

epigeal. The mode of germination in which the cotyledons appear above ground and carry on photosynthesis during the early stages of establishment.

epigynous. Describes a flower, or preferably the petals, sepals and stamens (or perianth and stamens) of a flower in which the ovary is inferior (figure 43(14), p. 554).

epigynous zone. A rim or cup of tissue on which the sepals, petals and stamens are borne in some flowers with inferior ovaries (figure 43(14), p. 554).

epiphyte. A plant which grows on another plant but does not derive any nutriment from it.

exocarp. The outer, skin-like layer of a fruit-wall in those fruits in which the wall is distinctly 3-layered.

farina. The flour-like wax present on the stem and leaves of many species of *Primula* and of a few other plants.

fastigiate. With all the branches more or less erect, giving the plant a narrow tower-like outline.

filament. The stalk of a stamen, bearing the anther at its tip (figure 43(3,4), p. 554).

filius. Used with authority names to distinguish between parent and offspring when both have given names to species, e.g. Linnaeus (C. Linnaeus, 1707–1788), Linnaeus filius (C. Linnaeus, 1741–1783), son of the former).

floret. A small flower, aggregated with others into a compact inflorescence.

follicle. A dry dehiscent fruit derived from a single free carpel and opening along one suture.

free. Not united to any other organ except by its basal attachment.

free-central. A form of placentation in which the ovules are attached to the central axis of a 1-celled ovary (figure 44(5), p. 555).

fruit. The structure into which the gynaecium is transformed during the ripening of the seeds; a *compound* fruit is derived from the gynaecia of more than one flower. The term 'fruit' is often extended to include structures which are derived in part from the receptacle (*Fragaria*), epigynous zone (*Malus*) or inflorescence-stalk (*Ficus*) as well as from the gynaecium.

funicle. The stalk of an ovule (figure 44(1), p. 555).

fusiform. Spindle-shaped; cylincric but tapered gradually at both ends.

gamete. A single sex-cell which fuses with one of the opposite sex during sexual reproduction.

gland-dotted. With minute patches of secretory tissue usually appearing as pits in the surface, as translucent dots when viewed against the light, or both.

glandular. (1) Of a hair: bearing at the tip a usually spherical knob of secretory tissue. (2) Of a tooth: similarly knobbed or swollen at the tip.

glaucous. Green strongly tinged with bluish grey; with a greyish waxy bloom.

graft-hybrid. A plant which, as a consequence of grafting, contains a mixture of tissues from 2 different species. Normally the tissues of 1 species are enclosed in a 'skin' of tissue from the other species.

gynaecium. The female organs (carpels) of a single flower, considered collectively, whether they are or united.

gynophore. The stalk which is present at the base of some ovaries (and the fruits developed from them).

half-inferior. Of an ovary; with its lower part inferior and its upper part superior (figure 43(15), p. 554).

haploid. Possessing in its normal vegetative cells only a single set of chromosomes.

hastate. With 2 acute, divergent lobes at the base, as in a mediaeval halberd (figure 41(12), p. 552).

haustorium. The organ with which a parasitic plant penetrates its host and draws nutriment from it.

herb. A plant in which the stems do not become woody, or, if somewhat woody at the base, do not persist from year to year.

herbaceous. Of a plant; possessing the qualities of a herb, as defined above.

heterostylic. Having flowers in which the length of the style relative to that of the stamens varies from one plant to another.

hilum. The scar-like mark on a seed indicating the point at which it was attached to the funicle.

hybrid. A plant produced by the crossing of parents belonging to 2 different named groups (e.g. genera, species, subspecies, etc.). An F1 hybrid is the primary product of such a cross. An F2 hybrid is a plant arising from a cross between 2 F1 hybrids (or from the self-pollination of an F1 hybrid).

hydathode. A water-secreting gland immersed in the tissue of a leaf near its margin.

hypocotyl. That part of the stem of a seedling which lies between the top of the radicle and the attachment of the cotyledon (s).

hypogeal. The mode of germination in which the cotyledons remain in the seed-coat and play no part in photosynthesis.

hypogynous. Describes a flower, or, preferably the petals, sepals and stamens (or perianth and stamens) of a flower in which the ovary is superior and the petals, sepals and stamens (or perianth and stamens) arise as individual whorls on the receptacle (figure 43(10–12), p. 554).

incised. With deep, narrow spaces between the teeth or lobes.

included. Not projecting beyond the organs which enclose it.

indefinite. More than 12 and possibly variable in number.

indehiscent. Without preformed lines of splitting, opening, if at all irregularly by decay.

inferior. Of an ovary; borne beneath the sepals, petals and stamens (or perianth and stamens) so that these appear to arise from its top (figure 43(13,14), p. 554).

inflorescence. A number of flowers which are sufficiently closely grouped together to form a structured unit (figure 42), p. 553).

infraspecific. Denotes any category below species level, such as subspecies, variety and form. To be distinguished from *subspecific*, which means relating to subspecies only.

integument. The covering of an ovule, later developing into the seed-coat. Some ovules have a single integument, others 2.

internode. The part of a stem between 2 successive nodes.

involucel. A whorl of united bracteoles borne below the flower in some Dipsacaceae.

involucre. A compact cluster or whorl of bracts around the stalk at or near the base of some flowers or inflorescences or around the base of a capitulum; sometimes reduced to a ring of hairs.

keel. A narrow ridge, suggestive of the keel of a boat, developed along the midrib (or rarely other veins) of a leaf, sepal or petal.

laciniate. With the margin deeply and irregularly divided into narrow and unequal teeth.

lanceolate. 3–4 times as long as wide and tapering more or less gradually towards the tip (figure 41(9), p. 552).

layer. To propagate by pegging down on the ground a branch from near the base of a shrub or tree, so as to induce the formation of adventitious roots.

leaflet. One of the leaf-like components of a compound leaf (figure 41(15), p. 552).

legume. The typical fruit of the Leguminosae, formed from a single carpel and opening down both sutures; legumes are, however, variable, some indehiscent, others breaking into 1-seeded segments, etc.

lenticel. A small, slightly raised interruption of the surface of the bark (or the corky outer layers of a fruit) through which air can penetrate to the inner tissues.

linear. Parallel-sided and many times longer than broad (figure 41(6), p. 552).

lip. A major division of the apical part of a bilaterally symmetric calyx or corolla in which the petals or sepals are united; there is normally an upper and lower lip, but either may be missing.

lyrate. Pinnatifid or pinnatisect, with a large terminal and small lateral lobes.

marginal. Of placentation; describing the placentation found in a free carpel which contains more than 1 ovule.

medifixed. Of a hair; lying parallel to the surface on which it is borne and attached to it by a stalk (usually short) at its mid-point.

mericarp. A carpel, usually 1-seeded, released by the break-up at maturity of a fruit formed from 2 or more joined carpels.

mesocarp. The central, often fleshy layer of a fruit-wall is distinctly 3-layered.

micropyle. A pore in the integument(s) of an ovule and later in the coat of a seed (figure 44(1), p. 555).

monocarpic. Flowering and fruiting once, then dying.

monoecious. With separate male and female flowers on the same plant; male flowers may contain non-functional carpels (and vice versa).

monopodial. A type of growth-pattern in which the terminal bud continues growth from year to year.

mucronate. Provided with a short, narrow point at the apex (figure 41(25), p. 552).

mycorrhiza. A symbiotic association between the roots of a green plant and a fungus.

nectary. A nectar-secreting gland.

neuter. Without either functional male or female parts.

node. The point at which 1 or more leaves or flower parts are attached to an axis.

nut. A 1-seeded indehiscent fruit with a woody or bony wall.

nutlet. A small nut, usually a component of an aggregate fruit.

obconical. Shaped like a cone, but attached at the narrow end.

oblanceolate. As lanceolate, but attached at the more gradually tapered end.

oblong. With more or less parallel sides and about 2–5 times as long as broad (figure 41(7), p. 552).

obovate. As ovate, but attached at the narrower end.

obovoid. As ovoid, but attached at the narrower end.

opposite. Describes 2 leaves, branches or flowers attached on opposite sides of the axis at the same node.

orthotropous. Describes an ovule which stands erect and straight (figure 44(1), p. 555).

ovary. The lower part of a carpel, containing the ovules(s) (ie. excluding style and stigma); the lower, ovule-containing part of a gynaecium in which the carpels are united (figure 43(2), p. 554).

ovate. With approximately the outline of a hen's egg (though not necessarily blunt-tipped) and attached at the broader end (figure 41(10), p. 552).

ovoid. As ovate, but applied to a solid body.

ovule. The small body from which a seed develops after pollination (figure 44, p. 555).

palmate. Describes a compound leaf composed of more than 3 leaflets, all arising from the same point, as in the leaf of *Aesculus*. also used to described similar venation in simple leaves (figure 41(19), p. 552).

palmatifid. Lobed in a palmate manner, with the incisions pointing to the place of attachment, but not reaching much more than halfway to it (figure 41(20), p. 552).

palmatisect. Deeply lobed in a palmate manner, with the incisions almost reaching the base (figure 41(21), p. 552).

panicle. A compound raceme, or any freely branched inflorescence of similar appearance (figure 42(3,5), p. 553).

papillose. Covered with small blunt protuberrances (papillae).

parietal. A form of placentation in which the placentas are borne on the inner surface of the walls of a 1-celled ovary, or rarely, in a similar manner in a septate ovary (figure 44(11), p. 555).

pectinate. With leaves, leaflets, or hairs in regular, eye-lash-like rows.

pedate. With a terminal lobe or leaflet, and on either side of it an axis curving outwards and backwards, bearing lobes or leaflets on the outer side of the curve (figure 41(22), p. 552).

peltate. Describes a leaf of other structure with the stalk attached other than at the margin (figure 41(27), p. 552).

perennial. Persisting for more than 2 years.

perfoliate. Describes a pair of stalkless opposite leaves of which the bases are united, or a single leaf in which the auricles are united so that the stem appears to pass through the leaf or leaves (figure 41(3,4), p. 552).

perianth. The calyx and corolla considered collectively, used especially when there is no clear differentiation between calyx and corolla; also used to denote a calyx or corolla when the other is absent (figure 43(2), p. 554).

perigynous. Describing a flower, or, preferably, the petals, sepals and stamens (or perianth and stamens) of a flower in which the ovary is superior and the petals, sepals and stamens (or perianth and stamens) are borne on the margins of a rim or cup which itself is borne on the receptacle below the ovary (it often appears as though the sepal, petals and stamens (or perianth and stamens) are united at their bases) (figure 43(11,12), p. 554).

perigynous zone. The rim or cup of tissue on which the sepals, petals and stamens) are borne in a perigynous flower (figure 43(11,12), p. 554).

petal. A member of the inner perianth-whorl (corolla) used mainly when this is clearly differentiated from the calyx. The petals usually function in display and often provide an alighting place for pollinators (figure 43(1), p. 554).

phyllode. A leaf-stalk taking on the funtion and, to a variable extent, the form of a leaf-blade.

pinnate. Describes a compound leaf in which distinct leaflets are arranged on either side of the axis (figure 41(15), p. 552). If these leaflets are themselves of a similar compound structure, the leaf is termed *bipinnate* (similarly *tripinnate*, etc.).

pinnatifid. Lobed in a pinnate manner, with the incisions reaching not much more than halfway to the axis (figure 41(16), p. 552).

pinnatisect. Deeply lobed in a pinnate manner, with the incisions almost reaching the axis (figure 41(17), p. 552).

pistillode. A sterile ovary in a male flower.

placenta. A part of the ovary, often in the form of a cushion or ridge, to which the ovules are attached.

placentation. The manner of arrangement of the placentas.

pollen-sac. One of the cavities in an anther in which pollen is formed; each anther normally contains 4 pollen-sacs, 2 on either side of the connective, those of each pair separated by a partition which shrivels at maturity.

polyploid. Possessing in the normal vegetative cells more than 2 sets of chromosomes.

proliferous. Giving rise to planlets or additional flowers on stems, leaves or in the inflorescence.

protandrous. With anthers beginning to shed their pollen before the stigmas of the same flower are receptive.

protogynous. With stigmas becoming receptive before the anther in the same flower shed their pollen.

pulvinus. A swollen region at the base of a leaflet, leaf-blade or leaf-stalk.

pyrene. A small nut-like body enclosing a seed, 1 or more of which, surrounded by fleshy tissue, make up the fruit of, for example, *Ilex.*

raceme. An inflorescence consisting of stalked flowers arranged on a single axis, the lower opening first (figure 42(2), p. 553).

radially symmetric. Capable of division into 2 similar halves along 2 or more planes of symmetry (figure 43(8), p. 554).

radicle. The root preformed in the seed and normally the first visible root of a seedling.

raphe. A perceptible ridge or stripe, at one end of which is the hilum, on some seeds.

receptacle. The tip of an axis to which the floral parts, or perigynous zone (when present), are attached (figure 43(1), p. 554).

reflexed. Bent sharply backwards from the base.

rhizome. A usually horizontal stem, situated underground or rarely on the surface, serving the purpose of food-storage or vegetative reproduction or both; roots or stems arise from some or all of its nodes.

rootstock. The compact mass of tissue from which arise the new shoots of a herbaceous perennial. It usually consists mainly of stem tissue, but is more compact than is generally understood by rhizome.

runner. A slender, above-ground stolon with very long internodes.

sagittate. With a backwardly directed basal lobe on each side, like an arrowhead (figure 41(11), p. 552).

samara. A winged, dry, indehiscent fruit or mericarp.

saprophytic. Dependent for its nutrition on soluble organic compounds in the soil. Saprophytic plants do not photosynthesise and lack chlorophyll; some plants, however, are *partially saprophytic* and combine the two modes of nutrition.

scale-leaf. A reduced leaf, usually not photosynthetic.

scape. A leafless flower-stalk or inflorescence-stalk arising usually from near ground level.

scarious. Dry and papery, often translucent.

schizocarp. A fruit which, at maturity, splits into its constituent mericarps.

scion. A branch cut from one plant to be grafted on the rooted stock of another.

seed. A reproductive body adapted for dispersal, developed from an ovule and consisting of a protective covering (the seed-coat), and embryo, and, usually, a food-reserve.

semi-parasite. A plant which obtains only part of its nutrition by parasitism.

sepal. A member of the outer perianth whorl (calyx) when 2 whorls are clearly differentiated as calyx and corolla, or when comparison with related plants shows that a corolla is absent. The sepals most often function in protection and support of other floral parts (figure 43(1), p. 554).

septum. An internal partition.

shrub. A woody plant with several stems or branches arising from near the base, and of smaller stature than a tree.

simple. Not divided into separate parts.

sinus. The gap or indentation between 2 lobes, auricle or teeth.

spathulate. With a narrow basal part, which towards the apex is gradually expanded into a broad, blunt blade.

spicate. Similar to a spike.

spike. An inflorescence or subdivision of and inflorescence, consisting of stalkless flowers arranged on a single axis (figure 42(1), p. 553).

spur. An appendage or prolongation, more or less cylindric, often at the base of an organ. The spur of a corolla or single petal or sepal is usually hollow and often contains nectar (figure 43(7), p. 554).

stamen. The male organ, producing pollen, generally consisting of anther borne on a filament (figure 43(1), p. 554).

staminode. An infertile stamen, often reduced or rudimentary or with a changed function.

stellate. Star-like, particularly of branched hairs.

stigma. The part of a style to which the pollen adheres, normally differing in texture from the rest or the style (figure 43(2), p. 554).

stipule. An appendage, usually 1 pair, found beside the base of the leaf-stalk in many flowering plants, sometimes falling early, leaving a scar, In some cases the 2 stipules are united; in others they are partly united to the leaf-stalk.

stock. A rooted plant, often with the upper parts removed, on to which a scion may be grafted.

stolon. A far-creeping, more or less slender, above-ground stem giving rise to a new plant at its tip and sometimes at intermediate nodes.

stoma. A microscopic ventilating pore in the surface of a leaf or other herbaceous part.

style. The usually slender, upper part of a carpel or gynaecium, bearing the stigma (figure 43(2), p. 554).

subtend(ed). Used of any structure (e.g. a flower) which occurs in the axil of another organ (e.g. a bract); in this case bract subtends the flower.

subulate. Narrowly cylindric, and somewhat tapered to the tip.

sucker. An erect shoot originating from a bud on a root or rhizome, sometimes at some distance from the parent plant.

superior. Of an ovary; borne at the morphological apex of the flower so that the petals, sepals and stamens (or perianth and stamens) arise on the receptacle below the ovary (figure 43(10–12), p. 554).

suture. A line marking an apparent junction of neighbouring parts.

sympodial. A type of growth-pattern in which the terminal bud ceases growth, further growth being carried on by a lateral bud.

tendril. A thread-like structure which by its coiling growth can attach a shoot to something else for support.

terete. Approximately circular in cross-section; not necessarily perfectly cylindric, but without grooves or ridges.

tetraploid. Possessing in its normal vegetative cells 4 similar sets of chromosome.

throat. The part of a calyx or corolla transitional between the tube and limb or lobes.

triploid. Possessing in its normal vegetative cells 3 similar sets of chromosomes.

tristylic. Having flowers of different plants with long, short or intermediate-length styles; the stamens of each flower are of 2 lengths which are not the same as the style-length of that flower.

truncate. As though with the tip or base cut off at right angles.

tuber. A swollen underground stem or root used for food storage.

tubercle. A small, blunt, wart-like protuberance.

turion. A specialised perennating bud in some aquatic plants. consisting of a short shoot covered in closely packed leaves, which persist through the winter at the bottom of the water.

umbel. An inflorescence in which the flower-stalks arise together form the top of an inflorescence-stalk; this is a *simple umbel* (figure 42(8), p. 553). In a *compound umbel* the several stalks arising from the top of the inflorescence-stalk terminate not in flowers but in secondary umbels.

undivided. Without major divisions or incisions, though not necessarily entire.

urceolate. Shaped like a pitcher or urn, hollow and contracted at or just below the mouth.

valve. The part of the fruit, covering the seeds, that falls from the septum during dehiscence (Cruciferae).

vascular bundle. A strand of conducting tissue, usually surrounded by softer tissue.

vein. A vascular strand, usually in leaves or floral parts and visible externally.

venation. The pattern formed by the veins in a leaf or other organ.

versatile. Of an anther; flexibly attached to the filament by its approximate mid-point so that a rocking motion is possible.

vessel. A microscopic water-conducting tube formed by a sequence of cells not separated by end-walls.

viviparous. Bearing young plants, bulbils or leafy buds which can take root; they can occur anywhere on the plant and may be interspersed with, or wholly replace, the flowers in an inflorescence.

whorl. A group of more than 2 leaves or floral organs inserted at the same node.

wing. A thin, flat extension of a fruit, seed, sepal or other organ.

xerophytic. Drought-tolerant. Can also describe the environment in which drought-tolerant plants live.

INDEX

Synonyms and names mentioned only in observations are printed in *italic* type